Fahrenheit-Celsius Conversion

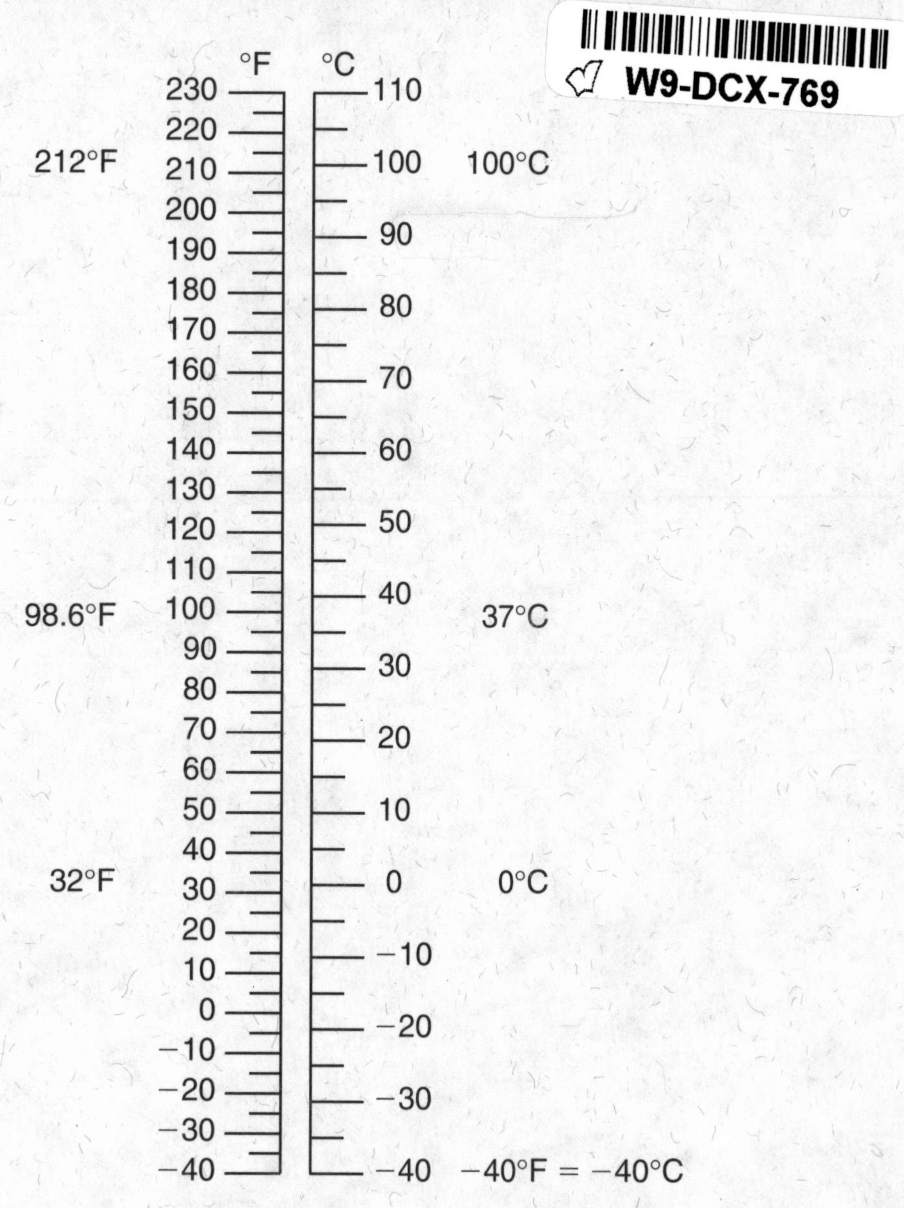

Temperature

°C = degrees Celsius

°F = degrees Fahrenheit

°C = 5/9(°F − 32)

°F = 1.8(°C) + 32

Kelvin (absolute temperature scale): Kelvins (K) = °C + 273.15

Biology

BURTON S. GUTTMAN

The Evergreen State College

Contributing Author

Johns W. Hopkins III

Washington University

Boston Burr Ridge, IL Dubuque, IA Madison, WI New York San Francisco St. Louis
Bangkok Bogotá Caracas Lisbon London Madrid
Mexico City Milan New Delhi Seoul Singapore Sydney Taipei Toronto

WCB/McGraw-Hill
A Division of The McGraw-Hill Companies

BIOLOGY

This book is printed on recycled, acid-free paper containing 10% postconsumer waste.

1 2 3 4 5 6 7 8 9 0 VNH/VNH 9 3 2 1 0 9 8

ISBN 0-697-22366-3

Vice president and editorial director: *Kevin T. Kane*
Publisher: *Michael D. Lange*
Sponsoring editor: *Patrick E. Reidy*
Senior development editor: *Connie Balius-Haakinson*
Marketing manager: *Lisa L. Gottschalk*
Senior project manager: *Peggy J. Selle*
Senior production supervisor: *Sandra Hahn*
Designer: *K. Wayne Harms*
Senior photo research coordinator: *Lori Hancock*
Art editor: *Jodi K. Banowetz*
Compositor: *GTS Graphics, Inc.*
Typeface: *10.5/12 Minion*
Printer: *Von Hoffmann Press, Inc.*

Cover design: *Kaye Farmer*
Cover photograph: *Anthony J.F. Griffiths*
Illustration rendering: *Page Two Incorporated*

The credits section for this book begins on page 1139 and is considered an extension of the copyright page.

Library of Congress Cataloging-in-Publication Data

Guttman, Burton S.
Biology / Burton S. Guttman. — 1st ed.
p. cm.
Includes bibliographical references and index.
ISBN 0-697-22366-3
1. Biology. I. Title.
QH308.2.G874 1999
570—dc21 98-20858
CIP

www.mhhe.com

To Erica, who has always been there, and
Lois, who will always be there.

Brief Contents

Contents

* These elements appear in every chapter.

Chapter 14 ~ Protein Synthesis and Cell Growth 273

Growth is primarily protein synthesis, and here we examine the transmission of information from DNA to protein and protein synthesis in the context of cell organization.

Chapter 15 ~ Cycles of Growth and Reproduction 292

Most organisms reproduce through a type of sexual cycle, a series of events that includes division and recombination of their chromosomes.

Chapter 16 ~ Mendelian Heredity 311

Sexual organisms inherit their characteristics in relatively simple patterns, as we explore in this chapter.

Chapter 17 ~ Microbial and Molecular Genetics 331

Our basic understanding of genetic processes has come from work with bacteria and phage, leading more recently to recombinant-DNA techniques.

Preface

This book has grown out of a personal journey of discovery. I have been trying for many years to understand biology as a coherent science with organizing principles, and Johns Hopkins and I have tried for most of our professional lives to help students develop their personal understanding of often complicated biological concepts. (Here I must explain that this book developed as a long-term collaboration with Johns W. Hopkins III, and it bears much of his contributions and influence. Johns's untimely illness took him out of the project as it was coming to a climax, but I must write this preface with a mixture of "we" and "I" to reflect our collaboration.) Pedagogically, the book relies on many years of teaching introductory biology to a wide variety of students in many contexts.

This project began long ago as an exploration in theoretical biology, a search for general principles in biology and general ways to understand organisms, but it soon became a project in pedagogy, based on the conviction that students can learn biology more effectively if the myriad details of the science are developed in a strong conceptual framework. When the project was pursued by the team at WCB/McGraw-Hill, the critical comments of nearly 100 reviewers told us that the book was indeed on the right track, with unique features that would make it welcome even in a crowded market. This preface explains the background to this project and the features that make *Biology* unique, innovative, and worthy to teach and learn from.

The trouble with biology is that it is full of facts. An unimaginable number of factual statements could be made about the few million species of organisms on earth. Someone once published an example of a college zoology exam from the pre-Darwinian era that required only the recitation of endless anatomical facts; the Darwinian paradigm changed that, and for a long time, biology was taught primarily as a collection of these facts organized around the principle of natural selection and the fact of evolution. Of course, students of biology must still learn many facts about the natural world, often fascinating facts that motivate them to continue their personal explorations. But as the science of biology matures, it should increasingly subsume facts under general principles and develop coherent general concepts. As our knowledge of molecular, cellular, and physiological processes has grown, that foundation emerges from the genetic conception of an organism: a structure that operates on the basis of information in its genome.

The beginnings of this foundation emerged with Norman Horowitz's (1959) conception of an organism as a structure that can reproduce itself and mutate. For a long time, I used this as a foundation for thinking about biology. But its emphasis gradually shifted as I realized that the definition was too limited—for instance, there are perfectly good organisms that for one reason or another cannot reproduce. A more satisfactory foundational statement is this: An organism is a structure whose organization and operation are governed by the information in a genome.

Fortuitously, this conception is consonant with the thinking of a computer age, since everyone now knows that a computer is a hardware device that operates on the basis of particular software instructions; biological theory was therefore strengthened by John von Neumann's (1951) theoretical demonstration that it is possible to create a self-reproducing automaton that would operate on the basis of its instruction tape and produce an identical automaton. As such automata reproduce, occasional errors (mutations) will creep into copies of the genome. It is then obvious that such systems, operating within an ecological framework where resources are necessarily limited, will undergo evolution through natural selection (Guttman, 1966). Thus, the entire Darwinian paradigm emerges from the genetic conception of an organism.

Biology is still full of facts, and students can only understand the science by learning many of them. This book, however, is based on the conviction that students of biology will benefit far more from an emphasis on concepts and principles developed within this genetic-evolutionary-ecological framework, as explained more fully below.

Four Unifying Conceptual Themes

This book is tied together by four primary themes. This conceptual framework gives students something solid on which to hang virtually all the specific facts and more limited principles we develop. Let's examine these ideas carefully.

1. ***Organisms are genetic systems.*** They operate and reproduce themselves on the basis of instructions encoded in their genomes. As I pointed out above, populations of organisms will experience mutation and evolution through natural selection, so the entire Darwinian framework that traditionally forms such an important foundation for biology emerges from this broader genetic framework. The genetic theme also makes sense of the principal cellular activities of synthesizing molecular structure under genetic instructions, using materials and energy derived from other phases of metabolism. The book's genetics theme, and the central importance of inheritance with variation, is symbolized by the cover photo of this book kindly contributed by Tony Griffiths of the University of British Columbia. Tony is also co-author of the first genetics book to stress a secondary theme of the present book—the use of genetic analysis to

dissect biological process. For a few places where this theme is emphasized, look at:
—Sections 2.7 and 2.8, for basic concepts about organisms
—Section 12.1, about how a genome instructs cells to operate
—Sections 15.6 to 15.8, about how a genome determines development

2. ***Organisms live in ecosystems, where they are adapted to particular ways of life and engage in complex interrelationships with other organisms and the environment.*** From the beginning, the genetic framework is immersed in an indispensable ecological framework. The book emphasizes the triumvirate of selection, adaptation, and ecological niche, which describes how organisms live, how they come to have their particular features, and how they interact with one another. For places where this theme is emphasized, look at:
—Section 2.9a, for basic ideas about ecosystems
—Section 7.10, about energy in ecosystems
—Section 11.7, for chemical communication among organisms in a community

3. ***Evolution, operating primarily through the process of natural selection, has produced the enormous variety of life and continues to operate today.*** This cornerstone of modern biology needs no explanation. However, instructors commonly complain that the theme gets lip-service by being relegated to an introductory chapter and a few later chapters specifically about evolution. I have tried to integrate the theme more consistently throughout the book, and the reader can judge how successful the integration has been by examining some representative sections such as:
—Section 2.9, about the concept of natural selection
—Section 4.13, about the evolution of proteins
—Section 14.5, which shows how cellular processes restrict evolution

4. ***Organisms function through molecular interactions.*** This theme emphasizes the basic fact that organisms are molecular machines and that they operate through molecular interactions. Scientific explanation is a matter of showing how phenomena fit into the causal structure of the universe, and much of the explanatory structure of biology shows how molecules push and pull on one another. I establish the concept early on that biomolecules are structures with unique shapes that interact with one another in unique ways, and then we keep returning to this general way of thinking through many examples.

   Now a general biology course is obviously no place to discuss the molecular details covered in a biochemistry course with students who understand organic reaction mechanisms. So this idea is developed almost entirely through pictorial examples. Thus, the book emphasizes molecular explanation without being strongly chemical. Although this sounds self-contradictory, it is basically simple. I continually stress the concept of biomolecules having specific, complementary shapes, and this is demonstrated through pictures that are little more than artists' conceptions, requiring little or no understanding of organic chemistry. For places where this theme is emphasized, look at:
—Section 8.6, about the structure of membranes
—Sections 8.12 and 8.13, about how membranes move molecules
—Sections 11.7 to 11.10, about proteins of the cytoskeleton

Three General and Pedagogical Themes

In addition to these conceptual themes, I have worked hard to develop certain features of pedagogical importance to students.

Science as stories of living history. Scientists know that our subject is a dynamic structure based on the investigations of many people. The facts and principles we teach were not handed down from some mystical source, yet they are often presented as if humans had nothing to do with their discovery. But students should know at least some history of discovery and should think of science as an ongoing human enterprise. If only there were room to tell all the great stories! But there isn't, and I have confined myself to some of the best. In any case, you will find many names of contemporary investigators in the text, and Section 1.5 explains why their biographies are not given.

Learning through exercises. Exercises interspersed throughout allow students to test their comprehension of ideas immediately after they have been presented and to go beyond mere memorization by applying concepts to new situations. Students need to check their understanding as soon as possible, to be sure they can understand the ideas to be developed next. These exercises emphasize the skill of problem-solving; many are quantitative and are used to develop a conception of sizes and rates in the biological world.

Telling students what they need to know when they need to know it. Teachers have realized for themselves that there is a time to explain concepts and a time to hold back. But many authors seem to believe that students can and should learn everything the instructor knows about a topic as soon as the topic is introduced. However, understanding develops slowly, with repetition, and it is easy to create an informational overload that just confuses students. Accordingly, I have tried to follow the principle that concepts and information should only be introduced at the time when students can use it, not before.

Chapter 4, for example, introduces the general principle of polymeric structure and explains the structures of polysaccharides, nucleic acids, and proteins; but this is not the time to start developing complicated concepts about the role of nucleic acids in the genetic apparatus. Let the first idea sink in and develop a bit. The genetic ideas will come in good time. (I know from experience that introducing the functions of different kinds of RNA at this point is useless to students because they have no basis for assimilating the information.) Along with the major polymers, I introduce lipids briefly, but lipids are not polymers. They just aren't. And talking about them as if they were confuses the issue of biological structure. Furthermore, students don't have to know about lipids until they encounter membranes in Chapter 8, so I delay most of the information about them until I can introduce it in a more meaningful biological context.

Intended Audience

This book is designed for the typical majors biology course taken by college students in their first or second year. It assumes relatively little prerequisite knowledge—principally that students have become aware of many natural phenomena during 18 years or so of living, including some experience with the world of organisms. It does make the assumption that students can think as well as memorize, so they are able to work out some ideas through exercises rather than just being told every point dogmatically. Since science is quantitative, students must know elementary algebraic manipulations and must be able to work with scientific notation (powers of 10), although this is explained in Appendix A. The book explains elementary chemical concepts as needed. Although a lot of organic molecules are displayed in the book for the sake of reference and concreteness, I actually treat most chemical structures as little more than unique shapes that interact with other unique shapes, so no knowledge of organic chemistry is required beyond the simple structural ideas in Chapter 3.

Unique Chapters and the Placement of Ideas

Organisms as Genetic Systems that Evolve within Ecosystems

Several specific features of the chapters set this book off from traditional books. First, please pay attention to Chapter 2. Although the book generally follows the now-traditional "micro-to-macro" approach, Chapter 2 provides the broad genetic-ecological-evolutionary context for the whole book. It defines an organism(and a virus) and provides some of the language for thinking about all the details and the specific concepts to be developed in the rest of the book, before we can deal with genetics, ecology, and evolution more extensively. The information comes too fast here, too condensed—of course, it does, and I don't expect students to really understand it all during a first go-around. But it sets a tone, an orientation. It says to students that they should start to look at organisms as genetic systems that evolve within ecosystems.

Understanding Basic Chemical Structure

To return to the treatment of biomolecular structure in Chapter 4, beginning students have trouble understanding the basic chemical structure of organisms. The key is understanding the principle of polymeric structure and learning the structures of the principal monomers and their polymers (polysaccharides, nucleic acids, and proteins). This principle is not usually emphasized enough. I discuss polysaccharides in some detail, because they won't be discussed in much detail elsewhere, and nucleic acids in outline.

Chapters 4 and 5 then devote a lot of time to proteins, and throughout the book we relentlessly emphasize proteins, proteins, proteins. The biological reality is that virtually everything an organism is or does depends upon its particular protein composition, and it is hard to emphasize enough that an organism is made of thousands of different proteins, each with its own function. I point out that amino acids and proteins have charges, because students should understand this and because it is necessary to understand electrophoresis, one of the most important modern techniques for analyzing structure. The story of protein structure is then developed through Sanger's work on insulin, Kendrew and Perutz's work on myoglobin and hemoglobin, and Anfinsen's work on ribonuclease. This story makes some critical points about protein structure; understanding biology depends upon the point that each protein consists of a specific sequence of amino acids with a unique shape. And it is best, I think, to make the point by telling how these discoveries were made.

Chapter 4 ends with a return to the genetic and evolutionary themes. Even though I have only sketched the genetic conception in general terms so far, I want to continue building on it at every opportunity; I want students to start thinking about the genetic reason that proteins have their particular structures, rather than simply thinking that proteins exist, that they come to be in some obscure or magical way. And I want to emphasize the importance of evolutionary thinking at every opportunity.

Let me emphasize another point here: As I introduce the idea of mutation as a basis for evolution, it is easy to say that most mutations are likely to be disadvantageous rather than advantageous. But the principle of polymeric construction gives us new insight here by showing that a mutation may only replace one amino acid in a protein out of hundreds, and in some cases the mutation might only replace a few atoms out of thousands. Thus, polymeric construction makes for great subtlety in evolutionary change. When I emphasize that evolution occurs through a kind of editing process, I show that it can be a very subtle editing.

Introduction to Enzymes

Chapter 5 introduces enzymes as classic examples of proteins and develops another general principle that will pervade the whole book: That proteins bind to ligands through weak interactions. The chapter introduces the idea of saturable (Michaelis-Menten) kinetics—not the mathematics, but rather the idea that M-M kinetics implies that proteins have a discrete number of binding sites. The idea is used later in discussing myoglobin and hemoglobin, and it will be used again in Chapter 8 to distinguish simple diffusion from facilitated diffusion. Chapter 5 also introduces the general idea of allostery, which will, of course, be used throughout the book.

Survey of Cell Structure

Chapter 6 is a traditional survey of cell structure, but not quite traditional. I think it is a mistake to try to tell students everything about each cellular organelle when they aren't prepared to understand the information. So this is only a light survey, emphasizing the overall organization of a cell and the size relationships of cellular components (especially by means of two special boxes). The details come in later chapters, when students can integrate information about the functions of each structure one by one into their developing conceptual framework.

General Concepts of Regulation

Chapter 11, The Dynamic Cell, introduces some important concepts that I return to several times later, especially general concepts of regulation. Here we meet the ideas of feedback, steady state, homeostasis, informational transduction, receptor proteins, signal ligands (the various "-mones"), and the general signal transduction pathway that includes G-proteins and protein kinases.

Mitosis and Meiosis as Distinct Processes

Biologists have long been divided into two armed theological camps on the mitosis-meiosis issue; I strongly believe they should be discussed at different times. The conceptual reason is that mitosis and meiosis are distinct processes with entirely different functions. Mitosis is a phase of the cell cycle, and it should be understood in that context along with cell growth and DNA replication; meiosis is a phase of the sexual cycle, and it should be understand in that context as part of the life cycles of most eucaryotic organisms. The pedagogical reason is that the processes are too similar, and trying to discuss them together would be like trying to teach a child about football and soccer simultaneously.

I think it is important for a modern biology book to put mitosis into its proper place in the cell cycle. Until recently, mitosis was all we knew about; the rest of the cycle was relegated to the anonymity of interphase. But now that so much is known about the cycle as a whole and its regulation, our view should be corrected. Students come from high school biology thinking that mitosis *is* the cell cycle, and in my own teaching I have to reverse the emphasis and put all the events in perspective. My students have asked why the modern textbooks we use fail to do this, and I don't have a good answer.

In introducing the general sexual cycle, I use the convenient terms haplontic, diplontic, and haplodiplontic. They replace a multiplicity of more complicated terms that have been used for certain taxonomic groups, and they are easy to grasp.

Instructors have conflicting ideas about how Mendelian genetics should be developed, but I feel strongly that it should be done after developing a clear view of meiosis and the sexual cycle. Then it is easy to show how Mendel's laws follow from the events of meiosis and random fertilization. Furthermore, I believe it is important that students should already have a conception of what genes are for, so the entire DNA-RNA-protein story comes first (Chapters 12–14).

Introduction to Plants and Animals

Chapter 32 is a unique introduction to plants and animals. Many classic biology textbooks of an older era treated the general functions of plants and animals together, and many instructors still prefer this approach. I believe plants and animals are different enough to justify the now-current separate approach, but it is still valuable to discuss their commonalities. Here we can consider general body form, water relationships, exchange of oxygen, nitrogen, and carbon dioxide, and some general considerations of size.

Other Features of the Book

The book incorporates several regular features to help pique students' interest and help them study.

The Art Program: I have attempted a closer integration of text with art by incorporating some drawings into the text. It was Johns's idea that this text art should be the equivalent of the simple drawings a good instructor continually makes on the blackboard while lecturing, although the illustrators have made far better pictures than most of us can sketch. The more elaborate numbered figures are the equivalent of the slides and transparencies we show during a lecture, and the publisher has gone all-out to create a new art program with real pedagogical value as well as the highest aesthetic qualities.

Some general themes of consistent formatting and color coding have been incorporated into the more conceptual figures. Thus, DNA is consistently blue and RNA consistently red. The parts of a eucaryotic cell have consistent colors, introduced on the general drawings in Chapter 6. Phospholipid bilayers are consistently blue and yellow. In physiological processes, hormones are consistently blue, other substances such as enzymes are green, organs and other structures are red-orange, conditions to be regulated are yellow-orange, and effects are violet. Below is an example of the consistency of color.

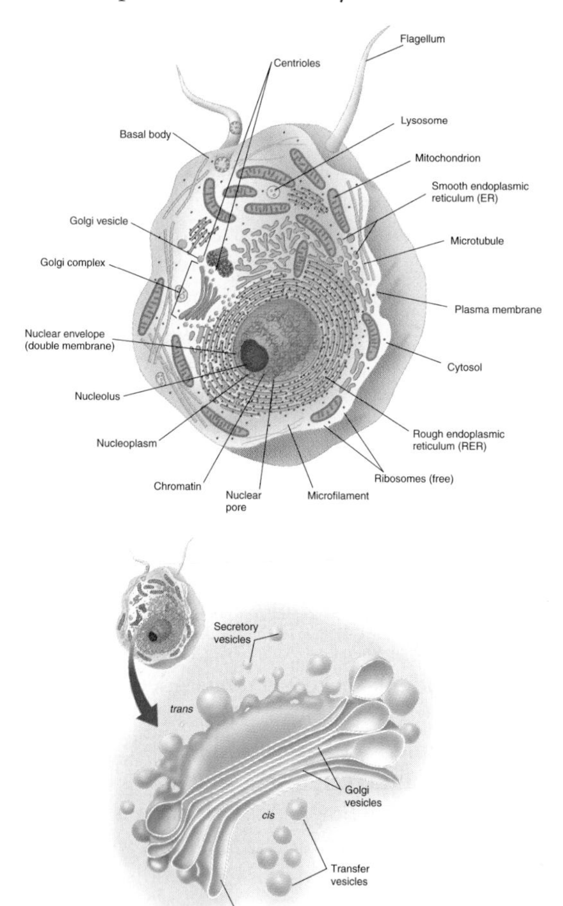

Key Concepts: The headings of sections are propositions that summarize key concepts, and these are listed at the beginning of each chapter.

Stories: Chapter 1 introduces the theme of science as story telling, and each chapter opens with a story to help put the subject into a context.

31 The Fungi

Armillaria mellea, the Honey Mushroom, is prized by many mushroom-hunters, though it makes some people ill. It is actually a plant pathogen, and its phosphorescent mycelium glows in the dark.

Key Concepts

31.1 The kingdom Fungi is monophyletic and defined by a few characteristics.
31.2 Most fungi form mycelia, made of long, thin hyphae that grow from a spore.
31.3 Basidiomycetes include most of the common mushrooms.
31.4 Ascomycetes develop spores inside sacs.
31.5 Some deuteromycetes are harmful to humans, while others are beneficial.
31.6 Zygomycetes are molds that reproduce through zygospores.
31.7 Many fungi are involved in mutualistic relationships with other organisms.

Clean a pound of fresh mushrooms, remove the stems, and cut the caps into thick slices. Saute them in 2 tbsp. of butter until they are barely tender. Then transfer them to a well-buttered baking dish, arranging them in neat, overlapping rows. In a saucepan, combine a cup of sour cream, 2 tbsp. of flour, salt and pepper to taste. Simmer the mixture until heated, but do not let it boil. Spoon the sauce over the mushrooms. Top with 1/2 cup grated cheddar cheese, 1/2 cup grated monterey jack cheese, and a dash of cayenne pepper. Bake for 15 minutes at 350°. Serves 4.

Many mushrooms make wonderful additions to our cuisine, but this is only one context in which fungi enter our lives. Wise mushroom hunters know how to distinguish the delicious ones from their close relatives that can kill painfully.

Similarly, some molds lend delightful tastes to cheeses, but others grow on our food and ruin it. Some fungi produce skin diseases, and others are terrible plant pests that can ruin crops. Fungi are well worth knowing about. Sometimes that knowledge literally makes the difference between life and death. Although the fungal kingdom embraces fewer than 80,000 named species, fungi are important both ecologically and in human affairs. We eat them and they eat us. They also eat (decompose) much of the detritus of our ecosystems. The study of fungi is **mycology**, and it practitioners are mycologists. Many amateur mycologists delight in the beauty and variety of mushrooms and their relatives, the challenge of identifying them and understanding their life histories, and the pleasant nuances some of them can add to our food. ■

650

Boxed Readings: Three types of boxed readings are used as a kind of extended footnote. *Concepts* summarize ideas or introduce some necessary physical and chemical concepts. *Methods* discuss experimental methods. *Sidebars* tell interesting stories that are too good to pass up but don't fit into the text.

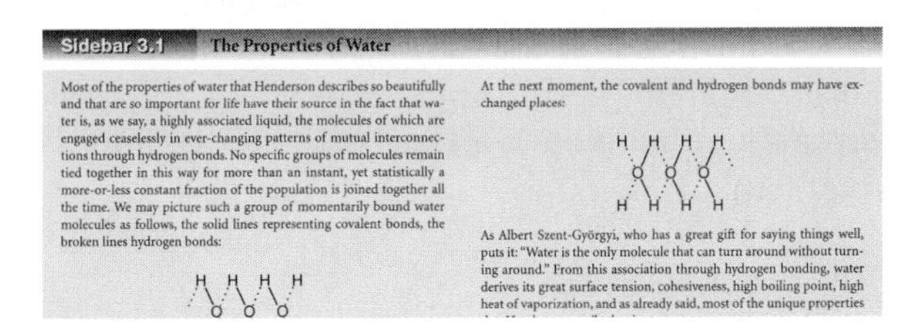

Sidebar 3.1 The Properties of Water

Most of the properties of water that Henderson describes so beautifully and that are so important for life have their source in the fact that water is, as we say, a highly associated liquid, the molecules of which are engaged ceaselessly in ever-changing patterns of mutual interconnections through hydrogen bonds. No specific groups of molecules remain tied together in this way for more than an instant, yet statistically a more-or-less constant fraction of the population is joined together all the time. We may picture such a group of momentarily bound water molecules as follows, the solid lines representing covalent bonds, the broken lines hydrogen bonds:

At the next moment, the covalent and hydrogen bonds may have exchanged places:

As Albert Szent-Györgyi, who has a great gift for saying things well, puts it: "Water is the only molecule that can turn around without turning around." From this association through hydrogen bonding, water derives its great surface tension, cohesiveness, high boiling point, high heat of vaporization, and as already said, most of the unique properties

Exercises: Students can test their understanding of concepts by working through exercises that emphasize the skill of problem-solving. These exercises are interspersed in each chapter and take students beyond memorization by applying concepts to new situations.

Internal Index: To provide more information about a topic, perhaps as a reminder, an Internal Index running through the book gives references to more extensive explanations of concepts. In general I have tried to present only the information needed to grasp a point without burying the idea in unnecessary information.

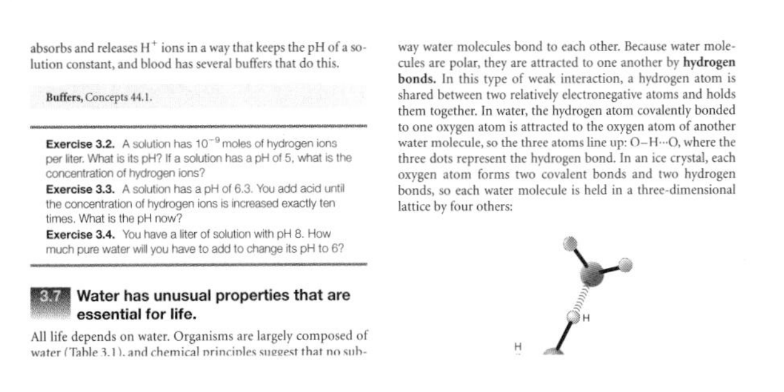

absorbs and releases H^+ ions in a way that keeps the pH of a solution constant, and blood has several buffers that do this.

Buffers, Concepts 44.1.

Exercise 3.2. A solution has 10^{-9} moles of hydrogen ions per liter. What is its pH? If a solution has a pH of 5, what is the concentration of hydrogen ions?
Exercise 3.3. A solution has a pH of 6.3. You add acid until the concentration of hydrogen ions is increased exactly ten times. What is the pH now?
Exercise 3.4. You have a liter of solution with pH 8. How much pure water will you have to add to change its pH to 6?

3.7 Water has unusual properties that are essential for life.

All life depends on water. Organisms are largely composed of water (Table 3.1), and chemical principles suggest that no sub-

way water molecules bond to each other. Because water molecules are polar, they are attracted to one another by **hydrogen bonds.** In this type of weak interaction, a hydrogen atom is shared between two relatively electronegative atoms and holds them together. In water, the hydrogen atom covalently bonded to one oxygen atom is attracted to the oxygen atom of another water molecule, so the three atoms line up: O–H···O, where the three dots represent the hydrogen bond. In an ice crystal, each oxygen atom forms two covalent bonds and two hydrogen bonds, so each water molecule is held in a three-dimensional lattice by four others:

Codas: I think of the chapters as tapestries of themes, analogous to pieces of music, so it is appropriate to conclude each chapter with a short coda that generally relates the subject to the major themes of the book.

Summary: Each chapter ends with summary statements that briefly recapitulate the principal ideas.

End-of-Chapter Review Questions: Students can test their understanding by answering these questions, almost all written by Gail Patt of Boston University. To properly answer the True-False questions, students must change a false statement to a true one, so they cannot simply guess T or F. The five concept questions in each chapter require students to state ideas in their own words. The answers to these questions are found in Appendix B.

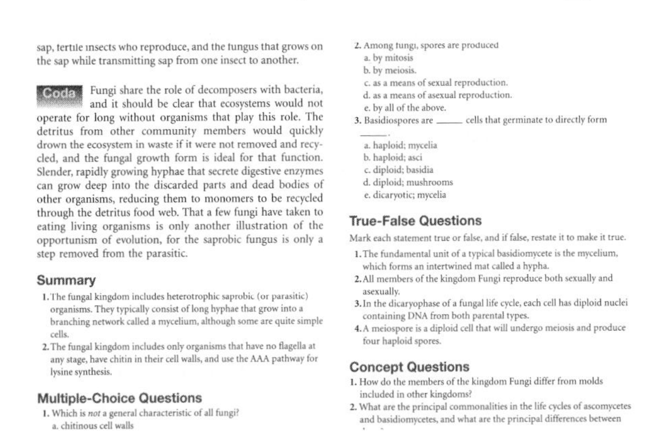

sap, fertile insects who reproduce, and the fungus that grows on the sap while transmitting sap from one insect to another.

Coda Fungi share the role of decomposers with bacteria, and it should be clear that ecosystems would not operate for long without organisms that play this role. The detritus from other community members would quickly drown the ecosystem in waste if it were not removed and recycled, and the fungal growth form is ideal for that function. Slender, rapidly growing hyphae that secrete digestive enzymes can grow deep into the discarded parts and dead bodies of other organisms, reducing them to monomers to be recycled through the detritus food web. That a few fungi have taken to eating living organisms is only another illustration of the opportunism of evolution, for the saprobic fungus is only a step removed from the parasitic.

Summary

1. The fungal kingdom includes heterotrophic saprobic (or parasitic) organisms. They typically consist of long hyphae that grow into a branching network called a mycelium, although some are quite simple cells.
2. The fungal kingdom includes only organisms that have no flagella at any stage, have chitin in their cell walls, and use the AAA pathway for lysine synthesis.

Multiple-Choice Questions

1. Which is *not* a general characteristic of all fungi?
   a. chitinous cell walls

2. Among fungi, spores are produced
   a. by mitosis
   b. by meiosis.
   c. as a means of sexual reproduction.
   d. as a means of asexual reproduction.
   e. by all of the above.
3. Basidiospores are _____ cells that germinate to directly form _____.
   a. haploid; mycelia
   b. haploid; asci
   c. diploid; basidia
   d. diploid; mushrooms
   e. dicaryotic; mycelia

True-False Questions

Mark each statement true or false, and if false, restate it to make it true.

1. The fundamental unit of a typical basidiomycete is the mycelium, which forms an intertwined mat called a hypha.
2. All members of the kingdom Fungi reproduce both sexually and asexually.
3. In the dicaryophase of a fungal life cycle, each cell has diploid nuclei containing DNA from both parental types.
4. A meiospore is a diploid cell that will undergo meiosis and produce four haploid spores.

Concept Questions

1. How do the members of the kingdom Fungi differ from molds included in other kingdoms?
2. What are the principal commonalities in the life cycles of ascomycetes and basidiomycetes, and what are the principal differences between

Supplements to this textbook

Student Study Guide

The *Student Study Guide* contains an average of 65 page-referenced questions per chapter, written by Iain Campbell, University of Pittsburgh. Also included is a section entitled *The Language of Biology,* written by the author of this textbook. It simplifies difficult and unfamiliar scientific terminology through the use of fill-in-the-blank questions. The textbook author also provides an overview of each chapter of *Biology.*

Essential Study Partner CD-ROM

This CD-ROM is an interactive student study tool packed with over 120 animations and more than 200 learning activities. From quizzes to interactive diagrams, your students will find that there has never been a more exciting way to study biology. A self-quizzing feature allows students to check their knowledge of a topic before moving on to a new module. Additional unit exams give students the opportunity to review an entire subject area. The quizzes and unit exams hyperlink students back to tutorial sections so they can easily review coverage for a more complete understanding. This CD-ROM tutorial supports and enhances the material presented in *Biology* and is offered free with the textbook.

Instructor's Manual with Test Item File

The *Instructor's Manual* prepared by Carla Barnwell and Melissa Michael, both of the University of Illinois, Urbana-Champaign, includes Chapter Overviews/Introductions, Extended Chapter/Lecture Outlines with figure references, Teaching Strategies, Discussion Activities, and Demonstration Activities.

The *Test Item File* by Donald G. Ruch and Ken Badger, both of Ball State University, is available in soft cover and on disk (Microtest) and includes over 4000 test questions in the forms of matching, multiple choice, true/false, and discussion.

Computerized Testing Software

A computerized test generator, Microtest, is available free to qualified adopters. The program is available in Windows and Macintosh formats and enables instructors to generate tests from questions in the test item file.

Transparencies

A boxed set of 300 full color transparency acetates feature useful images for classroom presentations.

Slide Set

A boxed set of 100 electron and photomicrograph slides of images found in the text are available to qualified adopters.

Visual Resource Library

The *Visual Resource Library* is a CD-ROM that contains hundreds of biological images from *Biology*. The CD-ROM contains an easy-to-use program to quickly view images and easily import them into PowerPoint to create multimedia presentations or use the already prepared PowerPoint presentations.

Text-Specific Web Site

http://www.mhhe.com/biosci/genbio/guttman/

The *Biology* web site allows students and instructors from all over the world to communicate. By visiting this site, students can access study quizzes, explore links to other relevant biology sites, and catch up on current information.

McGraw-Hill Learning Architecture

The McGraw-Hill Learning Architecture is a browser-based product that is a solution for delivering educational content over networked environments. The Learning Architecture connects students with each other as well as their instructor in an integrated environment. In addition to providing support and collaboration tools for users, such as built-in messaging and discussion lists, the Learning Architecture also manages all students on the server as well as the course material assigned to them. The benefits of this system to both instructors and students are tremendous.

Recommended Laboratory Manuals

***Biological Investigations: Form, Function, Diversity, and Process,* 5th edition,** by Warren D. Dolphin, introduces students to scientific investigation. This manual contains thirty-four labs that emphasize the scientific method. Students are introduced to critical thinking, and each lab contains all of the information needed to perform experiments. Students are also given alternative ways of exploring science via technology, such as Internet activities and CD-ROM products. (ISBN 0-697-36049-0)

Biology Laboratory Manual, 5th edition, is a full-color lab manual written by Darrell S. Vodopich of Baylor University and Randy Moore of the University of Louisville. It contains approximately fifty laboratory exercises. The lab manual is customizable by chapter in full color and is accompanied by a *Laboratory Resource Guide.* (ISBN 0-697-35356-7)

Additional Supplements and Technology Products From WCB/McGraw-Hill

How to Study Science, 2nd Edition

by Fred Drewes, Suffolk County Community College

This excellent workbook offers students helpful suggestions for meeting the considerable challenges of a college science course. It offers tips on how to take notes, how to get the most out of laboratories, and how to overcome science anxiety. The book's unique design helps students develop critical thinking skills while facilitating careful note taking. (ISBN 0-697-15905-1)

A Life Science Living Lexicon

by William N. Marchuk, Red Deer College

This portable, inexpensive reference helps introductory-level students quickly master the vocabulary of the life sciences. Not a dictionary, it carefully explains the rules of word construction and derivation, in addition to giving complete definitions of all important terms. (ISBN 0-697-12133-X)

Biology Study Cards

by Kent Van De Graaff, R. Ward Rhees, and Christopher H. Creek, Brigham Young University

This boxed set of 300 two-sided study cards provides a quick yet thorough visual synopsis of all key biological terms and concepts in the general biology curriculum. Each card features a masterful illustration, pronunciation guide, definition, and description in context. (ISBN 0-697-03069-5)

Critical Thinking Case Study Workbook

by Robert Allen

This ancillary includes 34 critical thinking case studies designed to immerse students in the "process of science" and challenge them to solve problems in the same way biologists do. The case studies are divided into three levels of difficulty (introductory, intermediate, and advanced) to afford instructors greater choice and flexibility. An answer key accompanies this workbook. (ISBN 0-697-34250-6)

The AIDS Booklet

by Frank D. Cox

This booklet describes how AIDS and related diseases are commonly spread so that readers can protect themselves and their friends against this debilitating and deadly disease. This booklet is updated quarterly to give readers the most current information.

Basic Chemistry for Biology

by Carolyn Chapman

This workbook is a self-paced introduction or review of the basic concepts of chemistry that are most useful in other areas of science. This pocket-sized tutorial covers organic and inorganic chemistry. (ISBN 0-697-36087-3)

The Internet Primer

by Fritz J. Erickson & John A. Vonk

This short, concise primer shows students and instructors how to access and use the Internet. The guide provides enough information to get started by describing the most critical elements of using the Internet.

The Dynamic Human CD-ROM Version 2.0

This guide to anatomy and physiology interactively illustrates the complex relationships between anatomical structures and their functions in the human body. Realistic, three-dimensional visuals are the premier feature of this exciting learning tool. The program covers each body system, demonstrating to the viewer the anatomy, physiology, histology, and clinical applications of each system. (ISBN 0-697-38935-9)

Virtual Physiology Laboratory CD-ROM

This CD-ROM features ten simulations of the most common and important animal-based experiments ordinarily performed in introductory lab courses. The program contains video, audio, and text to clarify complex physiological functions. (ISBN 0-697-37994-9)

Life Science Animations 3D Videotape

Forty-two animations of key biological processes are available on a videotape. The animations bring visual movement to biological processes that are difficult to understand on the text page. The videotape is narrated and animated in vibrant color with dynamic three-dimensional graphics. (ISBN 0-07-290652-9)

Life Science Living Lexicon CD-ROM

by William N. Marchuk, Red Deer College

A Life Science Living Lexicon CD-ROM contains a comprehensive collection of life science terms, including definitions of their roots, prefixes, and suffixes as well as audio pronunciations and illustrations. The Lexicon is student-interactive, providing quizzing and notetaking capabilities. It contains 4,500 terms, which can be broken down for study into the following categories: anatomy and physiology, botany, cell and molecular biology, genetics, ecology and evolution, and zoology. (ISBN 0-697-37993-0)

References for the Preface

Guttman, B. S. 1966. A resolution of Rosen's paradox for self-reproducing automata. Bull. Math. Biophys. **28**:191-194.

Horowitz, N. H. 1959. On defining 'life.' *The Origin of Life on Earth,* pp. 106-107. London: Pergamon Press.

von Neumann, J. 1951. The general and logical theory of automata. *Cerebral Mechanisms in Behavior.* New York: John Wiley & Sons.

Acknowledgments

The best thing that could have happened to this project was falling into the hands of the wonderful editorial people at WCB/McGraw-Hill. Kevin Kane believed in the project initially, took it on, and put it into the capable hands of Connie Haakinson, who has been the world's finest developmental editor. Connie has been instrumental in shaping this book, in both large and small ways, by marshalling the consultation of about a hundred expert reviewers, by weighing and discussing their ideas together with hers and mine, by critically reading text and figures, and by editing and coordinating thousands of details through innumerable revisions. Meanwhile, she has held my hand through the Internet and phone lines, laughed and cried with me, sworn and yelled with me, put up with my neuroses, and has become a friend and colleague. Michael Lange has wisely encouraged, criticized, and guided us around many potential pitfalls. Matt T. Lee helped us reshape the first draft manuscript by analyzing the content and suggesting ways to improve clarity. Kennie Harris, our copyeditor, pitted her red pen against my word-processor; as we fought over semicolons and writing style, her intelligent reading and relentless questioning uncovered inconsistencies and vagaries, and forced me to explain better, thus making an immeasurably tighter, more coherent book. Connie Mueller enthusiastically searched for photos that would create a beautiful, instructive book. Carlyn Iverson oversaw the whole art project, farmed out material to several expert initial illustrators, and then helped guide the final products through the final illustrators at Page Two. I would also like to express my gratitude to the production staff: Peggy Selle, project manager; Wayne Harms, designer; Jodi Banowetz, art coordinator; and Lori Hancock, photo coordinator.

I am deeply indebted to all the reviewers listed below. They have provided uncountable corrections and suggestions, sometimes providing just the shot of support and adrenaline needed to carry on, sometimes sending me back to the literature to check on facts. Some have provided excellent ideas, which often gave me the courage to do what the voice in the back of my head kept urging me to do. In addition to these consultants, I have been blessed with the patient and generous help of my friends and colleagues at Evergreen, especially Clyde Barlow, Michael Beug, Richard Cellarius, Jeff Kelly, Betty Kutter, Jack Longino, Nalini Nadkarni, Jim Neitzel, Janet Ott, John Perkins, Jude van Buren and Al Wiedemann. As usual, any remaining errors are my responsibility. I should also like to thank the faculty of the Kobe University of Commerce in Kobe, Japan, for their hospitality during a faculty exchange visit in the spring and summer of 1997, which gave me the leisure and freedom to do final editing and writing.

Barbara J. Abraham
Hampton University

Kenneth W. Andersen
Gannon University

J. David Archibald
San Diego State University

Steve Baker
Oxford College of Emory University

Ruth E. Beattie
University of Kentucky

Wayne M. Becker
University of Wisconsin—Madison

Arlene G. Billock
University of Southwestern Louisiana

Andrew R. Blaustein
Oregon State University

George M. Bleekman, Jr.
American River College

Joy Bonnema
Clarke College

Gary Borisy
University of Wisconsin—Madison

Dorothy M. Brecheisen
University of Northern Iowa

Donald P. Briskin
University of Illinois

Ruby L. Broadway
Dillard University

Pamela J. Bryer
Bowdoin College

Emily A. Buchholtz
Wellesley College

Linda K. Butler
University of Texas at Austin

Byron K. Butler
Yale University

Carolyn Chapman
Suffock County Community College

Joe Coelho
Western Illinois University

William Cohen
University of Kentucky

James J. Copi
Madonna University

David J. Cotter
Georgia College & State University

George W. Cox
San Diego State University

Charles Creutz
University of Toledo

Kenneth J. Curry
University of Southern Mississippi

David B. Czarnecki
Loras College

Garry Davies
University of Alaska Anchorage

Jean DeSaix
University of North Carolina at Chapel Hill

Randy DiDomenico
University of Colorado at Boulder

Ronald Dimock
Wake Forest University

Sharon Eversman
Montana State University

Bruce Felgenhauer
University of Southwestern Louisiana

Stephen George
Amherst College

Andrew Goliszek
North Carolina A&T State University

Sheldon R. Gordon
Oakland University

Glenn A. Gorelick
Citrus College

Joseph L. Graves, Jr.
Arizona State University West

Mary H. Gray
Purdue University

Jean G. Heitz
University of Wisconsin—Madison

Stanton Hoegerman
The College of William & Mary

James Philip Holland
Indiana University

Terry L. Hufford
University of Texas at San Antonio

Craig T. Jordan
University of Texas at San Antonio

J. Eric Juterbock
Ohio State University

Chris L. Kapicka
Northwest Nazarene College

Travis Knowles
Francis Marion University

Will Kopachik
Michigan State University

Charles Levy
Boston University

Lynn Lewis
Mary Washington College

Harvey B. Lillywhite
University of Florida

Ruby W. Littlepage
Camosun College

David Marcey
California Lutheran University

Tom McKinney
Mohave Community College

Michael Ray Meighan
University of California-Berkeley

Ralph R. Meyer
University of Cincinnati

Stephen A. Miller
College of the Ozarks

Herbert Monoson
Bradley University

Debra M. Moriarity
University of Alabama in Huntsville

James C. Munger
Boise State University

Murray Nabors
Colorado State University

Jane Noble-Harvey
University of Delaware

Ken Nuss
University of Northern Iowa

Laura J. Olsen
University of Michigan

Beryl Packer
Des Moines Area Community College

Glenn R. Parsons
University of Mississippi

Gail R. Patt
Boston University

Charles Paulson
Arizona International College

Barbara Pleasants
Iowa State University

James C. Pushnik
California State University—Chino

Arlyn E. Ristau
Hawkeye Community College

Rodney A. Rogers
Drake University

Donald G. Ruch
Ball State University

Mark Sanders
University of California—Davis

Diane Shakes
College of William & Mary

Thomas D. Sharkey
University of Wisconsin

Lisa A. Shimeld
Crafton Hills College

John Smarrelli
Loyola University

Kingsley R. Stern
California State University—Chico

Dennis Swanger
Eastern Oregon University

Robet H. Tamarin
University of Massachusetts Lowell

Roger E. Thibault
Bowling Green State University

Craig R. Tomlinson
U.S. BioGenix, Inc.

Dennis G. Trelka
Washington and Jefferson College

Judy Verbeke
University of Arizona

Alexander Werth
Hampden-Sydney College

Norman Williams
University of Iowa

Esther Wilson
University of Wisconsin—Parkside

Robert Winget
Brigham Young University Hawaii Campus

Vernon Wranosky
Colby Community College

To The Student

The purpose of this book is to provide you with methods and principles for understanding the fascinating, beautiful, marvelous biological world. More than anything else, I hope this book will change your perceptions of that world and deepen your enjoyment of it. There is a superstition afoot that scientific knowledge of a subject, especially quantitative knowledge, ruins one's aesthetic pleasure and direct experience. I think this is dead wrong. Some cynics would have us believe, for instance, that knowing about music or art in a technical sense interferes with the pleasure of experiencing them. My personal experience tells me that the opposite is true, that my experience of music and art is enriched by what little knowledge I have of them. Similarly, some people would have us believe that our enjoyment of nature is ruined by knowing something about geology or biology. Again, my personal experience belies this notion. My aesthetic pleasure in walking through a forest or watching birds in a marsh is heightened and enriched by my knowledge of biology and my sense of a deep ecological connection to my surroundings. In a plant's colors, I see the molecules of chlorophyll and anthocyanin; in a bird's flight, I see the interaction of actin and myosin and the effects of lift and drag. I know I share a common ancestry with the plants and animals around me, and I know that the molecules of my body are constantly being exchanged with them. I hope you will acquire some of this sense of the natural world.

Much of your life has been spent in the company of people called teachers. The society you have grown up in promotes the idea that these teachers should be able to teach you many ideas and skills.

If you buy this way of looking at the world, you will be handicapped for life. Many years of teaching have taught me that I can't do it. Not that I'm a poor teacher—I have a stack of evaluations by students telling me that I'm generally doing a good job. No, the trouble is that "to teach" is not an active verb. "To learn" is an active verb. You cannot learn through passive activities, the equivalent of someone opening your skull and pouring in the information. You only learn through personal action. Like anyone who teaches biology or other sciences in college, I know a lot about the subject, and essentially everything I know I taught myself. Oh, yes, like you, I have sat in many courses and have listened to many lectures, often excellent coherent lectures by fine instructors. But I did not really learn from those lectures, and neither will you.

I learned by reading carefully, by writing about my ideas, by thinking about them again and again, by solving problems requiring me to apply concepts, and by discussing ideas with others. And this is the primary way I try to teach at Evergreen. Yes, I do give lectures, and a good lecture can usually clarify points that the book left obscure; but I try to make them opportunities for dialogue, where students can ask questions derived from the books they should have read before class. Students learn primarily by writing about each topic, putting the concepts into their words; they learn by engaging in workshops where they discuss concepts with their peers and solve problems that require an application of principles; and they learn by themselves, in their own rooms, often in "study gangs" where they continue the dialogue and the problem-solving collaboratively.

To learn any organized subject, you must put the ideas into a coherent framework. This book provides that framework. You must work to understand it, to elaborate it for yourself, and to connect every new concept or fact to the framework, for only in this way will the information stay with you and become useful. In this way, you make the ideas meaningful to yourself. If you approach biology as a collection of largely unrelated facts, the ideas will quickly spill out of your mental space, and you might as well not have taken the course.

This book contains some exercises and questions. Do them. You will only learn by doing.

Problem Solving

The most important aspect of problem solving is understanding the physical situation. If you're not entirely sure what the problem refers to, make a simple sketch of the situation. You might be asked to think about the number of cells carrying water in a tree trunk, for instance, and it may help to draw an imaginary section of the trunk showing the cells in question with their dimensions, so you can determine the number of cells in a given area. Even the simplest sketch can clarify and guide your thinking.

The next most important technique in problem solving is called *dimensional analysis*. It takes various forms, all based on recognizing that the objects and processes of the world have dimensions, primarily length (L), mass (M), and time (T). Then an area is length squared (L^2) and a volume is length cubed (L^3). Velocity has dimensions such as miles/hr or cm/sec, or L/T. Density is the mass in a given volume and therefore has the dimensions of M/L^3. The most important units and conversion factors are given in the inside front cover. In thinking about physical situations, you must always include the dimensions of each unit and treat them just like numbers during all algebraic manipulations and calculations. Suppose, for instance, we know that 100 sheep are grazing in a field and we want to know the number of legs. The essential information you need is the equation 1 sheep = 4 legs. Suppose you divide both sides of "1 sheep = 4 legs" by 1 sheep:

$$\frac{1 \text{ sheep}}{1 \text{ sheep}} = \frac{4 \text{ legs}}{1 \text{ sheep}} = 1$$

We can multiply the quantity "100 sheep" by any number that is equal to 1, since multiplying by 1 doesn't change its value, so let's multiply it by the fraction 4 legs/1 sheep:

$$100 \cancel{\text{sheep}} \times \frac{4 \text{ legs}}{1 \cancel{\text{sheep}}} = 400 \text{ legs}$$

where "sheep" in the numerator cancels "sheep" in the denominator to give the desired answer. On the other hand, if you thought carelessly about the units, you might have multiplied

$$100 \text{ sheep} \times 1 \text{ sheep}/4 \text{ legs} = 25 \text{ sheep}^2/\text{leg}$$

Since the dimension "sheep^2/leg" doesn't have any obvious physical meaning, this result should tell you that you've made a mistake and have calculated the wrong quantity.

By multiplying and dividing dimensions, you can easily convert from one unit to another. For instance, to calculate the number of seconds in a year:

$$365 \cancel{\text{day}} \times 24 \cancel{\text{hr}}/\cancel{\text{day}} \times 60 \cancel{\text{min}}/\cancel{\text{hr}} \times 60 \text{ sec}/\cancel{\text{min}} = 3.15 \times 10^7 \text{ sec}$$

where you cancel units all along until you are left with only seconds. Notice that to make many calculations, you must be able to use powers of 10 correctly and convert correctly between metric units, as explained in Appendix A.

I hope all this talk about calculations and problem-solving doesn't detract from the basic message of this book. On the contrary, the ability to calculate sizes and rates should deepen your understanding of the biological world. Finally, I hope you will take a minute now and then to step off the academic treadmill that drives us every year, to reflect on what you are learning and take some pleasure in it. Good luck in your studies.

Burton Guttman
Olympia, WA
June, 1998

About The Author

Burton S. Guttman studied math, philosophy, physics, and zoology at the University of Minnesota, then spent a year as a Woodrow Wilson Fellow at Johns Hopkins, where he intended to study embryology. But the Institute of Molecular Biology at the University of Oregon called, and he finished his Ph.D. work there in 1963. After two years of postdoctoral work at the California Institute of Technology, he taught at the University of Kentucky for six years, developing his interest in educational pedagogy and the explanation of scientific research to students and non-scientists. He has been a member of the faculty of The Evergreen State College since 1972, where he teaches biology and is able to explore many other subjects in interdisciplinary programs of study. He continues to collaborate with Elizabeth Kutter, James Neitzel, and many undergraduate students on the molecular genetics of phage T4.

Evergreen was founded in 1968 to explore new directions in liberal arts education. Most of the work involves students and faculty members engaging in collaborative explorations of ideas by bringing several disciplines to bear on a central theme. The College emphasizes cooperative work rather than competition and makes students responsible for their own learning; lectures are deemphasized in favor of discussion and problem-solving through seminars, workshops, and writing.

Introduction

1

Key Concepts

1.1 Scientific exploration and story-telling satisfy our curiosity about the natural world.
1.2 Organisms have evolved as one stage in the continuing evolution of the universe.
1.3 Natural sciences develop a systematic understanding of the world based on observation and experimentation.
1.4 It is important to distinguish science from pseudoscience and nonscience.
1.5 Our study of biology will emphasize four major themes.
1.6 Biologists may use more types of explanations than those used in other sciences.
1.7 Complex structures are understood by referring to both their components and their emergent properties.
1.8 What is life?
1.9 Why should anyone study biology?

Chimpanzees in their natural habitat.

In silence broken only by occasional soft grunts, a band of chimpanzees moves through the underbrush of a central African forest. They head toward a stand of msulula trees, for the fruit is at the peak of ripeness and it is one of their favorites. Swinging agilely up into the trees, they virtually disappear behind the large leaves, only their hairy brown arms emerging to pluck the fruits. The chimps are silent now, feeding contentedly. Suddenly there is a commotion on a high branch. An older female, going after a batch of fruit, has come across a bird's nest, and she excitedly announces her discovery as she pounces on a newly hatched nestling and begins to eat it. Her cries attract a young male, who happens to be her son, and in the next few minutes the two of them consume the still-unhatched eggs in the nest, balancing their vegetable diet with more protein-rich food.

As the chimps continue to feed, another small group—some of them young females carrying their infants—approaches the trees, announcing their impending arrival with a series of pant-hoots, crying "Hoo, waa, hoo, waa" as they scamper along. Now as they reach the msulula trees, several males already feeding there take up the "hoo, waa" cry, acknowledging the new arrivals. The newcomers easily scale the trees and settle down to feed.

Meanwhile, two old females and two of their adolescent offspring climb down the trees and move off. Satisfied with their meal of fruit and a few caterpillars, they settle in a grassy patch to rest. One of the younger chimps makes a playful gesture toward the other, and suddenly they are chasing each other, rolling on the ground together, and laughing heartily as they play. Occasionally they pause, face each other, and egg each other on with a characteristic play-face, exposing their lower teeth; then they are off, chasing and laughing again.

Not far away, two older males are feeding in a different way. Huddled next to a termite nest, they break twigs from nearby bushes, stripping off the leaves to make suitable tools. Again and again, they dip these twigs into holes in the termite nest, wait a few seconds, then withdraw the twigs and eat the termites clinging to them. Their termite fishing is practiced and intelligent. If the end of a twig becomes bent, they bite it off. If a twig becomes unusable, they discard it in favor of a fresh one.

Anyone who watches chimps for a long time (as Jane Goodall has done) will see them engage in a marvelous variety of activities. While they live mostly on fruits and leaves, they supplement their diet with grubs and insects; sometimes they will kill an animal, even a fairly large one such as a baboon, a piglet, or a young bushbuck. Occasionally chimps will fight, and their fights may be severe enough to permanently injure or even kill one party. At intervals when a female is receptive, she will copulate with males, perhaps with several of them one after the other, and after a gestation period averaging 231 days, she will give birth. She will cuddle and carry her infant, nurse it, and gradually help it learn to secure its own food, protect itself against the dangers of the forest and threats from other chimps, and fit into chimpanzee society.

Chimpanzees exemplify, in the most obvious ways, what it is to be alive. But their activities not only sustain their species. They continually affect all the other living organisms around them (Figure 1.1). By eating msulula fruits and defecating the seeds, they help these trees reproduce and spread. By eating grubs and caterpillars, ants and termites, they reduce the populations of several kinds of insects. By ripping twigs and leaves off trees and bushes, they affect the growth of those plants, and the minute wounds they make are entry points for viruses and bacteria that may cause plant diseases. By favoring some kinds of plants over others for their food, they have subtle effects on the diversity of plants in the forest. A bird whose nest they have raided will need to lay another clutch of eggs, but it may be too late in that bird's breeding season to do so, and so the bird population is pushed a little closer to a critically small number. The urine and feces chimps void into the forest soil will become food for some bacteria and molds and inhibit the growth of others. And all this time, they are taking in oxygen, which is constantly being generated by the vegetation around them, and producing carbon dioxide, which is constantly being used by that vegetation for its own growth. ■

Figure 1.1
Chimpanzees live in a world of other organisms, and together they form a community characterized by complicated ecological relationships.

1.1 Scientific exploration and story-telling satisfy our curiosity about the natural world.

We tell this story about chimpanzees because humans are a story-telling species. We treasure our most gifted storytellers, for they take us to places we have never seen, worlds we may never visit, and into the minds and hearts of our fellow humans. Scientists are also storytellers. Their stories are about great adventures and what the venturers found in the new worlds they discovered: "When Charles Darwin was a young man, he had the opportunity to take a voyage around the world on a ship. . . ." "When the French wine industry was plagued with diseased wines, it called on Louis Pasteur to help. . . ." "In the 1930s, when Max Delbrück was working in Niels Bohr's lab, he got interested in basic questions of biology and. . . ." "Last week I drove out into the desert and noticed some lizards engaged in a really strange behavior, and I started to wonder what. . . ." Stories of science begin this way. Then they take on a different cast: "It turns out that what we once thought was an irritation caused by nervous stimulation is actually an infection by a new type of bacterium that we call. . . ." "Our hypothesis is that the cell membrane contains a protein that recognizes interleukin-2 and opens up to let calcium ions leak into the cytoplasm. . . ." "There are two species of grass here, adapted to soils of slightly different pH, and they serve as food for two different species of beetles because. . . ."

People explore and tell stories out of a sense of wonder (Figure 1.2). On the one hand, we experience pure *aesthetic wonder* and the joy of encountering the remarkable and the beautiful. We see birds with unbelievable colors streaking through the bush, insects of bizarre shapes, flowers whose colors and forms can entrance us for hours—and we naturally exclaim over these marvels, savor the experience, and try to

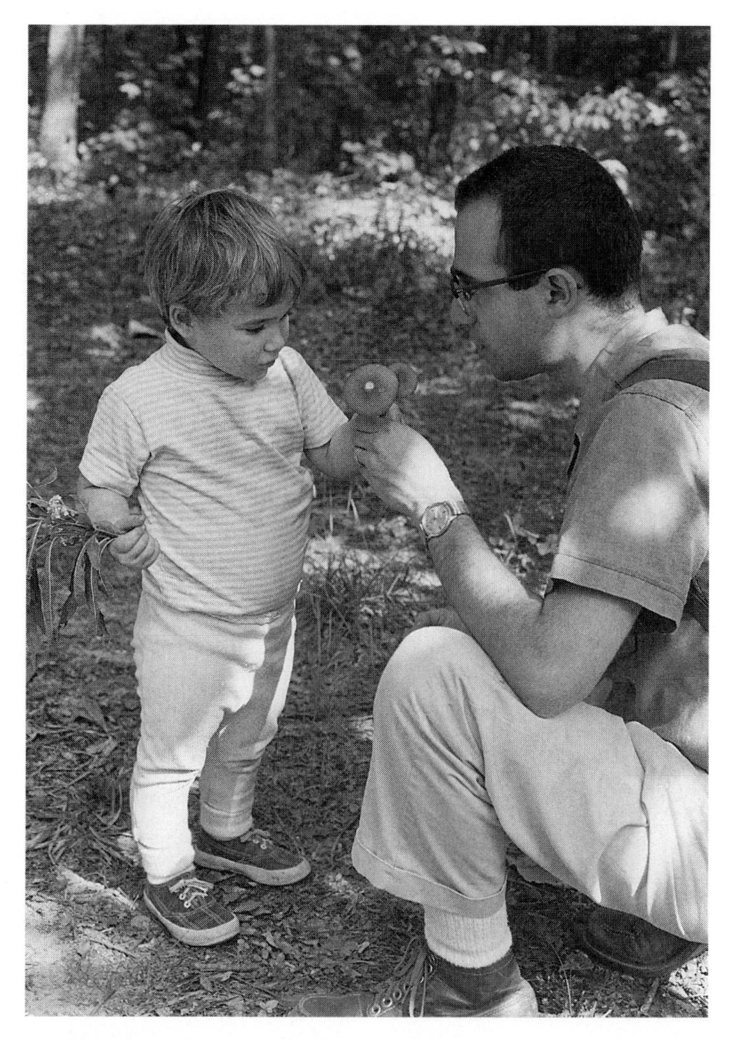

Figure 1.2
The human sense of wonder is natural in children. How can we preserve this sense and nurture it throughout a lifetime?

prolong it. This is pure aesthetic pleasure. The same experience of wonder and pleasure comes from fine music and art. And because the natural world presents unlimited subjects to admire and treasure, people who have experienced these marvels are extremely worried about the devastation of the environment and the possibility that many of its inhabitants could become extinct.

The natural world also inspires *scientific wonder,* or curiosity, which makes us ask questions in an effort to understand. While marveling at the delicate blues and greens of a bird's feathers, we ask, "What creates these colors? How are the feathers built, and how do they develop in just this shape? What function do the colors have in the lives of the birds? Are the colors protective, or do they send signals to prospective mates? How often are the feathers molted, and are the birds impaired during a molt?" Experiencing our own bodies and seeing ourselves as part of a larger community of life leads to more questions: "Where did I come from? How can a tiny egg grow into an adult with all the right parts in the right places? Why does something occasionally go wrong and produce a baby with defects? What should I eat to stay healthy? How will the chemicals in my food or in the air affect me? Why do I have blue eyes when my parents both have brown eyes? What causes the pain in my back? How can I see and feel and hear? How can all these trees make leaves with such distinctive shapes? Why do these insects live only on this plant and no other? Why do birds sing? How can they migrate south every fall and come back to the same place in the spring? How. . . ? Why. . . ?" Aesthetic wonder is content to experience. Scientific wonder demands explanations, and leads to endless questions.

The sciences are one result of human curiosity and exploration. They are part of the human attempt to understand this complex, beautiful, sometimes frightening world—a world we must understand if we are to live comfortably and successfully. Biology is the science that observes, explores, and attempts to explain the world's living organisms. Thus, to begin the story of biology that will occupy the rest of this book, we must first go back and outline the story of life's origins.

1.2 Organisms have evolved as one stage in the continuing evolution of the universe.

We understand organisms as structures that change in time, and we had best begin our story at the beginning of time itself, at the origin of the universe. According to the best physical evidence, the universe originated at least 13 billion years ago (Figure 1.3), with all matter concentrated at a single spot that was hot and dense beyond comprehension. This mass exploded, in the so-called "Big Bang," and began to expand rapidly and to cool. Within a few seconds the basic particles of ordinary matter—protons, neutrons, and electrons—had formed. Eventually these particles combined to create the smallest atoms, those of hydrogen and helium.

The universe has been evolving from the beginning of time. Evolution does not imply progress, only persistent change, and *cosmic evolution* has meant the formation of new stars and different types of stars. During the first few million years of the universe's existence, the first stars formed as clouds of hydrogen and helium gas condensed under their own gravitational attraction (Figure 1.4). As the stars grew more dense, their energy was converted into heat, until they became hot enough to glow and to radiate vast amounts of energy. Inside these stars, as in gigantic pressure cookers, hydrogen and helium nuclei combined to make heavier and heavier elements through a series of nuclear reactions. It is especially interesting that one series of nuclear reactions forms carbon, nitrogen, and oxygen atoms, the chief components of organisms. These stars churned out heavy elements for many millions of years until they ended their existence by blowing up, as novas or supernovas, spewing their contents out into space where they formed new stars. The resulting atoms may have resided in several new generations of stars as astronomical objects were born, lasted for millions or billions of years, and then were destroyed and their matter given to

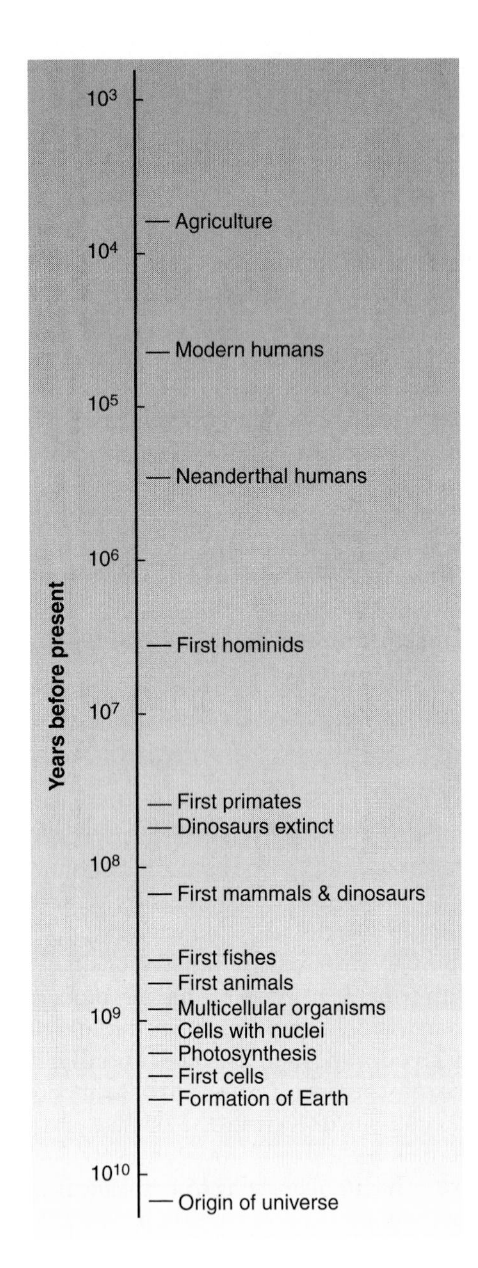

Figure 1.3

This time scale shows key biological events in the history of the universe and the earth. Notice that the scale is logarithmic, so more recent parts *(top)* are expanded to show details.

Figure 1.4

The great nebula in Orion is a mass of stars and interstellar gas from which new stars appear to be forming now, just as the first stars formed billions of years ago.

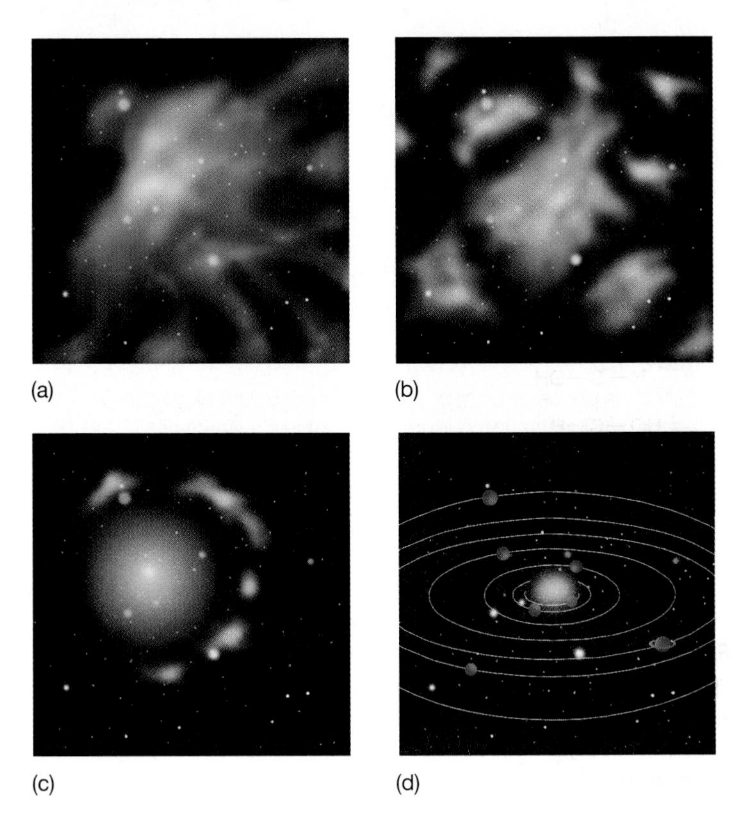

Figure 1.5

As a solar system begins to form, *(a)* a gas cloud collapses into fragments, which each become proto-stars. *(b)* A cloud condenses under its own gravitational force, and its central mass eventually begins to radiate energy and becomes a star. *(c)* Meanwhile, a considerable amount of material remains in the periphery; radiation from the star blows the surrounding gases into space, while the heavier remainder condenses into planets. *(d)* Planets like the earth that are made largely of heavy elements eventually form molten metallic cores with cool rocky plates floating on their surfaces.

something else. Some of them are the atoms of this paper and ink and those you are using to read and turn these pages.

Our existence depends on an accident. When our own sun formed, much of the heavy matter was left outside the sun where it condensed into planets, some with metal cores of iron and nickel and rocky surfaces with substantial amounts of water (Figure 1.5). Our earth was one of those planets, and it happens to lie in a narrow temperature zone where some water could remain liquid and an atmosphere could form. Here the conditions were right for life to begin.

Only large, complex, carbon-based molecules can form the highly ordered structures of organisms. These molecules—

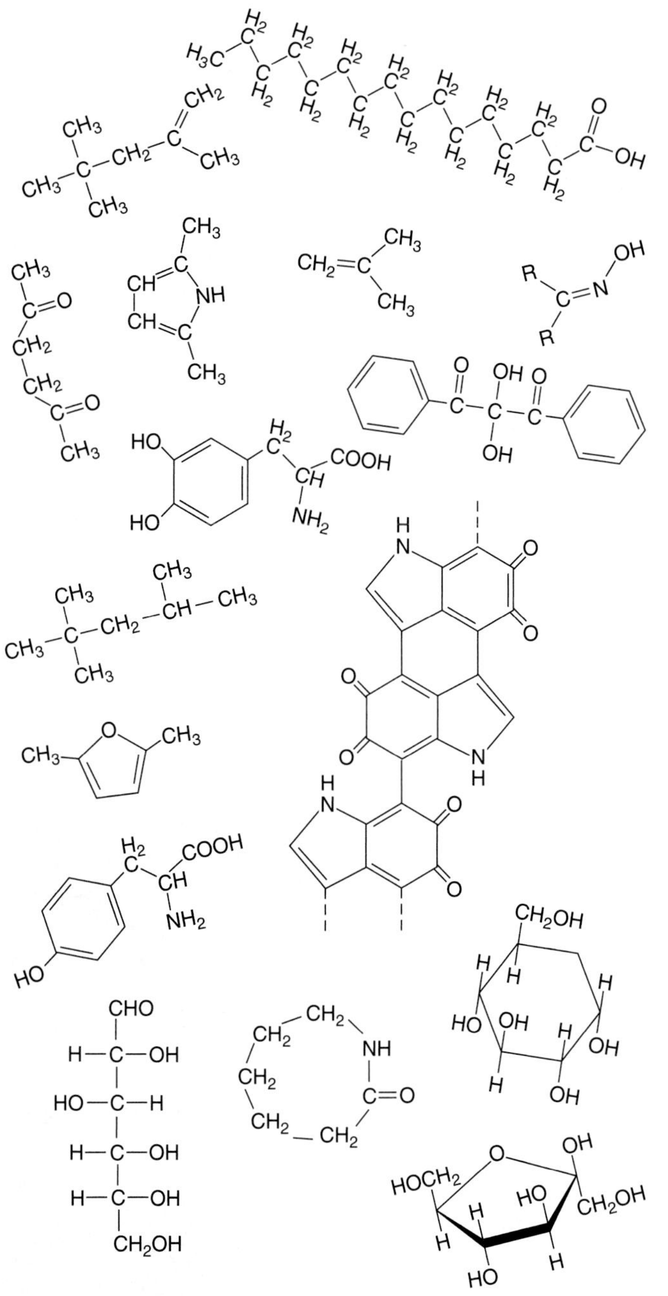

Figure 1.6

This illustration depicts a potpourri of carbon-based molecules, the kinds of molecules that constitute living organisms. Notice that they are made mostly of carbon (C) and hydrogen (H) atoms, along with some oxygen (O), nitrogen (N), phosphorus (P), and sulfur (S). The simple geometric shapes of some of the molecules are abbreviations for rings made of carbon atoms with hydrogen atoms attached.

called *organic* because they are typical of organisms—are made of carbon and hydrogen atoms, often combined with oxygen, nitrogen, and a few other elements (Figure 1.6). Hydrogen, oxygen, nitrogen, and carbon make up over 99 percent of the atoms in an organism. Experiments have shown that the building blocks of biological molecules form easily under the

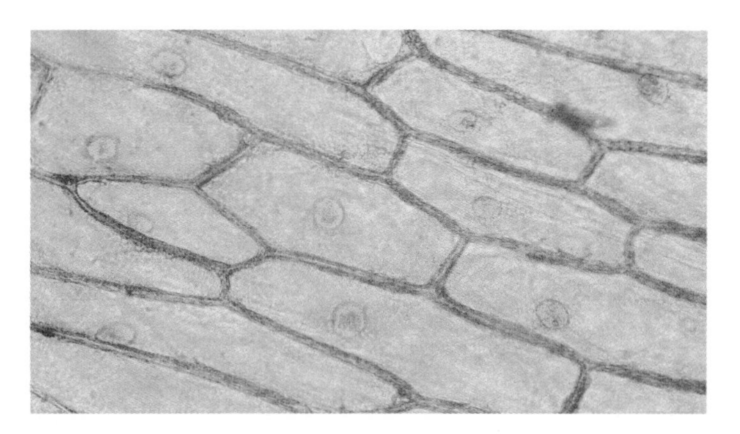

Figure 1.7

The cells of an onion skin share common features with those of all other organisms. They are made of characteristic carbon-based molecules and have distinct boundaries called membranes (supplemented by cell walls in this case) separating their internal structures from their surroundings.

conditions—vastly different from present conditions—that must have existed on the primitive earth. Once the simplest carbon-based molecules formed, a new kind of evolution began—a **biological evolution** (also called organic evolution) that led to the formation of living organisms and then, over vast times, to the millions of different kinds of organisms that have inhabited the earth up to the present day.

The transition from the first organic compounds to the first living organisms happened slowly over a long time. The earth formed about 4.5 billion years ago, and the first simple forms of life appeared about a billion years later. During this time, through complicated and still rather mysterious steps, the first small, carbon-based molecules combined into large molecules, which eventually associated to make the first simple **cells** (Figure 1.7). The formation of cells was a critical event, since they are the smallest units that exhibit the essential properties of a living organism, and every organism consists of one or more cells.

Our knowledge of life, climate, and geological events in the past comes from rocks containing **fossils,** the mineralized remnants of organisms or their traces, such as footprints and impressions (Figure 1.8). The rocks of the earth's crust are laid down in layers, or strata, one upon the other, so looking at the strata is like reading backwards in time. Quite reliable methods are now available for measuring the ages of these rocks, primarily based on the rate at which radioactive materials decay. Some of the most ancient rocks collected, from North America, Australia, South Africa, and elsewhere, show remnants of small, simple, single-celled organisms quite similar to certain bacteria living today (Figure 1.9). During the next billion years or so, the descendants of these bacteria diverged into many forms, including larger, more complex cells such as modern algae and protozoa. These, too, have left their record in the rocks. As we will discuss in Chapter 2, the fossil record is one of the primary reasons we believe all the diverse species on earth are the result of biological evolution.

(a)

(b)

Figure 1.8

A fossil is formed when an organism is covered with a sedimentary deposit, such as mud or lime, and the decaying body is gradually replaced by minerals, which preserve the original form quite precisely. *(a)* The fossil of an ancient fish. *(b)* A bed of ammonites, relatives of modern squid.

The story of cosmic and biological evolution explains the origin of the world as we see it. Now it is time to say more about the general nature of science as a way of exploring and explaining the world.

Exercise 1.1 Looking at Figure 1.6, describe the major features of organic molecules.

Exercise 1.2 Even though we have barely introduced the concept of cells, by looking at Figure 1.7 and using your general knowledge, identify some of their obvious features.

1.3 Natural sciences develop a systematic understanding of the world based on observation and experimentation.

The nature of science

Science begins with the observation that the world is full of interesting objects and regularly recurring events. People have been gathering scientific knowledge of the world for as long as they have been able to conceptualize and communicate, because science is fundamentally an extension of the common sense and practical work that humans have always used to survive. Even preliterate cultures have a lot of shared knowledge of their worlds. Ancient monuments made of rock attest to humans' recognition of astronomical regularities from thousands of years ago, and naturalists have sometimes been amazed to find that natives have quite complete catalogs, and very sophisticated knowledge, of the plants and animals around them. Although these people don't share the scientific community's understanding of the universe, some of their knowledge has the general characteristics of science.

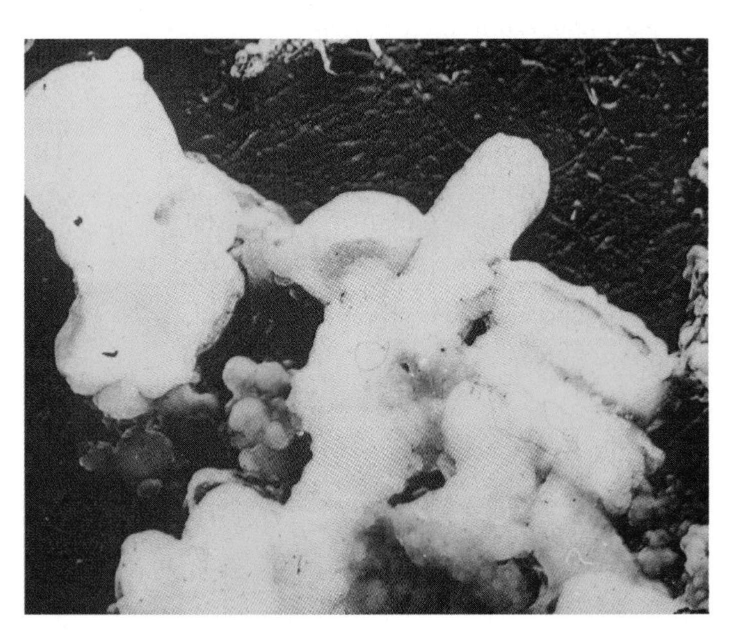

Figure 1.9

A photograph taken with an electron microscope shows primitive fossilized cells. (The coloring is artificial.)

A science begins by cataloging regularities in the universe and then trying to explain them. Events have causes. The universe has a causal structure, and scientific explanations show how phenomena fit this structure—how various factors push and pull and influence one another. For example, the long story of stellar evolution and the beginnings of biological evolution on earth is based on causal sequences: Atoms formed stars, which generated heavier elements through known sequences of reactions, so when new stars formed later, some were accompanied by planets that had the materials essential for life. Under the right conditions, those materials eventually assembled into simple organisms, and a process of biological evolution began. We outline this story to show how the origin of living organisms and their properties can be explained by means of a causal chain and by the intrinsic properties of the materials involved.

It is not easy to give a short definition of science; science may be best defined by listing its principal features:

- Science is **empirical:** Its information comes from observation and experiment, not just from thinking and imagining.
- Science is *systematic,* organized around a larger conceptual framework.
- Science is *open* and *progressive,* and has the potential to grow without limit.
- Science is *logical* and based on common sense. No matter how complex the instruments of a science may be, and no matter how much it depends on sophisticated mathematics, it uses basic human reasoning.
- Scientific explanations must be **intersubjectively testable.** That is, different people must be able to observe the same events and confirm or challenge one another's stories.
- Scientific explanations can only postulate physical objects or forces whose existence can be tested empirically; they cannot involve spirits or other nonphysical forces. This rule is called **naturalism.** So-called supernatural phenomena can either be investigated through experiment and observation, or they are outside the realm of science because they cannot be made empirically meaningful.
- Scientific explanations can *change.* When the authors of this book were in high school, humans had 48 chromosomes. This was an established fact of biology; all the textbooks said so. But with improved methods for examining chromosomes, it turned out that the correct number is 46. All the textbooks now say so. We don't expect this particular fact to change, but anything we accept today as a well-established fact or a satisfactory explanation could change tomorrow with better evidence or an improved theory.

Hypotheses

Scientific explanation depends upon both *hypotheses* and *theories.* Both are explanations, but they work at two quite different levels. A hypothesis is an educated guess, a conjecture offered to explain a phenomenon. We all use this common-sense way of dealing with the world. For instance, you get into your car one morning and it won't start. What's wrong? Is it out of gas? Is the battery dead? Is the starter defective? Is there some break in the electrical system? We immediately think of these hypothetical explanations, and one by one we test them. Notice that a hypothesis, *H,* is only fruitful if it predicts an observation, *O,* that could be made in a certain situation: If *H,* then *O.* We then set up the proper conditions and see whether we observe *O.* We can eliminate possible explanations with simple logic:

If the battery is dead, then the engine will not crank.
The engine *does* crank.
Therefore, the battery is *not* dead.

It is perfectly logical to reason: If *H,* then *O; O* is not true, therefore *H* is not true. However, it is tempting—but wrong—to reason this way:

If the battery is dead, then the engine will not crank.
The engine does not crank.
Therefore, the battery is dead.

In other words, we cannot reason: If *H,* then *O; O* is true, therefore *H* is true. This reasoning is illogical, because there could be many reasons why the engine doesn't turn over.

In dealing with singular practical situations like a car that won't start, we eliminate possible explanations and eventually come upon one that works (or we consult a local mechanic). With the more general statements of science, if we do not observe *O* as predicted by a certain hypothesis, that hypothesis is disconfirmed or invalidated. The hypothesis might have to be thrown out or at least modified. But even though each favorable observation helps confirm and strengthen a hypothesis, no observation *proves* the hypothesis. This odd situation arises because of the constraints of logic. No matter how much our confidence in a hypothesis is strengthened by observations consistent with it, an unlimited number of alternative hypotheses could predict the same observations (in principle), and we cannot test them all (or even think of them all). This means that the test of whether a statement is scientifically meaningful is not whether the statement can be proven true but whether it can be proven *false* and therefore rejected. Science isn't like mathematics, where one can prove that a theorem is true; science is much more like law, where one can only establish a fact or explanation beyond a reasonable doubt—and still leave room to reconsider the case later.

Unscrupulous people take advantage of the logic of science to argue that no one can prove they are doing any harm. A wealth of evidence, for instance, indicates that tobacco is a causative factor in cancer, heart disease, and other health problems. Nevertheless, spokespeople for the tobacco industry can argue quite logically that no one has proven tobacco is harmful. They can claim that many other harmful factors might actually be causing each ailment and that the high incidence of tobacco use among people with these diseases is just coincidental. In fact, they can even point to counterexamples, to many people who use tobacco, seem quite healthy, and live to old age.

Theories

A theory is quite different from a hypothesis. To nonscientists, the word "theory" implies something airy and unrealistic, in contrast to "facts," which are hard and indisputable. In science, however, a theory is a logical structure of ideas that not only explains a few observations but also makes sense of a large body of knowledge and explains how a part of the world

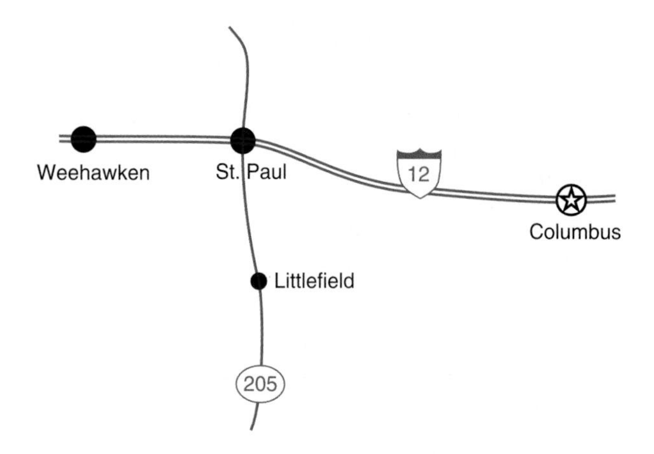

Figure 1.10

Like a map of a territory, a theory shows the relationships among otherwise isolated pieces of information.

operates. The British philosopher Stephen Toulmin's helpful conception is that a theory is very much like a map (Figure 1.10) in that the elementary facts of science are analogous to facts about a territory:

Weehawken is 15 miles due west of St. Paul.
Littlefield is 10 miles south of St. Paul on Highway 205.
Columbus is 23 miles east of St. Paul on Highway 12.

While a person can get around a territory simply by memorizing such facts, the human intellect naturally puts them together into a more comprehensive picture, a map. The map satisfies the natural desire to understand things as a whole.

Even an incomplete map is valuable, because it makes useful predictions and suggests other observations that could be made. In Figure 1.10, the territory between Columbus and Littlefield isn't described, but the map predicts that they should be about 30 miles apart. Someone who explores that region can verify that the map is realistic and improve it by providing more precise measurements; the exploration might yield quite unexpected facts and result in a much more useful map. (Maybe there is an undiscovered lake or river in the territory.) Similarly, a theory guides future work. Making unlimited observations about the world and piling up trivial knowledge is futile without first having a plan. A good theory directs observation and hypothesis formation in potentially productive directions, making exploration more fruitful and worthwhile.

What, then, is the idea of evolution? Is it a fact? Is it "merely" a theory? The concept of evolution actually has two faces—one fact, one theory. If we ask how all the organisms on Earth have reached their present forms, the answer is that they have evolved. This answer is based on such an enormous, coherent body of evidence that we must take it as fact. By contrast, the other face of evolution, the complex body of ideas about *how* evolution occurs, is a theory. Only this latter aspect is properly called the *theory of evolution,* though people who don't understand the subject commonly misuse the phrase. Evolutionary theory, like even the best-confirmed scientific theories, is constantly changing as biologists learn more.

Exercise 1.3 Examine these arguments and determine if they are logical.

If George Washington was assassinated, he is dead.
George Washington is dead.
Therefore, he was assassinated.

If my explanation for this strange phenomenon is correct, 0.73 amp of current will be flowing through this wire.
The current measures 0.72 amp.
Close enough. My explanation is correct!

Exercise 1.4 Over several decades, unusually high numbers of people living downwind from a nuclear facility develop cancer and other functional diseases. A spokesman for the facility says, "You can't prove that radiation from our plant caused any of these illnesses. We aren't to blame." What do you think about this contention?

1.4 It is important to distinguish science from pseudoscience and nonscience.

People love good stories, and in trying to make sense of our world they have told stories like these:

"The earth is actually flat. All the photographs they say were taken from space are faked."

"I saw this bright light, and suddenly a spaceship came down and landed. These strange creatures came out and took me inside, and that's all I can remember, but I have these scars where they must have operated on me."

"You're a Taurus, I can tell. All Tauruses are very strong-willed, and they keep getting themselves into romantic trouble. Your horoscope says you're going to have a chance to make a lot of money soon."

"Cancer comes from vitamin deficiencies, but I can cure it. These peach pits contain a vitamin called laetrile, and if you come to my sanitorium for several months of treatment, I guarantee to cure your cancer."

These stories belong to pseudoscience and superstition—a shadowy area that includes astrology and parapsychology as well as sensational and faddish beliefs in such theories as the Bermuda Triangle and a variety of irregular medical practices. It would be wonderful to be able to distinguish these stories from those of science, but there is no sharp dividing line between one and the other. The late Linus Pauling, for instance, a distinguished chemist and two-time winner of a Nobel Prize, became convinced that large doses of vitamin C are essential to normal health and will cure the common cold and other diseases. Though supported by some research, his claims remain highly controversial, on the borderline between established science and pseudoscience. This is only one of the more famous cases of problematic science; legitimate science continually pushes the

Sidebar 1.1 Creationism Is Not Science

In 1994, the American Institute of Biological Sciences (AIBS), an organization that promotes research and teaching in biology, adopted the following resolution.

The AIBS Executive Committee passed a resolution in 1972 deploring efforts by biblical literalists to interject creationism and religion into science courses. It is troubling that more than 20 years later, there is an urgent need to reaffirm AIBS's earlier position. Despite rulings by the Supreme Court declaring it unconstitutional to promote a religious perspective in public school education, such attempts by creationists continue in a variety of guises.

The theory of evolution is the only scientifically defensible explanation for the origin of life and development of species. A theory in science, such as the atomic theory in chemistry and the Newtonian and relativity theories in physics, is not a speculative hypothesis but a coherent body of explanatory statements supported by evidence. The theory of evolution has this status. The body of knowledge that supports the theory of evolution is ever growing; fossils continue to be discovered that fill gaps in the evolutionary tree, and recent DNA sequence data provide evidence that all living organisms are related to each other and to extinct species. These data, consistent with evolution, imply a common chemical and biological heritage for all living organisms and allow scientists to map branch points in the evolutionary tree.

Biologists may disagree about the details of the history and mechanisms of evolution. Such debate is a normal, healthy, and necessary part of scientific discourse and in no way negates the theory of evolution. As a community, biologists agree that evolution occurred and that the forces driving the evolutionary processes are still active today. This consensus is based on more than a century of scientific data gathering and analysis.

Because creationism is based almost solely on religious dogma stemming from faith rather than demonstrable facts, it does not lend itself to the scientific process. As a result, creationism should not be taught in any science course.

Therefore, AIBS reaffirms its 1972 resolution that explanations for the origin of life and the development of species that are not supportable on scientific grounds should not be taught as science.

AIBS. 1995. Creationism is not a science. *Bioscience* 45: 97. © 1995 American Institute of Biological Sciences.

edges of the unknown, where investigators argue rationally about which story describes reality.

A story is *pseudoscience* if people continue to believe it in the absence of sound evidence or in the face of strong counterevidence. Claims of astrologers, for instance, have been tested many times by unbiased observers, and the claims always fail, but people continue to believe in astrology anyway. *Superstitions* are hasty generalizations from insufficient observation: "I accidentally walked under a ladder once, and then I lost a lot of money. And one of my friends walked under a ladder, and he broke his arm right after." Or, "Well, my friends Jan and Kelly are both Tauruses, and they're very strong-willed." These stories are poorly confirmed hypotheses, which people continue to believe irrationally.

Pseudoscience often violates an important scientific guide, the **Principle of Parsimony,** which states that one ought to prefer the simplest satisfactory explanation. This is sometimes called Occam's Razor after the philosopher William of Occam (or Ockham, 1285?–1349?). Many people believe, for instance, that they have predictive dreams because they have had a dream experience that seemed to come true soon afterward. But statistical analysis shows that even with a very small probability that a dream and an event are linked just by coincidence, among millions of people, many each year will experience apparently predictive dreams. To believe that some people have psychic powers that allow them to predict events, we would have to believe in a whole complicated structure for the universe, entailing changes in physics and human biology that no one has yet demonstrated. Therefore, we ought to prefer the simple, statistical explanation.

Religious fundamentalists who do not believe in biological evolution because it contradicts the biblical story of creation have tried to support their beliefs with pseudoscientific arguments. Sidebar 1.1 presents a statement about the issue of "creationism," which is especially important to biologists.

It is important to see that these beliefs are *unscientific* and to distinguish them from the *nonscientific.* The *un*scientific imitates science but does it badly, with too strong a will to believe and too little objective testing and skepticism. The *non*scientific, on the other hand, is human activity that has little or nothing to do with science and is done for different reasons. Included in this category are the arts and humanities, building personal relationships, sports and games, and virtually everything else people do to add pleasure and meaning to their lives. Although elements of science and technology may be involved in nonscientific endeavors (such as the technical aspects of painting, photography, and theater), their fundamental purpose is quite different from the scientific enterprise of explaining natural phenomena and systematizing that knowledge.

Exercise 1.5 A student wrote an article about a petrified forest in which he reported that he had counted the rings in several trees and found them to be no more than a couple of hundred years old, so there was no reason to believe that the earth is thousands or even millions of years old. Criticize this argument.

1.5 Our study of biology will emphasize four major themes.

People who study and teach biology generally believe they are engaged in an exciting adventure. The purpose of this book is to invite students to come along on this adventure, to experience the excitement of discovery and share the stories of past discoveries. Most of the book tells the essential stories about how organisms work. Although these stories can be complicated, our primary goal is to make them as clear as possible.

No one comes to a college biology course ignorantly or naively. Previous courses and years of daily experience have already taught you a great deal about biology. You already know that you must eat a variety of foods containing nutrients your body needs for maintenance, repair, and growth, that you must take in oxygen to stay alive and produce carbon dioxide as a waste, and that your nervous system carries messages rapidly from point to point, enabling you to react quickly to new stimuli. The remainder of this chapter is designed to remind you of what you already know and to show how we will build on this foundation throughout the book as we develop four fundamental themes of modern biology. The themes introduced in Concepts 1.1 are woven throughout the book, not limited to certain chapters, and the key terms italicized here will help you recognize when each of these themes reappears later. These four themes form a theoretical basis for biology and provide a valuable organization for students who are just beginning to understand the science.

Along the way, we will also tell a number of fascinating stories of scientific discovery to show how science is done and to emphasize that science is a human activity. These stories are told too seldom; science is often presented as a collection of facts and theories that seem to emerge from nowhere. We prefer to show that science is done by people—not only a few prominent people from the past, such as Darwin and Newton, but also people alive today who are doing experiments and developing new ideas. Scientists associate notable accomplishments with the names of the people responsible—the Watson-Crick model, Batesian mimicry, and the Hardy-Weinberg Principle—and we will often mention people in this book. However, we will emphasize their contributions to biology, not their biographies. The important fact is not that someone is a professor of biology or biochemistry at a particular university, but that he or she did the critical experiment that tested X's hypothesis and provided the basis for later work by Y and Z. This is the view of the structure of science, and how it advances, that we want to develop.

We hope to show why scientists believe certain facts and theories rather than others. Obviously, however, time and space do not allow us to present the evidence supporting every conclusion. Often we will have to be content with explaining the conclusions a majority of scientists have reached after a great deal of work and study.

1.6 Biologists may use more types of explanations than those used in other sciences.

Having introduced the four main themes of biology, let's return to the matter of scientific explanation. In physics and chemistry, you will only encounter causal explanations. Biology, however, relies on at least three kinds of explanations. For example, take the question Why do birds build nests? Consider these answers:

- Birds build nests because the increasing day length stimulates certain brain centers to start producing hormones, which stimulate other brain centers to set in motion a series of behavior patterns built into the birds' nervous systems and encoded in their genes. One of these behaviors is building a nest of a specific architecture. This answer is based on *causation.*
- Building a nest is a functional behavior for birds; it provides a relatively secure place for them to raise their young. This answer is based on function or *survival value.*
- Because nest-building is a functional behavior, the genetic instructions for making a particular kind of nest have been gradually shaped over millions of years by selection of genes that provide practical behavior patterns. This answer is based on *evolutionary history.*

When we ask questions about biology, it is important to think about the kind of question we are asking and the kind of answer we want. Most often, people are searching for either a causal or a functional answer, and the distinction is important.

Another kind of answer is missing from our list. No serious biologist would say, "The birds know that if they build a nest, their little ones will be safe and happy, so they're making a nice home for them." This answer, which has been made purposely Mother Goose-ish, imputes purpose and foresight to the birds, and such an explanation, called a **teleological** explanation, is excluded from biology. Teleological explanations try to explain events in terms of the purposes they serve (Greek, *telos* = end, purpose), and they are excluded from science for several reasons. Teleology of a kind is acceptable in explaining human behavior because humans can anticipate the future and act rationally. Thus it would be perfectly acceptable to say that a human mother built a playpen for her child because she knew it would keep her child safe and happy. Similar explanations probably make sense in explaining the behavior of some other intelligent animals, as Jane Goodall has shown for chimpanzees, but if we extended this way of thinking to the whole biological world, we would be guilty of **anthropomorphism** (*anthropos* = human; *morph* = form): attributing human characteristics to nonhumans. A lot of teleological thinking stems from anthropomorphism.

Nevertheless, it is clear that organisms and their parts have functions, and Colin Pittendrigh coined the term **teleonomy** to mean the apparently goal-directed behavior of organisms. We will show later that teleonomy is inherent in certain kinds of systems, such as a computer running a program or a genetic system like an organism. Unfortunately, teleonomic systems seem to seduce people into a kind of sloppy teleological thinking. For instance, it is easy to anthropomorphize computers and react to them as if they were rational beings with their own goals instead of dumb machines that only do what they are programmed to do. The idea of evolution—an essential property of organisms—is especially vulnerable to teleological thinking. People tend to fall into traps such as imagining that evolution is directed toward a goal (like creating humans) or that organisms

Concepts 1.1 Four Themes of Biology

1. Organisms are genetic systems.

They operate and reproduce themselves by using information contained in their genes.

This concept is entering more and more into the thinking of modern people. Everyone knows that we *inherit* most of our features from our parents, and that the inheritance occurs through *genes,* which are made of *DNA.* The use of DNA evidence in some notorious court cases has made people aware that genes can be used for identification, and conversations often turn to the question of how much of human behavior is determined by our genes and how much by experience. Genetic inheritance is even the stuff of jokes: "I'm afraid I've inherited my dad's gene for making bad investments." Progress reports on the Human Genome Project in newspapers and magazines have brought the term *genome* into our lives. A genome is the set of all an organism's genes. Much of our study will be devoted to understanding the structure of a genome, how *information* is encoded in it, and how that information determines an organism's structure and behavior. By organizing biology around this central genetic concept, we can make sense of the whole.

For more discussion of this theme, look at:

- Sections 2.6 and 2.7 (basic concepts about organisms)
- Section 12.1 (how a genome instructs cells to operate)
- Section 15.6 (how a genome determines development)

2. Organisms live in ecosystems.

Within these systems, they adapt to particular ways of life and engage in complex interrelationships with other organisms and the environment.

As arguments heat up about using natural resources and preserving the natural environment, the concepts of *ecology* force themselves upon our consciousness. Although people take different sides of the political and economic debates, they are generally becoming aware that all organisms are members of *ecosystems,* where an *ecological community* is made of many species that interact in complex ways. Thus cutting down a forest, dumping acid wastes into a river, or driving a single species to extinction can have far-reaching consequences, affecting many other species—including humans. Here the idea of *niche* becomes critical, a niche being a species's particular way of life and the place where it lives. We commonly say that every species is *adapted* to its niche, that its characteristics just fit it for living there. The question of how adaptation occurs then leads to the fundamental concept of evolution through natural selection, our third theme.

For more discussion of this theme, look at:

- Section 2.8 (basic ideas about ecosystems)
- Section 7.9 (energy in ecosystems)
- Sections 27.8–27.11 (chemical communication between organisms in a community)

3. Evolution operates primarily through the process of natural selection.

It has produced the enormous variety of life and continues to operate today.

The concept of *biological evolution* pervades modern thought, even though many people still refuse to accept the idea for religious reasons. People are fascinated by dinosaurs and why they became extinct, and newspapers regularly report discoveries of fossils that cast new lights on human evolution. *Natural selection,* too, has become part of our thinking—which is not to say that people really understand the concept. Natural selection simply means that in each generation the organisms whose features make them best *adapted* for their ways of life are the ones that have the most offspring that survive and reproduce another generation, so their favorable features are passed along and gradually become accentuated. Natural selection only makes sense if organisms are genetic systems, creatures that can inherit their characteristics and change that inheritance. Thus there is *genetic variation* in a *population* of similar organisms if the population contains individuals with different genes and different features. The concept of natural selection, as first outlined by Charles Darwin and Alfred Russell Wallace in the nineteenth century, provides a powerful explanation for a great deal of biology. It explains how organisms become so well adapted to a particular niche in an ecosystem and, on a molecular scale, how biological molecules are fashioned to perform particular cellular tasks. The threefold concept of *selection, adaptation,* and *niche* will become an important part of our thinking.

For more discussion of this theme, look at:

- Section 2.8 (the concept of natural selection)
- Section 4.14 (the evolution of proteins)
- Section 14.5 (how cellular processes restrict evolution)

4. Organisms function through molecular interactions.

Organisms are made of complex *carbon-based molecules* that interact with one another because of the ways they fit together and the forces they exert on one another. You cut your hand, soak it in hydrogen peroxide, and watch oxygen bubbles form along the cut; the bubbles show the interaction between hydrogen peroxide and the molecules in the wound. A *chemical reaction* is initiated there. When you cut apples or potatoes, their surfaces turn dark quickly as they are exposed to the air; this chemical reaction is due to *enzymes* in these fruits and vegetables. Good cooks know that lemon juice can keep apples from turning dark because the acidic juice interferes with the interaction. They also know better than to put fresh pineapple in gelatin; the mixture won't gel because enzymes in the pineapple digest the gelatin. Similar enzymes in the mouth, stomach, and intestines digest our food.

A number of chemical concepts are central to modern biological thought. In general, organisms are made of *large molecules* (mostly *proteins*) with specific, distinctive shapes. Molecules interact by *binding* to one another through *chemical bonds* and by fitting their shapes together much like jigsaw-puzzle pieces. Two molecules that fit together in this way are said to be *complementary,* and every biological function depends on molecules interacting with one another in *shape-specific* ways. We will discuss several types of molecular interactions that keep occurring, in different forms, again and again.

For more discussion of this theme, look at:

- Section 8.6 (the structure of membranes)
- Sections 8.11–8.15 (how membranes move molecules)
- Sections 11.9–11.12 (proteins that shape cells and make them move)

foresee what characteristics they need and then somehow choose to acquire those features. Since organisms do acquire their characteristics through evolution, teleological thinking would explain why an organism has some function—but the explanation would be unscientific. But a teleo*logical* explanation for the properties of organisms can always be replaced by a teleo*nomic* explanation that addresses the function served and uses evolutionary history to explain how something serving that function could appear. For instance, a teleological (and wrong) explanation for giraffes having long legs and necks is that they kept trying to reach higher leaves on trees. A teleonomic explanation is that some primitive giraffes had genes for longer legs and necks, which allowed them access to a new source of food; continuous selection for genes of this kind over many generations eventually produced all long-legged, long-necked animals.

The difference between thinking teleologically, as nonscientists often do, and thinking teleonomically, as biologists should always do, is subtle but critical. Biologists often make casual statements that sound teleological but are just shorthand substitutes for the long teleonomic statement. If a biologist says, "This species of plant spreads its leaves out to catch the sunlight," he or she is not really suggesting that the plants can grow or behave with some purpose. This statement is a short way of saying, "Since it is advantageous for a plant to collect as much sunlight as possible, the ancestors of these plants that grew with their leaves spread out were more successful than those with a different growth form, so these plants were gradually shaped by evolution to grow in this form." That long, detailed statement would be thought odd by a group of people who think like modern biologists; everyone would understand the short, casual statement as an abbreviation for the other one.

In this book we will sometimes use the shorthand type of statement. Since our purpose is to teach you to think like a modern biologist, exercises from time to time will challenge you to give the proper teleonomic translation and encourage you to convert from a teleological thinker into a teleonomic thinker.

Exercise 1.6 The statement, "Birds have light bones so they can fly more easily" has a teleological flavor. Translate it into a teleonomic statement.

1.7 Complex structures are understood by referring to both their components and their emergent properties.

Throughout this book, we will show how biological processes depend upon molecular interactions. By elucidating these phenomena, the research programs of molecular biology and cell biology have provided deep, satisfying insights into biological structure and function. In showing how the structures of biological molecules explain biological functions, this research has created a revolution in biology over the past half-century and has led to the insights that underlie this book. Now explaining a complicated process or system in terms of its constituent parts is one kind of **reduction.** As physics and chemistry developed into such powerful explanatory forces, some scientists proposed reductionism as a scientific ideal; that is, they proposed that science should try to replace all the terms, concepts, and laws of chemistry with terms, concepts, and laws of physics, and that ultimately even biology could be reduced entirely to physics. Thus, if we knew everything about physics, we could predict and explain everything about biology as well. Reductionism is actually a complicated issue in the philosophy of science, and it could take different forms (such as reducing concepts or reducing laws). But even though molecular explanations are essential to biology, the concepts and laws of biology cannot be replaced completely by those of physics and chemistry. The language of biology simply cannot be translated entirely into a language of physics.

To understand biology, we will supplement molecular explanations with an emphasis on the properties of whole organisms and their ecological relationships because complex systems, including biological systems, have **emergent properties,** properties of a whole structure that go beyond those of its constituents. Both emphases—the partly reductionistic and the emergent—are important, and neither tells the whole story by itself. This point is not unique to biology. An automobile, for example, is a complicated assemblage of small parts, yet we explain its operation with the details of internal combustion engines, steering mechanisms, gear boxes, and brakes, each having parts that push and pull on one another in certain ways. But the automobile also has emergent properties—that is, as a whole, it will move and carry passengers, which no single part or collection of its parts can do. So we understand automobiles both through the details of their mechanical and electrical operation and also by the way they function as integrated wholes. The same is true for biology. Although biological systems are complicated and fascinating, they are just very complex chemical systems, and they can be analyzed and studied just like other, simpler systems.

1.8 What is life?

Since biology is "the science of life," students expect a biology course to teach them what life is. Unfortunately, the issue is complicated. One problem is that "life" is a noun, and by talking about it, we trick ourselves into thinking that a substance called life actually exists: "At last, I have isolated life; just add some of these shimmering crystals to anything, and it will come alive." Biologists of an earlier generation talked about "protoplasm," a supposed substance in every cell that confers all the properties of life. There is no such thing. Living organisms perform a wonderful variety of activities, but all of these activities depend on the interactions of many specific, identifiable substances, and none of these substances by itself could be called "life."

It is perfectly obvious, however, that a firm, green plant standing erect in the sunlight is quite different from one that has withered and turned brown—or that an eating, breathing, moving animal is in a very different condition from one decaying on the ground. Clearly there is a state of *living,* of being *alive,* that is actually quite easy to discern. Living plants and animals carry out **metabolism** (Figure 1.11), a complicated series of chemical processes. They take in raw materials from the environment and shape them into molecules of their own structure. Organisms extract energy from the environment, too, and only maintain themselves and grow if energy and materials are continually moving through them.

In this book we will use "living" or "alive" to mean a state of metabolizing, even though the definition gets us into some counterintuitive situations. For instance, some simple organisms—even small animals—can be dehydrated (dried out) or frozen to stop their metabolism, but when they are thawed or placed in water, their metabolism resumes and they function normally. We might have to say that they were dead and are now alive again, even though the idea of reversible death is rather strange. Instead, people who work with frozen or dehydrated organisms say that they are *viable,* meaning capable of metabolizing and growing in the right conditions. If water is the key ingredient in restoring metabolism, we could conclude that water is life, but that would be silly, to say the least!

Thus we still haven't defined "life." Defining "life," in fact, seems like a futile enterprise. And it certainly isn't a goal in biology; no biologists are engaged in a hot competition to learn "the secret of life," and no one who understands biology seriously believes there is such a thing. Instead, biologists study many specific biological processes and try to understand the lives of particular organisms, with an unspoken agreement that if "life" means anything at all, it is a *process,* or more accurately, a lot of complicated processes, all occurring together. We will outline these interrelated, interacting processes in subsequent chapters.

Exercise 1.7 Dr. Brown has advanced a new hypothesis to explain life. He believes all cells contain minute units called "bions." They are not made of ordinary matter, so it is not possible to detect them through any physical apparatus used for studying atoms and molecules. They are most like bits of electricity, but they do not have charges as electrons do, and they do not affect ordinary electrical devices. Dr. Brown believes the universe contains a limited (but very large) number of bions and that they enter the body of an organism at the time it comes into existence and leave when it dies. He knows the standard explanations for metabolism, heredity, and other biological processes but believes they are inadequate; he has a feeling that we must postulate these mysterious units in order to explain all the mysterious activities of organisms. Discuss Dr. Brown's hypothesis in the light of what you know about scientific hypotheses and theories.

1.9 Why should anyone study biology?

Biology is the hottest field of modern science. Perhaps we say this because we are biologists, and our colleagues in cosmology or high-energy physics might dispute our claim, but look what makes headlines these days (Figure 1.12). Every day, newspapers and television report some new and exciting discovery in basic biology or an application to health, medicine, or agriculture.

In just the last few years, molecular biologists have developed techniques for working with minute amounts of DNA, the stuff that genes are made of. They can determine the detailed structure of DNA molecules and apply their techniques to important problems such as identifying the defective genes in people with genetic diseases, treating cancers, and finding ways to stop infection by the AIDS virus.

The best reason we know to study biology is simply to satisfy the human senses of aesthetic and scientific wonder we mentioned earlier. As trained biologists look at the world, they see not only its surface features but also the intricacies of biological structure and the interrelations among different organisms. That heightened awareness enriches one's life, adding interest and enjoyment to daily activities. A second reason to study biology

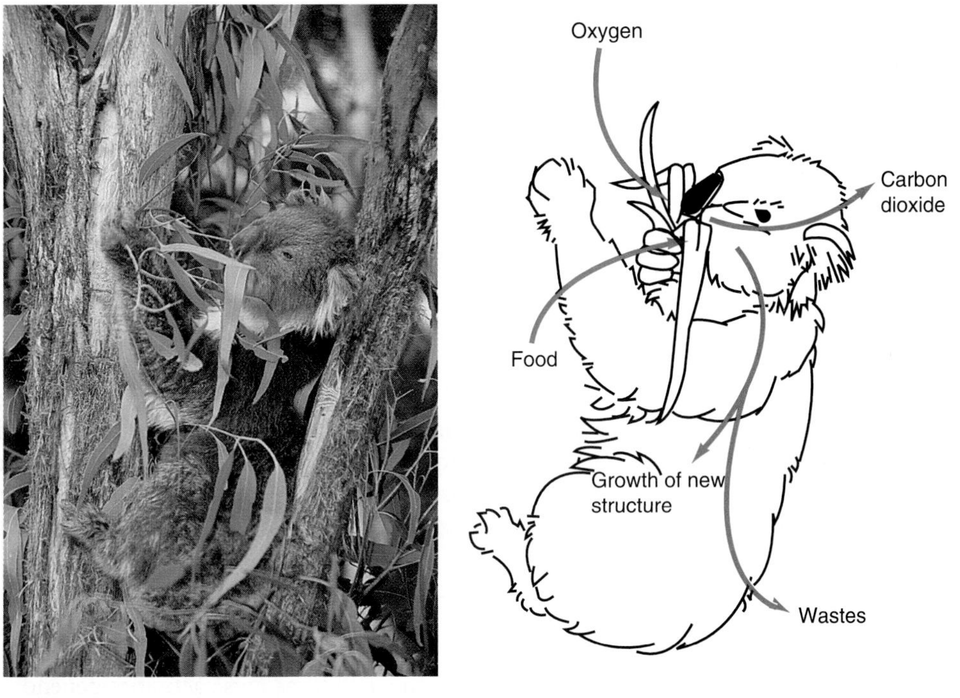

Figure 1.11

In the processes of metabolism, an organism extracts energy and raw materials from its environment, transforms these materials into its own structure, and eliminates wastes. By means of these chemical activities, the organism maintains its structure and even grows.

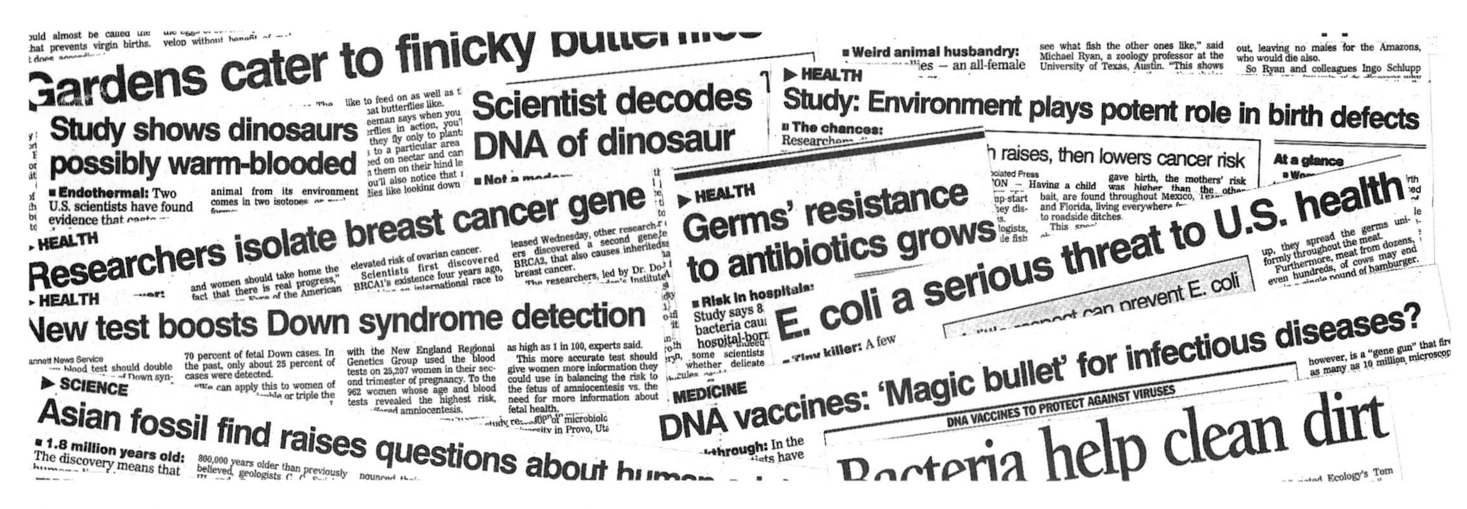

Figure 1.12
Advances in biology make major headlines these days.

is that the world is such an exciting, mysterious, and beautiful place that people not only want to know more about it, but *should* know more about it. Scientific knowledge about the world we live in ought to be part of a liberal education. Liberal education is *liberating* education, freeing people and giving them greater power over their lives. A knowledge of biology will enable you to make wise choices—about healthful living and the many problems informed citizens must face in our society as complex scientific issues become increasingly important.

Exciting advances have been made in biology recently. The worldwide community of biologists understands a great deal about life. At the same time, an enormous amount is yet to be learned, and the fields of biology are wide open for people who want an intellectual challenge. A few years ago, a special issue of the journal *Science* listed the following key fields that offer challenging careers in biology, with some bias toward applied research:

- Molecular biology: studying the three-dimensional structures of molecules to learn how they operate; making molecules with designed structures as model systems or to enhance the immune system.
- Genetics: deciphering the whole genetic structure of organisms, including humans; identifying specific genetic diseases in humans and correcting them with various genetic therapies.
- Immunology: working out the detailed biology of viruses such as the AIDS virus and developing methods to combat such viruses; determining precisely how the cells of the immune system operate and using this knowledge to treat immunological problems in humans; using immunology to develop new therapies.
- Differentiation and development: unlocking the complex mechanisms through which genes direct the development of a single cell into a large organism.
- Neuroscience: working out the details of brain function by applying and developing many new techniques; determining how specific brain centers control certain processes and precisely what is happening when we perceive something or perform an action.

The future offers enormously exciting and rewarding careers to people who train themselves well and have the imagination to ask fruitful questions combined with the determination to develop research programs over many years. You could be one of those people.

A different kind of opportunity faces people who are more interested in organisms living in natural habitats. The challenges are both to understand the ecology of these systems and to preserve them. While ecology has also been making great advances, it is intrinsically more difficult than other areas of biology because the systems involved are so large and the organisms in them interact in such complex ways. So much remains to be done in this field too. Furthermore, ecologists—and others who share their concerns—are working under the gun of increased environmental destruction. Natural areas for study are disappearing, and ecologists are being called upon not only to help preserve these areas but also to educate and persuade humanity as a whole to change its ways and avoid the ultimate worldwide ecological disaster.

In addition to basic research and its immediate applications, an increasing number of people will be needed in applied areas, especially the health sciences. As our knowledge of basic biological processes grows, so do the ways that knowledge applies to the improvement of health. Until about a century ago, physicians could do little to cure illness except prescribe a few folk remedies or perform some crude and dangerous surgery. Once they learned that many diseases are caused by microorganisms and once antibiotics became available, physicians acquired real power over infections. In the last few decades, knowledge of cell biology and genetics has provided greater understanding of metabolic disorders, cancer, and other serious problems, and the possibility for prevention and cures is promising. Better understanding of physiology and nutrition is pointing the way to more healthful living through exercise and moderate diet. Medical practice becomes more sophisticated every year, and health workers at all levels need up-to-date knowledge to do their work.

Coda Biology in an evolutionary framework will be one of the major themes of this book, for it is fundamental that all organisms are the result of biological evolution, which is one phase of the evolution of the universe. We will also continually emphasize the genetic nature of organisms and the fact that they operate through molecular interactions. The next chapter sets the stage for an evolutionary understanding of biology by outlining the process of evolution as it relates to ecology. The two themes are closely connected.

Summary

1. Humans tell stories to explain their experiences and to make sense of the phenomena they observe. Science also begins with stories of investigation and goes on to provide explanations of natural phenomena.
2. Science is inspired by aesthetic wonder at the beauty and complexity of the natural world and by scientific wonder, which asks questions and searches for explanations.
3. We understand organisms in the context of their history, as part of an evolving universe. About 4.5 billion years ago, the earth was formed by the condensation of interstellar gas and dust. On the primitive earth, organic molecules formed; these molecules are made primarily of carbon and hydrogen, often combined with oxygen and nitrogen. The first organic molecules gradually formed more complex molecules, until eventually the first cells were formed. From this beginning, biological, or organic, evolution proceeded to form the great variety of organisms we see today.
4. The history of evolution is recorded by fossils, the mineralized remnants of organisms or their traces. Fossils are preserved in layers (strata) of rocks whose ages can be determined, primarily by using the rate of decay of radioactive materials in them.
5. Science provides explanations for natural phenomena, principally by showing how events fit into the causal structure of the universe.
6. Science is characterized as empirical (based on observation and experiment), systematic, open and progressive, and logical. Scientific explanations must also be intersubjectively testable, so one observer can confirm or challenge the claims of another; and they must be naturalistic, postulating only physical objects and forces that can be investigated empirically.
7. A hypothesis is an educated guess that can be tested. If consequences predicted by the hypothesis are actually observed, the hypothetical explanation is supported and strengthened, but not proven. However, if the predicted consequences are not observed, the hypothesis can be ruled out.
8. A scientific theory is a general explanatory framework, analogous to the map of a region.
9. There is no sharp boundary between science and pseudoscience or superstition. Legitimate science, pushing the boundaries of the unknown, will always be controversial and problematic. However, pseudoscience is generally characterized by hasty generalizations, poorly confirmed hypotheses, and irrational beliefs that continue even in the face of contrary evidence.
10. It is important to distinguish the unscientific (pseudoscience) from the nonscientific, which includes human pursuits, such as the arts and humanities, whose purposes and methods are quite different from those of science.
11. Our course of study will emphasize four themes: (1) Organisms are genetic systems. (2) Organisms live in ecosystems. (3) Evolution, operating primarily through natural selection, is responsible for the variety of life we observe. (4) Organisms function through molecular interactions. These themes build on your knowledge from previous experience and education.
12. All sciences use explanations based on causation. In addition, biological explanations may refer to the function or survival value of a characteristic and to its evolutionary history.
13. Biology cannot use teleological explanations, which refer to an intentional purpose or end that a characteristic is supposed to serve, because that ascribes intelligence and foresight to most organisms. However, biology regularly uses teleonomic explanations, which explain how features with certain functions have arisen by referring to evolutionary history.
14. Reductionistic explanations in biology explain the operation of a system by referring to the structure and operation of its parts. Although such explanations are central to modern biology, organisms also have emergent properties that arise from their organization and go beyond the properties of their parts.
15. The term "life" is problematic. "Life" does not exist as a special substance or property. We will use the terms "alive" and "living" to mean that an organism is carrying out metabolism, a complex series of chemical processes in which the organism uses raw materials and energy from the environment to maintain itself and to grow.
16. The study of modern biology ought to be part of a liberal education that gives people greater understanding of their world and greater control over it. In addition, modern biology offers many opportunities for rewarding and exciting careers.

Key Terms

Multiple-Choice Questions

1. Which is *not* characteristic of a good scientific hypothesis?
   a. It postulates a function for a phenomenon.
   b. It predicts particular results from a set of initial conditions.
   c. It can be tested by experiment or observation.
   d. Its consequences can be tested by different investigators.
   e. It is naturalistic.
2. Which of the following are empirical?
   a. Implanting electrodes in an animal's brain to study its behavior.
   b. Sitting quietly in a forest and observing an animal's behavior.
   c. Thinking of an explanation for an animal's behavior.
   d. All three activities.
   e. *a* and *b* but not *c*.
3. Hypotheses are tested by
   a. repeating the initial observation many times.
   b. deducing their implications.
   c. confirming a hypothetical prediction.
   d. disproving all challenging hypotheses.
   e. finding a logical link to an existing theory.

4. Some mystics claim unique abilities to "channel" communication with the souls of ancient people. The principal trouble with such claims is that they are
   a. not empirical.
   b. superstitions.
   c. not intersubjectively testable.
   d. hypothetical.
   e. irrational.
5. All organic molecules
   a. contain rings of atoms.
   b. contain carbon atoms.
   c. were originally formed in ancient stars.
   d. were made through organic evolution.
   e. All of the above are true.
6. Which of these could be fossils?
   a. An animal's body replaced by minerals.
   b. The impression of a leaf in a rock.
   c. Footprints of human ancestors preserved in rock.
   d. *a* and *b,* but not *c.*
   e. All three are fossils.
7. Which of the following are nonscientific, rather than scientific or unscientific?
   a. Basing your decisions on astrology.
   b. Believing in creationism.
   c. Determining the acoustic properties of an opera house.
   d. Majoring in biology because you enjoy it.
   e. Determining the stresses in a building you are designing.
8. Which is the most accurate definition of a theory?
   a. A tentative causal statement about an observation.
   b. An initial observation of a natural process.
   c. An explanatory framework that integrates a group of hypotheses.
   d. A prediction of the outcome of an experiment.
   e. A hypothetical explanation for a natural event.
9. Which of the following statements is (are) teleological?
   a. The function of the heart is to pump blood.
   b. Many bacteria swim toward a source of sugar.
   c. Birds evolved strong wings as they strove to fly.
   d. *a* and *c* but not *b.*
   e. All three statements are teleological.
10. Emergent properties result from
   a. having many different molecules in a mixture.
   b. complex interactions among the diverse parts of a system.
   c. the sum of the individual properties of constituent elements.
   d. the emergence of organisms from the water onto the land.
   e. the appearance of unexplained properties in complex organisms.

True-False Questions

Mark each statement true or false, and if false, restate it to make it true.

1. Organic molecules contain hydrogen and oxygen.
2. Casual explanations are essential in physical science but of minor importance in biology.
3. Cosmic evolution was directed toward eventually producing the elements of living organisms so organic evolution could begin.
4. A hypothesis is a tentative explanation for an observed phenomenon.
5. Experiments bear out the consequences of my hypothesis, so the hypothesis must be correct.
6. Experiments contradict the consequences of my hypothesis, so the hypothesis must be false.
7. If someone claims only he and his disciples can observe a certain phenomenon, his claim is not scientific because it is not empirical.
8. Enjoying a Mozart symphony is an unscientific activity.
9. Teleological explanations are based on evolutionary theory.
10. If total reductionism were possible, we could know all the principles of biology by first determining all the principles of physics and chemistry.

Concept Questions

1. What distinguishes aesthetic from scientific wonder? Does one necessarily negate the other?
2. What is wrong with the following scenario? After several years of carefully observing the rising and setting sun, you exclaim, "My hypothesis is that the sun rises in the eastern sky and sets in the western sky."
3. Someone asserts that scientists and science are untrustworthy, arguing that many scientific theories are later proven false, so how do we know today's explanation won't also be proven false? How can you reply to this argument?
4. Explain the difference between a hypothesis and a theory.
5. How can scientific research lead to an objective understanding of the universe if scientists, like all other humans, are fallible?

Additional Reading

Collins, H. M., and Trevor Pinch. *The Golem: What Everyone Should Know About Science.* Cambridge University Press, New York, 1993. History and social aspects of science.

Hanson, Norwood R. *Perception and Discovery: An Introduction to Scientific Inquiry.* Freeman, Cooper & Company, San Francisco, 1969. A general view of science that emphasizes the nature of perception.

Medawar, P. B. *Advice to a Young Scientist.* Harper & Row, New York, 1979. Words from an elder scientist about doing science.

Roberts, Royston M. *Serendipity: Accidental Discoveries in Science.* John Wiley & Sons, New York, 1989.

Toulmin, Stephen. *The Philosophy of Science: An Introduction.* Harper Torchbooks, Harper & Row, New York, 1960 [1953].

———. *Foresight and Understanding: An Enquiry into the Aims of Science.* Harper Torchbooks, Harper & Row, New York, 1963 [1961].

van Lawick-Goodall, Jane. *In the Shadow of Man.* Houghton Mifflin Co., Boston, 1971.

Internet Resource

To further explore the content of this chapter, log on to the web site at:

http://www.mhhe.com/biosci/genbio/guttman/

PART I

Basic Principles

2 The Diversity of Organisms

An archerfish aims its watery missile at a spider.

Key Concepts

2.1 The biological world is extraordinarily diverse.

2.2 Species are shaped through selection.

2.3 The concept of evolution developed slowly with the growth of modern science.

2.4 Evidence from the fossil record and the study of comparative anatomy indicate that organisms are related by evolution.

2.5 Darwin and Wallace outlined the general process of natural selection.

2.6 An organism is a structure that functions on the basis of information carried in its genome.

2.7 An organism is like a self-reproducing automaton.

2.8 Evolution occurs through natural selection operating on populations in ecosystems.

2.9 Organisms measure out their lives in cycles of growth and reproduction.

2.10 Populations of organisms tend to grow but are held in check by environmental factors.

2.11 The genetic variability of populations is the basis for natural selection.

2.12 Organisms are opportunists that meander into certain strategies of survival.

2.13 Phylogenies describe the apparent course of evolution.

2.14 Biologists divide the world of organisms into kingdoms.

At the edge of an Indonesian lake, an archerfish quietly slips underneath a dragonfly perched on a blade of grass. Raising its snout to the surface, it suddenly spits a few drops of water at the insect, knocking it into the water where the fish quickly gobbles it up. Archerfish are known for their accuracy. Their survival depends on their ability to routinely shoot a stream with William-Tell accuracy at a target as much as a meter away. Because a light ray bends as it passes between the air and the water, the fish must even compensate for refraction. Yet young archerfish are not instructed by their parents. Each fish develops with a hunting routine wired into its nervous system and muscles, though presumably it fine-tunes the routine through practice. Then where does this accurate hunting ability come from?

Archerfish are only a representative. We might focus on any one of millions of species and ask how it has acquired its particular way of life. The answer will always be that the species has gradually evolved into its present form. Ancestors of the archerfish, sometime in the remote past, did not possess the skill of archery. The ability evolved over a very long time as these animals changed generation after generation, and this theme—*continuity with change*—will inform our entire study of biology. The evolutionary perspective is fundamental to biology, and it is essential to see every organism as the product of evolution. But to understand how evolution occurs, we must understand the fundamental nature of organisms. We will show that organisms naturally evolve because they are self-reproducing genetic systems, that each generation inherits its characteristics with occasional novelties, and that evolution depends on natural selection operating on accidental changes in heredity. In this way, organisms take advantage of opportunities for change and adapt to their ecological surroundings. ■

2.1 The biological world is extraordinarily diverse.

One of the most impressive features of the biological world is its diversity—the enormous number of different kinds, or *species*, of organisms we see around us (Figure 2.1). Humans are fascinated with the variety of organisms, and field guides that help us identify birds, flowers, and other organisms are very popular. A plate from such a field guide (Figure 2.2) shows a variety of birds that live in eastern North America. Although they all lead similar lives, each kind of bird is called a distinct species. Why? There are many "kinds" of humans, too, with quite different appearances (Figure 2.3), but we are all one species. How do biologists decide whether organisms belong to the same species or not?

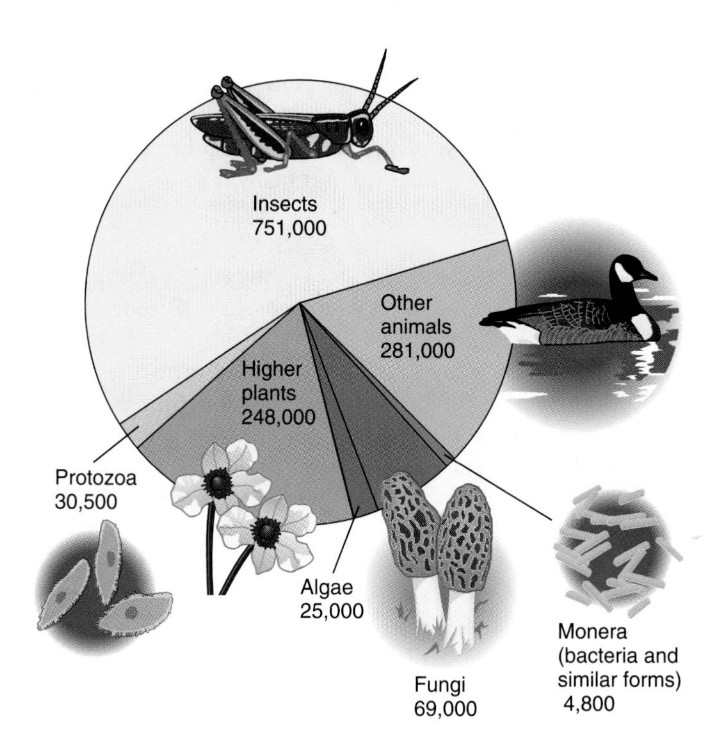

Figure 2.1

At present, over 1.4 million species of living organisms have been identified, divided by major groups as shown here. The known species are probably a small fraction of the actual numbers.

Figure 2.2

This plate from a popular field guide shows several species of North American warblers, arranged to help observers identify them in the field. (Plate 49 from *A Field Guide to the Birds,* 2d ed. by Roger Tory Peterson. Copyright © 1974 by Roger Tory Peterson. Reprinted by permission of Houghton-Mifflin Company. All rights reserved.)

Figure 2.3
All humans belong to one species, but the species has enormous variability. Even in a population of people who are closely related and have similar general features, it is easy to recognize individuals.

One central task in biology is to clearly define a species and establish criteria for determining the limits of each. Categorizing and naming organisms is known as **systematics** or **taxonomy.** After more than two centuries of work, systematists have identified well over a million species of animals, about 250,000 species of plants, and several thousand species of algae, fungi, and microorganisms. Many now-extinct species have also been named. Still, some authorities estimate that the number of living species yet to be identified far exceeds those we already know. Meanwhile, biologists are concerned that humans are rapidly destroying the natural areas where the majority of unknown species probably live.

Biology must explain the incredible diversity of life. *Why* are there so many kinds of organisms in the world, rather than just a few? Where did they all come from? The modern theory of evolution addresses these questions, and the study of evolution begins by identifying and systematizing the world of organisms. Systematics is also concerned with explaining the origins of species, showing their natural relationships, and explaining how they have evolved.

2.2 Species are shaped through selection.

In 1155 A.D., the Genji family of Japan finally achieved supremacy by defeating its rivals, the Heike, in a naval battle at Dan-no-ura. According to legend, the defeated Heike, following the only honorable course prescribed by samurai tradition, threw themselves into the sea in a mass suicide. They were immediately reincarnated as crabs, their shells sculpted with the face of a Japanese warrior (Figure 2.4). You don't have to believe the legend, or even have much imagination, to see the samurai's

Figure 2.4
The Heike crab got its name from a Japanese legend. The pattern on its shell is thought to resemble the face of a Samurai warrior.

face on the back of the Heike-gani, or Heike crab *(Dorippe japonica)*, and related crabs found in this region off southwestern Japan. The Heike crab was once called Oni-gani, *Oni* meaning a frightening demon that might be the spirit of a dead man, and though the species is edible, the Japanese will not eat them.

The English biologist Julian Huxley, who first told this story, pointed out that the resemblance to a Japanese warrior is far too specific and detailed to be accidental. Yet the explanation for the image on the crab's shell is quite simple. Long ago, some of the crabs must have borne shells on which an imaginative person could see the outlines of a human face. Superstitious Japanese fishermen very likely threw back any such crabs they pulled from a fishing net, keeping only the unsculpted crabs. Year after year, generation after generation, more and more surviving crabs were those with the clearest sculptures on their shells, until the whole species was gradually transformed into its present appearance.

So for centuries the Japanese selectively allowed crabs with a certain form to live and reproduce—unwittingly doing what a breeder of plants or animals does purposefully. Breeders, desiring cows that give more milk, select and breed only the best cows in their herd; farmers sow seed from only the finest plants of last year's crop. Through conscious selection, they shape their plants and animals into the most desirable forms.

Around 10,000 years ago, humans began to trap and tame a small wolf that they eventually bred into the dog. But how strange it is even to say "the dog," for dogs have developed into an amazing variety of forms (Figure 2.5), and anyone encountering a Saint Bernard and a Chihuahua in the wild would hardly be inclined to consider them the same species.

If humans can mold plants and animals by selecting breeding stock with desirable features, perhaps a similar process could account for the origin of all organisms, including humans. This was a critical insight that Charles Darwin and Alfred Russel Wallace, in the middle of the nineteenth century, brought to the problem of the origin of species. Darwin and Wallace argued that, just as breeders impose an *artificial*

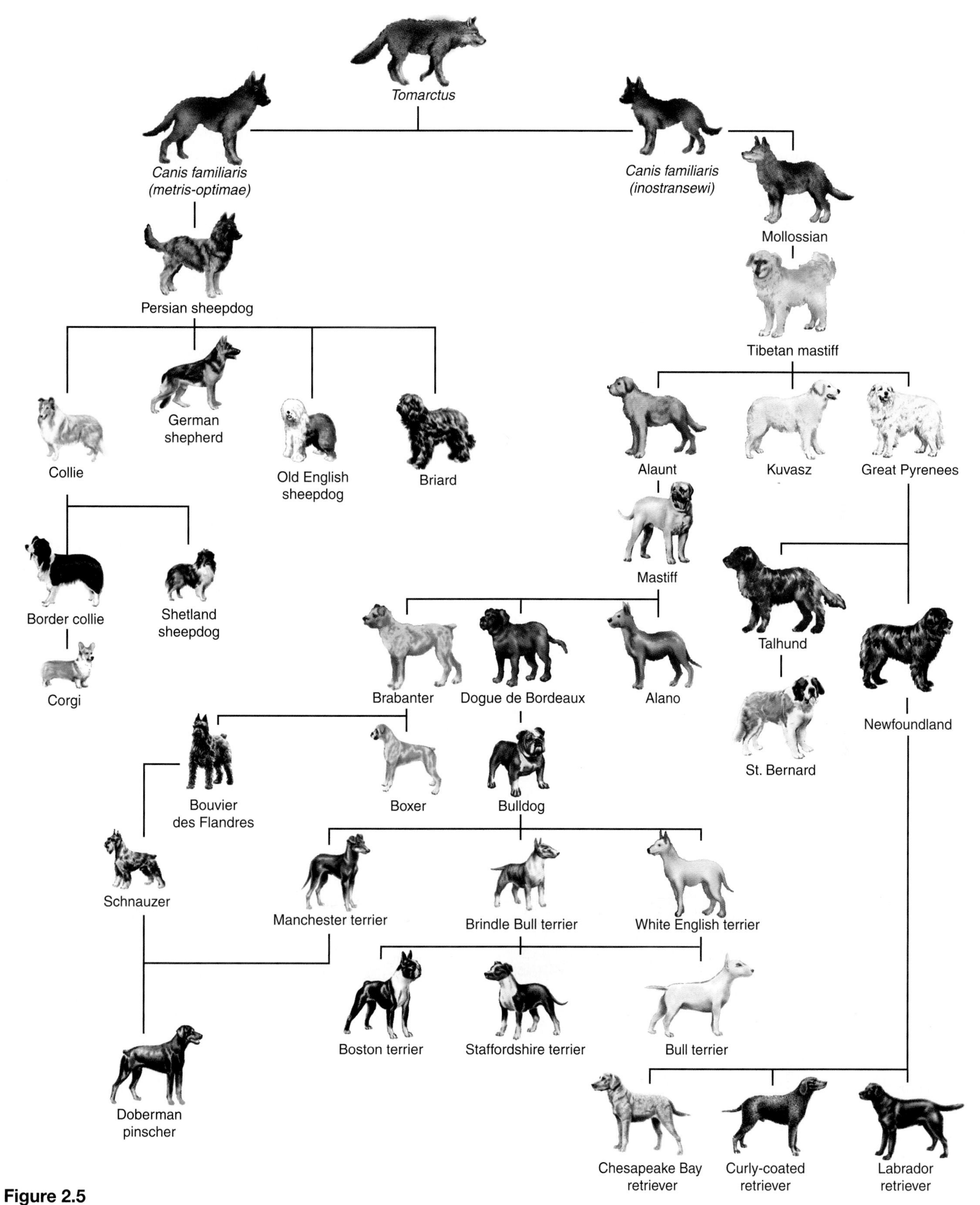

Figure 2.5

All these breeds of dogs belong to the same species. The breeds are arranged to show how they were derived from one another through selective breeding.

selection on their plants and animals, evolution occurs through *natural selection* imposed, without foresight or purpose, by the forces of nature. Darwin's book *On the Origin of Species* (1859) presented so much evidence and argued these new ideas so convincingly that most rational people were persuaded. To understand how these ideas developed, we must first trace the history of thought about the nature and origin of species.

2.3 The concept of evolution developed slowly with the growth of modern science.

Western civilization's conception of the earth and its place in the universe has long been dominated by a blend of Greek science and Judeo-Christian theology. People saw the earth as the center of a universe made of concentric spheres bearing the sun, moon, planets, and stars (Figure 2.6). No one had any comprehension of the vastness of space—or any reason to doubt the biblical story that the universe had been created by divine fiat about 4000 B.C., followed by the separate creation of each living thing. This picture, however, began to fall apart with the Renaissance, when the new view of the universe developed by Copernicus, Kepler, and Galileo showed the earth to be only one of several planets orbiting the sun, thus removing it from the central position human vanity had assigned it. New generations of astronomers started to portray the universe as a vast space far older than a few thousand years and occupied by many suns. But naturalists still believed that all living organisms had been created independently in a very short time. Not until the nineteenth century did serious challenges to the story of creation appear and people begin to accept that life has developed slowly, through biological evolution.

Religious tradition aside, two obstacles to thinking kept scientists from abandoning the traditional viewpoint. One is *essentialism*, a philosophy that goes back at least to Plato. Essentialism proposes that everything in the world has a distinct, unvarying *essence* underlying its outward features. All horses, for instance, possess an ideal of "horseness"—that is, while they may be short or tall, swaybacked or straight-backed, all horses are fundamentally the same below their surface appearances because they all share the same essence.

In biology, this way of thinking became *typology*. Biologists in the seventeenth and eighteenth centuries considered that every species conforms to an idealized, characteristic *type* and all members of the species are just like that type. Even if some organisms are really quite variable and hard to classify, the typological mind tries to ignore the variation among individuals and sorts them into neat species anyway (Figure 2.7).

Those early biologists who saw the world typologically thought the variations among individuals are trivial, and simply assumed that each species had been created separately. John Ray (1627–1705), one of the fathers of modern biological classification, reflected his contemporaries' thought when he said that two individuals belong to the same species if one is the ancestor of the other or if both are descended from a common ancestor. With this definition in mind, typological thinkers could not even conceive of the possibility of evolution (Figure 2.8). A question such as, "Could the wolf and the fox have had a common ancestor?" would make no sense to them. Such an ancestor must have been either a wolf or a fox. If it were a wolf, it could not have been the ancestor of a fox, and vice versa. So the question would never even arise.

A second reason the concept of evolution arose slowly was that people had little comprehension of the age of the earth.

Figure 2.6
At the time of Galileo and Copernicus, the universe was thought to be a series of spheres, with the earth at the center. Notice how the orbit of the sun is represented by a circle of light around the earth, with the orbits of the planets beyond it.

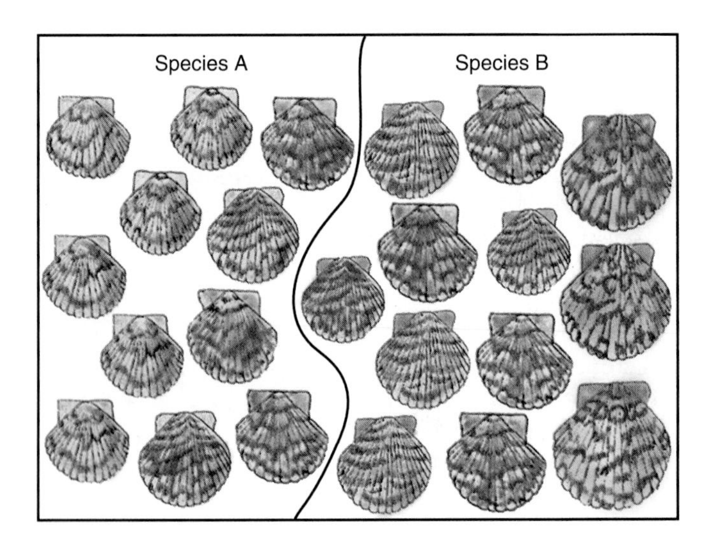

Figure 2.7
The individuals in a group of organisms may actually vary continuously, with no obvious sharp differences among them. Typological thinking, however, forces them into distinct categories, ignoring their complexity.

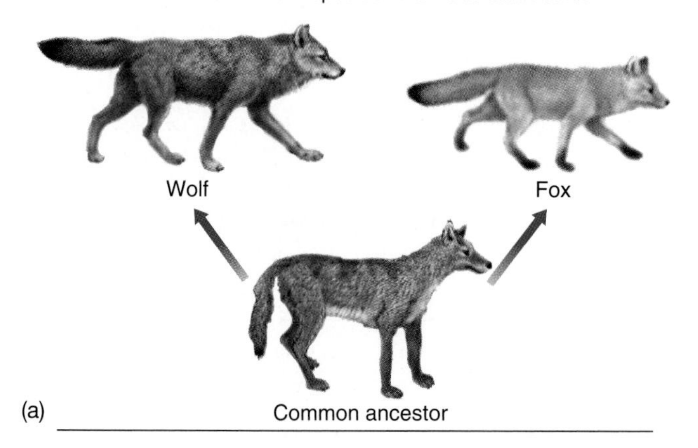

Figure 2.8

(a) If a species is defined by reference to contemporary organisms, such as the wolf and fox shown here, it is possible to imagine how one ancestral species became two. But if a species is defined by reference to its ancestors, as in *(b),* all individuals in a lineage must be the same species, and there is no way to even conceive of evolution.

Even if they could have imagined a change from one species to another, such a change would require a long time, and it was generally believed that the world had been created less than 6,000 years earlier. Eventually, however, a few innovators began to ask scientific questions about the age of the earth and to extend the time scale a little. For example, George Louis Leclerc, Comte de Buffon (1707–88), seriously considered the possibility that the earth and planets had been formed from very hot material, and he conducted experiments to test the rate at which metal balls cool off after being heated. From his data, making the false assumption that the sun's radiation has little effect on the earth's temperature, he calculated that the earth must be about 75,000 years old. He was wrong, of course, but at least he had started to ask the question empirically, through observation and experimentation.

Other scientists—among whom James Hutton (1726–97) stands out—began to realize that geological forces must have been shaping the earth's features for an exceedingly long time. They saw rocks deposited in layers, or strata, which must have been laid upon one another over eons, and they noticed many strata contained fossils that were obviously the remnants of ancient organisms. Some naturalists, seeing that many of these fossilized remains were clearly different from any living species, postulated that the earth had experienced a series of geological catastrophes that had wiped out all life on Earth or at least all life in a large area of it. Some of these *catastrophists* imagined that an entirely new creation had taken place after each of these events, perhaps with each new flora (plant life) and fauna (animal life) a little more advanced than those before. To explain some of their observations, though, the catastrophists had to postulate over twenty separate extinctions and creations, and the theory began to look ridiculous. Catastrophism was struck a final blow in 1830–33, when Charles Lyell's great work *Principles of Geology* laid the foundations for modern earth science by arguing most convincingly that geological forces are at work constantly and rather uniformly, a viewpoint that became known as *uniformitarianism.* Of course, the earth is occasionally rocked by devastating earthquakes, volcanic eruptions, and collisions with meteors, but none of these events destroy all life on the planet. It is clear now that, even as movements of the earth's crust are raising new mountain chains, the mountains are being worn down slowly by erosion and their rocks deposited in beds where plant and animal remains become buried and eventually fossilized. This uniformitarian view of geology prepared the way for thinking that organisms, too, are gradually changing.

2.4 Evidence from the fossil record and the study of comparative anatomy indicates that organisms are related by evolution.

Radically new ideas, such as evolution in biology or relativity in physics, develop only when more conventional theories can no longer explain observations. As nineteenth-century naturalists noted the detailed similarities among species, they became skeptical of the conventional explanations and were forced to consider some kind of evolutionary theory.

What kind of observations make us turn to evolution to explain the diversity of organisms? The fossil record provides the most persuasive evidence for evolution. Each layer of rock records the flora and fauna of a particular time, and even before the development of modern methods for determining their ages by means of radioactive elements that decay at known rates, it was obvious that the lower layers of rock are older than the upper layers. The fossils within these layers of rock show that the life of past eons differed from contemporary life not only in small ways but in large ways as well (Figure 2.9). The fossil record reveals that different kinds of organisms successively appear and disappear, one group after another. The first traces of life are the remnants of unicellular (single-celled) organisms. Some of the earliest animal fossils, from around 600 million years ago, were so different from any animals living today that we hardly know what to make of them; yet several million years later, they were replaced by more familiar types, such as clams,

Era	Period	Epoch	Major Events in Earth History
Cenozoic	Quaternary	Holocene — 10,000	
		Pleistocene — 1,600,000	Earliest humans; Modern horse evolves in North America, then dies out; Ice Ages; Grand Canyon carved; Cascade Range and Pacific Coast Ranges formed
	Tertiary	Pliocene — 5,300,000	
		Miocene — 23,700,000	Rapid spread and evolution of grazing mammals
		Oligocene — 36,600,000	Earliest elephants
		Eocene — 57,800,000	First primitive horses, rhinoceroses, and camels
		Paleocene — 66,400,000	First primates; Initial uplift and folding of Rocky Mountains
Mesozoic	Cretaceous	144,000,000	Extinction of dinosaurs; Great evolution and spread of flowering plants; Half of North America covered by seas; Initial uplift of Sierra Nevada
	Jurassic	208,000,000	First birds and mammals; Dinosaurs at their peak; Dinosaurs; Pangaea divided into Laurasia and Gondwanaland
	Triassic	245,000,000	Arid climates in much of western North America; Continents have converged to form Pangaea
Paleozoic	Permian	286,000,000	Mammal-like reptiles; Ice Ages in Southern Hemisphere; World climate much like today; Deserts in western United States
	Pennsylvanian	320,000,000	First reptiles; Large insects; Widespread swamps (coal source); Tropical climate in United States; Uplift and folding of Appalachian Mountains
	Mississippian	360,000,000	Sharks; Widespread flooding of North America; limestone deposited
	Devonian	408,000,000	First amphibians; First forests
	Silurian	438,000,000	First air-breathing animals (scorpions); first land plants; Trilobites; Deserts in eastern and central United States
	Ordovician	505,000,000	Trilobites at peak; First vertebrates (fish); Widespread flooding of North America by seas
	Cambrian	570,000,000	Ediacaran fauna; rapid diversification of animals
Precambrian	Proterozoic	2,500,000,000	Eucaryotic cells; Glaciation–probably worldwide; Many mountain systems uplifted and eroded
	Archean	3,800,000,000	First simple (procaryotic) cells
Hadean		4,500,000,000	Formation of the earth

Figure 2.9

The geological time scale shows the major events in the changing fauna and flora of the earth. Numbers show the approximate time since the beginning of each era, period, or epoch.

corals, and snails. All these animals were *invertebrates;* they lacked the bony internal skeleton of animals such as fish, birds, and mammals, which are called *vertebrates.* The first vertebrates do not appear in the fossil record until about 400 million years ago—200 million years after the earliest invertebrates! Figure 2.9 shows the successive changes in vertebrates: first primitive jawless fishes, later fishes with jaws, and still later, amphibians, the first terrestrial (land-dwelling) vertebrates. It took another 50 million years or so for the reptiles to arise, and they dominated the scene for a long time. Traces of birds and mammals date from about 200 million years ago, but neither group began to flourish until 60–70 million years ago, after most of the reptiles had disappeared.

Plants show the same kind of successive change. The most primitive fossil plants are simple algae. The first vascular plants—those with tubelike cells for transporting water—appeared around 400 million years ago. No conifers such as pine and fir trees existed until about 225 million years ago, and flowering plants first appeared 140 million years ago.

Only evolution can explain this sequential fossil history. If all species of organisms had been created at one time, their fossils would all be in the same layer, and we would not see the systematic appearance and disappearance of various kinds of organisms in the rock strata.

Clear evidence showing the course of evolution also comes from comparative studies of biological structure. Today, we can see some of the most compelling evidence for evolution in the structures of biological molecules, but nineteenth-century anatomists and naturalists had to confine their studies to large anatomical features, especially bones, teeth, and shells. Take vertebrate skulls, for example. A skull is not a single bone but an assemblage of about forty bony plates that grow together. All skulls show remarkable similarities, with systematic differences from one to another (Figure 2.10). Most of the bones in a fish skull, for instance, have counterparts in the skulls of amphibians, reptiles, and mammals, both modern and extinct, but their shapes and positions are gradually modified in later and later forms. Occasionally a new bone appears or an old one is lost. We can only understand such changes in the light of evolution—if each kind of organism descended from some other, with gradual modifications in its structure along the way.

Look at the situation from an engineering viewpoint. Organisms are obviously very functional. They are well-designed for the lives they lead. If each type had been created independently, designed by some cosmic engineer, there would be no reason to use the same structural elements for fish, snakes, mice, elephants, whales, birds, and bats, all of which have very different anatomical requirements. Yet all these varied animals are made of the same units, just modified from one group to another. Equivalent structural elements in different species are said to be **homologous** if they occupy similar positions and come from the same place in a developing embryo. In contrast, the wings of insects and birds are **analogous** because they serve the same function but are not made of the same structures. Homology argues for evolution—for continuity with modification. If the biological world were dominated by merely analogous structures, the evidence would lean more toward independent creation.

With careful study one can usually identify homologous structures on the basis of their relative positions. Thus we identify particular skull bones of various species as homologous and give them the same names (nasal, frontal, prefrontal) by tracing the bones from one species to another, taking into account their modifications (Figure 2.10). The functions of homologous bones may change, but in modern species they can be identified by their common embryological origins, since each one arises from a distinct place in the embryo.

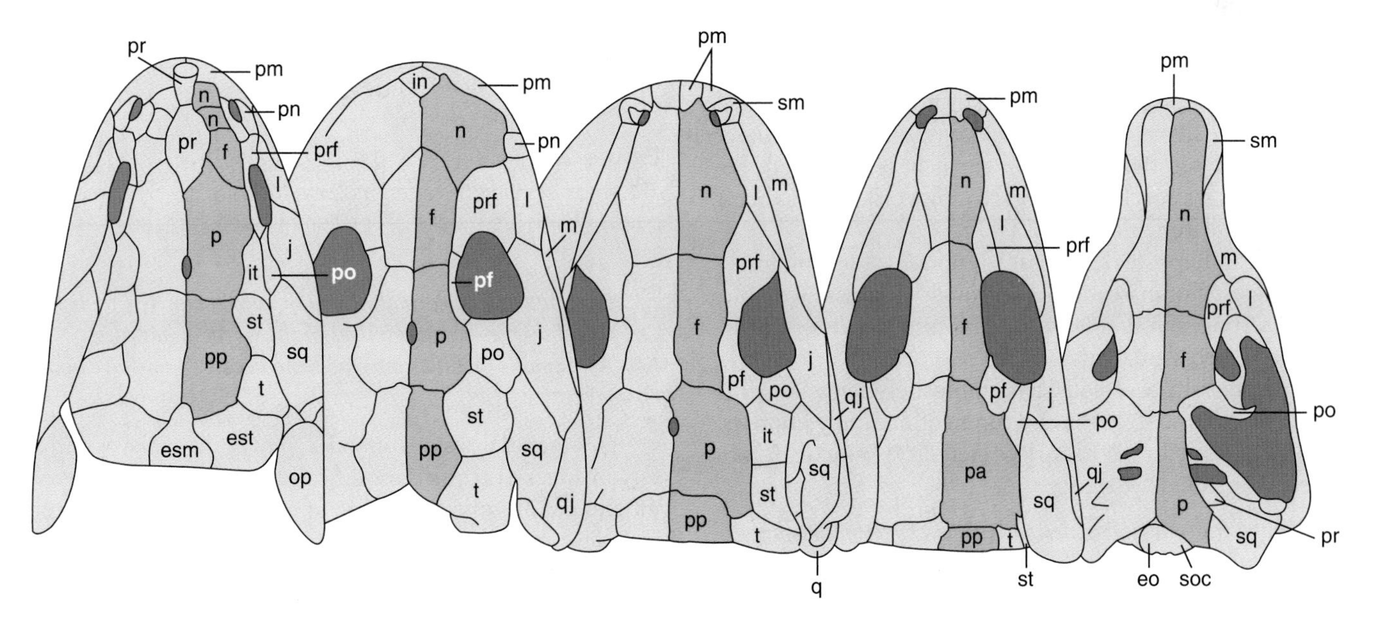

Figure 2.10

The skulls of several vertebrates share common structural features that show their common ancestry. Here several bones in each skull are identified, with four emphasized *(in orange)* for comparison: nasal *(n),* frontal *(f),* parietal *(p),* and postparietal *(pp).*

Vertebrate limb bones show engineering sense—or the lack of it—particularly well (Figure 2.11). Here we can see how homologous structures were modified for different functions, so the bones that originally made up the fin of a fish were gradually shaped into the different limbs of land animals. In some animals one bone is a mere vestige that has no function at all; it is still present only because the animal's ancestors had it and it hasn't been entirely lost yet. The existence of vestigial structures, such as the human appendix, points clearly to biological evolution.

We now see systematic changes in the fossil record and homology in structure as clear evidence for biological evolution. But even though scientists before Darwin also had this evidence, they did not propose that organisms had evolved from one another, principally because no one could explain how such evolution could happen. This is what Darwin supplied, in the concept of natural selection.

Exercise 2.1 Suppose one biologist says, "The successive changes we observe in the fossil record are best explained by evolution," while another argues, "I admit there is a kind of succession in the fossil record, but it can be explained just as well by a long series of creations, one after another." Do the general principles of scientific explanation presented in Chapter 1 require us to prefer one argument over the other?

2.5 Darwin and Wallace outlined the general process of natural selection.

While scientists of the late eighteenth and early nineteenth centuries knew about fossils and the anatomical similarities we now call homology, almost all stuck tenaciously to the idea of separate creation. That may seem peculiar, but old ideas die hard. Indeed, when Robert Chambers ventured to propose an evolutionary theory in 1834, he did so anonymously, and with good reason. His book, *The Vestiges of the Natural History of Creation,* was roundly attacked and castigated, even by people who later became fierce advocates for evolution. And although Chambers could see that evolution must have occurred, he could not provide a mechanism to account for it. Other theorists had the same difficulty.

In 1809 Jean Baptiste de Lamarck proposed a new philosophy of biology including evolution through a process we now know to be wrong. But it is worth mentioning because similar ideas, labeled "Neo-Lamarckian," keep reappearing in biology. Lamarck's philosophy postulated a harmony between an organism and its environment, so the environment naturally imposes itself upon heredity. In his history of heredity, François Jacob describes Lamarck's views thus:

> *. . . only the transmission to descendants of experience acquired by individuals appeared to account for the harmony between organisms and nature. Never had this idea been exploited so systematically and with so much detail, however—nor so confidently since Lamarck took for granted that an organ disappears because it is of no use. For him, whales and birds have no teeth because they do not need them. The mole lost the use of its eyes because it lives in the world of darkness. Acephalous molluscs have no head because they have no need for it.*[1]

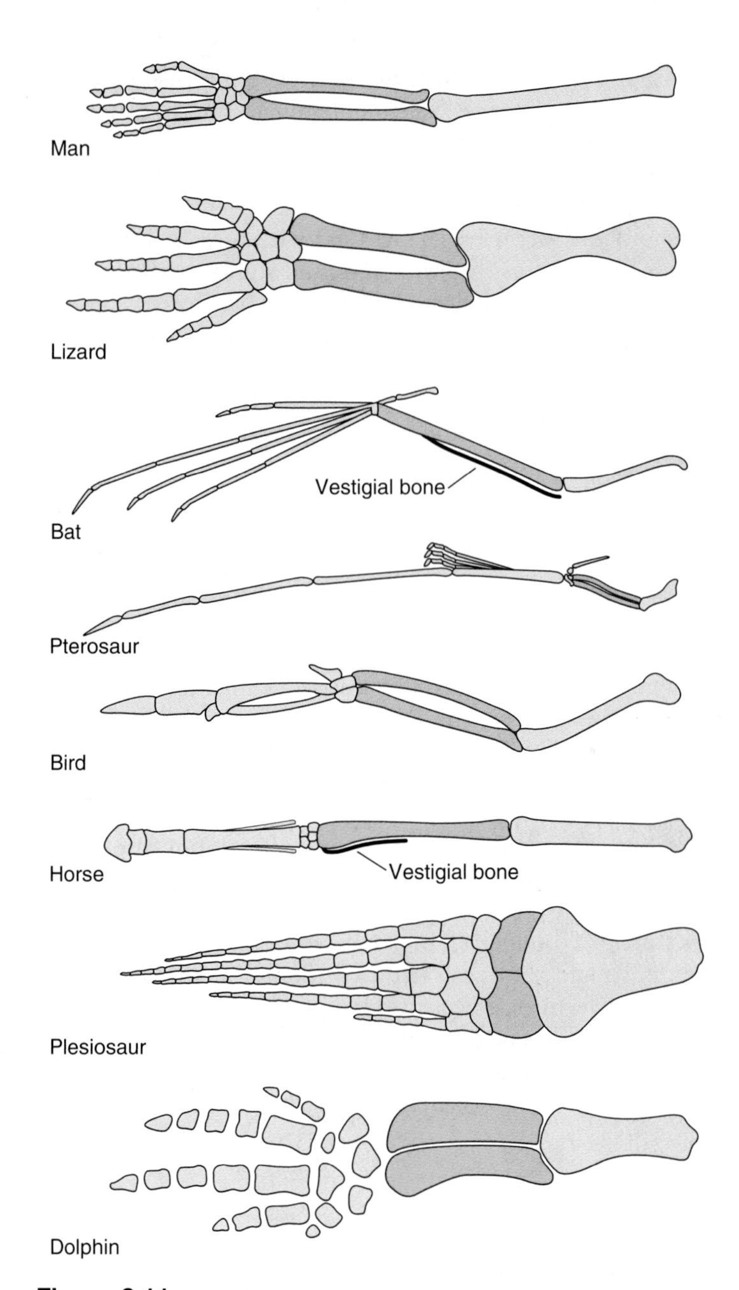

Figure 2.11
The limbs of different vertebrates are made of clearly homologous bones, which are modified for different functions in each species.

In a general way, Lamarck's viewpoint makes good sense. Organisms do acquire modifications in response to environmental conditions. A running animal develops strong running muscles, and an animal that continually rubs some part of its body grows protective calluses. Lamarck thought such acquired features were somehow incorporated into an organism's

[1]François Jacob, *The Logic of Life,* pp. 149–50. New York: Pantheon Books, 1973.

heredity, endowing its offspring with a greater tendency to have the same useful features; in the same vague way, an organism that didn't use something would simply lose it (Figure 2.12). While Lamarck's proposal had a certain appeal, it did not really *explain* evolution, as explanation is understood by modern science. That is, it did not provide even the outline of a real mechanism that could account for a hereditary change in response to the environment. It happened, too, that the prevailing intellectual climate in both France and England was inhospitable to Lamarck's philosophy and to all ideas about evolution, and they were not accepted.

Faced with evidence demanding an evolutionary theory, but with no satisfactory mechanism, Darwin and Wallace developed their concepts on the basis of their extensive experience as field naturalists. Rather than staying in museums and examining dead specimens, they traveled to distant lands and spent long periods in the field observing, experimenting, and thinking about the lives of the organisms they saw. Their field experience taught them to think about **populations** of organisms, all the individuals of one kind that live in one area. They saw that populations are highly *variable*—that is, the individuals in each population vary considerably in size, shape, coloration, and other features. They also saw that populations remain quite stable in size, even though organisms have an immense capacity for reproduction. Apparently most individuals die before they have a chance to mature and reproduce. From this realization it was only a short step to conclude that the organisms that do survive to produce another generation must be those most fit for living

(a)

(b)

(c)

Figure 2.12

The Lamarckian viewpoint is that organisms are in such harmony with their environment that their characteristics become modified in accordance with their needs. Thus, giraffes *(a)* acquire longer necks as they try to reach higher leaves on trees, while cave animals, like the fish and crayfish *(b, c)* lose their eyes because they have no need for them. Although this explanation sounds reasonable, it does not explain how these events could occur, as a modern explanation does.

in their habitats. The principle of natural selection that Darwin and Wallace recognized can be summarized in four points:

1. Every organism has the potential to produce more offspring than can survive.
2. There is always variation among individuals in a population; some of this variation is inherited, so the next generation inherits some of these variable features from their parents.
3. Specific variations may make an individual either more or less likely to survive and reproduce than individuals with different variations.
4. Variant traits that enhance survival and reproduction will be passed on to offspring and will be found in more of the population in each succeeding generation.

The principal problem Darwin faced in developing the idea of natural selection was that it depends on heredity, but in his time heredity was hardly understood at all. To understand what a species is and how species evolve, we must first understand what an individual organism is, and we turn to this question next.

Figure 2.13
Reproduction is one of the most universal and obvious features of the biological world.

2.6 An organism is a structure that functions on the basis of information carried in its genome.

It has taken biologists a long time to realize that the process of evolution reveals something quite fundamental about the nature of organisms. Evolution and natural selection are only possible because organisms can reproduce themselves using a hereditary mechanism that is very stable but is also able to change. Here is the fundamental paradox, or near-paradox, of biology: stability with just a touch of instability.

To explain evolution and the other properties of organisms, we will develop a genetic model. The term **model** is used in two senses in science, and here we use it to mean a mental construction that summarizes and represents the most critical features of a system, enabling us to think about its structure and operation in order to understand it. A **system** is any part of the universe we choose to isolate and study. A good model serves not only as a theoretical basis for a science but also as a guide to learning the concepts of a science; throughout this book we will show how a genetic model organizes the central ideas of biology and makes them coherent.

Organisms reproduce themselves (Figure 2.13). This is one of the most basic facts of biology. Reproduction necessarily includes *inheritance.* Sugar maple trees produce more sugar maples, St. Bernards produce St. Bernard puppies, and humans produce small humans. Offspring tend to have the same features as their parents. We say that organisms are *genetic systems* because "genetic," like "genesis" and "generate," refers to the origin of something, and organisms are systems that owe their origins to their parents, through the processes of reproduction and inheritance.

Saying that two things have the same features, the same structure, says something about the **information** they carry. Concepts 2.1 explores the concept of information that specifies its characteristics—Tom, rather than Dick or Harriet, or a blue flower with three petals rather than a yellow one with five. The information that specifies biological structure is **genetic information.**

Just as the information humans exchange is encoded in a physical form, such as printed letters on paper or an electronic pattern on a computer disk, genetic information is encoded in the form of carbon-based molecules that constitute an essential set of instructions called an organism's **genome.** When organisms reproduce, they pass on copies of their genomes to their offspring, and so offspring resemble their parents.

A genome is the collection of the organism's *genes,* and it is now grade-school knowledge that genes are made of molecules of deoxyribonucleic acid, or DNA (Figure 2.14). DNA has just the properties required of a genome. DNA is an *informational* molecule made of subunits that can be arranged in any sequence to carry genetic information, just as different sequences of letters convey different messages. DNA also has the ability to make copies of itself, or to *replicate.* Replication is critical because an organism must make exact copies of its genome to pass on to its offspring—or at least *nearly* exactly copies, since the ability to change slightly is also important. Organisms probably could not exist if DNA (or the chemically similar ribonucleic acid, RNA) did not exist.

Because reproduction is so fundamental, some biologists have proposed that an organism should be defined by its ability to reproduce. However, some organisms, such as mules and infertile people, can't reproduce, yet they are still organisms because they carry out metabolism and operate on the basis of instructions carried by their genomes. For this reason, the genetic model focuses on the genome rather than on reproduction and inheritance.

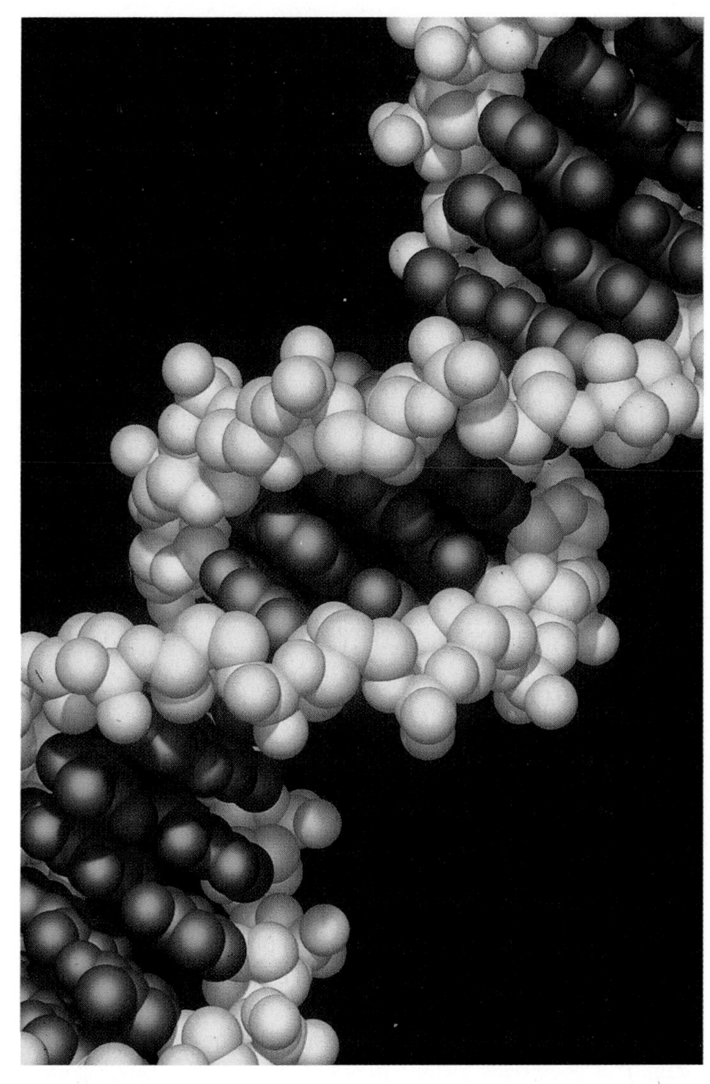

Figure 2.14

The genomes of all organisms are made of DNA (deoxyribonucleic acid), a molecule that carries information and can replicate itself.

2.7 An organism is like a self-reproducing automaton.

The second sense of "model" in science is an analogy. For example, when physicists early in this century realized that an atom has a relatively massive nucleus and also contains electrons, they made a model of atomic structure in which electrons revolve around the nucleus as planets revolve around the sun, and this is still a good model for many purposes. Our model to illustrate the operation of an organism and the concept of a genome is a computer. A computer can perform enormous numbers of operations quickly and reliably if it is given a program with detailed operating instructions. An organism operates in much the same way, as we can illustrate by examining a specialized computer, a universal constructor (Figure 2.15). Fitted with an array of tools (screwdrivers, wrenches, soldering guns, and so on), this machine can assemble small parts to make all kinds of devices. We only have to fill a series of bins with the necessary small parts and

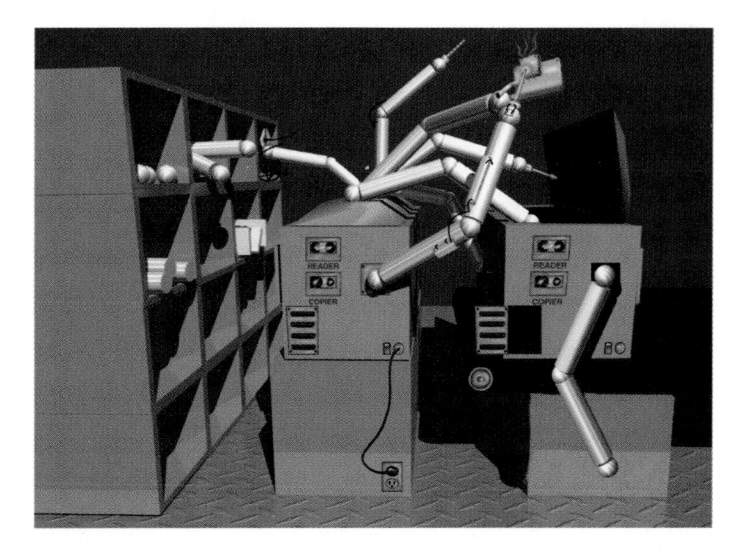

Figure 2.15

A universal constructor working to build a machine by assembling small parts according to instructions in its tape is like an organism that reproduces itself by following instructions in its genome. Notice that the instruction tape is being read in one slot and that there is another slot where a blank tape can be inserted for copying.

give the constructor a tape with operating instructions: "Take plate from bin #1. Drill 3-mm hole in position (7.5 cm, 6.4 cm). Take bolt from bin #35. Insert in hole. . . ," and so on, although clearly in machine language, not in English. The constructor is an automaton, a sophisticated robot capable of making a television set, or anything else, just as robot units already perform many of the steps in assembling automobiles and other devices.

Now let's change the instruction tape. Instead of directing the constructor to make a television set, suppose we tell it to make another universal constructor. The constructor will follow its tape and make another device exactly like itself. The last instructions on the tape say, "Put a blank tape in the tape copier. Make a copy of this instruction tape. Rewind the copy, put it into the new constructor, and start it running." When this is done, the new constructor will start to work just like the old one. It will make a third constructor. Meanwhile, the first constructor will start working again to make a fourth constructor. Pretty clearly, if this goes on for long, the world will be swarming with constructors—just as it would be swarming with oysters or sparrows or anything else that kept reproducing itself without limit.

The universal constructor, with a tape telling it to reproduce itself, is like an organism. Look what it does:

- It takes energy from its environment. (Maybe it is plugged into a source of electricity. Better yet, it could have panels for collecting solar energy.)
- It takes raw materials from its environment. (The small parts are manufactured elsewhere, but the constructor could be programmed to make its own.)
- It uses the energy to assemble the small parts into another structure like itself.
- Its entire operation is determined by the instructions on its tape. This tape is its genome.

Concepts 2.1 Information

Pick up the phone and dial 1-555-1212.
"Directory assistance," says the operator.
"Doe," you say, "John C."
"The number is 987-6543."
"Thank you."

You have just received some information. But what exactly have you gotten? Is the information a thing? Can you measure how much you have received? Before you called, you were very uncertain about John C. Doe's phone number. It might have been any of 10 million numbers. But now you have no uncertainty at all; you know which of 10 million numbers is the right one. By definition, you receive information when you reduce the number of possibilities in some range. For instance, you might get some information about the right phone number by eliminating certain prefixes, narrowing the range from 10 million to, say, 100,000, but you still wouldn't know what number to call. However, you get a great deal of information when you narrow the range to only one number. The information exchanged in a transaction is measured by the ratio of the *number of possibilities beforehand* to the *number of possibilities afterward.*

The game "twenty questions" makes an important point about measuring information. In this game, you are supposed to guess what someone is thinking of by asking only twenty questions that have "yes" or "no" answers. Since you can only get one of two answers, a wise strategy is to cut the range of possibilities roughly in half with each question. For instance, "Is it in the Western Hemisphere?" "Is it north of the equator?" "Is it west of the Mississippi?" and so on.

Computers also operate on a binary logic system by using electronic elements that are either "on" or "off." In computer language, reducing the amount of uncertainty—the range of possibilities—by a factor of two conveys one *bit* (an abbreviation for "binary digit") of information. So to determine the number of bits of information received, express the reduction in uncertainty as a power of 2. For instance, suppose one out of 1,000 possibilities is specified. Since 1,000 is about 2^{10}, about 10 bits of information have been received.

Information is usually carried in a *sequence* of symbols. It takes information to specify the sequence:

Please wipe your feet on the mat.

instead of the sequence:

phshr zaal t spuprn skjey ml ohg.

A small change in a sequence can make a big difference:

Please wipe your feet on the cat.

Technically, all these sequences carry about the same amount of information, even though their meanings are quite different.

Information is carried in organisms through strings of symbols too, but those symbols are the subunits of DNA molecules, which can be arranged in different sequences. Evolution consists of selecting organisms with the most functional sequences.

This is what a real organism does. Its genome contains a *genetic program* that dictates the organism's structure and, therefore, its basic operation. Real organisms are made of complex, carbon-based molecules, and their metabolism—all the processes of getting energy and transforming small parts—is chemical rather than mechanical and electronic.

To reinforce the key role of the genome, let's vary this story. The device shown below consists of a tape cassette plus the barest mechanical parts necessary to move around and find universal constructors.

When it finds a constructor, it ejects the constructor's tape (genome) and inserts its own. When the constructor reads this new tape, it receives instructions for making copy after copy of identical mobile tape cassettes, which go out into the world looking for more universal constructors. So the critical element of the constructor is its genome: different genome, different product. The mobile tape cassette that can subvert the constructor is a **virus.** It is essentially an independent genome that goes around using the resources of organisms for its own reproduction. You can see that viruses and organisms are quite different, though both are genetic systems and both are part of the world biologists study.

Viruses, Section 12.5 and Section 29.14.

The genetic model not only serves as a foundation for the science of biology but also clarifies the old bugaboo of teleology. As we pointed out in Chapter 1, organisms have an intrinsic characteristic called *teleonomy,* which means they appear to be purposeful. The evolutionary biologist Ernst Mayr has defined a process or behavior as teleonomic if it owes its goal-directedness to the operation of a program. A computer again is the perfect example, because a computer following a program has a kind of goal—to achieve the purpose of the program. The computer, of course, is not aware of having goal-directed behavior, and it did not choose to carry out the program. It is merely a dumb machine doing what it has been programmed to do. Because an organism follows the directions in its genome, it too is goal-directed. It has inherited its program from its ancestors, who shaped that program in the course of their evolution. Because its ancestors were successful, an organism that carries out the program it has inherited has a reasonable chance of being successful—that is, maintaining itself and reproducing. (François Jacob has written jokingly of the "dream" of every cell: to become two cells.) But the organism is no more aware of what it is doing than the computer is. It doesn't consciously strive to reproduce itself or to leave successful offspring. The organism just does what its genome dictates. Only we humans, observing the organism with a certain scientific objectivity, can identify its goal-directedness.

Exercise 2.2 Based on the model described in this section, list some ways in which a virus differs from an organism.

2.8 Evolution occurs through natural selection operating on populations in ecosystems.

The earth, with its varied climates and physical features, provides many different environments, including marshes, swamps, streams, lakes, hot springs, coral reefs, city parks, prairies, forests, backyard gardens, and deserts. Each of these is an **ecosystem.** It is a physical environment of water, soil, and air that supports a **community,** a collection of plants, animals, and microorganisms that live in the same area and are potentially able to interact. The members of a community share living spaces and interact in complex ways (Figure 2.16). They eat each other and are eaten. They afford shelter and are in turn sheltered. They provide stages for one another on which each one acts out the drama of its life. No organism could live without the others in its community.

Figure 2.16
The members of a community live together and interact in complicated ways—sharing or competing for living spaces and by consuming one another. A community is largely structured by the ways its members use one another for food. Plants capture sunlight and thus bring some of the light energy into the ecosystem; animals consume this energy as they eat plants and other animals. Each organism is a source of energy and nutrients for other organisms.

The relationships within an ecosystem—the interactions among the members of the community and between the community and its environment—are its **ecology,** and the same word is used for the science of these relationships. (*Eco-* comes from Greek *oikos* = house, because an ecosystem operates rather like a household in which people serve different functions.) Ecology is the *economy* of an ecosystem. Ecology depends on the chemical fact that *molecules contain energy.* We can see this most clearly in a fire, where the molecules of wood release their energy as heat while they are converted into gas molecules (Figure 2.17). The members of a biological community contain energy in their structures, and they share their energy because one serves as food for another. Plants, which capture the energy of sunlight as they grow, are the energy foundation of most ecosystems, and these plants then serve as food for animals:

$$\text{Sunlight} \rightarrow \text{Plants} \rightarrow \text{Animals}$$

In this way, the members of a community support one another

Figure 2.17
The heat of a fire is the energy that was held in the molecules of wood. This energy is released as the wood molecules are broken down into small molecules that escape as gas.

by passing along their energy and their molecules. This means, also, that their metabolisms all depend on one another and are intertwined. The metabolic waste products of one organism become food for another. Plants grow by incorporating molecules of carbon dioxide, including those you are now exhaling, and in turn, produce molecules of oxygen, including those you are now inhaling. Humans should have a deep sense that we are parts of ecosystems. We should also realize that the environment is not something remote, "out there," but something we are intimately connected to. The water, the food, and the components of air that pass through us blend seamlessly with the molecules of our bodies. Whatever happens to our environment happens to us.

Biology only makes sense if we recognize that organisms have acquired their forms and functions through evolution, and evolution only makes sense if we view it ecologically, understanding that each type of organism is suited for a particular role in an ecosystem. In the metaphor of the noted ecologist G. Evelyn Hutchinson, the drama of evolution is played out in the theatre of the ecosystem. The result of natural selection is **adaptation,** meaning that each kind of organism becomes shaped for one particular way of life in one environment. An organism's way of life is called its **ecological niche** in the community. Niche, adaptation, and selection are interrelated terms: A *niche* is the particular place and way of life of each species; *adaptation* is the evolutionary process by which a species is shaped to live that way; and the shaping occurs through *selection* of genetic differences.

Selection is necessarily a *populational* matter, because evolution is a change in a natural population, with all its variation. Organisms spread out geographically also, and in each place, because of random events and the particular features that are selected, they become adapted to local conditions and different niches. As a result, a vast variety of different kinds of organisms inhabit the world.

2.9 Organisms measure out their lives in cycles of growth and reproduction.

The lives of organisms can be drawn as *cycles* that repeat, generation after generation. Life cycles vary from the very simple to the extremely complex (Figure 2.18), depending largely on the way an organism reproduces. A simple, unicellular organism such as an amoeba moves about, feeding and growing, until it has roughly doubled in size. Then it divides into two smaller cells, and each of them repeats the process. This is **asexual reproduction,** and it is characterized by having no stage during which different cells combine to make a new organism. By contrast, most multicellular plants and animals engage in **sexual reproduction** in which mature individuals produce sperm and egg cells (or their equivalents) that combine, during fertilization, to form a cell called a *zygote,* a new individual. The zygote generally grows and matures into an adult that repeats the process. Thus a species may consist of organisms with radically different structures because they are at different stages of their life cycle. Although an oak tree looks nothing like one of its egg cells or like an acorn, all three are stages in the life cycle of an oak. Even though a human egg, a sperm, and an adult are remarkably different, collectively they constitute the human species.

What ties different stages together and makes them part of one species? First, and most obviously, *one stage can develop into another.* For example, an egg becomes part of an acorn, which develops into an oak tree. Second, all the forms in one life cycle are united because *they share the same genome.* Each genome contains a program of instructions that specifies growth, development, and a way of operating, and all the stages read and follow instructions from different parts of that program.

In viewing organisms as life cycles, we come close to the dominant conception of what a species is. (This conception only applies to sexual species, because the reproduction of asexual organisms is so different that the word "species" must mean something different in each case). We asked earlier why each kind of warbler shown in Figure 2.2 is considered a separate species while all humans are considered a single species. In a population of animals, males and females regularly combine their sperm and egg cells to produce a new generation of zygotes; these zygotes grow into adults that shuffle their sperm and eggs in new combinations, and this goes on endlessly. However, Yellow-throated Warblers share one set of intertwined life cycles, Canada Warblers another, Cape May Warblers another, and so on. Even though species are identified primarily on the basis of **morphology**—form, size, structure, and color—biologists commonly think of a species as one of these reproductive units, a group of organisms that reproduce only with one another and are isolated from all other similar units. Humans constitute a single species because all humans, in spite of their morphological differences, can reproduce with one another. We will argue later (Chapter 22) that this is a very good conception of a species, although it entails many difficulties. For now, it is best to think of a species naively as a "kind," defined by distinctive morphology.

2.10 Populations of organisms tend to grow but are held in check by environmental factors.

Every organism has an enormous potential for reproduction. An amoeba divides into two cells, each of which divides again to make four cells; the four produce 8, then 16, 32, 64, and so on. In only ten doublings, a single cell grows to over a thousand. A graph of the number of individuals against time would soon go off the paper:

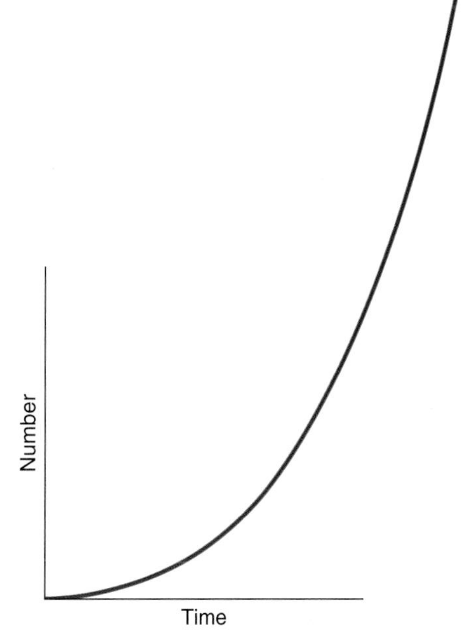

This pattern, called *exponential growth,* is typical of organisms growing in favorable surroundings—with no lack of food or space and no predators to eat them.

No population, however, can continue to grow exponentially for long, because eventually environmental limitations start to restrict growth. Every organism requires **resources,** such as living space, nutrients, and energy, that are necessarily limited. Organisms also become prey for other organisms, so

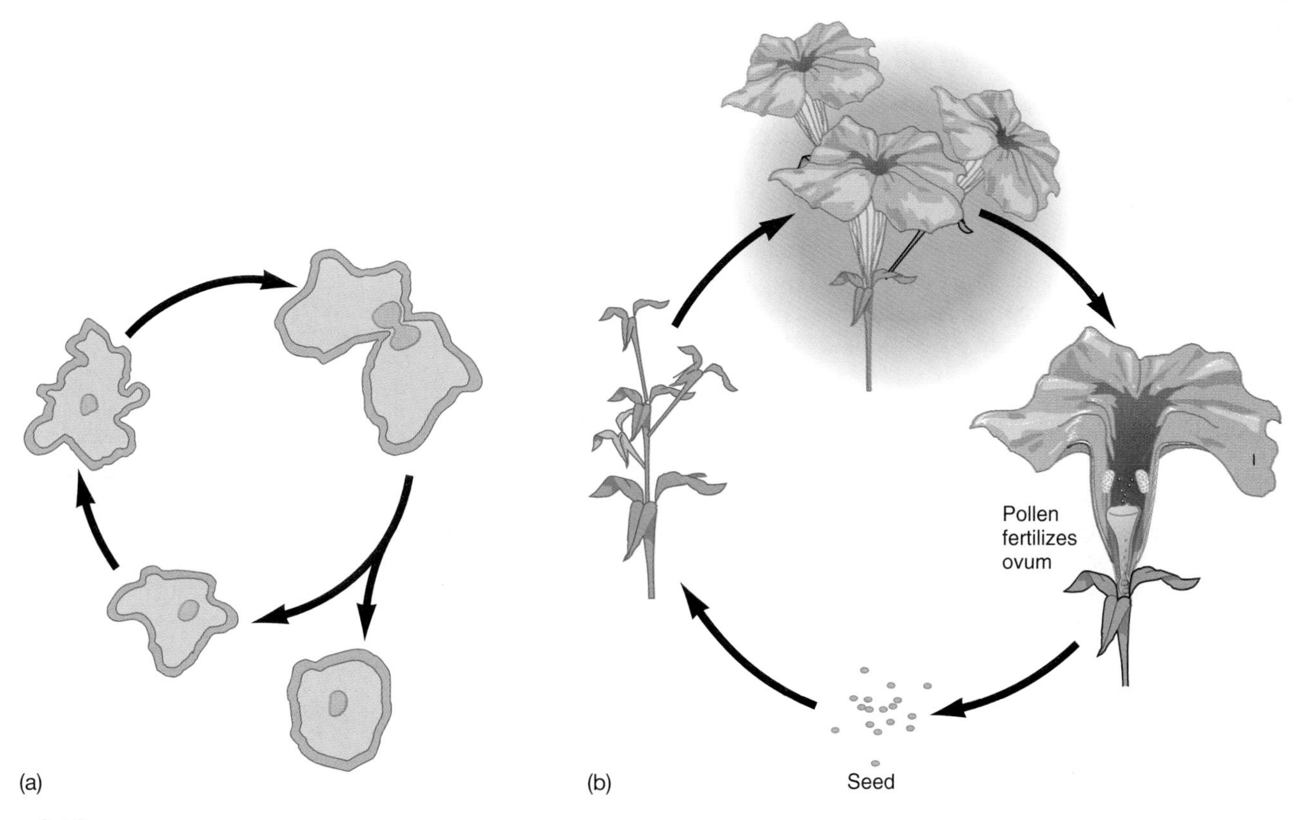

Figure 2.18

Some organisms' life cycles are very simple, while others are quite complex. *(a)* A single-celled organism such as an amoeba grows to about twice its original size and then divides in half; each life cycle is merely a repetition of this process. *(b)* A plant develops flowers in which eggs and sperm (in the pollen) form separately. Eventually a sperm fertilizes an egg, and the resulting cell (zygote) grows into an embryo that remains in a seed for a time. The seed then grows into a mature plant.

not all of them survive to reproduce. The growth curve of a real population therefore levels off at a maximum, the **carrying capacity** of the environment:

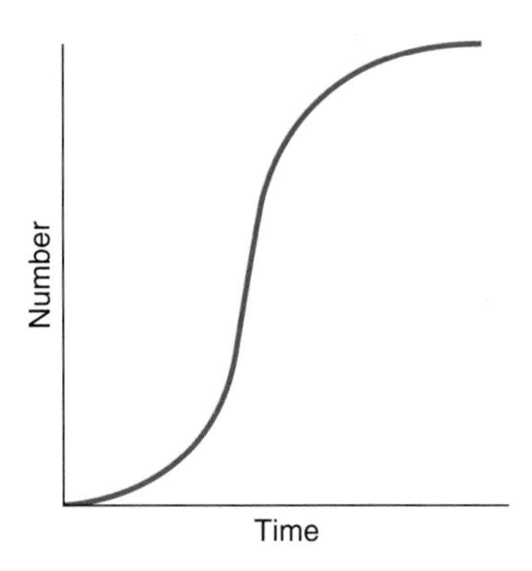

The population comes into a kind of balance with the other populations in its community and with the physical limitations of the ecosystem.

The fact that growth is inevitably limited, called the *Malthusian principle,* was proposed by Reverend Thomas Malthus (1766–1834). Malthus pointed out that the number of people increases faster than the food supply, so severe economic pressures, starvation, disease, and war are necessary to hold down the population to a number the planet can support. In this competition for space and food, only the fittest—the best-adapted—survive. This principle had a great influence on Darwin's thinking.

2.11 The genetic variability of populations is the basis for natural selection.

The poet John Greenleaf Whittier once said he would believe in evolution when he could watch an onion turn into a geranium, and the way people commonly (mis)understand evolution today is almost equally absurd. Evolution is a change in a population of organisms from generation to generation, not a change in an individual, though its result, of course, is to produce individuals that are different from their ancestors.

Based on the work of Darwin and Wallace, modern biology now recognizes that every population is extremely variable—just look at the variation among humans (see Figure 2.3). This variability stems from two sources, both explained by the genetic model.

Anyone who has copied recorded tapes or transferred a computer file has probably created a tape or a file with a glitch in it at some time or another. Anything containing information is corruptible—subject to errors and bits of noise—and this applies to genomes as well. Errors in a genome, called **mutations,** are small changes in the structure of the DNA molecules that constitute the genome. Since the information in the genome determines the structure of an organism, any little change in that information could result in an organism with a different structure. Mutation is the first source of variation in organisms. The second source of variation is sexual reproduction, during which genomes are randomly shuffled and rearranged, creating different combinations of genes that contain mutational differences. Each new individual gets half its genome from one parent and half from the other parent; you have half the information from your mother's genome and half from your father's, and you probably resemble each of them in some features. In these two ways, even closely related organisms become different from one another, and natural populations always harbor a lot of genetic variability (Figure 2.19).

Selection operates on this variability. The genetic differences among members of a population give some individuals more favorable features and others less favorable ones, so they have different abilities to compete, to adapt to new conditions, and ultimately, to reproduce. Since populations belong to communities, they change largely in response to conditions created by the other populations they interact with. Today, those other populations are often humans. Farmers spread insecticides over the land to control crop pests, but soon find that the susceptible insects have been replaced by resistant ones, so they have to turn to new chemicals. The widespread use of antibiotics has selected many strains of antibiotic-resistant bacteria, so pharmaceutical companies have to keep searching for new agents. In both cases, the chemicals kill all the susceptible organisms, but allow a few resistant mutants to multiply. In a population of identical individuals, however, there would be no resistant insects or bacteria to select.

For most organisms, few offspring can actually survive to reproduce another generation, and those that do are the ones with the most advantageous features. Comparing individuals with different genomes in the same species, the genetic type that leaves the greatest proportion of offspring in the next generation is assigned the greatest **fitness,** a measure of reproductive success. Natural selection means that individuals with the highest fitness are more likely to pass their features on, so more individuals in succeeding generations come to be like them:

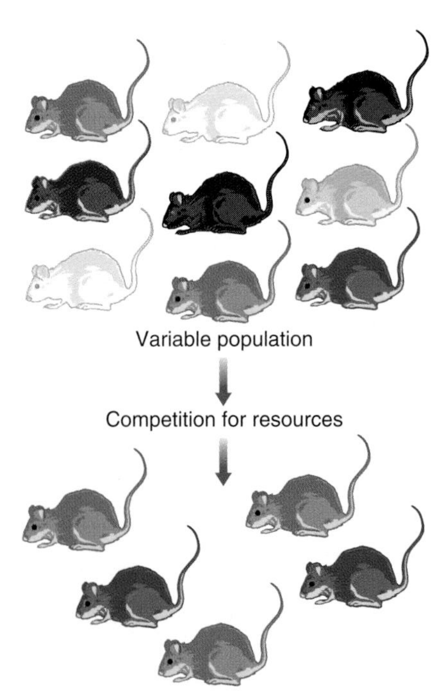

Figure 2.19

The members of every population will differ from one another significantly because they carry different mutations and different combinations of mutations, like this patch of pink-flowered fireweed plants in the midst of a population of red-flowered plants.

H. B. D. Kettlewell described a classic contemporary case of natural selection in the English peppered moth, *Biston betularia.* Across most of England, this moth was originally light in color and beautifully camouflaged on the lichen-covered bark of trees (Figure 2.20). But around heavily industrialized areas like Manchester and Birmingham, smoke and pollutants killed the lichens, and in 1848 a melanic (dark) form of the moth was first observed that is almost invisible on exposed, dark bark. Fifty years later, 99 percent of the moths around Manchester were melanic, and in more lightly polluted areas, the frequency of light and dark forms correlated very well with the abundance of light and dark trees. The pollution reached a peak in the 1950s, when Kettlewell did his studies; now, with greater industrial regulation, the pollution has been reduced, the lichens are reappearing, and with them, the light-colored moths. Each kind of moth stands out sharply on one background or the other, and Kettlewell's observations of birds feeding showed that the better-camouflaged moths escape their predators more successfully. In one experiment near Birmingham, Kettlewell captured 447 melanic and 137 light moths, marked and released them, and then later tried to recapture them to see how many had survived. He recaptured 123 (27.5 percent) of the melanics and only 18 (13.0 percent) of the light moths. Clearly, more of the melanic moths survived because they were less likely to be seen and eaten than the conspicuous light-colored moths. These survival rates provide a rough measure of fitness. In a pollution-damaged woods, if we assign the melanic moths a fitness of 1, the light-colored moths have a relative fitness of 0.47, because their survival rate is only 47 percent ($13.0/27.5 = 0.47$) as high as that of the melanics in this particular environment.

This *differential survival* rate among the moths leads to *differential reproduction:* Where the bark of trees is dark, most parents of the next generation will be the surviving dark moths rather than the light ones. The essence of natural selection is that different types of individuals within a population survive and reproduce at different rates.

As the American biologist George Wald has pointed out, natural selection is a kind of *editing* process, in which the best features of each organism are developed, while the less suitable features are discarded (Figure 2.21). As a result of mutation, selection, and such factors as the isolation of populations by geographic barriers, many different kinds of organisms have evolved, each adapted to a particular way of life and gradually becoming well-shaped to that existence. Through this subtle editing process, organisms become *designed* for their ways of life, and we can use the word "design" like this in a strictly scientific sense, without embarrassment or apology. We know now that evolution is also driven in part by random events that have little or no adaptive significance, and we will factor these events into the process as we develop it in later chapters.

Exercise 2.3 What is the relationship between the Malthusian principle and the process of natural selection?

Example 2.3 Suppose that in early spring a population of insects consists of equal numbers of two types, one slightly larger than the other. One month later, 60 percent of the individuals in the next generation are smaller, and 40 percent are larger. What is the fitness of the larger insects relative to the smaller ones? Solution: Suppose the initial population consisted of 100 individuals, 50 of each kind, and the next generation also had 100 individuals, 60 smaller and 40 larger. The relative survival rates would be $60/50 = 1.2$ for the smaller type and $40/50 = 0.8$ for the larger type; the relative fitness would be $0.8/1.2 = 2/3$. So the fitness of the larger type would be two-thirds that of the smaller type.

Exercise 2.4 Given the relative fitnesses for the two types of insects described in Example 2.3, predict their numbers after another month if the population is limited to 100 individuals.

Figure 2.20
When light and dark variants of the peppered moth are viewed on different backgrounds, it is easy to see how their colors provide good or poor camouflage in different habitats.

Exercise 2.5 "Natural selection is, fundamentally, just differential reproduction." Justify this statement or present an improved counterstatement.

2.12 Organisms are opportunists that are led into certain strategies of survival.

Looking at all the marvelous structures, behaviors, and other adaptations of organisms, we may often ask, "How could all this diversity ever have developed through evolution?" and "How did these organisms ever get to be so different?" The answers are the subjects of much of this book; some false answers are discussed in Concepts 2.2.

Even though some biologists object to the term, it is now common to say that a species follows a *strategy* for survival. This is really just the species's niche and behavior viewed from a certain human viewpoint, for each species grows and behaves in ways *analogous* to human behavior, as if it had chosen one way of life over another. A viable strategy for one species of plant is to grow in the forest shade; a viable strategy for another is to grow in the bright sunlight of a prairie. Of course, no species has "chosen" a strategy. Rather, its ancestors—little by little, generation after generation—merely wandered into a successful way of life through the action of random evolutionary forces.

Figure 2.21
Organisms evolve through gradual editing. The ancestors of giraffes were adapted to a way of life that required only moderately long necks and legs. They produced slightly variant offspring, some of whom were taller and were able to reach food higher in the trees. These were selected and their features perpetuated. So over many generations, giraffes were gradually shaped into a long-necked form.

Survival strategies evolve because all organisms are *opportunists.* We see again and again that when the opportunity arises for living a certain way, some kind of organism eventually takes advantage of it. The paleontologist George Gaylord Simpson commented,

> *. . . over and over again in the study of the history of life it appears that what can happen does happen. There is little*

Concepts 2.2 Pitfalls in Thinking About Evolution

Ever since the idea of evolution entered the consciousness of our society, it has been a source of much philosophical speculation and a good deal of muddled thinking. Students (and many others) who approach the subject are beset by at least three kinds of barriers to thinking clearly.

1. Rose-Colored Glasses

Since it is so obvious that organisms are well suited to their ways of life, people tend to believe that adaptations are perfect—that every feature must be an adaptation for some definite function. But there is no perfection in adaptation. Detailed studies often reveal the imperfections in generally functional systems. Among the variants in any population, some individuals are clearly better adapted than others. Furthermore, the belief that every structure and every behavior has evolved from some specific function leads people to tell "just so" stories that may sound convincing but have no basis in fact.

2. The Great Warehouse in the Sky

People often talk as if organisms can develop advantageous features by ordering what they need from what the writer Elaine Morgan has called a "Great Warehouse in the Sky." This leads them to ask, "If it's an advantage to have this feature, why don't all organisms have it?" The answer, of course, is that organisms only get their features by chance, and if accidental events haven't produced a particular characteristic, it simply won't be there.

3. Planning for Success

The most common impediment to thinking about evolution is imagining that organisms are evolving toward some future goal. This is the viewpoint of *teleology,* discussed in Chapter 1, and it has been the basis of great philosophical schemes, including one developed by the French theologian Pierre Theilhard de Chardin in a series of popular books. Such thinking is outside the pale of science and has no scientific evidence to support it. But in rejecting teleology, we must not reject *teleonomy.* All organisms are objects with functions, and they evolve so as to maintain and enhance those functions. It would be absurd to deny, for instance, that birds have wings so they can fly; flying is a functional behavior for birds, and wings have evolved to serve the function very well. However, birds were led by accidental events—chance and necessity—to the condition of flight. What we must deny is that birds have been in any way *striving* to evolve wings or that they could in any way *foretell the future* and change in a favorable direction. Adopting one viewpoint rather than another is a subtle balancing act that every student of biology should try to develop.

suggestion that what occurs must occur, that it was fated or that it follows some fixed plan, except simply as the expansion of life follows the opportunities that are presented.

The great genetic diversity in natural populations enables some individuals to move off on new evolutionary paths and develop new strategies whenever the opportunities arise. Nothing *makes* this happen. As the French biologist Jacques Monod emphasized, all of biology is the result of chance and necessity. *Chance* brings certain combinations of genes together at times and places where they can be used, and these genes determine a structure, a certain organism that operates by *necessity,* by following the laws of physics, chemistry, and biology. Now and then, the strategy for survival dictated by an organism's structure will be new and will lead it and its descendants along a new path. Then that strategy will be gradually refined in each generation by further chance events. Once pointed in a certain direction, a line of evolution survives only if the cosmic dice continue to roll in its favor. Species survive for various times, maintaining themselves and diversifying until their luck runs out. Many potential paths are never taken, because conditions do not become right. However, just by chance, a wonderful diversity of life has developed during the billions of years in which organisms have been evolving on earth.

Exercise 2.6 Explain carefully how the Heike crabs were shaped into their present form.

2.13 Phylogenies describe the apparent course of evolution.

When pre-Darwinian naturalists classified species, they were just making convenient catalogs. Since they believed each species had been created independently for its way of life, they saw no need to explain similarities and differences between species. But the evolutionary perspective raised the question, "Why do some organisms resemble one another more closely than others?" and simultaneously provided the answer: "Because similar organisms have evolved from common ancestors."

On the basis of the fossil record and comparative anatomy, modern systematists develop **phylogenies,** describing how species are related to one another. These relationships are commonly represented by *phylogenetic trees.* The tree in Figure 2.22 shows the relationships among the major groups of cephalopods—molluscs like squid, octopi, and the chambered nautilus. Their fossilized shells have left such an excellent record that systematists can be quite confident of this phylogeny. The tree branches at several points, where one species must have divided into two or more. Individual branches of a tree—sometimes called *lines of evolution*—often show a trend, such as an increase in size or a change in the manner of coiling.

2.14 Biologists divide the world of organisms into kingdoms.

Acquiring a biological sense includes coming to appreciate the diversity of organisms, as well as learning general principles that apply to all, or to most, organisms. Some of the most

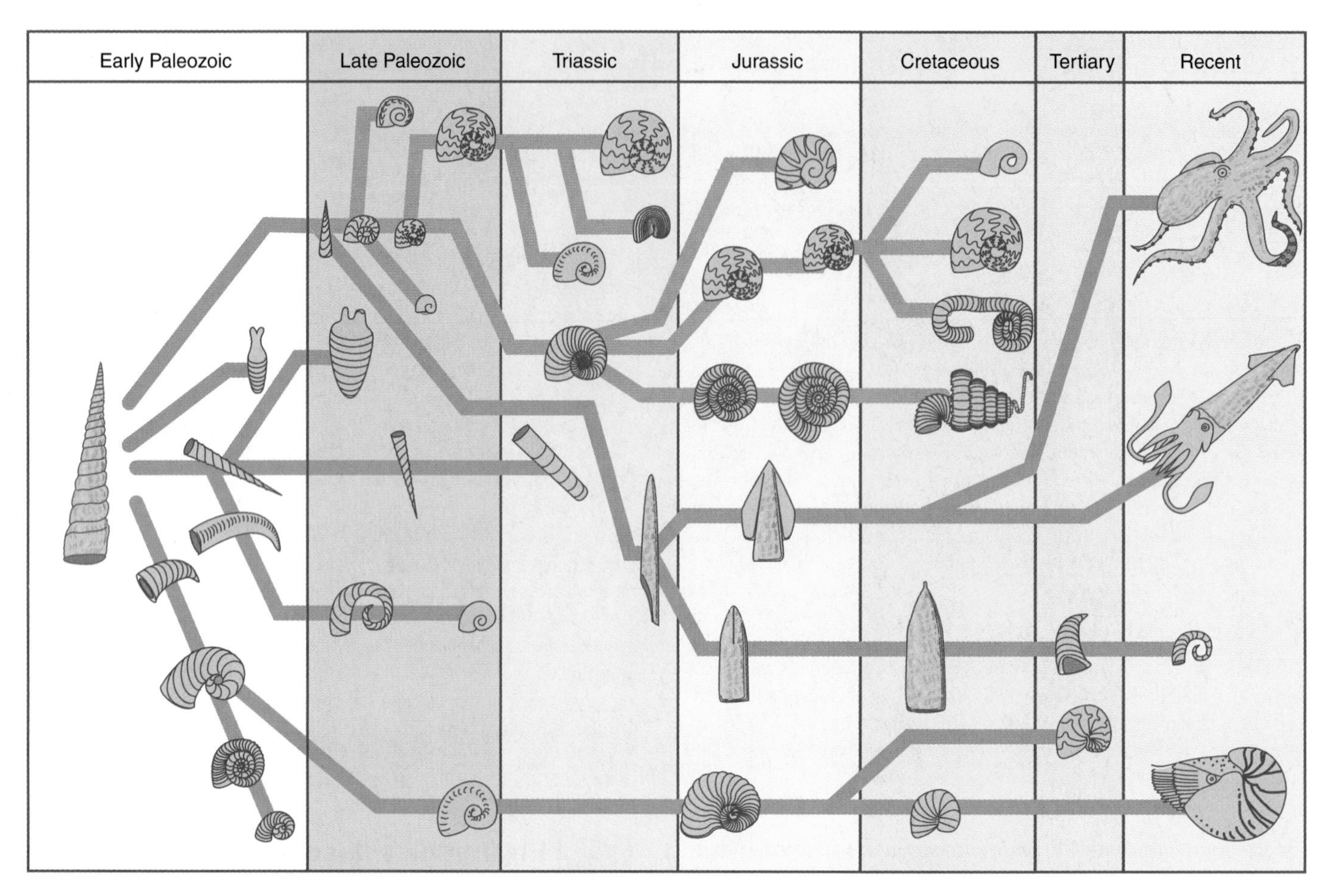

Figure 2.22

This diagram shows the probable phylogenetic relationships among the major groups of cephalopods, beginning with a single coiled form in the early Paleozoic era *(far left).* Some lines of evolution show a tendency toward greater coiling of the shell, while in the line leading to modern octopi and squid, the shell has become internal. The four modern types are the remnants of the thousands of species that have existed in the past.

fascinating creatures are microorganisms that can only be seen with the aid of a microscope, but long before this instrument was invented in the seventeenth century, naturalists had amassed a long list of obvious organisms—plants and animals visible to the naked eye. In ancient times, the world as a whole was divided into three kingdoms—animal, vegetable, and mineral—so the world of biology comprised the plant and animal kingdoms, Plantae and Animalia (Figure 2.23*a*). In 1867, however, the German biologist Ernst Haeckel recognized that this classification is simplistic, so he established a third kingdom, Protista, to include the microorganisms (mostly single-celled creatures) and others that he felt were neither plants nor animals (Figure 2.23*b*). Since then, many schemes have been proposed for dividing the complex, confusing array of organisms into categories that closely reflect their apparent phylogeny and that also meet other ideals of taxonomy. Within each kingdom, species are now classified formally in a hierarchy of categories, ranging from the broadest (kingdom) through phylum, class, order, family, and genus; this is discussed in Chapter 22. Classification is a difficult task, if only because it has been so hard to determine the evolutionary relationships among many prominent groups. Over the years, taxonomic schemes have had their advocates and critics, and have gone in and out of fashion.

A currently widely accepted taxonomy divides organisms into five kingdoms, as shown in Figures 2.23*c*. These kingdoms are not consistently easy to define, but it is useful to characterize them for our first look at biological diversity:

1. Monera: bacteria, whose cells are smaller (usually) and less organized than those of other organisms, as explained in section 6.5.
2. Protista: a very heterogeneous group containing algae, protozoans, and some molds; most species are microscopic, but some of the algae are huge and resemble plants.
3. Fungi: mushrooms and most of the molds.
4. Plantae: the plants.
5. Animalia: the animals.

A fairly recent proposal for a still broader classification divides all organisms into three *domains,* encompassing six kingdoms (Figure 2.23*d*). This scheme is based on detailed molecular analysis of cell structure, which shows that a few unusual

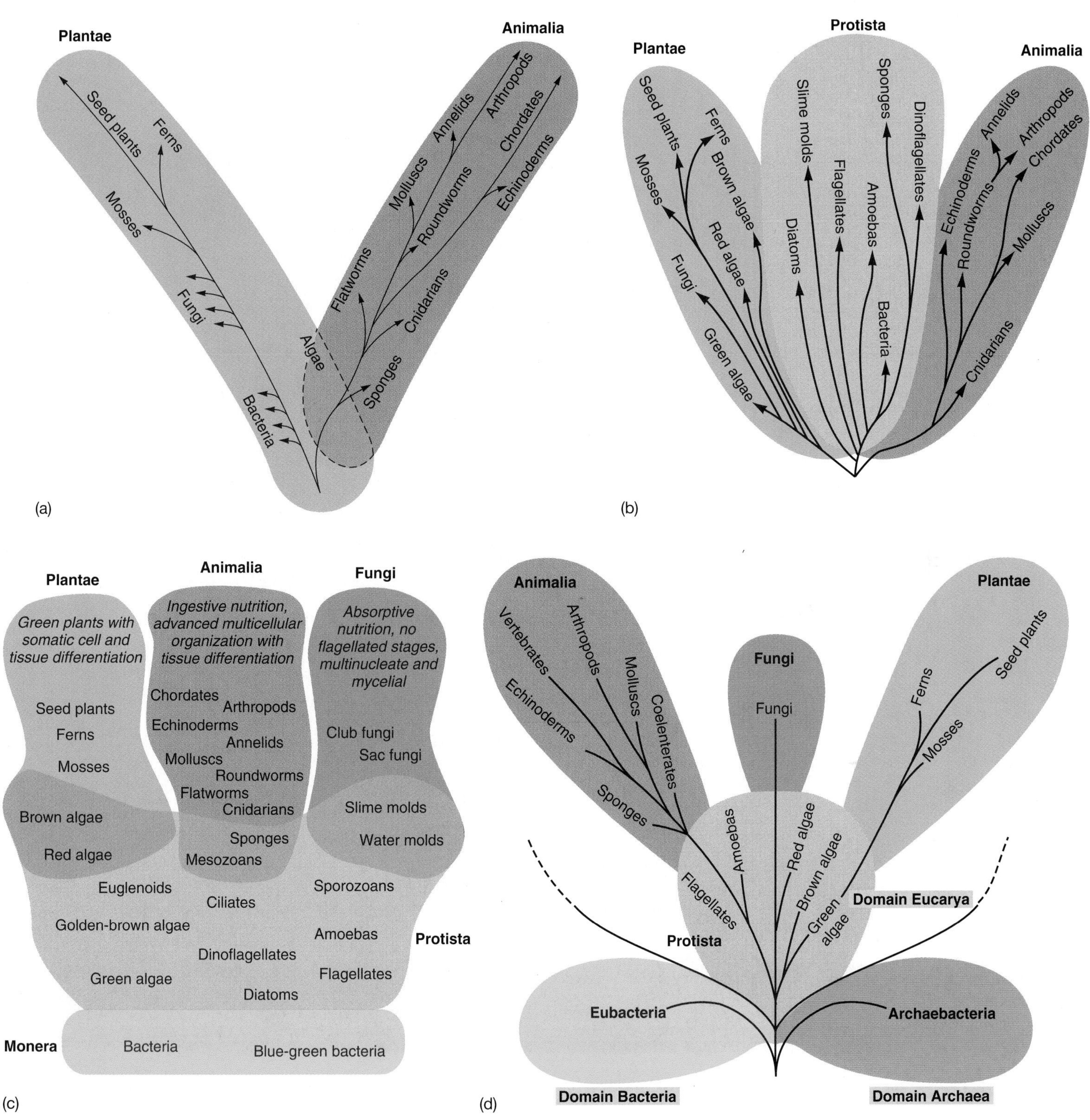

Figure 2.23

Biologists' views of the overall phylogeny of organisms and how they should be classified have changed over about the last 150 years. *(a)* The traditional view, from pre-Darwinian times, was that all organisms are either plants or animals. *(b)* In the 1860s problematic organisms, mostly single-celled, were placed in a third kingdom, Protista. *(c)* A scheme that divides all organisms into five kingdoms expressed the consensus of biologists as of a few years ago, even though it is difficult to set boundaries between Protista and other kingdoms. This scheme has sometimes been supplemented by placing Monera in a separate superkingdom, Procaryota, and the other kingdoms in a superkingdom Eucaryota, to recognize a major distinction in cell structure. *(d)* A new proposal advocates establishing three domains, as categories above the kingdoms, to recognize fundamental differences between archaebacteria and other bacteria and to note the differences of bacteria in general from all other organisms.

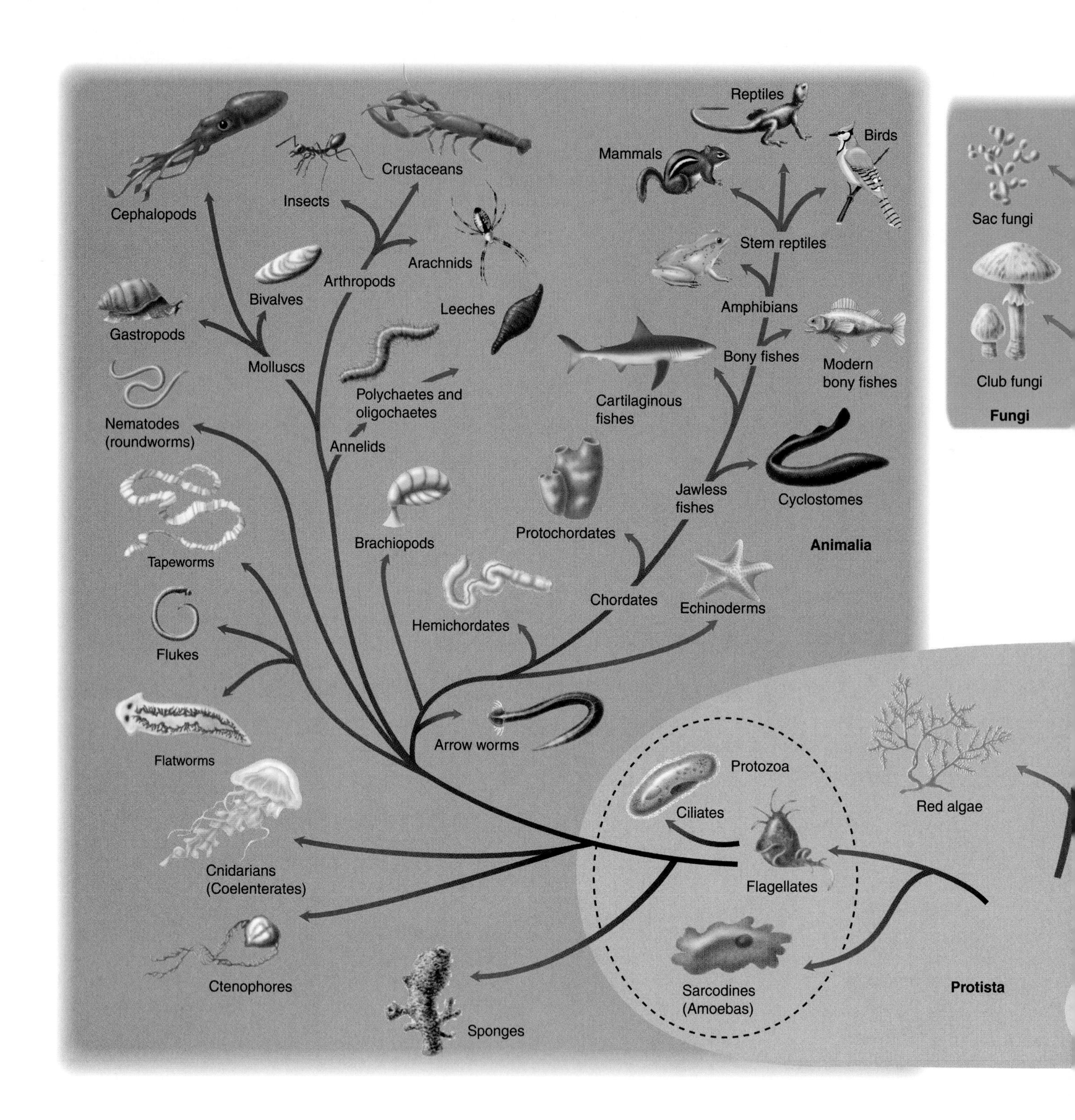

Figure 2.24

The principal groups of organisms are arranged to show their taxonomic relationships in this version of a modern classification.

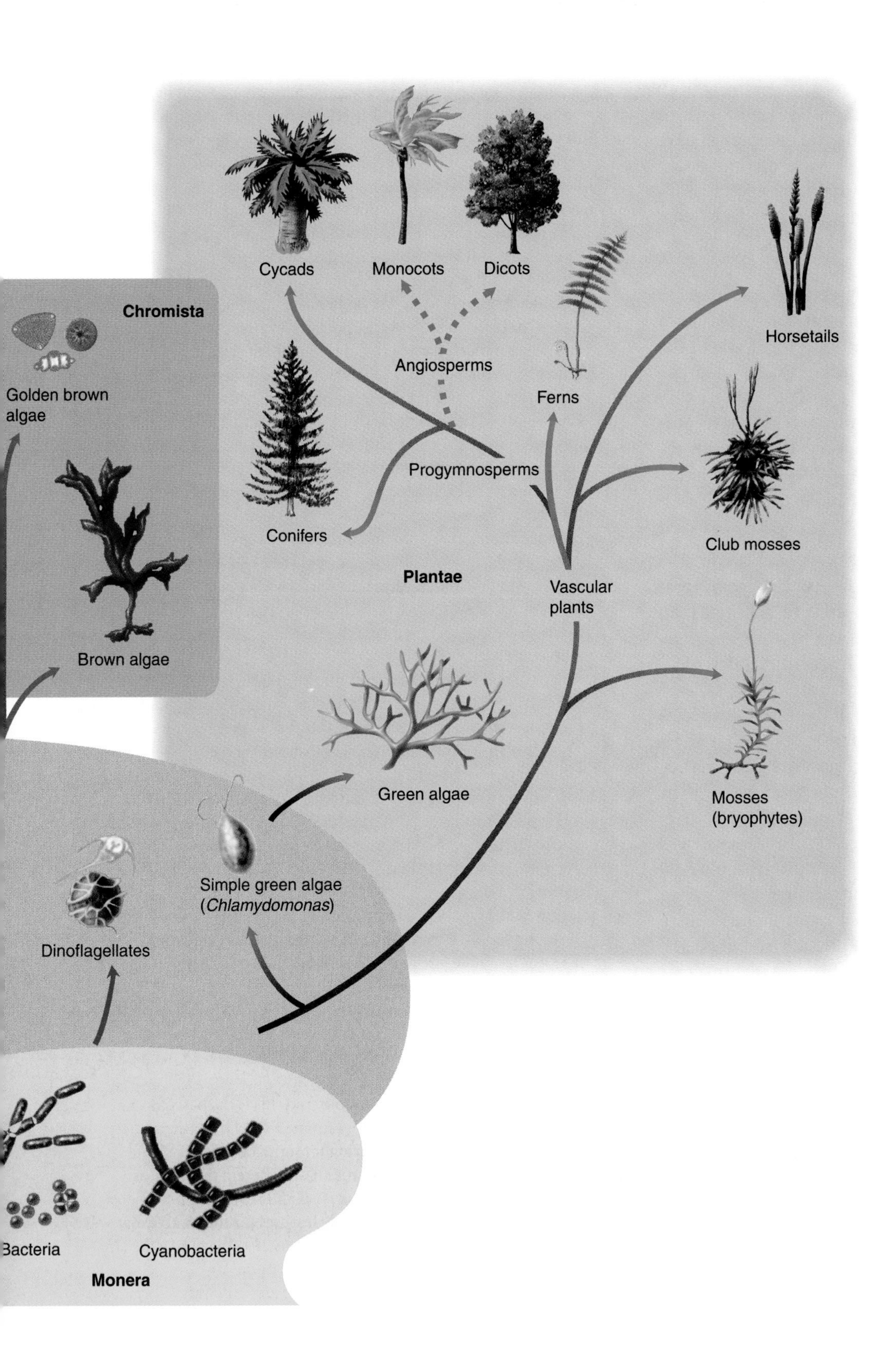
Cycads
Monocots
Dicots
Chromista
Horsetails
Angiosperms
Golden brown algae
Ferns
Progymnosperms
Conifers
Club mosses
Plantae
Vascular plants
Brown algae
Green algae
Mosses (bryophytes)
Simple green algae (*Chlamydomonas*)
Dinoflagellates
3acteria
Cyanobacteria
Monera

bacteria (archaebacteria) are fundamentally different from both the majority of bacteria (eubacteria) and all other organisms (eucaryotes). The distinctions are so subtle and complex that we will wait to explain the procaryote/eucaryote distinction in Chapter 6 and the other distinctions in Chapter 29. Many biologists now favor this domain concept because it reflects what appear to be very ancient, fundamental differences.

Figure 2.24, on page 40, shows the principal categories of organisms for the purpose of becoming generally acquainted with them, ignoring the subtle domain distinctions. The major problem in defining kingdoms centers around the limits of the very heterogeneous kingdom Protista; many of its plantlike members are closely related to the plants, while others are not. To reduce its heterogeneity, we remove one large group of protists to a kingdom called Chromista and move those closely related to plants into the plant kingdom. You may use this figure as a general reference to the world of life on Earth.

Coda Humans' interest in biology stems from the observation that we are surrounded by a great diversity of other organisms, and each culture has made up stories to explain where these creatures came from and how they are related. Modern biology offers a scientific story, well-supported by a wealth of evidence: Organisms have evolved into their present states over vast times. Biological evolution is only possible because organisms are genetic systems, that operate by following the instructions in their genomes. They reproduce themselves and pass on copies of these genomes to their offspring, but during this process they acquire slightly different genomes, so their characteristics vary. To survive and reproduce, each kind of organism must fit into an ecological niche, and those whose characteristics make them best suited for each niche will be most successful in reproducing; thus a process of natural selection goes on continually, shaping each kind of organism for its particular way of life. With this general framework in mind, a number of questions naturally arise: What is a genome, and how does it operate? If information is required to specify an organism's structure and operation, how does this happen? How is an organism built so that it can form its own structure and reproduce itself by following the genetic program in its genome? To answer such questions, we must develop a conception of biological organization, and we will begin to do that in the next chapter by introducing some basic concepts of chemistry.

Summary

1. The biological world encompasses enormous diversity, and a major goal of biology is to catalog and explain this diversity.
2. Humans mold organisms for their own purposes by artificially selecting genetic differences. Darwin and Wallace took account of this fact in developing the theory of evolution through natural selection.
3. Because of the persistence of old ideas, many years passed before the concept of evolution was developed and accepted.
4. Two major lines of evidence indicate that all existing organisms are related by evolution: The fossil record shows that different kinds of organisms successively appeared and then disappeared over millions of years; comparative anatomy reveals that related organisms have diverged from one another, or from a common ancestor, through modifications of their structures.
5. Structures are said to be homologous if they have the same embryonic origins and occupy similar positions in different species, even though they may be highly modified in their fully developed forms. Homology provides strong evidence for evolution.
6. Darwin and Wallace proposed that evolution occurs primarily because of natural selection. Populations are variable but limited in size, so only those individuals most fit for their ways of life survive and pass on their characteristics to their offspring.
7. Organisms may acquire features during their lifetimes that help them adjust to their environments. However, their offspring do not inherit such features, and evolution cannot be explained on this basis.
8. Organisms generally reproduce themselves, although some organisms, such as sterile hybrids, cannot. Reproduction entails heredity, because the offspring of each organism resemble their parents.
9. Inheritance means organisms carry genetic information that specifies their structures. This information is carried in the genome, which is always made of deoxyribonucleic acid (DNA). A genome is like an instruction tape that directs the operation of a computer, and an organism is like a self-reproducing computer.
10. A virus is an independent genome that uses functioning organisms to reproduce itself.
11. An ecosystem consists of a community of interacting organisms in a certain physical environment. The relationships among members of the community depend upon the fact that molecules contain energy. The members of the community transfer energy as one serves as food for another.
12. Evolution occurs principally through natural selection operating on populations in ecosystems. Each type of organism is adapted to a particular way of life in one ecological niche, and selection is the mechanism that shapes it for this way of life through evolution.
13. The lives of organisms can be represented as cycles, including a stage of reproduction. Even though the stages of an organism's life cycle may not resemble one another, they are united in that they all get their information from a common genome.
14. Populations of organisms tend to increase exponentially, without limit. However, organisms require resources from their environment, and since resources are limited, population size is held in check by the environment.
15. Populations are genetically variable, and this is the basis for natural selection. Variability arises from mutations and from the rearrangement of genomes during reproduction.
16. Organisms with different genomes have different fitnesses, as measured by different rates of survival and reproduction. The individuals with the most favorable features for their particular way of life are said to have the greatest fitness, because they leave the largest proportion of offspring in the next generation.
17. Random genetic events (chance) create different kinds of organisms whose characteristics are determined by the laws of physics, chemistry, and biology (necessity). Through the continued selection of favorable characteristics, organisms are directed into strategies for survival—certain ecological niches and patterns of behavior.
18. Evolution does not occur with any purpose or direction. Rather, organisms are led by random events into genetic opportunities, and natural selection takes advantage of these events to shape the organisms into particular strategies for survival.

19. Biologists try to establish phylogenies, which show the probable course of evolution and indicate how different species are related to one another.
20. Organisms are often classified into five kingdoms: Monera, Protista, Fungi, Plantae, and Animalia. Alternative classifications are also used, including one with three domains and seven kingdoms, based on complex characteristics of cell structure and metabolism.

Key Terms

systematics 20
taxonomy 20
homologous 25
analogous 25
population 27
model 28
system 28
information 28
genetic information 28
genome 28
virus 30
ecosystem 31
community 31
ecology 31
adaptation 32
ecological niche 32
asexual reproduction 32
sexual reproduction 32
morphology 33
resource 33
carrying capacity 34
mutation 34
fitness 34
phylogeny 37

Multiple-Choice Questions

1. Essentialism and typology are not easily reconciled with ideas of evolution because
   a. by definition, a fundamental essence is unchanging.
   b. similarities between separate species would be considered more important than their differences.
   c. differences between members of a species would be considered more important than their similarities.
   d. neither similarities nor differences among species are considered significant.
   e. the age of the earth is significant for evolution but not for essentialism.
2. A seventeenth-century naturalist who discovered a remote island populated with previously unknown creatures would first
   a. study the phylogeny of the inhabitants.
   b. prepare a taxonomy.
   c. examine homologies.
   d. look for evidence of artificial selection.
   e. look for evidence of natural selection.
3. Vestigial organs are considered evidence of evolution because
   a. their structure may show similarities to functional organs in other species.
   b. they arise from the same embryological source as functional organs in other species.
   c. it is unlikely that separately created species would contain useless parts.
   d. their presence fits into the history of the organ as seen in the fossil record.
   e. All of the above are true.
4. Structures in organisms of different species are homologous if they
   a. have a similar appearance.
   b. are found in many species within the same community.
   c. perform the same or similar functions.
   d. develop from the same embryonic structure and occupy comparable positions.
   e. contribute to fitness in the same way.
5. Which term is the most inclusive?
   a. niche
   b. ecosystem
   c. population
   d. community
   e. species
6. In the process of *natural selection,* the role of the environment is to
   a. stimulate the mutation of particular genes.
   b. create particular adaptations needed at that time.
   c. increase the reproductive rate of some members of a population.
   d. increase the rate of reproduction of all members of a population.
   e. enable the population to evolve toward a predetermined goal.
7. Viruses are not considered real organisms because they
   a. lack usable genetic information.
   b. are not capable of sexual reproduction.
   c. do not have their own mechanisms for exchanging matter and energy with the environment.
   d. do not evolve.
   e. are not subject to natural selection.
8. Carrying capacity always acts as a limiting factor, or brake, on
   a. fitness.
   b. exponential growth.
   c. evolution.
   d. adaptation.
   e. natural selection.
9. In the following list, ___ form(s) the least variable genetic group.
   a. the protists
   b. the bacteria
   c. any single species
   d. any single kingdom
   e. the organisms found in one ecosystem
10. Viruses are included among the
    a. monerans.
    b. archaebacteria.
    c. eubacteria.
    d. protists.
    e. None of the above.

True-False Questions

Mark each statement true or false, and if false, restate it to make it true.

1. Most of the species on Earth have been identified and named.
2. Phylogenetic trees attempt to classify organisms according to their ecological distribution.
3. Natural selection creates particular mutations that enable organisms to be better adapted to their surroundings.
4. If members of two different species exhibit several analogous sets of structures, it is likely that the species share a relatively recent common ancestor.
5. Vestigial organs are homologous to their functional ancestral structures.
6. Kettlewell's observations of peppered moths showed that, as industrial soot accumulated in the environment, the mutation rate of light-color genes to dark-color genes also increased.
7. A molecule composed of 20 identical subunits carries as much information as one composed of 20 different subunits.
8. Viruses contain both genetic information and a mechanism to supply the energy needed to express that information.
9. When the proportion of individuals in a population with a particular adaptation increases or decreases, the population is not evolving.

10. As long as the environment remains relatively stable, the survival strategy of a successful species will generally not change.

Concept Questions

1. In what ways did the acceptance of Darwin's ideas require advances in the science of geology?
2. Distinguish between the meanings of the following terms: population, community, ecosystem, niche.
3. Explain the difference between the terms *adaptation* and *fitness.* Which is the more significant for evolutionary change?
4. If an unending and easily obtainable supply of food, space, and mates were available, would natural selection be likely to occur? Explain.
5. Suppose one of our space probes discovered a planet that not only supported intelligent life but also appeared to have a geologic and environmental history very similar to that of Earth. Would you expect to find the same assortment of organisms there that we have on Earth?

Additional Reading

Dawkins, Richard. *The Selfish Gene.* Oxford University Press, New York, 1976.

———. *The Blind Watchmaker.* W. W. Norton & Co., New York, 1987. These two books by Dawkins, both eminently readable, present a modern view of the evolution of genetic systems.

Goodwin, Brian C. *How the Leopard Changed Its Spots: The Evolution of Complexity.* Charles Scribner's Sons, New York, 1994. About the evolution of morphology.

Gould, Stephen Jay. *Bully for Brontosaurus: Reflections in Natural History.* W. W. Norton, New York, 1991. One of Gould's popular collections of essays on natural history and evolution.

Huxley, Julian. *New Bottles for New Wine.* Chatto and Windus, London, 1957. Essays about evolution, population, and other important biological matters.

Irvine, William. *Apes, Angels, & Victorians: The Story of Darwin, Huxley, and Evolution.* McGraw-Hill Book Co., New York, 1955.

May, Robert M. "How Many Species Inhabit the Earth?" *Scientific American,* October 1992, p. 18. Anywhere from 3 million to 30 million may inhabit the earth, according to various estimates, and an accurate census may be crucial for efforts to preserve biodiversity.

Internet Resource

To further explore the content of this chapter, log on to the web site at:

http://www.mhhe.com/biosci/genbio/guttman/

The Chemistry of Biology

3

Key Concepts

3.1 Basic principles of chemistry are fundamental to biology.

3.2 Some atoms form molecules by sharing electrons.

3.3 Many compounds are made from combinations of positive and negative ions.

3.4 Atoms are rearranged in chemical reactions.

3.5 Acids and bases liberate or accept hydrogen ions.

3.6 Hydrogen ion concentrations are expressed as pH.

3.7 Water has unusual properties that are essential for life.

3.8 Organisms are composed of only about 25–30 elements.

3.9 Organisms are constructed of carbon-based molecules.

3.10 Organic molecules have distinctive shapes, and most can change shape without breaking bonds.

3.11 The solubility of an organic molecule in water is important biologically.

3.12 Different types of organic molecules are characterized by their functional groups.

The world is colored by chemicals, such as these dyes.

We live in a chemical world. Commercials try to sell us shampoo that is "pH balanced," while the foods on our shelves contain such mysterious additives as monosodium glutamate, diglycerides, sorbitan monostearate, and propylene glycol monoesters. Other messages from the media make us wonder if we are getting enough zinc or calcium in our diets, or if we are being exposed to PCBs and dioxins. It is a confusing, complicated world of oddly named materials that we apparently ought to understand—but how?

Our world is also chemical in another sense, for organisms are chemical machines that take in nutrients from the environment and transform these raw materials into their own structures. Organisms run on chemical reactions. **Metabolism,** the sum of all this chemical activity, consists of thousands of reactions in which incoming molecules are changed from one into another, including reactions that build biological structure.

In this chapter we will first develop some concepts of chemical structure and then examine the structure and special properties of water. Finally we will introduce organic molecules, those made mostly of carbon and hydrogen. This will prepare us for Chapter 4, which explores **biomolecules**—that is, organic molecules that occur naturally in organisms, especially huge organic molecules called *macromolecules* that form most of an organism's structure. As we investigate these topics, we will emphasize one of the principal themes of the book: that organisms function through molecular interactions and that molecules interact with one another because of their specific shapes. ■

3.1 Basic principles of chemistry are fundamental to biology.

We'll begin by developing some general principles of chemistry.

Matter is made of atoms.

All the materials of the world are made of **elements,** which are basic substances that cannot be broken down into simpler substances by a chemical process. Chemists have now identified more than one hundred elements, including hydrogen, oxygen, carbon, nitrogen, sulfur, and metals such as mercury, aluminum, iron, and copper. An **atom** is the smallest particle of an element that has any of the element's characteristic properties. Atoms are made of subatomic particles that do not have such properties: **electrons** with a negative electrical charge (−), **protons** with a positive electrical charge (+), and **neutrons** with no electrical charge. An atom's mass comes almost entirely from the protons and neutrons that form the **nucleus** in its center. (This nucleus is not to be confused with the nucleus of a biological cell.) Electrons surround the atomic nucleus. Because particles with the same charge repel each other, and those with opposite charges attract each other, each positive proton usually attracts a single negative electron, making an atom as a whole electrically neutral.

Atoms are very small—about 0.1 to 0.24 nanometers (nm) in diameter ($1 \text{ nm} = 10^{-9}$ meters). Remarkably, techniques have been developed for manipulating individual atoms and observing them with powerful microscopes (Figure 3.1). An atom's size means little, though, until it is compared with larger structures, such as the biomolecules that are made of many atoms.

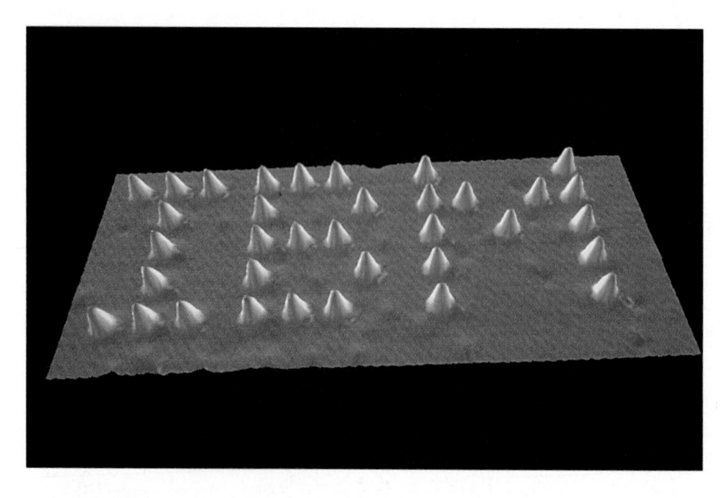

Figure 3.1

These gold atoms have been arranged by micromanipulation and are visualized by a powerful new microscopic technique.

Atoms form molecules; elements form compounds.

Most atoms combine with other atoms to make **molecules;** two atoms are said to form a **bond** when they are held in a stable arrangement at a specific distance. In drawings of molecules, the bonds between atoms are represented by lines, but the bond is just an energetically stable arrangement of electrons and atomic nuclei, most often made by a pair of electrons held between two nuclei. If two different elements combine in a stable molecular arrangement, they form a **compound,** a substance with well-defined properties that are quite different from those of its component elements. Hydrogen and oxygen, for instance, combine to make water, a compound whose molecules have two atoms of hydrogen bonded to one atom of oxygen. Water is obviously quite different from either of its elements. Similarly, ordinary salt, sodium chloride, is a combination of sodium, a soft metal, with chlorine, a noxious, greenish gas. The salt, too, is nothing like its elements.

As biologists, we are interested in chemistry primarily because *the distinctive shapes of biomolecules and their ability to change shapes are the keys to their function.* All molecules have distinctive shapes because of the various sizes of each atom and the ways atoms are bonded to one another. Many molecules, especially biomolecules, are able to change their shapes subtly by rotating bonds from one position to another without breaking them. This phenomenon explains a great deal of biology. The concept of a chemical bond is also extremely important; molecules are held together by quite strong bonds, while they interact with one another through quite weak bonds. Biomolecular interactions occur principally between **complementary molecules,** molecules whose shapes allow them to fit together like jigsaw puzzle pieces; they can interact in distinctive ways, and these interactions produce biological activities. As molecules interact, they often induce shape changes in each other. Mutually induced shape changes and other aspects of molecular form will be a continuing theme in this book, because the simple features of shape make the apparent complexity of biological function comprehensible.

Each element has a particular atomic number and mass.

The atoms of each element contain distinctive numbers of protons, electrons, and neutrons, which account for the properties of the element. Each element has an **atomic number,** the number of protons in each of its atoms, and an **atomic mass,** the total mass of protons and neutrons in each atom. Atomic masses are given in *atomic mass units (amu);* as a standard, atoms of the most common form of carbon are assigned a mass of exactly 12 amu. Then the **molecular mass** of a compound is the sum of the masses of the atoms in one molecule; the molecular mass of water (H_2O) is 18, the sum of the atomic mass of one oxygen atom (16) plus two hydrogen atoms with a mass of 1 each.

Most elements have isotopic variants.

All atoms of an element have the same atomic number, but their atomic masses may vary. For instance, all atoms of hydrogen have a single proton and thus an atomic number of 1. Most of them have no neutrons, so their atomic mass is also 1; this form is symbolized by ^1H:

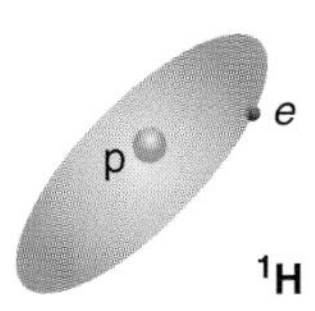

But rare forms of hydrogen called deuterium (^2H) and tritium (^3H) have, respectively, one neutron and two neutrons in addition to the proton:

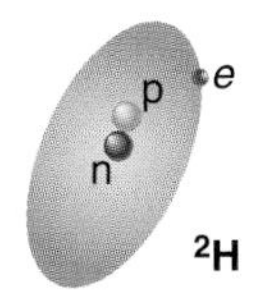

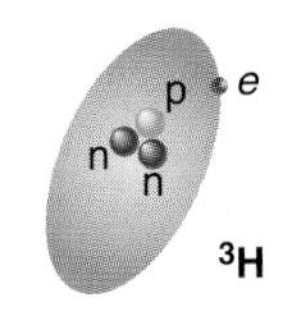

These three forms of hydrogen, which have the same number of protons but different numbers of neutrons, are called **isotopes.** Most elements exist in isotopic forms, so their atomic masses are averages of the various isotopes. Tritium is a radioactive isotope, or **radioisotope;** radioisotopes are unstable, and they regularly—and spontaneously—change into more stable atoms by giving off radiation, usually electrons. Since this radiation can be detected by various instruments, radioisotopes are important experimental tools in biology (see Methods 10.1).

The atoms of helium, the next heavier element, have two protons and two neutrons, so their atomic number is 2 and their atomic mass is 4. Next is lithium, with three protons and, most commonly, four neutrons. Oxygen has eight protons, and its most common isotope has eight neutrons, giving it an atomic mass of 16. Above atomic number 20, neutrons begin to outnumber protons more and more; the most common isotope of a very heavy element, uranium (atomic number 92), has an atomic mass of 238. (How many neutrons do the atoms of this isotope have?)

Each atom has a distinctive arrangement of electrons.

The chemical properties of elements depend largely on how the electrons are distributed in each of their atoms. This is best explained in relation to the **periodic table,** which was first devised in 1869 by the Russian chemist Dmitri Mendeleev and is a foundation of both modern chemistry and modern biology. The table arranges the elements in order of their atomic numbers and relates their properties to their atomic structures (Figure 3.2). Each horizontal row is a *period.* Each vertical column is a *family* of elements with similar properties. To understand these groupings, we must examine the electronic structure of atoms.

Atomic structure is easily visualized with a simple model that places the electrons around the atomic nucleus in a series of concentric shells representing energy levels with the lowest-energy level closest to the nucleus:

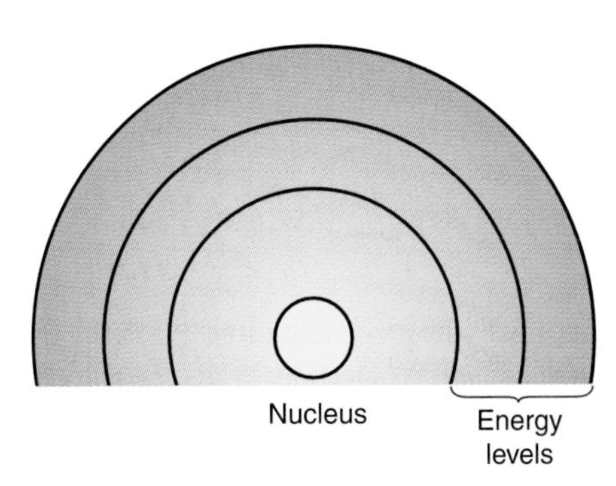

Each electron resides in one level and has the energy of that level. An electron goes into the lowest level available, but each level can only accommodate a limited number of electrons: two in the first level, eight in the second, and eighteen in the third. The periodic table is based on this electronic structure, and each row corresponds to one level. Atomic structure becomes rather complex beyond the third level, but to understand biology we only need to consider the smaller atoms that comprise most biological structure.

1 (Alkali metals)	2 (Alkaline earth metals)	3	4	5	6	7 (Halogens)	8 (Noble gases)
1 **H** 1.008							2 **He** 4.003
3 **Li** 6.941	4 **Be** 9.012	5 **B** 10.81	6 **C** 12.01	7 **N** 14.01	8 **O** 16.00	9 **F** 19.00	10 **Ne** 20.18
11 **Na** 22.99	12 **Mg** 24.31	13 **Al** 26.98	14 **Si** 28.09	15 **P** 30.97	16 **S** 32.07	17 **Cl** 35.45	18 **Ar** 39.95
19 **K** 39.10	20 **Ca** 40.08	31 **Ga** 69.72	32 **Ge** 72.59	33 **As** 74.92	34 **Se** 78.96	35 **Br** 79.90	36 **Kr** 83.80

Figure 3.2

In the periodic table, the elements are arranged in order of atomic number. (Here we show only the first four periods, representing the elements that are most important in biology.) Those in each column form a family with similar properties, including valence—that is, the number of bonds that each atom typically forms with other atoms.

The arrangement of electrons in its atoms determines many of an element's properties, including how it forms compounds. One way to explain atomic structure is by the "buildup" method: Starting with a hydrogen atom, with its single proton and single electron, we can imagine building larger atoms by adding more protons and neutrons to the nucleus. For every new proton, with its positive charge, we must add another electron, with its negative charge, to keep the atom neutral. The single electron of a hydrogen atom is in the first energy level. We make a helium atom(He) by adding a second proton to the nucleus and a second electron to the first level, filling the first level:

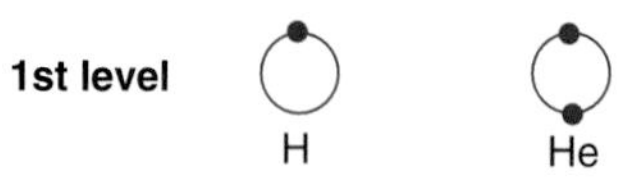

Notice that only hydrogen and helium are in the first period of the periodic table. The next electrons must go into the second level, and as we fill this level, one electron at a time, we make the atoms of the elements lithium (Li), beryllium (Be), boron (B), carbon (C), nitrogen (N), oxygen (O), fluorine (F), and neon (Ne):

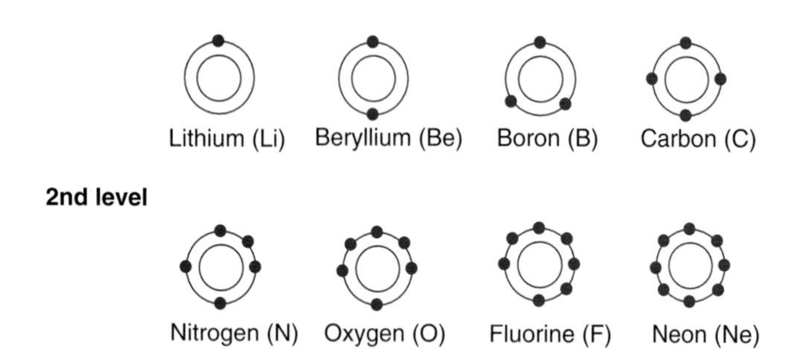

Filling the second energy level completes the second period of the table. Atoms of the next eight elements, from sodium (Na) to argon (Ar), are made in the same way, with the electrons going into the third energy level:

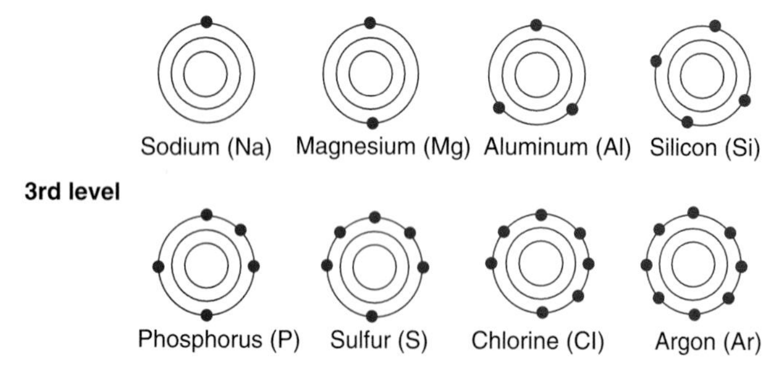

The elements in each family (each column of the table) have similar electronic structures and chemical properties. Moving to the right across the table, the elements in each family have atoms with one more electron in their highest occupied energy level. The elements in the right-hand column of the table (helium, neon, argon, krypton, and xenon) have one energy level filled. These elements are called *noble gases* because they are not chemically reactive and only form molecules through special chemical manipulations. Atoms of the other elements, with only partially filled energy levels, form bonds with one another by transferring or sharing some electrons. We will use the elements that form biomolecules as examples of how this bonding process works.

3.2 Some atoms form molecules by sharing electrons.

All biomolecules are basically made of only six elements—hydrogen, carbon, nitrogen, oxygen, phosphorus, and sulfur. These elements have small atoms, so their electrons are close to the atomic nuclei, and they make strong bonds with one another, forming stable molecules. To understand the structure of these basic atoms, we must use a slightly more sophisticated model of atomic structure and show how these atoms form strong **covalent bonds** by sharing electrons. Hydrogen, oxygen, nitrogen, and carbon are the smallest atoms that can form 1, 2, 3, and 4 covalent bonds, respectively, as shown by the structures of molecular hydrogen, methane, ammonia, and water.

Hydrogen

Each hydrogen atom has one electron, which resides in a small spherical region around the atomic nucleus, a space called an **orbital.** All the electrons in an atom reside in orbitals of various shapes and energies. Each orbital can contain a maximum of two electrons, so a hydrogen atom has room for one more in its orbital. Two atoms can share electrons by contributing one electron each to a single *molecular orbital* that encompasses both atoms. Two hydrogen atoms commonly create a molecule with the formula H_2 (diatomic or molecular hydrogen) by forming a molecular orbital that surrounds both nuclei and contains their two electrons; the sharing of these electrons creates a covalent bond between the atoms:

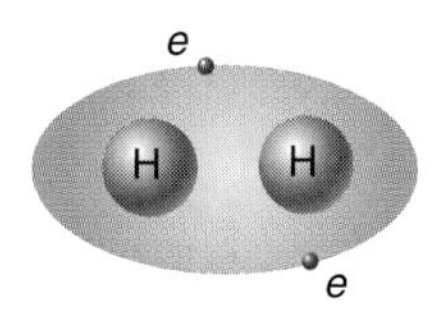

Methane

The main component of natural gas is methane, a colorless, odorless gas with the formula CH_4, which means each molecule of methane is made of one atom of carbon and four of hydrogen.

Carbon atoms have six electrons: two that fill the lowest energy level and four in the second level. Carbon atoms have four orbitals at the second level; they are club shaped and equally spaced, directed toward the four corners of a tetrahedron (a triangular pyramid):

The four electrons at this level are distributed according to a rule that each electron goes into a separate orbital if possible, so each orbital in a carbon atom has room for one more electron from another atom. Four hydrogen atoms, also with one electron each, can combine with the carbon atom to make a CH_4 molecule with four covalent bonds:

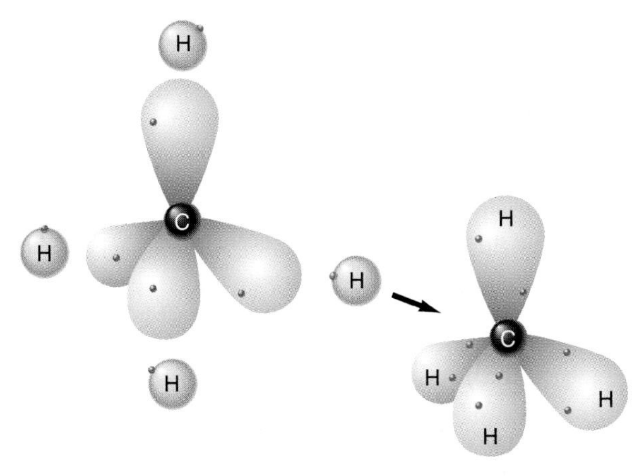

Methane (CH_4)

In each of the four bonds, one electron of the carbon is shared with the electron of a hydrogen atom in a single, sausage-shaped molecular orbital encompassing the two nuclei. Methane is the simplest organic molecule. Even the largest organic molecules are held together by strong covalent bonds like these.

Ammonia

By mentally adding one more proton and one more electron to a carbon atom, we form nitrogen. The additional electron fills one of the four orbitals at the second energy level, so only three are available for making bonds. A nitrogen atom therefore commonly makes only three bonds. It can combine with three hydrogen atoms to make a molecule of ammonia (NH_3), which is similar in chemical structure to methane:

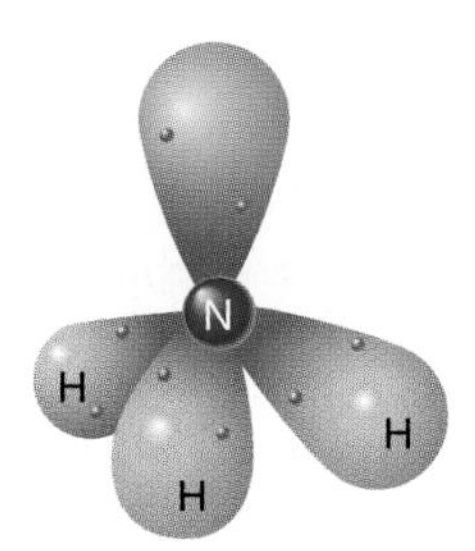

Ammonia (NH_3)

Water

To mentally make an oxygen atom from nitrogen, we again imagine adding an electron; this fills another orbital at the second energy level, leaving an oxygen atom with two orbitals filled and two left for bonding. An oxygen atom can therefore combine with two hydrogen atoms to make water (H_2O). The molecule is V-shaped, with an angle of about 105 degrees between the bonds:

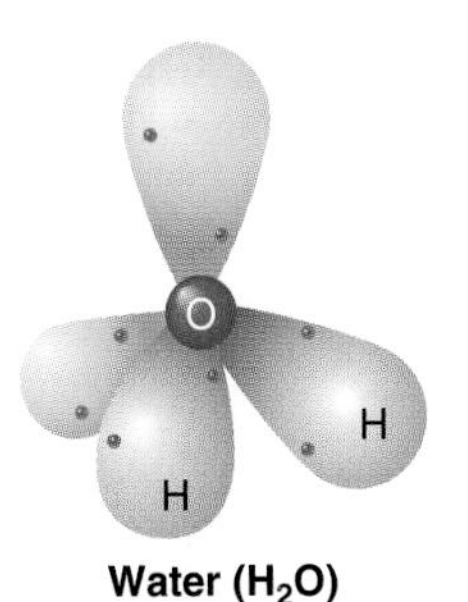

Water (H_2O)

In the bonds of methane, ammonia, and water molecules, the electrons are shared between the central atom and each hydrogen atom. But central atoms of increasing size, with increasing positive charges in their nuclei, have an increasing tendency to attract electrons; this tendency is called the **electronegativity** of the atom. Nitrogen is somewhat more electronegative than carbon, and oxygen is still more electronegative than nitrogen. If two atoms with equal electronegativities are bonded to each other, they share the electrons in the bond equally, but if one atom is much more electronegative than the other, it holds the electrons more tightly than the other one does, making a *polar* covalent bond. The oxygen atom of a water molecule, for example, pulls electrons partially away from the hydrogen atoms, making water a **polar molecule** with an uneven distribution of electrical charge. The O end of water is somewhat negative, and the end with the H atoms is somewhat positive, a situation represented by the lowercase Greek letter delta (δ) to show a small unequal distribution of charge, rather than the whole unit charge of an electron or proton:

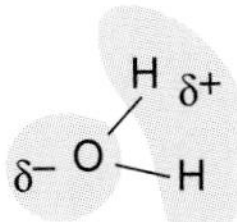

Since atoms are held together by sharing electrons, the polarity of the O–H bonds in water makes them relatively weak—that is, the oxygen atom holds the electrons so strongly that they are not able to hold onto the hydrogen atoms as tightly as carbon holds the hydrogen atoms in methane; in fact, one hydrogen atom in water may become unbonded and leave the rest of the molecule. When this happens, the hydrogen atom leaves its sole electron with the oxygen atom and becomes a **hydrogen ion,** denoted H^+. An **ion** is an atom or molecule that has an electric charge because it has unequal numbers of protons and electrons. Since the hydrogen atom consists of only one proton and one electron, a hydrogen ion is just a lone proton. The remainder of the water molecule has an extra electron and therefore a single negative charge; it is a **hydroxyl ion,** denoted OH^-:

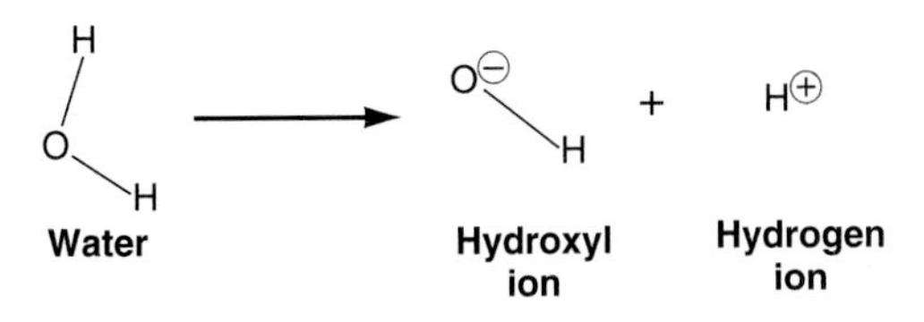

Let us emphasize that a hydrogen ion is simply a proton, for we will use the two terms interchangeably.

3.3 Many compounds are made from combinations of positive and negative ions.

If you have a sense of the structure of methane, ammonia, and water, you have the basis for understanding the huge biomolecules as well as the aqueous environment that surrounds them in all cells. Now let's extend this chemistry lesson by looking at fluorine, the next element after oxygen in the periodic table. Fluorine is even more electronegative than oxygen, so when fluorine and hydrogen atoms combine to form hydrogen fluoride (HF), the fluorine pulls the electron completely away from the hydrogen atom and becomes a fluoride ion, denoted F^-:

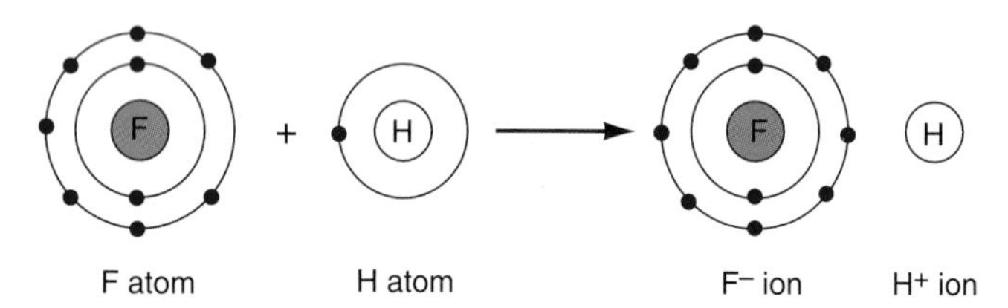

The fluoride and hydrogen ions tend to stay together, but there is no longer a covalent bond between them. Instead, they are held by an **ionic bond** due to the mutual attraction of positive and negative ions rather than a sharing of electrons. Ions with positive charges are called **cations,** and those with negative charges are called **anions.** (As a memory aid, think of the "t" of "cation" as a plus sign.) In fact, purely covalent and purely ionic bonds are extremes. The increasingly polar bonds in ammonia and water have more and more ionic character and less covalent character.

The capacity of an atom to combine with others is described as its **valence,** the number of electrons it can donate or accept. In the series of molecules methane, ammonia, water, and hydrogen fluoride, the central atom is always surrounded by eight electrons in four orbitals, which are either involved in bonds or are filled so they cannot participate in bonding. These four orbitals are called the *s* and *p* orbitals, but to avoid the complications of electronic structure, we will refer to them simply as the *valence orbitals* and the electrons in these orbitals as *valence electrons.* As we move to elements in higher periods of the table, the valence orbitals will be at higher energy levels, but it is principally the *s* and *p* orbitals at each level that are involved in forming bonds with other atoms. Most atoms form molecules by following the *octet rule:* Atoms form bonds by transferring or sharing electrons so that each atom is surrounded by eight valence electrons. The obvious exception to this rule is hydrogen, which has only one valence orbital and forms bonds so that this one is filled. Each atom is, in effect, achieving the electronic configuration of the nearest noble gas in the periodic table. Hydrogen achieves the configuration of helium, with its first energy level filled by two electrons.

To see how the octet rule applies, let's return to our partial periodic table (Figure 3.2) and consider the atoms in certain families of elements. The atoms of the elements in column 1, such as lithium, sodium, and potassium (called alkali metals), have one electron in their valence orbitals:

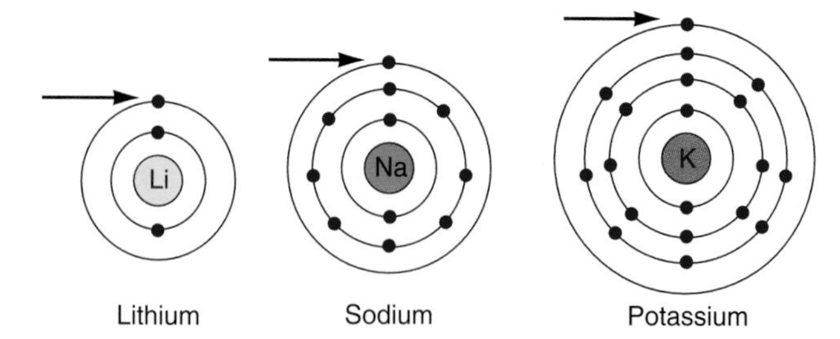

They tend to lose that electron and become cations: Li^+, Na^+, and K^+. On the other side of the table, column 7 contains the halogens (fluorine, chlorine, bromine, iodine) whose valence orbitals contain seven electrons and are only one electron shy of having an octet of electrons. The halogens all behave like fluorine, gaining an electron to become anions: fluoride (F^-), chloride (Cl^-), bromide (Br^-), and so on. (Notice that the names of monatomic anions take the *-ide* ending.) Just as the hydrogen and fluoride ions tend to stay joined by an ionic bond to make hydrogen fluoride, any alkali metal can combine with any halogen, forming such compounds as lithium fluoride (LiF), potassium chloride (KCl), and sodium iodide (NaI). Sodium, potassium, fluorine, chlorine, and others in these families have a valence of 1, and their ions are *monovalent cations* (Na^+, K^+) or *monovalent anions* (F^-, Cl^-).

In contrast, the atoms of elements in column II (beryllium, magnesium, calcium—the alkali earth metals) have two electrons in their valence orbitals. Magnesium and calcium atoms give up two electrons and become *divalent cations* (Mg^{2+} and Ca^{2+}); each of these atoms can combine with *two* atoms of a halogen, such as two chlorine atoms, to form compounds such as calcium iodide (CaI_2) or magnesium chloride ($MgCl_2$):

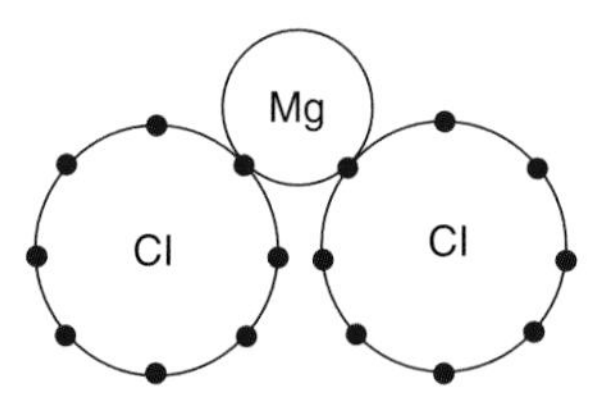

The atoms of elements in the sixth column, such as oxygen and sulfur, can *accept* two electrons, although free oxide and sulfide ions hardly exist. Elements with a valence of 2 can combine one for one, forming compounds such as calcium oxide (CaO) and magnesium sulfide (MgS). They can also engage in two-to-one exchanges, forming such compounds as sodium oxide (Na_2O) and potassium sulfide (K_2S). These examples show that the number of valence electrons in an atom determines its bonding properties, and elements with similar atomic structure have similar chemical properties.

Exercise 3.1 List several compounds of each of the four types that can be made by combining monovalent or divalent cations with monovalent or divalent anions.

Elements in the middle of the periodic table tend to share electrons in covalent bonds rather than losing or gaining electrons to become ions. We have already seen this in the carbon, nitrogen, and oxygen atoms, and in the polar bonds between oxygen and hydrogen, a situation intermediate between purely covalent and purely ionic bonding. Because of the complications of electronic structure and the different electronegativities of atoms, some atoms have different valences in different situations. Nitrogen, for instance, has a valence of 1 in N_2O, 2 in NO, 3 in HNO_2, 4 in NO_2, and 5 in HNO_3.

3.4 Atoms are rearranged in chemical reactions.

Heating and cooling are called physical processes because they may change the outward appearance of materials, but do not change their composition. Thus ice can be heated to liquid water and then boiled to steam, or steam can be cooled to ice again, but the material is still water (H_2O). However, passing an electric current through water in an appropriate apparatus can transform it into its components, hydrogen and oxygen gases; such a process, in which one kind of material is converted into another, is a **chemical reaction.** In some chemical reactions, elements combine to make a compound, as when heat is released by combining the carbon of charcoal briquets with oxygen from the atmosphere:

$$\underset{\text{Carbon}}{C} + \underset{\text{Oxygen}}{O_2} \rightarrow \underset{\text{Carbon dioxide}}{CO_2}$$

Salts are formed when an acid and a base neutralize each other:

$$\underset{\text{Hydrochloric acid}}{HCl} + \underset{\text{Sodium hydroxide}}{NaOH} \rightarrow \underset{\text{Sodium chloride}}{NaCl} + \underset{\text{Water}}{H_2O}$$

In other reactions, compounds decompose into their elements:

$$\underset{\text{Water}}{2\,H_2O} \rightarrow \underset{\text{Hydrogen}}{2\,H_2} + \underset{\text{Oxygen}}{O_2}$$

The number 2 is written in front of H_2O and H_2 to *balance* the equation, so all the material that reacts is accounted for among the products. This raises the question of how much material takes part in a reaction. Even the smallest amounts of material we can handle in a laboratory contain enormous numbers of atoms, and we use a unit of mass that reflects this. A **mole** (abbreviated *mol*) of any substance is about 6×10^{23} molecules (or atoms) of it. This number, known as *Avogadro's number,* has been determined experimentally; it is the number of atoms in X grams of a substance with an atomic (or molecular) mass X. For instance, because the mass of molecular hydrogen (H_2) is 2, a mole of hydrogen—6×10^{23} hydrogen molecules—weighs 2 grams. Because the atomic mass of carbon is 12, a mole of carbon weighs 12 grams; the molecular mass of water is 18, so a mole of water weighs 18 grams. A mole is a convenient unit because a mole of anything is a number of grams equal to its atomic or molecular mass. (For reference, twenty drops of water weigh about a gram, and an American nickel weighs 5 grams.)

The numbers in all chemical equations should be read as numbers of moles, such as, "One mole of carbon plus one mole of oxygen yields one mole of carbon dioxide." If we did not balance the equation for decomposition of water, there would be two moles of oxygen on the right side and only one on the left, which makes no sense physically. It would also be wrong to denote oxygen by O to achieve a balance, because oxygen consists of a pair of atoms bonded together strongly, and its proper formula is O_2. The numbers multiply the formulas, just as in algebra, and show that two moles of water decompose into two moles of hydrogen and one of oxygen.

3.5 Acids and bases liberate or accept hydrogen ions.

If you've thought of travelling around the world to see some famous monuments and works of art, you might be wise to get your tickets soon, before many of them are ruined. For a long time, increasing air pollution has been producing acid rain, which eats away at buildings and statues, especially those made of limestone or marble. The Taj Mahal, the Capitol building in Washington, D.C., and many famous statues and other irreplaceable works of art are gradually disappearing into thin air as their calcium carbonate structure is converted into carbon dioxide. To understand this phenomenon, we must first examine the nature of acids and the role of water as a solvent. Many substances can be used as solvents, but an **aqueous solution,** in which some material is dissolved in water, is the principal kind of interest in biology. From now on when we use the word "solution," we are referring to an aqueous solution.

To understand how small molecules behave in water, let's consider hydrogen fluoride. When dissolved in water, it dissociates into hydrogen and fluoride ions:

$$HF \rightarrow H^+ + F^-$$

A substance that gives up a hydrogen ion in solution is an **acid,** so HF is commonly called hydrofluoric acid. Hydrochloric acid (HCl) also dissociates into a hydrogen and a chloride ion. A substance that can combine with a hydrogen ion is a **base,** so the fluoride ion is a base because it can recombine with the hydrogen ion to form HF again. Another example of a base is ammonia, which can combine with a hydrogen ion to become an *ammonium ion:*

$$\underset{\text{Ammonia}}{NH_3} + \underset{\text{Hydrogen ion}}{H^+} \rightarrow \underset{\text{Ammonium ion}}{NH_4^+}$$

Water acts like an acid because it can dissociate into a hydrogen ion and a hydroxyl ion:

$$\underset{\text{Water}}{H_2O} \rightarrow \underset{\text{Hydrogen ion}}{H^+} + \underset{\text{Hydroxyl ion}}{OH^-}$$

In fact, the ion H^+ doesn't exist to any extent in water. Water also acts like a base, similar to ammonia, and combines with a hydrogen ion to form H_3O^+, a *hydronium ion:*

$$\underset{\text{Hydrogen ion}}{H^+} + \underset{\text{Water}}{H_2O} \rightarrow \underset{\text{Hydronium ion}}{H_3O^+}$$

Nevertheless, it is generally more convenient to refer to H^+, or protons, rather than to hydronium ions. Thus the action of acid rainwater on limestone or marble can be written:

$$CaCO_3 + 2\,H^+ \rightarrow Ca^{2+} + CO_2 + H_2O$$

Although each water molecule rarely dissociates, and the ions formed exist only briefly, this phenomenon is crucial for life. The concentration of hydrogen ions in a solution is critical because each species is adapted to living with a certain amount of acidity. In addition to its effect on human structures, acid rain is having devastating effects on ecosystems, killing forests and turning once-thriving lakes into sterile pools. A special measure has been devised to express hydrogen ion concentration, as we explain next.

3.6 Hydrogen ion concentrations are expressed as pH.

The concentration of a chemical solution is commonly expressed as **molarity,** the number of moles of solute (the substance dissolved) per liter of solution; a 1-molar (1 *M*) solution contains one mole per liter. Since the molecular mass of NaCl is 58, a 1.0 *M* solution of NaCl contains 58 grams per liter; the solution contains about 6×10^{23} Na^+ ions and an equal number of Cl^- ions. A 0.1 *M* solution of NaCl contains 5.8 g per liter: 6×10^{22} Na^+ ions and an equal number of Cl^- ions. Brackets around a chemical symbol mean the molar concentration of that substance, so we could write: $[Na^+] = [Cl^-] = 0.1\ M$.

Pure water at 25°C is found experimentally to have a hydrogen ion concentration, $[H^+]$, of $10^{-7}\ M$ and an identical concentration of hydroxyl ions, $[OH^-]$. The product of these concentrations remains constant:

$$[H^+][OH^-] = [10^{-7}][10^{-7}] = 10^{-14}$$

So as the concentration of one ion increases, that of the other must decrease. Other substances can add hydrogen or hydroxyl ions. Hydrochloric acid (HCl), for instance, contributes hydrogen ions, so a strong HCl solution could have a $[H^+]$ of $10^{-2}\ M$. Ammonia, on the other hand, combines with hydrogen ions and leaves an excess of hydroxyl ions, so an ammonia solution could have a $[H^+]$ of $10^{-10}\ M$. A solution with more H^+ ions than OH^- ions is **acidic,** and one with more OH^- ions than H^+ is **basic,** or **alkaline.** Because the two ions are in equal concentration in pure water, they balance each other, making water **neutral,** neither acidic nor basic.

The hydrogen ion concentration of a solution can vary over a wide range and is of great significance in both chemistry and biology. Because expressing these concentrations with negative exponents is cumbersome, we use a special measure called **pH,** which is defined as the negative logarithm (to the base 10) of the hydrogen ion concentration:

$$pH = -\log_{10}[H^+]$$

Logarithms are explained in Appendix B. For some reason, they scare people more than other mathematical ideas, but "logarithm" is really just another name for "exponent." In the simplest cases, we simply remove the minus sign from the exponent of the hydrogen ion concentration, and the remaining exponent is the pH. Thus, in a neutral solution, where $[H^+] = 10^{-7}\ M$, the pH is 7. In an acidic solution, if $[H^+]$ is $10^{-4}\ M$, the pH is 4. In an alkaline solution, if $[H^+] = 10^{-9}\ M$, the pH is 9. It should now be clear that an acidic solution has a pH *lower* than 7, and an alkaline solution has a pH *higher* than 7.

The **pH scale** shown in Figure 3.3 expresses the concentration of hydrogen ions in various substances. The scale runs from 0, where $[H^+] = 1\ M$ ($10^0\ M$), to 14 where $[H^+] = 10^{-14}$ *M*. It is important to understand that the pH scale is not linear. It is a logarithmic scale, and each unit means a hydrogen ion concentration ten times different from the next. A pH of 2 (0.01 *M*) means a $[H^+]$ ten times less than pH 1 (0.1 *M*) and ten times greater than pH 3 (0.001 *M*). It is instructive to note the pH values of various biological fluids on this scale. Human blood is just on the basic side of neutrality (pH 7.4), while the stomach contents during digestion are very acidic, as low as pH 2. Blood is an example of a buffered solution; a buffer is a substance that

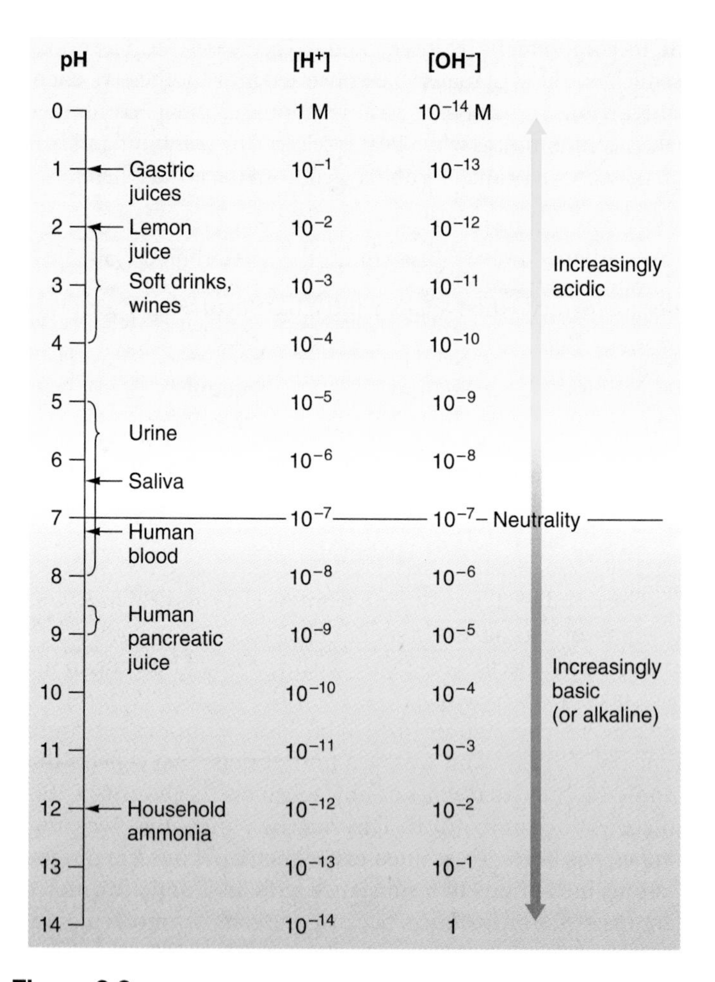

Figure 3.3

The pH scale measures the concentration of hydrogen ions $[H^+]$ in a solution. A low pH, such as 1, indicates a high concentration of hydrogen ions; a high pH, such as 14, indicates a low concentration. A change of one pH unit is an increase or decrease in $[H^+]$ by a factor of 10. pH 7 is neutrality, where the concentrations of hydrogen and hydroxyl ions are equal.

absorbs and releases H^+ ions in a way that keeps the pH of a solution constant, and blood has several buffers that do this.

Buffers, Concepts 44.1.

Exercise 3.2. A solution has 10^{-9} moles of hydrogen ions per liter. What is its pH? If a solution has a pH of 5, what is the concentration of hydrogen ions?

Exercise 3.3. A solution has a pH of 6.3. You add acid until the concentration of hydrogen ions is increased exactly ten times. What is the pH now?

Exercise 3.4. You have a liter of solution with pH 8. How much pure water will you have to add to change its pH to 6?

3.7 Water has unusual properties that are essential for life.

All life depends on water. Organisms are largely composed of water (Table 3.1), and chemical principles suggest that no substance could substitute for water in a biological system. In a fascinating book originally published in 1913, *The Fitness of the Environment,* Lawrence J. Henderson pointed out that it isn't enough for organisms to adapt to their environments; life couldn't exist unless the environment also had the right properties. Henderson showed how the inorganic world—at least on our planet—is suited to supporting life, and he emphasized the properties of water. It is worth reviewing his major points and seeing that all these properties stem from the structure of water.

Water crystals are among the most beautiful natural objects. Though a lump of ice is very commonplace, snowflakes are delicate, lacy hexagons with wonderfully varied symmetry (Figure 3.4). Ice crystals grow in this form because of the unique way water molecules bond to each other. Because water molecules are polar, they are attracted to one another by **hydrogen bonds.** In this type of weak interaction, a hydrogen atom is shared between two relatively electronegative atoms and holds them together. In water, the hydrogen atom covalently bonded to one oxygen atom is attracted to the oxygen atom of another water molecule, so the three atoms line up: O–H···O, where the three dots represent the hydrogen bond. In an ice crystal, each oxygen atom forms two covalent bonds and two hydrogen bonds, so each water molecule is held in a three-dimensional lattice by four others:

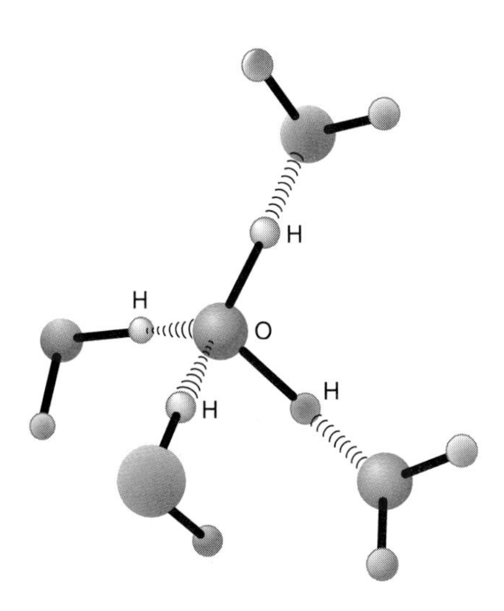

Although each hydrogen bond is weak, the combination of many such bonds keeps ice highly structured. Even in liquid water, molecules stay bonded in icelike clusters, whose size decreases with increasing temperature. One scientist, only partly joking, has referred to the oceans as single molecules because each water molecule is attracted so strongly to its neighbors. Sidebar 3.1 elaborates on this internal bonding of water.

Table 3.1 Water Content of Various Organisms and Tissues

Organism or Tissue	Percent Water (By Mass)
Bacterium	80
Paramecium	79
Mushroom	85–90
Alga	90–95
Jellyfish	96
Nudibranch (shell-less mollusc)	95
Sea urchin egg	77
Rat	65–70
Human (newborn)	66
Brain, kidney, muscle	80
Spinach	93
Typical green plant	75
Plant seeds	5–10
Plant leaves	50–90
Freshly felled timber	40–65

Figure 3.4
Snowflakes show the regular hexagonal structure of ice crystals.

Sidebar 3.1 The Properties of Water

Most of the properties of water that Henderson describes so beautifully and that are so important for life have their source in the fact that water is, as we say, a highly associated liquid, the molecules of which are engaged ceaselessly in ever-changing patterns of mutual interconnections through hydrogen bonds. No specific groups of molecules remain tied together in this way for more than an instant, yet statistically a more-or-less constant fraction of the population is joined together all the time. We may picture such a group of momentarily bound water molecules as follows, the solid lines representing covalent bonds, the broken lines hydrogen bonds:

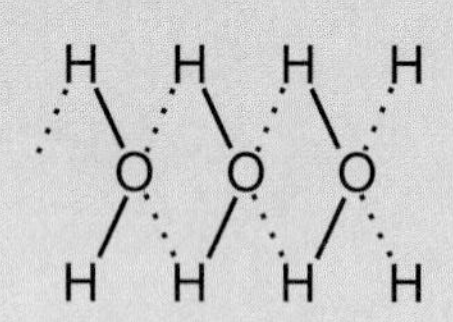

At the next moment, the covalent and hydrogen bonds may have exchanged places:

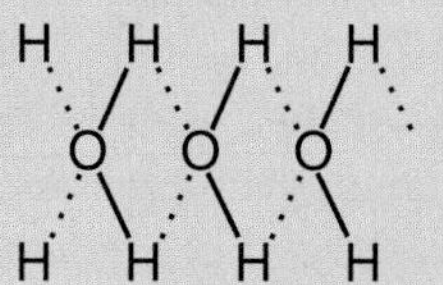

As Albert Szent-Györgyi, who has a great gift for saying things well, puts it: "Water is the only molecule that can turn around without turning around." From this association through hydrogen bonding, water derives its great surface tension, cohesiveness, high boiling point, high heat of vaporization, and as already said, most of the unique properties that Henderson ascribed to it.

Source: From George Wald, in the Introduction to Lawrence J. Henderson, *The Fitness of the Environment,* 1958, Beacon Press, Boston.

Hydrogen fluoride and ammonia molecules can form hydrogen bonds, just as water does. A hydrogen bond can form wherever oxygen, nitrogen, or fluorine atoms hold a hydrogen atom between them. As we will see, hydrogen bonds such as O–H···O and O–H···N are extremely common in biomolecules. Hydrogen bonding within water also gives it at least seven biologically important properties.

1. Water is **cohesive.** Because of hydrogen bonding, water molecules stick together. Cohesiveness is critical for plants, especially large trees. Plants have fine tubes leading from their roots to their leaves (Figure 3.5); water evaporates from the leaves at the top, and because the water molecules in the tubes stick together, more water is drawn in at the roots, so a constant stream of water flows through the plant. It is also significant that water is **adhesive,** able to form hydrogen bonds with many biomolecules, such as the cellulose in the walls of these plant tubes.

Cohesiveness also causes the surface of water to act like an elastic skin. Water is said to have a **surface tension,** because the molecules on its surface are held together by hydrogen bonding from within (but of course are not subjected to equal forces from above the surface). Surface tension allows small objects to rest atop a body of water, even if they're denser than water. An early type of compass was simply a magnetized needle gently placed in a container of water, where it floated and turned to point north. Some creatures, such as water striders (sometimes called water skippers), take advantage of surface tension by walking on water as if it were solid (Figure 3.6).

2. Water is an excellent solvent for many substances. The alchemists of the Middle Ages searched for a universal solvent, a liquid that would dissolve everything. (What would you keep it in, by the way?) There is no such solvent, of course, but water comes closer than anything else. Salts, such as sodium chloride, dissolve in water by dissociating as each ion becomes surrounded by the polar water molecules:

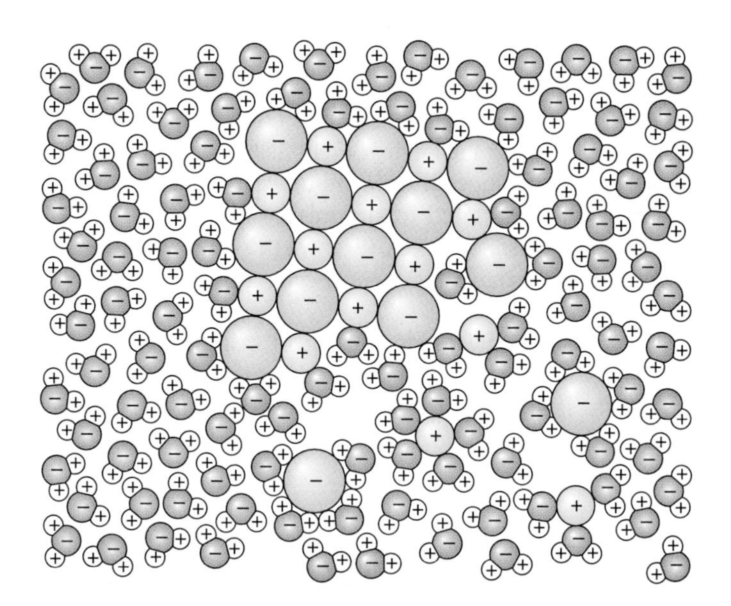

Shielded by a shell of water molecules, the ions stay in solution because they are no longer affected by attractive forces from other ions.

Water also dissolves many kinds of organic compounds. The chemical reactions of organic molecules in metabolism take place most easily in solution. Thus water, being such an excellent solvent, facilitates metabolism. The inside of a cell is mostly an aqueous solution called **cytosol** that contains many kinds of molecules and ions.

3. Water has a high *heat capacity,* the ability to hold heat. *Specific heat,* a measure of heat capacity, is the heat required to raise the temperature of 1 gram of a substance 1°C. By definition, the specific heat of water is 1 **calorie** (cal) per gram. (The term *Calorie,* with a capital "C," used to describe the energy in food, is 1,000 calories or 1 kilocalorie [kcal]). The heat capacities of most organic compounds are about 0.6 cal/gram or less.

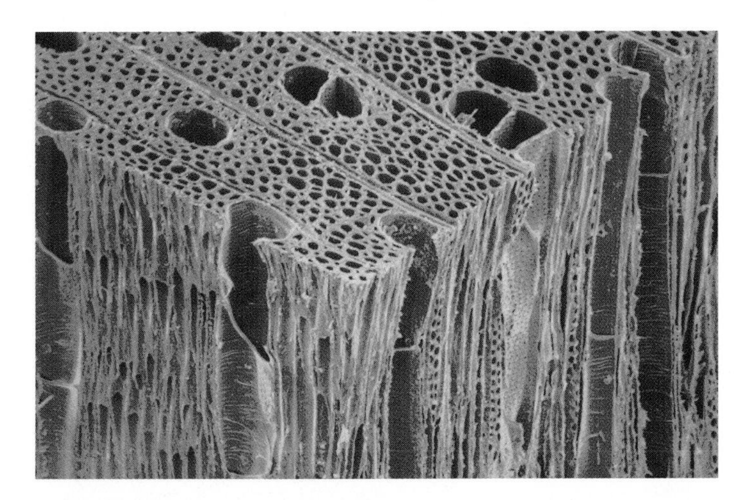

Figure 3.5
Plant stems contain fine tubes that conduct water from the roots to the top. This transport depends on the cohesion between water molecules in each tube. (The bright green color is artificial.)

Figure 3.6
The surface molecules in water are held by hydrogen bonding from within; this creates a surface tension, and the surface is almost like a skin. A needle will float on water if placed there gently, and small animals, like these water striders, can walk on the surface.

Water, with its high heat capacity, therefore changes temperature more slowly than do other compounds when gaining or losing energy. This resistance to sudden changes in temperature makes water an excellent medium for organisms to live in because they are adapted to narrow temperature ranges and may die if the temperature fluctuates widely.

The heat capacity of water stems directly from its hydrogen-bonded structure. Although hydrogen bonds are weak, their effect in combination is enormous. As heat is added to ice or to liquid water, the energy first goes into breaking hydrogen bonds, allowing the molecules to move about freely. Since temperature is a measure of the average kinetic energy of molecules—the rate at which they are moving—the temperature of water rises slowly with the addition of heat.

4. Water has a high *heat of vaporization,* the energy required to vaporize it. Because of the energy needed to break the hydrogen bonds holding a water molecule to its neighbors, more energy is required to evaporate liquid water than most other substances. It requires between 540 cal (at 100°C) and 596 cal (at 0°C) to evaporate a gram of water, but only 236 cal for a gram of alcohol (ethanol). Most substances require far less energy. This feature is significant biologically because evaporation of water is one way terrestrial (land) organisms cool themselves. When we perspire, the water that evaporates carries heat from our bodies. Evaporation from the leaves of a plant or a panting dog's nasal passages does the same thing (Figure 3.7).

5. Water has a high *heat of fusion,* the amount of energy that must be removed to freeze a substance, because when water freezes, each molecule must give up a lot of energy to form hydrogen bonds with its neighbors and settle into the growing ice crystal. For water, the heat of fusion is about 80 cal per gram, compared to around 20–40 cal/gram for most organic compounds. So it is much harder to freeze an organism containing water than it would be to freeze one made of an organic solvent. Furthermore, on a much larger scale, the temperature of the environment does not change drastically with the onset of winter or summer, since large amounts of water on the earth's surface freeze and thaw at those times. These water masses tend to keep the surface temperature more constant than it would be otherwise, modulating sudden shifts of temperature. This is why deserts, without the tempering effect of water, have wider swings of temperature between night and day than areas near lakes or oceans (Figure 3.8).

6. Water remains liquid over a wide temperature range, from 0°C to 100°C. Most other substances only remain liquid over a narrower range. Since the chemical reactions of metabolism depend on interactions between molecules moving about freely in liquid water, the limits of life are set by water's freezing and boiling points. This property of water makes possible a wide variety of habitats (niches) in which organisms can exist. Ice fish live in water at a temperature close to freezing, while bacteria and algae live in hot springs where the temperature is near the boiling point of water (Figure 3.9).

7. Ice floats. This apparently trivial fact is very important. Ice is less dense than liquid water because molecules in the open, lacy, hydrogen-bonded crystal lattice are held farther apart than in liquid water. When water freezes, it expands, has reduced density, and floats on the liquid. Therefore, when the temperature drops below freezing, ice forms an insulating layer on the surface of bodies of water, reducing the heat loss from them and enabling life to go on below. As the air warms up again, the surface ice melts. If ice sank, oceans and lakes might be packed from the bottom with ice, and many of them would not be able to thaw out, since they only receive energy from warm air and from sunlight that does not penetrate very far. The earth's climate would be very different—far colder than it is.

Exercise 3.5 Imagine a water molecule in ice, held in place among its neighboring molecules. As the molecule warms up, it jiggles in place faster and faster, until it can break free. Compare the actual situation, where the molecules are linked by hydrogen

(a)

(b)

Figure 3.7

These two organisms use the same method to keep cool on a hot day. *(a)* The plant transpires, continually evaporating water through its leaves. The water molecules that leave remove excess energy, and the organism remains relatively cool. *(b)* The dog pants to evaporate water from its mouth and nasal passages. Dogs have a special nasal gland that seems to supply water to the nasal surface where it can evaporate for cooling.

Figure 3.8

Water's high heat capacity, high heat of vaporization, and high heat of fusion combine to maintain relatively constant environmental conditions where water is abundant. Where it is scarce, as in this desert, the temperature changes much more drastically between night and day.

bonds, to an imaginary situation in which there is no hydrogen bonding, so the molecules have only weak attractions to one another. Then compare the energy a molecule must expend to break free, with and without hydrogen bonding.

Exercise 3.6 Explain why evaporation cools a body of water. *(Hint:* The temperature of an object measures the average kinetic energy of its atoms and molecules, and kinetic energy increases with velocity.)

3.8 Organisms are composed of only about 25–30 elements.

A copper wire is flexible and conducts electricity well; an aluminum tube is strong and lightweight; and a wafer of silicon has unusual electrical properties that make it ideal for constructing small electronic circuits. Since an object's properties depend on the materials it is made of, we should ask which chemical elements occur in organisms, and whether they explain how living

Figure 3.9
The enormous range over which organisms can live in water is shown by the fish in an icy Minnesota lake and algae that lend their colors to the hot springs at Yellowstone National Park.

organisms operate. Traces of all elements have been found in organisms, but only about 25–30 of them are known to have any biological function. Table 3.2 summarizes the principal elements in the human body. Figure 3.10 shows that the distribution of elements in organisms in general is remarkably similar to their distribution in the whole universe, with some exceptions. (Hydrogen, for instance, is relatively much more abundant in the universe than in organisms.) So organisms are made of the same elements found elsewhere; there are no uniquely biological elements. The uniqueness of living things doesn't come from the elements they are made of, but rather from the molecules those elements form and the way these molecules react with one another.

As we saw in Section 3.7, water usually constitutes more than half of an organism's total weight; this accounts for the large percentages of oxygen and hydrogen in Table 3.2. The weight of an organism after water is removed is its *dry weight.* If most of an organism is water, then what is the chemical nature of the rest? A simple, revealing way to analyze the chemical composition of a dried organism is to weigh it, incinerate it carefully, and then weigh the remains. During incineration, about 80–90 percent of the dry weight is lost as the organic constituents of the organism burn off and are converted into gases, mostly carbon dioxide and water. This experiment explains why the six major components of organic compounds—hydrogen, carbon, nitrogen, oxygen, sulfur, and phosphorus—constitute the bulk of an organism's substance other than water. (Students sometimes remember organic compounds by their symbols as CHNOPS molecules.)

The inorganic constituents are left as ash that makes up no more than 10–20 percent of the dry mass, and the larger amount only comes from organisms that have a lot of bone or shell. This ash contains less abundant elements, principally sodium, magnesium, potassium, calcium, and chlorine; these elements constitute the principal ions dissolved in an organism's water, and calcium is a constituent of supporting material such as bone. Some organisms contain substantial amounts of silicon or iodine. Selenium, fluorine, boron, and vanadium are only essential for special groups of organisms and seem to be generally unimportant for most. Still lower on the abundance scale are several **trace elements,** including the metals cobalt, nickel, manganese, iron, copper, zinc, and molybdenum. The

Table 3.2 Principal Elements in the Human Body

Element	Percent by Mass
Hydrogen	9.5
Carbon	18.5
Nitrogen	3.3
Oxygen	65.0
Sodium	0.2
Magnesium	0.1
Phosphorus	1.0
Sulfur	0.3
Chlorine	0.2
Potassium	0.4
Calcium	1.5
Chromium	Trace
Iron	Trace
Cobalt	Trace
Nickel	Trace
Copper	Trace
Zinc	Trace
Molybdenum	Trace
Iodine	Trace

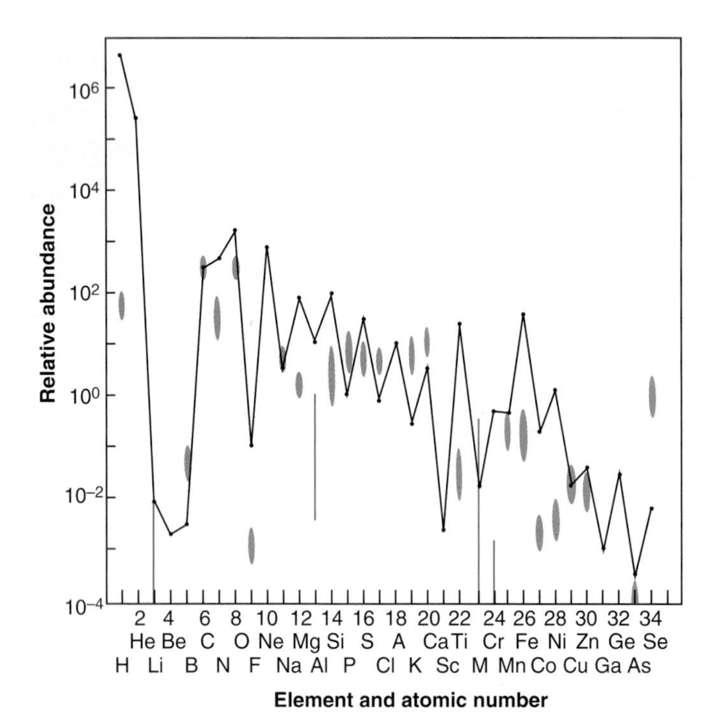

Figure 3.10

The abundance of elements in the universe (relative to 100 atoms of silicon) compared with their abundance in organisms per 1,000 total atoms (blue-green bars). Data for elements that are important biologically (such as carbon, oxygen, and nitrogen) are quite reliable, and these elements fall into narrow ranges. Data for less abundant elements are less certain. The abundance of elements with little or no known biological importance tends to vary a lot.

major role of these metals is to work with biological catalysts to speed up metabolic processes. Current research, especially in nutrition, keeps demonstrating that other elements have vital roles in metabolism, so the list of essential elements will probably keep growing. Perhaps some day selenium will be added to the list; though no illness is yet associated with a lack of selenium, it apparently has value, in small amounts, as an antioxidant that protects cells against some kinds of damage.

3.9 Organisms are constructed of carbon-based molecules.

Now we turn to the organic compounds, the CHNOPS compounds, that constitute the bulk of biological structure. By definition, organic compounds are based on carbon and hydrogen; only carbon oxides such as CO_2 and carbonates are considered inorganic. These compounds are obviously called "organic" because they are characteristic of organisms, and until the early nineteenth century, they were thought to be uniquely products of living organisms. It was a great triumph of chemistry when, in 1828, Friedrich Wöhler synthesized the organic compound urea from an inorganic compound, ammonium cyanate, and thus showed that there is no mysterious gap between the two kinds of matter.

Thousands of different organic molecules are possible because a carbon atom forms four covalent bonds, and they can be linked to one another and to atoms of other kinds in an enormous variety of combinations. To illustrate this, we'll begin with the **hydrocarbons,** compounds made only of carbon and hydrogen. In the simplest hydrocarbon, methane (CH_4), a carbon atom is bonded to four hydrogen atoms. The next larger hydrocarbon, ethane (C_2H_6), has two carbon atoms bonded to each other, and the three remaining bonds on each carbon atom are made to hydrogen atoms:

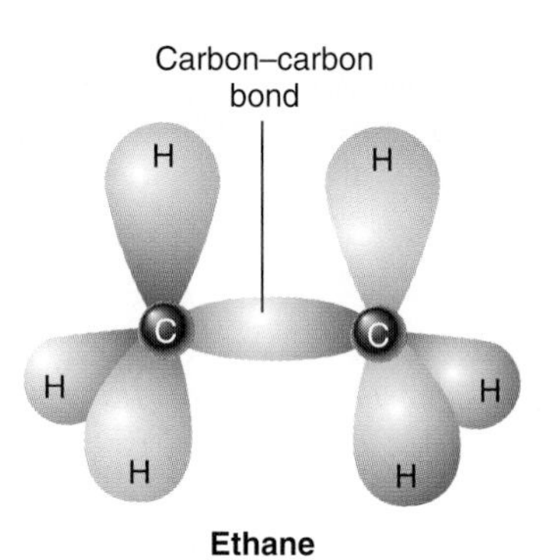

The covalent C–C bond is the basis of most organic molecules. Carbon atoms can be linked to one another in chains, which can become quite long because the C–C bond is very strong.

Ethane is the first of a series of hydrocarbons, each with one more carbon atom and two more hydrogen atoms. The next hydrocarbon in this series is propane (C_3H_8):

```
    H   H   H
    |   |   |
H — C — C — C — H
    |   |   |
    H   H   H
```

Propane

It is made (conceptually) by inserting a third carbon atom, with two hydrogens, between the two end carbons, which each retain their three hydrogens. This is the common pattern for larger organic molecules. Butane (C_4H_{10}) is next in this series, but it has two possible structures:

```
    H   H   H   H               H   H   H
    |   |   |   |               |   |   |
H — C — C — C — C — H       H — C — C — C — H
    |   |   |   |               |   |   |
    H   H   H   H               H   C   H
                                  / | \
                                 H  H  H
```

Butane

These two molecules, with the same composition but different structures, are **isomers.** They are different compounds, which can be separated by chemical methods. The one with the unbranched, so-called "straight" chain, is *n*-butane (*n* for *normal*), and the one with the branched chain is isobutane (*iso-* = equal, referring to the symmetrical shape of the molecule). Now we start to see why there are so many different kinds of organic molecules. Even with only four carbon atoms, two structures are possible, and as the number of carbon atoms increases, the number of possible isomers also increases substantially. For instance, the six carbon atoms of a hexane (C_6H_{14}) can be linked in five ways:

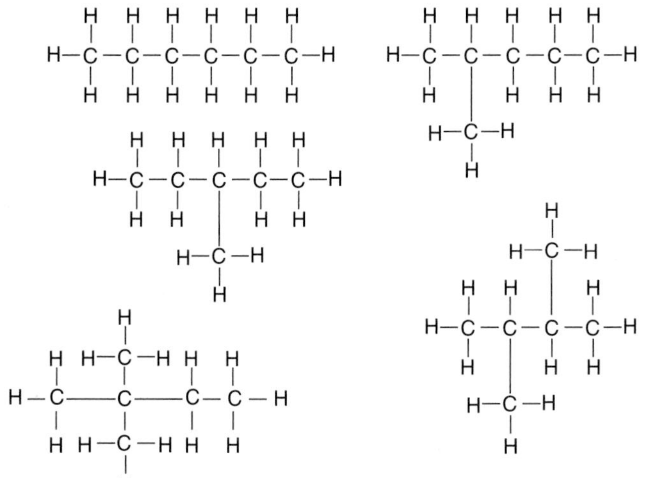

Hexane

The larger compounds in the hydrocarbon series are systematically named according to Greek roots to show the number of carbon atoms in a molecule. Thus *pent-*, *hex-*, *hept-*, and so on, which are used for geometric figures (pentagon, hexagon), are also used for these compounds (pentane, hexane, heptane) to show the number of carbon atoms in a molecule.

Notice that we can represent molecules in several ways. Molecular formulas, such as C_6H_{14}, simply tell the number of atoms of each element in a molecule; drawings that show how the atoms are bonded to one another are structural formulas. The structures of simple molecules can be shown compactly by a kind of abbreviated structural formula, such as $CH_3CH_2CH_3$ for propane and $CH_3CH_2CH_2CH_3$ for *n*-butane.

The hydrocarbons discussed so far are said to be **saturated** because each carbon atom carries the maximum number of hydrogen atoms, either two or three. Many other hydrocarbons are **unsaturated,** which means they have fewer than the maximum number of hydrogen atoms; in each molecule, at least two carbon atoms are connected by a **double bond,** in which they share *two* pairs of electrons. Double bonds are drawn as C=C, as in ethylene ($H_2C{=}CH_2$) or butene ($CH_3CH{=}CHCH_3$):

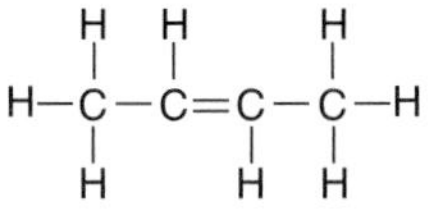

All the nutritional questions about saturated and unsaturated fats revolve around molecules like this. When unsaturated fats such as corn and safflower oils are made into margarine, a saturated fat, hydrogen atoms are added to their double bonds, a process called hydrogenation.

Carbon chains also form rings, commonly pentagons and hexagons as in cyclopentane and cyclohexane:

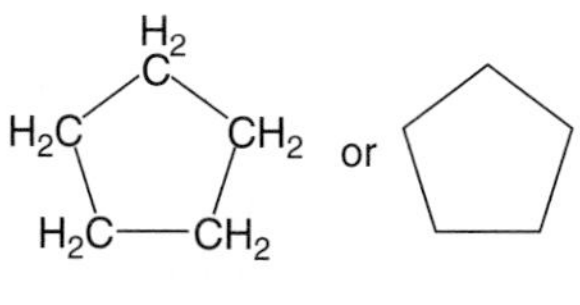

Cyclopentane 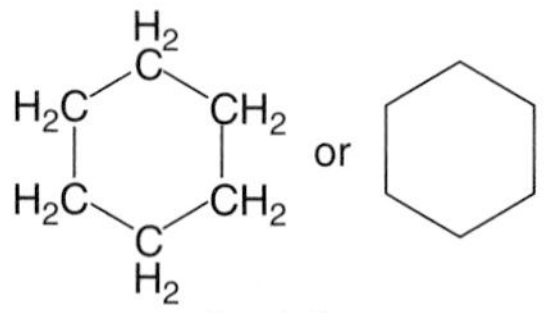 **Cyclohexane**

Many biomolecules contain rings with oxygen, nitrogen, or sulfur atoms as well as carbon, sometimes with complex shapes. Two molecules are drawn here in two equivalent ways to show a convention of organic chemistry: In the abbreviated drawings on the right, each vertex in a ring represents a carbon atom, usually with one or two hydrogen atoms attached, and only O, N, P, and S atoms are shown explicitly, along with H atoms bonded to them:

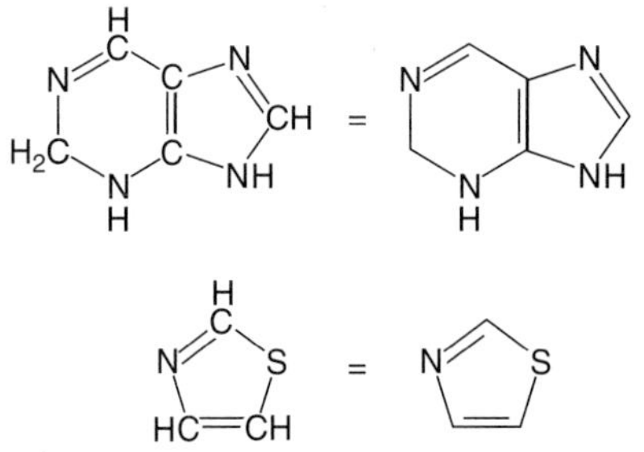

Some bonds are drawn as heavy or wedge-shaped lines to suggest perspective, because it is often important to see the three-dimensional relationships of the atoms:

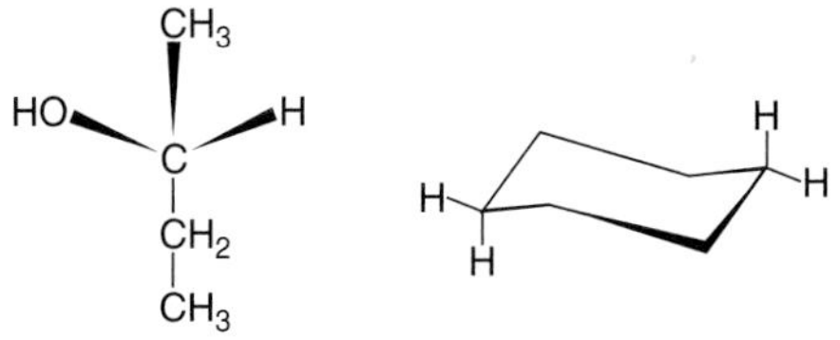

Also, the atoms in a chain or ring are numbered (N-1, C-2, C-3), using standard rules, to specify where other atoms may be attached:

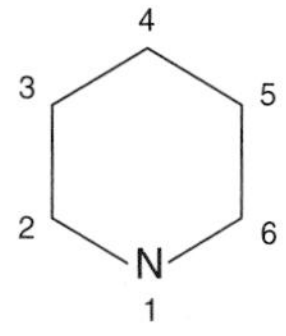

Exercise 3.7 *(a)* Draw a structural formula to show the following atoms properly connected. (Be sure the carbon atoms always make four bonds, nitrogen three bonds, and oxygen two.)

$$H_2C = CH(HCNH_2)CH_2COOH$$

(b) Draw abbreviated structures for the following molecules:

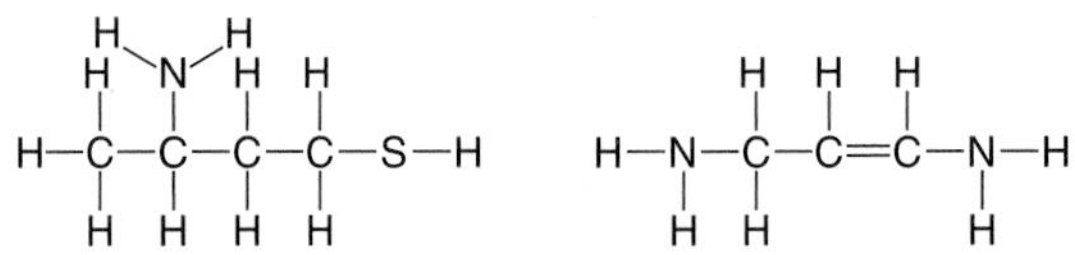

3.10 Organic molecules have distinctive shapes, and most can change shape without breaking bonds.

We began this chapter by emphasizing the significance of molecular shape. *All organic molecules have distinctive three-dimensional shapes.* Methane, for example, is a tetrahedron with the carbon atom in the center and a hydrogen atom at each of the four corners, and ethane is like two linked tetrahedrons:

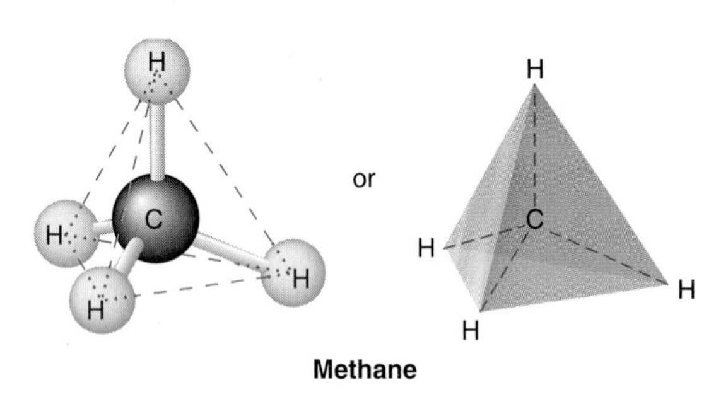

Methane

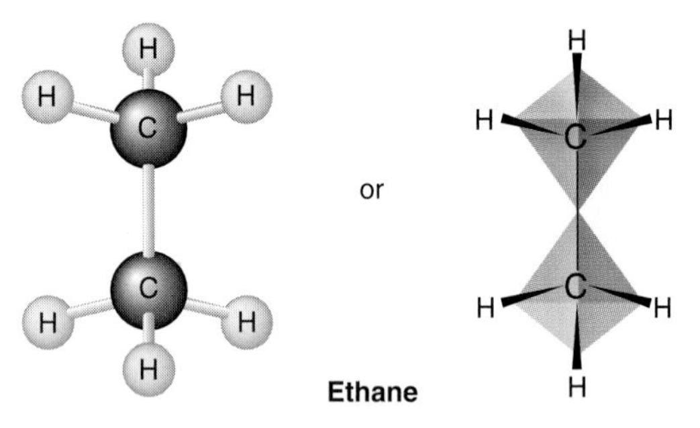

Ethane

Hydrocarbon chains have a zigzag shape, with the hydrogen atoms alternately forming a pair of knobs on opposite sides:

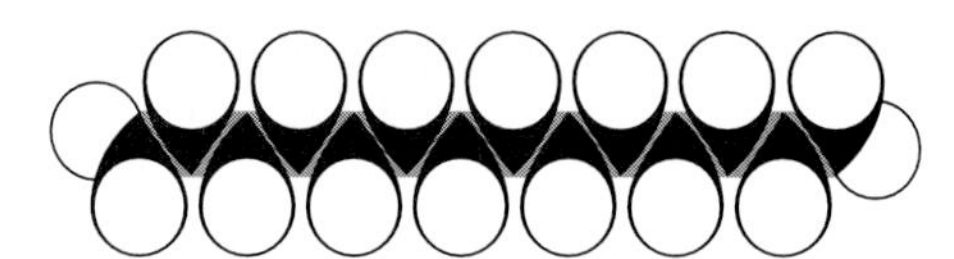

Therefore, hydrocarbons are often represented by zigzag lines:

H_3C–CH_2–CH_2–CH_2–CH_2–CH_3 or

Branched-chain molecules have increasingly complex forms as the chains become longer, and larger molecules have even more distinctive shapes. When students make organic molecules with ball-and-stick models, they inevitably notice that a propane molecule looks like a little animal:

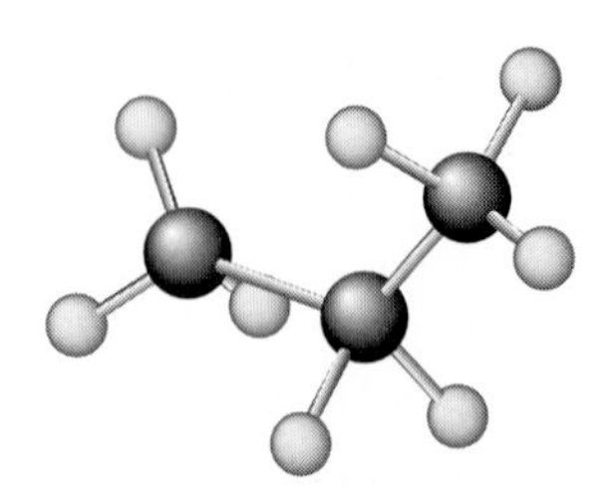

Compounds called *aromatic hydrocarbons* because of their strong odors illustrate a common feature of some organic structures: electrons occupying large orbitals. The simplest of these compounds is benzene (C_6H_6). Each of its carbon atoms is bonded to only one hydrogen atom, and the carbons form a ring that is conventionally drawn with alternating single and double bonds:

H C, HC, CH, HC, CH, C H or

This representation, however, is unrealistic. The electrons are not localized, or confined to single and double bonds between carbon atoms as shown but are really *delocalized*—that is, able to occupy large doughnut-shaped orbitals spread around the whole ring, a situation better represented as:

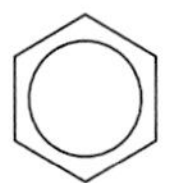

Aromatic molecules are flat, with all the atoms in the ring lying in a plane, in contrast to the zigzag forms of saturated molecules such as cyclohexane. In biomolecules like DNA, the aromatic constituents are stacked on top of one another like slices in a loaf of bread. Furthermore, these ring-shaped orbitals have different energies from those in other orbitals, so they interact with light differently (as explained in Section 10.3); the more complex aromatic materials are often distinctively colored because of the way they absorb light, and color is one of the most significant features of the biological world.

When the atoms in organic molecules are joined only by single bonds, they can generally rotate around these bonds, so the molecule's shapes are not only distinctive but also changeable. The cyclohexane molecule shows this very well; by simply rotating some atoms, without breaking the bonds between them, the molecule can switch between two distinct **conformations**—that is, different arrangements of the atoms in space—called the chair and boat forms:

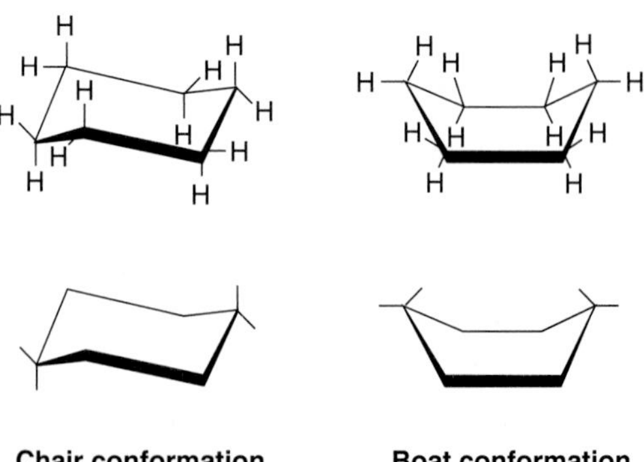

Although the chair conformation is the more stable, cyclohexane molecules are constantly switching back and forth from one conformation to the other. The ability of very large biomolecules

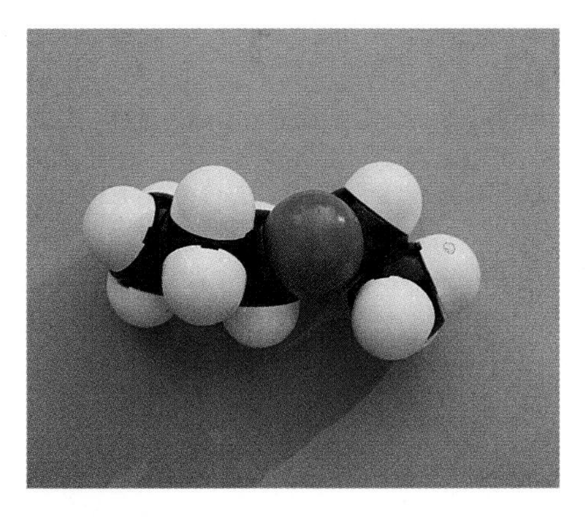

Figure 3.11
Three-dimensional models of organic molecules show their distinctive shapes.

to change conformations accounts for most of their biological activity. Figure 3.11 shows the shapes of a few organic molecules.

3.11 The solubility of an organic molecule in water is important biologically.

Oil and water don't mix, as we can see in a bottle of salad dressing or the oil slick on a puddle of water (Figure 3.12). Yet, as we showed earlier, many other substances dissolve in water perfectly well. What makes the difference? The general rule is, "Like dissolves like." Remember that water molecules are polar—the oxygen atom is somewhat negative, and the two hydrogen atoms somewhat positive. So other polar molecules dissolve in water because of mutual interactions with the polar water molecules, as do ions like those in a salt. Water-soluble molecules are **hydrophilic,** literally "water loving."

In contrast, nonpolar materials that are not water soluble, such as hydrocarbons, are **hydrophobic,** or "water fearing." The insolubility of nonpolar materials is explained by the *hydrophobic effect.* Water molecules attract one another strongly but have very little attraction for hydrocarbon molecules, so when hydrocarbons are mixed with water, they tend to disrupt the hydrogen-bonded water structure:

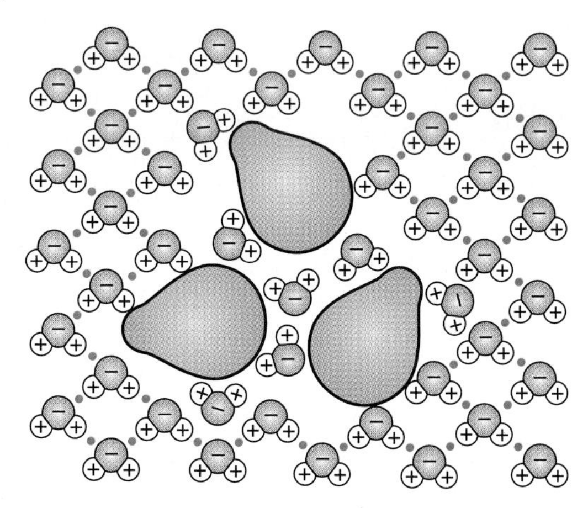

Water, however, tends to form as many hydrogen bonds as possible, forcing the hydrocarbon molecules aside into clusters

Figure 3.12
Oil and water don't mix. Oil (hydrocarbon) molecules are hydrophobic and do not form bonds with water molecules, so oil forms a distinct layer on the top of water.

where they have the least disruptive effect on the water structure. So the hydrocarbons form isolated droplets:

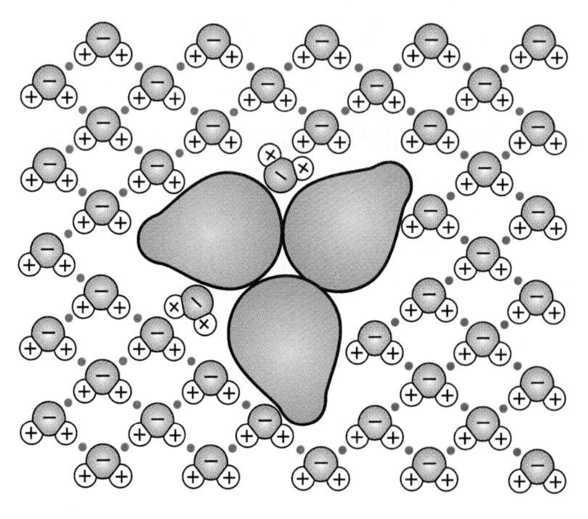

Within a droplet, the hydrocarbon molecules are held together by weak interactions called *van der Waals forces.* After shaking an oil and water mixture, you can see the oil separate into small droplets that gradually fuse into a single, separate phase.

The hydrophilic or hydrophobic character of an organic molecule determines much of its biological function. As we pointed out earlier, water-soluble organic molecules are compatible with the water contained in cells, where most chemical reactions of metabolism occur. On the other hand, the membranes that separate one aqueous compartment from another—and separate the cell from its environment—consist of hydrophobic molecules (described in Chapter 8). The affinity of a biomolecule for the water in cytosol or for the surrounding membranes determines where it resides and how it functions.

3.12 Different types of organic molecules are characterized by their functional groups.

Think of a simple hydrocarbon chain as a scaffolding to which other atoms can be attached. A few common groups of atoms, called **functional groups,** can change a molecule's character when added to a hydrocarbon. The principal types of organic compounds are characterized by the functional group(s) they carry, so an experienced chemist, looking at the formula of an organic molecule, knows what class it belongs to and many of its chemical characteristics. Here we'll review the most common classes of compounds and their functional groups.

Alcohols

An organic molecule with a **hydroxyl group** (–OH) is an alcohol, such as CH_3CH_2OH (ethanol) or $CH_3HCOHCH_3$ (isopropanol). Some biologically important alcohols have more than one hydroxyl group, such as glycerol:

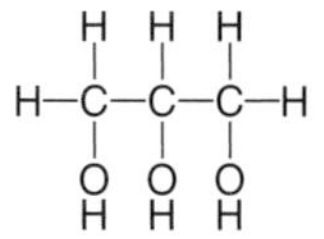

```
   H  H  H
   |  |  |
H—C—C—C—H
   |  |  |
   O  O  O
   H  H  H
```

Because a hydroxyl group is polar (it is two-thirds of a water molecule), short-chain alcohols are water-soluble; when the hydrocarbon chain is very long, a single polar group on one end has little influence in making the molecule hydrophilic.

Aldehydes and ketones

In a carbonyl group (–C=O) a carbon atom is double-bonded to an oxygen atom. Such a group on the end of a chain makes an aldehyde, such as $CH_3HC{=}O$ (acetaldehyde). A carbonyl on an internal position in a chain makes a ketone, such as acetone:

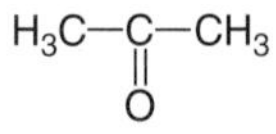

$$H_3C{-}\underset{\underset{\displaystyle O}{\|}}{C}{-}CH_3$$

Some classes of organic molecules carry more than one functional group. **Sugars,** for example, are aldehydes or ketones that also have one or more hydroxyl groups.

Organic acids

A carbon with a double bond to an oxygen atom and a single bond to a hydroxyl group is a **carboxyl group,** conventionally represented by –COOH, and molecules that bear carboxyl groups are organic acids. Vinegar is dilute acetic acid (CH_3COOH), and butyric acid ($CH_3CH_2CH_2COOH$) provides one of the disagreeable tastes and odors of rancid butter. The carboxyl group is acidic because it dissociates, releasing a hydrogen ion:

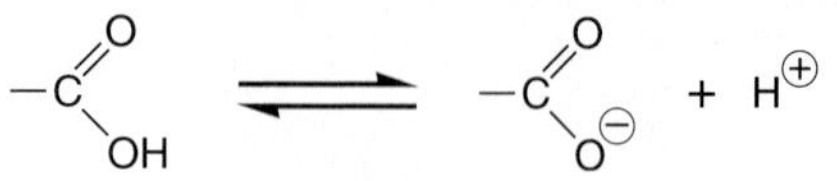

$$-C(=O)OH \rightleftharpoons -C(=O)O^{\ominus} + H^{\oplus}$$

Amines

Molecules that carry **amino groups** ($-NH_2$) are called amines; an example is $CH_3CH_2CH_2NH_2$ (propylamine). These are bases because an amino group acts very much like an ammonia molecule and can accept a hydrogen ion:

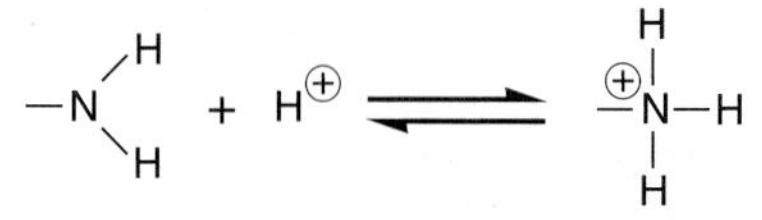

$$-NH_2 + H^{\oplus} \rightleftharpoons -\overset{\oplus}{N}H_3$$

Molecules that bear carboxyl or amino groups exist in both ionized and nonionized forms, the degree of ionization depending on the pH of the solution around them. In acidic solutions, with an excess of hydrogen ions, amino groups pick up hydrogen ions and are positively charged ($-NH_3^+$), but in alkaline solutions, they are not ionized ($-NH_2$). The opposite is true for carboxyl groups: They are not charged (–COOH) in acidic solutions, but lose their hydrogen ions and become negatively charged ($-COO^-$) in alkaline solutions.

Thiols

Molecules that carry a **sulfhydryl group** (–SH) are thiols, or mercaptans. Mercaptoethanol ($HO{-}CH_2CH_2SH$) is a thiol commonly used in biochemical research. We will see that thiols and their modifications are important in protein structure. These molecules are remarkable for their pungent odors. Simple hydrogen sulfide (H_2S) is the characteristic odor of rotten eggs; freshly chopped onions liberate propanethiol; and garlic livens our meals with its allyl mercaptan. On the other hand, you detect a skunk by its 3-methyl-1-butanethiol.

Phosphates

Organic compounds containing some form of phosphoric acid (H_3PO_4) are common and essential in biology. The **phosphate group** on an organic molecule can have different forms, depending on the pH of the medium, because two of its oxygen atoms can be either associated with a hydrogen, as –OH, or ionized, as $-O^-$:

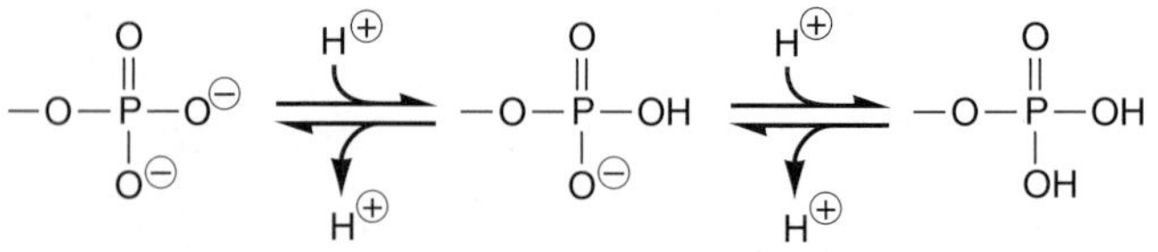

$$-O-P(=O)(O^{\ominus})-O^{\ominus} \underset{H^{\oplus}}{\overset{H^{\oplus}}{\rightleftharpoons}} -O-P(=O)(O^{\ominus})-OH \underset{H^{\oplus}}{\overset{H^{\oplus}}{\rightleftharpoons}} -O-P(=O)(OH)-OH$$

The functional groups described previously are all polar and make a hydrocarbon more soluble in water. However, molecules become more hydrophobic and less soluble in water if they carry nonpolar groups, most commonly small hydrocarbon chains such as the methyl ($-CH_3$) and ethyl ($-CH_2CH_3$) groups. Notice the word construction here: The ending *-yl* indicates a portion of a molecule (generally, the molecule with one hydrogen atom removed) considered as a group. Thus methane is transformed into a methyl group, and acetic acid with its hydroxyl group removed becomes an acetyl group ($CH_3C{=}O$).

Coda By introducing basic chemical concepts in this chapter, we have laid the foundation for understanding how organisms function through molecular interactions. Biology depends on the fact that atoms form molecules with distinctive shapes and distinctive abilities to form weak bonds (such as hydrogen bonds) with one another. We will show that their interactions with the still more distinctive shapes of large biomolecules are the foundations of all biological processes. In the next chapter, we will show that these very large biomolecules are made by assembling many small organic molecules like those we have described here. We have also stressed the importance of water, without which life as we know it would be impossible, and have seen how organic molecules of various kinds interact with water. The interactions between biomolecules and water are critical for biological functions, and a function may depend critically on a simple property such as pH, the hydrogen ion concentration in a solution. The cytosol—the fluid inside a cell—and other biological fluids all have their characteristic pH, and each type of organism is adapted to an environment with a certain pH. Chemical considerations impose limitations on the structure and operation of all organisms. Yet, within these limitations, organisms have evolved into the great variety that we hinted at in Chapter 2 and have developed the ability to inhabit a wide range of niches, even to live in conditions that seem quite inhospitable. In Chapter 4, we begin to show how this is possible—how huge biological molecules can be built from small organic molecules, how these molecules can be adapted for the whole range of biological activities, and how they can carry the vital genetic information.

Summary

1. An atom has a nucleus made of positively charged protons and neutrons with no charge, surrounded by a space containing negatively charged electrons. All atoms of one element have the same number of protons and electrons; this is the element's atomic number. However, there may be isotopes whose atoms have different numbers of neutrons. So an element's atomic mass (the total mass of protons and neutrons) may be an average of the masses of its isotopes. An atomic mass unit is defined as 1/12 the mass of the most common isotope of carbon.
2. Most atoms form bonds with one another to make molecules. Most elements combine with other elements to make compounds.
3. The periodic table arranges elements according to their atomic numbers and places them in families with similar properties. The atoms of all elements in one family have similar electronic structures. Electrons reside at different energy levels in an atom, and each row (period) of the table corresponds to filling one energy level.
4. The electrons in an atom occupy a region of space called an orbital. Each orbital can hold a maximum of two electrons, and two atoms can form a molecular orbital, encompassing both of their nuclei, by each contributing one electron to this orbital. This arrangement forms a covalent bond that holds the atoms together, creating a molecule.
5. In methane, a carbon atom forms four covalent bonds with hydrogen atoms; in ammonia, a nitrogen atom forms three covalent bonds with hydrogen atoms; and in water, an oxygen atom can only form two covalent bonds with hydrogen atoms. In this sequence, the large central atom becomes increasingly electronegative and holds electrons more tightly, thus weakening the covalent bonds. The bonds in a water molecule are weak enough for one hydrogen atom to occasionally pull away from the rest of the molecule as a hydrogen ion, a single proton. An ion is a charged atom or molecule, with unequal numbers of protons and electrons.
6. The principal orbitals that participate in bonding, called valence orbitals, can hold a total of eight electrons. Most atoms form molecules by transferring or sharing electrons so that each atom is surrounded by eight valence electrons.
7. Elements in the first two columns of the periodic table tend to lose one or two electrons and become cations. Elements in columns 7 and 6 of the periodic table tend to gain one or two electrons and become anions. Anionic and cationic elements commonly combine to form compounds in which their atoms are held together by ionic bonds.
8. An acid is a substance that liberates hydrogen ions, and a base is one that accepts them. Hydrochloric acid is an acid because it can dissociate into hydrogen and chloride ions. Ammonia is a base because a hydrogen ion can combine with its one filled orbital and share the "extra" electron there.
9. Changes in hydrogen ions are very significant in biology. The pH scale expresses the concentration of hydrogen ions in a solution. The scale runs from pH 0, indicating a high concentration of hydrogen ions, to pH 14, indicating a low concentration. At pH 7, the concentrations of hydrogen and hydroxyl ions are equal, and the solution is neutral.
10. All organisms consist largely of water, which has unique properties, including cohesiveness, high heat capacity, high heats of vaporization and fusion, excellent capabilities as a solvent, and a lower density when solid than when liquid. These properties, which are largely due to its hydrogen-bonded structure, are essential for life.
11. Organisms are made of relatively few elements. Because most of their mass is water, they contain a great deal of oxygen and hydrogen. The bulk of their dry weight consists of organic molecules, which are made chiefly of carbon, hydrogen, oxygen, and nitrogen with lesser amounts of phosphorus and sulfur. The inorganic components of organisms include potassium, sodium, chlorine, magnesium, and calcium, plus trace amounts of several metals, including iron, copper, zinc, and molybdenum.
12. Each type of organic molecule has a unique shape, and interactions between distinctively shaped biological molecules are the basis of most biological activities.
13. Many organic molecules can change their shapes, or conformations, by rotating atoms without breaking any bonds. Changes in conformation are critical in biological actions.
14. Some organic molecules are hydrophilic, or polar, with an affinity for water; others are hydrophobic, or nonpolar, with a tendency to avoid water. These distinct tendencies are important in structuring cells and in biological functions.

15. Different classes of organic molecules carry certain characteristic functional groups of atoms, especially hydroxyl, carbonyl, carboxyl, amino, sulfhydryl, and phosphate groups. Each type of functional group confers different properties, and the various types of organic compounds have distinctive biological functions.

Key Terms

metabolism 45
biomolecule 45
element 46
atom 46
electron 46
proton 46
neutron 46
nucleus 46
molecule 46
bond 46
compound 46
complementary molecule 46
atomic number 46
atomic mass 46
molecular mass 46
isotope 47
radioisotope 47
periodic table 47
covalent bond 48
orbital 48
electronegativity 49
polar molecule 49
hydrogen ion 49
ion 49
hydroxyl ion 49
ionic bond 50
cation 50
anion 50
valence 50
chemical reaction 51
mole 51
aqueous solution 51
acid 51
base 51
molarity 52
acidic 52
basic 52
alkaline 52
neutral 52
pH 52
pH scale 52
hydrogen bond 53
cohesive 54
adhesive 54
surface tension 54
cytosol 54
calorie 54
trace element 57
hydrocarbon 58
isomer 58
saturated 59
unsaturated 59
double bond 59
conformation 60
hydrophilic 61
hydrophobic 61
functional group 62
alcohol 62
hydroxyl group 62
sugar 62
organic acid 62
carboxyl group 62
amine 62
amino group 62
thiol 62
sulfhydryl group 62
phosphate group 62

Multiple-Choice Questions

1. An atom is to an element as a molecule is to a (an)
   a. ion.
   b. compound.
   c. isomer.
   d. isotope.
   e. mole.
2. The ability of an atom to form chemical bonds is primarily a function of the
   a. distribution of its protons.
   b. distribution of its neutrons.
   c. distribution of its electrons.
   d. difference between its atomic weight and its atomic number.
   e. stability of its nucleus.
3. Elements in one family in the periodic table have similar chemical properties because they have.
   a. nuclei of similar sizes.
   b. equal numbers of neutrons.
   c. the same number of electrons in their innermost orbital.
   d. the same pattern of electrons in their outermost orbital.
   e. the same number of orbitals.
4. Which is the best description of a covalent bond?
   a. the attractive force between two ions with opposite charges.
   b. the attractive force between two nonpolar molecules.
   c. the force created when two atoms with incomplete orbitals share a pair of electrons.
   d. the attraction between the nucleus of an atom and its electrons.
   e. the ability of orbitals to alternate between a localized and a delocalized configuration.
5. If water had a lower specific heat,
   a. seasonal climate changes would be greater.
   b. heavier insects could walk on water.
   c. water molecules would be more polar.
   d. it would not move from the roots to the leaves in plants.
   e. it would not form steam.
6. If the pH of a solution is 2,
   a. 2 percent of the water molecules are ionized.
   b. pOH is also 2.
   c. $[H^+] = 2\ M$.
   d. $[OH-] = 10^{12}\ M$.
   e. $[H^+] = 10^{-2}\ M$.
7. Hydrogen bonding between water molecules is directly responsible for
   a. surface tension.
   b. the property of cohesion.
   c. the fact that ice is less dense than water at 0°.
   d. water's solvent properties.
   e. all of the above.
8. Hydrogen bonds form
   a. between hydrogen atoms.
   b. when hydrogen is shared between two nonpolar molecules.
   c. when hydrogen ions are accepted by a base.
   d. when hydrogen ions are released by an acid.
   e. when hydrogen in one polar molecule is attracted to the electronegative portion of another molecule.
9. Water is an excellent solvent for solutes that are
   a. polar.
   b. ionized.
   c. acidic.
   d. basic.
   e. all of the above.
10. A hydrocarbon is said to be unsaturated if
    a. one end of the molecule is hydrophilic while the other end is hydrophobic.
    b. it has one or more double bonds between carbon atoms.
    c. it contains more than one functional group.
    d. each internal carbon atom is covalently bonded to two hydrogen atoms.
    e. its functional groups include at least one aromatic ring.

True-False Questions

Mark each statement true or false, and if false, restate it to make it true.

1. An atom that has one unfilled valence orbital may achieve a more stable configuration by losing electrons and becoming an anion or by gaining electrons and becoming a cation.
2. Since a mole is a unit of weight, one mole of any compound weighs one gram.
3. A carbon atom contains six protons in its nucleus, but has only four valence electrons.

4. Oxygen is more electronegative than fluorine because the oxygen nucleus has fewer protons than the fluorine nucleus.
5. A 1 *M* solution of HCl contains 6×10^{23} ions; 3×10^{23} are hydrogen ions, and 3×10^{23} are chloride ions.
6. If the pH of solution A is 3 and that of solution B is 6, there are a thousand times more free hydrogen ions per mole in solution A than in solution B.
7. While both ionic bonds and hydrogen bonds result from the attraction of oppositely charged atoms, only hydrogen bonds hold two negative atoms together.
8. When a molecule of water dissociates to form one hydroxyl ion and one hydrogen ion, the hydroxyl ion is left with an extra proton that has been released from the hydrogen ion.
9. Different isotopes of an organic compound will have different molecular shapes.
10. If two hydrocarbons differ in length but contain the same polar functional group at one end, the two compounds will not be equally polar.

Concept Questions

1. What property of water makes it useful as the principal ingredient in automobile radiator coolants? What is the significance of this property for organisms?
2. Is pure water an acid or a base?
3. Distinguish between a hydrocarbon and a functional group.
4. How do hydrophobic molecules differ from hydrophilic molecules?
5. Explain why short-chain alcohols (such as the ethanol in alcoholic beverages) are soluble in water but long-chain alcohols are not?

Additional Reading

Bloch, Konrad E. *Blondes in Venetian Paintings, The Nine-banded Armadillo, and Other Essays in Biochemistry.* Yale University Press, New Haven, 1994.

Dickerson, Richard, and Irving Geis. *Chemistry, Matter, and the Universe.* Benjamin/Cummings, Menlo Park, 1979. One of the best introductions to chemistry; now out of print.

Hoffmann, Roald, and Vivian Torrence. *Chemistry Imagined: Reflections on Science.* Smithsonian Institution Press, Washington, D.C., 1993.

Oxtoby, David W., and Norman H. Nachtrieb. *Principles of Modern Chemistry.* Saunders College Publishing, Fort Worth, 1996.

Simmonds, Richard J. *Chemistry of Biomolecules: An Introduction.* Royal Society of Chemistry, Cambridge (England), 1992.

Snyder, Carl H. *The Extraordinary Chemistry of Ordinary Things.* John Wiley and Sons, New York, 1995.

Solomons, T. W. Graham. *Organic Chemistry,* 6th edition. John Wiley & Sons, New York, 1996.

Internet Resource

To further explore the content of this chapter, log on to the web site at:

http://www.mhhe.com/biosci/genbio/guttman/

4 Polymers and Proteins

A computer-generated model of a protein shows that it is a large molecule, made of many atoms, with a distinctive shape.

Key Concepts

- **4.1** Most biological materials are macromolecules constructed as polymers.
- **4.2** Polysaccharides illustrate the concept of polymers as structural materials.
- **4.3** Nucleic acids are an organism's informational molecules.
- **4.4** Proteins, the most diverse biological polymers, have many functions.
- **4.5** Enzymes are catalysts for chemical reactions.
- **4.6** The three-dimensional structure of an enzyme makes it very specific in the reaction it catalyzes.
- **4.7** Proteins are polymers whose monomers are amino acids.
- **4.8** Amino acids are linked together by peptide bonds to form proteins.
- **4.9** Amino acids and proteins are commonly ionized.
- **4.10** The amino acids in a peptide are linked in a definite sequence known as the peptide's primary structure.
- **4.11** Regular coiling or folding of a polypeptide chain determines the polypeptide's secondary structure.
- **4.12** Much of our knowledge of protein structure comes from studies of hemoglobin and myoglobin.
- **4.13** The primary structure of a polypeptide uniquely determines its three-dimensional shape.
- **4.14** The structure of a protein is determined by the genome and shaped by evolution.

How do we construct a house? First we lay a foundation of concrete or perhaps layers of concrete blocks held together by cement. On top of this, we lay a plate of heavy timbers and then begin constructing a frame of lighter timbers upward, joists and studs with regular spacing. On top of the joists, we lay a flooring of plywood sheets, and on top of the plywood, a finished floor of hardwood strips, carpeting, asphalt tiles, linoleum, or ceramic tiles. We cover the studs with a base wall of sheetrock, made of gypsum compacted between heavy paper, and cover the sheetrock with a plaster finish that can be painted or maybe covered with wallpaper. So a modern house is made of several distinct structural materials, each with its own characteristics and function.

In trying to understand the structure of a living cell, we encounter a series of structural materials that are just as distinctive in their own way as the wood, concrete, gypsum, and asphalt of a house. Each has its own place and its own function. The main structural components of a cell are proteins, polysaccharides, nucleic acids, and lipids. Once you understand them, you could no more mistake one for the other than confuse a sheetrock panel with a two-by-four. The main impediment to understanding these materials is that we don't ordinarily handle pieces of protein or nucleic acid, whereas we have all handled wood and carpeting. In its basic structure, a cell is probably no more complex than a modern building, but its complexity is at a molecular level that lacks the familiarity of everyday objects. One of the major goals of science is to analyze such complexities and reduce them to simpler laws and generalizations. In this chapter, we will show that biological structure is based on simple underlying principles. Most of the chapter is devoted to proteins, because throughout the book we will emphasize their absolutely critical role in biological activities. ■

4.1 Most biological materials are macromolecules constructed as polymers.

The bulk of every organism consists of very large molecules, or **macromolecules,** that are analogous to the large, stable structural materials of a house. We characterize all biomolecules by their molecular mass, M_r. As with all other molecules, M_r is measured in atomic mass units (amu), but biochemists and biologists generally use the term **dalton (Da)** instead of amu, honoring John Dalton, a pioneer in the development of an atomic theory of chemistry. "Dalton" is used especially for large molecules and particles such as viruses; one **kilodalton (kDa),** a thousand daltons, is a more convenient unit for describing large particles. Thus we might say that a virus has a mass of 8×10^6 daltons or 8,000 kDa. Biologists generally choose a mass of about 1,000 daltons as an arbitrary dividing line between small molecules and macromolecules. Some macromolecules have masses of hundreds of thousands or even millions of daltons.

Organisms are constructed according to a simple principle: All macromolecules are made of small subunits covalently bonded in various combinations. The macromolecules are **polymers** (*poly-* = many; *-mer* = part), and their subunits are **monomers** (*mono-* = one). Although cells are made of thousands of different polymers, they only have to make about 30–40 different types of monomers, and this greatly simplifies the task of making their structure. This principle of polymeric construction is at the heart of biology.

Biological polymers include *homopolymers,* made of only a single kind of monomer:

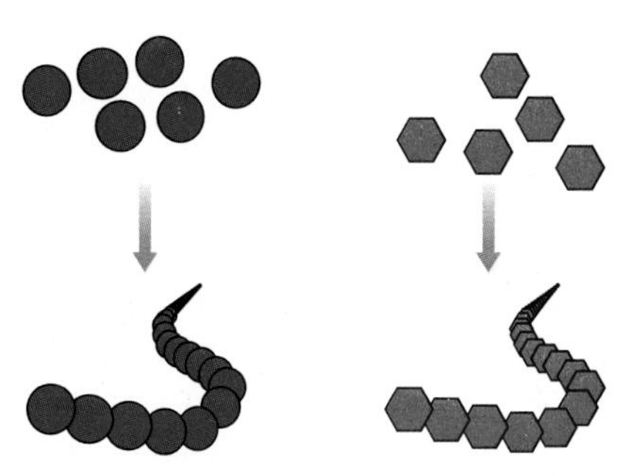

and *heteropolymers* made of different kinds of monomers.

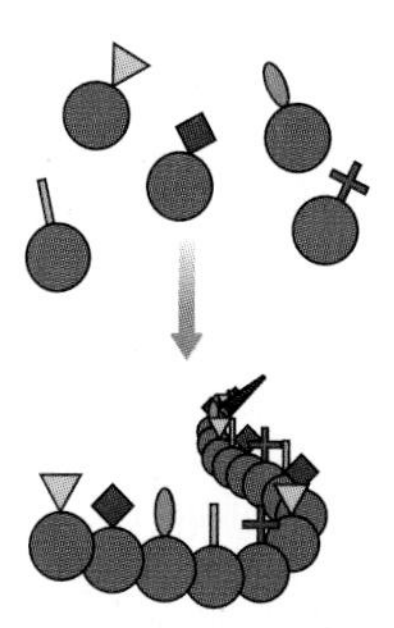

Although cells usually use very complicated processes for **polymerization,** the process of synthesizing polymers, in principle it is simple: Each subunit is linked to the next by removing a molecule of water, a process called **dehydration synthesis** or **condensation.** We can illustrate this with a generic polymer whose monomer, represented by a square, has –H at one end and –OH at the other, like this:

$$H\text{-}\blacksquare\text{-}OH$$

When two of these monomers bond chemically, a molecule of water is removed:

$$H\text{-}\blacksquare\text{-}OH + H\text{-}\blacksquare\text{-}OH \rightarrow H\text{-}\blacksquare\blacksquare\text{-}OH + H_2O$$

The process can continue in the same way to make a chain of hundreds or thousands of subunits:

$$H\text{-}\blacksquare\blacksquare\text{-}OH + H\text{-}\blacksquare\text{-}OH \rightarrow H\text{-}\blacksquare\blacksquare\blacksquare\text{-}OH + H_2O$$

The synthesis of biological polymers, however, never actually takes place in the simple way implied by these diagrams. Polymerization is an energy-requiring process that uses much of an organism's energy budget, and in later chapters we will demonstrate how energy is put into polymerization.

The reverse of polymerization is **hydrolysis** (*hydro-* = water; *lysis* = splitting) where water is added back to separate the subunits. Digestion is largely hydrolysis. We eat food made of polymers, which are attacked by acids and enzymes, and broken down into their monomers:

$$H\text{-}\blacksquare\blacksquare\text{-}OH + H_2O \rightarrow H\text{-}\blacksquare\text{-}OH + H\text{-}\blacksquare\text{-}OH$$

Our bodies then use these monomers to make polymers for our cells.

The three most important biological polymers are polysaccharides, nucleic acids, and proteins. They are shown here alongside their respective subunits:

Macromolecule (Polymer)	**Subunit (Monomer)**
Polysaccharide	Monosaccharide (sugar)
Nucleic acid	Nucleotide
Protein	Amino acid

These three types of polymer form the bulk of biological structure, along with *lipids,* biomolecules that are not polymeric in structure. A basic lipid is made by linking three long hydrocarbon molecules, called *fatty acids,* to a molecule of glycerol; so in a minor way, lipids reflect the general principle of constructing larger molecules from similar small ones, but lipids never achieve the size of polymers. We will explain lipids when we discuss biological membranes in Chapter 8.

Exercise 4.1 The formula of the sugar glucose is $C_6H_{12}O_6$. What is its molecular mass?

Exercise 4.2 Chemists have used the general principle of polymeric construction to make plastics, such as polyethylene and polystyrene. The useful plastic Teflon® is a polymer of tetrafluoroethylene, made by linking $CF_2{=}CF_2$ units into long chains. Is it a homo- or heteropolymer? Unlike the construction of biological polymers, water is not removed at each step; instead, the double bonds are replaced by single bonds as other bonds are made between the monomers. Sketch part of a Teflon® molecule.

Exercise 4.3 Suppose a polymer is made out of the monomer $HS{-}CH_2CH_2COOH$ by dehydration synthesis. Draw the repeating structure of the polymer.

4.2 Polysaccharides illustrate the concept of polymers as structural materials.

The function of a macromolecule is determined by its structure, which depends on the structures of its monomers. As we go along in this chapter, we will emphasize that the structure of each molecule suits it for a particular biological function. Let's begin with some important structural materials, the **polysaccharides.** They are predominantly homopolymers, and their monomers are *simple sugars,* or **monosaccharides.** Sugars and the larger molecules made by combining them are also called *carbohydrates* because the formula of a sugar is a multiple of CH_2O, as if it were a "hydrated carbon." A sugar is either an aldehyde or a ketone—that is, a compound with a carbonyl group (–C=O)—that also bears one or more hydroxyl groups (–OH). All names of sugars end in *-ose,* so the two main classes of sugars are aldoses (those with a carbonyl group at the end of the chain) and ketoses (those whose carbonyl group is not on the end of the chain). Sugars are also named for the number of carbons in the chain, as in C_3 (triose), C_4 (tetrose), C_5 (pentose), and C_6 (hexose). An aldohexose is therefore an aldehyde sugar with six carbon atoms:

```
     H    H    H    H    H     H
     |    |    |    |    |    /
H —  C —  C —  C —  C —  C — C
     |    |    |    |    |    \\
     OH   OH   OH   OH   OH    O
```

Chemists have found that two or more types of sugar are often made of the same atoms bonded to one another in almost exactly the same way, and yet they are distinct compounds with different physical properties. The molecules of these sugars are **stereoisomers,** which differ only in the arrangement of their atoms in space. If you examine the sugar drawn above, you'll see that some carbon atoms in it are *asymmetric*—that is, they are linked to four different groups, so the –H and –OH on these carbon atoms can be oriented in two ways. This structure gives sugars left- and right-handedness, or *chirality,* as explained in Concepts 4.1. With two choices at each position, several stereoisomers correspond to each molecular formula, as Figure 4.1 shows. All sugars have either the D or L configuration. Then there are two distinct D-tetroses, four D-pentoses, eight D-hexoses, and a parallel series of L sugars. Only a few of these sugars, however, have much biological importance. The four most common hexoses are:

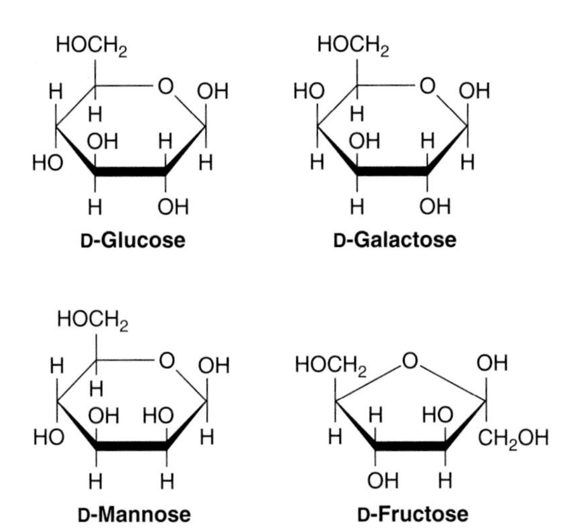

These molecules are drawn in the ring form that pentose and hexose sugars usually assume, although they can also open into the straight-chain form and reclose in a different conformation. Closing the ring makes another asymmetric carbon, so each sugar also has α and β isomers:

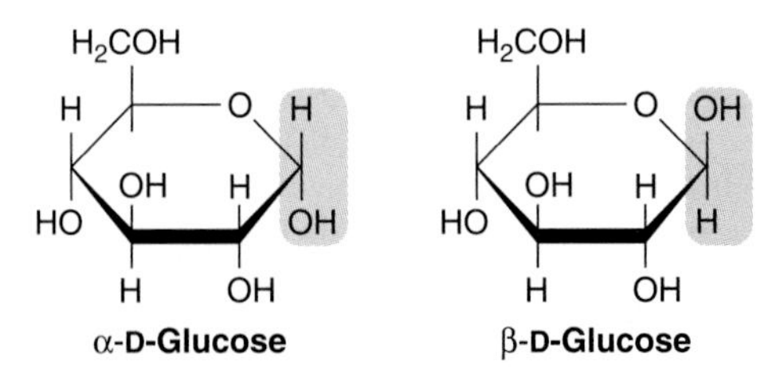

Simple sugars can be bonded in a *glycosidic linkage* by dehydration synthesis, removing –H from one sugar and –OH from the other to form water. Several common sugars are *disaccharides* made of two simple sugars, such as:

Concepts 4.1 Chirality and Stereoisomers

In the 1840s, Louis Pasteur started his doctoral studies by investigating the properties of tartaric acid. Chemists of the time knew that some crystals are able to polarize light; that is, when light vibrating in all directions passes through such a substance, the emerging light vibrates in only one plane. They also knew that solutions of some substances can rotate the plane in which polarized light vibrates. Pasteur discovered that tartaric acid crystals have two forms, which are mirror images of each other:

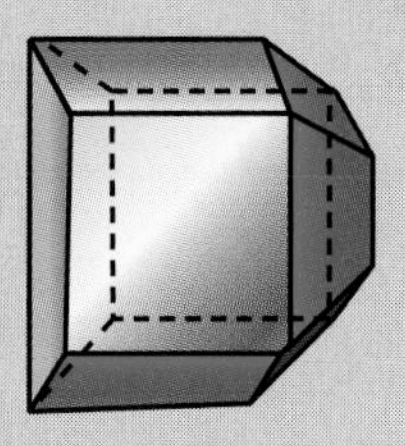

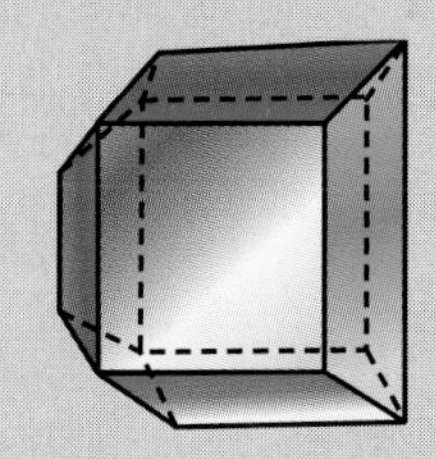

When solutions are made with only one type of crystal, one solution rotates plane-polarized light to the left and the other to the right. Pasteur had discovered what we now call the **chirality,** or "handedness," of some molecules (*cheir* = hand). Chiral structures are mirror images of each other. A molecule of one form cannot be superimposed on a molecule of the other form so that all parts of the two molecules coincide any more than you can superimpose your left hand onto your right hand:

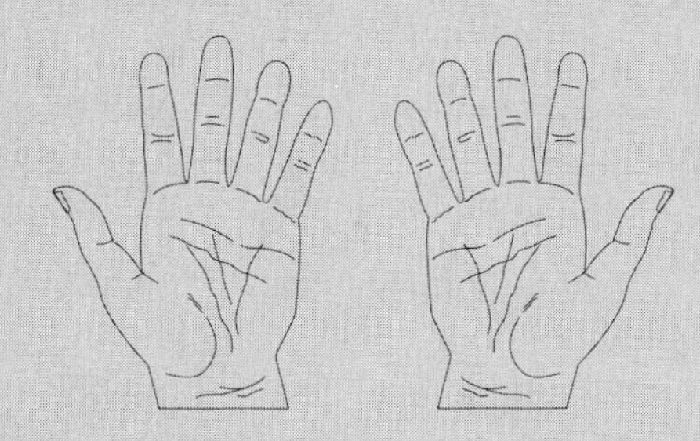

A few years later, Pasteur was studying the problems of molds and other contaminants in the French wine-making industry. Tartaric acid is a by-product of wine-making. Pasteur discovered that, as a mold grows on a mixture of the two forms of tartaric acid, it uses only the right-rotating form. Thus he found that a living organism can have an inherent preference for a subtle difference between molecules, and we now know this to be a most important feature of the structure and metabolism of organisms.

Chiral organic molecules occur if an asymmetric carbon atom is bonded to four *different* atoms or groups of atoms:

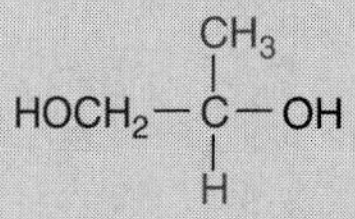

$$HOCH_2-\underset{H}{\overset{CH_3}{C}}-OH$$

There are two different arrangements of these four groups, which are mirror images of each other. Chiral molecules are also one kind of *stereoisomer,* which differ only in the arrangement of their atoms in space. (This contrasts with the isomers we introduced in Chapter 3 that have their atoms linked in different ways.) Organic chemistry books discuss the complications of identifying and naming stereoisomers. We will follow the conventions used in biochemistry by taking the glyceraldehyde molecule as a standard. One form is designated L (for *laevo* = left) and the other D (for *dextro* = right):

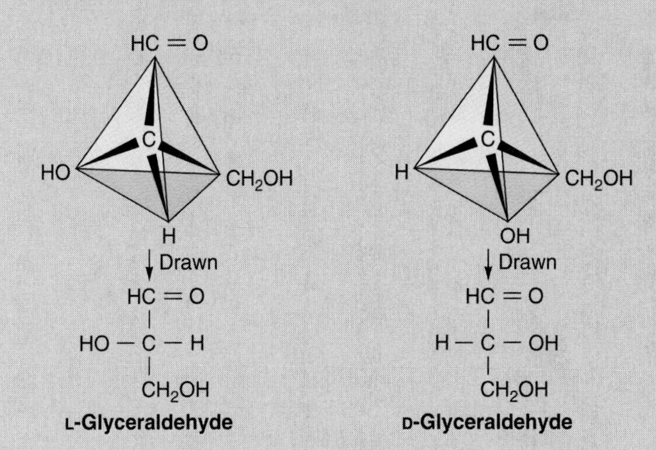

Larger molecules, such as sugars, are then related to this standard.

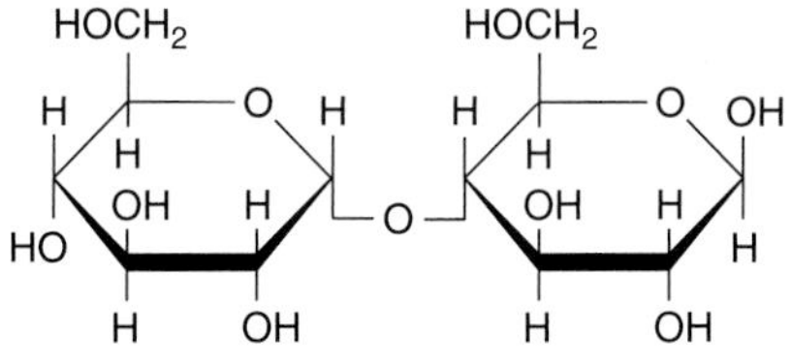

Maltose

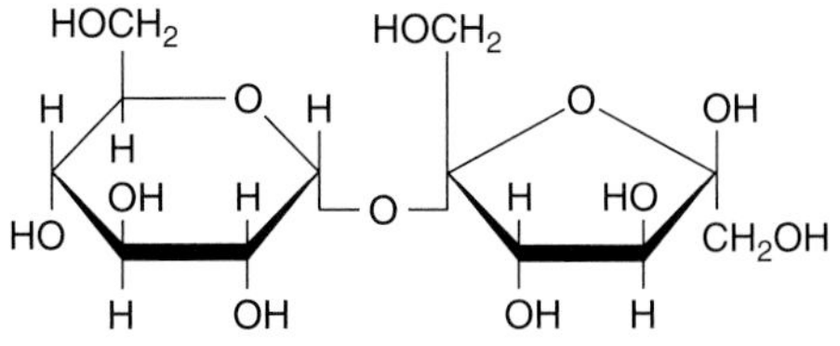

Sucrose

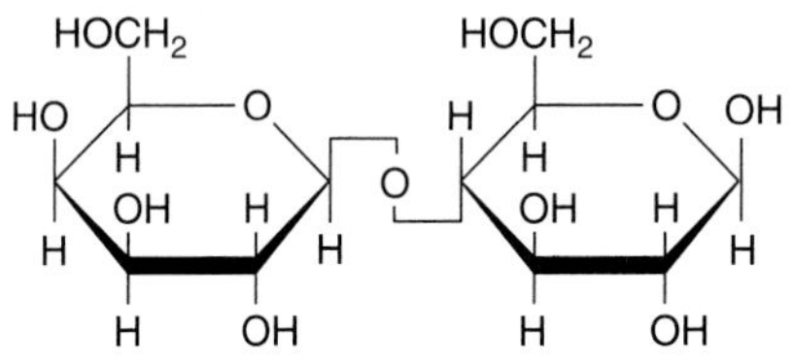

Lactose

Chains of two to ten sugars are called *oligosaccharides* (*oligos* = few), and the smaller oligosaccharides are called *complex sugars;* molecules of more than ten subunits are called *polysaccharides.*

Because each sugar can be linked at several positions, various kinds of glycosidic linkages are possible. These are designated by notations that show which carbon atoms are joined;

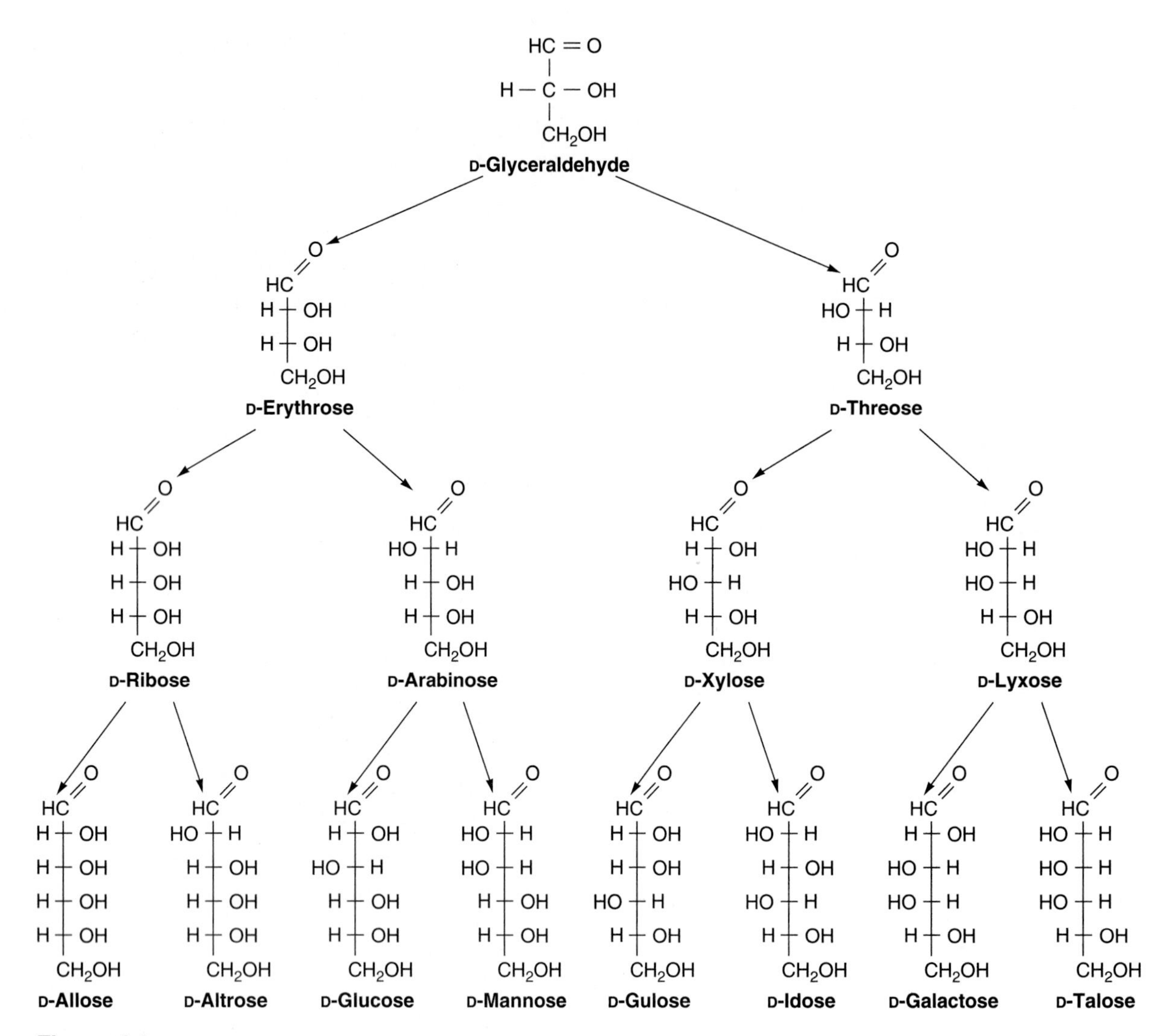

Figure 4.1
A sugar with only three carbon atoms (triose) has two optical isomers, D and L. The series of D sugars is shown here. Note that the addition of each –HCOH– group produces two more possible arrangements, so there are two D-tetroses, four D-pentoses, and eight D-hexoses.

(a) Cellulose

(b) Starch (amylopectin)

Figure 4.2
(a) Cellulose and *(b)* starch or glycogen, both polymers of glucose, are the most common biological polysaccharides. They differ only in having α linkages (in starch) or β linkages (in cellulose) between the glucose residues. Notice that starch also has some side-branches and tends to curl up into helices, while cellulose makes straight chains.

for instance, an $\alpha(1 \rightarrow 4)$ linkage connects C-1 of a sugar in the alpha configuration to C-4 of another sugar, and a $\beta(1 \rightarrow 3)$ linkage connects C-1 of a sugar in the beta configuration to C-3 of another sugar. The type of linkage is crucial in determining the nature of a polysaccharide. If you are eating mashed potatoes or rice on a paper plate, you are dealing with two very different glucose polymers, though they differ only in the linkage. The plate is made of D-glucose molecules joined by $\beta(1 \rightarrow 4)$ linkages to make **cellulose** (Figure 4.2*a*), which forms the tough fibers of wood and cotton. The **starch,** or amylopectin, in the potatoes or rice is made of the same monomers joined with $\alpha(1 \rightarrow 4)$ and $\alpha(1 \rightarrow 6)$ linkages (Figure 4.2*b*), and it stores energy. You will digest these starch molecules back to glucose, which your cells may use immediately for energy, but much of the glucose will probably be polymerized in your liver and muscles into **glycogen,** a polymer very similar to starch that is used for energy storage in animals. Because a hexose sugar can form a glycosidic linkage at three or four of its carbon atoms simultaneously, polysaccharides commonly form *branched* polymers. Starch and glycogen differ in the degree of $1 \rightarrow 6$ branching, and the chains take on a helical form. (One test for starch is to add iodine, which turns blue when combined with starch because the iodine atoms fit into the helices, where they have different optical properties.)

Cellulose, which is built into the walls of plant cells, is the most abundant polymer on Earth because of the enormous plant biomass, especially in the wood of trees. In addition to important homopolymers such as cellulose and starch, there are many kinds of polysaccharide heteropolymers. Animals produce a series of interesting materials called *mucopolysaccharides* with monomers of hexose sugars carrying sulfate and amino groups. Mucopolysaccharides form mucins, the slippery, viscous fluids that function to lubricate and protect. Mucins occur in the synovial fluids in the joints between bones, in the mucus of our digestive and respiratory tracts, and in the substance that snails and other invertebrates glide on (Figure 4.3). Saliva and the fluids in the eye are partly composed of mucopolysaccharides, as are major parts of bone and cartilage. One especially abundant mucopolysaccharide is *chitin,* which forms the cell walls of most fungi and the skeletons of arthropods, including the crustaceans, insects, spiders, and their relatives. In addition, plants polymerize a variety of unusual sugars into pectins, which are built into all plant tissues, as well as gums and other complex molecules that constitute large parts of their fruits. Several of these polysaccharides are discussed in a functional context in later chapters.

One of the first things to notice about sugar molecules, whether in monomeric or polymeric form, is that they have many hydroxyl groups. This is the structure of water, the principal component of an organism, and the sugars and their polymers bond strongly to water. Thus glucose is the principal energy-carrying molecule in our blood; sucrose and other sugars have a comparable role in plants; and polysaccharides commonly function through their interactions in an organism's aqueous internal environment.

Figure 4.3

Mucopolysaccharides are very common in nature. They are important components of structures such as ligaments and bone, as well as slippery fluids such as mucus that many animals use to lubricate their movements.

4.3 Nucleic acids are an organism's informational molecules.

Nucleic acids, a second class of biological polymers, will occupy much of our attention because they constitute the genome and the cellular machinery that translates information in the genome into cellular structures. The monomers of nucleic acids are **nucleotides,** which have three parts: a *sugar,* a ring-shaped, nitrogenous *base,* and a *phosphate:*

Base

Phosphate

Sugar (ribose)

(Missing in deoxyribose)

A nucleic acid is made by condensing nucleotides into a chain, making bonds from sugar to phosphate to sugar to phosphate:

Base Base Base Base Base

Sugar Sugar Sugar Sugar Sugar

P P P P P

Phosphate

Thus a nucleic acid has a sugar-phosphate backbone with the bases extending to the side. The two types are defined by the sugars in their nucleotides: **Deoxyribonucleic acids (DNA)** contain deoxyribose, and **ribonucleic acids (RNA)** contain ribose. Nucleic acids serve mostly as carriers of genetic information. As we will show in Chapters 12 and 13, nucleic acids are ideally suited for storing genetic information, because it is so easy to make replicas of them, so the information they contain is passed along from generation to generation.

4.4 Proteins, the most common biological polymers, have many functions.

Aside from water, the most abundant material in an organism is **protein,** which may comprise 80 percent of its dry mass. Every organism contains many types of protein that differ in their structure and function. Bacteria, the simplest organisms, are made of about 2,000–3,000 different kinds of proteins, while humans have perhaps ten to twenty times that many. A cell contains a few molecules of some proteins and many thousands of others, depending on the protein's function.

Proteins are so fundamental to biological processes that, whenever we think of a molecular or cellular function, it is natural to immediately ask what kind(s) of protein(s) are involved. Let's look at the variety of roles proteins play, so you can see why many different kinds are needed and why they are so fundamental to life.

1. *Proteins form structures.* Hair (including wool), fingernails, feathers and other surface structures of animals are mostly made of the protein keratin. The protein collagen may comprise 25–30 percent of the protein in some vertebrates, since it makes part of their bones, tendons, ligaments, and skin.
2. *Proteins effect movement.* Muscles contract because of the action of two proteins, actin and myosin, which pull on each other. Many other movements are mediated by fine tubules of the protein tubulin; some of these tubules form thin extensions called flagella and cilia that lash back and forth. Many microorganisms use cilia or flagella for movement, and ciliated cells on some tissue surfaces, such as those in the respiratory tract, constantly sweep along a layer of mucus that traps dirt and microorganisms.
3. *Proteins protect against foreign materials.* Proteins called antibodies are made by animals, especially birds and mammals, in response to potentially harmful invaders like bacteria and viruses. They help inactivate and destroy the invaders, and without them we would have little protection against disease.
4. *Proteins transport small molecules and ions through cell membranes.* Proteins of this type control the internal composition of a cell and what enters and leaves it.
5. *Proteins receive stimuli and transmit information.* All cells carry proteins that recognize molecules in their environment and allow the cell to respond—perhaps by moving away from noxious materials or by taking up useful nutrients. An animal's tissues and organs communicate with one another by sending specific chemical messengers through the bloodstream or the nervous system. These messengers attach to special receptor proteins on the cells that receive the message. Cells in the retina of the eye contain the protein rhodopsin, which detects light, while many proteins on the tongue and in the nose detect molecules in our food or in the air, giving us the sensations of taste and smell.
6. *Proteins communicate messages between the cells of an organism.* Cells of plants and animals frequently communicate via chemical messengers called *hormones,* and many hormones, including insulin and growth hormone, are proteins.
7. *Proteins carry and store material.* Hemoglobin, the red protein of blood, transports oxygen from the lungs to all parts of the body and carries carbon dioxide back to the lungs where it is exhaled. Ferritin, a protein in the liver, stores excess iron and releases it when needed.
8. *Proteins control or regulate processes.* Certain proteins can turn genes on and off or create chemical signals that control the many reactions of metabolism.
9. *Proteins catalyze chemical reactions.* These proteins, known as **enzymes,** increase the reaction rate. In this chapter we will introduce some concepts about protein structure and function as they relate to enzymes, and in Chapter 5 we will discuss enzymes themselves in more depth.

Exercise 4.4 A bacterial cell has a wet mass of 2×10^{-12} g. Eighty percent of it is water, and half the dry mass is protein. How many grams of protein does the cell contain?

Exercise 4.5 We can convert from grams to daltons with the relationship that Avogadro's number is 6×10^{23} daltons/gram. Convert the mass of protein from Exercise 4.4 into daltons of protein. If the average protein weighs 30,000 Da, how many protein molecules are in the cell?

4.5 Enzymes are catalysts for chemical reactions.

The thousands of chemical reactions of metabolism, will occur in test tubes as well as in cells, but they usually go much too slowly to sustain life. Increasing the temperature will increase reaction rates, but if you tried to speed up an organism's metabolism by raising its temperature, you would cook it. Instead, organisms use **catalysts.** A catalyst accelerates a chemical reaction without being permanently changed itself, so it can continue to act again and again. For instance, hydrogen peroxide breaks down slowly into water and oxygen, but a pinch of platinum dust will catalyze the reaction, causing hydrogen peroxide to liberate oxygen explosively. Because the platinum itself is not changed by this reaction, it can catalyze the reaction repeatedly. In the catalytic converter on the exhaust pipe of your car, heavy metals such as platinum and palladium catalyze the formation of carbon dioxide from oxygen and carbon monoxide, as well as the conversion of nitric oxide (NO) into molecular nitrogen and oxygen. This simple device reduces the tremendous pollution caused by motor vehicles by converting noxious products of combustion in the engine into ordinary components of the atmosphere.

Inorganic catalysts like platinum lack the specificity needed for biological purposes. Instead, organisms catalyze their many chemical reactions with proteins called *enzymes,* which are highly specific catalysts. Each enzyme can only operate on a particular molecule (or a few similar molecules) known as its **substrate:**

$$\text{Substrate} \xrightleftharpoons{\text{Catalyzed by a specific enzyme}} \text{Product}$$

4.6 The three-dimensional structure of an enzyme makes it very specific in the reaction it catalyzes.

The specificity of enzymes is an example of a general biological principle: Specific biological processes occur through the interaction of molecules that have *complementary shapes* (see Section 3.1), particular three-dimensional shapes that match one another like the shapes of interlocking jigsaw puzzle pieces. An enzyme interacts with its substrate at the **active site,** a pocket in the protein whose shape just fits the substrate's complementary shape. Emil Fischer, who first developed these ideas, pictured enzyme activity as an interaction between a lock and a key:

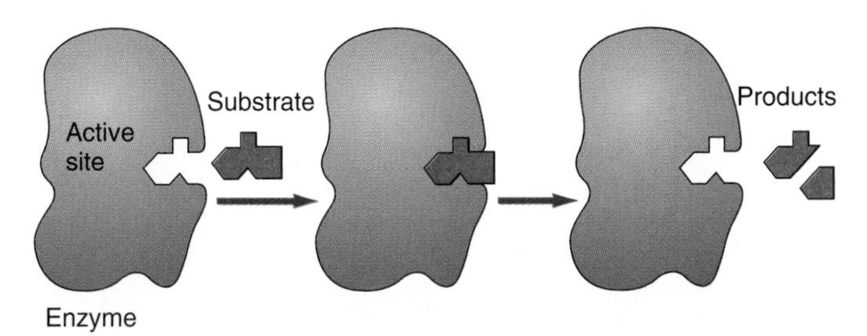

A substrate enters the active site like a key entering a lock and is converted into a product, which comes off the enzyme and leaves the active site ready to receive another substrate molecule. Modern research on enzymes has shown that the lock-and-key model is too simple, but it is a good way to begin understanding how enzymes work. In fact, both the enzyme and the substrate change their structure during the crucial events of catalysis in the active site, and of course, the product that leaves the active site may be quite different from the substrate that entered it.

The role of enzymes in metabolism is inseparable from metabolism itself, for nearly every reaction in living organisms is accelerated by enzymatic catalysis. Until recently, we could generalize that all enzymes are proteins, but now we know that some ribonucleic acids, called *ribozymes,* are also enzymes. (But, as we showed earlier, many proteins are not enzymes.)

Exercise 4.6 Plants conduct complex photosynthetic reactions in which they combine carbon dioxide with organic molecules to make sugars and other products. The process will not occur unless CO_2 is present. Is CO_2 a catalyst of this process? Why or why not?

4.7 Proteins are polymers whose monomers are amino acids.

All proteins are linear polymers consisting of long chains of **amino acids.** Although hundreds of different amino acids exist, all known proteins in all organisms use the same twenty amino acids as subunits (with only a few rare additions). All twenty have the same general structure: A central atom known as the alpha (α) carbon is bonded to an amino group ($-NH_2$), a carboxyl group ($-COOH$), a hydrogen atom, and a distinctive *side chain,* or R group:

$$H_2N-C_{\alpha}H(R)-C(=O)OH$$

The amino group gives the molecule half its name, and the carboxyl group makes the molecule an organic acid. It is the side chain, however, that determines the identity of a particular amino acid. In the simplest amino acid, glycine, the side chain is a single hydrogen atom, but in other amino acids, the side chain is more complex (Figure 4.4).

In all amino acids except glycine, the α-carbon atom is asymmetric, since four different groups are attached to it. Therefore there are D and L isomers of these compounds. Glycine does not have chiral forms because its α-carbon atom, being bonded to two hydrogen atoms, isn't asymmetric. Only L-amino acids form proteins. (D-amino acids, however, are found in some antibiotics produced by microorganisms and in the cell walls of bacteria.)

In Figure 4.4, the amino acids are grouped according to the nature of their side chains. Some amino acids, having nonpolar side chains, are relatively insoluble in water, but most amino acids have polar side chains and are hydrophilic. Some polar side chains are acidic or basic, and will therefore carry a negative or positive charge, respectively, in the pH range typical of cells. Notice that one molecule, proline, is unusual in being a secondary amino acid because its side chain is linked to its amino group, but the cellular apparatus that synthesizes proteins treats it like the typical amino acids. Notice also that the amino acids cysteine and methionine contain sulfur, which explains why most of the sulfur in most organisms is in their proteins.

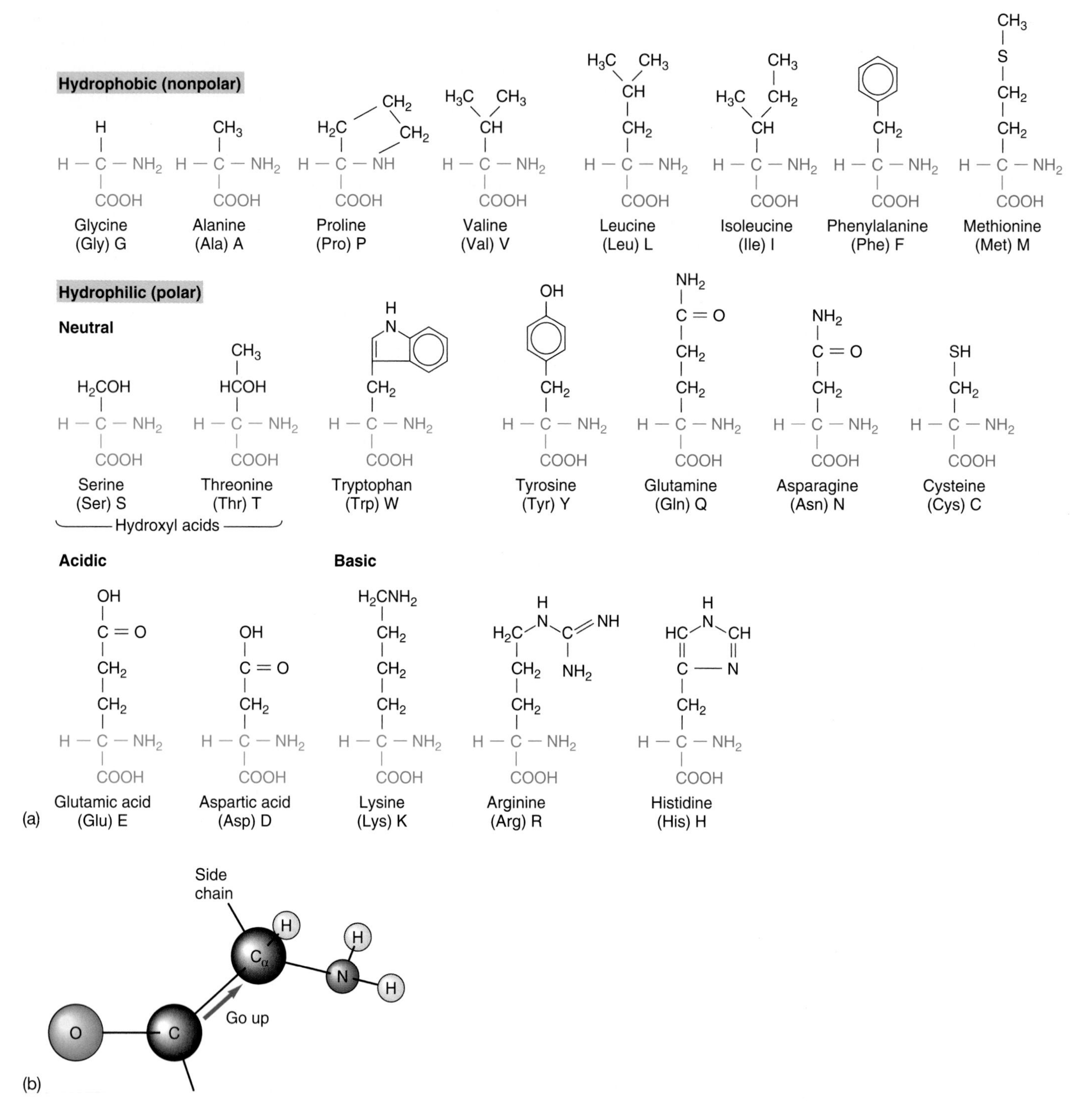

Figure 4.4

The twenty amino acids that commonly occur in proteins are shown in *(a).* Notice that each one has a standard three-letter abbreviation as well as a one-letter abbreviation. (The latter is used mostly by protein specialists.) All amino acids except glycine exist in either the D or L configuration; only the L isomers form proteins. *(b)* To determine the configuration, look from the carboxyl end toward the amino end and orient the molecule so the bonds to the α carbon go *upward.* The side chain will be on the left in L isomers and on the right in D isomers.

4.8 Amino acids are linked together by peptide bonds to form proteins.

Cells make proteins from amino acids by dehydration synthesis, just as in other biological polymerization processes. The actual process in cells is fairly complicated, but a simplified explanation is that water is removed from two amino acids (H from one, OH from the other), and they become covalently bonded through a *peptide linkage:*

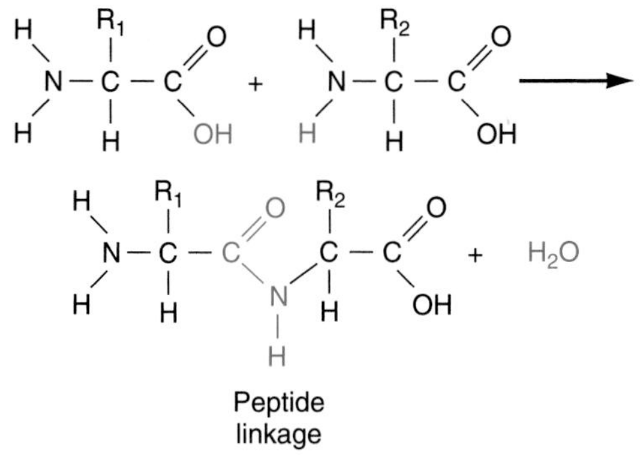

Peptide linkage

A chain of amino acids linked together this way is a *peptide.* The portion of an amino acid that remains in a peptide once the atoms of water have been removed is an *amino acid residue.* Thus every peptide consists of a covalently linked backbone, with the repeated sequence –N–C–C– formed by adjacent residues and the amino acid side chains extending from the backbone. Two amino acids form a dipeptide, three a tripeptide, and so on. A chain of a few amino acids is an *oligopeptide,* and a chain of many is a **polypeptide,** which is another name for a protein. (Some people limit the term "protein" to large polypeptides, but the distinction between large and small is quite arbitrary. Others prefer to use "protein" only for polypeptides that have assumed their proper functional forms and, perhaps, are combined with other polypeptides or with nonpeptide molecules that are essential for their function.)

Notice that a peptide chain has a direction—that is, the two ends are different. A peptide has a free amino group at one end, the *N-terminus* or amino terminus, and a free carboxyl group at the other end, the *C-terminus* or carboxy terminus:

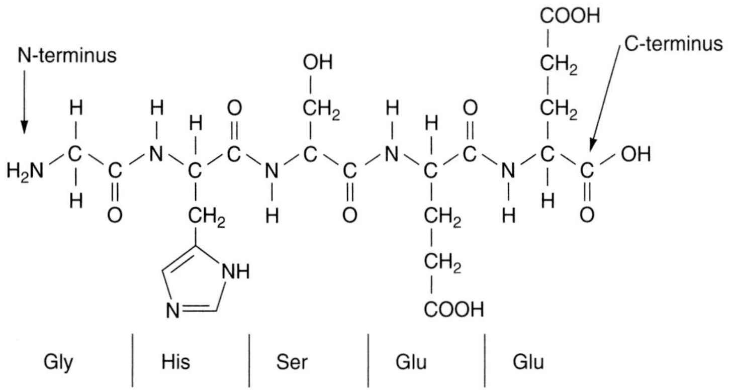

N-terminus C-terminus

Gly | His | Ser | Glu | Glu

By convention, the sequence of a peptide is always written from the N-terminus to the C-terminus, a direction often shown in diagrams by an arrow. This also turns out to be the direction in which cells synthesize proteins, starting with the N-terminal amino acid.

Exercise 4.7 The average mass of an amino acid residue in a protein is about 110 Da. If a protein contains 275 amino acids, what is its approximate mass?

Exercise 4.8 Draw the structures of the dipeptides Gly–Ser and Ser–Ala.

Exercise 4.9 Nucleic acids and proteins both have backbones with molecular groups extending to the side, but otherwise their structures are very different. Draw the basic structures of the two molecules. How do they differ?

Exercise 4.10 When you eat food containing the three principal biological polymers, they are hydrolyzed into their monomers in your stomach and intestines. What does that mean? What kinds of monomers does each kind of polymer yield?

Exercise 4.11 Some theorists have suggested that life on other planets might be based on silicon, instead of carbon, since silicon has many of the properties of carbon and can form large molecules of the same kind. However, the Si–Si bond is only about half as strong as the C–C bond. What does this fact imply about the ability to make a functional organism based on silicon?

4.9 Amino acids and proteins are commonly ionized.

At the pH typical of biological systems, close to neutrality, amino acids and peptides gain and lose protons (H^+) and become ionized. The amino group tends to become $-NH_3^+$, and the carboxyl group to become $-COO^-$, so the simplest amino acid, glycine, has both groups ionized at pH 7:

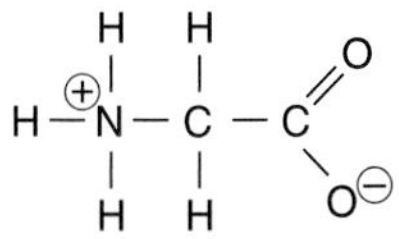

Since the two charges balance each other, the molecule has zero net charge; such doubly ionized molecules are called *double ions* or *zwitterions.* At a high concentration of hydrogen ions (low pH), a hydrogen ion binds to the ionized carboxyl group, and the molecule's net charge becomes positive. At low concentrations of hydrogen ions (high pH), both ionizable groups lose their protons, and the net charge is negative:

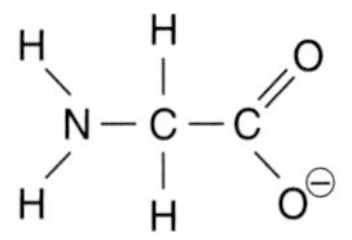

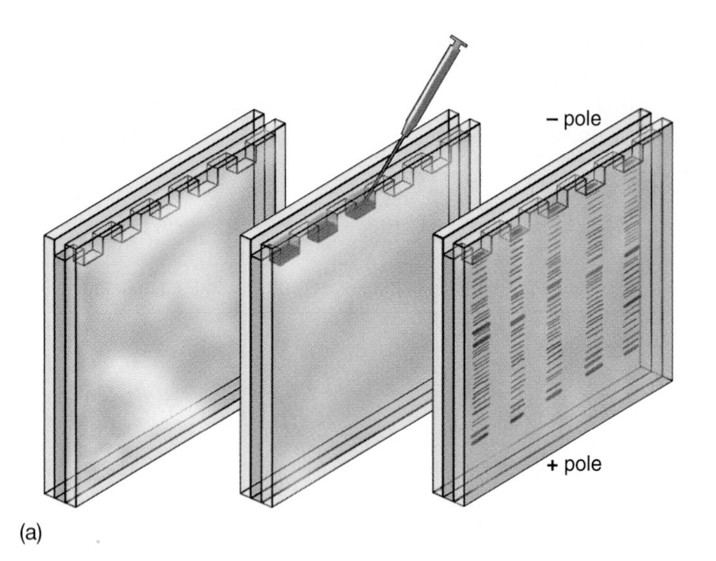

(a)

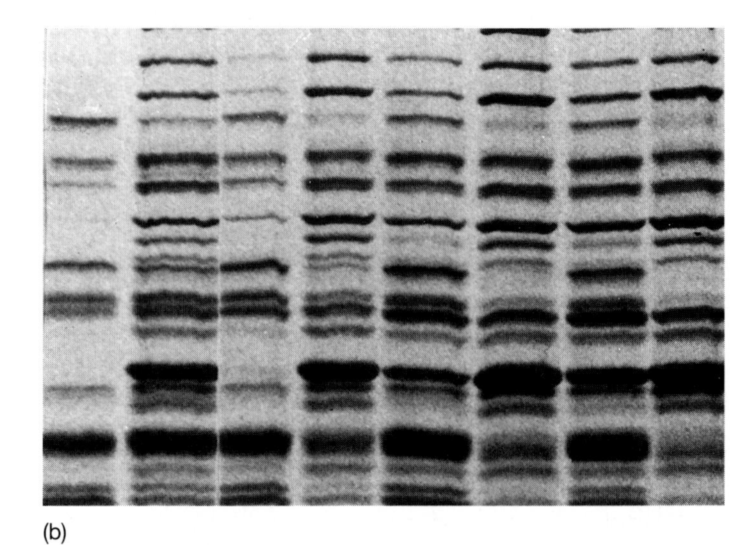

(b)

Figure 4.5

Polyacrylamide-gel electrophoresis (PAGE) is a powerful technique used in characterizing proteins and other macromolecules. A solution of the monomer acrylamide is mixed with chemicals that polymerize the monomers into polyacrylamide. This solution is poured into the space between two glass plates *(a)*, where it polymerizes into a thin slab of gel with the consistency of gelatin. A mixture of proteins to be separated is combined with a detergent, sodium dodecyl sulfate (SDS), to denature the proteins, so their shape is no longer important, and to coat them with overwhelming negative charges, so they all have the same charge. This mixture, along with a blue dye, is laid into a well at the top of the gel, and a strong electrical current is run through it. The negatively charged proteins are drawn toward the positive electrode, and the gel acts like a molecular sieve, allowing small molecules to move faster than large ones; this technique therefore separates proteins on the basis of their molecular weight. *(b)* When the gel is stained after electrophoresis, the proteins appear as a series of distinct bands, with those of highest molecular weight at the top of the gel. Each lane (column) is a different protein sample.

The amino and carboxyl groups of other amino acids behave like those in glycine, as do the amino group and carboxyl group at the ends of a peptide.

In addition, acidic and basic amino acids have ionizable side chains that gain and lose protons within characteristic pH ranges. The basic amino acids lysine and arginine lose the proton from the amino group of their side chain around pH 10–11, but at a lower pH, the side chain is positively charged. The opposite is true for the carboxyl groups on the side chains of acidic amino acids (glutamic and aspartic acids). They are negatively charged at a pH above 2–3, and only at very low pH do the side chains gain a proton and lose their charge.

Exercise 4.12 A polypeptide is made of amino acids, so why don't the internal amino and carboxyl groups of a peptide ionize, as well as those on the ends? (*Warning:* This is tricky question designed to make you look closely at polypeptide structure again.)

The ionization of organic acids creates an ambiguity in nomenclature that can be confusing at first. For example, when the side chain of glutamic acid is ionized, with a $-COO^-$ group, the molecule is a base and is named glutamate, the *-ate* ending being standard for the base forms of organic acids. Glutamate could be combined with an ion other than hydrogen—for instance, with Na^+ to make sodium glutamate. Organic acids shift continually between the acid and base forms in a cell, and since their particular form is generally unimportant, we will use the base name for glutamate and for other organic acids, such as pyruvic acid/pyruvate and lactic acid/lactate.

Electrophoresis is a widely used technique for purifying amino acids, peptides, and proteins by taking advantage of their ionization (Figure 4.5). Charged molecules migrate in an electric field at rates related to their electrical charge and size. In the conditions shown in the figure, the proteins are given a uniform charge and are separated from one another on the basis of their mass.

Exercise 4.13 Draw the structure of alanine at pH 2 and at pH 11.

4.10 The amino acids in a peptide are linked in a definite sequence known as the peptide's primary structure.

Around 1950, Frederick Sanger at Cambridge University in England established some basic facts about the structure of the protein hormone *insulin* in one of the most important investigations in biology. Sanger won a Nobel Prize for this work in 1958 and then went on to do comparable work with nucleic acids, which we will discuss in Chapter 14.

Sanger chose to use insulin because it is a small protein and because large amounts of quite pure material were readily available from drug firms supplying insulin to patients with diabetes, a disorder in which people can't make enough of their own insulin. Sanger knew that insulin, being a protein, is a polymer of amino acid subunits. He asked, "In what order, or sequence, are these amino acids arranged?" knowing that the answer would shed significant light on protein function. When he began his work, some people believed that a protein has no unique sequence of amino acids, but rather is a collection of molecules with similar, but not identical, sequences. Sanger showed that, without doubt, all molecules of each type of protein have the same unique sequence of amino acids. (The reason is that protein structure is genetically encoded in the genome, and Sanger's work was also critical in establishing this central biological concept.)

Sanger developed a way to identify the amino acid in the N-terminal position of a peptide by binding an appropriate chemical reagent to it. He found that two amino acids, glycine and phenylalanine, are at the amino end of insulin, and therefore concluded that insulin consists of two polypeptide chains. He confirmed this conjecture by treating the protein with mercaptoethanol, a compound that breaks **disulfide bridges,** a kind of covalent linkage that forms in proteins between two cysteine residues:

Disulfide bridge

$$O{=}C(-)\,H{-}C{-}CH_2{-}S{-}S{-}CH_2{-}C{-}H\,(N{-}H,\ C{=}O),\ H{-}N$$

$$\downarrow\ HS-CH_2-CH_2OH\text{, Mercaptoethanol}$$

$$O{=}C(-)\,H{-}C{-}CH_2{-}SH\quad HS{-}CH_2{-}C{-}H\,(N{-}H,\ C{=}O),\ H{-}N$$

Two sulfhydryl groups

The mercaptoethanol split insulin into two peptide chains, one with N-terminal glycine and the other with N-terminal phenylalanine. Two disulfide bridges, in fact, hold the two chains together, while a third bridge links two cysteines on the same chain.

Sanger then cut insulin into several small peptides with acid, which cuts proteins at random, and with digestive enzymes that only hydrolyze peptide bonds between specific amino acids. He separated these peptides by *paper chromatography* (Figure 4.6). This procedure separates molecules on the basis of their solubility in different solvents. The substances to be separated—peptides, in this case—are dissolved in a solvent and spotted on a piece of heavy filter paper. One edge of the paper is placed in a solvent, and as the solvent slowly seeps into the paper, it carries the peptides along at different rates. After several hours, the paper is removed from the solvent and dried. For better separation, the paper is rotated by 90 degrees, and the procedure is repeated with another solvent. The separated peptides are located by spraying the paper with ninhydrin, which turns amino acids blue. Each peptide is then hydrolyzed into its component amino acids, which are separated by chromatography. Spots of amino acids are identified by running other chromatographs with known amino acids to see how they move on the paper.

In this way, Sanger identified the amino acids and the N-terminal residues in a set of peptides cut randomly from the protein. This gave him a set of *overlapping* peptide sequences that he could piece together, like a puzzle, to determine the overall sequence of the protein. For instance, after finding the peptides Ile–Val, Val–Glu, and Glu–Glu, he could guess that the original protein contained the sequence Ile–Val–Glu–Glu. He then found a peptide with the sequence Gly–Ile–Val–Glu, where the glycine is the N-terminus of one chain, so he could infer that the chain begins Gly–Ile–Val–Glu–Glu. With this logic, he gradually reconstructed the entire sequence of both chains, as shown in Figure 4.7. This work demonstrated that a protein has a unique sequence of amino acids, which is now called the **primary structure** of a protein.

When the technique of protein sequencing was applied to insulins from several species of animals, it revealed that these proteins have quite similar amino acid sequences, with only a few differences (Figure 4.8, p. 79). This pattern makes sense if all these species evolved from a common ancestor but accumulated small differences as they diverged from one another. Comparisons of molecular structures like this help establish the phylogenetic relationships among different species.

Sanger's technique for determining the primary structure of a protein has been largely replaced by automated instruments that use the Edman degradation procedure developed by the Swedish chemist Pehr Edman; by removing amino acids one at a time from the N-terminus of the polypeptide and identifying them, this procedure can determine the sequence of up to sixty amino acids in a short time. Although Sanger took years to determine the structure of insulin, his pioneering work made today's techniques possible, and the amino acid sequences of

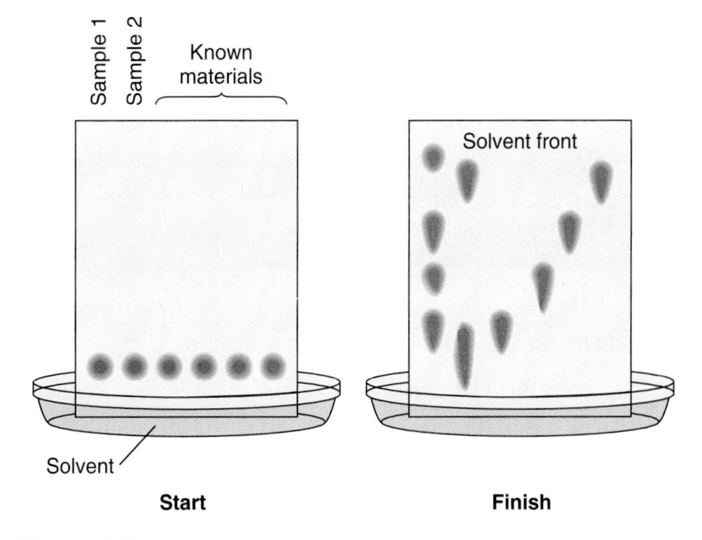

Figure 4.6

In this illustration of paper chromatography, amino acids move through the paper to different points because of their solubility in the solution.

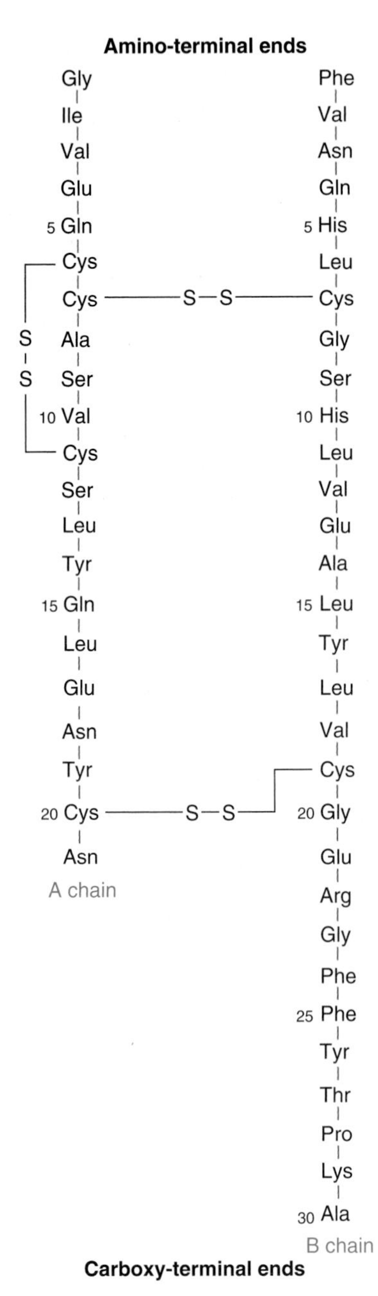

Figure 4.7

Porcine (pig) insulin is made of two polypeptide chains held together by disulfide bonds.

thousands of proteins are now known. Information from protein sequencing has led to two main conclusions:

1. The amino acid sequence of each kind of protein, its primary structure, is unique. All copies of the protein are identical, because of the way their sequence is genetically determined.
2. A protein may consist of more than one polypeptide chain. These chains may be held together by covalent bonds, usually disulfide bridges between cysteine residues in different chains, or by noncovalent forces such as hydrogen bonding and other weak attractive forces.

Exercise 4.14 If a peptide has a molecular mass close to 1,000 Da, how many amino acids does it have? After acid digestion, the following peptides are identifiable (using the one-letter abbreviations for amino acids): LK, AC, CE, GG, EL, CG, SC, GSC, GGG. The N-terminal residue is A. What is the sequence of the peptide?

4.11 Regular coiling or folding of a polypeptide chain determines the polypeptide's secondary structure.

We noted in Chapter 3 that organic molecules can assume different *conformations,* or shapes, without breaking any of the bonds between atoms. A polypeptide is a long, flexible molecule whose conformation can change extensively, but every protein actually *folds* into a particular conformation and is held in that shape by bonds, mostly weak bonds, between the various atomic groups in its backbone and side chains. These bonds stabilize the polypeptide into one of two general forms—either a long, thin **fibrous protein,** used mostly for structure, or a compact, rounded **globular protein.** These forms are shown in Figure 4.9, on pp. 80–81, by the fibrous proteins of hair (Keratin), silk (fibroin), and collagen, and by the globular forms of hemoglobin and myoglobin.

Around 1950, the chemists Linus Pauling and Robert Corey, working at Caltech, examined the structure of fibrous proteins by X-ray diffraction; in this method, a crystal of the protein is placed in an X-ray beam so its atoms diffract X rays onto a photographic plate in a complex pattern, which can be deciphered to show the structure of the crystal. Pauling and Corey found that several proteins have the form of the **alpha (α) helix,** shown in the structure of hair in Figure 4.9*a*. This shape, which is maintained largely by hydrogen bonds between the carbonyl (–C=O) and imino (–NH) groups of the backbone, is one example of a protein's **secondary structure.** Many fibrous proteins are little more than long α helices. Such proteins constitute large parts of animal structure, including the α-keratins of hair, horn, hooves, and porcupine quills; the myosins that make up much of the contractile apparatus of muscles; and the fibrin that forms blood clots.

If you pull out a hair and try to stretch it, it will stretch about a third longer than its initial length, then break. The usual explanation is that the hair stretches as hydrogen bonds are disrupted and the polypeptide chains open up from α helices into linear structures. Then it breaks when the chains have stretched as far as they can and covalent bonds are finally disrupted. But although many knowledgeable biochemists, including several Nobel Prize winners, have used this demonstration in their classes, no one has produced hard evidence that this is the correct explanation. Nevertheless, it's a neat thing to try—and the explanation may even be correct!

Pauling and Corey, again using X-ray diffraction, described a second conformation of proteins, the **beta (β)-pleated sheet** (Figure 4.9*b*), where several parallel chains are connected by

		1	2	3	4	5	6	7	8	9	10	1	2	3	4	5	6	7	8	9	20	1	2	3	4	5
Rabbit		Phe	Val	Asn	Gln	His	Leu	Cys	Gly	Ser	His	Leu	Val	Glu	Ala	Leu	Tyr	Leu	Val	Cys	Gly	Glu	Arg	Gly	Phe	Phe
Elephant		Phe	Val	Asn	Gln	His	Leu	Cys	Gly	Ser	His	Leu	Val	Glu	Ala	Leu	Tyr	Leu	Val	Cys	Gly	Glu	Arg	Gly	Phe	Phe
Rat, mouse		Phe	Val	Lys	Gln	His	Leu	Cys	Gly	Ser	His	Leu	Val	Glu	Ala	Leu	Tyr	Leu	Val	Cys	Gly	Glu	Arg	Gly	Phe	Phe
Dog, whale		Phe	Val	Asn	Gln	His	Leu	Cys	Gly	Ser	His	Leu	Val	Glu	Ala	Leu	Tyr	Leu	Val	Cys	Gly	Glu	Arg	Gly	Phe	Phe
Goat		Phe	Val	Asn	Gln	His	Leu	Cys	Gly	Ser	His	Leu	Val	Glu	Ala	Leu	Tyr	Leu	Val	Cys	Gly	Glu	Arg	Gly	Phe	Phe
Chicken		Ala	Ala	Asn	Gln	His	Leu	Cys	Gly	Ser	His	Leu	Val	Glu	Ala	Leu	Tyr	Leu	Val	Cys	Gly	Glu	Arg	Gly	Phe	Phe
Cod	Met	Ala	Pro	Pro	Gln	His	Leu	Cys	Gly	Ser	His	Leu	Val	Asp	Ala	Leu	Tyr	Leu	Val	Cys	Gly	Asp	Arg	Gly	Phe	Phe
Tuna	Val	Ala	Pro	Pro	Gln	His	Leu	Cys	Gly	Ser	His	Leu	Val	Asp	Ala	Leu	Tyr	Leu	Val	Cys	Gly	Asp	Arg	Gly	Phe	Phe

	6	7	8	9	30	1	2	3	4	5	6	7	8	9	40	1	2	3	4	5	6	7	8	9	50	1
Rabbit	Tyr	Thr	Pro	Lys	Ser	Gly	Ile	Val	Glu	Gln	Cys	Cys	Thr	Ser	Ile	Cys	Ser	Leu	Tyr	Gln	Leu	Glu	Asn	Tyr	Cys	Asn
Elephant	Tyr	Thr	Pro	Lys	Thr	Gly	Ile	Val	Glu	Gln	Cys	Cys	Thr	Gly	Val	Cys	Ser	Leu	Tyr	Gln	Leu	Glu	Asn	Tyr	Cys	Asn
Rat, mouse	Tyr	Thr	Pro	Met	Ser	Gly	Ile	Val	Asp	Gln	Cys	Cys	Thr	Ser	Ile	Cys	Ser	Leu	Tyr	Gln	Leu	Glu	Asn	Tyr	Cys	Asn
Dog, whale	Tyr	Thr	Pro	Lys	Ala	Gly	Ile	Val	Asp	Gln	Cys	Cys	Thr	Ser	Ile	Cys	Ser	Leu	Tyr	Gln	Leu	Glu	Asn	Tyr	Cys	Asn
Goat	Tyr	Thr	Pro	Lys	Ala	Gly	Ile	Val	Glu	Gln	Cys	Cys	Ala	Gly	Val	Cys	Ser	Leu	Tyr	Gln	Leu	Glu	Asn	Tyr	Cys	Asn
Chicken	Tyr	Ser	Pro	Lys	Ala	Gly	Ile	Val	Glu	Gln	Cys	Cys	His	Asn	Thr	Cys	Ser	Leu	Tyr	Gln	Leu	Glu	Asn	Tyr	Cys	Asn
Cod	Tyr	Asn	Pro	Lys	—	Gly	Ile	Val	Asp	Gln	Cys	Cys	His	Arg	Pro	Cys	Asp	Ile	Phe	Asp	Leu	Gln	Asn	Tyr	Cys	Asn
Tuna	Tyr	Asn	Pro	Lys	—	Gly	Ile	Val	Glu	Gln	Cys	Cys	His	Lys	Pro	Cys	Asn	Ile	Phe	Asp	Leu	Gln	Asn	Tyr	Cys	Asn

Figure 4.8

The sequences of the insulin β chain from different vertebrates show that the same protein in related species has very similar, but not identical, structures. The pattern of differences fits the phylogenetic relationships among the species, as indicated by their anatomy.

hydrogen bonds. Beta sheets are found in the β-keratins of birds' beaks and claws and in the scales of birds and reptiles. This is also the molecular structure of fibroin, the silk protein produced by silkworms and other insects.

Collagen is the chief protein of some of an animal's toughest structures—its bones, teeth, tendons, and skin. Collagen must be very strong to fulfill its structural functions, and purified collagen fibers look like steel cables, with a characteristic banded pattern (Figure 4.9*c*). Each collagen fiber is made of many smaller units, called tropocollagen, and a tropocollagen molecule consists of three polypeptides twisted around each other in a unique helical form.

4.12 Much of our knowledge of protein structure comes from studies of hemoglobin and myoglobin.

In the 1950s, X-ray diffraction was used to determine the structure of two closely related globular proteins: myoglobin from muscle (*myo-* = muscle) and hemoglobin from blood (*hemo-* = blood). Both proteins carry oxygen and become bright red when bound to it. Myoglobin has been a favorite material for study by biochemists and physiologists because it is easily purified in large quantities, especially from whale muscle, which was once plentiful. A whale must store oxygen in its muscles when it descends into the ocean for long periods (Figure 4.10, on p. 82), so whale meat is rich in myoglobin and is very dark red in color.

Hemoglobin captured the attention of scientists over a century ago. Its red color makes it easy to identify, and large amounts of pure hemoglobin can be obtained from blood, because red blood cells comprise nearly half the volume of blood, and hemoglobin comprises about half the blood cells' dry mass. As Figure 4.11, on page 83, shows, when blood is diluted in water, the cells break open (lyse), releasing their hemoglobin, which is then easily collected as quite pure crystals.

Myoglobin and hemoglobin are much less regular in structure than fibrous proteins, so X-ray diffraction analysis of their structure is difficult and requires extensive calculations. Nevertheless, Max Perutz and John Kendrew, with their associates, began X-ray analyses of hemoglobin and myoglobin, aided by computers. We will first look at the structure Kendrew deduced for myoglobin, which is the simpler protein.

Myoglobin Structure

Each myoglobin molecule is a single polypeptide chain of 153 amino acid residues (Figure 4.12, on p. 83), which folds up into the form shown in Figure 4.9*e*. Notice that about 75 percent of the chain is α helix; the helical structure is disrupted only where the molecule bends. Certain amino acid sequences tend to form α helices, and others favor β sheets, which also occur in globular proteins. A proline residue, for instance, doesn't fit either form, so every proline in the sequence interrupts the regular secondary structure, forcing the polypeptide chain into a

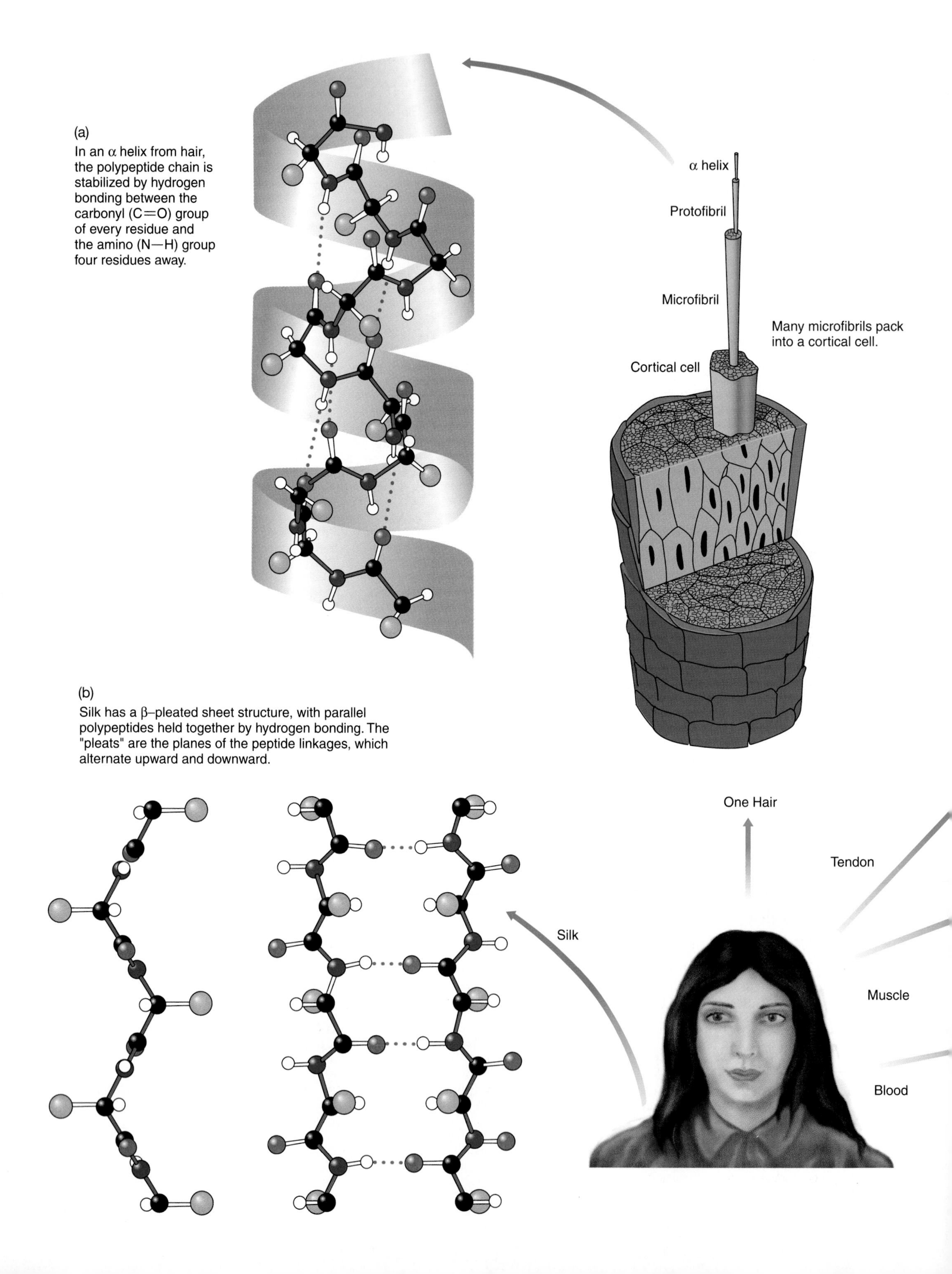

(a)
In an α helix from hair, the polypeptide chain is stabilized by hydrogen bonding between the carbonyl (C=O) group of every residue and the amino (N—H) group four residues away.
α helix
Protofibril
Microfibril
Many microfibrils pack into a cortical cell.
Cortical cell
(b)
Silk has a β–pleated sheet structure, with parallel polypeptides held together by hydrogen bonding. The "pleats" are the planes of the peptide linkages, which alternate upward and downward.
One Hair
Tendon
Silk
Muscle
Blood

Figure 4.9

The structures of five common proteins: *(a)* hair keratin; *(b)* silk fibroin; *(c)* collagen from a tendon or ligament; *(d)* hemoglobin from blood; *(e)* myoglobin from muscle.

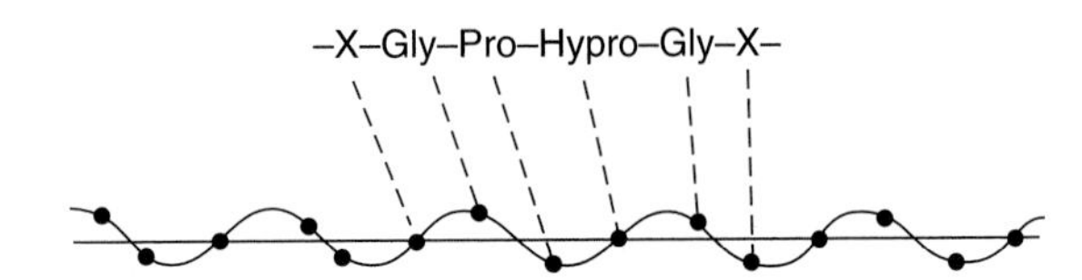

Polypeptide with typical repeating amino acid sequence.

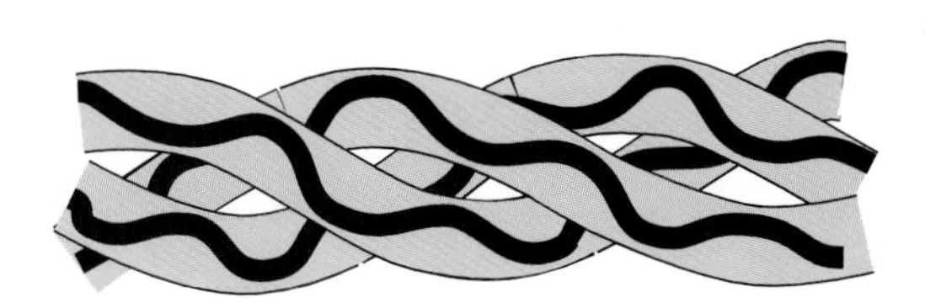

Each tropocollagen is a helix of three polypeptide chains.

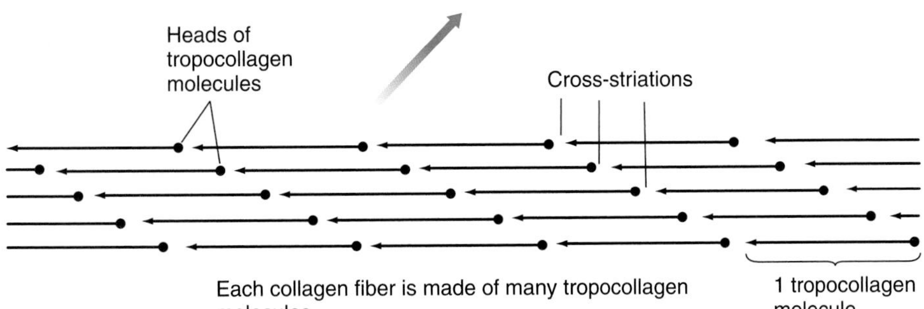

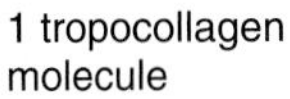

1 tropocollagen molecule

Each collagen fiber is made of many tropocollagen molecules.

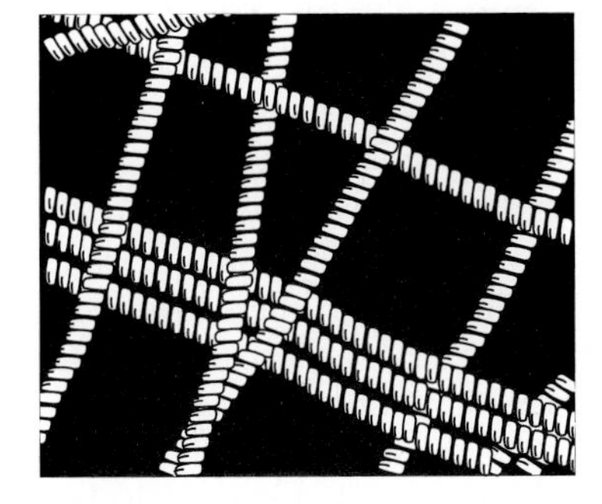

(c)
Isolated fibers of collagen, from a tendon or ligament, look like cables and have a distinctive repeating pattern.

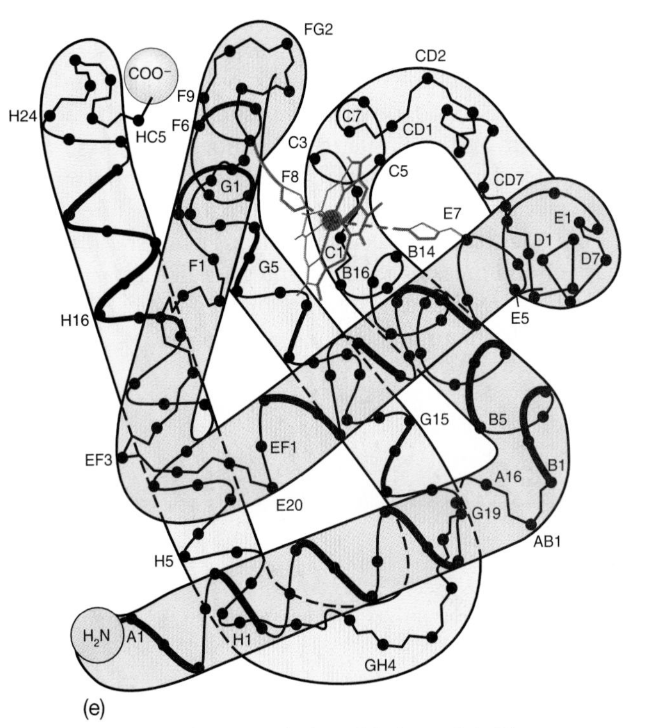

(e)
Myoglobin, from muscle, is a globular protein. The heavy line shows the backbone of the polypeptide; the side chains are omitted for simplicity. Trace the outline of the polypeptide backbone and notice the α-helical sections.

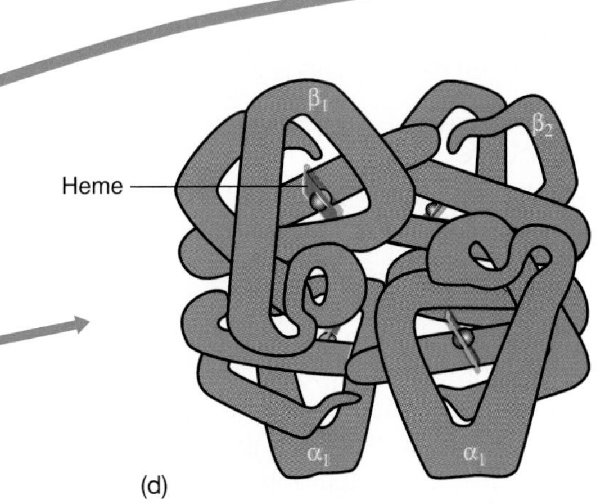

(d)
Hemoglobin, from blood, is made of four polypeptide chains—two α chains and two β—that are each very similar to one myoglobin molecule. The chains fit together compactly and are held in a specific quaternary structure by interactions between the amino acid side chains on their surfaces. Notice the positions of the four hemes, one on each globin chain.

Figure 4.10

Whales can stay under water for an hour at a time, in part because their muscles contain large amounts of myoglobin that store a considerable amount of oxygen.

different shape. Computer programs based on known protein structures can now predict the likely structure of new peptide sequences with some accuracy.

Interactions between the side chains of a globular protein hold it in a particular three-dimensional shape known as its **tertiary structure.** (It is hard to distinguish sharply between secondary and tertiary structures; a protein's overall shape is largely determined by secondary structures such as α helices and β sheets within it, and any other features not due to these forms are considered tertiary structures.) Polar and nonpolar side chains interact differently in forming tertiary structures. Around pH 7 the side chains of basic amino acids like lysine are positively charged, and those of acidic amino acids like aspartic acid are negatively charged. These opposite charges attract each other, holding the side chains together in a *salt bridge:*

HN — HC — $(CH_2)_4$ — $NH_3^{\oplus}$ $^{\ominus}O$ — C(=O) — CH_2 — CH (C=O, HN)

O=C

Lys Asp

Tyrosine and aspartic acid show one example of hydrogen bonding inside a protein:

NH — HC — CH_2 — (benzene ring) — OH ··· O — C(=O) — CH_2 — CH (C=O, HN)

Hydrogen bond

O=C

Tyr Asp

Other residues show examples of dipole interactions and hydrophobic interactions:

Hydrophobic interaction

NH — HC — CH_2 — CH(CH_3)(CH_3) H_3C(H_3C)HC — CH_2 — CH (C=O, HN)

O=C

Leu Leu

Dipole interaction

HN — HC — CH_2 — O—H   H—O — CH_2 — CH (C=O, HN)

O=C

Ser Ser

All these interactions stabilize the tertiary structure of a protein.

Because X-ray diffraction analysis is done by computer, the computer can be programmed to show the points where side chains interact. Kendrew found many such points between polar side chains, but far more between nonpolar side chains. The general rule is that globular proteins tend to fold up with their polar groups on the outside in contact with the surrounding water and almost all their nonpolar groups clustered together inside, forming a hydrophobic core that is largely responsible for holding the protein in shape (Figure 4.13). Although experts still debate the mechanism for the interactions of nonpolar groups in protein folding, the two most common explanations are those given earlier (see Section 3.11) for the behavior of hydrocarbons in water: (1) The nonpolar side chains, excluded from the surrounding water structure, are pushed together into small islands, and (2) the nonpolar groups are held together by van der Waals forces.

Each myoglobin molecule contains a small **heme** molecule, a member of an important family of compounds discussed in Concepts 4.2. Heme serves as a **prosthetic group**—that is, a nonpeptide molecule tightly bonded to the protein that is essential for its function. Heme is the part of myoglobin that binds oxygen, and without its heme group, the protein alone (the globin) is useless as an oxygen carrier. Heme molecules are flat, or planar, and each one fits into a nonpolar slot in its polypeptide chain.

Hemoglobin structure

In contrast to myoglobin, hemoglobin is made of four polypeptide chains—two alpha (α) chains and two beta (β) chains (Figure 4.9*d*). (Polypeptides are often designated by Greek letters; this has nothing to do with the α helix and β sheet.) The α and β chains have similar three-dimensional shapes and sizes, the α chain containing 141 amino acids and the β chain 146. Their amino acid sequences are also very similar, and a close look reveals that even where the sequences are not identical, amino acids in comparable positions have side chains with similar properties. Polar amino acids in one chain substitute for different polar amino acids in the other;

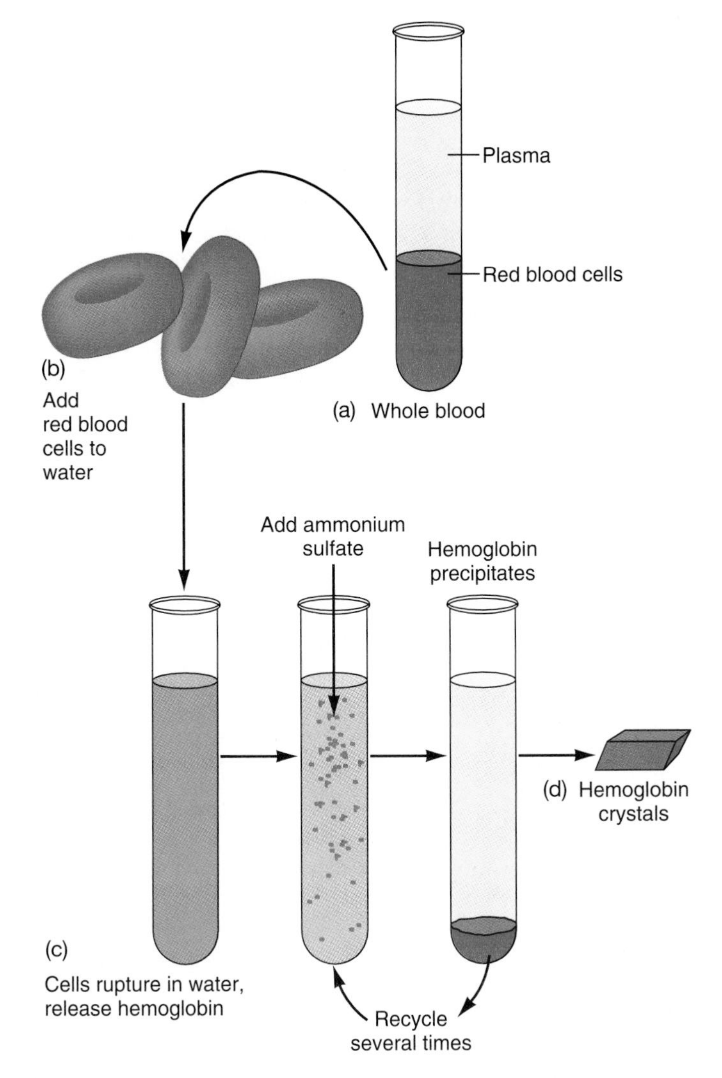

Figure 4.11

Hemoglobin can be extracted and purified from blood using the following procedure. *(a)* Red blood cells are collected in a centrifuge tube. Each red blood cell, or erythrocyte *(b)*, is a biconcave disc with a thin membrane, which ruptures when placed in pure water *(c)*. The hemoglobin released from the cells is collected by the procedure of *salting out:* when a salt, such as ammonium sulfate, is added to the solution, the protein precipitates in a crystalline form *(d)*. This hemoglobin can be purified by repeating the precipitation, eventually producing quite large crystals of pure hemoglobin.

nonpolar amino acids substitute for different nonpolar ones. Such comparisons indicate that a side chain's nature is more important than its actual identity in determining a protein's three-dimensional shape.

A protein made of two or more polypeptides, such as hemoglobin, has an additional degree of complexity beyond that of a single polypeptide; the **quaternary structure** of a protein is the form it has because of the assembly of its subunits. In such proteins, each polypeptide chain is a *protomer,* and the entire assemblage is a *multimer.* The protomers are held together through interactions among some of their surface groups. Many enzymes and other proteins function only as multimers.

Gly-Leu-Ser-Asp-Gly-Glu-Trp-Gln-Leu-Val-Leu-Asn-Val-Trp-Gly-
Lys-Val-Glu-Ala-Asp-Ile-Pro-Gly-His-Gly-Gln-Glu-Val-Leu-Ile-
Arg-Leu-Phe-Lys-Gly-His-Pro-Glu-Thr-Leu-Glu-Lys-Phe-Asp-Lys-
Phe-Lys-His-Leu-Lys-Ser-Glu-Asp-Glu-Met-Lys-Ala-Ser-Glu-Asp-
Leu-Lys-Lys-His-Gly-Ala-Thr-Val-Leu-Thr-Ala-Leu-Gly-Gly-Ile-
Leu-Lys-Lys-Lys-Gly-His-His-Glu-Ala-Glu-Ile-Lys-Pro-Leu-Ala-
Gln-Ser-His-Ala-Thr-Lys-His-Lys-Ile-Pro-Val-Lys-Tyr-Leu-Glu-
Phe-Ile-Ser-Glu-Cys-Ile-Ile-Gln-Val-Leu-Gln-Ser-Lys-His-Pro-
Gly-Asp-Phe-Gly-Ala-Asp-Ala-Gln-Gly-Ala-Met-Asn-Lys-Ala-Leu-
Glu-Leu-Phe-Arg-Lys-Asp-Met-Ala-Ser-Asn-Tyr-Lys-Glu-Leu-Gly-
Phe-Gln-Gly

Figure 4.12

A single polypeptide chain of whale myoglobin has 153 amino acid residues bonded in this primary structure.

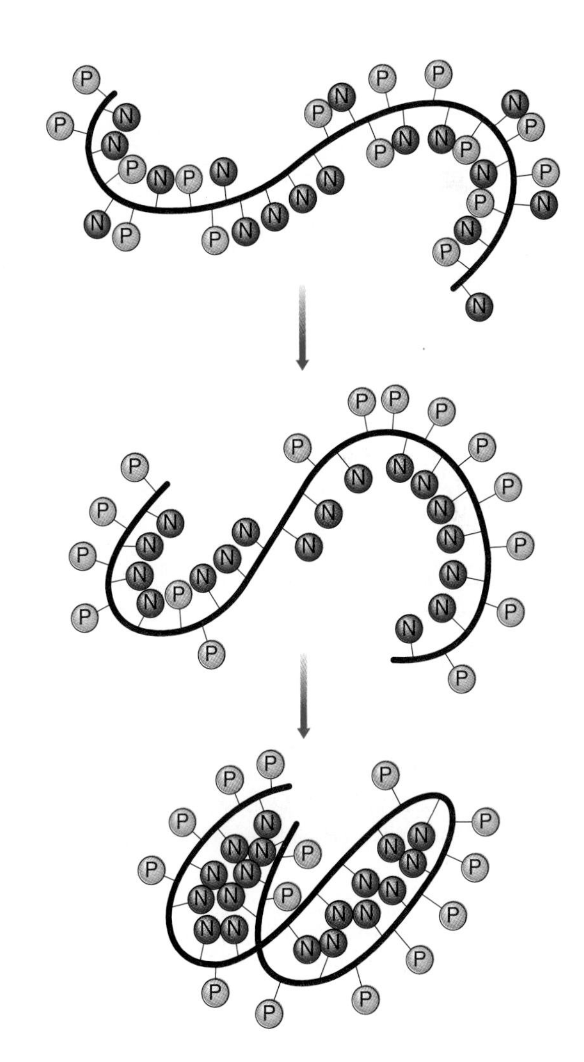

Figure 4.13

A polypeptide has both polar (P) and nonpolar (N) side chains. It tends to fold up so that the polar groups are on the outside in contact with the surrounding water and the nonpolar groups are on the inside, where they interact with each other to make a hydrophobic core.

Concepts 4.2 A Remarkable Family of Compounds

In the course of evolution, organisms occasionally evolve the ability to synthesize a compound that opens up great possibilities for further development, allowing future organisms to explore whole new ways of life. The *tetrapyrroles* are a prime example. Some ancient organism evolved the ability to assemble four pyrrole molecules into a tetrapyrrole; with specific side chains, it is called a *porphyrin:*

Pyrrole
4 Pyrroles
Me = Metal ion
R = Side chain
Porphyrin

The four nitrogen atoms of the porphyrin ring can hold a positive ion between them. Heme, for instance, has an iron in the middle. The iron ion is said to be *chelated* in the ring, for it is held there much as the chela of a lobster or crab (its large pincer) holds a bit of food:

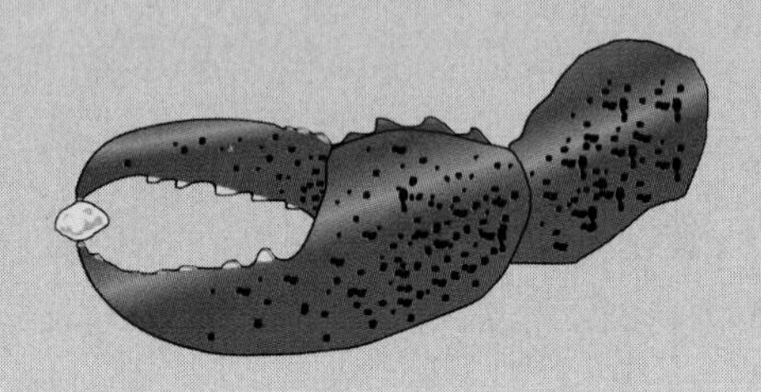

Many other metal ions can be put into a porphyrin ring, and some of them have biological uses. A copper porphyrin, turacin, is a red pigment in some birds' feathers, and a cobalt porphyrin is the core of vitamin B_{12} (cobalamin), which has various roles in metabolism. But two kinds of tetrapyrroles—chlorophylls and hemes—stand out above all others.

The *chlorophylls* are all magnesium tetrapyrroles with various side chains. Life depends on chlorophylls because the large, square molecule acts like an antenna that absorbs light. This enables organisms such as algae and plants to capture light energy in the process of photosynthesis, thus bringing energy into an ecosystem.

The *hemes* are versatile molecules. They are built into different kinds of proteins, which determine just how the heme iron atom will behave. It is already ringed on four sides by nitrogen atoms, and some proteins provide two more nitrogens, above and below the ring:

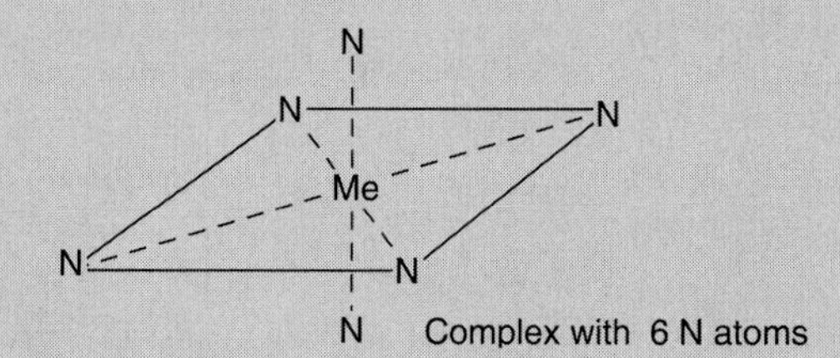

In the proteins called cytochromes, the iron atom shifts between the Fe^{2+} and Fe^{3+} states, and these proteins pass electrons from one to another in reactions that are critical to metabolism. Hemes are also components of peroxidases and catalases, enzymes that oxidize various substances with peroxide (H_2O_2).

In hemoglobins and myoglobins, on the other hand, the heme binds an oxygen molecule to carry and hold oxygen in our blood and muscles. These molecules probably make large animals like humans possible. Hemoglobin is also used in some plants that fix nitrogen (that is, convert N_2 to ammonia), where it holds oxygen molecules that would otherwise interfere with this process. All these functions became possible because of the evolution of one complex molecule.

Figure 4.14 summarizes the different levels of folding in globular proteins. This folding determines the protein's secondary structure (through the formation of α helices and β sheets), tertiary structure (through interactions between the side chains in a polypeptide), and quaternary structure (through the assembly of polypeptide subunits). All three levels combine to establish the overall shape of the protein. Globular proteins, especially large ones, are often made of compact globular regions called **domains** that fold up independently of other such regions. Domains are significant because they are generally distinct functional units; a protein often interacts with two or three other molecules, and it may have a distinct domain for each of these interactions. Some new proteins apparently have evolved by combining the domains of existing proteins in new ways.

4.13 The primary structure of a polypeptide uniquely determines its three-dimensional shape.

Around 1960, Christian Anfinsen investigated the question of how a protein folds into its natural shape. He was interested in ribonuclease (RNase), an enzyme that breaks the bonds linking the nucleotide monomers of ribonucleic acid. Earlier he had been involved in determining the sequence of amino acids in RNase, the second protein to be sequenced. Anfinsen showed that RNase becomes enzymatically inactive when treated with chemicals that gently disrupt interactions among its amino acid residues. Urea and guanidine hydrochloride, for instance, break the hydrogen bonds and other noncovalent interactions within the protein, and mercaptoethanol breaks disulfide bridges

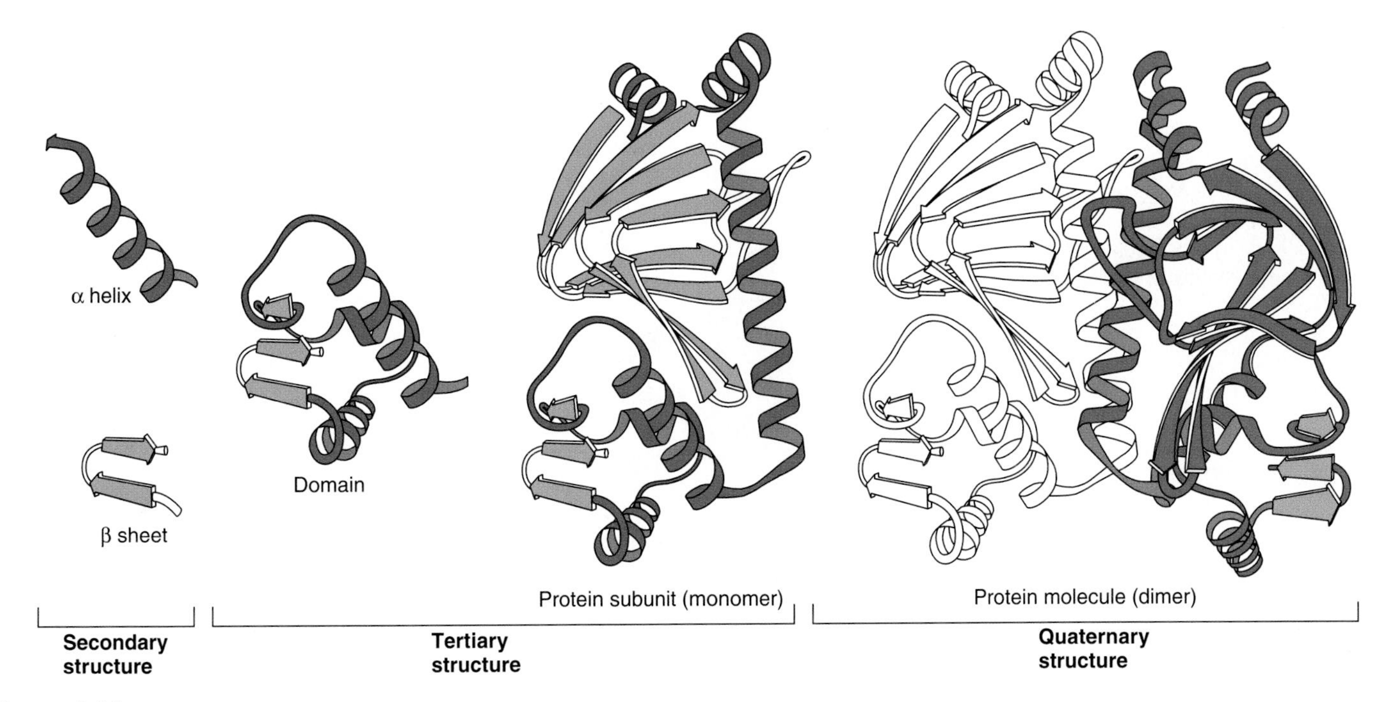

Figure 4.14

The three-dimensional structure of a protein is described by a hierarchy of different levels of folding, each level made of the structures just below it. The polypeptide, whose primary structure is not shown explicitly, is represented by a ribbon pointing toward the carboxy terminal end. Elementary structures are either α helices or β sheets; these are combined in compact local regions called domains, and a whole protein subunit is made of two or more domains. Two or more subunits (protomers) may combine to make a multimer. Notice how the terms "secondary," "tertiary," and "quaternary structure" are applied.

between cysteine residues. When the chemical agents disrupt these bonds, which normally hold the protein in a certain conformation, its shape changes. The protein is said to be *denatured* (Figure 4.15), and it is no longer active as an enzyme because it does not have a proper active site where substrate molecules can fit.

Proteins can also be denatured by heat, and in cooking an egg, you are denaturing its proteins. While you can't uncook an egg, Anfinsen found that RNase is reversibly denatured by the chemical agents; removing them *renatures* the RNase and restores its three-dimensional structure, including the arrangement of amino acids forming its active site, so it regains its enzymatic activity.

The demonstration that a protein can be inactivated and then can spontaneously return to its original shape and be active again led to an important conclusion: The sequence of amino acids in a protein determines its three-dimensional shape. As a protein is synthesized, it folds up spontaneously into its lowest-energy conformation, held there by many internal interactions. This is a protein's proper functional shape, the shape that has the right features to act as an enzyme or perform some other function. Most proteins spontaneously assume their proper conformation, but others need some assistance. While they are being synthesized, these latter proteins bind to *chaperonins*, other proteins that prevent them from folding improperly until their synthesis is complete.

Exercise 4.15 Amino acids can be linked into peptides in any sequence. Let's get a rough idea of how many proteins are possible. *(a)* Consider a dipeptide: There are 20 choices for the first amino acid and 20 choices for the second, resulting in $20 \times 20 = 400$ dipeptides. Now, how many different tripeptides (made of three amino acids) are there? *(b)* How many different proteins 100 amino acids long are possible? How many 300 amino acids long?

4.14 The structure of a protein is determined by the genome and shaped by evolution.

Why does a protein, such as myoglobin, have its particular shape? There are two ways to answer this question, and both are necessary for a full explanation. The functional answer is that this is a very good shape for a protein molecule whose function is to hold oxygen in muscle cells. The mechanistic answer is that the polypeptide folds up as it does because of the way the various amino acid groups push and pull on one another as the protein is being synthesized. Anfinsen's work shows that the shape is uniquely determined by the amino acid sequence—that is, if it had a different amino acid sequence, it would fold up into a different shape.

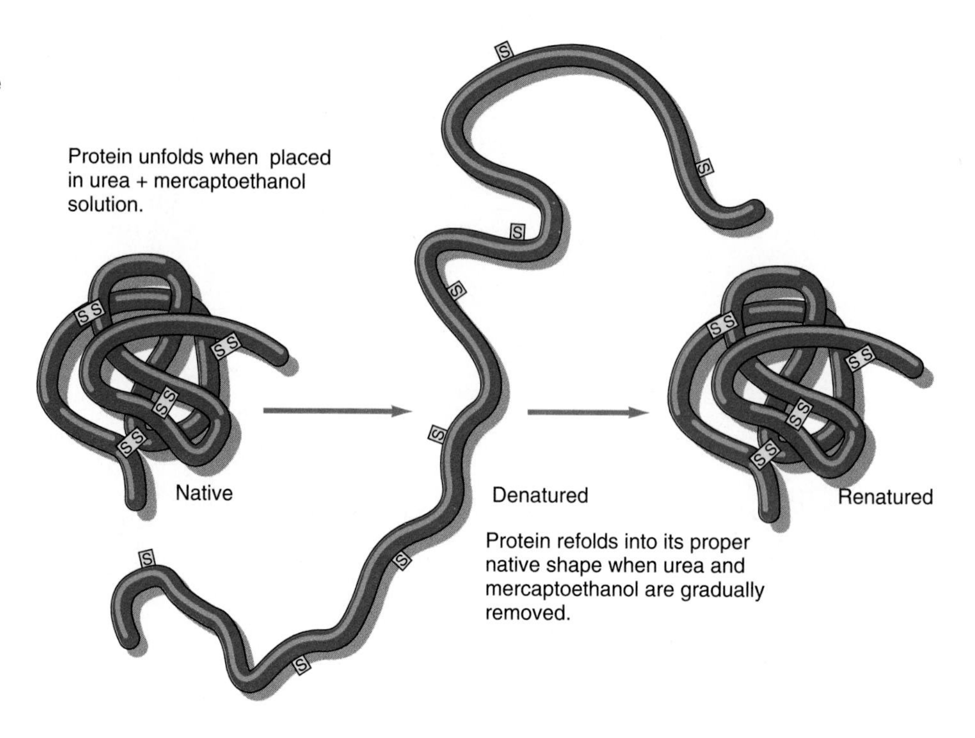

Figure 4.15
A globular protein, such as ribonuclease, can be denatured with reagents that disrupt its hydrogen bonds and disulfide bonds, allowing the protein to unfold and become inactive. When these reagents are removed gently, so disulfide bridges and other bonds can form again, virtually all the molecules return to their original shape and recover their activity.

So what causes the muscle cells to make a protein with a particular sequence? The genome of every cell in an organism contains the information for making all of that organism's proteins. Exercise 4.15 points out that the number of possible proteins is unimaginably large; there are 20^{100} possible small proteins of 100 amino acids, for instance, and vastly more possible proteins of medium and large size. Information is required to specify one object out of a range of possible objects (see Concepts 2.1), such as one particular protein, and therefore an enormous amount of information is required just to specify the proteins of one cell. The protein factories in a muscle cell follow molecular instructions that tell them to link certain amino acids in one particular sequence that forms myoglobin, as well as instructions for making thousands of other proteins.

Furthermore, instructions from the genome also direct the *amount* of each protein to be made, and this is why some muscle cells make more myoglobin than others. The dark meat of a chicken, for example, contains a lot of myoglobin, while the white meat contains very little. (The color of the tissue, incidentally, is also determined largely by its pattern of blood vessels.)

But why do cells carry a particular kind of information in the first place? This brings us to the functional answer to the original question: Muscle cells contain the information for making myoglobin because it is a protein with a very good shape for storing oxygen. If the genetic information were slightly different, muscle cells would make a molecule with a different shape that might not be as functional. Yet again we ask, "How do these cells happen to have the information for making such a functional protein?" The answer is that the information for making a functional myoglobin must have gradually *evolved* into its present form over many generations.

The evolution of a protein is basically an editing process. Remember that mutations are always occurring in genomes, so some individuals will have instructions for making myoglobin molecules with minor variations in their sequences. If any of those variations make the individuals better adapted to their ways of life, those individuals are more likely to be selected. In other words, they will tend to have more offspring and more successful offspring, and that particular change will become common in the population. Keep in mind, of course, that the evolution of a single protein is just one aspect of the whole organism's evolution.

Figure 4.8 shows that different species have slightly different instructions for making insulin, which are reflected in slight variations in the protein's amino acid sequence. The same is true of myoglobin. The functional explanation for these differences is that each species is adapted to different conditions. It is unlikely that *every* difference in amino acid sequence really makes a difference in the life of the animal, but at least *some* of those differences must be adaptive and must suit each species for its way of life.

When we edit something we have written, we make specific, intentional changes to improve the writing. But a pod of whales—or any other organism—can't make intentional changes in their genomes. The variations underlying evolution are the result of random mutations that occur by chance. It might seem that such mutations could only ruin the animal, not help it, and in fact many mutations—probably most of them—are deleterious. But mutations can also make small, subtle changes in the structure of a protein, and some of these may really be improvements. The key is that proteins are polymers and that a mutation can change a single amino acid at a time. For instance, the difference between alanine and serine is just a single oxygen atom, and the difference between aspartic and glutamic acids is just a single $-CH_2-$ group:

```
       H                    H
       |                    |
   H — C — H            H — C — O — H
       |                    |
H2N — C — COOH       H2N — C — COOH
       |                    |
       H                    H
    Alanine               Serine

                          COOH
                            |
                           CH2
     COOH                   |
       |                   CH2
      CH2                   |
       |             H2N — C — COOH
H2N — C — COOH              |
       |                    H
       H
  Aspartic acid        Glutamic acid
```

Alanine **Serine** **Aspartic acid** **Glutamic acid**

If a mutation instructs a cell to insert a serine instead of an alanine, or a glutamic acid instead of an aspartic acid, its effect will be to add only that one oxygen atom or $-CH_2-$ group to all the copies of a particular protein, leaving the thousands of other atoms in the protein as they are. Other changes are bigger, of course, but the potential for making such subtle changes in the structure of a protein allows proteins to be gently, gradually edited generation after generation so that their shapes are functional.

The ideas outlined here will become clearer as we see how genetic information is encoded and how it is used to make proteins. These are some of the most central concepts of biology, and it is important to keep them in mind in order to make sense of the structure and operation of organisms.

Coda The principle of polymeric structure underlies biological organization. Organisms only make (or obtain from their environments) twenty kinds of amino acids, five kinds of nucleotides, and a few kinds of sugars. They polymerize these monomers into all their proteins, nucleic acids, and polysaccharides. By making only a few other small molecules, they synthesize their lipids. If chemistry were different—if it were not possible to assemble many small units into huge molecules—organisms probably could not exist. Because amino acids and nucleotides can be arranged in different sequences, enormous numbers of different proteins and nucleic acids are possible. Each kind of protein folds into a molecule with a unique shape that may be suitable for conducting a certain biological process. Evolution depends upon a kind of experimentation in which random genetic events (usually mutations) create proteins with sometimes subtle differences, and the organisms whose proteins are best suited for their ways of life become more likely to survive and reproduce. In Chapter 5, we will look more closely at the actions of proteins.

Summary

1. Biological materials are primarily very large molecules called macromolecules. These are polymers made by combining many small molecules, or monomers. Homopolymers are made of only a single kind of monomer; the most important biological polymers are heteropolymers, which are made of several kinds of monomers.
2. Polysaccharides, whose monomers are sugars (monosaccharides), function as energy stores and structural elements.
3. Molecules that have four different atomic groups bonded to one carbon atom (an asymmetric carbon) can be arranged in two different ways, called optical isomers. Sugars have both L and D forms, and there are several isomers within each of these classes.
4. Nucleic acids, whose monomers are nucleotides, make the genome and the structures that translate its information.
5. The major polymers in all organisms are proteins, which function as structural components, effectors of movements, transporters that move ions and molecules across cell membranes, regulators of many functions, and enzymes.
6. All proteins are polymers whose monomers are amino acids. Twenty kinds of amino acids are commonly found in proteins. Amino acids (except glycine) have both L and D forms, but all amino acids in proteins are L isomers. Proteins may, in addition, contain small organic molecules of other kinds.
7. A protein is made by linking amino acids together with peptide bonds. A peptide linkage is formed by removing a water molecule. The portion of an amino acid that resides in a peptide is an amino acid residue.
8. Amino acids and proteins are ionized at typical biological pH. Their amino groups tend to combine with a hydrogen ion and become NH_3^+, while their carboxyl groups tend to lose a hydrogen ion and become COO^-. Some amino acids also have acidic or basic side chains, and these too become ionized.
9. The amino acids in a polypeptide are linked in a definite sequence, known as the polypeptide's primary structure. Every kind of protein has its own distinctive primary structure.
10. Once a polypeptide chain has been formed, it folds up into some kind of secondary structure, most commonly with stretches of α helix or β-pleated sheet. These structures are stabilized by hydrogen bonds between the N–H and C=O groups of the polypeptide backbone.
11. Some major structural proteins are long, fibrous molecules with α-helical or β-sheet forms or with different helical forms, like that of the common structural protein collagen.
12. Globular proteins, such as myoglobin and hemoglobin, have tertiary structures formed by further folding of the polypeptide backbone. These structures are largely stabilized by covalent disulfide bridges, by weak bonds between polar side chains, and by interactions between hydrophobic side chains in the interior of the protein. The protein folds up in such a way that these side chains avoid the water around the protein and associate with one another.
13. Globular proteins, especially large ones, have domains made of α helices and β sheets that form small, compact regions with limited functions, such as the ability to interact with one other kind of molecule.
14. The primary structure of a polypeptide uniquely determines its three-dimensional shape, and many polypeptides spontaneously assume their proper functional shape as they are being synthesized.
15. The primary structure of a protein is specified by information in the genome. Through evolution, each protein is shaped into an appropriate structure for its particular function.

Key Terms

macromolecule 67
dalton / kilodalton 67
polymer 67
monomer 67
polymerization 67
dehydration synthesis 67
condensation 67
hydrolysis 67
polysaccharides 68
monosaccharide 68
stereoisomer 68
chirality 69

cellulose 71
starch 71
glycogen 71
nucleic acid 71
nucleotide 71
deoxyribonucleic acid (DNA) 72
ribonucleic acid (RNA) 72
protein 72
catalyst 72
substrate 73
active site 73
amino acid 73
polypeptide 75
disulfide bridge 77
primary structure 77
fibrous protein 78
globular protein 78
alpha (α) helix 78
secondary structure 78
beta (β)-pleated sheet 78
tertiary structure 82
heme 82
prosthetic group 82
quaternary structure 83
domain 84

Multiple-Choice Questions

1. Which term includes all the others?
  a. polysaccharide
  b. nucleic acid
  c. protein
  d. homopolymer
  e. macromolecule

2. A polysaccharide made from 100 monosaccharides
  a. is a macromolecule.
  b. has been made by condensation reactions.
  c. has been made by dehydration synthesis.
  d. has resulted in the production of 99 molecules of water.
  e. is all of the above.

3. A ketopentose contains, or is,
  a. five glucose monomers.
  b. five glycosidic linkages.
  c. five ketone groups.
  d. a 5-carbon ketone monomer.
  e. five 5-carbon ketone monomers.

4. Which of the following statements is correct?
  a. All carbohydrates are composed of complex polysaccharides.
  b. Polypeptides are constructed of covalently linked protein residues.
  c. Nucleotides are formed from repeating monomers of nucleic acids.
  d. Polysaccharides are formed from repeating monosaccharide subunits.
  e. Oligopeptides are composed of repeating units of oligo-amino acids.

5. Ribose is a ______ while sucrose is a ______.
  a. pentose; hexose
  b. hexose; pentose
  c. monosaccharide; disaccharide
  d. polymer; monomer
  e. kind of starch; kind of sugar

6. Disulfide bridges are least significant in establishing a protein's ______ structure.
  a. primary
  b. secondary
  c. tertiary
  d. quaternary
  e. multimer

7. Which level(s) of structure establish a domain in a multimer?
  a. primary only
  b. secondary only
  c. tertiary only
  d. quaternary only
  e. all of the above

8. If a protein is reversibly denatured, which structural level can you be sure has remained intact?
  a. primary only
  b. secondary only
  c. tertiary only
  d. quaternary only
  e. none of the above

9. Both glycosidic and peptide linkages
  a. are formed by covalent bonds between carbon atoms.
  b. can only form linear polymers.
  c. result from condensation reactions.
  d. contain the configuration –C–O–C–.
  e. result from linking a carbonyl group and an acid group.

10. Which level of three-dimensional structure of a linear polypeptide results primarily from strong internal bonds between R groups?
  a. primary linear structure
  b. helical secondary structure
  c. pleated sheet
  d. tertiary structure
  e. quaternary structure

True-False Questions

Mark each statement as true or false, and if false, restate it to make it true.

1. Theoretically, at least, it is always possible to lengthen a linear polymer by adding another monomeric unit to one or both ends.
2. Peptide bonds and glycosidic linkages both use covalent bonds.
3. The most abundant protein in many vertebrate animals is cellulose.
4. The R group of amino acids participates in forming peptide linkages.
5. The higher the pH, the higher the concentration of amino acids in the form of zwitterions.
6. Naturally occurring macromolecules of starch, glycogen, and cellulose are homopolymers, but naturally occurring proteins are heteropolymers.
7. All enzymes are catalysts, but all catalysts are not enzymes.
8. The three-dimensional configurations known as β-pleated sheets and α helices result from the binding of a prosthetic group to a protein.
9. Chlorophyll, hemoglobin, and myoglobin are all tetrapyrroles with an iron-containing porphyrin ring.
10. Although different proteins such as insulin and hemoglobin have different structures, all insulin or all hemoglobin molecules are identical.

Concept Questions

1. Relate the processes of dehydration synthesis and hydrolysis to events in your daily life by giving an example of when each process is of primary importance.
2. Explain how glucose monomers can form branched as well as unbranched polysaccharides.
3. Suppose you wish to purchase a dietary amino acid supplement and must choose between a product containing a mixture of D and L isomers and another product containing only L isomers. If you prefer a product that has been derived from cellular sources as opposed to one that was synthesized in the laboratory, which one will you select? Explain.
4. Contrast the roles of amino and carboxyl groups on the one hand and R groups on the other hand with respect to the structure and function of a polypeptide.
5. What is the difference between a multimer and a domain?

Additional Reading

Creighton, Thomas E. *Proteins: Structures and Molecular Properties.* W. H. Freeman, New York, 1993.

Dickerson, Richard E., and Irving Geis. *The Structure and Action of Proteins.* Harper & Row, New York, 1969. A small book with excellent illustrations of protein structures.

Lehninger, Albert L., David L. Nelson, and Michael M. Cox. *Principles of Biochemistry.* Worth Publishers, New York, 1993.

Radetsky, Peter. "Kim's Coils." *Discover,* June 1995, p. 112. Peter Kim's research on coiled coils, a type of protein folding, revealed that the coils attach to each other by a series of snaps, rather than a zipperlike mechanism as commonly thought.

Rawn, J. David. *Biochemistry.* Neil Patterson Publishers, Burlington (NC), 1989.

Stryer, Lubert. *Biochemistry.* W. H. Freeman, New York, 1995.

Vollrath, Fritz. "Spider Webs and Silks." *Scientific American,* March 1992, p. 52. Structural engineering of spider webs and the molecular differences between silks.

Weinberg, Robert A. "The Molecules of Life." *Scientific American,* October 1985, p. 48. A general overview of molecular mechanisms underlying biological complexity.

Internet Resource

To further explore the content of this chapter, log on to the web site at:

http://www.mhhe.com/biosci/genbio/guttman/

5 Enzymes and the Dynamics of Proteins

Organisms engage in actions—sometimes very conspicuous actions—that are effected by proteins.

Key Concepts

A. Enzyme function

5.1 Enzymes work by lowering energy barriers.

5.2 Chemical reactions come to an equilibrium if left undisturbed.

5.3 An enzyme's structure enables it to catalyze a specific reaction.

5.4 Michaelis-Menten kinetics shows that an enzyme has a limited number of active sites.

5.5 Temperature affects the rate of enzyme-catalyzed reactions.

5.6 Each enzyme has an optimal pH.

5.7 Some enzymes require cofactors and coenzymes to function.

B. General protein function

5.8 Studies of enzyme inhibition help explain protein action and emphasize the importance of weak bonds between molecules.

5.9 Hemoglobin illustrates cooperative effects between the subunits of a protein.

5.10 The binding of proteins to other molecules is fundamental to biological processes.

5.11 Many proteins, including hemoglobin, show allosteric effects.

5.12 A small change in protein structure can have profound functional consequences.

5.13 The primary structure of a protein is a record of its evolutionary history.

5.14 Computer searches help identify the function of an unknown protein on the basis of its structure.

Raise your arm. Breathe deeply. Have a bite of food. Read a book. The business of being alive is about *doing,* and at every moment, we are doing many things—moving, breathing, digesting, thinking, talking. It is so natural for all living organisms, including humans, to be doing something all the time that it takes special effort to stop and ask *how* we can do all these things. That , however, is one of the central questions of biology.

In Chapter 4 we started to develop the concept that virtually everything an organism does depends on specific proteins, on their interactions with one another and with other molecules. When any biological function is performed, biologists assume the process is carried out by specialized proteins. This is a general way of biological thinking that students should learn too. In this chapter we will show that proteins typically work by binding at certain sites to shape-specific molecules that fit into those sites. We begin with enzymes, because they are proteins that are well understood and can serve as models of specific protein interactions. To make sense of these processes, we will emphasize that proteins do not interact with other molecules through strong covalent bonds but rather through weak, easily broken bonds; for this reason, the molecules can bind to each other briefly, come apart, and then bind again. A fresh look at myoglobin and hemoglobin will bring out some other general principles about protein function, especially the concept that proteins change their shape and activity when bound to other molecules. The chapter concludes by returning to the general theme of evolution, specifically the evolution of proteins; we will show how some hemoglobins have evolved and how the phylogenies of organisms are recorded in the amino acid sequences of their proteins. ■

A. Enzyme function

5.1 Enzymes work by lowering energy barriers.

Chemical reactions occur constantly in the world around us. For instance, iron combines with oxygen in air or water to form iron oxide (rust), and the hydrocarbons in gasoline combine with oxygen inside automobile engines to form carbon dioxide and water (and release useful energy). Then isn't it remarkable that so many materials do *not* react chemically with one another? We depend, after all, on most objects staying quite stable and lasting for years. Why do chemical reactions occur so easily between some substances and slowly or not at all between others? One main reason is that chemical reactions involve changes in energy, and most substances don't have enough energy to react. In Chapter 7 we will discuss energy in some detail, but here we will rely on what you already know about energy as we describe the mechanics of chemical reactions.

Metabolism consists of hundreds of chemical reactions. Here is one simple reaction that is occurring right now in your red blood cells:

$$\underset{\text{Carbon dioxide}}{CO_2} + \underset{\text{Water}}{H_2O} \rightarrow \underset{\text{Carbonic acid}}{H_2CO_3}$$

Carbon dioxide and water (the *reactants*) combine to form carbonic acid (the *product*). All the reactions of metabolism are catalyzed by enzymes, and the reactants are the *substrates* of the enzyme.

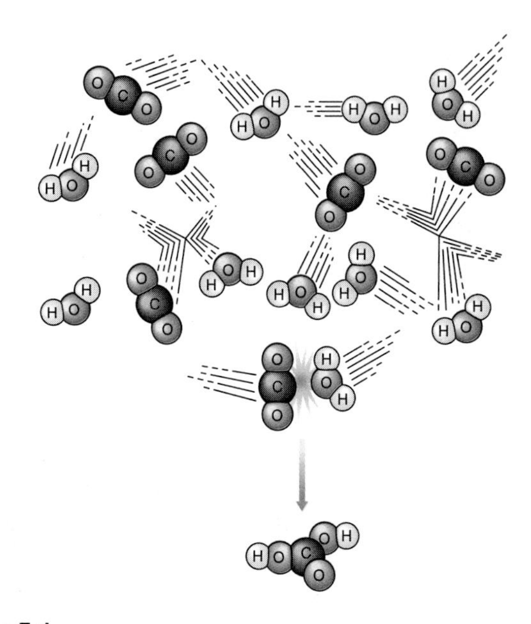

Figure 5.1
A carbon dioxide molecule and a water molecule can react to form carbonic acid when they collide in just the right way.

In a chemical reaction, certain bonds between atoms in the reactants are broken, and new bonds are formed. It takes energy to break bonds, but all molecules already have their own kinetic energy, the energy of their vibration and movement. A CO_2 molecule and a water molecule react only if they collide in just the right way and with sufficient energy to break their bonds and create a new arrangement of atoms (Figure 5.1). Reactants must therefore cross an *energy barrier* to be converted into products, as illustrated by a graph showing the energy of the molecules as the reaction proceeds:

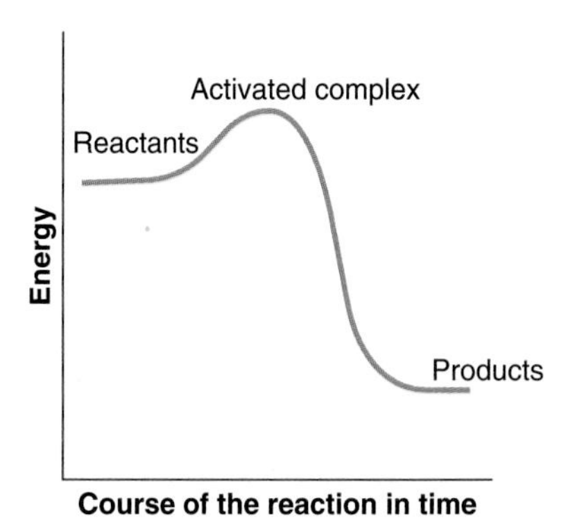

The reactants must have sufficient energy to temporarily form an **activated complex,** or **transition state,** a fleeting combination of atoms that has more energy than either the reactants or the products. The activated complex then changes into the products. The rate of the reaction depends on the rate at which molecules can cross the energy barrier. Thus the molecules in our tables, carpets, and other stable objects aren't reacting with one another largely because they don't have sufficient energy.

One way to increase the rate of a chemical reaction is to raise the temperature. Temperature measures the average *kinetic energy* of the atoms and molecules in any material—how rapidly they are vibrating and moving around. As the temperature increases, the molecules move faster, and more of them will have enough energy to cross the barrier. The temperature-dependence of a reaction is given by its Q_{10}, the factor by which the rate of the reaction increases when the temperature increases 10°C. The Q_{10} of many reactions is approximately 2—that is, their reaction rate doubles with every 10°C increase in temperature.

Some organisms do depend on an increase in temperature to boost their metabolism; for instance, insects, turtles, and snakes sun themselves to get warmer so they can move more quickly. But no organism can raise its metabolic rate high enough to sustain life simply by increasing its temperature. Therefore, metabolism always depends upon catalysts, specifically *enzymes.*

Enzymes and other catalysts accelerate chemical reactions by reducing the energy barrier. When reactant molecules are free in solution, they can only react if they meet in just the right orientation and with enough energy. But an enzyme holds its substrate molecules in the right orientation to react and exerts forces on them that cause the right bonds to break and form. In this way, an enzyme lowers the energy barrier that substrates must pass and increases their reaction rate:

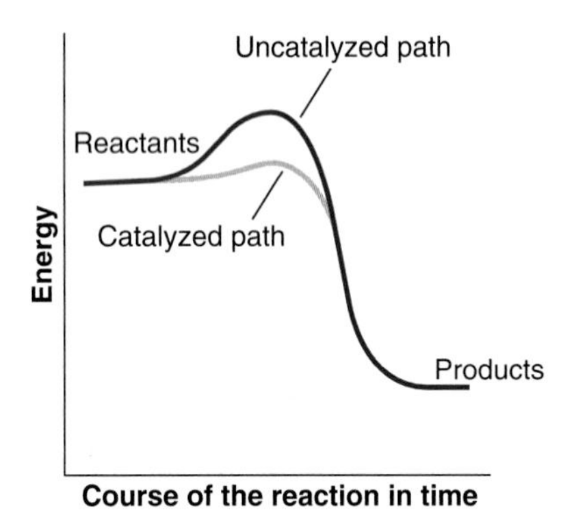

The activity of an enzyme is often expressed by its **turnover number,** the number of substrate molecules transformed per second by each enzyme molecule. To get a feeling for the enormous speed an enzyme can achieve, again consider the reaction of carbon dioxide with water: Inside red blood cells and in a few other places in the body, this reaction is catalyzed by the enzyme carbonic anhydrase, which eliminates the CO_2 produced in metabolism. Without the enzyme, the reaction is very slow, but in its presence, the rate increases enormously. The turnover number for carbonic anhydrase is 4×10^5, which means each enzyme molecule can convert 400,000 molecules of CO_2 and H_2O every second! Most enzymes are slower, with typical turnover numbers of 1,000–10,000, although a few enzymes have turnover numbers of less than one (Table 5.1). Still, the resulting enhancement of reaction rates is enormous.

Enzymes are commonly named by adding the suffix *-ase* to the name of the substrate. A protease, for instance, is an enzyme that digests proteins, while sucrase splits the disaccharide sucrose into two simple sugars. Some names describe the type of reaction the enzyme catalyzes; oxidases oxidize, and ligases ligate (join) two molecules. Still other enzymes, such as the digestive enzymes pepsin and trypsin, were named before this standard terminology was introduced, and they retain their old names. People who think biologically immediately respond to the *-ase* ending; they know it signifies an enzyme and understand that some kind of protein can perform the operation indicated by the enzyme's name.

Exercise 5.1 (True story.) An astronomer relaxing outside his observatory one day noticed a stream of ants moving along the wall and saw that they moved at different speeds in the sun and in the shade. How did he use the ants as a "thermometer" to determine the relative temperatures in these two places?

Exercise 5.2 An undergraduate student who knows some biology but very little chemistry overhears a graduate student talking about "carbamoyl transferase." What should he understand immediately about what the graduate student is discussing?

Exercise 5.3 From the names alone, what can you tell about the functions of the enzymes pyruvate oxidase and DNA polymerase?

Exercise 5.4 A typical biological reaction occurs at the rate of 10 micromoles of product per minute at 20°C. How fast will it occur at 40°C?

Table 5.1 Turnover Numbers of Typical Enzymes

Enzyme	Function	Turnover Number (Per Second)
Lysozyme	Cuts bacterial cell walls	5×10^{-1}
DNA polymerase I	Synthesizes DNA	1.5×10^1
Chymotrypsin	Cuts proteins	1×10^2
Penicillinase	Breaks down penicillin	2×10^3
Carbonic anhydrase	$CO_2 + H_2O \rightarrow H_2CO_3$	4×10^5
Catalase	$2 H_2O_2 \rightarrow 2 H_2O + O_2$	4×10^7

5.2 Chemical reactions come to an equilibrium if left undisturbed.

If carbon dioxide is mixed with water, the molecules will start to react to make carbonic acid. So if we mix a mole of CO_2 and a mole of H_2O, how long will it take for them to be completely converted into carbonic acid? The answer is "forever" because this reaction is reversible. Just as there is a certain chance that the CO_2 and H_2O molecules will collide and react, there is an equally certain chance that a molecule of carbonic acid will rearrange itself to form CO_2 and H_2O again. Consequently, the reaction will never go to completion. Instead, if we wait long enough, the reactants and products will reach a state of **equilibrium,** a point at which the concentrations of the various materials don't change. As we will now show, this is a *dynamic equilibrium,* which means the forward and reverse reactions are still occurring at equilibrium, but they *balance* each other so the overall conditions do not change.

Instructors have devised many analogies to illustrate equilibrium, and we will use the story of the Distressed Babysitter: A boy has been employed to take care of a little girl with a rascally streak. She owns some toy farm animals, and in an effort to keep her busy, the babysitter dumps the animals out on the floor and starts to stand them up. But the child, being roguish, starts to knock the animals down. Suddenly the two of them are in a race, the babysitter trying to stand the animals up as fast as he can and the child trying to knock them down as fast as she can (Figure 5.2). What condition will the animals be in once both youngsters are working as fast as possible?

Two factors will determine how fast each of them can operate. For the babysitter, the factors are (1) how fast he can move around the room to find animals and stand them up, and (2) the concentration of animals (number per square meter) lying down in the room. If he has just dumped the animals onto the floor, most of them will be lying down and he will be able to find them easily, but if most of the animals are standing up at some point, few will be lying down, and it will take him longer to find them. Similarly, the two factors for the child are (1) her speed in moving around the room and knocking animals down and (2) the concentration of upright animals.

Let's denote the rates at which the babysitter and the child can work as k_b and k_c, respectively, and the concentrations of upright and downed animals as [u] and [d], respectively. Then the rate at which the baby-sitter stands animals up will be:

$$\text{Rate}_b = k_b\,[d]$$

and the rate at which the child reverses his work will be:

$$\text{Rate}_c = k_c\,[u]$$

Equilibrium occurs when the two rates are equal—that is, when:

$$\text{Rate}_b = \text{Rate}_c, \text{ or } k_b\,[d] = k_c\,[u]$$

(a)

(b)

(c)

Figure 5.2

(a) A battle between two forces begins when the babysitter sets up a number of animals, initially causing the concentration of standing animals to increase. *(b)* The child begins to reverse the babysitter's work by knocking the animals over. *(c)* The babysitter and the child work at a steady pace, each undoing the other's work. The equilibrium concentrations of animals standing or lying down depend on the relative rates at which the two antagonists can work.

Rearranging this equation, we have:

$$\frac{[d]}{[u]} = \frac{k_c}{k_b} = K_{eq}$$

where K_{eq}, the **equilibrium constant,** is a measure of the relative rates of the babysitter and child. We see that the concentrations of the upright and downed animals reach a certain ratio that is measured by the equilibrium constant. If, for instance, the babysitter and child work at exactly the same rates, K_{eq} will be 1, and as many animals will be standing as are lying down. If the babysitter were able to work twice as fast as the child, twice as many would be standing as lying at equilibrium. But it is important to see that this is a *dynamic* equilibrium: The child and the babysitter are both working furiously, so each animal is being converted rapidly back and forth between the two states. However, the two processes balance each other.

The equilibrium constant of a chemical reaction is defined as a ratio:

$$\frac{\text{Concentration of products}}{\text{Concentration of reactants}}$$

It also predicts the direction of the reaction. If this ratio is greater than 1, the reaction will proceed in the forward direction (to the right as it is written); if it is less than 1, the reaction will go in the reverse direction. We will return to this idea later when we discuss energy in more detail.

One important point to remember is that catalysts, including enzymes, increase the rate of a reaction but they don't change the equilibrium because they increase the forward and reverse rates equally. Enzymes catalyze the reactions that build the macromolecules of an organism's structure, but they also catalyze those reactions just as rapidly in the reverse direction, contributing to the breakdown of biological structure. Enzymes alone therefore cannot carry out the processes of metabolism. Organisms also require a source of energy to drive reactions that have small equilibrium constants in the direction of synthesis. In Chapters 7, 9, and 10, we will discuss how this is done.

Exercise 5.5 The equilibrium constant for chemical reaction 1 is 10^3 and for reaction 2, it is 10^3. What do these numbers say about the relative concentrations of reactants and products in each case?

Exercise 5.6 Which of these statements about the distressed babysitter analogy is/are true . . . and why?

a. The babysitter is like an enzyme.
b. The child is like an enzyme.
c. The babysitter and child working together are like an enzyme.

Exercise 5.7 Suppose you have a container in which CO_2, water, and carbonic acid are all at their equilibrium concentrations, and you add a small amount of radioactive CO_2. Which of these statements is correct?

a. All the radioactivity will remain with the CO_2, because at equilibrium no more CO_2 can react with water.
b. Some radioactive carbonic acid will soon appear, because the chemical reaction is still taking place.

5.3 An enzyme's structure enables it to catalyze a specific reaction.

Biological processes are very specific. Each enzyme catalyzes only a single reaction with one substrate or sometimes the same kind of reaction with very similar substrates. Each biomolecule, such as an enzyme, has a specific three-dimensional shape and only interacts with other molecules that have suitably complementary shapes. Thus, carbonic anhydrase doesn't digest protein molecules, and a protease doesn't change the rate of carbonic acid formation.

We saw in Chapter 4 that an enzyme is so specific because its substrate fits into the *active site* somewhat like a key fits into a lock. The active site is where the chemical reaction actually takes place. Forces of attraction and repulsion between the atoms of the substrate and the enzyme distort some bonds of the substrate, lowering the reaction energy barrier. As an example, Figure 5.3 shows the interaction between lysozyme and its substrate. Lysozyme is an enzyme in tears, mucus, and egg white that cuts bonds in bacterial cell walls, destroying the bacteria. (It was discovered by Alexander Fleming, the microbiologist who later discovered penicillin. Fleming had a cold one day, and he let some of the mucus from his nose drip onto a culture dish with growing bacteria; lysozyme in the mucus destroyed them.) The substrate interacts specifically with the side chains of amino acid residues that form the active site. Some of these side chains enter into the reaction and are slightly changed while the reaction is occurring; they are restored when the reaction is done, so the enzyme is ready for the next substrate molecule.

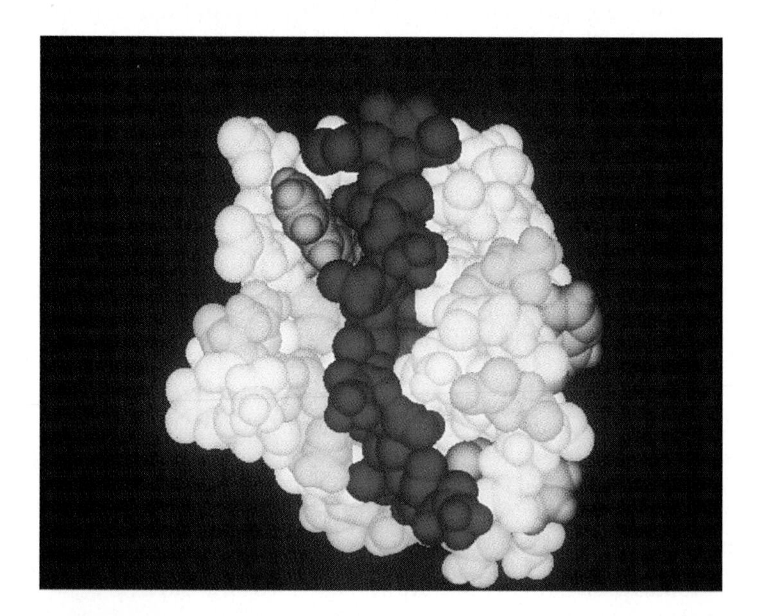

Figure 5.3

The enzyme lysozyme binds to its substrate, a chain of sugar molecules *(red)* in the bacterial cell wall, at its active site, a groove formed by the atoms of the protein (*white, yellow,* and *blue*). Then it catalyzes a chemical reaction in which the sugar chain is cleaved in the middle.

5.4 Michaelis-Menten kinetics shows that an enzyme has a limited number of active sites.

In 1913, Leonor Michaelis and Maude Menten developed a way to analyze enzyme-catalyzed reactions that is still used today. Using a constant amount of enzyme, they studied the kinetics of the reaction—that is, how the initial rate (velocity) of a reaction changes as the starting concentration of substrate, [S], increases (Figure 5.4). Plotting the initial reaction rate against [S] produces a curve characteristic of each enzyme. As [S] increases, the rate increases rapidly at first, but eventually levels off and doesn't increase further, no matter how much additional substrate is added.

Such curves tell a lot about enzymes and their function. First, the curve supports the view that each enzyme molecule usually has only one active site that can interact with substrate molecules (Figure 5.5). This is shown by the fact that the reaction rate levels off at a maximum rate, V_{max}. When all the sites are occupied, the enzymes are operating at maximum capacity, and we say the enzymes are *saturated.* Remember that each curve represents the rate for a constant amount of enzyme; adding more enzyme makes additional active sites available and leads to a new, higher V_{max}. As an analogy to this situation, imagine a yard where many spiders have spun webs to trap flies; each spider can only trap and consume one fly at a time. Now consider the rate of fly consumption as the number of flies increases. At first, the spiders will be able to handle the increasing numbers, consuming flies faster and faster. But eventually every spider will be busy all the time trapping and consuming flies. The spiders will be saturated with flies and will not be able to consume them any faster, no matter how many flies appear.

A second point is that every enzyme can be characterized by a **Michaelis constant,** K_m, defined as the *substrate concentration* at which the reaction rate is half of V_{max}.

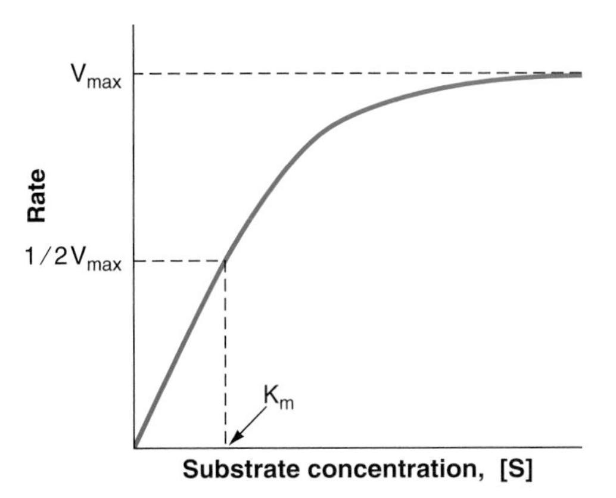

K_m measures the tightness of binding, or **affinity,** between enzyme and substrate; a high K_m indicates loose binding (low affinity), and a low K_m indicates tight binding (high affinity). If an enzyme has a high affinity for its substrate, it doesn't take much substrate to make the reaction proceed rapidly.

The same kind of kinetic analysis has been done with other proteins, such as the transport proteins that move ions and small molecules through membranes (Chapter 8). These proteins aren't enzymes, since they don't catalyze chemical reactions. Nevertheless, most of them show the same kinetics as enzymes, indicating that they too interact with ions or molecules at specific binding sites. Thus the Michaelis-Menten model pertains in general to many kinds of proteins.

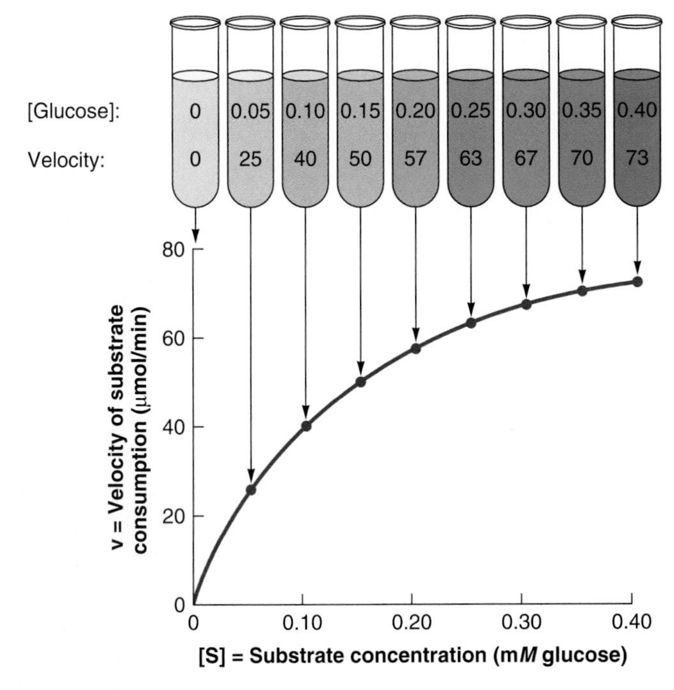

Figure 5.4

In an experiment devised to study the kinetics of an enzymatic reaction, each tube contains the same amount of enzyme but a different amount of substrate (glucose). The experimenter measures the rate at which substrate is removed or product is formed, and then plots the rate of the reaction against the initial substrate concentration. From *The World of the Cell* by W. M. Becker. Copyright © 1986 by The Benjamin/Cummings Publishing Company. Reprinted by permission.

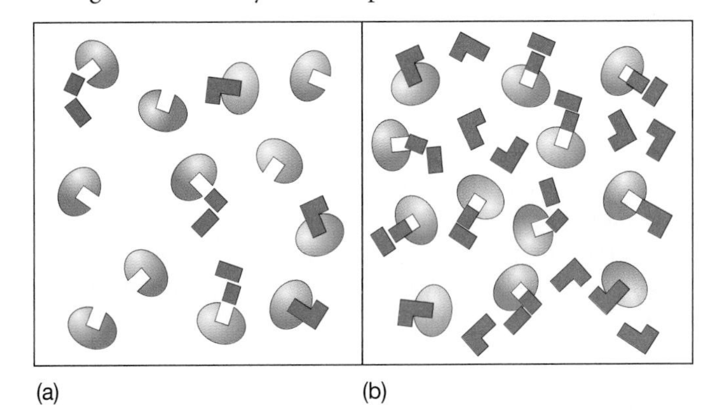

Figure 5.5

(a) Each enzyme molecule has an active site where it can bind a substrate molecule and transform it. As the concentration of substrate increases, the rate of the reaction increases because it is easier for an enzyme molecule to encounter a substrate molecule and bind to it. *(b)* At high substrate concentrations, however, all the active sites are bound to substrate, and the enzymes are saturated. Each enzyme is working as fast as it can, and the reaction cannot go faster, no matter how many substrate molecules there are.

5.5 Temperature affects the rate of enzyme-catalyzed reactions.

Every enzyme has an optimal temperature, a temperature at which it is most active. Below the optimum, raising the temperature increases enzyme activity by increasing the kinetic energy of molecules. The rate of most chemical reactions roughly doubles for every increase of 10°C. However, an enzyme must have a specific three-dimensional structure to function properly, and that structure is temperature-sensitive. As the temperature is raised above the optimum, enzymes become denatured, and their active sites can no longer bind substrates. Each enzyme's optimal temperature therefore represents a balance between increasing kinetic energy and protein denaturation.

It isn't surprising that organisms have evolved with enzymes whose optimal temperatures are suited for the conditions they live in. This is probably the simplest illustration of the principle that biomolecules are shaped by evolution to function in a specific ecological niche. Organisms that live at extreme temperatures, such as microorganisms in hot springs and fish in icy lakes, have enzymes capable of functioning at these extremes; those able to live at high temperatures have enzymes with amino acids that form more heat-stable bonds. Organisms whose body temperatures vary widely have enzymes that function well over a broad temperature range or enzymes with different optima that cover the whole temperature range:

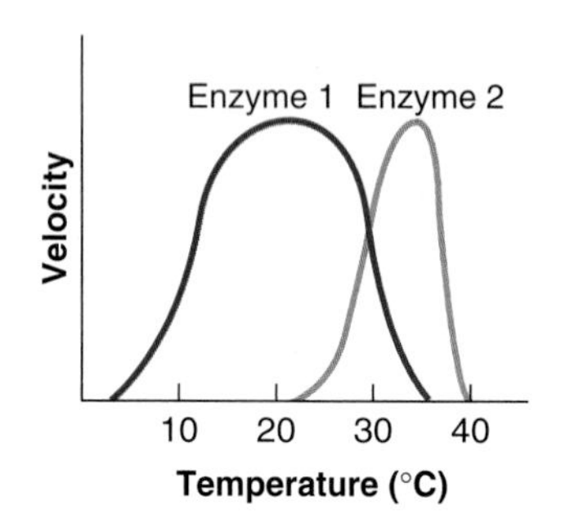

The temperatures at which similar species of bacteria live are closely correlated with the temperatures at which their enzymes are denatured (Table 5.2).

Human body temperature stays close to 37°C, and that is also the optimal temperature of most human enzymes. However, the enzymes that function in sperm production in human males operate best at a temperature slightly below 37°C; these enzymes occur in the testes, which are suspended outside the abdominal cavity in the scrotum where the body temperature is lower. In rare instances in which the testes fail to descend into the scrotum during development, their enzymes don't produce fertile sperm.

5.6 Each enzyme has an optimal pH.

The pH of the solution around an enzyme also affects its structure and activity, because hydrogen ions interact with the polar groups on proteins and often take part in the reactions they catalyze. The rate of enzyme-catalyzed reactions therefore depends on pH, and each enzyme has its own optimal pH. At extreme pH values, enzymes are generally denatured just as they are by high temperatures. Thus an organism's enzymes have been shaped by evolution to operate at a certain pH, as well as at a certain temperature. In humans, blood enzymes, such as carbonic anhydrase, have an optimum pH of 7.4, the pH of blood, whereas pepsin, the protease secreted by the stomach, has an optimum around 2, the pH of the stomach while digestion is taking place:

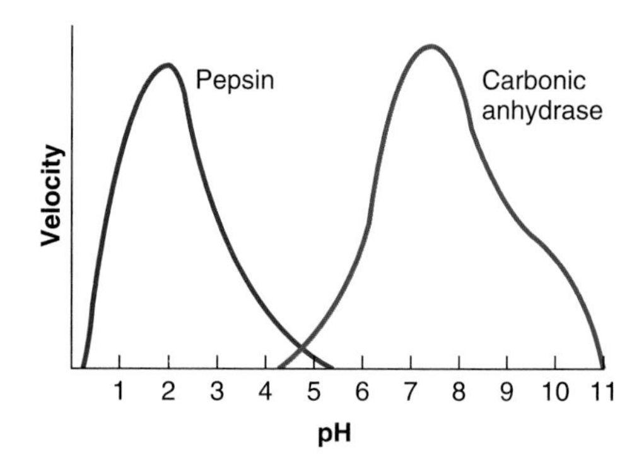

5.7 Some enzymes require cofactors and coenzymes to function.

The concentration of ions such as sodium, potassium, magnesium, and chloride may affect the rate of enzyme-catalyzed reactions because these ions interact with polar groups on an enzyme and subtly change its properties. Curves relating reaction rates to the concentration of these ions resemble the curves

Table 5.2 Temperature Effects on Bacteria and Their Enzymes

	Maximum Growth	Enzyme Denaturation Temperature (°C)		
Species	*Temp. (°C)*	*Indophenol Oxidase*	*Catalase*	*Succinic Dehydrogenase*
Bacillus mycoides	40	41	41	40
B. megatherium	46	48	50	47
B. alvei	46	51	50	53
B. vulgatus	55	56	56	50

relating rates to temperature or pH. For instance, an enzyme that carries out a complex reaction on DNA depends on magnesium ions:

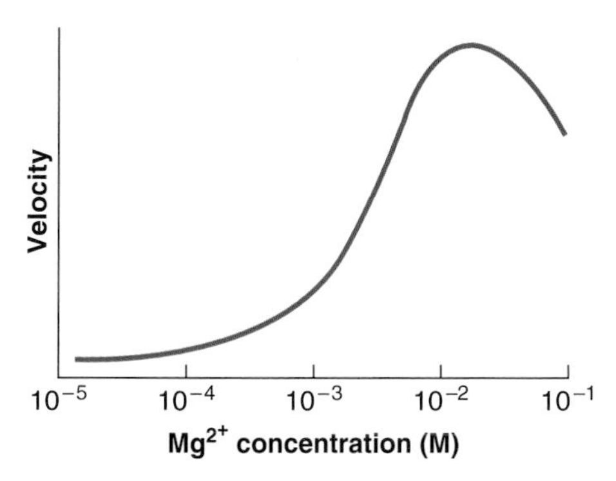

Several metal ions—including copper, zinc, nickel, and manganese—act as **enzyme cofactors,** being bound to the active site of the enzyme, where they participate in the chemical reaction. The ions are not interchangeable; an enzyme that uses such a cofactor requires a particular element, and without it the enzyme simply won't function. This is why these metals are among the trace elements commonly found in organisms and why they are essential for good health. Both plants and animals show distinct disease symptoms if they live in soil or eat food that is deficient in an essential element.

Some enzymes also function only with **coenzymes,** small, nonprotein organic molecules that are needed to make an enzyme complete:

$$\text{Inactive enzyme} + \text{Coenzyme} \rightarrow \text{Active enzyme}$$

Many metabolic reactions involve adding a small group of atoms, such as a methyl group ($-CH_3$), to a substrate or removing it from a substrate. The coenzyme participates in the reaction by carrying the group in or removing it. Many coenzymes are derived from **vitamins.** Vitamins are organic compounds that an organism requires in small amounts but cannot make for itself. They are a diverse group, best obtained by eating a variety of foods, and it is clearly important for one's diet to be varied enough to supply all of them. **Prosthetic groups,** such as the heme molecules bound to each polypeptide chain of hemoglobin, are similar to coenzymes but remain attached to the protein. As we discuss metabolism in later chapters, we will encounter several examples of coenzymes.

Exercise 5.8 Years ago, some people with backgrounds in chemistry and physics began doing biological research and tried to grow bacteria in media (mixtures of nutrients) according to laboratory recipes, using distilled water. The bacteria would not grow for them, although other people, familiar with growing bacteria, had no trouble using these recipes. Eventually, the newcomers discovered with dismay that the more experienced scientists had simply gotten their water from the faucet. Why did tap water work for media, but pure water did not?

B. General protein function

5.8 Studies of enzyme inhibition help explain protein action and emphasize the importance of weak bonds between molecules.

Enzymes are inhibited by many compounds, and studies of the way they are inhibited help us understand how they normally function. In particular, these studies emphasize the critical point that proteins typically interact with other molecules through weak bonds.

Many enzymes are subject to **competitive inhibition.** A competitive inhibitor is so similar in structure to the enzyme's normal substrate that it can occupy the active site and block entry by the substrate. Thus it *competes* with the substrate for entrance into a limited number of active sites. The competitor may be an alternate substrate or a molecule that is unable to enter into a chemical reaction as a substrate does. For instance, folic acid is an important coenzyme that carries one-carbon units such as methyl groups from one compound to another. Bacteria can synthesize folic acid, but humans cannot, and this difference is exploited in antibiotic therapy. One component of folic acid is para-aminobenzoic acid (PABA), and the antibiotic sulfanilamide is so similar to PABA that it acts as a competitive inhibitor of a key enzyme in folic acid synthesis, thus inhibiting bacterial growth:

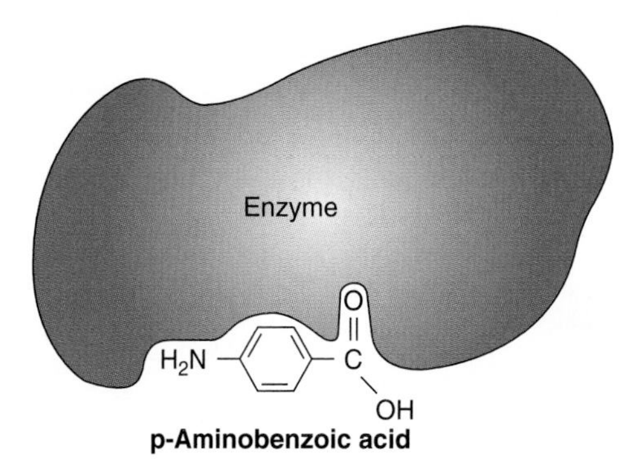

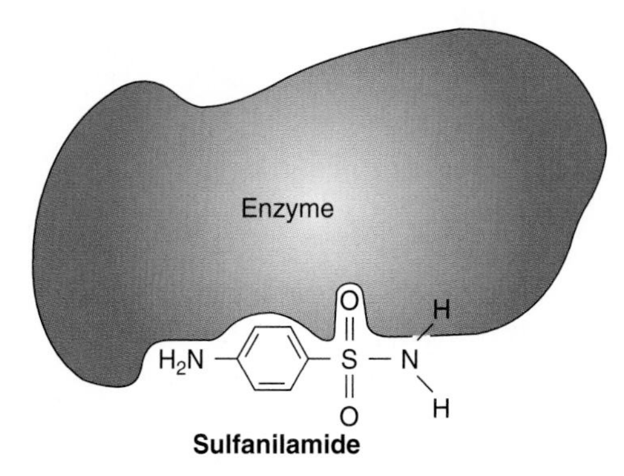

The antibiotic has virtually no effect on the metabolism of human cells, which use ready-made folic acid—one of the vitamins—from food.

Studies of enzyme kinetics make it clear that both the substrate and the inhibitor bind to amino acid side chains in the active site through weak interactions, such as hydrogen bonds and electrostatic attractions, which are easily broken. Enzyme molecules (E) and substrate molecules (S) are involved in a dynamic equilibrium; they combine to make a complex (ES), which can either go back to separate E and S molecules or forward to produce the enzyme plus the product (P):

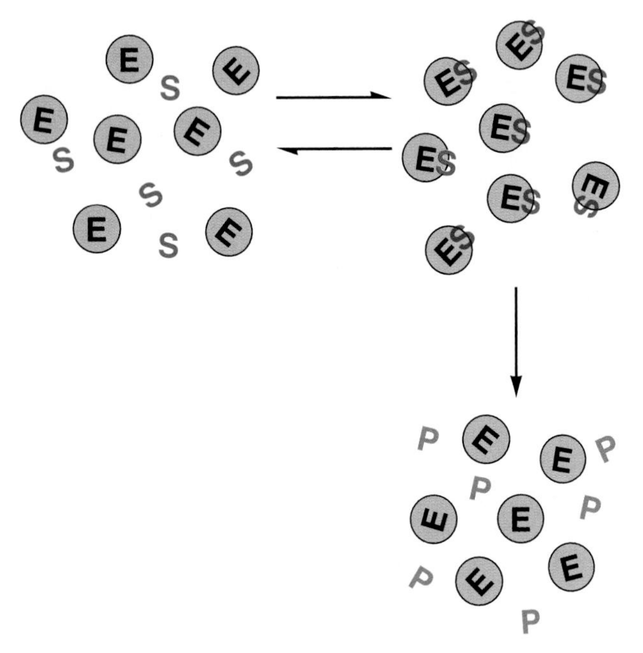

Similarly, if a competitive inhibitor (I) is added, it too forms only weak bonds with the active site, and there will be an equilibrium between inhibitors bound to the enzyme and those not bound, with competition between S and I molecules:

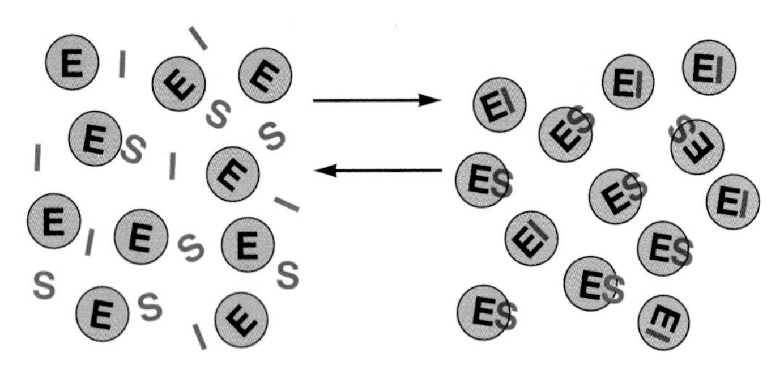

Which molecule enters a site depends on their relative concentrations; the one in higher concentration is more likely to bind to the site. Because the molecules are continually binding and dissociating, the inhibition can be overcome by raising the concentration of substrate, so substrate molecules will tend to displace inhibitor molecules from the active sites.

This shifting equilibrium is of enormous importance in biology. Throughout this book, we will emphasize that biological processes occur through biomolecular interactions, particularly those in which proteins bind to other molecules. It is important to develop a certain view of such interactions. Whenever we picture molecules binding to one another, we should visualize them in equilibrium between the bound and unbound states:

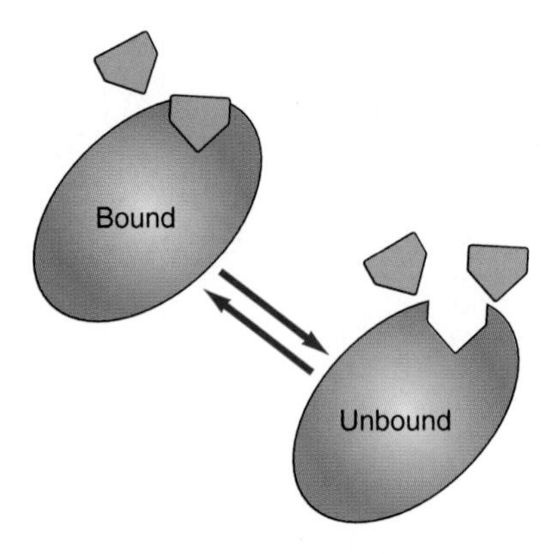

The medical profession often uses competitive inhibition to treat a certain type of alcohol poisoning. Alcoholics may resort to cheap and readily available substitutes for the grain alcohol (ethanol) found in most alcoholic beverages. Sometimes they use ethylene glycol, an unusual alcohol that is a component of some antifreezes. Ethylene glycol itself is not extremely toxic, but it is transformed into the poisonous compound oxalic acid by the action of two enzymes, one of which, alcohol dehydrogenase, is a rather nonspecific enzyme that can use both ethylene glycol and ethanol as substrates. The treatment for ethylene glycol poisoning is simply to administer large quantities of ethanol. Ethanol displaces ethylene glycol from the active sites of the dehydrogenases and thus prevents formation of the very toxic oxalic acid. The kidneys eliminate unchanged ethylene glycol from the body rather quickly. If all goes well, the patient merely wakes up with a painful hangover, having avoided fatal poisoning.

Enzymes can also be affected by *irreversible inhibitors,* which form a covalent bond with part of the enzyme's active site. This permanently destroys the enzyme's activity, because any method of removing the inhibitor will also break covalent bonds elsewhere in the protein. The action of poison gas, which killed or maimed thousands during World War I, was a classic example of irreversible inhibition (Figure 5.6). Nerve impulses are transmitted between nerve cells and muscle cells by a compound called acetylcholine, and excess acetylcholine must be removed by the enzyme acetylcholinesterase. The poison gas diisopropyl-phosphofluoridate (DIPF) forms a covalent bond with the amino acid serine in the active site, thus blocking access of acetylcholine to the active site:

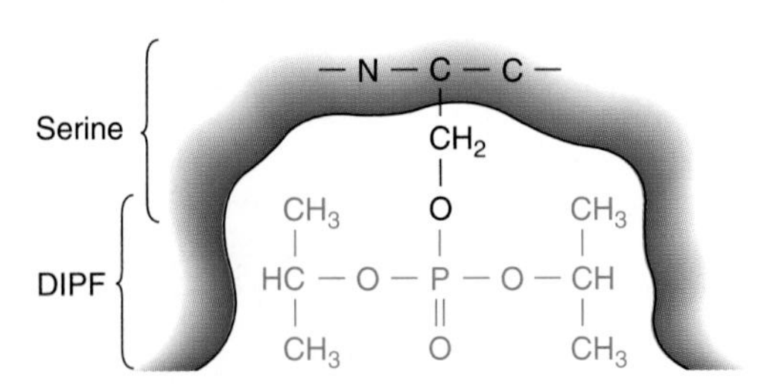

Figure 5.6
Many soldiers during World War I wore gas masks as protection against the poison gas diisopropyl-phosphofluoridate, which was used as a weapon.

When the enzyme is inhibited, the skeletal muscles go into a state of contraction, and death results from asphyxiation because the muscles needed for breathing are paralyzed.

Exercise 5.9 (Fairy tale.) Green flies are attracted to the leaves of the kiukiu plant, where they sit and suck the plant juices. Each leaf is only large enough for one fly. Blue flies are also attracted to the plant, and they compete with green flies.

a. If a plant has 100 leaves, describe its saturation as more and more green flies light on it.
b. Do you expect any one leaf to be constantly occupied by flies? What would you expect to observe if you watched one particular leaf for a while?
c. If a plant is saturated with green flies and a swarm of blue flies comes along, describe how the plant will change with varying numbers of green and blue flies.

Exercise 5.10 Valeric acid, which has a chain of five carbon atoms, is the substrate of a certain enzyme. The similar compound caproic acid has six carbon atoms and is not a substrate. However, caproic acid interferes with the action of the enzyme on valeric acid; the reaction rate decreases as caproic acid is added and increases as it is removed. What do you conclude about the interaction of caproic acid with the enzyme?

5.9 Hemoglobin illustrates cooperative effects between the subunits of a protein.

The proteins hemoglobin and myoglobin have provided significant insights into protein function as well as protein structure. We have already noted that hemoglobin carries oxygen from an animal's lungs or gills to all its tissues; that it consists of four globular polypeptide chains, two α and two β, with known amino acid sequences; and that each chain is bound to an iron-containing heme group. Myoglobin consists of a single heme group and a single polypeptide chain, whose amino acid sequence is similar to that of hemoglobin's chains. Here we are interested in how hemoglobin and myoglobin function as they carry oxygen. To make sense of this discussion, it is essential to have in mind the picture we have just developed of a protein interacting through weak bonds with other molecules, which continually bind and unbind.

To investigate the way oxygen binds to hemoglobin and myoglobin, we do an experiment very similar to the one used to study enzyme kinetics (see Figure 5.4). The experimental results are displayed as a **saturation curve** by plotting the fraction of protein molecules bound to O_2 against the concentration of O_2, which is generally given as the partial pressure of oxygen (Concepts 5.1). A saturation curve for myoglobin (Figure 5.7) resembles the curve for the rate of an enzymatically catalyzed reaction with Michaelis-Menten kinetics. This result is very enlightening, because it shows that the same principle underlies the action of both proteins: Myoglobin has a single binding site for oxygen, just as an enzyme has a single active site for binding its substrate. The curve shows that more binding sites are occupied at increasing concentrations of oxygen, until all the molecules are saturated.

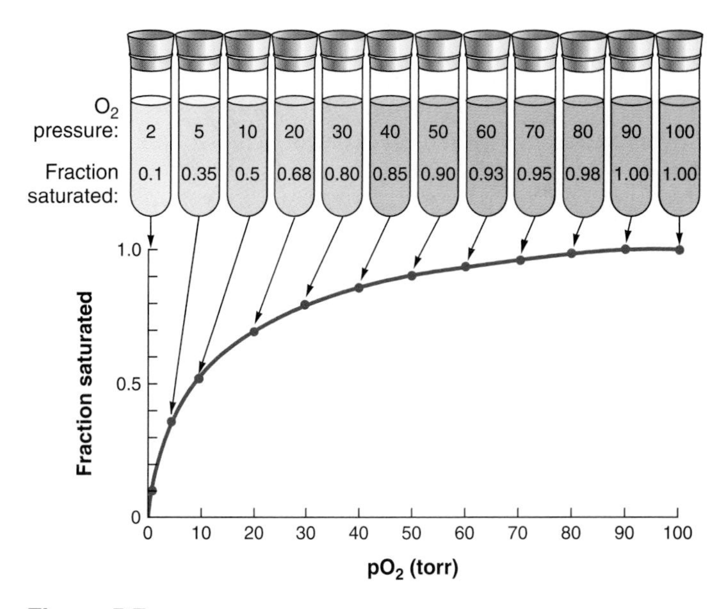

Figure 5.7
In an analysis of the oxygen saturation of myoglobin, each tube contains the same myoglobin solution, but the pressure of oxygen over each solution is different; the higher the pressure, the higher the concentration of oxygen in solution. The color of the protein changes when it is bound to oxygen, so we can determine how much of the myoglobin is bound or unbound in each tube with an appropriate optical instrument.

Concepts 5.1 Pressure

Pressure is the force per unit of area that the force is exerted on. When you blow up a balloon, the pressure inside increases because more molecules are being forced into a confined space, and each of them exerts a force when it strikes the walls of the balloon. Pressure also changes if the area to which the force is applied changes. For example, when a 50-kg (110-lb) woman puts her weight on one tennis shoe (with an area of about 200 cm^2), the pressure under her foot is a modest 0.25 kg/cm^2. But each of her steps on the narrow heel of a dress shoe (with an area of about 5 cm^2) exerts a pressure of 10 kg/cm^2.

We can apply this simplified view of pressure to the earth's atmosphere: We live at the bottom of an ocean of air that exerts a pressure of about 1 kg/cm^2 (14.7 lbs/in^2) on everything. Around 1643, Evangelista Torricelli demonstrated that the atmosphere can support a column of fluid, as in this diagram:

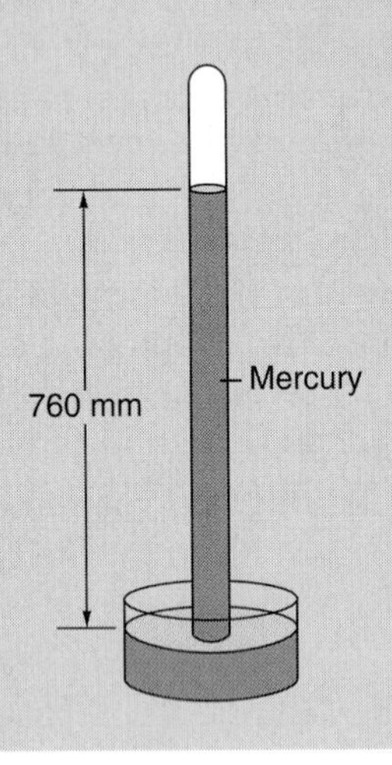

Here the tube has been filled with mercury and then inverted in a dish of mercury, leaving a vacuum at the top of the tube. This is a simple barometer, since the level in the tube rises and falls with atmospheric pressure. A pressure of 1 atmosphere by definition will hold up a column of 760 mm of mercury (or about 30 inches—the number usually used by meteorologists). A commonly used measure of pressure is a **torr,** which is 1/760 atmosphere.

In a mixture of gases, such as the earth's atmosphere, each gas exerts a pressure, called *partial pressure,* independently of the others. Thus, since about one-fifth of the atmosphere is oxygen, the partial pressure of atmospheric oxygen at sea level is about 150 torr. This is important biologically because every gas dissolves in a liquid in proportion to the pressure it exerts. The amount of O_2 or CO_2 dissolved in the blood is proportional to the pressure of that gas on the blood. Thus oxygen dissolves in blood in the lungs, where the oxygen pressure is high, and it comes out of solution in the tissues, where oxygen pressure is low. At sea level, the oxygen pressure is about 100 torr in a human lung and about 25 torr in the small blood vessels of the body, the capillaries. In muscles and other active tissues, the concentration of O_2 is very low, about 1–2 torr.

In contrast, the saturation curve for hemoglobin is sigmoid (S-shaped):

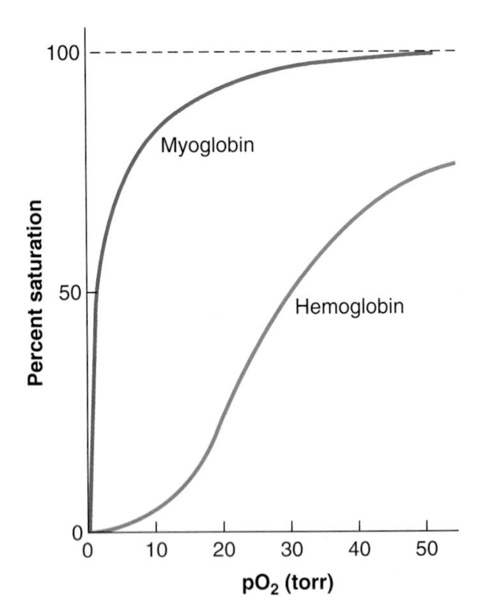

This curve shows that at relatively low concentrations of oxygen, where most molecules of myoglobin are already saturated with O_2, just a small fraction of hemoglobin molecules are saturated. Hemoglobin only becomes saturated at a higher concentration of oxygen than is required to saturate myoglobin. These curves indicate that myoglobin binds O_2 more tightly than hemoglobin does, which is to say that myoglobin has more *affinity* for O_2 than hemoglobin has. For a mammal's body to operate successfully (as it has evolved), myoglobin must have more affinity for oxygen than hemoglobin has. Myoglobin's greater affinity enables it to accept O_2 from hemoglobin and store it briefly before releasing it for use in metabolism:

O_2 in lungs → O_2 bound to hemoglobin → O_2 bound to myoglobin → O_2 used in metabolism

The sigmoid shape of the saturation curve of hemoglobin is the result of a **cooperative effect** between the protein's subunits. Each hemoglobin subunit can switch between two distinct conformations: a B conformation in which O_2 is *bound* to the heme and an N conformation in which O_2 is *not bound:*

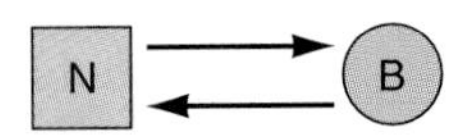

When an oxygen molecule binds to a heme group of one protein subunit, that subunit flips from the N shape into the B

shape. First, picture a molecule with all four units in the N conformation, without oxygen:

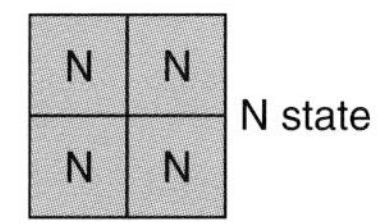

When the molecule encounters oxygen, one subunit will bind an O_2 molecule and switch into its B conformation:

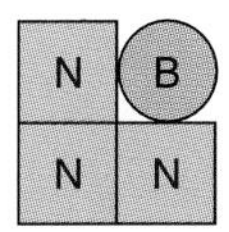

Because of the tight linkage among the four subunits, the shape of one subunit affects the shape of the subunits it touches. So the first subunit to change its shape by binding O_2 causes the other three subunits, in turn, to switch into their B conformation, where they bind O_2 more readily:

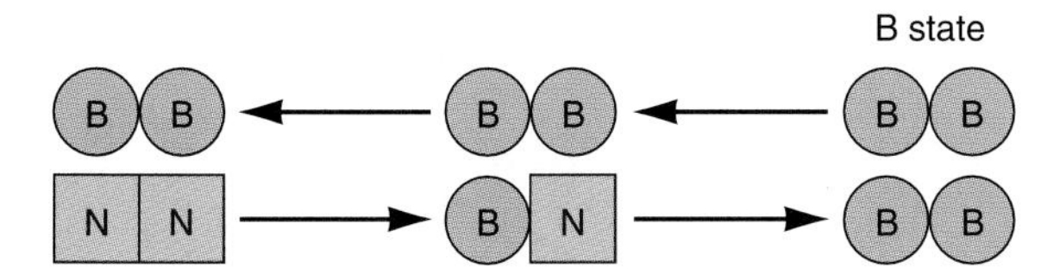

Binding thus becomes progressively easier until oxygen is bound to all four subunits. By contrast, myoglobin does not show a cooperative effect because it has only one polypeptide chain and a single heme group that can bind O_2.

Cooperativity makes hemoglobin an ideal molecule for carrying oxygen in the blood. In the lungs or gills, where the concentration of oxygen is high, all the hemoglobin molecules quickly become saturated with O_2. When these molecules are transported by the blood to the body tissues, where the concentration of O_2 is very low, they begin to unload O_2. As soon as one subunit releases its oxygen, the cooperative effect works in reverse and the other units unload theirs:

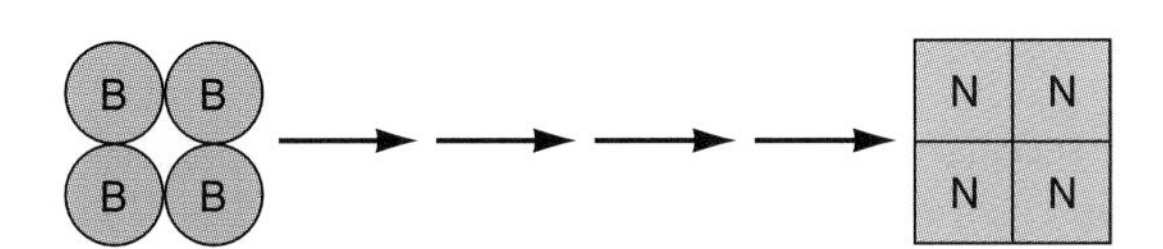

By comparing the saturation curves for hemoglobin and myoglobin, you can see that where the oxygen concentration is low, as it is in active tissues, hemoglobin releases much more of its oxygen than myoglobin does, so oxygen is transferred from hemoglobin to myoglobin.

Exercise 5.11 Suppose you isolate hemoglobin from two closely related mammals, one that lives at sea level and one that lives high in the mountains. Predict the relative shapes of the saturation curves for their hemoglobins.

Exercise 5.12 The hemoglobin in adult humans is mostly hemoglobin A (Hb A), but the hemoglobin in a fetus is a slightly different protein, hemoglobin F. Predict which hemoglobin, A or F, has a higher affinity for oxygen, and explain why this difference is functional.

5.10 The binding of proteins to other molecules is fundamental to biological processes.

This discussion of the operation of enzymes and of hemoglobin and myoglobin leads us to some general ideas about the functions of proteins and also to one of the most important principles of biology. Consider the features of an enzyme and how these features can be applied to proteins in general.

1. An enzyme interacts with another molecule, its substrate. In general, biological processes occur because a protein interacts with a molecule called a **ligand,** which is most often a small molecule but could even be a macromolecule. Oxygen is the principal ligand with which hemoglobin and myoglobin interact.
2. An enzyme-substrate interaction takes place in a specific place on the enzyme, its active site. In general, the site on a protein where it binds to a ligand is a **binding site:**

   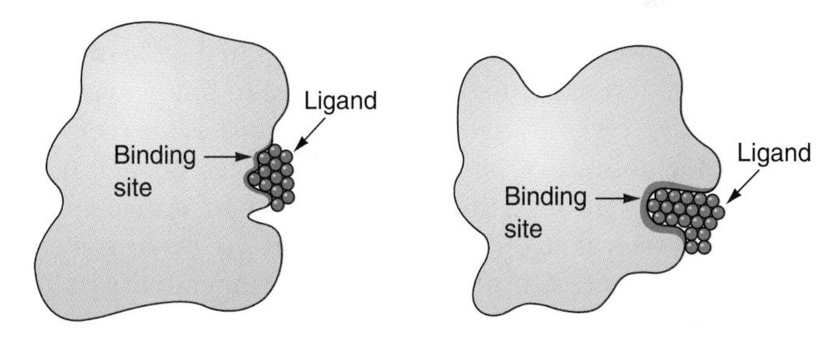

   In hemoglobin and myoglobin, the oxygen-binding sites are at the heme groups.
3. An enzyme only binds a specific molecule with a structure that just fits its active site. A competitive inhibitor interacts with an enzyme because it is very similar in structure to the substrate. In general, a protein only interacts with a ligand that specifically fits into its binding site. This interaction between molecules with shapes that fit together is called shape-specificity or **stereospecificity** (*stere-* = solid). It is similar to the interaction between interlocking jigsaw puzzle pieces, and molecules that interact in this way are said to be *complementary.*
4. The interaction between an enzyme and its substrate has a biological consequence: The substrate is converted into a product. In general, the interaction between a protein and its ligand has some other biological consequence, such as:
   - The protein simply holds the ligand and carries it for a while, as in the globins.
   - The protein transports the ligand across a cell membrane.
   - The protein prevents the ligand from having some other effect.
   - The ligand changes the protein's shape, thus either inhibiting the protein or activating it.

This list of possible interactions between a protein and a ligand is incomplete. We will expand it throughout this book. It is no exaggeration to say that *every biological process depends upon some interaction between a protein and a specific ligand.* This principle is one of the great generalizations of biology and is absolutely essential for understanding how organisms operate.

5.11 Many proteins, including hemoglobin, show allosteric effects.

The cooperativity exhibited by hemoglobin is one of a large class of actions called **allosteric effects.** An allosteric protein is one that has two or more binding sites with quite different shapes that can bind ligands with correspondingly different shapes (*allo-* = different; *-stere* = solid):

The concept of allostery is nicely illustrated by hemoglobin. The function of hemoglobin is not just to bind oxygen—it is to *release* oxygen to the tissues that require it for metabolism. Hemoglobin must be able to bind oxygen efficiently as the blood passes through the lungs or gills of an animal, but it would be useless if it did not release that O_2 while passing through the other tissues of the body. The cooperativity of hemoglobin makes it release most of its oxygen, but this is accentuated by at least two other molecules. One is a small molecule, 2,3-bisphosphoglyceric acid (BPG), which is produced by metabolizing cells:

```
 O    OH
  \\ /
   C          O
   |          ||
H — C — O — P — O⊖
   |          |
H — C — H     O⊖
   |
   O
   |
O = P — O⊖
   ||
   O⊖
```

2,3 bisphosphoglyceric acid (BPG)

Red blood cells contain about equimolar amounts of hemoglobin and BPG, and each hemoglobin molecule has a site between the two β chains where BPG binds (Figure 5.8). The binding of BPG shifts the O_2-binding curve of hemoglobin so the protein gives up its oxygen even more readily at low O_2 pressure. The significance of the BPG effect becomes most apparent at high elevations; within a few hours after moving to a high altitude, a normal person's BPG level in the red blood cells begins to rise,

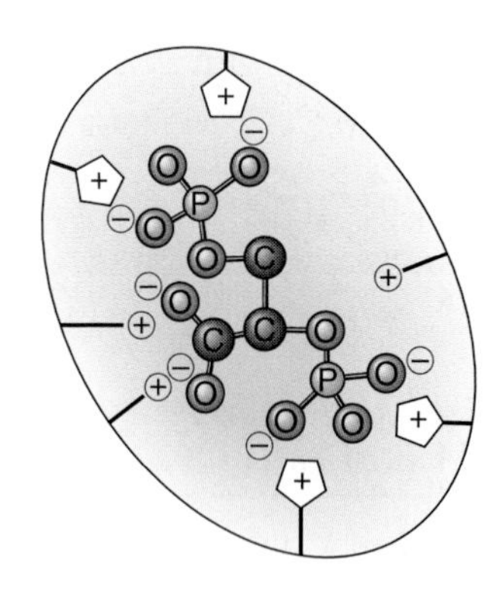

Figure 5.8

The hemoglobin molecule has a regulatory site formed by seven positively charged amino acid side chains. They form a pocket where BPG, with its corresponding negatively charged groups, can bind. When BPG occupies this site, it changes the interactions between the protein subunits.

so the hemoglobin has less affinity for oxygen and can more effectively release it where it is needed. Saturation curves show the dramatic effects of adding or removing BPG:

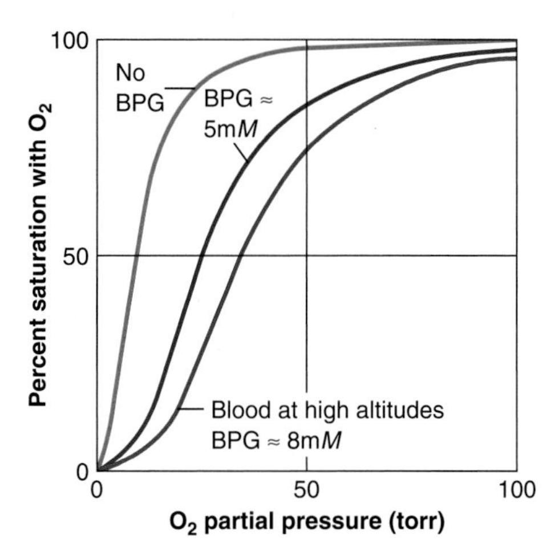

The place where BPG binds is called a **regulatory site** because it regulates the operation of hemoglobin. BPG acts as an *allosteric effector,* a ligand that changes the activity of an allosteric protein.

The second molecule that affects the conformation of hemoglobin dramatically is CO_2, although it has no binding site such as the BPG binding site or the active site of an enzyme. CO_2 is produced abundantly in the metabolizing tissues of the body. As the CO_2 concentration rises in the tissues and the blood passing through them, some CO_2 molecules bind to the amino-terminal residue of the hemoglobin:

$$R-NH_2 + CO_2 \longrightarrow R-N(H)-C(=O)-O^{\ominus} + H^{\oplus}$$

This subtle change is enough to shift the hemoglobin toward its N conformation, thus releasing more oxygen. O_2 and CO_2 have opposite effects on the hemoglobin structure; in the lungs or gills, where the oxygen concentration rises, oxygen tends to bind to the hemoglobin and shift it back into the B conformation, where CO_2 is released. Thus hemoglobin does double duty, transporting both gases, and by shifting from one conformation to the other, it does so very effectively.

5.12 A small change in protein structure can have profound functional consequences.

Hemoglobin has evolved into its present form because this structure is excellent for carrying O_2 and CO_2 in the blood. Even a small change in the basic structure of the protein can have enormous consequences. Many people, however, have mutant hemoglobins due to an accidental change (a mutation) in the genetic information that specifies the amino acid sequence of one hemoglobin subunit. In most cases, a single amino acid has simply been replaced by another. Some of these people have associated health problems, such as anemia, and among the most interesting and socially significant of these problems is *sickle-cell anemia.* Instead of normal adult hemoglobin A (Hb A), people with this condition have Hb S, in which the glutamic acid at position 6 in the β chain has been replaced by valine:

Val-His-Leu-Thr-Pro-Glu-Glu-Lys-Ser-Ala-Val-Thr-Ala-Leu-Trp-

↓

Val-His-Leu-Thr-Pro-Val-Glu-Lys-Ser-Ala-Val-Thr-Ala-Leu-Trp-

That slight change has profound effects. Humans, like most other animals and plants, have two copies of each gene, one inherited from each parent. Most people have two copies of the normal gene symbolized by Hb^A and produce only normal hemoglobin, Hb A. A small percentage of people have one Hb^A gene and one for hemoglobin S, Hb^S. These individuals have both kinds of hemoglobin in their red blood cells and are essentially healthy, but they are *carriers* who are able to transmit the mutant gene to their offspring. An even smaller percentage of people have two copies of the Hb^S, so they produce only Hb S and become very sick. Under conditions such as reduced oxygen concentration, cells containing Hb S change into elongated "sickled" forms (Figure 5.9*a*) because the protein crystallizes into long fibers when it becomes deoxygenated, rather than undergoing a subtle change in shape like Hb A. These sickled cells clog small blood vessels and cut off the oxygen supply to nearby tissues. Sickled cells are also destroyed more rapidly than normal red blood cells, leading to anemia. Modern medical treatments can help relieve the symptoms of sickle-cell anemia, but without treatment the condition can cause fever, dizziness, pain, pneumonia, rheumatism, and heart and kidney disease, generally ending with death at an early age.

Sickle-cell anemia is relatively common in Africa, southern Europe, and other malaria-ridden areas, because people who have one Hb^A gene and one Hb^S gene happen to be unusually resistant to malaria. The malaria parasite (a protist carried by mosquitoes), enters their red blood cells and begins to reproduce. Those cells then release more parasites, which infect vast numbers of other red blood cells. In this way, the malaria

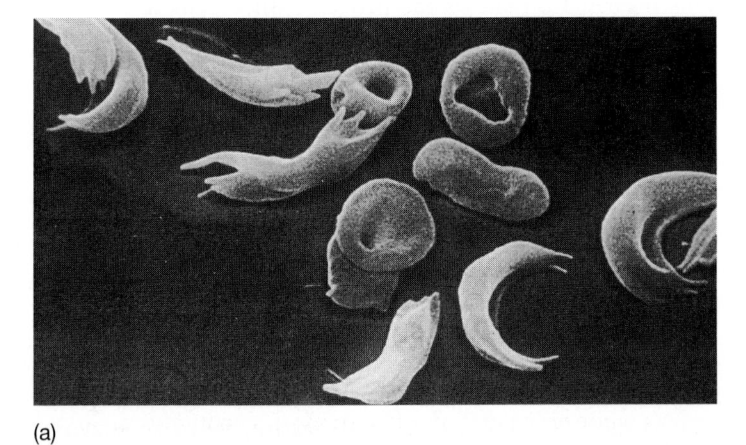
(a)

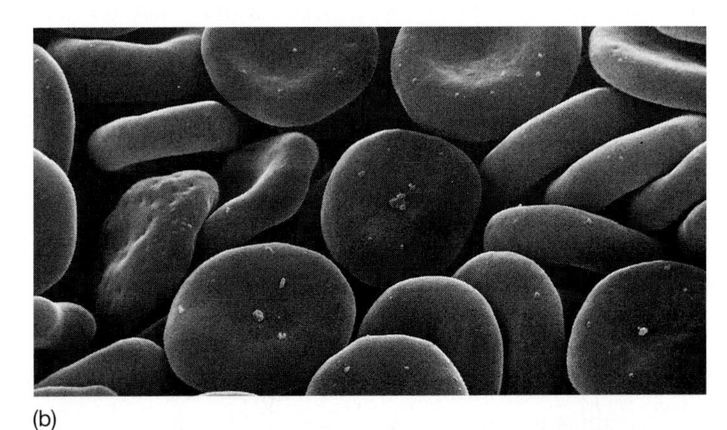
(b)

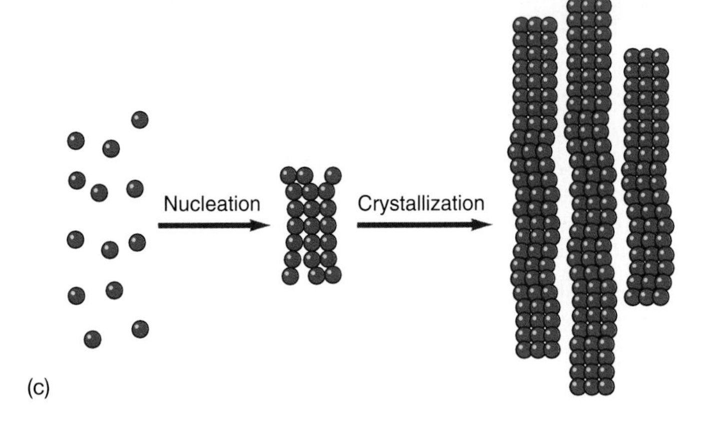

(c)

Figure 5.9
(a) Sickled red blood cells are clearly different from *(b)* normal red blood cells. *(c)* Sickling results from the crystallization of hemoglobin S into long fibers.

infection progresses. While growing inside red blood cells, the parasites lower the concentration of oxygen and cause the cells to become sickled. These misshapen cells are destroyed, along with their enclosed parasites, by scavenger cells that are part of the body's defense system. Consequently, where malaria is rampant, natural selection favors individuals with one copy of the Hb^{S} gene because they are resistant to the malaria parasite. Of course, individuals unfortunate enough to have two copies of Hb^{S} have sickle-cell anemia and are poorly adapted to any environment.

Sickle-cell anemia is a prime example of a "molecular disease," a disease caused by a change in the molecular structure of a gene product. This whole phenomenon results from a very small change in one gene: The information for the hydrophilic amino acid glutamic acid changes into information for the hydrophobic amino acid valine. The mutation only changes a small part of the protein while leaving most of it—and therefore most of its function—intact. The essential point is that this change in amino acid sequence creates a protein that is less soluble and undergoes rapid crystallization when it loses its oxygen, instead of the subtle conformational changes of normal hemoglobin.

The story of sickle-cell anemia also shows that a subtle change in a protein can have a tremendous effect, and such changes may be critical to evolution. Some amino acid replacements, for instance, can make a protein more stable at higher temperatures—or at very low temperatures. Thus an organism with such a protein can live in a different environment and occupy a new ecological niche, just as the sickle-cell mutation allows people to live in a malaria-ridden environment. In this way mutations can lead to evolution and perhaps to great changes in the structure of ecosystems.

5.13 The primary structure of a protein is a record of its evolutionary history.

Modern technology for sequencing proteins quickly has opened the door to comparative studies of proteins. This work has yielded some of the most important information we have about biological structure and function and about the course of evolution.

We explained in Section 2.4 that homologies between similar structures in different species are evidence for evolution. Proteins also show homology, sometimes more clearly than anatomical structures do. Figure 4.8 shows that the insulin molecules from many vertebrates have very similar amino acid sequences, with some small differences in each species. Proteins with very similar sequences in different species are considered homologous, even though they may take on different forms and functions during the course of evolution, just as bones and muscles do. Suppose we compare the amino acid sequences of one kind of protein in seven species (made very simple for illustration). The dots represent residues that are identical throughout, and each letter stands for a single amino acid change.

1 ....
2 ...........Q.............M......C...............D.........
3 ....E......................
4 ...........Q.............M......C.........................
5 ....Q..
6 ................................E.......K...............
7 ................................E.......G...............

Three proteins have the Q change, and three have the E change, suggesting that these substitutions occurred early in separate lines of evolution:

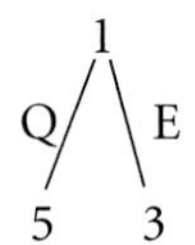

Two of the proteins with Q also have M and C; one of these also has D. Those with the E change may have acquired either K or G. So we may summarize the lines of phylogeny like this:

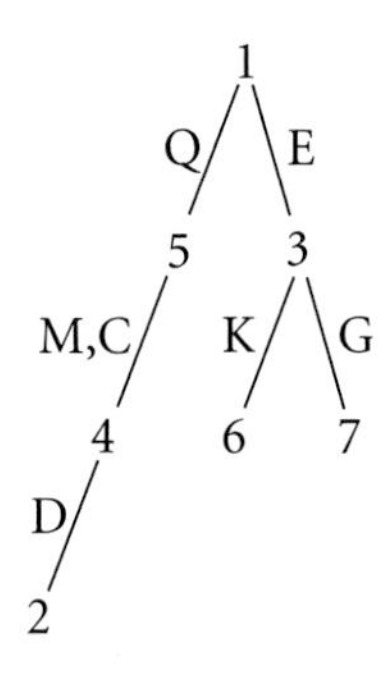

This information shows the probable phylogenetic relationships among all seven species, based on this protein. Real sequences, of course, are more complex and may be harder to rationalize. But phylogenies based on proteins have generally confirmed phylogenies based on anatomy, and comparisons of protein sequences from many species can refine phylogenies and clarify complex situations.

Homology takes on a different meaning when we find proteins with different functions but very similar sequences. A phylogenetic tree of different globin chains (Figure 5.10) shows that new proteins can become diversified to serve new functions. The homology between a myoglobin and a hemoglobin chain is different from the homology among all vertebrate insulins or all the egg white lysozymes of different birds. Furthermore, we can talk about partial homologies, where only portions of two molecules show homology. These different kinds of proteins evolve through genetic mechanisms that we will examine in subsequent chapters.

5.14 Computer searches help identify the function of an unknown protein on the basis of its structure.

As information about protein structure has accumulated and computers have become more powerful, comparative amino acid sequences have taken on new significance. We now know a

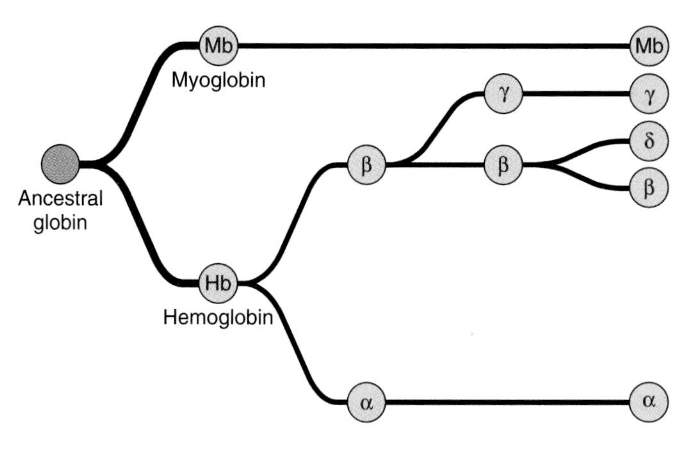

Figure 5.10
The family tree of different globins has been deduced by comparing their sequences. The predominant hemoglobin is made of α and β chains. Delta chains (δ) occur in Hb A_2, which is present in minor amounts in adults; gamma chains (γ) occur in fetal hemoglobin, Hb F.

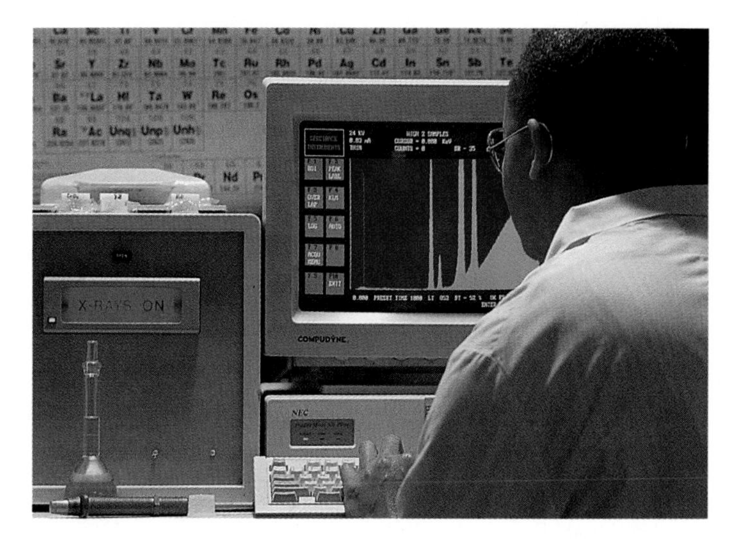

Figure 5.11
Biologists use computers more and more to study protein and nucleic acid structures. Through Internet connections, they can tap into enormous informational databases, and in collaboration with computer scientists, biologists are developing more powerful programs for analyzing this information and representing the structures of cells and their components.

great deal about the kinds of secondary and tertiary structure that a given amino acid sequence is likely to form. Some sequences easily form α helices; others are most common in β sheets. Sophisticated computer programs can now examine the sequence of a protein and predict its likely structure—and thereby indicate something about its function as well.

Enormous databases of protein sequences are growing daily, stored in central data banks (for instance, at the National Institutes of Health in Bethesda, Maryland, and the Japan International Protein Information Database [JIPID] in Tokyo), which scientists all over the world can tap into electronically via the Internet (Figure 5.11). Computer programs can compare the sequence of an otherwise unknown protein with all the sequences in the database and find those that show homologies to the newly sequenced protein. Such analyses may show that the protein has one domain characteristic of a certain kind of enzyme and a second domain that can probably bind a small molecule. This information will suggest the function of this protein and point to further experiments to characterize it. Thus biologists are working in an ever-expanding background of molecular detail, and their research can advance rapidly by making use of everything that is already known about protein structure.

The critical point underlying all this work is that structure and function are intimately tied to each other. Each protein molecule—indeed, each part of the molecule—has a specific structure that has been shaped by evolution to serve a particular function. Molecules with similar structures probably have similar or identical functions.

Coda The concept of chemical equilibrium has a key role in this chapter. Organisms operate through chemical reactions, and every reaction has an equilibrium point. Enzymes cause reactions to occur more rapidly, but they do not change this equilibrium. The interactions of enzymes with substrates (or more generally, proteins with molecules called ligands) entail weak bonds, so there are equilibria between the proteins associated with their ligands or dissociated from them. These weak interactions allow proteins to do their work. Hemoglobin, for instance, is only functional because it can associate with oxygen at one time and release it later. In a subsequent chapter we will see that muscles contract because two other proteins are able to associate briefly, while one pulls on the other, releases, and moves. Because the shape of a protein determines which ligands it interacts with and what it can do, proteins with similar structures have similar functions. Modern computers enable us to compare the amino acid sequences of newly discovered proteins with those whose functions are known and thus get clues to the functions of the new proteins. In Chapter 6, we will put proteins into a larger context by surveying cell structure to see where different functions are performed in typical cells.

Summary

1. Enzymes lower the energy barriers to chemical reactions by exerting specific forces on their substrates.
2. Chemical reactions come to an equilibrium if no external forces disturb them. Enzymes change the rates of reactions but don't change their equilibria.
3. An enzyme is very specific for a single type of substrate or, more rarely, a few substrates with similar structures.
4. Kinetic studies of enzymes indicate that each enzyme molecule has an active site where the substrate binds. As the concentration of substrate increases, the active sites become saturated, and the rate of the reaction approaches a maximum value.
5. An enzyme is characterized by its Michaelis constant, K_m, the substrate concentration for which the rate of the reaction is

half-maximal. The greater the affinity of an enzyme for its substrate, the smaller the K_m.

6. Temperature and pH affect the rate of enzyme-catalyzed reactions. Every enzyme has an optimal temperature and an optimal pH, which are close to the temperature and pH of the environment where that enzyme normally functions.
7. Some enzymes require cofactors and coenzymes for their function. Cofactors are often metal ions that bind to the active site, and the concentrations of other ions around the enzyme can also affect an enzyme's function. Coenzymes are small organic molecules that work with the enzyme; similar molecules that are covalently bonded to the protein are called prosthetic groups.
8. Enzymes can be inhibited by other compounds. Competitive inhibitors resemble the normal substrate of the enzyme and compete for access to the active site. The kinetics of their interaction with the enzyme confirm the general model of enzyme activity. Noncompetitive inhibitors irreversibly inactivate an enzyme by binding covalently to its active site.
9. Hemoglobin shows cooperative effects between its subunits. The subunits can undergo conformational changes between a state in which oxygen is bound (B) and another state in which it is not bound (N). They shift back and forth between the B and N states in a coordinated way. This cooperativity enables hemoglobin to pick up oxygen efficiently where it is abundant, as in the lungs or gills, and release it where oxygen is scarce, as in the tissues.
10. Allosteric proteins such as hemoglobin have two or more distinct binding sites where ligands with different shapes can bind. The binding of each ligand tends to stabilize the protein in one conformation or another, thereby regulating the protein's function.
11. A small change in protein structure can have profound functional consequences, as illustrated by sickle-cell anemia. A change of only one amino acid in the β chain of the protein gives it quite different properties, so it crystallizes into long fibers when the oxygen concentration is low. However, the mutated protein also inhibits the growth of the malaria organism, thus allowing people with this protein to adapt to different ecological conditions.

Key Terms

activated complex 91
transition state 91
turnover number 92
equilibrium 93
equilibrium constant 94
Michaelis constant 95
affinity 95
enzyme cofactor 97
coenzyme 97
vitamin 97
prosthetic group 97
competitive inhibition 97
saturation curve 99
torr 100
cooperative effect 100
ligand 101
binding site 101
stereospecificity 101
allosteric effects 102
regulatory site 102

Multiple-Choice Questions

Which alternative best completes the statement or answers the question?

1. If the Q_{10} of a process is 1, the rate of the process
   a. doubles when the temperature increases 10°C.
   b. doubles when the temperature decreases 10°C.
   c. changes by the same amount when the temperature increases or decreases by a given amount.
   d. is independent of increases in temperature.
   e. is below the measurable level.
2. Which of the following phrases best describes how enzymes affect chemical reactions?
   a. lower the activation energy
   b. raise the activation energy
   c. supply activation energy
   d. remove activation energy
   e. release activation energy
3. Which statement is correct?
   a. Enzymes generally catalyze reactions in the forward direction only.
   b. Enzymes theoretically can catalyze a particular reaction in both directions.
   c. Enzymes usually catalyze one kind of reaction in the forward direction and another kind of reaction in the reverse direction.
   d. Enzymes increase the ratio of substrate to product at equilibrium.
   e. Enzymes decrease the ratio of substrate to product at equilibrium.
4. Which factor(s) alters the rate of an enzyme-catalyzed reaction?
   a. increasing the concentration of enzyme while maintaining the same concentration of substrate
   b. increasing the concentration of substrate while maintaining the same concentration of enzyme
   c. adjusting temperature
   d. adjusting pH
   e. all of the above
5. What is the difference between a cofactor and a coenzyme?
   a. coenzymes are proteins; cofactors are not.
   b. coenzymes are not proteins; cofactors are protein in nature.
   c. coenzymes are organic molecules; cofactors are generally inorganic.
   d. coenzymes form prosthetic groups; cofactors do not.
   e. coenzymes activate enzymes; cofactors activate coenzymes.
6. At equilibrium, a system that includes an enzyme plus equal concentrations of its normal substrate and a competitive inhibitor will have
   a. all enzyme molecules bound to substrate.
   b. all enzyme molecules bound to inhibitor.
   c. half the enzyme bound to substrate and half bound to inhibitor.
   d. some enzyme bound to neither substrate nor inhibitor.
   e. all enzyme bound to either substrate or inhibitor but not in any fixed ratio.
7. Myoglobin does not exhibit a cooperative effect in binding oxygen because
   a. it is composed of one polypeptide chain.
   b. it already has a high affinity for oxygen.
   c. its primary structure inhibits cooperativity.
   d. oxygen binds irreversibly to the active site.
   e. unlike hemoglobin, it does not circulate in the blood.
8. Because of cooperativity, the binding of oxygen to
   a. one molecule of hemoglobin stimulates oxygen binding to other molecules of hemoglobin.
   b. hemoglobin prevents oxygen binding to myoglobin.
   c. hemoglobin stimulates hemoglobin's uptake of CO_2.
   d. one heme group stimulates binding of oxygen to the other heme groups within the same molecule.
   e. one or more heme groups prevents hemoglobin's assuming the B conformation.
9. Which of the following acts as a ligand?
   a. an enzyme
   b. the substrate
   c. the binding site

d. the active site
e. all of the above

10. Carbon monoxide binds to the active site of hemoglobin, displacing oxygen. Fortunately, if found in time, victims of carbon monoxide poisoning can be resuscitated by being given oxygen. This suggests that carbon monoxide is a(an)
    a. competitive inhibitor.
    b. irreversible inhibitor.
    c. allosteric effector.
    d. coenzyme.
    e. cofactor.

True-False Questions

Mark each statement true or false, and if false, restate it to make it true.

1. In general, covalent bonds link the active site of an enzyme to its substrate.
2. Although enzymes do not affect the equilibrium point of a reaction, they do affect the rate at which equilibrium is reached.
3. If enzyme A forms tighter bonds to its substrate than enzyme B, the Michaelis constant of enzyme A will be higher than that for enzyme B.
4. The optimal temperature of all enzymes in a single organism falls within a small range.
5. A competitive inhibitor binds to an enzyme's active site while an irreversible inhibitor binds to an enzyme's allosteric site.
6. In terms of enzyme kinetics, the addition of a competitive inhibitor to an enzyme-catalyzed reaction will alter the turnover rate but not the V_{max}.
7. At very high oxygen concentrations, myoglobin's saturation curve becomes sigmoid in shape.
8. People who have only one gene for Hb S are resistant to malaria because only half of their red blood cells sickle.
9. A person with two genes for Hb S will have anemia but not be resistant to malaria.
10. The more essential and efficient a cellular protein is, the slower its rate of evolutionary change.

Concept Questions

1. How does the concept of an *energy barrier* relate to that of an *activated state?*
2. Explain the physical mechanism by which enzymes lower the activation energy of a reaction.
3. What would be the advantage of not using fever-lowering drugs such as aspirin or ibuprofen to counteract a low-grade fever that resulted from a bacterial infection?
4. Explain why it is unlikely that an enzyme could be competitively inhibited and irreversibly inhibited by the same ligand.
5. With reference to hemoglobin, distinguish between cooperative effect and allosteric regulation.

Additional Reading

Cech, Thomas R. "RNA as an Enzyme." *Scientific American,* November 1986, p. 64. RNA can act as an enzyme by cutting, splicing, and assembling itself.

Doolittle, Russel F., and Peer Bork. "Evolutionarily Mobile Modules in Proteins." *Scientific American,* October 1993, p. 34. Certain modular protein structures may be transmitted horizontally across species lines.

Gutfreund, H. *Kinetics for the Life Sciences: Receptors, Transmitters and Catalysts.* Cambridge University Press, Cambridge and New York, 1995.

Karplus, Martin, and J. Andrew McCammon. "The Dynamics of Proteins." *Scientific American,* April 1986, p. 42. Proteins function by way of internal motions, which are being explored in computer simulations.

Lea, Peter J. (ed.). *Enzymes of Primary Metabolism.* Academic Press, London and San Diego, 1990.

Internet Resource

To further explore the content of this chapter, log on to the web site at:

http://www.mhhe.com/biosci/genbio/guttman/

6 Introduction to Cells

Light microscopes are important tools of modern biology.

Key Concepts

6.1 The cell theory has had a long history.
6.2 Cells are surrounded by a plasma membrane.
6.3 Most cells are very small.
6.4 Microscopy reveals cell structure.
6.5 The two major types of cells are procaryotic and eucaryotic.
6.6 Cells contain structures called organelles that have specialized functions.
6.7 A eucaryotic cell is compartmentalized by its nucleus and other membranous organelles.
6.8 Eucaryotic cell surfaces have characteristic features.
6.9 Cells in a tissue may be connected by several types of junctions.
6.10 Procaryotic cells also contain organelles.

The American humorist James Thurber recounts in his book *My Life and Hard Times* that he could not pass the college botany course "because all botany students had to spend several hours a week in a laboratory looking through a microscope at plant cells, and I could never see through a microscope. I never once saw a cell through a microscope. This used to enrage my instructor." All around the lab, students were seeing and drawing cells, but Thurber—who had suffered an eye injury in childhood—couldn't see anything. So he dropped the course and came back to try again the next year. He describes his next encounter with the instructor:

> *"Well," he said to me, cheerily, when we met in the first laboratory hour of the semester, "we're going to see cells this time, aren't we?"*
>
> *"Yes, sir," I said.*"

Thurber adjusted the microscope until finally he was able to see something and sketch it, but the drawing he made didn't look like a cell. Puzzled, his professor then looked through the microscope himself. Thurber describes his reaction:

> *His head snapped up. "That's your eye!" he shouted. "You've fixed the lens so that it reflects! You've drawn your eye!"*

We all have to learn to see, especially to see new, unfamiliar things. And, like James Thurber, every biology student today must learn to look through a microscope and see cells. This "rite of passage" is a necessary introduction to the concept that all organisms consist of cells, a concept that is central to our modern understanding of biological processes. After centuries of observation, with the aid of increasingly sophisticated microscopes and other instruments, biologists are coming to understand the intimate connection between the structure and function of cells. Wherever a function is to be performed, appropriate structures must be in place to do the job. This chapter provides an overview of cell structure and lays the groundwork for our discussion of many biological processes in later chapters. ■

6.1 The cell theory has had a long history.

Around 1665, the English scientist Robert Hooke[1] took on the duty of presenting public demonstrations—"three or four considerable experiments"—each week for the Royal Society of London. While searching for interesting new examples, he examined a thin slice of cork with a microscope he had built and noted that it was divided into small compartments. He called these compartments *cells* because they reminded him of monks' chambers, or cells, in a monastery (Figure 6.1). Hooke described his experiment in his book *Micrographia:*

> *I took a good clear piece of Cork, and with a Pen-knife sharpen'd as keen as a Razor, I . . . cut off from the former smooth surface an exceeding thin piece of it, and placing it on a black object Plate, because it was it self a white body, and casting the light on it with a deep plano-convex glass, I could exceeding plainly perceive it to be all perforated and porous, much like a Honey-comb . . . Our Microscope informs us that the substance of Cork is altogether fill'd with Air, and that Air is perfectly enclosed in little Boxes or Cells distinct from one another.*

Thus the term *cell* was introduced into biology. Because cork is dead matter, Hooke actually saw only the thickened walls of dead plant cells, and some scientists have criticized him for emphasizing these walls while neglecting the substances and processes within. Nevertheless, Hooke recognized that green plant cells were "fill'd with juices," and he suggested that these chambers contain minute but important functional structures.

Other microscopists of this era made amazing observations with their primitive instruments. Foremost among them was Antonie van Leeuwenhoek, a Dutch amateur who observed and made careful drawings of everything that came his way. Leeuwenhoek discovered **microorganisms,** organisms too small to be seen with the unaided eye (Figure 6.2), and he was the first to describe human cells, including sperm.

As microscopes gradually improved, particularly in the early nineteenth century, other observers who examined living material began to recognize a common structure among them all. René Dutrochet, for instance, wrote in 1824 that "all organic tissues are actually globular cells of exceeding smallness, which appear to be united only by simple adhesive forces." Today we use the word **tissue** to refer to a mass of similar cells, such as a muscle or the outer layer of a leaf.

Although we recognize the early work of researchers like Dutrochet, credit for recognizing the fundamental role of cells in biology is generally given to two German biologists. In 1838, Matthias J. Schleiden wrote a new botany textbook in which he

[1]Hooke didn't limit his talent to biology. Physicists know him as the experimenter who formulated Hooke's Law: *Stress is proportional to strain.* Inventors know him as the developer of the escapement mechanism in clocks and watches, only now being replaced by quartz technology. Historians know him as the astute politician who guided the Royal Society in its early years, helping make it one of the most prestigious organizations of scientists and an influential force in advising government policy-makers.

(a)

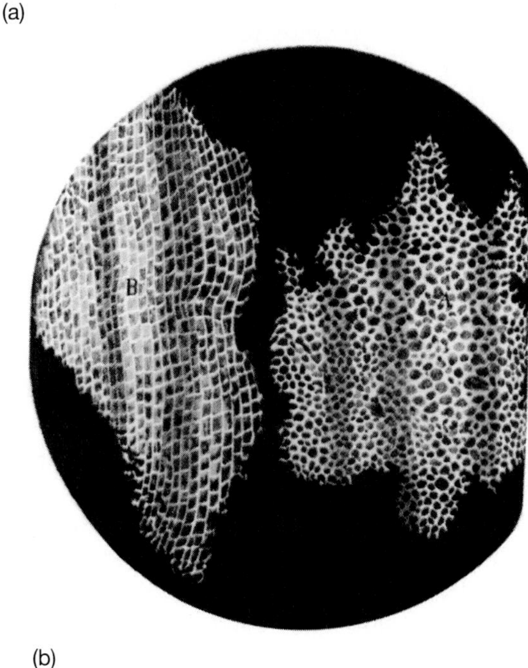

(b)

Figure 6.1

(a) Robert Hooke built a simple microscope with which he observed many minute structures for the first time. *(b)* His original drawing of a thin slice of cork shows its cellular structure.

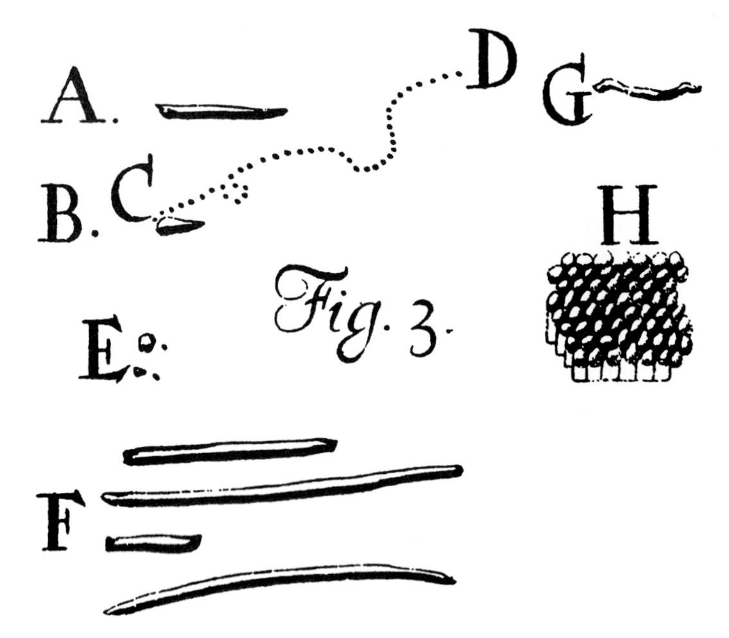

Figure 6.2

Some of Leeuwenhoek's original drawings, published in 1684, show the microorganisms, or "animalcules," he observed in pond water. These drawings are recognizable as bacteria, mostly rod-shaped, with packets of spherical cells at E and H.

championed the idea that all plants are made of cells, and the next year Theodor Schwann extended this view to animals. Schleiden and Schwann did not get the whole concept right in its modern form; for instance, they had rather odd misconceptions about the origins of new cells. But once they had shown how universal cellular structure is, other observers also began to see the components of tissues as cells and to advance the same theory. This was a major change in perception and thinking. When scientists discover totally new objects, they must form hypotheses about what they are looking at and gradually confirm them so others may see accurately. It took time for people to accept the cell hypothesis and then to really *see* that all tissues are made of cells.

The modern **cell theory** (or principle), which has been developing for over a century now, consists of these points:

1. All organisms are made of cells and the products of cells. **Unicellular** organisms consist of only one cell, although many algae and bacteria, which are basically unicellular, grow as regular filaments or clusters of cells. The most obvious organisms are **multicellular,** made of many cells. Some multicellular organisms have only a few hundred to a few thousand cells; humans have about 100 trillion (10^{14}).
2. The activities of a multicellular organism are largely the activities of its individual cells and the interactions among them. A cell is the smallest unit capable of carrying out fundamental biological processes, making it the basic functional unit of biology.
3. All cells, in all organisms, are very much alike, being composed of the same types of chemical compounds and functioning through similar chemical reactions.
4. All cells come from preexistent cells. This is the basic point of heredity: Each organism consists of cells that were derived from the cells of its parent(s).

The fourth point, which was established in 1858 by the German pathologist Rudolf Virchow, puts the cell theory into an evolutionary context. Ever since the first cells developed over three billion years ago, all cells—basically, all organisms—on the planet have been derived from one another by repeated cell division. Each of us developed from a single egg cell from our mother that was fertilized by a sperm cell from our father, and their cells, in turn, arose from innumerable generations of cell division, back to unicellular organisms and ultimately to some original cell, or cells.

6.2 Cells are surrounded by a plasma membrane.

With a little instruction, it is easy to recognize cells in a tissue by the boundaries between them (Figure 6.3). The boundary of every cell is a **plasma membrane.** Membranes are thin, flexible sheets of lipid and protein whose structure is explained in Chapter 8. Most cells contain a number of internal membranes, but all membranes are too thin to see in light micrographs (photographs taken with light microscopes). The boundaries of plant cells (Figure 6.3*b*) are more obvious because their plasma membranes are surrounded by heavy **cell walls** made of cellulose and other polymers. The plasma membrane controls passage of material in and out of the cell. It excludes some substances completely, allows in (or even pumps in) nutrients, oxygen, water, and other essential substances, and lets wastes leave.

6.3 Most cells are very small.

Cells and their components constitute a microscopic world measured in units that are rarely used for any other purpose. The scale of sizes in Figure 6.4 shows metric units that differ by factors of 10. The metric units used for measuring cells are a *micrometer* (μm), which is a thousandth of a millimeter, and a *nanometer* (nm), which is a thousandth of a micrometer. One-tenth of a nanometer is an Angstrom unit (Å), used for expressing the dimensions of atoms. Figure 6.4 also shows the smallest structures that can be seen clearly with the human eye, a light microscope, and an electron microscope. We will explain types of microscopy later in this chapter.

Most cells are tiny. Hooke, astonished by their size, wrote:

> *I . . . found that there were usually about threescore of these small Cells placed end-ways in the eighteenth part of an Inch in length, whence I concluded there must be neer eleven hundred of them, or somewhat more than a thousand in the length of an Inch, and therefore in a square Inch above a Million, or 1166400. and in a Cubick Inch, above twelve hundred Millions, or 1259712000. a thing almost incredible, did not our Microscope assure us of it by ocular demonstration.*

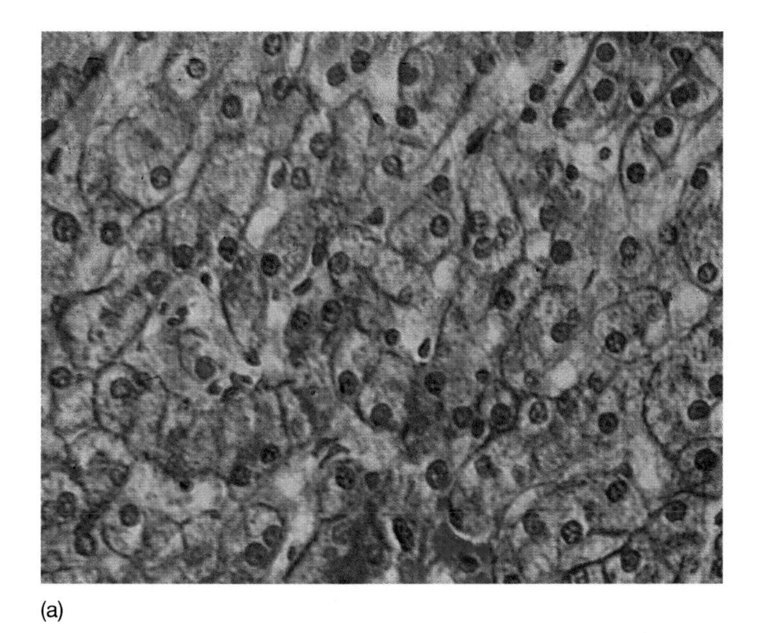

(a)

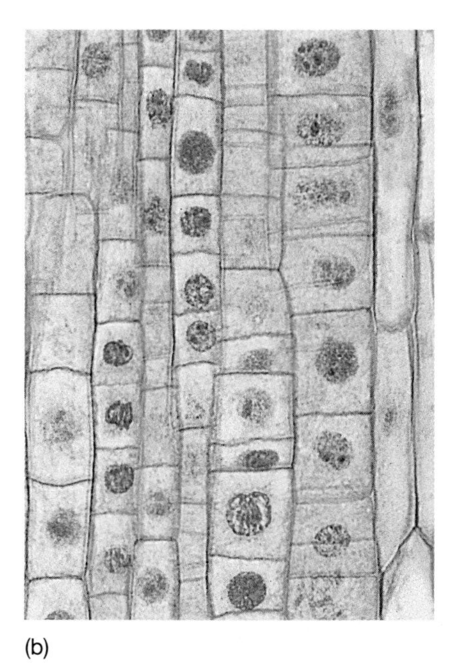

(b)

Figure 6.3
Light micrographs show that similar cells are packed together like building blocks to make a tissue. *(a)* Animal tissue; *(b)* plant tissue.

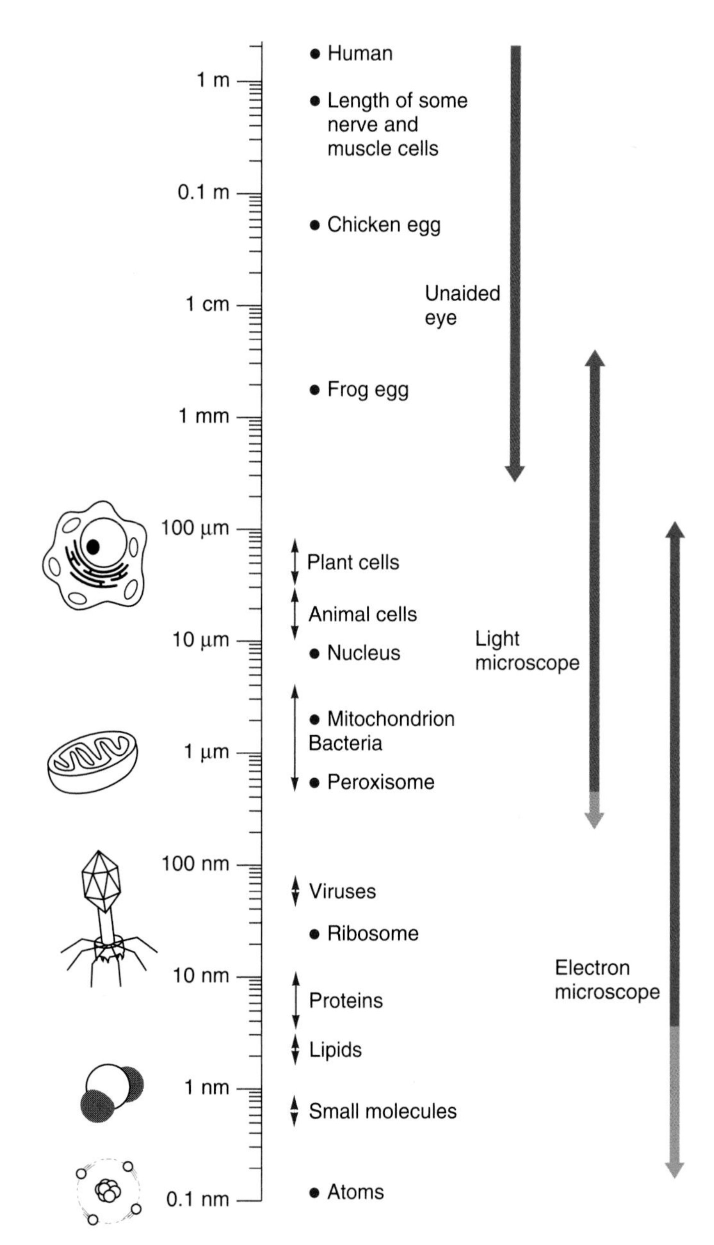

Figure 6.4
The units used for measuring cells and their microscopic components differ by factors of 10^3, so they are arranged here along a logarithmic scale, where each marked unit is 10 times larger than the next. The heavy arrows at the right show the limits that can be seen by the human eye and by microscopes, as explained in Section 6.4

With few exceptions, cells can only be seen with a microscope, and their dimensions are measured in micrometers. The smallest known cells are mycoplasmas, infectious agents about 0.1–0.3 µm in diameter. Typical bacterial cells are 0.5–1.0 µm wide, and the diameters of most plant and animal cells are about 10–50 times greater. A hundred cells, each 10 µm in diameter, could be lined up between a pair of millimeter marks on a ruler. Why aren't cells much larger?

Basically, the size of a cell is determined by its need for a favorable ratio of surface area to volume to support metabolism. Metabolism, which occurs throughout a cell's volume, requires materials such as glucose and oxygen, and produces wastes such as CO_2 and urea. Those substances can only enter and leave the cell through its surface, but as a cell grows, its volume changes with the cube of a dimension and increases much faster than its surface area, which changes only with the square of a dimension. Consider, for instance, a series of cubes measured in arbitrary units:

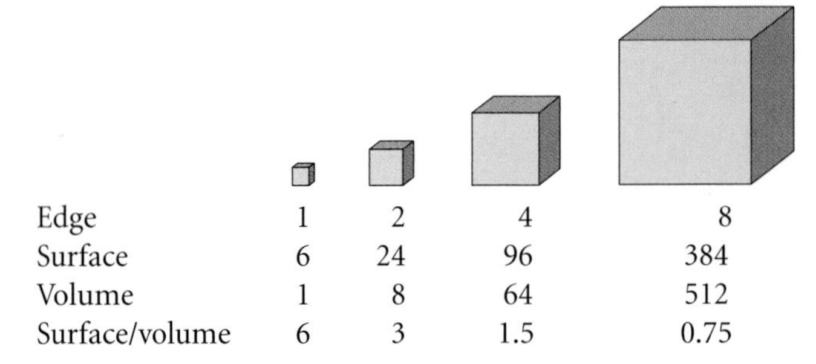

Edge	1	2	4	8
Surface	6	24	96	384
Volume	1	8	64	512
Surface/volume	6	3	1.5	0.75

Because the ratio of surface to volume decreases as the size of the cube increases, a cell with a large volume could not exchange matter with its surroundings fast enough to support metabolism. Thus only small cells—or the long, thin cells of many algae and molds—have a favorable surface/volume ratio. Some large cells effectively increase their surface area by having internal membranes that make connections to the outside, and other cells increase communication between their insides and outsides by stirring their contents and moving materials around, an activity easily observed in some plant cells.

Exercise 6.1 A large lecture hall has 2,000 seats and 10 exits. A small classroom has 40 seats and 2 exits. People can leave each room at the rate of 2 per second per exit. Which room can be emptied fastest? What does this have to do with the sizes of cells?

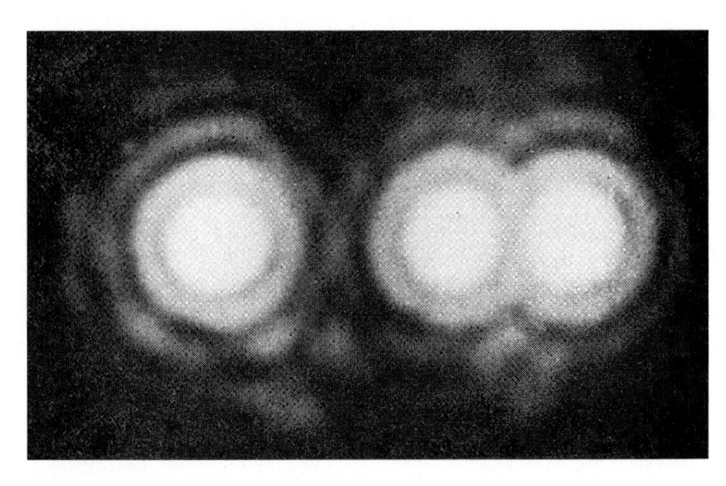

Figure 6.5

Light from a point source is not a point but rather a series of concentric diffraction rings. (Diffraction is a modification that light undergoes when it passes by the edge of an object or through a slit or hole.) If two points are very close, the rings overlap so the source appears to be a single broad point. A microscope is judged by its ability to resolve this pattern into distinct points.

Figure 6.6

A photograph is prepared for printing by a screening process that breaks it into dots of ink. No matter how much the print is enlarged (magnified), no further detail can be resolved.

6.4 Microscopy reveals cell structure.

A few cells can be seen without microscopes. A huge bird's egg is technically a single cell; the cells of some algae are macroscopic; and, surprisingly, even a human egg is visible without magnification. Nerve cells are no wider than others, but some of them have incredibly long extensions, so a cell in the human sciatic nerve can extend from the spinal cord into the foot—and you can imagine the lengths of certain nerve cells in a giraffe or a whale! Nevertheless, most cells and their structures can only be investigated with instruments much more powerful than a human eye.

We cannot see most cells or any of their internal structures with the unaided eye because our eyes lack sufficient **resolving power,** the ability to see two adjacent points as separate points. No matter how small a source of light may be, its image is not simply a point but rather a peak of light intensity surrounded by diffraction rings (Figure 6.5). Two points that are very close produce overlapping peaks, and if the points are too close, their peaks meld into one and appear to be a single point. A good microscope can still resolve them into separate peaks, so we judge a microscope by its power of resolution. It is important to remember that resolution differs from *magnification*—how large an object appears. Resolution provides real information, while magnification may simply make an object look bigger without providing additional structural detail. For instance, the printed photograph in Figure 6.6 is made of minute spots of ink; magnifying it makes the spots larger but does not reveal more detail, or resolution.

In either an eye or a microscope, light is focused by a lens. The **limit of resolution** of a lens is the minimum distance *(d)* between two points that the lens can resolve. This limit depends on the numerical aperture (NA) of the lens, as defined in Figure 6.7, and on the kind of radiation used for observation. Electromagnetic radiation, including visible light, is an oscillating wave (an electrical wave and a magnetic wave together) with varying wavelength (Figure 6.8). In the visible portion of the spectrum,

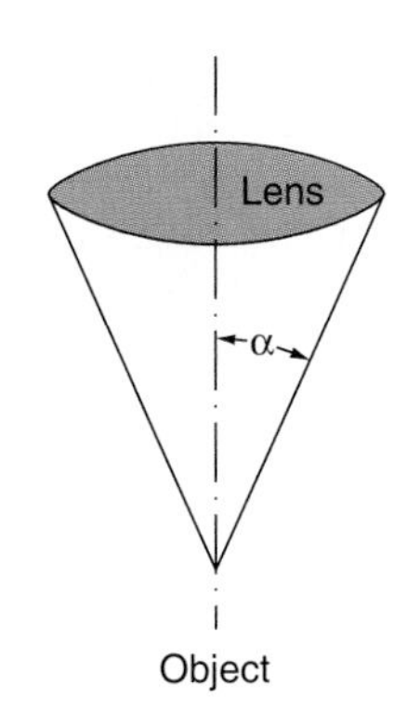

Figure 6.7

A lens receives a cone of light from a point on an object. The numerical aperture of the lens (NA) is equal to $n \sin \alpha$, where α is the half-angle of the cone of light entering the lens and n is the refractive index of the space between the lens and the object. That space may be filled with air, for which $n = 1$, or with a special oil with an index of refraction as high as 1.4.

red light has the longest wavelength and violet light the shortest; infrared and ultraviolet radiation are just beyond the visible limits. Thus if the light has wavelength λ (lambda):

$$\text{Limit of resolution} = d = 0.61\ \lambda/\text{NA}$$

Because of this relationship, a microscope that uses shorter-wavelength radiation has a smaller limit of resolution (greater resolving power) than one that uses longer-wavelength radiation. Some microscopes use ultraviolet (uv) light, whose wavelength is shorter than that of visible light, and quartz lenses, since ordinary glass absorbs ultraviolet light. Resolution is also improved by increasing the NA, either by increasing the angle of the cone of light the lens takes in or by surrounding the specimen with a substance that has a higher refractive index[2] (n) than air, such as oil with n of about 1.52. The most refined light microscopes (those that focus visible light; Figure 6.9*a*) have a theoretical limit of resolution of 0.2 μm, though their practical limit of resolution is no better than about 0.25–0.40 μm. (Note that 0.40 μm = 400 nm, the wavelength of violet light, the shortest-wavelength part of the visible spectrum.)

[2]The refractive index of a substance is the ratio of the speed of light passing through it to the speed of light in a vacuum. This ratio determines how much a light ray is bent in passing from one substance to another.

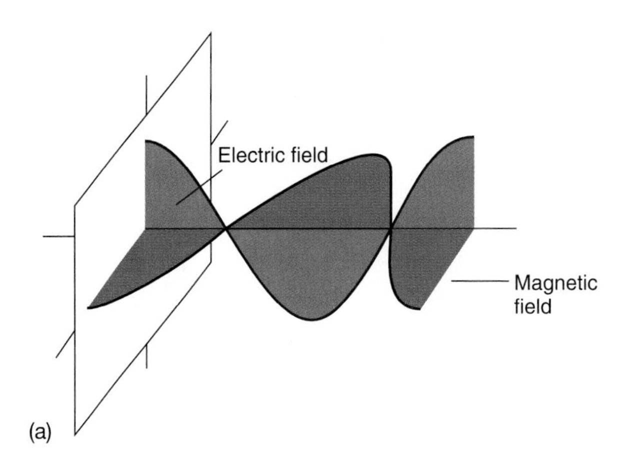

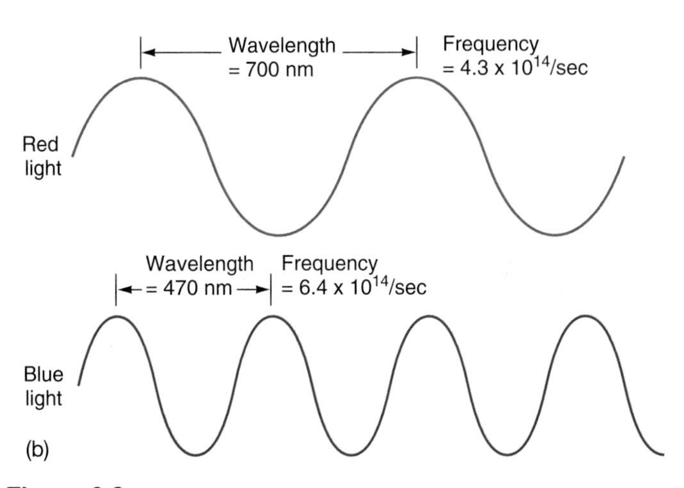

Figure 6.8

(*a*) Light is an oscillating electromagnetic wave. (*b*) Different colors of light have different wavelengths.

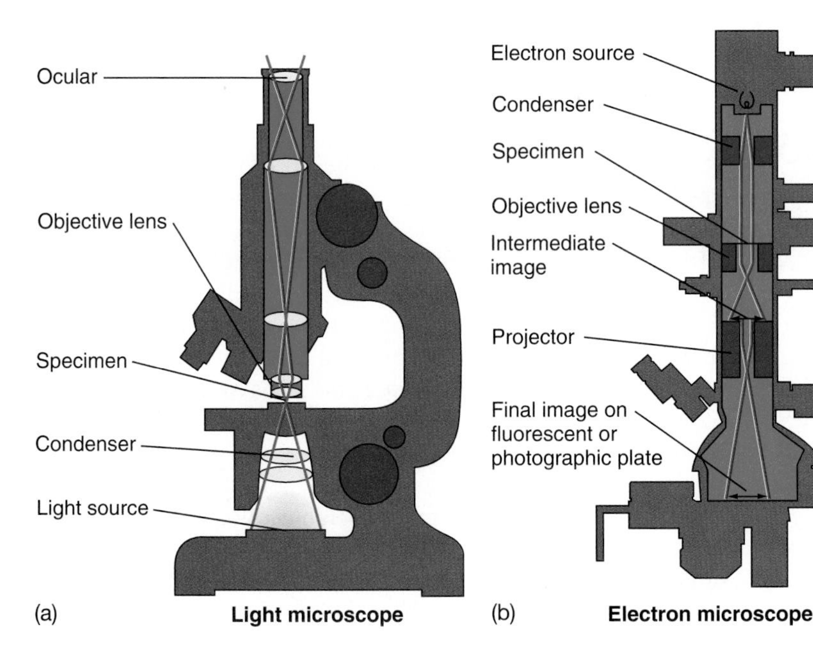

Figure 6.9

A modern compound light microscope is similar to an electron microscope. *(a)* In a light microscope, light from the source is focused on the specimen by a condenser. Light leaving the specimen is picked up by an objective lens and passed on to an ocular, or eyepiece. The magnification of the instrument is the product of the magnification achieved by the ocular (generally 10 times, or 10×) and that of the objective (from 10× to 100×). *(b)* In an electron microscope, a beam of electrons is substituted for light, and the lenses (*dark blue*) are electromagnets. The beam is focused on a fluorescent plate for viewing; a photographic plate can be placed in this position to take a picture of the specimen.

Methods 6.1 Preparing Specimens for Microscopy

A specimen to be examined by microscopy is first fixed to preserve its structure by soaking it in a preservative such as formaldehyde. The fixed tissue must then be cut into slices, or **sections,** thin enough to transmit light. But since it is hard to cut soft materials cleanly, the tissue is first *embedded* in something solid for slicing—commonly paraffin (for light microscopy) or polymerized plastics (for both light and electron microscopy). Because the tissue is largely water, these embedding substances will not penetrate it, so the water is removed by soaking the tissue in a series of nonaqueous solutions. The fixed, water-free specimen is then embedded either in warm paraffin, which soaks into the specimen and solidifies when cool, or in liquid plastic, which first penetrates and then polymerizes into a solid when heated or exposed to ultraviolet light.

For some simple light microscopy, only a steady hand and a razor blade are needed for sectioning, but generally the material must be sliced with a *microtome,* an instrument with a sharp blade that can be adjusted to make slices of different thicknesses. It's really just a very sophisticated (and expensive) meat slicer. The sections for electron microscopy must be so thin that a diamond knife is commonly used.

Finally, the thin slices are mounted for staining on a glass slide for light microscopy or a wire grid for electron microscopy. The challenge is to find stains that bind only to specific materials and make them stand out. Stains for light microscopy are brightly colored because they absorb visible light, but electrons are stopped by heavy atoms, not by colored substances, so specimens for electron microscopy are treated with heavy-metal compounds such as lead hydroxide, $Pb(OH)_2$; potassium permanganate, $KMnO_4$; phosphotungstic acid; or osmium tetroxide, OsO_4, which is also a good fixative. Electron micrographs (pictures taken with an electron microscope) are always black and white—black where heavy-metal stains absorb electrons, white where electrons pass through unimpeded. But whether the contrast is provided by colored stains or by heavy metals, when you look at a micrograph you are not seeing pristine cellular structures—you are simply seeing where the stains bind.

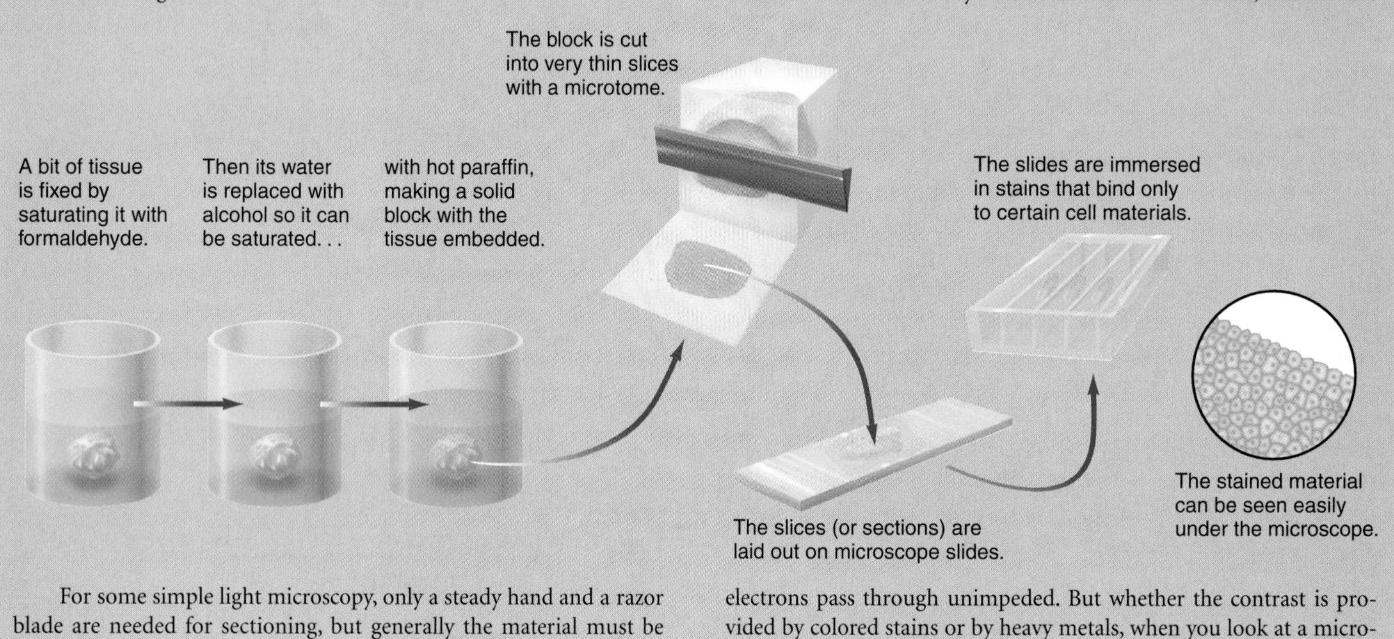

Exercise 6.2 While you are out watching birds one morning, you see a small bird that you can't identify immediately. Even with your binoculars, you can't tell if it has one dark line or two dark lines above its eye. What optical limitation are your binoculars exhibiting?

Microscopy depends on stains or special optical techniques.

We rarely see biological structures in their natural state with a microscope. If you look at fresh tissue with an ordinary light microscope, you will see very little because seeing depends upon *contrast* between objects, and—except for a few colored bodies, such as chloroplasts—cell structures are all made of similar, virtually colorless substances. For this reason, microscopy relies on creating contrast with **stains,** reagents that bind strongly to certain cellular materials and make them stand out in contrast to others. (Methods 6.1 describes some techniques used in microscopy.) Before stains can be applied, the tissue must be preserved with a **fixative** such as formaldehyde. Fixatives stabilize the macromolecules of cell structure in their natural positions, preserving native structures so they do not deteriorate during staining. Unfortunately, fixatives can also interact chemically with cell molecules to produce unnatural structures called **artifacts.** Only painstaking experimentation and the fact that different methods of preparation show the same structures give us confidence that we are looking at a close reflection of natural structures through a microscope.

Because fixation and staining inevitably create some distortion, and because fixed material is necessarily dead, microscopists often turn to *phase-contrast microscopes* and *interference microscopes* that can reveal the structures of unstained, living cells (Figure 6.10). These instruments depend upon the

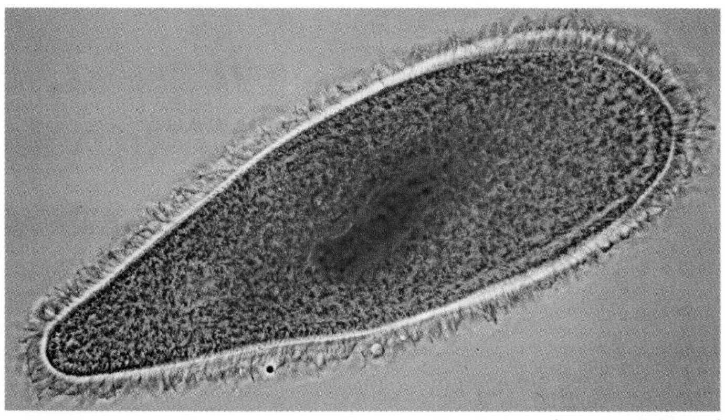

Figure 6.10

A phase-contrast micrograph of a ciliated protist, *Paramecium,* reveals some internal structures and makes the cilia on the cell surface stand out quite sharply. This specimen is unfixed and unstained. The structures are made visible by the optics of the microscope, which creates contrast from the interference of light rays that have been retarded to different degrees by the materials of the cell.

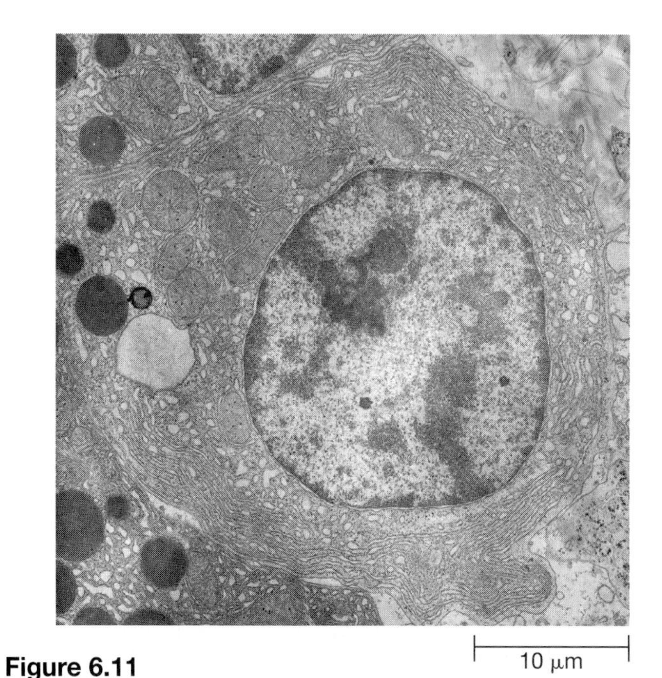

Figure 6.11

This electron micrograph of a pancreas cell shows its major structures.

fact that even colorless cellular structures bend and interfere Look, for instance, at an air bubble in clear glass or at the boundary between clear oil and clear water. These boundaries are visible because they bend light waves or because the two substances retard light differently. The optical systems of phase-contrast and interference microscopes enhance such contrasts. Although these microscopes have provided considerable insight into the operation of living cells, they are subject to the same limit of resolution as other light microscopes.

Electron microscopes enormously increase resolution and provide the modern picture of cell structure.

A critical advance in microscopy was the invention of the **electron microscope** in 1931 (see Figure 6.9*b*). An electron microscope uses beams of electrons with far shorter wavelengths than visible light and thus far greater resolving power. Electromagnets focus the electrons just as glass lenses focus light, and the images are visualized with fluorescent screens, photographic film, and electronic devices. *Transmission electron microscopy* (TEM), where the electron beam passes through the specimen, has a theoretical resolving power about 100 times greater than light microscopy and can resolve structures as small as 2–3 nm.

Biologists' view of cell structure was dramatically changed during the 1950s and 1960s by the pioneering studies by Keith Porter and George Palade, then at the Rockefeller University, and by Christian de Duve, who integrated information about purified cell components with the images from electron microscopy to clarify the nature of the components shown in micrographs. The electron micrographs of this time—mostly of rat or guinea pig liver—opened up a new view of cells (Figure 6.11). These pictures were so illuminating that biologists continue to think of them whenever they envision the general structure of a cell. Electron microscopy not only brought a clearer understanding of larger structures that had already been seen by light microscopy, such as nuclei and chloroplasts, but also revealed many previously unknown structures, such as internal membranes.

When you look at electron micrographs, remember that they are usually pictures of thin slices of material, and they show the boundaries of objects as distinctive lines. Figure 6.12 will help you relate what you see in a micrograph to actual cell structures. Notice that what appears to be a line is usually a section through a membrane. (The best-resolved electron micrographs show a membrane as a sandwich of two dark lines around a white space.)

Two techniques of electron microscopy provide a more three-dimensional view of objects. *Negative staining* shows the shapes of small particles by surrounding them with a puddle of a heavy-metal stain, so the portions not covered by the stain stand out in contrast as light areas:

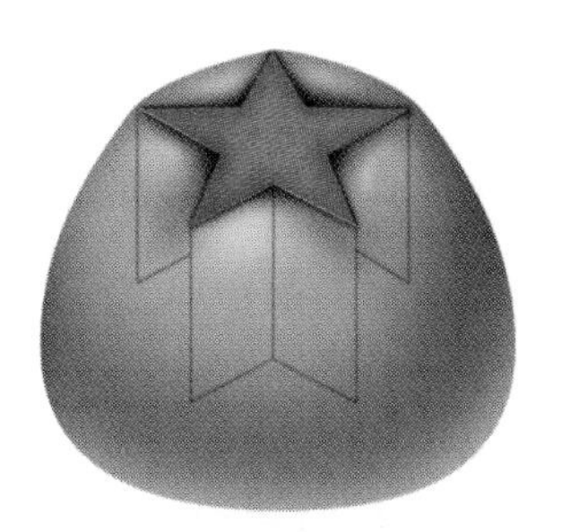

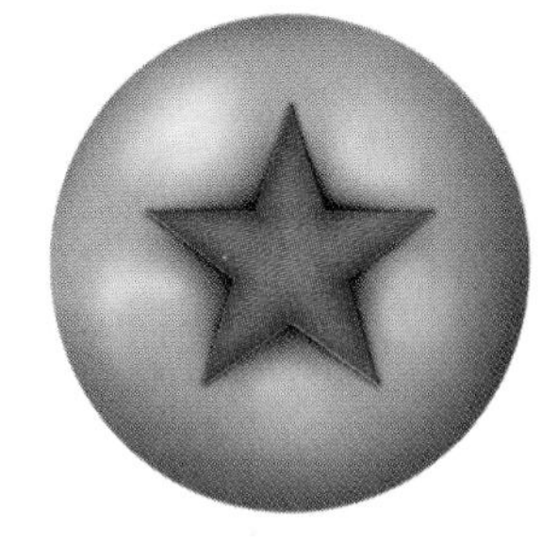

This gives a good sense of their form, especially when several similar objects can be seen in different positions. The second technique employs a *scanning electron microscope* (SEM), which plays an electron beam over the metal-coated surface of an object and detects the secondary electrons that are knocked out of

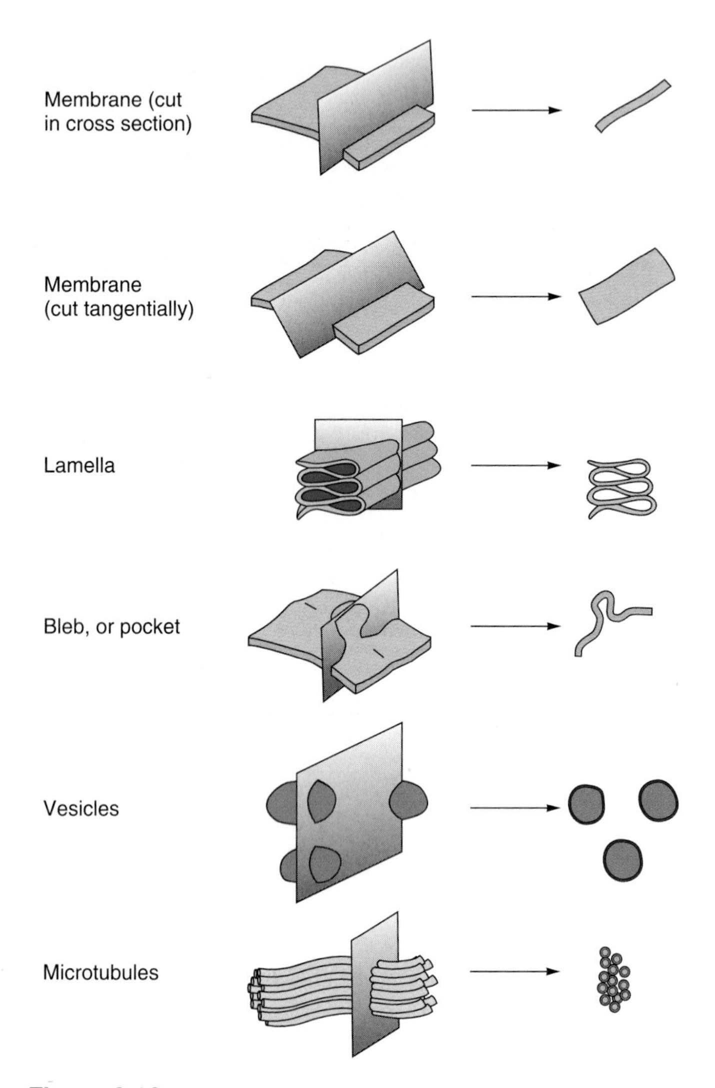

Figure 6.12

Cellular structures look different when sliced thin and viewed with a microscope. In the left column, some common cellular structures are shown in a three-dimensional view, and in the right column, as they appear through a microscope.

the specimen. Scanning electron micrographs show three-dimensional forms spectacularly (Figure 6.13). Although the SEM has lower resolving power than the classical transmission EM (a limit of resolution of about 3 nm at best), it produces revealing and elegant pictures of surface structures.

Exercise 6.3 To visualize how the three-dimensional structure of an object can be determined from two-dimensional sections, draw cross sections through an orange in two directions—across the sections and parallel to the sections.

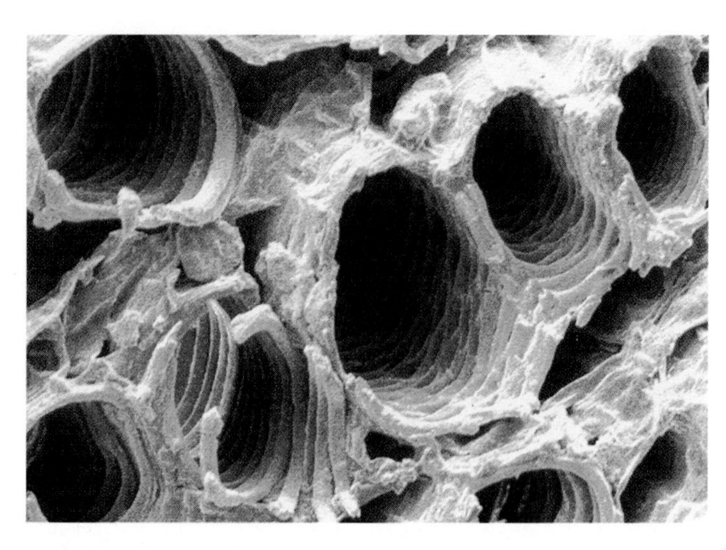

Figure 6.13

A scanning electron micrograph gives a feeling for the three-dimensional shapes of objects. These are tubules that conduct fluids through a leaf. The coloring is artificial.

6.5 The two major types of cells are procaryotic and eucaryotic.

Among the great variety of organisms outlined in Chapter 2, the bacteria stand out because they are so much smaller than other unicellular microorganisms, such as algae and protozoa. However, the most significant difference between bacteria and other cells is not size but structure. A high-power light micrograph shows that each plant or animal cell contains a large, central **nucleus** (Figure 6.14*a*). At this level of resolution, bacteria also have what appear to be nuclei (Figure 6.14*b*). But electron microscopy reveals that the nuclei of plant and animal cells (Figure 6.15*a*) are bounded by a pair of membranes that form a **nuclear envelope,** whereas bacteria have only a diffuse body with no surrounding envelope, called a **nucleoid,** rather than a nucleus (Figure 6.15*b*).

The nucleus houses the **chromosomes** (Figure 6.16), the complexes of deoxyribonucleic acid (DNA) and protein that carry the genetic information of the cell. The DNA of a bacterial cell, in contrast, is one long chromosome that is highly compacted to form the nucleoid. Based on this difference in nuclear structure, E. Chatton drew a major distinction between two types of cells and organisms. He termed the cells of bacteria **procaryotic** (*pro-* = before or primitive; *karyon* = kernel or nucleus); the cells of all other organisms—protists, chromists, fungi, plants, and animals—he called **eucaryotic** (*eu-* = true). (These words may also be spelled with a "k": prokaryotic, eukaryotic.) The organisms themselves are called **procaryotes** and **eucaryotes.** Table 6.1 shows several other features besides the differences in size and nuclear structure that distinguish these two types of cells and organisms.

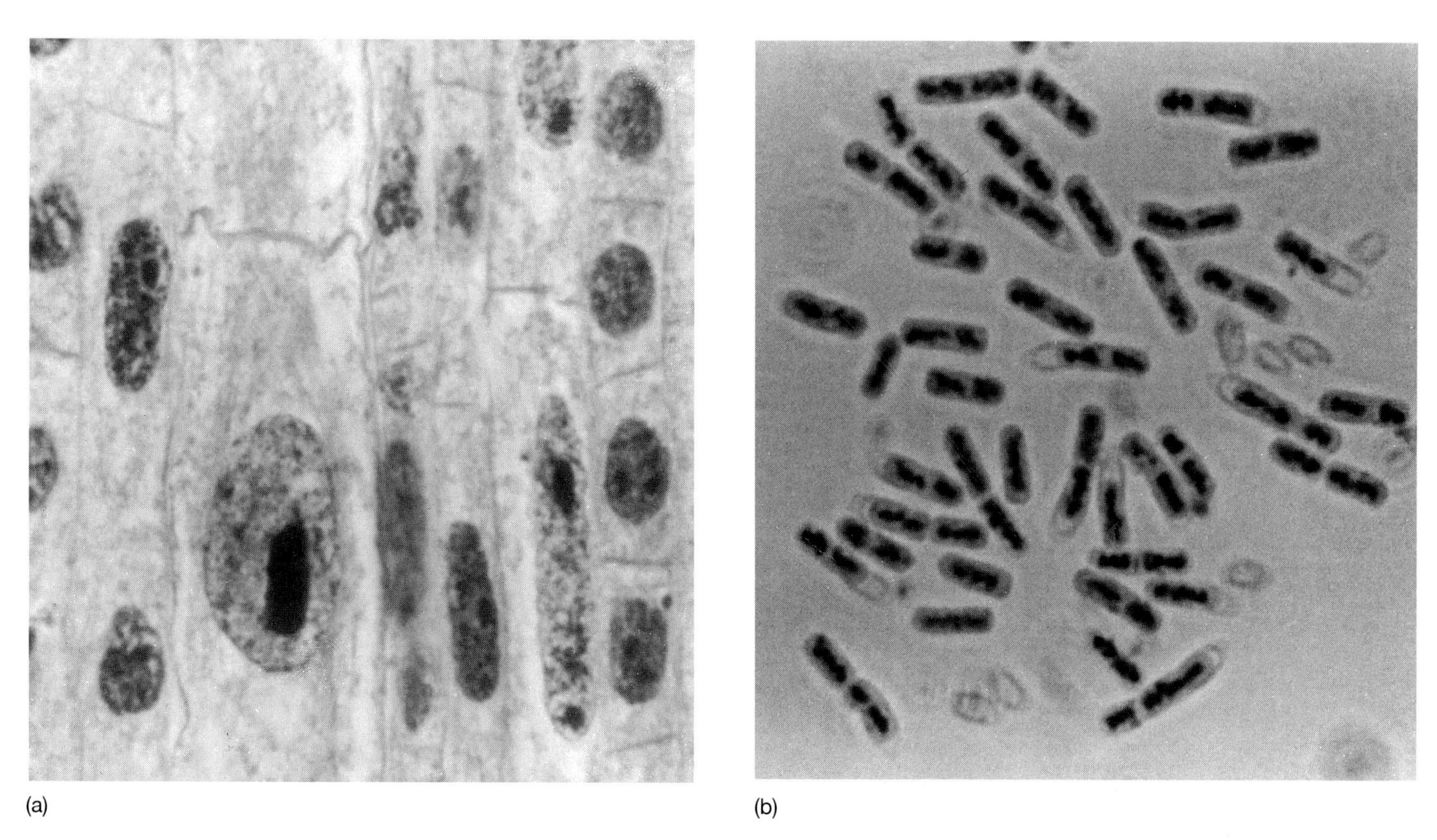
(a) (b)

Figure 6.14

(a) A high-power light micrograph of plant tissue shows that each cell contains a large, spherical nucleus. *(b)* Bacterial cells also show large, central nuclear regions, but these are not true nuclei.

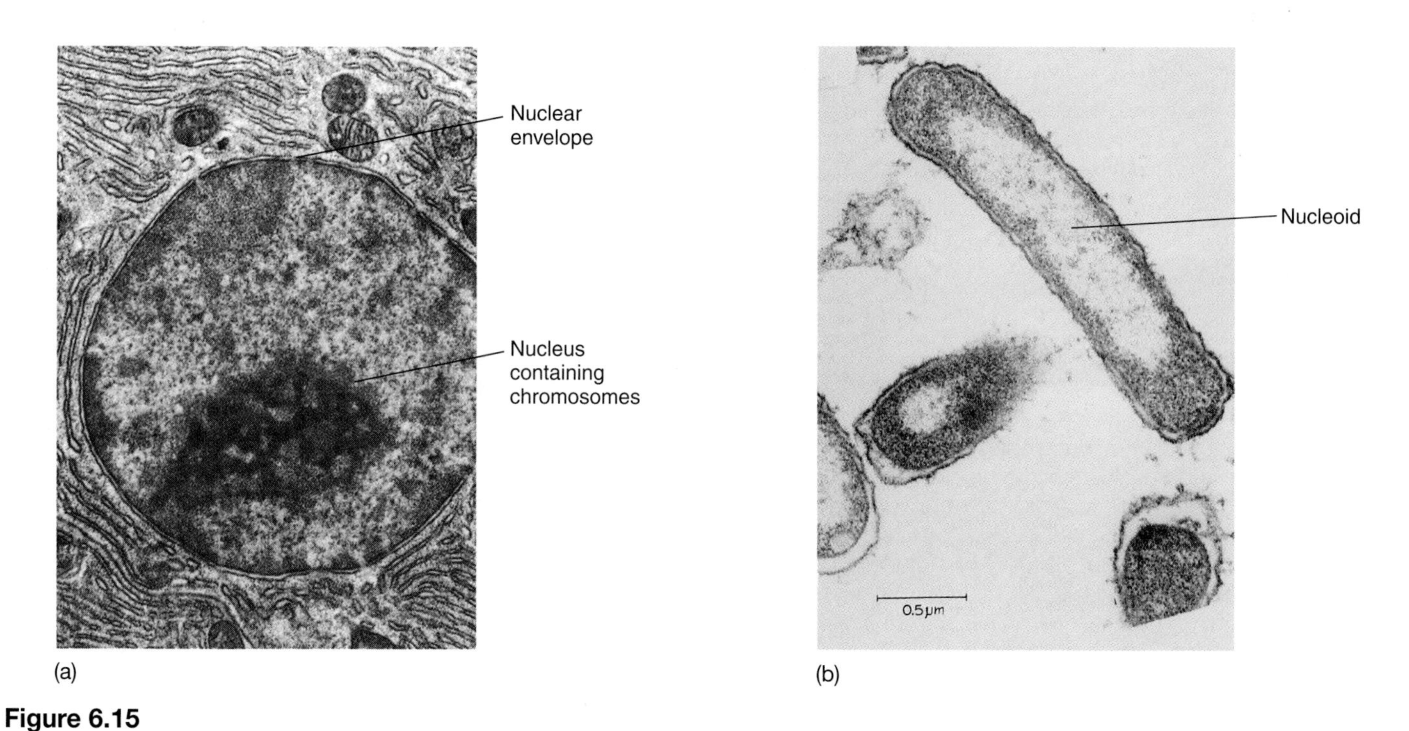

(a) (b)

Figure 6.15

(a) An electron micrograph of an animal cell shows that the nucleus is defined by a sharp membrane, the nuclear envelope. However, bacteria *(b)* have no such envelope; they have only nucleoids, which appear white in this picture, in the middle of the cytoplasm. Each bacterial cell is much smaller than the representative animal cell.

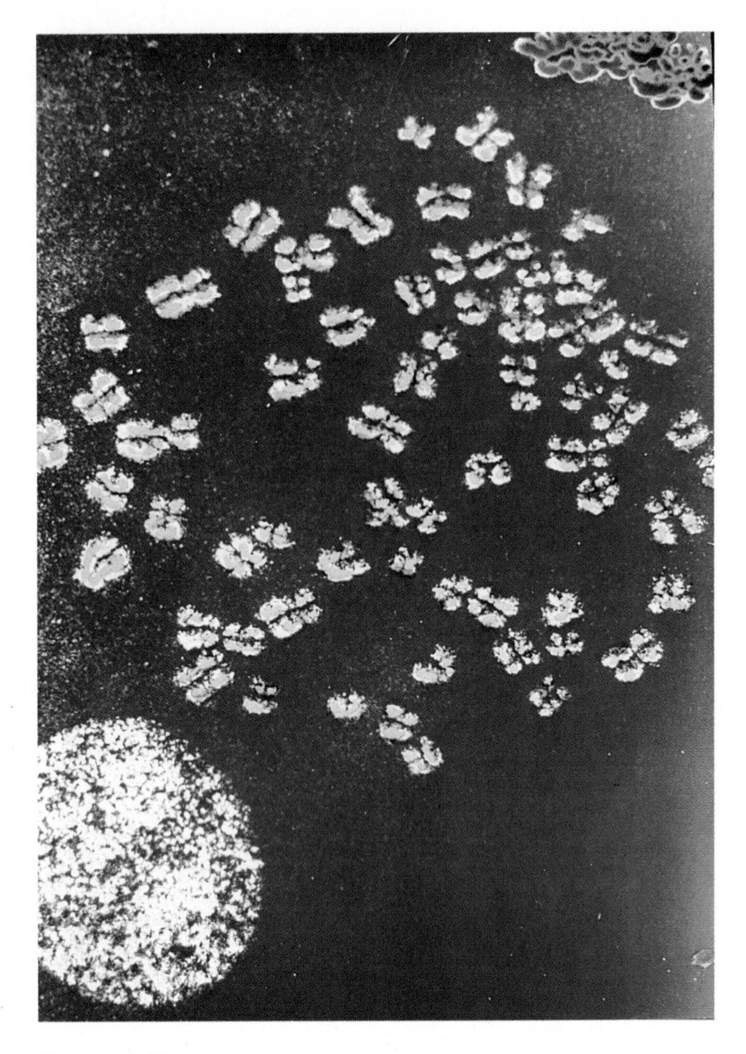

Figure 6.16
All the chromosomes from one nucleus are spread out here next to an intact nucleus. Chromosomes, the principal organelles inside a nucleus, are long nucleic acid molecules, combined with proteins, that contain most of the genes of a cell. (Mitochondria and chloroplasts also carry some genes.)

Exercise 6.4 Typical procaryotic cells seldom have diameters of more than 1 μm. Eucaryotic cells are typically about 10–50 μm in diameter. If a bacterium were a sphere with a radius of 1 μm and a eucaryotic cell were a sphere with a radius of 10 μm, what would be the relative volumes of the two cells? (Note: The volume of a sphere is $V = 4/3\ \pi r^3$, where r is the radius, but calculate only the *ratio* of the volumes.)

Exercise 6.5 Suppose a new organism discovered on Mars had cells about the same size as bacteria but with a definite nuclear envelope. Would you call it a procaryote or a eucaryote?

Table 6.1 Some Differences between Procaryotic and Eucaryotic Cells

Feature	Procaryotic	Eucaryotic
Typical diameter	< 1 μm	10–50 μm
Nucleus	Nucleoid	True nucleus
Intracellular compartments	Few or none	Extensive
Chromosomes	One (circular)	Several to many (linear)
Ribosomes	Small (70S*)	Large (80S*)

*S values measure the sizes of particles and molecules by their rate of sedimentation.

6.6 Cells contain structures called organelles that have specialized functions.

Cells are much more than membranous bags of biomolecules. They are highly organized and contain distinctive **organelles** ("little organs"), specialized parts that perform specific functions. (Procaryotic cells have fewer organelles than eucaryotic cells.) There are three kinds of organelles:

1. Structures made of internal membranes, which may be permanently or temporarily connected to one another. (Some biologists restrict the term "organelle" to these structures.)
2. Filaments and tubules made of specific proteins that form the *cytoskeleton* (*cyto-* or *-cyte* = cell), a complex that holds a cell in shape and makes it move.
3. Large complexes of enzymes and other proteins, sometimes including nucleic acids, that perform certain metabolic functions. One such complex is described in Sidebar 9.1.

Eucaryotic cells are highly **compartmentalized**—divided by internal membranes into compartments where specific metabolic processes are localized, often in the membranes themselves. The nuclear envelope divides a cell into two large compartments, the **nucleoplasm** inside and the **cytoplasm** outside, whose contents are quite different. Each compartment contains different proteins and smaller organelles that make compartments of their own. The **cytosol** is the aqueous fluid of the cytoplasm outside of membrane-bounded structures (Sidebar 6.1), and this term also applies to the fluid in procaryotic cells, which have no distinct nucleus. Procaryotic cells have little compartmentalization.

Sidebar 6.1 The Cytosol is Crowded

Although an electron microscope gives us a conception of the larger components of cells, no existing microscope can show the smallest molecules that fill the cytosol, the aqueous space around organelles. However, X-ray diffraction and other techniques have revealed the structure of proteins and small molecules, and we can also calculate how many molecules of each type occupy the cytosol. David Goodsell has used this information to create some pictures of the insides of cells, as shown here. This bacterial cell is about 2 μm long and about 0.5 μm (500 nm) in diameter. A square 100 nm wide is enlarged a million times to show all the molecules except water and salt ions (mostly potassium, magnesium, sodium, and chloride). Proteins of various shapes are shown in blue; most of them are enzymes. A bit of DNA, which fills the center of the cell, is shown in red. Much of the cytoplasm is occupied by molecules involved in protein synthesis: two types of RNA in orange interact with ribosomes in magenta—the factories where proteins are made. The smaller green molecules among the proteins are a variety of organic molecules such as amino acids, sugars, nucleotides, and coenzymes; most of them are molecules being transformed in the chemical reactions of metabolism. A volume the size of this square of cytoplasm and 1 nm deep also contains about 1.3 million water molecules.

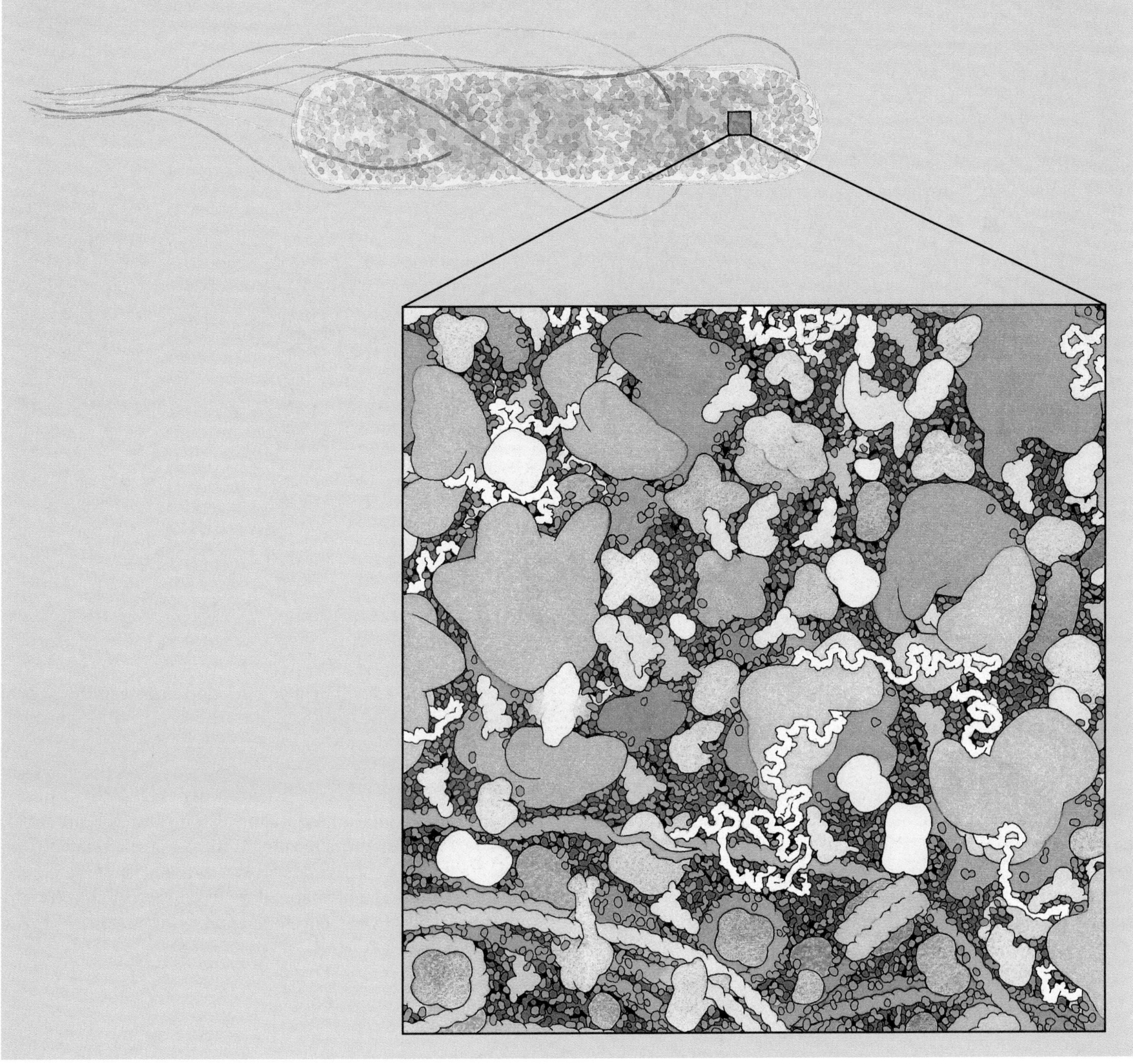

Sidebar 6.2 A Trip into a Cell

To get a better feeling for the size of a eucaryotic cell and its components, imagine that a whole cell is blown up by a factor of a million so that we can actually go inside and look around. Now the micrometers in which cells are measured will be meters (m), the nanometers will be millimeters (mm), and 10 nanometers will be a centimeter (cm). (Remember that a meter is about a yard and a centimeter is a little less than half an inch.) If a typical cell has dimensions of 20–40 μm, our cell is now a space that fills a good-sized lecture hall—a rectangular space at least 20 m on an edge in each direction. It would fill a basketball or tennis court and be about as high as the court is wide.

Let's step inside such a cell and look around. Most of the structures we see are made of, or surrounded by, membranes. These sheets are incredibly thin. Picture them as sheets of transparent plastic about 6–10 mm thick. On or in each sheet are many proteins, irregular blobs ranging in size from large pearls to typical marbles. Some of the proteins extend through the sheets; others are attached to one face or the other, either recessed into the plastic or more lightly attached. Each membrane is a mosaic of these protein blobs, which are actually free to move around in the membrane. If the proteins average a centimeter in diameter, there can be 10,000 of them *in each square meter* of the membrane! They have many different shapes and sizes, because they have many different functions.

The cytosol surrounding all these structures is filled with water molecules about 0.5 mm in diameter, and we are constantly being struck by them and by ions, such as sodium, potassium, and chloride, which are (appropriately) like grains of salt, about 0.1 mm in diameter. The cytosol is also thick with proteins, mostly enzymes—some alone and some in clusters—averaging about 1 cm in diameter, and with many other organic molecules 1–2 mm wide.

Suspended in the middle of this cell and occupying much of its volume is a round nucleus, a membranous bag about 5–10 m across. Peering inside the nucleus, we see chromosomes. The DNA of these chromosomes by itself is only a heavy string, about 2 mm thick, but it is covered with all kinds of proteins. We may find it hard to move around near the nucleus, for the space is filled with layers of endoplasmic reticulum separated by only a few centimeters, so we can hardly get between them. The ER membranes are covered with thousands of ribosomes, each about the size of a ping pong ball and each made of many smaller proteins.

Throughout the cell, we encounter mitochondria that look like huge loaves of bread, about a meter in diameter and several meters long; they are always changing shape a little, like fat worms. Other membranous bags—vesicles—are scattered through the cell. But much of the space not occupied by membranes is taken up with two kinds of long protein structures of the cytoskeleton: microtubules and microfilaments. The microtubules are tubes about 0.25 mm in diameter, but they are incredibly long and stretch throughout the cell, holding some membranes in shape. The microfilaments are only about 6–7 mm wide, rather like thin ropes. Bundles of them run here and there through the cell, also holding it in shape. We can see the bead-sized protein subunits in both microtubules and microfilaments.

One impression that should come out of this brief trip is the great contrast between sizes of objects. Even at this great enlargement, atoms and small molecules would be barely visible, and the basic macromolecules of the cell would be only a few millimeters in diameter, forming functional structures a few centimeters across. The only cellular structures we can see with a light microscope are those that have dimensions of several meters in this imaginary cell.

6.7 A eucaryotic cell is compartmentalized by its nucleus and other membranous organelles.

In the rest of this chapter we will outline the structure of idealized eucaryotic plant and animal cells and introduce their organelles (Figure 6.17). We will discuss the functions of these organelles in later chapters, but first it's useful to learn to identify them in electron micrographs and to get a general idea of their unique functions, structural features, and relationships to one another. Sidebar 6.2 will help you develop a sense of the sizes of cellular components.

The nucleus

The rather spherical nucleus is a prominent feature of a eucaryotic cell and often its largest organelle. Forming its boundary is the nuclear envelope which controls the passage of material between the nucleus and cytoplasm just as the plasma membrane regulates the passage of material between the cell and its surroundings. The nuclear envelope consists of two membranes,

Figure 6.17

(a) An idealized animal cell and *(b)* a plant cell summarize the principal components of eucaryotic cells. We will use these for guidance and orientation; they do not portray any actual cells.

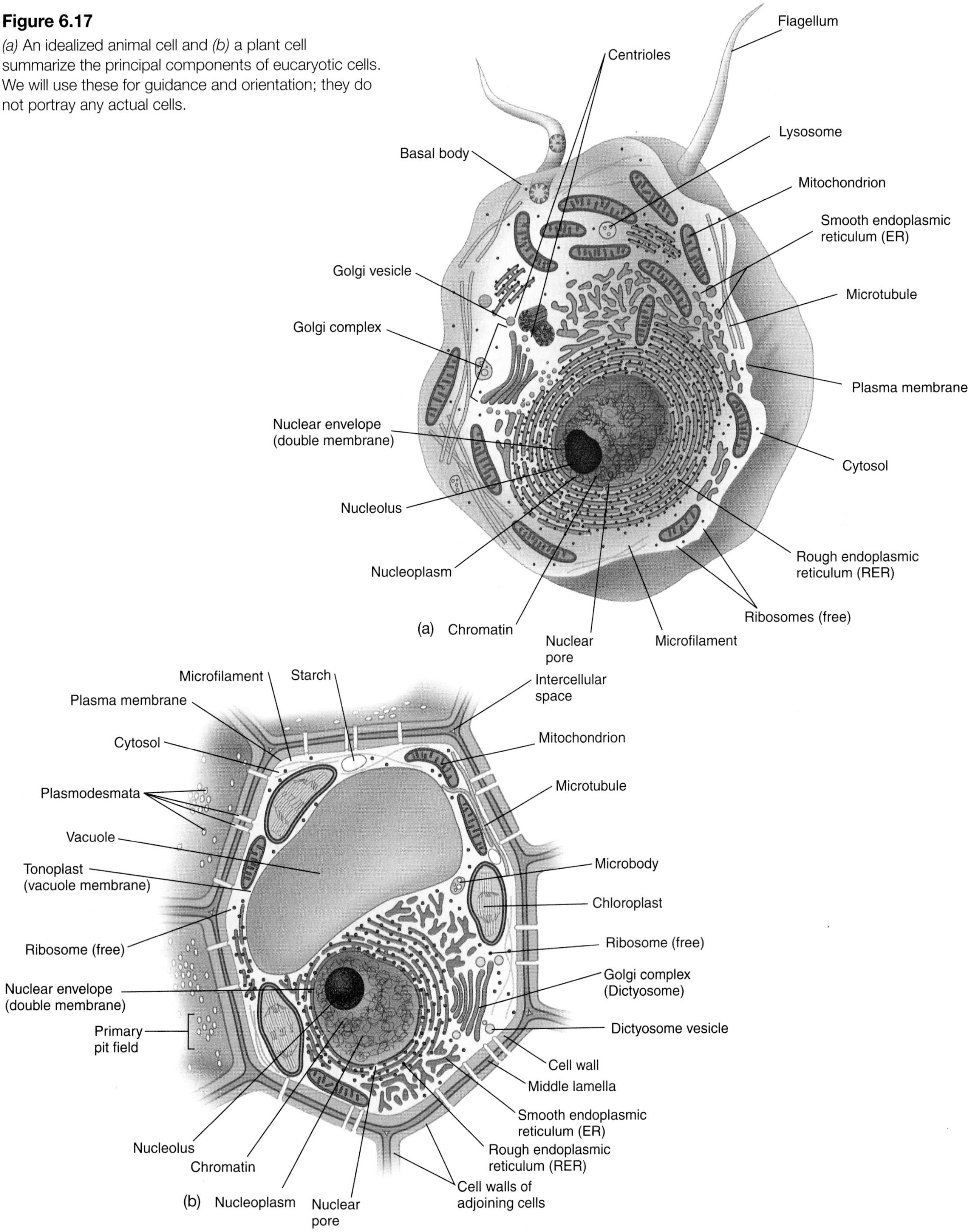

spaced about 5–15 nm apart except at many points where they are fused to make a pore or channel that connects the nucleoplasm and cytoplasm. Each pore is about 50 nm in diameter and surrounded by an octagonal complex of protein granules (Figure 6.18). The pore complex is made of an estimated 100 types of protein, including several that transport both proteins and nucleic acids into and out of the nucleus.

In light micrographs, often the most prominent object is a **nucleolus** ("little nucleus"; Figure 6.19) inside the nucleus. Nucleoli are the sites where the ribonucleic acid (RNA) of **ribosomes** is synthesized; ribosomes are particles made of protein and RNA, which are then exported to the cytoplasm to serve as the protein factories of cells. Although nucleoli have no surrounding membranes, they can be isolated in quantity and analyzed chemically because they are very dense organelles, the result of tight associations between their constituent macromolecules.

The endomembrane system

Much of the cytoplasm in a typical eucaryotic cell is occupied by the **endomembrane system,** made of internal membranes involved in synthesizing proteins and distributing them to other cellular compartments or exporting them from the cell as detailed in Chapter 14. The most prominent part of the system is the **endoplasmic reticulum (ER),** an extensive series of membranes called smooth ER or rough ER, depending on their appearance in electron micrographs. Rough ER membranes look rough because they are covered with ribosomes, whereas smooth ER membranes have no ribosomes attached. Since the bulk of a cell's dry mass is protein and proteins are so critical to an organism's activities, everything connected with protein synthesis is of central importance. The ribosome factories are abundant in all cells, and in eucaryotic cells many ribosomes are bound to ER membranes.

Typical rough ER (Figure 6.20) is a remarkable stack of membranes that lie parallel to one another and to the nuclear envelope, partially or entirely surrounding the nucleus. Detailed micrographs show that nuclear and ER membranes are connected at some points, but ER membranes contain some distinctive proteins not associated with nuclear functions. The ER membranes separate a space within themselves, the **lumen,** from the surrounding cytoplasm. Ribosomes attached to the ER membranes on their cytoplasmic face synthesize proteins that pass through the membranes into the ER lumen, where they may be chemically modified.

A general principle that links structure and function in cells is the membrane-metabolism principle: Where a metabolic process is carried out by the components of a membrane, a cell makes extensive membrane to provide for that process. The enormous layers of rough ER membranes are one instance of this principle; these are membranes where proteins are made (by attached ribosomes), and to synthesize a lot of protein, the cell requires a lot of membrane. This principle relates metabolism to visible structure—that is, cells may also make large amounts of

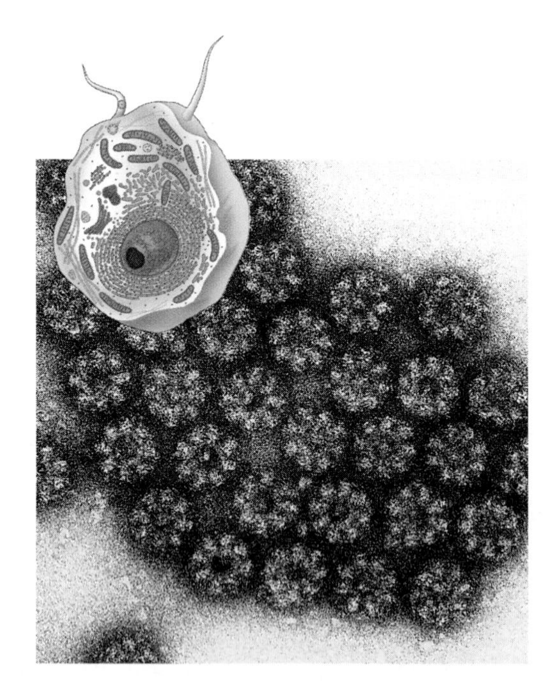

Figure 6.18
The nuclear envelope contains pores made by an octagonal complex of many proteins that regulate the passage of substances into and out of the nucleus.

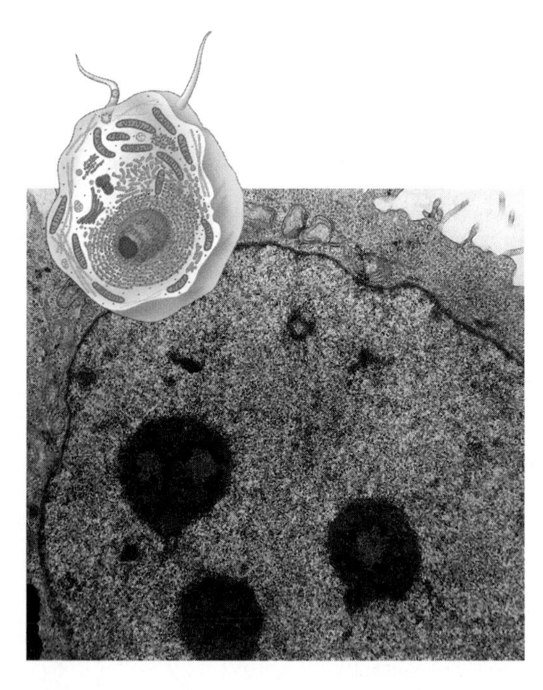

Figure 6.19
Nucleoli are prominent dark structures inside the nucleus. Here is where the RNA component of ribosomes is made.

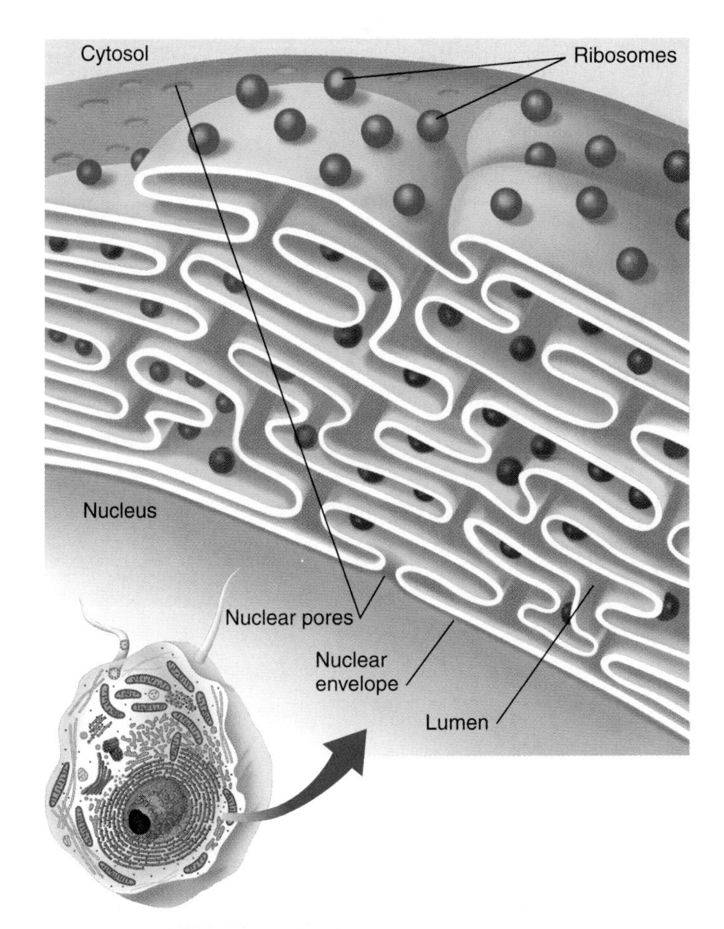

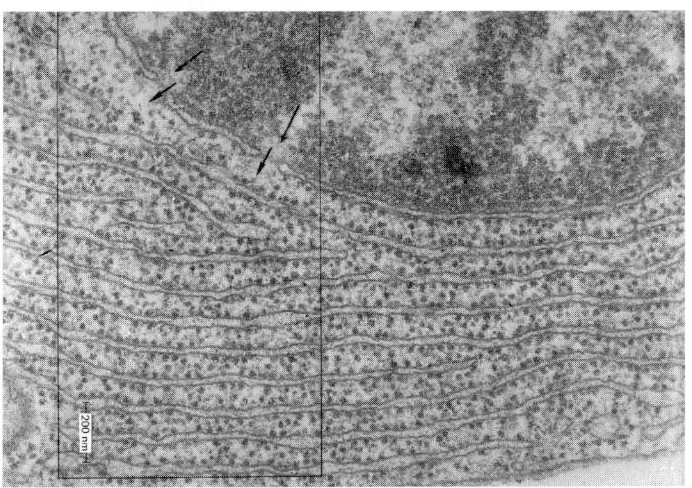

Figure 6.20

The rough endoplasmic reticulum typically consists of parallel membranes with ribosomes attached. Notice that the ribosomes are all on the cytoplasmic face of the membrane; the folds of membrane enclose a separate space, the lumen of the ER. The arrows show that the nucleus is connected through pores in the nuclear envelope to the cytoplasmic space.

an invisible, soluble enzyme in response to a metabolic need, but the visible membrane immediately reflects a metabolic function.

Smooth ER membranes form tubules and other irregular shapes (Figure 6.21). These membranes are the sites of lipid synthesis, and they are abundant in cells that produce nonprotein hormones. The smooth ER of liver cells contains enzymes that alter and detoxify drugs and other foreign materials. Biopsies from patients taking large doses of certain drugs show that they develop extensive smooth ER in their liver cells.

Exercise 6.6 Explain how the membrane-metabolism principle is illustrated by the presence of extensive smooth ER in liver cells from patients who are taking a lot of drugs.

Around 1900, Camillo Golgi described structures that readily took up a silver stain in the nerve cells of barn owls. These structures came to be called the **Golgi complex,** or Golgi apparatus. It took half a century before electron microscopy showed that the Golgi complex is a stack of distinctive, flattened membranous vesicles connected to branching tubules and associated with ER membranes (Figure 6.22). Golgi complexes, called *dictyosomes* in plants, are the packaging, sorting, and exporting centers of the cell. Proteins that accumulate in the ER lumen move on to the Golgi membranes where they are modified (for instance, by having sugar molecules added). Then they are sorted and directed to certain destinations in the cell or exported from the cell to be used elsewhere.

Mechanism of sorting and directing within a cell, Section 14.8.

Vesicles, vacuoles, and other bodies

The cytoplasm of eucaryotic cells contains a variety of small, rounded sacs called **vesicles,** which are bounded by single membranes. They have diverse functions. Some transport substances from one compartment to another, as from the ER to the Golgi membranes. Others contain proteins or other substances that have been packaged in the Golgi membranes and are being held for export from the cell.

Animal cells contain distinctive vesicles called **lysosomes** that are packed with many hydrolytic enzymes—that is, enzymes that can hydrolyze (digest) various polymers, thus reducing large molecules to small ones. The enzymes in lysosomes digest materials that a cell takes in from outside, and they even digest the cell's own components—for instance, destroying old cell structures or removing entire cells during some processes of development.

Peroxisomes are small vesicles containing specific enzymes, including oxidases that break down amino acids and fats. One by-product of these reactions is hydrogen peroxide (H_2O_2), which is very destructive. The enzyme catalase in the peroxisomes breaks down excess H_2O_2 into water plus oxygen, thus protecting the cell. Peroxisomes have quite diverse functions (and therefore diverse contents) in different cells. For instance, peroxisomes of white blood cells contain peroxidases that destroy invading bacteria with H_2O_2. In plants that store fats in their seeds, specialized peroxisomes help developing seedlings utilize these fats. Peroxisomes also aid in the metabolism of a photosynthetic by-product in the leaf cells of many species.

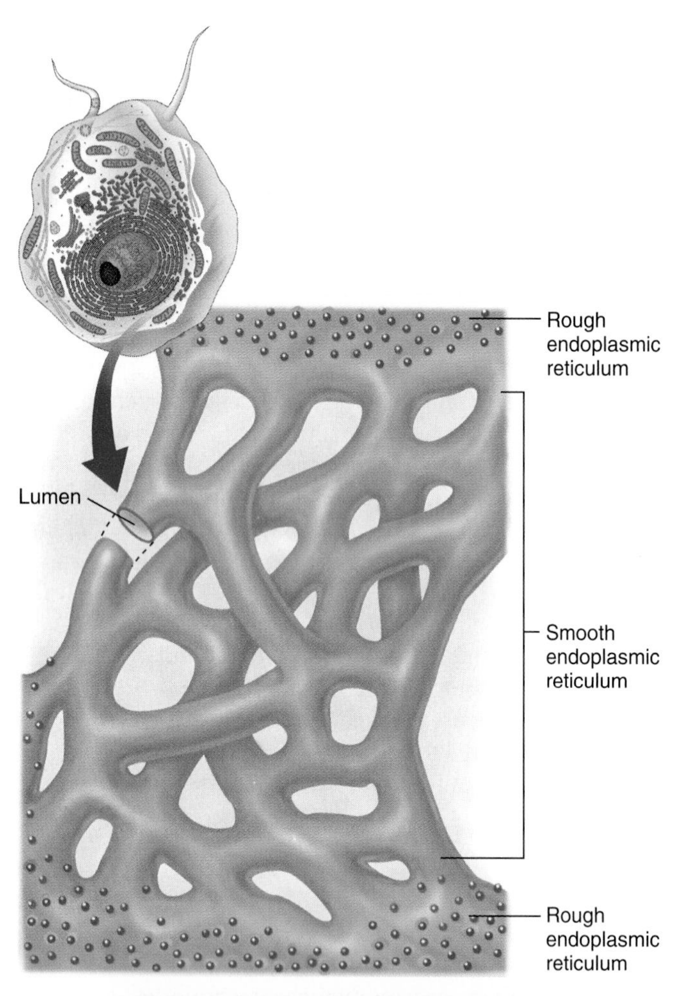

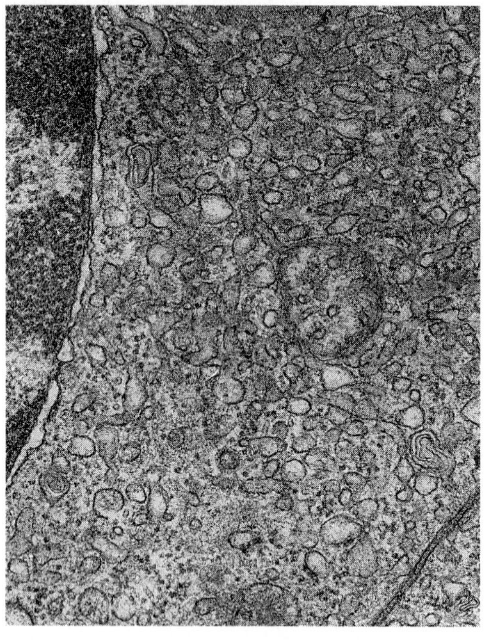

Figure 6.21

The cytoplasm of this gland cell contains extensive smooth ER. This cell produces large amounts of a hormone (adrenaline) that is synthesized in these membranes.

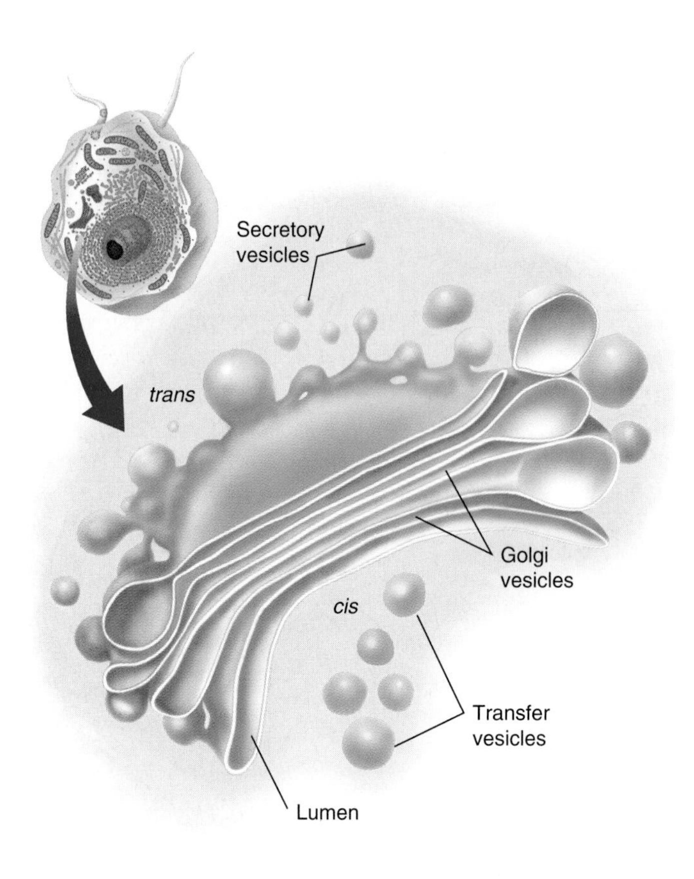

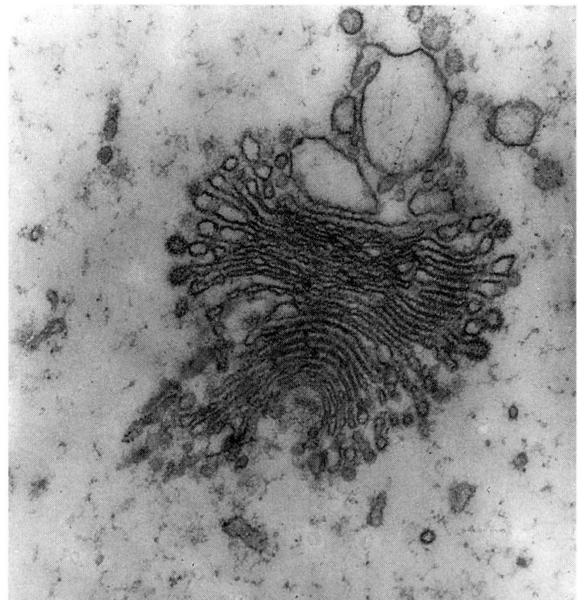

Figure 6.22

The Golgi complex consists of a set of membranes with very distinctive shapes. Here is where proteins are modified and prepared for distribution into various compartments.

Plant cells commonly have a large, prominent, central **vacuole** whose fluid contents, called *cell sap,* are quite different from the surrounding cytoplasm. As a plant cell matures, the vacuole often expands and may eventually occupy most of the cell's interior. Proteins embedded in the **tonoplast,** the membrane of the vacuole, regulate the passage of ions and molecules, and the vacuole has a crucial role in keeping a cell turgid—that is, swollen and distended. Plant cells are surrounded by a rigid wall and do not have lysosomes, but the vacuole serves for waste disposal. The tonoplast transports wastes and other substances destined for degradation into the vacuole, where enzymes digest them. The purple and red pigments (anthocyanins) of structures such as petals and fruits are also deposited in the vacuole. Substances in the cell sap may become concentrated enough to form crystals.

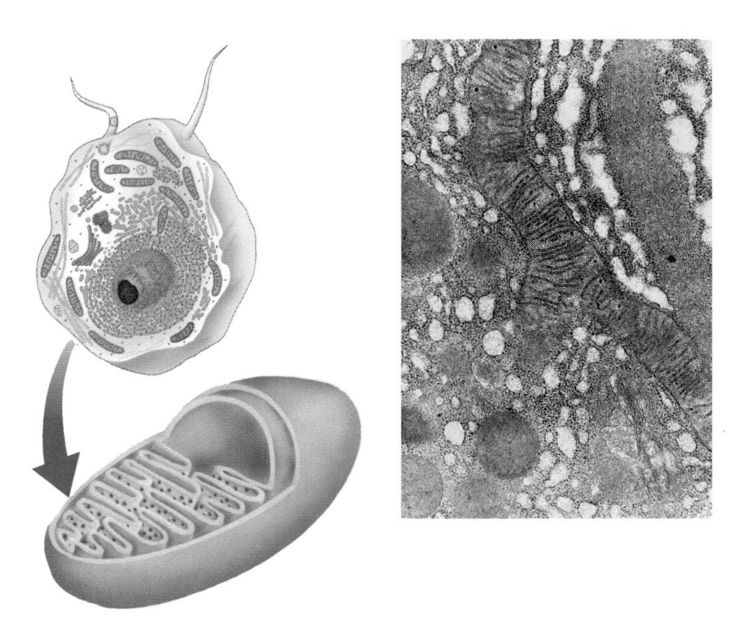

Figure 6.23
Mitochondria are long, thin bodies with internal cross-membranes (cristae). Most of the energy-yielding processes of the cell are carried out here.

Mitochondria and chloroplasts

Eucaryotic cells contain many **mitochondria** (sing., **mitochondrion**), the organelles that carry out the principal chemical reactions in which energy is extracted from foodstuffs (Figure 6.23). Mitochondria are especially concentrated in muscle cells, where they support movement, and in regions of other cells where a lot of energy is needed. Mitochondria are comparable in size to bacteria but are quite long and thin (*mitos* = thread; *chondros* = grain). They are made of two membranes; the inner membrane divides the organelle into two internal compartments and is thrown into many crosswise folds called **cristae.** Here is another instance of the membrane-metabolism principle: The inner membrane contains many of the enzymes and other proteins that carry out the reactions of energy metabolism, and the cristae greatly increase its area, thereby increasing the rate at which metabolism can be carried out.

Eucaryotes that conduct photosynthesis, such as plants and algae, contain **chloroplasts,** which capture light energy and store it in chemical forms. Chloroplasts are surrounded by two membranes, with a third, the *thylakoid membrane,* occupying much of the inside. The membrane-metabolism principle explains the structure of the thylakoid membrane, too: It contains the chlorophyll that absorbs light (and gives plants their green color), along with many of the proteins that carry out photosynthesis. The thylakoid membrane's extensive, highly folded structure provides a huge surface for capturing light and carrying out these unique reactions. Many plants and algae can be identified by their distinctively shaped chloroplasts (Figure 6.24). A chloroplast is one of a variety of small plant organelles called **plastids.** Some plastids are immature chloroplasts, and others are the sites of special phases of metabolism; *amyloplasts,* for instance, synthesize and store large amounts of starch.

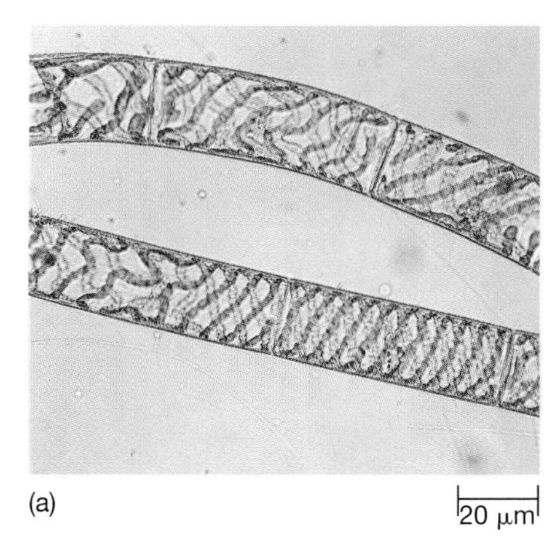

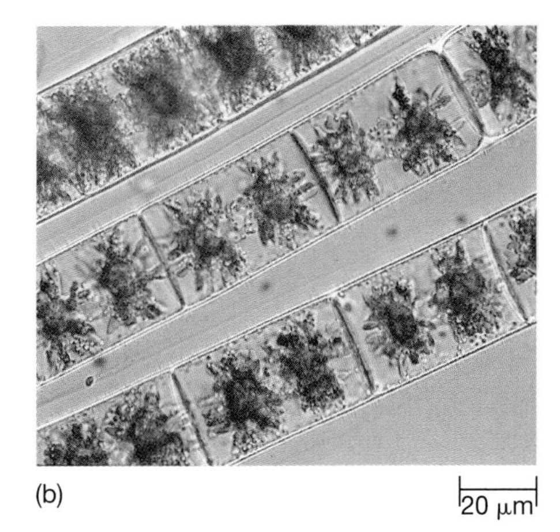

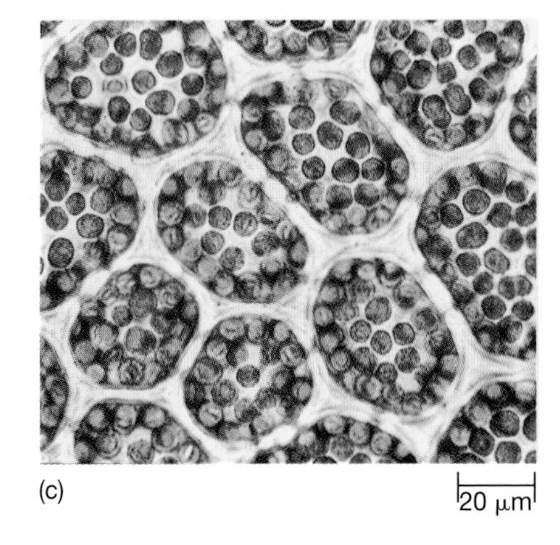

Figure 6.24
Chloroplasts are the centers of photosynthesis in plants and algae. They are generally green, due to the chlorophyll that captures light energy, but may also be rather yellow or brown. Their distinctive shapes are shown here by (a) *Spirogyra,* (b) *Zygnema,* and (c) *Rhizomnium.*

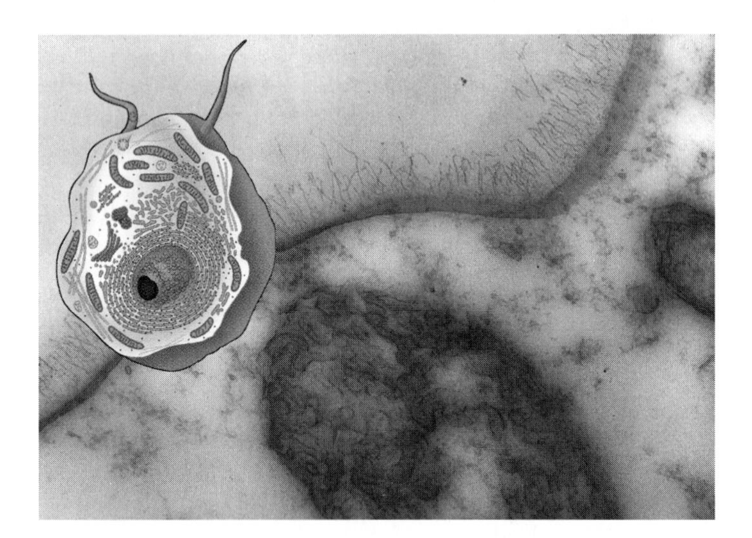

Figure 6.25
The glycocalyx is a fuzzy layer of oligosaccharide filaments on animal cell surfaces.

The cytoskeleton

Eucaryotic cells have unique shapes, and many of them change their shapes extensively, especially when they grow and divide. All these forms and movements are created by the **cytoskeleton,** a network of protein tubules and filaments running through the cytoplasm. Because the cytoskeleton and its activities require a carefully told story, we devote most of Chapter 11 to telling it.

6.8 Eucaryotic cell surfaces have characteristic features.

The surfaces of animal cells are covered with a **glycocalyx** of fine filaments on the outside of the plasma membrane (Figure 6.25). These are oligosaccharides (short chains of sugars) that tend to stick to one another, so the glycocalyx helps bind neighboring cells together. Because the sequences of these sugar chains are distinctive, they are probably tags that identify different types of cells, enabling them to recognize one another and interact appropriately as the animal develops.

The cells of plants, some fungi, and many protists are surrounded by a rigid *cell wall* outside the plasma membrane (Figure 6.26). The rigidity of this wall gives the cell its shape, even though the plasma membrane within it remains flexible. Plant cell walls are largely constructed of cellulose; some protists have walls composed of silica and other substances, that may be beautifully sculpted (Figure 6.27).

6.9 Cells in a tissue may be connected by several types of junctions.

Adjacent cells in animal tissues are packed closely, but still separated by a space of about 20–30 nm. This space is filled with extracellular fluid, which supplies the cells with nutrients and carries away their wastes. Yet adjacent cells are often connected by junctions, which we divide into two categories: **adhering junctions** and **communicating junctions** (Figure 6.28).

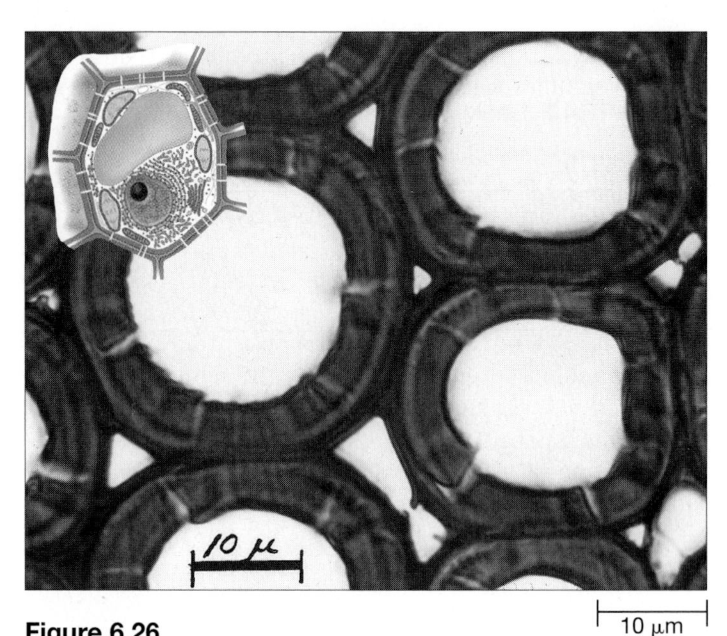

Figure 6.26
Plant cells have heavy walls that define their boundaries and keep them intact. These cell walls, containing cellulose and related polysaccharides, account for the solid, rigid nature of plant structures, especially wood.

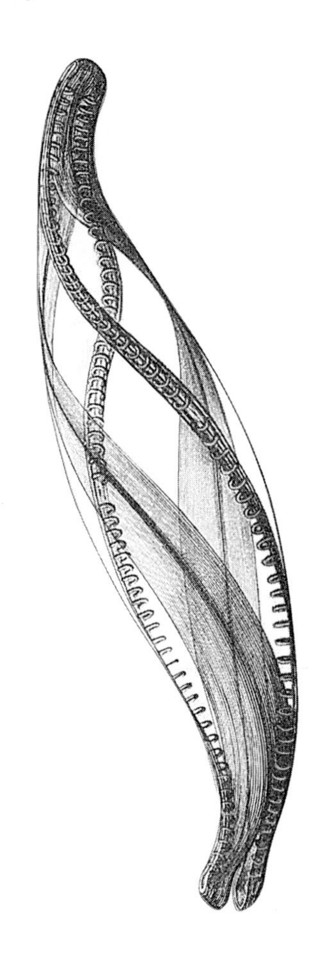

Figure 6.27
The cell wall of a diatom (a chromist) is a finely detailed, symmetrical sculpture of silica.

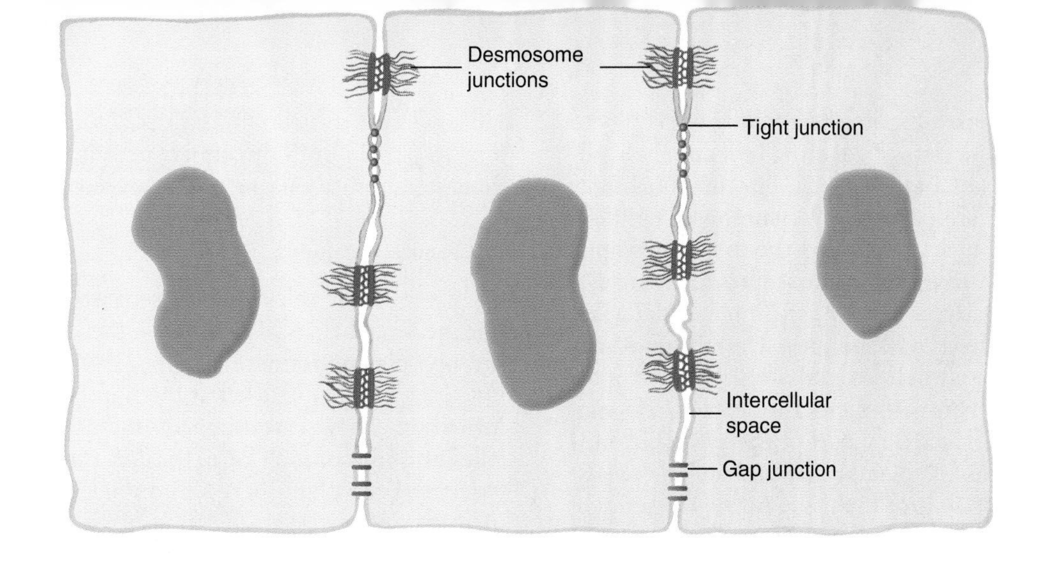

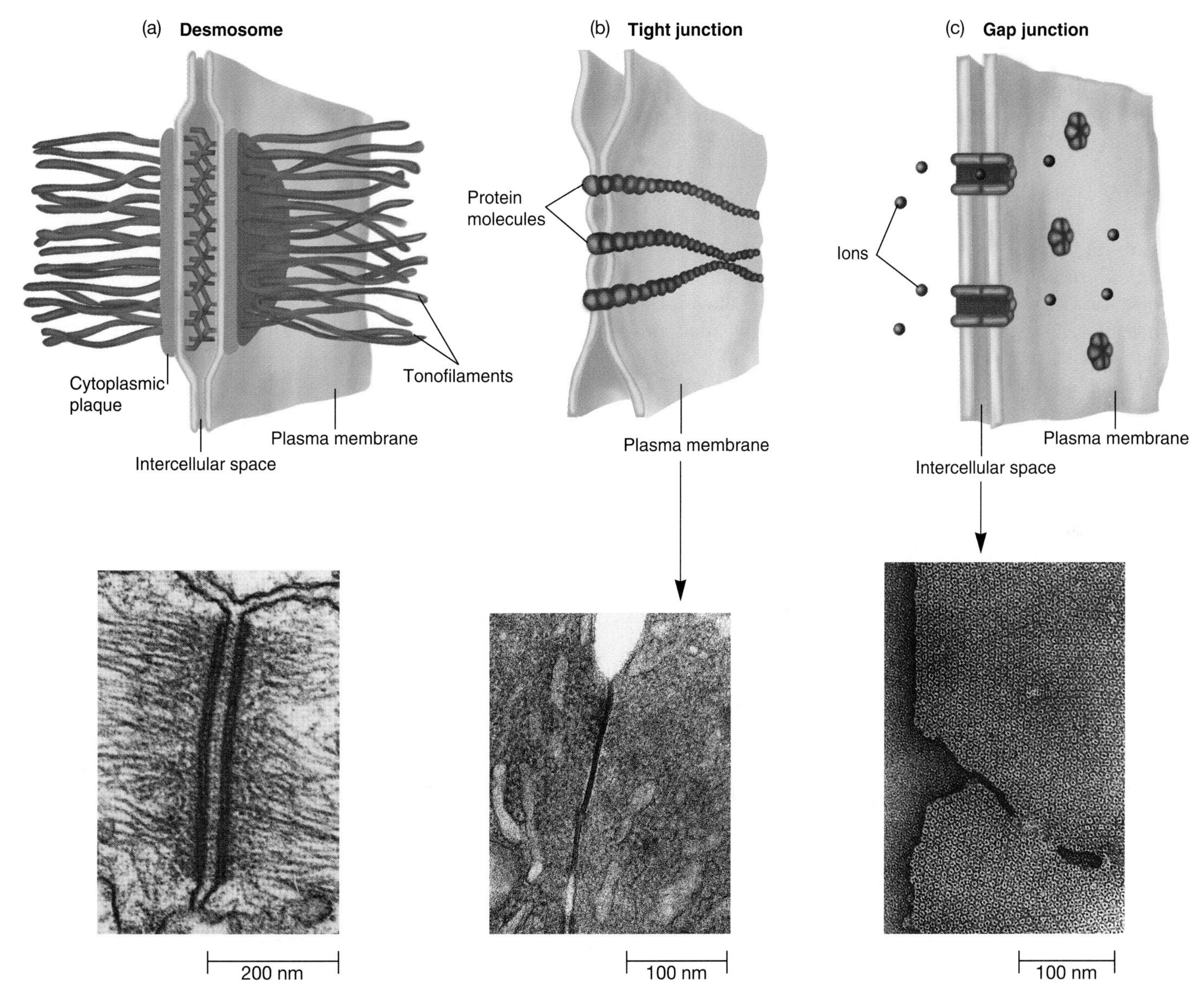

Figure 6.28

Adjacent cells in animal tissues are separated by narrow spaces filled with intercellular fluid, and are joined at adhering junctions, either desmosomes or tight junctions. *(a)* Spot desmosomes form between two half-desmosomes, one in each cell, connected by *inter*cellular filaments; they are anchor points for an *intra*cellular skeletal network that gives the tissue strength. *(b)* At tight junctions, the plasma membranes of the neighboring cells are held together by rows of proteins that join one membrane to the other with no intervening space. *(c)* Gap junctions contain channels between cells. Each channel is a hexagonal pipe made of the protein connexin, formed from a half-pipe in each cell. The channels are about 2 nm in diameter, allowing ions and small molecules, but not proteins, to pass through.

Adhering junctions link cells into sheets and give rigidity to organs. **Desmosomes,** one class of adhering junction, hold adjacent cells together like spot welds or bits of glue. They strengthen a tissue and are particularly common in epithelia (layers of surface cells) such as skin and the intestinal lining. Desmosomes are created when half-desmosomes form in adjacent cells, and the two halves join across the intercellular space through a complex of fibrils and granules. Desmosomes may form long belts that run entirely around a cell or small spots that hold cells together here and there. Inside each cell, spot desmosomes are interconnected by part of the cytoskeleton, a network of filaments (tonofilaments) that probably helps distribute stresses applied to part of the tissue through a series of linked cells to the tissue as a whole (Figure 6.28*a*).

Tight junctions, another type of adhering junction, are marked by a network of extraordinary ridgelike structures where the membranes of the two cells are held together intimately by connecting proteins. By forming rings around cells, tight junctions fuse each cell to its neighbors so tightly that extracellular fluid can't leak past. Therefore, to get across a layer of cells, materials must pass through the cells, not around them. For instance, tight junctions in the epithelial linings of some body cavities, such as the intestines and bladder, prevent the contents of these cavities from leaking out.

Communicating junctions, in contrast to adhering junctions, pass information from cell to cell, and some allow ions and small molecules to flow between the cytosol of adjacent cells. The prime examples are **gap junctions** made of protein pores that span the membranes of adjacent cells. Each pore is a multimer of six molecules of the protein connexin in one cell bound tightly to an identical multimer in the other cell, making an open pore about 2 nm wide. Molecules smaller than about 1,000 daltons can flow through these channels. For instance, gap junctions convey ions that stimulate contraction in heart muscle cells, thereby causing the heart to contract in a coordinated way.

The channels in gap junctions are controlled by the concentration of calcium ions around them. The usual concentration of free Ca^{2+} in the cytosol is 10^{-7} to 10^{-6} molar, and if it rises above this level, the channels close. Since the Ca^{2+} concentration in the extracellular fluid is about a thousand times higher, any injury to the plasma membrane will admit calcium to the cytosol and cause the channels to adjacent cells to close. Through this mechanism, cells in a tissue protect themselves by sealing off any that become injured.

We have just defined communicating junctions structurally, but from a *functional* viewpoint, many of them are known as **synapses,** regions where cells can send signals to one another. Gap junctions are also called *electrical synapses* because they have very low electrical resistance and provide conduits for electrical communication between cells. If a pulse of electricity is given to one cell with a microelectrode, it can be detected in an adjacent cell. *Chemical synapses,* are used by cells in the nervous system to transmit information to other cells by means of molecules called *neurotransmitters.* While gap junctions allow ions to pass in both directions, neurotransmitters pass in one direction, from a sending cell to a receiving cell.

Chemical Synapses, Section 41.7.

Many plant cells, though separated by the considerable thickness of their walls, still have their cytoplasms connected through **plasmodesmata** (Figure 6.29). These communicating junctions are channels left at the time of cell division and wall formation. They serve as conduits for the exchange of molecules and ions among cells, thus creating a network in which large parts of a plant share a common cytoplasm. This arrangement has interesting implications for plant biology that are explored in the chapters on plant physiology.

6.10 Procaryotic cells also contain organelles.

Procaryotic cells, small as they are, also have some distinct organelles. Most bacteria are held in characteristic shapes, usually rods or spheres, by a rigid wall (quite different in structure from plant cell walls) just outside a plasma membrane. As in eucaryotic cells, the bacterial cytoplasm is rich in ribosomes, the sites of protein synthesis, but in procaryotes the ribosomes are free-floating structures concentrated in the outer regions of the cell. The center of the cell is largely occupied by the nucleoid—or nucleoids, since rapidly growing cells may have more than one. In some bacteria, the plasma membrane has **mesosomes** that extend into the interior of the cell and connect to the nucleoid (Figure 6.30). When we discuss cell division in Chapter 13, we will show how mesosomes help separate the nucleoids of a dividing cell equally into the daughter cells. By definition, procaryotic cells have no envelope around the nucleoid that would make it a true nucleus. All other organelles typical of eucaryotic cells are absent in bacteria.

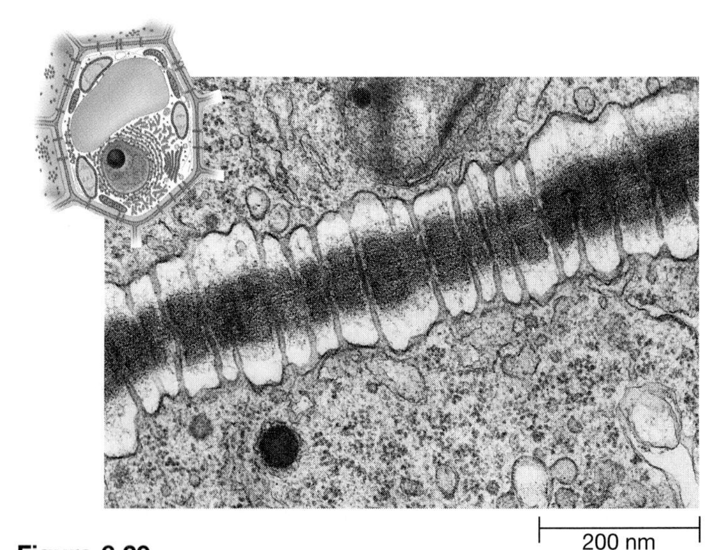

Figure 6.29

Plasmodesmata are open junctions, analogous to gap junctions, that are left between plant cells when new walls form after cell division.

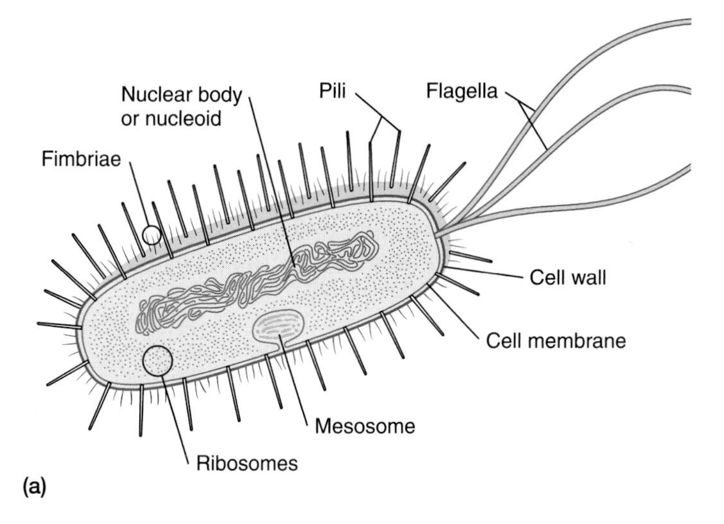

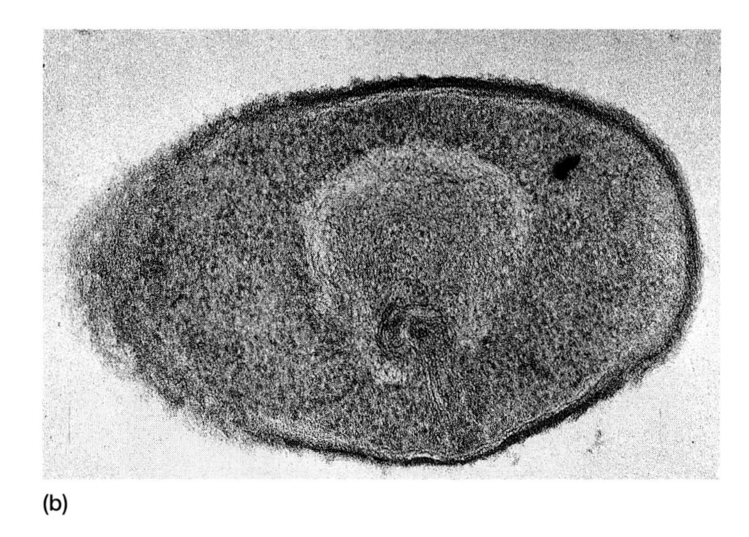

Figure 6.30

(a) A typical procaryotic cell has extensive surface structures, a nucleoid (DNA), but no internal membranes or very few, though mesosomes are common. *(b)* The mesosome of the bacteriium *Bacillus subtilis* is an extension of the plasma membrane connected to the nucleoid (light-colored fibrous material).

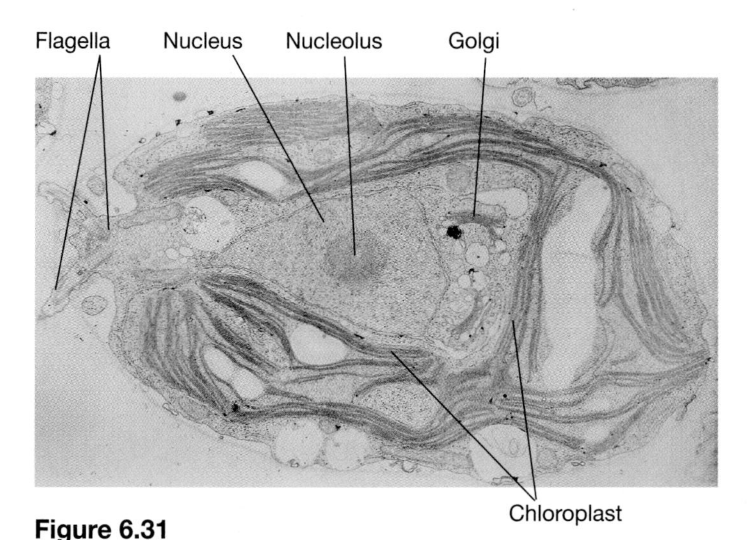

Figure 6.31

The green alga *Chlamydomonas* is a single eucaryotic cell, showing most of the features we discussed earlier in this chapter.

As yet another illustration of the membrane-metabolism principle, some procaryotes, especially photosynthetic bacteria, have extensive internal membranes that form a large apparatus for energy metabolism. Along with the plasma membrane, these internal membranes perform the functions of mitochondria and chloroplasts in eucaryotic cells; that is, they contain the proteins that extract and store energy and, in photosynthetic cells, capture light energy.

Coda Cells are the basic functional units of organisms, and it is essential to have a feeling for their general organization. The distinction between procaryotic and eucaryotic cells forms a significant dividing line in the evolutionary history of life on Earth, a line between organisms with relatively little functional organization and those with many specialized organelles. To review our discussion of cell structure, look at a simple algal cell (Figure 6.31). Each of its organelles is made of specialized proteins, often embedded in membranes, that perform their functions by interacting with one another and with various ligands. The information for making these proteins resides in the cell's genome, mostly in the nucleus except for small amounts in the mitochondria and chloroplasts. This information is transformed into actual protein structures in the organism's ribosomes. Also, the information has been shaped by evolution to give each protein a very functional structure, where the criterion is essentially ecological—whether the structure helps adapt the whole organism to its particular way of life, a life of swimming in a pond within a certain ecosystem where it stays alive by capturing the energy of light through photosynthesis. Photosynthesis occurs in the alga's large, cup-shaped chloroplast, and the organism moves through the pond by means of two long flagella, which are extensions of its cytoskeleton. Thus, by focusing on particular cellular structures, we can synthesize a functional view of the organism that relates its functions to its evolutionary history and ecological role.

Summary

1. The concept that organisms consist of cells took a long time to develop, but it is now a basic principle of biology. The cell theory entails four basic points:
   a. All organisms are made of cells and the products of cells.
   b. The activities of a multicellular organism are largely the activities of its individual cells and their interactions.
   c. All cells are made of similar chemical compounds, and they function because of similar chemical reactions.
   d. All cells come from preexistent cells.
2. A cell is surrounded by a plasma membrane that defines its boundaries and regulates what may enter or leave.
3. Cells are very small because they need a favorable surface/volume ratio to carry out their metabolic activities.
4. Cell structure is revealed primarily by microscopes, which are able to resolve very small objects by using either light or electron beams.
5. A microscope's limit of resolution is determined largely by the radiation it uses. The limit ranges from about 0.2 µm for a light microscope to a few nanometers for an electron microscope.
6. Materials that are to be examined by microscopy must be put through a regime of fixation, slicing into thin sections, and staining. Microscopy depends primarily on stains, which bind to certain cellular components and make them stand out in contrast to others. Microscopes may also use special optical techniques, such as light interference.
7. The two major types of cells are procaryotic (bacteria) and eucaryotic (all other organisms). Eucaryotic cells have a real nucleus surrounded by a distinct nuclear envelope; procaryotic cells have only a diffuse body called a nucleoid.
8. Cells contain organelles, structures with specialized functions, but eucaryotic cells have many more than procaryotic cells. Most organelles are membrane-bounded compartments, but some are filaments or tubules that shape cells and are responsible for their movement. Other organelles are complexes of proteins, sometimes including nucleic acids, that perform special metabolic functions. A eucaryotic cell is compartmentalized by its nucleus and other membranous organelles; procaryotic cells have relatively little compartmentalization.
9. The nucleus of a eucaryotic cell, defined by the nuclear envelope, contains chromosomes, which carry most of the genes. It also contains one or more nucleoli, where the RNA of ribosomes is made. Ribosomes are the organelles that synthesize proteins.
10. The endomembrane system consists of the endoplasmic reticulum (ER), Golgi complex, and vesicles that temporarily connect them. The ER encloses a separate space, the lumen.
11. Ribosomes on the cytoplasmic face of the rough ER synthesize proteins, which move into the lumen of the ER. Here they may be modified chemically before passing on to the Golgi complex, which modifies them further and directs them to various cellular compartments.
12. Much of cell structure is governed by the membrane-metabolism principle, which states that where a metabolic process is performed by proteins associated with a membrane, a cell makes large amounts of the membrane to provide for that process.
13. Eucaryotic cells contain a variety of vesicles with specialized functions. Some carry substances from one compartment to another. Lysosomes contain digestive enzymes that break down all kinds of materials brought into cells, as well as old cellular components. Peroxisomes contain enzymes that produce and use peroxide. Vacuoles are prominent in plant cells, where they may occupy much of the cell volume.
14. Mitochondria and chloroplasts are the principal sites of energy metabolism in cells. Mitochondria, which are found in all eucaryotic cells, contain metabolic systems that extract energy from foodstuffs. Chloroplasts, found only in photosynthetic eucaryotes, capture light and transform some of its energy into a useful chemical form.
15. Cells have specialized surface structures. The glycocalyx marks the external surface of animal cells with specific chemical structures and probably helps cells bind together and interact properly during development. The cells of plants, fungi, and many protists and chromists are surrounded by a rigid cell wall.
16. Eucaryotic cells in tissues are joined by various junctions. Adhering junctions, including desmosomes and tight junctions, provide strength to tissues. Communicating junctions, including gap junctions (or electrical synapses), chemical synapses, and plasmodesmata, allow ions or small molecules to flow between cells, largely for transmitting information.
17. Procaryotic cells are not usually divided into compartments, but they do contain some distinctive structures, such as the mesosome, an invagination from the plasma membrane to which the nucleoid is attached.

Key Terms

microorganism 109
tissue 109
cell theory 110
unicellular 110
multicellular 110
plasma membrane 110
cell wall 110
resolving power 112
limit of resolution 112
stain 114
fixative 114
artifact 114
section 114
electron microscope 115
nucleus 116
nuclear envelope 116
nucleoid 116
chromosome 116
procaryotic/procaryote 116
eucaryotic/eucaryote 116
organelle 118
compartmentalized 118
nucleoplasm 118
cytoplasm 118
cytosol 118
nucleolus 122
ribosome 122
endomembrane system 122
endoplasmic reticulum (ER) 122
lumen 122
Golgi complex 123
vesicle 123
lysosome 123
peroxisome 123
vacuole 125
tonoplast 125
mitochondria 125
cristae 125
chloroplast 125
plastid 125
cytoskeleton 126
glycocalyx 126
adhering junction 126
communicating junction 126
desmosome 128
tight junction 128
gap junction 128
synapse 128
plasmodesmata 128
mesosome 128

Multiple-Choice Questions

1. The cell theory specifically entails the idea that
   a. all modern cells are descended from ancestral cells.
   b. all modern cells are descended from one ancestral cell.
   c. all modern eucaryotic cells are descended from one ancestral procaryotic cell.
   d. all modern procaryotic cells are descended from viruses.
   e. all modern eucaryotic cells are descended from one ancestral eucaryotic cell.

2. In order to accurately visualize the internal structure of a living cell, you could
   a. increase the wavelength of radiation used by your microscope.
   b. use appropriate fixatives and stains to increase contrast.
   c. switch to electron microscopy.
   d. try phase-contrast microscopy.
   e. increase the preserving power of the fixative.
3. Which of the following can achieve the highest degree of resolution?
   a. ordinary light microscopy
   b. phase-contrast microscopy
   c. ultraviolet microscopy
   d. transmission electron microscopy
   e. scanning electron microscopy
4. Procaryotic cells are characteristic of
   a. bacteria.
   b. protists.
   c. plants and animals.
   d. fungi.
   e. more than one of the above.
5. Which is characteristic of eucaryotic cells but not procaryotic cells?
   a. plasma membrane
   b. nuclear envelope
   c. ribosomes
   d. several chromosomes
   e. *b, c,* and *d,* but not *a*
6. The cytosol is best described as
   a. a set of internal cellular membranes.
   b. a component found only in procaryotic cells.
   c. a viscous fluid that is the site of many chemical reactions.
   d. a collection of filamentous organelles.
   e. the highly structured internal membranes of mitochondria and chloroplasts.
7. Hydrolysis is the primary kind of chemical reaction in
   a. rough endoplasmic reticulum.
   b. smooth endoplasmic reticulum.
   c. the Golgi complex.
   d. lysosomes.
   e. the nucleus.
8. Which kind of cells would have the most extensive rough endoplasmic reticulum and Golgi complex?
   a. cells that produce and store lipid
   b. glandular cells that produce and secrete a protein hormone
   c. liver cells that detoxify drugs
   d. complex bacterial cells
   e. cells that are specialized to line body cavities
9. Which is a functional mismatch?
   a. nucleolus—RNA synthesis
   b. Golgi complex—packaging and sorting secretions
   c. lysosomes—synthesis of polysaccharides
   d. ribosomes—synthesis of proteins
   e. mitochondria—energy metabolism
10. Which would you expect to find as a characteristic of circulating red blood cells?
   a. tight junctions
   b. desmosomes
   c. plasmodesmata
   d. plastids
   e. none of the above

True-False Questions

Mark each statement true or false, and if false, restate it to make it true.

1. A rectangular cell that measures $2 \times 4 \times 8$ μm has twice the surface and four times the volume of a rectangular cell that measures $1 \times 2 \times 4$ μm.
2. A microscope that uses infrared radiation would have greater resolving power than a microscope using ultraviolet radiation.
3. Every cell is surrounded by a cell wall that regulates the inflow of nutrients and the outflow of metabolic wastes.
4. Phase-contrast microscopy is used to increase the resolving power of light microscopy.
5. The nucleolus in procaryotic cells is functionally analogous to the nucleus in eucaryotic cells.
6. The elements of the cytoskeleton are primarily composed of nucleic acids.
7. Ribosomes facing the lumen of the endoplasmic reticulum synthesize proteins.
8. Thylakoid membranes contain enzymes that are essential for photosynthesis.
9. The glycocalyx is composed of proteins that provide the cell with an external rigid skeleton.
10. Tight junctions enable nutrients and ions to travel easily between neighboring cells.

Concept Questions

1. Why does the cell membrane appear as a single line in a light micrograph but not in an electron micrograph?
2. Why are tissues fixed, sectioned, and stained before being viewed with an ordinary compound microscope?
3. Why are most cells so small?
4. List the major subdivisions of the endomembrane system and state the particular function(s) of each part.
5. Distinguish between the functions of tight junctions and gap junctions.

Additional Reading

Alberts, Bruce, Dennis Bray, Julian Lewis, Martin Raff, Keith Roberts, and James D. Watson. *Molecular Biology of the Cell,* 3d edition. Garland Publishing, New York & London, 1994.

de Duve, Christian. "Microbodies in the Living Cell." *Scientific American,* May 1983, p. 74. Organelles with different metabolic tasks.

——— *Blueprint for a Cell: The Nature and Origin of Life.* Neil Patterson, Burlington (NC), 1991. On the origin of life and the structure and origin of cells.

Fawcett, Don W. *An Atlas of Fine Structure: The Cell.* W.B. Saunders Company, Philadelphia and London, 1966.

Goodsell, David S. *The Machinery of Life.* Springer-Verlag, New York, 1993.

Jacobs, William P. "Caulerpa." *Scientific American,* December 1994, p. 66. This large algal plant is a single gigantic cell, challenging the belief that organisms must be multicellular to have great size and a complex, specialized form.

Jensen, William A., and Roderic B. Park. *Cell Ultrastructure.* Wadsworth Publishing Company, Belmont (CA), 1967.

Internet Resource

To further explore the content of this chapter, log on to the web site at:

http://www.mhhe.com/biosci/genbio/guttman/

7 Energy and Metabolism

A beaver feeding on aspen leaves shows that much of an organism's life is spent obtaining energy from its environment. The leaves grew by capturing the energy of sunlight, and the beaver maintains itself and grows by capturing some of the energy stored in the plant.

Key Concepts

A. Basic Chemical Concepts

- 7.1 Energy cannot be created or destroyed.
- 7.2 Chemical reactions entail changes in energy.
- 7.3 Some reactions occur spontaneously, and others do not.
- 7.4 Chemical reactions entail changes in entropy as well as in heat content.

B. Applications to Metabolism

- 7.5 The need to decrease entropy is a central problem of biology.
- 7.6 Organisms construct and maintain themselves through enzyme-catalyzed pathways.
- 7.7 Energy-consuming processes can be driven by coupling them to energy-yielding processes.
- 7.8 In biological reactions, free energy is carried primarily in ATP.
- 7.9 Ecosystems operate on a flow of energy that comes from the sun.
- 7.10 Useful energy can be obtained from oxidative reactions.
- 7.11 Two kinds of nucleotides are used as oxidizing and reducing agents.
- 7.12 ATP and NADPH provide the energy and reducing power needed for biosynthesis.

Small trees stand in the soft April sunlight, pushing out their leaves, gathering in the light, turning greener with every passing hour. Each day the sun irradiates the leaves a bit more strongly. As the leaves mature, insects that have been passing the cold in their pupas emerge and mate. Soon the leaves are dotted with small white eggs, and the eggs become caterpillars, which gnaw at the leaves, taking in some of the stored sunlight as they grow. Now small birds returning from the south find a feast of insects packed with the sun's energy, which powers them as they build their nests, lay their eggs, and start to raise their young. But other animals in the woodland need the sun's energy, too. A crow steals a nestling bird, lays it out over a limb, and picks out the rich flesh, dripping with blood and sunlight. A fox discovers an unwary thrush and turns its store of sunlight into fox energy. A hunter with gun ready comes upon a covey of quail, and his evening meal is quail meat well seasoned with sunlight.

When the Welsh poet Dylan Thomas wrote,

The force that through the green fuse drives the flower
Drives my green age; that blasts the roots of trees
Is my destroyer,

he was describing the sense that life is unified energetically, by dependence on a common force. Ultimately, that force is sunlight, and much of biology is the story of how sunlight is transformed into the energy stored in plant structure and then shared with the rest of the living world. All organisms need energy. They grow. They constantly convert materials from the environment into their own structure. They move, or at least their cell parts do. Our goal in this chapter is to explain how the energy transformations in chemical reactions can accomplish the work of living. Instead of a rigorous analysis, which is better left to more advanced courses, we will try to build on your intuitive understanding of energy and your knowledge from earlier science courses to show how energy relates to biology. ■

A. Basic Chemical Concepts

7.1 Energy cannot be created or destroyed.

Matter and energy are fundamental to science, and we deal with both in our everyday lives. *Matter* is easy to understand. It is anything that has mass, occupies space, and can be perceived in some way. *Energy,* defined as the capacity to do work, is more elusive, partly because we can only know energy exists somewhere if work is actually done. Suppose you lift a large rock off the ground (Figure 7.1*a*). You are doing work, and you can only do so by using energy stored in your body. In doing this work, you also give the rock some **potential energy** (strictly speaking, gravitational potential energy), energy that could be turned into work if the rock were dropped. This potential energy is proportional to the height of the rock above an arbitrary zero point such as the ground, so an object 20 meters off the ground has twice the potential energy of the same object 10 meters off the ground. As a falling rock gets closer to the ground, its potential energy continually decreases, but in picking up speed, the rock is acquiring greater **kinetic energy,** the energy of motion. As the rock hits the ground, all its energy is dissipated into movement of the soil, sound waves from the impact, and heat. In principle, the energy of the falling rock could also do useful work; the classic flour mill operates by using the kinetic energy of falling water to turn a grinding wheel (Figure 7.1*b*).

We see, then, that energy takes different forms. Because organisms run on chemical reactions, their energy must ultimately take a chemical form. Even biological movement is fundamentally chemical. Your muscles move because highly ordered proteins interact and pull on one another, using the energy derived from your food. Many chemical reactions, especially those characteristic of organisms, will not occur unless energy is added to drive them.

Energy arises because objects exert *forces* on one another. Physicists recognize four basic forces, and each of them creates a kind of potential energy that can manifest itself as movement, or work. Because of the *gravitational force,* objects have gravitational potential energy due to their position above the ground. The previous example of the falling rock illustrates how this can be converted into kinetic energy. Tapping the energy of a flowing river to move objects or generate electricity is another example. Nuclear energy exists because of the *strong force* that holds the nuclei of atoms together, and the *weak force* underlies the most common kind of radioactivity. The *electromagnetic force,* the force of attraction and repulsion between charged particles, is responsible for electrical energy. We commonly experience this force as a current of electrons in a wire that can be used to run appliances, and this force is also the basis of chemical energy. Thus the electrons and protons of atoms are held in position by the forces between them, and each kind of molecule has a characteristic chemical energy because of its arrangement of atoms. Electromagnetic events also

The person uses chemical energy derived from food to move muscles. Some of this energy is given to the rock.

As the rock falls, it *loses* potential energy and *acquires* kinetic energy.

As the rock is carried higher, it acquires greater potential energy

The faster the rock falls, the more kinetic energy it has. The lower it falls, the less potential energy it has. The sum of the two energies is a constant.

Rock on the ground has zero energy.

As the rock hits the ground, it has maximum kinetic energy, which is converted into kinetic energy of sound (motion of air) and heat (random motion of molecules).

(a)

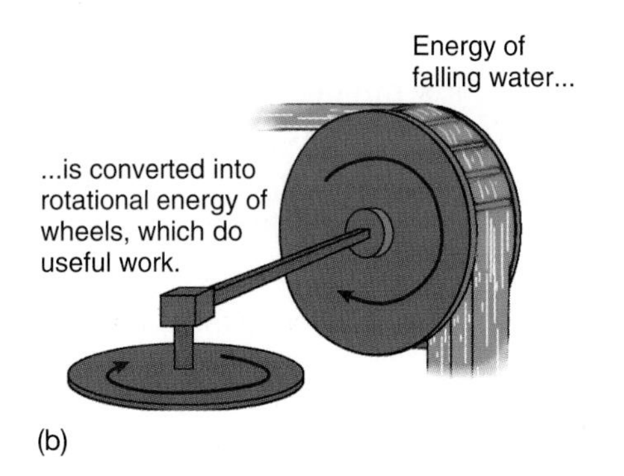

(b)

Figure 7.1

Objects have energy because of their positions and because of their motion. *(a)* If a rock is lifted, it acquires potential energy because of its height. If it falls, it acquires kinetic energy as it moves faster, and this energy is dissipated into heat and sound when the rock hits the ground. *(b)* The energy of a falling object, such as water, can also be converted into useful work, for instance turning a water wheel to run a mill. However, the total amount of energy is always the same, regardless of the form it takes.

produce radiation, which we experience as light, X rays, microwaves, and radio waves. Finally, all matter has *internal energy,* which we commonly detect as heat; it is the combined kinetic energy of atoms and molecules in random motion, and the faster these particles are moving, the warmer a substance is.

Because one force can produce another, one form of energy can be converted into another. The radiation in a microwave oven is converted into the internal energy of our food; energy in hydrocarbon molecules can be released by combustion to power an engine. However, in all these transformations the total amount of energy remains constant. Thermodynamics, the science of energy, begins with this principle, the *first law of thermodynamics,* also known as the law of conservation of energy: *Energy can be put into different forms but cannot be created or destroyed.* (Einstein's realization that matter and energy are interconvertible revolutionized physics, but living organisms don't engage in nuclear reactions that convert matter into energy.)

The connection between force and energy provides a way to measure energy. We can only tell that energy exists in a system if the system is able to do work, which means that a force must move matter through a distance. The work done—the energy expended—is equal to the force times the distance. Although energy in biological systems has traditionally been measured in calories, biologists increasingly use the standard physical unit of a **joule (J).** 1 J = 0.239 cal, or 1 cal = 4.184 J. These are very small quantities, and the energies of most chemical reactions are commonly given in the thousand-times larger units of kilojoules (kJ) and kilocalories (kcal) per mole. We will give both quantities for many biological processes.

Thermodynamics deals with the energy of a system, which you may recall from Chapter 2 is simply any part of the universe we choose to isolate and study: a single chemical reaction, a cell, an individual organism, an ecosystem in which many organisms live, the earth, or even the whole universe. The simplest systems are *isolated systems* that cannot exchange either matter or energy with their surroundings. We are more familiar with *closed systems* that can exchange energy but not matter; for instance, when you cook by applying heat to food in a tightly closed pot, you have a closed system in which the heat affects the desired chemical changes. Biological systems are especially interesting, though difficult to analyze, because they are *open systems;* matter and energy constantly flow through them, and this flow is necessary to maintain the organisms involved.

Exercise 7.1 A power company gets electricity from a dam on a river. The electricity is used to power a motor that lifts an elevator with a ceiling light. Show how one form of energy is changed into another at each step along the way.

7.2 Chemical reactions entail changes in energy.

As organisms carry out metabolism, they take energy from their surroundings and use it to build up their own structure. To understand metabolism, we must understand changes in energy as one substance is converted into another in a cell. As a model, we will use the formation of water from its elements:

$$H_2 + \frac{1}{2} O_2 \rightarrow H_2O + \text{Heat}$$

Mix a mole of hydrogen with half a mole of oxygen in a calorimeter (an instrument that measures heat production) and give the mixture a spark. The mixture explodes, the hydrogen and oxygen are converted to steam (water vapor), and the calorimeter shows that 58 kcal of heat has been released.

Energy, of course, is conserved here. To draw a balance sheet showing where the energy in this reaction comes from and where it goes, we begin by asking, "How can a molecule contain energy?" Of course, all matter carries some heat because its component atoms and molecules are in motion, and we showed in Chapter 5 that this energy is important in determining the *rate* of a chemical reaction. But the energy that concerns us here, the energy that determines whether a reaction can occur at all, involves the bonds between atoms. Consider two hydrogen atoms (Figure 7.2). Their protons carry positive charges and repel each other; their electrons, carrying negative charges, also repel each other. However, the protons and electrons of the two atoms attract one another. These forces of attraction and repulsion create energy in the interaction between the atoms. When the atoms are infinitely far apart, this energy is zero. As they get close together, the energy falls, but if they are pushed too close, the energy rises again. So there is an energy well, at a distance of 0.074 nm between the hydrogen nuclei, where the energy is lowest. As a general principle of physics, every system is most stable at its point of least energy. Thus a rock on the ground is more stable than an elevated rock because it would take energy

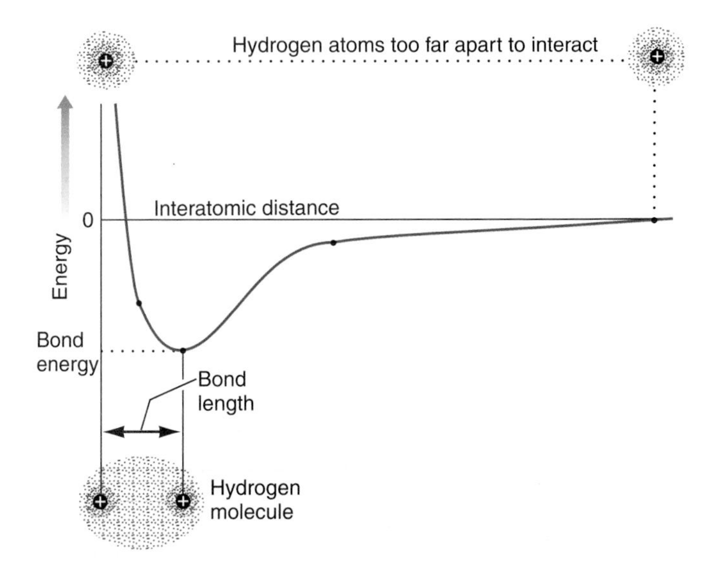

Figure 7.2
As two hydrogen atoms come closer together, their protons and electrons start to attract each other. The energy of interaction between them *(red curve)* decreases, and energy is required to separate them. If, however, they are pushed very close together, they start to repel each other. So they come to rest at a distance where the energy of their interaction is at a minimum, and in this position they form a stable molecule.

to lift the lower rock, and objects don't acquire energy spontaneously. Similarly, at the distance where their energy is least, the hydrogen atoms form a stable covalent bond. To break this bond—to separate the hydrogen atoms—we have to add about 103 kcal/mol, and this energy, by definition, is the **bond energy** of the hydrogen molecule. In this sense we say there is energy "in" a molecule, but notice that a *deficiency* of energy is what actually keeps the atoms together.

Since heat escaped during the formation of water, the water (steam) must now have less energy than the original gas mixture had. This experiment shows, in fact, that a mole of water contains 58 kcal less energy than an equivalent amount of hydrogen and oxygen. The difference is the *heat of formation* of water. During this reaction, the bonds between hydrogen atoms and between oxygen atoms break and are replaced by bonds between hydrogen and oxygen atoms in water (Figure 7.3). All the bonds in the water molecules must contain less energy than did the bonds in the original gas molecules. All bonds of a certain kind have about the same amount of energy, regardless of the molecules they are in, and we can use these values (Table 7.1) to write a simple balance sheet for the energy changes that occur in making water from hydrogen and oxygen.

Bonds Broken

One mole of H—H: 103 kcal
½ mole of O=O: 0.5 × 117 = 58.5 kcal
Total: 161.5 kcal put in

New Bonds Made

Two moles of O—H: 2 × 109 = 218 kcal released

The difference is 161.5 − 218 = −56.5 kcal, reassuringly close to the measured value of 58 kcal, since the calculation is based on *average* bond energies. The negative value indicates that net energy comes *out*, so water is shown at a lower point on the energy scale of Figure 7.3 than the gases it is made from.

What difference does it make if one compound has more energy or less than another? Just as matter tends to achieve its lowest-energy, most stable state, a chemical reaction tends to go from a higher to a lower energy state. Hydrogen and oxygen atoms are in a more stable condition in water than in pure hydrogen and oxygen molecules because more energy is needed to break all the bonds in water than to break all the bonds in the molecules of the pure elements. In sum, when atoms are bonded they are in a certain energy state, and the total energy of all the bonds involved determines whether the products of a reaction hold more or less energy than do the reactants.

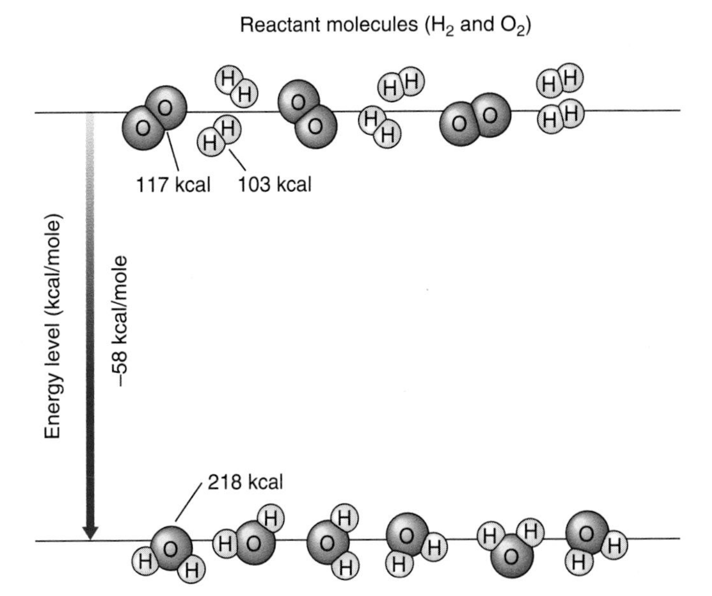

Figure 7.3

(Top) The bond between two oxygen atoms contains 117 kcal of energy per mole, and the bond between two hydrogen atoms contains 103 kcal of energy per mole. To convert hydrogen and oxygen to water, the H–H and O=O bonds must be broken. *(Bottom)* Then two O–H bonds are formed in making each water molecule, which collectively have less energy (218 kcal/mole) than the separate molecules of hydrogen and oxygen.

Strictly speaking, the energy we are discussing here is called **enthalpy,** or heat content, and is denoted by *H*. Thus, during a chemical reaction, the difference in enthalpy between the initial and final states of a system is:

$$\Delta H = H_{\text{of all products}} - H_{\text{of all reactants}}$$

where Δ (Greek delta) means "change in" or "difference." Reactions often produce heat, sometimes enough to cause an explosion. This means that energy is coming out, so the products of the reaction contain less energy than the reactants, and ΔH is negative ($\Delta H < 0$). On the other hand, heat is absorbed in some reactions, such as the formation of benzene from carbon and hydrogen. This means the resulting molecules have more energy than their components, so the change in enthalpy is positive ($\Delta H > 0$). This kind of reaction will only proceed if it is heated. So during chemical reactions, excess energy (heat) usually comes out, but sometimes energy must be put in to begin with.

Table 7.1 Average Bond Energies

Bond	Energy (kcal/mol)	Bond	Energy (kcal/mol)
H—H	103	N—N	37
C—C	80	N—H	92
C—H	98	O—O	34
C=C	145	O=O	117
C≡C	198	C—O	79
C=O	173	H—O	109

Exercise 7.2 Acetylene (HC≡CH) burns with oxygen:

$$2\,C_2H_2 + 5\,O_2 \rightarrow 4\,CO_2 + 2\,H_2O$$

producing so much heat that acetylene torches can melt steel. Use the numbers in Table 7.1 to calculate the energy produced by burning a mole of acetylene.

Exercise 7.3 The heat of formation of ethane (C_2H_6) is −20.2 kcal/mol, and that of ethylene (C_2H_4) is +12.6 kcal/mol. Which compound is more stable? In which case is energy put into the compound when it is made from its elements, and in which case is energy taken out?

7.3 Some reactions occur spontaneously, and others do not.

A *spontaneous process* occurs without any outside help, while a *nonspontaneous process* requires energy. A spontaneous process is one that goes *downhill* energetically, the natural way for things to flow, while a non-spontaneous process goes *uphill.* Since matter tends toward its lowest-energy state, we expect chemical reactions to naturally run downhill to a more stable condition. In other words, we expect natural, spontaneous chemical processes to give off heat. However, some processes violate this rule. Melting ice may be the most familiar exception; as ice melts spontaneously, the water absorbs heat, so we can cool cans of drinks by putting them into an ice bath. As another example, when ammonium chloride dissolves spontaneously in a beaker of water, the vessel gets cold as heat is taken out of the water. These are energy-consuming processes that occur naturally, but they seem to violate a law of nature. How is this possible? It happens because events are driven not only by the tendency to achieve a state of low energy but also by the tendency for things to become *disordered,* a tendency we will examine next.

7.4 Chemical reactions entail changes in entropy as well as in heat content.

We spend our lives trying to counteract the forces of disorder (Figure 7.4)—organizing our desks, arranging the clothes in our dresser drawers, and removing accumulations of trash and dirty dishes. Orderly piles of papers or clothes tend to fall into disordered heaps, not the other way around. And so it is with atoms and molecules. This drive toward disorder is measured by the **entropy** *(S)* of a system. A highly ordered system has low entropy, and a disordered one has high entropy. The *second law of thermodynamics* states: *In all natural processes, the entropy of an isolated system tends to increase to a maximum.*

The action of water illustrates the second law nicely. We place ice cubes on one side of a thin partition in an insulated tank (Figure 7.5) and warm water on the other side. Ice has low entropy and is highly ordered, with its molecules held in place by hydrogen bonds. Heat will pass from the warm water into the ice. Since heat is just the energy of random molecular motion, molecules in the ice vibrate and turn faster as they acquire more kinetic energy. Some eventually get enough energy to break the hydrogen bonds holding them in the ice crystal and become liquid. Water is clearly more disorganized than ice because its molecules move freely and randomly. Even in freshly melted ice, most of the molecules still form hydrogen-bonded "icebergs," but as the water continues warm up, more and more molecules break free and move independently.

Entropy introduces *direction* to a process, direction in time and sometimes direction in space. In our experimental tank,

Figure 7.4

Left to themselves, most things tend to become more disordered and mixed up—their entropy increases. Living organisms appear to be an exception to this rule, though nothing lives forever.

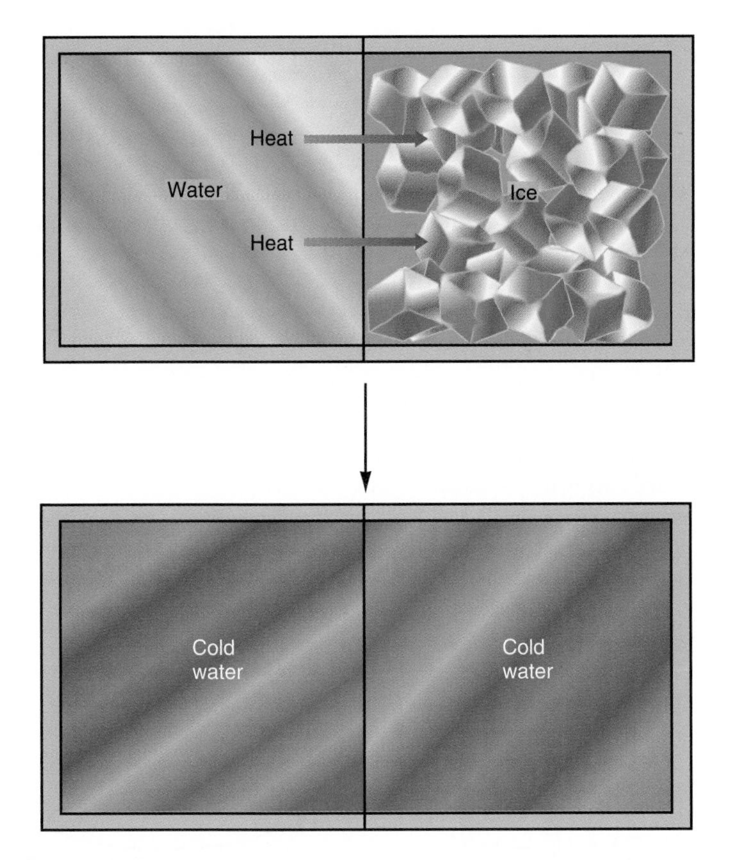

Figure 7.5

Ice has low entropy; all the water molecules in it are neatly lined up and vibrating relatively little in place. If the ice is placed in contact with liquid water, heat flows from the water into the ice. The ice warms and melts, and the whole system acquires higher entropy. Thus the system has a direction in space because heat flows only in one direction, and it has a direction in time, going from a more orderly to a less orderly condition.

heat only flows one way. The ice melts; the water does not spontaneously freeze. If someone showed us a film of this tank, having no cooling and heating system, with the water on one side of the partition freezing while the other side gets hot, we would be quite sure the film was being run backwards.

The first and second laws of thermodynamics come together through another quantity, the **free energy,** denoted by G:

$$G = H - TS$$

where T is temperature, the absolute or Kelvin temperature. The point at which all molecular motion ceases ($0K$) corresponds to $-273.16°C$. Remember that H is enthalpy (energy) and S is entropy (a measure of disorder). When a process occurs at a constant temperature, the change in free energy is:

$$\Delta G = \Delta H - T\,\Delta S$$

The laws of thermodynamics are satisfied if the free energy decreases in every process—that is, if ΔG is negative. Generally, ΔH is negative (energy decreases), and ΔS is positive (disorder increases). The minus sign in front of the $T\,\Delta S$ term then makes ΔG always negative. However, some processes are driven mostly by increasing entropy and others mostly by decreasing enthalpy (Figure 7.6). For instance, the hydrophobic interactions inside a typical globular protein are influenced more by entropy than by enthalpy, and since the $-T\,\Delta S$ term becomes more negative as the temperature increases (because of the minus sign), such proteins are more stable at higher temperatures (up to a point).

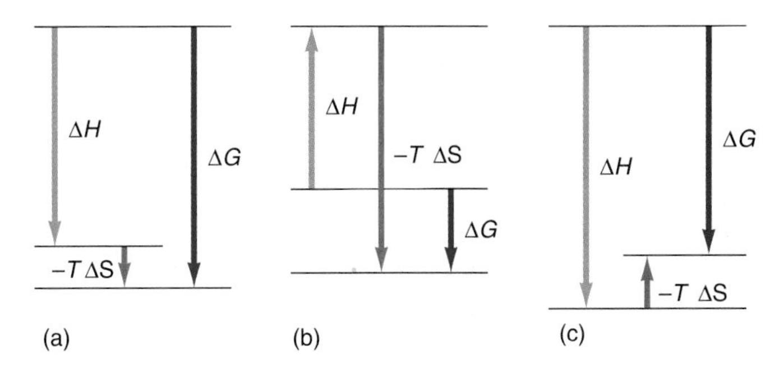

Figure 7.6

(a) In some reactions, ΔH is negative and ΔS is positive, so ΔG is negative because of their combination. *(b)* In others, ΔH is positive, but $T\,\Delta S$ is large enough to overcome the increase in enthalpy, so the reaction is driven by the entropy change. *(c)* In still others, ΔS is negative, but ΔH is also negative and is large enough to make ΔG negative; such reactions are enthalpy-driven.

The change in free energy, ΔG, determines whether a process is going uphill or downhill, because it takes account of both the tendency toward minimum energy and the tendency toward maximum disorder. When a real system is at equilibrium, its free energy is a minimum. In **exergonic** reactions (*ex-* = out; *ergon* = work), like the formation of water, the components lose free energy; these reactions occur naturally and spontaneously. The opposite processes, like the separation of hydrogen and oxygen gas from water, are **endergonic** reactions (*end-* = in) that can only happen if energy is put into the system. For instance, water can be dissociated into its elements with an electric current. Concepts 7.1 provides additional clarification of some thermodynamic concepts introduced in this section.

Now consider a typical biological reaction—say, making a dipeptide from two amino acids:

$$\text{Glycine} + \text{Alanine} \rightarrow \text{Dipeptide}$$

Although we have written the reaction as the formation of the dipeptide, the process can also go the other direction. Which direction does it actually tend to go spontaneously? To find out, we do the chemical equivalent of the previous experiment with ice and water (see Figure 7.5). We mix a mole of glycine, a mole of alanine, and a mole of dipeptide, add a catalyst to speed up the reaction (since catalysis has no effect on the equilibrium), and wait until the mixture comes to equilibrium (Figure 7.7). Then we measure the concentrations of the substances. If there is more dipeptide than glycine or alanine at equilibrium, the formation of the dipeptide is an exergonic, or downhill, reaction. In this case, however, the reaction goes in the opposite direction spontaneously: The dipeptide tends to hydrolyze into its components instead of being synthesized, so this is an endergonic, or uphill, reaction.

Concepts 7.1 Energy and Entropy

The concepts of thermodynamics are among the most subtle and complicated ideas in physics and chemistry, and it is very difficult to discuss them well enough in an introductory biology text for a truly deep, satisfying understanding. Students generally develop this understanding only when they take advanced courses in physical chemistry, and even well-educated scientists often don't understand the ideas very well. However, the following additional notes may be helpful at this stage.

Enthalpy, or **heat content** ***(H),*** is the total energy that may be available in a system, such as the total energy contained in a chemical compound.

Entropy ***(S)*** is not a form of energy; energy is measured in joules (or comparable units), but entropy has the dimensions of joules/Kelvin, and only the quantity $T \, \Delta S$ (that is, Kelvin × joules/Kelvin) is an energy. However, entropy may be thought of as a measure of the *quality* of energy. That is, energy in the form of electricity or certain chemical structures has low entropy and is "high-quality energy" that has the potential for doing a lot of work; in contrast, heat has high entropy and is a "low-quality energy" with much less potential for doing work.

Free energy ***(G)*** is the amount of energy available to do work. Considering the explanation of entropy given above, the equation $\Delta G = \Delta H - T \Delta S$ shows that the greater the entropy change in a system (the larger ΔS is), the less free energy is available. Thus, if a process releases energy in the form of heat, with a large ΔS, the available energy is largely wasted from the point of view of doing work.

Bond energy is the energy "in" a chemical bond—strictly speaking, the energy released when a given chemical bond is formed or the energy required to break a bond.

Synthesizing a dipeptide from two amino acids is just one of the many endergonic reactions organisms have to carry out to build their complex molecules. Since such reactions do not occur spontaneously, an organism must clearly obtain energy from somewhere and use it to force these reactions to occur. In the remainder of this chapter, we will see how this is done.

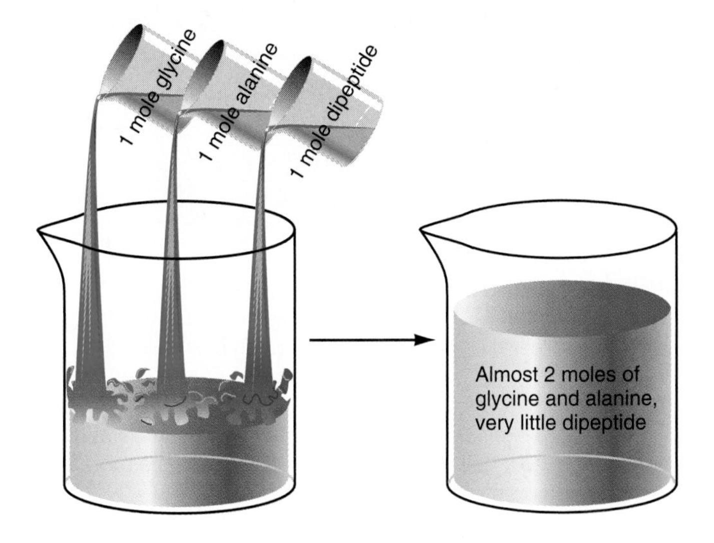

Figure 7.7

To determine the equilibrium point of a reaction, one mole each of glycine, alanine, and a glycine-alanine dipeptide are mixed, with a catalyst added to speed up the reaction. After the system has come to equilibrium, the concentrations of the three components are determined. In this case, there is very little of the dipeptide at equilibrium, showing that the reaction tends to go in the direction of dissociating the dipeptide into its component amino acids.

B. Applications to Metabolism

7.5 The need to decrease entropy is a central problem of biology.

All organism are organized and orderly—and that means low entropy. Obviously, an organism has lower entropy—and more free energy—than the materials it is made from. To prove that point, compare two different arrangements of the same molecules—a solution of amino acids and the same amino acids linked into identical protein molecules, all with the same sequence:

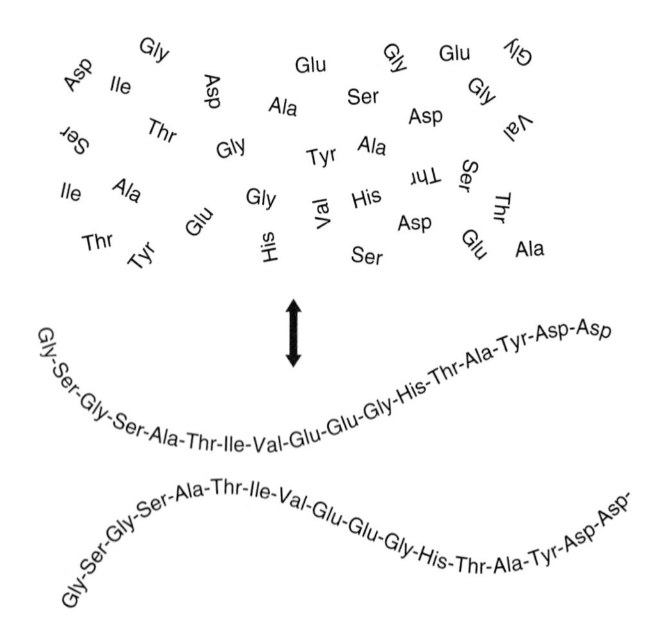

The free amino acids (at left) can move around into many positions, so the solution is disordered and has high entropy, but in the protein chain, the amino acids are constrained, ordered, and have much lower entropy. As an organism grows and builds these regular, organized molecules, it is building up free energy within itself. Thus, organisms seem to violate the second law of thermodynamics, but that's only an illusion. Actually, the second law says that an *isolated* system achieves minimum free energy, but an organism isn't isolated. Rather, an organism stays alive by constantly exchanging energy and matter with its environment—light, carbon dioxide, oxygen, and nutrients—and an organism that is cut off from these resources for long soon disintegrates into a muddle with rapidly increasing entropy. A system consisting of an organism alone just isn't realistic (Figure 7.8*a*). Instead, we must consider a system containing the organism plus its nutrients and its wastes, because an organism extracts free energy from its surroundings, leaving wastes with less energy (Figure 7.8*b*). In such a system, entropy increases and free energy decreases, in accordance with the laws of physics.

To make a room neat or arrange a deck of playing cards in its original sequence, you need: (1) information (about where things ought to go) and (2) energy. So it is with organisms. To build its organized form, an organism needs the information in its genome, which specifies what structures to make, and it needs energy. It must carry out endergonic reactions to build up structures with increased free energy. Next we will see how this is done.

7.6 Organisms construct and maintain themselves through enzyme-catalyzed pathways.

An organism is like a factory, a chemical factory whose product is more of its own structure. Metabolism is organized into streams of chemical activity known as **metabolic pathways,** which are analogous to the assembly lines of the factory. The first worker puts some small parts together and passes the unit on to the next person in line, who performs another operation on it. As each worker makes a small change and passes the unit on, the product gradually grows and assumes its ultimate shape. In an organism, the materials moving through a metabolic pathway are known as **metabolites** and are sometimes called *intermediary metabolites* because they are in an intermediate stage of assembly or breakdown. A pathway is a sequence of chemical reactions, each catalyzed by one enzyme (the analogue of an assembly line worker) and each making a small change in the metabolite, just as each worker makes a small change in the product (Figure 7.9). Some pathways break down molecules step-by-step, while others synthesize the monomers of cell structure and then polymerize these monomers into macromolecules. Some pathways consume energy, while others provide the energy needed to do all this work.

As an example of a metabolic pathway, Figure 7.10 shows the conversion of the amino acid aspartate, into another, threonine. First, the —COOH group of aspartate is converted to HC=O; next, HC=O is converted to H_2COH. Then a water molecule is removed, leaving a double bond between two carbon atoms, and water is added across the double bond in the other direction, making threonine. Notice that only a small change occurs in each reaction and that each step in the pathway is catalyzed by a different enzyme. However, we have purposely omitted two crucial details of these reactions; we will fill them in later to illustrate two important metabolic principles.

In a factory, materials usually move on a conveyor belt directly from one worker to the next. In an organism, some enzymes are built into enzyme complexes or membranes, where they apparently operate like assembly lines. But many other enzymes float about freely, neither organized nor attached to a

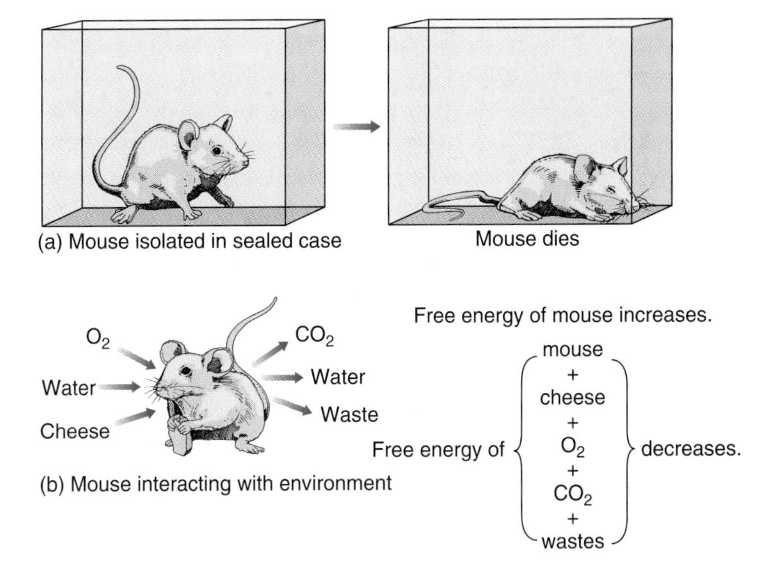

Figure 7.8
(a) An organism doesn't exist by itself; without exchanges of matter and energy with its environment, it would die. *(b)* An organism must take in nutrients (and, in many cases, energy from the sun) and produce wastes. Even though the organism itself acquires higher energy, it does this at the expense of its nutrients (or light). If we consider the organism + its nutrients + its wastes, the whole system shows a decrease in energy, in accordance with the laws of physics.

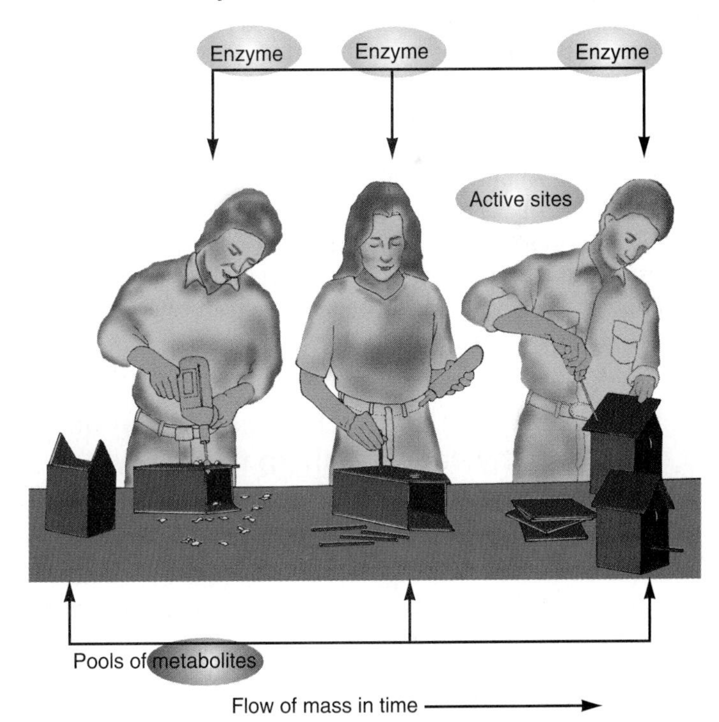

Figure 7.9
A metabolic pathway is like an assembly line. Each person (enzyme) takes the product (metabolite) into a small space made by his or her hands and tools (an active site) and makes one small change in it (chemical reaction). The altered product (a different metabolite) then passes on to the next person.

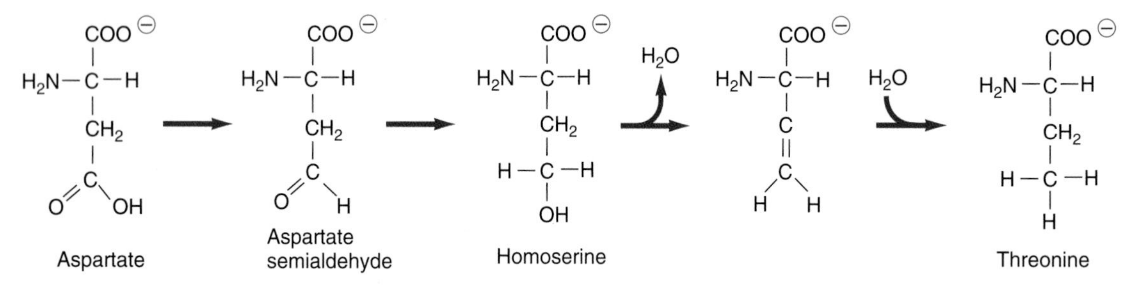

Figure 7.10

The amino acid aspartate is converted into a different amino acid, threonine, through a metabolic pathway consisting of four reactions. Some important features of this pathway are omitted here.

solid foundation. The metabolites, too, are all mixed together in the cytosol, and they move by random molecular motion (diffusion) from enzyme to enzyme. All the molecules of any metabolite form a **pool** in a cell—a pool of aspartate, a pool of homoserine, a pool of threonine, and so on. These pools are just abstractions, not physically separated collections of molecules, since hundreds or thousands of molecule types are actually mixed together in the cytosol (Figure 7.11). However, compartmentalization in eucaryotes confines some materials and raises their concentrations so that substrates can more easily encounter the enzymes that operate on them. Molecules may be added to a single pool from several sources and withdrawn in several ways, depending on how a metabolite is used.

Metabolism has two aspects, catabolism and biosynthesis (Figure 7.12). In **catabolism,** cells break down incoming food molecules into smaller molecules (*cata-* = down) and save some of their energy for use elsewhere. In **biosynthesis,** also called **anabolism** (*ana-* = up), cells build up smaller molecules into monomeric molecules such as amino acids and then polymerize these monomers into macromolecules. (Anabolic steroids, which have created much controversy in athletic circles, are hormones that can stimulate muscle formation [anabolism], but at the cost of serious health problems.) The question now is, "How can energy be derived from catabolism and used in endergonic synthetic reactions?"

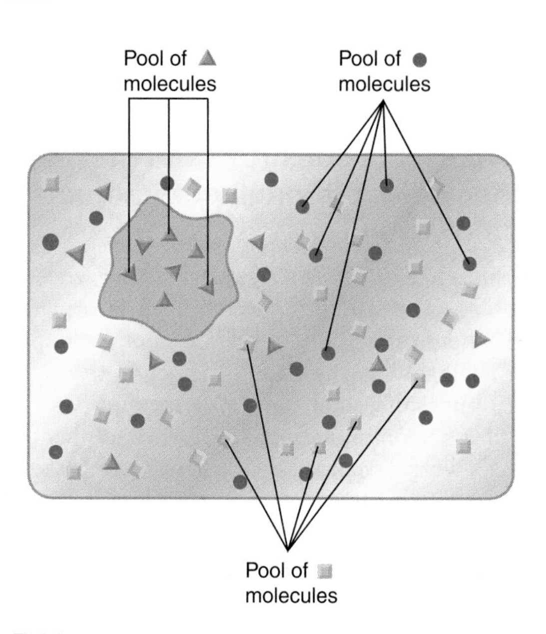

Figure 7.11

All the molecules of one kind in a cell constitute a pool. Physically, the pools are all mixed with one another, but conceptually, molecules move from one pool to another as enzymes transform them.

7.7 Energy-consuming processes can be driven by coupling them to energy-yielding processes.

The downhill reactions of catabolism *drive* the uphill reactions of biosynthesis. However, one reaction can only drive another if the two are linked or coupled to each other *by sharing a chemical component,* and such **coupled reactions** can occur in two general ways (Figure 7.13). The first way depends on the fact that the actual free energy change in a reaction decreases as the concentrations of the products decrease. Look at the general reaction A + B → C + D, where the letters represent different compounds. Even if the reaction is endergonic, and tends to go to the left rather than to the right, it can be *driven* to the right if something removes C or D (or both). Imagine two reactions in sequence:

$$A + B \rightarrow C + D \rightarrow E + F$$

where the first reaction is endergonic and the second exergonic. Because the second reaction occurs spontaneously, it will reduce the concentrations of C and D, and thus *pull* the first reaction forward. *As long as the second reaction releases more energy than the first one consumes,* the sum of the reactions has a negative free energy, so together they will take place spontaneously.

A second kind of coupling uses a **group transfer** reaction in which an atom or group of atoms is transferred from one molecule to another. We saw in Section 3.9 that a salt dissolves in water by dissociating into its ions. However, silver chloride (AgCl) barely dissolves in water—that is, the dissociation reaction between silver and chlorine ($AgCl \rightarrow Ag^+ + Cl^-$) is highly endergonic and does not happen spontaneously. However, silver ions combine with ammonia in an exergonic reaction to make a complex ion:

$$Ag^+ + NH_3 \rightarrow AgNH_3^+$$

This reaction yields so much free energy that it can drive the first one, and the two are coupled by transferring the silver ion

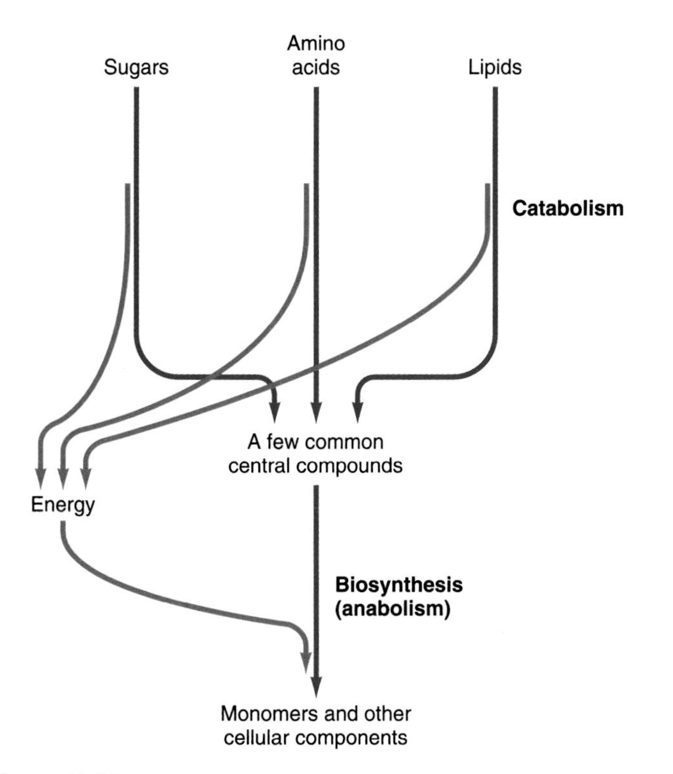

Figure 7.12
Most metabolic pathways are either catabolic, in which nutrients are broken down and some of their energy is captured, or anabolic (biosynthetic), in which the nutrients and their breakdown products are built up into cellular components, using some of the energy released in catabolism.

from the chloride to the ammonia. Intersecting curved arrows show reactions that are coupled in this way, and they make it easy to see the transfer of a group (G) from one substance to another. The reactions:

$$AG \rightarrow A + G$$
$$B + G \rightarrow BG$$

are represented by:

AG ╲╱ B → A, BG

The inorganic reaction of our example can be drawn:

AgCl ╲╱ NH_3 → Cl^-, $AgNH_3^+$

You should be able to see that the silver ion is being transferred. Living organisms drive many processes through the transfer of a few special groups that have particular advantages energetically, as we will explain next.

7.8 In biological reactions, free energy is carried primarily in ATP.

A *potential* is a measure of a system's ability to do work, and organisms make use of several kinds of potentials. A cell can acquire a *chemical potential,* for instance, by concentrating a particular molecule or ion in a compartment (Figure 7.14*a*). Since this orderly arrangement has low entropy, the concentrated molecules or ions tend to go to a high-entropy state by escaping

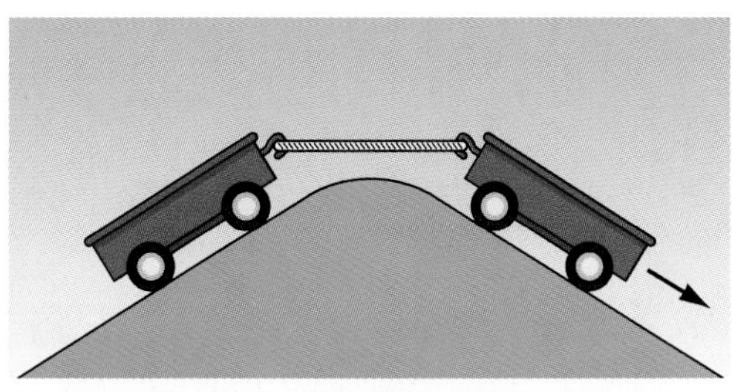

(a) Reaction that spontaneously runs downhill pulls reaction that must run uphill.

(b) Reaction that spontaneously releases a molecular group... ...can drive a reaction that must go uphill

...if the group is transferred from the first reaction to a reactant of the second reaction.

Figure 7.13
An endergonic biological process that requires an input of energy can be driven by coupling it to an exergonic process that yields even more energy. *(a)* By sharing a common intermediate (rope connecting the wagons), an exergonic reaction can pull an endergonic reaction. *(b)* An exergonic reaction can transfer a group of atoms from one molecule to another, thus driving the second to a higher energy level.

from the compartment and moving out into regions of lower concentration. As they move, they can be made to do work. By analogy, a reservoir filled with water has a high potential, and if the dam is opened, the escaping water can be made to do work. An *electrical potential,* or voltage, is created by building up a high concentration of electrons or ions in one place (Figure 7.13*b*); this system, too, can do work as the charged particles escape, perhaps in an electric current. A *group transfer potential* permits a chemical group to be transferred in a coupled reaction (Figure 7.13*c*). There is no absolute zero point of group transfer potential, just as there is no absolute zero point for gravitational potential energy, so as a standard measure of potential, we arbitrarily use the free energy of the reaction in which the group is transferred to water—in other words, when a compound containing the group is hydrolyzed.

Metabolic reactions can be driven by the transfer of many chemical groups, but nature has selected one group as a kind of universal energy currency—the **phosphoryl group:**

$$-\overset{\displaystyle O}{\overset{\|}{\underset{\underset{\displaystyle O^{\ominus}}{|}}{P}}}-O^{\ominus}$$

This group is customarily represented by —Ⓟ (This should not be confused with a phosphate, which is the phosphoryl group combined with another molecule; X with a phosphoryl group attached is X phosphate.) The ability of a compound to transfer this group is measured by its **phosphoryl-group transfer potential,** which is just the free energy of removing the phosphoryl through hydrolysis:

$$X—Ⓟ + H_2O \rightarrow X + H_3PO_4$$

The product, inorganic phosphate or phosphoric acid (H_3PO_4), is commonly abbreviated P_i. Some molecules have a low phosphoryl-group transfer potential, meaning that they have little ability to donate a phosphoryl group to something else. Other compounds can donate the group quite readily and have a high potential; cells use these compounds to carry free energy.

Remember that nucleotides, the monomers of nucleic acids, consist of a sugar combined with a nitrogenous base and a phosphate group. Cells build up a supply of nucleotides in their *triphosphate* form, carrying three phosphoryl groups rather than just one as stores of usable free energy. For metabolism in general, the most important of these is **adenosine triphosphate (ATP),** which is a universal energy carrier in all organisms:

Adenosine

Adenosine monophosphate (ADP)

Adenosine diphosphate (ADP)

Adenosine triphosphate (ATP)

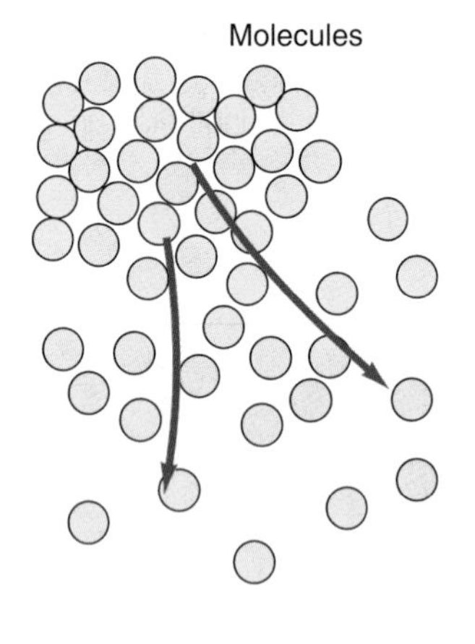

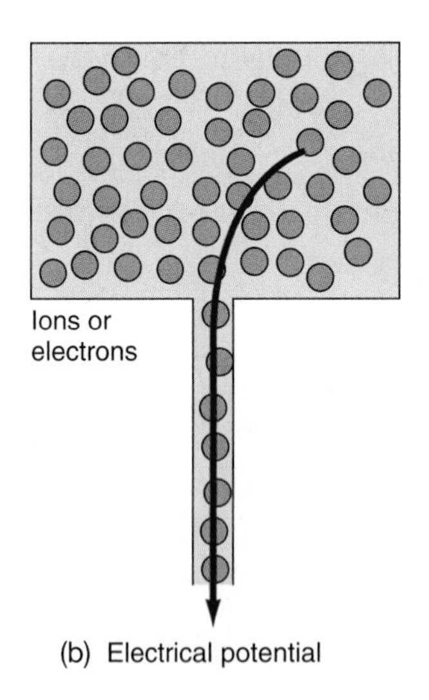

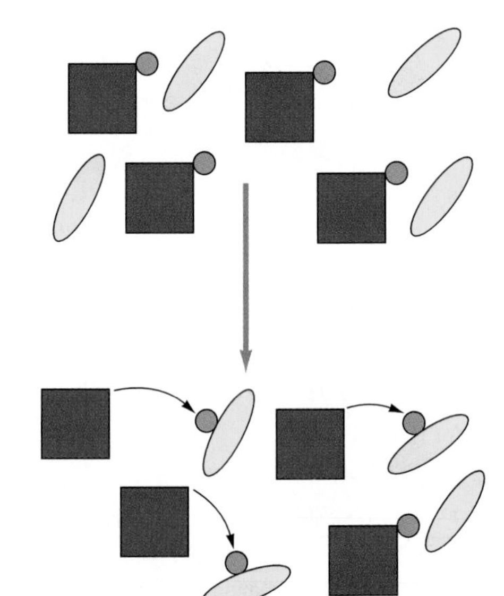

Figure 7.14
A system with a potential is able to do work. *(a)* A chemical potential measures the tendency of a highly concentrated substance to move and disperse. *(b)* An electrical potential measures the tendency of a high concentration of electrons or ions to move. *(c)* A group transfer potential measures the tendency of a molecule to transfer a group of atoms.

Adenosine with one phosphoryl group added is adenosine monophosphate (AMP); adding a second phosphoryl makes adenosine diphosphate (ADP), and adding a third makes ATP. The reaction in which ATP is hydrolyzed to ADP:

$$ATP + H_2O \rightarrow ADP + P_i$$

has a ΔG of -7.3 kcal/mol (30.5 kJ/mol) under so-called "standard" conditions, which means that all the components of the reaction are at a concentration of 1 molar. In contrast, so-called "low energy" compounds, such as glycerol phosphate, only give up about 2 kcal/mol when they are hydrolyzed, so ATP has a high group transfer potential. Depending on the actual concentrations of ATP and P_i inside a cell, the transfer of a phosphoryl group from ATP provides about 11–13 kcal of energy per mole, and this is enough to drive many endergonic reactions. Some phosphates have even higher potentials than ATP, such as creatine phosphate, which stores energy in muscle. In the past, the bonds between the phosphoryl groups of ATP were designated "high-energy bonds," but this is incorrect. The bond is not unusual; the critical point is the compound's ability to transfer a group to many other compounds.

ATP or a similar nucleotide is used in **phosphorylation** reactions, where it transfers its terminal phosphoryl group to another molecule, thereby effectively energizing, or activating, the second molecule. Suppose a cell must convert compound A to compound B, which has more free energy than A (Figure 7.15). First, ATP activates A by phosphorylating it to A-phosphate, which has even more free energy than B. Even though the formation of A-phosphate is endergonic, splitting ATP releases enough energy to drive the formation of A-phosphate uphill, and the whole coupled reaction is still exergonic. Then, in a second reaction, A-phosphate is easily hydrolyzed into B plus inorganic phosphate. One step in the pathway for converting aspartate to threonine is shown in Figure 7.16. In fact, aspartate can't be converted to aspartate semialdehyde directly because this is an endergonic reaction. Aspartate is first activated by ATP to aspartyl phosphate, and then the phosphate is removed. This is typical of a biosynthetic reaction driven by ATP.

ATP is the major source of free energy in cells, with similar nucleotides playing secondary roles, and its energy is used for three kinds of work (Figure 7.17). In addition to driving biosynthetic reactions, ATP supplies energy for transporting ions and molecules across cell membranes, an essential process for all organisms. By activating certain motor proteins, ATP also supplies the energy for motion, including small movements of cellular components, locomotion of single cells, and muscle contraction in animals.

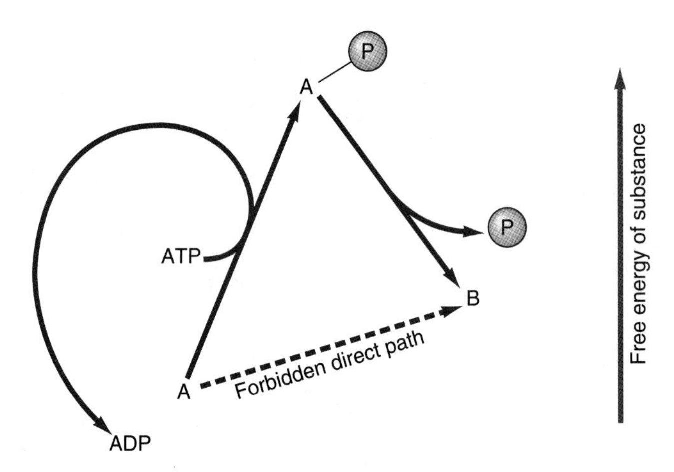

Figure 7.15

ATP may be used to convert compound A to compound B, which has more free energy than A. The phosphoryl group of ATP is added to A to make A-phosphate; this is an exergonic reaction because splitting ATP releases more free energy than is needed to make A-phosphate. The release of the phosphate then forms compound B.

Exercise 7.4 The phosphoryl-group transfer potentials (kcal/mol) for three compounds are: ATP (to ADP + P_i), 7.3; creatine phosphate, 10.3; and glucose 6-phosphate, 3.3. Which compounds can transfer phosphoryl groups spontaneously to which others?

7.9 Ecosystems operate on a flow of energy that comes from the sun.

Because ATP and similar molecules are the energy currency of organisms, cells build up a supply of them to be used wherever energy is needed. But by the first law of thermodynamics, "you can't get something for nothing," so where does the energy come from to make these highly energetic compounds? In the next section, we will provide a chemical answer to this question, but here we'll consider the ecological answer: This energy comes from the organism's environment, often from other organisms (Figure 7.18).

The ultimate source of energy for life on Earth is the sun. (Here we are ignoring some recently discovered marine ecosystems that operate entirely on chemical energy.) For this reason, we only expect to find life on planets that are close enough to a star to receive a lot of energy. Plants, algae, and some kinds of bacteria are **phototrophic** organisms, or **phototrophs** (*photo-* = light; *-trophy* = a mode of nourishment); they

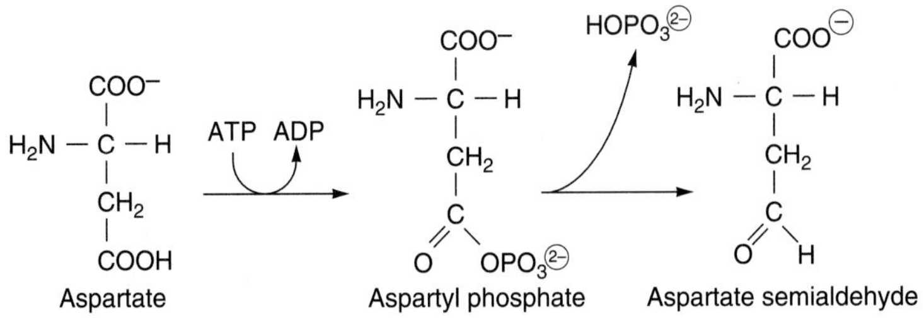

ATP ADP
Aspartate — Aspartyl phosphate — Aspartate semialdehyde
$HOPO_3^{2-}$

Figure 7.16

One step in the pathway of threonine synthesis actually requires activation of a metabolite (aspartate) by ATP.

Figure 7.17

The energy stored in ATP is used for three kinds of work: *(1)* activating molecules to drive endergonic reactions; *(2)* building up the chemical potential of molecules or ions in one cellular compartment by moving them to one side of a membrane; and *(3)* activating motor proteins, which carry out mechanical work.

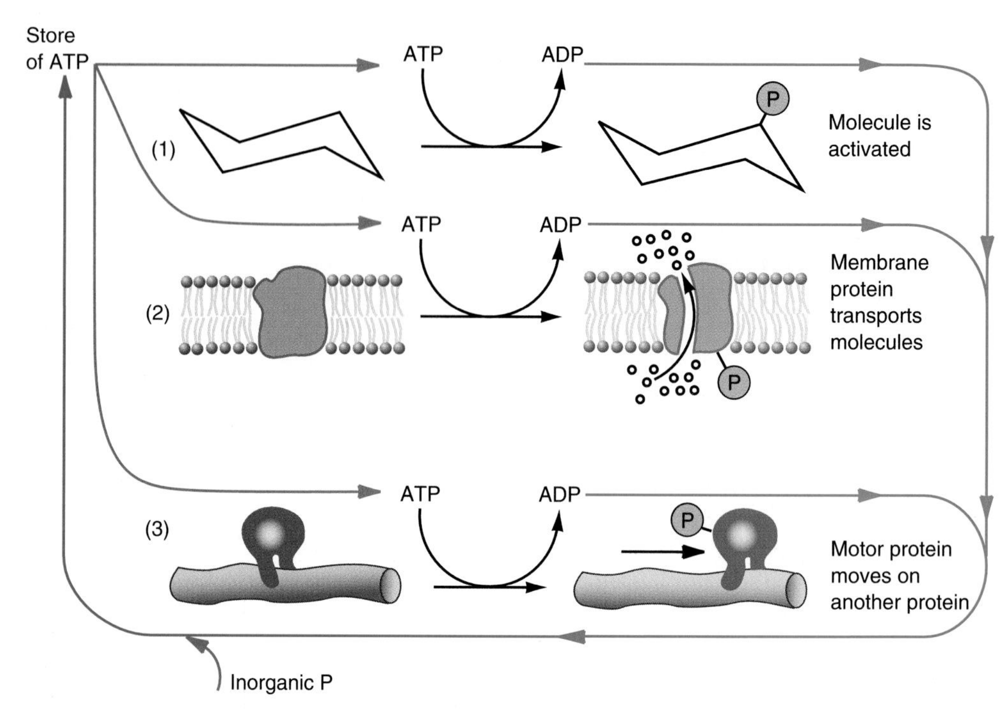

Figure 7.18

There are two kinds of organisms with respect to energy source. *(a)* Phototrophs capture some of the sun's energy and store it in their molecular structure. *(b)* Chemotrophs then eat other organisms and obtain some of that stored energy (or, more rarely, they extract energy from inorganic compounds). Organisms that pass their energy along from one to another like this constitute a food web.

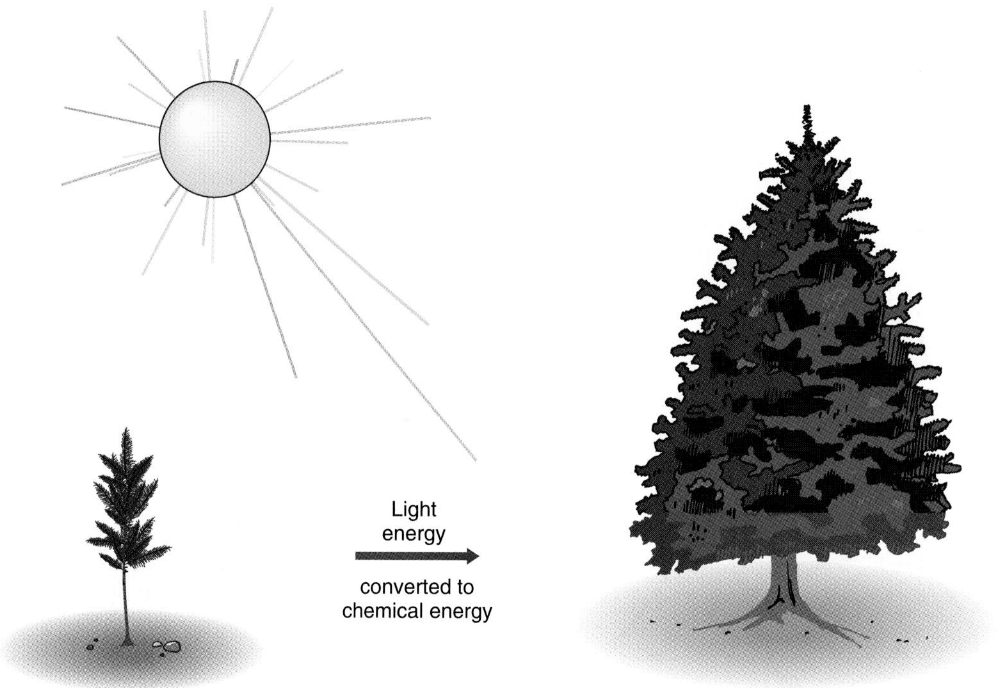

(a) Phototroph

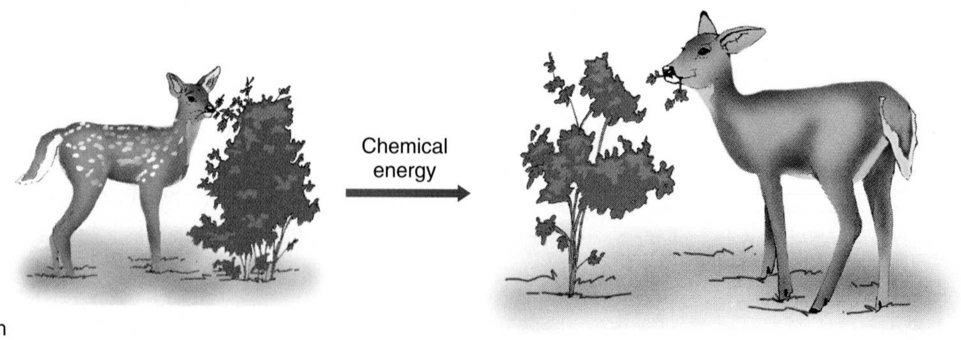

(b) Chemotroph

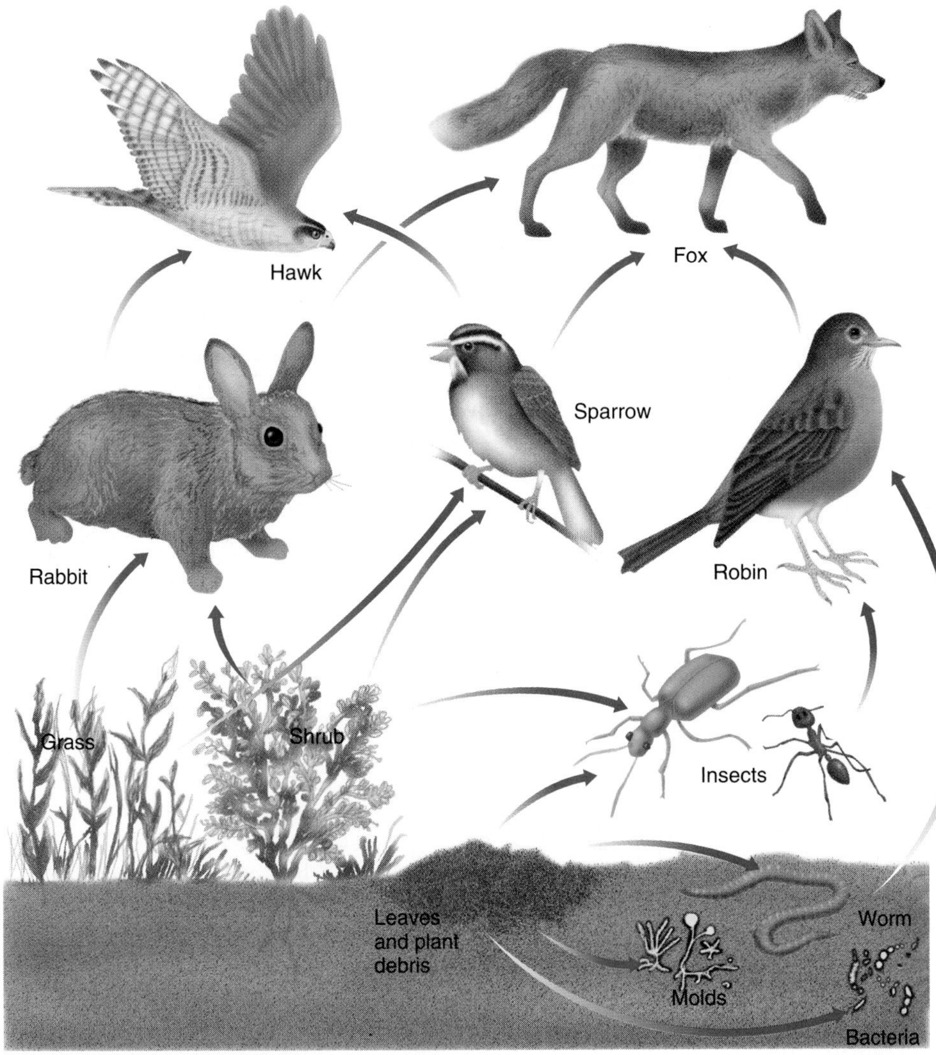

Figure 7.19
Organisms that eat one another form a simple food web. Real food webs are generally more complicated.

capture light energy in the process of **photosynthesis,** which is a light-driven synthesis, as the name implies. Phototrophs synthesize the organic molecules of their own structure from carbon dioxide and other simple inorganic materials. Energy—here derived from light—is needed to drive this synthesis, because these organic molecules have more energy than do the inorganic molecules. So as they grow, phototrophs are storing up energy in their structures. These energy-rich materials then serve as food for **chemotrophic** organisms, or **chemotrophs,** organisms such as animals whose energy source is chemical rather than light.

The story of the ecosystem has a simple plot: eat and be eaten. Organisms are related energetically as members of a *food chain,* a series of organisms that eat one another, although the ecosystem is best described as a complicated, tangled **food web** (Figure 7.19). Energy flows through this web, starting with phototrophs that act as **producers,** bringing energy into the system through photosynthesis (Figure 7.20). Producers store energy in their structure and are eaten by **primary consumers,** or **herbivores,** which use some of that stored energy for their own growth and reproduction. These, in turn, are eaten by **secondary consumers,** or **carnivores,** which use some of the energy in the herbivores' structure. Some carnivores may be eaten by still higher carnivores (called top carnivores). These distinctions are somewhat idealized, and many members of an ecosystem are **omnivores** that eat a mixture of plant and animal materials. Finally, every organism—producer or consumer—dies and ultimately becomes food for **decomposers,** the molds and bacteria that decay biomolecules as they grow and take their share of energy. An organism's position in a food web is called its *trophic level.*

The phototroph/chemotroph distinction refers to an organism's energy source. Organisms are also distinguished by their source of carbon, the major constituent of organic molecules. **Autotrophic** organisms (**autotrophs;** *auto-* = self) can make their own organic compounds from CO_2, water, and other inorganic materials, while **heterotrophic** organisms (**heterotrophs:** *hetero-* = different) can only live on the organic compounds already made by some other organism. Combining the energy source with the carbon source creates four categories

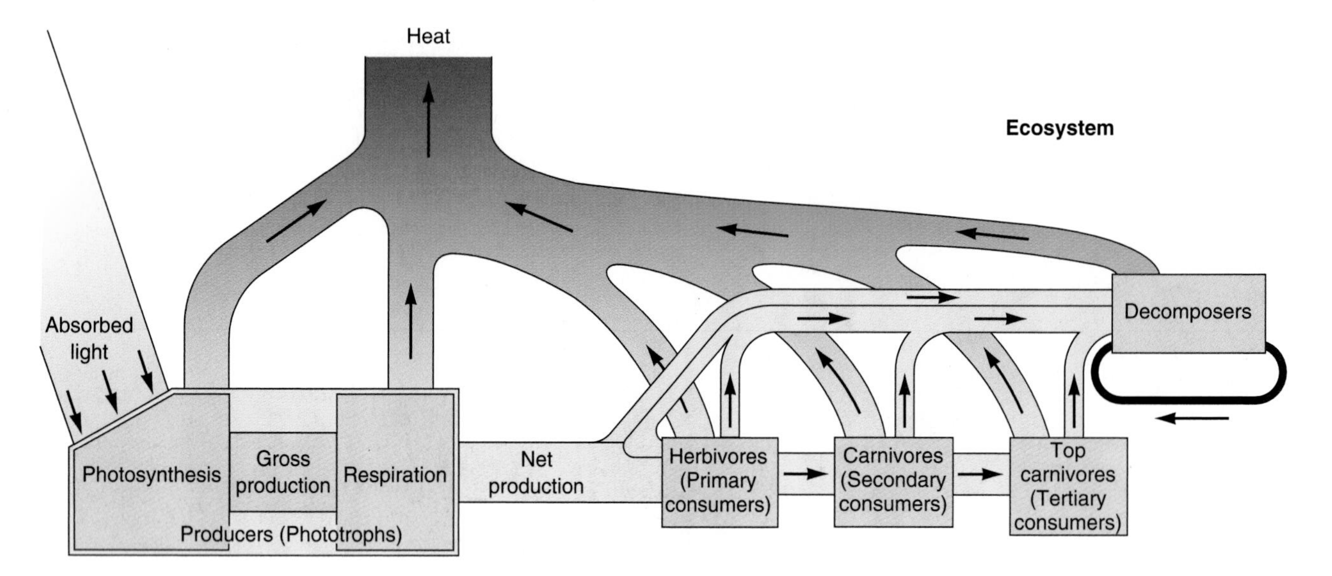

Figure 7.20

The energy that is initially captured from sunlight by the phototrophs (producers) of an ecosystem is passed on through various food webs. At each stage, some energy is lost as heat, and some is conserved in the structure of growing organisms. But though this energy supports several trophic levels, ultimately it all goes off as heat.

(Figure 7.21). Most of the ecosystem's producers are *photoautotrophs,* like plants, which use light for energy and CO_2 as their carbon source. However, some unusual bacteria are *chemoautotrophs,* which grow on CO_2 by extracting energy from inorganic materials. Most consumers are *chemoheterotrophs,* which extract energy from the organic molecules they consume as they transform these molecules into their own structure. Other bacteria are *photoheterotrophs,* which use organic molecules as the source of their material but get their energy from light.

Although metabolism occurs within each organism, when we adopt a broader ecological viewpoint, we see that the metabolisms of organisms are interconnected. That is, a by-product discarded by one can become an important nutrient for another. Thus it is quite realistic to say that a whole ecosystem has a metabolism, and we will see later that an ecosystem may be characterized as autotrophic or heterotrophic, depending on its stage of development and overall balance of activities.

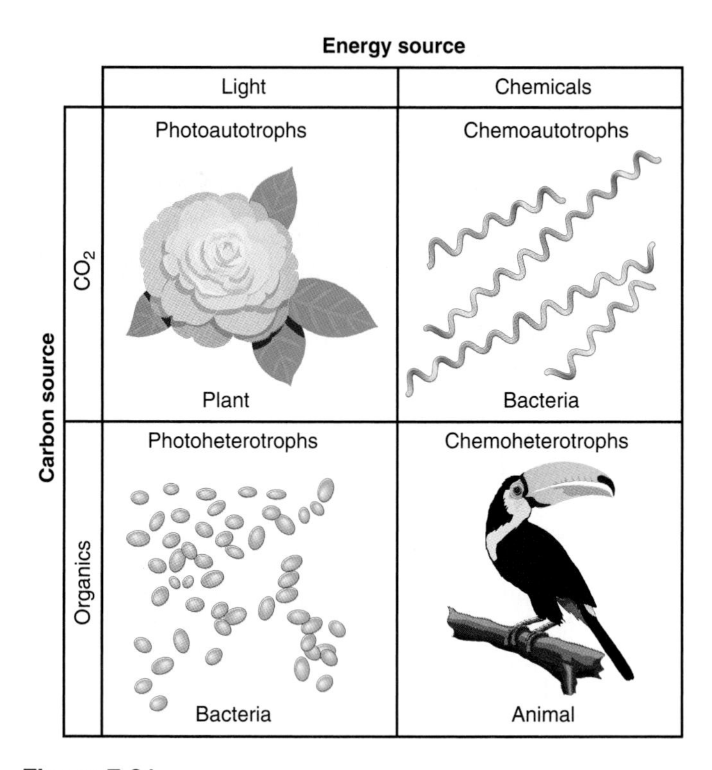

Figure 7.21

Organisms are distinguished as chemotrophs or phototrophs on the basis of their energy source, and as autotrophs or heterotrophs on the basis of their carbon source. These distinctions create four metabolic classes of organisms.

7.10 Useful energy can be obtained from oxidative reactions.

We have learned that organisms obtain energy from some environmental source and store it in compounds such as ATP. To complete the story, we will show how energy is extracted from compounds in the environment through processes of *oxidation* and *reduction,* the primary energy-yielding reactions of life.

Originally, oxidation meant combining something with oxygen (as in the rusting of iron: $Fe + O_2 \rightarrow FeO_2$) and reduction meant just the opposite, removing oxygen. Some reactions do involve oxygen itself, but there are now more general and useful definitions: A substance is **oxidized** when it gives up electrons (or hydrogen atoms) and **reduced** when it gains electrons (or hydrogen atoms). We include hydrogen atoms in these definitions because biological oxidation and reduction reactions most commonly entail the transfer of H atoms, rather than just electrons. If A gives an electron to B, A is oxidized and B is reduced. Since each substance is the agent that changes the other, A is the **reducing agent** and B is the **oxidizing agent.** The combined processes, called *redox reactions* or *oxidoreduction,* have to occur together, even if they are separated in space, because electrons have to come from somewhere and go somewhere. We can see an example of oxidoreduction by putting a zinc rod (or a galvanized nail) into a copper sulfate solution (Figure 7.22). Initially the zinc atoms are in the Zn^0 state, not oxidized at all, while the solution contains Cu^{2+} ions. But zinc ions (Zn^{2+}) soon start to appear in solution, and a thin layer of copper metal (Cu^0) is deposited on the rod. Each zinc atom loses two electrons and is oxidized to a zinc ion (Zn^{2+}). Each copper ion gains two electrons (e^-)and is reduced to a copper atom (Cu^0):

$$Zn^0 \rightarrow Zn^{2+} + 2\,e^-$$
$$Cu^{2+} + 2\,e^- \rightarrow Cu^0$$

The reaction is really one coupled process:

Zn^0 Cu^{2+}

Zn^{2+} Cu^0

where the transferred group is a pair of electrons. In this reaction, zinc is the reducing agent and copper is the oxidizing agent.

Would the opposite reaction occur between a copper rod and a zinc sulfate solution? No, it wouldn't. The reaction that actually occurs is exergonic, and the reverse would be endergonic. Zinc is simply a stronger reducing agent than copper. The strength of any substance as a reducing agent is indicated by its **reduction potential** (Table 7.2), expressed in *volts* and measured relative to a standard reaction in which hydrogen ions are converted into hydrogen gas. On the scale of reduction potentials, the substance with the algebraically lower potential is the stronger reducing agent. Since the reduction potential of zinc is −0.76 volts and that of copper is −0.34 volts, zinc will reduce copper.

Now we can understand why more reduced compounds have more energy. Remember that phototrophs store energy as they convert CO_2 to organic compounds; they can recover that energy for their own growth, and so can chemotrophs that eat the phototrophs. In an organic compound such as methane, the carbon atom is relatively reduced, so the electrons are held quite equally between carbon and hydrogen atoms:

H
H—:—C—:—H
H

There is a tension in this structure, as if the electrons were on rubber bands, being pulled by both atoms. In H_2O and CO_2, the products of burning methane with oxygen, the oxygen atoms hold the electrons tightly:

H—:O:—H

O::=C=::O

as if one of the rubber bands has been cut so the electrons are in a more "relaxed" state, a more stable state. Energy is released in the transition from the "tense" to the "relaxed" condition. If this energy is released rapidly, we feel it as the heat of a fire. Energy can also be released more slowly to fuel the biological processes of metabolism. Thus the story of chemical conversions in an organism or an ecosystem is largely a story of redox reactions.

To show whether a substance is being oxidized or reduced in a reaction, each element in a compound is assigned an

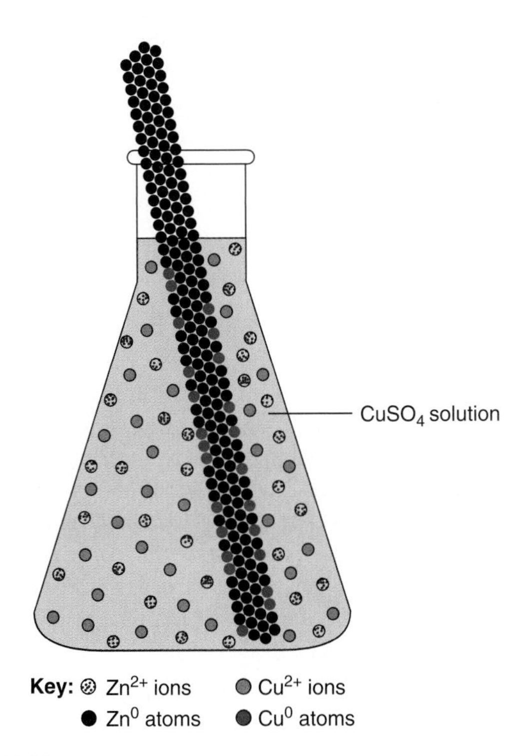

Figure 7.22
When a zinc rod is placed in a copper sulfate solution, the zinc atoms (Zn^0) are oxidized to Zn^{2+} ions while copper ions (Cu^{2+}) in the solution are reduced to a layer of copper metal (Cu^0) deposited on the rod. Each zinc atom transfers two electrons to a copper ion.

Table 7.2 Standard Reduction Potentials (Volts)

Reaction	Volts
$Mg^{2+} + 2e^- \rightarrow Mg$	−2.37
$Zn^{2+} + 2e^- \rightarrow Zn$	−0.76
$Fe^{2+} + 2e^- \rightarrow Fe$	−0.41
$Cu^{2+} + 2e^- \rightarrow Cu$	−0.34
$NAD^+ + H^+ + 2e^- \rightarrow NADH$	−0.320
$FAD + 2H^+ + 2e^- \rightarrow FADH_2$	−0.219
Pyruvate $+ 2H^+ + 2e^- \rightarrow$ Lactate	−0.185
$2H^+ + 2e^- \rightarrow H_2$	0.000
Fumarate $+ 2H^+ + 2e^- \rightarrow$ Succinate	+0.031
$NO_3^- + 2H^+ + 2e^- \rightarrow NO_2^- + H_2O$	+0.421
$\frac{1}{2}O_2 + 2H^+ + 2e^- \rightarrow H_2O$	+0.816
$Ag^{2+} + e^- \rightarrow Ag^+$	+1.99

oxidation state, a formal number that shows how many positive or negative charges it must carry to account for the charge of the whole molecule, which is usually zero. Although atoms generally share electrons, for the sake of assigning numbers, we imagine that the more electronegative atom takes the electrons all to itself. Hydrogen is always in the $+1$ state, and oxygen is always -2; other atoms then take the values necessary to make a total of zero. In CO_2, the two oxygen atoms together are $2 \times (-2) = -4$, so carbon must be $+4$ to balance. This is carbon in its most oxidized state. In sugar ($C_6H_{12}O_6$), C is reduced to an average oxidation state of zero, and in methane (CH_4), C is still more reduced to the -4 state. (Notice that reduction literally means a reduction in oxidation state, while oxidation means an increase in oxidation state.)

Redox reactions are also central to storing energy in ATP as it occurs in certain cellular structures—mitochondria and chloroplasts in eucaryotes, special cell membranes in procaryotes (Figure 7.23). The process depends on creating minute electric currents in the membranes of these structures and using some of their energy to make ATP, just as we tap the electric current from a battery to light a flashlight or power a radio. We get power from a dry cell battery by creating a circuit that carries electrons from the oxidation of the zinc shell around the outside to the reduction of manganese in a mixture inside. Similarly, in mitochondria, electrons are carried from reduced organic compounds to the final oxidizing agent, oxygen (which animals like us need to stay alive). In chloroplasts, electric currents originate from compounds that have become energized by absorbing light. Both systems synthesize ATP through a common mechanism, as we will show in Chapters 9 and 10.

7.11 Two kinds of nucleotides are used as oxidizing and reducing agents.

In addition to energy-carrying molecules like ATP, cells need oxidizing and reducing agents for many steps in metabolism. Most metabolic oxidations are **dehydrogenations** in which a pair of hydrogen atoms is *removed;* most metabolic reductions are **hydrogenations** in which a pair of hydrogen atoms is *added.* The enzymes for these reactions (dehydrogenases and reductases) use two kinds of coenzymes, which are also nucleotides.

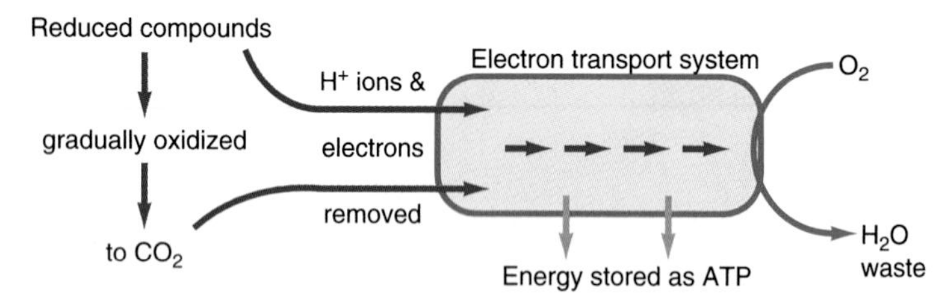

Figure 7.23

Cells run on oxidoreduction. They contain structures in which minute electric currents are created by the sequential transfer of electrons (or hydrogen atoms) from one molecule to another in a chain. The energy released in these transfers is stored in a chemical form, generally as ATP.

One of these coenzymes is a double nucleotide, **nicotinamide adenine dinucleotide,** or **NAD⁺** (Figure 7.24*a*). Notice that half of NAD^+ is adenosine monophosphate (AMP), and the other half is a nucleotide whose base is **nicotinamide:**

O
C — NH_2
N

Nicotinamide has a special capacity for oxidation and reduction. With its positively charged nitrogen atom, NAD^+ oxidizes a reduced organic compound (XH_2) by removing a pair of hydrogen atoms:

DH_2 + **NAD⁺** (H, $CONH_2$, N⁺) ⟶ D + **NADH** (H H, $CONH_2$, N) + H^+

One proton is released as a hydrogen ion, so one proton and two electrons join NAD^+, canceling its positive charge and making NADH.

The reverse of this reaction is a reduction of X to XH_2. While NAD^+ is the chief coenzyme for dehydrogenation in catabolism, reductions are commonly catalyzed by enzymes that use the coenzyme NAD^+ phosphate ($NADP^+$), which is simply NAD^+ with another phosphoryl group. As an illustration, look at two other steps in the conversion of aspartate to threonine that we couldn't show completely in Section 7.6. The conversions of the —COOH group to HC=O and then to H_2COH are both reductions; notice in Figure 7.25 that the molecules lose oxygen atoms and gain hydrogens, and both reactions require NADPH.

Many organisms, including animals, are unable to make the carbon-nitrogen ring of nicotinamide, so they must obtain it from a vitamin, nicotinic acid, commonly known as niacin. (Neither substance is related to the nicotine of tobacco.)

A second important coenzyme of oxidoreduction is **flavin-adenine dinucleotide (FAD),** whose **flavin** ring is an oxidizing agent (Figure 7.24*b*). FAD is usually attached as a prosthetic group on certain proteins, called *flavoproteins,* which are built into the cellular systems that oxidize organic molecules and synthesize ATP. In these reactions, FAD picks up a pair of hydrogen atoms and is reduced to $FADH_2$:

$$XH_2 + FAD \rightarrow X + FADH_2$$

Animals also lack the ability to synthesize the flavin ring, and they must obtain it in the form of vitamin B_2 (riboflavin).

Prosthetic groups, Section 4.12.

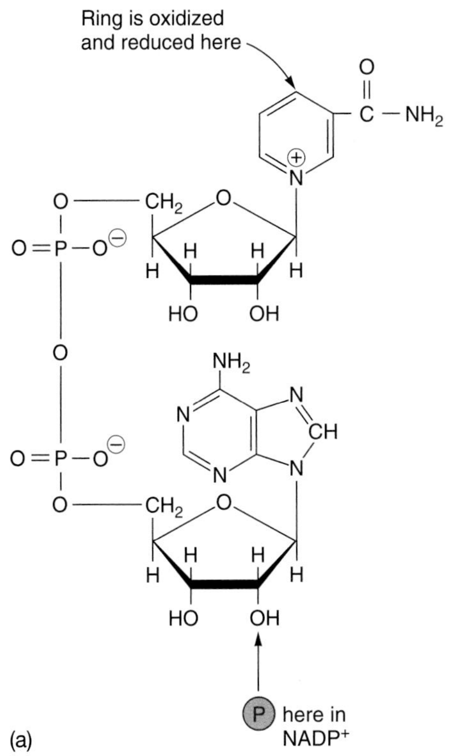

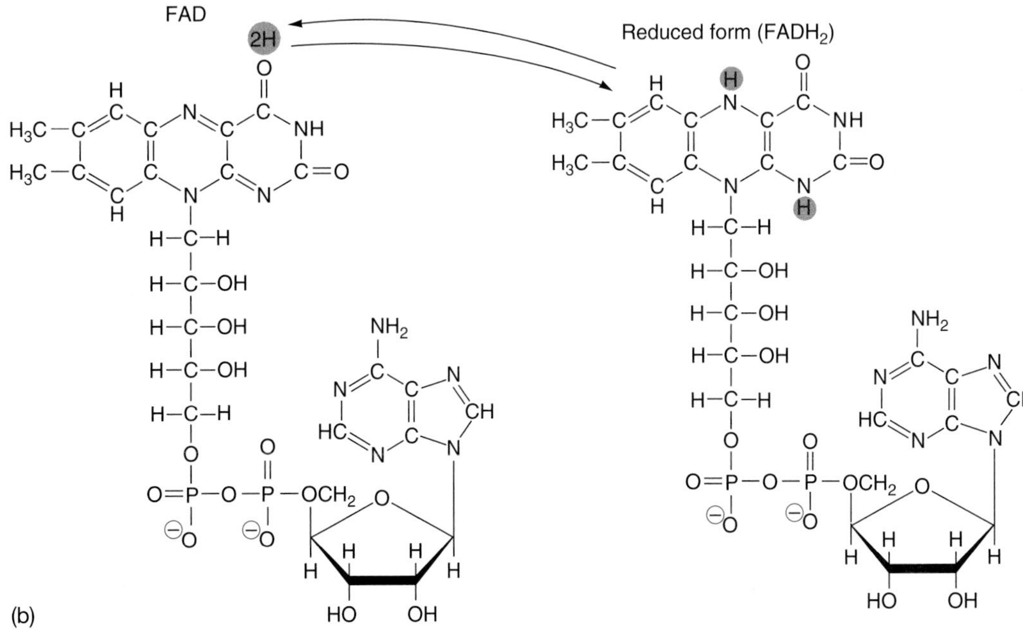

Figure 7.24

(a) The chief coenzyme of oxidation and reduction is nicotinamide-adenine dinucleotide (NAD^+). The coenzyme $NADP^+$ has another phosphate at position P. *(b)* The flavin ring is oxidized and reduced in the coenzyme flavin-adenine dinucleotide (FAD).

Figure 7.25

Two steps in the pathway of threonine synthesis are reductions and require NADPH as a source of hydrogen atoms.

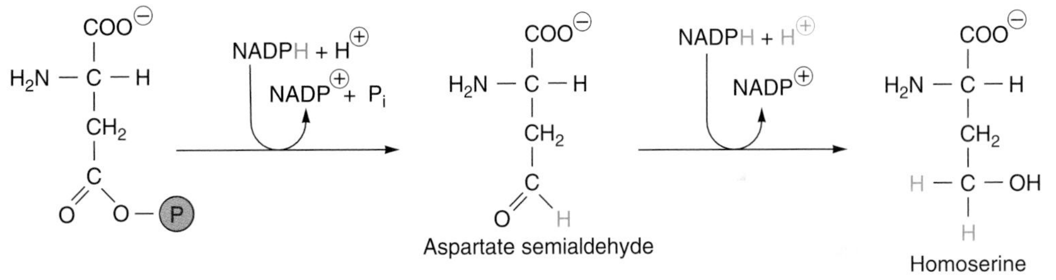

> **Exercise 7.5** Referring to Table 7.2, which of the following processes will occur spontaneously?
> a. Iron will reduce magnesium.
> b. $FADH_2$ will reduce pyruvate to lactate.
> c. NADH will reduce oxygen to water.
> d. Nitrate (NO_3^-) will reduce FAD.

7.12 ATP and NADPH provide the energy and reducing power needed for biosynthesis.

Now it is time to sum up the main points of this chapter into an overall picture of metabolism. Metabolism provides energy for biosynthesis and growth, for movement, and for transporting materials in and out of cells. Since organisms are made of polymers, biosynthesis means making monomers (and other monomer-sized molecules, such as coenzymes) and polymerizing them into polymers. The chemical reactions of biosynthesis are organized into metabolic pathways, each step catalyzed by a different enzyme.

The endergonic (uphill) reactions of biosynthesis are driven by compounds such as ATP that have high group transfer potentials. Since biomolecules are more reduced than the molecules they are made from, some steps in their synthesis require a reducing agent, generally NADPH. All organisms must have supplies of both ATP and NADPH—the one for energy, the other for reducing power. Both coenzymes are derived from oxidation, either the oxidation of reduced compounds (in chemotrophs) or the flow of high-energy electrons activated by light (in phototrophs).

A heterotroph lives on reduced monomers, such as sugars. During catabolism, it breaks these molecules down to smaller molecules while oxidizing them to release some of their energy, which is stored as ATP and NAD(P)H (Figure 7.26). Then new monomers are synthesized from some of the partially oxidized

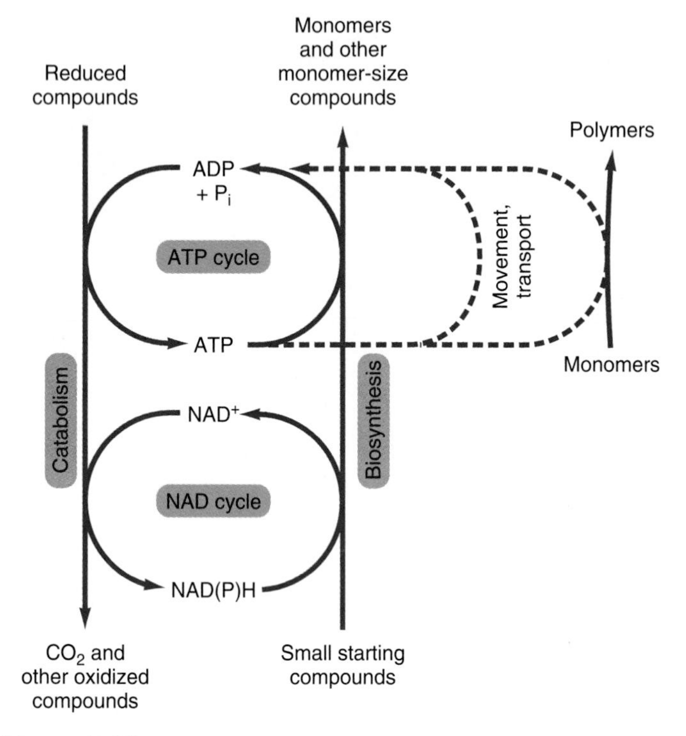

Figure 7.26

Heterotrophic metabolism consists of two main phases. In *catabolism,* reduced molecules are broken down to smaller molecules and oxidized to obtain energy. In *biosynthesis,* new monomers are synthesized from some of the partially oxidized molecules, and these monomers are polymerized into structural molecules.

molecules resulting from catabolism, and these monomers are finally polymerized into structural molecules. Autotrophs, which get their carbon atoms from CO_2, supplement these processes with a cycle of reactions in which CO_2 is reduced to sugars.

Coda When considering the energy requirements of organisms, their ecological foundations become most obvious. All organisms need energy, and the source of that energy must be their surroundings. For half the living world, that means light; for the other half, it means other organisms, or at least the materials they have produced. It's very likely that no planet or biosphere can maintain itself for very long unless some phototrophs evolve there. That is, only the phototrophs can tap the sun's virtually unlimited, inexhaustible energy supply, and practically every ecosystem depends on them. (Since we now know of ecosystems that operate entirely on chemical energy, it would be interesting to speculate on the nature of a planet inhabited only by chemotrophs. How could they survive on a large but limited supply of reduced inorganic compounds?)

Through evolution, every species has found a particular place in the energy web of its ecosystem and a particular way of life that supplies its energy. But regardless of their energy source, all organisms use the same basic metabolic mechanisms; they live by tapping the energy of certain compounds through oxidative reactions and then storing some of that energy in universal energy currencies (ATP) and universal redox currencies (NAD^+/NADH). The first organisms to evolve solutions to their energy problems were successful because they drew upon the potential of phosphates, which have the required chemical properties. Here again we see the fitness of the environment, as discussed in Chapter 3. If there were no such things as phosphates, organisms might not be able to exist. But our remote ancestral cells evolved proteins that could use phosphates, and they passed on instructions in their genomes for doing so. We function as we do because we have inherited that genetic and chemical capability. We make proteins that can oxidize reduced compounds from our environment and tap their energy, and we make proteins that can interact with organic phosphates in just the right way to store energy in these compounds. Thus we obtain and use the energy that keeps us alive.

Summary

1. Energy is the ability to do work. Energy can be transformed into different forms but cannot be created or destroyed.
2. Every molecule contains chemical energy in its bonds due to the forces between protons and electrons. Chemical reactions entail changes in energy (enthalpy or heat content).
3. A chemical reaction will occur spontaneously if it proceeds downhill energetically, but some reactions occur spontaneously even though the enthalpy of the products is greater than the enthalpy of the reactants.
4. Entropy is a measure of the disorder of a system. All isolated systems tend to increase in entropy (become more disordered). Chemical reactions entail changes in entropy as well as in enthalpy, and sometimes increasing entropy drives a reaction, even though its enthalpy increases. Free energy takes account of both enthalpy and entropy, and the requirement for decreasing free energy is really what determines the direction of a chemical reaction.
5. Organisms are highly organized and have lower entropy (and greater free energy) than the materials they are made from. To maintain this condition, organisms derive free energy from their environment.
6. Organisms construct and maintain themselves through metabolic pathways, which are sequences of reactions, each catalyzed by an enzyme.
7. Energy-consuming reactions can be driven by coupling them to energy-yielding processes. Two reactions can be coupled if they share a common component. An exergonic reaction can pull an endergonic reaction forward, or a chemical group can be transferred from an exergonic reaction to an endergonic one.
8. The most important group transferred in biological reactions is the phosphoryl group. The free energy of such a reaction is measured by a compound's phosphoryl-group transfer potential. Thus free energy is carried primarily in adenosine triphosphate (ATP).
9. Energy enters an ecosystem from sunlight through phototrophs and is passed on through a food web in which one chemotroph eats another. Some of the energy stored in the structure of one organism may be used to make the structure of another. The members of a food web are classified as producers, primary consumers, secondary consumers, and decomposers. Autotrophs obtain their carbon from CO_2, and heterotrophs obtain it from organic compounds.
10. A substance is oxidized when electrons or hydrogen atoms are removed from it and reduced when they are added to it. Organisms

obtain useful energy through oxidative reactions.

11. The nucleotides NAD^+ and FAD are used as oxidizing agents. NADPH is used as a reducing agent.
12. Organisms must maintain stores of ATP and NADPH to provide the energy and reducing power needed for biosynthesis. Heterotrophs obtain them through catabolic pathways in which reduced organic compounds are oxidized and broken down into smaller molecules, which are then built up in biosynthesis. Autotrophs carry out additional reactions in which they reduce CO_2 to organic compounds.

Key Terms

Multiple-Choice Questions

1. The chemical energy contained within a stable molecule, which can be used to do work,
   a. is a form of kinetic energy.
   b. is derived from the gravitational force.
   c. can be converted to heat.
   d. Answers *a, b,* and *c* are true.
   e. Answers *a* and *b,* but not *c* are true.
2. Which is not correct about anabolic reactions taken as a whole?
   a. ΔH is positive.
   b. ΔG is positive.
   c. $T\Delta S$ is negative.
   d. ΔS is negative.
   e. Disorder increases.
3. Chemical bond energy is best described as
   a. the energy necessary to form a bond.
   b. the energy necessary to break a bond.
   c. the heat of formation when a bond is made.
   d. *a* or *b* but not *c.*
   e. *b* or *c* but not *a.*
4. The heat of formation of a molecule is a change in
   a. entropy.
   b. enthalpy.
   c. potential energy.
   d. free energy.
   e. endergonic energy.
5. Assuming constant temperature, when a metabolic reaction approaches and then reaches equilibrium,
   a. the entropy level rises.
   b. the system increases in stability.
   c. the first and second laws of thermodynamics are satisfied.
   d. *a* and *b,* but not *c* are correct.
   e. *a, b,* and *c* are all correct.
6. If an isolated system at constant temperature undergoes a reaction and shows a decrease in both heat content and free energy, then
   a. the energy of the system has not been conserved.
   b. heat must have been converted to free energy.
   c. the order of the system has increased.
   d. the disorder of the system has increased.
   e. it has moved away from equilibrium.
7. When glucose is catabolized to carbon dioxide and water,
   a. the oxidation state of carbon is increased.
   b. ΔG is negative.
   c. the products (carbon dioxide and water) contain less free energy than the substrate (glucose).
   d. the products (carbon dioxide and water) have a greater entropy than the substrate (glucose).
   e. all of the above are correct.
8. In compounds such as ATP that have high phosphoryl-group transfer potentials, the chemical bond holding the terminal phosphoryl group is best described as
   a. stable and strong.
   b. unstable and easily broken.
   c. resistant to hydrolysis.
   d. an oxidizing agent.
   e. a reducing agent.
9. Growing organisms lower their internal entropy levels by
   a. excreting metabolic wastes containing high levels of free energy.
   b. engaging only in anabolic reactions.
   c. coupling exergonic reactions to those with a positive ΔG.
   d. increasing their body temperature.
   e. increasing the entropy of their environment.
10. The two kinds of chemical processes that usually couple spontaneous reactions to nonspontaneous reactions are phosphoryl-group transfers and
    a. redox reactions.
    b. hydrogenations and dehydrogenations.
    c. hydrolysis and dehydrolysis.
    d. *a* and *b,* but not *c.*
    e. *a, b,* and *c.*

True-False Questions

Mark each statement true or false, and if false, restate it to make it true.

1. Heat is a form of potential energy.
2. Energy, or work, is equal to force multiplied by distance, and is measured in calories or joules.
3. Chemical energy results from the strong nuclear force.
4. Energy is interconvertible with respect to kind (e.g., chemical, radiant, electrical, etc.) as well as between kinetic and potential forms.

5. Reactions that entail a positive change in free energy release energy to their surroundings.
6. The melted form of a compound has a higher entropy level than its solid counterpart but a lower entropy level than its vaporized state.
7. An exergonic reaction in which ΔG is -5 kcal/mol can drive an endergonic reaction in which ΔG is $+5$ kcal/mol or higher.
8. Given the compounds A and AH_2, compound A contains more energy than its oxidized partner AH_2.
9. NAD^+ is the reduced form of NADP.
10. The formation of ATP from ADP and P_i is an example of a reduction; the formation of ADP and P_i from ATP is an oxidation.

Concept Questions

1. Explain why virtually all metabolic reactions require an enzyme, but only some require NADH and/or ATP in addition to an enzyme.
2. What is meant by the phrase *coupled reactions,* and why is reaction coupling necessary?
3. Can a nonspontaneous reaction be changed into a spontaneous one by employing the appropriate enzyme? Explain.
4. Explain why every ecosystem must contain heterotrophs as well as autotrophs.
5. Compare and contrast the functions of ATP and NADH.

Additional Reading

Bergethon, P. R., and E. R. Simons. *Biophysical Chemistry: Molecules to Membranes.* Springer-Verlag, New York, 1990.

Smith, E. Brian. *Basic Chemical Thermodynamics,* 4th ed. Clarendon Press, Oxford, 1990.

Tomasi, Thomas E., and Teresa H. Horton (eds.). *Mammalian Energetics: Interdisciplinary Views of Metabolism and Reproduction.* Comstock Publ. Associates, Ithaca (NY), 1992.

Zotin, A. I. *Thermodynamic Bases of Biological Processes: Physiological Reactions and Adaptations.* W. de Gruyter, Berlin and New York, 1990.

Internet Resource

To further explore the content of this chapter, log on to the web site at

http://www.mhhe.com/biosci/benbio/guttman/

Membranes and Transport Processes

8

Key Concepts

A. Membrane Structure

8.1 Molecules move by diffusion.

8.2 Membranes allow some molecules to diffuse freely, while they inhibit the passage of others.

8.3 Cells have osmotic properties.

8.4 Lipids are insoluble in water.

8.5 Experiments with osmosis in cells show that membranes are made of lipids.

8.6 Phospholipids form bilayers, which are the structural basis of all membranes.

8.7 A membrane is a fluid mosaic of lipid and protein.

8.8 The fluidity of a membrane depends on the composition of its lipids.

8.9 Membrane proteins move laterally, but do not "flip-flop" across a membrane.

8.10 Studies of red blood cell membranes have increased our understanding of membrane structure.

B. Transport Mechanisms

8.11 Proteins transport some substances across membranes through facilitated diffusion.

8.12 Some substances are actively transported against a concentration gradient.

8.13 A gradient carries a potential and can do work.

8.14 Secondary active transport uses the energy of an established concentration gradient.

8.15 Membrane proteins can move substances by vectorial action.

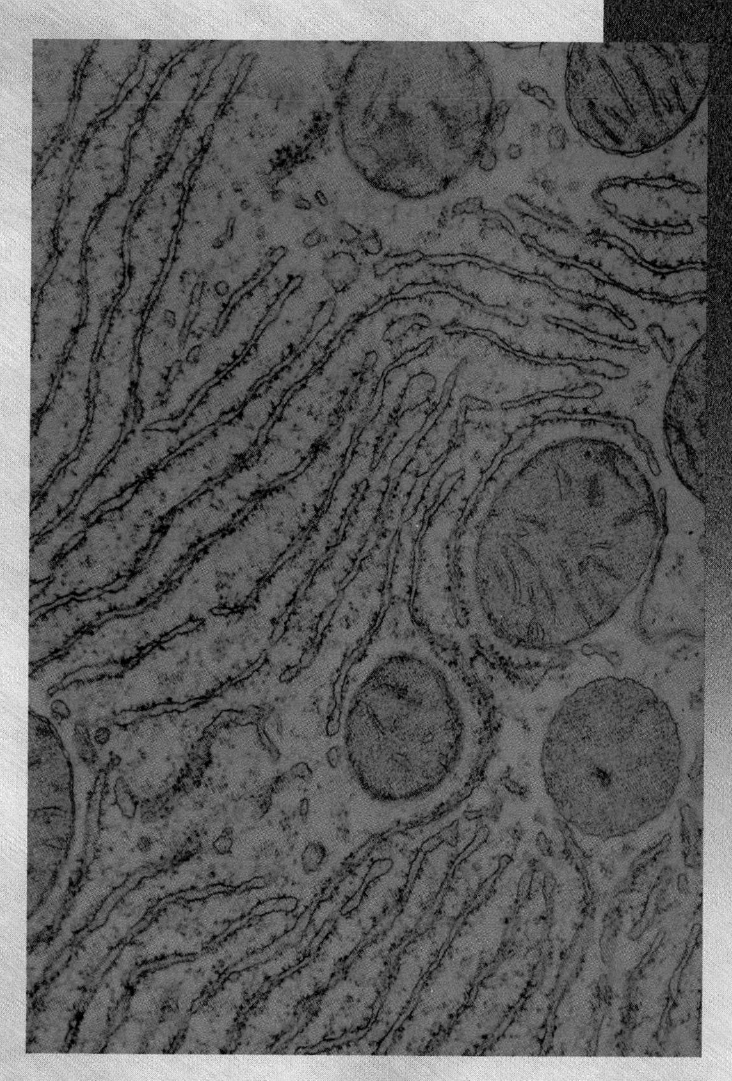

Electron microscopy reveals the many membranes in a typical eucaryotic cell.

In 1930, the American biologist Janet Plowe did a series of simple but revealing experiments. She manipulated cells with very fine needles and showed that the cell surface is made of a special structure that holds the body of the cell together. As Plowe put it, "There is an external layer, or plasmalemma, surrounding the protoplast [cell body], which, while fluid, is more elastic and more extensible than the remainder of the cytoplasm." In spite of investigations by Plowe and other experimenters, biologists continued to argue about the existence of plasma membranes for a long time, until electron micrographs laid all doubts to rest. When we look at cells with standard microscopic methods, we see all kinds of membranes. Every cell is bounded by a plasma membrane, and most eucaryotic organelles are bounded by other membranes: the nucleus, defined by its nuclear envelope; mitochondria and chloroplasts, with their own internal membranes; and the endomembrane system of the endoplasmic reticulum and Golgi membranes.

Contemporary biologists recognize that membranes do much more than hold cells and organelles together. Membranes are intricate mosaics of lipid and protein molecules that—simply by virtue of their structures—are able to conduct important metabolic processes themselves, including mobilizing energy. They can establish chemical and electrical potentials across themselves, and these potentials can be turned into the synthesis of ATP as well as the basic activities of animal nervous systems and even some movements of plants. In this chapter, after describing the lipid-protein membrane structure, we will consider some typical membrane proteins. We will then show how molecules are transported across membranes by various mechanisms, most of them based on the action of specific proteins. ■

A. Membrane Structure

8.1 Molecules move by diffusion.

The cytosol within a cell is extraordinarily different from the surrounding extracellular liquid and from the solutions within internal cellular compartments bounded by membranes. These differences are critical to the survival of the cell, and they are maintained by membranes, which regulate the passage of materials.

To understand how membranes operate, we must understand that molecules move constantly and rapidly. (In this discussion, we will use the term "molecule" to include ions of all kinds.) Every moving body has kinetic energy, and the collective kinetic energies of molecules are experienced as *heat,* or *thermal energy.* The rate at which molecules move depends on their temperature and size. At higher temperatures, molecules move faster and collide with one another with greater force. Calculations show that at 37°C, water molecules move at fantastic speeds, perhaps 2,500 km/hr (1,500 miles/hour), and a molecule of glucose, whose mass is ten times that of water, moves at about one-third that speed when dissolved in water. This movement occurs randomly in all directions, causing different kinds of molecules to intermingle as they collide with one another and with the walls of their containers. We call this random molecular movement **diffusion.**

Diffusion can be seen clearly by putting a concentrated dye solution on a block of gelatin (Figure 8.1). (Diffusion occurs in a gas or liquid too, but there the molecules are also mixed by stirring or convection; using gelatin eliminates these factors.) As the dye molecules spread out by moving randomly, the intensity of color shows that the dye forms a **concentration gradient,** a continuous gradation in the concentration of dye molecules from a region of high concentration to one of low concentration. Molecules diffuse randomly in all directions, but their net tendency is to diffuse *down* this gradient—from the region of higher concentration to the region of lower concentration:

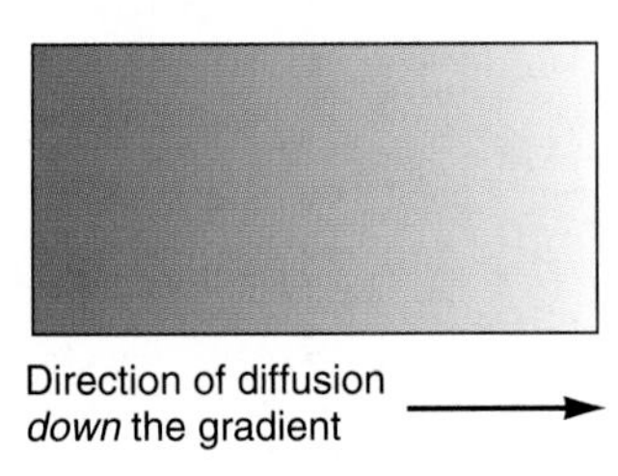

Eventually they come to an equilibrium and are uniformly distributed, so the gradient no longer exists.

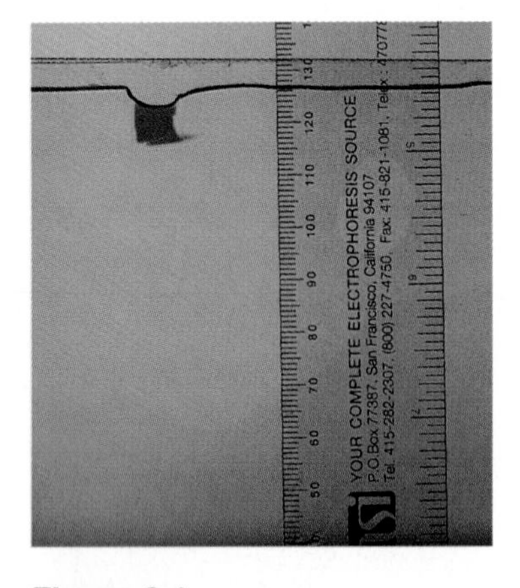

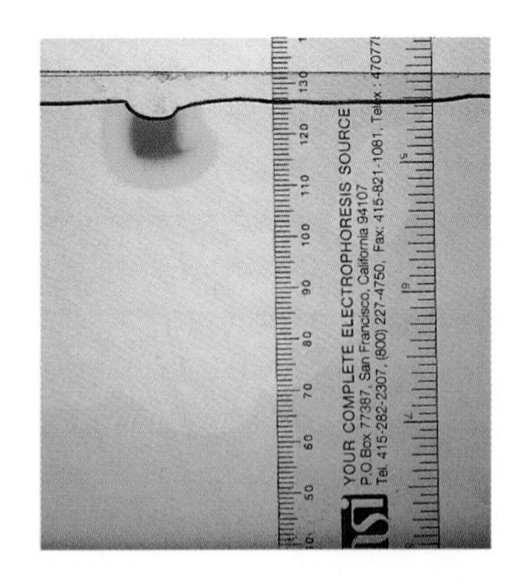

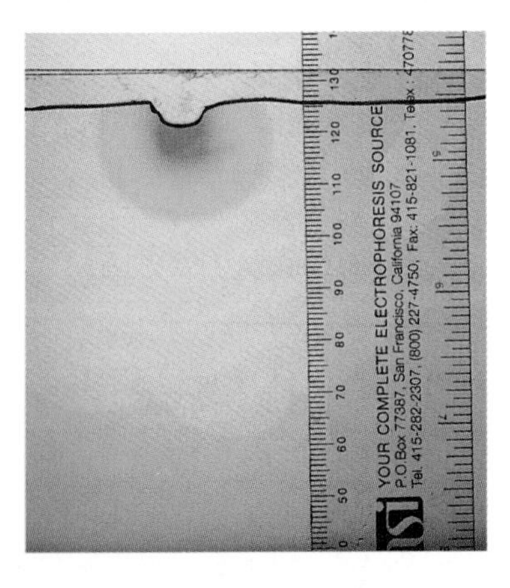

Figure 8.1

Random motion due to thermal energy will cause molecules to move throughout a container, between other molecules. Here the dye fluorescein diffuses through water and gelatin. At first the dye molecules form a gradient, with the highest concentration closest to their origin. Eventually the dye molecules will diffuse until they are uniformly distributed.

Exercise 8.1 In an ordinary thermometer, the column of mercury expands as it gets warmer. Explain why.

8.2 Membranes allow some molecules to diffuse freely, while they inhibit the passage of others.

Membranes restrict the passage of some materials and not others. A membrane is said to be *permeable* to molecules that can pass through it readily and *impermeable* to those that cannot. If a container is divided into two compartments by a permeable membrane, certain molecules can still diffuse through the membrane; there will be a *flux,* or flow, from one compartment to the other. The flux occurs in both directions, but if there is a gradient across the membrane, more molecules will move down the gradient than up, creating a *net flux* in one direction. Such an unequal flux will drive the system closer to equilibrium:

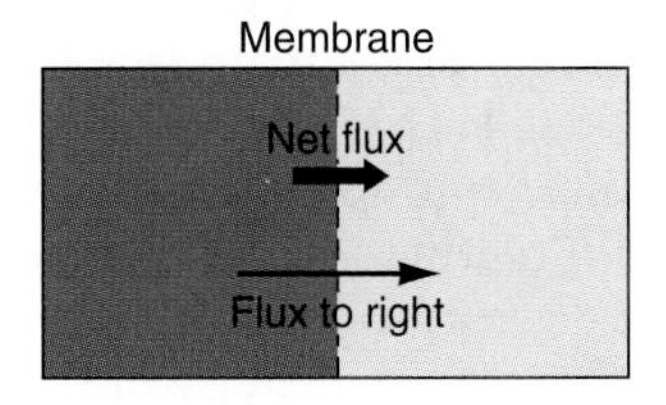

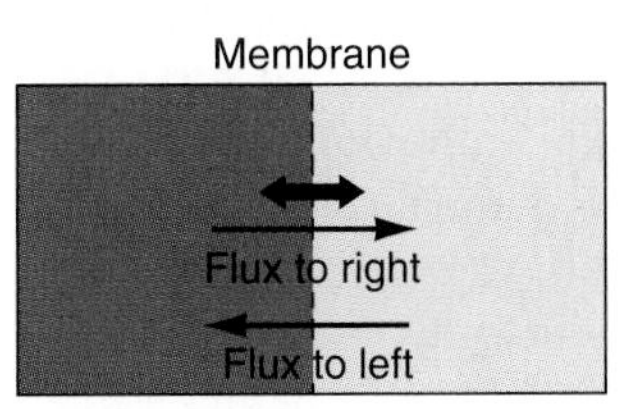

Flux to right = flux to left
Net flux = 0

The rate at which the system reaches equilibrium depends on four factors: (1) the temperature, since molecules with more thermal energy move faster; (2) the size of the molecules, since large molecules move more slowly than small ones; (3) the area of the membrane, since the greater the area, the greater the chance that a molecule will hit the membrane and penetrate it; and (4) the concentration difference across the membrane—that is, the steepness of the gradient. **Fick's Law of Diffusion** expresses the rate of diffusion *(R)* as a function of these factors:

$$R = DA\frac{\Delta c}{\Delta x}$$

D is a diffusion constant that is characteristic of the substance diffusing and the medium it is moving through, and *A* is the area through which diffusion occurs. The substance diffuses through a distance *x,* and its concentration at any point is *c,* so the concentration gradient is $\Delta c/\Delta x$; a steep gradient, for instance, means a large change in concentration over a short distance.

A **semipermeable membrane,** like cellophane, is permeable to small molecules such as water but not to larger molecules within a solution. If a semipermeable membrane separates two compartments with unequal concentrations of a solute to which it is impermeable, a net flux of water in one direction—called **osmosis**—will occur across the membrane. Consider a semipermeable membrane separating pure water from a solution of sucrose:

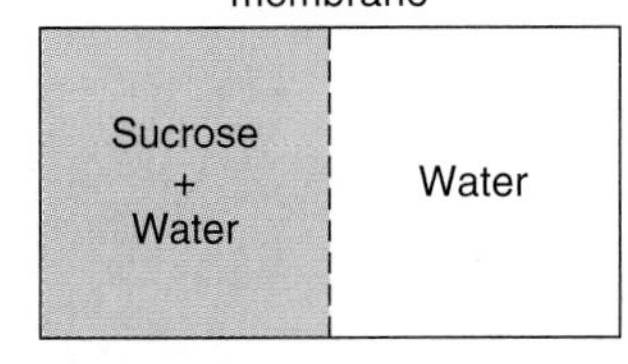

Although the strength of a solution is usually given by the concentration of solute, it makes more sense here to think of the concentration of *water.* Pure water is 55.55 *M* (molar) because the molecular mass of water is 18, so a liter of water, weighing 1,000 grams, contains 1,000/18 = 55.55 moles. Adding a solute to water will lower the water concentration; as a general rule, one solute molecule displaces one water molecule in dilute solutions, so in a 1 *M* solution (1 mole/liter) of sucrose, the water concentration is reduced to 54.55 *M:*

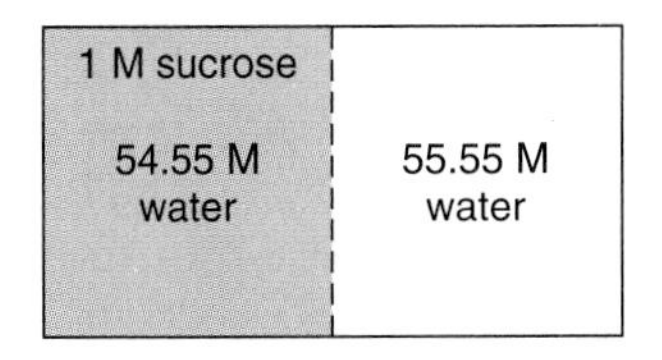

The water concentration is clearly higher in pure water than in the solution. Since sucrose molecules can't diffuse through the membrane, but water molecules can, there will be a net flux of water into the sucrose solution—down the water concentration gradient from a higher concentration of water to a lower concentration. This flow is osmosis.

As osmosis continues, the amount of water in the sucrose compartment will increase at the expense of the pure water compartment. If the compartment walls could move, the volumes would change to compensate for this net flux of water. If the compartments are fixed in place, however, pressure will build up in the solution until it is just high enough to oppose any further flux. At this point the system has come to an equilibrium. The pressure in the sucrose solution is defined as its **osmotic pressure.** With a pressure gauge attached, such a system becomes an *osmometer* that can measure the osmotic pressure (Figure 8.2).

The ability of a solution to develop an osmotic pressure is measured by its **osmolarity,** which is determined by the concentration of all particles in solution, regardless of their identity, size, or charge. A 1 *M* solution of sucrose in water is also, by definition, 1 osmolar (1 Osm). NaCl dissociates into Na^+ and Cl^- ions, so each NaCl molecule becomes two particles, and a 1 *M* NaCl solution is 2 Osm. Similarly, $MgCl_2$ dissociates into three particles—one Mg^{2+} ion and two Cl^- ions—so a 1 *M* $MgCl_2$ solution is 3 Osm (assuming it dissociates completely).

Osmosis in most biological systems occurs between compartments with different concentrations of solutes, not between compartments containing pure water. The *difference* in osmolarity determines the flow of water through semipermeable membranes, so water flows from a solution with low osmolarity (high water concentration) to a solution with high osmolarity

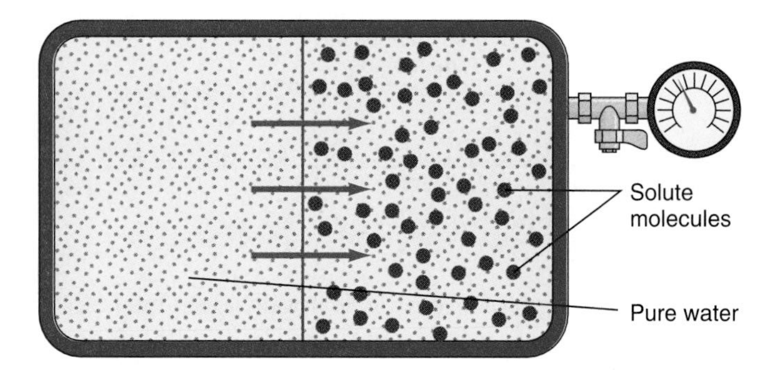

Figure 8.2

A simple osmometer can measure osmotic pressure. Osmosis will occur from the water into the solution, increasing the pressure in the solution side until it just opposes further osmosis. At this point, the system will be at equilibrium, and the osmotic pressure in the solution can be measured with the pressure gauge.

(low water concentration). Solutions that have the same osmolarity are **isosmotic** (*iso-* = equal). If two solutions have unequal osmotic pressures, the one with a higher osmolarity is **hyperosmotic** (*hyper-* = more or above), and the one with the lower osmolarity is **hypoosmotic** (*hypo-* = less or below). It should be clear that water will flow from a hypoosmotic solution to a hyperosmotic one.

Exercise 8.2 Which osmotic system will show the stronger flux of water: 2 *M* glucose against pure water or 1 *M* glucose against 4 *M* glucose?

Exercise 8.3 Which osmotic system will show the stronger flux of water: 2 *M* NaCl against pure water or 1 *M* $MgCl_2$ against 2 *M* NaCl?

8.3 Cells have osmotic properties.

Evidence for the structure of cell membranes comes in part from the fact that they show osmotic behavior, which means that their plasma membranes are semipermeable. However, we say that biological membranes are *selectively permeable,* because they do not simply discriminate on the basis of molecular size. Other factors allow certain molecules to diffuse through while blocking others. Water is in constant flux back and forth through cell membranes. The cytosol of a cell—being a solution of ions, metabolites, proteins, and other materials—has a higher osmolarity than pure water. In the human body, the fluids inside cells (intracellular) and outside cells (extracellular) have approximately the same osmolarity, about 0.3 Osm or 300 milliosmolar (mOsm).

Cells can be used as osmometers of a kind by observing their behavior in solutions of different osmolarity (Figure 8.3), but the strength of a solution measured in this way is called *tonicity* rather than osmolarity. The difference between osmolarity and tonicity is that osmolarity refers to all solutes in a solution, both those that pass through the membrane and those that do not, as measured by an ideal osmometer with a membrane that only admits water; tonicity depends only on solutes that do not penetrate the membrane, and is determined experimentally by observing the behavior of cells or tissues in a solution. When placed in an **isotonic** solution, cells neither shrink nor swell. In a more dilute solution, which is **hypotonic** to the cytosol (less than 300 mOsm, for human cells), cells swell as water flows into them. In a more concentrated solution, which is **hypertonic** to the cytosol (more than 300 mOsm, for human cells), water flows out of the cells, so they shrink, or *plasmolyze* (Figure 8.3).

Such experiments show that cell membranes are permeable to water molecules. Water, CO_2, and O_2 are among the few small molecules of biological importance that can diffuse freely through membranes, apparently between the membrane components. When we consider certain aspects of animal physiology in later chapters, it will be important to know that some membranes have specific protein channels for water, called *aquaporins.*

In hypotonic solutions, the cells of plants and microorganisms such as bacteria and algae tend to swell and develop considerable internal pressure, called **turgor pressure.** In fact, the cells would burst if they weren't surrounded by strong cell walls. But animal cells don't have such walls, so in a hypotonic solution, they take up water and swell until they burst, releasing their contents and leaving ruptured plasma membranes behind. (Recall from Chapter 4 that this is the first step in purifying the hemoglobin from red blood cells.) However, animal cells normally don't rupture because they are bathed in internal fluids whose overall solute concentrations do not differ strongly from those inside the cells.

Medical practitioners have to pay attention to the concentrations of the solutions they inject into their patients, since solutions that aren't isotonic will damage cells. Drugs to be injected intravenously are typically dissolved in a 0.9 percent saline (salt) solution, which is isotonic for red blood cells (RBCs) (Figure 8.4). The same principle applies in microscopy. Artifacts are minimized when cells are kept in isotonic solutions, often 0.28 percent sucrose.

Artifacts in cells, Section 6.4.

Exercise 8.4 How would you use osmosis to determine the concentration of water (or of total solutes) inside a cell?

Exercise 8.5 Fill in this table to show the relative concentrations of a solution and the cytosol and whether a cell swells or shrinks.

Hypertonic solution	___ concentrated than cytosol	Cell _____
Isotonic solution	Equal to cytosol	Cell undergoes no change
Hypotonic solution	___ concentrated than cytosol	Cell _____

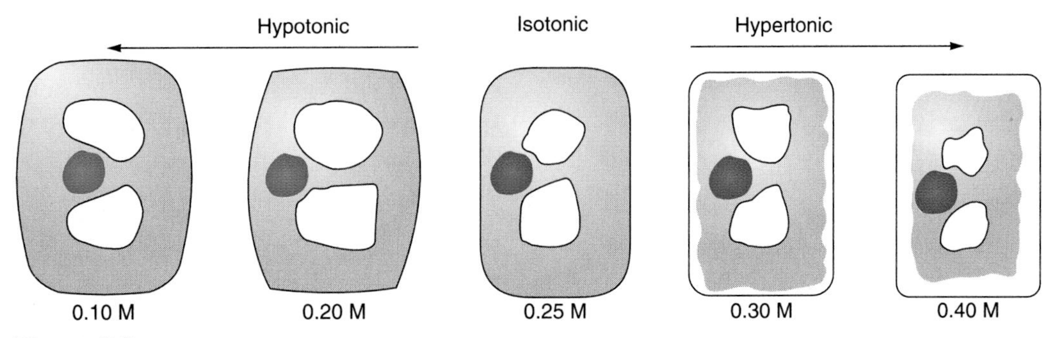

Figure 8.3

Cells placed in a solution with the same osmolarity as the cytosol neither swell nor shrink; such a solution is isotonic to the cytosol. Pure water, or any solution with a lower concentration of solute than in the cytosol, is hypotonic; in such a solution, the cells tend to take in water and swell. In a more concentrated solution, which is hypertonic to the cytosol, they tend to lose water and shrink (plasmolyze), so the plasma membrane pulls away from the cell wall.

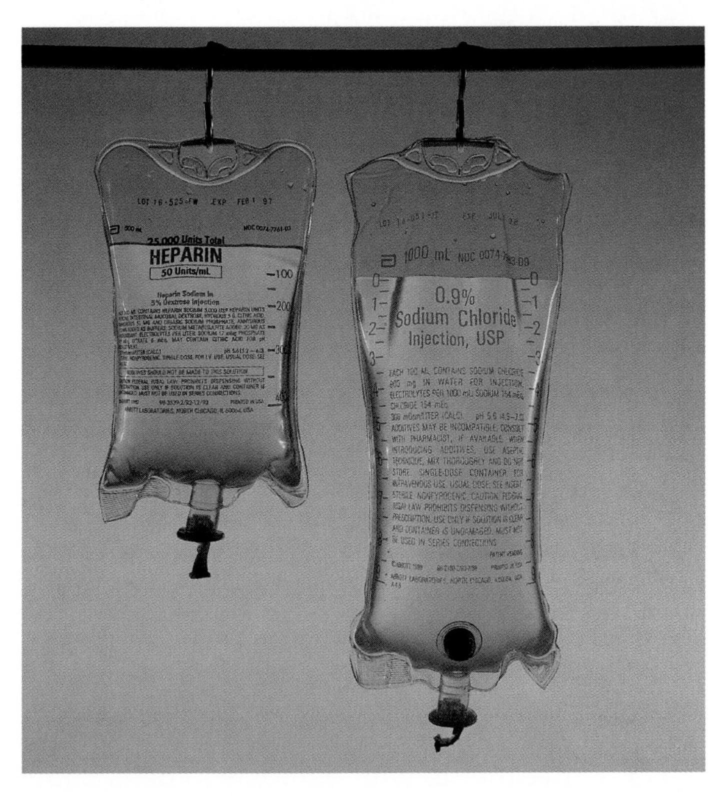

Figure 8.4

Solutions to be injected into the body are often made up in isotonic saline (sodium chloride).

8.4 Lipids are insoluble in water.

The foundation of a membrane is a double layer of **lipids** (*lipos* = fat), organic molecules that include fats, waxes, and several other hydrophobic substances that are quite insoluble in water. Most biologically important lipids are based on **fatty acids,** which are long-chain hydrocarbons with a carboxyl group at one end:

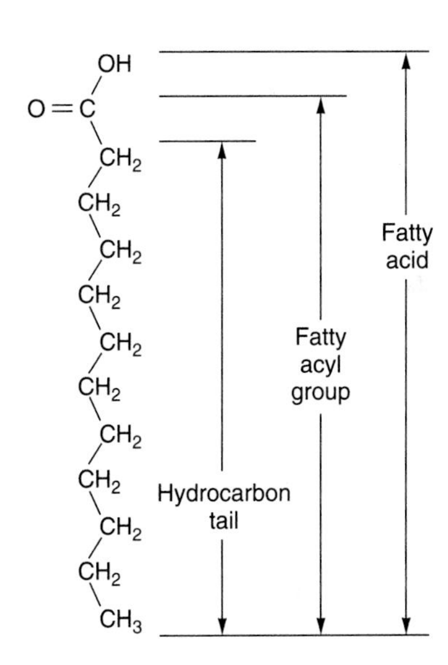

The longer the chain, the less soluble a fatty acid is in water, because the hydrocarbon chain makes the molecule hydrophobic. The carboxyl group is hydrophilic, however, so the molecule as a whole is said to be **amphipathic** (*amphi-* = both; *pathos* = feeling): It "loves water" at its carboxylic acid end and "fears water" at its hydrocarbon end. This characteristic underlies the action of *soaps,* which are sodium or potassium salts of fatty acids. When you wash your hands with soapy water, the soap molecules surround particles of dirt and droplets of oil with their hydrophobic ends, while their hydrophilic ends dissolve in the water so the dirt and oil can be rinsed away (Figure 8.5). For thousands of years, people have been making soap by boiling animal fat with wood ashes, which are rich in sodium and potassium hydroxides:

$$NaOH + CH_3CH_2CH_2CH_2\cdots CH_2COOH \rightarrow$$
$$CH_3CH_2CH_2CH_2\cdots CH_2COONa + H_2O$$

Table 8.1 lists some of the most important fatty acids. These molecules have an even number of carbon atoms, commonly 16 or 18, because organisms synthesize them by repeatedly linking

Figure 8.5
Oily, dirty material is surrounded by the nonpolar hydrophobic ends of a cloud of soap molecules, and the complex dissolves in water so it can be rinsed away.

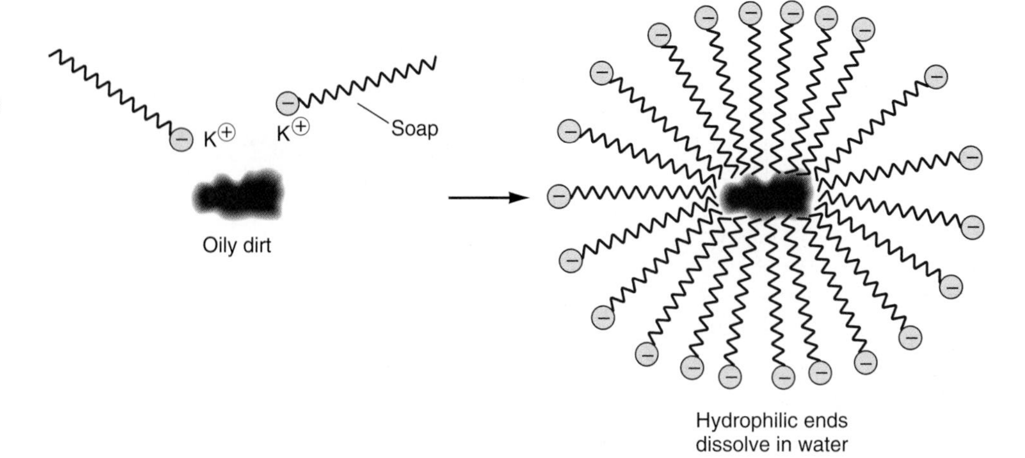

two-carbon molecules. (The fact that molecules in coal and oil have an even number of carbon atoms is important evidence that these deposits came from ancient organisms.)

Fatty acids are constituents of **triglycerides,** also known as **neutral fats.** A triglyceride consists of three fatty acid molecules linked to glycerol, a three-carbon compound with a hydroxyl group on each carbon (a polyalcohol); *ester linkages* between the fatty acids and the glycerol are formed, as in the synthesis of biological polymers, by removing water molecules:

Glycerol + **3 fatty acids**

$$H_2C{-}OH \quad H{-}O{-}\overset{O}{\overset{\|}{C}}{-}CH_2{-}CH_2{-}CH_2\cdots{-}CH_2{-}CH_2{-}CH_2{-}CH_3$$
$$HC{-}OH \quad H{-}O{-}\overset{O}{\overset{\|}{C}}{-}CH_2{-}CH_2{-}CH_2\cdots{-}CH_2{-}CH_2{-}CH_2{-}CH_3$$
$$H_2C{-}OH \quad H{-}O{-}\overset{O}{\overset{\|}{C}}{-}CH_2{-}CH_2{-}CH_2\cdots{-}CH_2{-}CH_2{-}CH_2{-}CH_3$$

Triglyceride

Ester linkages

$$H_2C{-}O{-}\overset{O}{\overset{\|}{C}}{-}CH_2{-}CH_2{-}CH_2\cdots{-}CH_2{-}CH_2{-}CH_2{-}CH_3$$
$$HC{-}O{-}\overset{O}{\overset{\|}{C}}{-}CH_2{-}CH_2{-}CH_2\cdots{-}CH_2{-}CH_2{-}CH_2{-}CH_3 + 3H_2O$$
$$H_2C{-}O{-}\overset{O}{\overset{\|}{C}}{-}CH_2{-}CH_2{-}CH_2\cdots{-}CH_2{-}CH_2{-}CH_2{-}CH_3$$

Although too much fat in the diet may pose a health risk for humans, some is essential. Notice in Table 8.1 that fatty acids such as palmitic and stearic are *saturated,* with each carbon atom bonded to as many hydrogen atoms as possible; others are *unsaturated,* with at least one C = C double bond. Unsaturated lipids are most typical of plants, which often store them as energy reserves—in seeds, for example. These lipids are liquid at room temperature, and we commonly extract them from seeds and use them for cooking oils, such as olive, corn, and safflower oils; Figure 8.6. These oils can be saturated in the process of *hydrogenation,* which substitutes hydrogen atoms for the double bonds, thus converting liquid oils into margarine. Animals store neutral fats in their fatty, or adipose, tissues; these deposits insulate against cold, cushion internal organs such as the kidneys from damage, and provide energy reserves. Animal fats generally contain more saturated than unsaturated fatty acids.

Table 8.1 Common Fatty Acids

Common Name	Structure
Palmitic	$CH_3(CH_2)_{14}COOH$
Stearic	$CH_3(CH_2)_{16}COOH$
Palmitoleic	$CH_3(CH_2)_5CH = CH(CH_2)_7COOH$
Oleic	$CH_3(CH_2)_7CH = CH(CH_2)_7COOH$
Vaccenic	$CH_3(CH_2)_5CH = CH(CH_2)_9COOH$
Linoleic	$CH_3(CH_2)_3(CH_2CH = CH)_2(CH_2)_7COOH$
Linolenic	$CH_3(CH_2CH = CH)_3(CH_2)_7COOH$

8.5 Experiments with osmosis in cells show that membranes are made of lipids.

It was clear by the middle of the nineteenth century that cells are the basic units of biological structure, and around 1855, Karl Nägeli concluded that they must have a distinct boundary, which he called a plasma membrane. He based this conclusion on the fact that dyes freely penetrate injured cells but not intact cells. Biologists continued to argue about the existence of plasma membranes for a long time, even though Janet Plowe's experiments, described at the beginning of this chapter, ought to have settled the issue.

In the 1890s, Ernst Overton studied the osmotic properties of plant cells and concluded that plasma membranes are made of lipids. Overton used plant roots, whose cells have delicate cytoplasmic extensions called root hairs. With a microscope, he

Figure 8.6

Lipids are an important part of the human diet, although an excess can lead to various health problems. Liquid cooking oils largely contain unsaturated fatty acids; they can be converted into saturated fatty acids, as in margarine, in the process of hydrogenation.

watched these cells plasmolyze when placed in a hypertonic sugar solution—that is, the plasma membrane and cytoplasm shrank and pulled away from the cell wall, as shown in Figure 8.3. Overton observed that in a sugar solution the cells stayed like this for a long time. In solutions of other materials, however, the cells shrank for a while and then returned to their normal size. Overton suggested that these other materials, in contrast to sugar, were able to move through the plasma membrane (though more slowly than water molecules) and equilibrate, until no osmotic pressure difference remained.

Overton used these observations to determine the rate at which a substance permeates the membrane, since the faster it enters the cell, the sooner the cytoplasm returns to its normal size. He found that the rate at which molecules enter depends on their molecular mass (smaller molecules enter faster) and on how easily they dissolve in a lipid. The solubility of a compound in lipid is measured by its oil-water *partition coefficient;* after mixing a substance with water and oil, the amount dissolved in the two liquids is measured, and the partition coefficient is the amount dissolved in oil divided by the amount dissolved in water (Figure 8.7). Remember that "like dissolves like"—hydrophobic substances dissolve in lipids and other hydrophobic substances, and hydrophilic substances dissolve in water and other hydrophilic substances. The higher the partition coefficient of any material, the faster it diffuses into a cell (Figure 8.8). Since substances that are more lipid-soluble penetrate the cell membrane faster, Overton came to the obvious but significant conclusion that the membrane is made of lipid. Modern analyses of different kinds of membranes show that they contain

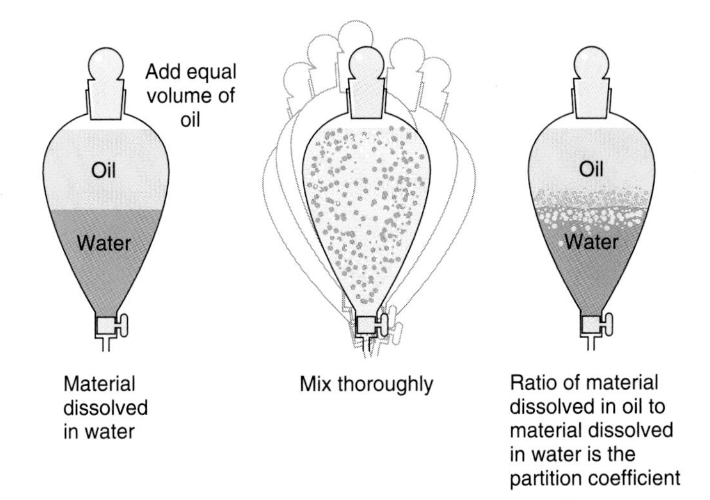

Figure 8.7

The partition coefficient of a substance between oil and water is measured by letting the substance equilibrate between the two solvents and then measuring its concentration in each solvent. The higher its partition coefficient, the more is dissolved in the oil.

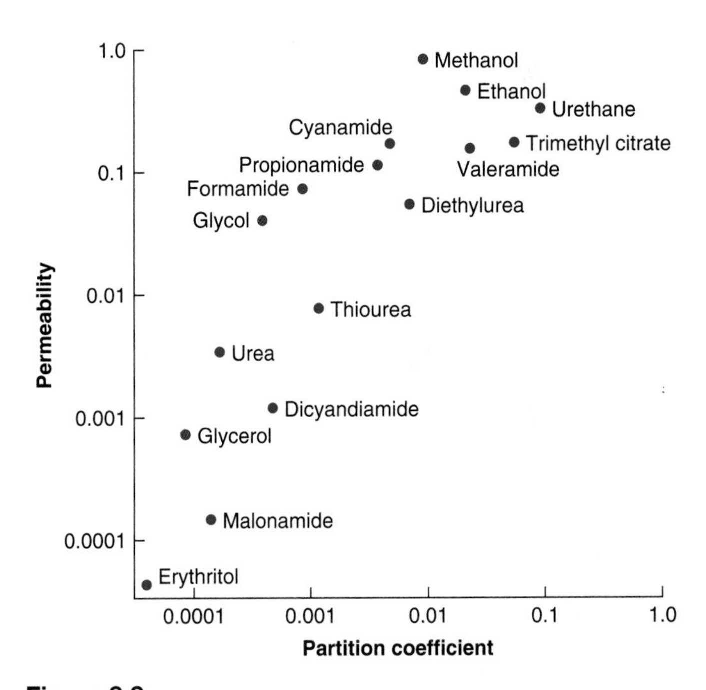

Figure 8.8

The higher a substance's partition coefficient between oil and water, the faster it permeates into cells. Since "like dissolves like," this is strong evidence that the cell membrane is made of lipids.

comparable amounts of lipid and protein, with the lipid/protein ratio varying from about 3:1 to 1:3, and that many membranes also contain small amounts of sugars (Table 8.2).

Most of the lipids in membranes are not triglycerides but **phospholipids,** which are like triglycerides with one fatty acid replaced by a phosphate linked to a small molecule, such as serine or choline (Figure 8.9). Phospholipids are named for this small molecule, as phosphatidyl serine or phosphatidyl choline.

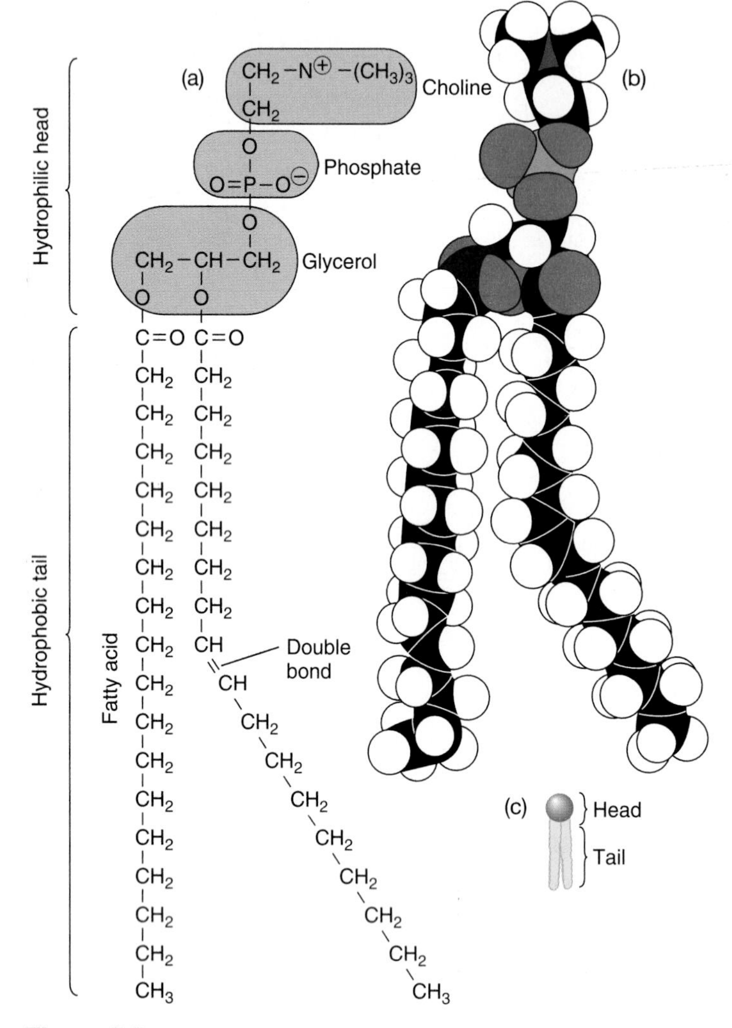

Figure 8.9

The structure of a phospholipid is represented here in three ways. *(a)* The three carbon atoms of glycerol are linked to two fatty acyl molecules and to a phosphate group, which in turn is attached to a small molecule such as serine, choline, ethanolamine, or another glycerol. *(b)* A chemical model shows the molecule's three-dimensional shape, using atomic models with standard colors: Carbon is black, hydrogen white, oxygen red, nitrogen blue, and phosphorus orange. *(c)* In drawings of membrane structure, a phospholipid will be represented by a cartoon symbol, the blue sphere representing the polar "head" (glycerol, phosphate, and the small attached molecule) and the yellow "tails" representing

Since the phosphate carries negative charges and the small molecule often carries a positive or negative charge, this region of the phospholipid molecule is strongly polar and hydrophilic. Because the rest of the phospholipid consists of two nonpolar hydrocarbon chains, the whole phospholipid is strongly amphipathic. As we will show next, this characteristic is the key to membrane structure.

Exercise 8.6 How quickly would you expect the following molecules to penetrate a cell in experiments like Overton's?

Table 8.2 Compositions of Selected Membranes (Percentages by Mass)

Membrane	Protein	Lipid	Carbohydrate
Myelin sheath	18	79	3
Plasma membranes			
Human red blood cell	49	43	8
Mouse liver	44	52	4
Amoeba	54	42	4
Inner mitochondrial	76	24	0
Spinach chloroplast	70	30	0

Give your reasons in each case: *(a)* pentane, $CH_3(CH_2)_3CH_3$; *(b)* malonate, $^-OOC{-}CH_2{-}COO^-$; *(c)* ATP^{4-} (ATP is normally charged like this at the near-neutral pH of most cells).

Exercise 8.7 Which substance of the following pairs would enter cells faster: *(a)* butane or butyl alcohol; *(b)* an aromatic molecule such as benzene or a similar molecule with two charged groups; *(c)* glycine or the dipeptide glycyl glycine?

8.6 Phospholipids form bilayers, which are the structural basis of all membranes.

The probable arrangement of lipids in a membrane was revealed in 1925 by the work of two Dutch scientists, E. Gorter and F. Grendel, who worked with the membranes of red blood cells. Mammalian RBCs have no membranes other than the plasma membrane, having lost all the others, including the nuclear envelope, as they mature. Gorter and Grendel carefully measured some RBCs and estimated their average surface area. Then they used chloroform to extract the lipids from a known number of cells and spread the lipids on the surface of a tray of water (Figure 8.10). Just as a film of oil spreads over a rain puddle, the phospholipids form a film, one molecule thick, on the trough, and a wire stretched across the trough compresses this film so the area it covers can be measured. Gorter and Grendel estimated that the lipid from one RBC covers about 200 μm^2, while the surface of the cell has about half that area. They therefore concluded that the cell is covered by a double layer, or **bilayer,** of lipid molecules with the hydrophobic tails inside, where they interact with one another, and the hydrophilic heads outside, where they interact with the surrounding water (Figure 8.11). In fact, Gorter and Grendel slightly underestimated the surface area of a red blood cell, but they also did not extract all of the membrane lipids; the two errors cancel out. Also, they ignored proteins in the membrane. Still, the experiment was brilliant in its direct simplicity, and the conclusion remains correct

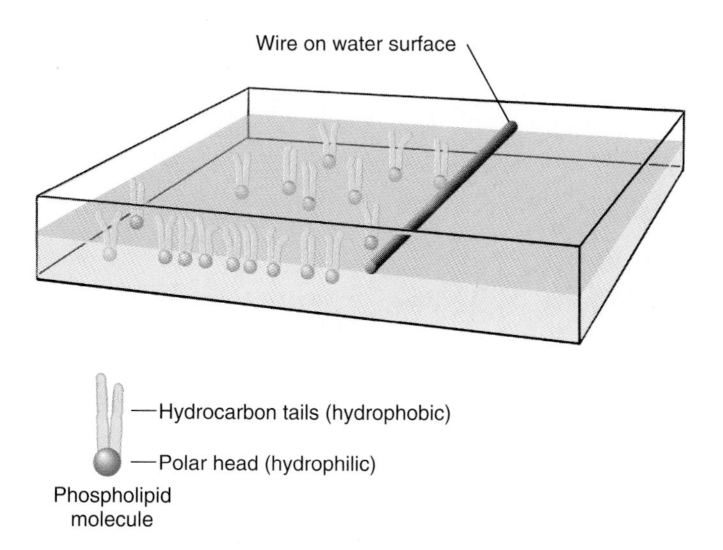

Figure 8.10
Phospholipid molecules spread out in a single layer on the surface of water. A wire placed at the water's surface can be moved to compress the molecules into a minimum area.

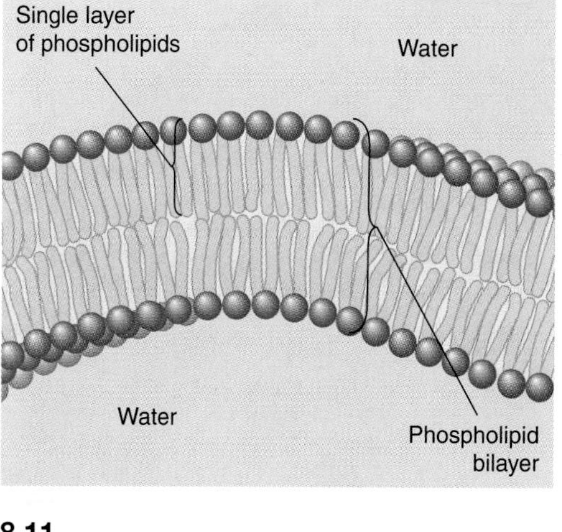

Figure 8.11
Phospholipids are arranged in a bilayer in water. The molecules associate so their hydrocarbon tails make a hydrophobic layer that excludes water and the polar groups are in contact with the surrounding water.

and forms the basis for our present understanding of membrane structure.

In 1935, two English scientists, James F. Danielli and Hugh Davson, also suggested that membrane lipids form a bilayer, without even referring to Gorter and Grendel's work, and they assumed that proteins coat the surfaces of the bilayer. Supporting evidence for the Danielli-Davson model came from studies on the myelin sheaths that insulate nerve fibers. These sheaths are formed by cells that wrap themselves around a nerve fiber many times, leaving many layers of closely packed membrane (Figure 8.12). Measurements by X-ray diffraction, and later by electron microscopy, showed that a single layer is about 7–8 nm thick, just the calculated thickness of a bilayer with protein on both sides. Although this model for membrane structure persisted for a long time, new techniques eventually proved that it was wrong.

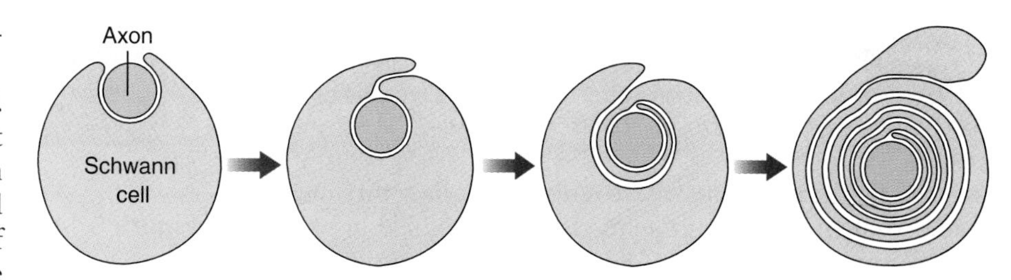

Figure 8.12
By wrapping around an axon (part of a cell in a nerve), a Schwann cell forms a myelin sheath. The cytoplasm is squeezed out of the sheath, leaving a spiral of closely packed membranes. This formation has been studied by various physical techniques, such as X-ray diffraction, to determine the structure of membranes.

8.7 A membrane is a fluid mosaic of lipid and protein.

In 1972, S. J. Singer and G. L. Nicolson described a model of membrane structure that accommodated new experimental results showing that globular proteins are embedded within the lipid layer. The evidence came especially from the **freeze-fracture** method of preparing material for microscopy (Figure 8.13). In this technique, a tissue is frozen in liquid nitrogen and then fractured by a sharp blow with a knife edge. It may then be "etched" by putting it in a vacuum to evaporate its water, so other molecules stand out more strongly. The fractured surface is coated with platinum, or some other heavy metal, to make it visible by electron microscopy. This technique reveals the insides of membranes as well as their surfaces, because the fracture lines generally go right down the center of a bilayer where the molecular interactions are weakest, splitting it into two halves with their associated proteins. A freeze-fracture picture of the myelin sheath shows only smooth layers, since these membranes are mostly lipid Figure 8.14*a*). More typical membranes containing a lot of protein, such as those from chloroplasts, show smooth layers with embedded particles, which are apparently globular protein molecules (Figure 8.14*b*).

Singer and Nicolson incorporated this information into their **fluid mosaic model** (Figure 8.15). They thought of a membrane as a sea of lipid containing protein icebergs, which float at various depths because of their interactions with the lipid bilayer. The important point is that many proteins lie *within* the lipid bilayer, not spread over its surface as in the Danielli-Davson model, and all current evidence supports this view.

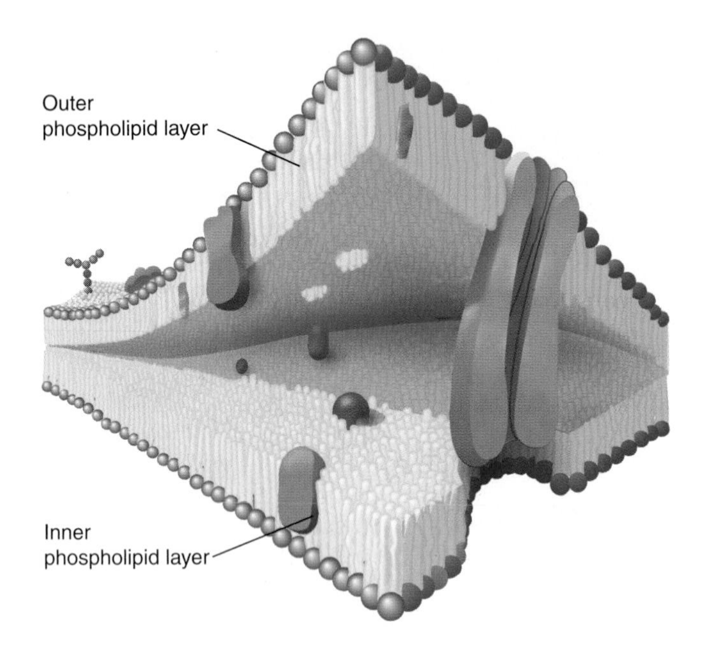

Figure 8.13

The freeze-fracture method splits membranes down the middle of the lipid bilayer, where the attraction between molecules is weakest. The structure may be enhanced by etching, putting the preparation into a vacuum to evaporate water in the membrane. It is then coated with a thin metal film for electron microscopy. Integral proteins of the membrane will be detectable as protrusions or cavities.

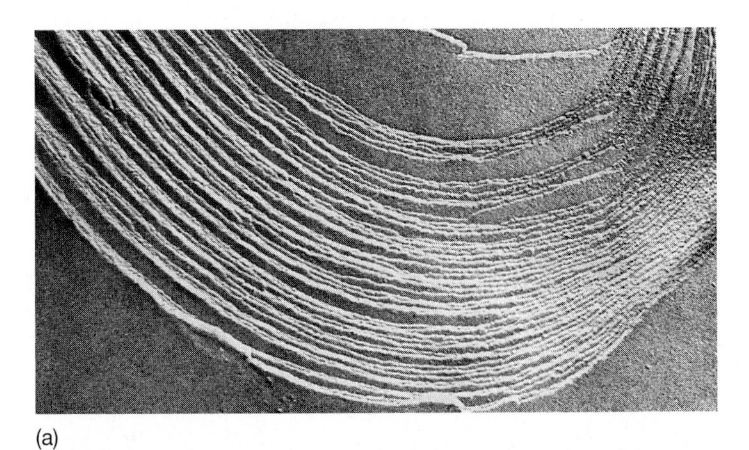

(a)

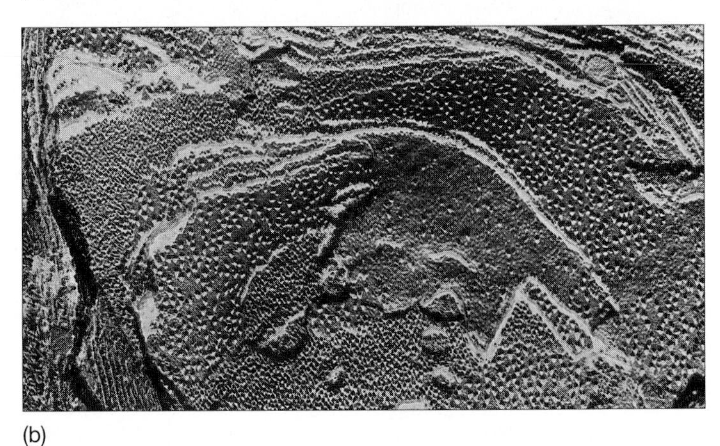

(b)

Figure 8.14

Freeze-fracturing shows that membranes differ in their protein composition. *(a)* Myelin sheath membranes appear as uniform layers with little obvious structure. *(b)* Chloroplast membranes, in contrast, show an internal structure full of small particles, which are presumed to be protein molecules, singly or in clusters.

The fluid mosaic model accounts for the fact that membranes contain globular proteins that are insoluble in water. Globular proteins in the cytosol have hydrophilic surfaces, but any protein embedded in the hydrophobic middle of the lipid bilayer must have a hydrophobic surface. **Integral proteins** form part of the membrane structure, and just where each of them lies in the membrane depends on the fit between hydrophobic and hydrophilic groups of the protein with similar regions of the lipid bilayer (Figure 8.16). An integral protein with a hydrophobic middle can span the bilayer completely as a **transmembrane protein,** or spanning protein. In contrast, **peripheral proteins** have hydrophilic surfaces and bind more loosely to one surface of the membrane.

Exercise 8.8 The following sketches show membrane proteins; the shaded portions have hydrophobic exteriors. Show how each of the three proteins will fit into the adjacent lipid bilayer:

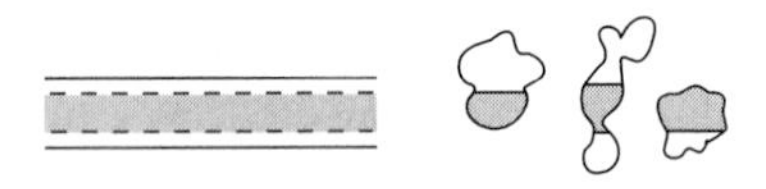

The fluid mosaic model also explicitly recognizes that membranes aren't static. In fact, it proposes that membrane molecules, at the typical temperatures of biological systems, will move around and diffuse past one another. A number of experiments show that membranes are quite fluid. An especially beautiful demonstration was performed in 1970 by Larry D. Frye and Michael A. Edidin. They took advantage of the fact that cells growing in tissue culture can be made to fuse and merge into a single cell, a *heterokaryon* that has two different nuclei (*hetero* = different; *karyon* = nucleus). Frye and Edidin used human and mouse cells. Since each type of cell has distinctive surface proteins, specific antibodies can be made against them by injecting other animals with these cells, as explained in Methods 8.1. These antibodies can then be used to label the cell surfaces. Frye and Edidin attached a red fluorescent dye to the antibodies against human proteins and a green dye to those against mouse proteins, so they could follow the movements of mouse and human membrane proteins just by watching the colored antibodies attached to them. Antibody-labeled cells were fused to form heterokaryons (Figure 8.17). Initially, the dyes were localized on separate halves of each heterokaryon. The two colors soon began intermingling, and after an hour, the dyes were completely mixed; this meant that the two types of protein had been able to diffuse in the lipid bilayer and mix

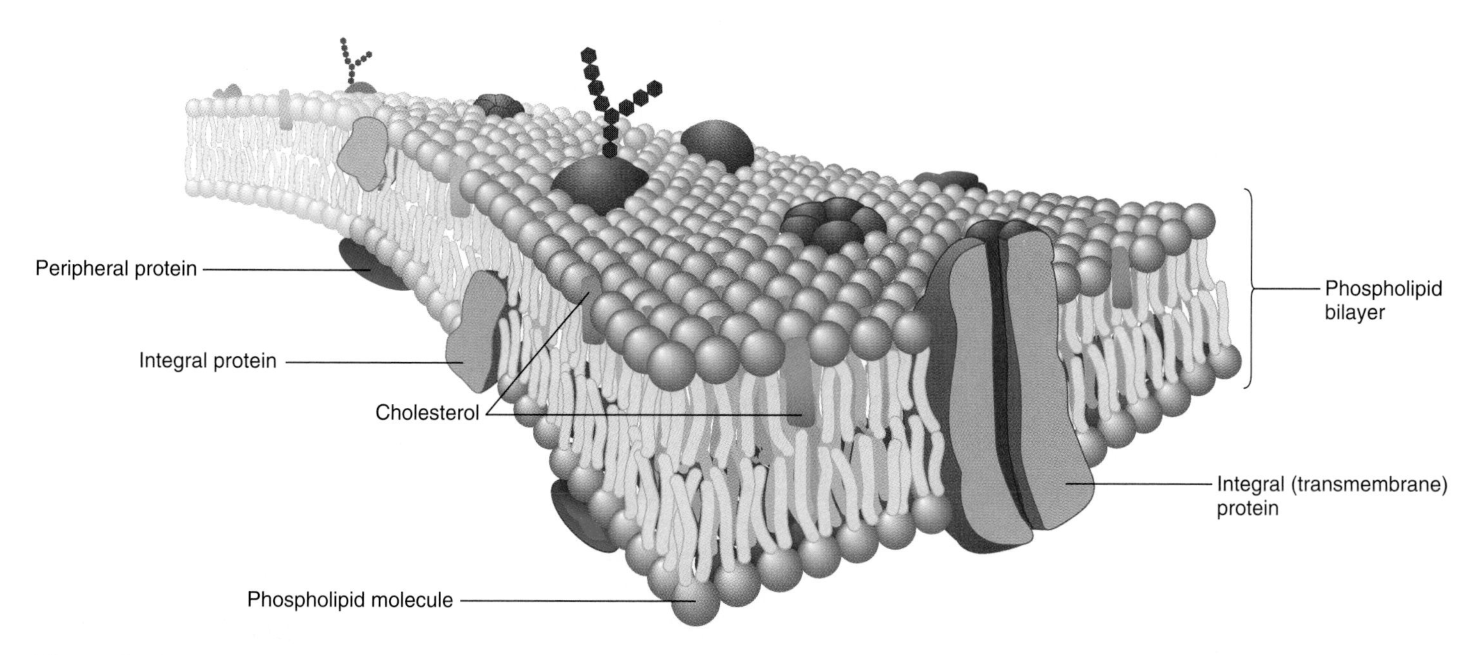

Figure 8.15

The fluid-mosaic model proposes that a membrane consists of a fluid phospholipid bilayer with proteins buried in it to various depths and positions, depending on the arrangement of hydrophilic and hydrophobic groups on their surfaces. All molecules are able to move around freely.

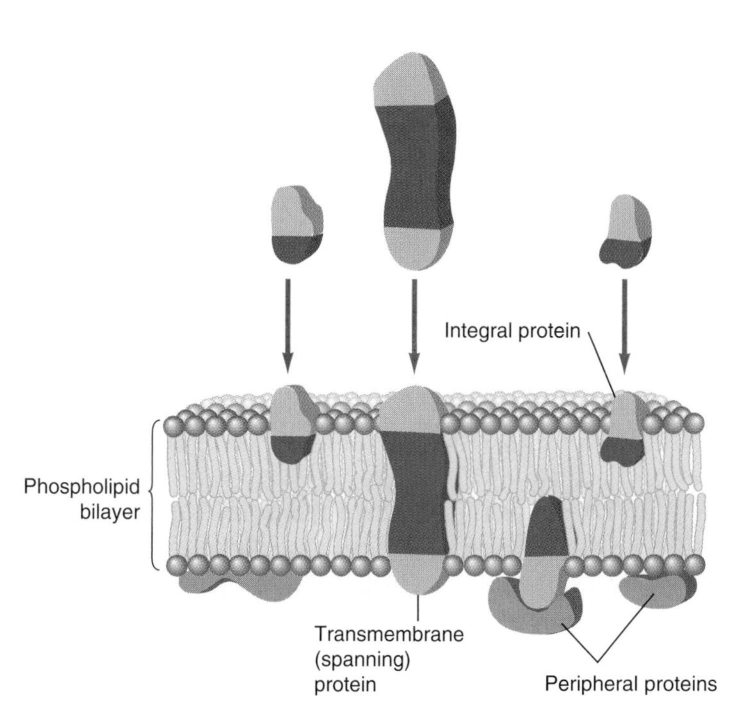

Figure 8.16

Integral proteins are at least partially buried in the lipid layer. Their positions are determined by the extent of hydrophobic (*dark green*) and hydrophilic (*light green*) regions on their surfaces. Among the most important integral proteins are transmembrane, or spanning, proteins. Peripheral proteins are attached more weakly on the surface.

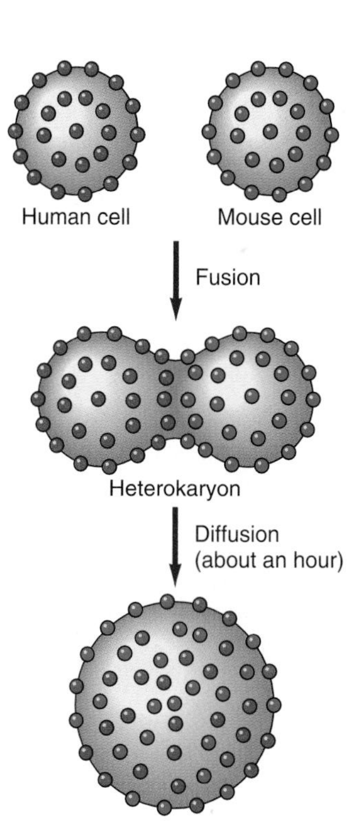

Figure 8.17

Frye and Edidin demonstrated that membranes are really fluid by labeling proteins on human cells with red fluorescent dyes and those on mouse cells with green dyes. The mixing of dyes after about an hour demonstrates how freely the proteins move around in the membrane.

Methods 8.1 Antibodies and Antigens

Proteins can be extraordinarily specific in their interactions with other molecules. Vertebrates, particularly birds and mammals, use this specificity to protect themselves against potential invaders by making **antibodies,** proteins that recognize and attack potentially infectious agents such as bacteria and viruses. Antibodies can bind to foreign materials, especially proteins and polysaccharides, known as **antigens.** By binding to antigens on potential pathogens or on the toxic materials they produce, antibodies inactivate these threats. Antibodies can also be used experimentally to find and identify molecules.

Every protein has distinctive little knobs and bumps called *antigenic determinants,* or *epitopes,* on its surface. An antibody is a Y-shaped protein with sites on its arms that will bind specifically to these epitopes. Each type of antibody is specific for one kind of epitope, and their specificity makes antibodies powerful experimental tools. For instance, to locate a particular type of protein in a cell, we use the protein as an antigen, injecting it repeatedly into an animal where it is normally not found. The animal then develops a high concentration of antibodies in its blood directed against this one protein.

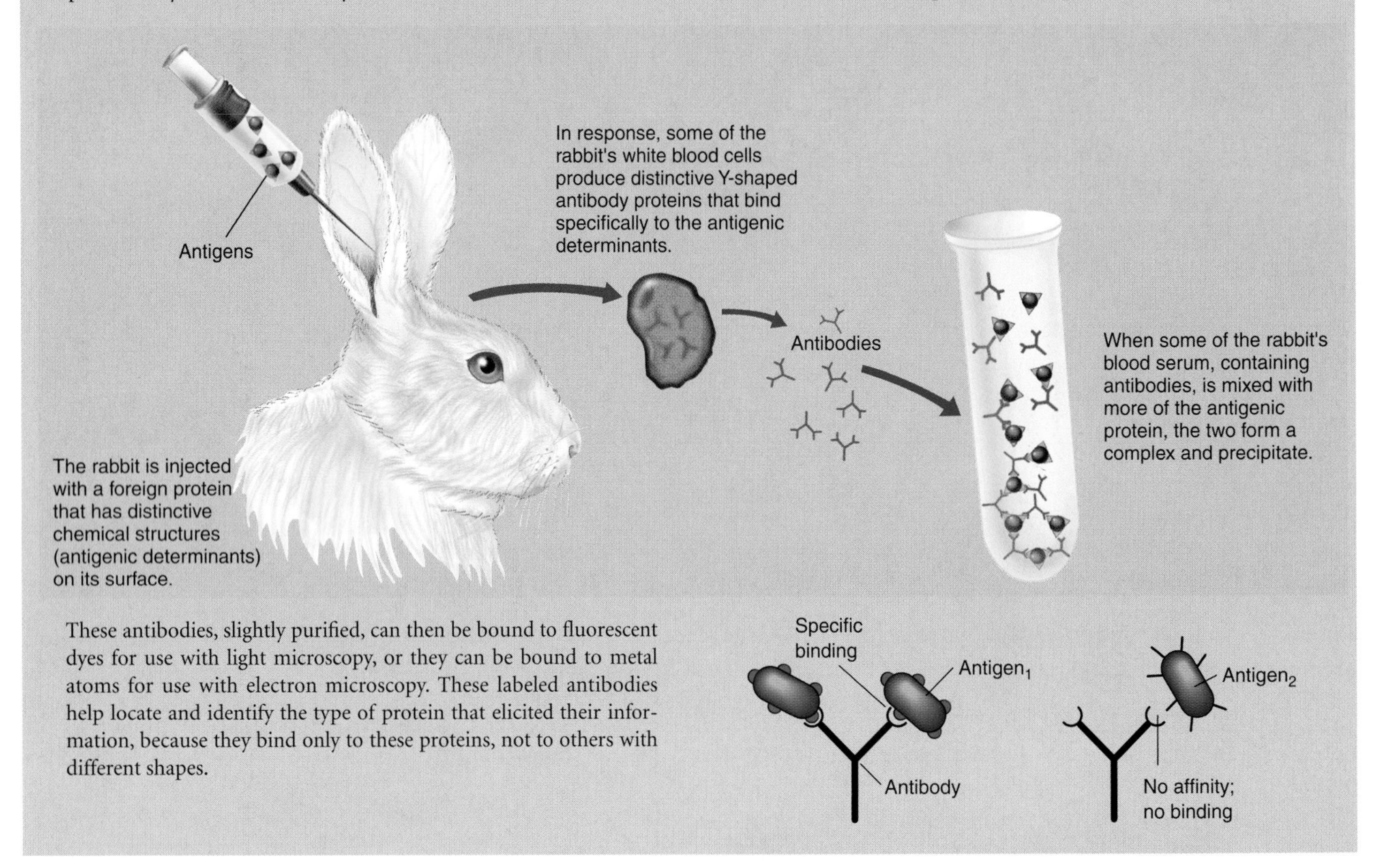

These antibodies, slightly purified, can then be bound to fluorescent dyes for use with light microscopy, or they can be bound to metal atoms for use with electron microscopy. These labeled antibodies help locate and identify the type of protein that elicited their information, because they bind only to these proteins, not to others with different shapes.

completely within that short time. Such rapid diffusion indicates that the phospholipid core of a membrane must be about the consistency of olive oil.

The fluid mosaic model is now very well confirmed and has replaced older ideas about membrane structure. Notice, however, that it is still based on the lipid bilayer structure. When phospholipids are mixed with water, they spontaneously form a bilayer. The fact that amphipathic lipid molecules form this structure must have been important in the early evolution of cells and in modern cells it means newly made phospholipids will assemble themselves spontaneously into membrane structures, without any additional genetic information to determine where they should go. We will return to this question of molecules assembling themselves into functional structures in Section 14.14.

8.8 The fluidity of a membrane depends on the composition of its lipids.

The hydrocarbon tails of membrane phospholipids line up in parallel and interact with one another. Because these interactions tend to restrict the mobility of the phospholipid molecules, they largely determine a membrane's fluidity. Two features of phospholipid tails—their length and their saturation—affect membrane fluidity by influencing these interactions. Phospholipids with long tails pack together more tightly, forming less fluid membranes (Figure 8.18*a*). In contrast, a double bond in an unsaturated hydrocarbon tail puts a kink in the tail, resulting in looser packing and greater membrane fluidity (Figure 8.18*b*).

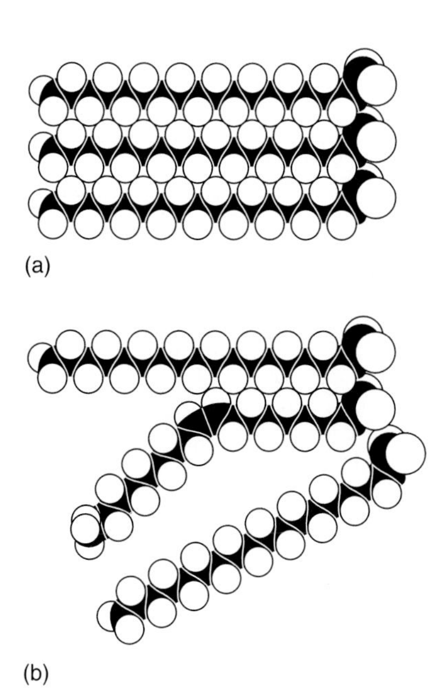

Figure 8.18

(a) Saturated and *(b)* unsaturated fatty acid chains pack differently in the membrane structure. The chains interact weakly through van der Waals forces, and the longer they are, the stronger the interaction. The double bonds in unsaturated fatty acid chains make kinks, so the molecules aren't able to interact as tightly as straight, saturated chains. Membranes made largely with unsaturated chains are more fluid (less viscous) than those formed by saturated chains of the same length.

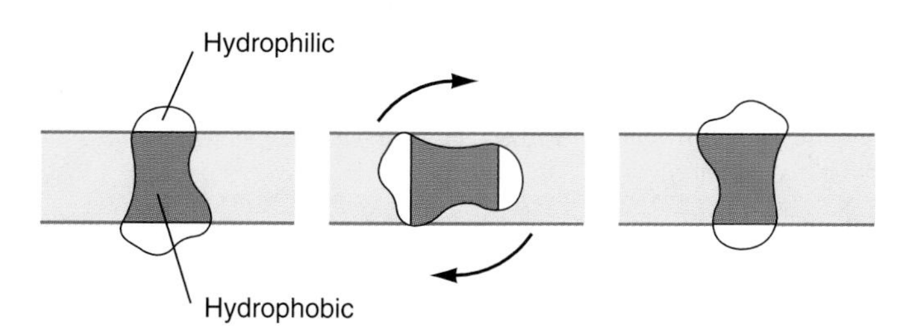

Figure 8.19

To flip-flop through a membrane, a protein would have to go through energetically unfavorable interactions between hydrophobic and hydrophilic molecules. Therefore, flip-flopping does not happen.

The fluidity of a biological membrane increases with temperature, and an analysis of membranes from a variety of organisms shows that their membrane lipids adapt them to the temperature of their environment. Cold-water fish, for example, tend to have membranes with short, unsaturated phospholipid tails, that keep their membranes fluid at the low temperatures in which they live. Some bacteria and other organisms are able to change the composition of their membrane lipids rapidly in response to sudden temperature changes.

In addition to phospholipids, many membranes contain sterols, such as cholesterol in animal membranes and stigmasterol in plant membranes:

H_3C CH_3 CH_3 CH_3 H_3C HO **Cholesterol (animals)**

CH_3 H_3C CH_3 CH_3 CH_3 H_3C HO **Stigmasterol (plants)**

A large fraction of animal cell membranes—up to 17 percent by weight in myelin—consists of cholesterol. Cholesterol molecules are embedded in the phospholipid bilayer, between hydrocarbon chains, where they disrupt interactions between saturated fatty acid chains but fit between unsaturated chains. The result is that cholesterol keeps a membrane consistently fluid over a broad range of temperatures.

8.9 Membrane proteins move laterally but do not "flip-flop" across a membrane.

A number of techniques have been developed to determine the orientation of proteins that span the bilayer of a membrane. One is to prepare antibodies specific for proteins on the outer surface of the membrane and to chemically link these antibodies to the protein ferritin. The iron atoms in ferritin make it a natural stain for electron microscopy. When ferritin-labeled antibodies are applied to broken membranes, electron micrographs show that they attach only to one side of the membrane, implying that each membrane protein always keeps the same orientation—outside parts remain outside, inside parts inside. In other words, the protein doesn't "flip-flop" and reverse its orientation. This is to be expected if we consider the energy necessary to change orientation, because during a reversal, hydrophilic groups on the protein would have to come in contact temporarily with the hydrophobic part of the membrane, and vice versa, an energetically unfavorable situation (Figure 8.19).

8.10 Studies of red blood cell membranes have increased our understanding of membrane structure.

The easily prepared plasma membranes of red blood cells, which Gorter and Grendel used to study membrane structure, have also been useful for studies of membrane proteins. To characterize these proteins—that is, to determine their sizes, structures, and functions—they must be separated from the lipids. This task at first proved difficult. In fact, the proteins could only be investigated after the discovery that certain non-ionic detergents, such as triton X-100,

$$H_3C-C(CH_3)_2-CH_2-C(CH_3)_2-C_6H_4-O-(CH_2CH_2)_{9-10}OH$$

will solubilize the proteins and separate them from their associated lipids. Once dissolved, the proteins can be separated by polyacrylamide-gel electrophoresis, which we introduced in Section 4.9. It is instructive to examine three types of proteins derived from RBC membranes ("ghosts"): glycophorin, anion channel protein, and a group of proteins that form an internal skeleton. All these proteins have specific roles in the RBC due to their structure and their arrangment in the plasma membrane.

Glycophorin is an integral protein that spans the membrane bilayer. Even though each RBC contains about 6×10^5 molecules of glycophorin, its function is unknown. Still, glycophorin illustrates important principles that govern the placement of proteins in membranes. Its amino acid sequence suggests that a nonpolar sequence of 23 amino acid residues in the middle of the chain fits into the hydrophobic middle of the membrane (Figure 8.20). Most of the remaining 108 amino acid residues are polar, with hydrophilic side chains; they are thought to predominate outside the bilayer, with 38 residues projecting into the cytoplasm and 70 facing the exterior. Many short chains of sugars (oligosaccharides) are attached to the exterior domain, giving the protein its name (*glyco-* = sugar; *-phor* = bear or carry). Proteins with sugars attached, or **glycoproteins,** account for the substantial amount of carbohydrate in some membranes.

The *anion exchange protein* has a chain of 930 amino acids, which spans the lipid bilayer a dozen times, forming a channel that allows passage of chloride (Cl^-) and bicarbonate (HCO_3^-) ions (Figure 8.20). The protein exchanges one chloride ion for one bicarbonate ion, and since bicarbonate is essentially a soluble form of CO_2, RBCs can pick up CO_2 from tissues and eliminate it in the lungs.

On the inner surface of the RBC membrane, some integral proteins are anchored to a skeleton of peripheral membrane proteins, including spectrin, ankyrin, actin, and band 4.1 protein. This lattice holds the cell in its characteristic shape, a biconcave disc (Figure 8.21), showing how proteins with specific shapes bind to one another and perform distinctive functions. People who have inherited the diseases spherocytosis or elliptocytosis are anemic because their RBCs are abnormally shaped, fragile, and shorter-lived than normal RBCs. They have defective spectrin or band 4.1 protein, showing again how a small change in genetic information translates into an abnormal protein that can produce severe biological effects.

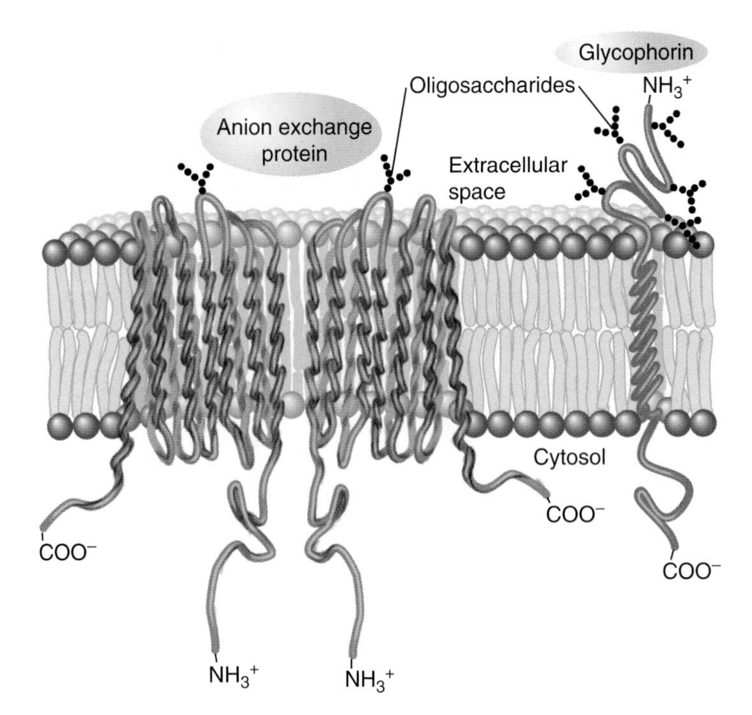

Figure 8.20
Two integral proteins of the red blood cell membrane are glycophorin and anion exchange protein. Glycophorin spans the membrane, with a predominantly nonpolar, α-helical region that fits into the lipid bilayer and two more polar regions that extend to the interior and exterior. Sugar chains (oligosaccharides) are attached on the exterior surface. The anion exchange protein is wound back and forth across the membrane twelve times, and two molecules form an anion channel through the bilayer that can exchange one Cl^- ion for one HCO_3^- ion (which is made from CO_2 and water). In this way, the protein helps to carry CO_2 from the tissues into the RBC and to remove the CO_2 in the lungs where it is eliminated.

All these proteins have specific roles in a cell by virtue of their structure and arrangement in or around membranes. The function of some proteins is to recognize the existence of small molecules in the extracellular fluid so the cell can respond to them. Many membrane proteins transport ions and small molecules across the membrane and control their fluxes. We take up this function next.

B. Transport Mechanisms

The plasma membrane maintains the proper internal composition of a cell by transporting many kinds of molecules and ions. Transport is accomplished by two types of specific **transport proteins.** *Carriers* are proteins that specifically transport a variety of molecules and ions, acting very much like enzymes. *Channels* are proteins with open pores that permit the passage of small ions.

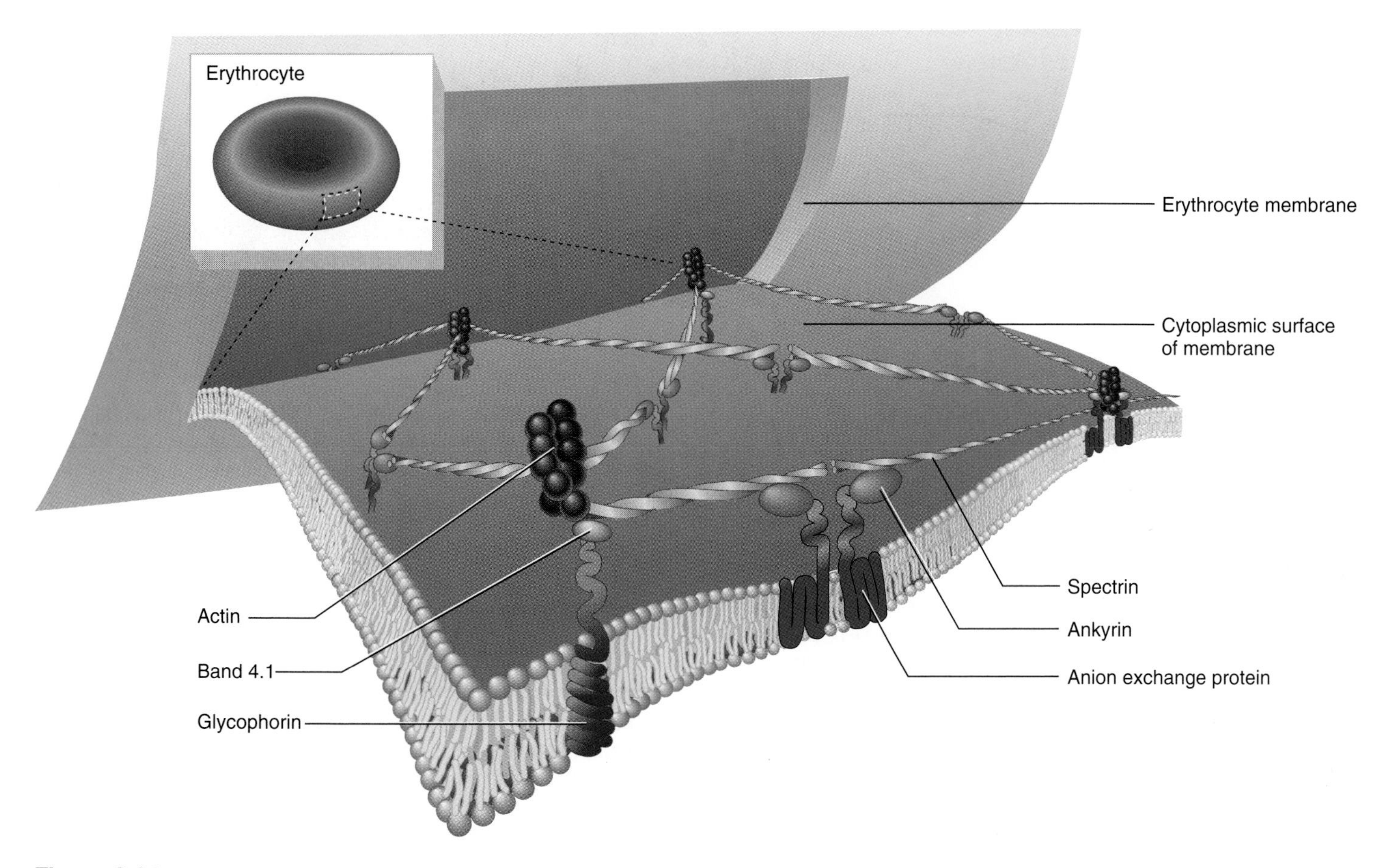

Figure 8.21

The structure of a protein network inside a red blood cell that holds it in its characteristic biconcave-disc shape.

8.11 Proteins transport many substances across membranes through facilitated diffusion.

While some small or lipid-soluble molecules can cross membranes by simple diffusion, most molecules that are large or polar—such as sugars, amino acids, and proteins—cannot. The same is true for all ions. To move from one side of the lipid bilayer to the other, these substances require the assistance of **carrier proteins,** also sometimes known as *permeases.*

Recall from Section 8.3 that a few materials, including H_2O, CO_2 and O_2, simply diffuse through membranes at rates that depend mainly on their concentration gradients. If molecules are moving by simple diffusion, the rate at which they get through depends only on the rate at which they strike the membrane, and this is proportional to their concentration:

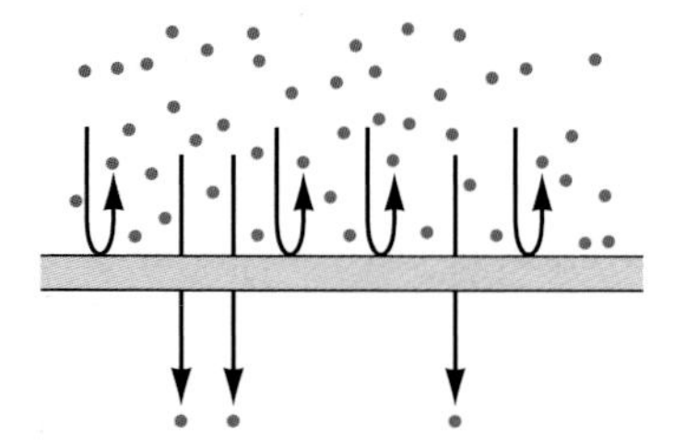

So there will be a net flux through the membrane as long as a concentration difference exists. Such a process *cannot be saturated;* as the following graph shows, the rate continues to increase as the concentration difference increases:

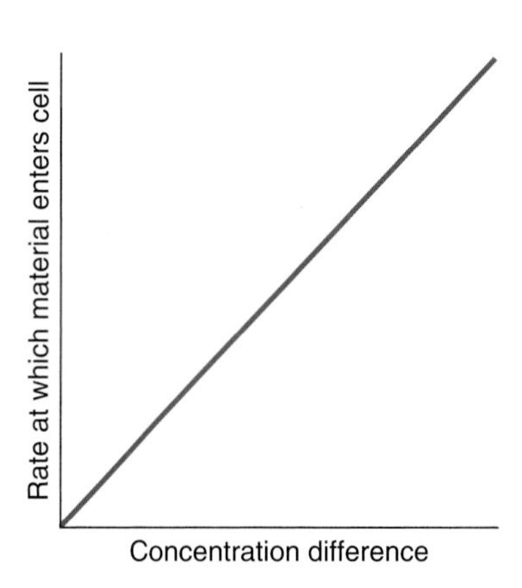

In contrast, we infer the existence of carriers from the fact that many larger molecules, such as sugars and amino acids, can only cross membranes through a different process that *can be saturated.* That is, the rate of transport is proportional to the concentration at low concentrations, but it reaches a maximum as the concentration increases:

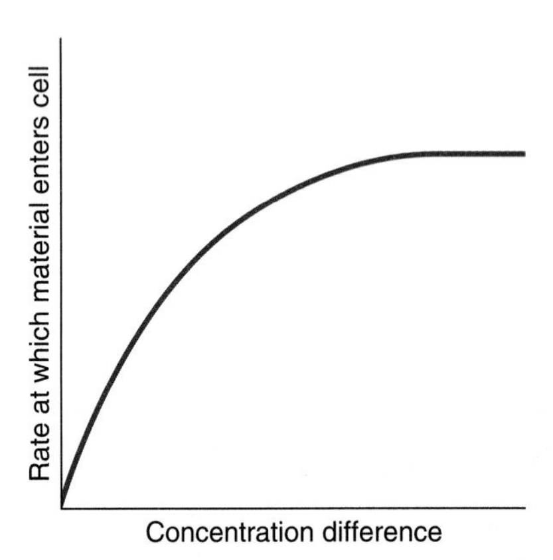

It is no coincidence that this curve resembles the rate curve of an enzyme with Michaelis-Menten kinetics (see Section 5.4); just as the number of active sites is limited by the number of enzyme molecules, a transport process can be saturated if each cell has only a limited number of sites where each substance can get through, so once all these sites are full, the rate of transport can't increase. Amino acids, sugars, and other metabolically important organic molecules can cross membranes through a type of protein-assisted transport called **facilitated diffusion.** Carriers in the membrane facilitate the passage of molecules that would be blocked by the hydrophobic interior of the membrane:

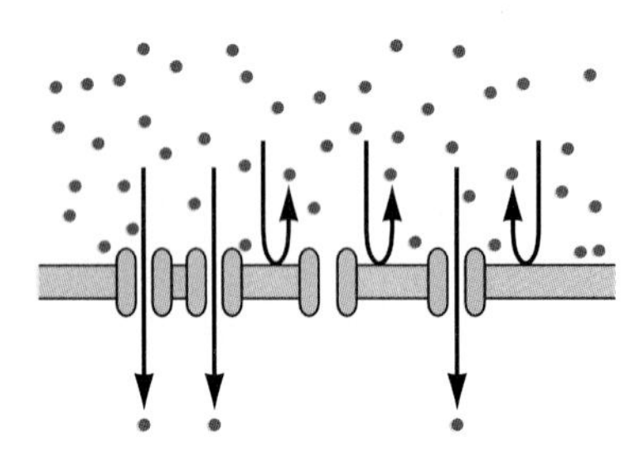

We use the general term *ligand* for the molecule each protein transports. Rather than simply binding to a site on a protein, ligands bind to carriers in such a way that they are able to move through a membrane. Facilitated diffusion is highly selective, since each carrier interacts stereospecifically with its ligands and can only transport the one ligand that has the right shape (or sometimes a few similar ligands). For this reason, every cell needs many types of carrier proteins to handle the many ligands in its environment.

The kinetics of facilitated diffusion indicate that permeases work much as enzymes do, but how can a protein move material through a membrane? Carriers like the anion exchange protein have several α helices spanning the membrane, forming a channel through which the ligands can move. Just as an enzyme changes its conformation while transforming its substrate, a carrier mediates facilitated diffusion by changing its conformation in a **ping-pong mechanism.** The protein constantly shifts back and forth between a conformation in which it is open to the outside ("ping") or open to the inside ("pong") (Figure 8.22). When a ligand enters the binding site of the protein on one side, the protein flips into its other conformation, allowing the ligand to leave on the other side.

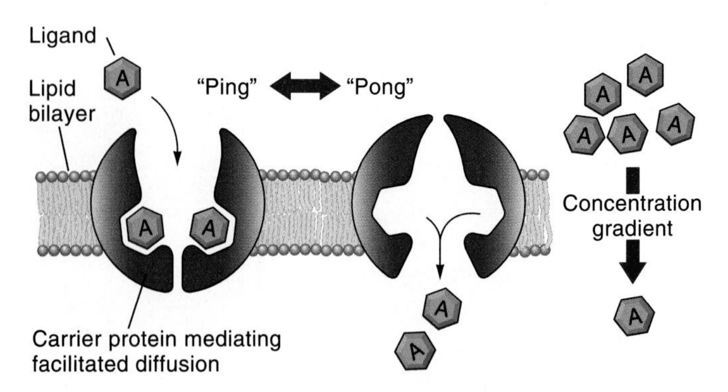

Figure 8.22

In this model for the action of a carrier, the protein forms a channel that will only accept one kind of ligand (or a few similar ligands). The protein is constantly undergoing transitions between a "ping" and a "pong" conformation, so it acts as a shuttle to move ligands through the membrane. However, the protein simply allows diffusion of the ligands down their concentration gradient because they are more concentrated on one side of the membrane; it cannot transport them actively against the gradient.

Exercise 8.9 Compare the interaction of enzymes with their substrate and the interaction of carriers with the ligands they transport. Explain why facilitated diffusion is a saturable process.

8.12 Some substances are actively transported against a concentration gradient.

Both simple and facilitated diffusion can only transport substances *down* a concentration gradient, from a higher to a lower concentration. However, a cell keeps many substances in its cytosol at concentrations that are either higher or lower than their concentrations outside the cell. To maintain these concentration differences, the cell must transport the substances *against* their gradients, from a lower to a higher concentration. Such transport processes require energy and are therefore known as **active transport.** The needed energy can be supplied in various ways, but it most commonly comes from ATP, the same energy source used to drive other endergonic processes. ATP can activate a carrier through phosphorylation, just as it activates a substrate by phosphorylating it (review Section 7.8). Phosphorylation shifts the protein into a different conformation, and in this way the ATP gives the protein the energy needed to transport a ligand.

One very well-known active transport system is the Na^+-K^+ exchange pump in the plasma membrane of all animal cells. This *sodium-potassium pump* maintains a high concentration of K^+ and a low concentration of Na^+ in the cytosol relative to the cell's surroundings. Since the carrier gets its energy by hydrolyzing ATP, it is known as the Na^+-K^+ ATPase (Figure 8.23). In each round of its activity, this protein uses the energy of one ATP molecule to transport three Na^+ ions outward and two K^+ ions

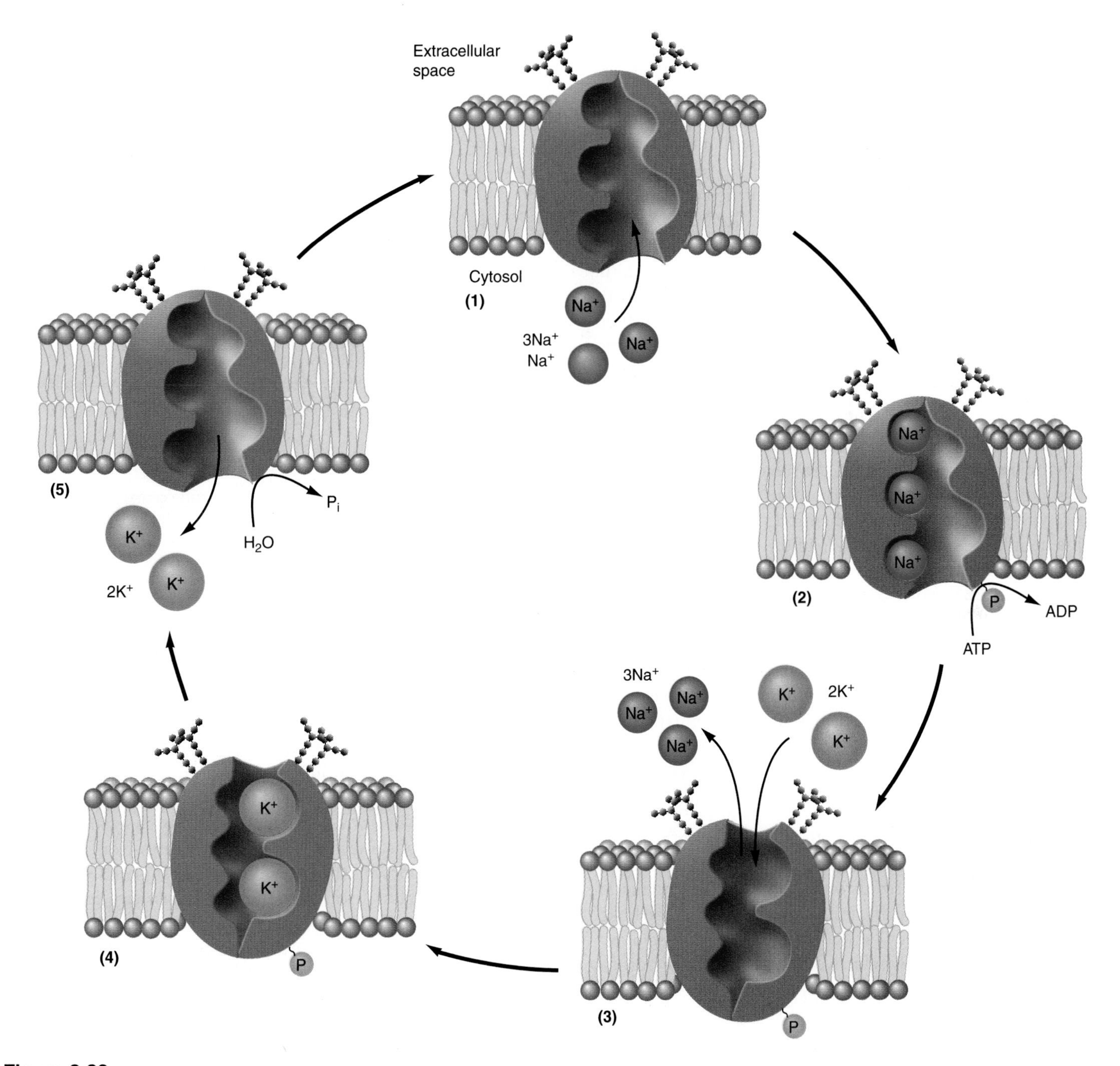

Figure 8.23

The Na^+-K^+ ATPase actively transports Na^+ and K^+ ions in opposite directions across the cell membrane. *(1)* Na^+ ions bind to their site from the inside. Phosphorylation of the protein *(2)* shifts its conformation so these ions are released to the outside *(3)*. K^+ ions bind to their site from the outside *(4),* and this stimulates removal of the phosphoryl group, allowing the protein to shift back into its original form so these ions are released inside *(5).*

inward. Like the carriers that function in facilitated diffusion, the protein has binding sites that are specific for the two kinds of ions and a channel that the ions move through. ATP phosphorylates the protein at a site on the cytoplasmic side, putting the protein into an active, energized conformation; removing this phosphoryl group causes the protein to return to its original conformation. Through its slight changes in shape, the protein moves each ion to the other side. Other proteins couple Na^+-transport to the transport of some other ion, such as H^+.

8.13 A gradient carries a potential and can do work.

In Chapter 7 we showed that a potential exists wherever there is an unequal distribution of matter. A rock held above the ground and water falling from a height have gravitational potential energy, which can be converted into work. Similarly, wherever molecules are more concentrated in one place than in another, a *chemical potential* exists, and it is capable of doing work as the molecules diffuse toward equilibrium. However, the energy in a chemical potential is very small. In the next section, we will show how cells can use the chemical potential of one substance to transport a second substance, but one could hardly build a practical machine to do any work with a chemical potential alone.

On the other hand, if the gradient is formed of ions rather than neutral molecules, it has an unequal distribution of electrical charge as well as matter, and the potential energy it carries is an **electrochemical potential.** Energy resulting from an unequal distribution of charge is far greater than that due to the unequal distribution of molecules alone. Such an electrical potential, carried by ions or electrons, is a **voltage,** and a voltage can do a great deal of work, as we prove every day as we use electrical devices.

Because of the Na^+-K^+ ATPase, eucaryotic cells have a high concentration of K^+ ions and a low concentration of Na^+ in their cytoplasm; in animal tissues, the surrounding extracellular fluid often has the opposite composition. For this reason and others, all cells have an electrochemical potential across their plasma membranes, with the inside of the cell electrically negative and the outside positive. Because of this potential, a cell must do additional work (beyond strictly chemical work) to transport ions across its plasma membrane and having built up an electrochemical potential, it is able to perform useful work. Electrochemical potentials in biological systems are the sources of all kinds of work, including the synthesis of ATP in mitochondria and chloroplasts and the operation of animal nervous systems.

Structure and function of nervous systems, Chapter 41.

In addition to carrier proteins, the membranes of plant and animal cells contain **channel proteins** that conduct ions. Channels form water-filled pores through the lipid bilayer that are just large enough to let specific ions pass. They cannot be linked to an energy source, so they only conduct ions passively (downhill), but each one is highly selective for one type of ion, usually K^+, Na^+, Ca^{2+}, or Cl^-. A channel can conduct a million ions per second, much faster than the fastest carrier protein. However, channels do not stay open all the time, and they are said to be *gated.* Each kind of channel has a gate that can be opened by one of three types of stimuli: a mechanical stress on the membrane, a ligand that binds specifically to the protein, or a change in the voltage across the membrane (Figure 8.24). All three kinds of channels are critical elements in the behavior of plants and animals, especially in animal nervous systems.

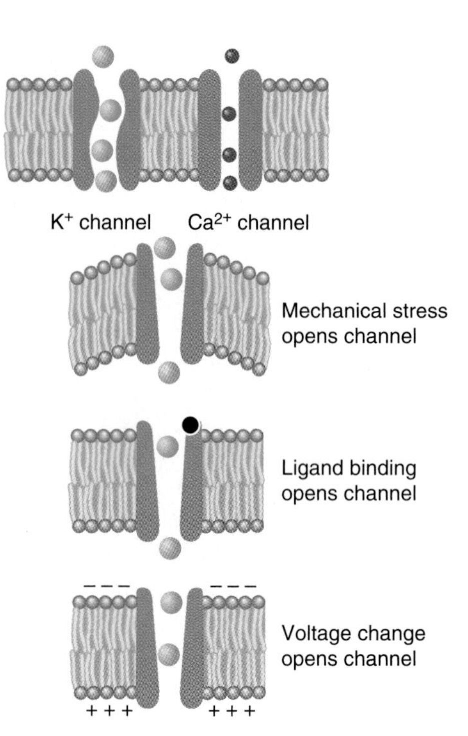

Figure 8.24
A channel protein allows one type of ion to diffuse through passively. The protein is gated, and each type opens in response to one of three stimuli: a mechanical stress, binding a specific ligand, or a change in the voltage across the membrane.

8.14 Secondary active transport uses the energy of an established concentration gradient.

Cells actively transport some substances by **cotransport** processes, which couple the movement of one substance to the movement of a second one. In general, energy from ATP is used to create a gradient of substance A, which naturally tends to diffuse down this gradient. But the diffusion of A is coupled to the transport of substance B, through a protein that simultaneously carries A and B. As long as the gradient of A is steeper than the gradient of B, B will be actively transported against its gradient. Cotransport mechanisms that use energy from ATP indirectly to transport another substance are designated **secondary active transport** to distinguish them from *primary active transport* in which ATP is used directly to phosphorylate a carrier. For example, cells that line an animal's intestine actively take up foodstuffs such as glucose and amino acids, using the energy stored in the Na^+ gradient that has already been established. Since sodium ions are more concentrated outside the cells, they tend to diffuse inward. A carrier protein simultaneously binds Na^+ and glucose (or Na^+ and an amino acid) and moves both ligands in together, even though the concentration of glucose (or of the amino acid) may be higher on the inside (Figure 8.25). Such cotransport of two materials in the same direction is called **symport.**

The cotransport of two substances in opposite directions, known as **antiport,** uses the same principle. For instance, the band-3 protein of red blood cells exchanges bicarbonate ions (HCO_3^-) from inside the cells with Cl^- ions from outside the cells (Figure 8.26). In this case, both ions are moving passively down their concentration gradients. Other antiport proteins, such as one that exchanges Ca^{2+} ions for Na^+ ions, operate via secondary active transport.

Carriers operate on two general principles. First, each protein is specific for the ligand it transports, and it undergoes

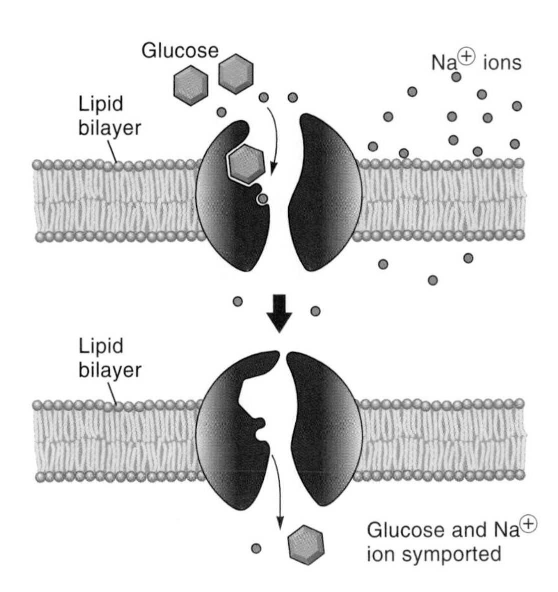

Figure 8.25

One symport process uses a carrier that simultaneously transports glucose and Na^+ ions. The gradient of Na^+ ions is developed by the Na^+-K^+ ATPase; the energy of the gradient is then used to carry glucose into the cell.

conformational changes while interacting with that ligand. Second, proteins that interact with two or more ligands can act allosterically, much as hemoglobin does when it binds to O_2 or BPG (Chapter 5). The Na^+-glucose symport protein, for instance, is an allosteric protein because it has a binding site for each ligand; binding at each site apparently changes the protein's shape, and it will only operate if both sites are occupied.

Allosteric proteins, Section 5.11.

8.15 Membrane proteins can move substances by vectorial action.

An important point about the action of membrane proteins was made in 1961 by the English biochemist Peter Mitchell. He recognized that all proteins of a given kind will be oriented in the same way across an asymmetrical membrane, and therefore will move their ligands in the same direction with respect to the membrane. This *vectorial flow* is the basis for the action of some carriers. (A vector is a quantity, such as force or velocity, that has a direction as well as a magnitude; it can therefore be represented by an arrow.) The vectorial principle depends on the asymmetry of membranes and on the general activity of a protein. A protein changes conformation when it takes a ligand into its binding site. As the protein then releases the ligand and reverts to its original conformation, the ligand moves in a specific direction (Figure 8.27). This action is unimportant if the proteins are floating freely in the cytosol. However, if they are integral proteins in a membrane, all oriented in one direction, they will move their ligands in the same direction with respect to the membrane.

Thus Mitchell pointed out that metabolism can have a vectorial aspect to it. A membrane-bound enzyme not only operates at

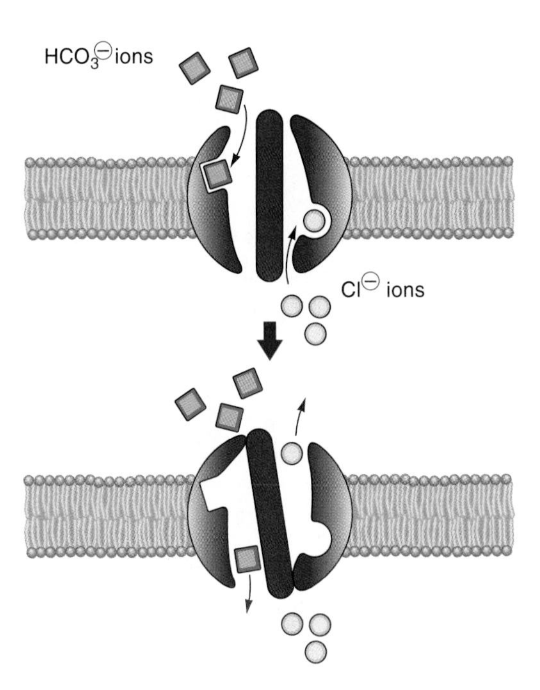

Figure 8.26

The anion exchange protein of red blood cells is an antiporter that carries HCO_3^- and Cl^- ions in opposite directions.

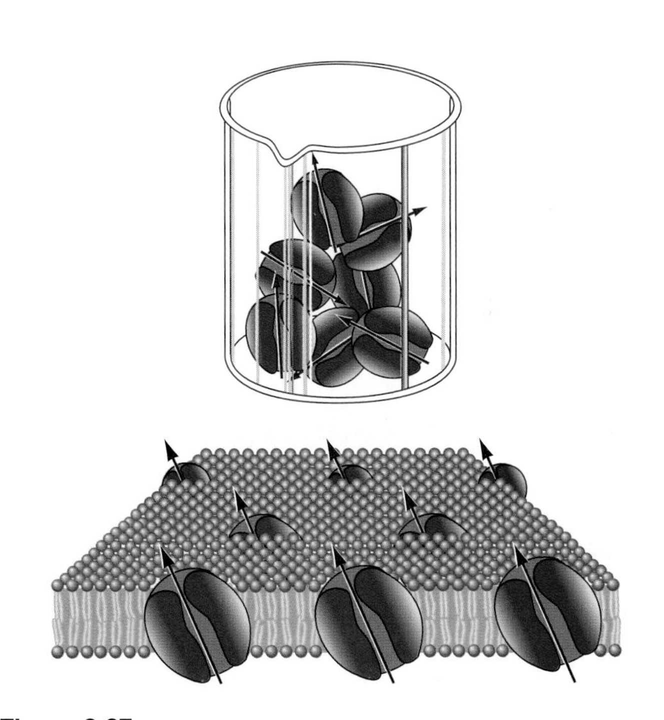

Figure 8.27

Proteins will act vectorially if they are arranged in a specific way in a membrane, so they move their substrates, or other small molecules, in one direction while interacting with them.

a certain rate but also in a certain *direction,* and it can simultaneously carry out a step in metabolism while moving a molecule across a membrane. This is the basis for **group translocation,** another type of active transport (Figure 8.28). For instance, a sugar

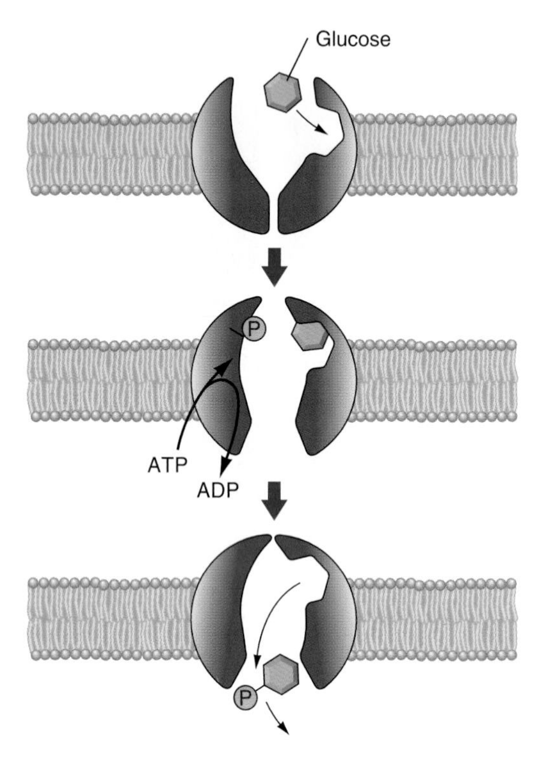

Figure 8.28

An enzyme that operates vectorially transports a glucose molecule into the cell while making a change in its structure—in this case, phosphorylating it.

can be transported into a cell by an enzyme that carries out a step in metabolism, altering the sugar while it is transported.

Figure 8.29 summarizes all the transport mechanisms we have discussed in this chapter.

Exercise 8.10 A bacterial cell can pick up amino acid Q from its growth medium and use it as a source of energy and carbon. The first step in the metabolism of Q is removing its amino group; sketch a vectorial process that simultaneously transports Q into the cell and starts to metabolize it.

Coda The significance of cell membranes in biology becomes more apparent every year. Though biologists once thought of cells as mere bags of enzymes, we now see their extensive internal organization, an organization largely due to membranes. Eucaryotic cells, of course, are full of specialized organelles that are bounded by membranes and often have internal membranes as well, and even procaryotic cells are organized internally by the properties of their plasma membranes. Membranes, being asymmetrical arrays of proteins, are capable of organizing a region by coupling an energy source to the transport of substances from one place to another. For this reason, membranes may be the most significant organizing structures in the biological world.

In the next two chapters, we get into the heart of metabolism, where membranes again have critical roles in obtaining energy. The first steps in the evolution of primitive biological systems apparently occurred in solutions, with little organization, but evolution must have accelerated enormously once the first cell membranes—and thus the first cells—formed. Membranes create boundaries, and thus individuality and privacy. They create organisms with their own distinct genomes, with individual characteristics not shared with their neighbors—organisms that face the world alone, succeeding or failing in the race to reproduce. Without such individuality, there can be no natural selection, no evolution.

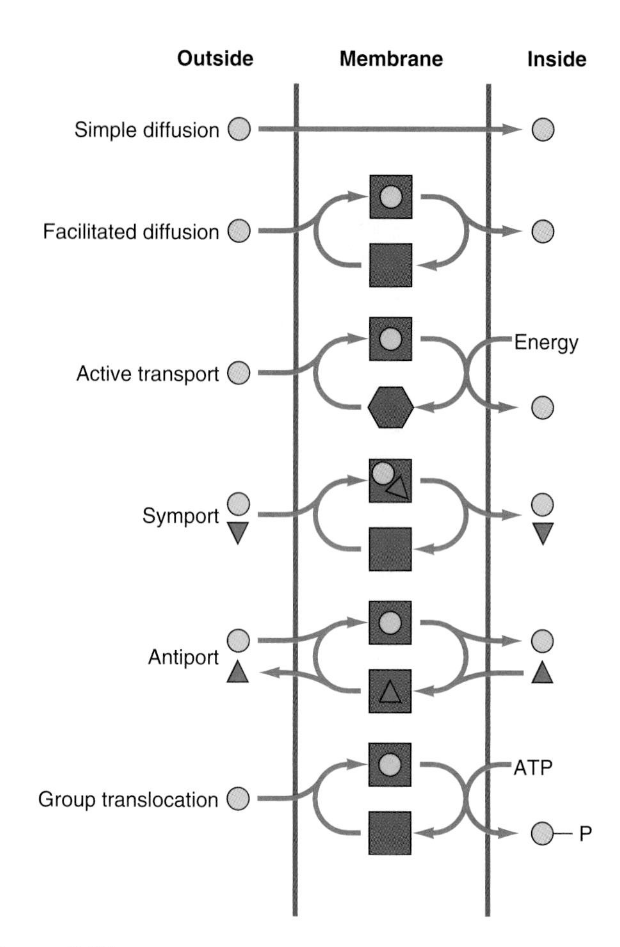

Figure 8.29

This schematic summarizes and compares the mechanisms of transport across membranes.

Summary

1. Molecules move by diffusion, due simply to their kinetic energy.
2. Membranes allow some molecules to diffuse freely across them, but inhibit the passage of others. A semipermeable membrane is permeable to water but impermeable to one or more solutes.
3. Osmosis occurs when a membrane separates solutions with different concentrations of materials and therefore different concentrations of water. Water diffuses from its higher concentration to its lower concentration.
4. Cells shrink in hypertonic solutions and swell in hypotonic ones.
5. Experiments with osmosis in cells show that membranes are made of lipids. Materials diffuse into cells at a rate proportional to their size and their solubility in lipids.

6. Lipids are insoluble in water. Most of them are based on fatty acids, which typically have 16 or 18 carbon atoms. Fatty acids are amphipathic molecules, with both a hydrophobic "tail" and a hydrophilic "head." A neutral fat (triglyceride) is made of three fatty acids linked to a glycerol molecule.
7. The major lipids in membranes are phospholipids, which are neutral fats in which one fatty acid has been replaced by a phosphate group linked to a small molecule such as choline.
8. Membrane lipids form bilayers with their hydrophobic tails together inside the bilayer and their hydrophilic heads in contact with the surrounding aqueous environment.
9. Membranes are fluid mosaics of lipid and protein. The proteins associate with the phospholipid bilayer on the basis of their affinity for the lipids. Experiments on fusing antibody-labeled cells show that plasma membranes are as fluid as olive oil.
10. Membrane fluidity depends on the lengths of hydrocarbon chains and the number of double bonds in them. These factors determine how tightly the lipids pack together.
11. Membrane proteins move laterally but cannot "flip-flop" across the membrane, since such a motion is energetically unfavorable.
12. The membrane proteins of red blood cells exemplify those of cells in general. They are characteristically embedded with their hydrophobic regions inside the membrane, often with several α helices traversing the membrane. Some of them carry extensive oligosaccharides on their outer faces.
13. Cell membranes allow some small molecules (O_2, CO_2, H_2O, lipids) to diffuse through freely; such transport by simple diffusion is not saturable. Many organic molecules are transported by facilitated diffusion, by means of a specific carrier protein; this process is saturable.
14. Some ligands are carried by active transport against a concentration gradient. This process requires an input of energy, usually from ATP. If the energy in ATP is supplied directly to the carrier, the process is called primary active transport.
15. While chemical gradients have the potential to do some work, a gradient of ions has an electrochemical potential and is capable of doing much more work. This is important in the generation of ATP and in animal nervous systems. Channel proteins conduct specific ions passively and are gated so they can be opened by various stimuli.
16. Secondary active transport processes use the energy of a concentration gradient established by primary active transport. Symport is the transport of two ligands in the same direction; antiport is the transport of two ligands in opposite directions. The energy in either case is derived from one ligand moving down its concentration gradient.
17. Membrane proteins move molecules by vectorial action, due to the specific orientation of the proteins in the membrane. By group translocation, molecules are transformed enzymatically as they are transported.

Key Terms

diffusion 154
concentration gradient 154
Fick's Law of Diffusion 155
semipermeable membrane 155
osmosis 155
osmotic pressure 155
osmolarity 155
isosmotic 156
hyperosmotic 156
hypoosmotic 156
isotonic 156
hypotonic 156
hypertonic 156
turgor pressure 156
lipid 157
fatty acid 157
amphipathic 157
triglyceride / neutral fat 158
phospholipid 159
bilayer 160
freeze-fracture 161
fluid mosaic model 161
integral protein 162
transmembrane protein 162
peripheral protein 162
antibody 164
antigen 164
glycoprotein 166
transport protein 166
carrier protein 167
facilitated diffusion 168
ping-pong mechanism 168
active transport 168
electrochemical potential 170
voltage 170
channel protein 170
cotransport 170
secondary active transport 170
symport 170
antiport 170
group translocation 171

Multiple-Choice Questions

1. Under uniform conditions, which will diffuse most rapidly?
   a. water
   b. glucose
   c. disaccharides
   d. ATP
   e. methane
2. Which has the least effect on the rate of diffusion of a solvent through a permeable membrane?
   a. temperature of the system
   b. size of the solvent molecule
   c. pH of the solvent
   d. membrane area
   e. steepness of the concentration gradient
3. Assume a compartment of fixed volume, divided by a semipermeable membrane. Which of the following systems will develop the highest osmotic pressure?
   a. 1 *M* glucose/pure water
   b. pure water/pure water
   c. 1 *M* NaCl/2 *M* NaCl
   d. pure water/2 *M* NaCl
   e. 1 *M* glucose/2 *M* NaCl
4. What is the osmolarity of a 1 *M* $NaHCO_3$ solution in which the solute dissociates into Na^+ and HCO_3^- ions?
   a. 1 Osm
   b. 2 Osm
   c. 3 Osm
   d. 4 Osm
   e. 5 Osm
5. Osmolarity refers to all solutes in an aqueous solution, while tonicity refers only to solutes that
   a. cannot pass through a semipermeable membrane.
   b. dissociate to form ions.
   c. do not dissociate completely.
   d. do not dissociate at all, such as oxygen or carbon dioxide.
   e. none of the above.
6. Which of the following is (are) amphipathic?
   a. soap
   b. phospholipids
   c. triglycerides
   d. *a* and *b*, but not *c*
   e. *a*, *b*, and *c*
7. Hydrolysis of 1 mole of a neutral fat requires the addition of ______ mole(s) of water, yielding ______ mole(s) of glycerol and ______ mole(s) of fatty acids.
   a. 1; 3; 3
   b. 3; 3; 3
   c. 3; 3; 1
   d. 3; 1; 3

8. Which of the following is believed to be the best description of the molecular structure of a plasma membrane?
   a. a lipid bilayer
   b. a lipid bilayer coated by proteins on both the inner and outer surfaces
   c. an equal mixture of randomly distributed lipid and protein
   d. a mosaic of proteins embedded in a sea of lipids
   e. a bilayer of phospholipids, cholesterol, and glycophorin
9. Amino acids and sugars cross the cell membrane by
   a. ligand transport.
   b. facilitated diffusion.
   c. phosphorylation.
   d. endocytosis.
   e. phagocytosis.
10. By which of the following are transport proteins usually activated?
   a. a sodium-potassium pump
   b. a cotransport process
   c. phosphorylation by ATP
   d. vectorial flow
   e. all of the above

True-False Questions

Mark each statement true or false, and if false, restate it to make it true.

1. When a molecule diffuses, each leg of its moment-to-moment trajectory is directed down its concentration gradient.
2. In an osmotic system containing two solutions that differ in osmolarity, water will flow toward the more hypoosmotic solution.
3. The cytosol of a cell has a lower osmolarity than pure water.
4. Cells placed in a hypertonic solution will develop turgor pressure or burst.
5. While a diffusible substance is flowing from an area of high concentration to a region of lower concentration, a concentration gradient exists.
6. Substances that diffuse easily through the plasma membrane have a low oil-water partition coefficient.
7. Transmembrane proteins have hydrophilic ends and a hydrophobic midregion.
8. Permeases function as carriers during active transport.
9. If the entry of a solute into a cell requires the addition of energy and does not change with increasing solute concentration, the solute is likely moving by active transport.
10. When a symport protein transports two or more ligands, the ligands bind one at a time and are transported sequentially.

Concept Questions

1. What is meant by a concentration gradient?
2. Compare the functions of transmembrane proteins and peripheral proteins.
3. How could one determine whether a solute was being transported by diffusion or active transport?
4. What are *gated* channel proteins, and why is gating necessary?
5. Compare and contrast primary active transport with secondary active transport and cotransport.

Additional Reading

Unwin, Nigel, and Richard Henderson. "The Structure of Proteins in Biological Membranes." *Scientific American,* February 1984, p. 78. How the structure of membrane proteins is analyzed.

Vance, Dennis E., and Jean E. Vance (eds.). *Biochemistry of Lipids, Lipoproteins, and Membranes.* Elsevier, New York, 1991. A comprehensive treatise on lipids, lipoproteins, and membrane structure.

Zimmer, Carl. "First cell." *Discover,* November 1995, p. 7. David Deamer believes life originated within a protective membrane made of lipids that form bubbles called liposomes in water. Such cells may have trapped RNA during the wet-dry cycle of primordial tidepools.

Internet Resource

To further explore the content of this chapter, log on to the web site at:

http://www.mhhe.com/biosci/genbio/guttman/

Cellular Respiration

9

Key Concepts

9.1 The biosphere operates on a carbon cycle.
9.2 Respiration produces NADH and ATP.
9.3 Overview: Respiration consists of two processes.
9.4 Glucose is oxidized to pyruvate.
9.5 Pyruvate is commonly oxidized to an acetyl group.
9.6 Mitochondria are the principal sites of respiration in eucaryotes.
9.7 The Krebs cycle is the core of metabolism.
9.8 The electron transport system synthesizes ATP.
9.9 A proton gradient across a membrane can be used to synthesize ATP.
9.10 A proton gradient itself can do work.
9.11 Many organisms obtain energy through fermentation.
9.12 Excess sugar can be made into fatty acids.
9.13 Some organisms use other types of respiration and inorganic energy sources.
9.14 Many compounds are catabolized into the central pathways.
9.15 Summary: Heterotrophic metabolism consists of five phases.

Children and other animals get a lot of their energy from carbohydrates.

We all know we have to eat to get the energy that sustains life and the materials for maintaining our bodies, but what does the body do with that food to get the energy out and put the matter into the right places? The relationship between putting food in the belly and having the energy to move around is pretty obscure. When a tired child is given a candy bar or ice cream, is it more than just the psychological boost that converts him back into a rambunctious bundle of energy? Why does indulging in too many pastries put fat on the waist and thighs? How is a football player's diet of beefsteak converted into muscle?

How our bodies use food is far from obvious, and in this chapter we lay the foundation for understanding that process, with emphasis on the metabolic mechanisms for obtaining energy. Metabolism includes all of an organism's chemical activities, although we noted in Chapter 7 that the metabolic activities of the organisms sharing an ecosystem are so intimately linked that the whole system can be said to have one gigantic metabolism. We also established in Chapter 7 that autotrophs and heterotrophs have different roles in an ecosystem. Autotrophs (mostly phototrophs) bring energy (usually from the sun) into the system as they reduce CO_2 to organic compounds, while heterotrophs, including humans, consume those organic compounds by oxidizing them to recover the energy they store. We use organic molecules partly for energy and partly for substance, for making our structures. So a secondary theme, to which we will return at the end of the chapter, is the conversion of unoxidized metabolites into structural molecules. ■

9.1 The biosphere operates on a carbon cycle.

Most organisms obtain energy from reduced compounds through the process of *respiration.* This word has meant different things at different times. It commonly refers to breathing—inhaling and exhaling—and a moment's reflection will show that breathing is intimately related to getting energy. If you start to exercise hard, for instance, you'll automatically start to breathe faster and more deeply (Figure 9.1). Eventually you may have to stop and catch your breath because you feel you have no more energy. Your hard-working muscles are expending a lot of energy, and to produce it, they somehow demand a lot of air. But what does breathing have to do with getting energy?

The relationship between air and energy first started to emerge in 1755 when Joseph Black showed that respiring animals give off a gas then known as "fixed air," which we now call carbon dioxide. In 1771 Joseph Priestley studied respiration and combustion by sealing small animals or burning candles into glass vessels (Figure 9.2 *a, b*). Priestley and his contemporaries believed in a substance called "phlogiston," which was supposed to be a material ingredient of fire. To Priestley, the animals and the candles in his experiments were both producing phlogiston. So when the animals died and the candles went out, Priestley concluded that they were doing the same thing—that they had "phlogisticated" the air. Next Priestley discovered that by sealing a sprig of mint or some other small plant in the vessel or by adding mercuric oxide to the vessel and focusing the sun's rays on it, the candles burned longer and the animals survived longer. He therefore inferred that the phlogiston had been removed from the air. (He did not call the substance mercuric oxide, by the way, since this is a modern chemical name and oxygen hadn't been discovered yet.) In 1780 Jan Ingenhousz found that light is needed to make a plant "dephlogisticate" the air and that in darkness plants "phlogisticate" it just as animals do. The observations of Priestley and Ingenhousz thus suggested a natural cycle of reactions in which plants and animals conduct opposite processes.

This cycle could only be understood with the concepts of a more modern chemistry. One of its founders, Antoine Lavoisier, established that the gas he named oxygen is removed from the air during respiration or combustion and replaced by carbon dioxide. Nicholas de Saussure found that when plants

Figure 9.1
An athlete can expend a tremendous amount of energy and must breathe hard during and after exercise to recover that energy from stored foodstuffs.

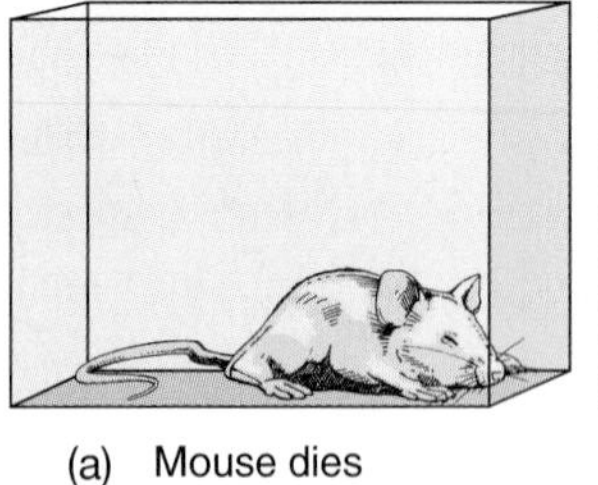

(a) Mouse dies

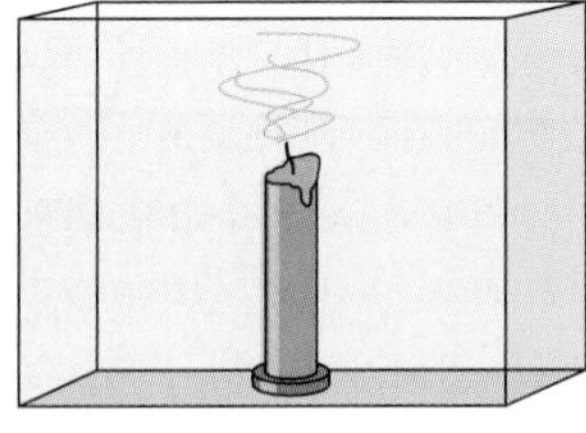

(b) Candle goes out

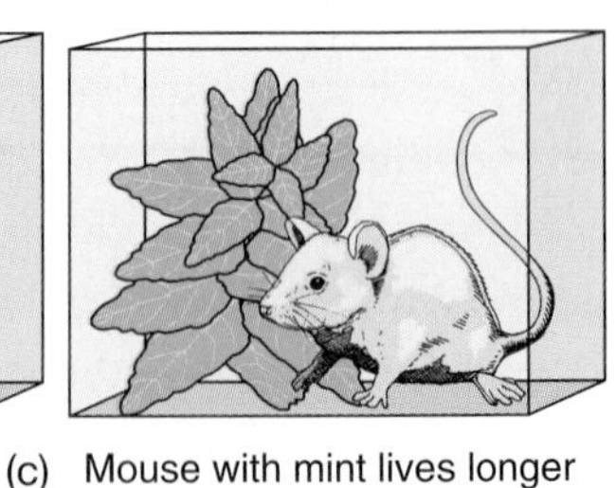

(c) Mouse with mint lives longer

Figure 9.2
Joseph Priestley's observations indicated that plants and animals perform opposite processes. The mouse dies from lack of oxygen *(a)* for the same reason the candle goes out *(b)*, but the mint can extend the life of either by producing extra oxygen *(c)*. However, the mint is only active while it is being illuminated.

are exposed to light in chambers containing measured amounts of CO_2, they grow by consuming the gas, and also that they produce oxygen and consume CO_2 at the same rate. de Saussure concluded that plants incorporate CO_2 as they grow, since those given no CO_2 don't increase beyond the mass of the seed from which they sprouted. As chemistry developed in the nineteenth century, the process of photosynthesis came to be described by the general equation:

$$6\,H_2O + 6\,CO_2 \rightarrow C_6H_{12}O_6 + 6\,O_2$$

where $C_6H_{12}O_6$ is a general formula for the sugars formed.

Lavoisier and Pierre de Laplace studied combustion and respiration quantitatively, using a chamber surrounded by ice as a calorimeter; the heat produced in the chamber by burning material or by the respiration of an animal is proportional to the amount of ice melted during the process (Figure 9.3). One experiment revealed that a small animal produced enough heat in 10 hours to melt 341 grams of ice. Lavoisier and Laplace also measured the CO_2 the animal produced and determined that 326.7 grams of ice are melted when that much CO_2 is produced by combustion. Because these values are so close, they argued that respiration is just a slow combustion. Lavoisier thought at first that carbon is carried through the blood to the lungs where it is slowly combined with oxygen, but he gradually developed a more realistic theory by moving the site of respiration from the lungs to the blood and finally to all the body tissues. Thus the view developed that during respiration every tissue is burning carbon compounds—again represented by sugar—according to the equation:

$$C_6H_{12}O_6 + 6\,O_2 \rightarrow 6\,H_2O + 6\,CO_2$$

This equation is the opposite of photosynthesis. Matters are not really so simple, because these two equations summarize many reactions, but they still show that the biosphere operates on a **carbon cycle** consisting largely of photosynthesis and respiration (Figure 9.4). Autotrophs, mostly photoautotrophs such as plants, synthesize organic compounds from CO_2 as they grow, while chemoheterotrophs such as animals oxidize those compounds back to CO_2.

Did you notice that the definition of respiration changed during this discussion? Even though "respiration" still means breathing, **cellular respiration** is quite different. It is a chemical process in which energy-rich materials are oxidized to release their energy. These are usually carbon compounds, which are oxidized to CO_2 and water. Concepts 9.1 summarizes key events that led to our current understanding of cellular respiration.

To illustrate cellular respiration, we will focus on the metabolism of *carbohydrates,* which are sugars or compounds that

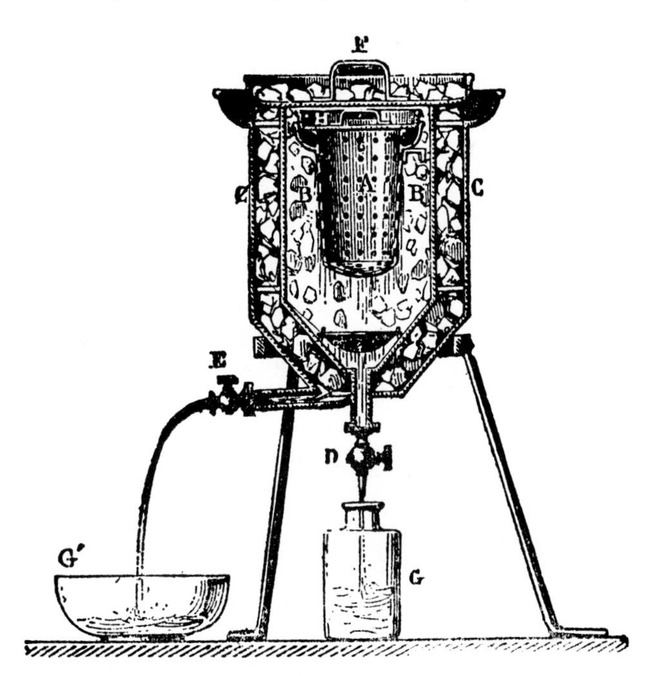

Figure 9.3
Lavoisier and Laplace used a simple calorimeter in which the animal was placed in a cage *(A),* and the heat of its body melted ice in chamber *B.* The water was collected in a container *(G)* and weighed. The outer layer of ice in *C* insulated the apparatus and prevented heat from entering. This instrument is preserved in the Conservatoire des Artes et Metier in Paris. It is described in the Memoirs of the French Academy, 1780, "Sur la Chaleur."

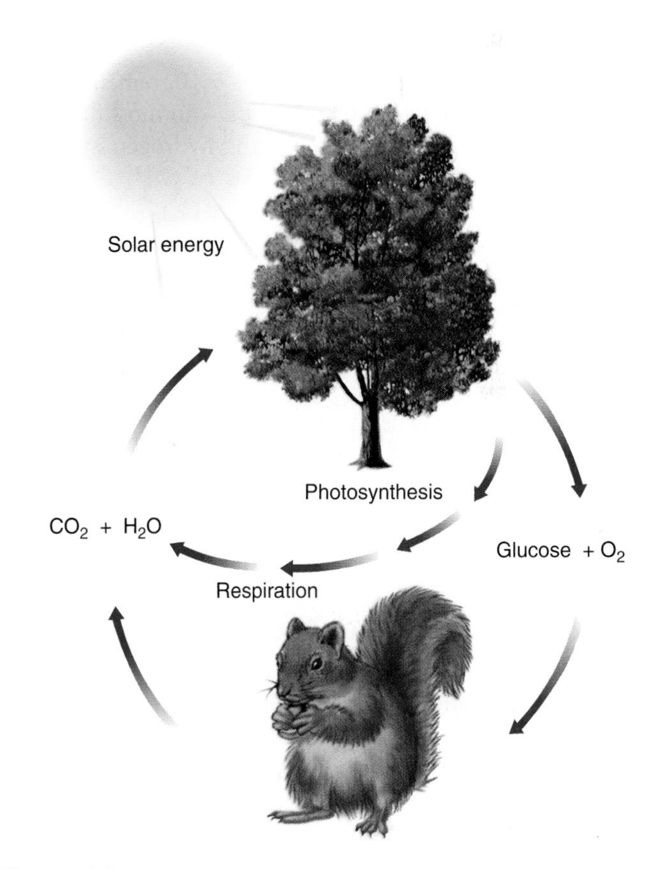

Figure 9.4
A general overview of the carbon cycle shows that it consists largely of two processes: photosynthesis, which converts CO_2 to organic molecules, and respiration, which converts organic molecules to CO_2. Notice that only phototrophs such as plants carry out photosynthesis, but that both phototrophs and chemotrophs carry out respiration.

Concepts 9.1 Key People and Events in Biological History

1755. Joseph Black: Animals produce "fixed air" (CO_2).

1771. Joseph Priestley: Burning candles and animals both "phlogisticate" the air (remove O_2), while plants "dephlogisticate" the air (renew O_2).

1780. Jan Ingenhousz: Light is needed to make a plant dephlogisticate the air.

1780. Antoine Lavoisier and Pierre de Laplace: Respiration and combustion are the same process.

1789. Antoine Lavoisier: Both respiration and combustion consume O_2 and produce CO_2.

1804. Nicholas de Saussure: Plants exposed to light grow by incorporating CO_2.

can be hydrolyzed to sugars. Cells commonly live on the simple sugar D-glucose, one of the isomers of hexose ($C_6H_{12}O_6$). During photosynthesis, plants manufacture hexoses, including glucose, which is stored as starch or transported as part of sucrose. Many microorganisms live on glucose, and animals carry it in their blood to nourish their cells. (If you feel weak and a little dizzy between meals, your blood glucose level may have fallen too low.) In fact, more of the biological world is made of glucose than any other substance, because it is the monomer of cellulose, which comprises a huge share of the world's biomass in plant structure, especially in the wood of standing forests. The organisms that eventually break down and recycle this wood must catabolize glucose.

To reemphasize the point of Section 7.10, remember that energy is commonly released by oxidation—in most biological situations by dehydrogenation, removing hydrogen atoms. In cellular respiration, the hydrogen atoms of $C_6H_{12}O_6$ are removed (and some oxygen is added), leaving CO_2. These hydrogen atoms usually combine with oxygen to make water. Remember, however, that oxidation does not necessarily involve oxygen; later we will show alternative kinds of respiration that don't require oxygen.

In respiration, sugar is oxidized to CO_2 and water, whereas in photosynthesis, essentially the opposite process occurs: CO_2 and water are reduced to sugar. Yet both respiration and photosynthesis entail very similar oxidative reactions that release energy, which is stored in compounds such as ATP. In the remainder of this chapter, we will see how this happens in respiration and then take up photosynthesis in Chapter 10.

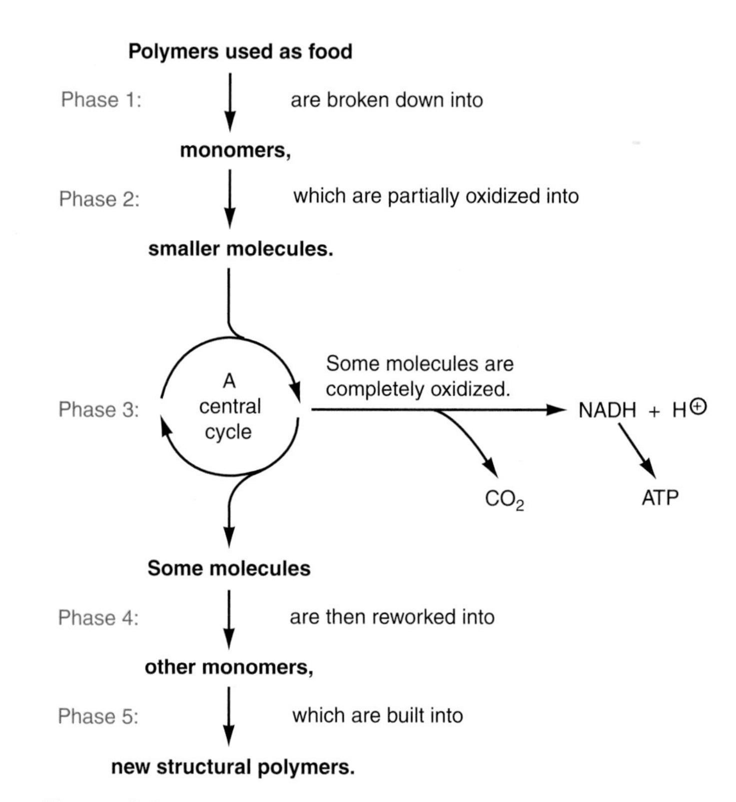

Figure 9.5

This overall view shows the central pathways of metabolism. Food molecules are mostly polymers, which must be digested into their monomers in phase 1. These monomers are broken down and oxidized *(2)*, some completely to CO_2 and H_2O *(3)*, some to other molecules that can be converted into new monomers *(4)* and polymerized into new cellular components *(5)*.

9.2 Respiration produces NADH and ATP.

Students who start to learn about metabolism sometimes feel ensnared in a tangled web, not knowing which way is up. This feeling comes from trying to learn everything at once, without first gaining an overall perspective. Our strategy for avoiding this confusion is to begin with a brief outline of metabolism and its function. We will then go over the same ideas twice more, each time in more depth, so that you can gradually fill in the specifics.

First, let's see how metabolism as a whole is organized in a chemoheterotroph. It is divisible into five phases, starting with a process of digestion, since food often consists of polymers that must be hydrolyzed into monomers for further metabolism (Figure 9.5). In phases 2 and 3 of this diagram, the reactions that convert organic molecules completely to CO_2 and H_2O constitute cellular respiration, the process that mobilizes energy for the endergonic reactions of phases 4 and 5 in which cellular structure is made.

Understanding respiration depends on grasping a few key concepts:

1. The function of the process is to make stores of ATP (for energy) and NADH (for reducing power). Remember that the coenzyme NAD^+ is the most common oxidizing agent in cells, and it is reduced to $NADH + H^+$. We will use "NADH" to represent the reducing power that accumulates in a cell, even though we know some oxidations entail the reduction of

FAD to $FADH_2$ and NADPH is really the coenzyme used for reduction in biosynthesis.

2. Energy is obtained from foodstuffs by oxidizing organic substrates while reducing NAD^+ to $NADH + H^+$. This solves the problem of storing up NADH.
3. A cell has a limited amount of NAD^+ or NADH. If every molecule were reduced to NADH, there wouldn't be any NAD^+ left for further metabolism. So there must be a way to oxidize NADH back to NAD^+. This could be done by using oxygen as an oxidizing agent and reducing it to water:

$$NADH + H^+ + \tfrac{1}{2}O_2 \rightarrow NAD^+ + H_2O$$

But this reaction would waste the large amount of energy stored in NADH. Instead, electrons from NADH are run through a respiratory **electron transport system (ETS).** This is a specialized set of membrane-bound proteins and coenzymes that conducts a flow of electrons and conserves some of their energy by synthesizing ATP. Oxygen is used as the terminal electron acceptor for the ETS.

4. Metabolism centers around an important compound, pyruvate, and around a cycle of reactions, the **Krebs cycle** or **citric acid cycle,** discovered by Sir Hans Krebs. Most catabolic pathways contribute molecules to this cycle, and many biosynthetic pathways begin with metabolites in the cycle. Pyruvate is also a center for many pathways.

9.3 Overview: Respiration consists of two processes.

It is useful to realize that respiration consists of two quite different processes: First the organic substrate (glucose) is oxidized completely to CO_2, with the reduction of NAD^+ to $NADH + H^+$. Then the NADH that has been generated is used to operate an electron transport system, which generates ATP.

The first process begins with glucose molecules whose carbon atoms are in a relatively reduced condition; they end up in their most oxidized state, in CO_2. The metabolites in this pathway are mostly oxidized by dehydrogenation, usually by simply removing a pair of hydrogen atoms, but some metabolites are oxidized by adding water (two hydrogen atoms and one oxygen) and then removing the two hydrogens. Thus the first stage of respiration can be summarized by:

$$C_6H_{12}O_6 + 6\,H_2O \rightarrow 6\,CO_2 + 24\,H$$

where 24 H represents the hydrogens held by NADH or $FADH_2$ or in solution as hydrogen ions.

In the second process, the 24 H are used to reduce an electron transport system that generates ATP:

$$24\,NADH \rightarrow 24\,NAD^{\oplus};\quad \text{Electron transport system};\quad 6\,O_2 \rightarrow 12\,H_2O$$

Here oxygen is the **terminal electron acceptor** that removes the hydrogen atoms. This process can be summarized by:

$$24\,H + 6\,O_2 \rightarrow 12\,H_2O$$

The sum of the two equations shows that sugar is completely oxidized to carbon dioxide and water:

$$C_6H_{12}O_6 + 6\,O_2 + 6\,H_2O \rightarrow 6\,CO_2 + 12\,H_2O$$

If this were pure chemistry, we would cancel six molecules of water on each side of the equation. But this is biology, and the water molecules are important, for they tell us something about how the energy is obtained.

With this outline as a guide, we will consider the stages of respiration in more detail and then discuss some variations on the process.

9.4 Glucose is oxidized to pyruvate.

Glucose has three possible fates in a cell, depending on intracellular conditions: to be stored, oxidized, or converted into cell structure. For now, we ignore the third fate, because we are focusing on the generation of energy in respiration. A cell with overabundant glucose can store some as starch (in plants) or as glycogen (in animals). Glycogen granules are visible in some animal cells, such as muscle tissue that has been fed well but not worked very hard. When runners "carbo load" before a race, they are building up a supply of glycogen. Eventually this stored glucose is oxidized, mostly through the pathway of **glycolysis** (*glyc-* = sugar, *-lysis* = splitting) (also called the Embden-Meyerhof pathway) shown in Figure 9.6. Although the details are complicated, the essence of glycolysis is splitting the 6-carbon (C_6) sugar into two 3-carbon (C_3) molecules and oxidizing them, while storing energy in NADH and ATP:

$$C_6\text{ (glucose)} \rightarrow 2\,C_3\text{ (reduced)} \rightarrow 2\,C_3\text{ (oxidized)};\quad 2\,NAD^{\oplus} \rightarrow 2\,NADH$$

As a result, each glucose molecule becomes oxidized to two molecules of pyruvate (pyruvic acid):

$$CH_3-C(=O)-COO^{\ominus}$$

Although glucose has a lot of energy, in the first and third steps of the pathway, it is activated even more with ATP—not just once, but twice. Here a cell invests some energy at the beginning with the promise of getting out more energy later. The whole phosphorylated molecule, which we can represent as Ⓟ–C_6–Ⓟ, is split in step 4 (and half of it is rearranged in step 5) to make two molecules of C_3–Ⓟ, glyceraldehyde 3-phosphate. Each of these molecules is oxidized in step 6 by NAD^+, while a molecule of inorganic phosphate is added to each carbon chain to make 1,3-bisphosphoglyceric acid (Ⓟ–C_3–Ⓟ). In reducing NAD^+ to $NADH + H^+$, some energy from the substrate has been stored.

1,3-bisphosphoglyceric acid has such a high phosphoryl transfer potential that it can transfer one phosphoryl group to ADP, thus making ATP directly (step 7). This mode of ATP synthesis in glycolysis is called *substrate-level phosphorylation* to distinguish it from the very different process of ATP synthesis in the mitochondria (explained in Section 9.9). Since the cell initially invested two ATPs to activate the glucose and one ATP is recovered from each half-glucose, the cell has recovered its investment. Yet the remaining phosphoglyceric acid still has a great deal of energy. It is rearranged in step 8 to make phospho*enol*-pyruvate (PEP), another compound with a high phosphoryl transfer potential; in step 9, PEP is converted to pyruvate as it phosphorylates another molecule of ADP to ATP.

A cell that carries out only these reactions obtains two extra ATPs for every glucose molecule. The maximum energy that could be obtained from glucose is 686 kcal/mol, so if each ATP could provide 11 kcal, these two ATPs carry only about 3.2 percent ($[2 \times 11]/686$) of the energy of glucose. Still, this is enough to let many cells live perfectly well. The glycolytic pathway also yields two molecules of NADH per glucose, and as we will show later, if the cell is carrying out a complete respiration, each NADH can be used in the ETS to make additional ATP.

Glycolysis finally produces pyruvate, which stands at the crossroads of metabolism; it has several possible fates in different organisms and circumstances. Next we will follow its preparation for the Krebs cycle.

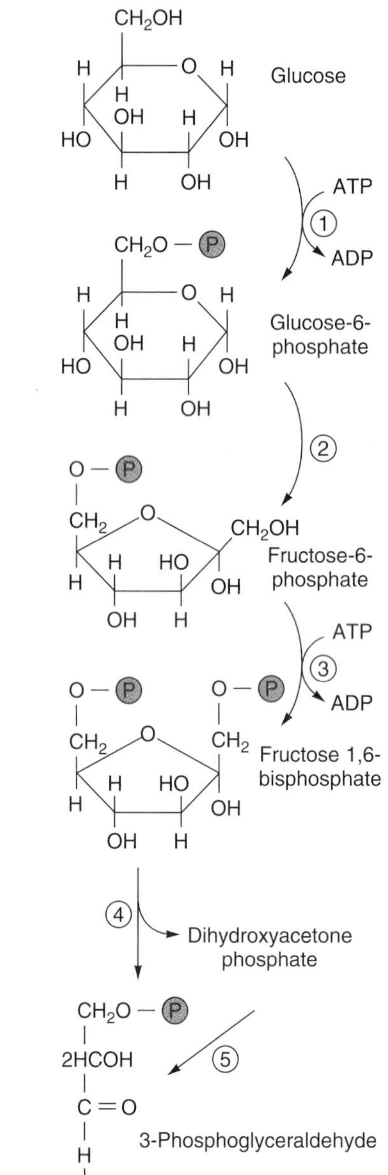

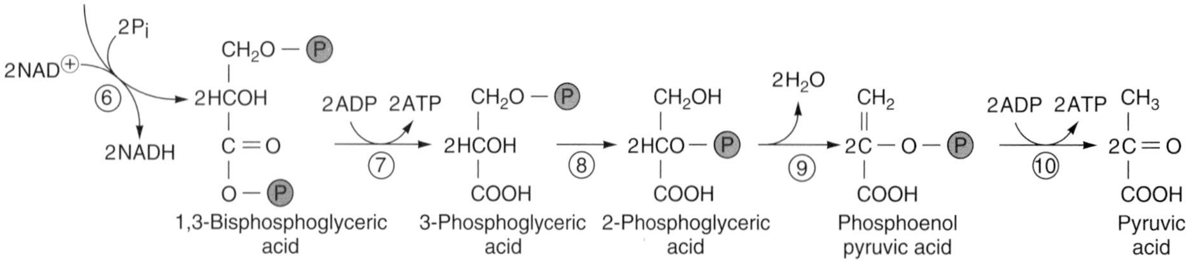

Figure 9.6
In glycolysis, glucose is split into smaller molecules and gradually oxidized as described in the text. At steps 7 and 10, ATP is formed directly.

9.5 Pyruvate is commonly oxidized to an acetyl group.

An interesting and complex enzyme, pyruvate dehydrogenase catalyzes reactions in which the carboxyl group of pyruvate ($-COO^-$) is removed as CO_2, leaving a C_2 acetyl group. (We will see several reactions in which a carboxyl group is removed as a CO_2 molecule; one is nearly as oxidized as the other.) If this reaction occurred outside a cell, the remaining 2-carbon molecule would be acetate (CH_3COO^-), but in metabolism the acetate (minus one oxygen atom) is attached as an acetyl group to a coenzyme called coenzyme A (CoA) to make acetyl-CoA:

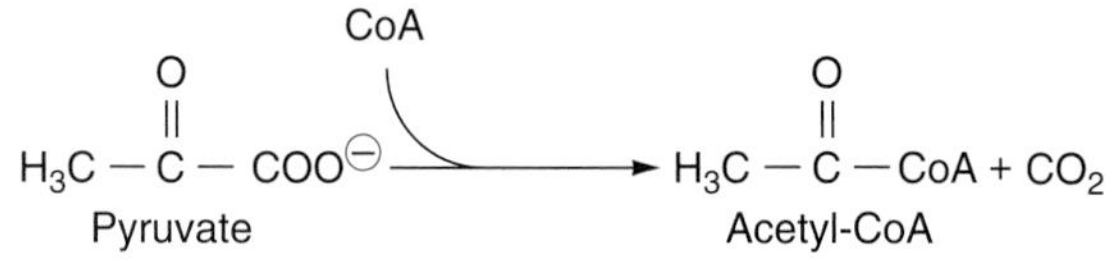

$$\underset{\text{Pyruvate}}{H_3C-\overset{O}{\overset{\|}{C}}-COO^{\ominus}} \xrightarrow{\text{CoA}} \underset{\text{Acetyl-CoA}}{H_3C-\overset{O}{\overset{\|}{C}}-CoA} + CO_2$$

Each coenzyme has a particular function in metabolism; a major role of coenzyme A is to carry C_2 groups. Acetyl-CoA has such a high group-transfer potential that it can easily transfer the acetyl group to other molecules, thus building up larger molecules from this C_2 fragment. In the main path of respiration, the acetyl group will go into the Krebs cycle.

Pyruvate dehydrogenase (discussed in Sidebar 9.1) requires four coenzymes, which animals obtain as vitamins.

Sidebar 9.1 A Protein Assembly Line

The pyruvate dehydrogenase that converts pyruvic acid to acetyl-CoA is actually a beautiful complex of three kinds of enzymes that carries out a sequence of reactions, like an assembly line:

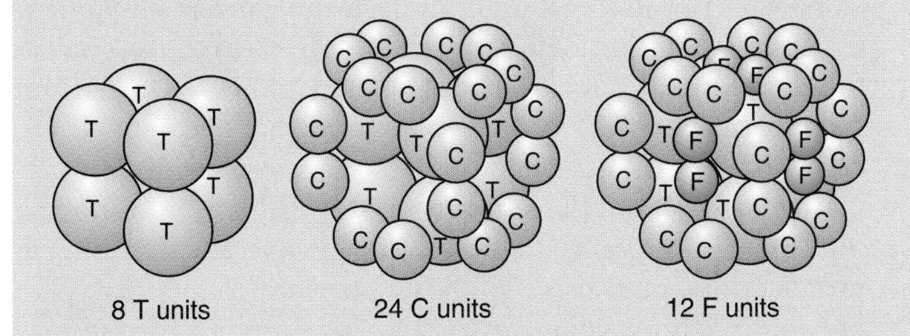

The complex, with a total mass of 4.6 million daltons, consists of 60 subunits of the three enzymes, which assemble themselves into a symmetrical crystal. The reactions themselves are shown as:

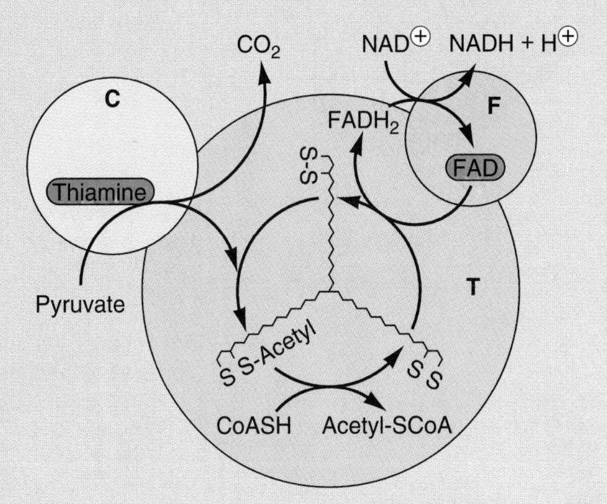

Pyruvate dehydrogenase works with four coenzymes, including some that are vitamins or are made from vitamins: thiamine (vitamin B_1); pantothenic acid, which is built into coenzyme A (CoA):

Sulfhydryl group
P
P
SH
Adenosine
Long chain of C and N atoms, including pantothenic acid

and riboflavin (vitamin B_2), a part of flavin-adenine dinucleotide (FAD). Once CO_2 has been removed from pyruvate, the remaining C_2 fragment is attached to lipoic acid, a long molecule with two sulfur atoms on one end; it swings the C_2 fragment from active site to another on the enzyme complex, like a conveyer belt. Finally, this 2-carbon fragment, now an acetyl group, combines with coenzyme A to make acetyl-CoA.

Since the enzyme plays such a key role in metabolism, it is obviously important to eat foods that supply enough of these vitamins every day.

9.6 Mitochondria are the principal sites of respiration in eucaryotes.

Glycolysis occurs in the cytosol in both procaryotic and eucaryotic cells. Procaryotic cells have very little compartmentalization of metabolism, except that some systems are located in membranes. In eucaryotic cells, however, the next phases of metabolism—pyruvate oxidation, the Krebs cycle, and electron transport—are confined to the mitochondria (Figure 9.7). Mitochondria are easily recognized in electron micrographs as elongated bodies of bacterial size—that is, about 0.5 μm wide and a few micrometers long. A mitochondrion is made of two membranes; the outer one is simple and quite permeable, but the inner one restricts the passage of materials and has extensive folds known as **cristae.** The composition of the **matrix space** inside the inner membrane is very different from the cytosol, since the inner membrane contains several transport proteins that regulate the passage of molecules such as ATP, ADP, and the intermediary metabolites of the Krebs cycle. All the Krebs-cycle enzymes are confined to the matrix and inner membrane, and the proteins involved in electron transport are built into the inner membrane.

9.7 The Krebs cycle is the core of metabolism.

Cycles of reactions are critical to metabolism. Generally, compound X is combined with A to make B. Then B is converted to C, C to D, and so on through a series of reactions that eventually yields a by-product, Y, and a new molecule of A:

X
A
B
Y
C

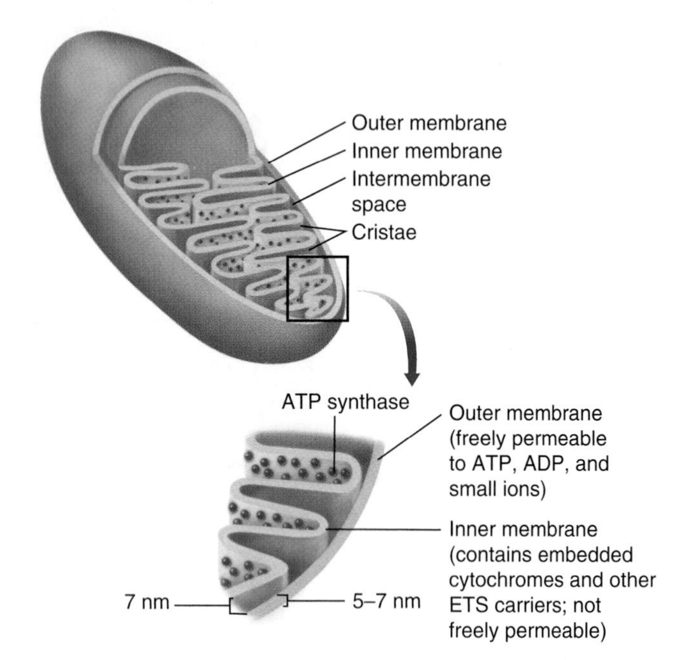

Figure 9.7

A mitochondrion has two membranes. The inner one, which has extensive folds called cristae, contains some enzymes of the Krebs cycle and components of the ETS. One of its faces is rich in enzyme complexes, the ATP synthase that forms ATP from ADP and P_i (small knobs). Other Krebs-cycle enzymes are confined to the matrix space.

Thus the whole system is really a kind of large-scale catalyst for converting X into Y. In the Krebs cycle, the C_2 acetyl group is combined with a C_4 molecule to make a C_6 molecule:

C_2 → C_4 + C_2 → C_6 → CO_2 + C_5 → CO_2 + C_4 → (back to C_4)

Two molecules of CO_2 are then removed, leaving a new C_4 molecule that can start the cycle all over again. So the cycle oxidizes the C_2 unit completely to two CO_2 molecules. Notice the fate of the six carbon atoms from the original glucose molecule: They were divided into two C_3 molecules; then one carbon atom from each of these was lost as CO_2 in converting pyruvate to an acetyl group; and now each acetyl group has been converted into two more CO_2 molecules. In this way, the glucose is completely oxidized to six CO_2 molecules.

This overview shows that the Krebs cycle is fundamentally quite simple. Now let's look at it more closely. As Figure 9.8 shows, the C_2 acetyl group of acetyl-CoA is linked to the C_4 molecule oxaloacetate to form the C_6 molecule citrate (citric acid). In the first few steps of the cycle, two carboxyl groups are removed as CO_2 molecules; the second of these reactions, in which α-ketoglutarate is converted to succinyl-CoA, is very similar to the reaction in which pyruvate is converted to acetyl-CoA. Another few steps restore the oxaloacetate, which combines with a new acetyl group to start the cycle again.

The metabolites in the cycle are oxidized at several points. Three reactions are dehydrogenations in which NAD^+ is reduced to $NADH + H^+$. Notice how neatly an oxidation is accomplished by adding a water molecule to convert fumarate to malate, and then removing the two hydrogens to convert malate to oxaloacetate. Each NADH can be used in the ETS to generate 2.5 molecules of ATP, so the three steps that generate NADH can yield 7.5 ATPs. At one point, succinate is oxidized to fumarate by FAD, and the reduced $FADH_2$ feeds electrons directly into the ETS to make 1.5 ATPs. Finally, one reaction directly produces one nucleoside triphosphate through substrate-level phosphorylation: guanosine triphosphate (GTP) in animals and ATP in plants. These are energetically equivalent, since the structure of the base doesn't affect the energy carried by the molecule.

Exercise 9.1 Complete oxidation of glucose to 6 CO_2 would yield 686 kcal/mol. Count all the ATPs that can be made in oxidizing glucose through glycolysis and the Krebs cycle, and determine what percentage of the energy of glucose is saved, using 11 kcal/mol as the energy transferred by one ATP. Don't forget the NADH made when pyruvate is converted to acetyl-CoA.

9.8 The electron transport system synthesizes ATP.

As a cell carries out oxidative metabolism, it accumulates NADH and $FADH_2$, whose energy can be used to convert ADP into ATP in the process of **oxidative phosphorylation.** The ETS, which carries out this final stage of respiration, is located in the inner membrane of mitochondria in eucaryotes and in the plasma membrane in procaryotes. We'll concentrate on mitochondria here to simplify the description.

The term "electron transport system" may be a bit confusing. When the ETS accepts hydrogen atoms from NADH and $FADH_2$, oxidizing them back to NAD^+ and FAD, it accepts both protons and electrons. Why, then, do we focus on the electrons alone? What happens to the protons? We will show that most components of the respiratory chain carry only the electrons, but some components operate vectorially by transferring the protons to one side of the membrane. In this way, the ETS creates a proton gradient, an electrochemical gradient that has enough energy to drive ATP synthesis. Finally, at the end, protons and electrons come back together and combine with oxygen to form water:

NADH → $NAD^{\oplus}$ + H; H → Protons → $H^{\oplus}$ gradient → ATP; H → Electrons → Respiratory ETS; O_2 → H_2O

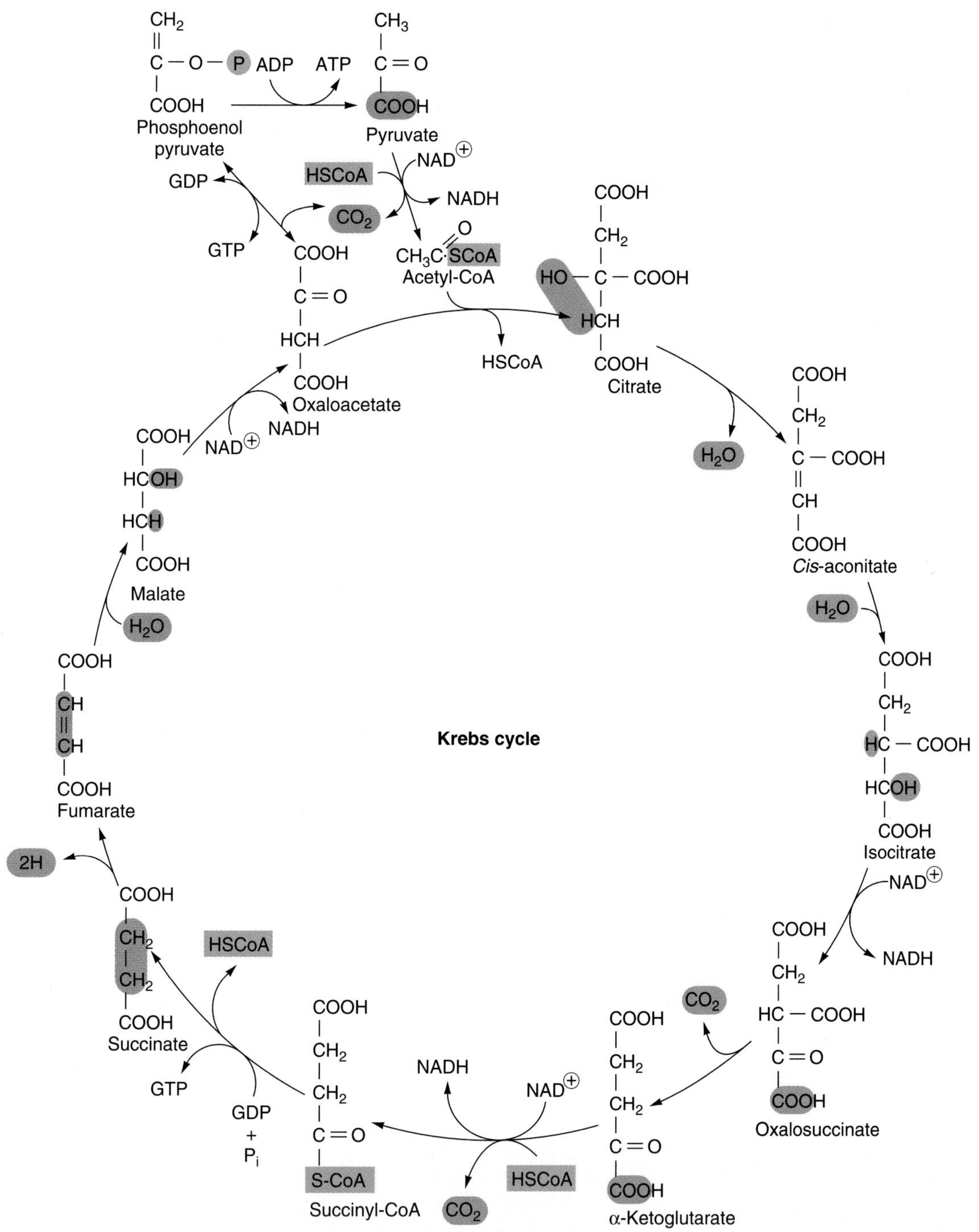

Figure 9.8

The Krebs cycle and some closely related reactions are the heart of metabolism. The cycle takes in a C_2 unit at the top (the acetyl group of acetyl-CoA), combines it with a C_4 molecule (oxaloacetate) to make a C_6 molecule (citrate), and then releases two molecules of CO_2 to regenerate another C_4 molecule. Several oxidation steps produce NADH or $FADH_2$, which may then be used to generate ATP, as explained later.

An ETS operates like a bucket brigade, a chain of people who pass buckets of water from hand to hand to put out a fire (Figure 9.9*a*). If the buckets are hydrogen atoms or electrons, the bucket carriers are "reduced" while they are holding a bucket and become "oxidized" as soon as they pass it on. Each person alternates rapidly between the "reduced" and "oxidized" states, just as the carriers of the ETS do (Figure 9.9*b*). But as the electrons pass from one carrier to the next, they are running downhill energetically—gradually losing energy. Much of their energy is used to build up the proton gradient, and much of the energy in this gradient will then be stored in the form of ATP.

The components of an electron transport system (described in Concepts 9.2) form four protein complexes that are embedded in the inner mitochondrial membrane like a mosaic (Figure 9.10), but are free to move laterally because membranes are so fluid. Electrons, donated primarily by NADH and secondarily by $FADH_2$, pass from one complex to another via mobile carriers (ubiquinone and cytochrome *c*) and finally to Complex IV, which becomes oxidized again by reducing oxygen to water:

NADH → Complex I → CoQ → Complex III → Cytochrome *c* → Complex IV → O_2

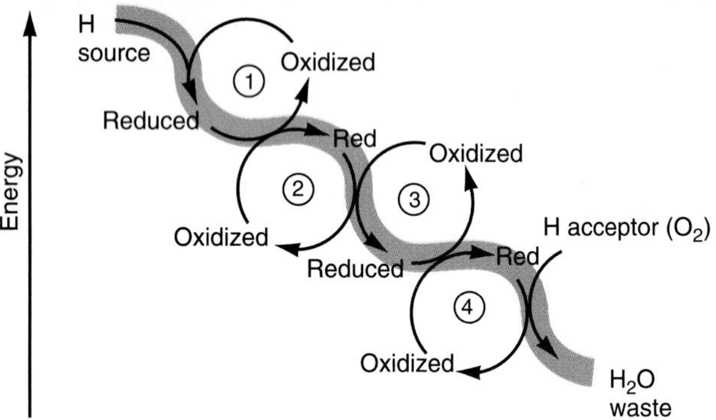

Figure 9.9

An electron transport system operates like a bucket brigade. Each electron carrier is alternately reduced and oxidized, as if the electron were a bucket being passed from person to person.

The earlier stages of respiration depend on this final step. Without oxygen, the whole system backs up. The ETS stays reduced, so it can't oxidize NADH back to NAD^+, thus stopping the Krebs cycle, which depends upon NAD^+. Of course, ATP synthesis through oxidative phosphorylation stops. So it is easy to see how the lack of oxygen can lead to a lack of energy and even to the death of an organism in a short time.

Next we will see how the energy in a proton gradient is converted into the energy of ATP.

Exercise 9.2 Cyanide and carbon monoxide both inhibit the activity of Complex IV. Explain why they are poisonous. (CO also interferes with hemoglobin, but that's another story.)

Figure 9.10

The components of an ETS can pass electrons from one to another because of their arrangement in the membrane. The complexes are made of the cytochromes and other components shown in Concepts 9.2. NADH + H^+ reduces Complex I; an additional complex (succinate-Q reductase) carries out the reduction of succinate to fumarate shown in the Krebs cycle, with the reduction of FAD to $FADH_2$. Complex III accepts electrons from both these two complexes via ubiquinone (UQ and UQH_2 in its oxidized and reduced forms), which is mobile in the lipids of the membrane. Complex III passes electrons via the mobile cytochrome *c* to Complex IV, which finally combines the protons and electrons with oxygen to make water. The Complexes I, III, and IV each transport three or four protons across the membrane for every pair of electrons they carry.

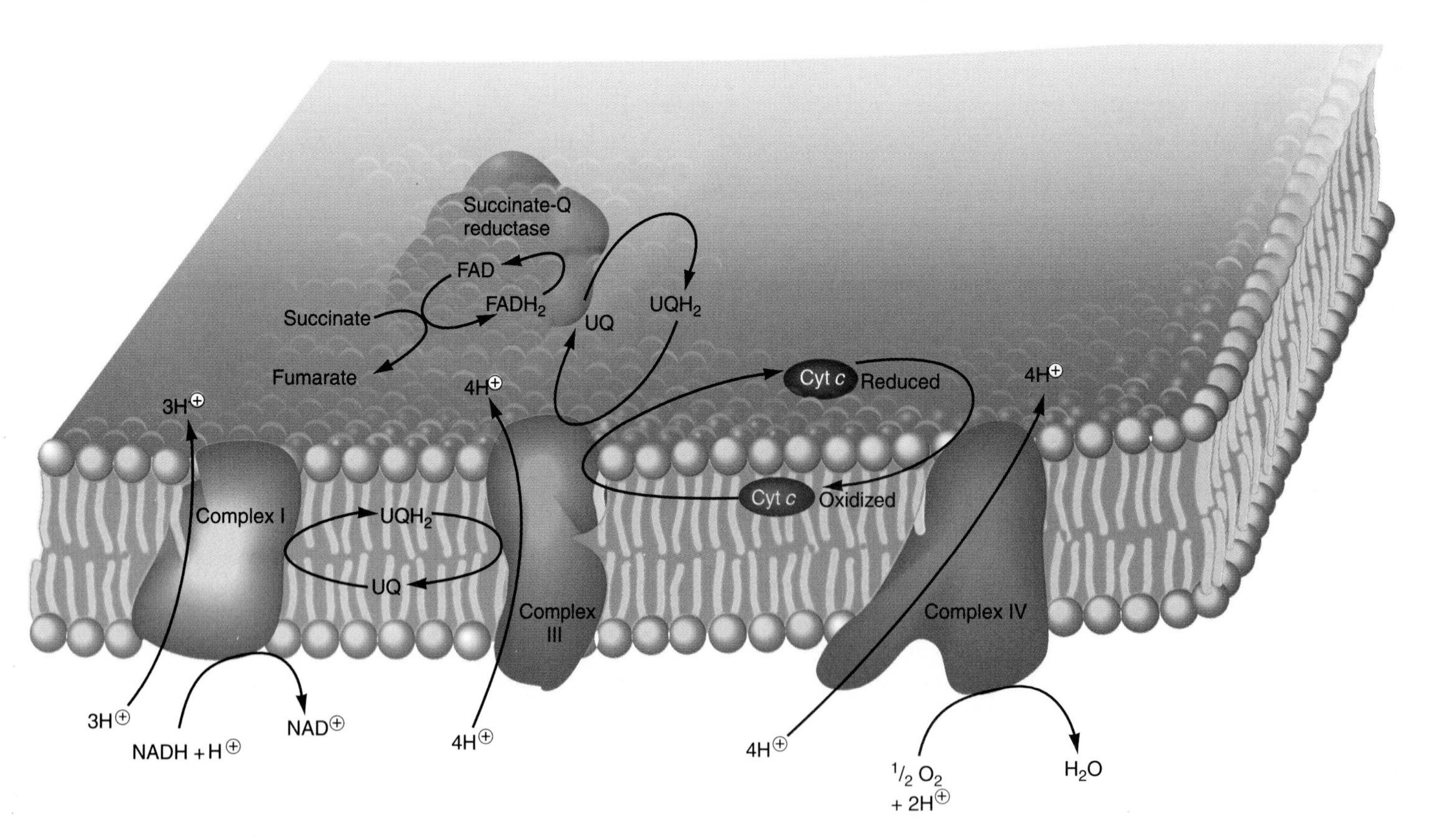

Concepts 9.2 Components of the ETS

1. **Quinones** are lipid-soluble, single- or double-ring molecules with two oxygen atoms that can be reduced and oxidized reversibly. Thus a quinone can accept a pair of hydrogen atoms and then pass them on to the next carrier. Whereas most ETS components only carry electrons, a quinone carries both protons and electrons, so it can shuttle protons to make the proton gradient across the membrane.
2. **Iron-sulfur proteins** are rather small proteins with a core of interconnected iron and sulfur atoms. The best known are *ferredoxins,* whose name reminds us that they contain iron (*ferrum* = iron) and conduct redox reactions as an iron atom is reversibly oxidized to Fe^{3+} and reduced to Fe^{2+}.
3. **Flavoproteins,** such as the one in the pyruvate dehydrogenase complex, carry a flavin prosthetic group, usually FAD. Because they also employ a heavy metal (molybdenum, manganese, iron, or copper), these elements are critical micronutrients.
4. The **cytochromes** are a large, varied group of proteins built around a heme group (see Concepts 4.2) whose iron atom can be reversibly oxidized and reduced, as in an iron-sulfur protein. A chain of cytochromes can pass an electron from one to the other. The numbered cytochromes in this chain represent cytochromes in general; when specific types are known, cytochromes are designated *a, b, c,* or *d* with subscript numbers and appear in diagrams as Cyt *a,* Cyt b_6, and so on.

Quionone

Iron-sulfer protein core

Chain of cytochromes showing the path of the electrons

9.9 A proton gradient across a membrane can be used to synthesize ATP.

The step-by-step transfer of electrons by an ETS leads to the synthesis of ATP by a remarkable mechanism that conserves some energy from the oxidation of organic molecules. Electron microscopy shows that the inner mitochondrial membrane has distinctive little knobs along its inside face; these are proteins that form an *ATP synthase,* which combines ADP with inorganic phosphate to make ATP. Since this is an endergonic process, it must be driven by a source of energy. The ETS provides that energy.

The work of Peter Mitchell explained how an ETS generates ATP. After biochemists with conventional ideas had failed for years to find compounds involved in ATP synthesis, Mitchell developed the radically different theory of **chemiosmotic coupling,** in which ATP is synthesized by a membrane system that creates a proton gradient. Every ETS is built into a closed membrane—the inner membrane of the mitochondrion, the thylakoid membrane of a chloroplast, or a bacterial plasma membrane. The ETS carriers in this membrane operate *vectorially* (see Section 8.15); that is, these carriers are oriented in the membrane so that as they pass electrons from one to another they also create a gradient of protons (H^+ ions) across the membrane (Figure 9.11). Mitchell identified this gradient as a **proton-motive force:** a combination of a chemical potential due to the gradient of protons plus an electrical potential (since each proton carries a positive charge). The proton-motive force can then do work, such as synthesizing ATP.

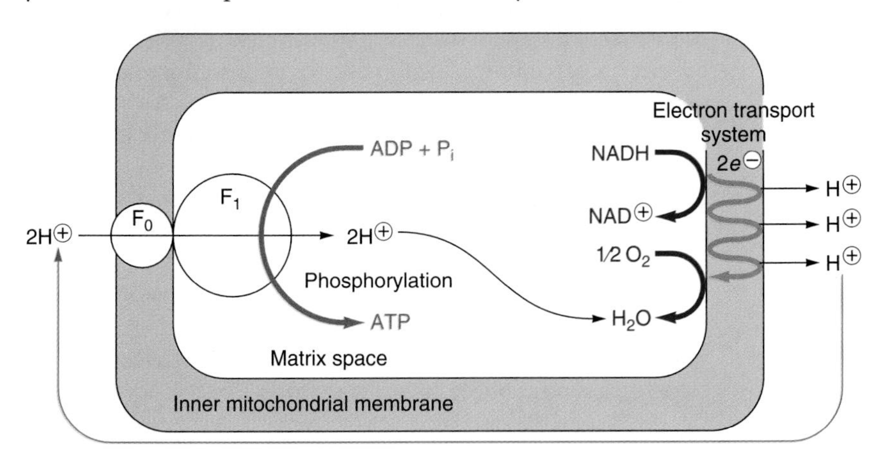

Figure 9.11

The arrangement of electron carriers, and their action, creates a high concentration of protons on one side of the membrane. The protons can only get back across the membrane through the mitochondrial ATP synthase system, which forms one molecule of ATP for every three protons that pass through it.

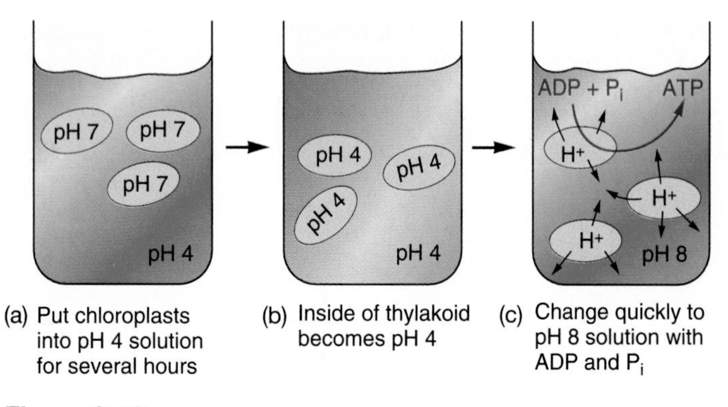

Figure 9.12

Jagendorf's experiment demonstrated clearly that Mitchell's hypothesis for ATP generation is correct. *(a)* Chloroplasts are put in a pH 4 solution for several hours. *(b)* After a high concentration of protons has accumulated inside the thylakoid membrane, the chloroplast generates ATP while these protons diffuse across the membrane *(c)*.

Mitchell's hypothesis is now well confirmed. Chloroplasts and mitochondria synthesize ATP through the same chemiosmotic mechanism, and Andre Jagendorf confirmed Mitchell's hypothesis with a remarkably simple, elegant experiment (Figure 9.12). He soaked chloroplasts in a solution at pH 4 for several hours, giving the space inside the thylakoid membrane a high concentration of protons. Then he transferred the chloroplasts to a pH 8 solution and observed a burst of ATP synthesis as protons moved down their concentration gradient from the inside to the outside of the membrane.

In mitochondria, each complex pumps three or four protons into the space between the inner and outer membranes for every pair of electrons it transports. So as electrons gradually give up their energy at each step in the ETS, falling to lower and lower energy levels, this energy is stored in the proton gradient across the inner membrane. The energy in this gradient is captured by letting the protons cross the membrane through the ATP synthases (see Figure 9.11), which carry out the reaction:

$$ADP + H_3PO_4 + \{H^+\}_{Outside} \rightarrow ATP + H_2O + \{H^+\}_{Inside}$$

The details of ATP synthesis are complicated and the subject of a lot of current research. The best estimates are that 2.5 ATPs are produced per proton/electron pair coming from NADH and 1.5 ATPs are produced per proton/electron pair coming from $FADH_2$.

9.10 A proton gradient itself can do work.

While the energy of the proton gradient can be stored in the form of ATP, it can also be used directly—for instance, to drive a symport mechanism just as a sodium-ion gradient does. The galactoside permease system of some bacteria is a proton-lactose symporter; it uses the energy of the proton gradient for active transport of the sugar lactose (Figure 9.13). The same proton gradient also supplies the motive force for bacterial flagella, which are protein rods that rotate within an anchoring ring in the membrane. Apparently the proteins that turn the flagellum shift back and forth between two conformations; they are activated as they combine with protons and are deactivated as they use their energy to pull the flagellum a fraction of a turn. Thus the flagellum turns much the way a group of people can make a children's merry-go-round turn—by spacing themselves around it so each one can repeatedly pull on a bar of the merry-go-round, contributing a bit of energy and maintaining its speed. Now imagine that each person is "activated" by binding a proton and that the ion is released as he or she pulls. The person will be activated, expend the energy by pulling, and then be activated again. This is roughly how the proton gradient makes the flagellum rotate.

* * *

Before continuing to the next section, let's pause and take stock. We have followed the atoms that were originally in a glucose molecule through their stages of oxidation: glycolysis to pyruvate; oxidation of pyruvate to acetyl-CoA; oxidation of each acetyl group to 2 CO_2 in the Krebs cycle, all producing NADH + H^+ as a by-product; and finally, oxidation of the NADH in the ETS, yielding ATP and, as a by-product, water.

This process should be clear to you, and if it is not, this is a good place to stop and review it. Next, we will enlarge on it. At two important points, the atoms moving through these pathways could take a detour, either into fermentation or into lipid synthesis. There are two additional considerations: Some organisms get their energy from metabolites other than carbohydrates and oxygen, and organic compounds such as amino acids can also be catabolized.

9.11 Many organisms obtain energy through fermentation.

If an organism metabolizes glucose completely to CO_2 and water, it is able to store 50–60 percent of the energy of the sugar in the form of ATP. To do this, it must be living in an **aerobic** environment (that is, with oxygen) and must use oxygen as a terminal electron acceptor. Many organisms, however, are adapted to **anaerobic** environments (without oxygen), and without oxygen they can't carry out a complete respiration as described here. One alternative is **fermentation.** Most fermentations use glycolysis, and even though glycolysis does not recover much of the energy of glucose, many organisms can use it to grow perfectly well because it produces both ATP and NADH. Remember, though, that it is essential for the NADH pool to be oxidized back to NAD^+, so metabolism can continue; in respiration, this is accomplished by using oxygen as a terminal electron acceptor, but how can it be done without oxygen? The

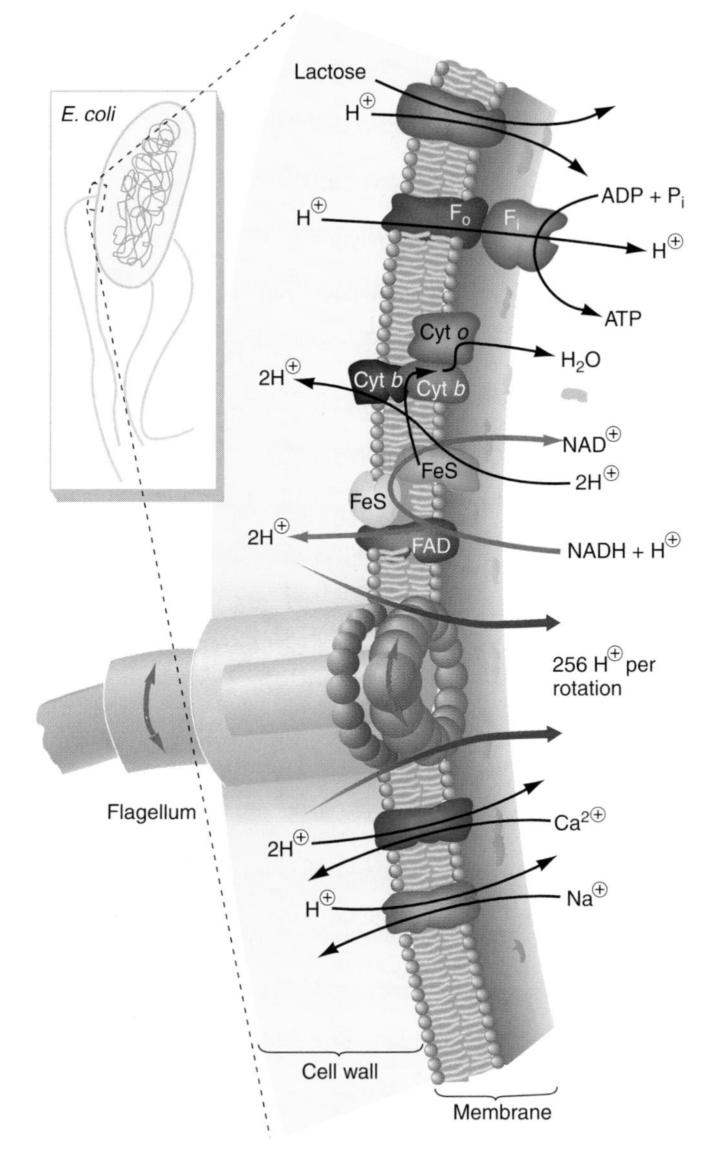

Figure 9.13

The electron transport system of the bacterium *E. coli* is similar to that of a mitochondrion and is arranged so its carriers expel protons from the cell as they are reduced and oxidized. The proton gradient may be used to make ATP through the ATP synthase, but it may also drive other systems, such as the permease that transports lactose and the mechanism driving the bacterial flagellum. The flagellum rotates in a ring of sixteen identical proteins, which are activated by protons. If each protein must be activated by one proton to rotate the flagellum one-sixteenth of a turn, 256 protons are required per revolution.

answer is that in fermentation an organic compound—most often pyruvate—replaces oxygen. In fact, we can now define respiration as oxidative metabolism in which the final oxidizing agent is inorganic, usually oxygen, whereas in fermentation the final oxidizing agent is an organic compound. In most fermentations, NADH + H^+ is oxidized to NAD^+ by reducing pyruvate to an end product, as shown in Figure 9.14. Some yeasts, for instance, convert the pyruvate to CO_2 and acetaldehyde, then reduce the acetaldehyde to ethanol (ethyl alcohol):

$$\underset{\textbf{Pyruvate}}{CH_3\overset{O}{\overset{\|}{C}} - COO^{\ominus}} \longrightarrow \underset{\textbf{Acetaldehyde}}{CO_2 + CH_3CHO} \xrightarrow{NADH + H^{\oplus} \;\to\; NAD^{\oplus}} \underset{\textbf{Ethanol}}{CH_3CH_2OH}$$

This blessed process is the basis for baking bread, since carbon dioxide is what makes dough rise (Figure 9.15). It is also the basis for making beer, wine, and other alcoholic drinks. Thus a humble organism provides us with food and drink just by growing under adverse conditions.

Other fermentations produce various by-products that we find either delectable or detestable, depending on our tastes and cultures. Some of them produce the bitter taste of spoiled milk, but we value others for the flavors they give to cheeses. For instance, propionic acid provides the distinctive taste of Swiss cheese, and the CO_2 produced in the fermentation makes the big holes in it. One of the simplest and most common fermentations produces lactate simply by reducing pyruvate directly:

$$\underset{\textbf{Pyruvate}}{\begin{array}{c} CH_3 \\ | \\ C{=}O \\ | \\ COO^{\ominus} \end{array}} \xrightarrow{NADH + H^{\oplus} \;\to\; NAD^{\oplus}} \underset{\textbf{Lactate}}{\begin{array}{c} CH_3 \\ | \\ H-C-OH \\ | \\ COO^{\ominus} \end{array}}$$

Lactate provides the sharp flavor of sauerkraut, sour cream, yogurt, authentic dill pickles (not the common commercial kind made with vinegar, which is acetic acid), and other delights. (Notice that lactate, a C_3 acidic compound, is quite different from lactose, a double sugar made of galactose and glucose.)

An important part of our economy depends on fermentation. Industries have cultivated many strains of bacteria and fungi for their ability to make the by-products of various fermentations, since this is often a far cheaper, more efficient way to produce these materials than purely chemical synthesis would be.

Animals depend on the lactate fermentation as a backup to respiration. Our cells use glucose as their main energy source, and when we are breathing normally and not working too hard, we oxidize glucose completely. A short burst of work will use up the ATP and some other energy reserves stored in the muscles, but these can be restored once the muscles are at rest again. Sustained exercise, on the other hand, makes different demands. Glucose cannot be oxidized completely if the lungs and circulatory system can't supply oxygen fast enough. Furthermore, the respiratory apparatus works rather slowly compared to glycolysis. Glycolysis, ending in the reduction of pyruvate to lactate, can supply the needed ATP quickly enough to sustain exercise for some time, even without enough oxygen to supply the muscles completely. In vertebrates, the lactate is carried to the liver where it is converted back to glucose.

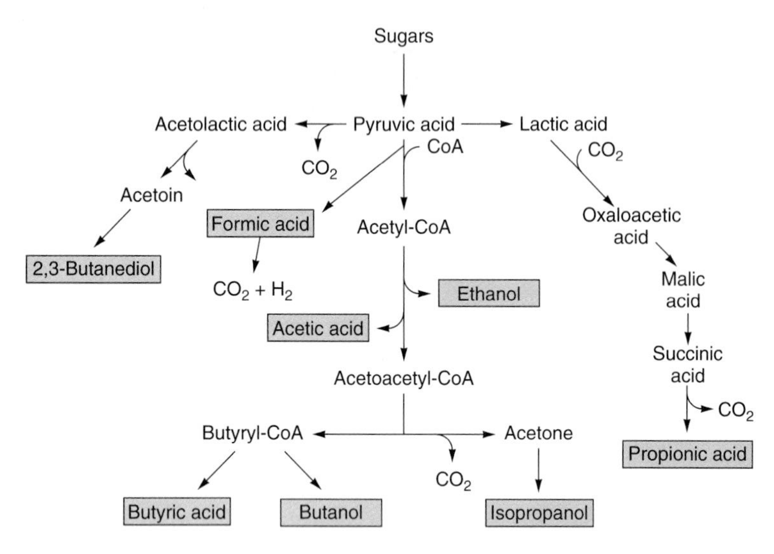

Figure 9.14

The major pathways of fermentation all start with pyruvic acid (pyruvate). Pyruvic acid can be oxidized in a variety of ways, sometimes with the production of CO_2. The oxidized molecule that remains is then used as a hydrogen acceptor for NADH + H^+, thus regenerating NAD^+ so that glycolysis can continue. The end products are excreted from the cell, so materials such as ethanol and various acids accumulate in the medium where fermenting organisms are growing. No organism uses all of these pathways, and most use only a single one.

Figure 9.15

Many important products are synthesized by bacteria and fungi in industrial fermenters, and even baking bread depends on yeasts fermenting sugar to produce the CO_2 that makes the bread rise.

The buildup of lactate in muscles creates the tired, burning feeling associated with muscle fatigue. You can experience this by holding a weight in your hand with your arm extended straight out in front of you. If you hold still, the constant contraction of your shoulder muscles will cut off blood circulation, and those muscles will develop that burning sensation. When you lower your arm, you can feel the renewed blood flow wash the acid away.

9.12 Excess sugar can be made into fatty acids.

Sometimes cells, especially animal cells, receive more carbohydrate than they can use immediately—perhaps as a result of taking in too much dessert and getting too little exercise. While excess glucose can be stored as glycogen, an animal's capacity for glycogen storage is limited. Fat is a better storage material anyway, because one gram of fat yields about 9 kcal (37.5 J) when oxidized, in contrast to about 4 kcal/gram (16.7 J/gram) for carbohydrate. So, primarily in liver cells and fat (adipose) cells, sugars are oxidized to pyruvate, then to acetyl-CoA, and the acetyl groups are put together end-to-end to make fatty acids, as shown in Figure 9.16. The acetyl units are reduced by NADPH as they are added, storing considerable energy in each fatty acid.

Since fatty acids are made of 2-carbon units, the resulting molecules have an even number of carbon atoms, commonly 16 or 18. Fat synthesis is not limited to animals, of course. Many algae store food materials as oils, and seeds often store considerable amounts of fat as they ripen; these become our sources of olive oil, corn oil, and other cooking oils.

Cells retrieve the energy stored in fat through essentially the reversal of the synthetic reactions. These reactions cut the fatty acids back into acetyl-CoA molecules that can be fed into the Krebs cycle. The pathways of synthesis and oxidation are not exact opposites. They use different coenzymes and occur in different places—synthesis on the endoplasmic reticulum and oxidation in the mitochondrial matrix. Using slightly different paths for the forward and reverse directions of a process is a standard strategy in cells, allowing each process to be controlled independently.

9.13 Some organisms use other types of respiration and inorganic energy sources.

A truly general biological viewpoint recognizes several types of respiration that use different electron donors and acceptors. Even though oxygen is the most common terminal electron acceptor, many kinds of bacteria use nitrate instead, reducing it to nitrite. A few bacteria, such as *Desulfovibrio,* even employ a sulfate respiration in which they reduce sulfate to sulfide. Since respiration is defined as a process in which the terminal oxidizing agent is inorganic, using nitrate and sulfate falls within this definition. None of these organisms are prominent in ecosystems, because they don't get as much energy from their food as do organisms that use oxygen. The energy that can be obtained from oxidation depends on the potential difference between the electron donor and acceptor. If electrons from NADH + H^+, with a potential of -0.320 volts, are used to reduce oxygen to water, at a potential of 0.816 volts, they can give up $0.816 - (-0.320)$

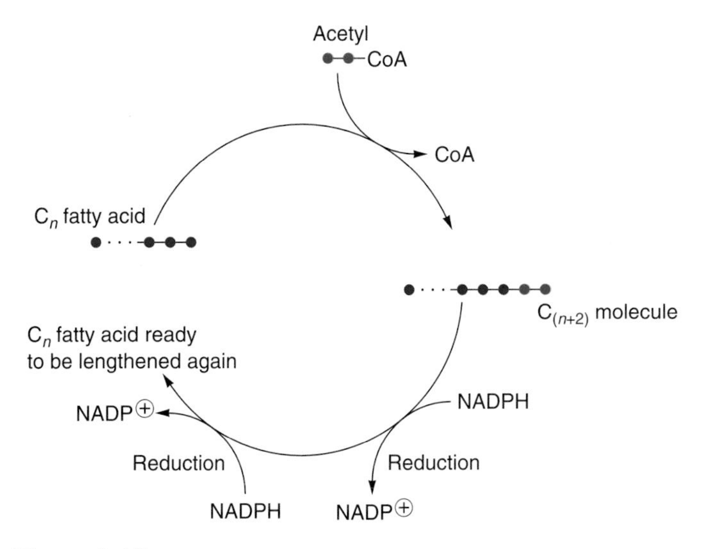

Figure 9.16

Fatty acids are synthesized from acetyl-CoA. A fatty acid made of *n* carbon atoms *(blue dots)* is made two carbon atoms longer *(green dots)* by the addition of a C_2 (acetyl) unit transferred from coenzyme A. The added acetyl group is then reduced by NADPH, and the resulting fatty acid is ready to receive another acetyl group.

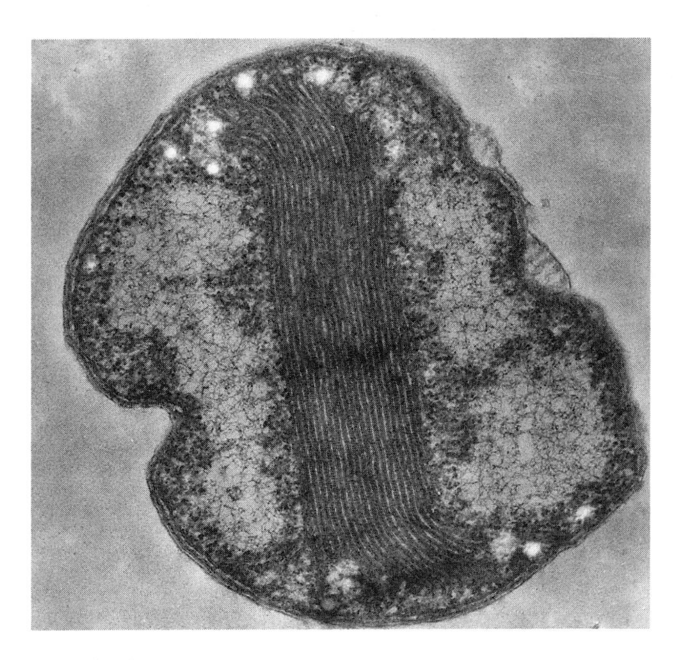

Figure 9.17

The bacterium *Nitrocystis gracilis* has extensive internal membranes in which respiration and ATP synthesis occur.

= 1.136 electron-volts of energy that can be turned into ATP. In contrast, if the electrons are used to reduce nitrate to nitrite, at only 0.421 volts, they give up only 0.841 electron-volts and can be used to make fewer ATPs.

It is tempting to ask, "If oxygen is so much better as a terminal electron acceptor, why do these bacteria use nitrate or sulfate?" That question contains a trace of teleological thinking, implying that an organism could survey the chemical world, choose a good strategy for survival, and evolve in that direction. Organisms are, indeed, opportunists. They use whatever resources are available, and it is better to grow inefficiently than not to grow at all. However, organisms can only evolve to the extent allowed by random, unpredictable genetic events. They cannot follow one strategy of growth and reproduction and then willy-nilly choose an alternative, no matter how advantageous it might be. Species that use unusual metabolic processes have found their own niches in ecosystems. There is no selective pressure to make them change, but a good deal of conservative selective pressure to keep them as they are.

Respiration merely uses oxygen, nitrate, or sulfate as sinks to remove spent electrons. In contrast, bacteria known as **chemoautotrophs** take advantage of a variety of inorganic chemical reactions that *yield* energy, instead of using organic compounds or sunlight as energy sources. They can oxidize inorganic compounds to get ATP and NADH, which are then used to reduce CO_2 to organic compounds, just as photosynthetic organisms do. Those in the genus *Thiobacillus,* for instance, use sulfur compounds, oxidizing H_2S and S to sulfate. *Hydrogenomonas* oxidizes H_2 to H_2O and *Ferrobacillus* oxidizes Fe^{2+} to Fe^{3+}. *Nitrosomonas* oxidizes ammonia to nitrite (NO_2^-), and *Nitrobacter* oxidizes the nitrite further to nitrate (NO_3^-). Chemoautotrophs grow slowly because these reactions don't yield a lot of energy; to increase their energy production, these bacteria develop huge internal membrane structures (Figure 9.17) that provide large amounts of the enzyme systems for generating ATP. (This is another instance of the membrane-metabolism principle established in Chapter 6.) The chemoautotrophs are not as successful as the dominant organisms on this planet, which use the much more energetic organic compounds or tap the energy of sunlight. Still, they have all found niches where they are able to eke out an existence and survive. Those that transform nitrogen and sulfur compounds play important roles in the cycling of materials in ecosystems.

Nitrogen and sulfur cycles, Section 28.25.

9.14 Many compounds are catabolized into the central pathways.

Our food supplies carbohydrates and fats, which we usually oxidize for their energy, but most of its bulk (aside from water) consists of proteins and nucleic acids. We use some of these compounds for their monomers; in fact, our bodies require eight of the twenty amino acids ready-made in food, since we can't make them ourselves. However, much of the protein and other compounds can be catabolized for their energy, and many chemotrophs depend on these organic constituents as energy sources. Figure 9.18 shows how amino acids are catabolized into metabolites in and around the Krebs cycle, using the same general mechanisms used for catabolism of carbohydrates and fats: large molecules split in two, the reduction of NAD^+ to NADH + H^+, and sometimes the formation of acetyl-CoA.

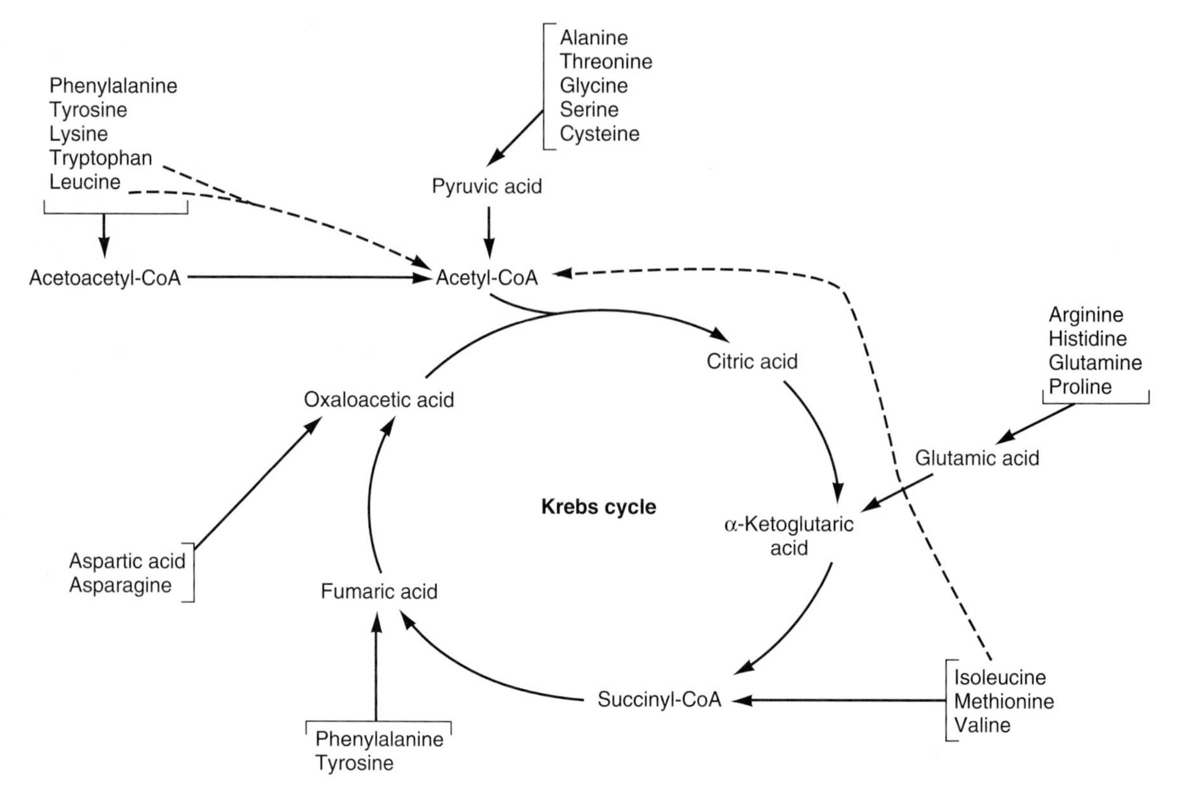

Figure 9.18

The amino acids in food that are not required for protein synthesis can be catabolized into the central metabolic pathways at several points so they can be used for energy.

When amino acids are catabolized, all their amino groups generally cannot be used, because most organisms simply don't need so much nitrogen. Instead, they are removed and eliminated as ammonia, urea, or uric acid.

Excretion of excess nitrogen, Section 46.1.

9.15 Summary: Heterotrophic metabolism consists of five phases.

For perspective, Figure 9.19 summarizes the pathways we have been examining. The picture begins with the digestion of the major food materials—proteins, polysaccharides, and lipids—into smaller molecules. Some macromolecules, such as starch and glycogen, are energy-storage materials. As we saw in Section 4.1, macromolecules are synthesized by removing a water molecule to make each bond; during digestion, they are broken apart again by adding water in the process of **hydrolysis:**

$$H{-}\blacksquare\blacksquare{-}OH + H_2O \rightarrow H{-}\blacksquare{-}OH + O{-}\blacksquare{-}OH$$

These reactions are carried out by **hydrolytic enzymes.**

The products of digestion enter the central metabolic pathways and are converted to a few key compounds: pyruvate, acetyl-CoA, and metabolites of the Krebs cycle. It is worthwhile to run through each phase of metabolism again, noting particularly where materials are oxidized while NAD^+ and FAD are reduced to NADH + H^+ and $FADH_2$. Some NADH is used to generate NADPH, which is reserved as a reducing agent for biosynthesis, but most NADH is used to synthesize ATP. In most cells, the bulk of ATP is synthesized in an electron transport system (ETS), a mosaic of electron carriers that operate vectorially, so they use the energy of reduced compounds such as NADH and $FADH_2$ to generate a gradient of protons. The energy in this gradient may be used to synthesize ATP or to perform work directly. Since NADH and $FADH_2$ reduce the ETS at two different points, each NADH can be used to generate 2.5 ATPs, and each $FADH_2$ only 1.5. With this diagram, one can calculate the energy that various oxidizable substrates can yield.

An ETS is always a closed membranous sac of some kind whose interior is often extensively folded to provide a large functional surface. In eucaryotes, this sac is the inner mitochondrial membrane; in procaryotes, it is the plasma membrane, which is extensively folded inward in some bacteria. In the next chapter, we will see that the electron transport systems used in photosynthesis are very similar to those used in respiration.

Metabolism is so complicated, in part, because a cell is simultaneously carrying out respiration and supplying molecules for biosynthesis. We have focused on respiration. We must reemphasize that many of the molecules moving through these pathways are not oxidized completely but are saved to build the structure of the organism. This is shown in Figure 9.20. Several of the metabolites in the central pathways also serve as precursors of new monomers. For this reason, the Krebs cycle and some pathways that serve it are called *amphibolic* pathways, because they have both catabolic and anabolic functions (*amphi-* = both). So molecules can be removed from the pools of these metabolites for biosynthesis of new monomers, which are then

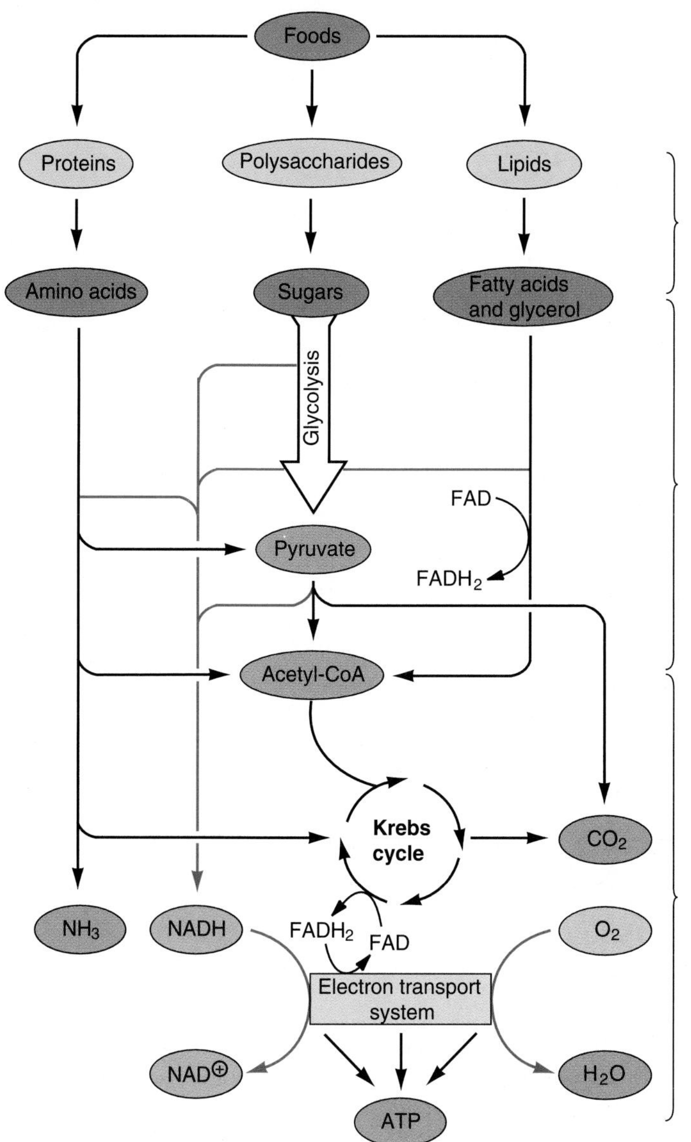

Figure 9.19

Here is a more detailed summary of the first three phases of metabolism as outlined in Figure 9.5. In phase 1, food molecules are hydroylized (digested) to their subunits. In phase 2, we can trace the path of incoming glucose (sugars) as it is partially oxidized through the glycolytic pathway to pyruvate, which is then decarboxylated (by removal of one CO_2 molecule) to acetyl-CoA. Phase 3 entails complete oxidation, in the Krebs cycle, to CO_2. Along the way, FAD and NAD^+ molecules are reduced to $FADH_2$ and NADH + H^+ (shown here as just NADH). These reduced molecules then contribute electrons to the electron transport chain that synthesizes ATP.

polymerized into new cell structure. All this chemical activity is directed toward growth, or at least maintenance of existing structure, and metabolism simultaneously provides energy and raw materials for growth.

Coda Some of our most ancient ancestors, perhaps two billion years ago, were simple cells that evolved the ability to oxidize organic compounds, using glycolysis and similar processes to carry out some kind of fermentation. These primitive cells could not have attained even that degree of organization unless they were genetic systems that had been shaped generation after generation through a gradual editing of their proteins. That editing included their genes, since the structure of each protein is determined by the structure of a particular gene. Little by little, each enzyme in each metabolic pathway became shaped to interact with its substrate and often with coenzymes such as NAD^+ and ATP, so the energy stored in those substrate molecules could be mobilized for cellular processes. All metabolism depends on enzymes that have been shaped by evolution to act upon specific substrates.

Those ancient cells were also members of a primitive ecosystem, one that must have included some phototrophic cells that were able to make organic molecules by using the energy of sunlight, so the ecosystem included a carbon cycle. The fermenting cells may have lived on organic compounds released by the breakdown of other cells. At the same time, some kinds of bacteria were evolving membrane-bound systems for making ATP through a chemiosmotic process, allowing them to use more of the energy stored in the available organic compounds, and it seems likely that some of these bacteria became incorporated into larger fermenting cells and became mitochondria, a story we take up in Chapter 30. By adapting to local conditions at each stage, cells evolved with the metabolic capabilities of modern eucaryotic cells that we have described here, including the cells of plants and animals. In the next chapter, we show how photosynthesis is carried by another kind of organelle, the chloroplast, and by similar bacterial cells.

Figure 9.20

Other metabolic processes, including phases 4 and 5 of Figure 9.5, begin with the central pathways, which should be considered amphibolic because they serve both catabolism and biosynthesis. It is useful to think of synthesis from the central pathways occurring for two distinct functions—energy storage and structural synthesis—though they overlap. Excess foodstuffs, such as glucose or amino acids, can be made into energy stores *(purple arrows),* either fatty acids and neutral fats or starch or glycogen. Structural materials are synthesized *(red arrows),* by converting certain central metabolites into amino acids and other monomers (phase 4), which are then polymerized into structural polymers (phase 5). Finally, under anaerobic conditions, pyruvate can be used for fermentation *(green arrow)* to reoxidize NADH to NAD^+.

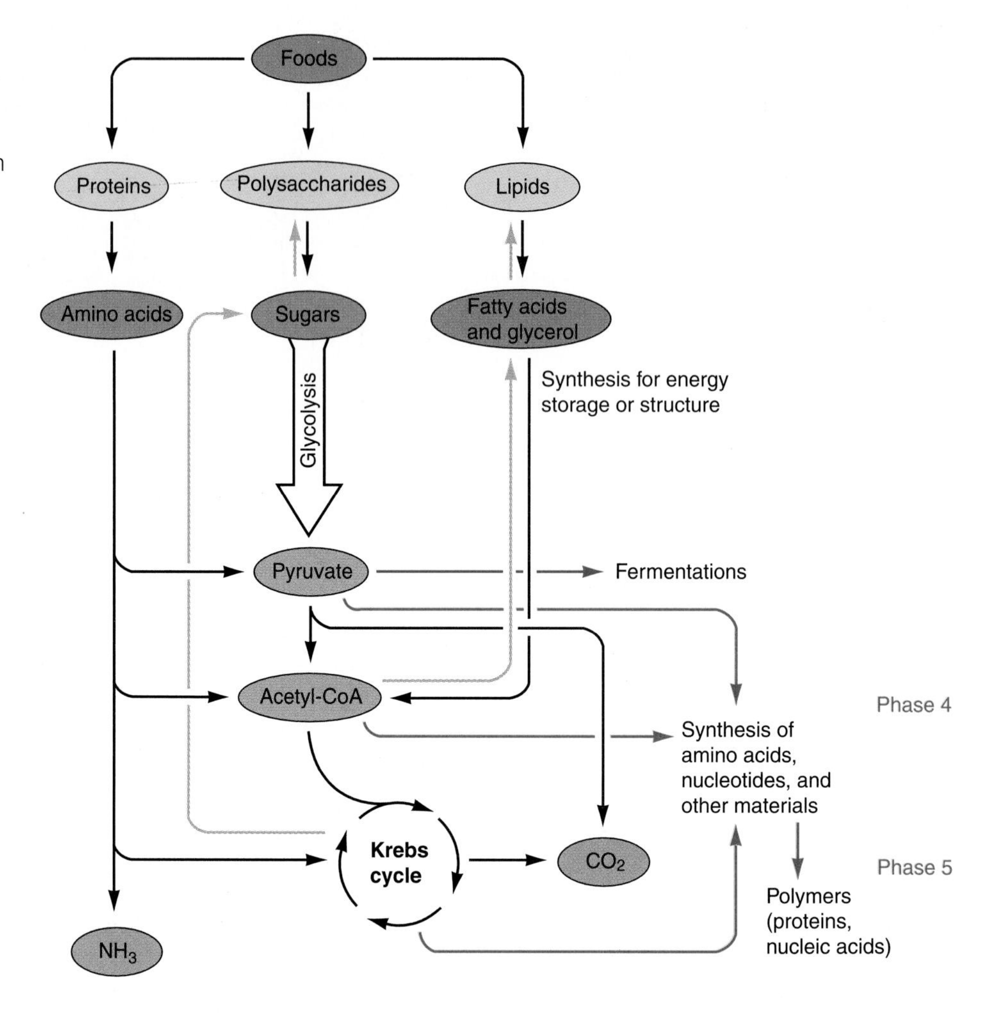

Summary

1. The biosphere operates on a carbon cycle, consisting mostly of two processes: photosynthesis, in which CO_2 is reduced to organic molecules by using light energy, and cellular respiration, in which organic molecules (especially glucose) are oxidized to CO_2.
2. The principal function of cellular respiration is to build up a store of reducing power (NADH) and energy to drive endergonic processes (ATP). To oxidize NADH back to NAD^+, hydrogen from NADH is used to reduce a respiratory electron transport system (ETS), which finally reduces O_2 to H_2O. The protons in this system create a proton gradient across a membrane, which is then used to generate ATP.
3. In the first part of respiration, glucose is entirely oxidized to CO_2, largely by means of a cycle of reactions, the Krebs or citric acid cycle. In the second part, the NADH and $FADH_2$ generated in the first part are used to synthesize ATP in the ETS.
4. In glycolysis, the C_6 sugar glucose is activated with two ATPs. It is then split into two C_3 molecules, each of which is oxidized in several steps to pyruvate. During the oxidation reactions, NAD^+ is reduced to $NADH + H^+$, and four molecules of ATP are made, for a net two ATP per glucose.
5. Pyruvate is commonly oxidized by an enzyme complex to an acetyl group, bound to coenzyme A, while the carboxyl group of pyruvate is removed as CO_2.
6. The further reactions of respiration in eucaryotic cells occur in mitochondria. Each mitochondrion has a relatively permeable outer membrane and an inner membrane that contains many enzymes of the Krebs cycle, the carriers of the electron transport system, and the ATP synthases that convert ADP into ATP.
7. In the Krebs cycle, a C_2 compound (acetyl group) is combined with a C_4 compound (oxaloacetate) to form a C_6 compound (citrate). Then two CO_2 molecules are removed, leaving a C_4 compound that is converted back into oxaloacetate. Meanwhile, the metabolites are oxidized with the production of NADH and $FADH_2$.
8. The respiratory electron transport system is made of complexes of cytochromes and other proteins, as well as quinones and iron-sulfur proteins. These components are reversibly oxidized and reduced as they pass electrons from one to another.
9. The ETS develops a proton gradient across its membrane, a proton-motive force. In the process of chemiosmotic coupling, the protons diffusing down their gradient are directed through ATP synthase enzymes, which convert ADP into ATP and store some of the energy of the gradient, energy that was originally obtained through oxidation of the organic substrates in the first part of respiration.
10. The energy of the proton gradient can be used for other processes, such as the active transport of molecules across a cell membrane or the rotation of bacterial flagella.
11. Organisms living in anaerobic environments, without oxygen, can obtain energy from glucose and other organic molecules through fermentation; instead of using O_2 as a terminal electron acceptor, they use an organic end product, generally pyruvate. The NADH produced during glycolysis is used to reduce pyruvate or a product of pyruvate

to substances such as lactate and ethanol, thus restoring $NADH + H^+$ to NAD^+.

12. If an organism has more carbohydrate than it can oxidize completely through the Krebs cycle, it may store some by synthesizing fatty acids; fat, which yields 9 kcal/gram upon oxidation, is a more efficient storage form than carbohydrate (4 kcal/gram). Fatty acids are made linking the acetyl groups from acetyl-CoA into chains, typically composed of 16–18 carbon atoms.
13. Some bacteria carry out other kinds of respiration in which they reduce nitrate to nitrite or sulfate to sulfide instead of reducing oxygen to water. Other bacteria use inorganic energy sources—such as sulfide, ammonia, iron, and hydrogen—instead of organic sources.
14. Amino acids and other organic substances can be oxidized into the central metabolic pathways using mechanisms similar to those for oxidizing glucose.
15. The central metabolic pathways, especially the Krebs cycle, are called amphibolic because they serve both a catabolic function (oxidizing substrates) and an anabolic function (producing molecules that can be synthesized into amino acids and other monomers for growth).

Key Terms

carbon cycle 177
cellular respiration 177
electron transport system (ETS) 179
Krebs cycle/citric acid cycle 179
terminal electron acceptor 179
glycolysis 179
cristae 181
matrix space 181
oxidative phosphorylation 182
quinones 185
iron-sulfur proteins 185
flavoproteins 185
cytochromes 185
chemiosmotic coupling 185
proton-motive force 185
aerobic 186
anaerobic 186
fermentation 186
chemoautotroph 189
hydrolysis 190
hydrolytic enzyme 190

Multiple-Choice Questions

1. ATP is produced by substrate-level phosphorylation
   a. only during glycolysis.
   b. only during the Krebs cycle.
   c. only during electron transport.
   d. during glycolysis and the Krebs cycle.
   e. during glycolysis, the Krebs cycle, and electron transport.
2. When glucose is catabolized to pyruvic acid, most of the energy of glucose
   a. is stored in ATP.
   b. is stored temporarily in NADH.
   c. is retained in pyruvic acid.
   d. is used to oxidize phosphoglyceraldehyde.
   e. is used to pump protons across the outer mitochondrial membrane.
3. Coenzyme A is not a significant reactant
   a. in the breakdown of glucose to pyruvic acid.
   b. during the Krebs cycle.
   c. during the conversion of pyruvic acid to citric acid.
   d. in the glycolytic pathway and the Krebs cycle.
   e. in the mitochondrion.
4. Carbon dioxide is mostly produced in
   a. the Embden-Myerhof pathway (glycolysis).
   b. the Krebs cycle.
   c. the electron transport system.
   d. glycolysis and the Krebs cycle.
   e. glycolysis, the Krebs cycle, and the electron transport system.
5. Assume that a molecule of glucose contains the radioactive isotope tritium (^3H) instead of hydrogen. After complete oxidation, the isotope would be found in
   a. sugar.
   b. carbon dioxide.
   c. oxygen.
   d. molecular hydrogen.
   e. water.
6. Assume that a molecule of glucose, synthesized using a radioactive isotope of oxygen, enters the glycolytic pathway. After completing that phase of cellular respiration, the radioactive marker would be found in
   a. carbon dioxide.
   b. water.
   c. pyruvic acid.
   d. ATP.
   e. NADH.
7. Compounds with a high enough phosphoryl transfer potential to generate ATP act
   a. only during glycolysis.
   b. only during the Krebs cycle.
   c. only during electron transport.
   d. during glycolysis and the Krebs cycle.
   e. in all phases of cell respiration.
8. The energy of the proton gradient established within the mitochondria drives the
   a. oxidation of NADH.
   b. reduction of NAD^+
   c. cleaving of glucose into two 3-carbon molecules.
   d. substrate phosphorylation of ATP.
   e. oxidative phosphorylation of ATP.
9. The pool of NAD^+ that is required for glycolysis can be maintained by oxidation of NADH during
   a. glycolysis.
   b. fermentation.
   c. electron transport.
   d. *a* and *b*, but not *c*.
   e. *b* and *c*, but not *a*.
10. Which of these metabolic pathways would you not expect to find in organisms that do not use oxygen as a terminal electron acceptor?
    a. Krebs cycle
    b. glycolysis
    c. synthesis of ATP
    d. synthesis of fatty acids
    e. all of the above

True-False Questions

Mark each statement true or false, and if false, restate it to make it true.

1. In the overall equation of cellular respiration ($C_6H_{12}O_6 + 6\ O_2 \rightarrow 6\ H_2O + 6\ CO_2$), the carbon of CO_2 comes from glucose, and the oxygen in CO_2 comes from water.
2. The carbon cycle couples the processes of cellular respiration and photosynthesis.
3. While photosynthesis is primarily a process in which carbon is oxidized, cellular respiration is a process in which carbon is reduced.
4. After completion of glycolysis and the citric acid cycle, the energy of glucose remains only in CO_2 and H_2O.
5. The oxidation of pyruvate to alcohol or lactate is known as fermentation.

6. The enzymes of the Krebs cycle are located on the external mitochondrial membrane, while the enzymes of the electron transfer system are found on the inner mitochondrial membrane.
7. Anaerobic organisms that utilize the pathways of glycolysis and fermentation do not produce ATP by means of oxidative phosphorylation.
8. As electrons travel from one component of the electron transport system to the next, they gain energy from the surrounding proton gradient.
9. In oxidative metabolism, the final oxidizing agent is organic, whereas in fermentation, the final oxidizing agent is inorganic.
10. As a result of the events of chemiosmosis, the oxygen we breathe is combined with carbon from sugar to form carbon dioxide.

Concept Questions

1. Predict what would occur if the enzyme-coenzyme complex that converted pyruvate to acetyl-CoA stopped functioning.
2. Trace the route of the hydrogen atoms through cellular respiration from glucose to water.
3. Demonstrate that glycolysis, the Krebs cycle, and the electron transport system are coupled processes.
4. Why do you breathe? Be very specific and account for the functions of oxygen, carbon dioxide, and the water vapor in your breath.
5. The reactions that begin with pyruvate and include the Krebs cycle have been called the final common pathway of metabolism. What is meant by the phrase *final common pathway,* and why is the Krebs cycle but not glycolysis included?

Additional Reading

Daviss, Bennett. "Power Lunch." *Discover,* March 1995, p. 58. The idea of batteries powered by microbial digestion.

Douce, R., D. A. Day, and T. Ap Rees (eds.). *Higher Plant Cell Respiration.* Springer-Verlag, Berlin and New York, 1985. A part of the encyclopedia of plant physiology; research on plant mitochondria and respiration.

Shulman, R. G. "NMR Spectroscopy of Living Cells." *Scientific American,* January 1983, p. 86. Methods for studying metabolic reactions in functioning tissue.

Internet Resource

To further explore the content of this chapter, log on to the web site at

http://www.mhhe.com/biosci/genbio/guttman/

Photosynthesis

10

Key Concepts

10.1 Photosynthesis in eucaryotes occurs in chloroplasts.

10.2 Molecules absorb light through activation of their electrons.

10.3 Chlorophylls are the major pigments used in photosynthesis.

10.4 Photosynthesis requires a reducing agent, which is generally water.

10.5 Two photosystems cooperate in plant photosynthesis.

10.6 Cyclic photophosphorylation creates only ATP.

10.7 Noncyclic photophosphorylation creates both ATP and NADPH.

10.8 CO_2 is reduced to organic compounds in the Calvin cycle.

10.9 Some plants use an alternative pathway for CO_2 fixation.

A diver with air tanks rests amid the plants that synthesize his oxygen.

In addressing current environmental issues, people tend to talk as though "the environment" were just an abstract concept, a place remote from humans and the other organisms that live in it. In fact, the environment is in intimate contact with us, and it seeps into us in countless ways. People go around every day casually breathing the air and eating nourishing foods, taking in the molecules of the environment for the energy and nutrients they supply. Perhaps we assume too confidently that we are only breathing and ingesting beneficial molecules, and we take for granted that the essential ones will always be available. Anyone planning to make a scuba dive or fight a forest fire plans to carry tanks of compressed air, but most of us take the air for granted, never thinking how much we depend on the world's plants to keep it rich in oxygen. Only when crops fail and people are threatened with starvation do we realize our critical dependence on plants for food. Both oxygen production and plant growth depend on photosynthesis.

In Chapter 9 we showed how cells obtain energy by oxidizing reduced organic compounds (typically glucose) in the process of respiration. These reduced compounds were originally formed through an input of energy, which initially comes from the sun. Now we will show how phototrophic organisms—plants, algae, and certain types of bacteria—capture light energy in photosynthesis and use it to build organic molecules from CO_2. Scientists often use the term "black box" to refer to a system with an unknown structure that is carrying out a particular process. The little box shown here that carries out photosynthesis is a "black box," and our purpose is to understand what is going on inside:

Light, CO_2, H_2O → Black box → Glucose → Other organic compounds; Black box → O_2

10.1 Photosynthesis in eucaryotes occurs in chloroplasts.

Although a green leaf looks uniformly green, even a low-power microscope reveals that the color is confined to numerous small bodies called **chloroplasts** (Figure 10.1). These are the sites of photosynthesis. Many protists have a few large, distinctively shaped chloroplasts (see Figure 6.24), while the chloroplasts of plant cells are typically disc-shaped, numerous, and small—about a micrometer thick and 2–4 μm wide. Cells located just below the upper surface of a leaf, where most photosynthesis occurs, may each have 25 to 50 chloroplasts, corresponding to at least 10^5 chloroplasts per square centimeter of leaf surface. Chloroplasts can move about through the cytosol, change their shapes, and sometimes even orient themselves in different positions in response to changing light intensities.

Each chloroplast is bounded by a double membrane. Most of the chloroplast's volume is occupied by an extensive third membrane, the **thylakoid membrane,** which divides the chloroplast into an internal space, the **lumen** and an external space, the **stroma.** The thylakoid membrane is folded into many disc-shaped vesicles in stacks called **grana** (sing., **granum**), which are connected by irregular sections of membrane. The division of the chloroplast into thylakoid and stroma reflects a division of the black box of photosynthesis into two parts, corresponding to two phases of photosynthesis. One part of the box

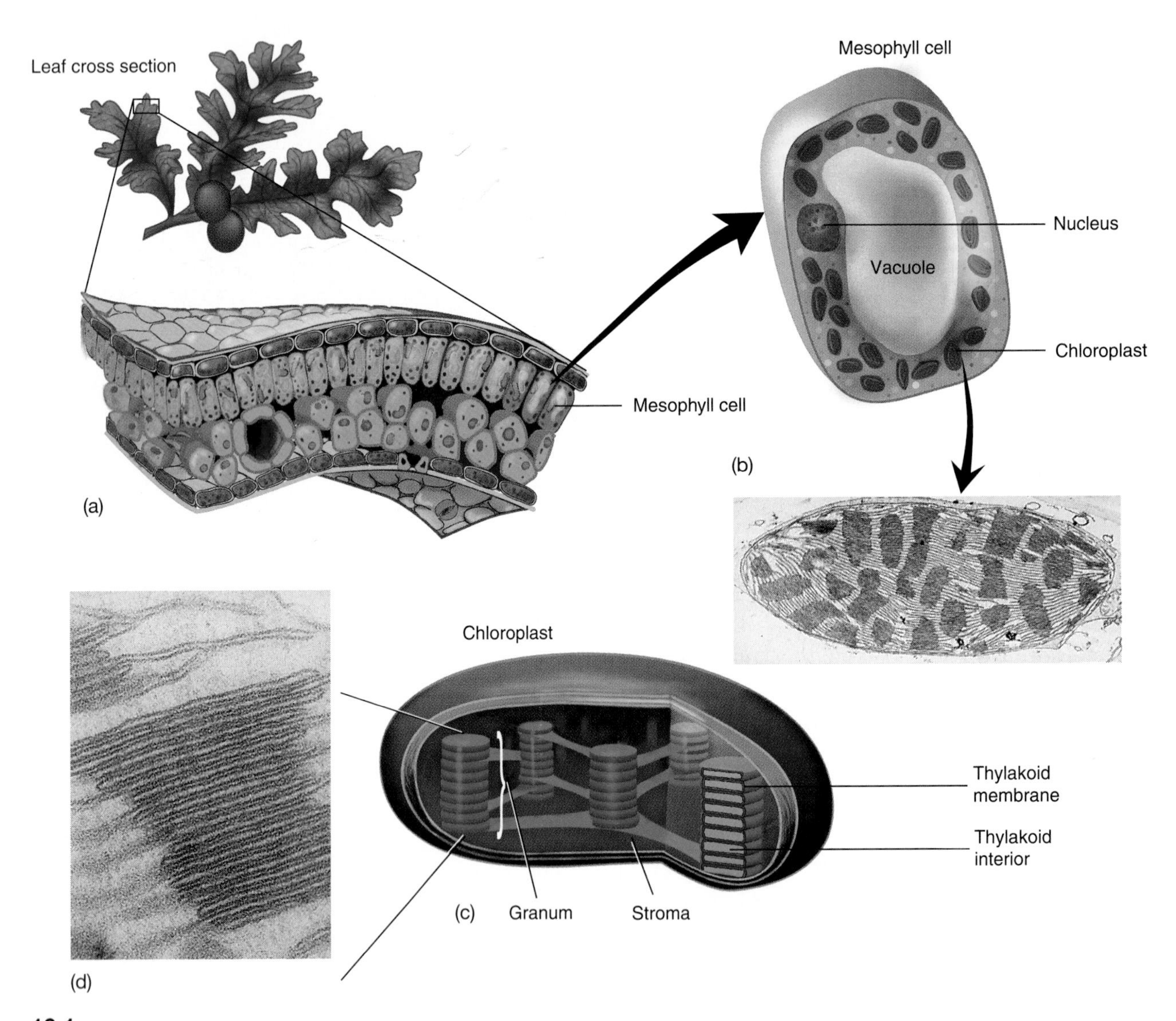

Figure 10.1
(a) Most plant chloroplasts are confined to mesophyll cells in the middle layers of leaves. *(b)* An enlarged view of a mesophyll cell shows many chloroplasts in the cytoplasm around the cell's periphery. *(c)* Each chloroplast is disc-shaped and contains an extensive thylakoid membrane that forms a number of grana, stacks of flattened vesicles shown at higher magnification in *(d)*. Other parts of the thylakoid membrane connect grana as shown in the drawing.

contains **light-dependent reactions** in which light energy is absorbed and stored; the other contains **light-independent reactions** in which the energy acquired in the first box is used to reduce CO_2 to organic molecules, a process called *CO_2 fixation.* The thylakoid membrane, which provides the whole leaf with an enormous area for absorbing light and capturing some of its energy, contains the chlorophyll and other molecules that conduct the light-dependent reactions. This system stores energy in the form of ATP and NADPH. (Remember that $NADP^+$ is simply NAD^+ with an extra phosphate. Cells use NADPH as a coenzyme for reduction and NAD^+ for oxidation.) The stroma around the thylakoid membrane is full of enzymes that carry out the light-independent phase. So the black box has this structure:

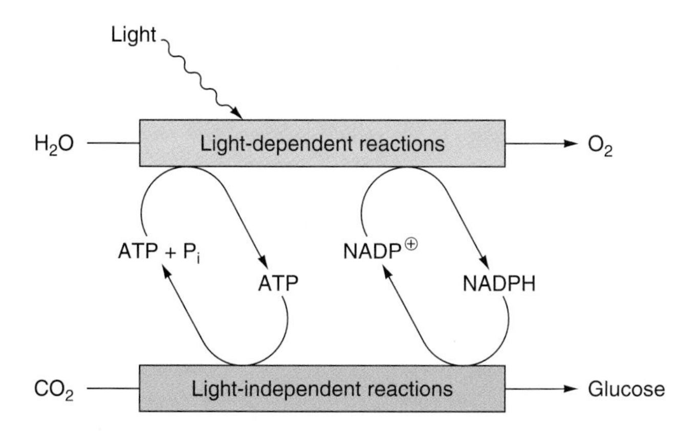

ATP provides energy for reducing CO_2 to organic compounds, and NADPH is the reducing agent. In plants, the hydrogen for reduction (in the form of separated protons and electrons) comes from water, leaving O_2 as a by-product.

What is the product of photosynthesis? Ultimately, of course, it is the whole plant or other phototroph, or even the entire food web that relies on phototrophs, but let's confine the question to chloroplasts and the cells they reside in. We will show later that the first product of CO_2 reduction is glyceraldehyde-3-phosphate, a triose phosphate (that is, a phosphorylated 3-carbon sugar) that is also an intermediate metabolite of glycolysis. Inside plant chloroplasts, triose phosphates can be converted into hexose phosphates and stored as starch through a pathway that is essentially the reversal of glycolysis. Alternatively, triose phosphates are exported from the chloroplast to the cytoplasm, where they can either be metabolized through the central metabolic pathways or converted into the double sugar sucrose, which is transported to other parts of the plant. In one sense, therefore, the end product of photosynthesis is triose phosphate. On the other hand, triose phosphates are commonly converted into hexose sugars, in starch or sucrose, so we will represent the product of photosynthesis as hexose ($C_6H_{12}O_6$), because it is conventional and makes it easier to compare photosynthesis with respiration.

The two phases of photosynthesis are remarkably parallel to the two phases of respiration, but they run in opposite directions. In the first phase of respiration, glucose is oxidized to CO_2, with the reduction of NAD^+ to NADH; in the second phase, NADH is oxidized in an electron transport system (ETS) to make ATP, using the reduction of oxygen to water as a final electron (or hydrogen) acceptor. In the light-dependent reactions of photosynthesis, water donates protons and electrons to an ETS that reduces $NADP^+$, which is functionally the opposite of the respiratory scheme:

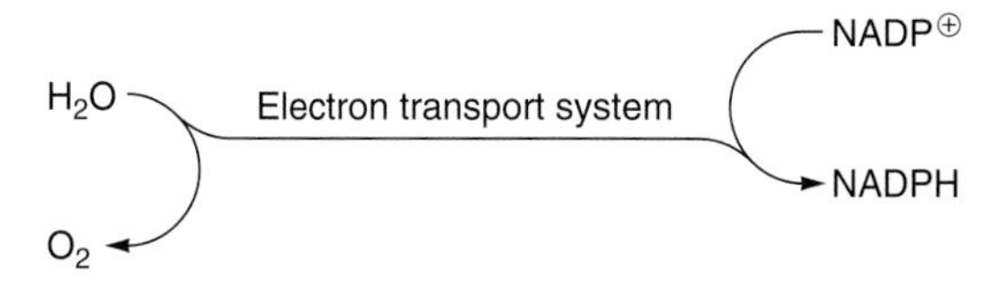

Then in the light-independent reactions, the reduction of CO_2 to sugar is the opposite of respiration, though it uses entirely different pathways—not just the respiratory pathways in reverse.

While most phototrophs are eucaryotes, the procaryotic phototrophs are also very important ecologically. Cyanobacteria (blue-green bacteria) are common in water, in soil, and on the bark of trees, where they often form black films that contrast with the greenish film of eucaryotic algae. A variety of green, purple, red, and brown bacteria inhabit muddy, often sulfurous waters where they carry out a photosynthesis that differs from eucaryotic photosynthesis: The bacteria draw electrons not from water but from other oxidizable inorganic substances such as H_2S or, in some cases, organic compounds. The phototrophic bacteria lack chloroplasts; instead, they develop beautiful internal sacs and lamellae (Figure 10.2) that are functionally equivalent to the thylakoid membrane of chloroplasts.

Phototrophic bacteria, Chapter 29.

Exercise 10.1 What general principle, introduced in Chapter 6, is illustrated by the extensive internal membranes of chloroplasts and phototrophic bacteria?

10.2 Molecules absorb light through activation of their electrons.

Why are leaves green? One answer is, "Because they aren't red and blue," and that is not a flippant reply. Natural, white light is a mixture of all colors. Substances called *pigments* absorb part of the light and thus subtract out certain colors, allowing the remaining colors to be visible in the transmitted or reflected light. Leaves absorb red and blue light, so what we see is mostly green light. But how does a pigment absorb light?

As we pointed out in Chapter 6, light is one type of electromagnetic (EM) radiation, a form of energy that behaves like both a particle and an oscillating wave, straining the human ability to represent natural phenomena realistically. As a particle, each unit of EM radiation is called a photon. It travels at a velocity, *c*, which is 3×10^8 m/sec in a vacuum; the radiation

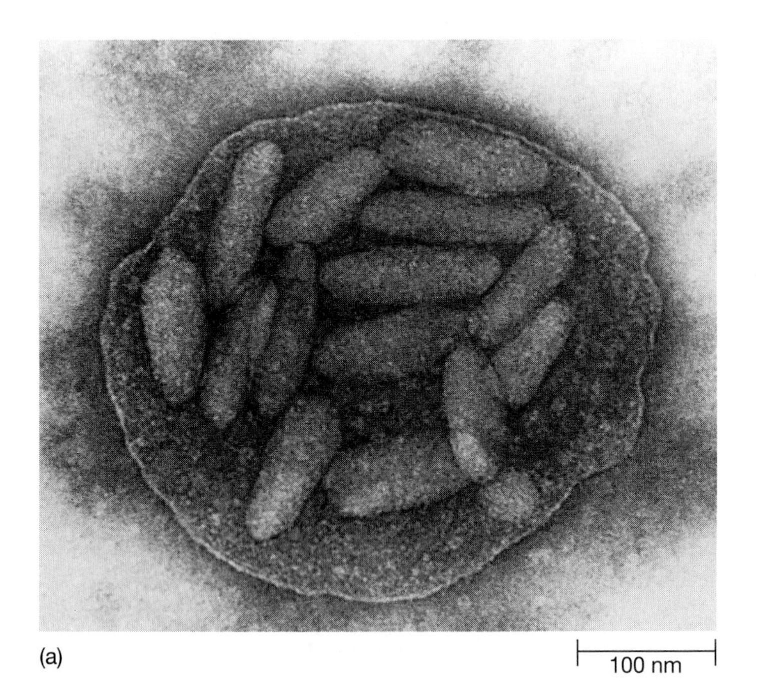

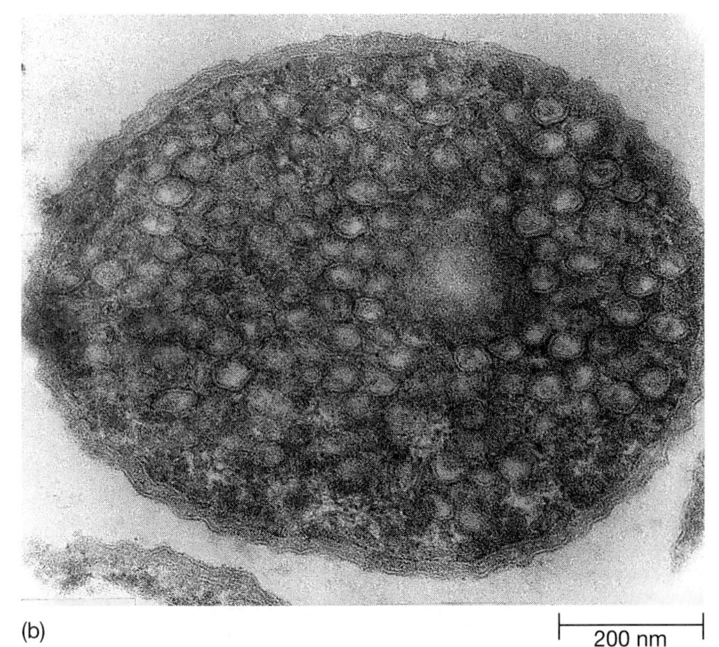

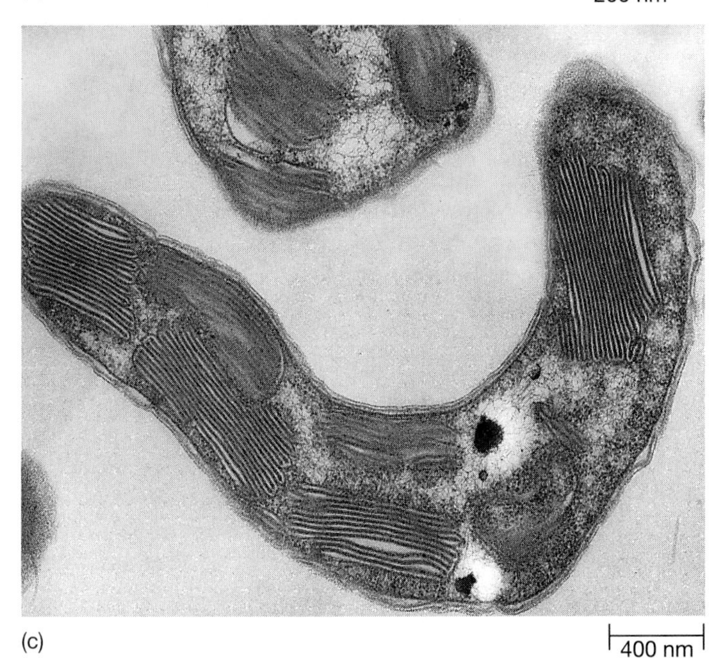

Figure 10.2

The photosynthetic membranes of bacteria occupy large portions of the cytoplasm. *(a)* The green bacterium *Chloropseudomonas ethylicum* has several large vesicles; *(b)* the cytoplasm of the purple bacterium *Chromatium D* is filled with small vesicles; *(c)* the purple bacterium *Ectothiorhodospira mobilis* has stacks of folded membranes.

also vibrates with a frequency of ν (Greek nu) vibrations per second and has a wavelength of λ (Greek lambda). The frequency and wavelength are related inversely—as one increases, the other decreases, as shown by the equation $c = \lambda\nu$. A photon may be considered a packet, or quantum, of energy. However, all photons do not carry the same energy. The energy of EM radiation is proportional to its frequency, and the EM spectrum (Figure 10.3) ranges from very high-energy, short-wave gamma rays and X rays through visible light of intermediate wavelengths, to low-energy, long-wavelength radio waves.

EM radiation of different energies interacts with matter in characteristic ways. Microwaves, with wavelengths in the range of 1 mm to 1 m, make molecules move and vibrate faster, so we can use them to cook food. Infrared radiation, with wavelengths of 2–15 μm, has only enough energy to make the bonds between atoms stretch and bend; this also gives molecules more energy, and we commonly experience infrared radiation as heat. Radiation in the visible light range (about 380–760 nm) and in the ultraviolet range (about 200–380 nm) has just enough energy to make electrons jump from one level to another in an atom or molecule (Figure 10.4). Remember that the electrons in an atom reside in orbitals that have specific energy levels, and each electron must be in one level or another. Ordinarily, the electrons in a molecule are in the lowest energy levels available to them, in their *ground states;* at usual biological temperatures, light is the only source of energy that can move an electron to a higher level. An electron moves in a quantum jump (a single jump all the way) by absorbing a quantum of light energy precisely equal to the energy difference between two levels. For example, an electron at a level of 2.0 volts of potential energy can jump to the 3.1-volt level only if it absorbs a light quantum of precisely 1.1 volts. It can't pick up 0.5 volts in one event and then 0.6 volts in a second event, nor can it pick up 1.2 volts. An electron can also absorb a quantum of very high-energy radiation, causing *ionization,* because the electron is knocked completely out of an atom or molecule, leaving an ion.

White light is a mixture of photons with all the energies in the visible range. Molecules irradiated by white light absorb those quanta that allow electrons to make the energy transi-

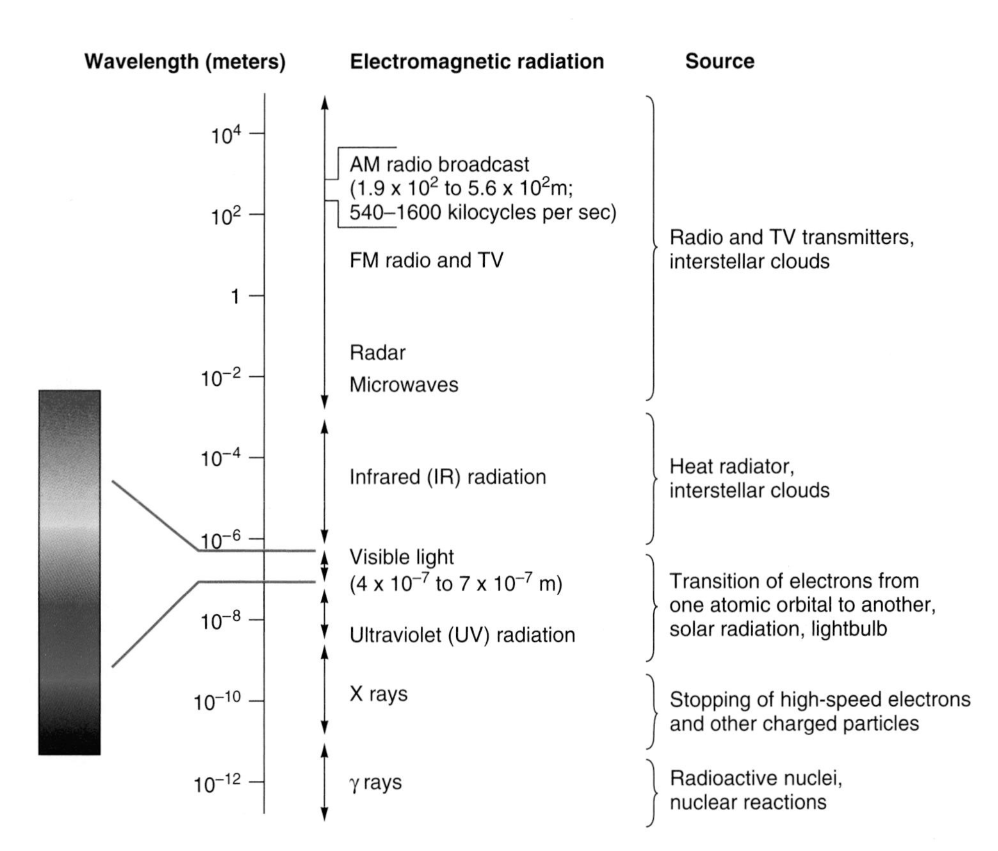

Figure 10.3
Electromagnetic radiation ranges from very long-wave (low-energy) radio waves through visible light to very short-wave (high-energy) gamma and X rays.

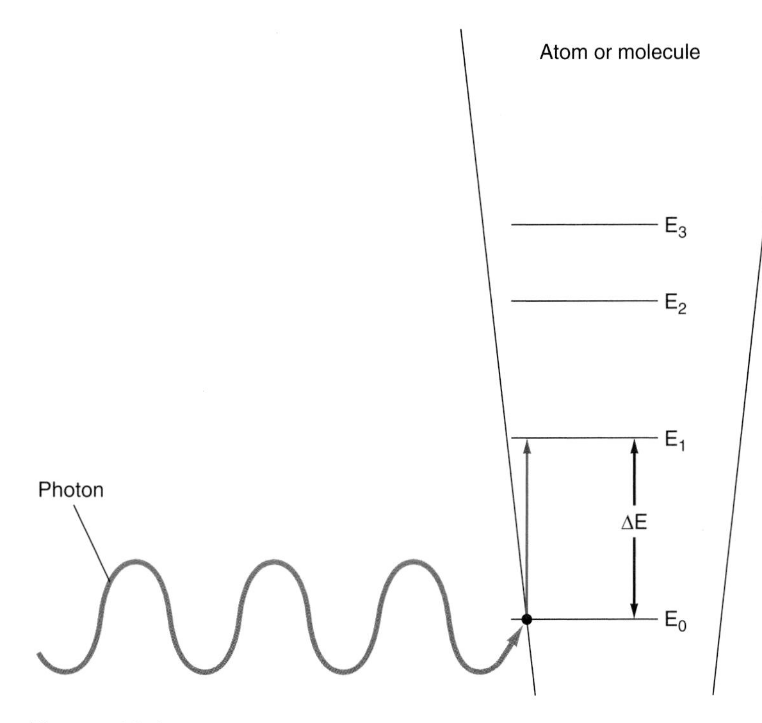

Figure 10.4
A packet of light (a photon) travels as a wave and carries a quantum of energy; the amount of energy is proportional to its frequency of vibration. Every atom has certain energy levels where electrons can reside. An electron can jump to a higher level by absorbing a photon that has precisely the energy needed to raise it to the higher level.

tions available to them. Since every type of atom or molecule has slightly different energy levels, every substance absorbs light with a characteristic series of wavelengths, as displayed in an **absorption spectrum** that shows the light it absorbs at each wavelength (Figure 10.5). For instance, the absorption spectrum of chlorophyll *a* has strong peaks at 680–700 nm (red light) and at 430 nm (blue light). Since chlorophyll absorbs light of those two wavelengths, most of the remaining light passing through the pigment is the green light we see. Absorption spectra are so characteristic that they are routinely used to identify substances.

An excited electron stays at a higher energy level for only a very short time, about 10^{-15}–10^{-9} second. During that time, the molecule is *excited,* meaning that it has additional energy and a greater ability to engage in chemical reactions. The excited electron has moved farther from its nucleus where, for instance, an oxidizing agent can remove it more easily. This is just what happens in photosynthesis, as an excited electron in chlorophyll is drawn away by a nearby molecule and started on its path through a system of electron carriers that makes ATP.

Incidentally, if an excited electron is not removed while it is at a higher level, it falls back to its ground level and must give up its extra energy as heat or as emitted light. Many materials emit light of a slightly longer wavelength than the light they absorb, a phenomenon called *fluorescence.* (Some of the energy of the excited electron goes into

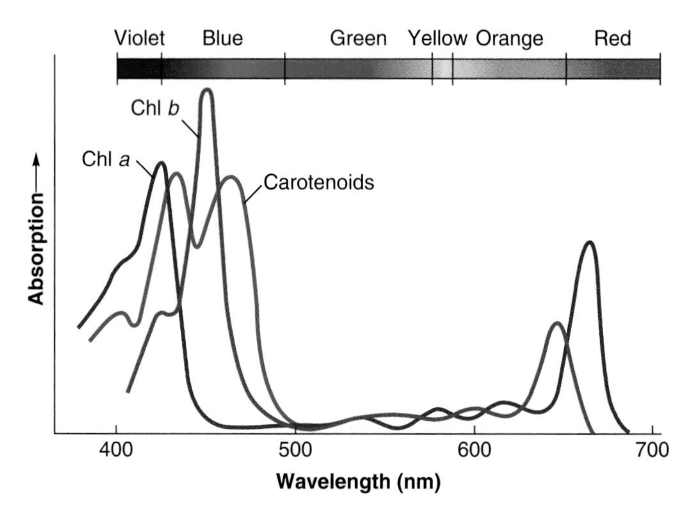

Figure 10.5

This graph shows the absorption spectra of three compounds that are important in photosynthesis. Each absorption peak indicates that the compound absorbs photons of a particular wavelength—that is, energy. The higher the peak, the more strongly the light of that color is absorbed. Chlorophyll *a* is the only pigment that is central to photosynthesis.

increasing the kinetic energy of the molecule, so the electron first drops to a slightly lower level before it falls to its ground level, emitting a lower-energy photon of longer wavelength.) Fluorescent materials are very useful in the laboratory because they can be attached to biological structures, causing them to glow brilliantly under the microscope when irradiated with ultraviolet light.

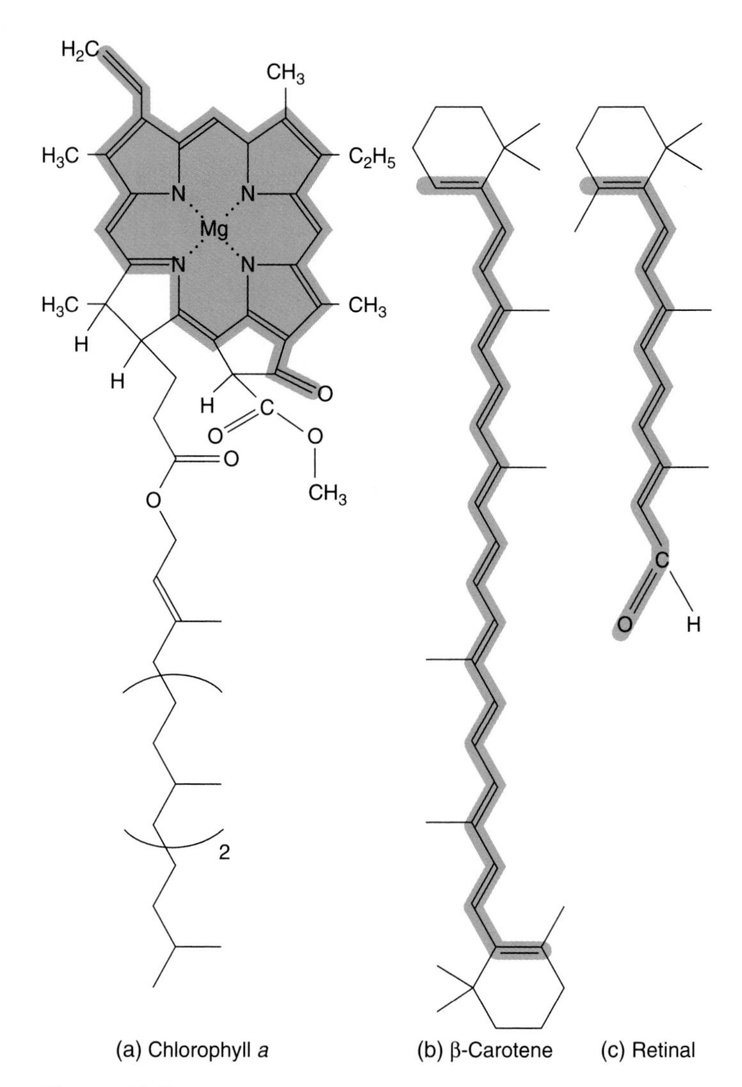

Figure 10.6

The pigments *(a)* chlorophyll *a, (b)* β-carotene, and *(c)* retinal all have large orbitals with delocalized electrons *(colored regions).*

10.3 Chlorophylls are the major pigments used in photosynthesis.

Most of the light that drives photosynthesis is absorbed by **chlorophylls** (Figure 10.6*a*). This group of similar pigments belongs to the remarkable class of tetrapyrrole compounds discussed in Concepts 4.2. Chlorophylls have a magnesium atom chelated in the center of their porphyrin ring, a long tail of the alcohol phytol, and various small chemical groups attached around the ring. All eucaryotic phototrophs have chlorophyll *a* (abbreviated Chl *a*) plus smaller amounts of either Chl *b* (in green algae and plants) or Chl *c* (in chromists such as golden-brown and brown algae). The phototrophic bacteria have similar bacteriochlorophylls.

All phototrophs also use secondary pigments, especially **carotenoids,** which include *carotenes* (Figure 10.6*b*) and more oxidized and modified molecules called *xanthophylls.* These spectacular yellow, orange, red, and purple pigments account for the colors of phototrophic bacteria, tomatoes, and carrots—whence their name—and (with other pigments) form the brilliant hues of autumn leaves, after the predominant chlorophyll has disappeared. While chlorophyll absorbs red and blue light, carotenoids absorb maximally in the blue-violet range, so we see the yellow-orange-red light they don't absorb.

Notice that carotenoids are very similar to *retinal* (Figure 10.6*c*), the major visual pigment that absorbs light in the eye. Retinal, in fact, is synthesized from the group of vitamin A compounds found in animal sources, such as fish oils, or from the β-carotene that is abundant in carrots and tomatoes, so you can see why good vision depends on eating some of these foods. We conventionally draw the structures of all these compounds, as well as other pigments, with alternating single and double bonds between carbon atoms, but just as in a benzene ring, this formal picture is a poor representation of the actual arrangement of electrons. It is more accurate to say that the electrons, rather than being confined to local orbitals, are *delocalized* and occupy a single large orbital extending over the whole region represented by alternating single and double bonds. Electrons in such an orbital absorb light in the visible part of the electromagnetic spectrum. But it is precisely because such molecules are used in our eyes to absorb light that this light *is* visible and that some substances appear colored to us.

Although an absorption spectrum shows which wavelengths of light a pigment absorbs, it doesn't show whether that light energy is used for a chemical process. In contrast, an **action spectrum** shows the rate of a particular process at each wavelength of light and tells whether light of a particular wavelength is actually used. The action spectrum for photosynthesis (Figure 10.7) is broadly similar to the absorption spectra of the chlorophylls plus carotenoids, showing that light absorbed by both kinds of pigments contributes to photosynthesis. Other experiments show that Chl *a* is the chief photosynthetic pigment, and that carotenoids and other chlorophylls absorb light of other wavelengths and pass the energy on to Chl *a*.

The German plant physiologist Theodor Engelmann demonstrated the action spectrum of photosynthesis in 1882 with a simple but clever experiment (Figure 10.8*a*). He focused a spectrum of visible light on an algal filament *(Cladophora)* with bacteria that require oxygen. The bacteria swam to the regions of highest oxygen concentration, in the blue and red regions of the spectrum, indicating that photosynthesis is occurring in these places. In a second experiment, he focused very fine beams of light on a different alga, *Spirogyra*, and showed that the chloroplast is the locus of photosynthesis, since only light focused there generates oxygen (Figure 10.8*b*).

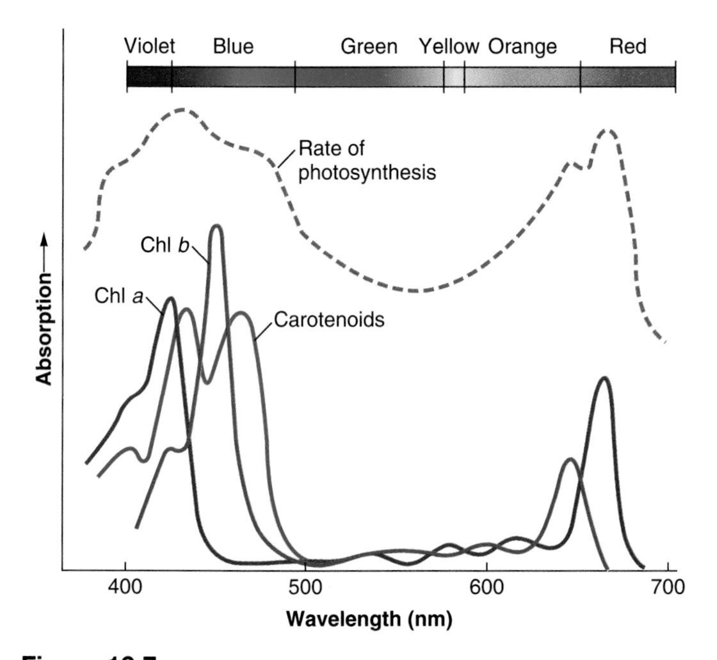

Figure 10.7

An action spectrum shows the relative rates of photosynthesis in chloroplasts irradiated with light of specific wavelengths.

Sometimes, when studying a process like photosynthesis, we find an absorption peak that can't be identified with a known molecule. Then the molecule responsible for the peak is simply labeled P, for pigment, along with the wavelength of the peak, such as P680 or P700. These peaks pose intriguing mysteries and have set some investigators to work trying to identify the molecules involved. The molecules responsible for peaks at 680 and 700 nm have been identified and found to have key roles in photosynthesis.

Exercise 10.2 In the action spectrum for photosynthesis (Figure 10.7), identify the pigments that most likely contribute to each of the peaks and shoulders. (A shoulder is a bulge in the spectrum that isn't distinct enough to be a separate peak.)

10.4 Photosynthesis requires a reducing agent, which is generally water.

As early as the beginning of the nineteenth century, it was clear that photosynthesis evolves oxygen. Chemists generally assumed that CO_2 is somehow split into oxygen and carbon, which is then made into organic molecules. This idea should not have persisted after the 1890s, when Theodor Engelmann showed that the purple sulfur bacteria carry out a photosynthesis

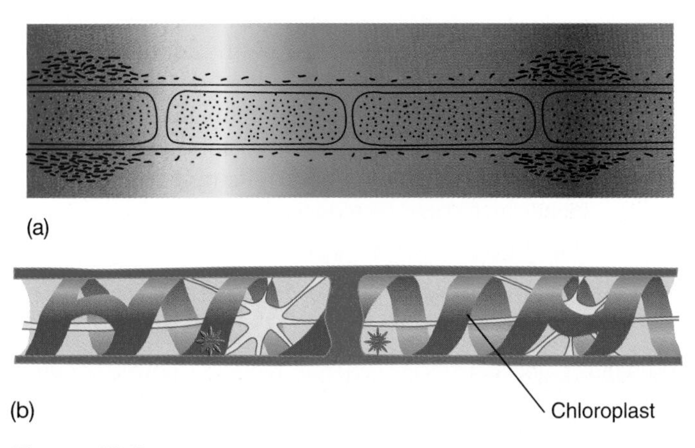

Figure 10.8

Engelmann demonstrated that only certain wavelengths of light are effective in photosynthesis. *(a)* With a fine optical device developed by his friend Carl Zeiss, he focused a light spectrum on a strand of the alga *Cladophora,* whose cells contain a large, continuous chloroplast. Oxygen-requiring bacteria in the surrounding water were attracted to the red and blue regions of the spectrum, showing that photosynthesis—which produces oxygen—was occurring there. *(b)* To show that the chloroplast is the site of oxygen production, Engelmann used another device developed by Zeiss to shine minute spots of light on a strand of *Spirogyra.* The bacteria were only attracted to spots of light on the chloroplast.

without evolving oxygen. However, it did persist until 1931, when C. B. van Niel, then a graduate student at Stanford University, confirmed that these bacteria depend on light and that they oxidize hydrogen sulfide (H_2S) and other reduced sulfur compounds to elemental sulfur and sulfate, instead of producing oxygen. For instance, they produce sulfur in a photosynthesis summarized by:

$$6\,CO_2 + 12\,H_2S \rightarrow C_6H_{12}O_6 + 12\,S + 6\,H_2O$$

The sulfur accumulates as visible particles inside these cells. Van Niel also found that some bacteria use hydrogen gas and leave no residue:

$$6\,CO_2 + 12\,H_2 \rightarrow C_6H_{12}O_6 + 6\,H_2O$$

He therefore proposed that there are several kinds of photosynthesis, all having one common feature: They use light to split a reduced compound, H_2X. The X is a by-product, and the H atoms combine with CO_2 to make carbohydrate:

$$6\,CO_2 + 12\,H_2X \rightarrow C_6H_{12}O_6 + 6\,H_2 + 12\,X$$

Then the proper equation for green plant photosynthesis becomes:

$$6\,CO_2 + 12\,H_2O \rightarrow C_6H_{12}O_6 + 6\,H_2O + 6\,O_2$$

Notice that this is precisely the opposite of the summary equation for respiration. Again, we don't cancel water molecules on both sides because the equation tells us something about the biology of the process—that water is split by light to provide a reducing agent. Proof that oxygen comes from water was obtained a few years later when water labeled with a heavy isotope of oxygen ($H_2{}^{18}O$) became available. This isotope served as a tracer, described in Methods 10.1, to show the fate of the oxygen atoms in water.

In 1937 Robin Hill showed that illuminated chloroplasts produce oxygen if they are simply given an appropriate electron acceptor. He used ferric compounds, containing iron atoms in the Fe^{3+} state, which could be reduced in the process to ferrous (Fe^{2+}) compounds:

$$2\,H_2O + 4\,Fe^{3+} \rightarrow O_2 + 4\,H^+ + 4\,Fe^{2+}$$

This process, now called the **Hill reaction,** demonstrated that illuminated chloroplasts can generate reducing power through the light-driven splitting *(photolysis)* of water. In photosynthesis, however, the electron (and hydrogen) acceptor is $NADP^+$. $NADP^+$ is reduced to NADPH, which is then used to reduce CO_2 to carbohydrate. Next we will see how NADPH is formed, along with ATP, in the light-dependent reactions.

Exercise 10.3 Many dyes have different colors in their oxidized and reduced states. For example, methylene blue is blue when oxidized and colorless when reduced, and its reduction potential is appropriate for being reduced by chloroplasts. Outline how you would set up an experiment, with isolated chloroplasts, to demonstrate the Hill reaction using methylene blue.

Exercise 10.4 Heavy oxygen (^{18}O) is indeed useful as a tracer, since atoms and molecules with different masses can be identified with an instrument called a mass spectrometer. Outline how you would determine experimentally whether the oxygen produced in plant photosynthesis comes from CO_2 or H_2O.

10.5 Two photosystems cooperate in plant photosynthesis.

The light-dependent reactions of photosynthesis, as they occur in a plant, consist of three main events:

1. *The primary event of photosynthesis.* A chlorophyll molecule absorbs a photon of light, thus exciting one of its electrons.
2. *Charge separation.* A primary electron acceptor removes the excited electron, so it becomes negative and leaves a positively charged chlorophyll molecule. By creating separate positive and negative centers, the system has effectively made a minute battery and an energized electron capable of moving from one center to the other.
3. *Photophosphorylation.* The electron travels through an electron transport system in the thylakoid membrane, very much like the respiratory ETS, which generates ATP through the same chemiosmotic mechanism. ATP synthesis in photosynthesis is called **photophosphorylation.** The electron can take a cyclic pathway, which generates only ATP, or a noncyclic pathway, which also reduces $NADP^+$ to NADPH, using water as the ultimate reducing agent.

The photosynthetic pigments and proteins of the thylakoid membrane are organized into complexes called **photosystems: PS I and PS II.** The evidence for these systems first came from the work of Robert A. Emerson and his colleagues around 1943. While measuring the efficiency of photosynthesis with monochromatic light, they found that efficiency dropped quite sharply at the far-red end of the spectrum (700 nm and longer). They noticed, however, that if chloroplasts irradiated with far-red light were also given light of a shorter wavelength, photosynthesis was restored to its full efficiency. This is the *Emerson enhancement effect.* It occurs because photosynthesis requires the interaction of *two* light reactions. They can both be driven by light of a shorter wavelength (less than 680 nm), but only one can be driven by far-red light. We now know that PS I absorbs most of the far-red light, while PS II absorbs most of the shorter-wavelength light.

Photosystems I and II themselves are complexes of proteins, chlorophylls, and other materials, and they can be separated from the thylakoid membrane with detergents. Each photosystem is organized around a **photosynthetic unit** of about 200–300 chlorophyll molecules, packed together in a single, large, light-collecting structure; this is called an **antenna complex** because it functions in photosynthesis rather the way an antenna functions for a radio or television set. Any molecule in the photosynthetic unit can absorb a photon of light, which excites one of its electrons briefly. The chlorophyll molecules in

Methods 10.1 The Use of Tracers in Experiments

In his novel *A Study in Scarlet,* Sir Arthur Conan Doyle had Sherlock Holmes follow a murderer across half of London with the aid of a sharp-nosed hound. Having unwittingly stepped in a bit of creosote, the man left a perfect trail (until he crossed the path of the carts carrying creosote from the factory). The creosote was a **tracer,** a label attached to his feet that allowed him to be followed.

Biological research often depends on our ability to trace the path of one component through a system. A scientist investigating the movement of birds, for example, can trap them and attach metal bands to their legs. If a bird is recaptured later, or found dead, the band reveals something about its habits. The movements of small mammals through their meadow home can be traced by attaching tiny radio transmitters to their bodies and following them with a directional antenna. This provides valuable information, since the animals' behavior seems to be unaffected by their little burdens.

The most commonly used tracers, however, are radioactive isotopes *(radioisotopes).* These by-products of nuclear technology give biologists excellent tools for following the transformations of materials in metabolism. We can now buy many kinds of compounds labeled with radioisotopes, especially ^{14}C (carbon-14), ^{3}H (tritium), ^{32}P (phosphorus-32), and ^{35}S (sulfur-35). With various instruments for detecting radiation, we can locate the tracers and determine how much material is present. (The ^{18}O used in photosynthesis research is a heavy isotope of oxygen, but not a radioisotope.)

For instance, humans and other vertebrates have only one use for iodine: as part of the hormone thyroxine, which is made in the thyroid gland in the throat. The rate at which someone is synthesizing thyroxine can therefore be determined by feeding a person a small amount of the radioisotope ^{131}I (iodine-131) diluted with unlabeled ("cold") iodine (^{127}I). Then a Geiger counter held over the throat can detect the radioactivity of the thyroid gland, which shows how actively it is working. Incidentally, since the radiation from ^{131}I is strong and damaging, it can sometimes be used to destroy a controlled amount of tissue in a person with an overactive thyroid gland. Although radioisotopes are damaging, they are useful because one can detect such minute amounts of radioactive material—much less than can be found by chemical methods.

Most modern work with radioactive tracers employs specific labeled compounds. For instance, to follow the synthesis of protein, we can give cells radioactive amino acids, take samples periodically, and determine their radioactivity. We can also determine the fate of particular atoms this way. For example, glucose in which only the number-6 carbon atom (C-6) is labeled with ^{14}C can be used as a carbon source for growing cells. Then we can isolate some of their amino acids and, through chemical analysis, determine which atoms are radioactive; those atoms were once the C-6 atoms of glucose. In this way, we can determine precisely how one metabolite is transformed into another in a metabolic pathway.

Autoradiography is often used with radioisotopes. After some material is labeled, it is covered with photographic emulsion or laid against X-ray film and kept in the dark for a few days. The emissions from the radioisotope expose the film nearby, and upon developing the film, one can see where the labeled material is located. Isotopes with relatively low-energy emissions, such as tritium, are most useful because their electrons are absorbed by the film close to the source. Autoradiography is especially useful with gel electrophoresis, because it can reveal bands of labeled proteins or nucleic acids.

the complex are arranged so they can transfer energy to one another. Thus no matter where an electron is first excited, its energy can be passed along rapidly to other molecules in a random path, until it encounters a **reaction center,** a unique place where the energy can be trapped (Figure 10.9). The reaction-center molecules are identified as P700 in PS I and P680 in PS II; each of them is a special pair of chlorophyll molecules associated with certain proteins and electron carriers. As we show next, electron transfer begins at the reaction center and can take either a cyclic or a noncyclic path.

10.6 Cyclic photophosphorylation creates only ATP.

The process of **cyclic photophosphorylation** (Figure 10.10), which is restricted to PS I, is a mechanism for converting light energy to chemical energy. When a photon of light excites an electron of a chlorophyll molecule, the energy is passed on to the reaction center, P700, designated P700* in its excited state. P700* can reduce the iron-sulfur protein *ferredoxin* (Fd) using an intermediate ferredoxin-reducing system (FRS). The reduction potential of P700 in its ground state is 0.4 volts, while that of the FRS is about −0.6 volts, a difference of 1 volt. But a photon of red light carries 1.8 electron-volts of energy, so P700* has

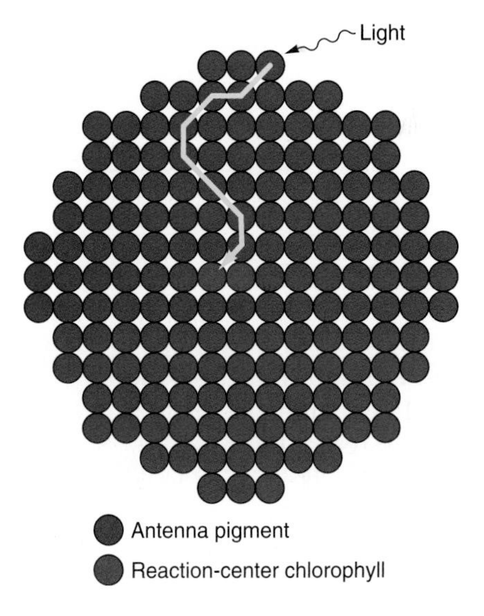

Figure 10.9

Chlorophyll molecules *(green circles)* are arranged so that the energy of an excited electron in one molecule can pass directly to those in neighboring molecules, with no loss of energy. An electron is excited in one molecule that absorbs a photon of light; it travels a random path *(yellow)* until it is drained off from a reaction center *(red circles)* to begin its movement through an electron transfer system.

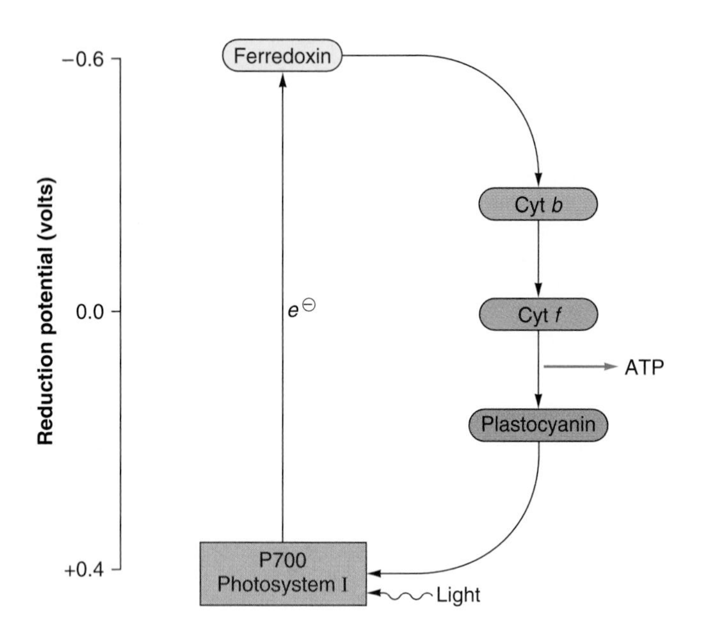

Figure 10.10

In the cyclic pathway, a photon of light is absorbed by the antenna complex of photosystem I and transferred to a reaction center, P700. The excited P700* then has enough energy to reduce a chain of carriers, including a complex of cytochromes *b* and *f* and plastocyanin, a small copper protein; P700 becomes temporarily oxidized to $P700^+$, and the electron reduces it to P700 again at the end of its cyclic path. Through a chemiosmotic mechanism like that of mitochondria, one ADP is phosphorylated to ATP.

more than enough energy to reduce the FRS, which in turn reduces ferredoxin:

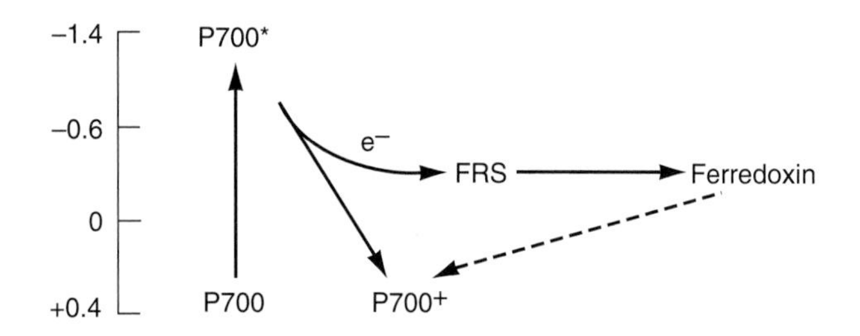

At this point, the system has an oxidized $P700^+$ and a reduced ferredoxin. The electron then passes back to $P700^+$ on a cyclic pathway (Figure 10.10), through an electron transport system that generates ATP by chemiosmosis.

Light effectively acts like a pump to raise electrons to a high enough potential to initiate the whole process, by giving P700 the energy to reduce the chain of electron carriers. As electrons pass between the two cytochromes, protons are transported from the outside of the thylakoid membrane to the inside (lumen), creating a proton gradient that is used to synthesize ATP. This system is therefore very similar to the oxidative phosphorylation system of mitochondria, including a similar ATP synthase, but here the protons are pumped to the inside of the membrane instead of the outside. The cyclic pathway thus conserves some of the energy of the absorbed light in the structure of ATP. The cyclic pathway does not generate any NADPH, but it is valuable because more ATP than NADPH is needed to convert CO_2 to sugar.

10.7 Noncyclic photophosphorylation creates both ATP and NADPH.

In addition to producing ATP, chloroplasts must produce a store of reducing power in the form of NADPH, but the reduction of $NADP^+$ to NADPH requires a noncyclic pathway and the absorption of a second photon of light by photosystem II. To understand this system, let's return to the point at which photosystem I, after absorbing a photon of light, has initiated a series of electron transfers. Once ferredoxin (Fd) has been reduced, it can either pass electrons along the cyclic path or reduce $NADP^+$ to NADPH:

$$2\ Fd_{red} + H^+ + NADP^+ \rightarrow 2\ Fd_{ox} + NADPH$$

But electrons that have been transferred to $NADP^+$ cannot get back to $P700^+$, which remains oxidized and unable to function again. Reducing $P700^+$ requires an electron from an external source and the energy of a second photon of light, and PS II supplies both.

P680, the reaction-center molecule in PS II, absorbs a photon of light and is activated. Like P700, P680 is oxidized (to $P680^+$) as it donates an electron to a noncyclic pathway of electron carriers (Figure 10.11). The resulting process is known as the *Z scheme* because it looks like a Z when drawn out with each component at its proper reduction potential. This electron finally reduces $P700^+$ back to its original state. Meanwhile, electron transport through the noncyclic pathway, from one electron carrier to another (Figure 10.12), ferries protons across the thylakoid membrane, adding to the proton gradient that drives ATP synthesis. ATP synthesis through this pathway is called **noncyclic photophosphorylation.**

The $P680^+$ in PS II is still oxidized. The power to reduce it back to P680 comes from water, which is split by an enzyme:

$$H_2O \rightarrow 2\ H^+ + 2\ e^- + \tfrac{1}{2}O_2$$

The electrons liberated from water return $P680^+$ to its original state, while the protons contribute to the proton gradient. Oxygen is released as a by-product.

To summarize the light reactions, chlorophyll molecules of PS I absorb light and pass the energy of excited electrons to the reaction center, P700, which then becomes oxidized to $P700^+$ by reducing the electron transport chain. Electrons may pass through a cyclic pathway, back to $P700^+$, generating only ATP. Or electrons may reduce $NADP^+$ to NADPH, requiring other electrons to reduce $P700^+$ again. These electrons come from the photolysis of water, using the energy of a second photon of light absorbed in PS II. As electrons pass from PS II to PS I, some of their energy is used to generate ATP.

For contrast, Figure 10.13 shows the photosynthetic apparatus of purple sulfur bacteria, which produce both ATP and NADH with only a single photosystem. The pathway of

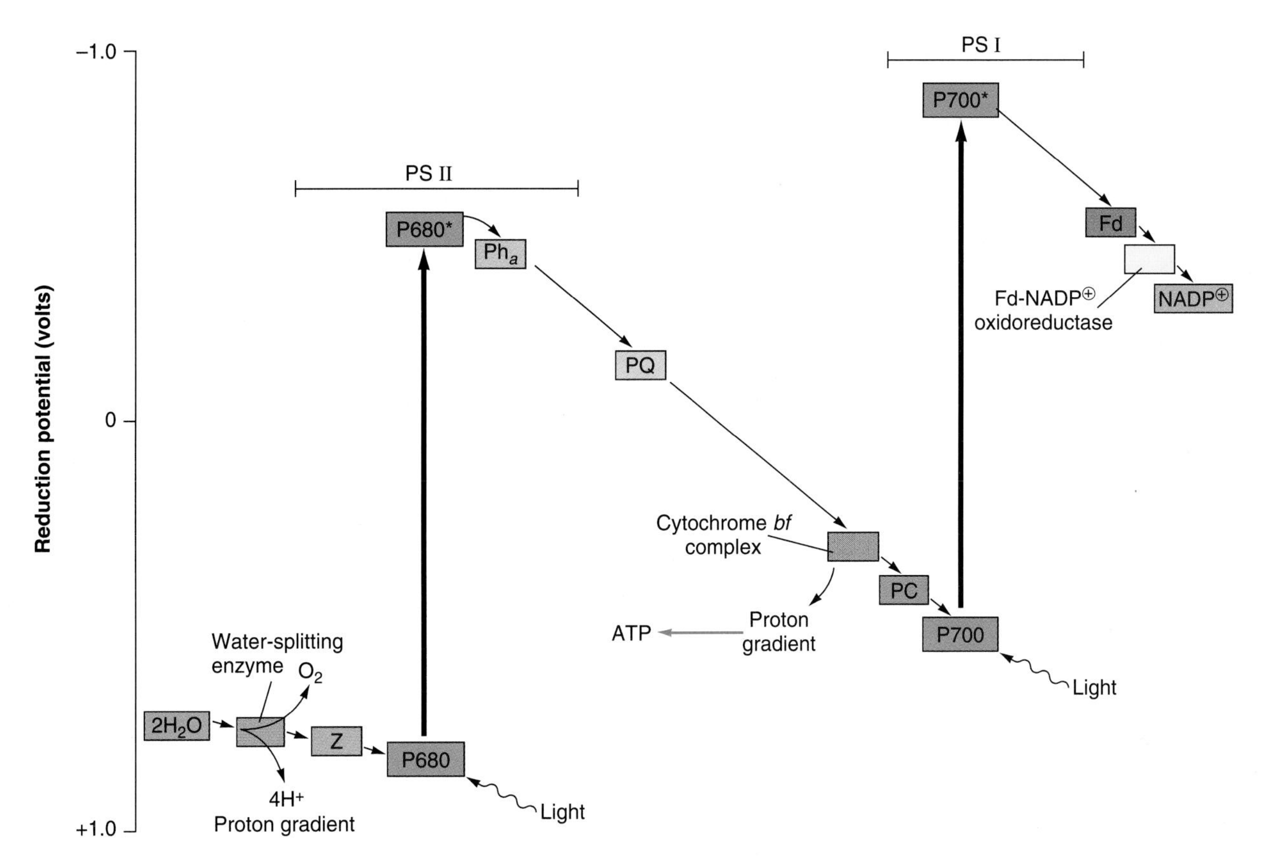

Figure 10.11
In noncyclic photophosphorylation, two electrons are transferred to reduce one $NADP^+$. The energy of a photon absorbed by PS I is used to pump an electron to the potential of a system containing ferredoxin (Fd), which can then reduce $NADP^+$ to NADPH. This leaves PS I oxidized. A second photon absorbed by PS II is used to split water, raising an electron to the potential where it can reduce phaeophytin a (Ph_a). This electron then runs down the chain through plastoquinones (PQ), cytochromes *b* and *f*, and plastocyanin (PC) to reduce P700. PS II is then reduced by electrons from water.

electron transport is essentially cyclic, with the creation of a chemiosmotic potential that generates ATP. Apparently, however, the component (X) that accepts the excited electron from the reaction center does not have a high enough potential to reduce NAD^+ directly, so NADH is formed through a partial reversal of electron transport. To supply the missing electrons, reducing power comes from compounds such as H_2S, leaving sulfur or other oxidized materials as byproducts. The green bacteria have a similar system, but the initial electron acceptor is at a high enough potential to reduce $NADP^+$ directly.

Types of bacterial photosynthesis and their evolution, Section 29.4.

Exercise 10.5 Write an equation for the reduction of $NAD(P)^+$ to $NAD(P)H + H^+$, being sure to use the correct number of electrons. Refer back to Chapter 7 if necessary. How many photons must PS I absorb to reduce one molecule of $NADP^+$? How many photons must then be absorbed by PS II to reduce PS I?

10.8 CO_2 is reduced to organic compounds in the Calvin cycle.

We have now described the apparatus in half of the black box of photosynthesis—the light-dependent reactions that generate ATP and reducing power with the energy of light. These are the truly unique reactions of photosynthesis. The light-independent reactions in the other half of the black box, which reduce CO_2 to organic compounds, are used by all autotrophs, including the chemoautotrophs that generate ATP and NADPH by oxidizing inorganic materials. Now we'll outline these reactions.

When the radioisotope carbon-14 (^{14}C) became available for research just after World War II, many biochemists put it to use as a tracer. Among them, Melvin Calvin, Andrew Benson, and James Bassham tried to determine the pathway of carbon dioxide reduction to sugar. They grew illuminated algae *(Chlorella)* with ordinary carbon dioxide, then added radioactive CO_2 ($^{14}CO_2$) and quickly dropped the cells into alcohol to stop their metabolism (Figure 10.14*a*). The labeled metabolites were then separated by paper chromatography (Figure 10.14*b*). Although the reactions of CO_2 incorporation are incredibly fast,

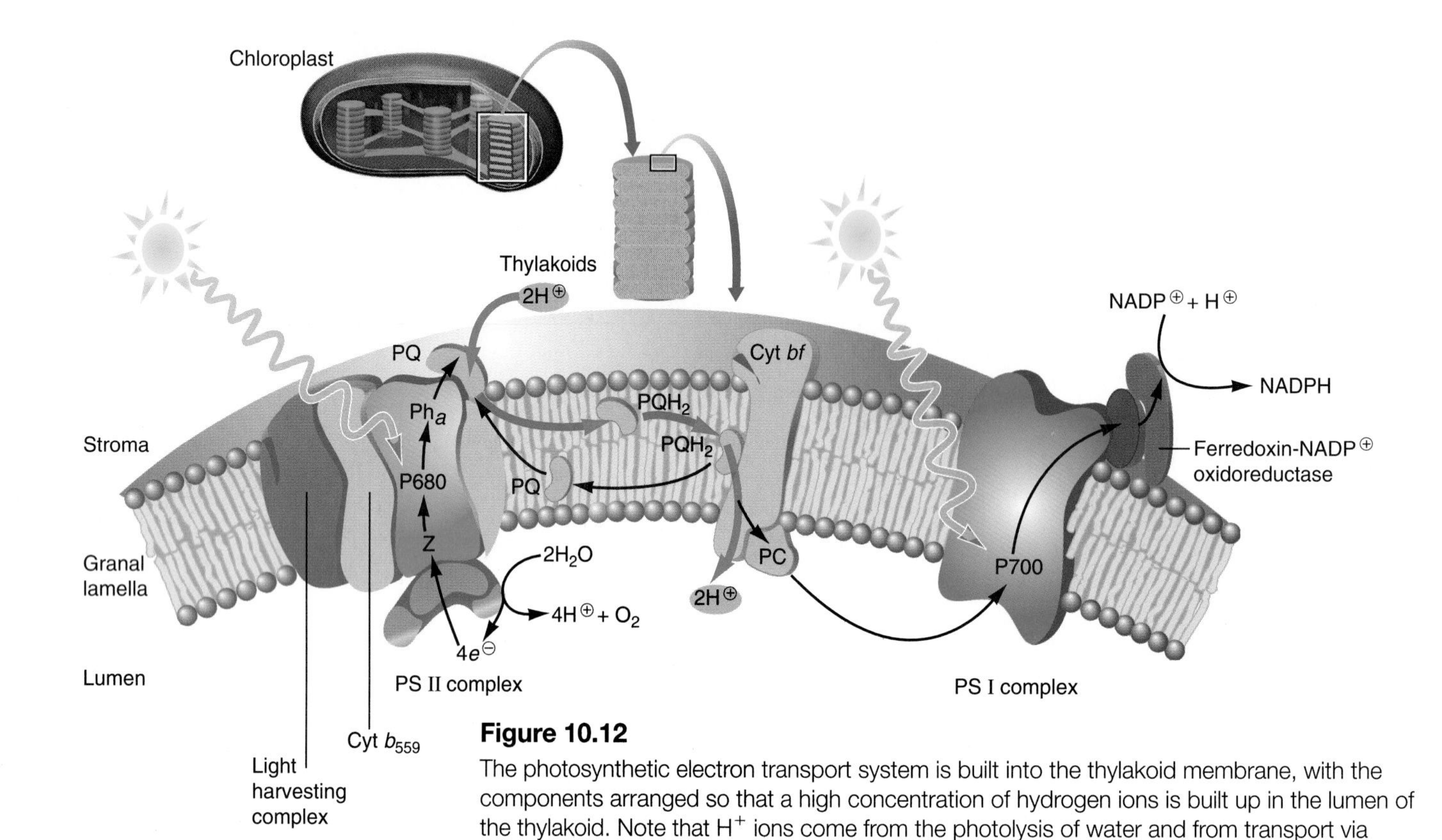

Figure 10.12

The photosynthetic electron transport system is built into the thylakoid membrane, with the components arranged so that a high concentration of hydrogen ions is built up in the lumen of the thylakoid. Note that H^+ ions come from the photolysis of water and from transport via plastoquinone (PQ) and the cytochrome *bf* complex. The return of three H^+ through the ATP synthetase system phosphorylates one ADP to ATP.

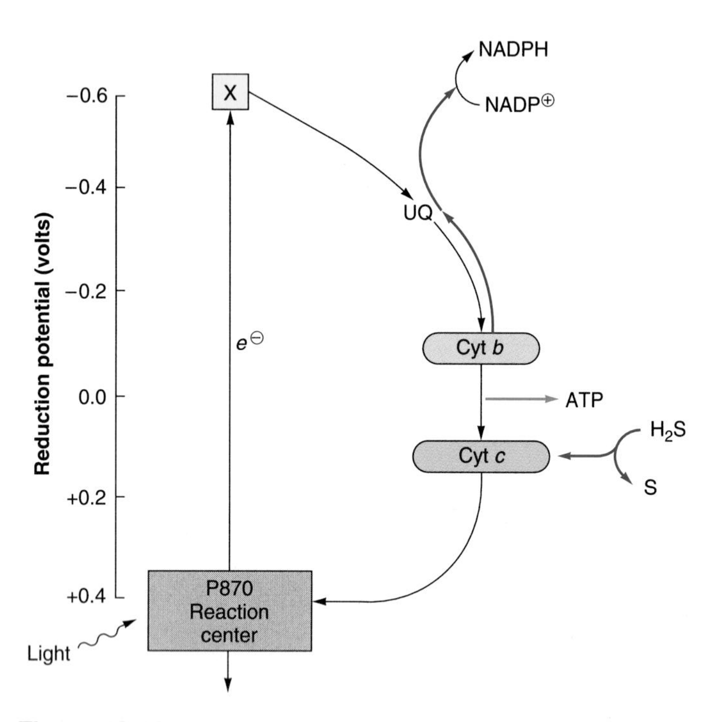

Figure 10.13

Some purple bacteria use this path of electron transport and photophosphorylation. The path is essentially cyclic, but some electrons are sent in a reverse path to reduce $NADP^+$ to NADPH, and H_2S or other reduced compounds supply electrons withdrawn from the system in this way.

they can be studied using this method. If the cells are stopped after the tracer has been incorporated for only a short time, the most highly radioactive metabolites should be the ones made first, those in the earliest part of the pathway. With longer exposure to the tracer, the later metabolites should be labeled. Furthermore, the details of ^{14}C distribution in each compound reflect the way carbon atoms are shuffled as one metabolite is converted into the next; that is, chemical analysis of the radioactive compounds shows *which* atoms in each compound are labeled, and this information can be used to deduce how one compound is converted into another. From these experiments, Calvin and his associates determined that CO_2 is incorporated through a complex cycle of reactions, now called the **Calvin cycle** or the **photosynthetic carbon reduction (PCR)[1] cycle.** Incidentally, when CO_2 is dissolved in water and goes into cells, it enters a complex equilibrium among dissolved CO_2, bicarbonate ion (HCO_3^-), and carbonic acid (H_2CO_3). The enzymes involved in CO_2 reduction may actually be using bicarbonate ions, but we will continue to symbolize the molecules as CO_2.

The key reaction of the Calvin cycle (Figure 10.15) begins with ribulose 5-phosphate, a phosphorylated C_5 sugar. One mol-

[1]Unfortunately, the abbreviation PCR for this cycle could be confused with the same acronym now commonly used for an important technique of molecular genetics, the polymerase chain reaction. However, the context should make clear which meaning is intended.

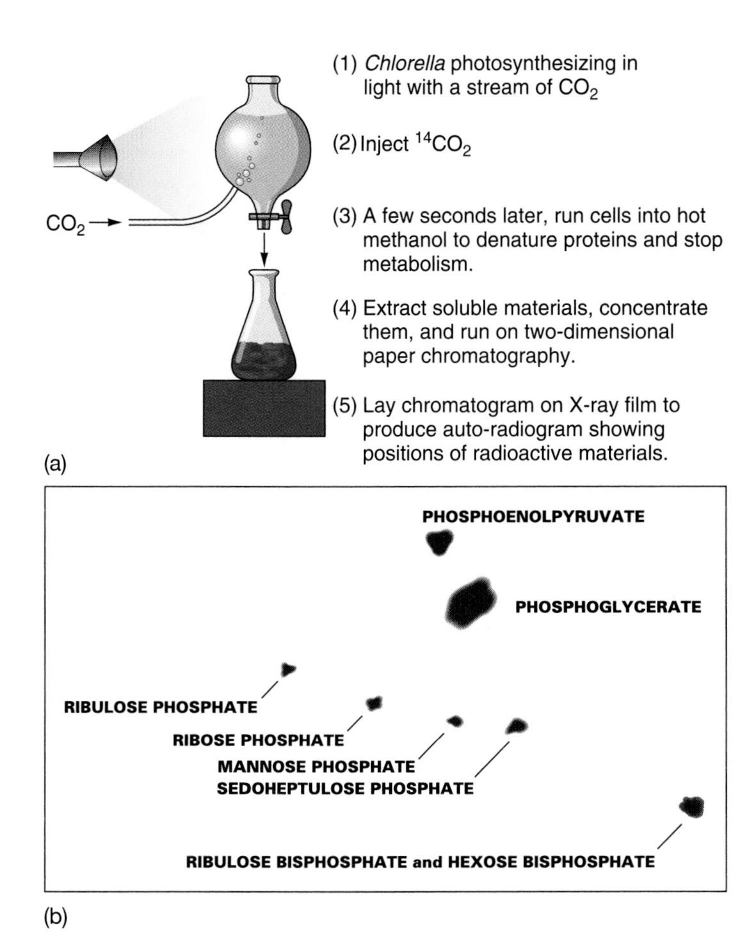

Figure 10.14

(a) Calvin and his associates grew *Chlorella* in an apparatus that allowed them to give very brief flashes of light and to introduce labeled CO_2 rapidly. The algae were then dumped quickly into alcohol to kill them, and their metabolites were separated by paper chromatography. *(b)* The dried paper was laid over a sheet of X-ray film, so spots containing radioactive atoms showed up as dark spots on the film, and the compound in each spot was identified. The conversion of one metabolite into another could be inferred from the changes in labeling by ^{14}C.

ecule of ATP is used to phosphorylate it to ribulose bisphosphate, which then accepts a CO_2 molecule; the resulting C_6 molecule is immediately split into two molecules of 3-phosphoglyceric acid, a C_3 compound. (This reaction is catalyzed by another remarkable enzyme, called rubisco, described in Concepts 10.1.) An additional ATP and one NADPH are used to reduce each of these to glyceraldehyde 3-phosphate, two of which can be made into one molecule of glucose. However, the process is cyclical. As Figure 10.16 shows, for every three molecules of CO_2 that enter the cycle, six molecules of glyceraldehyde 3-phosphate circulate through the cycle. Only one out of six molecules in the glyceraldehyde 3-phosphate pool is withdrawn for other metabolism. Five other molecules ($5 \times C_3 = 15$ carbon atoms) are reshuffled to make three new molecules of ribulose bisphosphate ($3 \times C_5 = 15$ carbon atoms) to begin the cycle again.

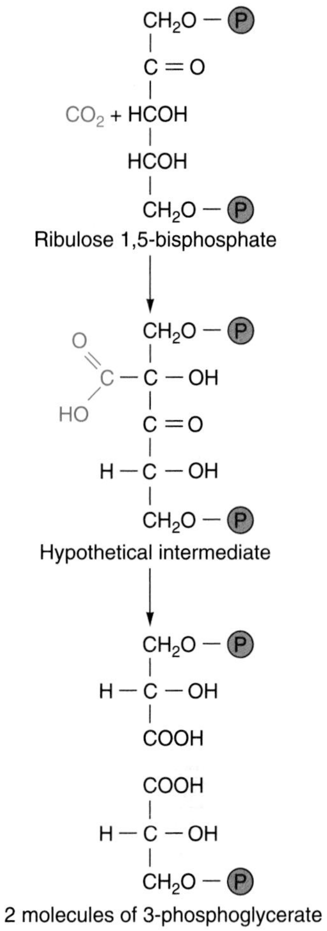

CO_2 + Ribulose 1,5-bisphosphate (CH_2O-P, $C=O$, $HCOH$, $HCOH$, CH_2O-P)

↓

Hypothetical intermediate

↓

2 molecules of 3-phosphoglycerate

Figure 10.15

Carbon dioxide is fixed in a reaction with ribulose 1,5-bisphosphate.

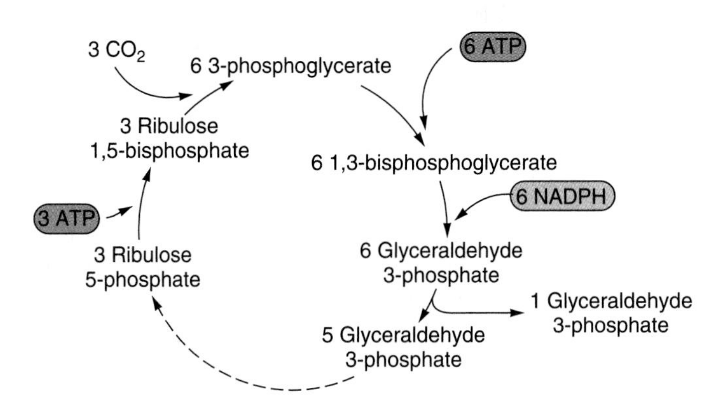

Figure 10.16

In the Calvin cycle outlined here, complex reactions (not shown) convert five molecules of glyceraldehyde 3-phosphate to three of ribulose 5-phosphate.

When all the reactions are added up, the synthesis of one molecule of sugar requires 6 CO_2, 18 ATP, and 12 NADPH. That seems like a lot of energy, but a phototroph has an enormous potential source of that energy in sunlight. The efficiency of photosynthesis is actually quite high. Since the

Concepts 10.1 The Most Abundant Enzyme in the World

The enzyme that catalyzes the initial step of CO_2 reduction to glucose is ribulose 1,5-bisphosphate carboxylase-oxygenase, affectionately called *rubisco* (as a play on "Nabisco"). It may be the most abundant enzyme in the world, because it accounts for half of the protein in chloroplasts and half of the soluble protein in leaves. The whole enzyme (560,000 Da) is a symmetrical complex of eight large subunits (56,000 Da) and eight small subunits (14,000 Da) arranged as shown here:

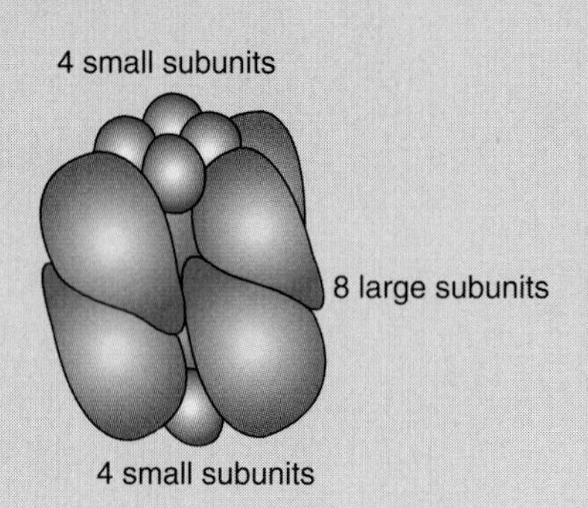

Each large subunit has an active site, making eight active sites per complex; the chloroplast stroma, where the enzyme resides, is therefore rich in sites for CO_2 reduction.

The "oxygenase" in the name of this enzyme refers to a second reaction it catalyzes: combining ribulose bisphosphate with O_2 instead of CO_2 and splitting the molecule into a C_2 and a C_3 compound. The C_2 compound is then oxidized to CO_2 without generating any ATP, apparently wasting energy. CO_2 and O_2 compete for rubisco, and their relative concentrations determine the extent of the carboxylase or oxygenase reactions. The latter is the first reaction of a process called **photorespiration,** which takes in oxygen and releases CO_2, and whose function is unknown. The oxygenase reaction, with subsequent photorespiration, may simply be a defect in rubisco. In any case, photorespiration raises the cost of fixing on CO_2 from 3 ATP to about 5, depending on the temperature and CO_2 concentration, thus lowering the efficiency of photosynthesis by a few percent and significantly reducing plant growth. Agronomists are interested in finding ways to control or eliminate photorespiration, thereby increasing crop yields. Photorespiration has been a significant selective factor in leading some plants to evolve another pathway for incorporation CO_2, as explained in Section 10.9.

free energy of formation of hexose is 686 kcal/mol, reducing one CO_2 to the level of hexose requires 686/6 = 114 kcal/mol, and this is done with 2 NADPH (that is, one-sixth of the 12 required per hexose). As you could calculate in Exercise 10.5, it takes two electrons from PS I to reduce each $NADP^+$, requiring two photons; these must be replaced from PS II, which requires two more photons. Thus four moles[2] of photons must be absorbed to make each mole of NADPH, or a total of eight moles of photons to reduce one mole of CO_2. One ATP is produced along with each NADPH, but we must add in one more ATP—hence, one more photon—per CO_2, for a total of nine. The energy of 600-nm light is 47.6 kcal/mol, so 9 moles of light provide 428 kcal. Therefore, 114/428, or about 26 percent, of the light energy is stored in the sugar.

Glyceraldehyde 3-phosphate, the immediate product of the Calvin cycle, is an intermediate of glycolysis, so it can be converted into sugar or can go into the central metabolic pathways for conversion into amino acids and other cellular materials. However, these photosynthetic reactions only supply energy and materials for the cells that are actually photosynthesizing. A whole plant also contains nonphotosynthetic parts, such as roots, and to supply their needs, the photosynthetic cells produce the disaccharide sucrose, which is carried away in the plant sap. (This is why sap is sweet and why it is such a fine nutrient for many animals.) These sugars may be used in respiration in other tissues, or they may be stored as starch. Ultimately, they may nourish a microorganism or an animal and then a whole food web.

[2]It may seem odd to talk about a mole of photons, but if a mole is defined as an Avogadro's number (6×10^{23}), one could have a mole of anything. An Avogadro's number of photons is an einstein.

The reducing power generated by the light reactions can be used for more than just CO_2 fixation. NADPH and reduced ferredoxin are also used (1) to reduce sulfate to sulfide and nitrate to ammonia for synthesis of amino acids; (2) for lipid biosynthesis; and (3) in cyanobacteria for nitrogen fixation, the reduction of free nitrogen to ammonia, which is so essential to the economy of every ecosystem.

10.9 Some plants use an alternative pathway for CO_2 fixation.

Since the Calvin cycle is so widespread, M. D. Hatch and C. R. Slack's demonstration of a new pathway of CO_2 fixation came as a surprise in the 1960s. The pathway is used by at least one hundred species of plants, primarily those growing in hot, dry conditions. Photosynthesis that uses the Calvin cycle generates a C_3 product, glyceraldehyde 3-phosphate, and is called **C_3 photosynthesis;** the Hatch-Slack pathway generates C_4 compounds first and is called **C_4 photosynthesis** (Figure 10.17*a*). C_4 photosynthesis is a mechanism for pumping CO_2 into the inner tissues of the plant. It begins in the mesophyll cells that constitute the bulk of leaf tissue, using metabolites we have met before in glycolysis and the Krebs cycle. Carbon dioxide is first combined with phospho-*enol*-pyruvate (PEP), one of the highly energetic intermediates of glycolysis, to make oxaloacetate, and the oxaloacetate is reduced to malate. Most plants that conduct C_4 photosynthesis have a so-called *kranz anatomy* (German: *kranz* = wreath or ring), with a layer of bundle-sheath cells surrounding the veins that carry materials in and out of the leaf (Figure 10.17*b*). The mesophyll cells transport the malate they have made into the bundle-sheath cells, where it is decarboxylated; the resulting CO_2 is used to operate the standard Calvin cycle.

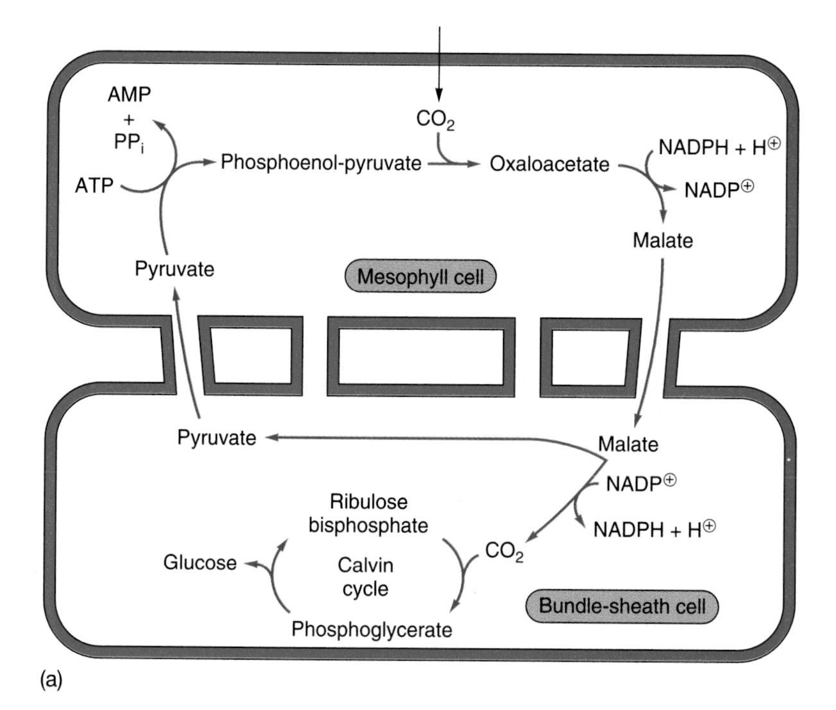

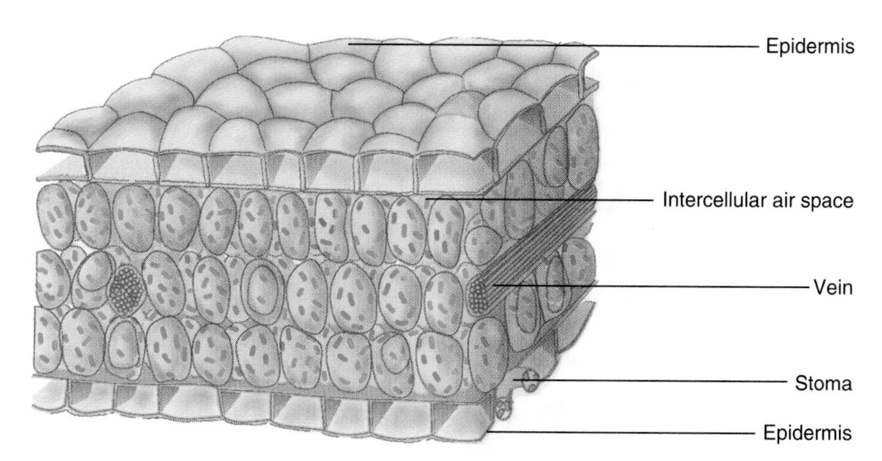

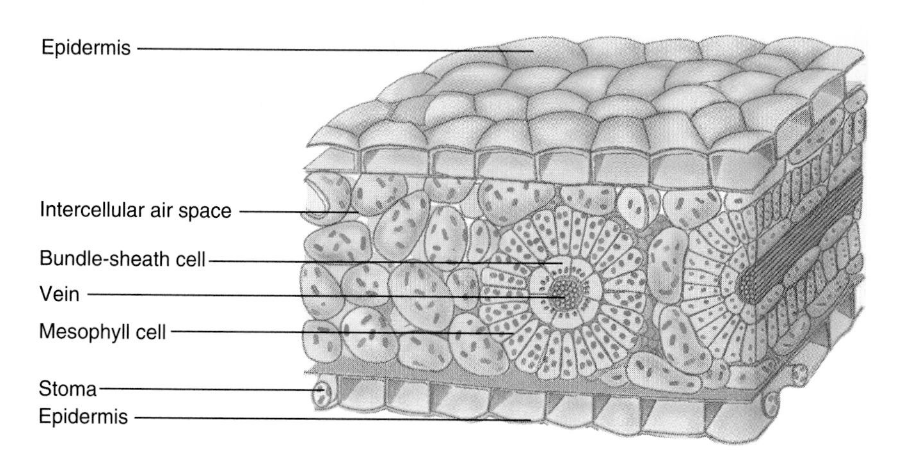

Figure 10.17

(a) C_4 photosynthesis begins in mesophyll cells where CO_2 is combined with phospho-*enol*-pyruvate to make oxaloacetate, which is converted into malate. Malate is transported into bundle-sheath cells, where the CO_2 is removed and used in ordinary C_3 photosynthesis. Pyruvate can cycle back to the mesophyll. *(b)* Many C_4 plants have kranz anatomy in their leaves, with mesophyll cells in a "wreath" (German *kranz*) around the bundle-sheath cells.

Figure 10.18

Sugar cane is a prime example of a C_4 plant. Its photosynthetic apparatus functions well in hot climates, where a C_3 plant often might not grow well because photorespiration would outpace photosynthesis.

Although this C_4 photosynthesis seems like a waste of energy, it actually allows C_4 plants, such as corn and sugar cane, to grow very well in hot climates (Figure 10.18). As explained in Concepts 10.1, the rubisco enzyme catalyzes an alternative oxygenase reaction, leading to photorespiration, and the extent of photosynthesis or photorespiration is governed by two variables:

1. As the temperature increases, the rate of the oxygenase reaction increases faster than the rate of the carboxylase reaction.
2. As the ratio of CO_2 or O_2 increases, photosynthesis is favored over photorespiration.

All vascular plants exchange gases through pores in their leaves known as *stomata,* which admit enough CO_2 to support photosynthesis while allowing O_2 to escape. But plants also maintain a continuous flow of water, which evaporates through the stomata, and when water is limited, the stomata close as a water-conservation mechanism. This closure, however, cuts off the supply of CO_2 while letting the O_2 concentration increase due to the light-dependent reactions of photosynthesis. The higher level of O_2 stimulates photorespiration and wastes energy. Plants that live at moderate temperatures function well in spite of photorespiration, though their cost of fixing one CO_2 can be raised from 3 ATPs to about 5, depending on the temperature and CO_2 concentration. C_4 plants avoid the problem of photorespiration. The carboxylase that converts PEP to oxaloacetate is not inhibited by oxygen. The mesophyll cells therefore make excellent use of the CO_2 they can obtain with limited stomatal opening, and they transfer this CO_2 to the bundle-sheath cells where the CO_2/O_2 ratio can be kept high and where classical C_3 photosynthesis can go on uninhibited.

This story has a historical aspect. Geologic evidence shows that, about 60 million years ago, the CO_2 concentration in the atmosphere was very low, perhaps as low as 160 parts per million (ppm) compared with 360 ppm today. Because the ratio of CO_2 to O_2 determines the extent of the carboxylase and oxygenase reactions of rubisco, plants depending on rubisco were at a real disadvantage in that atmosphere. Between 60 and 30 million years ago, many of them independently evolved a kind of C_4 photosynthesis—several types occur that differ in small ways from the mechanisms described here. Some plants today have an intermediate C_3-C_4 metabolism and can use both mechanisms, but once a plant has switched to C_4 metabolism, it is committed to that method.

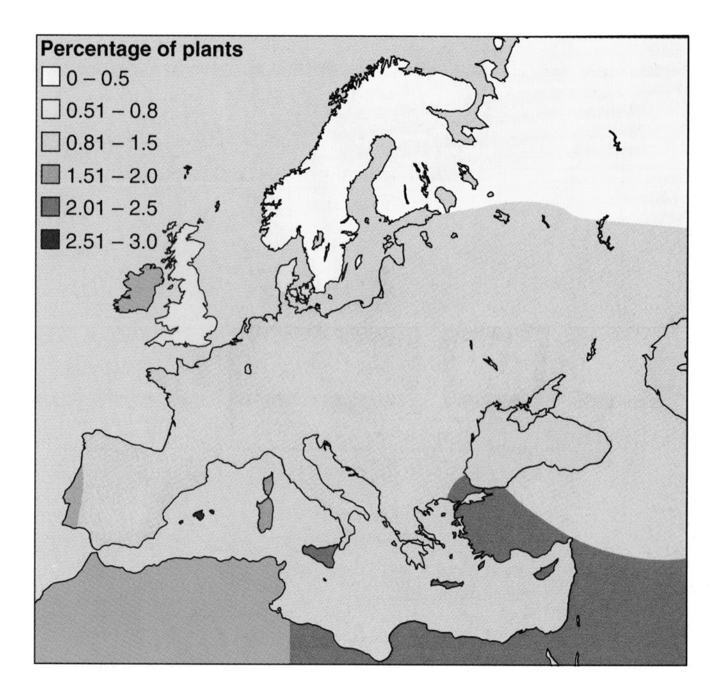

Figure 10.19

The percentage of angiosperms (flowering plants) in Europe that use the C_4 mechanism increases toward the south. This type of photosynthesis is correlated with increasing temperature and aridity.

Because of the cost of the pump mechanism, the Hatch-Slack mechanism requires 5 ATPs, rather than 3, to reduce one CO_2; this is comparable to the cost of photorespiration in C_3 photosynthesis. But photorespiration varies with temperature, while the cost of the C_4 pump does not. Therefore, C_3 photosynthesis is more efficient than C_4 at low temperatures, and C_4 is more efficient at high temperatures. The crossover temperature—the point at which the two costs are equal—depends on the CO_2 concentration in the atmosphere. When C_4 evolved, the crossover temperature was down around 20°C. In today's atmosphere, it is about 28°C, and if the CO_2 in the atmosphere doubles, it will be 40°C.

C_4 plants are growing with plenty of sunlight. The C_4 mechanism is an excellent adaptation for them, because under circumstances that would cause most plants to stop photosynthesis and growth (such as limited water), they are able to continue. This pathway adapts them very well to their way of life. Studies in North America, Australia, and Europe (Figure 10.19) have shown that the percentage of C_4 plants increases with the minimum temperature during the growing season and with greater aridity (dryness).

Many plants use a variation on C_4 metabolism called **crassulacean acid metabolism (CAM)** because it is common in cacti and similar fleshy-leaved plants of the family Crassulaceae, such as the kalanchoës often grown as house plants. The plant opens its stomata at night, taking in CO_2 and fixing it with a C_4 process to make malic acid and some related acids. During the day, the plant closes its stomata, and the accumulated acids are decarboxylated to make CO_2, which is incorporated via the common C_3 pathway. CAM differs from other C_4 photosynthesis in two ways: (1) It is correlated with a daily stomatal cycle, and (2) each cell accumulates the C_4 acids in its vacuole and then converts them with C_3 photosynthesis in its chloroplasts, rather than transporting these acids to another cell. CAM is typical of succulent plants from very arid regions, especially where a great temperature change occurs from day to night, as in many deserts. While other kinds of C_4 photosynthesis are primarily adaptations to heat and low CO_2 concentrations, CAM is a water-conserving process because the stomata are only opened at night when relatively little water evaporates. But between 20,000 and 30,000 species of plants living in a wide variety of habitats use CAM in some form, and it is a facultative process—one that can be induced by environmental circumstances—so some plants only switch to CAM when they become water-limited or salt-stressed.

Exercise 10.6 Human activities are producing a general warming of the atmosphere and an increase in its CO_2 concentration. Given the information presented here about the relative efficiencies of C_3 and C_4 photosynthesis, what would you predict about the survival and evolution of plants in the near future?

Coda It is hard to overemphasize the significance of photosynthesis in biology. The metabolism of virtually every ecosystem begins with photosynthesis, and it is essential for maintaining the atmospheric oxygen we depend on. There is good reason to believe that the most primitive organisms obtained their energy by oxidizing organic molecules in their environment, but such chemotrophy would have come to an end after a while as these energy sources were used up. Only the evolution of the first phototrophs saved the day by tapping an enduring energy source. Furthermore, those phototrophs evolved the first electron transport systems, some of which were converted in later evolution to be used for chemotrophy (a story we tell in Chapter 29).

Historically, therefore, photosynthesis underlies the oxidative metabolism carried out by our mitochondria. The evolution of such different metabolic systems demonstrates the flexibility and potential resourcefulness of a genetic system. Having evolved the proteins to carry out one process efficiently (photosynthesis, in this case), the system has the potential to modify these structures for related processes, perhaps opening up whole new possibilities for evolution and ecological niches. In the next chapter, where we develop a more complete picture of cellular structure and function, we will again see how the evolution of a fundamental mechanism or structure—such as a pathway for responding to an external signal or a protein useful for holding a cell in shape—can be adapted for a wide range of functions.

Summary

1. Photosynthesis in eucaryotes occurs in chloroplasts. The internal thylakoid membrane contains the molecules that absorb light and carry out electron transport in a light-dependent process. This generates ATP and NADPH.
2. The stroma around the thylakoid membrane contains enzymes that carry out a light-independent process, using ATP and NADPH to reduce carbon dioxide to sugar.
3. Molecules absorb light through activation of their electrons. An electron can absorb a photon that contains exactly enough energy to carry the electron to a higher orbital. The absorption spectrum of a compound shows which wavelengths of light it absorbs.
4. Chlorophylls are the major light-absorbing pigments. They have a large ring with delocalized electrons that absorbs red and blue light. Carotenoids also absorb some light for photosynthesis.
5. In photosynthesis, $NADP^+$ is reduced to NADPH, and a proton gradient is created that drives the synthesis of ATP. In plants, the reducing agent for this process is water. Some photosynthetic bacteria use other reducing agents, such as sulfide or hydrogen.
6. Experiments using different colors of light show that two photosystems must cooperate in plant photosynthesis. Each photosystem is a complex of proteins and pigments embedded in the thylakoid membrane.
7. A simple form of electron transport in photosynthesis uses a cyclic pathway, beginning with the absorption of light by chlorophyll. The cyclic photophosphorylation effected by this process creates only ATP, but no NADPH.
8. Plants have a noncyclic pathway that creates both ATP and NADPH. The Z scheme of photosynthesis shows how photosystem I absorbs light and reduces $NADP^+$ to NADPH, leaving the system oxidized. It is reduced again by electrons derived from the splitting (photolysis) of water, and the energy for this process is derived from the absorption of light by photosystem II.
9. Sugar is made from CO_2 in the Calvin cycle. The key reaction is the activation of ribulose-5-phosphate to ribulose bisphosphate by ATP, so it is able to combine with CO_2. A complicated series of reactions, requiring NADPH and additional ATP, then generates a carbohydrate product (glyceraldehyde 3-phosphate) and regenerates ribulose bisphosphate. Six turns of this cycle yield a molecule of sugar.
10. Some plants that are adapted to high temperatures with relatively little water use an alternative pathway for CO_2 fixation called C_4 photosynthesis. In their mesophyll cells they combine CO_2 with phospho-*enol*-pyruvate to make oxaloacetate, which is converted into malate. Malate is transported into bundle-sheath cells, where the CO_2 is removed and used in ordinary C_3 photosynthesis. This process circumvents photorespiration and uses enzymes with a high affinity for CO_2.
11. In crassulacean acid metabolism (CAM), a variation of C_4 photosynthesis, plants open their stomata at night, take in CO_2, and use it to synthesize malate and related acids, which are stored. Then, in the daytime, the stomata close, the acids are decarboxylated, and the resulting CO_2 is used in the C_3 pathway of photosynthesis.

Key Terms

chloroplast 196
thylakoid membrane 196
lumen 196
stroma 196
grana/granum 196
light-dependent reactions 197
light-independent reactions 197
absorption spectrum 199
chlorophyll 200
carotenoid 200
action spectrum 201
Hill reaction 202
photophosphorylation 202
photosystems I and II (PS I, PS II) 202
photosynthetic unit 202
antenna complex 202
tracer 203
autoradiography 203
reaction center 203
cyclic photo-phosphorylation 203
noncyclic photo-phosphorylation 204
Calvin cycle 206
photosynthetic carbon reduction (PCR) cycle 206
photorespiration 208
C_3 photosynthesis 208
C_4 photosynthesis 208
crassulacean acid metabolism (CAM) 211

Multiple-Choice Questions

1. CO_2 fixation occurs
   a. within the thylakoid membranes.
   b. inside the lumen.
   c. within the stroma.
   d. in the grana.
   e. within the membranes connecting the grana.
2. ATP and NADPH are produced
   a. in the Calvin cycle.
   b. in the light-dependent reactions.
   c. within the stroma.
   d. within the matrix.
   e. between the two membranes that form the boundary of the chloroplast.
3. If you know the absorption spectrum of a pigment, you can be reasonably certain of
   a. the pigment's color.
   b. the wavelength of the visible light it reflects.
   c. the wavelength of the visible light it absorbs.
   d. the energy of the visible light it reflects and absorbs.
   e. all of the above.
4. Which of the following can supply electrons to reduce the carbon in CO_2 to sugar?
   a. H_2S
   b. water
   c. light
   d. *a* and *b*, but not *c*
   e. *b* and *c*, but not *a*
5. The excited electrons that are removed from chlorophyll may be used to reduce
   a. NADP.
   b. H_2O.
   c. O_2.
   d. ATP.
   e. glucose.
6. ATP is produced by means of ______ phosphorylation carried out within ______.
   a. oxidative / an ETS
   b. oxidative / a photosystem
   c. photophosphorylation / chlorophyll
   d. substrate-level / chlorophyll
   e. photophosphorylation / an ETS
7. During cyclic photophosphorylation, which is (are) reduced using excited electrons from chlorophyll?
   a. an ETS
   b. NADP
   c. P680
   d. water
   e. all of the above
8. Which does *not* occur during cyclic photophosphorylation?
   a. oxidation of chlorophyll
   b. production of NADH
   c. production of ATP
   d. an ETS
   e. reduction and subsequent oxidation of ferredoxin
9. In the Z scheme, $P680^+$ is reduced with electrons directly from
   a. H_2O.
   b. P700.
   c. NADPH.
   d. ferredoxin.
   e. an ETS.
10. Rubisco can be defined as
    a. the final sugar to leave the Calvin cycle.
    b. the first sugar to be formed in the Calvin cycle.
    c. an enzyme that adds CO_2 to a 5-carbon sugar.
    d. the compound that accepts CO_2 in the Calvin cycle.
    e. the enzyme that forms 6-carbon sugars from 3-carbon sugars.

True-False Questions

Mark each statement true or false, and if false, restate it to make it true.

1. The longer the wavelength of electromagnetic radiation, the higher its energy.
2. In the absence of other interactions, an electron that has been boosted to a higher energy level will either fall back to its ground state immediately or remain in its new orbital indefinitely.
3. If a pigment is responsible for a specific process, the action spectrum of the process will be similar to the absorption spectrum of the pigment.
4. Both photosystems I and II can carry out cyclic photophosphorylation, but only photosystem I can engage in noncyclic photophosphorylation.
5. Since photosystem I absorbs light of longer wavelengths than photosystem II, photosystem I absorbs quanta at a higher energy state.
6. Oxidized chlorophyll in photosystem II can be reduced by electrons from water, whereas oxidized chlorophyll in photosystem I can be reduced by electrons from photosystem II or by the return of its own electrons.
7. During photosynthesis in bacterial autotrophs, reducing power comes from water while energy comes from light.
8. The Calvin cycle requires the input of ATP, NADPH, and CO_2, in addition to the necessary enzymes and the 5-carbon sugar, ribulose 5-phosphate.
9. Several of the intermediate molecules of the Calvin cycle are the same or similar to intermediates of the Krebs cycle.
10. For every six molecules of CO_2 that enter the Calvin cycle, two molecules of glyceraldehyde 3-phosphate will leave the cycle to form one molecule of glucose.

Concept Questions

1. Contrast the structure of a chloroplast with that of a mitochondrion.
2. Where and when is a proton-motive force generated during photosynthesis?
3. What is the advantage to a plant of undergoing both cyclic and noncyclic photophosphorylation?
4. What are the products of the light-dependent and light-independent phases of photosynthesis?
5. Explain why C_3 photosynthesis is more efficient than C_4 photosynthesis in temperate regions, while the reverse is true in hot, arid regions.

Additional Reading

Barber, J. (ed.) *The Photosystems: Structure, Function, and Molecular Biology.* Elsevier Co., New York, 1992.

Govindjee, and William J. Coleman. "How Plants Make Oxygen." *Scientific American,* February 1990, p. 50. A biochemical mechanism enables phototrophs to use solar energy to split water molecules into oxygen gas, protons, and electrons.

Kirk, John T. O. *Light and Photosynthesis in Aquatic Ecosystems,* 2d ed. Cambridge University Press, New York, 1994.

Lawlor, D. W. *Photosynthesis: Molecular, Physiological, and Environmental Processes,* 2d ed. John Wiley & Sons, New York, 1993.

Tobin, Alyson K. (ed.) *Plant Organelles: Compartmentation of Metabolism in Photosynthetic Cells.* Cambridge University Press, New York, 1992. Seminar series of the Society for Experimental Biology (Great Britain).

Youvan, Douglas C., and Barry L. Marrs. "Molecular Mechanisms of Photosynthesis." *Scientific American,* June 1987, p. 42. Modern techniques for analyzing the photosynthetic apparatus.

Internet Resource

To further explore the content of this chapter, log on to the web site at:

http://www.mhhe.com/biosci/genbio/guttman/

11 The Dynamic Cell

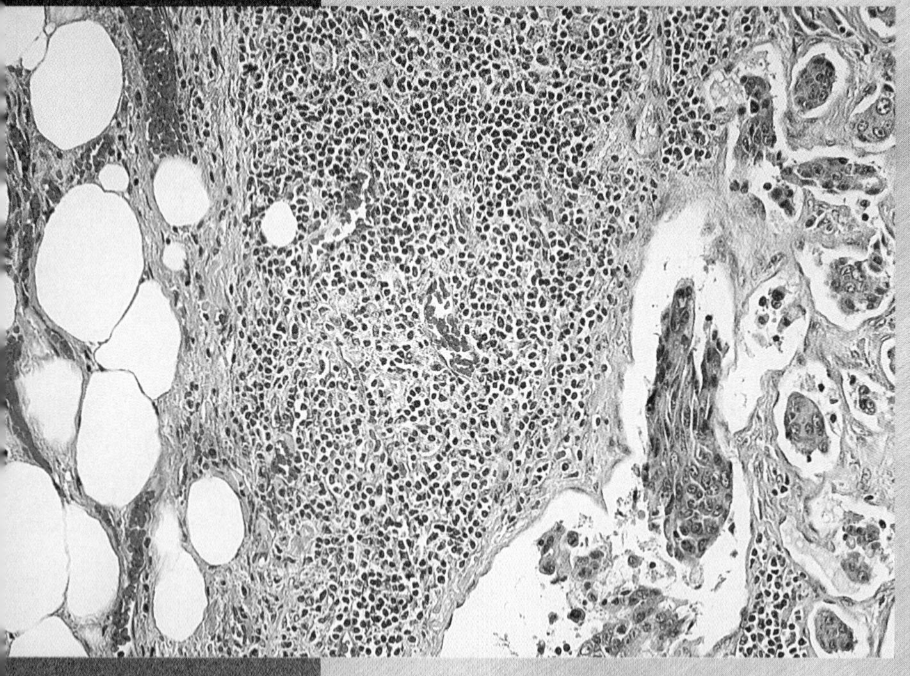

The large cells occupying spaces on the right side of the picture are cancer (adenocarcinoma) cells that have invaded a lymph node containing the many normal small lymphocyte cells that occupy most of the micrograph.

Key Concepts

A. How Cells Regulate Their Growth

11.1 Cell growth means both synthesizing new biomolecules and dividing to form more cells.

11.2 Cells grow by assimilating materials from their environment through biosynthetic pathways.

11.3 Organisms maintain themselves in steady-state conditions through homeostatic mechanisms.

11.4 Allosteric proteins are general informational transducers.

11.5 Allosteric enzymes at critical points regulate the activity of metabolic pathways.

11.6 Cells can recognize and respond to external ligands.

11.7 Signal ligands carry information at different levels of activity.

11.8 Eucaryotic cells respond to external signals through a common transduction pathway.

B. How Cells Move and Change Their Form

11.9 Actin filaments effect many cell movements.

11.10 Microtubules shape cells and are used for movement.

11.11 Microtubule structures are organized by special centers.

11.12 Objects move on microtubules by means of specialized motors.

11.13 Cilia and flagella are movable bundles of microtubules.

11.14 Bacterial flagella are made of flagellin, a different globular protein.

11.15 Materials can be moved across membranes by bulk transport.

The doctor sits down with her patient, but before she can speak, the look on her face tells the story. The news is bad, as bad as it can be. The patient has cancer, and it has metastasized—spread from its place of origin to many tissues throughout the body. In each site, cancer cells are growing, spreading, overwhelming their more civilized neighbors.

The concept of a cell that we have developed so far is incomplete and even misleading. It leaves the impression that cells are simply building blocks that form the larger structures of plants and animals, and then remain in place, never moving or changing. But of course that isn't true. Cells synthesize themselves and grow. They constantly change shape, move, and interact with each other in ways that even the electron microscope cannot capture. Cancer is probably the most dramatic way—but not the only way—that cells call attention to their dynamic nature. Cells often move around. Some of our cells normally move through the blood and into other tissues, protecting us, but in an abnormal cancerous state, a cell's ability to move becomes deadly. Most tissue cells grow slowly and remain in place because of complex inner controls; a cancerous cell has lost those controls.

In this chapter, we will develop a more dynamic view of cells. We will examine cell growth and the biosynthetic pathways responsible for that growth. Cells and organisms, whether growing or not, regulate their complex activities and maintain a so-called steady state and an internal condition called homeostasis. To understand this, we will develop the general principles of communication circuits, which include regulation through negative feedback, the process of information transfer called transduction, and the roles of allosteric proteins. Finally, we will discuss the organization and structure of the cytoskeleton, which shapes cells and allows them to move. ■

A. How cells regulate their growth

11.1 Cell growth means both synthesizing new biomolecules and dividing to form more cells.

Cells maintain themselves and grow, but what does growth mean? A cell is growing when it makes more of its own structure, but even though cells are about 80 percent water, just adding water—perhaps by swelling in a hypotonic solution—would not be considered growth. Fundamentally, growth means synthesizing the characteristic biomolecules of cell structure—proteins, nucleic acids, polysaccharides, and lipids. Remember that most biomolecules are polymers, and their monomers are made in extensions of the metabolic pathways outlined in Chapter 9. The cellular stores of energy, mostly in ATP and NADPH, are largely used to synthesize new biomolecules. In this chapter we will explore growth and growth regulation in general, leaving the specific mechanisms that synthesize proteins and nucleic acids for the following chapters.

Although many of the cells in our bodies are constantly growing, replacing old tissues, this process is hidden from us. It is easy, though, to observe growth in unicellular organisms by putting bacteria or common baker's yeast into a **nutrient medium,** a mixture of the materials the cells need for growth, such as the simple medium shown in Table 11.1. A bacterial cell in such a medium increases in length by synthesizing more of its biomolecules. After an hour or so, it rather quickly undergoes **binary fission,** dividing down the middle to become two cells (Figure 11.1*a*). A baker's yeast grows by putting out a bud, which then grows until it is the size of the original cell (Figure 11.1*b*). In both cases, one cell becomes two. So growth means

Table 11.1 Recipe for a Growth Medium

Component	Amount
Na_2HPO_4	7.0 g
KH_2PO_4	3.0 g
NaCl	0.5 g
NH_4Cl	1.0 g
Glucose	6.0 g
$MgSO_4$	10^{-3} M
$CaCl_2$	10^{-4} M
$FeCl_3$	10^{-6} M
$ZnCl_2$	10^{-6} M
Water	1,000 ml

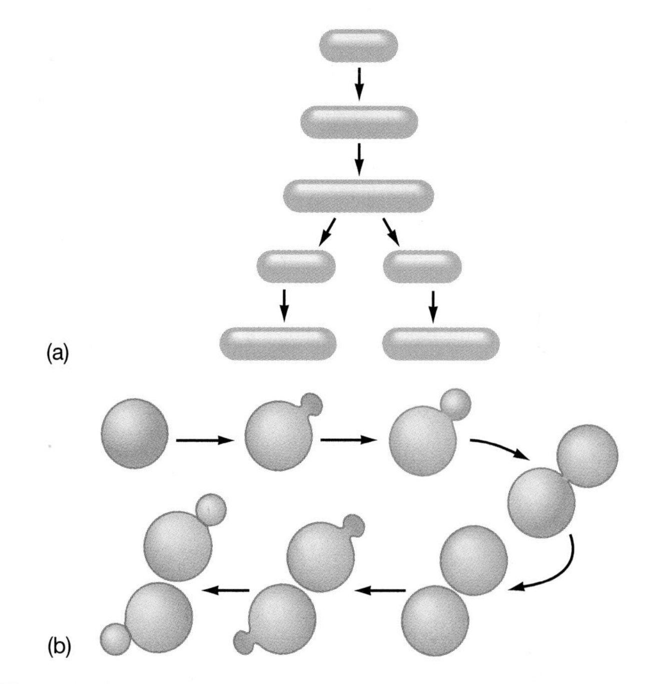

Figure 11.1
(a) A bacterial cell grows by adding new material until it is twice its original size and then dividing in two (binary fission). *(b)* A budding yeast cell grows by putting out a small bud, which grows to full size and separates from its parent.

two things: Cells get larger, and they *proliferate*—increase in number. As we outlined in Chapter 2, the number of cells doubles periodically under ideal conditions, in a pattern of *exponential growth* that can produce a huge population in a short time. All cells grow this way, though eucaryotic cells take longer, usually several hours, to double in size and divide.

A multicellular organism also grows by cell proliferation. Plants actually grow a great deal through elongation of some cells, rather than through further division. In animals, nerve cells and certain muscle cells do not divide after reaching a mature state, though animals grow partly by the enlargement of muscle cells. However, cells in tissues such as the skin and intestinal lining of animals or the root cap of plants are continually being lost and replaced by new cells through cell division. Even organisms that are not growing are replacing old molecules. We can observe the growth of plant and animal cells in a

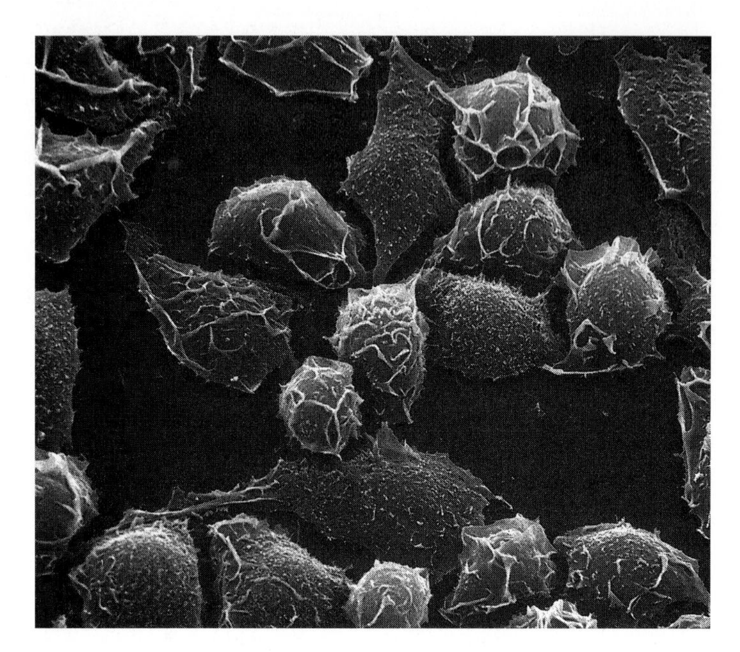

Figure 11.2

Animal cells growing in cell culture (here, human white blood cells called macrophages) have a generally rounded shape, in contrast to the more rectangular shape that they assume in a tissue. They move by pushing out thin extensions.

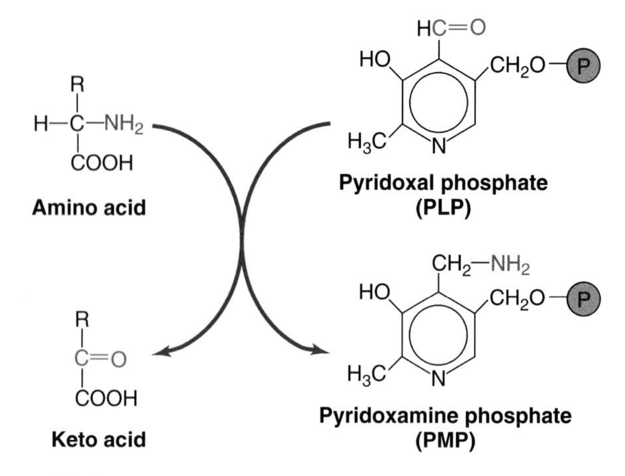

Figure 11.3

Transaminations are carried out by vitamin B_6, which is converted back and forth between its pyridoxal and pyridoxamine forms as it replaces amino groups *(red)* with keto groups *(green),* thus converting a keto acid into an amino acid or vice versa.

cell culture by putting bits of tissue in nutrient media similar to those used for growing bacteria (though richer in nutrients and more complex). The cells separate in the culture flask and often move about (Figure 11.2), and they, too, grow exponentially by repeated division.

The exponential growth of organisms is like the growth of money earning compound interest—that is, a percentage of the capital is added to the capital periodically, so interest is always being paid on a larger amount. In biological growth, each cell is a factory that makes more cellular material, and each one will eventually produce a copy of itself. As with money, the more you have, the more you get. Each cell contains enzymes, which operate metabolic pathways, which produce more amino acids, which form more enzymes, which operate its metabolic pathways faster, and so on without end. Thus, the enzymes in every new bit of cell mass produce more cell mass, just as every dollar your savings account earns goes to work to make more dollars. Growth, an organism's principal activity, is therefore a circular process.

Exercise 11.1 How will each element in the nutrients listed in Table 11.1 be used to make cellular materials?

11.2 Cells grow by assimilating materials from their environment through biosynthetic pathways.

Organisms are said to **assimilate** materials from their surroundings into their structure. Bacteria in a nutrient medium, for instance, take up raw materials such as glucose and convert them into their own biomolecules. The carbon, hydrogen, and oxygen atoms in glucose combine with the nitrogen atoms of ammonia, the sulfur atoms of sulfate, and phosphate from the medium. The metabolic pathways of the cell build these atoms into monomers such as amino acids, nucleotides, fatty acids, and sugars; we call these *end products* of each pathway because generally their only fate is to be built into the molecules of cell structure (proteins, nucleic acids, lipids, and polysaccharides), not transformed into other small molecules:

$C_6H_{12}O_6$ $NH_4^{\oplus}$ $SO_4^{2\ominus}$ $PO_4^{3\ominus}$ → Metabolic pathways → Nucleic acids, Polysaccharides, Lipids, Protein

Cells also assimilate dissolved sodium, potassium, magnesium, calcium, and chloride ions into their cytoplasm, and they incorporate heavy metal ions into their enzymes, including iron, zinc, nickel, molybdenum, and manganese.

Biosynthesis begins with the metabolites of the central metabolic pathways discussed in Chapters 9 and 10, especially glycolysis and the Krebs cycle. Several of these metabolites are keto acids, which have a keto group (C=O) in addition to a carboxyl group (–COOH); keto acids are converted into amino acids through reactions called **transaminations.** Enzymes transfer amino and keto groups from one acid to another by using as a shuttle the coenzyme pyridoxal, which we obtain as vitamin B_6 (Figure 11.3). (This gives us good reason to make sure we have enough vitamin B_6 in our diets.) Thus, pyruvate is converted into alanine:

$$CH_3\text{–}C(=O)\text{–}COO^{\ominus} \longrightarrow CH_3\text{–}CH(NH_2)\text{–}COO^{\ominus}$$

Oxaloacetate is converted into aspartate:

$$\mathrm{COOH{-}CH_2{-}C(=O){-}COO^{\ominus}} \longrightarrow \mathrm{COOH{-}CH_2{-}CH(NH_2){-}COO^{\ominus}}$$

α-ketoglutarate is converted into glutamate:

$$\mathrm{COOH{-}CH_2{-}CH_2{-}C(=O){-}COO^{\ominus}} \longrightarrow \mathrm{COOH{-}CH_2{-}CH_2{-}CH(NH_2){-}COO^{\ominus}}$$

As Figure 11.4 shows, these three amino acids are the gateways to pathways used to synthesize most of the other amino acids, as well as the purine and pyrimidine nucleotides of nucleic acids. (All these reactions, of course, require specific enzymes, which make up much of the protein in a cell.) These metabolic maps allow us to trace how the atoms in glucose—and other compounds that are catabolized in the central pathways—are assimilated into the proteins and other structures of a cell.

Animals differ from most other organisms because they lack the enzymes for many biosynthetic reactions, so they must obtain eight of the twenty amino acids—the *essential* amino acids—from their food. Remember, too, that animals and some other organisms are unable to synthesize many coenzymes, so vitamins that supply these substances or their precursors must be included in the diet.

Essential amino acids and nutrition, Section 47.18.

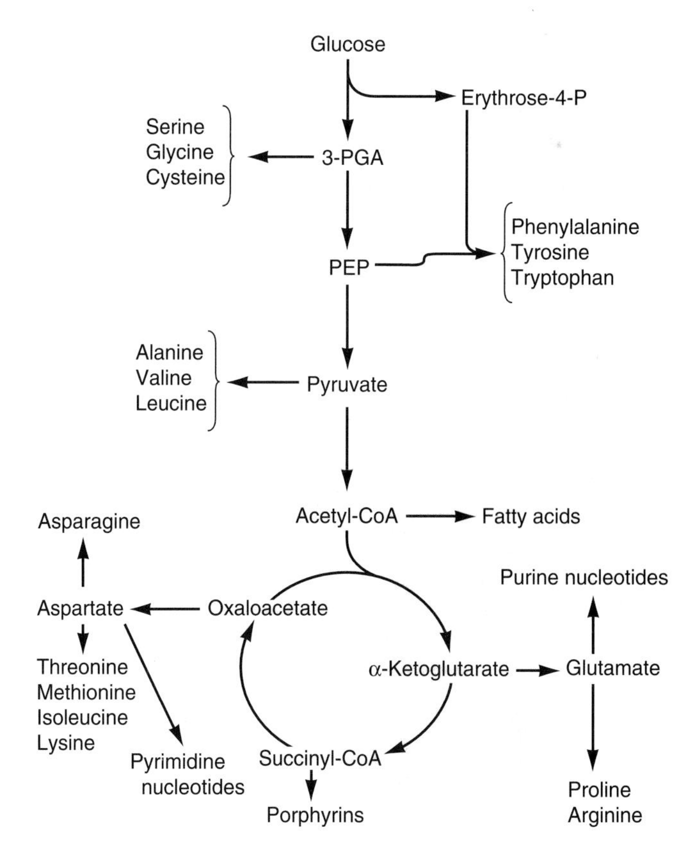

Figure 11.4

Cells synthesize the monomers that make the bulk of their structure through the major biosynthetic pathways shown here, so carbon atoms that originate in glucose could become part of amino acids, nucleotides, or other molecules. These pathways are very efficient because they simultaneously synthesize several amino acids, with relatively short, specialized branches. Notice that the amino acids alanine, aspartate, and glutamate, once formed, are gateways to several other molecules.

Exercise 11.2 (Teleonomy exercise.) Translate the following teleological statement into a teleonomic one: Organisms have their biosynthetic pathways organized in a branching pattern—instead of having a separate biosynthetic pathway for each amino acid—so they can make their components efficiently.

11.3 Organisms maintain themselves in steady-state conditions through homeostatic mechanisms.

We humans are subject to a lot of stresses. The temperatures of July and January make us sweat and shiver, but our body temperature remains the same. We sometimes eat too much salt or sugar, drink too much or too little water, eat strange mixtures of foods, overwork, and get too much or too little exercise. Yet our bodies retain essentially the same composition, at least over the short term. (Too many of these stresses over long times, of course, can lead to disease processes.)

Humans illustrate a remarkable fact of life—that an organism's structure and composition can remain virtually the same even though it continuously takes in nutrients and produces wastes. The situation looks like an equilibrium. Remember that if a mixture of materials is in a dynamic equilibrium, their concentrations all remain constant even though individual molecules are constantly shifting back and forth. Something like this happens in an organism, but the condition certainly is not an equilibrium. (An organism at equilibrium is dead.) This is a **steady state,** in which materials constantly flow *one way* through the system: Incoming nutrients are converted into more of the organism's structure or into wastes, with no reverse flow. So an organism is like a river that maintains the same level and shape, even though water is constantly flowing through it. Like the river, metabolic flow through an organism is *balanced:* The rates of all reactions are so tuned to one another that the concentrations of metabolites remain quite constant, even though no one molecule remains in any pool for long. The organism's structure—proteins, nucleic acids, lipids—also stays constant, even though some molecules are being added or

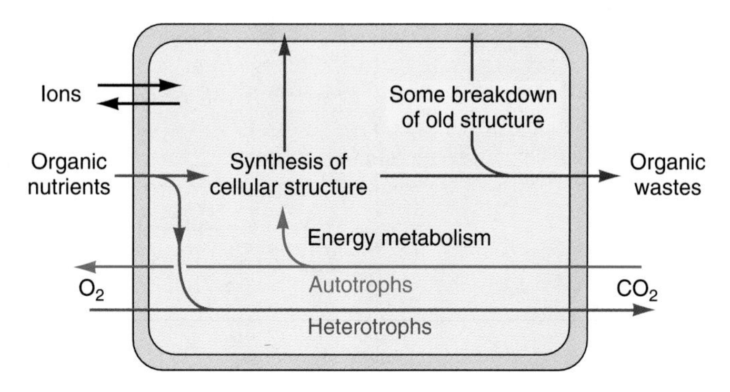

Figure 11.5

An organism maintains a steady state by taking in nutrients, synthesizing more of its own components from them, breaking down old components, and excreting wastes. The concentrations of all metabolites and structural components remain essentially constant. If the organism is growing, its growth is balanced, so the relative amounts of all components remain constant even though the total mass is increasing. The principal difference between autotrophs *(green arrows)* and heterotrophs *(red arrows)* lies in their energy metabolism and carbon sources.

replaced (Figure 11.5). In a growing organism, all the newly added materials remain in balance with one another.

Dynamic equilibrium, Section 5.2.

Organisms have evolved mechanisms for survival in the face of harsh and constantly changing environmental conditions. Even if its surroundings change, every organism, to one degree or another, maintains quite a constant internal state called **homeostasis.** That is, each cell must keep its internal conditions constant within a narrow range; failing to do so would be fatal. Each enzyme is adapted to operate best in a cytosol with certain concentrations of ions such as H^+, K^+, Cl^-, and Mg^{2+}. A cell controls these concentrations, and other factors, by adjusting the flow of materials across its plasma membrane and by regulating the rates at which materials flow through its metabolic pathways and are polymerized into cell structures. It slows down or accelerates each action to maintain a balanced, steady-state flow.

Enzyme activity in relation to ion concentrations, Section 5.7.

Homeostasis is achieved by regulating certain variables, such as temperature or the concentration of an ion. Regulation always depends on closed circuits that employ negative feedback. **Feedback** is a process in which information from the output of a device is sent back to control the device:

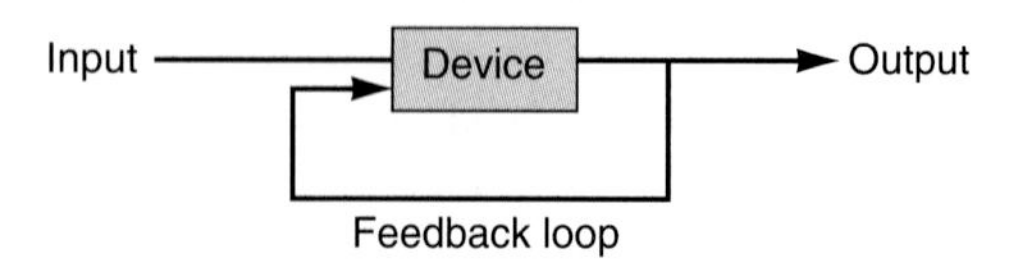

In *negative* feedback, the information is used to keep the regulated variable close to a certain desired point. Temperature regulation in a house illustrates negative feedback very well:

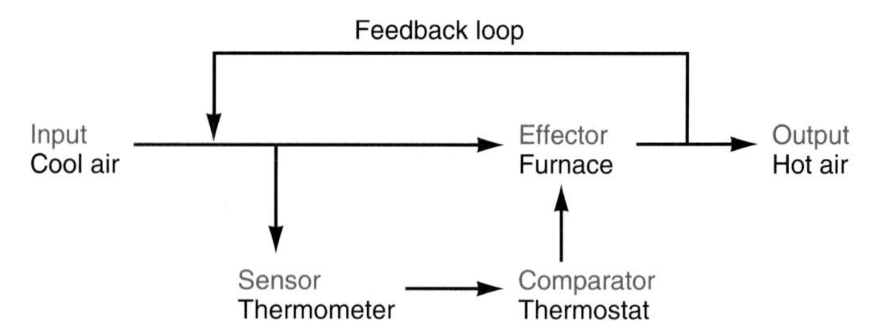

The temperature in the house is the *controlled variable.* Every circuit has a *sensor* that monitors the variable from moment to moment; in the house, this is a thermometer that measures the temperature. The sensor communicates with a *comparator,* a thermostat, which compares the actual temperature with a *set point,* the desired temperature set on the thermostat. If the temperature is close to the set point, nothing happens. If the air is colder than the set point, the thermostat sends an activating signal to an *effector,* the furnace, whose output, hot air, raises the temperature to the set point. This is a feedback system because information about the output of the system is fed back into the system, and it is negative feedback because it keeps the temperature close to the set point. It depends on a communication circuit joining the sensor, comparator, and effector.

Negative feedback is fundamental to physiology. Homeostatic systems in our bodies monitor conditions from moment to moment, both internally and externally, and send out signals to correct any problems (Figure 11.6). They keep our temperature at 37°C and our blood pH at 7.4, maintain set concentrations of CO_2, glucose, calcium, and other materials in the blood, and govern many other processes. These homeostatic mechanisms operate on fundamentally *chemical* information. Plants and animals can also detect and respond to factors such as light, temperature, gravitation, and sound, but these too involve complicated processes that depend upon chemical interactions.

A regulatory circuit requires at least one **transducer,** a device that converts one form of energy or information into another. Some transducers operate strictly on energy, such as a motor that transduces electricity into mechanical energy. Generally, though, the term is reserved for devices that convert information as well as energy. The thermometer in our house-heating circuit transduces thermal energy into an electrical signal or a point on a visible scale; a photoelectric tube transduces light into electricity; and the receptor cells of the eye transduce light signals into nerve impulses. Organisms have some transducers, such as muscles, that only convert energy, and many others that convert information. In this discussion we will be concerned mostly with informational transducers, and we will emphasize the chemical nature of all biological transducers.

A control system with negative feedback leads to homeostasis and stability. In contrast, *positive feedback* creates amplification and leads to instability—that is, a small change leads to

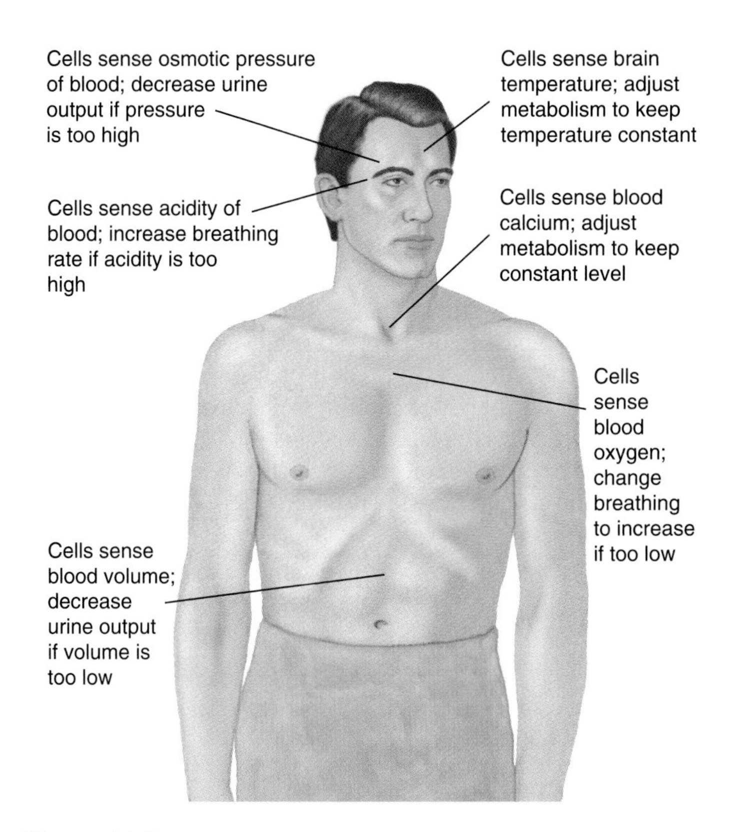

Figure 11.6

Animals are able to monitor their internal conditions from moment to moment and maintain quite constant homeostatic conditions by means of specialized cells in critical locations that create negative feedback systems.

increasingly greater effects. A circuit containing a microphone, amplifier, and speaker can deafen you at a concert because of positive feedback if the microphone gets near the speaker. The loud squeal comes from a weak initial sound produced as the speaker is amplified again and again by the circuit. This is a positive-feedback effect. A biological case of positive feedback occurs during childbirth. As the baby in the uterus (womb) grows, the uterine muscles respond to stretching by signaling the pituitary gland to release the hormone oxytocin. Oxytocin signals the uterine muscles to contract further, which in turn brings about production of more oxytocin in a cycle that eventually culminates with the baby's birth. A small contraction and a little hormone lead to lots of hormone and forceful contractions. Because a positive-feedback system is inherently unstable, it can only be used to produce a short-term effect, such as birth; long-term regulation requires negative feedback.

Exercise 11.3 You are driving along a freeway at a constant 55 mph, despite changes in the slope and condition of the road. Using the general regulatory circuit model, explain how your own regulatory circuit allows you to do this. (And don't say you use the cruise control, unless you can explain precisely how *that* works.)

Exercise 11.4 A certain cold-water fish maintains a high level of glycerol in its blood as a kind of antifreeze. It has a hormone (H) that regulates this glycerol level. If the glycerol concentration falls below a set level, the fish's liver cells release H, which stimulates fat cells to release glycerol until the concentration is back to normal. Identify the components of this process in terms of a general regulatory circuit.

11.4 Allosteric proteins are general informational transducers.

In Section 5.11 we explained the essential properties of an *allosteric protein:* It can bind two different molecules, or **ligands,** with quite distinct shapes, at sites with correspondingly distinct shapes, so that the interaction at one site affects the other site (Figure 11.7). One ligand is commonly a small molecule or ion; the other may be a different small molecule or a macromolecule such as another protein or a nucleic acid. Allosteric proteins show two features that we have emphasized before: First, ligands bind to proteins through weak bonds, so the interaction between the two molecules is readily reversible; second, a ligand interacts stereospecifically with a protein and tends to stabilize the protein in one conformation.

Allosteric proteins are vitally important in biology because they are informational transducers. We will illustrate this point with allosteric enzymes that regulate metabolic pathways and

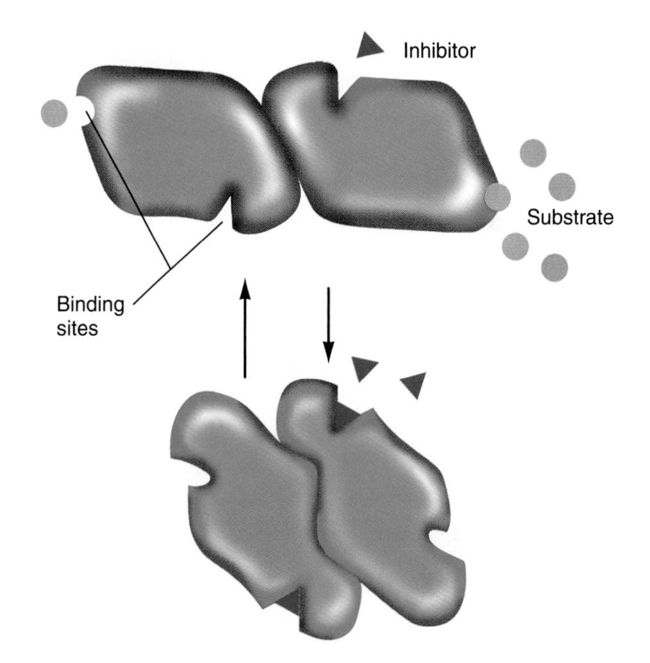

Figure 11.7

An allosteric protein (here made of two identical subunits) has two dissimilar binding sites, which are specific for different ligands: in this case, substrates *(circles)* and inhibitors *(triangles).* It can also shift back and forth between two (or more) conformations: in one conformation, substrates can bind but inhibitors cannot; in the other conformation, the reverse is true. Binding each ligand tends to stabilize the protein in one conformation or the other.

with receptor proteins that change cellular activities in response to chemical signals. In Chapter 18 we will show how allosteric proteins turn genes on or off in response to chemical signals.

11.5 Allosteric enzymes at critical points regulate the activity of metabolic pathways.

To maintain a steady state, a cell must coordinate its hundreds of metabolic reactions so they stay in step with one another. Cells accomplish this coordination by having allosteric enzymes at key points in pathways, enzymes whose activities are regulated by specific metabolites. Most commonly, the enzyme is inhibited by the end product of its pathway, and the process is called **end-product inhibition** or **feedback inhibition** (Figure 11.8). Consider a general biosynthetic pathway such as:

$$A \xrightarrow{\text{Enzyme 1}} B \xrightarrow{\text{Enzyme 2}} C \xrightarrow{\text{Enzyme 3}} D \xrightarrow{\text{Enzyme 4}} E$$

Enzyme 1 responds to the concentration of the end product (E) and is inhibited when this concentration becomes too great, so it reduces the flow of material through the pathway. For example, the amino acid threonine is converted into the amino acid isoleucine through a multistep pathway. Threonine and isoleucine have quite different structures, and several enzymes are needed to convert one into the other. The first enzyme in the pathway (threonine deaminase) is allosterically inhibited by isoleucine. The enzyme has two conformations. In one conformation, it is a good enzyme that converts threonine into the next metabolite. In its alternative conformation, the enzyme exposes a regulatory site that can bind isoleucine, while its active site is not available to threonine, so it has no enzymatic activity. As the isoleucine concentration rises, more enzyme molecules bind isoleucine and are stabilized in their nonfunctional form, thus shutting off the production of isoleucine. When the concentration of isoleucine falls, isoleucine molecules will tend to come off the enzyme molecules, which revert to their active conformation and increase the reaction rate.

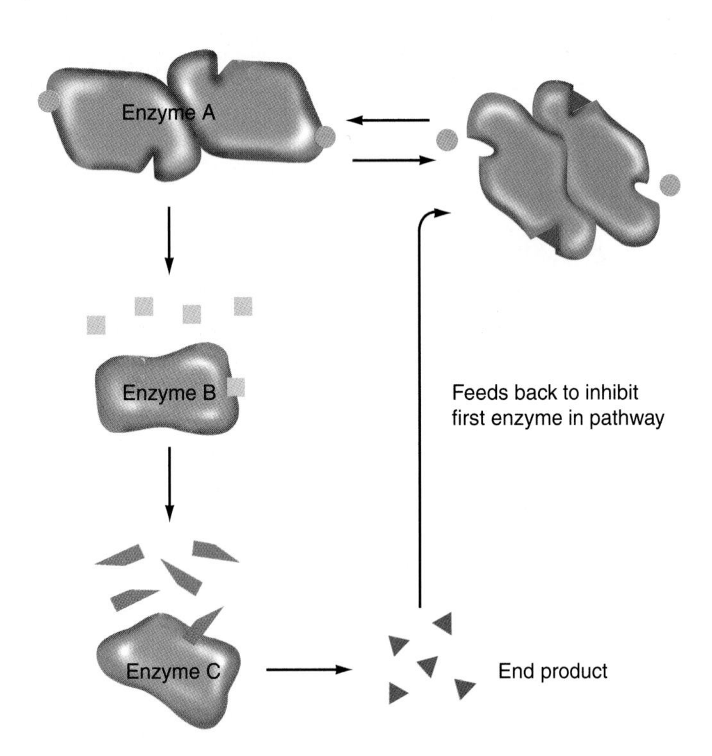

Figure 11.8
Feedback inhibition occurs in a biosynthetic pathway as the concentration of the end product rises. Enzyme A is an allosteric protein with both an active site and a regulatory site that can bind molecules of the end product. In one conformation, it is a good enzyme that operates on its substrate (such as threonine). When bound to the end product (such as leucine), the protein is stabilized in an alternative conformation in which it has no enzymatic activity. Thus, the concentration of the end product determines how many enzyme molecules are active and how many are inhibited.

Exercise 11.5 Allosteric effects depend on the binding of a ligand (L) to a protein (P), forming a complex (C). This interaction is characterized by an equilibrium constant, $K = [C]/[L][P]$, where the brackets, as usual, mean the concentration of each component. If the amount of protein in a cell is constant, use the idea of an equilibrium constant to explain why the amount of complex increases if there is more ligand and decreases if there is less.

Regulatory circuits of this kind show how remarkably well a system can be designed by evolution, because feedback controls in cells are placed for maximum efficiency, just where a good engineer would place them (Figure 11.9). In a branched biosynthetic pathway, each end product inhibits the first enzyme specific to its synthesis. Sometimes different end products inhibit the same early reaction; a cell may have two distinct enzymes to catalyze the same reaction, each one inhibited by a different end product. Such excellent design testifies to the power of natural selection operating over a long time; these regulatory systems were probably developed during the early evolution of cells, when those that accidentally developed the most efficient

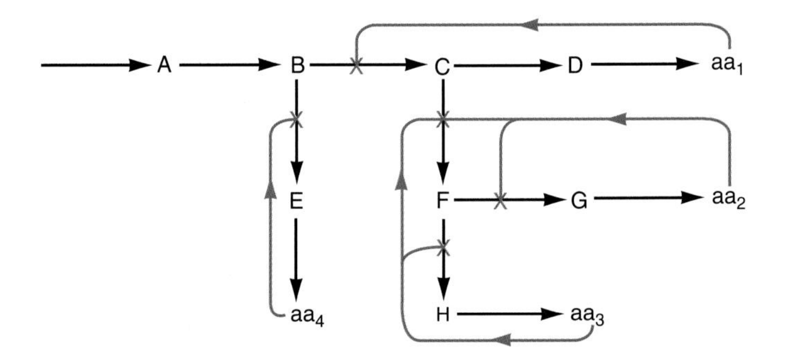

Figure 11.9
A hypothetical pathway that synthesizes four amino acids (aa_1–aa_4) shows several points of regulation. Each amino acid feeds back to inhibit the first reaction specific to its synthesis. Sometimes two or more amino acids inhibit the same reaction, as in a situation where aa_2 and aa_3 both inhibit the conversion of C to F. Here there may be two enzymes, one sensitive to aa_2 and one to aa_3.

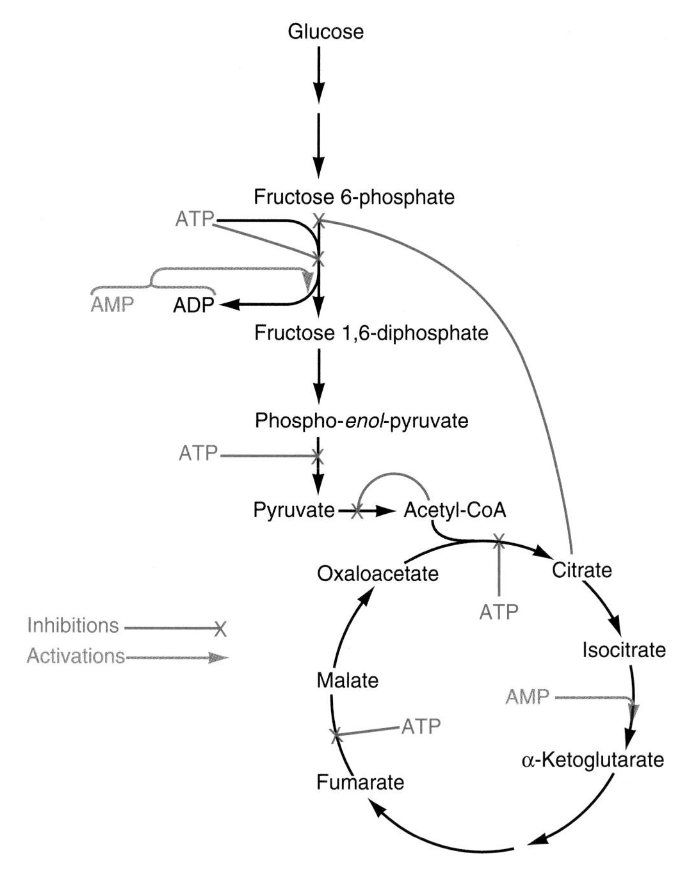

Figure 11.10

The major pathways are regulated at a few key points. Inhibitions are marked by a red X, activations by an orange arrow. Notice particularly that when the ATP pool is large, it feeds back to inhibit several reactions that tend to generate ATP. When the AMP and ADP pools are large, they activate these reactions.

controls were able to reproduce faster and more efficiently than those with less functional controls.

Regulatory ligands may also *activate* allosteric enzymes, rather than inhibit them. The main pathways that ultimately produce ATP are regulated at several points by ATP, ADP, and AMP (Figure 11.10). The pools of these three compounds rise and fall; as ATP is used, it is converted to ADP and AMP, while the latter two are converted back to ATP during respiration. ATP tends to turn off its own synthesis by allosterically inhibiting some key enzymes. Conversely, ADP and AMP allosterically activate certain enzymes. Cellular respiration therefore decreases when the ATP pool is large; when the ATP pool is small, the pools of ADP and AMP are large, and they stimulate respiration.

Exercise 11.6 Explain why selection has favored the placement of regulatory enzymes at the first metabolic reaction specific to the end product being regulated, rather than at an earlier or later reaction.

Exercise 11.7 Following is a set of biosynthetic pathways that produce three amino acids (aa_1, aa_2, and aa_3). Which enzymes should be inhibited by each end product?

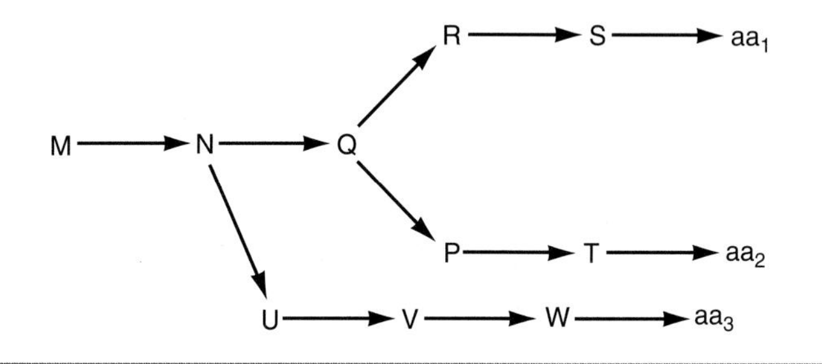

11.6 Cells can recognize and respond to external ligands.

If you take a walk in the woods on a pleasant summer morning, the local plants will tempt you to supplement your breakfast with their fruits. Should you pick a wild apple or nibble some berries? The answer will depend on what you know about your local flora, and perhaps on a little judicious sampling. Of course, you can safely eat familiar fruits like blackberries or strawberries, but you may notice birds eating bright red berries that you can't identify without a good field guide, and you'll wonder whether you could eat them, too. While we don't recommend random sampling of unknown fruits, if you were really hungry, you might be able to distinguish safe ones from dangerous ones by the difference between a sweet and a bitter taste.

Microorganisms swimming about in the water or animals wandering across the land don't carry field guides to safe foods, so they must rely on a sense akin to taste. To maintain their integrity, organisms need information about their external environment as well as about their internal conditions. This information may be critical. The environment poses opportunities, such as new sources of energy, as well as threats from destructive agents. Even simple unicellular organisms detect external chemicals and respond appropriately; they move toward sources of food and avoid dangerous materials like strong acids. Any environmental factor an organism recognizes and responds to is a **stimulus,** and the most basic stimuli are chemical.

Flagellated bacteria swim in response to chemical stimuli. If we put a fine tube filled with a nutrient like glucose into a bacterial culture, the cells swarm around the mouth of the tube (Figure 11.11). This movement in response to a chemical is **chemotaxis** (*-taxis* = turning), and when the bacteria are moving toward an attractant such as glucose that they can use, it is *positive chemotaxis.* Bacteria can perceive that the sugar concentration increases as they move toward the tube and decreases as they move away, and an internal mechanism makes them move in the positive direction. On the other hand, they show *negative chemotaxis* by retreating from a tube containing a repellant, such as a noxious alcohol.

Bacterial flagellum structure, Section 11.11.

Organisms detect each kind of external ligand by means of a distinctive **receptor protein,** an allosteric protein whose binding site on the external surface of the plasma membrane is stereospecific for some ligand. Its only function is to detect the

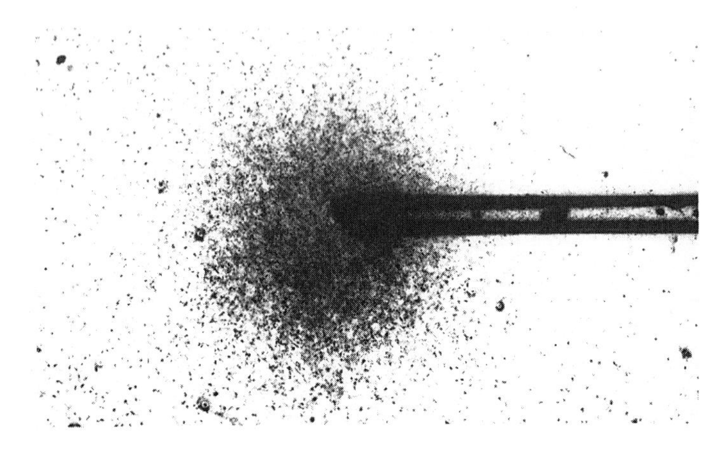

Figure 11.11
Bacteria *(Escherichia coli)* are attracted to a glucose solution in a capillary tube.

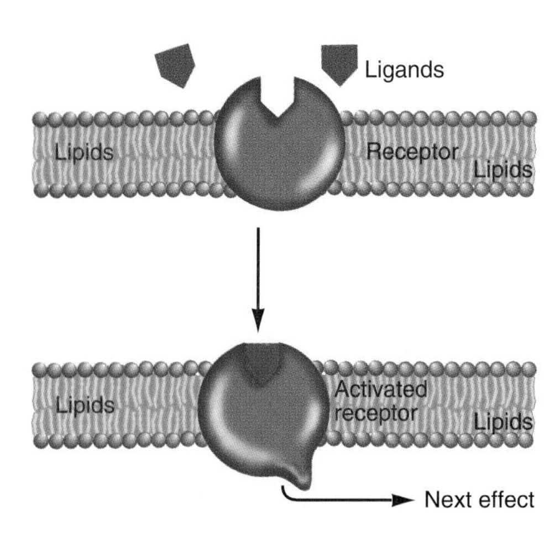

Figure 11.12
A receptor can recognize the presence of a distinctive ligand by binding to it. This interaction changes the conformation of the receptor, setting a new chain of events in motion.

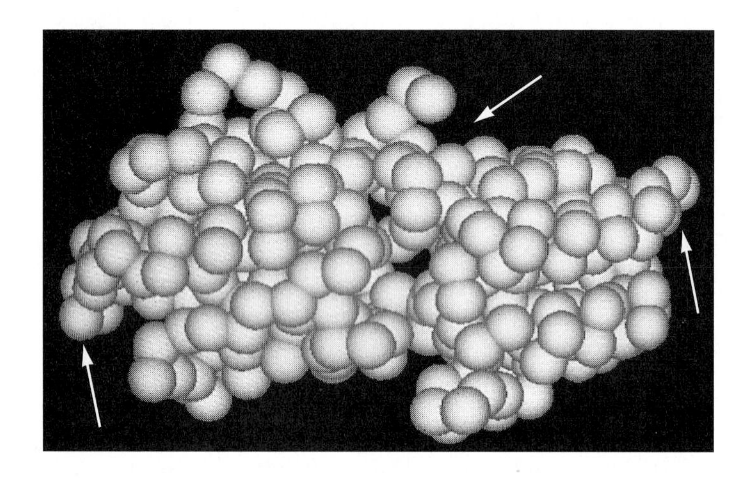

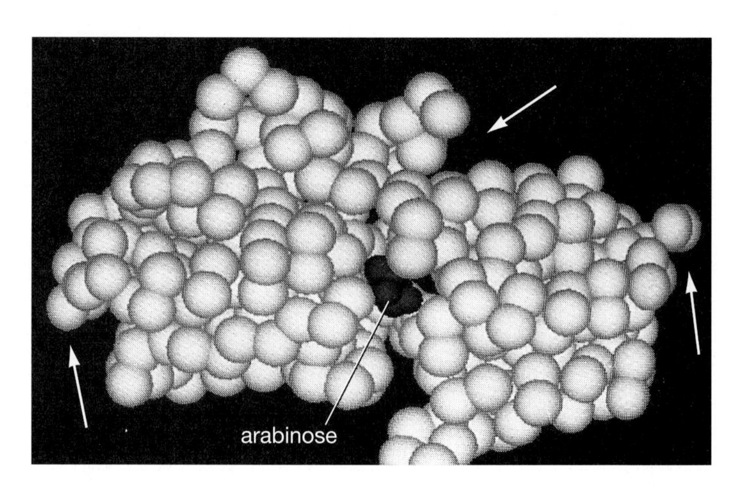

Figure 11.13
The arabinose-binding protein on the surface of an *E. coli* cell detects arabinose by binding to it and changing its conformation. The change in shape is subtle, but the arrows point to just three of the places where one can see clear differences between the two conformations by comparing clusters of atoms.

ligand. When the ligand binds, the receptor changes its shape, setting other events in motion (Figure 11.12). Thus the mere presence of the ligand—detected by the receptor—produces a significant effect, such as chemotaxis. Biologists now commonly say the receptor *recognizes* a ligand that binds to it or allows an organism to recognize the ligand.

Bacteria have a **chemoreceptor** for each type of ligand they can recognize. For instance, the arabinose-binding protein shown in Figure 11.13 has a site stereospecific for the sugar arabinose. When bound to arabinose, the chemoreceptor changes its shape subtly, starting a string of changes in other proteins that direct the movements of the flagella. Each cell has several of these sensory systems because each chemoreceptor is very specific. Distinct receptors recognize glucose, fructose, maltose, and other sugars. One chemoreceptor detects both L-aspartic and L-glutamic acids, which are very similar. Another chemoreceptor detects several alcohols (methanol, ethanol, *n*-propanol, and so on) to which the cells respond negatively.

An animal's senses of smell and taste also start with stereospecific chemoreceptors that detect many external ligands. These senses guide us to food, which usually has a sweet or pleasing taste, and warn us away from harmful materials, which tend to have bitter tastes and noxious odors. Chemoreceptor proteins on specialized receptor cells in our tongues and nasal passages relay their information to other parts of the nervous system, which operates through its own specialized ligands and receptors to pass chemical signals from one nerve cell to another.

Exercise 11.8 Receptors that recognize external ligands are integral proteins of the cell membrane. Their binding sites for their ligands are on the outside, but they trigger changes inside the cell. Think of two distinct ways in which the external binding and conformational change might initiate the next event. *Hint:* Consider changes in quaternary protein structure for at least one mechanism.

11.7 Signal ligands carry information at different levels of activity.

Organisms use chemical signals and chemoreceptors to carry information between the parts of a cell, between the cells of a multicellular organism, and between individuals of the same or different species. These communication systems use special **signal ligands,** which are generally unusual molecules that are not catabolized for energy and materials. They include:

- Hormones, which carry signals between the cells of a multicellular organism
- Pheromones, which carry signals between organisms of a species
- Alarmones, which carry signals within a single cell

The *-mone* ending has been extended to two other types of molecules called *allomones* and *kairomones,* which have ecological functions that we will explain shortly. Strictly speaking, allomones and kairomones don't all act as signal ligands, but it is useful to explain them in this context.

The first signal ligands to be discovered were **hormones,** materials produced by one type of cell in a multicellular organism to elicit a distinctive response in other types of cells (Figure 11.14). Hormones are essential agents of homeostasis. The hormone insulin, whose structure was discussed in Chapter 4, regulates the glucose concentration in the blood of many vertebrates. Beta cells in the pancreas monitor the amount of glucose in the blood circulating around them, using glucose receptors on their surfaces. As the blood glucose concentration rises, the beta cells release insulin into the blood. Many cells throughout the body have insulin receptors on their surfaces, and the binding of insulin to these receptors causes these cells to remove glucose from the blood and store or metabolize it. Other hormones do the opposite, signaling cells to release more glucose into the blood. Plant hormones may signal certain cells to start growing faster or to slow their growth. Hormonal systems become extremely complex; one hormone may signal one type of cell to release a second hormone that may signal another type of cell to release still a third hormone.

Individuals of a single species signal one another with **pheromones** (Figure 11.15). Males and females of a species, for example, use sex pheromones to find and recognize one another, and to stimulate mating behavior. Recruiting or aggregating pheromones signal individuals to gather together; thus the massive numbers of ladybird beetles that congregate in the fall find one another through pheromones. Other pheromones

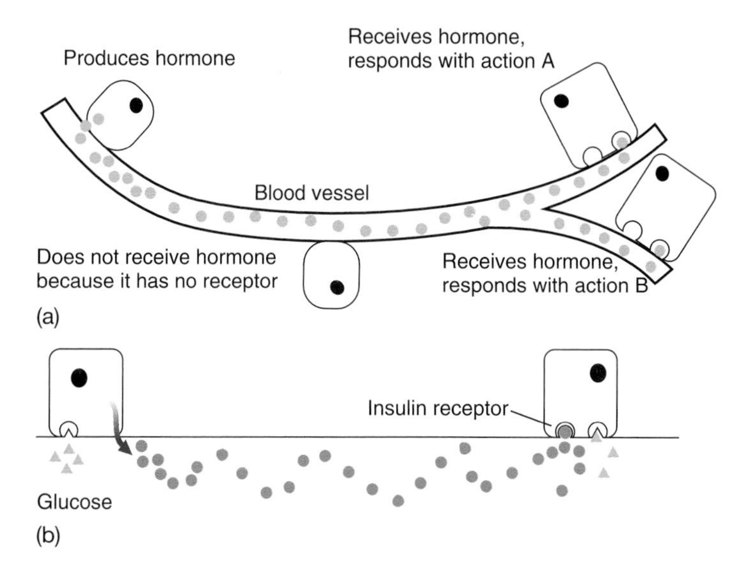

Figure 11.14

(a) A hormone is a signal ligand produced by one kind of cell that has a distinctive effect on other cells in the same organism. The hormone does not affect cells that have no receptors for it; cells that do have receptors may respond in different, distinctive ways. *(b)* Here a beta cell in the pancreas detects too much glucose in the blood; in response, it releases the hormone insulin, which signals other cells to remove glucose from the bloodstream.

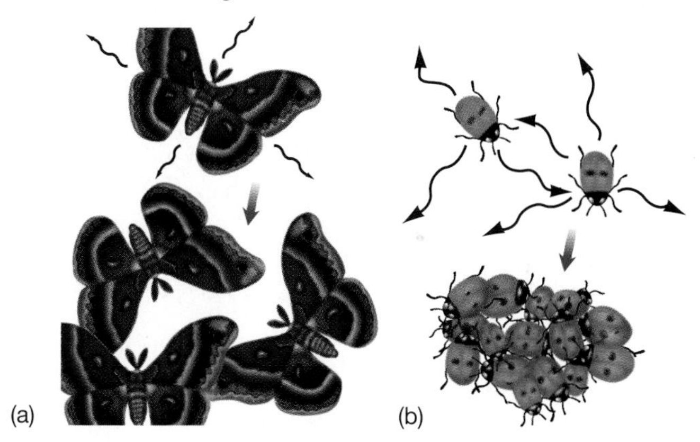

Figure 11.15

Pheromones are signal ligands produced by one individual to send a specific message to other individuals of the same species. *(a)* A female *Cecropia* moth releases a sex pheromone, which attracts males to her. *(b)* Ladybird beetles release an aggregation pheromone, which signals many beetles to congregate in one place.

have special functions. A queen honeybee, for example, produces a pheromone that prevents any other female in the hive from maturing into a queen, thus preserving her own dominance over the hive.

The full story of queen dominance, Section 50.8.

Intracellular signal ligands called **alarmones** trigger specific responses inside cells when certain threatening conditions arise. We will show in Chapter 18 that bacteria set off one kind of alarmone when their energy source is exhausted and another when something interferes with protein synthesis.

In addition to these signal ligands, many comparable compounds have roles in ecological communities. These compounds can hardly be called signal ligands because some are inhibitors whose message would only be, "Die!" or "Stop growing!" For instance, the allomones mentioned previously are produced by members of one species and have detrimental effects on other species. Antibiotics are prime examples; they are produced by some kinds of fungi, algae, and bacteria to kill or inhibit other microorganisms. In an evolutionary sense, organisms have been most inventive in developing agents of chemical warfare that injure potential competing species.

Other compounds, known as kairomones, benefit the individual that detects them, not the one that produces them. Thus if insects of one species communicate with one another by means of a distinctive pheromone, a small mammal may prey on them by detecting this pheromone and homing in on it, so the insect pheromone is also a kairomone for the mammal. We explore allomones and kairomones under the heading of "Community Structure" in Chapter 27.

Eucaryotic cells respond to external signals through a common transduction pathway.

Eucaryotic cells receive signals from many sources and respond by controlling many processes, but they use a remarkably small number of molecular devices to do so. We expect this for evolutionary reasons: Once a primitive cell evolved a control system for one function, it would have been much easier to adapt it for similar functions than to evolve a totally new system. Because we will meet it again and again in later chapters, we introduce a common pathway with this structure:

External ligand —*Binds to*→ Receptor protein —*Activates*→ G protein —*Produces*→

Second messenger —*Activates*→ Protein kinase —*Activates*→ Enzyme

This system transduces many kinds of external signals into internal responses, using three important molecular devices: G-proteins, second messengers, and protein phosphorylation. In studying this system, students encounter a forest-and-trees problem: If you read the following explanation too closely, you may get lost in the details. You must understand that the function of this transduction pathway is to convert an external signal into an internal effect. Then, keep the general structure of the pathway in mind and read the following for general concepts, not for details.

G-proteins

A signal begins when an external ligand binds to a membrane receptor, a transmembrane protein that conducts the external signal into the cytoplasm to effect a second process (Figure 11.16). This receptor typically interacts with a guanine-nucleotide binding protein, or **G-protein,** in the cytoplasmic face of the cell membrane. G-proteins bind the nucleotides guanosine diphosphate (GDP) and guanosine triphosphate (GTP), which are used here as signals, not for the energy they carry. When the external ligand binds to the receptor, the G-protein is activated. G-proteins are thus general transducers that convert an external stimulus into an internal signal.

First and second messengers

Most hormones remain outside the cells that respond to them, even though their effects are internal. In the 1960s, Earl Sutherland proposed that a hormone acts like a *first messenger,* speeding from cell to cell, which stimulates its target cell to produce an intracellular *second messenger* that initiates the cell's specific action. The second messenger Sutherland discovered is **cyclic AMP** (**cAMP,** adenosine 3′, 5′-monophosphate), in which the phosphate group is attached in a ring structure to two points on the ribose. Cyclic AMP is made from ATP by the enzyme adenylate cyclase:

ATP

Adenylate cyclase → **Pyrophosphate (PP_i)**

Cyclic AMP

Adenylate cyclase is commonly activated by a G-protein. Another enzyme, phosphodiesterase, inactivates cAMP by converting it into AMP. We will encounter cyclic AMP in many contexts, because it is a widespread signal ligand that has several distinct roles.

Calcium ions (Ca^{2+}) are also second messengers. They are effective because the Ca^{2+} concentration in cytosol is usually very low, around 10^{-7} *M,* so suddenly increasing this level can serve as a signal. Cells store Ca^{2+} in calcium-sequestering membrane compartments and release it into the cytosol through a complicated pathway that also uses G-proteins. Once released, Ca^{2+} frequently acts by binding to intracellular proteins such as *calmodulin,* which initiate various processes (Figure 11.17).

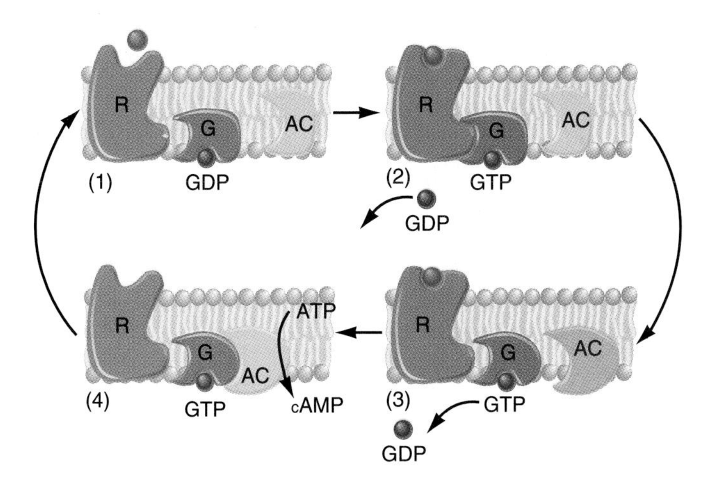

Figure 11.16
The action of G-protein-linked receptors. *(1)* A ligand binds to a receptor (R) and activates it. *(2)* The G-protein *(purple),* initially bound to GDP, binds to the active receptor. *(3)* GTP replaces GDP on the G-protein. *(4)* G-protein-GTP binds to adenylate cyclase (AC) and activates it, so it converts ATP to cAMP. Hydrolysis of GTP to GDP then restores the system to its original condition.

Protein phosphorylation

The third transducing device illustrates one of the most important general principles to emerge in cell biology: Some proteins in a cell are activated or inactivated by the addition and removal of small, simple chemical groups. Just controlling these conversions can control vital functions. Most often, the activity of a protein is changed by phosphorylation (adding a phosphoryl group), and the enzymes that phosphorylate other proteins are called **protein kinases:**

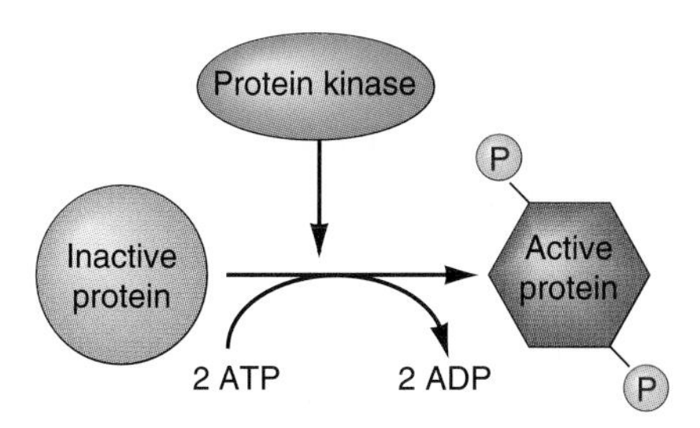

Eucaryotic cells carry an assortment of specialized protein kinases to regulate their many activities; each kinase phosphorylates its target protein on either a serine or a tyrosine residue, and each is controlled through a different path, commonly by one of the second messengers. Other enzymes, called protein phosphatases, remove the phosphoryl groups from specific proteins to reverse the effect.

B. How cells move and change their form

Eucaryotic cells often have distinctive shapes, and they may also be highly flexible. Their surfaces spread out and then retract. Their cytoplasm is constantly being stirred around, and some of

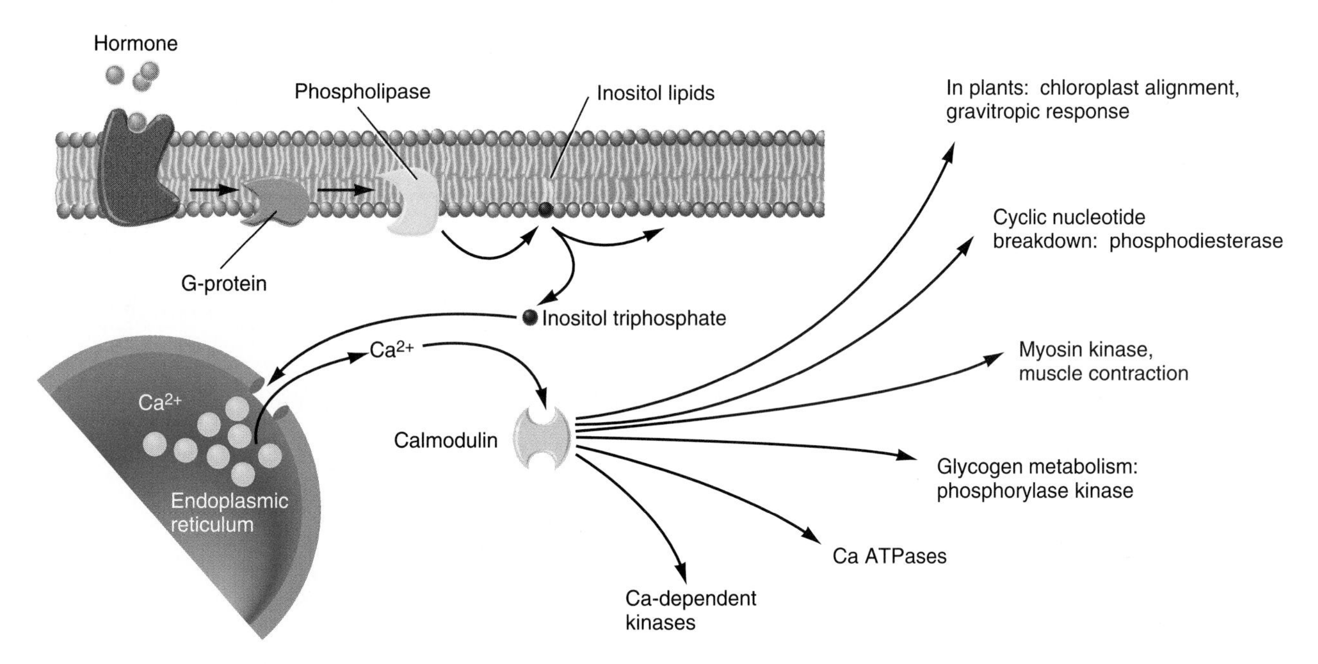

Figure 11.17
Some ligands bind to receptors linked to G-proteins that activate phospholipases, which then release inositol triphosphate (IP_3) from the cell membrane. IP_3 opens calcium channels, allowing Ca^{2+} ions to flow into the cytoplasm. The calcium ions often act by activating calmodulin, which initiates other actions specific to each cell.

their organelles move through the cytoplasm, often very quickly. Cells divide in two, a process that changes their shapes drastically and temporarily reorganizes their structure. The cells themselves often move around as well. Some cells travel actively through an animal's body, between the stationary cells, while in a developing embryo, masses of cells migrate to shape new organs. Free-living cells in soil and water propel themselves either by amoeboid movement—stretching out pseudopods and flowing from place to place—or by means of flagella and cilia (Figure 11.18).

A eucaryotic cell's shape and most of its movements are controlled by its **cytoskeleton,** a fibrous framework that stretches throughout the cell. The cytoskeleton is made of three elements: **actin filaments** (also called **microfilaments**), **microtubules,** and **intermediate filaments.** By forming complexes with other specific proteins, these versatile elements perform many functions.

The involvement of actin filaments or microtubules in various cellular processes has been demonstrated by very useful experimental logic: If a drug specifically disrupts or inhibits a structure S, any process inhibited by the drug must depend on S. Actin filaments were first identified as structures inhibited by cytochalasin B, a fungal antibiotic, whereas microtubules are inhibited by colchicine, or colcemid, a drug made by the autumn crocus, which is used to treat gout and some kinds of cancer. Microtubules are also affected by the alkaloids vincristine and vinblastine, which are anticancer agents. We can often infer that a process is carried out by actin filaments or by microtubules if it is inhibited by one of these drugs.

The complex and often beautiful structures of the cytoskeleton illustrate the principle that *specific cellular structures are made of specific proteins.* Each structure has its particular form and function because it is made of distinctive proteins with the requisite properties. This should be one of the major take-home lessons of this survey; though the details may be forgettable, remember that wherever a cell employs part of the cytoskeleton to perform a certain function, special proteins interact to form molecular complexes that do the job. Here and elsewhere we will emphasize this point by showing some of these complexes and naming the proteins.

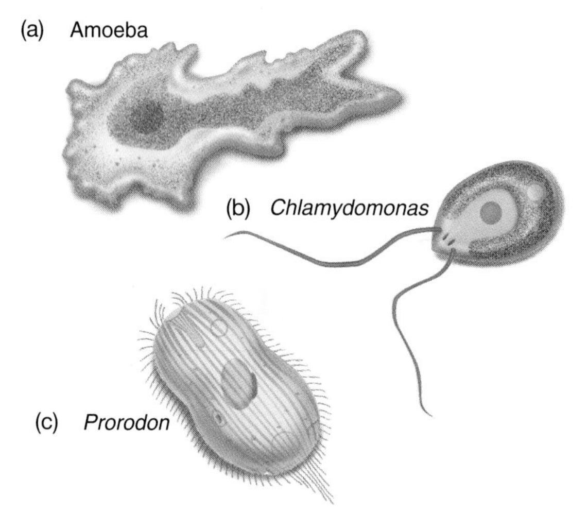

Figure 11.18

(a) Amoebas are common single-celled organisms that move by extending temporary "arms," called pseudopods, and flowing into them. *(b)* Many protists, like this *Chlamydomonas* cell, swim by means of whiplike flagella, while *(c)* ciliates, such as *Prorodon,* propel themselves with the many short cilia on their surfaces.

Exercise 11.9 Eucaryotic cell division entails two separate processes: separating chromosomes (mitosis) and dividing the cell in two (cytokinesis). The former is inhibited by colchicine, and the latter (in animals) is inhibited by cytochalasin B. What do these facts imply about the mechanisms of the two processes?

11.9 Actin filaments effect many cell movements.

Much of the cytoskeleton consists of 6-nm-wide microfilaments of the protein **actin,** which may constitute 5–20 percent of the cellular protein. Actin filaments are responsible for many dynamic cellular processes, including muscle contraction and movements during embryonic development in animals. Fluorescent antibodies against actin bind to these filaments and reveal them as a spectacular, glowing network extending throughout the cell, clearly holding it in shape (Figure 11.19). Actin filaments also form the **cell cortex,** a layer just under the cell surface, which holds the cell in shape and accounts for much of its movement. Some animal cells, such as those lining the intestine, have minute, fingerlike **microvilli** projecting from one face and creating a huge surface for the absorption of food molecules or for other processes (Figure 11.20). Each microvillus contains a bundle of actin filaments, anchored to the plasma membrane by the protein α-actinin.

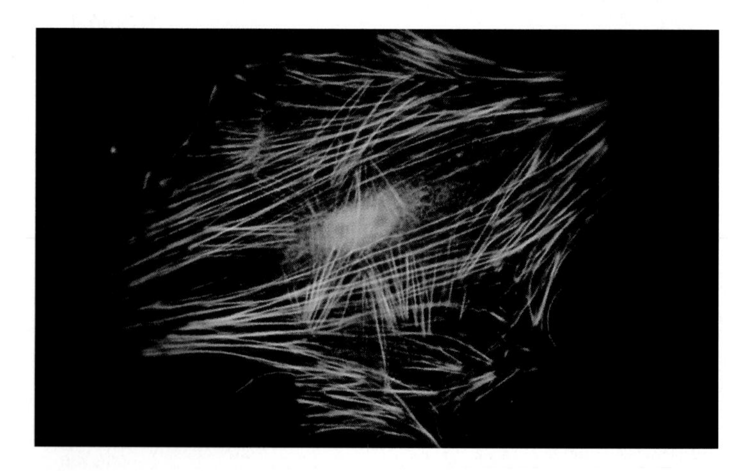

Figure 11.19

Fluorescent antibodies against actin reveal the cytoskeleton of an animal cell as a sheaf of filaments throughout the cytoplasm that holds the cell in its shape.

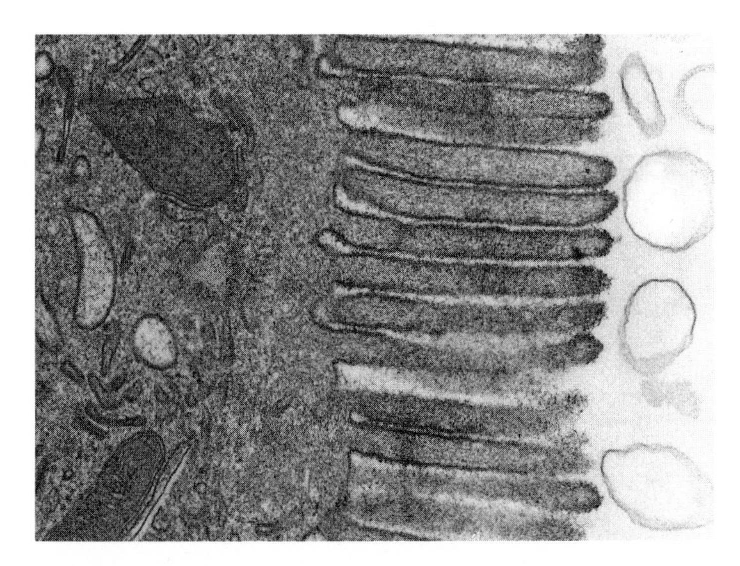

Figure 11.20
The microvilli of some animal cells are extensions held in shape by organized actin filaments.

Animal cells in cell cultures contact the substratum (the plastic culture flask) at many *adhesion plaques,* where actin filaments attach to the plasma membrane. Adhesion plaques are complexes of at least six proteins besides actin, including some proteins of the extracellular matrix (Figure 11.21). Actin filaments can change their shape and pull the cell along by means of these attachments.

Actin filaments are dynamic structures that can quickly change their form, and this is one way they effect cellular movements. Actin is composed of 45-kDa subunits called G (globular) actin, which polymerize into long, double-stranded filaments of F (fibrous) actin (Figure 11.22*a*). These filaments have a distinct polarity, with a plus and a minus end (though now that the molecules can be seen well by electron microscopy, the ends are being called "barbed" and "pointed"). Subunits attach more rapidly to the plus end than to the minus end, and actin filaments change their form when subunits are added or removed in a process called **treadmilling** (Figure 11.22*b*). Subunits attach to the plus end while others detach from the minus end, and in this way the whole filament moves in the plus direction while remaining the same length. Treadmilling actin filaments beneath the surface of a cell can move the entire cell, and they cause the cell to push out thin sheets and small spikes with which it probes the environment (Figure 11.23).

Actin filaments also produce movement by interacting with filaments of **myosin,** a long, fibrous protein whose globular end is an ATPase (Figure 11.24). Myosin and actin are organized so that myosin molecules pull on actin molecules by activating this ATPase. The globular head of a myosin molecule, activated by the hydrolysis of ATP, attaches to one actin subunit and pulls a little through a slight change in conformation, so the myosin filament slides a short distance past the actin filament (Figure 11.25). In effect, the myosin walks along the actin much as a centipede walks—by attaching, contracting, and releasing its legs in a regular rhythm. Actin-myosin complexes are responsible for the movement called **cyclosis** or **cytoplasmic streaming** in many plant cells, where the cytoplasm moves vigorously in a circular pattern (Figure 11.26). We will see these interactions again in the highly organized arrangements of actin and myosin in animal muscle (Section 44.6).

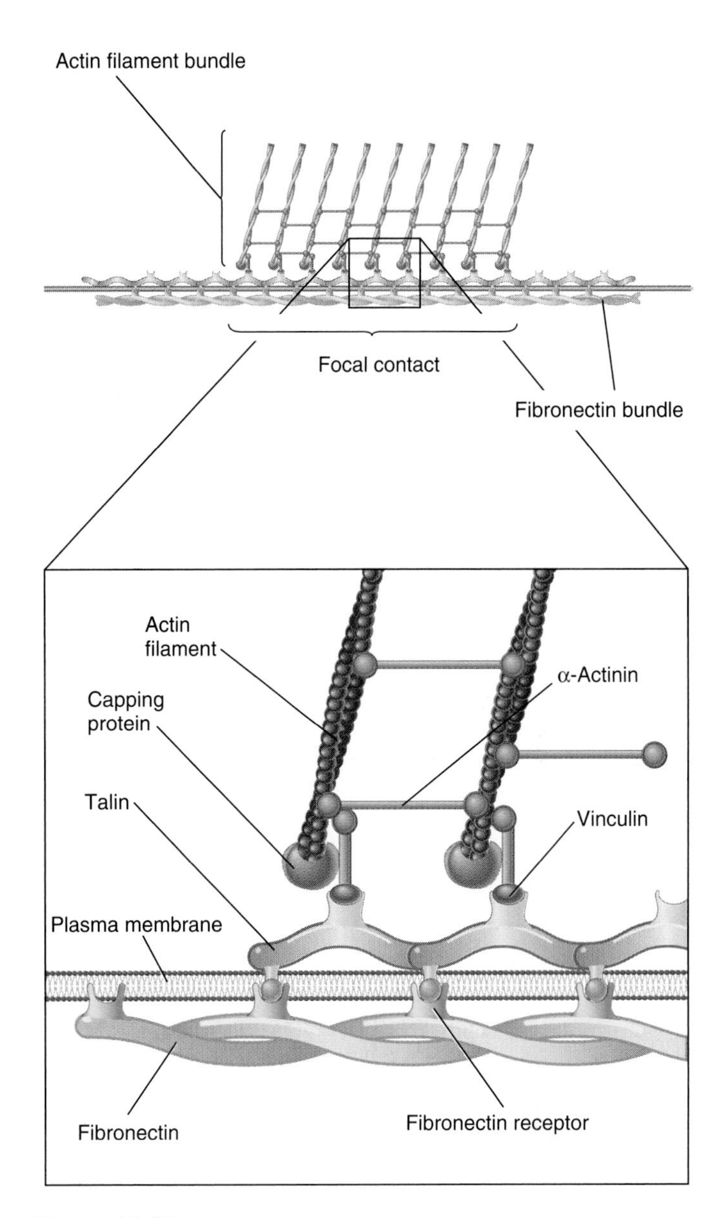

Figure 11.21
In an adhesion plaque, actin filaments are bound to the cell surface, so they can push or pull on it, by at least four kinds of intracellular proteins and two more (fibronectin and fibronectin receptor) that form an extracellular matrix. This possible structure is quite diagrammatic, because the structures of all the proteins involved are not really known.

Because actin filaments can rapidly depolymerize from F to G actin and then polymerize again, parts of the cytoskeleton can change to serve different functions. In animal cells, for instance, much of the cytoskeleton breaks down just before cell

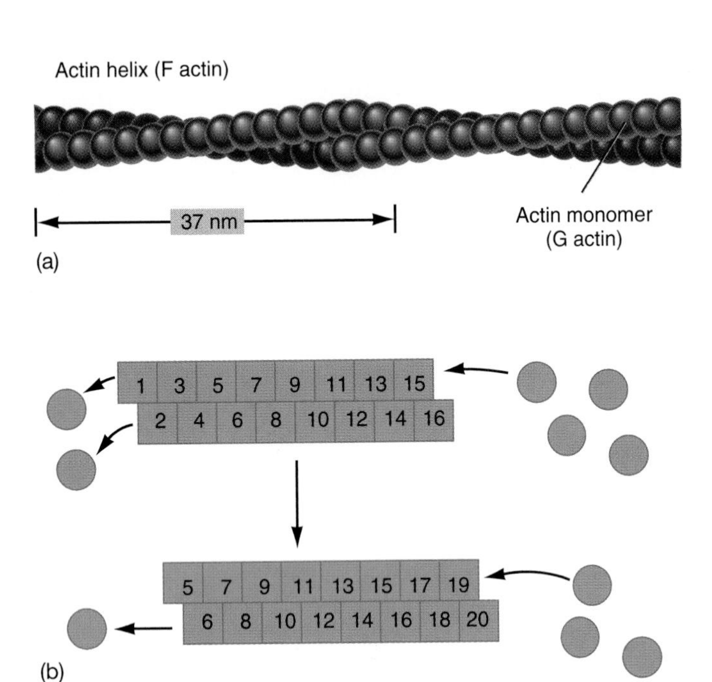

Figure 11.22
(a) An actin filament (F actin) is a double helix of globular subunits (G actin). *(b)* The filament assembles, or disassembles, rather quickly, but subunits add more rapidly to the plus end than to the minus end, so the whole filament treadmills in the plus direction.

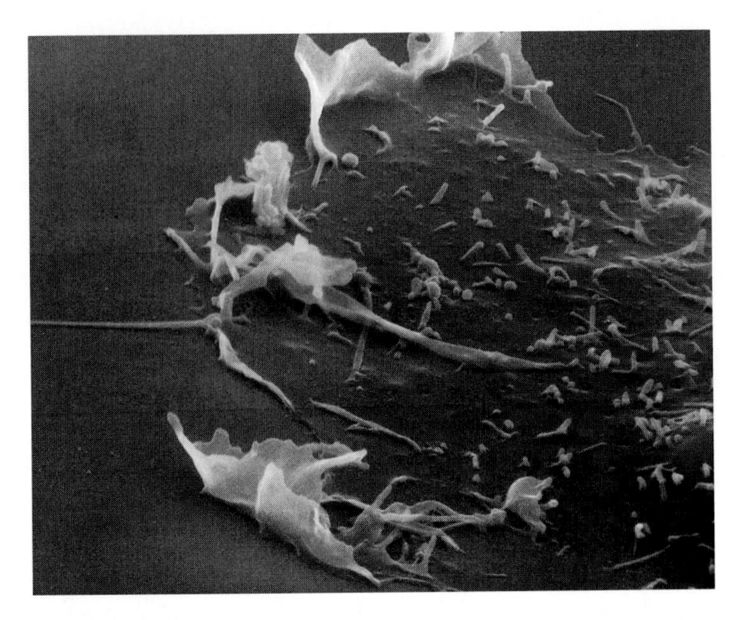

Figure 11.23
A tissue-culture cell moves by extending a broad, thin foot, which is pushed along by a layer of actin filaments that extend by treadmilling.

division, so the cell relaxes and becomes round; then some filaments assemble in a contractile ring that closes, like a purse string, to pinch the cell in two (Figure 11.27).

The movement of cells such as free-living amoebas and many white blood cells in animals is an old biological mystery. This amoeboid movement is explained by the interactions of actin filaments with some accessory proteins: Filamin binds the filaments into a stiff gel, like gelatin that has set in the refrigerator, and gelsolin disperses the filaments into a liquid sol, like liquid gelatin. In amoeboid movement, the pseudopods have stiff cortical layers, and a stream of cytosol seems to flow through the middle of a pseudopod, changing from a sol at the rear of the cell into a gel as it reaches the tip of the pseudopod (Figure 11.28).

11.10 Microtubules shape cells and are used for movement.

Microtubules were discovered in 1963 when Keith R. Porter and Myron Ledbetter introduced the fixative glutaraldehyde into electron microscopy. They discovered that what had looked like structureless cytoplasm contains a multitude of small tubules, 25 nm in diameter with a central 14-nm hole. Long microtubules may extend for great distances through a cell. They are built of two very similar proteins, α- and β-tubulin (50 kDa each), which form an α-β dimer. A microtubule is a helix made of 13 parallel columns of dimers, just as 13 staves might make a barrel:

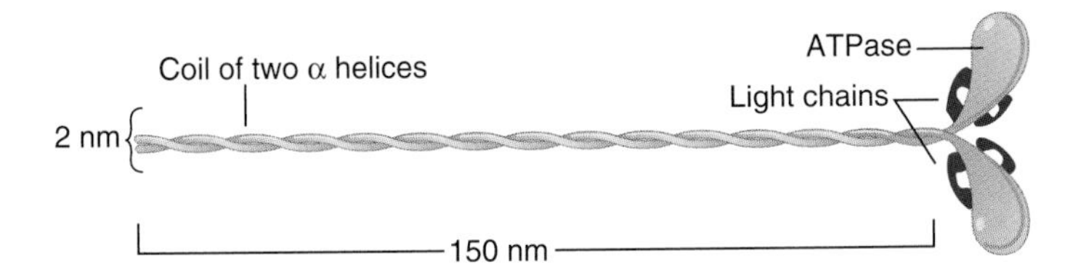

Figure 11.24
A myosin molecule is a long filament made of two α helices, with a globular head that has ATPase activity. Small (light chain) proteins bind to the head.

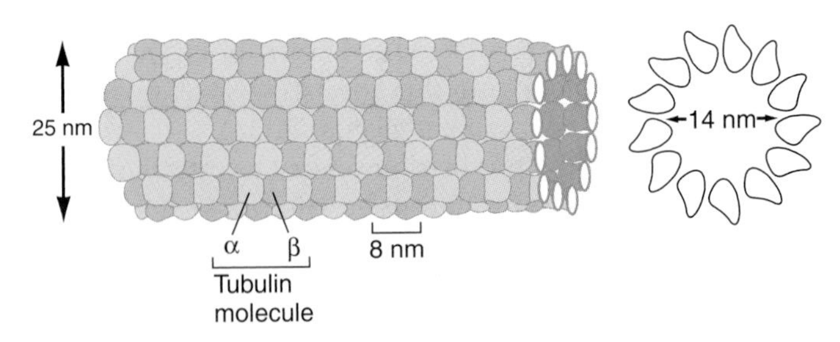

Tubulin dimers assemble spontaneously into microtubules, though the assembly is enhanced by Mg^{2+} ions and high temperature, and inhibited by Ca^{2+} and low temperature.

Like actin filaments, microtubules change form by the addition or removal of protein subunits. Microtubules have a

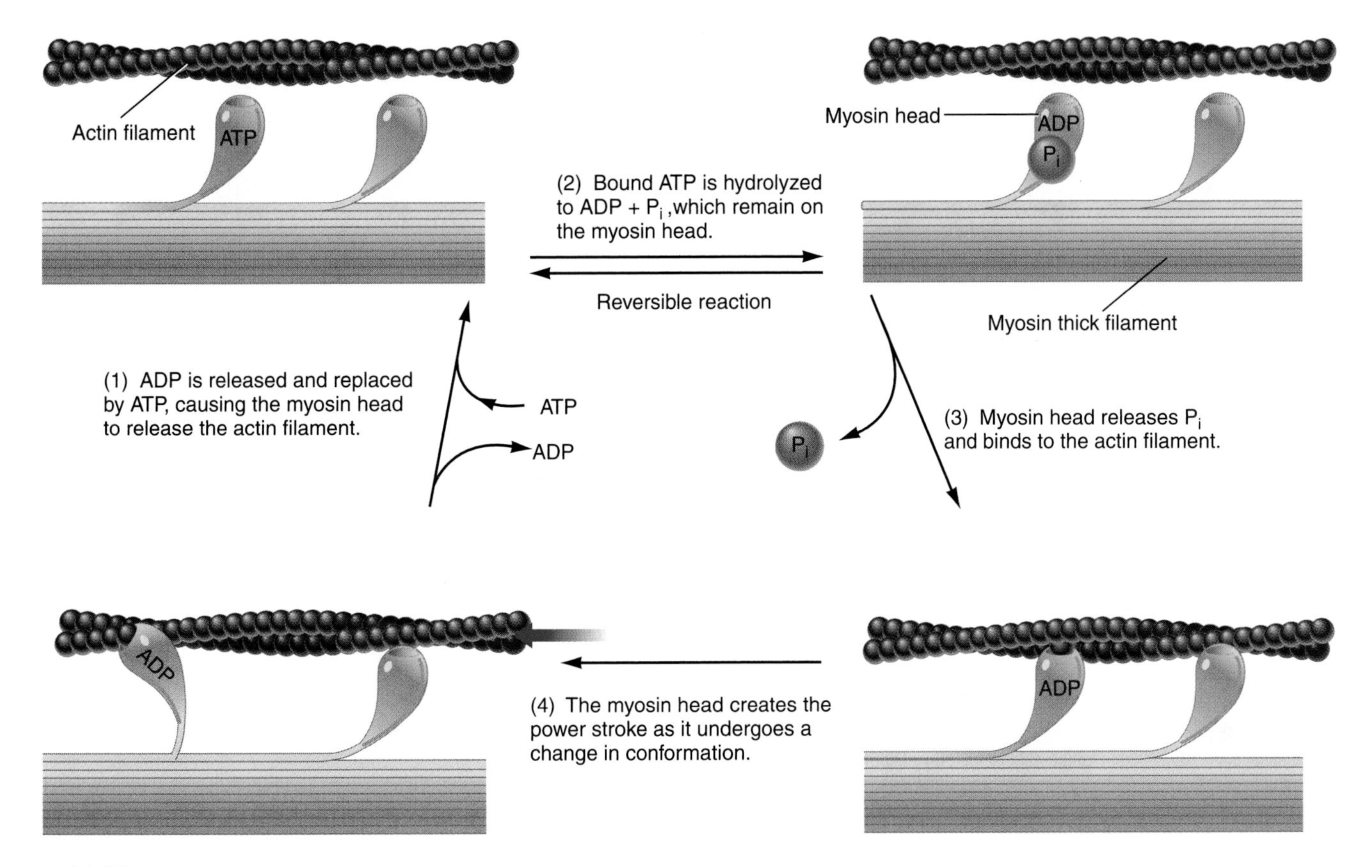

Figure 11.25

Myosin can pull on an actin filament through a cycle of attachment, change of conformation, and reattachment. The energy for movement comes from hydrolysis of ATP.

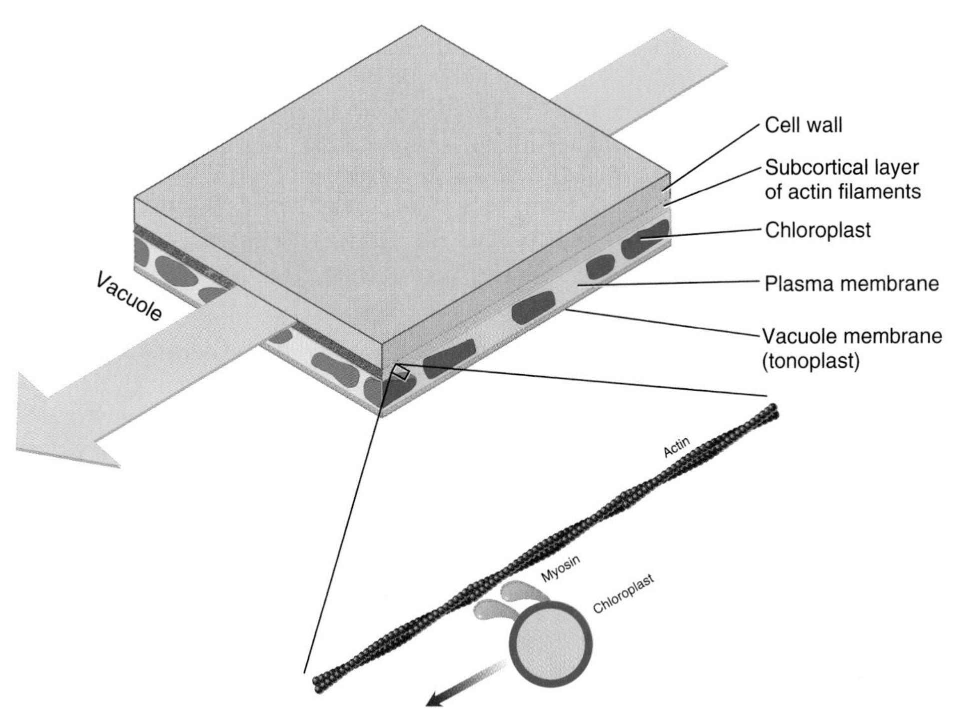

Figure 11.26

Plant cells commonly move their contents around by cytoplasmic streaming. The movement derives from myosin molecules bound to organelles, such as chloroplasts, that pull themselves along the actin filaments of the cell cortex much as myosin filaments pull along actin filaments in muscle.

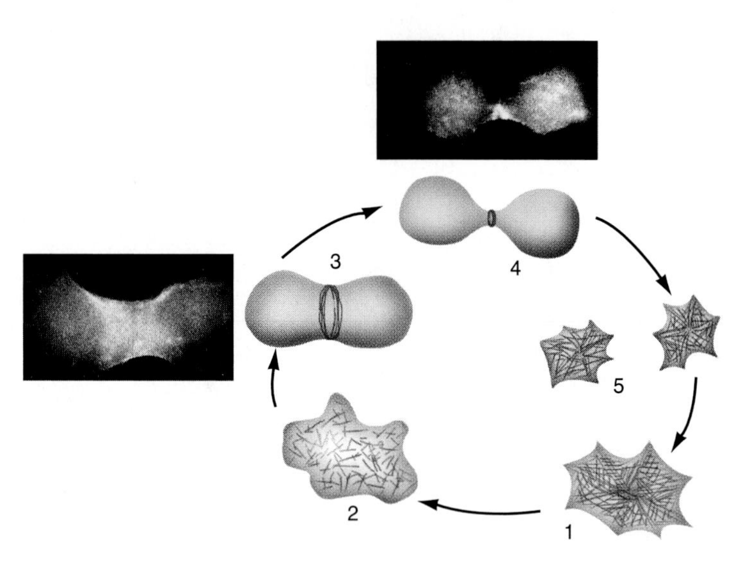

Figure 11.27

An animal cell that is not dividing is held in a characteristic shape by its cytoskeleton *(1),* here shown only by its actin microfilaments. As it prepares for cell division *(2),* much of its cytoskeleton breaks down. Some of the actin reassembles as a contractile ring around the middle of the cell *(3).* This ring contracts like a purse string to divide the cell in two *(4).* After division, the cytoskeleton starts to reform into its characteristic shape *(5).* The micrographs show the form of the actin purse string at two stages of contraction.

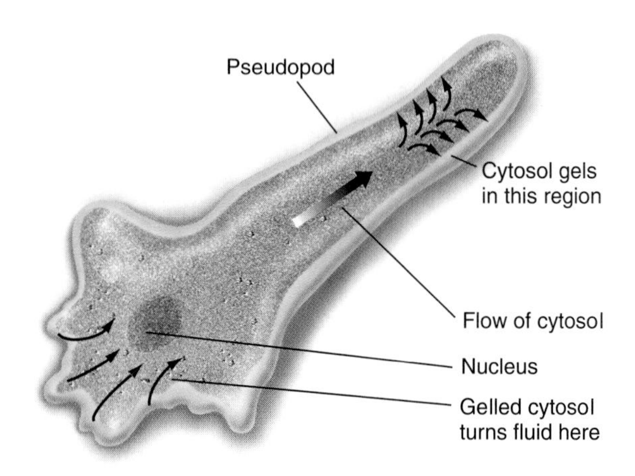

Figure 11.28

Amoeboid movement appears to depend upon the aggregation of actin filaments into an organized cortex near the front of a pseudopod and its disaggregation into a fluid form at the rear.

minus end to which dimers attach slowly and a plus end to which dimers attach more rapidly. A general principle of cytoplasmic organization is that the minus end is protected by a **microtubule-organizing center (MTOC),** so it is anchored in place and doesn't change, while the plus ends are free to grow longer or shorter.

Microtubules are the elements of all the following structures and processes:

- They supplement the actin filament cytoskeleton to maintain the shapes of many cells.
- They form the motile skeletal structure of cilia and flagella.
- They form the core of organelles called centrioles and basal bodies, which serve as microtubule-organizing centers.
- They form structures that some protozoans use for movement.
- They are the spindle fibers of the mitotic apparatus, which separates chromosomes during cell division.
- They organize the cell wall between daughter cells in plant cell division.
- They help develop and maintain axons, the long, thin extensions of nerve cells that carry nerve impulses over long distances. They also help move vesicles of material through the axon.
- They help move many materials in cells, such as releasing secretion granules during exocytosis and moving pigment granules in pigment cells (chromatocytes).

Cell division and the mitotic apparatus, Section 13.8.

Some microtubular structures are quite remarkable, as shown in Figure 11.29. Notice that they may hold a cell in a certain shape by forming arrays just under the plasma membrane. The densely packed microtubules in structures such as axopods and axostyles, which some protozoans use for movement, probably effect movement by sliding past one another, as they do in cilia and flagella.

As mentioned previously, the drugs colchicine and vinblastine inhibit microtubule-mediated processes by binding to tubulin and blocking normal microtubule assembly. These drugs are effective as anticancer agents, probably because they disrupt cell division by interfering with the mitotic apparatus. On the other hand, they may preferentially attack cancer cells because they inhibit several cellular activities, and cancer cells are generally much more active than normal cells.

11.11 Microtubule structures are organized by special centers.

Just as actin filaments depolymerize at some times and repolymerize at other times, perhaps in different form, many microtubule complexes shift around as cells change their activities. MTOCs orient microtubules and initiate their polymerization from tubulin dimers. A diffuse region near the nucleus, the **centrosome,** serves as an MTOC. In the cells of animals and some protists (but not in those of plants), it contains a **centriole,** a beautiful structure that seems to be the same as the **basal body** at the base of every cilium and flagellum (Figure 11.30). Centrioles or basal bodies are built with ninefold symmetry around nine short triplets of microtubules. The centrosome (with or without a centriole) organizes a loose cytoskeleton of microtubules in nondividing cells, and then organizes the mitotic

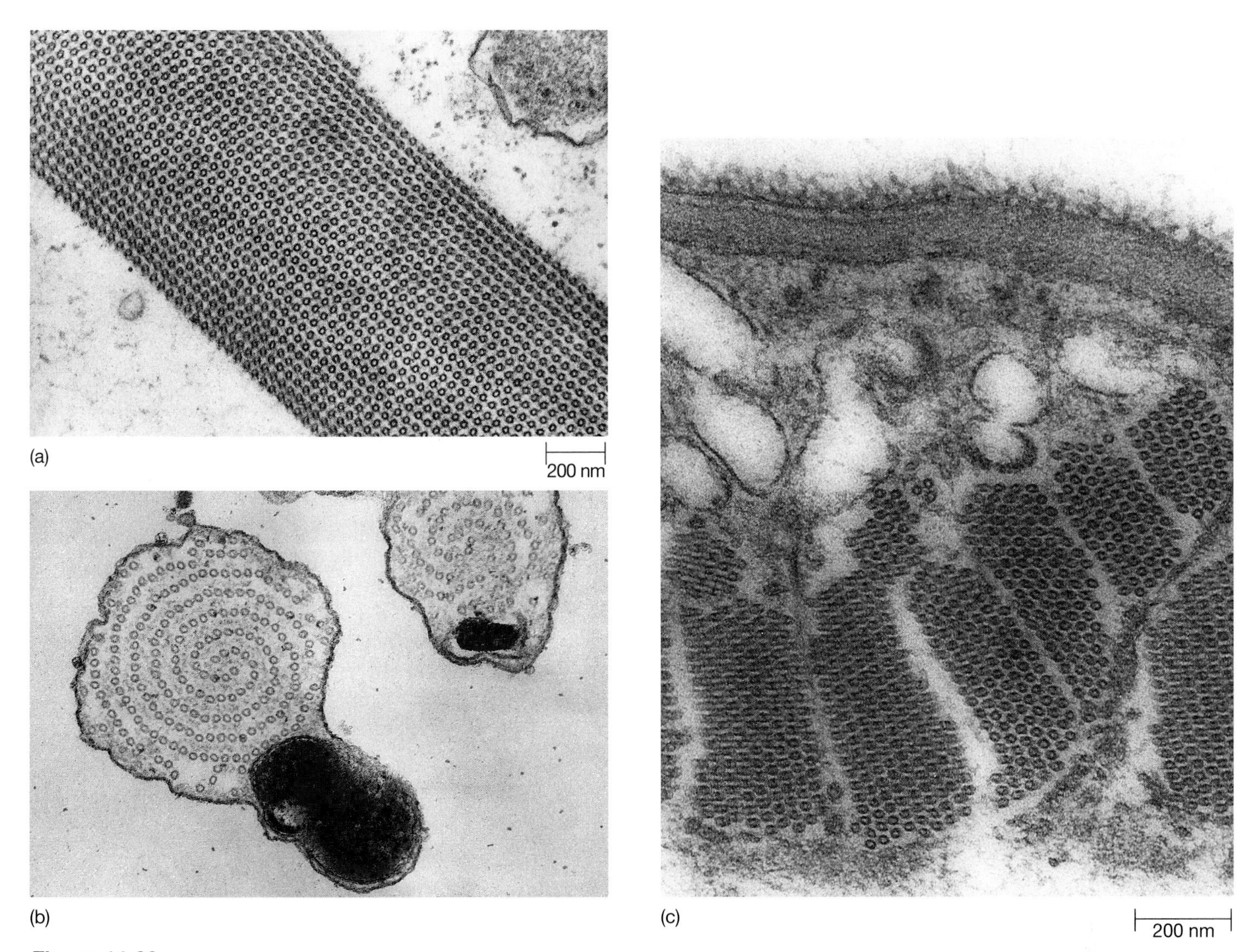

Figure 11.29

A sampling of cellular structures made of microtubules includes: *(a)* the axostyle of a ciliate, *Saccinobaculus,* which moves by contorting this apparatus; *(b)* concentric spirals of microtubules in the pseudopods of the heliozoan *Actinosphaerium;* and *(c)* bundles of microtubules under the surface of a flagellate, *Diplodinium.*

spindle, which separates the chromosomes when a cell divides (Figure 11.31). (Plant cells show that a centriole isn't essential, since their mitotic apparatus forms without a centriole or any other visible structure in the centrosome.) Basal bodies help organize the microtubules of cilia and flagella. It is thought that the centriole originated as a basal body in the most primitive flagellated cells and later became an organizer of the mitotic spindle in some organisms.

11.12 Objects move on microtubules by means of specialized motors.

It is one thing to arrange microtubules in a cell in a functional way and quite another for the cell to use this structure for movement. How is it done? Some movement comes from the assembly and disassembly of microtubules, as we noted previously, but this doesn't account for the faster movements associated with microtubules. The general principle is that cells use **molecular motors,** specialized microtubule-associated ATPases that hydrolyze ATP and use the energy to move along the microtubule. A motor acts rather like the ATPase of myosin. That is, it goes through a cycle of conformational changes as it repeatedly attaches to a tubulin monomer, pulls, and releases. Each attachment-pull-release cycle requires a molecule of ATP.

The microtubule motors form two families: *dyneins* and *kinesins* (Figure 11.32). Dyneins move from the plus end to the minus end of a microtubule, and kinesins do the reverse. These remarkable little devices can be used to anchor organelles to microtubules and to move them at speeds of up to 50 μm per minute. Motors are essential for moving chromosomes during

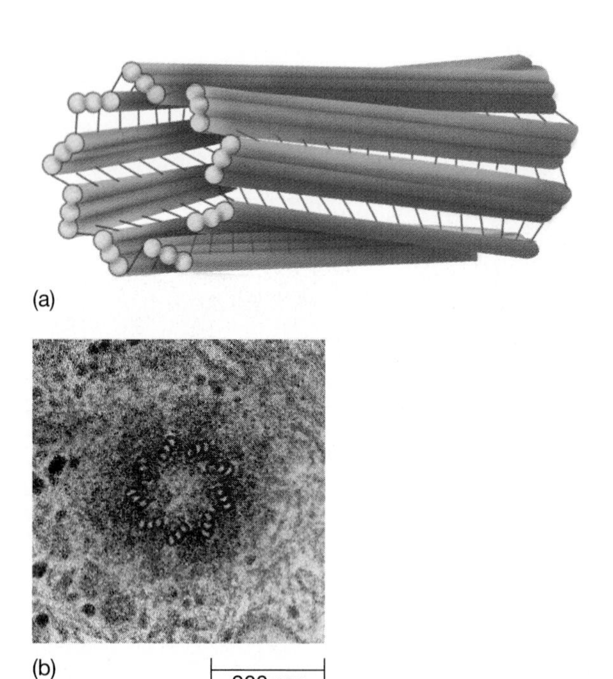

Figure 11.30

(a) A centriole is made of nine triplets of microtubules. *(b)* The electron micrograph showing a cross section through the structure has been enhanced photographically to show its ninefold symmetry more clearly.

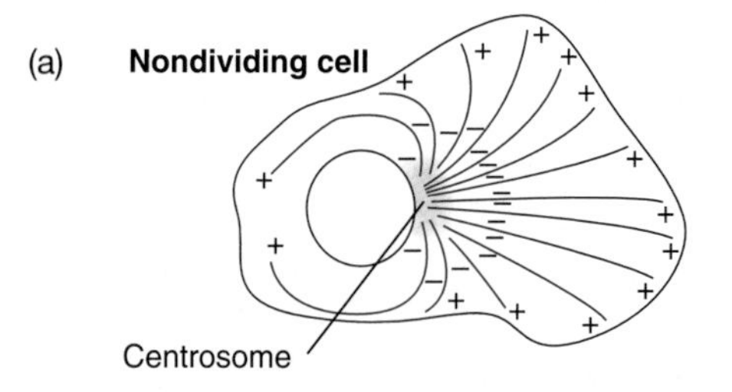

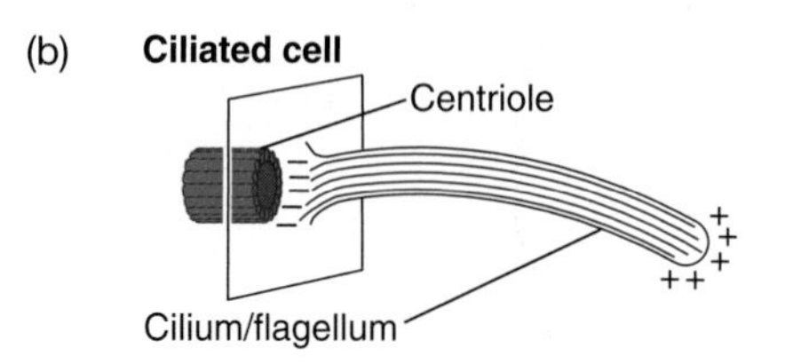

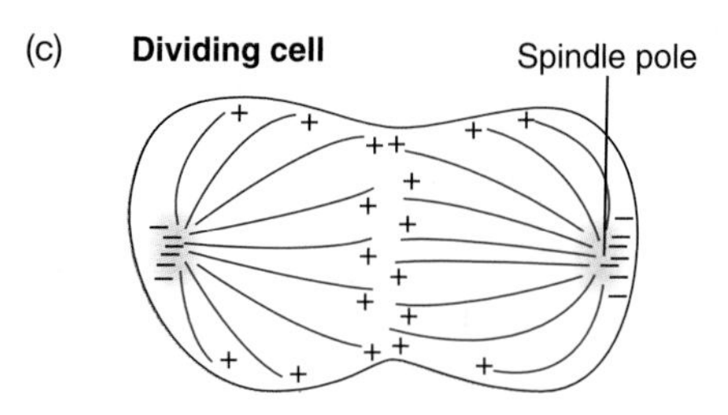

Figure 11.31

The centrosome (which contains a centriole except in plant cells) is an organizing center (MTOC) for various microtubule structures: the cytoskeleton in a nondividing cell *(a),* a cilium or flagellum *(b),* or the spindle apparatus that separates chromosomes in a dividing cell *(c).* Microtubules grow from it from their minus ends. The basal body serves the same function for cilia or flagella.

cell division, as we will show in Section 13.8, and for the actions of cilia and flagella, as we will show next.

11.13 Cilia and flagella are movable bundles of microtubules.

Many eucaryotic cells bear extensions of the cell surface called **kinetids** that are used for rapid movement. Short kinetids are **cilia;** long ones are **flagella.** Sperm cells typically have one flagellum, many algae have two, and some protozoans have many. Other protozoans may be covered with hundreds of cilia, and the external surfaces of certain animals, such as flatworms, are also covered with cilia. In vertebrates, many surfaces, as in the nasal passages and genital tract, are rich in ciliated cells that sweep along a steady stream of mucus, trapping and removing bits of dirt and microorganisms.

Cilia and flagella have the ninefold symmetry of basal bodies and centrioles, but with nine doublets of microtubules, instead of triplets, plus two single tubules in the center, making a characteristic 9 + 2 pattern (Figure 11.33). (Some cells such as the rods in vertebrate eyes are built around modified cilia having the circle of microtubules without the central pair.) The doublets are linked by the protein nexin and by other interconnecting filaments to make a framework about 200 nm in diameter called an *axoneme.* Each doublet bears an arm of dynein, the motor that drives kinetid movement and makes the doublets slide past each other, much as myosin and actin fibrils do

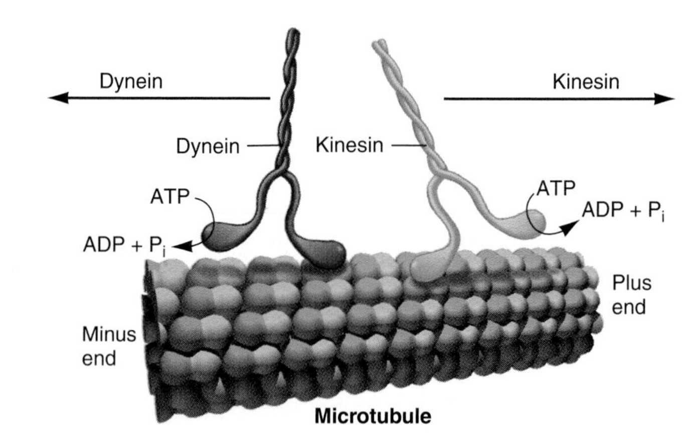

Figure 11.32

Dynein and kinesin are two types of protein motors that characteristically move along a microtubule toward the minus end or the plus end, respectively. They are both powered by ATP.

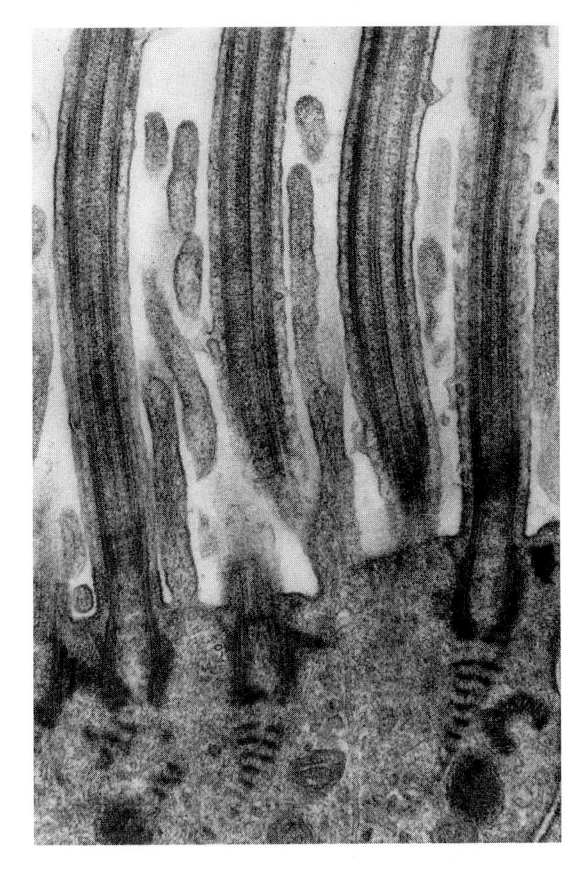

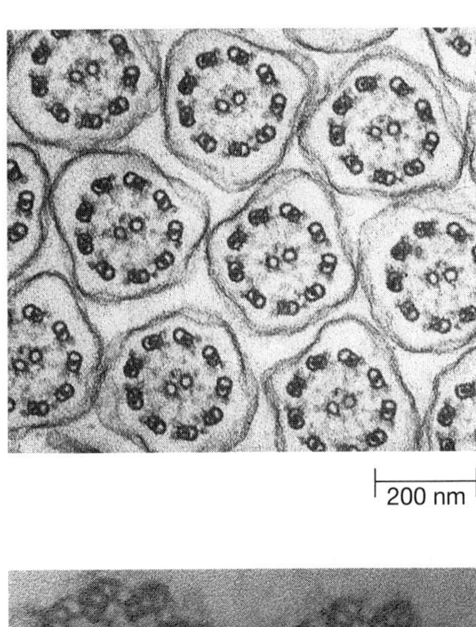

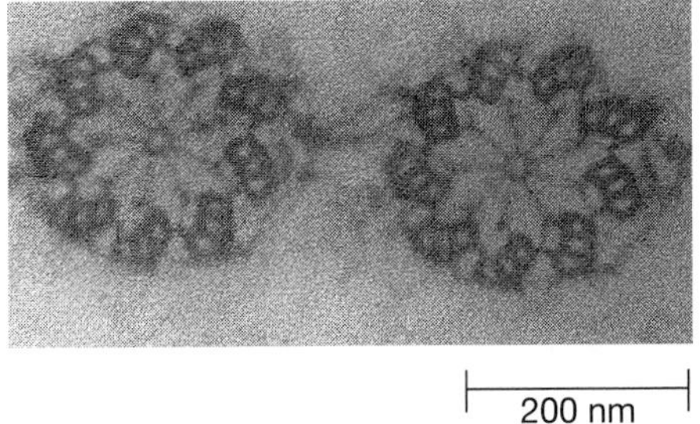

Figure 11.33

A longitudinal section through a field of cilia *(left)* shows that each one contains an axoneme made of several parallel microtubules. A cross section *(top, right)* shows that an axoneme consists of nine doublets of microtubules with two more in the middle. Notice that the doublets are not complete, separate microtubules but share several subunits. Neighboring doublets are linked by the protein nexin, and the "ears" on each doublet are molecules of the motor protein dynein, which pulls each microtubule doublet past its neighbors and provides the energy for movement. *(Bottom, right)* A cross section through the basal bodies shows that they have the same structure as centrioles.

(Figure 11.34). The ATP needed for this action is supplied by mitochondria closely associated with the axonemes.

Exercise 11.10 Kartagener syndrome in humans is due to a nonfunctional dynein. What defect will these people have at a cellular level? Why do they have chronic bronchitis and sinusitis (inflammations of the respiratory tract)? Would you expect affected males to be fertile? Why?

11.14 Bacterial flagella are made of flagellin, a different globular protein.

Although sharing the same name, the procaryotic flagellum is very different from eucaryotic flagella. It is a long, naked helix of the protein flagellin that protrudes from the bacterial wall and membrane where its hooked end is anchored into a system of rings (Figure 11.35). Rather than sliding, the bacterial flagellum rotates inside the rings. Both the flagellin helix and the protein rings that surround it must have rotational symmetry. The flagellum can only move by one structure pulling on the other, which necessarily produces rotation. The energy for this movement comes directly from the proton gradient across the cell membrane, as shown in Figure 9.13.

11.15 Materials can be moved across membranes by bulk transport.

In addition to transporting specific molecules and ions across intact membranes by simple diffusion and protein-mediated transport, cells can move larger amounts of material in and out by means of vesicles. In **exocytosis,** a vesicle inside the cell fuses with the plasma membrane and disgorges its contents; in **endocytosis,** a vesicle from the plasma membrane invaginates and pinches off (Figure 11.36). Through exocytosis, cells export materials that they have synthesized in the ER-Golgi system, including enzymes that function outside the cell and the components of extracellular structures such as cell walls and bone.

Cells import bulk material in at least two distinct ways—endocytosis and **phagocytosis,** or "cell eating" (Greek, *phagein* = to eat). In phagocytosis, cells engulf relatively larger bits of material through movements of the cell membrane directed by

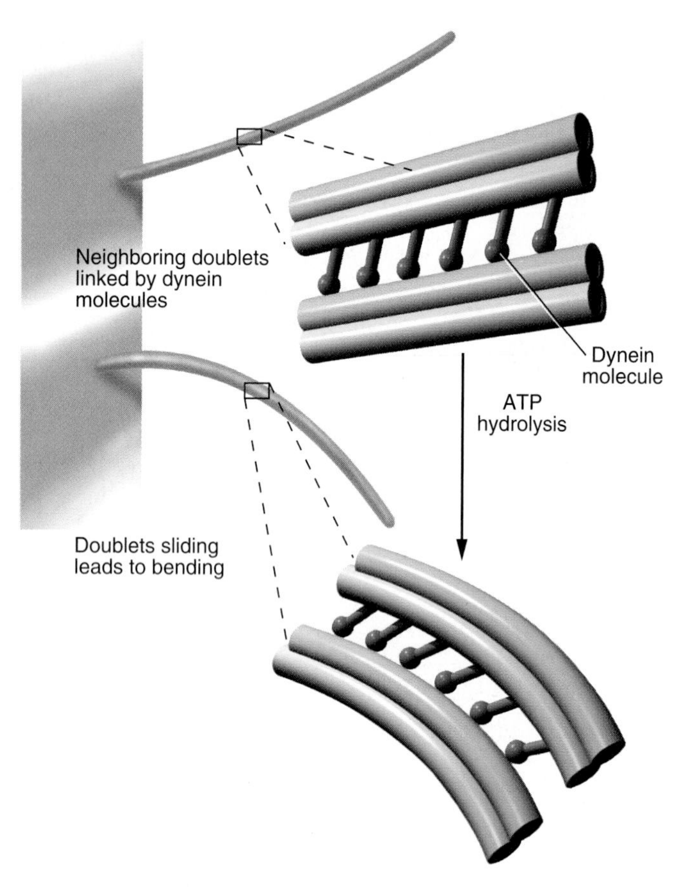

Figure 11.34

A cilium bends as the dynein of one microtubule doublet "walks" along a neighboring doublet in an action very much like the action of myosin to actin.

actin filaments, as if the cell were creeping around the object (Figure 11.37). Many protista, such as amoebas, feed this way, as do phagocytic white blood cells in animals, which ingest and destroy invaders such as bacteria.

Endocytosis, in contrast, begins with **coated pits,** depressions in the cell surface formed by networks of the protein clathrin. These pits regularly invaginate, round up, and pinch off to make a **coated vesicle,** which is covered by a regular cage of clathrin (Figure 11.38). Coated vesicles bring materials into the cell and shuttle materials from one cellular compartment to another. In the process of *receptor-mediated endocytosis,* receptors on animal cell surfaces bind proteins in the blood that carry nutrients such as iron or lipids (Figure 11.39), and coated vesicles bring these receptor-protein complexes into the cytoplasm. Coated vesicles then carry their contents to **endosomes,** which are larger vesicles that lie beneath the cell surface (Figure 11.40). Receptor proteins and clathrin networks are recycled to the cell surface for reuse. Then other vesicles deliver hydrolytic enzymes from the Golgi complex to the endosomes, converting them into **lysosomes.** Lysosomal enzymes, in concert, are able to break down all kinds of polymers to their monomers and generally reduce large biological structures to small molecules, so they attack the materials that have been brought in by endocytosis. The small molecules pass through the lysosomal membrane and enter the cell's metabolic pathways. Since coated vesicles always enclose a certain amount of fluid when they close, endocytosis has also been called **pinocytosis,** or "cell drinking" (Greek, *pinein* = to drink).

Lysosomes are also used in *autophagy* (literally, "eating self"), in which normal cells continually break down some of their components and recycle the molecules. Lysosomes may dispose of an old chloroplast, for example, by fusing with it and digesting it. Autophagy is a way to recycle selected portions of a

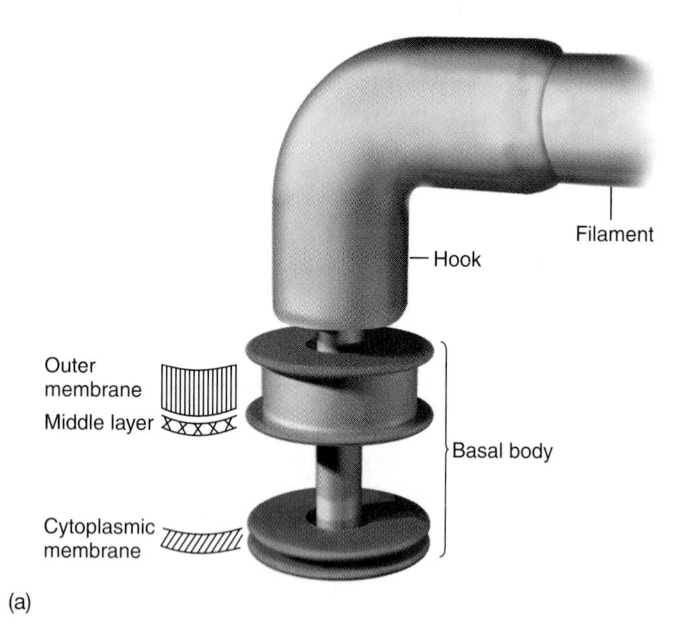

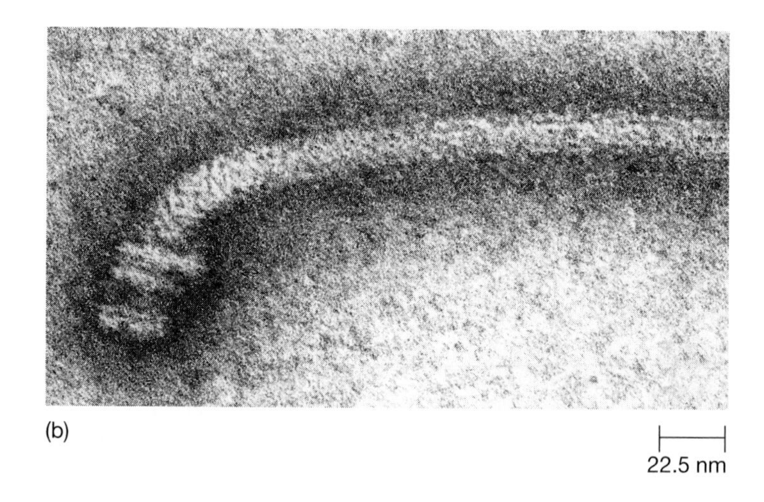

Figure 11.35

(a) The bacterial flagellum is made of a helix of small, globular proteins (flagellin) similar to tubulin or actin. It is anchored in two rings of proteins clearly visible in the electron micrograph *(b),* that fit into the cell wall and cell membrane.

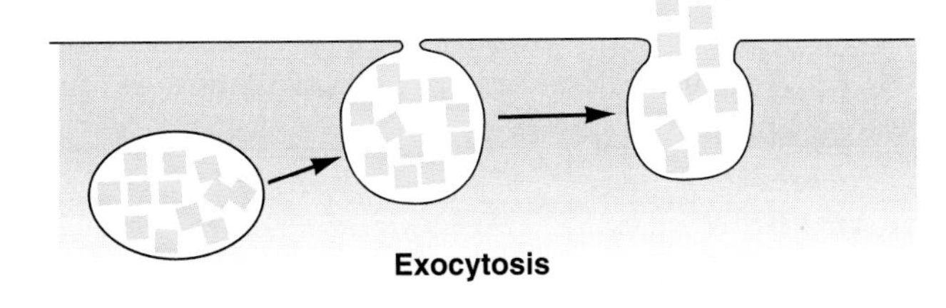

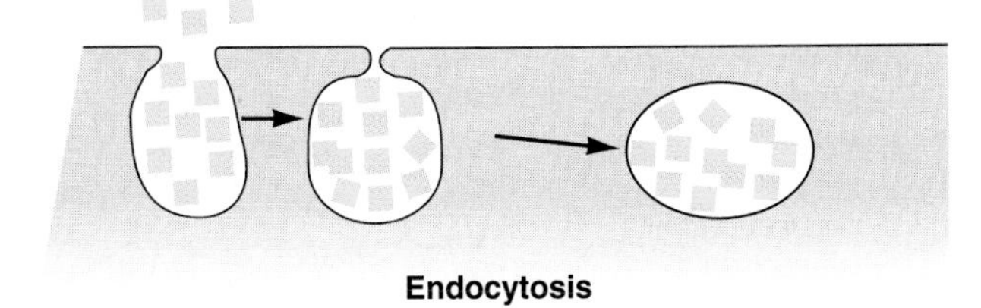

Figure 11.36
Exocytosis and endocytosis are opposite processes that move vesicles of material out of cells or into their cytoplasm.

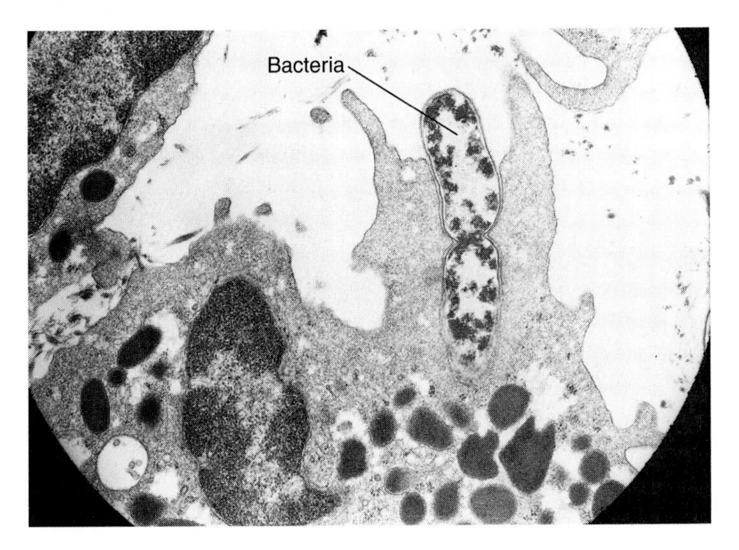

Figure 11.37
A white blood cell phagocytizes a pair of bacteria.

cell without destroying the whole cell. The versatile lysosomes are also used to remove whole cells and large structures; when a tadpole becomes an adult frog, lysosomes break down the tail tissue, and the growing frog uses some of the old tissue. Lysosomes are activated during embryonic development wherever structures must be broken down and reshaped; the hands and feet of a human embryo initially grow with webbing between the fingers and toes, and lysosomes break down the webs to sculpt out the digits. Lysosomes are also abundant in a uterus after birth, when it must be reduced from its great bulk of pregnancy to its small, nonpregnant size.

Exercise 11.11 Trace the likely fate of one amino acid residue and one fatty acid molecule in a lipoprotein in the blood after the lipoprotein binds to a cell.

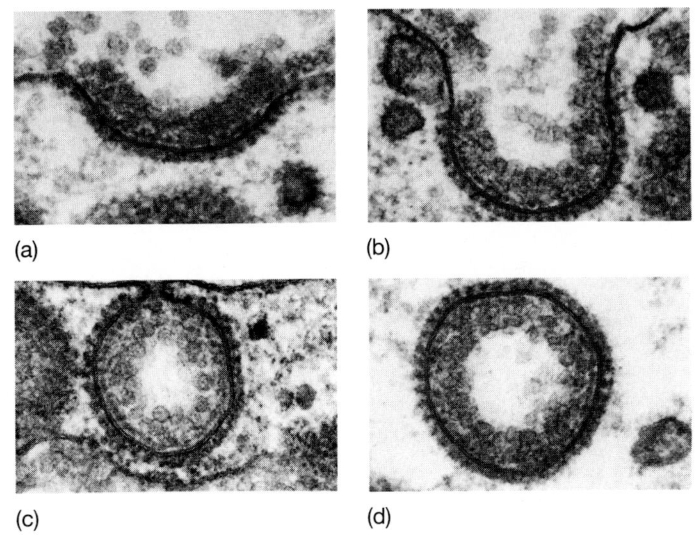

Figure 11.38
Coated vesicles form through the invagination of coated pits covered with regular networks of the protein clathrin. The clathrin is visible in all four micrographs here as a regularly spaced series of particles on the cytoplasmic side of the plasma membrane (of an egg cell.) *(a)* A coated pit develops in the plasma membrane, and some darkly staining material accumulates in it on the exterior face. The pit deepens *(b)* as the clathrin forms a closed cage around it and the plasma membrane closes in behind the pit *(c)*. Finally, the pit closes off as a coated vesicle *(d)*, which carries its enclosed material deeper into the cytoplasm.

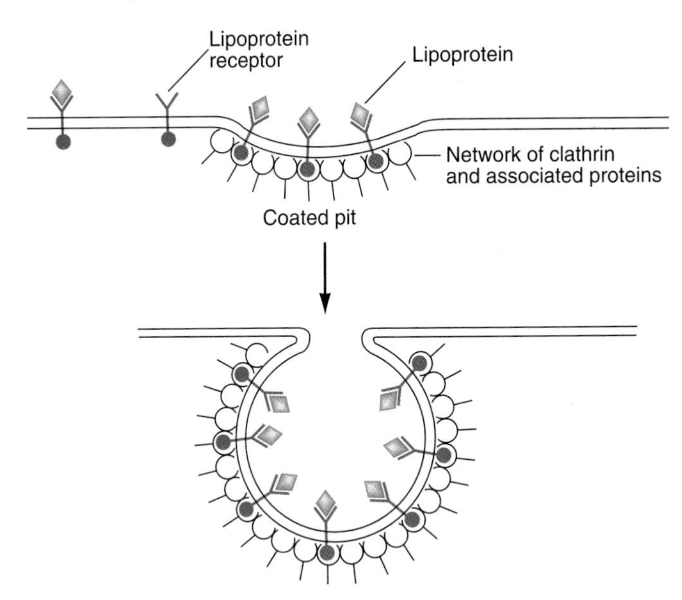

Figure 11.39
Animal cells may have receptors for lipoproteins, which carry lipids in the bloodstream. When these molecules bind to lipoprotein receptors in coated pit areas, they are carried inside the cells by endocytosis, and in this way the lipids are delivered to the cell.

Coda We have now completed an introduction to basic cell processes by bringing together a number of disconnected facts presented earlier into a view of cells that begins to approach reality. As the chapter title "The Dynamic

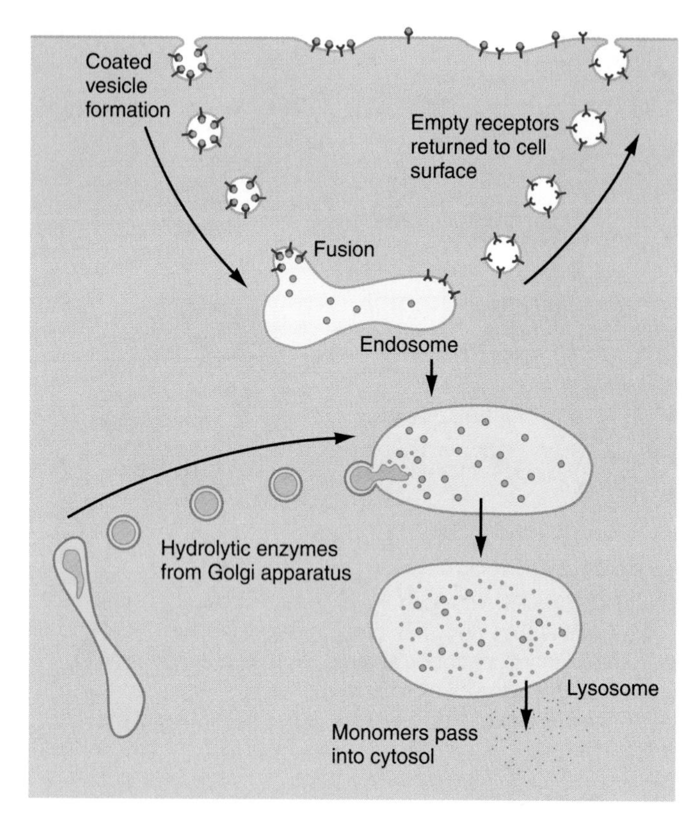

Figure 11.40

Lysosomes form as vesicles carried into the cytoplasm by endocytosis fuse with endosomes. Other vesicles, carrying digestive enzymes that have been packaged in the Golgi membranes, deposit their contents in the endosomes, converting them into lysosomes. These enzymes hydrolyze incoming material to monomers, which enter the cytoplasm. Receptors carried in by coated vesicles, and their clathrin coats, are recycled to the cell surface.

Cell" implies, cells are busy: They metabolize, grow, and move about. They produce signals that other cells respond to, and they in turn respond to signals from other cells and the environment. Fundamental to all of this is the fourth major theme of this book: *organisms function through molecular interactions.* Biosynthetic pathways with their enzymes depend on hundreds of quite specific interactions, including those between enzyme and substrate; many enzymes are allosteric proteins, and the regulatory function they play is fundamentally one of molecules interacting with each other. The way hormones and other ligands fit their receptors is another illustration of molecular interactions. Growth, in general, involves fitting molecules together in quite precise ways. Cell movements, by means of microtubules or actin filaments, depend on molecules interacting with one another, and it is instructive to consider movement from this perspective.

The structures and processes described in this chapter have become efficient and functional because organisms are genetic systems whose structures are shaped and designed by natural selection. The large molecules, especially proteins, that are needed for growth and other processes originate in a cell's genome; their structures, and of course their shapes, are determined by the cell's genes. Those genes have been gradually edited so the proteins they encode have specific functions in the life of every organism. Part II of this book explains how this genetic apparatus works.

Summary

1. Cells grow by synthesizing their components through biosynthetic pathways, which arise principally in the central metabolic pathways, especially the Krebs cycle.
2. Populations of cells tend to grow exponentially, which means the amount of new material made during each time period is proportional to the amount already present. Exponential growth results because each cell or cellular component is engaged in making more cells or components.
3. Cells regulate themselves so as to stay in a steady-state condition, with optimal concentrations of all their components. This constant internal condition is called homeostasis, and there are many regulatory mechanisms for maintaining it.
4. Allosteric proteins have two or more distinct binding sites where other molecules (ligands) can bind. The protein has two different conformations, with different activities; one may be active and the other inactive. Binding to specific ligands shifts the protein back and forth from one conformation to the other.
5. Allosteric proteins regulate metabolic pathways. Allosteric regulatory enzymes at key points in pathways are inhibited or activated in response to the concentrations of end products or other metabolites.
6. Receptor proteins recognize and bind to ligands stereospecifically. Chemoreceptors recognize external ligands, which may represent sources of energy or potential threats.
7. Signal ligands carry information at different levels of activity. Hormones carry information between different cells in a multicellular organism, and pheromones carry information between individuals of the same species; alarmones signal intracellular conditions that require a broad response. Allomones and kairomones are important ligands in ecological relationships.
8. Cells move and change their form actively by means of the cytoskeleton, a protein complex consisting of actin filaments (microfilaments), microtubules, and intermediate filaments.
9. Actin filaments effect many kinds of cell movement. They have a polarity and can treadmill by adding subunits at the plus end while releasing them at the minus end. They also form stereospecific functional complexes with many other proteins. Actin filaments interact particularly with myosin, which pulls on actin through a cycle of conformational changes powered by ATP.
10. Actin filaments form the cell cortex and structures such as the microvilli on intestinal cells. They form an extensive apparatus that holds cells in shape, and this network breaks down during cell division into other structures that help divide the cell in two.
11. Microtubules shape cells in many ways, as extensive bundles below the cell surface. They also contribute to cell movement and constitute the mitotic apparatus that separates chromosomes during cell division.
12. Microtubular structures are organized by special microtubule-organizing centers (MTOCs), which include the centrosome, basal bodies, and in some cells, the centriole.
13. Dyneins and kinesins are ATPases that function as molecular motors. They are used to anchor organelles to microtubules and to move them along microtubules in a defined direction.
14. The microtubules of cilia and flagella are organized in a characteristic 9 + 2 pattern, and they slide past one another by means of dynein motors.

15. Bacterial flagella consist of long helices of flagellin. They rotate within rings in the plasma membrane and cell wall, powered by a proton gradient across the membrane.
16. Bulk materials can be moved across membranes by exocytosis, which releases them from vesicles to the cell surface, and by endocytosis, which carries them from the surface into vesicles.
17. Endocytosis occurs by means of coated vesicles, which form at the cell surface around materials bound to receptors and carry these materials into endosomes in the cytoplasm.
18. Endosomes are transformed into lysosomes, which contain many digestive enzymes that can reduce large molecules to monomers. Lysosomes digest materials from outside, as well as some cellular organelles and sometimes whole cells.

Key Terms

nutrient medium 215
binary fission 215
cell culture 216
assimilate 216
transamination 216
steady state 217
homeostasis 218
feedback 218
transducer 218
ligand 219
end-product inhibition/ feedback inhibition 220
stimulus 221
chemotaxis 221
receptor protein 221
chemoreceptor 222
signal ligand 223
hormone 223
pheromone 223
alarmone 223
G-protein 224
cyclic AMP (cAMP) 224
protein kinase 225
cytoskeleton 226
actin filament 226
microfilament 226
microtubule 226
intermediate filament 226
actin 226
cell cortex 226
microvilli 226
treadmilling 227
myosin 227
cyclosis/cytoplasmic streaming 227
microtubule-organizing center (MTOC) 230
centrosome 230
centriole 230
basal body 230
molecular motor 231
kinetid 232
cilium/cilia 232
flagellum/flagella 232
exocytosis 233
endocytosis 233
phagocytosis 233
coated pit 234
coated vesicle 234
endosome 234
lysosome 234
pinocytosis 234

Multiple-Choice Questions

1. Which phrase best defines homeostasis?
   a. a dynamic steady state in which reactants flow in a one-way path
   b. an equilibrium condition that results from reversible, enzymatic reactions
   c. unchanging concentration of substrates and products
   d. the ability to undergo reversible reactions
   e. the ability to grow and divide
2. Which is an example of homeostasis in humans?
   a. increased rate of breathing when the activity level rises
   b. diverting blood to deeper blood vessels in order to minimize heat loss when ambient temperature falls
   c. increasing metabolic rate and heat production the body temperature falls
   d. *a* and *b*, but not *c*
   e. *a*, *b*, and *c*
3. Which is the best example of negative feedback?
   a. If no policeman is in sight, you keep increasing the speed of your auto, but when you see a policeman, you slow down to the speed limit.
   b. You decide that, as long as your income keeps growing, you will have more children.
   c. As long as your bank balance remains above zero, you keep purchasing whatever you want, but when your funds are depleted, you do not go into debt.
   d. You work as many hours as you can to save a specific amount of money; when you reach your goal, you work only as many hours as you must to maintain your bank balance.
   e. In even-numbered years, you have a roommate; in odd-numbered years, you live alone.
4. Which statement about allosteric proteins is correct?
   a. Some signal ligands bind to allosteric proteins by means of weak bonds while other signal ligands bind strongly.
   b. Most allosteric proteins freely oscillate between two different shapes as they rapidly bind and release each ligand in turn.
   c. Informational binding of ligands by an allosteric protein occurs because both ligands bind to the same site on the protein.
   d. When one ligand binds to an informational allosteric protein, the other ligand is unlikely to bind because its shape no longer fits its binding site.
   e. Ligands that bind to allosteric proteins can only inhibit the activity of the protein.
5. An antibiotic produced by a fungus or bacterium is best classified as a (an)
   a. hormone.
   b. pheromone.
   c. allomone.
   d. alarmone.
   e. allosteric protein.
6. Which is a second messenger?
   a. cAMP
   b. Ca^{2+}
   c. a hormone such as insulin
   d. *a* and *b*, but not *c*
   e. *a*, *b*, and *c*
7. Which is the primary transducing mechanism by which an extracellular stimulus is converted to an intracellular signal?
   a. G-proteins
   b. calmodulin
   c. cAMP
   d. Ca^{2+}
   e. protein kinase
8. Suppose a particular external stimulus caused a specific intracellular response. Which is the most likely order in which signal transducers would work?
   a. G-protein—cAMP—protein kinase
   b. protein kinase—cAMP —G-protein
   c. cAMP—protein kinase—G-protein
   d. protein kinase—G-protein—cAMP
   e. All the above are equally likely.
9. Microtubules are part of all of the following except
   a. the mitotic spindle.
   b. cilia.
   c. flagella.
   d. centrioles.
   e. microvilli.
10. Microfilaments function in all these ways except
    a. forming the mitotic spindle.
    b. establishing the shape of a cell.

c. amoeboid motion.
d. ciliary action.
e. formation of adhesion plaques.

True-False Questions

Mark each statement true or false, and if false, restate it to make it true.

1. Growing cells need transaminase enzymes to convert keto acids from their central pathways into nucleic acids.
2. Bacteria cannot grow in a medium without sulfur because they need this element to synthesize their nucleic acids.
3. An allosteric protein can bind two kinds of ligands at the same binding site, where they have different effects on the protein's activity.
4. If compound A is regulated by negative feedback, a decrease in the level of A will result in an increase in the rate of its production.
5. End-product inhibition generally occurs when the metabolic product of a pathway inhibits one of the final enzymes in the pathway.
6. Our sensations of taste and smell arise from the binding of environmental allosteric proteins to ligands in the plasma membrane of a receptor cell.
7. G-proteins are structural constituents of the plasma membrane, whereas cAMP is a mobile compound in the cytosol.
8. Once activated, second messengers such as cAMP or Ca^{2+} ions act as receptors.
9. The effects of certain inhibitors allow us to determine that microtubules are the primary structures involved in cytoplasmic streaming, or cyclosis.
10. Materials brought into a cell by exocytosis end up in mesosomes where they are digested by hydrogenase enzymes.

Concept Questions

1. Explain how a homeostatic steady state differs from a chemical equilibrium.
2. Explain why positive feedback does not lead to a homeostatic steady state.
3. Persons of normal weight are believed to have a normal negative feedback between food intake and metabolic rate. A decrease in food intake is followed by a decrease in metabolic rate, while an increase in food leads to an increase in metabolic rate. Why is this type of regulation called *negative?*
4. Compare and contrast hormones, pheromones, and alarmones.
5. Compare and contrast the interaction of tubulin with microtubular organizing centers and molecular motors.

Additional Reading

Alberts, Bruce, Dennis Bray, Julian Lewis, Martin Raff, Keith Roberts, and James D. Watson. *Molecular Biology of the Cell.* 3d ed. Garland Publishing, New York & London, 1994.

Bretscher, Mark S. "How Animal Cells Move." *Scientific American,* December 1987, p. 72. Animal cells move by bringing pieces of the outer membrane into the cytoplasm and then recycling them to the surface in a directed way.

Carafoli, Ernesto, and John T. Penniston. "The Calcium Signal." *Scientific American,* November 1985, p. 70. Calcium ions as second messengers in cells.

Carraway, K. L., and C.A.C. Carraway (eds.). *The Cytoskeleton: A Practical Approach.* IRL Press at Oxford University Press, Oxford and New York, 1992.

Dautry-Varsat, Alice, and Harvey F. Lodish. "How Receptors Bring Proteins and Particles into Cells." *Scientific American,* May 1984, p. 52. More information about receptor-mediated endocytosis.

Glover, David M., Cayetano Gonzalez, and Jordan W. Raff. "The Centrosome." *Scientific American,* June 1993, p. 32. More detail about the structure that directs assembly of the cytoskeleton.

Stossel, Thomas P. "The Machinery of Cell Crawling." *Scientific American,* September 1994. The role of the cytoskeleton in cellular movement.

Taubes, Gary. "Conversations in a Cell." *Discover,* February 1996, p. 48. Steven Schreiber and Jerry Crabtree have discovered how to fabricate signal pathways between cells, called dimerizers. Control of the pathways has been applied to genetic engineering and genetic-based therapies for diseases such as sickle-cell anemia.

Internet Resource

To further explore the content of this chapter, log on to the web site at:

http://www.mhhe.com/biosci/genbio/guttman/

PART II

The Structure and Function of the Genome

12 The Structure of the Genome

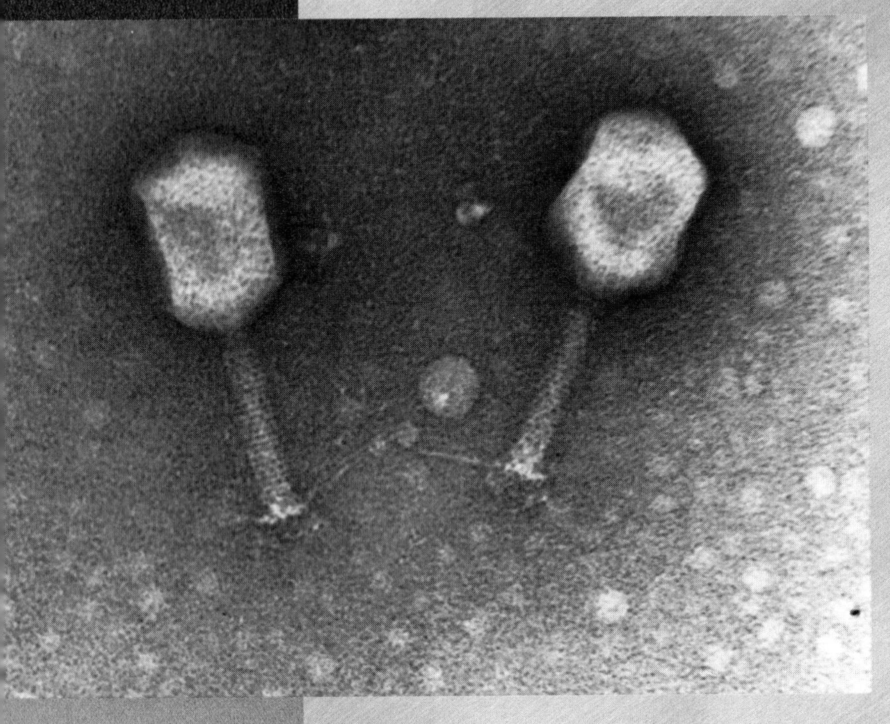

Electron micrograph of the bacteriophage (bacterial virus) T4.

Key Concepts

12.1 A cell operates on the basis of instructions in its genome.

12.2 Modern biological thought has been shaped by a Mendelian outlook.

12.3 Mutants extend the conception of a gene.

12.4 Genes control the steps in metabolism.

12.5 Transformation experiments pointed to the genetic role of DNA.

12.6 Bacteriophage are valuable tools for studying the molecular mechanisms of heredity.

12.7 The genome is DNA.

12.8 Two strands of DNA typically make a double helix.

12.9 DNA can replicate through specific base-pairing.

12.10 Nucleic acids encode information for the amino acid sequences of proteins.

12.11 Mutations are changes in nucleic acid sequences.

In 1943, the Austrian physicist Erwin Schrödinger published a series of lectures under the title *What Is Life?* Though ignored by traditional biologists, the little book became the inspiration for a growing movement that eventually became molecular biology. In his publication Schrödinger, one of the revolutionaries who had invented quantum mechanics in the 1920s, emphasized the importance of genes, which must specify and control the entire development of a complex organism from zygote to adulthood. He based his thinking on the ideas of the young physicist Max Delbrück. Delbrück conceived of genes as complex molecules that could exist in alternate states, jumping from one to another in the process of mutation, much as atoms can jump from one quantum state to another. After World War II, with the inspiration of Schrödinger's book and the experimental leadership of Delbrück and others, another little band of revolutionaries created the genetic view of life we are developing in this book. That is the story we begin in this chapter.

We have already outlined the idea that the information for forming each of an organism's structures—and thus performing each of its functions—lies in the genome, its DNA. Now it is time to examine DNA structure and see how DNA molecules can encode genetic information. After introducing the Mendelian outlook on the biological world, we will see how mutants can demonstrate that genes of certain kinds exist and how mutants were used to develop the concept that genes control the steps in metabolism. Then we will show how viruses, particularly bacterial viruses, are especially useful in genetic experimentation and how they revealed that DNA is the hereditary material. Finally, we will outline DNA structure and explain how it can replicate and carry genetic information. Just as we traced the gradual change in the meaning of "respiration" in Chapter 9, we begin by using the words "mutant" and "gene," which we all understand rather vaguely, and then show how advances in research gradually sharpened and defined these terms. ■

12.1 A cell operates on the basis of instructions in its genome.

Organisms resemble their parents. This simple, commonplace observation is the beginning of genetic biology, and it implies that organisms inherit something from their parents that gives them the same characteristics. That something is their genome. As we pointed out in Chapter 2, the genetic information in a genome provides a *genetic program* specifying the structure of the whole organism and how it is to operate:

Genome → Structures of biomolecules → Specific functions

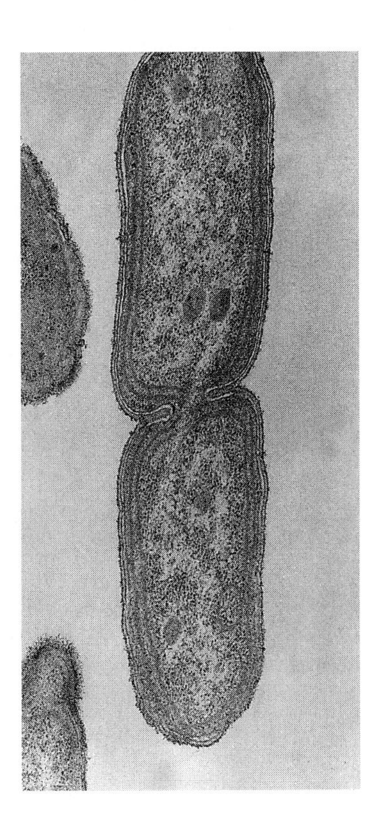

Figure 12.1
When a cell divides, both resulting daughter cells usually have the characteristics of the original cell because they have identical copies of the genome, which determines their characteristics.

Every cell is made of a few thousand different types of proteins, and each protein, if it is to be functional, must have just the right sequence of amino acids. As the example of sickle-cell hemoglobin in Section 5.12 shows, changing even a single amino acid out of hundreds can disrupt a protein's function. Genetic information in a cell's genome specifies the structures of all a cell's proteins, and to specify a structure is to specify a function too. The structure of an enzyme, for example, determines exactly what it can do—what chemical reaction it catalyzes, what other ligands can bind to it, and their effects. An organism creates its own form by "reading out" the instructions in its genetic program that determine exactly what structures to make and when and where to make them.

Sickle-cell hemoglobin, Section 5.12.

Although inheritance is most obvious in complex organisms like humans, it is apparent even in unicellular organisms that reproduce by simple cell division. When a cell divides, the two resulting **daughter cells** preserve the structure of the parent cell (Figure 12.1), and generally all the cells that may be formed by generations of division have the same chemical composition and operate the same way. Each cell passes on the information in its genome to its descendants as the genome forms new copies, or replicas, of itself in the process of **replication:**

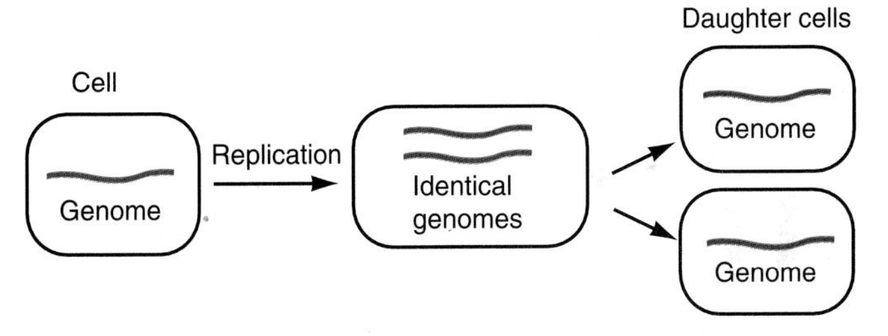

Replication provides two identical genomes, so both daughter cells receive a copy. Since all the cells in a multicellular organism arise from repeated divisions of the original zygote (fertilized egg), they all can have (and with minor exceptions apparently do have) the same genome. Replication, like the synthesis of proteins, must occur with high fidelity—that is, with very little change or loss.

Now we will elaborate on this general model for the operation of a genome and see what evidence there is to support it.

12.2 Modern biological thought has been shaped by a Mendelian outlook.

A little garden in an Austrian monastery has the honor of being the birthplace of modern thought about heredity, for it was there that a priest, Gregor Mendel, carried out experiments from 1857 to 1868 that laid the foundation for twentieth-century genetic research. Mendel had been raised and ordained in the monastery of Brünn (now Brno, Czech Republic), and sent to study science at the University of Vienna. Shortly after returning to Brünn to teach in 1853, he began experiments on inheritance, using the common garden pea. This work led to a paper, published in 1866, in which he demonstrated that heredity is governed by a few simple rules and that, rather than being complex and mysterious, the process of heredity could be understood quite easily. But Mendel's work was ignored, and he eventually gave it up as he became more involved with the business of the monastery. Mendel died in 1884, virtually unknown in science. However, in 1900, Carl Correns, Hugo de Vries, and Erich von Tschermak rediscovered Mendel's rules through their own experiments, recognized Mendel's original discovery,[1] and proclaimed it to the world, leaving us with the adjective "Mendelian" to describe the pattern of inheritance characteristic of all sexually reproducing organisms.

Twentieth-century biology grew up in a Mendelian world. For nearly a century now, all our thinking about biology has been shaped by Mendel's most critical and most basic discoveries: *Organisms carry units of heredity,* which we now call genes, and *discrete genes determine discrete characteristics.* He showed that the peas he worked with had distinct "factors" that determined such features as the shape of the pea or the color of its flowers. Although Mendel made all his discoveries and worked out the basic patterns of heredity without knowing about DNA, chromosomes, or the chromosomal dances called mitosis and meiosis, it is most instructive to discuss Mendelian inheritance in the light of modern knowledge of these things, as we will do in Chapter 16. Meanwhile, in this chapter details such as specific patterns of heredity, dominant and recessive genes, and the interactions among genes needn't concern us, as long as we recognize the most essential Mendelian idea and realize that investigators into genetics have spent much of this century trying to understand just what these mysterious genes are.

12.3 Mutants extend the conception of a gene.

During the early years of this century, Thomas Hunt Morgan's laboratory at Columbia University was an exciting place. Following the new knowledge of Mendelian rules of inheritance, Morgan and his students pioneered modern genetics through their brilliant research with a common fruit fly, *Drosophila melanogaster.* This little animal is an ideal subject for genetic work; notice in Figure 12.2 that the fly has many complex traits such as veins in its wings and tiny bristles covering its body, each of which is an inherited feature. Furthermore, it breeds rapidly and prolifically. A female *Drosophila* lays many eggs, and each of them develops into a small, white larva that feeds and grows for several days before becoming a pupa with a hard outer capsule. Soon the pupa is transformed into an adult fly, just as a caterpillar becomes a butterfly within its cocoon. This whole reproductive cycle takes only about two weeks, so it is easy to raise large numbers and many generations of flies in a relatively short time.

Figure 12.2
A fruit fly, *Drosophila,* is a complex animal, in spite of its small size. Its many traits are specified genetically.

The flies Morgan and his students began with all looked alike, having red eyes, tan bodies, bristles, and so on, as shown in Figure 12.2. These so-called **wild-type** flies are used as a standard for comparison with variant flies. Some variants always appear in any population—flies with white eyes or purple eyes, with black bodies, or with wings that are curly or distorted in some other way (Figure 12.3). Where do the variants come from? One hypothesis is that each is caused by a stress in the fly's environment as it develops. Many chemicals, for instance, can produce grossly distorted animals, and even giving fly larvae a heat shock can make them develop into defective adults.

The hypothesis about stress can be tested by breeding the variant flies while keeping environmental conditions as constant as possible. If a dumpy-winged fly is mated with a wild-type fly, their offspring all appear wild-type (a pattern of inheritance Mendel had already noted), but if these offspring are then mated with one another, flies with the same dumpy wings appear in the next generation. And if these dumpy-winged flies breed with one another, they produce nothing but flies with the same dumpy wings. Most of the variant traits show this pattern of inheritance. Each variant **breeds true**—that is, when it mates with another variant of the same type, all its offspring have the same traits. These results show that the variants are not caused by environmental stresses (which would produce all kinds of variants haphazardly) but that some flies, called **mutants,** have experienced **mutations,** stable changes in their heredity. (The adjective "mutant" describes both genomes and individuals, and the event that originally produced a mutant genome is also called a mutation.) Even though a genome replicates with quite high fidelity, the process isn't error-free. Mistakes occur occasionally during replication, and a mutation is just such an error. Damages to the genome from radiation and chemicals also produce mutations. But inheritable changes in the genome are caused only by certain kinds of radiation and chemicals, and their effects can be distinguished from the nonheritable changes caused by other environmental factors.

[1]Mendel's work had been ignored, but not really lost. He had communicated his results to the Swiss-German botanist Karl von Nägeli, who was not impressed by them, perhaps because they did not fit with the theory of heredity he himself was developing. In any case, Nägeli knew about Mendel's work, and Correns was his student.

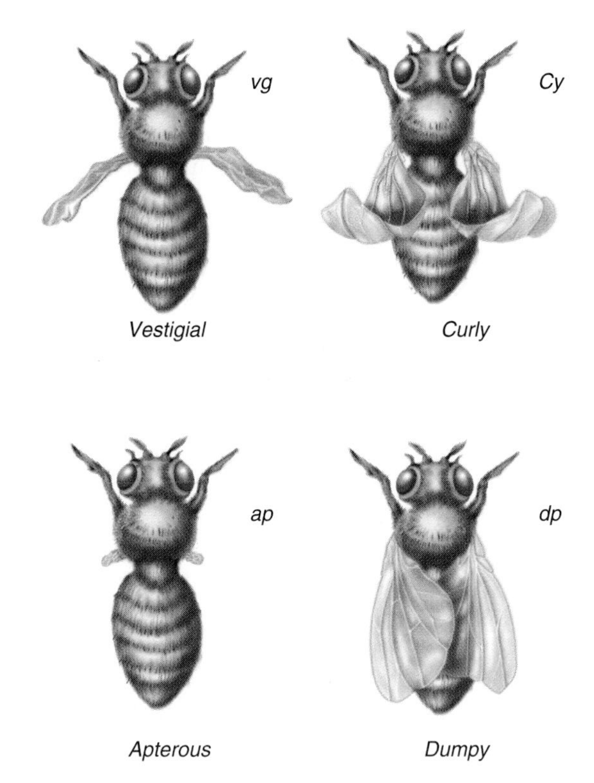

Figure 12.3

A variety of mutant fruit flies show how accidental changes in a genome can alter an organism's developmental program and thus its form.

The genome of a wild-type fly carries a program for normal wing development. Mutations can change this program, and a fly that inherits a mutant genome will therefore develop abnormal wings. A fly with dumpy wings, however, still has normal eyes, bristles, and body color. One with white eyes still has normal traits in all other respects. These observations confirm Mendel's principle that the genome must contain many *distinct* pieces of information, each specifying only a small part of the fly's anatomy. Each of these pieces of information is what we call a **gene.** Morgan's work extended the developing conception of a gene by showing that each gene is capable of mutating independently. We will develop a better definition later, but these simple observations carry two important concepts:

- A gene is a unit that specifies a particular trait.
- A gene is a unit that can mutate independently.

Notice, incidentally, that saying "a gene specifies a trait" is not the same as saying "each trait is specified by one gene." We will see that heredity is much more complicated than this and that each feature of a complex organism is actually specified by several genes.

12.4 Genes control the steps in metabolism.

During the 1940s, George W. Beadle and Edward L. Tatum linked genetics and metabolism with their experiments on the red bread mold *Neurospora crassa.* This simple organism is a suitable subject for studying gene function for two reasons. First, it grows on a nutrient medium made only of water, glucose, salts, and the vitamin biotin. It is hard to analyze the metabolism of an animal such as *Drosophila* that feeds on complex foods, but since this mold makes all its biomolecules from so few nutrients, it is quite easy to analyze its metabolism. Second, it is also easy to find mutants of *Neurospora.* Wild-type *Neurospora* makes all its cellular materials (except biotin) from glucose and salts, and an organism that can make all its components from such simple ingredients is a **prototroph.** A mutant that cannot make a certain cellular component, such as an amino acid or a coenzyme, is an **auxotroph,** and it will only grow if that missing substance is added to its growth medium.

To study the relationship between genes and metabolism, Beadle and Tatum collected auxotrophic mutants of *Neurospora.* They first found mutants that would only grow if their medium was supplemented with yeast extract, a rich mixture of amino acids, vitamins, and other nutrients. Further analysis showed that each mutant generally required only a single nutrient or, more rarely, a few nutrients with very similar chemical structures. They decided to analyze mutants that required only the amino acid arginine. These arginine auxotrophs are called *arg* mutants. Let us be quite clear about their growth requirement: The molds need arginine to make their proteins, and since *arg* mutants can't make arginine for themselves, they can only grow if it is supplied in their growth medium.

Beadle and Tatum found that the *arg* mutants fall into three groups, which we will designate *argA, argB,* and *argC,* based on their distinctive nutritional patterns. All these mutants will grow on a medium supplemented with arginine, but the *argA* mutants will also grow on media containing either ornithine or citrulline, two other amino acids, instead of arginine. The *argB* mutants will grow on arginine or citrulline, but not on ornithine; and the *argC* mutants require arginine specifically. Beadle and Tatum realized that these results only make sense if there is a biosynthetic pathway for arginine and if the mutants have defects in this pathway. They postulated a pathway like this:

$$\text{Precursor} \xrightarrow{\text{Enzyme 1}} \text{Ornithine} \xrightarrow{\text{Enzyme 2}} \text{Citrulline} \xrightarrow{\text{Enzyme 3}} \text{Arginine}$$

As we will show in a moment, the experimental results can only be explained if ornithine and citrulline are intermediary metabolites in the pathway, in this sequence. (We don't know what might come before ornithine.) There must also be three enzymes, one for each reaction shown, and the fact that a mutation can block the pathway means *a mutant gene produces a defective enzyme.* Most important, *a mutation in one gene only affects one enzyme.*

Now suppose we try to feed each mutant the various metabolites (Figure 12.4). A mold can only grow if it has arginine or if it can transform another metabolite into arginine by using the intervening enzymes. Thus the *argA* mutants also grow if they are given ornithine or citrulline because they can take up these compounds from the growth medium and convert them into arginine, using enzymes 2 and 3, which are specified by their

argA mutant

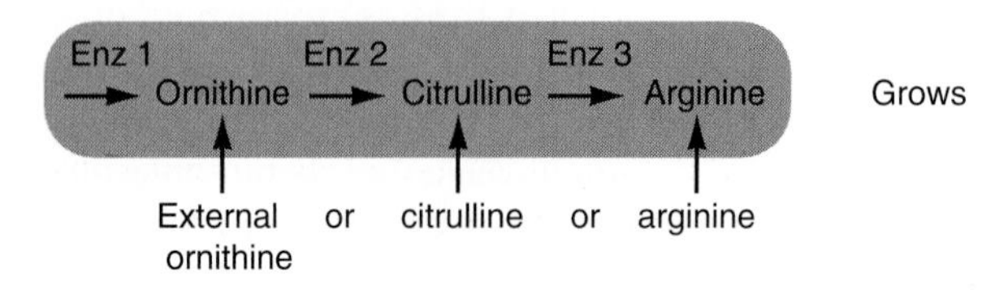

argB mutant

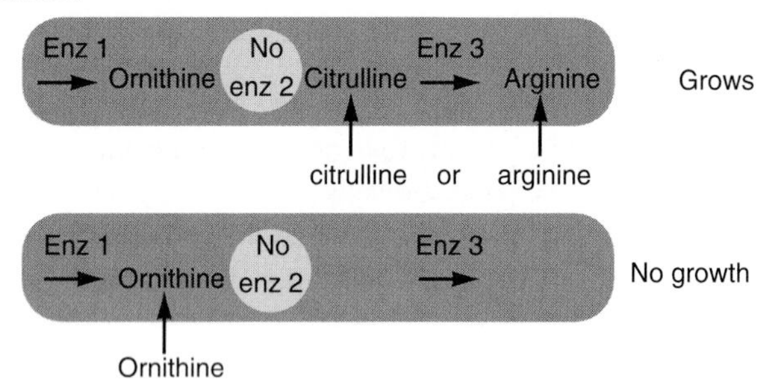

argC mutant

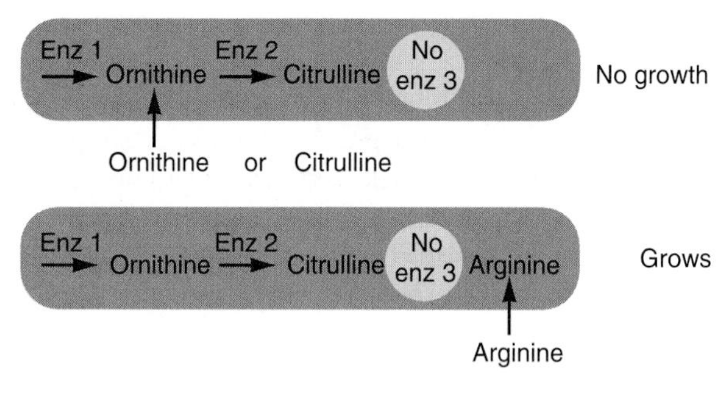

Figure 12.4

The three types of *arg* mutants are distinguished by their ability, or inability, to grow in media containing the different metabolites.

good (wild-type) genes. If *argA* mutants had defects in enzymes 2 or 3, they couldn't do this; therefore, they must have defects in enzyme 1 (or in an earlier enzyme). The *argB* mutants can grow on citrulline, but not on ornithine; thus they can convert citrulline to arginine with enzyme 3, but they must have a defect in enzyme 2, because they can't convert ornithine into citrulline. Finally, the *argC* mutants must have defects in enzyme 3, since they cannot convert either ornithine or citrulline into arginine.

This simple genetic experiment shows that, without having to do any chemical analysis, we can infer the sequence of enzymes and metabolites in a metabolic pathway by assuming that a mutation in each gene blocks a different enzyme. A simple rule underlies this analysis: A mutant can grow on a compound that comes *after* its mutational block in the pathway, but not on one that comes *before* that block. This explains the pattern of nutritional requirements.

Beadle and Tatum were thus forced to a monumental conclusion: *The function of each gene is to determine the structure of a single enzyme.* This concept, the "one gene, one enzyme" principle, is a cornerstone of modern biology. However, we now generalize this principle by saying "one gene, one polypeptide" because some proteins are made of more than one polypeptide and many proteins are not enzymes. This principle says that each gene specifies a distinct protein. Most proteins are enzymes that carry out metabolism, and it is through metabolism, of course, that the organism synthesizes its structure.

Beadle and Tatum's work was anticipated early in the twentieth century by the English physician Sir Archibald Garrod, who studied several inherited human abnormalities that we now know are biochemical deficiencies related to metabolism of the amino acid phenylalanine. He concluded that genes, leading to the synthesis of enzymes, were involved in these deficiencies, which he called "inborn errors of metabolism." While the molecular basis for most human disorders is unknown, several well-known conditions result from simple deficiencies in known enzymes. For instance, phenylketonuria (PKU) is a metabolic disorder caused by the inability to metabolize phenylalanine properly (Figure 12.5). Just as *Neurospora* mutants accumulate the metabolite that is normally converted by their missing enzyme, phenylketonuric people accumulate phenylpyruvate, a breakdown product of phenylalanine, which leads to severe mental retardation. Fortunately, PKU can be detected easily at birth by a simple test on a drop of the infant's blood, and a child who has the disorder can be maintained very well, without brain damage, on a diet with restricted amounts of phenylalanine. The test for PKU is now required by law in most states.

Figure 12.5 also shows the enzymatic defects in several other disorders. In each case, an unusual intermediary metabolite, normally present in only minute amounts, accumulates because of the enzymatic blockage. Usually, however, we know little about why the blockage leads to its peculiar physiological consequences. A classic example is Lesch-Nyhan syndrome, manifested in mentally deficient children with spastic muscle movements and an uncontrollable desire to mutilate themselves; they will bite their lips and fingers, and must be restrained to prevent serious injury to themselves and others. Yet, at its root, the syndrome is due to a lack of one enzyme in the pathway for the metabolism of purines (one component of nucleic acids); these children cannot properly dispose of excess purines from the breakdown of nucleic acids, and they produce excess uric acid, among other materials. Just why the buildup of these metabolites produces such bizarre behavior is unknown.

Given the relationship between genes and metabolism, a major question still remains: What *is* the genome? We will address this question in the next section.

Exercise 12.1 Mutants that cannot synthesize the amino acid tryptophan fall into two classes: *trpX* mutants can grow on compound B, a precursor of tryptophan, while *trpY* mutants cannot. Draw the positions of the X and Y enzymes in the pathway for tryptophan synthesis.

Exercise 12.2 Wild-type *E. coli* cells are able to accumulate the sugar fructose by facilitated diffusion. A certain mutant lacks this ability. *(a)* What structure in these cells is most likely inactivated by this mutation? *(b)* Does this observation

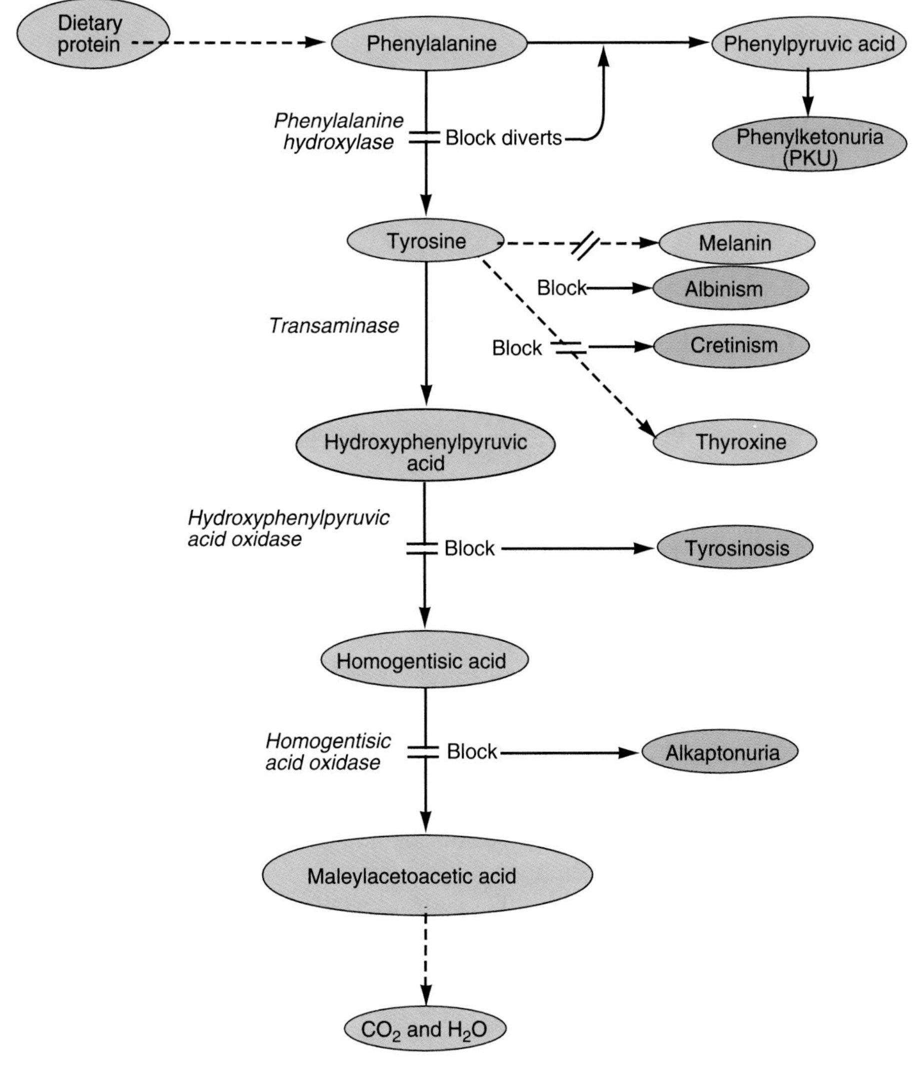

Figure 12.5
Enzymatic blocks in several of the steps for metabolism of the amino acids phenylalanine and tyrosine produce various inherited disorders *(pink ellipses)*.

contradict or support the "one gene, one polypeptide" principle? Does it contradict or support the narrower "one gene, one enzyme" form of the principle?

12.5 Transformation experiments pointed to the genetic role of DNA.

Once the elementary ideas of heredity started to make an impact on the thinking of biologists, it became obvious that cells must carry some distinctive substance, some chemical, that carries genetic information. However, biologists could only speculate about the nature of this substance. The most obvious candidate was protein, which early biochemists knew was a complex material, probably able to carry information, but some biologists thought genes might be made of nucleic acid, a substance that is abundant in cell nuclei. Nucleic acid was discovered in 1869 by Friedrich Miescher, who extracted it from pus cells that he bravely collected from discarded hospital bandages. "Nuclein," as he called it, was something of a chemical novelty to him because it contains both nitrogen and phosphorus. Some years later it became clear that there are two kinds of nucleic acid, RNA and DNA, although not much was known about either one—or about protein, for that matter. Yet different bits of evidence pointed to genetic roles for both nucleic acid and protein. In a famous book that laid the foundation for much of contemporary thought about cell biology, *The Cell in Development and Heredity,* E. B. Wilson went back and forth, from one edition to another, changing his mind about which substance had the honor of being the genetic material. It wasn't until about 1952 that it finally became clear to everyone that nucleic acid carries the genetic information, but the story of this recognition had begun 24 years earlier.

In 1928, Frederick Griffith made a perplexing observation while experimenting with *Streptococcus pneumoniae* (pneumococcus), a bacterium responsible for a type of pneumonia (Figure 12.6). The smooth (S) strain of this organism is surrounded by a glistening capsule and is pathogenic (disease-causing), while the rough (R) strain lacks capsules and is not pathogenic. Griffith found that mice died of a severe infection within a few hours after being injected with live S cells, but survived injections of live R cells. Heat-killed S cells were also harmless. However, the mice died after being injected with a mixture of heat-killed S cells and live R cells, and live S cells could be recovered from them. Even though he couldn't explain this observation, Griffith said that there must have been a **transformation** of R cells into S cells in this experiment. (R and S strains differ in several characteristics, so S cells could not have arisen simply through mutation.)

In 1943, Oswald Avery, Colin MacLeod, and Maclin McCarty set out to determine just what the transforming agent is. They too used mixtures of heat-killed S cells and live R cells, but they systematically destroyed components from the S cells with digestive enzymes. Destroying proteins, lipids, and polysaccharides had no effect on transformation, but destroying DNA eliminated it. The fact that DNA can transform one type of cell into another should have elicited great excitement and convinced everyone that DNA is the stuff genes are made of. Indeed, some people who knew about this research did recognize its implications, but most members of the biological community weren't yet ready to accept this conclusion.

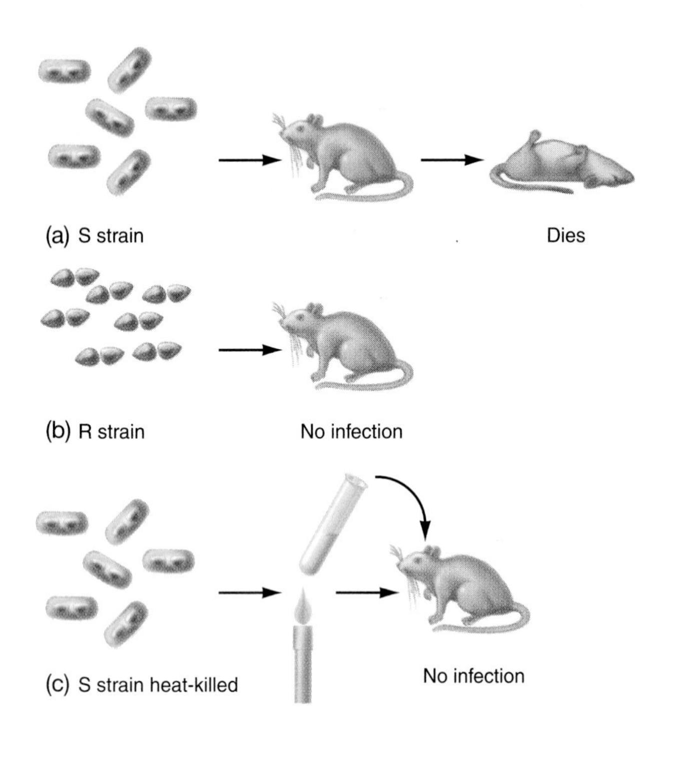

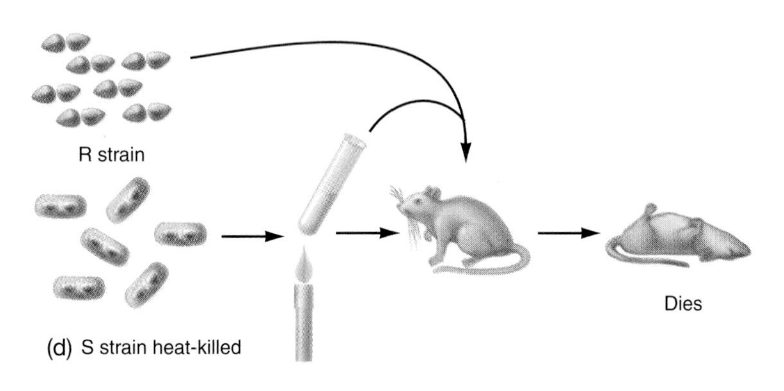

Figure 12.6

Griffiths knew that a smooth (S) strain of pneumococcus is pathogenic *(a)* but a rough (R) strain is not *(b)*. He then showed that heat-killed S bacteria are harmless *(c)*, but that some kind of transforming material taken from heat-killed S cells can change R bacteria into S bacteria *(d)*.

12.6 Bacteriophage are valuable tools for studying the molecular mechanisms of heredity.

During the 1930s, the physicist Niels Bohr and his colleagues often discussed the nature of life in their Copenhagen laboratory, and one of Bohr's students, Max Delbrück, was inspired to turn his attention to biology. Convinced that understanding biology lay in understanding heredity, Delbrück searched for a simple genetic system that could be analyzed easily. He focused his attention on viruses that grow in bacteria, called **bacteriophage** (literally, "bacterium eater"), or **phage,** for short. In 1939, he

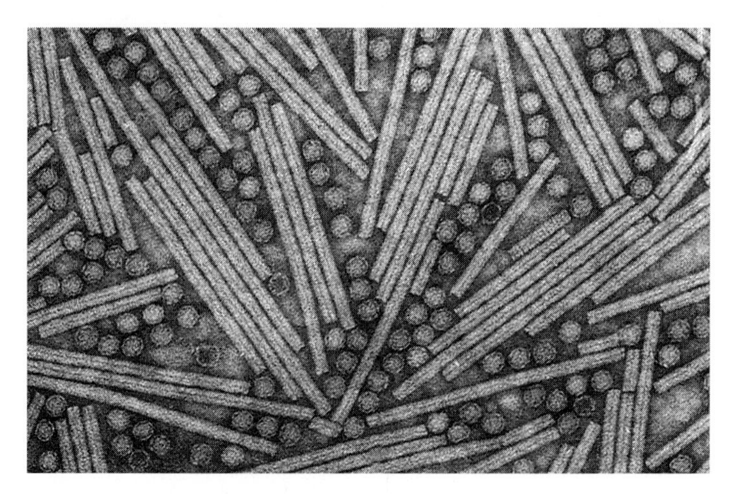

Figure 12.7

Viruses are not cells, but take the form of minute infectious particles called virions. Viruses are often named for the diseases they cause. The round virions here are turnip yellow mosaic virus, and the rods are tobacco mosaic virus.

went to Caltech (in Pasadena, California), and with Emory Ellis, developed basic methods for working with phage. During the 1940s, Delbrück, Salvador Luria, Alfred D. Hershey, and their students, known informally as the American phage group, began studies that led to fundamental insights into the molecular basis of biology. To understand this work, we must consider the multiplication of phage, which is representative of viruses in general.

We introduced the concept of a virus in Section 2.8 but didn't include viruses in the portfolio of organisms in Chapter 2 because they are a different kind of entity, quite distinct from organisms. The genetic model developed in Section 2.8 shows that a virus is essentially a little genome that goes around using cells for its own replication. Viruses are not cells, but rather, minuscule particles called **virions,** much smaller than even bacterial cells (Figure 12.7; Concepts 12.1). Viruses are parasites that reproduce rapidly, but only inside a cell that serves as their *host,* and each virus is specific for a particular type of host. We humans are plagued by animal pathogens like influenza, poxviruses, and polioviruses, while plants are subject to viral diseases with such names as "yellows," "mosaic," and "crinkle."

The original phage workers studied viruses that grow in *Escherichia coli,* a normally harmless, rod-shaped bacterium that lives in the intestines of all mammals, including humans, and is therefore abundant in sewage-contaminated waters. To grow phage, we must first know how to grow bacteria, and *E. coli* cells grow quickly in a nutrient medium made of inorganic nutrients and sugar. Also, we must be able to determine the number of cells in a culture at any moment; we do not talk about the *concentration* of bacteria or viruses, but rather about their **titer,** which expresses the concentration of anything that must be measured in a biological (rather than chemical) way. It is easy to determine the titer of bacteria by counting them in a calibrated chamber under the microscope, but the number of viable (live) cells can only be determined by the procedure shown in Figure 12.8. We *dilute* a sample by a known amount in some of the growth medium and then *plate* the diluted culture, which means spreading small amounts on a layer of nutrient medium in shallow glass dishes called **petri plates.** This growth medium is made semisolid by adding **agar,** a polysaccharide derived from seaweed, which gives it the consistency of jello. The bacteria can't move around easily on an agar medium, so every cell stays where it is deposited and grows there. All the cells derived from one cell asexually, by repeated division, constitute a **clone,** and they all stay together on the plate, making a visible **colony**—a small, circular mass of cells—in 12 hours or so. Since each viable bacterium originally deposited on the plate grows into one colony, counting the colonies determines the titer of viable bacteria in the original culture.

Example 12.1 Suppose we dilute a culture by a factor of 100 and then plate 0.1 ml from the dilution tube. If 350 colonies grow on the plate, the tenth-milliliter sample must have contained 350 viable cells. Therefore, the dilution tube it came from must have contained 3,500 bacteria per milliliter, and the culture must have contained $3{,}500 \times 100 = 3.5 \times 10^5$ live bacteria per milliliter. (Note: Appendix II explains arithmetic with the powers-of-10 notation.)

Example 12.2 A culture of bacteria is diluted by a factor of 10^4, two samples of 0.1 ml each are spread on plates, and 187 and 210 colonies then grow on these plates. What was the titer of bacteria in the original culture?

The average count of the two plates is about 200 colonies, so 1 ml of the solution that was plated contained $10 \times 200 = 2 \times 10^3$ viable cells. Since the plated solution was a 10^4-fold dilution, the original culture contained $2 \times 10^3 \times 10^4 = 2 \times 10^7$ viable bacteria per milliliter.

Exercise 12.3 A culture of bacteria is diluted by a factor of 10^6. Samples of 0.1 ml are plated, and an average of 95 colonies grow on the plates. What is the titer of live bacteria in the original culture?

The viruses (phage) that can grow in *E. coli* are also abundant in contaminated water and can be isolated as shown in

Concepts 12.1 Virions, Viruses, and Life

Students often ask, "What's the difference between a virus and a virion?" For comparison, let's ask, "What's the difference between a bacterium and a cell?" A bacterium is an organism, and a cell is a form it takes. A virus is a different kind of biological entity (not an organism), and a virion is the form it takes. It is not quite right to say that a virus is made of virions; a virus is *made of* protein and nucleic acid. Someone might say his home is a house or has the form of a house, just as someone else might say his home is an apartment or a houseboat, or has the form of an apartment or houseboat. The home is *made of* lumber, concrete, and roofing material.

The distinction between virus and virion rests in part on the concept we established in Chapter 2 that one must conceive of an organism (even though a virus is not an organism) as a life cycle. "Human" refers to a life cycle, and it takes different forms during that cycle—egg, sperm, baby, adult. "Poliovirus" is also a life cycle, and it takes different forms during that cycle, the most obvious form being a small ball of protein with RNA inside, which we call a virion; but poliovirus in the midst of infecting a cell is still poliovirus.

People often get exercised over the question of whether viruses are alive or not. This is purely a semantic issue, a question of how we use words, and not a serious biological issue. Since we have said (in Chapter 1) that being alive can be defined as a condition of having metabolism, maybe we can say that viruses are alive during the intracellular phase of their multiplication cycle but not when they are particles outside cells.

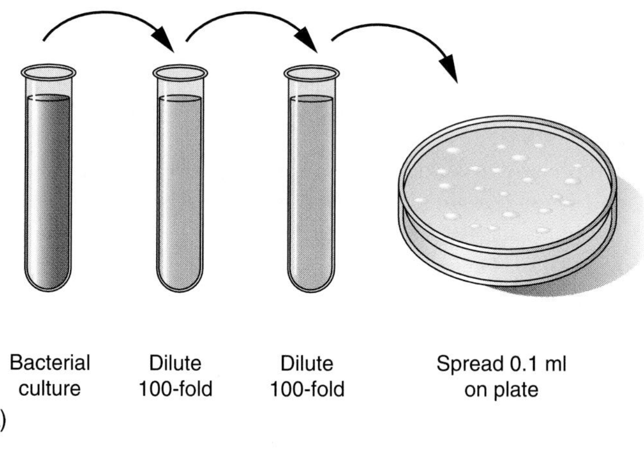

(a)

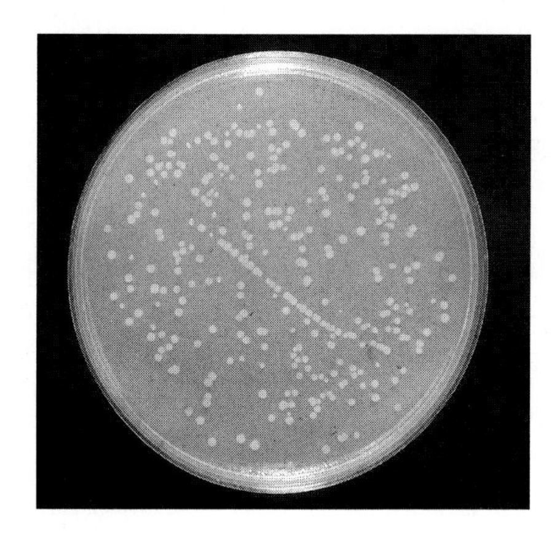
(b)

Figure 12.8
(a) A bacterial culture, containing millions or billions of cells per milliliter, is diluted serially: A sample of the culture is diluted into a tube of sterile medium, then a sample from this tube is diluted into a second tube, and so on. Finally, a sample from the last tube is spread on agar medium in a petri plate. *(b)* After incubation for several hours, each cell deposited on the plate grows into a colony.

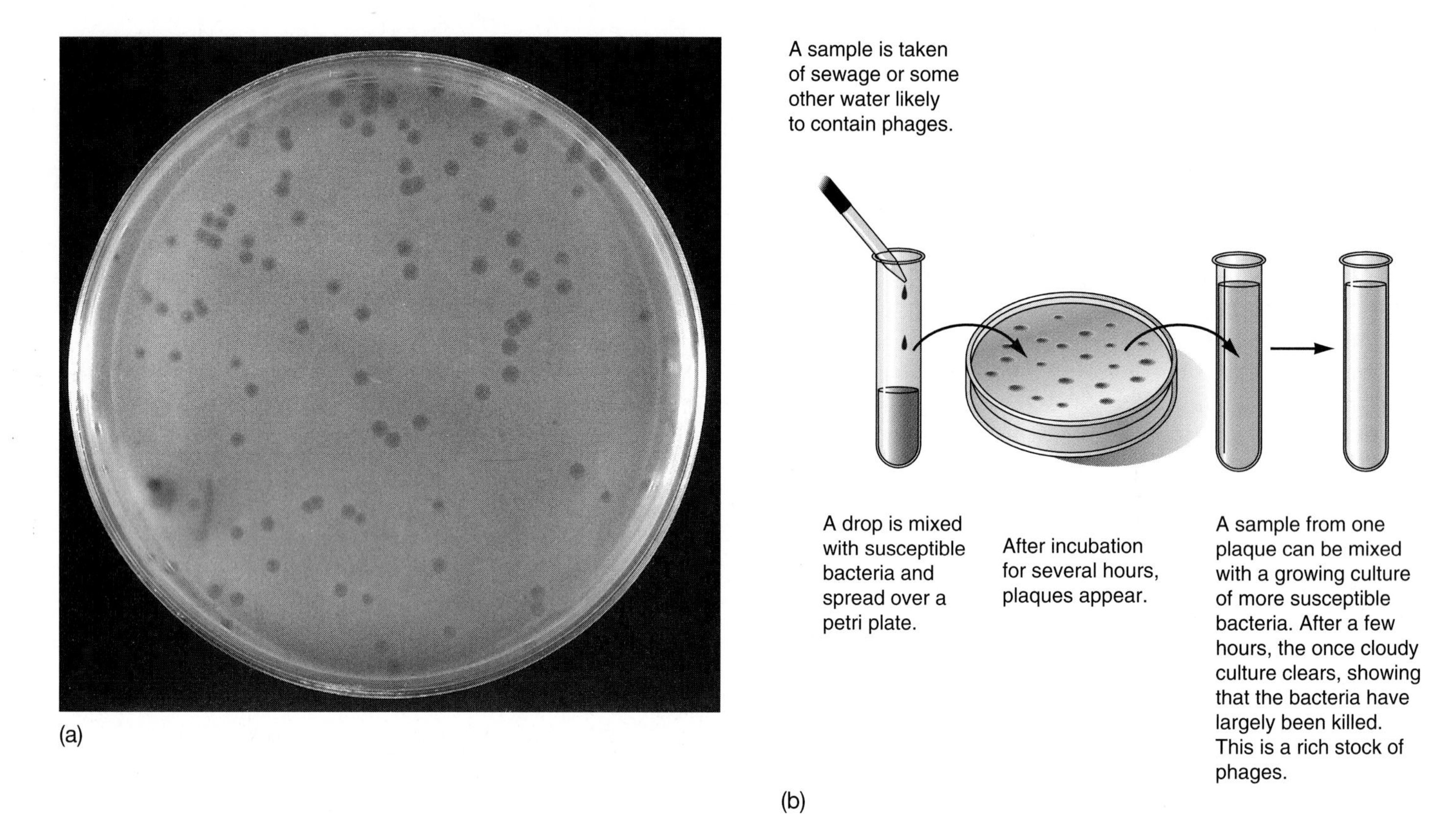

Figure 12.9

(a) One type of bacteriophage can be isolated from a new source by spreading a sample on a petri plate with a lot of bacteria, so phage in the sample can form distinct round plaques in a lawn of bacteria *(b).* Each plaque is a focus of infection where bacteria have been killed. The material in a plaque, which is rich in phage, is removed to a tube where the phage are allowed to grow on fresh cells to make a concentrated suspension of phage.

Figure 12.9. A culture with about 10^8 bacteria per milliliter is turbid, or cloudy, but within a few hours after a drop of water containing phage is added, the culture suddenly becomes clear. The phage break the bacteria open, or **lyse** them, and the cleared medium then contains many phage along with broken fragments of bacteria. Each type of phage can be isolated by plating a bit of the cleared medium on a petri plate along with millions of host bacteria—so many that they grow into a continuous layer (a "lawn") instead of separate colonies. Wherever a phage falls on this layer, it starts an infection that can be seen in a few hours as a clear circular area of cell lysis called a **plaque.** With this plating method, we can also count phage because, when appropriately diluted, each phage makes one plaque.

Various types of phage are designated by letters and numbers, as T2, T4, SPO1, or S13, and they make plaques of distinctive sizes and forms. Electron microscopy shows that each type of phage virion has a characteristic size and shape, such as a sphere or a polyhedron with a tail (Figure 12.10). The fact that each type of virus has a distinct form, which it maintains as it reproduces, tells us that viruses too must carry genetic information, making them ideal subjects for genetic studies, as Delbrück and his colleagues realized.

Exercise 12.4. A stock of phage is diluted serially by four factors of 100—in other words, by $100 \times 100 \times 100 \times 100$, or 10^8. Three 0.1-ml samples from the last tube are plated, and they yield 320, 300, and 310 plaques. What is the titer of the phage in the stock?

12.7 The genome is DNA.

Purified phage can be kept in a refrigerator for years, neither increasing nor decreasing in number, but when they are incubated with their host cells, they multiply in a *single step* of growth (Figure 12.11). Delbrück and Ellis discovered this by mixing phage T4 with bacteria and then plating samples at intervals to look for plaques. For about 23 minutes, all the samples yielded the same number of plaques (black curve on the graph), but then the number rose quickly and leveled off at about 200 times its original value. Delbrück and Ellis immediately realized what this pattern of growth means: Phage multiply *inside* the intact bacteria, so their increased numbers only become apparent when they lyse their host cells about 23–30 minutes after infection and burst out. Electron microscopy later supported this

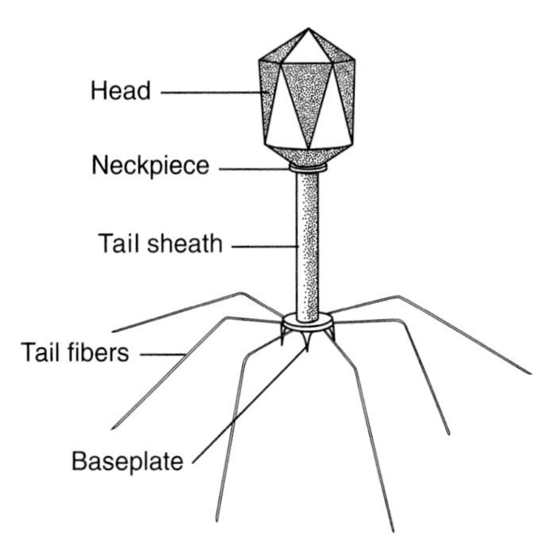

Figure 12.10

A virion of phage T2 or T4 is a complex structure with subunits (head, tailpieces, and tail fibers) made of many distinct proteins. The head encloses a genome of nucleic acid (DNA).

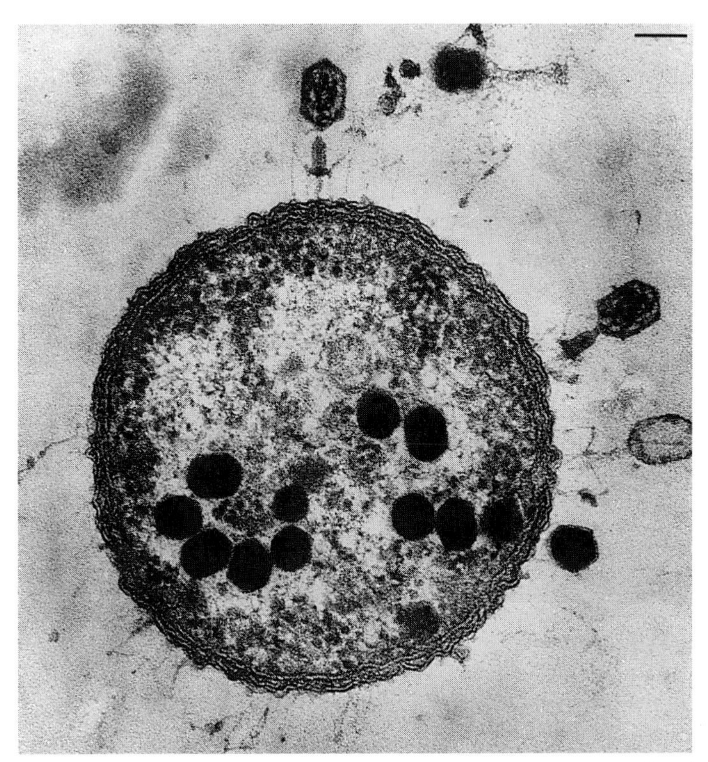

Figure 12.12

An electron micrograph of phage T2 infecting bacteria shows the protein part of a few infecting virions attached to the cell surface. The phage DNA from each virion has been injected into the cell, where it cannot be seen, but the DNA has replicated extensively, and the large black particles inside the cell are new phage heads packed with phage DNA.

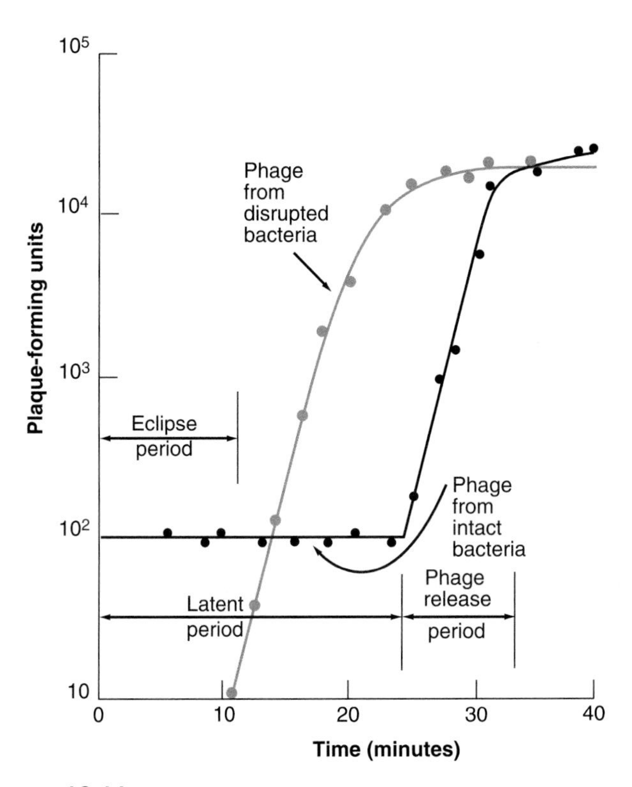

Figure 12.11

Phage T4 multiplies in a single step of growth. The black curve shows the plaques formed when infected cells are allowed to burst spontaneously; infected cells remain intact for about 23 minutes and then suddenly start to lyse. The colored curve shows numbers of intracellular phage, as revealed by lysing the cells artificially at various times; this curve shows that virions accumulate rapidly inside the cells before lysis begins.

view by showing that phage initially adsorb to the outside of their host cells (*ad*sorption means adhesion of molecules to a surface), and tailed phage like T4 attach by the tips of their tails (Figure 12.12). Micrographs taken late after infection show lysed cells releasing hundreds of new phage.

To find out what happens inside a cell during phage multiplication, A. H. Doermann broke open infected cells artificially and measured the number of phage. He thus obtained the colored curve in Figure 12.11, showing a dramatic rise in the number of intracellular phage until the cells start to burst. However, the graphs show that during the first 11 minutes after infection, there are no phage in the cell at all! As soon as they start an infection, the phage go into a noninfectious *eclipse period.* What happens during this time? To find out, Alfred D. Hershey and Martha Chase did a landmark experiment in 1951.

T2 phage are about half protein and half DNA. Hershey and Chase realized they could separate the roles of these two components by labeling them with radioactive tracers. Protein contains sulfur but no phosphorus, so phage produced in cells growing in a medium with radioactive ^{35}S (in sulfate) will only incorporate the ^{35}S label into their proteins. DNA, on the other hand, contains phosphorus but no sulfur, so phage produced in bacteria growing in a medium with ^{32}P (in phosphate) will have

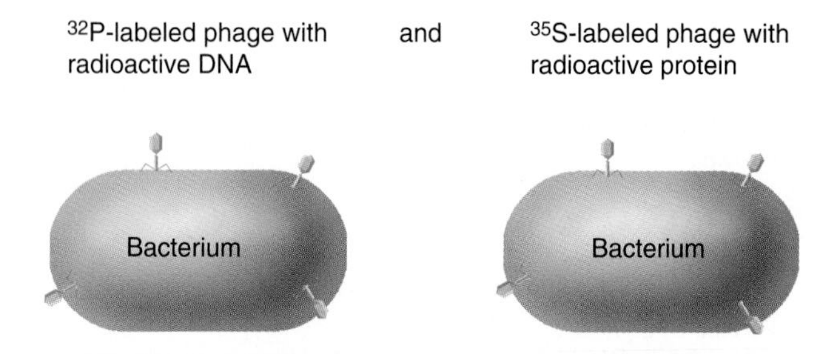

are used to infect separate cultures of bacteria. Shortly after infection, the cells are agitated in a small, high-speed blender. This does little damage to the cells, but it strips off the phage coats (capsids) attached to their surfaces. The two infected cultures are then spun in a centrifuge at speeds high enough to sediment the cells into a pellet at the bottom of the tubes. The stripped phage coats remain in suspension in the supernatant.

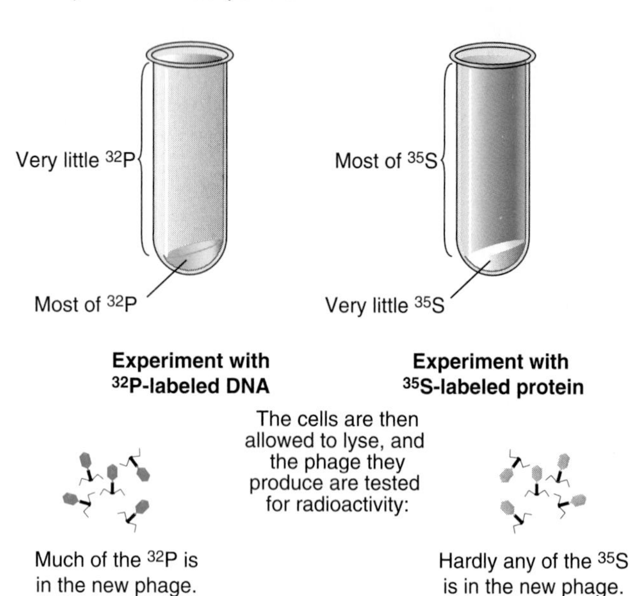

Explanation: Since most of the ^{32}P remained with the bacteria, the original labeled DNA was inside them, and much of this DNA was transferred to the new phage. Most of the ^{35}S was not with the bacterial cells because the phage protein remained outside the cells and was removed.

Figure 12.13

A composite version of the Hershey-Chase experiment shows how radioactive tracers were used to distinguish the roles of DNA and protein during a phage infection.

labeled DNA. Figure 12.13 diagrams the most revealing parts of the Hershey-Chase experiment. It shows, first, that only the DNA gets into the cell during a phage infection, while the protein is left attached to the cell surface where it can be sheared off by a blender. Second, the viral DNA passes into the next generation of phage, but the protein does not. The conclusion is clear: *The genome is DNA. DNA carries the genetic information.*

Figure 12.14 summarizes a cycle of phage multiplication and explains the eclipse phase. An infecting phage transfers its DNA into a bacterial cell. This DNA is the phage genome, which carries a genetic program for taking over the cell and converting it into a factory for manufacturing more phage. The genome

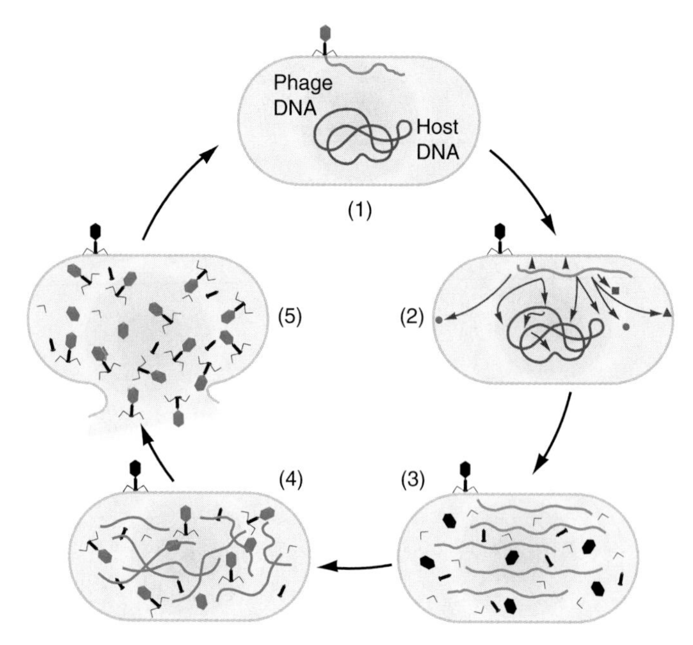

Figure 12.14

(1) A phage adsorbs to a bacterial cell and transfers its DNA into the cytoplasm. This DNA is the phage genome, which carries a genetic program for taking over the cell and converting it into a factory for manufacturing more phage. *(2)* The synthesis of new phage proteins begins. Some of these proteins stop bacterial protein synthesis while others inactivate bacterial DNA and even digest it. *(3, 4)* Specific phage-encoded enzymes begin to replicate the phage DNA. The phage DNA also directs the synthesis of new structural proteins, which assemble themselves into new phage particles with DNA packed into their heads. *(5)* When the cell is filled with many new phage virions, it lyses and releases these phage.

encodes many new phage proteins, which interfere with some normal cellular functions, destroy the cellular genome, replicate the phage DNA, and form components of new phage particles. During the eclipse period, the cell contains separate DNA and protein molecules but no intact phage. About halfway through the growth cycle, these components start to assemble themselves into new phage virions, which accumulate inside the cell. The virions are released when the cell lyses. Although viruses that grow in plant and animal cells behave similarly, they have different ways of invading their host cells, each cell produces much larger numbers of virions, and an infection cycle usually takes hours or days, rather than minutes. Every kind of organism probably has its own viruses. They have evolved along with their hosts and are common members of every ecological community.

Virus-infected cells are important model systems because the viral genome replaces the cellular genome and introduces a small set of new genes. An *E. coli* cell has an estimated 4,000 genes. Phage T4 has about 300 genes, and some very small viruses have only 7–10. Virus-infected cells and normal cells carry out essentially the same molecular processes, but the small viral genome is easier to analyze and to manipulate (using techniques that we will discuss in later chapters). Much of our knowledge of cellular processes has come from studying virus multiplication.

The next question is, how can DNA replicate and carry this genetic information?

Exercise 12.5. A culture of bacteria is infected with 10^8 phage per milliliter. After the cells have lysed, there are 1.5×10^{10} phage per milliliter. Assuming each phage infects one bacterium, how many phage on the average were produced in each cell?

12.8 Two strands of DNA typically make a double helix.

By 1951, it was clear that DNA is a polymer and that its monomers are four kinds of nucleotides, each consisting of a phosphate, the sugar deoxyribose, and one of four kinds of cyclic, nitrogenous bases. Two of the bases, having only a single ring of C and N atoms, are **pyrimidines:**

Thymine (Thy) **Cytosine (Cyt)**

The other two bases, with a double-ring structure, are **purines:**

Adenine (Ade) **Guanine (Gua)**

The nucleo*sides* of these bases (that is, the base-sugar combinations, without phosphates) are named, respectively, thymidine, cytidine, adenosine, and guanosine, abbreviated T, C, A, and G. To be formal, each of these names should be prefixed with *deoxy-*, since the sugar in each nucleoside is deoxyribose, and they should be abbreviated dT, dC, dA, and dG. But we will only do this if necessary to avoid confusion with RNA.

A nucleic acid molecule consists of many nucleotides linked to form a sugar-phosphate backbone so each phosphate connects the 5′ carbon of one sugar to the 3′ carbon of the next; the molecule therefore has a 5′→3′ polarity:

5′–3′ direction

The prime marks on these numbers show that they refer to the carbon atoms of the sugars. Atoms in the bases take numbers without prime marks. Nucleic acid sequences are conventionally written in the 5′→3′ direction, from left to right whenever possible.

For a long time, people were misled by the belief that DNA is made of a quartet of A, T, G, and C units repeated over and over. Since such a regular molecule could not carry information, the idea that DNA is the genetic material couldn't be taken seriously. But by 1951, Erwin Chargaff had shown that DNA preparations from different organisms vary in composition, so DNA molecules must actually be quite variable and capable of carrying information, as the results of the Hershey-Chase experiment demand.

Also in 1951, Francis H.C. Crick and James D. Watson, one of Luria's students, began to investigate DNA structure at Cambridge University, hoping that its structure would explain its function. They based their work largely on X-ray diffraction photographs taken by Rosalind Franklin and Maurice Wilkins (Figure 12.15). Chargaff's work also gave them an important

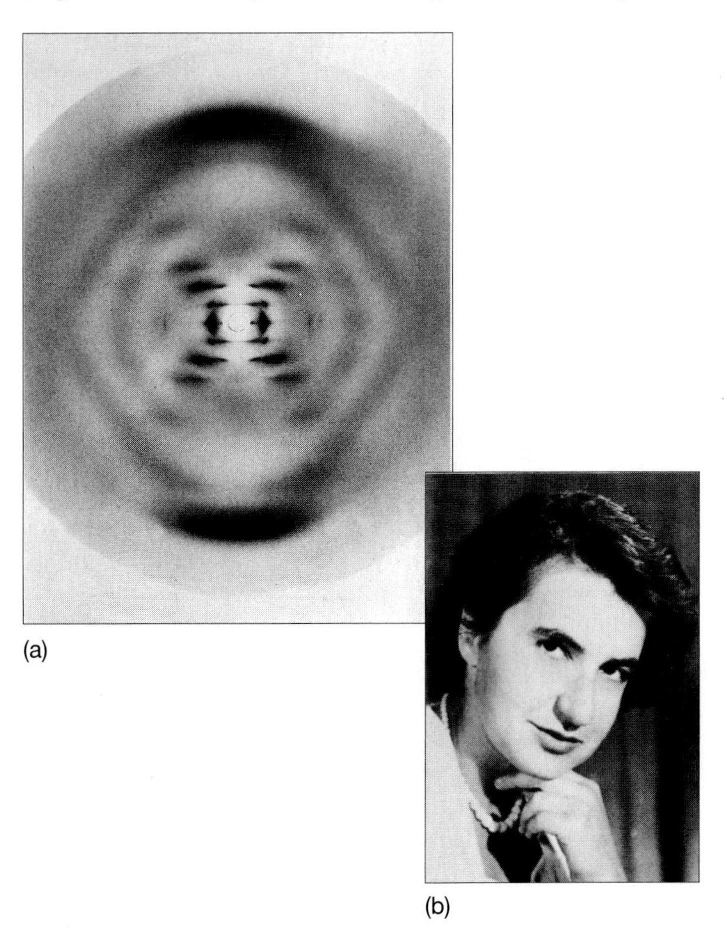
(a) (b)

Figure 12.15

(a) An X-ray diffraction photograph of DNA, made by Maurice Wilkins, shows a regular pattern of spots. This pattern can be interpreted to give the form and dimensions of a DNA molecule. Photographs like this one, made by Rosalind Franklin and Wilkins, were critical in helping Watson and Crick determine the DNA structure. *(b)* Franklin died in 1958, too early to share in the Nobel Prize awarded to the other three scientists.

clue: Every DNA sample contains equal numbers of adenine and thymine molecules and equal numbers of guanine and cytosine molecules—that is, A=T and G=C.

By making physical molecular models with this information in mind, Watson and Crick hit upon a now-famous DNA structure (Figure 12.16). DNA, they said, consists of a double helix of two polynucleotide strands with opposite 5′→3′ polarities. The Watson-Crick model places the sugar-phosphate backbones on the outside of the helix, because DNA is truly an acid whose phosphates lose hydrogen ions, become negatively charged, and repel each other. The molecule resembles a spiral staircase whose frame is the sugar-phosphate backbones and whose steps are the bases. Each step is a pair of bases held together by hydrogen bonds, always one pyrimidine and one purine: either T bonded to A or C bonded to G, thus explaining Chargaff's data:

CH_3 Deoxyribose Deoxyribose

Thymine **Adenine**

Deoxyribose Deoxyribose

Cytosine **Guanine**

These two combinations are the most stable because these bases are *complementary* to each other; their shapes fit together and form hydrogen bonds. Adenine is the complement of thymine, and guanine is the complement of cytosine. Furthermore, one whole strand of DNA is complementary to the other; they have opposite shapes that fit together neatly.

Exercise 12.6 One component of a nucleotide carries electrical charges in the pH range of cytosol. Which component is this, and are the charges positive or negative? If a protein molecule binds strongly to DNA, what kind of charges is it likely to have?

Exercise 12.7 Shortly before Watson and Crick published their model of DNA structure, Linus Pauling, having recently determined the α-helix and β-sheet structures of proteins, wrote a model for a double-stranded DNA helix in which the sugar-phosphate backbones are close together on the *inside.* How could Watson and Crick see immediately that this structure is wrong? (*Hint:* Think about Exercise 12.6 again.)

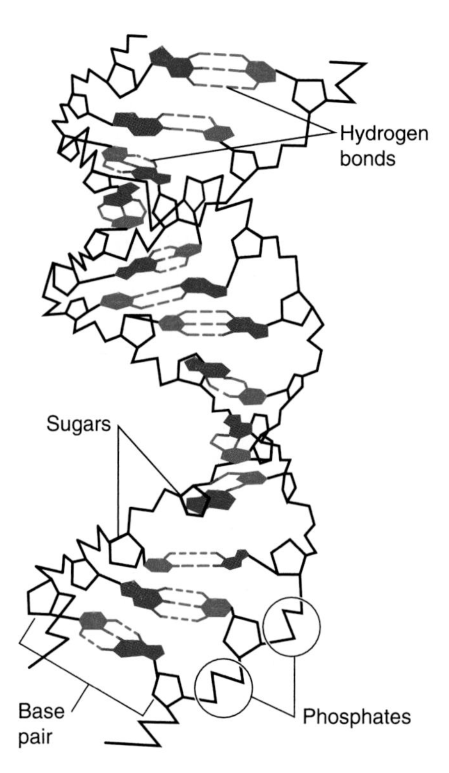

Figure 12.16

DNA has the form of a double helix. The backbones on the outside are chains of deoxyribose sugars *(pentagons)* linked by phosphates. Each base pair is made of a pyrimidine base (*green hexagons*) connected by hydrogen bonds *(red lines)* to a purine base *(blue hexagon + pentagon shapes).*

12.9 DNA can replicate through specific base-pairing.

In publishing their model, Watson and Crick commented wryly, "It has not escaped our notice that the specific pairing we have postulated immediately suggests a possible copying mechanism for the genetic material." A double-stranded DNA molecule can replicate by obeying the Watson-Crick pairing rules (Figure 12.17; Sidebar 12.1). The two polynucleotide strands are held together only by weak hydrogen bonds. If the double-stranded molecule is "unzipped" from one end, the nucleotides in the polymer are free to hydrogen-bond to their complementary nucleotides, which are abundant in the cytosol where they are being made by biosynthetic pathways. If the first three nucleotides in one strand are A–C–G, they will bind in sequence to a free T, G, and C. As the free nucleotides form hydrogen bonds with the intact strand, they are covalently bonded to each other, forming a new strand that begins with the complementary sequence T–G–C:

A C G G T C A → A C G G T C A
T G C → T G C C

The process continues along each strand as the helix unwinds, until a new strand of complementary nucleotides has been

Sidebar 12.1 Do Nucleic Acids Replicate Themselves?

Biologists always say "DNA replicates," and we talk about how "self-replication" occurs. Yet we know that in fact DNA *is replicated* in a complicated process by several enzymes working coordinately, as explained in Chapter 13. If you put a DNA molecule into a solution with some of its nucleotide components, you would wait a very long time before the single molecule replicated to make two. Are biologists chronically ignoring reality?

To answer this question, we must revisit the concept of evolution. As biological systems were slowly evolving out of simple organic compounds billions of years ago, the formation of functional nucleic acids was a key event because organisms are fundamentally genetic systems requiring genomes, which are always nucleic acids. As we explain in Chapter 29, the critical functions were probably carried out at first by ribonuceic acid (RNA) molecules, which only differ chemically in small ways from DNA and still comprise the apparatus in all cells that converts the information in DNA into actual cellular structure. The whole genetic apparatus was only able to evolve because those primitive nucleic acids had an intrinsic ability to replicate themselves. The enzymatic apparatus that eventually took over replication and made it fast and efficient did not exist yet; it, too, was slowly evolving. But if nucleic acids had not had this inherent property of self-replication, biological systems probably could never have evolved. Our way of speaking about replication reflects this recognition. We continue to speak about nucleic acids *replicating*, rather than *being replicated*, because it is still the property of the nucleic acid that permits self-replication to occur. We relegate the enzymatic apparatus to the role of a helper in the process, albeit an essential helper.

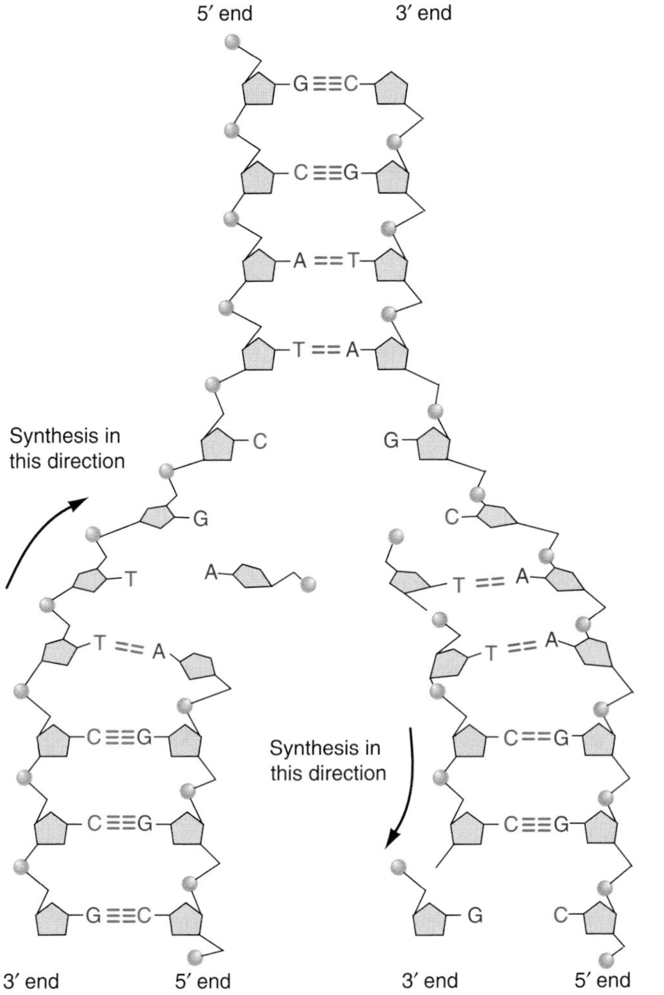

Figure 12.17

This is a general conception of DNA replication, according to the Watson-Crick model. The strands of the helix separate from each other; free nucleotides from the cytosol pair with the exposed bases and are linked into new DNA strands by enzymes (not shown here).

formed on each of the old strands and two identical double-stranded molecules have replaced the original molecule. Notice that DNA replication is a polymerization, and it is catalyzed by a DNA polymerase enzyme.

Mechanism of DNA replication, Section 13.4.

Exercise 12.8. Here is one end of a DNA molecule:

3′–A–A–T–G–C–C–T–A–C–C–A–G–T–5′
5′–T–T–A–C–G–G–A–T–G–G–T–C–A–3′

Separate it into its strands, and then replicate each strand with new nucleotides. Be sure the two molecules you end up with are identical to the original, and designate the polarity of their strands.

12.10 Nucleic acids encode information for the amino acid sequences of proteins.

Now we can understand how a genome specifies the rest of the organism. The Hershey-Chase experiment established that a genome specifies the structures of proteins, because only phage DNA enters a cell, but phage virions, made of both DNA and protein, come out. Therefore, the phage DNA, the genome, must specify the structures of those phage proteins. The Beadle-Tatum experiments established that genes determine protein structure, a gene being a unit that specifies the structure of one polypeptide. The genomes of all cells are very long DNA molecules (a bacterial genome consists of about 4 million nucleotide pairs and is over a millimeter in length), but they can be divided functionally into a few thousand genes, each of which is a short sequence of nucleotides that specifies one polypeptide.

Proteins and nucleic acids are both heteropolymers, made of several different monomers. They contain information or, in other words, it takes information to specify them—that is, to specify which of the four nucleotides to put in each position of a nucleic acid molecule or which of the 20 amino acids to put in each position of a protein. Once such a polymer is formed, it contains information. Just as the three symbols of Morse code (dot, dash, space) can encode a message that can be translated into the 26 letters plus punctuation of English, a nucleic acid genome can encode a genetic message with its four "letters," the four nucleotides of a single strand. This message can be translated into the 20-amino-acid language of a protein, and this is exactly what happens in a cell: DNA sequences (genes) encode amino acid sequences (proteins). And the words *encode* and *translate* are standardly used to describe the informational relationship between a genome and its protein products. In Chapter 14, we elaborate on the question of how the genetic message, encoded in a nucleic acid molecule, can specify the amino acid sequence of a protein.

Let us emphasize that the genome simply specifies the *primary structure,* the amino acid sequence, of each protein. As Anfinsen's experiments with ribonuclease showed, once a protein with a certain primary structure has been made, the molecule folds up into the right three-dimensional shape, its lowest-energy state. (We now know that many proteins need enzymes called chaperonins to help them fold into their lowest-energy form, but the protein's primary structure still determines its shape.) Enzymes, which are protein (or, rarely, RNA) products of the genome, then synthesize all the nonprotein components of a cell. Furthermore, proteins and other structural materials assemble themselves into larger structures with the right shapes for their particular biological activities. This assembly happens spontaneously, with little or no additional information needed from the genome.

Anfinsen's experiments on protein structure, Section 4.11.

Exercise 12.9 Explain why a homopolymer such as cellulose, made of only one kind of monomer, cannot carry information.

Exercise 12.10 The sequence of nucleotides in DNA specifies the sequence of amino acids in a protein. If a *single* nucleotide specified one amino acid, the DNA could only specify four amino acids because there are four nucleotides. If a *pair* of nucleotides (such as AA, AC, AG, AT, and so on) could specify an amino acid, how many amino acids could be determined? Since there are 20 amino acids, what is the shortest nucleotide sequence that could specify one of them? This question anticipates a point we will expand upon in Chapter 14.

12.11 Mutations are changes in nucleic acid sequences.

We can see now what a mutation is. If the sequence of bases is a message, any change in the sequence can change the message, and such a change is therefore a mutation. Some mutations will have no visible effect ("silent" mutations); others will be **nonsense mutations** that make gibberish out of a message; and still others will be **missense mutations** that change the meaning of a message. Think of genetic messages as printed messages and mutations as typographical errors. For instance, the engagement announcement that read, "He will receive his B.S. degree in Wildwife Management" suffered from a missense mutation. A small change can obviously make a big difference. Structurally, mutations can take several forms. For example, a mutation could be a **substitution,** like "w" for "l," or an **insertion** of some snrgaflzg@hojp#i extra material. A mutation could also result from **deletion** of an important. Another interesting mutation is an **inversion** in which a piece of DNA is accidentally removed and put back the wrong way. A **frameshift** mutationc ausest hes pacingb etweenw ordst ob es hifted so the message can't be read properly. A **duplication** is another obvious kinnother obvious kind of error. We will discuss the causes of mutations in Chapter 17.

Coda This chapter brings us to the heart of modern biology and begins to give meaning to one of the four themes around which this book is organized: *Organisms are genetic systems.* We have shown how model organisms, such as fruit flies, molds, and bacteria, have been used to elucidate the nature of heredity and the connection between genes and metabolism. The chapter also uses viruses to show that the genetic material is DNA, and we will emphasize viruses in later chapters as important tools for exploring the genetic apparatus. We have shown how the structure of DNA enables it to carry genetic information and to perform the important task of replication. The two complementary strands of DNA separate, and each lends its structure to the synthesis of a new complementary chain; the result, of course, is two new, identical DNA molecules where one had existed before. Thus the genome of one cell produces the genomes of two daughter cells, and in this way the genetic heritage is passed from generation to generation. We have introduced mutations, informational changes in genes that were essential in originally defining genes and are constantly used to explore the nature of genes. The genetic nature of organisms emerges even more strongly as we show that nucleic acids encode information for the assembly of amino acids into the correct sequence to manufacture every protein needed for organisms to function.

Summary

1. A cell operates on the basis of instructions in its genome. These instructions form a genetic program that specifies how the cell is to be constructed and how it will operate.
2. Gregor Mendel's work demonstrated that an organism's characteristics are determined by discrete units, which we now call genes. Mendel thus set the scene for twentieth-century research on heredity, which became focused on the question of what a gene is.
3. The existence of genes was demonstrated by the discovery of mutants, each of which usually affects one particular trait of the organism. The existence of many distinct mutations shows that there must be many distinct genes.
4. Genes control the steps of metabolism. Mutants that are defective in different enzymatic steps show that each gene carries the information for one enzyme. In general, a gene carries the information for one

polypeptide chain, since not all proteins are enzymes and some proteins are made of more than one polypeptide.

5. Simple blocks in metabolic pathways, due to the lack of an enzyme, can have severe physiological effects. Many inherited human disorders, such as phenylketonuria, are caused by enzymatic deficiencies.
6. The question of the chemical nature of the genome was answered by transformation experiments in which it was shown that DNA can carry genetic information from bacteria of one type into those of another type.
7. Bacterial viruses (bacteriophage, or phage) are infectious agents that multiply in bacterial cells. Viruses consist of minute particles, or virions, made of nucleic acid and protein. A phage infects a cell by injecting its nucleic acid, which takes over the cell, replicates rapidly, and makes new phage proteins, eventually resulting in lysis of the cell and release of new virions.
8. The use of radioactively labeled phage by Hershey and Chase demonstrated that the genome is DNA, since only DNA gets into a cell and only DNA is passed along to the next generation of phage.
9. Watson and Crick developed the proper model for DNA structure. It shows two strands of DNA forming a double helix held together by hydrogen-bonding between complementary nucleotides (A with T, G with C).
10. The Watson-Crick model explained how DNA can replicate itself through specific base-pairing. All that is needed is for new nucleotides, made by the metabolic apparatus of the cell, to line up with the nucleotides of an existing strand and be polymerized into a new molecule by an appropriate enzyme.
11. The Watson-Crick model also showed how, in general, nucleic acids can encode the information for the amino acid sequences of proteins. A DNA molecule must carry a coded message for protein structure, but it need only specify the primary structure of a protein, its amino acid sequence, since once this is done, the protein will fold up into its functional shape.
12. Mutations can be understood as simple changes in nucleic acid sequences. These can take such forms as changing one nucleotide pair to another, deleting or adding nucleotides, or inverting a sequence of nucleotides.

Key Terms

daughter cells 241
replication 241
wild-type 242
breed true 242
mutant 242
mutation 242
gene 243
prototroph 243
auxotroph 243
transformation 245
bacteriophage/phage 246
virion 246
titer 246
petri plate 247
agar 247
clone 247
colony 247
lyse 248
plaque 248
pyrimidine 251
purine 251
nonsense mutation 254
missense mutation 254
substitution 254
insertion 254
deletion 254
inversion 254
frameshift 254
duplication 254

Multiple-Choice Questions

1. Which of the following statements is *not* correct?
   a. Mutant organisms have experienced changes in their DNA.
   b. Variations in a genome caused by environmental chemicals are not true mutations because the chemicals were environmental agents.
   c. Some mutations arise as a result of errors during DNA replication.
   d. If a variation breeds true, the variation will always appear when two similarly mutant organisms are mated.
   e. All wild-type organisms of a single species are similar in appearance.
2. The "one gene, one polypeptide" principle derives from work with
   a. bacterial transformation.
   b. viral infection of bacteria.
   c. nutritional mutants in a mold.
   d. analysis of the chemical structure of DNA.
   e. radioactively labeled bacteriophage.
3. Assume that mice are infected with one or two strains of *Streptococcus pneumoniae.* In which of the following cases would the mice develop a severe infection and die?
   a. inject with live S strain
   b. inject with live S and live R strains
   c. inject with heat-killed R and live S
   d. inject with live R and heat-killed S
   e. all of the above
4. The experiments of Avery, MacLeod, and McCarty confirmed the results of previous experiments
   a. analyzing auxotrophic mutations in bread mold.
   b. analyzing plaque formation by bacteriophage.
   c. studying viral infection.
   d. analyzing bacterial transformation.
   e. involving chemical analysis of DNA.
5. One of the following is *not* characteristic of viruses. Which is it?
   a. able to parasitize cells
   b. able to reproduce rapidly in a sterile chemical medium
   c. can infect only specific host cells
   d. contain nucleic acids
   e. can undertake metabolic reactions within a host
6. Although viruses are not alive in the strict sense, they do carry genetic information because they can
   a. reproduce within a host cell.
   b. display the same size and shape generation after generation.
   c. contain DNA or RNA.
   d. produce plaques of a particular size and shape.
   e. do all of the above.
7. If all the hydrogen in a phage virion were replaced by radioactive tritium (^3H), and the labeled phage infected a bacterium, where would the label be found?
   a. All would be inside the bacterium.
   b. All would be outside the bacterium.
   c. About equal amounts would be found inside and outside the bacterium.
   d. Most would be inside, but a small amount would be outside.
   e. Most would be outside, but a small amount would be inside.
8. Hershey and Chase concluded that DNA carried genetic information because
   a. DNA was the transforming principle.
   b. transformation could be eliminated by digesting the ineffective DNA.
   c. complete virions were not present during the eclipse period.
   d. each auxotrophic mutant carried one mutation.
   e. the DNA of phage T2 is passed on to the next generation.
9. Which of the following does not occur when a double helix of DNA replicates?
   a. Covalent bonds in the backbone of the helix break.
   b. Hydrogen bonds between bases break.

c. Adenine pairs with thymine, and cytosine with guanine.
d. A new complementary strand forms alongside each old strand of the double helix.
e. DNA polymerase catalyzes the polymerization of complementary strands.

10. In a double helix of DNA,
a. the backbones have opposite polarity.
b. bases on one strand are attached to bases on the other strand with hydrogen bonds.
c. a complementary pair of bases always includes one purine and one pyrimidine.
d. *a* and *b,* but not *c* are correct.
e. *a, b,* and *c* are correct.

True-False Questions

Mark each statement true or false, and if false, restate it to make it true.

1. Any particular genome has two major attributes—the ability to replicate and the ability to direct the synthesis of proteins.
2. If, in an environment that remains uniform, a mating between two flies carrying the same variation produces offspring that also exhibit the same variation, the cause of the variation must be genetic.
3. Wild-type *Neurospora* are nutritional auxotrophs.
4. If a strain of *Neurospora crassa* is auxotrophic for arginine, the enzyme that converts citrulline to arginine is not functioning properly.
5. Bacteriophage can only metabolize when inside a host bacterium, but animal and plant viruses can cause diseases even before they enter their host cells.
6. A nucleotide is a phosphorylated nucleoside.
7. Thymine, cytosine, and guanine are purine bases, while adenine is a pyrimidine.
8. DNA is said to have a $5'-3'$ polarity, which means that a phosphate group is bonded to the $5'$ carbon within the nitrogenous base and another phosphate group is bonded to the $3'$ carbon of the base.
9. According to Chargaff's data: $(A + G)/(T + C) = 1$.
10. In contrast to most other biological reactions, DNA replication does not require enzymes.

Concept Questions

1. Assume that three different auxotrophic strains of an organism grow if supplied with a minimal medium plus the hypothetical coenzyme X. Strain 1 specifically requires the coenzyme. Strain 2 requires either the coenzyme or compound Z. The third strain can grow if coenzyme X or either of compounds Y or Z is added to minimal medium. What information can be deduced from this data regarding the synthesis of coenzyme X?
2. Explain the difference between a bacterial colony and a bacteriophage plaque.
3. What fundamental difference between polypeptides and polynucleotides ensured the success of the Hershey-Chase experiments?
4. Why are virions not obtainable during the eclipse period?
5. What data support the base-pairing relationships in a double helix of DNA, and why are other pairings unlikely?

Additional Reading

Crick, Francis. *What Mad Pursuit: A Personal View of Scientific Discovery.* Basic Books, New York, 1988. Crick's autobiography and history of contemporary science.

Judson, Horace Freeland. *The Eighth Day of Creation.* Touchstone Books, New York, 1979. A history of genetics and molecular biology.

Luria, Salvador E. *A Slot Machine, A Broken Test Tube: An Autobiography.* Harper & Row, New York, 1984. Luria's autobiographical account of his early research in microbial genetics.

Sayre, Anne. *Rosalind Franklin and DNA.* W. W. Norton & Co., New York, 1975. A biography of Rosalind Franklin that contrasts with the picture Watson paints in his account.

Watson, James D. *The Double Helix: A Personal Account of the Discovery of the Structure of DNA.* Atheneum Press, New York, 1980 (1968). Watson's account of the discovery of DNA.

Internet Resource

To further explore the content of this chapter, log on to the web site at:

http://www.mhhe.com/biosci/genbio/guttman/

DNA Replication and the Cell Cycle

13

Key Concepts

13.1 Genes are located on chromosomes.

13.2 Chromosomes are DNA-protein complexes.

13.3 DNA replicates semiconservatively.

13.4 DNA is replicated by a protein complex that operates in different ways on the two strands of a double helix.

13.5 DNA synthesis requires energy furnished by nucleoside triphosphates.

13.6 All growing cells proceed through a regular cycle.

13.7 The cell cycle is driven by the synthesis and degradation of special proteins.

13.8 Mitosis is a mechanism for dividing chromosomes into two identical sets.

13.9 Procaryotic and eucaryotic chromosomes are replicated in different ways.

In our children, we see the continuity of ourselves, our families, and our species.

In the musical play *Baby*, by Richard Maltby and David Shire, a pregnant young woman named Lizzie feels her baby move for the first time. It is the most remarkable event in her life, and suddenly she feels herself connected to the entire chain of life that has gone before and will come after, as she sings:

So this is the tale my mother told me,
That tale that was much too dull to hold me,
And this is the surge and the rush she said would show
Our story goes on.

Oh, I was young, I forgot that things outlive me.
My goal was the kick that life could give me,
And now, like a joke, something moves to let me know
Our story goes on.

And all these things I feel and more
My mother's mother felt, and hers before.
A chain of life begun upon the shore
Of some primordial sea
Has stretched through time
To reach to me.

And now I can feel the chain extending
My child is next, in a line that has no ending.
And here am I, feeling life that her child will feel
When I'm all gone.

Yes, all that was is part of me
As I am part of what's to be,
And thus it is our story goes on, and on, and on,
And on, and on.[1]

The continuity Lizzie feels is a continuity of cells generating cells, generation after generation. Her cells were derived from one of her mother's cells, and hers from her mother before, back through uncountable generations to the first cell. The principle that all cells come from preexistent cells is part of the cell theory. And one of the most important events in this story is the replication of DNA as each cell prepares to divide, a replication that is essential if each cell is to have the memory—the genome—needed to ensure the continuity of generations. In this chapter we tell the story of DNA replication and cell reproduction, a story that never ends. ■

13.1 Genes are located on chromosomes.

Genetics began to develop after 1900, based on Mendel's demonstration that factors of heredity (genes) carry the information for specific characteristics. Investigators began to ask about the physical structure of genes and where they are located. For some time microscopists had known about chromosomes, the long, thin bodies that become visible in a eucaryotic cell when it starts to divide (Figure 13.1). During cell division, the chromosomes replicate and then go through an elaborate ballet, which separates them so that each daughter cell receives one copy of each. The function of the chromosomes and the meaning of their movements were a mystery until 1902, when Theodor Boveri and William S. Sutton independently pointed out that chromosomes behave exactly as the hypothetical genetic "factors" postulated by Mendel should behave. The precise separation of chromosomes during cell division, for example, is exactly parallel to the separation of genes into daughter cells. Sutton and Boveri therefore inferred that genes are located on chromosomes, and evidence gradually accumulated to support this conclusion.

Since genes are located on chromosomes and the chromosomes of eucaryotes are in the cell nucleus, it follows that the nucleus plays a critical role in heredity. Joachim Hämmerling confirmed this point most elegantly with the experiment shown in Figure 13.2. One of the most beautiful algae in the ocean is

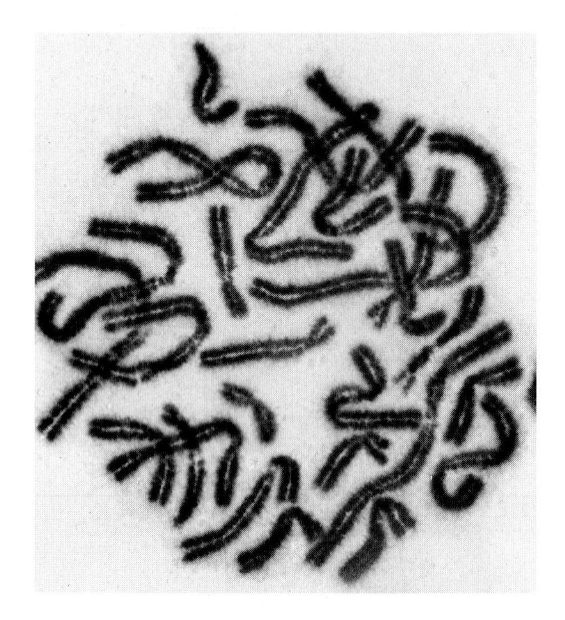

Figure 13.1
The chromosomes of a eucaryotic cell reside in its nucleus and contain almost all the genes in the cell, except for small amounts of DNA that carry a few genes in mitochondria and chloroplasts.

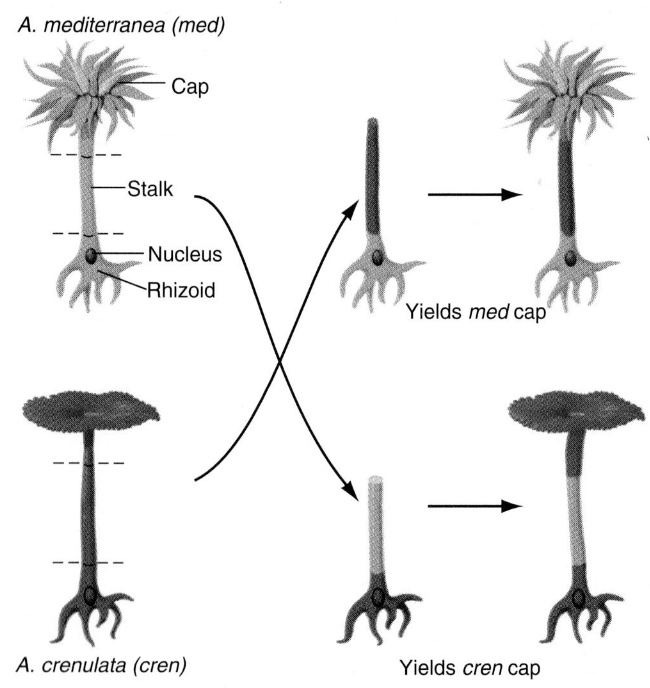

Figure 13.2
Hämmerling's experiments with grafted segments of *Acetabularia* demonstrated that the information for cap structure must come from the nucleus.

[1] *The Story Goes On,* by David Shire and Richard Maltby, Jr. © 1984 WB Music Corp., Progeny Music, Warner-Tamerlane Publishing Corp., and Lang Pond Music. All Rights Reserved. Used by Permission. WARNER BROS. PUBLICATIONS, U.S. INC., Miami, FL 33014.

Acetabularia, an exceptionally long, single-celled organism with a delicate umbrella at one end and a nucleus located in the "root" end a few centimeters away. Hämmerling used two species, *A. crenulata* and *A. mediterranea,* which have distinctively shaped caps. After cutting cells of each species into pieces, he grafted the stalk of one species onto the root of the other. These hybrid cells eventually regenerated new caps, and in every case, the shape of the cap was characteristic of the species that contributed the root piece containing the nucleus. Hämmerling's experiments showed that in regenerating, each cell follows the instructions of a genetic program that specifies the shape of its cap, a program located in its nucleus.

Procaryotic cells, by definition, don't have well-defined nuclei bounded by an envelope. Their nucleoids, which are visible in electron micrographs as central light patches within the otherwise dark cytoplasm, are long, highly condensed DNA molecules that look rather like dense coils of yarn (Figure 13.3).

13.2 Chromosomes are DNA-protein complexes.

How large are genomes? As we saw in Chapter 12, the genomes of all organisms are made of DNA, a polymer of nucleotides, which generally takes the form of a double helix. The unit of a double-stranded DNA molecule is a nucleotide pair, commonly called a *base pair (bp);* however, genomes are so huge that they are more conveniently measured in units of *kilobase pairs (kbp),* where 1 kbp = 1,000 bp. One turn of the DNA helix encompasses 10 bp and is 3.4 nm long, so a kbp is 100×3.4 nm = 340 nm long, or 0.34 μm. Since a nucleotide pair weighs 614 daltons (A–T and G–C pairs are almost identical in mass), one kbp weighs $1{,}000 \times 614 = 6.14 \times 10^5$ daltons, or 1.8 million daltons/μm for double-stranded DNA. So a single-stranded nucleic acid weighs half that much, close to a million daltons per micrometer. Table 13.1 gives the sizes of some representative genomes.

A eucaryotic genome is divided into several chromosomes, each of them a long, double-stranded DNA molecule with attached proteins. The 3.6×10^{12} daltons of DNA in a human body cell has a total length of 1.8 meters, but it is divided among 46 chromosomes. Since all this DNA is packed into a nucleus only 10 μm in diameter, about 100,000 times smaller than its length, the DNA is wound up and severely compressed to fit into the nucleus. (This is equivalent to packing a DNA molecule nearly 20 football fields long into the volume of a marble.)

Procaryotic cells have only one chromosome each, though it is usually in an intermediate stage of replication and is partially doubled. This chromosome is a long, circular DNA molecule with no free ends, associated with proteins. The chromosome of *E. coli* has about 3,800 kbp and is about 1.2 mm long, close to 1,000 times longer than the cell, and it is wound back and forth on itself many times to make a compact nuclear body.

When eucaryotic cells are prepared for light microscopy, staining with basic dyes reveals colored material called **chromatin** (*chroma* = color) throughout the nucleus. This is the DNA of the cell combined with proteins, mostly structural

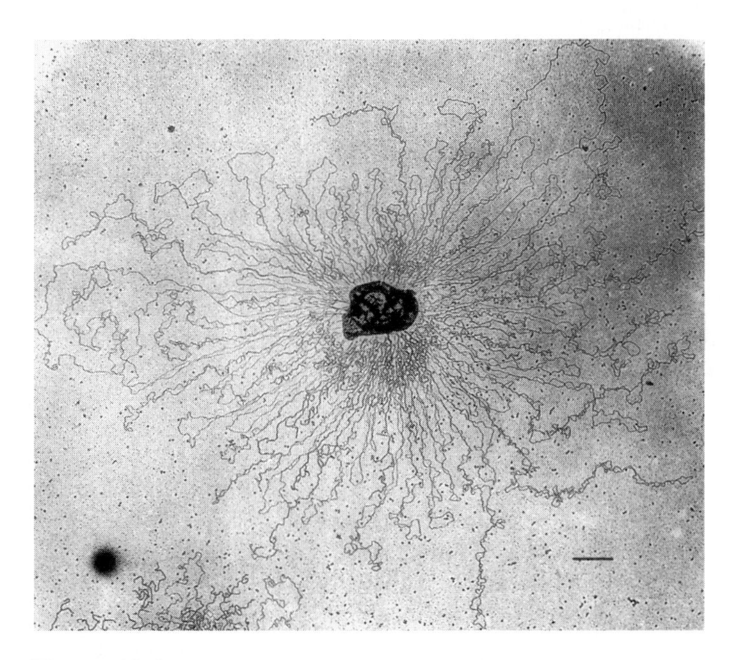

Figure 13.3

When a bacterial cell *(dark body in the center)* is ruptured gently, its chromosome is revealed as one extremely long, highly tangled DNA molecule. Before the cell was ruptured, this DNA formed the light nuclear body visible in Figure 6.15.

Table 13.1 Sizes of Representative Genomes*

Organism or Virus	Nucleotide Pairs	Length
Phage φX174	5.5×10^3	1.7 μm
Phage T4	1.69×10^5	50 μm
E. coli	3.8×10^6	1.2 mm
Yeast	2×10^7	6 mm
C. elegans (roundworm)	1×10^8	3 cm
Drosophila	1.7×10^8	5 cm
Human	3×10^9	90 cm
Frog	3.5×10^9	1.1 m
Lily	9×10^{10}	27 m

* Amount of DNA per haploid genome for sexual species. φX174 has single-stranded DNA.

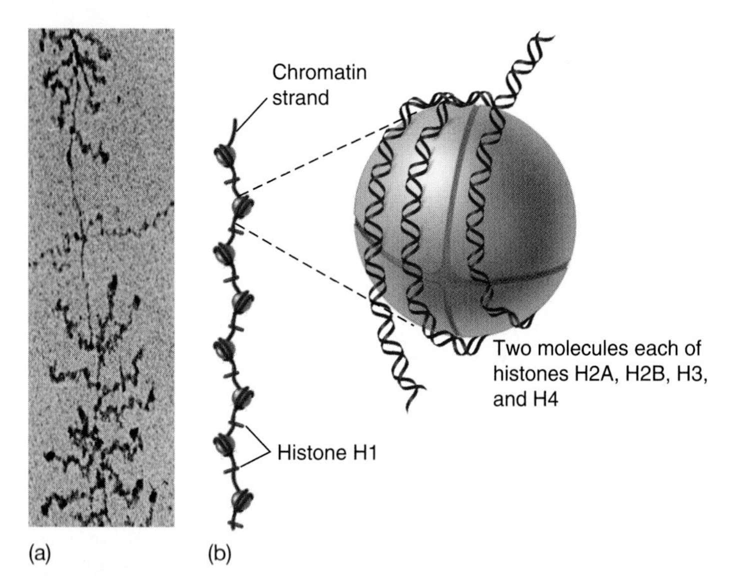

Figure 13.4

(a) An electron micrograph of chromatin shows it as a string with regularly spaced beads. *(b)* Analysis shows that each bead is an octamer made of two molecules each of histones H2A, H2B, H3, and H4, and a long, double-stranded helix of DNA is wrapped around histone beads to form nucleosomes. Histone H1 binds to the spacer DNA between them.

proteins called **histones.** Histones are rich in basic, positively charged amino acids (lysine and arginine), which bind to the negatively charged phosphate groups of DNA. Except for sperm cells, which have unique chromosomal proteins, the histone proteins of all eucaryotic cells are of just five types (H1, H2A, H2B, H3, and H4), regardless of species, cell type, or stage of development. Electron micrographs of extracted chromatin fibers (Figure 13.4*a*) show beads about 11 nm in diameter, uniformly spaced about 14 nm apart. These beads are **nucleosomes,** made of DNA wrapped around histone complexes. Nucleosomes reduce a chromatin fiber to one-seventh the length of an extended DNA molecule.

Bacterial DNA is combined with a histonelike protein called HU. Although the organization of bacterial chromosomes is not known as well as that of eucaryotic chromosomes, there is evidence that HU forms structures similar to nucleosomes with the DNA.

The chromosomes of a nondividing cell cannot be seen with a light microscope. They first become visible early in cell division, when chromatin strands become short and compact by coiling up on themselves (Figure 13.5); the compaction is apparently caused by an increased salt concentration and is stabilized by the binding of histone protein H1. At that time it becomes clear that each chromosome consists of two half-chromosomes called **chromatids** attached to each other at a unique point, the **centromere.** A centromere is simply a special region of DNA. To

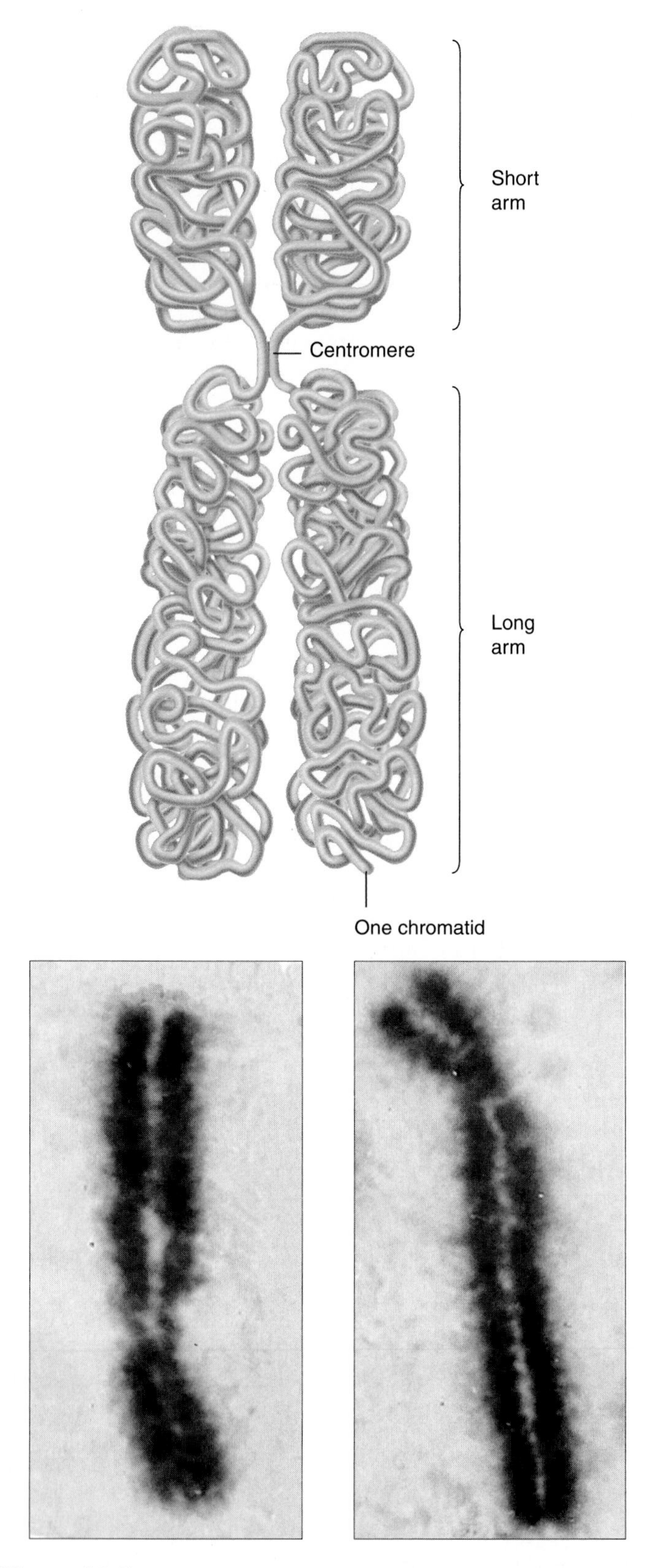

Figure 13.5

When a chromosome is visible during cell division, it consists of two chromatids joined at the centromere. The chromatin of each chromatid is highly condensed because the nucleosomes coil up into cylinders.

understand chromosomes, it is essential to see that they change in a regular way during the cell cycle. We will refer to a chromosome consisting of two chromatids as a *duplex chromosome;* then, during cell division, the chromatids of each chromosome will separate into the two daughter cells, and each of them will be a chromosome in its own right, which we will call a *simplex chromosome.* If the cell is going to divide again, each chromosome must replicate itself and will again become a duplex chromosome made of two chromatids:

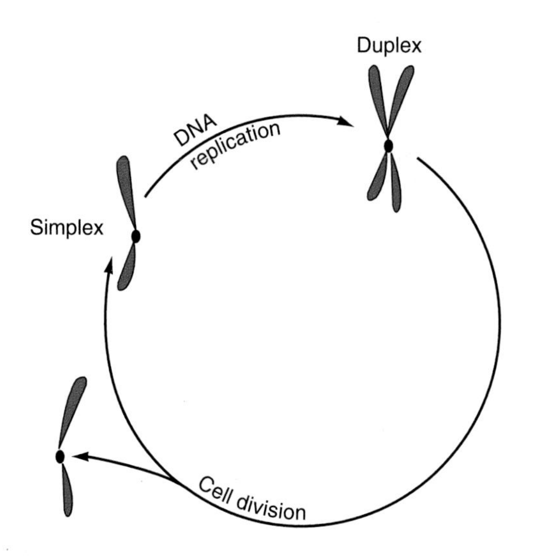

Both duplex and simplex structures are chromosomes, but a chromosome changes as a cell goes through a period of DNA replication and then divides in two. Next we will see how DNA is replicated.

13.3 DNA replicates semiconservatively.

The base complementarity of the Watson-Crick model immediately suggested that DNA should replicate as shown in Figure 13.6*a,* with each strand remaining intact and specifying a new complementary strand. This mode is called **semiconservative** because each strand is conserved—kept intact—even though the double helix is not. However, other modes of DNA replication are theoretically possible, albeit less likely. For example, DNA might undergo **conservative** replication (Figure 13.6*b*), in which each old strand could direct the synthesis of its complement, with the two old strands staying together as the original helix while the two new strands make a whole new helix. DNA might also replicate in more complicated ways, perhaps by breaking up into smaller units, as shown in Figure 13.6*c.*

In 1958, Matthew Meselson and Franklin W. Stahl found a superb way to test the prediction of the Watson-Crick model regarding replication. It is an excellent example of scientific reasoning (see Sidebar 13.1), because the various modes of replication predict different outcomes from the experiment. Working with Jerome Vinogard, Meselson and Stahl centrifuged a dense

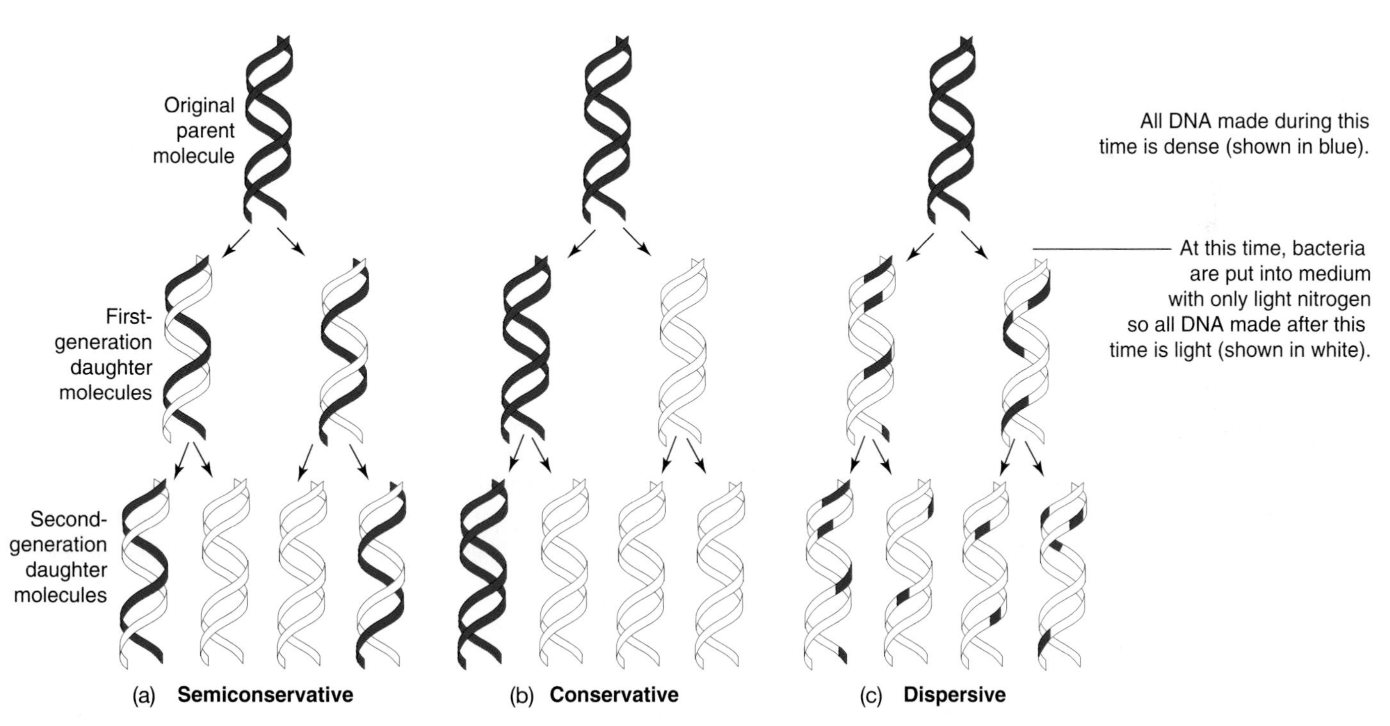

Figure 13.6

(a) If DNA does replicate semiconservatively, the original material of a DNA molecule *(blue)* should become distributed like this in two rounds of replication. If the colored strands are labeled, then after one round of replication, both molecules should be half-labeled. After another round, half the molecules should be half-labeled and half should be unlabeled. *(b)* In conservative replication, the two old strands stay together after replication, and a new molecule is made of the two new strands. *(c)* A third possibility is dispersive replication. DNA might replicate by breaking up into smaller pieces that could re-form new molecules rather irregularly, so the label becomes mixed in each generation.

Sidebar 13.1 The Logic of Scientific Experiments

Science is a creative business, and it does not have a single method or a single logic, no matter what traditional textbooks may say. In Section 1.3 we outlined one important kind of reasoning in science, the logic of the hypothetical. That is, to be meaningful, a theory, a hypothesis, a hunch must make a difference empirically—it must have testable consequences. Hypotheses in science take the form, "If this hypothesis (H) is right, we should be able to make a predictable observation (O): if H, then O." If we do not make the predicted observation O, H cannot be right. Remember, though, that the hypothesis cannot be verified in this way—it can only be disproved.

The late physicist John R. Platt pointed out a particularly powerful form of hypothetical reasoning, which he called strong inference. Rather than setting up a single hypothesis, an investigator sets up alternative hypotheses, each of which predicts a different outcome of a particular experiment. That is:

If H_1, then O_1.
If H_2, then O_2.
If H_3, then O_3.

and more, if possible. Next the investigator does a critical experiment designed to make all of the outcomes ($O_1, O_2, O_3, \ldots$) possible. There can only be one outcome—let's say O_2 in this case. Then the investigator can reason:

O_1 is not observed, so H_1 is not true.
O_3 is not observed, so H_3 is not true.
O_2 is observed, so H_2 is probably correct.

Again, the experiment does not *prove* H_2 is true. Reasoning that way would be contrary to elementary logic. But H_2 has withstood a powerful test, especially if the hypothesis and experiment are quantitative and the results agree closely with the prediction. This reasoning is exactly what Meselson and Stahl used to test the replication prediction of the Watson-Crick model.

solution of cesium chloride (CsCl) for a few days at top speed in an ultracentrifuge, an instrument that can spin samples up to 60,000 revolutions per minute. They found that the salt ions form a **density gradient,** a solution whose density is low at the top of the tube and increases steadily toward the bottom. Large molecules in the solution will find an equilibrium position where their buoyant density equals that of the gradient. All DNA has a density of about 1.7 g/ml, but there are small differences between DNAs from different sources. So if a mixture of DNAs is centrifuged in a CsCl solution, the molecules in the mixture separate from one another because each type moves to its own position in the gradient.

Cells make their DNA bases by assimilating whatever nitrogen is in their growth medium. Knowing this, Meselson and Stahl made bacteria with dense DNA by growing them for several generations in a medium containing the heavy isotope ^{15}N instead of ordinary nitrogen (^{14}N). When these bacteria were transferred into an ordinary medium containing ^{14}N, they started making DNA of normal density. Meselson and Stahl followed the change in density over two generations after the transfer by taking samples periodically, extracting the DNA, and centrifuging it in CsCl.

Figure 13.6*a* shows how the density of the DNA should change if it replicates semiconservatively. The original molecule, synthesized in medium containing ^{15}N, is all dense. In the first round of replication in ^{14}N-containing medium, it should separate into two strands that each acquire a light partner strand, thus making hybrid molecules of intermediate density. After a second round of replication, the original dense strands should still be combined with light strands, and an equal number of molecules should be made exclusively of light strands. Figure 13.7 shows what Meselson and Stahl actually saw. Initially, all the DNA is dense. After one round of replication (generation 1), it is all half-dense, halfway between the densities of DNA made entirely of ^{15}N and DNA made entirely of ^{14}N. This is the right density for DNA made of a dense strand combined with a light one. After a second round of replication (generation 2), half the DNA is still half-dense and half of it is light. These results are precisely what the Watson-Crick model predicts, supporting the semiconservative mode of replication.

Meselson and Stahl then had to show that the units being separated in the CsCl gradient were double-stranded molecules with each strand either all heavy or all light. Their evidence for this took advantage of a useful property of DNA: Since the double helix is held together by weak hydrogen bonds, heat of 80–100°C will separate the strands by disrupting, or *melting,* these bonds (Figure 13.8). Just as water molecules break their hydrogen bonds and move about freely as they acquire more energy, the DNA strands break their hold on each other and separate into single strands. Meselson and Stahl melted some of their half-dense DNA and showed that it separates into one dense fraction and one light fraction, as it should. They concluded that each strand of half-dense DNA consists of either totally dense or totally light nucleotides, not a mixture of dense and light nucleotides in one strand. This work provided a definitive confirmation of the Watson-Crick model.

Exercise 13.1 If DNA replicated conservatively (as shown in Figure 13.6*b*), what would Meselson and Stahl have observed after one round of replication in ^{14}N-containing medium? After two rounds?

Exercise 13.2 DNA could replicate by each strand breaking into segments (as shown in Figure 13.6*c*). How could Meselson and Stahl's experiments have distinguished this mode of replication from the others?

Exercise 13.3 After Meselson and Stahl did their work on bacterial DNA, J. Herbert Taylor tried a comparable experiment with eucaryotic cells. By labeling chromosomes with radioactive thymidine, he could locate radioactive DNA by autoradiography (see Methods 10.1). Then the chromosomes were allowed to replicate in a nonradioactive medium. After one round of replication, both chromatids in each chromosome were radioactive. What was the appearance of the chromosomes after a second round of replication?

13.4 DNA is replicated by a protein complex that operates in different ways on the two strands of a double helix.

DNA replication turns out to be a complicated process, even though base-pair complementarity is at the heart of it, as Watson and Crick surmised. It is carried out by a **replisome,** a protein complex built around the enzyme **DNA polymerase,** which synthesizes new DNA strands that are complementary to existing strands. Each existing strand acts as a **template** for a new one, because it forces a complementary series of nucleotides to line up just as a drawing template forces your pencil to follow a prescribed line (Figure 13.9). Replication requires the four types of nucleotides (A, C, G, and T) in their triphosphate form: dATP, dCTP, dGTP, and dTTP, where the "d" reminds us that these are *deoxy*ribonucleotides. (The cell, of course, is constantly synthesizing these compounds.) One by one, nucleoside triphosphates bind to their complements in the

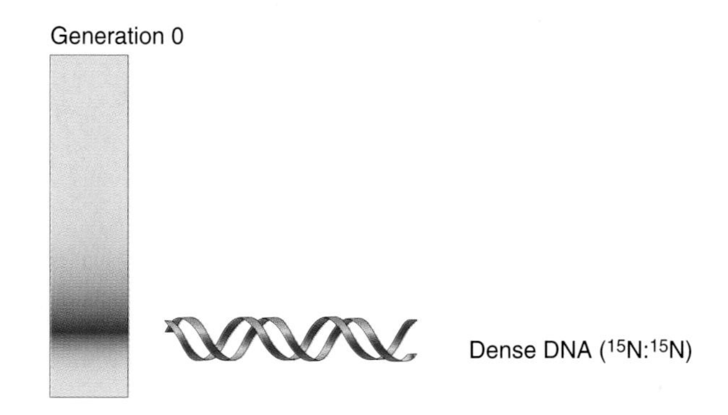

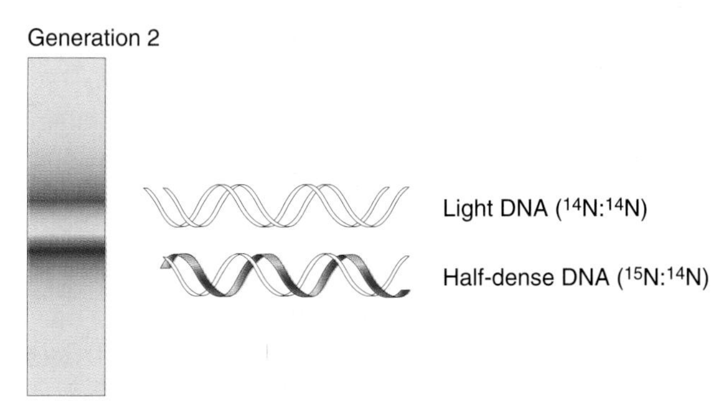

Figure 13.7

Meselson and Stahl started with cells containing fully dense DNA. After replicating once in light medium, all the DNA was half-dense. After replicating again, half of it was half-dense and half was light. Compare this with Figure 13.6*a.*

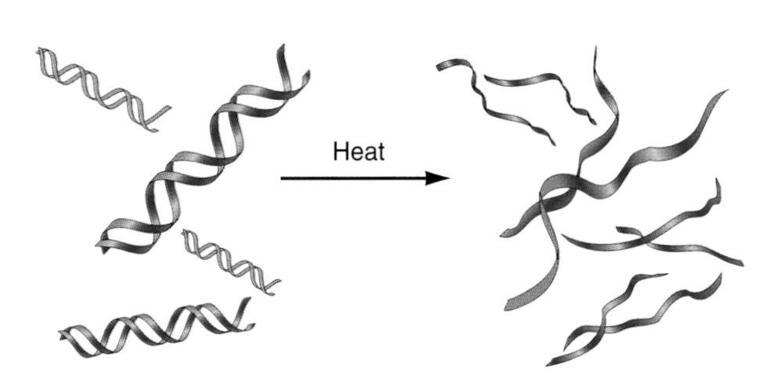

Figure 13.8

When a DNA molecule is heated, the weak hydrogen bonds that hold its strands together are disrupted, or melted, so it separates into two single-stranded molecules.

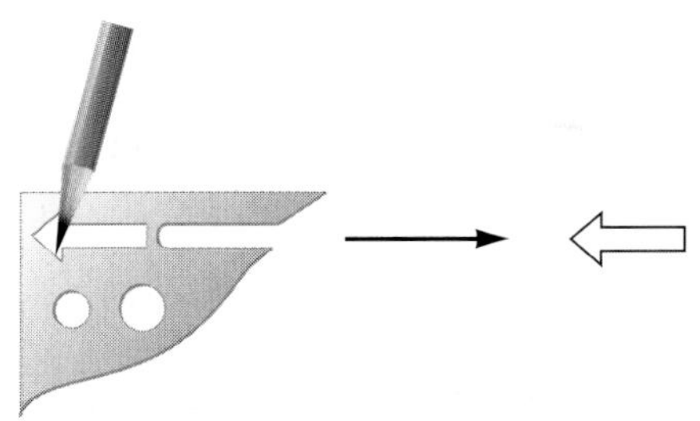

(a) A drawing template forces the pencil to make a complementary shape of line.

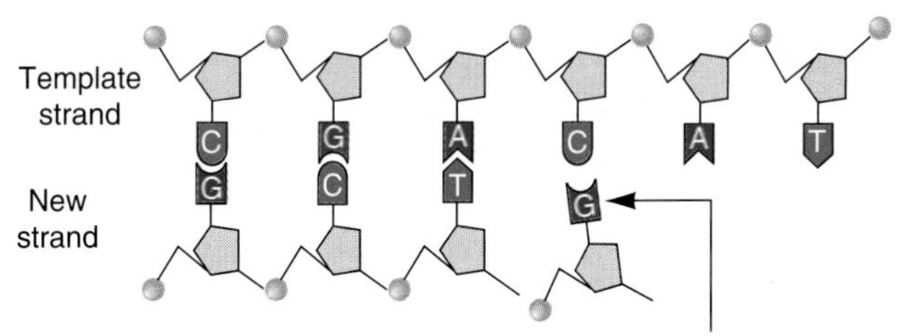

(b) The DNA template forces a complementary strand of nucleotides to fit in.

Only G will fit in this position. A, C, and T will be rejected because they have the wrong shapes.

Figure 13.9

Just as a drawing template determines the shape of a line *(a),* the nucleotide sequence of a DNA strand determines the sequence of a new strand by only allowing the insertion of complementary nucleotides *(b).*

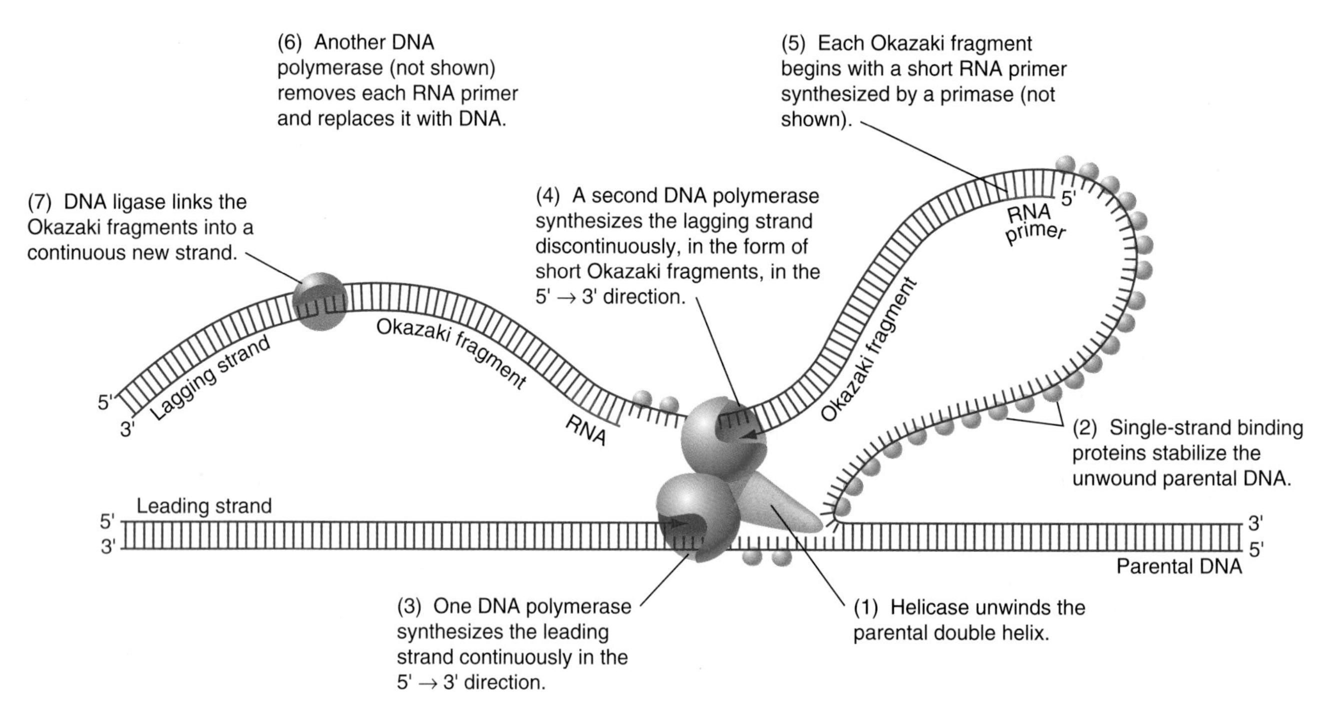

Figure 13.10

A model for the actual mechanism of DNA replication depends on a replisome made of several proteins, including a helicase that unwinds the double helix into single strands. The single-strand-binding proteins keep the template DNA from going back into its double-stranded form. One DNA polymerase synthesizes the leading strand continuously by reading one template strand in the 3′ → 5′ direction. The second polymerase synthesizes the lagging strand in short pieces (Okazaki fragments) by replicating short stretches of template DNA. The fragments are then linked by DNA ligase to make a continuous DNA strand.

intact, template strand. The polymerase moves along, linking these nucleotides into a single new strand.

If this were all DNA replication entailed, it would be simple, but the replisome contains several proteins that carry out additional functions (Figure 13.10). First, replication occurs on a single-stranded template, so the double helix must be unwound. This process requires both energy and an enzyme called helicase that turns the parental helix, opening it into separate strands. As the DNA helix divides into two single strands, it forms a *replication fork,* a Y-shaped region where the replisome acts. Since single DNA strands tend to re-form a double-stranded structure, replication also requires single-strand binding proteins, which stabilize the single-stranded molecules.

A DNA polymerase can only operate in the 5′ → 3′ direction while "reading" along a template strand in the 3′ → 5′ direction. It uses a nucleoside triphosphate, bearing phosphates on the 5′ carbon to form a phosphodiester linkage to the free 3′ hydroxyl group on the end of the DNA strand it is synthesizing. Since the two strands of a double helix have opposite polarities, it would seem that a replisome could only synthesize one strand in each replication fork. However, two DNA polymerases in each replisome, oriented in opposite directions, actually synthesize both strands by working in quite different ways. The replisome moves along one template strand in the 3′→ 5′ direction, replicating it in a continuous process to form what is called the **leading strand.** The other polymerase synthesizes the second new strand, the **lagging strand,** in short pieces averaging about 1,000 nucleotides long; these pieces are called **Okazaki fragments** because they were discovered by Reiji Okazaki. Rather than replicating continuously, the replisome uncovers a stretch of the second template strand and then replicates it, uncovers another stretch and replicates it, and so on.

DNA polymerase cannot simply start to polymerize nucleotides on a bare single strand. It is only able to elongate a strand that has already been started, called a *primer,* and the replisome contains an additional enzyme, a *primase,* that starts replication by making a short RNA molecule. The DNA polymerase then takes over and elongates this RNA primer with DNA. The leading strand only contains such a primer at its start, but each Okazaki fragment begins with one; a different kind of DNA polymerase replaces each of these RNA segments with DNA, and finally, the enzyme DNA ligase links the Okazaki fragments into a continuous strand.

13.5 DNA synthesis requires energy furnished by nucleoside triphosphates.

The four nucleoside triphosphates (dATP, dCTP, dGTP, and dTTP) not only contribute the monomers of new DNA strands but also the energy needed to form the phosphodiester bonds between neighboring deoxyriboses. Each nucleoside triphosphate fits into a little nook on the existing DNA molecule, as shown in Figure 13.9. Its base is hydrogen-bonded to a base of the template strand; this is the recognition event that makes replication precise. Its two terminal phosphates *(pink)* are split off as inorganic pyrophosphate, represented by PP_i:

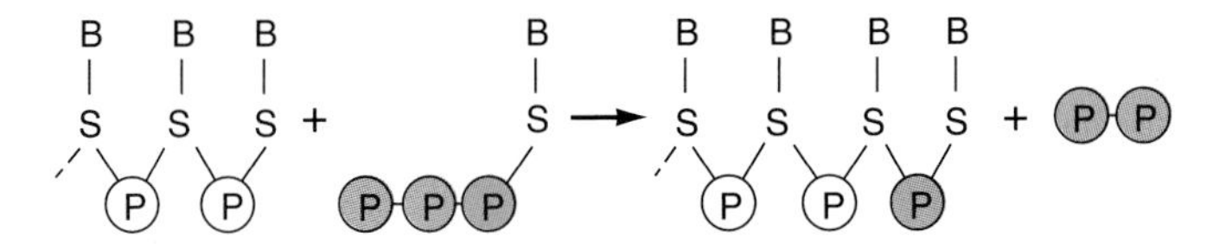

This event supplies the energy needed to link the remaining phosphate to the 3′ end of the growing chain, just as hydrolyzing ATP supplies the energy for many endergonic reactions in metabolism.

The fate of the pyrophosphate is extremely important. If it were left intact, DNA synthesis would be reversible, and an organism might disassemble its genome instead of replicating it. But cells have enzymes (pyrophosphatases) that hydrolyze PP_i into two molecules of phosphate, P_i, so synthesis cannot be reversed easily and organisms continue to grow. It is remarkable that the growth and maintenance of an organism can depend on such a simple process.

Having outlined DNA replication, our next step is to put this important process into a cell and show its place in cellular growth and division.

Exercise 13.4 The *E. coli* chromosome contains 3.8×10^6 nucleotide pairs. If two replisomes working in opposite directions replicate it in 40 minutes, how many nucleotides is each DNA polymerase adding per second?

13.6 All growing cells proceed through a regular cycle.

As they grow and divide, all cells go through a series of regular events called the **cell cycle** (Figure 13.11). Procaryotes and eucaryotes have comparable cell cycles, but the eucaryotic cycle is more clearly marked, so we begin there.

Every growing eucaryotic cell periodically goes through a nuclear division called **mitosis,** in which the set of chromosomes, having been duplicated, divides into two identical sets. Mitosis varies quite a bit in different microorganisms, but in the "classical" mitosis of plant and animal cells, the nuclear envelope breaks down while the chromosomes condense into tight bundles and are separated into two daughter nuclei. Mitosis is usually followed by **cytokinesis,** a division of the whole cell into daughter cells, but not always, for in cells called *coenocytes,* the cytoplasm divides irregularly or only occasionally while nuclear division proceeds. Coenocytes, which are common in fungi and algae, therefore have more than one nucleus. (Most of the muscles in the human body also have this structure, but they are called *syncytia* and are formed by *fusion* of cells, not by division.)

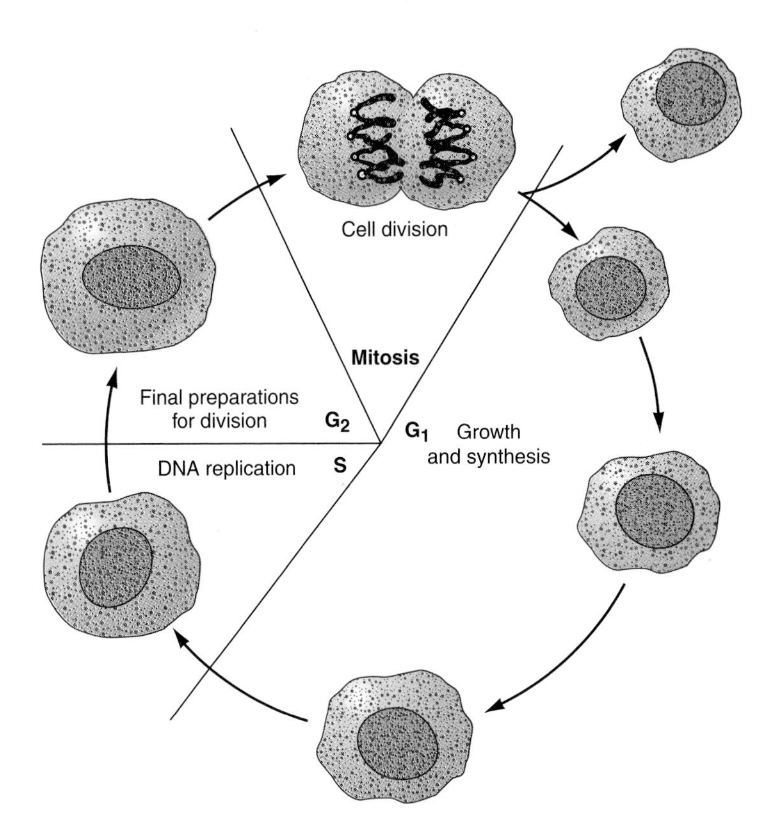

Figure 13.11
The general eucaryotic cell cycle consists of four phases. S is the time of DNA replication in preparation for mitosis. The cell spends most of its time in G_1 and G_2, during which most growth and preparation for division (mitosis) take place.

Before a nucleus can divide, its chromosomes must be replicated. In addition to mitosis (M), the cell cycle includes a synthetic period, **S,** when the DNA is replicated. Between S and M are two periods, $\mathbf{G_1}$ and $\mathbf{G_2}$, that were originally called "gaps" in the cycle, even though they account for most of the time and are periods of intense activity, including most of cell growth and all the routine functions cells usually carry out. The whole nonmitotic period ($G_1 + S + G_2$) is called **interphase.** Cycle times vary considerably, from a typical length of about 10–20 hours to many days. In mammalian cells, the S phase typically takes seven hours, G_2 three hours, and M about an hour. The greatest variation comes in G_1, which may last only a few minutes during the initial cleavages of a fertilized egg but as long as several hours or even several days in the cells of a mature organism. Some mature cells that will never divide again except in unusual circumstances—including many of the differentiated cells in plant and animal tissues—are considered to be in a perpetual G_1 condition called G_0.

13.7 The cell cycle is driven by the synthesis and degradation of special proteins.

As you might expect, a process as complicated as the cell cycle depends on an elaborate series of events, which cell biologists are just now beginning to understand. Information about these events has emerged from studies of a few selected organisms, mostly budding yeast *(Saccharomyces cerevisiae)* and fission yeast *(Schizosaccharomyces pombe),* using methods of genetic analysis that we will discuss later. The key to all genetic analysis is to find a mutant organism that is unable to do something and then study it to determine its defect. This is the main way experimenters find previously unknown genes, proteins, and functions. Genes and proteins discovered by using mutants are named with letters and numbers that suggest their functions. Mutants defective in the *c*ell *d*ivision *c*ycle, for instance, are designated *cdc* mutants, and each mutant may reveal the existence of a new protein involved in this process. The protein encoded by each gene is designated by the symbol for that gene (Cdc, for example), but with a capital letter and not italicized.

The cell cycle is driven by an engine that turns once per cycle, changing the intracellular level of certain critical proteins that act as *checkpoints.* Controls on the cycle have evolved to ensure that the key events of cell growth and division do not happen unless previous critical events have occurred. It would be disastrous, for instance, if a cell could start to go through mitosis before its chromosomes were all duplicated or before it had grown large enough, so the cell cycle is marked by at least two checkpoints—one for the S phase and one for mitosis. To pass these checkpoints, the cell must make a series of proteins, especially *S-phase promoting factor (SPF)* and *M-phase* (or *mitosis*) *promoting factor (MPF).* Each of these (and other cell-regulatory proteins) is a combination of two subunits: a cyclin and a protein kinase (Figure 13.12). Here is an application of the principle we established in Chapter 11: Cells contain many specialized protein kinases, each one phosphorylating a different protein and each controlled in a different way. Phosphorylating a protein changes its conformation and activity, and by regulating these protein kinases, cells regulate their many operations. MPF induces mitosis by phosphorylating other key proteins. Chromosomes are stabilized into compact bodies during mitosis because MPF phosphorylates histone protein H1, which then binds to the chromosomes. Furthermore, the activity of MPF itself is controlled by phosphorylating and dephosphorylating it, so the whole cell cycle engine is regulated by a series of other proteins identified by *cdc* mutants.

At the end of mitosis, a cell enters G_1, and if it is to go through another cell cycle, it must pass a unique point called *Start,* which signals the beginning of the S phase. To pass Start, the cell must have a sufficient level of SPF, and this requires the Cdk2 protein and the previous synthesis of other proteins that effectively measure cell size, so DNA synthesis only begins in a cell that has grown large enough. In budding yeast, a newly budded daughter cell is much smaller than its mother cell, and if it were allowed to divide without adequate growth, the cells would get smaller and smaller. A series of proteins monitor the cell as it grows during G_1 and delay Start until it reaches a certain threshold size. Passing Start then initiates DNA replication itself. Some cells cannot make DNA because they lack the enzymes for synthesizing its key components, thymidine and dTMP. (Thymidine is unique to DNA; RNA contains uridine instead.) The synthesis of these enzymes appears to trigger replication, because once they have been made, DNA synthesis can begin.

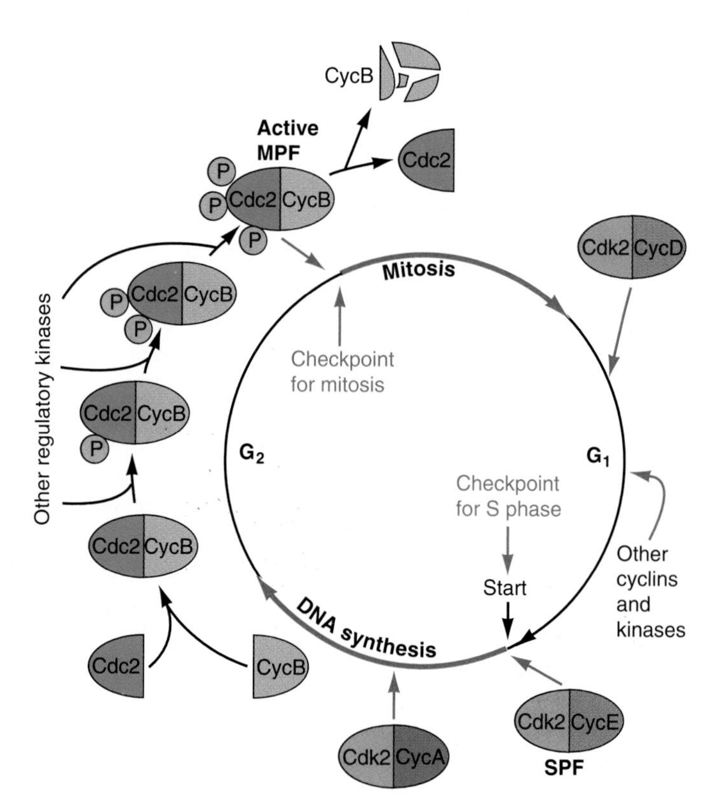

Figure 13.12

The cell cycle is driven by cyclins and cyclin-dependent kinases (Cdks, including one called Cdc2), forming complexes that move the cell past certain checkpoints. Several cyclins and Cdks form during G_1, and finally Cdk2 and cyclin E form S-phase promoting factor (SPF), which initiates the S phase. M-phase promoting factor (MPF) is made of cyclin B and Cdc2, and is activated by phosphorylation. Synthesis of Cdc2 begins during interphase, so the level of MPF rises until mitosis starts; Cdc2 is broken down during mitosis.

After completing the S phase, a cell enters G_2 and prepares to enter mitosis. If cells in G_2 are given agents that cause them to make faulty proteins, they won't divide properly, indicating that mitotic proteins are made during this time. Then, to initiate mitosis, a cell must pass that checkpoint by having sufficient MPF. The MPF level is low during most of interphase, and mitosis begins when it is high. MPF is made of cyclin B and Cdc2. As Figure 13.12 shows, cyclin B is made throughout interphase and the early stages of mitosis, but it is suddenly degraded toward the end of metaphase (the middle of mitosis), and this event signals the beginning of interphase.

The cell cycle engine may continue to turn even if other parts of the cycle are inhibited. For instance, DNA synthesis, which normally precedes mitosis, can be inhibited in fertilized frog eggs, but MPF continues to be activated and deactivated. The nuclear envelope breaks down as if mitosis were about to occur and then re-forms again as if mitosis were over, but the cell doesn't divide.

Exercise 13.5 Frog eggs can be induced experimentally to make MPF, and their cytoplasm can then be extracted and injected into other cells in interphase. What effect do you think this has on the injected cells?

13.8 Mitosis is a mechanism for dividing chromosomes into two identical sets.

Mitosis is designed (by evolution) to ensure that each daughter cell ends up with one complete copy of the genome. During the S period, each chromosome, which was simplex during G_1, has become a duplex chromosome made of a pair of identical chromatids; mitosis simply separates the chromatids to make two identical sets of simplex chromosomes. Though mitosis is a continuous process, it is classically divided into five phases, as shown here by mitosis in annual ryegrass (*Lolium multiflorum*):

The interphase nucleus is compact and dark, with chromatin forming an undifferentiated, colored smear; individual chromosomes are not yet visible.

During **prophase,** the nucleus enlarges, and the chromosomes begin to condense enough to be seen as separate bodies. The centrosome, the principal microtubule-organizing center of an interphase cell (Section 11.11), divides into two centers at the beginning of prophase. These centers start to migrate toward opposite poles of the cell, and as they separate, they form and orient the **mitotic spindle** between them. The spindle, which will later separate the chromatids of each chromosome, is a dense collection of microtubules; they were named *spindle fibers* before their structure was known. The mitotic spindle enlarges throughout prophase.

Prophase is followed by a period of frantic chromosomal activity called **prometaphase,** when the nuclear envelope breaks down and the chromosomes are attaching to the mitotic spindle. The nuclear envelope is made of two layers of membrane, lined on the inside by a scaffolding, the *nuclear lamina,* made of intermediate-filament proteins called lamins. When the lamins are phosphorylated, either by MPF or some other kinase controlled by MPF, the lamina comes apart, and the nuclear envelope breaks down into small vesicles with attached lamins, as do the endoplasmic reticulum membranes. While this is going on, each chromosome attaches to the mitotic spindle by structures on its centromere called **kinetochores,** one on each chromatid. Each chromosome sweeps back and forth across the cell as it captures microtubules from the two poles and is pulled one way and the other.

As the cell enters **metaphase,** the chromosomes settle down with their centromeres in the middle of the cell, more or less in a central plane called the *metaphase plate.* They are now maximally condensed, and each one clearly consists of two chromatids.

Suddenly, at the onset of **anaphase,** the connections that have been holding the chromatids together at the centromeres break apart, owing to a change associated with the destruction of cyclin B. Each duplex chromosome separates into its chromatids—now simplex chromosomes in their own right—

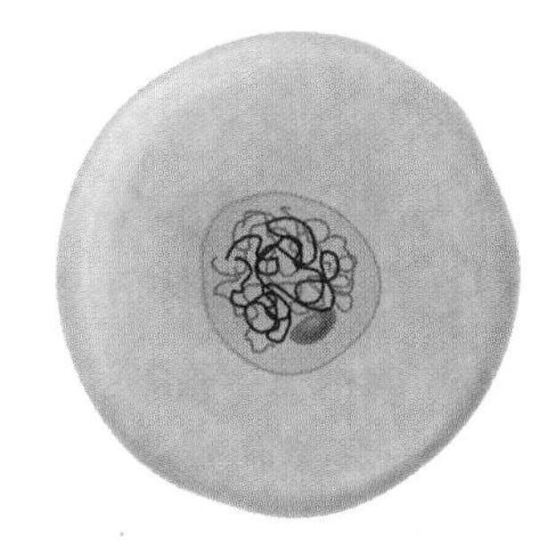
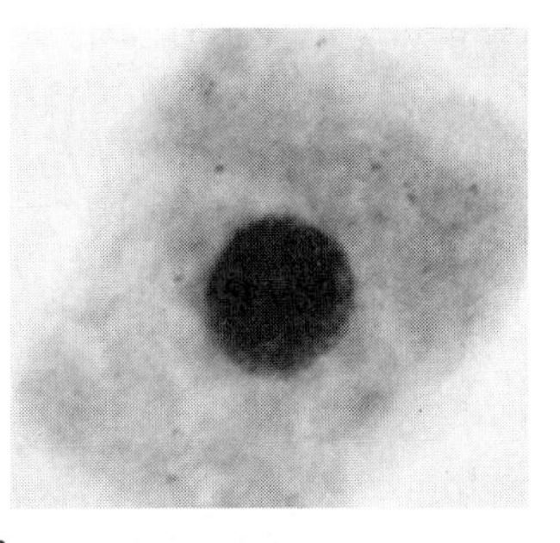
Interphase

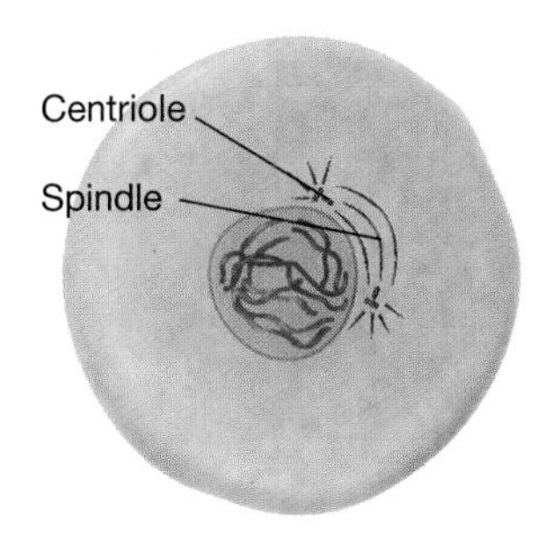

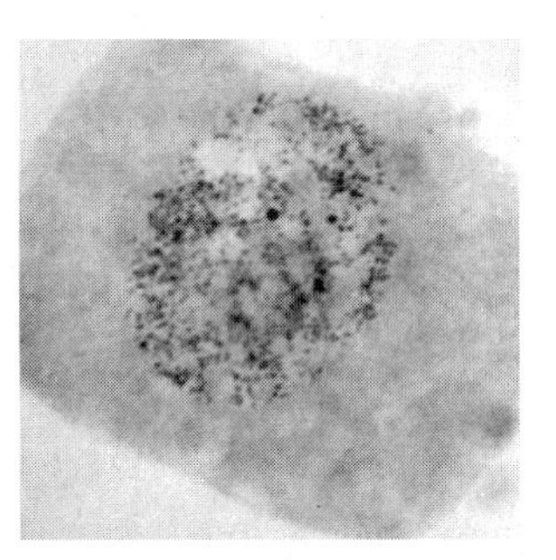
Prophase

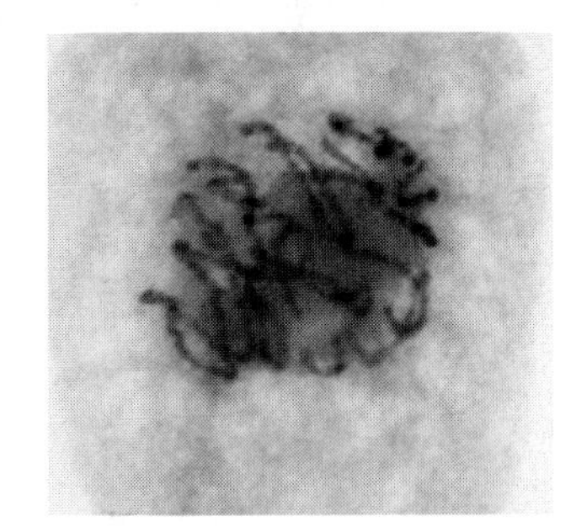
Prometaphase

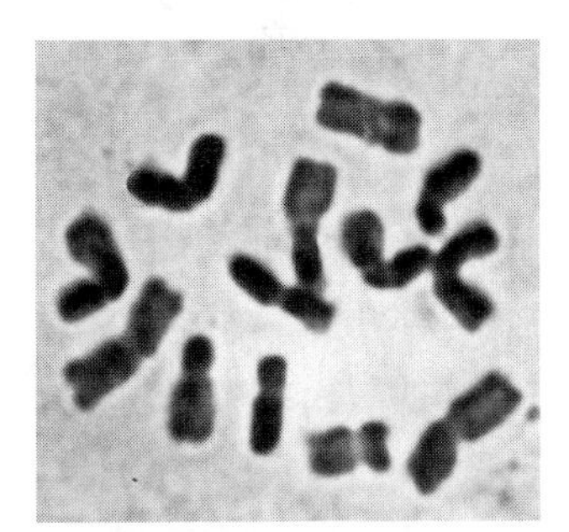
Metaphase

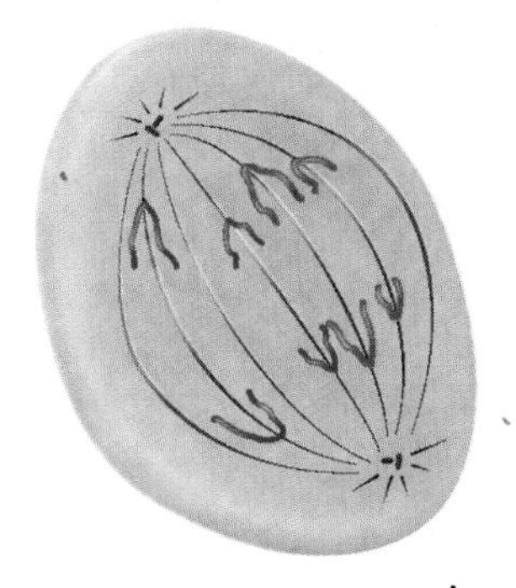
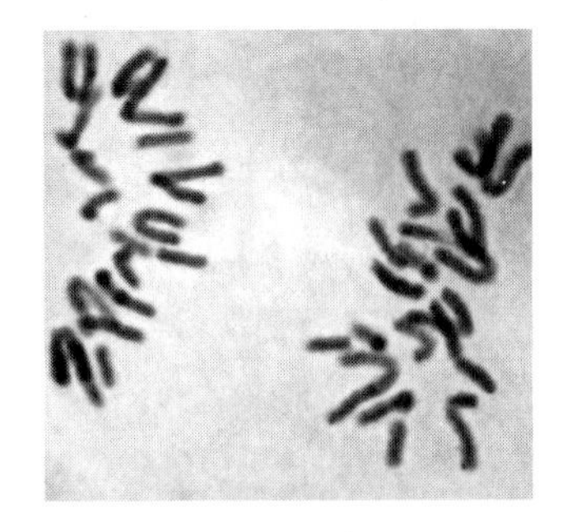
Anaphase

which move comparatively quickly toward opposite poles. As the chromatids are pulled through the cytoplasm, their arms drag behind to form V or J shapes.

Telophase begins when the chromosomes arrive at their poles. They start to uncoil and assume their interphase condition. New nuclear envelopes begin to form around each set, as the membrane vesicles first condense around individual chromosomes and then fuse into a single membrane. The spindle microtubules disappear and are replaced by interphase microtubules, and the next interphase begins.

The mitotic spindle is able to move chromosomes because of its structure. By metaphase, it contains three types of microtubules (Figure 13.13*a*): Astral microtubules lead toward the cell membrane, kinetochore microtubules attach to the chromosomes, and polar microtubules extending from each centrosome overlap in the middle of the cell. For the chromatids to be pulled to opposite poles, each chromosome must capture at least one fiber from each pole. During prometaphase, chromosomes are "fishing" for microtubules, and by metaphase, the average chromosome is attached to several of them (the number varies from one to 40 in different species). Early in mitosis, the spindle microtubules are turning over rapidly, much faster than interphase microtubules, as tubulin subunits are being constantly added at their plus ends and removed from their minus ends in the centrosomes. Once a microtubule is attached to a kinetochore, it becomes more stable, and its subunits turn over more slowly until anaphase begins.

The anaphase movement of chromosomes toward the poles results from two processes. First, the cell is elongating as its poles are pushed apart by the interaction of polar microtubules from opposite poles pushing against each other. Second, the kinetochore microtubules shorten as they depolymerize—slowly from the centrosome end and rapidly from the kinetochore end. How then can the chromosomes stay attached to structures that are constantly depolymerizing? Just as you stay on a treadmill by walking, the chromosomes stay on the chromosomes with a motor protein, a kind of kinesin built into the kinetochores, which continually walks toward the minus end. As long as the kinesin can move as fast as the microtubule disassembles, the chromosome will stay attached (Figure 13.14).

Microtubule motor proteins, Section 11.12.

As a result of these complicated movements, each daughter nucleus gets one chromatid of each original chromosome, and the two nuclei remain genetically identical. It makes no difference how many chromosomes are in the nucleus; each chromosome divides independently. These events have been observed in plant and animal cells for about a century, and constitute a classical mitosis. The process of mitosis itself has evolved over billions of years. Primitive cells probably grew and divided rather haphazardly, producing daughter cells that did not have complete functional genomes, so there was great selective pressure for mechanisms to evolve that could regulate the cell cycle and make mitosis as error-free as possible.

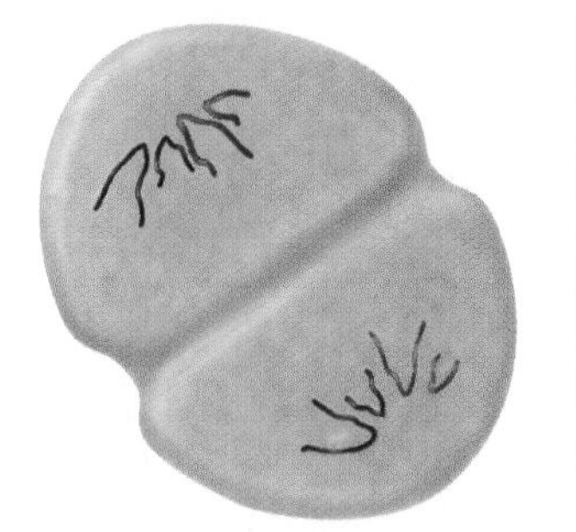
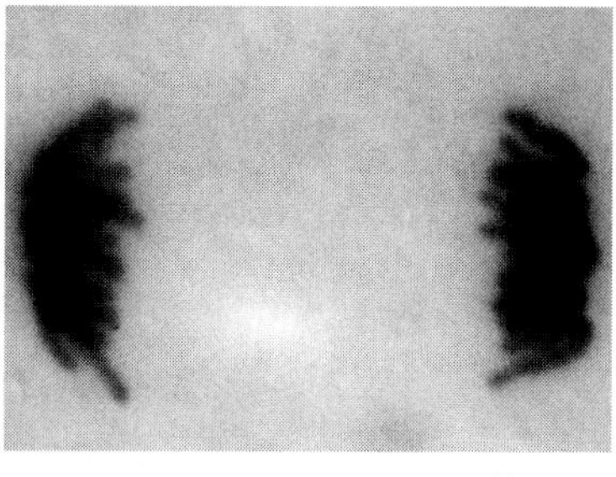

Telophase

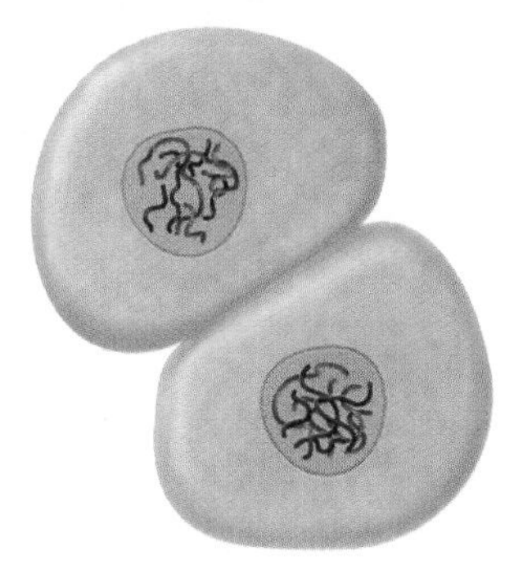
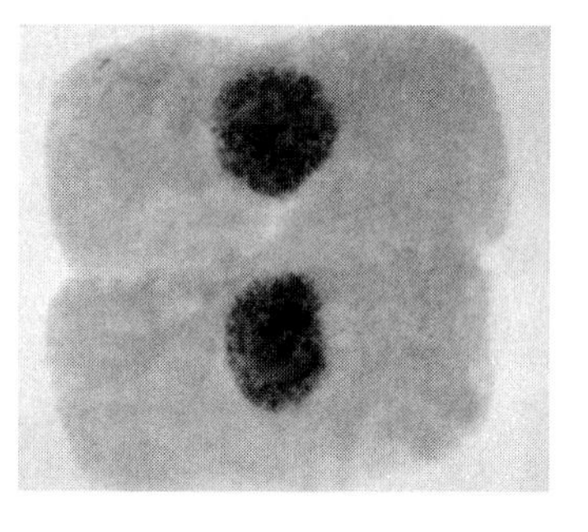

Interphase

Mitosis in primitive eucaryotes, Chapter 29.

Exercise 13.6 Using short pieces of wire or string to represent chromosomes, show a cell with only two simplex chromosomes, one long and one short. Carry this cell through one complete cycle, being sure to replicate the chromosomes during the S phase and divide them during mitosis.

Exercise 13.7 During anaphase, the chromosomes move comparatively quickly—about 1 μm per minute. But you can show that this is actually quite slow by determining how long it would take a chromosome to move 1 cm.
(*Note:* 1,440 min = 1 day.)

Exercise 13.8 Suppose the spindle microtubules shorten due to removal of tubulin subunits (dimers) from one end. The subunits are about 3 nm long, and there are 13 subunits in one turn of the microtubule helix. How fast must protomers be removed from a fiber to achieve a speed of 1 μm/min?

13.9 Procaryotic and eucaryotic chromosomes are replicated in different ways.

The cell cycle in procaryotes, though similar to that of eucaryotes, is much simpler. *Procaryotes do not engage in mitosis,* so they have no mitotic apparatus. The single chromosome is a circular DNA molecule attached inside the plasma membrane, sometimes to the mesosome, an extension of the plasma membrane (see Figure 6.30). The circular DNA molecule replicates during much of the cell cycle by means of two replication forks moving

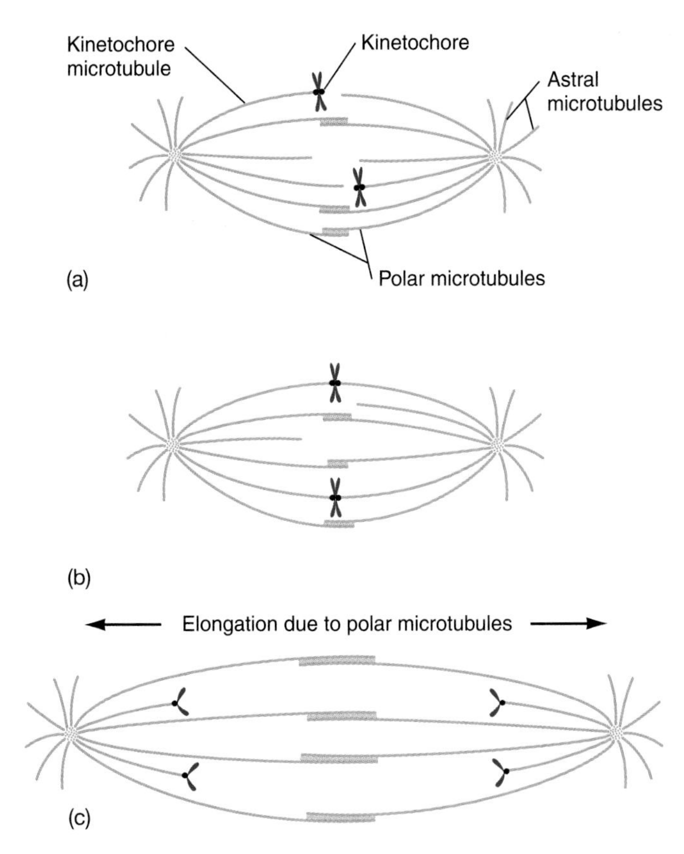

Figure 13.13

(a) Microtubules grow out of the centrosomes and elongate rapidly at their plus (cytoplasmic) ends while being disassembled at their minus ends in the centrosomes. Kinetochores on chromosomes capture some microtubules and stabilize them. Polar microtubules from opposite centrosomes overlap and interact. (If astral microtubules have a distinct function, it is unknown.) *(b)* By metaphase, each chromosome has captured at least one microtubule from each centromere and is pulled to the middle of the cell. *(c)* At anaphase, chromatids are drawn to the centromeres by the shortening of kinetochore microtubules while polar microtubules push against each other to move the centromeres apart.

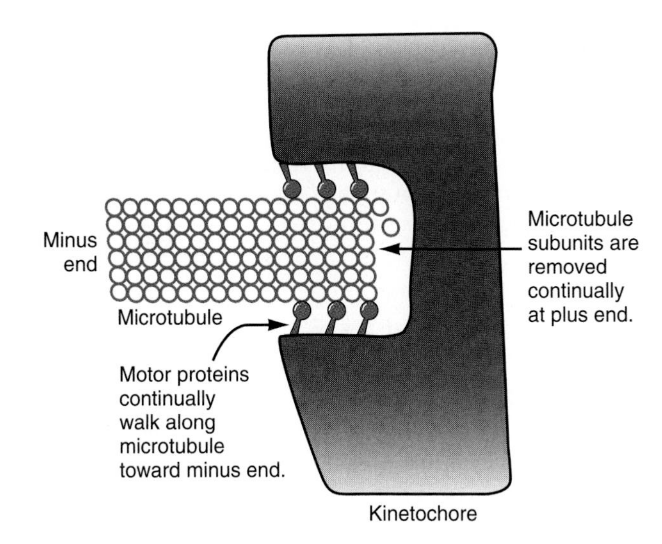

Figure 13.14

During anaphase, each spindle microtubule is being rapidly disassembled at its plus end. The kinetochore remains attached to the changing microtubule by means of kinesin motor proteins that continually walk toward the minus end.

in opposite directions around the circle. Bacterial cells such as *E. coli* take 40 minutes to complete a round of replication.

Figure 13.15 shows a general picture of the bacterial cell cycle. As the loop of DNA replicates, the daughter molecules attach to the plasma membrane and are pulled apart by elongation of the cell as new cell membrane (and the wall around it) grows between the attachment points. The cell divides after a round of replication is completed. Slowly growing cells have a gap corresponding to G_1 before a new round of replication starts. Rapidly growing cells are synthesizing DNA continually and can divide in less than 40 minutes by using more than two replication forks simultaneously, so they replicate their DNA faster in a rather complicated pattern.

If a round of replication takes 40 minutes in *E. coli,* and the average eucaryotic chromosome contains 50 times as much DNA as a bacterial chromosome, the S period in eucaryotic cells would be at least 2,000 minutes, or over 30 hours. But in fact,

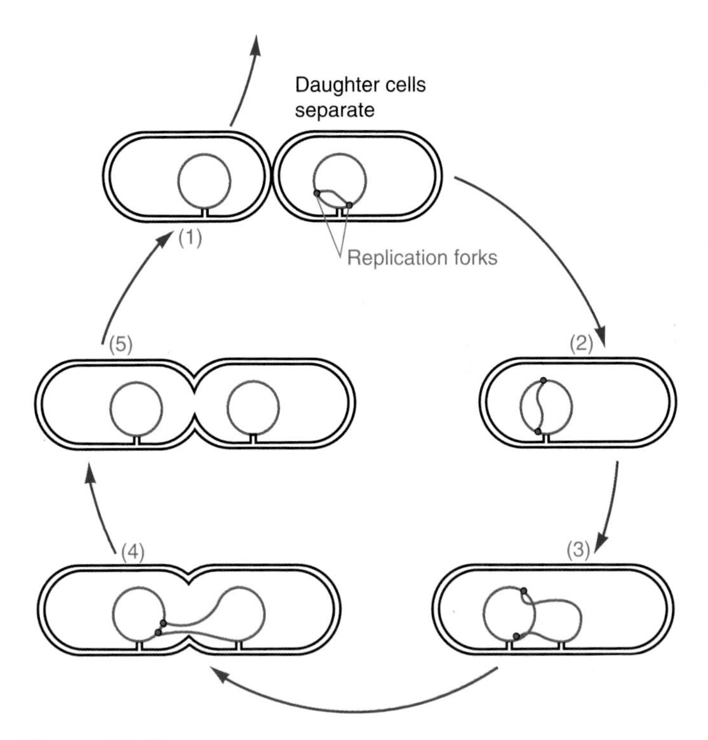

Figure 13.15

The bacterial cell cycle entails one round of DNA replication. Each chromosome is a circle, attached at one point on the plasma membrane (perhaps to mesosomes). *(1)* In a cell that has just completed a round of replication and has finished dividing, replication begins again. *(2)* The replicated regions become longer as the cell elongates, and *(3)* both daughter molecules acquire attachments to the cell membrane. *(4)* As replication is nearly complete, the daughter molecules are largely separated into the two halves of the cell by continued cell elongation, and cell division begins again. *(5)* Finally, replication is complete; each half of the cell has its own genome, and cell division separates them as independent cells.

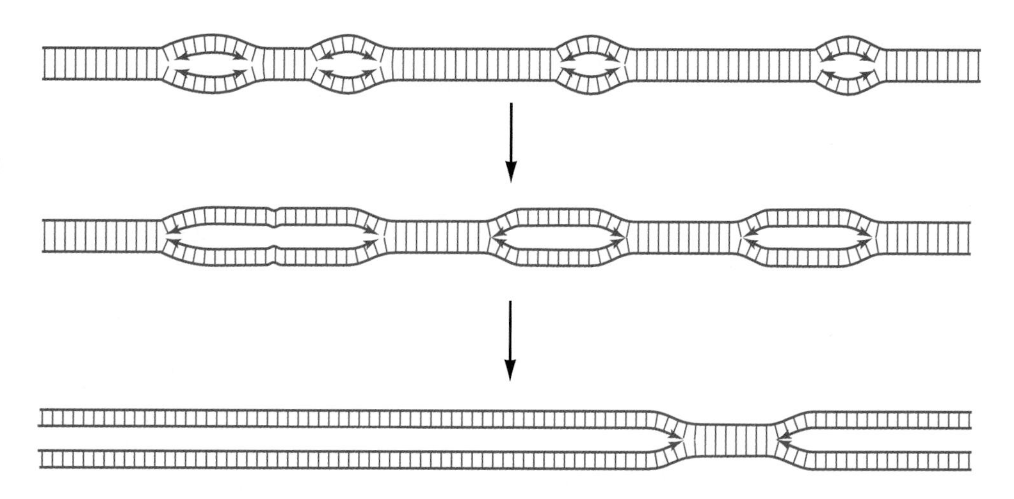

Figure 13.16
Replicating eucaryotic chromosomes are divided into many tandem replication units of varying lengths. Within each unit, replication forks proceed from the middle in both directions, on both DNA strands, until they meet the neighboring units.

replication occurs much faster because each chromosome is divided into thousands of replication units, averaging about 30 μm long, that work simultaneously. Replication begins in the center of each unit and proceeds in both directions until it meets the neighboring units (Figure 13.16). A cell can regulate the number of replication units it employs simultaneously. While an early embryo is dividing into cells, one division follows another with incredible speed. With all the replication units operating at once, the S phase in some embryos may be only an hour, and in *Drosophila* the entire genome may be replicated in only three or four minutes! But in adult tissue where the S period ranges from about 6 to 20 hours, fewer units operate simultaneously. Regulation at this level can provide a clock mechanism to control the timing of growth.

Exercise 13.9 If the *Drosophila* genome has 5,000 replication units, each including about 30,000 nucleotide pairs and using two replisomes, how fast does each polymerase have to work to replicate this DNA in 4 minutes?

Coda In this chapter we continue to develop a more complete picture of cells from the genetic perspective as expressed by the theme that organisms are genetic systems. One can just as well say *cells are genetic systems,* as we explain here. Chapter 11 described cells as dynamic; Chapter 12 showed that DNA carries genetic information and can be replicated and passed on to daughter cells when cells divide. In this chapter we bring the ideas of those two chapters together and give the molecular details of replication, noting that in eucaryotic cells DNA is complexed with protein to form chromosomes. Continuity of genetic information requires not only synthesis of more DNA, but also correct partitioning of that DNA into daughter cells by mitosis during which daughter cells each normally get equivalent sets of chromosomes. Growing cells proceed through a very orderly cycle of growth and division, which we have outlined here.

This chapter also develops another of the book's unifying themes: *Organisms function through molecular interactions.* A few examples of these interactions illustrate this concept. Eucaryotic chromosomes are examples of interactions between DNA and chromosomal proteins; various molecular forces hold these macromolecules together to form the chromosome. Chromosomes, of course, move by interactions among the parts of microtubules. Replication at its root is a matter of fitting together molecules with shapes that match—DNA strands, nucleoside triphosphates, and replicating enzymes. Control of a cell cycle means producing and degrading proteins that fit the shapes of other molecules, and control occurs as the molecules interact. It is helpful to visualize all these processes occurring as molecules fit together like pieces of a puzzle. For instance, imagine the shape of a DNA strand, the shape of a nucleoside triphosphate, and the shape of a replicating enzyme. Where and how do they interact, and what are the consequences of the interactions? Don't worry about all the details; just consider the overall processes. This approach is used by one of the most exciting fields of research in biology, in which molecular interactions of biological interest are visualized using new and powerful techniques of computer imaging.

Summary

1. Genes are located on chromosomes, which reside in the nucleus of eucaryotic cells. Most of the information for cell structure comes from the nucleus.
2. Eucaryotic chromosomes are built of DNA complexed with histone proteins, forming nucleosomes. This structure shortens the DNA. A chromosome becomes still shorter as the nucleosomes wind up together to make tightly compacted structures.
3. Each eucaryotic chromosome changes regularly as a cell proceeds through the cell cycle. Immediately after cell division, a chromosome is a simplex structure consisting of one long DNA molecule combined with proteins; during DNA replication, it is converted into a duplex structure consisting of two chromatids, which are DNA molecules (plus protein) identical to the original molecule.
4. Procaryotic chromosomes are made of DNA combined with HU protein. Each cell has one circular chromosome, which forms a nucleoid (nuclear body).
5. The Meselson-Stahl experiment showed that the strands of each DNA helix separate, and a new strand is constructed on each old strand, by complementary base-pairing, to form two identical helices.

6. DNA replication is effected by a replisome complex containing two molecules of DNA polymerase. One polymerase synthesizes the leading strand continuously by reading along one of the template DNA strands in the $3' \rightarrow 5'$ direction. The other DNA polymerase replicates the lagging strand in short fragments (Okazaki fragments) on the opposite template strand. Adjacent fragments are ligated together to form the complete lagging strand.
7. As each nucleoside triphosphate (dATP, dCTP, dGTP, dTTP) is added to a growing polynucleotide, the terminal pyrophosphate group is removed and split into two molecules of phosphate. These reactions provide the energy for polymerization and keep the system directed toward synthesis rather than breakdown of DNA.
8. A eucaryotic cell cycle consists of a period of growth (G_1), a period of DNA synthesis (S), a period of preparation for cell division (G_2), and mitosis (M). Mitosis is generally followed by cell division; when it is not, a cell with several nuclei in a single cytoplasm (coenocyte) results.
9. The cell cycle is driven by a series of cyclin-dependent kinases regulated by cyclins and required to pass key checkpoints. Mitosis promoting factor (MPF) consists of the protein Cdc2 combined with cyclin B, which is synthesized during interphase and degraded during mitosis. S-phase promoting factor (SPF) is required to pass Start and initiate DNA replication; it is only made after the cell has grown large enough for another division. Other kinases and cyclins monitor other phases of the cycle.
10. Cyclin-dependent kinases promote events in the cell cycle by phosphorylating certain proteins, and the kinases are themselves regulated by other kinases.
11. Mitosis is divisible into five phases: (1) prophase, when the chromosomes become compact and visible while a mitotic spindle forms; (2) prometaphase, when the nuclear envelope breaks down and chromosomes attach to the mitotic spindle; (3) metaphase, when the chromosomes line up in the middle of the cell; (4) anaphase, when the chromatids of each chromosome separate from each other; and (5) telophase, when nuclear envelopes re-form and chromosomes become extended again.
12. In mitosis, chromosomes attach to very dynamic spindle microtubules that, at first, are constantly adding and removing tubulin subunits. The kinetochores of each chromosome capture microtubules, so by metaphase, each chromatid is attached to one or more microtubules from one pole. Attachment to a kinetochore stabilizes a microtubule. The anaphase movements of chromosomes toward the poles are effected by axial microtubules from opposite poles pushing against each other, thus elongating the cell, and by depolymerization of the kinetochore microtubules.
13. Procaryotic cells go through a comparable cycle, which does not include events comparable to mitosis. The circular chromosome is replicated during most of the cycle, and the daughter chromosomes are separated as the cell elongates.
14. A eucaryotic chromosome has thousands of replication units where replication may occur simultaneously. The units eventually meet one another, resulting in a complete replicated chromosome made of two chromatids.

Key Terms

chromatin 259
histone 260
nucleosome 260
chromatid 260
centromere 260
semiconservative 261
conservative 261
density gradient 262
replisome 263
DNA polymerase 263
template 263
leading strand 264
lagging strand 264
Okazaki fragment 264
cell cycle 265
mitosis 265
cytokinesis 265
S period 265
G_1 period 265
G_2 period 265
interphase 265
prophase 267
mitotic spindle 267
prometaphase 267
kinetochore 267
metaphase 267
anaphase 267
telophase 268

Multiple-Choice Questions

1. Which of the following statements is correct?
a. Any animal has more DNA than any plant.
b. Humans have more DNA than other animals.
c. All animals have the same amount of DNA.
d. Eucaryotes have more DNA in each cell than procaryotes.
e. Procaryotes and viruses have the same amount of DNA.

2. Which structure includes all the others?
a. histone proteins
b. chromatin
c. nucleosome
d. DNA
e. centromere

3. The primary function of nucleosomes is to
a. assist DNA replication.
b. assist protein synthesis.
c. compact DNA within the nucleus.
d. unwind DNA.
e. stabilize DNA in the single-stranded state.

4. Which of the following is found in both simplex and duplex chromosomes?
a. only one chromatid
b. two chromatids
c. only one centromere
d. two centromeres
e. only one strand of DNA

5. The Meselson-Stahl experiments
a. proved that DNA replicates conservatively.
b. disproved that DNA replicates conservatively.
c. proved that DNA replicates semiconservatively.
d. disproved that DNA replicates semiconservatively.
e. Both *a* and *c* are correct.

6. Which is *not* part of a replisome?
a. proteins
b. DNA
c. RNA
d. enzymes
e. None of the above are part of a replisome.

7. DNA polymerase catalyzes the formation of ___ bonds.
a. hydrogen
b. ionic
c. phosphodiester
d. base-pairing
e. weak

8. The phrase *leading strand* refers to
a. a new strand of DNA synthesized continuously.
b. the template from which DNA is synthesized continuously.
c. a new strand of DNA that is synthesized discontinuously.
d. the template from which DNA is synthesized discontinuously.
e. the moving replication fork.

9. The energy for lengthening the DNA backbone is derived from
   a. ATP.
   b. DNA polymerase.
   c. Okazaki fragments.
   d. splitting triphosphates into pyrophosphates.
   e. joining pyrophosphates into ATP.
10. Which does *not* occur during interphase?
   a. DNA synthesis
   b. growth
   c. cytokinesis
   d. protein synthesis
   e. cellular respiration

True-False Questions

Mark each statement true or false, and if false, restate it to make it true.

1. Hämmerling's experiments with *Acetabularia* demonstrated the semiconservative nature of DNA replication.
2. It is possible to study the method of DNA replication in bacteria with density gradient centrifugation by using bacteria that have been growing and dividing in a medium containing radioactive nitrogen.
3. A replisome is composed of all four types of nucleotides: dATP, dCTP, dGTP, and dTTP.
4. At each replication fork, one strand of DNA replicates continuously to produce the leading strand while the other strand replicates discontinuously to produce the lagging strand.
5. The energy needed to synthesize DNA is derived from the breakdown of phosphodiester bonds in the backbone.
6. A eucaryotic chromosome contains two chromatids during all phases of the cell cycle except G_2 and the early stages of mitosis.
7. Barring the use of specific inhibitors, any cell that reaches G_1 in the cell cycle will undergo mitosis.
8. The eucaryotic cell cycle is regulated by the synthesis of Cdc2 during interphase and its destruction during metaphase.
9. The primary function of a kinetochore is to join the two chromatids of a duplex chromosome.
10. During anaphase, chromosomes are pulled toward opposite poles of the cell as polar microtubules and kinetochore microtubules simultaneously push and pull on the chromatids.

Concept Questions

1. Why do you think J. Herbert Taylor used radioactive thymidine and autoradiography rather than heavy nitrogen and density gradient centrifugation when analyzing DNA replication in eucaryotic cells?
2. In addition to DNA polymerase, what three other proteins are part of replisomes, and what is the function of each?
3. Would Okazaki fragments and primase be essential if DNA polymerase could synthesize DNA in both the $5' \rightarrow 3'$ and the $3' \rightarrow 5'$ directions? Explain.
4. What is MPF, and how does its action regulate the cell cycle?
5. Contrast the pattern of DNA replication in bacteria with that in eucaryotes.

Additional Reading

Campisi, Judith, et al. (eds.). *Perspectives on Cellular Regulation: From Bacteria to Cancer (Essays in Honor of Arthur B. Pardee).* Wiley-Liss, New York, 1991. A collection of papers about the cell cycle and the regulation of growth.

Hughes, P., E. Fanning, and M. Kohiyama (eds.). *DNA Replication: The Regulatory Mechanisms.* Springer-Verlag, Berlin and New York, 1992.

Kornberg, Arthur, and Tania A. Baker. *DNA Replication,* 2d ed. W.H. Freeman, New York, 1992. An advanced, detailed treatment of the subject.

Wang, James C. "DNA Topoisomerases." *Scientific American,* July 1982, p. 94. DNA topoisomerases are enzymes that convert rings of DNA from one topological form to another.

Internet Resource

To further study the content of this chapter, log on to the web site at:

http://www.mhhe.com/biosci/genbio/guttman/

Protein Synthesis and Cell Growth

14

Key Concepts

- **14.1** DNA specifies RNA, and RNA specifies protein.
- **14.2** Unstable messenger RNA carries genetic information.
- **14.3** Transfer RNA adaptors carry amino acids to the mRNA templates.
- **14.4** The genetic code is systematic and redundant.
- **14.5** The mechanism of information transfer in cells restricts the process of evolution.
- **14.6** Transcription and translation are coupled in procaryotes but separated in eucaryotes.
- **14.7** Cells sort their proteins into compartments by using intrinsic protein structures.
- **14.8** The Golgi complex packages, sorts, and exports materials.
- **14.9** Secretion entails the synthesis and outward flow of membranes.
- **14.10** Eucaryotic genes contain noncoding sequences.
- **14.11** Coding sequences can be used to make more than one kind of protein.
- **14.12** The nucleolus is the site of ribosome assembly.
- **14.13** Mitochondria and chloroplasts contain their own genetic apparatus.
- **14.14** Biological structures conserve their own patterns during growth and determine the form of new structures.

Each person is made of a distinctive set of proteins encoded in his or her genome.

What makes you the unique person you are? A psychologist or sociologist would attribute the development of your personality to a particular cultural background, a certain family life, and a wealth of unique experiences. A biologist, however, while not denying any of those influences, would focus on your proteins. You came into the world with a genetic inheritance, a mixture of genetic information from your father and mother that mostly encodes thousands of proteins. Look at yourself in a mirror. What you see is basically protein. Your hair consists of strands of protein, colored by certain pigments made by enzymes. Your eyes have irises of a distinctive color made by other enzymes. Your skin is a heavy, waterproof protein layer shaped into your unique facial form by an underlying pattern of muscle, bone, and cartilage, and whatever parts of that underlying structure are not themselves protein have been made by proteins—that is, by certain enzymes. Your face resembles your parents' faces because you share with them a series of developmental proteins that instructed your cells to form their structures in the common patterns that shaped your parents' faces. Psychologists have acquired very good evidence that much of your personality is also inherited, so it is not simply the result of your experiences. This too indicates a pattern of protein expression in your nervous system. You really are what your proteins are.

Most of the genetic information in an organism's genome determines the amino acid sequences of its proteins, and much of metabolism is devoted to reading and translating the information carried in DNA sequences. So protein synthesis is the most central process in a cell. Electron micrographs of typical cells, both procaryotic and eucaryotic, show that much of the cytoplasm is occupied by ribosomes, the factories where protein is made, and in eucaryotic cells, by membranes that further process new proteins. As you grow, or simply maintain your structure, you are making new cells, new proteins. This is the story we will tell in this chapter. The story has three major points: how the information in DNA is transcribed into RNA, how the information in RNA is in turn translated into protein, and how a cell sorts proteins to their proper destinations. ■

14.1 DNA specifies RNA, and RNA specifies protein.

Shortly after Watson and Crick published their paper describing DNA structure, Crick proposed two basic theses about the use of genetic information. He called his first point the "Central Dogma": *Genetic information stored in DNA provides the information for protein synthesis, and information does not flow in the reverse path from protein to DNA.* This is now a central concept of biology. Remember that a cell encodes each of its many kinds of protein in a separate DNA sequence, and we have defined a gene (Chapter 12) as a portion of the genome that encodes a single polypeptide. The whole genome contains many genes, marked off by signals that specify their ends and control their use (Figure 14.1).

Crick called his second basic thesis the "sequence hypothesis." It states that, *since DNA is a sequence of nucleotides and a protein is a sequence of amino acids, the DNA sequence forms a* coded message *that can be translated into the protein sequence.* However, biochemical experiments had shown long before that protein synthesis depends upon ribonucleic acid (RNA) synthesis. RNA is an informational intermediary between DNA and protein. DNA first passes its information to RNA in a process called **transcription** because the DNA and RNA sequences both use the same nucleotide "language." This process is just like

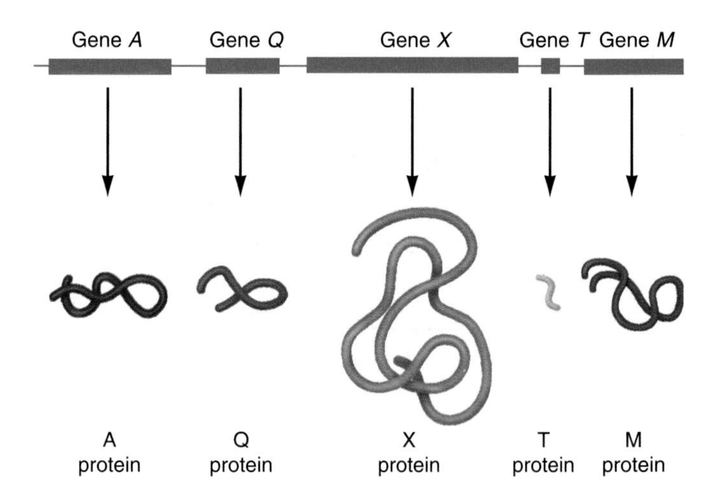

Figure 14.1
A genome consists of distinct genes, each of which is a coded message for one type of polypeptide. Each of these messages must be translated into protein structure.

transcribing handwriting into typing, symbol by symbol. The second step, where RNA passes its information on to protein, is **translation,** because the information is changed from the language of nucleic acids into the language of proteins, whose "words" are amino acids:

Replication DNA —**Transcription** (RNA synthesis)→ RNA —**Translation** (Protein synthesis)→ Protein

RNA is very similar to DNA, except that its sugar is ribose instead of deoxyribose, and the base uracil replaces thymine:

Uracil

The nucleoside of uracil is uridine (U). Since uracil and thymine are so similar, they form identical hydrogen bonds with adenine. Even though most RNA molecules are single-stranded, they can fold back on themselves to form double-stranded structures such as hairpin loops that observe the Watson-Crick bonding rules: G bonds with C, and A bonds with U (as if it were T):

A—U—C—C—A—G≡C—C—G—U—A—C

DNA and RNA strands can also bind to each other through complementary base-pairing. Just as a DNA strand serves as a template for making a complementary DNA strand in replication, it acts as a template for making a complementary RNA strand during transcription. RNA synthesis, like DNA synthesis, requires the four nucleotides in their triphosphate form—ATP, CTP, GTP, and UTP—and it is catalyzed by the enzyme **RNA polymerase** (Figure 14.2). An RNA polymerase temporarily opens up the DNA helix. In a process similar to replication, each DNA nucleotide binds to its complementary ribonucleoside triphosphate from the cytosol, and the RNA polymerase moves along the DNA, catalyzing the union of these ribonucleotides into an RNA strand. As in DNA synthesis, the polymerase operates in only one direction, making an RNA molecule by adding nucleotides to the free 3′-end of the growing strand as it copies from a template DNA strand in the 3′ → 5′ direction. Thus if a DNA sequence begins:

3′–A–C–C–G–T–C–T–G–A–C–G–5′
5′–T–G–G–C–A–G–A–C–T–G–C–3′

and if its upper strand is used as a template, it specifies an RNA that begins 5′–U–G–G–C–A–G–A–C–U–G–C–3′. This is transcription. The RNA polymerase is sometimes called a "transcriptase," and the RNA molecule it produces is a **transcript.** It is important to see that an RNA transcript does not exist in the cell until it is made in this process; information is transferred to the RNA *as it is synthesized.*

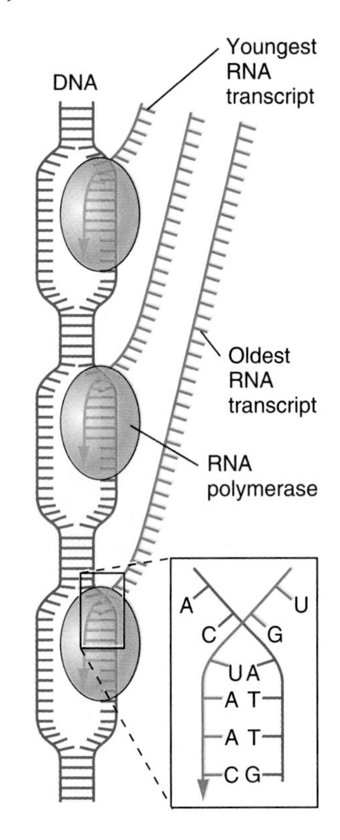

Figure 14.2
An RNA polymerase reads along the DNA, using it as a template for inserting complementary ribonucleotides and polymerizing them into an RNA molecule.

Since DNA, RNA, and protein are all linear molecules (unbranched chains of monomers), Crick hypothesized that the RNA transcribed from a specific gene and the polypeptide translated from that RNA are all **colinear** with one another. This means information is transferred from a sequence of deoxyribonucleotides (dN) to a sequence of ribonucleotides (rN) and then to a sequence of amino acids (aa) by reading *directly* from one to the other:

dN — dN — dN — dN — dN — dN — dN — dN — dN — dN
↓ ↓ ↓ ↓
rN — rN — rN — rN — rN — rN — rN — rN — rN
↓ ↓ ↓ ↓
aa — aa — aa — aa — aa — aa — aa — aa — aa

Transcription occurs quite directly, through Watson-Crick base-pairing. But amino acids don't resemble nucleotides, and transferring information from an RNA sequence into protein structure requires a complex apparatus, which we will describe shortly. Since nucleotides encode amino acids, a group of nucleotides that stands for one amino acid is known as a **codon.** A single nucleotide, however, can't encode a single amino acid because there are 20 kinds of amino acids but only four kinds of nucleotides. Even two nucleotides aren't enough to encode all the amino (as by letting AA mean Gly, AC mean Ser, and so on) because there are only 16 such combinations. (There are 4 ways to choose the first nucleotide and 4 ways to choose the second, so $4 \times 4 = 16$.) To encode 20 amino acids, a codon must be at least three nucleotides long, since there are 64 such triplets ($4 \times 4 \times 4$). Experiments we will describe later show that amino acids are, in fact, specified by codons of three nucleotides. The first three nucleotides in a gene specify the first amino acid in a protein; the next three specify the next amino acid; and so on:

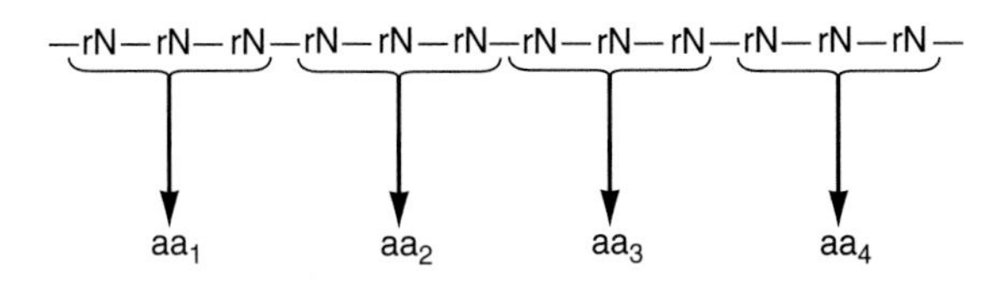

Exercise 14.1 Convince yourself (if you aren't already convinced) that 16 sequences (permutations) of four nucleotides are possible, taken two at a time. Make a square of four columns and four rows, and write A, T, G, and C at the tops of the columns and also to the left of the rows. Fill in the 16 two-letter combinations.

Exercise 14.2 Similarly, if you need convincing that there are 64 sequences of four nucleotides, taken three at a time, try this: Print A, T, G, and C down the left side of a sheet of paper with lots of space between them. From each letter, draw four lines, and put A, T, G, or C at the end of each line. From each of *these* letters, draw four more lines and print the letters again. You should now be able to see 64 paths from a first letter to a third.

Students generally ask several important questions at this point:

1. Is the same strand of the DNA used as a template throughout the genome? Some genes are transcribed from one strand and some from the other, but any one gene is read only from one strand; the information in the complementary strand in that region is not used (with rare exceptions).

2. What determines which strand an RNA polymerase will copy and where it will begin copying? Near each gene (or each small group of genes) is a site called a **promoter,** a specific control signal that determines where transcription begins. RNA polymerases bind preferentially to promoters, and this binding orients the enzyme so it can only start to copy the correct strand in the right direction. Other control sites in the DNA, in combination with control proteins, also regulate the activities of RNA polymerases. A polymerase stops synthesizing each transcript at sites called **terminators,** DNA sequences that form hairpin loops of the kind shown earlier.

3. If only one strand is copied, what function does the other strand have? It is, at least, a reservoir of information for repair synthesis; DNA is always being damaged by chemical reagents and radiation, and certain enzymes move along the genome like railroad repair crews, finding damaged nucleotides and replacing them. These enzymes need the information in the undamaged strand to insert the correct nucleotides. The double-stranded structure may also help control transcription by keeping the template strand covered except when an RNA polymerase opens the double helix.

We can summarize the "central dogma" and the "sequence hypothesis" as Crick's Law:

A sequence of nucleotides in DNA encodes a colinear sequence of amino acids in a protein, but there is no reverse information transfer from protein to DNA.

Exercise 14.3 (Science fiction story.) The organisms living on the planet Tau Centauri C have proteins with only 12 types of amino acids. What is the most likely size of a codon in their DNA? Suppose that, instead of 12 types of amino acids, they have 80. What is the most likely size of their codons?

Exercise 14.4 The following segment of a DNA molecule encodes the beginning of a protein:

3′–T–A–C–C–C–G–G–T–A–C–G–T–T–C–G–G–A–A–A–T–C–5′
5′–A–T–G–G–G–C–C–A–T–G–C–A–A–G–C–C–T–T–T–A–G–3′

Transcribe an RNA molecule from it, starting at the left end and using the *upper* strand as a template. Then divide both the DNA and the RNA into codons. Remember that RNA uses U instead of T. Save this transcript for Exercise 14.5.

14.2 Unstable messenger RNA carries genetic information.

Ribosomes, the factories that synthesize proteins, are large complexes of RNA and protein. Procaryotic ribosomes are a little smaller than eucaryotic ribosomes, but both consist of two

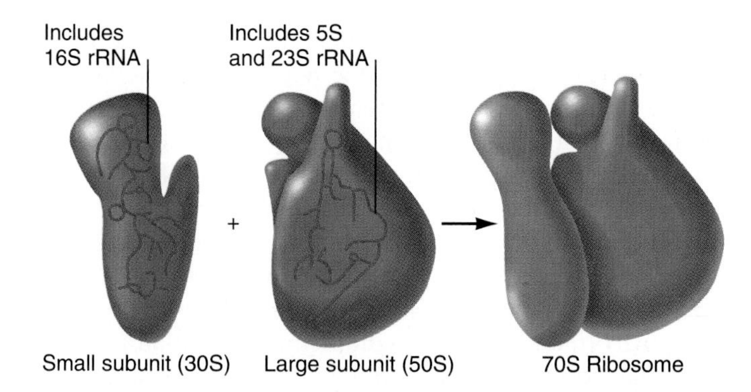

Figure 14.3

A ribosome consists of two subunits. Each one is a cluster of structural proteins attached to a framework of ribosomal RNA. Ribosomes and their component macromolecules are often isolated and characterized by centrifugation; their sedimentation rate is expressed in terms of the sedimentation constant (S). Larger particles fall faster than small ones in a solution, but S values are only roughly proportional to molecular weight, and they are not additive.

subunits made of many proteins bound to structural **ribosomal RNA (rRNA)** molecules (Figure 14.3). About 80 percent of the RNA in a bacterial cell is rRNA. For years it was assumed that the transcripts from each gene are built into the ribosomes, where they are translated into proteins. However, François Jacob, Sydney Brenner, and Matthew Meselson had reason to doubt this. They pointed out that the RNAs carrying information for different kinds of proteins should be very different, but all ribosomal RNA seems to be about the same, even in different species. Furthermore, because many cells can quickly change the proteins they make, the RNA templates for protein synthesis must be quite unstable in a chemical sense; they must be made rapidly, used for a time to direct protein synthesis, and then broken down and replaced by other molecules. In other words, template RNA molecules should **turn over** rapidly, much as the inventory in a store turns over rapidly if business is good. However, rRNA molecules are quite stable; once made, they are only broken down slowly, if at all. Therefore, rRNA cannot be the template.

For these and other reasons, Brenner, Jacob, and Meselson postulated in 1961 that rRNA is just a structural component of the ribosome and that the actual messages for protein synthesis are carried in molecules they called **messenger RNA (mRNA).** Messengers should be short-lived molecules that bind to the ribosomes, where they are translated into a few molecules of the protein they encode and then are destroyed. These investigators then showed that the information for making new proteins after phage infection is carried by unstable mRNA molecules, and soon afterward other research showed that unstable messenger RNAs are the templates in uninfected bacteria and in eucaryotic cells.

A ribosome can make any protein specified by the mRNA it translates, much as a videocassette recorder can produce any image on a television screen that is specified by the tape it plays.

Because a ribosome associates temporarily with an mRNA molecule, it can translate different messengers, one after the other:

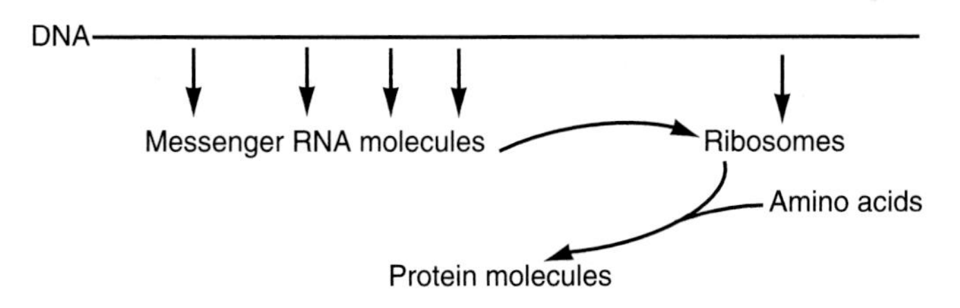

Notice that protein synthesis requires an input of information from mRNA—and an input of material in the form of amino acids.

14.3 Transfer RNA adaptors carry amino acids to the mRNA templates.

Crick pointed out in 1955 that nucleotide shapes are not complementary to amino acids, and that an amino acid alone is too small to line up with a codon of three nucleotides. He therefore postulated that cells contain adaptor molecules—probably another kind of RNA—that carry an amino acid on one end and have a shape on the other end that is complementary to an RNA codon. So there was great excitement in 1957 when Mahlon Hoagland, Paul Zamecnik, and their associates discovered the small RNA molecules now known as **transfer RNA (tRNA),** which have just the properties Crick had predicted (Figure 14.4). Cells have at least one type of tRNA for each type of amino acid, and each one has an **anticodon,** a sequence of three nucleotides on one end that is complementary to the mRNA codon for that amino acid.

Cells also have enzymes called amino-acyl tRNA synthetases—one kind for each kind of amino acid—that recognize each amino acid and combine it with the proper tRNA (Figure 14.5). Protein synthesis requires an input of energy too, so first the enzyme activates the amino acid with ATP:

$$\text{Amino acid} + \text{ATP} \rightarrow \text{Amino-acyl AMP} + \text{PP}_i$$

The enzyme links the activated amino acid to the tRNA to make an amino-acyl tRNA, which can then bind to the mRNA. Notice that in protein synthesis, as in DNA replication, a pyrophosphate molecule (PP_i) is released as one step in synthesizing the polymer. Pyrophosphatases then split pyrophosphate in half, effectively keeping the process of protein synthesis from being reversed, since forming pyrophosphate and ATP again would be endergonic reactions.

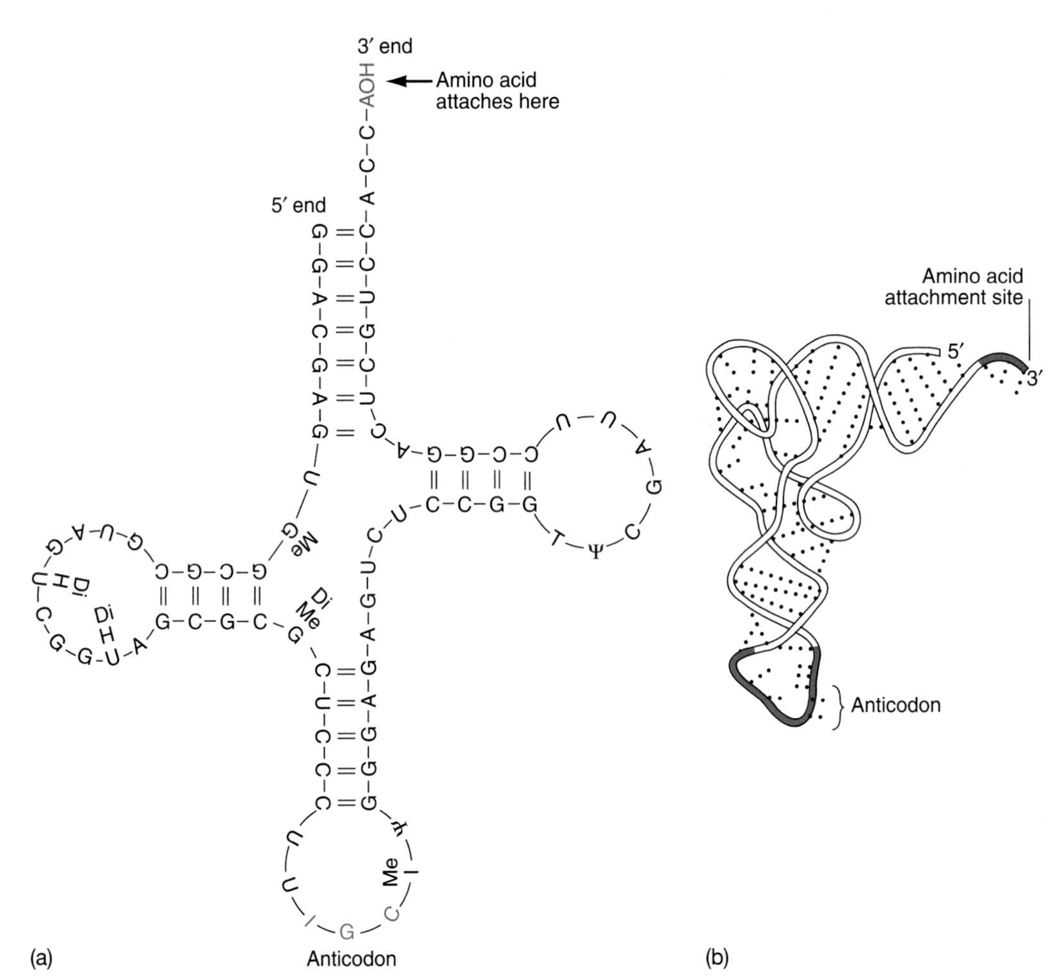

Figure 14.4

A transfer RNA is a relatively small molecule (about 70–90 nucleotides) with internal base-pairing that makes a cloverleaf structure *(a),* although this is distorted in the three-dimensional form of the molecule *(b).* An amino acid can be bound on one end to make an amino-acyl tRNA, and an anticodon on one loop is complementary to a codon on the mRNA.

Figure 14.5
For each of the 20 amino acids, cells have an enzyme that combines that amino acid with its appropriate tRNA to make an amino-acyl tRNA. The alanyl-tRNA synthetase, for example, only recognizes alanine and the tRNA that should carry alanine, and it combines them.

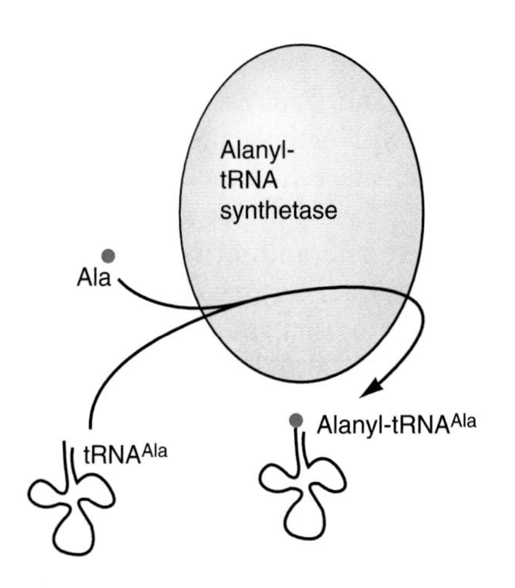

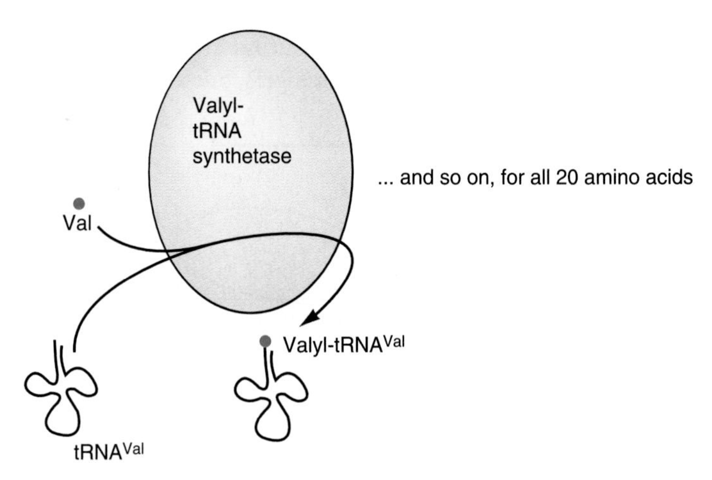

Figure 14.6 shows how the mRNA-ribosome complex binds amino-acyl tRNA molecules, one at a time. As the ribosome feeds the mRNA along, codon by codon, it assembles the *nascent* (that is, growing) protein, amino acid by amino acid. Thus the sequence of amino acids is determined by the sequence of the mRNA. As each codon of the mRNA appears in a particular site on the ribosome, an amino-acyl tRNA with the complementary anticodon binds into place. Then the latest amino acid is linked to the one just before it through a peptide linkage. This process continues until the entire gene copy on the mRNA has been translated into a polypeptide.

Ribosomes are constantly recycled. Several ribosomes bind to a single mRNA to make a **polyribosome,** and each ribosome moves along, translating the messenger sequence into a protein. As a ribosome comes to the end of the messenger, it dissociates into its subunits, which are recycled to form new ribosomes (Figure 14.7).

14.4 The genetic code is systematic and redundant.

In 1961 Marshall Nirenberg identified the first word of the genetic code and a general method for cracking the rest of the code. He developed a system for making proteins *in vitro* (literally, "in glass," in a cell-free system, as contrasted with *in vivo,* in an intact cell). To a cell extract containing ribosomes, tRNAs, and other cellular components, Nirenberg added artificial mRNA molecules with known or partially known sequences. To his great delight, a messenger made only of U (polyuridine) would only stimulate the synthesis of polyphenylalanine. In other words, the message U–U–U–U–U–U–U–U– . . . was translated Phe–Phe–Phe– . . . , which means UUU is the codon for phenylalanine. By refining this approach over the next few years, investigators in Nirenberg's laboratory and several other laboratories elucidated the whole code, which is summarized in Table 14.1.

The code is quite remarkable. Since the 64 triplets must specify only 20 amino acids, 44 of them might have turned out to be meaningless; however, nearly all 64 *do* code for amino acids. This means the code is redundant, or degenerate, in that several codons can specify the same amino acid. Moreover, the code is systematically redundant. Reading the mRNA in the $5' \rightarrow 3'$ direction, the first two bases carry most of the information, and sometimes the third base doesn't matter. For instance, UCX always means serine, regardless of what X is. In some cases it only matters whether the third base is a purine or a pyrimidine: UUU and UUC both mean phenylalanine, while UUA and UUG both mean leucine. When such different but closely related codons encode one amino acid, a single type of tRNA is usually able to translate them—that is, a cell doesn't need a different species of tRNA for each codon. Crick explained this by the *wobble hypothesis,* which says the base of the anticodon that binds to the third base of the codon is not fixed in place but can wobble slightly. This freedom allows U in the wobble position of the tRNA to recognize both A and G on the mRNA, and G to recognize both U and C.

Three triplets—UAA, UAG, and UGA—do not code for amino acids. Though commonly called "nonsense" triplets, they are actually **termination codons** used at the end of a gene to terminate protein synthesis. The codon AUG stands for methionine and has a special role as an *initiation* codon. It is the first codon in (almost) every gene, and in eubacteria, the special tRNA that binds to it carries a modified amino acid, N-formylmethionine, whose amino group is blocked so it can only go into the N-terminal position:

$$\underbrace{H-C(=O)}_{\text{Formyl group}}-N(H)-C(H)(CH_2-CH_2-S-CH_3)-C(=O)-OH$$

In eucaryotes and archaebacteria, ordinary methionine is the first amino acid in each polypeptide.

One of the delights of biology is the inventiveness of cells, and a particular surprise was the discovery that *E. coli,* and presumably other bacteria, can encode a twenty-first amino acid.

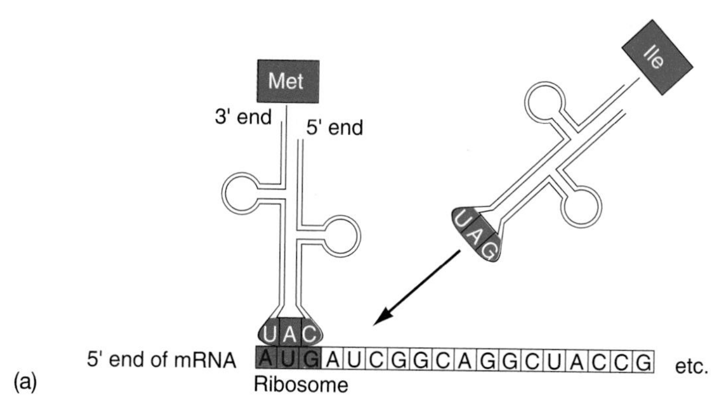

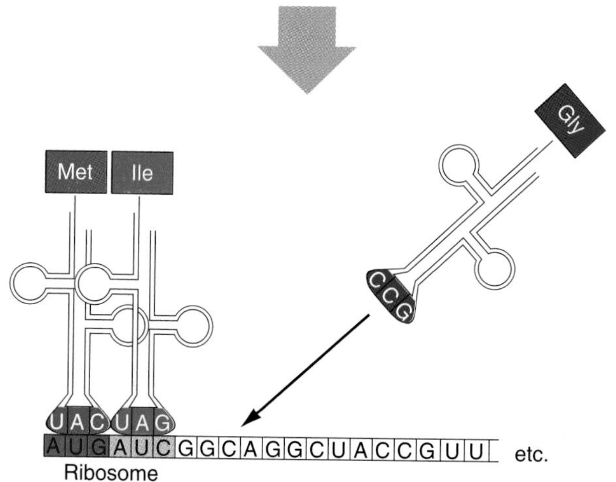

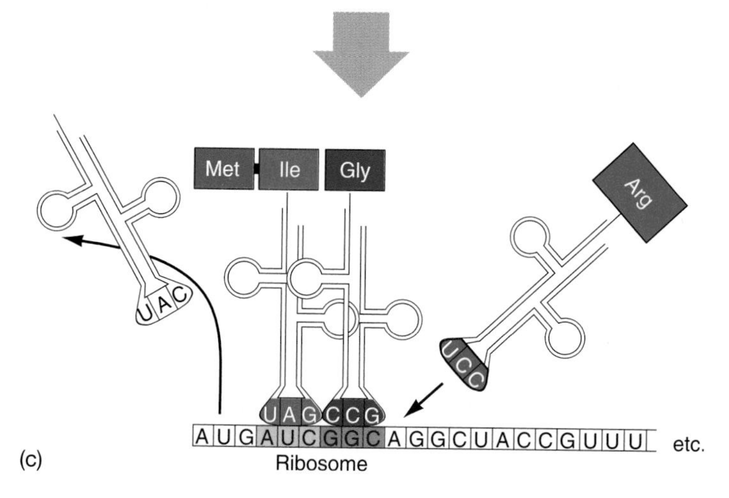

Figure 14.6

Protein synthesis occurs as a messenger RNA, attached to a ribosome, is translated one codon at a time. *(a)* The first codon to be translated, near the 5′ end of the mRNA, specifies the N-terminal amino acid of the protein, because only the charged tRNA for that amino acid will bind to this codon on the mRNA. *(b)* Then the second amino-acyl tRNA binds to the second codon of the messenger. *(c)* When the first two amino-acyl tRNAs are in position on the messenger, enzymes of the ribosome make the peptide linkage between these two amino acids, removing the first amino acid from its tRNA. This dipeptide remains bound to the second tRNA, as a dipeptidyl-tRNA. The messenger then moves along to bring the third codon into position, and the ribosome repeats the process, adding the third amino acid to the dipeptide to make a tripeptide. The peptidyl-tRNA will grow one amino acid at a time until the polypeptide is complete.

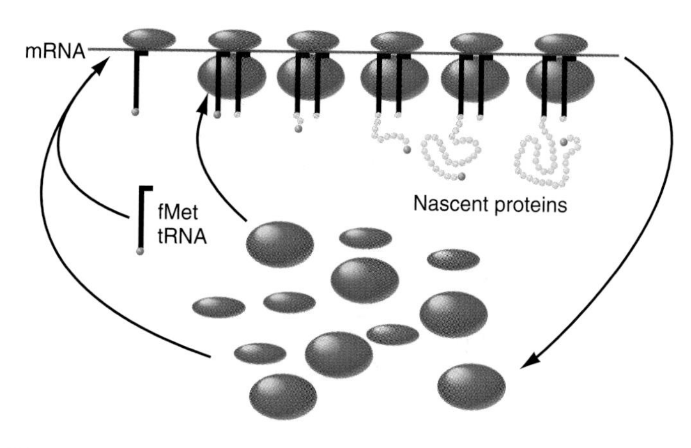

Figure 14.7

Several ribosomes translate an mRNA simultaneously, as a polyribosome. As a ribosome comes to the end of the messenger, it dissociates into subunits that recycle into new functional ribosomes.

Table 14.1 The RNA Code

First Base (5′-End)	Second Base: U	C	A	G
U	U UU Phe C A UU G	U C UC Ser A G	U UA Tyr C A Stop UA G Stop	U UG Cys C UGA Stop UGG Trp
C	Leu U C CU A G	U C CC Pro A G	U CA His C A CA Gln G	U C CG A G Arg
A	U AUC Ile A AUG Met	U C AC Thr A G	U AA Asn C A AA Lys G	A AG G U AG Ser C
G	U C GU Val A G	U C GC Ala A G	U GA Asp C A GA Glu G	U C GG Gly A G

In a specific context, the stop signal UGA encodes selenocysteine, a cysteine containing selenium instead of sulfur. The mRNA near this codon must have a particular sequence, which makes it form a looped structure, and allows a special tRNA for selenocysteine to bind to UGA. We will show in Chapter 48 that this obscure fact has important implications for the growth of HIV, the virus that causes AIDS.

Exercise 14.5 Using the standard code, translate the messenger made in Exercise 14.4. As you translate the molecule, do you find any ambiguity—in other words, a place where you are not sure which amino acid to insert? Although an amino acid may be specified by more than one codon, does one codon ever specify more than one amino acid?

Exercise 14.6 The most common mutations are *transitions*, in which an A–T pair is replaced by a G–C pair, or vice versa, with purine replacing purine and pyrimidine replacing pyrimidine. Which of the codons can most easily mutate into one of the stop signals? What would happen if such a mutation occurred somewhere in the middle of a gene?

Exercise 14.7 In the DNA shown in Exercise 14.4, a mutation changes the codon TCG (on the upper strand; AGC on the lower strand) so the T–A pair becomes a C–G pair. Write the sequence of the mutated messenger RNA and translate it.

Exercise 14.8 What is the difference between a terminator and a termination codon?

14.5 The mechanism of information transfer in cells restricts the process of evolution.

Nearly a century before Darwin proposed the idea of natural selection, Jean Baptiste de Lamarck had devised a different explanation of evolution. As we discussed in Chapter 2, he saw that animals become modified by their ways of life; for instance, they develop calluses over parts of the skin that are regularly irritated, and they develop strong muscles when they run to catch food or to escape becoming food. Lamarck therefore proposed that the harmony between organisms and their environments allows successive generations to inherit the acquired characteristics that allowed their parents to survive.

According to Crick's Law, however, information is only transferred from nucleic acid to protein, never in the reverse direction. So because there is simply no cellular mechanism for a reverse transfer, a mechanism of evolution like that proposed by Lamarck is simply not possible. An animal might, indeed, develop strong muscles or protective pads made of protein, but those environmentally determined changes cannot become encoded in its genes. At best, its offspring could have a propensity to develop the same structures.

Evolution, therefore, only occurs through Darwinian mechanisms (or some unknown mechanism compatible with Crick's Law). Random mutations in DNA produce changes in protein structure and therefore in patterns of growth, development, and behavior. Individuals whose mutations make them better adapted to their ways of living produce more offspring; thus those mutant genes appear in the population with greater frequency.

14.6 Transcription and translation are coupled in procaryotes but separated in eucaryotes.

Figure 14.8 compares the processes of transcription and translation in procaryotes and eucaryotes. Notice first that all cells synthesize three types of RNA—messenger, transfer, and ribosomal. This means we must broaden the concept of a gene; in addition to a DNA sequence that specifies the structure of a polypeptide, *the DNA sequences that specify stable RNAs are also genes.* These transfer-RNA and ribosomal-RNA genes are transcribed just like those that encode proteins, but their RNA transcripts are stable components of the cell and are not translated into proteins. Messenger RNA, on the other hand, is usually unstable.

In eucaryotes, transcription occurs on chromosomes confined to the nucleus, while ribosomes are in the cytoplasm, so messenger and transfer RNAs must be transported out of the nucleus to interact with ribosomes. In procaryotes, however, ribosomes are in direct contact with the DNA, and the processes of transcription and translation occur together. Figure 14.9 is an electron micrograph showing the simultaneous transcription and translation of a stretch of bacterial DNA; RNA polymerases are synthesizing mRNAs, which in turn are covered with ribosomes translating them into proteins. All these ribosomes, one after the other, are synthesizing exactly the same proteins as they move along the mRNAs. In bacteria, in fact, each mRNA generally carries the information for a series of genes that are transcribed together, so the ribosomes repeatedly translate the same series of proteins, one after another, from each transcript.

Now it is time for an instructive calculation, which you should perform before reading further:

Exercise 14.9 In *E. coli* at 37°C, mRNA is synthesized at a rate of about 55 nucleotides/sec, and proteins are synthesized at a rate of about 17 amino acids/sec. If a typical protein is made of 300 amino acids, how long does it take to transcribe its messenger and to synthesize one protein molecule?

This calculation should show that it takes less than 20 seconds for an RNA polymerase molecule to transcribe an average-sized gene and about the same time for a ribosome to translate the RNA into an average-sized protein. Each mRNA is translated several times, but soon after it has been completed and released from the DNA, ribonucleases start to attack it and degrade it to nucleotides. In bacteria, mRNA molecules turn over with a half-life of about 3 minutes, which means half the mRNA molecules that exist at any instant will be broken down to nucleotides 3 minutes later. In contrast, many eucaryotic cells make much more stable mRNAs with half-lives of several hours, which continue to be translated into vast numbers of identical protein molecules.

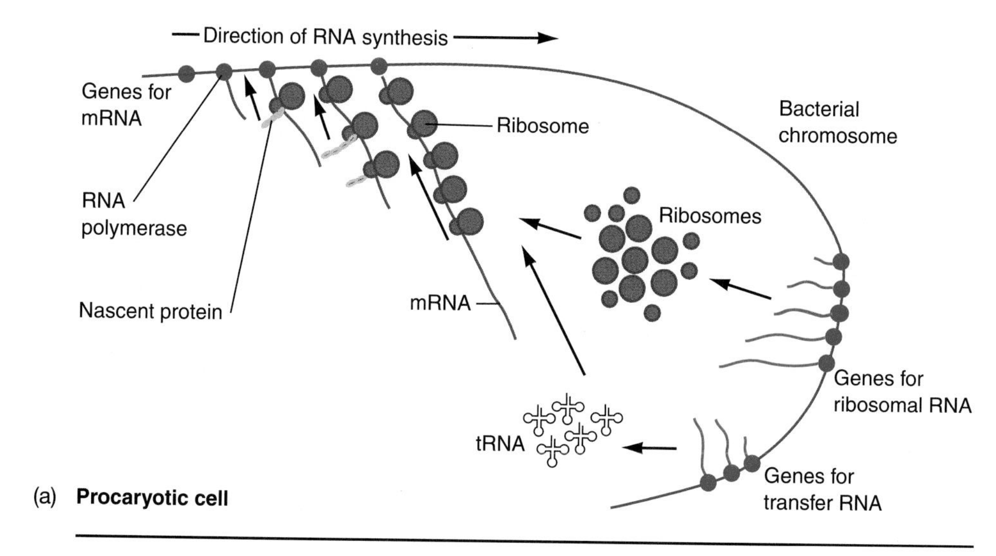

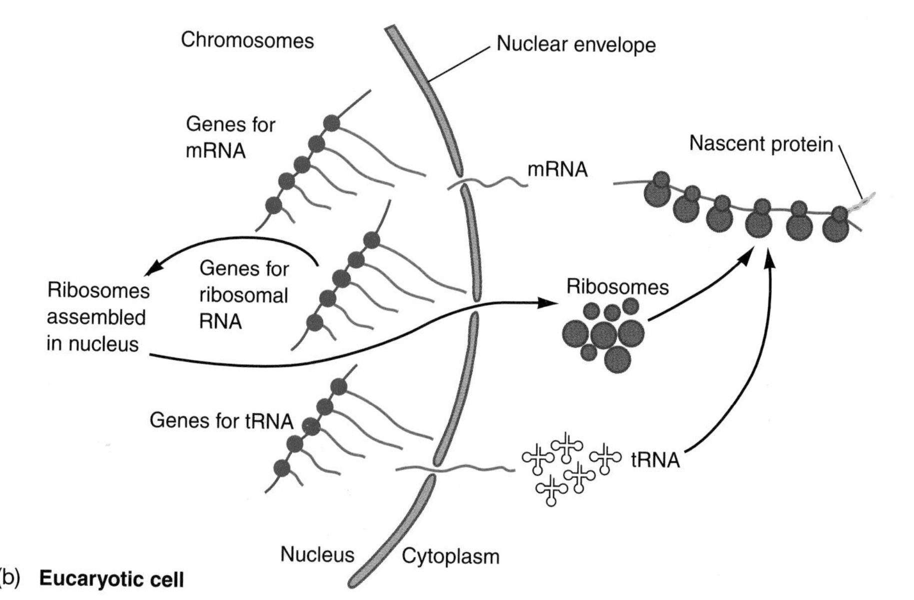

Figure 14.8

(a) In procaryotes, transcription and translation are coupled; ribosomes attach to an mRNA before it is completed and begin to translate it into protein. *(b)* In eucaryotes, transcription occurs in the nucleus; mRNA transcripts are transported into the cytoplasm, where they bind to ribosomes that translate them into protein. Notice that in both cases some genes are transcribed into stable tRNA and rRNA molecules, in addition to the genes that encode proteins through mRNA.

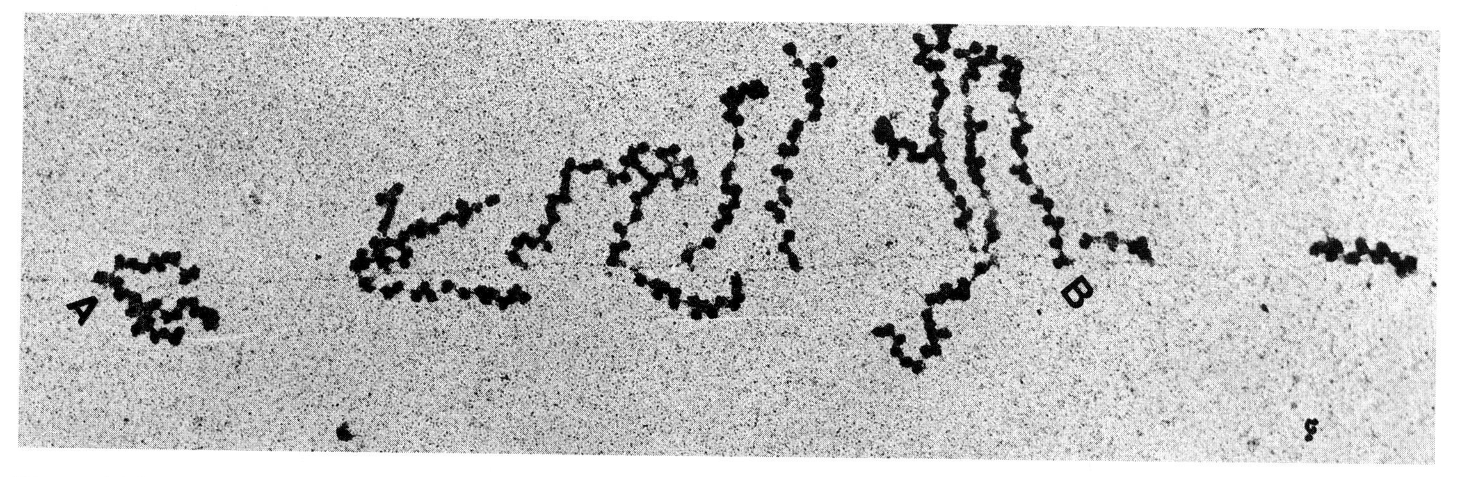

Figure 14.9

This electron micrograph shows simultaneous transcription and translation of a region of a bacterial genome. Several RNA polymerases are working along the same stretch of DNA, each synthesizing a messenger RNA. The nascent messengers are covered with ribosomes (large black particles), so are hardly visible, but judging their length by the number of attached ribosomes, we can see that the messengers are elongating as the RNA polymerases move along the DNA from point A to point B. The nascent proteins are too small to be seen.

Exercise 14.10 A bacterial mRNA encodes proteins A, B, and C, in that order. Ten ribosomes attach to this messenger, and the first one synthesizes protein A, then protein B, and then protein C. What do the other nine ribosomes synthesize?

14.7 Cells sort their proteins into compartments by using intrinsic protein structures.

The endoplasmic reticulum (ER), a system of interconnected membranes in the cytoplasm of eucaryotic cells, plays an important role in protein synthesis and in sorting proteins into cellular compartments. It is most prominent in cells that are making materials to be secreted, including enzymes or hormones that function outside the cell. The ER membranes separate the cytosol from the space inside the ER, the **lumen,** which is continuous with the space between the two membranes of the nuclear envelope (Figure 14.10). As mRNA molecules exit the nucleus through the pores in the nuclear envelope connecting the nucleoplasm with the cytosol, they encounter the ribosomes, where they are translated into proteins.

Cells, especially eucaryotic cells, are highly organized, and every protein must reach its proper functional location. Proteins made in eucaryotic cells can be destined for one of at least three places: the cytosol, a membrane-bounded organelle, or outside the cell. A long-time mystery of cell biology has been the sorting problem: How can a cell sort its proteins and send them to their proper destinations? The answer lies in mechanisms that recognize **signal peptides,** distinctive sequences in the proteins themselves, generally on their N-terminal ends where synthesis begins. (Some proteins have *signal patches,* distinctive formations that appear only after the complete protein has folded into its tertiary structure.) Proteins without a signal peptide remain in the cytosol where they are made. Proteins destined for the mitochondria or the nucleus are attracted by their distinctive signal peptides to sites on the mitochondrial or nuclear envelope, which transport them inside (Figure 14.11*a*).

Other proteins are sorted out during their synthesis on the ER membranes and subsequently in the Golgi membranes. As such a protein is made on a ribosome, proteins that recognize its signal peptide guide the ribosome complex to the ER (Figure 14.11*b*) where the nascent protein is then guided through the ER membrane *as it is synthesized.* Integral membrane proteins spontaneously assume their proper orientations, perhaps by weaving themselves back and forth through the membrane (Figure 14.11*c*). Proteins to be secreted or eventually moved into other cellular compartments wind up in the lumen of the ER, where enzymes remove the signal peptide and perhaps process the protein into its functional form. Many of these proteins are glycoproteins, and enzymes in the ER attach parts of their sugar chains, which may be used later to guide each protein to its proper place.

The pathway of protein secretion was worked out by George Palade and his associates, who studied the secretion of digestive enzymes from glandular (acinar) cells in the mammalian pancreas. Cells synthesize these enzymes in an inactive form called a **zymogen,** because active enzymes would digest one another and the cell containing them. (Zymogens are only activated in the environment of the intestine, where they digest food macromolecules.) Moving through the endomembrane system of the cytoplasm, zymogens are finally carried to the plasma membrane in small vesicles. Most other secreted materials use a similar pathway.

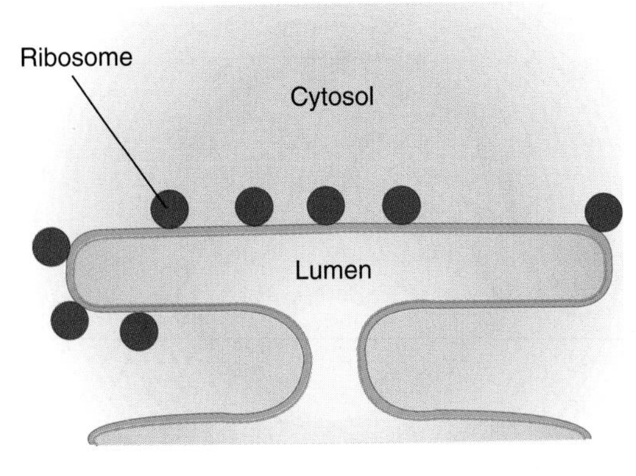

Figure 14.10
The endoplasmic reticulum (ER) separates the cytosol on one side, including ribosomes, from the lumen inside its folds.

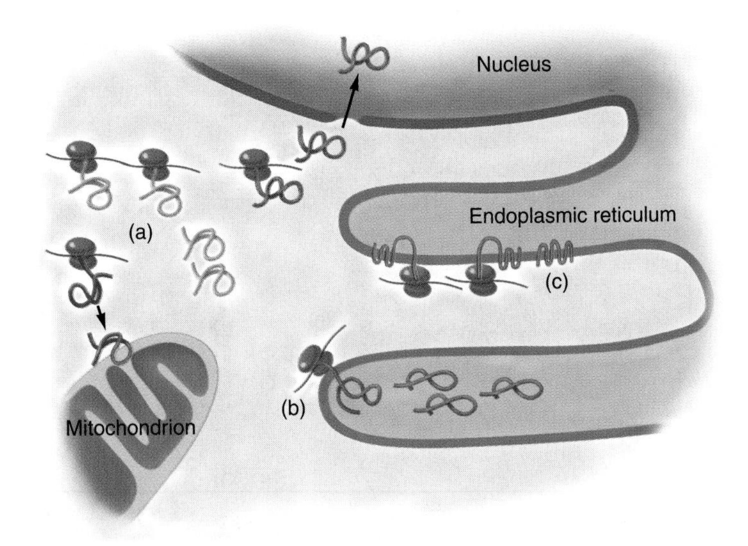

Figure 14.11
Proteins synthesized in the cytoplasm have different fates. *(a)* Proteins made on ribosomes in the cytosol either remain there or are guided by their signal peptides into other compartments, such as the mitochondria or nucleus. *(b)* Some proteins are made on the ER membranes; the N-terminus of the nascent protein is a signal peptide that guides the ribosome to the ER membrane, so it moves into the lumen as it is made. *(c)* Integral membrane proteins are made on the ER membranes and are guided into position by their amino acid sequences.

Anatomy of the pancreas and intestine, Section 47.6.

The experiments that revealed this pathway illustrate the **pulse-chase technique,** the biochemical equivalent of dropping a dye marker into flowing water and watching its movement. Cells are pulse-labeled by letting them metabolize for a very short time with a radioactive ("hot") compound. Then they are given an excess of the same type of compound in a nonradioactive ("cold") form to "chase" the unused hot compound out of the cells. This procedure leaves the cells with a small amount of material labeled with a tracer, which is located by taking samples at various times and analyzing them by a method that reveals radioactivity.

Palade pulse-labeled the cells in a slice of pancreatic tissue with radioactive amino acids, which the cells incorporate into the proteins they synthesize, mostly zymogen. With autoradiography and electron microscopy, he followed the fate of the labeled zymogen (Figure 14.12). A few minutes after labeling, the radioactivity appeared in the lumen of the ER, and in a few more minutes, in **transitional vesicles** located at the edge of the ER. These vesicles pinch off from ER membranes and move toward the Golgi complex, where the radioactivity appeared next—and where the story of secretion continues.

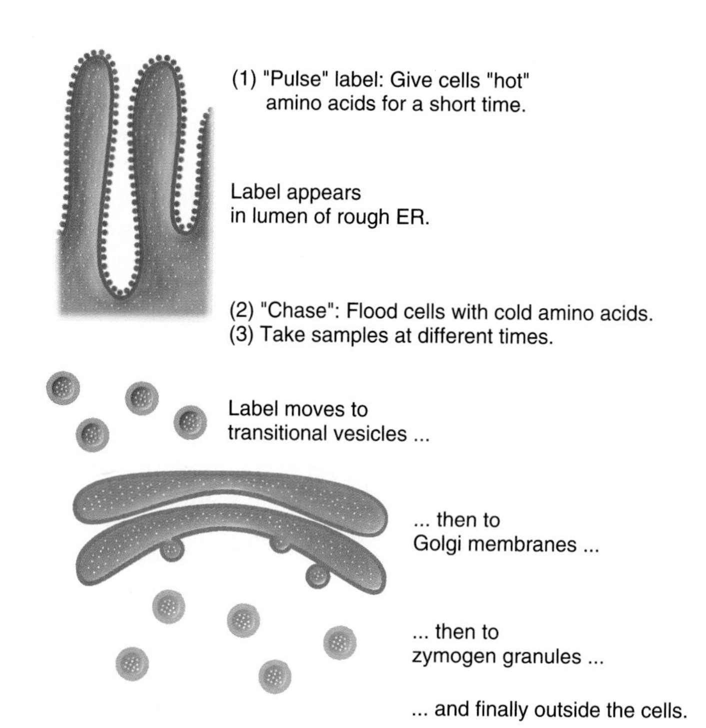

Figure 14.12
A pulse-chase experiment shows the fate of zymogen, enzyme precursors destined to be exported from the cell. The radioactivity appears first in the rough ER, then in transitional vesicles, and then in the Golgi membranes before it finally appears in zymogen granules near the cell surface.

14.8 The Golgi complex packages, sorts, and exports materials.

We are still following the fate of zymogen as it is processed in the endomembrane system. After radioactive zymogen has passed through the ER membranes, transitional vesicles carry it into the Golgi complex, a general center for packaging and sorting materials. As Figure 14.13 shows, the transitional vesicles merge into the *cis* face of the Golgi membranes, and other vesicles carrying processed zymogen emerge from the *trans* face. (The terms *cis* and *trans* are used in several contexts in chemistry and biology; they mean, roughly, "on the same side" and "on opposite sides," respectively.) Electron microscopy shows that zymogen in the vesicles that emerge from the *trans* face becomes more densely packed as the vesicles move toward the secreting surface of each pancreatic cell, where they stay until a signal triggers the cell to release them. Then the vesicles fuse with the plasma membrane and expel their contents by exocytosis.

Proteins passing through the Golgi complex are changed in various ways. Golgi enzymes add the bulk of the sugar chains to glycoproteins, and some of these sugars are distinctive molecular "tags" used to sort proteins to their final destinations, just as airport scanners read the bar codes on suitcase tags to direct them to the correct flights (we hope!). Hydrolytic enzymes destined for the lysosomes, where they digest materials imported into the cell, are tagged with the sugar mannose 6-phosphate. Receptors specific for this sugar reside in the membranes of the vesicles that transport these enzymes from the Golgi complex to the lysosomes (Figure 14.14). Through such specific interactions, cells organize their components so they can function properly.

Coated vesicles, Section 11.15.

As well as adding sugars, enzymes in the Golgi complex can change the structure of proteins themselves. Many proteins are

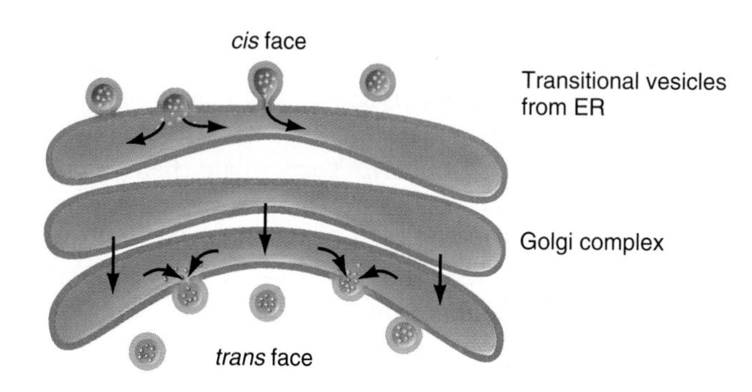

Figure 14.13
Materials move into the Golgi membranes as they fuse with vesicles from the endoplasmic reticulum on the *cis* face. The materials are processed in various ways in the Golgi membranes and then exported through other vesicles that form on the *trans* face.

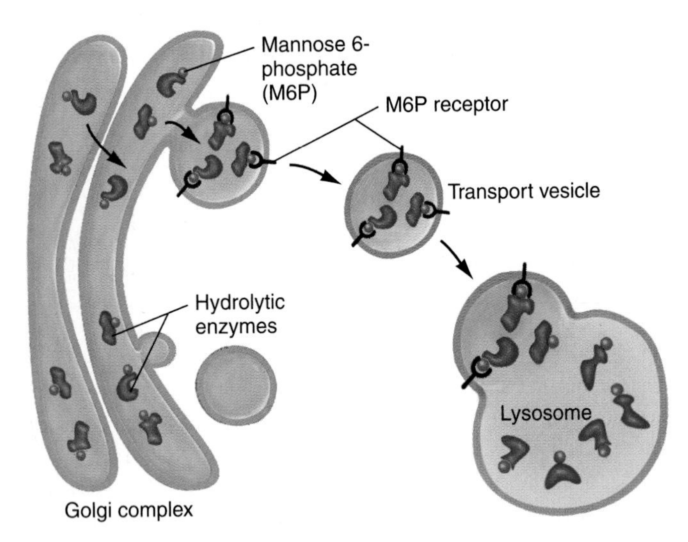

Figure 14.14
Enzymes destined for the lysosomes have the sugar mannose 6-phosphate attached. Receptors for this sugar in one type of coated vesicle are used to shuttle these enzymes to the lysosomes.

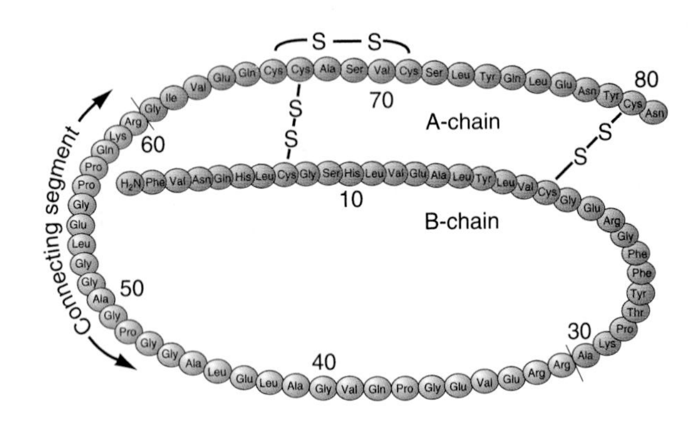

Figure 14.15
Insulin is made as a single polypeptide, which is a proprotein because it needs to be processed into a functional form. The internal C (connecting) peptide is removed to give the protein its final form of two separate chains joined by disulfide bridges.

first synthesized in a *proprotein* form, with extra peptide sequences not present in the final, functional molecule. Although the hormone insulin, for example, consists of two separate polypeptide chains, it is synthesized as a single chain, a *proinsulin* molecule (Figure 14.15). Converting enzymes in the Golgi complex then remove the internal C peptide from proinsulin, converting it into insulin. One or a few N-terminal residues are commonly removed from most proteins.

14.9 Secretion entails the synthesis and outward flow of membranes.

Figure 14.16 summarizes the pathway that is used for protein and nonprotein secretions alike. During secretion, membrane material itself continually moves outward from the ER, through

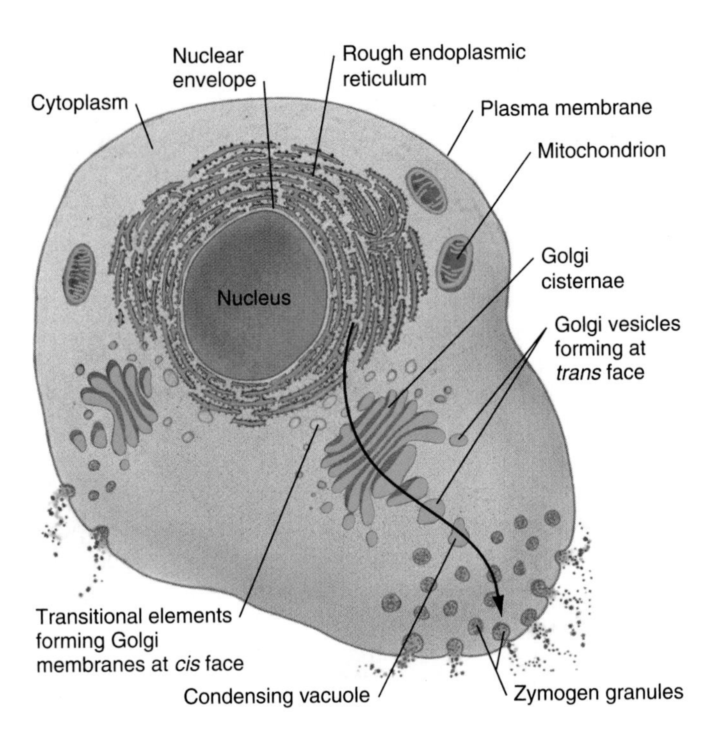

Figure 14.16
In the general pathway for synthesis of secreted proteins, the proteins pass from their place of synthesis in the rough ER through a series of membranous compartments to the cell surface, where they are secreted.

the Golgi complex, and into the plasma membrane. If a cell is not growing, excess membrane material would accumulate in the plasma membrane, were it not for the coated vesicles we discussed in Section 11.15. As these vesicles transport material from the cell surface into the cytoplasm, they are also recycling membrane components, thus keeping membrane synthesis in balance with other cellular processes. Membranes grow mostly in the ER where the enzymes of phospholipid synthesis are located and where their integral proteins are made.

Exercise 14.11 A group of cells that are synthesizing glycoproteins is divided into two batches. One is given a pulse label of the sugar galactose, and the other is pulse-labeled with another sugar, fucose. Both labels are chased out with cold sugars. Cells labeled with galactose show the label in the rough ER lumen and in the Golgi complex. Cells labeled with fucose show no label in the ER; it first appears on material in the Golgi complex. Both labels stay with the completed protein. Explain these results, and tell what they imply about the structure of the oligosaccharide chains on the protein.

14.10 Eucaryotic genes contain noncoding sequences.

The initial RNA transcripts in eucaryotic cells are much larger than necessary for coding purposes. Even though it takes only 1–2 kilobases of RNA to encode proteins of 30,000–60,000

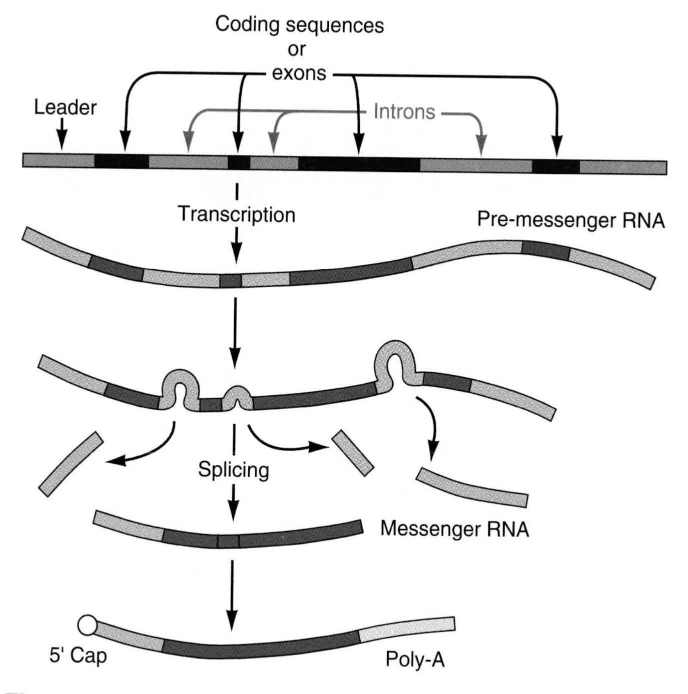

Figure 14.17

Most eucaryotic genes are interrupted by introns (intervening sequences) that do not encode the gene's protein product. The entire region is transcribed into a long piece of pre-messenger RNA. This is processed into the messenger by enzymes that remove the intron sequences and splice the coding sequences together. The 5′ cap and poly-A tail apparently protect the ends of the messenger from degrading enzymes.

daltons, typical RNA transcripts are 5–50 kilobases long. These initial transcripts, known as *pre-messenger RNA* or *heterogeneous nuclear RNA,* don't last long. They are soon reduced to much smaller mRNA molecules containing only the sequence that encodes the primary structure of a protein. How this happens was revealed using methods developed for quickly determining the sequences of DNA molecules (Methods 14.1). When the results finally emerged, they gave everyone an exciting surprise.

It is easy to translate a DNA or RNA sequence mentally. Knowing the genetic code, we can read along a nucleic acid sequence and write down the amino acid sequence it encodes. But when the sequences of eucaryotic genes were first determined, it became evident that they contain much more than the sequences that encode their proteins. Along with the DNA sequences that *do* encode protein structure, the **expressed sequences (exons),** eucaryotic genes generally contain inserted stretches called **intervening sequences (introns)** that do *not* encode protein. In the mouse gene for the hemoglobin β chain, for instance, the 432-bp sequence that encodes the amino acid sequence of the protein is interrupted by two introns of 116 bp and 646 bp. Other genes may be even more fragmented; a sequence of 500–1,000 bp that encodes a moderate-sized protein may be interrupted by several introns to make a total **transcription unit** (introns plus exons) up to 50,000 bp long. Such sequences are more intron than exon.

RNA polymerase transcribes a whole transcriptional unit, both introns and exons, into a pre-mRNA transcript (Figure 14.17). Other enzymes add an unusual nucleotide cap to the 5′ end of this transcript and a poly-A tail of 100–300 adenosine nucleotides to the 3′ end. Both additions protect the RNA against degradation, and the cap is required for translation. Special enzymes then clip out all the intervening sequences and splice the remaining coding sequences together into an mRNA with one uninterrupted sequence that encodes a protein.

Why are eucaryotic genes so complex? The intron-exon structure is probably valuable to eucaryotic cells for complicated reasons related to both development and evolution. In many cases, each exon encodes a single *domain* of a protein, which often has a specific function, such as binding to one kind of ligand (review Section 4.12). If exons from different genes are combined in new ways, they can rather quickly (on an evolutionary timescale) form new genes that encode proteins with novel functions. Thus the intron-exon organization may be preserved primarily for the genetic novelty it can create. This organization also allows organisms to use the information in DNA efficiently during development, as we show in the next section.

14.11 Coding sequences can be used to make more than one kind of protein.

In procaryotic cells, with their simple gene structure, a DNA sequence is transcribed and translated directly into a protein. The DNA and protein carry the same information. But eucaryotic cells, with their complex intron-exon gene structure and processing of the initial transcript, are able to use the information in one DNA sequence in alternative ways. This makes for efficient information storage, since one region of DNA can be used for two or more functions.

The muscle protein tropomyosin in *Drosophila* illustrates alternative splicing of the four exons in an RNA transcript. Adult muscle tissue uses all four exons (1-2-3-4), but in embryonic cells, the transcript is spliced to eliminate exon 3, so the protein is translated from an mRNA with the exon sequence 1-2-4 (Figure 14.18). In different tissues of a mammal, the same mechanism provides varying amounts of the enzyme α-amylase, which digests starch. There are two different 5′ exons, one used to make the mRNA in salivary glands and the other used in the liver. The salivary glands produce saliva containing large amounts of α-amylase, which is made in their cells about 100 times faster than in liver cells, apparently because of the difference in structure of the mRNAs.

The 3′-poly-A sequence can also be added at different points to produce different mRNAs (Figure 14.19). In the thyroid gland, mRNA with one set of exons produces the hormone calcitonin, which regulates calcium metabolism; in this instance, the poly-A tail is attached just after the sequence for most of the calcitonin molecule. But in certain nerve cells the

Methods 14.1 Determining DNA Sequences

Biologists are currently making great advances with methods for rapidly sequencing DNA molecules. Now we can easily determine sequences of hundreds and even thousands of nucleotides to see precisely how a genome is structured and to locate all its genes and control sites. In fact, it has become easier to read a DNA molecule and determine the sequence of amino acids it encodes than to determine the sequence of amino acids in the protein directly.

Two methods of DNA sequencing are currently in use. A technique devised by A. M. Maxam and Walter Gilbert uses reagents that selectively destroy the four kinds of nucleotides, thus creating DNA fragments of different lengths, each ending in a known nucleotide. Another method, invented by Frederick Sanger, uses DNA synthesis that terminates randomly at known nucleotides. We'll illustrate the Sanger method here for a long, single-stranded molecule (we must work with one strand at a time) containing the sequence –TGTCGAC–, so the complementary sequence –ACAGCTG– will be made when it is replicated. We will set up four tubes where replication will occur; note that the reaction mixture in each tube will contain millions of molecules of each kind. DNA replication must begin with a short primer DNA molecule. We use primers containing ^{32}P-labeled nucleotides, so we can detect the resulting DNA molecules, and their sequence is carefully chosen so they will bind near the region to be sequenced. Then, when these primers are elongated by DNA polymerase, the region to be sequenced will be replicated. DNA synthesis goes on in the four tubes simultaneously. Each tube has adequate amounts of dATP, dCTP, dGTP, and dTTP to support a lot of DNA synthesis, but many of the resulting molecules stop at different points because we add a small amount of a dideoxy-nucleotide (ddNTP). Since a ddNTP has no oxygen atom on the 3′-carbon of the sugar, it can only make one phosphate linkage to a polynucleotide, rather than two as ordinary nucleotides do. So when a dideoxy-nucleotide is incorporated into a chain (linked by its 5′-carbon), the chain ends there:

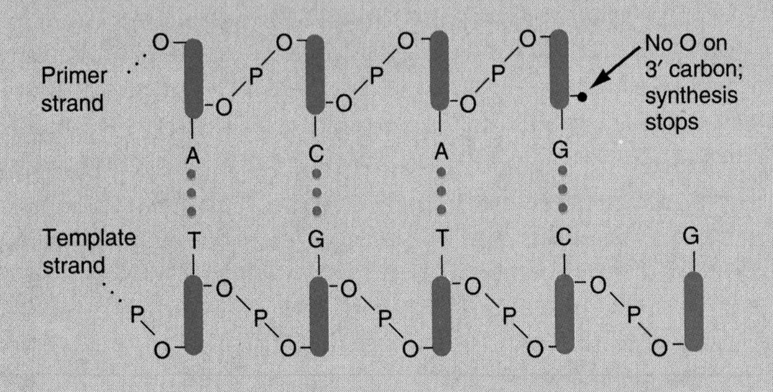

One of the four tubes has a small amount of ddTTP added, a second has ddCTP, the third ddATP, and the fourth ddGTP. To use the tube with ddGTP as an example, there is enough ordinary dGTP in the reaction mixture, so G is inserted almost everywhere it is needed, but each molecule will incorporate a ddG somewhere and terminate. Molecules will grow to different lengths, each one stopping randomly at a G, at two points in the sequence of our example. Synthesis in the other three tubes stops randomly at a T, C, or A. Each tube then contains labeled molecules that start at the same point and terminate at all possible points within the region of interest. We separate the molecules in each tube by size with gel electrophoresis, where the smallest travel fastest. Molecules of each size form bands that are revealed through autoradiography by laying the dried gel on X-ray film. Reading from the smallest molecules upward, the sequence of the DNA in the 5′ → 3′ direction can be read directly because each band contains molecules one nucleotide longer than the band just below it, and we can see which nucleotide is added to make the next longer molecule (see illustration).

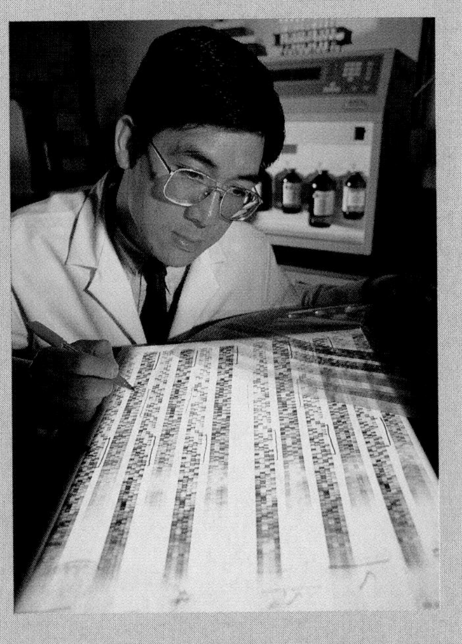

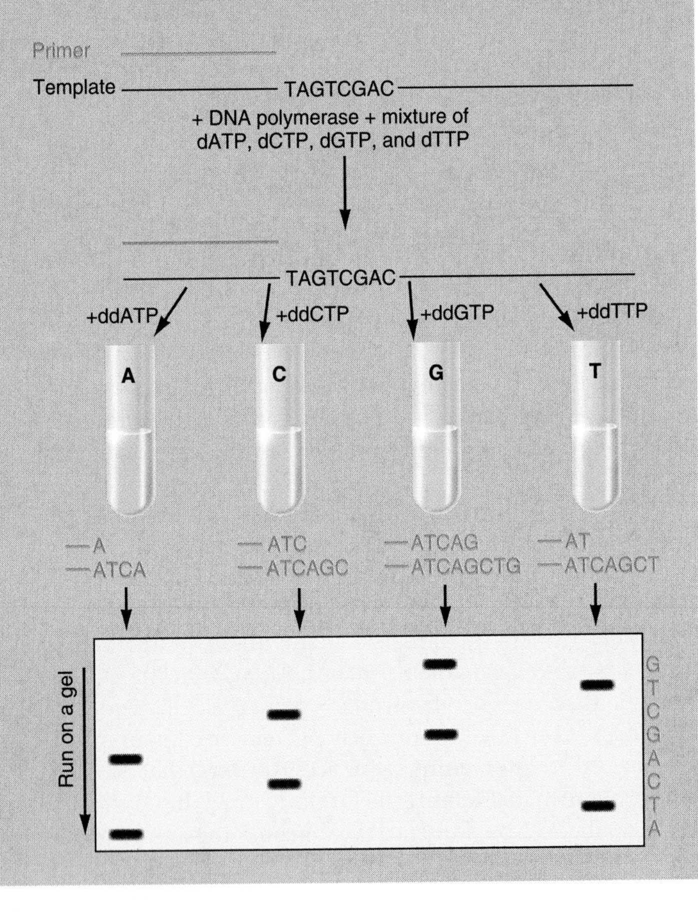

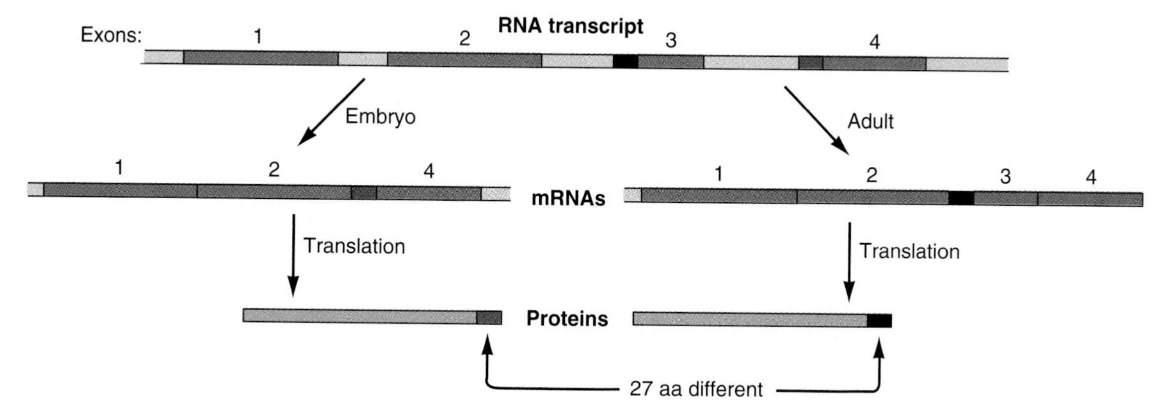

Figure 14.18

Alternative ways of splicing an RNA transcript can produce different proteins. The muscle protein tropomyosin in *Drosophila* can be made in embryonic cells from exons 1, 2, and 4; a similar protein is made in adult cells from all four exons.

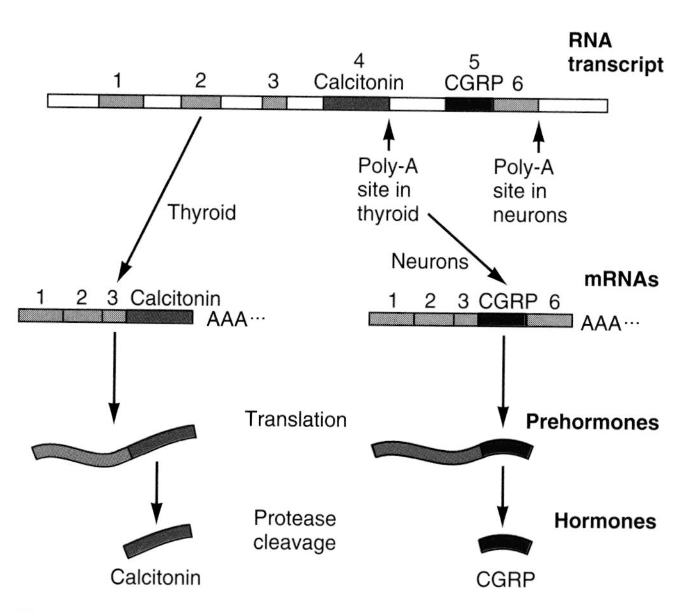

Figure 14.19

Information in a DNA sequence can be used in different ways if the poly-A sequence is added to the RNA transcript differently. A sequence is processed in thyroid glands to yield an mRNA for the hormone calcitonin; it is processed differently in some nerve cells to yield an mRNA for a different protein.

splicing occurs differently, and the poly-A tail is added further along, producing the mRNA for a calcitonin-gene-related peptide. Its function is unknown, although it may be involved in taste reception. Here again the same information is used in two ways for quite different functions.

14.12 The nucleolus is the site of ribosome assembly.

The prominent nucleolus of a eucaryotic cell is the site where the ribosomes are assembled. Nucleoli are dense aggregates of protein and RNA around a central DNA molecule, but with no surrounding membrane. They develop around the **nucleolar organizer** on chromosomes where the genes for ribosomal RNA (rRNA) are located. A typical cell has a pair of chromosomes with nucleolar organizers and therefore two nucleoli, but there is much variation, with some cells having many nucleoli. For example, amphibian oocytes, the cells that are about to become eggs, may contain thousands. These extra nucleoli allow them to manufacture many ribosomes quickly in preparation for the vast amount of protein they will synthesize if they become fertilized.

Remember that ribosomes and their components are often characterized by their sedimentation constant (S; see Figure 14.3). Eucaryotic ribosomes are large 80S particles, made of a 40S and a 60S subunit. Each subunit is composed of many protein molecules attached to a specific rRNA: the 40S subunit has an 18S RNA, and the 60S unit has 28S, 5S, and 5.8S rRNAs. The smaller ribosomes of procaryotes are 70S particles composed of 30S and 50S subunits with smaller rRNAs.

To study events in the nucleoli, Oscar L. Miller and Barbara R. Beatty spread out the nucleolar organizer DNA and took the beautiful electron micrograph in Figure 14.20*a*. It shows many rRNA transcription units, each being transcribed simultaneously by about 100 RNA polymerase molecules. A nucleolar organizer has 100–200 identical transcription units in tandem, each unit encoding the 18S, 5.8S, and 28S RNA molecules. Each RNA polymerase in the picture is transcribing a 45S precursor RNA, which grows longer as the polymerase moves along. Excess sequences are cut out of each 45S RNA molecule, leaving only functional rRNA molecules. Meanwhile, ribosomal proteins, made in the cytoplasm, are guided by their own signal peptides to the nucleus, where they associate with rRNA molecules in the nucleolus to make ribosomal subunits (Figure 14.20*b*). Then the completed subunits move back into the cytoplasm to begin their work of protein synthesis.

14.13 Mitochondria and chloroplasts contain their own genetic apparatus.

The whole eucaryotic genome is not carried in the nucleus; mitochondria and chloroplasts also contain portions of it. Each one carries its own internal genetic system, including 5 to 10 copies of a single, closed circle of DNA without histones, rather like a bacterial genome. Internal RNA polymerases transcribe

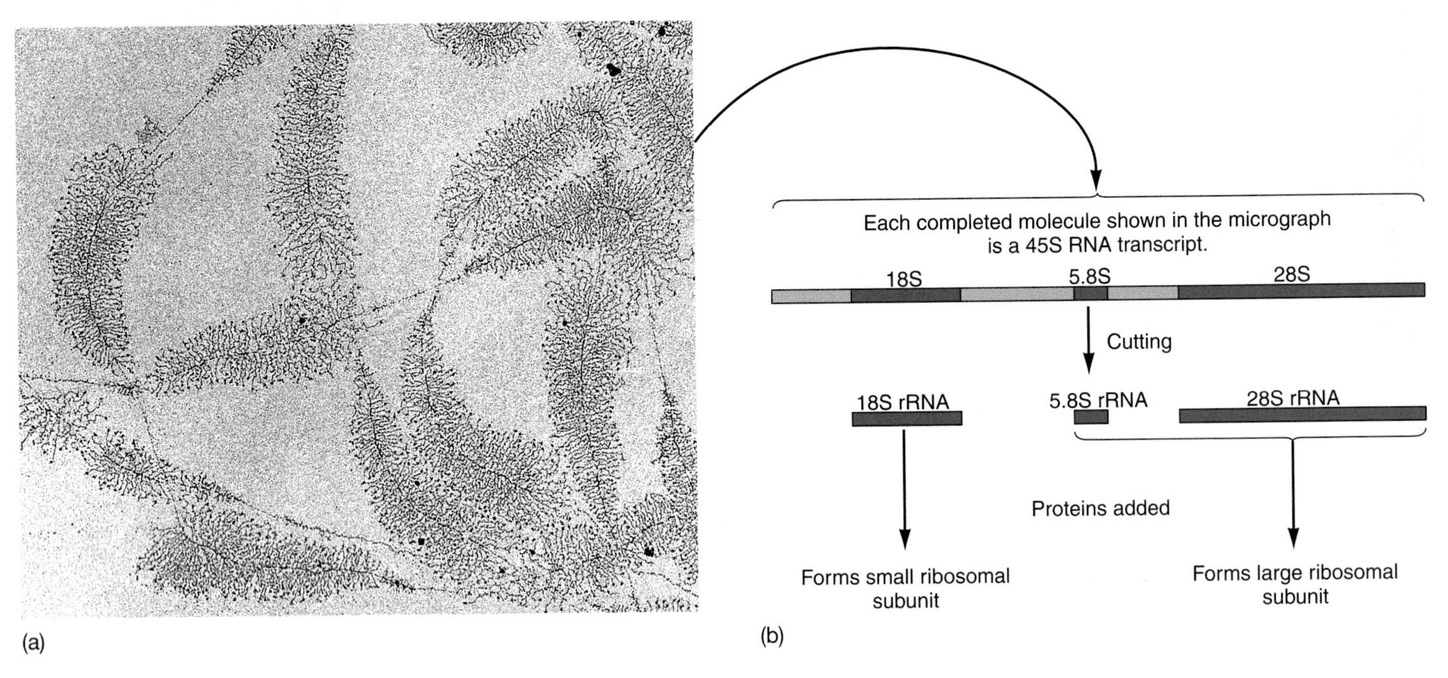

Figure 14.20

The nucleolar organizer region consists of many tandem transcription units, one next to the other, that are transcribed simultaneously by RNA polymerases. *(a)* Each transcription unit in this micrograph looks like a pine tree because transcripts of increasing length are being synthesized on it. *(b)* Each completed 45S RNA molecule is then cut into stable rRNA molecules, which are combined with proteins to make ribosomal subunits.

this DNA into mRNAs, which are translated within the mitochondrion or chloroplast by each organelle's own distinctive ribosomes and tRNAs. These semiautonomous organelles grow by expanding their membranes and dividing, and motion pictures of live cells show that mitochondria and chloroplasts continually change shape, fuse with one another, bud, and divide.

Frederick Sanger and his group determined the complete sequence of human mitochondrial DNA. Its 16,569 nucleotide pairs encode two types of rRNA, 22 tRNAs, and 13 proteins. The mitochondrion encodes only a few of its own proteins; the mitochondrial enzyme cytochrome oxidase is made of several polypeptide subunits, some encoded by mitochondrial genes and some by nuclear genes. Mitochondria use a genetic code that is very slightly different from the universal code in Table 14.1, and mitochondria employ different tRNAs.

Evidence indicates that the chromosomes of mitochondria and chloroplasts are remnants of larger genomes, and that both organelles arose in ancient times, as independent bacteria developed a symbiotic relationship with larger cells and started to live inside them. These ideas are explored further in Chapter 30.

14.14 Biological structures conserve their own patterns during growth and determine the form of new structures.

An organism's genome carries the information for constructing the whole organism. But now we must recognize that the patterns of some structures are actually carried in previously existing structures rather than being directly determined by the genome. For example, while the genome provides the information for membrane proteins and for the enzymes that make membrane lipids, it cannot specify the correct orientation of protein and lipids in a functioning membrane. If membrane components are isolated and then put back together, they form a symmetrical membrane that doesn't work properly. That is, its transport proteins are oriented symmetrically, so they shuffle molecules equally—and uselessly—in both directions, and some of its receptors for external ligands are also useless because they are internal. In a functioning cell, molecules properly orient themselves in the membrane because they are made and put in place by cellular structures such as ribosomes that insert proteins into the ER membranes. This leads us to an important generalization: *Biological structures conserve their intrinsic patterns during growth by determining the arrangement of new units as they are added.*

The information that resides in existing cellular structures, beyond the information in a genome, is called **epigenetic information,** and it is also essential for normal cell functioning. We showed earlier that each protein must be guided to the right place in a cell, where it can function properly, and this requires an intact ER and Golgi system, with the right proteins already in place in each membrane, ready to sort out the newly made proteins. Tracy Sonneborn and his colleagues demonstrated the importance of epigenetic information by using delicate microsurgical techniques to remove and reverse small sections of cilia from ciliated protists (Figure 14.21). When the cells divided,

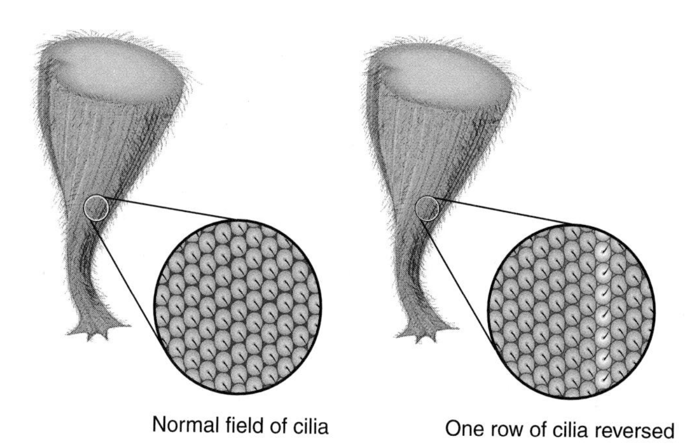

Figure 14.21

The surface of a ciliate is made of many regular rows of cilia. If one row is removed and turned around, all the cells derived from the altered cell retain the same altered pattern.

each daughter cell had the reversed row, and this feature persisted through many later cell divisions. This result means that the orientation of new cilia is determined by the existing pattern; in other words, the existing biological structure engenders more of the same structure.

Epigenetic information makes an enormous contribution to the proper development of an animal embryo, as we will show in Chapter 21. An egg is not just a random mixture of materials that begin to organize themselves properly under instructions from genes. In fact, the egg cytoplasm is highly organized, and it receives specific inputs of epigenetic information from surrounding maternal tissues as it develops.

As a general rule, molecules will crystallize rapidly into a regular structure if they are given a *nucleating agent* as a foundation. You can grow many beautiful crystals, for instance, by making a highly saturated solution of a salt and then adding a tiny nucleating crystal on which growth can begin. Bacterial flagella assemble in the same way. A solution of flagellin subunits condenses very slowly into flagellar rods, but if fragments of broken flagella are added as nucleating agents, condensation occurs explosively. Fragments of normal flagella will condense flagellin into normal rods, and fragments of "curly" flagella, from mutant bacteria, will condense flagellin subunits into the curly form. Again, we see that the form of an existing structure determines the form of newly synthesized structures.

The need for epigenetic information in cell organization complicates the *Jurassic Park* scenario—the movie in which a complete living dinosaur is made from an intact dinosaur genome. Even if one could find an intact dinosaur genome somewhere, it would have to be transcribed on the proper developmental schedule (a problem discussed in Chapter 21), and each protein would have to be sorted in the right way. The epigenetic information in functioning dinosaur cells—now obviously lost—may have been quite different from that in modern reptilian cells.

Coda The central theme running through this chapter is Crick's Law, which states that a sequence of nucleotides in DNA encodes a colinear sequence of amino acids in a protein. In addition, the whole chapter reflects three of the four central themes of this book. First, it gives substance to the theme that organisms are genetic systems. Second, because Crick's Law also specifies the direction in which information flows—from DNA to protein, with messenger RNA as an intermediary—it has profound implications for evolution, another of the book's themes. Natural selection, the process leading to evolution, now gains additional meaning. Random mutations in DNA change the structure of proteins, and these proteins in turn affect an organism's growth, behavior, and all the other characteristics that influence its survival and reproduction. Natural selection picks out the most favorable mutant types, those that most enhance survival and reproduction.

Another of this book's unifying themes is the principle that organisms function through molecular interactions. RNA synthesis, protein synthesis, and the other processes discussed in this chapter are examples of interacting molecules—three-dimensional shapes fitting into complementary three-dimensional shapes. We see this in the interactions between amino acids, the tRNAs that carry them, the enzymes that link the two, and the binding of tRNAs with mRNAs. We also see it in the proteins that sort newly made proteins by recognizing their signal peptides or certain sugars attached to them. Finally, this chapter has emphasized that all the information in an organism is not in its genome; existing structures engender more of the same, and this too comes about because of the way newly made molecules interact with the molecules already in place.

Summary

1. DNA specifies the structure of RNA, and RNA specifies the structure of protein. Each gene, which is the DNA sequence that encodes a single polypeptide, passes its information to a molecule of RNA.
2. RNA is transcribed on one strand of the DNA by the enzyme RNA polymerase. The transcript is made with each DNA base specifying its complementary RNA base by following the rule that A pairs with U (uracil) and G with C.
3. The RNA transcript is then translated into protein. Each set of three nucleotides forms a codon that specifies one amino acid, and information is transferred colinearly from DNA to RNA to protein.
4. The RNA transcripts that serve as templates for protein synthesis are actually unstable messenger RNAs (mRNA).
5. Cells also contain stable RNAs: ribosomal RNA (rRNA), which is a part of the ribosome structure, and transfer RNA (tRNA), which carries the amino acids for protein synthesis.
6. Each mRNA binds to a ribosome, or more generally to several ribosomes, where its information is translated into protein.
7. Transfer RNA adaptors carry amino acids to the mRNA templates. Each tRNA carries one kind of amino acid and binds to the mRNA-ribosome complex because the anticodon on one end binds to the codon on the mRNA. A second tRNA then binds to this complex, and the ribosome forms a peptide bond between the two amino acids. The mRNA moves through the ribosome, calling each amino-acyl tRNA, in turn, into the proper position; the amino acids are then polymerized, one by one, into a polypeptide.

8. The genetic code is systematic and redundant. The 64 triplets of bases code for amino acids in a systematic way. The first two bases, reading from 5′ to 3′, carry most of the information. Three triplets form stop signals at the ends of genes and one, AUG, is an initiation codon that, in eubacteria, specifies the special amino acid N-formylmethionine, which begins every polypeptide. Eucaryotic and archaebacterial polypeptides begin with the amino acid methionine.
9. The mechanism of information transfer restricts the process of evolution. Since information is transferred only from nucleic acid to protein, changes that occur in protein structures during the lifetime of an organism cannot be incorporated into the genome. Evolution must occur through the selection of accidentally occurring mutations that confer some advantage on the organism carrying them.
10. Transcription and translation are coupled in procaryotes but separated in eucaryotes. In eucaryotic cells, RNA is transcribed inside the nucleus and then transported to the cytoplasm, where proteins are made. But in procaryotes, the mRNAs are being translated by ribosomes while they are still incomplete and are being synthesized on the DNA.
11. Proteins reside in different places in cells. They are brought to their proper positions either because they are made there or by means of special features of their structures. Some proteins are synthesized by ribosomes in the cytoplasm, and they either remain there or are moved to such places as the mitochondria or the nucleus by signal peptides that are part of their structures.
12. Other proteins are made by ribosomes attached to the endoplasmic reticulum. These proteins are sorted into various compartments as they pass through the ER and Golgi membranes, largely by means of signal peptides on the proteins themselves. Thus proteins that are supposed to function outside the cell are exported in special vesicles, while those that are supposed to end up in lysosomes are carried there by other vesicles containing proteins that recognize their specific structures.
13. As materials are moved from the ER membranes to the outside of the cell, the components of the membranes involved are also being synthesized and moved outward. Thus membranes are constantly growing; in particular, more materials are always being added to the plasma membrane. Their components are probably recycled back into the cytoplasm by coated vesicles.
14. Eucaryotic genes have very complex structures, with expressed sequences (exons) interrupted by intervening sequences (introns). An entire transcription unit, containing both introns and exons, is transcribed into pre-messenger RNA. Then the introns are spliced out, and the remaining messenger RNA is exported from the nucleus and translated into protein.
15. A single DNA sequence can be used to encode two or more different proteins through differential splicing of exons and differential addition of the poly-A tail to a transcript.
16. The nucleolus is the site of ribosome assembly. The nucleolar organizer region contains many tandem units that are transcribed into long RNA molecules. Then the stable ribosomal RNAs are removed and combined with proteins to make ribosomes.
17. Mitochondria and chloroplasts contain their own genetic apparatus. They contain small, circular DNAs, rather like bacterial DNAs, which encode some proteins of these organelles, as well as special rRNAs and tRNAs.
18. Biological structures conserve their own patterns during growth and determine the form of new structures of the same kind. All the information for biological structure is not contained in a genome; some structure is determined by the existing structure, which makes newly added molecules assume their proper form and orientation in a cell.

Key Terms

transcription 274
translation 274
RNA polymerase 275
transcript 275
colinear 275
codon 275
promoter 276
terminator 276
ribosomal RNA (rRNA) 276
turn over 276
messenger RNA (mRNA) 276
transfer RNA (tRNA) 277
anticodon 277
polyribosome 278
termination codon 278
lumen 282
signal peptide 282
zymogen 282
pulse-chase technique 283
transitional vesicle 283
expressed sequence/exon 285
intervening sequence/intron 285
transcription unit 285
nucleolar organizer 287
epigenetic information 288

Multiple-Choice Questions

1. All varieties of RNA that function in protein synthesis are
   a. transcribed from a DNA template.
   b. stable and long-lived.
   c. able to bind with and form stable relationships with several polypeptides.
   d. double-stranded.
   e. possessed of a codon or anticodon.
2. Lamarckian theories of inheritance are refuted by the central dogma because
   a. information cannot flow from protein to DNA.
   b. phenotypic change does not generally cause mutation within gametes.
   c. a specific environmental change does not cause a specific mutation.
   d. *a* and *b*, but not *c*
   e. *a*, *b*, and *c*
3. Identify the polypeptide that would be produced as a result of transcribing and translating the following single strand of DNA:
   3′ T–A–C–T–C–C–A–C–C–T–T–T–C–A–G 5′
   a. val – gly – gly – lys – leu
   b. leu – lys – gly – gly – val
   c. met – leu – lys – gly – val
   d. met – arg – trp – lys – val
   e. val – arg – lys – try – met
4. What would change in the polypeptide if the strand of DNA used in the previous problem underwent a mutation in which the third C from the left changed to T, as follows:
   3′ T–A–C–T–C–T–A–C–C–T–T–T–C–A–G 5′
   a. There would be no change.
   b. The polypeptide would be one amino acid shorter.
   c. The polypeptide would be one amino acid longer.
   d. One amino acid would substitute for another.
   e. Translation would terminate within this stretch.
5. Using the strand of DNA shown in the previous problem, what would be the effect of a mutation that resulted in a change of the second C from the right to a T, as follows:
   3′ T–A–C–T–C–T–A–C–T–C–A–G 5′
   a. There would be no change.
   b. The polypeptide would be one amino acid shorter.
   c. The polypeptide would be one amino acid longer.
   d. One amino acid would substitute for another.
   e. Translation would terminate in this stretch.
6. Which molecule or part of a molecule is properly matched to its function?
   a. DNA polymerase—transcription

b. RNA polymerase—translation
c. amino-acyl tRNA synthetase—DNA replication
d. promoter—RNA decay
e. None of the above pairs are properly matched.

7. The wobble mechanism results in fewer different ______ than ______.
   a. anticodons; codons
   b. codons; anticodons
   c. amino acids; codons
   d. amino acids; anticodons
   e. anticodons; amino acids
8. A protein that lacks a signal peptide sequence will probably reside in the
   a. mitochondria.
   b. Golgi complex.
   c. cytosol.
   d. nuclear membrane.
   e. plasma membrane.
9. In order for a length of eucaryotic mRNA to be translated, the pre-mRNA transcript is modified in several ways, including
   a. intron removal.
   b. addition of a cap on the 5′ end.
   c. addition of a poly-A tail to the 3′ end.
   d. *a* and *b*, but not *c*.
   e. *a*, *b*, and *c*.
10. Which is *not* a difference between functional eucaryotic rRNA and mRNA?
   a. Both are produced in the nucleus, but only mRNA moves into the cytosol.
   b. There are many copies of the genes for rRNA, but not of the genes for each mRNA.
   c. rRNA genes are transcribed but not translated; genes for mRNA are both transcribed and translated.
   d. mRNA, but not rRNA, is modified by the addition of a cap and a tail.
   e. There are many more varieties of mRNA than rRNA.

True-False Questions

Mark each statement true or false, and if false, restate it to make it true.

1. To qualify as a gene, a region of DNA must be both transcribed and translated.
2. The promoter precisely aligns the ribosome on mRNA during translation.
3. Amino-acyl tRNA synthetase joins mRNA to tRNA, liberating pyrophosphate in the process.
4. Of the 64 possible codons, 60 code for amino acids, 1 is an initiator, and 3 are termination codons that signal the end of a gene.
5. Base substitutions in the first position of a codon generally result in an altered polypeptide, whereas substitutions in the third position often do not.
6. If one knows the sequence of amino acids in a polypeptide, it is easy to deduce the order of bases in DNA.
7. If one knows the linear sequence of bases in a stretch of active DNA, it is easy to deduce the order of bases in the anticodons that will subsequently be aligned during translation.
8. The wobble hypothesis states that one mRNA anticodon may base-pair with several tRNA anticodons.
9. A secretory protein will appear in these structures or regions in the following order: lumen of ER, transitional vesicles, *cis* face of Golgi, *trans* face of Golgi, plasma membrane.
10. In eucaryotes, transcriptional units of DNA are produced by removing the introns and splicing the exons together.

Concept Questions

1. What does the word *colinear* mean, and what is its significance in protein synthesis?
2. Relate the concepts of a redundant code and wobble.
3. In pulse-chase experiments (see Section 14.7), an intact cell is first given a radioactive metabolite, followed immediately by a much larger amount of a nonradioactive *chaser.* Explain why *(a)* the cell must be intact; *(b)* the time between the pulse and the chase must be brief; and *(c)* the amount of nonradioactive chaser greatly exceeds that of the radioisotope.
4. A quantity of a short length of single-stranded DNA is divided among four test tubes. You are given the physical equipment to sequence the DNA by the Sanger method and are instructed to list the chemical compounds you will need, indicating which will be used in all of the test tubes and which will be used only in an individual tube. Prepare your list.
5. Distinguish between genetic and epigenetic information.

Additional Reading

Darnell, James E. "The Processing of RNA." *Scientific American,* October 1983, p. 90. Further information about RNA processing in eucaryotic cells.

Grivell, Leslie A. "Mitochondrial DNA." *Scientific American,* March 1983, p. 78. The organization of mitochondrial genes.

Rothman, James E., and Felix T. Wieland. "Protein Sorting by Transport Vesicles." *Science,* April 1996, p. 227. Recent advances in understanding the machinery of vesicle transport have established general principles that underlie a broad variety of physiological processes, including cell surface growth, the biogenesis of distinct intracellular organelles, endocytosis, and the controlled release of hormones and neurotransmitters.

Internet Resource

To further explore the content of this chapter, log on to the web site at:

http://www.mhhe.com/biosci/genbio/guttman/

15 Cycles of Growth and Reproduction

A family of Summer Tanagers is engaged in extending the life cycle of their species for another generation.

Key Concepts

15.1 The lives of organisms are characterized by rhythms of growth and activity.

15.2 It is an advantage for an organism to be a sexual diploid.

15.3 Organisms exhibit three kinds of sexual cycles.

15.4 Meiosis divides a diploid cell into haploid cells.

15.5 Spermatogenesis and oogenesis entail different patterns of cell division.

15.6 A cycle of growth implies a cycle of gene regulation.

15.7 Some life cycles comprise distinct morphological phases.

John Hammond lay in his bed in the July heat, covered with blankets and shivering as if he were in an arctic blizzard. In his blood, millions of minute parasites were breaking out of his red cells, causing his temperature to drop, wracking his body with a phase of malaria. Yet he knew that, in only a few hours, he would be reacting in just the opposite way, burning with fever as the parasites entered another phase of their life cycle.

Hammond and his wife were in the midst of their own life cycles, and downstairs their two small children, oblivious to their father's misery, played with their cat and her four kittens. Above the kitchen sink, several pots of flowers bloomed and rested periodically. So in this one household, many life cycles of growth and reproduction were being played out at once. How many others were hidden? Were mice reproducing in the walls, and minute insects multiplying in the backs of the cupboards? The refrigerator probably harbored some mold growing on food, and many microorganisms flourished in the soil of the flowerpots. Everywhere, bacteria were continually reproducing, perhaps to be washed away or flushed away, in a never-ending chronicle.

Malaria is now generally preventable and controllable, in spite of the spread of drug-resistant strains of the parasite, but John Hammond's story still emphasizes that organisms live their lives in cycles. Having explored the cycle of cell growth and division, we are now ready to develop the concept of a sexual cycle of reproduction, especially as it occurs in plants and animals. An understanding of this cycle will make it easy to understand in Chapter 16 how genes are inherited in simple *patterns* in all plants, animals, and fungi. Thinking in terms of cycles also reminds us that, instead of remaining static, unchanging structures, organisms continually change form as they progress through their lives. This viewpoint is basic to a proper ecological and evolutionary understanding of biology, because evolution has shaped each species to go through a series of life stages that may bear little resemblance to one another and may involve quite different ecological relationships at different times. Life cycles are often closely tied to the natural cycles of the earth—the daily cycle of light and dark, the tides, the annual cycle of seasons—and this is where we begin. ■

15.1 The lives of organisms are characterized by rhythms of growth and activity.

Around the coral islands of the South Pacific, at dawn one week after the November full moon, the water suddenly begins to look like noodle soup. The palolo worms are breeding. Hidden in their coral crevices beneath the surface, the worms have been growing for a year, developing sperm and eggs in their tails (Figure 15.1). And now, all at once, these tails break off and rise to the surface, where they soon release their sex cells and turn the water milky. The head ends remain below, ready to repeat the cycle exactly a year later. You could set your calendar by the palolos, and Ralph Buchsbaum has written about them:

> *The natives of the Samoan and other islands are familiar with the habits of the palolos. They consider them a great delicacy and look forward to their breeding season. When the day arrives, they scoop them up in buckets and prepare a great feast, gorging themselves just as we do on Thanksgiving day, knowing that there will not be another treat like it until exactly the same day of the next year. Actually, there is a small "crop" of swarming palolos a week after the October full moon, but it is too small to interest the natives.*

All organisms are creatures of time. Their lives are series of events, often cycles of events. Rocks can roll downhill whenever they get loose, winds can always blow, but birds would not last long if they migrated south before they had raised their summer broods, and a farmer cannot reap his harvest before he has sown the seeds and tended his crop. We all go through cycles in our lives—some too subtle to notice and others so obvious we take them for granted. Time is important to us, and we adjust our lives to it.

The palolos' yearly cycle is adjusted to the relentless movements of the earth and moon and the resulting cycles of seasons and tides. The annual breeding cycles of plants and animals—the blooming of spring and summer flowers, the pairing of males and females—are adaptations to such cycles (Figure 15.2). These intricate patterns of growth and behavior must be triggered by built-in mechanisms that somehow count the tides or perceive other signals such as changes in temperature and day length.

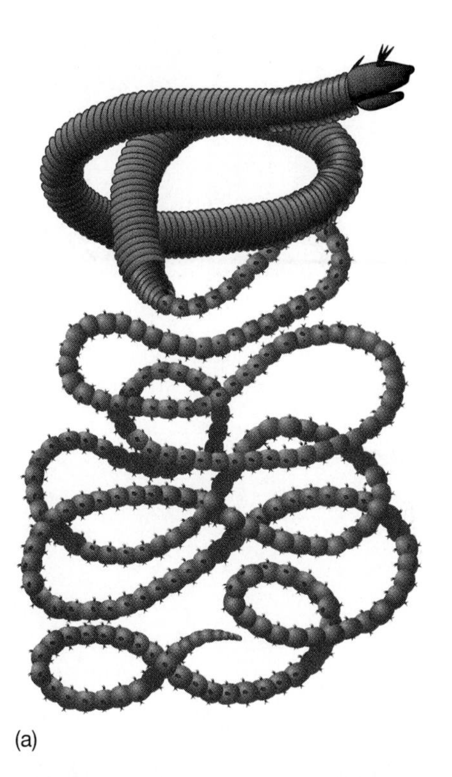

(a)

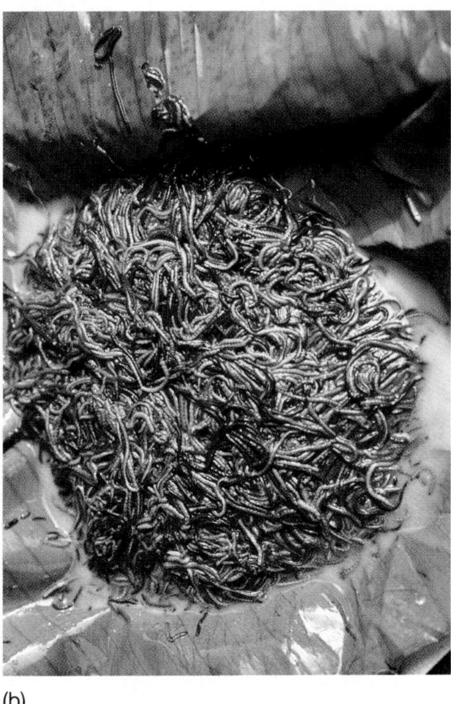

(b)

Figure 15.1

(a) A palolo worm *(Eunice viridis)* has a short, thick body with a long, segmented tail containing the reproductive organs. *(b)* A leafy basket of worms ready to serve at a Samoan feast.

Figure 15.2

Plants and animals often renew their reproductive activities in the spring. The warmth and increasing day length stimulate new hormonal activities; plants start to grow or flower, while animals commonly begin to mate.

Figure 15.3

Grunions are small, silvery fish that mate on a regular tidal cycle on beaches along the California coast.

Figure 15.4

Male fiddler crabs *(Uca)* are distinguished by one large claw, or chela, which can inflict a nasty pinch. The crab's body, just behind its chela, has two eyestalks containing tiny glands that are partially responsible for a regular daily cycle of color change.

If you live near the California coast, you can watch another remarkable cycle in action any time between March and August, precisely at high tide on the second, third, and fourth nights after the highest tides of each lunar month, when silvery little fish called grunions come onto the beach by the thousands to mate. Between waves, the females wriggle tail-first into the soft sand just above the water line and deposit their eggs (Figure 15.3). The nearest males then fertilize the eggs by covering them with milt, their sperm-bearing fluid. The eggs develop in their protected place under the sand and are washed out into the ocean to mature about two weeks later, during the next cycle of high tides.

Another kind of cycle correlated with lunar or tidal changes is the regular estrus cycle of many female mammals. Similarly, the brown alga *Dictyola* releases its sperm and eggs every 14–15 days, with the tides. In some cases, we know about the mechanisms behind these cycles. The human menstrual cycle, for instance, is controlled by several hormonal signals exchanged between the ovaries and the brain, and the length of the cycle depends largely on the time it takes for an egg to grow and for the surrounding cells to produce their hormones.

Some of the most obvious and fascinating cycles in nature are correlated with the 24-hour day. Many plants open and close their leaves and flowers in synchrony with the day and night hours, leading Darwin to suggest this as a mechanism for conserving heat. Along many Atlantic coast beaches and mud flats roam bands of little fiddler crabs, the males waving great, threatening claws at one another (Figure 15.4). These crabs turn dark in the middle of the day and light at night, as pigment cells in their skin expand and contract. The change is controlled by hormones released from nerve cells, especially from one tiny blue pearl of a gland in their eyestalks. Even if the crabs are kept in total darkness, this cycle continues indefinitely.

All animals wake and sleep on a regular daily schedule. As these words are being written, a Song Sparrow who comes at the same time every morning is foraging for its food under the apple tree outside—a fact that says something about the daily cycle of this writer as well. Rats and mice also show a 24-hour cycle of activity. However, if they are kept in constant light, their cycles continue but become only *approximately* 24 hours in length and so are called **circadian rhythms** (*circa-* = approximately; *dia* = day). A rat living in constant light may show a peak of activity, say, every $25\frac{1}{2}$ hours, so the peak comes later every day. But if the rat is then given a regular light signal every 24 hours, such as at sunrise, its natural rhythm adjusts to coincide with the day; the circadian rhythm is thus said to be *entrained* by the light cycle.

Many rhythms are created by internal hormonal and nervous mechanisms. Yet, despite years of research on these so-called biological clocks, their fundamental mechanisms are just barely coming to light. One theoretical model suggests that the information-transfer system from DNA to RNA to enzyme to metabolite, with all its feedback controls, is intrinsically cyclical and could be the basis for internal clocks. But the clock mechanism must be unusual, because it is affected very little by temperature, whereas typical metabolic processes double in rate for every 10°C increase. And, whatever the mechanism is, the internal clock can be adjusted and correlated with the larger external rhythms of the world.

The events of reproduction should be seen against the background of cyclic activities by which all organisms measure out their lives. Organisms don't stay the same. They move on from stage to stage, and many events only occur at certain times in an individual's life history.

15.2 It is an advantage for an organism to be a sexual diploid.

Simple organisms like bacteria only grow clonally—that is, by repeated cell divisions—and they can only evolve slowly. Evolution depends on mutation and on rearrangements of genes in new combinations, but mutation rates (the probability per generation that a certain kind of mutation will occur) are low. True, we find a lot of mutants in a population of bacteria, but only because we can grow such enormous numbers of bacteria that mutants inevitably occur among them. Some simple changes, such as becoming resistant to an antibiotic, may require only one mutation, but evolution generally requires a series of mutations and rearrangements of the genome, one change adding to another by chance until they have made new structures and functions. Suppose an organism carries genes *X, Y,* and *Z;* then suppose it would gain a valuable new function if it acquired mutant forms of these genes, *x, y,* and *z.* If the organism only grows clonally, one cell may acquire mutation *x,* but it may take a long time for one of its mutant progeny to acquire a second advantageous mutation *(y)* to go with the first one, and still more time for a cell with *x* and *y* to acquire *z.* (Meanwhile, it may be a disadvantage to have one of these mutations without the others.) So it takes many generations for strictly clonal organisms to change much, although each generation time is very short; in fact, bacteria survive and evolve by force of numbers because they reproduce so rapidly. (Some pseudosexual processes in bacteria, discussed in Chapter 17, do allow them to transfer and combine their genes, but irregularly and in limited ways.)

Suppose we wanted to design an organism with reproductive and genetic advantages over all clonal organisms. We would give it two features. First, to guard against deleterious mutations, it should have two copies of every gene, so if one is inactivated by mutation, it still has the other. Second, in every generation we would arrange for two organisms to pool and mix copies of their genomes, to make novel combinations of genes. Then if each one has a potentially valuable feature that the other lacks (for instance, one carries *x* and the other carries *y*), the two might combine their good points to make a better organism. Organisms with these characteristics obviously have evolutionary advantages over their simpler cousins.

As a matter of fact, humans are the kind of "ideal" organism we have just designed, as are almost all plants, animals, fungi, and many microorganisms. The first favorable feature, having two copies of each gene, is *diploidy.* A cell that has only one copy of its genome is **haploid** (*haplos* = simple), and one with two copies is **diploid** (*diplos* = double). The second feature is *sex.* In every generation, sexual organisms can combine sperm and eggs (or the equivalents) to make offspring with new combinations of genes. Because diploid, sexual species are so much more adaptable than haploid, asexual species, they have evolved—more rapidly than asexual organisms could—into a great variety of forms that can live in a wide range of conditions. Most species in each major branch of the eucaryotic family tree are sexual and have prominent diploid stages.

Sexual and asexual organisms also live on different time scales. In addition to the short growth cycles of their individual cells, sexual organisms measure out their lives by the longer time span of a **sexual cycle** (Figure 15.5). The sexual cycle alternates between haploid and diploid phases—haplophase and diplophase. Diploid cells form when two haploid cells called **gametes** (typically an egg and a sperm) fuse with each other in the process of **fertilization** or **syngamy** (*syn-* = together, *-gamy* = marriage). Haploid cells form again when a diploid cell goes through **meiosis,** a type of nuclear division that reduces the number of chromosomes by half. This kind of division is essential in a sexual cycle, because if fertilization continued generation after generation without a compensating division to reduce the diploid set back to a haploid set, the number of chromosomes would keep doubling every generation, and this obviously couldn't go on for long.

The question has been asked, "If sexual reproduction is so wonderful, why haven't all the organisms on Earth become sexual?" This is the question we warned against in Concepts 2.2. Organisms can only evolve to the extent allowed by unpredictable, random events. There is no Cosmic Warehouse to which organisms can send orders in time of need: "Have noticed that certain organisms have become sexual and seem to be doing very well.

Figure 15.5

A general sexual cycle is an alternation between haploid and diploid phases. It is drawn here to simulate reality by showing that the cycle is really a spiral through time; a diploid organism, such as an adult human, does not turn back into haploid cells, but simply produces another generation of haploid cells.

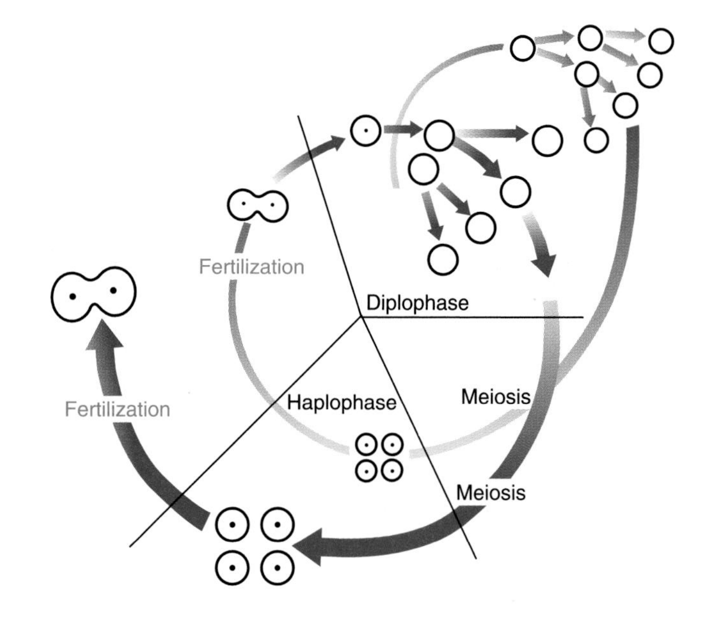

Figure 15.6

Humans are typical of organisms in which the diplophase dominates. Each sex makes haploid gametes, which fuse to begin a new diploid generation.

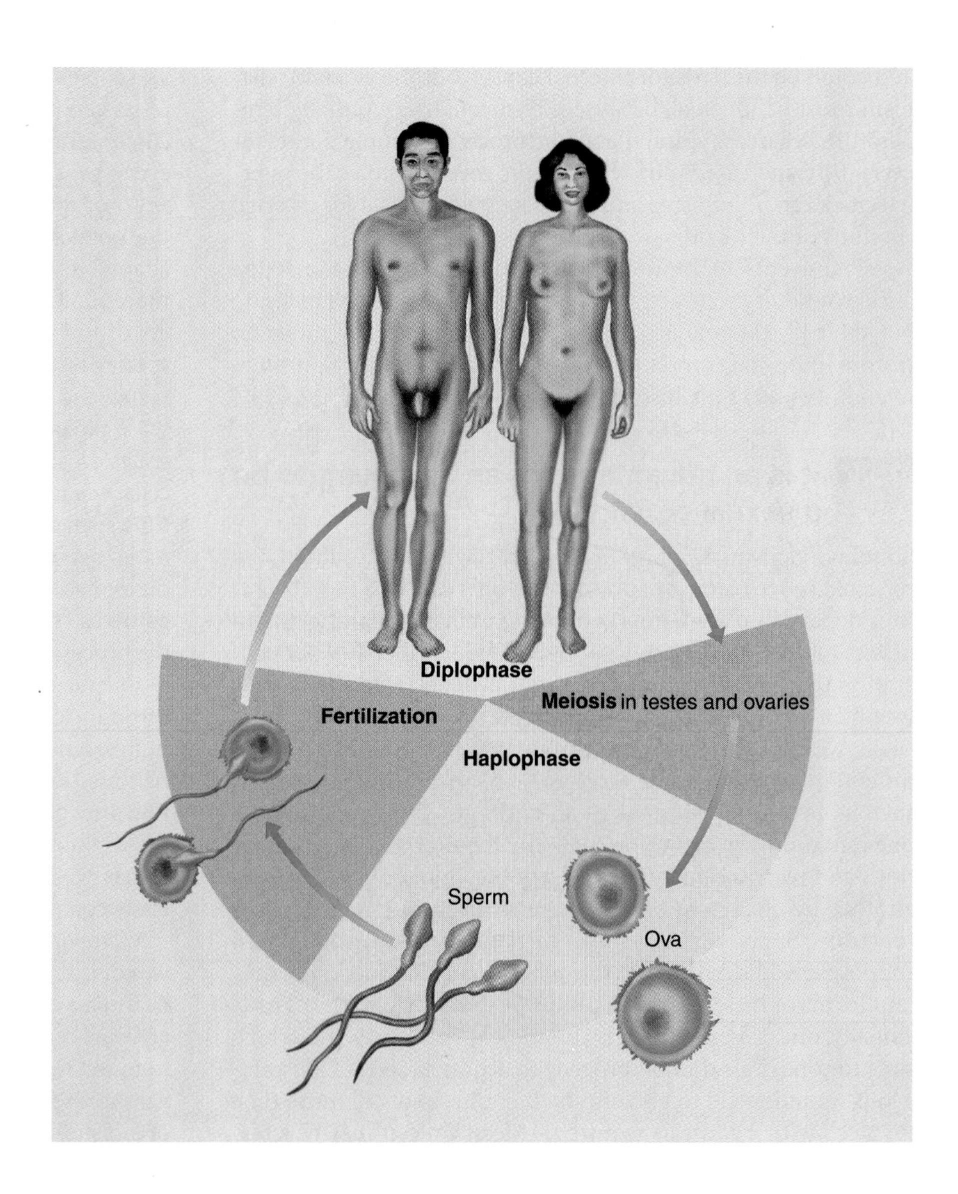

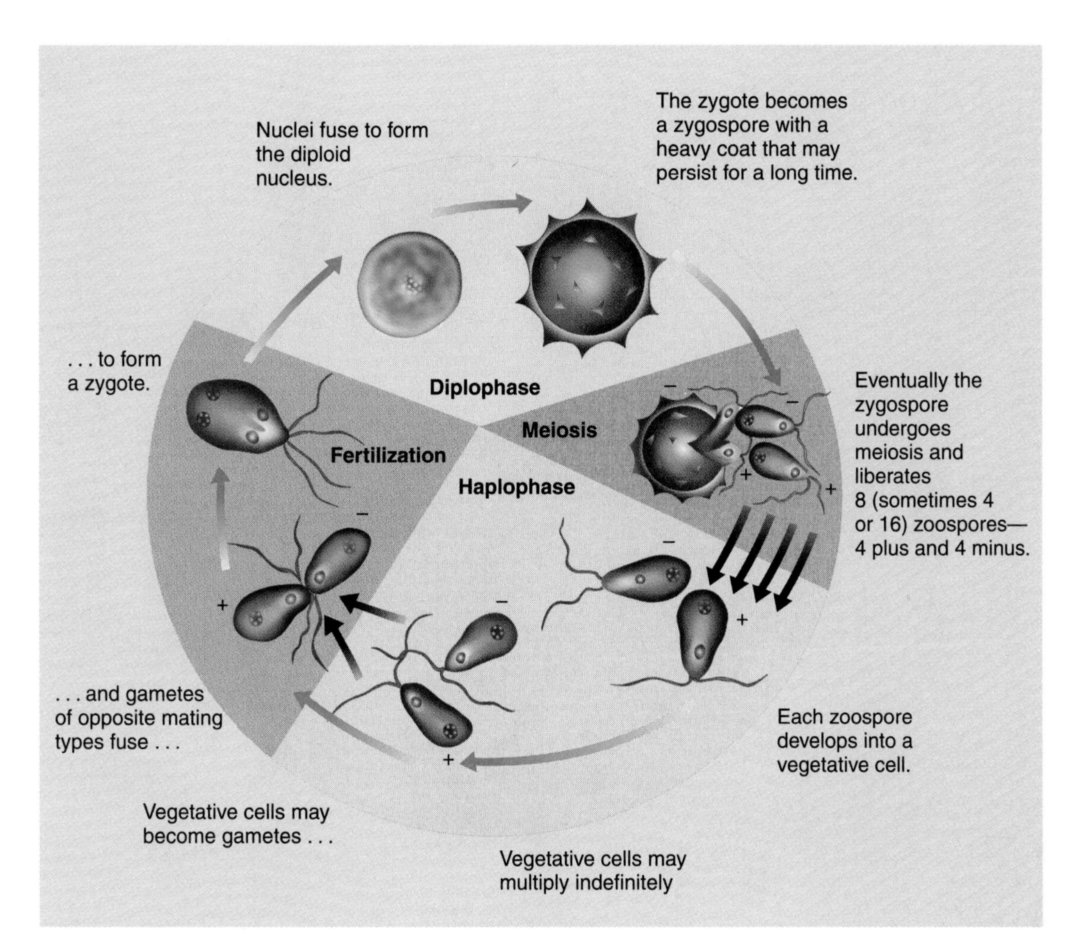

Figure 15.7
In the life cycle of the green alga *Chlamydomonas,* the haplophase dominates. The life cycle includes only a brief diplophase—a zygote that eventually undergoes meiosis to produce more haploid cells.

Please send full kit of sexual parts, including efficient apparatus for meiosis, with full instructions for installation." Anything as complex as a sexual cycle must have evolved over a very long time. We do not know how many times it has evolved—perhaps only once. In any case, it does not happen easily.

15.3 Organisms exhibit three kinds of sexual cycles.

You've known about sex for a long time. Now let's examine "the facts of life" more carefully and see exactly what is happening during a typical sexual cycle in humans (Figure 15.6). Our bodies (muscle, nerve, blood, liver, and so on) are made of **somatic cells** (*soma* = body), which are diploid and grow by mitosis. Since daughter cells produced by mitosis have the same chromosome sets as the parent cell, all our somatic cells have essentially the same genetic complement. On the other hand, our gametes, or **germ cells** (sperm and eggs) are haploid and are made by meiosis in organs called **gonads;** *this is the only place where meiosis occurs.* Male gonads, or **testes,** produce **sperm;** female gonads, or **ovaries,** produce **eggs (ova).** Haploid egg and sperm cells combine to form a new diploid cell, the **zygote,** which develops into another adult.

This is the reproductive cycle we are all familiar with. It is not, however, the only kind of sexual cycle. Ordinary mitotic growth, as in our somatic cells, is **vegetative growth,** and in animals, it only occurs in the diplophase, so it is called a *diplontic cycle.* But in other kinds of organisms, vegetative growth may occur in the haplophase (a *haplontic cycle*) or in both the haploid and diploid stages, a *haplodiplontic cycle.* Sidebar 15.1 explores some other variations on sexual reproduction.

The little green alga *Chlamydomonas* illustrates a haplontic cycle (Figure 15.7). The flagellated cells whipping about in pond water are all haploid, and their vegetative growth is simple proliferation, each cell dividing in two by mitosis. Sexual reproduction only enters their life cycle in certain circumstances—for instance, during nitrogen depletion—when some cells become gametes, and two of them fuse to make a diploid zygote. Because the gametes all look identical and differ only in subtle chemical ways, they are called different **mating types** (not different sexes), either plus (+) or minus (−), and syngamy occurs when a + cell fuses with a − cell. But the resulting zygote doesn't divide mitotically and grow, as an animal zygote does. Instead, it develops a tough coat and becomes a **spore.** A spore is a single cell that is either a stage in reproduction or is specially resistant to damage and can survive unfavorable conditions, such as drying. Inside the *Chlamydomonas* spore, the nucleus undergoes meiosis plus one or two rounds of mitosis in the haploid condition to make eight (in some species, 16) new haploid cells, half + and half −, which eventually break out and become another generation of vegetative cells.

In the haplodiplontic cycle of a common yeast like *Saccharomyces,* both haploid and diploid cells can grow vegetatively (Figure 15.8). Haploid yeast cells grow and proliferate by

Figure 15.8
Yeast can grow vegetatively, by mitosis, as either haploid or diploid cells.

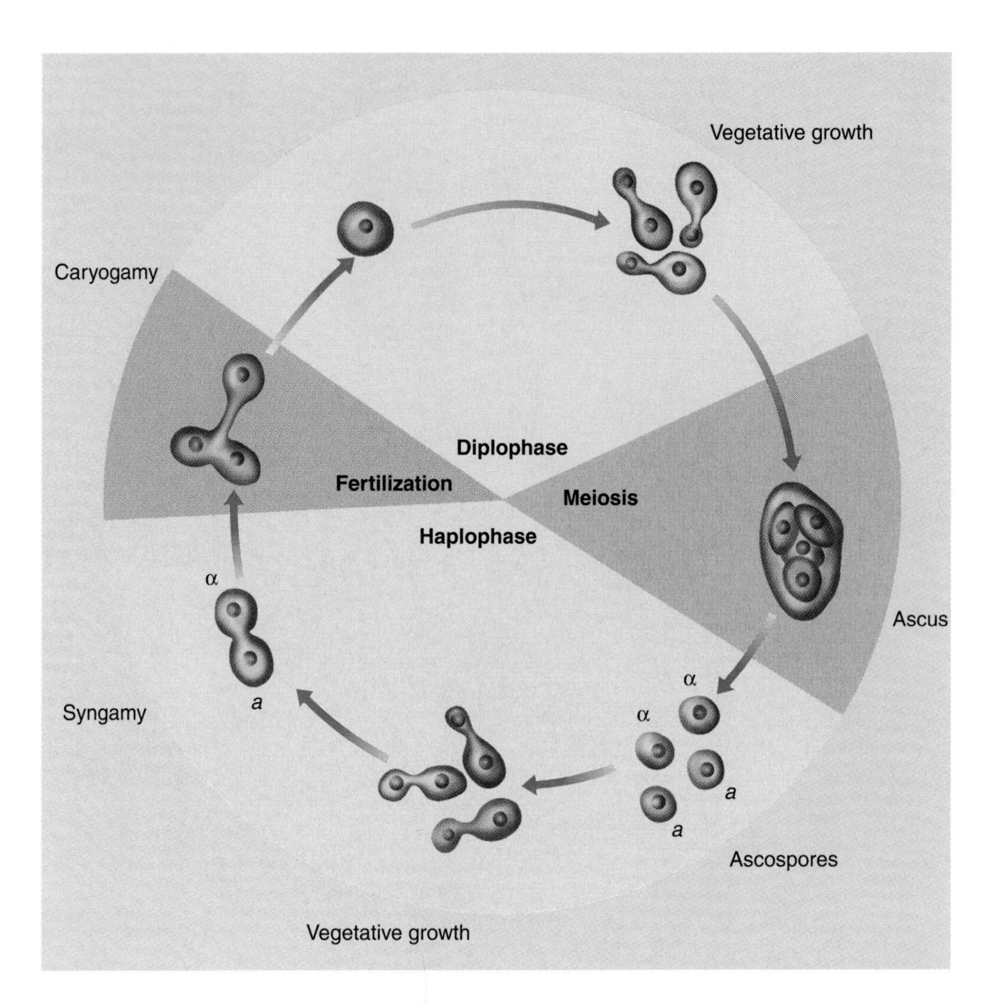

mitosis (and budding cell division; Section 11.1). Yeast can go on like this indefinitely, but in the right nutritional conditions, two cells of opposite mating types, designated *a* and α (alpha) in yeast, fuse to make a diploid cell that hardly looks any different. Diploid cells can also grow indefinitely by mitosis, but occasionally they undergo meiosis to make four haploid spores (two *a* and two α) that can divide and grow as before.

Familiar land plants also have haplodiplontic cycles, including some growth in a haploid phase—called the **gametophyte** because it produces gametes—and some growth in a diploid phase, the **sporophyte,** which produces spores. The familiar leafy structure of a fern, for instance, is diploid, and it grew from a small haploid structure buried in the soil. In Chapter 33 we will tell the story of plant evolution, whose main theme is the increasing dominance of the sporophyte in the later plant groups, apparently because of the advantages a diploid has over a haploid.

Figure 15.9 shows another distinction often made between life cycles. In **isogamy** the gametes are all the same size, like the gametes of *Chlamydomonas.* In **anisogamy** one gamete is clearly larger than the other. And in **oogamy** one gamete is a motile sperm while the other is a nonmotile ovum. Another general trend in evolution is toward oogamy, with a large ovum and a small motile sperm. Isogamy has the advantage that both gametes are motile and perhaps better able to find one another; oogamy provides one large cell that stores nutrients and an apparatus for development, and is also a large, stationary "target" for the motile cell to find.

Exercise 15.1 Just for fun, speculate about what human society would be like if humans had a haplontic cycle or a haplodiplontic cycle.

15.4 Meiosis divides a diploid cell into haploid cells.

To study chromosomes, it is useful to create a **karyotype,** a picture of a set of chromosomes that shows them all in order of size so one can determine, for instance, whether the set is diploid or haploid. To produce a karyotype, dividing cells are treated with the drug colchicine, which disrupts the mitotic spindle and stops mitosis in metaphase, when the chromosomes are maximally condensed. The chromosomes are then spread out on a microscope slide and photographed (Figure 15.10*a*), and the pictures are cut out and arranged as in *(b).* Karyotyping is easily done and is now routinely used in medical practice to detect possible genetic abnormalities.

Meiosis is a mechanism for dividing chromosome sets in half. Diploid cells of each species have a characteristic num-

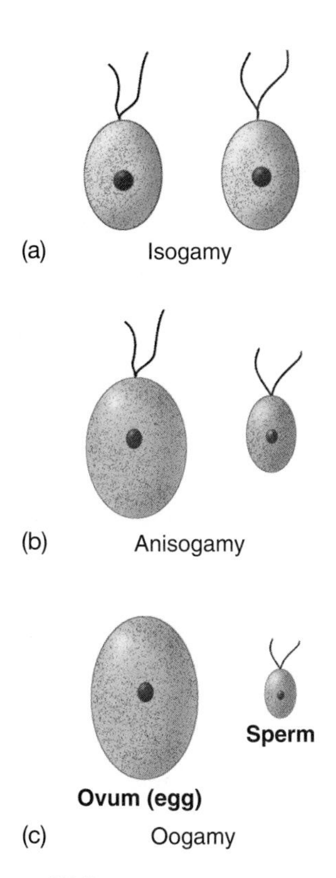

Figure 15.9

(a) Isogamy, *(b)* anisogamy, and *(c)* oogamy are three variations on sexual reproduction. In oogamy, the gametes are different enough to be called an ovum (or egg) and sperm.

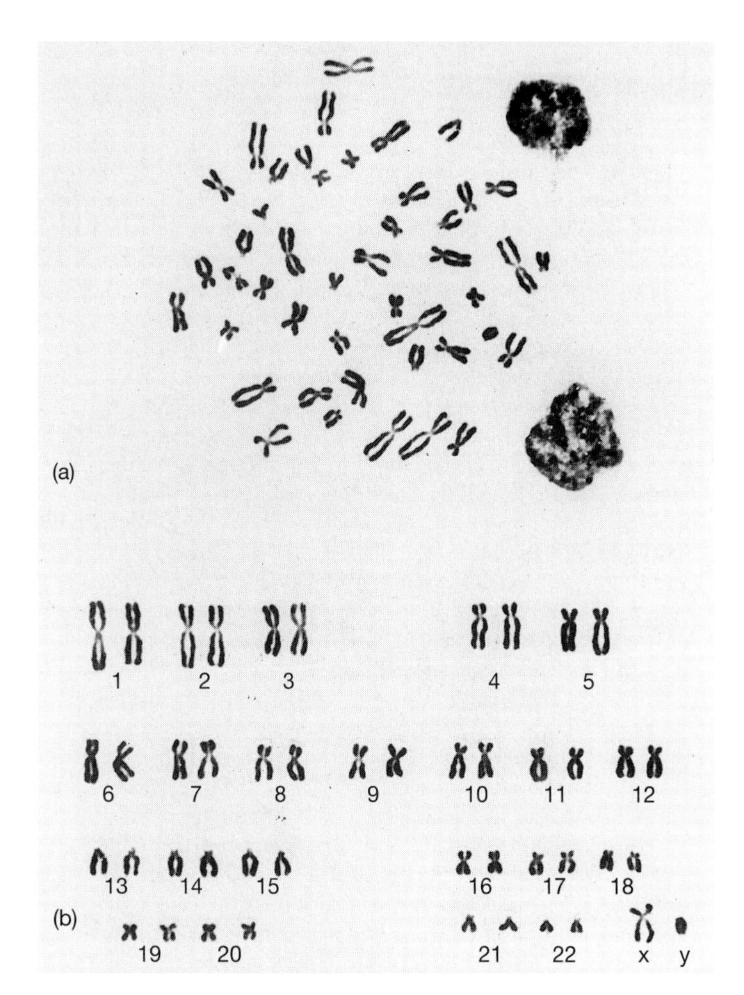

Figure 15.10

(a) To produce a karyotype, mitosis is halted at metaphase, and the chromosomes are spread out for viewing. *(b)* Photographs of individual chromosomes are cut out and lined up in order of size.

ber of chromosomes, $2n$, where n is the haploid number. For instance, *Drosophila* has 8 chromosomes ($n = 4$), and humans have 46 ($n = 23$). In making a karyotype, chromosomes are matched up into n pairs of **homologous chromosomes,** or **homologs,** both having their centromeres in the same position, the same overall length, and the same arm lengths. (Techniques that reveal distinctive bands on chromosomes aid in identifying homologs, which also have identical banding patterns.) The only nonmatching chromosomes are certain sex chromosomes in animals, generally named X and Y, that distinguish the two sexes. Female mammals, for instance, have two X chromosomes, and males have one X and one Y.

Banding patterns on chromosomes, Section 19.5.

Meiosis must clearly be very precise. A set of chromosomes is like a chess set. It is no good having a collection of just any 32 chess pieces, for each player must have precisely one king, one queen, two rooks, and so on. Likewise, if a diploid set has 10 pairs of chromosomes, it is no good dividing it randomly into any two sets of 10 chromosomes each because each haploid set must have only one chromosome #1, one #2, and so on. A **euploid** cell (*eu-* = true or good) contains either a complete haploid set or an integral multiple of such a set ($2n$, $3n$, and so on). A cell with any deviation, such as an extra chromosome or a missing chromosome, is **aneuploid.** An aneuploid set with double copies of some parts of the genome and no copy of other parts is generally not viable, and the mechanism of meiosis has evolved to separate a diploid set back into haploid, euploid sets like those of the parents, with the right balance of genes.

In general, meiosis begins with a diploid cell that has gone through an S phase, so its chromosomes are doubled, but it then goes through *two* successive divisions, meiosis I and II, without further DNA replication, to make four haploid products. Let us emphasize that *a cell entering meiosis must have passed through an S phase.* During this phase of DNA replication, each chromosome becomes a duplex structure consisting of two sister chromatids, and meiosis cannot proceed unless the chromosomes have this structure.

We illustrate meiosis here with micrographs of sperm formation in the salamander *Amphiuma,* which has a diploid number of 28, so the micrographs show 14 pairs of homologs. The drawings show meiosis in a cell with only four chromosomes—two pairs of homologs. The imaginary organism in which this meiosis is occurring probably had a mother and father or their

Sidebar 15.1 Variations on a Sexual Theme

Humans tend to think that it's natural for each species to have two sexes, male and female, but some organisms would find the idea totally foreign (if they could think about it). Many plants and animals, for instance, simply aren't differentiated into two sexes. Certain worms, among other animals, are *hermaphrodites,* named after the mythical Hermaphroditus, the child of the gods Hermes and Aphrodite, who had both male and female organs. Bearing both ovaries and testes, hermaphrodites sometimes fertilize themselves, but more often, one animal fertilizes the eggs of another. (Cross-fertilization promotes more genetic variation, which is necessary for evolution.) Similarly, most flowering plants are hermaphroditic, with flowers containing both male and female parts. If the flowers are specialized and produce either pollen or eggs but not both, they may be borne on a single *monoecious* plant (literally, "one house"), or on *dioecious* plants ("two houses") that are either male or female and bear one kind of flower or the other. Again, the rule is cross-fertilization, and hermaphroditic flowers are often constructed so they can't fertilize themselves.

In some animal species that have separate males and females, an individual changes its sex regularly during its life. In shallow water along the Atlantic coast of North America, we find stacks of a dozen or so somewhat flattened molluscs, each clinging tightly to the one below. These are *Crepidula fornicata,* the slipper shell. The small animals at the top of the stack are males, the large ones at the bottom are females, and those in between are changing from male to female as they grow. Some oysters change sex many times in their lives. Each animal begins as a male, but after shedding its sperm, it changes into a female; then, after spawning in this form, it becomes a male again. This cycle continues throughout the animal's life, at a rate influenced by nutrients and temperature. For example, oysters living in cold water may only change sex every other year, while those living in warm water may change a few times during a single year. The sex of other animals, including turtles, is determined by the temperature at which they develop.

Some organisms develop by *parthenogenesis,* which literally means "virgin birth"—in other words, without fertilization. In some species an unfertilized egg develops into a male, in some into a female, and in others into either sex under the influence of different chemical determinants. In each hive of honeybees, for instance, a diploid queen produces ordinary haploid eggs, and periodically she copulates with a male bee and receives his sperm in a little storage sac. While laying her eggs, she can either lay unfertilized haploid eggs, which develop into male drones, or allow sperm to escape and fertilize the eggs, and these develop into female workers.

Artificial parthenogenesis can be induced in frog eggs with certain chemical baths or by pricking them with a needle, so they develop into haploid females. The natural counterpart, *pseudogamy,* occurs in many species of worms, where a sperm activates an egg to begin developing, but the two do not actually fuse and combine their chromosomes.

Females can also develop parthenogenetically from eggs that become diploid without fertilization, in either of two ways. In *apomixis,* meiosis simply doesn't occur—the chromosome number in the developing ovum is never halved. In *automixis,* meiosis occurs, but then two sets of separated chromosomes rejoin and the ovum starts to develop.

Insects provide prime examples of female-producing, parthenogenetic animals. Many species of aphids, for instance, produce several generations of females during the summer. Eventually, at the summer's end, a stimulus such as a reduced food supply or a drop in temperature induces a return to bisexuality. In some species, all the females begin to produce both male and female offspring; in others, some females produce only males and others only females. This last generation then produces diploid zygotes that overwinter and hatch in the spring. Other species of insects never have any males.

equivalents, even though there are many variations on sexuality (see Sidebar 15.1). To remind us of this fact, we will consider the colored chromosomes maternal and the black ones paternal.

Early prophase I

The chromosomes are just becoming visible with a light microscope. They appear to be single, but as in mitosis, the chromosomes are already double. In mitosis every chromosome acts independently, but in meiosis the chromosomes engage in **synapsis,** a process in which homologs attract one another and pair up. (The mechanism that makes homologs synapse so precisely with each other is still largely unknown.) Each pair is called a **bivalent** (pronounced bih′-vuh-lent) if it is considered a combination of two chromosomes or a **tetrad** if it is considered a combination of four chromatids, but the two terms refer to the same thing. Homologous chromosomes are already beginning to synapse with each other at this early stage. Homologous chromosomes synapse very precisely, so their matching parts come together, and the distinctive knobs and constrictions on each chromosome align with the homologous features on the other.

Early prophase I

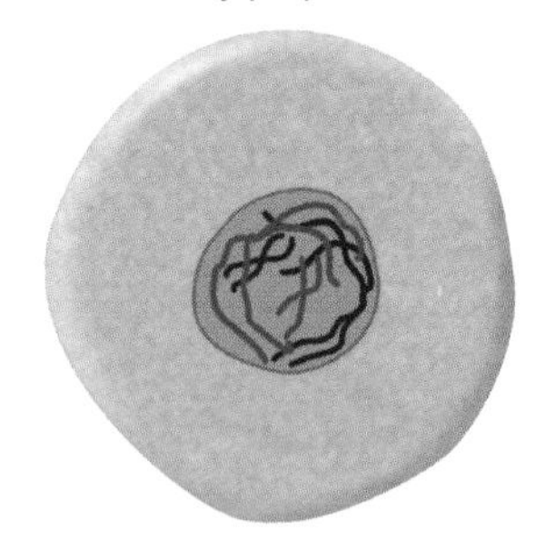

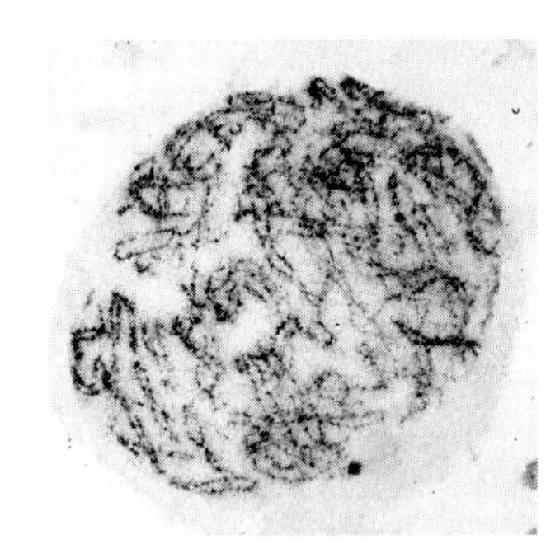

Late prophase I

The chromosomes are now more compact and visibly duplex, with two chromatids each. Previously, the chromosomes have been held in synapsis by strong forces, but now, homologous chromosomes appear to be pulling away from each other. They are still strongly attracted and held together at points where homologous chromatids cross or intertwine, called **chiasmata** (sing., **chiasma;** compare with the Greek letter chi, χ). We will show later that the chromatids are exchanging segments in the chiasmata, thus creating new combinations of genes and increasing the diversity of genomes. During these early stages, the nuclear envelope has been breaking down, a spindle has been forming, and the chromosomes are attaching to it, as they do in mitosis.

Late prophase I

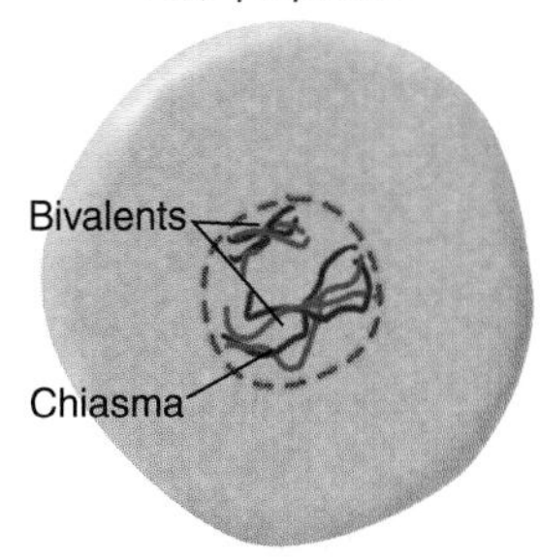

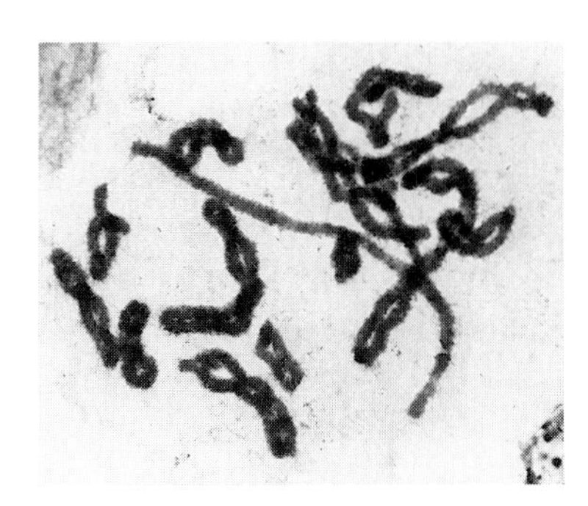

Metaphase I

The nuclear envelope has broken down, and the spindle is complete. The bivalents orient themselves on the metaphase plate with their centromeres actively pulling away from each other; the chromosomes appear to be under considerable tension and act as if they are only held together by the chiasmata.

Metaphase I

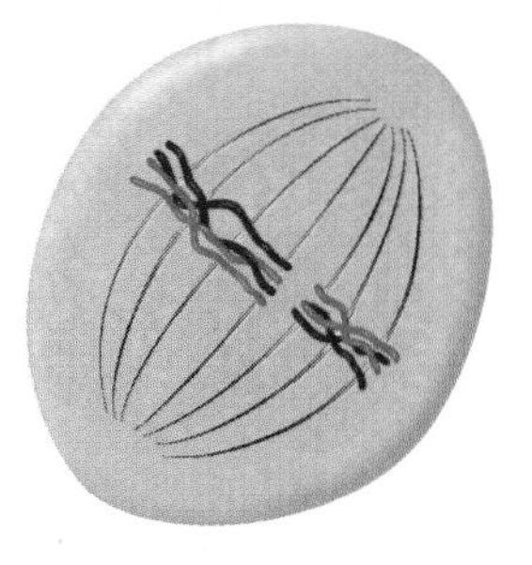

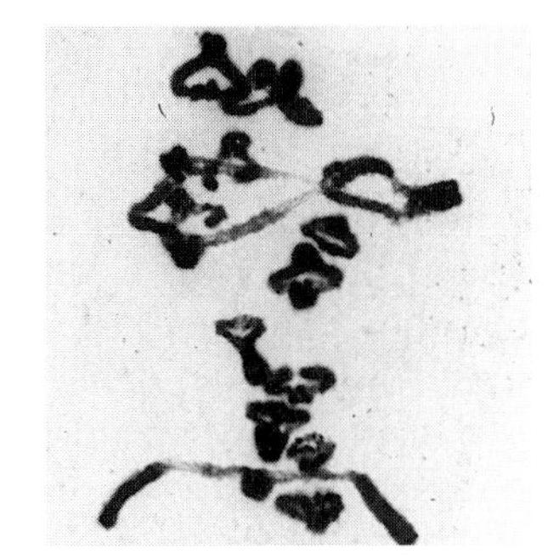

Anaphase I

The homologs in each pair separate completely and start to move toward opposite poles of the cell. Now notice that *sister chromatids do not separate from each other during anaphase I.* The homologs move to opposite poles with their centromeres intact. This contrasts with mitosis, in which the centromeres divide, so sister chromatids do separate.

Anaphase I

Telophase I-Prophase II

There is an interphase or interkinesis period, generally brief, between the first and second meiotic divisions. During this time, the nuclear envelope usually returns, and the chromosomes disperse again, but no DNA replication occurs. Though each daughter cell is now haploid, its chromosomes are still duplex, as if it had just gone through an S period, so *the second meiotic division is essentially a haploid mitosis.* As prophase II begins, the chromatids start to condense again, and the nuclear envelope disappears.

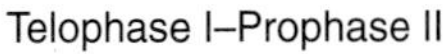

Telophase I–Prophase II

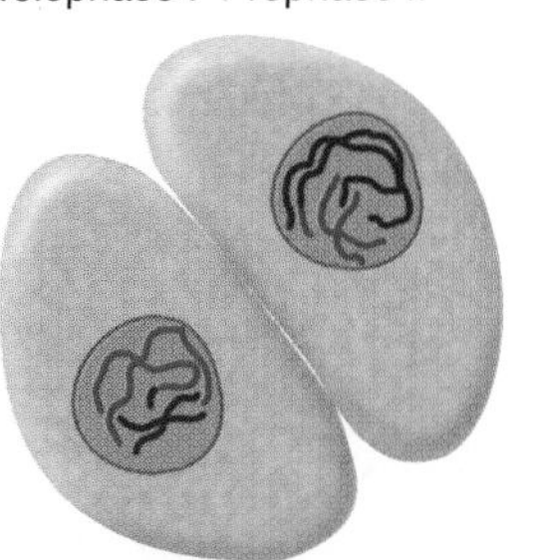

Metaphase II

The chromosomes are again very short and thick; their chromatids are visibly separated and appear to be pulling away from each other. The chromosomes line up as in mitosis with their centromeres along the metaphase plate.

Metaphase II

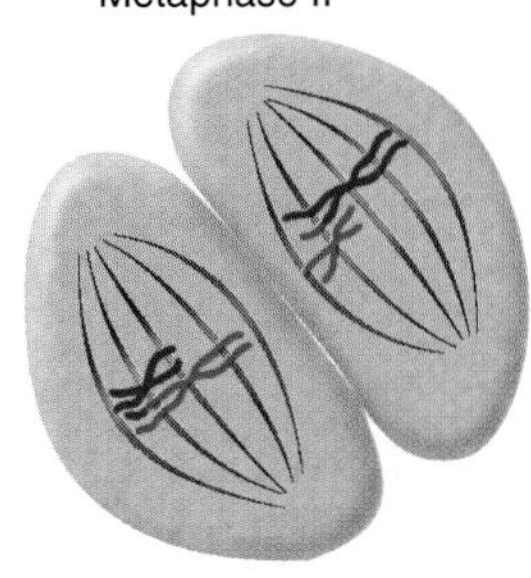

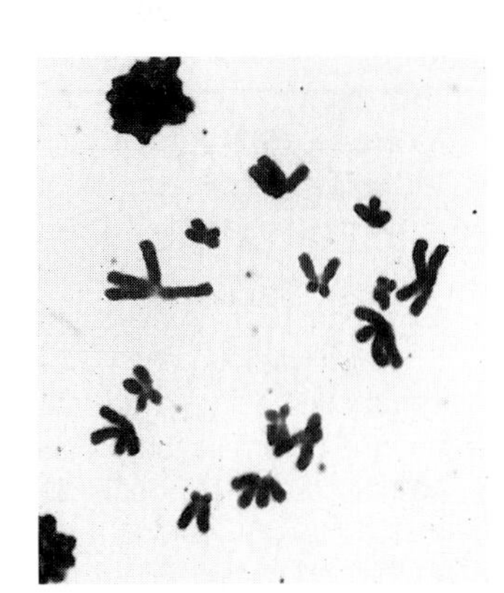

Anaphase II

The centromeres finally divide, and sister chromatids move toward opposite poles.

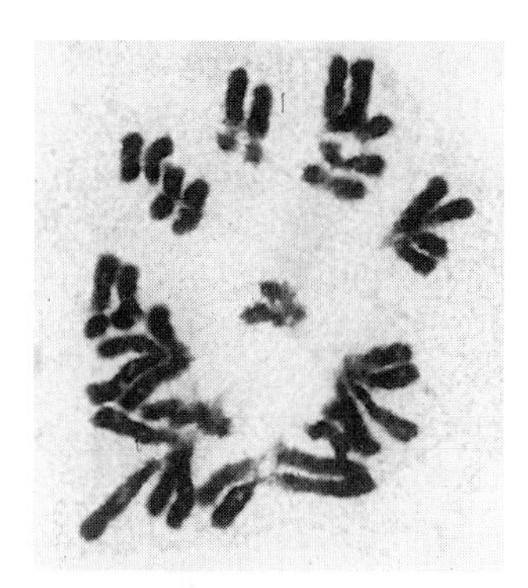

Telophase II

Nuclear envelopes re-form, and the chromatids—now independent chromosomes—return to their interphase condition. The original diploid cell has been divided into four essentially identical haploid cells.

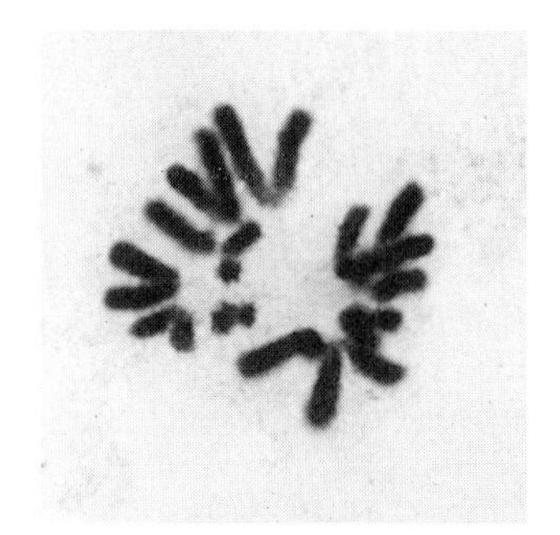

The chromosome sets of the four resulting haploid cells look identical, but they may be quite different genetically. Homologous chromosomes of the original diploid cell may have subtle mutational differences between their genes, and since each pair of chromosomes separates independently during anaphase, different haploid cells may have quite different combinations of genes derived from the two parents of this diploid organism. Meiosis is therefore a major source of genetic variation within a population. An additional source of variation lies in the formation and resolution of chiasmata, because in this process, homologous chromatids regularly **cross over,** exchanging segments of their DNA, as shown by the exchange of colors in the long chromosomes in the previous drawings. Crossing over makes new combinations of genes on a single chromosome. We will explore the genetic consequences of these meiotic events in Chapter 16.

Precise pairing of homologs during prophase I of meiosis is essential for proper segregation of chromosomes at anaphase. Occasionally *polyploid* individuals occur that have more than a diploid number of chromosomes, such as triploids ($3n$) or tetraploids ($4n$), because the chromosomes failed to segregate properly during the formation of one gamete (nondisjuction; see Figure 19.10). Such individuals may be quite viable; triploid *Drosophila* and polyploids of many plants are known. But triploids are commonly sterile because only two chromosomes of each set of three homologs can synapse and segregate regularly, while the third chromosome goes randomly to one daughter cell or the other, resulting in aneuploid, inviable offspring. However, polyploids with even numbers of chromosome sets, such as tetraploids or hexaploids ($6n$), are generally fertile and may be more vigorous than their diploid ancestors.

Polyploidy in evolution, Section 24.9.

Exercise 15.2 Compare mitosis and meiosis with respect to: *(a)* division of centromeres, *(b)* whether the chromosomes move independently, *(c)* haploid or diploid condition of the dividing cell, and *(d)* times of DNA replication.

Exercise 15.3 A cell that is about to undergo meiosis contains two long chromosomes and two short chromosomes, with the maternal chromosomes marked red and the paternal marked blue. Draw the chromosomes in the cell, or cells, at the beginning of meiosis, at the end of the first meiotic division, and at the end of the second meiotic division. At the end of each division, consider whether the chromosomes could occur in different combinations from those you drew first, and draw all the possible combinations.

Exercise 15.4 Denote homologous chromosomes by A and a, B and b, and so on. Suppose, in a species with $n = 2$, a diploid organism is made from gametes with the maternal chromosomes A,B and the paternal chromosomes a,b. When this diploid undergoes meiosis, what combinations of chromosomes can its gametes have? Suppose, in a species with $n = 3$, the original gametes carried A,B,C and a,b,c. What new combinations could then be made in meiosis?

Exercise 15.5 Exercise 15.4 shows that four distinct haploid combinations are possible if $n = 2$, and eight combinations are possible if $n = 3$. You should be able to state a general formula for the number of distinct haploid cells with n pairs of homologs. Given that $n = 23$ for humans and it is virtually certain that everyone has some genetic difference between every pair of homologs, how many distinct sperm or egg cells can a man or a woman make? How many genetically different children could every couple create? (We are ignoring additional differences that can result from crossing over.)

15.5 Spermatogenesis and oogenesis entail different patterns of cell division.

As previously stated, meiosis in a plant or animal produces haploid gametes, either sperm or eggs. The cells may have different names in different organisms, but the nuclei of one sex are always called sperm nuclei. The formation of sperm, **spermatogenesis,** is quite simple. In an animal, a testis is generally made of fine tubes, and diploid cells on the inner wall of a tube enlarge, move inward, and become *spermatogonia.* As shown in Figure 15.11*a,* each spermatogonium divides once mitotically

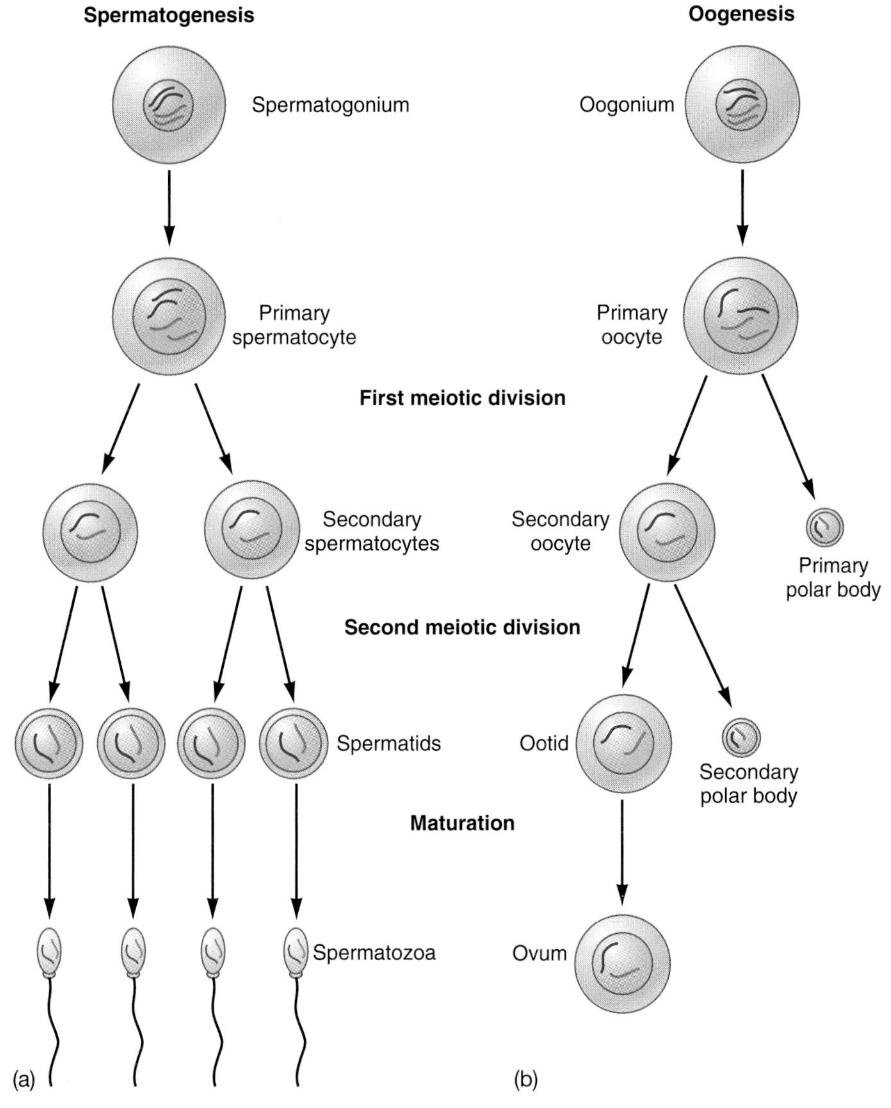

Figure 15.11

(a) Spermatogenesis and *(b)* oogenesis are similar, but oogenesis only produces one mature product instead of four. Notice that the cell entering meiosis is a *primary* spermatocyte or oocyte; after the first meiotic division, it is a *secondary* spermatocyte or oocyte. Both spermatozoa and polar bodies are actually much smaller in comparison with the ovum.

and then by meiosis into *spermatids.* Each spermatid then matures into a sperm, or spermatozoon (pl., spermatozoa).

A sperm is a single cell with an unusual form. As a vehicle for carrying genetic information to the next generation, it is mostly a nucleus, bearing chromosomes, with a long tail powered by ATP generated in the mitochondria packed into the neckpiece (Figure 15.12). (These mitochondria, however, do not enter the zygote along with the sperm nucleus. So mitochondrial genes are inherited exclusively in a maternal line, from mother to daughter.)

Egg formation, or **oogenesis** (Figure 15.11*b*), begins with single cells on the surface of an ovary and differs from spermatogenesis in an important way: Each *primary oocyte,* the cell that undergoes meiosis, produces only one ovum, compared with the four spermatids that spermatogenesis yields. A maturing primary oocyte accumulates *yolk,* a mixture of carbohydrates, lipids, and proteins that can nourish a developing embryo for a time. (The embryos of some animals, such as birds, develop within a shell and require a great deal of yolk because they will have no other source of food until they hatch.) The oocyte accumulates numerous ribosomes and messenger RNA molecules, forming an apparatus for protein synthesis during the first stages of embryological development. Oogenesis has evolved so that almost all of the oocyte cytoplasm is preserved through highly asymmetrical meiotic cell divisions. In each division, one set of chromosomes is removed with just a little cytoplasm as a **polar body** that soon degenerates. The timing of cell division and fertilization varies a bit. In some animals, the oocyte goes through the second meiotic division, producing a second polar body, before it is fertilized; in humans and other animals, the oocyte undergoes the first meiotic division but will not undergo the second division, to complete meiosis, unless it is fertilized by a sperm.

A sexually mature human male can produce about 50,000 sperm per minute. A healthy young man releases at least 300 million sperm in each ejaculate, and therefore vast numbers in a lifetime of sexual activity. But while a baby girl is born with about two million primary oocytes (a million in each ovary), very few actually mature during her lifetime. Oddly enough, all the oocytes she carries began the earliest stages of meiosis, and then stopped, while she herself was still a fetus inside her own mother. Beginning at the time of puberty, about 13 years later, the oocytes start to mature at the rate of one or two every 28 days, or about 13 per year. So in a reproductive lifetime of 35 years, a human female will only mature a few hundred eggs. The rest of the two million simply never develop.

The gametes of flowering plants develop in a comparable way, except for a technical difficulty in describing them. An idealized flower includes an *ovary,* which produce eggs, and an **anther,** which ultimately produces sperm, so these are the equivalent of the female and male gonads (Figure 15.13). This is sensible if we define a gonad as the organ in which meiosis occurs, but the products of meiosis in the ovary and anther are actually spores, not gametes. Nevertheless, all four products of meiosis in the anthers develop into **pollen grains** containing sperm nuclei. In the ovary, meiosis produces three small haploid cells, which degenerate, and one large spore cell (megaspore), which divides a few times more and produces an egg; this elimination of three of the four products of meiosis parallels the elimination of polar bodies in animals. Further

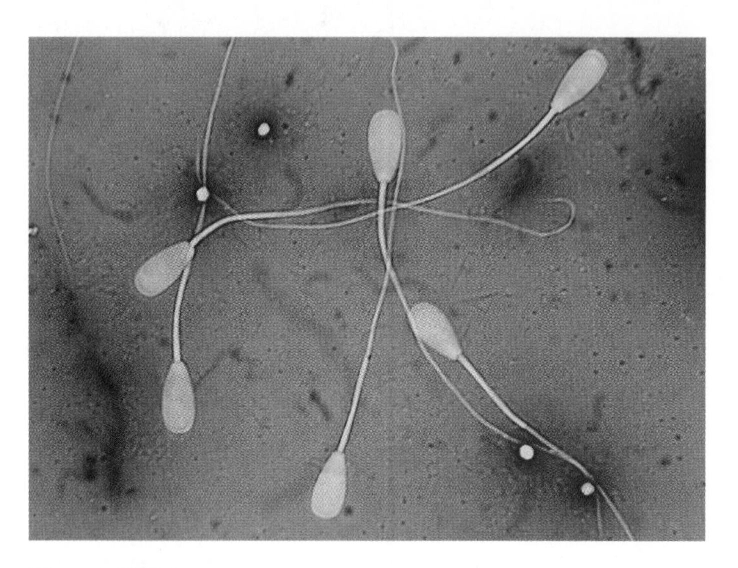

Figure 15.12
The spermatozoa of animals (here, from a bull) have a large head, containing the nucleus with its chromosomes, and one or more propulsive tails.

developments vary from one plant group to another, thus this story is best told in the context of plant evolution in Chapter 33.

Exercise 15.6 Why is it valuable for oocytes in animal cells to form polar bodies with only a small portion of their cytoplasm and for a similar process in plants to eliminate three cells and leave only one to produce an egg?
Exercise 15.7 For each of the following cells or organs, tell whether the cells can go through mitosis only, meiosis only, both mitosis and meiosis, or neither: liver, sperm, flower petals, ovary, skin, brain, roots, testis, leaves, ovum, bone marrow.

15.6 A cycle of growth implies a cycle of gene regulation.

A growth cycle depends on different genes turning on and off in a regular sequence. If the same genes were being expressed at all times, an organism would never change much, so even a simple cell cycle requires different genes acting at different times. The

Figure 15.13
In flowering plants, meiosis in both megasporocytes and microsporocytes produces four haploid cells. In an ovary, only a single large megaspore survives (compare with oogenesis in animals), and its nucleus (with some variations among plant groups) divides into eight nuclei, one of which is the egg nucleus. All four microspores become pollen grains (again notice the parallel with spermatogenesis in animals). The pollen grains are shown with the tubes that grow from them through the flower tissue, each tube carrying a sperm nucleus that can combine with an egg nucleus to form a new diploid plant.

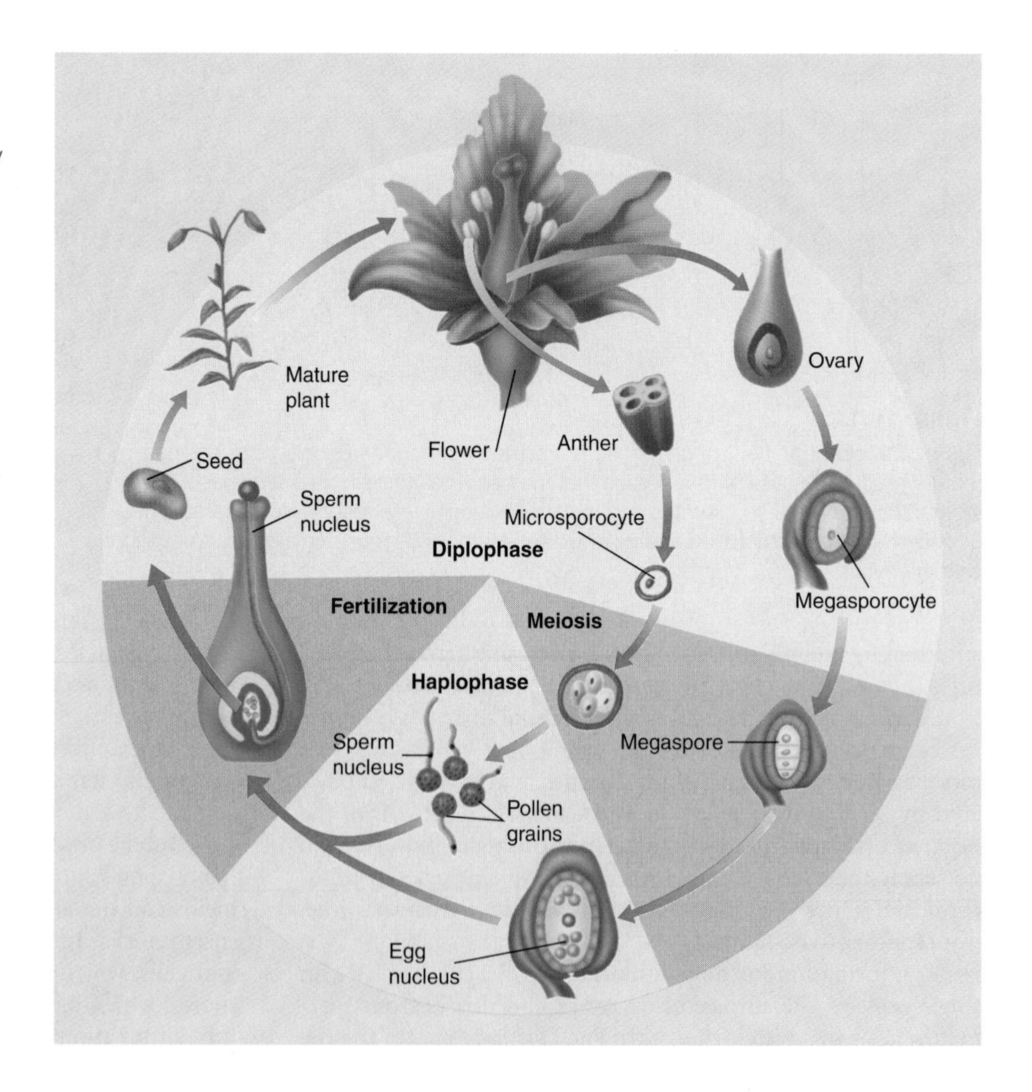

genes for synthesizing most cytoplasmic components are expressed during the G_1 period while the cell is growing a lot. Genes for DNA synthesis are turned on at the beginning of the S period and turned off at its end, when still other genes, encoding materials needed for mitosis, are switched on. These sequential changes are the most obvious expressions of a genetic program in which one event follows another.

In vegetative cells that just grow and divide, cycle after cycle, the cycle of genes being switched on and off repeats itself monotonously. But even some simple procaryotic cells have more complex cycles. Bacteria that only grow vegetatively can easily be wiped out by a sudden change in their environment, such as a drought, but other bacteria have evolved life cycles that allow them to weather such times. For instance, in aging cultures of bacteria such as *Bacillus* and *Clostridium,* each cell walls off one nuclear body in a protective coating to make an **endospore** (Figure 15.14). Endospores may be tough enough to last for years, enduring many stresses, and then to start growing vegetatively again when they fall into water with the right nutrients. Some of these bacteria are serious pathogens. Anthrax, for instance, a disease common in sheep, is caused by *Bacillus anthracis,* whose seemingly indestructible spores can stay viable for years, even in a field where no animals have been grazing for a long time or on the wool from an infected animal. The disease is passed to animals who graze in the field, or to weavers and other people who handle the wool, through cuts on their hands. Clostridial spores may lie dormant in soil for a long time until a sharp instrument carries them into, say, the warm, moist flesh of a foot, where they start to grow and to produce diseases like tetanus and gas gangrene. Organisms whose life cycles include such suspended states have a different relationship to time, for they may endure for long periods until conditions are right for them to start growing again.

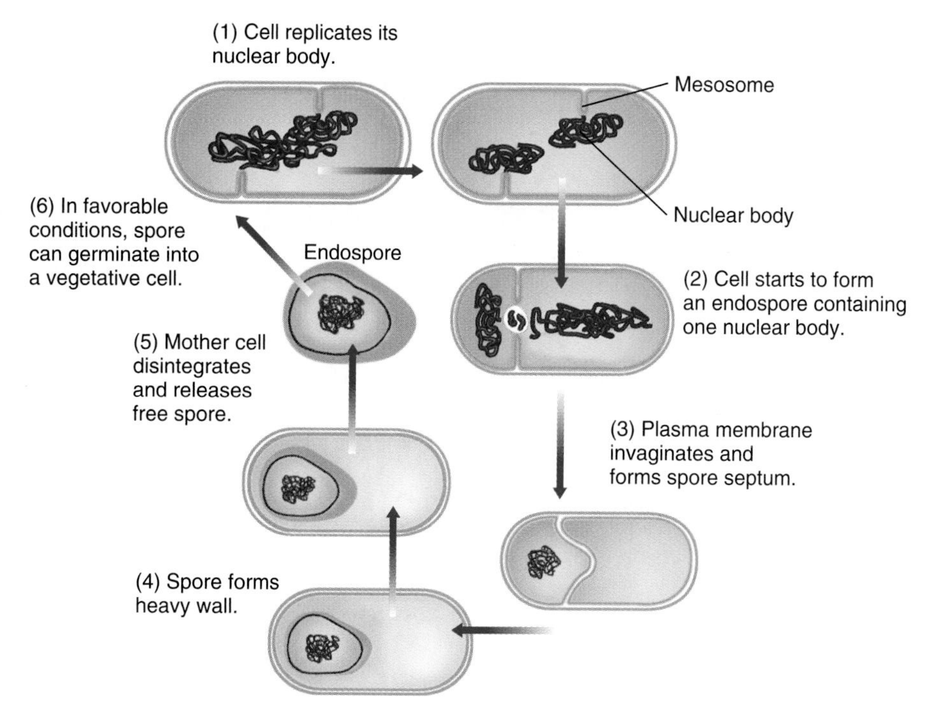

Figure 15.14

Some bacteria, such as *Bacillus,* have a special phase in their life cycles in which they can form rugged endospores that can survive adverse conditions.

The stalked bacterium *Caulobacter* varies its life cycle by **differentiation,** the production of different cell types. In each cycle, a cell differentiates into a stalked cell and a flagellated cell (Figure 15.15). Only stalked cells divide, and one daughter of the division is a new flagellated cell, which swims off and changes into a new stalked cell. You can see the ecological advantage of this pattern of growth, allowing a stalked cell to remain attached where it has adequate nutrients (or it would not be able to grow) while producing a mobile flagellated cell that can swim off and seek another nutrient-rich spot. Since the stalk at one end consists of different proteins from those in the flagellum that develops at the other end, two different sets of genes must be turned on in the two cells. These genes must be precisely regulated to generate the rhythm of the organism's life cycle. *Caulobacter* illustrates a simple version of the cycles we turn to next, in which a zygote develops into an adult with many kinds of cells.

The growth of a short bacterial cell into a longer one is relatively simple, while that of a *Caulobacter* flagellated cell into a double cell and back to a new flagellated cell is more complex. However, the development of a zygote into a multicellular organism, with all its organs and tissues, is so complicated that we are just beginning to understand it. We discuss this process in Chapters 20 and 21, but let us pause for a moment and consider just what it means for a human.

A human zygote is one cell, barely visible to the naked eye. As it develops into an embryo and eventually grows to adulthood, it becomes a structure of about 10^{14} cells of perhaps a hundred named types, each type specialized for a different function and therefore making distinctive proteins. Some of these cells are able to make massive amounts of collagen, which forms bones, ligaments, and tendons, and then to lay down calcium phosphate crystals on the collagen to form hardened bones. Other cells grow into muscles, filled with complexes of actin and myosin, that connect the bones and move them. The muscles also become connected and coordinated with nerve cells linked by long, wirelike extensions, which communicate with one another through specialized informational ligands. Cells roll themselves into a network of tubules—arteries, capillaries and veins—connected to a powerful, muscular heart, and the blood that develops in this system becomes rich in red cells filled with hemoglobin. An intestine develops through the

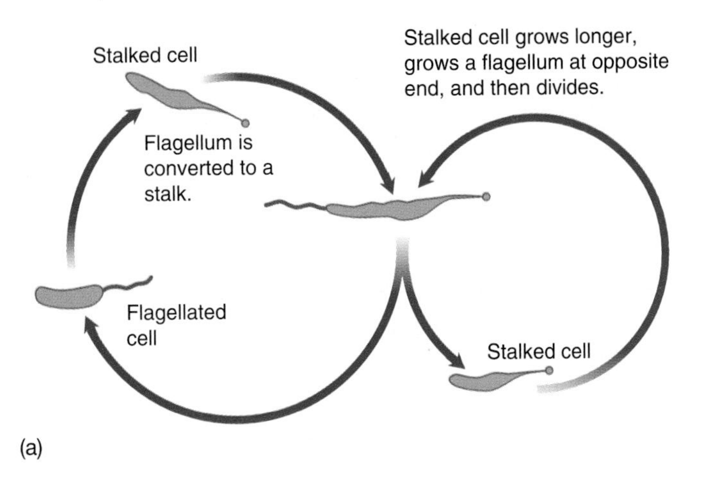

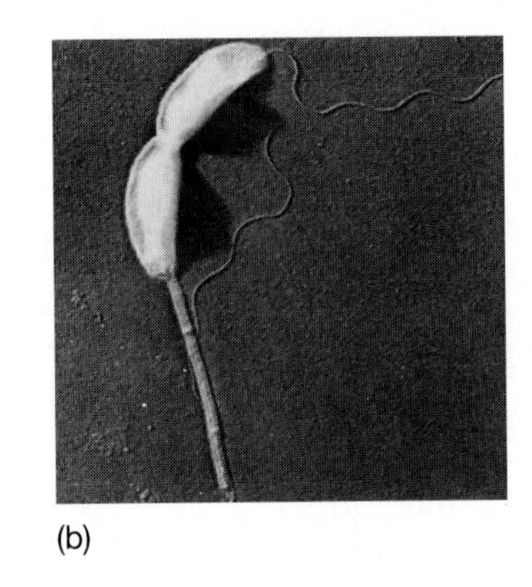

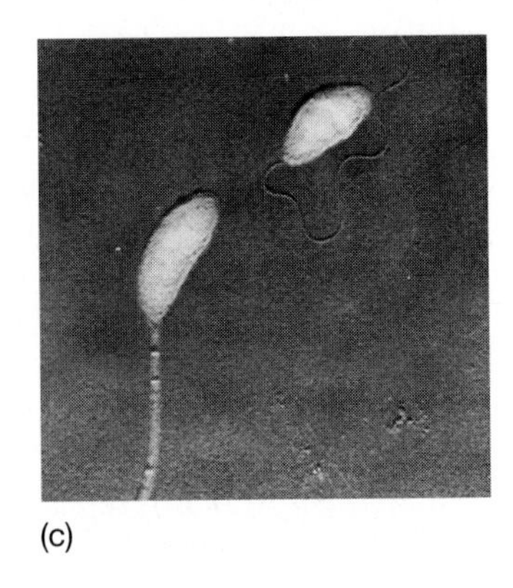

Figure 15.15
(a) The bacterium *Caulobacter* has a more complex life cycle than most procaryotes, with an alternation between attached stalked cells and motile flagellated cells. Electron micrographs show a cell partway through division *(b)* and one that has just divided into a stalked cell and a flagellated cell *(c)*.

middle of the body, where specialized cells produce a variety of enzymes that can digest food into usable molecular units. A leathery skin grows over the whole body, a waterproof layer of cells filled with keratins, sprouting hairs made of keratin. Each type of cell uses a portion of the human genome to produce its special materials and achieve its unique form. Understanding how this is done remains one of the great challenges of biology.

Development and differentiation, Chapters 20 and 21.

15.7 Some life cycles comprise distinct morphological phases.

A complicated plant or animal certainly changes as it develops, but it changes continuously and is obviously just one organism that is simply growing up. In contrast, the life cycles of other organisms include several discrete phases with distinctive forms and different ways of life. Insects are obvious examples. Butterflies, for instance, lay eggs that grow into wormlike *larvae,* go through a *pupa* stage, and eventually become winged adults (Figure 15.16). Parasites, too, commonly develop in a series of very distinct stages. Let's look at two examples.

Puccinia graminis, the wheat rust, is a notorious scourge of one of our most valuable crops. Though it has been recognized since ancient times, it still plagues us. Agricultural scientists try to fight it primarily by breeding resistant wheat strains, but the rust, a fungus, can evolve at least as fast as its host, so the battle never ends. The rust has a peculiar and complex life cycle, with quite distinctive phases shown in Figure 15.17. Notice that barberry plants are intermediate hosts for the fungus, and barberry has to be controlled to protect wheat. In one phase of the cycle, the fungus becomes a *dikaryon,* where each cell carries two haploid nuclei that go through mitosis simultaneously, maintaining the dikaryotic condition, but never fuse into a diploid nucleus.

Figure 15.16
Insects that undergo complete metamorphosis, such as the monarch butterfly, develop from zygotes into larvae, pupae, and then adults. Monarchs live in intimate association with common milkweed plants, which their larvae eat.

Some of the liver flukes—worms that invade the livers of mammals—have comparably complex life cycles (Figure 15.18). These parasites also use intermediate hosts. The stages of development in such worms are so morphologically distinct from one another that they were often identified as different

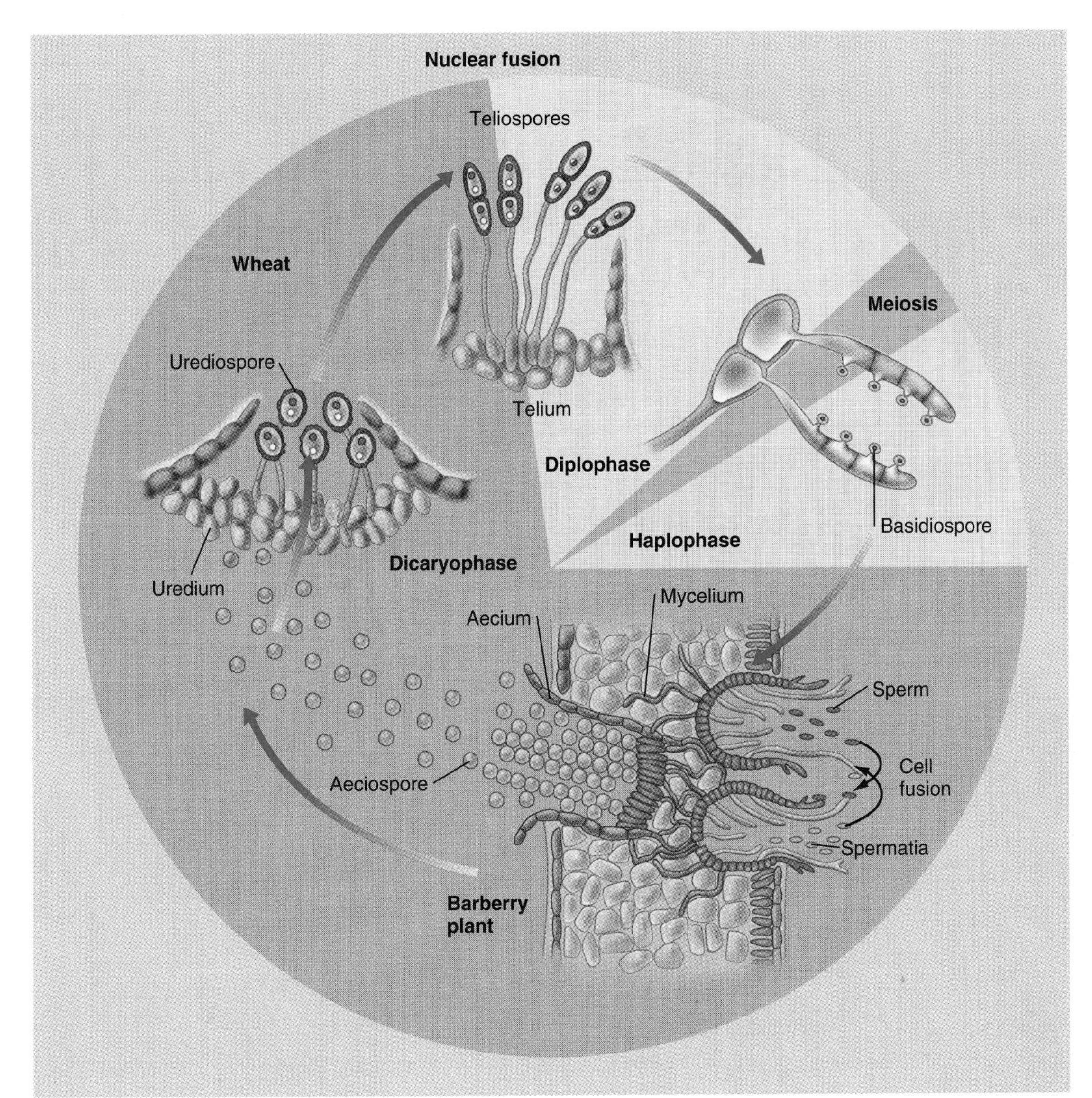

Figure 15.17
Wheat rust goes through a complex life cycle during which it lives on two hosts (wheat and barberry) and has several different morphological stages, yet all these stages are expressions of the same genome. Each stage produces spores that develop into the next stage. Notice that the mold has only a brief diplophase followed by meiosis, and that in the aecium, uredium, and telium stages, the cells are dikaryons containing two separate haploid nuclei.

species until their life cycles were determined. All these stages carry the same genome, but for all the resemblance between the stages, there might as well be several different genomes. Yet these differences between stages are simply expressions of different genes at different times.

Every organism (and every virus too, as shown in Chapter 12) goes through a distinctive cycle as it reproduces, a cycle created as the program in its genome is read out sequentially. The life of every organism can only be understood in the context of its life cycle. To understand its physiology or its ecological relationships with other organisms, it must be seen as a changing system that develops under certain conditions by following the instructions in its genome.

Exercise 15.8 Without knowing the details of a complex life cycle such as occurs in one of the parasites just described, you should be able to comment on the ecological significance of having morphologically distinct stages.

Coda In Moliere's play *The Would-be Gentleman,* the principal character is astonished to learn that all his life he has been speaking prose. It may come as a similar revelation to you that all your life you have been experiencing cycles, daily and yearly cycles of activity, and that the lives of all of us are part of a cycle as well. Not, of course, that each of us will be reborn after death and grow to adulthood again, but in every generation, each person will begin with the fertilization of an egg by a sperm, and will then develop into an embryo, a baby, and an adult. And in each adult, the process of meiosis will repeat itself to produce more eggs or sperm. We have also come to think of this kind of sexual cycle as natural, as typical, and it may come as another surprise to learn that other organisms have sexual cycles that repeat the same essential events (meiosis, fertilization) but place growth and other activities in the haploid phase, rather than in the diploid, or in both the haploid and diploid phases. In fact, as we pointed out in Sidebar 15.1, even the sexual cycles of many animals are incredibly

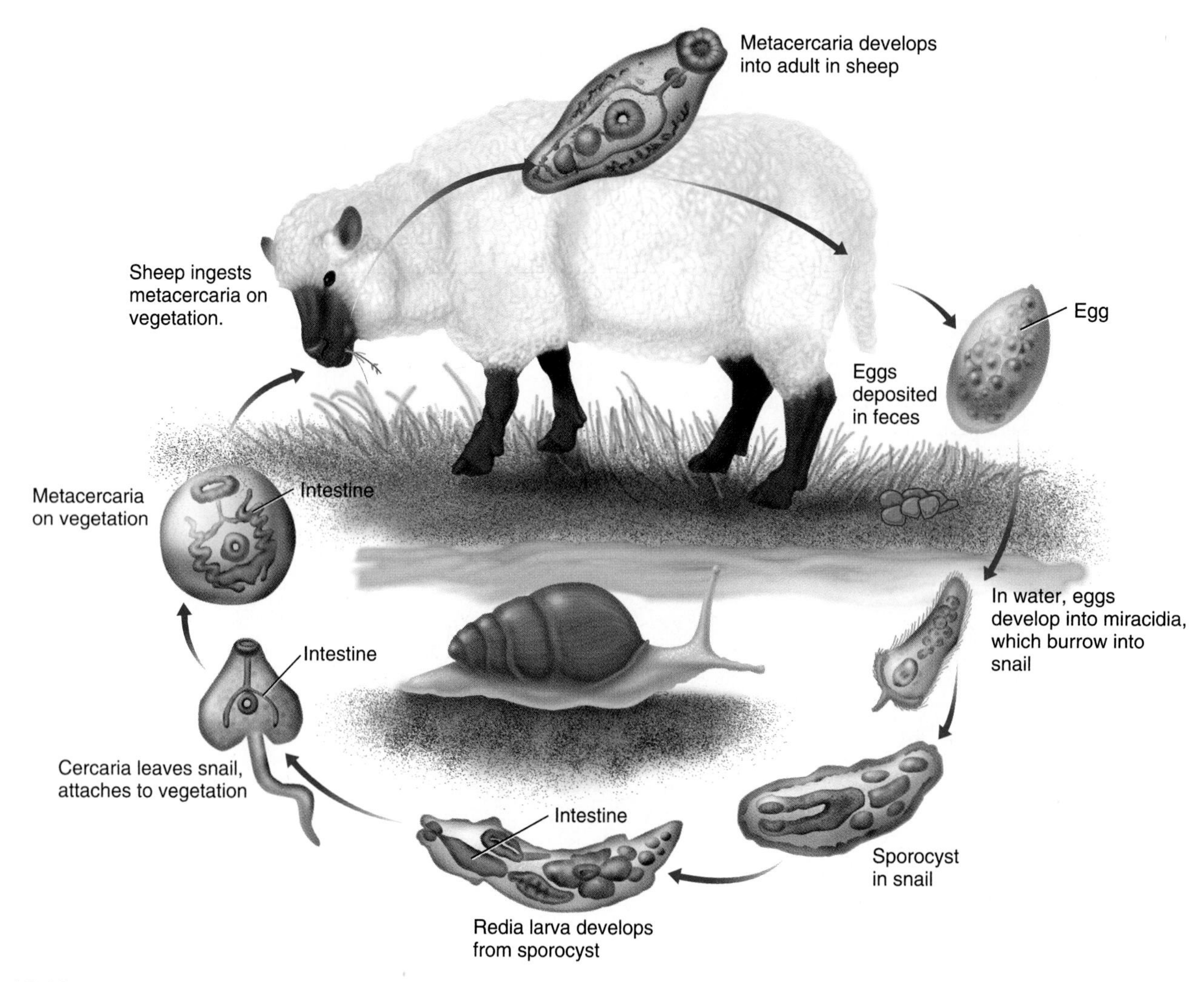

Figure 15.18

A liver fluke goes through several distinct morphological stages, which cycle between sheep and snails. The distinctive names miracidia, redia, cercaria, and metacercaria reflect their identification as distinct organisms before it became clear that they are all stages in the life cycle of one organism and are distinct expressions of the same genome.

different from ours. All these cycles are mechanisms for producing generation after generation of organisms with complete genomes, genomes that direct their development, determine their structure, and influence how they will operate. We have shown that these regular cycles must result from genetic programs that turn genes on and off in sequence.

The complicated mechanisms of meiosis and fertilization evolved over a long time, yet sexual reproduction also made the evolution of these and other intricate structures and processes easier. Remember one of the principal points we established in Chapter 2: Populations are highly variable. Sexual reproduction takes advantage of that variability by bringing together many combinations of genes with different mutations, encoding variant proteins. Since organisms operate through molecular interactions, especially between different proteins, a critical feature of evolution is making new combinations of proteins, which may be superior at performing ongoing functions or may even perform new functions. Sexual reproduction produces this recombination regularly, every generation.

Summary

1. All organisms go through cycles of growth, development, and reproduction.
2. Most eucaryotes have a cycle of sexual reproduction, in which they regularly exchange and recombine portions of their genomes with one another. Eucaryotes are predominantly diploids that have two copies of each chromosome.
3. The sexual cycle is an alternation between a haploid and a diploid phase. The diploid state is achieved through fertilization or syngamy, in which two haploid gametes fuse; the haploid state is achieved through meiosis.
4. There are three general kinds of sexual cycles, because vegetative growth, through mitosis, can occur during the diplophase, the haplophase, or both.
   a. In most animals, the haploid stage is brief, consisting only of sperm and egg cells; the life cycle is dominated by the diploid animal.

b. In some microorganisms, the products of meiosis are simply haploid cells that start to grow vegetatively by mitosis, and there is only a very brief diploid stage.
c. Certain organisms, such as yeasts, can grow in both their haploid and diploid phases. In primitive plants, both haplophases and diplophases are prominent; the diplophase becomes more dominant in more advanced plants.

5. Meiosis is similar to mitosis, but it can occur only in diploid cells (or those with greater multiples of the haploid number).
6. In the first meiotic division, homologous chromosomes (those with the same structure and genetic complement) pair, or synapse, with each other. Then the two chromosomes in each pair separate intact, without separation of their chromatids, and move to opposite poles. In the second meiotic division, which is just like mitosis in a haploid cell, the chromatids of each chromosome separate.
7. Meiosis generally yields four haploid cells from every diploid cell. In plants and animals, the products of meiosis are gametes that can combine again during fertilization, but meiosis in male structures (testes or anthers, for instance) produces four sperm (or pollen grains), while meiosis in ovaries produces only one egg.
8. Meiosis is a significant event in the evolution of a species because it creates genetic variation. This occurs in two ways. First, since chromosomes separate independently at anaphase I, different gametes can be made with all possible combinations of chromosomes, which may be carrying different mutations. Second, homologous chromatids may cross over during synapsis and exchange segments, thus creating new combinations of genes on each chromosome.
9. Even some simple organisms show a certain amount of development and differentiation, the production of morphologically distinct cells.
10. All cycles of growth and development depend on certain genes being switched on and off in a regular way, according to the fixed instructions of each organism's genetic program, so that appropriate proteins are made at the right times and places. This must happen even in a cell cycle.
11. The use of different portions of the genome, as in a genetic program, is much more obvious in cycles that entail a complex development from a zygote to an adult. As the cells of the organism divide clonally, a series of decisions are made that determine what kind of cell each one will ultimately become.
12. The use of different portions of the genome is also clearly shown by organisms that pass through different morphological stages during their life cycles.

Key Terms

circadian rhythm 295
haploid 295
diploid 295
sexual cycle 295
gamete 295
fertilization 295
syngamy 295
meiosis 295
somatic cell 297
germ cell 297
gonad 297
testis/testes 297
sperm/spermatozoon 297
ovary 297
eggs/ova 297
zygote 297
vegetative growth 297
mating type 297
spore 297
gametophyte 298
sporophyte 298
isogamy 298
anisogamy 298
oogamy 298
karyotype 298
homologous chromosomes/ homologs 299
euploid 299
aneuploid 299
synapsis 300
bivalent 300
tetrad 300
chiasma/chiasmata 301
cross over 302
spermatogenesis 302
oogenesis 303
polar body 303
anther 303
pollen grain 303
endospore 305
differentiation 305

Multiple-Choice Questions

1. Sexual cycles must include
a. a diplophase.
b. a haplophase.
c. a zygote.
d. meiosis.
e. all of the above.

2. Which is haploid?
a. cells of the ovary
b. cells of the testes
c. somatic cells
d. the zygote
e. sperm

3. Among animals, plants, and microorganisms, meiosis is not found in
a. zygotes.
b. spores.
c. somatic cells.
d. haploid cells.
e. all of the above.

4. Vegetative growth is synonymous with
a. mitosis.
b. meiosis.
c. isogamy.
d. oogamy.
e. anisogamy.

5. A karyotype is a
a. set of chromosomes located on a mitotic spindle.
b. pictorial representation of a set of chromosomes.
c. chromosomally visible genetic abnormality.
d. drug used to detect genetic abnormalities.
e. haplontic life cycle.

6. In which respects are a pair of homologous chromosomes alike?
a. location of the centromere
b. banding pattern
c. overall length
d. length of each arm from the centromere
e. all of the above

7. The term *aneuploid* refers to a cell with
a. one or a few extra or missing chromosomes.
b. more than two sets of homologs.
c. a single haploid set of chromosomes.
d. two sets of homologs.
e. a normal karyotype.

8. Meiotic and mitotic cell cycles differ in
a. the stage when DNA is synthesized.
b. the number of chromatids per chromosome in prophase.
c. whether or not synapsis takes place.
d. the number of times DNA replicates per cycle.
e. whether or not chromosomes become oriented on a spindle.

9. Which is not a difference between mitosis and meiosis I?
a. separation of sister chromatids
b. number of chromatids per chromosome in prophase
c. formation of chiasmata

d. division of centromeres
e. formation of tetrads

10. Genetic variation arises during oogenesis as a direct result of
    a. chiasmata formation and crossing over during meiosis.
    b. the number of patterns in which tetrads can align on the meiotic spindle.
    c. the formation of only one ovum from each oocyte.
    d. *a* and *b*, but not *c*
    e. *a*, *b*, and *c*

True-False Questions

Mark each statement true or false, and if false, restate it to make it true.

1. In meiosis, chiasma formation occurs only in prophase I, not in prophase II.
2. In a diplontic cycle, meiosis occurs at the end of the diplophase, and in a haplontic cycle, meiosis occurs at the end of the haplophase.
3. Organisms with haplodiplontic cycles engage in fertilization after both the haplophase and the diplophase.
4. Cells replicate their DNA before going through mitosis, and those going through meiosis replicate their DNA before each meiotic division.
5. After meiotic anaphase I, each chromosome consists of one chromatid.
6. The chromosomal activity of the second meiotic division resembles mitosis in a haploid cell.
7. The cytoplasmic events in spermatogenesis and oogenesis are similar, but the nuclear events are quite different.
8. A flower ovary is to a human ovary as an anther is to a sperm.
9. In males, the four sperm that result from a single meiotic event are cytoplasmically and genetically alike.
10. If an organism has morphologically distinct stages in its life cycle, each stage must have its own special genome.

Concept Questions

1. If you compare meiosis I with mitosis, which stages are the most different?
2. Suppose you are given many unlabeled drawings of dividing cells in metaphase and asked to identify which show meiosis, which show mitosis I, and which show mitosis II. You are told that all are from diploid organisms, but you are not told what the haploid number is for each specimen. What criteria would you use, and how would each one help in your diagnosis?
3. Compare the timing of meiosis, fertilization, and vegetative growth in haplontic and diplontic life cycles.
4. If organisms produce gametes that are designated by + and − mating types, is it likely that the species is isogamous or oogamous? Explain.
5. What events of meiosis lead to an increase in genetic variability within a species? Are new genes formed by these activities?

Additional Reading

Binkley, Sue. *The Clockwork Sparrow: Time, Clocks, and Calendars in Biological Organisms.* Prentice-Hall, Englewood Cliffs, NJ, 1990.

Cole, Charles J. "Unisexual Lizards." *Scientific American,* January 1984, p. 94. Whiptail lizards reproduce by parthenogenesis: virgin birth.

John, Bernard. *Meiosis.* Cambridge University Press, Cambridge and New York, 1990.

Internet Resource

To further explore the content of this chapter, log on to the web site at:

http://www.mhhe.com/biosci/genbio/guttman/

Mendelian Heredity

16

Key Concepts

16.1 Heredity is determined by discrete, conserved "factors."

16.2 Regular inheritance depends upon stability of the chromosome set.

16.3 The chance of two events happening together is the product of the chances that they will happen independently.

16.4 Not all genes exhibit dominance.

16.5 Genetic crosses show only a few simple patterns.

16.6 Genotypes can be determined with testcrosses.

16.7 Two genes may be inherited quite independently of each other.

16.8 Interactions between genes can produce unexpected ratios of phenotypes.

16.9 Genes are linked to one another.

16.10 Many genes have multiple effects.

16.11 Any gene can have more than two alleles.

16.12 Special chromosomes often determine an organism's sex.

16.13 Genes on sex chromosomes show a distinctive pattern of inheritance.

Modern genetics tells us that the color patterns of animals are inherited like other characteristics, not produced by environmental influences.

The book of Genesis records that when Jacob was working for his father-in-law Laban, he set spotted and striped sticks in front of the sheep and goats to influence them to produce young with similar markings, which were to become his own by contract. The story shows the concern of primitive farmers and herdsmen for improving their plants and animals, as well as their superstitions about influences on heredity. In spite of these early limitations, they were gradually able to cultivate more nutritious grains, cattle that gave more and better milk, faster horses, and dogs specialized for different kinds of work. Myths and folklore show that people have also been deeply concerned with their own inheritance; all kinds of rituals have been passed along for ensuring the birth of either a boy or a girl. Monstrous births have inspired wonder and fear, as well as spells and blessings for producing normal babies.

Today, people concerned about genetic diseases they may be carrying consult genetic counselors for advice, rather than applying to the local shaman, but they are only able to get sound advice because of the understanding of inheritance that has developed in the last century and because of techniques for analyzing hidden genetic factors that have been developed very recently. Knowing the rather simple rules that govern the inheritance of genes and the traits they produce, we do not have to learn different rules of inheritance for each characteristic. The goal of this chapter is to develop the principles that govern most situations, so we can determine the probability that various combinations of characteristics, and certain disorders, will occur. We will apply these principles to human heredity in Chapter 19. ■

16.1 Heredity is determined by discrete, conserved "factors."

Heredity in humans is obvious (Figure 16.1). We can easily identify subtle features, such as a certain shape of nose or ear, that link parents with their children. We know of families of tall people or short people, families whose members seem to live long lives and others who tend to get gray hair at an early age. However, the rules that govern inheritance are not nearly as obvious as the traits themselves, and were unknown until the ground-breaking work of Gregor Mendel came to light around 1900.

Mendel was successful because he took a new approach to the study of inheritance. First, he studied only a few selected characteristics that showed either one definite form or another, rejecting any trait that had, as he said, a "more or less" character. He therefore chose flower colors because they were either purple or white, never violet, and he chose pea shapes because they were unambiguously either round or wrinkled. Mendel realized he could not follow the inheritance of characters that cannot be described plainly.

Second, Mendel studied each characteristic independently. He did not allow data on the inheritance of flower color to be confused by information about the inheritance of seed shape or seed color. The idea that each characteristic is determined by separate factors is now fundamental to modern genetic thought.

Mendel began with **purebred** strains of peas, plant varieties that had been inbred (bred only with themselves) for several generations, so they consistently showed only one form of each trait. Each of his experiments was a **cross,** or hybridization, of two such strains—that is, breeding them with each other to make hybrid plants. In the experiment illustrated in Figure 16.2, Mendel crossed purebred yellow-seeded plants with purebred green-seeded plants by removing the pollen-bearing parts of each plant and dusting the plant with pollen from the other

Figure 16.1
The inheritance of traits in a family is usually quite obvious.

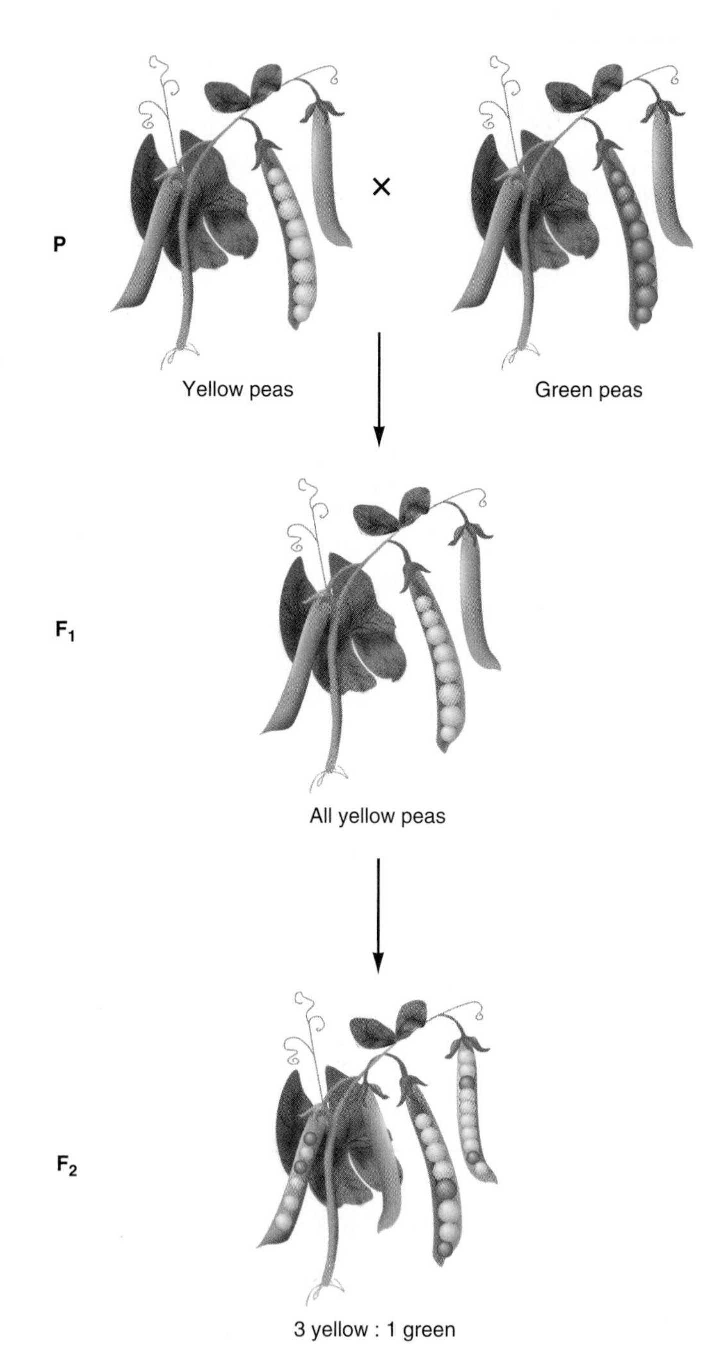

Figure 16.2
In one of Mendel's experiments on inheritance in peas, he studied seed color by crossing purebred yellow and green strains to make an F_1 generation, which had all yellow seeds. When these seeds were planted and the resulting plants were crossed, the F_2 generation had a ratio of three yellow seeds to one green seed.

strain. Though the resulting peas might have been yellow-green intermediates, or a mixture of green and yellow, they all turned out to be yellow. It was as if the green color had simply disappeared in this generation, the **first filial generation** (*filius* = brother), abbreviated $\mathbf{F_1}$. Mendel then planted the F_1 seeds and crossed the resulting plants with one another to make a **second filial generation** ($\mathbf{F_2}$). Remarkably, the green color that had disappeared in the F_1 plants appeared again; some of the F_2 seeds were yellow, and some were green. The other features Mendel studied showed the same pattern. Thus, when a purebred strain with purple flowers was crossed with a purebred strain with white flowers, all the F_1 plants had purple flowers, but the white color reappeared in the F_2 generation.

Another of Mendel's innovations was to *count* the plants (or seeds) with each characteristic, for he believed heredity could only be unraveled through quantitative, rather than merely descriptive, work. Furthermore, he looked at his numbers statistically. Following the inheritance of seed color, he obtained 6,022 yellow and 2,001 green seeds in the F_2 generation. Working with flower colors, he obtained 705 purple and 224 white plants in the F_2. These numbers look meaningless by themselves, and in similar situations, Mendel's predecessors had thrown up their hands and concluded that nothing could be concluded. Mendel, however, saw that both sets of numbers have a ratio very close to 3 to 1, and this observation led him to two simple explanations.

First, although he did not use these modern-day terms, Mendel assumed that the hereditary factors, whatever they are, are *conserved* from generation to generation—that is, the factors are not destroyed, even if their effects cannot be seen. This principle explains how green seeds or white flowers can reappear in the F_2 generation after disappearing in the F_1. The concept of genetic conservation is a foundation of our understanding of inheritance, and it is a shame Darwin didn't know about Mendel's work, because it provided the answer to a major objection that was raised against the idea of natural selection. Having a vague notion that features are somehow blended during inheritance, critics argued that if a superior variant appeared its good features would be diluted and lost. Mendel's work, however, showed that hereditary factors are conserved and can be recombined in future generations to produce other individuals that, like their ancestors, have superior features.

Second, Mendel assumed that every plant carries two hereditary factors for each characteristic, and that when a plant has two different factors, one of them is **dominant,** meaning its effect is visible, while the other is **recessive,** meaning its effect is hidden. In the garden pea, yellow seed color is dominant and green is recessive; in flower color, purple is dominant to white. This feature of heredity is the basis of one common system for symbolizing hereditary factors. A capital letter is used to represent the dominant factor, and a lowercase letter is used for the corresponding recessive factor. For instance, *Y* stands for the factor that specifies yellow seed color, and *y* stands for the one that determines green. Today, we realize these two factors are alternative forms of a single gene that determines seed color, and we call them **alleles,** or *allelomorphs,* of each other (*morph* = form; *allelon* = of each other).

The inheritance of pea color illustrates Mendel's hypothesis (Figure 16.3). One parent of each plant provides the pollen (containing sperm), and the other provides an egg. If both parents contribute equally to heredity, each plant must have two factors for each trait. (Today we say such plants are diploid.) The purebred yellow-seeded plants must have two copies of the yellow-determining allele *YY,* while the purebred green-seeded plants have two copies of the green-determining allele (*yy*). Individuals with a pair of identical alleles for a certain gene are **homozygotes,** or are said to be **homozygous.** *YY* is the

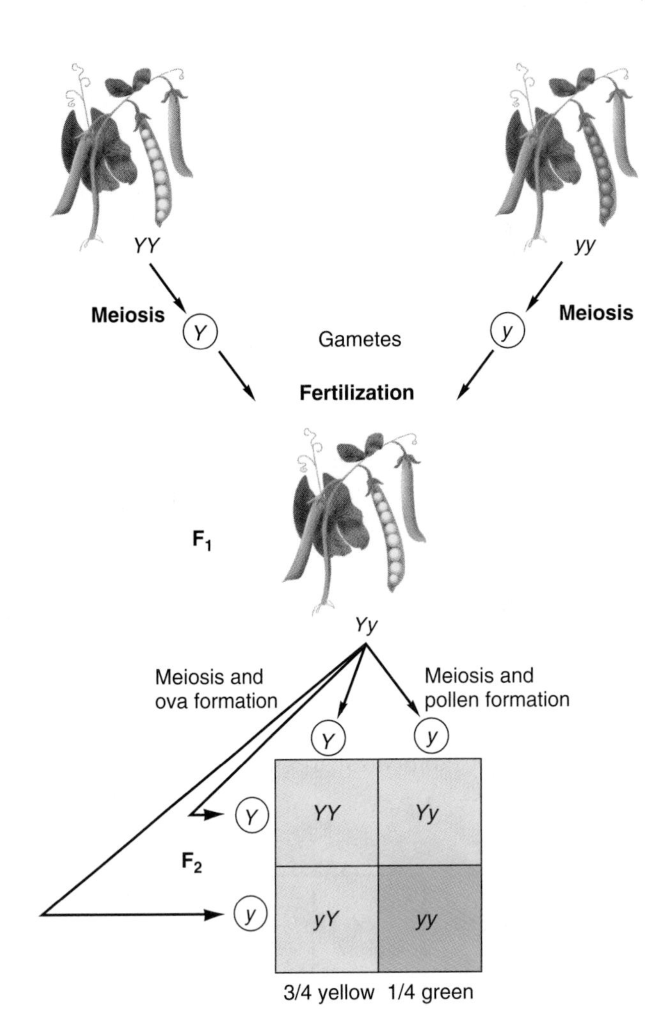

Figure 16.3

The inheritance of seed color is explained if the purebred parental strains carry two factors that determine color and are either *YY* (yellow) or *yy* (green). In meiosis, one parent produces gametes that carry *Y*, while the other produces gametes that carry *y*, so the F_1 generation is all hybrid *(Yy)*. Since *Y* is dominant to *y*, they have yellow seeds. Each of these plants produces half *Y*-bearing and half *y*-bearing gametes, so random mating of the *Yy* plants produces a ratio of 1 *YY* to 2 *Yy* to 1 *yy*, which translates into a ratio of three yellow peas to one green.

homozygous dominant combination, and *yy* is the homozygous recessive.

While somatic cells must have two factors, germ cells can have only one. Each sperm or egg can receive only one of the two factors that determine each trait, so each individual in the next generation again has only two factors, one from each parent. Otherwise, as we noted in Chapter 15, the number of factors—like the number of chromosomes—would double each generation. The factors, of course, are genes on chromosomes, which are reduced from two sets to one during meiosis. Remember, however, that Mendel knew nothing about chromosomes or meiosis.

When the parental plants are crossed, each gamete carrying a *Y* factor combines with one carrying a *y* factor, so all the F_1 plants—regardless of which parent produced the pollen and which the eggs—are hybrid and carry the factors *Yy*. Individuals carrying different alleles are said to be **heterozygous,** or are **heterozygotes.** For some reason, the *Y* factor is dominant, so all the F_1 seeds are yellow.

These seeds grow into pea plants whose somatic cells carry both factors, but when they form germ cells, each sperm or egg gets only one of the two factors: Half the gametes get the *Y* allele, and half get the *y* (Figure 16.3). Mendel said the two factors **segregate,** or separate, from each other during the formation of germ cells, though he knew nothing of the actual mechanism. We describe this principle in modern terms as *Mendel's First Law (Law of Segregation):*

> *The alleles of each gene segregate from each other during the first division of meiosis, so half the gametes carry one of the two alleles and the other half carry the other allele.*

Now suppose the sperm and eggs of the F_1 generation combine at random to form the F_2 generation. The results, as Mendel could see, are exactly like the results of flipping two coins simultaneously. Since each coin can land heads (H) or tails (T), the possible combinations are HH, HT, TH, and TT. Similarly, random fertilization can produce *YY*, *Yy*, *yY*, and *yy*, as illustrated in this Punnett square:

Pollen \ Egg	*Y*	*y*
Y	*YY* Yellow	*Yy* Yellow
y	*Yy* Yellow	*yy* Green

Half the eggs carry *Y*, and half *y*; the same is true for pollen. A large population of eggs and pollen therefore produces four combinations with equal frequency: one-quarter *YY* zygotes; one-half *Yy* or *yY*, which are the same; and one quarter *yy*. The first three combinations have at least one dominant *Y* factor, and these seeds will be yellow. The fourth has two recessive *y* factors, making green seeds. The ratio should be three yellow peas to one green, which is what Mendel observed.

An organism's observable characteristics, its **phenotype,** often will not reveal all the information encoded in its genome, its **genotype.** Because of dominance, a heterozygote may show exactly the same phenotype as a homozygous dominant individual. Some of the yellow peas in these experiments have the genotype *YY*, and some have the genotype *Yy*. The ratio of different genotypes (in this case, 1:2:1) is often different from the ratio of different phenotypes (here, 3:1). Another reason the phenotype may not reveal the genotype is that the environment in which an organism develops also has a role in determining the phenotype, as explained in Sidebar 16.1. Notice that in specifying an organism's genotype we only note the one gene, or a few genes, in question. We don't try to specify its entire genotype by describing each of the thousands of genes it carries. The most commonly confused terms in genetics are summarized in Concepts 16.1.

Mendel's ability to reason from his observations to this simple, elegant explanation of heredity represents a major triumph of human intelligence. But notice how important it was for him to realize that patterns of inheritance depend on such random events as the combination of a sperm with an egg. It was this insight that forced him to think statistically and to recognize that he could hardly expect to obtain phenotypic ratios of precisely 3 to 1.

16.2 Regular inheritance depends upon stability of the chromosome set.

We have just seen that the pattern of Mendelian inheritance follows directly from the events of the sexual cycle—meiosis and fertilization. It also depends on the fact that a chromosome set is a specific arrangement of genes that usually does not change.

Every sexually reproducing organism has a characteristic set of chromosomes, with a specific number (*n*) in its gametes

Concepts 16.1 Genetic Confusables

Gene or allele: The word *gene* refers to one hereditary unit, usually carrying the information for one kind of polypeptide. A diploid organism has (generally) two copies of each gene. An allele is a form of a gene; alleles result from mutation and encode variants of the same polypeptide. Although no individual normally can have more than two different alleles of a gene, many may exist within a population if different mutations have occurred in the gene.

Homozygous or heterozygous: An individual carrying two identical alleles of one gene is homozygous for that allele. If it carries two different alleles of a gene, it is heterozygous.

Genotype or phenotype: Genotype describes the information in a genome generally specifying which alleles of a gene the organism carries. Phenotype is the visible expression of that information, as seen in the organism's form, the proteins it can make, how it grows, and so on.

Sidebar 16.1 Heredity and Environment

People often argue about whether heredity or environment—"nature" or "nurture"—is most important in determining an organism's characteristics, especially the traits of humans. The disputants tend to argue as if the truth is all on one side or all on the other, whereas in reality, both factors act to different degrees in different situations. Still, when thinking about heredity, it is easy to forget the effect of the environment on how every organism grows. An organism's traits are always partly determined by its environment, which influences how its genetic information will be expressed. Identical twin babies separated at birth can grow into very different adults if one is raised with good nutrition in healthful surroundings by loving parents while the other is deprived of these influences. Similarly, a plant growing in poor soil with barely adequate water will certainly not be the same as one with identical genes growing in a very supportive environment. The genotype reacts to a particular environment. The range of possible forms it might express is the *norm of reaction,* and the environment determines precisely where the phenotype falls within that range.

Every population is variable. Even if we plant a single package of commercial beans, which have been bred for uniformity, the next generation of beans will vary in size. A graph of the number of beans of each size is a *normal distribution curve:*

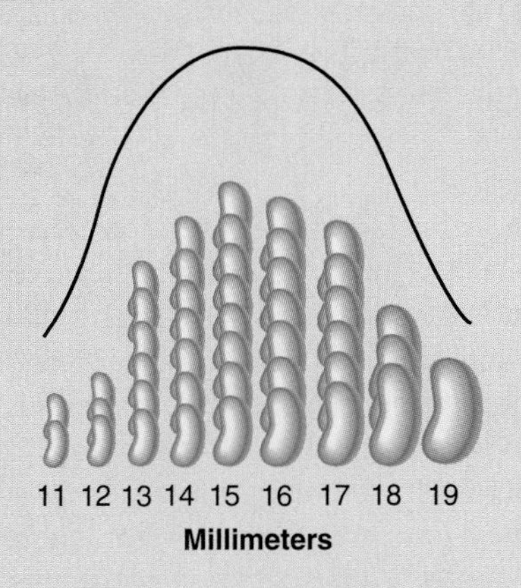

Most beans cluster around a mean (average) of, say, 15 mm, while the graph tapers off sharply on both sides. The width and shape of the distribution curve are described by its *variance.* That is, if the mean length of the beans is $\overline{x}$ and the length of each individual bean is x, we take the difference ($\overline{x} - x$) for each bean, sum the squares of these numbers, and divide by the number of beans. The resulting number is the variance. Uniform populations have small variances (narrow curves), and more diverse populations have large variances (broad curves).

The variability comes from both heredity and environment. Slight differences in the genomes of the beans make them grow to different sizes, and variations in the amounts of water, sunlight, and nutrients make some grow larger than others. We can reduce the genetic variance of a population by controlling both genetic and environmental factors, but the variance will never be zero.

From a variable strain of beans, we can try to breed lines of small beans and large beans by repeated selection (see illustration). Several generations of selective breeding will, indeed, create two distinct strains, but we cannot push the beans beyond the limits of their genetic potential. Selection, whether natural or artificial, can only make use of genetic variance. We could induce enormous variation in the tail lengths of mice by cutting their tails off at various points with a carving knife, but the variation would be environmental, and the next generation of mice would all have tails of approximately the standard length. (Remember the implications of Crick's Central Dogma; there is no reverse flow of information from protein [phenotype] to DNA [genotype].) The more genotypic variance a population holds, however, the more raw material there is for evolution to work on, and the rate of evolution in a population is proportional to its genotypic variance.

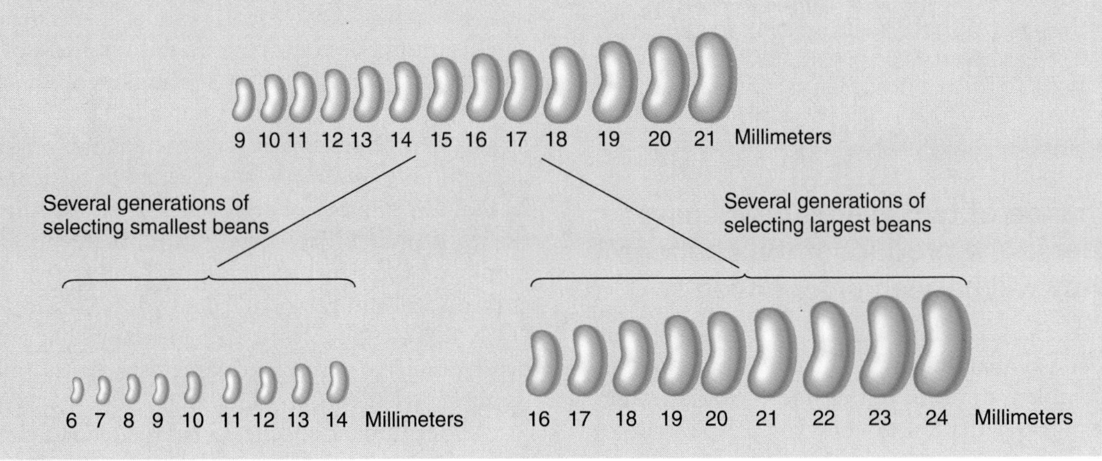

and twice that number ($2n$) in its somatic cells. (Only sex chromosomes, discussed in Section 16.12, may make some variation in the diploid set.) The chromosomes of each organism have a relatively stable arrangement of genes, so the two homologous chromosomes of each kind are essentially the same, though rarely identical in every gene. For example, somewhere in the pea genome (for which $n = 7$) is a gene that determines seed color. There may be many such genes, but breeding experiments clearly show there is at least one. In fact, the seed-color gene Mendel studied is known to be located at a certain position on chromosome #1 (Figure 16.4). Except in very rare instances, this gene occurs in the same position on *every* copy of chromosome #1, in every plant. Because there are two alleles of the gene, one specifying yellow seeds *(Y)* and the other green seeds *(y)*, some #1 chromosomes carry the *Y* allele and others carry *y*.

Another gene that determines the height of the plant is on chromosome #4, and one that determines whether the pods are green or yellow is on chromosome #5. These genes do not move around at random either, although chromosomes #4 and #5 in different individuals may have different alleles of these genes.

When homologous chromosomes separate in meiosis, the genes they carry are segregated, so each gamete receives only one allele of each gene. But if genes could move easily from one chromosome to another, some gametes might receive two alleles of a single gene while others received none. Such instability would make heredity intolerably chaotic. Although events like this actually happen, they are so rare that we can ignore them in establishing the usual patterns of heredity.

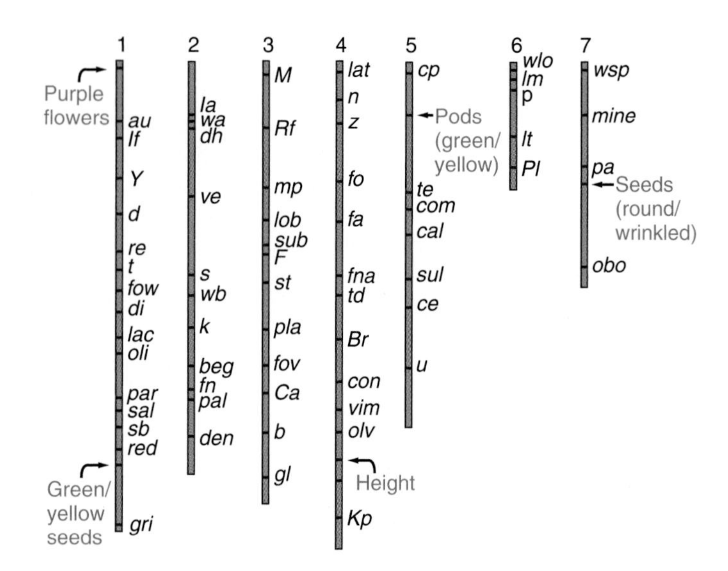

Figure 16.4

The chromosome set of each species, such as this haploid set of pea chromosomes, is quite stable. Each abbreviation is the name of an identified gene, which remains at a point on one chromosome, except for very rare chromosomal rearrangements. This stability, combined with the normal precision of meiosis, is what creates viable haploid sets that can combine to produce viable diploid sets.

Exercise 16.1 You can make model chromosomes from pieces of wire or string. Make a maternal set of one long and one short chromosome from one color of wire and a similar paternal set from another color. Then combine the sets in fertilization and carry them through DNA replication (add another wire of the matching color to each chromosome) and meiosis. Be sure to leave each chromosome intact during the first meiotic division and divide it during the second. Repeat this exercise until you hardly have to think about it.

16.3 The chance of two events happening together is the product of the chances that they will happen independently.

It is convenient to analyze simple genetic problems by using Punnett squares and similar drawings, but as genetic situations become more complex, drawing out all the possibilities becomes very clumsy. It is better to learn how to think in terms of probabilities and their combinations.

Suppose the chance that it will rain today is 20 percent (0.2), and the chance that you will get a letter from a friend today is 10 percent (0.1). What is the chance that it will rain and you will *also* get a letter? The probability of these (or any) two events happening together is the *product* of their individual probabilities—in this case, $0.2 \times 0.1 = 0.02$, or 2 percent.

Figure 16.5 illustrates the situation very simply. Let the area of a square represent the whole universe of possible events; each side of the square has a length of 1 arbitrary unit. Divide the square horizontally into 20-percent and 80-percent segments, representing the chances of rain and no rain, respectively. Now divide the square vertically into 10-percent and 90-percent segments, representing the chances of getting a letter and not getting a letter, respectively. The square is now divided into four rectangles. Each rectangle represents the simultaneous occurrence of two events, and its area is the product of the lengths of its sides. The sum of all four areas, of course, is 1.0. For example, the area of the rectangle representing the chance of getting neither rain nor a letter is $0.8 \times 0.9 = 0.72$, which tells us that 72 percent of the time, neither of these events will happen. This notion can be extended to calculate the probability that any number of events will occur together. So if you estimate a probability of one-third that you will find some money on the street, then the probability that it will not rain and you will not get a letter and you will find some money is $1/3 \times 0.72 = 0.24$. Notice, incidentally, that probabilities can be expressed as percentages, fractions, or decimals, whichever is most convenient.

The arithmetic of genetics is a matter of combining elementary probabilities. Thus, in the example of pea color, the probability of getting a green pea is the probability of getting a *y* egg (which is 1/2) times the probability of getting a *y* pollen (also 1/2), and $1/2 \times 1/2 = 1/4$. This way of thinking can be extended easily to more complicated situations, keeping in mind one more rule: If an event can occur in two or more ways, one must *add* the probabilities of each way, as the following exercise illustrates.

Exercise 16.2 *(a)* What is the probability of rolling a 12 with a pair of dice? *(b)* What is the probability of rolling an 11? Since 11 can be rolled in two ways (5 + 6 or 6 + 5), you must add the two separate probabilities to get the total. *(c)* What is the probability of rolling a 7? (Again, you must add several probabilities for mutually exclusive events.)

16.4 Not all genes exhibit dominance.

In his breeding experiments with peas, Mendel examined several characteristics that showed simple dominance of one allele over another. But all genes aren't like this. In the common garden zinnia, for example, a cross of purebred red-flowered plants with purebred white-flowered plants produces F_1 plants with pink flowers (Figure 16.6). If the F_1 flowers are allowed to pollinate themselves, the F_2 generation has a mixture of one-quarter red, one-half pink, and one-quarter white.

These results simply mean that red flower color shows *incomplete dominance* to white in zinnias. Zinnias have a gene *(R)* encoding an enzyme that makes a red flower pigment. An *RR* cell in a flower petal has two copies of the gene, gets two units of the enzyme, and therefore makes a full red pigment. A mutant gene *(r)* produces a defective enzyme, so homozygous *rr* plants cannot produce any red pigment and have white flowers. An *Rr* flower cell has only one *R* gene, produces only one unit of enzyme, and therefore has a diluted red, or pink, color. A single dose of enzyme in some plants is enough to produce the same phenotype as two doses, while in zinnias it produces a different phenotype. We will see other cases later in which two alleles contribute more or less equally to a phenotype.

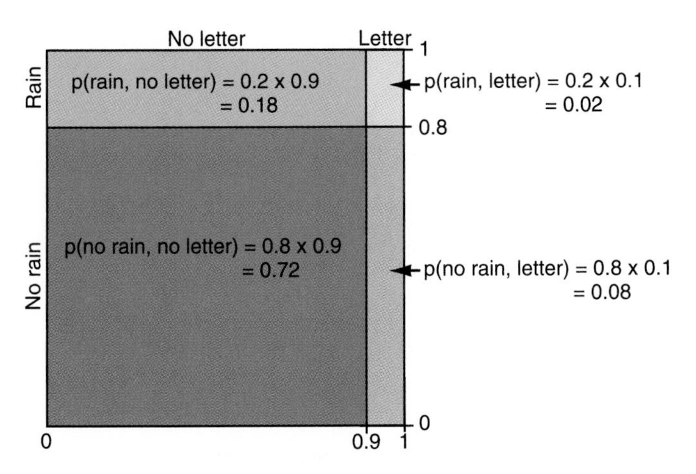

Figure 16.5
The probability of rain versus no rain is represented by dividing this square horizontally, while the probability of receiving a letter versus receiving no letter is shown by dividing it vertically. The probability of two events occurring together is proportional to the area of a rectangle where the two events intersect, which is the product of the individual probabilities.

A cross between two heterozygotes, say *Aa*, always produces offspring in a genotypic ratio of 1:2:1 (for *AA*, *Aa* and *aa*), but dominance usually produces a phenotypic ratio of 3:1. You should now be able to recognize this phenotypic ratio as meaning that the characteristic under consideration shows simple dominance and that the parents are both heterozygotes.

Exercise 16.3 Most humans can taste the bitter compound phenylthiocarbamide (PTC). Tasting is determined by a dominant allele *(T)*; nontasting depends on a recessive allele *(t)*. What are the phenotypes of *TT, Tt,* and *tt?*

Exercise 16.4 Two mice with black fur are mated with each other several times and produce 24 offspring, 19 with black fur and 5 with cinnamon-colored fur. What can you conclude about the parents and about the inheritance of fur color? (*Hint:* Given 24 offspring, what are the expected numbers of each phenotype if a single gene determines fur color? What must the genotypes of the parents be to get such a ratio?)

Exercise 16.5 In fruit flies, gene *b* determines black body color, and its dominant allele *B* determines wild-type body color. If a black fly is mated with a homozygous wild-type fly, what kind of offspring will they have? What will an F_2 generation look like?

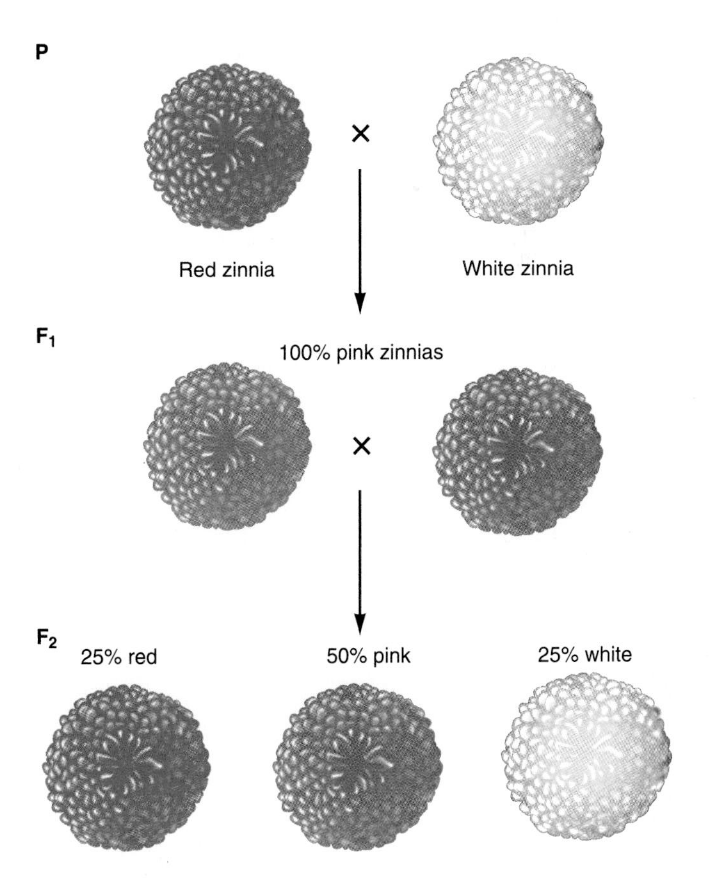

Figure 16.6
When red and white zinnias are crossed, the first generation of hybrids is all pink. The next generation is one-quarter red, one-half pink, and one-quarter white.

16.5 Genetic crosses show only a few simple patterns.

It has been said that the purpose of thinking is so you don't have to think. In other words, if you reason out a situation beforehand, you can react quickly when it arises, instead of thinking about it anew each time. People who are familiar with genetics know that only a few basic situations are possible, and analyzing them in general makes it easy to think about each case in particular.

Table 16.1 shows that there are only six possible matings involving a pair of alleles, *A* and *a*. All basic genetic situations are one of these six. For instance, the crosses *AA* × *AA* and *aa* × *aa* each yield only offspring of the same genotype. Instead of struggling with every new problem, it is easier to learn these situations and then recognize them when you encounter what appears to be a new problem.

Exercise 16.6 Convince yourself that the ratios of progeny in Table 16.1 are correct by working each situation out carefully.

16.6 Genotypes can be determined with testcrosses.

In peas, purple flower color, due to allele *P*, is dominant to white color, due to allele *p*. Pea plants with purple flowers could therefore be either homozygous *(PP)* or heterozygous *(Pp)*. When part of a genotype is unknown or unspecified, it is represented by a hyphen, so a purple plant has the genotype *P*-. How can we tell its actual genotype?

First, assume the plant is *PP*. It can only make gametes bearing *P*, so if it is bred with a white plant, which can make only *p*-bearing gametes, all the offspring will be heterozygous *(Pp)* and therefore purple. Second, assume the plant is *Pp*. Half its gametes should bear the *P* allele, and half the *p* allele. If it is bred with a white plant, half the offspring should be *Pp* and therefore purple, while the other half should be *pp* and therefore white. Thus the way to determine the genotype of an individual with a dominant phenotype is to mate it with a homozygous recessive individual. Such a mating, called a **testcross,** reveals the unknown genotype (Figure 16.7). A homozygote produces only offspring with the dominant phenotype, while a heterozygote produces equal numbers of dominant and recessive phenotypes. This is a very useful test.

Table 16.1 Possible Matings with a Pair of Alleles

Mating	Progeny
AA × *AA*	All *AA*
AA × *Aa*	1/2 *AA*, 1/2 *Aa*
AA × *aa*	All *Aa*
Aa × *Aa*	1/4 *AA*, 1/2 *Aa*, 1/4 *aa*
Aa × *aa*	1/2 *Aa*, 1/2 *aa*
aa × *aa*	All *aa*

Exercise 16.7 A female mouse with black fur is mated with a cinnamon-colored male. Approximately half their offspring are black, and half are cinnamon. What is the female's genotype?

16.7 Two genes may be inherited quite independently of each other.

In Mendel's experiments with peas, he looked at several traits at once, although he kept them separate in his analyses. He saw that the cross of a purple-flowered and a white-flowered plant produced an F_2 generation that was three-quarters purple and one-quarter white, while a parent with round seeds and another with wrinkled seeds produced an F_2 generation with three-quarters round seeds and one-quarter wrinkled seeds. Each trait was clearly inherited by itself in the same, simple way, and each one showed dominance independently of the others. We now know that these traits are inherited independently because their genes are on different chromosomes. (Many genes for other traits are actually linked on each chromosome, so they are inherited together. In this section we will only consider independent traits.)

To calculate the probability that a *combination* of traits will occur in a cross, we have to follow Mendel's basic rule: Since the traits are inherited independently, *think* about them independently. To illustrate this point, consider seed color and seed shape in peas (Figure 16.8). We begin by crossing a parent that is homozygous for yellow seeds *(YY)* and round seeds *(RR)* with another parent homozygous for green seeds *(yy)* and wrinkled seeds *(rr)*. All plants in the F_1 generation have both dominant traits,

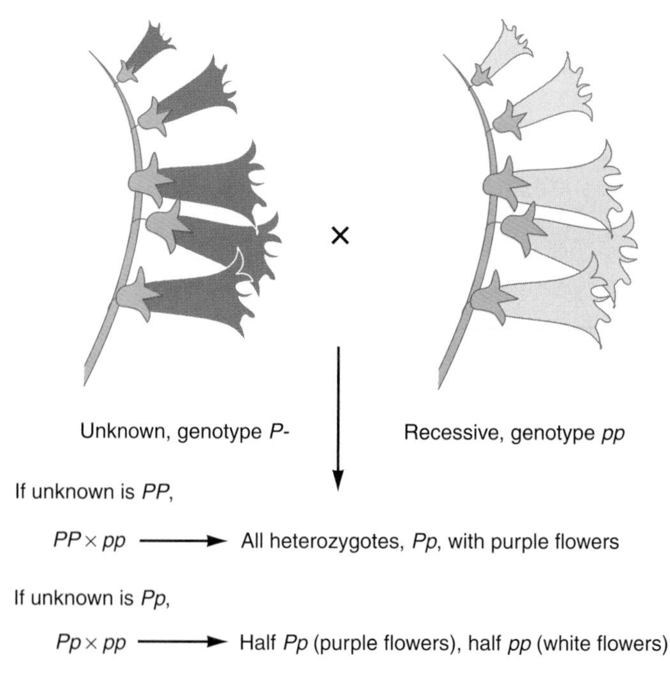

Figure 16.7
A testcross distinguishes between homozygous and heterozygous dominant individuals on the basis of their offspring.

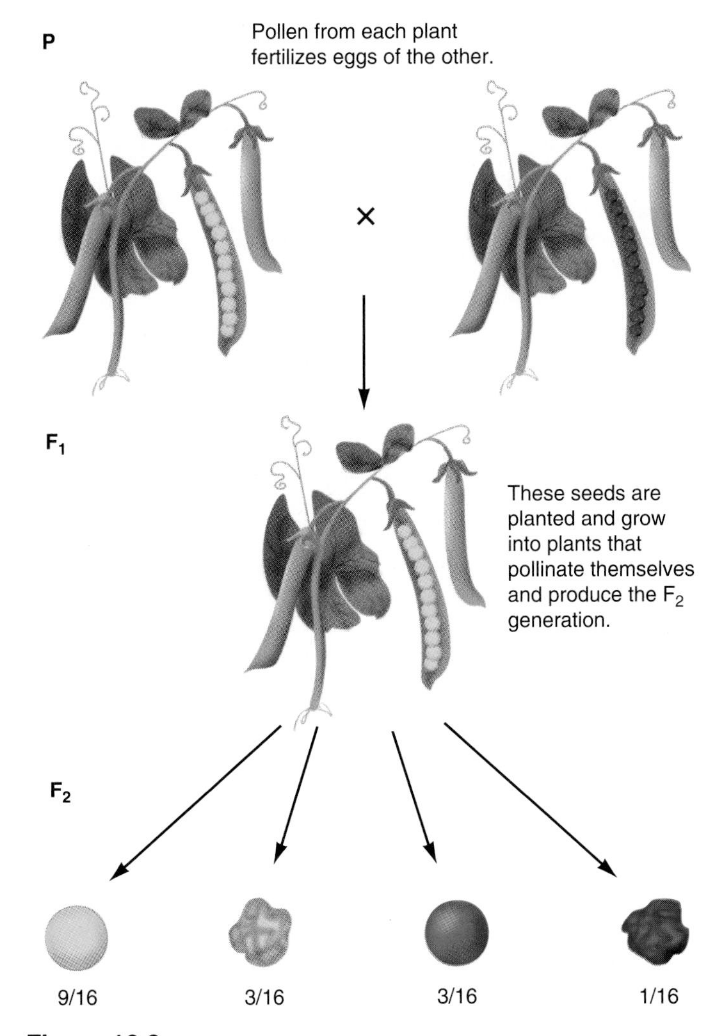

Figure 16.8
In a *dihybrid* cross, two characteristics, determined by different genes, are inherited with a regularity that is only slightly more complicated than the inheritance of a single characteristic. Beginning with homozygous yellow, round peas and homozygous green, wrinkled peas, the F_1 generation is all yellow, round. In the F_2 generation, there is a ratio of 9:3:3:1 of four different combinations.

producing yellow, round seeds. Crossing these F_1 plants with one another produces an F_2 generation with a ratio of 9/16 round and yellow, 3/16 wrinkled and yellow, 3/16 round and green, and 1/16 wrinkled and green. How can we understand this ratio?

These numbers are to be expected if the genes for the two traits are on different chromosomes, like this:

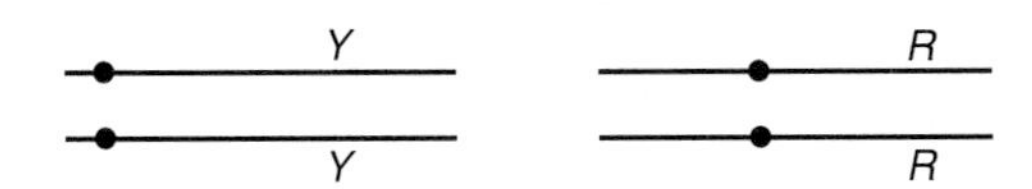

We begin with parental plants of genotypes *YY RR* and *yy rr*. When meiosis occurs in these plants to form gametes, each gamete gets one chromosome bearing a gene for seed color and one bearing a gene for seed shape. Because one parent can only make *Y R* gametes, and the other only *y r*, all the F_1 individuals must be *Yy Rr*. To determine the phenotype of such an individual (seed), we think about each trait independently, since in this simple situation the genes act independently: The seed color is yellow (seed shape does not affect the color), and the seed shape is round (seed color does not affect the shape). Individuals heterozygous for two genes are said to be *dihybrid.*

Now, after these seeds grow into mature plants and start to make their own gametes, what happens in meiosis? At metaphase I, the homologs in each chromosome pair are oriented at random on the metaphase plate, so there is no way to predict which way any homolog will move. Also, the separation of one pair doesn't influence the separation of any other pair, as shown in Figure 16.9. Therefore, the chromosomes can separate in either of two ways. Sometimes chromosomes bearing *Y* and *R* move to one pole while those with *y* and *r* move to the other, but it is just as likely that *Y* will go with *r* and *y* will go with *R*. This important point, which Mendel discovered, is known as *Mendel's Second Law (Independent Segregation):*

> *The alleles for genes on different chromosomes assort independently of one another in meiosis.*

The formation of eggs and sperm through meiosis is just the kind of large-scale dice game that can be analyzed by means of probabilities. With alleles of only a single gene segregating, there are only two kinds of gametes, and the probability of getting either one is one-half. With alleles of two genes, *R* and *Y*, there are four kinds of gametes—*Y R*, *Y r*, *y R*, and *y r*—so the probability of getting any particular one is one-quarter. Figure 16.10 shows the possible combinations of gametes in the F_2, using the Punnett square method. Again, to determine the phenotype of each individual, we must consider the two traits independently. Of the 16 combinations, nine have at least one *Y* allele and one *R* allele (*Y- R-*), so they have yellow, round seeds. Three of the 16 are *Y-* but are homozygous recessive for seed shape *(rr)* so they are yellow and wrinkled. Another three of the 16 are *yy R-*, so they are green, round seeds. Only one of the 16 has the genotype *yy rr*, which makes for green, wrinkled seeds. The phenotypic ratio 9:3:3:1 is a classical indication of a mating between two individuals who are both heterozygous for two genes on different chromosomes, with simple dominance.

By using probabilities alone, we can analyze this situation more quickly and with no chance of making an error in filling out a Punnett square. First consider seed color alone. Three-quarters of all seeds are *Y-*, making them yellow, and one-quarter are *yy*, making them green. Then consider seed shape: three-quarters of the seeds are *R-*, making them round, and one-quarter are *rr*, making them wrinkled. Now, to calculate the probabilities of each combination of features, we multiply these probabilities and find that the chances of getting seeds with particular phenotypes are:

Yellow and round, $3/4 \times 3/4 = 9/16$
Yellow and wrinkled, $3/4 \times 1/4 = 3/16$
Green and round, $1/4 \times 3/4 = 3/16$
Green and wrinkled, $1/4 \times 1/4 = 1/16$

This way of analyzing a problem is fast and accurate if you think clearly about each part of the problem. But until you get used to thinking genetically, it is advisable to draw out the Punnett square and also calculate the probabilities.

Exercise 16.8 Using only two homologous pairs ($2n = 4$), how many different kinds of gametes can be made if the homologs of both chromosomes are different? With three homologous pairs ($2n = 6$), how many different kinds of gametes can be made if all chromosomes are different? What if $2n = 8$? What is the general formula for any value of n?

Exercise 16.9 If a purebred pea plant bearing yellow, wrinkled seeds is crossed with one bearing green, round seeds, how will the seeds of the F_1 generation appear? Does this cross produce different results from the one discussed above? What will the F_2 generation look like?

Exercise 16.10 Though eye color in humans is really determined by several genes in ways that are not entirely understood yet, assume for purposes of analysis that one gene *(B)* determines brown eye color and is dominant to blue color, determined by *b*. A brown-eyed woman, whose father has brown eyes and whose mother has blue eyes, marries a blue-eyed man. What color eyes can their children have, and in what proportions?

Analysis: In many situations, the genotypes of individuals are not given explicitly, so the first step is to deduce those genotypes from other information. This brown-eyed woman could be either *BB* or *Bb*. But her blue-eyed mother must have given her a *b* allele, so she is heterozygous. We know the husband is *bb,* from his phenotype. So this is a $Bb \times bb$ mating; half the children will be *Bb,* brown-eyed, and half will be *bb,* blue-eyed.

Exercise 16.11 Assume that the allele for brown hair *(Br)* is dominant to that for blonde hair *(br)*. (Here we use a two-letter symbol, *Br* or *br*, to represent a single gene.) If the parents from the previous exercise are both heterozygous for brown hair, what combinations of hair and eye colors can their children have, and in what proportions?

Exercise 16.12 Another kind of gene notation uses a symbol with a superscript plus sign to signify a wild-type allele and the symbol without a plus sign for a mutant. In fruit flies, a mutation causing purple eyes *(pr)* is recessive to the wild-type allele that produces red eyes (pr^+). Ebony body color (*e*) is recessive to normal tan body color (e^+). If two flies that are heterozygous for both genes are mated, what characteristics will their offspring have, and in what proportions?

Exercise 16.13 A fruit fly with wild-type eyes and wild-type body color is mated with a fly with purple eyes and ebony body color. The offspring all have red eyes, but half have tan bodies and half have ebony. What is the genotype of the fly with the wild-type appearance?

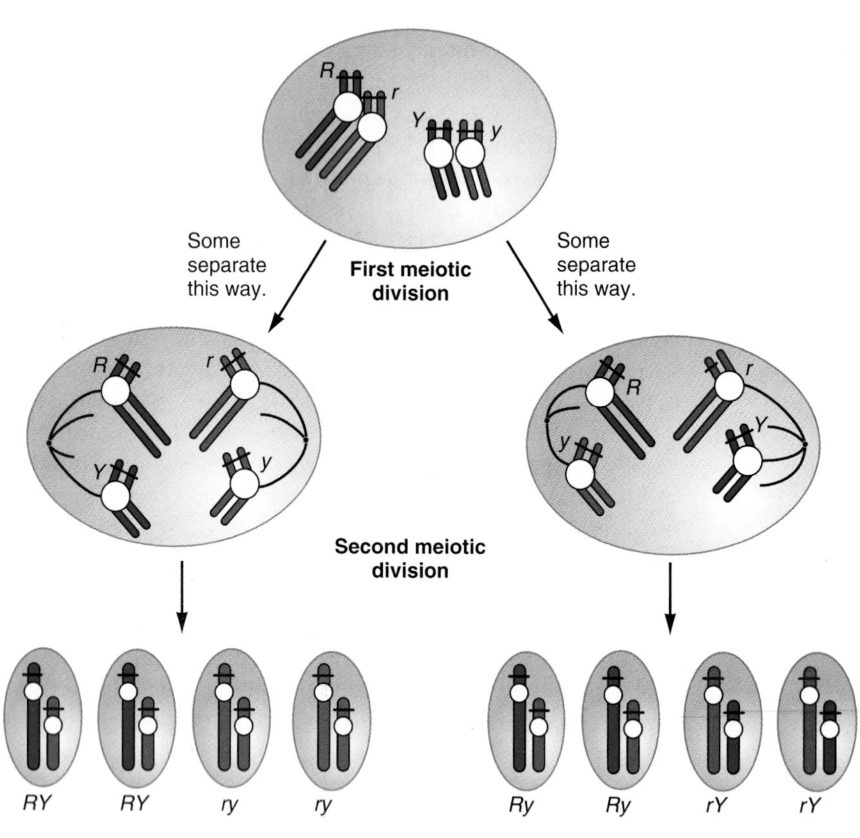

Figure 16.9

Suppose there are two pairs of chromosomes, one bearing alleles *R* or *r*, the other bearing alleles *Y* or *y*. In half the meioses *(left arrow at first meiotic division), R* segregates with *Y* and *r* with *y*; in the other half *(right arrow), R* segregates with *y* and *r* with *Y*, yielding four kinds of gametes in equal numbers.

16.8 Interactions between genes can produce unexpected ratios of phenotypes.

So far, we have considered crosses involving two genes that affect quite different characteristics, such as seed color and seed shape. But genes, we must remember, often carry the information for enzymes, and two genes may encode enzymes for a single pathway. In mice, gene *B* produces black fur, and the homozygote *bb* has brown fur. Gene *C* affects fur color, and a *cc* homozygote is white. If we cross mice with the genotypes *BB CC* and *bb cc*, the F_1 individuals are all *Bb Cc* and are brown. If we cross these offspring to produce an F_2 generation, we expect a 9:3:3:1 ratio of four genotypes, but the actual ratio is 9 black to 3 brown to 4 white.

Of the 16 F_2 combinations, nine are *B- C-* and therefore black. Three are *bb C-* and therefore brown. Three are *B- cc* and one is *bb cc,* but the latter four are white because the genotype *cc* produces white fur regardless of which *B* alleles are present. This phenomenon, in which one gene suppresses the effect of another, is called **epistasis,** a term meaning "standing upon." In this example, the recessive allele of the *C* gene "stands upon" the *B* gene and covers up the effects the *B* gene exerts by itself, so this effect is called *recessive epistasis.* A similar epistasis appears with genes that

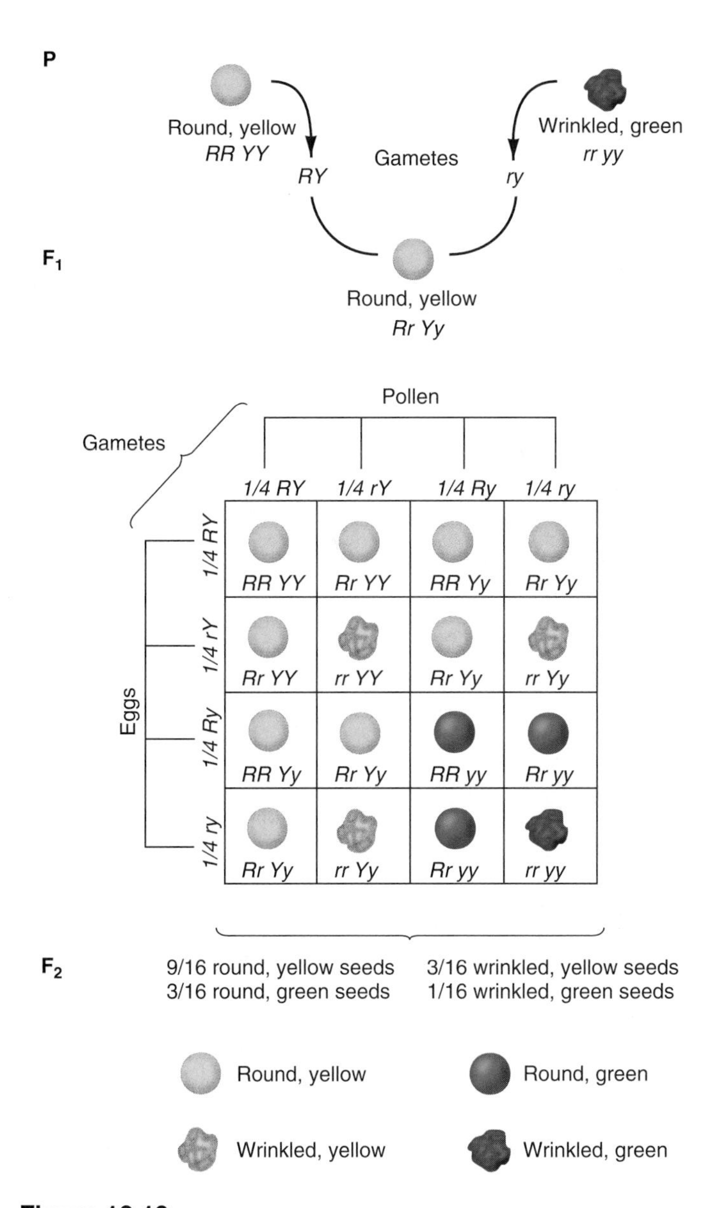

Figure 16.10

The genes for seed shape and seed color are inherited independently. Each parent makes equal numbers of the four types of gametes; the 16 combinations of gametes are easily seen by filling in the Punnett square.

encode enzymes in a single pathway (Figure 16.11). Suppose the recessive alleles of genes *M* and *N* produce the same phenotype when homozygous. Parents with the genotypes *MM nn* and *mm NN* produce F_1 individuals that are all *Mm Nn*. But instead of the 9:3:3:1 ratio in the F_2 generation, the ratio is 9:7. Nine of 16 genotypes are *M- N-*, so they have at least one dominant allele of each gene, and this provides enough of each enzyme to make a normal end product. The other seven have two recessive alleles of a gene, lack at least one enzyme, and therefore have an inoperative pathway. These examples show that the ratios in dihybrid crosses can provide important clues to the nature of the two genes and their possible interactions.

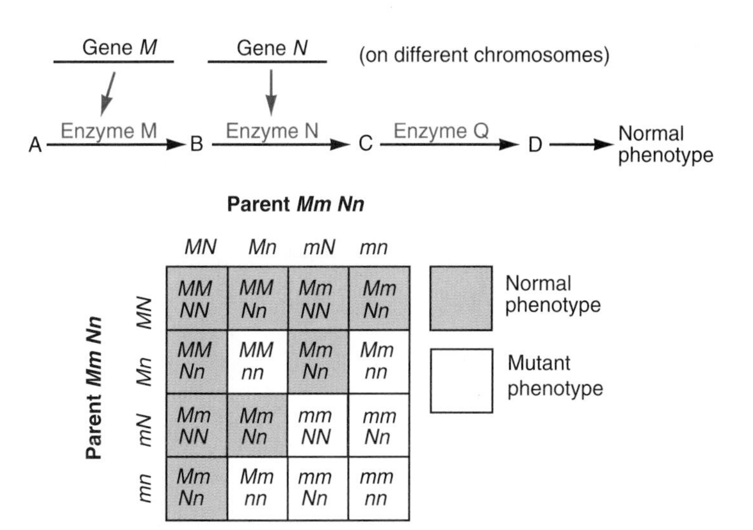

Figure 16.11

One kind of epistasis results from two genes, inherited independently, that encode enzymes in the same metabolic pathway. The normal end product, and its visible consequences, will be produced in 9 of 16 genotypes, which have at least one normal copy of each gene. The other 7 genotypes have two recessive alleles for at least one gene and therefore lack one enzyme or the other.

16.9 Genes are linked to one another.

When two genes with simple dominance are on different chromosomes, they are inherited independently, and the four phenotypes appear in the classical ratio of 9:3:3:1. But after Thomas Hunt Morgan and his students had accumulated a number of *Drosophila* mutants, they discovered exceptions to this pattern, owing to **gene linkage:** the existence of two genes on the same chromosome. Around 1913, A. H. Sturtevant, one of Morgan's students, reasoned that if genes are located at definite points on chromosomes, it should be possible to locate them—that is, to draw a **linkage map** of a chromosome.

To see how a linkage map is made, consider a cross between flies with vestigial wings, due to the *vg* gene, and flies with black bodies, due to the *b* gene (Figure 16.12). Since these mutations are recessive, the parents have the genotypes $vgvg\,b^+b^+$ and $vg^+vg^+\,bb$. (Notice again that the superscript plus sign indicates wild-type.) The F_1 flies are all $vg^+vg\,b^+b$, and because of dominance in both genes, they have wild-type wings and wild-type coloration.

However, instead of a 9:3:3:1 ratio in the F_2, about half the flies are wild-type for both traits, almost a quarter have black bodies, nearly another quarter have vestigial wings, and only about 1 percent have both black bodies and vestigial wings. These confusing numbers arise because the mutations *b* and *vg* are linked (Figure 16.13). The black flies and vestigial flies, having the original combination of alleles, are called **parental types.** The wild-type and double mutant flies, having different combinations of alleles, are called **recombinant types.** The original parental combinations, $vg^+\,b$ or $vg\,b^+$, tend to stay together, since chromosomes are quite stable. Most of the

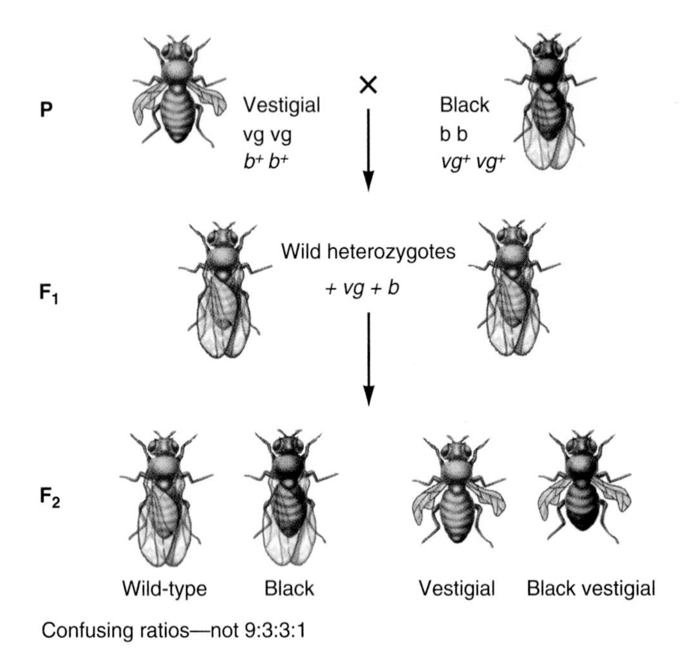

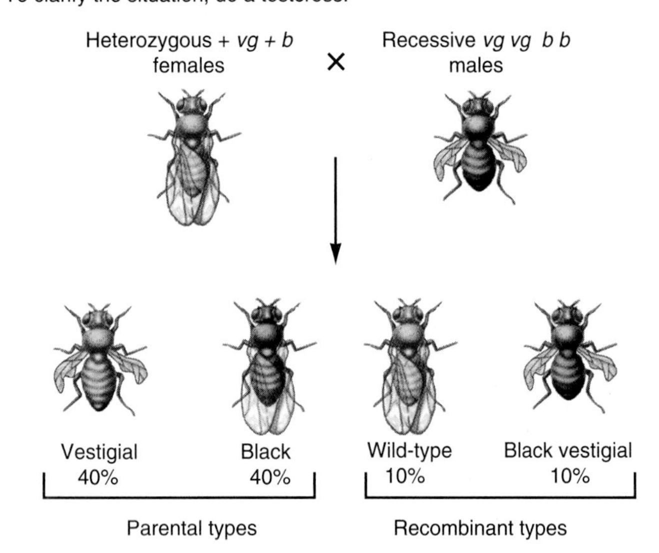

Figure 16.12

A cross between vestigial-winged and black-bodied flies yields very odd numbers in the F_2. The numbers from a testcross make more sense. They show that 20 percent of the chromosomes of the female flies are recombinant.

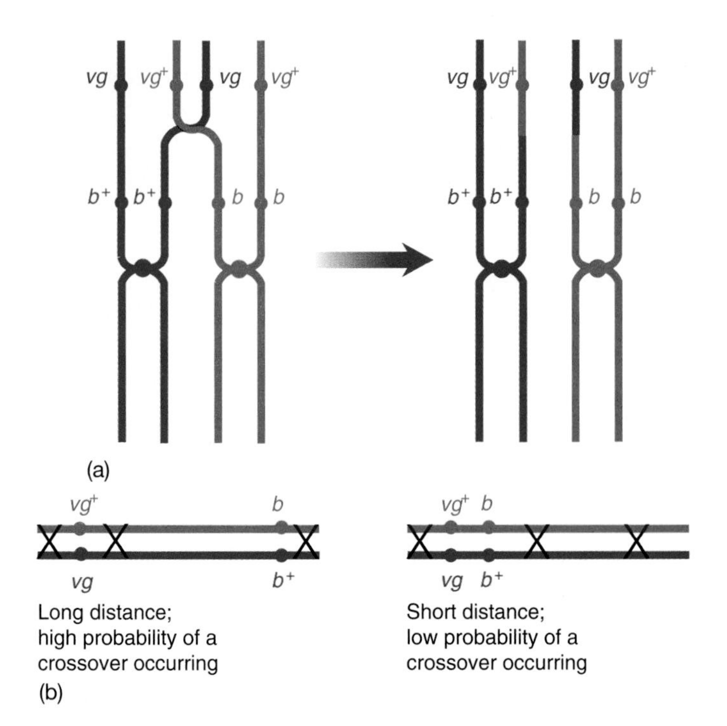

Figure 16.13

(a) Linked genes recombine because of crossovers that occur at random between them. *(b)* The probability of a crossover is proportional to the distance between the genes.

wild-type flies in the F_2 generation result simply from dominance in heterozygotes. But the rare black-vestigial flies must have two chromosomes with the combination *vg b*. Where did they come from?

In meiosis, homologous chromosomes pair tightly in prophase I, with their chromatids twisted around one another in *chiasmata.* In this state, the chromatids often **cross over:** They break at corresponding points and exchange homologous segments. Crossing over accounts for recombination between linked genes. It occurs regularly in meiosis, although for some reason it doesn't occur in male *Drosophila.* To show how crossing over explains the appearance of recombinants, we will do a testcross, mating females of the F_1 generation to *vgvg bb* males. The result is that 40 percent of the offspring have vestigial wings, 40 percent have black bodies, 10 percent have both vestigial wings and black bodies, and 10 percent are wild-type. As Figure 16.13 shows, the recombinants (vg^+ b^+ and *vg b*) arise from crossing over, and they are in the minority because crossovers don't occur between the two genes in every tetrad.

Now we can understand Sturtevant's idea for mapping genes. Crossovers occur at random locations. But, as Sturtevant realized, the chance of a crossover occurring between two genes should be roughly proportional to the distance between them. The space between the genes is a kind of target in which crossovers may fall randomly, and the bigger the target, the more likely it is that a crossover will hit it.

We can tell how often a crossover occurs simply by counting the recombinant individuals that result from a mating. In the testcross, 20 percent of the flies are recombinant, which means 20 percent of all the chromosomes that came through meiosis (in the females) had experienced a crossover between the *b* and *vg* genes. Thus the **frequency of recombination,** denoted by *R*, is 20 percent, or 0.2. This is the first piece of information we need to draw a map of this chromosome. The distances between genes are measured in **map units,** where 1 percent recombination is arbitrarily called 1 map unit (rather like arbitrarily using a scale of 1/4″ to 1 foot for an architectural drawing). So we place the *b* and *vg* genes 20 map units apart.

A third gene, *cn* for cinnabar eyes, is linked to *b* and *vg*. We cross cinnabar-eye flies with vestigial-wing flies to get an F_1 generation in which all the individuals are phenotypically wild-type. Then we use the females in a testcross with cinnabar-vestigial males. This cross yields 45 percent cinnabar-eye, 45 percent vestigial-wing, 5 percent both cinnabar-eye and vestigial-wing, and 5 percent wild-type. So here the frequency of recombination is only 10 percent (5 percent + 5 percent), indicating that the *cn* and *vg* genes are 10 map units apart. However, these two measurements aren't sufficient to draw a map of the chromosome, because *vg* could be between *b* and *cn*:

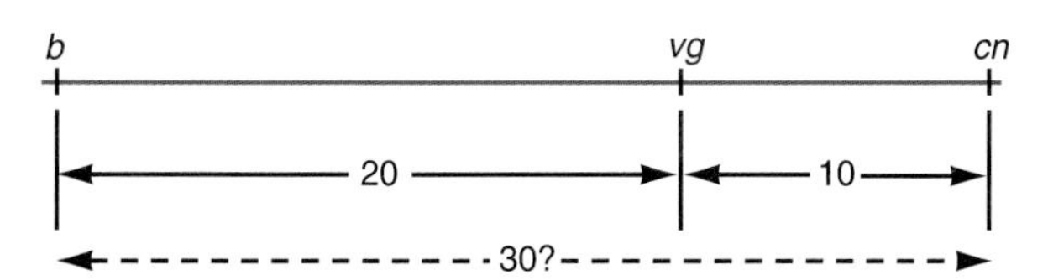

Or *vg* could be to the right of *cn*:

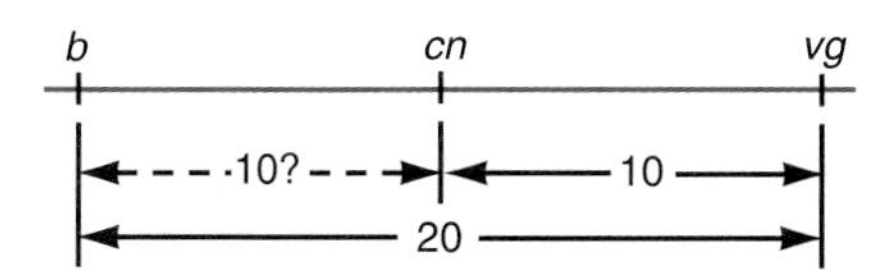

A third cross, between cinnabar-eye and black-body flies, settles the question; the frequency of recombination between *cn* and *b* is also about 10 percent, so the genes map in the order *b–cn–vg*.

We have just described the standard way to locate genes on chromosomes. A mutation used to map a gene is called a **marker,** and after establishing the relative positions of a few markers, we gradually add more and more to fill out the map by crossing flies bearing each new marker with flies carrying markers that are already mapped. However, the simple rule that R is a good measure of distance is only true when R is small. To convert R into map distances, we really need a **mapping function** like the one graphed here:

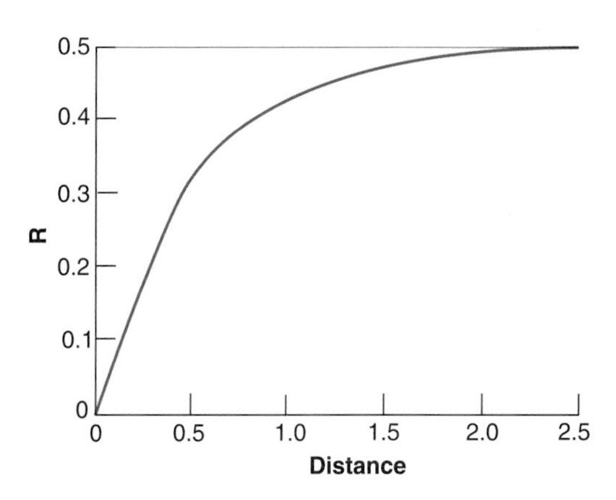

Such functions are derived theoretically, but experimental evidence shows that they provide a good relationship between R and physical distances. When R is small, it is indeed proportional to distance. However, as the distance between markers increases, R falls off to a maximum of 0.5, because at greater distances, two or more crossovers are likely to occur together, and an even number of crossovers cancel one another. All the markers that are demonstrably linked to one another form a *linkage group,* but since markers that are very far apart on the same chromosome show 50 percent recombination, a cross between them cannot demonstrate their linkage. Therefore, we can only show that they are in the same linkage group by showing their linkage to markers that lie between them. After doing several crosses between different markers, one can arrange the markers on a good, linear map. Ambiguities work themselves out as more and more markers are used. Methods 16.1 explains a method that establishes gene sequences with more certainty.

Exercise 16.14 Draw the pattern of recombination between two markers and convince yourself that an odd number of crossovers between the markers produces recombination, and an even number does not.

Methods 16.1 Three-Factor Crosses for Mapping

Although we illustrate the mapping of linked genes with a series of crosses in which only two genes are involved each time, the more certain, and often simpler, way to determine the sequence of genes is with a three-factor cross in which the parents differ in all three genes. Consider the offspring of a parent carrying chromosomes marked *ABC* and another carrying chromosomes with the alleles *abc*. For simplicity, let's assume the genes are linked in this sequence, with *B* in the middle. We create the heterozygote *ABC*/*abc* and perform a testcross, so we can easily identify the genotypes of all the resulting progeny from their phenotypes. Most of the progeny will carry the parental chromosomes, either *ABC* or *abc*. There are six possible recombinants:

ABc, *abC* } Have had a crossover between the *B* and *C* genes.

aBC, *Abc* } Have had a crossover between the *A* and *B* genes.

AbC, *aBc* } Have had crossovers both between *A* and *B* and between *B* and *C*.

Which of these recombinants do you think will be *least* abundant? The probability of a single crossover anywhere is low, so the probability of having two crossovers together must be even lower. Therefore, the third classes will occur least often; since these are the types in which the *B* allele has been changed (relative to the parental combinations), we infer that the *B* gene is in the middle. We assumed this was true, for illustration, but if we had not known the sequence initially, the results of the cross would have determined the middle gene.

16.10 Many genes have multiple effects.

Discussions of genetics commonly refer to "the gene for eye color in humans" or "the gene for comb shape in roosters." We use these shorthand designations so often that people forget the biological reality. In fact, there is no gene for eye color in humans or for the number of bristles in *Drosophila* or for any other similarly complex feature. The product of a gene is a protein, typically an enzyme. So it makes good sense to talk about "a gene for glutamic dehydrogenase" or "a gene for tyrosine deaminase" because these products can be directly identified with their genes. In a simple unicellular organism, there may be little more to say about the gene. But in a developing multicellular organism, a gene ought to be described with more information as, say, "coding for tyrosine deaminase, which is expressed particularly in the pigment cells of the retina, producing a brown eye color." Because of the state of knowledge in developmental biology, we generally can't give such designations yet. They are ideals to work toward.

When thinking about the actions of most genes in most organisms, two things must be kept in mind. First, no complex trait is likely to be produced by a single gene. The absence of a gene product can certainly eliminate or alter a feature, as many mutations show, but a normal feature, such as a bristle on a fruit fly or the color of a mouse's fur, must result from the coordinated action of several genes. Second, many genes—probably most genes—don't have a single effect and a single function. Genes that have multiple effects are said to be **pleiotropic** (Figure 16.14). We tend to notice the most obvious effect of a mutation and to name the genes accordingly, but it is easy to be misled by a name and to assume that the function of the gene is to produce the named trait. We generally find other effects of a mutation if we look for them. Few genes are so restricted in their actions that they only produce a single, simple phenotype.

Pleiotropy becomes especially important when the adaptive value of an allele is taken into account, as we do in Chapter 22. A new allele that arises by mutation generally has multiple effects, some obvious and some quite subtle. Some effects will probably confer an adaptive advantage, others a disadvantage. The overall balance of these effects then determines whether the mutation is likely to be eliminated from the population or perpetuated. The effects that are most obvious to humans may not be the most important in determining a mutation's adaptive value in fitting the organism to its particular way of life.

16.11 Any gene can have more than two alleles.

The alleles of a gene arise through mutation. Mutations in several *Drosophila* genes, for example, can produce purple, white, and other eye colors instead of the normal red eyes. Each mutant allele encodes a different defective protein, so the resulting eye tissue has an abnormal combination of pigments. A single gene can also have more than one allele, because some mutations can produce different degrees or kinds of activity, instead of just creating a defective protein.

In rabbits, a gene that determines fur color has four alleles (Figure 16.15). The wild-type allele (c^+) is dominant to the others, so homozygotes or heterozygotes for c^+ have normal, dark gray fur (agouti). The allele c^a is recessive to the other alleles, and $c^a\,c^a$ individuals are albinos. The c^h allele, either by itself or with c^a, produces the Himalayan pattern with pigment on the ears, feet, and tip of the nose. The c^{ch} allele produces chinchilla fur when homozygous, but $c^{ch}\,c^h$ or $c^{ch}\,c^a$ individuals are light gray.

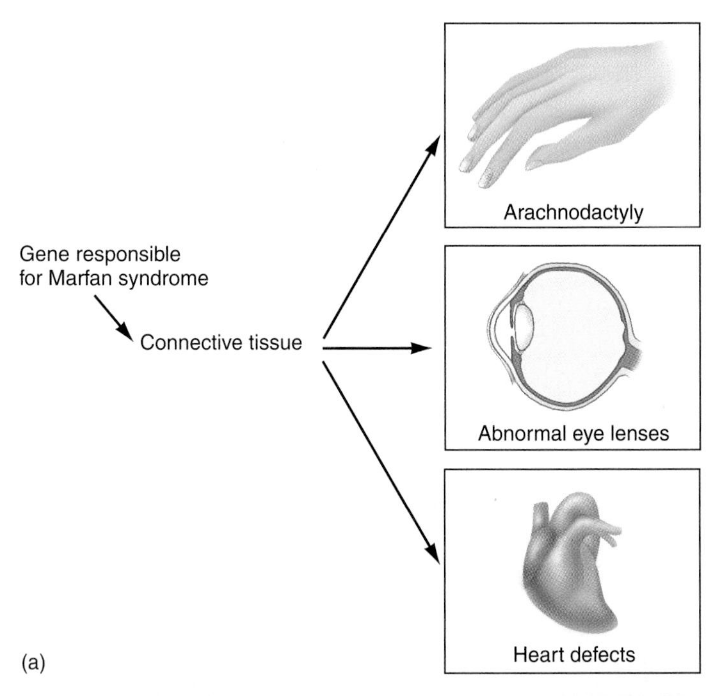

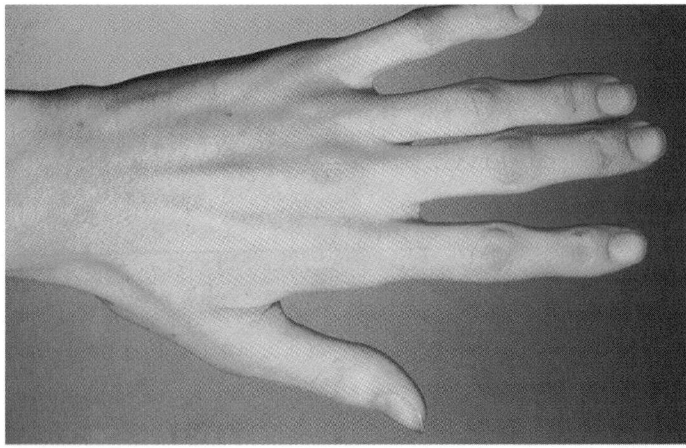

Figure 16.14

(a) Marfan syndrome is caused by a pleiotropic mutation that results in abnormal connective tissue (cartilage, bone, and sheets that wrap around segments of muscle). *(b)* The hand of a person with Marfan syndrome has a characteristic arachnodactylous ("spider-fingered") form.

Figure 16.15

Four phenotypes of rabbit fur that are due to multiple alleles. The Himalayan pattern results from expression of the genes for fur color only in the coolest regions of the skin, showing that a temperature-dependent process is involved.

Exercise 16.15 In mice, the degree of brown fur color is determined by a gene *(c)* that has a series of alleles: c^k, c^d, c^r, and c^a. The genotype *c*- has 100 percent color, and other genotypes have these percentages:

c^kc^k	c^kc^d	c^kc^r	c^kc^a	c^dc^r	c^dc^a	c^rc^r	c^rc^a	c^ac^a
80%	60%	45%	40%	25%	20%	10%	5%	0%

Assume that the enzyme made by each allele produces a certain percentage of color by itself and that the activities are additive in each case. From these percentages, determine the activity of each allele by itself.

16.12 Special chromosomes often determine an organism's sex.

In Sidebar 15.1, we give just a glimpse of the many forms of sexual reproduction, pointing out that in some species, sex is determined by environmental factors, such as temperature, or that sex changes regularly as part of a life cycle. An organism's gender is commonly determined genetically like any other trait. The simplest sexual organisms have only *mating types,* which differ primarily in the chemical structures of their cell surfaces, allowing their gametes to recognize each other and fuse. Although mating type is apparently determined by only one or a few genes, the distinct sex differences among many animals and plants may be due to many genes.

Special **sex chromosomes** determine gender in some plants and most animals. Humans, for instance, have somatic cells with 22 pairs of "ordinary" chromosomes, or **autosomes.** In addition, they have a pair of sex chromosomes; a woman has two X chromosomes, and a man has one X and one Y chromosome (Figure 16.16). During meiosis, therefore, a woman can only make one kind of egg, bearing an X chromosome. While the X and Y chromosomes share only a small region of homology,

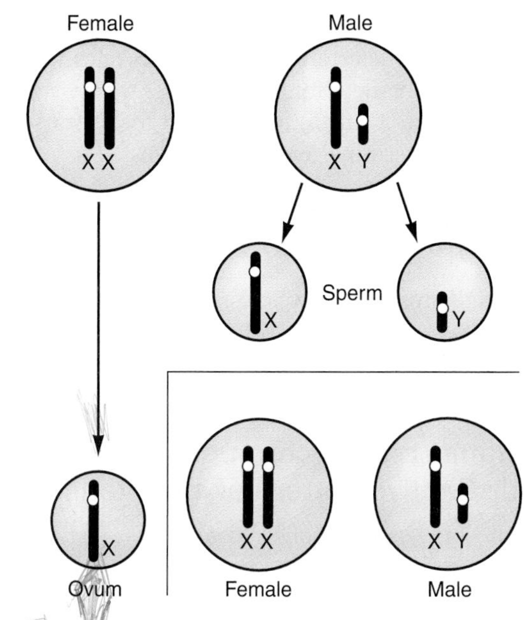

Figure 16.16

In the XY system of sex determination, a male is XY and a female is XX. A simplified Punnett square shows that since the male produces X sperm and Y sperm, it is the sperm that determines the sex of the offspring.

they behave like homologs during meiosis, so a man can make two kinds of sperm, half with an X chromosome and half with a Y. Then an egg fertilized by an X-bearing sperm is XX, or female, and one fertilized by a Y-bearing sperm is XY, or male. This is why the sperm determines the sex of the child.

All mammals have this XY system, as do fruit flies. In insects such as bugs and grasshoppers, sex is determined somewhat differently: The females are XX, but the males have only one X chromosome and are designated XO, the O indicating that there is no homolog. In all chromosomal sex-determination systems, one sex (here, the female) makes only one kind of gamete and is said to be *homogametic,* while the other (the male) makes two kinds of gametes and is *heterogametic.* In animals where the female is heterogametic, the chromosomes are customarily designated Z and W instead of X and Y. Male butterflies and some male birds and fishes are ZZ, while females have a small W chromosome and are ZW. In chickens and other birds, males are ZZ and females ZO. So all four possible versions of this sex-determination system occur and work equally well. Complex factors that no one understands yet determine differentiation between the two sexes. In humans, a small region of the Y chromosome, called SRY, determines maleness by converting the gonads into testes instead of ovaries; thus any zygote with a Y chromosome develops into a male, while a zygote without one becomes a female. In insects such as *Drosophila,* on the other hand, gender is determined by a balance between autosomal and X-linked genes. Each X chromosome has one-and-a-half units of female-determining genes, and each autosomal set has one unit of male-determining genes. This causes an XX fly to be female and an XY fly to be male, but for reasons different from those that determine gender in humans.

Exercise 16.16 Many stories from history tell of a king who divorced his wife (or worse) because she was "unable to present him with a son." Now you should be able to comment on the question of who is to blame, if "blame" can be assigned, in such a case.

16.13 Genes on sex chromosomes show a distinctive pattern of inheritance.

Morgan and his students collected many *Drosophila* mutants like those shown in Figure 12.3. One day they discovered a male fly with white eyes. When they mated it with a red-eyed female, all the flies in the F_1 generation had red eyes. These flies were then mated among themselves to produce an F_2 generation with a 3:1 ratio of red-eyed to white-eyed flies. This phenotypic ratio is to be expected when both parents are heterozygous for a characteristic that shows simple dominance, but in this case all the white-eyed flies were male.

The mutation producing white eyes *(w)* is inherited so peculiarly because it is on the X chromosome. To keep track of X chromosomes and their mutations, we'll denote a chromosome bearing *w* as X^w and a chromosome bearing the wild-type allele as X^+. The original white-eyed male must have been X^wY, because a male has only one X chromosome. The uniformly red eyes of the F_1 generation show that w is recessive to w^+, and an X^wY male has white eyes because it has no w^+ allele to mask its *w* allele, as there would be in a heterozygous female. This is the first important feature of an **X-linked,** or sex-linked, trait: Even if it is recessive, it appears in all males that carry the allele for it.

Figure 16.17 shows that sperm made by a white-eyed male carry either X^w or Y and fertilize X^+ eggs. So the offspring are either X^+Y (normal males) or X^+X^w (normal females). In the F_2 generation, the females are either X^+X^+ or X^+X^w and are all normal. The males get their single X chromosome from their mothers; half of them get X^+ and are normal (X^+Y), while the other half get X^w and are white-eyed (X^+Y^w). These white-eyed males are a quarter of the whole F_2 population.

A heterozygote for a recessive allele is also called a **carrier** for the trait that particular allele determines, so an X^wX^+ female is a carrier of white eyes. While not expressing the trait herself, she passes it to half her male offspring. This is another characteristic of an X-linked trait, one of the features

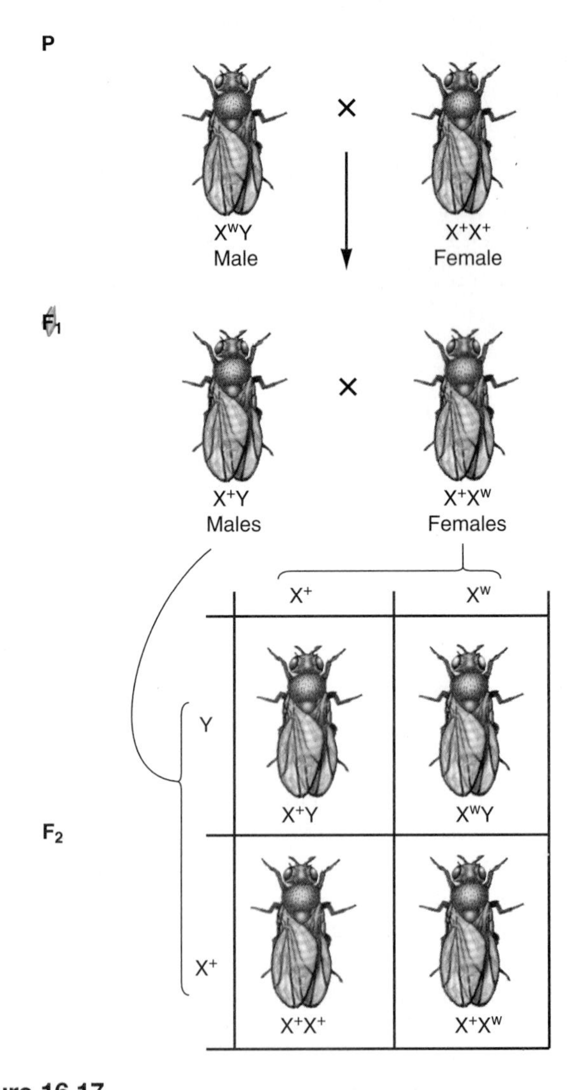

Figure 16.17
Inheritance of the X-linked allele for white eye color in *Drosophila* shows a pattern typical of X-linked characteristics.

to look for when trying to determine if a gene is on the X chromosome.

Inheritance in humans and domestic animals is often analyzed by drawing a **pedigree,** in which males are represented by squares and females by circles:

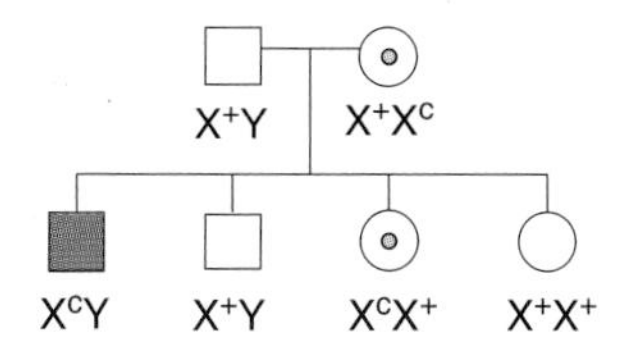

A line connecting parents shows a marriage or mating, and the individuals in each generation appear on the same horizontal line. The pedigree shown here is for a form of red-green color blindness, one of the most common X-linked traits in humans. A person with the trait is indicated by a colored symbol, and a carrier by a dot in the symbol. Notice that no female is affected, but half of them are likely to be carriers, and each one can be expected to pass on the trait to half her sons. Half her daughters are also likely to become carriers themselves. Of course, no male is a carrier.

The most famous case of X-linked inheritance is the hemophilia, a deficiency in blood-clotting, that has passed through the royal families of Europe for several generations (Figure 16.18). The original mutation must have occurred in Queen Victoria or her mother. Some royal families may still be transmitting the gene, even though a number of males have died of its effects over the years.

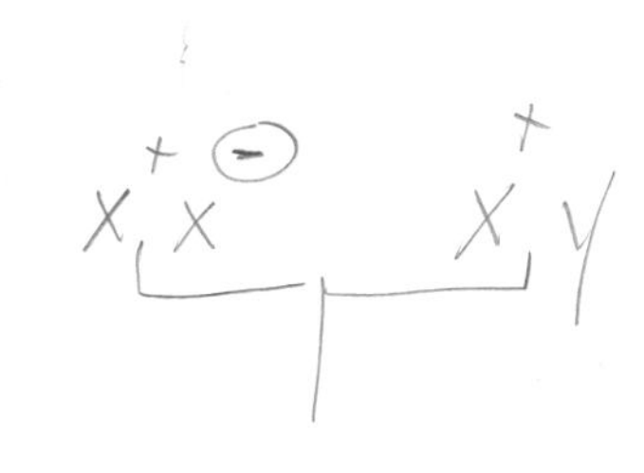

(a)

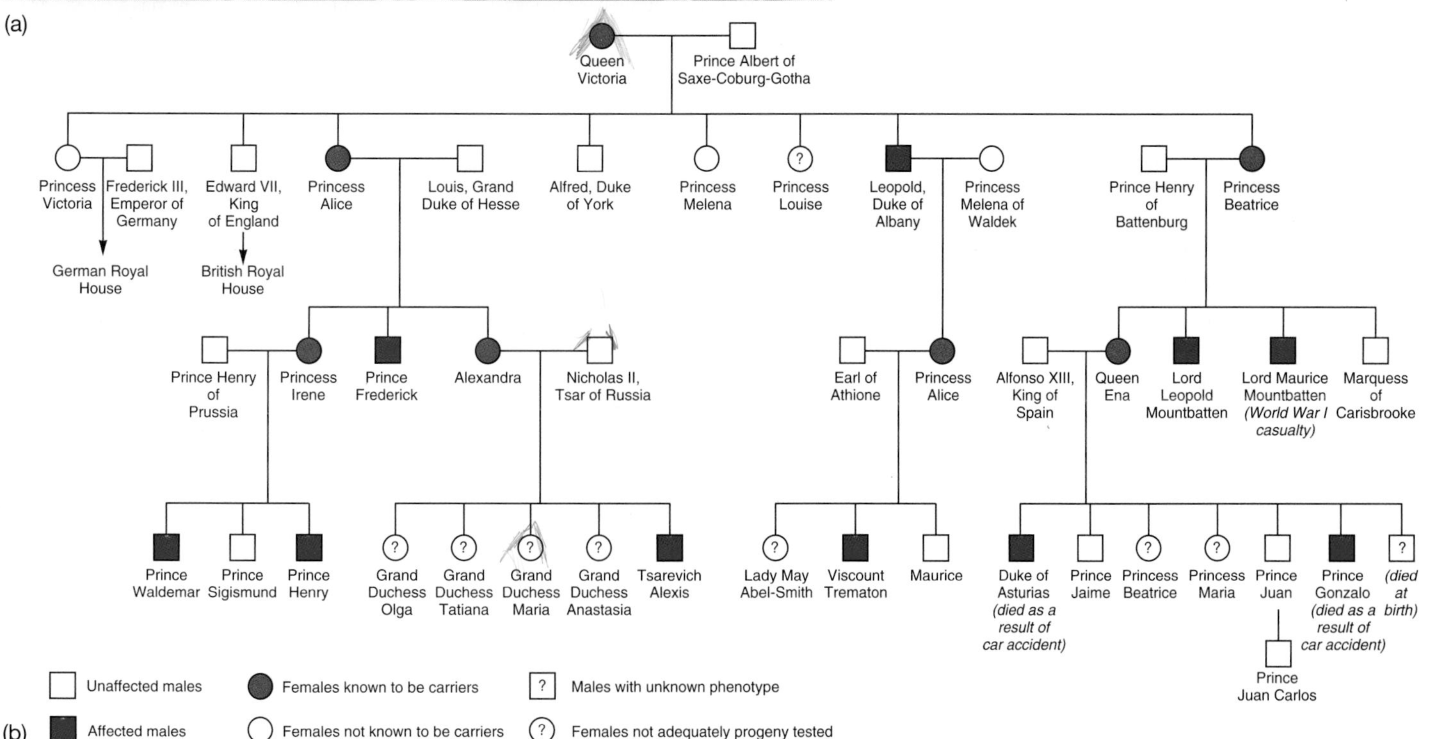

Figure 16.18

(a) Britain's Queen Victoria transmitted the gene for hemophilia to at least three of her children. *(b)* This pedigree shows how the trait then spread through some of the royal houses of Europe.

Exercise 16.17 Assume that an autosomal recessive allele (*b*) produces blond hair in humans. If two people with brown hair and normal vision have a blond, color-blind son, what is the genotype of the mother? Of the father?

Exercise 16.18 Cats show an interesting example of X-linked inheritance. A recessive allele (*l*) produces yellow fur, and its dominant form (*L*) produces black fur. A heterozygous *L l* female has calico fur, with mixed patches of yellow and black. (*a*) A calico cat has a litter of eight kittens: one yellow male, two black males, two yellow females, and three calico females. What color is their father? (*b*) A black cat has a litter of five: two black males, a black female, and two calico females. What can you tell about the paternity of these kittens?

Coda When describing meiosis in Chapter 15, we promised that understanding that process would make the patterns of heredity understandable. This chapter makes good on the promise. Mendel's laws of inheritance are virtual descriptions of the events of meiosis with emphasis on the alleles that may reside on homologous chromosomes. Homologs separate; alleles segregate. Each pair of homologs separates independently; unlinked genes are inherited independently. Homologs exchange segments as they cross over in chiasmata; linked genes can recombine.

In describing these regularities, we are simply showing how sexual organisms have evolved mechanisms that ensure that each generation inherits complete, functional genomes from its parents, with only rare exceptions. Since an organism's genome provides basic information for its structure and operation, this regularity is essential. It should also be clear that novelty can arise at several points in the process: Different combinations of chromosomes arise from the random events of mitosis; different kinds of gametes can combine in fertilization; and new combinations of linked genes can be made through crossing over. In addition, of course, each gene can mutate to produce new alleles (as witness the different alleles of the rabbit fur-color gene), and as we will show later, chromosomes can exchange and rearrange their parts. All these mechanisms provide the variability on which evolution depends, and even subtle differences between genomes can make significant differences in an organism's fitness, permitting it to adapt to quite different ecological conditions.

Summary

1. In organisms that go through a regular sexual cycle, traits are inherited in simple patterns that follow directly from the random events of meiosis and fertilization.
2. A diploid organism has two sets of chromosomes. In meiosis, homologous chromosomes pair up and then separate, so each gamete gets a regular haploid set of chromosomes, with only one of each kind.
3. A gene may have two or more forms, called alleles, which determine different forms of a particular trait. An organism with identical alleles for a gene is homozygous, and one with different alleles is heterozygous.
4. When meiosis occurs in a heterozygote, chromosomes carrying the two different alleles of each gene are segregated into different gametes (Mendel's First Law).
5. Gametes combine at random in fertilization. Therefore, gametes carrying alleles *A* and *a* will make equal numbers of four combinations: *AA*, *Aa*, *aA*, and *aa*. If one allele is not dominant to another, three phenotypes will occur in the ratio of 1:2:1. If one allele is dominant, there will be two phenotypes in the ratio of 3:1.
6. Homozygotes and heterozygotes for a trait that shows dominance can be distinguished with a testcross. The individual with the unknown genotype is crossed with a homozygous recessive. If all the offspring are of the dominant phenotype, the individual being tested must be homozygous dominant; if the offspring are half of the dominant and half of the recessive phenotype, the individual being tested must be heterozygous dominant.
7. Traits whose genes lie on different chromosomes are inherited separately; each pair of alleles will segregate and recombine independently of the others (Mendel's Second Law). Therefore, the proportion of individuals who should have any combination of traits is predicted from the product of the proportion who should have each trait separately. For example, if the probability of inheriting trait A is three-quarters and the probability of inheriting trait B is one-quarter, the probability of having A and B together is three-sixteenths. ($3/4 \times 1/4 = 3/16$.)
8. Traits whose genes are linked on the same chromosome tend to be inherited together. An original combination of alleles is only disrupted by crossing over between homologous chromosomes in early meiosis, so some chromatids exchange homologous segments.
9. The frequency of recombination between two linked markers (mutations used to identify a gene) is a measure of the distance between the locations of the genes; the greater the distance, the greater the chance of a recombination. Numbers derived from appropriate crosses can then be used to draw a map of a chromosome.
10. Special chromosomes generally determine sex. In mammals, each male has an X and a Y chromosome, and each female has two Xs in addition to a set of autosomes.
11. In mammals, the sperm determines the sex of each offspring, since each female produces only eggs with an X chromosome and males produce two kinds of sperm, half with an X chromosome and half with a Y.
12. Traits determined by recessive alleles on the X chromosome (such as hemophilia and color blindness) are inherited in distinctive patterns. A male exhibits the trait even though he has only one copy of the allele. A heterozygous female does not show the trait but, on the average, passes it to half her male offspring and makes carriers of half her female offspring.

Key Terms

purebred 312
cross 312
first filial generation/F_1 313
second filial generation/F_2 313
dominant 313
recessive 313
allele 313
homozygote/homozygous 313
heterozygote/heterozygous 314
segregate 314
phenotype 314
genotype 314
testcross 318
epistasis 320
gene linkage 321
linkage map 321
parental type 321
recombinant type 321
cross over 322
frequency of recombination 322
map unit 322

marker 323
mapping function 323
pleiotropic 324
sex chromosome 325
autosome 325
X-linked 326
carrier 326
pedigree 327

Multiple-Choice Questions

1. Which definition is correct?
   a. A dominant allele is expressed only if homozygous.
   b. A dominant allele is expressed only in a heterozygote.
   c. A recessive allele is expressed only in a homozygote.
   d. A recessive allele is expressed only in a heterozygote.
   e. A dominant allele is more likely to be passed to future generations than a recessive allele.
2. If a geneticist refers to a *purebreeding strain,* she means a group of organisms that
   a. are homozygous for a particular trait.
   b. are heterozygous for a particular trait.
   c. result from mating two heterozygotes.
   d. are homozygous for all their phenotypic traits.
   e. have parents exhibiting two different forms of the trait.
3. A cross of two phenotypically different purebred parents results in a phenotypically uniform F_1 generation and an F_2 generation containing three phenotypically different kinds of offspring. Out of 100 F_2 individuals, 25 resemble one parent, 25 resemble the other parent, and 50 resemble neither. Which hypothesis best explains the inheritance of the trait involved?
   a. one pair of alleles; one allele is completely dominant to the other
   b. two pairs of independently assorting alleles
   c. one pair of alleles showing incomplete dominance
   d. one locus that includes three different alleles
   e. three pairs of independently assorting alleles
4. What is the phenotype of a fruit fly whose genotype is $e^+e^+ \; pr^+pr$ (where e = ebony body color and pr = purple eyes; wild-type = tan body and red eyes)?
   a. ebony body; purple eyes
   b. ebony body; red eyes
   c. tan body; purple eyes
   d. tan body; red eyes
   e. ebony and tan body; one eye red; one eye purple
5. Which of these statements is correct?
   a. Two organisms with identical genotypes may not have identical phenotypes.
   b. Two organisms with identical phenotypes may have different genotypes.
   c. Many pairs of alleles may interact to produce one phenotypic characteristic.
   d. One pair of alleles may have several phenotypic effects.
   e. All of these statements are correct.
6. If genes *C* and *D* are on the same pair of homologous chromosomes and exhibit recombination, a cross of two individuals heterozygous for both genes will result in ___ phenotypic classes.
   a. one
   b. two
   c. three
   d. four
   e. none of the above
7. Assume that the genes for hair color and curliness are autosomal and unlinked. Dark hair is dominant to light hair, and curly hair is incompletely dominant to straight hair (the heterozygote has wavy hair). If one parent has light, curly hair and the other has light, wavy hair, which phenotypes could be present in their children?
   a. light or dark; curly, straight, or wavy
   b. all light; all wavy
   c. light or dark; wavy or straight
   d. only dark; curly, straight, or wavy
   e. only light; curly, wavy, or straight
8. An individual who is trihybrid for unlinked genes *A*, *B*, and *C* can produce ___ genetically different types of gametes.
   a. 2
   b. 4
   c. 6
   d. 8
   e. 12
9. Which of the following phenotypic ratios would you expect in a mating of a female carrier of hemophilia and a normal male?
   a. all normal daughters; all normal sons
   b. all normal daughters; all sons with hemophilia
   c. all daughters with hemophilia; all sons normal
   d. hemophila in half of daughters and sons
   e. all normal daughters; half normal sons
10. In domestic cats, the fur pattern called calico or tortoiseshell results from the heterozygous condition of an X-linked coat color gene. There are two alleles of this gene. The dominant allele specifies black coat color, while the recessive allele specifies yellow. Which of the following statements is correct?
    a. Calico cats are normally female.
    b. Male as well as female cats may be phenotypically black or yellow.
    c. If a calico female is mated with a black male, the litter will include black, yellow, and calico kittens of both sexes.
    d. *a* and *b* but not *c*.
    e. *a*, *b*, and *c*.

True-False Questions

Mark each statement true or false, and if false, restate it to make it true.

1. In general, testcrosses are not necessary in cases where alleles are incompletely dominant.
2. Independent assortment occurs as a result of crossing over during meiotic prophase I.
3. Segregation of alleles occurs as a result of crossing over during meiotic prophase I.
4. Recombination of linked genes occurs as a result of crossing over during meiotic prophase I.
5. If a particular gene is known to have more than two alleles, some diploid individuals may contain only one allele, some may contain two, and some may contain more than two.
6. If gene *A* has a lower frequency of recombination with gene *B* than gene *C*, gene *A* is probably closer to *B* than *C* on the chromosome.
7. If one observes 9:7 or 15:1 phenotypic ratios among F_2 members of standard genetic crosses, the traits under study probably result from the activity of genes having two allelic forms.
8. In a typical dihybrid cross (no linkage, true dominance and recessiveness within each allelic pair), four genotypically different classes will be formed.
9. In mammals, sex is chromosomally regulated so that the females are heterogametic and the males are homogametic.
10. Since females carry two alleles for X-linked alleles while males have only one, recessive traits such as hemophilia and color blindness will be found more frequently in females.

Concept Questions

1. How does the term *allele* differ from the term *gene*?
2. Suppose you are a genetically informed farmer with a few spotted cows. More than anything else, you want your herd to be purebred with respect to the spotted trait. Rumor has it that the allele for spotted fur is recessive. Do you hope the rumor is true or false? Explain.
3. Explain the difference between dominance and epistasis.
4. If the F_2 generation of a cross yields a phenotypic ratio of 12:3:1, what can you deduce about the numbers and relationships between alleles and genes?
5. What phenotypes would you expect for both parents of a color-blind female?

Additional Reading

Ayala, Francisco J., and John A. Kiger. *Modern Genetics.* The Benjamin/Cummings Publishing Co., Menlo Park (CA), 1980.

Bailey, Jill. *Genetics and Evolution: The Molecules of Inheritance.* Oxford University Press, New York, 1995.

Gelbart. *An Introduction to Genetic Analysis,* 5th ed. W.H. Freeman & Company, New York, 1993.

Griffiths, Anthony J.F., Jeffrey H. Miller, David T. Suzuki, Richard C. Lewontin, William M. Keller, Evelyn Fox. *A Feeling for the Organism: The Life and Work of Barbara McClintock.* W.H. Freeman and Co., New York, 1983.

Sapienza, Carmen. "Parental Imprinting of Genes." *Scientific American,* October 1990, p. 52. Even when fathers and mothers contribute identical genes to their offspring, the genes may have different effects. Sex-specific gene imprints can influence normal development and trigger diseases.

Internet Resource

To further explore the content of this chapter, log on to the web site at:

http://www.mhhe.com/biosci/genbio/guttman/

Microbial and Molecular Genetics

17

Key Concepts

17.1 A set of mutations defines a genetic map of a genome.

17.2 Mutations are alterations in nucleic acid structure.

17.3 It is easy to find many kinds of mutants for genetic analysis.

17.4 A simple phage cross illustrates the principles of gene mapping.

17.5 Genes are defined by complementation tests.

17.6 A gene is really colinear with its protein product.

17.7 A messenger is read by threes without commas.

17.8 Bacteria have a pseudosexual mechanism.

17.9 Donor cells transfer genes in their linkage sequence.

17.10 A zoo of little genetic entities broadens our conception of the biological world.

17.11 Viruses can promote genetic exchange.

17.12 Bacteria restrict the growth of foreign entities like viruses.

17.13 DNA molecules can be combined at will by using restriction enzymes.

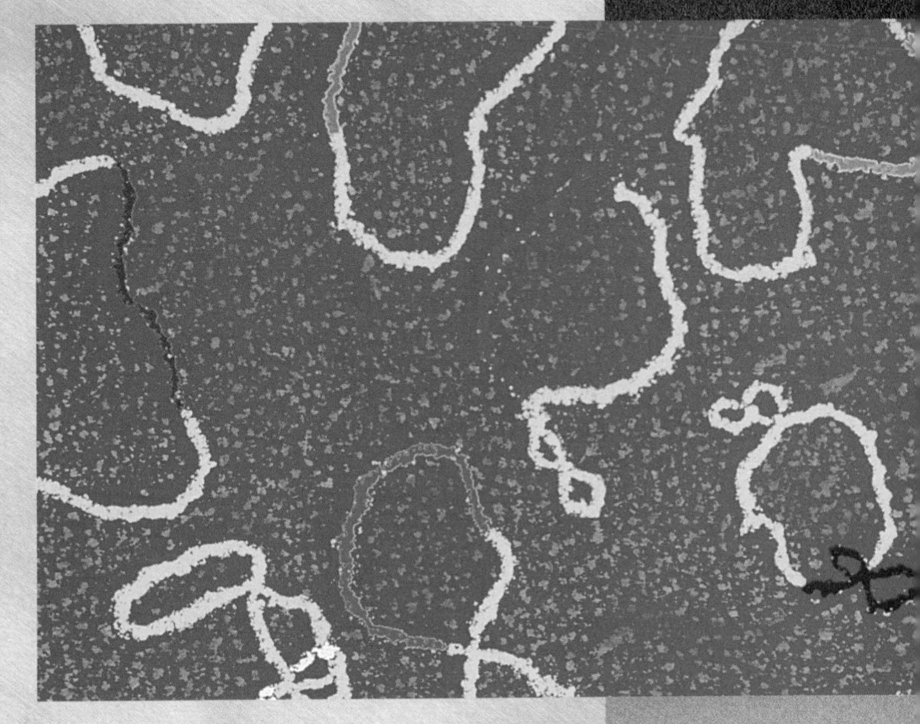

Small ring-shaped DNA molecules called plasmids have been artificially colored to show that segments of foreign DNA have been inserted in them through recombinant-DNA methods.

Aldous Huxley, in his book *Brave New World,* imagined a future society in which people would be engineered for various roles in society, and where the very idea of natural birth was obscene and embarrassing. Huxley's engineers used crude chemical methods. He could not have envisioned the subtle power of genetic engineering that has developed 60 years later.

We now live in a society with all the potential to be a "brave new world"—a society in which people skilled in the use of a few simple tools are learning to locate and identify new genes, then manipulate them almost at will to create whatever new combinations they desire. Genetic engineers have tailored cells with mixed DNAs of bacteria, yeasts, viruses, plants, and humans. The new technology is revolutionizing basic biological research while promising to fulfill old medical and agricultural dreams of curing humankind's ailments and creating improved sources of food. For a time, some entrepreneurs saw genetic technology as the road to fortune, and genetic engineering companies became hot items on the stock exchange. Now, though the enthusiasm for investing in such technology has cooled somewhat, large corporations still bank heavily on this research. Today's world is a place in which more and more students will find employment, success, and possibly fame, along with the satisfaction of helping improve the human condition; it is also a world in which they will have to struggle with increasingly complicated moral and social issues.

Every science builds on earlier science, and recombinant-DNA technology has been built upon the methods and concepts of classical microbial genetics. We will begin this chapter with an analysis of relatively simple research methods, showing how to find mutants and use them for genetic analysis. Then we will show how genes are exchanged between bacteria by viruses and similar genetic entities called plasmids, and we will conclude by introducing the basic principles of recombinant-DNA technology. ■

17.1 A set of mutations defines a genetic map of a genome.

What would it be like to know everything about a genome? It would mean knowing the complete nucleotide sequence of the DNA, precisely where each gene begins and ends, what each gene encodes, and how the information is used. No organism uses all the information in its genome all the time; we say that a gene is **expressed** when its product (usually a protein, less often a stable RNA like tRNA or rRNA) is actually produced. It is also critical to know what factors control when and where each gene is expressed.

In fact, we *can* know all these things through a program of research called **genetic analysis.** This research is providing quite complete knowledge of the genomes of bacteria such as *E. coli* and certain viruses, and it is being extended to a few plants and animals. Genetic analysis starts by determining that a gene exists and is located in a certain place on a chromosome relative to other genes. Ultimately, the goal is to develop a **genetic map** of the genome that shows the locations of all genes, with their punctuation or control elements and a description of what each gene does (Figure 17.1). This work is basically an extension of the chromosome mapping introduced in Chapter 16.

Since it is now easy to determine DNA sequences (see Methods 14.1), you might think that just learning the whole DNA sequence of a genome would identify the genes and tell us everything we want to know about them. However, the DNA sequence doesn't tell how the information is *used.* We can search for possible genes in a DNA sequence because we know the sequence is read by codons of three bases and we know the genetic code (Table 14.1). But where to start reading? Here is a DNA sequence, purposely made repetitive:

3′-GTAGTAGTAGTAGTAGTAGTAGTAGTAGTAGTAGTAGTA . . . 5′
5′-CATCATCATCATCATCATCATCATCATCATCATCATCAT . . . 3′

The message in this sequence is translated by codons, and it can be read with a *reading-frame* mechanism that moves along in the direction of mRNA synthesis and translation, the 5′ → 3′ direction, marking off three bases at a time. But there are no "commas" marking off the codons, so we don't know whether to choose one of the three reading frames on the lower strand—CAT, ATC, or TCA—or one of the three on the upper strand, reading now right to left—ATG, TGA, or GAT. Nothing in the DNA sequence tells us which of the six possible reading frames is correct, and we don't even know which strand to read. Ah, but we know the code! We know that genes usually—usually—start with the RNA codon AUG, which encodes methionine or formylmethionine, so we search for the DNA equivalent, 5′ATG, which *could* be the start of a gene. Of course, ATG could also be in the middle of a gene, since methionine could appear anywhere in a polypeptide, so we look (or have a computer look) for a reading frame starting with ATG that encodes a fairly long sequence of amino acids before coming to one of the termination codons, TAA, TAG, or TGA. Such a sequence is called an **open reading frame (ORF),** and an ORF *could* be a gene. Other clues help identify genes, such as common control sequences at the beginnings of genes, but there is no way to tell if an ORF of any length is actually used as a gene without strong experimental evidence.

This evidence, and information about the actual function of a gene, only comes from genetic experiments using mutants. The old irony remains: We can only know a gene exists by finding a mutant that abolishes or alters the gene's function. You may recall from Chapter 16 that a mutation used to map and identify a gene is a **marker,** and it is critical to understand that a marker provides an *experimental definition* of a gene. We then locate each gene by performing a series of genetic crosses between individuals

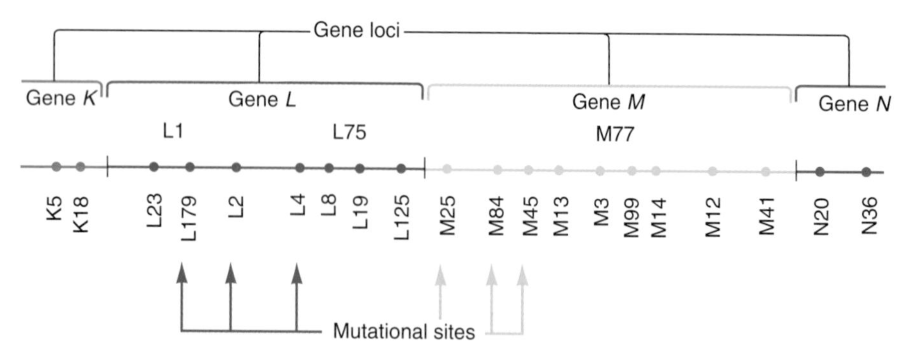

Figure 17.1

A genetic map shows the positions of all the genes, control elements, etc., in the genome. This imaginary map depicts a small region of a genome, including genes designated *K, L, M,* and *N,* which are defined by mutations that occur in them. Mutants are first assigned arbitrary numbers, since we know nothing about them. The locations of their mutations are determined by genetic crosses; the position of a point mutation is its *site,* and a few mutations that appear to be deletions are represented by bars that cover the deleted region. Then genes are defined by mutations that lie within them on the basis of complementation tests, explained in Section 17.5. The position of a gene or other functional element is its *locus* (pl., *loci*).

carrying these markers, as described in Section 16.9. With animals such as fruit flies and mice, we often find only one allele, other than wild-type, for a gene. However, it is easier to find mutant viruses and microorganisms, and they may provide many alleles of a gene, allowing us to get quite detailed information about it.

Since DNA is a linear polymer, a genetic map should be linear, as in Figure 17.1. The smallest possible mutation, a change in a single nucleotide pair, is called a **point mutation** and is represented on the map by a point, called its **site.** A gene can encompass many mutational sites, and these sites establish the location, or **locus,** of the gene.

At the risk of emphasizing what may be obvious, the map is not the same as the genome it represents. The map is just a picture, and it may be distorted, just as a flat map of the earth necessarily distorts its round surface. For instance, a genome that has a deletion mutation is actually shorter than an unmutated genome, but on the map we still show where the deleted DNA is missing.

17.2 Mutations are alterations in nucleic acid structure.

Although heredity depends on reasonable fidelity in replicating a genome and passing copies on to the next generation, mutation is also essential for the existence of organisms. Without genetic changes, there could be no evolution, and organisms would never become adapted to particular ways of life. Mutations occur spontaneously in every organism or virus at low rates, typically about 10^{-6} to 10^{-8} per generation for any one gene. This means, in every generation, the chance that a given gene will mutate is very slim—anywhere from one per million to one per hundred million. Mutations may also be induced by agents called **mutagens,** such as X rays, ultraviolet (UV) light, and many chemical agents, all of which are used in genetic experimentation. In addition, many apparently spontaneous mutations probably result from unknown mutagens in the environment. It is important to understand that all mutations, whether spontaneous or induced by a known mutagen, occur at random sites. Mutagens increase the rate of mutation, but we cannot use them to produce mutations at any chosen sites.

The chemical effects of some mutagens are well known. X rays are one kind of *ionizing radiation,* radiation carrying so much energy that it knocks electrons completely out of atoms and molecules, leaving ions. Such radiation produces free radicals, molecules with a single unpaired electron that are able to engage in chemical reactions potentially damaging to a cell. In addition to point mutations, X rays often break chromosomes and cause gross genetic damage and rearrangements of the DNA, including deletions, in which two breaks produced close together heal with the middle piece missing.

When DNA absorbs UV light, neighboring pyrimidine bases can become chemically rearranged and covalently linked into a pyrimidine dimer (Figure 17.2). Because they stop DNA replication, these UV damages can be lethal. UV light is often used to sterilize materials in a laboratory. But cells contain repair enzymes that cut out the damaged region, replace it with a new segment of polynucleotide using the intact strand as a template, and reseal the molecule with covalent bonds. These enzymes often save a cell that would otherwise die. Other enzymes allow DNA replication to proceed past a pyrimidine dimer, but with errors; these errors can be mutations.

Chemical mutagens can alter DNA in many ways. Nitrous acid (HNO_2), for instance, mutagenizes by removing amino groups from some nucleotide bases, converting cytosine to uracil and adenine to hypoxanthine:

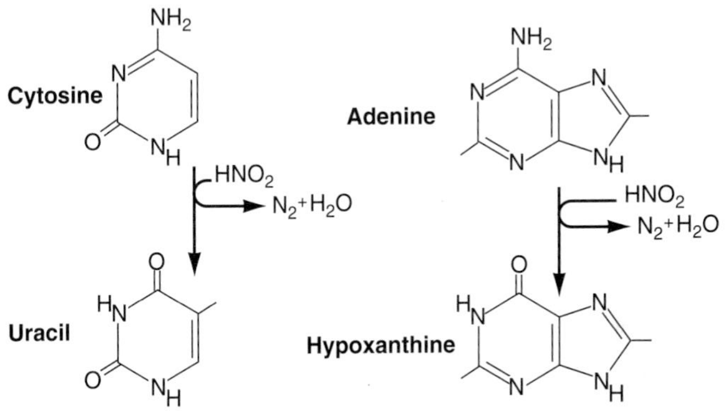

The first change can replace a G–C pair with an A–T pair when the DNA replicates (Figure 17.3), and the second can do the reverse. Both changes will generally show up as mutations. This agent deserves our attention because nitrous acid is formed in the acid conditions of the stomach from sodium nitrite, a common food preservative. (Read the label on the next package of cold cuts you buy).

Mutagenesis is valuable experimentally, but it is of even more concern in human life. Among the huge amounts of chemicals that modern industry and agriculture put into the air, water, and soil are many that pose an enduring mutagenic threat to the human genome, as well as more immediate dangers to particular individuals. Mutagens are usually also carcinogens, which cause cancers, and teratogens, which cause deformed fetuses. The dangers posed by such agents are discussed in Chapter 20.

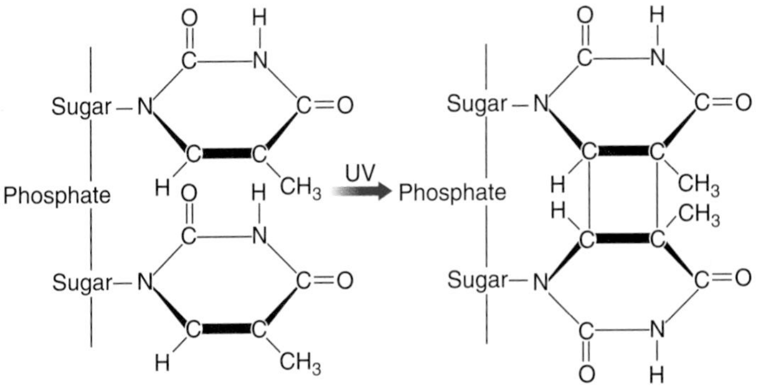

Figure 17.2

A pair of pyrimidines in DNA is converted into a dimer when they absorb a photon of UV light. This dimer distorts the DNA.

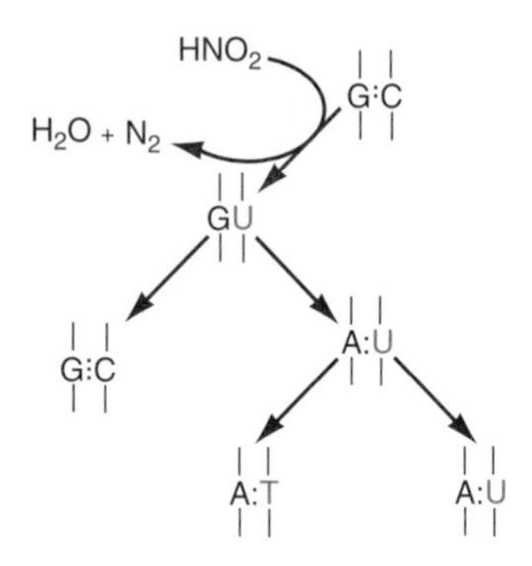

Figure 17.3

Mutations often occur as DNA containing an altered nucleotide replicates. Here a G–C pair, with the cytosine converted into a uracil, is converted into an A–T pair after two rounds of replication. The strand containing the U can continue to produce identical mutant molecules indefinitely, probably until repair enzymes discover the U and correct it.

Exercise 17.1 We said in the preceding section that a change from one base pair to another will *generally* be a mutation. By considering the genetic code (Table 14.1), explain why many such changes will *not* be mutations.

17.3 It is easy to find many kinds of mutants for genetic analysis.

A general rule for making progress in science is to attack the simplest problems first, using simple experimental systems. Our understanding of the genetic apparatus and the expression of genetic information was developed through the study of bacteria, particularly *E. coli,* and the viruses (phage) that infect them. This now-classical work shaped basic methods of analysis in biology, providing techniques and information that could be used later to unravel much more complicated cellular, developmental, and physiological processes.

An old recipe for rabbit stew begins, "First, catch a rabbit." To explore the structure of a genome, we need to catch some mutants. This just requires a little ingenuity. Most bacteria, for instance, are sensitive to antibiotics like streptomycin. We can find streptomycin-resistant mutants simply by plating a lot of bacteria on nutrient agar containing a little streptomycin, where the sensitive cells are killed and only the rare streptomycin-resistant mutants make colonies. This method, like nearly all genetic manipulations of microorganisms, uses the principle of selection to find mutants with specific characteristics.

Auxotrophic mutants—those that cannot synthesize one or more of their own components—are valuable for genetic analysis. Suppose we want mutants that cannot synthesize their own arginine. To produce mutations, we first subject a culture to UV light or to a chemical mutagen. Since mutations are rare, even after mutagenesis, we have to select the few mutants that have characteristics we want (Figure 17.4). After growing the culture for a while with arginine, so auxotrophic mutants can grow, we transfer the cells to a medium without arginine and

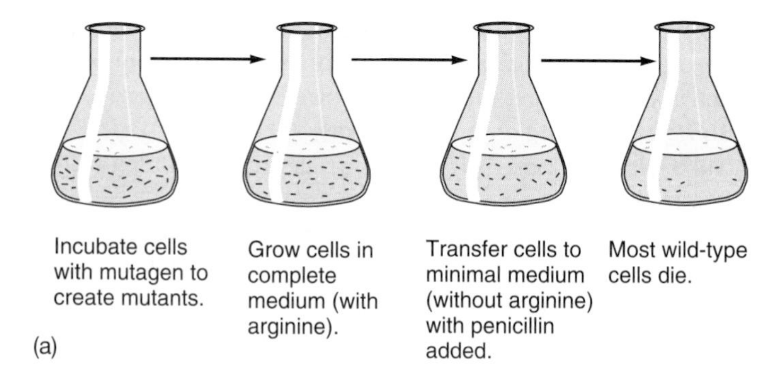

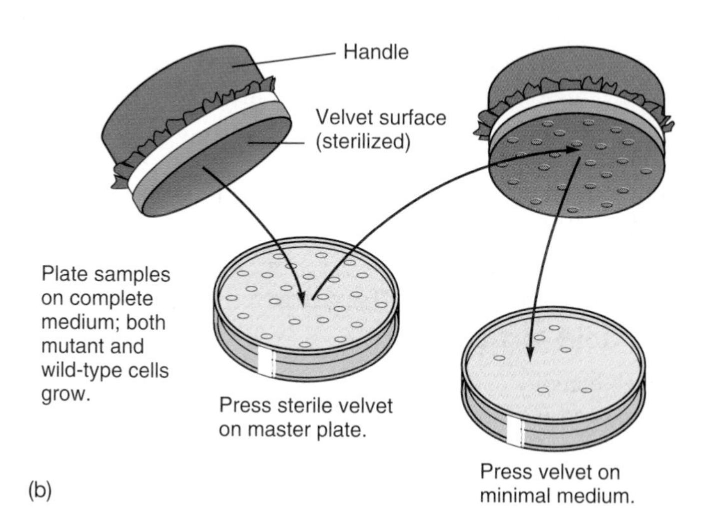

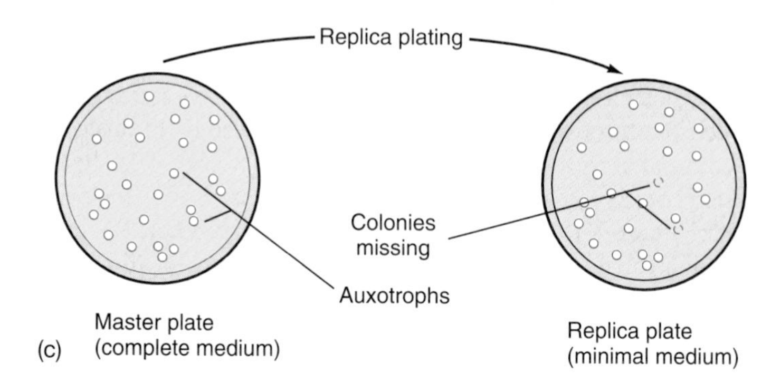

Figure 17.4

(a) A classical way to catch bacterial mutants is illustrated by mutagenizing cells and growing them in a medium containing arginine, where auxotrophs can grow. Then they are transferred to a medium with penicillin and no arginine. Only wild-type cells (prototrophs) grow, and most of them are killed by the penicillin, enriching the culture for auxotrophs. *(b)* Some cells are spread on plates with arginine. *(c)* Replica plating onto plates without arginine reveals the mutants that cannot grow without this amino acid.

add the antibiotic penicillin, which interferes with the synthesis of bacterial cell wall structure; this causes growing bacteria to develop weak walls and lyse spontaneously because of their high turgor pressure. The arginine auxotrophs survive because, lacking arginine, they can't grow, but they remain viable and able to metabolize for a long time. To find them, we plate samples of bacteria on medium containing arginine, where all the bacteria grow, and then use the replica plating method developed by Joshua Lederberg to transfer cells from each colony to plates without arginine. Those that can't grow there are probably arginine auxotrophs.

Auxotrophic mutants, Section 12.4.

Some of the most interesting mutants have **lethal mutations,** defects in essential genes that produce an inviable organism. But how could an organism with a lethal mutation ever be studied? The answer is to use mutants with **conditional lethal mutations,** which are only lethal under certain circumstances. Auxotrophic mutants, for instance, are conditional lethals because they only fail to grow (and eventually die) in the absence of the required nutrient; given the nutrient, they survive and grow. **Temperature-sensitive *(ts)*** mutants are very useful conditional lethals; they grow normally at a low temperature (say, 30°C) but not at a high temperature (say, 42°C) (Figure 17.5*a*). A *ts* mutation can occur in any gene because a simple change in the amino acid sequence of a protein can greatly affect the protein's stability. Some mutations make a protein unfold into a useless shape at all temperatures, but *ts* mutations produce proteins that are functional at low temperatures and only unfold at high temperatures. Interestingly, because of the nuances of protein structure, there are also cold-sensitive *(cs)* mutations that have the opposite effect, leaving the protein relatively normal only at a high temperature.

Bacteriophage are favored subjects for genetic exploration because it is also easy to find and analyze phage mutants. Some of the most valuable phage mutants are **host-dependent** (Figure 17.5*b*). You must understand that we identify bacteria by species, such as *Escherichia coli,* but there are many variants or *strains* of each species that differ in specific traits. Biologists have isolated many strains of useful bacteria and keep them in sometimes huge strain collections for future use. Host-dependent phage mutants reproduce easily in *permissive strains* of bacteria, where their defect is corrected in some way or where the mutant gene isn't needed; their defects can be studied by infecting *restrictive strains,* in which they do not reproduce, in order to determine what function they lack.

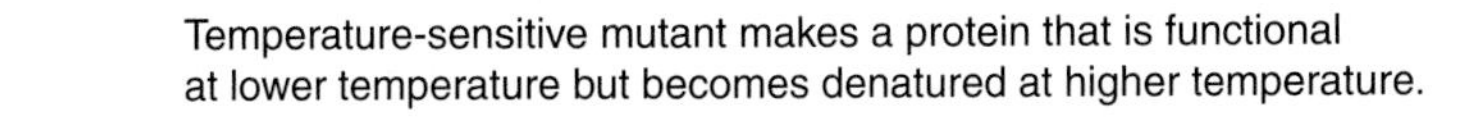

(a) Temperature-sensitive mutant makes a protein that is functional at lower temperature but becomes denatured at higher temperature.

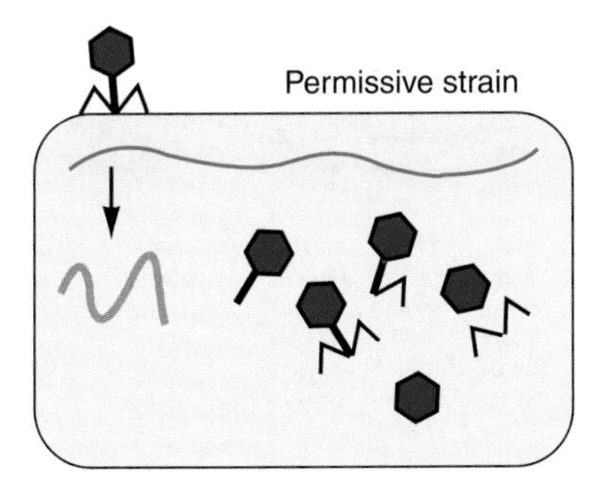

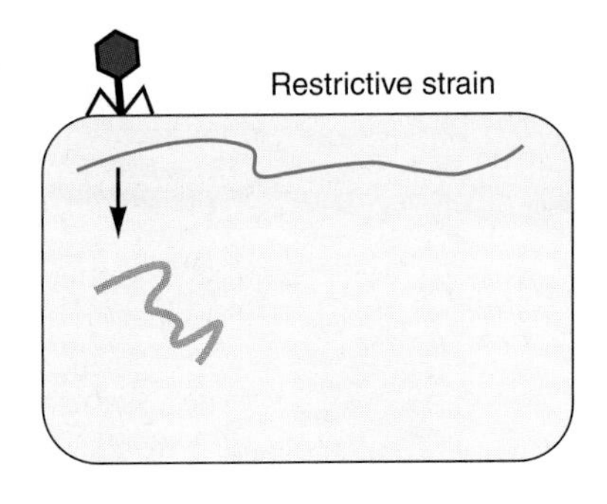

(b) Host-dependent mutant makes faulty protein that is not needed for growth in the permissive strain, but growth is aborted in the restrictive strain for want of this protein.

Figure 17.5

Two types of conditional lethal mutants are used in research. *(a)* Temperature-sensitive mutants have a protein that only functions at low temperatures, so the phage do not grow at an elevated temperature. *(b)* Host-dependent mutants can only grow in certain strains of bacteria.

17.4 A simple phage cross illustrates the principles of gene mapping.

Having established the general idea of gene mapping, we will show how genetic markers can be mapped in a simple cross with phage and why phage are such valuable genetic tools. Genetic mapping is a kind of microscope for examining genomes, and it becomes more valuable as we increase its "resolving power"—its ability to distinguish two points that are close together. The resolution of a light microscope is determined by the quality of its lenses and the wavelength of the light; a genetic microscope is limited by the smallest measurable frequency of recombination, R. Since R is the number of recombinants divided by the total number of progeny, and the smallest measurable number of recombinants is 1, R can only be made smaller by increasing the number of progeny. If we could only examine 100 progeny, we couldn't measure a value of R less than 1/100 or 0.01, which corresponds to one map unit; we could not tell if

Methods 17.1 Deletion Mapping

A number of *rII* mutants are **deletion mutants,** which are missing small segments of the genome. Point mutants can back-mutate to wild-type, but deletion mutants never do because it would be nearly impossible to add back the right sequence of nucleotides by accident. A deletion can't be represented as a single site on a map but must be shown as a bar that may overlap the sites of several point mutations. If the nucleotides of a gene are designated *a, b, c,* and so on, then each point mutation is a change in one nucleotide, but the deletion is the loss of a whole series of nucleotides:

abcdefghijklmnop . . .
abcdefghijklmnop . . . —Point mutations
abcdefghijklmnop . . .
*abc**defghijk**lmnop . . .* —Deletion

The *rII* mutants allowed Benzer to ask a simple yes-or-no question: *Can any recombination at all occur between two given mutations?* Suppose we cross two phage-carrying deletions that do not overlap, such as one missing *defghi* and one missing *mnopqrst.* The DNA of these mutants can combine to make a wild-type genome, since each one has the nucleotides the other is missing. However, mutants with overlapping deletions, such as *defghi* and *hijkl,* are both missing some of the same nucleotides (*h* and *i*, in this example), so their DNA cannot recombine to make a wild-type phage. Therefore, if a cross between two deletion mutants produces even a single wild-type recombinant, their deletions must not overlap, but if the cross produces no wild-type recombinants, we infer that the deletions do overlap. Since a deletion is a change in a linear sequence of nucleotides, it must be possible to place all deletions in an unambiguous sequence, the sequence in which the string of missing letters would appear in a dictionary. Suppose we cross six deletion mutants with one another:

Results of crosses:

	A	B	C	D	E	F
A	0	0	0	+	+	+
B	0	0	0	0	0	+
C	0	0	0	0	+	+
D	+	0	0	0	0	0
E	+	0	+	0	0	+
F	+	+	+	0	+	0

Pairs that can produce wild-type recombinants (+ in the table) don't overlap; those that don't recombine (0 in the table) do overlap. We can infer that these deletions overlap one another in a pattern like this:

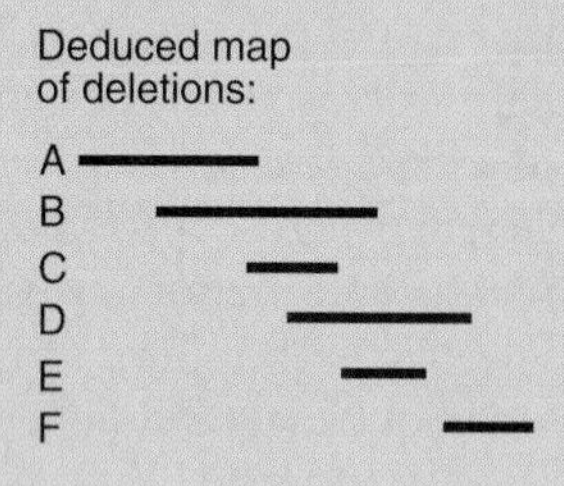

Benzer was able to place a large set of deletions in an unambiguous sequence, as a first step in creating a detailed map of a small genetic region.

Having established the positions of several deletions, Benzer used this information to quickly map new mutations, including point mutations. To see how, consider this question: A businessman whose store is on Main Street is angry because the city has torn up the street between 3rd and 10th Avenues, disrupting traffic in front of his store. Last year he had the same problem when Main was torn up between 9th and 15th. Where is his store located?

The store is obviously between 9th and 10th Avenues, where the two "deletions" of the street overlap. Suppose, then, we have established the order of five deletions and we cross phage bearing a point mutation (X) to all five deletion mutants:

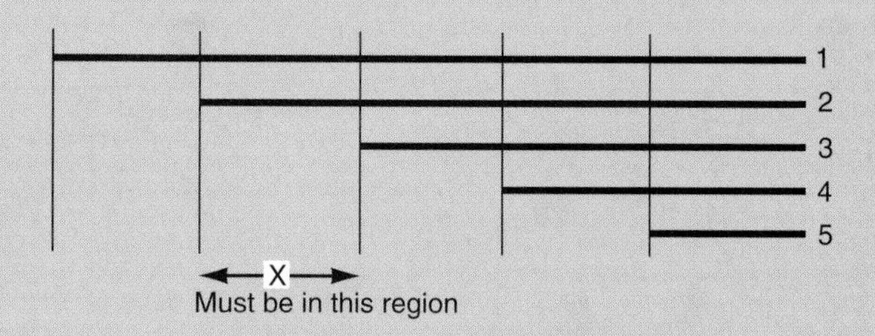

If we get wild-type recombinants when crossed with deletions 3, 4, and 5, but not with 1 and 2, deletions 1 and 2 overlap X and the others do not; X therefore lies between the left end of 2 and the left end of 3. Having established the approximate location of X, we can locate it more precisely using a series of nested deletions that define shorter and shorter regions:

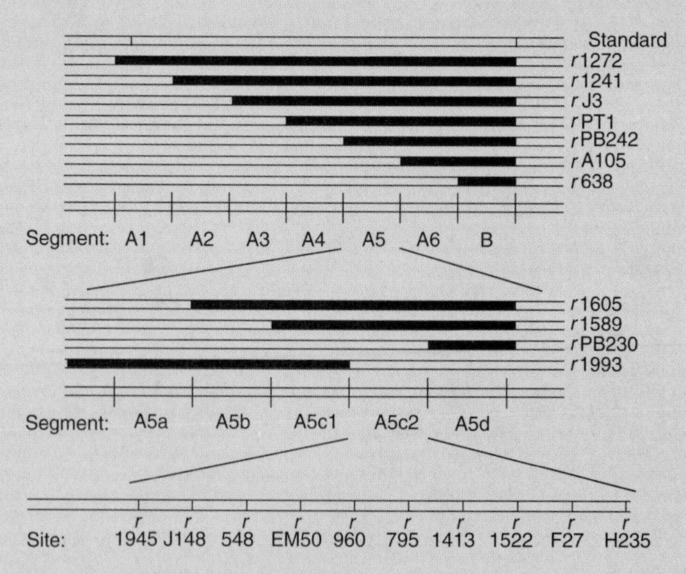

By this method, Benzer located hundreds of point mutations at many sites in the *rII* region, showing that changes at many sites in the

(continued)

genome—probably at every site—can be mutations. Deletion mapping provides a rapid mapping method and a detailed map. Such experiments demonstrate that genetic analysis is a powerful tool, which can disclose information that not even the most powerful microscope can provide.

Exercise 17.A Six deletions are crossed with one another in pairs; the results are shown in the accompanying table. (+ means that they give wild-type recombinants, 0 that they do not.):

	1	2	3	4	5	6
1	0	0	0	0	+	0
2	0	0	+	0	+	0
3	0	+	0	+	+	+
4	0	0	+	0	+	+
5	+	+	+	+	0	0
6	0	0	+	+	0	0

Arrange the deletions in proper order, assigning them arbitrary lengths just to make them fit one another. (*Hint:* Establish the relationship between any two deletions and add others one at a time, modifying the arrangement to fit each new piece of information.)

Exercise 17.B The total length of the T4 map, obtained by adding all the short distances between markers, is about 2,000 map units. The phage genome contains about 2×10^5 base pairs. How many bp are there per map unit? If two mutations are 0.01 map units apart, how close are they (in bp)?

From Exercise 17.B, there are 100 bp per map unit, so if two mutations are only 0.01 map units apart, they are 1 nucleotide apart—that is, right next to each other.

two markers were closer than that. If, however, we could examine 10,000 progeny, we might measure an R value as small as 0.0001.

In the late 1950s, Seymour Benzer developed some fundamental insights into genomic structure, along with new methods of analysis. Benzer found that he could work with large numbers of phage, and thus measure small frequencies of recombination, by using the *rII* (read "r two") mutants of phage T4. The *rII* mutants are conditional lethals that will grow in what we will call strain C (*can* grow) of *E. coli* but not in strain N (*no* growth). In addition to the method described in Methods 17.1, we can illustrate mapping with this genetic system by using two *rII* mutants, *a* and *b* (Figure 17.6). We start with parents that are $a\ b^+$ and $a^+\ b$, and cross them in strain C (where they can grow) by simultaneously infecting C bacteria with both types, using enough phage that the average bacterium gets at least a few phage of each kind. The DNA molecules of these phage mix in the cytoplasm, where they can interact and recombine, as we will explain shortly. When the cells lyse, they produce phage of four possible genotypes: $a\ b^+$ and $a^+\ b$ (the parental types) and $a^+\ b^+$ and $a\ b$ (the recombinant types).

To examine the progeny phage, we must allow them to form plaques on petri plates, as shown in Figure 12.9. We do this by mixing a diluted phage sample with

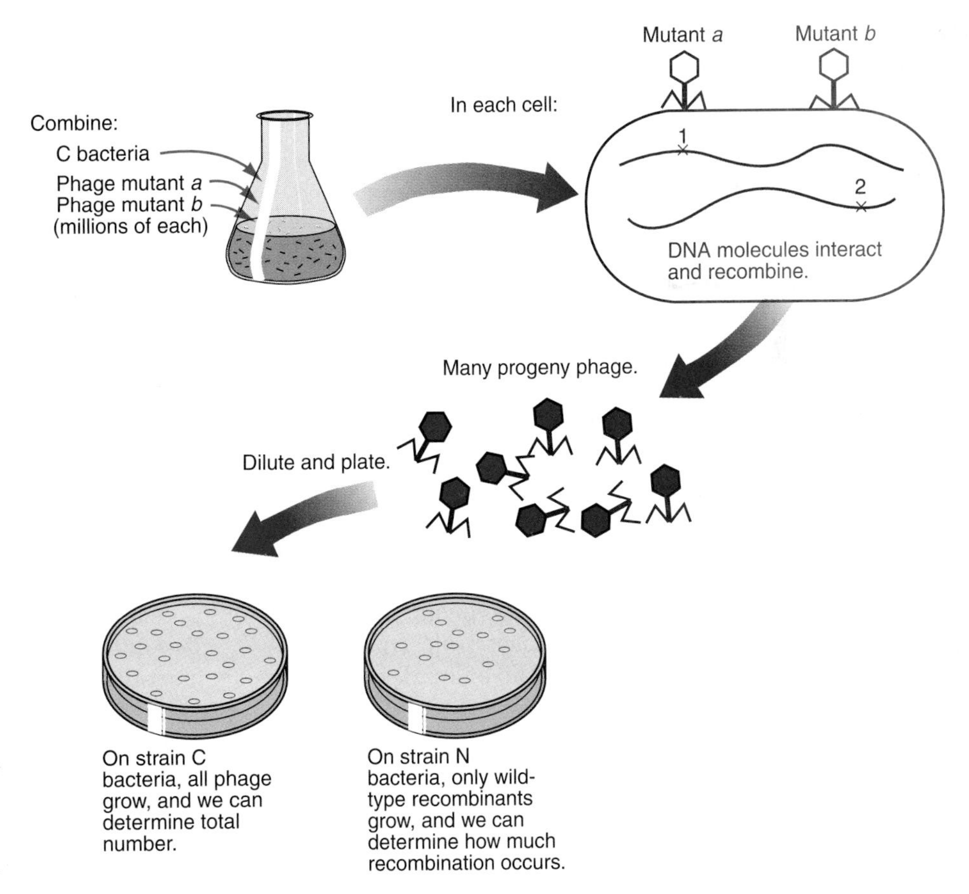

Figure 17.6
In a typical phage cross, two *rII* mutants are crossed by simultaneously infecting strain C bacteria, where both mutants and wild-type phage can grow, so their DNAs can interact and recombine. When the infected cells lyse, we dilute a sample and plate in two ways. By plating on strain C, we can count the total number of phage; by plating on strain N, we can measure the number of wild-type recombinants. The ratio of recombinants to the total is the frequency of recombination.

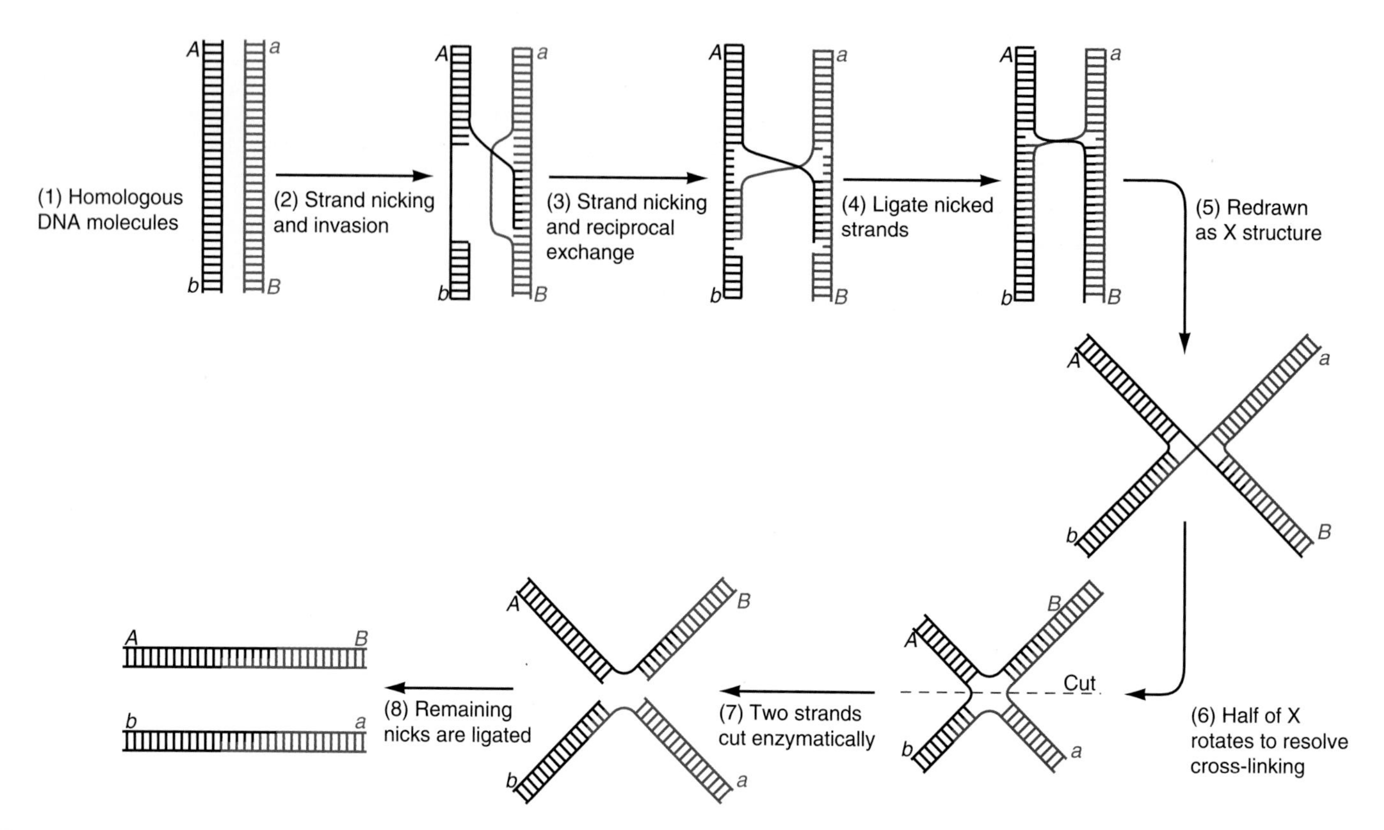

Figure 17.7

Recombination between two DNA molecules probably takes different forms in different situations. This is one simple model for DNA molecules carrying markers *A b* and *a B*. One type of enzyme nicks a strand of one molecule and promotes invasion of the other molecule. A second enzyme is a DNA ligase that synthesizes new covalent bonds between nucleotides to heal a nick. The cross-linked molecules assume an X shape, which has been observed by electron microscopy; half of this structure rotates 180 degrees and the X is resolved by enzymatic cutting of two strands to produce molecules that are hybrids of the original molecules.

millions of plating bacteria on which they grow. We can choose to plate on strain C, where all the phage can grow, and thus count the total number of phage produced, or we can choose to plate on strain N, preventing the growth of *rII* mutants and thus counting only the few wild-type ($a^+ b^+$) recombinants. (This technique is so sensitive that it could detect a single recombinant in 10^6 phage, if recombination ever occurred at such a low rate.) If, for instance, a total of 10^{10} phage/ml are produced in this experiment and 4×10^8 phage/ml are wild-type, the frequency of recombination for this pair of markers is $(4 \times 10^8)/10^{10} = 0.04$.

What is actually going on as the phage DNA molecules interact with one another? Although the chemistry of recombination is complicated and still being intensely investigated, we know it is conducted by specialized enzymes that allow two DNA molecules with similar sequences to exchange parts (Figure 17.7). During the process, the DNA molecules form *heteroduplex* structures, in which a segment of double helix is made of one strand from one parent and one strand from the other parent. Recombination is probably similar in all organisms, though the enzymatic details may differ. Eucaryotic chromosomes have complex protein structures that hold homologous chromatids together in chiasmata and promote recombination.

17.5 Genes are defined by complementation tests.

Mapping experiments locate a series of *mutations,* but they say nothing about the limits of *genes.* The *rII* mutations, for instance, all map in a continuous sequence, but how many genes do they represent, and where are the boundaries between genes?

We use **complementation tests** to answer these questions. Two mutants *complement* each other if each one can supply a function the other lacks. Suppose the *rII* region consists of two genes, *A* and *B,* that are both essential. A gene, by definition, encodes one polypeptide, so a mutation in either gene will produce a faulty protein and thereby stop growth. Neither an *A*-gene mutant nor a *B*-gene mutant by itself can grow on strain N. If, however, N cells are infected simultaneously with these two mutants (so the products of both phage DNAs are in the same cytoplasm), the *B* mutant can supply a normal A protein and the *A* mutant can supply a normal B protein (Figure 17.8). Consequently, the phage will multiply. If, however, both mutations were in the *A* gene or both were in the *B* gene, neither genome could supply an essential protein, and no phage would be produced. The test therefore separates mutations into genes.

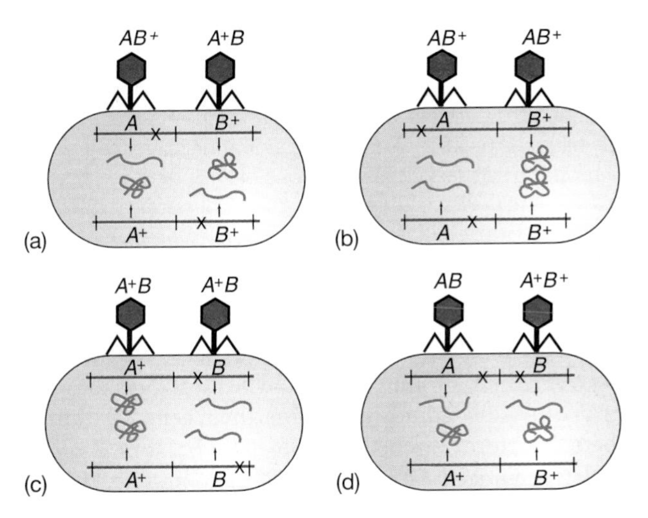

Figure 17.8

A complementation test defines the limits of two genes. *(a)* If two mutations are in different genes, each genome has a good copy of one gene; these complement each other and growth can occur. *(b, c)* If the mutations are in the same gene, there are two bad copies of that gene and no growth can occur. *(d)* As a control, cells are infected with wild-type phage and with a phage carrying both mutations in one genome. The phage should grow in this situation whether the mutations are in the same gene or not; this control checks that the mutant genes do not somehow block expression of the wild-type genes.

Benzer's complementation tests separated the *rII* mutants into two groups. Mutants that map in the left part of the region do not complement one another, and collectively they define the *rIIA* gene; those that map on the right define the *rIIB* gene. Any mutant in one group can complement any mutant in the other. Mutations that define each gene are not mixed with each other along the map; they form two neat, discrete regions, exactly as they should if our picture of genetic organization is correct. In principle, a complementation test can define genes in any organism or virus.

17.6 A gene is really colinear with its protein product.

Crick's Law says that the sequence of codons in a gene is the same as the sequence of the amino acids they specify. One of the clearest experimental demonstrations of this fact came in 1964 from the work of Charles Yanofsky and his colleagues with the *E. coli* enzyme tryptophan synthetase, the last enzyme in the pathway of tryptophan biosynthesis.

The Yanofsky group collected many tryptophan auxotrophs, bacteria that require tryptophan for growth, and mapped their mutations (using methods we describe later). Using complementation tests, they identified mutations in the *A* gene that encode one polypeptide of tryptophan synthetase, and created a map of several mutations in the *A* gene. By also determining the amino acid sequences of the wild-type protein and of each mutant form, they could compare the amino acid replacements in the proteins with the mutational sites in the gene (Figure 17.9). The location of each mutation correlates perfectly with the position of a single amino acid replacement in the protein. This was an elegant, direct demonstration of the colinearity between gene and protein.

Two of the *A*-gene mutants had different replacements of the same amino acid. In mutant *A23*, the glycine at position 210 is replaced by arginine; in mutant *A46*, it is replaced by glutamic acid. The two mutational sites must therefore be in the same codon. The genetic code (Table 14.1) shows that the mutations affect the first two nucleotides of the codon for position 210, and you can easily figure out what substitutions must have occurred in the DNA to produce these replacements.

17.7 A messenger is read by threes without commas.

The experimental evidence we have just examined shows that a DNA molecule and the protein it encodes are really related in the simple, direct way we outlined in Chapter 14. Biochemical work by Nirenberg and others also identified the three-base codons that encode each amino acid (see Section 14.4), but this biochemical work built upon earlier experiments by Crick, Brenner, Leslie Barnett, and R. J. Watts-Tobin who used the *rII* system for an elegant genetic analysis that showed the genetic code is triplet and "commaless." Crick and his colleagues used *rII* mutants produced by the dye *proflavine,* which induces short additions or deletions of one or a few nucleotide pairs. These experiments only make sense if the specific coding mechanism they postulated is correct. They proposed that the RNA message is translated by a reading-frame mechanism, as we explained earlier, that marks off three bases at a time and associates the correct amino acid with each triplet of bases. (We now know this happens as proteins are made because each transfer RNA associates with three bases in the messenger RNA, but this experiment was crucial in demonstrating that something like this must occur.) There are no "commas" or spaces in the message,

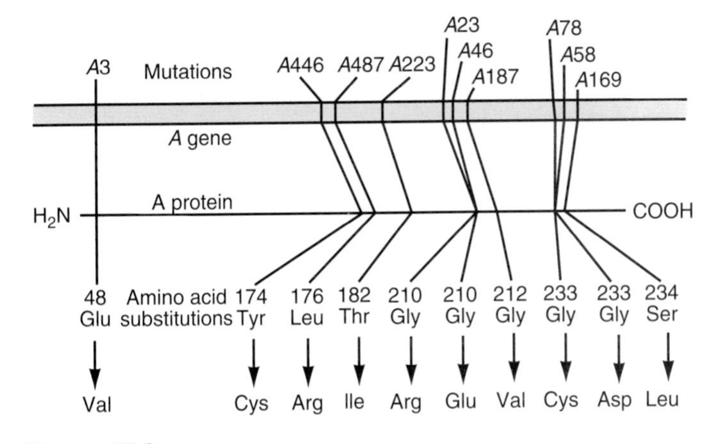

Figure 17.9

Yanofsky's results show that the sequence of mutations in a gene is colinear with the sequence of changes in the amino acids they encode.

so if the reading frame is accidentally shifted by one or two bases, it will read the wrong triplets and make the wrong polypeptide. Suppose the message, in triplet words, is properly read:

THE BIG RED DOG RAN AND BIT THE TAN CAT BUT ITS LEG WAS TOO FAT AND . . .

With a proflavine-induced deletion of one nucleotide pair, the resulting mRNA would be read:

THE BIG RED DOG RNA NDB ITT HET ANC ATB UTI TSL
↑
EGW AST OOF ATA ND . . .

This is a **frameshift mutation** because the reading frame is shifted out of phase, one base to the right, and the message is read incorrectly after the point of the mutation.

Now suppose a second mutation inserts a base close to the deletion. This frameshift mutation shifts the reading frame back to the left by one base and puts it into the proper phase:

THE BIG RED DOG RNA NDB ITT **X**HE TAN CAT BUT ITS
↑ ↑
LEG WAS TOO FAT AND . . .

There is still a short, mistranslated segment, but in some parts of a protein a few wrong amino acids won't matter. This second change is called a **suppressor mutation** because it cancels the effect of the first mutation. Without knowing what change has occurred in the DNA, every mutation can be classified arbitrarily as left-shift or right-shift (L or R) by simply noting that, in general, any L mutation should suppress any R mutation to yield the wild-type phenotype.

The Crick group collected proflavine-induced mutations in a segment of the *rIIB* gene. They started with one mutation, arbitrarily designated an L type, and collected several suppressors of it, which must all be R types. Then they separated out each of these mutations and found suppressors of *them;* these suppressors of R mutations must naturally be L mutations. In this way, they collected many mutations that could be designated (still arbitrarily) L or R. In agreement with their prediction, they found that any R mutation can suppress any L mutation when they are combined (if their sites are close enough), but a double mutant made with two Ls or two Rs is still mutant.

The critical test was to combine three L or three R mutations. If the code is really triplet, a triple mutant should behave like wild-type, and it does. For instance, three additions close together should look like this:

THE BIX GRE DDO GRA NAX NDB ITT HXE TAN CAT BUT
↑ ↑ ↑
ITS LEG WAS TOO FAT AND . . .

The code could actually be sextuplet, or some higher multiple of three, and it took Nirenberg's later work to confirm that the code is triplet. Yet it was this elegant genetic analysis that prepared the ground for the biochemical analysis and showed the power of clever genetic work.

Exercise 17.2 Proflavine can induce an addition or deletion of more than one base pair. By experimenting with triplet words, convince yourself that adding 1, 4, or 7 bp has the same effect as deleting 2, 5, or 8, and that deletions of 1, 4, or 7 have the same effect as additions of 2, 5, or 8.

17.8 Bacteria have a pseudosexual mechanism.

In 1945, Edward Tatum and Joshua Lederberg recognized that bacteria, being such simple cells, would be ideal tools for genetic studies if only they could recombine their genes. Tatum and Lederberg therefore searched for a sexual process in *E. coli.* Realizing that recombination in bacteria must be rare, if it happens at all, they set out to select for the few cells that have experienced recombination, as if they were rare mutants. They tried to recombine two auxotrophic mutants to make a prototroph: one strain that required threonine and leucine (thr^- leu^- met^+ thi^+) and another that required methionine and thiamine (thr^+ leu^+ met^- thi^-). Only thr^+ leu^+ met^+ thi^+ recombinants could grow on minimal medium containing none of these substances, and sure enough, a few recombinants appeared in a mixture of many parent cells (Figure 17.10). Other experiments demon-

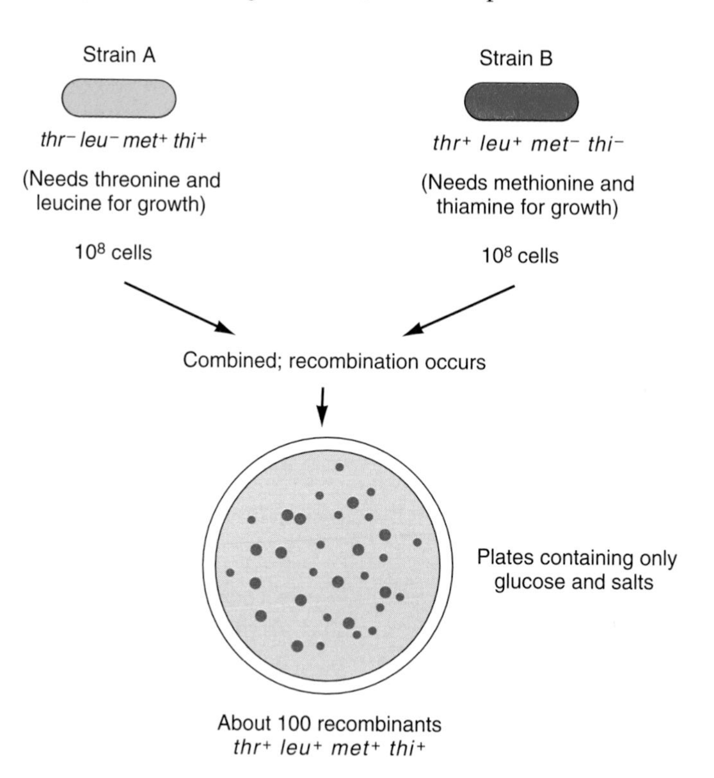

Figure 17.10
Genetic recombination in bacteria was discovered by mixing two auxotrophic parental strains with the genotypes *thr leu* and *met thi.* A few wild-type prototrophic bacteria were found; these could only have arisen through recombination, since the probability of any bacterium mutating back to the wild-type condition in two genes is essentially zero.

strated that recombination occurs when two parental cells come into intimate contact in the process of **conjugation.**

We now know that conjugation only occurs because some bacteria carry extra genetic elements called **plasmids,** which have become crucial for modern genetic work. Plasmids are rather small, circular DNA molecules, of about 10,000–100,000 bp, contrasted with a bacterial chromosome of about 4 million bp (Figure 17.11). Plasmids resemble viruses, as both are small, semi-independent genomes, and in Section 17.10 we will show that the two kinds of entities share many properties and can't be clearly distinguished from each other. The principal difference is that plasmids, unlike viruses, never exist as particles outside a cell. Plasmids maintain themselves inside their host cells by self-replication, but some kinds of viruses can enter a state in which they do the same. Plasmids—and many benign viruses—do their hosts no particular harm and may confer several benefits. Even the smallest plasmid has several genes, and collectively, plasmids carry quite a variety of genes.

Many plasmids can initiate conjugation and transfer copies of themselves into other cells. The **F factor** (F for fertility) is a transferable plasmid responsible for most conjugation in *E. coli* (Figure 17.12). One of the proteins it encodes is pilin, which forms long, thin extensions called **pili** (sing., **pilus**) on the cell surface. An F^+ cell, which carries an F factor, makes pili that stick to an F^- cell, which has no F factor, and the two cells are drawn together into intimate contact. The F factor then replicates and transfers a copy of itself into the F^- cell, thus converting the F^- cell into an F^+ cell. In this way some plasmids spread themselves through a population of bacteria like a disease. (F^- bacteria remain in nature for various reasons, including the rarity of F^+ strains.)

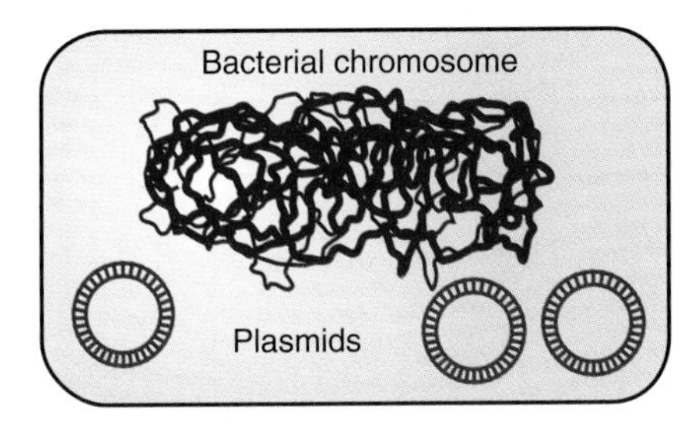

Figure 17.11
Plasmids are small, circular DNA molecules that can replicate themselves inside host cells. Thus they maintain themselves as the bacteria multiply, generally doing their hosts no harm and often conferring valuable traits, such as resistance to antibiotics.

The F factor belongs to a subclass of plasmids known as **episomes** (Figure 17.13) that lead a double life. An episome can be an independent DNA molecule in the cytoplasm of a bacterium, or it can integrate itself into the bacterial chromosome. Integration occurs through crossing over. Recombination enzymes cut the two DNA molecules and link the plasmid and chromosomal DNAs into a single, covalently bonded molecule:

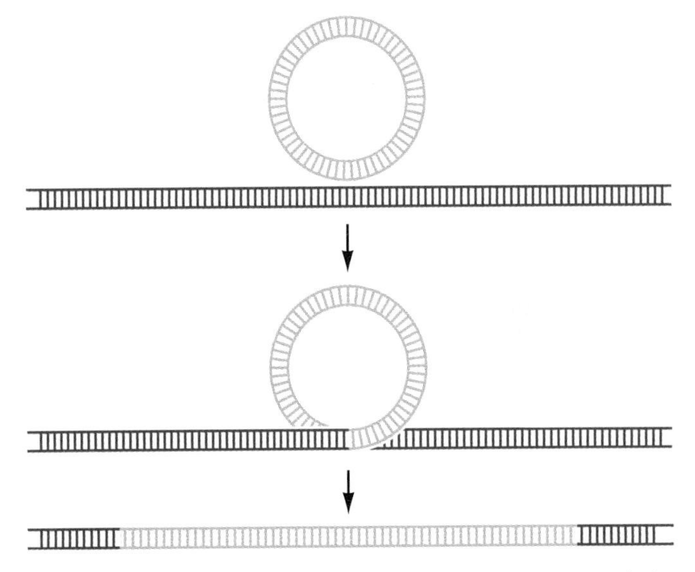

Then each time the chromosome replicates, the integrated plasmid DNA is also replicated, and in this way the plasmid ensures its maintenance in the cell.

Cells with an integrated F factor are called **Hfr strains** (for High frequency of recombination) because during conjugation they can regularly transfer their genes to an F^- cell. When an Hfr cell encounters an F^- cell, the F factor is stimulated to transfer a copy of its DNA (Figure 17.14), but since that DNA is continuous with the chromosomal DNA, a copy of the chromosome (or at least a substantial part of it) is transferred into the F^- cell. Through recombination, some genes from the Hfr cell then replace some of the F^- genes. Notice that bacterial conjugation is not a two-way exchange of genes. The Hfr parent (sometimes called a male strain) donates DNA, while the F^- (sometimes called a female) is a recipient. Recombination

Figure 17.12
The F factor is a plasmid that can transfer itself from one cell to another. An F^+ cell, which carries an F factor, makes protein extensions called pili that can make contact with F^- cells containing no F factor. The two cells are drawn together as the pilus depolymerizes; then the F factor replicates a copy of itself and sends it into the F^- cell, converting it to F^+.

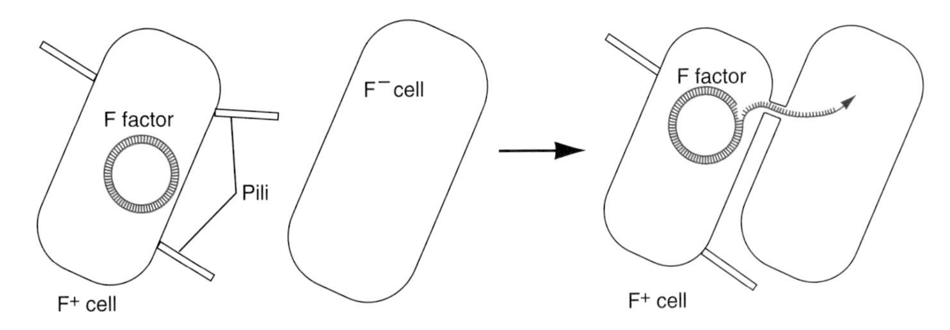

actually occurs in the recipient cell. The F^- cell generally is not converted to F^+ because it does not receive the entire F factor.

17.9 Donor cells transfer genes in their linkage sequence.

Before much was known about plasmids, François Jacob and Elie Wollman performed important experiments that provided insights into conjugation and recombination. They crossed Hfr and F^- strains differing in several genes and tried to follow the transfer of the Hfr chromosome by interrupting conjugation at intervals by whipping the conjugating bacteria around in a blender to break the weak connections between them. Then they tested the F^- cells to determine which genes they had received from the Hfr donors. The experiment showed that the bacterial genes occur in a definite sequence, the sequence in which they are transferred into the F^- cells. Time-of-transfer experiments are routinely used to map the bacterial chromosome, and the standard map uses minutes rather than recombination frequencies as map units (Figure 17.15). The *E. coli* map, developed through this and other methods to be described shortly, now contains over 1,000 gene loci.

Jacob and Wollman found that each Hfr strain transfers its genes in a different sequence, but that they can all be combined into a single *circular* sequence. Thus, before anyone had distinctly seen the bacterial chromosome with an electron microscope, genetic analysis had established that it is a circle. Each Hfr strain begins transferring from the point where its F factor is integrated and transfers its genes in sequence until mating is interrupted.

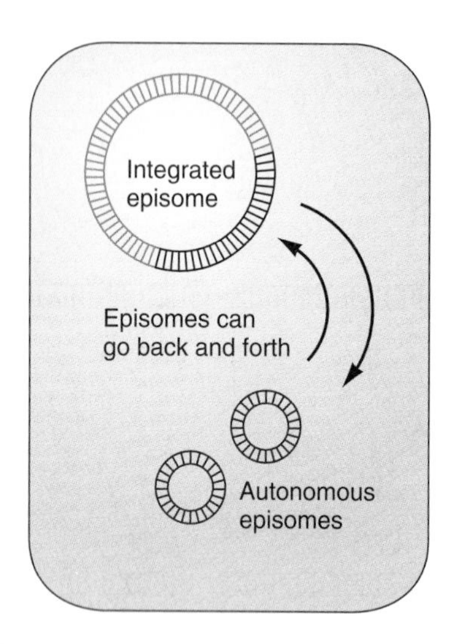

Figure 17.13
Episomes are genetic elements that can exist either autonomously in the host cytoplasm or integrated into the host cell genome.

Exercise 17.3 Four Hfr strains are found that transfer various genes in different sequences, as follows:

Hfr 1: *ala thi his proA gal leu lac*
Hfr 2: *his thi ala proC ara metB pyr*
Hfr 3: *leu lac pyr tyr metB ara ala*
Hfr 4: *metB tyr lac leu gal val proA*

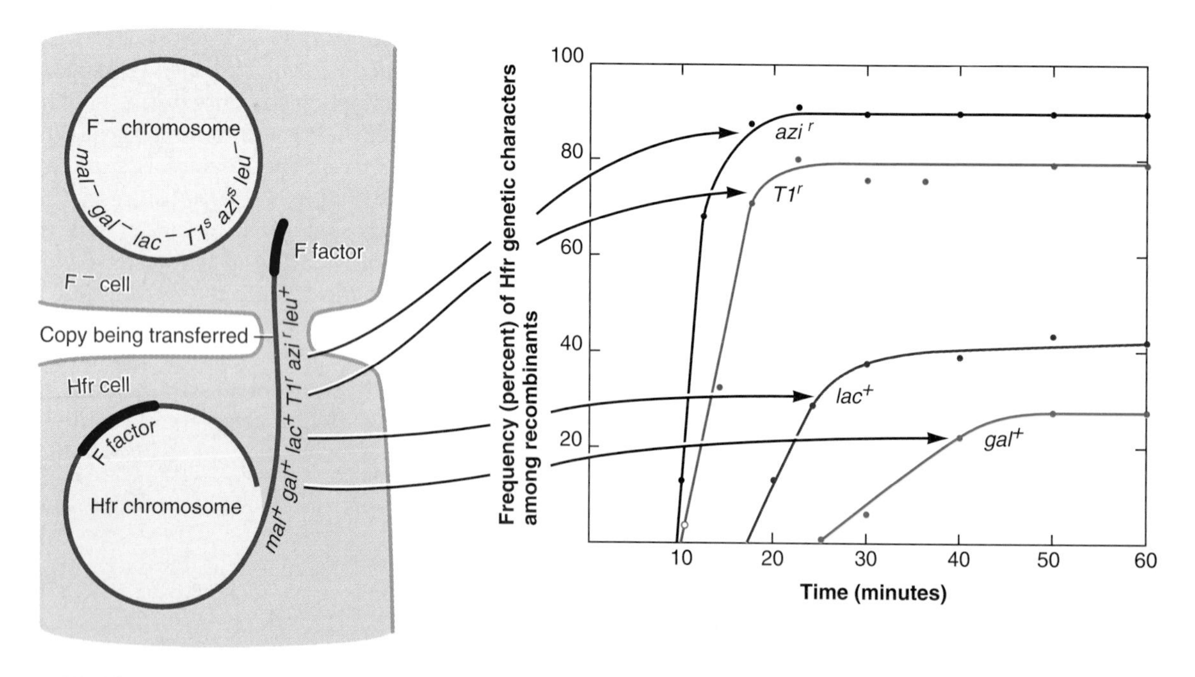

Figure 17.14
When an Hfr cell conjugates with an F^- cell, the F factor again starts to transfer a copy of itself, but since the F DNA is covalently linked to the cellular DNA, a copy of the Hfr genes is transferred into the F^- cell. The genes enter the F^- cell in their linkage order, and each gene first appears at a specific time: the *azi* (azide resistance) gene 9 minutes after conjugation begins, the *T1* gene at 10 minutes, the *lac* gene at 17 minutes, and so on. Recombination between Hfr and F^- chromosomes occurs in the F^- cell after the transfer has occurred.

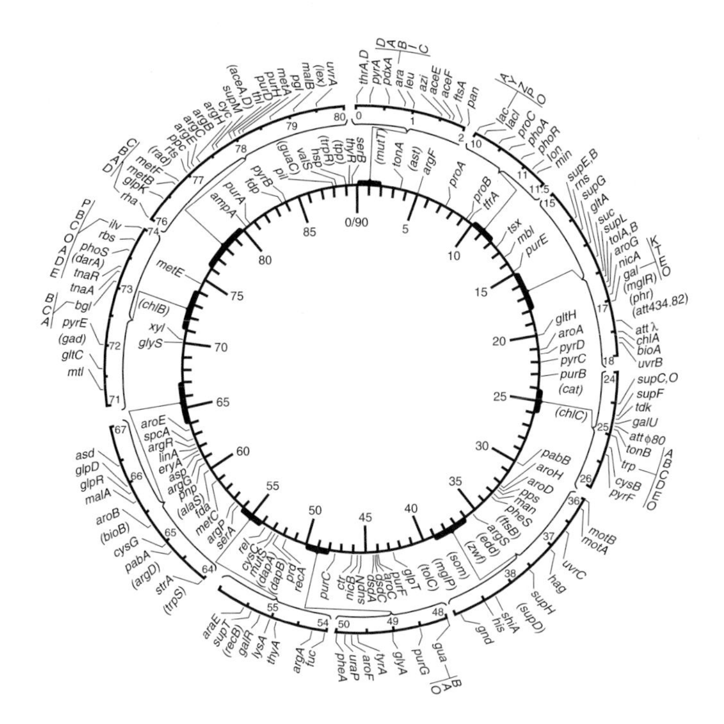

Figure 17.15

By using time of entry in place of map units, a map of the bacterial genome can be established. The three-letter name of each gene indicates its function, and a fourth (capital) letter distinguishes different genes with related functions. For instance, all the *ilv* genes (near 74 min on the map) encode enzymes for the synthesis of the amino acids isoleucine and valine; the *arg* genes located at several places encode enzymes for the synthesis of arginine. This is a rather simple version of the *E. coli* map; a detailed contemporary map shows so many genes and so much information that it must be printed on a large sheet of paper.

(Not all genes were examined in each strain.) Assuming that all these strains have the same gene sequence, draw a genetic map and note the location of the F factor in each strain. (*Hint:* Place the markers of one Hfr strain on part of a circular map; then add markers from another Hfr. You'll see how they all fit together.)

17.10 A zoo of little genetic entities broadens our conception of the biological world.

The biological world is full of amazing things, and biologists have uncovered a remarkable microcosm of small genomes that are behind a variety of important phenomena, especially some nasty diseases that often invade our lives. These entities have evolved so many ways to replicate themselves that it is hard to fit them neatly into categories, which is why we hedge on drawing hard lines among viruses, plasmids, and episomes.

Remember that viruses are minute genetic entities that are not cellular like organisms and can only multiply within a cell by subverting the cell's genetic apparatus to the production of viral genomes and proteins. Each virus particle, a *virion,* consists of a nucleic acid genome protected by a protein coat, a *capsid,* as shown by two representative viruses:

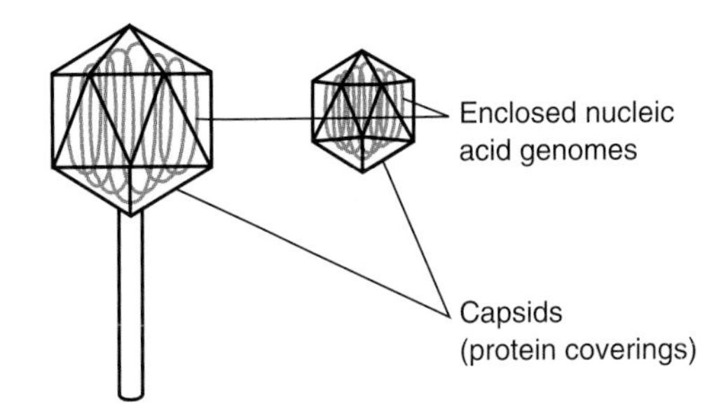

We have used phage as model viruses. Many, perhaps most, viruses reproduce in a **lytic cycle,** the process exemplified by a phage like T4: First, the phage genome is injected into a bacterial cell, where it subverts the cell's normal genetic apparatus. After about half an hour, the cell is filled with new phage particles and then lyses. But other viruses reproduce differently. Phage λ (lambda), for instance, which also infects *E. coli,* sometimes goes through a lytic cycle and kills its host while multiplying rapidly. At other times, the λ DNA does not start to replicate after injection. Instead, it forms a circular molecule that acts just like an episome, even though phage λ is a virus, not a plasmid. The λ genome integrates at one point on the bacterial chromosome, between the *gal* genes (for catabolism of galactose) and the *bio* genes (for synthesis of the coenzyme biotin):

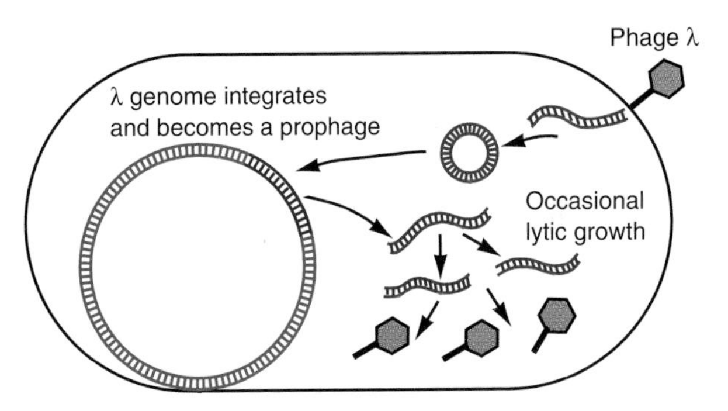

Once integrated into the chromosome, the phage genome has gone into a noninfectious **prophage** state. A cell that harbors a prophage is said to be **lysogenic.** In spite of the virus genomes they carry, lysogenic cells grow and multiply like any others, and since the prophage is replicated every time the bacterial chromosome replicates, the lysogenic condition is quite stable, like the Hfr state of having an integrated F factor. The whole phenomenon is called **lysogeny,** and phage that can establish lysogeny (can *lysogenize*) are called **temperate phage.**

Lysogeny occurs because the phage establishes a *control system* that keeps all its genes quiescent (except those for the control system itself), so it doesn't enter a lytic cycle. Every once in a while, however, in perhaps one cell in 10,000, the control mechanism keeping the prophage quiescent is disrupted. Then a prophage excises itself from the chromosome, through a reverse crossover, and starts to replicate itself and grow lytically, so this one cell lyses and releases a burst of new phage. (Lysogenic means "capable of lysing.") Prophage can also be induced, with ultraviolet light or certain other agents, to leave the chromosome and enter the lytic cycle. The prophage does not have to be integrated to be controlled, and the prophages of other temperate phage, such as P1, simply inhabit the cytoplasm like plasmids:

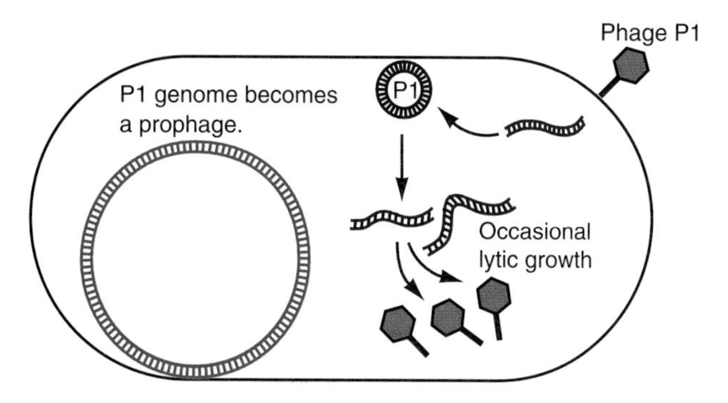

Temperate phage are of great interest for quite practical reasons, because some bacteria only cause serious diseases if they are lysogenic for certain phage. *Corynebacterium diphtheriae,* which commonly inhabits the human throat along with a "normal flora" of other bacteria, only produces diphtheria and a characteristic toxin (poison) if it carries prophage β. *Clostridium botulinum* only causes food poisoning with its botulinus toxin if it harbors the phage DEβ, and only lysogenic strains of *Streptococcus* can cause scarlet fever. Occasional outbreaks of intestinal infections are caused by *E. coli* strains carrying lambdoid prophages with genes for intestinal toxins; they caused illness and some deaths among customers of Jack-in-the-Box restaurants in the United States in 1994 and several outbreaks in Japan in 1996.

Other viruses that can only multiply in a lytic cycle replicate their DNA in quite a different way. The minute virus ϕX174 (Greek letter phi, pronounced "fee") has a tiny, circular, *single-stranded* DNA genome. When this DNA enters a cell, it uses a special DNA polymerase to set up a *replicative form (RF),* a circular, double-stranded molecule similar to a plasmid:

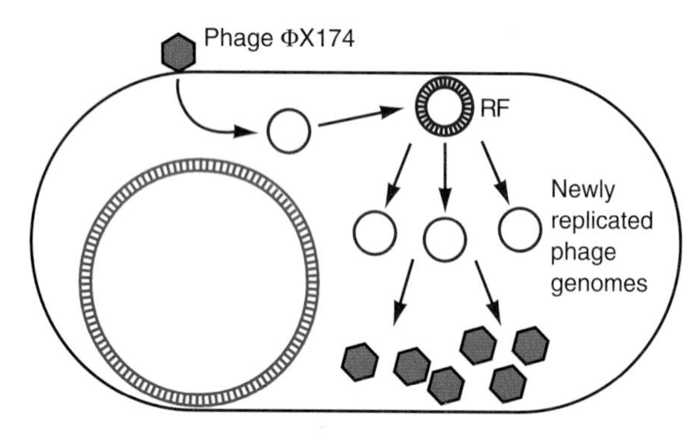

Using a second unique enzyme, the RF then replicates hundreds of new single-stranded genomes, which get wrapped up in viral proteins to make new virions. Phage of a different family, such as M13, have long, thin virions that look like the pili of an F^+ bacterium and actually infect by attaching to a pilus, rather than to the bacterial surface. M13 also replicates its single-stranded DNA with an RF, but it reproduces without lysing its host cell; as its RF continually casts off new, single-stranded genomes, they combine with capsid proteins at the cell surface, and the cell continually extrudes virions without any damage to itself except the drain on its energy:

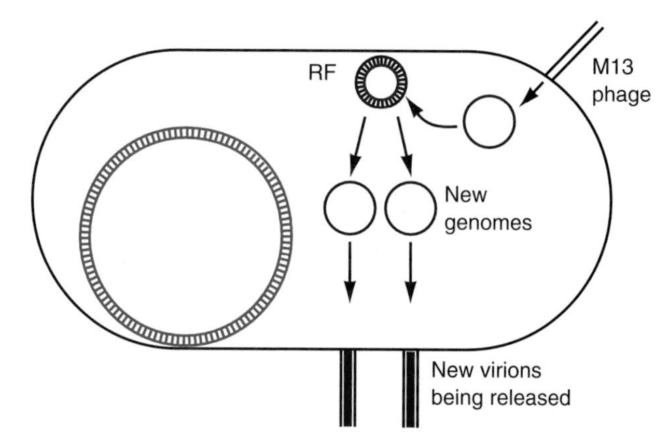

Here, then, is a virus that behaves much like a plasmid. Cholera, an often-fatal intestinal disease, is caused by the bacterium *Vibrio cholerae,* but this organism is only pathogenic if it carries a virus similar to M13 that has the genes for the cholera toxin.

The F factor is a typical *transmissible plasmid,* one that can pass copies of itself into other cells, and we have seen that it is also an episome because it can integrate itself into the host chromosome. Genes carried by plasmids confer all kinds of abilities. Sidebar 17.1 describes the *resistance factors (R factors),* plasmids that create major health problems because their genes confer resistance to antibiotics. Some plasmids are *virulence factors* whose genes make their hosts pathogenic, like some of the temperate phages listed above; virulence factors include the plasmids that make the bacteria responsible for intestinal infections ("Montezuma's Revenge") that are the bane of travelers. Other plasmids produce *colicins,* toxic agents that kill certain bacteria. *E. coli* strain K30, for instance, produces a colicin, E1; K30 is immune to E1, but the colicin kills other strains of *E. coli.* The E1 genome is a plasmid, which usually remains quiescent in its host cell, but sometimes E1 DNA starts to replicate rapidly and produce its colicin, which is released upon cell lysis. So E1 behaves like a very defective temperate phage, producing a protein that kills cells but cannot infect them and reproduce:

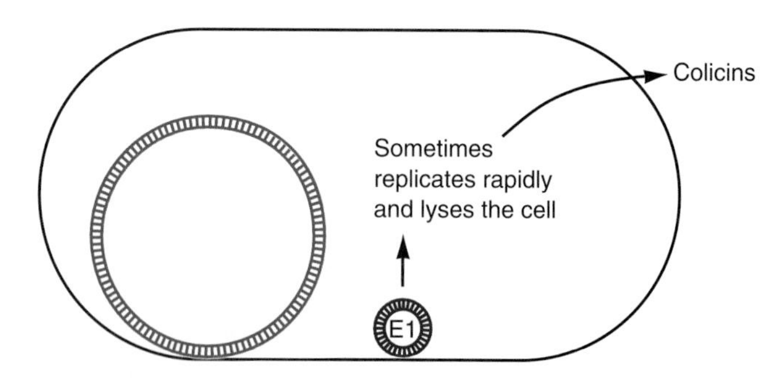

All these genetic elements are made more varied by the presence of *transposons,* small DNA elements that can transpose themselves from place to place and insert themselves into plasmids or into the genomes of organisms. Transposons encode enzymes (transposases) that promote recombination of the transposons themselves with other DNA molecules:

Sidebar 17.1 R Factors and Antibiotic Resistance

An important kind of plasmid was discovered in Japan in 1955 when a Japanese woman contracted a form of dysentery caused by *Shigella* bacteria. Such infections are usually controlled easily by antibiotics, but these particular bacteria were resistant to four antibiotics: sulfanilamide, streptomycin, chloramphenicol, and tetracycline. This was amazing, since antibiotic resistance was rare at the time, and simultaneous resistance to four different drugs was unheard of. Yet, in subsequent years, dysentery in Japan became very intractable to drug treatment as multiply-resistant strains of *Shigella* became more common.

Tsumoto Watanabe found that multiple resistance is conferred by plasmids called **R factors** (R for "resistance") that behave very much like F. Many such plasmids are now known that confer resistance to several drugs. Like F, they can spread rapidly through a population, transferring themselves from a resistant to a sensitive cell by conjugation, and they can even transfer between different species of bacteria. In Japan, the proportion of drug-resistant *Shigella* strains increased from 0.2 percent in 1953 to 58 percent by 1965, and most of them carried R factors. One Japanese study found that 84 percent of all *E. coli* and 90 percent of *Proteus* (another gut bacterium) in hospital patients were resistant to the same antibiotics. The spread of R-bearing bacteria has resulted in outbreaks of multiply antibiotic-resistant infections all over the world.

An R factor can carry more than 10 different antibiotic-resistance genes at once. R-bearing bacteria have been reported in marine fish, and a comparison of bacteria in the Stout River in England sampled in 1970 and 1974 revealed that, while the number of bacteria remained the same, antibiotic resistance had doubled. It is clear that R factors are being strongly selected by widespread use of antibiotics. The rapid spread of R factors even across generic boundaries promises drug-resistant outbreaks of other diseases. Studies at Birmingham Accident Hospital showed that drug-resistant pseudomonads (another type of bacteria) isolated from infected burn victims probably acquired R factors from the patients' own intestinal bacteria.

The problem is basically ecological. Vast reservoirs of bacteria carrying R factors exist in the environment because of the extensive use of antibiotics as growth promoters in feed for chickens, pigs, cattle, and other livestock, which amounts to about 15 million pounds a year. Apparently, when low-grade infections are eliminated by such treatment, animals grow faster. However, antibiotic treatment rapidly selects for drug resistance, so the antibiotics lose their effectiveness, and their levels in feed have to be continually increased to maintain growth. As a result, the incidence of R-bearing cells in livestock has increased dramatically. In Great Britain, the potential hazard of R factors spreading from farm animals to bacteria in humans was documented in the Swann Report, which recommended a drastic cutback in the use of antibiotics as growth promoters. However, the economic returns of antibiotic sales to farmers are enormous—over $270 million in the United States in 1983—and in North America, pharmaceutical companies have resisted the pressures to stop selling their products for feed to farm animals.

Although the antibiotic and agriculture industries claim their practices do not select for resistant bacteria and pose no threat to human health, a definitive demonstration of the real hazards came in 1984. It started in February of 1983 when Michael Osterholm of the Minnesota Health Department called the Centers for Disease Control (CDC) in Atlanta to report an unusual outbreak of gastrointestinal disease in the Minneapolis-St. Paul area, caused by *Salmonella newporti.* The bacteria all carried the same plasmid, making them resistant to ampicillin, carbenicillin, and tetracycline. Other infections by the same bacteria were reported in South Dakota by the state epidemiologist. The bacteria were eventually traced to a feedlot that routinely added chlorotetracycline to its animal feed. Beef from the lot had then been shipped to supermarkets where the Minnesota victims bought their meat.

The story of R factors shows what can happen when short-term economic gain overshadows the possible long-term consequences. Now R-bearing strains of syphilis and gonorrhea bacteria promise to add to the already heavy medical load of venereal disease because these infections can no longer be treated with common antibiotics such as penicillin. Some microbiologists are warning that the era of general antibiotic therapy is coming to an end, as more and more bacteria become resistant. Fighting infectious disease in the future may depend much more on people developing good natural immunity through more healthful regimes of living, rather than depending on physicians to cure them.

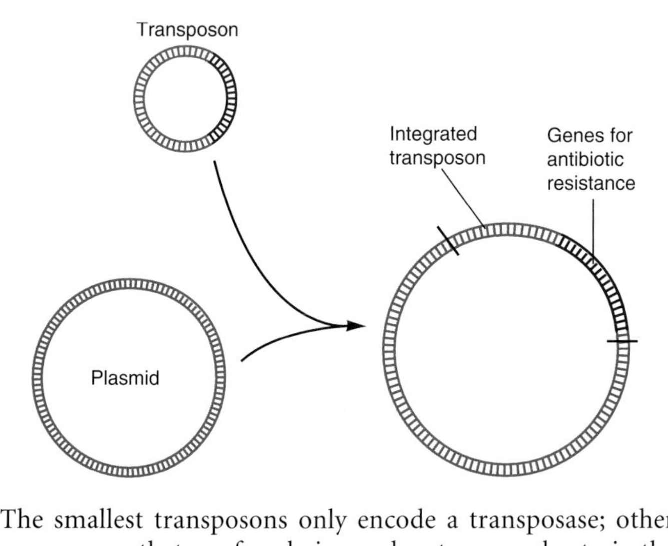

The smallest transposons only encode a transposase; others carry genes that confer obvious advantages on bacteria that carry them. A transposon may include genes for novel metabolic pathways (such as the catabolism of unusual compounds like camphor or toluene) or for resistance to antibiotics, so various plasmids can become resistance factors by incorporating transposons that carry these genes.

Bacteria may also carry *insertion sequences,* DNA sequences of a few hundred to a few thousand bp that can randomly insert themselves into the genome and thus inactivate any gene they land in. The enigmatic bacteriophage mu (for "mutator") is a temperate phage that can only reproduce by first inserting itself into a bacterial genome, producing a mutation as it does so.

Transposable elements and plasmids are not confined to bacteria. Yeasts and other fungi carry their own distinctive plasmids. The geneticist Barbara McClintock spent much of her productive life studying similar elements in maize that she called "controlling elements"; they are able to transpose themselves from one place to another in the maize chromosomes, activating and inactivating genes as they go, and often operating in pairs composed of a *receptor* element that moves in response to molecular signals from a *regulator* element. Similar transferable genetic elements have been found in *Drosophila* and in yeast; they may occur in all genomes.

Biologists are just starting to appreciate the full potential of these genetic elements. They have been implicated in rearranging chromosomes by breaking and recombining them. They may be generally responsible for a degree of chaos and instability in genomes, perhaps creating large-scale changes that allow the evolution of quite novel organisms. Finally—in contrast to the ordinary *vertical transmission* of genes from parents to offspring—they add the possibility of *horizontal transmission* of genetic information between unrelated organisms, as from one bacterium to another via a plasmid or a virus, as we show in the next section. For instance, similarities between certain genes of two different species of bacteria, or between genes of an organism and a virus, suggest that copies were moved from one creature to another, and this adds an entirely new dimension to the mechanisms of evolution.

17.11 Viruses can promote genetic exchange.

In 1952 Lederberg's student Norton Zinder looked for recombination in *Salmonella typhimurium,* a close relative of *E. coli.* He mixed two auxotrophic strains, LT-2 and LT-22, and to his delight, found prototrophic recombinants. However, he also discovered that genetic exchange could occur even if the cells were not in contact, so they couldn't be conjugating, and that the cells were not being transformed by free DNA. He then demonstrated that LT-22 was lysogenic for a phage, P22, and that P22 virions were actually carrying some of the host DNA into the LT-2 cells. Zinder called this new mode of genetic exchange **transduction.**

Phage such as P22 are *generalized transducers* that can carry any bacterial genes, as well as any other piece of DNA, from one cell to another (Figure 17.16*a*). Phage-encoded enzymes generally cut the DNA of the host cell into pieces, and as new phage particles are made, they may incorporate these host DNA molecules instead of phage genomes, so they can transfer this DNA into other cells. Other phage, such as λ, are *specialized transducers* that can only carry a few bacterial genes near their point of integration in the bacterial chromosome (Figure 17.16*b*). When λ prophage are induced to enter a lytic cycle, the phage genome normally reverses the steps by which it was integrated, but occasionally the prophage is excised incorrectly: It carries along some bacterial genes while leaving some phage genes behind. Phage λ virions, for instance, may pick up *gal* genes from the cell they are made in and carry them into other cells. Because transducing phage carry small pieces of the bacterial genome, transduction is commonly used for detailed mapping of small regions.

Having outlined the natural phenomena that promote genetic exchange and recombination in bacteria, phage, and plasmids, we will now introduce some powerful experimental methods that have been developed for manipulating genomes.

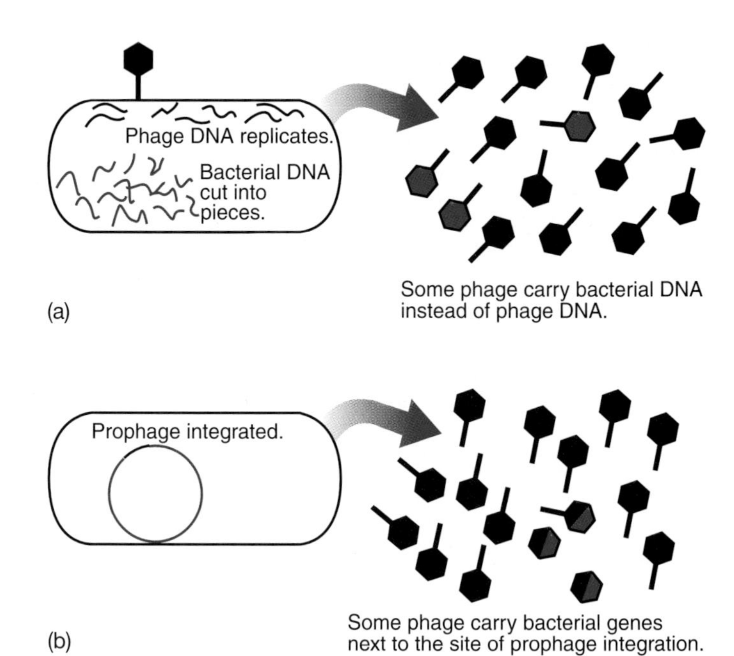

Figure 17.16

Transducing phage can be made in two ways. *(a)* The bacterial DNA is broken into pieces, some of which are accidentally incorporated into newly formed phage virions. *(b)* Temperate phage that integrate into the host genome as prophages may accidentally carry along some host DNA if the prophage excises from the host chromosome incorrectly, picking up some of the host genes on either side of its integration point.

17.12 Bacteria restrict the growth of foreign entities like viruses.

Around 1961, Werner Arber and Daisy Dussoix tried to find out why phage λ that have been grown in one strain of bacteria often grow very poorly in other strains. They discovered that when the phage DNA enters a type of cell different from the one it was made in, it is attacked and degraded by enzymes now called **restriction endonucleases.** We now know that every bacterial strain protects itself by making distinctive enzymes that degrade foreign DNAs, such as phage DNA. (*Endo*nuclease means that the enzyme breaks phosphodiester bonds in the interior of the DNA molecule, not at its ends.) Each enzyme recognizes and cuts a particular sequence of nucleotides. These sequences are always *palindromic*—that is, both strands have the same sequence in the $5' \rightarrow 3'$ direction, like this:

G·A·A·T·T·C
C·T·T·A·A·G

(A palindrome is a sentence that reads the same backwards and forwards, such as Adam's introduction to Eve, "Madam, I'm Adam," or Napoleon's lament, "Able was I ere I saw Elba.")

The enzyme cuts this sequence in the same way on each strand:

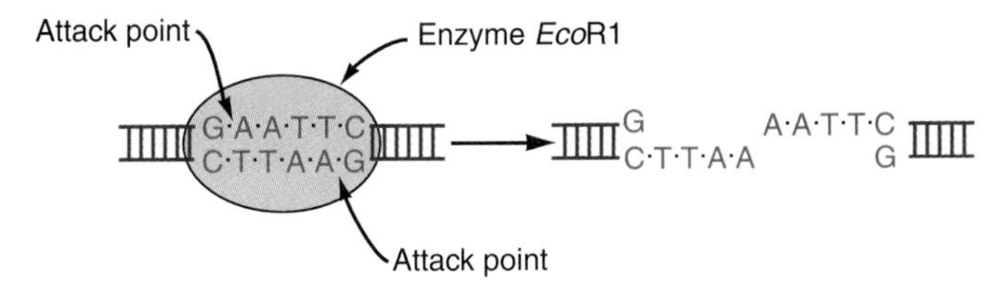

This action reduces the foreign DNA to fragments and protects the cell from invaders.

Why doesn't a cell destroy its own DNA with such enzymes? Each cell also has modification enzymes that recognize the same palindromic sequences and protect them by adding methyl groups, so they are no longer substrates for the nucleases. (The few phage made in one strain that survive in another are the lucky ones that get methylated before they are degraded.) Once a DNA has been methylated, it is protected only against the endonucleases of the same bacterial strain, and it is helpless in other strains.

These facts might have been only a minor footnote in the annals of molecular biology if a few investigators had not recognized the potential of restriction endonucleases for research.

17.13 DNA molecules can be combined at will by using restriction enzymes.

All DNA molecules that have been cut with the same restriction endonuclease have identical ends. Most endonucleases leave short, complementary, single-stranded ends of 2–4 bases that are "sticky" for one another—that is, such complementary sequences can hydrogen-bond to one another, like the ends cut by *Eco*R1:

Around 1972, Annie Chang, Paul Berg, and Seymour Cohen realized that any two pieces of DNA that have been cut with the same enzyme can be recombined with each other. This insight was the beginning of the **recombinant-DNA methodology** that has revolutionized investigations in cell and molecular biology. In principle, the method can be used to put any desired gene, from any organism, into a simple system where it can be grown and studied—for instance, putting a mammalian or plant gene into *E. coli* or a yeast.

The general method is to combine a DNA molecule of interest—what we will call "target DNA"—with a **vector,** a plasmid or viral DNA molecule that can carry the target DNA into a recipient cell called a **host.** Many host-vector systems have been developed for different purposes; Figure 17.17 shows one classic method. First, we extract some of the target DNA, say mouse DNA, and "restrict" it—cut it into fragments with one type of endonuclease, in this case *Sal*I[1]. This reduces the DNA to fragments with identical ends. We also grow and purify some of the vector, in this case the plasmid pBR322. Remember that plasmids commonly carry genes for antibiotic resistance, and pBR322 has the useful genes tet^R and amp^R that confer resistance to the antibiotics tetracycline and ampicillin. (The genes encode enzymes that destroy these antibiotics.) *Sal*I will cut pBR322 DNA at only a single site, within the tet^R gene, thus opening the molecules into a linear form with the same sticky ends as the mouse DNA. We then mix the mouse and plasmid DNAs with the enzyme DNA ligase, which *ligates* DNA: It seals single-stranded breaks by making new covalent bonds between nucleotides and making longer molecules from shorter ones, such as joining the Okazaki fragments during DNA replication. As this mix incubates for several hours, two kinds of plasmids will form. Some will close on themselves, and the enzyme will ligate them back into ordinary plasmids, with both their tet^R and amp^R genes intact. Others will incorporate a piece of mouse DNA and will be ligated into slightly larger plasmids with a functional amp^R gene but no tet^R gene, since the latter is interrupted by the mouse DNA.

When *E. coli* cells are treated briefly with Mg^{2+} and Ca^{2+} ions at the right temperature, they become **competent,** which means their cell membranes become permeable to DNA. We add the plasmid DNA mixture to competent host cells, and some cells become *transformed* by taking up a plasmid molecule. Because all the plasmids carry an intact amp^R gene, we plate the cells on agar containing ampicillin. Only cells now transformed with a plasmid will grow; the others die because they are sensitive to ampicillin. Now we need to distinguish cells carrying a recombinant plasmid, containing mouse DNA, from those carrying a nonrecombinant plasmid, so we replica-plate these colonies onto agar containing both ampicillin and tetracycline. Bacteria that have incorporated a recombinant plasmid have no tet^R gene and cannot grow here, so we can identify colonies that grow on ampicillin but not tetracycline as those that have incorporated some of the mouse DNA. Each of these colonies is a **clone,** derived from a single bacterium, and the process of getting a specific gene into a host cell is called cloning. A set of these clones, containing many different fragments of mouse DNA, is a **gene library.** Since we did not select for any specific mouse gene beforehand, this is a *shotgun experiment* because, like firing a shotgun wildly, we could have incorporated every mouse gene into some clone. The basic technique has been modified for working with selected genes of interest.

Recombinant-DNA technology opened the door to many new lines of basic research. Genes from plants and animals, which might be difficult to study in their complicated native settings, can be cloned into bacteria, other microorganisms, or cells in tissue culture where it is relatively easy to examine them. Furthermore, many biologists have foreseen practical

[1] Restriction endonucleases are named for the bacterium that produces them. For instance, *Sal*I is isolated from ***S**treptomyces **al**bus*.

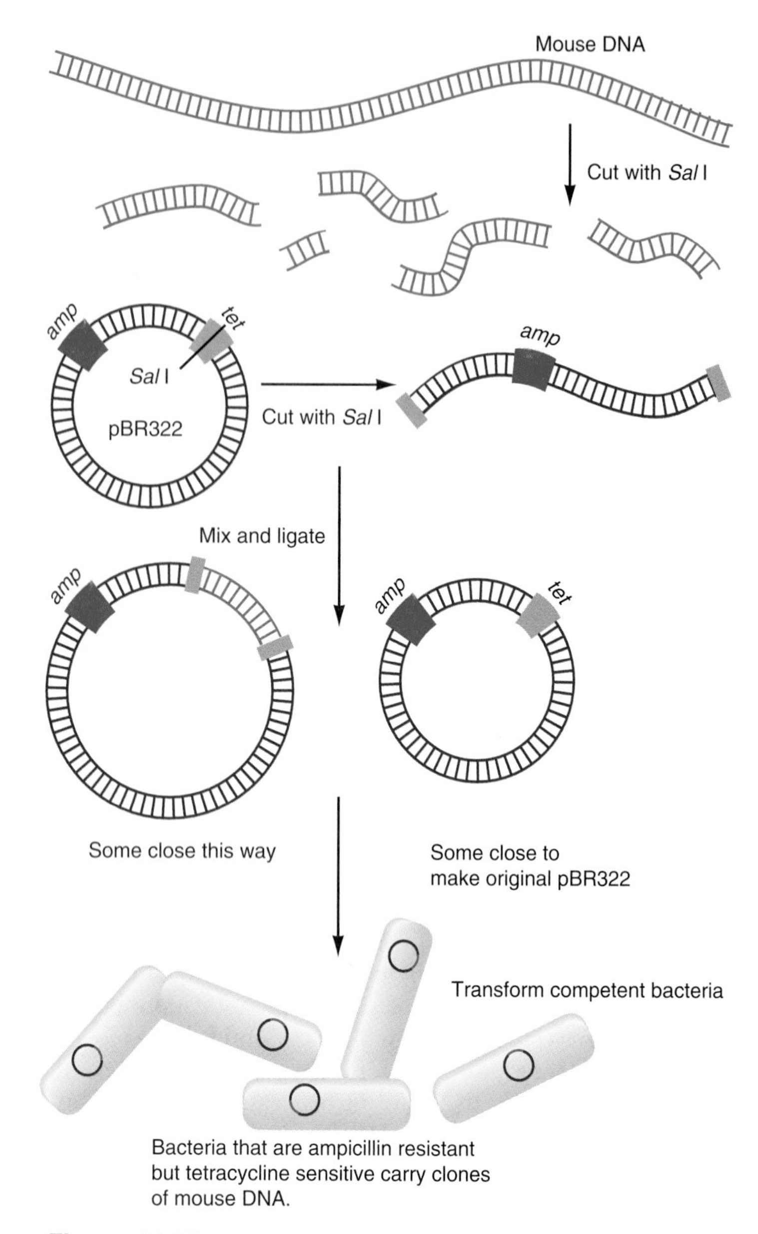

Figure 17.17

A basic recombinant-DNA method can be used to insert random pieces of mouse DNA *(red)* into plasmids *(blue)* and clone them. This shotgun experiment yields a library of clones that can contain any segment of the mouse genome.

applications of the methods, especially in domesticated plants and animals, to speed up and enhance traditional agricultural techniques. After all, humans have been changing these plants and animals for thousands of years through artificial selection to make all the varieties we know today. Agricultural breeding is not a new concept, and there is no obvious reason not to use the newest methods of breeding to achieve desirable ends—though many will argue. These same techniques might also be used for identifying human genes and even curing human genetic diseases.

To understand further developments in recombinant-DNA techniques, we must discuss genetic regulation, which we will do in Chapter 18.

Exercise 17.4 Dr. Smith wants to use the plasmid shown here to clone some plant DNA:

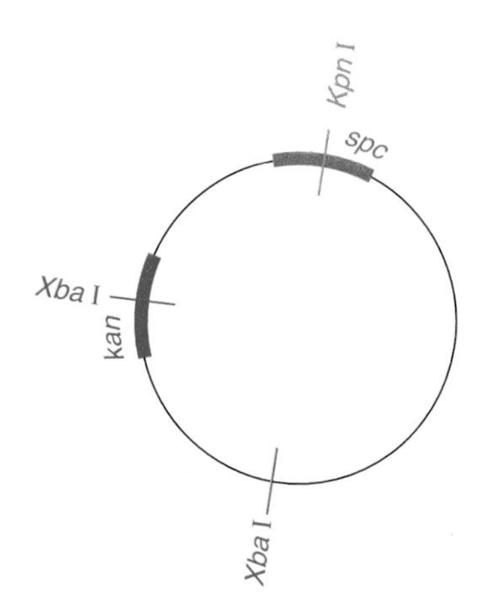

Notice that the plasmid has a *kan* gene that confers resistance to kanamycin and an *spc* gene that confers resistance to spectinomycin. Smith also knows that the restriction enzymes *Xba* I and *Kpn* I will make cuts as shown, and he can use either of them to cut the foreign DNA. Which enzyme should he use to cut his DNA, and what regime of antibiotics should he use to find bacteria with plasmids carrying certain plant genes?

Coda This chapter further amplifies the genetic theme of the book. Since organisms are genetic systems, it is appropriate to use genetic methods for investigating them. Indeed, it is essential. As we pointed out at the beginning of the chapter, asking strictly chemical questions—about the nucleotide sequence of a DNA molecule, for instance—yields chemical answers that are essential for a full biological understanding, but it fails to provide essential biological information. It fails to identify genes, their products, and their functions. The biological methods described here provide that essential information.

In exploring the genetic apparatus of bacteria, viruses, and some eucaryotes, biologists have uncovered a world of fascinating genetic entities that live in cells and replicate in various ways. The genes they carry give their hosts the ability to occupy different ecological niches, perhaps to live as pathogens or to grow in the presence of antibiotics. They may also carry genetic material from one organism to another, promoting horizontal transmission of genetic information instead of the more usual vertical transmission. These genetic elements create another factor in evolution and ecology, though its implications are far from clear yet.

The entire recombinant-DNA technology that has developed from classical molecular genetics depends upon specific biomolecular interactions. Each restriction endonuclease recognizes and cuts a specific DNA sequence; the resulting single-stranded segments then join because they are complementary and recognize each other. Other additions and refinements to this technology depend upon similar specificities of enzymes and nucleic acids, and it is only because biomolecular interactions *are* so specific that we can identify, analyze, and manipulate each element of a genome.

Summary

1. A set of mutations is used to create a genetic map of the genome of any organism or virus. Ideally, such a map will show the location and extent of all genes and sequences that control gene expression. Mutations used to map and analyze a genome are called markers.
2. Mutations are caused by changes in nucleic acid structure. These changes are commonly caused by ultraviolet or other radiation, or by specific chemical agents that change DNA bases, thereby altering DNA sequences.
3. Mutants for genetic analysis can be selected in various ways. The most useful mutants are conditional lethals, which are only lethal in certain circumstances—for instance, at high temperatures or (for phage) in certain strains of bacteria.
4. A phage cross is performed by infecting cells simultaneously with two different phage mutants so their genomes can interact and form recombinant genomes, with different combinations of markers, through crossing over. The distance between two markers is measured by the frequency of recombination, which equals the number of new recombinant phage divided by the total number of phage produced.
5. Mutations can be mapped rapidly by using deletions. Deletions can be placed in an unambiguous order by the way they recombine, or fail to recombine, with each other. Then point mutations can be located relative to the ends of the deletions by the same logic.
6. Genes are defined experimentally by complementation tests. Two mutations complement each other if they produce a wild-type phenotype when combined in different genomes in one cell. This means they lie in different genes, so each genome has a good copy of one gene, and each good gene can produce one needed gene product.
7. Experiments that compare the sequence of mutational sites in a gene with the sequence of amino acid changes in the protein it encodes show that a gene is really colinear with its protein product.
8. Experiments combining different frameshift mutations demonstrated that a messenger RNA is read by threes with a commaless reading-frame mechanism.
9. Bacteria such as *E. coli* can transfer genetic material from cell to cell during conjugation, so they may be used for genetic studies. Conjugative genetic transfer is due to plasmids, which are extrachromosomal genetic elements that inhabit bacteria and sometimes may be integrated into the bacterial genome.
10. Plasmids such as the F factor transfer a copy of themselves from one cell (the donor or F^+ cell) to another (the recipient or F^-) by promoting conjugation in which the two cells join tightly. In Hfr cells, the F factor is integrated into the bacterial chromosome, so when an Hfr cell conjugates, the F factor transfers a series of bacterial genes.
11. Interrupted mating experiments demonstrate that donor genes are transferred in the sequence in which they are linked on the bacterial chromosome. These experiments are therefore used to map the genome of the donor cell.
12. Although phage such as T4 can only replicate lytically, by destroying their host cells, some phage replicate in a noninfectious form. They become prophages that generally remain integrated in the host chromosome, expressing only those genes that form a control system. Cells carrying prophages are said to be lysogenic.
13. Cells carry a great variety of genetic elements categorized as plasmids, episomes, transposons, and viruses. They replicate in many ways, both as part of the cellular genome and independently. Many can transfer themselves from cell to cell, or can be transferred, and they confer many new properties on their hosts.
14. Phage can promote genetic exchange, called transduction, by carrying host genes from cells in which they have grown into other cells. Some phage can carry any host gene by incorporating pieces of the host genome instead of their own genomes. Others can only transduce a few genes near the sites where their prophages integrate; these transducing phage are made by faulty excision of the prophage DNA.
15. Bacteria produce enzymes called restriction endonucleases that restrict the growth of foreign DNAs, such as viral genomes. Each endonuclease attacks a specific, generally palindromic DNA sequence. Bacteria protect their own DNA from endonuclease attack by adding methyl groups to the critical sequences.
16. DNA molecules can be combined at will using these restriction enzymes. All DNA molecules that have been cut with one type of enzyme have the same "sticky" ends, with the same short, complementary sequences. Such molecules can be joined in any desired combination. DNA molecules from any source can thus be inserted in plasmids, so they can be transferred into bacteria or yeast, manipulated, and studied.

Key Terms

express/gene expression 332
genetic analysis 332
genetic map 332
open reading frame (ORF) 332
marker 332
point mutation 333
site 333
locus 333
mutagen 333
lethal mutation 335
conditional lethal mutation 335
temperature-sensitive (*ts*) 335
host-dependent 335
deletion mutants 336
complementation test 338
frameshift mutation 340
suppressor mutation 340
conjugation 341
plasmid 341
F factor 341
pilus/pili 341
episome 341
Hfr strain 341
lytic cycle 343
prophage 343
lysogenic/lysogeny 343
temperate phage 343
R factor 345
transduction 346
restriction endonuclease 346
recombinant-DNA methodology 347
vector 347
host 347
competent 347
clone 347
gene library 347

Multiple-Choice Questions

1. Antibiotics such as penicillin are used to select nutritional auxotrophs because the antibiotic
   a. kills the nutritionally mutant cells.
   b. kills the bacteria that are not auxotrophs.
   c. inhibits the effect of the missing nutrient.
   d. allows the growth of only the wild-type cells.
   e. keeps the level of environmental contaminants low.
2. Three *rII* mutations are mapped. The frequency of recombination between markers *a* and *b* is 0.003. Between markers *a* and *c*, the frequency of recombination is 0.005. Which is closer to marker *a*?

a. *b*
b. *c*
c. Cannot determine because recombination frequency is not related to distance between markers.
d. Cannot determine because they may be part of the same gene.
e. Both *b* and *c* are equally close to *a*.

3. If wild-type recombinants are observed when crossing two *rII* mutants,
   a. the mutations are contained within one deleted length of DNA.
   b. complementation has taken place.
   c. the mutations are at different sites.
   d. *a* and *b*, but not *c*.
   e. *b* and *c*, but not *a*.
4. Which type of mutation is likely to have the greatest effect on phenotype?
   a. deletion or insertion of a codon at the start of a gene
   b. deletion or insertion of a codon at the end of a gene
   c. substitution of one base pair with another
   d. frameshift mutation at the start of a gene
   e. frameshift mutation at the end of a gene
5. To isolate *E. coli* cells that are auxotrophic for methionine and sensitive to penicillin from penicillin-sensitive prototrophs, which growth medium would be used?
   a. complete medium
   b. complete medium plus penicillin
   c. minimal medium plus penicillin
   d. minimal medium plus methionine
   e. minimal medium plus methionine and penicillin
6. Which is *not* correct about conjugation between F^+ and F^- cells?
   a. The F^- cell is converted to F^+.
   b. The F^+ remains F^+.
   c. The F^+ cell contains a plasmid that codes for pilin.
   d. Once F^- cells have become F^+, they can conjugate with Hfr cells.
   e. Chromosomal genes from the F^+ cell do not transfer during conjugation.
7. Which is characteristic of lytic phage life cycles, but not lysogenic phage life cycles?
   a. Most of the phage genes are active.
   b. Bacteria infected by lytic phage are always destroyed immediately.
   c. Phage proteins are generally not synthesized by the host cell, but host proteins continue to be made.
   d. The phage enters an infective prophage state.
   e. Recombination can occur between phages.
8. When biologists refer to the dramatic increase in antibiotic resistance, they mean the increase in the proportion of antibiotic-resistant to antibiotic-sensitive
   a. human cells.
   b. human beings.
   c. viruses that infect human cells directly.
   d. viruses that infect bacteria.
   e. bacteria that infect humans.
9. When genetic recombination occurs following viral transduction, the genes of one ______ replace those of ______ .
   a. virus; another virus
   b. virus; a bacterium
   c. bacterium; another virus
   d. bacterium; another bacterium
   e. plasmid; a transposon
10. When making a recombinant plasmid, it is useful to use a restriction enzyme that cuts the plasmid DNA within an antibiotic-resistance gene because that will specifically enable you to determine whether the
    a. antibiotic-resistance gene comes from the vector or the host.
    b. plasmid has incorporated the target DNA.
    c. host cell has incorporated a plasmid.
    d. plasmid has been infected by a virus.
    e. restriction enzyme has worked properly.

True-False Questions

Mark each statement true or false, and if false, restate it to make it true.

1. A DNA sequence leading from an initiation codon to a termination codon is called an open reading frame and is identical to a gene.
2. Most mutagens cause a mutation by deleting bases from DNA.
3. If two mutants complement each other, the mutations are in different genes.
4. The deletion or insertion of any number of nucleotide pairs in a gene results in a frameshift mutation.
5. In a cross between two auxotrophic mutants, wild-type recombinants are selected by plating on minimal medium.
6. Both viruses and plasmids can exist as genetic elements inside cells and as independent particles outside cells.
7. Episomes are genetic elements of bacteria that can only exist integrated into the bacterial chromosome.
8. When a prophage is excised from a bacterial chromosome, it always takes some bacterial genes with it.
9. Each type of plasmid such as an R factor or virulence factor is restricted to a particular species of bacterium and cannot confer its properties on other species of bacteria.
10. Restriction enzymes are viral products that destroy bacteria by cutting up bacterial DNA.

Concept Questions

1. Explain why knowing the complete DNA sequence of a genome does not automatically produce a genetic map.
2. Distinguish between recombination and complementation experiments. What kind of information do they yield?
3. Lederberg and Tatum crossed two strains of bacteria that each carried two auxotrophic mutations. How could they be sure the wild-type bacteria they recovered were recombinants and not simply reverse mutants?
4. Explain the differences among plasmids, episomes, and temperate phage. In what ways are they alike?
5. Has the increased use of antibiotics increased the rate at which bacteria mutate from sensitivity to resistance for various antibiotics? If not, why do we now find so many more resistant bacteria?

Additional Reading

Caldwell, Mark. "Prokaryotes at the Gate." *Discover,* August 1994, p. 44. A report on antibiotic resistance in bacteria. One source at the Centers for Disease Control reports that organisms resistant to every known antibiotic now exist.

Fedoroff, Nina V. "Transposable Genetic Elements in Maize." *Scientific American,* June 1984, p. 84. Mobile genes occur in bacteria, plants, and animals.

Stahl, Franklin W. "Genetic Recombination." *Scientific American,* February 1987, p. 90. How chromosomes trade parts and reshuffle their genetic information.

Internet Resource

To further explore the content of this chapter, log on to the web site at:

http://www.mhhe.com/biosci/genbio/guttman/

Gene Regulation and Genetic Engineering

18

Key Concepts

A. Fundamentals of Gene Regulation

18.1 In procaryotes, genes with related functions tend to be located together in blocks and regulated together.

18.2 The regulation of genes for lactose metabolism is a classical model system.

18.3 Genetic regulatory circuits fit the general model of communication and regulation.

18.4 Genes may also be regulated by positive mechanisms.

18.5 Biosynthetic genes may be regulated by repressors.

18.6 Alarmones regulate still larger blocks of genes.

18.7 Promoters, the DNA sequences to which RNA polymerase binds, also regulate gene expression.

18.8 Eucaryotic genes are regulated primarily by combinations of promoters and enhancers.

18.9 Some genes are directly regulated by steroid hormones.

18.10 Chromosome structure may regulate gene transcription.

B. Applications to Recombinant-DNA Work

18.11 Recombinant-DNA research employs a few basic methods.

18.12 Foreign genes can be cloned in plasmids that permit their expression, and can be used for research and practical applications.

18.13 Genes can be cloned in plants by means of a natural bacteria-plant-plasmid system.

18.14 Animal cloning has been limited by unknown physiological problems.

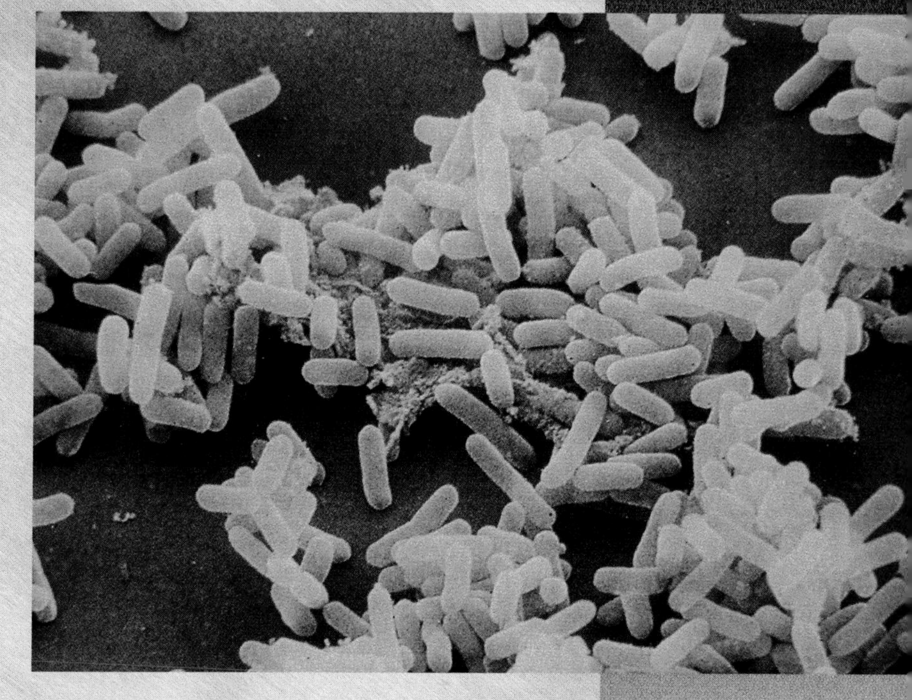

A scanning electron micrograph of *Escherichia coli* cells, artificially colored.

When France was an occupied territory during World War II, Jacques Monod divided his time between fighting the war as a member of the underground and doing microbiology experiments at the Institut Pasteur. Monod was using protozoa called ciliates to study growth when his advisor, Andre Lwoff, suggested he use bacteria instead. Assured that *E. coli* was not pathogenic, Monod grew a culture of these bacteria on a medium containing both glucose and lactose, and obtained an odd result: The bacteria grew for a few hours, paused for a while, and then continued growing. Puzzled, Monod asked Lwoff what this could mean. Lwoff said he didn't know, but maybe it had something to do with "adaptive enzymes." When the war was over, Monod continued his experiments on "enzyme adaptation," an old term that refers to the change in the enzymatic composition of cells grown in different conditions. By 1961, Monod and his colleagues had found the key to the puzzle, and in 1965, he and François Jacob were awarded the Nobel Prize for their work.

A bacterial cell has about 2,000–3,000 genes, and complex eucaryotes probably have about 50,000–100,000. But all these genes are not used—we say *expressed*—all the time. An organism expresses different genes (makes different proteins) at different times in its life cycle, as we emphasized in Chapter 15. Organisms have evolved mechanisms for *regulating* the expression of their genes in various circumstances. In free-living cells such as bacteria and yeast, many genes encode proteins needed for metabolizing various substances under special circumstances; these cells can cope with changes in their environment by switching over rapidly to make new proteins. Differentiated cells of multicellular organisms are distinctive because each has its own special proteins; although all the cells in the organism have the same genes, their expression is precisely regulated. In this chapter, we will explore genetic regulation and then discuss applications of this information in making genetically engineered plants and animals. ■

A. Fundamentals of Gene Regulation

In procaryotes, genes with related functions tend to be located together in blocks and regulated together.

Our basic conceptions of gene regulation came from studies in bacteria, particularly *E. coli.* As biologists identified bacterial genes, an interesting pattern emerged: Genes that encode functionally related proteins tend to be located next to each other. For instance, the 11 genes encoding the series of enzymes that synthesize the amino acid histidine all lie in a single block (Figure 18.1). The sequence of the genes, in fact, is almost the same as the sequence of the enzymes in the pathway.

Gene regulation prevents a cell from wasting energy by making unnecessary components. For example, if a bacterial cell is in a medium with plenty of histidine, or any other cell component, there is no point in its making more. Bacteria have evolved regulatory mechanisms at two levels that adjust the activity of their biosynthetic pathways to the availability of nutrients in the environment (Figure 18.2). In one of these mechanisms, feedback inhibition (see Section 11.5), the end product of a pathway inhibits enzymes that already exist in the cell. In the other, control mechanisms *repress* genes for biosynthetic enzymes when a cell already contains enough of the materials those enzymes synthesize. In this context, **repression** means keeping the genes from being expressed. Furthermore, by growing cells with different amounts of histidine, we can show that the enzymes are *regulated coordinately,* so the amounts of all these enzymes remain proportional to one another. Coordinate repression is common in bacteria, and it implies that a single mechanism controls all these genes together.

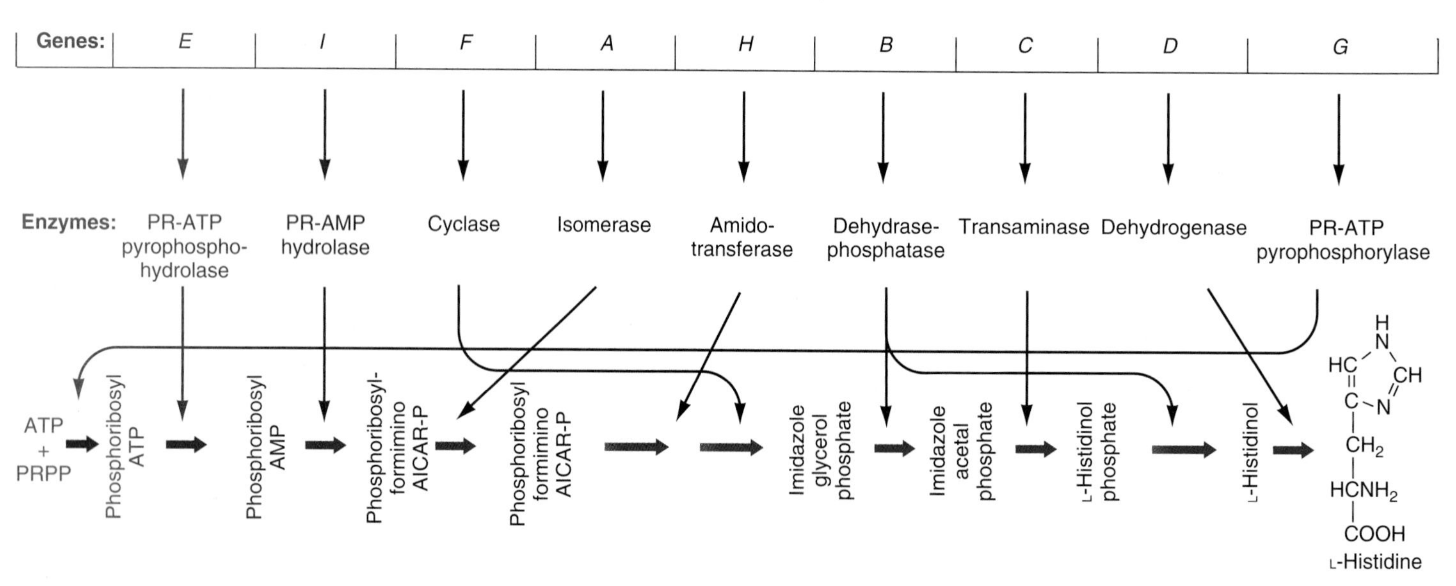

Figure 18.1

The genes for the enzymes that synthesize histidine occur in one block, and even their order is the same as the order in which the enzymes act in histidine biosynthesis.

Perhaps it is obvious that a well-adapted organism should make only the proteins it needs at any moment, but it is important to understand why. Inserting each amino acid into a protein costs the cell three ATPs, in addition to the energy needed to synthesize the amino acid. Cells making unneeded proteins grow and multiply more slowly than their better-regulated competitors, and since the faster-growing cells have a selective advantage in any environment, those with the most functional controls win out. The expression of a gene could be controlled at either the point of mRNA synthesis (transcription) or the point of protein synthesis (translation). On the whole, transcriptional controls are favored. This, too, is adaptive, since it wastes energy to make an RNA molecule that will not be used. In a number of cases, however, controls are exerted over translation and also over protein activity.

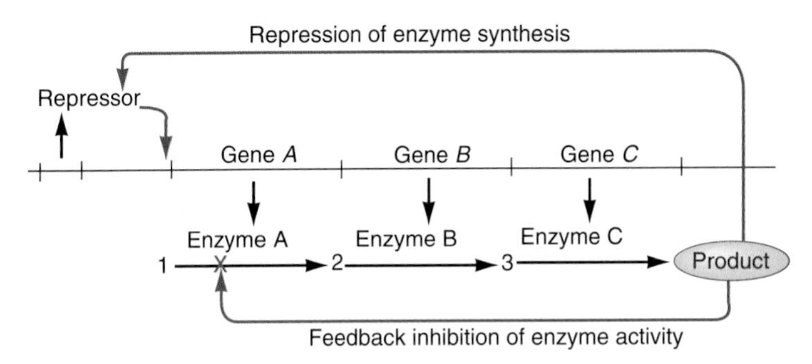

Figure 18.2
Regulation operates at two levels. An excess of the end product of a pathway *(yellow ellipse)* feeds back to inhibit the activity of existing enzymes and also to repress the synthesis of additional enzymes.

18.2 The regulation of genes for lactose metabolism is a classical model system.

All cells have **housekeeping proteins,** proteins needed for the essential metabolic functions they perform at all times. In *E. coli* these proteins include the enzymes for glucose metabolism (the glycolytic pathway), and *E. coli* cells use glucose in preference to any other energy source. They can also grow on the disaccharide sugar lactose by making an enzyme, β-galactosidase, that cuts lactose into two simple sugars, galactose and glucose:

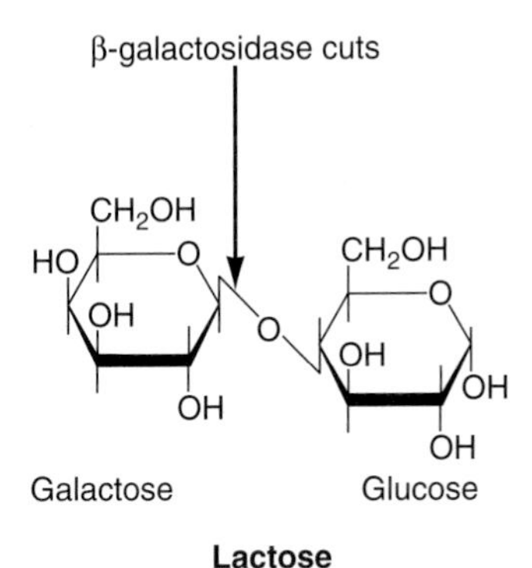

Lactose, however, is a relatively rare sugar, found most commonly in milk, so it is advantageous for the bacteria to make the enzyme only when its substrate, lactose, is present and glucose is absent. When grown with glucose, bacteria make very little β-galactosidase, but as soon as lactose is substituted for the glucose, they begin to make β-galactosidase at a high rate (Figure 18.3). If the lactose is removed, they quickly stop making the enzyme. We therefore say that β-galactosidase is an **inducible enzyme,** and lactose acts as an **inducer.** This process is essentially the opposite of gene repression, although the mechanisms of the two processes are similar. (To study the regulation of this enzyme, we actually use an artificial inducer called IPTG [isopropyl-thiogalactoside], which is structurally similar to lactose but is not metabolized. IPTG also becomes important in recombinant-DNA methods, as we will explain later.)

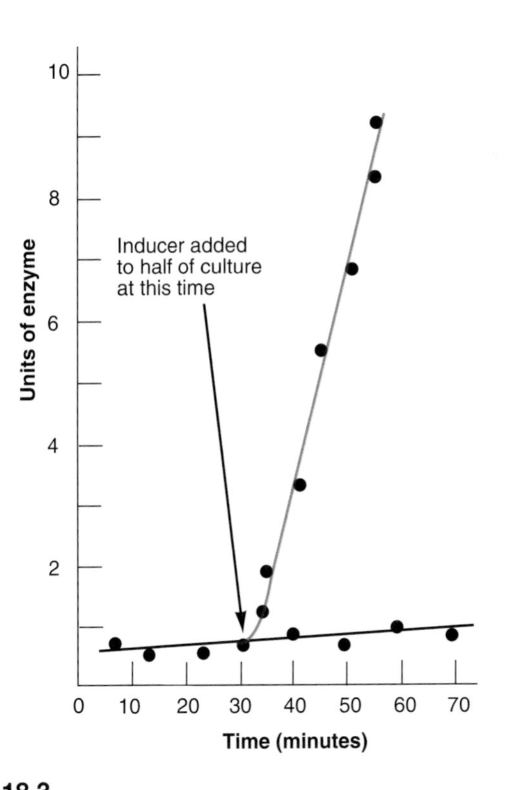

Figure 18.3
β-galactosidase is an inducible enzyme. The rate of enzyme synthesis increases by a factor of about 1,000 when lactose is added to the growth medium.

Around 1961, François Jacob and Jacques Monod made clever use of mutants to determine how the genes for lactose metabolism are regulated. Some mutants can't make β-galactosidase. Their mutations identify a gene, *lacZ,* that encodes this enzyme. A second protein, galactoside permease, transports lactose across the cell membrane and is encoded by the *lacY* gene. These two genes lie right next to each other, and a third gene, *lacA,* whose role is obscure, lies next to *Y*. Genes like these three are sometimes called *structural genes* because they

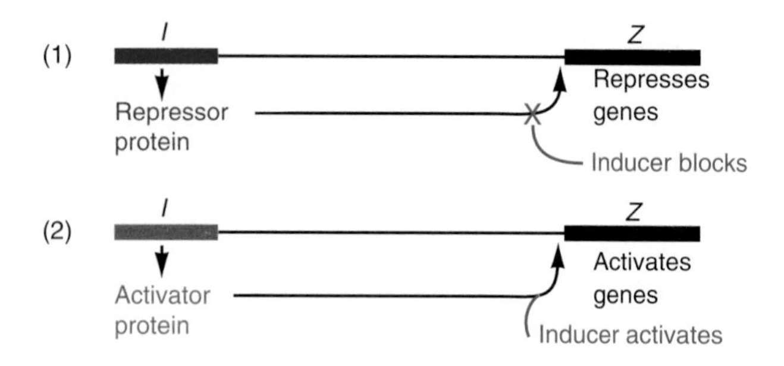

Figure 18.4

We can imagine two models of gene regulation. In model 1, a repressor protein prevents expression of the genes unless it is combined with a small inducer molecule (in this case, lactose). In model 2, an activator protein, which must combine with the lactose to be effective, stimulates expression of the genes.

encode protein structures rather than being part of a regulatory mechanism. Furthermore, the proteins they encode are all regulated together, like the histidine enzymes.

Other mutants, designated *lacI,* have a defective control mechanism and make all three *lac* proteins at high rates even when there is no inducer. The *lacI* mutations identify another gene close to *Z*. In conjunction with Arthur Pardee, Jacob and Monod set up experiments to test two alternative models for the action of the *I* gene (Figure 18.4). In model 1, *I* encodes a **repressor protein** that keeps the *Z*, *Y*, and *A* genes silent. The role of an inducer is to inactivate the repressor, so the genes are turned on. In model 2, *I* encodes an **activator protein** that combines with an inducer such as lactose to turn the *Z*, *Y*, and *A* genes on. Pardee, Jacob, and Monod then tested these models with some ingenious experiments. The central question is whether I^- or I^+ is the dominant allele. Model 1 says that a mutant repressor (from an I^+ gene) cannot turn the genes off, but it should be recessive to a normal repressor (from an I^+ gene). Model 2 says that a mutant activator protein works even in the absence of lactose, so it should be dominant to a normal inducer protein.

The most revealing Pardee-Jacob-Monod experiment is shown in Figure 18.5. Donor (Hfr) cells with the genotype I^+ Z^+ are mated with recipient (F^-) cells that are I^- Z^-, in the absence of inducer, so neither strain can make any enzyme. However, immediately after mating begins, and for about two hours, the recipient cells make a lot of enzyme. This result is consistent with model 1, the repressor model. As soon as the Hfr chromosome enters the F^- cytoplasm, its Z^+ gene is in a cytoplasm with no repressor protein, so it begins to express itself. It takes a couple of hours for the incoming I^+ gene to make enough functional repressor to block the expression of the Z^+ gene. This newly introduced I^+ gene changes the character of the cell, meaning it is dominant to the I^- gene that encodes a defective repressor.

This genetic experiment revealed the existence of a repressor and the other elements of this control system. The repressor is an allosteric protein with one binding site for DNA and one for the inducer, which can exist in two conformations. In the absence of inducer, it binds tightly to a site on the DNA, called the **operator, O,** next to the *Z* gene (Figure 18.6). By binding or not binding the repressor, the operator regulates one or more nearby genes. Genes regulated by an operator constitute an **operon,** and the *Z-Y-A* block is known as the *lac* operon. When the repressor protein is bound to the operator, the *lac* operon is turned off because RNA polymerase cannot transcribe the operon; if an inducer binds to the repressor protein, the repressor changes conformation so it can no longer bind to the operator. With the repressor removed, the operon is turned on because RNA polymerase is free to transcribe the structural genes *Z*, *Y*, and *A*. The whole operon is transcribed as a single, long mRNA encoding all three proteins. The direction of transcription is called **downstream** and the opposite direction **upstream,** so the operator site is upstream of the genes it controls.

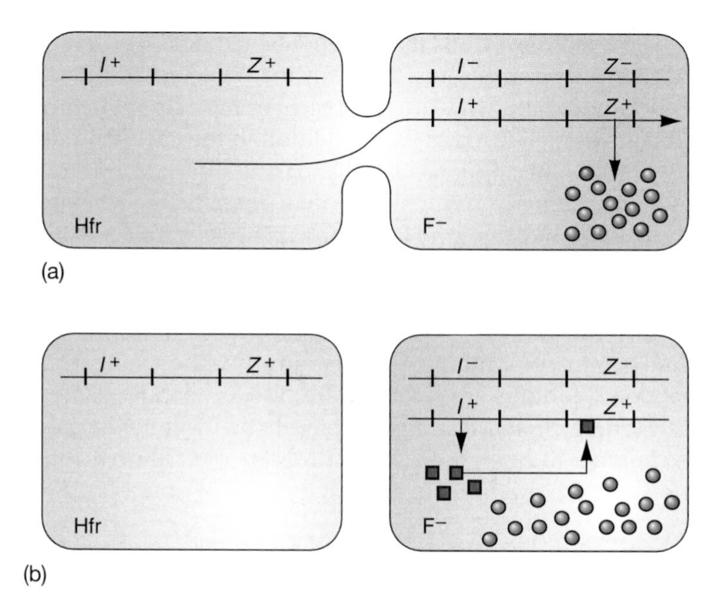

Figure 18.5

In the Pardee-Jacob-Monod experiment, two strains of bacteria are mated that cannot make β-galactosidase by themselves under the conditions of the experiment. *(a)* For a while after mating, β-galactosidase *(green spheres)* is synthesized at a high level. *(b)* By two hours after mating, enzyme synthesis is turned off. This result is consistent with the repressor model for gene regulation: Enzyme synthesis begins when the Z^+ gene enters the cell and stops as repressor *(red squares)* is made from the incoming I^+ gene.

18.3 Genetic regulatory circuits fit the general model of communication and regulation.

Using the *lac* system as an example, we can show how gene regulation fits the general model of a regulatory circuit, as outlined in Chapter 11, with its *sensor* and *effector*. Remember that in a home heating system designed to maintain a set-point temperature, the sensor is a thermometer in the thermostat, and the effector is the furnace. Likewise, a genetic regulatory circuit maintains homeostasis by keeping a metabolite at a set level or

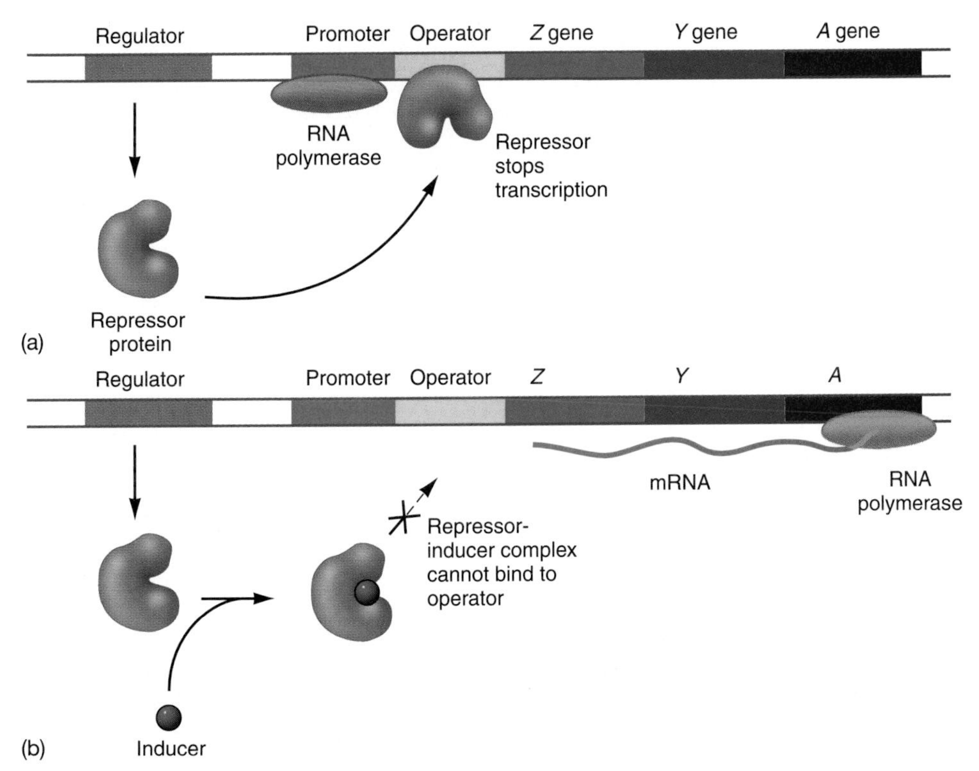

Figure 18.6
An operon is one or more genes regulated by an operator. *(a)* The *Z-Y-A* genes regulated by the operator are kept "off" (not transcribed) as long as repressor protein is bound to the operator. *(b)* When the repressor protein binds to an inducer, it undergoes an allosteric shift so it can no longer bind to the operator. Then the repressor comes off the DNA, and RNA polymerase molecules can transcribe the region into messenger RNA.

responding appropriately to a new signal by using a sensor protein that senses an environmental factor and responds to it. In the *lac* system, the sensor is the repressor protein, which responds to the presence of lactose. The same protein acts as an effector by turning the *lac* genes on in response to the signal from the sensor. In other circuits, the effector may be a separate protein.

The common genetic regulatory pattern is:

- An allosteric regulatory protein combines with a small signal ligand, thus changing the protein's shape and its ability to bind to a control site somewhere on the DNA.
- In response to the change at this DNA site, a neighboring gene (or block of genes) is turned on or off. This means that either messenger RNA can be transcribed from the gene(s) or that transcription is inhibited.

Cells use at least three variations on this kind of regulatory circuit (Figure 18.7):

- Negative repression, or *induction,* as in the *lac* system just described.
- Positive repression, as in amino acid biosynthesis systems (Section 18.5).
- Positive activation, as in the arabinose system (Section 18.4).

Positive and negative regulatory systems of this type can work alone, or they can be combined to perform more complicated regulatory tasks.

To get a clear picture of genetic regulation, you must see that regulatory proteins *diffuse throughout the cell* and bind to any control site they recognize. Genes on the same DNA molecule are in the *cis* position, and those on different DNA molecules are in the *trans* position. A regulatory protein can act both *cis* and *trans* to the gene that encodes it, but the control sites where the protein binds, such as the *lac* operator, can only regulate neighboring genes in the *cis* position.

To determine experimentally how particular genes are regulated, we make cells called **merodiploids** (*mero-* = part or partial) with two copies of the genes in question, one copy in the bacterial chromosome and one in a plasmid like an F factor. Using different combinations of mutations, we can predict how much protein a cell should make with different regimes of regulation and then test these predictions with the merodiploids, as in the following exercise.

Exercise 18.1 We have constructed a number of merodiploids for the *lac* operon, using several regulatory mutants (Table 18.1). As usual, there are wild-type and defective alleles of the *Z* and *Y* genes. I^- produces an inactive repressor; I^s produces a "super-repressor" that has no affinity for inducer; and O^c is an operator that cannot bind repressor. Imagine that an uninduced cell normally makes 0.1 units of β-galactosidase or permease, and from each induced (nonrepressed) gene, it gets 100 units of protein. Determine how much protein each cell type should make, with and without inducer.

Example 18.1 Let's analyze the second case in Table 18.1. The cell has one functional and one defective *I* gene. However, the I^+-gene product, the repressor, can bind to any *lac* operator (and both operators are normal). A functional repressor is dominant over a defective one, so the cell will exhibit normal regulation overall. Note, though,

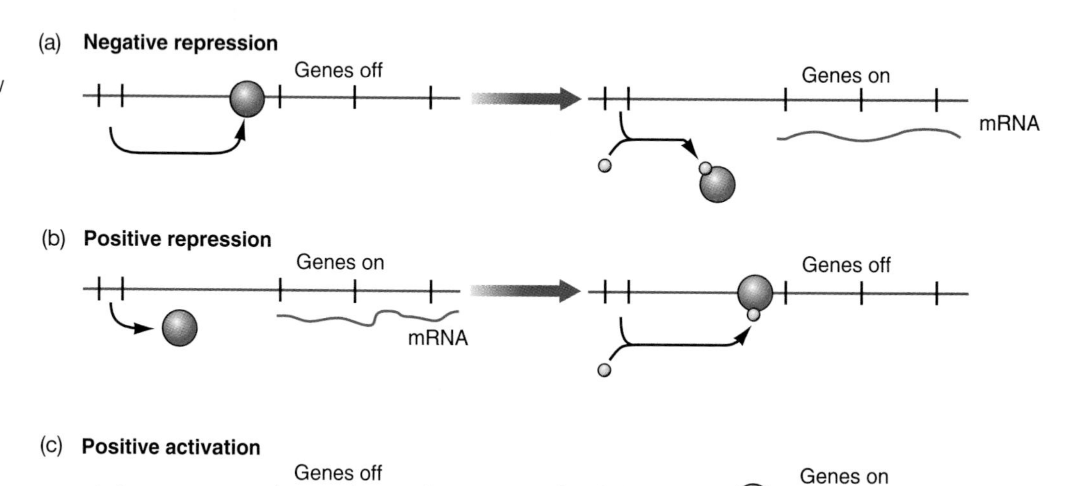

Figure 18.7
Genetic regulation takes three principal forms. Each case shows how the regulator protein *(red ball)* responds to a signal ligand *(small yellow ball)*. *(a)* Negative repression: A repressor protein turns genes off *unless* it is antagonized by a signal ligand. *(b)* Positive repression: A repressor protein only turns genes *off* when combined with a signal ligand. *(c)* Positive activation: An activator protein only turns genes *on* when combined with a signal ligand.

Table 18.1

	Uninduced		Induced	
Strain	*β-galactosidase*	*Permease*	*β-galactosidase*	*Permease*
$I^+O^+Z^+Y^+$ / $I^+O^+Z^+Y^+$				
$I^-O^+Z^+Y^-$ / $I^+O^+Z^-Y^+$	0.1	0.1	100	100
$I^-O^cZ^+Y^-$ / $I^+O^+Z^+Y^+$				
$I^sO^+Z^+Y^+$ / $I^+O^+Z^+Y^-$				
$I^-O^+Z^+Y^+$ / $I^sO^+Z^-Y^+$				

that there is only one functional *Z* gene and one functional *Y* gene. Therefore, we expect only 0.1 units of each protein in the uninduced cell and only 100 units of each in the induced cell.

18.4 Genes may also be regulated by positive mechanisms.

After the *lac* system's structure had become clear, Ellis Englesberg and his colleagues determined how the sugar arabinose regulates genes that encode three enzymes for arabinose metabolism (Figure 18.8). In this case, experiments of the Pardee-Jacob-Monod type gave a different result from that of the *lac* system. These genes are controlled positively, by an activator protein, rather than being controlled negatively by a repressor. The *C* gene encodes an activator that is inactive unless arabinose is present. Arabinose binds to the activator (much as lactose binds to the *lac* repressor), and the altered activator binds to the upstream initiator site *(I)*, activating it so RNA polymerase can transcribe the structural genes of the operon.

Exercise 18.2 In a positive regulatory system of the arabinose type, merodiploids should show different dominance relationships from those in a negative regulatory system of the lactose type. Suppose you make a merodiploid with a C^+ gene (allele) and a C^c gene, the latter encoding a protein that acts as an activator even when arabinose is absent. Which allele should be dominant?

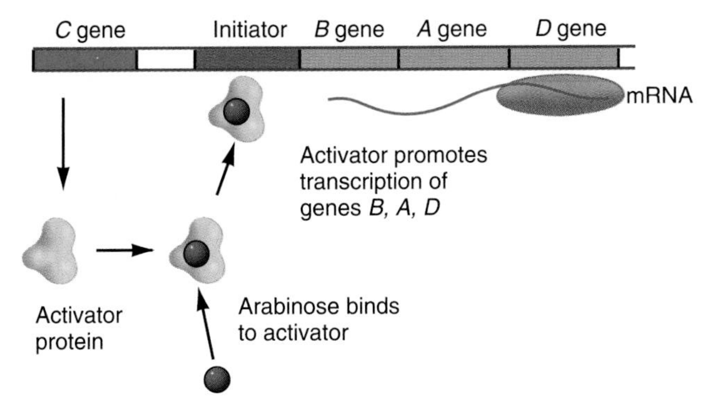

Figure 18.8

The arabinose genes of *E. coli* are regulated by positive activation. The *C* gene encodes an activator protein that can combine with arabinose. In its combined form, it is able to bind to the initiator site and activate transcription of the *B-A-D* block.

18.5 Biosynthetic genes may be regulated by repressors.

Biosynthetic enzymes, such as those for histidine biosynthesis, can be regulated as an operon by another kind of repressor protein: A *positive* repressor only has an affinity for its operator when the appropriate ligand is *bound* to it, in contrast to the *lac* repressor, which binds to an operator site when the regulatory ligand is *not bound* to it (Figure 18.9). In this way, the operon responds to a high concentration of its end product. We must emphasize again that this feedback control mechanism is quite different from end-product inhibition of the first enzyme in a pathway (see Figure 18.2 and Section 11.5). Repression stops the synthesis of new proteins; inhibition regulates the activity of existing proteins.

Because regulatory proteins diffuse through the cytoplasm, a single type of protein can coordinately control the expression of several operons. The genes for arginine biosynthesis in *E. coli,* for example, are located in five distinct operons, widely separated on the genome, but a single repressor protein encoded by one regulator gene controls them all (Figure 18.10). A set of coordinately regulated operons is a **regulon.**

Exercise 18.3 While studying the biosynthesis of histidine in a previously unknown bacterium, you find that the genes for the biosynthetic enzymes map in four distinct blocks *(A-B, C-D-E, F-G-H,* and *I-J),* but they are all regulated coordinately by histidine. Among the many mutants you collect, three are especially interesting. All three mutants can make all the enzymes, but #1 cannot regulate the F, G, and H enzymes, #2 cannot regulate the A and B enzymes, and #3 cannot regulate any of the enzymes. Interpret these results by specifying the most likely location of each mutation.

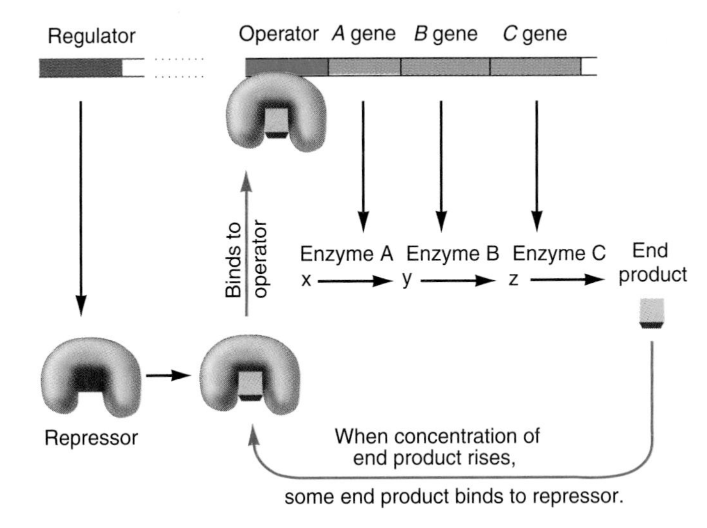

Figure 18.9

Biosynthetic operons are commonly regulated by positive repression. The repressor protein only blocks transcription when it is bound to the end product, and this only happens if there is an excess of the end product.

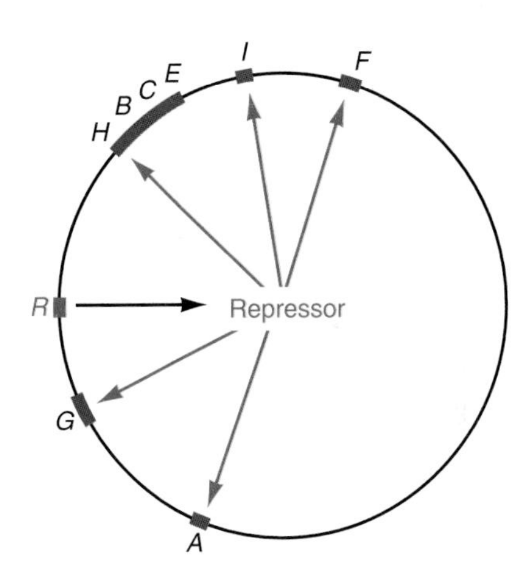

Figure 18.10

The arginine regulon consists of separated operons that are all under control of the same repressor protein.

18.6 Alarmones regulate still larger blocks of genes.

The common pattern of several regulatory mechanisms should now be clear: An allosteric sensor protein detects the presence of a critical ligand by binding to it and changing conformation, so the protein turns genes on or off in response to that ligand. Such proteins can regulate one gene, a block of related genes (an operon), or several blocks of genes (a regulon). They can also regulate a large set of genes in response to emergencies when triggered by ligands called **alarmones.**

An obstruction to protein synthesis is a serious emergency in a cell. When this happens in bacteria, the ribosomes make an unusual nucleotide, **guanosine tetraphosphate (ppGpp):**

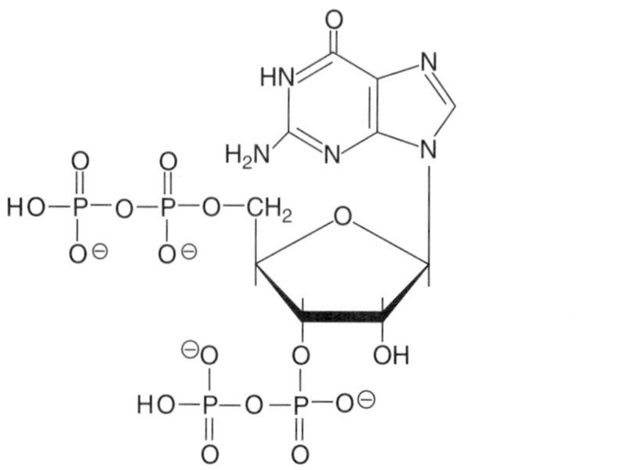

This nucleotide acts as an alarmone by immediately stopping transcription of the tRNA and rRNA genes and stimulating the operons for amino acid biosynthesis. Both responses to ppGpp are functional: They halt production of the apparatus for protein synthesis, which is useless if the whole process is stopped, and they stimulate the production of amino acids when the lack of a certain amino acid might be causing the blockage.

Another potential calamity in a cell is the lack of an energy source; this event is signaled by the alarmone **cyclic AMP (cAMP).** Most bacteria always have the glycolytic pathway enzymes and are set to use glucose as their default energy and carbon source. A cell with adequate glucose has very little cAMP. But if the glucose concentration falls, the enzyme adenyl cyclase starts to convert ATP into cAMP (Figure 18.11). A **cAMP acceptor protein (CAP),** which is usually inactive, becomes activated when bound to cAMP, and it turns on genes that may be able to supply energy. The *lac* operon, for instance, contains an upstream site for binding active CAP-cAMP, and the operon can only be induced if CAP-cAMP is bound to this site (Figure 18.12). So if both lactose and glucose are available, the cAMP level remains low, and glucose is used preferentially. The *lac* operon will only be induced if the cell has lactose but no glucose. This and other global alarm systems keep a cell from making proteins it doesn't need but permit a rapid response to changing conditions. This system explains Monod's early results; the bacteria grew on glucose until they exhausted the supply, then paused in their growth while switching over to use lactose.

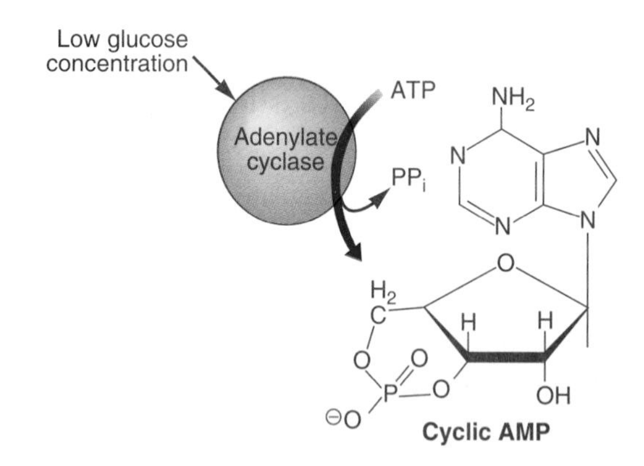

Figure 18.11

Cyclic AMP (cAMP) in bacteria is synthesized by the enzyme adenylate cyclase in response to a low level of glucose.

Exercise 18.4 While studying regulation by cyclic AMP, you discover a peculiar mutant. All its systems seem to respond normally to the presence and absence of glucose, except the *lac* operon. This operon makes *lac* proteins at rather low levels in response to lactose, but the level is not affected by adding or removing glucose. What's wrong with this mutant?

18.7 Promoters, the DNA sequences to which RNA polymerase binds, also regulate gene expression.

Some genes are regulated transcriptionally, without the complicated mechanisms of repressors, activators, and alarmones, but simply by the properties of their promoters. Promoters are the regions of DNA to which RNA polymerase binds initially, so the polymerase is oriented to transcribe the proper strand in the right direction. Procaryotic promoters are about 70 bp long, just about the length of the RNA polymerase complex, and are located just upstream of the genes to be transcribed (Figure 18.13). Comparing the sequences of many promoters from different genes and organisms shows that their similarities can be summarized by means of a **consensus sequence,** a kind of statistical "most typical" sequence (Figure 18.14). Few promoters have the actual consensus sequence, but their sequences are all slight variations on it. The *strength* of a promoter—its effectiveness in promoting transcription—depends on its sequence; the more its sequence resembles the consensus sequence, the stronger the promoter. As Figure 18.14 shows, procaryotic promoters always have a **TATA box,** also called a *Pribnow box* for its discoverer, David Pribnow. This short sequence of thymines and adenines is important for binding the RNA polymerase to the DNA.

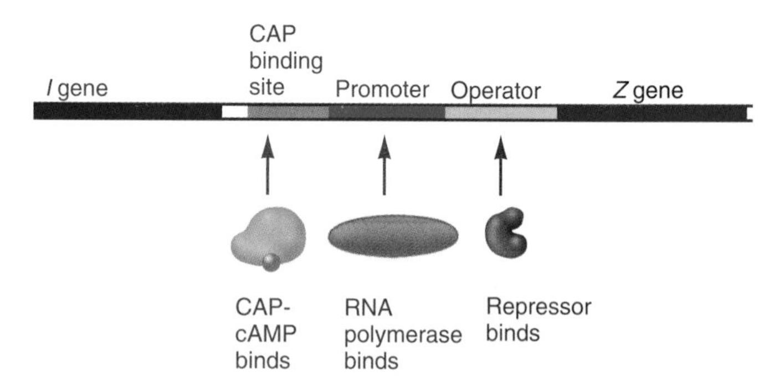

Figure 18.12

The regulatory region of the *lac* operon contains: an operator where the repressor protein binds; an upstream promoter site where RNA polymerase binds to initiate transcription; and a CAP-binding site, where the CAP can bind if it is bound to cAMP. CAP-cAMP is needed for efficient transcription, even if lactose is present.

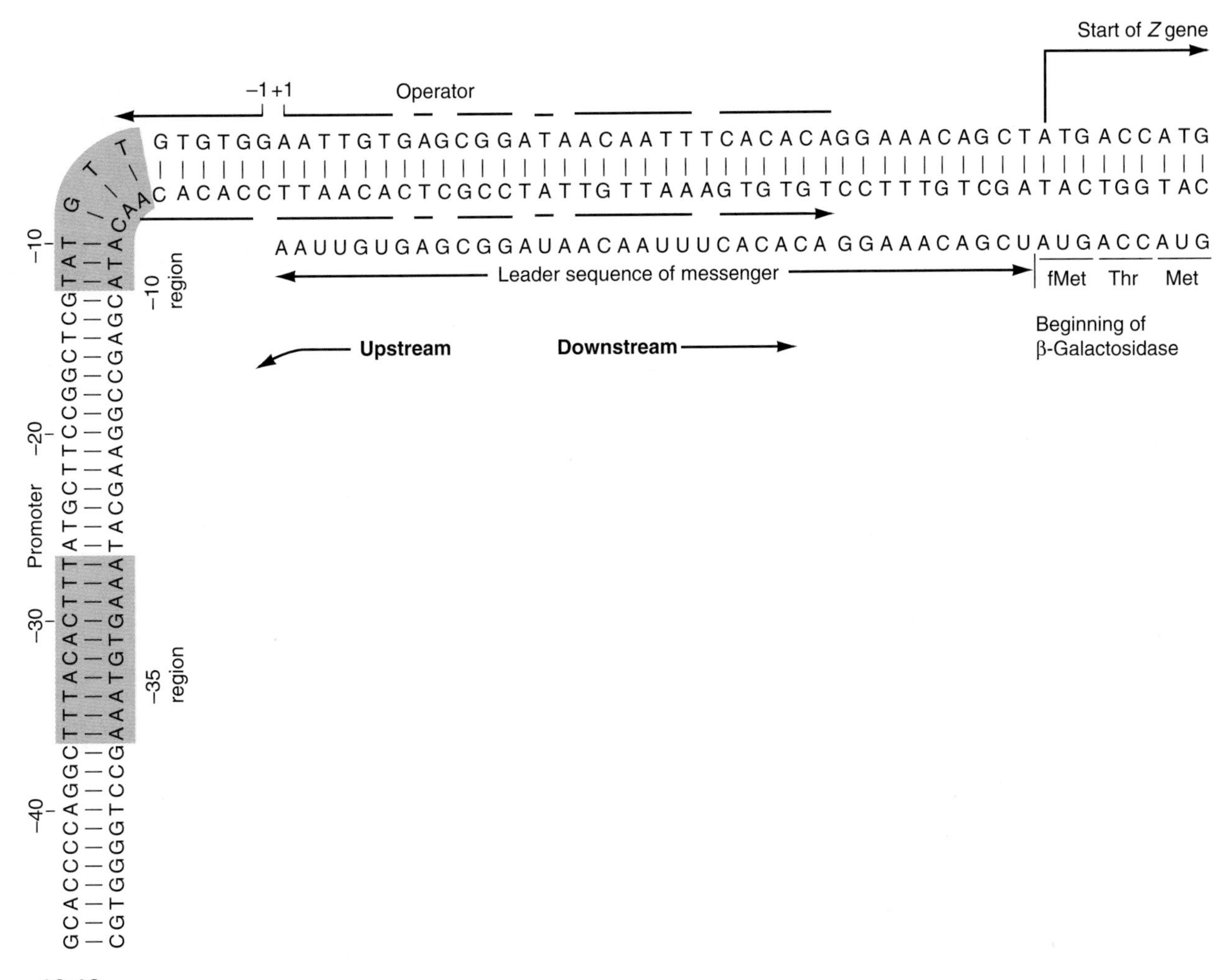

Figure 18.13

The DNA sequence of the regulatory region of the *E. coli lac* operon shows where proteins actually bind to the DNA. +1 is the first nucleotide that is transcribed; nucleotides to the left of it (upstream) are numbered from −1. The RNA polymerase primarily recognizes the −10 region, or TATA box, and the −35 region, but the polymerase is such a large protein complex that it covers most of this region when it first binds. The operator is a region of about 35 nucleotide pairs whose symmetry is shown by the antiparallel arrows. Notice that the messenger RNA is transcribed with a long leader sequence before the start of the *Z* gene.

Gene	−35 region		−10 region		Initiation site	
lac	CCCAGGC	TTTACACTTT	ATGCTTCCGGCTCG	TATGTT	GTGTGGA	ATTGTGA
trp	ATGAGCTG	TTGACAATT	AATCATCGAACTAG	TTAACT	AGTACGC	AAGTTCA
araBAD	ATCCTAC	CTGACGCTT	TTTATCGCAACTCTC	TACTGT	TTCTCC	ATACCCGT
araC	CGTGATTA	TAGACACTT	TTGTTACGCGTTTT	TGTCAT	GGCTTTG	GTCCCGC
tRNA^{Trp}	ACGTAACA	CTTTACAGC	GGCGCGTCATTTGA	TATGAT	GCGCCCC	GCTTCCC
bioB	TAATCGAC	TTGTAAACC	AAATTGAAAAGATT	TAGGTT	TACAAGT	CTACACC
Consensus sequences:		TTGACAATT		TATAAT		

Figure 18.14

Promoter sequences in the DNA of selected bacterial and phage genes are compared by aligning their +1, −10, and −35 regions. +1 is the first base to be transcribed; the −10 region is the TATA box. Note that not all sequences in the −10 and −35 regions are exactly the same. The consensus sequence is a statistical "best fit" showing the base that most often appears in each position.

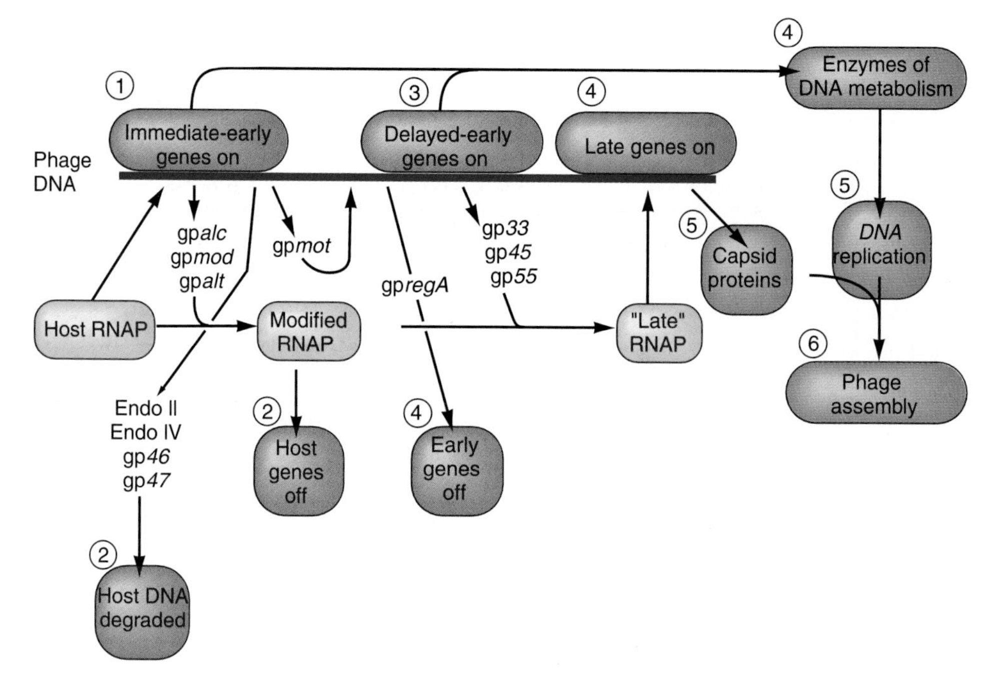

Figure 18.15

The transcription of phage T4 DNA in a well-regulated sequence is determined by several T4-encoded proteins, generally named gp (gene product) with the number or name of the gene. Each regulatory protein changes the host RNA polymerase (RNAP) and directs it toward different promoters. Circled numbers show the approximate timing of events.

Changing the affinity of RNA polymerase for various promoters is one way genes can be regulated using the properties of their promoters. The transcription of phage T4 genes shows this very nicely (Figure 18.15). The bacterial RNA polymerase attaches to a set of very strong *early promoters,* for which it has high affinity. Among the proteins encoded by these early transcripts are some that change the polymerase so it starts to transcribe from *middle promoters;* proteins encoded by these middle genes then change the polymerase again so it transcribes from *late promoters.*

Some proteins are also regulated by post-transcriptional and translational mechanisms. Messenger RNAs, particularly in eucaryotic cells, have different stabilities, and the amount of protein synthesized from a messenger can be limited by breaking the messenger down rapidly. Protein synthesis can also be regulated by factors that bind to a messenger and inhibit its translation. In virus-infected cells, for instance, some viral proteins bind to their own mRNAs and thus stop further translation that would otherwise produce excess protein.

Exercise 18.5 Suppose the following sequences have been observed in the −35 position of some bacterial promoters. Write a consensus sequence for this position.

TTGACA, TCGACA, TTAATA, TTGACT, CTGACA, TTAACA, TAGACA, TTGACG, ATGACA

Exercise 18.6 The *lacI* gene, which encodes the *lac* repressor protein, is not itself regulated by any other protein, and this is true for regulatory proteins in general. What, then, allows these genes to be expressed at the proper level?

Exercise 18.7 Phage Z has only two types of genes, early and late. Immediately upon infecting a cell, its early genes are all transcribed by the cellular RNA polymerase. Around 7 minutes after infection, its late genes are all turned on, and the early genes are no longer transcribed. A single phage gene, *reg,* is responsible for the transformation. What regulatory mechanism is used here, and what does the product of the *reg* gene probably do?

18.8 Eucaryotic genes are regulated primarily by combinations of promoters and enhancers.

Initially it was thought that gene regulation in eucaryotes would be similar to that in procaryotes, and it is true that some eucaryotic cells, such as yeasts, live very much like bacteria. For example, they have the same needs to respond to nutritional conditions, and they can turn genes on and off with some of the same general regulatory mechanisms, such as repressor proteins that control the expression of single genes. However, eucaryotes have also evolved mechanisms not found in procaryotes to address their principal problems of gene regulation in the specialized, differentiated cells of plant, animal, and fungal tissues. These cells have genes for two kinds of proteins. In addition to the housekeeping proteins required in all, or almost all, cells (enzymes for standard metabolic pathways, structural proteins, and so on), eucaryotes make **tissue-specific proteins** in certain types of cells, such as myoglobin and contractile proteins in muscle or hemoglobin in red blood cells. The primary questions about eucaryotic gene regulation are about genes for these specialized proteins.

Research on eucaryotic genes, conducted largely with recombinant-DNA methods, indicates that they are regulated primarily by two elements: promoters and enhancers (Figure 18.16). A typical eucaryotic promoter is a region of about 100 bp, slightly longer than a procaryotic promoter, containing critical sequences called **upstream promoter elements (UPEs).**

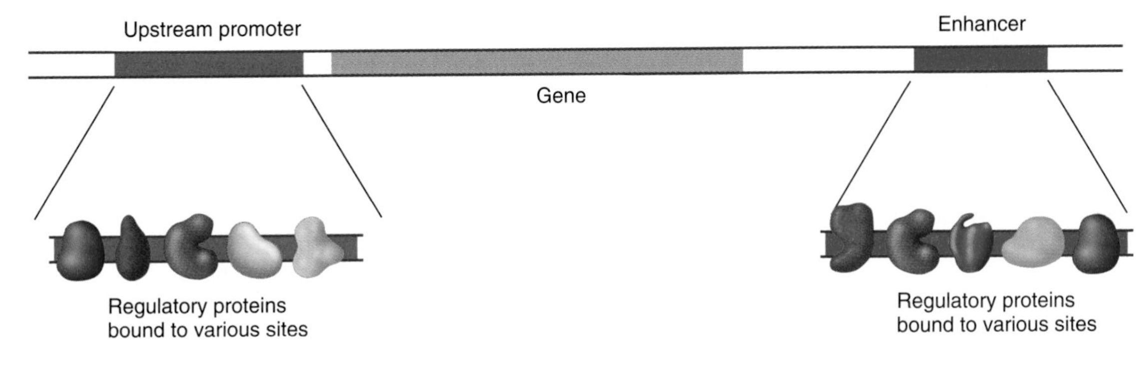

Figure 18.16
Typical genes of eucaryotic cells have a promoter, consisting of about 100 nucleotide pairs just upstream of the RNA start site, and an enhancer, which may be some distance away on either side of the gene. Both control elements have sites to which specific regulatory proteins can bind.

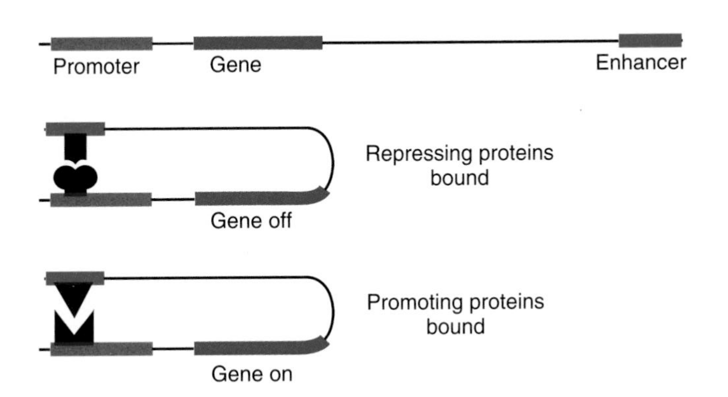

Figure 18.17
An enhancer can interact with a promoter because the DNA between them loops out, so that proteins bound to these elements can come into contact. The enhancer can bind different proteins, which have either a positive or negative effect on the promoter. How strongly the gene is transcribed depends on the balance of their interactions.

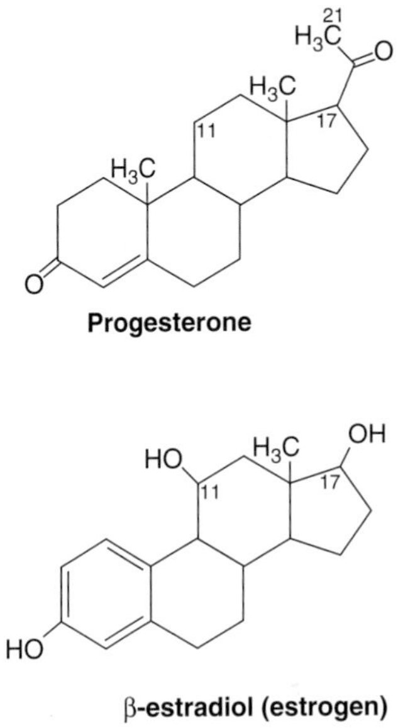

Progesterone

β-estradiol (estrogen)

Figure 18.18
Steroids are lipid-soluble compounds with the basic ring structure of cholesterol. Many of them are hormones, like progesterone and estrogen shown here.

Changing the sequence of a UPE abolishes transcription of the gene. Like the *lac* promoter region, which has critical sequences for binding CAP protein and the *lac* repressor, a UPE seems to be a site to which specific regulatory proteins can bind.

Enhancers are regulatory sites that can enhance transcription of a gene and may lie either upstream or downstream of the gene. Each enhancer contains a series of short binding sites for stimulatory and inhibitory proteins, and its effect on a gene depends on the balance of proteins bound to it. The existence of enhancers came as a surprise because, unlike UPEs, they may be thousands of nucleotide pairs distant from the gene they regulate. How can an element so far away from a gene have any effect on it? As Figure 18.17 shows, the effect depends on the flexibility of DNA and its ability to bend around, bringing two sequences and their associated proteins into contact. (In fact, two elements that are too close cannot interact because the DNA can't bend so sharply. A distance of about 500 nucleotide pairs seems to be optimal.) The bound proteins affect transcription by interacting with RNA polymerase molecules to enhance their binding to the gene's promoter. Enhancer-like elements that inhibit transcription have been called **silencers.**

A plant or animal seems to have a rather small set of regulatory proteins that can bind to UPEs and enhancers. But although there are only a few dozen different proteins they can be combined in an enormous number of ways, and each combination, in principle, can determine a different state of gene expression. Genes at a higher regulatory level—"master" control genes—may then select a combination of proteins for each tissue, which in turn select the genes to be turned on and off. How these master genes are regulated as cells differentiate is a question we will take up in Chapter 21.

18.9 Some genes are directly regulated by steroid hormones.

Steroid hormones (Figure 18.18) differ from other hormones in that they don't bind to cell surface receptors (see Section 41.4). Instead, steroids, being lipid-soluble, pass through cell membranes easily and regulate gene expression by binding to cytoplasmic receptor proteins. Animals have a whole family of steroid receptor proteins, each one specific for a different hormone and each having a DNA-binding site specific for enhancer elements (Figure 18.19). Each steroid selectively affects a small set of genes in a particular tissue because of the specificity of its receptor proteins for particular enhancers. In chickens, for instance, the hormone estrogen stimulates cells of the oviducts (the tubes through which the eggs descend) to produce ovalbumin and other egg proteins.

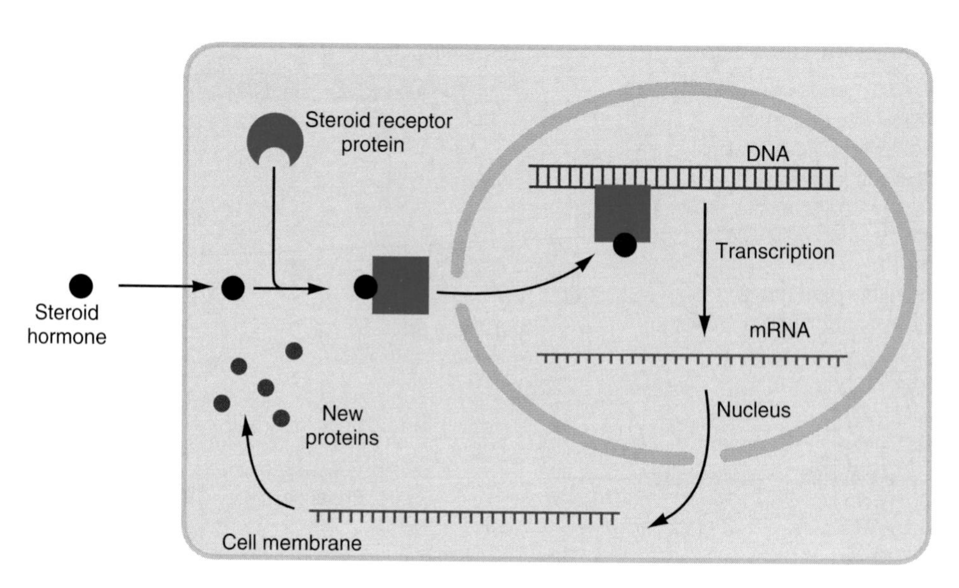

Figure 18.19

Steroid hormones generally act like inducers. The hormone enters a cell and binds to an intracellular receptor protein, thus activating the protein so it promotes the transcription of certain genes.

18.10 Chromosome structure may regulate gene transcription.

A eucaryotic chromosome is a DNA molecule complexed with histone proteins, largely in nucleosomes. This structure can be opened up or compacted so as to regulate the expression of its genes.

Chromosomes are normally visible with a light microscope only during mitosis or meiosis, when they supercoil on themselves and become very compact. However, microscopists saw long ago that some chromosome regions remain highly compacted and therefore visible during interphase. They called these segments **heterochromatin,** in contrast to the **euchromatin** in the rest of the chromosome. Heterochromatic regions are scattered about, but are prominent at the chromosome ends and around the centromere. Early genetic experiments indicated that heterochromatin contains very few genes and may repress genes that are brought near it. Occasionally, a segment of a chromosome is inverted through a pair of breaks followed by repair in the wrong orientation (Figure 18.20). When one break is near a heterochromatic region, a new gene comes under its influence, and the gene's action may be repressed. This is a classic example of a **position effect,** which shows that a gene's activity is determined by its chromosomal environment as well as by its own structure.

Little or no RNA is transcribed from heterochromatin. Indeed, compacting DNA into a heterochromatic structure appears to be an important mechanism for keeping it silent. In Chapter 19 we will discuss permanently heterochromatized X chromosomes that are never transcribed. Their genes are never expressed, and a heterozygous female has a mosaic of tissues, each derived from a cell in which the genes of one X chromosome or the other have been silenced.

These observations about chromosome structure are amplified by experiments on two kinds of specialized chromosomes—the *lampbrush chromosomes* of many animals' oocytes and the giant chromosomes of some flies. A developing ovum stores a great deal of material during the oocyte stage, and the ova of animals contain all the mRNA needed to carry an embryo through an early developmental stage and enough ribosomes to last until a late larval stage. So a great deal of RNA is being transcribed from some genes in the oocyte, and oocyte chromosomes often take on an unusual form with thousands of thin, paired chromatin loops, reminiscent of an old-fashioned lampbrush (Figure 18.21). Experiments by Joseph G. Gall and H. G. Callan showed that each chromosome consists of two double-stranded DNA molecules

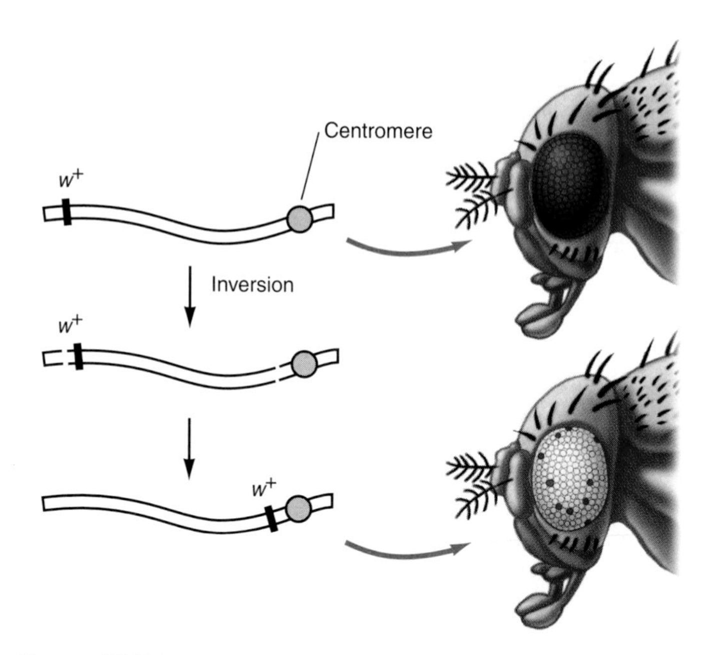

Figure 18.20

The *Drosophila* gene w^+ specifies a red eye. An inversion can move the gene close to heterochromatin around the centromere, where it specifies a white eye with flecks of red color, showing that the gene is being largely repressed but expressed irregularly in some cells.

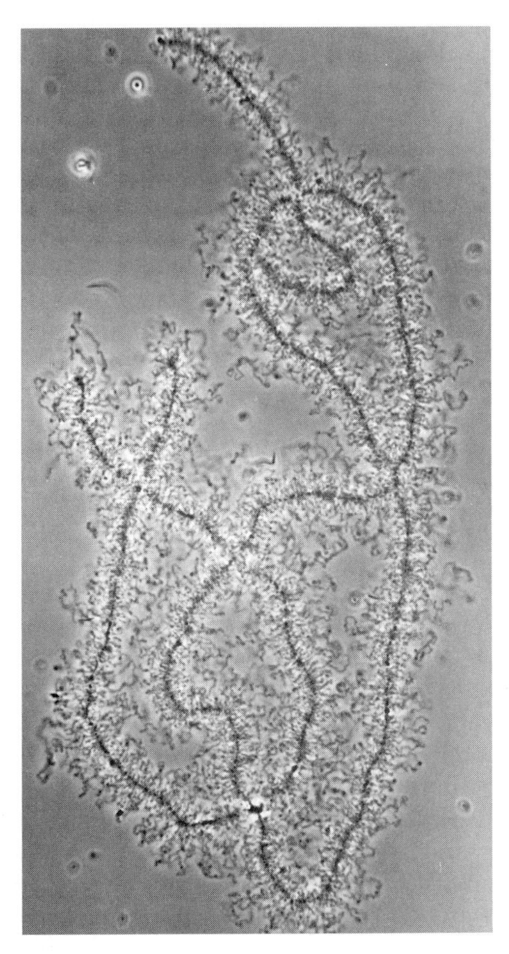

Figure 18.21

A lampbrush chromosome has thousands of loops of extended DNA that are being transcribed into RNA molecules. Here two chromosomes can be identified by their centromeres in the middle of the photograph.

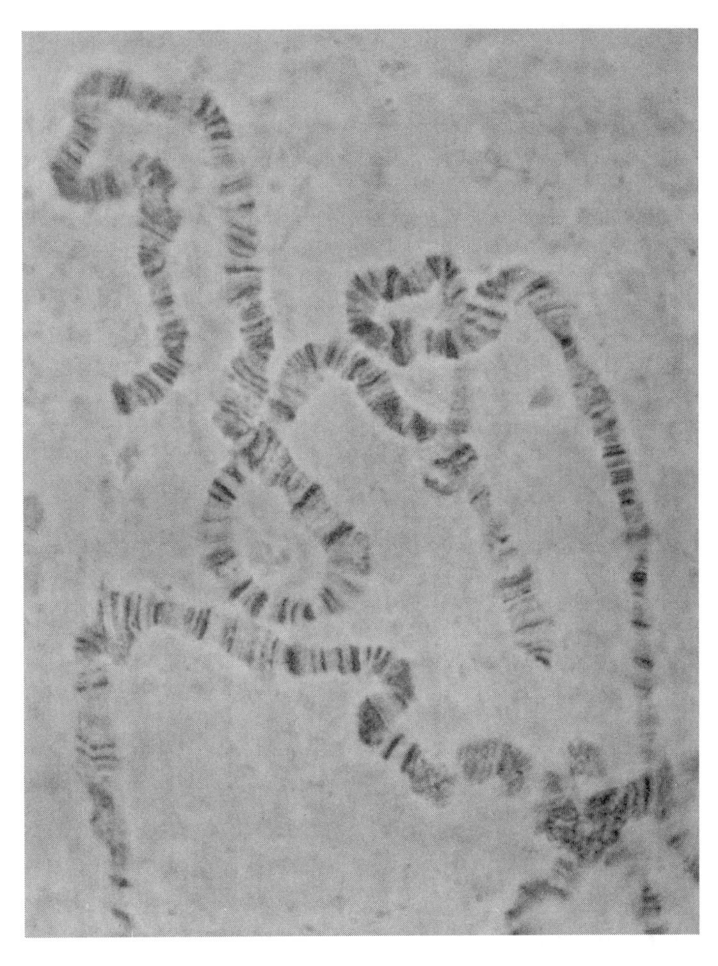

Figure 18.22

The giant salivary chromosomes of *Drosophila* are characterized by their distinctive banding patterns.

that alternately loop out opposite each other. Each loop is at least one transcription unit, covered with RNA strands being transcribed from it.

In the larvae of some flies, including *Drosophila,* certain tissues such as the salivary glands have chromosomes so large that they can be seen in the interphase nucleus (Figure 18.22). Their characteristic bands are used to identify gene locations. These **polytene** ("many-threaded") **chromosomes** consist of over a thousand tandem DNA molecules made by repeated replication without separating from one another. Polytene chromosomes appear only in some embryonic tissues that make large amounts of certain proteins. The salivary glands, for instance, synthesize enzymes to digest all the food the larvae eat, as well as a protein that forms the pupa case. If only a few types of protein are being synthesized in each cell type, one might expect to see that only a few genes are being transcribed from their polytene chromosomes. This is just what we see. The DNA in most bands remains compact and inactive, and little or no RNA synthesis can be detected there. But a few bands are opened out into **chromosomal puffs** (Figure 18.23), looking like clusters of DNA loops, where there is extensive RNA transcription. Apparently one type of protein is encoded at each puff; in some tissues these proteins have been separated chromatographically and correlated one-to-one with identified puffs. Furthermore, the polytene chromosomes in each tissue type exhibit a distinct pattern of puffs, correlated with the different kinds of proteins they are making.

One of the clearest cases of induction by steroids is in insect larvae whose molting is controlled by the steroid hormone ecdysone. The hormone induces protein synthesis in specific cells, including those of salivary glands, where its effects can be seen as puffing on polytene chromosomes. Figure 18.24 shows a sequence of puffing, in which ecdysone turns on one set of genes first, and then the products of these genes apparently induce puffing and transcription of the other bands.

Exercise 18.8 Think again about the observation that puffs appear in different chromosomal regions in a definite sequence. This implies a somewhat more complex regulatory circuit that can be made by linking the simple regulatory mechanisms we have already discussed. Sketch out one example of such a circuit.

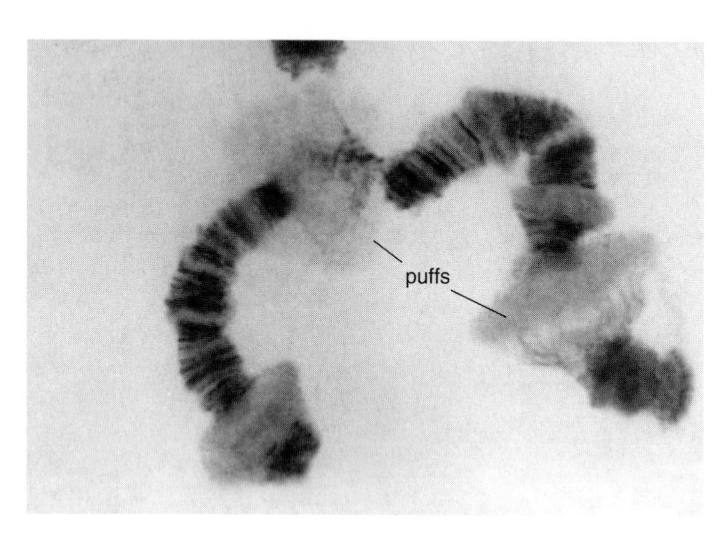

Figure 18.23

Puffs are regions of a chromosome where DNA loops out and is apparently being transcribed.

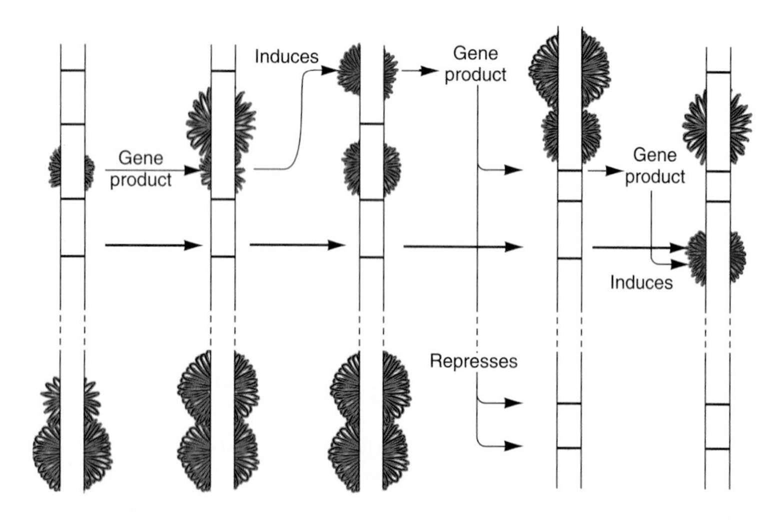

Figure 18.24

During the development of some insects, chromosome puffs occur in a definite sequence. Products of the first genes that puff appear to induce puffing in the next genes.

B. Applications to recombinant-DNA work

Having seen how genes are regulated naturally, we can turn to the regulation of genes in artificial systems produced by recombinant-DNA technology for the purpose of studying genetic regulation and for biotechnology applications. First we will discuss getting a target gene into a system where it can be expressed and studied. Then we will tackle the much greater problem of tailoring such a gene into some other organism—a plant, an animal, even a human—to change its characteristics.

18.11 Recombinant-DNA research employs a few basic methods.

The enzymatic processes that underlie recombining DNA molecules must be supplemented with other methods and processes.

Melting and annealing nucleic acids

Many methods of analysis depend upon the ability of single-stranded nucleic acids, both DNA and RNA, to bind to their complementary sequences. We showed in Chapter 13 that double-stranded DNA molecules become single-stranded at high temperatures; the same thing happens at a high pH because the hydrogen bonds that hold DNA in a double helix are replaced by hydrogen bonds with the excess hydroxyl ions in the solution. We can make a mixture of single-stranded DNA molecules with different sequences and combine them with RNA molecules having a sequence complementary to some of them, at a pH where double-stranded molecules can form. Then the RNA molecules will diffuse about and bind only to their complementary DNAs:

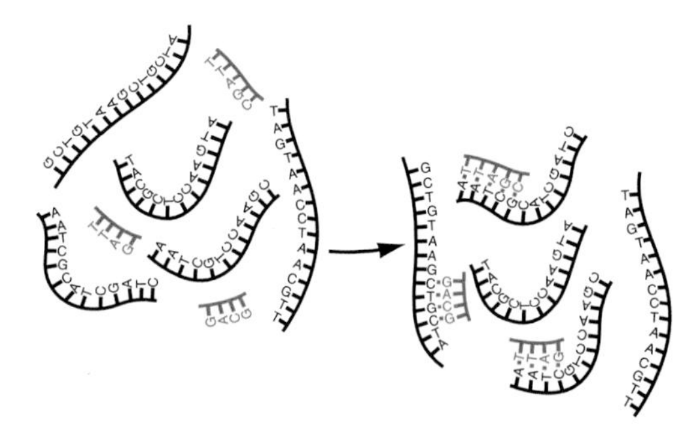

Both DNA and RNA molecules with complementary sequences can find each other this way.

The polymerase chain reaction

Occasionally someone invents a simple method that finds enormous new applications, improves everyone's work, and makes them wonder, "Now why didn't I think of that?" Such is the polymerase chain reaction (PCR) technique invented in 1983 by Kary B. Mullis, which enables investigators to make large numbers of identical DNA molecules from a minute DNA sample. In addition to its role in molecular genetics research, PCR has revolutionized forensic medicine by allowing criminal investigators to amplify an insignificantly small sample at a crime scene into valuable evidence (see Sidebar 19.2).

The PCR method depends on the fact that DNA replication must start from a small *primer* sequence: A DNA polymerase cannot start to assemble new nucleotides onto a bare template strand, but rather begins by elongating a short second strand. Primers can now be made with an instrument that chemically synthesizes short DNA molecules (usually around 20 nucleotides long) with any desired sequence. In the PCR method, we want to amplify a region that we will call "target DNA." We may not know its sequence—sometimes the goal is simply to get enough DNA for sequencing—but we must at least know short sequences at each end of the target region. Then we make primers with complementary sequences. Calling one strand of the DNA plus and the other minus, we make a plus-strand primer for one end and a minus-strand primer for the other end.

The method also depends on the fact that some bacteria can live at very high temperatures, as in boiling hot springs, so all their enzymes are very resistant to heating. We use the DNA polymerase from *Thermus aquaticus,* known as Taq polymerase, or the polymerase from some other high-temperature organism. In addition to the polymerase, the reaction mixture contains at least one DNA molecule with the target region, the primers just described, and the four kinds of nucleoside triphosphates (dATP, dCTP, dGTP, dTTP) for making new DNA. Finally, we need a simple machine called a temperature cycler, which quickly moves small tubes (in which the PCR reaction occurs) from one temperature bath to another or rapidly changes the temperature around the tubes. The cycler puts the tubes through a cycle of three processes:

Melting. At a temperature of about 95°C, double-stranded DNA molecules melt apart into single strands.
Annealing. At about 70°C, primers anneal onto the single strands.
Replication. At about 60°C, the Taq polymerases extend the primers and replicate the DNA.

The reaction works like this:

1. Melt the target DNA into single strands, then move the tube to annealing temperature and allow a plus primer to bind to the minus strand while a minus primer binds to the plus strand:

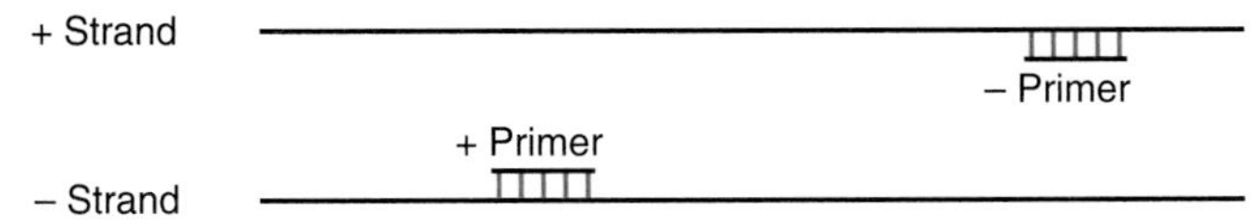

2. Leave the tube at replication temperature for a few minutes, so polymerases can complete a new plus strand and a new minus strand. Synthesis will go beyond the target region on each strand, but notice that it *begins* at defined points, with the primers we have made:

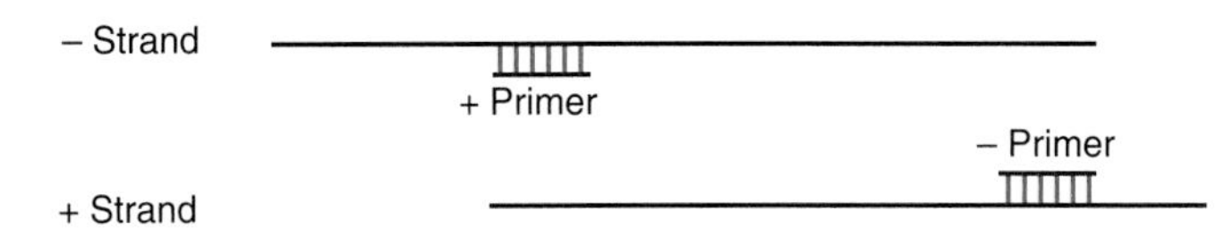

3. Melt all the double-stranded molecules, then return to annealing temperature. Primers now bind to the new plus and minus strands:

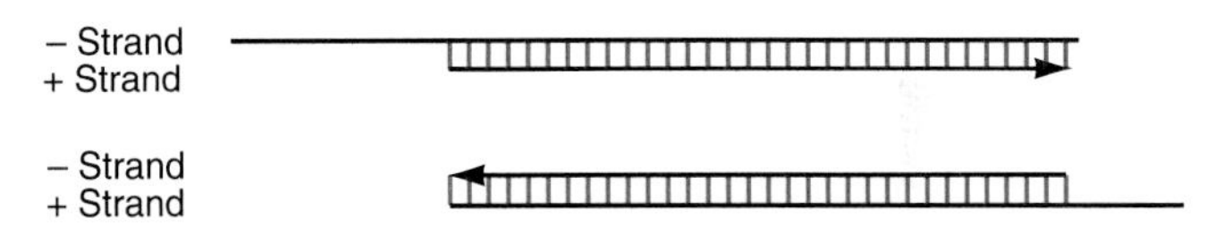

(Primers also bind to the original strands, but we will ignore them because from now on, synthesis on those molecules will become insignificant.) Synthesis continues from each primer, and now each of the new strands also *ends* at a point defined by the opposite primer:

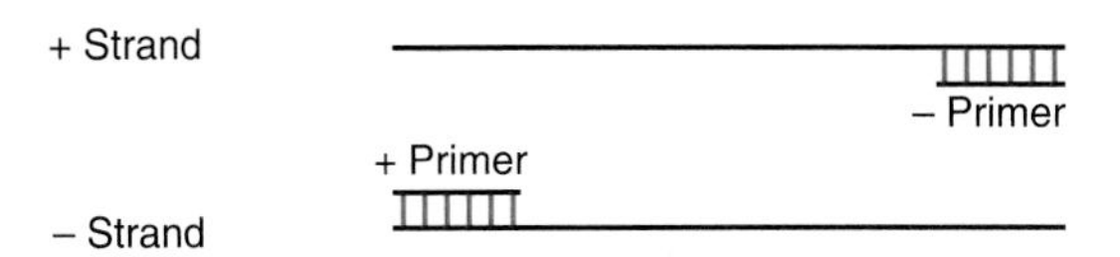

4. In the next cycle, the newest strands now begin and end at the primers, so they contain only the target region:

From now on, each cycle of replication will double the number of such strands. Ten cycles will produce $2^{10} = 1{,}024$ molecules if each strand is replicated at each step. In 20 cycles, there can be over a million molecules; in 30 cycles, over a billion. Each cycle takes only a few minutes, so in a few hours the technique can make enough DNA with a chosen sequence to isolate for use in further experiments.

Agarose gel electrophoresis of nucleic acids

Just as proteins can be separated by electrophoresis in a polyacrylamide gel, fragments of nucleic acids are separated by electrophoresis in gels made of agarose, a material similar to the agar used for culturing microorganisms (Figure 18.25). The sizes of the fragments are measured against standards of known size. A fragment of interest can be cut out of the gel for further analysis, perhaps to be sequenced.

Southern blotting

Most restriction enzymes cut a whole mammalian genome into many fragments. To find the few fragments that carry a particular gene, we use a powerful technique invented by Edward Southern and therefore called **Southern blotting** (Figure 18.26). (Stereotypical scientists are pretty humorless, but real molecular biologists have whimsically named two variants on the technique "northern blotting" and "western

Figure 18.25

Nucleic acids can be separated in agarose. *(a)* A thin slab of agarose gel is cast with a row of slits, which can be filled with small samples of DNA, perhaps PCR products or DNA that has been cut with a restriction enzyme. *(b)* Because DNA carries negative charges, the fragments move toward the anode in an electrical field and separate quickly, with the smallest fragments moving fastest. To visualize the DNA, the gel is soaked in a solution of ethidium bromide, a dye that attaches to DNA molecules and fluoresces red-orange in ultraviolet light.

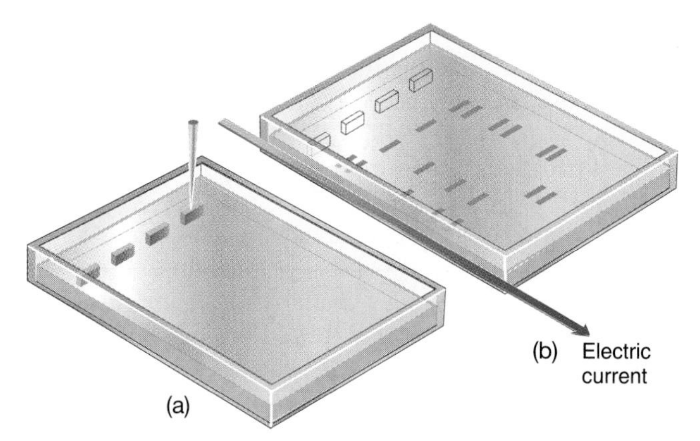

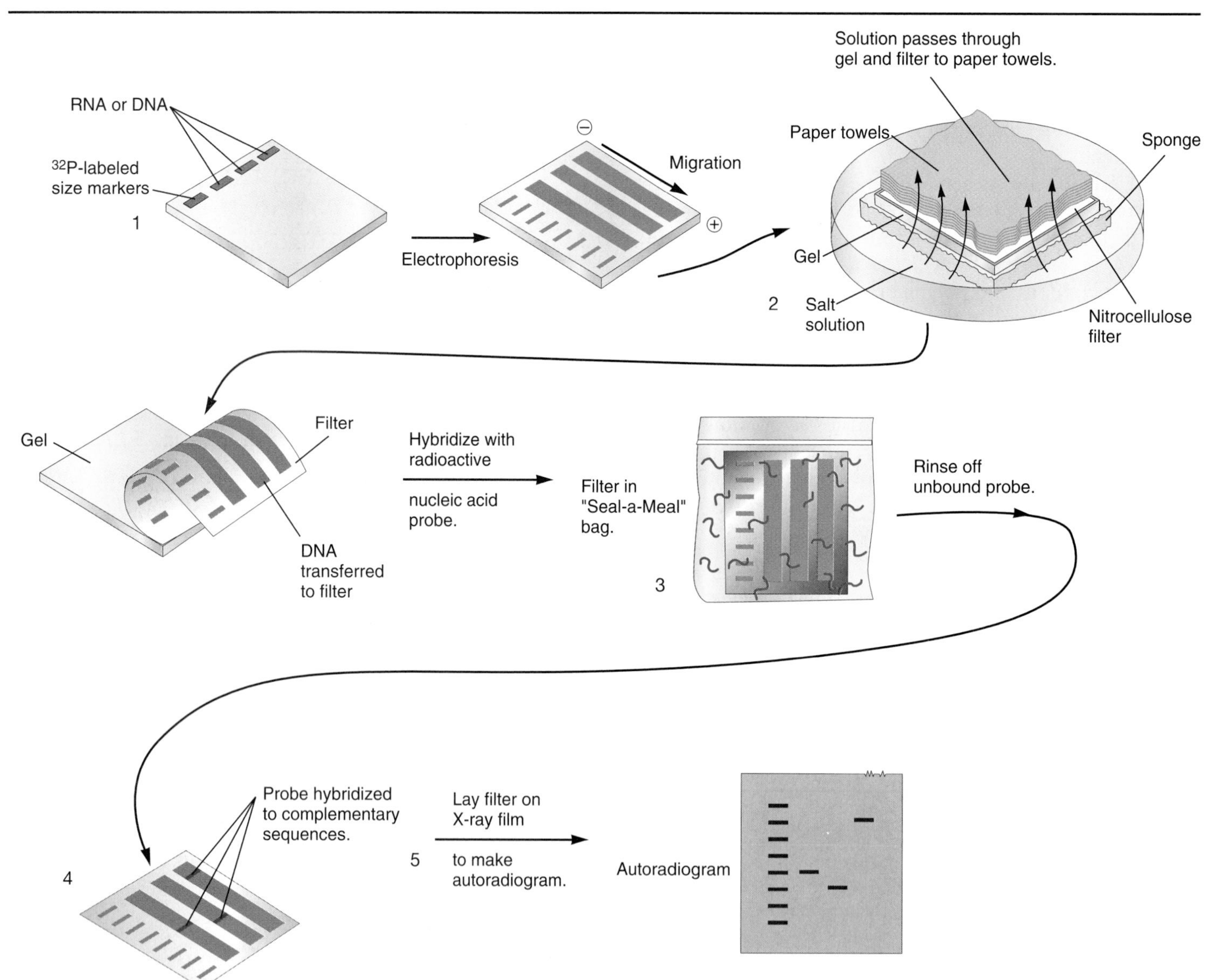

Figure 18.26

The Southern blotting technique requires several steps. *(1)* DNA fragments are separated in a gel by electrophoresis. *(2)* A nitrocellulose filter is pressed against the gel; when liquid is drawn through with a stack of paper towels, the DNA is extracted from the gel and sticks to the sheet. *(3)* The DNA is denatured with alkali to make it single-stranded, and the sheet is agitated for several hours in a plastic bag containing radioactive RNA probes that can bind to the DNA. *(4)* Unbound RNA is rinsed off the sheet, leaving RNA bound only to complementary DNA. *(5)* The dried sheet is pressed against X-ray film to make an autoradiogram.

blotting.") Suppose we want to find the gene for the mouse α hemoglobin chain. Fragments of mouse DNA that has been cut with a restriction enzyme are separated on a gel and then transferred to a nitrocellulose filter. From mouse cells that are making hemoglobin, we extract the corresponding messenger RNA (as discussed later) and make it radioactive; this RNA is to serve as a **probe,** a specific nucleic acid used to find complementary nucleic acids. The nitrocellulose filter is sealed in a bag with a solution containing the labeled probe, so the probe has an opportunity to bind to DNA. The filter is then dried and laid on X-ray film to make an autoradiogram. Because the hemoglobin mRNA probe binds only to complementary DNA, just a few bands containing parts of the hemoglobin gene show up as radioactive; we say that these bands "light up."

This general method will work with any combination of nucleic acids in the gel and as a probe. Thus a piece of DNA that has already been identified as having an interesting sequence can be used as a probe to find other DNA fragments with the same sequence or to find sequences adjacent to it in the genome.

Isolating specific mRNA

Many methods require the mRNA for a specific protein to be used as a probe for the corresponding gene. Naturally, we extract RNA from cells that should be rich in the particular messenger we want, so for hemoglobin, we use cells that are maturing into red blood cells. Eucaryotic messengers can be separated from other kinds of RNA because they have a poly-A tail, so they will bind to complementary poly-T molecules attached to a solid base. Finding specific mRNAs requires advanced techniques that we won't discuss here.

In some techniques, the RNA itself is not useful; for instance, RNA molecules cannot be cloned. However, DNA molecules with the sequence of the RNA can be made with an enzyme called reverse transcriptase, which is made by retroviruses (see Figure 18.34). These viruses make DNA copies of their RNA genomes, and we employ their enzyme to make **complementary DNA (cDNA)** copies of useful RNA molecules (Figure 18.27).

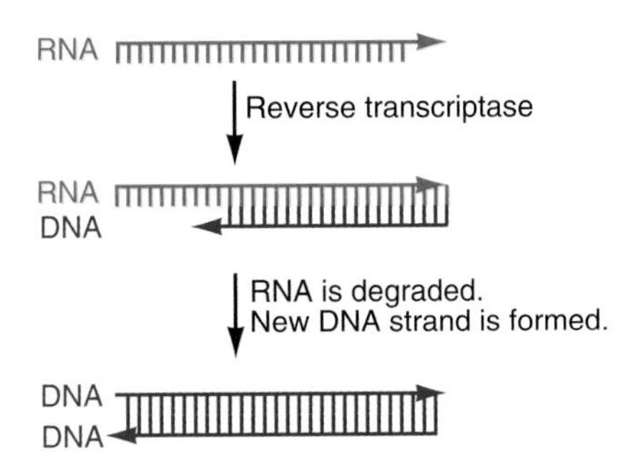

Figure 18.27

Reverse transcriptase from a retrovirus will make a DNA copy (complementary DNA, or cDNA) of an RNA molecule.

Exercise 18.9 Sometimes we need to make a probe for a gene whose sequence is unknown, but we know at least part of the amino acid sequence of the gene's protein product. Consult Table 14.1 (the genetic code) to help think about this question: What kind of amino acid sequence would you look for in making an artificial DNA to be used as a probe? Why?

18.12 Foreign genes can be cloned in plasmids that permit their expression and can be used for research and practical applications.

As we showed in Chapter 17, it is relatively simple now to clone target DNA into unicellular organisms such as bacteria or yeast. It is more difficult to find a single gene of interest among the many clones in a gene library, but one method, using a form of Southern blotting, is shown in Figure 18.28. As described in Section 18.11, we can find (or synthesize) a probe for the gene. Bacteria carrying a gene library are spotted on petri plates, where they grow into colonies. Their DNA, transferred to a nitrocellulose sheet, is then incubated with a radioactive probe, thus identifying clones that probably carry the target DNA.

The kind of vector originally used for cloning, such as the workhorse pBR322, is only good for obtaining a desired gene from its source and converting it into a form that can be manipulated further. Such vectors lack the control elements needed for gene expression, and the next step is to clone the gene into an **expression vector** whose promoters and control elements allow the gene to be transcribed under controlled conditions. Many of these vectors have now been tailored, each with its own use. Some practical ones, the pUC and pET vectors (called "puck" and "pet," of course), share several useful characteristics (Figure 18.29). Since the *lac* operon is so well known, it was natural to use it for controlling the expression of cloned genes. A typical expression vector carries the *lacI* gene encoding the repressor, the *lac* promoter region, and the *lacZ* gene encoding β-galactosidase. This β-galactosidase gene is an example of a **reporter gene,** a gene that can be detected easily and is used to show ("report") that a cell carries the desired DNA. At the start of the *lacZ* gene, the plasmid carries a **polylinker,** a short stretch of DNA containing the sites cut by several useful restriction endonucleases. The polylinker allows us to choose the best restriction enzyme for cloning the target gene. The vector also contains an antibiotic resistance gene to be used for selection. If a desirable gene is cloned into one of these vectors and the vectors are taken up by bacteria, we can distinguish different kinds of bacteria by plating on agar with the antibiotic, the *lac* inducer IPTG, and a substrate called X-gal (Figure 18.30). Cells that don't get a plasmid cannot grow because they are sensitive to the antibiotic. The bacteria that do grow will be blue if they have a functional β-galactosidase and can hydrolyze the X-gal, or white if they have an insert in the *lacZ* gene and cannot hydrolyze X-gal. We pick the white colonies and grow them for further analysis.

Many genes can be cloned with vectors of this kind. Other genes, however, encode proteins that will inhibit or kill any

Figure 18.28

One method of screening a gene library for a gene of interest begins by spotting a series of bacteria (carrying inserts) on petri plates. After they have grown, a nitrocellulose sheet is laid over each plate, and the cells that stick to it are gently lysed in place so their DNA sticks to the sheet. The DNA is denatured into single strands with a strong basic solution and incubated for several hours with a solution containing a radioactive probe. This probe binds only to complementary DNA. The sheet is rinsed to remove accidentally bound probes, dried, and laid on X-ray film for autoradiography. If a colony "lights up"—shows that its DNA binds the probe—it probably carries the gene of interest.

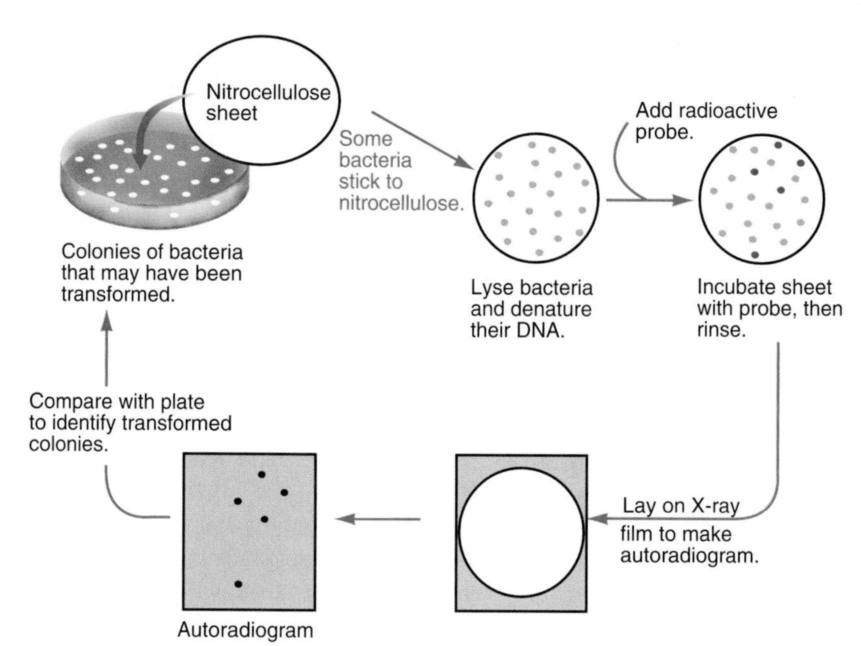

Figure 18.29

A commonly used type of expression vector, such as a pET vector, carries the *lac* regulation system, including the *lacZ* gene for β-galactosidase. Just upstream of *lacZ* is a polylinker sequence with the cut sites for several restriction endonucleases (*Eco*R1, *Sac*I, and so on).

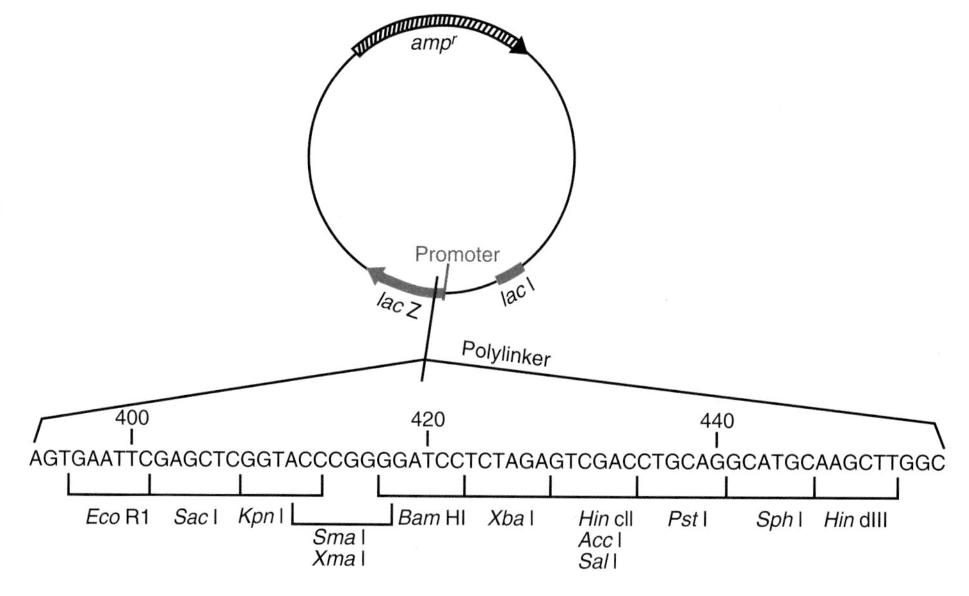

bacterium that expresses them, and special plasmids have been devised to ensure that such genes are tightly controlled. For instance, the gene may be cloned downstream of a promoter recognized only by the distinctive RNA polymerase of a particular phage. The gene for this polymerase is only introduced after the cells have grown, and even if the gene product kills the cells, they will be able to make enough of it for biochemical analysis.

Bacterial cloning systems are excellent for certain purposes, and they have the advantage—important for industrial production—that a cloned gene can be expressed at a high level in a dense culture of cells, to produce enormous amounts of one protein rather easily and cheaply. The method has been used to produce small peptide hormones, starting in 1977 with somatostatin, a peptide of only 14 amino acids that regulates the effects of growth hormone on metabolism. The gene for this hormone was not even cloned—instead, a sequence of 42 nucleotides encoding the right amino acid sequence was made artificially and inserted into the *lacZ* gene in such a way that the hormone could be removed from the resulting hybrid protein by chemical cleavage. A few years later, human growth hormone (HGH) was produced in large quantities from a clone. Previously, tiny amounts of HGH had been isolated at great cost from pituitary glands removed at post-mortem examinations. In addition to the expense, the method had a tragic side effect when three patients who had received extracted HGH contracted Kreuzfeld-Jacob disease, which is caused by a virus transmitted through brain tissue. Now HGH is available cheaply, safely, and in sufficient quantity from pharmaceutical companies. Insulin, too, has been produced in cloned cells, replacing the older method of extracting it from animals in the slaughterhouse.

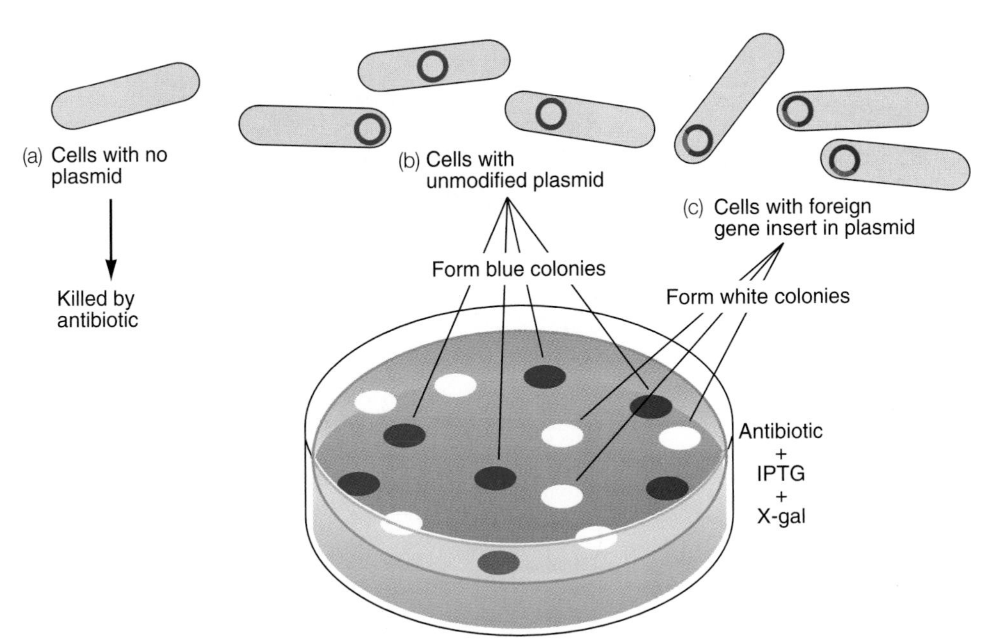

Figure 18.30
After cloning an insert into an expression vector, different bacteria can be distinguished by plating on a medium containing a selective antibiotic, IPTG, and X-gal. *(a)* Only bacteria that have taken up a vector can grow; the rest are killed by the antibiotic. *(b)* Bacteria with an unmodified plasmid have a functional, inducible *lacZ* gene; IPTG will induce production of β-galactosidase, which hydrolyzes the X-gal into a blue dye. *(c)* Bacteria carrying a plasmid with a foreign gene inserted into the *lacZ* gene will also grow, but they cannot make a functional β-galactosidase, cannot hydrolyze X-gal, and have colonies that are white.

Bacteria, however, lack the complex apparatus for expressing a typical eucaryotic gene, splicing out its introns, and perhaps modifying it through the ER and Golgi membranes. So to study many eucaryotic genes, one needs a eucaryotic expression system. Yeasts carry a number of useful plasmids that have been tailored for cloning. Genes have also been cloned in a tissue culture of insect cells by using a large insect virus, baculovirus, that infects arthropods and grows in a culture of insect cells. A gene of interest is first cloned into a special vector tailored so it can recombine with the baculovirus DNA and thus introduce that gene into the insect cells. A third useful system employs vaccinia virus in mammalian cells.

After a gene has been cloned and expressed, the challenge remains to create a **transgenic organism,** generally a plant or animal carrying a foreign gene that is expressed in such a way that the organism's characteristics are actually changed. We will discuss this process next.

18.13 Genes can be cloned in plants by means of a natural bacteria-plant-plasmid system.

In addition to the intrinsic interest in plant genes shared by many biologists, plants have been a principal target for genetic engineering because of their fundamental importance in agriculture. The problem is to get foreign DNA across the plant cell wall so it can be expressed. Some experimenters have used a literal "shotgun" approach by coating tiny metal particles with DNA and shooting them into plant tissues, but the most common method of plant engineering takes advantage of a natural genetic engineer, the cause of crown gall. The soil bacterium *Agrobacterium tumifaciens* produces galls (round tumors or growths) on plants just above the soil line, where the bacteria can insinuate themselves into an injury (Figure 18.31). The actual cause of gall formation is a large (200 kb) plasmid, the **Ti** (tumor-inducing) **plasmid.** Ti carries a number of important genes within a region called T-DNA, which is transferred into plant cells rather the way the F factor transfers copies of itself into other bacteria. T-DNA then inserts itself at random somewhere in the plant chromosomes and transforms the plant cell so it grows into a tumor through the expression of several genes. Six or seven *vir* genes in the T-DNA encode proteins that detect the wounded plant cells and help transfer the T-DNA into the plant cells, while enzymes encoded by other genes synthesize plant hormones (auxins) that stimulate growth of the tumor itself. Another set of genes makes the cells produce unusual amino acids called opines, which the tumor secretes; agrobacteria in the surrounding soil then use the opines as a source of carbon and nitrogen, using other Ti-encoded enzymes, and opines also induce conjugational transfer of the Ti plasmid between bacteria. So Ti is a very sophisticated little parasite that uses plant tissues to feed the bacteria that carry it and to promote its own replication.

To transform plants with new genes, Ti derivatives called *disarmed vectors* have been developed that lack the tumor-inducing genes but retain the genes necessary for transfer and integration of the T-DNA into the plant genome. The target DNA is cloned into a bacterial plasmid that can recombine, in *Agrobacterium,* with the modified Ti plasmid, so the target gene integrated into the T-DNA is transferred to the plant cells. It is easier with plants than with animals to pass the critical step from one cell with an integrated gene to an entire organism that can express the gene. A modified Ti plasmid may be used to transform plant cells in tissue culture, and these cells can then be grown into entirely new plants, some of which will carry the desired genes. Genes have also been introduced into germinating seeds; the resulting mature plants are then bred.

Growing plants from single cells, Section 21.1.

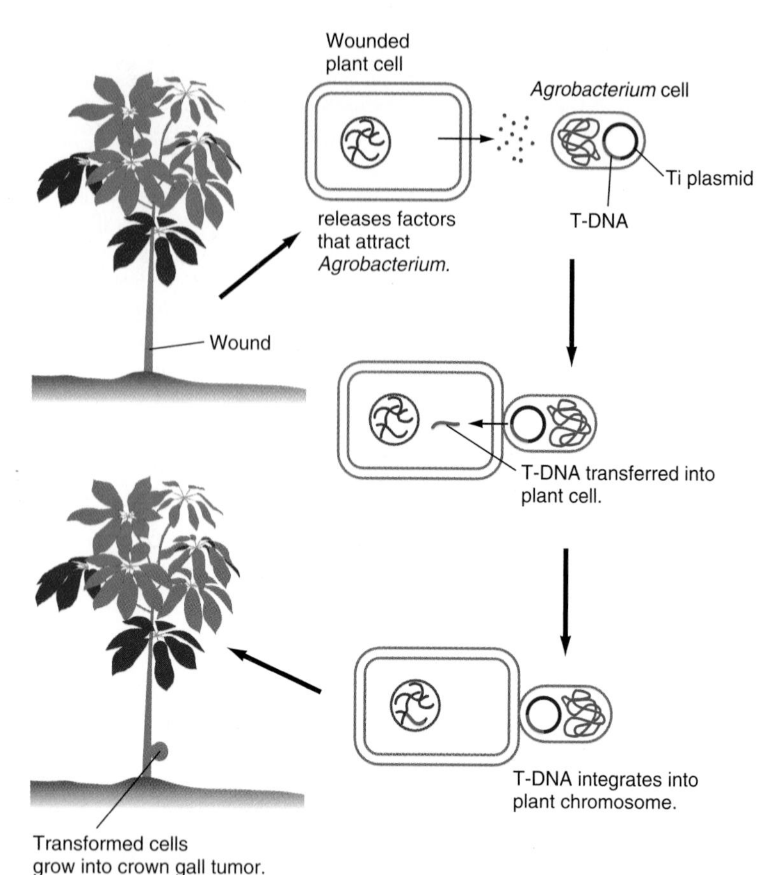

Figure 18.31

Crown galls are caused by the Ti plasmid carried by *Agrobacterium tumifaciens.* The bacteria infect wounds in the plant. In a process similar to bacterial conjugation, they transfer part of the Ti plasmid, the T-DNA, into some plant cells. T-DNA contains genes that induce these plant cells to grow into tumors.

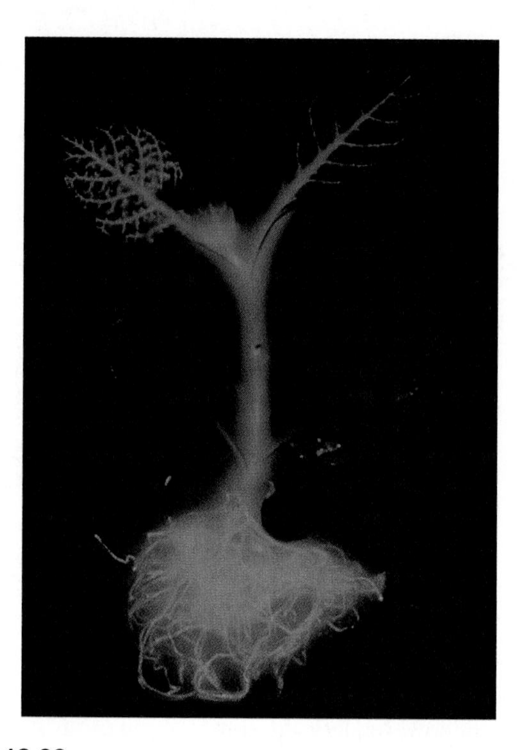

Figure 18.32

A tobacco plant glows like a firefly because the luciferase gene has been transformed into its cells.

Given these and other methods for transforming plants with new genes, what kinds of genes have been transferred and what are the results? An early and spectacular success was expression in tobacco of the firefly gene for the enzyme luciferase, which is responsible for the insect's glow (Figure 18.32). This experiment was much more than a showy tour de force; the luciferase gene is a spectacular reporter gene here, showing that a foreign gene can, indeed, be expressed in the plant tissue. This system is useful for studying regulatory mechanisms, because any desired upstream control element can be cloned next to the luciferase gene, which then serves as a reporter to show the effects of the regulatory sequence.

Resistance to herbicides is one obviously desirable trait in cultivated plants because it allows farmers to apply herbicides to weeds without killing the food crop itself. Herbicides can act at many distinct places, including mitochondria, chloroplasts, protein synthesis sites, and membrane processes, and if the mode of action of any herbicide is known, one can try to introduce genes for proteins that are resistant to the herbicide or will detoxify it. Some investigators have tried to introduce genes that confer resistance to insects. Many infestations by lepidoptera (moths and butterflies) are fought with a deadly and effective toxin made by the bacterium *Bacillus thuringiensis,* sold commercially under the name B.T. Plants have been engineered with this toxin or with other proteins that confer some resistance to insect infestation. A plant could also be made resistant by giving it an enzyme system for producing an allomone against its chief insect enemies, though this approach apparently has not been tried yet.

Allomones, Section 27.9.

One of the first applications of *Agrobacterium* transformation was aimed at producing tomatoes that wouldn't ripen too early, thus being less likely to become soft and bruised before reaching the grocery store. Tomato ripening requires several enzymes and is induced by the hormone ethylene in a positive feedback loop, so a little ethylene induces a fruit to produce still more. The successful strategy introduced *antisense DNA* for the hormone: The hormone is synthesized by a series of enzymes, and DNA is introduced that can be transcribed into an RNA complementary to mRNA for one of these enzymes (Figure 18.33). This anti-messenger RNA hybridizes with the cellular mRNAs and blocks their translation, so the fruit does not produce ethylene and remains relatively hard. The tomatoes can then be ripened with ethylene at their destination.

Ripening Fruit, Section 39.9.

Another desirable trait for a plant to acquire is its own ability to fix nitrogen—that is, to reduce N_2 to NH_3. A great deal of research has focused on this problem, which we discuss in Chapter 40.

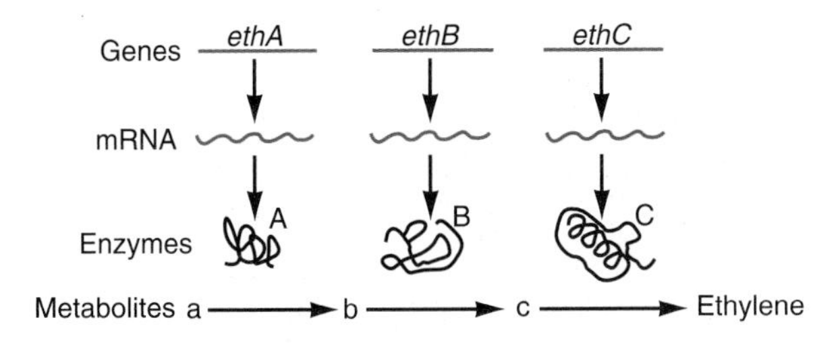

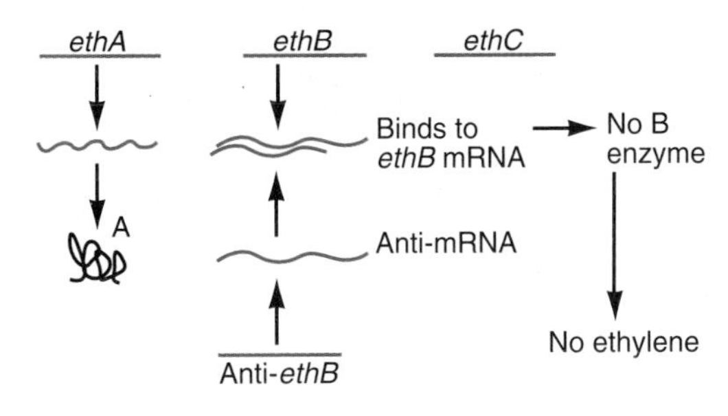

Figure 18.33

Tomato cells normally express genes whose enzyme products produce ethylene. A DNA sequence and promoter is cloned in so its transcript is complementary to the mRNA for one of these genes. The RNAs hybridize, thus blocking the expression of one gene, so ethylene is not produced.

18.14 Animal cloning has been limited by unknown physiological problems.

Transforming animals with new genes presents different kinds of challenges; for instance, whole animals cannot (so far) be grown from a few cells in tissue culture. Two general approaches to creating transgenic animals have been successful: using viral vectors and microinjecting DNA into embryos. Modified viruses would seem like natural vehicles for introducing transgenes, especially some animal viruses known to integrate into their hosts' chromosomes in the manner of prophages of temperate phage. The polyoma viruses, for instance, have small, circular, double-stranded DNA genomes and are associated with transformative changes in animal tissue; some of them are known to induce tumor formation by integrating into their hosts' genomes. A cell with a viral genome integrated into one of its chromosomes loses its normal cellular controls and grows wildly. The virus that causes human warts is a member of this group; thus a wart consists of skin cells growing in an abnormal lump rather than in the usual thin layer. **Retroviruses** grow similarly (Figure 18.34). These are RNA viruses, but after infecting a cell, the viral genome is copied into a DNA form, which inserts itself into a chromosome. In principle, any virus of this kind could be tailored to carry another gene with appropriate control sequences, so it could be expressed as desired in some tissue. "One-round" retrovirus vectors have been made by

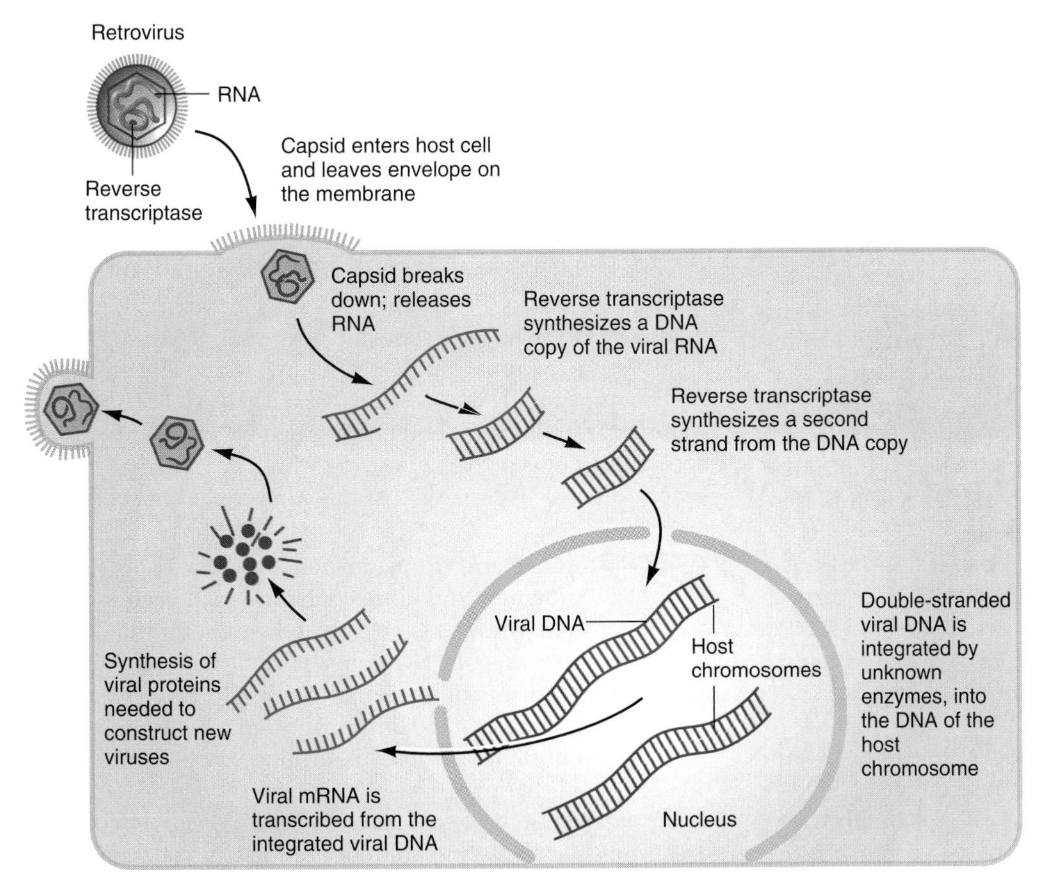

Figure 18.34

A retrovirus has an RNA genome but integrates a DNA copy of the genome into the chromosomes of its host. After the virus enters a host cell, its genome is uncoated; the enzyme reverse transcriptase makes a DNA copy of the genome, which then integrates as a provirus at some random site in one of the chromosomes. RNA genomes of a normal retrovirus are replicated from the integrated DNA, but a modified retrovirus could be used to carry other genes into the cell.

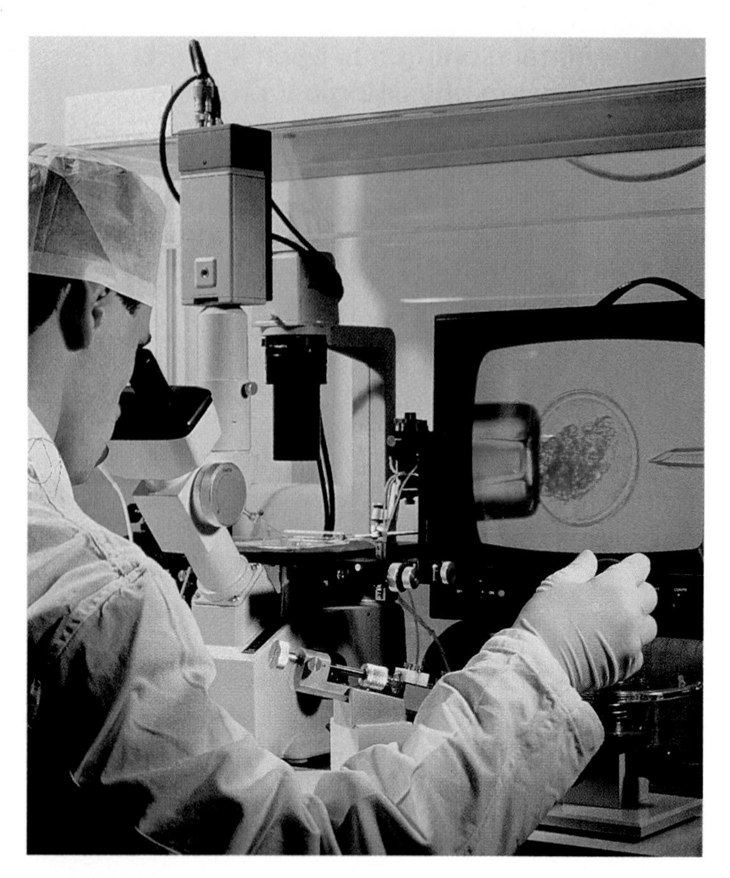

Figure 18.35
Using a micromanipulator attached to a microscope, cell biologists can operate on cells with very fine tools. An extremely thin needle, like a hypodermic syringe, can be used to inject DNA directly into the nucleus of a cell.

Figure 18.36
A transgenic mouse *(right)* carrying genes for human or rat growth hormone is shown next to its normal relative.

removing genes essential for viral reproduction and inserting a transgene; theoretically, such vectors can insert the transgene into a chromosome but cannot reproduce. This kind of manipulation, however, entails a constant danger—creating a virus with unpredictable properties that could replicate wildly.

For this reason, many experimenters have turned to microinjection (Figure 18.35). Micromanipulators attached to a light microscope allow subtle movements of minute tools, including tiny needles. Properly tailored DNA can be injected right into the nucleus of an animal embryo, where it may integrate into chromosomal DNA. Although efficiency is low, it is often enough to make this method successful.

Animal breeders have tried many kinds of transgene transfers in domesticated animals, with varying results. Mice carry a single gene, *Mx1*, that makes them resistant to influenza viruses. The gene was cloned and transferred into embryo pigs to confer genetic resistance on them, but none of the grown pigs expressed the transgene protein.

Richard Palmiter and his associates developed one of the first successful transgenic animals, a mouse dubbed "Supermouse." They cloned genes for human or rat growth hormone into the mouse's liver cells under control of the promoter for a gene that responds to certain ions. Supplying extra zinc turned the gene on, causing the transgenic animals to grow to about double their normal size (Figure 18.36). Transgenic sheep were later created with genes for extra growth hormone meant to enhance their development. However, even though the sheep had higher levels of growth hormone in their blood, none of them grew any faster. In fact, most of them had health problems such as diabetes and pneumonia, and died at an early age. Perhaps the most successful results of genetic technology so far are dairy cows injected with synthetically made bovine growth hormone to stimulate them to produce more milk. They do, indeed, produce more milk, but like the sheep, they have health problems and are more susceptible to infections. In response, dairymen feed them higher doses of antibiotics, which lead to additional problems (see Sidebar 17.1). The public has reacted quite negatively to the idea of drinking milk from these cows, partly as an informed response to the possible dangers of taking in excess antibiotics and hormones, partly out of uninformed fear of the unknown. Responses of this kind have cast considerable doubt on the advisability of producing transgenic animals. Experimental results so far have shown that we don't yet know enough about intricate animal physiology to understand and anticipate the effects of adding one isolated, foreign gene to a complicated genome. Still, the most potentially exciting, and controversial, applications of genetic engineering would be in curing human genetic diseases. We will discuss this work in the next chapter.

Coda An organism operates on the basis of its genetic program; thus close control of each gene is necessary. A gene must only be expressed at the right time and place, and its product must be made in the right amount. We have already seen that organisms must be able to respond appropriately to environmental stimuli, and one of the most common and important responses is turning certain genes on or off. The first genetic regulatory systems to be elucidated, such as the *lac* operon, have this function. More generally, complex organisms like plants and animals need to express many genes only in certain differentiated cells, giving each part of the organism its own structure and function.

Genetic regulatory mechanisms illustrate most strongly the principle of operation through molecular interactions. The regulatory proteins are mostly allosteric proteins, which bind both to sites on DNA and to controlling ligands. Regulation may be positive or negative, and genes may be repressed or induced. Some mechanisms seem very complicated, but remember that each one has evolved opportunistically—not through rational design but merely as each organism followed the genetic opportunities that opened up to it through random events. There is no way to predict what kind of mechanism will control a particular gene; the organism uses whatever has evolved as long as it works. A new question now arising is whether humans can understand the intricacies of these control mechanisms well enough in the context of a plant or animal's physiology to tinker with them for some perceived "improvement."

Summary

1. In procaryotes, genes with related functions tend to be located together in blocks and regulated together.
2. The regulation of the genes for lactose metabolism was worked out by using mutants that are defective in regulation.
3. An operon consists of a block of genes controlled by an operator. In the *lac* operon, the genes are regulated negatively by a repressor protein that binds to the operator and prevents transcription. An inducer, such as lactose, binds to the repressor (which is an allosteric protein) and changes its conformation so it can no longer bind to the operator.
4. Genes may also be regulated by positive mechanisms. An activator protein promotes transcription. The arabinose activator, for instance, is only active when arabinose is bound to it.
5. Biosynthetic genes may be regulated by repressors; these repressors only bind to their operators when bound to the end product of the biosynthetic pathway. Several blocks of genes can be regulated by a single repressor.
6. Still larger blocks of genes are regulated by alarmones. Guanosine tetraphosphate (ppGpp) is made in response to a block in protein synthesis; cyclic AMP (cAMP) is made in response to the absence of an energy source. These alarmones are able to turn blocks of genes on or off.
7. Promoters, the DNA sequences to which RNA polymerase binds, also regulate gene expression. The expression of phage genes, for instance, may be determined largely by having promoters of different types and strengths, as well as proteins that modify the RNA polymerases so they recognize the correct promoters.
8. Eucaryotic genes are regulated primarily by combinations of promoters and enhancers. These sequences can bind specific regulatory proteins, and the activity of a gene is determined by the combination of proteins bound to the promoters and enhancers that control it.
9. Some genes are directly regulated by steroid hormones, which bind to specific receptor proteins inside a cell.
10. Chromosome structure may regulate gene transcription. Chromosomes may be compacted into tight structures that cannot be transcribed, or specific regions may be opened up so the genes located there may be transcribed. Heterochromatic regions are permanently or temporarily shut down, so none of their genes can be expressed, and genes translocated near heterochromatin may be shut off.

Key Terms

repression 352
housekeeping protein 353
inducible enzyme 353
inducer 353
repressor protein 354
activator protein 354
operator 354
operon 354
downstream 354
upstream 354
merodiploid 355
regulon 357
alarmone 357
guanosine tetraphosphate (ppGpp) 358
cyclic AMP (cAMP) 358
cAMP acceptor protein (CAP) 358
consensus sequence 358
TATA box 358
tissue-specific protein 360
upstream promoter element (UPE) 360
enhancer 361
silencer 361
heterochromatin 362
euchromatin 362
position effect 362
polytene chromosome 363
chromosomal puff 363
Southern blotting 365
probe 367
complementary DNA (cDNA) 367
expression vector 367
reporter gene 367
polylinker 367
transgenic organism 369
Ti plasmid 369
retrovirus 371

Multiple-Choice Questions

1. In the regulation of genes for lactose metabolism, lactose functions
   a. as an inducer.
   b. as the substrate of the enzyme encoded by a structural gene.
   c. as an inducible enzyme.
   d. *a* and *b*, but not *c*.
   e. *b* and *c*, but not *a*.
2. A bacterial mutant that cannot synthesize β-galactosidase, but can produce galactoside permease and repressor protein normally, probably carries a mutation in the ______ gene.
   a. *lacZ*
   b. *lacY*
   c. *lacA*
   d. *lacI*
   e. *lacO*
3. If lactose-metabolizing enzymes are produced whether or not lactose is present, a mutant ______ gene is likely the cause.
   a. *lacZ*
   b. *lacY*
   c. *lacA*
   d. *lacI*
   e. *lac* promoter
4. Which sequences of DNA encode a diffusible product?
   a. *lacO*
   b. *lacI*
   c. *lacZ*
   d. *a* and *b*, but not *c*
   e. *b* and *c*, but not *a*
5. Which of the following statements about the *lac* operon is *not* correct?
   a. The operator is upstream of its structural genes.
   b. The information carried by the *lacI*, *O*, *Z*, *Y*, and *A* genes is transcribed as one length of mRNA.
   c. The *lacI* gene produces a diffusible protein.
   d. The *lacI* gene product acts as an active repressor when lactose is not present.
   e. If lactose is not present, structural genes of the *lac* operon are not transcribed.

6. An *E. coli* cell that produces the *lac* operon gene products whether or not lactose is present is converted into a merodiploid by the insertion of a plasmid. Afterwards, the lactose-metabolizing enzymes are produced only when lactose is present. Which of the following is a necessary and sufficient explanation of these observations?
   a. The plasmid contains a wild-type *lacI* allele.
   b. The plasmid contains a wild-type *lacZ* allele.
   c. The plasmid did not contain any structural genes of the *lac* operon, but it did contain a wild-type *lac* operator.
   d. When the plasmid was introduced, wild-type *lacZ* and *Y* gene product was introduced by error.
   e. None of the above would account for the observations.
7. The TATA or Pribnow box has all the following characteristics except:
   a. It is both transcribed and translated.
   b. It provides a binding site for RNA polymerase.
   c. It is upstream of its structural genes.
   d. It is part of a consensus sequence.
   e. It is found within procaryotic promoters.
8. Pre-transcriptional gene control includes all except
   a. enhancers and silencers.
   b. strong and weak promoters.
   c. UPEs.
   d. alarmones.
   e. unstable mRNA.
9. Chromosomal puffs indicate regions where
   a. heterochromatin is found.
   b. DNA is being actively transcribed.
   c. operon regulation is inhibited.
   d. steroid binding is inhibited.
   e. DNA replication is taking place.
10. Which of the following must be part of the DNA of expression vectors but need not be included in transfer vectors?
   a. a restriction site
   b. reverse transcriptase
   c. Taq polymerase
   d. a promoter
   e. primers

True-False Questions

Mark each statement true or false, and if false, restate it to make it true.

1. An operator can act only in the *trans* position.
2. The *lacI* gene can act in either the *cis* or *trans* position because it is upstream of the operator.
3. The three structural genes of the *lac* operon are regulated in a coordinated way because the operator binds to a diffusible product.
4. The primary difference between the repressor proteins of the *lac* and histidine-synthesis operons is that the histidine repressor must be bound to a ligand to be active, while the *lac* repressor is active alone.
5. Alarmones are regions of DNA that alter the transcriptional activity of RNA polymerase.
6. Chromosomal regions that stain heavily are called heterochromatic, and are indicative of a high level of transcription.
7. In order to replicate DNA by PCR, one must combine a DNA primer, all four nucleoside triphosphates, RNA polymerase, and the target DNA in a thermal cycler.
8. In order to identify a particular gene in a Southern blot, a probe made of either radioactive RNA or radioactive DNA must be used.
9. Transgenic plants have been constructed using a bacterium called Ti as the vector.
10. In order for a transgenic animal to express the transgene in a medically or economically useful manner, an appropriate promoter must lie upstream of the gene.

Concept Questions

1. After reading this chapter, how would you answer Jacques Monod's question about the on-off-on growth pattern of *E. coli* grown on a culture containing both glucose and lactose?
2. Would you expect most regulatory pathways in bacteria to be reversible or irreversible? What about eucaryotes? Explain.
3. How do promoters differ from enhancers and silencers?
4. Compare the action of alarmones with that of steroid hormones.
5. If you want to produce a human protein in a bacterial cloning system, why would you choose to work with human cDNA rather than human genomic DNA?

Additional Reading

Anderson, W. French. "Gene Therapy." *Scientific American,* September 1995, p. 96. A review of the current situation and hopes for the future.

Chilton, Mary-Dell. "A Vector for Introducing New Genes into Plants." *Scientific American,* June 1983, p. 50. Using the Ti plasmid for genetic engineering.

Grunstein, Michael. "Histones as Regulators of Genes." *Scientific American,* October 1992, p. 40. Histones are not just structural proteins but vital participants in gene expression and repression.

Hiss, Tony. "How Now, Drugged Cow?" *Harper's Magazine,* October 1994, p. 80. Biotechnology, with bovine growth hormone, comes to rural Vermont.

Lacy, Paul E. "Treating Diabetes with Transplanted Cells." *Scientific American,* July 1995, p. 40. The main obstacle is overcoming interference by the immune system.

Levine, Joseph S., and David T. Suzuki. *The Secret of Life: Redesigning the Living World.* WGBH Boston, Boston, 1993. A companion to the PBS television series, discussing social aspects of molecular biology and genetic engineering.

Lyon, Jeff, and Peter Gorner. *Altered Fates: Gene Therapy and the Retooling of Human Life.* W. W. Norton, New York, 1995.

McKnight, Steven L., and Keith R. Yamamoto (eds.). *Transcriptional Regulation.* Cold Spring Harbor Laboratory Press, Plainview (NY), 1992. A series of papers about gene regulation.

Mullis, Kary B. "The Unusual Origin of the Polymerase Chain Reaction." *Scientific American,* April 1990, p. 56. How the PCR technique was conceived during a moonlit drive through the mountains of California.

Ptashne, Mark, Alexander D. Johnson, and Carl O. Pabo. "A Genetic Switch in a Bacterial Virus." *Scientific American,* November 1982, p. 128. How a repressor operates.

Tjian, Robert. "Molecular Machines That Control Genes." *Scientific American,* February 1995, p. 38. More information about gene regulation in eucaryotes.

Welsh, Michael J., and Alan E. Smith. "Cystic Fibrosis." *Scientific American,* December 1995, p. 36. How the genetic flaw leads to the disorder.

Internet Resource

To further explore the content of this chapter, log onto the web site at:

http://www.mhhe.com/biosci/genbio/guttman/

19

Human Genetics

Key Concepts

19.1 Pedigrees indicate the mode of inheritance of a characteristic.

19.2 Quantitative characteristics may result from the action of many genes.

19.3 Humans have ABO and other blood group antigens.

19.4 Rh factors are associated with hemolytic disease of the newborn.

19.5 Human chromosomes are distinguished by their banding patterns.

19.6 Chromosomes sometimes fail to separate in meiosis.

19.7 Aneuploids for sex chromosomes are relatively common.

19.8 Human genes can be localized to chromosomes by making hybrid cells.

19.9 Human genes can be localized to chromosomes by linkage disequilibrium.

19.10 Human genes can be identified and located with recombinant-DNA methods.

19.11 Genetic disorders might be cured through gene therapy.

19.12 Information about genetics has moral consequences.

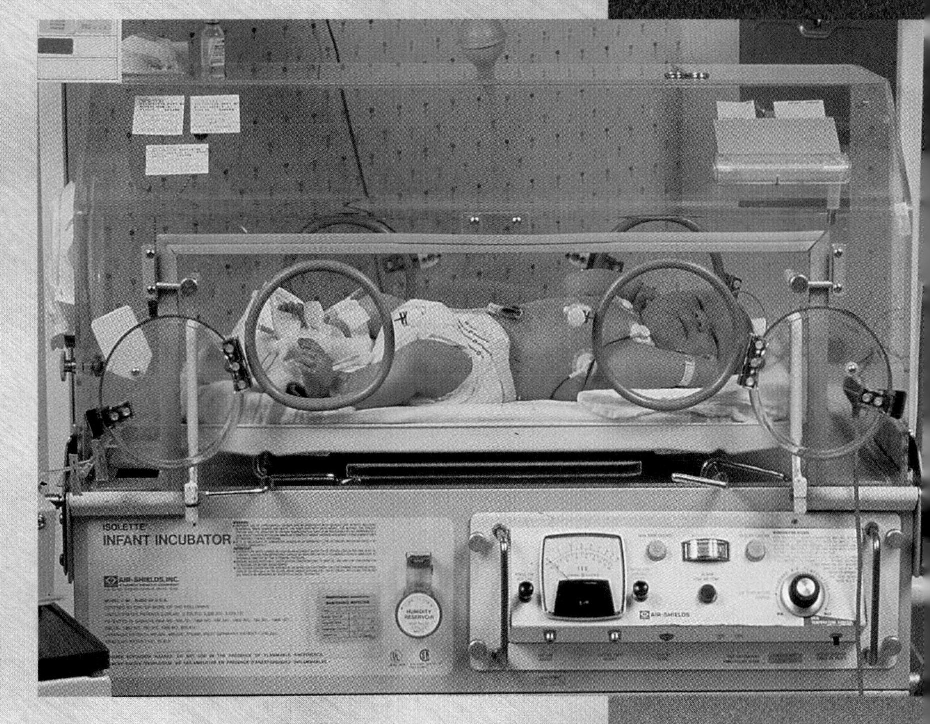

Babies born with genetic abnormalities require special care.

Teddy seems like a normal four-year-old boy. He is bright, alert, interested in everything, loving, and apparently happy. But Teddy has a terminal illness that is likely to end his life early. He often has difficulty breathing and has a vaporizer in his room to raise the humidity and help him breathe. Mucus accumulates in his lungs. Medicines help him cough it up, and his parents help drain the mucus two or three times a day by laying him across their laps and clapping him across the chest. Teddy's sweat is abnormally salty, his sense of taste and smell unusually sensitive.

Teddy's disease is called cystic fibrosis (CF). It was recognized in 1936, and in 1965 was shown to be a recessive hereditary condition. CF alleles are relatively common in people of European descent, producing one CF child in 1,000–1,500 live births in the U.S. Its symptoms vary considerably. Due to increased awareness, affected babies are now commonly identified by their salty sweat. Accumulation of mucus in the lungs, with accompanying bacterial infections, is universal with CF patients, who generally die in young adulthood as a result of these infections. The pancreas is often affected; its ducts become clogged with mucus, so it cannot secrete its digestive enzymes. CF patients therefore have enormous appetites, brought on by their bodies' efforts to obtain adequate nutrition, but much of their food remains undigested and emerges as unusually large amounts of foul-smelling feces. So CF patients usually take supplements of digestive enzymes as well as several vitamins.

Modern human genetics, as we explain in this chapter, is largely directed toward identifying mutant genes that cause such disorders, understanding the disorders, and searching for a cure. ■

19.1 Pedigrees indicate the mode of inheritance of a characteristic.

Well over 6,000 human genes have now been identified on the basis of mutations occurring in them, and many of the genes have been mapped to known loci on the chromosomes. After identifying a genetically determined characteristic, we first want to know its mode of inheritance—whether it is autosomal or sex-linked, dominant or recessive. We can generally determine this from a **pedigree,** which shows the phenotypes (and possibly the genotypes) of everyone in a family tree. Males are shown by squares, females by circles, and a marriage or mating by a horizontal line, ■—● or a double line if the couple are closely related: ■═● The offspring of a marriage, collectively known as a **sibship,** are shown on the next line below, from left to right in the order of their birth, so:

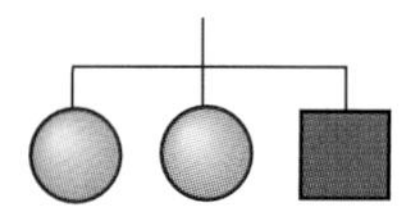

means a family of three, born in the order girl, girl, boy. Fraternal twins, who develop from two different eggs, and identical twins, who come from a single egg that divides, are shown by lines leading to the sibship line like this:

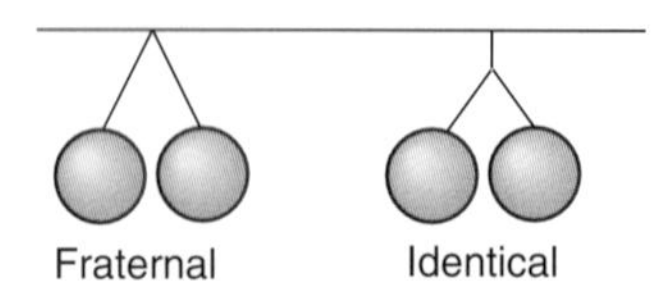

An aborted or stillborn fetus of unknown sex may be shown by:

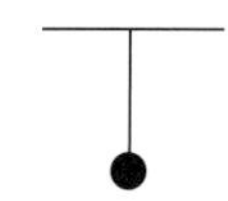

The generations are numbered successively I, II, III, etc., and individuals within each generation are designated by Arabic numbers. Anyone who exhibits the characteristic of interest is indicated by shading the symbol in some way, and a heterozygous carrier for the feature (called an obligate

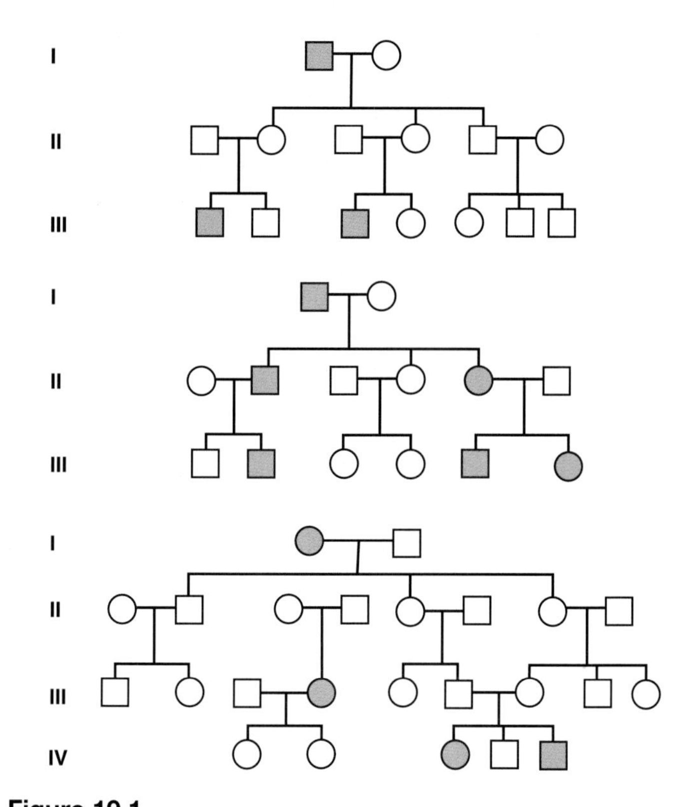

Figure 19.1
Three pedigrees showing different patterns of inheritance.

heterozygote) may be shown by a dot in the middle of the symbol:

The typical pedigrees in Figure 19.1 illustrate ways of thinking about how a trait is inherited. The first question is whether the trait could be an X-linked recessive. Remember that such a characteristic appears almost exclusively in males, but females who get the allele from their fathers carry it in recessive form. Pedigree *(a)* shows this pattern. The fact that only males are affected is the first clue; the second clue is that affected males pass the trait to their grandsons, through their daughters. Only half the grandsons are affected, on the average, because their mothers have two X chromosomes and only one carries the allele for the trait.

X-linked inheritance, Section 16.13.

When tracing relatively rare conditions in a family, we always assume that people marrying into the family do not carry an allele for the condition. This is a safe assumption, except in small communities whose inhabitants may be highly inbred, such as rather isolated villages, or areas inhabited by long-established religious sects, such as the Hutterites and Mennonites in the United States.

Figure 19.2

A characteristic such as height is distributed according to a normal distribution curve. The largest number of people have the average height, with smaller and smaller numbers toward the extremes of shortness and tallness.

If pedigree analysis rules out an X-linked trait, the trait is autosomal, but it may be either dominant or recessive. The principal clue is the frequency of affected individuals. A dominant trait will obviously show up in every generation, as in pedigree *(b)*; a recessive will appear rarely, as in pedigree *(c)*, since it requires the marriage of two heterozygotes, which doesn't happen often. Unfortunately, the analysis of pedigrees is complicated by traits that don't appear as regularly as we expect; some traits, for instance, are said to have reduced **penetrance,** which means complicating factors keep them from appearing in all the individuals who have the appropriate genotype.

19.2 Quantitative characteristics may result from the action of many genes.

Some human characteristics are clearly due to a single gene, and their inheritance can be analyzed by means of pedigrees. For instance, the *T* gene determines the ability to taste a compound called phenylthiocarbamide (PTC). To tasters, who have the dominant allele *(T-)*, the compound is bitter; nontasters *(tt)* detect nothing. Other characteristics are clearly influenced by a number of genes, perhaps in a complicated way. Human eye color is a good example; it is a gross simplification to talk about a "gene" that determines brown eyes when it is dominant and blue eyes when recessive, for the color is really made by the interactions of several gene products.

Quantitative features like height appear to be so complex that one might wonder if they are genetically determined at all, but a casual look at the populace shows there must be a strong genetic component in determining height. Tallness or shortness very clearly runs in families, and the people of native tribes have characteristic average heights, from very tall Watusi to very short pygmies. Mendel, you'll recall, rejected quantitative features because there was no simple way to determine the phenotype of each individual unambiguously. That is, a person is not just either tall or short; height varies smoothly over a considerable range and in fact conforms to a *normal distribution* (Figure 19.2). This is the typical curve showing the number of people at each value of a continuous variable, and it is generated by intelligence tests and course examination results as well as measurements of size. How can we understand the inheritance of a quantitative characteristic?

Let's develop a model (Figure 19.3). Suppose people have only two genes that influence height, *A* and *B*, and that each copy of a dominant allele produces one unit of height. So the tallest person would have genotype *AA BB*, and the shortest, *aa bb*. As shown in Figure 19.3*a*, the possible combinations of alleles create five phenotypic classes: There is only one way to get all four dominant alleles or all four recessive alleles, four ways to get either one or three dominant alleles, and six ways to get two dominant alleles. Thus the middle height will be the most common, and the extremes will be the rarest. Now if we suppose, instead, that height is determined by three genes, *A*, *B*, and *C*, six phenotypes are possible, with frequencies of 1, 5, 10, 10, 5, and 1. As Figure 19.3 shows, each time another gene affecting height is added, the distribution becomes more like the normal distribution. (You may recognize the number of combinations in each class as the coefficients of a binomial distribution.) The actual distribution of heights can be understood, then, if height is a **polygenic trait,** the result of many genes, each making a small contribution to height. There may be no gene whose sole function is to make a person taller, but if many genes make small contributions to growth and if people carry different alleles for them, the population will show the continuous distribution of height that we observe.

The same general model fits other quantitative characteristics, including skin color and the results of intelligence tests when applied to any large population. Estimates of how much intelligence is influenced by genetic or by environmental factors vary substantially, but there is certainly good evidence for a considerable genetic component. Like height, intelligence must be a polygenic trait, and the normal distribution of intelligence can be understood on the basis of small contributions by many genes.

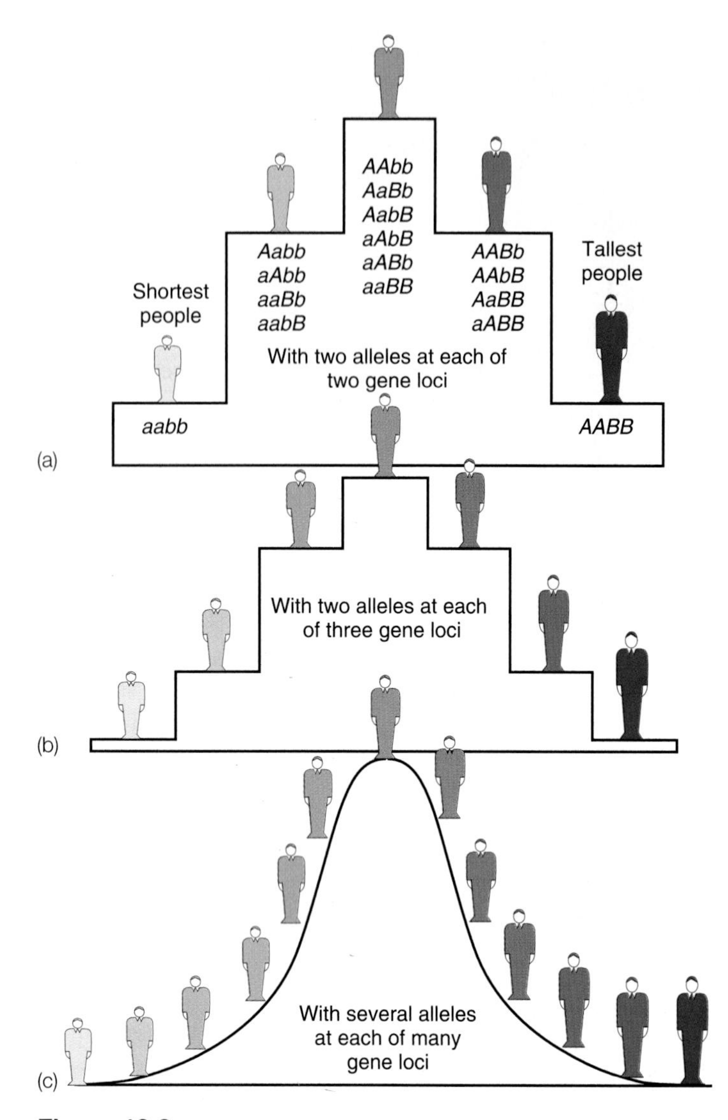

Figure 19.3

If a quantitative trait is determined by two genes, and each dominant allele contributes one unit of the feature, five phenotypic classes will be possible in the ratios shown *(a).* If the trait is determined by three genes, six classes will be possible. The relative numbers get closer and closer to a normal distribution as the number of genes increases *(b, c).* Thus a character such as height, with a continuous distribution, can be explained as the result of many genes.

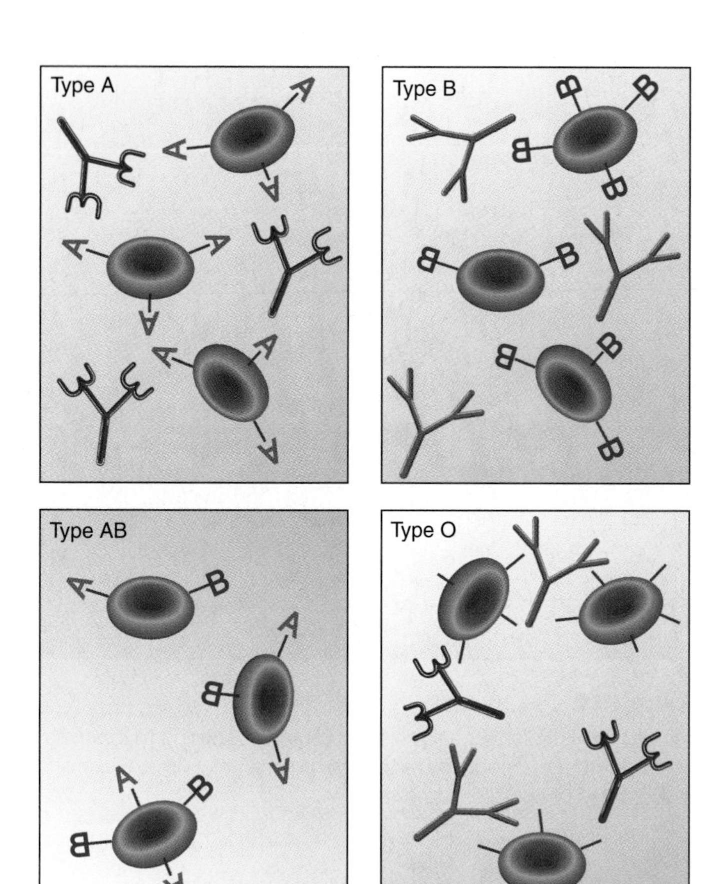

Figure 19.4

Red blood cells carry various antigenic specificities designated A and B. People normally cannot make antibodies (Y-shaped molecules) against their own antigens, but each person's blood serum does contain antibodies against antigens not carried on his or her own cells. Thus a person with type A blood has only anti-B antibodies, and vice versa.

19.3 Humans have ABO and other blood group antigens.

A wounded soldier lies on the battlefield losing massive amounts of blood and will soon die, while another who has just died lies nearby with liters of blood going to waste. What could be simpler than revitalizing the dying man with the dead man's blood? History records many cases of early attempts to transfuse blood; sometimes they worked, but more often they failed. As medicine became more rational and physicians developed the techniques for transfusing blood from one person to another, they discovered that some combinations are incompatible and lead to clumping and destruction of blood cells, tissue breakdown, fever, and sometimes death. In 1900, Karl Landsteiner observed that blood cells appear to have two possible specificities, denoted A and B, and he classified people into corresponding blood groups (Figure 19.4). Some individuals have type A blood and others type B, while still others have both (type AB) or neither (type O). We now know that A and B designate two types of glycoprotein antigens found on the surfaces of red blood cells (Sidebar 19.1).

Antigens and antibodies, Methods 8.1.

Everybody has antibodies directed against those antigens that are *not* found on their own blood cells: A person with type A blood has anti-B antibodies, and one with type B blood has anti-A antibodies. Blood cells clump, or **agglutinate,** when they

Sidebar 19-1 The ABO Blood Group Substances

The substances responsible for the ABO blood groups are oligosaccharides on surface glycoproteins of red blood cells. These sugar chains are made of four kinds of sugars: galactose (Gal), fucose (Fuc), *N*-acetylglucosamine (GNAc), and *N*-acetylgalactosamine (GalNAc). These monomers can be linked in many combinations to make distinctive antigens, but some humans lack one of the enzymes that assemble these chains, and therefore they make only partial chains.

The base of the ABO chain has the structure Gal-GNAc-Gal-GalNAc:

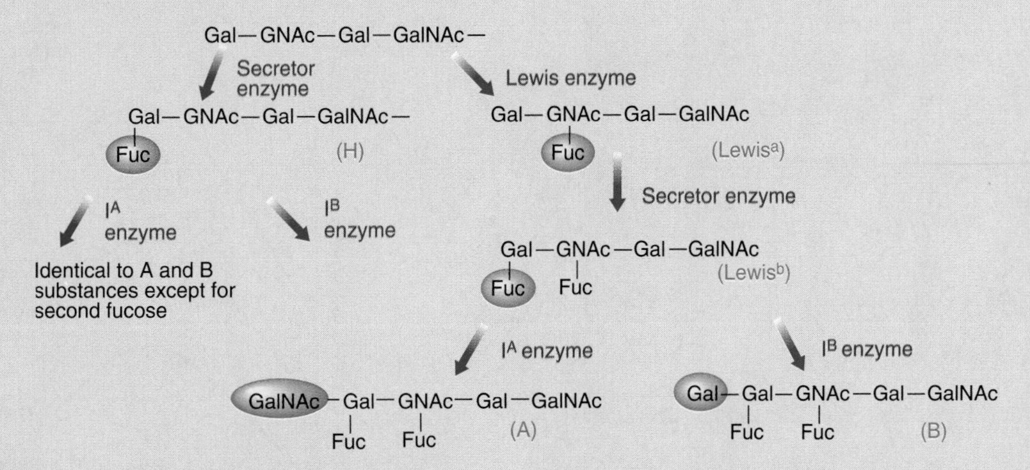

An enzyme encoded by the *Lewis (Le)* gene adds a fucose to this chain to make the Lewis[a] substance, but *le/le* homozygotes can't add this sugar. Next, the Secretor enzyme, encoded by the *Se* gene, adds a second fucose to make the Lewis[b] substance. Finally, the *I* gene enzyme adds a terminal sugar to any chain that has the second fucose residue, but there are two active forms of the enzyme. The I^A allele encodes an enzyme that adds another GalNAc, whereas the I^B allele encodes an enzyme that adds another Gal. People with blood type O lack the *I*-gene enzyme, so their red blood cells have only the Lewis[b] substance, which is not antigenic.

are mixed with antibodies against the antigens they bear. Since relatively few antibodies come into a recipient with transfused blood, and they are quickly diluted, the problem arises when the recipient has antibodies directed against the donor's cells. People with type AB blood, having neither anti-A nor anti-B antibodies, can *accept* blood of any type, while people with type O blood, having no A or B antigens on their blood cells, can *donate* to anyone (Figure 19.5). In practice, however, blood samples from the donor and the recipient are always mixed beforehand to avoid unforeseen problems.

Why do we have antibodies against blood cells at all? Most of us have never actually been exposed to the antigens that would induce the formation of these antibodies in the way we are exposed to other agents that induce immunity. One explanation is that the sugar chains on cell surface glycoproteins are similar to sugar chains on intestinal bacteria, so the antibodies we make against these bacteria also recognize the blood-cell glycoproteins.

Although the A and B blood types were first thought to result from two different genes, it soon became clear that a single gene, *I*, with three alleles, specifies the ABO antigens. I^A and I^B specify types A and B, respectively, and are **codominant,** which means they are expressed equally in an I^AI^B heterozygote to produce type AB, and they are both dominant to the recessive allele *i*, which specifies no blood type. With this information, you should be able to answer the following questions.

Exercise 19.1 What are the phenotypes of people with these genotypes? *(a)* I^AI^A *(b)* I^Ai *(c)* ii *(d)* I^AI^B *(e)* I^Bi

Exercise 19.2 What are the possible phenotypes of children from the following marriages? *(a)* $I^AI^B \times I^Ai$ *(b)* $I^Ai \times I^Bi$ *(c)* $I^Bi \times I^Bi$ *(d)* $ii \times I^AI^B$

Exercise 19.3 Fill in all the possible blood types of offspring from the marriages represented in this table, with the blood types of the mothers listed along the top and those of the fathers down the side.

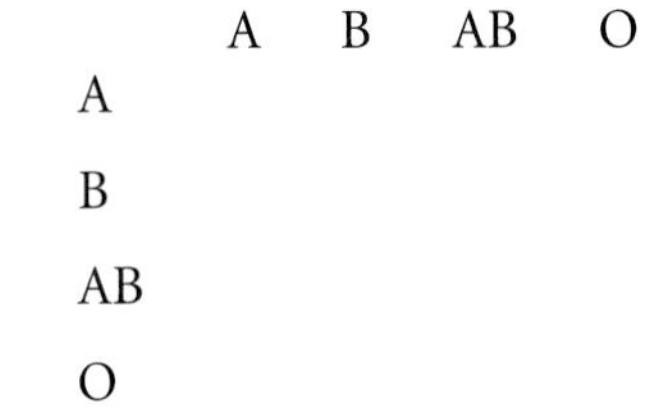

	A	B	AB	O
A				
B				
AB				
O				

Exercise 19.4 A woman with blood type A bears a child with blood type AB and accuses a man with type O of being its father. If you were the judge, would you hold the accused responsible for support of the child? (We should emphasize that blood typing can only be used to rule out certain possible fathers; it can never prove that a particular man is the father.)

The ABO system and the Rh system, which we discuss in the next section, are the most important blood type systems clinically, but other blood types have also been discovered,

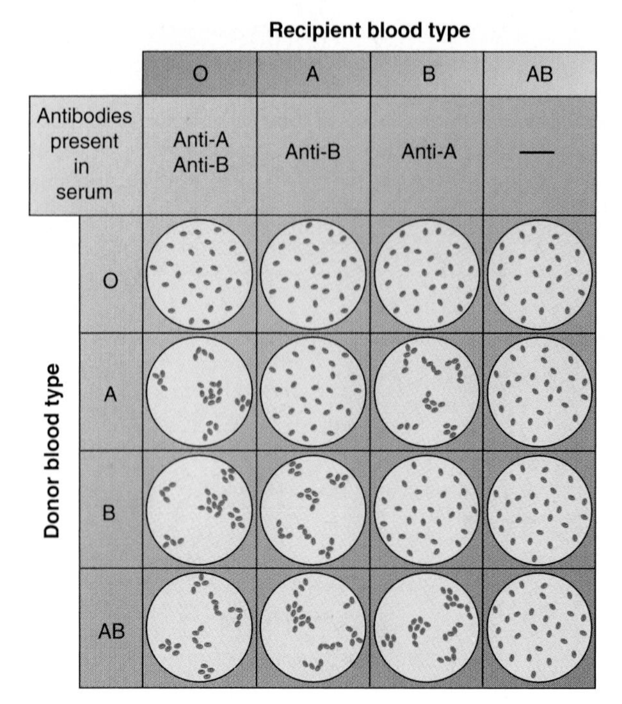

Figure 19.5

Incompatible transfusions result in the clumping of blood cells. Transfusions between people with identical blood types always work (unless complicated by some incompatibility outside the ABO system). People of type AB are universal recipients because they have no antibodies against either type. People of type O are universal donors because their cells do not bear either antigen.

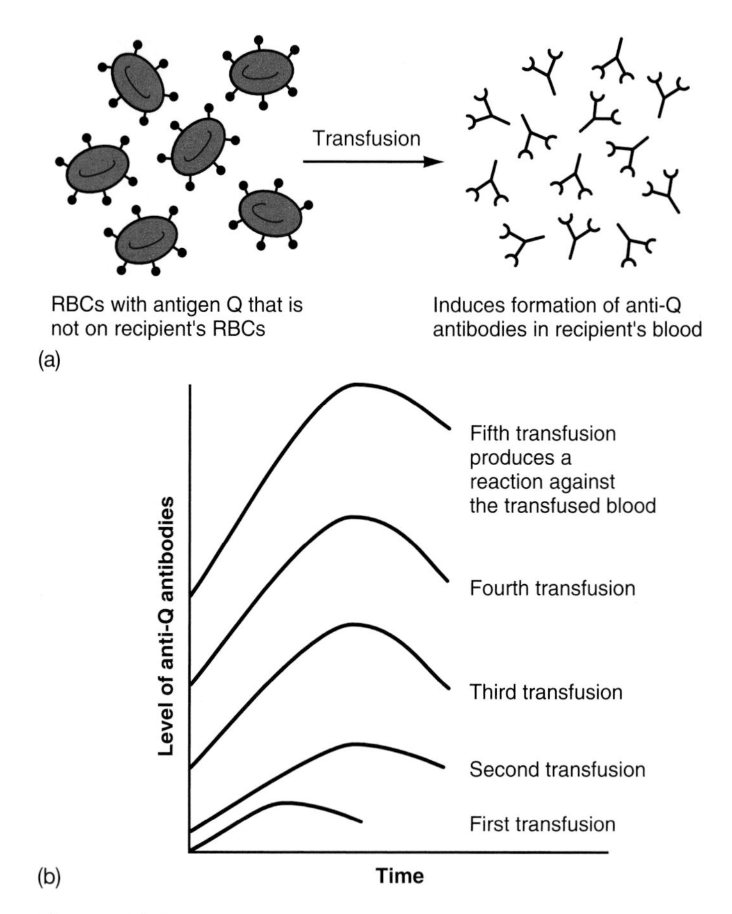

Figure 19.6

A patient with a previously unknown blood type, q, receives several blood transfusions from people who have the more common type, Q. *(a)* With each transfusion, his immune system is stimulated to make more and more antibodies against Q. *(b)* Eventually, the anti-Q concentration in his blood serum rises to the point where he reacts against another transfusion of Q blood. Physicians discover the Q-q blood-type difference while trying to determine the cause of the reaction.

showing how heterogeneous humans are. New blood types are usually found when a person develops an immune reaction after receiving several blood transfusions, indicating that he or she has a blood antigen different from those of the general population (Figure 19.6). For instance, the Duffy blood type was found when a hemophiliac by that surname reacted to some blood after a series of transfusions. The antigen is found in 65 percent of people who have been tested, and two alleles, Fy^a and Fy^b, are known.

The MN antigens form another well-known blood type system, but people normally don't carry antibodies against these antigens, so incompatibilities only arise after repeated transfusions. The M and N types, like A and B, are due to a pair of codominant alleles; every individual is either M, N, or MN. A related antigen results from a closely linked gene, *S*, which has a recessive allele, *s*.

These blood groups usually cause few clinical problems. However, the rhesus factors, which we take up next, can cause serious incompatibilities associated with childbirth.

19.4 Rh factors are associated with hemolytic disease of the newborn.

Before blood typing became common, some families experienced a tragic pattern of births. Their first one or two children would generally be born normal and healthy, but subsequent children were stillborn or became sick shortly after birth and often died. The sick or dead infants were usually swollen with excess fluid or had yellowish skin (jaundice); some had unusual numbers of immature red blood cells, or erythroblasts, a condition called **erythroblastosis fetalis.** All these disorders are now called **hemolytic disease of the newborn (HDN).**

The cause of HDN came to light in 1940, when Landsteiner and A.S. Wiener injected rabbits with the blood of rhesus monkeys and found the rabbits produced antibodies that agglutinate not only the monkeys' red blood cells but also the red cells of about 85 percent of humans. The common antigen in human and rhesus blood is called the **rhesus antigen (Rh).** In fact, a group of three closely linked genes encode the rhesus antigens, but for simplicity, everyone is said to be rhesus positive (Rh^+) or rhesus negative (Rh^-).

HDN results from incompatibility between an Rh^- mother and Rh^+ fetuses who get their Rh antigens from an

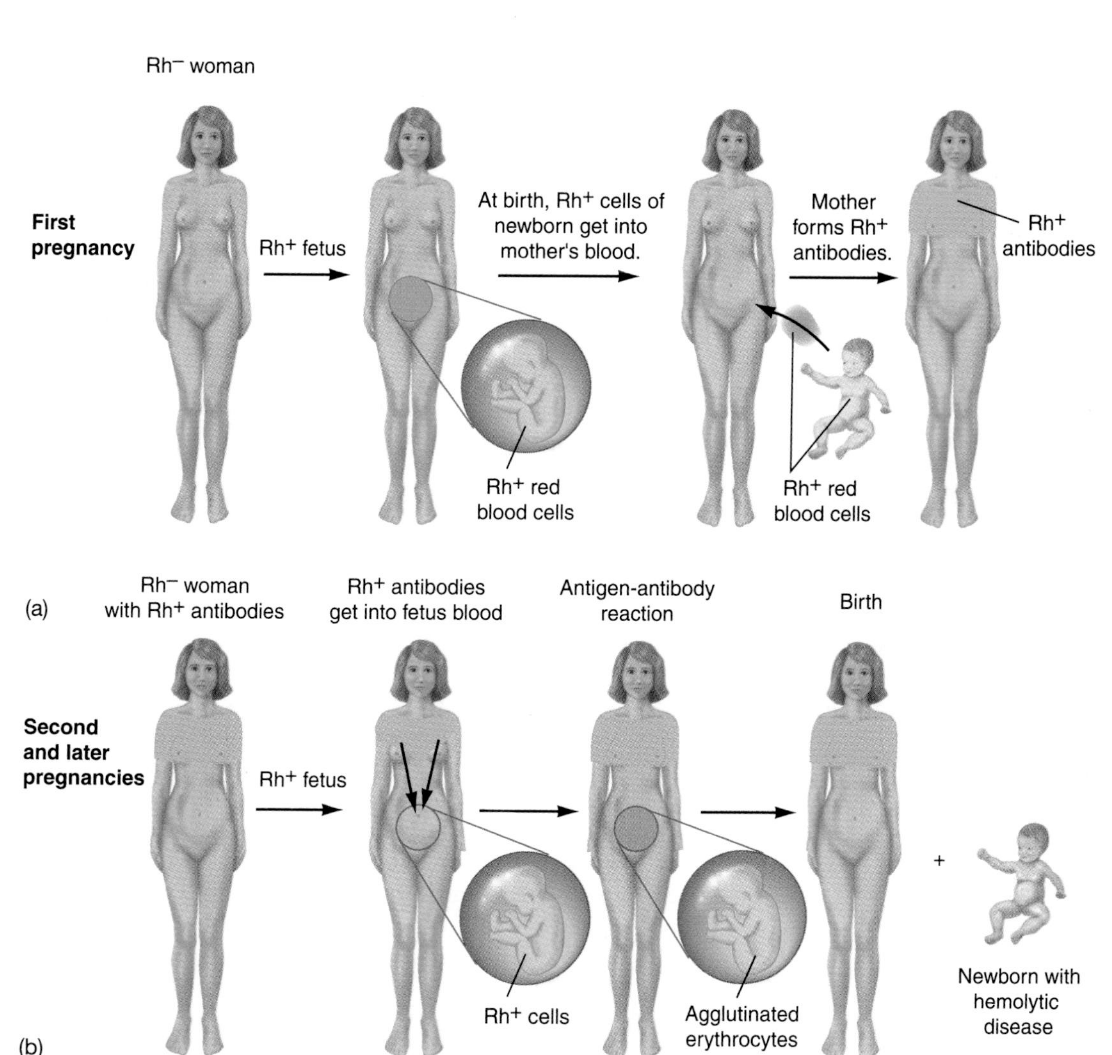

Figure 19.7

(a) Hemolytic disease of the newborn (HDN) develops when an Rh^- woman bears an Rh^+ baby and receives some of the baby's blood. She begins to make antibodies against the Rh factor, so in subsequent pregnancies *(b),* some of her anti-Rh antibodies can get into the fetus, where they make red blood cells clump together (agglutinate); this either kills the fetus or produces a baby with HDN. (Source: *Human Genetics,* 2/e by Novitski, ©1982. Adapted by permission of Prentice-Hall, Inc., Upper Saddle River, NJ.)

Rh^+ father (Figure 19.7). Even though a fetus is nourished by a transfer of materials from its mother's bloodstream through the placenta, the two blood circulations are quite separate. But although blood cells normally do not cross the placental membranes, some maternal antibodies do. A mother may occasionally get some of the baby's blood into her own bloodstream during pregnancy, through a little cross-bleeding, but she is most likely to be exposed during birth, when there is normally considerable bleeding. This is when an Rh^- mother giving birth to an Rh^+ child starts to become immunized against her baby's Rh antigens. Then, in a second or later pregnancy, the anti-Rh antibodies she has developed can pass into the fetus's circulation, react against its red blood cells, and produce HDN.

Because most people are Rh^+, couples are usually Rh-compatible; the rate of HDN without medical intervention is about one in 200 births. Incompatible couples can now avoid the problem in several ways. **Amniocentesis,** a procedure in which amniotic fluid is withdrawn from a pregnant woman for analysis (Figure 19.8), can detect a fetus at risk because the fluid will contain excessive amounts of bilirubin, a breakdown product of red blood cells. Such babies can be delivered prematurely, or if born full-term, they can be saved by a massive transfusion to replace all the Rh^+ blood with Rh^- blood. Moreover, the whole problem can generally be avoided altogether by preventing an Rh^- mother from producing anti-Rh antibodies. If such a woman gives birth to an Rh^+ child, she is given an injection of anti-Rh antibodies within 72 hours[1] after birth. These antibodies then combine with any fetal red cells that may have gotten into her blood, coating them so they can't start to immunize her. This procedure, which must be repeated with each pregnancy, is very effective, and in clinical trials it has reduced the fraction of women reacting against their Rh^+ children's blood from 7 percent to 1 percent.

[1]The 72-hour limit was originally chosen for a nonmedical reason, as Edward Novitski notes in his book *Human Genetics.* The clinical trials on suppressing antibody formation were originally made with male inmates of Sing Sing Prison in New York. The investigators wanted to simulate birth conditions by injecting the suppressing antibodies some time after injecting Rh^+ red blood cells, and they decided that 72 hours between their visits to the prison would minimize the danger of the inmates incorporating their schedule into an escape plan or using it as an excuse to riot.

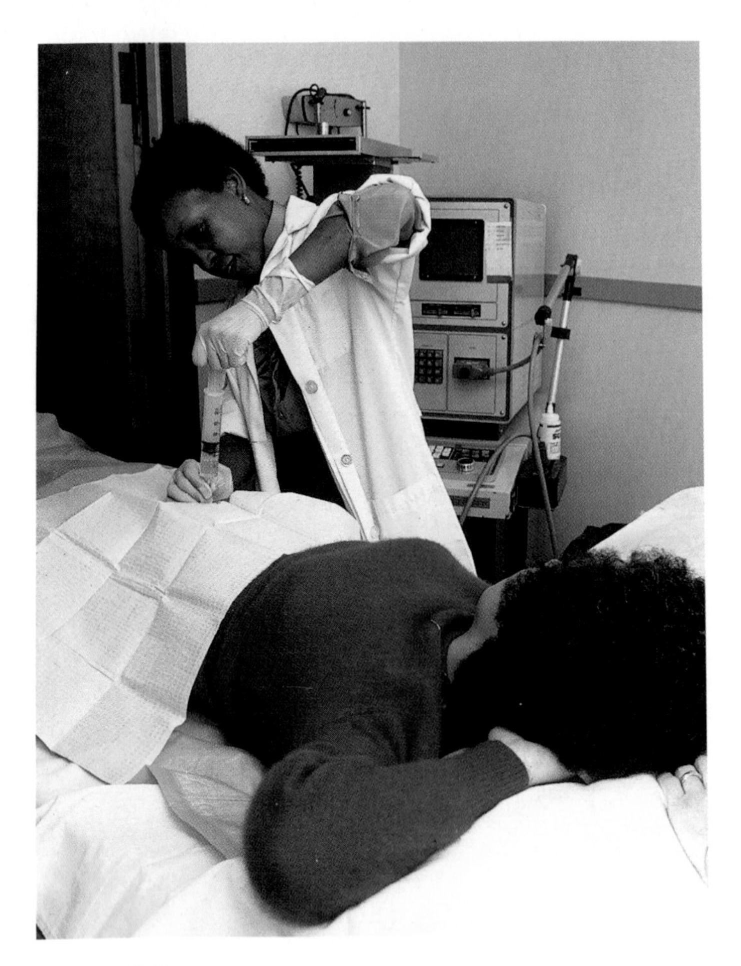

Figure 19.8

In amniocentesis, a sample of amniotic fluid from the sac surrounding the fetus is drawn off for analysis. Biochemical tests for certain genetic abnormalities can be performed on it, and the fluid contains fetal cells that can be karyotyped to look for chromosomal abnormalities.

Exercise 19.5 The Kell blood group was discovered in a family by that name when one of their children was born with HDN, even though mother, father, and baby were all Rh^+. The mother's blood had reacted against red cells from both her husband and her baby. What factor in this family's particular genetic makeup would explain the mother's reaction?

19.5 Human chromosomes are distinguished by their banding patterns.

The human karyotype, as revealed by the methods described in Section 15.4, has been well known for many years. However, some of the chromosomes visualized in the classical way can be confused with one another, and it is hard to see small changes that might occur in their structure through mutation. Fortunately, chromosomes acquire distinctive banding patterns (Figure 19.9) when they are treated with a fluorescent dye, either quinacrine mustard (Q banding) or Giemsa stain (G banding). Chromosomes treated in this way can be identified unambiguously, and it is then easy to identify the ones that have had significant alterations, such as large deletions or the transfer of part of one chromosome to another (a translocation). Furthermore, these bands can be used to locate genes of interest or to specify abnormalities, using a standard two-digit numbering system established by an international conference in Paris in 1971. By convention, the short arm of each chromosome is called p and the long arm, q. Since the number of genes in the whole haploid set has been variously estimated at 10,000–50,000, the average chromosome may have a thousand genes. Each gene therefore occupies only a minute part of even the narrowest band. We will discuss methods for locating a gene later in this chapter.

19.6 Chromosomes sometimes fail to separate in meiosis.

One class of serious genetic problems arises from occasional failures of meiosis. The separation of chromosomes or chromatids during a meiotic anaphase is called **disjunction,** so a failure to separate is termed **nondisjunction** (Figure 19.10). Normal development requires a certain gene balance that is provided by two copies of each chromosome (except for the sex chromosomes). Nondisjunction produces gametes with extra or missing chromosomes, and zygotes formed from such gametes are **aneuploid,** having an abnormal number of chromosomes. Usually they are either **monosomic,** with only one copy of a particular chromosome, or **trisomic,** with three copies. Most of these zygotes, including all monosomics in humans, cannot develop at all beyond the early stages, and so are aborted spontaneously. Of all human conceptions, 15–20 percent are aborted spontaneously, and at least half of these have a recognizable aneuploidy, including trisomies of almost any autosome. Only humans with trisomies of chromosomes 13, 18, and 21 can live for varying periods of time after birth.

The most common result of nondisjunction in humans is **Down syndrome,** formerly called mongolism. A syndrome is a set of abnormalities that typically occur together. People with Down syndrome have subnormal intelligence and a characteristic facial appearance, including a folding of the eyelid reminiscent of the epicanthic fold of Asiatic people (but not really the same). A karyotype of their chromosomes shows that they have three copies of chromosome 21 (Figure 19.11*a*), so the condition is also known as **trisomy-21.** With intensive early therapy, Down children are now doing much better than anyone had previously thought possible; they can generally learn to become somewhat independent and to hold down simple jobs as adults.

The nondisjunction that produces Down syndrome and other trisomies becomes more likely as the mother ages. Mothers younger than 30 have 0.04 percent Down babies; this rate increases to 0.11 percent for ages 30–35, 0.33 percent for ages 35–40, 1.25 percent for ages 40–45, and 3.15 percent for moth-

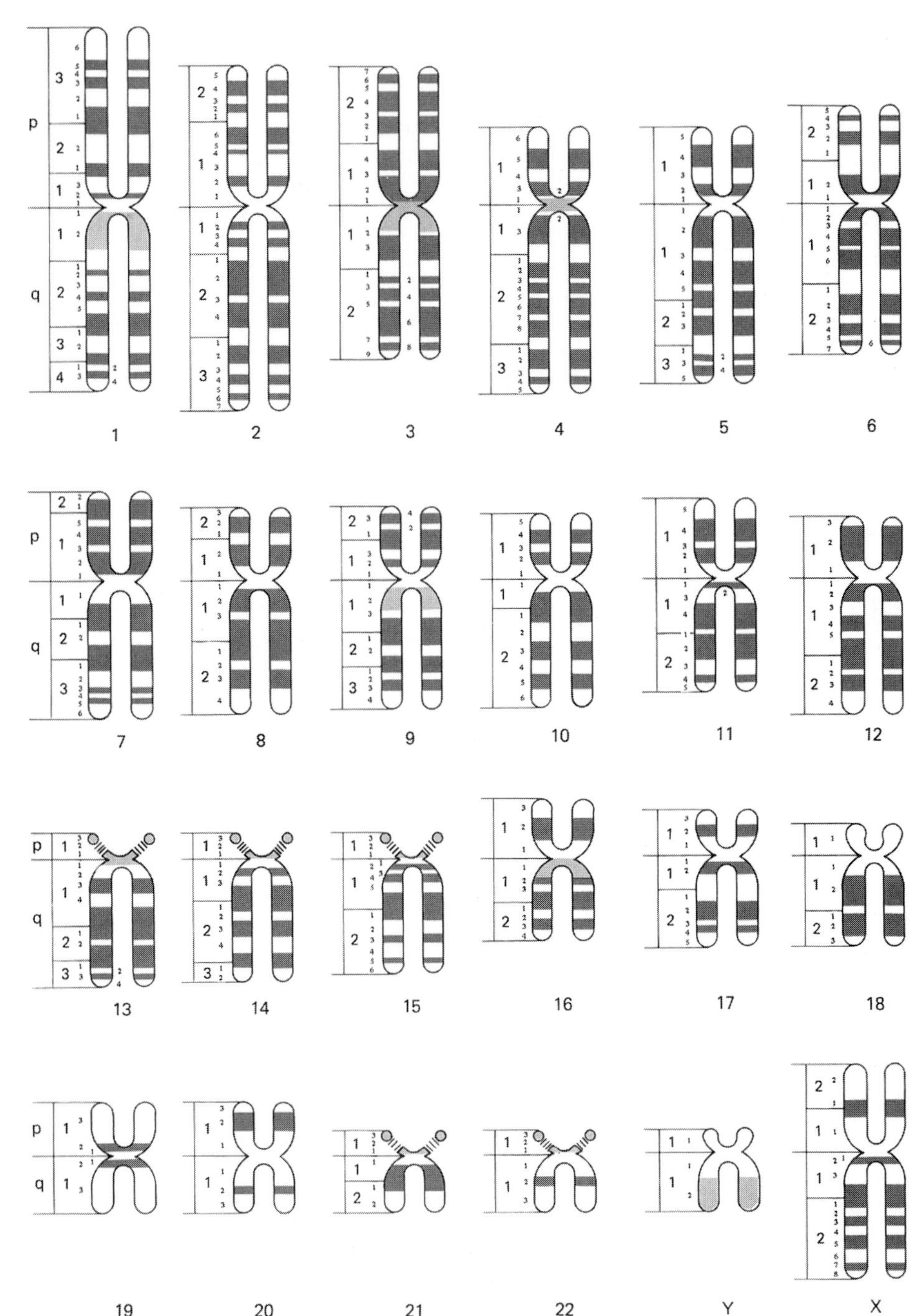

Figure 19.9

This standard human karyotype shows the positions and numbers of the bands. The short arms are designated p, and the long arms, q. The bands are specified by two digits, so the light band at the end of the long arm of chromosome 5 is 5q35. (Source: *Birth Defects: Orig. Art. Ser.,* ed. By D. bergsma, Vol. VIII (7), 1972, published by Williams & Wilkins Co., Baltimore, for The March of Dimes Birth Defects Foundation.)

ers over 45. This correlation is unexplained, except that it appears to be related to the extremely extended period of gametogenesis in women. Even before a baby girl is born, her ovaries have developed, and all the eggs have gone through the early stages of meiosis and then stopped. The eggs resume development one by one years later when she becomes sexually mature and her ovulatory cycle begins. As the process is prolonged with age, there is simply more time for something to go wrong in meiosis.

However, about 4 percent of Down syndrome cases come from families with an inherited abnormality due to a rearrangement of certain chromosomes. Rarely, a chromosome set shows a **translocation,** in which a piece of one chromosome becomes attached to another. Some people carry a 14/21 translocation, meaning that part of chromosome 21 is attached to chromosome 14 (Figure 19.12). People with only one other chromosome 21 are perfectly normal, even though they have only 45 chromosomes, but their abnormal chromosomes can separate in six ways during meiosis. Assuming that one of the resulting gametes combines with an ordinary gamete, carrying one copy each of chromosomes 14 and 21, three of the six types of zygotes will die, one will be normal, one will become a carrier like the parent, and one will become a person with Down syndrome.

19.7 Aneuploids for sex chromosomes are relatively common.

While most fetuses with autosomal trisomies die before birth, infants with trisomy-13 and trisomy-18 have occasionally lived for a short time after birth. People with unusual numbers of sex chromosomes are much more common. They generally look quite normal despite some abnormalities that turn up in a clinical examination. For instance, about one girl in 3,000 has **Turner syndrome.** These individuals are short, have a broad chest with widely spaced nipples, may have a somewhat webbed neck, and fail to mature sexually even though they have female genitals (Figure 19.13*a*). Karyotyping shows such a person to be 45 XO, which means that she has 45 chromosomes, including only one X, because one of her parents produced a gamete with no sex chromosome at all (Figure 19.13*b*). Some other women have just the opposite condition—three or more X chromosomes. They are mentally retarded and may or may not be fertile.

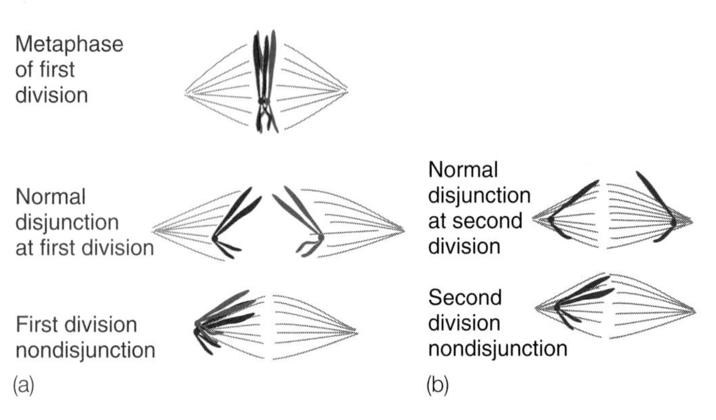

Figure 19.10
Nondisjunction is a failure of chromosomes to separate properly at a meiotic anaphase. *(a)* At the first meiotic division, nondisjunction means both homologs go to one daughter cell. *(b)* At the second meiotic division, nondisjunction means the chromatids of a chromosome fail to separate. Nondisjunction at one division can produce gametes bearing two copies of one chromosome or no copies; if it happens at both divisions, it can produce gametes with four copies of a chromosome.

About one young man in 400 is found to have **Klinefelter syndrome.** These men are generally tall, mentally retarded, sexually underdeveloped, and frequently have some female breast development (Figure 19.14). Karyotyping shows these men to have genotypes such as 47 XXY, 48 XXXY, or even 49 XXXXY. This syndrome is also the result of nondisjunction.

The fact that people with these syndromes are relatively normal raises questions about gene balance in ordinary people. Since normal males and females have different numbers of X chromosomes, there must be a mechanism to compensate for the difference in the number of copies of X-chromosome genes and still maintain a genetic balance. In 1961, Mary Lyon and Liane Russell independently presented a model for compensation of X-linked genes in which each noted that heterozygous females often show a variegated phenotype; calico cats, for instance, are heterozygotes marked by randomly distributed patches of black and yellow fur (Figure 19.15). Lyon and Russell proposed that each cell uses only one of its two X chromosomes, so in every cell of a developing female embryo, one X chromosome is inactivated at random, and the same chromosome remains inactive in all the cells derived from that embryonic cell (Figure 19.16). In the heterozygous cat, the X chromosome with the black-fur allele is

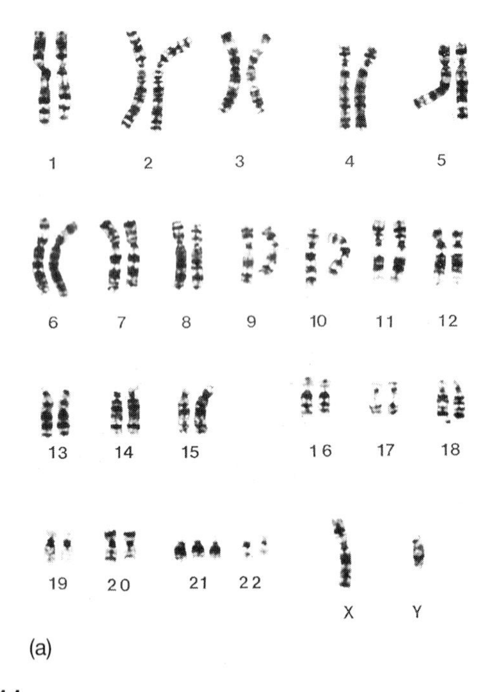

Figure 19.11
(a) Down syndrome is due to trisomy for chromosome 21, as shown by the karyotype. *(b)* Down-syndrome children have certain typical physical characteristics and slight to severe mental retardation.

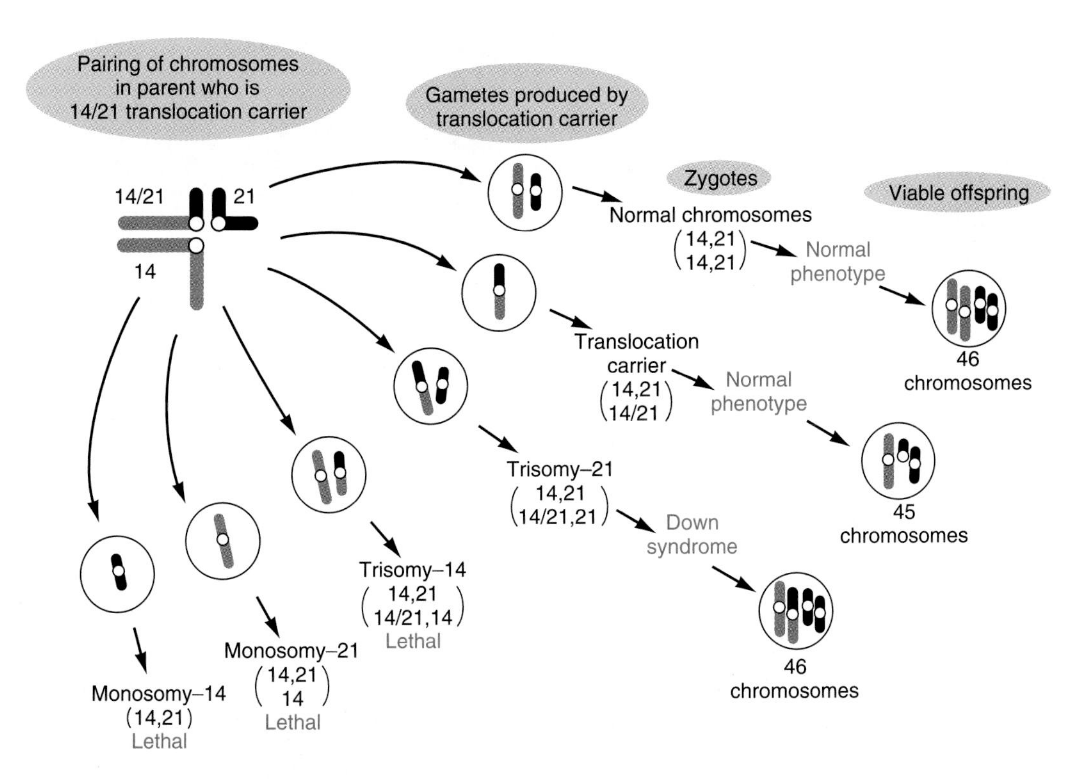

Figure 19.12

People with a 14/21 translocation (that is, a portion of chromosome 14 attached to chromosome 21) carry the potential for producing children with Down syndrome. Segregation of the chromosomes is rather complicated and can happen in six ways. Given a fertilization with a normal 14, 21 gamete, half the combinations are lethal, and one of the six leads to Down syndrome.

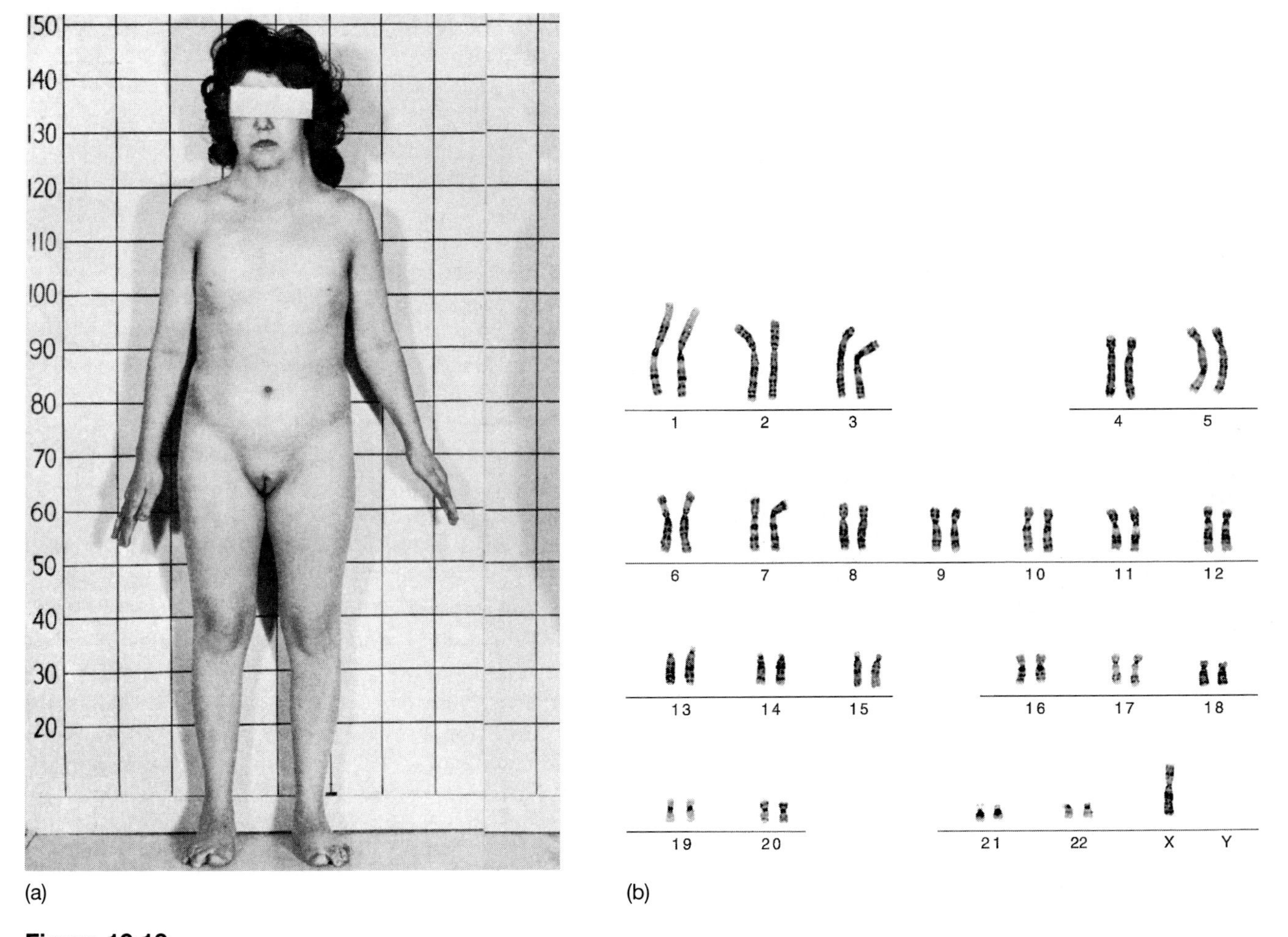

Figure 19.13

(a) This young woman has Turner syndrome (45, X) as shown by her karyotype *(b)*.

Figure 19.14

(a) This young man originally showed the typical appearance of Klinefelter syndrome, with a somewhat feminized body, including breast development. *(b)* After breast reduction surgery and three years of therapy with the masculinizing hormone testosterone, his body assumed a more typical masculine form.

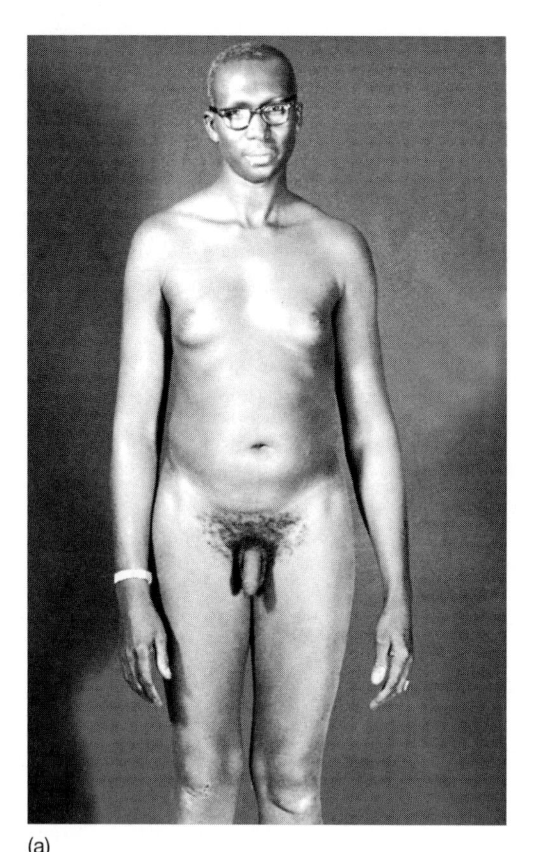

(a)

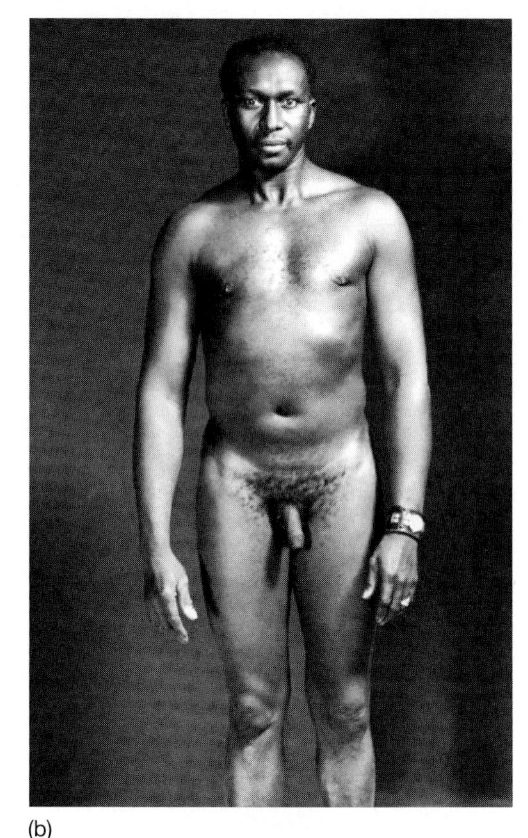

(b)

Figure 19.15

A calico cat has distinct patches of black and yellow fur. She is a heterozygote carrying alleles for these fur colors on her X chromosomes, and in each patch of skin, only one of these genes is expressed.

turned off in some skin cells, which then grow into yellow fur patches, while the chromosome with the yellow-fur allele is turned off in other skin cells, which develop into black fur patches. Although this pattern of X-inactivation is most obvious in the fur of female cats and mice, every female mammal is a mosaic of two cell types, and any allelic difference between her X chromosomes may show up as a phenotypic difference.

Lyon and Russell hypothesized that an X chromosome is turned off by being condensed into a tight bundle. We showed in Section 18.10 that genes in highly condensed heterochromatic regions are not expressed, and the inactive X chromosome is effectively heterochromatized. Compact X chromosomes, which can be seen in the cells of normal females, are called sex chromatin or **Barr bodies,** after Murray Barr, who discovered them (Figure 19.17). A cell typically has one less Barr body than the number of X chromosomes—that is, one X is left functional and the rest are condensed. So a normal woman has one Barr body in each cell, a woman with Turner syndrome has none, and women with extra X chromosomes have two, three, or even four. Men normally have no Barr bodies, but those with Klinefelter syndrome show one, two, or more, depending on the number of X chromosomes in their cells. The fact that X-aneuploids are, on the whole, so normal shows that the mechanism of X-chromosome inactivation works quite well, leaving only one functional X chromosome, but there is still enough of a gene imbalance in aneuploids to create some abnormalities.

A few men have the genotype XYY. When they were discovered, Patricia Jacobs and her colleagues pointed out that an un-

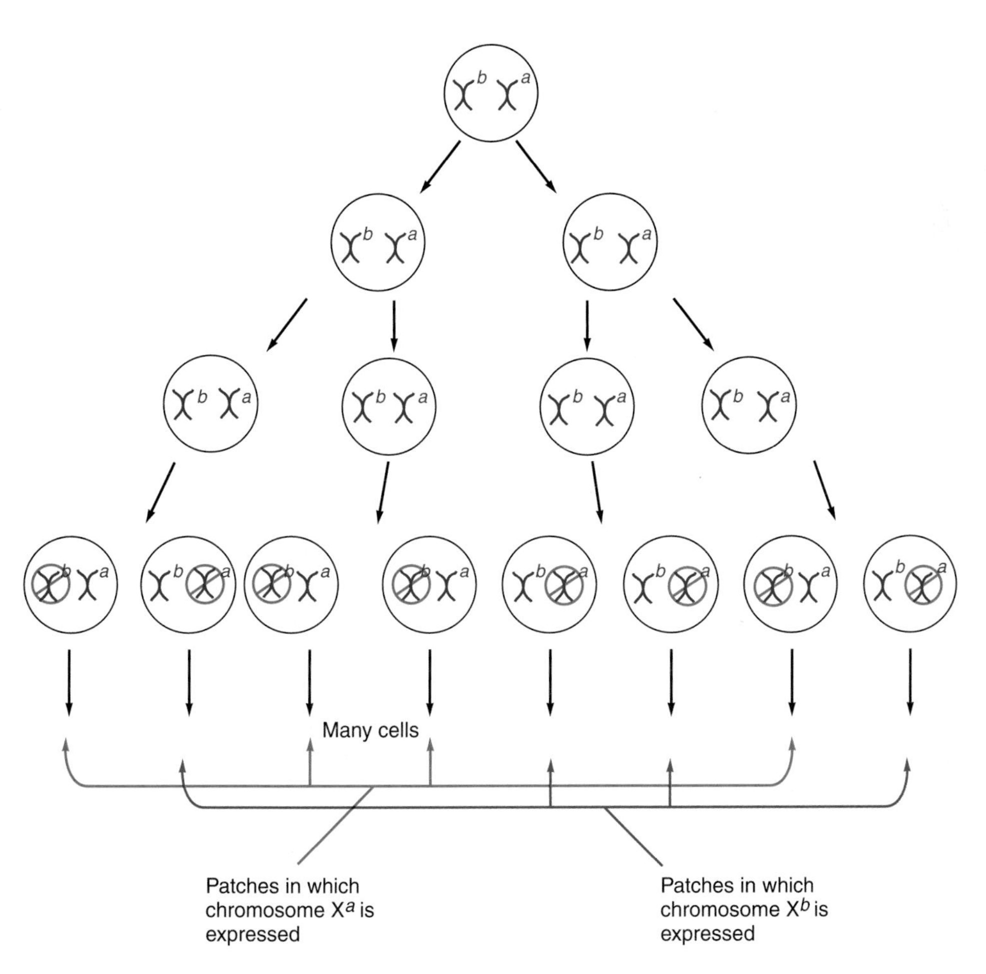

Figure 19.16
In a female embryo with two X chromosomes, one X is randomly selected in each cell and made into a compact structure that does not express its genes. The same chromosome remains inactivated in the entire clone of cells derived from this embryonic cell, so a heterozygote will have patches of cells with different phenotypes.

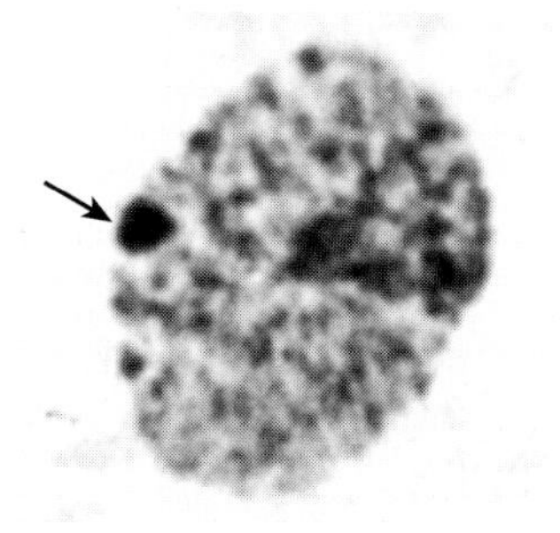

Figure 19.17
A Barr body *(arrow)* is visible at the edge of the nucleus in cells scraped from the inside of the cheeks and properly stained.

usually high proportion of the men in penal institutions are XYY, although most of them are not in prison and live quite ordinary lives. Some critics claimed that genes on the Y chromosome tend to create aggressive and antisocial behavior, and in the midst of the first excitement over this discovery, several XYY men on trial for murder used their genotypes as a legal defense, claiming they had been compelled by powerful genetic forces. One man in Australia was acquitted on this basis, and others had their sentences reduced. However, the relationship between the Y chromosome and human behavior is more complicated and problematical than it once appeared. XYY men tend to be tall, to have somewhat reduced intelligence, and to be emotionally immature and rather compulsive, showing the kind of behavior that is likely to get them into trouble. But since most XYY men behave normally and have intelligence in the normal range, the correlation between this genotype and unusual behavior remains unclear.

Exercise 19.6 A woman with normal vision has a color-blind daughter with Turner syndrome. Explain this.

Exercise 19.7 A condition called anhidrotic ectothermal dysplasia results from an X-linked allele. Men with this allele have no teeth and no sweat glands in their skin. Heterozygous women have no teeth in portions of their jaws and large islands of skin lacking sweat glands. Why?

19.8 Human genes can be localized to chromosomes by making hybrid cells.

X-linked genes are relatively easy to identify in humans because they are inherited in such distinctive patterns, but most genes are autosomal. Since humans aren't bred and mated like fruit flies or mice, the classical techniques for mapping human genes are generally limited to unusual situations. First, one parent must be homozygous recessive for two linked genes *(aa bb)*, and the other parent heterozygous for both *(Aa Bb)*. Then they either must have a large number of children, or we must find several such families and sum the results from all their children. With each child, such parents are effecting a testcross that measures the linkage between *A* and *B*, and if data from enough offspring can be obtained, the two genes can be mapped. This method has been used to show that the ABO blood group locus is about 10 map units from a gene for nail-patella syndrome, which produces abnormal connective tissue (ligaments, bones, and similar materials).

The principal reason for wanting to locate many human genes precisely is that most genes are identified by the disorders they produce when mutated, disorders that are sometimes terribly debilitating and cause untold misery. Human geneticists try to locate and identify these genes in order to identify the proteins they encode and study the physiological problems they produce. This research can reveal ways to correct the problems and perhaps even correct the mutated gene.

Although we can't ethically manipulate people, we can manipulate their cells just like those of any other organism. Some genes can be localized by making hybrid cells, fusing somatic human cells in tissue culture with those of other animals. (As we saw in Section 8.7, Frye and Edidin used this method to demonstrate the fluidity of membranes.) The nuclei also combine, and at first the two sets of chromosomes go through the cell cycle together, so the hybrid cells have a full set of both human and mouse chromosomes. The chromosomes of one species or the other tend to be lost as the cells continue to grow in culture, perhaps because the chromosomes of each species are going through a cell cycle on a different schedule and the cell divides before the slower chromosomes are prepared to disjoin. When human cells are fused to mouse cells, the human chromosomes are lost (Figure 19.18), but when mouse cells are fused with hamster cells, the mouse chromosomes are lost.

These odd circumstances, which were discovered quite accidentally, make it possible to assign many human genes to particular chromosomes. Mouse cells lacking the enzyme thymidine kinase (TK) have been used to locate the human gene for this enzyme (Figure 19.19*a*). After hybridizing mouse and human cells, human chromosomes are lost at random, so we can find clones with various combinations of human chromosomes and test each one for the ability to synthesize TK. As Figure 19.19*b* shows, only clones that carry human chromosome 17 make TK, thus localizing the TK gene to this chromosome.

In addition to losing entire chromosomes, mouse-human hybrid cells lose fragments of human chromosomes at unusually high rates, or they translocate human segments to mouse chromosomes. Cell lines with these chromosomal abnormalities are even more valuable for localizing human genes than those in which entire human chromosomes have been lost. Cells that make TK and retain only parts of chromosome 17 can be used to localize the TK gene still more narrowly. Many hybrid cell lines, containing various bits of the human genome, are now kept in laboratories for use in locating human genes.

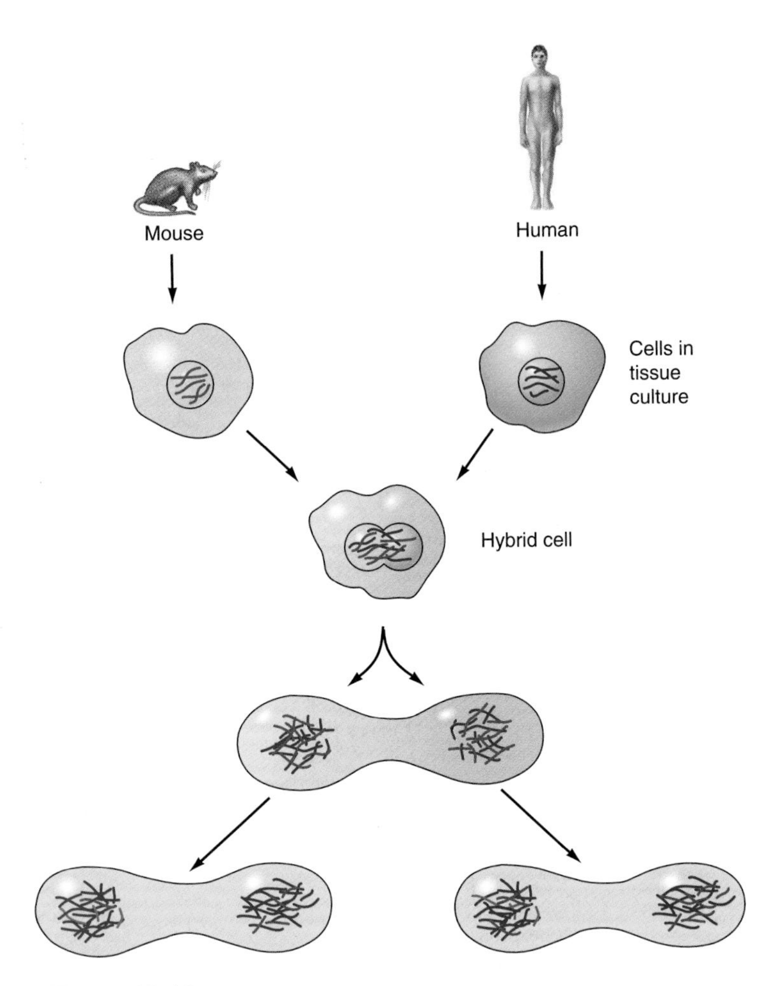

Figure 19.18
Hybrid cells can be made by combining mouse and human cells in tissue culture. The initial hybrid has a full set of both mouse and human chromosomes, but as cells multiply, they gradually lose human chromosomes at random.

19.9 Human genes can be localized to chromosomes by linkage disequilibrium.

Not all interesting genes can be located with cell hybrids, and even if hybrids help, it is important to have additional evidence for the position of a gene locus. A second method comes from studying families with particular traits, including genetic disorders.

Suppose that in the human population, 90 percent of the copies of a particular chromosome carry allele *A*, and 10 percent carry allele *a*. In this case we say that the **allelic frequency** is 0.9 for the *A* allele and 0.1 for the *a* allele. Suppose, also, that several generations ago, someone with an *a* allele acquired a mutation, *b*, in gene *B* near the *a* locus, and that *bb* homozygotes have a genetic disorder. As this *a b* chromosome was inherited over many generations, spreading the *b* mutation through the population, some *A b* chromosomes were made by crossing over (Figure 19.20). If recombination were to go on for a long time, the *b* allele would reach a **linkage equilibrium**—that is, it would be randomly distributed, linked to *A* in 90 percent of the chromosomes and to *a* in

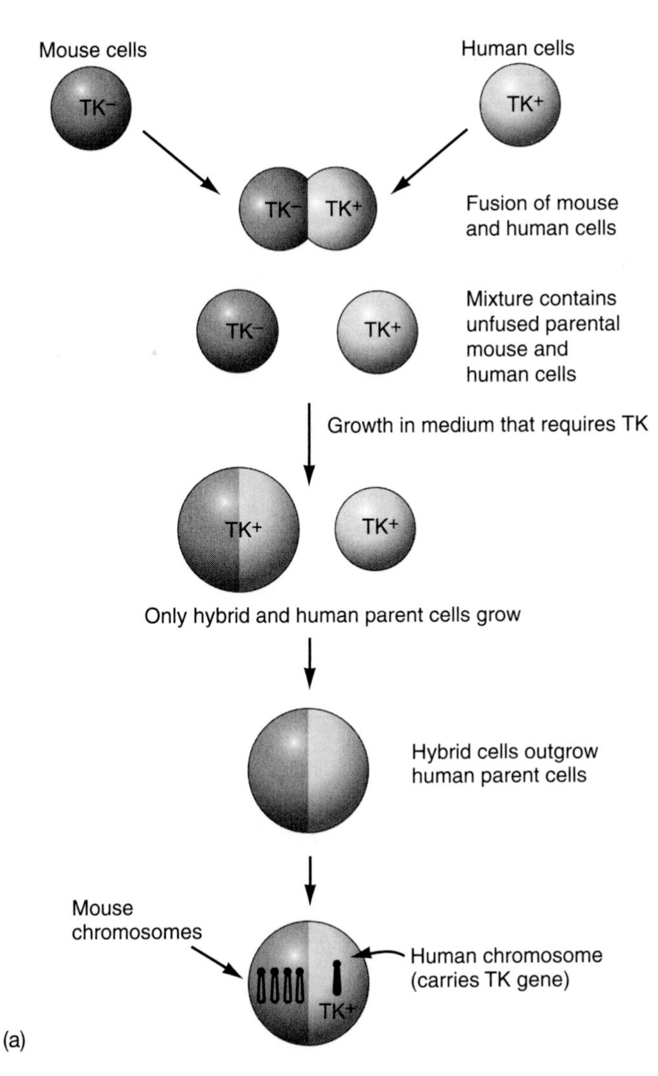

Mouse/human clone	Human chromosomes persisting in the clone	Presence of thymidine kinase
A	5,9,12,21	
B	3,4,17,21	
C	5,6,14,17,22	
D	3,4,9,18,22	
E	1,2,6,7,20	
F	1,9,17,18,20	

(b)

Figure 19.19

(a) Human cells are hybridized with mouse cells deficient in thymidine kinase. Various clones that are isolated later are tested for the presence of the enzyme and for the human chromosomes that remain. *(b)* The pattern shows that the gene for this enzyme is on chromosome 17.

only 10 percent. However, if there isn't enough time to achieve equilibrium, *b* will remain linked more often to *a* than to *A*, and there will be a **linkage disequilibrium.** So if we find an unusually high frequency of the *a* allele among victims of the disorder (*bb* individuals) and in obligate heterozygotes (*Bb* carriers of the disorder), we can conclude that *b* is linked to *a*. If we know what chromosome has the *A* locus, we can infer that it also has the *B* locus.

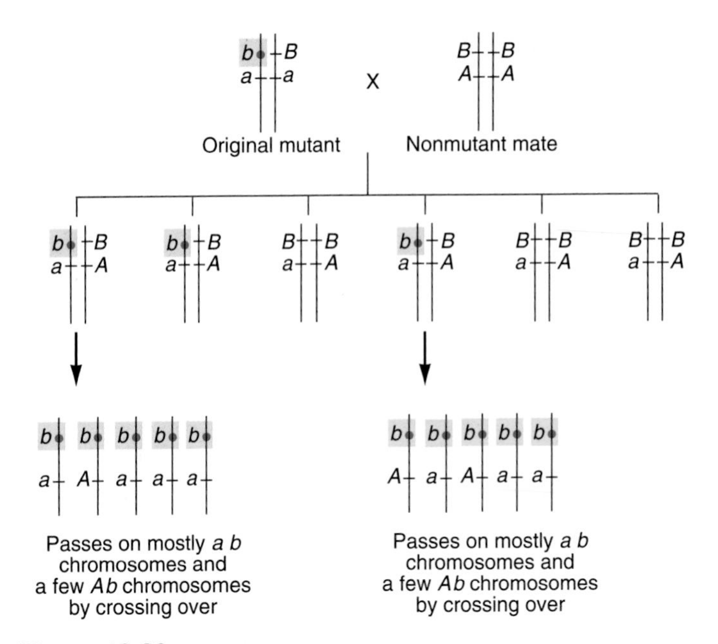

Figure 19.20

Given enough time, a rare allele, *b,* will be combined at random with the alleles of a nearby locus, alleles *A* or *a*. However, if not many generations have passed since the *b* allele appeared, it will be linked more often to the allele to which it was originally linked (here, *a*). This is a linkage disequilibrium.

Analysis of linkage disequilibrium has been an important technique in mapping the genes for genetic disorders when the functional basis of the disorder is unknown. However, it requires the occurrence of some gene on the same chromosome that can be used as a marker. Such genes are rare. Fortunately, we can use another kind of genetic marker that may have no functional consequence.

Mutations are always occurring in human chromosomes, and even if they occur in noncoding DNA where they produce no phenotypic change—between genes and in introns—they can be detected by changes in the patterns of cuts made by restriction endonucleases, because these enzymes locate specific DNA sequences. Suppose we collect DNA samples from a number of unrelated people and cut this DNA with an endonuclease—say, *Sma*I. We separate the fragments on a gel, and then, as described in Section 18.11, probe them with a labeled messenger RNA for the β hemoglobin chain (Figure 19.21). The result is that different people have different patterns of labeled DNA fragments, not because their hemoglobins are different but because they have different *Sma*I sites in and around the β hemoglobin gene. These differences are known as **restriction fragment length polymorphisms** (**RFLPs,** pronounced "rif-lips"). (A population is polymorphic when its members have more than one form of some character, such as black, brown, blond, and red hair.) Investigators have now identified about 1,000 RFLPs in the human genome, and they are routinely used for mapping genes. They are also used in police work for so-called DNA fingerprinting as evidence that a suspect committed or did not commit a crime (Sidebar 19.2).

Sidebar 19.2 Identifying People

Police investigations received an enormous boost with the discovery that people have distinctive fingerprints, and they were boosted further with the discovery of blood groups that can often be identified in bits of evidence from a crime scene. Now, as a result of recombinant-DNA research, the law has a still more powerful forensic tool that makes it possible to characterize each person by his or her unique genome. The British molecular biologist Alec Jeffreys discovered an important class of RFLPs called *VNTRs (variable number of tandem repeats)*, which are now used routinely in forensic work.

A RFLP results from a small change in the DNA sequence that alters the cut site of a restriction endonuclease. The human genome is full of sites where VNTRs, short sequences 15 to 100 nucleotides long, are repeated anywhere from about 4 to 40 times. These VNTRs are very distinctive. If a DNA sample is cut with a restriction enzyme that does *not* cut within the repeated sequence itself, many DNA fragments of different sizes will be produced because of the variable numbers of repeats attached to each one. Then if a probe is available for the repeat, Southern blotting will produce a distinctive banding pattern, now called a *DNA fingerprint:*

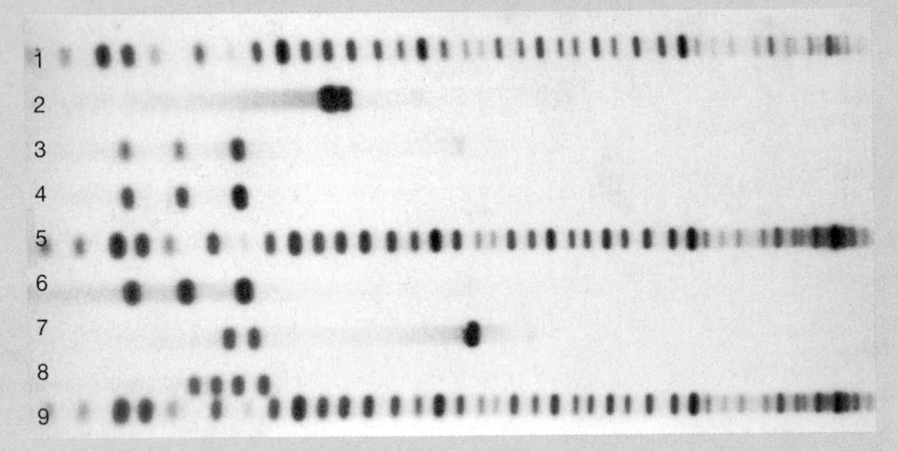

The minute amount of DNA in the follicle cells on a single hair or in a droplet of blood can be amplified by PCR and used in this method. The gel shown here was used in a case of two men being held as suspects in attacking and raping a young woman. DNA from suspect A's blood is in lane 2, and DNA from suspect B's blood in lane 4. Lanes 1, 5, and 9 are viral DNA used to establish sizes of fragments. Lanes 3 and 6 are DNA from semen samples recovered from the victim's clothing and vaginal canal; lane 7 is from the victim's own blood. Lane 8 is a control sample. It is easy to see which man is clearly innocent and which one is probably guilty. The technique has also been applied to identifying fragmentary human remains, based on the patterns of their close relatives, such as the remains of political victims who had "disappeared" in Argentina while the country was in the grip of a repressive militarist regime.

19.10 Human genes can be identified and located with recombinant-DNA methods.

While many people remain skeptical and cautious about some applications of recombinant-DNA techniques because of their possible dangers or misuse, these techniques have already shown their potential for improving the human condition by being used to identify and study genes responsible for genetic disorders. Cystic fibrosis, which we introduced at the beginning of this chapter, is a prime example of a genetic disease whose gene was located through painstaking application of these methods. In 1985 the underlying problem was identified as defective regulation of chloride-ion channels in cell membranes, due to a protein called the *cystic fibrosis transmembrane conductance regulator (CFTR)*. However, the gene that encodes this protein was unknown.

Identification of the CFTR gene, by a large group of investigators at the University of Toronto and the University of Michigan, was a major triumph of molecular genetics. The CFTR mutation had already been located on chromosome 7 through studies of families with the gene, showing that it is in linkage disequilibrium with *met,* a so-called oncogene known to be on chromosome 7. By 1988, the Toronto group had localized the CF mutation between *met* and RFLP number D7S8, in the 7q31 band (Figure 19.22).

With the CFTR mutation thus localized, the Toronto and Michigan groups constructed a *restriction map* of this 500,000-bp region of chromosome 7. Such a map shows the locations of the sites cut by several restriction enzymes, and these sites can be used in somewhat the same way that mutational sites are used in recombination experiments to obtain a detailed picture of a small region of the genome. After completing the restriction map of the region, the investigators located the CFTR gene itself with probes, as described in Section 18.11. They isolated RNA transcripts from cells that normally produce the CFTR protein, including sweat-gland and pancreas cells; from these transcripts, they made many complementary-DNA (cDNA) clones, testing each one laboriously for its ability to hybridize with cloned fragments in the mapped region. This method localized the gene for the CFTR protein, a large gene composed of 24 coding sequences separated by introns.

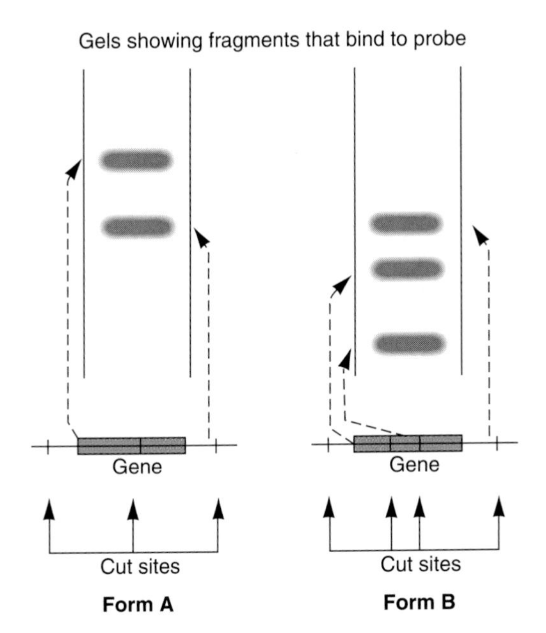

Figure 19.21
A restriction fragment length polymorphism is discovered by cutting the DNAs of different people with one restriction enzyme, separating the fragments, and probing for one gene of interest (see Section 18.11). Forms A and B, found in different people, show that they differ in the locations of cut sites for this enzyme.

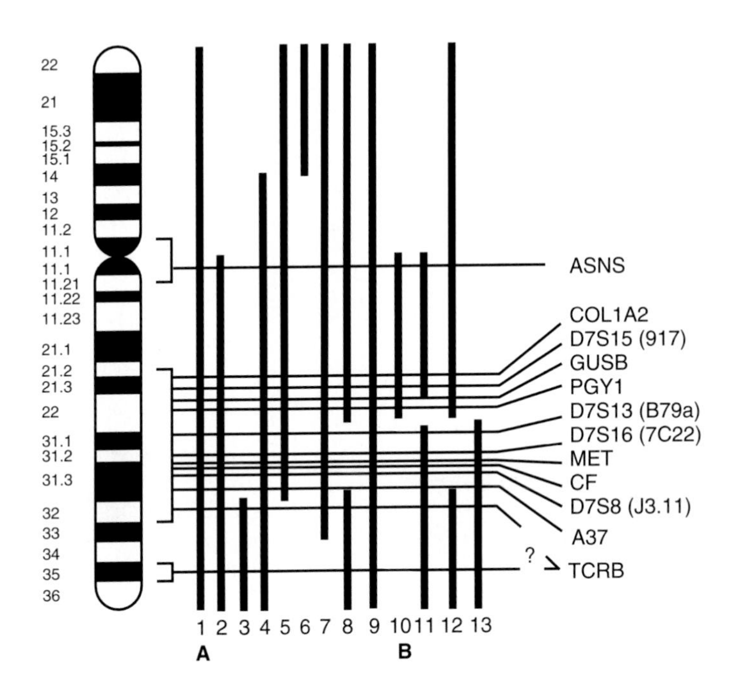

Figure 19.22
A map of human chromosome 7 shows the location of markers (listed along the right side) close to CF; most of these are RFLPs. The vertical bars show the extents of human chromosome 7 fragments among somatic cell hybrids. A1–9 are from human-mouse hybrids, and B10–13 are from human-hamster hybrids.

The nucleotide sequence of the CFTR gene was then determined and translated (theoretically) into a protein sequence. It encodes a protein of 1,480 amino acids. Some good computer models can predict the most likely secondary and tertiary structures of a polypeptide. The CFTR protein is predicted to fold up with 12 transmembrane helices, so it would fit neatly into a membrane and could conduct ions across the membrane. Another domain of this protein, one that doesn't fit into a membrane, has the structure of domains that are known to bind ATP. All this is theoretical, of course. But now that the protein has been identified in this way, investigators can find the actual protein, study its properties, and perhaps use the information gained to alleviate the disorder.

The cDNA clones obtained from CF patients, for comparison with the normal clones, revealed only one consistent difference: a deletion of three nucleotides that remove residue 508, a phenylalanine. Previous studies had indicated that about 85 percent of CF patients of northern European descent probably carry this mutation, so that single mutational event, the removal of just three nucleotides and one amino acid, accounts for the enormous difficulties suffered by generations of CF patients.

A total of 25 people wrote the three papers describing this work in the journal *Science* in September, 1989, with at least a dozen more given credit for their technical assistance. Even though our description of their work barely outlines all these investigators had to do, we have gone into enough detail here to make two points. First, this kind of research requires an extraordinary amount of effort and painstaking, time-consuming attention to detail. Second, it can be done. The techniques for determining the location and sequence of an unknown gene, and the protein it encodes, are all available and are being continuously refined and improved. Methods for manipulating cells and chromosomes, cloning DNA, and determining nucleotide sequences raise new ethical problems (see Section 19.12), but they have already demonstrated their potential for alleviating human suffering.

19.11 Genetic disorders might be cured through gene therapy.

Recombinant-DNA technology has always held the promise of correcting human genetic defects by substituting a normal gene for a defective one. Once the CFTR gene was identified and its function confirmed, medical investigators turned to possible applications of this knowledge through gene therapy. However, instead of "gene therapy," we will use the term *gene transfer,* as a number of researchers do to keep patients aware that the current studies are *not* therapeutic but may some day lead to an effective treatment. A first step in this research was to determine whether a normal CFTR gene could be transformed into isolated cells, and this was accomplished by making vaccinia virus (Figure 19.23) into part of a vector system for controlling the expression of a cloned gene: The virus was modified with a gene for the phage T7 RNA polymerase, and the CFTR gene was then cloned in a plasmid downstream of a promoter recognized only by that polymerase. When the virus and the plasmid were introduced into cells from a CF patient that lacked the critical ion transport regulator, these cells acquired normal ionic regulation.

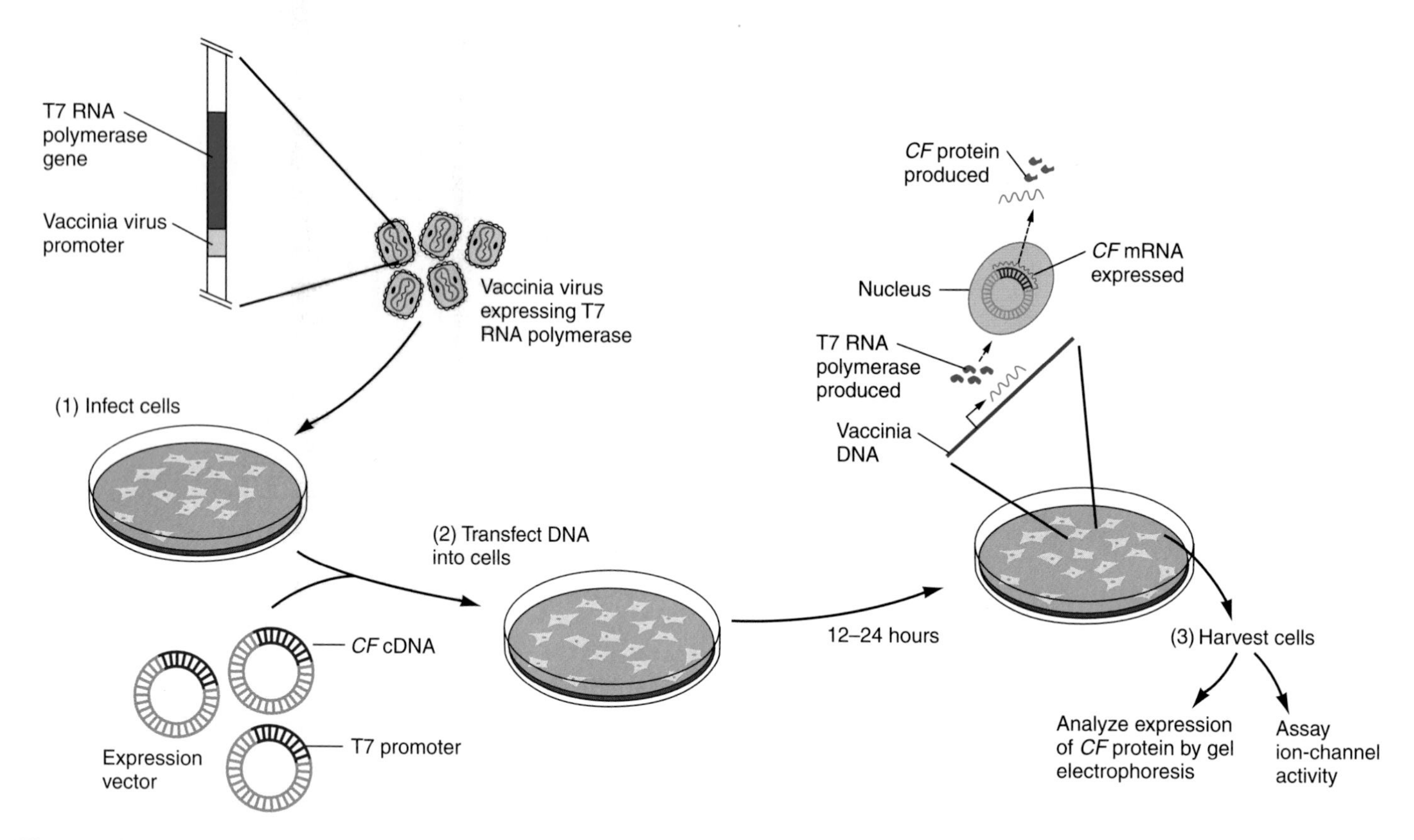

Figure 19.23

The CFTR defect can be corrected in individual cells by means of a vaccinia virus that has been altered to carry the gene for bacteriophage T7 RNA polymerase. *(1)* Cells isolated from a CF patient are grown in tissue culture and first infected with the vaccinia virus, which expresses the T7 polymerase at a high level when introduced into cells. *(2)* An expression vector is introduced, carrying the normal CFTR gene downstream of a T7 promoter, so it is only transcribed by this special polymerase. *(3)* When the cells are collected and tested, they show normal ion-transport activity as a result of expression the wild-type CFTR gene.

As of 1996, nine CF Gene Therapy Centers were carrying out gene transfer studies with CF adults in the U.S., plus a few centers elsewhere, with a total of over 70 patients—more gene transfer tests than for any other disease. Dozens of vectors are being examined for efficacy, but none have all the characteristics desired in a gene transfer vector. Some gene transfers are performed by bronchoscope, a tube for examining the lungs, used in this application to introduce material into them. Other centers prefer to treat the nose or sinuses, because they are easier to reach and mistakes are not as irrevocable.

As shown first by Ronald Crystal and his colleagues at Cornell University, adenovirus can deliver the CFTR gene to lung cells in humans, where it is expressed. Because the virus does not integrate into the human genome, repeated dosing is required, but unfortunately, subsequent doses result in lowered transfer efficiencies and also produce inflammatory reactions. To counteract this problem, more and more of the adenoviral DNA has been removed, with the aim of producing a vector with no viral gene products except the few needed for packaging and expressing the CFTR DNA. It is hoped that such a vector will not be recognized by the host immune system, and that repeated dosing will be as efficacious as the first.

A small adeno-associated virus (AAV) does not produce the inflammatory reaction that adenovirus vectors do, but it does not transfer the gene as well either. An improvement in efficiency has been seen when CF cells in tissue culture are treated with UV or gamma radiation, or with various alkylating agents. But even without these treatments, AAV retains a degree of popularity because, unlike adenovirus, it does not cause disease. The results of the first AAV gene transfer trials are just being evaluated at this writing. Retrovirus carrying CFTR has not worked in slowly dividing cell populations like those of the lung; however, retroviral technology is being refined and improved, and may become useful in the future.

The question remains as to whether expressing a normal CFTR protein will rid the CF patient of bacterial infections in the lung, which cause more than 90 percent of the morbidity and mortality. Work by Jeff Smith from the University of Iowa suggests it might. Smith examined a protein in lung secretions whose function is to kill bacteria. The protein is inactive in the presence of high salt, the condition characteristic of CF patients, but once the salt concentration is lowered as it would be by an active CFTR, the factor becomes active and is able to kill bacteria again.

Many other methods are being developed to treat the known genetic diseases. For instance, disorders associated with abnormal blood cells might be treated by transforming those cells in tissue culture and then introducing them into a patient's bone marrow, where they find a natural home. It is all but certain that some of these methods will succeed and will become part of medical practice within the next few years.

19.12 Information about genetics has moral consequences.

Knowledge has consequences. People who are ignorant of events or powerless to affect them can hardly be blamed for their actions, but as people learn more about the world, and especially as they acquire more ability to control and change the world, they also acquire a moral responsibility to use that knowledge and power wisely. Nowhere is this becoming more obvious than in human genetics.

Sickle-cell anemia, a disease we described in Chapter 5, affects about 1 percent of the African-American population. Heterozygotes for the Hb S mutation, who constitute 14 percent of the black U.S. population, are virtually normal, but homozygotes become very sick and generally die young. Assuming no homozygote lives long enough to become a parent, people with sickle-cell anemia must be the offspring of two heterozygous carriers. These carriers can now be detected with a simple test and advised that if they marry another carrier, a quarter of their children on the average will be affected and half their children will be carriers. With this information, carriers are now able to make informed, though certainly no less difficult, decisions as to whether or not to have children.

Several disorders due to recessive genes are unusually common among Ashkenazic Jews, the majority of Jews of eastern European descent. The best-known of these disorders is Tay-Sachs disease, a form of mental retardation that results from abnormal metabolism of certain complex lipids. It affects one in 3,600 infants among Ashkenazim, and about 3 percent of the Ashkenazic population are carriers. Tay-Sachs carriers can be identified with a biochemical test, but sociological studies in Jewish communities have shown how the possibility of gaining knowledge of one's genotype opens a series of problems that never existed before. Testing is voluntary, but people who don't want to be tested may be viewed as antisocial or irresponsible. Those who have not been tested—or have learned that they are carriers—may be shunned by prospective marriage partners.

While genetic diseases appear infrequently compared with problems such as heart disease and cancer, they still cause their share of human misery. More and more tests are being devised for detecting carriers in a high-risk population, since carriers, with only one copy of the mutant allele, are a much larger percentage of the population than the afflicted homozygotes who have two copies. Yet, for all the obvious benefits of these new detection techniques, they force us to confront some disturbing choices. What do we say to people who may be at risk? Do we say, "You must consent to this test to find out if you are a carrier?" That position would violate certain human rights—if nothing else, the right to not know. This is a new and difficult position for humanity to be in. In the past, when most information about our genomes was unavailable, the problem didn't arise. Suddenly, modern science has put us in a position where we can know, and we must then ask if we want to know. So far, this decision has been left up to each individual; no state or government has forced any of its citizens to submit to genetic testing to find heterozygotes. But any government could move in this direction, justifying its action in the name of the common good.

Taking the situation a step further, what do we say to someone who has been tested voluntarily and finds that he or she is a carrier? Do we say, "You cannot have children?" So far, no government has taken the step of preventing carriers of any disease from having children. But suddenly individuals are faced with a new kind of moral decision, which may affect not only their own personal happiness but also the future of humanity. And each addition to our knowledge of human genetics opens a new opportunity for the state to step in and make a decision for its citizens.

The abilities to test and to manipulate in new ways raise other questions. Now that we can learn about the condition of a fetus through amniocentesis, parents can make informed decisions about terminating a pregnancy if the child is likely to be deformed. Surveys show that the majority of Americans favor allowing people the choice of abortion, but now the question arises about conditions that justify an abortion. Should it be done if karyotyping shows that the child will have a severe deformity and will probably die early? What about Down syndrome, recognizing that many Down children are quite functional and live happy, fulfilling lives? What if karyotyping shows that the child will be a boy, and the couple wants a girl? And who should have the right to make such decisions: the parents? the physician? or the government?

The power of recombinant-DNA methods, as shown in the case of cystic fibrosis, raises another set of difficult questions. Suppose one method of gene transfer is successful and we acquire the ability to correct CF through gene therapy. Should we? To most people, the answer would be a resounding "Yes," but some ethicists have maintained that gene therapy should be used to correct the defect in individuals, but not to change the genomes they pass on to their offspring and thus change future generations. Is this a reasonable guideline? If it became possible to correct cystic fibrosis or Tay-Sachs or sickle-cell anemia for all time, why shouldn't it be done? On the other hand, if we begin correcting such obvious defects, we start down a slippery slope; what other characteristic might next be deemed a "defect" in need of correction by enthusiastic genetic engineers? Will genes for misshapen noses or a tendency toward excess body fat or a low IQ then need to be eliminated from the human gene pool? And, again, who will decide?

Many biologists are currently engaged in one phase or another of the Human Genome Project whose aim is to identify every element of the human genome. Portions of the genome have even been distributed to centers that engage high school biology students in sequencing small regions. As a result, many laboratories have been reporting the discoveries of genes associated with human traits and disorders. There is little doubt that this information has the potential for a great deal of good; we will learn how these genes and their products act, gaining insights into basic cellular functions as well as disease processes, and we will acquire the ability to test people for alleles they may carry that predispose them to develop certain diseases. However, in a society whose health-care system is driven by a massive insurance industry operating for profit, the same information can be devastating. Will each person have the right to keep his or her genotype secret, or will it be made known to insurance companies? Will people be denied insurance if they carry genes that predispose them to conditions that will be expensive to treat, or will they be forced to pay premiums they cannot afford? Questions like these open up the possibility that people could be penalized and discriminated against for their genes. How will our society handle these issues of ethics and economics? Or will this growing knowledge force our society to totally revise its health-care system?

Today the need is greater than ever before for educated, informed citizens to take an interest in these questions and in the actions, or potential actions, of their government. If a government agency allows an experiment in gene therapy to proceed, is it acting in the best interest of humanity or only in the interests of a company seeking profit or a biologist seeking fame? The citizens of an increasingly technological world must be able to think clearly about such matters.

Although there are no simple answers to the ethical questions now being raised, this much is certain: Once humanity has opened the doors to greater knowledge, there is no going back. Once we know something, we cannot pretend we don't know. We cannot cry "Halt!" and call for a moratorium on knowledge. Science is too public a process. The ways of achieving greater knowledge are too well known; someone is inevitably going to learn the next thing, and it is probably better that everyone can know it. The only general solution appears to be greater knowledge, not less, and an informed citizenry ensuring that this knowledge is used as wisely as possible.

Coda The fundamental principle that organisms are genetic systems comes home strongly when we consider humans. The very existence of a Human Genome Project places the genome at the center of our attention and emphasizes that our characteristics depend fundamentally on information encoded in our DNA. We are also acutely aware of differences among ourselves; the enormous genetic diversity in human populations has resulted from mutations and chromosomal rearrangements, and this diversity is one aspect of evolution. Genetic diversity is the raw material on which evolution operates, and in *Homo sapiens,* evolution has produced people of many types who were isolated in many corners of the earth before our species became so mobile. Some of those differences were apparently adaptations to differences in the human ecological niche—for example, darker skin color is associated with more tropical regions and lighter color with more temperate regions. Ecological factors associated with other differences, such as blood types, are obscure at best, although we have seen the subtle effect of sickle-cell hemoglobin on resistance to malaria. When we consider the molecular basis for our characteristics, we have only to look at cystic fibrosis to see how a seemingly trivial change in a protein—one amino acid out of nearly 1,500—can abolish its normal function. The interactions between a protein and its substrate or another ligand depend on delicate, precise adjustments shaped over millions of years. Mutations quickly put them awry. In the human genome, such mutations have caused premature death and much distress. Now humanity faces the question of whether we will wisely use our growing power to correct these mutations.

Summary

1. We can often determine the mode of inheritance of a characteristic from pedigrees. X-linked traits are inherited most distinctively, since they primarily affect males but are passed on through females. Whether an autosomal characteristic is dominant or recessive is usually determined from the frequency of affected individuals.
2. Humans have many blood group antigens that may cause problems of incompatibility in blood transfusions. In the ABO system, people may have A, B, or AB antigens on their red blood cells; their blood contains antibodies against the antigens that their red blood cells do not have. Rare antigens are sometimes encountered when people become immunized by a series of transfusions.
3. Rh-factor incompatibilities can cause hemolytic disease of the newborn. Generally an Rh$^-$ mother becomes immunized against her Rh$^+$ baby, and her antibodies then react against the blood of future fetuses. Hemolytic disease of the newborn can now be prevented in several ways.
4. If homologous chromosomes fail to separate (disjoin) in meiosis, gametes may form with either extra chromosomes or with a missing chromosome. Some genetic diseases, such as Down syndrome, are the result of trisomy for one chromosome.
5. Human chromosomes are distinguished by karyotyping and by distinctive banding patterns created by various reagents. A standard system allows any position on a chromosome to be designated unambiguously.
6. Quantitative characteristics may result from the action of many genes. If each copy of a dominant allele adds to a particular feature, such as height, then as the number of relevant genes increases, the distribution of that feature in a population will more closely approximate a normal distribution. Presumably, the effects of all these alleles add to one another.
7. Aneuploids for sex chromosomes, resulting from nondisjunction of X or Y chromosomes, are relatively common. Varying numbers of X chromosomes are tolerated because of a mechanism for randomly inactivating all but one of them in each cell and converting them into Barr bodies.
8. Human genes can be localized to specific chromosomes by making hybrids with cells of mice or hamsters. These hybrids lose chromosomes or parts of chromosomes, and from the phenotypes of cells retaining different combinations of chromosomes, we can deduce the location of some genes.
9. A human gene can be localized to a specific chromosome by finding a linkage disequilibrium. If all genes were in equilibrium, the gene for a rare condition would be found linked to each allele of a nearby gene in proportion to the frequencies of those alleles in the population. If the rare gene is found more often with one particular allele, this shows the two are closely linked.
10. Human genes can also be identified and located with recombinant-DNA methods. This combination of techniques includes developing a restriction map, sequencing DNA, and using RNA probes to identify fragments of DNA with complementary sequences.

11. Wild-type human genes have been introduced experimentally into defective tissue-culture cells and into patients with genetic disorders. These genes can be expressed to correct the cellular defects, and this method has been used with some success in patients.
12. Information about genetics has moral consequences. Growing human knowledge of genetics is posing difficult moral questions. Increasingly, questions arise about who will decide whether an individual must submit to genetic tests, or who will be responsible for determining whether a person shall have children that may be affected by a genetic disease. These questions demand an informed public that is sufficiently educated to make such serious decisions.

Key Terms

pedigree 376
sibship 376
penetrance 377
polygenic trait 377
agglutinate 378
codominant 379
erythroblastosis fetalis 380
hemolytic disease of the newborn (HDN) 380
rhesus antigen (Rh) 380
amniocentesis 381
disjunction 382
nondisjunction 382
aneuploid 382
monosomic 382
trisomic 382
Down syndrome 382
trisomy-21 382
translocation 383
Turner syndrome 384
Klinefelter syndrome 384
Barr body 386
allelic frequency 388
linkage equilibrium 388
linkage disequilibrium 389
restriction fragment length polymorphism (RFLP) 389

Multiple-Choice Questions

1. Red-green color-blindness appears in significantly fewer women than men, but even in families with several affected males, the trait does not appear in every generation. From this evidence, you could logically deduce that red-green color-blindness is caused by a ___ allele.
   a. dominant autosomal
   b. recessive autosomal
   c. dominant X-linked
   d. recessive X-linked
   e. None of the above is a logical choice.
2. If all individuals known to be homozygous for a recessive allele do not exhibit the mutant phenotype, the mutant allele is said to be
   a. dominant.
   b. recessive.
   c. polygenic.
   d. incompletely dominant.
   e. incompletely penetrant.
3. A trait caused by a dominant autosomal allele will generally result in a pedigree that shows
   a. roughly equal numbers of affected males and females.
   b. affected individuals having at least one affected parent.
   c. affected individuals in every generation.
   d. *a* and *b*, but not *c*.
   e. *a*, *b*, and *c*.
4. If a pedigree shows that males, but not females, are affected by a particular characteristic, that all affected males have affected fathers, and that all sons of affected fathers are similarly affected, the responsible allele could be
   a. Y-linked and recessive.
   b. autosomal and dominant, with severely reduced penetrance in females.
   c. X-linked and dominant.
   d. *a* and *b* but not *c*.
   e. *b* and *c*, but not *a*.
5. If a pedigree indicates that every affected individual has or had two unaffected parents, the allele responsible for the condition is probably a(n)
   a. autosomal dominant.
   b. autosomal recessive.
   c. X-linked dominant.
   d. X-linked recessive.
   e. autosomal or X-linked recessive.
6. Which statement is correct?
   a. Transfusion of type O blood into a type AB recipient will stimulate the production of an anti-O antibody.
   b. Transfusion of Rh-negative blood into an Rh-positive recipient will stimulate the production of an anti-Rh antibody in the recipient.
   c. Transfusion of O-negative blood into an AB-positive recipient will stimulate the production of anti-B and anti-Rh antibodies in the recipient.
   d. Transfusion of O-negative blood into a B-positive recipient will cause agglutination of the donor red blood cells.
   e. Transfusion of O-positive blood into a B-negative recipient will not cause agglutination, but may trigger anti-Rh antibody production in the recipient.
7. One of four siblings is blood group A, the second is B, the third is O, and the fourth is AB. Their parents must have blood types ___.
   a. A and B.
   b. AB and O.
   c. AB and AB.
   d. O and O.
   e. A and AB.
8. Banding patterns in human chromosomes enable geneticists to
   a. find point mutations in genes.
   b. distinguish members of homologous pairs.
   c. determine the precise location of single genes.
   d. establish which characteristics are determined by polygenes.
   e. analyze RFLPs.
9. Some calico cats have large, distinct patches of black fur and yellow fur, while others, called tortoiseshell, have a less bold pattern with small interspersed areas of yellow and black fur. Assuming both patterns result from the same basic cause, which hypothesis best explains these observations?
   a. The bold-patterned cats are males, while the tortoiseshells are females.
   b. The bold-patterned cats are females, while the tortoiseshells are male.
   c. The earlier the X-inactivation in the embryo, the bolder the pattern.
   d. The later the X-inactivation in the embryo, the less bold the pattern.
   e. The bold-patterned cats are homozygous; the tortoiseshells are heterozygous.
10. The alleles for hemophilia and anhidric ectothermal dysplasia occupy different loci on the X chromosome. Among the offspring of a normal woman heterozygous for both conditions and a male with hemophilia you would expect
    a. all males to have hemophilia; all females to be dysplastic.
    b. all males to be dysplastic; all females to have hemophilia.
    c. some males and some females to be normal for both conditions.
    d. no dysplastic females.
    e. no females with hemophilia.

True-False Questions

Mark each statement true or false, and if false, restate it to make it true.

1. Pedigree analysis of human families is used in place of controlled matings to determine whether a mutant allele is X-linked or autosomal.
2. Inherited traits that show up in each generation of an affected family are probably caused by sex-linked recessive alleles.
3. The ABO blood group alleles are a good example of a polygenic system involving three genes.
4. Rhesus antigen got its name when it was found that all routine transfusions from a rhesus monkey donor to a human recipient elicited the production of a human antibody.
5. Each band in a chromosome stained with Giemsa stain or quinacrine mustard represents one gene.
6. Down syndrome that results from aneuploidy is probably a result of random non-disjunction during gamete production in a parent, while translocation Down syndrome can be inherited generation after generation within a family.
7. A Barr body can be formed by inactivation of any of the autosomes.
8. DNA fingerprinting relies on analyzing RFLPs found within a few particular structural genes.
9. Since RFLPs are genetic markers produced by restriction endonucleases, the use of different restriction endonucleases will result in different patterns of RFLPs.
10. Although the mutant CFTR protein is present and functions as a chloride channel in the plasma membrane of all body cells, the transfer and expression of the normal allele in lung cells is sufficient to cure the disease.

Concept Questions

1. Suppose an X-linked recessive allele is so common in an isolated population that unaffected females are generally heterozygotes. Would pedigrees of affected families look like those for the typical X-linked allele? Explain.
2. A recessive X-linked allele causes hemophilia. A study of 100 female heterozygotes reveals a wide range of clotting times, despite the fact that all the women have the same genotype. Explain.
3. How do geneticists account for the difference in lethality of aneuploidy in sex chromosomes versus aneuploidy in most other pairs of chromosomes?
4. Explain why DNA fingerprinting has been used more frequently to prove innocence than to prove guilt.
5. Since the CFTR protein was isolated before the gene for cystic fibrosis had been mapped, why couldn't researchers simply sequence the protein and use that information to construct and transfer a normal allele into CF patients?

Additional Reading

Berdanier, Carolyn D., and James L. Hargrove (eds.). *Nutrition and Gene Expression.* CRC Press, Boca Raton (FL) 1993.

Lawn, Richard M., and Gordon A. Vehar. "The Molecular Genetics of Hemophilia." *Scientific American,* March 1986, p. 48. Hemophiliacs bleed because a defective gene deprives them of a key blood-clotting protein. The protein has now been made artificially by isolating the normal gene and then inserting it into cultured cells.

Mange, Elaine J., and Arthur P. Mange. *Basic Human Genetics.* Sinauer Associates, Sunderland (MA) 1994.

Marion, Robert. *Was George Washington Really the Father of Our Country? A Clinical Geneticist Looks at World History.* Addison-Wesley Publishing Co., Reading (MA) 1994. The influence of genetics on history, from George III to John F. Kennedy.

Necia, Grant C. (ed.). *The Human Genome Project.* Los Alamos National Laboratory, Los Alamos (NM) 1992.

Patterson, David. "The Causes of Down Syndrome." *Scientific American,* August 1987, p. 52. The genes responsible for many of the pathologies associated with the disorder are being identified and mapped to sites on chromosome 21.

Verma, Inder M. "Gene Therapy." *Scientific American,* November 1990, p. 68. Treatment of disease by introducing healthy genes into the body is becoming feasible.

Internet Resource

To further explore the content of this chapter, log onto the web site at:

http://www.mhhe.com/biosci/genbio/guttman/

20

Developmental Biology I: Morphogenesis and the Control of Growth

Key Concepts

A. Early Embryonic Development

20.1 Fertilization activates an egg and initiates development.

20.2 Animal embryos are shaped by large-scale movements of cells.

20.3 Reptile, bird, and mammal embryos develop extraembryonic membrane systems.

B. Factors That Regulate Morphogenesis and Growth

20.4 Patterns of cell division determine the form of many tissues.

20.5 Differential adhesion can determine the arrangement of some tissues.

20.6 Cells bind to an extracellular matrix and migrate on it.

20.7 Contact inhibition normally restricts the division and movement of cells.

20.8 Specific proteins regulate the growth of tissues.

20.9 Sheets of tissue may be shaped by several factors.

C. Cancer and the Regulation of Growth

20.10 Cancer cells escape restraints on reproduction and invade areas of normal tissue.

20.11 Cancer, like evolution, involves natural selection.

20.12 Cancers are derived from single cells through DNA modification.

20.13 Several independent events are needed to change a normal cell into a fully cancerous cell.

20.14 Both tumor initiators and tumor promoters contribute to the transition from normal cells to malignant ones.

20.15 Some viruses can transform normal cells into tumor-producing cells.

20.16 Oncogenes are derived from cellular proto-oncogenes whose products are components of signal-transduction pathways.

20.17 Environmental factors cause most cancers.

20.18 More than one mutation is necessary to cause cancer in an animal, yet one oncogene can transform a normal cell.

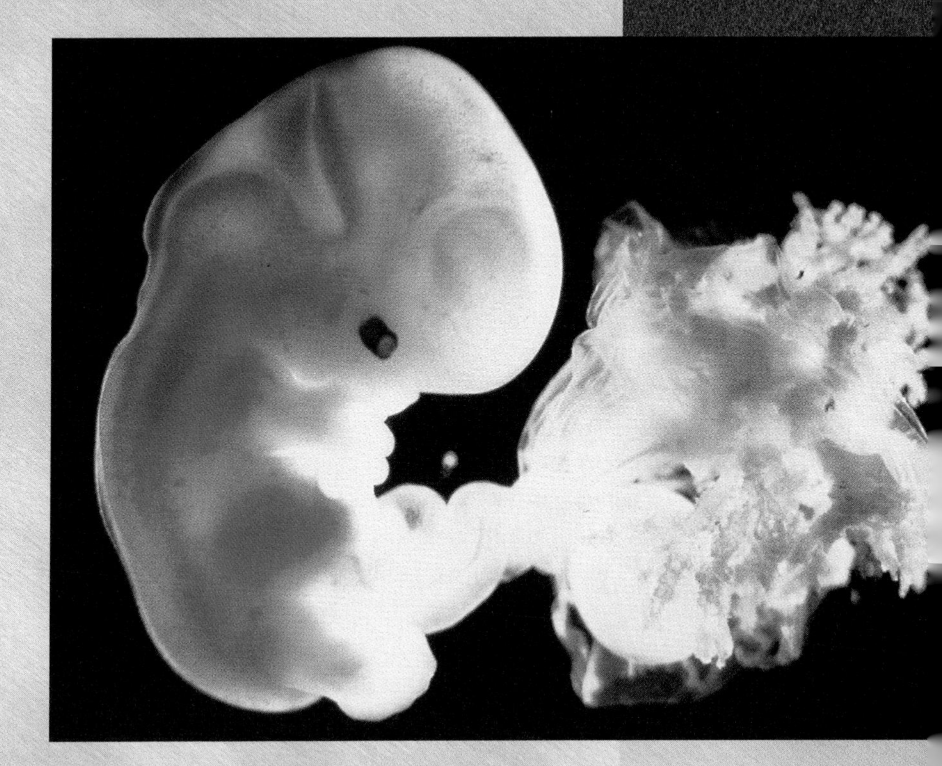

A human fetus at four weeks has a large head, with regions of the brain visible inside.

As expectant parents await the birth of their baby, they commonly experience an unspoken fear that their child will not be normal. Yet it is remarkable that these fears are so seldom borne out. Somehow a minute fertilized egg manages to grow into a small human with its eyes and nose and fingers and toes in the right number, in the right places. We humans, like all organisms, carry genetic programs that determine development, and the program seldom goes wrong. Biologists who study development cannot yet explain how the program works, but we can at least describe the process and point to some of its mechanisms.

Developmental biology is among the great frontiers of biology. This science grew out of the classical science of embryology, which described the anatomy of development and elucidated some of its mechanisms. Development entails two distinct processes:

- Every organism grows in a life cycle, and most show a distinct change in form, or **morphogenesis,** during the cycle.
- Development is accompanied by **differentiation,** the formation of cells with distinctive features and specialized functions.

Morphogenesis and differentiation imply a schedule of gene regulation, with various genes being turned on and off at certain times and places. The idea that a developing organism is following a genetic program shapes all our thinking about development. The genome carries information that causes some cells to grow into a sheet of tissue that rolls into a blood vessel and other cells to form a sphere that becomes an eye. The genome also contains information that allows some cells in an eye to become the retina, others to become a clear lens, and still others to become the muscles that shape the lens. To understand these processes, developmental biologists are now using many modern techniques, especially those of molecular genetics, and important new ideas are constantly emerging. We will explore our current knowledge of morphogenesis in this chapter, paying special attention to the case of cancer, in which development is distorted; we will then discuss differentiation in Chapter 21. ■

A. Early embryonic development

20.1 Fertilization activates an egg and initiates development.

Embryonic development begins when a sperm fuses with an egg and sets a new series of reactions in motion. The female cell preparing for fertilization is an oocyte going through meiosis, and it may be fertilized at various times in this process; starfish oocytes, for instance, are generally fertilized during the first metaphase, while most mammalian oocytes are fertilized during the second metaphase. Outside its plasma membrane, the egg has a **vitelline envelope,** a fibrous mat containing structures necessary for the species-specific binding of the proper sperm. The whole egg is often surrounded by a thick layer of egg jelly, made of glycoproteins that may attract or activate sperm. The egg jelly of a sea urchin, *Arbacia,* contains a short peptide called *resact* that diffuses into the surrounding water and guides sperm of the same species toward the egg by chemotaxis.

Chemotaxis, Section 11.6.

On making contact with the egg jelly, a sperm experiences an influx of Ca^{2+} ions, which stimulate the acrosomal material in its tip to release digestive enzymes (Figure 20.1). These enzymes break down the egg jelly. Simultaneously, a thin acrosomal process pushes forward carrying the protein *bindin,* a species-specific recognition protein that matches equally specific receptors on the vitelline membrane and ensures that only sperm of the right species will be able to fertilize the egg. We introduced the concept of specific recognition in Chapter 11, and cell-cell recognition is a principal theme in development. When a sperm has passed this recognition test, the vitelline envelope lyses where the sperm tip is attached, preparing the way for the plasma membranes of the sperm and egg to fuse. The surface of the egg has minute extensions called *microvilli,* which rise in a fertilization cone toward the sperm. As the two plasma membranes come in contact, they fuse, and the fertilization cone engulfs the sperm nucleus, pulling it inward. At this point, a stimulus activates the egg, using the familiar signal-transduction pathway of G-proteins (guanine-nucleotide-binding proteins), inositol phosphates, and protein kinases. Hydrogen ions flow out of the cell, raising the cytosolic pH, and the cytosolic Ca^{2+} concentration increases dramatically as the inositol phosphate pathway releases Ca^{2+} ions from their internal stores.

Common signal-transduction pathway, Section 11.8.

Ionic changes in the cytosol then initiate the next events, especially changes in the sperm and egg nuclei. There is a sudden burst of respiration and oxygen consumption. In sea urchins, for example, the sperm nuclear envelope disperses into vesicles, and the highly condensed sperm chromosomes start to decondense, all within two minutes. Then the sperm and egg nuclei migrate toward each other and eventually fuse to make a diploid zygote nucleus. New protein synthesis begins, and by about 20 minutes after fertilization, DNA replication has started. Mitosis begins about an hour after fertilization. In mammals, nuclear fusion takes about 12 hours, but the chromosomes have been replicating while still in separate nuclei, and when they meet, they immediately form a mitotic spindle and align on it, starting the first cleavage of the zygote.

Several sperm may be attracted to an ovum and attempt to fertilize it (Figure 20.2). Fertilization by more than one sperm,

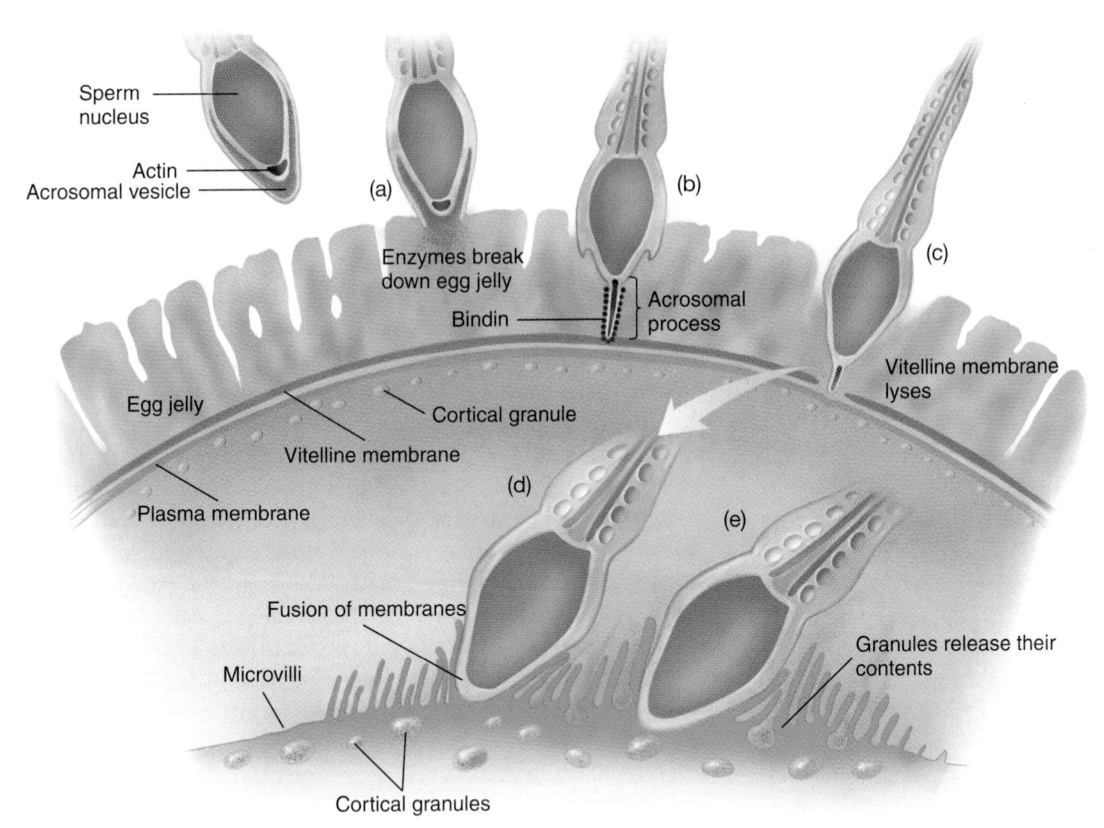

Figure 20.1

(a) When a sperm contacts the egg jelly, its acrosomal membrane opens, releasing enzymes that break down the jelly. *(b)* The acrosomal process pushes forward, carrying bindin proteins that make contact with complementary recognition proteins on the vitelline membrane. *(c)* The membrane lyses, allowing the sperm head to contact the plasma membrane. *(d)* Microvilli from the egg cell surface rise toward the sperm and the membranes of the sperm and egg fuse. *(e)* Granules in the egg cortex release their contents in a wave that expands over the surface, to block the entrance of other sperm.

known as *polyspermy,* occurs routinely in some animals (mollusks, birds, and reptiles), and yet they develop normally. Generally, however, animals have mechanisms to prevent polyspermy. In sea urchins, for example, a fast reaction occurs within seconds of the first sperm penetrating: The membrane potential suddenly rises from −70 mV (millivolts) to about +20 mV, the same kind of change that occurs in a nerve cell as it carries signals. This change lasts for about a minute and blocks other sperm. A second, slower mechanism is the **cortical granule reaction** (Figure 20.3), a visible change that advances across the egg's surface from the point of sperm penetration. The calcium ions released into the cytosol cause thousands of cortical granules lying just below the plasma membrane, in the egg cortex, to fuse with the membrane and pour their contents into the space between the plasma and vitelline membranes. Materials from these granules cause the vitelline membrane to rise and form a fertilization membrane, which releases any attached sperm and prevents the attachment of others.

Action potentials in nerve cells, Section 41.10.

Some eggs have been activated artificially with an electric shock, a needle prick, or immersion in certain chemical baths. Such eggs develop *parthenogenetically,* meaning without chromosomes from a male, and go through at least the early stages of development before aborting. Some adult (haploid) frogs have even been produced this way, but mammalian embryos stop developing early.

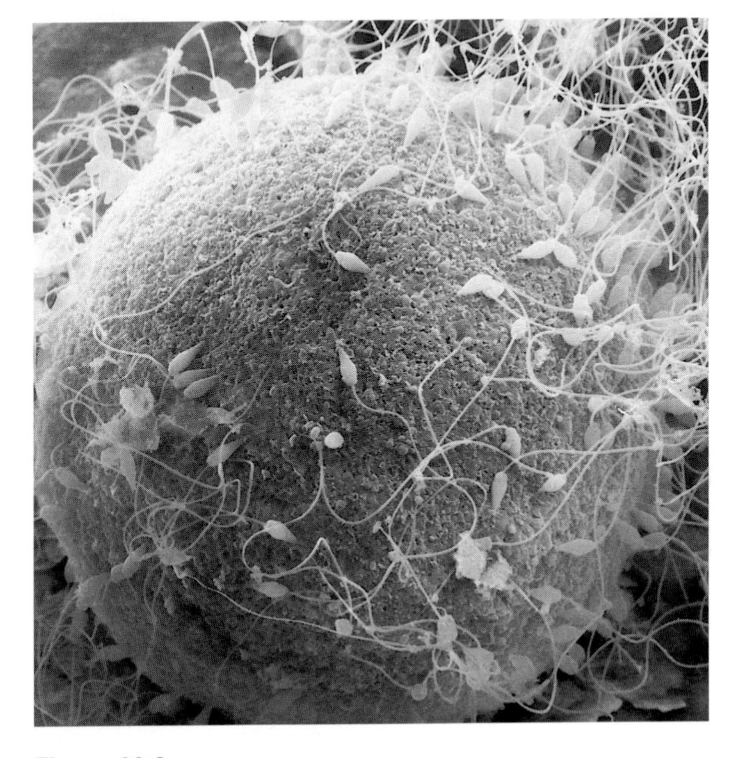

Figure 20.2

A number of sea urchin sperm stick to the surface of an egg, although only one will fertilize it.

20.2 Animal embryos are shaped by large-scale movements of cells.

Every embryo, whether plant or animal, begins with a single cell that grows into a rather round cluster of cells, but mature plants and animals do not end up round. Each organism has at least one *axis of polarity*, a line along which its form changes in some significant way. In a typical plant, the axis is along a main stem from its leafy shoots to its root, and in a typical animal, the axis goes from its head to its tail (Concepts 20.1). Establishing this polarity is one of the most important transformations in embryonic development (Figure 20.4).

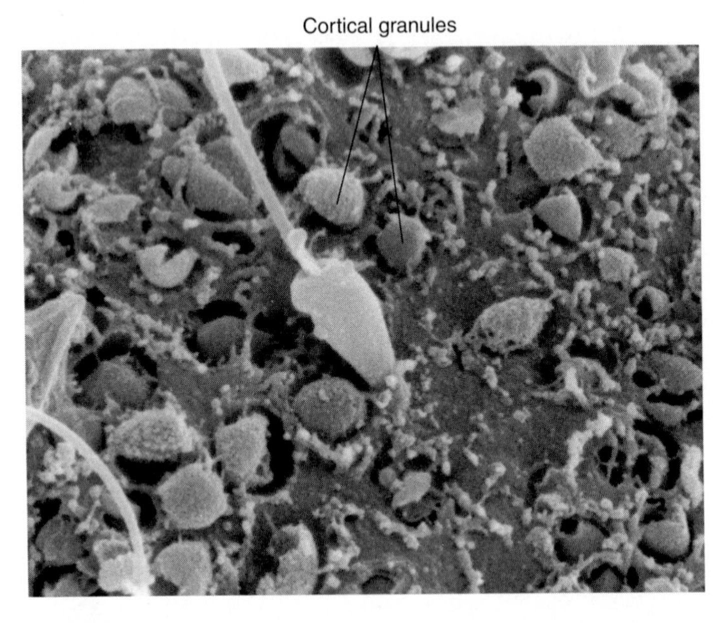

Figure 20.3

A single sperm penetrates the egg and fertilizes it. The cortical reaction spreads away from the point of penetration as cortical granules release their contents.

Frog embryos have long been favorite subjects for embryological research (Figure 20.5). Half the egg, the **animal hemisphere,** is marked by darkly pigmented cortical cytoplasm (the layer just under the cytoplasmic membrane), and the other half, the **vegetal hemisphere,** has light-colored, yolky cytoplasm. When a sperm penetrates the egg cytoplasm during fertilization, it induces the cortical cytoplasm to rotate about 30 degrees relative to the inner cytoplasm toward the point of sperm entry; this movement leaves a region, the *gray crescent,* opposite the point of sperm entry and creates a dorsal-ventral (back-belly) axis on the embryo. The gray crescent will eventually be at the posterior end. The zygote then begins to cleave into smaller cells called **blastomeres:** first two, then four, eight, and so on. The process is called **cleavage** because the zygote is just being divided, not increasing in size. (Nevertheless, each blastomere goes rapidly through a cell cycle, including an S period of DNA replication, before it can divide.) Eventually the blastomeres form a **morula,** resembling a raspberry, and then a **blastula,** a hollow ball of cells enclosing a fluid-filled space, the **blastocoel.**

The blastula then develops an indentation, the **blastopore,** just below the gray crescent near the equator between the animal and vegetal halves. This event signals the beginning of **gastrulation,** a complex movement of cells in which the blastula is transformed into a **gastrula.** A blastula is like a basketball or soccer ball, and the gastrula is effectively the result of pushing your fist into one side of the ball to collapse it, though to be more realistic, imagine that the rubber closes in around your wrist. In a great sweeping motion, cells in both animal and vegetal regions of the blastula surface migrate toward the

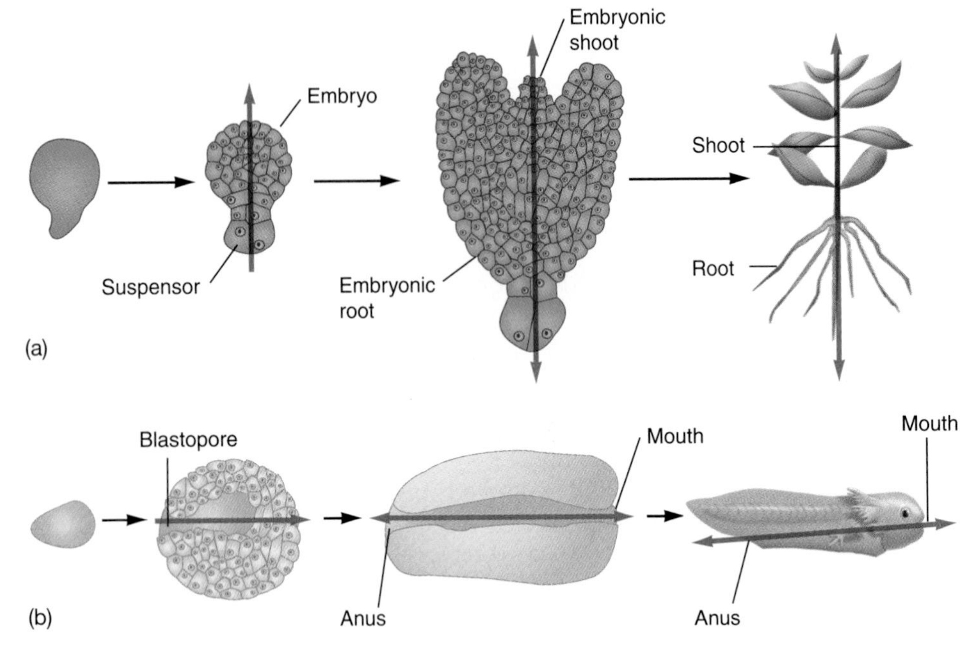

Figure 20.4

Round early embryos grow so as to establish axes of polarity. *(a)* The round plant embryo develops a primitive shoot and root that grow upward and downward, establishing a vertical axis. *(b)* The animal blastula develops an indentation on one side (the blastopore) where gastrulation begins, thus establishing the intestinal tube that runs through its middle and its head-to-tail axis.

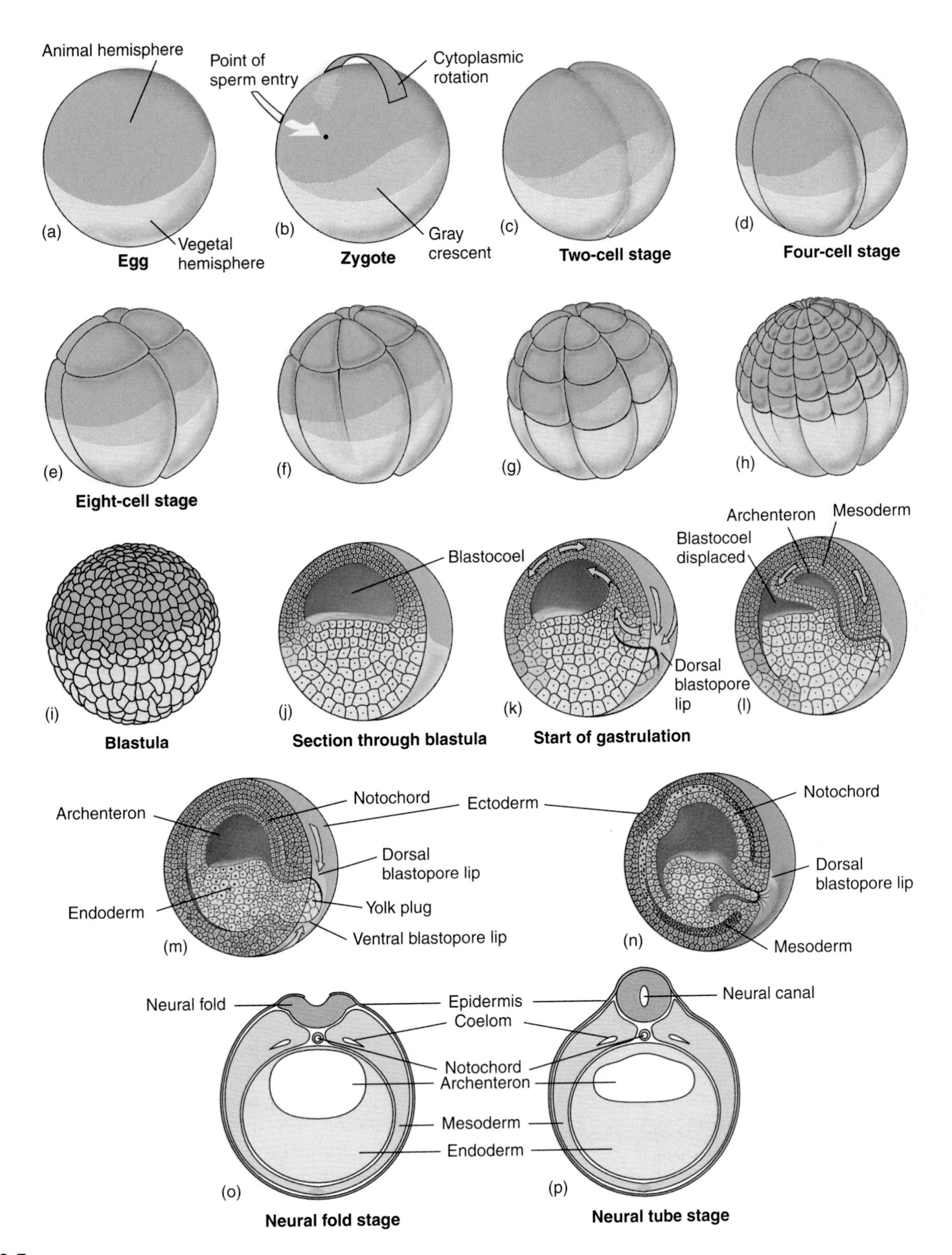

Figure 20.5

A frog egg *(a)* has a pigmented animal half and a light vegetal half. Penetration of the sperm *(b)* causes a rotation of the cortical cytoplasm relative to the interior cytoplasm, exposing the gray crescent. Cleavage of the embryo into many cells (blastomeres) *(c–h)* then produces a hollow blastula *(i)* with smaller cells in the animal hemisphere and larger, yolky cells in the vegetal hemisphere, so the blastocoel *(j)* is confined to the animal hemisphere. Gastrulation *(k–m)* entails massive cell movements *(yellow arrows)*; cells sweep inward at the blastopore, mostly from the animal hemisphere, creating two germ layers above: an outer ectoderm *(blue)* and an inner mesoderm *(red)*. Meanwhile, endoderm *(yellow)* that has formed below will spread out to become the lining of the archenteron, the primitive gut tube. Further development of the primary germ layers is most obvious in cross sections *(o, p)*. The nervous system forms as a tube (see also Figure 20.8); much of the mesoderm will form muscle and bones; and a split in the mesoderm creates the coelom, or body cavity.

Concepts 20.1 The Forms of Animals

Because this chapter about development comes before chapters about plant and animal anatomy, you may need some guidance in understanding the developing body. First, animals and plants are made of *tissues*, which are masses and sheets of cells with similar anatomy and functions. Tissues take many forms, such as blocks of rather cubical cells lined up neatly in the liver or more irregular cells pressed together in some glands. An *epithelium* is a sheet of cells that usually forms the boundary of a structure. Second, most animals have the general form of a tube within a tube:

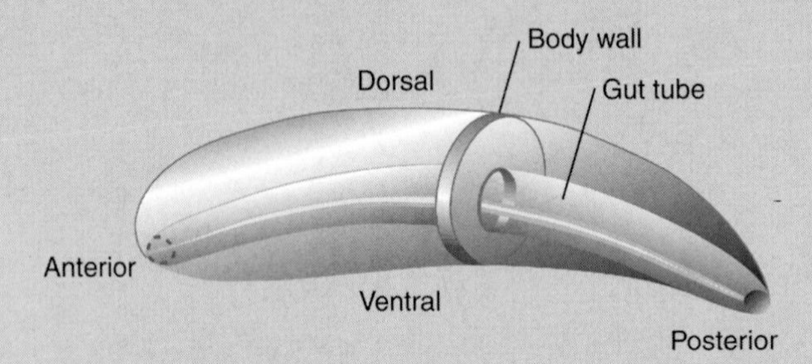

The outer tube is the body wall, and the inner tube is the digestive tract or gut, with a mouth at the *anterior* (head) end and an anus at the *posterior* (tail) end. In vertebrates, the central nervous system (brain and spinal cord) runs down the middle of the back and is largely enclosed by the vertebral column (backbone). The term *dorsal* refers to the back, and *ventral* refers to the belly.

blastopore, push through it, spread out inside the blastocoel, and end up inverted against the inner surface of the cells that remain outside. This movement obliterates the blastocoel and replaces it with the **archenteron** (literally, "ancient gut"), the cavity formed inside the gastrula that opens to the outside through the blastopore. (In the ball analogy, this is the space around your hand.) The archenteron will become the animal's gut (digestive tract.) In frogs and other lower vertebrates, the blastopore will become the anus, and later an indentation at the other end of the embryo will break through into the archenteron to form the mouth.

Starting with gastrulation, the cells that proliferated during cleavage become organized into distinct tissues. Most animals develop from three primary tissue layers (or germ layers) formed early in embryonic development: **ectoderm** on the outside, **endoderm** on the inside, lining the archenteron, and **mesoderm** in between:

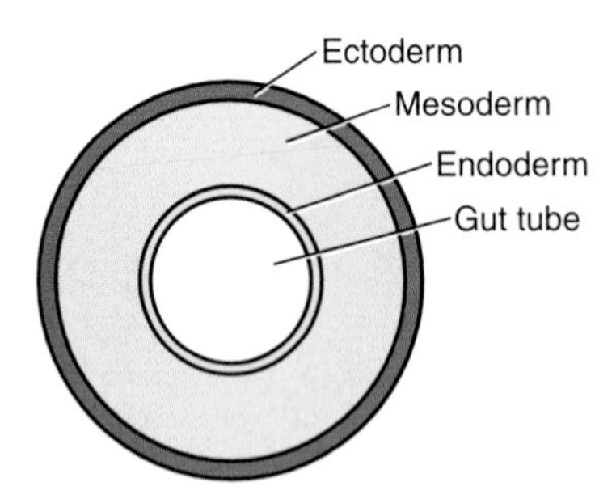

Gastrulation starts the formation of these layers. Each layer eventually develops into a distinctive set of differentiated tissues (Figure 20.6). Ectoderm, formed from cells on the outside of the gastrula in what was the animal hemisphere, develops into the outer layer of skin (epidermis) and the nervous system. Endoderm forms from the yolky vegetal cells, which are essentially overrun by cells of the mid-zone where animal and vegetal halves meet; the endoderm lies in the middle of the embryo and becomes the lining of the gut (digestive tube) and its extensions into the liver, pancreas, and lungs. Mesoderm, derived from cells that migrated through the blastopore from the mid-zone, forms the bulk of bones, muscles, and parts of organs such as the heart and kidneys.

The embryo now consists of two types of cells that behave very differently (Figure 20.7). **Epithelial cells** adhere to one another tightly and form sheets or tubes; **mesenchymal cells** are loosely organized, moving and acting more independently. Most of the mesoderm consists of mesenchyme, which will eventually condense around endodermal tubes to make the bulk of the internal organs. We will see later that cell-cell interactions between the endoderm of the tubes and the surrounding mesenchyme are essential for development.

After gastrulation, a second great tissue movement, **neurulation,** creates the foundation of the central nervous system (the brain and spinal cord). As Figure 20.8 shows, ridges begin to rise at the left and right edges of the central dorsal surface, the **neural plate.** These ridges roll in from both sides while the plate sinks inward, and when they meet and fuse with each other, they convert the neural plate into a hollow **neural tube.** The broad anterior end of the tube becomes the brain, the narrow posterior region the spinal cord. Notice that, as the tube forms, masses of **neural crest cells** break free from the ridges and remain outside the tube; later they will migrate throughout the embryo and form pigment cells, many of the nerve cells, part of the adrenal glands, and part of the skeleton. Although these tissues seem very dissimilar, they share significant features because of their common origin in the neural crest.

Meanwhile, the mesoderm has been spreading out below the developing neural tube. Even before neurulation, some mesodermal cells have condensed into the **notochord,** a thin rod that supports the body of vertebrates and is largely obliterated by the backbone in higher vertebrates. At each side of the notochord, mesoderm forms a row of blocks, or **somites,** that will produce the backbone, limbs, and the muscles of the body wall. Somites continue to form as the embryo elongates.

Exercise 20.1 Identify places on the surface of a frog blastula where you could put a bit of dye and have the dye end up inside the gastrula. If the dye is on the outer surface of the surface cells, in what cavity of the embryo will the dye be?

Exercise 20.2 Most animals have an elongated body with a gut tube running through it. In vertebrates, the blastopore becomes the anus. But other kinds of animals develop in an alternative way and produce the same adult body form. What is the alternative?

20.3 Reptile, bird, and mammal embryos develop extraembryonic membrane systems.

As embryos develop, they require water and a way to get rid of wastes. Those that develop in aqueous environments can easily exchange materials with the surrounding water, but adaptation to terrestrial life requires further evolution of development. Reptiles, the first vertebrates to become fully terrestrial, adapted to land by evolving a structure, the **amniote egg,** that preserves an aqueous environment for the embryo (Figure 20.9). Embryos of the *amniotes*—reptiles, birds, and mammals—develop *extraembryonic membranes* outside the embryo itself; here, "membrane" means a tissue made of one or more layers of cells, not a thin layer of lipid and protein surrounding an individual cell. These membranes include the **amnion,** which forms a fluid-filled amniotic cavity around the embryo, the **chorion** through which oxygen and CO_2 are exchanged, and the **allantoic membrane** enclosing the **allantois,** which forms a sac where nitrogenous wastes are deposited.

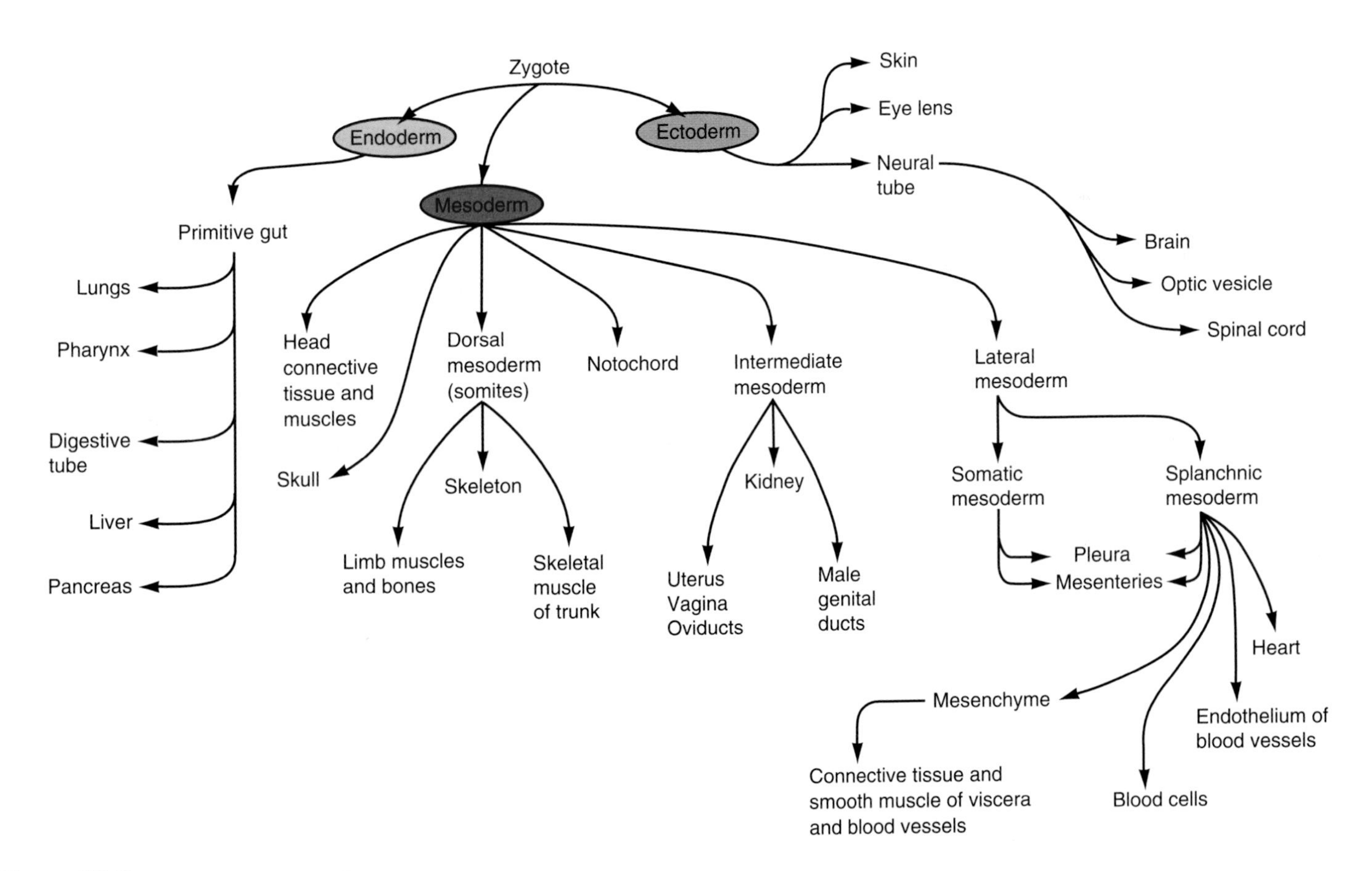

Figure 20.6
Tissues in a vertebrate differentiate along these lines from the three primary tissue layers of the embryo.

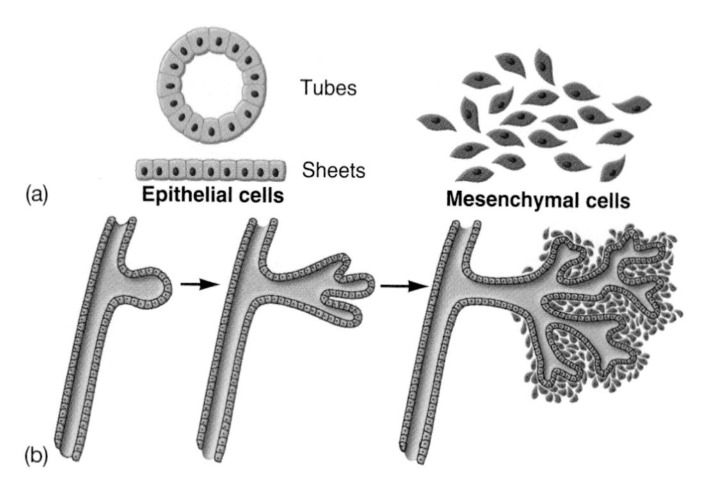

Figure 20.7

(a) Embryonic cells are either epithelial or mesenchymal. *(b)* Some structures arise from the epithelial archenteron as tubular outpocketings, which then become highly branched and surrounded by mesenchyme.

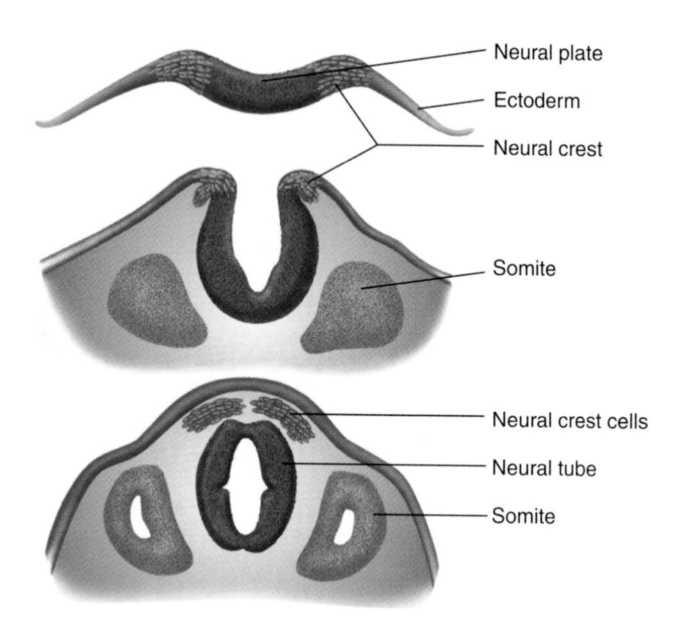

Figure 20.8

The nervous system arises from a tube that invaginates down the back of the embryo, as shown in cross sections at various times. Notice the position of neural crest cells, which spread out through the embryo to form other structures.

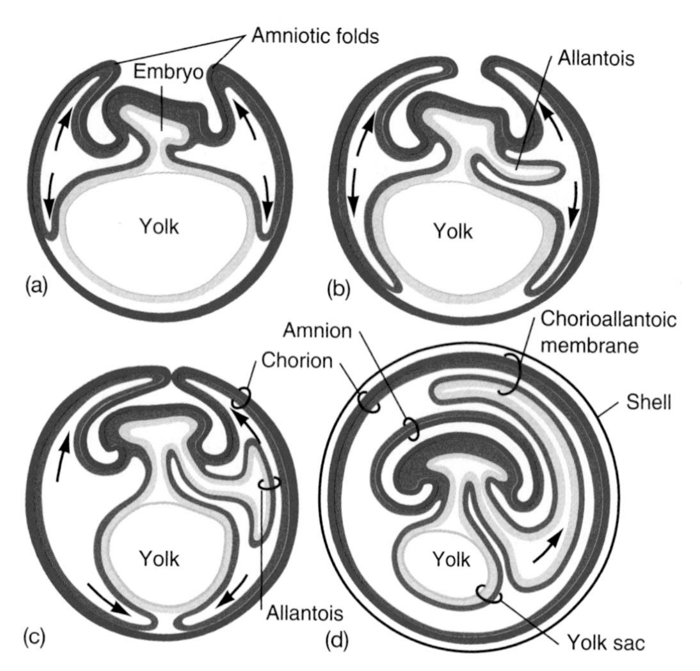

Figure 20.9

(a) Early in its development, the bird embryo extends amniotic folds made of mesodermal tissues *(red)* and ectodermal tissues *(blue)*. The yolk is enclosed in a yolk sac of endodermal tissue *(yellow)*. *(b)* The allantois, made of endodermal and mesodermal tissues, grows out dorsoventrally. *(c)* As the embryo uses the nutrients of the yolk, the membranes continue to grow, and the amniotic folds meet and fuse. *(d)* The embryo is now enveloped in an amnion. Its chorion and allantois form a chorioallantoic membrane that serves gas exchange, and wastes are stored in the allantois.

The transition from the freely growing amphibian embryo (as exemplified by frogs and salamanders) to the enclosed reptile embryo required considerable modification of the embryo itself. The strategy of enveloping the embryo in a closed egg to retain water makes it necessary to provide food as well, and thus reptiles and birds store a lot of yolk in their eggs to nurture the developing embryo. A sea urchin or frog egg has *holoblastic* cleavage, meaning that the entire zygote divides into blastomeres; a reptile or bird embryo has *meroblastic* cleavage, meaning that only a small part of the egg, on one side of the large yolk, becomes the embryo proper. This spot, the **blastodisc,** is equivalent to the frog blastula. It divides into blastomeres and soon splits into two layers, the **epiblast** above and the **hypoblast** below (Figure 20.10). Then a linear invagination, the **primitive streak,** forms down the middle of the epiblast, and gastrulation occurs as epiblast cells migrate into this streak from both sides. These cells move much more independently than do the sheets of tissue that migrate during gastrulation in frogs.

Cells that will become the endoderm displace the hypoblast cells toward the periphery of the embryo; other cells form layers of mesoderm between the endoderm and the remaining epiblast; the latter becomes the ectoderm. Afterward, the neural tube develops in the ectoderm along the axis of the embryo, much as it does in a frog embryo, and an anterior fold of tissue that will become the head begins to rise and separate from the underlying yolk. The embryo is thus spread out over the yolk, and as it consumes the yolk, it gradually rolls up into the adult tubular form—a round

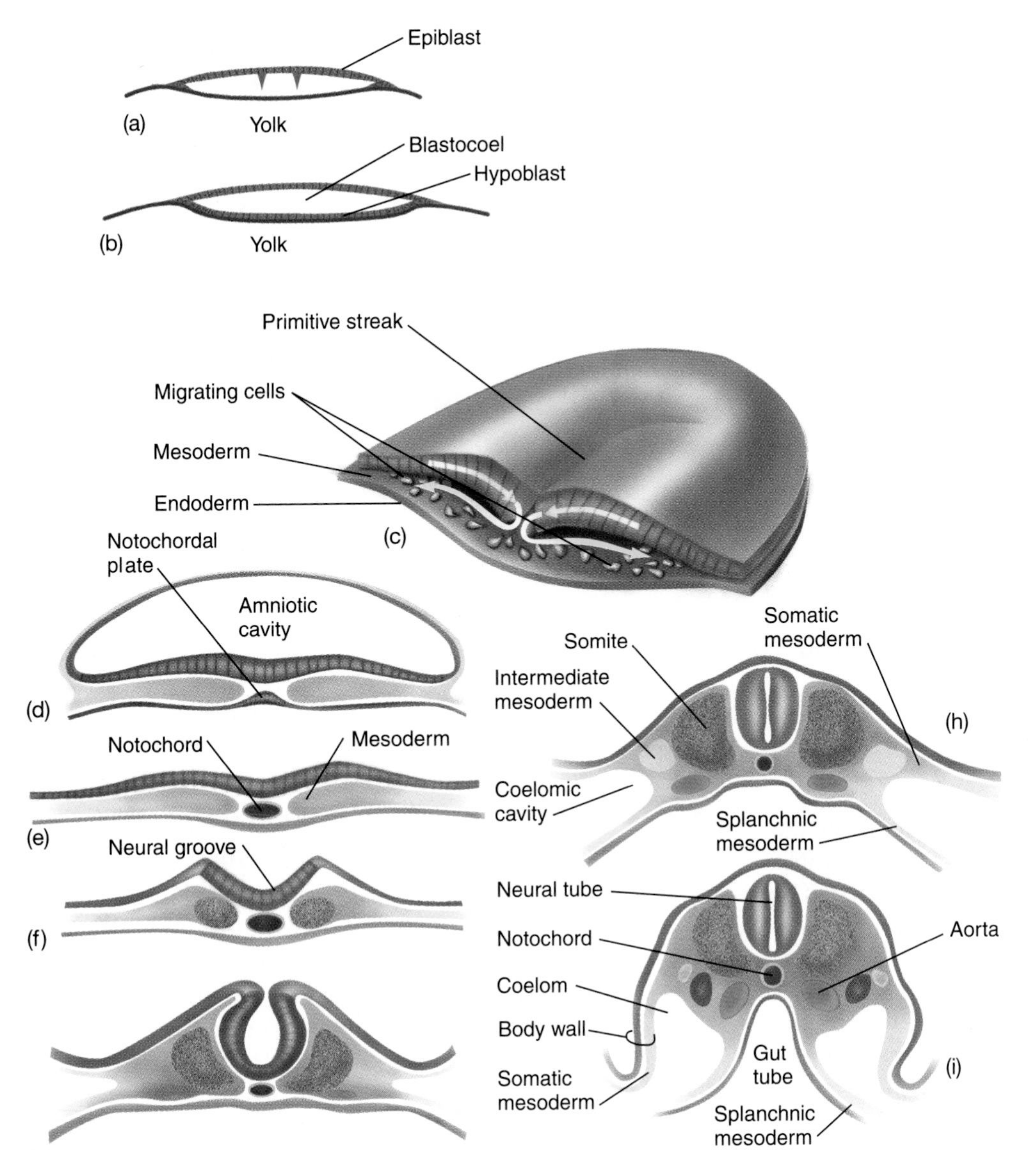

Figure 20.10

(a) A bird embryo develops as a flat disc on the yolk surface. *(b)* Cells leave the upper (epiblast) layer to form the lower hypoblast. *(c)* Gastrulation occurs by the movements of cells inward through the primitive streak. *(d–f)* The mesoderm spreads out laterally into regions whose fates are shown in Figure 20.6. Embryonic mesoderm divides into the somites (muscles and skeleton) and intermediate mesoderm (kidneys and genital system). *(g, h)* At the edge, the mesoderm splits into two layers. Splanchnic (visceral) mesoderm forms visceral components such as the heart, blood vessels, and blood; somatic (body) mesoderm forms membranes that suspend the intestines. These two layers enclose a space which eventually closes to form the coelom *(i)* as the body becomes round. Note the simultaneous formation of the neural tube, gut tube, and aorta. Meanwhile, extraembryonic membranes have been developing around the embryo from sheets of ectoderm, mesoderm, and endoderm.

body wall around a central gut. Meanwhile, the extraembryonic membranes develop and come to occupy much of the egg (see Figure 20.9). The **chorioallantoic membrane,** made by fusion of the chorion and the allantoic membrane, becomes rich in blood vessels and is the embryo's organ for gas exchange.

Mammals evolved from reptiles, and mammalian development is a modification of the reptile-bird pattern (Figure 20.11). The zygote develops into a ball called a **blastocyst,** with an **inner cell mass** from which the embryo proper and some extraembryonic membranes develop, and an outer layer of cells, the **trophoblast,** from which the chorion will grow. After the inner cell mass divides into layers of epiblast above and hypoblast below, gastrulation takes place just as it does in birds, with cells migrating into the primitive streak from both sides. Mammalian eggs have little yolk, and the membranes that evolved in reptiles are used to make the intimate connection between the embryo's and the mother's blood vessels, which supply both food and oxygen to the embryo (though the bloodstreams themselves do not mix). The amnion and chorion develop much as in birds, but in mammals, the chorion and allantois combine with tissues from the wall of the uterus to make the embryonic part of the *placenta,* the organ through which the embryo will be nourished and will exchange gases and wastes.

Further development of a human embryo, Section 51.10.

Exercise 20.3 Take a frog embryo as a model and show the principal modifications that must be made to convert it into a bird embryo.

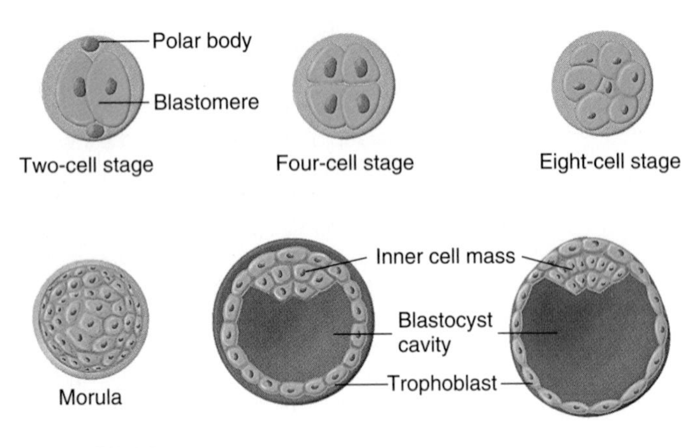

Figure 20.11

Early stages in development of a mammalian (human) embryo are similar to those in other vertebrates, but the embryo will develop from the inner cell mass, while the trophoblast becomes the placenta.

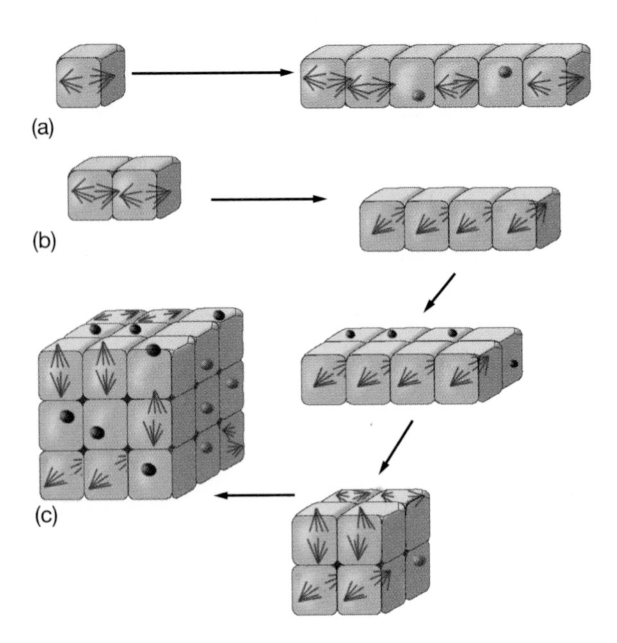

Figure 20.12

(a) A series of cells that divide only in one plane form a filament. *(b)* Dividing in two planes, they form a sheet, and *(c)* dividing in three planes, they form a block.

B. Factors that regulate morphogenesis and growth

20.4 Patterns of cell division determine the form of many tissues.

One major question in developmental biology concerns the morphogenetic forces that mold the embryo into its proper form. Although some tissues are molded into specific shapes by complicated forces, they may also be shaped by two simple factors: cell division in different planes, discussed here, and cell adhesion with different strengths, discussed in Section 20.5.

Just as many architectural structures can be made by piling bricks in various ways, biological structure can be made largely by putting cellular bricks together in columns, sheets, or blocks. Such formations depend on controlling the orientation of the cell-division axis as a tissue grows. Cells that always keep their division axes oriented in the same direction will grow into filaments (Figure 20.12*a*), as in algal filaments or at the growing tip of a plant root. Cells that always divide in one plane make a thin sheet of cells that can form a surface (Figure 20.12*b*), such as the outer layers of a leaf or an animal's skin. Cells that divide in all three directions of space will form a solid block (Figure 20.12*c*).

We don't know what factors orient cell division to produce different shapes, but the results are clearly visible. Animal cells can move about with some freedom and create new forms, as seen in the massive reorganization that takes place during gastrulation. But plant cells are fixed in place by their own cell walls, so plant morphology is determined by controlling the planes of cell division. For this reason alone, bacteria, algae, and plant tissues grow in characteristic patterns (Figure 20.13). Plant structures are often cylinders that grow by cell division and elongation in the axial (lengthwise) direction, sometimes followed by radial cell division to form a thicker cylinder.

20.5 Differential adhesion can determine the arrangement of some tissues.

Differences in the adhesiveness of cell surfaces are especially important in animal tissues. During embryonic development, animal cells move around actively and assemble themselves into structures partly because of the way specific surface molecules cause them to adhere to one another. For instance, cells with such molecules only in a ring around their middles (Figure 20.14) will form a sheet one cell thick. Evidence indicates that many animal cells stick together in certain forms simply because of differences in their adhesiveness.

In 1907, Henry V. Wilson did pioneering experiments to find out if isolated cells would form tissues. By squeezing living sponges through fine-mesh cloth in seawater, he separated them into single cells. These cells began to move around by amoeboid action like unicellular organisms, and as they contacted each other, they stuck together, gradually forming larger and larger clumps. After several days, they began to form typical sponge structures, and in a few weeks, they had reformed into functioning sponges. Furthermore, a mixture of cells from a red species and a purple species sorted themselves out into red and purple masses, and eventually formed two functioning sponges. These experiments suggested that reaggregating cells recognize one another through their surface properties.

Knowing that simple organisms can reaggregate properly from their cells, Aron A. Moscona tried comparable experiments with chick embryos. He dissociated various tissues into individual cells with the proteolytic enzyme trypsin and an agent that removes calcium ions. Moscona found that these cells, too, will gradually move around and adhere to each other,

(a)

(b)

Figure 20.13

(a) Dreparnaldria is a filamentous alga whose cells divide only in a single plane. *(b) Ulva* has the form of thin sheets with its cells dividing in two planes.

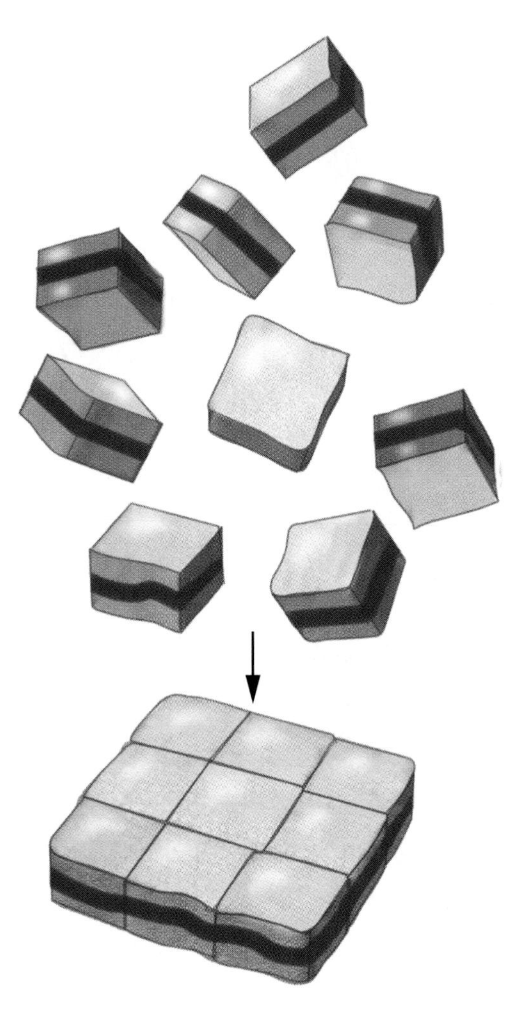

Figure 20.14

Cells will stick together in a sheet if they have sticky patches confined to a band. The proteins could be kept in place by tight junctions or a system of desmosomes and internal fibrils.

making small clumps and then large clumps. If two or three tissue types are dispersed together, they first form clumps of mixed cells, which soon sort themselves out to make well-organized structures. Cells dispersed from an embryonic organ like a kidney will reaggregate to make tissues similar to the structures of the original organ.

These experiments demonstrate that something intrinsic to the cells gives them the information for making proper tissues, which suggests that an organ of any size and complexity could assemble itself if its component cells were simply able to adhere to one another properly. Malcolm S. Steinberg has argued convincingly that some organs form in the right pattern simply because their cells adhere to one another with different strengths. Suppose cells of type A bind to each other more strongly than do those of type B, and that A–B bonds have an intermediate strength. If the two types are then allowed to sort themselves out, they should form a mass consisting of A cells in the center surrounded by a layer of B cells. (You can convince yourself this is true by imagining how the cells will interact as they move around and touch one another.) Now suppose C-type cells bind more strongly than B but more weakly than A. We would expect that, in a mixture of A and C, A would end up in the center with C surrounding; in a mixture of B and C, the C cells would settle in the middle. This is exactly the way embryonic tissues assort themselves (Figure 20.15), showing how this rather simple explanation can account for an apparently complex phenomenon.

Cell adhesion depends on glycoproteins (proteins with sugars attached) known as **cell adhesion molecules (CAMs).** CAMs are either *cadherins,* which require calcium ions, or *immunoglobulin CAMs* that resemble antibody molecules and are independent of calcium ions. CAMs are specific for different tissues; in mammals, N-cadherin (neural cadherin) functions on nerve and heart cells, E-cadherin on epithelial cells and on all early embryonic cells, and P-cadherin on placental cells and epithelia. Both classes of CAMs are transmembrane proteins anchored in the plasma membrane with extracellular domains that can bind specifically to CAMs of other cells, thus joining the cells together (Figure 20.16). Small proteins (catenins) link the cytoplasmic ends of cadherin molecules to the actin filaments of the cytoskeleton, forming a continuous intracellular and extracellular network able to transmit forces that shape and move cells. Immunoglobulin CAMs have extracellular regions similar in structure to antibody molecules (see Section 48.7). By sticking to each other as cadherins do, these CAMs can bind cells more strongly, or they can be modified to decrease cell adhesion by the addition of sialic acid, a negatively charged sugar containing D-mannose and pyruvic acid. Sialic acid attracts positive ions and water, thus making the cell surface rather slippery, so cells with a lot of sialic acid on their surfaces tend to repulse one another rather than to adhere (Figure 20.17).

Actin's role in cell shape and movement, Chapter 11.

As development proceeds, cells make and break connections with each other, and the timing and tightness of these connections in part involves interactions among the various types of CAMs in the membranes of different cells. Neural crest cells, for example, are held together by N-cadherin in the neural crest region; they lose the cadherin while migrating to their destinations and then acquire it again when forming their characteristic adult structures. Each cell type may have a distinctive set of recognition proteins, which make it adhere to its neighbors or repulse its neighbors as necessary in order to move it into its correct position in the developing embryo.

Exercise 20.4 Cells of types P and Q bind to one another in all combinations, but Q cells bind very tightly to other Q cells and less tightly to P cells. What kind of structure will these cells form?

Exercise 20.5 M cells bind tightly to one another, and N cells bind tightly to one another—but M and N cells do not bind to each other at all. If these cells are mixed together, how will they associate?

20.6 Cells bind to an extracellular matrix and migrate on it.

Tissues of plants and animals consist of more than the cells themselves (see Section 32.8). In addition to the junctional proteins that weld them together (see Section 6.9), animal cells secrete an **extracellular matrix** of proteins and polysaccharides. In connective tissues such as bone and cartilage, this matrix constitutes the bulk of the tissue; in epithelia, the matrix is only a thin **basement membrane.** The matrix of animal cells is complex:

- The matrix consists largely of a network of collagen fibrils.
- The collagen is supplemented by *proteoglycans,*

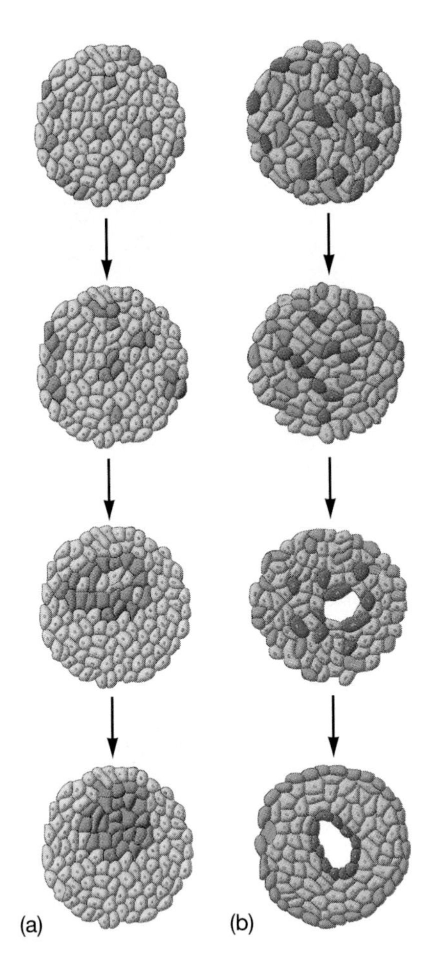

Figure 20.15
If cells from different embryonic tissues are disaggregated and mixed, they will assemble themselves into recognizable tissues. In *(a),* epidermal cells surround neural tissue; this will occur if neural cells adhere to one another more tightly than do epidermal cells. In *(b),* mesodermal cells segregate in the middle; they will do this if they adhere to one another with a strength intermediate between that of the other cells.

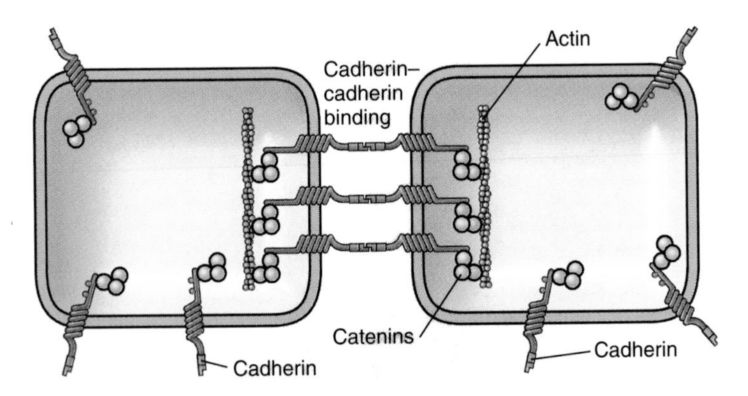

Figure 20.16
Cadherins are bound into the plasma membrane, and they can bind to cytoplasmic proteins such as the actin of the cytoskeleton. The external domains of the cadherins have complementary structures, so they bind to each other and hold cells together, at the same time organizing the actin molecules.

mucopolysaccharides linked to fibrous proteins (Figure 20.18*a*).

- On top of these fibers are adhesive proteins called laminin, fibronectin, and tenascin.
- Cells of the tissue attach to the adhesive proteins by means of *integrins.*

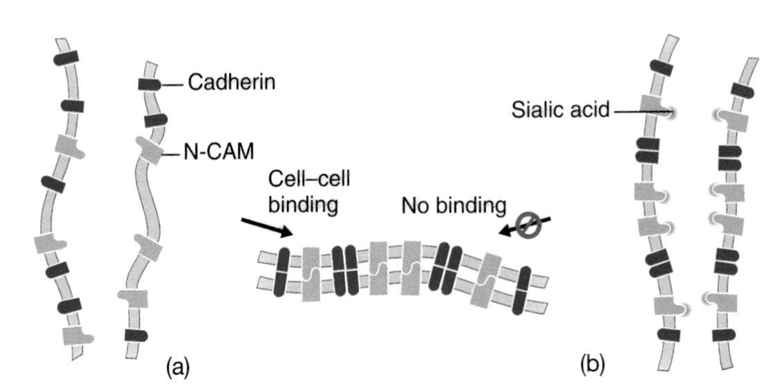

Figure 20.17

(a) Neural cell adhesion molecules (N-CAM) recognize each other and can bring cells close together so they can be linked more strongly by cadherins or other molecules. *(b)* N-CAM molecules can carry varying amounts of sialic acid, which prevents their binding to each other. Cells can strengthen or weaken their affinity for one another by changing the amount of sialic acid.

Integrins are a family of plasma membrane proteins that attach specifically to the proteins of the matrix: some to laminin, others to fibronectin, and so on. They get their name because they integrate the intracellular cytoskeleton to the extracellular cell-matrix connections, just as cadherins link the cytoskeleton to cell-cell connections (Figure 20.18*b*). Plasma membranes are rich in integrins, but each molecule binds quite weakly, allowing a cell to move a little on its matrix much as a centipede moves with some of its feet while resting on others. During mitosis, cells normally become quite spherical as they lose these contacts that have been holding them in their characteristic shapes.

Embryogenesis depends on cellular movements—gastrulation, neurulation, the migration of neural crest cells, and the migration of cells that will become blood cells and germ cells—sometimes over considerable distances. These movements must be guided. Some cells are guided by chemotaxis; white blood cells to be converted into specialized cells called T lymphocytes migrate into the embryonic thymus gland because they are "homing" on characteristic thymus proteins. Other cells migrate along protein gradients established on an extracellular matrix; prospective mesodermal cells and neural crest cells migrate along a base of fibronectin because their surface integrins remain bound to the underlying fibronectin. Here we see the importance of integrins tying the cell surface to the cytoskeleton, so as the integrins search out the right path, they guide the cell's movements.

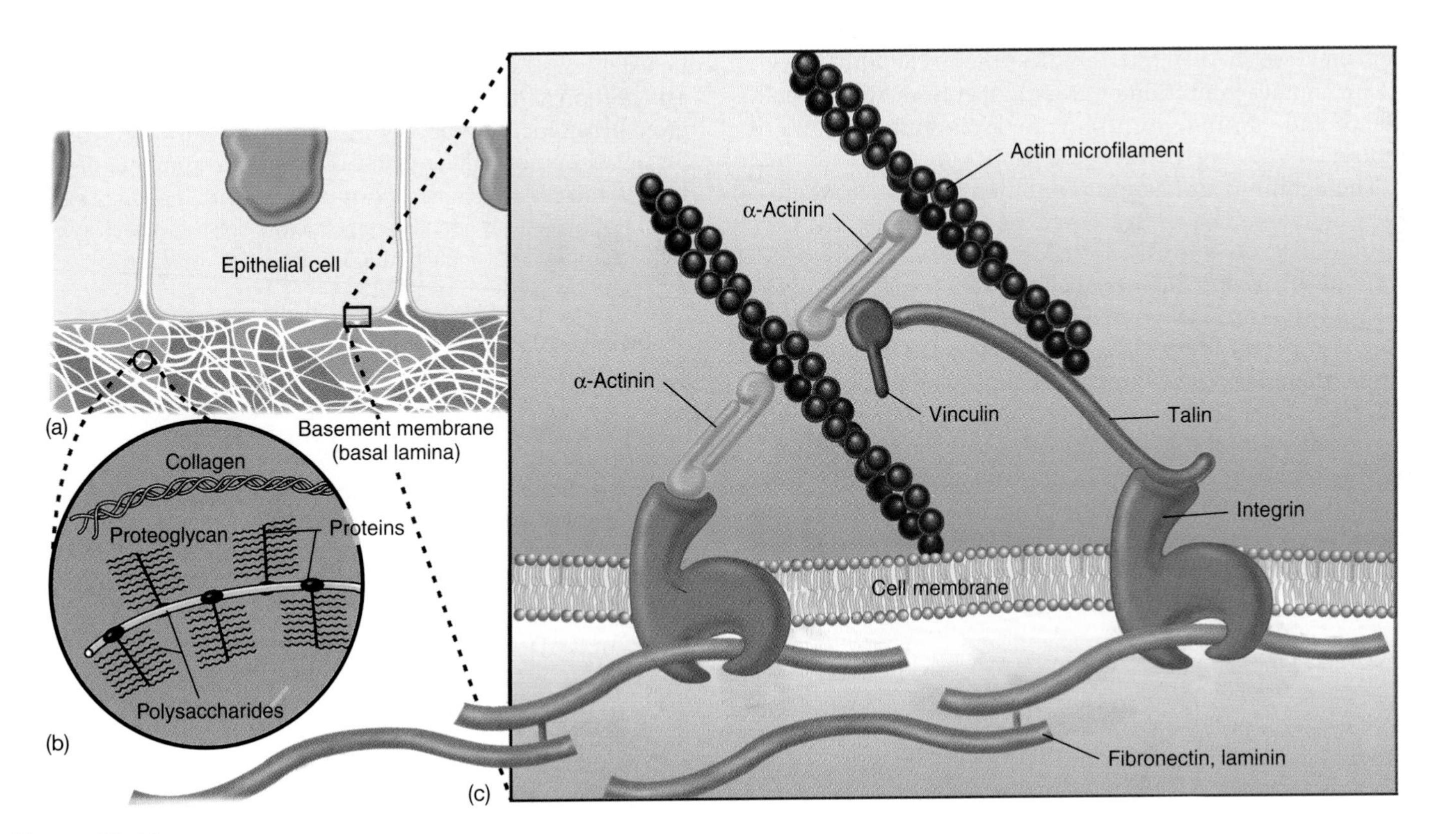

Figure 20.18

Cells in a tissue *(a)* are held together in part by an extracellular matrix, which is a network *(b)* of collagen fibers, proteoglycans, and other proteins such as laminin and fibronectin. *(c)* The matrix is attached to the cell by integrins, which are transmembrane proteins that link adhesive proteins of the matrix to intracellular actin microfilaments.

20.7 Contact inhibition normally restricts the division and movement of cells.

Cell migration is partially regulated by **contact inhibition,** the repression of cell division and movement when cells are in contact. As the cells in a tissue come into contact, they exchange signals that regulate their activities. Animal cells in tissue culture move around vigorously, extending a thin "foot," or *lamellipodium,* at their forward edge. The lamellipodia of two cells suddenly stop their movements when they come in contact. They may form lamellipodia on another side and move off in other directions, but as cells become more crowded in a culture, they inhibit one another from all sides and eventually settle down and stop migrating. Contact inhibition can be regulated. For example, the migration of embryonic cells may be guided in this way: As neural crest cells lose their N-cadherins when they start to migrate, they also lose contact inhibition and move off in different directions. Eventually, migrating cells reach their destinations, reacquire contact inhibition, and stay in place.

Cell contact may also inhibit mitosis. Tissue-culture cells normally grow and divide on an approximately 20-hour cycle, but as they become packed together, their rate of mitosis falls until they finally stop dividing entirely. Contact inhibition of both division and movement keeps the normal cells in every tissue and organ confined to their proper limits rather than growing indefinitely into nonfunctional lumps. Although the movements of embryonic cells are inhibited when they are in their proper sites, they generally experience no contact inhibition of growth; therefore, they remain in place and continue to proliferate as an animal grows to adult size.

The significance of contact inhibition is best seen where it fails. Though tumor cells are packed together tightly, they continue to grow and divide wildly because, among other things, they have lost their contact inhibition mechanisms. A wart, for instance, is a common kind of benign tumor. It is the result of a specific warts virus that infects some skin cells and changes their pattern of growth. Normal skin cells grow in a thin layer, under control by contact inhibition. If the skin is cut, the cells along the edge of the cut lose their inhibition and begin to grow and move again until they have repaired the wound. But cells altered by the warts virus are insensitive to inhibitory signals from other cells, so they continue to grow into a clump.

Exercise 20.6 Each integrin is made of two subunits, α and β. Several types of each subunit are known, and each α-β combination binds best to a different matrix protein. Explain how this structure produces diverse specificity with a minimum of genetic information.

Exercise 20.7 A small cluster of embryonic cells remain in place but continue to undergo mitosis. Are they contact inhibited?

20.8 Specific proteins regulate the growth of tissues.

Cell proliferation, or growth, is an inherent part of development. The cell cycle, as described in Chapter 13, is subject to regulation by external **growth factors.** In Chapter 39, we will show how several small organic molecules (plant growth factors or hormones) regulate the growth and proliferation of plant tissues. Animals have a number of regulatory protein growth factors. In vertebrates these include a nerve growth factor (NGF) that stimulates certain nerve cells, a brain-derived growth factor that stimulates other nerve cells, and an epidermal growth factor (EGF) that stimulates epidermal and several other types of cells. Some factors stimulate specialized cells; for instance, a long list of proteins called *cytokines* control growth and interactions among cells of the immune system. Some growth factors, like EGF, are **mitogens** that induce growth and mitosis. By binding to cell surface receptors, they turn on genes that promote cell division through protein kinases, including the cyclins and cell division kinases (Cdks) that drive the cell cycle. Cell growth is regulated by complicated systems that check and balance one another, usually ensuring normal cell proliferation. Several genes that encode proteins of the system are what we will describe later as *proto-oncogenes* (see Section 20.15), genes that, when mutated, promote the unregulated cell growth of cancer.

NGF and some other growth factors, however, are not mitogens. When taken up by nerve cells in the sympathetic nervous system, NGF assures their survival and guides their characteristic extensions (axons) toward their target cells. Without NGF, this portion of the nervous system does not grow. Specialized growth factors are probably needed to support other tissues. Since tissues normally contain only minute amounts of these factors, they have been hard to study, but there is good reason to think that relatively few different factors exist and each type of cell is stimulated by a particular combination of them.

Exercise 20.8 Plant proteins known as *lectins* are nonspecific mitogens that provoke nondividing animal cells into growth and mitosis. Phytohemagglutinin (PHA) from kidney beans, for instance, is a mitogen that also agglutinates certain blood cells—makes them stick together. Concanavalin A (Con A), a protein in jack beans, agglutinates tumor cells by binding between their cell surface oligosaccharides. Con A also agglutinates normal cells, but only if they are in the midst of mitosis. What does this fact about Con A indicate about the surface structures of normal cells and tumor cells?

20.9 Sheets of tissue may be shaped by several factors.

In addition to cell adhesion, much of morphogenesis can be understood as the result of three other factors: how cells elongate and contract, differential cell division, and the deposition of extracellular fibers that form a supporting matrix.

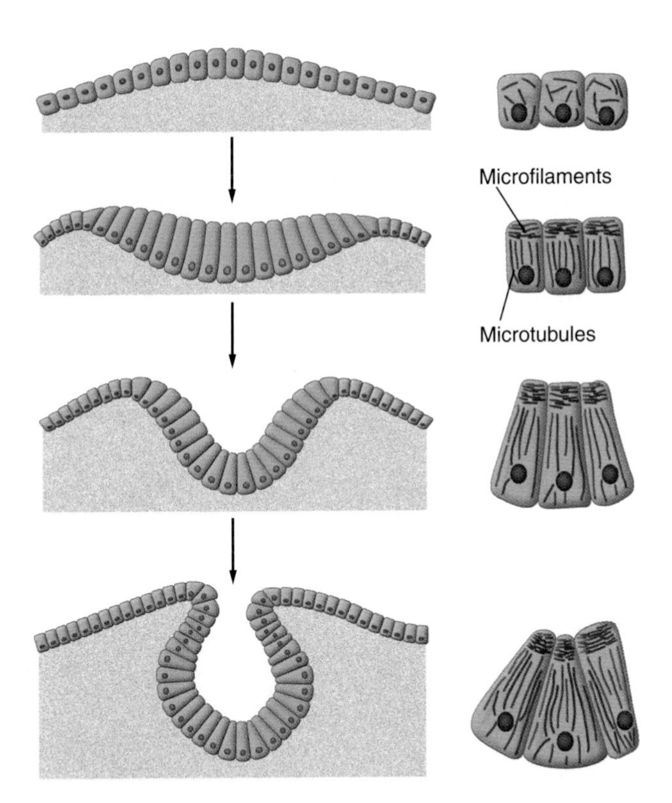

Figure 20.19

The neural tube is shaped by two processes. Neural plate cells become elongated by the growth of microtubules in one direction; then microfilaments at one end contract to draw the cell into a wedge shape and roll up the whole layer of tissue.

As the sheet of neural plate cells rolls into the neural tube (Figure 20.19), the cells in the center of the plate elongate due to the extensive microtubules running lengthwise through them. Then bundles of actin filaments at upper ends of these cells contract quite suddenly, throwing each cell into a wedge shape. This coordinated movement in the whole sheet makes it bulge downward and assume its characteristic form. These cellular changes are coordinated in part because the contractile belts of actin and myosin fibers are linked from cell to cell. Consequently, the contraction of one cell generates a tension that extends to its neighbors, and eventually the whole sheet is contracting in the same way.

Linkage of the cytoskeleton among cells, Section 6.9.

The same kind of movement, with additional factors, accounts for the shaping of sheets of epithelium into tubes in such organs as the lungs, salivary glands, and pancreas. Recall that these organs form around tubes that branch off the gut tube (see Figure 20.7) and then become highly folded and branched as they mature (Figure 20.20). Some of the folding and branch formation is caused by changes in cell shape, as in neurulation, but at least two other factors are involved: (1) Cells divide fastest at the tip of a tubule, so the accelerated cell proliferation in this region elongates the tubule. (2) At the same time, clefts form

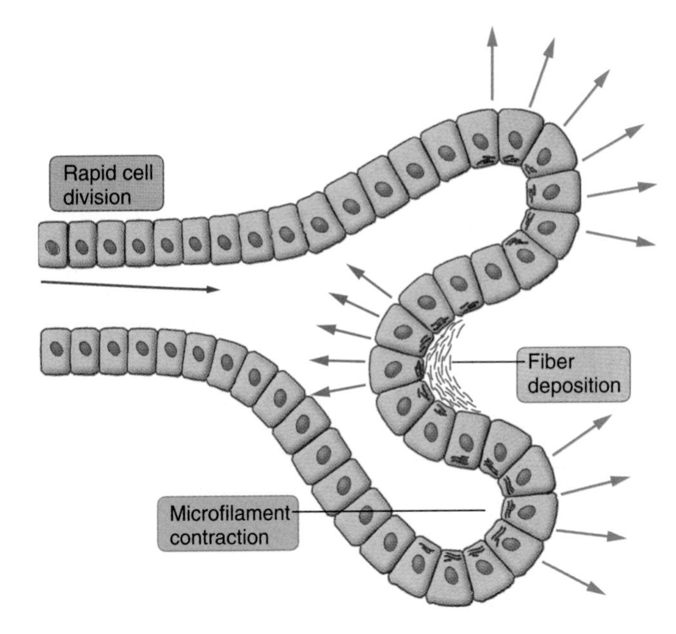

Figure 20.20

Tubules form as a result of at least three processes. Rapid cell division in some regions extends the tubules outward. Invaginations (infoldings) occur where supports of extracellular fibers are laid down, and the invaginations may be created by the combination of forces, including microfilament contraction, that create folding of tissues, as in neural tube formation.

between branches of the tubule at points where large amounts of extracellular matrix fibers are being laid down, a matrix that will eventually support the whole branched structure.

The more profound questions about morphogenesis are yet to be asked and answered. We still don't know why forces such as rapid mitosis or contraction of microfilaments are set in motion just when and where they are, or why an extracellular matrix is laid down only in certain places. So while the forces themselves are relatively simple and many can be understood in terms of known mechanisms, their regulation remains a mystery.

Exercise 20.9 Draw a row of several adjoining cells. Then show how a combination of microtubules and microfilaments in the right places could make the row bulge in the middle and contract at the ends to form a shape like the lens of the eye.

C. Cancer and the regulation of growth

A normal embryo develops its form through tightly regulated patterns of cell division and movement. To successfully complete the internal genetic programs determined by their genomes, cells must have functional internal controls, but they must also be able to receive signals from neighboring cells and respond to them appropriately. If any of these controls and signal systems fail, the embryo may become badly deformed, sometimes developing into a mass of tissue that hardly resembles an organism at all. Defects in these controls and signal systems later in life can also produce abnormal growths, including the serious, destructive growths called cancers. We will discuss cancer here because of its intrinsic interest in human health and the light it sheds on normal growth and development.

Cancer is a major (and often preventable) cause of death in all countries, accounting for about one death in five in Europe and North America. Although heart diseases cause more deaths in these regions and infectious disease and malnutrition are more devastating in many other parts of the world, cancer remains a major problem everywhere. Studies of the causes and progression of cancer have alleviated much human suffering, while also enormously increasing our understanding of normal cell growth. Here we will introduce some ways of thinking about cancer, about how it develops and how it can be prevented.

20.10 Cancer cells escape restraints on reproduction and invade areas of normal tissue.

One abnormal cell that begins to reproduce without the controls that normally keep growth in check will grow into a mass of abnormal cells called a **tumor,** or **neoplasm.** Some tumors simply grow in place without causing any damage, except for devouring more than their share of nutrients and perhaps creating a mechanical blockage in the body as they enlarge. Such tumors are **benign.** Their cells remain differentiated and may not differ much from those of normal tissue. Benign tumors may remain quite small unless they become vascularized through growth of a branch of the circulatory system. More serious tumors are those whose cells are able to invade other tissues; these are cancers, or **malignant** tumors (Figure 20.21). All tumor cells have lost some contact inhibition of growth, but cancer cells become dedifferentiated, and, like embryonic cells, have different surface molecules and have also lost contact inhibition of movement. Cancer cells tend to break away from their parent tissue and migrate through blood vessels and body cavities. By making enzymes that degrade parts of normal tissues, they are able to set up invading colonies of proliferating cells in tissues far away; this process, called **metastasis,** is the main reason cancers are so hard to fight. Once a cancer has metastasized, it is very hard to locate and surgically remove every colony. Then it must be treated with radiation therapy or chemotherapy (using chemical agents that inhibit or destroy the cells), or in more experimental procedures, by using the body's natural immune defenses.

Many cancerous cells superficially resemble the normal cells from which they are derived. Cancers called **carcinomas,** for instance, grow from epithelia, and their cells resemble normal epithelial cells. About 80 percent of all human cancers are carcinomas, which develop in the skin, mouth, throat, stomach, colon, rectum, lung, breast, uterus, ovary, and prostate gland. **Sarcomas** are cancers that originate in muscle or connective tissue, and **leukemias** are derived from white blood cells.

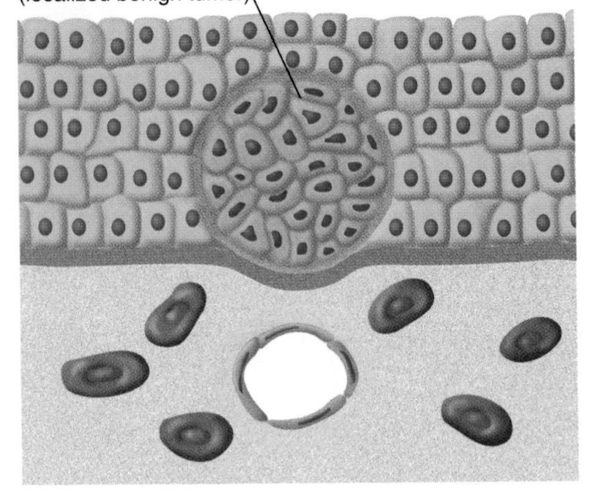

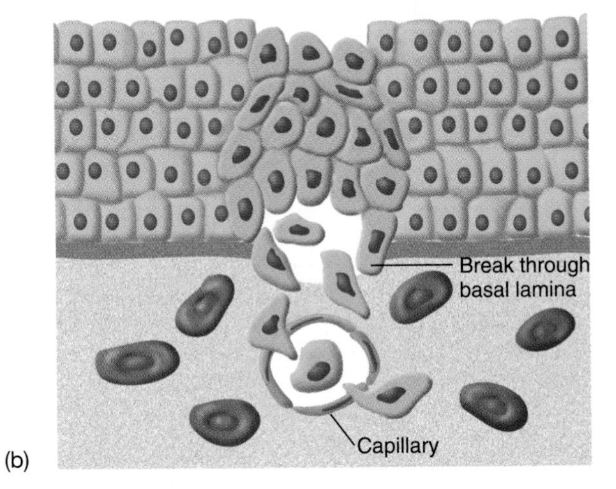

Figure 20.21
A tumor is a mass of cells that grows without normal controls. *(a)* Benign tumors simply grow in place. *(b)* Malignant tumors metastasize as cells escape into the bloodstream or body cavities, migrate to other tissues, and invade them.

20.11 Cancer, like evolution, involves natural selection.

Understanding cancer depends on the ideas of natural selection, since cancer cells compete with other cells much as organisms compete with one another in biological evolution. Normal tissues in an animal coexist with one another like the organisms in an ecosystem, although they are not in competition and have control mechanisms that keep them growing harmoniously. Cancer cells destroy this harmony by competing with their neighbors. If Darwin had observed the progression of cancer in an animal, he might have seen the principles of competition acting in a population of cells, just as in a population of organisms. The factors that are important in natural selection among individual organisms also apply to competition among individual cells: the rate of mutation to alternative genomes, the number of cells involved, the rate at which the cells reproduce, and the selective advantage mutants have over normal cells. In evolution, of course, the result is a degree of success—an enduring ecosystem and its component species; in cancer, the result is a failure—the death of the organism.

20.12 Cancers are derived from single cells through DNA modification.

Several lines of evidence lead to the conclusion that cancers are **monoclonal,** derived from a *single altered cell* that proliferates wildly. One kind of evidence comes from the fact that an X chromosome is randomly turned off early in development in every cell of a female mammal, and the same chromosome remains inactivated in all the cells of a clone. In a woman who is heterozygous for an X-linked gene, all the cells of a cancer have the same X chromosome inactivated, indicating that they grew from a single cell.

X-chromosome inactivation, Section 19.7.

Even more rigorous proof of the monoclonal origin of cancer comes from DNA analysis. Chronic myelogenous leukemia results from a translocation between chromosomes 9 and 22. Sequencing the DNA to determine the precise location of the chromosome break and reunion shows that all the leukemic cells in a patient have precisely the same sequence where the chromosome fragments join, meaning that they all came from a single normal cell that experienced the translocation. Leukemic cells from other individuals with the disease have different sites of breakage in the same general region.

All the experimental evidence indicates that the initial cell of a cancer, the "founder cell," arises from an alteration in its DNA. A cancer might also result from altered *expression,* rather than a change in the DNA itself, such as inactivation of part of the genome much as an X chromosome is converted into a Barr body. However, such changes in expression are rare. *Most cancers arise from mutation of somatic cells.*

Sir Percivall Pott observed in 1775 that young men in London who worked as chimney sweeps suffered abnormally high rates of cancer of the scrotum. In an era when personal cleanliness left much to be desired, he correctly attributed the cancers to the accumulation of chimney soot and tars on their skin. Since then, many chemical compounds have been shown to be cancer-causing agents, or **carcinogens** (Methods 20.1). One

Methods 20.1 The Ames Test

Investigators have identified and studied many carcinogens, using colonies of mice, rats, and other animals as test subjects. To test a compound or agent for its carcinogenic potential in an animal, one must have enough animals susceptible to cancer to give a statistically meaningful result. One must also control concentrations, times of exposure, methods of application, the genetic background of the animals, and many other variables. However, Bruce Ames devised a much simpler test to screen for potential carcinogens by measuring their effect on mutation rate. The test uses mutants of the bacterium *Salmonella* that have a defective gene for one enzyme for biosynthesis of the amino acid histidine (*his* mutants); only bacteria that have mutated back to the his^+ state can grow on a medium without histidine. A suspected mutagen is added to petri plates containing a growth medium lacking histidine. Bacteria are spread on these plates, and if the substance is an effective mutagen, colonies of his^+ bacteria will appear at a rate greater than on the control plates. To more nearly mimic the behavior of the test compound in an intact animal, the plates are supplemented with an extract of rat liver taken from rats that have been given drugs to induce the formation of liver enzymes that alter these drugs. Theoretically, these enzymes have evolved because they are able to inactivate toxic substances, and they convert the compound being tested into a variety of related substances. Ironically, these supposedly protective enzymes sometimes convert materials that are not themselves carcinogenic (that is, mutagenic) into carcinogens, instead of the other way around. Liver extract with these enzymes is used in the Ames test to reconstruct what presumably happens in the body.

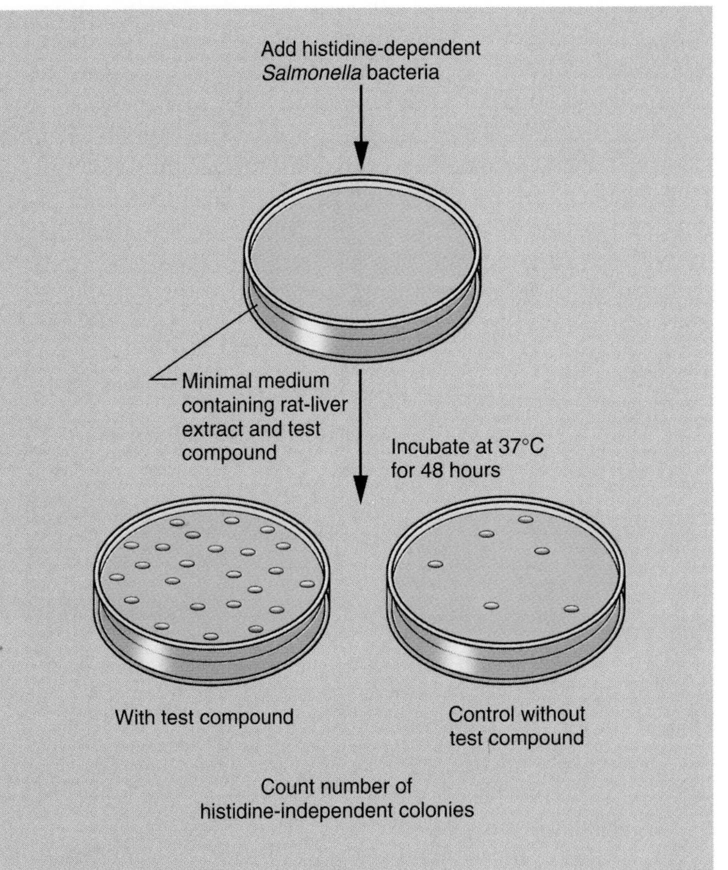

The Ames test is a highly efficient way to screen compounds for their mutagenicity. Obviously, working with vast numbers of bacteria and short experimental times is far superior to working with large colonies of experimental animals and testing periods that could span years. Ames reasoned that mutagens identified in his test would also be carcinogens, and the test does show a very high correlation between an agent's mutagenic ability and its carcinogenic ability as measured by more traditional and direct methods, such as causing tumors in mice. Many industrial products and by-products have been identified as mutagens by this test, and in laboratories all over the world, the test is a standard procedure in screening compounds for possible carcinogenicity.

highly carcinogenic chemical (2-naphthylamine) caused bladder cancer in virtually all the men who had worked with it in a British chemical factory in the early 1900s. We now see that carcinogenic compounds are the kind of molecules that either interact directly with DNA or are converted into forms that interact with DNA. And, most important, they are *mutagens,* compounds that cause mutations. Many studies show that most carcinogens are also mutagens. Radiation, such as X-rays and ultraviolet light, is both a mutagen and a carcinogen.

20.13 Several independent events are needed to change a normal cell into a fully cancerous cell.

Thinking about cancer, by laymen and experts alike, was misled for many years by the reasonable, but wrong, assumption that a single event could turn a normal cell into a cancer cell. Searching for *the* cause of cancer is futile because a normal cell and its progeny undergo not one but several independent events before becoming cancerous; thus one must search for several causes. Carcinogenesis should be viewed as a progression of events.

This view is supported by the age-old observation that people are more likely to develop cancer as they grow old. Figure 20.22 shows the age at which English women have first been diagnosed with colon cancer. If women of all ages were equally likely to develop this cancer, the curve would have been a horizontal line; instead, it begins to rise steeply around age 40. Analysis of this curve for colon and other cancers leads to estimates that five or six independent events lie behind most cancers. And of course the probability of accumulating all these events increases with time. That is, if it takes five events to develop a cancer and the average person sustains one event per 10 years, clearly someone who has lived for 50 years is more likely to develop cancer than one who has lived for only 20 years.

Because several events are needed to produce a cancer, histologists (biologists who study tissues) can often recognize intermediate stages between normal cells and fully cancerous cells by microscopy. This is the basis for the well-known Pap smear test, a technique developed by George Papanicolaou for detecting early stages of cervical cancer (Figure 20.23). Normal

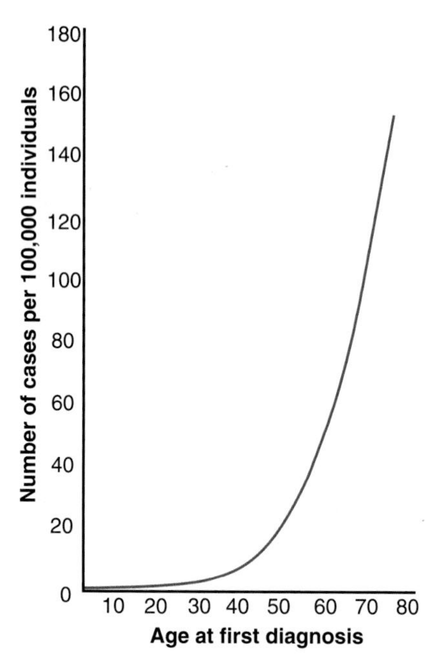

Figure 20.22

The number of newly diagnosed cases of colon cancer among women in England and Wales increases as a function of their age. If cancer were caused by single events that could occur at any age, the graph would be flat. The incidence clearly rises dramatically with age, showing that cancer is caused by a sequence of several events; mathematical analysis suggests that five or six events are necessary to produce a cancer.

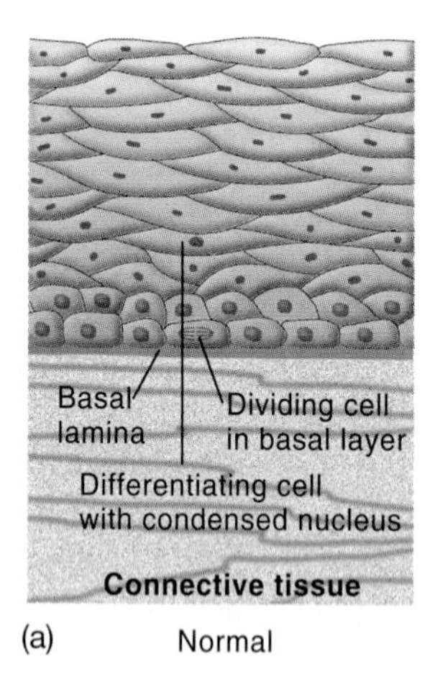

(a) Normal

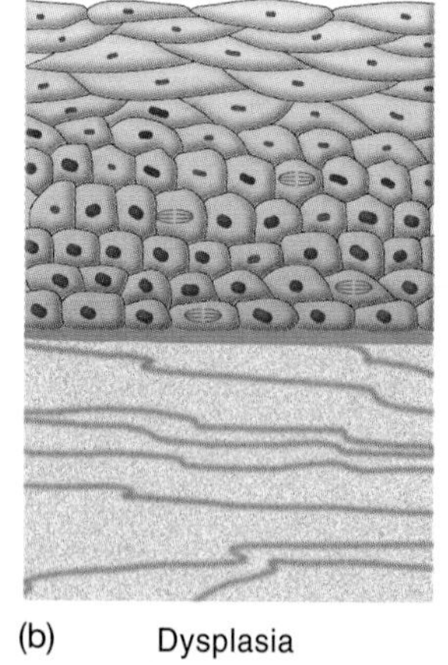

(b) Dysplasia

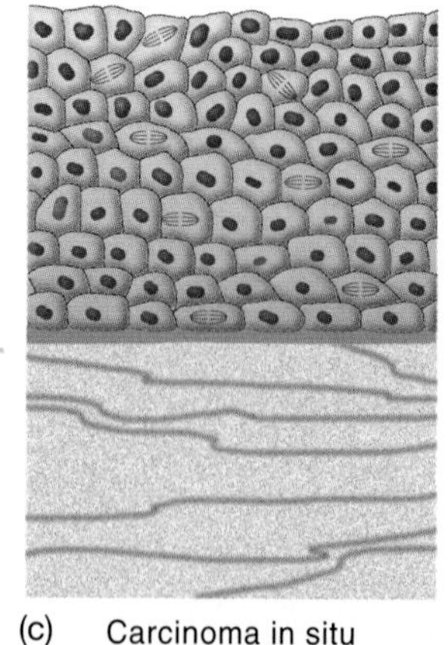

(c) Carcinoma in situ

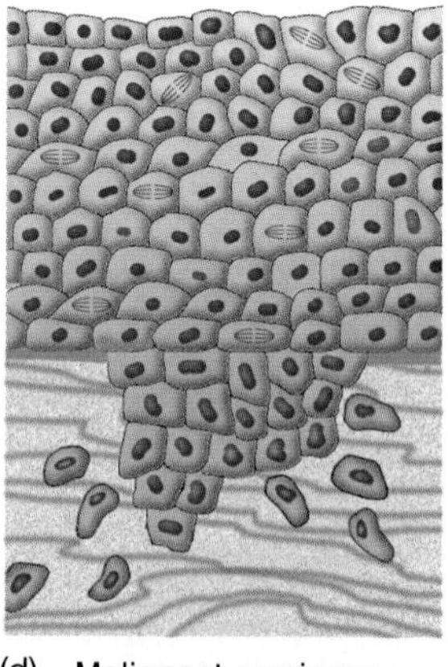

(d) Malignant carcinoma

Figure 20.23

In the Papanicolaou technique (Pap smear), cells are scraped from the uterine cervix and examined by microscopy for signs of carcinoma, whose stages of development are typical of other carcinomas. *(a)* Normal cells are well differentiated and have small nuclei. *(b)* In the stage known as dysplasia, many cells are differentiating and have larger nuclei. *(c)* In a carcinoma remaining in place (in situ), all the cells are proliferating and undifferentiated. *(d)* A malignant carcinoma begins when cells cross the basal lamina and start to move into other tissues.

epithelial cells scraped from the cervix have small nuclei and rather large cytoplasmic volumes, but those in the early stages of progression toward cancer have a different shape, with large nuclei. To the experienced observer, they are clearly abnormal; although obviously related to normal cervical cells, they have taken on a different identity. Later, more abnormal cells, with more pronounced abnormalities, will be seen. Still later, fully cancerous cells appear that have large nuclei and a small amount of cytoplasm, and that are no longer firmly attached to the underlying tissue. Recall the principles of natural selection: At each stage, the modified cells have a competitive advantage over their neighbors. The first modified cells have a slight advantage over normal cells, and each subsequent modification produces a still more successful cell, until the progeny outgrow normal cells as well as abnormal cells in the previous stage of change.

20.14 Both tumor initiators and tumor promoters contribute to the transition from normal cells to malignant ones.

Carcinogens can act as initiators or promoters. **Tumor initiators** are agents that alter DNA and initiate the long progression toward cancer. Coal tars and tobacco smoke contain compounds such as benzopyrene that are tumor initiators and may cause skin cancers. If applied to the skin, they alter the DNA in some cells, but skin cancer may not appear for years, if ever. Often, years or even decades pass between exposure to a tumor initiator and the detection of cancer. Other agents, called **tumor promoters,** enhance tumor formation during this time, generally by increasing cell proliferation. Initiated cells divide frequently under the influence of the promoter, producing many more initiated cells, which are subject to still further mutation. Each event carries the cells along the path to the cancerous state. Cancer only develops if a tumor initiator acts before a promoter does and if the promoter is intense enough:

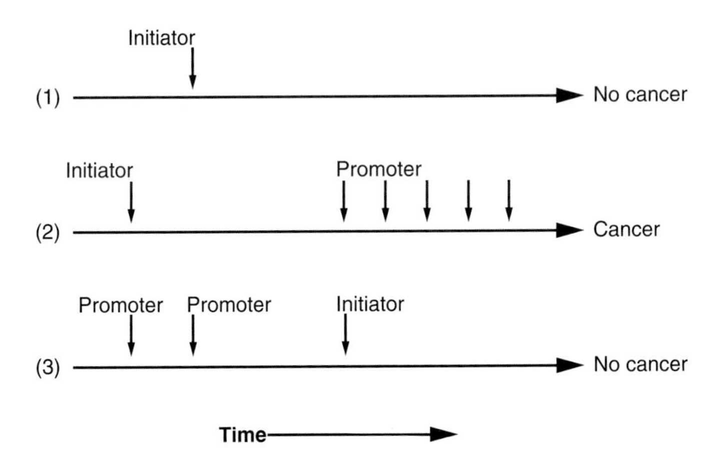

Promoters hasten the appearance of cancer if applied to the skin following initiation, and the more promoter that is applied, the sooner the cancer appears.

In sum, the progression toward cancer begins with modification of the DNA of a single cell, which then outcompetes its neighbors in growth. Its progeny, carrying the modified genome, in turn undergo successive mutations and periods of proliferation until finally, fully malignant cells result. The scenario is straightforward:

$$\text{Mutation} \rightarrow \text{Proliferation} \rightarrow \text{More mutation} \rightarrow \text{More proliferation} \rightarrow \ldots \rightarrow \text{Cancer}$$

Clearly, any modification of a cell that favors its proliferation leaves it in a preferred position to become malignant. Next we will see that the relevant mutations occur mostly in genes of the pathways that transduce intercellular signals and regulate cell growth.

20.15 Some viruses can transform normal cells into tumor-producing cells.

In 1910, Peyton Rous, a young medical school graduate, noticed a chicken carcass in a poultry market that was riddled with cancerous growths (sarcomas), and he began to search for the agent that might have caused them. During the late nineteenth century, techniques had been developed for distinguishing between two agents of infectious diseases, bacteria and viruses (see Section 29.9); bacteria were defined as agents able to be captured in a ceramic filter, whereas viruses are small enough to pass through such a filter. Rous was able to initiate cancer in a healthy chicken by injecting extracts from a chicken sarcoma. Since the agent responsible passed through filters, it was identified as a virus, now known as Rous sarcoma virus (RSV). Rous and his colleagues spent decades investigating viruses as causative agents of cancer, at first with little support from the scientific community but later as recognized pioneers. This work earned Rous the Nobel Prize for Medicine in 1966.

RSV is a *retrovirus* (review Figure 18.34). Its single-stranded RNA genome is copied into DNA and incorporated into its host's chromosome. Under the right conditions, the viral DNA will remain a stable part of the genome and will be replicated with it for generations. Normal fibroblast cells (embryonic connective tissue cells) grow in a culture medium and proliferate, with regulation by contact inhibition between neighboring cells, forming a single layer of cells attached to the bottom of a culture flask. However, cells infected with a tumor virus such as RSV round up, proliferate rapidly, lose contact inhibition, grow beyond a single layer, and are said to be **transformed.** (Don't confuse this sense of "transformation" with the transformation of a bacterial cell when it takes up DNA.) When injected into an appropriate animal, these transformed cells form tumors. RSV and other viruses that are able to transform cells are **tumor viruses,** and they play an important role in cancer studies. They transform cells by incorporating tumor-promoting genes called **oncogenes** (*oncos* = mass, meaning a tumor) into the cells' genomes. RSV carries the RNA for one oncogene, called *src.*

The DNA of normal vertebrate cells carries a region very similar to the RSV *src* gene—not identical, but close. This region in the normal cell can mutate into an oncogene and is therefore called a **proto-oncogene,** c-*src* ("c" for cell), while the modified counterpart in the virus is the v-*src* gene, the oncogene. All oncogenes have their proto-oncogene counterparts in vertebrate cells. According to the current model, a virus becomes a tumor virus by accidentally incorporating a proto-oncogene such as c-*src*, which then undergoes a mutation and becomes the v-*src* oncogene:

Proto-oncogene (c-*src* in a normal cell) —Copy of c-*src* transferred to retrovirus and mutated→ Oncogene (v-*src* in retrovirus [RSV])

(Retroviruses have evidently picked up oncogenes from the processed mRNA copy of the cellular gene, since the viral oncogenes lack the introns of the cellular copies.) Although an oncogene acquired by a tumor virus is not essential for the virus's multiplication, the virus can introduce the oncogene into a normal cell. This cell and its progeny are then transformed and display the phenotypes of cancerous cells.

Oncogenes also appear in cells that are not infected by viruses. A normal cellular proto-oncogene may be converted into an oncogene directly by undergoing a mutation, by mutation of its control factors, or by a change in its location through a chromosomal rearrangement. These changes can lead to an altered protein or to synthesis of a protein in the wrong amount or in an uncontrolled way. The obvious question now is, What are these mysterious genes that can become oncogenes? The answer is that they are growth-regulating genes that encode proteins of the signal-transduction pathway.

20.16 Oncogenes are derived from cellular proto-oncogenes whose products are components of signal-transduction pathways.

Oncogenes have been identified in several kinds of cancer. There are probably fewer than a hundred types of oncogenes, and the same ones appear over and over again. With few exceptions, proto-oncogenes normally encode proteins of the signal-transduction pathway that communicates growth-regulating signals between cells (Figure 20.24), including membrane receptors, G-proteins, inositol phosphate enzymes, and protein kinases. Proto-oncogenes encode proteins of all these types. The corresponding oncogene can interfere with any step of a

(a) **Normal genes (proto-oncogenes) encode proteins of signal transduction pathway.**

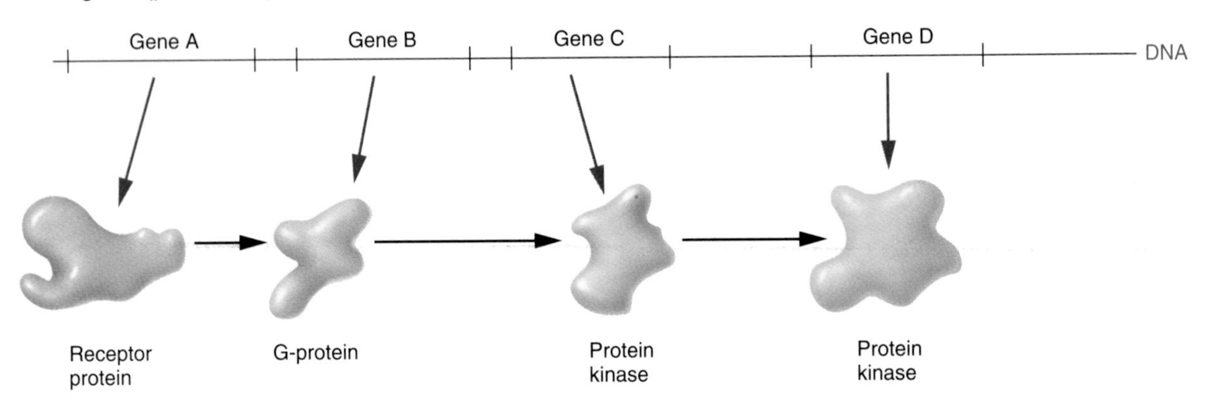

(b) **One mutant gene creates an abnormal signal transduction pathway.**

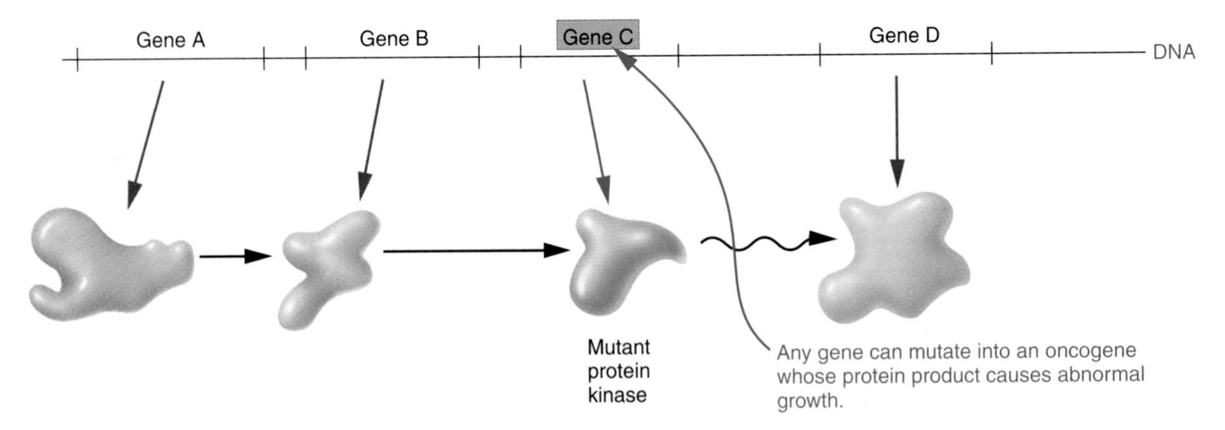

Figure 20.24
(a) Normal eucaryotic cells have several genes that encode proteins of signal transduction pathways. *(b)* Any of these genes must be considered a proto-oncogene because it could mutate into an oncogene.

cell-signaling pathway, creating an "outlaw" cell that ignores normal contact inhibition signals from its neighbors, outproliferates normal cells, and even destroys them. The c-*src* gene, for instance, encodes a protein kinase, and the oncogene *src* produces an altered protein kinase. The *abl, kit,* and *yes* oncogenes encode other protein kinases, and *ras* encodes a G-protein. Normal G-proteins are activated by binding GTP and have a GTPase activity that degrades the bound GTP and inactivates the protein, but the Ras protein lacks GTPase activity, so it is activated but never inactivated. The result is uncontrolled cell growth. About half the carcinomas of the lung and colon and more than 90 percent of pancreatic carcinomas are associated with the *ras* oncogene.

General signal-transduction pathway, Section 11.8.

Oncogenes may also arise from the genes for growth factors or growth-factor receptors. The receptor for epidermal growth factor (EGF) spans the plasma membrane, with an extracellular binding site for EGF and a cytoplasmic site that has tyrosine kinase activity. Binding to EGF activates the kinase, which then phosphorylates the tyrosine residues of certain proteins, thus regulating their activity. The *erbB* oncogene encodes an abnormal receptor that lacks the EGF binding site but has the kinase constantly activated (Figure 20.25); this oncogene is associated with stomach, breast, and ovarian adenocarcinomas.

Oncogenes are opposed by **tumor suppressor genes** whose protein products stop unregulated cell proliferation. One of the most important of these genes, *p53* (its protein product has a molecular weight of 53 kDa), encodes a protein that induces transcription of a smaller regulatory gene; the product of the regulatory gene, in turn, binds to one of the cyclin-Cdk2 complexes needed to pass the checkpoint at the beginning of the S phase of the cell cycle. A high level of the *p53* protein therefore keeps a cell from continuing through its cycle. The absence of *p53* has been implicated in a large percentage of cancers. Most people have two normal *p53* genes, and both must be inactivated to allow growth of a cancer. Some families pass on a propensity to develop certain kinds of cancer because they already have a mutated copy of one of these genes, and a heterozygote for such a mutant gene only has to experience a mutation in the one good copy to lose an important block to cancer development.

20.17 Environmental factors cause most cancers.

Epidemiological studies, which investigate the distribution of disease in relationship to environmental factors, have shown that the incidence of many kinds of cancer varies quite widely from country to country (Table 20.1). Amazingly, the rates of specific types of cancer in high-incidence regions can be ten times or even 100 times those in low-incidence regions. Are these differences due to environmental factors, including diet, or are they due to genetic differences between populations of these countries? To answer this question, studies have tracked the change in cancer rates in groups of people who have moved from one country to another. If they retain the rate in the country of origin, this would be evidence that the population has a genetic propensity to develop this type of cancer. Emphatically, however, they develop cancer at rates similar to those of other people living in the new region. Ovarian cancer, for instance,

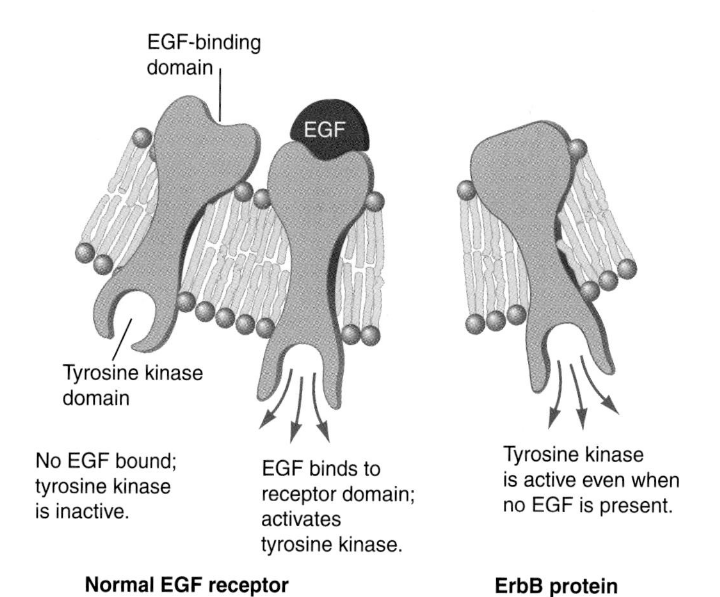

Figure 20.25
The normal cellular receptor for epidermal growth factor (EFG) becomes a protein tyrosine kinase when bound to EGF. The *erbB* oncogene produces a smaller protein that constantly has tyrosine kinase activity and constantly signals cell division. (Source: Albert L. Lehninger, et al., *Principles of Biochemistry,* 2nd edition, 1993, Worth Publishing.)

Table 20.1 Variation in Incidence of Common Cancers*

Type of Cancer	High Incidence	Low Incidence
Colon	Connecticut, USA (whites): 34	Tianjin, China: 1.3
Stomach	Nagasaki, Japan: 82	Kuwait (Kuwaitis): 3.7
Lung	New Orleans, USA (blacks): 110	Madras, India: 5.8
Liver	Shanghai, China: 34	Nova Scotia, Canada: 0.7
Breast	Hawaii (Hawaiians): 94	Israel (non-Jews): 3.0
Melanoma	Queensland, Australia: 31	Osaka, Japan: 0.2

*New cases/year per 100,000 population.

is prevalent in Western countries but rare in the Far East, especially in Japan. Yet the ovarian cancer rate among Japanese women who migrate to the United States approaches that of the general U.S. population, with a delay of about 20 years. Although genetic tendencies to develop cancer exist and can be identified in some families, genetic predisposition is of minor importance compared to environmental factors. Analyses of epidemiological data lead to the conclusion that 80–90 percent of cancer would be avoidable if we could identify and remove its environmental causes.

A person can't simply identify a country where cancer is rare, move there, and expect to lead a charmed existence free of the threat of cancer. Factors balance out. A country with a low incidence of one kind of cancer usually has a higher incidence of other kinds, so all countries have similar total cancer rates. And, as with all diseases, prevention is better than cure. Some factors can be eliminated, at least in principle: Cigarette smoking is highly correlated with lung cancer and cancer of the lip, and also associated with cancer of the larynx, esophagus, pancreas, kidney, and bladder. But people can choose to use or not use cigarettes or chewing tobacco, even though these products are addictive and their victims may need special help to kick the habit. Colon and rectal cancer is low in Japan, Southeast Asia, and sub-Saharan Africa, but high in Western countries. Low incidence is partly correlated with low-fiber diets, and people can certainly change their eating habits. Several studies have shown connections between occupations and cancers. Woodworkers, especially carpenters and cabinetmakers, have excess stomach cancers, suggesting that the tumors are caused by wood particles. Stomach cancer rates are high in asbestos workers, coal miners, rubber workers, farmers in Japan, and nickel refiners in Russia. Metalworkers have an excess of respiratory cancer and aluminum workers an excess of pancreatic cancer. Men who work around oil and gasoline have an excess of urinary bladder cancer. Those who work with phenoxy herbicides and chlorophenols, which are used in agriculture and forestry for weed control, show an excess of soft tissue tumors and lymphomas. It is likely that our workplaces can be made safer. Many cancers are caused by pollutants that are poured into our air, water, and soil by industries, agriculture, automobiles, and other elements of modern life; these factors, too, can be controlled if an educated public demands it. Furthermore, one function of the immune system (see Chapter 48) is to destroy neoplastic cells before they become dangerous, and maintaining good immunity through other healthy practices is a major defense mechanism.

The frequency of cancer has been rising in the United States and elsewhere due to increasing pollution and an aging population. It is now likely that one out of three people in the United States will have at least one encounter with cancer during their lifetime. In 1993 in the United States, 1,170,000 new cases of cancer were identified, and the deaths of 528,000 people were attributed to it. A few types account for the majority of deaths: lung cancer (28 percent), colon and rectal cancer (11 percent), breast cancer (9 percent), leukemias and lymphomas (9 percent), prostate (7 percent), pancreas (5 percent), ovary (3 percent), stomach (3 percent), bladder (2 percent), and central nervous system and eye (2 percent). For reasons that aren't understood, the incidence of a few kinds of cancer is decreasing; for instance, cancer of the stomach has decreased enormously in the United States since the 1930s.

20.18 More than one mutation is necessary to cause cancer in an animal, yet one oncogene can transform a normal cell.

We showed earlier that cancer only develops after a number of mutational events have occurred; that is, one mutation is not sufficient. Yet we have also seen that an oncogene, resulting from mutation of a normal cellular gene, can transform a normal cell to a cancerous state. Clearly these two bits of information are at odds. How can several mutations be required in one case but a single mutation be sufficient in another? The explanation is quite complex and still incomplete. Part of the ambiguity lies in the fact that oncogenes are assayed by their effect on cultured cells, usually fibroblasts, and the assay involves easily observed modifications that distinguish transformed cells from normal cells. Full-blown cancer in an animal is far more complicated because it involves all the interactions and regulatory systems of an animal, including enormous numbers of diverse cells and redundant mechanisms that can replace one system with another. In an animal, a single altered component is usually not catastrophic—not enough to lead to cancer.

Recall that cell proliferation in cancer can be caused either directly by activating oncogenes or indirectly by inactivating tumor suppressor genes that normally keep proliferation in check. Cancer progression in an animal entails an intricate interplay between processes that enhance cell proliferation and others that block it. So cancer requires the effects of a number of mutations. Some produce oncogenes, which enhance proliferation. Other mutations disable genes that normally suppress proliferation, thus eliminating a defense against cancer. On the whole, individual cells in tissue culture receive different influences from cells in their normal places in the body.

As cell biology research has progressed in the past few decades, we have gained remarkable insights into the regulation of cell growth and abnormalities in growth. This has been accompanied by applied research into the causes of cancers and specific cures. Several types of cancer that meant certain death only a few years ago can now be cured with high probability. There is good reason to be optimistic about cures for all cancers, but humanity will probably never be free of cancer. As Bruce Ames has pointed out, many carcinogens (and other disease-inducing compounds) are natural components of our food. Nevertheless, by being aware of the many factors involved in carcinogenesis, we can at least minimize our chances of contracting the disease.

Coda Embryology has always had a special place in biology because of the complexity, the very mysteriousness, of its subject. To many people, the general problems of development are still the central questions of biology. How can any creature engage in such incredible processes and perform them correctly time after time? Classical embryologists were probably more inclined than other biologists to become vitalists in their old age—that is, to adopt a philosophy that organisms are endowed with enigmatic properties beyond the ken of ordinary physical science. If a scientist has gained little understanding of development after studying it for a lifetime, it must be easy to feel that the process cannot be understood by the methods of science. Indeed, an embryologist dying of a cancer, which is, after all, a derangement of development, must experience a particularly strong sense of irony and helplessness. But now we have started to gain insights into the mysteries. Detailed studies of cell movement, of cell adhesion and surface molecules, and of the informational interactions among cells are beginning to show how ordinary physical interactions among biomolecules can effect the large-scale changes we observe. And behind these processes lies the ever-present genetic program. In the next chapter, we will gain some insights into the way such a program plays out to cause differentiation, including some of the morphological changes we have described here.

Summary

1. Developmental biology deals with questions about morphogenesis, or the development of specific form, and differentiation, the process by which each type of cell develops its distinctive features.
2. Both morphogenesis and differentiation require the differential expression of particular genes in each type of cell.
3. Fertilization entails a series of interactions between the sperm and the layers of the egg surface, including a recognition step that generally prevents mixing of the gametes from different species. Then the sperm and egg cytoplasms fuse, the nuclei are combined, and mitosis begins.
4. A typical animal embryo begins with cleavage of the zygote into small cells called blastomeres; they eventually form a hollow ball, a blastula. The blastula then becomes a gastrula as part of the cell layer pushes inside through a pore, the blastopore, and ends up inverted against the cells remaining outside.
5. The archenteron, the cavity formed by gastrulation, is the primitive gut of the animal. In vertebrates, the blastopore becomes the anus; later, the mouth will break through at the other end of the archenteron. Cells remaining on the embryo's surface form the ectoderm, which will become the skin and nervous system. Cells of the archenteron form endoderm, which will become the lining of the gut and organs branching from it. Masses of cells in between form mesoderm, which will become the bulk of internal organs and muscles.
6. The central nervous system (brain and spinal cord) forms from a neural tube that rolls up along the back of the embryo.
7. Several internal organs develop as branches from the gut tube that become surrounded by mesenchyme (loose mesodermal connective tissue).
8. The evolution of the amniote egg has allowed reptiles and birds to adapt to terrestrial life. The embryo develops from a small disc on the surface of the nutrient-rich yolk, and only gradually rolls up into a rounded body. Extraembryonic membranes grow from the embryo. The amnion surrounds the embryo and keeps it enclosed in fluid. The chorion forms a breathing organ, and wastes collect in the allantois.
9. A mammalian embryo grows from an inner cell mass; an outer layer of cells, the trophoblast, forms the extraembryonic membranes. The chorion and allantois combine to form the placenta, which will exchange gases, wastes, and nutrients with the mother's blood.
10. We can understand a great deal of tissue and organ shaping through simple mechanisms. Restricting the planes of cell division can produce filaments (division in one plane), sheets (division in two planes), and masses (division in all three planes).
11. Cells stick together in characteristic forms because of differential adhesion due to specific surface molecules. Dispersed embryonic cells will re-form their tissues, and they will arrange themselves properly just because of the strength with which they bind to one another. Differential strengths of adhesion between cells account for the patterns with which they form layered structures, such as limbs with ectoderm outside and mesoderm inside.
12. Cell adhesion depends on glycoproteins called cell adhesion molecules (CAMs). CAMs are either cadherins, which require calcium ions, or immunoglobulin CAMs that resemble antibody molecules and are independent of calcium ions. Several types of CAMs are characteristic of different tissues. They form specific connections between cells, and they may appear and disappear on a schedule that permits cells to move independently at times and stick together at other times.
13. The cells in a tissue are linked to an extracellular matrix of collagen, proteoglycans, and adhesive proteins by means of integrin proteins. Embryonic cells migrate on such a matrix to reach their proper positions for further development.
14. Normal differentiated cells inhibit one another's growth and movement through signals received by surface receptors.
15. The growth (and other activities) of tissues is often stimulated by specific growth factors that bind to surface receptors, which stimulate cell division or guide cell growth.
16. Cell sheets and masses may be shaped by elongation of cells by microtubules and local constriction by microfilaments. Bulges and clefts may be produced by more rapid mitosis in some regions, combined with the deposition of supporting extracellular fibers in other places.
17. Cancers are major causes of human mortality. Benign neoplasms are made of differentiated cells that grow in place and cause relatively little damage, except mechanical interference. The cells of malignant neoplasms are able to metastasize and invade other tissues, where they do great damage.
18. Carcinomas grow from epithelia, sarcomas from muscle or connective tissue, and leukemias from white blood cells.
19. Cancers grow because of natural selection among cells within an animal, just as evolution progresses because of natural selection among different organisms in a population. Cancer cells are uncontrolled and able to grow faster than normal cells. Cancer cells thus have a selective advantage over their normal neighbors, so they take over and destroy a tissue.
20. Cells are usually only released from contact inhibition in unusual situations, such as cuts. Transformed cells, which have been altered by mutation or by infection with a tumor virus, also have altered cell surfaces. Tumor cells become like embryonic cells, which also are not inhibited by contact.
21. Cancers arise from single cells whose genomes have been altered. Most neoplastic cells result from mutations caused by environmental mutagens, which are also carcinogens. These agents are routinely identified by the Ames test, based on their ability to cause mutations in bacteria.
22. Several independent events must occur to produce a cancer. Cells that have experienced some of these events may have altered forms and may be detected as precancerous cells by procedures such as the Pap test for cervical carcinoma.

23. Cancer is caused by a series of events over many years, not by a single event. A tumor is generally initiated by agents that cause changes in DNA. But the altered cell only develops into a tumor if it is changed later by other agents, called tumor promoters, which enhance tumor formation, generally by increasing cell proliferation.
24. Epidemiological studies show that cancers are caused by environmental factors, including diet, pollutants, and other factors associated with various industries. Eliminating these factors would significantly reduce the incidence of cancer. Different types of cancer are particularly common in certain societies and among workers in various industries.
25. Cancer cells have lost normal controls over their proliferation. Some of these alterations stimulate cells to go through their cell cycles more rapidly; others change the fates of cells that would normally stop dividing or die. Direct stimulation stems from changes in specific genes called oncogenes.
26. Tumor viruses carry oncogenes, and when they infect cells, the cells are transformed so they grow abnormally. Oncogenes encode proteins involved in cell signaling, such as G-proteins, growth-factor receptors, or protein kinases that normally transduce signals between cells. Oncogenes arise by mutation of proto-oncogenes, which encode the normal proteins.
27. It is still not clear why a cell can be transformed by a single mutation to create oncogenes while a cancer is only created by several independent events. The key may be the difference between individual cells in tissue culture compared with cells in their normal locations in the body.

Key Terms

morphogenesis 398
differentiation 398
vitelline envelope 398
cortical granule reaction 399
animal hemisphere 400
vegetal hemisphere 400
blastomere 400
cleavage 400
morula 400
blastula 400
blastocoel 400
blastopore 400
gastrulation 400
gastrula 400
archenteron 402
ectoderm 402
endoderm 402
mesoderm 402
epithelial cell 402
mesenchymal cell 402
neurulation 402
neural plate 402
neural tube 402
neural crest cell 402
notochord 403
somite 403
amniote egg 403
amnion 403
chorion 403
allantoic membrane 403
allantois 403
blastodisc 404
epiblast 404
hypoblast 404
primitive streak 404
chorioallantoic membrane 405
blastocyst 405
inner cell mass 405
trophoblast 405
cell adhesion molecule (CAM) 407
extracellular matrix 408
basement membrane 408
contact inhibition 410
growth factor 410
mitogen 410
tumor (neoplasm) 412
benign 412
malignant 412
metastasis 412
carcinoma 412
sarcoma 412
leukemia 412
monoclonal 413
carcinogen 413
tumor initiator 415
tumor promoter 415
transform 415
tumor virus 415
oncogene 415
proto-oncogene 415
tumor suppressor gene 417

Multiple-Choice Questions

1. All except ______ occur before the actual fusion of the sperm and egg nuclei.
   a. gray crescent formation
   b. cortical reaction
   c. conversion of the vitelline membrane into a fertilization membrane
   d. release of acrosomal material
   e. recognition of bindin by receptors on the vitelline membrane
2. The blastopore is the site of entry into the
   a. morula.
   b. blastocoel.
   c. archenteron.
   d. vitelline membrane.
   e. gray crescent.
3. Which of the following cellular layers includes cells that actually become part of the embryonic body?
   a. amnion
   b. chorion
   c. trophoblast
   d. allantoic sac
   e. epiblast
4. In plants, ______ is the least important factor determining the shape of an organ, whereas in animals, it is of great importance.
   a. the orientation of mitotic spindles
   b. the plane of cell division
   c. the number of planes of cell division
   d. differential cell adhesion
   e. the action of growth factors
5. The lens of the eye and the neural tube form in a similar way: The surface sheet of epithelial cells rolls inward to form a subsurface vesicle or tube. In both cases, the dominant morphogenetic force is likely to be
   a. contact inhibition.
   b. contraction of cytoskeletal proteins at one end of each of the cells involved.
   c. migration of lens or tube-forming cells on a fibronectin lattice.
   d. loss of cadherins between adjacent cells.
   e. change in orientation of the mitotic spindle in the affected cells.
6. Metastatic neoplasms are
   a. composed of malignant cells.
   b. one example of benign tumors.
   c. governed by strong forces of contact inhibition.
   d. generally restricted to their site of formation.
   e. composed of highly differentiated cells.
7. The cells of a young embryo and a malignant tumor share all the following characteristics except
   a. little or no contact inhibition.
   b. cells that are not highly differentiated.
   c. alternation between aggregated and mesenchymal states.
   d. they are monoclonal.
   e. they arise by somatic cell mutation.
8. Many sequences of events occur in the development of a cancer. Which one of the following is not a correct sequence?
   a. conversion of a benign neoplasm into a metastatic neoplasm
   b. action of tumor initiator; then action of tumor promoter
   c. transformation of a sarcoma into a carcinoma
   d. mutation of a proto-oncogene into an oncogene
   e. wild-type tumor suppressor gene altered to mutant form
9. Which is *not* correct about proto-oncogenes and oncogenes?
   a. If a proto-oncogene does not mutate, it will not lead to the production of cancer.

b. Proto-oncogenes are often regions of DNA that code for growth regulators or their membrane receptors.
c. Proto-oncogenes are often regions of DNA that code for G-proteins or protein kinases.
d. The transformation of a proto-oncogene into an oncogene requires the activity of a tumor virus.
e. An oncogene can result from mutation of a tumor suppressor gene.

10. Which of the following is associated with the transformation of a normal cell into a tumor cell?
    a. mutation of a proto-oncogene
    b. mutation of tumor suppressor gene
    c. infection by a virus carrying an oncogene
    d. environmental mutagens
    e. all of the above

True-False Questions

Mark each statement true or false, and if false, restate it to make it true.

1. The sea urchin oocyte is covered first by a jelly layer, next by its plasma membrane, and most peripherally by a vitelline envelope.
2. The correct order of early embryonic stages is blastula, morula, and then gastrula.
3. During gastrulation, the blastocoel is transformed into the archenteron.
4. Ectoderm gives rise to the outer covering of the embryo plus the neural tube and neural crests.
5. Ectodermal and endodermal cells tend to exhibit the wandering activity typical of mesenchyme, whereas mesodermal cells generally form flat epithelial sheets.
6. The extraembryonic membranes include the amnion, chorion, allantoic membrane, and vitelline membrane.
7. Ectoderm forms from surface cells that do not enter the interior of the embryo through the blastopore or the primitive streak.
8. Adhesive proteins such as fibronectin and laminin are cadherins located on the surface of migrating cells.
9. People born with one mutant tumor suppressor gene, such as the *p53* gene, are more likely to develop cancer, or more likely to develop it at an earlier age, than those born with two wild-type alleles.
10. Although many factors have been implicated in the development of cancer, inherited genetic predispositions have been clearly shown to be the most significant.

Concept Questions

1. Contrast the meaning of the terms morphogenesis and differentiation.
2. Amphibian embryos have an animal hemisphere and a vegetal hemisphere. Do bird embryos? Explain.
3. Distinguish between these sets of terms: *(a)* ectoderm, mesoderm, and endoderm; *(b)* epiblast and hypoblast.
4. As the neural tube forms, cells of the neural crest become mesenchymal. They migrate throughout the body, eventually forming many discrete masses. These differentiate into a variety of structures, including facial bones and the inner portion of the adrenal gland. Using the concepts of cell adhesion and matrix adhesion, outline a hypothetical sequence of events to account for the motility of these cells and their ability to form a variety of discrete structures.
5. Since we know that tumor viruses exist and some babies are born with tumors, can we "catch" or inherit cancer in the same sense that we "catch" a communicable disease such as flu or inherit a particular blood type? Explain.

Additional Reading

Cavenee, Webster K., and Raymond L. White. "The Genetic Basis of Cancer." *Scientific American,* March 1995, p. 50. More details about carcinogenesis as discussed in this chapter.

Cohen, Leonard A. "Diet and Cancer." *Scientific American,* November 1987, p. 42. Recommendations aimed at reducing the incidence of cancers associated with nutrition.

Croce, Carlo M., and George Klein. "Chromosome Translocations and Human Cancer." *Scientific American,* March 1985, p. 54. When chromosomes in a cell of the immune system recombine their DNA, they may activate cancer-causing genes.

Edelman, Gerald M. "Cell-adhesion Molecules: A Molecular Basis for Animal Form." *Scientific American,* April 1984, p. 118. The role of CAMs in development.

Goodman, Corey S., and Michael J. Bastiani. "How Embryonic Nerve Cells Recognize One Another." *Scientific American,* December 1984, p. 58. Developing neurons seek one another out and interconnect with high specificity.

Grimes, Gary W., and Karl J. Aufderheide. *Cellular Aspects of Pattern Formation: The Problem of Assembly.* Karger, Basel and New York, 1991. An essay on morphogenesis and the formation of patterns during development.

Hennig, W. (ed.). *Early Embryonic Development of Animals.* Springer-Verlag, Berlin and New York, 1992.

Hunter, Tony. "The Proteins of Oncogenes." *Scientific American,* August 1984, p. 70. Further information about the proteins normally encoded by these genes.

Hynes, Richard O. "Fibronectins." *Scientific American,* June 1986, p. 42. These adhesive proteins hold cells in position and guide their migration.

Liota, Lance A. "Cancer Cell Invasion and Metastasis." *Scientific American,* February 1992, p. 34. How cancer cells spread throughout the body.

Sachs, Leo. "Growth, Differentiation and the Reversal of Malignancy." *Scientific American,* January 1986, p. 40. Specific proteins regulate the growth of normal white blood cells and their differentiation into nondividing forms. Leukemic cells can also be made to differentiate, suggesting new approaches to cancer treatment.

Sharon, Nathan, and Halina Lis. "Carbohydrates in Cell Recognition." *Scientific American,* January 1993, p. 74. Carbohydrates form many combinations from a few components, and nature has selected them for use in many processes of recognition.

Weinberg, Robert A. "A Molecular Basis of Cancer." *Scientific American,* November 1983, p. 126. Further information about oncogenes.

Internet Resource

To further explore the content of this chapter, log on to the web site at:

http://www.mhhe.com/biosci/genbio/guttman/

21 Developmental Biology II: Differentiation

A seven-day-old chick embryo has a large head with a prominent eye.

Key Concepts

21.1 Differentiation does not generally entail a loss of DNA.

21.2 The fates of some embryonic cells are determined from the beginning of development.

21.3 In other embryos, cells are only determined later in development by various factors.

21.4 Cells become progressively determined as they divide.

21.5 Some cells remain stem cells indefinitely.

21.6 Certain tissues induce the differentiation of other tissues.

21.7 Inductive interactions may be instructive or permissive.

21.8 Differentiation may be determined by hierarchies of genes that regulate other genes.

21.9 Some differentiating tissues appear to be regulated by a clock mechanism.

21.10 Many cells differentiate on the basis of information about their positions in the embryo.

21.11 Series of homeotic genes determine the basic body plans of all animals.

21.12 Homeotic genes in plants have effects parallel to those in animals.

Scientists do not believe in miracles. The essence of the scientific enterprise is showing how phenomena can be explained by fitting them into the causal structure of the universe. But if anything has struck biologists as miraculous, it is the development of a plant or animal from a single cell into its mature form. Generations of embryologists have watched the zygote of a chick or pig gradually bulge, fold, and roll into a recognizable animal shape—have seen it acquire a brain and limbs and a beating heart with blood cells visibly moving through it. They have watched, drawn, named, cataloged—but have been unable to explain—this complicated process. Such an event that cannot be explained seems miraculous.

Science, to be fruitful, must be done at the right time and place. People can't even ask questions sensibly without the tools needed to answer them, and for most of the past century, the science of embryology has been struggling without those tools. But the end of the twentieth century is a particularly exciting time for biologists who have watched and wondered about development. Since development is the expression of instructions in a genome, embryology can now benefit from genetic methods developed for identifying genes, reading DNA sequences, isolating and characterizing proteins, and following intricate molecular interactions. Now we can begin to tell the emerging story of the progressive determination of cell fates and the differentiation of specialized types of cells. Regional differences in an embryo, which eventually produce the right organs in each part, arise from factors that turn certain genes on and off in each cell. In this chapter we will see how genes are activated in a time-dependent way and how they are regulated on the basis of information about each cell's position in the developing embryo. ■

21.1 Differentiation does not generally entail a loss of DNA.

A zygote is **totipotent,** meaning it has the potential to develop into every kind of adult tissue. Soon after fertilization, a zygote divides into smaller cells whose developmental potential becomes more and more limited until each one can only become one kind of cell. In principle, cells could differentiate by losing the DNA encoding proteins that aren't used. That is, the genes might be arranged on chromosomes so that each cell type could retain a different set, or unused genes could somehow be cut out and eliminated. But some elegant and painstaking experiments have ruled out this model for some cases and made it seem unlikely overall.

In 1952, Robert W. Briggs and Thomas J. King first performed the kind of experiment outlined in Figure 21.1. One cell of a late-embryo frog is sucked into a small-bore pipette, which generally breaks the plasma membrane but leaves the nucleus intact. The nucleus is then injected into a frog egg whose own nucleus has been removed with a needle. Briggs and King found that many of these eggs could then develop at least as far as the late embryonic stages, and in subsequent experiments, John B. Gurdon was able to raise some of them to adults. To be sure the frogs were developing from the implanted nucleus rather than from a remnant of the old one, Gurdon employed a mutant strain of the South African clawed toad *Xenopus laevis,* a strain that has only one nucleolus per cell. When Gurdon implanted nuclei from a one-nucleolus strain into the eggs of a two-nucleolus strain (after inactivating the egg nucleus with ultraviolet light), the developing toads, as expected, had only a single nucleolus, demonstrating that their nuclei were derived from the implanted nucleus.

These experiments show that many nuclei from even very late developmental stages still retain the whole complement of genes. It would be rash to claim that every adult nucleus retains a full set of chromosomes, and it is clear that lymphocytes, which make antibodies, become specialized by rearranging some genes and losing DNA (see Section 48.10). But most adult nuclei probably contain all the information present in the original zygote nucleus; in principle, almost any cell could throw off secondary modifications of its chromosomes—such as the heterochromatic inactivation of one X chromosome in mammalian females—and dedifferentiate into embryonic cells.

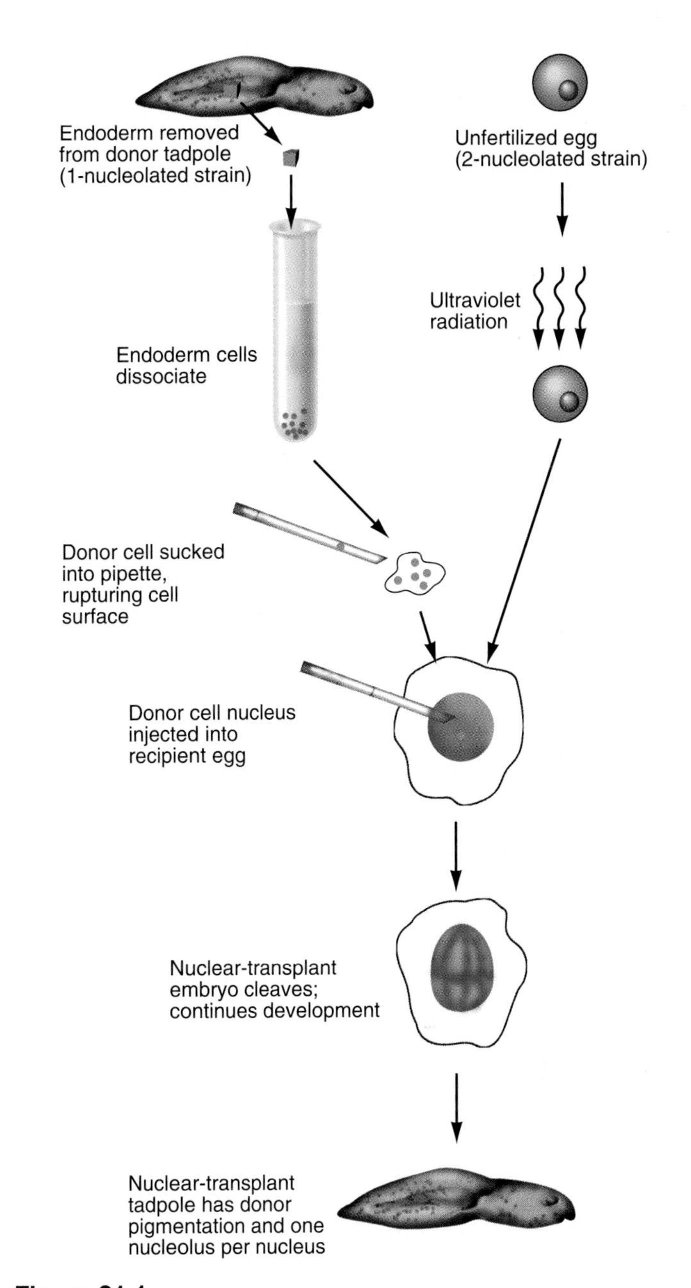

Figure 21.1

In the nuclear transplantation experiment, the original egg nucleus is either removed with a needle or destroyed with UV irradiation; then it is replaced with a nucleus from an embryonic cell. The number of nucleoli is used as a marker to show that the transplant tadpole gets its genome from the donor cell.

Most cells have the potential for producing normally functioning cells of all kinds.

Frederick C. Steward did an equivalent demonstration for plant tissues by isolating single mature carrot cells and letting them develop into normal carrot plants (Figure 21.2). This experiment, too, shows that each differentiated cell has an intact genome but only expresses certain genes. Steward's work, incidentally, has now been developed into a small industry for producing desired plants, often by starting with cells transformed by cloned genes.

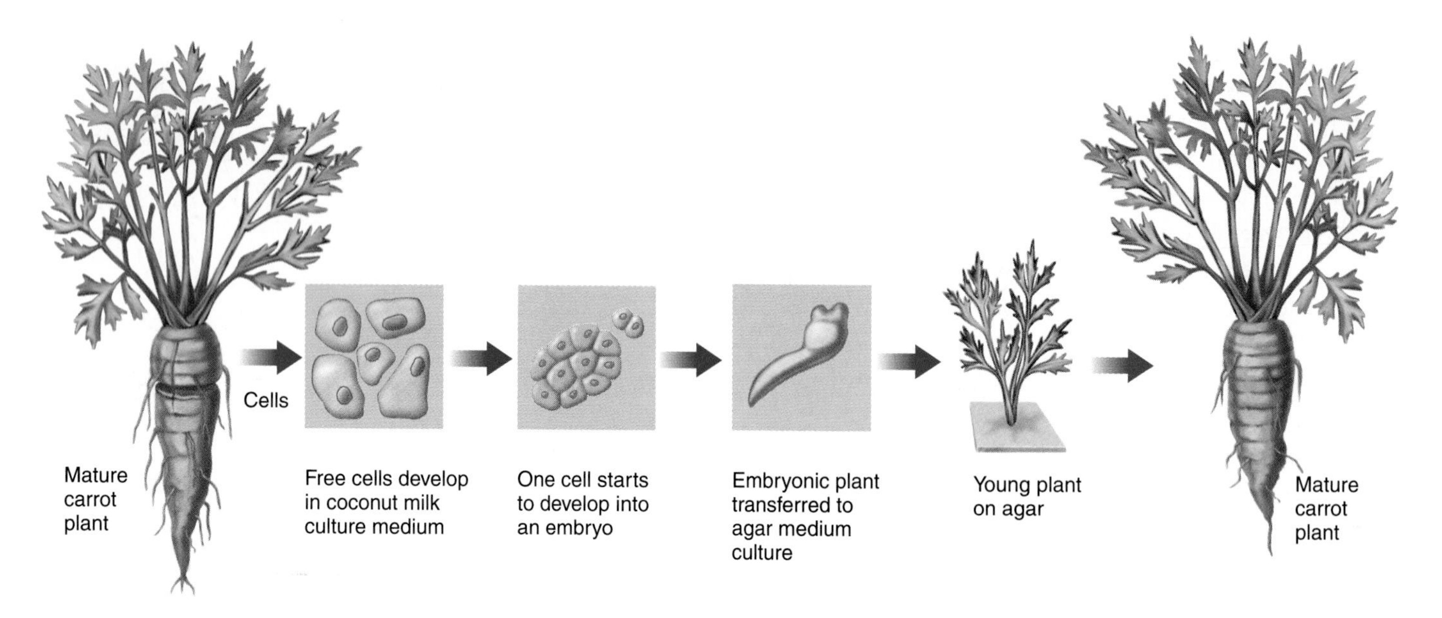

Figure 21.2
A whole carrot plant can be grown from a differentiated root phloem cell, showing that the cell's genome contains a complete set of genes.

Nuclear transplantation is another kind of "cloning," quite different of course from propagating a foreign gene in a microorganism. It underlies the fantasy that has caught the public imagination about making multiple copies of people. Science fiction writers have assumed that if frogs could be cloned, so could people. The complicated moral question arose again after the cloning of sheep and some primates in 1997. We can imagine sound reasons for wanting to clone domesticated animals, but the motivation to clone humans is far from obvious. Doing so, for any purpose, would raise enormous ethical problems.

Exercise 21.1 Suppose you wanted to test experimentally whether certain differentiated cells in an organism are still totipotent. You try an experiment of the Gurdon or Steward type and find that, again and again, cells will only develop into undifferentiated clumps. What will you conclude? *Note:* This is not so much a question about development as about the methods and logic of science.

21.2 The fates of some embryonic cells are determined from the beginning of development.

Before an embryonic cell becomes differentiated by synthesizing its characteristic proteins and acquiring its distinctive form, it becomes **determined,** which means its fate is sealed through an internal molecular decision. This event starts the cell on a single path of differentiation so that normally it can only become one kind of cell. Embryonic cells become determined at various times in different species, generally very early in development and, in some embryos, even in the early blastomeres.

E.B. Wilson demonstrated determination in 1892 in a classic study of the annelid worm *Nereis* (Figure 21.3). *Nereis* eggs, like those of many other invertebrates, cleave in a distinctive pattern called spiral cleavage (see Section 34.10), creating layers of small cells, or micromeres, in the animal half and large cells, or macromeres, in the vegetal half. In this embryo, Wilson could follow each blastomere to its ultimate fate. The top three quartets of micromeres become the ectoderm, the remaining

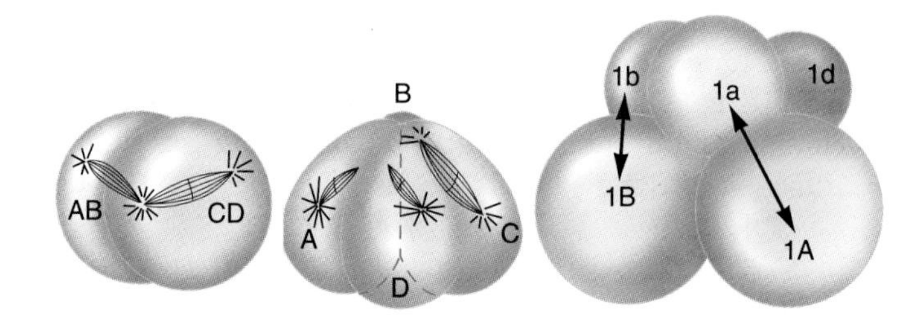

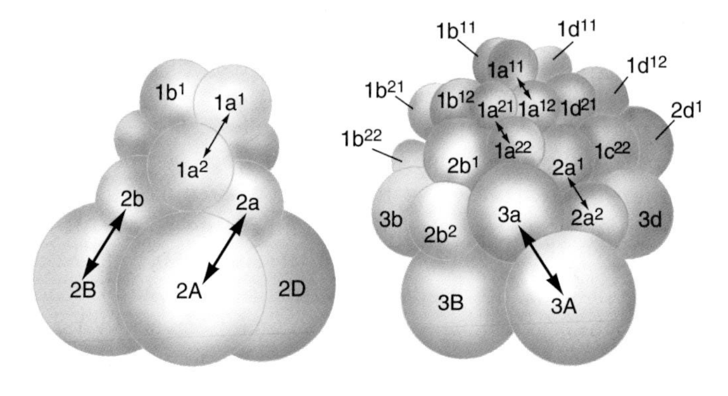

Figure 21.3
The *Nereis* embryo cleaves into a series of micromeres (small cells) and macromeres (large cells). Each cell, uniquely identified by a system of numbers and letters, is destined to develop into a specific structure, and removing one cell produces an embryo with specific parts missing.

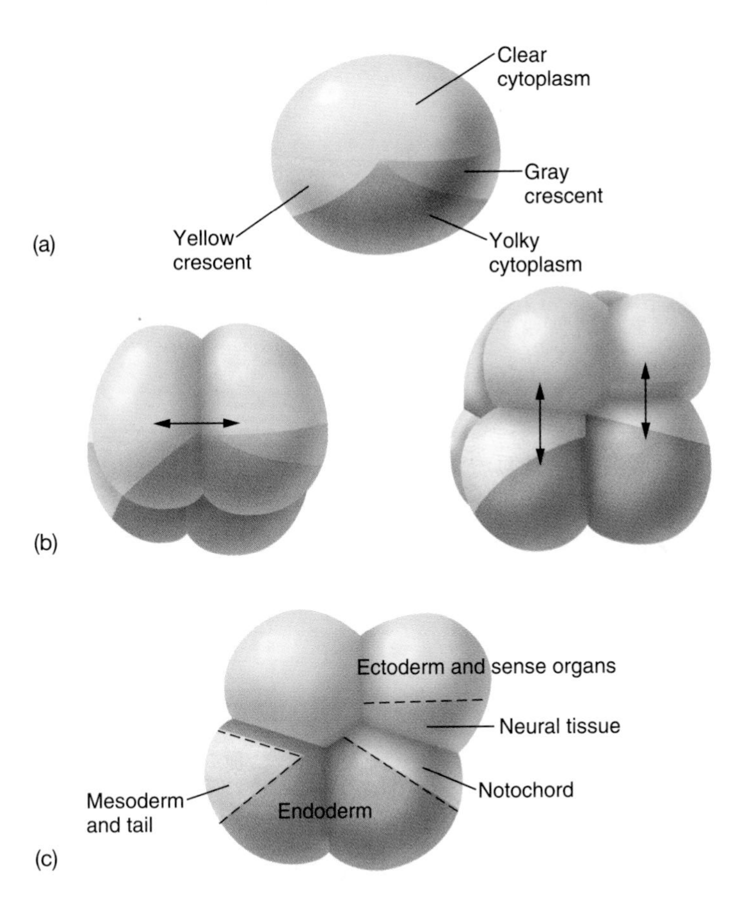

Figure 21.4
(a) A *Styela* egg becomes divided into distinctive regions at the time of fertilization. These are parceled out to the early blastomeres *(b)*, and the fate of each region is determined at this time *(c)*: The light gray cytoplasm becomes notochord and neural tissue; the clear cytoplasm in the animal half develops into ectoderm and sense organs; and the yellow-crescent cytoplasm becomes the mesoderm, including muscles and larval tail.

micromeres form the mesoderm, and the macromeres form the endoderm. Furthermore, each blastomere develops into a specific segment of an organ in the larval worm. The fate of each cell is sealed very early, and removing a single blastomere leaves a defective larva, since the remaining cells can't adjust to supply the missing tissue. Such an egg shows **mosaic development,** where each blastomere has as distinctive a role in the formation of the embryo as each tile in a mosaic picture.

In some embryos (maybe all embryos, to some extent), even the fates of regions of an undivided egg are determined. For instance, in 1905 E.G. Conklin studied the development of *Styela,* a chordate called a sea squirt. Even before the first cleavage, its zygote is divided into five regions with different pigmentation. Conklin demonstrated that these regions have different fates (Figure 21.4). Since the fate of each region is determined, we can draw a **fate map,** showing what each region will become. Cells destined to become a particular structure are designated **presumptive**—presumptive epidermis, presumptive muscle, and so on.

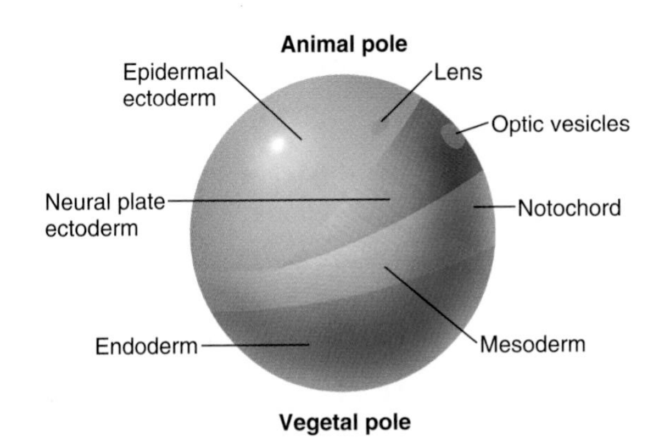

Figure 21.5
A fate map of an amphibian embryo at the beginning of gastrulation shows the structures each region will normally develop into.

The mechanisms of determination span a spectrum of processes. At one end of the spectrum are **autonomous processes,** as in mosaic development, where the development of each cell is determined entirely by its own internal information. At the other end are **cell-interactive processes** in which a cell's developmental path is determined by factors in its environment, mostly by neighboring cells. The cell-interactive mechanism prevails at a few developmental points in the roundworm *Caenorhabditis elegans* (Methods 21.1). At the second cleavage, two of the cells can switch fates if their positions are switched. A group of neighboring cells influences whichever of the two cells they touch to form a particular type of muscle. It makes no difference which of these two cells is in which position—they are equivalent before surrounding cells influence them. The specific influence is a protein, and the gene encoding it has significant roles in development in many species.

Exercise 21.2 Suppose you take embryos of an animal at the 32-cell stage and remove one specific cell from each embryo. When the animals are fully developed, they are all missing a prominent cluster of nerve cells. At the 32-cell stage, is that one cell determined? Is it differentiated?

21.3 In other embryos, cells are only determined later in development by various factors.

In contrast to the blastomeres in animals with mosaic development, the early embryonic cells of frogs and other vertebrates are not determined yet. Such embryos are said to have **regulative development,** because the blastomeres can change their developmental paths for a time. Although such an embryo isn't naturally marked, one can construct a fate map for it by using bits of charcoal dust or vital dyes that stain cells without hurting them. Figure 21.5 shows a fate map of an early gastrula frog embryo.

A classic experiment shows regulative development in an early-gastrula frog embryo (Figure 21.6). When a little patch

Methods 21.1 Choosing the Right Organism: The Roundworm *Caenorhabditis elegans*

Much of our detailed knowledge of biology has been provided through work with only a few organisms that happen to be excellent model experimental systems. Research on *E. coli*, *Drosophila*, and mice has taught us the basics of modern genetics, and developmental biology is acquiring its own favorite organisms. *Drosophila* is one, the mouse another; a third, *Caenorhabditis elegans* (see-no-rab-di'tis) a roundworm, is the result of a search led by Sydney Brenner, a pioneer in molecular genetics who set out in the late 1960s and early 1970s to find an animal that could serve developmental biologists as *E. coli* had served basic molecular genetics. Brenner and his colleagues realized they needed an organism that would be easy to work with in the laboratory. It had to have identifiable mutants, so its genetics could be worked out, and it had to make studies of cell division and cell movement accessible. They settled on *C. elegans* as an ideal experimental system, and this tiny animal has definitely lived up to Brenner's expectations.

C. elegans is a nematode with a very short life cycle (about three days at a temperature of 25°C). Only about a millimeter long, in nature it is a free-living soil organism.

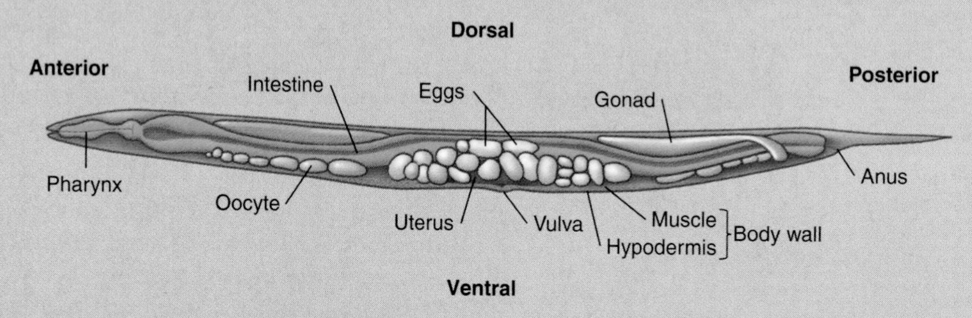

The worm has only 959 cells, compared to the 10^{14} cells in a human, and since it is transparent, all these cells can be identified and studied. In fact, the whole sequence of cell divisions in *C. elegans* is now known. Enormous numbers of worms can be raised in the laboratory either in liquid suspension or on the surfaces of petri plates. All the worm needs to eat is bacteria. Its small genome (about 10^8 base pairs, half the size of *Drosophila*) contains about 15,000 genes arranged on 5 autosomal chromosomes and an X sex chromosome. Genetic experiments with *C. elegans* are straightforward. Most organisms are hermaphrodites (XX), but males (XO) arise from time to time. Normally, reproduction takes place through self-fertilization by hermaphrodites, but sperm from males can also be used to fertilize hermaphrodites' eggs. Because large numbers of worms can be grown in a short time, it is easy to select mutants, even rare ones.

Many clever tricks have been used to follow development in *C. elegans*. Single cells, or even cell nuclei, can be destroyed with focused laser beams. RNA, DNA, proteins, and small molecules, including inhibitors, can be injected into cells or extracellular fluid. Genes defined by the abundant mutants can be cloned and their products identified, and the genes can be moved to different locations in the genome. Experiments are only limited by the imagination of the scientist.

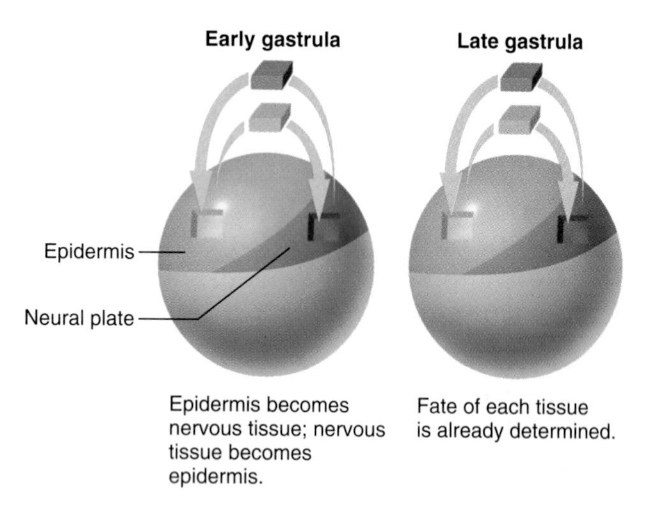

Figure 21.6

When early gastrula tissues in a frog embryo are exchanged, they regulate and become the same as the tissues around them. But in a later gastrula, the fates of cells have been determined, and they remain as they were before.

of presumptive epidermis is cut out and exchanged with a patch of presumptive nervous tissues, their fates change, and the cells in each patch uniformly develop into the same tissues as all the cells around them. If, however, the tissues are exchanged somewhat later, after the cells have been determined, a patch of nerve cells develops in the middle of the epidermis while a patch of epidermis develops in the middle of the nervous tissue.

Identical twins testify to the flexibility of regulative development, for each of them has grown from half of an embryo whose blastomeres separated after the first cleavage or perhaps a little later. (Conjoined twins develop from embryos that separate incompletely at an early stage.) Even more remarkable are some mice that Beatrice Mintz produced (Figure 21.7). She dissociated very early blastulas into separate blastomeres and mixed the cells from two embryos; the cells move around, adjust to each other, and develop into one perfectly normal mouse. Such mice have four parents. They are called *allophenic* because they may show two different phenotypes (such as coat colors).

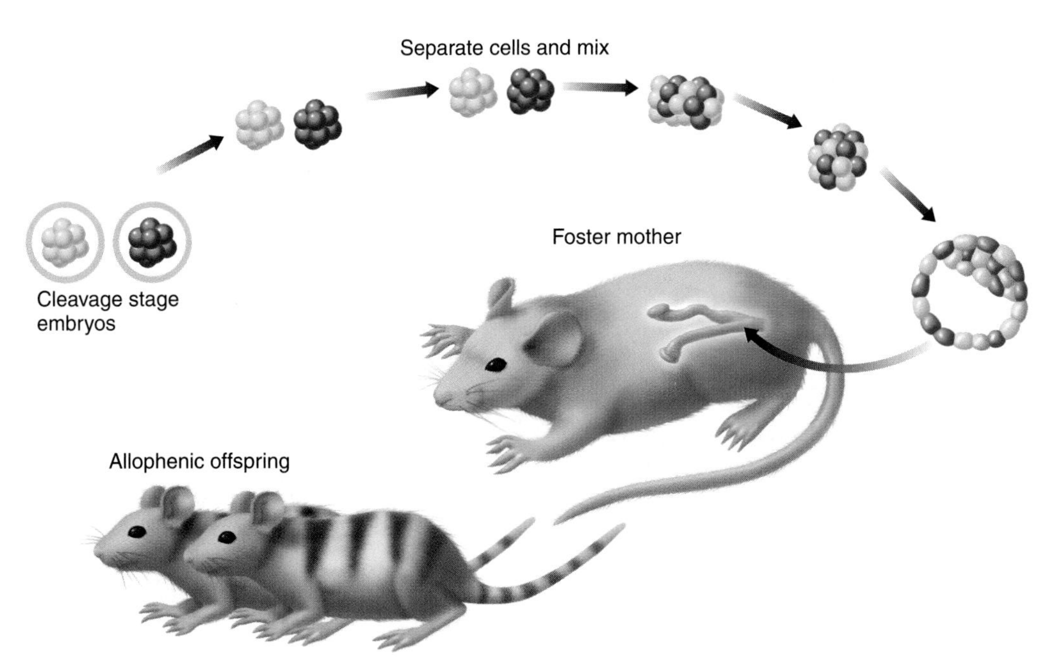

Figure 21.7
Allophenic mice with black and white stripes are produced by combining embryos of black-furred mice and white-furred mice.

Exercise 21.3 You have some embryos of a different animal (not the one used in Exercise 21.2) also at the 32-cell stage. You carefully exchange cells 7 and 14 in each embryo, and they all develop into normal adults. Does this embryo show mosaic development or regulative development?

21.4 Cells become progressively determined as they divide.

The experiment in Figure 21.6 shows that in an early embryo with regulative development interchanged cells can adjust to become like those surrounding them, but shortly afterward, this becomes impossible, for their fates have been sealed. Such experiments indicate that the fates of cells in the early stages of development become progressively restricted as they divide until they eventually have only a single fate. The totipotent zygote contains a genetic program for producing all cell types, but the cells it produces gradually begin to follow subprograms that limit them more and more. Figure 21.8 shows that as vertebrate mesodermal cells develop, some become **stem cells,** in this case pluripotent stem cells able to develop into any of the blood cell types. In further divisions, these developing cells pass through choice points where they are directed into one path or another, until eventually each stem cell is totally committed to becoming a specific kind of blood cell. These cells differentiate under the stimulation of several proteins called colony-stimulating factors (CSFs) that are produced by other cells. Thus erythropoietin made by kidney cells stimulates the development of erythrocytes (red blood cells), and different CSFs, made by blood cells and others, stimulate the development of other blood cells.

Relatively few cell divisions are needed to create the mass of an adult animal, even a large one. For instance, about 45–50 divisions would make the approximately 10^{14} cells of an adult human. As the zygote divides, its blastomeres are becoming more and more restricted, and eventually most of them become determined as cells of a single type. How many cells actually become determined as each type? Beatrice Mintz used the allophenic mice she developed (Figure 21.7) to answer this question. She observed that when blastulas from white and brown mice are mixed, the resulting mice have a series of stripes extending from the midline of the back. The maximum number of stripes in any mouse is 34 (17 on each side). The pigment cells that create the fur color (melanoblasts) are derived from the neural crest (Figure 21.9), and after the neural tube has formed, they spread out along the center of the back. Mintz's analysis suggests there are only 34 such cells. Each one then grows into a subclone that moves out through the skin to color one strip of fur. This experiment shows that a small number of cells have been set aside, or *allocated,* in the early embryo to develop into these particular adult cells.

Exercise 21.4 Describe the differentiation of blood cells by comparing the process to the programs and subprograms used by a computer.

21.5 Some cells remain stem cells indefinitely.

While many cells become fully differentiated, others remain stem cells throughout an organism's lifetime. Typically, as stem cells grow and proliferate, one daughter cell at each division remains a stem cell while the other differentiates (Figure 21.10). In the skin, for instance, stem cells in the lower layer of epidermis continually proliferate as the top layers of

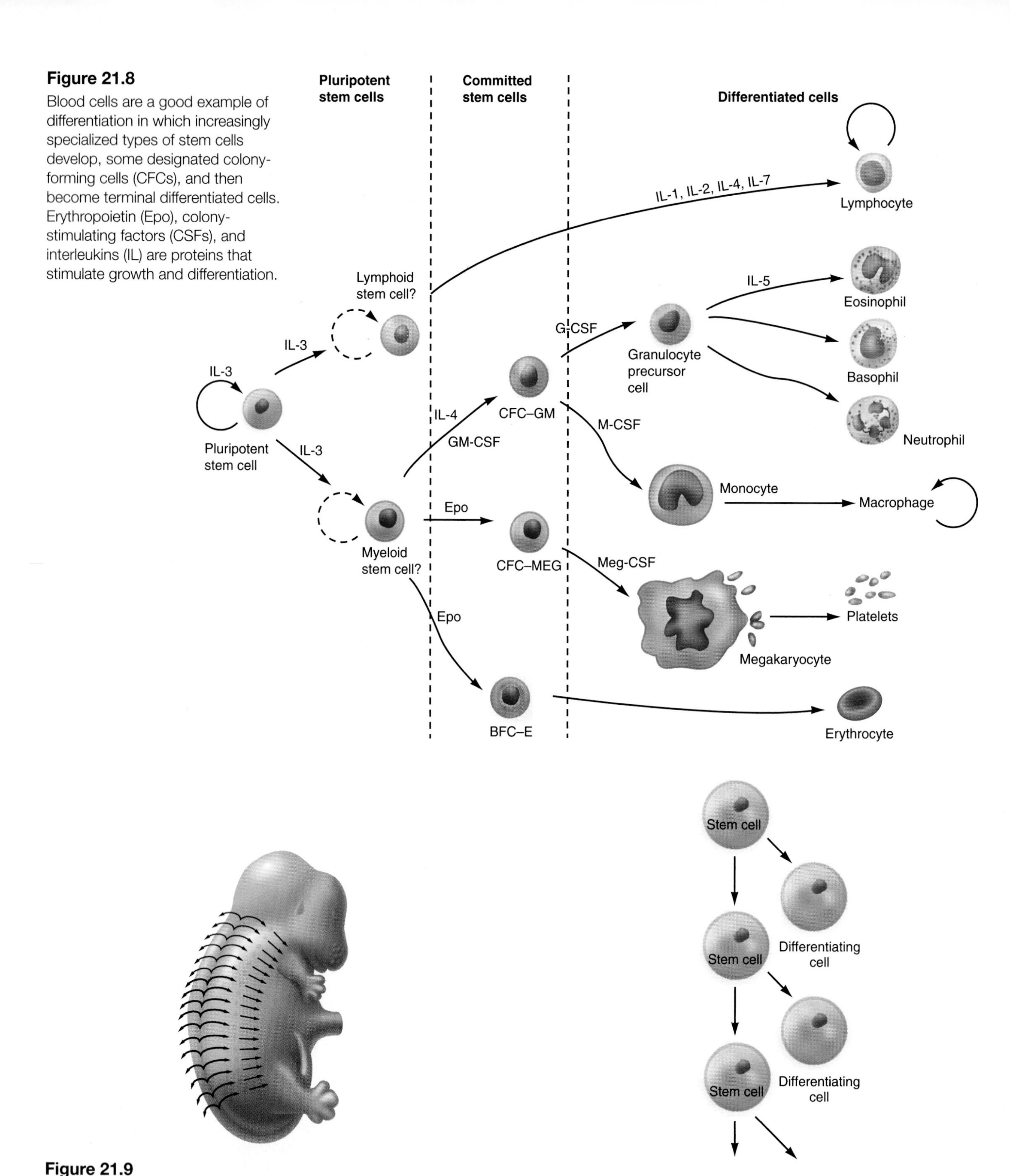

Figure 21.8

Blood cells are a good example of differentiation in which increasingly specialized types of stem cells develop, some designated colony-forming cells (CFCs), and then become terminal differentiated cells. Erythropoietin (Epo), colony-stimulating factors (CSFs), and interleukins (IL) are proteins that stimulate growth and differentiation.

Figure 21.9

Pigment cells derived from neural crest tissue distribute themselves along the back of the embryo mouse. Experiments with allophenic mice suggest that only 17 pigment cells are located on each side.

Figure 21.10

A common pattern of development: At each division, one daughter cell remains a stem cell while the other starts to differentiate.

skin wear off. The lower cell at each division remains a stem cell while the upper one loses its water, begins to fill up with the protein keratin, and eventually becomes a tough, protective scale.

Plant tissues also show this pattern of development, particularly in woody plants. As plant tissues differentiate, some stem cells remain in **meristems,** layers of tissue that never fully differentiate. Meristems give plants the capacity to grow indefinitely without ever reaching a maximum size, in contrast to most animals. In the stems of woody plants, a meristematic layer called the vascular cambium lies between the xylem and phloem, the tissues that conduct water and nutrients (Figure 21.11). The stem grows wider as the cambium cells proliferate, with the innermost daughter cells developing into xylem, the outermost cells becoming phloem, and those in the middle remaining meristematic.

Cells that are in the last stage of differentiation—such as nerve cells, blood cells, and upper epidermal cells—may never divide again. However, other specialized cells, including those in the liver, retain the capacity to divide. Adult liver cells proliferate very little, but up to 60 percent of a rat's liver or 10–20 percent of a human's liver can be removed surgically and the remainder will still regenerate to make a normal-sized organ. These cells demonstrate their capacity for growth when they are suddenly released from inhibitory influences.

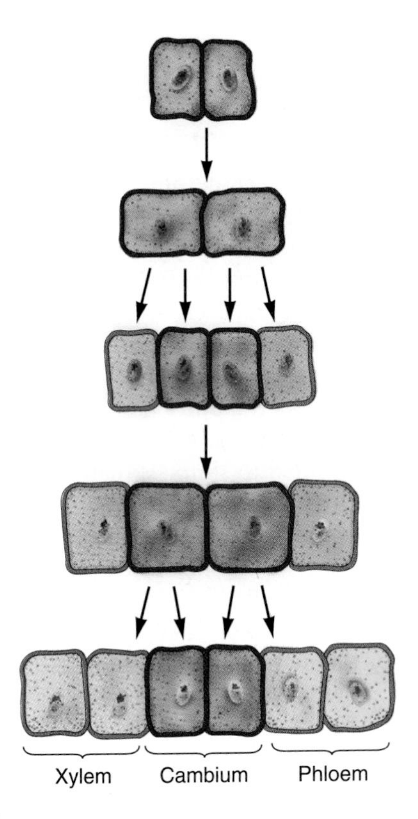

Figure 21.11

In the vascular cambium, dividing cells produce xylem tissue to one side and phloem to the other, but the central cells remain stem cells indefinitely.

Exercise 21.5 If a large part of a rat's liver is removed and the animal's circulatory system is connected to that of a rat with a normal liver, both livers start to grow. What does this experiment show about the factor(s) that stimulate growth?

21.6 Certain tissues induce the differentiation of other tissues.

Many cases of differentiation depend on **embryonic induction,** a process in which a tissue induces neighboring cells to develop in a certain direction through chemical signals called *inducers.* These compounds should not be confused with the well-known inducers that regulate gene expression in simple cells, even though they may turn out to operate similarly. Embryonic induction was discovered by Hans Spemann during the 1920s and 1930s. First he repeated experiments that had been done by others before him, separating the two blastomeres created by the first cleavage of an amphibian embryo. Usually only one of the two blastomeres developed into a normal embryo, but sometimes both blastomeres did. Spemann found that the critical difference is in the position of the first cleavage plane. As we saw in Chapter 20, an amphibian zygote develops a *gray crescent* opposite the point where the sperm nucleus enters the egg, and if the first cleavage cuts through the gray crescent so that both halves receive some of this patch, both can develop.

The gray crescent is particularly interesting because the dorsal lip of the blastopore develops from it. With his attention focused on this region, Spemann and his student Hilde Mangold did a landmark experiment in 1924 (Figure 21.12). They removed the dorsal lip of the blastopore from early gastrula newt embryos and transplanted them into the blastocoel in the early gastrulas of a different newt species. The host embryos developed a second nervous system and head, and some of them even became a pair of whole, conjoined-twin embryos. Because the two species of newts had different coloring, it was clear that the extra head and other structures were primarily host tissue. Thus the donor tissue didn't just grow as a parasite at the side of the host; rather, the donor tissue *induced* some host tissue to develop into structures that it would not otherwise become. Spemann and Mangold called the dorsal lip material the **organizer** because it can direct embryonic growth.

We now know that the dorsal lip of the blastopore acquires its ability to organize development through events initiated at fertilization. The rotation of cytoplasm that occurs when a sperm nucleus enters an egg activates components in the vegetal-pole cytoplasm that will eventually determine the characters of animal-pole cells. A patch of vegetal cells, the *Nieuwkoop center,* produces proteins (now identified as the products of known genes) that induce formation of the Spemann-Mangold center above them. So even though induction by the organizer has been called "primary" induction, the organizer itself is organized by preceding activities.

The dorsal-lip cells are destined to sweep through the blastopore and form the dorsal surface of the archenteron, an embryonic tissue called *chordamesoderm* because it will become

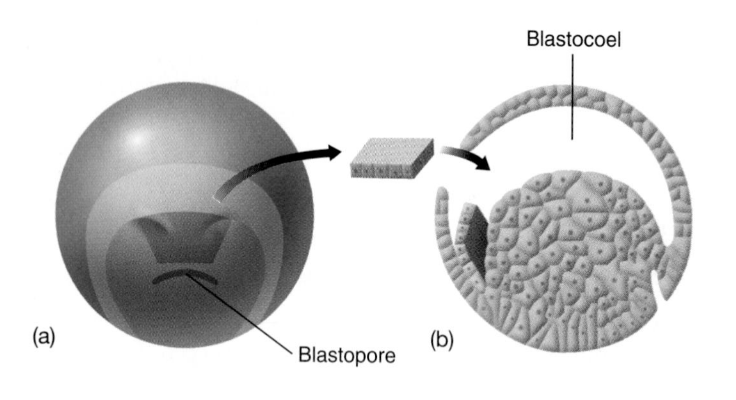

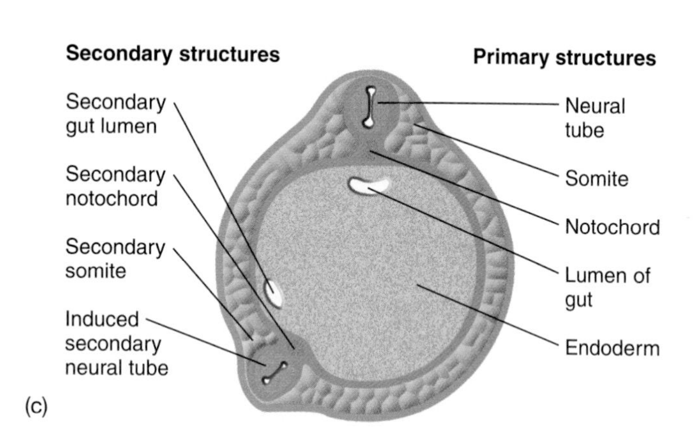

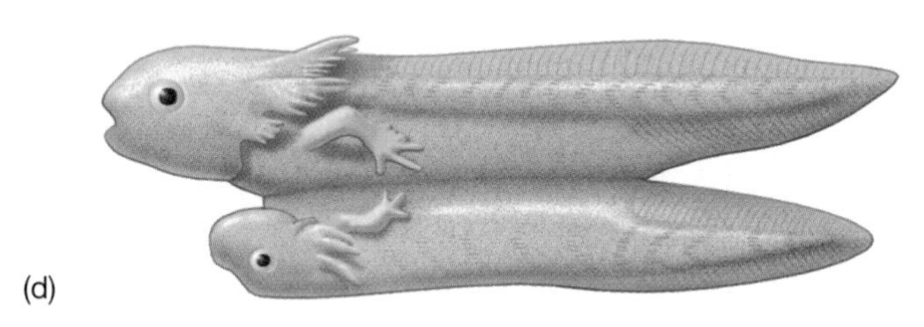

Figure 21.12
If the dorsal lip of the blastopore from one embryo *(a)* is transplanted into another *(b),* it induces the formation of secondary structures *(c),* and sometimes virtually a whole embryo conjoined to the main embryo *(d).*

the notochord and mesoderm. Chordamesoderm lies under the dorsal ectoderm and induces the central portion of ectoderm to form a neural plate, which rolls itself into the neural tube. Embryonic induction presumably occurs where the two kinds of cells are in contact, through a soluble chemical inducer that diffuses over from chordamesoderm cells to activate the ectoderm cells. Evidence for this mechanism comes from experiments in which induction still occurs when chordamesoderm is separated from ectoderm by a filter whose fine pores allow materials to diffuse but prevent direct cellular contact. The inducer is probably a combination of proteins produced by the Nieuwkoop center to induce the Spemann-Mangold center.

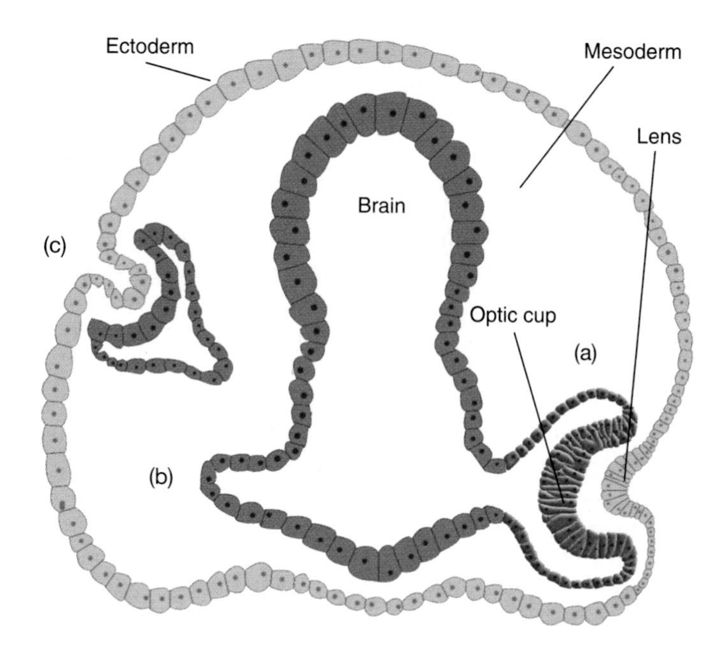

Figure 21.13
An optic cup grows out of the brain and induces the overlying ectoderm to develop into a lens (at *a*). The lens does not develop where the cup is removed *(b),* but it does develop where the cup is transplanted *(c).*

And, as we might expect from knowing about signal-transduction pathways, both protein kinases and cyclic AMP are involved in the inductive response.

21.7 Inductive interactions may be instructive or permissive.

The induction of neural ectoderm by chordamesoderm is often referred to as *primary induction,* since it involves an interaction between primary tissue layers. At later stages and other places in the embryo, a series of *secondary inductions* occur, showing that two kinds of cell-cell interactions occur during development. In an **instructive process,** an inducing tissue directs a target tissue to develop in a certain way; in a **permissive process,** an inducing tissue simply allows the other tissue to differentiate by expressing information it already possesses. Permissive interactions seem to be limited to simple situations in which a tissue requires a base of extracellular proteins such as fibronectin and laminin to develop, and these are supplied by a neighboring tissue. We will therefore focus on instructive interactions.

One powerfully instructive tissue is a vertebrate optic cup. Figure 21.13 shows how an optic cup grows out of the primitive brain region of the neural tube and induces the overlying epithelium (the single layer of cells that will generally become skin) to develop into a lens. If an optic cup is removed, no lens forms. This in itself suggests that the epithelium doesn't "know" it is to become a lens until the optic cup instructs it, but the definitive experiment is to place the severed optic cup under the epithelium somewhere else, and there it induces formation of a lens. However, an optic cup won't induce lens formation in every tissue; the target tissue can only receive the instruction

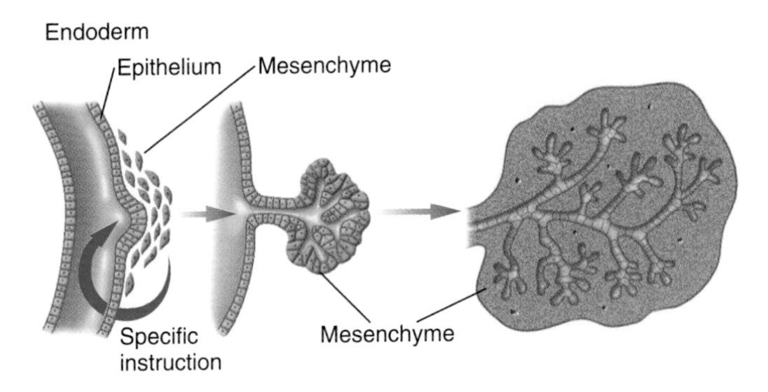

Figure 21.14

Mesenchyme *(pink)* forms around the tubular buds of endoderm *(yellow-brown)* from the intestine, and in each case the mesenchyme instructs the tube to become a particular organ. The endoderm can come from anywhere in the tube; the specificity lies in the mesenchyme.

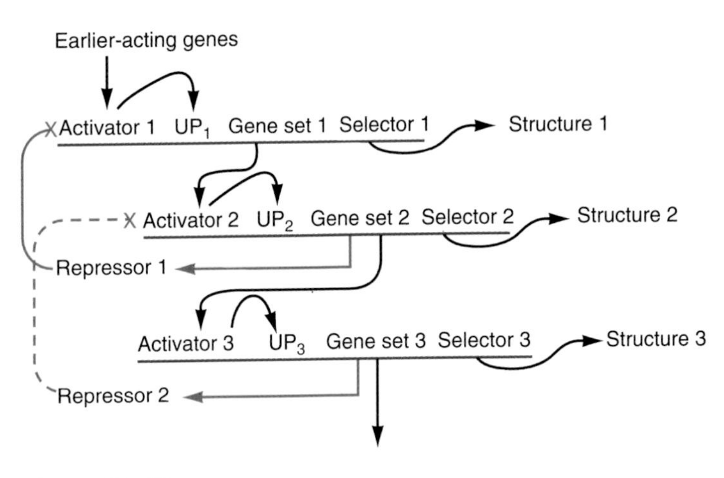

Figure 21.15

One general model for gene regulation in differentiation can be made by combining the regulatory elements of simple systems. Each set of genes, in its turn, can either activate or repress another set *(red lines)*. The product of a selector gene regulates a large series of other genes, selecting the formation of one structure rather than another.

and react properly if it is **competent**—that is, not restricted along some other line of differentiation. Ectoderm apparently is not competent to become a lens unless it has previously received specific signals from the neural plate; if it has, the optic cup performs the final determination into lens tissue.

Following the general rule that mesenchyme instructs epithelium, the epithelia of endoderm or ectoderm are commonly instructed by the nearby mesenchyme. The epidermis of a chick, for instance, develops into characteristic feathers and scales under instructions from the particular mesenchyme that underlies it. Lungs, pancreas, and liver develop from tubes that branch out of the central (endodermal) gut tube; in each case, this tube becomes increasingly branched and finally develops the definitive structure of its organ. Mesenchyme surrounding each region of the gut tube becomes part of the organ and carries the specificity that determines the organ's form (Figure 21.14). Presumably, in each interaction, one or more specific proteins pass from the mesenchyme to the epithelium, but these have not been identified yet.

Exercise 21.6 Consider the following situations: *(a)* The skin of a bird does not grow feathers unless it is in contact with mesoderm. If the mesoderm comes from the back, the skin develops back feathers, and if the mesoderm comes from the head, the skin develops head feathers. *(b)* The skin of a lizard does not grow scales unless it is in contact with mesoderm. Regardless of where the mesoderm comes from, head skin develops head scales and back skin develops back scales. Which of these situations shows an instructive process and which a permissive process?

21.8 Differentiation may be determined by hierarchies of genes that regulate other genes.

When Jacob and Monod proposed the operon model for gene regulation (see Section 18.2), they pointed out that it could be generalized in unlimited ways, and could be expanded into models for differentiation. If bacteria can use regulator genes to control blocks of genes through operators, more complex cells can use other genes to regulate the regulators. Many specific models have been made, and can be made, to account for differentiation, but a general model we can use for discussion is shown in Figure 21.15. The essential point is that genes can be linked in *regulatory circuits,* in which genes turn one another on and off through specific signals. Names such as "activator," "selector," and "sensor" have been proposed for the effective genes.

In this general model, which can be varied endlessly, one set of genes can activate another and repress a third. Differentiation then operates according to a binary logic: Each set of genes is either on or off. A series of selector genes can work one after the other, either in space or time, by using the appropriate logical connections. In each part of a developing embryo, different selector genes can be turned on whose function is to control genes for structural proteins, enzymes, and the like—the materials characteristic of each differentiated cell. Such a series can produce different structures in each part of the embryo or in a single developing region at different times.

As explained in Section 18.8, eucaryotic genes are regulated primarily by upstream promoters and by enhancers both upstream and downstream from each gene. Regulatory proteins bind to these regions in various combinations and interact so as to promote or inhibit transcription from a gene (Figure 21.16). All the eucaryotic genes known to function in genetic regulation encode proteins of this kind. We will see shortly how this model plays out in specific cases.

21.9 Some differentiating tissues appear to be regulated by a clock mechanism.

Instructions for differentiation can be selected by time, position, or both. In the following sections, we will take up differentiation on the basis of position and show how genes can be

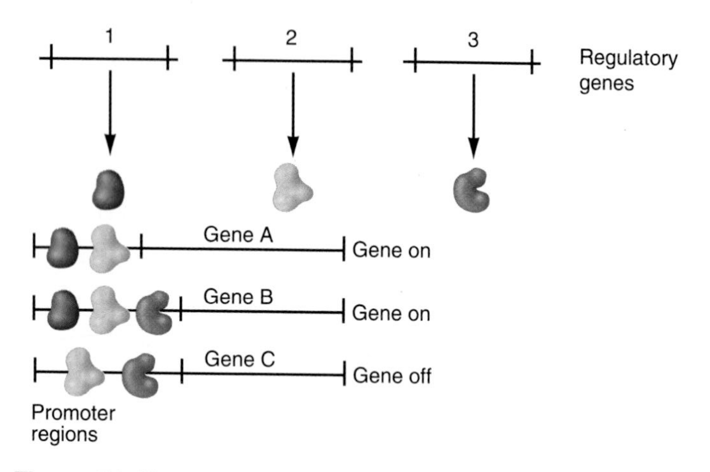

Figure 21.16

Many regulatory genes encode proteins that act as transcription factors by binding to regulatory regions, such as promoters and enhancers. Different combinations of these factors either promote or inhibit transcription of each gene.

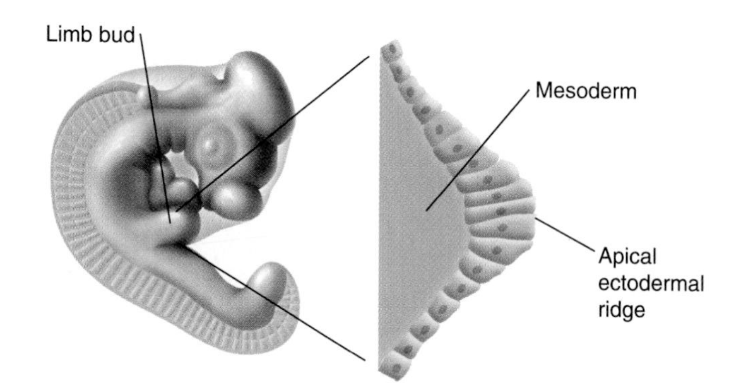

Figure 21.17

The chick limb bud has an apical crest made of a cap of ridged ectoderm with underlying mesoderm.

turned on sequentially in response to spatial cues; in this section we will show how they can be turned on in a temporal sequence, as in the growth of chick limbs.

The limbs of quadrupeds, all having the same general bone pattern, grow from limb buds that gradually elongate as they grow out of the body (Figure 21.17). A limb bud is made of a mesodermal core beneath a cap of ectoderm with a distinctive ridge, and differentiation involves an interaction between the two tissues. The ridge itself is essential; no limb develops if it is removed, and grafting on an extra ridge produces a divided double limb. Yet it is the mesoderm underneath that gives instructions for the form of the limb. In a chick embryo, if a bit of terminal mesoderm from a leg bud is inserted just under the ridge of a wing bud, the resulting limb is mostly wing with a normal foot at its end. The information for this foot structure obviously comes from the leg mesoderm.

Fascinating experiments conducted by J.H. Lewis, D. Summerbell, and Lewis Wolpert have shown generally how the limb differentiates. When they cut off the tip of a limb bud (including its mesoderm) and grafted on a tip from a younger limb bud, the result was a limb with duplicated basal bones, such as humerus and radius-ulna (Figure 21.18). The opposite graft, replacing the tip of a young bud with a tip from an older one, produced a limb with missing middle bones. These experiments suggest that the terminal ridge and the mesoderm immediately under it use a clock mechanism to specify the fate of each region of mesoderm. The terminal region first instructs the mesoderm behind it to become humerus, then instructs the next region to become radius-ulna, the next to become wrist, and the last to become phalanges (finger or toe bones). But it specifies each structure at a certain time, working on a clock that is always ticking. At a particular age, the limb-bud tip must send instructive materials to the nearby mesoderm. Then the clock ticks. New genes are turned on, and they give the mesoderm a different instruction. The clock ticks again, and another set of genes is turned on. The tip of an older limb has already specified the basal regions of the limb behind it. If the tip is replaced by a younger tip whose internal clock is set at an earlier time, it simply repeats the process, specifying a new humerus and a new radius-ulna. If the tip of a young limb, with only a humerus, is replaced by an older tip with a later clock, the new tip has already used its instructions for basal regions, so it is set to specify wrist or phalanges, leaving a limb with no middle bones.

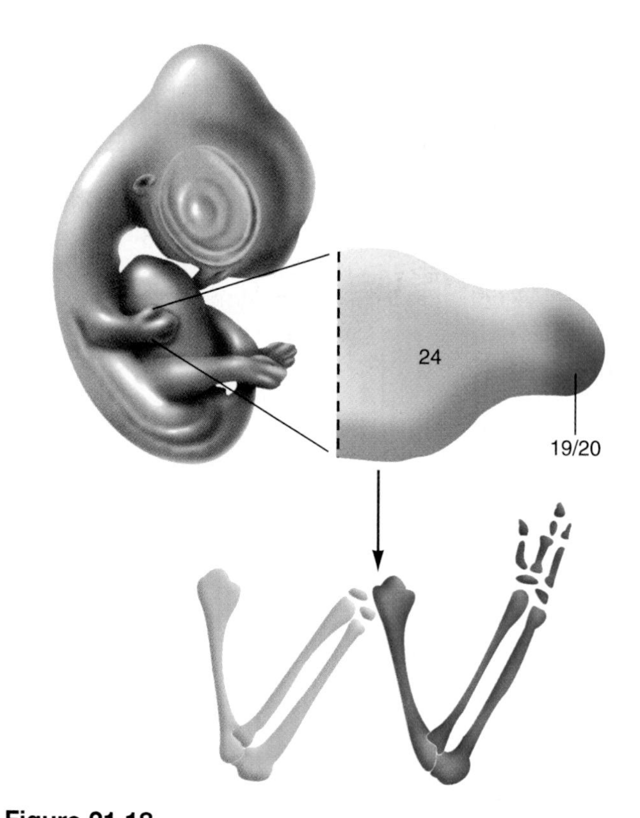

Figure 21.18

An apical crest of stage 19/20, which has not yet specified the humerus, is grafted to a limb of stage 24, which has already had both humerus and radius-ulna specified. The resulting limb has a duplication of several bones.

Clocklike mechanisms of this sort may be very important in development. While we don't know the genetic basis for any of them yet, they are consistent with the model in Figure 21.15.

Exercise 21.7 Here is a regulatory circuit with two blocks of genes:

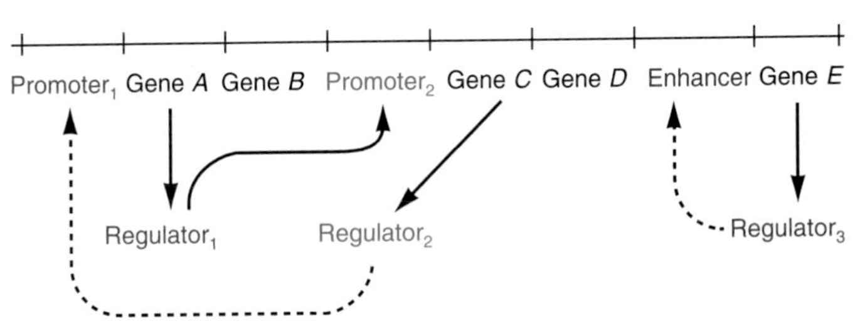

Regulator$_1$ encodes a protein that binds to Promoter$_2$ and prevents transcription of genes *C* and *D*. Regulator$_2$ encodes a protein that binds to Promoter$_1$ and prevents the transcription of genes *A* and *B*. The product of gene *E* binds to the enhancer and promotes transcription of genes *C* and *D*. Suppose genes *A* and *B* are expressed initially, and then an outside event promotes transcription of gene *E*. How will the expression of these genes change? Show how a regulatory circuit of this kind could be used in a clock mechanism to turn a series of genes on and off one after the other.

21.10 Many cells differentiate on the basis of information about their positions in the embryo.

Some of the most intensely investigated systems show how cells differentiate on the basis of **positional information** that specifies each cell's position in the body. Some tissues receive instructions from their neighbors through induction, but more general information about position is needed. Without it, the genes that specify head structures could be turned on in the stomach, for example, and an embryo could not develop.

The early determination of blastomeres in a mosaic embryo, and even of the undivided zygote, means that position-determining substances, called **localized determinants,** are already distributed in the egg itself. So the blastomeres become differentiated because these determinants get partitioned into different cells as the zygote is cleaved, and the determinants locally induce or repress different blocks of genes. This mechanism sets each blastomere on a particular course of differentiation. In effect, a cell knows its position because it responds to materials that have been placed in specific positions in the egg cytoplasm. An egg can be structured in this way because it matures in an ovary, surrounded by specialized cells that can contribute their own localized determinants.

There is good reason to think that these hypothetical materials are in the relatively structured cortex (outer layer) of an egg. E. Newton Harvey found that mild centrifugation of sea urchin eggs leaves the cortex intact but moves other components around, so mitochondria, yolk materials, and other structures form separate layers. Yet these eggs could still develop into normal larvae. On the other hand, eggs subjected to extreme centrifugation that disrupted the structure of the cortex did not develop normally. Thus normal development depends on a highly structured egg, with specific materials localized in each cortical region.

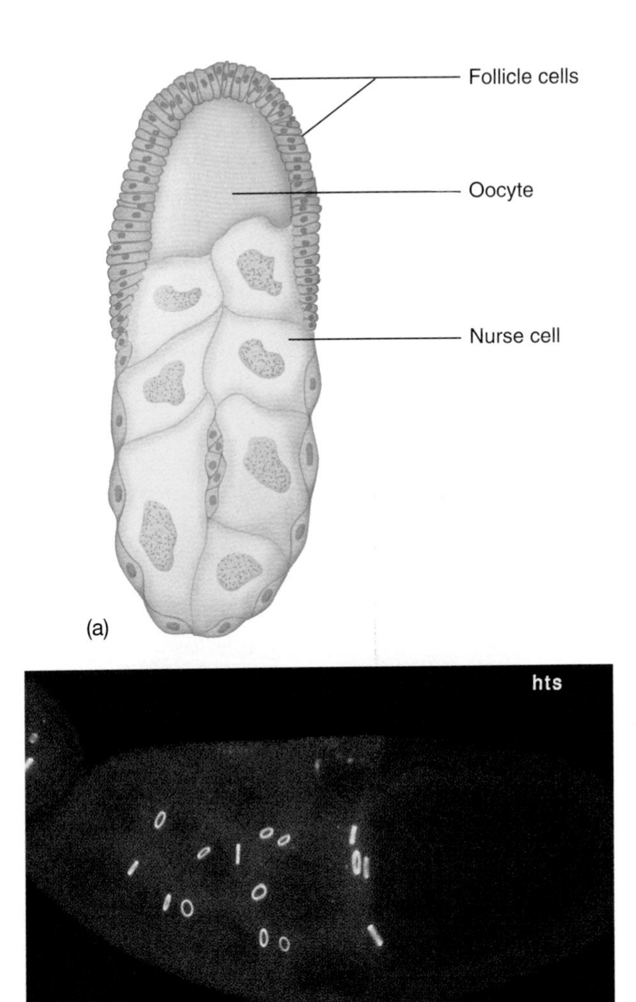

Figure 21.19
(a) A maturing oocyte is surrounded by nurse cells, which dump portions of their cytoplasm into the oocyte through connecting ring canals. *(b)* The ring canals are revealed by special staining in a photomicrograph of an egg follicle.

The role of genes in development has been highly analyzed in the fruit fly *Drosophila melanogaster,* the geneticist's workhorse, and it is important to understand its embryology. In a *Drosophila* ovary, an oogonium divides into 16 cells that share a single cytoplasm through ring canals. One cell then undergoes meiosis to become the oocyte, and the other 15 grow into polyploid *nurse cells* surrounding the oocyte (Figure 21.19). The oocyte enlarges as the nurse cells deposit some of their cytoplasm into it through the ring canals. Then it is released into the oviduct, fertilized by a sperm, and expelled to develop outside the female's body.

The zygote's nucleus first undergoes rapid mitotic divisions to produce a syncytium with hundreds of nuclei. Then these nuclei migrate to the surface of the embryo, and the cytoplasm around them divides into separate cells. Cells in the posterior end contain distinctive ribonucleoprotein particles called polar granules, and these cells become the entire germ plasm, from which the next generation of gametes will be made. The other 6,000 cells become the fly's somatic (body) cells. The embryo develops into a larva through gastrulation and other cellular movements that create tissue layers and embryonic organs. Most obviously, by about 10 hours after fertilization, the larva is divided into distinct segments typical of arthropods: three to five head segments (variously defined); T1–T3 in the thorax; and A1–A11 in the abdomen (Figure 21.20). The anterior and posterior halves of each segment may form different structures. Research has been directed toward understanding the genetic basis of this segmentation.

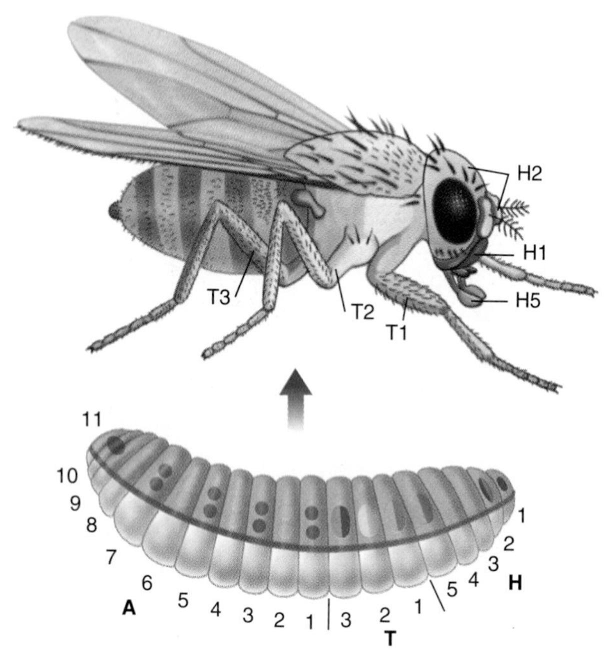

Figure 21.20
An adult fruit fly develops from three to five segments that form the head (H), three in the thorax (T) and eleven in the abdomen (A). The segments are clearly visible in the embryo. (Source: A. Garcia-Bellido, et al., *Scientific American,* 241:90-98, 1979.)

The larva feeds and grows for several days, then becomes a pupa that develops into an adult. While still a larva, it has been laying down patches of cells called *imaginal discs* (pronounced with a hard "g"); an adult fly is an *imago* (ih-may'go or ih-mah'go). Each imaginal disc will develop into one of the external adult structures: two wings, six legs, two antennae, and so on. Each imaginal disc resides in the larva in the position where the adult organ will form.

It is clear that the egg already carries localized determinants in its cytoplasm. Figure 21.21 shows that if a little of the posterior cytoplasm is allowed to leak out, information for the tail end is lost, and an embryo with no abdomen develops. If material from the posterior end is injected into the anterior, an embryo develops with two posterior ends. These polarities built into the embryo reflect the activities of the nurse cells. By injecting their own specific materials into the egg cytoplasm, they create asymmetries that make each position in the embryo unique. Every cell in a fruit-fly embryo is given an "address" along two axes: dorsal-ventral (that is, back-belly) and anterior-posterior (head-tail). A cell takes its cue for differentiation from this information. The asymmetries in the egg's cytoplasm translate into turning on certain genes, and these, in turn, induce other genes until each region becomes differentiated and makes its own unique structures. ("If my address is *x,y*, then I must be a cell of the upper posterior part of the second thoracic segment, and I should make wing proteins.")

Development in *Drosophila* is being analyzed more and more frequently by genetic methods, starting with mutants defective in a certain developmental step. Remember the central idea behind genetic analysis: Studies of a mutant gene are designed to determine what the normal gene does, and finding a mutant identifies a function that might be impossible to

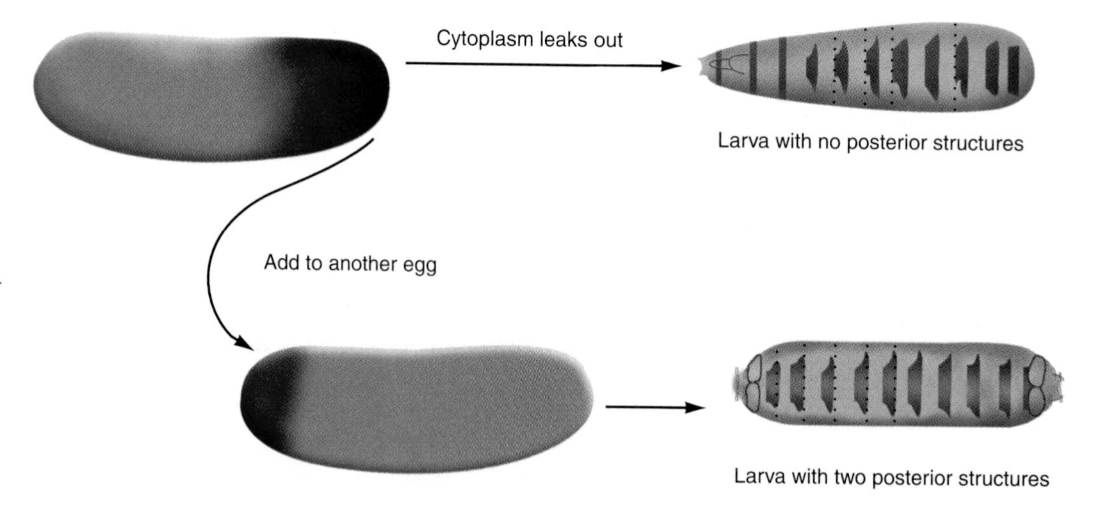

Figure 21.21
Information about the anterior-posterior position in a *Drosophila* embryo is built into the egg. If material is removed from one end, the embryo develops without structures characteristic of that end. If material from the posterior end is injected into the anterior, an embryo develops with two posterior ends.

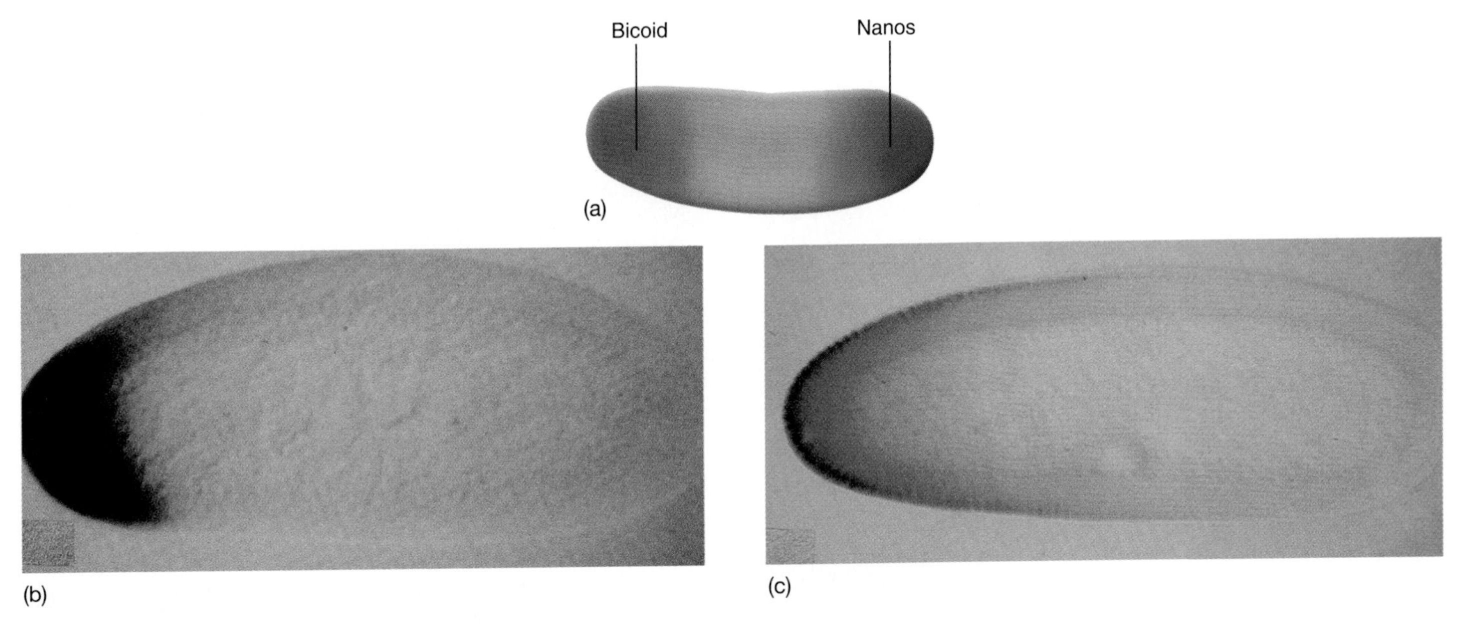

Figure 21.22
(a) In a *Drosophila* embryo, an anterior-determining gradient is set up by the Bicoid protein and a posterior-determining gradient by the Nanos protein. (A protein is named for the gene that specifies it, using a capital letter and no italics.) The photographs show embryos stained to show *(b) bicoid* messenger RNA and *(c)* Bicoid protein.

discover with any other method. Because the genetic analysis of *Drosophila* is so advanced, many genes that provide positional information are known. We will now describe the action of several genes and their proteins—not because these details are to be memorized, but to show paths of development quite concretely.

The anterior-posterior axis is set up initially by two genes, *bicoid (bcd)* and *nanos(nos)* (Figure 21.22), which are expressed in the nurse cells surrounding the developing egg. They inject the *bicoid*-gene mRNA into the anterior end of the egg and the *nanos*-gene mRNA into the posterior end. These messengers are translated in the egg, and their protein products diffuse through the egg, creating concentration gradients along the anterior-posterior axis.

The gradients of Bicoid and Nanos proteins initiate a cascade in which a hierarchy of several genes is turned on, genes with wonderful names that define each of the segments (Figure 21.23). First a series of *gap genes* are expressed; they are so named because mutations in these genes produce large gaps in the anterior-posterior structure, and they cannot grow beyond a larval stage. The *knirps* gene (German, "dwarf") acts in a posterior region, and the *krüppel* gene (German, "cripple") acts in an anterior block. Gap genes, in turn, activate a series of *pair-rule genes,* which lay down the basic division of the body into segments. Mutations in pair-rule genes can delete several segments; for instance, *even-skipped (eve)* mutants are missing all the even-numbered segments, and *fushi tarazu (ftz;* Japanese, "too few segments") mutants are missing all the odd-numbered segments. Finally, a series of *segment-polarity genes* are turned on that specify anterior and posterior halves of a segment. When one of these genes is mutated, as in *gooseberry,* the larva has segments with anterior and posterior halves that mirror each other, instead of being distinctive. These segment-determining genes appear to encode transcription factors, as specified by the general model we presented earlier. Figure 21.24 shows a few of the interactions among some genes.

With the body divided into segments defined by these three sets of genes, a series of **homeotic genes** is activated. The homeotic mutants that define these genes are fascinating because they grossly disrupt development, generally by converting a structure into a different one that normally occurs in some other part of the animal. The *Antennapedia* complex determines structures in the anterior half of the body; the *Antennapedia* mutation *(Ant)* itself substitutes a pair of legs for the antennae that normally grow on the fly's head (Figure 21.25). These genes have all the characteristics of selectors. A homeotic gene determines the choice a cell makes between different fates; a homeotic mutant abolishes one of the fates and replaces it with another.

The *bithorax* complex determines the form of segments from the middle of the thorax (segment T2) through the abdomen. It contains three genes and several regulatory regions that determine the expression of these genes. The complex operates as if segment T2 is a "ground state" or "default setting"—the condition all segments will have unless changed by the expression of certain genes, as happens naturally in all the segments posterior to T2. For instance, the products of three genes—*bithorax (bi), anterobithorax (abx),* and *postbithorax (pbx)*—regulate sites that convert the T2 state into T3. A fly that is homozygous for mutations in all three genes has four wings, since T3 then reverts completely to the T2 state (Figure 21.26).

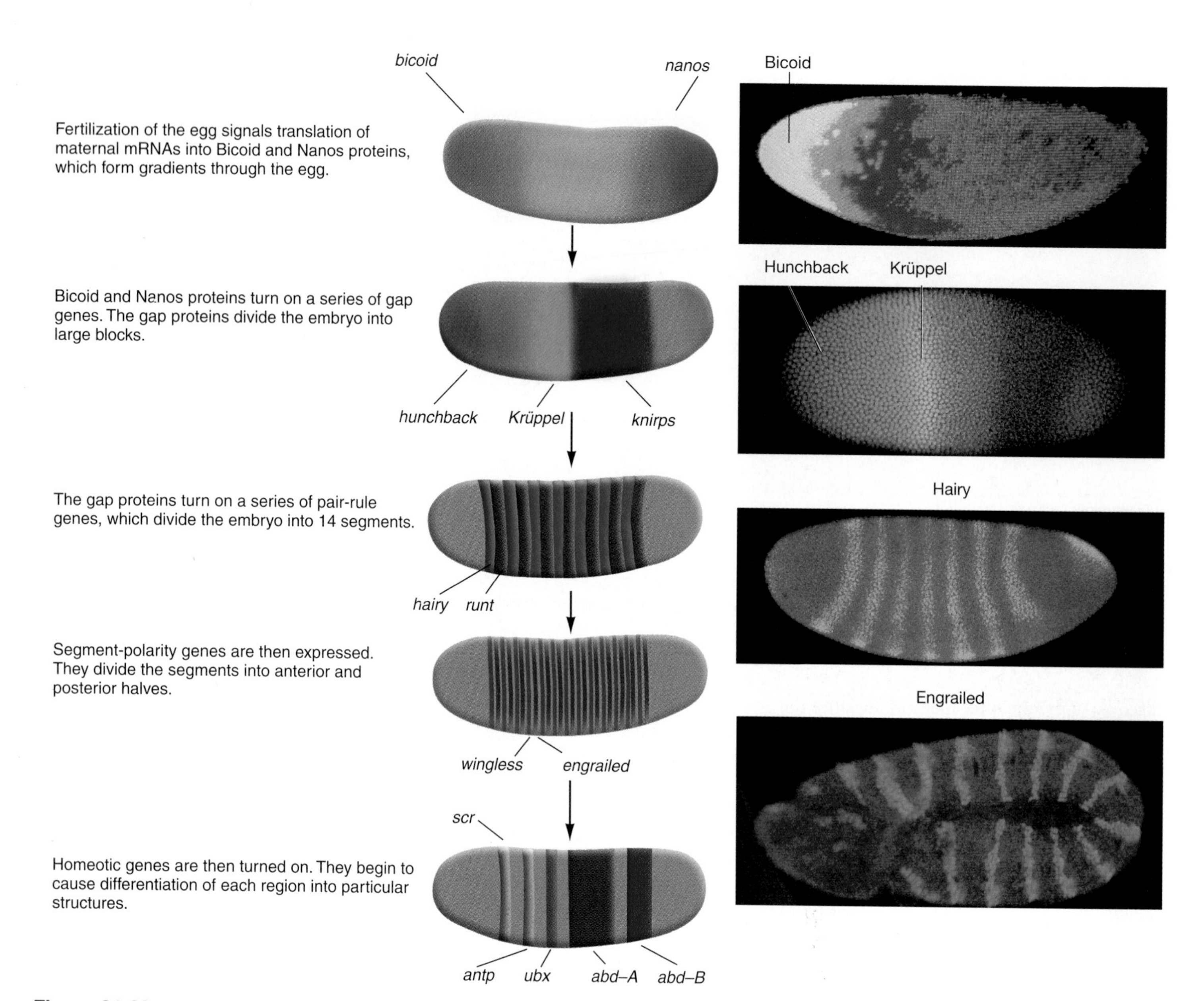

Figure 21.23

Each cell in a *Drosophila* embryo is given an address along an anterior-posterior axis through the sequential expression of a series of genes, which are defined by mutants. The cells are then given instructions to differentiate into particular structures. The micrographs, made by Christiane Nüsslein-Volhard and Sean Carroll, show the location of antibodies, labeled with fluorescent dyes, directed against each embryonic protein. (From *Introduction to Genetic Analysis,* 4th edition by J.F. Griffiths, et al. © 1996, 1993, 1989, 1986, 1981, and 1976 by W.H. Freeman and Company. Used with permission.)

The *bithoraxoid (bxd)* region converts the T3 state into A1, so a homozygous *bxd bxd* fly has legs on A1, just as if it were T3. Similarly, a series of other genes or regulatory sites determines each successive posterior segment.

Exercise 21.8 The shell of the common freshwater snail *Lymnaea* coils either dextrally (to the right, as seen from the tip) or sinistrally (to the left). The coiling direction is determined by one gene, a dominant *D* for dextral coiling and a recessive *d* for sinistral. When dextral and sinistral snails are crossed, the offspring have the shell form of the parent that produces the egg—the "mother," if you will, although snails are hermaphrodites. That is, the form of an embryo isn't determined by its own genotype but by that of its "mother." Explain this situation on the basis of a principle established in the preceding section.

21.11 Series of homeotic genes determine the basic body plans of all animals.

Homeotic mutants have been known for a long time, but when the DNA sequences of their genes were determined, they turned out to hold a remarkable surprise: a common sequence of 180 nucleotide pairs, called a **homeobox.** The homeobox encodes a

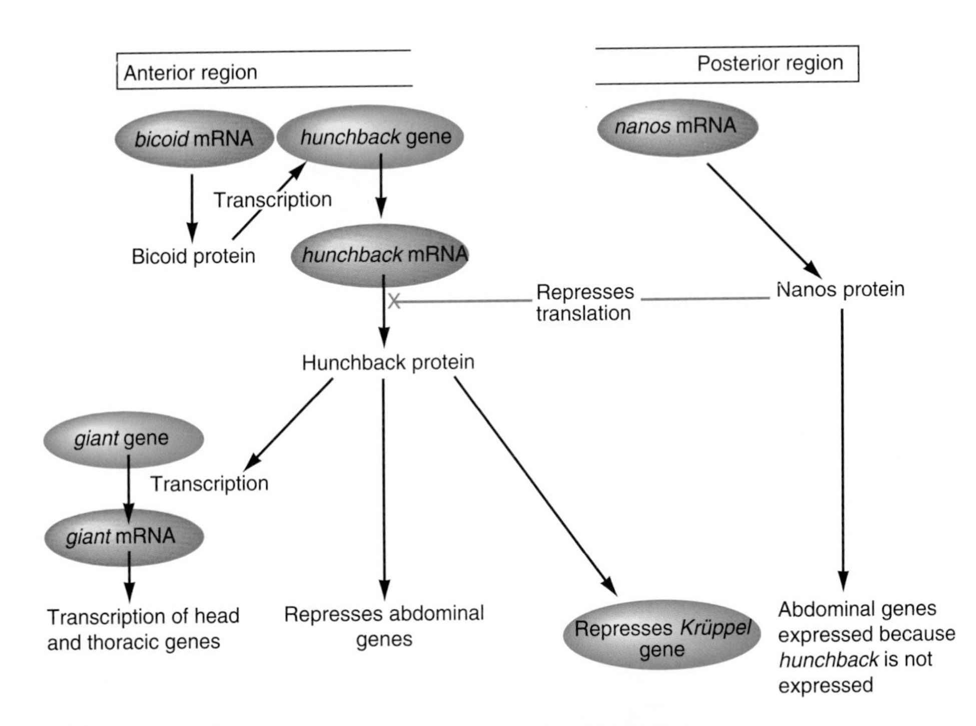

Figure 21.24

The Bicoid protein is active in the anterior part of the embryo, where it specifies transcription of the *hunchback* gene. In the posterior region, however, the Nanos protein prevents translation of the *hunchback* mRNA, so Hunchback protein is not made. High levels of the Hunchback protein, in turn, induce transcription of the *giant* gene; where the level of Hunchback is lower, the *Krüppel* gene is turned on.

Figure 21.25

An *Antennapedia* mutant has a pair of legs on its head instead of antennae.

Figure 21.26

This four-winged fly was made by a combination of mutations in the bithorax complex that converts the third thoracic segment into a copy of the second segment.

sequence of 60 amino acids, the **homeodomain,** in the gene's protein product, which enables the protein to bind specifically to a DNA sequence. So all the homeotic genes encode proteins that bind to regulatory sequences near other genes and turn them on or off. The details of this regulation are complicated and still not understood.

After discovering homeoboxes in *Drosophila,* investigators asked whether similar sequences appear in the genomes of other animals. As expected, other insects have them, but unexpectedly, they also appear in the DNA of quite different animals: in *Caenorhabditis elegans,* in the African toad *Xenopus laevis,* in chickens, and in mammals such as mice and humans. Genes containing homeoboxes are now called *Hox* genes, and the mammalian *Hox* genes are remarkably similar to those of insects, except that mammals have four clusters of *Hox* genes instead of one (Figure 21.27). The sequence of *Drosophila* genes

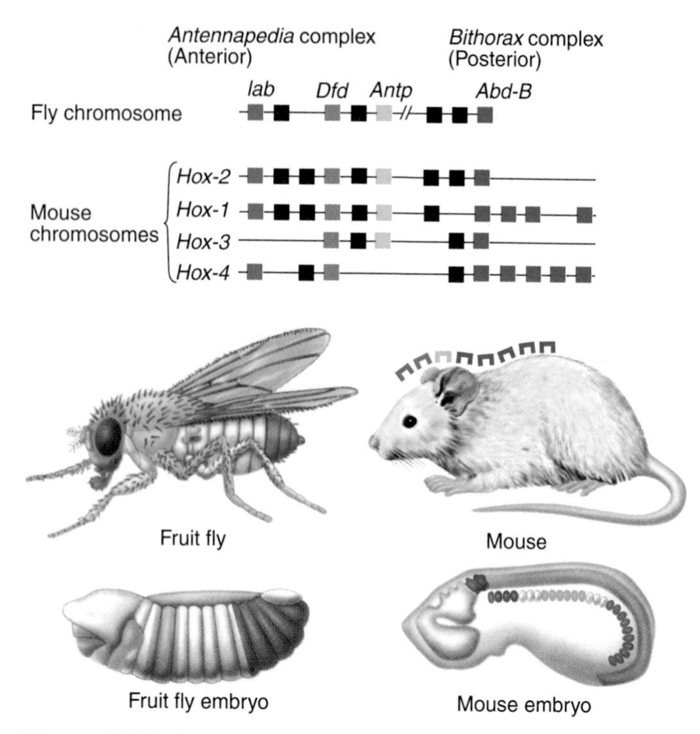

Figure 21.27

Genes of the *Antennapedia* and *Bithorax* complexes of *Drosophila,* which lie on a single chromosome, determine body structures in sequence from the head to the tail. They parallel the mammalian (mouse) *Hox* genes, arrayed on four chromosomes, which also determine structures along the head-to-tail axis. All these genes share homeoboxes. Those that are quite similar in structure are shown in the same colors. (Source: *From Egg to Adult,* 1992, Howard Hughes Medical Institute.)

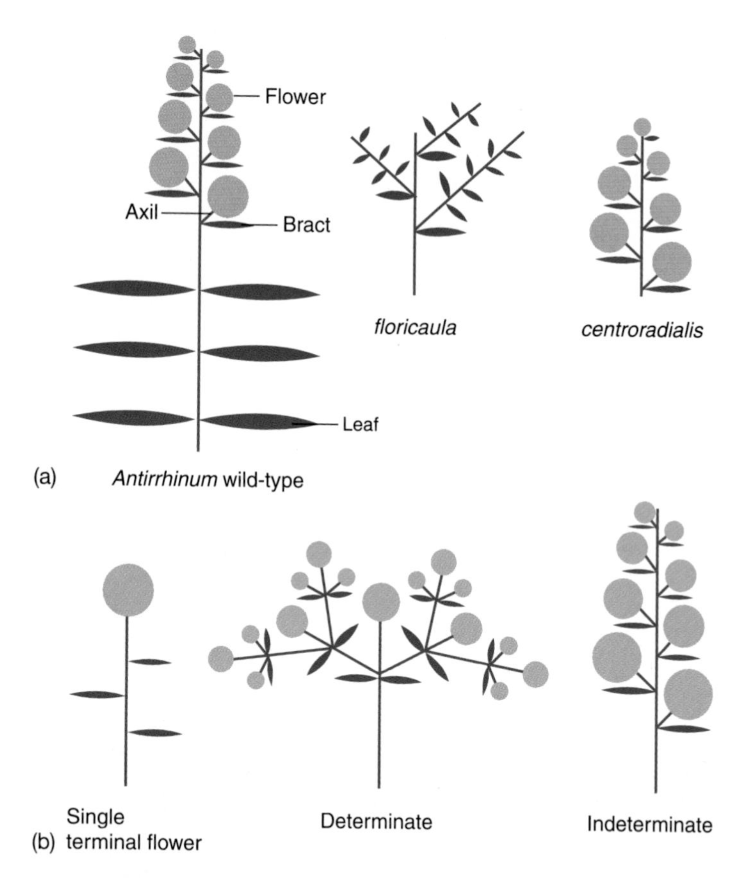

Figure 21.28

(a) Snapdragons normally grow from an inflorescence meristem that produces bracts, and one flower develops in the axil of each bract. In the *floricaula* mutant, another inflorescence meristem replaces the floral meristem. In the *centroradialis* mutant, a terminal flower develops. *(b)* The three classical types of inflorescences can be understood through the successive actions of homeotic genes like wild-type and mutant snapdragon genes.

in the *Antennapedia* and *Bithorax* complexes parallels the mammalian *Hox* genes, preserving the mapping between the gene sequence and the part of the body each gene affects; that is, the gene sequence runs from head-determination at one end to tail-determination at the other end. Insects and mammals have not had a common ancestor for at least 600 million years and perhaps as long as a billion years. Yet the mammalian and insect genes have remained so remarkably similar, in spite of their long evolutionary separation, that one can sometimes substitute for another. For example, *Drosophila* mutants without a functional *labial* gene can be rescued by introducing the chicken gene *Hoxb-1* so they develop normally.

These discoveries point to a deep commonality in the basic developmental paths of animals. The structure of the homeotic genes means that basic developmental mechanisms evolved in some of the earliest animals and have been used in more or less the same way to create all the varied body plans of the animal phyla (see Chapters 34 and 35). So animals have been able to form such a variety of functional body plans because their ancestors evolved a basic, flexible genetic mechanism for making an elongated body with a head and tail that could be modified in many ways. We will return to this theme at the end of Chapter 24.

21.12 Homeotic genes in plants have effects parallel to those in animals.

Although homeotic genes have been studied most extensively in animals, we should expect to find them in any organism that shows significant differentiation of its parts. Indeed, homeotic mutants in plants are starting to reveal how the growing plant makes molecular decisions between developing organs such as leaves and flowers, just as *Drosophila* mutants show how the decision is made between body segments. We can illustrate this with studies of mutants in the snapdragon, *Antirrhinum.*

Remember that plants grow from tissues called meristems in which stem cells retain the ability to multiply indefinitely. In *Antirrhinum,* the shoot rising above the soil grows from an *apical meristem,* which lays down a central stem and several pairs of leaves (Figure 21.28*a*). Then the apex undergoes a transition and becomes an *inflorescence meristem,* which produces a spiral arrangement of small leaves called *bracts.* (An inflorescence is a group of flowers.) In the angle (axil) between each bract and the stem, a *floral meristem* develops that grows into a flower with its petals and reproductive organs. So the shoot develops through temporal changes, similar to the changes in the growing chick limb, as the apex is converted from one kind of meristem to another. Two kinds of homeotic mutants can change this pattern. In the *floricaula (flo)* mutant, the meristem becomes an inflorescence meristem but never undergoes a transition to a floral meristem; instead, the shoot growing from the axil of each bract lays down another length of stem with bracts, and more stems of the same kind can grow in turn from the axils of these bracts. So the plant never flowers. In a mutant of the opposite type, *centroradialis,* the inflorescence shoot eventually ends in a terminal flower, which never develops in the wild-type.

Analysis of the protein encoded by the *flo* gene shows that it has some characteristics of transcription factors, though it does not show close similarities to any other protein. So, like the analogous genes of *Drosophila, flo* probably induces the transcription of a new set of genes needed to form a flower instead of a shoot with bracts.

The discovery of homeotic plant genes opens the door to thinking about the evolution of plant anatomy. Plants form inflorescences of three types (Figure 21.28*b*), and we can now see that the differences among them can be understood through these homeotic genes. The single terminal flower requires a gene like *centroradialis* to act early. The determinate type requires a gene of the mutant *floricaula* type to act for a while, followed by activation of another gene similar to *centroradialis.* Research of this kind is so fascinating and exciting partly because of the connections it makes between development and evolution.

Coda Nothing illustrates the genetic nature of organisms better than the process of development. It is easy to say that a bacterium or a yeast carries a genetic program, but we see its expression only in rather subtle changes in the forms of cells. But the genetic program, or developmental program, of a plant or animal becomes much more obvious as great changes occur over time. Inquiring into the mechanisms that carry out that program, we find a maze of intermolecular interactions as genes are turned on and off, as proteins are made at some times and places and suppressed at others. When all the molecular details of even a simple developmental episode are eventually worked out, they will surely have to be displayed on a very large piece of paper. But already the information available about *Hox* genes and others points to deep, basic developmental processes that were laid down in the most primitive plants and animals. We are starting to glimpse ways for small changes in these genes to manifest themselves in very significant changes in morphology, and they give us new ways to think about evolution. Some evolutionary theorists have contended that the great variety of body plans among animals, for instance, must have resulted from large-scale mutational changes. Now we can see that ordinary, small changes in critical developmental genes can have large-scale effect, and we can start to see how organisms suited to vastly different ecological niches might have arisen. The world where developmental and evolutionary thought meet is now bright with exciting possibilities.

Summary

1. A zygote is totipotent, meaning that it can develop into all adult tissues. As it divides, its cells become limited more and more to forming specific structures.
2. Even when a cell has become differentiated, it retains all its DNA. Transplantation of nuclei from relatively differentiated cells into enucleated eggs shows that, in general, no genetic material is lost as development proceeds. This is also shown by the growth of whole plants from single differentiated plant cells. Differentiation must be understood as the activation or repression of genes, not as the loss of genes.
3. Before a cell differentiates, it becomes determined. The fates of embryonic cells in some animals are totally determined at a very early stage (mosaic development), whereas those of other animals can regulate themselves and change their fates until a later stage. The fate of each region of an embryo is often shown by a fate map.
4. Some cells develop autonomously, on the basis of their own information. Others develop on the basis of information from neighboring cells with which they interact.
5. During development, some lines of cells become pluripotent stem cells, which are partly differentiated and become more restricted as they develop. As blastomeres divide, their fates become more narrowly restricted, and they pass through various stem cell stages. Eventually each one becomes completely determined—and then becomes a fully differentiated cell.
6. Embryos contain cells that have been set aside to develop into particular adult cell types. Analysis of allophenic mice (those made by combining two early blastulas with different genotypes) indicates that the number of cells allocated for a particular adult cell type can be very small.
7. Many tissues in a mature plant or animal are made of stem cells that do not fully differentiate. They continue to divide, and at each division one daughter cell remains a stem cell, and the other differentiates completely. Plant tissues of this kind are called meristems.
8. Some tissues differentiate because of induction, a process by which they receive information from other cells. In amphibians, the dorsal lip of the blastopore is an organizer that determines the development of all the tissues around it. The chordamesoderm induces the ectoderm above it to become a neural plate.
9. There are many cases of secondary induction during development. Induction may be instructive, in which the inducing tissue specifies the fate of the receptive tissue, or permissive, in which the inducing tissue simply supports development determined in some other way. In the development of limbs, the overlying ectodermal ridge has a permissive role, while the mesoderm instructs the development of limb structures, apparently through a clock mechanism.
10. Differentiation may be understood with an elaboration of the basic gene-regulation circuit, in which the products of some genes can turn blocks of other genes on or off. Such regulators may be combined in circuits of any complexity. Selector genes control large blocks of other genes responsible for forming specific structures, and they choose between such blocks.
11. Some tissues differentiate by means of a clock mechanism, which turns on different sets of genes in sequence. This process can use a

basic gene-regulation circuit because it takes some time for each set of genes to act and to induce the next set.

12. Cells may also differentiate on the basis of spatial information. A series of alternative developmental pathways may be specified by the distribution of materials called localized determinants in the egg cortex. These substances are injected into the developing egg by other cells in the ovary and are parceled out to each cell during cell division.
13. In a *Drosophila* embryo, an anterior-posterior axis is established by two proteins, one made in the head end and the other in the tail end. After this axis is established, a hierarchy of genes begins to act. Gap genes determine large segments of the body. Pair-rule genes determine each of the embryo's segments, and segment-polarity genes determine the halves of each segment.
14. After the segment structure of the embryo is established, a series of homeotic genes are turned on. These genes are defined by mutations that convert one segment or structure into another. Homeotic genes of the *Antennapedia* and *Bithorax* complexes specify structures in sequence from the anterior to the posterior poles of the body.
15. Homeotic genes are characterized by homeobox sequences, which specify homeodomains of proteins. Homeodomains are regions where the protein can bind to DNA. These proteins appear to be transcription factors; they form transcription complexes that regulate the expression of genes for each body structure.
16. Homeotic genes with structures very similar to those of *Drosophila* are found in the chromosomes of many other animals, including mammals. Each series of genes specifies a series of structures from the head to the tail of the animal. Such homologies across the animal kingdom point to an ancient mechanism of differentiation that underlies all animal body plans.
17. Homeotic genes also regulate plant development. Mutations that change one type of growth into another suggest mechanisms that can account for the variety of plant forms.

Key Terms

totipotent 423
determined 424
mosaic development 425
fate map 425
presumptive 425
autonomous process 425
cell-interactive process 425
regulative development 425
stem cell 425
meristem 429
embryonic induction 429
organizer 429
instructive process 430
permissive process 430
competent 431
positional information 433
localized determinant 433
homeotic gene 435
homeobox 436
homeodomain 437

Multiple-Choice Questions

1. The experiments of Briggs, King, and Gurdon supported the hypothesis that, in amphibians,
   a. genetic material is lost during development.
   b. cell nuclei from early developmental stages are unable to support development.
   c. whole cells taken from late developmental donor stages can be substituted for cells in recipient embryos.
   d. a complete complement of DNA is retained in the nuclei of older embryos.
   e. DNA taken from a zygote can substitute for DNA removed from a tadpole.
2. The fate of each cell in early vertebrate embryos, such as the frog 32-cell embryos,
   a. is determined at the time the egg cell is produced.
   b. is determined at fertilization.
   c. is similar to cells in other mosaic embryos.
   d. cannot be predicted with any degree of accuracy.
   e. is more restricted than at the zygote stage.
3. The natural ability to form identical twins or triplets is characteristic of
   a. all embryos.
   b. mosaic embryos.
   c. embryos that are regulative.
   d. zygotes that are totipotent.
   e. only humans.
4. If a patch of presumptive neural tissue is removed from a frog blastula and switched with a patch of presumptive epidermis from the same blastula,
   a. both patches will develop into epidermis.
   b. both patches will develop into neural tissue.
   c. both patches will die.
   d. each patch will develop according to its original fate.
   e. the presumptive neural tissue will become epidermis; the presumptive epidermis will become neural tissue.
5. If the dorsal lip of the blastopore is removed from a frog gastrula and transplanted into the region of presumptive endoderm in a young gastrula, it will induce the development of
   a. neural tissue.
   b. epidermis.
   c. a gray crescent.
   d. a partial tadpole, conjoined to the original.
   e. endoderm.
6. Which statement about amphibian development is *not* correct?
   a. The Niewkoop center induces the formation of the dorsal lip of the blastopore.
   b. The dorsal lip of the blastopore develops into chordamesoderm.
   c. The chordamesoderm induces the neural plate to form the notochord.
   d. In the absence of chordamesoderm, the neural plate would not form.
   e. The fate of prospective ectoderm includes both neural tissue and epidermis.
7. Epithelial ectoderm from the head that has been induced by the neural plate can
   a. form overlying skin.
   b. form a lens if acted upon by an optic cup.
   c. form a lens elsewhere in the body if transplanted along with an optic cup.
   d. *a* and *b*, but not *c*.
   e. *a*, *b*, and *c*.
8. What would you expect to develop if a chick leg bud core was replaced with a chick wing bud core, and the original epidermal ridge was retained?
   a. A normal wing would develop.
   b. A normal leg would develop.
   c. Two limbs would develop—one wing and one leg.
   d. One limb would develop, but half of it would be like a leg and half like a wing.
   e. Nothing would develop after such treatment.
9. If each pair of imaginal discs develops into different and discrete structures, which of the following genes must have been active and effective?
   a. *bicoid* and *nanos*
   b. segment-polarity
   c. pair-rule

d. homeotic
e. all of the above

10. A fruit fly larva that developed with a normal head end and a normal tail end, but an incomplete middle section likely has a mutation of ______ genes.
a. *bicoid* and *nanos*
b. segment-polarity
c. homeotic
d. pair-rule
e. gap

True-False Questions

Mark each statement true or false, and if false, restate it to make it true.

1. A mature unfertilized egg is totipotent, but that unlimited potential is lost under normal conditions at fertilization.
2. The cloning experiments performed by Briggs and King involved the transplantation of cells from older amphibian embryos into zygotes.
3. Animal species that exhibit highly mosaic development undergo determination at earlier stages than do species that exhibit regulative development.
4. Mosaic development and regulative development are the two patterns seen in animal embryology, and all embryos develop according to one of these two methods.
5. During amphibian development, the gray crescent becomes the dorsal lip of the blastopore, which then acts as the organizer discovered by Spemann and Mangold.
6. If mesodermal tissue A always induces the formation of the same structure when grafted under ectodermal tissue anywhere on the body, then the ectodermal tissue must be competent everywhere, and the interaction is permissive.
7. As a *Drosophila* larva feeds and grows, its body is gradually reshaped to form an adult fly.
8. In fruit flies, the homeotic genes activate the products of the *bicoid* and *nanos* genes.
9. There is little similarity in the DNA base sequence of homeotic genes from one organism to the next.
10. During *Drosophila* development, the homeotic genes do not act until each of the body segments has been determined and has acquired an anterior-posterior polarity.

Concept Questions

1. How does determination differ from differentiation? Are differentiated cells determined, and vice versa?
2. What functional characteristics distinguish a stem cell from any cell that is capable of cell division?
3. Explain why induction is characteristic of regulative development but not of highly mosaic development.
4. Compare the experiments shown in Figures 21.6 and 21.12. Why do conjoined twins containing chordamesoderm and a nervous system develop after the experimental procedures in Figure 21.12 but not in Figure 21.6?
5. Where are the genes that initiate the anterior-posterior axis in *Drosophila* embryos? Where is the site of transcription? Of translation?

Additional Reading

De Robertis, Eddy M., Guillermo Oliver, and Christopher V.E. Wright. "Homeobox Genes and the Vertebrate Body Plan." *Scientific American,* July 1990, p. 46. More information about the way this family of related genes determines the shape of the body.

Fletcher, Carol. "A Garden of Mutants." *Discover,* August 1995, p. 48. Elliot Meyerowitz studies *Arabidopsis* mutations that cause abnormalities in flower organs. He is seeking clues to the mutation process that led to the sudden appearance of flowers 150 million years ago.

McGinnis, William, and Michael Kuziora. "The Molecular Architects of Body Design." *Scientific American,* February 1994, p. 36. Further information about homeotic genes and the basic mechanisms that create body shape in animals.

Meyerowitz, Elliot M. "The Genetics of Flower Development." *Scientific American,* November 1994, p. 40. A story of homeotic genes in plant development.

Internet Resource

To further explore the content of this chapter, log on to the web site at:
http://www.mhhe.com/biosci/genbio/guttman/

Evolution

22 Classification and Evolutionary History

A well-organized collection of specimens is a taxonomist's workshop.

Key Concepts

22.1 Evolution may be either divergent or convergent.

22.2 Species are initially defined and classified on the basis of their morphology.

22.3 Natural populations are genetically heterogeneous.

22.4 Species are designated by a two-name system.

22.5 Species are classified through a hierarchy of taxa.

22.6 A classification should reflect phylogeny.

22.7 Information about DNA structure is becoming an important taxonomic tool.

22.8 Taxonomists have taken three general approaches to their work.

22.9 The biological species concept poses difficulties in practice.

22.10 The terms "higher" and "lower" organisms, though misnomers, are difficult to avoid.

Over 2,300 years ago, the philosopher Aristotle wrote,

> *The course of exposition must be first to state the attributes common to whole groups of animals and then to attempt to give their explanation. Many groups present common attributes, that is to say, in some cases absolutely identical affections, and absolutely identical organs—feet, feathers, scales, and the like—while in other groups the affections and organs are only so far identical as that they are analogous. For instance, some groups have lungs, others have no lung but an organ analogous to a lung in its place; some have blood, others have no blood but a fluid analogous to blood, and with the same office. To treat of the common attributes in connection with each individual group would involve, as already suggested, useless iteration. For many groups have common attributes.*

Here we see one of the ablest minds of ancient times wrestling with the issues of biological diversity. Seeking to categorize species with common features, he distinguishes between what today we call *homology* and *analogy* (see Section 2.4), recognizing that species should not be grouped together if they share mere analogies, but if a group is defined by homologous features, a single description applies to all of its members.

The main element lacking in Aristotle's thought is the idea of evolution. Taxonomists today wrestle with the same kinds of issues, but they are guided by the principle that a classification of a group of organisms should reflect its evolutionary history, its *phylogeny.* Classifications are not arbitrary, and the science of taxonomy is not mere stamp collecting. Nevertheless, it has often been difficult to identify the course of evolution and to fit species into the relatively simple classification schemes the human mind tries to devise. Furthermore, we are now in a period of intense debate among taxonomists; the field is undergoing a more profound examination of its fundamental philosophy than at any time since taxonomic thought was revolutionized by the concept of evolution over a century ago. The principal viewpoints that now divide taxonomists make for different classifications of organisms, as we will show later. ■

22.1 Evolution may be either divergent or convergent.

Evolution is generally studied under three headings: *Microevolution* means changes in the genetic makeup of a population; *speciation* means the formation of new species; and *macroevolution* means the large-scale changes we observe over geological time as new kinds of organisms arise, change form, and become extinct. The genetic composition of some populations fluctuates with changes in climate and other environmental factors; some biologists call these fluctuations evolution, but others do not.

Taxonomy summarizes macroevolution, showing how various species are related to one another. We showed in Chapter 2 how the course of evolution is depicted by phylogenetic trees linking existing species with known or hypothetical ancestors. The overall trends in most phylogenies are toward diversity and *divergence,* with one type of organism evolving into an increasing variety of diverse forms. This is the pattern we would expect, if only because evolution depends upon accidental, unpredictable events: mutation, recombination, and selection for adaptation to diverse habitats. It is highly unlikely that any group of organisms would reverse their evolutionary history and take on the form of an ancestor, because they are responding to current environmental pressures, and their genomes are experiencing new, unique mutations and recombinations. Some organisms appear to have evolved in what some call a "degenerate" way—changing into relatively simple forms from relatively complex ancestors—but they have not reversed their evolution. Such evolution could only be considered a reversal if evolution as a whole is seen as *progressing,* from simpler to more complex forms. But evolution does not produce progress; it just produces differences. We will return to this issue at the end of the chapter.

In contrast to divergent evolution, species that evolve into similar ways of life commonly develop similar forms, even though they come from quite different ancestors. This pattern of **convergent evolution** is exemplified by the similar forms of some streamlined swimming vertebrates (Figure 22.1). The ancestors of these animals were quite diverse—a primitive fish, a four-legged land reptile, a terrestrial bird, and a four-legged land mammal—yet the aquatic animals look very much alike. On the other hand, the internal structures of such animals are generally not very similar. Figure 22.2 shows the wing structures of three flying vertebrates; they are independent responses to the same opportunity—to occupy niches involving flight—but in spite of similar outward appearance, each one emphasizes quite different bones of the vertebrate limb. For this reason, these wings are only partly homologous, and the differences among them distinguish the three groups from one another. When these vertebrate wings are compared with the wings of insects, the relationship is clearly analogous.

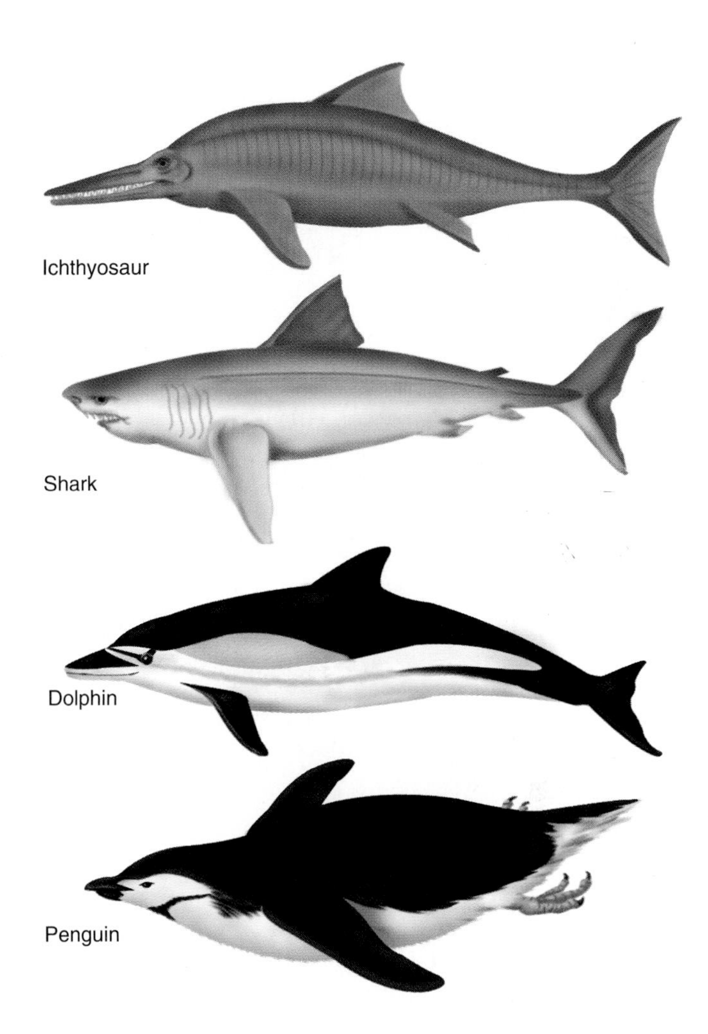

Figure 22.1
Four types of swimming vertebrates show convergent evolution in their body forms. Ichthyosaurs were Mesozoic reptiles. Sharks are an ancient group of fishes. Dolphins, like their relatives the whales, evolved from some primitive (but unknown) eutherian mammal, and penguins evolved from ancient (but unknown) birds that were probably able to fly rather than swim.

22.2 Species are initially defined and classified on the basis of their morphology.

Every science begins by systematizing the objects it tries to explain. As chemistry became a mature science during the nineteenth century, it had to account for the many kinds of substances that exist in the world, and Mendeleev's insightful

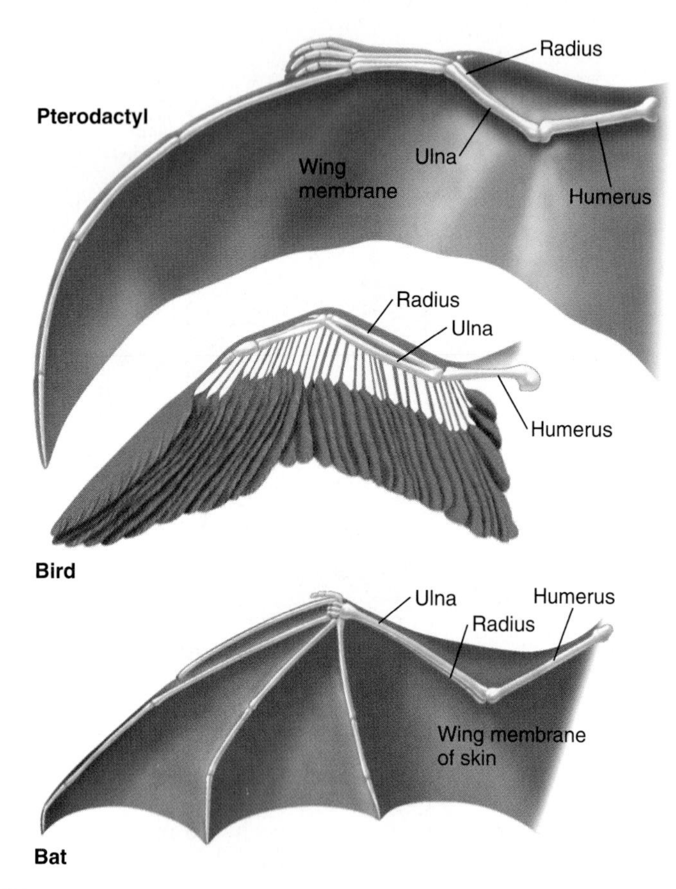

Figure 22.2

The wings of three vertebrates show convergent evolution. Their overall shapes are similar, even though they are based on different bone arrangements and have different surface structures (feathers and skin).

Figure 22.3

A variety of molluscs illustrate the problems of separating individuals into species on the basis of their morphology.

invention of the periodic table, which systematized the elements and their properties, set the stage for the explanatory power of atomic theory. Similarly, biology begins by identifying and systematizing the enormous variety of living things, and accounting for that variety by trying to reconstruct the course of evolution. Because evolution is largely about the origin of new species, we must define a species. Although we have been using the term casually to mean a "kind" of organism, this usage is obviously too vague. Just what is a species?

Figure 22.3 shows several kinds of mollusc shells. Even an unsophisticated eye can see that they fall into groups of individuals that resemble one another and are quite distinct from other groups. Systematists, or taxonomists, perceive these resemblances with a more practiced eye, paying attention to significant details and often using sophisticated statistical methods to determine whether two apparently distinct groups can be considered legitimately different. Classification begins by sorting individuals into groups that can be called *morphological species,* groups of organisms with similar or identical features that clearly distinguish them from other comparable groups. But two independent observers, confronting the variability of the biological world, often divide a series of organisms quite differently. They may emphasize different features or have different opinions about the importance of gaps and similarities in the series. So we naturally ask if there is a more objective way to define a species and some basis besides morphology for distinguishing species. To address the matter of defining a species, let's consider some instructive examples.

On almost any flat beach along the southern Atlantic coast of North America you will find pockets of little clams called coquinas at the water's edge (Figure 22.4). As each wave washes over them, they open to receive the fresh water and then squirm and burrow a bit into the sand. In only a few minutes you could collect many clams with a variety of patterns on their shells. Many populations include two or more obviously different forms, or **morphs,** that are unrelated to age or sex. If species were only defined by morphology, the coquinas might be classified into three or four different species, but all these forms are just different morphs of one species, *Donax variabilis.* Similarly, the Eastern Screech-owls *(Otus asio)* of North America contain two color morphs (Figure 22.5): reddish-brown and gray individuals in the same population and even the same brood, just as humans have different eye or hair colors in a single family.

Anyone who becomes serious about studying birds in eastern North America soon discovers that the flycatchers of the genus *Empidonax* pose a great problem in identification (Figure 22.6). These small, grayish birds are very difficult to distinguish from one another except by voice and habitat, and a silent bird during migration, which could be moving from one habitat to another, must be written off as an "unidentified empid." Yet, the Acadian Alder, Least, and Willow Flycatchers are all considered distinct species.

How can birds as similar as the flycatchers be considered different species when the two types of screech-owl are the same species? Even though the idea of a species begins with

Figure 22.4

The Coquina clams, *Donax variabilis,* are a single, highly polymorphic species. A population includes individuals of several distinct morphs, including different patterns of banding and coloration.

Figure 22.5

The Eastern Screech-owls of North America have both red and gray morphs.

Acadian
Deciduous woods and wooded swamps, especially beeches

Alder
Alder swamps and wet thickets, usually near water

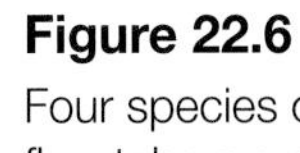

Figure 22.6

Four species of *Empidonax* flycatchers are practically identical morphologically and are best distinguished by their voices and habitats.

Least
Open woods, orchards, farms

Willow
Thickets, brushy pastures, old orchards, willows

morphology, some other criterion must be used to define and identify species. The dominant viewpoint, which we will adopt in this discussion, is embodied in the **biological species concept:** A species is a group of organisms that can breed with one another but not with members of other similar groups. All the different coquinas belong to the same species; they all reproduce with one another, and their forms have about the same significance as different eye colors among humans. The rufous and gray Screech-owls also mate with one another and have mixed broods of offspring. The *Empidonax* flycatchers, on the other hand, remain separate reproductively; they do not hybridize.

The biological species concept obviously applies only to sexually reproducing organisms, and it may have to be modified somewhat to take account of differences between plants and animals. To amplify this definition, we return to the idea of a phylogenetic tree. Each line of the tree represents a species (Figure 22.7*a*); the tree shows new species appearing where a line branches and species becoming extinct where lines end. Let's take a closer look at a line representing some sexual species (Figure 22.7*b*). Using the classical symbols for male (♂) and female ♀), lines representing gametes connect every individual to its mother and father in the previous generation and to its offspring in the next. The species is a network of criss-crossing lines. The question of whether an individual belongs to the species is decided quite objectively by seeing what it breeds with—so any individual that is part of the network is a member of the species. Applying the biological species concept takes systematics out of the laboratory and museum and into the field to supplement morphological distinctions with information about behavior, habitats, and breeding patterns. This kind of research often provides good reasons for splitting a morphological species in two or lumping two or more together into a single species.

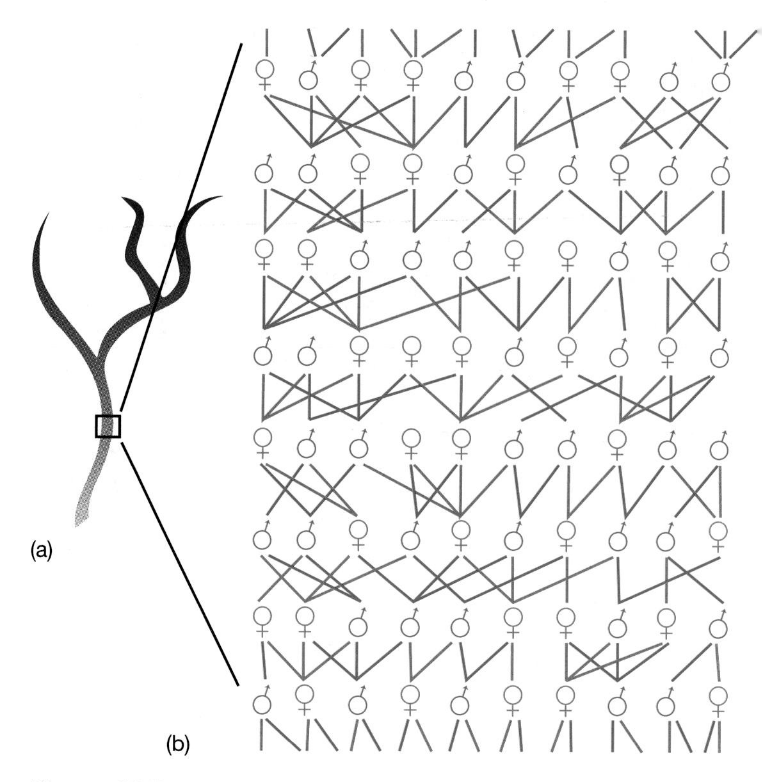

Figure 22.7

(a) A phylogenetic tree shows the course of evolution. If one of the lines of such a tree for sexual species is expanded *(b),* we see that the species consists of generation after generation of individuals connected to their parents and offspring by gamete lines.

Another way to describe Figure 22.7 is to say that all the individuals in a species share a *gene pool,* an imaginary pool where all members of the species deposit their gametes; new individuals are made by drawing one sperm and one egg out of the pool. The members of a species have very similar genomes because they are all drawing parts of their genomes from this common pool. We will expand on this idea in Chapter 23.

In contrast, a group of asexual organisms can only be called a distinct species by virtue of their common features. In this endlessly branching pattern, individuals will gradually diverge from one another:

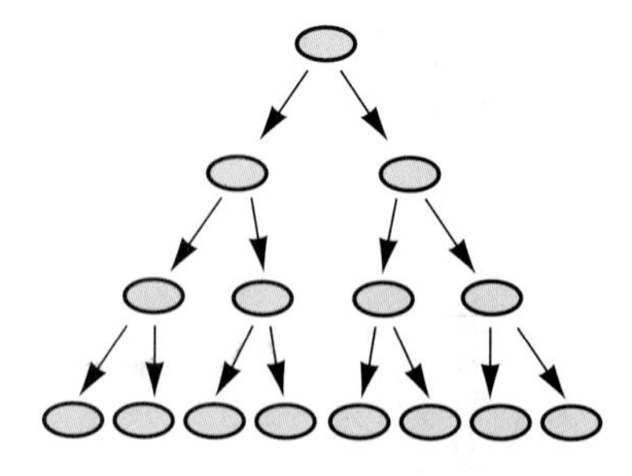

There is no firm criterion for including or not including an asexual individual in a species, and an asexual "species" is merely a convenient category, a collection of independent but very similar individuals.

Figure 22.8 shows what happens during an idealized process of speciation, when one species diverges into two, as shown by two distinct criss-crossing patterns. Once the two patterns have separated, every individual belongs to one or the other; no gamete line crosses from one branch to the other, and there are now two reproductively isolated species. In Chapter 24 we will show that it is sometimes hard to apply the biological species concept unequivocally because speciation never happens as quickly or as simply as the drawing suggests; populations are often at an intermediate stage of evolution where they exchange *some* gametes, and quite different situations may arise in different parts of a species' range. Also, some organisms, especially plants, often engage in regular hybridization between what are otherwise considered good species by morphological criteria. These events provide interesting complications, but the basic concept of a species as a reproductive unit is straightforward.

The biological species concept leads to other difficulties, as we will see in Section 22.5. Nevertheless, a hallmark of modern biological thought is recognizing that a species is not just an arbitrary collection of organisms but is a group with objective boundaries defined by reproduction. This is one of our legacies from Darwin. The question of defining a species is not primarily interesting because biologists want to catalog the world's organisms, but because the idea of a reproductively independent unit is so important. Unless such a unit—a species defined biologically—becomes genetically independent of other groups, it can't go on to produce other, more diverse groups in another evolutionary step, and it may even be unable to become a stable link in an ecosystem.

22.3 Natural populations are genetically heterogeneous.

The genetic term "wild-type allele" may convey the impression that one form of each gene can legitimately be considered normal and that other alleles of the gene are rare, unnatural, or defective. This isn't true at all. As species such as the coquinas and screech-owls show, natural populations actually harbor a great deal of variation, with different alleles of many genes and many different chromosomal arrangements. If this genetic heterogeneity produces obvious differences in form, we say the population is **polymorphic** (*poly-* = many; *-morph* = form). Polymorphism is common. Human populations are obviously very polymorphic, displaying different nose and ear shapes, different body forms, and several distinct colors of eyes, hair, and skin.

Even if genetic polymorphism is not apparent phenotypically, every population that has been studied reveals its genetic variability in polymorphisms of protein structure. Richard C. Lewontin, for instance, found that many "laboratory" mutants of *Drosophila* occur in wild populations. Using electrophoresis, he and J. L. Hubby then analyzed the proteins from individual flies and showed that wild populations harbor allelic forms of many enzymes. Small mutational differences between alleles often make little or no difference in the function of the proteins they encode, but the variant proteins have different charges and migrate differently in an electric field. Lewontin and Hubby separated the proteins of many flies and looked for particular enzymes by using stains that depend on the chemical reaction each enzyme catalyzes (Figure 22.9). In this way, they found a lot of variant forms of several enzymes. Further research on *Drosophila* shows that populations typically have two or more alleles at about 30 percent of their gene loci.

Genetic diversity in a species commonly shows itself geographically in differences between local populations, commonly called **demes.** If a certain characteristic of the species changes gradually across its range, the demes form a **cline.** For instance, populations of white clover distributed across Europe show a north-south cline of cyanide content (Figure 22.10). Similarly, the American Song Sparrows are small, medium-dark birds in the East and Midwest, look sun-bleached in the southwestern deserts, and become larger and rustier moving into the Pacific Northwest and along the Pacific coast to British Columbia and Alaska. Taxonomists have divided this species formally into **subspecies,** or **geographic races,** with distinct characteristics (Figure 22.11). In some organisms, subspecies may be distinguished by subtle differences in features such as size and color, while in other

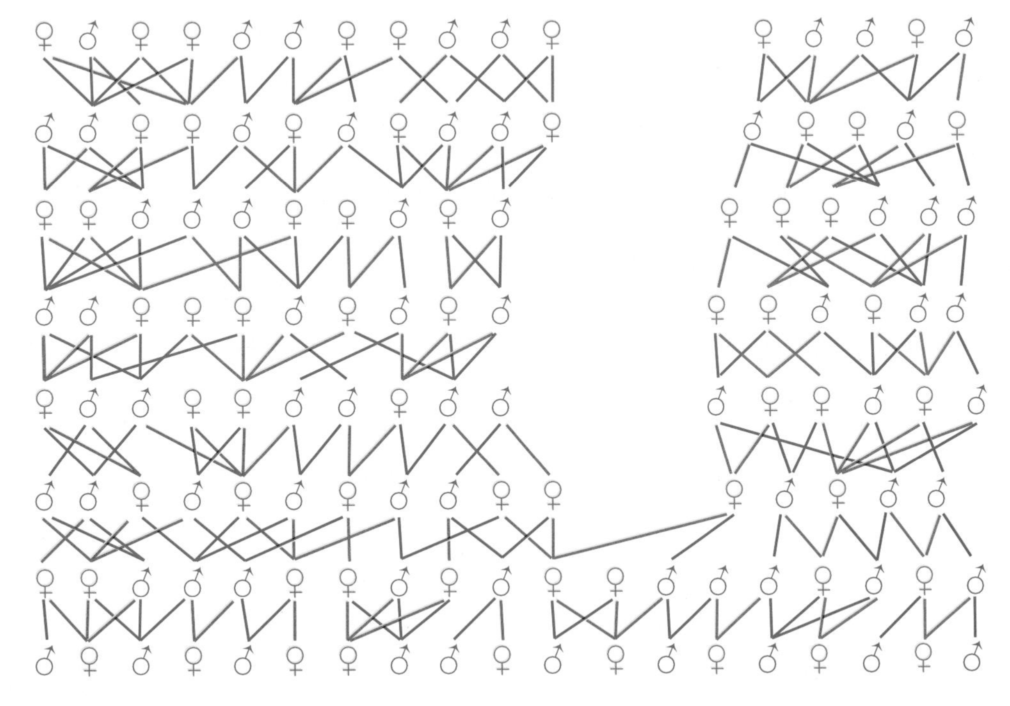

Figure 22.8

In a simple (and unrealistically rapid) episode of speciation, a single pattern of individuals connected by gamete lines separates into two distinct patterns, often because the two populations have been separated geographically. Each branch is now a separate species that does not exchange gametes (or genes) with the other species.

organisms, subspecies look so different that they would be considered different species (and often have been) if they did not interbreed. Named subspecies may reflect real discontinuities between neighboring populations, but the division of other species into subspecies is about as arbitrary as our division of the color spectrum into six colors. (People in some other cultures divide the spectrum quite differently.)

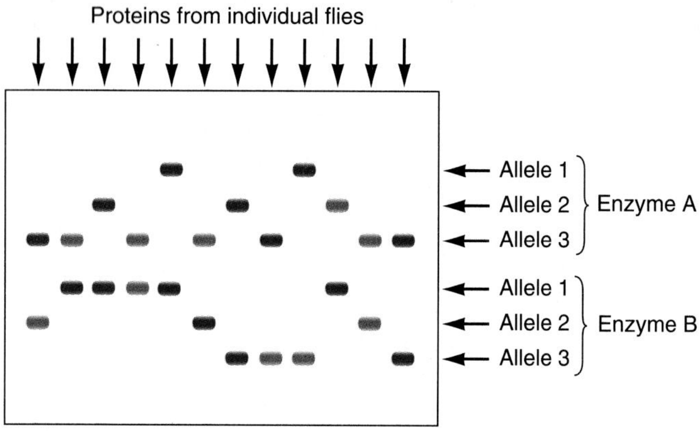

Figure 22.9
The proteins of individual flies are separated by electrophoresis and stained to locate specific enzymes. This reveals many different patterns and shows the genetic variation for each enzyme in the population.

22.4 Species are designated by a two-name system.

When a new species is defined, it must be named. Modern systematic biology began with the work of Carolus Linnaeus (Karl von Linné), who devised what we now call the *Linnaean system* of nomenclature and classification. Formal taxonomy begins with the tenth edition of Linnaeus's work, published in 1758. In this system, a species is one of a hierarchy of categories called **taxa** (sing., **taxon**) in which organisms are grouped on the basis of their similarities. The standard taxa in ascending rank are species, genus, family, order, class, phylum, and kingdom; taxa of other ranks such as tribe and cohort are sometimes inserted, and the prefixes super-, sub-, and infra- can be added to make other ranks. A taxon consisting of the most closely related species is a **genus** (pl., **genera**). Just as people have a family name that identifies the group they belong to and a given name that distinguishes them from others in the family, the scientific name of each species is a **binomial** (*bi-* = two; *nomen* = name) consisting of two parts: The first is the generic name, and the second is the trivial name that

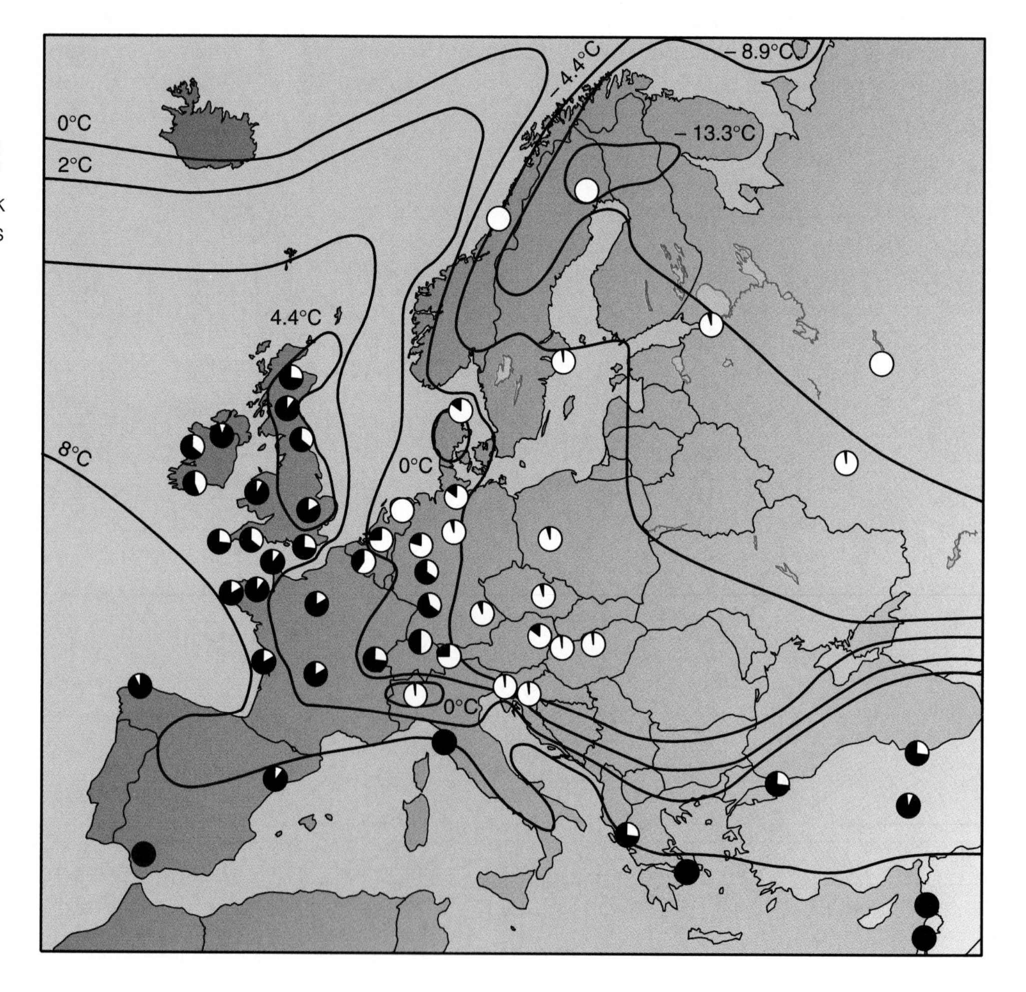

Figure 22.10
White clover populations in Europe include both cyanogenic (cyanide-producing) and non-cyanogenic plants. Populations show a cline from north to south in the frequency of cyanogenic plants, as indicated by the black sector in each circle. The lines are isotherms of the average January temperature.

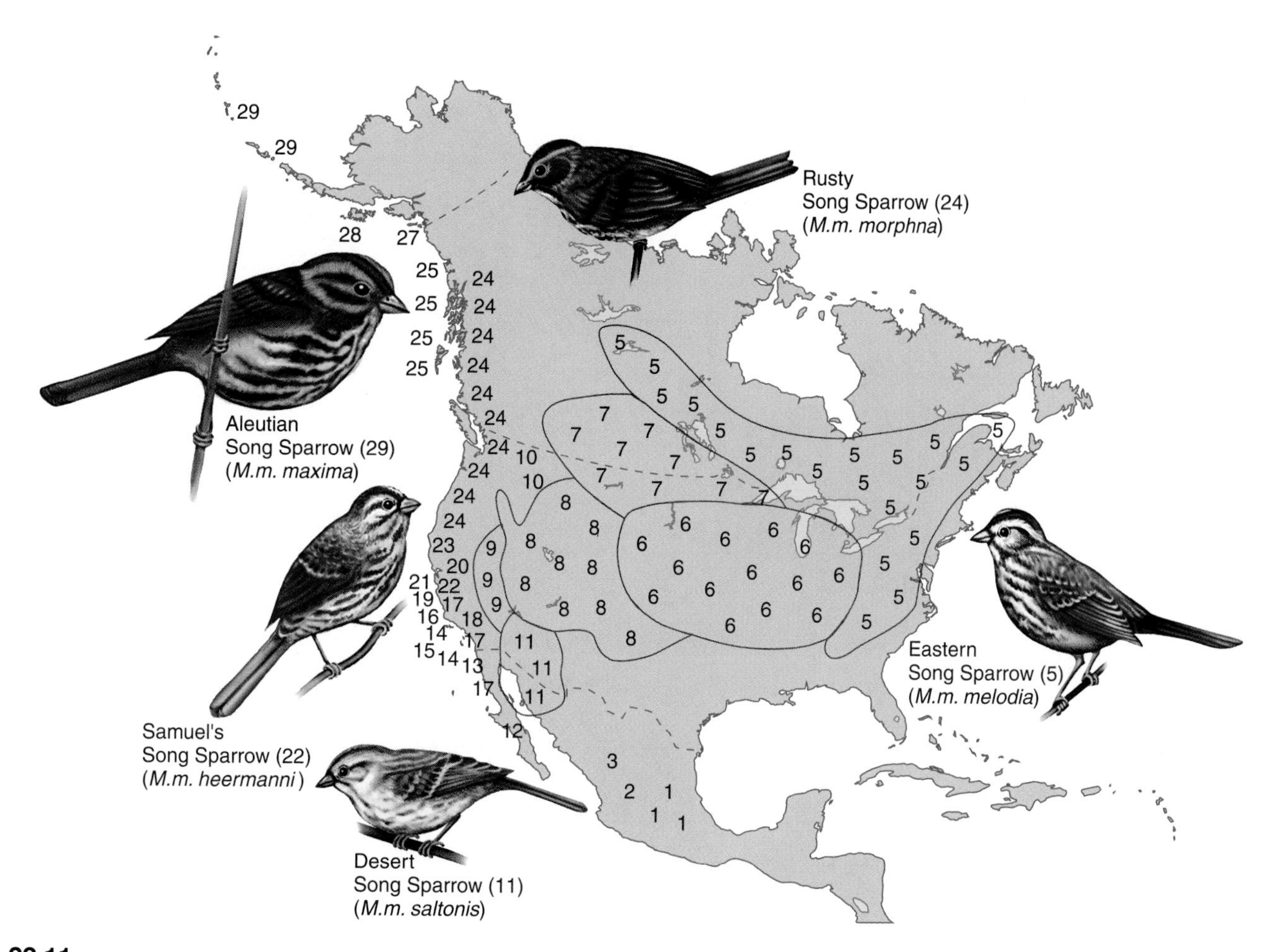

Figure 22.11

Song Sparrows are divided into several subspecies, or geographic races, all named *Melospiza melodia* with a third name to designate the subspecies. In its coloring and markings, each population shows its adaptation to local conditions.

designates the particular species within the genus. For instance, several species of similar cats are placed in the genus *Felis.* The domestic cat is *Felis catus;* the jaguar is *Felis onca;* the cougar is *Felis concolor;* and the African wildcat is *Felis libyca.* The scientific name is formed from Greek and Latin roots; it is always italicized or underlined, and the name of the genus is always capitalized. It may refer to a certain characteristic of the species or to a place where it is found, although some species are named after a person some taxonomist wanted to honor.

As we explained in Chapter 2, pre-Darwinian taxonomists thought that all of the members of a species conform to a single, ideal type and that variations within a species are insignificant or else represent imperfections. A remnant of that idea remains in modern systematics. Taxonomists often revise the classifications established by earlier workers, but there must be a system for keeping names straight; whenever a species is named, one specimen in a museum somewhere is designated the *type specimen* of that species, and in revisions the name of a species always stays with that specimen. If two species are lumped into a single one, the Law of Priority applies: The earliest name has priority in the new arrangement.

After more than two centuries of work, taxonomists have identified well over a million species of animals, perhaps 400,000 species of plants, and several thousand species of microorganisms. Many other extinct species have been named, and authorities estimate that many living species are yet to be identified, since more are discovered and named every year. Some biologists estimate that between 3 and 30 million contemporary species exist, the majority of them in the tropics.

Exercise 22.1 In 1867, Schmedlak established a species, *Phaecus graellsii;* in 1870, Grimaldi established another species, *Phaecus albus.* In 1987, Smith combined them into a single species. What is its name?

22.5 Species are classified through a hierarchy of taxa.

Once we see that very similar species can be put into a genus, the human eye for form and similarity lets us see that genera can be grouped into a hierarchy of larger and larger categories

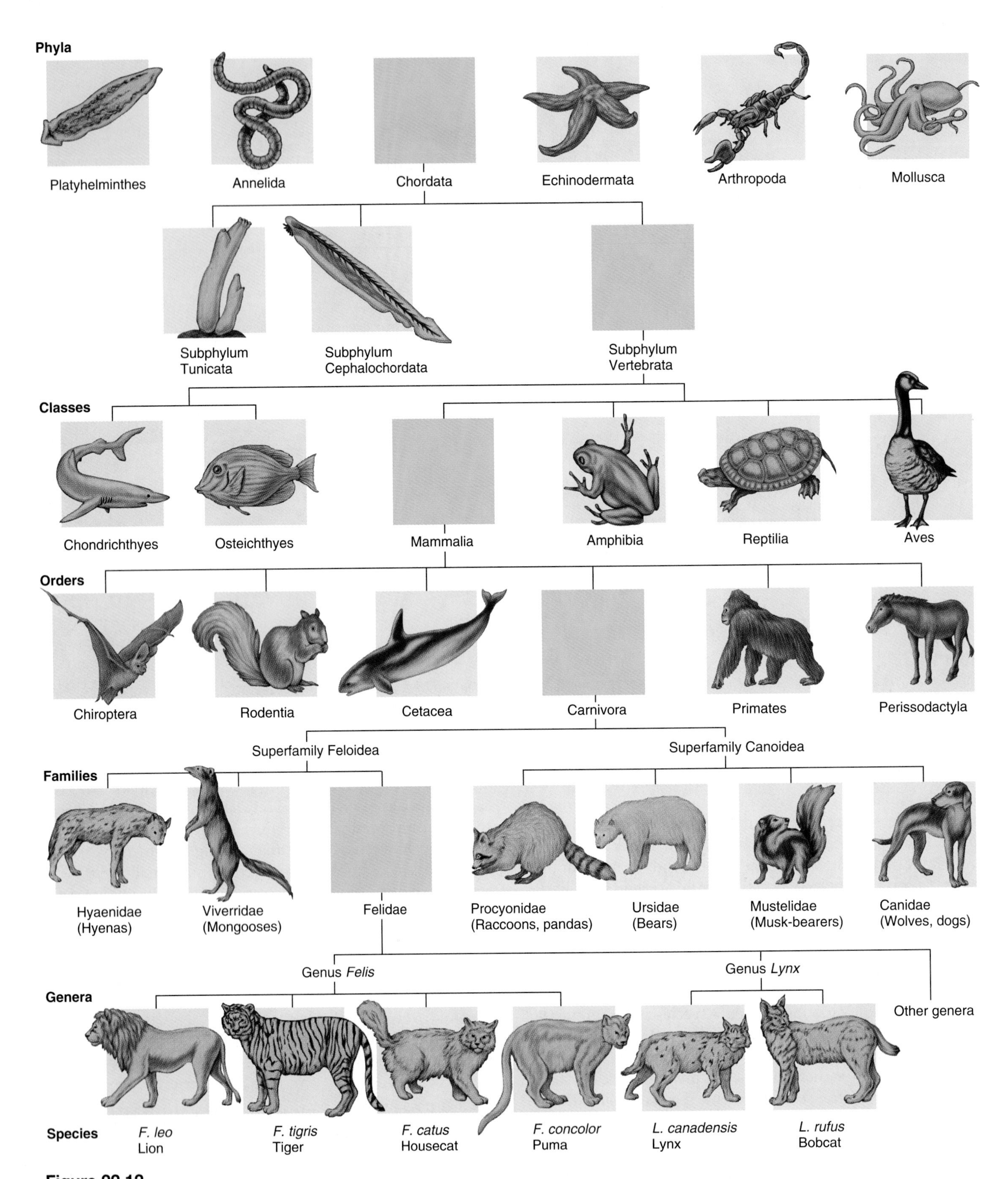

Figure 22.12
A classification places species in more and more encompassing taxa to show how closely they are related to one another.

(Figure 22.12). Several related genera are placed in a **family.** For example, the genus *Felis* and other genera of cats such as *Panthera* and *Lynx* all form the family Felidae. The ending *-idae* (pronounced "ih-dee"), the classical Latin way to indicate a family, denotes families of animals; botanists use the ending *-aceae* (pronounced "ay-see-ee") for plant family names—for instance, the pines and their close relatives belong to the family Pinaceae. Families are convenient taxa because we so often think of the members of a family as one kind of organism; thus rose, maple, violet, bear, deer, alligator, owl, and many other common names refer to families. In fact, for less familiar organisms, we often reduce the family's technical name to a common one, so nesticids and dinopids are spiders of the families Nesticidae and Dinopidae.

Continuing to classify cats, we observe that cats eat meat and share anatomical features such as clawed feet, similar dental patterns, and similar bone shapes. They share many of these features with members of other families, such as the bears (Ursidae) and the doglike animals (Canidae), so these families are all combined into a larger taxon, the **order** Carnivora. Carnivora and several other orders of animals in turn share more general characteristics, including hair on their bodies and females that produce milk, so they are combined into the **class** Mammalia, the mammals.

Mammals have a vertebral column, or backbone, and several other features that they share with the classes Aves (birds), Reptilia (reptiles), Amphibia (frogs and salamanders), and two or three different classes of fishes. All these classes constitute the **subphylum** Vertebrata (vertebrates). The vertebrates, combined with other groups of small animals that have a rod of supporting tissue called a notochord along their backs, make up the **phylum** Chordata. Chordata and many other phyla of animals form the **kingdom** Animalia, one of from five to eight kingdoms of organisms that are now commonly recognized in different systems.

Current kingdom concepts, Section 2.14.

This kind of hierarchy shows different degrees of similarity among species. Two species in the same genus have more in common than two that are in different genera but in the same family. As Aristotle recognized, classifying a species as a mammal means that it has hair, a certain skeletal structure, and suckles its young, so it isn't necessary to say these things about each species of mammal. This comprehensive system of classification introduces order to what would otherwise be a bewildering variety of organisms, and it allows us to study their differences and similarities systematically.

Next we must ask *why* organisms can be placed in taxa that share certain features. We will show that all the members of a genus are similar because they have a single common ancestor and that all the members of a family are also similar because they, too, have a single common ancestor. This principle relates classification to evolutionary history.

Exercise 22.2 Recite the sequence of taxa from largest (kingdom) to smallest (species). (There are mnemonics to help you remember, but it is probably easier to just say the sequence to yourself a few times.)

22.6 A classification should reflect phylogeny.

Biologists before Darwin had already arranged species into Linnaean hierarchies. For them, taxonomy was merely cataloging similar organisms. Evolutionary thought, however, provides a theoretical basis for taxonomy, and modern systematists try to define taxa containing only species whose similarity is derived from a common ancestor. Ideally , a classification should reflect phylogeny, the course of evolution. That isn't always easy.

Figure 22.13*a,b* shows a classification of contemporary organisms with their presumed phylogeny. All we know directly about these species is what we see now; we have no direct evidence about their past except perhaps for a few fossils. Based on their morphology and chemical structure, we arrange them into genera, families, and orders, always having in mind (explicitly or implicitly) the phylogenetic tree going back in time. That tree is our inference about the course of evolution. We try to make each taxon **monophyletic,** so that it consists only of organisms derived from one branch of the tree. In other words, all the species in one branch should be derived from a single ancestral species. Figure 22.13*c* shows how we might be fooled by appearances. Species that are not all derived from one immediate ancestor might evolve convergently, and the superficial resemblance among them could mislead us into making **polyphyletic** taxa, combining species that actually come from different branches of the tree. Systematists are always striving to eliminate such taxa by finding more certain indicators of the course of evolution. Many groups of organisms are still classified in different ways by different authorities, each classification reflecting an informed opinion about the importance of various characteristics and about the course of evolution. There will probably always be differences of opinion, and no one classification will ever win universal approval.

Figure 22.13 shows what you would see if you were living at any time in the past. Each branch of the tree, no matter how large, represents one species, and no matter where in time you cut the phylogenetic tree you would only see individual organisms that could be classified into species. In other words, each line could be expanded into the pattern of Figure 22.7; you would see the individuals and could draw the gamete lines connecting them. But you could never *observe* genera, families, and orders because they are creations of our imagination, invented for our convenience. When we combine species into a single genus or family, we infer that they came from one branch of the tree, but before that branch divided, it was only a single species.

A taxon may also be **paraphyletic** if it includes some, *but not all,* of the descendants of a single ancestor. Traditional modern taxonomies often include paraphyletic groups, which are

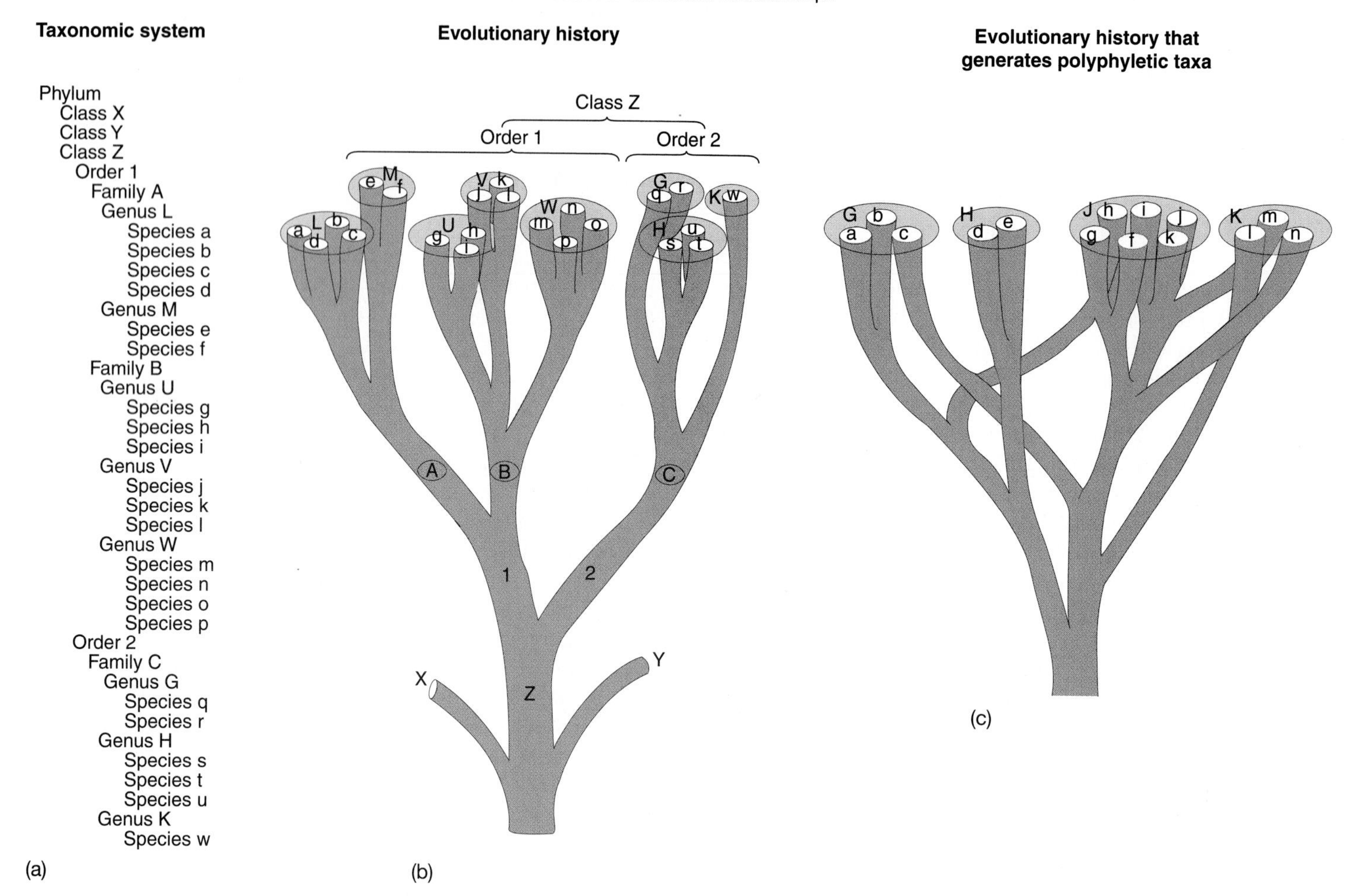

Figure 22.13

Every classification reflects a hypothetical phylogenetic tree. A taxonomic system *(a)* is based on the belief that the species involved are related by a certain evolutionary history *(b)*. If this phylogenetic tree is accurate, all the taxa are monophyletic because each one includes only the descendants of a single immediate ancestor. The most closely related species are included in genera, related genera are included in families, and so on. However, we might be deceived into making polyphyletic taxa that include the descendants of two or more immediate ancestors if a species from one branch becomes very similar to those in a different branch *(c)*.

acceptable to some taxonomists but anathema to others, as we will discuss later.

22.7 Information about DNA structure is becoming an important taxonomic tool.

While trying to sort out the phylogenetic history of a group of organisms, evolutionary biologists have been making increased use of information about DNA structure. Mitochondrial DNA is a small molecule inherited through a maternal line, since sperm do not contribute mitochondria to a zygote, so it makes a convenient phylogenetic tracer. Minor changes in this DNA, which may have little or no effect on protein structure, create differences in the location of restriction-endonuclease cut sites, so the DNA of each species can be characterized by its pattern of restriction fragments.

Restriction-fragment length polymorphisms, Section 19.9.

DNA-sequence homology has been particularly developed by several ornithologists. The underlying assumption in applying such information to taxonomy is that DNA sequences are more conservative than the developmental sequences they encode; that is, relatively small changes in DNA sequence can produce relatively large differences in the forms of flowers, leaves, claws, bones, or muscles, perhaps fooling us into separating species that look quite different but are actually closely related.

Developmental genes and evolutionary novelties, Sections 21.11, 21.12, 24.11.

The genomes of two species can be compared directly by amplifying certain regions with the polymerase chain reaction

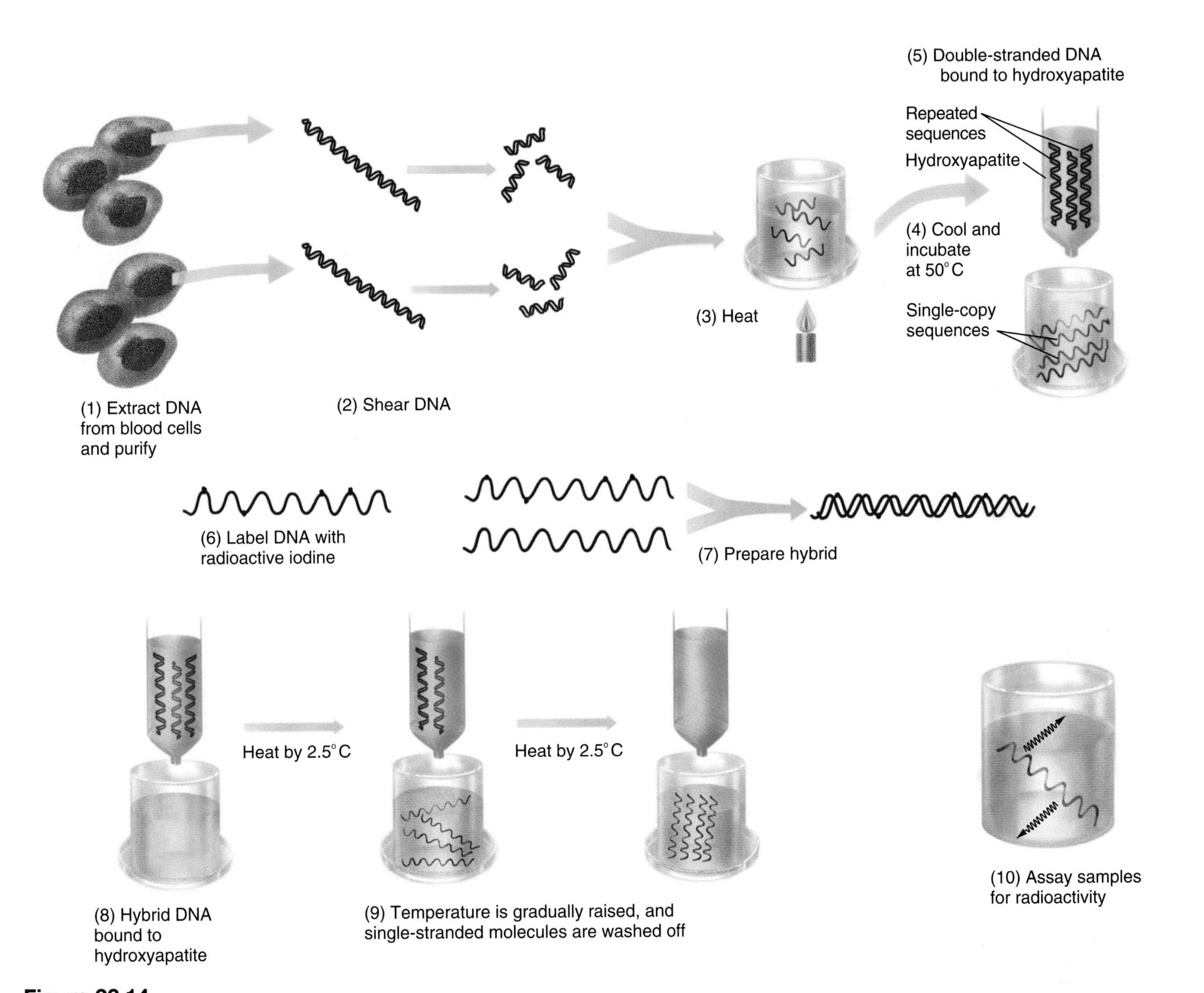

Figure 22.14

The general method of evaluating phylogenetic relationships on the basis of DNA hybridization is traced here. DNAs purified from organisms of two different species *(1)* are sheared into short pieces *(2)* and heated to melt the double-stranded molecules into single strands *(3);* then they are cooled so they can form double-stranded molecules again *(4).* The DNA that hybridizes rapidly consists of repeated sequences of no taxonomic interest, and this is eliminated by binding to a hydroxyapatite column *(5).* The remaining single-stranded DNAs, which do not bind to the column, presumably encode genes of taxonomic interest; they are labeled with radioactive iodine *(6),* mixed, and hybridized at 60°C for 120 hours *(7).* The hybrid DNA is bound to hydroxyapatite *(8),* and its temperature is raised slowly *(9),* while the single-stranded molecules are washed off the column and measured by their radioactivity *(10),* to determine the average melting temperature of the hybrids. A higher melting temperature shows that the DNAs bind tightly to each other, so the species are closely related. (Source: Sibley and Ahlquist, *Scientific American,* 254 (2):84, February 1986.)

(PCR) and comparing the sequences directly. The similarities of two genomes can also be determined rather easily by hybridization experiments (Figure 22.14). For taxonomic work, double-stranded DNA (almost entirely nuclear DNA) is extracted from each species and separated into single strands by heating. Single-stranded molecules, from whatever sources, will hybridize to one another if they have complementary, or nearly complementary, sequences. After labeling with radioactive iodine, the DNAs of two species are mixed, heated, and allowed to hybridize with each other for a long time at a moderate temperature. A significant fraction of DNA in complex eucaryotes consists of very common repeated sequences that have no taxonomic value, but these molecules are eliminated when the DNA is cooled rather quickly, because the repeated sequences reanneal first and may be trapped on the mineral hydroxyapatite, which preferentially binds double-stranded molecules. The remaining DNA contains unique sequences. The average melting temperature of the hybrid DNA is then determined. The DNAs of two species with very similar sequences will form hybrid molecules that stick together strongly and melt only at a high temperature, but if the sequences of the two are not so similar, they will form poorly hybridized molecules that melt at lower temperatures.

PCR and nucleic acid hybridization, Section 18.11.

Charles Sibley and his colleagues have systematically hybridized DNA samples from many bird species. Although their work has shown that the established taxonomy of birds was quite good on the whole, it also held some surprises and suggested needed revisions in the traditional classification. For example, it confirmed a relationship that some ornithologists had suspected on the basis of anatomy by showing that the New World vultures are actually more closely related to storks than to the Old World vultures they resemble superficially. Sibley and his coworkers then used the DNA hybridization data to establish a "molecular clock" indicating the approximate time that various lineages separated from each other (though this idea and the very idea of a reliable molecular clock are quite controversial). They place lineages that separated an estimated 90–100 million years ago in different orders, those that separated 80–90 million years ago in different suborders, and so on. Since similarities in DNA sequences are apparently preserved in spite of differences in morphology, these molecular studies will become even more important in future taxonomic work.

Exercise 22.3 Four species—A, B, C, and D—are studied by hybridizing their DNAs. The table below gives the average melting temperatures of the hybrid DNAs. From this information, draw a simple tree showing roughly how closely related the species are to one another.

	A	**B**	**C**	**D**
A	92	81	83	81
B		92	85	90
C			92	85
D				92

22.8 Taxonomists have taken three general approaches to their work.

Faced with a complicated array of species, a taxonomist must find an arrangement that makes sense. During most of this century, taxonomists have used a general approach that we can now characterize as *evolutionary taxonomy.* It will be easiest to characterize this approach after describing two newer alternatives, *phenetic* and *cladistic* grouping methods. They represent two distinctive philosophies about how judgments should be made in taxonomic work, as illustrated in Figure 22.15. Suppose species X, Y, and Z have evolved from a common ancestor, A. Y and Z remain much more similar to each other than either is to X, but X and Y are actually more closely related because they have a more recent common ancestor, B. A **phenetic** classification emphasizes their appearance alone (*pheno-*, from Greek, *phainaien-*, to

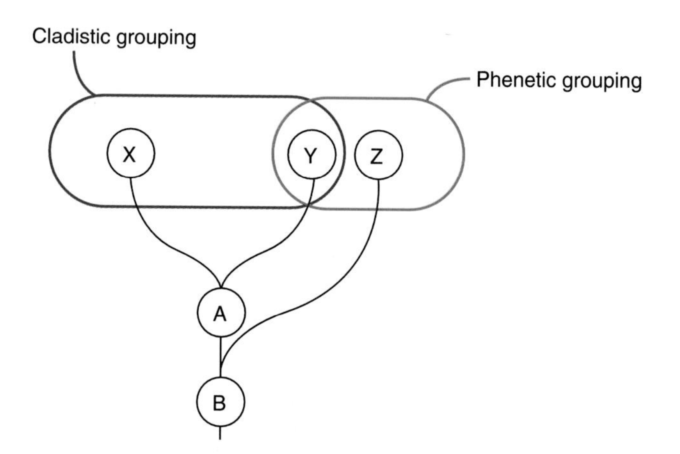

Figure 22.15

The phenetic approach to taxonomy classifies species exclusively on the basis of their appearance, so it would combine Y and Z into one taxon because they resemble each other more than either one resembles X. The cladistic approach is based exclusively on phylogenetic relationships, so it would combine X and Y into one taxon because they have a more recent common ancestor (A).

show) and unites Y and Z in one taxon because of their greater similarity, leaving X in a parallel taxon. A **cladistic** classification emphasizes their evolutionary history (Greek, *klados* = branch, referring to the branching pattern of a phylogenetic tree) and therefore unites X and Y, leaving Z in a second taxon.

Although pure phenetics has been largely abandoned by taxonomists, the phenetic viewpoint should be understood. This stance emphasizes that we have nothing to work with but the appearance of the organisms at hand and no way to know *a priori* which features are most indicative of evolutionary trends, so relationships can be uncovered most objectively by looking at many characteristics simultaneously and classifying organisms into taxa that share common features. Species that share the most features belong in the same genus; those with somewhat fewer features in common belong in the same family; and so on. A taxonomist following this approach uses the computerized methods of **numerical taxonomy** by making a table with many specific features, noting whether each pair of species shares the feature or not. All features are given equal weight. In principle, this method is supposed to produce greater objectivity by removing subjective judgments about the importance of different characteristics. The computer then compares species with one another and draws a matrix (Figure 22.16*a*) showing the fraction of these features that every pair of species has in common. From this matrix, the computer draws a branching tree of relationship, or *phenogram* (Figure 22.16*b*), which can be translated directly into a classification.

The phenetic method, unfortunately, creates several difficulties. Different computer programs that have been devised to analyze data phenetically sometimes give different results, so since these programs do not always manipulate data in the same way, the supposed objectivity of the method is often lost. But

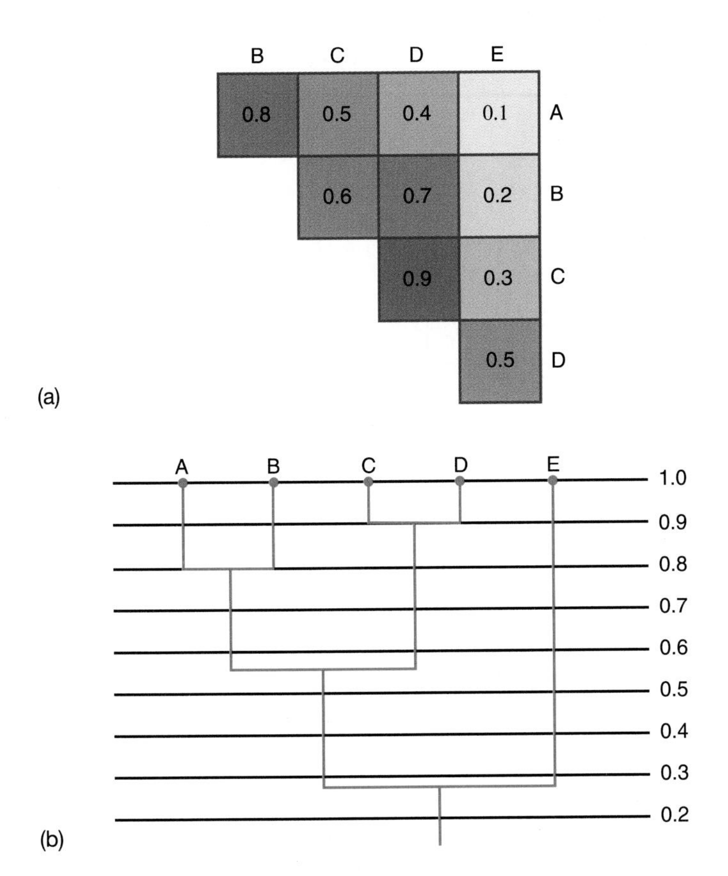

Figure 22.16

(a) A matrix shows the degrees of similarity among a group of species on the basis of many specific characteristics. *(b)* A phenogram derived from the matrix shows the inferred relationship among the species. The phenogram shows, for instance, that A and B are 80 percent (0.8) related and that C and D are 90 percent (0.9) related; the other degrees of relationship come from calculations not shown here.

the greatest problem is that this method sometimes produces taxa that are demonstrably polyphyletic—made of species with superficially similar features. It is a mistake to give all characteristics equal weight, because underlying anatomical features such as bone and muscle structures show relationships more strongly and reliably than do superficial features such as color.

In contrast, cladistics seeks to identify the branching pattern of the phylogenetic tree relating a series of species by identifying features with the greatest evolutionary significance. This method, introduced by the German taxonomist Willi Hennig in 1950, has become increasingly influential. Its goal is to produce testable hypotheses about the relationships among a series of organisms.

Cladistics depends on distinguishing primitive from derived characteristics. Every evolutionary lineage began with some species whose characteristics are primitive, by definition. Vertebrates, for instance, arose from some kind of ancestral chordate, an animal that lacked a backbone but had a notochord that was replaced by the backbone in later evolution. Having a notochord is therefore a **shared primitive characteristic,** and it defines the ancestral group from which all vertebrates arose. As vertebrates evolved, they diverged into several distinctive groups that are defined by **shared derived characteristics.** Mammals share the characteristics of hairy skins, three small bones in the middle ear, and milk production by females. Birds share the characteristics of feathered skins, light hollow bones, and forelimbs modified into wings. These distinctions are so obvious that they make good examples, but distinguishing primitive from derived characteristics is not always a simple task. Sometimes the distinction can be made by comparing the members of a taxon to those of some other taxon known to be primitive; this is one basis for considering the notochord a primitive feature, since vertebrates can be compared with chordates that are not vertebrates. Embryological development can also be useful; here, again, we see a structure like a notochord in an early-stage vertebrate embryo being replaced by a backbone. Fossils may also be very important clues, since they show the characteristics of ancestral organisms.

To return to the situation depicted in Figure 22.15, suppose species X and Y share derived characteristic 1 (perhaps a group of shared characteristics), which species Z does not have. Then we may infer that a branch of the phylogenetic tree leading to X and Y diverged from the branch leading to Z and that X and Y have a common ancestor, A, in which characteristic 1 first appeared. From this information, we draw a partial phylogenetic tree, called a **cladogram:**

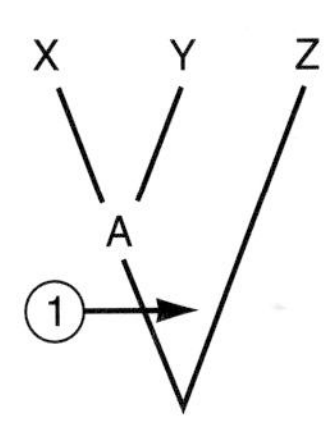

A monophyletic branch, such as the one including X, Y, and A, is called a **clade,** and the branch including Z is a *sister* clade to it (just as the term "sister" is used for entities such as chromosomes or cells that have divided from each other). Species X, Y, and Z share some other characteristic(s) 2, which they do not share with species W. From this, we infer another ancestor, B, from which X, Y, and Z were derived, and we draw a larger cladogram including species W:

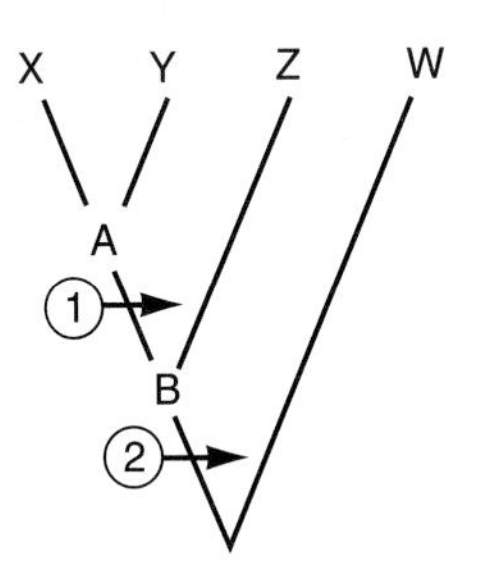

The cladogram can be transformed directly into a classification, but the result may not look like a traditional Linnaean classification (Figure 22.17).

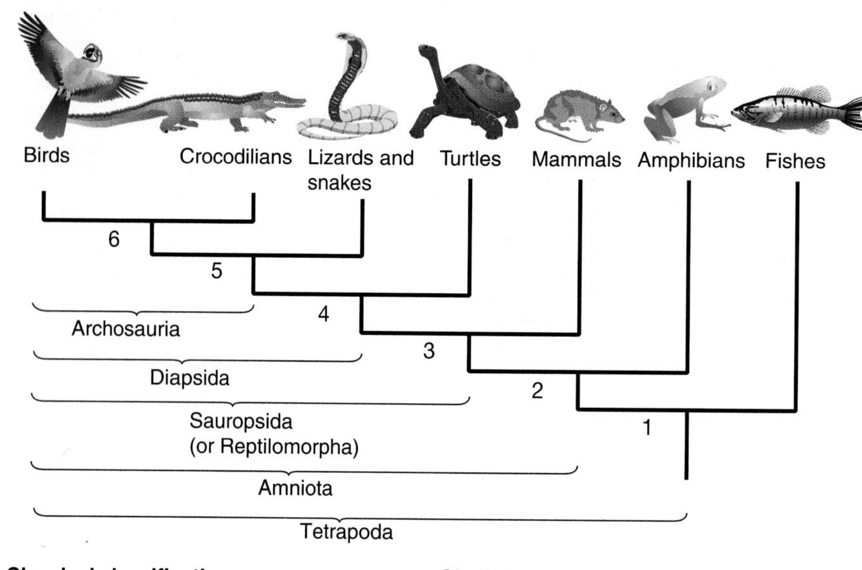

Figure 22.17
A classical classification of the vertebrates divides the subphylum Vertebrata into several classes of equal rank. Some of these taxa are paraphyletic, because they do not include all the descendants of a single ancestor; for instance, Reptilia does not include the birds and mammals, which evolved from reptiles. To understand the cladistic classification, follow the cladogram showing the vertebrate lineage: *(1)* Fishes diverge from other vertebrates (Tetrapoda). *(2)* Amphibians diverge from all remaining vertebrates (Amniota), so they are set aside as Lissamphibia. *(3)* Mammals diverge from all remaining vertebrates (Reptilomorpha), so they are set aside as Mammalia. *(4)* Turtles diverge from all remaining vertebrates (Diapsida), so they are set aside as Anapsida. *(5)* Snakes and lizards (Lepidosaura) are set aside in contrast to Archosauria. *(6)* Archosauria is divided into Crocodilia and Aves.

Classical classification

Phylum Chordata
- Subphylum Vertebrata
  - Class Pisces (fishes)
  - Class Amphibia (amphibians)
  - Class Reptilia (turtles, crocodilians, snakes, lizards, *Sphenodon*)
  - Class Aves (birds)
  - Class Mammalia (mammals)

Cladistic classification

Phylum Chordata
- Vertebrata
  - Tetrapoda
    - Lissamphibia (amphibians)
    - Amniota
      - Mammalia (mammals)
      - Reptilomorpha
        - Anapsida (turtles)
        - Diapsida
          - Lepidosaura (snakes, lizards, *Sphenodon*)
          - Archosauria
            - Crocodilia (crocodilians)
            - Aves (birds)

In contrast to cladistic taxonomy, evolutionary taxonomy, as practiced by many workers today, uses a combination of methods, neither strictly phenetic nor strictly cladistic, though it is closest to the latter. Evolutionary taxonomy uses a larger range of information than does pure cladistics, weights certain characteristics that are considered more indicative of phylogeny than others, and places greater emphasis on fossil data. Adherents of this method believe a classification should be based on more than the branching pattern alone and should also take into account the overall similarity or dissimilarity among species. The difference in emphasis is illustrated by Figure 22.18. Humans are closely related to the great apes (chimpanzees, gorillas, and orangutans), but on the basis of the branching pattern alone, they are much more closely related to chimps and gorillas than to the orangutan, and a purely cladistic taxonomy would place humans, chimps, and gorillas in the same taxon. However, humans have obviously diverged much more from the great apes than any of them have diverged from one another, so the evolutionary classification places humans in the family Hominidae and the great apes in the family Pongidae. Pongidae, however, is a paraphyletic taxon and not acceptable to a cladist.

In establishing relationships among the simplest organisms, which have no substantial fossil record, it may be difficult to distinguish between shared primitive and shared derived characteristics. Taxonomists have sought fundamental structural and metabolic characteristics that are so complex they have probably evolved only once. One such anatomical feature is the basal body, or centriole, with its complex 9 + 2 arrangement of microtubules; useful metabolic features include whole metabolic pathways or complexes of enzymes that are regulated in specific ways. Such features provide useful guidelines to the course of evolution.

Although systematists continue to identify phylogenetic relationships among organisms, based on increasingly sophisticated methods, several groups of organisms with "uncertain affinities" remain. Their study remains a delightful challenge, full of puzzles and controversies. The catalog of the diversity of the biological world is far from finished; it still offers unknown opportunities for the biological explorer.

22.9 The biological species concept poses difficulties in practice.

Having celebrated the biological species concept for its central place in modern biological thought, we must now take a closer look at some of its difficulties. One problem concerns intermediate stages in speciation, when populations are partially separated but still exchanging genes regularly. Other situations pose a conflict between the biological species concept and the desire to create only monophyletic taxa. Two recent studies will illustrate this difficulty.

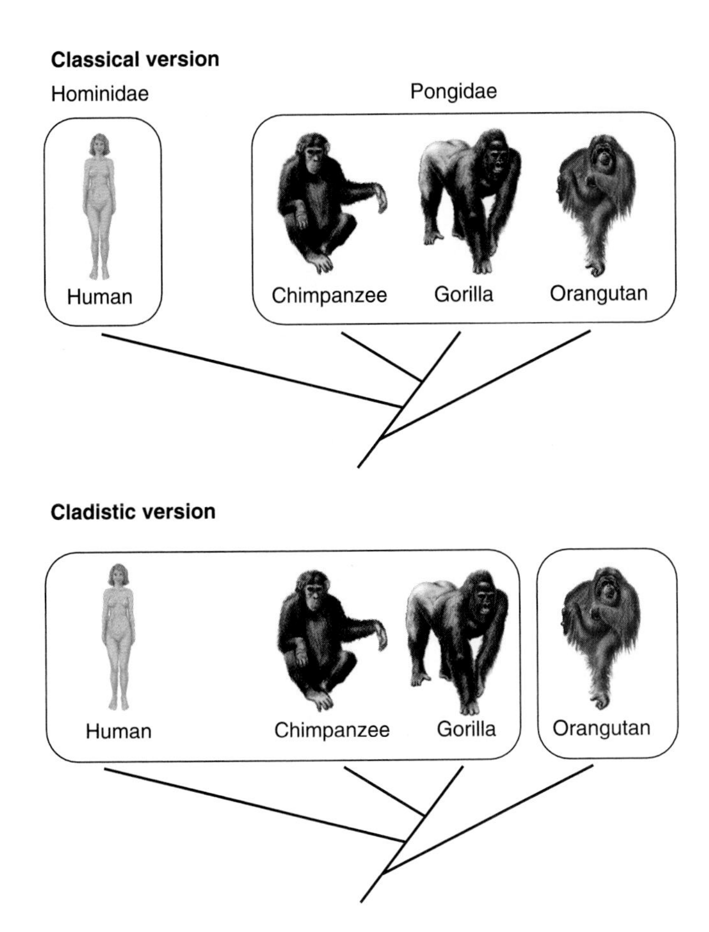

Figure 22.18
A classical classification, taking common features into account, places humans in one family and the great apes in another, while a cladistic classification places humans, chimps, and gorillas in the same category because they share more recent common ancestors.

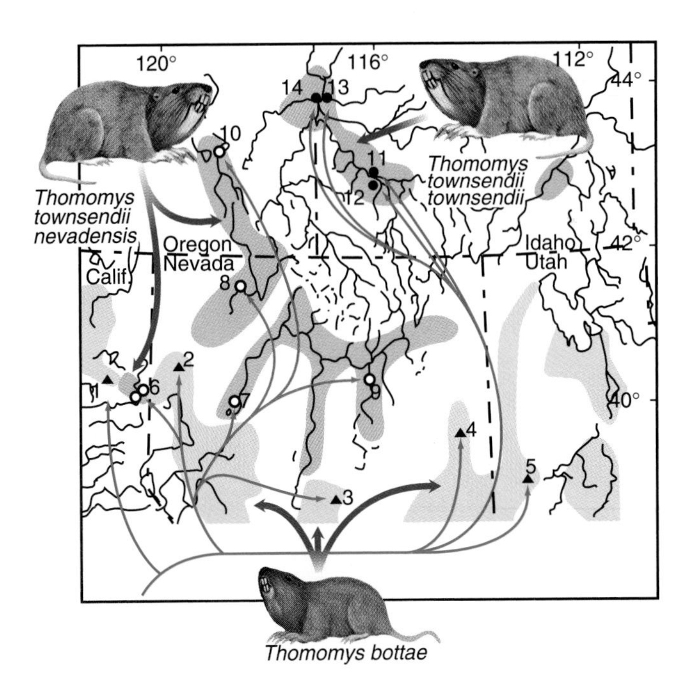

Figure 22.19
Two species of pocket gophers of the genus *Thomomys* are related in complicated ways. The range of *T. bottae* is shown in green, and the range of *T. townsendii* (with two named subspecies) in brown. Mitochondrial-DNA data show that *T. bottae* populations have a single origin *(green arrows),* but *T. townsendii* originated from three separate branches of the phylogenetic tree (locations 11-12-13-14, 7-8-9-10, and 6), making *T. townsendii* a polyphyletic species.

James Patton has used DNA similarities to study speciation in American pocket gophers of the genus *Thomomys* (Figure 22.19). Townsend Pocket Gophers, *T. townsendii,* are large animals that inhabit river valleys in Nevada and Idaho; they evolved from Valley Pocket Gophers, *T. bottae,* a group of smaller, more variable animals that occupy mountains, valleys, and deserts to the south. The interesting and disturbing feature that emerges from Patton's work is that populations of *townsendii* have arisen independently at least three times from separate populations of *bottae.* Thus *T. townsendii* is polyphyletic at the species level, yet its various populations are morphologically very similar and are classified as a single species.

Stephen Freeman has applied DNA methods to the North American orioles (Figure 22.20). The DNA evidence shows that *Icterus galbula,* the Baltimore Oriole, and *I. gularis,* the Altamira Oriole, are sister species, and they form a sister group with *I. bullockii,* Bullock's Oriole, and *I. pustulatus,* the Streak-backed Oriole. If nature were simple, one would expect hybridization to occur only between sister species, as a lingering artifact of incomplete separation. Nature, however, is not always simple; in fact, the Baltimore and Bullock's Orioles hybridize extensively where they meet in the Midwest, showing that groups that are not closely related may still interbreed and appear to be one species. For a while, Baltimore and Bullock's Orioles were considered a single species, called the Northern Oriole, but they have now been split back into distinct species on the official list. This is a formal matter, a matter of list-keeping, but if the Northern Oriole had been retained as a species because of the continued interbreeding it would be a polyphyletic species.

Because of difficulties like this, many taxonomists have abandoned the biological species concept in favor of a **phylogenetic species concept** of some kind. There are several versions of this concept, but they all regard a species as the smallest monophyletic group of organisms identifiable as a distinct group on the basis of morphology, without consideration of interbreeding. However, this species concept entails other kinds of difficulties. For instance, applying it to a comparison of mitochondrial DNA sequences produces the following genealogy:

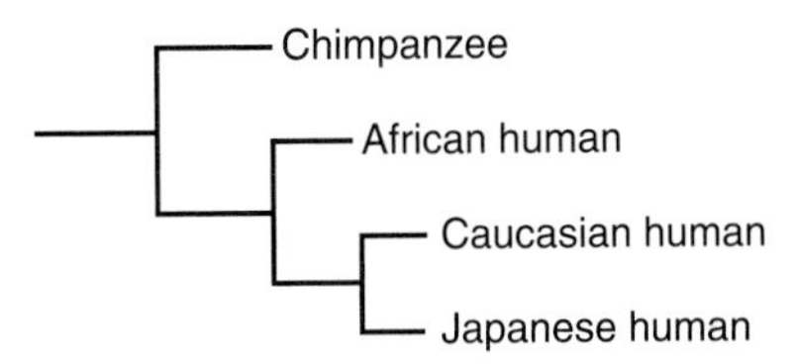

Thus, according to the phylogenetic species concept, the three groups of humans would have to be recognized as separate

Figure 22.20
The relationships among four species of North American orioles (genus *Icterus*) have been determined by analyzing their mitochondrial DNA. The Baltimore and Bullock's Orioles, though not as closely related as some other pairs of species, hybridize where they meet in the middle of the continent.

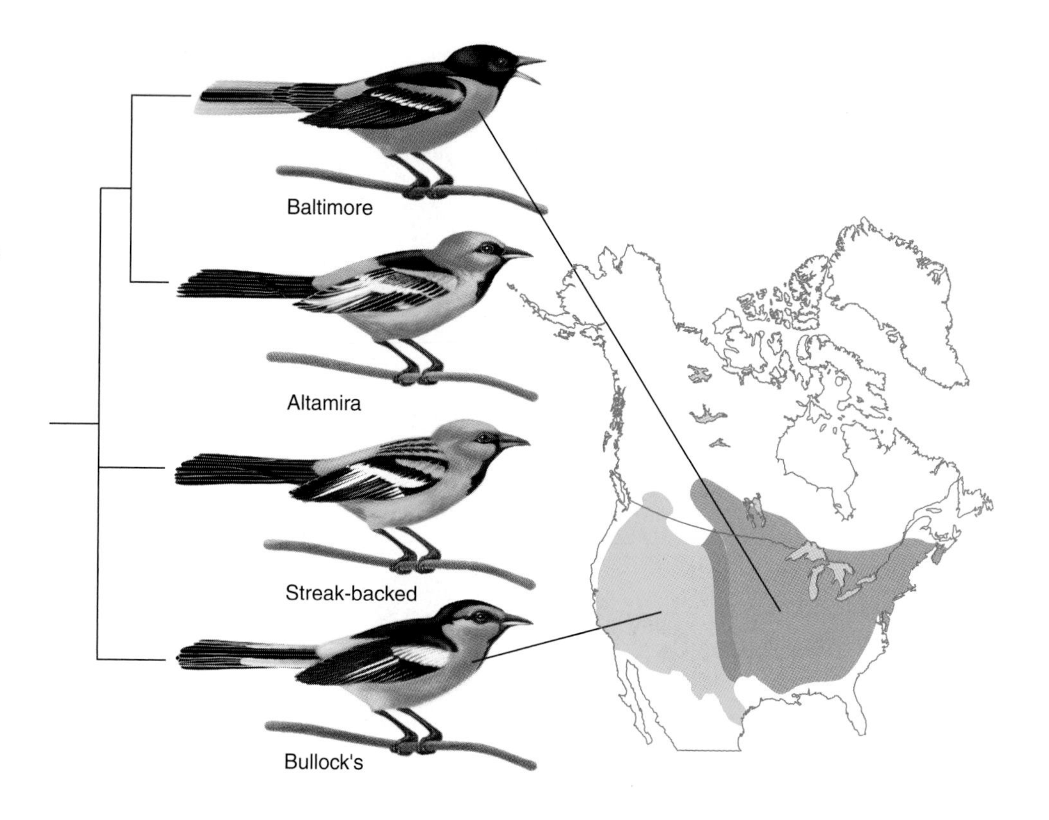

Concepts 22.1 Ancestors and Contemporaries

It is important to notice that biologists allow themselves a certain agreed-upon carelessness in describing the course of evolution. We commonly say that some green algae are the apparent ancestors of land plants or that certain fishes were the ancestors of amphibians. But it only takes a moment to realize that no group of living organisms can possibly be the ancestors of any other group of living organisms. A statement like the first example really means that the *ancestors* of these green algae were also the *ancestors* of the plants. In speaking as we do, we are tacitly assuming that algae like those we see today were living in the distant past, but to avoid such a cumbersome discussion, we use a kind of shorthand.

species, since they are monophyletic and morphologically distinct. Without a criterion such as interbreeding, groups now considered subspecies would have to be elevated to the rank of species.

Dissatisfaction with both the biological and phylogenetic species concepts is leading taxonomists and evolutionary biologists to search for new ways of thinking about the idea of a species. Nevertheless, it is important to see that a group of organisms that can exchange genes with one another, whatever it may be called, remains a significant biological entity. Concepts 22.1 clarifies a customary shortcut used by biologists in discussing evolution.

22.10 The terms "higher" and "lower" organisms, though misnomers, are difficult to avoid.

People often refer to creatures such as worms and insects as "lower animals," while they consider themselves and perhaps some other vertebrates "higher animals." Biologists should try to avoid that trap. It reflects a way of thinking about the world that began with Aristotle, who conceived of a "ladder of Nature," reaching from the simplest and most humble organisms step-by-step to the highest and noblest—humans, obviously. In his idealized conception of the world, every organism could be placed on the ladder, and there must be some creature to occupy every possible step. Medieval scholars codified this idea as the Great Chain of Being (Figure 22.21), extending it past humans to the angels and, ultimately, to God. The Great Chain of Being not only described the apparent physical and biological order of nature but also established a conception of rights. The position of humans clearly made us subject to the will and laws of God, but we, in turn, had a perfect right of dominion over all the lower forms and could use them as we pleased.

We still suffer from this misconception. It causes us to see the rest of the biological world from our position near the "top" of the ladder, and it supports the arrogance that destroys the natural world. However, modern biologists try to avoid language that reflects this old view of the world. They perceive a phylogenetic tree of the organisms on Earth as multibranched, not linear, with no species at the top. (In fact, in most classifications of mammals, the primates, including humans, are placed quite early in the sequence of orders, rather than at the end.) There is no real justification for considering any one type of mammal or bird to be "higher" than any other member of the same class, or for placing humans and other vertebrates collectively above insects, molluscs, or plants.

An alternative distinction is between *ancient* and *derived* lineages. Consider these relationships:

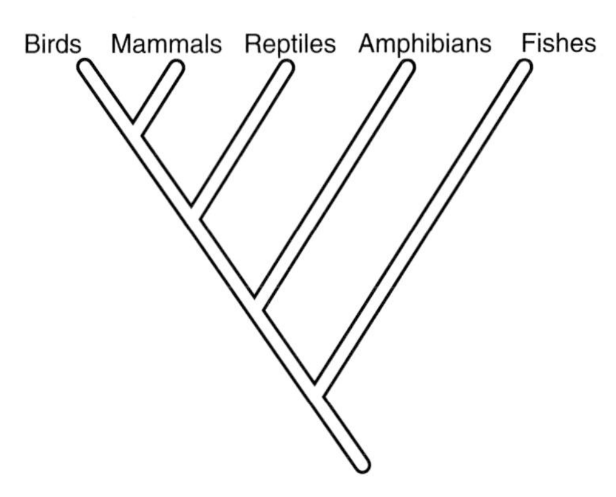

The first vertebrates to evolve were fishes; amphibians branched off later from some fish ancestor, reptiles still later from an amphibian ancestor, and birds and mammals even later from different reptilian ancestors. So according to their times of origin, fishes are more ancient, while birds and mammals are more derived. This distinction refers only to facts about the time when a lineage originated and avoids the "higher/lower" judgment.

Still, the "higher/lower" language reflects something that cannot be denied: There is a difference between the mammals and birds, with their complex structures and dominant roles in ecosystems, and the protozoans or worms. The terms *simple* and *complex,* while vague and perhaps not totally justified, are sometimes more suitable. Mammals are clearly more complex than worms, though whether mammals are more complex than fish is a more difficult question. Since mammals evolved after fish, we are tempted to say they are more complex for that reason alone. This temptation, however, reflects the modern prejudice that evolution is always progressive, so whatever came later must somehow be better. As we emphasized earlier, although organisms do tend to evolve from the simple to the more complex, the changes cannot be said to represent progress. Each species, after all, is simply adapting to the demands and opportunities presented by its environment, and there is nothing intrinsically progressive in these adaptations.

At least one other possible criterion for differences in evolutionary status might be characterized as plastic versus restricted. Compare the evolutionary potential of a simple worm with that of a mammal. With their unspecialized bodies, some kinds of worms could potentially evolve into a wide variety of other animals with quite different body plans. A mammal, however, is much more restricted; in later evolution, its descendants can only become other kinds of mammals. As organisms become more complex, they probably also become more restricted in possibilities for evolution.

Figure 22.21

In a classical representation of the Great Chain of Being, all creatures are assigned to positions in a continuous "ladder of life."

Coda All our thinking about the biological world entails evolution in some measure, and taxonomy or systematics is our way of cataloging the course of evolution on Earth. Systematics, however, goes well beyond mere cataloging to question the nature of the entities—the species—being cataloged and how they have arisen. When looking at populations realistically, we are struck by their variety. And on reflection, it is clear that this is only to be expected, because if organisms are genetic systems that depend on replication and inheritance of genomes, a lot of accidents must occur during these processes, and we must expect to see a lot of variant genomes. This variability creates problems of classification, which would be distressing if our sole purpose were to create a catalog. But our purpose is to understand how real organisms operate, how they change and evolve, so the variation is delightful, enlightening, and challenging. Perhaps our aesthetic delight in the variability of the natural world will motivate us to work for preservation of the world as its diversity is threatened. The patterns of variation can help us understand more clearly just how diversity arises in natural populations. And we can be challenged to learn what all the variants are and why they exist, for variation must often (or perhaps always?) have some ecological basis. Since organisms live in ecosystems, we expect enzymatic and morphological differences to make some difference in their lives and in their ecological relationships. We will explore some of these questions in the next chapter.

Summary

1. Species are first defined and classified on the basis of their morphology. But while this criterion can indicate groups that are likely to be species, biologists generally adopt the biological species concept, which uses reproductive isolation as the ultimate criterion.
2. Natural populations are polymorphic, containing different alleles and combinations of genes. Polymorphism may only be apparent in different forms of proteins or may result in visibly different forms within a population.
3. A species is given a binomial (two-name) designation, the first name being its genus and the second name its specific designation within the genus.
4. Each species is classified in a hierarchy of more and more inclusive taxa, in this sequence: genus, family, order, class, phylum, and kingdom. A hierarchy of this kind is called a taxonomy.
5. One aim of systematic work is to create a taxonomy that reflects phylogeny (evolutionary history) as closely as possible. This means, in part, creating only monophyletic taxa, which include only species derived from a single immediate ancestor, rather than polyphyletic taxa, which include species derived from two or more immediate ancestors.
6. The biological species concept runs into various difficulties in practice. Groups that are incompletely separated from each other, so they exchange some genes or exchange genes in irregular ways, create one difficulty. Recent studies show that a single group whose members exchange genes and act like one species may actually be derived polyphyletically from different ancestral groups.
7. Some taxonomists use a phenetic approach, which creates taxa exclusively on the basis of similarities in appearance. Others use a cladistic approach, which attempts to reconstruct phylogenetic trees and classifies organisms on the basis of their ancestry. Classical evolutionary taxonomy uses a combination of the two.
8. Some taxonomic work is being advanced using molecular criteria, especially similarities between the DNAs of different species as determined by sequencing or hybridization.
9. Classifying organisms as "higher" and "lower" is generally unrealistic; "ancient" compared with "derived," "complex" compared with "simple," or "plastic" compared with "restricted" may be more realistic distinctions.

Key Terms

convergent evolution 445
morph 446
biological species concept 448
polymorphic 449
deme 449
cline 449
subspecies 449
geographic race 449
taxa/taxon 450
genus/genera 450
binomial 450
family 453
order 453
class 453
subphylum 453
phylum 453
kingdom 453
monophyletic 453
polyphyletic 453
paraphyletic 453
phenetic 456
cladistic 456
numerical taxonomy 456
shared primitive characteristic 457
shared derived characteristic 457
cladogram 457
clade 457
phylogenetic species concept 459

Multiple-Choice Questions

1. Organs or structures that are homologous must exhibit the same
   a. appearance.
   b. function.
   c. genetic ancestry.
   d. *a* and *b*.
   e. *a*, *b*, and *c*.
2. Adaptive genetic variations within a single species are included within the meaning of the term
   a. taxonomic.
   b. convergent
   c. microevolutionary.
   d. analogous.
   e. phylogenetic.
3. A gene pool includes
   a. all the alleles shared by two species that have evolved convergently.
   b. all the genes that are important in the construction of homologous structures.
   c. all the wild-type alleles in a species.
   d. only those genes that are polymorphic.
   e. all the allelic variations found within a group of interbreeding organisms.
4. Species A, B, C, D, and E have a common ancestor. A, B, and C are assigned to family X; C and D to family Y. Family X is
   a. monophyletic.
   b. paraphyletic.
   c. polyphyletic.
   d. *a* and *b* but not *c*.
   e. *b* and *c* but not *a*.
5. Which term includes all the others?
   a. deme
   b. cline
   c. race

d. species
e. population

6. Which is a taxon?
   a. kingdom
   b. phylum
   c. order
   d. family
   e. All are taxa.
7. Which is the most accurate statement about a clade?
   a. A clade is equivalent to a species.
   b. A clade is the same as a genus.
   c. It is a taxon.
   d. It is a monophyletic lineage.
   e. All members of a clade can interbreed.
8. Taxonomists who accept the biological species concept will most readily agree upon the classification of
   a. a genus of living organisms.
   b. a family of extinct organisms.
   c. asexually reproducing species of living organisms.
   d. sexually reproducing species of extinct organisms.
   e. sexually reproducing species of living organisms.
9. The greatest similarity will be evident in the genotypes of members of the same
   a. species.
   b. genus.
   c. family.
   d. order.
   e. phylum.
10. The characteristics shared between members of the same ___ are more ancient than those shared by members of the same phylum.
    a. species
    b. genus
    c. family
    d. order
    e. kingdom

True-False Questions

Mark each statement true or false, and if false, restate it to make it true.

1. The evolution of an ancestral mammalian species into all the mammals we see today is an example of evolutionary convergence.
2. The difference in proportions of blood groups A, B, AB, and O among human populations is an example of macroevolution.
3. The proposed evolution of birds from ancestral dinosaurs is characterized as macroevolution.
4. Phylogenetic data are the most important evidence in defining species according to the biological species concept.
5. The concept of a morphological species is of greater importance in classifying asexually reproducing organisms than sexually reproducing organisms.
6. If a taxon includes two species that evolved convergently, the taxon is probably polyphyletic.
7. Modern humans are included in a polyphyletic taxon called *Homo sapiens.*
8. The lower the melting point of hybrid DNA, the more closely related are the parental species.
9. A binomial includes the genus name and the species name.
10. The ability to distinguish analogies from homologies is more significant for cladistic methods than for phenetic methods.

Concept Questions

1. Contrast the meaning of *phylogeny* with that of *taxonomy.*
2. Explain why reversals of evolutionary pathways are highly unlikely.
3. Why is it incorrect to think of evolution as progressive, i.e., from lowest or simplest to highest or most complex?
4. Suppose you are classifying automobiles using cladistic methods. Assume that all modern models of gasoline-powered autos evolved from horse-drawn carriages. Models that are no longer produced are considered to be extinct; surviving models will eventually be replaced by electric cars. Name one shared primitive characteristic of all autos and one shared derived characteristic.
5. How does the morphological species concept differ from that of a biological species or a phylogenetic species?

Additional reading

Ax, Peter. *The Phylogenetic System: The Systematization of Organisms on the Basis of Their Phylogenesis.* Translated by R. P. S. Jefferies. John Wiley, Chichester [West Sussex] and New York, 1987.

Eldredge, Niles, and Joel Cracraft. *Phylogenetic Patterns and the Evolutionary Process: Method and Theory in Comparative Biology.* Columbia University Press, New York, 1980. A study of evolution and the basis for classification.

Nelson, Gareth J., and Norman Platnick. *Systematics and Biogeography: Cladistics and Vicariance.* Columbia University Press, New York, 1981.

Quicke, Donald L. J. *Principles and Techniques of Contemporary Taxonomy.* Blackie Academic & Professional, London and New York, 1993.

Sibley, Charles G., and Jon E. Ahlquist. "Reconstructing Bird Phylogeny by Comparing DNA's." *Scientific American,* February 1986, p. 82. The technique of determining avian phylogeny through studies of DNA.

Wiley, E. O. *Phylogenetics: The Theory and Practice of Phylogenetic Systematics.* John Wiley, New York, 1981. An introduction to systematics and cladistic analysis.

Internet Resource

To further explore the content of this chapter, log on to the web site at:

http://www.mhhe.com/biosci/genbio/guttman/

23 Population Genetics

South American natives have a very high frequency of blood type O and no type B.

Key Concepts

23.1 Organisms in an ideal Mendelian population mate at random.

23.2 The genetic composition of a Mendelian population remains constant in the absence of mutation and selection.

23.3 The value of q may be determined from the frequency of homozygous recessives.

23.4 Selection can change gene frequencies.

23.5 Mutation is the source of all genetic change.

23.6 Gene frequencies may change rapidly in small populations.

23.7 Heterozygotes are sometimes more fit than either homozygote.

23.8 Some regimes of selection are notoriously ineffective.

23.9 Every organism is an integrated gene complex.

23.10 Genes can be rearranged by chromosomal inversions and translocations.

23.11 Fitness changes with changing environmental conditions.

23.12 Fitness generally changes geographically.

People who are migrating generally take their important possessions with them, but as humans spread out across the globe, many did not take along their genes for blood type B. Over the course of millions of years, humans evolved in the Old World—Africa, Europe, and Asia—and then very recently, perhaps 15,000 years ago, they moved into the Americas across the Bering Strait, into the Pacific Islands, and to Australia. The migrating hunter-gatherers of this time took with them all the knowledge, skills, and possessions they had gradually acquired, but curiously, many left that B gene behind. The alleles for the A and O blood types are the most common almost everywhere, while the B allele is mainly concentrated in Central Asia; however, Native Americans, Polynesians, and Australian aborigines have no B alleles at all. The map of human blood types displays a neat picture of human evolution by showing one kind of change a species can experience in a rather short time.

This chapter is about microevolution, the genetic changes that occur within a species. First we will develop a little of the now well-established theory of population genetics, which explains just how the genetic structure of a population changes. (This is the most highly developed mathematical theory in biology, although we will barely scratch its surface mathematically.) Starting with the observation from Chapter 22 that all natural populations have a lot of variation, we will show that an idealized population will achieve a genetic equilibrium in which the frequencies of many genes do not change. Then we will introduce some factors that contribute to evolution, such as mutation and selection. Finally we will examine geographic variation in natural populations and its implications for the other aspects of evolution. This theory has significant implications for the genetics of human populations and for policies directed toward improving human health. ■

23.1 Organisms in an ideal Mendelian population mate at random.

We showed in Chapter 22 that natural populations are generally polymorphic, at least at the protein level. Although the variation in populations comes from both genetic and environmental sources, natural selection can only operate on genetic variation, and only genetic variation plays a role in evolution. The more genetically variable a population is, the more raw material it contains for selection to work on. Sir Ronald Fisher, a pioneer in population genetics, called this principle the Fundamental Theorem of Natural Selection: *The rate of evolution in a population is proportional to its genetic variance.*[1] For this reason, we begin by examining some general features of genetic variability in Mendelian populations, which are populations of sexual, interbreeding individuals whose heredity follows Mendelian principles.

Heredity and Environment, Sidebar 16.1.

We expect natural populations to be polymorphic simply because mutations are always occurring in them. Of course, natural selection might operate by eliminating mutational differences, but populations actually tend to retain that genetic diversity because natural selection acts quite subtly to maintain a balance of different alleles in the population. In some cases, alleles are clearly retained because they confer advantages in different situations. Through selection, species remain adapted to their environments, and the microevolution that keeps them adapted depends upon their genetic diversity. As the populations of some contemporary species become very small, they are in danger of extinction because they are too homogeneous and don't harbor the genetic potential for adaptation to changing conditions. Later in this chapter we will see why some species are so polymorphic.

A species is a group of closely related individuals spread over a certain area, and its members can be unified or isolated to various degrees. At one extreme, the entire species may be one population in which any individual can make contact and mate with any other individual of the opposite sex (Figure 23.1*a*). The species is more likely divided into many local populations, or demes, but with a certain amount of genetic mixing, or *gene flow,* among them. At the other extreme, some or all of the demes may be isolated from one another, so little or no gene flow occurs (Figure 23.1*d*). We will see that evolution depends largely on populations being partially or completely isolated from one another, but we will start to analyze the genetic structure of populations with a large, undivided population of the first type.

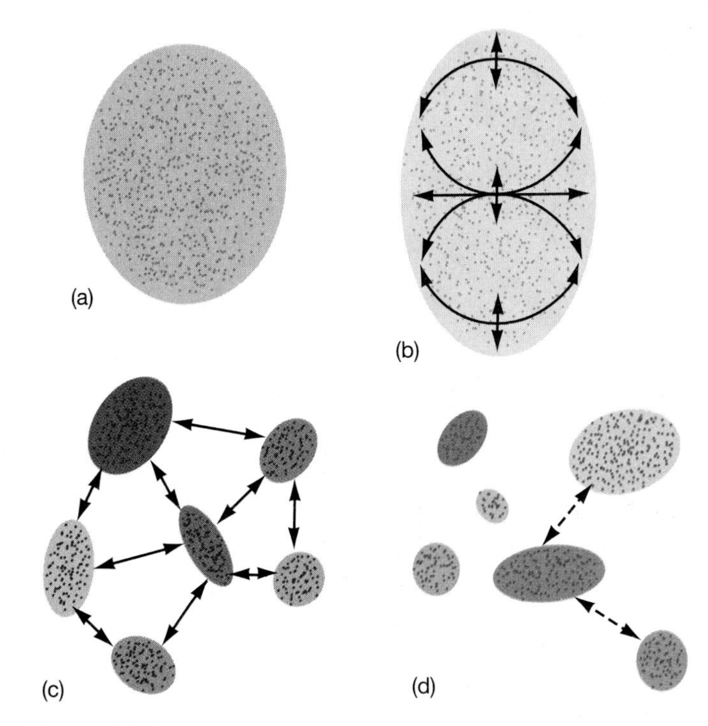

Figure 23.1
Various population structures are represented by individuals *(dots)* within geographic limits *(ellipses);* the arrows indicate gene flow. *(a)* A large population can have enough mobility of individuals or gametes for any two individuals of opposite sex to mate, so the concept of gene flow does not apply. *(b)* In a large population with local mobility and free mating, individuals at opposite ends of the area can be effectively isolated from one another and may differ genetically, but gene flow continues between regions. *(c)* A population can be broken up into local demes that are still able to exchange genes with one another. *(d)* Local populations may be so isolated that little or no gene flow occurs between them.

The theory of population genetics begins with an idealized situation: a large population of sexually reproducing diploids that is totally isolated, so no genes enter or leave it, and exhibits **random mating,** which means that any male and female can mate with each other (Figure 23.2*a*). Alternatively, if individuals show preferences for prospective mates with similar features, they exhibit **assortative mating** (Figure 23.2*b*). Humans, for instance, mate assortatively if they tend to marry those with the same eye color, comparable intelligence, or some other shared feature. For illustration, let's consider a population of mice, 70 percent with black fur and 30 percent with white, which nevertheless mate at random (Figure 23.3). The probability of picking out a black mouse at random is 0.7 and that of picking out a white mouse is 0.3. So the probability of two black mice forming a mating pair is $0.7 \times 0.7 = 0.49$, or 49 percent. Similarly, $0.3 \times 0.3 = 0.09$ (9 percent) of the matings should be between two white mice. The probability of a black male and a white female mating is $0.7 \times 0.3 = 0.21$ (21 percent), and the probability of a black female mating with a white male is the same, so these matings account for 42 percent. The combined probabilities (49 percent + 42 percent + 9

[1] Given a series of measurements, such as the height of individuals in a population, the standard deviation measures how broadly the measurements are spread around the mean. Variance is the square of standard deviation.

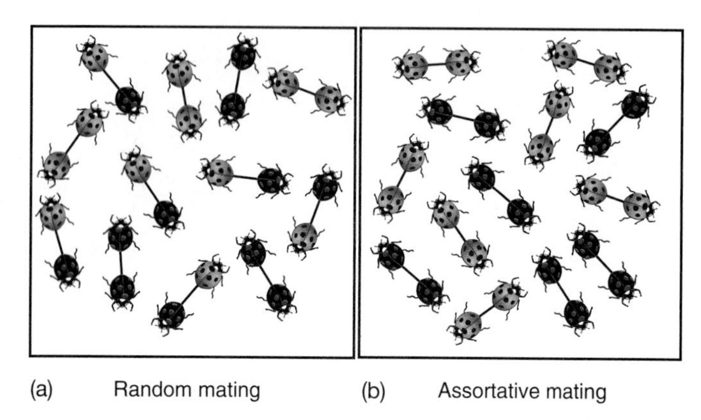

Figure 23.2
(a) Random mating means that any male and any female in a population can mate, with no preference for characteristics. With assortative mating *(b),* individuals tend to select their mates and mate primarily with those that have similar features.

percent) account for 100 percent of the possibilities. Genetic events in a population like this can be analyzed quite straightforwardly, as we will show next.

Exercise 23.1 In a small town, 10 percent of the people are red-haired and 90 percent are blond. If people marry at random, calculate the probabilities of marriages between two red-haired people, between two blonds, and between one red-haired person and one blond person.

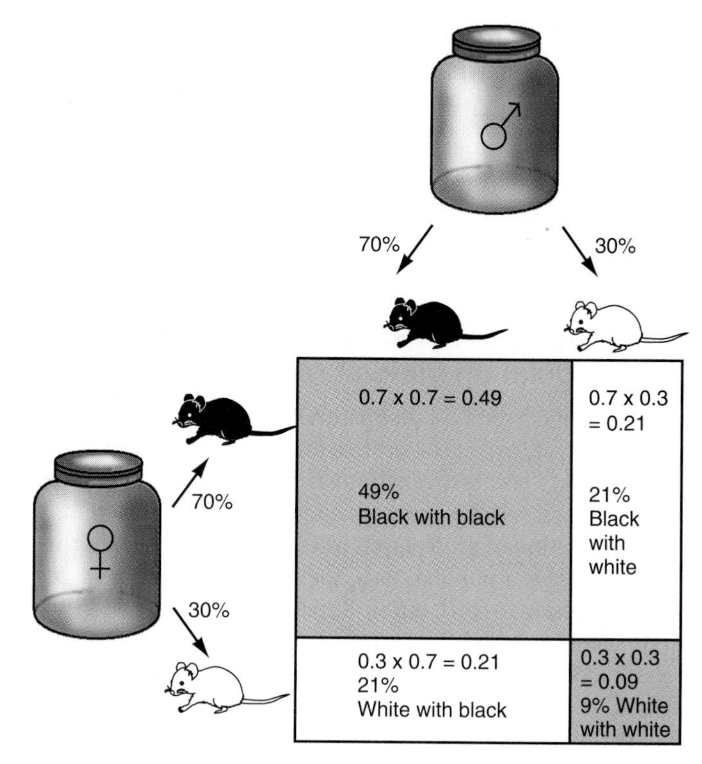

Figure 23.3
To analyze random mating of mice, we imagine that all the males are placed in one big jar and all the females in another. The probability of any combination is then determined by multiplying the probabilities of picking out a male and a female of each kind.

23.2 The genetic composition of a Mendelian population remains constant in the absence of mutation and selection.

The members of a Mendelian population are said to share a **gene pool,** a metaphorical common space where all their genes are combined, mixed, and then reassorted to create the next generation. With random mating, no female chooses the sperm that will fertilize her eggs and no male chooses the eggs. Any sperm can fertilize any egg to make a new individual by drawing gametes out of the pool randomly, subject only to the laws of probability.

In 1908, the English mathematician G. H. Hardy and the German biologist Wilhelm Weinberg independently discovered that an idealized Mendelian population will come to an *equilibrium* for each gene locus and will not change thereafter. This fact, now known as the **Hardy-Weinberg Principle,** is the foundation of population genetics. The Hardy-Weinberg analysis assumes:

- The population is diploid and sexually reproducing.
- The population is large enough for statistics to be valid.
- Mating occurs at random.
- No mutation takes place.
- No genotype has a selective advantage over any other, and all members of the population survive and reach reproductive age.
- There is no influx of genes from other populations.

Notice that since no mutation or selection takes place, such an idealized Mendelian population *is not evolving.* Analyzing this simple situation allows us to see the effects of factors such as mutation and selection.

The Hardy-Weinberg Principle is worth exploring in some detail for a couple of reasons. First, it assists in such diverse studies as the effects of mutagens in the environment, historical migrations of human populations, evolution of island bird populations, and effects of mutant conditions of medical importance in humans. When the ABO blood types were discovered, for instance, it was assumed that types A and B are produced by nonallelic genes, but Hardy-Weinberg analysis showed they must be alleles of a single gene. (The one-gene and two-gene hypotheses make quite different predictions about the frequency of AB people, and the data for several populations fit the one-gene prediction.) We showed in Chapter 19 how the whole idea of equilibrium or disequilibrium with respect to a particular allele underlies our ability to locate important genes, such as the CFTR gene whose mutation causes cystic fibrosis.

Second, many critical studies of population and evolution stem from examining the assumptions behind the Hardy-Weinberg Principle. If the distribution of genotypes in a population doesn't remain the same from generation to generation, at least one assumption must not be valid for this population. Is mutation occurring? Does one genotype have an advantage over others? Are individuals mating in a nonrandom way? Are individuals with different genotypes coming in from neighboring populations? Is the population so small that one genotype becomes more common just by chance? These kinds of questions underlie a systematic study of the reasons for change in a population, including human populations. Far from being just a matter of esoteric theoretical mathematics, Hardy-Weinberg equilibria are at the heart of exciting contemporary studies in anthropology, medicine, evolution, and human affairs.

Linkage disequilibrium and mapping human genes,
Section 19.9.

We will illustrate the Hardy-Weinberg analysis with a convenient human trait, the MN blood types, because the alleles are codominant, because there seems to be no selection for either M or N blood type, and because human populations conform perfectly to the Hardy-Weinberg principle for this trait. M and N types are encoded by two alleles, *M* and *N*, of a single gene. The abundance of each allele in a population is given by its **allelic frequency:** The frequency of the *M* allele, denoted by *p*, is the fraction of all these genes in a population that are *M*, and the frequency of the *N* allele, denoted by *q*, is the fraction that are *N*. Allelic frequencies can vary from 0 to 1; if there were equal numbers of the two alleles in a population, *p* and *q* would both be 0.5. Even though neither M nor N types have any selective advantage, populations don't necessarily have equal numbers of the two alleles, and the frequencies of *M* and *N* in different human populations vary quite a lot. If a gene has only two alleles, every copy must be one or the other, and $p + q = 1$. If one allele is dominant, its frequency *(p)* is often quite large, while the frequency of the recessive allele *(q)* may be quite small. (Incidentally, students who are used to problems in Mendelian genetics, where probabilities such as 1/2, 1/4, and 3/4 are common, sometimes want to apply these numbers in population genetics. Resist that temptation. Population genetics requires a different kind of reasoning, which we will develop here, and these simple fractions rarely come into it.)

Since every individual in a population carries two genes at each locus, the number of genes is twice the number of individuals. A population with a total of n_T individuals will contain three subpopulations with different genotypes: n_1 *MM* individuals have two *M* alleles each; n_2 *MN* individuals have one *M* allele and one *N* allele; and n_3 *NN* individuals have two *N* alleles each. Then the frequency of the *M* allele is:

$$p = \frac{2n_1 + n_2}{2n_1 + 2n_2 + 2n_3} = \frac{2n_1 + n_2}{2n_T}$$

and the frequency of the *N* allele is:

$$q = \frac{n_2 + 2n_3}{2n_1 + 2n_2 + 2n_3} = \frac{n_2 + 2n_3}{2n_T}$$

Example 23.1 A population contains 150 *MM*, 300 *MN*, and 100 *NN* individuals. Calculate the frequencies of the *M* and *N* alleles. We use the above equations, noting that $n_T = 550$. The frequency of *M* is $p = ([2 \times 150] + 300)/1{,}100 = 600/1{,}100 = 0.545$. The frequency of *N* is $q = (300 + [2 \times 100])/1{,}100 = 500/1{,}100 = 0.455$. Checking our arithmetic, $0.545 + 0.455 = 1.0$, so we've accounted for all the genes. Notice that in the numerators of these fractions we have simply calculated that 600 of the 1,100 genes at this locus in the population are *M* and 500 are *N*.

We think about breeding in this population with the same logic we used in choosing pairs of black and white mice, because in both cases mating occurs at random. It is as if all the sperm are put into one jar and all the eggs into another, and then a sperm and an egg are drawn at random to make a zygote (Figure 23.4). The probability of choosing a gamete carrying *M* is *p*, and the probability of choosing one carrying *N* is *q*. Then, as we established in Section 16.3, the probability of getting a

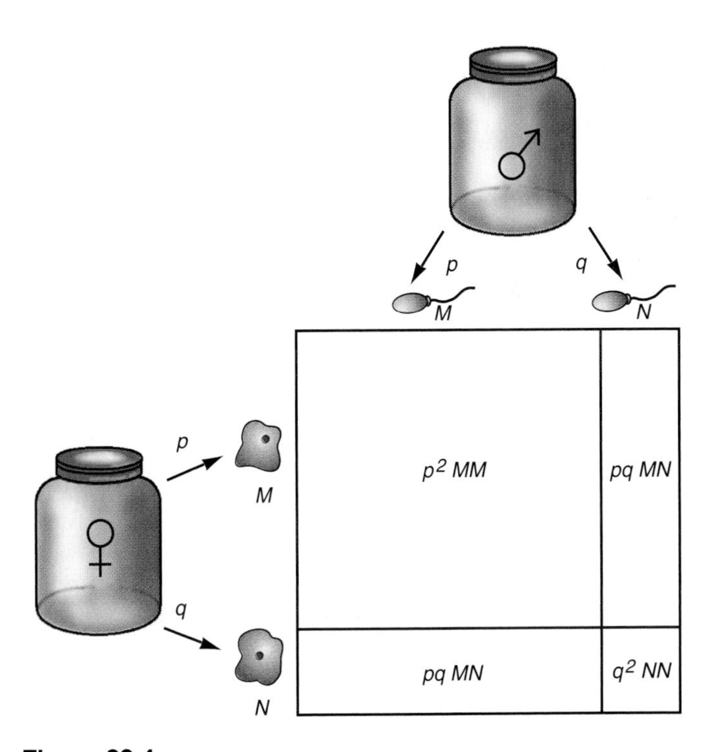

Figure 23.4
To analyze the genetic effects of random mating in a population with two alleles of a gene (here, the *M* and *N* genes for blood types), we imagine that all the eggs are placed in one big jar and all the sperm in another. The probability of each genotype is determined by multiplying the probabilities of picking out eggs and sperm of each type.

zygote of each kind is calculated by multiplying these probabilities for gametes:

1. The probability of forming an *MM* zygote is just $p \times p = p^2$.
2. The probability of forming an *MN* zygote is $(p \times q) + (q \times p) = 2pq$. (This is the probability of getting an *M* sperm with an *N* egg or an *N* sperm with an *M* egg, so it is just two times the same elementary probability.)
3. The probability of forming an *NN* zygote is $q \times q = q^2$.

The sum of the frequencies of all three genotypes is $p^2 + 2pq + q^2$, which, by simple algebra, is equal to $(p + q)^2$. Since $p + q = 1$, $p^2 + 2pq + q^2$ also is equal to 1 (as it has to be, since a probability of 1 means certainty, and every individual *must* be one of the three genotypes).

Exercise 23.2 A population of 1,000 individuals has 360 *MM*, 480 *MN*, and 160 *NN*. What are the allelic frequencies?

In the population in Exercise 23.2, $p = 0.6$ and $q = 0.4$. What will be the frequencies of the three genotypes in the next generation? The frequency of *MM* individuals will be $0.6^2 = 0.36$. The frequency of *MN* individuals will be $2(0.6)(0.4) = 0.48$. And the frequency of *NN* individuals will be $0.4^2 = 0.16$. If that generation consists of 1,000 individuals again, we can multiply these frequencies by 1,000 to get 360 *MM*, 480 *MN*, and 160 *NN* individuals, *which is exactly the composition of the previous generation.* What we have just shown numerically, for one case, is that a randomly mating population reaches an equilibrium at a ratio of $p^2 : 2pq : q^2$ for the three genotypes at a single locus and then *does not change.* You could prove the same thing in general with a little algebra, by taking all the possible matings into account. In only one generation of random mating, a population comes to a Hardy-Weinberg equilibrium, and as long as conditions stay the same, the population will remain at this equilibrium.

> *The Hardy-Weinberg Principle: In the absence of mutation and selection, a large, randomly mating population will reach an equilibrium ratio of* $p^2 : 2pq : q^2$ *for the three genotypes at one locus.*

Notice that the Hardy-Weinberg Principle applies to each gene locus independently, so a population may be at equilibrium for one gene while the frequencies of other genes are changing.

Exercise 23.3 Calculate the gene frequencies in a population with 500 *MM*, 250 *MN*, and 250 *NN* individuals.
Exercise 23.4 Using the gene frequencies calculated in Exercise 23.3, determine the number of individuals of each genotype at equilibrium in a total of 1,000. Is the initial population (in Exercise 23.3) at equilibrium?
Exercise 23.5 A population of field mice carries two alleles, *R* and *r*. If there are 100 *RR*, 500 *Rr*, and 400 *rr* mice, calculate the frequencies of the two alleles. Is this population at Hardy-Weinberg equilibrium? If not, calculate the equilibrium numbers of each genotype, assuming the total population remains 1,000.

23.3 The value of *q* may be determined from the frequency of homozygous recessives.

For illustration, we used a gene locus with no dominance, where we could distinguish *MM* and *MN* individuals. But how can we determine the frequencies of alleles when dominance gives both homozygous dominants and heterozygotes the same phenotype? We can recognize and count all the homozygous recessives. Their frequency is q^2. Therefore, the square root of that frequency is q, and $p = 1 - q$. Suppose, for instance, 1 percent of a population (0.01) has a certain recessive genetic condition. Since $q^2 = 0.01$, $q = \sqrt{0.01} = 0.1$. That is, 10 percent of the alleles at this locus in the population are recessives, and 90 percent are dominants.

Exercise 23.6 Suppose in one population the frequency of nontasters for PTC (a recessive trait) is one person in 10,000. What is the frequency of the recessive allele?
Exercise 23.7 Spina bifida is a genetic disorder caused by a recessive allele. If 26 children per 100,000 are born with this disorder in Japan and 426 per 100,000 are born with it in Ireland, what are the frequencies of the allele in these populations?

23.4 Selection can change gene frequencies.

Given the basic theory for a population at equilibrium, we can ask about the factors that create change. Hardy-Weinberg analysis assumes there is no selection for any genotype, but for most traits, the genotypes *AA*, *Aa*, and *aa* have different reproductive potential or **fitness** (sometimes called Darwinian fitness). The fitness of a genotype is a measure of its contribution to the gene pool of the next generation, compared with that of other genotypes. Fitness is always a *relative* measure; there is no such thing as an *absolute* measure of fitness (because a more fit genotype could always appear), so the best-adapted genotype is assigned a fitness of 1. Quite commonly, with complete dominance at a particular locus, both *AA* and *Aa* individuals have a fitness of 1, but homozygous recessives are at a disadvantage and have a lower fitness (Figure 23.5). In a flowering plant, different genotypes for flower color may have different fitnesses because one color is more attractive to pollinators, while genotypes that affect the length of an insect's leg may have subtle effects on the insect's ability to run or to hold onto its food. Then in each generation, instead of the three genotypes being at their classical ratio, p^2 *AA* : $2pq$ *Aa* : q^2 *aa*, the *aa* individuals will actually occur at a lower frequency $q^2(1 - s)$. This expression is the definition of s, the **coefficient of selection** against *aa*. The fitness of this genotype is then defined as $(1 - s)$.

To illustrate the idea of selection, consider again a population of 1,000 individuals, with 360 *AA*, 480 *Aa*, and 160 *aa*, for which $p = 0.6$ and $q = 0.4$. Suppose after one more generation the numbers of the three genotypes are 375, 500, and 125, respectively, so the total population is still 1,000. To determine the fitness of each genotype, we must find out how much they increase or decrease relative to their expected numbers in the absence of selection. Both *AA* and *Aa* have increased by about 4 percent, but instead of the expected 160 *aa* individuals, there are only 125; since $125/160 = 0.78$, there are only 78 percent as many *aa* individuals as expected. Then, assigning the *AA* and *Aa* individuals a fitness of 1.00, the fitness of *aa* is only $0.78/1.04 = 0.75$; in other words, the *aa* individuals are only 75 percent as fit as those with the other genotypes. The coefficient of selection against *aa* is $1 - 0.75 = 0.25$.

People still tend to have dramatic and romantic ideals about fitness. Slogans such as "the struggle for existence" or "nature red in tooth and claw" call up images of bloody battles between predators and their prey, where greater fitness means the ability to run faster or to win a battle to the death. In reality, fitness depends mostly on subtle factors, such as small changes in protein that confer different metabolic abilities, the ability to live at particular temperatures or oxygen pressures, and small changes in form. People also tend to pass along the myth that adaptations are of no value unless they are fully developed; an eye, for instance, is said to be of no value unless it is fully formed, so it is assumed there could be no selective value in acquiring any of the minute changes necessary to form an eye little by little over a long time. Modern studies of evolution show that this is not true. Peter and Rosemary Grant have conducted extensive studies on the finches of the Galápagos Islands off the coast of Ecuador, the subfamily Geospizinae, often called Darwin's finches because they had such a great influence on Darwin's thinking when he visited the islands as a young man aboard the *Beagle* in 1835–36. These islands are subject to severe changes ranging from very wet to very dry conditions. The Grants have shown that in very dry years the best-adapted ground finches are those with the largest, strongest bills, enabling them to open the large seeds that become most common in dry conditions. However, the average difference in bill dimensions between birds that survive the drought and those that die is only about half a millimeter, out of a total length of about 10 mm. Thus these data show that a subtle difference in form can make an enormous difference in fitness.

Figure 23.5
Different genotypes have different reproductive potential, or fitness. Typically, with dominance of one allele over another, the homozygous dominant and the heterozygote both have the same fitness, denoted by 1, while the homozygous recessive has a lower fitness, $1 - s$. The coefficient of selection, *(s)* is measured by a departure from the frequencies predicted by the Hardy-Weinberg equilibrium.

Similarly, Craig Benkman and Anna Lindholm studied Red Crossbills, finches whose bill tips cross and make an excellent tool for removing the seeds from Western Hemlock cones, on which they thrive (Figure 23.6). They cut off the bill tips of seven captive crossbills (an operation that doesn't hurt the birds since the bills have no nerves) to find out whether this highly

Figure 23.6
Red Crossbills feed very efficiently on the seeds of Western Hemlock because their crossed bill tips make an efficient tool for seed extraction.

specialized adaptation has adaptive value only in its fully developed form. The birds with uncrossed bills could remove seeds from dry, open hemlock cones, but were helpless with closed cones. However, as their tips started to grow back and become slightly crossed, they were able to start extracting seeds from closed cones as before, and the birds became more proficient as their bills grew longer and more crossed. This experiment showed that a mutation in an ancestral population that produced even a slight bit of crossing must have been advantageous, giving those birds superiority in occupying a distinct ecological niche.

23.5 Mutation is the source of all genetic change.

The frequency of an allele can also change by mutation. **Mutation rate** is defined as the probability per individual per generation that a gene locus will mutate, and measurements in organisms ranging from bacteria to mice show that these rates are very small, usually 10^{-5}–10^{-8}. The highest rate in this range would mean that every time 100,000 copies of a gene are replicated, one of them on the average will mutate. All genes are not equally likely to mutate because they are of different sizes, but let's assume an average mutation rate of 10^{-5}; then, if the human genome contains 100,000 genes, one mutation would occur every time it is replicated to produce a new sperm or egg, and each of us would carry, on the average, one or two new mutations. (Mutations would also occur when our somatic, or body, cells reproduce; this type may be significant in disorders such as cancer, but they do not contribute to the gene pool or to evolution of the species.)

Because mutations occur so infrequently, persistent mutation to a certain allele can only change its frequency quite slowly. In microorganisms that reproduce rapidly, such as bacteria, mutation can make a significant contribution to a population, though this contribution may only become apparent after some selective agent has taken effect. Every time a bacterial cell reproduces, there is a small chance that one daughter cell will become resistant to an antibiotic, and such resistant mutants may accumulate; the existence of these mutants only becomes significant, however, if the population is subjected to the antibiotic, which selects them by wiping out all the sensitive cells. On the other hand, mutation is a two-way phenomenon. If a mutation is only a minor change in a DNA molecule, such as replacement of a G–C pair by an A–T pair, then *back-mutation* or *reversion* to the original state will occur at approximately the same rate as a forward mutation, and the overall effect on a population will be minimal. As this graph shows, an allele may simultaneously tend to increase in frequency by mutation and decrease by selection:

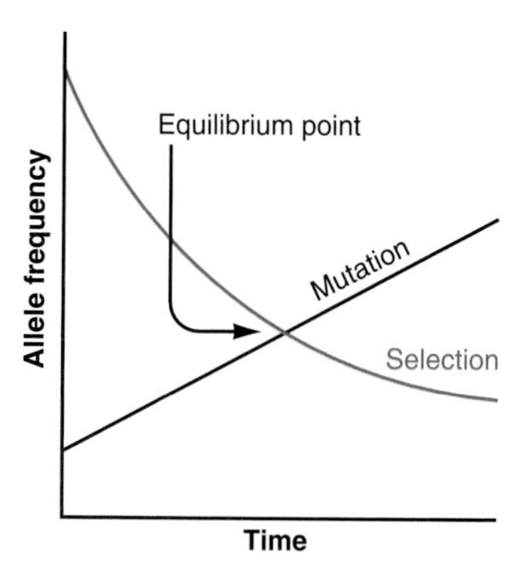

The frequency of an allele reaches an equilibrium level where the lines cross, determined jointly by the rates of selection and mutation; the situation is analogous to a leaky tub of water that is filled from a faucet, so its level is determined by the rates of inflow and outflow.

In spite of these considerations, *mutation is ultimately the source of all genetic variation and therefore the foundation for evolution.* A single mutation increases genetic diversity and thus increases the potential for future evolution, but by itself it is most likely useless or detrimental. An organism, after all, is a complex and finely tuned system, and a random change in its genome is not likely to improve it. Still, we showed in Section 4.14 that a mutation can make a very subtle change in protein structure, and organisms bearing different alleles, making slightly different proteins, bear the potential for evolutionary innovation.

Now we must return to the themes of *sexual reproduction* and *recombination* raised in Chapter 15. Sexual reproduction creates different combinations of alleles, and the combination of two or more genetic novelties creates the greatest potential for evolution (Figure 23.7). Furthermore, an important source

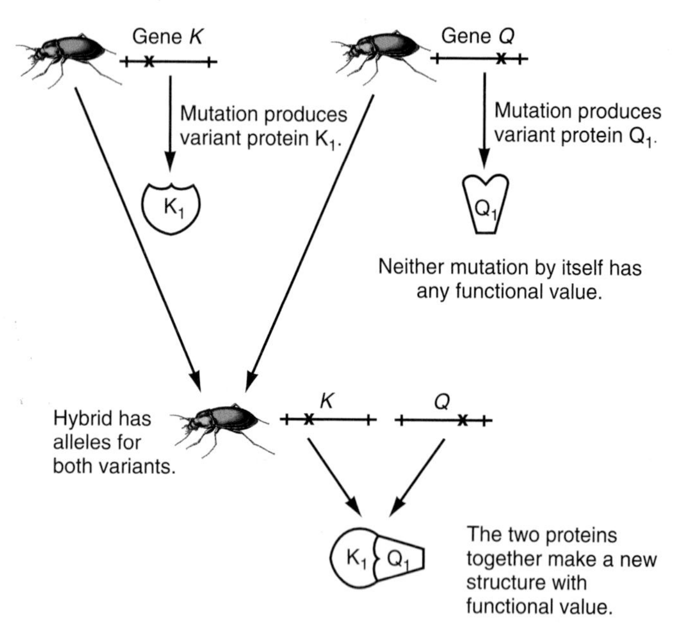

Figure 23.7
Genetic novelties in different organisms may be useless or detrimental, but when combined through sexual reproduction, they can form either an improved structure or a novel structure with a new function.

Concepts 23.1 Gene Duplication and Evolution

Evolution occurs through experimentation, and organisms have a greater chance of success as random events lead them to engage in more genetic experiments. One common pattern of evolution, a strategy that produces more genes for experimentation, is *duplication and diversification.* Gene duplications must be quite common, and the illustration shows how a genome can duplicate a gene and then diversify one of the copies. One copy of the gene is kept in essentially its original form by selection, because it encodes a protein with an important function. But the forces of evolution can play with the other copy. Mutation after mutation can accumulate in this copy, generally without hurting the organism that carries it. The mutant protein encoded by this gene may be quite useless, but it may also evolve to take on a totally new function, and it will start to undergo its own selective editing to shape it for that function.

Evidence for this process of evolution comes from comparing the structures of many proteins. For instance, vertebrate myoglobin and the various hemoglobins were clearly derived in this way; comparisons of their sequences suggest that they diverged as shown in Figure 5.10. Furthermore, sequencing the DNA of the hemoglobin genes has revealed *pseudogenes,* which do not produce functional proteins but may be in the process of evolution. Comparisons of proteins with quite different functions show that they must have arisen through duplication and divergence. Look at the structures of the hormones in Figure 47.14. The homology between glucagon and secretin is particularly striking.

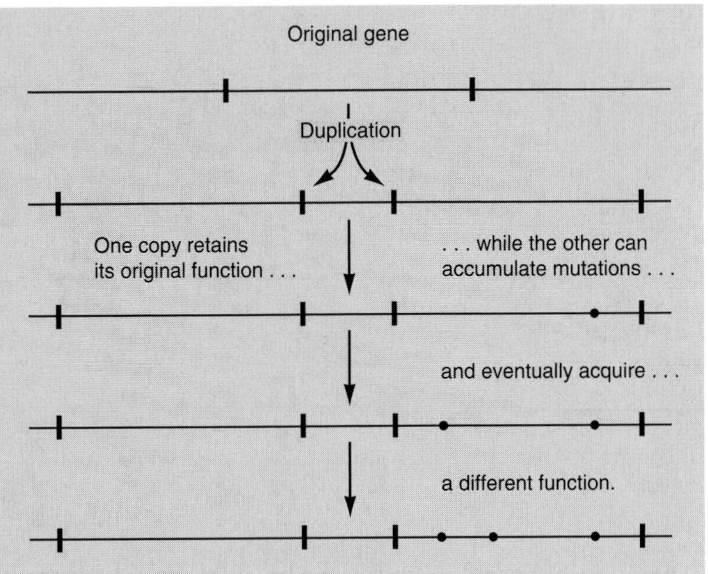

of genetic novelty is an extensive change in chromosome structure, producing a *duplication,* which can then change as shown in Concepts 23.1.

Exercise 23.8 A population of 10,000 beetles consists of 9,000 with two black spots on the back and 1,000 with four black spots. The difference is due to a single gene with two alleles and complete dominance. A year later, there are 9,100 with two spots and only 900 with four spots. Calculate the initial frequencies of the two alleles, the number of individuals expected in the next generation if there were no selection, and the coefficient of selection against the recessive allele.

Exercise 23.9 Blister beetles, which live in the Sonoran Desert of Arizona, exhibit assortative mating. The beetles differ in size, and when they pair off and copulate, tail to tail, they choose partners of similar size. Assuming size is determined genetically and does not simply reflect environmental effects or differences in age, what are some possible effects of this behavior on the genetic structure of the population?

23.6 Gene frequencies may change rapidly in small populations.

The Hardy-Weinberg Principle applies to a population that is large enough for random mating to occur, so the genotypes of the next generation are made in proportion to the allelic frequencies of the current generation. Natural populations may be very small, however, especially those that are somewhat isolated as shown in Figure 23.1*c* and *d,* and small populations can behave very differently from large ones.

If you play a coin-flipping game, in the long run you will turn up as many heads as tails, but in the short run you will probably have streaks in which one side or the other occurs more often than expected. If the coin is honest, no such streak will last forever, but it may last long enough to make the loser quit in discouragement. In a large population, the genetic "coins" are being flipped so much that the Hardy-Weinberg Principle applies: The frequencies of all alleles remain constant, in the absence of mutation and selection. In a small population, as Sewall Wright first pointed out, the gene pool may not behave this way, because each generation is made by only a few flips of a genetic coin. Suppose a population has equal numbers of two alleles for a particular gene locus: $p = q = 0.5$. If only a few sperm and eggs are drawn to make the next generation, they may not be a representative sample, so they deviate from the expected 50 percent of each allele. Perhaps in the next generation the frequencies will change to 0.48 and 0.52. Then in the following generation, since the frequencies are already somewhat lopsided, they may change a little more—say, to 0.46 and 0.54.

This genetic experiment can be done with a computer that simulates the evolution of a small population, starting with equal frequencies of two alleles. Figure 23.8 shows the results of two typical computer runs; the frequencies of the alleles drift rapidly in one direction or the other until, after several generations, one allele is eliminated (its frequency becomes zero) while the other is said to be *fixed* (its frequency becomes 1). In natural populations, such a rapid change in gene frequency is

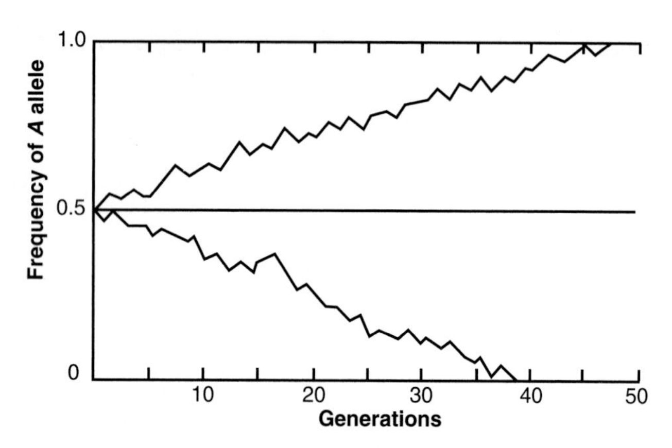

Figure 23.8

Genetic drift can occur in a small population in which, by chance, gametes area combined in a nonrandom way. Two computer runs, using a model of such a population, show how allelic frequencies can drift in either direction.

known as **genetic drift.** Genetic drift may be an important factor in speciation, as we will discuss in Chapter 24.

23.7 Heterozygotes are sometimes more fit than either homozygote.

For a long time, breeders have known that when they cross two strains of a plant or animal that aren't closely related, the resulting hybrid often shows clear superiority to both of its parents in growth, physiology, or general vigor, a phenomenon called **heterosis** or **hybrid vigor.** Inbreeding—generation after generation of breeding among close relatives—produces homozygosity at many gene loci, but each inbred population becomes homozygous for different alleles. A hybrid between distantly related strains, in contrast, is heterozygous at many loci, and heterozygosity often confers superiority.

Why should heterozygotes be superior? One reason is that some enzymes have variants called **isoenzymes,** slightly different forms of a single enzyme. Sometimes one form is best in one tissue and another is best in a different tissue, or each form may be best at a certain temperature or at a certain developmental stage. An inbred line may have only one form of enzyme, but the hybrid gets the ability to make both forms, giving it the biochemical versatility to grow better than either parent. In fact, the heterozygote may be still better off because isoenzymes are frequently multimers, proteins made of two or more subunits:

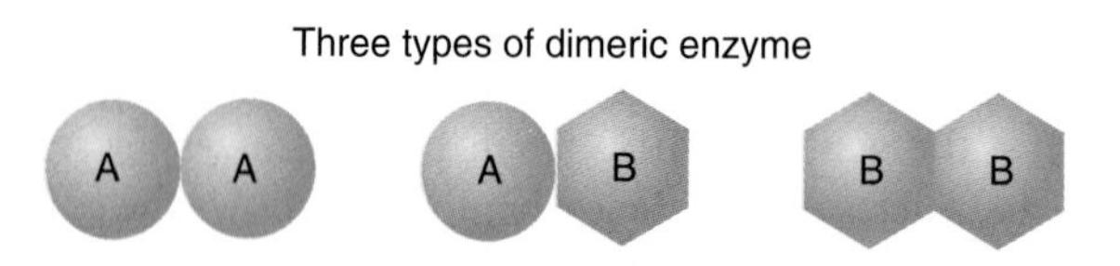

Suppose one strain can only make the A subunit, and therefore only the dimeric enzyme AA. The other strain can only make the enzyme BB. But the heterozygote can make AA, BB, and AB, so it has three forms that may function effectively in different situations.

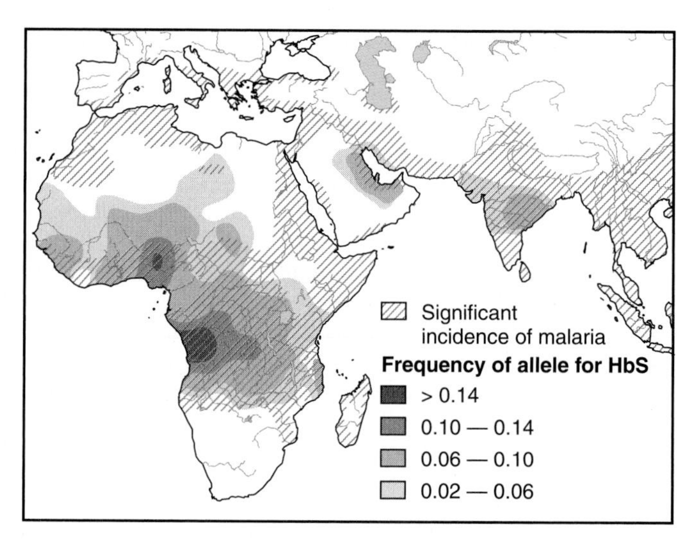

Figure 23.9

Malaria occurs over a number of tropical and subtropical areas in Asia and Africa. Sickle-cell anemia is common over the same general areas because people who are heterozygous for the HbS gene are more resistant to malaria than people with other genotypes.

Sickle-cell anemia in humans provides a classic illustration of heterozygote superiority. The critical mutation, affecting the β chain of hemoglobin, converts Hb A into Hb S. Calling the alleles of the β gene *A* and *S*, the *AA* homozygotes are normal, and the *SS* homozygotes tend to die young of sickle-cell anemia. In tropical regions where malaria is rampant, however, the *S* allele makes *SA* heterozygotes more fit than *AA* homozygotes (Figure 23.9). A. C. Allison found that in some East African populations the frequency of the *S* allele is about 0.2, and the relative fitnesses of genotypes *AA, SA,* and *SS* are 0.8, 1.0, and 0.24, respectively. About three-quarters of the *SS* individuals die before they can reproduce, but heterozygotes have an advantage of 25 percent over *AA* homozygotes.

Hemoglobin, sickle-cell anemia, and malaria, Section 5.12.

23.8 Some regimes of selection are notoriously ineffective.

Looking at a population in which the homozygous recessives have substantially lower fitness, it is tempting to think that the recessive allele will soon disappear because it is being selected against quite strongly. Not so. Allelic frequencies don't change so easily. Take an extreme example: a recessive human disease that is always fatal in childhood, so affected individuals never have a chance to reproduce. The fitness of the recessive allele is zero, and $s = 1$. One would expect the gene to disappear from the population rapidly, but it doesn't. The reason is that most copies of the recessive allele are carried by heterozygotes, who are not affected by it. If $q = 0.01$, for instance, $q^2 = 0.0001$, and only one person in 10,000 shows the disease and is eliminated; but the frequency of carriers is $2pq = 2 \times 0.99 \times 0.01 = 0.02$, so 2 percent of the population carries the deleterious allele.

One can calculate the number of generations (g) required to reduce the frequency of the recessive allele from q_0 to q_1. The relationship is:

$$g = \frac{1}{q_1} - \frac{1}{q_0}$$

If the disease occurs in one person in 10,000, the frequency of the recessive allele is 0.01. How long will it take to reduce q to 0.001 so the disease only appears in one person in a million? The equation says this requires $1/0.001 - 1/0.01 = 1{,}000 - 100 = 900$ generations. At 20–25 years per human generation, it would take a long time to effect this change. To reduce q to 1/10,000 would require another 9,000 generations, and humanity would still not be entirely free of the disease. When the allele frequency falls so low, mutation becomes a significant factor; the human population will never be rid of the allele because it will be created by mutation as fast as it is eliminated by selection.

Population genetics thus shows that the most effective way to eliminate a genetic disease like this from a human population is to identify the carriers and persuade them not to pass on their genes to another generation. In traditional societies where marriages were arranged, parents and matchmakers knew nothing of genetics, but they knew that certain families carried a tendency to have babies with known diseases, so they avoided arranging marriages between those families. Of course, this practice reduced the number of children with the disease but perpetuated the carriers.

Exercise 23.10 A student said, "Now I understand why it is so difficult to eliminate a deleterious gene from a population. Most copies of the gene are carried in heterozygotes, and the malaria example shows that heterozygotes always have a selective advantage over either homozygote. Heterozygosity is really the same thing as heterosis, isn't it?" The student is confused, but it is a common confusion. There are two incorrect statements here; find them and clarify the confusion.

23.9 Every organism is an integrated gene complex.

Discussing the inheritance of individual gene loci and calculating the fitnesses of different genotypes is misleading because it is far too simple. Certain obvious environmental factors, such as malarial parasites, may emphasize the contribution of one gene, but every organism is really a complex formed by at least a few thousand genes that must all function together. Unfortunately, we know too little about how a complex of genes interacts to create intricate structures, but consider the simple matter of a metabolic pathway: It is useless—and maybe even harmful—for an organism to have one enzyme in a pathway unless it has all the others and unless they are all adapted to working well together.

A simple example illustrates the importance of genes working together to make a functional whole. A certain species of plant (which we'll make haploid, for simplicity) has a gene with two alleles, L and l, that determine large or small leaves, respectively. It can also have either heavy or thin stems, as determined by the alleles S or s of another gene. These genes produce four different forms, or morphs. The morphs with large leaves and heavy stems or with small leaves and thin stems are well-adapted plants:

But the morph with large leaves and thin stems keeps falling over of its own weight and rotting, while the one with small leaves and large stems has so little leaf surface for photosynthesis that it can hardly support its own bulk, and it makes few seeds with low viability:

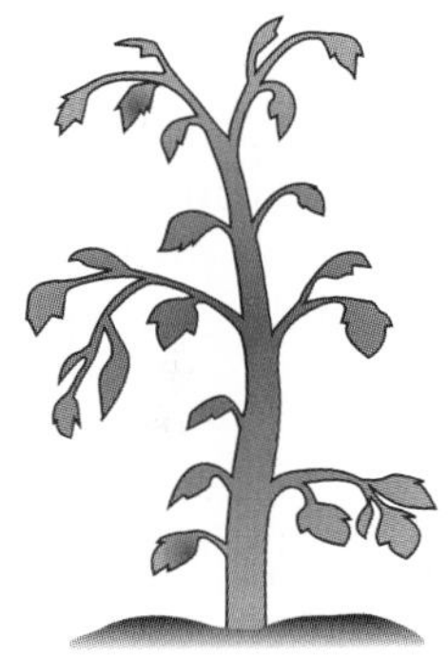

This example shows that not every combination of genes will work. Only certain combinations make the integrated gene complex of a viable organism. Furthermore, a population of these plants tends to evolve in an interesting, predictable direction. If the L/l and S/s loci are located on different chromosomes or are very far apart on the same one, the frequency of recombination between them is about 50 percent:

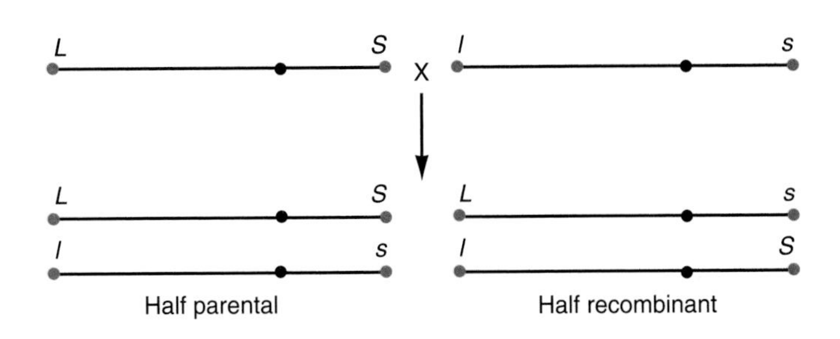

Therefore, in every generation, half the offspring of a healthy plant will be inviable, and the population will lose half its numbers every generation just for genetic reasons. This genetic burden creates a strong selective pressure for any rearrangement of the genome that places these genes close together and only in the combinations *LS* and *ls:*

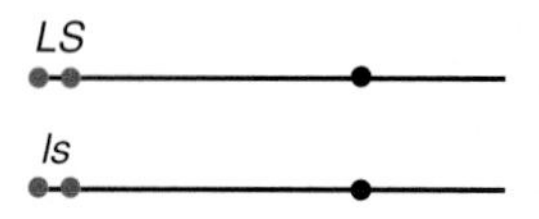

Before we explore the mechanisms that can do this, let's be quite clear about the result: Genes that must function together are brought together on a single chromosome with tight linkage, and are kept together, creating what C. D. Darlington and Kirtley Mather called a **supergene** (Figure 23.10). Successful organisms must create and inherit useful supergenes.

In Chapter 17 we showed that genes with related functions in viruses and bacteria tend to be closely linked. That is partially because tightly linked genes can be regulated together—but even unlinked genes can be regulated coordinately. Tight linkage brings together those genes whose functional products must act together, thus forming coadapted complexes of genes.

23.10 Genes can be rearranged by chromosomal inversions and translocations.

Chromosomes often suffer large structural damages and rearrangements. In contrast to small changes in a DNA sequence, these events remove or add large parts of a chromosome, creating large-scale changes in the linkages of genes. While deletions of portions of the genome are generally deleterious, duplications are also a source of evolutionary novelty (see Concepts 23.1). Two other kinds of changes, inversions and translocations, also have important evolutionary consequences.

A **translocation** occurs when part of a chromosome breaks off and attaches to a nonhomologous chromosome. This brings together genes that were formerly unlinked and breaks up old linkage groups. In a polymorphic population, however, where some individuals carry one gene arrangement and some the other, heterozygotes for the translocation can produce gametes with irregular sets of genes and therefore offspring that are inviable because they get too many copies of some genes and too few of others. The story of a common human translocation between chromosomes 14 and 21 is told in Section 19.6. This translocation causes a hereditary form of Down syndrome, and those who carry the translocation produce many inviable offspring.

In an **inversion,** a segment of a chromosome is cut out and reversed:

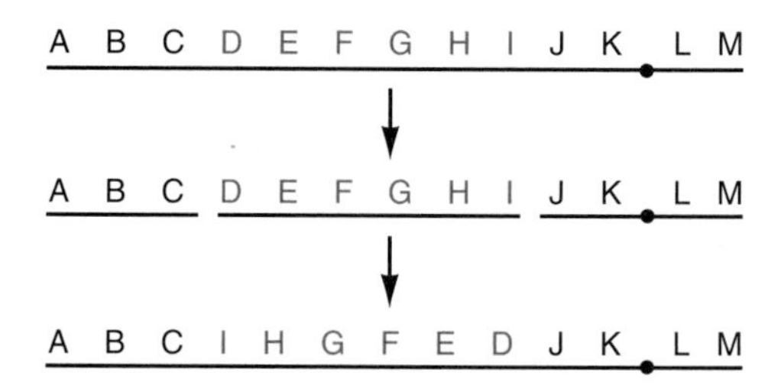

Unless one cut is within a gene, both wild-type (the original type) and inverted chromosomes carry the same information, and both sequences may be equally viable. But inversions inhibit recombination by crossing over. The reasons are a bit complicated, and they depend upon just where the inversion is relative to the centromere of the chromosome; the essential point is that in an inversion heterozygote—an individual carrying homologs with different gene sequences—crossing over between the chromosomes can produce chromosomes with duplicated and missing genes or mere fragments of chromosomes. Gametes that receive such defective chromosomes are inviable, so only chromosomes with the original parental combinations of genes can form viable gametes.

Why are these chromosomal rearrangements important? First, they create new gene combinations. If it is advantageous for two genes to become tightly linked, translocations and inversions are random events that can link them. Genomes with rearrangements that bring together related genes will be selected, since individuals who carry them have higher overall fitness. Second, once a certain combination of genes has been brought together, inversions and translocations tend to keep them together. If a population is polymorphic for an inverted arrangement in a certain chromosome, many individuals will be heterozygous for those arrangements. For the reasons just given, the viable gametes they produce will almost always have only the parental combinations, and only rarely a viable recombinant. The recombination-suppressing effects of inversions make them very important in natural populations.

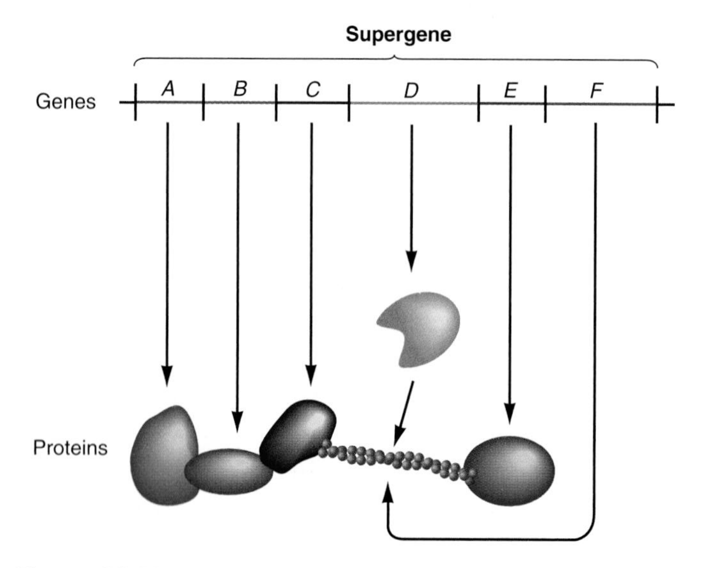

Figure 23.10

A supergene is a complex of genes that are closely linked and work together.

23.11 Fitness changes with changing environmental conditions.

We are now able to picture a natural population as a variable group of individuals that may differ significantly from one another yet have similarities and patterns of interbreeding that make them recognizable as members of a species. Every genotype represented in a population has a certain fitness, but fitness is not an absolute value that will never change. Fitness is measured relative to particular environmental conditions, and the one certainty about any natural environment is that it will eventually change.

Polymorphism gives a population greater potential to maintain itself in spite of conditions that change geographically and over time. Even in a small area, differences in local habitats may require different characteristics, and organisms must also adapt to changing weather conditions and other events. A population shows **balanced polymorphism** when it maintains the genes for two or more forms because selection favors each form in a different situation.

The value of polymorphism in different habitats emerged from a classic study of the British land snail *Cepaea nemoralis* by A. J. Cain and P. M. Sheppard. The snail populations are highly variable, having base colors of yellow, pink, and brown overlaid with various banding patterns and different colors on the shell lip. The base colors are due to three alleles at a single locus, with brown dominant to pink and pink to yellow. Another locus determines banding (the unbanded condition being dominant), and at least one more locus determines the banding pattern. Why all these different forms? Cain and Sheppard showed that they are associated with patches of different habitats. In woods with a carpet of brown leaves, the unbanded brown and unbanded or one-banded pink snails are particularly common, while in hedgerows and rough green areas, the banded yellow snails are abundant. The critical factor in the regional distribution of morphs is their visibility, especially their visibility to the Song Thrush, *Turdus philomelos,* which preys on them. Thrushes bring snails to "thrush anvils," large rocks where they break open the shells to get at the soft body inside. This habit makes it easy to study predation, because the broken shells left around a rock show what kinds of snails have been eaten (Figure 23.11). The thrushes obviously eat the more visible snails in each patch, and visibility changes with time—for instance, as the background changes from winter brown to the green of spring.

Thrushes are clearly a major selective agent in determining the genetic composition of the snail population, but the population survives very well in spite of the thrushes because of their balanced polymorphism. Snails of a single color and pattern

Figure 23.11

The snail *Cepaea nemoralis* has several distinct color morphs, which are camouflaged on different natural backgrounds. Birds such as the Song Thrush hunt the snails and break their shells open on rocks called "thrush anvils," where the fragments of shell show what kinds of snails the birds eat in each habitat.

might survive precariously in a restricted habitat, but that way of life would be dangerous because the habitat patches are small and ephemeral. The species actually adapts to a much broader habitat by producing individuals that are camouflaged against different backgrounds. By maintaining a variable gene pool, the snails buy survival in a varied environment, but they pay a genetic price by producing individuals with the wrong patterns in each habitat; this has been called a *genetic load* the population must bear.

One caveat: Different forms of a gene may confer the same fitness on an organism under the same environmental conditions. Much of the variation in populations is *selectively neutral,* so two or more alleles of a gene may be maintained in a population simply because there is no selection against any of them. Human blood groups such as M and N appear to be selectively neutral, though the ABO types may not be; type O people are slightly more susceptible to stomach and duodenal ulcers than type A, and the reverse is true for stomach cancer. But *every* difference in biological structure does not necessarily make a functional difference. Some of the changes that have occurred in the amino acid sequences of proteins over the eons are too subtle to have any significant effect on protein structure and function; they appear to be maintained in populations at random. However, any of these changes *could* have a distinct effect on fitness if circumstances should change.

One advantage of diploidy is that a population can carry some recessive mutations without harm and that occasionally two such mutations could come together and produce a superior individual, even though either mutation is deleterious when homozygous. This situation at its extreme, called **balanced lethality,** is illustrated by *Drosophila. Curly* (wing) and *plum* (eye) mutations are both lethal by themselves, but the recombinant *curly plum* survives (Figure 23.12). We can't explain such situations because we don't yet understand the developmental process in which these two genes interact. However, it is fair to say that the *curly plum* genotype constitutes a supergene, a set of genes that function well together; in a population carrying the two mutations, we would expect selection for individuals in which the gene loci are closely linked, as in the example of plants with different alleles for stem and leaf sizes.

Exercise 23.11 The Grants' work on Darwin's finches (see Section 23.4) has shown that dry years favor ground finches with the largest, strongest bills, while in very wet years selection favors individuals in the same species with smaller bills, which are best able to open small seeds. Explain what this situation implies about the genetic composition of the finch population.

23.12 Fitness generally changes geographically.

The organisms occupying a certain niche are, by definition, well adapted to that way of life, but the features required for occupying the niche may shift over a species's geographic range. (Some biologists might say that the niche is changing geographically, but if the species continues to play essentially the same role in its ecosystem throughout, we prefer to describe this as a single niche with different genetic requirements.)

In the 1940s, Theodosius Dobzhansky and his colleagues studied genetic polymorphism in wild populations of *Drosophila persimilis* and *D. pseudoobscura* in the southwestern United States and Mexico. After finding that these populations carry many different chromosome inversions, they performed a detailed study of 27 inversion rearrangements in the third chromosome. Figure 23.13 shows the gradients of genotypes that

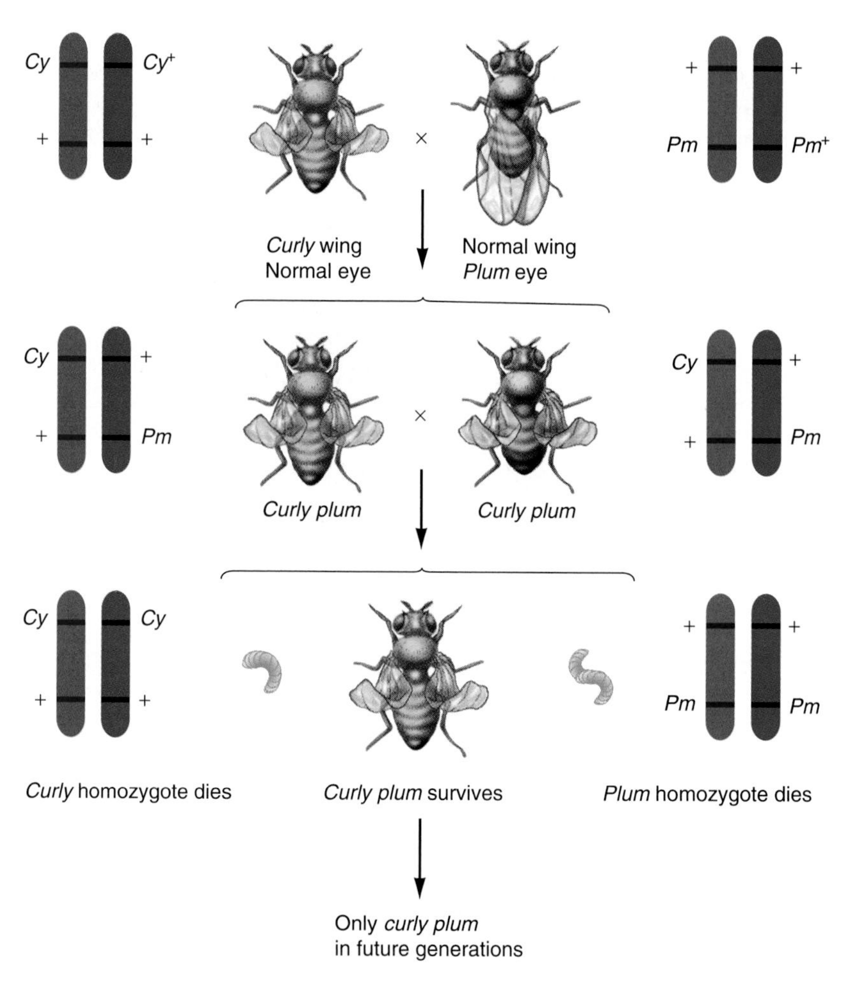

Figure 23.12
The mutations *curly* (wing) and *plum* (eye) are lethal when homozygous, but the *curly plum* heterozygote survives. This balanced lethality is still unexplained, but it is an example of a general phenomenon that makes heterozygotes for certain traits more fit than either homozygote.

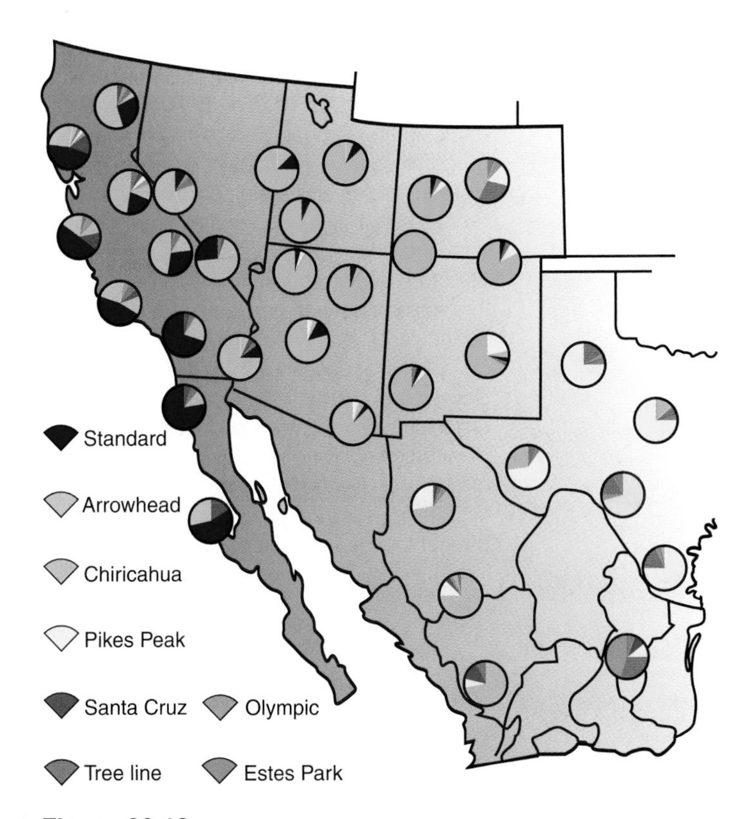

Figure 23.13

The frequencies of several inversions of the third chromosome of *Drosophila pseudoobscura* vary in a regular way across the southwestern United States and Mexico.

Dobzhansky and Carl Epling found. For unknown reasons, the Standard chromosome has a high fitness in California, but Arrowhead is much better adapted farther to the east, Pikes Peak has some advantage still farther east, and Chiricahua provides better adaptation in Central Mexico. The frequencies of these chromosomes change gradually rather than abruptly, presumably because the critical environmental variables also change gradually.

The mere fact that so many inversions coexist in the species is remarkable, and the geographic distribution of inversions shows that each chromosomal difference can make a difference. One might think that any sequence of genes would be as good as any other, but this isn't so. In Section 18.10 we showed how the expression of a gene may be influenced by its position, and the gene arrangements in *Drosophila* are obviously critical. Each type of chromosome is an integrated gene complex, maintained by the suppressing effects that inversions have on recombination, and some types are more valuable than others in each habitat.

As the seasons change, so do the fitnesses of certain gene complexes, as shown by the frequencies of four inversion types of *D. pseudoobscura* at Piñon Flats, California:

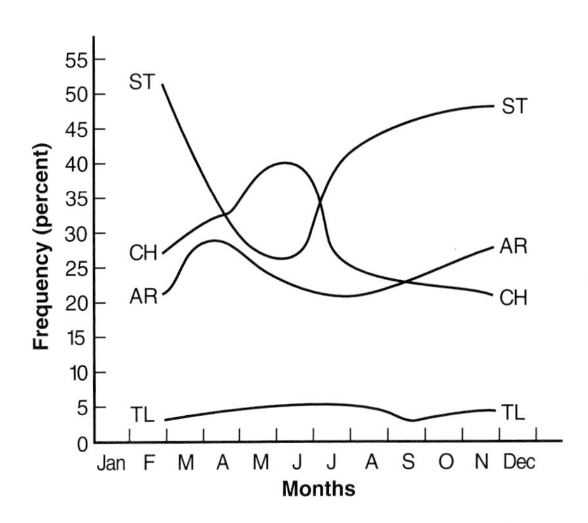

In this graph, ST = Standard, AR = Arrowhead, CH = Chiricahua, and TL = Tree Line. The Standard type is most common overall, but the Arrowhead and Chiricahua gene complexes provide better adaptations during the spring, for their frequencies increase at that time while that of the Standard gene complex decreases. In summer, the frequency of the Standard type rises again as the Arrowhead and Chiricahua types decline. (Some laboratory observations show that Standard is more advantageous in crowded populations, which may develop during the summer.) The relative fitness of each genotype clearly varies with time, and—as with the snails—the population as a whole survives by maintaining several forms that are adapted to different conditions.

These studies often showed the superiority of heterozygotes, with clear indications that flies heterozygous for a pair of inversions were more fit than either homozygote. This increased fitness is another reason populations maintain several different inversions.

John A. Moore's studies of leopard frogs also showed geographical change in phenotype. The frogs are all designated *Rana pipiens* but may actually be distinct species spread across eastern North America. These frogs are clearly adapted to the average temperature where they breed. Moore found that frog eggs taken from the northern part of the range can tolerate a temperature of 5°C and can develop in temperatures up to about 28°C, whereas those from Texas, Florida, and Mexico can tolerate nothing lower than 10–12°C and can develop up to 32–35°C. These differences in temperature tolerance must reflect distinct gene complexes that adapt each population to local temperature conditions.

Distinct features, such as color, size, and other aspects of form, may vary along different geographic gradients. Thus a species might show a cline in size along a north-south line and a cline in color along an east-west line. Each of these differences shows an independent response to a different environmental pressure.

Coda We began our study of biology with the idea that organisms are genetic systems that evolve through variations in their genomes, with the environment selecting out those variants that can reproduce most successfully in a

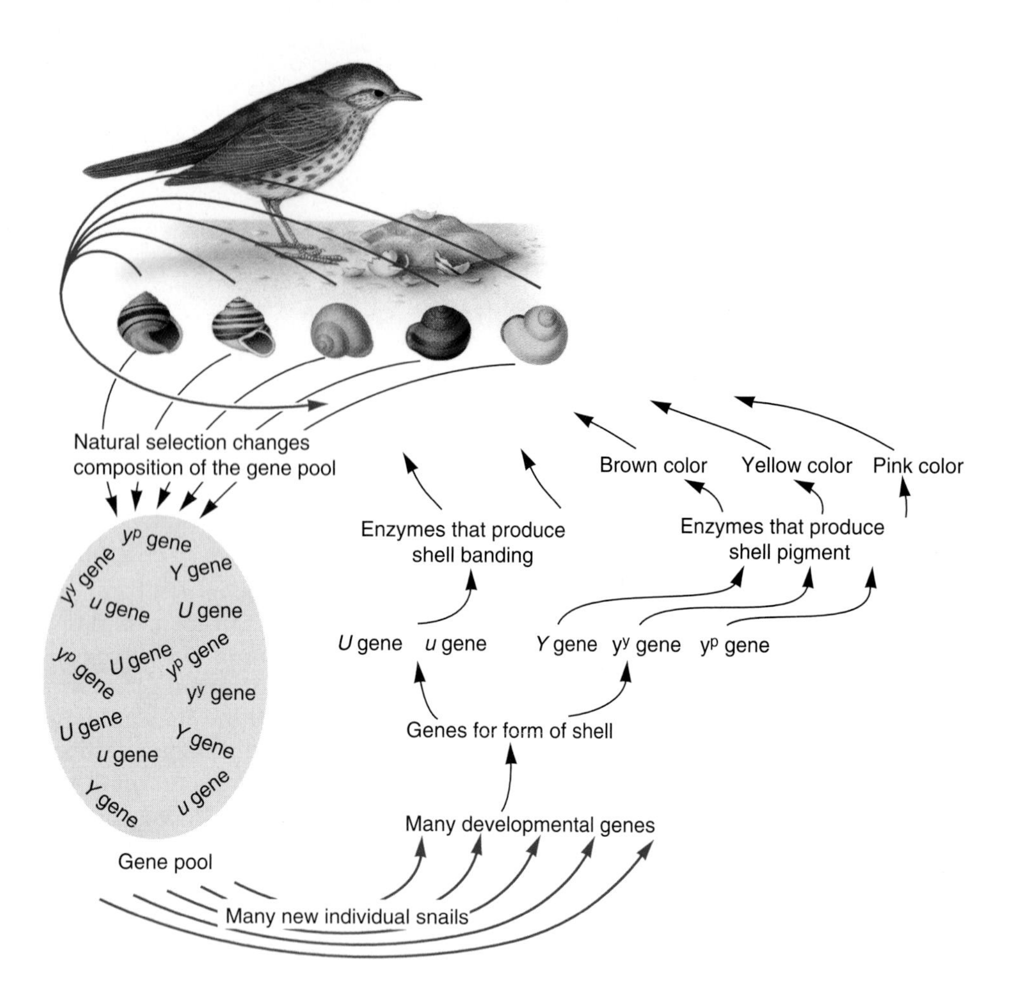

Figure 23.14

Another look at the *Cepaea* snails shows the causal pathways that underlie evolution. Their variation began with mutation and recombination in individuals, creating different alleles and allele combinations. Since each allele encodes a different protein, each individual forms particular protein structure due to specific molecular interactions between molecules. Such molecular differences result in variations in complex processes, such as pathways of development, eventually producing organisms with obvious phenotypic differences such as coloration and patterns of behavior. Because of their interactions with their environment, snails with different phenotypes have different fitnesses.

particular way of life. This chapter amplifies that idea. Evolution occurs fundamentally because genomes can change, and the change translates into effects at various levels. Let's look again at the *Cepaea* snails (Figure 23.14). Their variation arises from mutation and recombination, producing different proteins and thus different pathways of development; the resulting organisms have phenotypic differences. Natural selection then acts upon these distinct individuals as they interact with one another and with their environment. Populations harbor significant amounts of variability, often correlated with environmental factors. Populations vary geographically and with habitat. In the snail populations, we can easily see how forms are correlated with fitness in different environments; in other cases, we may not be able to tell just what environmental factors have selected particular variations (such as the colors of Song Sparrows), but since organisms can only survive if they are well-adapted to their environments, the variation almost surely reflects some adaptive value. Some tools for understanding how a population will change are embedded in population genetics theory, which enables us to calculate the effects of given rates of mutation and selection. An overall view of biological processes, from the molecular to the ecological relationships of an entire organism, ought to inform our thinking about evolution.

Summary

1. All populations are variable because of a combination of genetic and environmental factors. Natural selection, however, acts only on the genetic component of variability.
2. The theory of population genetics is developed for a large, randomly breeding population that is not disturbed by such events as mutation and selection.
3. In an ideal population, if there are two alleles, *A* and *a*, at a gene locus, the Hardy-Weinberg Principle states that the allelic frequencies will not change, and after one generation of mating, the genotypes *AA*, *Aa*, and *aa* will occur in the ratio $p^2 : 2pq : q^2$, where p and q are the frequencies of the *A* and *a* alleles, respectively.
4. The frequencies of alleles can be determined from the frequencies of the genotypes. If an allele shows total dominance so that *AA* and *Aa* individuals are identical, the frequency of the recessive allele is determined from the square root of the fraction of *aa* individuals.
5. Real populations change markedly because of mutation and selection. Since most mutants are at a selective disadvantage, it is common for the frequency of an allele to increase through mutation and to decrease through selection, coming to an equilibrium because of these forces.
6. Simply selecting against an allele by eliminating all homozygotes is very ineffective in eliminating the allele from the population, since most copies of the allele are carried by heterozygotes.
7. In many cases, the heterozygotes are more fit than either homozygote. They may have a hybrid vigor because they combine valuable features

of the homozygotes; for example, they may have a greater variety of isoenzymes that are valuable in different circumstances.

8. Each organism is a highly integrated complex made of many genes that must work together. Genes may be organized through recombination into supergenes, sets of closely linked genes that form functional complexes.
9. Chromosomes may be rearranged by translocations, in which segments move from one chromosome to a nonhomologous chromosome, and by inversions, in which a segment of a chromosome is cut out and inverted. Particular combinations of genes made through inversion are maintained because crossovers between them lead to inviable gametes in heterozygotes.
10. Populations carry variously rearranged chromosomes, and the adaptive value of a chromosome may change considerably in different places and climates, as well as over time.
11. Although every genotype can be assigned a relative fitness, that fitness may shift geographically. Populations in different areas have slightly different genotypes because each is adapted to local conditions.

Key Terms

random mating 465
assortative mating 465
gene pool 466
Hardy-Weinberg Principle 466
allelic frequency 467
fitness 468
coefficient of selection 468
mutation rate 470
genetic drift 472
heterosis 472
hybrid vigor 472
isoenzyme 472
supergene 474
translocation 474
inversion 474
balanced polymorphism 475
balanced lethality 476

Multiple-Choice Questions

1. If a population is in Hardy-Weinberg equilibrium for particular alleles,
   a. the population must be polymorphic at those loci.
   b. 50 percent of the individuals must be heterozygotes.
   c. the values for p and q will remain the same in the next generation.
   d. $p = 0.5$ and $q = 0.5$.
   e. evolutionary change can be expected.
2. When a population is in a state of Hardy-Weinberg equilibrium for a particular gene, which of the following relationships must be true?
   a. $p + q = 1$
   b. $p^2 + 2pq + q^2 = 1$
   c. The ratio of homozygotes to heterozygotes will be 50 : 50.
   d. *a* and *b* but not *c*
   e. *a*, *b*, and *c*
3. If 18 percent of a population in Hardy-Weinberg equilibrium has blood group MN and 81 percent has blood group M, q must equal
   a. 0.18.
   b. 0.81.
   c. 0.09.
   d. 0.01.
   e. 0.1.
4. For this problem, assume the following: Only one pair of alleles determines brown or blond hair color; all offspring of a brown × blond mating have brown hair; in Scandinavia, blonds are in the majority; in Italy, brown hair is more frequent. Which of the following statements is false:
   a. The values for p and q will be different for Scandinavia and Italy.
   b. In Scandinavia, the frequency of the blond allele is p, but in Italy, it is q.
   c. The value of $2pq$ in Italy is not the same as in Scandinavia.
   d. Each population may or may not be in equilibrium at this locus.
   e. If mating is strongly assortative for this locus, selection pressure is strong.
5. The frequency of an allele may remain constant if
   a. the rates of forward mutation and reversion are equal.
   b. the rate of mutation is balanced by selection.
   c. its frequency is fixed as a result of genetic drift.
   d. it is not subject to selection or mutation.
   e. All of the above are correct.
6. Which of the following conditions is the most conducive to a change in frequency of the alleles at one locus?
   a. genetic drift within a small population
   b. heterosis
   c. selection for the heterozygote
   d. a population in Hardy-Weinberg equilibrium
   e. a stable Mendelian population in a stable environment
7. All other things being equal, selection can most rapidly change the frequency of
   a. recessive alleles.
   b. dominant alleles.
   c. duplicated alleles.
   d. mutant alleles.
   e. heterozygotes.
8. A supergene is
   a. a gene that produces a greater amount of its product.
   b. an operator gene.
   c. a regulatory gene.
   d. a tightly linked collection of genes.
   e. a locus that is not polymorphic.
9. Which of the following chromosomal rearrangements is associated with suppression of crossing over during gametogenesis?
   a. deletion
   b. inversion
   c. translocation
   d. *a* and *b*, but not *c*
   e. *b* and *c*, but not *a*
10. When coupled with mutation, which of these chromosomal rearrangements increases the size and variability of the genome?
   a. inversion
   b. translocation
   c. deletion
   d. duplication
   e. None of the above increases the size and variability of the genome.

True-False Questions

Mark each statement true or false, and if false, restate it to make it true.

1. Mendelian populations do not undergo evolutionary change.
2. A Mendelian population is composed of a group of sexually reproducing organisms that share a gene pool and have unobstructed gene flow.
3. If a population is at Hardy-Weinberg equilibrium for one locus, it cannot be undergoing evolution at any other locus.
4. If the frequency of a given allele in a population decreases by 5 percent in one generation, the coefficient of selection against that allele is 0.95.
5. According to the Fundamental Theorem of Natural Selection, a species can change faster if its genetic and environmental variance increases.

6. A species occupying a homogeneous environment is more likely to be polymorphic than a species occupying a heterogeneous environment.
7. Mutation rates are generally so low that mutation by itself cannot provide the driving force for rapid evolutionary change.
8. If the rate of forward mutation (from A_1 to A_2) is twice the rate of back-mutation (from A_2 to A_1), eventually A_2 will be eliminated.
9. Translocations occur when a segment of a chromosome breaks off and attaches to its homolog.
10. Beneficial supergenes that are created by chromosomal inversions are more likely to be maintained within a population than are individual beneficial mutations.

Concept Questions

1. Suppose you study a natural population and try to determine the allelic frequencies for two gene loci, one with complete dominance and one in which the heterozygotes can be identified phenotypically. In which case can you determine the frequencies most surely?
2. Explain why genetic drift can occur in a small population but not in a large one.
3. The case of sickle-cell anemia is often cited as an example of balanced polymorphism with a genetic load. Explain.
4. If every mutation causes a change in the structure of a particular protein or RNA molecule, how is it possible for a mutation to be selectively neutral?
5. Which would you expect to be more polymorphic—a geographically isolated island species or a continental species that inhabits a large range? Explain.

Additional reading

Briggs, D., and S. M. Walters. *Plant Variation and Evolution.* McGraw-Hill, New York, 1969.

Crow, James F. *Basic Concepts in Population, Quantitative and Evolutionary Genetics.* W. H. Freeman, New York, 1986.

Gillespie, John H. *The Causes of Molecular Evolution.* Oxford University Press, New York, 1991.

Kimura, Motoo, and Naoyuki Takahata (eds.). *Population Genetics, Molecular Evolution, and the Neutral Theory: Selected Papers.* University of Chicago Press, Chicago, 1994. A collection of papers about the theory that much of evolution is selectively neutral.

O'Brien, Stephen J., David E. Wildt, and Mitchell Bush. "The Cheetah in Genetic Peril." *Scientific American,* May 1986, p. 84. The world's fastest land animal is in a race for continued survival. An ancient population bottleneck has resulted in genetic uniformity and has made the species extremely vulnerable to ecological change.

Real, Leslie A. (ed.). *Ecological Genetics.* Princeton University Press, Princeton (NJ), 1994. A collection of papers assessing current directions in population (ecological) genetics.

Schierwater, B., et al. (eds.). *Molecular Ecology and Evolution: Approaches and Applications.* Birkhauser, Basel and Boston, 1994. Papers about molecular aspects of population genetics.

Internet Resource

To further explore the content of this chapter, log on to the web site at:

http://www.mhhe.com/biosci/genbio/guttman/

24

Mechanisms of Evolution

Key Concepts

A. Major Features of Evolution

24.1 All evolution occurs within communities.

24.2 Evolution fundamentally depends on selection.

24.3 Speciation and extinction are major features of evolution.

24.4 Different organisms may evolve in quite different patterns.

B. Mechanisms of Speciation

24.5 Speciation occurs primarily through geographic isolation.

24.6 Isolating mechanisms may operate before or after mating.

24.7 Species diverge by taking advantage of new opportunities.

24.8 Both gene flow and selection influence speciation.

24.9 Plant evolution frequently involves hybridization and polyploidy.

C. Aspects of Macroevolution

24.10 The evolution of higher taxa does not require special mechanisms.

24.11 Anatomical differences may evolve through simple, systematic changes in development.

Plants of the genus *Lithops* resemble pebbles.

As you walk across a rocky desert, you must be careful where you place your feet. A close look will show that a particular "rock" is actually a plant like *Lithops,* which escapes being eaten by looking like the pebbles around it, and another might be a grasshopper, *Eremocharis,* looking like a weathered stone fragment as it blends with its surroundings. As you walk through a woods, it is easy to assume that the bark of the trees is just that, yet some pieces may be garden carpet moths, *Xanthorrhoe,* and others hawkmoths, *Xanthopan.* It is hard to see them even in photographs, yet the camouflage only works for *Xanthorrhoe* if it lies horizontally on the tree and only for *Xanthopan* if it lies vertically. That brown piece of a stump might actually be a nightjar or nighthawk sitting absolutely still, and what your eye has passed off as just another brown leaf is actually a well-disguised butterfly. These are only a few of the stories of camouflage and imitation that Julian Huxley tells, out of the innumerable stories of evolution and adaptation that could be told. By now it should be second nature to see these as stories of natural selection, working generation after generation to edit each species into a form suited for survival and more certain reproductive success.

These stories of evolution are also stories of ecology. In the metaphor of the noted ecologist G. Evelyn Hutchinson, the evolutionary play takes place in the ecological theatre, whose stage we set in outline in Chapter 2. Having examined the genetic apparatus and patterns of inheritance, we are now ready to watch the play more closely. We have established the essential idea that genetic diversity stems from mutation and recombination, and that natural selection operates on this diversity. The main additional point we will develop here is that evolution consists largely of speciation through geographic isolation, through the evolution of certain reproductive isolating mechanisms. ■

A. Major features of evolution

24.1 All evolution occurs within communities.

Evolution takes place in populations that already occupy roles in biological communities, and each species changes partly in response to other species. The evolution of two interacting species in a community is called **coevolution;** a flowering plant and the insect or bird that pollinates it are said to have coevolved, because each has adjusted its form, physiology, and behavior to the other, and each has taken advantage of the other so they can coexist (Sidebar 24.1). Interactions between only two species are most easily cataloged (see Section 27.4), but coevolution involves much more than these binary interactions. All evolution is coevolution. Because species belong to complex communities, none can change except in relation to the others. Evolutionary theory is about the changes in a species—and in the whole community—during the long periods over which evolution must be measured.

Charles Darwin originated a story that illustrates the interrelatedness of species in a community. Bumblebees, which collect nectar from flowers with their long tongues, are well adapted to pollinating red clover, so clover proliferates in England because of the bumblebee population. Bumblebee nests are raided by field mice, so where there are more mice, there are fewer bees. Fewer mice live around towns because the townsfolk keep cats that reduce the mouse population, so there are more bees and therefore more clover (Figure 24.1). Furthermore, red clover is the chief food of cattle, and since beef is the staple food of the British navy, the strength of the navy could be attributed to cats or, as Thomas Huxley said, to "old maids," who are the principal keepers of cats.

The story, though partly in jest, is by no means far-fetched. Its human connection makes it especially instructive, because we humans generally act as if we aren't members of any ecological community, pursuing our own interests without regard to their effects on the natural world. But some of those effects—logging, filling wetlands, polluting the air and water—have been disastrous and are likely to drive many species to extinction.

24.2 Evolution fundamentally depends on selection.

As we have already established in previous chapters, the basis for evolution is *genetic variation* coupled with *selection.* Variation comes from mutation, chromosomal rearrangements, and the reassortment of chromosomes during meiosis and fertilization. This genetic variability in populations leads to phenotypic variability, which is the material on which selection operates. Natural selection, generation after generation, allows

Figure 24.1

Bees fertilize clover, but their nests are raided by mice. Cats prey on mice, and so the more cats there are in a region, the more the clover is likely to flourish.

Sidebar 24.1 Pollination and Coevolution

Most modern flowering plants require the transfer of pollen from one flower to another for reproduction. (Self-pollination has usually been selected against, since it reduces the variation in a population.) Some species are simply pollinated by the wind, but many depend on animals, most commonly insects and birds. No animal, of course, goes about pollinating plants out of some kind of unconscious generosity; pollinators seek the nectar produced by special glands (nectaries) in the flower, and they only incidentally carry pollen as they move about.

The pollinator's behavior is guided, in part, by visual features of the flower indicating where the nectaries are, cues that can only be understood by asking what each kind of animal can see. With respect to color, birds have eyes like ours and apparently see much as we do. The eyes of some insects, however, are sensitive to different parts of the spectrum; if dishes of sugar water are placed on backgrounds of different colors, honeybees respond primarily to three colors, called bee's red, bee's blue, and bee's yellow:

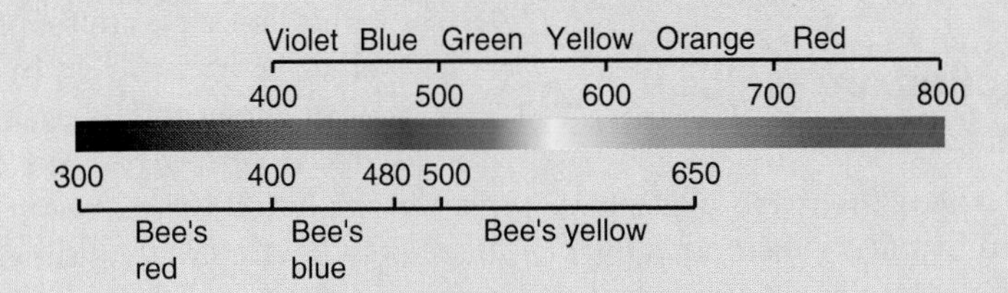

They are apparently unable to see the red that vertebrates see. These facts alone explain a great deal about flower colors. Flowers pollinated by bees tend to be yellow and violet while those pollinated by birds tend to be red and orange. The Americas, where hummingbirds are important pollinators, have many red and orange flowers. In Europe, though, there are no hummingbirds or their equivalent from the Old World tropics, and also few red or orange flowers.

Bee-pollinated flowers often have an interesting pattern: a solid background of purple with a few streaks or spots of yellow, or the opposite combination. These streaks, called *nectar guides,* point to the region an insect must enter to obtain nectar and, just incidentally, to pollinate the flower:

Flowers with nectar guides, pollinated by bees

Some flowers have evolved a form that especially fits their pollinators. The brilliant yellow flower of Scotch Broom, for instance, accommodate bumblebees perfectly. Before the bumblebee enters, the flower has one form, with the features of a set spring; as the insect moves into place, it trips the spring, and the flower changes shape, bringing its long stamens (male pollen-bearing parts) and style (female part) into contact with the bee's back. In this position, the pistil can pick up some pollen deposited by another flower while the stamens deposit more pollen, and as the bee darts from flower to flower, it pollinates the whole population of plants.

These remarkable features of flowers and their pollinators have been shaped by selection over a long time because each species benefits by enhancing the association.

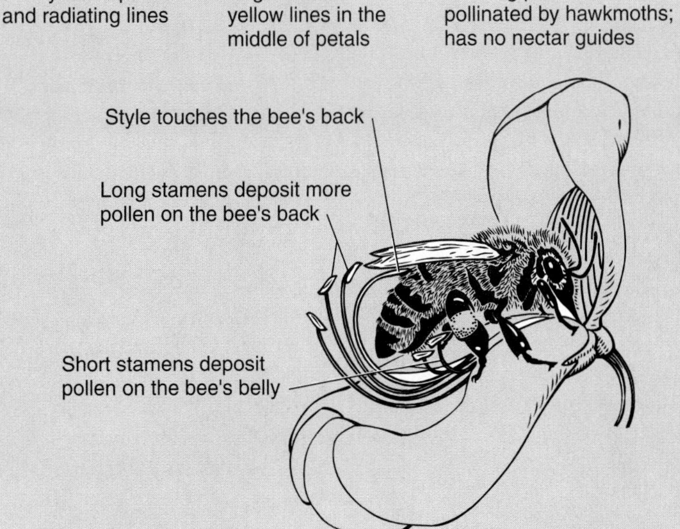

individuals with the most favorable combinations of genes to reproduce and eliminates those with the least favorable. As a result, an organism's structures gradually become shaped by evolution to perform their functions very well.

As George Wald has pointed out, and as the Heike crab story illustrates (see Chapter 2), evolution operates through a kind of *editing* process, rather like refining a piece of writing as it goes through successive drafts. So, too, are organisms shaped into functional forms by small changes generation after generation. Now we must show how putting natural populations into real situations, and especially spreading them geographically across the earth, explains the dominant processes of evolution.

24.3 Speciation and extinction are major features of evolution.

Fossils provide a record of the organisms that were living at each time in the past, as determined by dating methods outlined in Methods 24.1. A century ago, before many fossils had been discovered and while the fossil record was still relatively sparse, evolutionary biologists were misled into thinking that evolution takes the form of simple, straight-line changes in each group of organisms (a pattern they called orthogenesis). Horses, for instance, seemed to have changed steadily from the little, five-toed *Hyracotherium* into the big, one-toed *Equus* (Figure 24.2). But in fact, a larger collection of fossils has shown that horse evolution followed a much more complicated, more haphazard course (Figure 24.3). The phylogenetic tree for horses is typical. It shows three kinds of events. First, one species (one line in the tree) often divides into two or more species, in the process of **speciation.** At the branch point, organisms that had been breeding with one another divide into groups that do not interbreed. Second, most species lines end in **extinction,** the death of the species. Third, a line often changes gradually in one direction, indicating **phyletic evolution,** a consistent change such as a gradual increase in size or a gradual change in the shape of a body part. A fourth feature, **stasis,** is no evolution at all, and is represented by lines that continue straight on until they branch or end.

We outlined speciation in Chapter 22 as a division of one species into two or more. Now we must clear up an ambiguity regarding species of the past. Figure 24.3 includes single lines that represent an apparent change from one species to another, as from *Hippidium* to *Onohippidium* to *Hyperhippidium.* On the face of it, this looks like another kind of speciation, in which one species changes successively by phyletic evolution into forms worthy of different names, even different genera. However, because we have so few relevant fossils, we don't know whether phyletic evolution was actually occurring. It is just as likely that these horses were evolving through repeated episodes in which single species divided again and again, but that the fossils needed to show this are missing. With a rich fossil record, we could probably determine what was happening; however, we would also find it difficult or impossible to draw lines separating these fossils into different species because the horses were changing gradually and continuously, generation after generation. Since the existing fossils preserve only a tiny fraction of the individuals who ever lived, they conveniently divide the continuously changing populations of the past into segments; these artificial species, which replace one another in time, have been called *paleospecies* to distinguish them from contemporaneous species. But we won't describe the formation of paleospecies as "speciation."

If evolution is fundamentally due to natural selection, acting on variation, different patterns of evolution must result

Methods 24.1 Dating the Past

The rocks in which we find fossils were first laid down as sediments, where sand, silt, and clay were gradually deposited in layers, or strata. As the organisms buried in these strata decayed and dissolved, their forms were gradually replaced by minerals. Thus their anatomy was preserved, especially their hard parts such as shells and bones. How can geologists and paleontologists determine the age of an interesting stratum? Strata have been deposited on top of one another, so in an undisturbed landscape the oldest would be the deepest, but rocks aren't laid down at a uniform rate, so there is no simple clock inherent in a rock formation. Furthermore, some of the most ancient rocks have been lifted close to the surface or turned upside down, so the depth of a stratum is no real indication of its age. There must be a more reliable way.

The best methods depend on radioactive decay. Radioactive "parent" isotopes decay into stable "daughter" products, and in a rock, both isotopes are trapped in place. The age of the rock is determined by measuring the amounts of the two isotopes with an instrument called a mass spectrometer. Thus uranium 235 and uranium 238 decay into lead 206 and lead 207, respectively. The younger a rock is, the more uranium and the less lead it will have. The ratio of each uranium isotope to its lead daughter isotope determines its age.

Just as populations grow exponentially, radioisotopes decay exponentially. Each radioisotope has a characteristic half-life, the time required for half of it to decay. ^{14}C, for instance, has a half-life of 5,570 years, so after this time, only half the atoms in a sample of ^{14}C will remain. After 11,140 years, only a quarter will remain, and so on. Suppose N_0 atoms of a radioisotope are present when a rock is deposited. After t years, N atoms will be left, and the rest will have decayed into D atoms of a stable isotope: $N_0 \rightarrow N + D$. This regime of change translates into the decay equation $N = N_0 e^{-kt}$, where k is a known constant for each kind of radioisotope. Since $D = N_0 - N$, a little algebraic manipulation shows that $D/N = e^{kt} - 1$, or $ln(D/N + 1) = kt$, where *ln* means natural logarithm.

Several radioisotopes are used for rocks of different ages. The ages of the most ancient rocks can be measured through the decay of rubidium to strontium, $^{87}Rb \rightarrow {}^{87}Sr$, for which $k = 1.39 \times 10^{-11}$ years^{-1}. You can use this information to calculate the age of rocks for which the ratio $^{87}Sr/^{87}Rb = 0.0247$. (The answer is 1.75 billion years.)

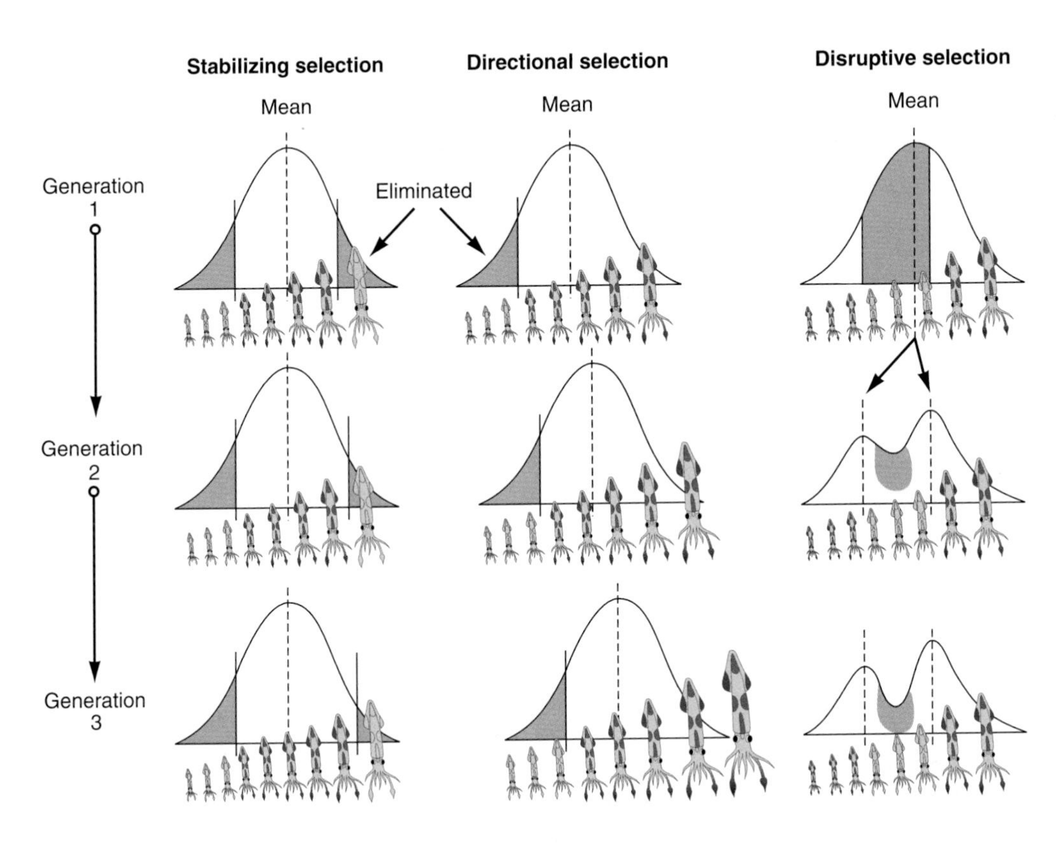

Figure 24.4

Selection can change the genetic variability of a population in three ways. Stabilizing selection keeps a population from changing; directional selection moves the population in one direction; and disruptive selection divides it into two or more populations.

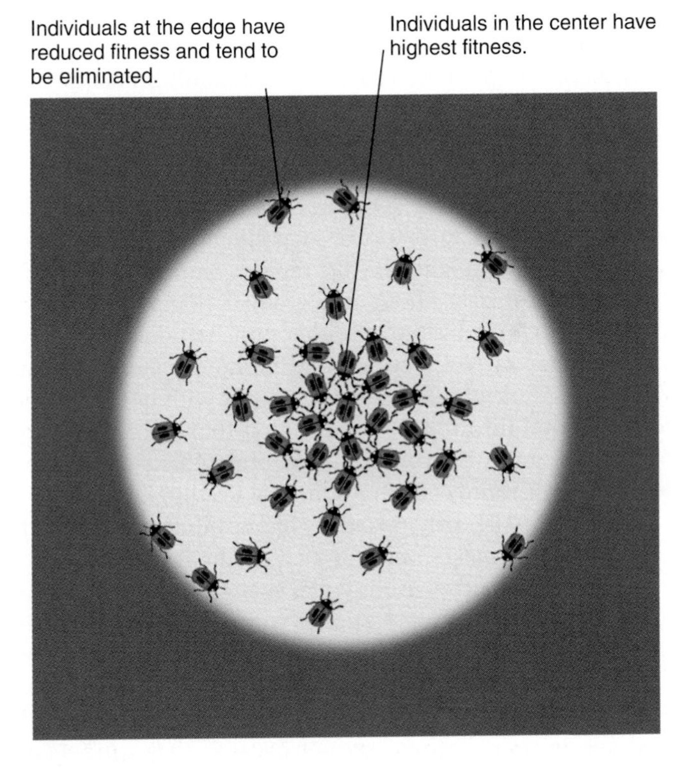

Figure 24.5

A population's niche may be modeled as a pool of light. Organisms that are best adapted to the niche live at the center of the pool of light, those more poorly adapted at its edge.

they are trying to live in the twilight at the edge of the niche, and they tend to be selected out and eliminated.

Phyletic evolution and directional selection

Stabilizing selection probably keeps a species adapted to its ecological niche in a particular environment, as long as that environment is stable. If each species were continually becoming better adapted through natural selection, one might expect the highest extinction rates among species that have lived for only short times; extinction should be less common among older species, which presumably have had the benefits of greater selection. However, Leigh van Valen demonstrated that this is apparently not the case among some groups of invertebrates, which became extinct at a constant rate, regardless of how long they had existed.

If natural selection isn't making organisms better adapted all the time, what is it doing? Van Valen's answer is his **Red Queen hypothesis,** named for the queen in Lewis Carroll's *Through the Looking-Glass* who tells Alice that in her country it takes all the running you can do just to stay in the same place. By this hypothesis, an ever-changing environment constantly challenges each species just to keep pace and remain well adapted. Applying this idea to the model presented in Figure 24.5, the pool of light in which each species lives is moving slowly, and the population living in it continually finds itself struggling genetically to keep up through directional selection. Extinction is a failure to keep up with changing environmental

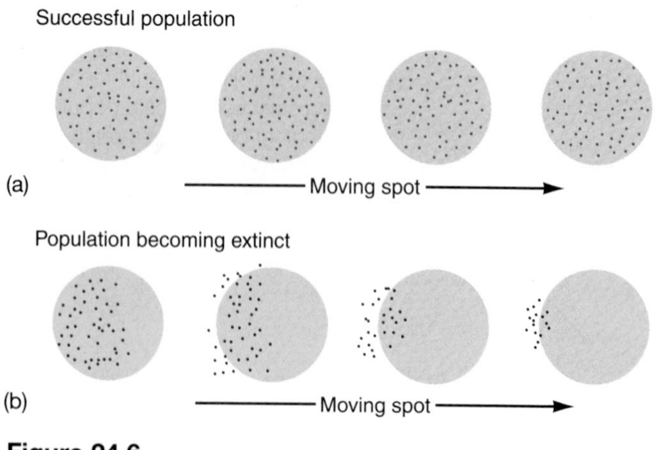

Figure 24.6

As the environment slowly changes, the niche requirements of a species change, and the population must change genetically to keep up with it. Here one population *(a)* keeps up with environmental demands and survives, while another *(b)* cannot change fast enough, falls behind, and becomes extinct.

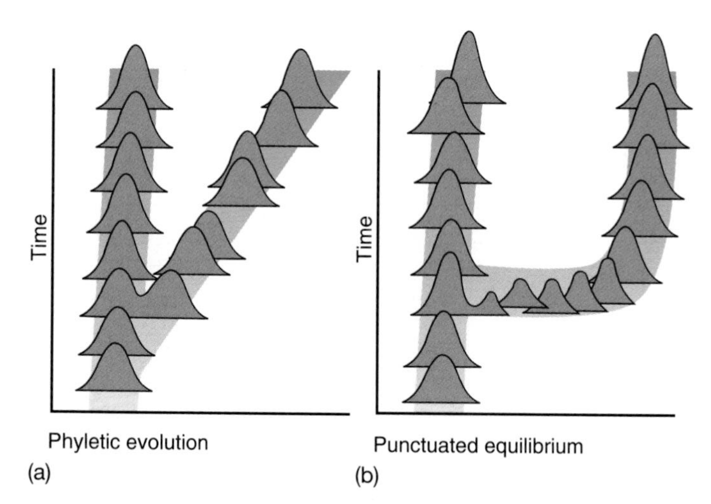

Figure 24.7

Two models have been proposed for the evolution of new forms. *(a)* In phyletic evolution, a species can change gradually. *(b)* In punctuated equilibrium, each species stays the same for a long time and eventually becomes extinct, but it occasionally produces new species in relatively sudden bursts.

conditions, the moving pool of light. As the population falls behind the pool of light, its fitness declines, and eventually the population dies out. This calls up an image of a crowd of insects trying to keep up with the spot of light moving slowly over a wall. They survive only as long as they stay within the spot of light; if they fall behind it, they become extinct (Figure 24.6).

We have seen that selection can only change a population within the limits imposed by its genetic variability. Although mutation and recombination continually add to genetic variability, their effects may not be adaptive, or the resulting changes may happen too slowly.

In summary, the long-term picture of evolution is that novel types continually arise through speciation, as we will describe shortly, and persist for a time, perhaps changing slowly to meet the demands of a changing environment or perhaps remaining very much the same through stabilizing selection. Eventually each species fails to meet the demands of the environment and becomes extinct, freeing up the resources it was using for some other species.

24.4 Different organisms may evolve in quite different patterns.

Over long periods of geological time—10 million to 100 million years—many groups show consistent trends in their evolution: A group of plants becomes larger while their flowers become smaller, or the shells of certain molluscs become more and more intricately coiled. Consistent changes in one direction could result either from phyletic evolution or from repeated speciation. Looking at the fossil record, some students of evolution have held that gradual, phyletic changes do not occur. Rather than seeing a phylogenetic tree with gently curving branches that indicate phyletic changes (Figure 24.7*a*), they see the branches emerging quite suddenly, as geological time goes, existing for a while in a state of stasis (equilibrium) and then becoming extinct (Figure 24.7*b*). In this model, termed **punctuated equilibrium,** evolution consists only of speciation and extinction, with little change in a species between its origin and its extinction. Proponents of punctuated equilibrium maintain that we interpret some evolution as phyletic because we focus on a general trend and don't see the many short branches.

Some paleontologists have provided good evidence for instances of punctuated equilibrium, such as the evolution of the molluscs shown in Figure 24.8. The abrupt speciations shown here occurred mostly in simultaneous bursts associated with changes in the rock formations, suggesting that times of environmental change might provoke rapid evolution. Other series of fossils, in contrast, show only gradual changes, at varying rates but with no indication of either punctuation or stasis. For instance, after carefully evaluating the known hominid (human) fossils, four physical anthropologists with quite different views of hominid evolution concluded that the fossils show continuous, gradual evolution, not punctuated equilibrium (see paper by Cronin et al. in Additional Reading).

So some fossil series apparently fit one model, and some the other—and this makes sense in view of the haphazardness of genetic variation and environmental change. Perhaps in a gradually changing environment, species with the potential to adapt will do so and will show phyletic evolution. Other species, adapted to one set of conditions, may be kept in stasis by stabilizing selection for long periods. If conditions change quickly, some species may become extinct because they can't adapt, while in other species, one population may survive by adapting to the new conditions, diverging in form so much that it is described as a new species (Figure 24.9).

The question of punctuated equilibrium revolves around the claim that speciation occurs too rapidly to result from a

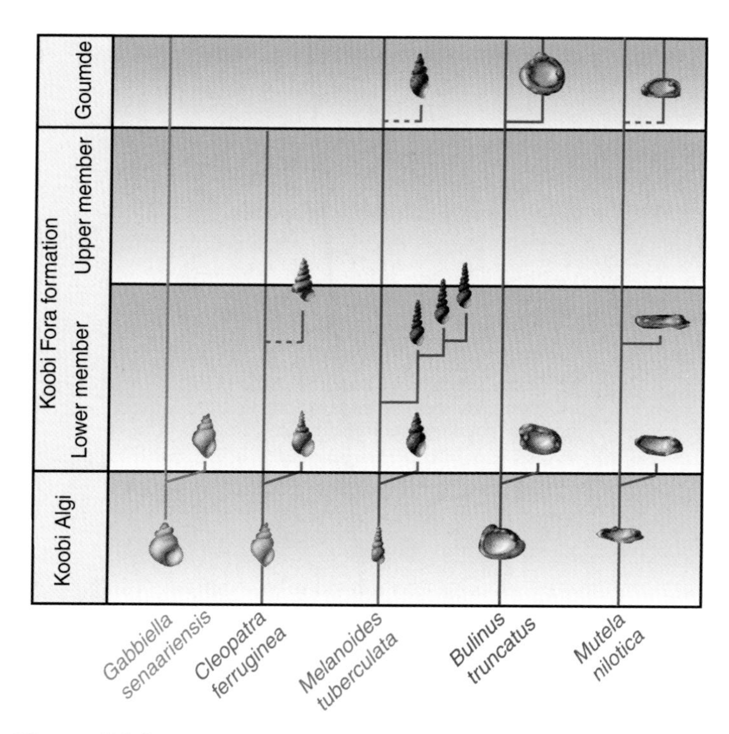

Figure 24.8

Evidence for punctuated equilibrium in molluscs of the Turkana Basin in East Africa. Each vertical line shows the lineage of one species. Dotted lines show speciations that are not well documented. Transitions between rock beds *(horizontal lines)* show changes in the environment,and these tend to be times when several speciation events occur simultaneously, especially during the change from the Koobi Algi to the Koobi Fora formations. The drawings are not all to the same scale. Source: P. G. Williamson, *A Century After Darwin,* edited by J. Maynard Smith, 1995, W. H. Freeman and Company.

slow accumulation of small differences through mutation, recombination, and selection. But just how rapid is "rapid"? The punctuated speciation events in the geological record have actually taken place over about 50,000 years—rapid from the viewpoint of a geologist, but not from the viewpoint of a population biologist. Speciation, as we shall describe it here, could easily happen in this time.

G. L. Stebbins and Francisco J. Ayala made an interesting calculation of how a gradual selection of small genetic changes can add up. As an example of gradual selection, they took the data of W. W. Anderson, who kept populations of *Drosophila* at different temperatures (16°C and 27°C); he found that after 12 years, those kept at the lower temperature were about 10 percent larger than the others. The flies changed slowly, at a rate of only about 8×10^{-4} of the average size per generation. (In other words, each generation, the flies increased by less than 1/1,000 of their size.) Stebbins and Ayala applied these data to one of the most spectacular changes in the fossil record, the increase in human brain size from approximately 900 cm^3 500,000 years ago to 1,400 cm^3 75,000 years ago. Could it have been accomplished in 425,000 years through gradual changes, at the rate at which the *Drosophila* populations changed in size? In fact, at that rate it would have taken only 540 generations, or

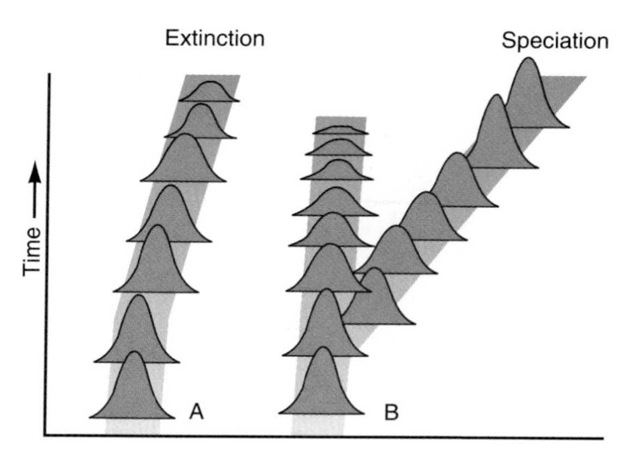

Figure 24.9

Organisms can respond in at least two ways to rapidly changing conditions. Species A cannot adapt and eventually becomes extinct. In species B, one population has the genetic variation necessary for adaptation, and it survives; it is different enough from the original B to be called a new species.

about 13,500 years! So there is good reason to think that the accumulation of small differences can account for both phyletic evolution and the apparently rapid changes of speciation, at least in some cases.

B. Mechanisms of speciation

24.5 Speciation occurs primarily through geographic isolation.

Among the most common birds of the North American woods are Northern Flickers, large, brown woodpeckers that feed on the ground and, when startled, fly off with a roller-coaster motion, flashing a patch of white rump feathers. If you live in the East, you will see flickers with bright yellow feathers under their wings and tails, and black mustaches adorning the faces of the males; in the West, your flickers will have salmon-red feathers and red mustaches (Figure 24.10). These birds were once called two different species, the yellow-shafted and the red-shafted flickers. However, the two types are not really so distinct, for intermediate forms occur quite often, especially in the middle of the continent: birds with orange feathers, sometimes sporting one red and one black mustache, or with some other combination of features. In other words, **gene flow,** the passage of genes from one population to another, occurs regularly among the flickers. The red-shafted and yellow-shafted forms are not distinct species but geographic variants of a single species.

In fact, species like the flickers that show great geographic variation provide the best evidence for the generally accepted view that speciation occurs largely through **geographic isolation** (Figure 24.11). Speciation begins with a geographically variable species in which a population becomes isolated. The population may cross an existing barrier, as by moving from the mainland to an island, or barriers may arise between

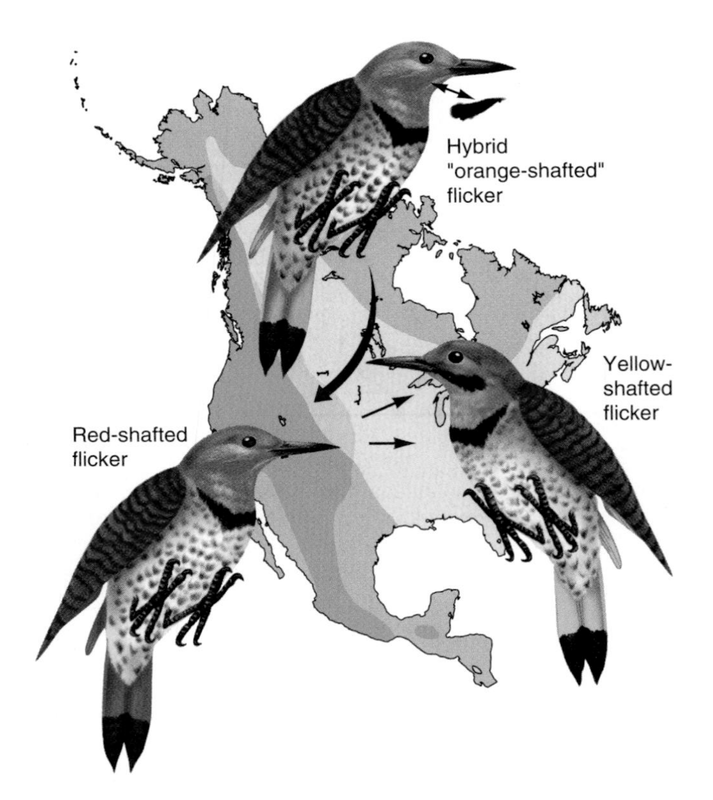

Figure 24.10

The Northern Flickers are divided into two large subspecies, called yellow-shafted and red-shafted, which occupy essentially separate ranges except for some overlap through the Great Plains. Hybrids with intermediate features, such as orange feathers, occur where the subspecies meet.

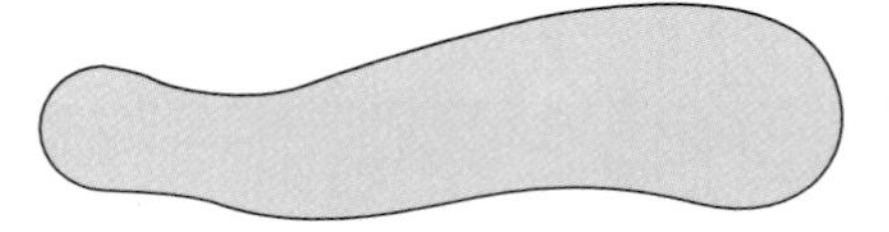

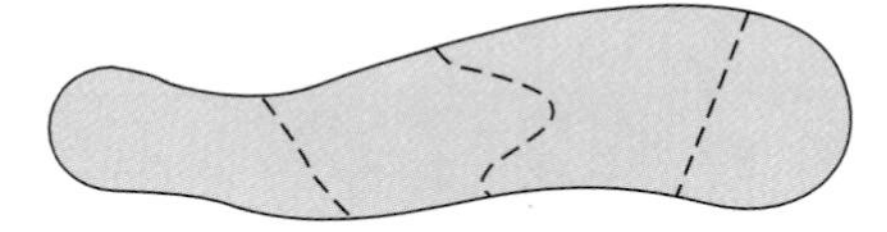

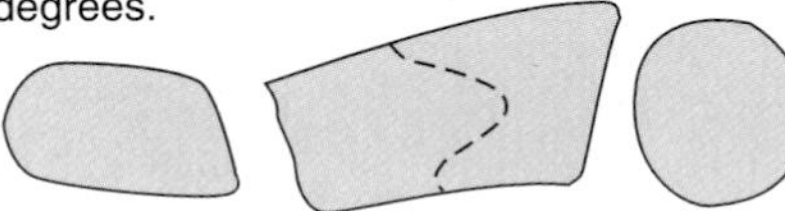

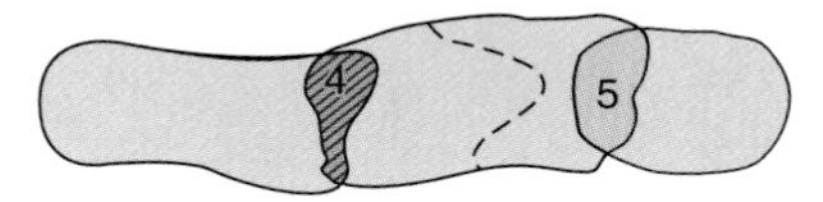

Figure 24.11

(1–3) Geographic speciation occurs as a species with a large range becomes divided into isolated populations. When the populations come together again, they may remain distinct, reproductively isolated species *(4),* or they may interbreed again *(5),* showing that they are still subspecies.

populations; a period of glaciation, for instance, might isolate them in pockets. While separated, members of the isolated populations undergo changes that make them distinct species. This model comes primarily from studies of animal species, especially birds and insects, and was developed most clearly by the ornithologist Ernst Mayr. While the same general processes presumably operate in all sexually reproducing species, some special forces operate in plant speciation, as discussed in Section 24.9.

We showed in Chapter 22 that geographic variation is common; as a species spreads out across a wide range, its populations may acquire many differences and perhaps become so distinct that they are called distinct species. By the biological species concept, however, morphological differences are not the defining feature of a species. Two populations are only separate species if they are *reproductively isolated* and do not interbreed with each other. It is only meaningful, of course, to talk about reproductive isolation if the populations in question are positioned so they *could* interbreed. If the ranges of two populations overlap, the populations are **sympatric** (*sym-* = together; Greek, *patra* = fatherland, thus homeland); if their ranges are adjacent, the populations are **parapatric** (*para-* = next to); if their ranges are separate, the populations are **allopatric** (*allo-* = different):

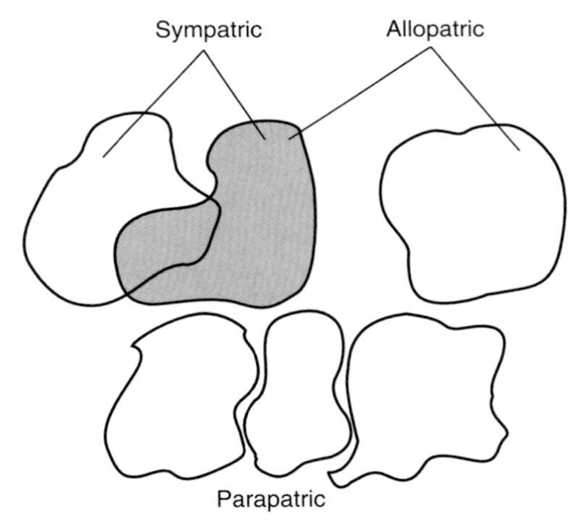

Reproductive isolation can only be tested with populations that are sympatric or at least parapatric; with allopatric populations, there is no way to ask the question.

Closely related allopatric populations, however, may be in the process of becoming different species. While populations are separated by geographic barriers, such as mountain ranges and bodies of water, they acquire significant morphological differences and also develop **reproductive isolating mechanisms,**

Concepts 24.1 Species Confusables

Species: As used here, a group encompassing all the individuals that are actually or potentially capable of breeding with one another.

Subspecies or **geographic race:** A geographically limited portion of a species that is morphologically distinct from other portions of the species.

Superspecies: A group of closely related species that are largely or entirely allopatric, so their ability to interbreed with one another cannot be determined.

Semispecies or **allospecies:** The kind of species that comprises a superspecies.

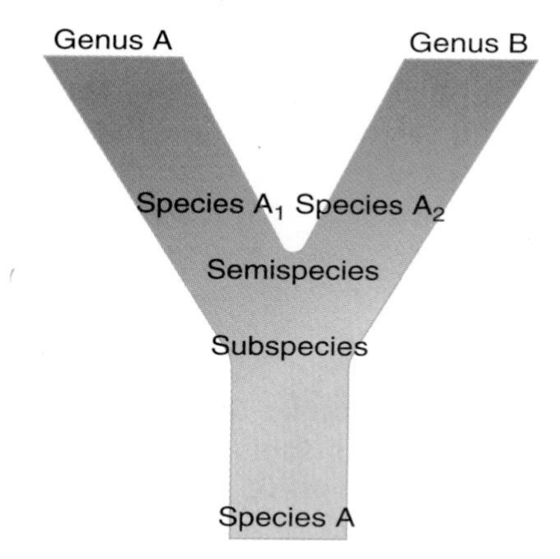

Figure 24.12
In the process of speciation, a single species divides into two or more species. At the intermediate semispecies stage, the taxonomic status of the groups will be uncertain.

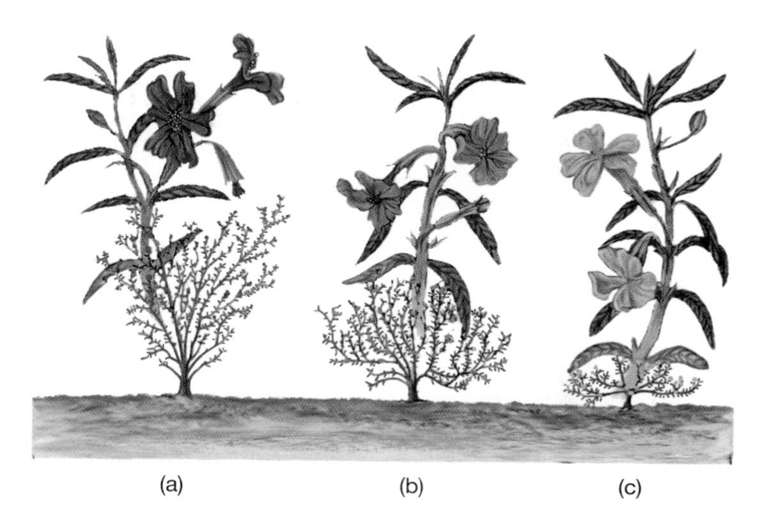

Figure 24.13
Monkey flowers of the genus *Diplacus* show complicated patterns of hybridization or lack of hybridization. In some areas, they appear to be distinct species; in other areas, they appear to interbreed and to be variants of a single species.

features that prevent interbreeding if the populations eventually come back into contact. No matter how different the populations are morphologically in this allopatric stage, their species status is indeterminate because we cannot tell if they interbreed. The populations are best called **semispecies** or **allospecies,** and the whole group of semispecies is a **superspecies** (Concepts 24.1). After an extended period of isolation, the semispecies may start to expand their ranges and become partially sympatric again. Then their species status can be determined. If they still interbreed, they are subspecies of one large species; if they do not, they are separate species. The Northern Flickers are at this intermediate stage in evolution, and the two types are still subspecies.

Speciation happens slowly, and as populations diverge from one another they will go through biologically interesting stages that are hard to classify. At first, they may still exchange genes and must be considered subspecies; later perhaps, they become semispecies, and finally species (Figure 24.12). But these terms describe arbitrary stages in a continuous and complex process.

It is impossible to predict just how populations will change as they spread into different areas. The flickers present a simple case of an intermediate stage of speciation. But some situations are not so simple. Traveling from the southern California coast up into the mountains, you will encounter several kinds of plants known as monkey flowers, genus *Diplacus.* Many of them are easy to identify and name with the aid of a field guide, but one group may give you some problems. The moist coastal areas support tall, red-flowered shrubs named *Diplacus puniceus* (Figure 24.13*a*). The drier foothills are home to shorter, bushier plants that usually have orange flowers, called *Diplacus longiflorus* (Figure 24.13*b*). Still higher, in the very dry mountains, are very short plants with yellow flowers, named *Diplacus calycinus* (Figure 24.13*c*). As the names imply, these plants are considered three distinct species. However, over a wide range, *puniceus* and *longiflorus* grade into one another, so they appear to be mere varieties of a single species, and populations of plants with the form of *longiflorus* can be found with flowers that are red, orange, or yellow. Yet there are places in the Santa Ana Mountains where *puniceus* and *longiflorus* grow together and are clearly distinct, with no interbreeding. Are these distinct species or not? Complicated relationships between populations may develop during evolution, and different degrees of separation may evolve in different places. The difficulty of applying names to such organisms is quite unimportant compared to the lessons they hold about evolution.

The monkey flowers and other complex situations reinforce the point we made in Section 22.9: The biological species concept, with its criterion of reproductive isolation, may run into difficulties because populations can change in complicated ways.

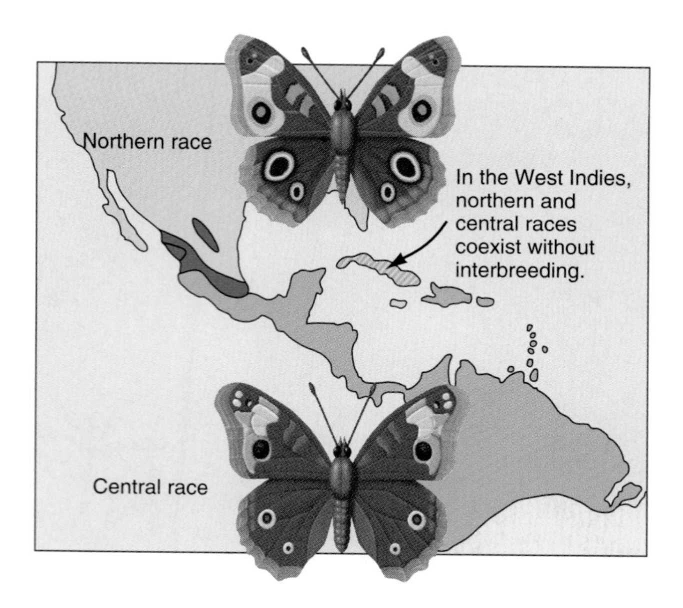

Figure 24.14

Adjacent populations of the Buckeye butterfly *Junonia lavinia* interbreed throughout their range, and the Northern and Central races interbreed in Mexico. However, the races invaded the West Indies from the north and the south, where they remain reproductively isolated, showing that they diverged genetically as they spread out geographically.

Sometimes a species cannot be delimited neatly because we are trying to apply a simple, idealized concept where it may not fit. The Buckeye *(Junonia lavinia),* a common American butterfly (Figure 24.14), illustrates the difficult situation of a ring of populations. Buckeye populations interbreed with one another around the Gulf Coast from Florida through Texas, Mexico, and Central and South America as the members of a species should. But where the ring closes in the West Indies, populations from Florida meet those from South America, and they will not mate with each other. Suddenly they are different. We don't really know how to apply names to these butterflies. It is as if a worm had curled around, met its own tail, and treated it as something foreign.

Two newly formed species can only become sympatric again if they come to occupy niches that do not overlap strongly. When two species that were formerly one come into contact again, they will compete for resources. (The whole matter of interspecific competition is discussed in Sections 27.5 and 27.6.) One natural response to competition is niche differentiation, in which competing species reduce competition by adapting to slightly different niches. Niche differentiation occasionally takes the form of **character displacement,** which means that two species are more dissimilar where they are sympatric than where they are allopatric. For example, two species of Asiatic nuthatches, *Sitta neumayer* and *S. tephronata,* have very similar plumage markings and bill sizes where they occur alone, but where their ranges overlap, their plumage patterns and bill sizes are quite different (Figure 24.15). The plumage differences

Figure 24.15

Where the nuthatches *Sitta neumayer* and *S. tephronata* meet in central Asia, their bills are quite different in length, showing specialization for eating food of different sizes. Where they occupy separate parts of their ranges, their bills are the same length.

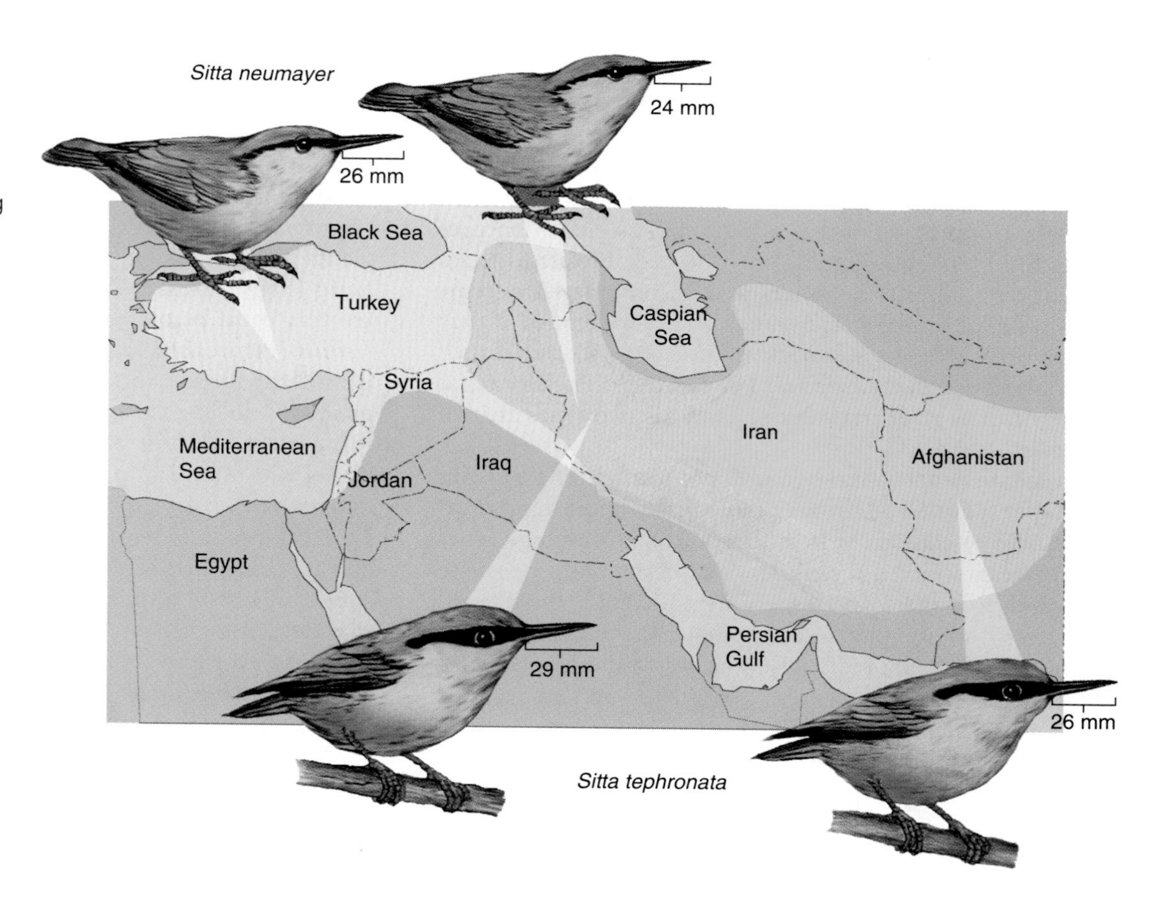

are most likely related to reproduction, enabling individuals to more surely recognize mates of their own species. The different bill lengths are adaptations for dividing food resources, since the size of a bird's bill fits the size of its food.

In the next sections we will examine two kinds of mechanisms that play important roles in speciation: those that operate in isolated populations to permit more rapid evolution into new forms and those that keep populations from interbreeding when they come into contact.

24.6 Isolating mechanisms may operate before or after mating.

Reproductive isolation means that two species cannot or do not produce viable offspring together—"under natural conditions." Some species that are perfectly well isolated in their natural habitats will occasionally produce viable hybrids in captivity, such as "ligers," and "tigons," which are hybrids between a lion and a tiger. In the wild, however, different patterns of reproductive behavior generally prevent hybridization.

Sexual reproduction can be very complicated (see Chapter 51), and incompatibility at any stage could result in reproductive isolation. Such mechanisms operate either before mating occurs or afterward. The difference is important. If two individuals mate and produce zygotes that die or reproduce poorly, they have both wasted much of their reproductive potential, perhaps all of it. So there is strong selective pressure to stop hybridization before mating can occur. The following are reproductive isolating mechanisms; the first four take place at the premating stage.

1. *Habitat isolation.* Two species may occupy such different habitats that they don't come into contact and thus never have a chance to hybridize. The Red Oak *(Quercus coccinea)* is adapted to swamps and wet bottomlands; the Black Oak *(Quercus velutina)* lives in drier, well-drained upland regions. Hybrids between these species are sometimes found in intermediate habitats, showing that they do not have physiological incompatibilities to interbreeding and are only kept apart by their ecological specialization.
2. *Temporal isolation.* Two species may breed at different times, so there is little chance of hybridization. Several cases are known in which two closely related types of plants release their pollen at different times, so there is little or no chance of cross-fertilization of one by the other.
3. *Behavioral isolation.* The elaborate courting and mating rituals of many animals (discussed in Chapter 51) help ensure that the wrong individuals don't mate. These behaviors are genetically encoded, so the final act of mating only occurs between individuals who share genes that give them compatible behavior patterns.
4. *Structural isolation.* The reproductive structures of two species may be incompatible. The genitals of animals may not fit together properly so a male can't transfer his sperm effectively to a female of the other species. In plants whose flowers are pollinated by animals, two species of plants may become isolated by acquiring different flower colors that attract different pollinators, or the flowers may develop different shapes, so they become specialized for pollination by different species of insects.

The next four mechanisms operate after mating.

5. *Gametic incompatibility.* The gametes of the species may fail to function together. For instance, the sperm of one species may not be able to fertilize the egg of another.
6. *Hybrid inviability.* A hybrid zygote may die because it is weak or deficient in some other way. A zygote with two different chromosome sets may fail to go through mitosis properly, or the developing embryo may receive incompatible instructions from the genetic programs of the maternal and paternal chromosomes, so it eventually aborts.
7. *Hybrid sterility.* The hybrid may develop into a sterile adult. Sterility generally results from complications in meiosis whereby different chromosome sets are unable to synapse with each other and form viable gametes. Recall that in a heterozygote for different inversions, the inevitable crossovers will give half the gametes improper chromosome sets. Translocations can lead to similar difficulties. The chromosomes of even closely related species may be different enough to make meiosis in a hybrid very difficult.
8. *Hybrid disadvantage.* Even if the hybrids are viable, their offspring may be inviable or have much lower fitness than the nonhybrid offspring of each species alone. Hybrids are often at a disadvantage because each species is adapted (or is becoming adapted) to a different way of life, so hybrids, with a mixture of gene complexes, generally are not well adapted to either way of life. Imagine two populations that are incompletely isolated and are becoming adapted to different niches. In each population some individuals ("hybridizers") have genes that tend to promote mating with the other population (Figure 24.16). If intermediates between the two populations have reduced fitness because they are not well adapted to either niche, the hybridizers will be putting their genes into a reproductive dead end, and genes that dispose their carriers to interbreed will be gradually eliminated from both species. The genes that remain in each population will tend to discourage interbreeding. Exceptions to this tendency, however, are quite common among plants, and successful hybridization often leads to the formation of new plant species (see Section 24.9).

Exercise 24.1 K. F. Koopman kept cages with a mixed population of *Drosophila persimilis* and *D. pseudoobscura* at 16°C. These species normally do not hybridize, but at a lower temperature the sexual barrier is reduced, and some hybrids (identified in this experiment by genetic markers) do occur. Koopman artificially gave these hybrids a fitness of zero by removing them, and he found that over a few generations the number of hybrid flies decreased sharply, practically to zero. Explain what happened in this experiment.

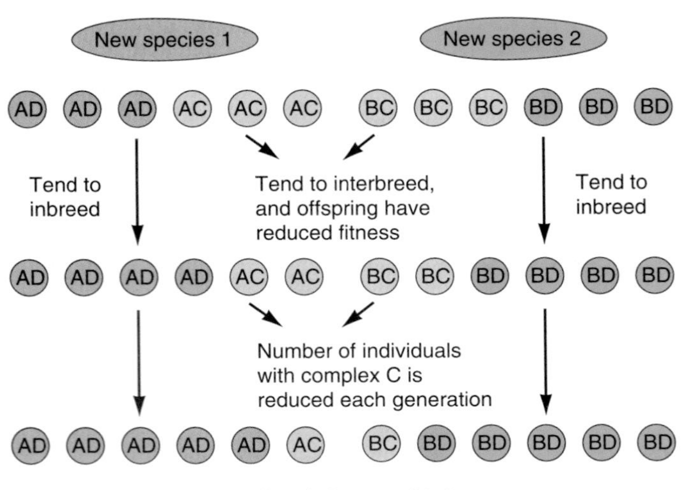

Figure 24.16

If two populations are becoming adapted to different niches (gene complexes A and B), individuals who still tend to mate with those of the opposite population (gene complex C, rather than complex D) produce intermediates whose genotypes are not well adapted to either niche and have lower fitness. Then gene complex C also has reduced fitness and it will tend to be replaced by complex D, which promotes inbreeding.

24.7 Species diverge by taking advantage of new opportunities.

When the first Europeans visited the Sandwich Islands—what we now call the Hawaiian Islands—they found seabirds familiar to any sailor, but most of the land birds were very different from anything they had encountered in Europe or the Americas. Most of them are unusual little red and yellow-green birds with an amazing variety of bills. They are now classified in a single, unique family called Drepanididae, and are known as Hawaiian honeycreepers.

The diversity of honeycreepers is unusual in contrast to the typical inhabitants of large landmasses. Each continent has many families of birds, and each family contains species that occupy very similar niches. For instance, sparrows have short, conical bills and are adapted for eating seeds; woodpeckers are adapted for digging into wood to catch burrowing insects; warblers are small, nervous, insect-eaters with thin bills; and thrushes have moderately heavy bills that are used for eating fruit, insects, and other small invertebrates.

Even though the internal anatomy of the Hawaiian honeycreepers shows that they are closely related, they occupy very different niches and show enormous external differences, especially in their bills (Figure 24.17). The chunky yellow and brown Chloridops uses (or *used,* for unfortunately many of these species are extinct now) its tough, massive bill for crushing the seeds of the naio plant. The Ou looks like a parrot and feeds on fruits just as parrots do in the American tropics, using its hooked bill to scoop out the insides of the ripe ieie fruit. The Koa Finches and Laysan Finch also fed on fruits, and the koas could split the tough twigs of the koa tree and eat grubs that lived inside them. The Akialoa feeds like a woodpecker, probing its long bill into crevices in trees and sometimes peeling off bits of bark to find grubs and insects. The mamos used their long, curved bills to suck the nectar out of deep flowers, while the Iiwi takes some nectar but prefers to eat the caterpillars off flowers. The Apapane's narrow bill is suited for its diet of insects combined with a bit of nectar. As a whole, the Hawaiian honeycreepers seem to have found most of the ways in which land birds can live, and the members of a single family have diverged enough to occupy the kinds of niches that are occupied by whole families of birds elsewhere.

As far as land birds are concerned, the Hawaiian Islands are quite isolated, well over 3,000 kilometers from North America. Migrating land birds occasionally wander or are blown off course by storms, and at some time in the past, perhaps only a few million years ago, a few birds that were ancestors of the honeycreepers, probably birds like our modern tanagers, must have wandered to Hawaii. They found few other birds competing for such foods as insects and nectar, and they adapted to using these resources. As the founding colony multiplied, it spread out over the islands where small populations became isolated from one another. Starting with slightly different founder populations, and given isolation and time, these differences became accentuated. Thus a colony on one island developed into seed-eaters while those on another island developed bills suited for eating insects. There is some reason to think that the founders were the red and black types that live primarily on nectar, that some of them evolved into the yellow-green insect-eaters, and that others developed later into the fruit-eaters. Figure 24.18 shows how this general trend can occur in island populations of any species.

The Hawaiian honeycreepers are *opportunists*—that is, the founders were faced with many opportunities for making a living, and their descendants eventually evolved ways to take advantage of them. But, as we pointed out in Chapter 2, all organisms are opportunists. By means of genetic variations that confer particular advantages, they seize the opportunities that arise for living in a new way.

The pattern of evolution in the honeycreepers is called an **adaptive radiation,** because one original population radiated out in several different biological directions (perhaps also different geographic directions) as its subpopulations took advantage of new opportunities and adapted to different ways of life (Figure 24.19). A general way of life to which a species can become adapted is called an **adaptive zone.** In the case of the honeycreepers, we may characterize insect-eating, nectar-eating, and fruit-eating as three general adaptive zones, with species in each zone specialized in their own particular niches. Given the right conditions, every species has the genetic potential to divide into more species, thus creating greater diversity. The term "adaptive radiation" is applied both to small episodes of evolution, as in the honeycreepers, and to large episodes, such as the divergence of the many basic types of birds or mammals. Verne Grant identified an adaptive radiation in one plant

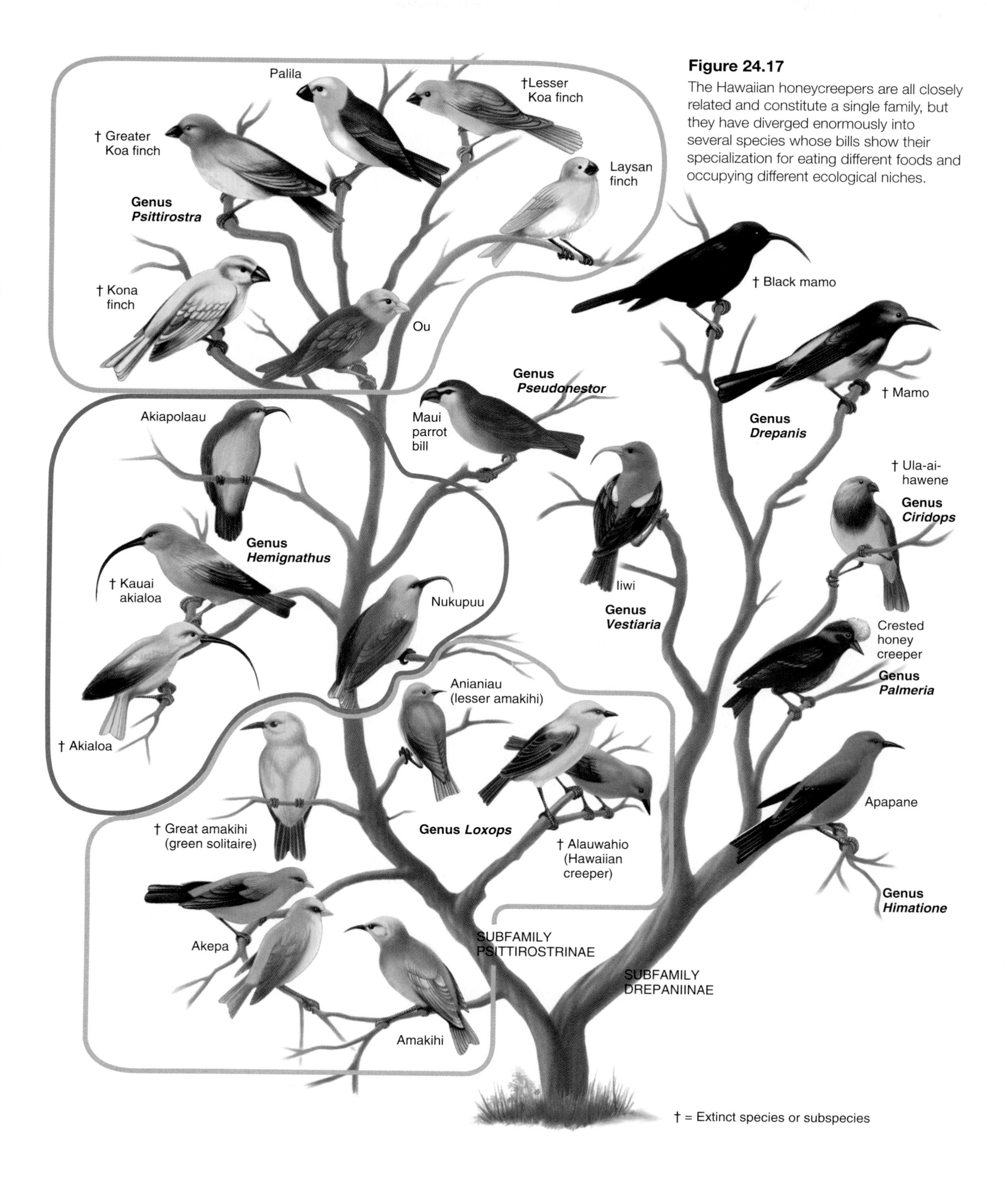

Figure 24.17
The Hawaiian honeycreepers are all closely related and constitute a single family, but they have diverged enormously into several species whose bills show their specialization for eating different foods and occupying different ecological niches.

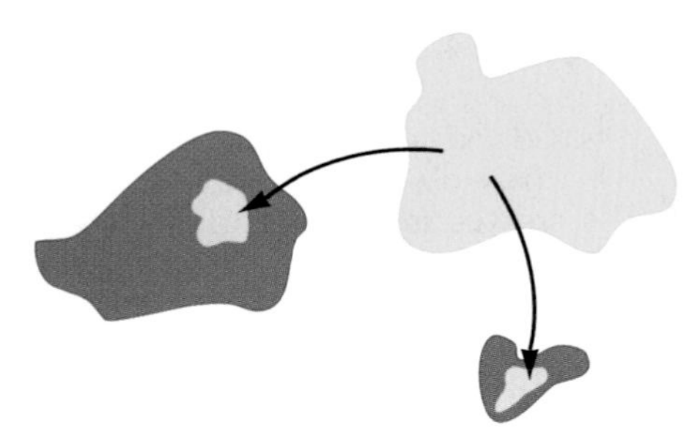

(1) Initial colony colonizes other islands.

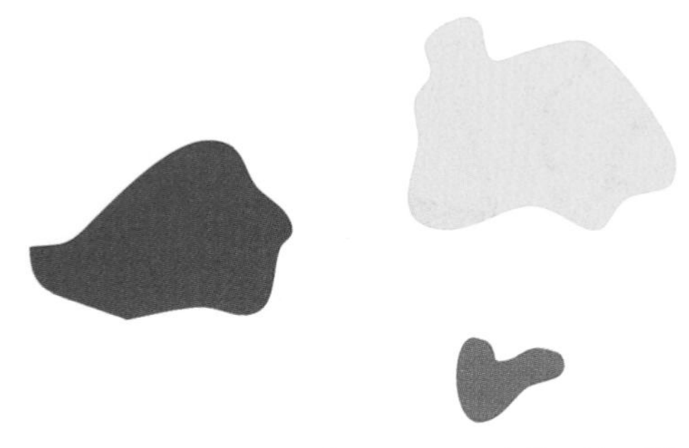

(2) Isolated populations become specialized to different niches and acquire reproductive isolating mechanisms.

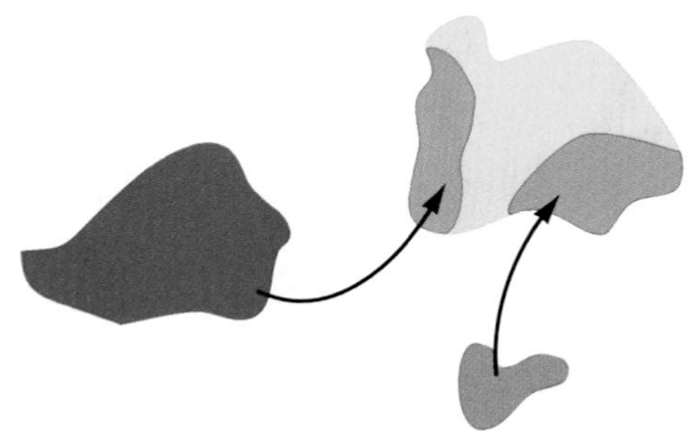

(3) Populations can expand their ranges, become sympatric, and remain distinct species.

Figure 24.18

Speciation on a series of islands may occur as local populations become isolated and acquire niche differentiation and reproductive isolating mechanisms.

family that produced flowers of sundry forms, each adapted to a different mode of pollination (Figure 24.20). All the diverse species we now see have arisen in this way over a long time.

During the transition times when species are dividing and radiating out into new adaptive zones, populations may be most likely to fail. Let's remember the point made in Section 23.4: Even a small change in a structure can make an important difference in fitness. We must not imagine that one of the specialized bills of a contemporary honeycreeper was useless until it acquired its modern form, nor were plants inviable while their flowers were changing into the forms we now see. Still, a rapid transition would seem the most advantageous. To become a successful new species, a population might have to cross a nonadaptive zone or a zone already occupied. Figure 24.21 illustrates this,

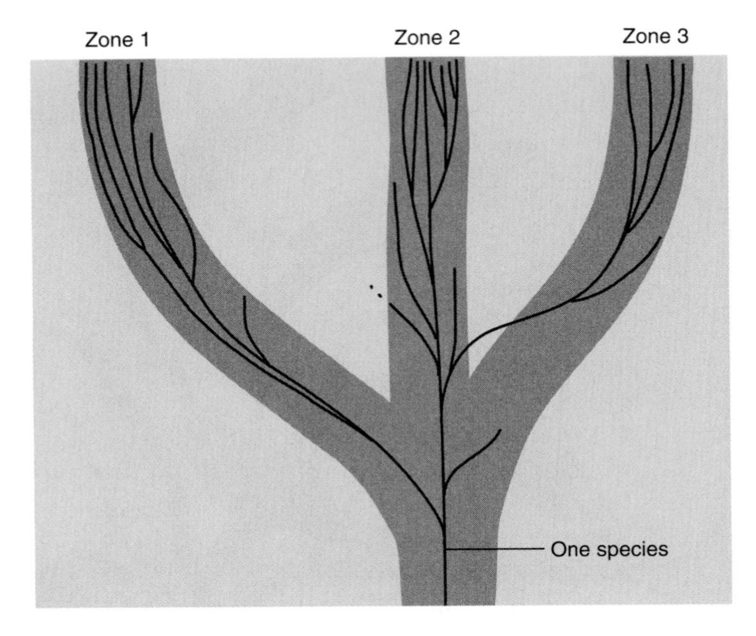

Figure 24.19

One species, or a few closely related ones, initially occupy one adaptive zone. Adaptive radiation occurs when one or more species change, perhaps quite rapidly, to occupy new adaptive zones. Here one such episode is shown briefly crossing a nonadaptive zone.

using the niche-spotlight model again: A population initially shares a patch of light with its parent population. A short distance away are unoccupied patches in different adaptive zones, but to reach them the population must cross a darker zone.

It may therefore be significant that new species are frequently founded by small populations in which *genetic drift* may occur. Genetic drift might carry a population quickly from one patch to another. The genetic makeup of a small population can change much faster by drift than by selection, and possibly in directions different from those in which selective forces would lead it. Isolated subspecies (semispecies) could acquire different habitat and niche preferences through genetic drift, and these small populations might drift to genotypes that are quite different from the typical genotype of their parent population. A special case of genetic drift is the **founder effect:** An isolated semispecies is founded by a few individuals whose genotypes are quite untypical of the main population, so the new population begins with an unusual average genotype and evolves further from there.

To provide some perspective on population structure and speciation, let's review the situations laid out in Figure 23.1, using the analysis of William Stansfield (Table 24.1).

24.8 Both gene flow and selection influence speciation.

Species such as the Song Sparrows, discussed in Section 22.3, show that as a species changes gradually over its range, populations at opposite ends of the range may become very different. These differences are generally due to selection, perhaps augmented by some genetic drift, with each local population

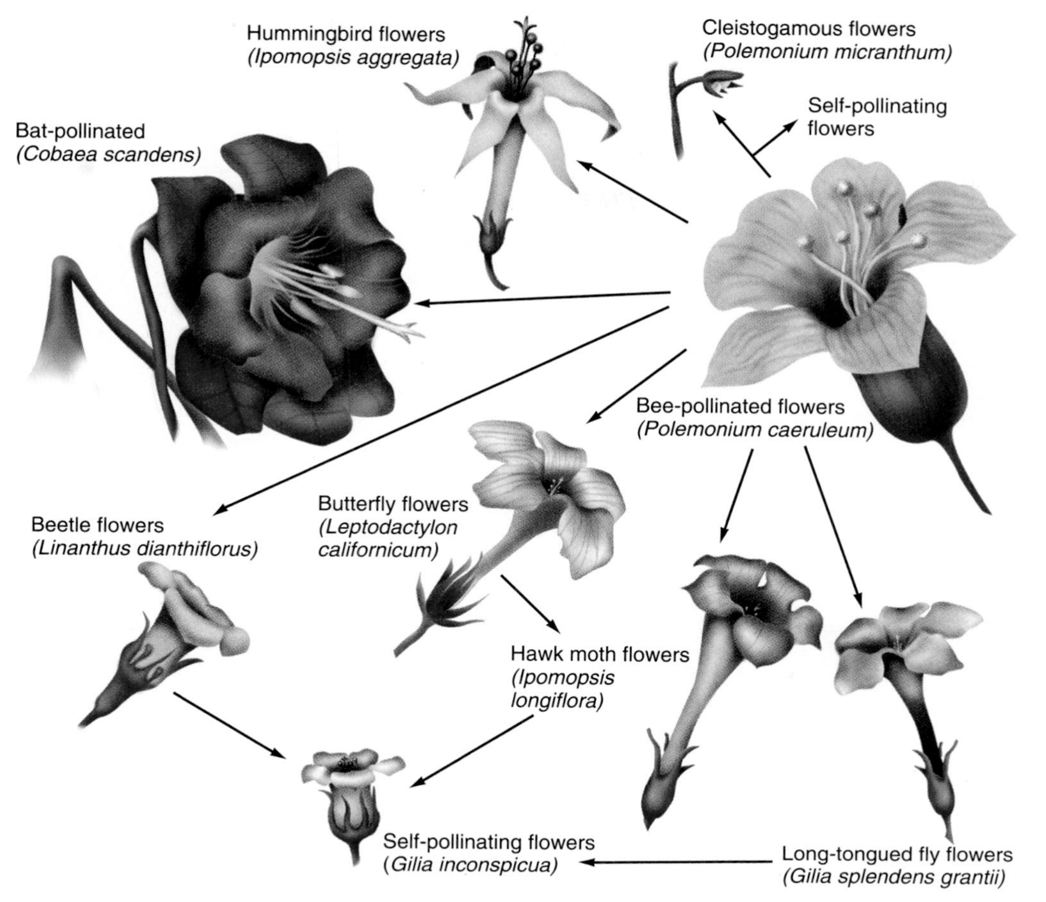

Figure 24.20
Plants of the family Polemoniaceae have apparently undergone radiation in several directions to produce flowers of many types, each specialized for pollination in a different way.

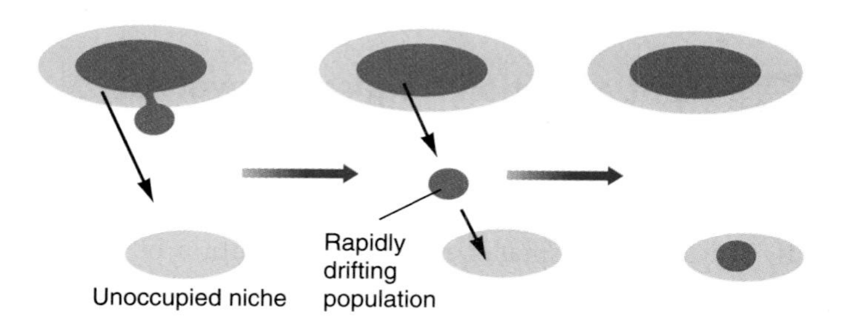

Figure 24.21
A small population, coming from a large species, may be able to change rapidly, through genetic drift, to occupy another pool nearby.

adapted to its own conditions. Many species consist of parapatric local populations (demes) that interbreed with one another to a degree, so at least some gene flow may continue between them. Two important questions have been asked about such species: First, how much gene flow actually occurs between neighboring demes across the range? Second, if there are no obvious geographical barriers to gene flow, and yet strong differences still develop, is there anything to prevent the species from breaking up into separate species? In other words, are geographic barriers always necessary for speciation?

Paul Ehrlich and Peter Raven have argued that individuals within a species move around much less than is often thought. Even though animals are generally very mobile, they tend to be stay-at-homes, rarely wandering very far, and returning to the same breeding areas year after year. Moreover, an individual that does wander far from its own deme may not contribute its genes to the gene pool of another deme. Such an animal may not be accepted by the local society or may be at a disadvantage in breeding compared with the residents. For instance, male butterflies usually emerge from their cocoons slightly before the females; they patrol the area searching for females and copulate with them as soon as they emerge, so virgin females are rare. A female only mates once, so a male entering a deme from outside has little opportunity to contribute to the deme genetically. Ehrlich and Raven also point out that even when plants are pollinated by wind dispersal, the success of pollination falls off very rapidly with distance; most pollen falls a few meters away, and hardly any goes more than a few tens of meters.

Therefore, relatively little gene flow may occur between neighboring demes of a species. Demes may keep essentially the same form and genotype because of selection—because they all occupy essentially the same niche in their local communities. Yet the difference between demes, which must also be due to local selection, can be striking. In the 1950s, E. B. Ford and W. H. Dowdeswell studied populations of the Meadow Brown butterfly, *Maniola jurtina,* and noted the variation in the number of spots on the butterflies' wings. Throughout its range in Eurasia and North Africa, the species shows a series of distinct "spot stabilizations," particular numbers of spots that are characteristic of the butterflies in a part of the range. Two small, distinct stabilizations are found in Devon and Cornwall in southwestern England (Figure 24.22). The boundary between them is unusually abrupt, sometimes spanning no more than a few meters, yet there is no physical barrier of any kind. In fact, while the boundary between the two stabilizations fluctuated eastward and westward over a distance of about 65 kilometers in the 13 years it was studied, it never broke down. Presumably it is

Table 24.1 Population Structure and Evolutionary Flexibility

Factors		Large Population	Intermediate-Sized, Partly Subdivided Population	Small, Isolated Population
Number of mutations		Many	Variable	Few or none
Rates of response to selection		High at intermediate gene frequencies; otherwise low	Fluctuating; depends on amount of gene flow	Negligible except in early stages of isolation
Genetic drift		Negligible	Occurs locally or periodically	Important
Gene flow		None	Important	None
Rate of Evolution Under:	Static environment	Slow or static	Fluctuating	Rapid approach to equilibrium; then slow or zero
	Slowly changing environment	Slow, progressive adaptation	Fluctuating, but generally progressive adaptation	Probable extinction
	Rapidly changing environment	Extinction or reduction to small isolated population or intermediate, partly subdivided population	Rapid, with maximal opportunity for origin of new forms	Certain extinction

maintained by some kind of selection and by a lack of gene flow between neighboring populations.

Thus a deme at the edge of a range may already be effectively isolated from the rest of the species, even without major geographic barriers. With gene flow already so small, it has been argued, such a deme could become quite different with sufficient selective pressure, and could develop reproductive isolating mechanisms. In this way, a series of parapatric demes could develop into separate species.

To find out how feasible parapatric speciation is, John A. Endler set up boxes with experimental populations of *Drosophila* in which he controlled fitness by artificial selection. He could arbitrarily assign any desired fitness to a given number of body bristles, for instance, by eliminating a certain fraction of flies with this characteristic from a population. Then he gave this feature a high fitness at one end of a population box and zero fitness to the population at the other end, with a gradient of fitness in between. He found that a strong cline (in number of bristles, in this example) can easily develop through the range, even though considerable gene flow continued to occur between neighboring regions. This situation may be common in nature, where the variables that determine fitness change smoothly across a geographic area, and neighboring populations continue to exchange genes. Endler's experiment suggests that parapatric speciation may be possible in wild populations, even with substantial gene flow between demes.

24.9 Plant evolution frequently involves hybridization and polyploidy.

For reasons that are not yet understood, hybridization is an important mechanism in plant evolution, even though it apparently plays a much smaller role in the evolution of animals. Edgar Anderson has emphasized the importance in plant evolution of **introgression,** a process in which some genes of one species work their way into the genome of another through hybridization. The F_1 hybrids between two species generally are less viable and fertile than their parents, but the progeny made by repeated backcrossing to one parent are intermediate varieties, with only part of the genome from the other parent, and they may be quite hardy and well-adapted to existing conditions. Introgression results from this sequence of hybridization, backcrossing, and selection of certain backcross types, and it is an important source of new variability in plant species. It tends to reduce sympatric species back to semispecies status, and sometimes leads to the emergence of new types.

In introgression, a few genes from one species might be introduced into another, but it is more likely that a whole chro-

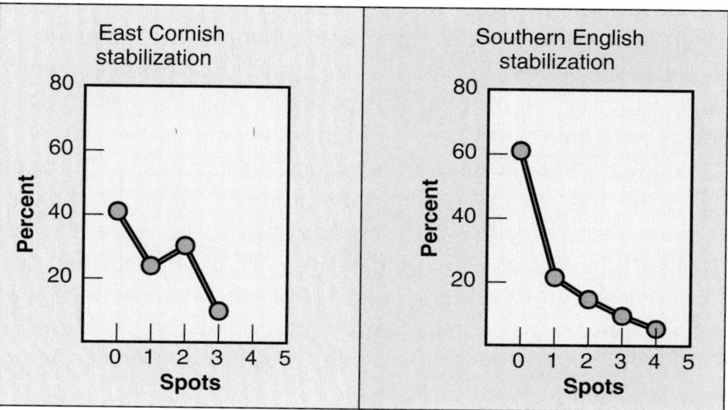

Figure 24.22

Two populations of the Meadow Brown butterfly, *Maniola jurtina* have different distributions of numbers of wing spots, as shown in the graphs. Where they meet in Cornwall, the eastern population remains distinct from the western population, even though there is no physical boundary between them. The populations are stabilized by some kind of internal mechanism.

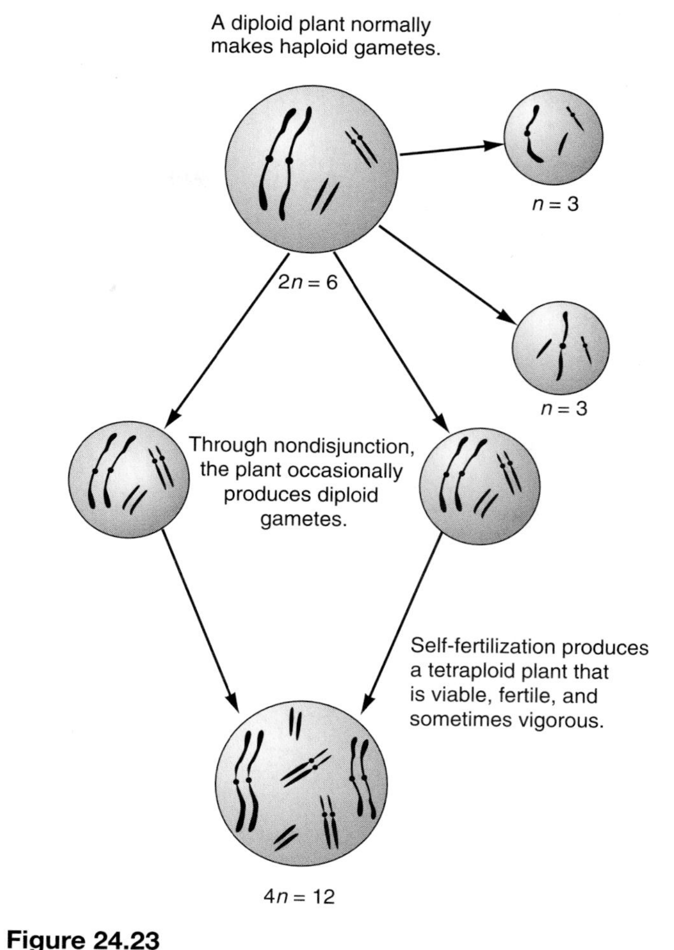

Figure 24.23

Autopolyploidy occurs in a diploid plant when nondisjunction produces diploid gametes. When they combine through self-fertilization, they create a tetraploid.

mosome or a segment of a chromosome replaces its homolog in the other species. A whole chromosome is an adapted gene complex, and its introduction into a different but closely related species may produce a new, more vigorous type with the ability to exploit a new environment and expand its range.

Introgression is prominent in some of the irises, or flags, which are so abundant and diverse in the lowlands of southern Louisiana. Around 1938, Herbert P. Riley found hybridization between the elegant blue *Iris giganticaerulea* and the brilliant orange *Iris fulva*. The two species are sympatric, but *I. giganticaerulea* is adapted to the waterlogged soil of marshes while *I. fulva* grows in the drier soil of banks and woods. The two only hybridize where the habitats have been broken down by human interference, and there Riley found a number of populations that show various degrees of hybridization, as measured by seven characteristics that mark one parental type or the other. Some populations appeared to be basically *giganticaerulea* with various amounts of the *fulva* genome resulting from continuous backcrossing.

New plant species may also develop more directly by the creation of **polyploid** individuals that have more than a diploid number of chromosomes. **Autopolyploidy** results from an abnormality in meiosis that creates diploid ($2n$) gametes (Figure 24.23). This sometimes happens in a plant that can pollinate itself. Self-fertilization, which is rare in animals, permits an unusual $3n$ *(triploid)* or $4n$ *(tetraploid)* zygote to be made, and the apparently greater plasticity of plant development allows many of these zygotes to grow into perfectly good, fertile plants. In fact, polyploids are generally larger than their parents and produce larger fruits; many of our domestic plants, which are cultivated for these features, are polyploids. Triploids are usually sterile because their three sets of chromosomes are unable to synapse properly and then separate to produce normal, viable gametes. (Triploids such as bananas are valued by humans because they have no large seeds. And, amazingly, some aquaculturists have now been able to breed triploid oysters, which don't spoil their culinary value by making large egg masses.) In a tetraploid, the

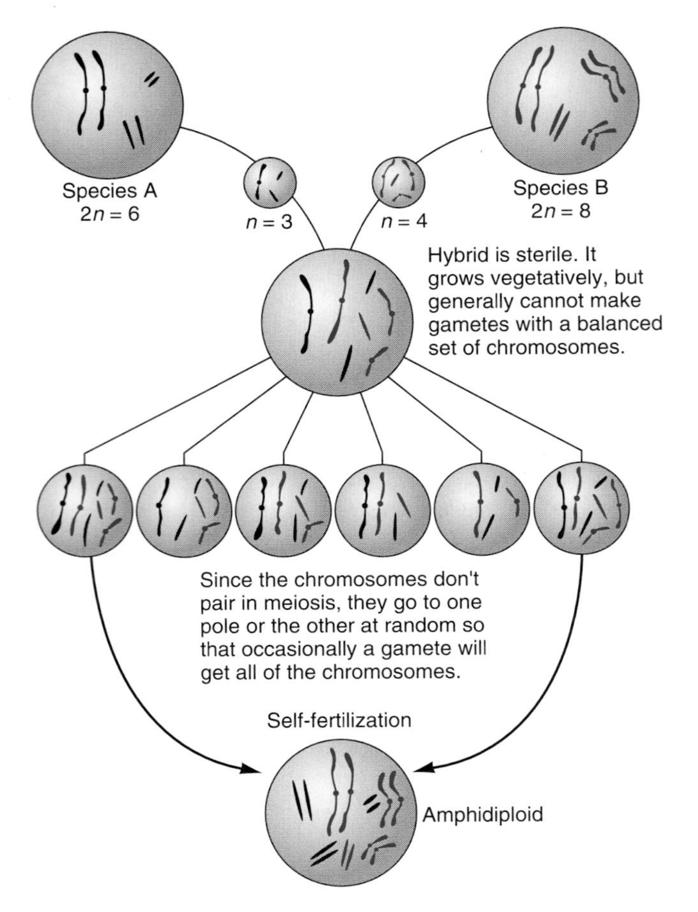

Figure 24.24

Allopolyploidy occurs in two stages. First, hybridization produces a plant that can only reproduce vegetatively because it is effectively haploid, with only one chromosome of each kind. Then some gametes combine through self-fertilization to produce a fertile diploid plant.

four homologous chromosomes generally seem to avoid irregular synapsis, so such plants tend to be viable and fertile.

Polyploidy may also result from a breakdown of interspecies barriers. In **allopolyploidy** (or amphiploidy), two chromosome sets come from different but closely related species. This may begin with haploid pollen from one species fertilizing haploid ova from another. The hybrids are sterile because the chromosomes from one set cannot synapse with those of the other set in meiosis, but because plants often reproduce vegetatively (asexually) so well, hybrids may be able to propagate themselves and compete effectively. If the chromosome set is doubled—for instance, through self-fertilization or some interference with mitotic division—an amphidiploid is produced. It will be fertile because each of its chromosomes has a homolog to synapse with (Figure 24.24).

An *auto*polyploid may have essentially the same characteristics and niche requirements as its parent, so it will have difficulty surviving, except under human cultivation, because it must compete with an already well-adapted type. Simply being larger is not always an advantage. An *allo*polyploid, on the other hand, may be more successful in nature because it combines the characteristics of both parents. It may be superior because of heterosis (hybrid vigor), and it may be suited to a slightly different niche.

Heterosis, Section 23.7.

It has been estimated that nearly half the known species of flowering plants have originated through polyploidy. One documented example is the salt-marsh grass *Spartina. S. maritima* ($2n = 60$) grows along European coastal marshes, while *S. alterniflora* ($2n = 62$) is found along the North American coast. The American species was accidentally introduced in Britain around 1800, and it started to grow in patches mixed with its European cousin. In about 1870, a sterile hybrid between them was identified and named *S. townsendii.* It is diploid and also has 62 chromosomes (indicating some minor chromosomal alteration from the expected number of 61), but can only reproduce asexually by extending rhizoids (runners). Then, around 1890, one of these plants apparently changed into a fertile allotetraploid named *S. anglica,* which has 122 chromosomes. *S. anglica* is a very vigorous grass that is now spreading around the coasts of Britain and France, replacing its parental species.

Modern bread wheat, *Triticum aestivum,* also arose through hybridization and allopolyploidy (Figure 24.25). Because of their distinctive forms, wheat chromosomes can be traced back to those of wild grasses, involving two episodes of hybridization followed by duplication. The ancestral wheats have been preserved; einkorn wheat is now grown especially on poor soils in Europe, and emmer wheat contains proteins that make the excellent gluten needed for pasta.

These regimes of hybridization and selection can produce *sympatric speciation,* without the need for geographic isolation. New species might also arise sympatrically through strong selection for adaptations to different habitats. In Britain, a variant of the grass *Agrostis tenuis* is genetically lead-resistant and is able to live on soils contaminated by lead-rich mine tailings. The original, lead-intolerant populations often grow only a few meters away on uncontaminated soil. Hybrids between the two types grow poorly on both types of soil, so the lead-resistant plants seem to constitute a new species that has arisen within the range of its parent species, perhaps because of only one or a few genetic changes.

C. Aspects of macroevolution

24.10 The evolution of higher taxa does not require special mechanisms.

Although speciation creates novelty and diversity in the biological world, some theorists have argued that it creates only relatively small differences, and that some other process is needed to account for the major differences between orders, classes, or phyla. The fallacy in this argument lies in believing that taxa established by taxonomists really exist. Even if a species can be defined quite objectively, that does not mean the abstractions "amphibian," "bird," and "mammal" are equally real. Each of these categories consists of many species. Each species goes its separate way, grad-

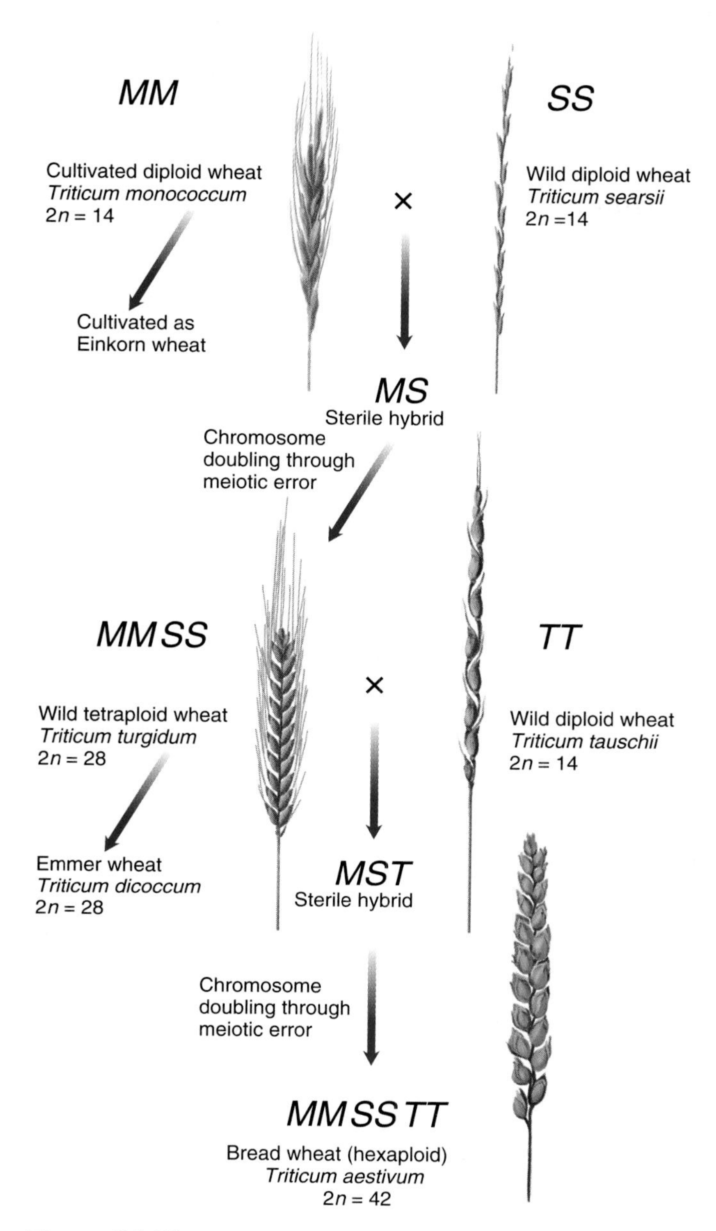

Figure 24.25

Bread wheat is the product of hybridization between ancestral species, whose haploid chromosome sets are represented by the large capital letters *M, S,* and *T. (Top)* Wild wheats with 14 chromosomes each hybridized, and about 11,000 years ago in Western Asia, the resulting sterile plant produced the ancestor of emmer wheat through nondisjunction. Then, around 8,000 years ago, a second episode of hybridization and chromosome duplication produced bread wheat *(bottom).*

ually acquiring differences, and some eventually produce new species so different from their ancestors and contemporaries that we choose to describe them by totally different words.

The point can be made by looking at just one situation, the origin of birds. *Archaeopteryx* (see Figure 35.39), the earliest known bird, clearly had feathers like modern birds, but it was apparently just a small, rather odd, thecodont dinosaur. It happened to evolve a gene complex that made its scales grow in a new and unusual pattern, as feathers, which it probably used for insulation and as nets for catching insects (gliding or flight probably developed only in its descendants). If *Archaeopteryx* had been the only species of its kind, it would undoubtedly just be considered an unusual reptile. We would not have invented the word "bird," and we certainly would never have erected a whole class just for *Archaeopteryx.* It happened, however, that *Archaeopteryx* had the potential to open a new adaptive zone, as its descendants have shown by their subsequent adaptive radiations, so about 9,000 species of birds now exist, each specialized for its own way of life. Yet each bird species met its environment independently; if it divided into new species, it did so in the ways we have described, and one clade diverged from another through relatively small, gradual steps. Taxa, then, are merely categories we devise for classifying what we see as distinctive groupings of species. No new mechanism need be postulated to account for their evolution.

24.11 Anatomical differences may evolve through simple, systematic changes in development.

The history of organisms preserved in the fossil record shows a huge variety of species that have undergone vast changes over billions of years. This is the essence of evolution—that a small ancestral horse with five toes grows into a large horse with only one, or that an ancestral flowering plant diverges into species with the variety of forms we see today. How could these great transformations in form have occurred? Or, to ask the question in a genetic context, how can genomes change to effect these great changes in form?

Though biologists do not know enough to answer this question in detail yet, some general answers are within our grasp. At the nexus between development and evolution, we must try to outline how changes in development can result in large differences in form. Some of our thinking about this problem rests on a way of picturing form that was devised by the great biologist D'Arcy Thompson, whose book *On Growth and Form* changed biologists' perspectives in its various editions from 1917 to 1942. Among his contributions, Thompson showed how we can account for changes in form by simply laying a grid over the drawing of a structure and treating the drawing as if it were made of stretchable and compressible rubber. To begin, draw a grid over a circle. If the grid is compressed in one direction, the circle becomes an ellipse:

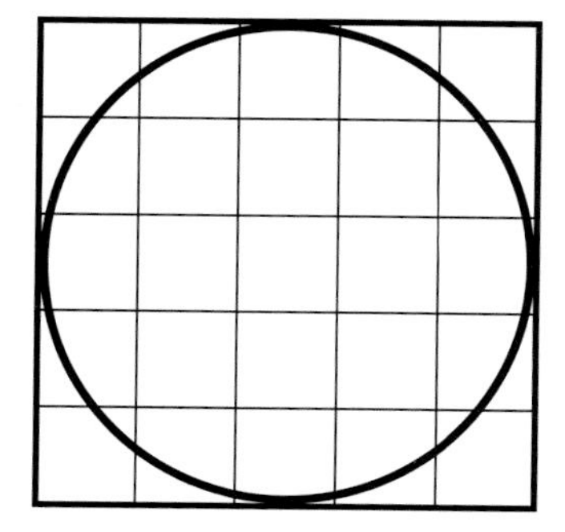

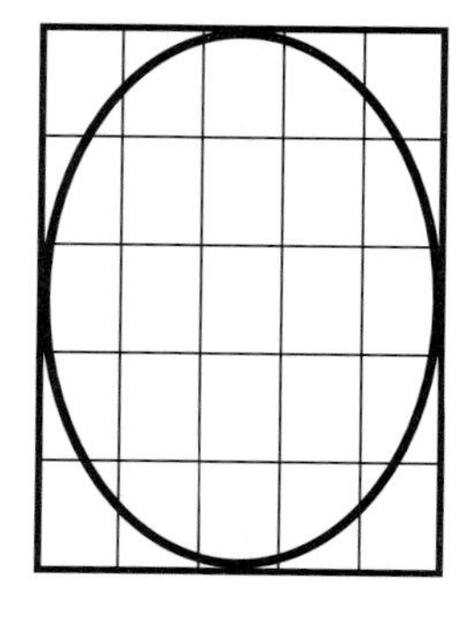

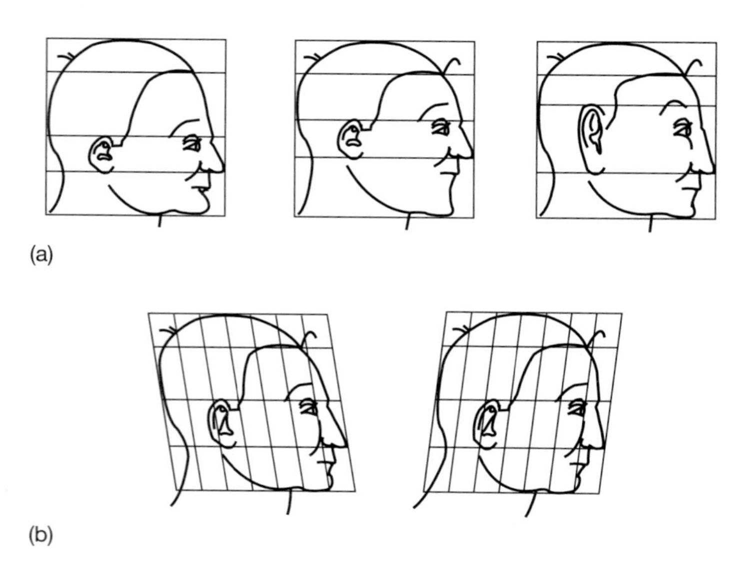

Figure 24.26

The German artist Albrecht Dürer showed how different faces can be created by changing a coordinate system superimposed on them: *(a)* expanding and contracting regions in the vertical direction; *(b)* changing the angles between lines in a grid.

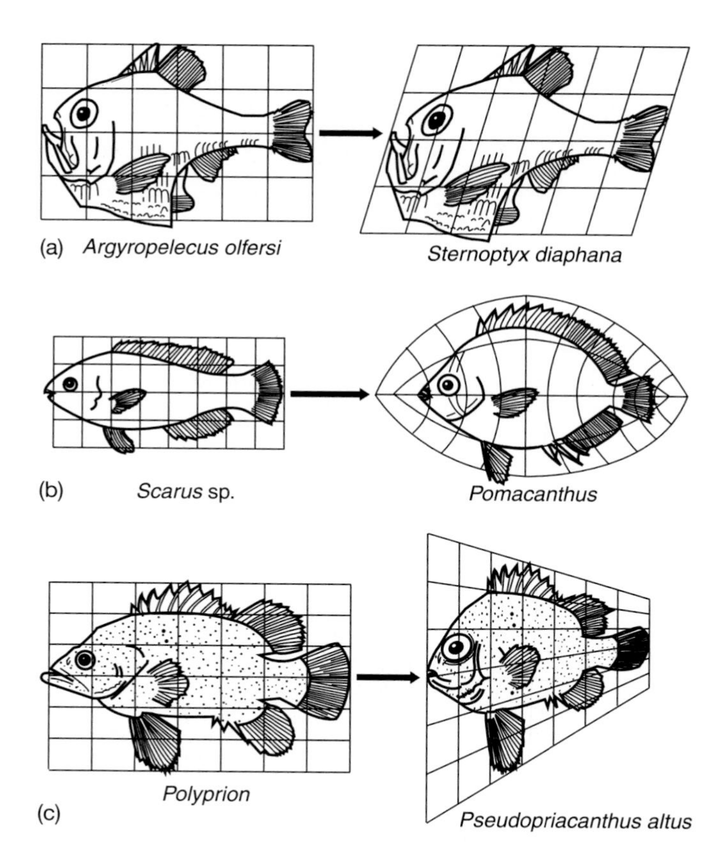

Figure 24.27

Evolutionary changes in related organisms can be described by systematic changes in coordinate grids, although this analysis is not meant to imply that these particular species have actually evolved in this way. *(a)* One species of hatchetfish could be converted into another by changing the 90° angles of a grid into 70° angles. *(b) Scarus* could be converted into the related *Pomecanthus* by deforming a linear grid into curves; D'Arcy Thompson noted that the color bands of *Pomecanthus* actually follow these curved coordinates. *(c)* Another pair of fishes seem to be related in outline by the triangular transformation of a grid.

Thompson applied this simple method to the forms of plants and animals, following the German artist Albrecht Dürer (1471–1528), who showed in his *Treatise on Proportion* how different human forms can be made by stretching and bending a single basic form (Figure 24.26). Thompson overlaid a regular grid on a drawing of one species and then imposed similar grids on drawings of related species, making the grid lines pass through comparable structures. From Thompson's many examples, we have chosen just two to show here: three transformations of fish (Figure 24.27) and the transformation of a human skull into those of two other primates (Figure 24.28). The general result is quite obvious: The forms of closely related organisms are often regular transformations of one another.

These transformations may be very interesting from an artistic viewpoint, but what do they have to do with actual evolution? A plant or animal develops its form through growth, and each region or structure may grow at its own rate as determined by the genome. Start with an animal in which all parts are growing at what we may call basal rates. If a change in gene expression causes one region to grow faster, the whole structure will be transformed, just as Thompson's drawings show. There are many possible transformations of this kind, all called **heterochrony** (*hetero-* = different, *chronos* = time), meaning a change in the timing or rate of a developmental process, relative to those of an ancestor. Heterochrony can result from stopping a developmental process early or continuing it beyond its normal time; in the growth of limb bones, for instance, the first would create shorter limbs and the second longer limbs, compared with the ancestor. Or the whole period of growth might be extended, resulting in a larger animal, as in the evolution of a small horse into a large one.

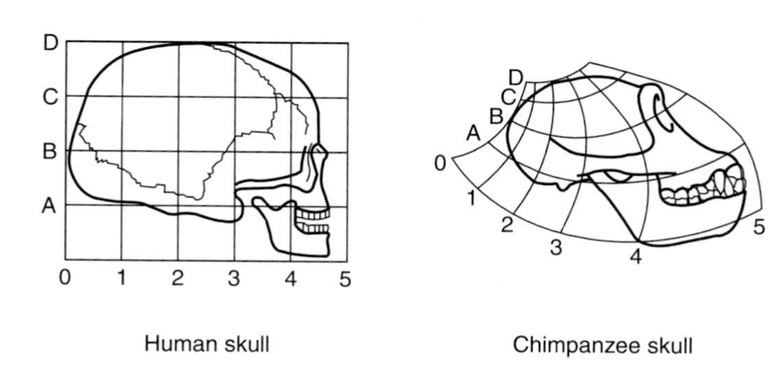

Figure 24.28

A rectangular grid laid over the human skull shows how it differs in a rather simple way from a chimpanzee skull. Thus the chimpanzee-like skull of a primitive hominid (human ancestor) is changed into the modern human skull by expanding the rear of the skull (as the brain becomes much larger) and rotating the top of the skull forward.

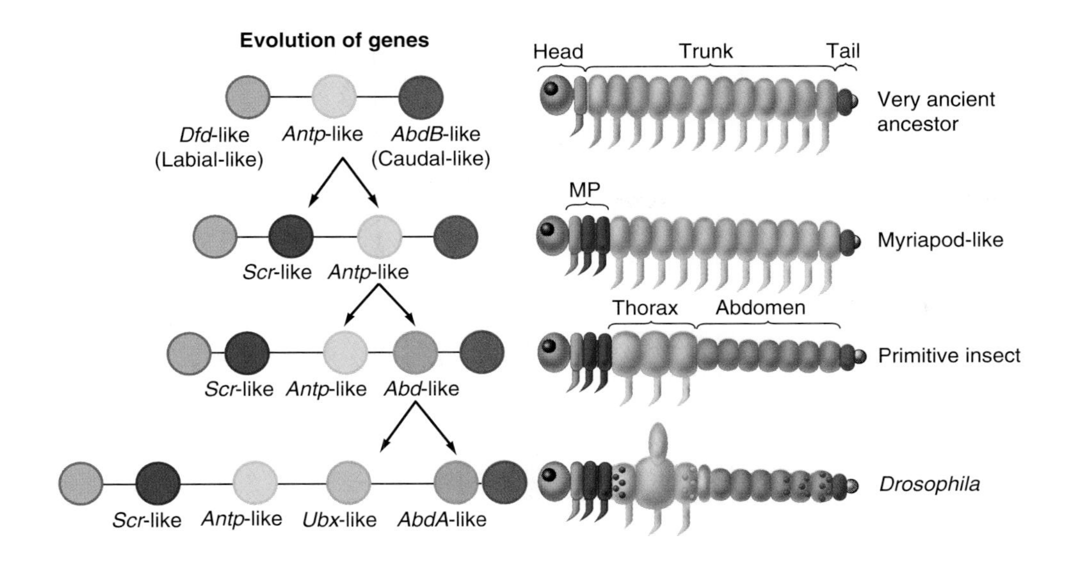

Figure 24.29
A series of duplications and transformations in certain genes that regulate development can explain the evolution of body form from a primitive arthropod with many undifferentiated segments to a winged insect such as *Drosophila.* Genes are generally described as *like* those identified and named in *Drosophila,* but not necessarily identical. It is proposed that the ancestral homeotic genes such as Antennapedia *(Antp)* duplicated to produce the gene called Sex-combs-reduced *(Scr)* that is responsible for the maxillipeds (MP) of myriapods (centipedes and millipedes). Further gene duplications produced homeotic genes governing the thorax and abdomen of a primitive insect and then the winged thorax of a fly. Other changes in genes affecting head and tail development (*Dfd*-like and *AbdB*-like) are not shown explicitly.

As developmental biologists learn how sequences of genes act to produce an adult body form, evolutionary biologists find a new basis for thinking about how new forms have evolved. It is too early to specify just how a heterochronic change might occur, but consider that if limb bones in chick development are laid down by a clock mechanism (see Section 21.9), any change that accelerates or slows that clock will affect the limb's form. Other genes set up concentration gradients of their protein products (see Section 21.10); changing the amount of one protein relative to another could then change the relative development of the embryo along an anterior-posterior or a dorsal-ventral axis.

Similarly, we can now think about the evolution of the basic arthropod body plan into an insect's body on the basis of knowledge about the sequence of *Drosophila* genes that produce differentiation of the body segments. M. Akam and his colleagues have shown how this probably occurred through a series of duplications and differentiations of the critical homeotic genes (Figure 24.29). As we learn more about the action of genes in development, it should become possible to explain other morphological transformations observed in evolution.

Coda Evolution through natural selection is one of the principal organizing concepts of biology. We have depended on it as an explanatory principle throughout this book, since the organisms we see today could not have acquired their particular forms in any other way. Yet natural selection takes different forms, and there is more to evolution than natural selection. Evolution occurs fundamentally because organisms are genetic systems whose genomes can change, and they change at various levels: The process begins with mutation and recombination in individuals, creating variability, and populations harbor significant amounts of variability. The result in each individual will be changes in structures, which depend upon specific intermolecular interactions; such molecular changes result in changes in more complex processes, such as a pathway of development, producing organisms with different forms and patterns of behavior. Natural selection will work upon these distinct individuals as they interact with one another and with their environment. At the same time, larger forces come into play, such as geographic or climatic events that isolate some populations from others, so they undergo independent evolution. Overall, evolution is dominated by randomness and opportunism. Mutations occur randomly and unpredictably. So do the chance events that bring alleles together in different combinations and in different environments. From time to time, these rolls of the cosmic and genetic dice produce new opportunities for novelty that organisms follow for a while. We see the results all around us; in fact, we are one of those results. It would be fun to stay around for another hundred million years or so and see what the cosmic dice have in store next.

Summary

1. All evolution occurs within communities. All evolution is coevolution, in which two or more species evolve together. The members of a community are very interdependent, affecting one another in subtle ways so that a change in any one member affects the entire community.
2. In outline, the forces of evolution are simple and straightforward. All populations are diverse because of mutation and recombination. Some of the genotypes in a population have higher fitness than others, which means organisms of those types tend to reproduce themselves more successfully and pass on copies of their genomes. Species diverge when isolated subgroups have changed genetically so much that they can no longer interbreed with one another.

3. The geological record shows that species typically last from about 100,000 years to a few million years before becoming extinct. The record also shows a great deal of speciation, in which one species divides into two or more.
4. Some lines of evolution show a consistent change in one direction (phyletic evolution); others show no evident change for long times (stasis).
5. Selection may favor the average types of a population (known as stabilizing selection), one extreme (directional selection), or two different types (disruptive selection).
6. Stabilizing selection keeps a population unchanged, with the average genotype in the population adapted to a particular niche. If the environment changes, the niche requirements may shift, and the population must change genetically through directional selection to meet the new requirements.
7. Local populations diverge from one another as they spread across a region. Speciation seems to occur most commonly when populations become geographically isolated for a time, during which they acquire genetic differences that make them unable to interbreed with other populations when their ranges expand again.
8. Reproductive isolation can occur through premating and postmating mechanisms—that is, before or after fertilization. If hybrids between two well-adapted types of organisms are inferior to their parents, there will be strong selection for premating mechanisms.
9. Genetic drift—a random genetic change that occurs in small populations—may account for rapid evolution in isolated populations.
10. As speciation takes place, populations are often at an intermediate stage (semispecies of a superspecies) where there is some gene flow (hybridization) between them. However, selection appears to be such a powerful force that it can override gene flow between neighboring populations; thus populations may be stabilized with quite distinctive features even if they are not reproductively isolated.
11. New plant species frequently arise through hybridization and backcrossing to one parent. Hybrid plants are often quite viable and may even be superior to their parents in some circumstances. Hybrids may be sterile because their chromosome sets cannot synapse, but fertile strains can then arise through polyploidy.
12. Special processes are not needed to account for higher taxa, since these categories are merely human abstractions. Once they are reproductively isolated, species tend to diverge into increasingly different forms.
13. Novel forms may arise through heterochrony, a process in which parts of an organism change their rate of growth or time of growth. Novelties may also arise through small changes in the homeotic genes that regulate the patterns of body form.

Key Terms

coevolution 482
speciation 484
extinction 484
phyletic evolution 484
stasis 484
stabilizing selection 485
directional selection 485
disruptive selection 485
Red Queen hypothesis 487
punctuated equilibrium 488
gene flow 489
geographic isolation 489
sympatric 490
parapatric 490
allopatric 490
reproductive isolating mechanisms 490
semispecies 491
allospecies 491
superspecies 491
character displacement 492
adaptive radiation 494
adaptive zone 494
founder effect 496
introgression 498
polyploid 499
autopolyploidy 499
allopolyploidy 500
heterochrony 502

Multiple-Choice Questions

1. When compared to drab-colored plants, those with colorful flowers are more likely to
   a. require wind for pollination.
   b. exist within a complex ecosystem.
   c. depend on coevolution with particular bird or insect species.
   d. have evolved by selection.
   e. evolve following geographic isolation.
2. Phyletic evolution generally results from ______ selection.
   a. stabilizing
   b. directional
   c. disruptive
   d. either stabilizing or disruptive
   e. a breakdown in the forces of
3. Starting from a parental stock, adaptive radiation to a variety of habitats results from ______ selection.
   a. stabilizing
   b. directional
   c. disruptive
   d. stabilizing or directional
   e. a breakdown in the forces of
4. In the long term, all species are most likely to undergo
   a. extinction.
   b. disruptive selection.
   c. stabilizing selection.
   d. directional selection.
   e. further speciation.
5. The survival rate for babies that weigh between 5 and 8 pounds at birth has always been higher than for heavier or lighter newborns. If birth weight is under some degree of genetic control, this is an example of
   a. disruptive selection.
   b. directional selection.
   c. stabilizing selection.
   d. phyletic evolution.
   e. gene flow.
6. Explanations of the fossil data that stress punctuated equilibrium give more importance to ______ than explanations stressing phyletic evolution.
   a. natural selection
   b. the slow pace of evolution
   c. adaptation
   d. extinction
   e. genetics
7. In the natural environment, gene flow occurs freely
   a. between members of different genera.
   b. between members of different species within the same genus.
   c. between members of the same species.
   d. between members of two or more isolated populations of the same species.
   e. All of the above are correct.
8. Which of the following constitutes postmating reproductive isolation?
   a. Males are isolated from females by geography.
   b. Courtship behavior of males does not elicit the proper response in females.
   c. External genitals of females and males prevent mating.
   d. Males of one population do not recognize females of another.
   e. An interspecific embryo is unable to develop to term.

9. A deme is best defined as
   a. an entire species.
   b. a population that has some gene flow with parapatric populations.
   c. members isolated from a parent population by geography.
   d. a population established by a limited number of founding members.
   e. a species that results by phyletic gradualism.
10. The golden hamster (a rodent often used in biomedical research or owned as a pet) is believed to have arisen as a separate species following one mating between a male and female of two other hamster species. Which of the following mechanisms might explain this event of speciation?
    a. adaptive radiation
    b. allopolyploidy
    c. autopolyploidy
    d. genetic drift
    e. punctuated equilibrium

True-False Questions

Mark each statement true or false, and if false, restate it to make it true.

1. Coevolution means division of a parent species into two or more species, occupying different adaptive zones.
2. According to the Red Queen hypothesis, every species is racing as fast as possible genetically to become better adapted to its ecological niche.
3. A species in a fairly stable environment is shaped largely by directional selection.
4. Since the half-life of ^{14}C is 5,570 years, determining the ratio of ^{14}C to ^{12}C is an accurate dating method for fossil remains that are more than 100,000 years old.
5. A more complete fossil record would allow us to distinguish phyletic evolution from repeated speciation.
6. Reproductive isolating mechanisms are likely to evolve more slowly in isolated populations than in contiguous populations.
7. If a group of semispecies become sympatric by expanding their ranges and continue to interbreed with one another, we would probably classify them as allospecies of a superspecies.
8. Two closely related, sympatric species usually occupy the same niche and are very similar morphologically.
9. In small populations, directional selection is the mechanism that promotes speciation fastest.
10. If a hybrid plant is infertile but viable, it could become a new, fertile species through introgression.

Concept Questions

1. Explain why extinction is the likely fate of any species, even if it is well adapted to its environment.
2. What kinds of environmental conditions foster stabilizing selection? What about directional or disruptive selection?
3. Suppose two populations have become somewhat distinct (morphologically, behaviorally, or ecologically) during a period of isolation but still interbreed when they become sympatric. If the hybrids between these populations have lower fitness than the parental types, what will eventually happen to these populations?
4. Explain why the species status of allopatric populations cannot be determined, based on the biological species concept.
5. Explain the likely role of heterochrony in divergent evolution.

Additional Reading

Alvarez, Walter, and Frank Asaro. "What Caused the Mass Extinction? An Extraterrestrial Impact." *Scientific American,* October 1990, p. 78;

Courtillot, Vincent E. "What Caused the Mass Extinction? A Volcanic Eruption." *Scientific American,* p. 85. Two papers that debate the event that resulted in a mass extinction 65 million years ago.

Cronin, J. E., N. T. Boaz, C. B. Stringer, and Y. Rak. "Tempo and Mode in Hominid Evolution." *Nature,* 292, pp. 113–22, 1981.

Farrington, Benjamin. *What Darwin Really Said.* Schocken Books, New York, 1966.

Gillespie, John H. *The Causes of Molecular Evolution.* Oxford University Press, New York, 1991.

Gould, Stephen Jay. *Dinosaur in a Haystack: Reflections in Natural History.* Harmony Books, New York, 1995. More of Gould's essays originally published in *Natural History* magazine.

Mangelsdorf, Paul C. "The Origin of Corn." *Scientific American,* August 1986, p. 80. Modern corn, the author argues, had not one ancestor but two: It is derived from a cross between a primitive corn and a perennial form of the wild grass teosinte.

Sheppard, P. M. *Natural Selection and Heredity,* 4th ed. Hutchinson and Co., London, 1975.

Stanley, Steven M. "Mass Extinctions in the Ocean." *Scientific American,* June 1984, p. 64. During brief intervals over the past 700 million years, many marine animals and plants have died out. Geologic evidence now suggests that most of the mass extinctions were caused by cooling of the sea.

Stebbins, G. Ledyard, and Francisco J. Ayala. "The Evolution of Darwinism." *Scientific American,* July 1985, p. 72. Recent developments in molecular biology and new interpretations of the fossil record are gradually altering and adding to the synthetic theory of evolution.

Weiner, Jonathan. *The Beak of the Finch: Evolution in Our Time.* Alfred E. Knopf, New York, 1994. A superb exposition of contemporary research on evolution, centering on the Grants' work on the Galápagos finches.

Zimmer, Carl. "Hypersea Invasion." *Discover,* October 1995, p. 76. Paleontologists Mark and Dianna McMenamin explain the mystery of how life on land spawned much more diversity than life in the sea in much less time, by viewing the fluids within the whole of land life as one interconnected "hypersea."

Internet Resource

To further explore the content of this chapter, log on to the web site at:

http://www.mhhe.com/biosci/genbio/guttman/

Ecology

25 The Biosphere

A quiet lake might be someone's home, although today it is likely to be modified by civilization.

Key Concepts

25.1 Energy from the sun drives most activities on the earth's surface.

25.2 The sun and the earth's rotation create prevailing wind patterns.

25.3 Ocean currents flow in horizontal surface patterns and in vertical mixing currents.

25.4 Oceans support several kinds of communities.

25.5 Freshwater habitats are among the richest and most varied on earth.

25.6 Precipitation falls in a zonal pattern across the continents.

25.7 Several major terrestrial biomes cover the earth.

25.8 Terrestrial ecosystems are determined by the relationships among climate, vegetation, and soil.

25.9 Soil is formed by chemical alteration of weathered rock and organic complexes.

25.10 The world is divisible into several biogeographic realms.

25.11 The earth's surface is made of slowly moving tectonic plates.

Where do you live? Most people would respond by giving an address—a house or apartment number on a city street or perhaps a rural route number. Such an answer reflects the governmental structure of our urbanized society. But a biologist might interpret the question differently: What are your latitude and longitude? What natural ecosystem did your city or farm replace? What type of soil do you have, and what kinds of plants grow best in it? Where does your water come from—a lake, a mountain stream, an aquifer? What is the climate like? What direction do the prevailing winds come from? How many sunny days and how many centimeters of precipitation occur annually?

Many of us are generally so far removed from the natural world that these questions don't occur to us. They may seem silly or irrelevant to our lives. But they are precisely the questions we must ask to understand the biological world. Political boundaries are irrelevant; limits of temperature, sunlight, and precipitation are very relevant.

Though we live on a lithosphere of rock and breathe an atmosphere of air, our general address is the **biosphere,** the part of the planet where all organisms live. The biosphere forms the setting for life, but the word "biosphere" is also used to include the organisms themselves—the sum of the earth's ecosystems. The biosphere is dynamic, changing in obvious ways from moment to moment and in more subtle ways that only become apparent on a geological time scale. Throughout this book we have emphasized that biology and biological processes must be understood in a broad ecological and evolutionary context. In this chapter we will examine the overall structure of the biosphere, especially the energy that drives it and the large geographical and geological forces that shape it. ■

25.1 Energy from the sun drives most activities on the earth's surface.

Most of the biosphere is powered by the energy of sunlight absorbed through photosynthesis, as we noted in Section 7.9. Exceptions are chemoautotrophic bacteria and the inhabitants of some deep-sea communities, which get their energy by oxidizing inorganic compounds, as we explain in Section 25.4. (Concepts 25.1 summarizes words that distinguish the places where organisms may live.) The drama of the biosphere has a simple plot: eat and be eaten (Figure 25.1). An ecosystem is structured largely by the *trophic,* or feeding, relationships among its community members, which can be described as a tangled **food web.** To review the distinctions introduced in Section 7.9: The plants, algae, and bacteria that obtain their energy from light are **phototrophs,** and the organisms that get their energy from chemical sources are **chemotrophs.** Based on their source of carbon, we also distinguish **autotrophs,** which can make their own organic compounds from CO_2 and other inorganic materials, from **heterotrophs,** which use pre-existing organic compounds. Looked at another way, the members of a community can be placed in somewhat idealized **trophic levels** reflecting their positions in a food web. Phototrophs are **producers** that bring energy into the system by capturing and storing light energy. **Primary consumers,** or **herbivores,** then eat the producers. **Secondary consumers,** or **carnivores,** in turn, eat the herbivores. In fact, many species are **omnivores** that feed at more than one level. All organisms produce wastes and eventually die. Dead organisms and their solid wastes, collectively called **detritus,** are eaten by **detritivores,** and all the organic materials in an ecosystem are eventually decayed by **decomposers.** In all our thinking about ecology, it is essential to keep in mind this fundamental trophic structure of an ecosystem.

How much energy actually passes through the biosphere to power all this activity? Each square meter of the earth's surface is irradiated by light at an average rate of 20 kilocal/min or 1.4 kilowatts (kW), a quantity called the *solar flux* or *solar constant.* Over the whole earth, this amounts to $1{,}750 \times 10^{11}$ kW of power. The following numbers are all in units of 10^{11} kW. Only a small part of the light energy striking the earth ever gets into the earth's organisms, but the physical activities of the oceans and atmosphere, which set limits for the existence of organisms, use enormous amounts of energy.

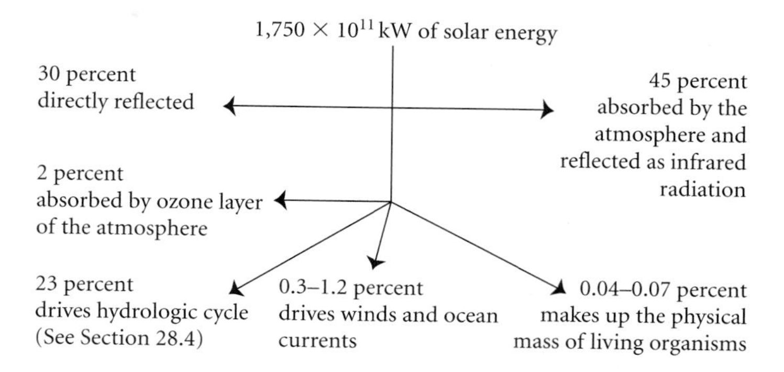

Only 0.04–0.07 percent of the energy of sunlight is incorporated into living organisms, but it supports a constantly changing biomass of about 1.8×10^{18} grams. Sunlight also drives powerful forces—the forces of wind, waves, water flows, and ocean currents—that are constantly bombarding and shaping the biosphere. Let's see how some of these forces shape the earth's ecosystems.

25.2 The sun and the earth's rotation create prevailing wind patterns.

The great sea of air enveloping the earth constantly absorbs energy from sunlight, especially by way of the heated water and ground below. The atmosphere, however, is not heated uniformly everywhere. Tropical latitudes receive much more energy per unit area than temperate latitudes because sunlight strikes the tropics most directly:

Concepts 25.1 Ecological Confusables: Places Where Organisms Live

Biogeographic realm: One of the six large divisions of the terrestrial part of the biosphere, corresponding roughly to the continents and having distinctive floras and faunas that differ considerably from one another.

Biome: A terrestrial community type, such as a temperate forest, tropical rain forest, or desert.

Biosphere: The part of the earth and its atmosphere in which organisms live, including the organisms themselves.

Community: All the organisms that live and interact together in the same area.

Ecosphere: Synonymous with "Biosphere."

Ecosystem: A community of interacting organisms and their physical environment.

Habitat: A general description of the place where a certain type of organism lives.

Zone: One of the regions of an aquatic habitat, defined primarily by depth of the water and amount of light.

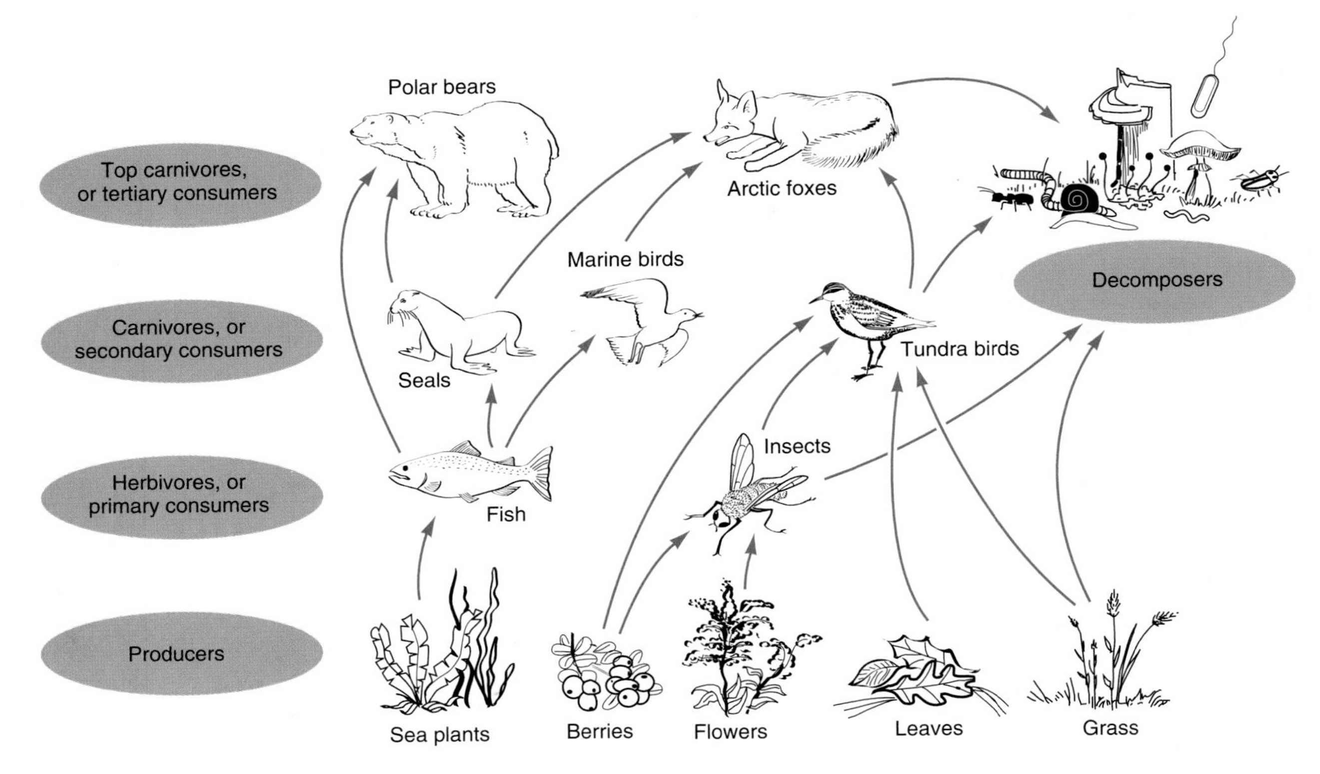

Figure 25.1
The members of a community form a food web in which species are said to occupy various trophic levels. The concept of a food chain or web originated in the 1920s with Charles Elton, who described this food web on Bear Island off the northern coast of Norway. A full description of the community would involve many species and would be far more detailed than this.

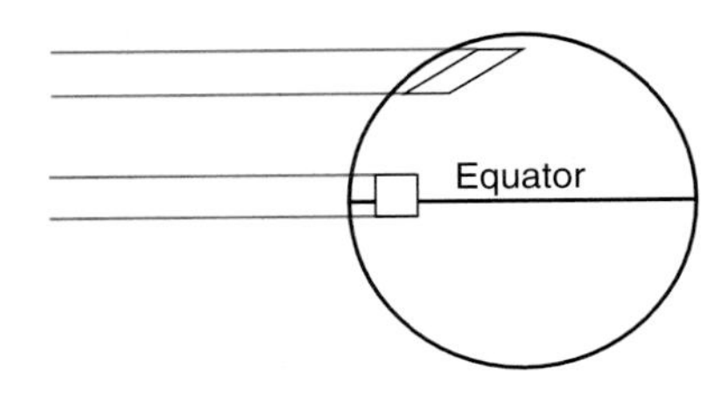

So the tropical air gets hotter than elsewhere. It expands, becomes less dense, and therefore rises. As this warm air rises into the upper atmosphere, it cools and deposits the water it was carrying as rainfall (Figure 25.2*a*), so the tropics experience a great deal of precipitation. The rising tropical air moves away from the equator—in other words, toward the poles—carrying heat with it, while cooler air at the surface moves in from the poles to replace the warm air. These movements create rotating cells of air, called Hadley cells after George Hadley who first proposed this model of the atmosphere in 1753. The atmosphere is actually divided into three cells in the Northern Hemisphere and three more in the Southern Hemisphere (Figure 25.2*b*). There is low pressure under the columns of rising air and high pressure where the air descends, as it does in the subtropics and near the poles.

The earth constantly rotates eastward as the meridional winds blow. Since the atmosphere is carried along with the earth, every bit of air has momentum in the direction of the earth's rotation; air at the equator has the most momentum (because it is farthest from the earth's axis and therefore rotates in a large circle), while air closer to the poles has less momentum (because it is rotating in smaller circles):

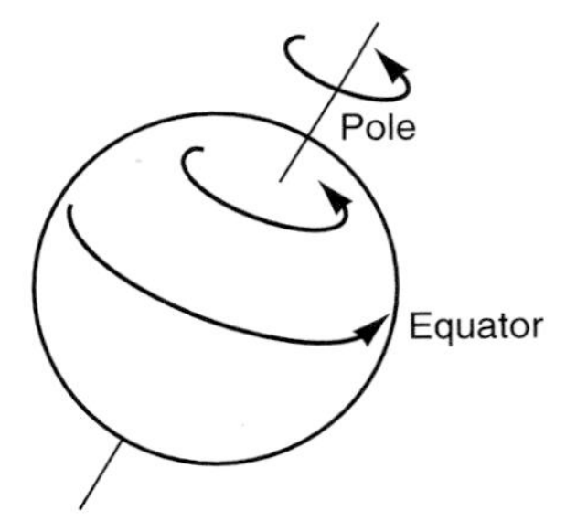

Air moving toward the equator therefore doesn't have enough momentum to keep moving with the earth below. It falls behind, toward the west, creating westward winds moving from temperate latitudes toward the equator:

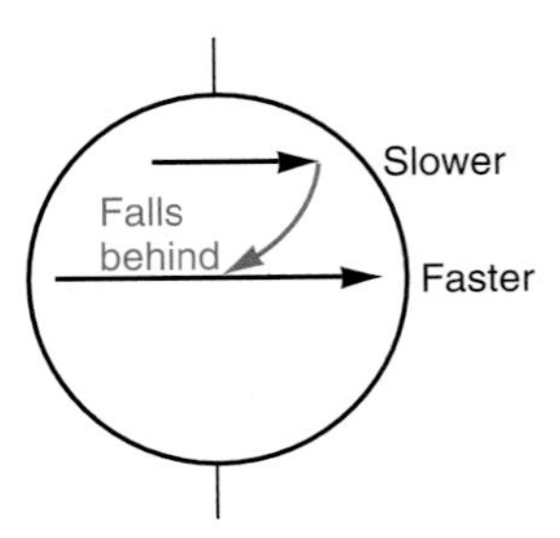

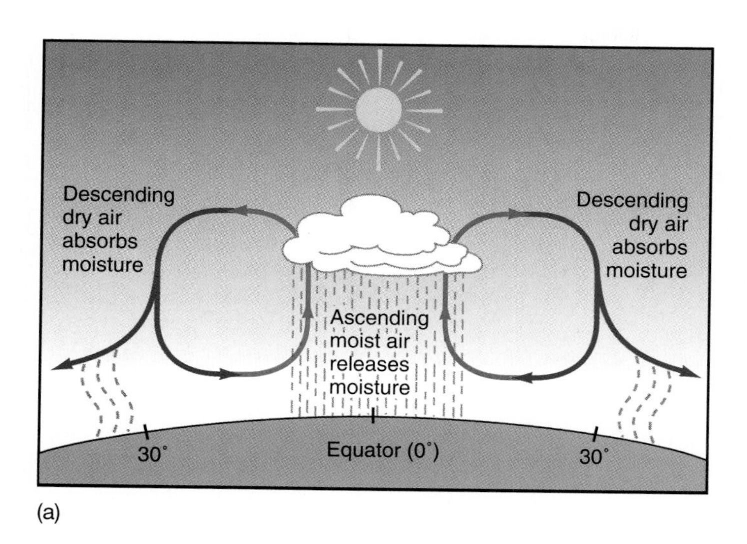

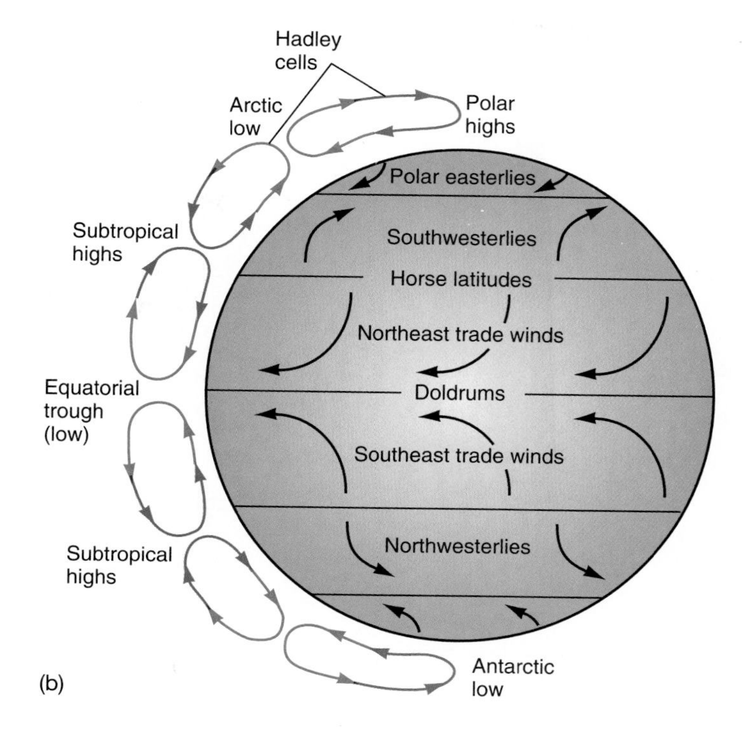

Figure 25.2

(a) Since the earth receives more sunlight at the equator than at any other latitude, heated air rises there *(red arrows)* and cools. Because cool air holds less water vapor than warm air, the excess water is released as rain, creating a humid tropical zone. The cool, drier air moves toward the poles and continues to cool. This cooler, denser air descends again around 30° latitude north and south *(blue arrows),* where it absorbs water vapor and creates arid subtropical zones. *(b)* These patterns of rising and descending air create three circulating cells of air extending around the earth in each hemisphere, producing subtropical and polar regions of high pressure where the air descends *(red arrows)* and equatorial and arctic regions of low pressure where it rises *(blue arrows).* Falling air spreads both north and south from 30°, completing a subtropical cell in each hemisphere. Air in a temperate-zone cell between about 30° and 60° accumulates water as it moves poleward and deposits this water as precipitation around 60° latitude. A third cell in the polar regions carries cool, dry air toward the poles and back toward the equator. Since the earth is rotating, air at low latitudes is moving faster than air at high latitudes. So air moving poleward moves toward the east, creating the southwesterly winds in the Northern Hemisphere and northwesterly winds in the Southern Hemisphere; air moving toward the equator moves toward the west, creating the northeast and southeast trade winds of the subtropics.

Conversely, air moving away from the equator continues to move faster than the earth below it, creating an eastward wind.

The force creating these eastward and westward winds is called the *Coriolis force.* The Coriolis force also adds a special twist (pun intended) to air moving from a high-pressure to a low-pressure region. In the Northern Hemisphere, a low-pressure center acquires a counterclockwise movement of air around it (a *cyclone*), while a high-pressure center acquires a clockwise movement around it (an *anticyclone*).[1] These directions are reversed in the Southern Hemisphere.

All these factors create the overall pattern of the earth's air circulation (Figure 25.2*b*): Prevailing westerly winds over much of the temperate latitudes, tropical easterly winds, and subtropical high-pressure regions in between. Because the earth's axis is tilted relative to its orbit, different latitudes face the sun directly at different times of the year, causing the wind patterns to change in a seasonal cycle. The winds, combined with flows of ocean currents and precipitation, create the climates we live in. Climate, let us note, is different from weather; it is a long-term average of weather as determined by temperature and precipitation. Most North Americans and Europeans live in warm, generally humid temperate climates that are getting warmer year by year, even though they experience freezing cold spells and snow at times.

25.3 Ocean currents flow in horizontal surface patterns and in vertical mixing currents.

The factors that create winds also produce water currents in the oceans (Figure 25.3). Water, however, being much denser and more viscous than air, circulates vertically much more slowly. Since the sun only heats the top 100 meters of the oceans, they are stratified, with a lot of circulation near the surface and much less below.

In the upper levels, ocean currents flow in a pattern comparable to the air pattern, but the water is deflected as it strikes the

[1]Mnemonic: Picture a clock high on the wall over a low counter. High pressure = clockwise; low pressure = counterclockwise.

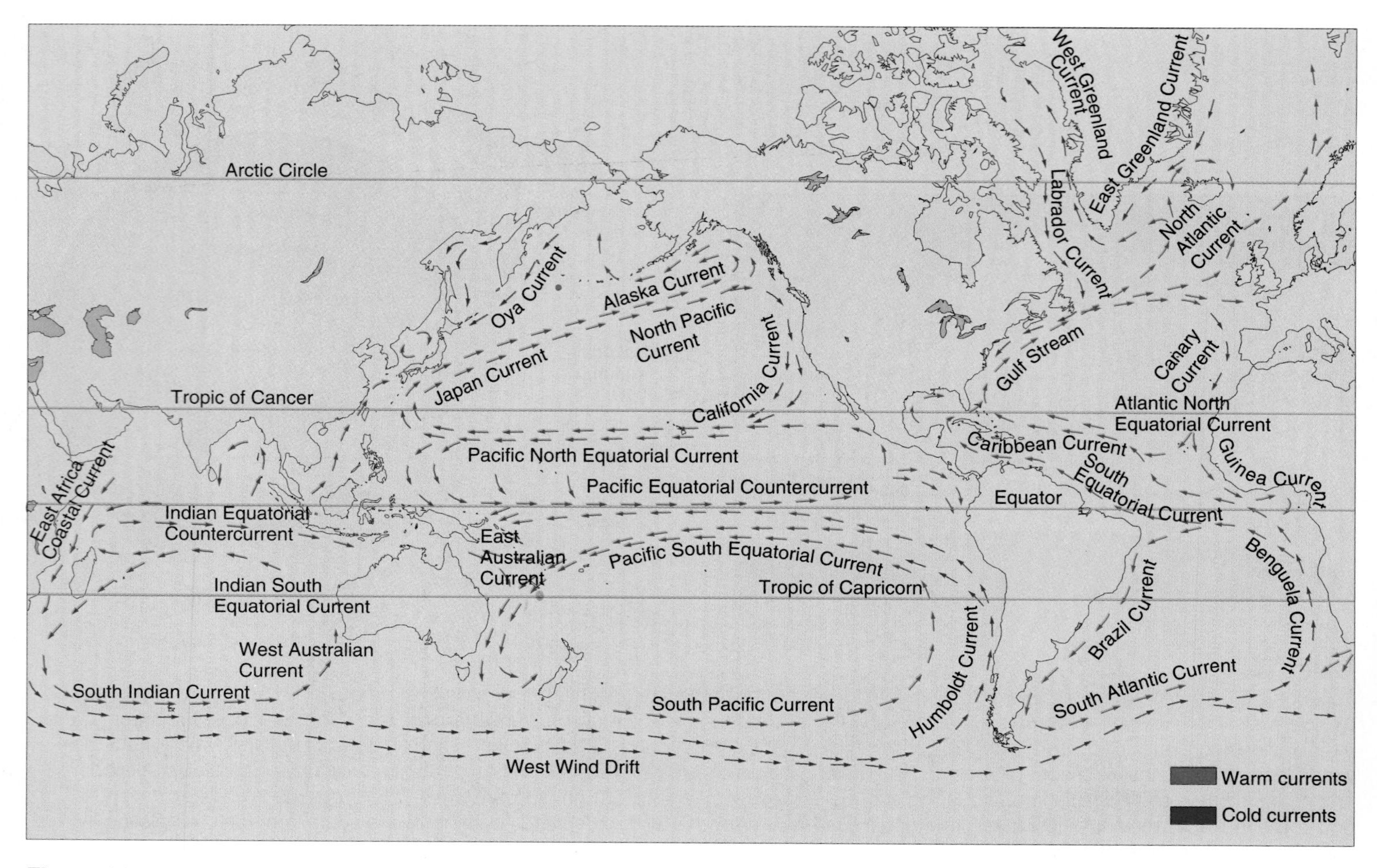

Figure 25.3
Surface currents of the oceans tend to form large circular patterns.

continents, throwing the currents into a series of circular patterns, or *gyres*. Since water changes temperature slowly, these massive water currents act as reservoirs of heat or cold, and therefore have great influences on world climate. Thus the Gulf Stream that flows northeast across the Atlantic Ocean carries enough tropical heat to make northwestern Europe much warmer than it would be otherwise; Moscow and Goose Bay, Labrador, are at latitudes comparable to Paris and London, but they are much colder. Similarly, along the western coasts of the Americas, the California current from the north and the Humboldt current from the south cool the land nearby considerably.

The oceans also experience a massive vertical movement of water, due largely to differences in the density of water at different temperatures. We saw in Chapter 3 that the hydrogen-bonded structure of water gives it unusual properties; when water freezes into ice, its density is only 0.92 g/ml, and its density increases to a maximum of exactly 1 g/ml at 4°C. The density of water also depends on the materials dissolved in it, and the water of the oceans is a solution of many mineral salts. Where this water freezes, it leaves its salts behind, thus increasing the density of the nearby liquid water. These factors of temperature and composition, along with winds blowing over the water's surface, create a phenomenon called *overturn*, which mixes the water vertically. We will examine overturn in a lake in Section 25.5. In the earth's oceans, overturn occurs massively in only a few places (Figure 25.4): There is a massive downward flow in the North Atlantic Ocean, between Greenland and Europe, a secondary downflow along the edge of the Antarctic ice mass, and a third in the Indian Ocean. The cold water falling into the ocean depths must be replaced by surface water, which comes primarily from the Pacific Ocean. This vertical flow is the only way oxygen from the surface gets into the ocean depths, and without it there would be little life in the oceans except near the surface.

25.4 Oceans support several kinds of communities.

Open oceanic communities

The oceans that cover almost three-quarters of the earth's surface are cold, largely dark bodies averaging 3 kilometers in depth. Only about 10 percent of the known species occur in the oceans, with the rest occupying terrestrial and freshwater habitats, but oceanic species represent a much greater variety

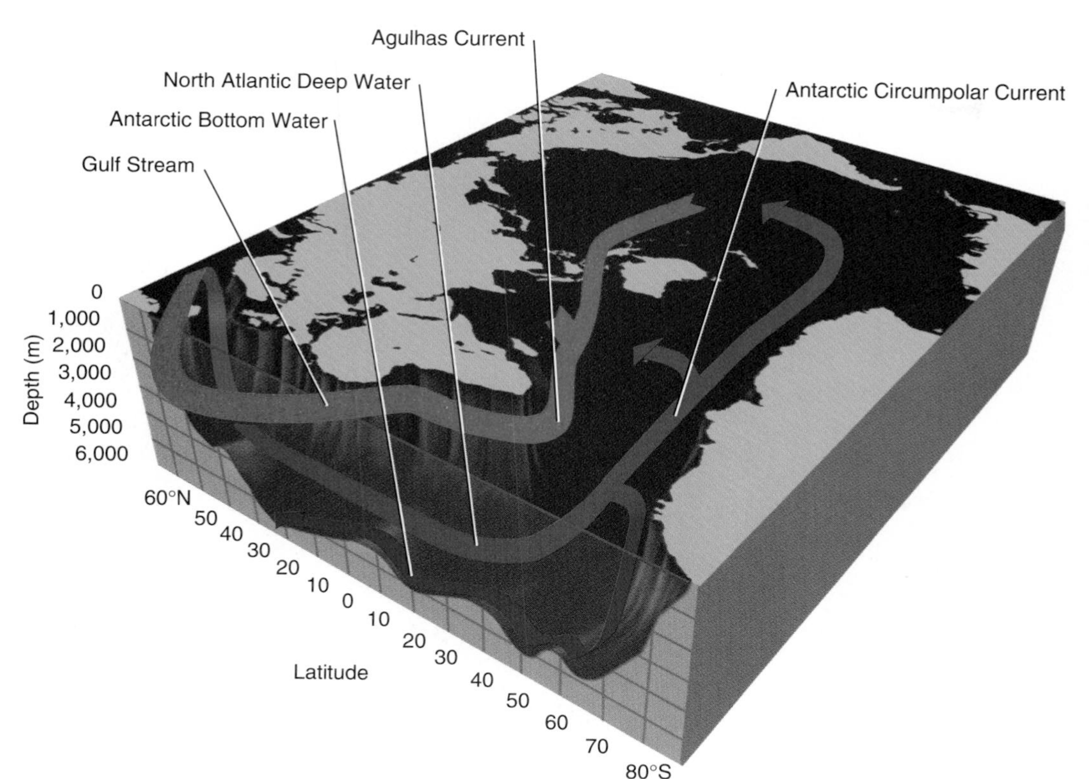

Figure 25.4

Surface water in the oceans becomes dense enough to fall to the ocean floor in two major places: the North Atlantic and the edge of Antarctica. The North Atlantic Deep Water moves southward, and most of it joins the Antarctic Circumpolar Current, which rises as it moves eastward. The Atlantic currents are replenished with warm water from the Pacific and Indian Oceans through the Agulhas Current, contributing to the Gulf Stream.

of phyla than are found on land. Oceanic organisms occupy two major regions: the **neritic province** in shallow waters over the continental shelves and the **oceanic province** of the open seas. Habitats within these provinces are defined by the depth of the water, the shape of the ocean bottom, and the penetration of light, as shown in Figure 25.5. Organisms that live on the ocean bottom are **benthic,** and the open water is divided by depth into a few **pelagic zones** (Greek, *pelagos* = sea). The surface (epipelagic) zone, to a depth of 200 meters, is the region of the ocean where enough light penetrates to support photosynthesis. The diverse community of this zone consists of **plankton,** the minute organisms that are tossed about at the mercy of the waves and currents (Figure 25.6), and **nekton,** animals that are able to swim independently. Plankton provide food for the nekton. Much of the plankton consists of algae, the **phytoplankton,** which are very significant in the world's economy since they carry out about two-fifths of all the photosynthesis on earth. **Zooplankton** is made of minute animals, including the larvae of larger animals, that feed on the phytoplankton.

The organisms of every ecosystem produce waste material, including detritus, the particulate breakdown product of organisms. Detritus from the epipelagic zone feeds the lower marine communities, and the organisms that thrive below are those that are able to feed on particles of detritus, including bacteria that decompose it. Recent explorations of the abyssopelagic zone, from 4,000 to 6,000 meters, have suggested that it is much more diverse biologically than had been thought. This is a world of bizarre animals with odd shapes, and many abyssal animals are luminescent or carry patches of luminescent

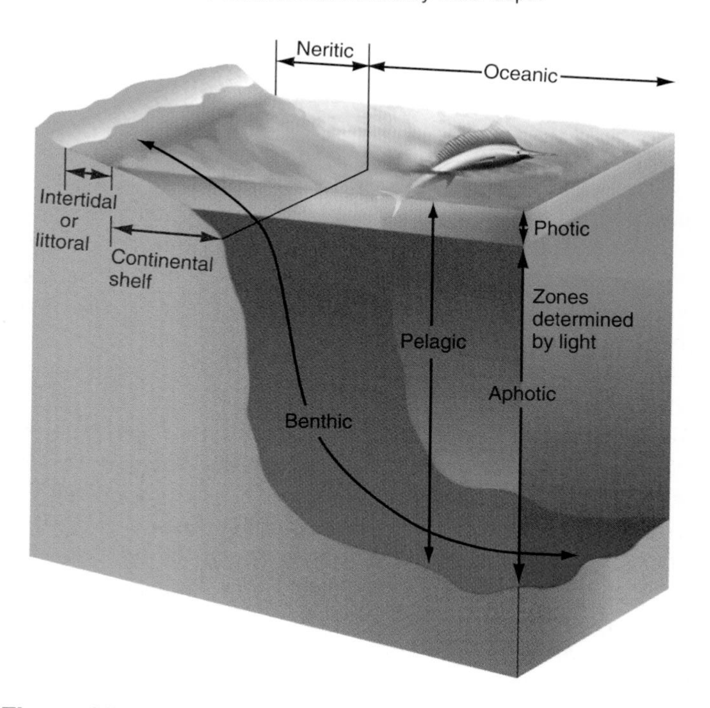

Figure 25.5

Marine habitats are determined by three criteria: *(1)* The oceanic province over the open seas contrasts with the neritic province over the continental shelf, including a littoral, or intertidal, zone between high- and low-tide lines. *(2)* The benthic zone is the ocean bottom; all the water above it is pelagic. *(3)* The photic zone is the surface region that light penetrates; below it is the aphotic zone.

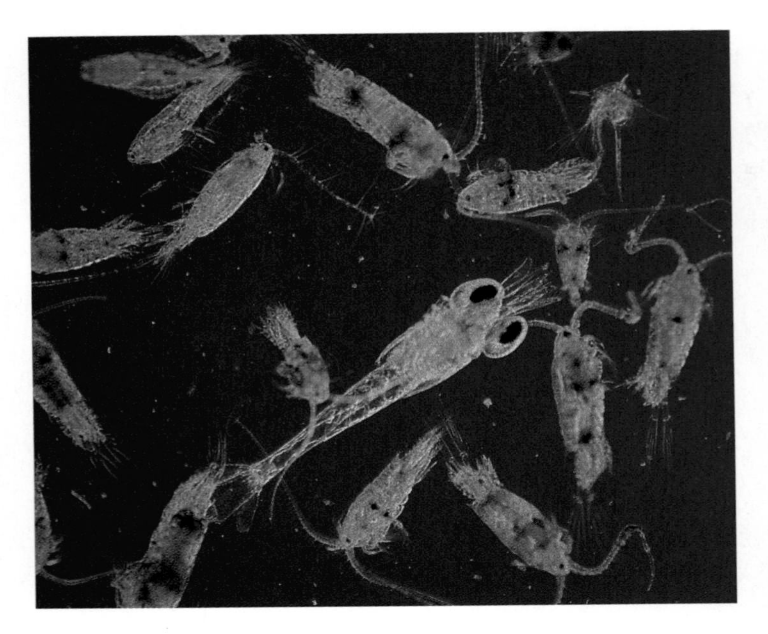

Figure 25.6

A net dragged through the ocean surface yields a sample of plankton, organisms form the base of the oceanic food web. This sample includes microscopic algae (phytoplankton) and small animals (zooplankton). Plankton are so small that they drift or swim weakly; animals strong enough to swim independently are called nekton.

Figure 25.7

A variety of fantastic fishes inhabit the abyssal zone of the ocean.

bacteria, so they produce light to signal one another or attract prey (Figure 25.7).

A remarkable abyssal community that is independent of the sun's energy and photosynthesis has been discovered around hydrothermal vents, where scalding water (350°C) emerges from the ocean floor. Chemoautotrophic bacteria use hydrogen sulfide (H_2S) in the water escaping from the vents for their energy, and they form the base of a food web that includes a variety of other organisms, such as the giant beardworms (phylum Pogonophora) shown in Figure 25.8.

Neritic communities

Most marine species inhabit the neritic province, even though its area is small. Where the ocean pounds the land, survival favors organisms that can anchor themselves securely against the waves. Here we find a very diverse fauna and flora, including kelps and other algae, barnacles and mussels that have literally glued themselves to the rocks, sea anemones attached securely in holes and crevices, sea stars that hold on by their many tube feet, and marine worms burrowed into the sediment (Figure 25.9). A part of the neritic zone called the **littoral** or **intertidal zone** is exposed at low tide, and its inhabitants must be able to withstand exposure to the air; they also become prey to a special class of predators, such as shorebirds that hunt on exposed mud or rocks. The littoral zone includes tide pools, depressions in the rocks that remain filled with water when the tide recedes (Figure 25.10), which are great places to study a marvelous variety of marine life.

Figure 25.8

A community based on the energy of hydrogen sulfide forms around deep-sea vents. Giant pogonophoran worms are among the unique members of this community.

Estuaries

An **estuary** is a particularly fertile type of neritic habitat consisting of a partly enclosed body of water where rivers empty into the ocean (Figure 25.11). Here the water is **brackish,** with a salinity intermediate between that of seawater and freshwater. Estuaries support an extraordinarily varied flora and fauna

Figure 25.9

Some organisms typical of the neritic zone include sea stars (starfish), purple sea urchins, and many kinds of molluscs attached to the rocks or burrowing in the mud.

Figure 25.10

Tide pools, such as this one in Olympic National Park in Washington state, provide extraordinarily varied samples of the marine environment.

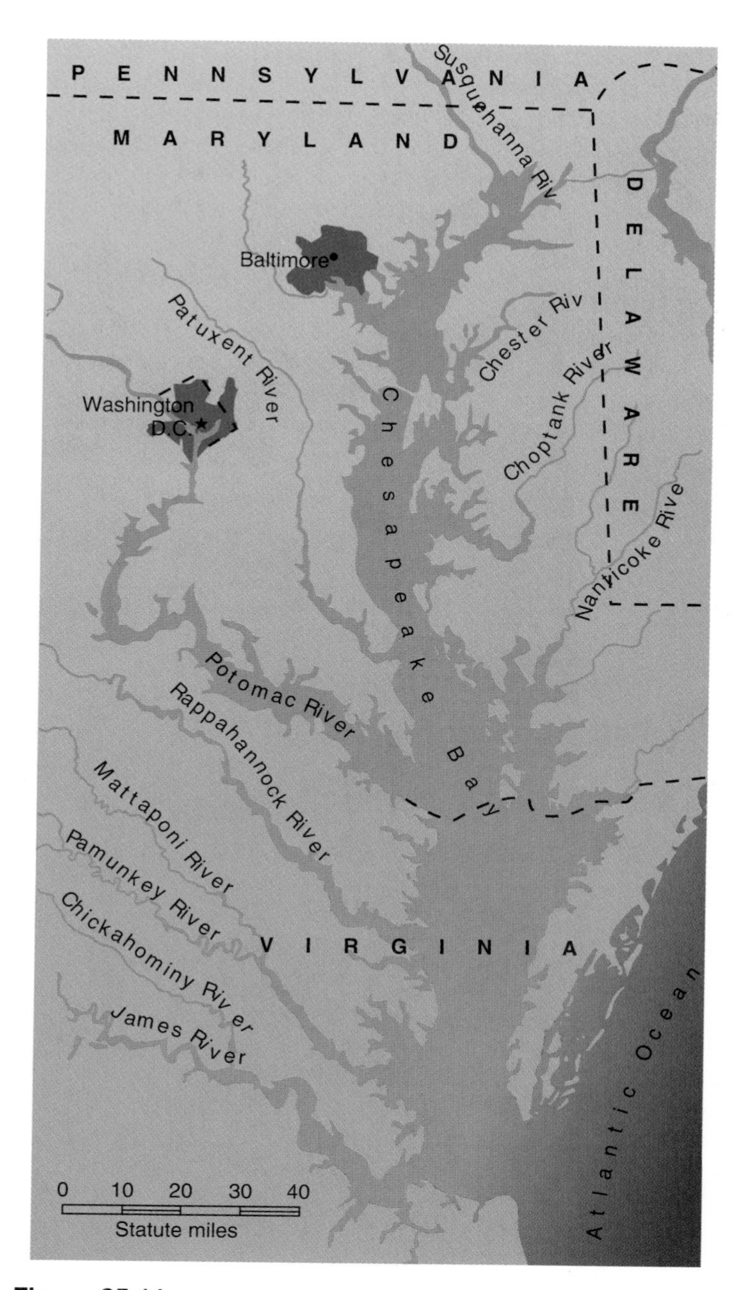

Figure 25.11

Chesapeake Bay that receives the water of several large rivers, with over 11,300 kilometers of shoreline. It provides large areas of littoral habitats, and its waters are spawning grounds for many kinds of fish and shellfish. But it is a classic example of a once productive environment that is at severe risk of being polluted by human activities, including industrial wastes and runoff from farms.

adapted to this condition: algae, plants, and the animals that feed on them. Estuaries have a higher concentration of nutrients than any other marine surface waters because they are fed from three sources: nutrients carried in by the surrounding rivers, currents carried in from the ocean, and abundant nitrogen-fixing bacteria in the estuarine mud. These nutrients support luxurious plant communities in the intertidal zone around the edges of the estuary, plants in the estuarine bottom, and an especially rich phytoplankton. On this foundation of producers, estuarine marshes and open waters support great populations of molluscs, crustaceans, and fish. Estuaries such as Chesapeake Bay therefore have enormous economic value, but they are also highly vulnerable to pollution carried in by rivers and by runoff from the surrounding land.

Coral reefs

Coral reefs support some of the greatest species diversity of all the ecosystems in the seas, and they provide a nice contrast to the more familiar terrestrial ecosystems. Three types of reefs occur within a narrow tropical or semitropical zone where the water remains at an average temperature of 27°C (Figure 25.12).

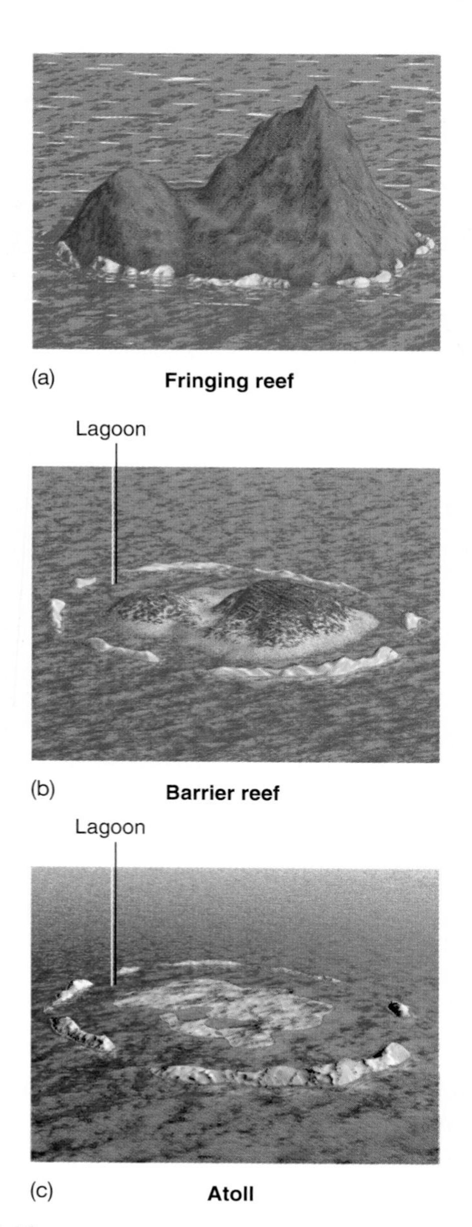

Figure 25.12
Coral reefs grow in three forms: fringing, barrier, and atoll.

This rarely occurs more than 22° latitude north and south of the equator. A *fringing reef* is one that runs along the edge of an island or continent, close to the land. A *barrier reef* stands some distance from the land, separated from it by a lagoon; for example, the Great Barrier Reef of northern Australia is separated from the mainland by a lagoon about 50–150 km wide and up to 40 meters deep. An *atoll* is a circular formation around a sunken island. All three types are built of the rocky skeletons of corals, so the physical foundation of the reef community differs from others in that its major structures are built by animals, not by plants.

Corals must grow on a solid foundation, but they live only in shallow waters near the surface of the reef. Since some reefs rise hundreds of meters above the ocean floor, the question arises as to how reefs could have developed. They surely could not have been built upward from corals growing deep in the ocean. The answer, given by Charles Darwin, is that all reefs started out as fringing reefs, in shallow water. The very deep reefs then developed over time, building upward as the ocean floor gradually sank or as the sea level slowly rose. The coral grows fast enough to more than keep up with geological changes, leaving communities that are perpetually in shallow water.

Reef communities are strongly dominated by a marvelous assortment of brilliant animals. The corals are cnidarians that secrete a calcium carbonate skeleton and grow in thick, often beautifully structured colonies (Figure 25.13). But the reef community also necessarily includes producers. Cyanobacteria abound in the tidal zone along the nearby land, green algae in the lagoon between land and reef, and coralline red algae—which also secrete calcium carbonate—are mixed among the corals. All these algae are the foundation of the reef food chain, and the coralline algae help glue the reef together. Furthermore,

Figure 25.13
Coral reefs are highly varied communities.

in among the tissues of the corals, and living symbiotically with them, are tiny algae known as *zooxanthellae* that use nitrogenous animal wastes for their growth, add oxygen to the water, and also promote secretion of calcium by the corals. The corals themselves are carnivores that live mostly on zooplankton.

Just as a forest creates a multitude of niches and crannies in which various organisms grow, the reef builders produce a fertile environment that houses a variety of plants and animals. Chief among these are annelid worms living in tubes in the substratum and molluscs such as clams. Both are filter feeders that secrete mucous layers in which they trap their food. In addition to the particles of organic matter and tiny organisms they extract from the water, they also absorb dissolved proteins and amino acids, and probably other dissolved monomers. Such organic compounds are abundant in the fertile waters of the reef, and it has been estimated that they may supply 10 percent of the food of reef animals.

Echinoderms, such as sea stars and sea urchins, wander through the reef community, preying on other animals. Many crustaceans also inhabit the reef community, the smaller being food for larger animals and the larger ones themselves feeding on medium-size animals and on plankton. And then there are the fish, many of them exotic creatures with brilliant colors and unusual shapes (Figure 25.14). They browse among the coral, feeding on vegetation and plankton. The reef also houses carnivorous fish—barracuda, groupers, moray eels, and other large predators that constitute the top layers of the food web.

The reef community is interesting partly because its members have fossilized so well. The calcium carbonate skeletons of its chief architects turn into limestone, and these largely unaltered skeletons, along with the shells of other reef inhabitants, have left a revealing fossil record at least 600 million years old. Evidence shows that previous coral reef communities were quite different from modern ones. They have gradually changed as new types of animals and plants evolved and became dominant for a time, providing excellent examples of the species turnover and extinction discussed in Chapter 28. We will see that communities may change substantially over time. Geological forces are always changing the face of the earth, while astronomical changes alter its climatic patterns. So new types of organisms will continue to evolve and take advantage of these changes, moving in and becoming dominant whenever the opportunity arises.

Figure 25.14
Coral reefs are home to many kinds of colorful fishes.

25.5 Freshwater habitats are among the richest and most varied on earth.

Freshwater habitats—ponds, lakes, bogs, marshes, swamps, and rivers—support an exceptional variety of life. Freshwater animals include: many kinds of fish and amphibians; a diverse assemblage of insects, which supply food for other animals as their zygotes grow into immature stages (nymphs) and then into adults; many kinds of snails and clams; some cnidarians such as *Hydra;* and even a few sponges. The energy base of aquatic communities consists of phytoplankton of many kinds of algae, as well as larger plants. Ponds and lakes are open bodies of water with rooted plants that provide food and shelter for animals and, of course, oxygen. These are generally classified as *submersed plants,* which remain below the water surface, *floating-leaved plants* such as water lilies whose leaves float at the surface, and *emergent plants,* which are anchored in the bottom and extend above the surface of the water (Figure 25.15).

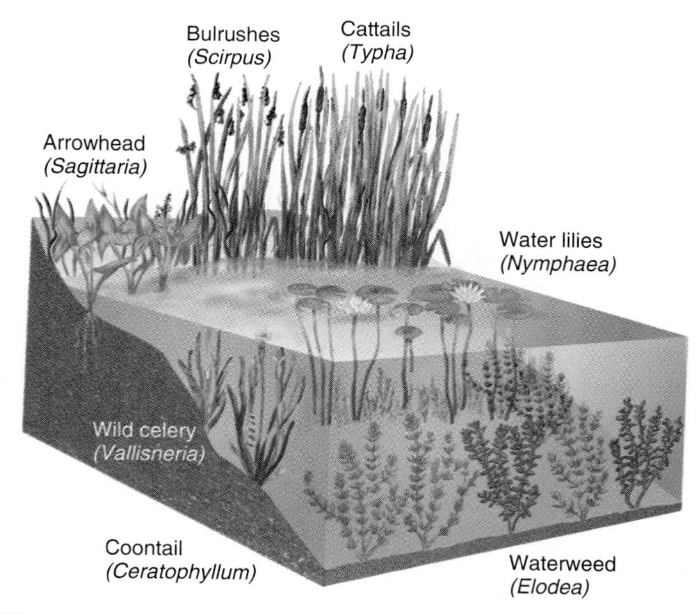

Figure 25.15
Freshwater plants rooted in the bottom include submersed plants, such as coontail *(Ceratophyllum),* waterweed *(Elodea* or *Anacharis),* and wild celery *(Vallisneria);* floating-leaved plants such as water lilies *(Nymphaea);* and emergent plants such as cattails *(Typha),* bulrushes *(Scirpus),* and arrowheads *(Sagittaria).*

Figure 25.16

A freshwater lake may support a rich community with a complicated food web.

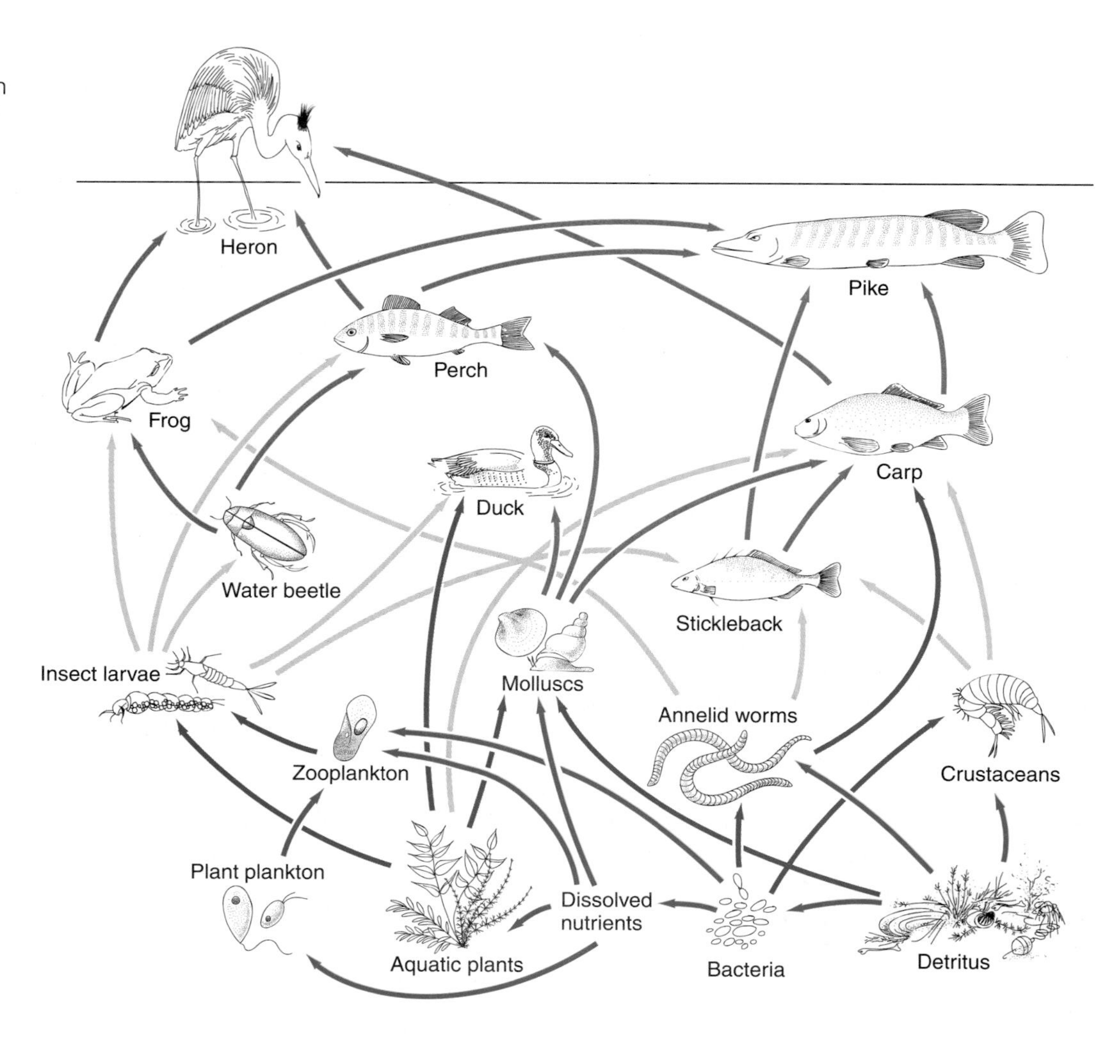

Two categories of ponds and lakes are distinguished by their nutrient levels. **Oligotrophic** lakes are nutrient-poor because the streams and groundwater running into them carry few nutrients. They are typically deep, and their lower depths are rich in oxygen. In contrast, **eutrophic** lakes have abundant nutrients, both inorganic and organic, and support an abundant, varied phytoplankton and many attached plants. Lake and pond habitats are divisible into three zones analogous to the zones of the ocean: a *littoral zone* at the edge; a **limnetic zone** where light penetrates and supports phytoplankton and thus a community of other organisms (Figure 25.16); and a **profundal zone** that is too deep to be lit.

Large lakes also show a dynamic called **thermal stratification,** another consequence of the fact that the density of water is maximum at 4°C. In winter, water that cools to 4°C sinks to the bottom of a lake, leaving colder water near the surface, where it freezes:

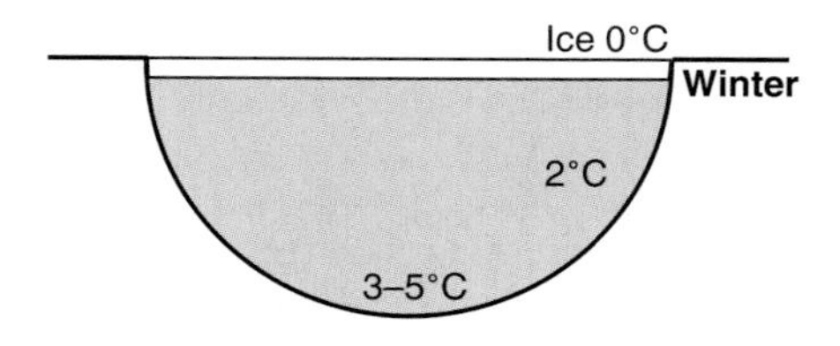

Below the ice, where the temperature varies between 0 and 4°C, many plants and animals survive quite well. When spring arrives, water from the melting ice warms to 4°C, and a layer of this temperature forms at the surface. Winds that did not affect the surface when it was frozen now push this dense water to one side of the lake, where it sinks lower into the surrounding water because of its weight, displacing lower layers of water to the surface; this phenomenon, called **spring overturn,** carries oxygen-rich water into the depths of the lake and brings up nutrients from the depths that support renewed biological activity near the surface as the temperature rises:

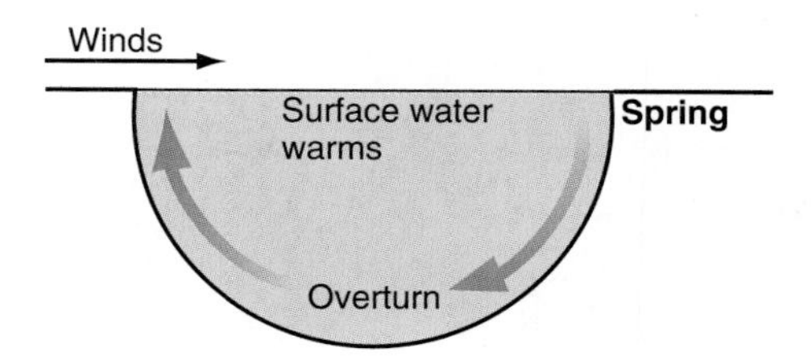

In summer, the surface water becomes quite warm to a considerable depth; the temperature of the upper water decreases gradually with depth and then drops sharply in a layer called the **thermocline:**

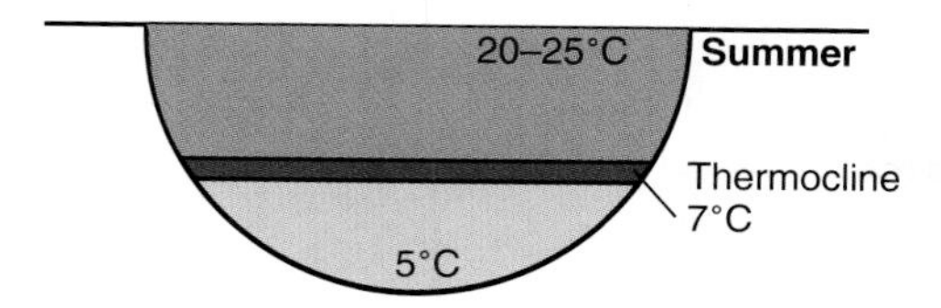

As autumn turns into winter, the surface temperature again falls to 4°, and at this point the lake experiences a **fall overturn,** again mixing the lake water and bringing nutrient-rich water to the surface:

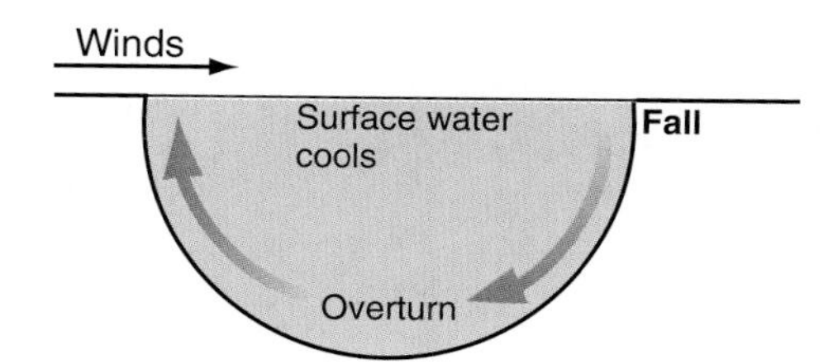

During the summer in a eutrophic lake, aerobic decomposers in the region below the thermocline use up the available oxygen and make these deep waters stagnant. The lowest waters are only refreshed when the lake becomes mixed during the fall overturn. As Edward Birge noted in 1904, such a lake virtually becomes two lakes, one superimposed on the other. The lake above the thermocline is warm, and its water is stirred by the wind; below the thermocline is a cold lake with denser, more viscous water virtually lacking in oxygen, where decomposition has been going on and the resulting gases accumulate. The upper lake supports the kind of community shown in Figure 25.16. The lower lake supports a different characteristic fauna, including red worms such as *Tubifex,* the larvae of small flies (midges) such as *Chironomus* and *Chaoborus,* some small clams, and microscopic crustaceans (see Figure 35.13).

Most natural lakes are oligotrophic but tend to become eutrophic over time as sediment builds up in them. They are highly susceptible to rapid eutrophication, which happens when they are enriched with nutrients in the runoff from fertilized fields and the waste streams of cities. A classic case of eutrophication was the transformation of Lake Erie, a lake once valued for its clear water, abounding in fish such as sturgeon, whitefish, and pike. As phosphate and nitrate pollutants flowed into the lake from the surrounding farms and cities, they supported the growth of microorganisms and the development of algae blooms, mats of filamentous algae that sometimes covered square kilometers of the lake surface. The decomposition of these algae during the hot summers depleted the lake of oxygen and killed many species of fish. The commercially valuable fish were replaced by other species, such as carp and catfish, and the lake's natural ecosystem was badly disrupted. While recent measures to clean up pollution in Lake Erie have been remarkably successful, the lake remains quite different from its pre-polluted condition.

Figure 25.17
The flora of a bog includes sphagnum moss, blueberries, leatherleaf, and carnivorous plants such as the pitcher plant.

In contrast to lakes and ponds, *bogs* are small, poorly drained wetlands with quite acidic water that few plants and animals can tolerate. They are dominated by *Sphagnum* mosses, which grow well in these waters (Figure 25.17). Decomposition in a bog is slow, so the partially decomposed plant tissue accumulates as *peat,* often to a great depth, giving bogs a thick, spongy base in which large animals—and people—can literally get bogged down. In this environment most plants have a hard time taking up nitrogen, so the flora is quite restricted to hardy evergreens, sedges, and a few other plants that can tolerate acidic, nutrient-poor conditions. These include evergreen shrubs such as rhododendrons and mountain laurel, and especially leatherleaf, a plant with small, leathery, greenish-brown leaves whose thickly intertwined roots help hold the spongy base of the bog together. Blueberries are also common here, as are the peculiar insectivorous plants such as sundews and pitcher plants, which trap and digest insects (see Section 40.15 and Figure 40.17).

Marshes and swamps are superficially similar to one another but distinguished by two criteria—water flow and vegetation. Water moves through marshes, although sometimes very slowly, and they are dominated by soft-stemmed

Figure 25.18
A North American marsh is a mosaic of tall grasses, reeds, and cattails. It harbors such animals as muskrats, weasels, raccoons, ducks, rails, and blackbirds.

herbaceous plants, especially reeds, sedges, and cattails. In swamps, water does not circulate, and the dominant plants are trees and shrubs. A typical North American marsh is a diverse blend of tall grasses, reeds, and cattails where birds such as ducks, wrens, rails, and blackbirds nest. These marsh plants are the primary producers of the ecosystem, supplemented by phytoplankton; the plants serve as food for raccoons, mice and muskrats, which are preyed upon by mink (Figure 25.18). The tiny pothole marshes scattered throughout the prairies of North America harbor up to 140 ducks per square mile and are the principal breeding grounds for ducks. One of the great ongoing conservation battles is the struggle to preserve these marshes in the face of agricultural interests trying to convert them into farmland.

The best-known North American swamps are vast northern wetlands dominated by trees such as black and white spruces, larch, and northern white cedar. Further south, swamps are characterized by red maple, ashes, yellow birch, and black gum; those of the deep south are dominated by cypress (Figure 25.19); and western swamps typically contain western hemlock and red alder. The ecology of swamps is similar to that of forests, except that they are much wetter, and their lower levels are characterized by plants adapted to those conditions, such as skunk cabbage.

Although open freshwater bodies constitute only about 2 percent of the earth's surface, vast areas of the earth were natural wetlands before human interference. Wetlands are fascinating places. One of the tragedies of our time is the rapid loss of these fruitful, highly productive areas that are crucial for the survival of many species, including some of our most beautiful and interesting plants and animals. Wetlands are strongly connected to and influenced by the surrounding land. Organic material from

Figure 25.19
Cypress swamps are characteristic wetlands of the southern United States.

healthy communities constantly falls and seeps into the water, enriching it, but where terrestrial ecosystems have been disrupted, the water ecosystems also suffer. Where a surrounding forest has been clear-cut, silt and debris wash into a river, clog it, and kill its natural fauna and flora, in part by blocking the sunlight and reducing photosynthesis; also, pollutants that accumulate on land wash into the water ecosystem and poison it.

The loss of wetlands is tragic not only because it eliminates fertile ecosystems and threatens the extinction of many species, but also because wetlands are natural reservoirs of groundwater. The widespread floods in midwestern regions of the United States during the summer of 1993 focused attention on the role of wetlands in flood control. In the Pacific Northwest, areas west of the Cascade Range are victims of deforestation in the mountains above them and the filling of former wetlands; so extensive flooding is almost guaranteed each winter and spring as mountain streams, carrying excess water no longer held by the forests, cascade down into lowlands deprived of the marshes that once absorbed this water. Flooding in urbanized areas that still retain their wetlands is much less severe than in those that have filled in their wetlands and built on them; by allowing unlimited "development" of wetlands, we pay dearly in real economic terms.

Exercise 25.1 Bogs are quite deficient in nutrients. Which trees are best able to grow in bogs—deciduous trees that shed their leaves every autumn or evergreens that retain their leaves? Why?

25.6 Precipitation falls in a zonal pattern across the continents.

Superimposed on the flows of air and ocean currents is a cycle of water evaporating in one place and precipitating in another as rain, snow, or hail. This flow, which we will explore in Section 28.4, has an enormous influence on soil types and vegetation. Because warm water evaporates much faster than cold water, water vapor comes largely from warm ocean water. Precipitation falls when water-saturated air rises and cools until it can no longer hold all its vapor. Therefore, precipitation is usually greatest where air currents rise over warm water and over land bathed by warm water; precipitation is least over cold water and land bathed by it, and over the interiors of continents far from water.

If there were no continents, rainfall would be distributed over the earth in several uniform belts of heavy precipitation and dry areas (Figure 25.20*a*). Inserting two landmasses similar to the real continents (Eurasia-Africa on one side of the earth, the Americas on the other) distorts these bands, creating the pattern of climates that we see, including an S-shaped dry zone across both the Northern and Southern Hemispheres (Figure 25.20*b, c*). The topography of the land, particularly its mountains, modifies this pattern. A cross section through western North America along the thirty-ninth parallel shows how the mountain ranges influence rainfall and vegetation (Figure 25.21). Winds moving eastward off the Pacific Ocean are forced upward on the western mountain slopes, where they cool and deposit their moisture abundantly, nurturing great forests. Having lost most of their moisture, the winds leave the eastern slopes of these mountains in **rain shadows,** which are drier and support different types of vegetation.

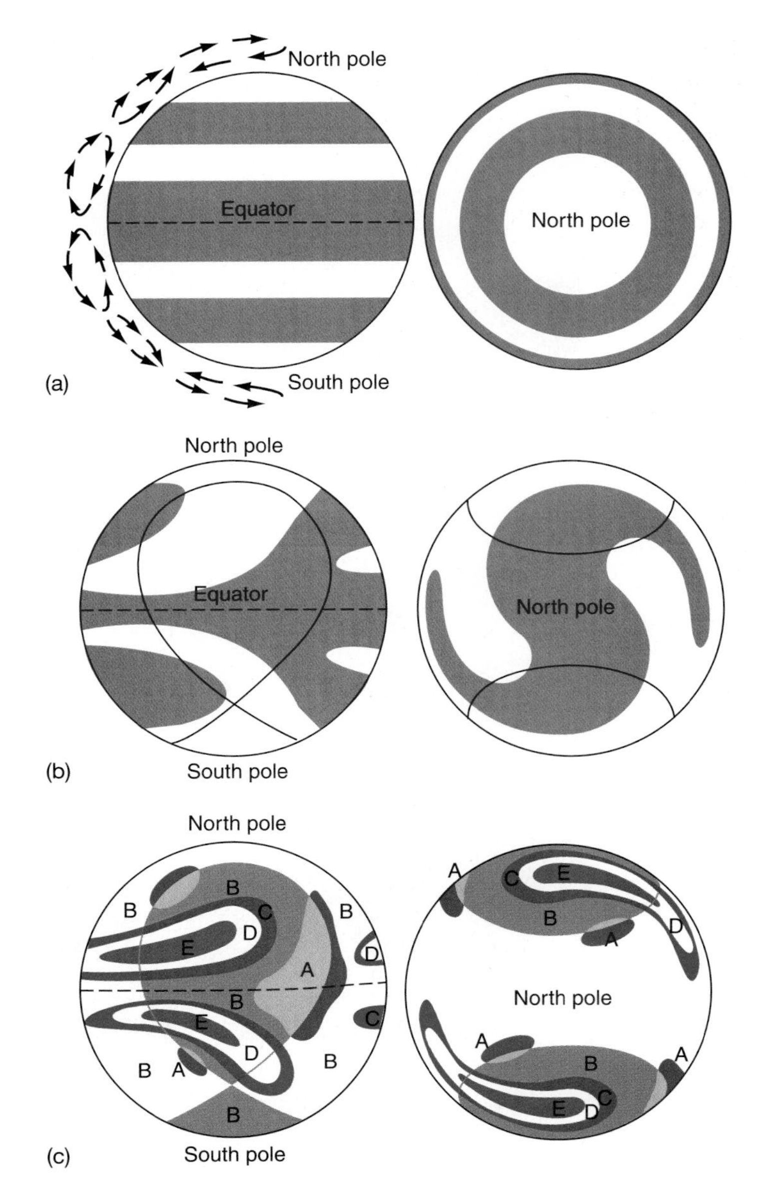

Figure 25.20

(a) On an earth without irregular continents and oceans, rainfall would be distributed in uniform belts of heavy precipitation *(color)* and dry areas *(white)*. *(b)* Two imaginary continents distort these belts. This creates the climates shown in *(c)*: superhumid (A), humid (B), subhumid (C), semiarid (D), and arid (E).

25.7 Several major terrestrial biomes cover the earth.

Most of the world's organisms live on land, and we now turn our attention to the characteristics of terrestrial ecosystems. Figure 25.22 shows the earth's major terrestrial ecosystem types, or biomes, which are defined mainly by their vegetation.

Tropical rain forest

These forests occupy extensive areas around the world where there is heavy rainfall (usually 2.5–4 meters per year) and a temperature between 20° and 28°C that changes very little with the seasons (Figure 25.23). The major rain forests are in the Amazon Basin, the Congo Basin, and parts of India, Burma, and Indonesia. They are very humid, with heavy rainfall. These forests are built of an enormous variety of tall trees whose canopies form three to five continuous layers as shown in Figure 25.23*b*. Giant emergent trees reach heights of 40–50 m, over an almost continuous main canopy of about 25–35 m. These trees capture 70–80 percent of the incident sunlight, leaving a dark, humid region below. An understory of smaller trees grows in this lower region, but vegetation near the ground is quite sparse.

The tropical rain forests are the richest biomes on earth, having an average biomass of 45 kg of plant and animal material per square meter, but as much as 80 kg in some places. They support a far greater variety of all organisms than do the temperate regions; up to 80 species of trees per acre have been recorded in places, compared with 4–8 species in temperate forests. Tropical forest communities—including spectacular, brilliantly colored birds and insects—are interrelated in exceedingly intricate food webs that biologists are just starting to understand. The nourishing fruits produced by canopy trees are an important resource for large birds such as toucans and hornbills, for monkeys, and for bats, which are dominant mammals in the tropics. These forests harbor some of the largest hawks

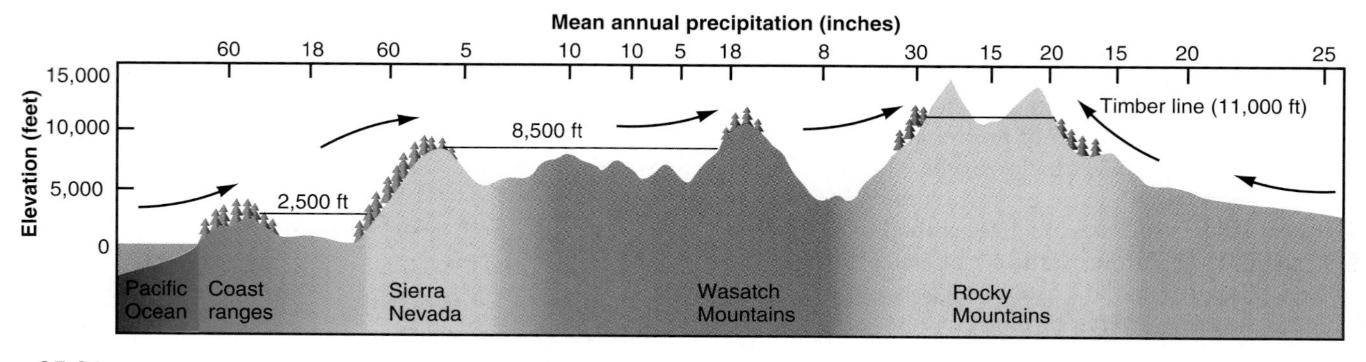

Figure 25.21

A profile through western North America shows that precipitation is heaviest on the western slopes of the mountains, where clouds deposit their water as they rise and cool, while the eastern slopes are in rain shadows.

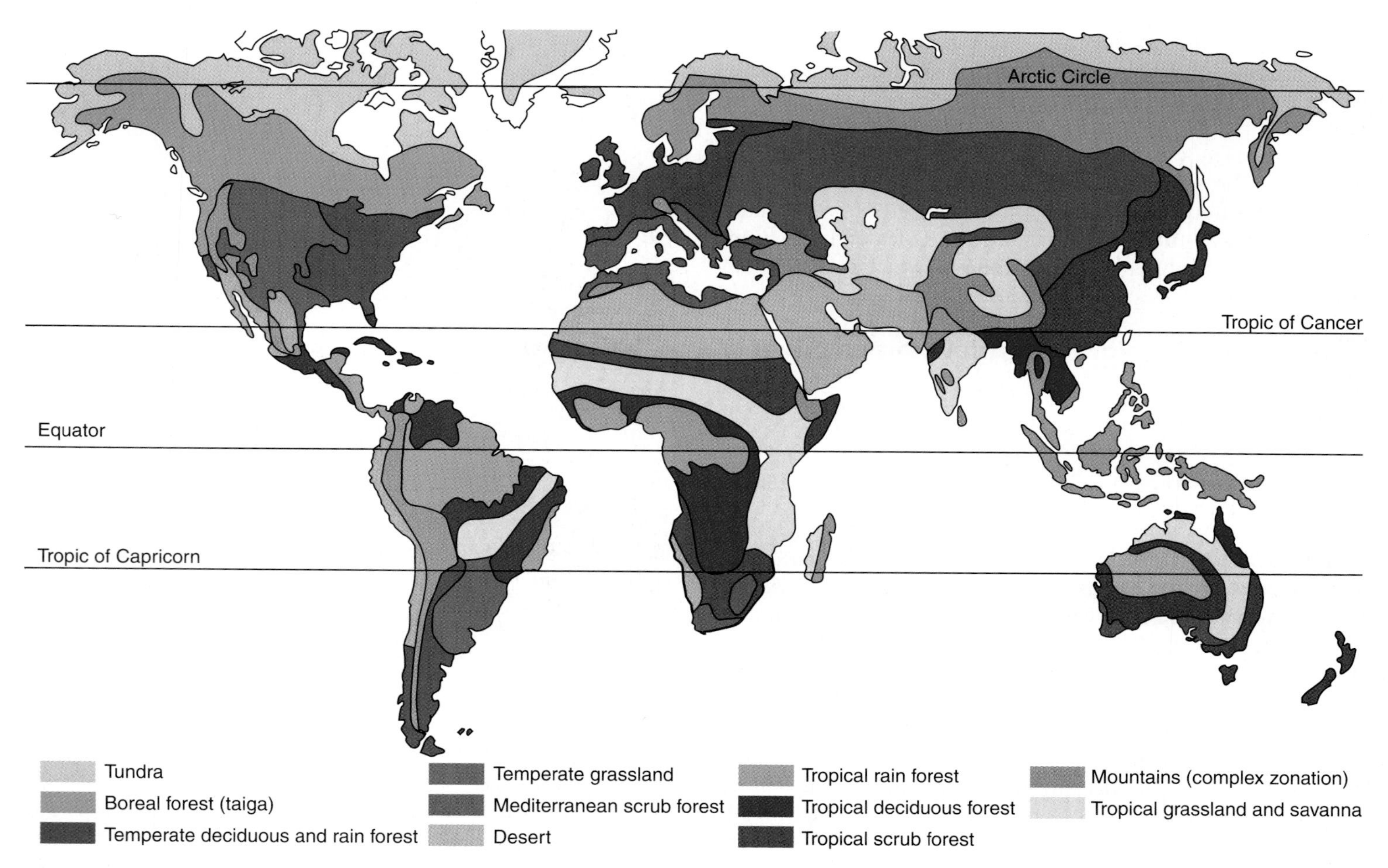

Figure 25.22

The continents are divided into several major biomes.

(a)

35 30 25 20 15 10 5 Meters

30 20 10 0 Meters

(b)

Figure 25.23

(a) A tropical rain forest habitat. *(b)* Tropical rain forests have three to five layers of vegetation: an understory near the ground, a low canopy layer 5–10 m up, another layer at about 15 m, a continuous canopy of treetops around 20 m, and emergent trees that reach to 35 or 40 m. The trees in each layer are adapted to different levels of light.

and eagles in the world, as well as top carnivores such as the clouded leopard in Southeast Asia and jaguar and ocelots in South America.

In spite of their richness in biomass and species diversity, the soil of rain forests is quite poor in nutrients. A massive amount of litter falls on the forest floor where it is rapidly broken down by termites and various decomposers, and in the high tropical heat it is quickly degraded chemically and recycled through the plants into new growth. Plants grow rapidly in this environment, but they also die younger, at maximum ages of 300–400 years compared to the trees of temperate forests, which may live well over a thousand years. The greatest threat to the rain forests comes from increased human activity, especially logging and clearing for agriculture. Multinational logging companies that have devastated large areas of Southeast Asia are now setting their sights on South American forests. Unfortunately, once a rain forest has been clear-cut, rain washes nutrients out of the soil rapidly, and it could take hundreds of years for a forest to regain its original condition. Clear-cut areas can be used agriculturally for a few years, but their soils are soon depleted. Some societies, as in Central America and the island of Madagascar, depend upon "slash-and-burn" agriculture; that is, they cut and burn an area, use it to grow crops until the soil is exhausted, and then move on to another patch. This practice becomes more damaging as these populations continue to grow, and it clearly cannot be sustained for long.

Tropical seasonal forest

In contrast to the rain forests, which have abundant water year around, areas farther from the equator in Latin America, Africa, India, Southeast Asia, and Australia experience periodic drought. Though these regions are warm all year, they have extensive rainfall (to about 100 cm) during the wet season, which is interrupted by a dry season. Tropical seasonal forests are more open than the rain forests, less diverse, and usually have fewer layers of vegetation. Teak, mountain ebony, and drought-resistant evergreens are typical trees; the northern Australian forest is characterized by eucalyptus. Typical animals are langur monkeys in India, elephants and okapi in Africa, and koalas and cockatoos in Australia.

Savanna

Savanna is a tropical grassland comparable to the tropical seasonal forest, in that it is uniformly warm and alternates between wet and dry seasons. Its inhabitants must therefore be adapted to growth and reproduction during the wet season, and to survival during the dry. Typical savanna consists of very tall, meter-high grasses with a few scattered trees; the flat-topped acacia trees give the African savanna its characteristic look, and the baobab tree is adapted to the dry conditions by its ability to store water in its swollen trunk. In the savanna live the animals that seem most typically African—the extensive herds of antelope, zebras, elephants, and giraffes, which are eaten by wild dogs, hyenas, and lions.

Mediterranean scrub forest

Also known as chaparral in North America, Mediterranean scrub forest occurs in such places as the Pacific coasts of North and South America, along the Mediterranean Sea, and in southern Africa. These areas receive cool, moist air from the oceans but are still relatively dry, with 25–75 cm of rainfall per year concentrated in a wet winter. Their summers are hot and dry. Their major vegetation consists of broad-leaved evergreens, such as live oak and eucalyptus, some evergreen conifers, some deciduous trees (those that shed their leaves during cold or dry seasons) such as Lombardy poplar, and a variety of shrubs, grasses, and succulents that are adapted to dry conditions (Figure 25.24). In western Asia, grasses that became domesticated as wheats and barleys originated in this kind of biome. Mediterranean forest is subject to periodic fires, even without human intervention, because of the hot, dry summer conditions and because many of the typical plants of the biome, such as lavenders and sage, produce volatile oils and burn easily. Animals that feed on this vegetation include guanaco in South America, mule deer in California, red deer and wild boar in Europe, and duiker and hyrax in South Africa.

Temperate grassland

Extensive inland areas of most continents (except Europe), where the winters are cold and the summers hot and dry, are occupied by prairies and steppes whose dominant plants are grasses (Figure 25.25*a*). Grasslands exist mostly on flat or rolling land, on slightly alkaline soil rich in organic matter. Precipitation in these areas varies from about 25 to 95 cm, concentrated in wet seasons that are often interspersed with droughts. Grasslands are not a single ecosystem, and before European settlement, North America held six major types characterized primarily by the length of the dominant grass (Figure 25.25*b*). Wet seasons on the grasslands are typically rather short, and grasses have adapted to these conditions by producing thick root masses that run deep into the soil. Roots constitute most of a grass's mass. This mode of growth allows the plant to reach water quickly and makes the plant particularly resistant to fire. Roots also carry the plants' major stores of food. The sod may keep trees from getting a foothold and thus prevent their growth, but grasslands apparently maintain their character primarily because of periodic fires, hard winter freezes, and grazing by large herbivores. Managers of American wildlife refuges on the plains have found that they can only prevent these areas from turning into forests by allowing herbivores (now including cattle) to graze and by using periodic controlled burns. Fire kills small trees, but grasses are adapted to it. Fire kills accumulated litter, releasing nutrients for further plant growth, and with the litter removed, the soil warms earlier in spring, so grasses start to grow earlier and grow taller. The grasses are supplemented by an array of often brilliantly flowered annual and perennial forbs (herbaceous plants other than grass). The annuals thrive—mostly in slightly disturbed areas where they can gain a foothold—by setting seeds quickly and overwintering

Figure 25.24

A Mediterranean scrub habitat at Mt. Tamalpais Park in California (where it is called "chaparral") includes broad-leaved evergreens such as Live Oak and drought-resistant shrubs such as Manzanita and Ceanothus.

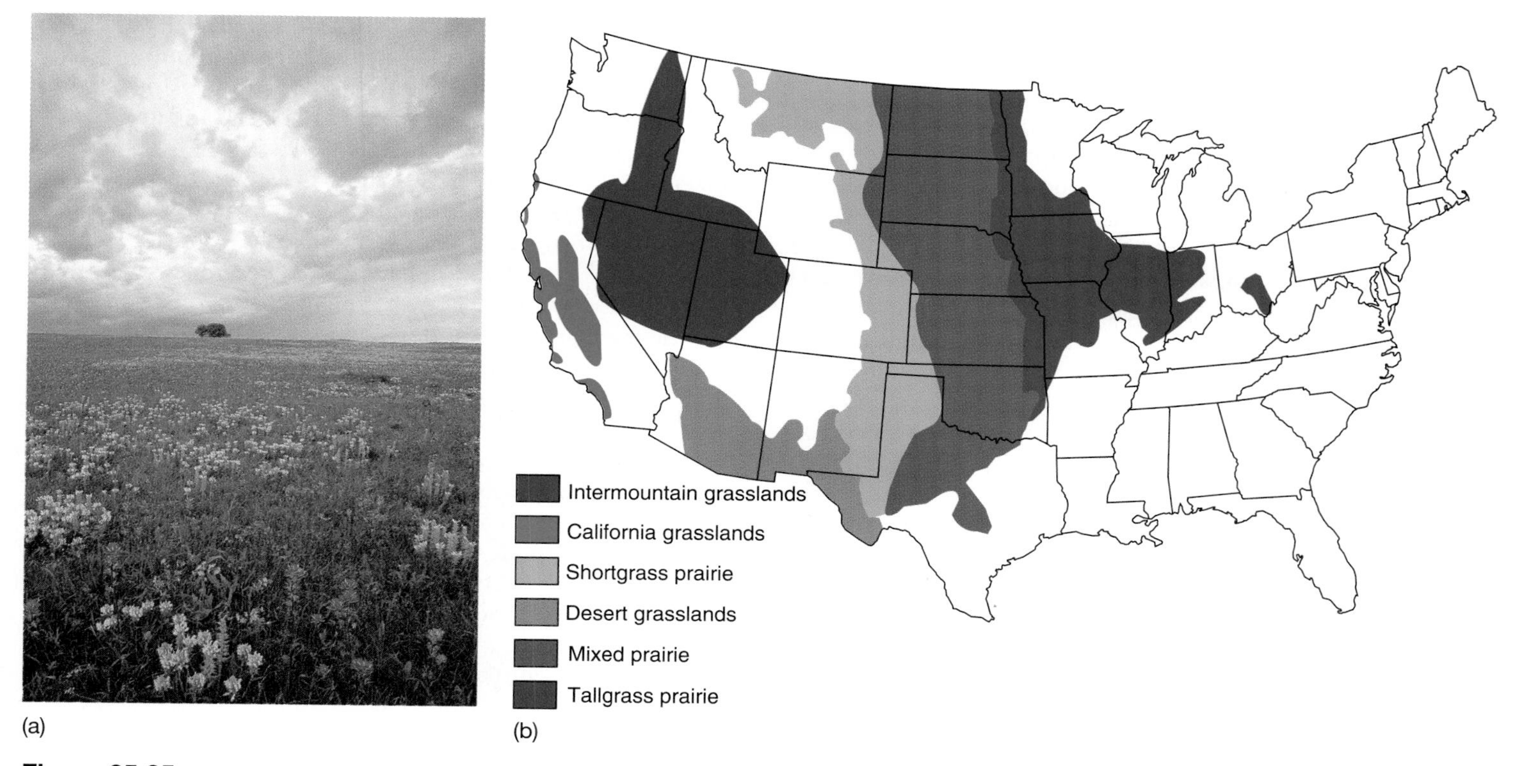

Figure 25.25
(a) A grassland habitat in Missouri contains lush grasses mixed with forbs such as Indian Paintbrush and Phlox. *(b)* North America has six distinctive types of grassland.

as seeds. The perennials thrive with many of the grasses' adaptations such as deep root systems, as well as leaves adapted for dry conditions. Grasslands have been taken over extensively for agriculture, but once the protective sod has been removed, the underlying soil is subject to rapid erosion by winds and rain.

Temperate grasslands are the homes of great herds of grazing animals, such as the pronghorns and bison of North America or the wild horses that once roamed the Asian steppes. Where there are no trees to hide in, the ability to run fast becomes an important adaptation. The grasslands of other continents harbor large, flightless birds such as ostriches and emus that also escape their predators by running. Many other grassland animals have adopted the burrowing habit. Grasslands commonly harbor small burrowing herbivores such as mice, voles, prairie dogs, pocket gophers, and rabbits. Consequently, they are also the homes of large predators, such as coyotes and lions, but many of the predators also live in burrows: burrowing owls, weasels, ferrets, badgers, and foxes. Burrowing is an obvious way of hiding, especially to escape from fires and from hot, dry summers and cold winters. Invertebrate herbivores like grasshoppers and caterpillars are important members of grassland communities.

Deserts

Deserts are regions that receive less than 25 cm of rainfall annually and may go without any rainfall at all for years (Figure 25.26). They have large areas of open land, with their plants

Figure 25.26
The Sonoran desert of the southwestern United States and northern Mexico is typical of the desert biome. It is dominated by Saguaro, which has a central role in the ecosystem as a shelter and food source for many birds and other vertebrates.

often concentrated in old riverbeds. Due to the scarcity of water, which tends to stabilize the temperature, deserts undergo dramatic shifts from the cold of night to the heat of day. Desert plants, including many kinds of cacti, are typical *xerophytes* (*xer-* = dry; *-phyte* = plant), plants adapted to dry conditions. Some desert plants grow extensive roots that can tap water sources deep underground, while others obtain their water from dew on the surface. Xerophytes can grow and reproduce rapidly during the brief rainy periods that most deserts experience, when the area becomes radiant with the color of their flowers. Many desert plants photosynthesize through crassulacean acid metabolism (CAM, see Section 10.9), opening their stomates only at night when the desert is cool. The animals of the desert are primarily arthropods (insects, scorpions, spiders), reptiles such as snakes and lizards, and some specialized birds and mammals. Many desert invertebrates, such as woodlice and millipedes, feed on plant debris; they are preyed upon by beetles, lizards, and some birds. These animals, in turn, are preyed upon by foxes, coyotes, wolves, and several members of the cat family, including the mountain lion, caracel, and cheetah. Small mammals such as gerbils, jerboas, and ground squirrels escape the heat by living in labyrinthine burrows during the day and becoming active at night, using a strategy parallel to CAM photosynthesis in plants. Many of these mammals feed on seeds; they are relatively abundant in the desert because desert plants often survive the driest periods as seeds; they only germinate and grow during the brief wet episodes.

Xerophytic adaptations, Section 38.9.

Semiarid scrub

The biome known as semiarid scrub is similar to desert, but receives more rainfall and has heavier vegetation. It supports little animal life, and its plants are thorny scrubs adapted to regions that receive limited or irregular rainfall and have unendurably hot summers. The scrub plants are often *sclerophylls,* with small, leathery leaves, such as creosote bushes and tamarisks. As in deserts, annuals survive in this biome as seeds, which sprout during the brief wet periods and grow quickly into plants that set new seeds. Snakes and lizards are common in the scrublands. Herbivores such as rabbits, gerbils, and gazelles manage to eke out a living on the sparse plants.

Temperate forests

Where the summers are warm and the winters cold, and where precipitation is relatively uniform throughout the year, the temperate forest biome predominates. These forests are generally moist, with an average annual rainfall of around 100 cm that supports the lush vegetation (Figure 25.27). Several types of these forests exist. Japanese forests are broad-leaved evergreens. In North America, the forests range from Pacific-Coast conifers to mixed coniferous-deciduous woods in the Great Lakes region to broad-leaved, deciduous trees in the East—oak, maple, beech, elm, birch, and hickory. These forests always harbor a thick underlayer of shrubs and a multitude of herbaceous plants close to the ground. The litter of this vegetation creates a thick layer of decaying biological material, or **humus,** that supports an abundant invertebrate fauna. Forest plants support such herbivores as deer, squirrels, rabbits, and rodents; they, in turn, support carnivores such as hawks, owls, bears, wolves, foxes, and bobcats, animals that are becoming rare as the vast forests are broken up into small patches.

Such forests were the original biomes of much of Europe and eastern North America before the intrusion of civilization, and since most cities and towns are built in these regions, these are the familiar woods of home to many people. Areas that are allowed to revert to wilderness quickly grow into these woods, and large areas of the eastern United States are coming to resemble the condition the first European settlers found. Other formerly wooded areas, however, continue to be divided and converted into housing developments and shopping centers, leaving smaller and smaller areas of woodland. A small patch of forest is not simply a smaller version of a large patch, primarily because of edge effects. The small patch has a high ratio of edge to area, but disruptions intrude at the edges. In North America, for example, several species of birds have been parasitized for a long time by cowbirds, which lay their eggs in the nests of other species. Cowbirds inhabit open areas, not dense woods. The species they have traditionally parasitized have come to an equilibrium with the cowbirds and have been able to evolve defenses, but as the forest patches become smaller, the cowbirds have started to prey on naive species, and they pose a severe threat. Furthermore, a small patch of forest may be too small to sustain even one pair of animals; woodland birds establish territories, and while a large patch of forest can support several

Figure 25.27

This temperate deciduous forest habitat of eastern North America is dominated by Beech and Red Maple. Although the tree canopy is quite thick, the lower level of the forest is relatively open, leaving room for a variety of shrubs and herbaceous plants on the forest floor.

territories, the patches left by development may be too small for even one.

Brood parasitism, Section 27.14.
Territoriality, Section 26.10.

Taiga (boreal forest)

Across Alaska, Canada, Scandinavia, Russia, and Siberia lies a great circumpolar band of coniferous forest called taiga (pronounced "tie′guh," from the Russian for "marshy pine forest"). Very little land lies at a comparable latitude in the Southern Hemisphere, but the southeastern coast of Australia has the same type of forest. The trees are mostly pines, spruce, and fir, with occasional intrusions of broad-leaved species like aspen, birch, and alder. These plants are adapted to long, cold winters with a short, moist, cool growing season; there may only be 30 days when the temperature reaches 10°C or more. Evergreens are able to begin photosynthesis and growth as soon as the sunlight and rising temperatures of spring permit; narrow-leaved evergreens are able to resist the cold, dry conditions of winter. The thick layer of needles that accumulates under conifers decays slowly because it remains cold and wet and oxygen penetrates it slowly. The polyphenolic compounds in these needles make them distasteful to worms and other animals that break down detritus and mix the soil in other biomes, so the taiga's soil is relatively poor in nutrients and strongly layered.

The taiga's most characteristic herbivores are deer; more species of deer live in this biome than anywhere else. Moose, hares, porcupines, and many rodents also live here (Figure 25.28). Beavers inhabit several biomes, but they are abundant in the taiga, where they have built extensive dams that flood forest areas and contribute to the wet soil conditions. The herbivores support such carnivores as weasels, bears, lynx, wolves, and wolverines, some of which are becoming scarce in all but the more remote forests. Most of the taiga's birds are migrants that live there only during the short summer, but several species of grouse, woodpeckers, and owls are resident. Insects are also abundant in the forests. Many of the taiga mammals are particularly large; we will show in Chapter 32 that large size is especially advantageous in a cold climate.

For a long time, the taiga remained relatively aloof from human activities that were devastating other biomes, but its riches are too tempting to keep it safe for long. Its huge conifers have great value to a civilization hungry for lumber and wood pulp, and the activities of loggers have ruined vast areas. Much of Siberia is still quite pristine, but multinational logging companies are already moving in. The taiga has also been used for immense mining operations, including iron, gold, and diamonds. Pollution moving northward from the highly industrialized regions of North America and Europe now threatens these great forests; they are sensitive to acid rain, and their acidic soils cannot neutralize this rain, which drains off into rivers. Hydroelectric engineering projects such as James Bay in Canada threaten to drown huge areas by damming rivers.

Figure 25.28
The northern coniferous forest, or taiga, is dominated by spruce and fir. Its dominant animals include Snowshoe Hare, Gray Wolf, and Moose.

Acid rain, Section 28.5.

Tundra

To the north of the taiga, south of the eternal polar ice caps, lies the tundra, an extremely cold, dry region with an annual precipitation of about 15 cm, mostly snow. During the summer growing season, lasting a brief 3–4 months, only the surface of the ground melts, and below a meter or so, the soil is in a continually frozen condition called *permafrost.* The major vegetation consists of low-lying grasses, mosses, lichens, and dwarfed shrubs such as birches and willows, which can grow and reproduce during the very brief summer (Figure 25.29). Barren-grounds caribou (reindeer) live in great herds that migrate from the northern edge of the taiga across the tundra, feeding on the temporary vegetation. The herds are followed by wolves, which pick off the old and sick animals. Muskoxen are permanent tundra residents that protect themselves against predators by forming a tight circle when threatened. Polar bears live primarily on seals, and arctic fox live on nestling birds, eggs, and small rodents such as lemmings, which are abundant and occasionally become superabundant, breaking out in massive numbers. Birds such as ptarmigan and snowy owls are common residents, and the small lakes of the tundra are important breeding grounds for many migratory birds, especially shorebirds and ducks. During the summer, the area abounds with hordes of mosquitoes, blackflies, and other insects, which emerge from eggs that have over-wintered. The principal human threat to the tundra comes from oil exploration, which despoils huge areas and blocks the migration of the animal herds with pipelines.

(a)

(b)

Figure 25.29

(a) A moose crosses a tundra near Mt. McKinley in Alaska, a habitat briefly made lush by the growth of colorful low shrubs such as bilberry and bearberry *(b)*.

Mountains

The sequence of biomes from tropical forests to tundra, correlated with latitude, is replicated with altitude as one ascends a typical mountain range (Figure 25.30). Mountains such as Mt. Kilimanjaro in Africa or Mt. Everest in Asia have their bases in tropical forests. These give way to a temperate deciduous forest above 1,000 m and to coniferous forest above 2,000 m. An alpine scrub resembling taiga appears around 3,400 m, and above 4,500 m is a permanent snowline similar to the polar regions. The principal difference between the high mountains and the Arctic is that the mountains receive quite direct sunlight during the day and may become very warm, instead of remaining frozen at all times. The fauna of each region often mimics the fauna of the corresponding lowland regions, such as ibex or mountain goats in the high mountains that correspond to caribou or muskoxen in the Arctic. However, the high mountain animals are partially limited by the greatly reduced oxygen pressure, which requires special adaptations in their circulatory and respiratory systems. High altitude plants, too, may mimic those of the tundra in their stumpy, dwarfed morphology.

25.8 Terrestrial ecosystems are determined by the relationships among climate, vegetation, and soil.

With the survey of biomes in mind, we can ask about the factors that create such diversity. The major geographic factors that determine the character of a terrestrial ecosystem are related rather simply: The worldwide patterns of precipitation and temperature determine climate; the climate determines the major types of vegetation in any region; and the combination of climate and vegetation is the major determinant of soil type. The earth's biomes are defined mainly by their vegetation, but the regions of the map in Figure 25.22 could largely be superimposed on climatic regions and regions of soil types as well, for the three factors follow one another very closely.

Climate is a combination of temperature and precipitation. The interaction between these factors determines different biomes, as shown in Figure 25.31. Around the poles, the low temperature is the major determinant of biome type. In the Northern Hemisphere, low temperatures create three broad biomes that run without interruption around the earth: the polar ice cap, a region of perpetual ice and frost of limited, specialized biological interest; the tundra; and the taiga. Over the rest of the world, the temperature is more moderate, and precipitation becomes a more important factor. Climates can be distinguished here in various ways, but it is convenient to recognize six types, with their associated major vegetation:

	Climate	Vegetation
Moist	Superhumid	Rain forest
	Humid	Forest
	Subhumid moist	Tall-grass prairie
Dry	Subhumid dry	Mid-grass prairie
	Semiarid	Short-grass prairie (steppe)
	Arid	Desert plants

This relationship is drawn out more graphically in Figure 25.31. The correlation between climate and vegetation type is quite obvious, though it is difficult to explain why grasses (rather than forests) grow in moderately dry temperate zones while forests (rather than grasses) grow in humid temperate zones. Part of the explanation lies in the close relationship among climate, vegetation type, and soil, as we explore next.

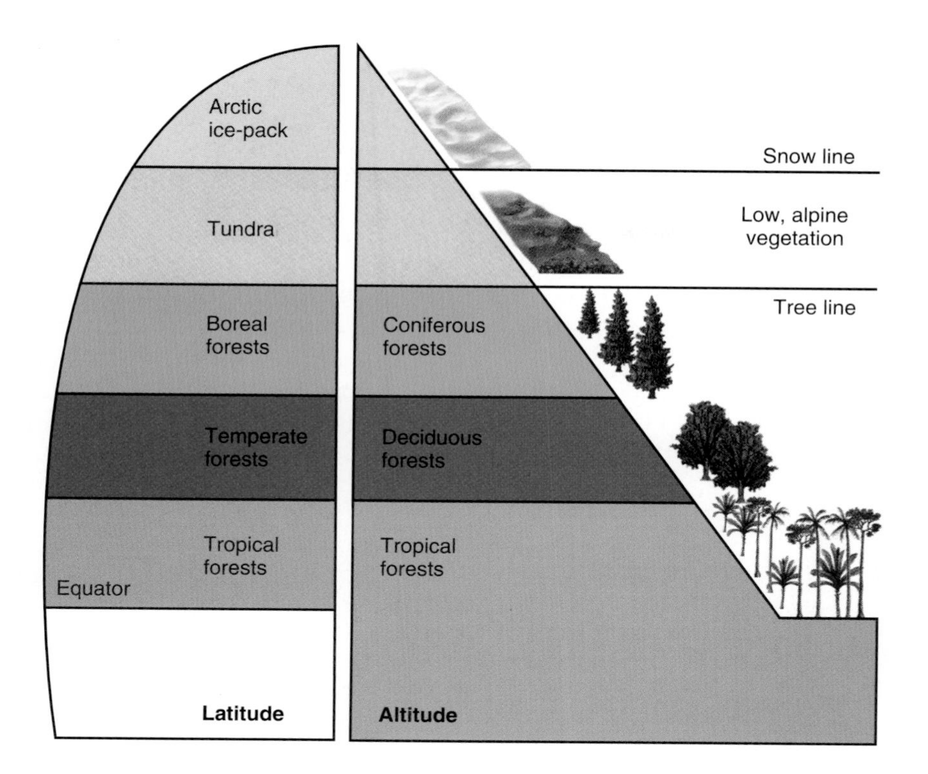

Figure 25.30

The zones of vegetation at increasing elevations on a mountain parallel the biomes at increasing latitudes on the earth's surface.

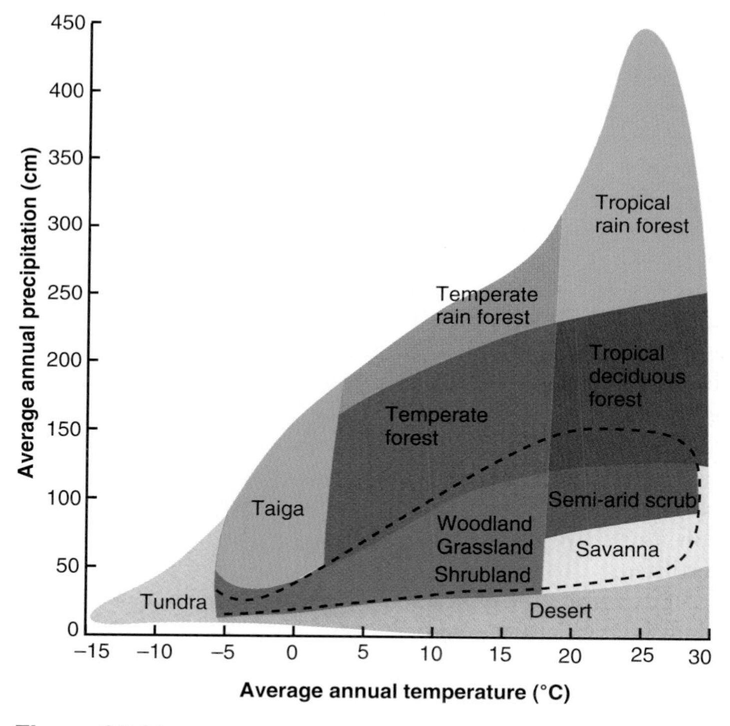

Figure 25.31

A classification correlates the major biomes with temperature and precipitation. The dashed line encloses a range of climates in which either grasslands or woody plants may constitute the prevailing vegetation of an area, depending on the seasonality of rainfall. This is Robert H. Whittaker's classification, but it is substantially the same as classifications that treat climatic factors in more complicated ways.

25.9 Soil is formed by chemical alteration of weathered rock and organic complexes.

We are wedded to water, an essential component of all organisms, but we are nurtured by the soil. Our earliest ancestors knew this. They made annual sacrifices to the Earth Mother and prayed for abundant crops; the Greeks told of the giant Antaeus who derived his great strength from his mother Gaea, the Earth.

Few organisms live on bare rock. Rain, wind, and organisms themselves—through acids excreted by lichens and others—break the parent bedrock into a thin layer of soil whose characteristics determine what can live in it and above it; reciprocally, the organisms of an ecosystem determine some characteristics of the soil.

Bedrock was laid down by the cooling of molten magma from the earth's core to form igneous rock, and by the deposition of layer upon layer of sediment from water and wind to form sedimentary rock. Some of these rocks have been altered into metamorphic rock by heat and by the enormous pressure of other rock masses. As erosion removes overlying rock, the initially solid rock beneath expands and cracks. Water fills these cracks, enlarging them as it freezes and thaws or as crystals of dissolved materials grow out of solution. Gradually these physical weathering processes reduce the rock to small chips (Figure 25.32).

Water, particularly rainwater, is the most effective agent in further altering rock chips. Rain is not simply H_2O. Atmospheric CO_2 dissolved in rainwater makes it a dilute carbonic acid solution containing H^+ and HCO_3^- (bicarbonate) ions.

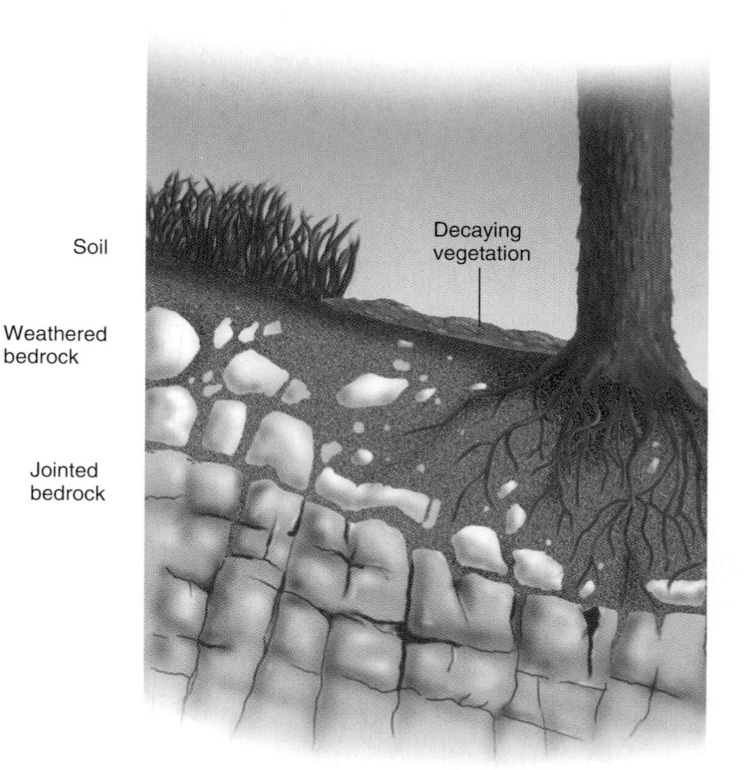

Figure 25.32

As underlying bedrock is transformed into sand and soil, large rock fragments are broken into smaller fragments by mechanical weathering, such as the freezing of water in cracks, and by chemical weathering, caused primarily by the action of substances dissolved in the water.

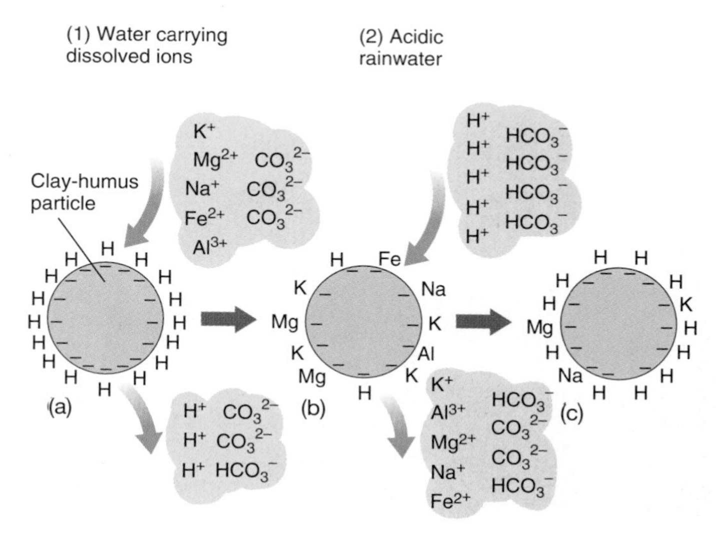

Figure 25.33

The hydrogen ions initially bound to clay-humus particles *(a)* are replaced by mineral ions from the breakdown of minerals *(b)*. Acidic rainfall then displaces the mineral ions, leaching them into the lower soil levels, so the particles remain acidic, and upper levels of the soil are poor in mineral ions *(c)*.

Even this weak acid is strong enough to alter the bedrock chemically. We discuss soil chemistry in more detail in Chapter 40, in the context of plant nutrition. For now, the essential point is that the parent rock reacts with the acidic water to release cations (positive ions) such as K^+, Ca^{2+}, and Fe^{2+} along with clay minerals, which have negative charges. A soil supporting plants accumulates humus, a black or brown complex of decaying organic material. Clay particles combine with humus to form complexes that can hold the cations.

A young soil is full of ions such as potassium, magnesium, and iron that have been released from the parent rock. These cations, bound weakly to the negative sites on the clay-humus complexes, are available as plant nutrients, but they can be displaced from the complexes by the hydrogen ions in rainwater (especially strongly acidic rain) percolating through the soil (Figure 25.33). Rainwater leaches mineral ions out of the upper soil and carries them to lower levels. Such a soil is poor in mineral ions and very acidic. The acidity reaches a limit, partly because the accumulated hydrogen ions break down the clay minerals even further, releasing more mineral ions.

The most extreme case of leaching is the process of **podzolization** in regions such as the western mountain chains of North America, where heavy rainfalls rapidly leach minerals out of the upper soil levels. In cross section, the soil shows distinct layers, or **horizons** (Figure 25.34). The upper horizons (called the A horizons) include dark layers at the very top underlaid by a layer that is light in color because the dark minerals have been leached out and deposited in the lower (B) horizons. The soil itself is called a **podzol**. It supports the growth of plants that have modest requirements for these minerals, especially the conifers of the mountains and the great forest belts of northern Eurasia and America. These trees bring up the ions they need from lower levels through their roots, but since their structures are not rich in minerals, the humus they deposit (from needles, branches, bark, and so on) is not rich in them either. Few minerals are therefore released when the humus decays. Moreover, in the cool climates of these regions, soil fungi and bacteria don't work fast as decay agents, so a deep humus layer accumulates. Although their natural vegetation is well adapted to podzols, they have only moderate value for farming and must be fertilized.

In contrast, the typical soil of the American and Asian grasslands is a black earth, or **chernozem**. Here the temperature and rainfall are both moderate, evaporation proceeds quickly, and leaching is just balanced by release of minerals from above. The native plants of these regions are grasses whose fine, deep roots decay quickly, leaving a deep humus layer, with abundant minerals. Chernozems are fine agricultural soils, particularly for domesticated grasses such as wheat.

A typical soil of wet tropical biomes is a **latosol,** or **oxisol,** a very deep, loose, red soil formed by high precipitation and high temperature. The high precipitation of the tropics causes

Figure 25.34

A section through soil shows an A horizon consisting of a dark upper layer with grass and a light layer below from which minerals have been leached. The lower B horizons, which are not set off by a sharp boundary here, include a brown layer where iron oxides have accumulated.

extensive leaching in latosols, while the high temperature promotes rapid decay, releasing minerals quickly and producing little humus, so the soil has few nutrients. Silicates break down faster than the other minerals at these temperatures, leaving insoluble reddish iron and aluminum oxides in the A horizons. In contrast, the typical desert soil—a **sierozem,** or **aridosol**—is formed by the breakdown of parent rock under high temperatures, but with little rain to leach the minerals away. Water penetrates the soil only a short distance and then is drawn back up by evaporation and by plant roots; this slight movement of water carries calcium carbonate down to the B horizons, where it may form such a hard layer that plant roots can't penetrate any deeper. The light desert vegetation leaves little humus, but the soil is rich in mineral nutrients and can be valuable agriculturally with the addition of water and nitrogen.

In addition to plants and physical factors, the soil of every ecosystem is conditioned by the other organisms living there. Animals mix and stir the soil very effectively. Earthworms, for example, are great soil-makers, for they burrow deep, cutting across horizons, opening the soil and carrying materials from one level to another. As worms digest their food, they grind soil particles into smaller pieces. Their excrement is rich in acids, minerals, and organic compounds.

The soil ecosystem, Section 40.5.

Plants, especially those with deep roots, continually reach lower into the soil and bring up nutrient elements. Much of what they bring to the surface and incorporate into their structure falls to the ground as litter, which decays into humus and is leached back into the soil, to be recycled again through the roots. This creates a nutrient cycle for every element, and we can trace its movements, sometimes against the larger background of chemical changes in the whole biosphere.

Nutrient cycles in plants, Section 28.3.

25.10 The world is divisible into several biogeographic realms.

Biogeographers—biologists who look at the earth's flora and fauna from a broad, geographic perspective—have seen that the earth can be divided not only into large biomes but also into regions called **biogeographic realms** (Figure 25.35). Each realm has its distinctive plants and animals, and their boundaries correspond approximately to the limits of the continents, though not exactly.

Figure 25.36 shows representatives of several animal groups that occupy similar niches in each realm. The similarities among the realms are obvious, aside from a few empty boxes showing that some forms simply did not evolve in those areas. Several continents support some long-snouted anteaters, some huge rhinoceros-like forms, some llama-giraffe forms, some gazelles, and so on. These are not necessarily closely related to one another; the Australian mammals, especially, are mostly marsupials or monotremes that are quite distantly related to placental mammals (see Section 35.14). These similarities show *convergent* or *parallel evolution.* Animals with quite different ancestry, on the different continents, come to resemble one another strongly because they evolve into forms with the features needed to occupy each kind of niche. All are responding to similar opportunities and demands in each realm. There is always a place for small animals like mice that can eat seeds and vegetation. There is always a place for fleet-footed grazing or browsing animals such as deer or gazelles. There is always a place for large carnivores, such as cats, wolves, and bears.

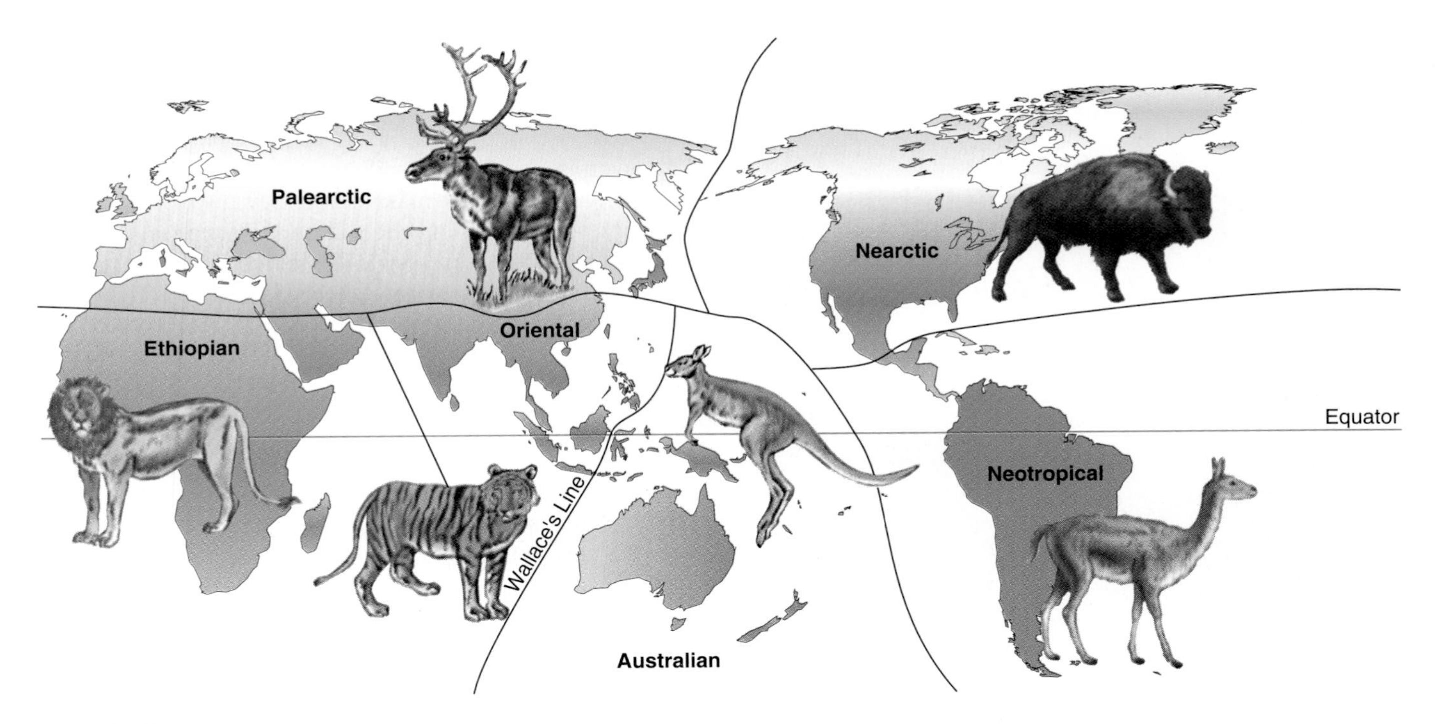

Figure 25.35
The earth can be divided into six biogeographic realms that cover major landmasses. The flora and fauna of each realm have distinctive features, which indicate that they have evolved independently of those in other realms. Wallace's Line was established by Alfred R. Wallace as the boundary between the Oriental and Australian realms; islands to the west of the line were once part of the Asian continent and have an Oriental fauna.

Convergent evolution, Section 22.1.

This biogeographic comparison leads to some questions: Why do the animals in each category often come from quite diverse ancestors? And why do the biogeographic realms have their rather peculiar boundaries? The answers lie partly in the isolation of the continents from one another. In looking at the mechanisms of evolution, we saw that diversity results from species being separated from one another long enough to become reproductively isolated and evolve in their own directions (see Section 24.5). If there were only one continent, the earth's flora and fauna would certainly be much less diverse. But the isolation or partial isolation of the different landmasses has allowed different organisms to evolve in similar niches on each continent.

This fact about evolution still does not explain why India and Southeast Asia have such different flora and fauna when they are mere bumps on the Asian continent. It also makes a mystery of the curious distribution of some fossil groups and some primitive living groups. Why, for instance, did mesosaurs, a type of aquatic Mesozoic reptile, inhabit only such distant places as Africa and South America? Why are birds such as emus, ostriches, and rheas found only in Australia, Africa, and South America? Why are lungfishes, which live in freshwater, found only in the tropical areas of those three continents but not in temperate areas or in Asia? The answers to these questions lie in the earth's geological history, as explained in the next section.

25.11 The earth's surface is made of slowly moving tectonic plates.

The science of geology has undergone a revolution in the last few decades. Early in this century, the German meteorologist Alfred Wegener saw that the coastlines of North America and Europe almost fit together, as do those of South America and Africa. He proposed that all the continents were once part of a single landmass, a supercontinent he called Pangaea, and that Pangaea broke up and began spreading slowly over the earth's surface. Wegener's theory of **continental drift** was derided for years until, in the 1960s, geologists found a number of lines of evidence to support his ideas. First, they showed that the coastline features of some continents can indeed be matched precisely to those on other continents. Second, the distribution of fossils and the patterns of magnetism in ancient rocks indicate that the continents were in different positions in the past. Third, the earth's surface is indeed moving, as Wegener suggested. There had to be some mechanism to account for continental drift, and Wegener's ideas were not accepted until the theory of **plate tectonics** became established.

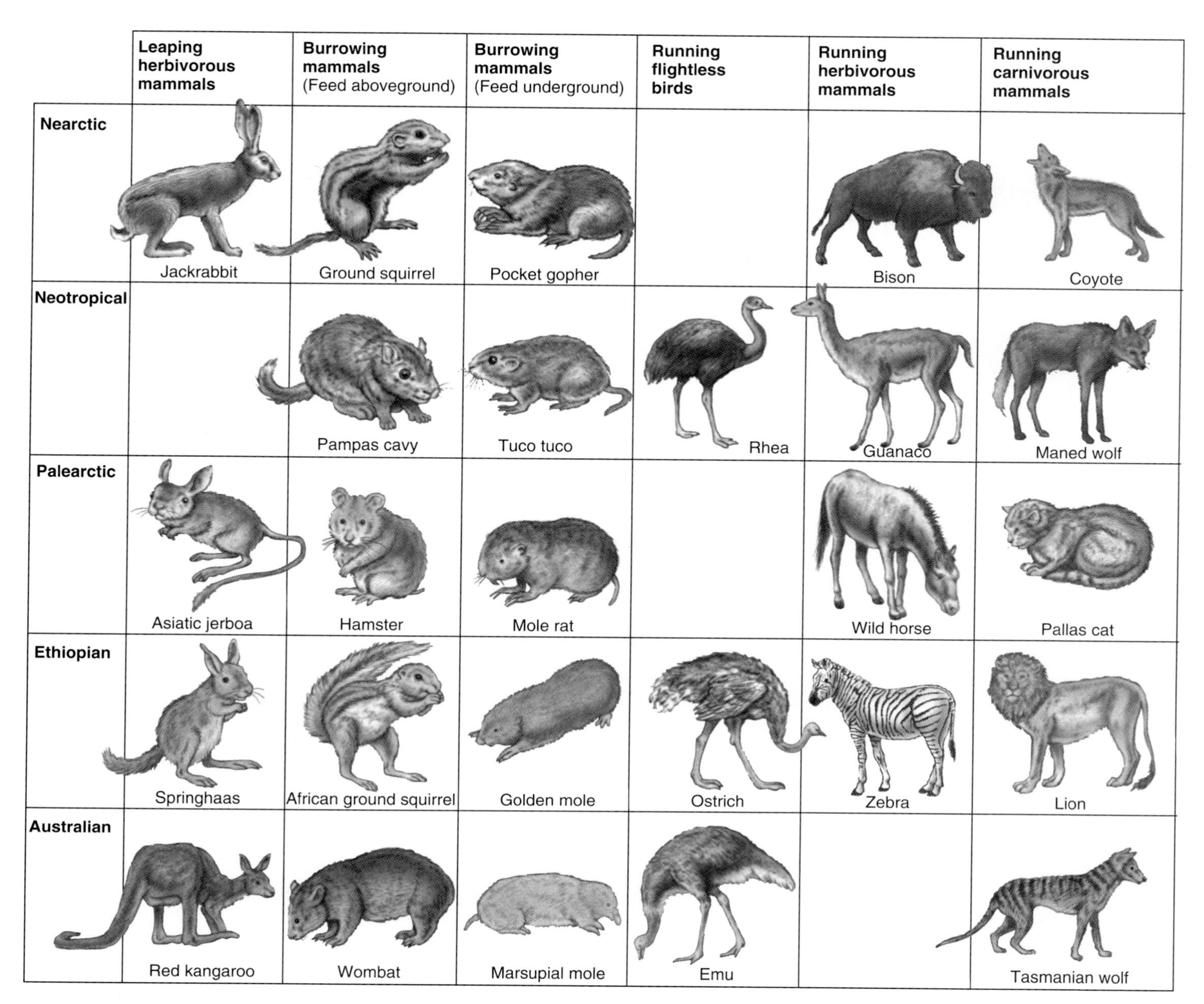

Figure 25.36

A comparison of prominent types of animals in different biogeographic realms shows that each ream generally has representatives of each type because these categories represent adaptations to general ways of life that have a place in many terrestrial ecosystems. Source: Peter Farb and Editors of *Life Ecology,* (Life Nature Library), 1963, Time-Life.

It is now clear that the earth's crust is divided into a number of *tectonic plates,* masses of rock that are literally floating on the melted rock (magma) below (Figure 25.37). Underlying geological forces continually move these plates, so over millions of years, some of them have moved apart and others have moved closer together. As two plates move apart, the magma wells up to create new ocean floor between them. The Atlantic Ocean, for instance, has been spreading in this way, separating North America and Europe. Where two plates collide, one pushes under the other, so the upper plate ripples and buckles into mountain ranges. The Himalayan Mountains, for example, are still being raised, as the Indian subcontinent, resting on a separate plate, presses northward into Asia. In other places where two plates come together, the two rock masses create enormous friction as they try to slip past each other, as at the San Andreas fault where the Pacific Plate carrying part of California meets the North American Plate. When the buildup of vast potential energy overcomes the friction, the plates suddenly slip past each other a short distance, causing an earthquake.

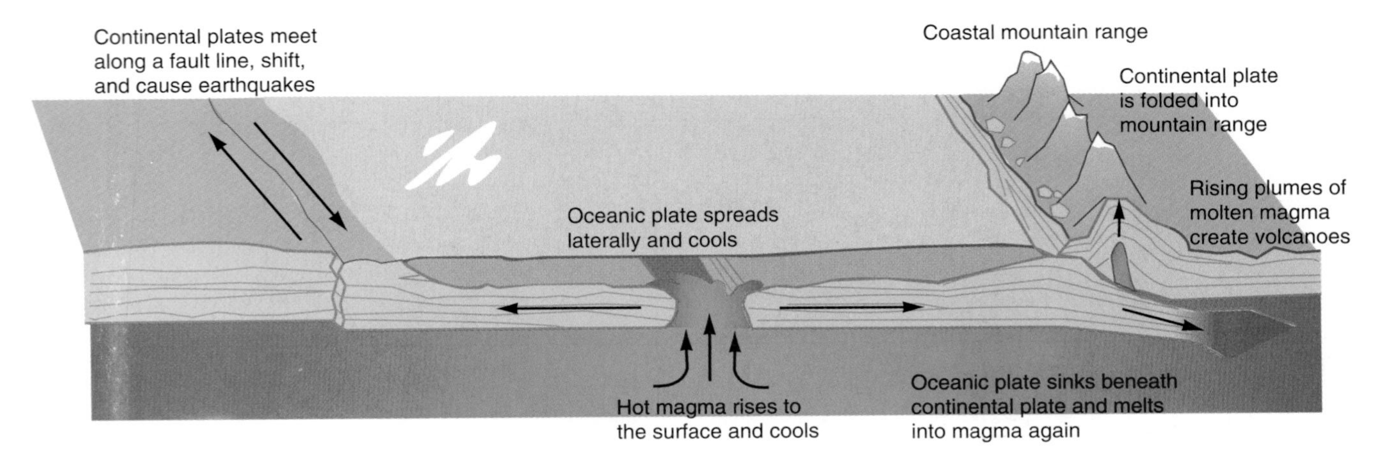

Figure 25.37

The earth's surface is divided into several solid tectonic plates floating on the fluid magma beneath them. Where two plates slowly grind past each other, they build up tension, which is released occasionally in the form of earthquakes. Where one plate pushes beneath another, mountains, including volcanoes, are raised. Where two plates gradually separate, they make rifts in ocean bottoms where fresh magma wells up and cools, thus enlarging the plate.

The geological record indicates that Pangaea was still intact around 250 million years ago, during the Permian period when the reptiles were becoming diversified and spreading over the earth (Figure 25.38). The landmass started to divide into a northern continent, Laurasia, and a southern one, Gondwana, about 220–210 million years ago. By 65 million years ago at the end of the Cretaceous period, Gondwana was well divided into the southern continents and India, which eventually collided with Asia about 50 million years ago. Eurasia and North America separated only 45 million years ago, and South America swung around to connect with North America at the Isthmus of Panama quite recently, about 2 million years ago.

In light of this continental history, the distribution of the world's flora and fauna makes more sense. India and Southeast Asia have different plants and animals because the two areas were separated for so long. South America, Africa, and Australia have so much in common because they were all parts of Gondwana, but they developed their own distinctive life forms after their separation 100 million years ago. This was a time when some of the major ancestral groups of mammals were just diverging, including the ancestral marsupials, which became the dominant mammals of Australia.

Coda The earth without organisms would be extraordinarily different—a mere ball of rock washed by water, with a very different kind of atmosphere. While evolving on this planet, organisms have obviously changed dramatically over 3 billion years or more, and they have changed the planet, too. Seen from above, the vast forests and grasslands present an extraordinary contrast to the nearly lifeless deserts and mountaintops. Organisms have shaped different environments for themselves in response to the abiotic factors available in each place—light, heat, water, and mineral resources. Looking at different biomes, we can see again how the environment must be able to support certain kinds of organisms: how generally high temperatures with a high input of energy and water are needed to support large forests, while low temperatures or little water only allow the sparse vegetation of tundra or desert. As we study the animals of each ecosystem, it becomes clear how they have evolved to take advantage of the opportunities made available by abiotic factors combined with the vegetation of a region, how they are able to fit into certain niches that reappear wherever a certain biome appears. This convergent evolution speaks again to the structural basis of biological functions, showing that where a function is to be served, similar structures will commonly evolve to serve it.

Summary

1. Ecosystems are largely structured by food webs in which matter and energy, transformed into biomass through photosynthesis, are passed to higher trophic levels as organisms consume one another. These activities use less than 0.03 percent of the solar energy received on the earth's surface.
2. The basic structure of an ecosystem depends largely on closely interrelated physical factors such as wind, temperature, moisture, and soil type.
3. A pattern of winds is created over the earth's surface because the earth is warmest at the equator and is rotating eastward. So warm air flows toward the poles and cooler air toward the equator, with tropical easterly winds and westerly winds in the temperate zones.
4. A similar set of factors creates extensive warm or cold ocean currents, which affect climates by carrying massive volumes of warm or cold water near land. Vertical currents carry surface water into the depths of the oceans.
5. The oceans present three major habitats: the neritic zone in the shallow waters along the continental edges, the surface layers of the open seas, and a deep-sea zone in the oceanic depths. Estuaries, partly enclosed bodies of water where the tide meets a river current, are particularly rich neritic habitats. Coral reefs are diverse ecosystems largely dominated and formed by animals, but still dependent on various algae for their energy.

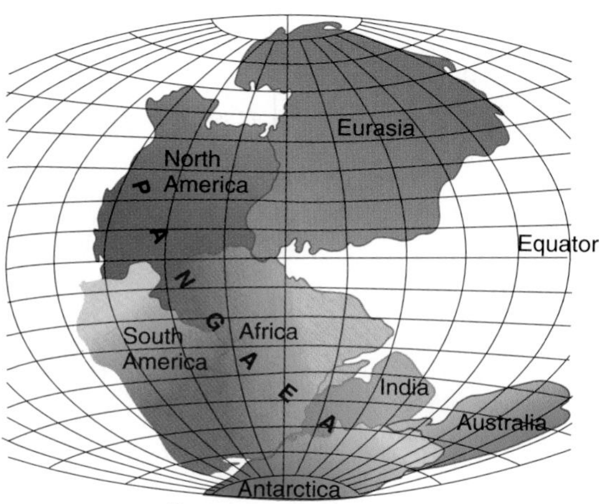

Pangaea:
Late Paleozoic, 250 MYA

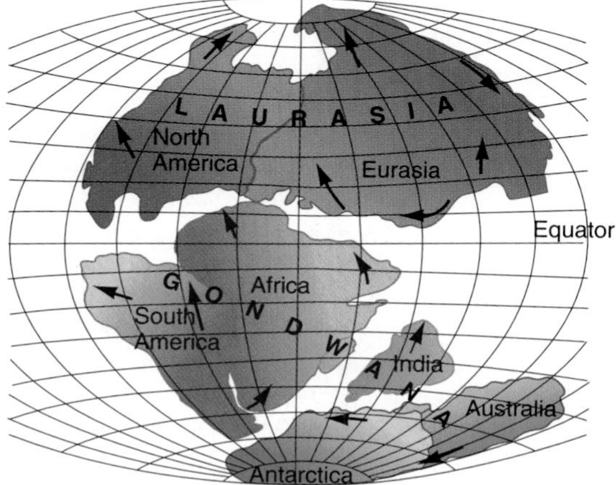

Laurasia and Gondwana:
Mesozoic, 210 MYA

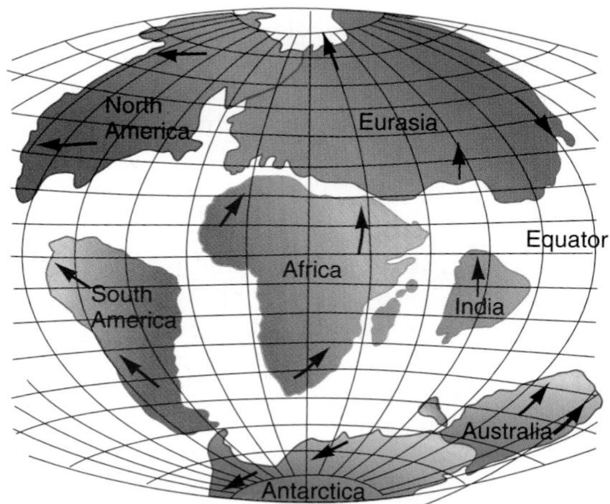

Most modern continents had formed by the end of the Mesozoic, 65 MYA

Figure 25.38

All the modern landmasses were in a single large continent, Pangaea, during the late Paleozoic (Permian) period. Shortly afterward, Pangaea started dividing into two large masses, Laurasia and Gondwana, though Gondwana was simultaneously dividing into smaller continents. These continents have been slowly changing positions as shown ever since.

6. In the open ocean, we distinguish plankton, the minute organisms that are moved by the waves and currents, from nekton, which are large and strong enough to swim independently.
7. The organisms of every ecosystem produce waste material called detritus, and the abyssal zone of the oceans is fed by the detritus of the surface zone.
8. Freshwater habitats, in ponds, lakes, bogs, marshes, swamps, and rivers, support a rich variety of life. Lakes have a littoral zone at their edges, a limnetic zone where light penetrates and supports phytoplankton, and a profundal zone that is too deep to be lit. Large lakes also show a thermal stratification of water layers with different temperatures.
9. Swamps, marshes, and other wetlands harbor a great variety of organisms and serve as reservoirs and filters of groundwater.
10. Water evaporates from the earth's surface, largely from the oceans, and falls as precipitation, mostly over land. Rainfall is distributed in characteristic patterns over the continents. It increases on the windward sides of mountain ranges and is reduced on the leeward (sheltered) sides.
11. The combination of temperature and precipitation produces certain climates, including circumpolar zones of ice cap, tundra, and taiga where cold temperatures predominate.
12. Each climatic zone produces a certain biome, or terrestrial ecosystem, defined fundamentally by its vegetation. The combination of climate and vegetation creates characteristic soils in each zone.
13. Soil is derived from the breakdown of bedrock into small fragments, which form clay-humus complexes that bind mineral ions. Slightly acidic water leaches minerals from upper to lower levels. Each type of soil supports certain plant types, and these partially determine the character of the soil.
14. The earth's surface is divided into a number of tectonic plates that are slowly moving because of underlying geological forces.
15. Each landmass has acquired a distinctive flora and fauna during its isolation from the others, and these are now seen as a number of biogeographic realms. Similar niches in these realms are filled with similar organisms that have evolved convergently.

Key Terms

podzolization 530
horizon 530
podzol 530
chernozem 530
latosol 530
oxisol 530
sierozem 531
aridosol 531
biogeographic realm 531
continental drift 532
plate tectonics 532

Multiple-Choice Questions

1. Which term includes all the others?
   a. habitat
   b. biome
   c. community
   d. biogeographic realm
   e. ecosystem
2. Most of the solar energy that reaches the earth's atmosphere is
   a. absorbed by green land plants.
   b. absorbed by photosynthetic aquatic plants.
   c. reflected back into space.
   d. used to heat the atmosphere.
   e. used to heat the ozone layer.
3. Which of the following is correct?
   a. In the tropics north of the equator, prevailing winds flow from west to east.
   b. In the 48 contiguous United States, prevailing winds flow from east to west.
   c. The primary zones of high air pressure are in the tropics and at the poles.
   d. In ocean currents, water moves as a uniform surface-to-bottom column of fluid.
   e. The earth's rotation and solar energy produce both prevailing winds and ocean currents.
4. Which best describes the epipelagic zone?
   a. sparsely inhabited zone near the ocean floor
   b. plankton-dwelling surface zone in the oceanic province
   c. zone of abyssal chemoautotrophic organisms
   d. intertidal region near the brackish water of estuaries
   e. coral reefs and reef-dwelling organisms
5. A description of an estuary would include all of the following except
   a. brackish water
   b. benthic zone
   c. rich in nutrients and phytoplankton
   d. found at latitudes with little or no seasonal change in temperature
   e. subject to pollution from runoff from surrounding land
6. During spring and fall overturn in a lake,
   a. nutrient-rich water sinks to the bottom.
   b. oxygen-rich water sinks toward the bottom.
   c. a thermocline develops.
   d. *a* and *b*, but not *c*.
   e. *a*, *b*, and *c*.
7. Which of these biomes is a uniformly warm grassland?
   a. savanna
   b. chaparral
   c. taiga
   d. tundra
   e. none of the above
8. Temperature is more important than precipitation in regulating the variety of flora and fauna in the
   a. tropical rain forest.
   b. coral reef.
   c. taiga.
   d. temperate forests.
   e. deserts.
9. Amount and pattern of precipitation is more important than temperature fluctuation in regulating the variety of plants and animals in the
   a. tundra.
   b. taiga.
   c. desert.
   d. polar regions.
   e. limnetic zone.
10. Evidence for continental drift comes from all except
    a. shapes of continental land masses.
    b. fossil remains within rocks on now-separate continents.
    c. upwelling of magma in regions where two plates move apart.
    d. increasing elevation of the Himalayas.
    e. even worldwide distribution of all terrestrial species.

True-False Questions

Mark each statement true or false, and if false, restate it to make it true.

1. Terrestrial ecosystems are subdivided into biomes, while aquatic ecosystems are subdivided into zones.
2. The overall pattern of air circulation constitutes the weather, but the day-to-day condition of the atmosphere at any one place constitutes the climate.
3. The primary producers in reef communities are chemoautotrophic bacteria.
4. Within the ocean, the open ocean habitats are richer in energy, nutrients, and organisms than the intertidal zones or the coral reefs.
5. The limnetic zone bears the same relationship to the epipelagic zone as the profundal zone does to the abyssal zone.
6. Eutrophic lakes are nutrient-rich and oxygen-rich while oligotrophic lakes are nutrient-poor and oxygen-poor.
7. More species live in temperate forests than in any other biome.
8. Reindeer, polar bears, dwarfed shrubs, and permafrost are all characteristic of the taiga.
9. Coniferous and deciduous trees, bobcats, owls, and a thick layer of humus are all characteristics of the temperate forest.
10. Because they were once united in the continent of Pangaea, the fauna and flora of Australia, southeast Asia, and India show striking similarities.

Concept Questions

1. Describe the components of a coral reef community.
2. Would freshwater communities in the tropics, subtropics, temperate, and polar regions be sustained year-round if water at 0°C had a greater density than water at 4°C? Explain.
3. Despite the fact that tropical rain forests are rich in numbers of species and in biomass, the soil is poorer than in tropical forests. How can this be?
4. Which aquatic and terrestrial ecosystems contain the greatest diversity of species? Which contain the least diversity? Do the same factors determine species richness on land as in water?
5. What is meant by convergent evolution? What role does convergent evolution play in determining the variety of plants and animals inhabiting any ecosystem?

Additional Reading

Allen, Durward L. *The Life of Prairies and Plains.* Our Living World of Nature Series. McGraw-Hill Book Co., New York, 1967.

Amos, William H. *The Life of the Seashore.* Our Living World of Nature Series. McGraw-Hill Book Co., New York, 1966.

———. *The Life of the Pond.* Our Living World of Nature Series. McGraw-Hill Book Co., New York, 1967.

Amos, William H., and Stephen H. Amos (eds.). *Atlantic and Gulf Coasts: The Audubon Society Nature Guides.* Alfred A. Knopf, New York, 1985.

Berreby, David. "Running on Tundra." *Discover,* June 1996, p. 74. The tundra is much the same all around the top of the earth, but an area around Kuparuk River in northern Alaska has been intensely studied since 1975. A rubbery, nutrient-poor wetland extends only a foot or two above permafrost. Researchers are studying the plant and animal life there.

Brown, Loren (ed.). *Grasslands: The Audubon Society Nature Guides.* Alfred A. Knopf, New York, 1985.

Coker, Robert E. *Streams, Lakes, Ponds.* Harper Torchbooks, Harper and Row, New York, 1968 [1954].

Hardy, Alister C. *The Open Sea. Its Natural History: The World of Plankton.* Houghton Mifflin Co., Boston, 1956.

McCormick, Jack. *The Life of the Forest.* Our Living World of Nature Series. McGraw-Hill Book Co., New York, 1966.

Niering, William A. *The Life of the Marsh.* Our Living World of Nature Series. McGraw-Hill Book Co., New York, 1966.

Schultz, Jurgen. *The Ecozones of the World: The Ecological Divisions of the Geosphere.* Springer-Verlag, Berlin and New York, 1995. A comprehensive overview of life zones and the biosphere.

Sutton, Ann, and Myron Sutton. *The Life of the Desert.* Our Living World of Nature Series. McGraw-Hill Book Co., New York, 1966.

Thorson, Gunnar. *Life in the Sea.* McGraw-Hill Book Co., New York, 1971.

United States Department of Agriculture. *Climate and Man: Yearbook of Agriculture 1941.* United States Government Printing Office, Washington, D.C., 1941. An old book that is still an excellent basic reference.

Usinger, Robert L. *The Life of Rivers and Streams.* Our Living World of Nature Series. McGraw-Hill Book Co., New York, 1967.

Internet Resource

To further explore the content of this chapter, log on to the web site at:

http://www.mhhe.com/biosci/genbio/guttman/

26 Population Structure and Dynamics

In normal conditions, a female rat nurtures her young and protects them until they can fend for themselves.

Key Concepts

26.1 Competition in communities occurs both within and between species.

26.2 All populations are dispersed in specific habitats over a certain geographic range.

26.3 Populations tend to grow exponentially.

26.4 Populations of many organisms tend to remain quite stable in size.

26.5 All populations are limited to a maximum size.

26.6 Populations may be limited by resources or by predation.

26.7 Some factors that limit population size depend on population density.

26.8 Density-dependent factors include nutrient limitation and physiological pressures associated with high density.

26.9 Members of a population tend to adopt strategies that maximize their use of resources and minimize competition.

26.10 Territories help allocate resources and control populations.

26.11 Survivorship curves show different patterns of reproduction.

26.12 Different kinds of selection produce two extremes within a spectrum of lifestyles.

26.13 The human population is growing far beyond its ecological limits.

John Calhoun once kept Norway rats in specially constructed pens to observe the effects of overcrowding. The pens were large enough to support about 48 rats comfortably, but initial populations of 32 and 56 individuals both grew to about 80. These rats underwent severe changes in physiology, and their social structure fell apart. Females largely abandoned their normal nest-building behavior and built only poor substitutes from the material provided. Their pups (young) left these nests too early to survive alone. Mortality among females and infants increased radically, infant mortality rising to 96 percent. The rats could move among four rooms, and infant mortality was lower in the two end rooms that acquired smaller populations. Although enough infants matured to keep the population fairly constant at 80, Calhoun believes the colonies might have died out in time due to reproductive failure, for when he later selected the healthiest rats to begin new colonies without overcrowding, they produced fewer litters than normal, and none of their offspring survived to maturity.

Many rats in Calhoun's colonies adopted bizarre behaviors. Cannibalism appeared among males, while some rats resigned completely from the social order and wandered about the pens like sleepwalkers, oblivious to all the other rats and taking no interest in sex. Some of the male rats, which Calhoun called probers, became hyperactive and excessively interested in sex with both genders. They abandoned the normal pattern of courtship and became unusually aggressive in attempting to mate with females.

It is hard to read about Calhoun's experiments without reflecting on their implications for human populations. The parallels with crowded urban societies—ridden with violence, socially dysfunctional, having high levels of stress—are too obvious. Is this what lies in store for humanity as our population increases? ■

26.1 Competition in communities occurs both within and between species.

One level of ecology is a story about populations. To be sure, organisms interact with one another as individuals—the individual fox eats a single hare, and the individual plant is outcompeted by a slightly better plant. But the population description provides a statistical summary of the overall effects of fox populations on hare populations and the dynamics of competition within plant populations. Ecology is also about resources, such as nutrients and space: how they are distributed, and how different organisms compete for them, divide them, and serve as resources for one another.

Populations and communities are both shaped largely by limited resources and competition for them. Darwin and Wallace observed that organisms have enormous potential for reproduction, but populations tend to remain about the same size; therefore, organisms are in competition and most do not survive. Competition for limited resources is both **intraspecific,** between members of the same species, and **interspecific,** between members of different species. These two kinds of competition have quite different effects. Intraspecific competition, which we discuss in this chapter, shapes the species; interspecific competition, discussed in Chapter 27, shapes an ecological community. In this chapter we will consider how populations grow, how their growth and structure are affected by the limitation of resources, and what factors may keep their numbers stable. We will show that different types of organisms have evolved a spectrum of living strategies from so-called opportunistic to equilibrium species, so their populations grow in different ways and respond differently to environmental change.

26.2 All populations are dispersed in specific habitats over a certain geographic range.

A population is a group of individuals of the same species in a given area—perhaps an entire species, perhaps only a small part of one. Its boundaries are often somewhat fuzzy, since they may shift geographically with time and since individuals frequently move in and out of an area. Ecologists may define a population rather arbitrarily as the individuals in an area that they are able to study.

Every species occupies a certain **geographic range,** an area where its members are likely to be found. But within that range, individuals select only a limited **habitat** that meets their requirements. Muskrats and eastern gray squirrels both occur over a large part of eastern North America; their geographic ranges overlap strongly, but muskrats live only in marshes while gray squirrels reside in deciduous forests. Thus a population is distributed, or *dispersed,* over its geographic range in accordance with the distribution of its resources. Individuals may be dispersed quite *randomly, uniformly,* or *clustered* in small areas of high concentration (Figure 26.1). Because resources are rarely distributed uniformly, the environment has a certain

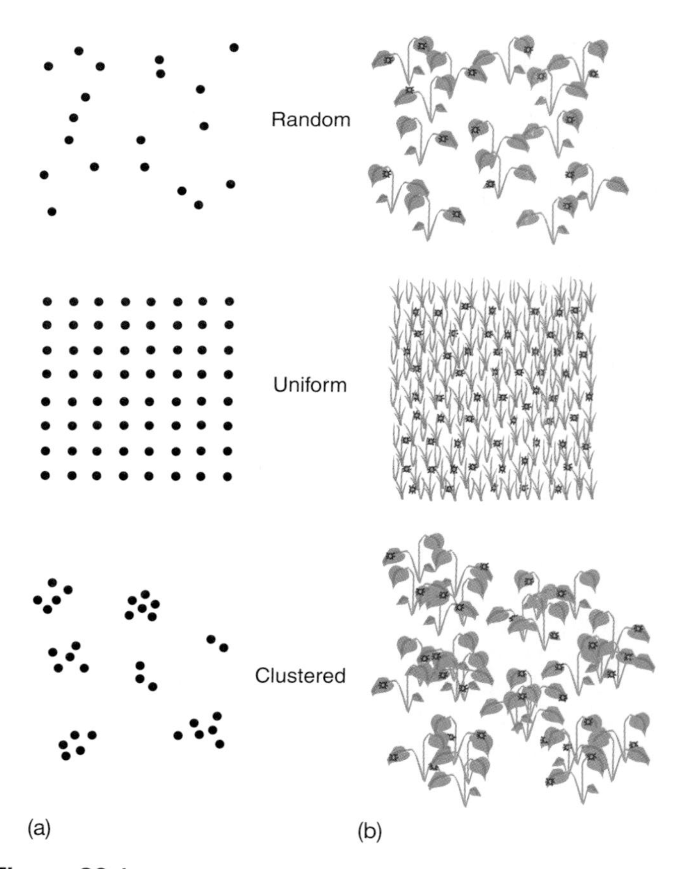

Figure 26.1

(a) The members of a population may be dispersed over their geographic range randomly, quite uniformly, or in clusters. *(b)* Their distribution is determined by the distribution of their resources, which is typically quite patchy.

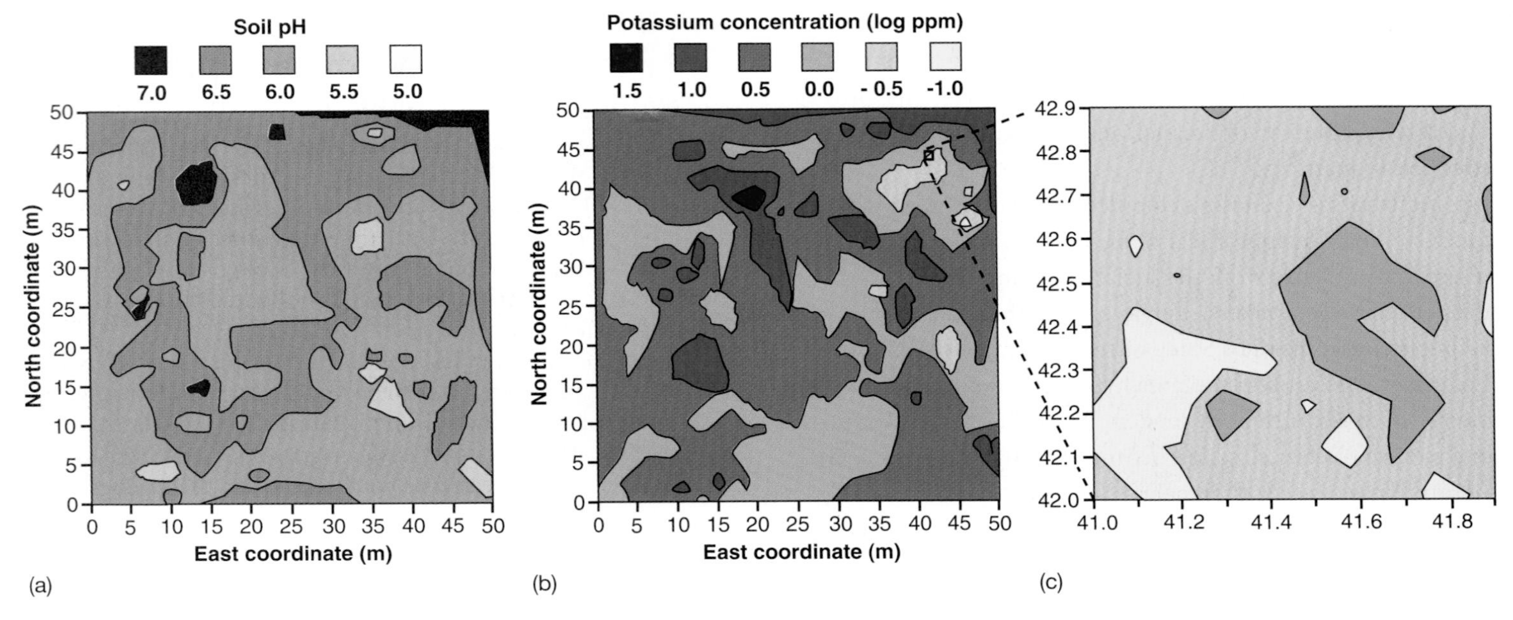

Figure 26.2

The environment is patchy. A small tract of forest floor (50 by 50 m) at Mont St. Hillaire, Quebec, was mapped for its concentration of *(a)* soil pH and *(b)* potassium. *(c)* Even one randomly chosen 1 × 1 m plot shows patchiness in potassium concentration. Source: Grant and Horn, *Molds, Molecules and Metazoa*, 1992, Princeton University Press. Source: Grant and Horn, *Molds, Molecules and Metazoa, 1992, Princeton University Press.*

patchiness to it. Because of differences in soil, temperature, topography, and other physical factors, plants tend to grow in patches (Figure 26.2), as do corals and other sessile animals of aquatic ecosystems. In each patch, plants adapted to different physical conditions interact with one another. Thus, if plant species A is inhibited by species B and enhanced by species C, it will only grow well in certain patches of land determined by the requirements of B and C. Insect species D, which lives only on A, will experience this patchiness by finding A only in those limited places. The size of patches is expressed by the **graininess** of the environment, which depends on the organism's size and ability to move. Different species may experience the same environment as **fine-grained,** made of small patches it can move through easily, or **coarse-grained,** made of large patches that confine it. Decaying spruce logs in a forest, for example, may provide habitats for both a bird and a slug (Figure 26.3), but the bird experiences the forest as fine-grained because it can move quickly from one such log to another, while the slug experiences it as coarse-grained because it may hardly be able to move from one log and may spend its entire life in a single grain.

Figure 26.3

A bird and a slug experience the same environment differently. Because the bird can move quickly and easily, each log and plant in the forest floor is a small grain to it, whereas the slug moves slowly and experiences each structure as a large grain.

Uniform dispersion may result from antagonism between individuals. Many plant species, for instance, produce substances that inhibit the growth of their own seedlings. In the deserts of the southwestern United States, the rubber-producing shrub guayule, *Parthenium argentatum,* releases cinnamic acid into the soil around it, and the creosote bush, *Larrea divaricata* (Figure 26.4), releases various pungent substances, as its common name implies. These materials inhibit the growth of plants of the same species, so the population spaces itself out quite uniformly, presumably as an adaptation for using the limited water and other resources of the desert. Since the same toxic materials severely limit the growth of other species in the region, these plants also impose a structure on the whole plant community.

Plants may grow in a clustered distribution because they require soil conditions that are only met in patches here and there. Similarly, animals may be clustered because of their social interactions—because they breed or hunt together. They

Figure 26.4
Many plants, such as the creosote bushes shown here, produce inhibitors from their roots that keep other plants of the same species from growing nearby, perhaps by inhibiting germination of seeds or by killing young plants. In this way, the species disperses itself quite uniformly.

Figure 26.5
Some animals, such as these albatrosses, reproduce in dense colonies, but at other times are dispersed widely over their whole geographic range.

may also spread out uniformly because of territoriality, breeding behavior in which they establish and defend territories for themselves (see Section 26.10). Dispersion often changes with time; seabirds such as shearwaters and albatrosses associate in dense colonies during the breeding season and then disperse over an enormous area of ocean during the rest of the year (Figure 26.5).

We measure the structure and changes in populations not only by population size but also by **density,** the number of individuals per unit area or volume. **Absolute density** taken over the entire range means much less than **ecological density,** the density within the limited habitat the population actually occupies. Measuring population size and density is a challenge. A mammalogist studying the herds of a large mammal may be able to collect excellent data by flying a small plane over the range and simply counting and photographing the subjects, but an investigator studying the mice in a meadow is hard-put to see these animals, much less count them. **Mark-recapture** is one commonly used method. First, a number of individuals are trapped, marked with bands or tags, and released. After giving the animals time to disperse themselves through the population, the traps are reset. Among the newly trapped animals will be some that were marked, and the fraction of these among the total trapped should equal the fraction of all marked individuals in the whole population size. Suppose 50 mice are initially marked and released. Then another 50 mice are trapped, and five of them turn out to be marked. If these mice are a random sample of the whole population, 10 percent of the population must be marked, so the 50 marked mice are 10 percent of the whole, which must be 500 mice. This method of estimation has its limitations, of course, but it can be improved by statistical procedures that make it more reliable for populations of various sizes.

Exercise 26.1 Identify at least two assumptions that underlie the mark-recapture method. How would an estimate of population size change if these assumptions are not valid?
Exercise 26.2 A biologist captured 200 ground beetles in the scrub habitat on top of a cliff, marked them, and released them. A few days later he captured 50 beetles and found that 5 were marked. What is the best estimate of the population size? How would your answer change if you were told that Sage Sparrows, which eat these beetles, are common in this habitat?

26.3 Populations tend to grow exponentially.

All natural populations tend to grow, since organisms have such an enormous potential for reproduction. But the capacity for growth isn't always obvious, because we are surrounded by long-established populations that have reached stable sizes and are not increasing any more. (Sidebar 26.1 describes some truly huge populations.) Sometimes, however, near-disasters force us to realize how rapidly organisms can reproduce, as when a population of tent caterpillars or other insects gets out of control and threatens to decimate a forest. Over several years, even casual observers of wildlife can see dramatic growth in numbers when a species is introduced into a new area that it finds congenial. The Ring-Necked Pheasant, English Sparrow, European Starling, and Rock Dove (the common city pigeon) were among many species introduced into North America by European settlers in misguided attempts to bring their familiar species and favorite game birds with them to the new world. These species spread rapidly across the continent and have achieved—or are achieving—stable populations, often to the detriment of native birds.

Simple unicellular organisms such as bacteria and yeast (and also growing cells in a cell culture) grow by repeatedly

Sidebar 26.1 The Enormous Numbers of Some Populations

The sizes of real populations sometimes strain the imagination. The nineteenth-century American naturalist John James Audubon spent part of the fall of 1813 in Kentucky, and he later described a flock of Passenger Pigeons that flew over the town of Henderson:

> The air was literally filled with pigeons; the light of noonday was obscured as by an eclipse; the dung fell in spots not unlike melting flakes of snow. . . . The people were all in arms. . . . For a week or more, the population fed on no other flesh than that of the pigeons. . . . The atmosphere, during this time, was strongly impregnated with the peculiar odour which emanates from the species. . . . Let us take a column of one mile in breadth, which is far below the average size, and suppose it passing over us without interruption for three hours, at the rate mentioned above of one mile in the minute. This will give us a parallelogram of 180 miles by 1, covering 180 square miles. Allowing 2 pigeons to the square yard, we have 1,115,136,000 pigeons in the flock.[1]

Five years earlier, Alexander Wilson estimated that a flock he had seen, also in Kentucky, numbered more than 2¼ billion birds. But due to unrestrained hunting, the numbers of passenger pigeons decreased rapidly during the nineteenth century. In 1892, Charles Bendire observed that the species might become extinct within a few years and pointed out that "from constant and unremitting persecution on their breeding-grounds they have changed their habits somewhat, the majority no longer breeding in colonies, but scattering over the country and breeding in isolated pairs." Apparently the survivors had been selected for their genetically weak social tendencies, since the downfall of the species was the great sociability that led it to form enormous flocks that were easily shot and netted. Unfortunately, even abandoning their colonial behavior could not save the birds. The last passenger pigeon died in the Cincinnati Zoological Gardens in 1914.

Another example of a large natural population is a colony of American driver ants, which may have more than 20 million workers, together weighing about 20 kg. One population in Maryland contained an estimated 12 million workers, spread over 10 acres in 73 nests, with a total mass of about 10 kg. This was not the only species of ant that inhabited the area, so ants alone must have constituted a huge biomass.

[1] *Ornithological Biography* (Edinburgh: A. Black, 1839).

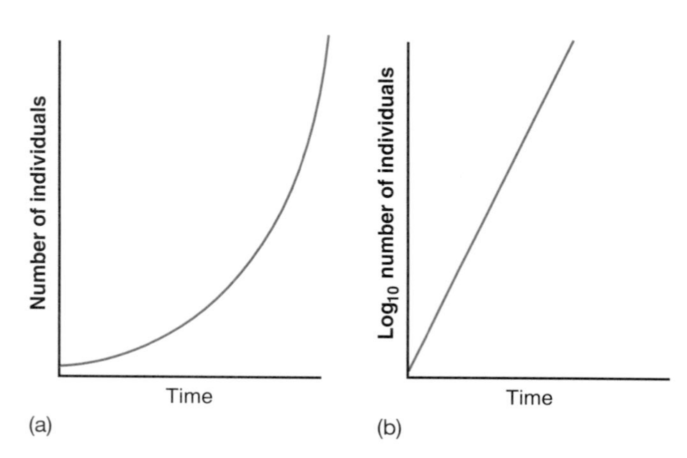

Figure 26.6

(a) An exponential growth curve rises faster and faster as time goes on, showing that the numbers added in each time interval keep increasing. *(b)* When the population size is graphed on a logarithmic scale, the curve is converted to a line. This graphical form is easier to analyze, since the slope of the line is the growth rate.

dividing in two, so their population doubles in size periodically. A population of N cells grows to $2N$, to $4N$, to $8N$, and so on. As we pointed out in Chapter 2, this pattern of *exponential growth* can produce large numbers in a short time (Figure 26.6; see also Sidebar 26.2). In general, populations of sexually reproducing organisms increase in the same way, although far below the rate at which bacteria can grow under optimal conditions. For instance, 8 pheasants were introduced in 1937 on Protection Island off the coast of Washington state; with no predators to inhibit their growth, the population grew to nearly 1,900 birds in 6 years (Figure 26.7). This episode ended artificially when the military occupied the island in 1942 and shot all the pheasants.

In a growing population of size N, we measure the number of individuals at regular time intervals, Δt. One definition of exponential growth is that the size *difference* between any two generations, ΔN, is proportional to N. Look at these series of numbers and at the differences between them:

Population:	1	2	4	8	16	32	64	128
Differences:	1	2	4	8	16	32	64	

As the population size doubles, the differences also double. Taking N at the end of each generation, the ratio $\Delta N/N$ is always 1/2. From this observation, we can see that a general way to express exponential growth is:

$$\frac{\Delta N}{N} = r\Delta t$$

r is called the **intrinsic growth rate** of the population. When Δt is measured in days, the maximum value of r for the ciliate *Paramecium* is approximately 1; for many insects, it is about 0.01; for small mammals such as mice, it is about 0.001; and for humans, it is about 0.0003. These numbers show how much faster small organisms can reproduce than large ones. A real population rarely doubles each generation; it is more likely to just gain a few individuals each year, perhaps like this:

10,000   10,100   10,201   10,303   10,405

Here the population grows by 1/100 of its size each year, so $r = 0.01$/year.

Sidebar 26.2 The Potential of Exponential Growth

A simple, but absurd example will show how much a population can grow just by doubling periodically. A single bacterium weighs 10^{-12} grams. The mass of the earth is 6×10^{24} grams. If a single bacterium could be maintained with enough nutrient medium to let it grow to a population with the earth's mass, doubling every half hour, how long would it take?

We can do the calculation roughly, considering that 10^3 is approximately 2^{10}. The cell must increase 6×10^{36} times, which is about 2^{122}. That is, the population must double about 122 times, but at a half-hour per doubling, this will take only 61 hours—just over 2.5 days!

This explains why people who use microorganisms as experimental tools can do their work so quickly, and also why even a few cells that start an infection in a susceptible person can become an enormous threat. It also should help you understand the frightening potential of the human population explosion now going on (see Section 23.13).

Since the growth rate is the increase in a population during any time interval, $\Delta N/\Delta t$, the growth equation is more commonly written:

$$\frac{\Delta N}{\Delta t} = rN$$

which shows that the growth rate is proportional to the size of the population. In other words, large populations grow faster than small populations of the same organism. Money grows this way when invested at compound interest. The interest it returns (every day or every quarter) is added to the capital, so interest is always being paid on a larger amount. Similarly, in each generation, a growing population produces more organisms that become the breeders in the next generation.

The intrinsic growth rate is really the difference between **natality,** or birth rate, b, and **mortality,** or death rate, d: $r = b - d$. In a population of N individuals, the birth rate is the number of births during a certain period divided by N, and the death rate is the number of deaths divided by N. As long as b is greater than d, the population is growing. For instance, in a population of 1,000 mice, there may be 300 births in one month, so the birth rate is 300/1,000 = 0.3 per month; but if 280 mice die, the population's death rate is 280/1,000 = 0.28 per month. (We assume that birth and death go on simultaneously, so the population doesn't increase by 300 at one time and then suddenly decrease by 280.) Overall, the population actually increases by 20 mice per month, and $r = 0.30 - 0.28 = 0.02$.

Another useful measure of growth is the ratio of numbers from one year to the next, the **net reproduction rate,** $R = N_{x+1}/N_x$. (The subscript applied to each N symbol is an *index,* which counts generations. Starting at any generation x, the next generations are $x + 1$, $x + 2$, and so on.) For a population of 10,000 that increases by about 100 individuals per year, $R = 1.01$. As long as R is greater than 1, the population is growing exponentially. R is a conceptually simple measure because it says directly that the average individual in one generation is replaced by R individuals in the next generation. If R remains constant, it is easy to predict what N will be after n generations:

$$N_n = R^n N$$

It is instructive, and rather startling, to see that even if R is only slightly larger or smaller than 1, a population can change size dramatically. If R is only 1.01, a population will grow by 2.7 times in 100 generations and by 21,000 times in 1,000 generations! On the other hand, if $R = 0.99$, the population will decrease to $0.000043N$ in 1,000 generations. The fact that many natural populations do *not* change so drastically means that some factors generally keep R very close to 1.00.

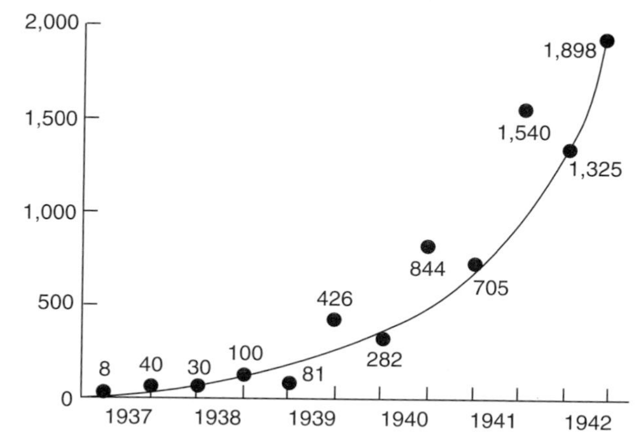

Figure 26.7
An island pheasant population grew exponentially over six years in the absence of limiting factors such as predation. The population was counted in spring and fall each year.

Exercise 26.3 A population of 10,000 individuals is growing with $R = 1.005$. How many individuals will there be after 10 generations?

Exercise 26.4 Determine which of the following series are growing exponentially by calculating $\Delta N/N$ to see if this ratio is constant:

a. 3 6 12 24 48 96
b. 4 6 8 10 12 14
c. 4 5 6.25 7.81 9.77 12.21

26.4 Populations of many organisms tend to remain quite stable in size.

We can see exponential growth in laboratory populations and under artificial conditions like those on Protection Island, but how do natural populations behave? Gilbert White, one of the

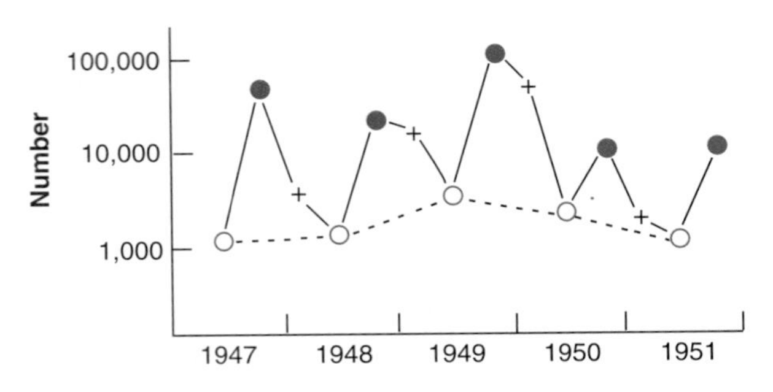

Figure 26.8

Over several years, a population of a grasshopper *(Chorthippus brunneus)* produced enormous numbers of eggs each summer *(green circles)*; much smaller numbers of young grasshoppers, or nymphs, survived *(+ signs)*, and the population of adults *(open circles)* remained at a stable, low level, which is presumably all that its environment can sustain.

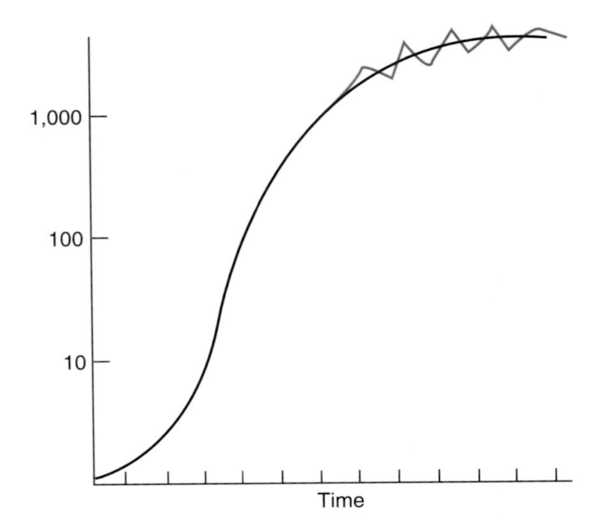

Figure 26.9

An idealized logistic growth curve is shown in black. A real population often fluctuates around an equilibrium level, as shown by the curve in color.

first naturalist-ecologists, began to study a classic case in the village of Selbourne in southern England in 1778. White noted that, year after year, just eight pairs of swifts lived in the village, nesting in the church and the thatched roofs of cottages. When the swift population was examined again in 1983, the area had changed considerably, its thatched cottages replaced by more modern houses, and yet the population had only increased to 12 pairs of swifts. This case is often cited to show the stability of a bird population over many generations, and other censuses of birds and mammals have also shown that populations tend to remain stable if the environment isn't seriously disturbed.

At this point, let us introduce two warnings. First, populations can change dramatically in the short term over a single generation. Figure 26.8 shows a typical situation, in which a population rises about a hundredfold early in the season when many young are produced and then falls to a small, stable level as most of the animals die. The chief factor holding the population down is predation, which we will discuss in Section 26.6 and in Chapter 27. For now, we must simply keep in mind that all organisms except top carnivores are preyed upon by others, and this is a major consideration in population stability.

Second, many studies have been done on birds and mammals, but focusing on them as examples tends to distort the picture. We will show Section 26.12 that different organisms have quite different modes of life; the populations of most birds and mammals tend to remain constant, but other kinds of organisms may experience sharp increases and decreases in numbers.

26.5 All populations are limited to a maximum size.

All natural populations stop growing at a certain maximum size, known as the **carrying capacity** of the environment and denoted by K. (Some populations actually grow beyond this size and then crash. We won't consider this phenomenon here, except to keep it in mind when thinking about human populations.) The carrying capacity is incorporated into the growth equation by rewriting it as follows:

$$\frac{\Delta N}{\Delta t} = rN(1 - \frac{N}{K})$$

This is called the **logistic growth** equation. You can see that as N increases, N/K gets closer to 1, so the growth rate keeps getting smaller. When $N = K$, N/K is exactly 1, and therefore the growth rate becomes zero. A graph of the logistic growth equation is described as *sigmoid* or *S-shaped,* and it levels off at K (Figure 26.9). Populations of small animals and protozoans grown in laboratories generally exhibit sigmoid growth curves, except that the equilibrium level often fluctuates rather than remaining steady (colored curve in the figure).

V. C. Wynne-Edwards devised one hypothesis to explain the limitation of population size; he conceived of a kind of group selection favoring populations that develop mechanisms to keep their sizes within the limits of their resources. Ordinary individual selection results from a competition between individuals; group selection, if it exists, is competition between groups. At first glance, this is an attractive idea: The populations that survive are those that, for their own good, limit intraspecific competition and come to live in greater harmony with their environment. However, most population biologists give this hypothesis little credence. The major difficulty is that a population limitation scheme is easily thwarted by "cheaters." Suppose all the individuals in some population have a special mechanism that limits their reproduction as their density increases. Eventually a mutant will appear that isn't limited by this mechanism, and mutants will then out-compete the others because they will tend to use up the resources and reproduce more rapidly. So the density-dependent factors that limit populations must either be unavoidable or must be selected for their benefits to individuals, not for the benefit of the species as a whole.

26.6 Populations may be limited by resources or by predation.

If populations aren't limited "for their own good" by intrinsic factors, what does limit them? Populations could be limited in three ways: by random catastrophes, such as weather; by limited resources; or by predators feeding on them, where predation is used in the broadest sense to include herbivory—herbivores eating phototrophs. Mice eat plants and are eaten by hawks, owls, foxes, and other predators. What is most important in limiting their numbers—not enough food, too many predators, or some other random factor? For a long time, the classic viewpoint in ecology has been that populations are limited primarily "from below"—that is, by the resources available to them. Many populations clearly reach a limit because their members are competing for limited resources. Competition takes the form of either *exploitation* or *interference.* If a population of mice lives on seeds, each mouse exploits the resource independently, and as a result, some mice get less food than they need. On the other hand, plants competing for appropriate soil to grow in may interfere with one another's growth by trying to occupy the same space, and animals that establish territories can interfere with one another as they compete for space. In animal species where males compete actively for females, two males engaged in combat are obviously interfering with each other. Whatever form it takes, intraspecific competition for resources ultimately affects the ability of individuals to survive and to reproduce. Mice that don't get enough food either die or have reduced fecundity (the rate of offspring production per female) due to ill health. A plant that is crowded out of a small patch of soil will die and fail to reproduce. Thus competition for resources may well limit population size. On the other hand, several ecologists have proposed that many populations are limited primarily "from above"—by predators feeding on them.

In a classic and controversial paper in 1960, Nelson Hairston, Frederick Smith, and Lawrence Slobodkin argued strongly for a compromise position: that herbivores are limited by predation, but the other trophic levels (producers, carnivores, and decomposers) are limited by food. The argument is that producers (plants) are not limited by herbivores because herbivores obviously do not affect them severely; the world is green, and there is plenty of vegetation around. Neither are they limited by weather and catastrophe. Therefore, they must be limited by resources. In fact, unnatural situations have sometimes shown how different vegetation can be when it is limited by an abnormally high herbivore population. Around 1907, the Kaibab Plateau near the Grand Canyon was converted into a national park, and its caretakers tried to protect the deer population by eliminating predators and prohibiting hunting. At the same time, cattle grazing and fires were stopped, thus making more food available to the deer. The deer population increased from about 20,000 in 1912 to 100,000 in 1924, followed by a crash in which 60 percent of the herd starved to death during two winters. The population explosion may have been due in part to the deer having access to additional food as well as to removal of predators, but the result was that the deer herd devastated its food supply. Since herbivores normally leave a great deal of edible vegetation, this case argues that vegetation is generally not limited by predation on it.

Continuing, Hairston and his associates argued that herbivores are not food-limited, since there is obviously plenty of food available; they are also not catastrophe-limited, so they must be limited by predation. Similarly, they argued that carnivores are food-limited, because there are no predators above them to limit their populations. Decomposers are food-limited because, by definition, they are the organisms that consume organic debris; decomposer populations might be limited by predation, but then organic debris would accumulate and would have to be consumed by something else. However, organic debris does not accumulate, and the "something else" would be a decomposer by definition.

Other investigators have leveled various criticisms at this neat thesis. We will see in Section 27.12 that predation does take its toll on some populations, perhaps enough to limit their size severely, but at other times predators appear to have little effect. With our current knowledge, it seems difficult or impossible to generalize about all cases, even at a single trophic level. For now, in thinking about the factors that limit populations, it is important to recognize that populations may be limited both from above and from below. We will always keep this view in mind when thinking about population size.

26.7 Some factors that limit population size depend on population density.

The factors that limit population growth are divided into those that are independent of population density and those that become more effective as population density rises. **Density-independent factors** affect large and small concentrations of organisms equally. For instance, harsh weather might kill 90 percent of the individuals in a particular population, whether there are 30 or 3,000. A severe fire kills organisms without regard for their density. In contrast, **density-dependent factors** affect a crowded population much more than a sparse one. As a population grows, various factors that reflect its density could start to change the organisms' behavior or physiology so that growth eventually stops (Figure 26.10). Ecologists have waged extensive arguments about the importance of density-dependent and density-independent factors, and both types operate to various degrees on different kinds of organisms. H. G. Andrewartha and L. C. Birch have argued that weather, partly through its effects on vegetation, is the major factor limiting insect populations, but that weather has much less effect on mammals or on plants adapted to withstand harsh conditions. Yet the weather is not an entirely density-independent factor, even for insects. A swarm of bees, for instance, can withstand low temperatures better than a single bee because they form a ball and keep each other warm with their metabolic heat. Conversely, all the members of a sparse population may be able to find shelter, whereas more individuals in a dense population may be forced into exposed conditions that kill them.

Density-dependent factors are of greater biological interest, and we will focus on them in thinking about the limitation of

Figure 26.10

As a population grows more dense, its members put pressure on one another and on the population's resources. These factors then inhibit further growth of the population in proportion to its density.

(1) When the density of a population is low, the individuals put relatively little pressure on one another and on their environment.

(2) As the density increases, the members of the population make greater demands on environmental resources and interact more strongly with one another.

(3) The increased density activates factors that tend to limit reproduction, such as stress and scarcity of food, thus stopping further growth.

population size. However, we must be skeptical about the importance of such controls in real populations. Many ecologists have noted, with Andrewartha and Birch, that unpredictable factors often disrupt habitats and severely affect populations. A fire breaks out. A flood rips through a lowland. A herd (or swarm) of herbivores reduces a lush field to stubble. Several years of drought beset an area. Events like these make a lot of thinking about population size irrelevant, because a natural population may not grow large enough or last long enough for the predictable biological factors to matter. This is another controversy within ecology, and we will take it up again later in connection with the issues of interspecific competition and ecosystem stability.

Exercise 26.5 Hawks search for small rodents in a meadow by flying back and forth over the area. Explain why such predation is a density-dependent factor that limits the rodent populations.

26.8 Density-dependent factors include nutrient limitation and physiological pressures associated with high density.

Many studies of growing animal populations have shown decreasing fecundity as the density of the population rises. In studies of the water flea *Daphnia,* Peter Frank and his associates found that at a density of only one animal per milliliter, each female produced up to four viable young per day on the average for several weeks, but with 32 animals per milliliter, each female bred for only a short time and produced far fewer than one offspring per day. Population biologists have documented density-dependent limitations by limited resources, by accumulated wastes, and perhaps by stress factors associated with high density.

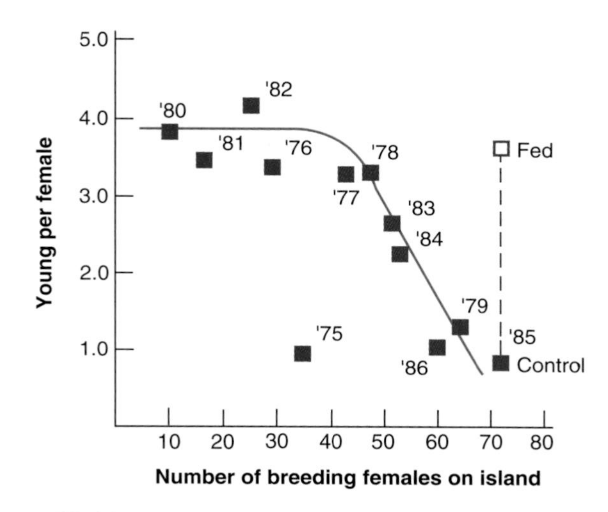

Figure 26.11

The fecundity of female Song Sparrows on an island is related to their population density. Notice that the population density fluctuated (numbers are years from 1975 to 1986) and did not simply increase, but fecundity remained close to 4 young per female when the population was about 40 or less, and dropped substantially above that density. Feeding the birds in 1985 raised their fecundity, indicating that nutrition was the limiting factor at high densities.

Nutrients can limit population growth.

Food is the most obvious resource that can reduce survival and fecundity. As crowded individuals find it harder to get adequate nourishment, they may die before they can mature and reproduce, or they may produce fewer, less viable offspring. In either case, the population size will fall. P. Arcese and J. N. M. Smith found that the population density of Song Sparrows on an island was closely related to the fecundity of females (Figure 26.11). They also showed experimentally that the sparrows' fe-

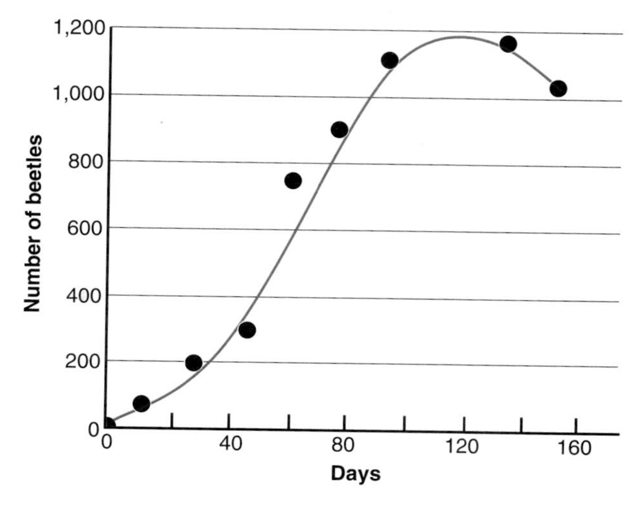

Figure 26.12

A population of flour beetles, *Tribolium confusum,* shows a typical population growth curve (number of beetles at all stages of growth in 32 grams of flour).

cundity was limited by food; when they provided extra food for the population, its fecundity rose to the low-density level. Similarly, G. D. Aumann and John T. Emlen showed that populations of the Meadow Vole (a small rodent) are limited by the sodium content of the soil. Areas rich in sodium had over 2,500 voles per hectare, while those with lower concentrations had no more than 575 per hectare. Furthermore, voles given sodium chloride produced more offspring, showing how this factor influences breeding.

Waste accumulation and other stress factors can limit population growth.

Density-dependent factors are clearly at work in some populations of microorganisms. As a culture of bacteria grows, it produces wastes such as acids that decrease the pH of the culture medium, so that the bacteria can't grow. Bacteria also become oxygen-starved as the culture becomes more dense. The more crowded the culture becomes, the more conditions depart from the optimum, until finally they become bad enough to stop growth.

Dense populations impose stresses on their members. Classical experiments by R. N. Chapman and by Thomas Park on flour beetles (*Tribolium confusum;* Figure 26.12) show how these stresses can limit their population density. *Tribolium* beetles spend their entire lives in flour, where they eat, defecate, mate, and die. Chapman raised beetle colonies, which he moved periodically to fresh flour, and found that their populations grew with a classical S-shaped logistic curve to a density of about 4.4 beetles per gram of flour, regardless of the size of the entire colony. This density, in other words, is the carrying capacity of this particular environment, and the beetles cannot exceed this limit because of a combination of environmental factors. Chapman attributed the stabilization to cannibalism of eggs by adults, because as a colony becomes more crowded, the rate of cannibalism increases.

Park continued these experiments by keeping some beetle colonies in one batch of flour, which the beetles fouled as they defecated and died in it. He found that females in the old flour laid fewer eggs than those in a comparable colony provided with fresh flour; he eventually showed that their egg-laying is inhibited by ethylquinone, which male beetles produce in special glands. Males are especially prone to release ethylquinone when another beetle interrupts them during mating, so they produce more inhibitor as the population density increases. Ethylquinone appears to be a population-control **pheromone,** pheromones being substances that carry signals between members of a population. Some pheromones, such as ethylquinone, seem to be specific inhibitors of reproduction. As another example, mice produce a material in their urine that inhibits ovulation; it accumulates under crowded conditions and so reduces the birth rate in response to population density. Thus some populations do appear to limit their growth by specific regulatory compounds, as Wynne-Edwards hypothesized.

Pheromones, Section 11.7.

In vertebrates with well-developed nervous and endocrine systems, the stresses of dense populations may affect reproduction through the **stress syndrome** (Figure 26.13), which was described by Hans Selye. Many of the factors that produce stress—injury, fear, anxiety, new environments, overcrowding, and others—activate the adrenal glands to produce their hormones. One hormonal effect is to inhibit the inflammatory reaction that fights infections, making animals under stress from population pressures more susceptible to disease, and reducing their numbers somewhat. Stress also inhibits the growth of reproductive organs. John Christian observed that as populations of house mice grew more dense, their adrenal glands increased in size while their reproductive organs shrank. With more

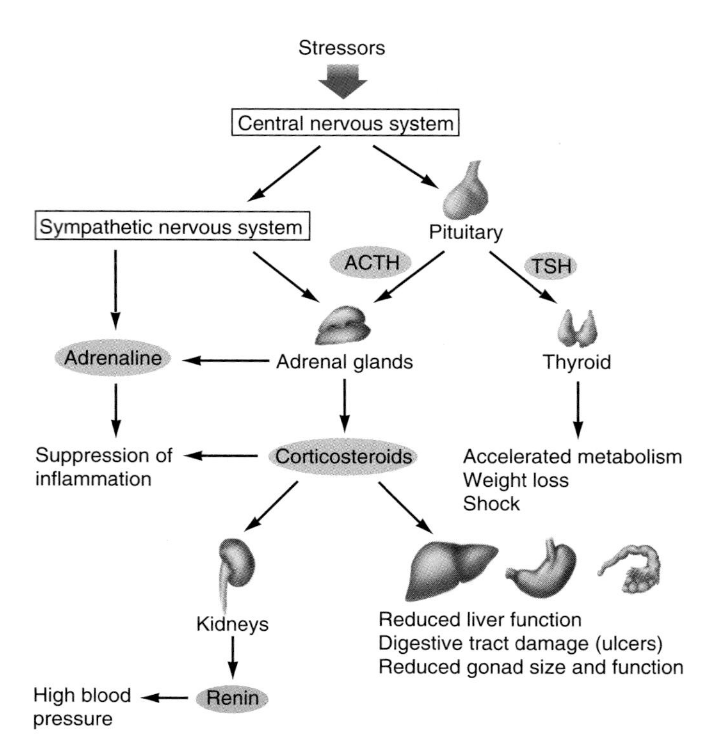

Figure 26.13

The stress syndrome in mammals is activated by many factors (stressors), including crowded living conditions. In humans, it may be brought on by the many complex factors associated with modern living conditions. Overproduction of adrenal hormones (corticosteroids) inhibits reproduction and interferes with the immune system and with inflammation, making it more difficult to fight off infections.

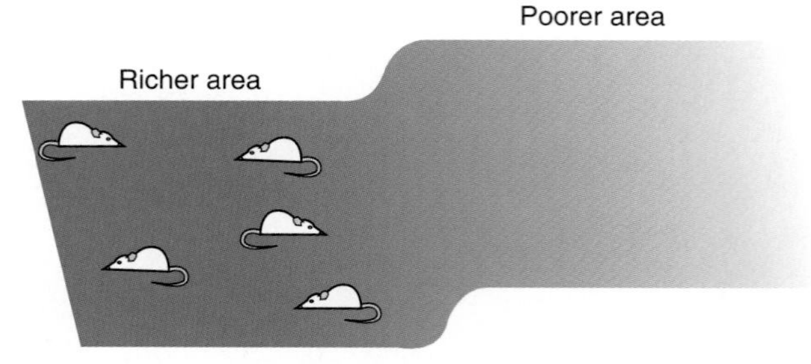

Mice initially tend to go into the richer area, where food is more abundant.

At a certain density, there is so much competition in the richer area that the average mouse gets just as much food in the poorer area, so mice then begin to occupy this area.

Figure 26.14

When animals have a choice of two habitats that differ in some resource, they should distribute themselves so as to maximize their access to the resource.

crowding, females stopped lactating, and some didn't bear any young at all. The growth and maturation of young mice in the population was also inhibited.

Inflammation and defense against disease, Section 48.2.

26.9 Members of a population tend to adopt strategies that maximize their use of resources and minimize competition.

Every organism has a limited amount of energy. It can allocate that energy in various ways for growth, maintenance, and reproduction, but allocating more energy to one activity reduces the energy available for others. This simple truth is called the **Principle of Allocation.** The choice it makes about allocation of energy has largely been made for each organism by its ancestors as they evolved a certain strategy of growth and reproduction. Each individual, following the instructions in its genome, is perpetuating that strategy with some variation. Its success in allocating its energy is one measure of its fitness, and we expect organisms to be growing and behaving quite efficiently. These patterns of growth and behavior will be different for each species, but their efficiency of allocation is sometimes quite remarkable.

Intraspecific competition, competition among members of the same species, certainly limits population sizes, but it is also precisely the pressure on a population that creates natural selection for factors and mechanisms that maximize the fitness of individuals, including factors that reduce competition among them. In addition to selection for optimal ways of growing and reproducing, organisms are selected for behavior patterns that tend to optimize their use of resources.

Suppose a population of animals has access to a range of habitats with different qualities of a resource, such as more or less food. How do they distribute themselves, and does their distribution give them the maximum possible payoff for their efforts? To analyze the situation, let's suppose a region has only two habitats, one richer in resources than the other, and let's imagine that the animals fill the area gradually (Figure 26.14). Assuming the animals can detect this difference in resources, the first animals will naturally go to the richer habitat and gain maximum rewards for their efforts, but as more animals enter the richer area, they compete with one another and thus reduce one another's gains. At what point does it become better for some animals to occupy the poorer area, where they will have to work harder for the resource but will have fewer competitors? The payoff to animals in the richer area will diminish to the point where it is equal to the payoff to the first animal in the poorer area, and at this point animals in both areas will have equal payoffs as their numbers increase. As long as the animals are free to move and seek resources, they should establish an **ideal free distribution,** the distribution of organisms who are free to optimize their access to resources, and their densities in the two areas will be proportional to the relative richness of the areas.

This theory assumes that animals are able to detect the richness of their environment and respond so as to maximize their payoff. So it is of interest to ask whether animals actually behave in a way predicted by the theory. Remarkably, some do! Manfred Milinski demonstrated this with stickleback fish in a neat, simple experiment: He kept six fish in a tank and fed them minute crustaceans *(Daphnia)* at one end twice as fast as at the other end, making one end richer in food (Figure 26.15). The ideal free distribution of the fish would be for four fish to feed at the rich end and two at the poor end, and indeed, this is how they distributed themselves. When Milinski reversed the feeding pattern, the fish redistributed themselves accordingly. Observations on natural populations have shown other cases of animals that conform quite well with their ideal free distributions.

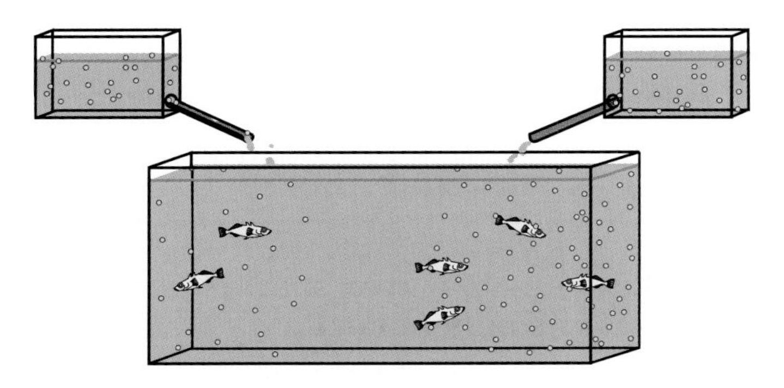

Figure 26.15

Sticklebacks distribute themselves in an ideal free manner when one end of their tank is made twice as rich in food as the other. When the food distribution is reversed, the fish reverse their own distribution.

Exercise 26.6 Two adjacent areas support populations of 1,000 insects per m^2 and 200 insects per m^2. A population of thrashers (ground-feeding birds) lives in these areas and eats the insects. Predict the relative densities of thrashers in the two areas.

26.10 Territories help allocate resources and control populations.

An important alternative strategy for dividing resources is **territoriality,** which means claiming an area for one's own and defending it actively, rather than simply roaming freely through a habitat. By defending a territory, many animals retain control over a set of resources, and territoriality is a way of limiting the number of individuals who try to share those resources. Individuals without a territory don't breed (except for cases in which outside males have been found to copulate with the female of a territorial pair), so each area is occupied by approximately the same number of individuals year after year. Figure 26.16 shows the territories established by a population of Song Sparrows in successive years. Even though the shapes of the territories varied from year to year, their sizes remained quite constant, so when the population increased, the territories limited the number of birds that could live in the area each year. Territoriality is therefore a population-control mechanism.

Because an animal must expend energy to defend an area's resources, territoriality is advantageous only if the animal obtains more energy than it expends. Frank Gill and Larry Wolf analyzed a prime case of resource defense in Golden-winged Sunbirds of Kenya, which drink flower nectar; the birds defend the feeding territories of mint flowers *(Leonotis)* (Figure 26.17), and remarkably, each bird defends an area containing just 1,600 flowers, which supply a bird's energy requirements for one day. The birds expend more energy in feeding (4.0 kJ/hr) than in perching (1.7 kJ/hr), and a lot of energy defending a territory

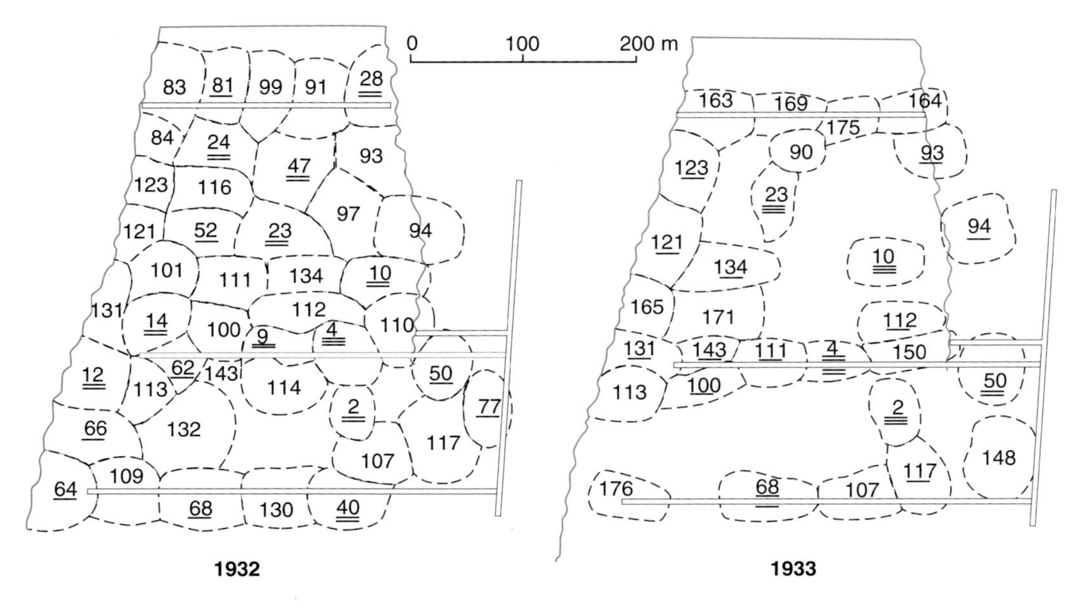

Figure 26.16

Song Sparrow territories in a small area of Columbus, Ohio, remained quite stable in size and distribution over several years, even though the number of breeding pairs varied. Lines under the numbers indicated additional years, after the first, that a bird continued to occupy a territory. These observations come from work by Margaret Nice, who began as an amateur naturalist and eventually produced one of the most important studies of bird territories ever done.

Figure 26.17
Golden-winged Sunbirds feed on *Leonotis* flowers and defend a territory of flowers.

(12.5 kJ/hr), but they can spend less time feeding if they have access to flowers of better quality. If the flowers yield 2 μl of nectar each, a bird can get enough nectar in only 4 hours, expending 4 hr × 4 kJ/hr = 16 kJ of energy; but if the flowers only yield 1 μl each, a bird must feed for 8 hours and expend 8 hr × 4 kJ/hr = 32 kJ. If a bird spends only 20 minutes a day keeping nectar thieves away from its territory, it will expend only 1/3 hr × 12.5 kJ/hr = 4.16 kJ, thus gaining almost 10 kJ for its efforts. Gill and Wolf concluded that resource defense is worthwhile in this case. However, suppose high-quality flowers contained 6 μl each, and nectar thieves only reduced them to 4 μl. The sunbirds could get enough nectar in 1.33 hours from rich flowers and in 2 hours from the poorer flowers. In this case, it is easy to calculate that the sunbirds would lose more energy defending a territory than they would gain.

Graham Pyke then asked, "Why 1,600 flowers, when large tracts of flowers are available?" He conjectured that the strategy the birds are following could be designed to optimize any of several factors: for instance, to maximize the energy they obtain per day or to maximize the time they can spend perching, presumably where they are relatively safe from predators. They might also be minimizing their daily expenditure of energy, because every activity probably wears the body out a little, and a less active bird is more likely to live long enough to breed. Pyke used data obtained from field observations and laboratory experiments, with a rather complex theoretical model, and found that the last hypothesis—minimizing energy expenditure—not only predicts that each bird should defend about 1,600 flowers, but also accurately predicts the time it should spend in each activity.

This story shows the power of good theoretical biology and also demonstrates that animals behave in a manner that optimizes their payoff in one way or another. Then the question naturally arises, "How does an animal get to be so smart?" The sunbirds, after all, didn't use ecological theory and a calculator to determine how to behave. Pyke's work implies that their behavior has been selected because it maximizes their ability to breed; to put it differently, birds that behave as these sunbirds do have been selected over many generations because they maximize their ability to breed. Of course, we still don't know what mechanisms could be built into an animal's nervous system to make it seek resources in accordance with an ideal free distribution.

Exercise 26.7 Determine how much energy the sunbirds would expend feeding on flowers containing 6 μl and those containing 4 μl. Compare these values with the energy expended in 20 minutes of territorial defense.

26.11 Survivorship curves show different patterns of reproduction.

The name of the survival game is *reproduction.* Remember that population size depends on a balance between the rates of natality and mortality. For a species as a whole, the number of new offspring produced is less important than the number that actually survive to reproductive age and the number of reproductively successful offspring that they, in turn, produce. The natality rate is determined by an organism's genetic program: Each plant produces a characteristic number of seeds, each female insect or bird lays a certain number of eggs, and so on. These are not necessarily the maximum numbers that each organism is able to produce. Each population allocates some of its limited energy to reproduction. It survives by producing just enough offspring to replace itself in the next generation, but its best strategy for survival isn't necessarily to produce enormous numbers.

The ornithologist David Lack noted that the females of each species of bird lay only a limited number of eggs, even though they are physiologically able to lay many more. He reasoned that each species has evolved to lay an optimal number of eggs determined by the cost of raising its young to adulthood. Even though laying more eggs increases reproductive potential, a pair of birds can supply only so much food and protection, and with a larger brood, fewer young are likely to survive. The clutch size, he argued, is the best compromise between these forces. A number of studies have supported Lack's hypothesis, although some investigators have found that the typical clutch

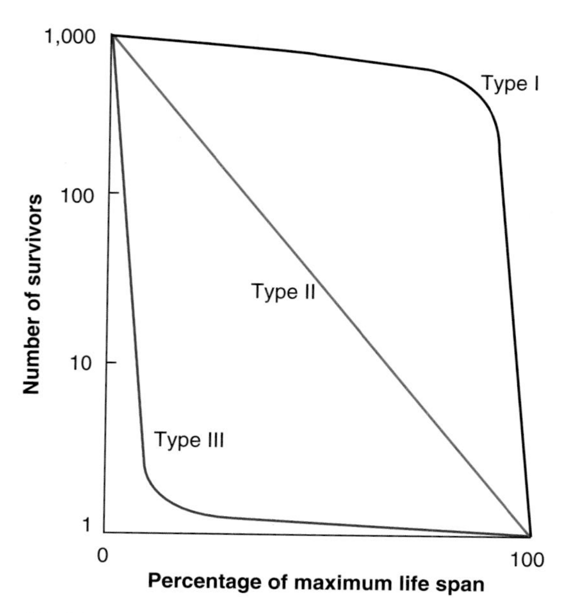

Figure 26.18

Survivorship curves show the fraction of individuals that survive to each age. Types I, II, and III are convenient ideals.

size is somewhat smaller than the maximum number of young that could be raised, suggesting that other factors must be taken into account, such as the cost of egg-laying to the female.

Each species is shaped by natural selection, generation after generation, into a successful pattern of growth and behavior. Different kinds of organisms have evolved quite different reproductive strategies, as shown by their **survivorship curves** (Figure 26.18). Suppose we mark the members of a **cohort,** all the individuals in a population born at one time, and determine how many of them survive to each age. The curves for different organisms form a spectrum. Industrialized human populations, for instance, experience a slight decrease shortly after birth due to a small infant mortality, but then the curve virtually levels off, since those who survive infancy generally survive to the age of about 60–80 years and then start to die off more rapidly. This pattern, called type I survivorship, reflects the reproductive strategy of mammals and birds: to limit the number of offspring and put a lot of energy into providing them with food and protection. Some plants show the same strategy by producing large seeds packed with food. In contrast, most invertebrates and many plants put their energy into producing large numbers of offspring that have little built-in protection and are left to fend for themselves. The majority of them die quickly (type III survivorship). Most seeds don't germinate, and most invertebrate eggs become food for other animals, but such species are successful because a few of the offspring survive, grow to maturity, and reproduce the next generation. Between these two extremes are many possible variations, representing species with quite steady mortality rates (type II survivorship). These distinct strategies obviously all work, but for different reasons and in different ecological situations.

26.12 Different kinds of selection produce two extremes within a spectrum of lifestyles.

The extremes of reproductive mode are correlated with two general modes of life, called opportunistic and equilibrium, which are at opposite ends of a spectrum. One example of an **opportunistic species** would be an insect species that inhabits relatively short-lived grassy areas that tend to be replaced by areas of broad-leaved plants. When the insects find a suitable grassy environment, they reproduce rapidly and expand throughout the area, exploiting its resources. They may only have a few years to live there. Then they disperse to find other areas, where they repeat the process. On the other hand, a typical **equilibrium species** might be a small mammal that lives in the much more stable environment of an ancient forest. This population has lived there for many generations and is at an equilibrium size, which only fluctuates a little with the vagaries of weather and other random forces. In contrast to the insects, the mammals are in sharp competition with other animals for limited resources.

These two situations are described by reference to the logistic growth equation (see Section 26.5); the insect has been selected for mechanisms that produce a high growth rate, r, while the mammal has been selected for mechanisms that maximize its steady-state population, K. Robert H. MacArthur and Edward O. Wilson therefore designated these two types of selection ***r*-selection** and ***K*-selection.** The concept—like virtually any scientific idea—has been criticized by other theorists, but it is useful as long as we recognize that the behavior and reproduction of a species are never shaped entirely by one force or the other.

Table 26.1 summarizes some features that are generally correlated with the two regimes of selection. Opportunistic species that do well in rapidly changing climatic and ecological conditions are the result of r selection. They are short-lived organisms that reproduce rapidly and may reproduce only once in a lifetime, perhaps in outbreaks like the insects mentioned previously. K-selection tends to produce equilibrium species that occupy more stable habitats and do well in relatively constant, predictable conditions. They are longer-lived organisms that reproduce several times, have low net reproductive rates, and invest large portions of their reproductive energies in each of a few offspring. Notice also alongside "Mortality" in the table that density-dependent and density-independent factors tend to have quite different effects on opportunistic and equilibrium species.

26.13 The human population is growing far beyond its ecological limits.

If success can be measured by the growth of a population, the human species takes the prize. Figure 26.19 shows how human numbers have grown over time. During most of history, the high birth rate was offset by almost equal infant mortality. The

Table 26.1 Factors Correlated with Different Modes of Life

	Opportunistic Species	Equilibrium Species
Selection	*r*-selection	*K*-selection
Climate	Variable or unpredictable	Fairly constant and predictable
Mortality	Often catastrophic; density independent	More directed; density dependent
Survivorship	Often type III	Usually types I and II
Population	Variable size, nonequilibrium, usually below carrying capacity of environment	Fairly constant size, at or near carrying capacity of environment
Length of Life	Short, usually less than 1 year	Longer, usually more than 1 year
Selection Favors	(1) Rapid development (2) High maximal rate of increase (3) Early reproduction (4) Small body size (5) Single reproduction	(1) Slower development (2) Greater competitive ability (3) Delayed reproduction (4) Larger body size (5) Repeated reproduction

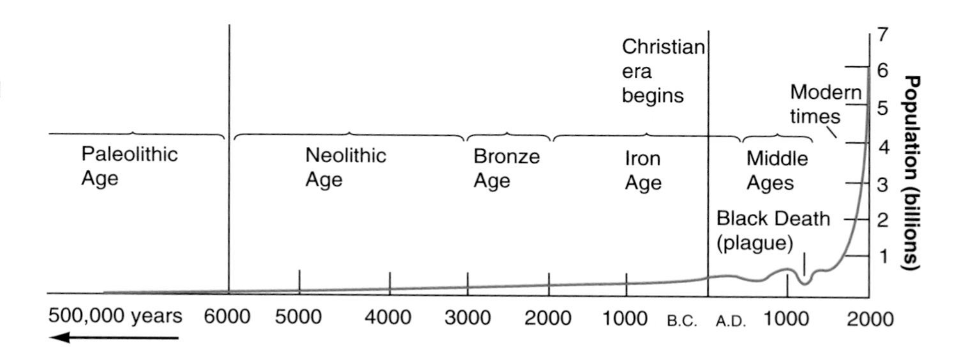

Figure 26.19
The human population grew very slowly until a few centuries ago, when it began to increase rapidly and exponentially. Early population sizes are based on estimates, more recent sizes on actual data.

population was further held in check by natural forces such as disease and famines. Indeed, the plagues of medieval and Renaissance times in Europe were such disasters that slight depressions in the growth curve show their effect on the whole human population. Most of the growth in the human population followed the industrial revolution, fueled by the fruits of civilization. Modern medicine and agriculture have steadily reduced death rates with little compensating change in the birth rate in many parts of the world. This rapid growth is aptly called a **population explosion.** Other populations have undergone similar explosions from time to time, temporarily exceeding the carrying capacity of their environments and then crashing, perhaps because of widespread disease or starvation.

The status of each population is displayed clearly by its **age structure distribution** (Figure 26.20), which shows the number of people in each age bracket. Notice that, every five years, each rectangle will replace the one just above it, so the future of the population can be predicted from its present structure, as long as the pattern of mortality holds steady. A population can be divided into prereproductive, reproductive, and postreproductive individuals, and whether the population increases, decreases, or remains the same size depends on the number of individuals in each category. We can see this on the graphs. The number of individuals who will soon be reproductive is determined by the number who are now prereproductive. The distribution of an expanding population is much broader at the bottom than at the top, indicating a large population of prereproductive individuals who will soon be reproducing (Figure 26.20*a*). In contrast, the age-structure distribution of a stable population has a uniform width from bottom to top (Figure 26.20*b*). A population is in decline if it has fewer prereproductive individuals than the number now reproducing, for there will soon be even fewer who can reproduce.

The human population explosion is not going on uniformly around the world. The relatively stable populations of some European countries have approximately the same number in each age group up to old age, while the rapidly expanding populations of many Third World countries have pyramidal age-structure distributions. These expanding populations continue to have enormous birth rates, even though their death rates have been falling. Without a compensating change in reproductive habits, they will continue to grow.

As the human population continues to grow, it puts enormous pressure on the world's ecosystems. Even the relatively slow growth in the United States means continued destruction of wild areas, the cutting of old-growth forests to provide lumber for new construction, and an industrial system that continually spews out toxic wastes capable of destroying ecosystems hundreds of miles away and poisoning our air and water. The worldwide economy encourages the destruction of vast forests, both tropical and temperate, hour after hour removing more of this essential phototrophic part of the biosphere. Such growth

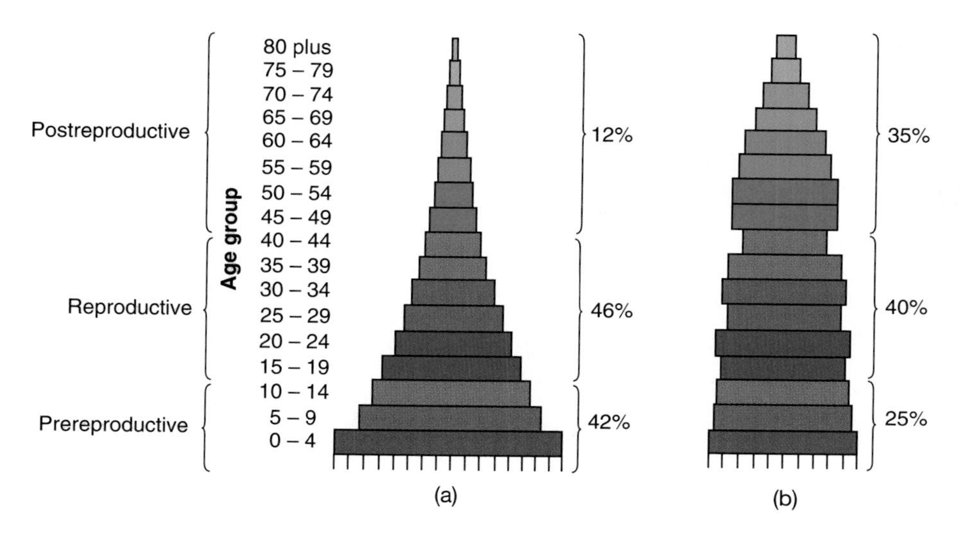

Figure 26.20
In an age-structure graph, the area of each block is proportional to the number of individuals in each age range. Individuals now in the prereproductive period will gradually move into their reproductive phase, so the graph for an expanding population, as in Central America *(a),* is quite triangular. The graph of a relatively stable population, as in a northern European country *(b),* has a more rectangular base.

and destruction cannot go on much longer if humans expect to survive. Just how long we can continue to live as we now do is a question every thinking person must wrestle with.

Coda Organisms reproduce, and the whole complex matter of fitness, natural selection, and evolution is about organisms reproducing more surely than their competitors. Populations have the potential to increase enormously, but ecological limitations keep them in check while selecting for individuals that are best able to survive and reproduce in their particular niches. If you have read any significant part of this book so far, all this is old hat. Until very recently, it would be hard to say that any species had ever become "too good" at reproducing itself. Undoubtedly there have been many episodes of a population temporarily growing beyond its carrying capacity, perhaps by devastating a forest as we have seen insects do now and then, but natural forces have always brought that population under control again, and the world has repaired itself and gone on. Now we are witnessing a new phenomenon in which a single species—humans—is growing worldwide at such an enormous pace, and with such a cruel impact on the world's resources, that it literally threatens the existence of life on Earth as we know it. If the combination of ruthless timber corporations and people with slash-and-burn economies wipes out the world's forests, what are we going to do for oxygen? What will the world be like if Siberia and Amazonia are turned into deserts? These questions become more serious every day we fail to address them.

As we reflect on population structure and growth, we are constantly brought back to the picture of John Calhoun's rats, with which we introduced this chapter. Calhoun and others commenting on his work have conjectured that there may be close parallels between his rat colonies and human populations, especially in overcrowded cities. If human and rat populations react similarly to overcrowding, we can see some obvious lessons for the structuring of human populations and human societies. Human population growth has implications not only for the biosphere as a whole but for the future of our species.

Summary

1. A population occupies a certain geographic range, but it is restricted to a specific habitat within that range. Individuals may be distributed randomly or in clusters.
2. Populations tend to grow exponentially, which means the increase in size between generations is a constant proportion of the population. A population that grows this way could become enormous in a short time, but all real populations grow only to a certain maximum size determined by environmental factors. Population growth can be measured in various ways, but it is always determined by a balance between natality (births, in the broadest sense) and mortality (deaths).
3. Each species is adapted to a certain pattern of growth, and the possible patterns fall on a spectrum shown by different survivorship curves. At one extreme, the organisms produce relatively small numbers of offspring, most of which survive to breed; at the other extreme, they produce enormous numbers of zygotes, most of which die early. Curves intermediate between these extremes are also found.
4. In spite of their potential, populations of many organisms tend to remain stable in size.
5. Population size can be limited by both density-independent factors, such as weather, that tend to reduce the population without regard to its density, and by density-dependent factors, such as intrinsic population mechanisms, that tend to decrease reproduction as the density rises.
6. Members of a population compete with one another for resources (intraspecific competition), and this is a major factor in limiting fecundity and in survivorship.
7. Because intraspecific competition limits survivorship, it is an important factor in natural selection, which maximizes the fitness of a population.
8. The two extremes of growth patterns are correlated with other features, making two modes of life. Equilibrium species have relatively long lifetimes and produce small numbers of offspring; they live in stable habitats and are not much disturbed by climatic conditions. Selection for this way of life is called *K*-selection. An opportunistic species, in contrast, is more short-lived and produces large numbers of offspring; it occupies more transient habitats and is affected by climatic conditions. This way of life is produced by *r*-selection. Between these two extremes is a whole range of intermediate modes of life.

9. Members of a population tend to conform to their ideal free distribution: They adopt the strategies of organisms that are free to maximize their use of resources. Such strategies tend to minimize competition. Territorial behavior, in which individuals defend territories and resources, is a common strategy of this type.
10. Physiological pressures that decrease fertility may be created by high density. For instance, chemical factors (pheromones) may build up in denser populations and decrease fertility. As the food supply decreases and as individuals interact with one another more strongly, stresses may be set up through a generalized stress syndrome that decreases fertility and increases mortality. For example, when rat populations are kept in confinement, they grow so dense that individuals start to act quite perversely, and one result is to lower their reproductive rate.
11. The human population is growing very rapidly and threatens to exceed the limits imposed by the earth's resources.

Key Terms

intraspecific competition 539
interspecific competition 539
geographic range 539
habitat 539
patchiness 540
graininess 540
fine-grained 540
coarse-grained 540
density 541
absolute density 541
ecological density 541
mark-recapture 541
intrinsic growth rate 542
natality 543
mortality 543
net reproduction rate 543
carrying capacity 544
logistic growth 544
density-independent factor 545
density-dependent factor 545
pheromone 547
stress syndrome 547
Principle of Allocation 548
ideal free distribution 548
territoriality 549
survivorship curve 551
cohort 551
opportunistic species 551
equilibrium species 551
r-selection 551
K-selection 551
population explosion 552
age-structure distribution 552

Multiple-Choice Questions

1. A population contains or includes
   a. several species within a limited area.
   b. several species plus abiotic factors within a limited area.
   c. all interacting members of related species.
   d. members of a single species that occupy a limited area.
   e. all members of a single species worldwide.
2. Which is a primary cause of a clustered distribution within a habitat?
   a. highly territorial population
   b. clustered resources
   c. release of toxins that affect members of the same species
   d. more-or-less random distribution
   e. sexual reproduction
3. Which statement is true about exponential growth of a population from generation to generation?
   a. Size increases by the same proportionate amount each generation.
   b. Size increases by the same number of individuals each generation.
   c. Size increases by a decreasing proportionate amount each generation.
   d. Rate of increase doubles with each generation.
   e. Rate of increase must be exactly whole integers—i.e., double, triple, quadruple.
4. Which is correct about the net reproduction rate of a particular population?
   a. It is the same as the intrinsic growth rate for the same time period.
   b. If the value is >1, growth is exponential.
   c. If the value is 0, growth is negative.
   d. *a* and *b*, but not *c*.
   e. *a*, *b*, and *c*.
5. If a population grows according to the logistic growth equation, the intrinsic growth rate
   a. is always exponential.
   b. is never exponential.
   c. increases over time.
   d. decreases over time.
   e. initially decreases, but then is constant.
6. Real populations are limited in size by all except
   a. predators.
   b. limited nutrient supply.
   c. group selection for optimum size.
   d. territoriality.
   e. competition.
7. Which of the following is least likely to be a density-dependent limiting factor?
   a. weather
   b. nutrients
   c. concentration of metabolic wastes
   d. stress
   e. infectious disease
8. Compared to animals with type I survivorship curves, those with type III survivorship patterns generally
   a. expend more energy per zygote but produce fewer of them.
   b. expend more energy to raise a zygote to reproductive age.
   c. have longer life spans if they survive to adulthood.
   d. have complex and specialized reproductive organs.
   e. are statistically more likely not to survive to adulthood.
9. Opportunistic species generally have all the following characteristics except
   a. large body size.
   b. *r*-selection.
   c. high mortality, especially after one breeding season.
   d. population size often regulated by density-independent factors.
   e. widely fluctuating population size.
10. Equilibrium species generally have all the following characteristics except
    a. they occupy a stable habitat.
    b. population size tends to be regulated by catastrophic events.
    c. *K*-selection.
    d. their life span includes more than one breeding season.
    e. type I or II survivorship patterns.

True-False Questions

Mark each statement true or false, and if false, restate it to make it true.

1. The terms *geographic range* and *habitat* can be used interchangeably because their underlying meaning is the same.
2. Uniform distribution of a population within a habitat generally results from a random dispersal pattern.
3. In mark-recapture experiments, the most critical value is the proportion of individuals that are initially caught and marked compared with those initially caught and not marked.

4. As a general rule, the smaller the organism, the faster the intrinsic growth rate.
5. If population size remains constant from generation to generation, the intrinsic growth rate is 0, and the net reproduction rate is 1.00.
6. Intraspecific competition can be either exploitative or random.
7. Population size of decomposers and carnivores is generally limited by the available food supply and not by predation.
8. The logistic curve flattens out when a population approaches the point of exponential growth.
9. The most successful populations are those that maximize the number of offspring produced each breeding season.
10. High natality is characteristic of opportunistic species.

Concept Questions

1. Is competition inevitable?
2. Explain the fallacy in the hypothesis that optimum population size is regulated by group selection.
3. Contrast the concept of an ideal free distribution with that of territoriality. Would you expect both strategies to be used by the same organism at the same or different times?
4. Explain why population growth will still continue even if an exploding population adopts serious and effective programs to limit the number of births.
5. Are human societies more like opportunistic species or equilibrium species? Does their behavior resemble that of r-selected organisms or K-selected ones?

Additional Reading

Beddington, John R., and Robert M. May. "The Harvesting of Interacting Species in a Natural Ecosystem." *Scientific American,* November 1982, p. 62. The problem of utilizing a biological resource without extinguishing the species.

Begon, Michael, John L. Harper, and Colin R. Townsend. *Ecology: Individuals, Populations and Communities,* 2d ed. Blackwell Scientific Publications, Boston, 1990.

Bongaarts, John. "Can the Growing Human Population Feed Itself? *Scientific American,* March 1994, p. 18. Some economic and ecological arguments about population growth.

Clutton-Brock, T. H. "Reproductive Success in Red Deer." *Scientific American,* February 1985, p. 86. A 12-year study of more than 1,000 of these gregarious animals in a nature reserve of the U.K. has revealed the determinants of the lifetime breeding success among red deer stags and hinds.

Dasgupta, Partha. "Population, Poverty and the Local Environment." *Scientific American,* February 1995, p. 26. Economic aspects of human population growth and ecology.

Diamond, Jared. "Easter's End." *Discover,* August 1995, p. 62. Easter Island turned from a tropical paradise with a civilization capable of making and transporting giant stone statues into a barren wasteland with a small, isolated population. Ancient bones and pollen grains give clues to its desecration and the natives' descent into cannibalism and starvation.

Homer-Dixon, Thomas F., Jeffrey H. Boutwell, and George W. Rathjens. "Environmental Change and Violent Conflict." *Scientific American,* February 1993, p. 38. Civil and international strife is growing out of human population growth and competition for resources.

Ryker, Lee C. "Acoustic and Chemical Signals in the Life Cycle of a Beetle." *Scientific American,* June 1984, p. 112. How the Douglas-fir beetle attracts a mate, repels intruders, and regulates the density of its population.

Wilson, Edward O., and William H. Brossert. *A Primer of Population Biology.* Sinauer Associates, Inc., Stamford (CT), 1971.

Zimmer, Carl. "How to Make a Desert." *Discover,* February 1995, p. 50. A study of the Jordana Basin in New Mexico and the general subject of desertification.

Internet Resource

To further explore the content of this chapter, log on to the web site at:

http://www.mhhe.com/biosci/genbio/guttman/

27 The Structure of Biological Communities

A moose feeds in its typical habitat in a quiet northern lake.

Key Concepts

A. General Features of Communities

27.1 A community is not a superorganism.

27.2 Species abundance varies within a community and from the tropics to the poles.

27.3 Community structure may be dominated by keystone species.

B. Niches and Relationships in the Community

27.4 Relationships within a community are complicated but may be analyzed into many binary relationships.

27.5 The concept of an ecological niche is problematic.

27.6 Two species cannot occupy the same niche—or can they?

27.7 Niche differentiation is commonly determined by subtle chemical factors.

C. Chemical Interactions in the Community

27.8 The members of a community are in perpetual "arms races" with one another.

27.9 Many organisms use allomones in chemical warfare against other species.

27.10 Some organisms create intolerable conditions for others.

27.11 Many species react to kairomones produced by other species.

D. Predation, Symbiosis, and Camouflage

27.12 Predation is an essential activity in every community.

27.13 Predator and prey populations may change in cycles.

27.14 Many organisms engage in symbiotic relationships.

27.15 Many animals are camouflaged by their forms and colors.

27.16 Some animals are protected by warning coloration.

27.17 Mimics may survive by imitating warning coloration.

Early in this century, moose colonized Isle Royale, a 210-square-mile island in Lake Superior near the Canadian shore, probably by walking across the frozen ice from Canada one winter. Without any predators to constrain them, the moose multiplied rapidly. By the mid-1930s, the herd was estimated at between 1,000 and 3,000 individuals, far exceeding the carrying capacity of the island. Then, after the moose had consumed all the low vegetation on the island, the population crashed, due to starvation, to a level below the carrying capacity. Within a few years, the vegetation grew back, and the herd expanded again, repeating the scenario with another population crash in the late 1940s. In 1949, timber wolves crossed the ice to the island.

Instead of adding to the problems of the moose population, the wolves had a beneficial effect on the whole ecosystem. Increasing to about 20–25 individuals, they have kept the moose population between 600 and 1,000, somewhat below the carrying capacity. The island again has abundant vegetation, enough to nourish the moose as well as a community of other animals.

The Isle Royale story is a classic example of community structure. Every population has such an enormous potential for reproduction that a single population of animals with a limited amount of food is just not a stable situation, leading instead to an endless cycle of population explosions and crashes. We have emphasized that the evolution of each species is shaped not only by interactions with abiotic factors but also by interactions with all the other organisms that constitute the biological part of an ecosystem, a community. A real community is made of several species at different trophic levels that eventually achieve a certain balance. This is the theatre in which the evolutionary play goes on. In this chapter, we will discuss some factors that structure communities and strongly influence the course of evolution, particularly stressing the concept of an ecological niche and the role of competition in the evolution of species. ■

A. General features of communities

27.1 A community is not a superorganism.

Because communities appear to stay organized—over short times, at least—it is tempting to view a community as a kind of huge "superorganism" with self-correcting internal control mechanisms, a viewpoint proposed by Frederick C. Clements, a pioneer in modern ecology. Just as an individual organism has parts that are well-designed by evolution for their specific tasks, Clements reasoned, the members of a community might have evolved to perform specific roles in an integrated complex. Just as an organism maintains a relatively constant internal condition of homeostasis in the face of destructive stresses, a community might be able to maintain constant conditions within itself. And certainly a community often looks like such a gigantic organism with all its species living together in a kind of harmony. We might even think that this apparent harmony is the result of a "community evolution," because all members of a community would certainly benefit if the whole system were evolving in the direction of stability and cooperation.

These ideas, however, contradict a basic concept of biology: Organisms are opportunists who will take advantage of any evolutionary opportunity that becomes available, and they will evolve in any way that perpetuates them, at the expense of any other organisms. Furthermore, the idea of community evolution appears to be unnecessary. We will show that relationships among members of a community can be understood in terms of mechanisms that have selective value to individual species, so there is no need to postulate complex mechanisms that favor the community as a whole. A general scientific principle (the principle of parsimony) advocates not invoking complicated explanations if simpler ones will do, and that principle certainly applies to this issue.

If communities were always tightly integrated associations, we would expect to find specific combinations of species that fit together well. And indeed, biologists have identified a number of distinct **plant associations,** certain groups of plants that live together, as well as characteristic groups of associated animals. The temperate deciduous forest biome has much the same characteristics everywhere, but in eastern North America, for instance, local conditions and the history of each region produce patches of communities such as jack pine-paper birch forest, burr-oak forest, aspen-birch-pine-cherry forest, and so on. Ecologists once believed that relatively few types of these associations exist and that each was an integrated community invariably found under certain physical conditions, but that view has turned out to be too simple.

Studies indicate that each plant species is distributed independently of others, in accordance with the gradual changes in environmental factors across a large area. Robert H. Whittaker, studying the distribution of plants along the slopes of the Great Smoky Mountains, found each species to be distributed in its own way along the gradients of altitude, temperature, humidity, and other variables, quite independently of other species (Figure 27.1). In some places a particular group of plants predominate together, giving the impression of a specific association, but in other areas, only one of these plants predominates. As the environment gradually changes, so does community structure, and frequently, no sharp discontinuities separate one plant association from another. The named plant associations and communities have merely been identified by casual observation.

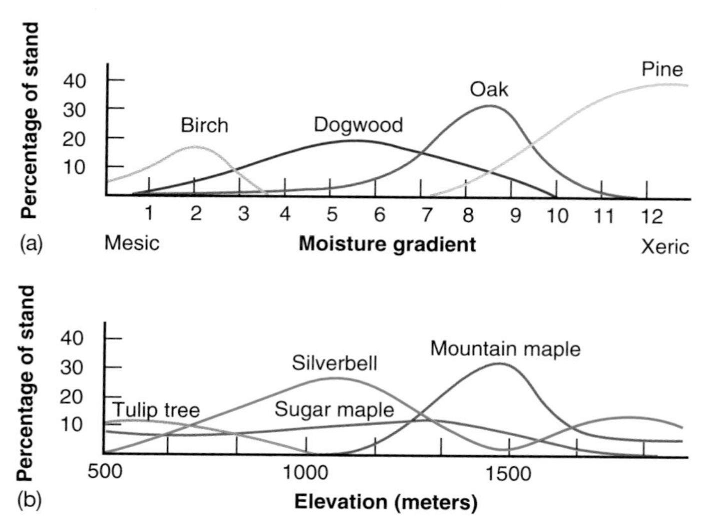

Figure 27.1

Two graphs show how trees in the Great Smoky Mountains of Tennessee are distributed along physical gradients: *(a)* along a gradient from moderately wet (mesic) to dry (xeric) conditions and *(b)* in relation to elevation.

Exercise 27.1 What kind of distribution profile would Whittaker have seen if plants did, in fact, occur only in distinct associations instead of as independent distributions of each species?

27.2 Species abundance varies within a community and from the tropics to the poles.

One of the most striking features of an ecological community is its diversity. Anyone who steps into a woodlot, a field, or a pond with open eyes and perhaps some good identification guides will notice many kinds of organisms in even a small area. And if you take a few samples into a laboratory equipped with good microscopes, the diversity becomes astonishing. It is easy to get lost for hours studying the jungle of algae, protozoans, and minute animals in a drop of pondwater or a spadeful of rich topsoil. Suppose, though, instead of just looking at all these organisms, we count the number of each species in a particular community. The resulting profile will look something like Figure 27.2*a*, containing very few individuals of many rare species and large numbers of a few common, dominant species. Such a distribution is generated by the so-called broken-stick model (Figure 27.2*b*). This is the distribution that would be generated if the resources of a habitat, represented by a stick, were divided sequentially, first by two species randomly "breaking the stick in two," then by others breaking the shorter sticks again and again as they take their share of the resources. This pattern may reflect the sequence in which species come to occupy a habitat, or their ability to competitively exclude other species, or both.

Superimposed on this pattern of diversity is another that the British ecologist Wallace Arthur has called nature's most impressive pattern: Going from the polar regions toward the tropics, species richness increases regularly. This gradient in diversity is very obvious to anyone who watches birds (Figure 27.3), but it has also been demonstrated in many other organisms, including trees, mammals, North American lizards, and small crustaceans called calanoid copepods in the Pacific and Arctic Oceans. Many observers have commented on this pattern. What are its causes? Ecologists have proposed several hypotheses, and while they've arrived at no fully satisfactory explanation, two main factors seem to be involved:

1. *History: opportunities for speciation.* Speciation occurs primarily when one portion of a species becomes geographically isolated in a refuge where it acquires reproductive isolating mechanisms. It has been argued that, because of the relatively harsh, uncertain climatic conditions at high latitudes compared with more favorable and constant conditions in tropical regions, isolated populations are more likely to survive in the tropics. Thus there will be a gradient

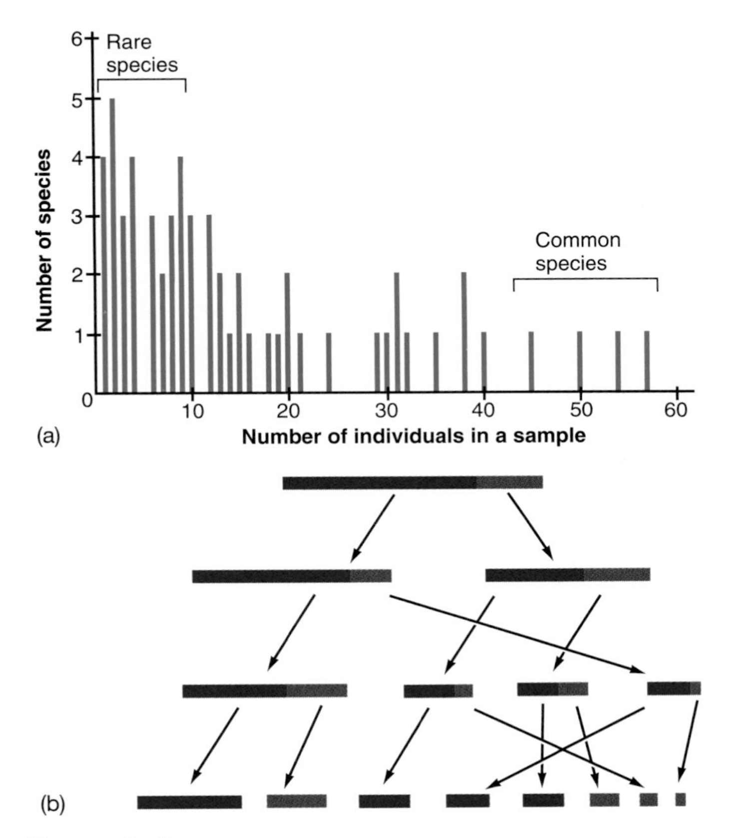

Figure 27.2

All members of a community are not equally abundant. *(a)* A sample taken by any standard random method (such as digging up a plot of soil and sifting out all the organisms) yields an abundance distribution with many rare species (only one or a few individuals in this sample) and a few common species (45 or more individuals in this sample). *(b)* This kind of distribution could result from "breaking a stick" repeatedly, where the stick represents resources.

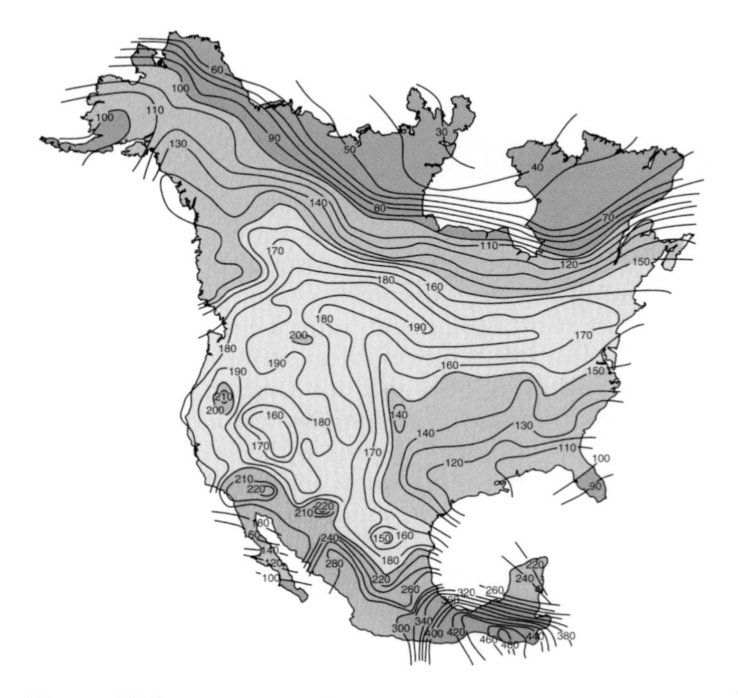

Figure 27.3

The abundance of bird species (number of species breeding in each area) increases steadily from the polar to the tropical regions.

in the number of species potentially available to occupy the available habitats.

2. *Climate: stability and energy availability.* The more favorable tropical climate and the availability of more energy there not only support greater speciation but also support more species once they have been formed. Species richness of both trees and vertebrates increases regularly as a function of the rate of *evapotranspiration* in a community, which in turn is a measure of solar radiation and temperature; this rate increases toward the tropics. The relatively stable tropical climate allows species to specialize to narrower niches and yet survive, whereas extreme specialists in more temperate or polar regions are likely to be wiped out by climatic fluctuations. Interestingly, greater richness of plant species *not* correlated with high *productivity* of a habitat—that is, the efficiency with which the plants turn light energy into their biomass.

27.3 Community structure may be dominated by keystone species.

All of the species in a community are probably never equally important in maintaining its structure. Robert Paine has suggested that many communities contain a **keystone species** whose role is so central that removing it would drastically upset the community. Paine studied littoral communities on the rocky shore of the Pacific coast, communities dominated by a single predator, the sea star *Pisaster ochraceous* (Figure 27.4). To test a hypothesis about the role of predation in community diversity, Paine removed *Pisaster* from experimental areas and found that a formerly diverse community of 15 species shifted toward one of only 8 species, dominated by the blue mussel *Mytilus californianus.* The conclusion from this experiment is that *Pisaster* keeps the populations of all its prey species so low that they are not in competition for space, and as a result, the diversity of the ecosystem is quite high.

Other communities may have comparable keystone species. Kangaroo rats in the Chihuahuan Desert largely determine the structure of the plant community by their preference for large seeds. When these rats were excluded from experimental plots, grasses with large seeds became so much more abundant that they interfered with the feeding of some birds that eat seeds on the ground, and the area became transformed into a grassland. Similarly, elephants have been labeled a keystone species in Africa because they regularly destroy shrubs and small trees, thus maintaining the habitat as a grassland instead of a forest.

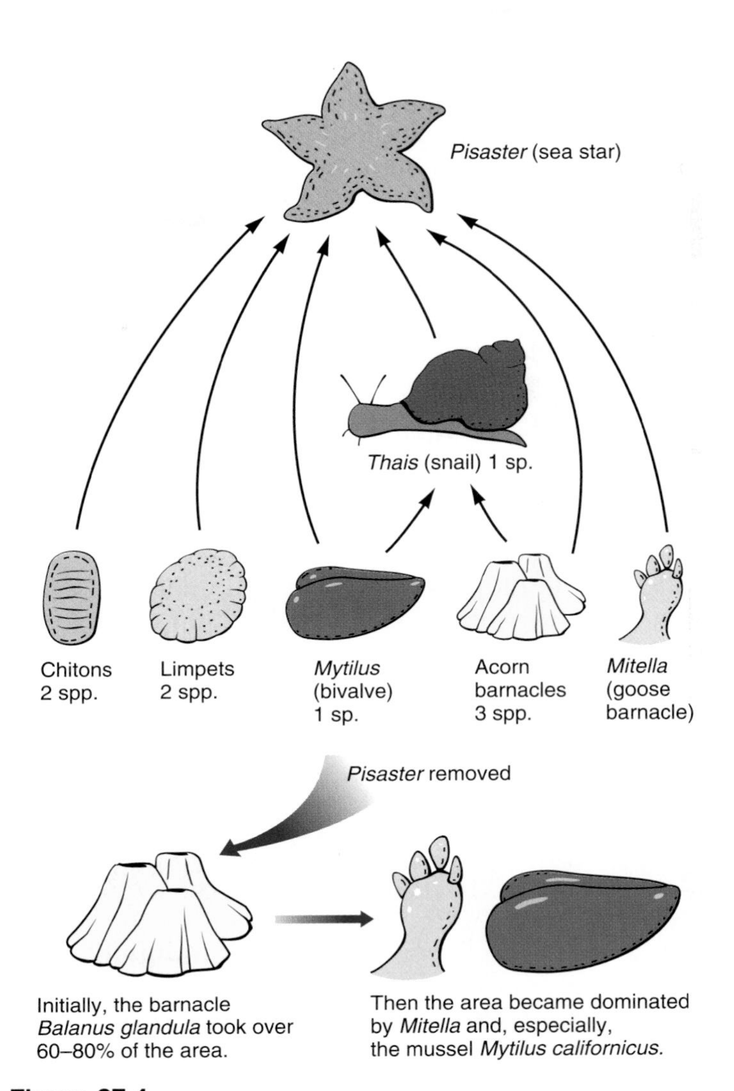

Figure 27.4

A rocky shore community is dominated by the top carnivore, *Pisaster.* When this species was removed, the community changed drastically to one of only 8 species dominated by *Mytilus* mussels.

B. Niches and relationships in the community

27.4 Relationships within a community are complicated but may be analyzed into many binary relationships.

We have been emphasizing that a community is structured fundamentally by its trophic relationships, which involve all the community members in a food web. Even the relationships among a few closely interacting species in a community can be complicated, and the relationships among thousands of species may defy description. For the sake of analysis, these interactions can be reduced to binary relationships: interactions between two species. Of course, ecologists know full well that focusing on binary relationships is too simple, and they have also delineated many interactions involving several species. The effect of species A on species B may be positive, neutral, or negative, and all the factors combine to create the six types of relationships listed in Table 27.1. Neutralism means the lack of any direct interaction between two species, so there is no effect on the community's structure. Amensalism, a relationship that is detrimental to one species and neutral to the other, may not even exist in the long run. Any situation of this kind would probably end soon because the adversely affected species would either evolve a way to resist the detrimental effects or die out.

The four remaining relationships that underlie community structure are predation/parasitism, commensalism, mutualism, and competition. Species A may live on or around species B, so they are primarily competing for space and resources; or species A may eat species B, putting selective pressure on the predator to become better at obtaining food and on the prey to better escape its fate. Commensalism and mutualism are part of a spectrum of intimate relationships entailing various degrees of harm or benefit to one species or the other. In this chapter we will discuss these relationships, emphasizing the chemical factors that structure certain species interactions.

27.5 The concept of an ecological niche is problematic.

Chapter 26 addressed *intra*specific competition, among members of the same species. *Inter*specific competition, on the other hand, stems from two species vying for the same resources, and to discuss it we must deal with the concept of an *ecological niche.* Although we have been using the word "niche" rather informally in earlier chapters, it is rather hard to define. Derived from the Latin *nidus,* meaning "nest," the word implies the space an organism occupies—but a habitat is also the physical space a species occupies. To distinguish these two terms, biologists tend to follow Eugene Odum's statement that a niche is a species's *occupation,* a description of how it makes a living and obtains its resources, while its habitat is its *address.* Ecologists have offered various other definitions of a niche, and it is instructive to compare them (Concepts 27.1). Some authors clearly restrict themselves to the space an organism occupies, in the broadest sense of "space," while others add a component of the organism's behavior and its role in the ecosystem. The ecologist Charles Elton expressed that viewpoint when he wrote, "When an ecologist says, 'There goes a badger,' he should include in his thoughts some definite idea of the animal's place in the community to which it belongs, just as if he had said, 'There goes the vicar.'"

The conception of niche most commonly used today comes from G. Evelyn Hutchinson, who pictured a niche as an *n*-dimensional space where each dimension is one variable affecting the life of the species. Consider just three factors that could limit a population—for a plant, say the temperature range it tolerates, the soil phosphate concentration in which it can grow, and its requirements for water (humidity). These factors interact to some extent, and in combination they define a growth range, which can be represented by a volume in three dimensions (Figure 27.5). This imaginary volume is the species's niche. A drawing like Figure 27.5 can only show three variables at a time, but in reality there must be many: the limits of its geographical range in latitude and longitude; the height above ground where an animal feeds; the range of sodium, zinc, phosphate, and other materials it requires or tolerates; and so on. The niche is an irregular volume in the space defined by all relevant factors. Notice, though, that this definition of niche does not easily accommodate a description of a species's behavior and way of life.

Each species's niche depends on its interactions with other species in its community. The volume a species could occupy in the absence of any competitors is its **fundamental niche,** but in reality it may be forced by competition to occupy a more limited volume, its **realized niche.** A. G. Tansley's classic 1927 experiment illustrates this difference. Two closely related British

Table 27.1 Binary Relationships in a Community

Effect of A on B	Effect of B on A	Relationship
+	−	Predation, parasitism
+	0	Commensalism
+	+	Mutualism
0	−	Amensalism
0	0	Neutralism
−	−	Competition

Concepts 27.1 Some Views of Niche

Joseph Grinnell (1917, 1928) said that niche is the functional role and position of an organism in its community—essentially a behavioral role.

Charles Elton (1927) stated that an animal's niche is "its place in the biotic environment, its relations to food and enemies, and the status of an organism in its community."

Lee R. Dice (1952) defined "niche" as the ecologic position that a species occupies in a particular ecosystem. The term therefore includes "a consideration of the habitat that the species concerned occupies for shelter, for breeding sites, and for other activities, the food that it eats, and all the other features of the ecosystem that it utilizes. The term does not include, except indirectly, any consideration of the functions that the species serves in the community."

G. L. Clarke (1954) distinguished "functional niche" from "place niche." He noted that different species of plants and animals fulfill different functions in the ecological complex and that the same functional niche may be filled by quite different species in different geographical regions.

Eugene Odum (1959) distinguished niche from habitat. For him, niche is an organism's "profession," whereas habitat is its "address." Niche is the "the position or status of an organism within the community and ecosystem resulting from the organism's structural adaptations, physiological responses, and specific behavior."

G. Evelyn Hutchinson (1957) said niche is defined by an *n*-dimensional region that specifies the total range of conditions in which the organism is able to live and reproduce—such factors as temperature, humidity, salinity, and so on. This does not include behavioral factors.

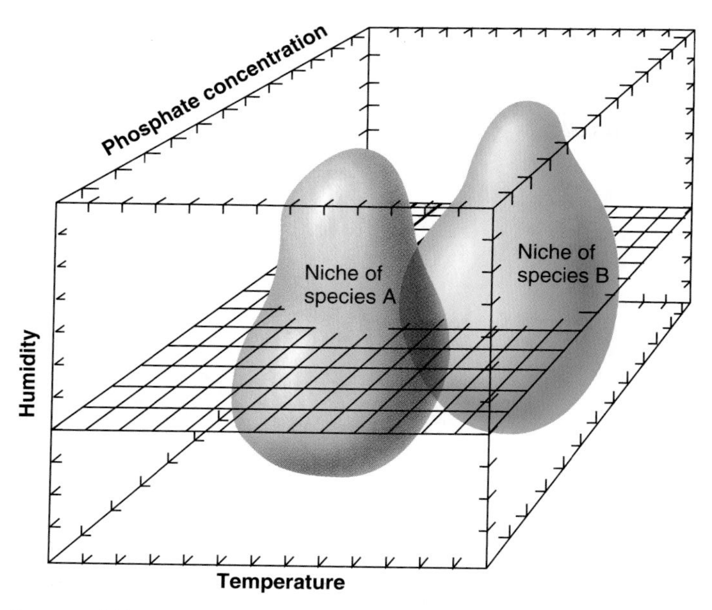

Figure 27.5

A niche may be understood as a space defined by *n* variables (dimensions) where a species lives. The niches shown here are only partially defined, by means of three dimensions, because no more than three can be pictured at one time. Two niches may partially overlap when the species have some of the same requirements and share some resources.

plants called bedstraws grow best in soils with different pH ranges: *Galium hercynium* in acidic soils and *G. pumilum* in basic soils. Tansley found that each species by itself would grow on both kinds of soils, but when planted together, *G. hercynium* invariably outcompeted *pumilum* in acidic soil, while the reverse was true in basic soil. This experiment shows that the fundamental niches of both species are quite broad in the pH dimension, but when grown together, each shows its superior adaptation to a specific pH range, and they restrict each other to a more limited niche, the realized niche.

27.6 Two species cannot occupy the same niche—or can they?

Given a conception of an ecological niche, let's return to the question of interspecific competition and ask to what extent species are actually in competition in nature. It has long been common wisdom in biology that interspecific competition must be short-lived and limited because if two species had identical or strongly overlapping niches so that they were competing for the same resources in the same habitat, either one species would outcompete the other or the species would evolve so as to have separate niches and no longer be in competition. Strong evidence for this concept came from experiments in the 1930s by the Russian ecologist G. F. Gause. He set up cultures in which protozoans fed on bacteria and yeast, which in turn fed on oatmeal. When raised by themselves, the larger, slower-growing *Paramecium caudatum* and the smaller, faster-growing *Paramecium aurelia* each showed classical growth curves. In mixed culture, however, *P. aurelia* always won out while the *P. caudatum* population diminished to almost nothing. In contrast, when Gause grew *P. aurelia* with *P. bursaria,* the populations of both species reached about half the levels they would have achieved in isolation, with *bursaria* living on bacteria suspended in the top half of the culture tube and *aurelia* living on yeast in the bottom half. Thus both survived by finding separate niches, even in this simple situation.

Based on this work, Gause stated the general principle that two or more species cannot continue to occupy the same niche indefinitely. The principle has been expressed before by other biologists, and Garrett Hardin finally formulated this concept as the **competitive exclusion principle:** *Completely competing species cannot coexist.* In other words, potentially competing species can only coexist through **niche differentiation**—evolution of one or more species so their realized niches don't overlap strongly and they are using different resources. Potential competitors can differentiate their niches by *partitioning* resources, either using different resources in the same space or dividing the space. This happened in the mixed culture of *P. aurelia* with *P. bursaria,* and it happens in natural situations. For example, Robert H. MacArthur reported in 1958 on the feeding patterns of five species of warblers living in the same area in New England (Figure 27.6). Although the species are very similar and apparently eat the same food, they divide the space by hunting in different parts of the tree canopy. On a much smaller scale, potentially competitive carnivorous beetles divide the space on leaves, where they feed on herbivorous insects (Figure 27.7). By evolving different behavior patterns, they divide a limited resource so they all get enough to maintain themselves. Two species can also share resources through temporal division; flying insects are hunted by swifts and swallows during the day, by nighthawks and their relatives around dusk, and by bats at night.

The competitive exclusion principle has been taken as a cornerstone of ecology, largely because it makes such good sense and is supported by experimental evidence, but the principle is actually more problematic than it might seem. It is virtually certain that two species cannot coexist if they are in total competition, but ecologists have tried to find out how much interspecific competition actually exists. In clarifying the question, competition has been said to take one of two forms: *Resource,* or *scramble,* competition occurs when organisms use common resources that are limited; *interference,* or *contest,* competition occurs when organisms seeking a resource harm one another in the process. Examples of resource competition abound. J. H. Connell's studies of two species of barnacles in Scotland revealed a classic example of interference competition. *Chthamalus stellatus* and *Balanus balanoides* live together on the same rocky Atlantic shores, but adult *Chthamalus* generally live higher in the intertidal zone. Connell observed that young *Chthamalus* settle in the *Balanus* zone but are killed by smothering, crushing, or undercutting. In spite of this aggressive competition, the distribution of the two species is partly determined by physiological factors, for *Chthamalus* is more resistant to desiccation and is therefore able to live in the more exposed upper zone.

Figure 27.6

Five species of warblers (genus *Dendroica*) that feed on bud worms in spruce forests divide the resources by foraging in different areas. MacArthur divided the tree space into 15 regions; the shaded areas show where he observed each species feeding at least half the time.

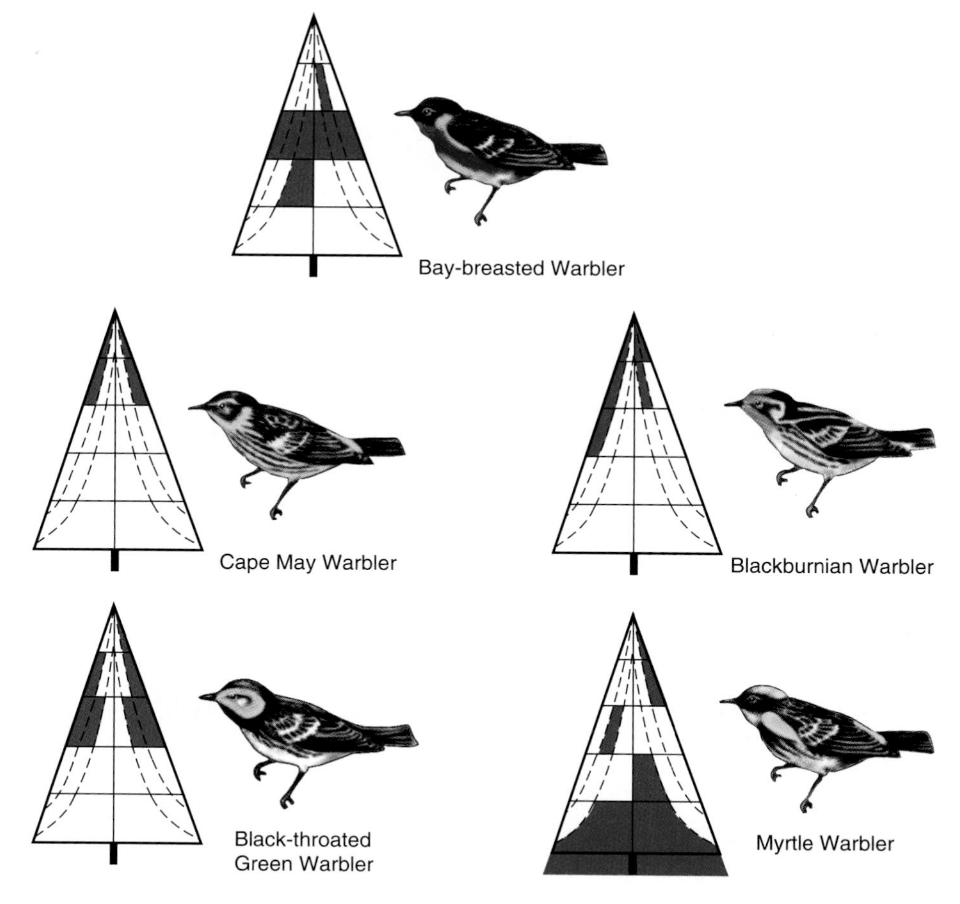

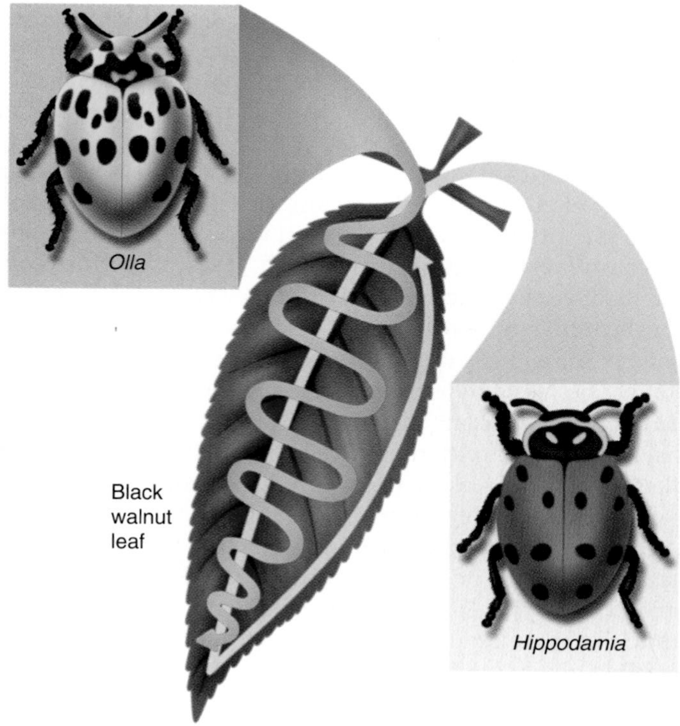

Figure 27.7

Carnivorous beetles divide the resources on a leaf (herbivorous insects) by using different foraging patterns.

There is no simple answer to the question of the amount of competition in natural ecosystems. Many species may have overlapping fundamental niches but restrict one another to limited, realized niches through competition, like the bedstraws Tansley studied. To assess the extent of such competition, many experimental studies have been performed in the field by removing one species from a community to see whether the niches of other species expand. These experiments show considerable differences in the amount of interspecific competition between plants, invertebrates, and vertebrates in terrestrial, marine, and freshwater habitats. The greatest amount has been found in marine habitats, where 90 percent of the vertebrates and over two-thirds of the plants showed evidence of competition. On the whole, however, evidence of competition was found in only about one-quarter to one-third of the situations examined.

The degree of competition also varies with trophic levels. Remember the argument of Hairston and his colleagues (see Section 26.6) that herbivores are limited by predation, not by the lack of resources; that would mean herbivores are often not in competition with one another, and some observations indicate this is true, particularly among herbivorous insects, which constitute one-quarter of all animal species. For instance, widespread plants sometimes have insects feeding on them in different ways in different areas, so in each area some modes are not used; the availability of unoccupied niches for herbivorous insects indicates a lack of competition that might have forced some species into these niches. D. R. Strong, Jr., studied 13 species of tropical leaf-mining beetles that use the same food and live in rolled-up leaves of *Heliconia* plants, but he could only find evidence that the niche of one species was weakly segregated from the niches of the others. These beetles apparently require exactly the same resources, but they live together without any aggression, either within or between species. These herbivorous insects, living in a rich tropical forest, may never reach large enough populations to actually be in competition, because they are exploiting such a large resource that their food supply isn't limited and predation keeps their numbers in check. But studies of other herbivores often show strong niche differentiation; for example, L. H. Emmons showed that nine species of African squirrels in an evergreen rain forest had clearly separate niches as determined by habitat type, the height of the vegetation, preferred food types, and the size of their food.

The competitive exclusion principle is in danger of being held on faith merely because it makes such good sense. It is the kind of scientific law that can't be proven true because there is no way to examine every situation where it might apply. On the other hand, it is almost impossible to prove the principle false because there are so many variables that could define a niche; a report of competition without niche differentiation might be explained away by arguing that *some* difference existed but simply was not found.

It should be clear that moving away from competition is an important factor in microevolution and speciation. Niche differentiation is one expected outcome of any actual or potential competition, and it must always occur in speciation. In geographic (allopatric) speciation, two populations that were once part of a single species acquire reproductive isolating mechanisms and are able to remain separate when they again become sympatric. However, the newly sympatric populations cannot remain in competition for precisely the same niche if both are to survive. So, along with reproductive isolation, the newly separated species must acquire some niche differentiation. The *Dendroica* warblers, for example, belong to a single clade; they must have separated through geographic speciation, but they could only become sympatric by occupying different niches. In their sharing of resources we are apparently seeing what J. H. Connell has called the ghost of competition past: Species once in competition have evolved with different niches. It is also possible that the species are still in competition for the same fundamental niche and are restricting one another's realized niches, so if any one species disappears, the niches of the survivors will expand.

Allopatry, sympatry, and geographic speciation, Section 24.5.

Exercise 27.2 Three species of similar snails live in a hemlock-cedar forest. They have different overall sizes, and their mouths average 5 mm, 3 mm, and 1.5 mm in width. How do you explain this situation? What is the effect of having such different-sized mouths?

27.7 Niche differentiation is commonly determined by subtle chemical factors.

One of the most elementary observations about communities is that members of each species occur only in certain patches within an ecosystem. This is our naive expectation, but what really determines where each species will occur? Start with a simple question: Why does an animal eat only certain plants? Few animals, of course, are able to make a *choice* in the human sense; their "choice" is simply an unconscious pattern of behavior whose evolution has been shaped by many factors. The important factors are often subtle and chemical. For instance, food choice has been studied extensively in the relationships among four species of fruit flies *(Drosophila)* feeding on 15 species of yeasts that grow on five species of cactus in the Sonoran desert of the southwestern United States and northern Mexico. Where the cacti are injured, bacteria, and later yeasts, move in and create pockets of rot that attract the fruit flies; one species of fly also lives on soil that has been soaked with juice from rotting cacti. The flies are remarkably specific in their choice of cactus, as shown in Figure 27.8. Subtle, specific factors keep the niches of the four fruit flies separate from one another. The flies effectively divide the available resources, so all four species survive without competition. Notice that casual observations of the flies and cacti would never have revealed any of this; only careful chemical analysis could show what is going on here.

C. Chemical interactions in the community

27.8 The members of a community are in perpetual "arms races" with one another.

The world is constantly changing. The forces of evolution are constantly at work, and each species may be changing in small ways to adapt to changing conditions. The members of a community must continually adapt to one another, each species acquiring adaptations that allow it to gain an upper hand, or at least to maintain itself. In discussing the course of evolution, we used the metaphor of the Red Queen for a species running as fast as it can to stay in the same place. To use a modern metaphor, all species are in an arms race with one another, a race that no one ever wins for long. Much of this chapter is about the status of specific races, as examples of mechanisms that a species can acquire to better exploit a resource or to reduce its chances of being a resource for a predator.

Adaptations may be morphological, behavioral, or biochemical. Morphological features—the elements of an organism's form—are the most obvious. Behavioral adaptations include stereotyped behavior patterns and a repertoire of rapid, automatic responses to certain stimuli (introduced in Chapter 49). The ability to learn, too, is an adaptation for dealing with short-term changes. Even plants exhibit adaptive behaviors, such as by curling up their leaves in dry conditions.

Among the biochemical adaptations, Robert H. Whittaker and Paul P. Feeny described a large and ecologically important class of **allelochemic** interactions. *Allelon* means "of one

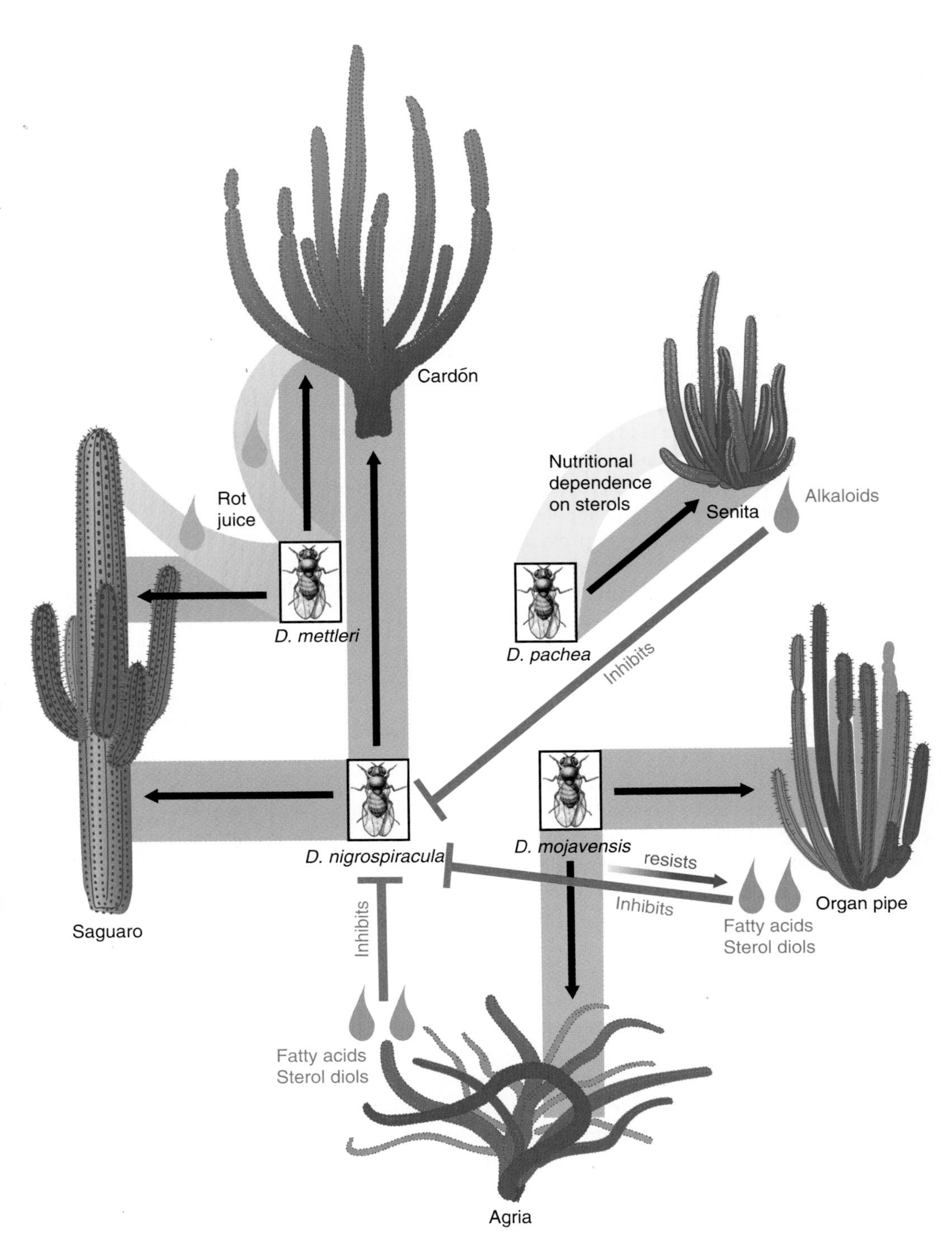

Figure 27.8
In the Sonoran desert, four species of *Drosophila* divide a cactus resource through chemical interactions. Each type of cactus produces a distinctive set of volatile compounds, mostly pungent alcohols, acetates, and acids, and these chemicals attract each type of fly to its cactus host. However, these volatile compounds, while attractive, do not determine the needs of each fly species. *D. pachea* is restricted to Senita because the other four species of cactus are nutritionally deficient: They lack sterols that the flies require but cannot make for themselves, and without these sterols females are infertile and larvae do not develop. *D. mojavensis* and *D. nigrospiracula* cannot live on Senita because they are intolerant of the alkaloids this cactus produces; *nigrospiracula* is also intolerant of fatty acids and sterol diols produced by Agria and Organ Pipe, so it is restricted to Saguaro and Cardon. Since *mojavensis* tolerates the materials produced by Agria and Organ Pipe, it lives on these two species, free from competition by *nigrospiracula.* *D. mettleri* avoids competition with the other fruit flies primarily through its behavior; although the adults live on Saguaro and Cardon along with *nigrospiracula,* the larvae escape competition with other species by living in soil soaked with Saguaro and Cardon rot juice.

another," so these are interactions in which members of a community engage in a kind of chemical warfare with one another called **allelopathy.** We discuss these interactions in the next sections.

By evolving some new structure or behavior or chemistry, a species may enhance its position in the community for a while, but eventually some other species will evolve another mechanism that improves its position, and so it goes. In the long run, the total of all the wins and losses is always zero, and no species ever quits the game a winner.

27.9 Many organisms use allomones in chemical warfare against other species.

Why are radishes so bitter, and why do onions make your eyes water? These plants didn't evolve as a favor to humans, to add interesting nuances to our cuisine. (Our modern types have been bred to accentuate these characteristics, but farmers began with the plants' strongly flavored ancestors.) These agents are repellents, the plants' natural defenses against predators.

Many species make substances called **allomones** whose sole functions are to hurt, inhibit, or repel other species. Some allomones, such as the venoms of wasps and bees, are used for defense or to subdue prey. Others enhance competition by excluding other species from a space or suppressing their growth. Still others are repellents that make an organism unattractive as food, such as skunk repellents or the nerve poison tetrodotoxin made by puffer fish, goby fish, and some amphibians. Allomones are best explained with stories of other examples.

The drug digitalis, a steroid glycoside derived from foxglove *(Digitalis purpurea)*, is known as a cardiac glycoside because it can produce heart attacks in animals, and physicians use it in controlled doses to treat heart disease in humans. An animal that eats foxglove can get a big dose of this allomone, and if it doesn't die of a heart attack, the one exposure will probably teach it to avoid foxglove. Grazing animals avoid buttercups, containing protoanemonin, and larkspurs, containing delphinine; these agents poison the nervous system, and they taste so bad that animals learn to avoid them. Animals that eat plants of the genus *Hypericum*, like St. John's wort, obtain hypericin, which causes skin irritations and extreme sensitivity to light. Again, most animals learn to avoid those plants, but *Chrysolina* beetles have evolved enzymes that detoxify hypericin, so they can feed on a resource that is not available to other animals.

Conifers and certain other plants have evolved a sneaky and very effective allomone defense by producing **phytoecdysones,** which mimic ecdysones, the hormones that regulate insect metamorphosis. An insect larva feeding on these plants is doomed to never develop into an adult. Since these insects can't reproduce, selection strongly favors the members of an insect population with a propensity to graze on other plants.

Small invertebrates often contain substances that make them irritating and distasteful. Meloid beetles, for instance, have cantharidin in their blood, a disagreeable substance also known as Spanish fly and long thought to be an aphrodisiac. When the beetle's leg is pinched, a drop of hemolymph (blood) oozes out, giving the would-be predator a warning taste of what it is about to ingest. Some grasshoppers produce both cardiac glycosides and histamine, which causes pain and inflammation, and ants produce the irritant formic acid.

Some animals have special chemical defense organs. The millipede *Apheloria corrugata* produces and stores mandelonitrile in one chamber of a gland, while a second chamber contains an enzyme that converts mandelonitrile into benzaldehyde and hydrogen cyanide (Figure 27.9). When attacked, the millipede squeezes the contents of the first chamber into the second and releases a droplet of the mixture, generating a shroud of protective cyanide fumes for half an hour or more. This mechanism keeps the defensive agent from poisoning its producer, since the material is made in an inactive form and then activated and released only when needed.

Allomones would not be able to deter predation unless some animals had evolved the ability to learn. The best defense teaches a lesson that doesn't have to be repeated. If a mouse couldn't learn that some grasshoppers taste terrible, it would go on killing them even if it didn't eat them, and the grasshoppers would still be at high risk to mouse predation. Other defense mechanisms, which we will discuss later, also show the importance of learning.

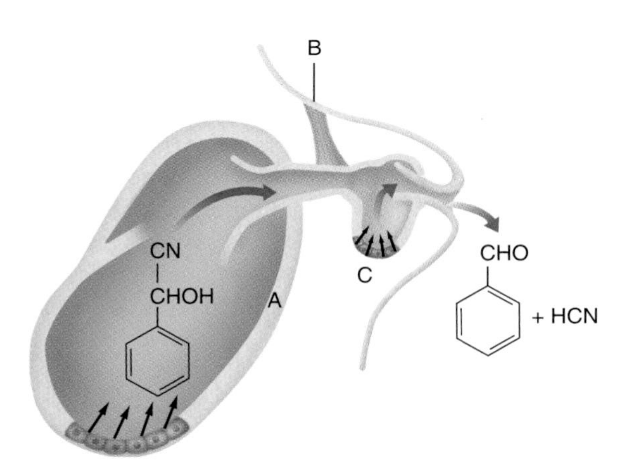

Figure 27.9

In the "reactor" gland of the millipede *Apheloria corrugata,* mandelonitrile is stored in the inner chamber (A). A muscle (B) opens a valve to the outer chamber (C) where an enzyme releases hydrogen cyanide gas.

27.10 Some organisms create intolerable conditions for others.

Plants commonly take over a space aggressively, sometimes by physical means, but more often with chemical adaptations. When light is the limiting factor in growth, as it often is among plants, some species acquire the physiological adaptations needed for growth in shade (see Section 38.8), or they may survive by growing taller, thus trapping light and also shading out lower-growing competitors. A species can claim a large share of living space by producing allomones that inhibit prospective competitors. The Black Walnut, *Juglans nigra,* releases juglone from its leaves, which washes into the soil and inhibits the growth of many plants. Plants that have evolved a tolerance to juglone can grow in the zone around the tree, but the walnut still regulates the plant community in its vicinity.

When environmental conditions are relatively stable, the species in a community seem to hold one another in check so none gets a strong enough hand to eliminate all competitors, but in a disturbed environment, where a secondary succession is beginning, an aggressive plant can take over. Home gardeners can spend a lot of energy fighting rapidly spreading vines and grasses, such as morning glory and bamboo. In some western areas of the United States, where the land has been overgrazed and the native grasses eliminated, cheat grass—a European import—has taken over, filling the once-rich range with prickly, inedible bristles that are highly flammable and crowd out the native shrubs.

Successions in ecosystems, Section 28.9.

In secondary successions in abandoned midwestern fields, one of the first species to move in is the common sunflower, *Helianthus annuus*, which produces allomones that inhibit the growth of other species. But the grasses *Aristida oligantha* and *Sorghum halepense* can tolerate these substances, and they invade the region to become dominant members of a second stage of the succession. *Aristida* and *Sorghum* produce their own allomones, primarily phenolic acids, which inhibit the growth of nitrogen-fixing bacteria and cyanobacteria, as well as the seedlings of other plants. But they themselves, being tolerant of these conditions and of the materials they make, continue to dominate and inhibit invasion of the region by other species. Many comparable situations probably exist in nature.

Fungi and certain bacteria make **antibiotics,** metabolic inhibitors that arrest the growth of other microorganisms in the soil and water around them. Naturally, the bacteria living near an antibiotic-producer have evolved enzymes and other defenses that enable them to resist these antibiotics. We can select antibiotic-resistant mutants in the laboratory, but they have been evolving in natural communities for billions of years.

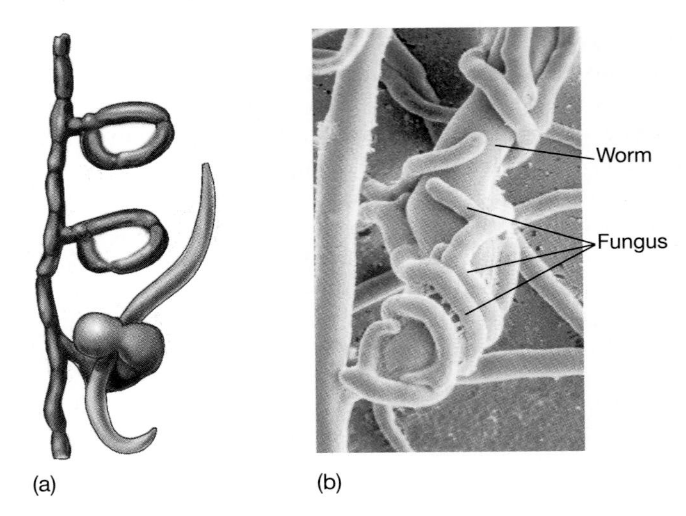

Figure 27.10
(a) One variety of fungus makes loops that expand when a nematode slips into them. The fungus then grows into the worm, digesting and absorbing its tissues. *(b)* A small roundworm (nematode) has been trapped by fungi.

27.11 Many species react to kairomones produced by other species.

In the chemical warfare of a community, a species's chemical adaptations are sometimes turned back against it by another species. Any distinctive compound—perhaps a pheromone, an allomone, or just a metabolic by-product—marks the species that produces it and can be used as a **kairomone:** a substance that gives a selective advantage to a species that *receives* it, in contrast to an allomone, which gives a selective advantage to the species that produces it. If a predator, for instance, locates its prey by means of chemicals that the prey species produces for other functions, those chemicals are kairomones. Many mammals and other animals track their food by smell, a behavior we take advantage of by using dogs to track game or people.

Some interesting fungi can trap nematode worms by growing in loops, which expand when stimulated by a worm that slips into the loop (Figure 27.10). The fungi only form these loops when stimulated by the presence of nematodes. Fungi that have been grown in isolation from nematodes make no loops, but they develop loops when bathed in water in which nematodes have lived, thus showing that they are responding to a kairomone produced by the worms.

The detection of kairomones works both ways. A prey species may have a certain odor to a predator, but the predator also has its own odor to the prey, allowing prey species to defend themselves. For this reason, humans learned very early to hunt their food from a downwind direction. Whelks (large marine snails) can distinguish herbivorous from carnivorous starfish by their chemical features, and the mere touch of a predatory starfish initiates a violent behavior pattern: The whelk jumps away, retracts its siphons and tentacles, and rolls away from the threat. Another example of growth influenced by kairomones is seen in certain rotifers. The rotifer *Asplanchna* preys on another rotifer, *Brachionus*. When growing in water with no predators, *Brachionus* develops with no particular protection, but *Asplanchna* releases a kairomone that is detected by young *Brachionus*, inducing that species to develop long, protective spines.

D. Predation, symbiosis, and camouflage

27.12 Predation is an essential activity in every community.

Communities are based on the trophic or predatory relationship, "A eats B." At each trophic level, the animals that eat are the predators, and those that are eaten are the prey, even if the predator is an herbivore and its prey is vegetation.

Hawks, owls, wolves, and wildcats are the victims of old prejudices about predation. These predators not only eat animals that people thought they had exclusive rights to, but in their ignorance, people assumed that predators should be eliminated because they damage the prey populations. (No one seemed to ask how the prey species had survived for so long before humans came along to protect them.) In fact, in any natural, established relationship, the predator is unlikely to jeopardize its prey populations in the long run. Predators can operate as density-dependent factors in limiting prey populations, for at least two reasons. First, prey can use a *refuge* in a density-dependent way (Figure 27.11). When a prey population is small, individuals can hide from predators to some extent, but in a larger population, many individuals are forced out into exposed places. Second, predators respond to changes in prey populations in at least three ways:

1. *Numerical response.* As the prey population increases, the predator population also increases.
2. *Functional response.* As one prey species increases, the predator concentrates more on that species. Many predators

apparently form a *search image* (see Section 27.15), a set of stimuli characteristic of one type of prey, which they use in hunting, and when one species becomes more abundant, they are more likely to hunt for it.

3. *Aggregative response.* When one prey population increases, predators are likely to move into the area and concentrate on it.

A predator's effect on its prey is a controversial matter, and the relationship varies considerably. Again, the best way to explore this subject is through stories about specific cases. In some situations, predators clearly do not hurt their prey populations and may even keep a prey population healthier than it would be otherwise. Paul L. Errington and his students showed this in their studies of Iowa muskrat populations. They observed that the muskrats maintained quite a stable population of 9,000 individuals in the fall on a 375-acre marsh they shared with about 30 muskrat-hungry minks. But the muskrat population was hardly affected by its predators—and certainly not adversely. In Errington's words:

> *The predation centered upon overproduced young; upon the restless, the strangers and those physically handicapped by injuries or weakness; upon animals evicted by droughts, floods, or social tensions; and upon what is identifiable as the more biologically expendable parts of the population.*

Studies of wolves have led to similar conclusions. L. David Mech studied the Isle Royale wolves and recorded 131 episodes in which they hunted moose. Fifty-four of the moose escaped before the wolves could even get close. Only six of the 77 moose confronted by wolves were overcome. Through continuous hunting, the wolves were able to kill about one moose every three days, providing them with 4.5–6.0 kg of meat per day. The wolves mostly caught the more expendable individuals—the very young or old or sickly. Considering the behavioral dimension of predation, this makes particularly good sense, especially when the predator can learn and can react intelligently to its environment. No wolf will rush into the midst of a herd of healthy moose if it has the option of picking off an old or sick animal at the edge of the herd. Indeed, the ability to learn and to behave intelligently has evolved largely for coping with this kind of situation.

Other studies confirm that predation (up to a point) has a beneficial effect on the prey population. This effect can be understood in terms of the *recruitment* of a population, the difference between birth and death at any time (Figure 27.12). Recruitment is severely reduced when a population approaches its carrying capacity, but if a population is held to a lower level by predation, its recruitment rises. Lawrence Slobodkin and his colleagues maintained laboratory populations of small invertebrates such as the cnidarian *Hydra* and the crustacean *Daphnia,* with the experimenters acting as predators by removing individuals at various rates. They found that as the intensity of predation increased, the prey reached greater ecological efficiency. Heavy predation thus increases the efficiency of the food web. Similarly, David Lack found that as fishermen take bigger fish out of an area, the fish population increases because more of the younger fish live to maturity, begin to breed earlier, and lay more eggs.

In contrast, other predators severely limit prey populations in restricted areas. Oliver Pearson, for instance, reported on a grassland and weed ecosystem in California where carnivorous

Figure 27.11
The members of a dense population are more likely to be exposed *(a),* whereas those in a sparse population can often find refuges where they are safe from predators *(b).* This is illustrated here by snails preyed on by Snail Kites,

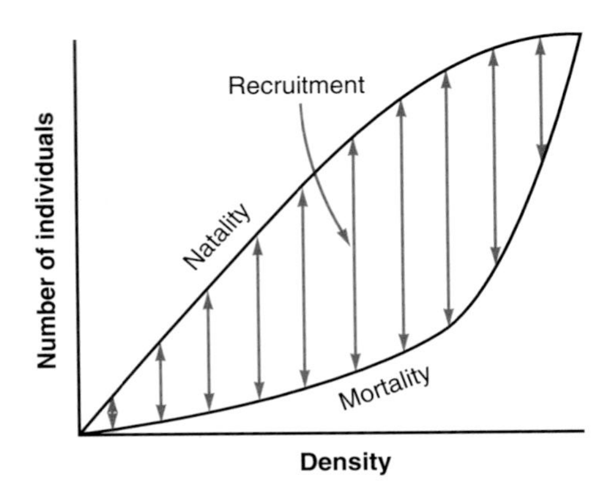

Figure 27.12
Recruitment, the difference between natality and mortality, is highest when a population is held below the carrying capacity of its environment.

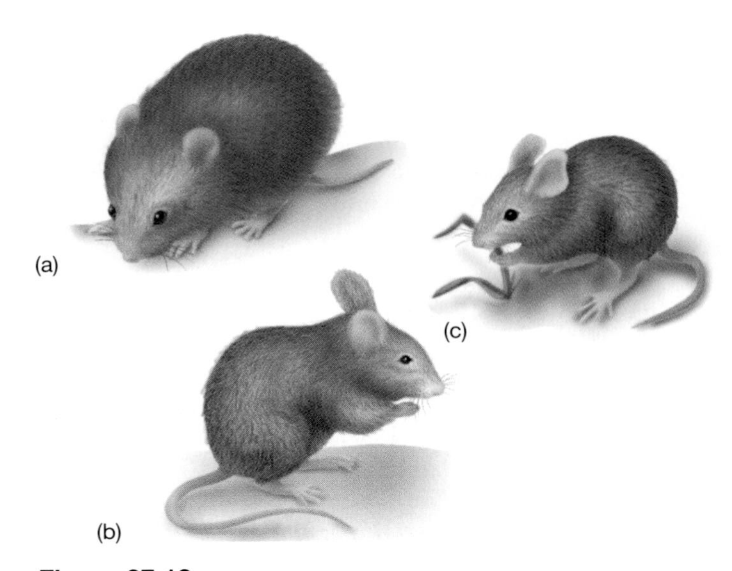

Figure 27.13
Three common herbivores in a grassland ecosystem are *(a)* voles *(Microtus), (b)* house mice *(Mus),* and *(c)* meadow mice *(Reithrodontomys).*

mammals, including gray foxes, raccoons, skunks, and feral cats, preyed heavily on populations of small rodents (Figure 27.13). The carnivores ate 88 percent of the voles *(Microtus),* and as these became scarce, they turned increasingly to mice. They then ate 33 percent of the *Reithrodontomys* and 7 percent of the *Mus* populations. The population of *Microtus,* which the carnivores clearly preferred, was virtually annihilated by predation, both by the mammals and by birds such as hawks and owls that passed through the area. Pearson noted, however, that the community could not have supported the voles and mice unless their numbers had been reduced by predation. We'll demonstrate this in the following calculations.

Exercise 27.3 Only 17.3 million kcal per acre of plant food was available at the beginning of the study. With an ecological efficiency (the ratio of prey biomass to the biomass of its food) of 10 percent, how much energy could these plants have supplied to the rodents?

Exercise 27.4 If all the rodents had been allowed to survive, the standing crops would have had to provide 1.4 million kcal/acre for the *Microtus,* 876,000 kcal/acre for the *Mus,* and 82,000 kcal/acre for the *Reithrodontomys.* Add these numbers and determine whether the plant crop could have supported all the rodents for a year.

Exercise 27.5 The carnivores removed a substantial proportion of the rodents: 55 percent of the available energy in the rodent populations. Calculate how much energy was left in the rodent populations. How much of the plant crop was required to maintain the remaining rodents?

The results of these exercises show that the heavy predation was essential to continued growth of the plants. Because so many rodents were eaten, 7 percent of the seed crop was saved to produce a new crop of plants.

A third situation shows that predators of very small animals may be even more effective than larger carnivores in controlling prey populations. The California citrus crop was plagued in the late 1880s by the cottony-cushion scale insect, *Icerya purchasi,* which feeds on tree sap. Since the species originated in Australia, an Australian predator was imported to control it, the ladybird beetle *Rodalia cardinalis.* The beetle proved extremely effective in controlling the insect population, sparing the crop further damage. Similar predator remedies have been successful in controlling other insect pests—sometimes with carnivorous beetles, sometimes with wasps that lay their eggs in the prey insect.

One conclusion from these varied situations is that the effects of predators on their prey may be primarily a function of different patterns of behavior. Large predators like wolves, which pursue large prey like deer, simply cannot approach and catch their prey at will. Cats preying on mice, on the other hand, have an easier time of it. As Pearson noted, "The consistent success of their vigils beside runways makes *Microtus*-catching seem absurdly easy." Similarly, some insect predators hardly have to work at all for their food. As Paul Colinvaux observed, a beetle "can walk up to the passive scale insects, chew them up at its leisure, and move on to the next one which sits there waiting to be eaten."

27.13 Predator and prey populations may change in cycles.

Predation is a critical factor in keeping a population below the limits set by availability of food or other resources. So we should expect the sizes of predator and prey populations to be closely correlated: the more predators, the fewer prey; the fewer prey, the fewer predators; and so on. The theorists Alfred J. Lotka and V. Volterra expressed this simple mutual relationship by a set of equations, now named for them, which predict that predator and prey populations should rise and fall in cycles (Figure 27.14). These cycles are slightly out of phase with each other,

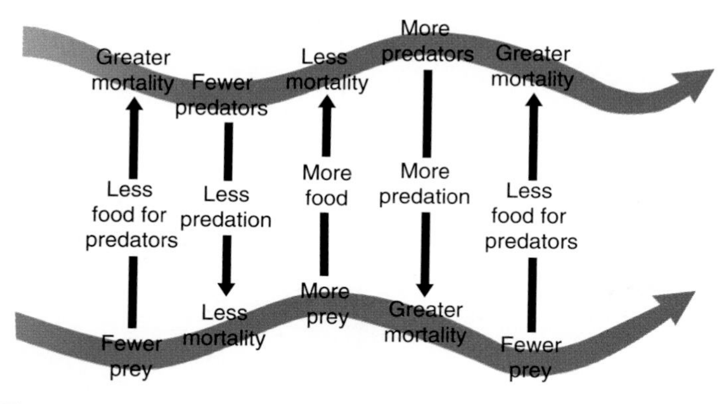

Figure 27.14

Interactions between predator and prey populations can produce a cyclic change in the size of both populations.

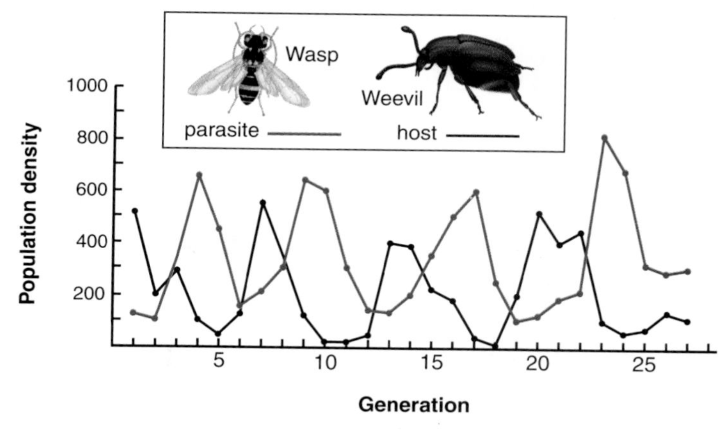

Figure 27.15

Since a parasite is just a special kind of predator, the interactions between a parasite and its host can be used to demonstrate predator-prey interactions. A cycle in population density occurs in the Azuki Bean Weevil *(Callosobruchus chinensis)* and the wasp *Heterospilus prosopidis,* a parasite on the larval weevils.

since each population takes a little time to react to a change in the other.

Interestingly, one set of data indicating that such cycles really occur in nature comes from records kept by the Hudson Bay Company on purchases of animal pelts over a 200-year period. These numbers reflect the population densities of each species, and the data on lynx and snowshoe hare pelts show the alternating rise and fall of these two species that the Lotka-Volterra equations predict. Lloyd Keith has now shown that what appeared to be a hare-lynx cycle is actually a hare-willow cycle; the hare population, acting as a predator on willows, rises and falls as predicted by the Lotka-Volterra equations, and although the lynx population changes in response to the hare population, it is not the driving force in this interaction. Figure 27.15 shows the fluctuations in experimental populations of a bean weevil and its parasite; the curves are highly correlated, apparently reflecting the kind of predator-prey relationship embodied in the Lotka-Volterra equations. Other investigators, however, have not found periodic behavior in experimental predator-prey situations, and field studies show that some populations undergo cyclic changes that are not at all associated with predation. Clearly, much remains to be learned about predation and population dynamics.

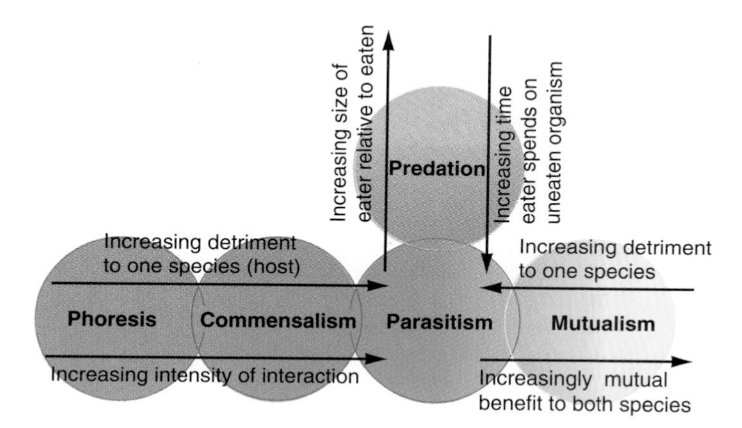

Figure 27.16

The different types of symbiotic relationships between species grade into one another in a spectrum.

27.14 Many organisms engage in symbiotic relationships.

In the midst of biological arms races, some species make temporary "truces" with one another, while others establish parasitic relationships. A whole series of intimate associations between species, broadly known as **symbiosis,** is common in communities. In common language, symbiosis always implies a cooperative relationship that is beneficial to both parties, but ecologists use the term *mutualism* for this association. Depending on the strength of the interaction and the degree of benefit to one species or the other, a symbiotic relationship can be categorized into one of four types along a continuum from simple coexistence to parasitism or mutualism (Figure 27.16).

At one end of the spectrum is **phoresis** (literally, "carrying"), which includes the loosest sort of interactions in which a larger animal (the *host*) carries around a smaller one (the *phoront*). But the larger is not hurt, and the smaller gets only the benefits of mobility and a place to live. The benign bacteria that live on our skin are good examples of phoronts.

Phoresis grades into **commensalism,** in which the smaller organism shares the host's food as well as its living space. For instance, the remarkable ciliated protozoan *Ellobiophyra* padlocks itself to the gills of a clam, *Donax vittatus* (Figure 27.17). Although the ciliates share the stream of food the clam feeds on, they don't take enough food to damage the hosts. If they did, this would be a parasitic relationship. Rather, the two species are simply eating "at the same table" (*co-* = together; *mensa* = table). Commensal relationships are very common. Sea anemones normally prey on fish by stinging them with their nematocysts, but some fishes live commensally among anemones

Figure 27.17
Commensalism is exemplified by the ciliate *Ellobiophyra donacis* attached to the gills of a clam. Source: E. Chatton and A. Lwoff, "*Ellobiophyra donacis*/Ch. et Lw., peritriche vivant surles branchies de l'acephate Donas vittatus da Costa," in *Bull. Biol. Belg.*, 63, 321 (1929).

and share their food. With a heavy mucous coating on their bodies, these fish are not troubled by nematocysts and can swim among the forest of anemone tentacles where they are safe from predation by other fish.

Commensalism grades into **parasitism,** a relationship in which a smaller organism, the *parasite,* lives on a larger host and depends metabolically on it. Parasitism differs from commensalism in that the parasite damages the host, whereas a commensal does not. A relationship on the borderline is the association between the pearlfish, *Carapus bermudens,* and some echinoderms such as sea stars and sea cucumbers (Figure 27.18). The fish, attracted to a kairomone the sea star releases, pushes its tail into the host's anal opening, forces its way inside, and lies within the intestine, where it feeds. It apparently does the starfish no particular harm. This relationship is referred to as *endocommensalism* (*endo-* = inside), but it is very nearly parasitism. However, the fish can also inhabit a sea cucumber, and in this case it devours the host's intestine, eventually kills it, and then continues to live inside the body. This is clearly parasitism, and it might even be called predation.

Figure 27.18
This pearlfish is about to twist its way into the digestive tract of a sea cucumber, where it lives as a commensal or parasite.

Parasite-host relationships are at one end of another spectrum (see Figure 27.16) that has predator-prey relationships at the other end. A tapeworm and a dog are certainly parasite and host; a wolf and a rabbit are clearly predator and prey. But how do we describe the relationship in which a mosquito draws blood from a human, a lamprey slowly eats a trout, or a snapping turtle bites off a fish's fin? The common denominator is that one organism gains mass and energy at the expense of another, and particular names for these activities are not very important.

Some of the most interesting ecology stories are told about true parasitism, in which the host is damaged. The chronic and natural condition of many organisms, including humans, is to be parasitized all their lives. Modern people, living in clean conditions and eating food free of parasites, may find it hard to realize that until very recently the natural condition of humans was to host a variety of **ectoparasites** (on the outside of the body), such as ticks, fleas, and lice, and **endoparasites** (on the inside), such as worms. Through modern medicine and sanitation we manage to keep ourselves reasonably free of them, but most animals (and many less-fortunate people) harbor a wide assortment of parasites (Figure 27.19). An **infection** is a parasitic relationship that has gotten out of balance, or in which some new, more virulent strain of parasite circumvents the host's defenses. Infection can also occur when a host lets down its defenses temporarily, as when it is ill or injured.

Aspects of disease caused by bacteria, Chapter 29.

Brood parasitism in birds is another unusual form of symbiosis. American cowbirds, European cuckoos, and some other birds make no nests of their own but instead lay their eggs in the nests of other species, which then raise their offspring for them. The relationship between the parasitic and the host nestlings varies; sometimes the parasite is larger and stronger than its nest-mates, and it may push them out. In other cases there is a certain amount of reciprocity, with the parasitic bird helping

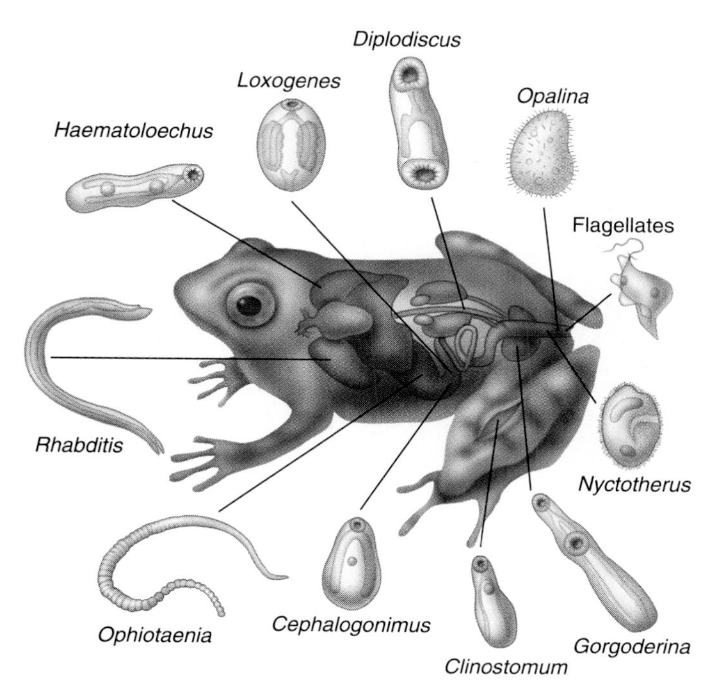

Figure 27.19
Most animals, like this frog, normally host a variety of parasites: fleas, flies, ticks, and mites externally; worms internally; and protozoans, fungi, bacteria, and viruses all over.

keep the host nestlings free of small parasites. As a result of its breeding habits, the common European cuckoo is going through the first stages of a kind of speciation. The species is dividing into distinct breeding lines, known as *gentes,* each specialized for parasitizing a different host and each laying an egg that almost perfectly matches its host's eggs (Figure 27.20). The gentes are becoming isolated from one another, perhaps undergoing a sympatric speciation, because each female lays her eggs in the nests of the species that raised her.

Parasitism grades into **mutualism,** a relationship from which both parties gain benefits and enhance their fitness. Hermit crabs, which live in snail shells, often carry anemones around on their backs (Figure 27.21). The anemone protects the crab with its stinging tentacles while feeding on scraps from the crab's meals. The bacteria living in our large intestines might be regarded as commensals with us, but the relationship is more like mutualism because they provide us with some vitamins and bulk for our feces. Other animals that feed largely on cellulose are quite dependent on their intestinal flora. Given the enormous amount of cellulose in the biosphere, it might seem strange that enzymes that digest cellulose are so rare in nature and that herbivores have not developed their own, but there has been no particular selective pressure for the evolution of such enzymes in herbivores, since they universally use bacteria and protozoa that have the necessary enzymes. A cow feeds on cellulose in grass and is able to use its component sugar, glucose,

Figure 27.20
European cuckoos deposit their eggs in the nests of several other species. The various gentes have eggs that closely resemble the eggs of their hosts. Source: James Fisher and Roger Peterson, *The World of Birds,* 1964, Bantam Doubleday.

Figure 27.21
Mutualism is illustrated by anemones and hermit crabs. The crab gets protection while the anemone gets mobility and food.

Figure 27.23
African crocodiles allow Egyptian Plovers to walk into their mouths to feed.

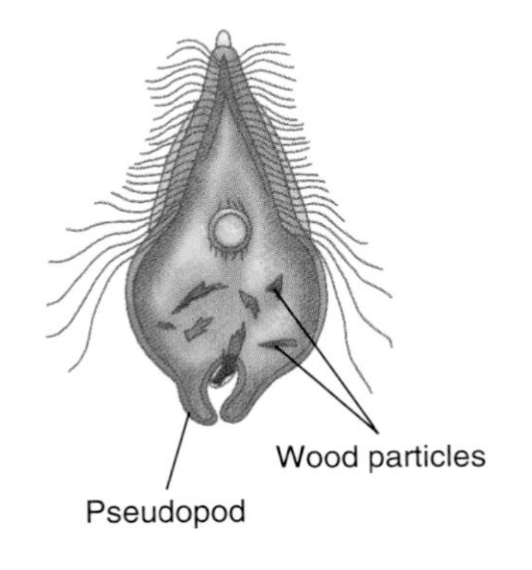

Figure 27.22
The flagellate *Trichonympha* lives in the intestines of termites. It engulfs and digests wood chips and thus provides food for both itself and the termite.

Figure 27.24
Spanish hogfish swim into the mouths of barracuda, where they feed and clean.

only because of microorganisms with appropriate enzymes in its specially adapted stomach (rumen). The notorious wood-eating termites cannot digest wood, but they chew and swallow wood chips, which are engulfed by flagellated protozoans in their intestines (Figure 27.22). The glucose these protozoans release from the wood feeds both themselves and their hosts. Microbial mutualists commonly supply their hosts with other nutrients, such as fatty acids, essential amino acids, and vitamins.

Another classic example of mutualism is the lichen relationship (see Chapter 31, Figure 31.12). Lichens grow in distinctive shapes and are named as if they were single species, but each is actually a specific association between a fungus and an alga. The two organisms form a minor ecosystem in themselves, because the alga uses photosynthesis to make organic compounds that partially feed the fungus. The fungus, for its part, provides water, minerals, protection, and a microenvironment suitable for the growth of the alga.

A **cleaning symbiosis** is a novel relationship in which one species gets some of its food by cleaning another. An African rhinoceros, for instance, lets tickbirds live on it. The birds devour ticks from the folds in the rhino's skin—an obvious benefit for both species—and the birds' alarm calls also warn their host of approaching animals. Although African crocodiles eat many kinds of birds, they let Egyptian Plovers walk into their gaping mouths to pick leeches from their gums (Figure 27.23). There are many examples of one fish cleaning another, because all kinds of small organisms grow on a fish's skin and in its mouth where other fish may eat them. The barracuda *(Sphyraena barracuda)* eats many small fish, but it allows the Spanish hogfish *(Bodianus rufus)* to swim into its open mouth to clean it (Figure 27.24). Other fishes patiently hold still while smaller fish clean encrusting organisms from their skin. The fish that are cleaned seem to seek out cleaners and to clearly benefit by the process; the cleaners always have distinctive patterns of markings and color, which are undoubtedly recognized by the potential predators as sign stimuli that inhibit eating behavior and promote tolerant actions in the larger fish.

Sign stimuli, Section 49.4.

27.15 Many animals are camouflaged by their forms and colors.

Predation has been a major factor in shaping the evolution of organisms' characteristics. One obvious protective mechanism is to become invisible, or as close to invisible as possible. The peppered moths *(Biston)* shown in Figure 2.20 are excellent examples of **cryptic coloration,** which camouflages the organism. Many insects have evolved to look like parts of plants (Figure 27.25). Birds often have nondescript brown and white plumage that blends with the dry grasses of their nests (Figure 27.26); species in which only the female incubates the eggs usually exhibit sexual dimorphism, the male plumage being quite colorful while the female is very dull and camouflaging. Sometimes coloration is combined with concealing behavior; an American Bittern, a small heron, blends into the reeds when it points its beak straight up in the air, as it does when threatened.

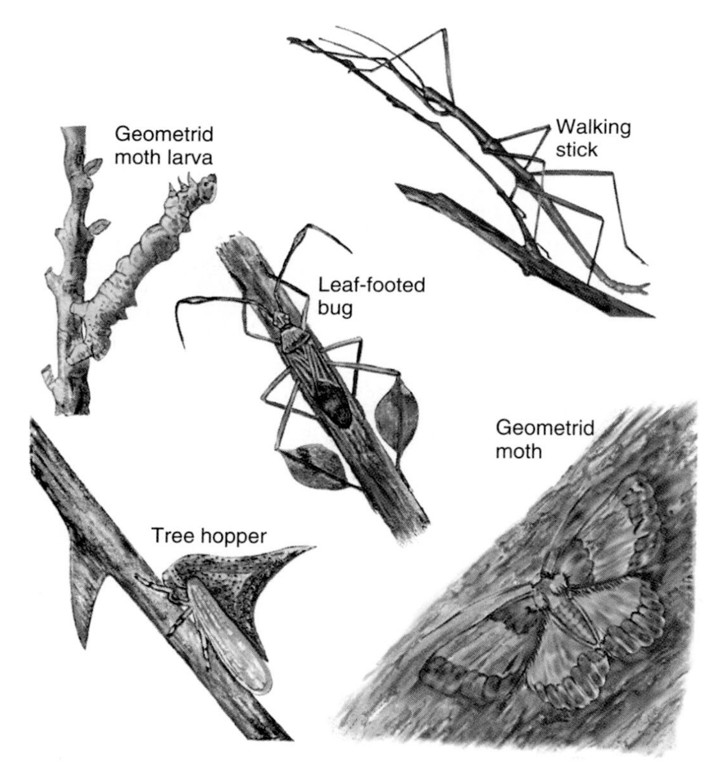

Figure 27.25

A variety of insects are camouflaged by their close resemblance to plant structures. Geometrid-moth larvae and walking sticks make excellent imitation twigs, tree hoppers look like thorns, and many species, such as leaf-footed bugs, look like leaves. The adult geometrid moth is hard to see against the bark of trees on which it commonly rests.

Figure 27.26

A female American Woodcock sitting on her nest is hard to see because her streaked, brown pattern blends so well with the surroundings.

A leopard's spots and a zebra's stripes illustrate **disruptive coloration** (Figure 27.27). While these markings stand out sharply when seen up close against a plain background, they disguise the animals beautifully when viewed against a broken background (Figure 27.27*b*). While moving through the underbrush dappled by spots of sunlight, these animals are very hard to see. Their colors make them more visible to us, with our color vision, than they are to animals who see only shades of gray.

Some animal morphologies are apparently effective because predators acquire **search images,** patterns they focus on in hunting and look for again and again. This behavior can account for the bright, often brilliant colors of some animals. Anyone seeing the gorgeous *Morpho* butterflies of the American tropics displayed in a collection (Figure 27.28) must wonder how they can survive with colors that make them so obvious to predators, but these butterflies flash their colors only when flying, and when they land and fold their wings, just the dull, cryptic parts are visible. A predator pursuing such a butterfly forms a search image that includes the brilliant color, so it is confused when the butterfly suddenly blends into the background. Similarly, the naturalist Edwin Way Teale has pointed out that Yellow-rumped Warblers are quite obvious when flying, but when they land, their wings cover the yellow patch and the birds seem to disappear (Figure 27.29). This effect could severely reduce a predator's success.

Experiments by A. T. Pietrewicz and A. C. Kamil support the idea that predators form search images. They showed Blue Jays some slides of cryptic backgrounds and some including cryptically patterned moths. The birds were rewarded for correctly pecking only at slides that showed moths. When presented with a series of slides showing only one morph (form) of a moth, the birds became quite good at seeing the moths in later trials, apparently by forming accurate search images, but when moths of different morphs were mixed, the birds didn't

(a)

(b)

Figure 27.27

(a) A real zebra *(right)* blends into a broken background because its stripes disrupt the edges of its body; an imaginary zebra with the stripes going the wrong direction *(left)* is quite visible because the stripes emphasize its edges. To see this, back away from the page until one of the images disappears. *(b)* The strong markings of a large cat camouflage it in the dappled light of a jungle or grassland setting.

improve. The use of search images by predators may be a factor in maintaining different morphs in prey populations, since predators that see different forms may have a hard time forming a useful search image.

27.16 Some animals are protected by warning coloration.

Some brightly colored animals seem to flaunt themselves in front of prospective predators, as though daring the predators to eat them. If camouflage is such an excellent defense mechanism, why hasn't it evolved universally? In some species, the bright colors have been selected for their role in communication within the species—to attract the opposite sex perhaps. Sometimes, though, the animals are conveying a message to potential predators, warning them to stay away through **aposematic coloration** (*apo-* = away; *sema-* = signal). A skunk can afford to be obvious. It tells prospective enemies that they can avoid a lot of trouble by learning to recognize the striking black and white pattern of its fur, and leaving it alone. Brightly colored insects may be conveying the same message, and in most cases they can afford to do so only because they are distasteful and because their predators can learn to avoid them.

Aposematic coloration becomes more effective through **Müllerian mimicry,** a phenomenon in which a number of species evolve a common color pattern that warns predators of the threat they pose. Thus the pattern of black and yellow stripes common to many types of wasps and bees (Figure 27.30)

Figure 27.28

The brilliant colors of *Morpho* butterflies are very noticeable and attractive during flight, but as soon as a butterfly alights, it folds its wings so the colors disappear, a move that probably confuses predators.

Figure 27.29

Yellow-rumped Warblers flash their color patches while flying, but with their wings folded they look quite different.

Figure 27.30
Müllerian mimicry is illustrated by the familiar yellow and black patterns common to a number of bees, yellow jackets, and wasps. Syrphid flies, on the other hand, have evolved the same pattern as a case of Batesian mimicry.

is advantageous to all of them because predators need only learn to avoid a single pattern, rather than attacking each species separately and learning many distinct warning signals. A painful encounter with one kind of bee helps protect all those that resemble it, whether of the same species or not.

Exercise 27.6 Thinking too carelessly about Müllerian mimicry, one might get the impression that all the species that come to resemble one another have evolved into a common form for the benefit of the predators that prey on them, so these predators only have to learn one pattern. Explain Müllerian mimicry to yourself in a sensible way, so it is clear that natural selection has operated to increase the fitness of the prey species.

27.17 Mimics may survive by imitating warning coloration.

In contrast to Müllerian mimicry, **Batesian mimicry** occurs when a harmless or palatable species called a **mimic** survives through a kind of false advertising by evolving to resemble a dangerous or distasteful **model.** A classic example is the brilliant black and orange monarch butterfly, *Danaus plexippus,* and the unrelated viceroy butterfly, *Limenitis archippus* (Figure 27.31). Monarchs feed on milkweed, *Asclepias.* Most animals avoid this plant because it contains noxious cardiac glycosides, but monarch larvae not only feed on milkweed but actually incorporate the glycosides into their tissues, thus becoming distasteful to predators. The monarch's aposematic coloring warns predators to stay away, and it is mimicked by the viceroy, which, though perfectly edible, is also avoided by birds that have learned to avoid the monarch.

Studies conducted by Lincoln and Jane Brower show that predators must learn these avoidance behaviors. They found that Blue Jays who have never seen either monarch or viceroy butterflies before will readily feed on viceroys. At first they will also attack monarchs, but after a few bites, they spit them out and even vomit up anything they have swallowed. Then, having learned to avoid monarchs, these birds also avoid viceroys.

Batesian mimicry poses an interesting ecological problem that can be analyzed with the game theory illustrated in Section 49.2. The mimic and model species are engaged in a contest. This kind of mimicry is clearly an advantage to the mimic species and a disadvantage to the model, for if a predator encounters mimics first, it will keep feeding on similar animals and may have to kill several models before it changes its habits. Thus it is advantageous for models to change so they don't resemble mimics. Game theory predicts that this regime of mimicry also will not work for either species if the mimics are too common; that is, aposematic coloration does little or no good if predators are rewarded with good food too often. One of Jane Brower's experiments is instructive on this point. She fed starlings mealworms, some of which had been made bitter with quinine (the models) and some with no quinine (the mimics). Both models and mimics were painted with green bands, while other unaltered mealworms with orange bands were offered as an alternative food source. Brower found that after a few experiences with bitter mealworms, the birds learned to feed only on those with orange bands and to avoid all green-banded worms.

(a)

(b)

Figure 27.31
Batesian mimicry is illustrated by the distasteful monarch butterfly *(a),* which is mimicked by the nontoxic viceroy *(b).*

The mimics were still largely protected even when they constituted 60 percent of the model-mimic population, but not at higher levels. Thus Batesian mimicry works quite well if the proportion of mimics is not too high and the predators have an alternative food source.

Coda In this chapter we have shown some of the interesting ecological and behavioral relationships among the species that share a community. These relationships create the community structure. They obviously have been shaped, and continue to be shaped, by evolution through natural selection, but nowhere have we had to postulate any mechanism more complex than natural selection that favors individuals or closely related groups of individuals. There has been no need to postulate anything like "community evolution," and certainly many examples of community structure show species in evolutionary arms races, competing with one another rather than acting cooperatively for some kind of supposed benefit to the community as a whole. Some examples of mutualism show species cooperating, but only because they have chanced into mutually beneficial interactions.

All the features being shaped by natural selection are the result of specific molecular interactions. The many chemical interactions in a community are the most obvious examples, cases in which a species has evolved special enzymes to produce an allomone or in which an animal's lack of a certain enzyme restricts it to certain foods. Yet the cases of camouflage and mimicry are also the results of molecular interactions. In probing into the course of development and differentiation in Chapters 20 and 21, we began to see how proteins interact to activate or repress sequences of genes whose ultimate result is a certain body form. At the end of Chapter 24 we began to get some insight into the evolution of different forms through changes in regulatory genes or in the timing of their expression. Though we are far from understanding the detailed course of evolution that has created any of the adaptations discussed here, this chapter should help us appreciate that each structure or behavior has resulted from the ecological shaping of molecular interactions.

Summary

1. A community consists of organisms that live together and interact, but its component species adapt to maximize their own fitness and are distributed in accordance with their own requirements. A community is not a "superorganism."
2. Communities are always diverse, but their component species are never equal in abundance. Their abundance commonly conforms to a "broken-stick" model. The diversity of communities increases in a gradient from the polar to the tropical regions.
3. Some communities are structured by keystone species, which have such a strong influence on other species that their removal changes the community structure significantly.
4. The complex relationships in a community can generally be simplified by considering possible relationships between pairs of species. Among the most important of these is interspecific competition, in which both species are adversely affected because they are rivals for the same resources. A spectrum of interactions including symbiosis and predation (or parasitism) also structures a community.
5. An ecological niche is difficult to define, but it is most generally considered to be the combination of a species's habitat and its way of life: the sum of all the conditions it requires. Many defining features of a niche can be represented by a volume in n dimensions.
6. The niches of different species may overlap, but Gause's Principle states that two species cannot continue to occupy the same niche space indefinitely. A number of observations suggest that if two species are potential competitors for the same niche, either they will evolve to occupy slightly different niches, or one will outcompete the

other. Competition, in general, should lead to the evolution of one or both species in a direction that lessens the competition. It is not definite, however, that Gause's Principle always holds true.

7. Subtle chemical factors may define a species's niche and may be responsible for limiting potentially competing species to different, non-overlapping niches.
8. Predation is an essential activity in every community, but it has variable effects. Many prey populations, especially those of large animals, appear to be damaged little by their predators, which prey primarily on the old, the diseased, and those who are outside the social structure. However, some prey species are apparently kept to very low levels by their predators.
9. The Lotka-Volterra theory predicts that predator and prey populations should vary cyclically, and in some situations there is evidence that they do.
10. Symbiotic relationships, in which two species live together, are very common. Phoresis, in which one species carries another around, and commensalism, in which the two species share resources and space without apparent harm to either, are quite common.
11. Commensalism grades into parasitism, a relationship in which one species, the host, is clearly harmed by the parasite. This is a very common situation, and most species of plants and animals are probably parasitized to some extent. Many species also live mutualistically, with each one contributing something and gaining something.
12. Organisms have evolved many adaptations for avoiding competition, thwarting their competitors and predators, and generally maintaining their positions in the community through various adaptations. Some of these are morphological, involving the development of specific structures. Others entail the evolution of particular behaviors.
13. Many adaptations are chemical. Among the ecologically most important are allelochemicals: the allomones that adversely affect other species and the kairomones used by other species against those that make them.
14. Allomones harm or inhibit other species. Allomones may be poisons or merely noxious or distasteful substances; some even interfere with the growth of competing species. On the other hand, any species may evolve countermechanisms for avoiding the effects of allomones.
15. Animals often avoid predation—or become more effective predators—by evolving forms and colors that disguise them or make them blend into the background.
16. Animals that can be dangerous sometimes evolve aposematic coloration that warns other species to avoid them, and similar species may take advantage of the behavior patterns of predators by evolving similar patterns (Müllerian mimicry), so predators that learn one pattern will avoid all of them. Sometimes harmless species take advantage of this system by evolving to resemble the more noxious or dangerous species (Batesian mimicry).

Key Terms

plant association 557
keystone species 559
fundamental niche 560
realized niche 556
competitive exclusion principle 561
niche differentiation 561
allelochemic 563
allelopathy 564
allomone 565
phytoecdysone 565
antibiotic 566
kairomone 566
symbiosis 569
phoresis 569
commensalism 569
parasitism 570
ectoparasite 570
endoparasite 570
infection 570
brood parasitism 570
mutualism 571
cleaning symbiosis 572
cryptic coloration 573
disruptive coloration 573
search image 573
aposematic coloration 574
Müllerian mimicry 574
Batesian mimicry 575
mimic 575
model 575

Multiple-Choice Questions

1. Within any particular community, the keystone species
   a. is usually autotrophic.
   b. may be a predator.
   c. contains the most members.
   d. is the most recent arrival.
   e. is in the greatest danger of extinction.
2. Among the binary relationships in a community, the opposite of mutualism is
   a. competition.
   b. commensalism.
   c. predation.
   d. amensalism.
   e. neutralism.
3. A white-tailed deer might be described as a medium- to large-sized herbivore that, along with raccoons, foxes, owls, hawks, and black bear, lives in North American temperate, second-growth deciduous forests. This description includes elements of the deer's
   a. habitat.
   b. niche.
   c. community.
   d. *a* and *b*.
   e. *a*, *b*, and *c*.
4. A description of an organism's niche generally includes all of the following except the organism's
   a. competitors.
   b. place in the food web.
   c. limiting environmental factors.
   d. growth rate.
   e. symbiotic and mutualistic associations.
5. Which is least likely to occur if two sympatric species share the same niche?
   a. The niche is divided into two or more distinct regions.
   b. The niche's resources are divided by behavioral means.
   c. Both species will limit their numbers to achieve stability.
   d. Traits evolve that increasingly differentiate the species physically.
   e. One species will not survive in that location.
6. Which of the following is the best functional description of an allomone?
   a. a density-dependent strategy to kill competitors
   b. a behavioral alteration to limit reproduction
   c. an anatomical adaptation to encourage niche partitioning
   d. a biochemical repellent that discourages predation
   e. an intraspecific sex attractant
7. The Lotka-Volterra equations predict all of the following except
   a. the number of predators is regulated by the number of prey.
   b. the number of prey is regulated by the number of predators.
   c. negative feedback mechanisms regulate the population size of both predator and prey.
   d. predator and prey numbers reach maximum levels at different times.

e. predatory relationships take precedence over all other environmental effects.

8. Which term describes the relationship in which one member of an interspecific pair benefits nutritionally while the other member neither benefits nor is harmed?
   a. phoresis
   b. commensalism
   c. parasitism
   d. predation
   e. mutualism
9. The kind of symbiosis exemplified by a lichen is best labeled
   a. parasitism.
   b. commensalism.
   c. mutualism.
   d. phoresis.
   e. predation.
10. Which of these surface patterns is used as a warning?
   a. cryptic coloration
   b. disruptive coloration
   c. aposematic coloration
   d. *a* and *b*, but not *c*
   e. *a*, *b*, and *c*

True-False Questions

Mark each statement true or false, and if false, restate it to make it true.

1. Tropical communities generally contain more individuals but fewer species overall than temperate communities.
2. A keystone species is the most abundant producer in a community.
3. A species's realized niche is generally more inclusive than its fundamental niche because the realized niche does not take interspecific competitors into account.
4. Competitive exclusion does not generally occur in an underpopulated community.
5. Niche differentiation will not occur as long as resources are sufficiently plentiful for two or more competing species.
6. Either niche differentiation or reproductive isolation is required for speciation.
7. Allomones function as intraspecific sexual attractants.
8. Production of antibiotics by some fungi and bacteria is an adaptation to enhance disease resistance and survival of humans.
9. Predation is a density-independent regulator for population size because as the prey population decreases, more refuge space becomes available for them, and their predators tend to hunt elsewhere.
10. Batesian mimicry works best if the model species outnumbers the mimics, while the relative numbers of participants has little or no effect on the efficiency of Müllerian mimics.

Concept Questions

1. Compare the rate of niche differentiation for a species with high levels of genetic variability as opposed to a species with high levels of allelic uniformity.
2. Distinguish predation from parasitism.
3. What kind of observations does the Lotka-Volterra equation attempt to explain?
4. Contrast cryptic coloration with aposematic coloration. Does the success of these strategies depend only on adaptations of the prey?
5. Distinguish Batesian from Müllerian mimics by *(a)* evaluating the role of convergent evolution in their adaptations, and *(b)* deciding whether one or both are truth-tellers or liars.

Additional Reading

Bannister, P. *Introduction to Physiological Plant Ecology.* Blackwell Scientific Publications, Oxford, 1976.

Bergerud, Arthur T. "Prey Switching in a Simple Ecosystem." *Scientific American,* December 1983, p. 130. How lynx adapt to changes in their prey populations.

Childress, James J., Horst Felbeck, and George N. Somero. "Symbiosis in the Deep Sea." *Scientific American,* May 1987, p. 114. The remarkable density of life at deep-sea hydrothermal vents is explained by the mutually beneficial symbiosis of invertebrate animals and sulfide-oxidizing bacteria that colonize their cells.

Colinvaux, Paul. *Why Big Fierce Animals Are Rare.* Princeton University Press, Princeton (NJ), 1978.

Gilbert, Lawrence E. "The Coevolution of a Butterfly and a Vine." *Scientific American,* August 1982, p. 110. An interesting story of a community interaction.

Handel, Steven N., and Andrew J. Beattie. "Seed Dispersal by Ants." *Scientific American,* August 1990, p. 76. Thousands of plant species rely on ants to disperse their seeds. With special food lures and other adaptations, a plant can induce the insects to carry away its seeds without harming them.

Krebs, Charles J. *Ecology,* 4th ed. Harper Collins, New York, 1994.

Morse, Douglass H. "Milkweeds and Their Visitors." *Scientific American,* July 1985, p. 112. The insects and other animals that frequent milkweed form a model community for the study of interactions among species. The animals come to forage, but a few of them also serve the needs of the plant.

Neal, Ernest. *Woodland Ecology.* Harvard University Press, Cambridge (MA), 1960.

Pietsch, Theodore W., and David B. Grobecker. "Frogfishes." *Scientific American,* June 1990, p. 96. Masters of aggressive mimicry, these voracious carnivores can gulp prey faster than any other vertebrate predator.

Prestwich, Glenn D. "The Chemical Defenses of Termites." *Scientific American,* August 1983, p. 78. Another example of chemical ecology.

Rosenthal, Gerald A. "A Seed-eating Beetle's Adaptations to a Poisonous Seed." *Scientific American,* November 1983, p. 164. A story of chemical adaptation.

Shuttlesworth, Dorothy. *Natural Partnerships.* Doubleday and Co., Garden City (NY), 1969.

Tumlinson, James H., W. Joe Lewis, and Louise E. M. Vet. "How Parasitic Wasps Find Their Hosts." *Scientific American,* March 1993, p. 46. The wasps find their hosts through kairomones from the plants on which they feed. The information might be useful in developing pesticide-free pest control.

Wickler, Wolfgang. *Mimicry in Plants and Animals.* McGraw-Hill Book Co., New York, 1968.

Internet Resource

To further explore the content of this chapter, log on to the web site at:

http://www.mhhe.com/biosci/genbio/guttman/

The Dynamics of Ecosystems

28

Key Concepts

A. The Cycling of Materials Through Ecosystems

28.1 Plants use only a small percentage of the available light energy for growth.

28.2 Higher trophic levels also have low conversion efficiencies.

28.3 Components of an ecosystem are always turning over, sometimes rapidly.

28.4 Water continuously cycles through the biosphere.

28.5 Organisms create cycles of elements through the biosphere and through inorganic compartments.

28.6 A detritus food web is a major part of every ecosystem.

28.7 A terrestrial ecosystem continuously loses nutrients to its surroundings.

28.8 Pollutants can accumulate in communities, with potentially disastrous results.

B. Ecological Succession

28.9 Communities tend to replace one another in successions.

28.10 Seral stages show definite trends in structure and productivity.

28.11 Communities tend to become more diverse in successions.

28.12 More diverse communities appear to be more stable.

28.13 Primary and secondary successions begin from different conditions.

28.14 The species in a community may continually replace one another.

An elephant in an African acacia forest.

The viral disease rinderpest found its way into Africa in 1895, with devastating effects on the typical animals of the grasslands. Ninety-five percent of the ungulates (cattle, gazelles, antelope) died. Since these animals had been grazing on the grasses and browsing on the trees and bushes of the area, these plants increased in number dramatically. Lions that had been feeding on the ungulates found food scarce, and in their hunger many became man-eaters, thus reducing the human population. The native humans had been preventing the growth of trees through controlled burning. With fewer people, fewer fires were set. The combination of reduced browsing and fewer fires allowed trees to spread from their formerly restricted areas, and by the early 1900s what had been an open savanna was a dense woodland. By 1920, however, the surviving wild ungulates had become resistant to rinderpest, and vaccines had been developed to protect domestic cattle. In the closed woodland, fewer tree seeds sprouted, and in combination with the increasing herds of ungulates, the woodland opened up again, creating the savanna ecosystems we now picture—open grasslands dominated by flat-topped acacia trees. None of these trees, however, are older than about 80 years, and they are now dying of old age, posing a problem for managers of the parklands: Tourists expect to see typical savannas with their umbrella-like trees, so to satisfy them, the managers must now try to maintain an ecosystem that is really in transition and not "natural." But what is natural? The woodland that would spread if there were no humans and no fires? A balance between woodland and veldt? Is periodic disease natural? How do humans decide what is natural in a world that is constantly changing?

In Chapter 27, we examined the general structure of communities and the interactions between their component species. The community of an ecosystem can hardly be separated from its abiotic (physical) part, but in this chapter we will focus more on the interactions of the community with its environment and on long-term changes in communities. We begin by revisiting a point that has already been informing our discussions: the structure of a food web and the flow of energy and matter through an ecosystem as organisms consume one another. ■

A. The cycling of materials through ecosystems

All ecosystems engage in enormous chemical transformations, the sum of the metabolic activities of their component organisms plus chemical reactions in the environment. These transformations are driven by a one-way flow of *energy* into the ecosystem and out again in the form of heat. *Matter,* by contrast, moves through ecosystems (and the biosphere as a whole) in vast cycles, as water moves through various compartments and as carbon and other elements are converted from one compound to another. The ecosystem is structured into a food web; energy and matter enter, become part of the organisms in the ecosystem temporarily, and are then lost or recycled. The ecosystem is also transformed by all this activity, and it slowly changes: Species replace one another as one ecological community replaces another in a succession. These linear flows and cycles are basic to an understanding of ecosystem structure.

28.1 Plants use only a small percentage of the available light energy for growth.

Ecologists sometimes refer to the biomass of an ecosystem as a "standing crop," and community members do treat one another like crops, since those at each trophic level harvest those in the level just below (Figure 28.1). With few exceptions, ecosystems obtain their energy as phototrophs transform sunlight into chemical energy while converting CO_2 and water into the organic molecules of their own structure. Matter moves through the community from one trophic level to another as energy is lost through metabolism. The detritivores and decomposers of the community serve as food for small animals, so the ecosystem gets its energy both from primary production fueled by light energy and from the recycling of dead organic matter.

Whenever energy is transformed, it is important to ask about the efficiency of conversion from one state to another. How efficiently is light energy converted into the chemical energy of phototrophs? There are two distinct answers: The **gross primary productivity** is the amount of incoming light energy captured in photosynthesis, but phototrophs use some of this energy for their own maintenance, leaving a smaller amount as **net primary productivity,** the amount of energy actually incorporated into the biomass of the phototrophs.

Studies on a natural field ecosystem (an old farm field now growing wild) and on a cornfield have shown that plants have a gross primary productivity of only 1.3–1.6 percent, so they only convert a small amount of sunlight into chemical energy. Plants

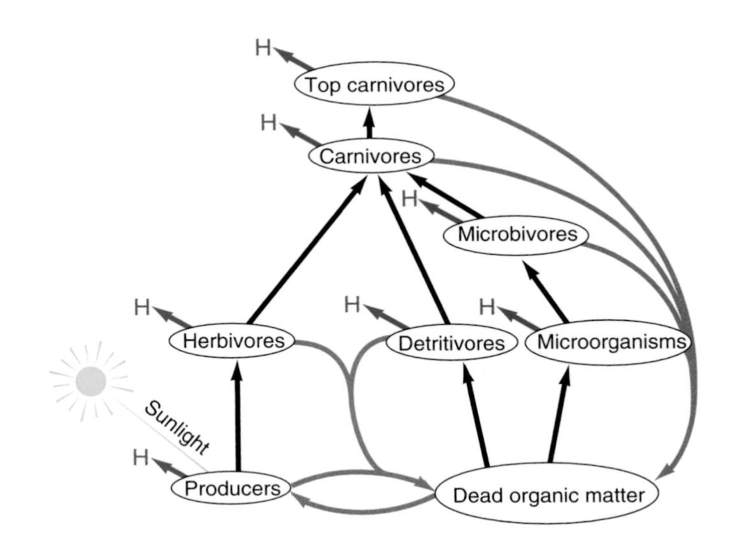

Figure 28.1

Energy and matter both move through a food web to higher trophic levels. Energy transformation is ultimately linear, ending with the conversion to heat energy (H). Matter, however, is continually recycled; the food web rests both on production by photosynthesis and recycling of wastes by detritivores and decomposers (microorganisms). Recycled organic matter carrying chemical energy may pass through the system several times before its energy is finally dissipated as heat.

Table 28.1 Net Primary Productivity and Biomasses of Ecosystems

Ecosystem	Area (10^6 km^2)	Mean Biomass (kg/m^2)	Net Primary Productivity (g/m^2/yr)
Tropical rain forest	17.0	45	2,200
Tropical seasonal forest	7.5	35	1,600
Temperate evergreen forest	5.0	35	1,300
Temperate deciduous forest	7.0	30	1,200
Boreal forest	12.0	20	800
Woodland and shrubland	8.5	6	700
Temperate grassland	9.0	1.6	600
Savanna	15.0	4	900
Tundra and alpine	8.0	0.6	140
Desert and semiarid scrub	18.0	0.7	90
Cultivated land	14.0	0.7	650
Rock, sand, and ice	24.0	0.02	3
Swamp and marsh	2.0	15	2,000
Lake and stream	2.0	0.02	250
Total continental	***149***	***12.3***	***773***
Open ocean	332.0	0.003	125
Upwelling zones	0.4	0.02	500
Continental shelf	26.6	0.01	360
Algal beds and reefs	0.6	2	2,500
Estuaries	1.4	1	1,500
Total marine	***361***	***0.01***	***152***
Total	***510***	***3.6***	***333***

use about a quarter of the captured light energy for their own maintenance, leaving only 1.0–1.2 percent as net productivity. Additionally, about 40 percent of the incident solar energy is used for *transpiration,* the movement of water through the plant and its loss via evaporation; this energy does not count as productivity, but is essential for the plant. Similar measurements on aquatic ecosystems have given gross primary productivities of 0.1–0.4 percent. However, water reflects and absorbs a substantial amount of light, and when this is taken into account, the productivity rises to about 1–2 percent of the light that actually enters the water and is available to the plants.

These productivity figures can be used to estimate the rate of growth—that is, the rate at which biomass is produced, starting with the data about incident radiation on the earth from Section 25.1. Follow the calculations through the following exercises.

Exercise 28.1 If we take 0.05 percent of the sunlight incident on the earth as the amount used in net primary productivity, this amounts to about 10^{14} watts (W). 1 W = 1 joule (J) per second, and there are 3.15×10^7 sec/yr. How many joules of energy are used in net primary production per year?

Exercise 28.2 One gram of biomass contains about 1.6×10^4 J. How many grams of biomass are formed each year by 3.15×10^{21} J?

Estimates from other sources give comparable values, on the order of 10^{17} grams per year. Table 28.1, which summarizes data for the world's ecosystems separately, gives about 1.7×10^{17} grams/yr. It is important to understand that there is a good deal of uncertainty in data of this kind regarding the earth, and any of these numbers could be off by a factor of 2 or so, but it is satisfying to see that different methods of calculating give similar numbers.

Actual productivities of different ecosystems vary enormously. Tropical rain forests and some other rich areas have net productivities of about 35,000 kJ/meter^2/year, compared with only 1,200 kJ/meter^2/year in some deserts.

Exercise 28.3 Humans are only one of millions of species on the earth, but it is interesting to see how much energy we consume just to stay alive. About 5.5×10^9 people now live on the planet. On the average, each one uses about 2.5×10^6 calories per day. 1 cal = 4.184 J. How much of the earth's net primary productivity do we use?

28.2 Higher trophic levels also have low conversion efficiencies.

Just as phototrophs use some of the energy captured in photosynthesis for their own maintenance, heterotrophs use much of their food for respiration. The energy obtained by oxidizing this material to CO_2 and water powers their activities and then goes back into the universe as heat. Furthermore, some of an animal's food passes through its intestine without being used. Thus relatively little biomass and energy actually get transferred to each higher trophic level as secondary productivity, tertiary productivity, and so on. Birds and mammals, being homeotherms ("warm-blooded"), convert only about 1–3 percent of their food into biomass because they respire so much of it for heat to keep their temperatures constant. Poikilotherms ("cold-blooded" animals) have higher efficiencies—typically about 10 percent for fish, 25–35 percent for some invertebrates, and even 50–60 percent for some insects. Bacteria grown on glucose convert about 30 percent of it into their own biomass.

The overall efficiency of conversion from one trophic level to another is shown dramatically by placing the biomasses at each trophic level above one another to form a pyramid (Figure 28.2). As expected, an ecosystem usually has a large mass of phototrophs, a smaller mass of herbivores living on them, and a still smaller mass of carnivores living on the herbivores. Studies of productivity in natural ecosystems show that, as a rule of thumb, biomass decreases by an order of magnitude (a factor of 10) at each trophic level. A 1942 study of a Minnesota lake showed that the biomass of herbivores was 13.5 percent of the autotroph biomass and that the carnivores converted 20 percent of the herbivores into their own biomass (Figure 28.3). Notice that the carnivores used 60 percent of their food intake for respiration (1.8/3), compared with only 30 percent for the herbivores (4.5/15). After all, carnivores have to expend much more of their energy to get food; chasing a rabbit takes more energy than grazing on grass.

Note, however, that the pyramid drawn for plankton in the oceans looks inverted, the zooplankton outweighing the phytoplankton on which it feeds. How is this possible? The answer is that the pyramid reflects only the mass of material at any instant, but not the rates of conversion between trophic levels. The inverted pyramid of plankton makes sense if the zooplankton consumes phytoplankton relatively quickly but is itself only consumed by higher carnivores relatively slowly. So it is more meaningful to draw a pyramid using the rate of energy input into each level, and such a diagram must be a real pyramid showing a decrease from level to level.

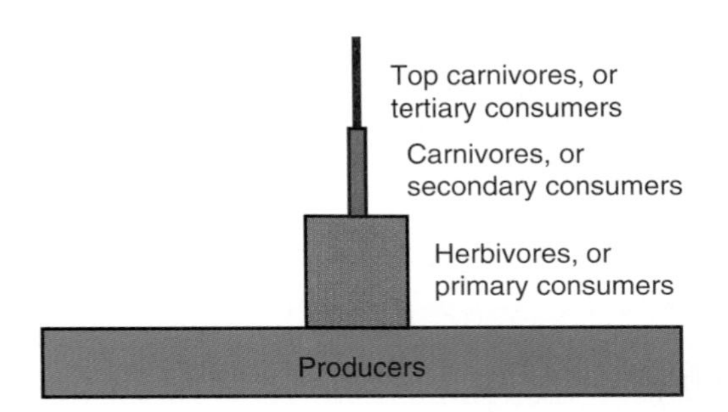

Figure 28.2
Each rectangle is proportional to the biomass or energy content at one trophic level. The resulting pyramids show how much mass and energy is lost with each trophic transformation.

Exercise 28.4 A plot of land is used to grow grass to support sheep, which will then support 100 people. Approximately how many people could be supported if the land were used, instead, to raise wheat or other grains for the people to consume directly?

28.3 Components of an ecosystem are always turning over, sometimes rapidly.

Organisms are dynamic. They continuously lose old molecules from their structures and replace them with new ones, sometimes very rapidly. This process is called **turnover,** and the materials themselves are said to be turning over, just as the inventory in a store or the patients in a hospital "turn over." It is important to see that the ecosystem *as a whole* is also turning over, since the dynamic condition of individual organisms extends to the whole system. Metabolism doesn't stop at the boundary of each organism, so it is sensible to talk about the metabolism of a whole ecosystem.

Let's first establish a way to think about turnover. Suppose a college has 10,000 students and everyone graduates in 4 years. The 10,000 students are the college's *store,* and their **residence time** is 4 years. It is then evident that to maintain a constant store, the college must graduate and admit an equal number of students per year:

$$\frac{10{,}000 \text{ students}}{4 \text{ years}} = 2{,}500 \text{ students/year}$$

This is the college's **turnover rate.**

Complex systems generally have different *compartments,* or subsystems, and materials flow from one compartment to

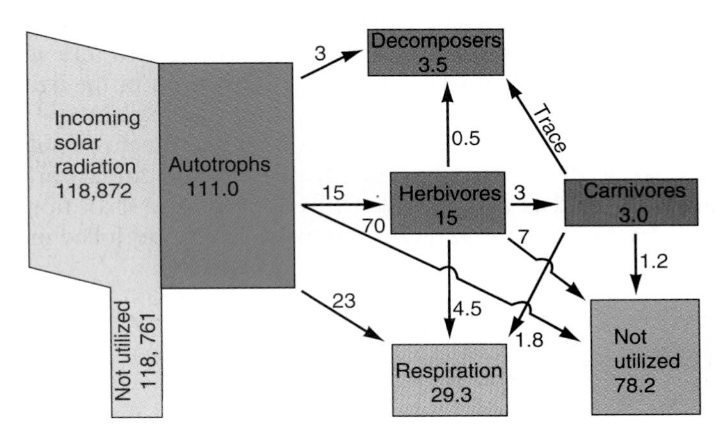

Figure 28.3
Cedar Bog Lake, Minnesota, supports a community like that shown in Figure 25.16 in which the energy flow from one trophic level to another has been measured. All numbers are calories per square centimeter per year. The energy in each trophic level is proportional to its biomass.

another. The compartments in a forest ecosystem may consist of the soil, the water washing through the soil, the long-lasting wood of the trees, and the short-lived leaves, flowers, and fruit, as well as comparable compartments in all the animals and microorganisms. In analyzing complex physical systems, we must deal with *average* residence times and turnover rates, since molecules move in and out of any compartment at random and at different individual rates.

Exercise 28.5 A record store with a stock of 40,000 compact discs sells an average of 250 discs per day. What is the average residence time of a disc in the store?

Many studies of turnover in ecosystems support two conclusions: Turnover is faster in summer than in winter and faster in the water than on land. In the early 1960s, F. H. Rigler studied the movement of phosphate, which cycles between biological structure and various soluble inorganic and organic compartments in the environment. Rigler used the radioisotope ^{32}P as a tracer for phosphate metabolism in several lakes in Ontario. He found that during the summer, when metabolism is rapid, the residence times of dissolved organic and inorganic phosphates in these lakes are only a few minutes, but during the winter, their residence times increased to 10^3–10^4 minutes in a typical lake. (A day is 1,440 minutes.) Similar studies by Lawrence Pomeroy indicate that phosphate in a salt-marsh creek and a river has a residence time of about one hour during the summer and about ten hours during the winter. In open coastal and sea waters, residence times are much longer, ranging from about 4 to over 50 hours in the summer, and much longer in winter.

The components of terrestrial ecosystems also turn over, but with residence times of years rather than hours. In certain forest ecosystems, calcium in the canopy (leaves, fruit, and flowers) turns over in about 0.4–1.5 years, while calcium in the wood turns over in about 6–20 years. Nutrients taken up from the soil largely accumulate under the trees as a ground litter of leaves, seeds, branches, and other debris, which turns over at a range of rates in different systems (Figure 28.4). In tropical forests, where very little material accumulates, calcium in the litter may turn over in 0.2 years, while in some northern forests, it turns over in 10–35 years. However, even though the exchange may be slow, material continues to move from one compartment of the system to another. Studies focusing on five nutrients in forests—nitrogen, calcium, potassium, magnesium, and phosphorus—have shown that ground flora (nonwoody plants) return as litter essentially all of the nutrients they take up from the soil; trees return 65–80 percent as litter and incorporate the rest into wood, where it still turns over slowly. Decomposition of the litter eventually releases virtually all its nutrients into the soil.

Using the numbers derived earlier, we can calculate some turnover times and residence times for ecosystems and for the biosphere as a whole.

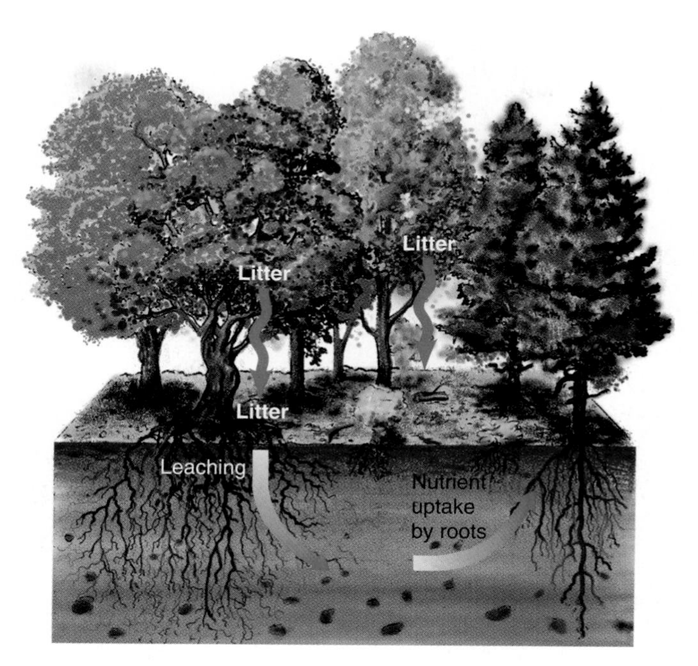

Figure 28.4
Nutrients cycle from the soil into vegetation. The litter that falls to the ground then decays; some nutrients leach into the soil, but they can all be recycled through the roots back into the plants.

Example 28.1 If the mass of the biosphere is 1.8×10^{12} metric tons and if 0.17×10^{12} metric tons of new material is synthesized per year, what is the average residence time of a molecule in the biosphere? Answer: 1.8×10^{12} tons divided by 0.17×10^{12} tons per year = 10.6 years.

Exercise 28.6 Since the biosphere is so diverse, a number for the residence time in the biosphere as a whole doesn't mean much, but from the data in Table 28.1 you can do similar calculations for each ecosystem. Calculate average residence times for molecules in a tropical rain forest, a temperate deciduous forest, and a desert.

28.4 Water continuously cycles through the biosphere.

Water is a major component of the earth and forms most of the mass of all organisms. The global interactions of water with the rest of the earth and with the organisms that live on it strongly influence the world's biological communities. The global flow of water called the **hydrologic cycle** (Figure 28.5) involves such enormous amounts of material that one can hardly comprehend the numbers. For example, about 4.2×10^{17} kilograms of water falls as precipitation over the earth's surface each year. For comparison, the reservoir of Grand Coulee dam on the Columbia River has a capacity of about 440 billion cubic feet, or 12 trillion kilograms; it would have to be filled about 35,000 times over to equal 4.2×10^{17} kilograms. Water flows over Niagara Falls at the rate of 1.7×10^{14} kg/year, so it would take about

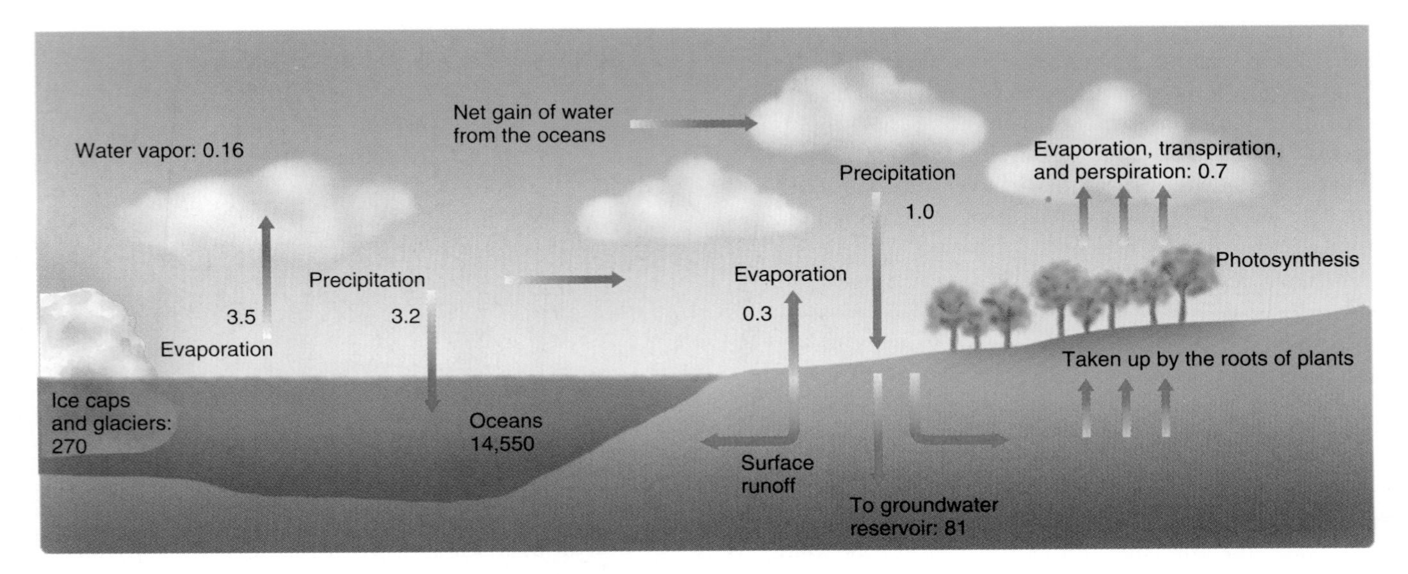

Figure 28.5
The hydrologic cycle entails enormous movements of water. The reservoirs here are in units of 10^{17} kilograms, and the flows are in units of 10^{17} kilograms per year.

2,470 identical rivers to deliver 4.2×10^{17} kg/year. But it is a mistake to think we can only use such numbers if we can picture them or comprehend them emotionally. We can use them to make practical calculations about transformations on the earth, especially the environmental effects of human activities, and that is all that is necessary.

Though about 95 percent of the earth's water is bound up chemically in rocks, the 5 percent that is free to circulate through the hydrologic cycle is 1.5 billion cubic kilometers. Ninety-seven percent of this circulating water is in the oceans, and most of the remainder is bound up for relatively long times in glaciers, including the polar ice caps.

A lifeless earth would still have a hydrologic cycle consisting of evaporation from the planet's surface and an equal amount of precipitation. However, much of the earth is covered by organisms that transpire enormous amounts of water, and the hydrologic cycle is driven largely by **evapotranspiration,** the combination of evaporation and transpiration. The resulting water vapor in the atmosphere eventually precipitates. While there is less precipitation than evaporation over the oceans, this is compensated for by more precipitation over the land, with a continuous runoff from streams and rivers back into the oceans. Of course, the whole cycle is powered by the sun, since none of this could happen without the sun's energy, which sustains plants, causes evaporation, and drives global air movements.

For a terrestrial ecosystem, the most significant part of the hydrologic cycle is precipitation and the percolation of water down through the soil, through the roots of plants, and back into the atmosphere through transpiration. This flow keeps plants alive and growing. In fact, even though plants divert some water into their own structure, a far greater amount just cycles through them (Figure 28.6). To produce a metric ton (1,000 kg) of fresh vegetation (a food crop, for instance), about

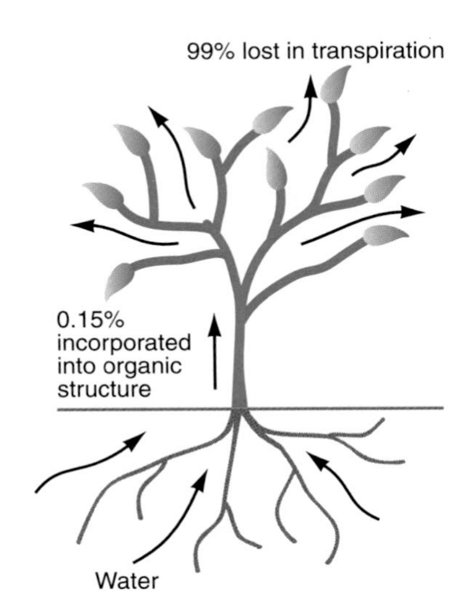

Figure 28.6
Most of the water that enters a plant doesn't become part of its structure; it simply flows through the plant and is lost by transpiration.

100 metric tons of water must move through the growing plants in a season. The resulting vegetation is about 75 percent water, and about three-fifths of the remaining 25 percent consists of atoms that were once in water and are now converted into organic materials. This means only about 0.15 percent of the water flowing through plants is converted into molecules of their structure.

The animals in an ecosystem participate in the hydrologic cycle as they eat plants, drink water, perspire, urinate, and otherwise exchange water with their surroundings. These activities are physiologically important, but on a global scale they are trivial compared with the enormous flows in other parts of the cycle.

As shown in Section 25.9, water is also a primary agent in soil formation. It is one of the major agents of weathering, the process that breaks down large rock masses into fragments, and surface water seeping through the soil transforms soil chemically.

28.5 Organisms create cycles of elements through the biosphere and through inorganic compartments.

Ecosystems are open, and matter is always flowing through them. Each ecosystem exchanges material with the biosphere as a whole and with inorganic compartments, such as the atmosphere, oceans, and bedrock. This creates a series of **biogeochemical cycles.** The *carbon cycle* outlined in Section 9.1 is just the most obvious one; also important are the nitrogen and sulfur cycles.

The nitrogen cycle

After the carbon cycle, the most significant elemental cycle in the biosphere is the **nitrogen cycle** (Figure 28.7), since nitrogen is a major component of biomolecules (proteins and nucleic acids). The atmosphere is about four-fifths nitrogen gas (N_2), and vast amounts of nitrogen reside in biological compartments, moving through them and the surrounding soil and water.

Nitrogen enters food webs through **nitrogen fixation,** the conversion of free N_2 to ammonia, NH_3. Most nitrogen is fixed by symbiotic associations of the roots of certain plants with rhizobia—bacteria of the genus *Rhizobium*—or with a curious organism named *Frankia* that is probably an actinomycete. Nitrogen is also fixed by some free-living bacteria in soil and water, especially *Azotobacter, Clostridium,* and cyanobacteria. Many other bacteria conduct the opposite process, **denitrification,** by converting nitrate (NO_3^-) and nitrite (NO_2^-) into atmospheric nitrogen.

Rhizobial associations and nitrogen fixation, Section 40.8.

Many bacteria interconvert nitrogen compounds to obtain energy. Some that live in anaerobic environments (without oxygen) carry out a nitrate respiration by reducing nitrate to nitrite. *Nitrosomonas* is a chemoautotroph that lives by oxidizing ammonia to nitrite, and *Nitrobacter* gets its energy by further oxidizing nitrite to nitrate.

The limited information now available indicates that the nitrogen cycle as a whole is in balance. However, recent human activity is adding major components. In addition to the many nitrogenous pollutants that humans add to the world unintentionally, the industrial fixation of nitrogen to make ammonia and urea, mostly for fertilizers, adds about 27 million metric tons of nitrogen per year to the soil. This is a large factor compared to the rest of the nitrogen-cycle components, and no one knows yet just what long-range effects it may have on the whole cycle.

The sulfur cycle

Another significant elemental cycle is the **sulfur cycle.** As a minor component of biomolecules, especially proteins, sulfur cycles through the biosphere via food webs. Atmospheric sulfur is primarily H_2S, which comes from three sources: sea-spray aerosols (44 million metric tons per year), respiration by sulfate-reducing bacteria (estimated at between 33 and 230 million metric tons per year), and volcanic activity (a minor contribution). Modern industrial pollutants now add substantially to atmospheric sulfur. The biologically interesting transformations of sulfur are oxidations and reductions carried out mostly by bacteria as part of their energy metabolism (Figure 28.8). Many bacteria grow heterotrophically while reducing sulfates and sulfur to sulfides under anaerobic conditions. Bacteria such as *Desulfovibrio* get their energy through an unusual form of respiration in which they too reduce sulfate to sulfide.

Accumulated sulfides inhibit the growth of most organisms. *Beggiatoa* and some other bacteria can oxidize sulfide back to sulfur, but this process requires free oxygen, which is unavailable in the anaerobic environments where most sulfides accumulate. Still, if enough light is available to support the growth of phototrophic bacteria, they can oxidize sulfide and sulfur back to sulfate and eliminate sulfide from anaerobic environments. These bacteria grow only in the photic zone of a body of water, close enough to the surface for light to penetrate but far enough below the surface to be anaerobic. In mud, where these phototrophs commonly grow, that zone is only a few millimeters deep. Although oxygen and photosynthesis can generally clear sulfide from the surface regions of a lake, it tends to accumulate deeper in the water, which accounts for the sterile depths of bodies such as the Black Sea.

A lot of sulfur is locked in rock deposits, especially inorganic sediments. Iron and calcium form insoluble sulfide and sulfate minerals, which release sulfur only by slow weathering. These sediments may also tie up metals such as copper, zinc, and cobalt, making them inaccessible to organisms. On the other hand, sulfides and sulfates may displace phosphates from mineral deposits, and since the growth of so many organisms is limited by the available phosphate, raising its concentration may enhance their growth enormously. We noted in Chapter 25 that lakes may become eutrophic when phosphates from human activities promote a prolific growth of algae. The sulfur cycle therefore has complicated effects on other nutrient cycles.

Today, burning sulfur-bearing fossil fuels (coal and oil) dumps large amounts of sulfur dioxide (SO_2) into the atmosphere. Some plants can take in SO_2 directly through their leaves, so in this way pollution does add some sulfur to ecosystems. However, a lot of atmospheric SO_2, in a light-driven reaction with oxygen, forms sulfur trioxide (SO_3), and SO_3 combines with water to make sulfuric acid, H_2SO_4:

$$SO_2 + \tfrac{1}{2}O_2 \rightarrow SO_3$$
$$SO_3 + H_2O \rightarrow H_2SO_4 \rightarrow H^+ + HSO_4^-$$

The resulting sulfuric acid makes a much more acid rainfall (pH 5 or less) than ordinary rain containing carbonic acid (pH about 5.6). When the pH of rain falls to the range of 2.1 – 2.8,

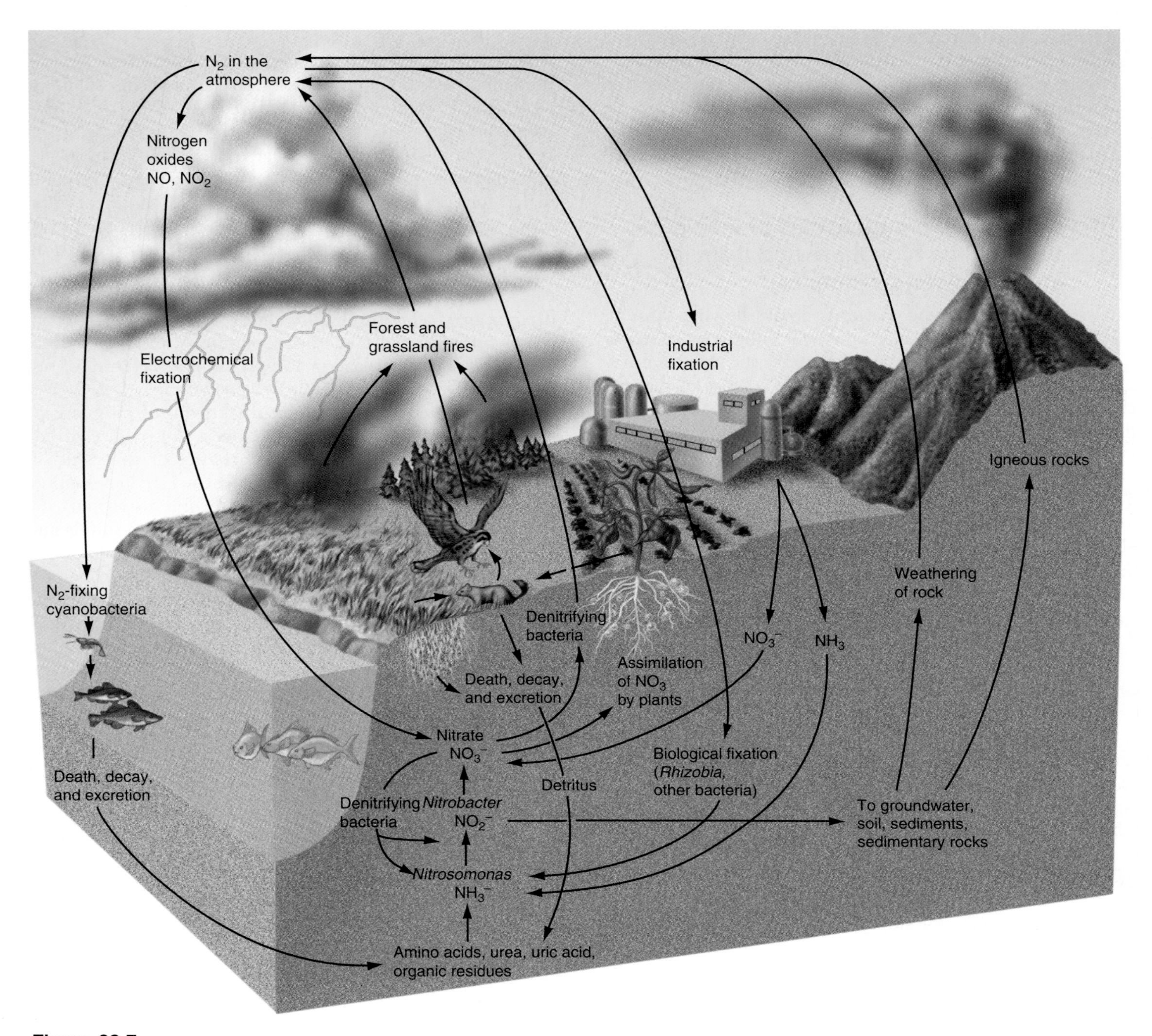

Figure 28.7

In the nitrogen cycle, atmospheric nitrogen (N_2) is converted to various nitrogen compounds in the soil; these nitrogen atoms are incorporated into plants and move through the food web. Other processes covert nitrogen compounds back to N_2.

as it has in both Europe and the United States, ecosystems are being soaked in only slightly diluted sulfuric acid, which breaks down the cell membranes of plants, leaches nutrients from the soil, and otherwise makes nutrients unavailable for plant metabolism. Acid rain is highly damaging to terrestrial and aquatic ecosystems. This form of pollution has become an especially serious concern in the northeastern United States and southeastern Canada, which get the bulk of the sulfur pollutants from the heavily industrialized Ohio River valley, and for some regions of Europe, where poorly controlled industries have polluted large areas and killed extensive forests.

28.6 A detritus food web is a major part of every ecosystem.

Some of the biomass of living organisms in an ecosystem is always being converted into **necromass** that takes the form of detritus—dead organic matter (Figure 28.9). Plants die and shed

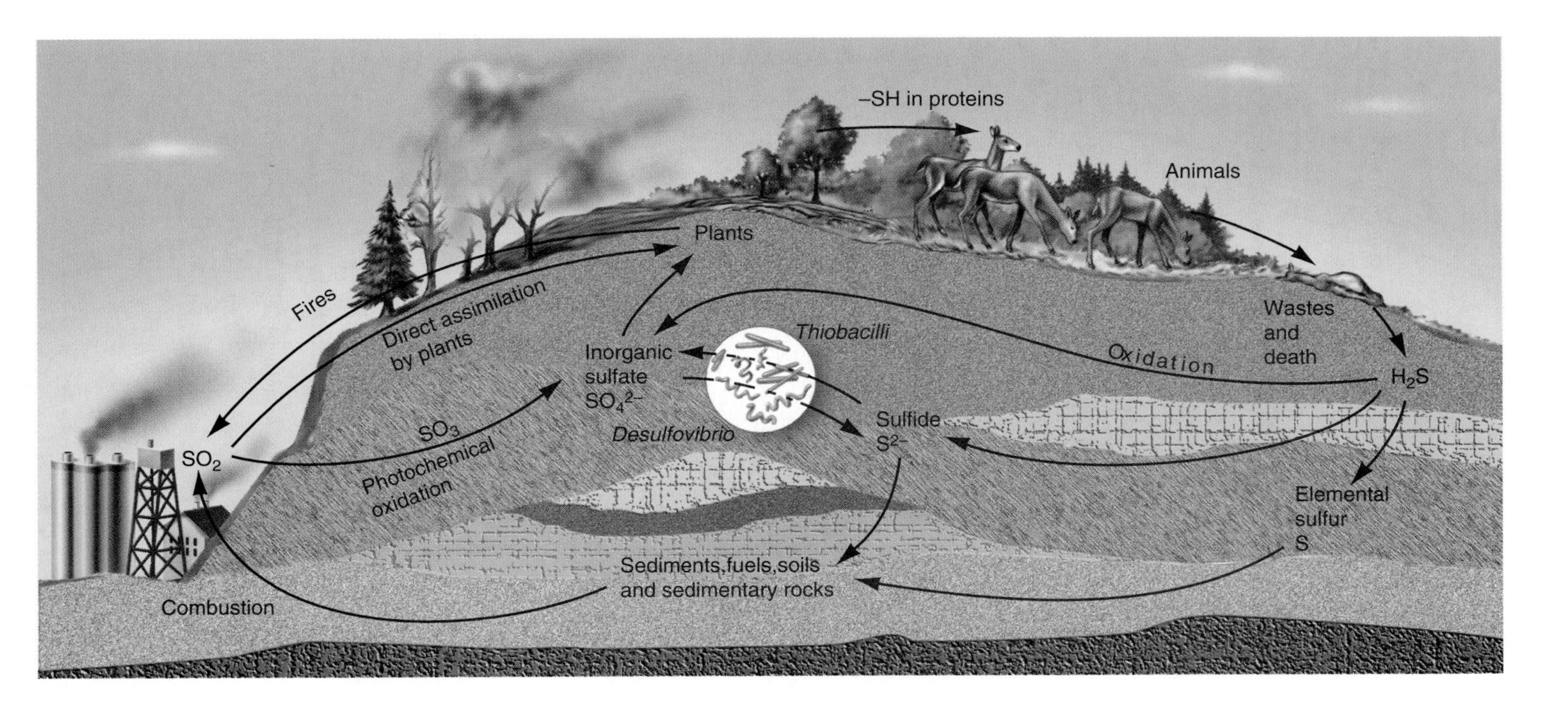

Figure 28.8
The sulfur cycle involves transformations similar to those in the nitrogen cycle, including several conversions by soil bacteria.

Figure 28.9
The biomass of every ecosystem is continually converted to necromass as the organisms of the community deposit their wastes and die.

their parts. Animals drop their feces, shed their skin, scales, hair, and feathers, and eventually die. All this material accumulates as detritus. Unfortunately, since detritus disappears from sight, we sometimes ignore its importance in the economy of an ecosystem. That is a great mistake. In fact, the food web of a community stands on two bases: live plant material and detritus. Detritus contains enough matter and energy to support a substantial food web, which recycles the necromass back into the community. Part of the necromass is consumed by a microflora, consisting of bacterial and fungal decomposers, and by a fauna composed of many kinds of animals, some specialized for eating detritus (detritivores). The community of animals supported by detritus includes some that then eat the decomposers and smaller detritivores (Figure 28.10). By tunneling through the detritus and shredding it, many animals make larger surface areas where bacteria and fungi can grow; by breaking into plant cells, they make the cellular contents available. Perhaps their most important effect is to mix pieces of detritus with their intestinal flora (microorganisms living in their guts), so even undigested material that comes out in their feces has been subjected to microbial activity. Many members of the detritivore community rely on mutualisms, such as microorganisms living in their intestines where they digest plant components that the animals themselves cannot digest. Some of the most fascinating stories of ecology have been told about the decomposers and detritivores.

The bulk of detritus is plant material. In spite of the many organisms that can attack it, plant detritus disappears more slowly than animal material because it is locked in plant cells with tough walls that are hard to digest. A bit of plant detritus becomes a small ecosystem that is colonized in turn by a succession

Figure 28.10
Many kinds of animals act as detritivores. They are classified into a few ecological types by their size, because the width of an animal's mouth determines the size of material it eats.

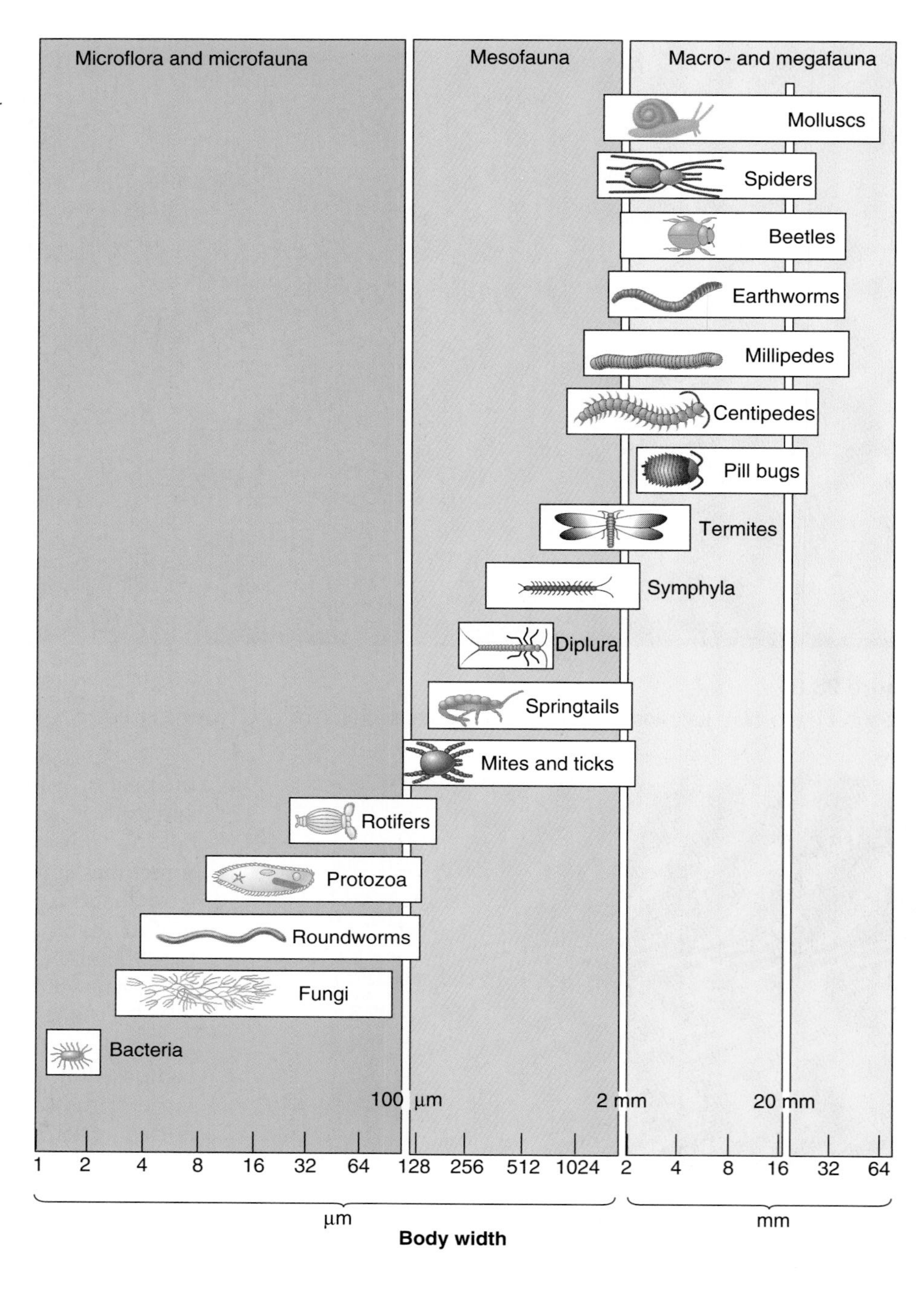

of opportunistic bacteria and fungi, each specialized for digesting different materials. (Ecological successions in general are discussed in Section 28.9.) Some bacteria and yeasts immediately use the readily available sugars and starch. They are followed by fungi that can digest somewhat more resistant plant materials, such as gums, and then by other fungi that attack cellulose, lignin, and other highly resistant compounds.

If you walk around woods and fields, you'll seldom see a dead animal. Or, if you do see a small carcass and then come back a few hours later, it will often be gone. The ecosystem is very efficiently "sanitized" by a crew of scavengers that range from vultures, ravens, and hyenas to specialized insects. Figure 28.11 shows the fate of a dead mouse found by *Nicrophorus* beetles; within minutes, they have buried the body, removed its hair, and transformed it into a little ball of flesh hardly recognizable as an animal.

Even though humans may flinch from the idea, feces are rich in nutrients and constitute an important resource for a number of organisms. They vary in nutritional quality, however. The fecal material of carnivores is relatively poor, containing only the least digestible substances, and there isn't much of it; it is probably decayed by the microflora. In contrast, the dung of herbivores is abundant and quite rich, so specialized insects, such as dung beetles, can use it to support their larvae. A herd of elephants or other herbivores, for instance, drops massive fecal patties, and during the rainy season in Africa, swarms of dung

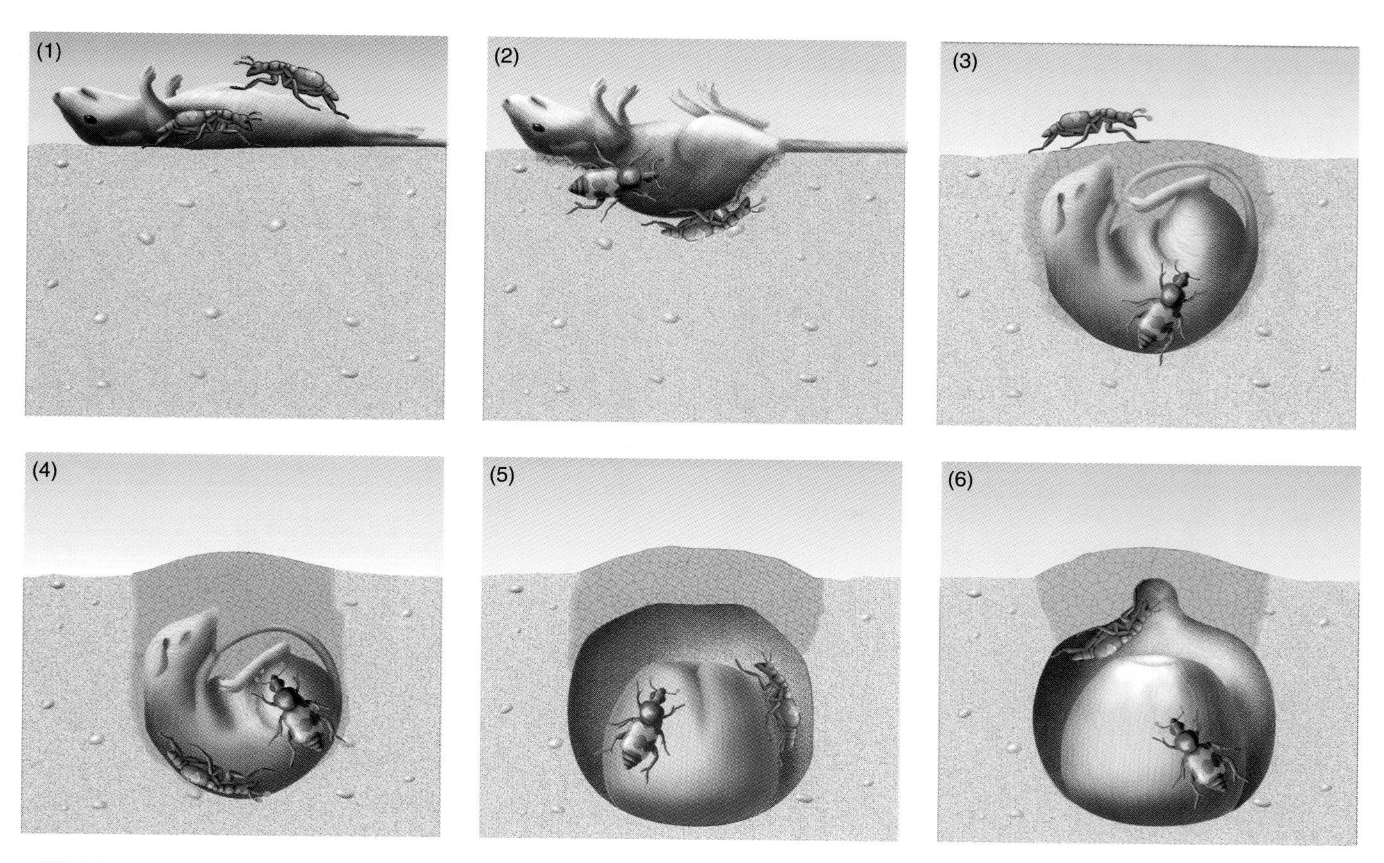

Figure 28.11

A pair of *Nicrophorus* beetles buries a mouse carcass by gradually excavating the soil under it. When the mouse is covered with soil, the beetles remove its hair and sometimes bits of flesh, converting it into a compact ball of flesh. Then they copulate, and the female lays eggs nearby. The parents (sometimes only the female) stay in the chamber until the eggs develop into larvae. The parents form a conical depression on top of the flesh-ball where they regurgitate bits of digested meat; they call the larvae to the feeding place by stridulation (making a noise by rubbing their legs together) and continue to care for the offspring until they pupate, after which the parents dig through the soil and fly away. Source: Milne and Milne, "Social Behavior of Burying Beetles" in *Scientific American,* 1976.

beetles descend on these feces. Each beetle tears off a wad, shapes it into a ball, and rolls it away. The beetle lays one egg in each ball, where its larva feeds on the dung and eventually emerges at maturity (Figure 28.12). To the ancient Egyptians, the image of a scarab beetle rolling its huge ball of dung suggested a god pushing the sun across the skies, so the beetle-headed god Kheper became the personification of the rising sun (Figure 28.13). The new beetle emerging from the dung also made the scarab a symbol of resurrection, of life emerging from death.

Some animals can profitably eat their own feces or those of their neighbors, recycling the material to get its nutrients. One example is the millipede *Apheloria montana,* which lives on leaf litter and thrives by eating its own feces, apparently because the action of its gut flora makes nutrients available that it could not get from the original material. A. J. Maclachlin and his coworkers uncovered a complex story of interactions among some detritivores in an English bog. The larvae of a small fly, *Chironomus lugubris,* eat particles of peat that have been made more nutritious by bacteria growing on them. The flies' feces are also enriched by the growth of fungi and other bacteria, but they are

Figure 28.12

A dung beetle rolls its ball of dung. A beetle larva will feed on this material and then emerge as an adult beetle.

Figure 28.13

An Egyptian wall painting in the tomb of Ramses VI shows many human figures around the scarab-beetle god Kheper, who is rolling the sun above him as a beetle rolls its ball of dung.

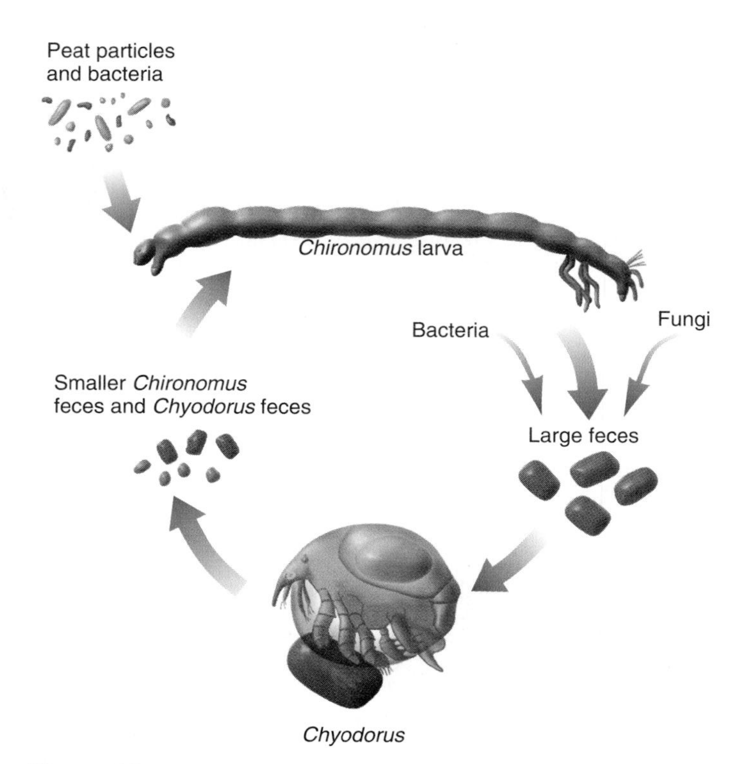

Figure 28.14

Larvae of the fly *Chironomus lugubris* eat particles of peat. Their feces, enriched by the growth of microorganisms, are suitable food for the cladoceran *Chyodorus sphaericus.* These partially eaten feces, mixed with those of *Chyodorus,* are then eaten by *Chironomus* larvae.

too large and tough for the *Chironomus* larvae to consume themselves. However, a cladoceran (*Chyodorus sphaericus,* a small crustacean), seems to depend upon these feces for its food. *Chyodorus* grasps a chironomid pellet and rotates it while grazing on it, gradually breaking it down to smaller particles. These particles, mixed with chyodorid feces, then serve as food for *Chironomus,* so the two species form a kind of mutualistic association, and both benefit (Figure 28.14).

28.7 A terrestrial ecosystem continuously loses nutrients to its surroundings.

Since the materials in an ecosystem are in a dynamic state, they move through the system as well as recycling within it. A classic study of the flow of nutrients through one system was conducted in the Hubbard Brook Experimental Forest in New Hampshire, an area of several small watersheds with very tight bedrock, so nothing can escape by flowing down through the underlying rock. Herbert Bormann, Gene Likens, and their colleagues monitored the flow of water and measured the input and output of nutrients for several years. They found significant annual losses, ranging from about 9 kg/hectare for calcium to 0.6 g/hectare for potassium. (A hectare is about 2.5 acres.) They also noted gains; chloride accumulated slightly, and the accumulation of nitrogen reflected the growth of the forest.

Even salts in the air contribute to an ecosystem. Chloride and magnesium ions, for instance, can be carried in airborne droplets of seawater. However, most of these materials must come from the bedrock, and it was estimated that 800 kg of rock are weathered per hectare per year in the Hubbard Brook watersheds.

Although enormous masses of rock underlie any terrestrial ecosystem, this kind of weathering can't go on forever. Mountains and hills are wearing down gradually. Everything living on them merely participates in the geological flow, and a second Hubbard Brook study demonstrated how much the ecosystem slows down the outflow of nutrients. After a few years of observation, one of the watersheds was clear-cut, and the new growth of plants was inhibited with an herbicide. In this area, the total runoff of nutrients increased about eight-fold: The outflow of calcium ions increased about nine times, of potassium about 20 times, and of nitrate—which had been accumulating previously—about 50 times. In addition, these losses accompanied much faster runoff of water, since plants normally retard runoff and give the water time to percolate into the rich soil of the forest, which acts a reservoir. In an undisturbed forest, about 40 percent of the input water escapes by evapotranspiration; in a clear-cut forest, most of it runs off as surface water. This study indicates the importance of ecosystems in the overall geological changes on the planet, and it has great implications for forestry and other practices involving human interference with the environment. We noted in Section 25.5 how clear-cutting forests and filling wetlands, which both hold water like sponges, lead to floods that ravage lowlands.

28.8 Pollutants can accumulate in communities, with potentially disastrous results.

The American writer Rachel Carson shook a complacent society with her 1962 book *Silent Spring.* She described how, after World War II, the growth of the chemical industry supported industrial societies in what appeared on the surface to be an increasingly good life—a world of fast cars, new toys made of brightly colored plastics, and neat suburban homes with well-kept lawns and gardens free of insect pests. Carson woke people to the dark side of this life by showing that DDT and other insecticides were responsible for reproductive failure in many birds and for heavy mortality in fish and other animals. Her work eventually led to the banning of these insecticides.

DDT is one of the halogenated hydrocarbons whose molecules are carbon ring structures with chlorine or bromine atoms attached in various combinations. One class of these compounds, the polychlorinated biphenyls (PCBs), have industrial uses, such as for noninflammable coolants in electrical systems, and are often manufactured for use as potent pesticides. PCBs share several biologically important characteristics. They are extremely stable. Even in the open environment, exposed to sunlight and a variety of microorganisms, they last for long times; DDT, for instance, has a half-life of 10–15 years. They are very mobile, and are carried long distances by wind and water. Although these compounds are poorly soluble in water, the massive quantities of water moving through the biosphere carry large amounts into the environment throughout the world.

These compounds are lipid-soluble and accumulate in the lipid deposits of organisms, especially in animals. Although an animal may live in water that has an extremely low level of a PCB, the material builds up in its lipid deposits, a process called **bioaccumulation.** This is simply a matter of the partition coefficient of the compound (see Figure 8.8). Even more serious, though, is the process of **biomagnification** in which the compound accumulates to higher concentrations at higher trophic levels (Figure 28.15). As one animal eats another, the contaminants accumulated in the lipids of the prey are transferred to the lipids of the predator. Thus a material at nominally sublethal levels in organisms low in the food chain can be magnified to dangerous levels in organisms high in the food chain.

As we pointed out in Chapter 20, cancers are caused by environmental factors, and the increasing cancer rates worldwide in the past 50 years or so are undoubtedly a side effect of the increasing chemical pollution of the world environment. PCBs are not only potent pesticides but also strong carcinogens and teratogens, and the effects of their accumulation in human tissues are becoming obvious. The pollutants accumulated in lipid deposits are flushed out when those lipids are mobilized for some use, as for the synthesis of milk, which is rich in fats. Mother's milk is now so highly contaminated that women must seriously consider whether the other health benefits associated with nursing are worth the additional risk from pollutants. This problem is not confined to women in industrialized countries.

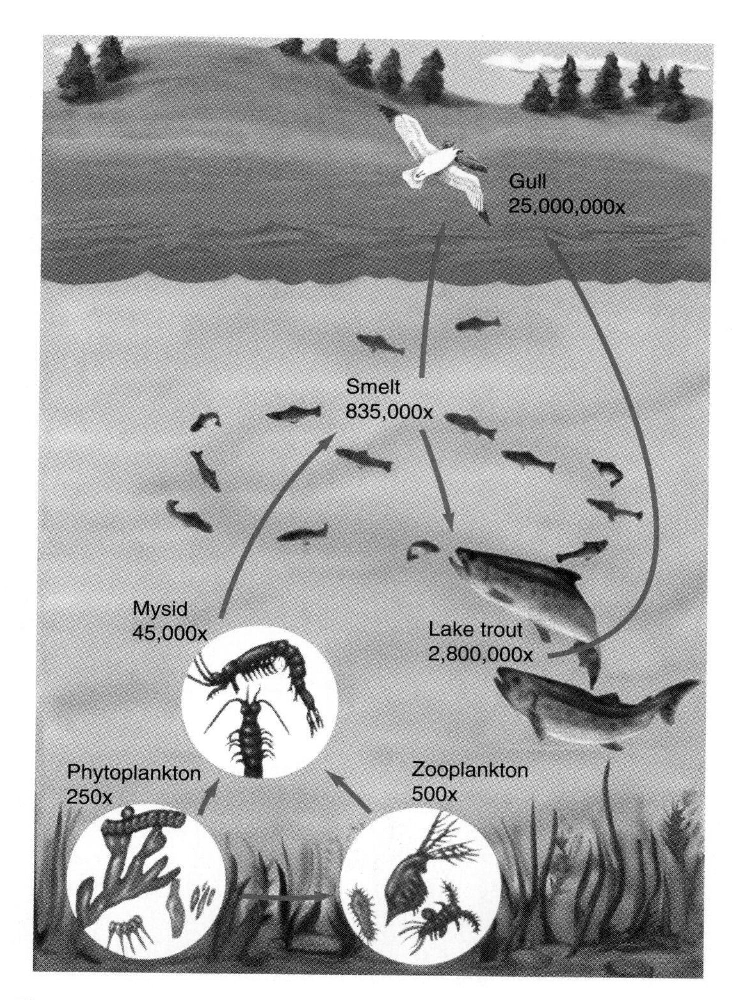

Figure 28.15
Polychlorinated biphenyls (PCBs) and other pollutants accumulate in organisms. Data from Lake Ontario show how microorganisms acquire them from sediments, and their concentration is magnified as they work their way through a food web.

Women of Arctic tribes like the Eskimos and Inuits, who live on fatty animal foods such as seals and fish, have some of the highest contaminant levels in their breast milk—so high that some samples of their milk brought into laboratories for analysis must be treated as toxic waste!

The actions of these pollutants in causing cancers and birth defects have been known for some time, but now an even more frightening effect has emerged. Beginning in the 1960s, investigators started to catalog a number of apparently unconnected disruptions: Mink in Michigan stopped reproducing; Herring Gull chicks at Lake Ontario died in the egg or hatched with deformities; Florida alligators developed as deformed intersexes, instead of as normal males or females, and the males developed undersized penises; and several European scientists noted sharp declines in human sperm counts. As investigators looked into these and many other curious and worrisome events, one causal factor emerged: pollutants acting as endocrine disruptors. The

Concepts 28.1 An Old Field Becomes a Forest

What happens if an old farm is allowed to return to its natural state? Farmers are constantly fighting the growth of weeds in their fields, and for a year or two after a farm is abandoned, it is a haven for small, annual plants: ragweed, chickweed, pigweed, and horseweed. These plants can grow in the hard ground, among the stubbles of the last harvest, with no protection from the hot sun. By making a little shade, the plants create a microclimate in a zone only a few centimeters above the ground that holds moisture in and protects the soil from erosion by rain and wind. Many insects thrive in this microhabitat. Grasshoppers, leaf hoppers, and aphids eat the plants and drain their juices, while predators such as spiders, ladybird beetles, praying mantises, and wasps live on the herbivorous insects. In turn, sparrows, meadowlarks, killdeer, quail, and other birds feed on seeds and all of the insects, and field mice and rabbits move in, providing food for hawks, owls, snakes, and carnivorous mammals.

principal regulators of reproductive physiology are steroid hormones, and because various manufactured organic compounds can bind to steroid hormone receptors, the effects on reproduction and development can be devastating.

In July 1991, a group of scientists concerned about this issue met at Wingspread Conference Center in Racine, Wisconsin, and issued a consensus statement summarizing the scientific evidence. It is clear, they said, that many human-made chemicals now in the biosphere are endocrine disruptors; the long list includes DDT and its degradation products, PCBs, dioxins, furans, EBDC fungicides, kepone, lindane, and other hexachlorocyclohexanes, and various organo-metal compounds. Their known effects already include interference with reproduction in many animal species as well as in humans, increased cancer rates, and other physiological disruptions. The Wingspread statement called for expanding the research on the problem, testing products for hormonal activity, and addressing the general lack of awareness in the scientific and public health communities.

B. Ecological succession

28.9 Communities tend to replace one another in successions.

With materials constantly moving through it, no ecosystem is likely to stay the same for long, although some undoubtedly last for centuries, or even millennia. Ecosystems change in two ways. In **ecological succession,** one community gradually replaces another, as shown in Concepts 28.1. In **species turnover,** some members of the community disappear and are replaced by new species, even though the community as a whole remains essentially as before.

Ecosystems are shaped by physical events, such as fire (Sidebar 28.1) and geological disturbances, and by interactions among community members. Through their activities, the organisms of a community inevitably change one another's environments. One change may *facilitate* another, as when the first plants that move into a new area prepare a soil that larger plants can take hold in. In this way, a community prepares the

Seeds dropped by the annual plants sprout the next year to produce more of the same plants, but among the new seeds being blown in by the wind or carried by animals are those of biennials such as mullein, which complete their life cycle in two years, and perennials such as goldenrod, asters, and grasses. Even if the perennials die back each winter, they maintain a foothold in the soil and outcompete the annuals. These plants form the base for new communities of animals, such as insects specialized for living on each species of plant, birds like bobolinks and blackbirds, and carnivores like foxes and weasels.

Small trees and shrubs eventually start to crowd out the perennials. The shade they cast is inhospitable to many of the herbaceous plants, and their larger, woodier bulk enables them to occupy wider areas and survive more surely over the winter. The trees vary from place to place, but commonly in eastern North America the old field is soon filled with a pine forest. Now the old fauna of the field is replaced with different animals. The meadow mice are replaced by forest mice. The meadow birds are replaced by chickadees, cardinals, flycatchers, towhees, warblers, blue jays, and woodpeckers. Within a couple of decades, the trees will have grown into a thick canopy, creating shade that excludes many plants and a thick carpet of needles that suppresses others.

Pines exclude their own seedlings, but they make conditions hospitable to hardwoods such as oaks and hickories. Gradually these trees move in, crowd out the pines, drop their leaves year after year, and create a new environment. They form a new forest that may persist for centuries.

way for new species to replace some of its members and, eventually, for an entirely different community to develop. Sometimes a change *inhibits* the growth of a species, as when a plant changes the pH of the soil around it, inhibiting the growth of other plants. Organisms also show *tolerance,* meaning that a species tolerates the conditions it finds or creates.

Terrestrial ecosystems are generally identified by their plants, which form the base of a community, but their fauna has critical effects on the plant community structure. Herbivores can determine the composition of a forest just by preferring certain food plants to others. Birds can affect the plant cover by their preferences for certain seeds, by their nesting or feeding habits (digging nests in trees, for instance), or by their habit of hiding acorns or other large seeds, many of which sprout rather than being eaten. The story of the rabbit plague in Australia is relevant here. European settlers introduced rabbits into Australia in 1859, but having no natural predators, the rabbits soon overran the land, consuming native plants. Finally, in 1951, the disease myxomatosis was introduced to hold the rabbit population in check; in just a few months, the population fell spectacularly, to about 1 percent of its original level. Trees and shrubs that had been held in check by rabbit grazing became much more common. As expected, evolution of the myxomatosis virus and the rabbits has been going on ever since. The rabbits that were not killed initially were more resistant to the virus, and their offspring have been growing still more resistant since then, while less virulent strains of the virus have been evolving.

The series of plant types that replace one another in succession is called a **sere,** and any particular plant community is a **seral stage.** Some ecologists have long theorized that a sere culminates in a **climax community,** a plant association that lasts indefinitely and will never be replaced by a more stable type. For a long time, it was commonly believed that all ecosystems would have climax communities in the absence of human interference.

Sidebar 28.1 Fire As an Ecological Agent

Humans and nature frequently come into conflict over the matter of fire. Fire in residential areas whips through brush and trees, destroying houses. Forest fires destroy trees designated for logging, as well as resorts and vacation homes. Since about 1940, the friendly Smokey the Bear has promoted a war against forest fires throughout the United States; the timber industry and Forest Service spend untold time and money preventing and fighting fires. Yet fire is a natural event in most ecosystems. Lightning strikes regularly start fires, and organisms have adapted to this fact of life.

As the early settlers noted, the grasslands of the American great plains burned regularly, yet they were lush ecosystems whose plants rebounded soon after the fire. The roots of prairie grasses can generally survive fire and grow new shoots quickly. In fact, this region has remained a grassland primarily because of periodic fires, combined with the effects of herds of large mammals, especially bison. Ecologists now see fire as a natural event that is essential to the health of the prairie ecosystem. Where fires have been suppressed, the grasslands tend to turn into forests; wildlife refuges, trying to maintain the native ecosystems, now use controlled burns to suppress tree growth. Similarly, the bunchgrass steppe of the eastern Oregon and Washington high country recovers quickly from a natural burn, which races through the aboveground vegetation without injuring roots. With the suppression of fire, these areas are being invaded extensively by cheat grass, a European import that makes poor food for cattle and other herbivores. Periodic burning would gradually eradicate cheat and restore the native grasses, but agricultural interests oppose this option.

Native forests, too, were subject to occasional fires, but trees are adapted to low-intensity burns. Such fires have been a factor in determining the distribution of certain species, such as the Red Pine of North America, which is resistant to fires of low to moderate intensity. Many conifers have adapted to fire by keeping their cones closed until a fire stimulates them to open and scatter seeds that replenish the forest. The occasional fires of natural forests burned off detritus periodically, so little loose fuel accumulated; therefore, each burn was relatively cool, compared with a modern forest fire fueled by huge buildups of detritus. Fire swept through the old natural forests quickly, killing some young trees but doing little damage to old trees with thick bark. In a "managed" forest, the trees are even-aged and equally susceptible to damage; fire is suppressed, but when a fire does inevitably break out, it is devastating. Modern ecologists know that Smokey the Bear is a friend of the timber industry and people who build homes in the forest, but not of the natural ecosystem.

In fact, a number of apparent climax communities have been cataloged, mapped, and named for their dominant plant species. The dominant species are usually trees, but not always; two or three types of prairies seem to be climax communities through the center of North America.

Recently this view of what is natural has been challenged more and more by the recognition that disruptive events, including fire, are the rule rather than the exception. Some large ecosystems may be in a kind of equilibrium condition between periodic disruptions and regrowth, as exemplified by Balsam Fir *(Abies balsamea)* forests in the northeastern United States (Figure 28.16). Other ecosystems may not achieve even this degree of stability. As Douglas Sprugel has pointed out, humans have short lives compared with many trees and with the time scale on which climatic change occurs; what we take to be stable, natural ecosystems may be only our short-term view of the world. The story of the African savanna at the opening of this chapter is one example. The old-growth forests of the Pacific Northwest provide another. These magnificent forests are dominated by huge trees such as Douglas-fir *(Pseudotsuga menziesii)* that may live for over 1,000 years, yet analysis of large areas has shown that most of the trees are only about 600–700 years old. The explanation is that during the very warm period called the "medieval optimum" between A.D. 1000 and 1300, the Pacific Northwest was apparently ravished by a number of huge fires. On a longer time scale, pollen analysis shows that these forests dominated by Douglas-fir did not even exist until about 6,000 years ago. This historical perspective in no way diminishes the importance of preserving the remaining contemporary forests, but it shows that humans must take a longer-term view in thinking about naturalness and stability.

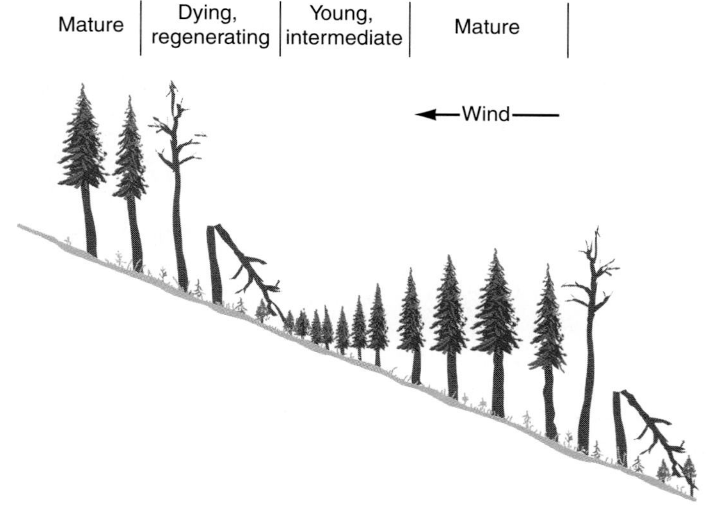

Figure 28.16

Balsam Fir forests show a regular wavelike pattern of growth and death, with a cycle time of about 60–80 years. (Prevailing winds knock down old trees.) If an ecosystem is large enough, the area as a whole may be in a kind of equilibrium condition, even though any smaller habitat changes periodically.

28.10 Seral stages show definite trends in structure and productivity.

As an ecosystem moves from one seral stage to another, its productivity and metabolism change in certain directions. One important trend is toward a balance between photosynthesis and respiration. In the early stages of succession in the field

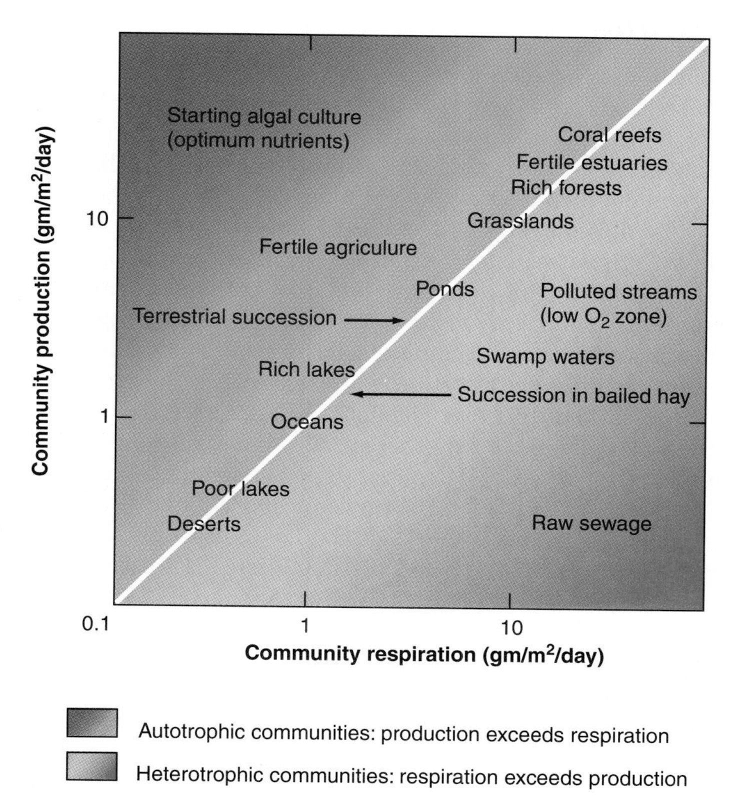

Figure 28.17
Various communities are represented by their relative rates of respiration and photosynthesis (as measured by net primary production).

described in Concepts 28.1, the ecosystem obtains a lot of energy through photosynthesis by small plants, which produce more organic material than is consumed. To use Eugene Odum's terminology, the community is **autotrophic,** since the ratio of photosynthesis to respiration is considerably greater than 1 (Figure 28.17). As an ecosystem matures, it tends to become more **heterotrophic,** with more respiration than photosynthesis, because later communities tend to be dominated by large plants, which have an intrinsic balance between respiration and photosynthesis. In general, most communities have a photosynthesis/respiration ratio close to 1. It is interesting that many kinds of stable ecosystems achieve this balance, even though earlier seral stages may be less balanced. (Some exceptions to the rule are cave ecosystems, where no photosynthesis occurs at all, and streams that get many of their nutrients from imported detritus.)

A second trend is toward greater biomass but lower productivity. For instance, a terrestrial ecosystem in an early seral stage consists mostly of herbaceous plants that have little biomass but photosynthesize and grow rapidly (that is, with high productivity). As the system ages, it acquires much more biomass, such as the great woody bulk of trees; however, since wood is not photosynthetic, the ratio of gross productivity to biomass falls. F. A. Bazzaz, who studied old field successions, found that the first plants synthesize organic molecules at a rate of about 20 to 40 mg of CO_2 per square decimeter of leaf surface per hour (1 decimeter [dm] = 10 cm). They are replaced by plants with rates close to 20 mg, then by trees with rates between 10 and 20 mg, and finally by trees predominantly in the range of about 4–12 mg. The net productivity of the entire system necessarily falls with maturity, because the system approaches the maximum biomass its nutrients can support.

Later plants also tend to be more shade-tolerant. The pioneers in an old field must grow in bright sunlight, but as the community becomes dominated by trees, species are selected that can grow well in the shade.

Sun plants and shade plants, Section 38.8.

28.11 Communities tend to become more diverse in successions.

A new community begins with only a few species that can gain a foothold in harsh conditions, but as seral stages replace one another, they create more and more possible niches. As these niches are filled, the community becomes more diverse. In fact, diversity increases quickly during the early stages of succession and then levels off and perhaps falls slightly. At the same time, the community's biomass increases. This trend is most apparent in the transition from a community of small herbaceous plants into a forest, but it may not be so apparent in a prairie or desert.

Later communities also last longer than earlier ones. After a farm field is abandoned, it takes only a few years for perennial plants to become well established, about 10 years for shrubs and small trees to become prominent, and perhaps 20–50 years for a typical pine forest to develop a canopy. A climax forest dominated by hardwoods, which replaces the forest of pines, may last for centuries and is clearly very resistant to change. Thus the community appears to become more stable as it grows older. This leads us to thinking about the general matter of community stability.

28.12 More diverse communities appear to be more stable.

Observations about trends in the evolution of a community lead to an interesting controversy in ecology. G. Evelyn Hutchinson and Robert H. MacArthur proposed a simple, convincing argument. If a community has only a few species of organisms and thus few trophic links, they said, it is unstable because an accident affecting one species will have enormous repercussions on the others. However, in more diverse communities with more trophic links, more checks and balances exist that can minimize the effects of a change in one species. An accident that reduces one population of herbivores, for instance, won't have much effect if several other herbivores use the resources and serve as food for carnivores. It is like the difference between a play with many roles and one with few roles. In a play with many characters, one role might be cut out or absorbed by some of the other characters without changing the essential structure, but in a play with only a few roles, each one is more important structurally and it is hard to eliminate any of them.

The argument over stability and diversity has important implications for conservation. As human activities threaten many species with extinction, conservationists have argued that it is important to maintain diversity in ecosystems in order to preserve their stability. But what do we mean by stability? Various definitions have been used, but for the sake of argument we will use stability to mean the tendency of an ecosystem to return to its original state if it is disturbed. An ecosystem is *resistant* if it is not easily disturbed in the first place, but *resilient* if it returns quickly to an equilibrium condition after being disturbed. One school of thought maintains that a more diverse ecosystem ought to be more stable because it is more likely to contain species that can thrive during an environmental disturbance, thus maintaining the integrity of the whole system; the alternative viewpoint is that all the members of a community have similar capabilities, so having more species doesn't make for greater stability. In fact, some ecologists have developed computer models of communities that make just the opposite prediction: that more diverse communities should be *less* stable.

What does the experimental and observational evidence say? Nelson Hairston and his colleagues set up simple experimental ecosystems with bacteria providing the base, protozoans such as *Paramecium* eating the bacteria, and carnivorous protozoans such as *Didinium* eating the paramecia. They obtained evidence that systems with three species of bacteria are more stable than systems with only one species, supporting the view that diversity creates stability. Ecologists in the field have identified some instances in which the more complex system appears to be the more stable, as in the rich Malaysian forests. David Tilman and John A. Downing studied the stability of Minnesota grasslands over several years, including a period of severe drought. The ecosystems were dominated by a few species of grass, including both C_3 and C_4 photosynthesizers, and they varied the species richness of small plots by controlling the nitrogen content of the soil. Tilman and Downing found a clear increase in stability, as measured by drought resistance, as the diversity of their plots increased. Another group of investigators at Imperial College in Berkshire, UK, set up experimental ecosystems with 9, 15, or 31 species each and demonstrated that plant productivity increased as diversity increased. Thus accumulated evidence supports the Hutchinson-MacArthur view of community structure and bolsters the conservationist position, though the argument for preserving ecosystems and species diversity can be made more strongly on economic, moral, and aesthetic grounds.

28.13 Primary and secondary successions begin from different conditions.

It is customary to distinguish between two kinds of succession. The story told in Concepts 28.1 is about a **secondary succession,** the most common kind, which begins with an ecosystem that has been disturbed—perhaps by fire, floods, or human activities.

Primary successions, which are rare, begin with a lifeless or nearly lifeless physical environment, such as bare rock and ash left after a volcanic eruption or glaciation. Inevitably, lichens, mosses, and other pioneer species move in and cling to the soil and rock. They gradually build up a thin organic layer in which the seeds of grasses and other herbs can sprout and grow, making the soil deeper, richer, and more moist. Eventually small shrubs and then trees can invade the area.

On the morning of 18 May 1980, the peak of Mt. St. Helens in Washington state blew off in a tremendous eruption, devastating over 600 km^2 of forested and clear-cut land. Though the eruption was a tragedy in human terms, it provided an opportunity to study the revegetation of a barren soil. The surrounding forest was a typical assortment of Pacific Silver Fir, Noble Fir, Douglas-fir, and Western Hemlock, with clear-cut land dominated by Fireweed, Pearly Everlasting, Vine Maple, and Pacific Blackberry. Areas along the local streams (riparian zones) were dominated by Red Alder, Cottonwood, and Willows. The eruption blew huge trees over like dominoes and spread ash over a broad area. Earthquakes and the rising magma weakened the north wall of the crater and created an avalanche of 2.8 billion m^3 of debris, spread over 60 km^2 to an average depth of 45 m.

During the first year after the eruption, the landscape was almost barren (Figure 28.18*a*). A few herbaceous plants (fireweed, thistle, and lupine) and a few willows had survived the landslide because they could grow from plant fragments. One significant factor in determining the survival of plants was the position of their dormant buds, and at first the only survivors were plants such as fireweed whose dormant buds lie in a protected position just below the soil surface. Although the normal forest of the area is coniferous, the first trees to move in were mostly deciduous. Red Alders, though not numerous, were especially vigorous, for several reasons: They are able to fix nitrogen, a nutrient largely missing from the debris; they are relatively resistant to browsing; and their wind-dispersed seeds germinate well and grow rapidly. Gradually more species moved into the area (Figure 28.18*b*), but recovery was slow, and in 1989 only 20 percent of the ground was covered by plants.

An interesting footnote to the Mt. St. Helens recovery occurred through efforts of the Soil Conservation Service to stabilize the soil against further erosion. Large areas were sprayed with seeds of exotic (nonnative) herbaceous legumes and grasses that grow quickly and can hold soil together. The effort failed. Areas with exotic species lost more conifers and gained fewer native species than natural areas, primarily because of one legume, *Lotus corniculatus,* which outcompeted the desirable natives. Still more disastrous, areas of rich grasses developed enormous populations of mice (250 per hectare, compared with a normal 10 per hectare), which killed off the established conifers by nibbling their bark. Over half the conifers had been lost by the time natural predators, such as hawks and coyotes, moved in and started to reduce the rodent population. The story teaches some lessons about human intervention in ecosystems. On the other hand, one purpose of ecological research is to gain an understanding that can be applied in the future. As natural disasters and human activities threaten the world's ecosystems, the knowledge needed to preserve them can perhaps be derived from past failures.

(a)

(b)

Figure 28.18

(a) After the eruption of Mt. St. Helens in Washington state, a large area was devastated and stripped bare of live vegetation. *(b)* A few years later, the area was becoming revegetated, starting with herbaceous plants, such as fireweed and lupine, and small willows and Red Alders.

28.14 The species in a community may continually replace one another.

Even in a relatively stable, long-lasting ecosystem, one species may be replacing another; each niche could be filled by different species at different times. The community structure shifts because, as conditions change, competitors for a resource may come along that are slightly better adapted to the new conditions and can outcompete the resident species. This phenomenon can also be explained using the analogy of a play: One production of a play can go on as one actor replaces another again and again without disrupting its structure. Eventually, the entire cast may change, but it is still the same play.

The best observations on species turnover have been made on islands, where species are easier to isolate and study. Robert MacArthur and E. O. Wilson suggested that the species on an island would become extinct at a rate proportional to the number of species already present, and that the island would be colonized by immigration at a rate inversely proportional to the number of species already present. They predicted that extinction and immigration should balance at a steady-state level, a level that every island should maintain, even though species continue to replace one another. To test this theory, Wilson and Daniel Simberloff studied four tiny mangrove islands in the Florida Keys whose main inhabitants are insects and other arthropods. After counting the number of species on each island, they surrounded each island with a big plastic tent and killed its inhabitants by fumigation. They then removed the tents and observed the islands being recolonized by new arthropods. As shown in Figure 28.19, the number of species on each island rose to its previous level and stayed relatively steady. However, the species that repopulated each island weren't identical to the original ones, and the species composition of each island changed slowly during the course of the experiment.

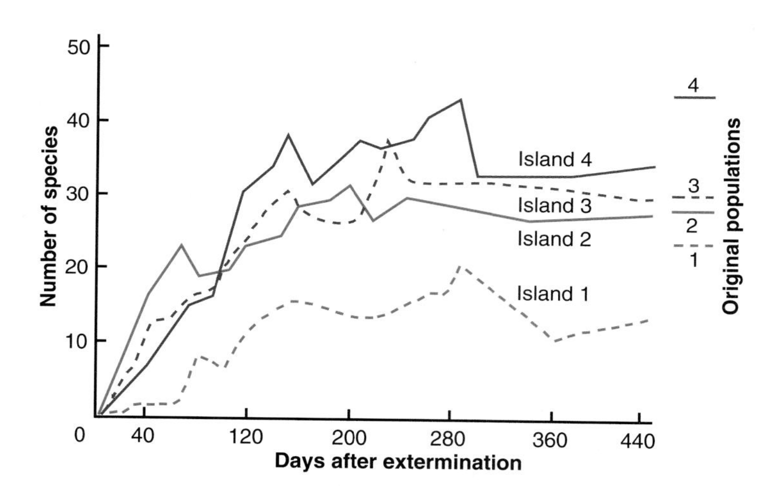

Figure 28.19

After the fauna of four small islands were exterminated, the islands became repopulated with numbers of species comparable to their original levels (right-hand scale), but with not all the same species.

Coda Understanding an ecosystem is probably the broadest view one can take of biology, as long as it is coupled with an evolutionary perspective. The evolutionary and ecological themes of this book are like two intertwined musical themes continually working in counterpoint to create the symphony of life. It should be clear by now that the symphony can only be played on fundamentally self-reproducing genetic instruments with their delicate balance of stability and transformation—preservation of information and mutation. No other known systems have the capability of maintaining themselves and adapting to a relatively stable environment in the short term while holding the potential for long-term evolution in response to much greater changes.

Looking at chemical transformations in the ecosystem again emphasizes the fundamentally chemical nature of organisms. We can never forget that we are dealing with instruments made of protein and nucleic acid, and that if these two molecules did not exist, organisms very likely could not exist either. Thus the great cycles of the ecosystem and biosphere—cycles of carbon, nitrogen, and sulfur—come to have a twofold molecular meaning: They are energy-transforming cycles of oxidation and reduction, and they are structure-transforming cycles in which the fundamental building blocks of organisms are broken down and remade. These transformations depend upon a maze of enzymes, each carrying out one brief molecular interaction in the vast chain of events, each one shaped for its function by the process of natural selection repeating endlessly in the background, creating the rhythm of the whole piece.

Summary

1. The organisms in an ecosystem occupy different trophic levels, which may be described by a pyramid showing the biomass or the amount of energy at each level. At each trophic level, most of the available matter is converted into wastes, and most of the energy goes off as heat, leaving relatively small amounts of matter and energy in the organisms themselves.
2. Plants use only small amounts of the available light energy for growth, storing about 1–2 percent in their biomass.
3. Organisms at higher trophic levels also convert relatively small amounts (10–20 percent) of their food into their own biomass.
4. The way materials move from one compartment of a system to another can be described by their turnover rate. The size of the compartment divided by the turnover rate equals the average residence time in the compartment.
5. The whole biosphere depends on a hydrologic (water) cycle of vast proportions involving evaporation, precipitation, percolation through the ground, and runoff. An important part of the cycle is the passage of water though plants and its transpiration back into the atmosphere.
6. Organisms participate in biogeochemical cycles. In addition to the water cycle, specific mineral nutrients, such as carbon, sulfur, and nitrogen, cycle through the biosphere and through inorganic reservoirs as they are transformed by various organisms.
7. Nitrogen cycles primarily as a component of protein, the major organic material in organisms. Special bacteria engage in nitrogen fixation, converting atmospheric nitrogen into ammonia; others interconvert ammonia, nitrate, and nitrite to get their energy; and some denitrifying bacteria change these compounds into nitrogen gas again.
8. The sulfur cycle is created largely by bacteria that oxidize and reduce various sulfur compounds to obtain their energy.
9. Components of the ecosystem are always turning over, moving from one compartment to another, sometimes rapidly. Turnover occurs faster during the summer than during the winter and much faster in aquatic than in terrestrial ecosystems.
10. The organisms of an ecosystem are constantly creating detritus, and the ecosystem is built on a detritus food web as well as on the food web that begins with producers. The detritus food web begins with a microflora (bacteria and molds) that feed on detritus; a variety of microfauna then eat the microflora and one another.
11. A terrestrial ecosystem continually loses nutrients to its surroundings. Enormous amounts of mineral nutrients may flow through certain ecosystems and are continually being washed out of the bedrock and soil with the flow of water from the system. However, the biological community slows this outflow considerably, compared with a system in which geological processes are operating alone.
12. Pollutants accumulate in organisms, mostly in fat deposits, and their concentration is magnified in a food web. Many types of manufactured organic molecules are endocrine disruptors, carcinogens, or substances that have other physiological effects.
13. Communities tend to replace one another in successions, as the organisms in each community create physical conditions that inhibit the growth of the existing community and are favorable for other types. A primary succession begins with a lifeless environment, a situation that seldom arises; a secondary succession begins with an existing community, such as an old farm field.
14. The first community in an area tends to be essentially autotrophic; later it achieves a balance between autotrophy and heterotrophy, or between photosynthesis and respiration. Communities also tend to become more diverse and to last longer through successions. Evidence indicates that more diverse communities have greater stability and are more resistant to disruption.
15. Later seral stages tend to have greater biomass but lower productivity, correlated with lower rates of photosynthesis.
16. In addition to large changes in community structure over time, the individual species in a community may continually replace one another, although the community retains essentially the same overall structure.

Key Terms

gross primary productivity 580
net primary productivity 580
turnover 582
residence time 582
turnover rate 582
hydrologic cycle 583
evapotranspiration 584
biogeochemical cycle 585
nitrogen cycle 585
nitrogen fixation 585
denitrification 585
sulfur cycle 585
necromass 586
bioaccumulation 591
biomagnification 591
ecological succession 592
species turnover 592
sere 593
seral stage 593
climax community 593
autotrophic 595
heterotrophic 595
secondary succession 596
primary succession 596

Multiple-Choice Questions

1. Which statement is the most accurate?
   a. Without phototrophs, matter would immediately cease to cycle within an ecosystem.
   b. Without detritivores and decomposers, energy would immediately cease to flow through the biosphere.
   c. Autotrophs and heterotrophs pass most of their input energy to the next organisms in the food web.
   d. Most of the solar energy that strikes phototrophs does not enter the biosphere.
   e. All of these statements are correct.
2. The energy differential between the gross primary productivity and the net primary productivity of an ecosystem is
   a. used up in the process of making the biomass of producers.
   b. contained in the biomass of the producers.
   c. found in the radiation reflected back into space.
   d. used for transpiration in the hydrologic cycle.
   e. the amount of energy lost when herbivores eat producers.

3. Turnover rate is equal to
   a. the average residence time of a component in a community.
   b. biomass divided by average residence time.
   c. average residence time multiplied by average temperature.
   d. the maximum time an element remains in the biosphere.
   e. the ratio of time of the biotic to the abiotic stages in a biogeochemical cycle.
4. In the phrase *acid rain,* the acid of interest is
   a. carbonic.
   b. hydrochloric.
   c. nitric.
   d. citric.
   e. sulfuric.
5. When a pollutant such as DDT is added to a community, the pollutant is
   a. primarily concentrated in the producers, rather than in the consumers.
   b. spread equally among all the organisms at all trophic levels.
   c. spread equally among the trophic levels.
   d. found equally in the aquatic and terrestrial parts of the community.
   e. concentrated in the organisms at the highest trophic level.
6. Which is not a trend in seral successions?
   a. increase in the proportion of woody parts of plants
   b. increase in the ratio of respiration to photosynthesis
   c. increase in the proportion of leaves to bark and wood
   d. increase in the height and size of trees
   e. decrease in the proportion of herbaceous plants compared to other plants
7. If community A consists of small plants that are intolerant of shade while community B includes larger, more shade-tolerant producers, it is logical to suppose that
   a. community A comes earlier in a succession.
   b. community A comes later in a succession.
   c. A and B represent different levels in the canopy of a single forest ecosystem.
   d. community A probably includes larger carnivores than community B.
   e. community A is probably more stable and diverse than community B.
8. Which of the following best exemplifies the starting point for a secondary succession?
   a. the new land created from volcanic activity
   b. the site of a recent tornado or hurricane
   c. an island community into which a new organism has been released
   d. the open ocean
   e. a decaying residential area turned back into a wetland
9. If the animal species in a stable, isolated community are counted and observed once per decade for five decades,
   a. the same numbers of the same species will be found at each count.
   b. the number of different species will probably remain constant.
   c. the island will show evidence of a primary succession.
   d. the total number of organisms will remain constant.
   e. the gross primary productivity will remain constant.
10. Within a single isolated community, the rate of species extinction is believed to vary with the
   a. number of species that have newly immigrated to the community.
   b. ratio of producers to consumers.
   c. ratio of primary to secondary consumers.
   d. number of species present.
   e. number of individuals present.

True-False Questions

Mark each statement true or false, and if false, restate it to make it true.

1. Matter cycles through the biosphere, while energy flows in a one-way path, entering as heat and leaving as heat.
2. The biomass of the secondary carnivores in an ecosystem is usually the same as that of the primary carnivores.
3. In the stages of succession, the ratio of photosynthesis to respiration remains constant.
4. Nitrogen fixation and denitrification are opposing processes that are generally carried out by different organisms.
5. Bacteria as well as some industrial processes fix nitrogen by using urea to generate molecular nitrogen.
6. Regions of ponds and lakes with high levels of sulfides generally have low levels of photosynthetic organisms.
7. The biotic parts of both the sulfur and nitrogen cycles include some organisms that oxidize and others that reduce the component sulfur and nitrogen compounds.
8. The necromass within a community is recycled by the cooperative activities of plants and bacteria.
9. Older communities are characterized by greater productivity as a result of an increasing biomass.
10. A community that is the most resistant to change is not necessarily the most resilient.

Concept Questions

1. Describe the hydrologic cycle if living organisms did not exist. What effect do organisms have on the cycle?
2. Contrast aquatic with terrestrial ecosystems with respect to molecular turnover and species turnover.
3. Correlate the proportion of *r*- to *K*-selected species in the early and late stages of successions.
4. Explain why lipid-soluble pollutants are subject to bioaccumulation, but water-soluble agents are not.
5. In many overpopulated tropical regions, forests are clear-cut to create new farmland. However, the process is counterproductive. Not only are normal forest products no longer available, but the soil rapidly becomes unfit for agriculture. What has happened?

Additional Readings

Baskin, Yvonne. "Forests in the Gas." *Discover,* October 1994, p. 116. The level of carbon dioxide in the atmosphere will double by the year 2050, producing an unknown effect on the world's forests and fields. Studies on the effects of CO_2 on various ecosystems.

Chauvin, Remy. *The World of an Insect.* McGraw-Hill Book Co., New York, 1967.

Colburn, Theo, Dianne Dumanoski, and John P. Myers. *Our Stolen Future.* Dutton, New York, 1996. The story of the endocrine disruptors and their dangers.

Harris, Larry D. *The Fragmented Forest.* University of Chicago Press, Chicago, 1984.

Holloway, Marguerite. "Sustaining the Amazon." *Scientific American,* July, 1993, p. 76. A review of the problems of balancing economic development with conserving the region's ecology.

Huggett, Richard J. *Geoecology: An Evolutionary Approach.* Routledge, London and New York, 1995.

Kusler, Jon A., William J. Mitsch, and Joseph S. Larson. "Wetlands." *Scientific American,* January 1994. A discussion of the importance of conserving these havens of biodiversity.

Morgan, Sally. *Ecology and Environment: The Cycles of Life.* Oxford University Press, New York, 1995.

Perry, Donald R. "The Canopy of the Tropical Rain Forest." *Scientific American,* November 1984, p. 138. Largely unexplored, the rain forest is home to one of the most diverse plant and animal communities on the earth. A new way of reaching the canopy allows close observation of its ecology.

Repetto, Robert. "Deforestation in the Tropics." *Scientific Amercian,* April 1990, p. 36. Government policies that encourage exploitation, in particular excessive logging and clearing for ranches and farms, are largely to blame for the accelerating destruction of tropical forests.

———"Accounting for Environmental Assets." *Scientific American,* June 1992, p. 64. The importance of accounting for the depreciation of forests, fisheries, and other natureal resources by development.

Revelle, Roger. "Carbon Dioxide and World Climate." *Scientific American,* August 1982, p. 35. The effects of increasing carbon dioxide in the atmosphere.

Shaw, Robert W. "Air Pollution by Particles." *Scientific American,* August 1987, p. 96. Acidic particles in the atmosphere are known to reduce visibility and damage materials. Ingenious methods have now demonstrated that the main source of the particles is the combustion of fossil fuels.

Internet Resource

To further explore the content of this chapter, log on to the web site at:

http://www.mhhe.com/biosci/genbio/guttman

Diversity of Life

29 Early Evolution and the Procaryotes

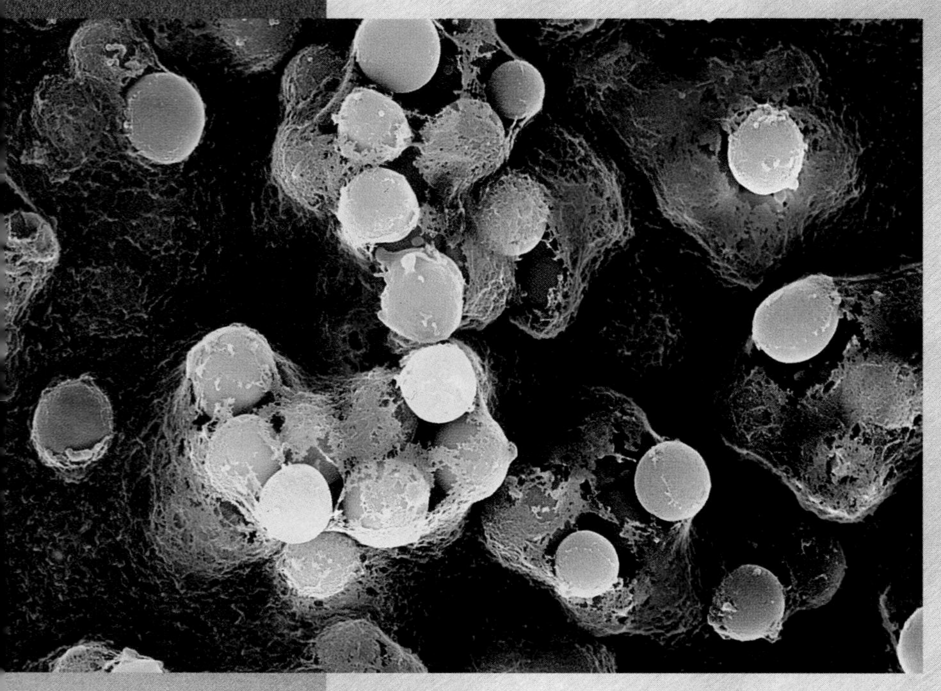

A scanning electron micrograph shows cells of *Streptococcus pyrogens* (artificially colored yellow), which commonly causes sore throats.

Key Concepts

A. The Origin of Biological Systems

29.1 Elements are formed through the natural evolution of stars.

29.2 Organic molecules form in the reducing atmosphere of primitive planets.

29.3 The evolution of a functioning genome is problematic.

29.4 We can reconstruct a likely course of procaryote evolution.

B. The Kingdom Monera

29.5 Procaryotic cells have no true nucleus and are usually very small.

29.6 Bacteria may be spheres, rods, spirals, or long filaments.

29.7 Bacteria are classified by cell shape, metabolism, and reaction to the Gram stain.

C. The Ecology and Uses of Bacteria

29.8 Bacteria have critical roles in every ecosystem.

29.9 Many bacteria cause infectious diseases.

29.10 Pathogens produce disease through invasion and toxin production.

29.11 Infectious agents are transmitted to new hosts from reservoirs of infection.

29.12 Some very small bacteria are intracellular pathogens.

29.13 Many bacteria are used in industrial processes.

D. Addendum: The Viruses

29.14 Viruses are not organisms.

29.15 Virions have simple, regular structures.

29.16 Viruses multiply in a common pattern.

26.17 Viruses have DNA or RNA genomes.

29.18 Viroids are unusual agents.

With this chapter we begin a survey of the major groups of organisms, both living and extinct. Because such a survey only makes sense as a story about evolution, we start at the beginning of evolution and trace some ideas about how simple organisms might have arisen. The "origin of life" has always been a knotty problem. While it is still far from being solved, the modern biological perspective we developed in previous chapters at least makes it easier to think about. Rather than conceiving of a quality called "life," which appeared as a whole at a particular time, we will ask how several features of organisms—membranes, metabolic pathways, genetic apparatus—might have evolved. Furthermore, biologists are no longer bound by the anthropocentric perspective that considers life unique to this planet. Although we have no evidence for organisms anywhere except on Earth, a cosmic view suggests that planets supporting living organisms may be relatively common in the universe. We therefore look for ways that primitive biological structures might develop as planets are forming. Formulating hypotheses about origins is speculative, of course, but it is sound speculation.

The rest of the chapter discusses the properties of procaryotes and the features used to classify them. We will close with some aspects of bacterial ecology, to develop an appreciation of the roles of procaryotes in ecosystems and human affairs. ■

A. The origin of biological systems

29.1 Elements are formed through the natural evolution of stars.

According to contemporary cosmology, as outlined in Chapter 1, the universe has been undergoing a cosmic evolution since its origin about 13 billion years ago. As stars go through their own cycles of birth and death, they build up the atomic nuclei of heavier elements from hydrogen and helium nuclei. These elements form rocky planets like Earth and also compose the organic molecules of which organisms are made.

Stars with planets must be quite common in the universe, even though direct astronomical evidence for them is meager. Chemical reactions that form primitive biomolecules can take place while a planetary system is still forming, so these molecules may be deposited on all primitive planets. Not all planets are suitable places for life, of course. A planet must be in just the right temperature zone around its sun, so that water—which appears essential for biological activity—remains liquid on its surface. The star must last for at least a few billion years before flaring into a nova, since fully functional organisms take a long time to evolve. Even with these limitations, however, one can calculate that a galaxy like ours, the Milky Way galaxy, must hold vast numbers of planets suitable for maintaining life. Let's trace the long chain of chemical reactions that naturally occur on such planets and eventually lead to organisms.

29.2 Organic molecules form in the reducing atmosphere of primitive planets.

The first speculations about the origin of life assumed that the earth's atmosphere has always been much as it is now. If that were true, however, molecular oxygen in the atmosphere would have attacked and oxidized any simple organic compounds that might have formed. In the 1930s, A. I. Oparin pointed out that the primitive Earth must have had a reducing—rather than an oxidizing—atmosphere, made mostly of hydrogen, methane, ammonia, nitrogen, and water. Oxides of carbon, nitrogen, and other elements were probably present, but molecular oxygen was scarce. Developing an argument that was also presented by J. B. S. Haldane in 1929, Oparin proposed that the energy of ultraviolet light and lightning discharges could have turned the gases in the hypothetical reducing atmosphere into a "primordial soup" of organic molecules.

In 1953 Stanley Miller tested the Haldane-Oparin hypothesis while he was a graduate student in Harold Urey's laboratory. Using the apparatus shown in Figure 29.1, he tried to reconstruct primitive conditions by passing electrical sparks through a mixture similar to the hypothetical primitive atmosphere. Remarkably, Miller found that after several days the mixture in this simple apparatus formed quite a variety of organic compounds, including the common amino acids. Miller's experiment has been repeated many times with various initial mixtures, always with similar results. The gases in the mixture form simple organic molecules like formaldehyde ($H_2C{=}O$), formic acid (HCOOH), hydrogen cyanide (HCN), and cyanoacetylene ($CH{\equiv}C{-}C{\equiv}N$). These react further to make a great variety of organic molecules: amino acids, including the 20 commonly found in organisms; fatty acids ranging from C_{12} to C_{20}; all the purine and pyrimidine bases of nucleic acids; and metabolites like acetate, lactate, succinate, and propionate.

Once formed, amino acids seem to polymerize rather easily into peptides. In fact, some small peptides are formed in the reconstruction experiments simulating a primitive atmosphere. Investigators have also suggested various ways that polymerization might be enhanced, as by certain clay minerals that can hold monomers on their surface at the right spacing for

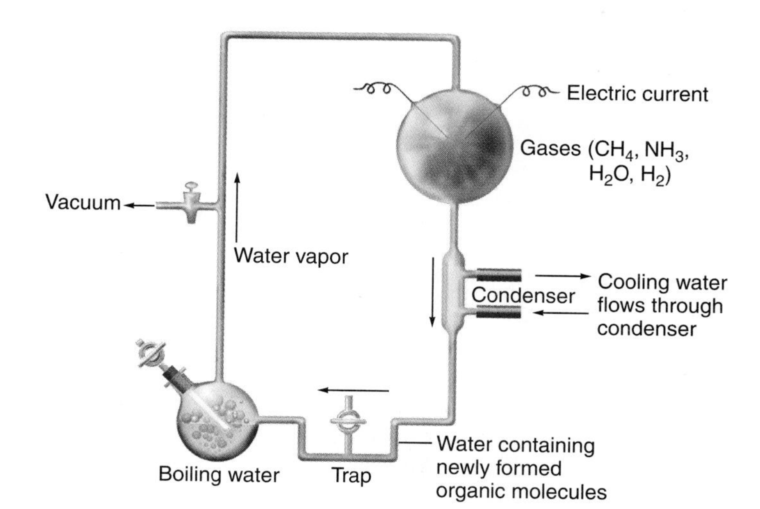

Figure 29.1

Stanley Miller's apparatus was designed to simulate the formation of organic molecules in the primitive atmosphere. The gases circulate while they are subjected to electrical discharge, simulating lightning, and organic molecules accumulate in the U-shaped trap where they can be drawn off.

polymerization. Simple proteins might be formed on a primitive planet at the edge of a pond or puddle where organic molecules wash up onto a rocky base and are concentrated by evaporation. Sidney Fox found that simply heating dry mixtures of amino acids produces so-called "proteinoids," small proteins (3–10 kDa) with many properties of modern proteins, including the ability to catalyze chemical reactions.

Fox and his colleagues later found that hot, salty solutions of peptides cool and form *microspheres,* stable nonliving structures with many properties of modern procaryotic cells and about the same size (about 2 μm in diameter). Microspheres even show some of the osmotic properties and selective permeability of modern cells, and changing the pH of the surrounding medium can induce them to form buds, rather like yeast cells. This is not to imply that microspheres are cells—only that cell-like objects can form under hypothetical primitive planetary conditions. These experiments, coupled with the observation that lipids are produced in primitive reaction mixtures and that lipids and proteins assemble rather easily into membrane structures, suggest that the formation of cells may not have been such a formidable barrier in the evolution of life.

The formation of cells is a critical event that must occur early in evolution, for two reasons. First, further evolution requires that metabolism be confined to small, enclosed spaces where high concentrations of metabolites and enzymes can develop. Second, genetic systems must be confined to cells to take advantage of mutational novelties; evolution depends on competition among individuals (cells or organisms), and this could not happen if each genetic novelty were shared with a large system.

We can't be sure just when the first functioning cells appeared, but the fossil record in the Onverwacht sediments of South Africa shows that simple cells already existed at least 3.4 billion years ago. These cells were procaryotes, some similar to our modern types. Excellent procaryotic fossils are found in rocks formed from 3.4 to 2 billion years ago in the Gunflint iron formation of Ontario.

Dating rock strata, Methods 24.1.

29.3 The evolution of a functioning genome is problematic.

The most critical feature of an organism, of course, is its genome. Without it, there could be no biological evolution, since evolution depends on the selection of individuals with variant genomes. So primitive biological systems could not have been shaped by natural selection unless they were specified by a genome. Although nucleic acid molecules do not form in primitive mixtures as easily as proteins do, they have been produced under simulated primitive conditions, and at least small polynucleotides must have formed on the primitive earth along with other organic compounds.

A genome has to replicate and direct the synthesis of proteins. Replication is an inherent property of nucleic acids, even though it must have been slow at first with no enzymes—or very poor ones—to catalyze the process. The problem lies in the second requirement. How could nucleic acids get *control* over protein structure and come to specify that structure?

The best we can offer are intelligent conjectures. Although modern cellular genomes are all made of DNA, the first nucleic acids to become functional were probably RNA, and for some time an "RNA world" was probably a critical stage in the evolution of functioning organisms. RNA molecules have an inherent ability to interact with one other through base-pairing, so they can replicate as DNA does. They can also fold into complicated forms (as tRNAs do) and act as *ribozymes*—that is, RNA molecules that catalyze chemical reactions just as protein enzymes do. Some modern pre-mRNA molecules, for example, are able to splice themselves and excise their own introns. During the RNA-world stage of evolution, many of the functions now performed by proteins must have been performed by RNA molecules. Francis Crick and Leslie Orgel have suggested that two kinds of RNA might have evolved early on (Figure 29.2). Some had the potential to become genomes (protogenomes) because their base sequences allowed them to replicate with particular ease. An RNA with an alternating A–U or G–C sequence can replicate efficiently by pairing with itself internally. It can bend back on itself:

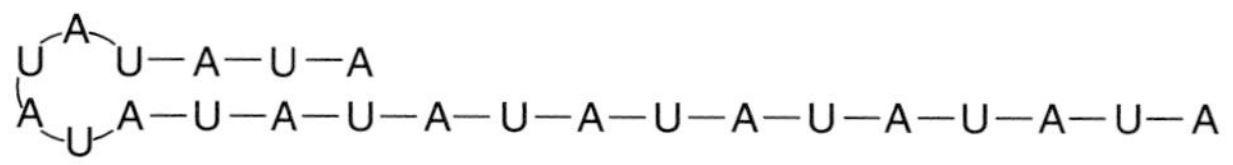

```
 A
U  U—A—U—A
A  A—U—A—U—A—U—A—U—A—U—A—U—A—U—A
 U
```

and can be extended into a longer molecule with the same sequence:

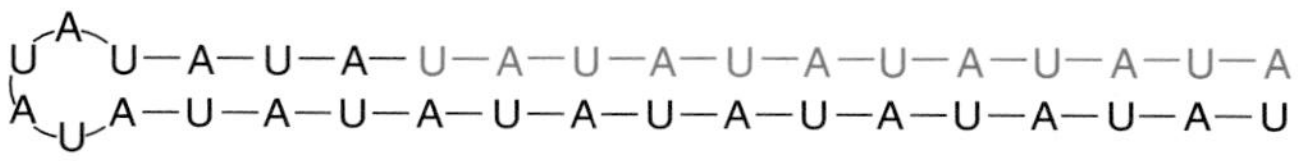

```
 A
U  U—A—U—A—U—A—U—A—U—A—U—A—U—A—U—A
A  A—U—A—U—A—U—A—U—A—U—A—U—A—U—A—U
 U
```

Other RNAs, having an affinity for specific amino acids, then served as primitive transfer RNAs (proto-tRNA) by helping line up amino acids along a protogenome, thus catalyzing a simple protein synthesis. Of course, the evolution of these primitive systems into efficient systems of the modern type must have been a long, slow process.

29.4 We can reconstruct a likely course of procaryote evolution.

Modern procaryotes represent an early period of evolution. Taken together with fossil evidence, they can be used to piece together a reasonable picture of how various forms of metabolism probably evolved (Figure 29.3). Concepts 29.1 reviews the terms for different modes of metabolism.

The earliest organisms must have been heterotrophs that fed on preformed organic compounds.

In thinking about early evolution, it is tempting to imagine that the first organisms were autotrophs that could grow on CO_2 and other inorganic compounds, maybe using light for energy.

Figure 29.2
A primitive genetic system might evolve in an RNA world in this manner.

(1) In the primitive soup, certain polynucleotides arise with sequences that...

can be extended easily by internal replication and can be replicated easily.

These polynucleotides proliferate more rapidly than others and form a pool of **protogenomes**.

Gly

Ala

(2) Other polynucleotides occur that do not replicate themselves as well, but have some affinities for amino acids. These form a pool of **proto-tRNA**.

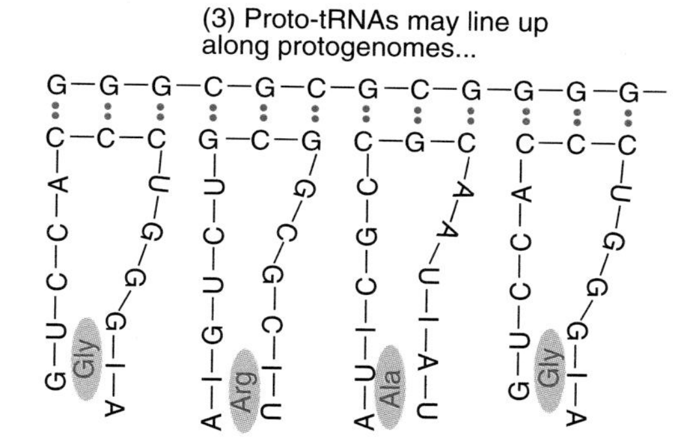

and catalyze the synthesis of certain polypeptides:

Gly — Arg — Ala — Gly —

(4) Gradually, protogenomes are selected that encode useful polypeptides, especially those with polymerase activity that can speed replication and protein synthesis.

Concepts 29.1 **Modes of Metabolism: A Reference**

Autotroph: Makes its organic molecules from CO_2.
Heterotroph: Lives on organic molecules made by other organisms.
Chemotroph: Obtains its energy by oxidizing reduced compounds.
Phototroph: Obtains its energy from light.

These can be combined in four ways, as shown in Figure 7.2. For instance,

Photoheterotroph: Obtains energy from light, but uses organic compounds as carbon sources.
Chemoautotroph: Obtains energy by oxidizing inorganic compounds, and converts such compounds into organics.

However, autotrophs—especially photoautotrophs—need complex metabolic machinery, machinery that the first organisms could not have had. In fact, the first cells must have been heterotrophs that fed on the organic compounds in the primordial soup, a veritable Garden of Eden that provided plentiful food while the cells evolved other features, such as a genetic apparatus. These cells evolved pathways for mobilizing energy through fermentation, probably including the glycolytic pathway, producing ATP by substrate-level phosphorylation, with no electron transport system.

Glycolysis and substrate-level phosphorylation, Section 9.4.

The first crisis in Eden was most likely the depletion of a critical monomer, such as an amino acid. Then the only organisms that could survive were those that happened to have an enzyme for converting some other compound into the depleted one. Such organisms were then selected by their ability to grow, but they eventually exhausted the precursor as well. Then a mutant was selected that could make the second compound from a third. Through this process of natural selection, pathways for

Figure 29.3

The principal metabolic types of procaryotes probably evolved in this sequence.

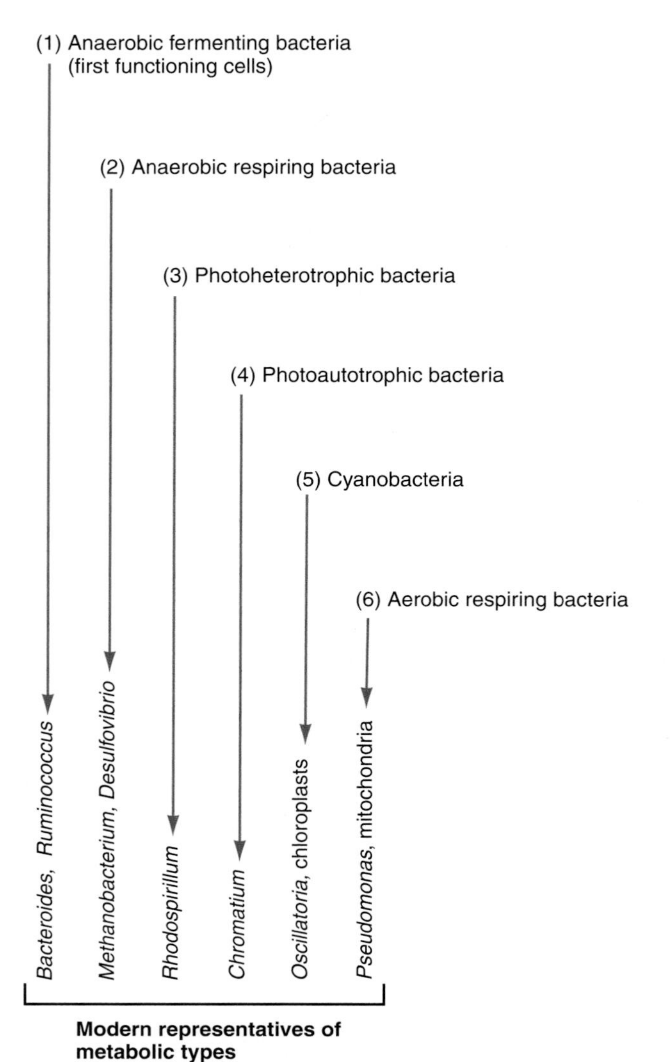

biosynthesis evolved, and organisms gradually became more independent (Figure 29.4). After a long time, autotrophs evolved that could synthesize all their organic components from CO_2, NH_3, water, and other inorganic compounds.

Photosynthesis evolved with the ability to make porphyrins, including chlorophyll.

The next steps in evolution built upon the ability to make the porphyrin ring (see Concepts 4.2). Iron-porphyrins (hemes) made possible a chain of cytochromes, an electron transport system for synthesizing ATP through respiration. Remember that cellular respiration means oxidizing a substrate, with an inorganic compound—not necessarily oxygen—as the terminal electron acceptor. In the absence of free oxygen, respiration must have been anaerobic at first, using terminal acceptors such as nitrate, sulfate, and CO_2.

Isotopic evidence indicates that autotrophs existed by the time the Onverwacht sediments were laid down; a sudden change in the ratios of carbon isotopes (^{12}C or ^{13}C) in these rocks probably means that autotrophic cells were suddenly assimilating CO_2 from the atmosphere, with their enzymes discriminating against ^{13}C in favor of ^{12}C. The first autotrophs were probably phototrophs, since photosynthesis was made possible by the evolution of chlorophylls, which are magnesium-porphyrins.

Plant photosynthesis produces oxygen as a by-product and is known as **oxygenic** photosynthesis. This advanced form requires two photosystems and must have taken a long time to evolve. Geologic evidence shows that free oxygen only appeared in the atmosphere around 2.7 billion years ago and did not accumulate to any extent until about 2.0–1.8 billion years ago. Before then, phototrophs used various forms of **anoxygenic** photosynthesis, which does not produce oxygen as a by-product. The first phototrophs (which we will call stage 1) probably used a form of photosystem I just to oxidize organic substrates. The modern survivors that retain stage-1 photosynthesis are green

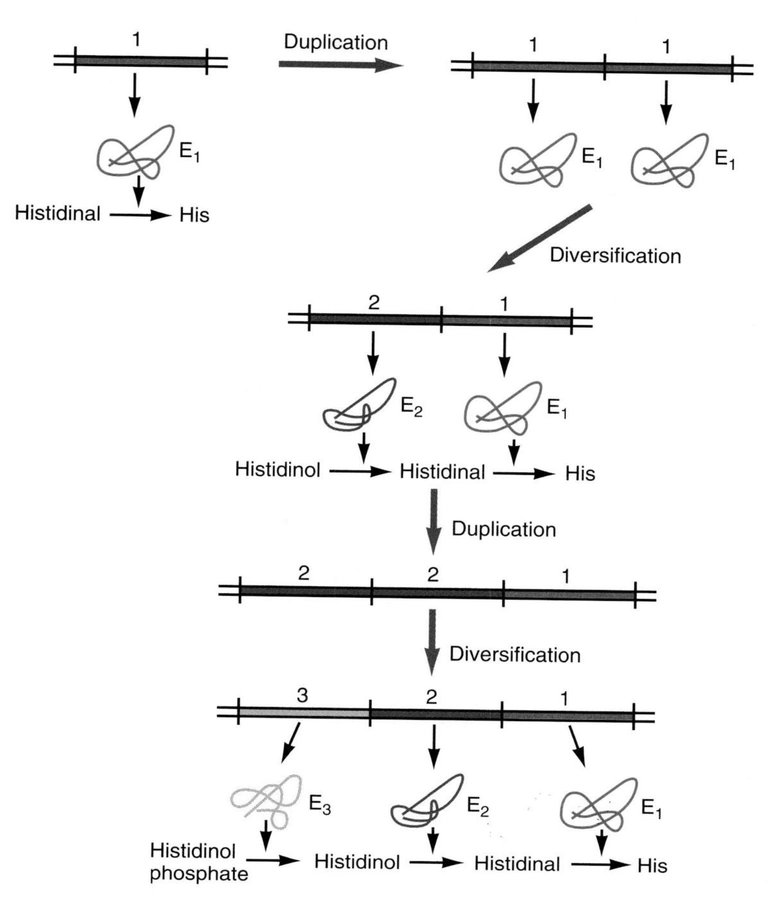

Figure 29.4

Metabolic pathways probably evolved backwards, with the last enzyme in the pathway appearing first, through gene duplication and diversification. The primordial soup contained all the amino acids required for protein synthesis, including histidine. When the histidine started to run out, cells were selected that could synthesize histidine from a similar compound (here, histidinal). These cells proliferated until the histidinal ran out; then other cells were selected that could synthesize histidinal from another compound.

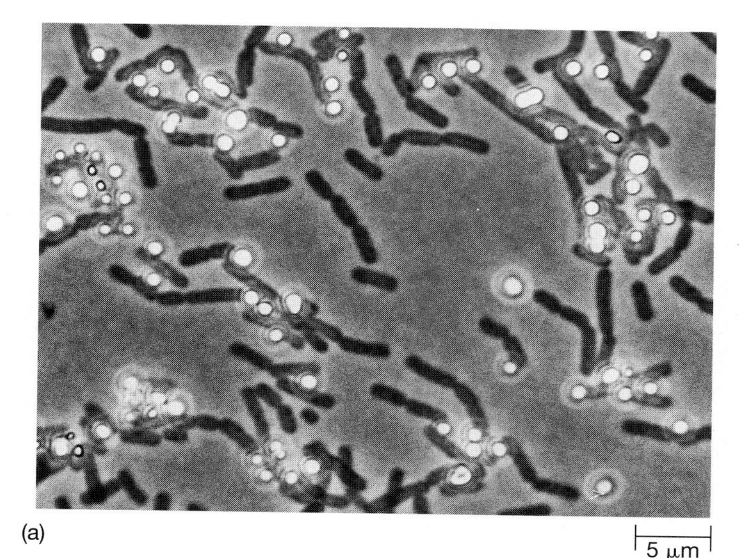

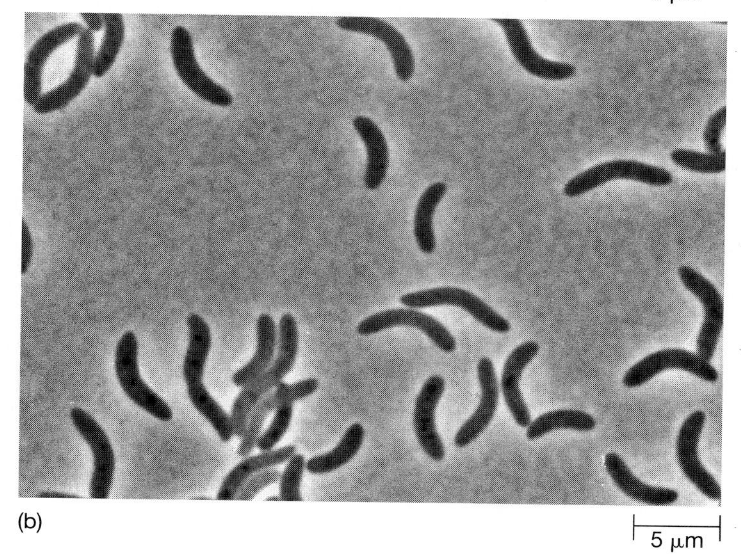

Figure 29.5

(a) The green sulfur bacterium *Chlorobium limicola* and *(b)* the purple nonsulfur bacterium *Rhodospirillum fulvum* are modern representatives of a relatively primitive form of photosynthesis. *Chlorobium* is photographed with phase-contrast optics and shows large extracellular globules of elemental sulfur as well as tufts of flagella.

sulfur bacteria such as *Chlorobium* and purple nonsulfur bacteria such as *Rhodospirillum* (Figure 29.5). *Chlorobium* assimilates CO_2 by reversing the tricarboxylic acid cycle instead of using the Calvin-Benson (PCR) cycle for reducing CO_2 to sugars. The PCR cycle is so complex that it required more time to evolve. *Rhodospirillum* and its relatives can grow as photoheterotrophs. They use light energy to produce ATP and reduced ferredoxin by means of cyclic photophosphorylation. With this stored energy, they use acetate and longer-chain acids as carbon sources.

Photosystems I and II, Sections 10.6 and 10.7.
Calvin-Benson cycle, Section 10.8.

When the PCR cycle finally did evolve, it was used in stage-2 photosynthesis, as exemplified by modern purple sulfur bacteria such as *Chromatium* and *Thiospirillum* (Figure 29.6). These bacteria can only grow anaerobically in the light, using H_2S as an electron (hydrogen) donor for photosynthesis instead of H_2O. Their intracellular deposits of little sulfur grains—a by-product of photosynthesis, analogous to the oxygen left by plants—make them quite distinctive. Sulfur is only an intermediate product, however, and it is oxidized further to sulfate. Some of these bacteria are large procaryotes, 3–5 μm wide and 20–50 μm long, whose cytoplasm is almost filled with extensive, generally spherical thylakoids.

Alternative forms of photosynthesis, Section 10.4.

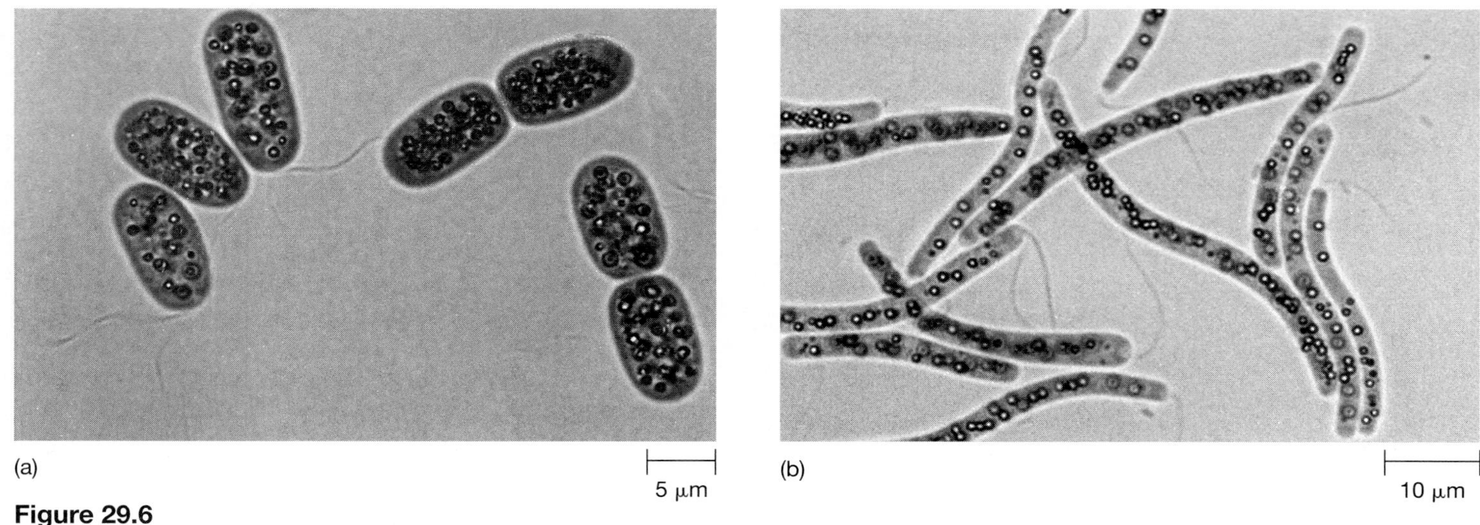

Figure 29.6

The purple sulfur bacteria *(a) Chromatium okenii* and *(b) Thiospirillum jenense* grow anaerobically and photosynthesize by using sulfur compounds as electron sources. *Thiospirillum* shows bright internal sulfur globules and flagellar tufts.

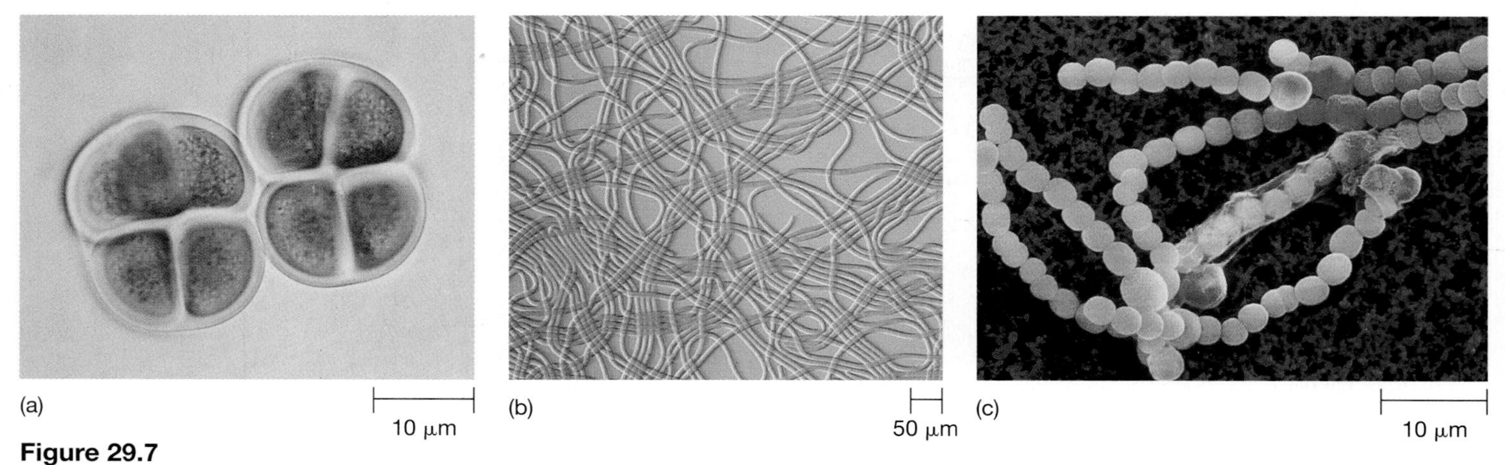

Figure 29.7

Cyanobacteria are usually single cells or chains of cells. *(a) Chroococcus* species; *(b) Oscillatoria tenuis; (c) Anabaena* species (SEM, artificial colors) showing some heterocysts (large cells, colored slightly orange) in which nitrogen fixation occurs.

With the development of photosystem II, oxygenic photosynthesis finally evolved (stage 3), as in modern **cyanobacteria** (blue-green bacteria; Figure 29.7). Cyanobacteria differ in several important ways from the other photosynthetic procaryotes. Instead of the bacteriochlorophylls of the purple and green bacteria, they have the same chlorophyll *a* as eucaryotic phototrophs, and like them, they produce oxygen as a by-product of photosynthesis. The striking colors of cyanobacteria come from supplementary light-gathering pigments called *phycobilins.* In combination with proteins, these pigments form large, diffuse granules called *phycobilisomes,* which cover the scattered photosynthetic membranes rather like ribosomes on an endoplasmic reticulum (Figure 29.8).

The production of oxygen opened the door to aerobic respiration.

Oxygenic photosynthesis began to shift the atmosphere from reducing to oxidizing. The increasing oxygen concentration provided a challenge and an opportunity. Since oxygen is potentially destructive to cellular structures, organisms could only survive by evolving ways to handle oxygen or by taking refuge in anaerobic environments (such as deep lakes and sulfurous mud). The opportunity lay in using O_2 as a terminal electron acceptor and reducing it to water, since this allows cells to get more energy from their electron transport systems than they can obtain with other terminal acceptors. The eventual evolution of cytochrome enzymes that could reduce oxygen to water was a great event in Earth's history, for it created modern aerobic respiration.

The evolution of aerobic respiration may have been the critical event that made possible the evolution of eucaryotic cells. Since the DNA genome is vulnerable to destruction by oxygen, it was advantageous to sequester it in a nucleus and to confine oxygen utilization and production to separate organelles, the mitochondria and chloroplasts. Furthermore, when cells evolved mitochondria and chloroplasts, they could become larger and produce ATP in local regions of the

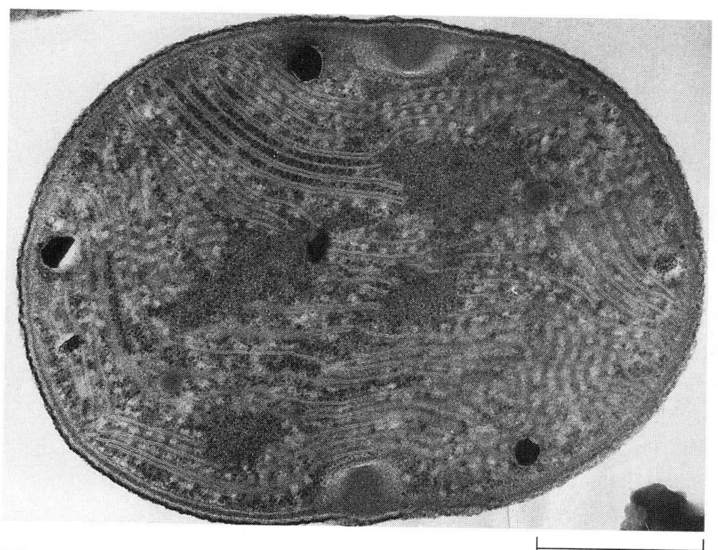

Figure 29.8
An electron micrograph of the cyanobacterium *Gleocapsa alpicola,* shows its photosynthetic membranes (thylakoids) covered with many phycobilisomes, the fluffy white particles.

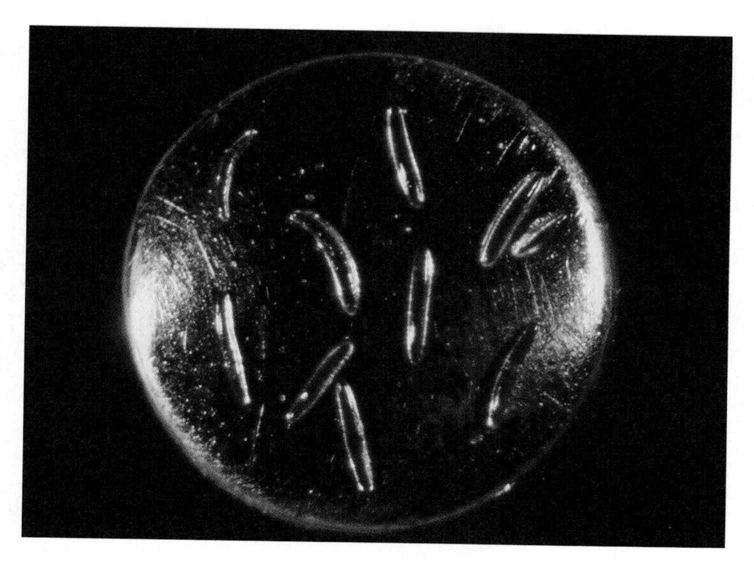

Figure 29.9
Epulopiscium, huge marine bacteria, are shown here on the head of a pin.

cytoplasm. Fossils of the first clear eucaryotic organisms are about 1.3 billion years old; we tell the story of early eucaryotic evolution in Chapter 30.

B. The kingdom Monera

29.5 Procaryotic cells have no true nucleus and are usually very small.

The thousands of known species of procaryotes (bacteria) occupy a wide range of environments—soil, hot springs with temperatures over 90°C, arctic ice shelves, the depths of the ocean, salt mines, plant roots, and the surfaces of larger organisms. Although some are pathogenic (causing disease), procaryotes as a group provide enormous benefits to humanity and to the whole biosphere. Bacteria are the principal decomposers of ecosystems. They build and transform soil, decompose waste products and detritus, and oxidize or reduce many compounds. Some fix nitrogen (N_2) into ammonia that can be used by plants and then by other organisms. Bacteria transform milk into cheese and cider into vinegar. Some antibiotics are bacterial products, and many industrial processes use bacteria. The bacteria inhabiting our body openings and intestines synthesize some essential vitamins such as vitamin K and prevent pathogens from colonizing those places.

All procaryotes except archaebacteria (Sidebar 29.1) constitute the kingdom Monera (Latin, *monos* = single). Most procaryotes are single cells with diameters of 1 μm or less, giving them a high surface-to-volume ratio. The procaryotic condition is not identical with small size, however; a few unusual species are 10–30 μm in diameter, and monster bacteria about 300 μm long were recently discovered (Figure 29.9). Procaryotic chromosomes are compacted into *nucleoids,* not surrounded by a nuclear envelope, and they have no endoplasmic reticulum or mitochondria. Their flagella, if present, are simple protein rods, not 9 + 2 complexes of microtubules. Each cell bears a single chromosome that is usually circular.

Procaryotic nucleoids, Section 6.5.

In addition, most procaryotes have a **cell wall** surrounding the plasma membrane. Since plant cells also have cell walls, this common feature led earlier taxonomists to place bacteria in the plant kingdom, but the two types of cell walls are not at all alike. Bacterial cell walls are built of a **murein,** also called **peptidoglycan,** a remarkable polymer made by bonding two kinds of monomers—amino acids and sugars—into a two-dimensional network (Figure 29.10). Mureins are the largest covalently bonded molecules in the world, gigantic bag-shaped molecules that completely enclose the cell.

Bacterial cell walls are quite rigid but rather porous, and they give bacteria their characteristic shapes. A cross-linked murein is strong enough to withstand the great turgor pressure created by the osmosis of water into a cell. Antibiotics such as penicillin and ampicillin kill by disrupting this structure; they inhibit enzymes that add new sugar-peptide units to the murein during growth, so the molecule becomes weakened, and the cell eventually bursts. We can use such antibiotics to fight infections because animal cells, lacking mureins, are not affected by them. The murein is also attacked by the enzyme lysozyme, which is abundant in egg white and in tears, saliva, and other secretions; this enzyme protects us against bacterial infections by lysing the invading cells. When lysozyme is added to bacteria in a medium with the same osmotic pressure as the bacterial cytoplasm, the cells assume a spherical

Sidebar 29.1 Kingdoms or Domains? The Place of Archaebacteria

As we explain in Chapter 2, the ancient division of the living world into plant and animal kingdoms was improved by the establishment of a third kingdom, Protista, for problematic organisms. The classification we use in this book recognizes a kingdom Monera for the procaryotes and five kingdoms for eucaryotic organisms: Protista, Chromista, Fungi, Plantae, and Animalia. The fundamental distinction between cell types may be reflected by establishing superkingdoms—Procaryota and Eucaryota.

Carl Woese and others have argued that some procaryotes, which Woese called **archaebacteria,** are remarkably different from both other procaryotes and the eucaryotes. Archaebacteria appear to be the remnants—refugees, almost—of ancient conditions (*archae-* = ancient). They include methanogens, which produce methane as a by-product of metabolism; some halophilic bacteria that live only in very salty water; and thermoacidophiles that live in hot, highly acidic conditions. Their distinctiveness appeared in studies of ribosomal RNA structure and some enzymes of the genetic apparatus, but their unusual features include lipids with ether (instead of ester) linkages and some peculiar coenzymes (Table 29.A). Notice that archaebacteria share characteristics both with other procaryotes and with eucaryotes. To reflect these features, Woese and others propose to establish three domains: Archaea, Procarya (Monera with archaebacteria removed), and Eucarya. The archaebacteria are certainly unusual, but whether they are unusual enough to justify such a classification is a decision the biological community must make.

The hypothetical scheme for the early stages of cellular evolution illustrated here shows how ancient archaebacteria may have been converted into eucaryotes by developing true nuclei and other organelles, as discussed in Section 30.7. Progenotes were the hypothetical first cells with functional genomes. Urcaryotes (German, *ur* = ancient or original), the ancestors of eucaryotes, may have shared many features with archaebacteria.

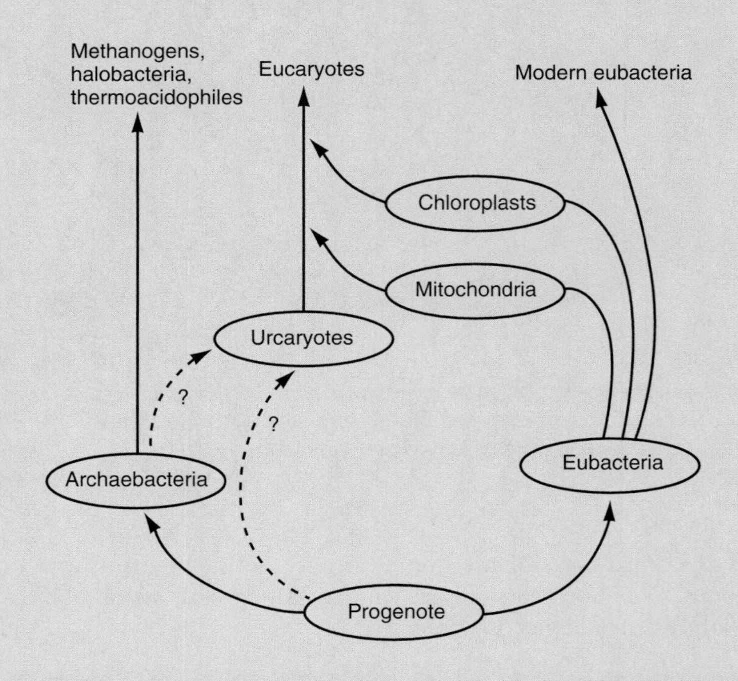

Table 29.A Features of Major Types of Organisms

Feature	Archaebacteria	Eubacteria	Eucaryotes
Typical cell size	About 1 μm	About 1 μm	10–50 μm
Nuclear envelope	Absent	Absent	Present
Cell wall	Various, but no mureins	Various, all with mureins	None in animals; various in other types, but no mureins
Membrane lipids	Ether-linked, branched	Ester-linked, straight	Ester-linked, straight
Initiator amino acid	Methionine	Formyl-methionine	Methionine
Ribosome subunits	30S, 50S	30S, 50S	40S, 60S
Lengths of ribosomal RNAs, bp	1,500, 2,900	1,500, 2,900	1,800, 3,500
Translation-elongation factor	Reacts with diphtheria toxin	Does not react with diphtheria toxin	Reacts with diphtheria toxin
Reaction to chloramphenicol	Insensitive	Sensitive	Insensitive

shape, no longer restrained by their walls. Such cells, called *spheroplasts* or *protoplasts,* are useful for certain experiments.

29.6 Bacteria may be spheres, rods, spirals, or long filaments.

When the French microbiologist Casimir Devaine first saw the rod-shaped organisms that cause anthrax, he called them "bacteria," from the Greek word *baktron,* meaning a rod. This general term still applies to all procaryotes, even though they aren't all rod-shaped. The common bacterial forms are small spheres, or **cocci** (pronounced "cock-sye"; sing., **coccus**); rods that curve, taper, or twist; and spirals (**spirilla;** sing, **spirillum**) (Figure 29.11*a*). These shapes often provide both the formal and informal names of many bacteria. Thus, diplococci are pairs of cocci (*diplo-* = double), streptococci are chains (*strepto-* = chain), and staphylococci form random clumps (*staphyle* = a bunch of grapes) (Figure 29.11*b*).

Most bacteria, if they are able to move at all, use flagella made of flagellin. In contrast, **spirochetes** are soft-walled, flexi-

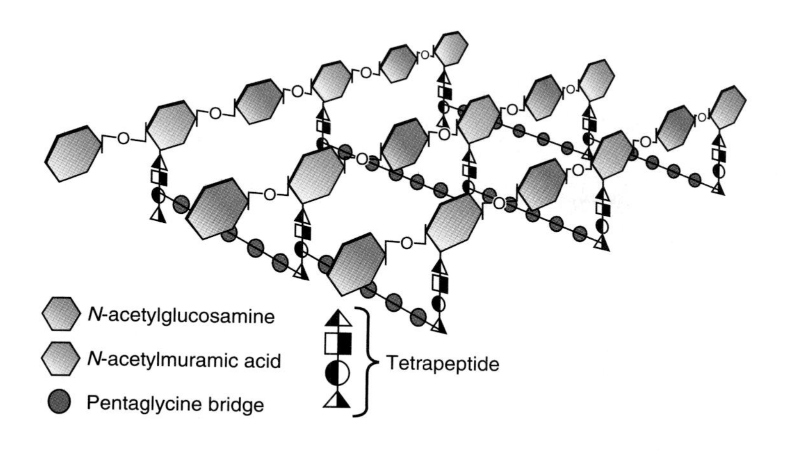

Figure 29.10

A murein, or peptidoglycan is made of two types of sugars, *N*-acetylglucosamine and *N*-acetylmuramic acid, alternating in a polymer chain and cross-linked by amino acids. Generally, a tetrapeptide, of four amino acids, connects the *N*-acetylmuramic acid residue; and the tetrapeptides are linked by a pentaglycine bridge made of five glycine residues. Such a covalently bonded sheet can be extended indefinitely and is actually curved around on itself to make a closed bag surrounding the whole cell.

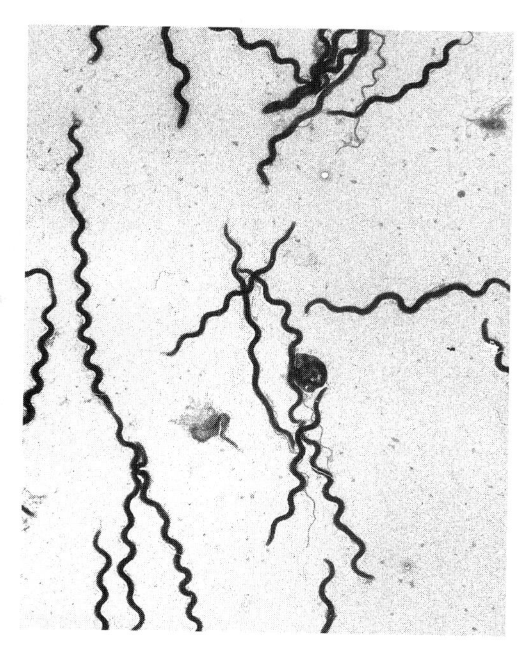

Figure 29.12

Spirochetes are common soil and water bacteria.

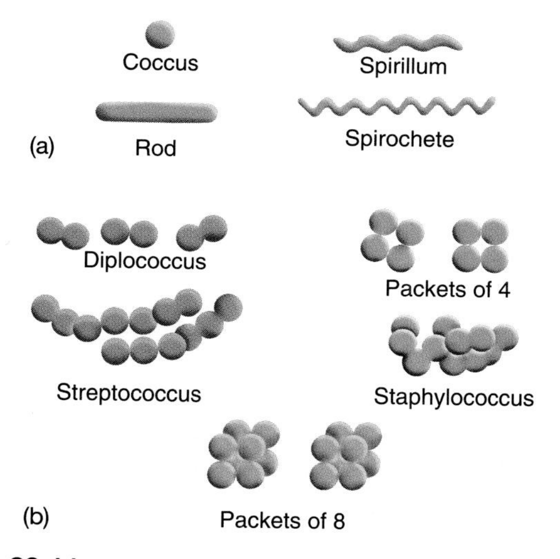

Figure 29.11

(a) Bacteria commonly have the forms of cocci (spheres), rods, or spirals of various kinds. *(b)* The cells, especially cocci, often associate with one another in several ways such as pairs, chains, random clusters, or packets of four or eight.

ble procaryotes that zip along through water like tiny corkscrews (Figure 29.12). They propel themselves by contracting an axial filament made of fibers similar to bacterial flagella that are attached at the ends of the cell and overlap in the middle. Some spirochetes are 500 μm long, yet they are thin enough (0.1–0.6 μm in diameter) to pass through filters that remove most other bacteria. Most spirochetes are soil and water organisms, but some are pathogens, such as *Treponema pallidum*, the causative agent of syphilis.

Bacterial flagella, Section 11.14.

Another group of procaryotes with flexible cell walls, the **slime bacteria** or **myxobacteria,** move with a gliding mechanism. They are common in soils, on rotting materials, and in animal dung, and are especially remarkable for living on other microorganisms such as green algae and cyanobacteria, which they actively kill with secreted enzymes and antibiotics. Myxobacteria aggregate into mushroom-like clusters, which may then become quite large and beautiful. In some species, certain cells in the cluster become spores that are tough enough to weather dry conditions; in other species, the whole mass may encyst until it can start to grow again (Figure 29.13).

Other unusual bacteria, found frequently on the surfaces of aquatic plants and animals, project stalks or blunt extensions (prosthecas), while stalked bacteria, many forming heavy filaments of slime, are common in contaminated water (Figure 29.14). For instance, *Gallionella* is an iron bacterium that grows profusely in iron-rich water and deposits ferric hydroxide in its stalk.

Mycobacteria include the agents that cause tuberculosis and leprosy, which have been sources of fear and misery throughout human history. Their irregular, branched cells have unusual cell walls containing complex lipids. The related **actinomycetes** form long, branched filaments with many nucleoids, but they are coenocytic because they lack cross-walls or have only incomplete walls (Figure 29.15). A representative actinomycete, *Streptomyces,* reproduces by forming spores, each containing a chromosome, at the end of each filament. Actinomycetes create the characteristic smell of fresh soil, and many species are cultivated commercially to produce important antibiotics, such as streptomycin, that kill other procaryotes.

Figure 29.13
The myxobacterium *Chondromyces* can reproduce as vegetative cells that divide like other bacteria, or the cells can form a fruiting body, which forms myxospores that germinate into vegetative cells.

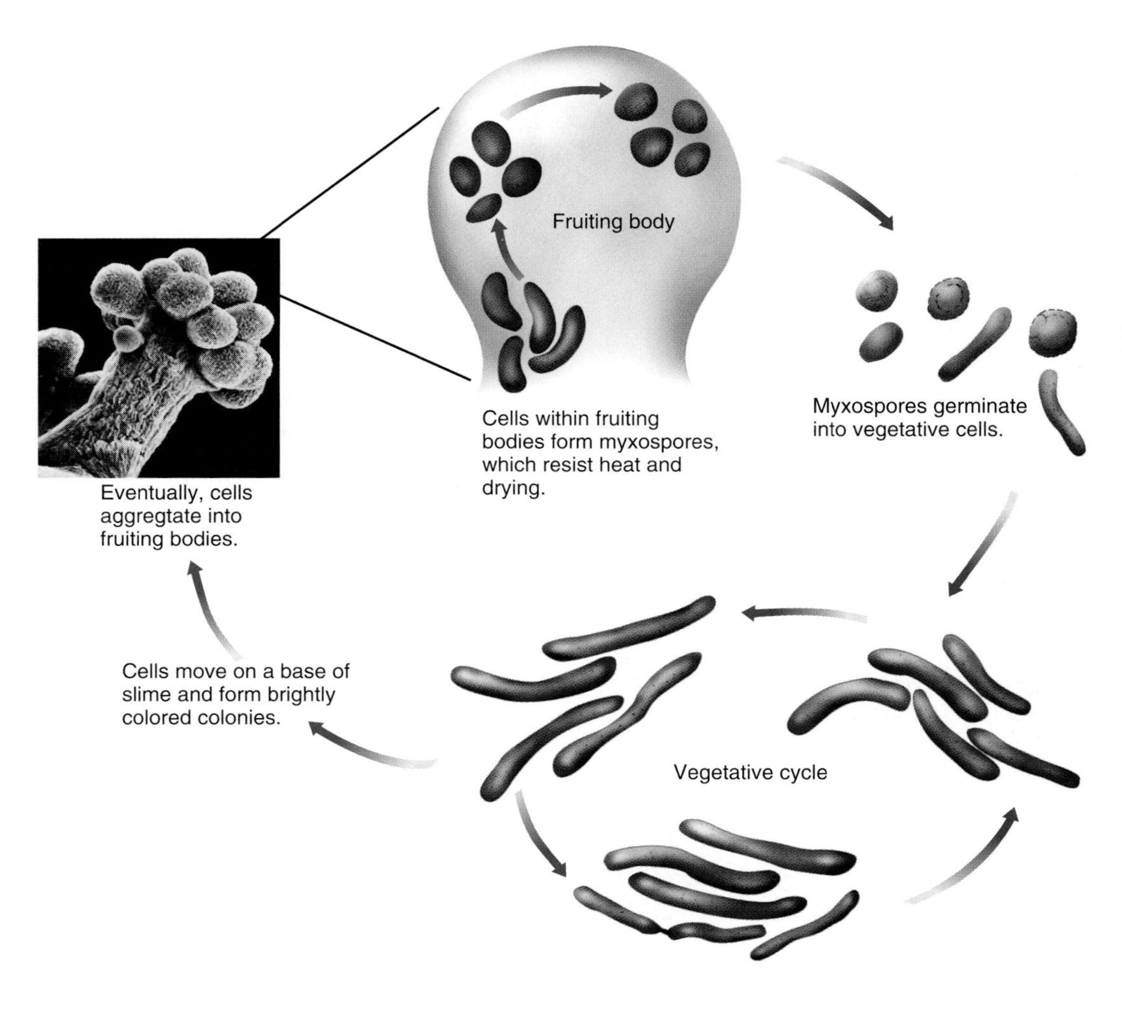

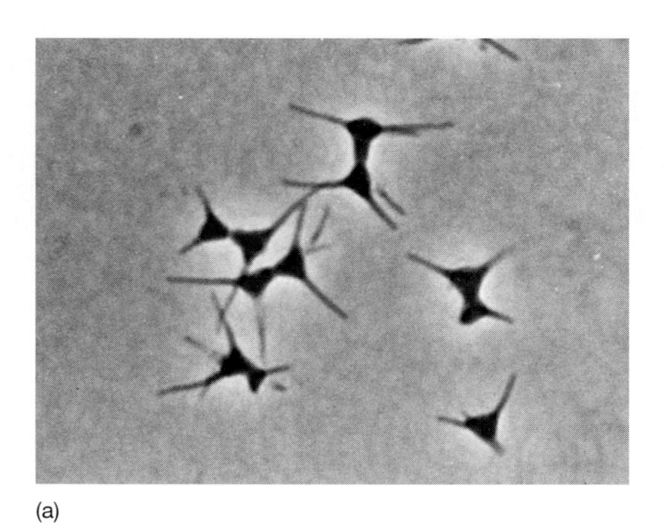

(a)

(b)

Figure 29.14
Some bacteria have unusual shapes. *(a) Ancalomicrobium* is a prosthecate bacterium; a prostheca is a narrow extension of the cell. *(b) Hyphomicrobium* is a stalked organism.

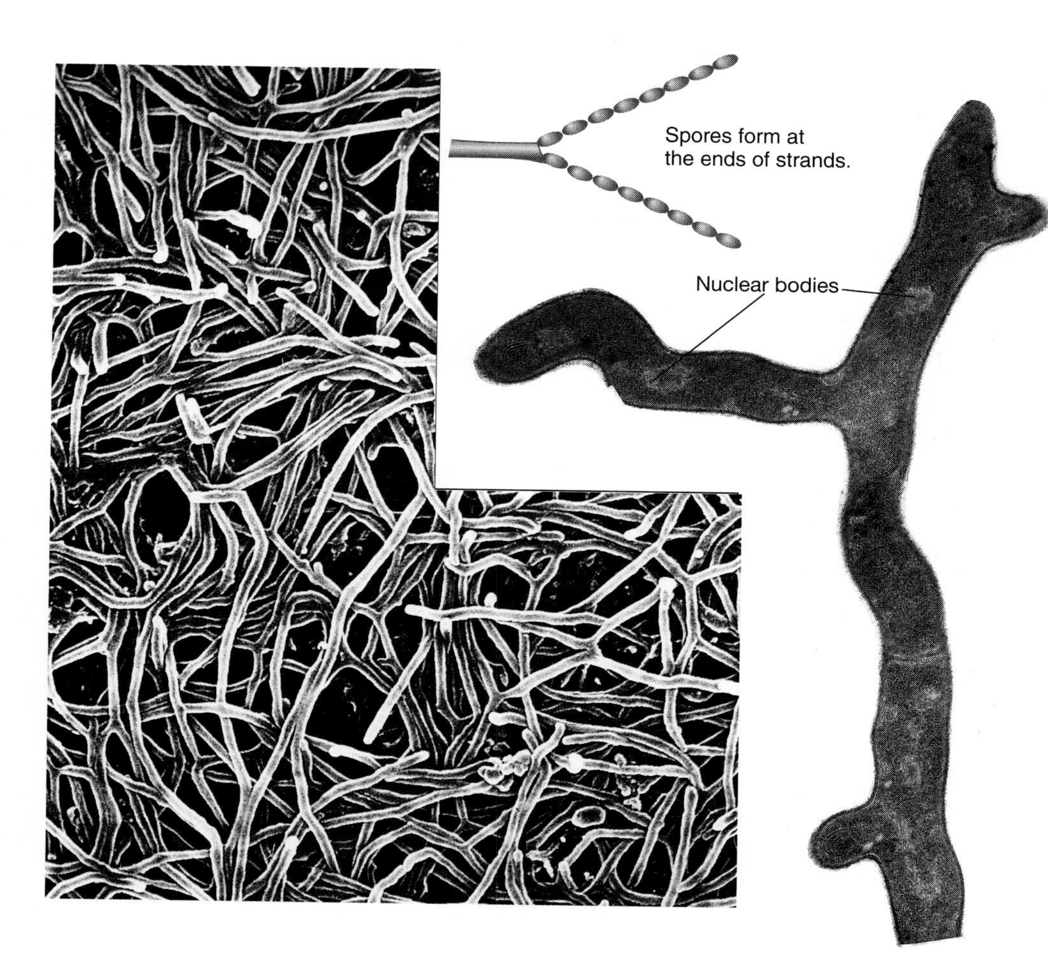

Figure 29.15
The bacterium *Streptomyces* grows as a mass of long filaments. An electron micrograph shows its coenocytic structure, with many nuclear bodies (nucleoids) only irregularly separated by cell walls. Chains of spores often form at the ends of strands.

In addition to these large cells, the procaryotes include **mycoplasmas,** the smallest of cells, with diameters of only 0.1–0.3 μm. Because they lack cell walls, mycoplasmas can grow into irregular, elongated forms (Figure 29.16). Since the volume of a sphere increases as the cube of its radius, these cells have only a hundredth to a thousandth the volume of more typical bacteria. Still, they contain a DNA genome and all the essential enzymes needed for complete metabolism. It is possible to estimate the volume needed for a membrane enclosing a genome, a few ribosomes, and single copies of the necessary enzymes, with enough water to function. This calculated volume is close to the actual volume of a mycoplasma, making them the smallest possible cells. While some mycoplasmas are harmless inhabitants of soil and vegetation, others are the agents of plant or animal diseases such as pleuropneumonia, a lung infection of cattle.

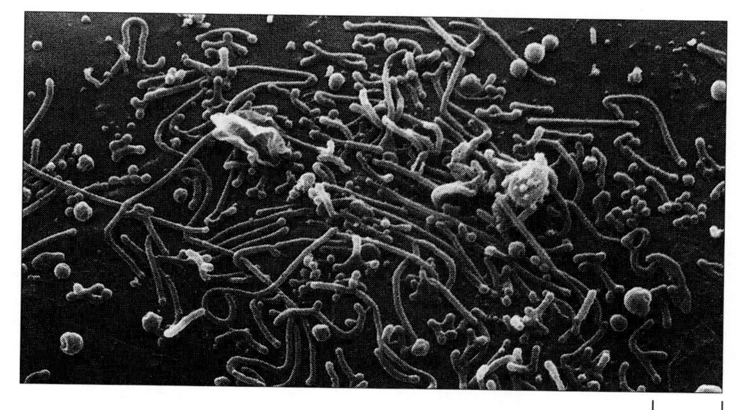

Figure 29.16
Mycoplasmas, the smallest known cells, have the minimal apparatus needed for metabolism and reproduction. They can assume a variety of irregular shapes.

29.7 Bacteria are classified by cell shape, metabolism, and reaction to the Gram stain.

Earlier classifications of procaryotes attempted to reflect their phylogeny, but it is now generally agreed that those arrangements were artificial; they included polyphyletic taxa and jumped to phylogenetic conclusions that couldn't be supported by good evidence. The best consensus on procaryotic taxonomy, in *Bergey's Manual of Determinative Microbiology,* avoids this mistake by simply using a phenetic classification with several groups that appear to be natural (Table 29.1), without implying any phylogeny.

Table 29.1 defines the major groups primarily by three criteria. The first is cell morphology, as outlined in Section 29.6.

Table 29.1 Major Groups of Procaryotes

I. Cocci
  A. Gram-positive cocci
  B. Gram-negative cocci
    1. Aerobic gram-negative cocci and coccobacilli
    2. Anaerobic gram-negative cocci

II. Rods and related coccal forms
  A. Gram-positive bacteria
    1. Endospore-forming rods and cocci
    2. Non-spore–forming rods
    3. Coryneform bacteria and actinomycetes
  B. Gram-negative bacteria
    1. Aerobic rods and cocci
    2. Facultatively anaerobic rods
    3. Strictly anaerobic bacteria
    4. Aerobic chemoautotrophic bacteria
    5. Sheathed bacteria
    6. Archaebacteria

III. Curved rods and flexible cells
    1. Gram-negative, aerobic spirilla and curved bacteria
    2. Gram-negative, anaerobic curved bacteria
    3. Spirochetes

IV. Special groups
  A. Phototrophs
    1. Anaerobic, anoxygenic phototrophs
    2. Aerobic, oxygenic phototrophs (cyanobacteria)
  B. Small, parasitic bacteria
    1. Rickettsias and chlamydias
    2. Mycoplasmas
  C. Bacteria with unusual forms or cell walls
    1. Gliding bacteria
    2. Bacteria with appendages, prosthecate bacteria, budding bacteria

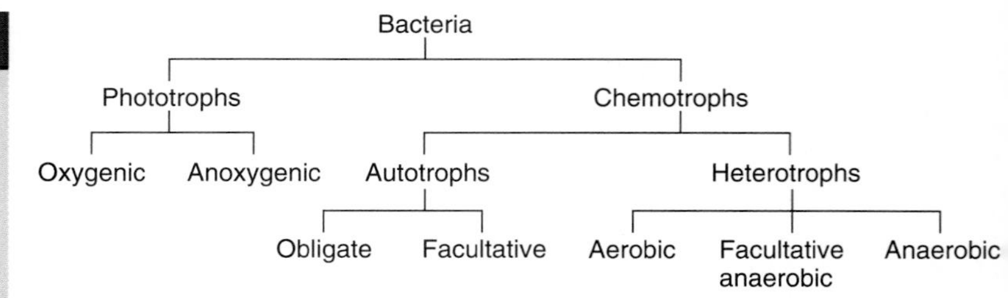

Figure 29.17
Bacteria can be classified into several metabolic categories.

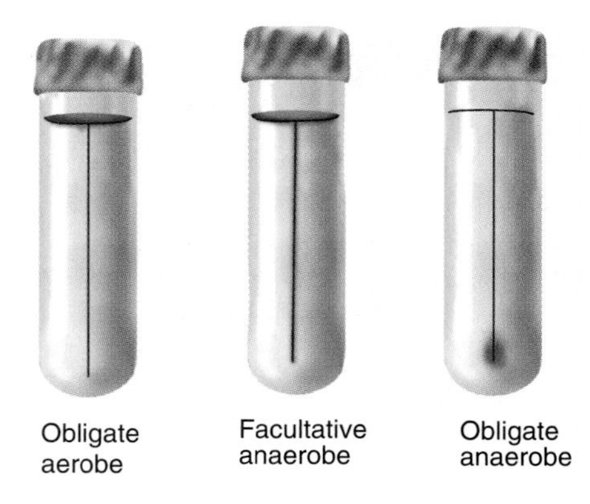

Figure 29.18
A nutrient agar medium is poured into deep tubes, and bacteria are introduced by stabbing the agar with a thin needle that has bacteria on its end. Obligate aerobes grow only at the top, obligate anaerobes grow only at the bottom, and facultative anaerobes grow throughout the tube, showing their ability to live both with and without oxygen.

The second is metabolic pattern; that is, phototrophs are separated from chemotrophs, autotrophs from heterotrophs, and oxygenic phototrophs from anoxygenic, as shown in Figure 29.17. Metabolic types are also distinguished as **obligate** or **facultative;** an organism's way of life is obligate if it is obliged to live that way and facultative if it can live in other ways. Obligate anaerobes, for instance, can only live in environments with no trace of oxygen, while facultative anaerobes can survive either with or without oxygen (Figure 29.18). The majority of known bacteria are aerobes, but many are facultative anaerobes. The group containing obligate anaerobes is probably polyphyletic.

A third criterion for classifying bacteria is their reaction to the **Gram stain,** devised in 1886 by Hans Christian Gram. This simple, quick procedure distinguishes gram-positive bacteria, which retain the stain, from gram-negative bacteria, which do not (Figure 29.19). Merely seeing that a bacterium is a gram-positive rod or a gram-negative diplococcus tells a bacteriologist a great deal and sometimes identifies its genus. Electron microscopy reveals that gram-positive bacteria have a simpler structure than gram-negative bacteria (Figure 29.20).

C. The ecology and uses of bacteria

The variety of bacteria is incredible, yet their differences are subtle. Walking through a zoo, it is easy to be impressed by the variety of animals because they are all so large and their differences are so obvious. If we were all reduced in size by a factor of a billion or so, the bacterial zoo would be just as astonishing. But just by looking at bacteria through microscopes, it is hard to appreciate them and easy to overlook their importance to the rest of the living world. Still more subtly, their varied ways of metabolism provide an additional source of fascination. So to develop a greater appreciation for bacteria, we will focus on

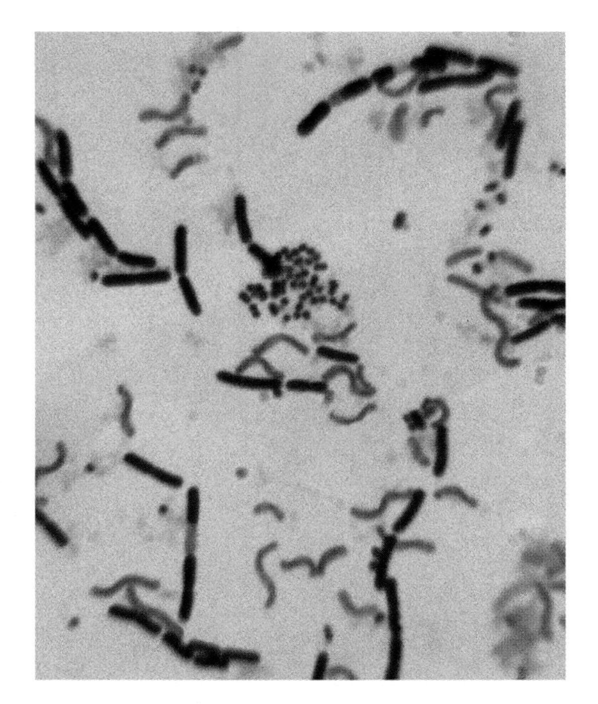

Figure 29.19
A mixed culture of bacteria that have been gram-stained shows the difference between purple, gram-positive cells (large, straight rods and a cluster of small, round cocci) and pink, gram-negative cells (thin, curved rods).

their ecological relationships and activities, including those most important to humans.

29.8 Bacteria have critical roles in every ecosystem.

The biosphere is suffused by a world of bacteria living on every surface, in every crevice, outside and often inside other organisms. Their activities are crucial ecologically. Bacteria are most significant as decomposers, organisms that reduce wastes and detritus to simpler materials. Without their activities, an ecosystem would choke on its own wastes. Bacteria (and molds) produce extracellular hydrolytic enzymes that attack the polymers of these wastes, releasing monomers on which they, and others that share their living space, can grow. Bacteria in turn become food for small animals; thus some of their mass is recycled back into the food chain.

The detritus food web, Section 28.6.

Bacteria that live on unusual transformations of inorganic compounds also play important roles in the cycling of materials

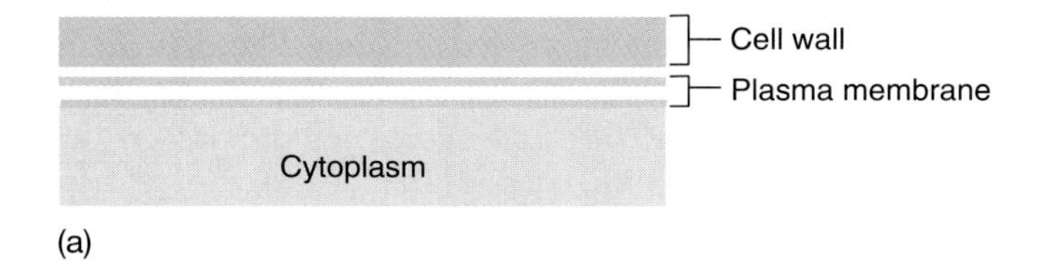

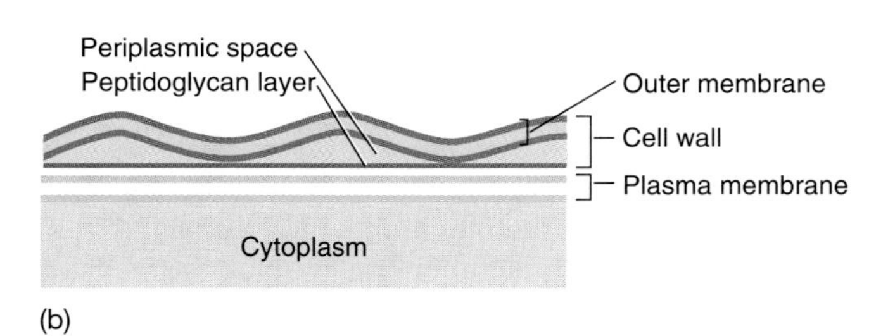

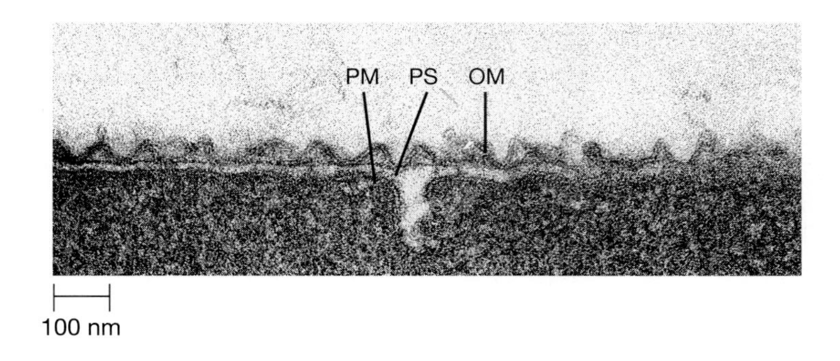

Figure 29.20
(a) The wall of a gram-positive cell is a thickened layer outside the plasma membrane (PM on micrograph). *(b)* Gram-negative cells have an additional outer layer (OM on micrograph), a lipid-protein membrane that has quite a different composition from the inner plasma membrane and is separated from it by a periplasmic space (PS on micrograph).

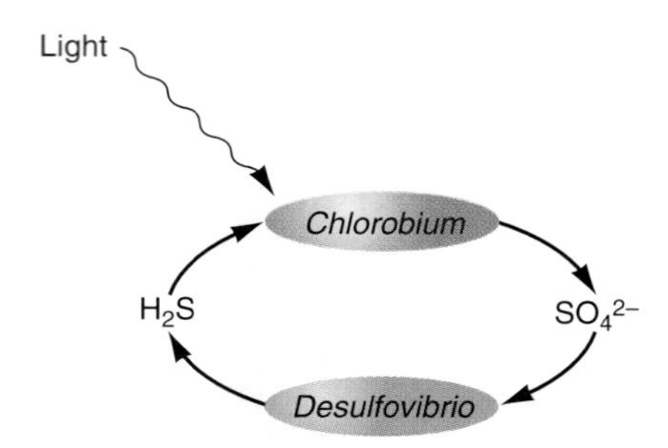

Figure 29.21
The phototrophic bacterium *Chlorobium* and the sulfate-reducing bacterium *Desulfovibrio* engage in a simple ecological relationship based on sulfur metabolism.

in ecosystems, performing critical steps in the nitrogen and sulfur cycles. Sometimes two or more species of bacteria cooperate fortuitously to degrade an unusual organic compound stepwise by combining their enzymatic capabilities, although each one by itself is unable to grow on the compound. Figure 29.21 shows a simple cycle between a phototrophic bacterium, *Chlorobium,* and a sulfate-reducing bacterium, *Desulfovibrio. Chlorobium* brings energy into the system and oxidizes H_2S to sulfate, while *Desulfovibrio* reduces the sulfate back to sulfide.

Bacteria in nitrogen and sulfur cycles, Section 28.5.

A similar relationship is established between *Thiobacillus ferrooxidans* and *Beijerinckia lacticogenes* (Figure 29.22). The thiobacillus is a chemoautotroph that reduces CO_2 to organic compounds, some of which are metabolized by *Beijerinckia;* the latter fixes nitrogen, producing nitrogenous compounds that are used by the thiobacillus. Through this mutualism, the two organisms grow together much better than either could grow alone. Such associations are particularly useful in extracting metals from low-grade ore through leaching: dissolving the metal compound out of rock, usually in an acid solution. Some industrial leaching processes are terribly dangerous and destructive; for instance, cyanide can be used to extract gold, but with disastrous effects on humans and the environment. However, organisms such as thiobacilli can produce just the right conditions for leaching certain metal ores relatively harmlessly.

29.9 Many bacteria cause infectious diseases.

Illness is a normal part of life. The **etiology** of a disease—its underlying cause or causes—is either functional or infectious. **Functional disease,** such as heart disease, results from a malfunction in the organism itself. **Infectious disease** is due to some other organism or virus that grows as a parasite, obtaining its nourishment through ectoparasitism on the surface of the host or through endoparasitism somewhere within its intestinal tract or other cavities or even within its tissues. Diseases are caused by bacteria, protozoans, fungi, worms, and viruses.

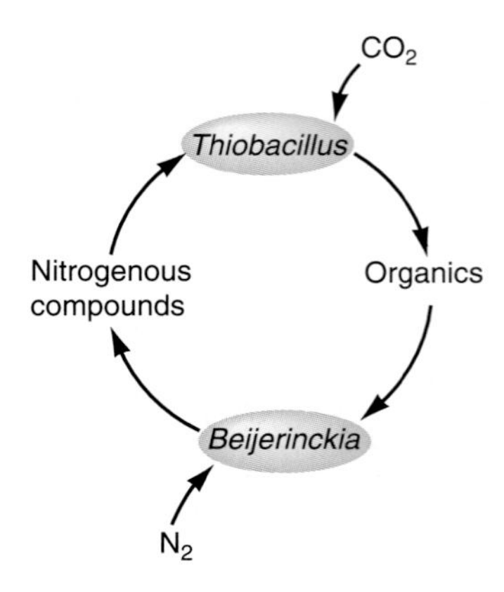

Figure 29.22
The chemoautotrophic bacterium *Thiobacillus ferrooxidans* and the nitrogen-fixing bacterium *Beijerinckia lacticogenes* have a mutualistic relationship.

Some small bacteria, such as *Bdellovibrio,* can even grow inside larger bacteria (Figure 29.23).[1]

A triumph of nineteenth-century medicine was the establishment of the germ theory of disease, the idea that many diseases are the result of a "germ," or infectious agent. Until then, disease had been attributed to all kinds of fanciful causes, and even though many diseases were known to be contagious, the basis for the contagion was a mystery. We owe our modern understanding largely to the research of Louis Pasteur. In 1835, when the cell theory was just being developed, Charles Cagniard-Latour and Theodor Schwann showed independently that fermentation in beer and wine is caused by the growth of yeast, a microorganism. Their findings earned them the ridicule of chemists, who could only laugh at the idea that small organisms could conduct what they saw as a purely chemical process. Pasteur's meticulous experiments, however, left no doubt that transferring yeasts and bacteria from one culture of plant juices or milk to another also transfers fermentative activity. He then showed that wines become "sick"—sour or bitter or stringy—because they are contaminated with undesirable bacteria. To counter such contamination, he invented the process now called **pasteurization,** in which the wine (or other food, such as milk) is heated briefly, just enough to kill the responsible organisms while leaving the food unharmed.

Pasteur then realized that animal diseases might also result from infection by microorganisms, and in the 1860s he demonstrated that infectious agents were indeed responsible for silkworm diseases, which at the time were ravaging the French silk industry. About a decade later, painstaking investigations conducted both by Pasteur and by Robert Koch showed that bacteria cause anthrax, a disease that can strike many mammals, including humans. Koch found that anthrax bacteria *(Bacillus anthracis)* form protective spores that may endure for a long time, so animals can be infected merely by grazing in certain pastures.

[1]The classic verse, attributed to various authors, is: "Big fleas have little fleas / Upon their backs to bite 'em, / And little fleas have lesser ones, / And so ad infinitum."

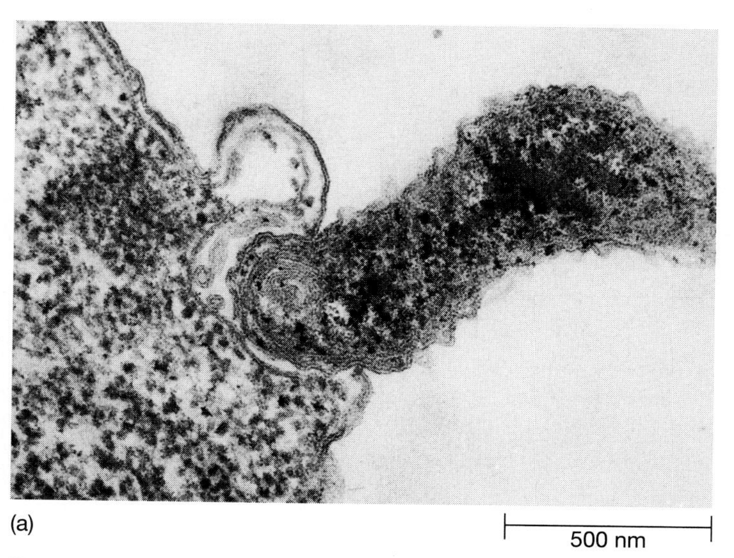
(a) 500 nm

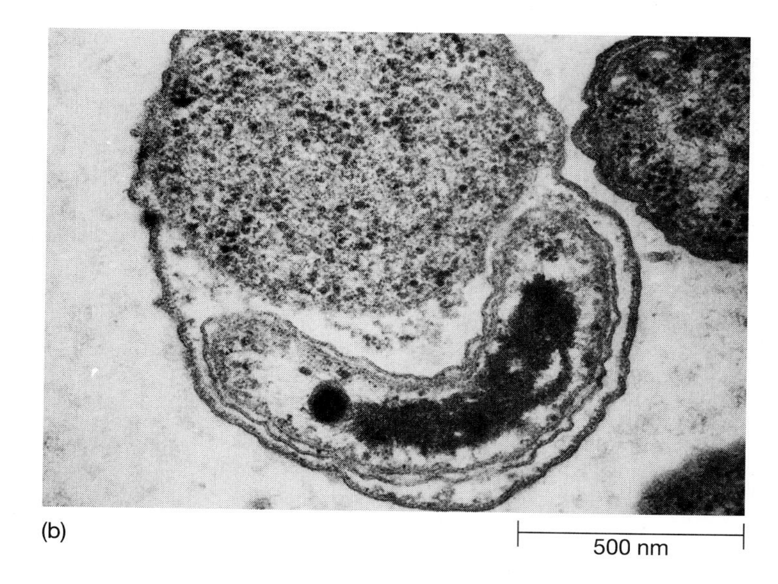
(b) 500 nm

Figure 29.23

Bdellovibrio infects larger bacteria by boring through their cell walls and membranes *(a),* to grow inside. *(b)* A *Bdellovibrio* lies within the cytoplasm of its host.

Koch devised a set of four conditions, now known as **Koch's Postulates,** that must be satisfied before a disease can be positively attributed to a certain organism. First, the organism must be recovered from animals that have the disease. Second, it must be grown in pure culture. Third, when healthy animals are inoculated with this culture, they must contract the disease. Fourth, the organism must again be recoverable from an inoculated animal. Using this method, the etiologic agents of many diseases were identified during the next decades.

Around 1864, the English surgeon Joseph Lister realized that infectious agents might be responsible for the high rate of sepsis (decay and death of tissues) associated with surgery, and he instituted procedures such as the sterilization of surgical instruments and the use of disinfectants.[2] These first steps toward modern surgical techniques naturally did much to reduce post-surgical complications and death. But old ideas die hard. One pathologist has recorded that, even at the beginning of the twentieth century, some surgeons in Dublin were still sharpening their scalpels on the leather soles of their boots before an operation to show their contempt for the germ theory of disease.

29.10 Pathogens produce disease through invasion and toxin production.

Pathogens are either *invasive,* causing disease by invading the host and growing in its tissues, or *toxigenic,* causing disease by producing **toxins,** or poisons. An extremely invasive pathogen is *Clostridium perfringens,* the cause of gas gangrene. Clostridia form spores that may lie dormant in the soil for a long time until they are carried into a wound, especially a deep puncture wound (from the proverbial rusty nail your mother warned you about). In the warm, moist tissue, they proliferate and make several extracellular enzymes: Collagenases attack the collagen fibers holding the tissue together; a hemolysin attacks red blood cells; a lecithinase destroys the lecithin (phosphatidyl choline) of cell membranes; and other proteases digest cell proteins. More bacteria then slip between the loosened and destroyed cells, invading the tissue further and living on the digested flesh. If not treated quickly, they can easily turn a once healthy foot into a rotting, blackened mess. *Bacillus anthracis* is also an extremely invasive organism; an animal that dies of anthrax has massive numbers of the bacteria growing in its bloodstream.

In contrast, *Clostridium tetani,* the agent of tetanus, does its damage almost exclusively with a toxin, a neurotoxin that inhibits transmission of information at some synapses in the central nervous system. Similarly, *Clostridium botulinum,* the agent of botulism, produces the botulinus toxin that paralyzes the neuromuscular junctions where nerve endings contact muscles; it is hardly invasive at all, since it does its damage while growing in improperly preserved foods, which people then eat.

29.11 Infectious agents are transmitted to new hosts from reservoirs of infection.

The growth of parasitic organisms in a larger host, with the accompanying destruction of tissues and production of toxic byproducts, produces an infection. Most parasites are restricted to a single host species or to a few closely related species.

Every organism is a potential habitat for others. Every surface, every nook and pocket of a multicellular organism is a specialized ecological niche where something else can grow. Eventually, most organisms achieve a certain equilibrium with one another, so the little hangers-on do the larger organism no

[2] *Sterilization* is the killing and removal of all microorganisms, spores, and viruses, particularly on materials to be used for medical procedures or biological experiments. *Disinfection* is the destruction, inhibition, and removal of microorganisms likely to cause infections or other undesirable effects, especially from large objects and spaces. *Pasteurization* is a process of briefly heating milk and other liquids just enough to eliminate organisms that cause disease and spoilage, leaving many living organisms.

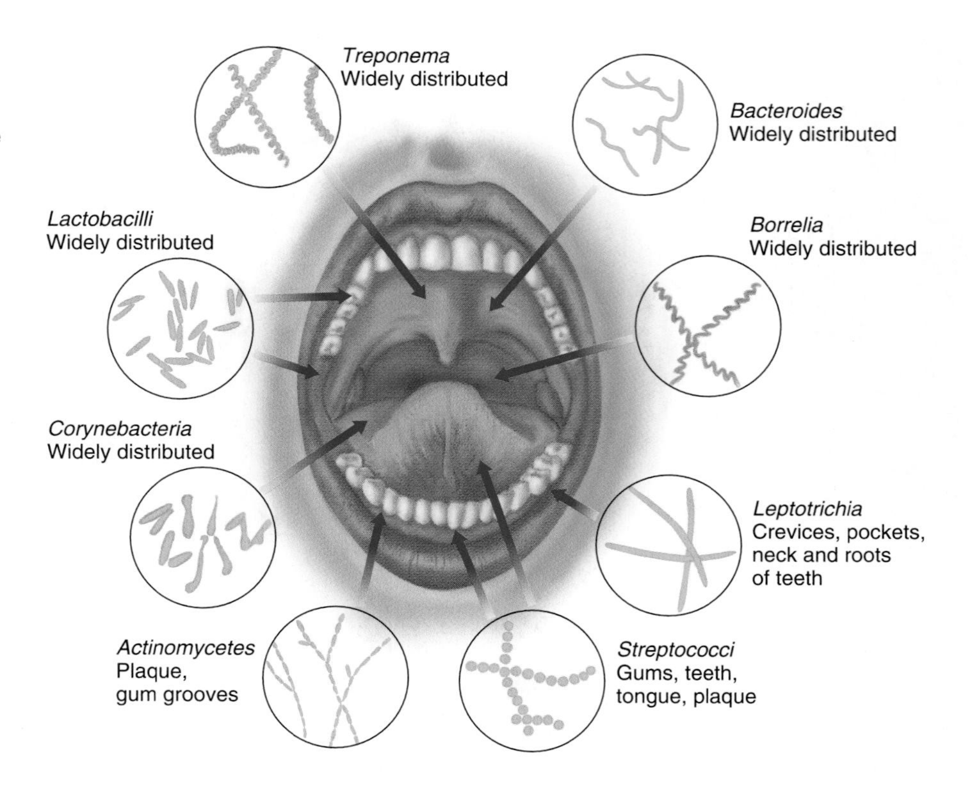

Figure 29.24

The normal mouth flora includes many kinds of bacteria and some fungi. Every surface and crevice forms an ecological niche for one or more species.

particular harm (commensalism) and may even help it (mutualism). Throughout our mouths, in the crevices of the gums, over tooth surfaces, in the pits of the tongue, and across the mucous membrane of the pharynx (where the nose and mouth join the throat), we all carry a rich flora of pneumococci, micrococci, neisseriae, lactobacilli, corynebacteria, spirochetes, actinomycetes, and eucaryotic yeasts such as *Candida* (Figure 29.24) (Since bacteria were once considered plants, an array of bacteria is called a "flora.") *Streptococcus salivarius* clings to the surface tissues of the mouth, while its relative, *S. mutans,* clings to tooth enamel and is the major agent of dental caries (tooth decay).

Commensalism and symbiosis, Section 27.14.

Large animals like humans have other rich floras over the entire skin, in the large intestine, and in the urogenital canals. Not only are these bacteria generally harmless, but they may even be essential and beneficial because they occupy spaces that pathogens might otherwise occupy, and their acid by-products inhibit the growth of other organisms. We are adapted to living with bacteria, as they are to living with us. Each area of the body has its own characteristic normal flora as well as characteristic pathogens that can initiate disease there. Our associations with microorganisms work perfectly well as long as they stay in their place; however, every organism is potentially an opportunist, and if the occasion arises—either genetically or environmentally—for a microorganism to grow faster at the expense of its host, it will do so. Some components of the normal flora may start infections if they get into an area where they don't normally live. For example, *Escherichia coli,* which normally inhabits the large intestine harmlessly, is the most common cause of urinary tract infections in humans, and the bacteria are almost always derived from the infected person's own intestinal flora. We rely on defense mechanisms, such as our immune system, to keep pathogens from entering places where they can start infections or to kill them if they do enter.

Every pathogen has a **reservoir of infection,** a place where it normally resides but may not cause any disease. The reservoir of infection for human disease may be in other people, in other animals, or in the soil. The agent may be transmitted to the host by direct contact, as from touching, kissing, or sexual intercourse; through contaminated objects or food; via airborne particles, like droplets released from the nose and mouth during sneezing and coughing; from infected large animals; and by insects or other invertebrate **vectors,** animals that can transmit pathogens from one host to another (Figure 29.25).

Rabies, for example, can be transmitted directly from an animal reservoir to humans. Though rabid dogs are the best-known vectors, the extensive reservoir includes foxes, wolves, coyotes, skunks, jackals, mongooses, raccoons, cats, and bats. The rabies virus, which can only survive in birds and mammals, is transmitted directly from one host to another through a bite wound, and one effect of the virus is to induce vicious biting behavior in its victims.

The most notorious vectorially transmitted disease is probably plague, caused by the bacterium *Yersinia pestis* (named for Alexander Yersin, who identified it as the causative agent in 1894). The *Y. pestis* reservoir is primarily in wild rodents, although several other kinds of mammals may be infected. Outbreaks of *Y. pestis,* called sylvatic plagues, are transmitted primarily by rat fleas and recur periodically among wild rodents. These fleas apparently carry the bacteria through human

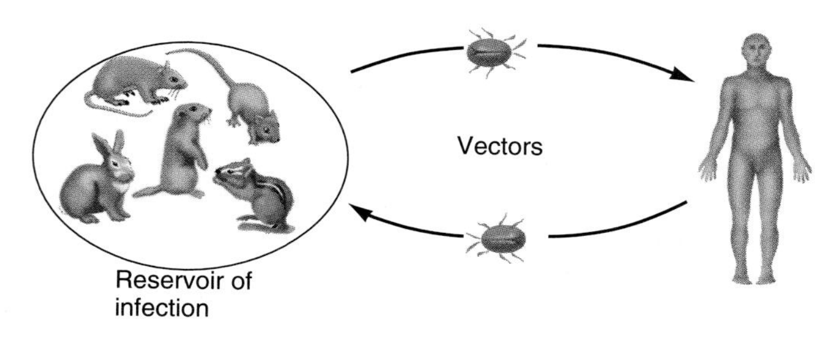

Figure 29.25

An infectious agent resides in a reservoir, including infected hosts, from which it is transferred to another host where it creates an infection. The transfer often requires a specific animal vector.

populations, sometimes with disastrous results. The Great Plague, beginning in 542 A.D., killed an estimated 100 million people in Europe and Asia over half a century, and the Black Death of the fourteenth century is believed to have wiped out a third of Europe's population. The last major epidemic of plague started in central Asia in 1871. Occasional outbreaks have occurred since then, but the disease can be controlled easily with modern antibiotic therapy. Without therapy, the pneumonic form of plague, a lung infection that is spread from person to person, is almost always fatal, and the bubonic form, which produces painfully swollen lymph nodes (buboes), is 70–90 percent fatal.

Though we are focusing on bacteria here, it should be clear that many diseases caused by organisms or viruses are spread via bites from insects, ticks, or other arthropods. Mosquitoes transmit malaria protozoa and yellow-fever viruses; the tsetse fly carries African sleeping sickness protozoa. Each pathogen has a reservoir among humans, other mammals, or sometimes birds, and it uses the arthropod as a vector. Generally, the pathogen has no other way to get into its host.

Plants, like animals, are subject to bacterial infections. Fire blight, for instance, is a bacterial disease that kills fruit trees so fast it leaves them looking as if they had been burned. Many soft rots of vegetables and fruits are bacterial infections, and crown gall tumors of plants are initiated by the bacterium *Agrobacterium tumefaciens.* Plant pathogens may have a hard time penetrating the tough, waxy walls of their hosts, but nematode worms, which open holes in the plant epidermis, are notorious vectors.

***Agrobacterium* infections,** Section 18.13.

29.12 Some very small bacteria are intracellular pathogens.

At the low end of the size scale, some particularly small bacteria have developed special ways of living. The **rickettsias** (Figure 29.26) are named for their discoverer, Howard T. Ricketts, who died in Mexico in 1910 from typhus, a rickettsial disease he was studying. Although these bacteria are true cells, typically about 0.3 μm by 1.0 μm, they are intracellular parasites that can only reproduce within a host cell because they lack certain metabolic machinery. Usually they infect the cells of arthropods, especially fleas, lice, ticks, and mites, which transmit them to humans and other animals through bites. Reproducing quickly in an animal's body, they cause such diseases as Rocky Mountain spotted fever, Lyme disease, Q fever, and epidemic typhus fever, which has been responsible for terrible plagues throughout human history.

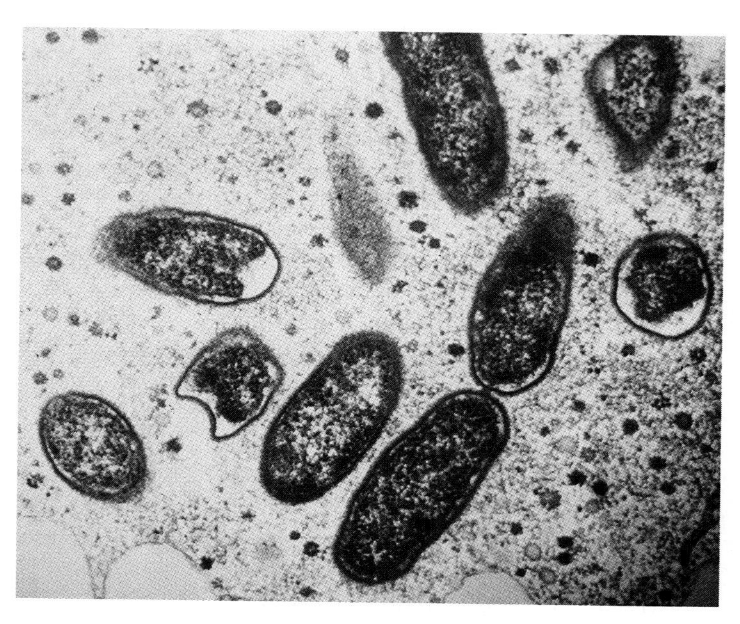

Figure 29.26

Rickettsias are small parasitic bacteria that are responsible for diseases such as typhus and Lyme disease. These rickettsial cells are growing inside a kidney cell.

Chlamydias are also very small intracellular pathogens. They depend even more completely on their host cells than do rickettsias, because they cannot metabolize glucose to make their own ATP. Chlamydias are responsible for the eye infection trachoma, a frequent cause of blindness. *Chlamydia psittaci* causes psittacosis, a kind of pneumonia transmitted by parrots and other birds, and other chlamydias are responsible for lymphogranuloma venereum, one of the lesser-known sexually transmitted diseases (STDs) of humans.

29.13 Many bacteria are used in industrial processes.

From the beginning of civilization, humans have taken advantage of microbial activities to enhance their food and drink. Natural yeasts growing on fruits fermented their juices and provided alcoholic drinks that became the basis of religious rituals and provided temporary relief from the burdens of life. Later, some prehistoric cook discovered that the same material that ferments fruit, when mixed with grain and water, could make the dough rise, and so bread was invented.

Microorganisms that produce ethanol provide a substrate for others that produce vinegar (French, *vinaigre* = sour wine). Wine must be kept anaerobic to prevent the growth of bacteria such as *Acetobacter,* which converts the ethanol into acetic acid:

$$CH_3CH_2OH + O_2 \rightarrow CH_3COOH + H_2O$$

The bacteria require oxygen and have the unusual ability (for procaryotes) to produce cellulose, which forms a mat on the surface of the wine and thus keeps the bacteria near the air.

Turning fruit juices into wines makes them more durable, for if wines are kept in tightly closed vessels, they will not change further. Before the invention of refrigeration, food storage was a real problem. One ancient solution was salting and smoking—hence, we have ham, bacon, and salt herring. Another solution, used on vegetables like cabbage and cucumbers, was pickling, a preservation technique that yields sauerkraut and pickles and relies on the action of lactate-producing bacteria (see Section 9.11). These bacteria have long been used to convert milk into buttermilk, yogurt, and kefir, and they play a major role in the formation of most cheeses. Even butter manufacturing is partly a bacterial process, since the milk must first be soured slightly by streptococci so the butterfat can be separated during churning. As a by-product of their fermentation, these bacteria produce acetoin, which oxidizes spontaneously to diacetyl, giving butter its characteristic taste.

Linen manufacturing, which uses fibers from flax, also depends on the growth of bacteria. Cellulose fibers are held together by pectin, and in the ancient process known as "retting," the fibers are released from the pectin. To break down this cement, the plants are first soaked in water. The aerobic microorganisms that first grow on the plant tissues use up the available oxygen, creating an environment for anaerobic butyric-acid bacteria whose pectinases hydrolyze the pectin. These bacteria, of course, are merely acting as decomposers, but people have learned to take advantage of the process.

Many similar stories could be told. In addition to the production of antibiotics by actinomycetes, bacteria with unusual metabolic capabilities have been used to manufacture commercial materials such as acetone. Other bacteria are used to break down noxious materials. Sewage treatment plants have long depended on bacteria to reduce wastes to gases (Figure 29.27), and genetic engineering is now producing strains of bacteria that can break down all kinds of pollutants. In 1990, such engineered bacteria were applied to oil spills in the Gulf of Mexico, where they effectively cleaned up potentially disastrous situations. Microbiologists tend to believe in E. F. Gale's Principle of Microbial Infallibility: Some organism can be found (or created, with current technology) to grow on any given organic

(a)

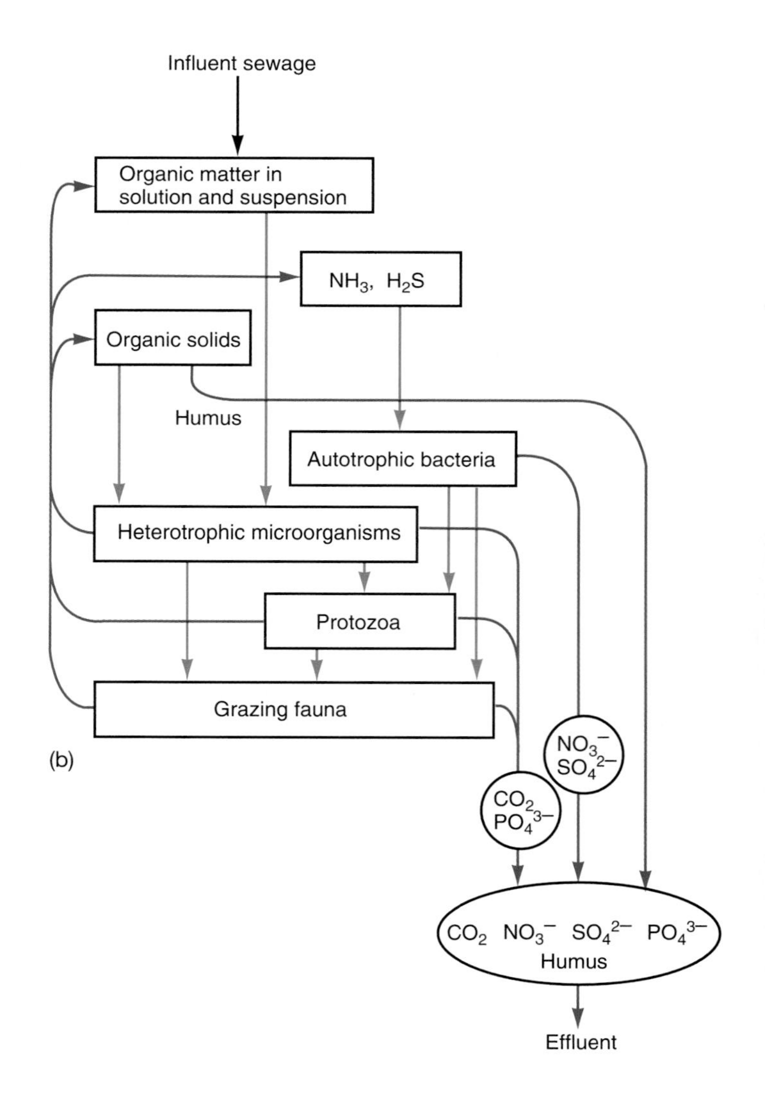

Figure 29.27

(a) One type of modern sewage plants uses a trickling filter, in which sewage is sprayed over a bed of crushed rock about 6 feet thick and then trickles down through the rock. *(b)* Bacteria growing in a film on the rock initiate a food chain *(red arrows)* by assimilating organic materials. Protozoa, which are very abundant and active in the film, eat the bacteria, and are themselves eaten by small invertebrates (grazing fauna). Wastes from the food chain are continually recycled *(blue arrows).* In this way the sewage is mostly converted into harmless organisms or into gases and inorganic materials that leave in the effluent *(green arrows).*

compound. To people who understand microbiology, the most obvious way to get rid of industrial pollutants is to grow something on them and convert the pollutants into the ordinary biomolecules of the cleansing organisms.

D. Addendum: the viruses

Viruses are such important elements in the biological world and have been so critical in shaping our understanding of it that we discuss them in several contexts throughout the book. Section 2.7, pp. 29–30, explains the concept of a virus. Sections 12.6–12.7, pp. 246–251, introduce bacteriophage and their key historical role in elucidating the importance of DNA; 17.3–17.5, pp. 334–338, explain how phage are used to map and define genes; 17.10–17.11, pp. 343–346, show phage as genetic elements, including the phenomenon of lysogeny, and how phage can carry genes from cell to cell. Section 18.17, p. 360, shows gene regulation in phage T4. Section 18.14, pp. 371–372, and 19.11, pp. 391–393, introduce viruses, including retroviruses, as agents for gene cloning and gene therapy; 20.15–20.16 introduces tumor viruses and their significance in cancer. Section 48.12, pp. 1033–1035, discusses HIV and AIDS in relation to the immune system. Thus, viruses are our constant companions in studying biology.

Viruses are agents that infect organisms, and all kinds of organisms probably have their own viruses. But viruses are not organisms. Rather, they are particles with genetic properties that use organisms as vehicles for their reproduction. It is appropriate to discuss viruses in this chapter because they are minute entities that cause diseases similar to those caused by bacteria and have often been confused with bacteria. But a clear understanding of microbiology depends on appreciating the differences between viruses and bacteria.

We'll ignore the question of whether viruses are alive or not because it is a rather fruitless semantic issue, but biologists cannot ignore the existence of viruses, and all of us should understand their place in nature. Although we discuss viruses in several places in this book, here we will discuss some of their general properties.

As we saw in Section 2.6, an organism consists of one or more cells that operate by following the instructions in their genomes. A virus is just an independent genome (nucleic acid) enclosed in a protective protein covering that allows it to survive outside cells and to invade a functioning cell, which becomes its host. The virus forces the cell's genetic apparatus to read the instructions in the viral genome instead of those in the cellular genome. Since the viral genome encodes directions for making more viruses, the virus subverts a functioning cell and converts it into a virus factory.

29.14 Viruses are not organisms.

The late French virologist Andre Lwoff pointed out that viruses differ from organisms by several criteria:

1. An organism is always a cell or assemblage of cells. No virus has such a structure. At one stage in its cycle of multiplication, a virus takes the form of particles called **virions,** each consisting of a nucleic acid genome enclosed in a protein covering, or **capsid.** The virion contains only one kind of nucleic acid, either DNA or RNA, whereas every cell uses both kinds. Viruses reproduce solely by using the information in this one nucleic acid, whereas organisms, including parasites, reproduce through an integrated action of all their constituents.
2. Viruses do not grow as cells do, by enlarging and dividing, nor do they reproduce as an organism does, either sexually or asexually. Instead, infected cells synthesize new virions in much the way that a factory manufactures its products.
3. Viral genomes do not contain the information for an apparatus to generate energy, nor for a protein-synthesizing apparatus of ribosomes, transfer RNAs, and other factors. A virus is thus totally dependent on its host cell for its energy and for translating its genome into proteins.

29.15 Virions have simple, regular structures.

The nucleic acid and capsid of a virion together form a *nucleocapsid* (Figure 29.28). Some nucleocapsids are *enveloped*—that is, surrounded by a membrane. Envelopes are derived from a

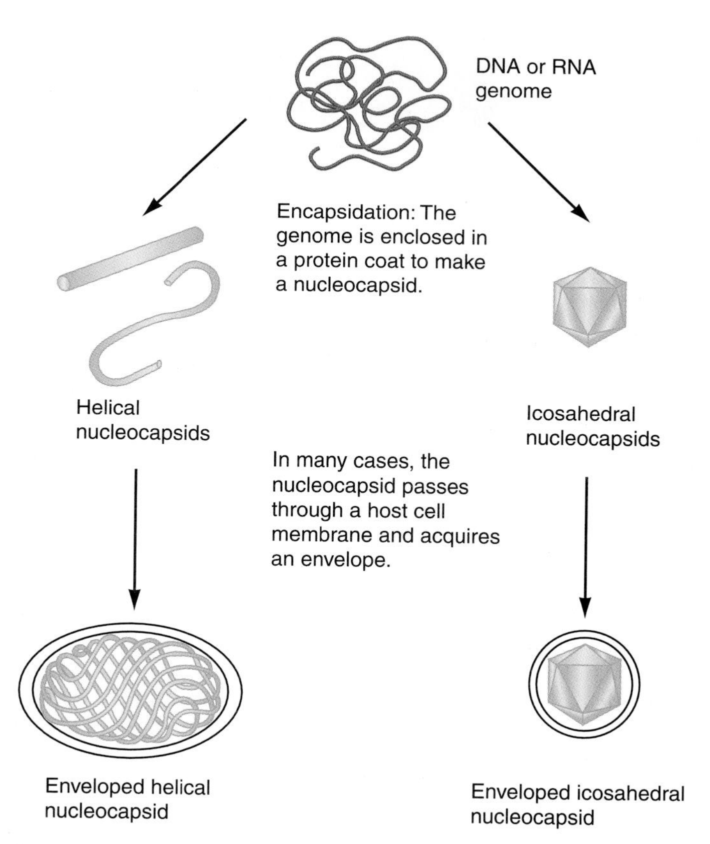

Figure 29.28
A nucleocapsid is a nucleic acid genome enclosed in a protein capsid; it is either helical or icosahedral in form. Some virions consist of the nucleocapsid alone; others have a surrounding envelope made of a modified cell membrane.

Figure 29.29
A helical nucleocapsid is made of identical protein subunits that enclose the nucleic acid in a groove. The structure grows until the entire nucleic acid is enclosed.

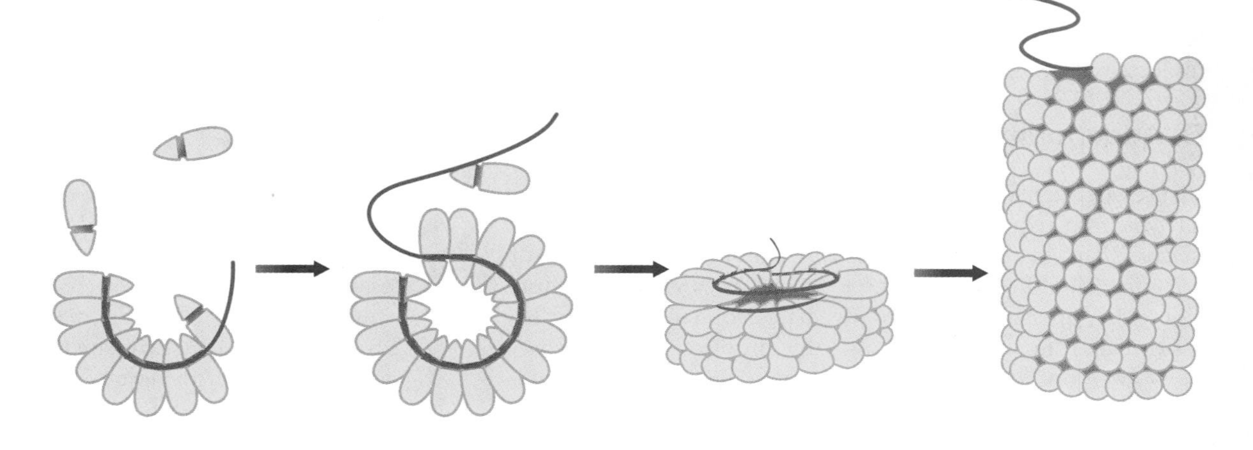

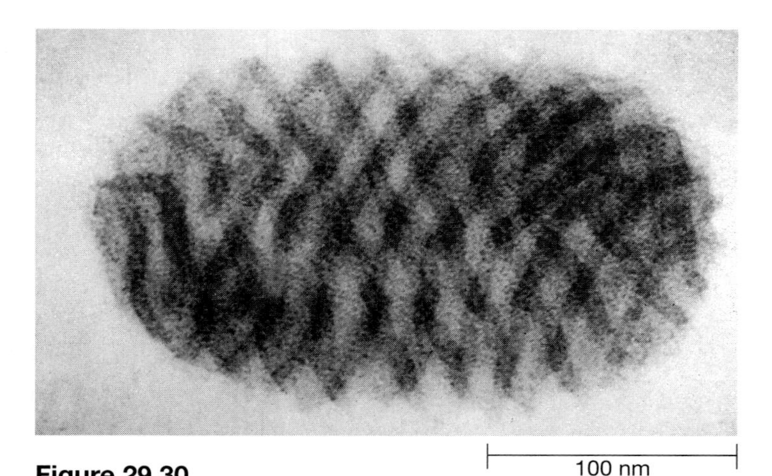

Figure 29.30
The large virion of Orf virus consists of a long helical nucleocapsid wound back and forth inside an envelope.

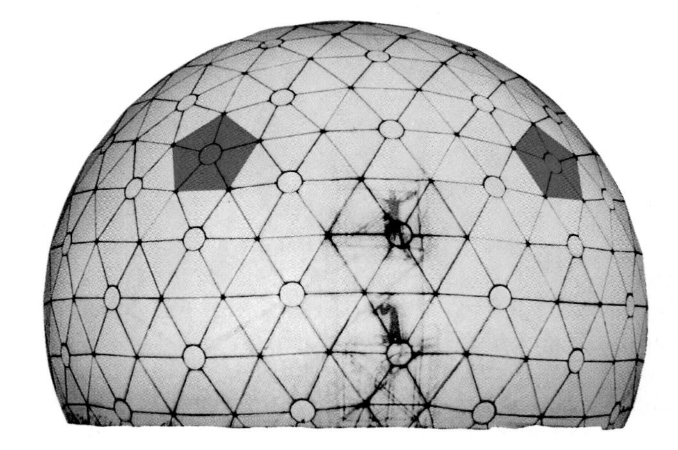

Figure 29.31
A geodesic dome is made of triangular subunits in a hexagonal pattern, but with a few centers made of five subunits each (two of them marked in color).

membrane of the host cell, but this feature does not give a virion the properties of a cell.

A protective capsid can be assembled around a nucleic acid to make either a helical or a spherical structure. A helical virion is made by stacking identical protein subunits that enclose the nucleic acid in an internal groove (Figure 29.29), and some viruses have such a nucleocapsid enclosed in an envelope (Figure 29.30). A spherical virion is made of protein subunits that form a shell around a core of nucleic acid. In each case, the size of the virion reflects the size of the nucleic acid. However, all spherical capsids are actually icosahedrons: solids with 20 identical triangular faces. In fact, their architecture is that of the geodesic dome invented by R. Buckminster Fuller (Figure 29.31). Fuller discovered that one can build a dome from triangular subunits connected hexagonally by inserting several pentagonal centers at intervals. A complete sphere requires 12 pentagons, and this is how spherical viruses are built. The smallest such virus has only 12 pentagonal centers, each made of five protein subunits. Larger viruses contain these 12 centers plus varying numbers (depending on the size of the virion) of hexagons made of six protein subunits (Figure 29.32).

29.16 Viruses multiply in a common pattern.

Every type of virus is specific for one type of host cell, and although viruses differ in the timing of their growth cycles and their specific effects, viral infections follow a common pattern. First the viral genome is replicated to make many new copies, and then these genomes are wrapped up inside capsid proteins to make new virions.

Viruses have varied modes of infection. Bacterial viruses attach to the cell surface, and their nucleic acids are transported through the cell membrane. Many animal viruses enter their host cells through a variation of the old Trojan Horse trick: The cells phagocytize the virions as if they were benign organic particles, but once inside the cell, the virus takes over. Since plant

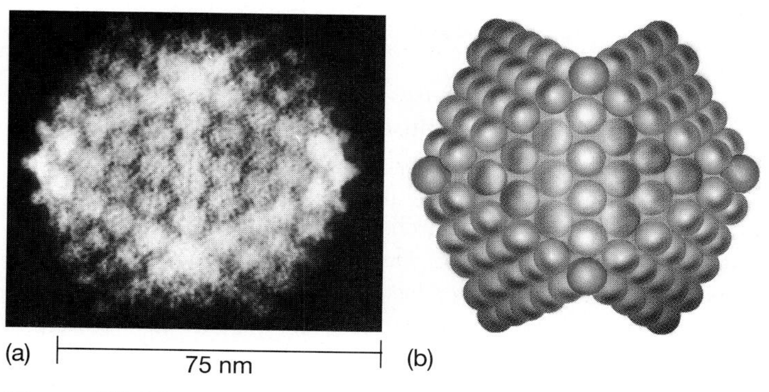

Figure 29.32

(a) Electron microscopy reveals the icosahedral structure of an adenovirus capsid. *(b)* Each vertex *(red units)* has fivefold symmetry, and the others have sixfold symmetry.

cells are enclosed in protective walls, plant viruses often need vectors, such as nematode worms or insects, to break through these walls and transmit the viruses into their host cells. Regardless of the mode of infection, viruses exhibit a common general pattern of multiplication. Viral genomes first express *early genes* that turn off some of the host's activities, usually stopping transcription of the host genome and translation of host messengers. Then, within a few minutes in bacterial viruses or a few hours in plant and animal viruses, the viral genome begins to replicate by means of unique new viral enzymes.

As replication begins, *late genes* of the viral genome are expressed; they encode new capsid proteins, which then combine with viral genomes to make new nucleocapsids. In some viruses, the capsid consists of many kinds of protein subunits, which are assembled through complex pathways (Figure 29.33). Assembly is always *self*-assembly; that is, the protein subunits, once made, simply come together into the right combinations because of their intrinsic structures, with only a few special enzymes to accelerate assembly. It is as if all the pieces of a car were made so they would automatically fit themselves together to make the whole vehicle.

Eventually, in most cases, the cell falls apart and liberates the hundreds or thousands of virions that have accumulated within. These virions, in turn, can infect other cells and repeat the cycle of infection. If the virus has enveloped virions, these envelopes are acquired from the membranes of the host cell. As shown in Figure 48.16, new viral proteins have been added to some cell membranes, and an animal's immune system can recognize such modified cells and destroy them, thus stopping the infection.

29.17 Viruses have DNA or RNA genomes.

Viruses are classified on the basis of several factors. Virions can have either a DNA or RNA genome, which may be double-stranded or single-stranded, linear or circular. The nucleocapsid may be helical or spherical, and may be enveloped or naked. Many bacterial viruses combine an icosahedral head with a helical tail. Some of the classes defined by these characteristics include only rather obscure viruses, such as helical RNA viruses that cause many kinds of plant diseases.

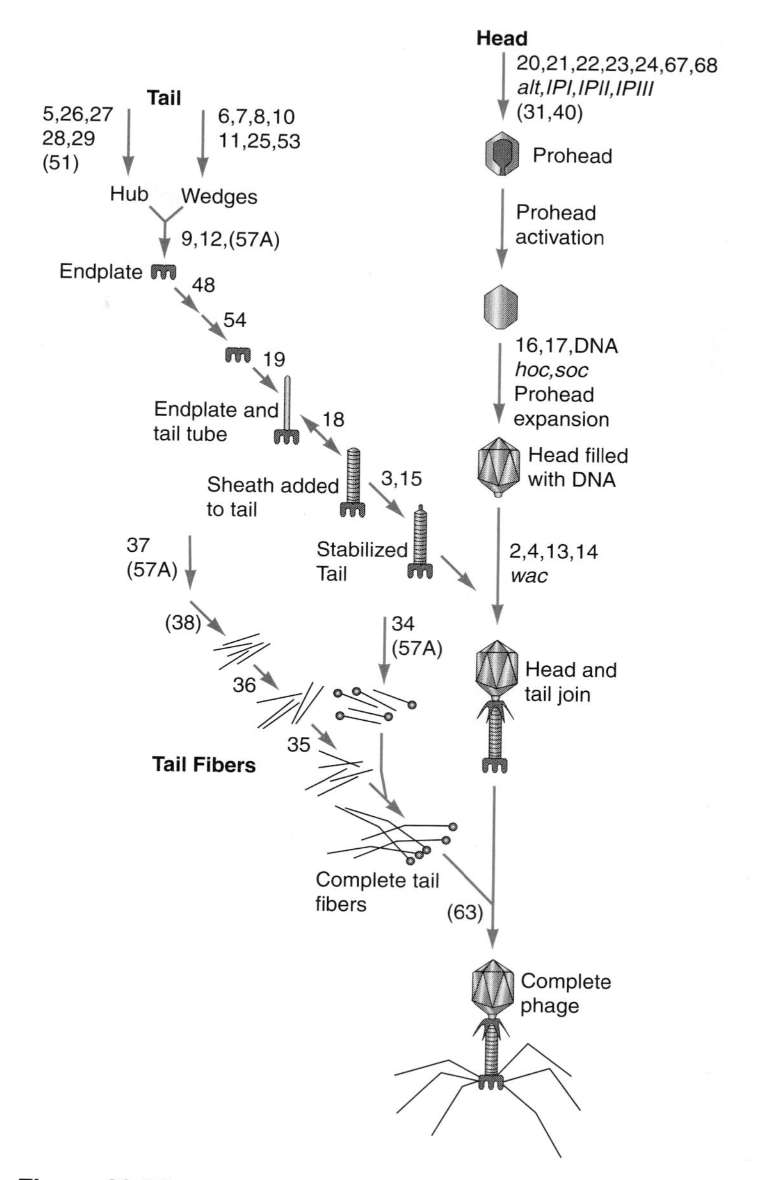

Figure 29.33

Virions of phage T4 are assembled through multistep pathways. Genes involved in each step are given numbers or short names such as *alt, hoc,* and *wac;* genes in parentheses encode enzymes for certain steps. One series of proteins forms a head, which is then filled with a DNA genome. The endplate is made by two pathways, one making a central hub from five proteins and the other making wedges of seven proteins; then six wedges combine with a hub to make an endplate. A tail tube forms on the endplate, and a sheath forms around the tail tube. A head and a tail then combine; tail fibers, which have been made via another pathway, join the tail to make a complete phage virion.

Deoxyviruses have DNA genomes. Those with naked icosahedral capsids include many minute bacteriophages; papilloma and polyoma viruses, including some that cause tumors;

adenoviruses that cause respiratory diseases and tumors; and large insect viruses. The enveloped icosahedral viruses are mostly herpes viruses. Deoxyviruses with naked helical capsids include bacteriophages that multiply without lysing their host cells. Enveloped helical deoxyviruses are pox viruses that infect humans and other animals.

Riboviruses are those with RNA genomes. Many riboviruses with naked icosahedral capsids are plant viruses. Animal viruses with this structure include rhinoviruses that cause colds, polioviruses, and the virus of foot-and-mouth disease. Enveloped icosahedral riboviruses are mostly arboviruses transmitted by insect vectors, including those that cause encephalitis, yellow fever, and dengue fever. The naked ribohelical viruses are almost all plant viruses. The enveloped helical riboviruses cause some well-known diseases such as influenza, mumps, measles, distemper, and rabies.

Since genetic information in cells is carried in DNA and transferred to RNA, the viruses with RNA genomes require special explanation. A few viruses (primarily reoviruses) have double-stranded RNA (dsRNA), and in this case replication and transcription are quite straightforward; using a specific replicase and transcriptase, the dsRNA is replicated as if it were DNA, and one strand is transcribed into mRNA. But most RNA viruses have single-stranded genomes (ssRNA), and these can be either *plus strands* with nucleotide sequences identical to mRNA or *minus strands* whose nucleotide sequences are complementary to mRNA. Let's first consider viruses such as polio and tobacco mosaic that have plus genomes. When a plus-strand viral genome enters a cell, it can serve as a mRNA directly; usually it combines with ribosomes and is translated directly into a polyprotein, a large polypeptide that is later cut into functional proteins. Some of these proteins are replicases that can convert the incoming ssRNA into a dsRNA called a *replicative form* (RF); from the RF, a virus-encoded transcriptase then copies new plus strands that serve as both mRNAs and new genomes.

In contrast, the myxo- and paramyxoviruses such as those that cause influenza, mumps, and measles have minus-strand genomes. When they infect, a transcriptase associated with the virion transcribes mRNAs from the genome that are translated into proteins. Some of these proteins are replicases that convert the minus RNA into RFs, from which new minus strands are replicated.

The viral world is so diverse that one constantly meets viruses with surprising properties; interactions between viruses and their hosts run the gamut from the most virulent to the most benign. Many viruses, for instance, coexist with their hosts in a way that doesn't necessarily kill the host cells. Viruses, and related viruslike elements, may even do their hosts a certain amount of good, since a cell carrying one type of virus in a lysogenic state (see Section 17.10) may be resistant to infection by other types. Some bacterial viruses destroy their hosts within minutes as they multiply massively, but many helical deoxyviruses multiply in their bacterial hosts indefinitely and quite benignly by extruding new virions harmlessly from the cell surface. Many plant viruses kill by disfiguring leaves and clogging the plant's vascular tissue; others create interesting color patterns in tulips.

Among the human viruses are poxes that produce nothing worse than localized skin pustules and a lot of itching. Rhinoviruses can create a miserable cold, and polioviruses can destroy critical nerve cells, thus paralyzing the body. Hepatitis B and Epstein-Barr virus cause chronic infections that are virtually without symptoms for years. Herpes simplex type I probably infects many children and persists in a latent form in ganglia in the nervous system; years later, a stress may activate the virus so it produces cold sores. The extreme of virulence is probably represented by the Marburg viruses, including the infamous ebola virus, which produce hemorrhagic fevers. They usually infect certain African monkeys but occasionally erupt in the human population; these viruses are so incredibly virulent that they infect every tissue of the body, changing the victim's personality as they destroy brain cells, turning the intestines into mush, destroying the intestinal lining so it is sloughed off with the feces, and ending in massive hemorrhaging throughout the body as the victim collapses in a mass of blood, dead cells, and viruses.

29.18 Viroids are unusual agents.

A few plant diseases—potato spindle-tuber disease is the best known—have been traced to particles called **viroids** that are even simpler than viruses. A viroid is merely a circular, single-stranded RNA molecule of 250 to 370 nucleotides, much smaller than the smallest viral genome. After being transmitted from one plant to another mechanically or through pollen or ovules, viroids may multiply massively in the new host's cells, mostly in the nucleoli; they don't act as mRNAs to direct protein synthesis, and it is not known how they cause disease. In fact, the same viroid may have little effect on one host but produce a severe disease in another. Viroids show that nucleic acids, with their intrinsic property of replication, may reproduce in surprising ways, often at the expense of other biological systems.

Coda

We have had a long and bumpy ride, we Earthlings, from a humble beginning "on the shore of some primordial sea." All our achievements are based on a few "inventions" developed by our most remote ancestors. Metabolism of some sort evolved early and continued to become more complex and diverse with the evolution of more complicated and versatile macromolecules. The cell, with its membrane isolating and concentrating metabolic activities, was an essential invention, as was the genome. What would we be without our genomes? From those beginnings have come an amazing variety of simple procaryotic cells that are able to live in the most diverse environments and conduct chemical reactions that rival the capabilities of organic chemists. Some of those environments are inside other organisms, so a few procaryotes are identified with the diseases they cause, but their broad roles in ecosystems are far greater and far more biologically significant than their pathologies.

As we eucaryotes search the procaryote lineages, trying to identify our own origin, we find it is not simple. The evidence indicates that primitive eucaryotes were mongrels with affinities to both archaebacteria and the more ordinary monera; our mitochondria were apparently derived from one kind of ancient bacterium and the chloroplasts of photosynthetic eucaryotes from other bacteria, as we will explore in the next chapter. Thus ancient eucaryotes derived their capabilities from multiple sources, from devices and inventions refined by independent evolutions. Perhaps our evolutionary potential, manifested in the diversity of eucaryotes, derives from this inheritance of a combination of biological inventions.

Summary

1. The chemical elements are formed through nuclear processes in stars.
2. Primitive planets have reducing atmospheres in which organic compounds can form from inorganic compounds such as CO_2, H_2O, and NH_3.
3. Simple, cell-like objects can be made from small proteins. The formation of cells that could isolate and confine metabolic processes must have occurred early in the evolution of life.
4. The evolution of a functioning genome is problematic. Both proteins and RNA molecules can form in primitive atmospheres, and the early genetic apparatus was probably made entirely of RNA. However, it is not easy to explain how RNA (and ultimately DNA) acquired control over protein synthesis.
5. The first organisms must have been heterotrophs that fed on the organic compounds in the "primordial soup."
6. Photosynthesis was a later development, which depended on the ability to make porphyrins, since chlorophyll is a magnesium-porphyrin. The first phototrophic organisms used the systems represented by some modern bacteria, which do not produce oxygen as a by-product.
7. Eventually organisms similar to modern cyanobacteria evolved, producing oxygen as a by-product of photosynthesis. This opened the door to aerobic respiration.
8. Procaryotic cells have no nuclear envelope. Most are very small (usually a micrometer or less in diameter), although a few are as large as the largest eucaryotic cells.
9. Most bacteria have a cell wall that includes a huge, bag-shaped molecule (called a murein or peptidoglycan) that encloses the entire cell and resists the high turgor pressure from within. This polymer is made of polysaccharide chains linked by short chains of amino acids.
10. Typical bacteria have the form of spheres, rods, or spirals. They are classified on the basis of these shapes, by their modes of metabolism, and by their reaction to the Gram stain, which divides them into gram-positive or gram-negative types.
11. Bacteria are important in every ecosystem. Most are decomposers that reduce wastes and detritus to simpler materials. As they oxidize or reduce various compounds to obtain energy for growth, they form important links in the nitrogen and sulfur cycles. Sometimes pairs of bacterial species engage in cyclic exchanges of materials.
12. A long series of investigations, beginning with Pasteur, demonstrated that many bacteria cause infectious diseases. To show that a particular organism is the etiologic agent of a disease, an investigator must satisfy Koch's Postulates.
13. Pathogens can cause disease by invading and destroying tissues or by producing toxins.
14. Infectious agents reside in a reservoir of infection in a particular population of organisms, from which they are transmitted to new hosts. Many vectors, usually insects or other arthropods, carry pathogens from one host to another.
15. Rickettsias and chlamydias are very small bacteria that grow as intracellular pathogens.
16. Bacteria are used in many industrial processes. They are cultivated to produce antibiotics and common foods such as cheese, yogurt, sauerkraut, and vinegar; they are also used to break down wastes and pollutants in sewage treatment plants and other sites.

Key Terms

Multiple-Choice Questions

1. The earth's original atmosphere is thought to have been rich in methane and ammonia, but poor in
   a. H_2.
   b. N_2.
   c. O_2.
   d. gases.
   e. all of the above.
2. In an RNA world, RNA serves all the following functions except
   a. catalysis.
   b. target of selection.
   c. replicating genome.
   d. template for DNA.
   e. adaptor molecule during translation.
3. The first phototrophs that evolved probably
   a. were oxygenic.
   b. used the Calvin-Benson cycle.
   c. used photosynthesis to oxidize organic compounds.
   d. used H_2S as an electron donor.
   e. had both cyclic and noncyclic photophosphorylation.
4. Which of these events or processes probably evolved before the appearance of oxygenic photosynthesis?
   a. photoheterotrophy
   b. anaerobic respiration
   c. the Calvin-Benson cycle
   d. *a* and *b*, but not *c*
   e. *a*, *b*, and *c*
5. The most direct and immediate cause of the shift from a reducing to an oxidizing atmosphere was the evolution of
   a. photoheterotrophy.
   b. the porphyrin ring.
   c. the appearance of CO_2 in the atmosphere.
   d. photosystem I.
   e. photosystem II.
6. Which statement is correct?
   a. Most of the organisms alive today are procaryotes.

b. Most procaryotes are pathogenic.
c. Most modern procaryotes have been forced to live in anaerobic habitats.
d. Most ancient procaryotes evolved into modern eucaryotes.
e. Most of the procaryotes are classified as archaebacteria.

7. One species of soft-walled bacteria that moves by means of a corkscrew action and causes the STD known as syphilis is classified as
   a. mycoplasmas.
   b. myxobacteria.
   c. spirochetes.
   d. cocci.
   e. archaebacteria.
8. Mycoplasmas contain all of the following except
   a. a genome.
   b. ribosomes.
   c. transcriptional apparatus.
   d. translational apparatus.
   e. a cell wall.
9. All of the following ecological activities or lifestyles can be ascribed to bacteria except
   a. oxygen generation.
   b. decomposition of detritus.
   c. parasitism.
   d. recycling of elements such as nitrogen and sulfur.
   e. production of functional diseases.
10. Several toxigenic bacteria harm the host by
   a. hydrolyzing critical enzymes within the host cell.
   b. digesting the intercellular material that holds host cells together.
   c. disrupting cell-to-cell communication within the host.
   d. invading and lysing red blood cells.
   e. destroying lipid components of the cell membrane.

True-False Questions

Mark each statement true or false, and if false, restate it to make it true.

1. Haldane and Oparin proposed that the gases in Earth's original atmosphere were converted to an organic soup by the energy of visible light.
2. The first organisms were probably unicellular autotrophs.
3. Photoautotrophs can synthesize sugar using inorganic sources of carbon and sunlight as an energy source.
4. Chemoheterotrophs synthesize sugar by ingesting inorganic sources of carbon and using the energy released from oxidation of reduced compounds.
5. Modern, oxygenic photosynthesis evolved in a stepwise manner that began with a cyclic photosynthetic pathway and matured with the addition of a noncyclic pathway and the Calvin-Benson cycle.
6. Like plants, all modern phototrophic bacteria contain and use chlorophyll *a*.
7. Bacterial cell walls are composed of a polysaccharide similar to cellulose.
8. A protoplast is the nucleoid region of a bacterial cell.
9. Streptococci are clumps of spherical bacteria, and staphylococci are chains of such cells.
10. Mycoplasmas are among the largest bacteria, and the actinomycetes are among the smallest.

Concept Questions

1. Why is the appearance of membranelike structures considered a landmark in the evolution of cells?
2. Assuming susceptible strains of organisms, why is penicillin an effective treatment for bacterial diseases but not viral diseases, and why doesn't the antibiotic kill our own cells?
3. How does pasteurization differ from sterilization?
4. List the conditions that must be satisfied in order to attribute a particular disease to a particular pathogen. Can you think of circumstances that would make it difficult or even impossible to meet all conditions?
5. How can you explain the observation that human pathogens transmitted by means of an insect vector often include other mammals, but not fish, amphibians, or reptiles, as part of their reservoir of infection?

Additional Reading

Brierley, Corale L. "Microbiological Mining." *Scientific American,* August 1982, p. 44. The role of bacteria in leaching metals from low-grade ore.

Burnet, Macfarlane. *Natural History of Infectious Diseases.* Cambridge University Press, Cambridge, 1962.

Cairns-Smith, A. G. "The First Organisms." *Scientific American,* June 1985, p. 90. A theory that the very first systems able to evolve through natural selection may have been crystals of clay.

De Duve, Christian. *Vital Dust: Life as a Cosmic Imperative.* Basic Books, New York, 1995. The origin and evolution of life and cells.

Dixon, Bernard. *Power Unseen: How Microbes Rule the World.* W.H. Freeman, New York, 1994. A popular explanation of the microbial world.

Ewald, Paul W. "The Evolution of Virulence." *Scientific American,* April 1993, p. 56. Why some pathogens become extremely virulent and others cause only minor infections.

Garrett, Laurie. *The Coming Plague: Newly Emerging Diseases in a World Out of Balance.* Farrar, Straus and Giroux, New York, 1994. An important popular exposition of how emerging diseases are likely to affect human health.

Ourisson, Guy, Pierre Albrecht, and Michel Rohmer. "The Microbial Origin of Fossil Fuels." *Scientific American,* August 1984, p. 44. Chemical analysis of the most varied organic sediments, including coal and petroleum, reveals a surprising commonality: All derive much of their organic matter from once unknown microbial lipids.

Postgate, J. R. *Microbes and Man,* 2d ed. Penguin Books, Harmondsworth, UK, 1986. An introduction to microbiology.

Prescott, Lansing M., John P. Harley, and Donald A. Klein. *Microbiology,* 3d ed. Wm. C. Brown Publishers, Dubuque (IA), 1996.

Rosebury, Theodore. *Life on Man.* Viking Press, New York, 1969.

Scientific American, October 1994. A special issue on the origin, distribution, and evolution of life.

Zinnser, Hans. *Rats, Lice and History.* Little, Brown Co., Boston, 1934. A classic exposition on the role of infectious disease in human history.

Internet Resource

To further explore the content of this chapter, log on to the web site at:

http://www.mhhe.com/biosci/genbio/guttman/

30

The Evolution of Eucaryotes

Key Concepts

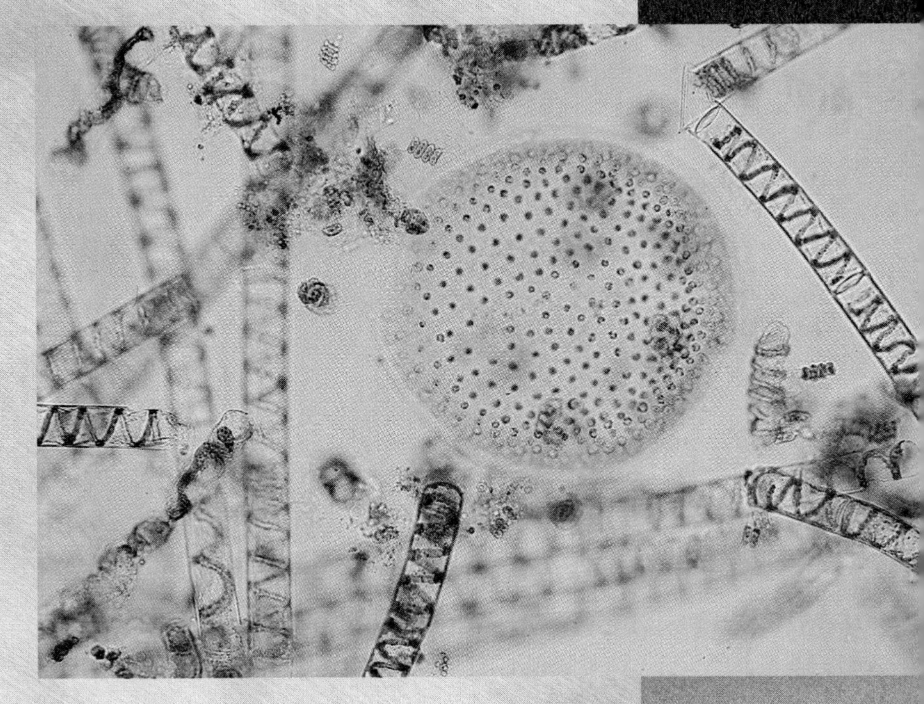

A good light microscope reveals the world of algae and other simple eucaryotes in a drop of pond water.

You place a drop of brownish green water on a slide, lay it on the microscope stage, adjust the instrument, and focus. Suddenly, strands of brilliant green appear, and you are lost in a jungle of huge green filaments, leaflike splotches, patches of brown—all within the space of a few millimeters. There are beasts in the jungle, too. Oddly shaped cells swim by—some clear and ghostly, some enclosing brilliant spheres of green, red, or amber, some flashing long whips, others moving mysteriously with no sign of a propellant. Occasionally a roundworm slithers across the stage, but this is clearly an animal, a huge intruder. You can ignore it for now, because the smaller creatures in the water are fascinating enough to occupy you for hours, or for a lifetime.

You have entered the world of the primitive eucaryotes, a beautiful, marvelously varied world. These are the algae, an amazing assortment of filaments, of rounded cells that live alone or in symmetrical packets, of geometrically precise cells that might have been designed on an engineer's drawing board. These are the amoebas, shapeless blobs that seem to flow from place to place or reach out their pseudopods ("false feet") to grasp a bit of food. These are the flagellates—green, golden, yellow, or colorless—that swim with fine flagellar lashes. These are white molds that draw nutrients from the water or from the cytoplasm of other creatures, and brilliantly colored slime molds that creep over the moist vegetation of a forest. And these are also parasites that invade the bloodstreams of animals and reduce them to helpless misery as surely as any bacterium or virus. The eucaryotic family tree begins with protists, and in this chapter we will study their modern descendants. By examining the great variety of protists and the subtle details of their structure and their lives, we will gain insights into the likely course of early eucaryotic evolution. Along the way, we will also develop some ideas about the probable evolution of the mitotic apparatus. ■

A. Problems of eucaryotic taxonomy

30.1 The classification of many eucaryotes is problematic.

At the end of Chapter 2, we reviewed some of the history of classifying organisms into kingdoms and showed that the first break with the ancient dichotomy of plants and animals was the establishment of a kingdom Protista to include various hard-to-classify organisms, mostly microorganisms. Different biologists have delimited Protista in various ways, but it has been a kind of grab bag of organisms that do not fit elsewhere. Protista became more homogeneous when microbiologists realized the importance of Chatton's distinction between procaryotes and eucaryotes and separated the procaryotes in their own kingdom. After being limited to eucaryotes, Protista was made still more homogeneous when the fungi were removed as a distinct kingdom (discussed in Chapter 31), although the definition and boundaries of the kingdom Fungi remain a problem that we will discuss later in this chapter. Still, Protista is a grab bag, a collection of leftovers that includes many groups of uncertain affinities, surely representing several clades. (Remember that a clade is a distinct monophyletic line of evolution stemming from one ancestral species.) Many investigators are now trying to identify well-defined clades among these eucaryotes, using modern molecular methods to sort out the phylogeny and taxonomy of these diverse, problematic, and often poorly known organisms. The purpose of this chapter is to review the major groups and identify their probably affinities.

The subjects of this chapter range from minute, unicellular creatures to kelps and seaweeds many meters long that look like perfectly good plants, except that they are brown instead of green. What shall we call all these creatures? The problems of terminology discussed in Section 22.10 now come back to haunt us, for words such as "lower" or "primitive" seem appropriate for some but not for all. Although no single term is exactly right, we will refer to them here as the *basal eucaryotes* because most of them occupy the region at the base of the eucaryotic family tree, although some have clearly evolved far from this base in their own distinctive directions. We only need this term for the beginnings of this discussion. Before long, we will separate some of the basal eucaryotes into the plant kingdom and others into a kingdom called Chromista. We will show why some organisms that appear to be fungi should be included in the Chromista. And we will leave a number of groups in a small kingdom Protista, which remains heterogeneous but at least more limited.

30.2 Basal eucaryotes include three general nutritional types of organisms.

The most typical basal eucaryotes are microscopic, consisting of rather round cells that grow singly, as filaments of cells, or as colonies of several cells cemented together. In contrast, other prominent organisms are large, brownish and reddish algae that look like plants (and have even been included in the plant kingdom by some biologists); these eucaryotes are multicellular, but unlike typical land plants, their cells are not differentiated for distinct functions, except for certain reproductive cells. It is useful to consider three informal categories of basal eucaryotes, based largely on nutrition: protozoa, and molds, and algae. Notice that we are only using these as very general English words, much as one might say "worm" or "flower." They imply nothing about phylogenetic relationships.

Protozoa

When all organisms were force-fit into either the plant kingdom or the animal kingdom, a phylum of animals called Protozoa was defined for all the unicellular, nonphotosynthetic organisms that feed rather like animals—that is, by ingesting bits of food. It is still convenient to refer to these creatures informally as protozoans, although they are only distantly related to one another (Figure 30.1). *Amoebas* are often quite shapeless and move by extending rounded pseudopods, which they also use to engulf bits of food. *Flagellates* move by means of long flagella, often only two but sometimes many. *Ciliates* are single cells that propel themselves with fine cilia.

Basic mechanisms of cellular movement, Section 11.9.

Protozoans can get their nutrients in two ways. **Osmiotrophic** organisms absorb monomer-sized organic molecules through their cell surfaces, while **phagotrophic** or **holotrophic** organisms engulf bits of solid food into cavities such as gullets or food vacuoles where digestion occurs, and then eject the undigestible remains (Figure 30.2). The most aggressive protozoans even ingest cells of about their own size. (Holotrophy is also the animal way of life, since most animals take in large molecules or pieces of food and reduce them to monomer-sized molecules.)

Molds

Molds commonly have thin, elongated cells and live as osmiotrophs, generally **saprobes** or **saprophytes** (*sapros* = rotten), which means they decompose the detritus of an ecosystem and absorb nutrients from the water and soil; other molds are parasites that absorb nutrients from the cells of other organisms. Molds produce their own extracellular digestive enzymes and pick up the released monomers. (The mycologist Clyde M. Christensen wrote that the difference between the digestive processes of humans and molds is that molds digest their bread before they eat it, while we eat it first and then try to digest it.)

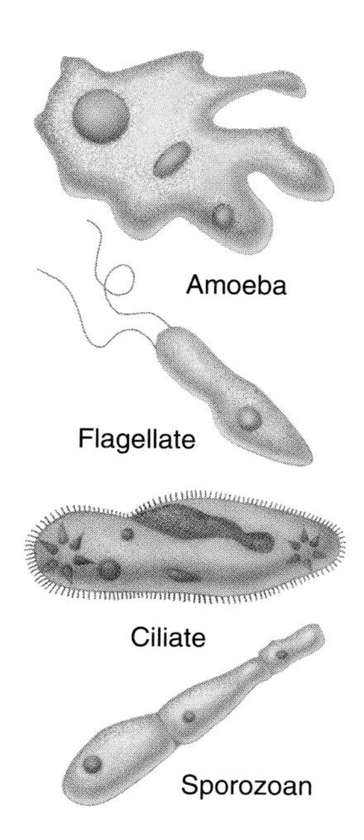

Figure 30.1
Protozoa are classified informally into four groups. Amoebas, flagellates, and ciliates are motile and usually free-living. Sporozoans are nonmotile parasites that have motile stages in their life cycles; they constitute two or three groups that are not closely related to one another.

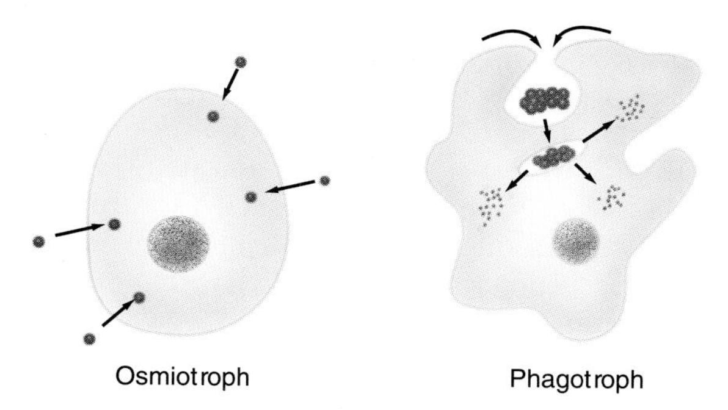

Figure 30.2
Some organisms are osmiotrophs that simply transport monomer-sized nutrients through their cell membranes. They may produce extracellular enzymes to digest solid materials, so they are common agents of decay. Phagotrophs pick up bits of food in cavities for digestion, so digestion is often intracellular. The most aggressive phagotrophs pick on cells of about their own size for their food.

Many types of molds are included in the kingdom Fungi (see Chapter 31), but we will show that certain organisms called water molds do not belong in that kingdom and are closely related to some other basal eucaryotes discussed in this chapter. The basal eucaryotes also include several kinds of slime molds, often brightly colored creatures that creep over wood and other decaying vegetation.

Algae

Most basal eucaryotes are relatively simple, undifferentiated, photosynthetic organisms known as **algae** (sing., **alga;** Figure 30.3). They are commonly grass-green, red, golden-brown, yellow-green, and brown; they include unicellular organisms, filaments of cells, clusters of cells and colonies, thin sheets of cells, and large seaweeds.

Many nonphotosynthetic protists probably evolved from algae that lost their chloroplasts. Chloroplasts are rather easily lost. Occasionally when a phototrophic cell divides, one daughter cell accidentally fails to get any chloroplasts and is said to be **apochromatic** (*apo-* = away from; *chrom-* = color). These cells can only survive if they have a metabolic apparatus for using other forms of nutrition and living as osmiotrophs or holotrophs. The basal eucaryotes include many apochromatic organisms. Many algae are flagellates or amoebas, and some of them were undoubtedly ancestral to flagellated or amoeboid protozoans. Many of the water molds appear to be apochromatic descendants of certain golden-brown or yellow-green algae.

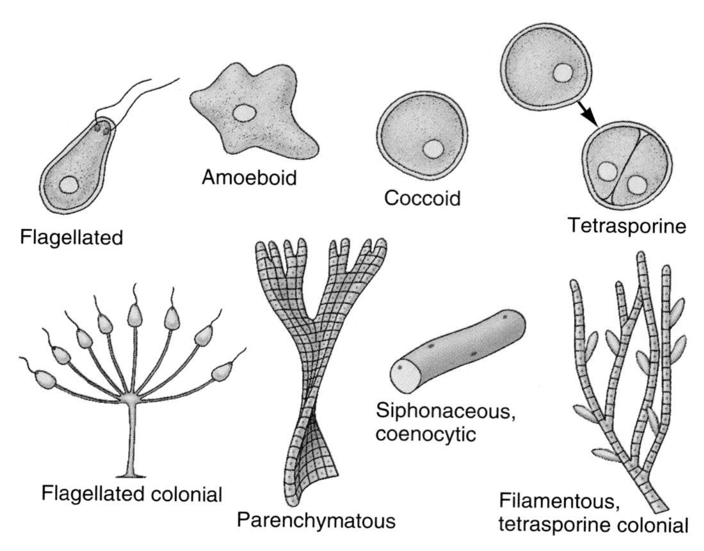

Figure 30.3
Algae take many forms. Motile algae may be flagellated or amoeboid, and some flagellates grow as colonies. Single, nonmotile cells are called coccoid, and similar cells that continue to multiply to form packets or colonies are called tetrasporine or tetrasporal. Tetrasporine algae may take the form of filaments. Other filamentous algae are siphonaceous or coenocytic, having many nuclei within a cytoplasm not divided into individual cells. Some large algae are parenchymatous, made of many boxlike cells.

Table 30.1 Major Features of Different Eucaryotic Clades

Feature	Green Clade	Brown Clade	Red Clade
Chlorophylls	Chl *a* and *b*	Chl *a* and *c*	Chl *a*
Carotenoids	β-carotene, lutein, violaxanthin, neoxanthin, astaxanthin	β-carotene, fucoxanthin, diatoxanthin, diadinoxanthin	
Thylakoids	Single	Triple stacks	Like cyanobacteria
Glucans for energy storage	α:1–4 (starch)	β:1–3 (laminarin, paramylum, and leucosin)	α:1–4 (glycogen)
Flagella	Two; both whiplash	One whiplash; one tinsel	None
Lysine biosynthesis	DAP	DAP	Unknown
Cell walls	Cellulose	Cellulose	Chitin

30.3 Several shared features provide guidelines to the phylogeny of eucaryotes.

Without any evidence to the contrary, we assume that the eucaryotic condition—possession of a membrane-bounded nucleus—evolved only once, so all eucaryotes arose from one ancestor and are members of a single clade. The basal eucaryotes then evolved into several distinct clades, some of which we now identify as fungi, plants, and animals. To understand the diverse groups of basal eucaryotes, it is important to identify those shared primitive and derived features that indicate lineages of evolution. A series of features (summarized in Table 30.1) point to the existence of three major clades among basal eucaryotes, what we will call the green, brown, and red clades. Following Thomas Cavalier-Smith and some other biologists, we consider the green clade to be the plant kingdom, and that is where we will place the green algae. We will also follow Cavalier-Smith in making the brown clade into a separate kingdom Chromista, just as the Danish botanist T. Christensen separated them as a group called Chromophyta. The red clade includes only the red algae, which we leave in the kingdom Protista; we will discuss evidence that the red algae are related to the true fungi (see Chapter 31), and we will see that the fungi are closely related to the animals.

Chloroplast structure and metabolism

Despite their basic similarity, the chloroplasts of various algae differ in significant ways and are important shared derived features. The non-green colors of algae come from supplementary pigments, notably the carotenoids listed in Table 30.1, which are yellow, orange, red, purple, and brown.

The green-clade algae have chloroplasts with single thylakoids (Figure 30.4*a*). They share similar carotenoids, and their Chl *a* is supplemented by Chl *b*. In the chloroplasts of brown-clade organisms, the thylakoids tend to occur in bundles of three (Figure 30.4*b*); they share different characteristic carotenoids, and although some have only chl *a*, most also have chl *c* instead of chl *b*. The brown-clade phototrophs are also remarkable in having their chloroplasts in the lumen of the rough endoplasmic reticulum, instead of in the cytosol.

Algal chloroplasts produce food reserves that are often stored in the *pyrenoids* visible in Figure 30.4. However, the green-clade algae make the familiar starch of plants, a polymer of glucose with α:1–4 linkages, while the brown-clade organisms make starchlike polymers of glucose (named laminarin, paramylum, and leucosin) with β:1–3 linkages and some β:1–6 branches. Many brown-clade algae also store oils.

Flagellar structures

Although eucaryotic flagella and cilia all have the same basic 9 + 2 structure, some differences in their surface structures correlate with other phylogenetic indicators. (Even though the mature, reproductive stages of many groups have no flagella at all, their gametes may be flagellated.) The two main types of flagella—**tinsel** and **whiplash**—are shown in Figure 30.5. The green-clade algae generally have a pair of whiplash flagella, and some have more than one pair. The brown-clade organisms typically have one whiplash flagellum (often pointed backward) and one tinsel flagellum (often pointed forward). In some groups, only the distinctive tinsel flagellum remains. The distinction in flagellation correlates very well with the other phylogenetic indicators.

Flagellar structure, Section 11.13.

Metabolic pathways

Many kinds of biochemical details, such as the way certain enzymes are regulated, offer additional information about phylogeny. Metabolic pathways are generally universal, because

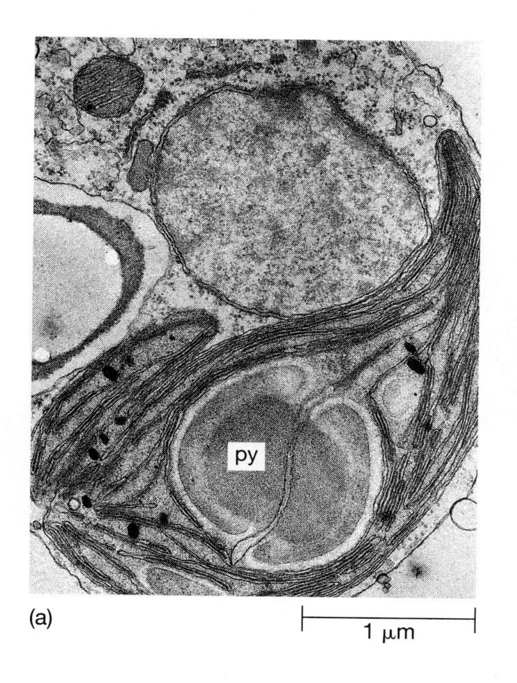

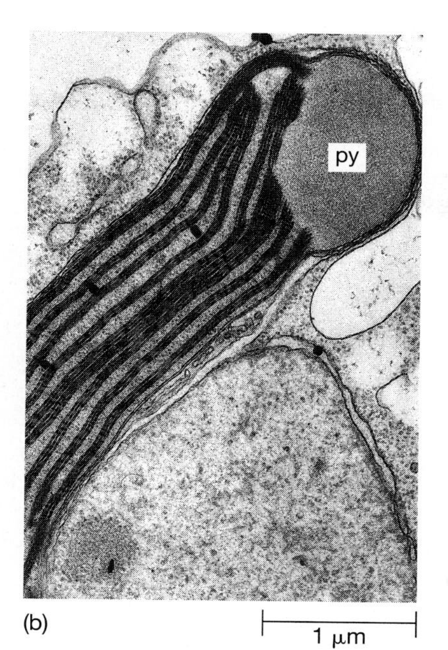

Figure 30.4

(a) A chloroplast of the green alga *Enteromorpha intestinalis* shows the irregular arrangement of thylakoids surrounding a pyrenoid *(py),* where starch accumulates. *(b)* A chloroplast of the brown alga *Pylaiella littoralis* shows the triple bundles of thylakoids that run into the large pyrenoid *(py).*

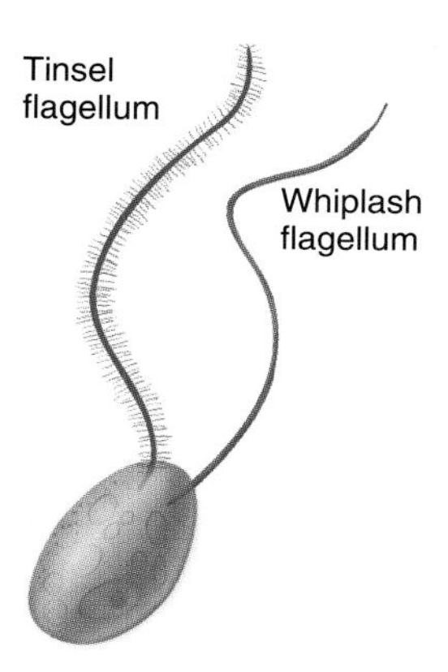

Figure 30.5

The flagella of eucaryotes may be either tinsel or whiplash.

once an adequate basic process has evolved, there is no selective pressure to change it. But lysine is an exception. Bacteria synthesize lysine through a metabolic pathway that includes diaminopimelic acid (DAP), a common component of their cell walls. Henry Vogel discovered that most eucaryotes that can make their own lysine also use the DAP pathway, but a few groups use a different pathway that includes α-aminoadipic acid (AAA) (Figure 30.6). The AAA club is quite exclusive. Its members include all the true fungi—those fungi that have cell walls made of chitin—plus two rather enigmatic groups that we will discuss later: a small group of algae called euglenoid flagellates and a group of molds called chytrids. Primitive eucaryotes probably used the DAP pathway; some eucaryotes lost the ability to make lysine entirely, and some of them later acquired the AAA pathway.

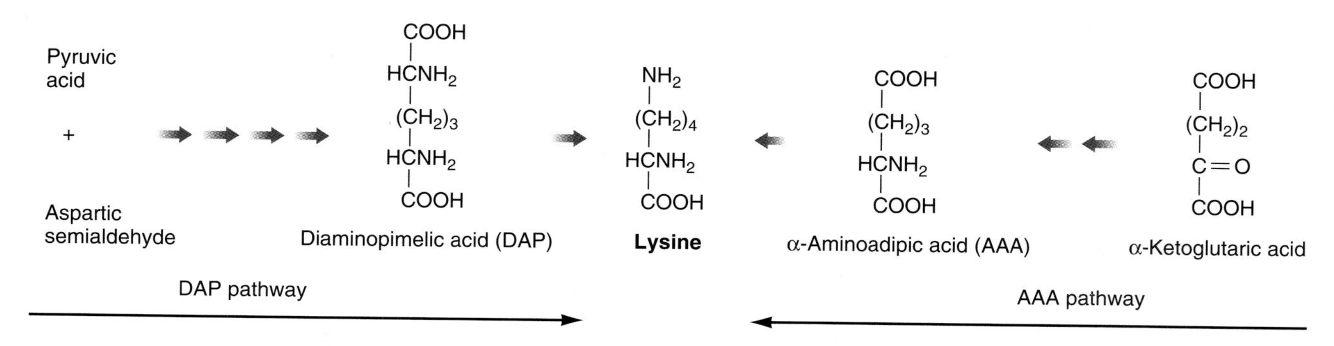

Figure 30.6

Organisms that can synthesize their own lysine use either of two pathways: via DAP or via AAA.

B. A survey of the basal eucaryotes

30.4 The chromists take many forms.

The brown-clade organisms, which we shall now call chromists, include unicellular and colonial algae, large seaweeds, and water molds.

Chrysophytes and xanthophytes

Some of the simplest chromists are single cells with golden, brownish, or yellow-green chloroplasts and a pair of flagella for movement (Figure 30.7). They are classified as **golden-brown algae** (phylum Chrysophyta) or **yellow-green algae** (phylum Xanthophyta). Some flagellates form small packets or colonies of several cells that stay cemented together. Single cells with no flagella are **coccoid,** and similar cells that stay cemented together in packets embedded in mucilage have acquired the unfortunate name **tetrasporine,** even though the cells are rarely in groups of four and are not spores of any kind. Many algae are filaments, often highly branched filaments (Figure 30.8), although this form is not common among chromists. Filamentous algae may also have a **siphonaceous** form: Like siphons, they are long tubes not divided into cellular segments, so several nuclei reside in a single cytoplasm (a coenocytic condition). Chromists also include **rhizopodial** algae, or amoebas (Figure 30.9), so-called because of their branching pseudopods (*rhizo* = root; *pod* = foot).

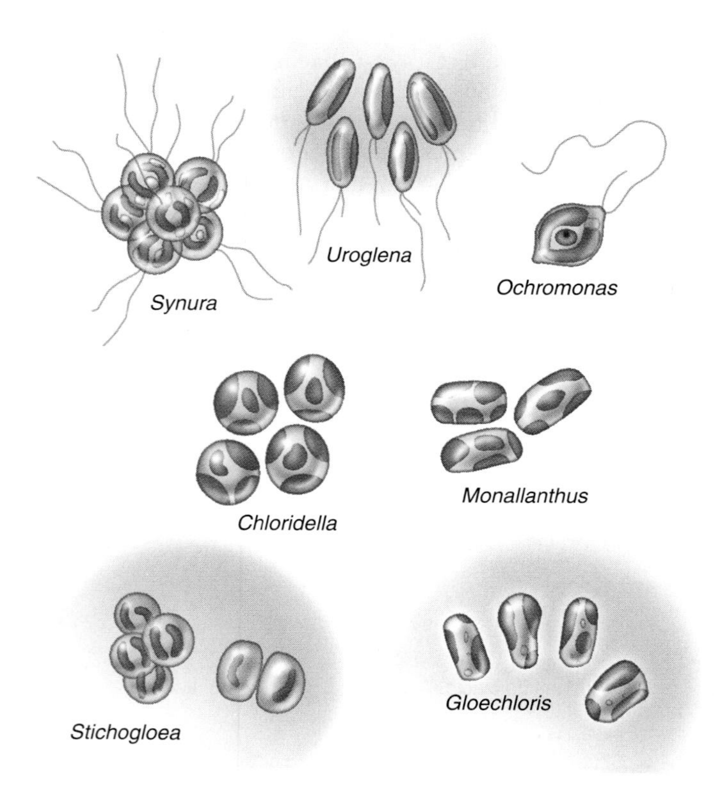

Figure 30.7

Some of the most common chromist algae are flagellate, coccoid, and tetrasporal.

Chrysophytes and xanthophytes are common members of both freshwater and marine plankton. They grow vegetatively by cell division. Colonial organisms grow into larger clusters of cells and then divide; filamentous algae may proliferate by breaking into short segments. Sexual reproduction, also

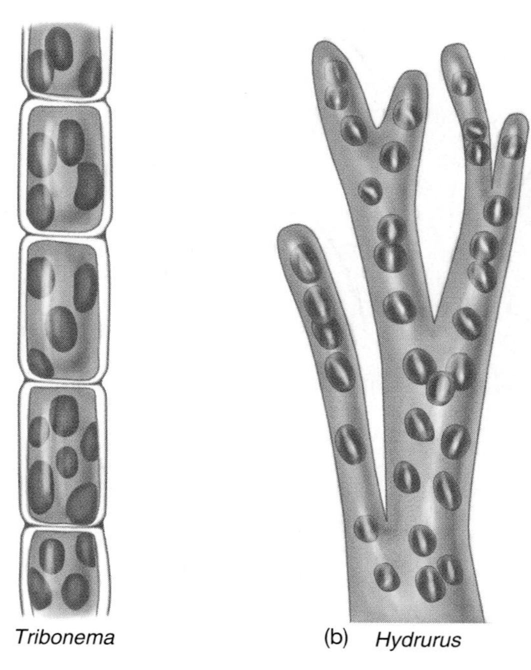

Figure 30.8

(a) Tribonema is a typical filamentous golden-brown alga.
(b) Hydrurus is a siphonaceous (coenocytic) alga having many nuclei in an undivided cytoplasm.

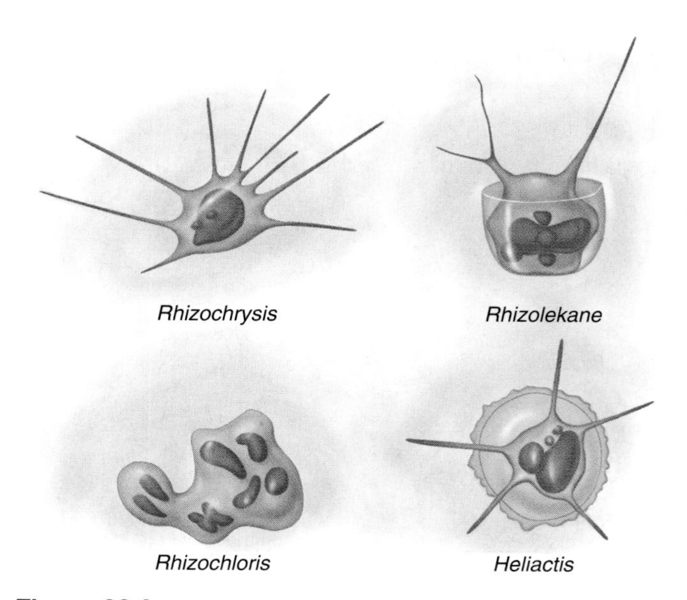

Figure 30.9

Some golden-brown algae are rhizopodial. Genera such as *Rhizolekane* and *Heliactis* include organisms with clear silica shells, reminiscent of similar protozoans with beautifully sculptured shells, as shown in Figure 30.19*d, e,* and *f.*

common among algae, is most often isogamous, with gametes that are morphologically identical; however, some species reproduce with motile zoospores. Freshwater species commonly overwinter by forming resistant cysts.

Diatoms

The elegance of form in chromists algae reaches a peak among the **diatoms** (phylum Bacillariophyta), whose magnificently sculptured shells are built of hydrated silica (silicon dioxide, the chief component of glass) in an organic matrix (Figure 30.10). Massive deposits of ancient diatoms form diatomaceous earth, which is valued as an absorptive material and for the abrasive properties of these tiny shells. The shells of diatoms overlap like the two halves of a box. In many species, as the cells divide, each new half is laid down inside the old one, so the cell volume gets smaller at each division. Eventually the diatoms get so small they must go through a cycle of sexual reproduction to restore larger cells (Figure 30.11). The zoospores formed during sexual reproduction are the only flagellated stage of a diatom's life cycle.

Although diatoms lack any obvious means of locomotion, such as flagella or cilia, many species are motile. They move in response to all kinds of mechanical, chemical, and light stimuli, and are particularly phototactic—that is, they tend to move toward a light source. In each half-shell, motile diatoms have slits (called raphes) that are associated with nearby bundles of contractile fibrils. These fibrils move dehydrated crystalline bodies into the raphe, where they take up water and expand into twisting fibrils. The fibrils move along the raphe until they adhere to some object; then they contract, and if the object is large enough, the diatom can move along it by means of these fibrils, leaving a trail reminiscent of a snail's trail.

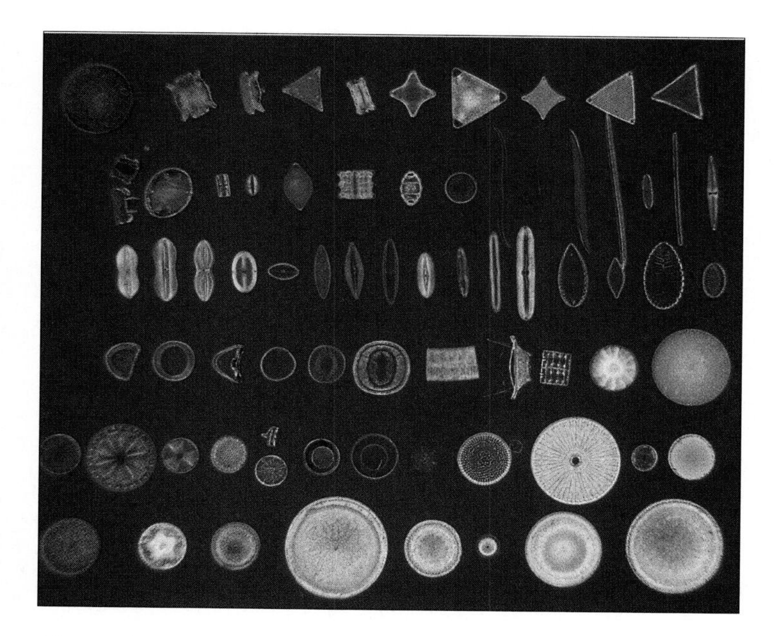

Figure 30.10
Diatoms are chromist algae whose intricately sculptured, symmetrical shells are made of silica.

Dinoflagellates

Among the most interesting chromists are golden-brown to greenish-brown organisms called **dinoflagellates** (phylum Pyrrophyta; Figure 30.12). Most are unicellular, distinctively shaped, covered with a cellulose armor divided into plates, and split by two grooves that contain flagella—one groove encircling the middle and the other perpendicular to it. The action of these flagella causes the organism to spin like a top and gives the

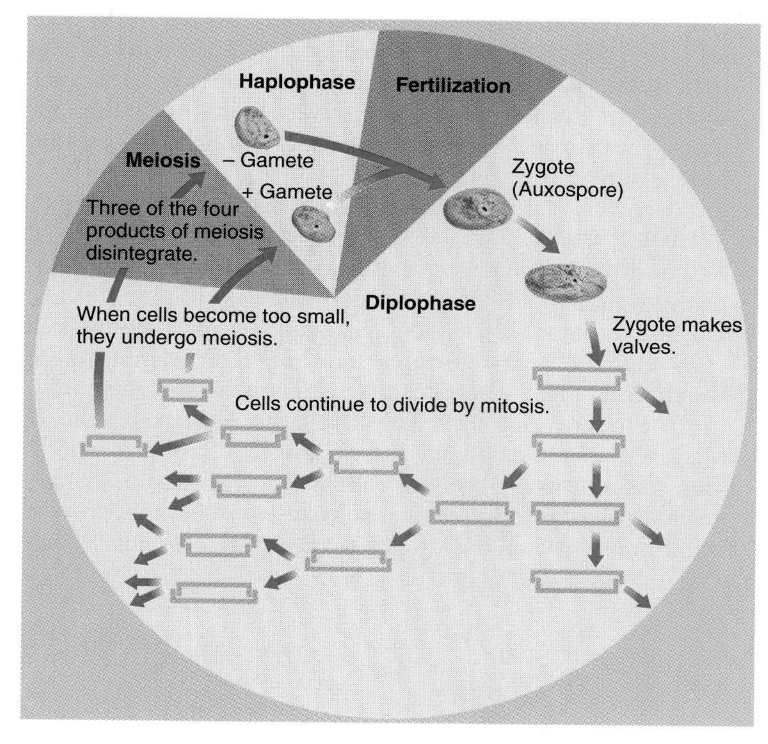

Figure 30.11
During the cell division cycle of a diatom, new shells form inside the old ones, so the cells get smaller with each division and must eventually restore their size with a round of sexual reproduction.

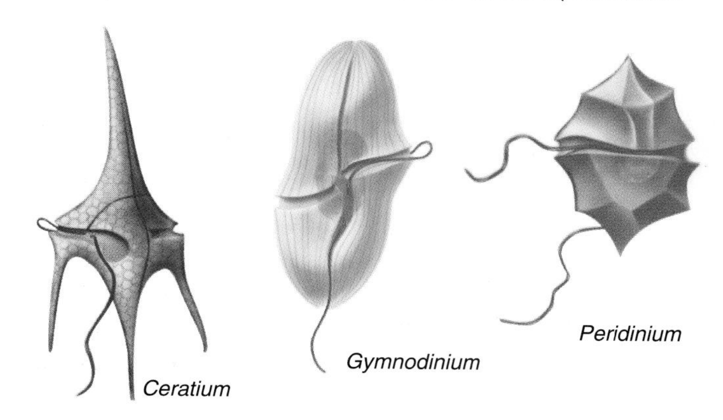

Figure 30.12
Dinoflagellates are distinguished by having a surface divided into plates and two flagella, one encircling and one trailing. Most are yellowish to golden-brown, but others, such as *Gymnodinium,* contain a red pigment and occasionally cause red tides, which poison some animals.

group its name (Greek, *dinos* = whirling). The dinoflagellates sometimes attract our attention quite dramatically. If you go swimming in the ocean on a dark night in late summer, you may see the water flash brilliantly as you agitate it; you are stimulating dinoflagellates such as *Noctiluca* to turn the energy of ATP into light by phosphorescence.

Other dinoflagellates can be very dangerous. Species such as *Gonyaulax* reproduce so massively in the warm summer waters that they create a red tide. As they grow, they produce a powerful vertebrate nerve poison that accumulates in the bodies of the suspension-feeding shellfish, especially mussels and oysters, that eat them. Although the toxin does not affect the molluscs, it can kill fish and other vertebrates, including humans, that consume the dinoflagellates or the molluscs that have fed on them.

Water molds

The **oomycetes** (phylum Oomyceta) are simple fungi that include water molds and downy mildews (Figure 30.13). Oomycetes have flagellated reproductive cells, cellulose cell walls, and a few other features that show their relationship to the chromist algae. These characteristics distinguish them from the true fungi, which never have flagella and whose cell walls are made of chitin, the polymer in the exoskeletons of arthropods. Oomycetes probably evolved from some siphonaceous yellow-green algae. Most oomycetes are obscure saprobes that grow in water or moist soil, but downy mildews like *Plasmopara* and *Peronospora* are plant pathogens on important crops. The mildew *Plasmopara viticolia* threatened to destroy the great vineyards of France during the 1870s. *Phytophthora infestans* is the agent of late blight of potatoes, a disease that practically destroyed the Irish potato crop during the 1840s, leading to widespread famine in which over a million people died and millions more were forced to emigrate to the United States and elsewhere. Such events illustrate again the enormous impact microorganisms can have on human affairs.

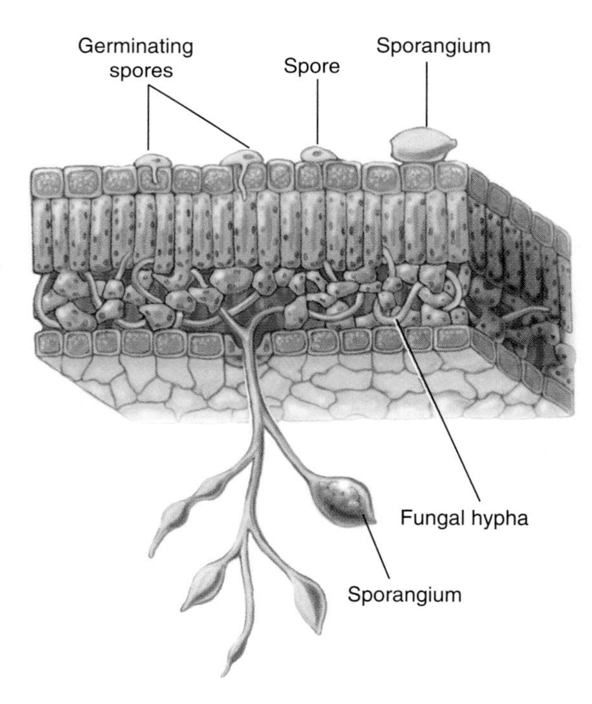

Figure 30.13

A representative oomycete grows as a long, thin hypha that reaches into dead or living plant tissue (here, a leaf) and extracts nutrients, thus acting as an agent of decay or as a parasite. It reproduces by means of spores produced in the sporangium.

Brown algae

The most conspicuous chromist algae are the **brown algae** (phylum Phaeophyta), mostly multicellular organisms that include the prominent brown seaweeds called kelps that dominate rocky coasts (Figure 30.14). While falling in the size range of typical plants, most brown algae show little or no differentiation of their tissues, other than reproductive cells. The general term for such undifferentiated multicellular organisms is **thallus** (pl., **thalli**), or they may be described as thalloid. Their tissues are made of thin-walled, boxlike cells called **parenchyma:**[1]

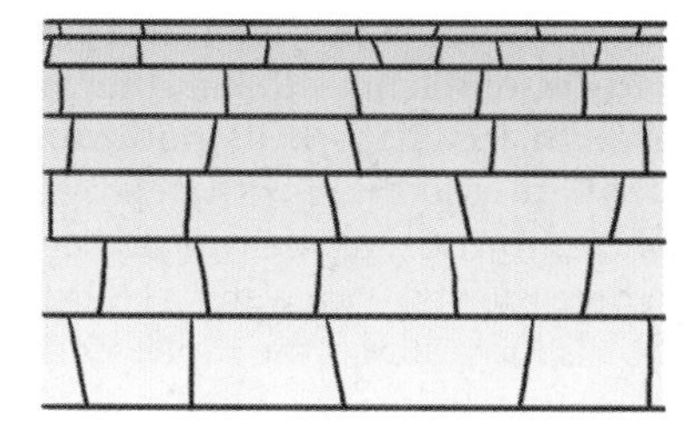

The 1,500 species of brown algae have evolved along several lines; they may be long and filamentous, finely branched, or flat and leaflike. Some kelps take the form of large sheets and cylinders of parenchymal cells.

In their similarities of form and life cycles, brown algae and plants show some remarkable parallel or convergent evolution. Many assume plantlike forms. *Laminaria* has a rootlike structure (called a holdfast) for attaching to the ocean floor, a stemlike portion called a stipe, and a flat, leafy portion. Seaweeds commonly grow air bladders that keep them afloat. The stipes of some kelps are differentiated into tissues like those of plants, including an outer epidermis. They may have vascular (conducting) tissue made of tubes that transport carbohydrates from the upper regions, where the most active photosynthesis occurs, to the portions buried deep in the water. However, no brown algae have reached quite the complexity of the vascular plants with distinct leaves, roots, and stems, nor have they evolved the woody tissue (xylem) that supports most land plants.

A second important parallel between evolution in brown algae and plants is the tendency for mature diploid phases of the life cycle to become dominant. Brown algae alternate between haploid (gametophyte) and diploid (sporophyte) phases. In

[1] Some botanists prefer "pseudoparenchyma" because the cells do not develop from an embryo as one type of differentiated tissue.

most groups, these two phases are morphologically identical and equally prominent, but in some complex algae such as *Fucus,* the sporophyte has become large and dominant while the gametophyte has become more obscure and may even be absent. As we discussed in Chapter 15, evolution tends to take this direction because of the genetic flexibility of a diploid, with its much greater potential for genetic novelty and rapid evolution. Watch for exactly the same trend in the evolution of plants.

Brown algae and other seaweeds are important in the diets of people who live near seashores. *Laminaria* is commonly used in soups, particularly in Japan and Korea. These algae are rich in certain minerals, especially iodine, but since much of their mass consists of indigestible polysaccharides, algae cannot form the basis of a diet. The polysaccharides are extracted and used as thickeners in processed foods such as puddings and ice cream; the polysaccharide agar, we have seen, is used to make semisolid growth media for microorganisms.

Figure 30.14

The kelp *Fucus vesiculosus,* commonly called bladder wrock, is typical of the large plantlike brown algae. This species grows along rocky ocean coasts.

30.5 Red algae have some primitive features.

Nearly 4,000 species of **red algae** constitute the phylum Rhodophyta. They are mostly marine and are especially abundant in tropical coastal waters, but many species live in freshwater and soil. These often spectacular-looking organisms (Figure 30.15) range from single cells to lacy, interwoven filaments and broad sheets marked by distinctive red-purple colors from their phycobilin pigments. These pigments are built into distinctive particles (phycobilisomes) covering their photosynthetic membranes, a structure they share with cyanobacteria (Figure 30.16; compare with Figure 29.8). Later we will consider the implications of this shared structure for the evolution of the protists.

Red-algae cells are surrounded by rather heavy walls with an inner layer of cellulose and an outer layer of pectic materials, like the pectins of fruit. Coralline red algae, whose walls contain calcium deposits, are important components of coral reefs. While the bulk of a coral reef is built of the skeletons of corals, which are animals, the coralline red algae also built into reefs are critical to the ecosystem's overall nutrition.

After cell division in red algae, the new wall does not close off completely but leaves a connection, like a plasmodesma in plant cells. Then a peculiar thing happens. The nucleus of one cell divides again, and one of the daughter nuclei migrates to the new wall and pushes its way through to make a secondary pit connection between the two cells (Figure 30.17). This would hardly be worth mentioning except that similar events (described in Chapter 31) occur in the true fungi, indicating that

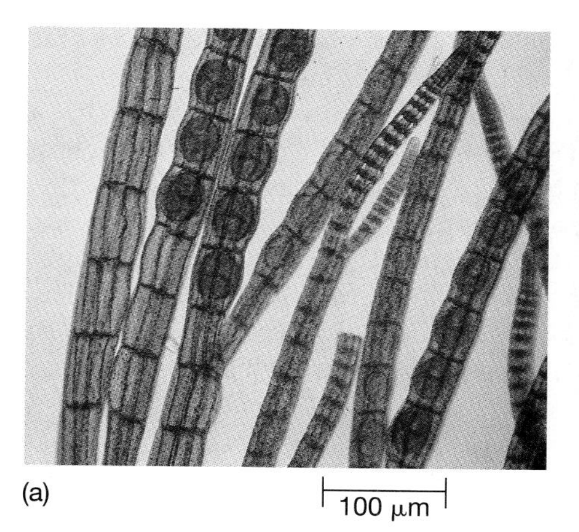

(a) 100 μm

(b)

(c)

Figure 30.15

Red algae take many forms. *(a) Polysiphonia* is a finely filamentous alga. *(b) Bossiella* is a coralline form that incorporates calcium carbonate into its structure. *(c)* Dulse, *Palmaria palmata,* is a large, leafy alga.

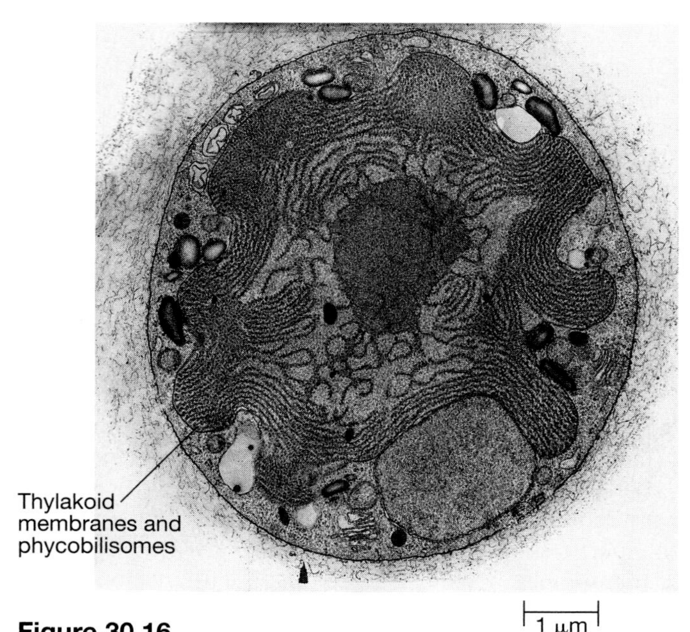

Figure 30.16

A cross section through the red alga *Porphyridium cruentum* shows its large chloroplast containing many thylakoid membranes. Phycobilisomes are fine particles that cover the membranes much as ribosomes cover the membranes of the endoplasmic reticulum.

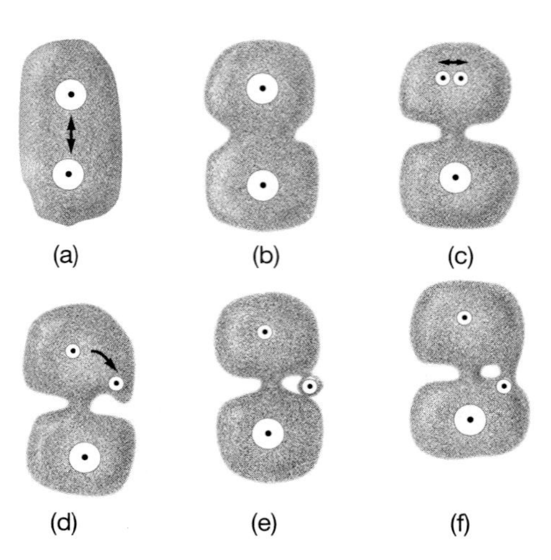

Figure 30.17

The cells of red algae are usually connected to each other at two points. *(a,b)* A primary pit connection forms as the cell divides. *(c–f)* A secondary pit connection forms as one nucleus divides again, and one nucleus migrates back through the cell wall.

red algae and fungi are probably closely related. This conclusion is supported by the similarities between the complex life cycles of many red algae (Figure 30.18) and those of rust fungi (see Figure 15.17).

30.6 Several types of protozoans probably evolved from algae.

Many protozoans are unicellular flagellates that are virtually identical to colored algae, especially some chromists. It seems clear that many evolved from apochromatic algae. Others may represent clades that were never photosynthetic. Some flagellates, including those that live symbiotically in the guts of termites (see Section 27.14) have hundreds of flagella.

Many of the colorless amoebas, phylum Sarcodina, undoubtedly arose from the similar rhizopodial algae of the Chromista. These heterogeneous organisms, though united by their pseudopods, are probably not all closely related. *Amoeba proteus* is a common freshwater example (Figure 30.19*a*). Other amoebas have rather formidable-looking pseudopods that resemble flagella (Figure 30.19*b*). The largest groups of Sarcodina have some kind of shell or skeleton. Foraminiferans (Figure 30.19*c*), which are abundant in warm oceans, secrete calcium carbonate shells that often look like microscopic snail shells. As they die, their empty shells settle into thick layers of mud at the ocean bottom, and in time, this material turns into limestone and chalk; the famous white cliffs of Dover are chalk beds made largely of ancient foraminifera. Because their shells are so distinctive and preserve so well, geologists use them extensively to determine the ages of sediments.

The heliozoans and radiolarians (Figure 30.19*d* and *e*) make beautiful, symmetrical skeletons of silica, suggesting a relationship with the diatoms and other chromist algae that metabolize silica so well. Although they appear rigid and delicate, the pseudopods of heliozoans contain spiral microtubule complexes (see Figure 11.29*b*) that enable them to move and engulf food as heavier pseudopods do. Shelled amoebas (Figure 30.19*f*) are very similar to some rhizopodial algae shown in Figure 30.9.

Amoebas look so different from flagellates, and move with such a different mechanism, that it seems hard to believe the two groups are closely related, but some of them apparently are. (Some biologists would combine them as a phylum called Sarcomastigophora.) One indication of the close relationship between these groups is the existence of *amoeboflagellates* that can switch back and forth between amoeboid and flagellated forms under the influence of environmental conditions.

The **sporozoa** are all parasites with complex life cycles, a diverse group now divided into at least two phyla, Apicomplexa and Cnidosporidia. Though the mature stages are nonmotile, some kinds of sporozoans reproduce with flagellated gametes, and some with amoeboid gametes, suggesting that the group is derived from both flagellated and amoeboid ancestors. The Apicomplexa include *Plasmodium* species, the agents of malaria in humans. *Plasmodium* goes through several stages as it cycles from person to person through *Anopheles* mosquitoes (Figure 30.20).

Ciliates (phylum Ciliophora) are a large, diverse group of organisms that propel themselves with the coordinated movements of the cilia that cover their surfaces (Figure 30.21). They probably arose from multiflagellated cells through shortening

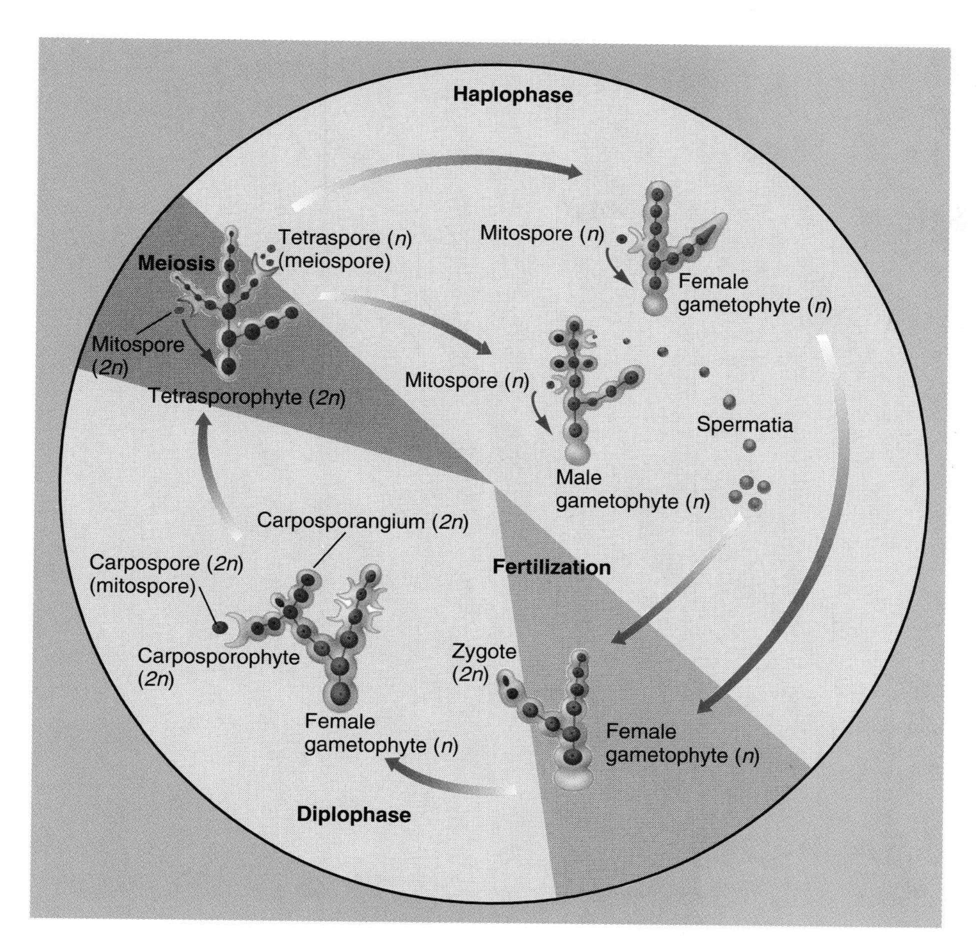

Figure 30.18

Life cycles of red algae are typically complex, having several distinct structural stages connected by different kinds of spores and gametes.

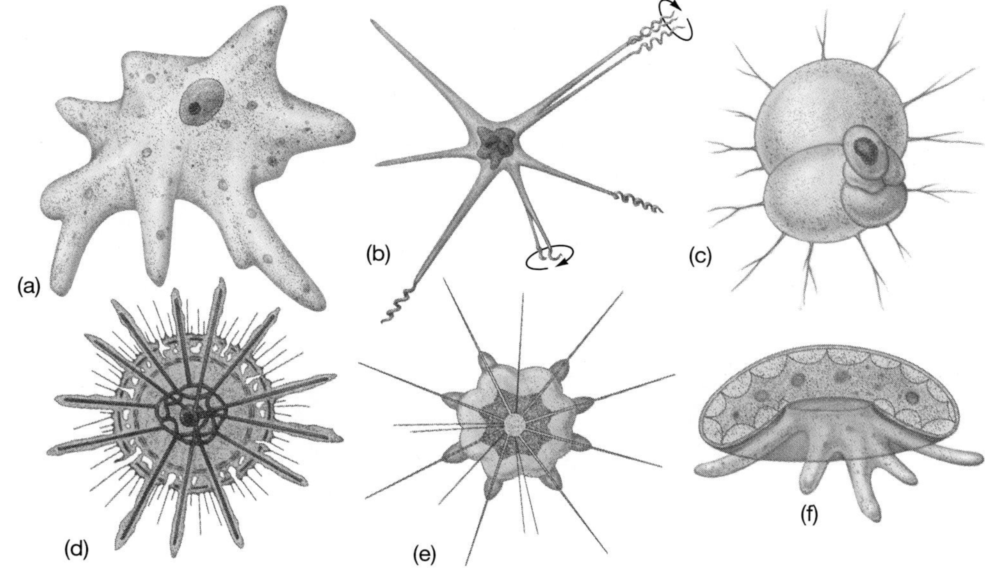

Figure 30.19

Sarcodina are protozoans that move by amoeboid action. Heliozoans and radiolarians have beautifully sculptured silica shells: *(a) Amoeba proteus,* a common freshwater organism; *(b)* a mayorellid amoeba with pseudopods that resemble flagella; *(c)* a foraminiferan with a shell of calcium carbonate; *(d) Actinosphaerium,* a heliozoan; *(e) Acanthometrium,* a radiolarian; *(f) Arcella,* a shelled amoeba.

Figure 30.20

The malaria parasite *Plasmodium* has a complex life cycle. *(1)* When a female *Anopheles* mosquito feeds on the blood of an animal, she releases sporozoites into the bloodstream. *(2)* Sporozoites enter the liver and divide many times to produce thousands of merozoites from a single liver cell. *(3)* Merozoites enter red blood cells (RBCs) and continue to divide, producing more merozoites that infect other RBCs. *(4)* Eventually, some merozoites form sexual forms (gametocytes), which are picked up by other mosquitoes. *(5)* In the mosquito's intestine, each female gametocyte produces one egg, and each male gametocyte divides into several sperm. Fertilization produces a zygote, which migrates into the salivary gland and divides into many sporozoites.

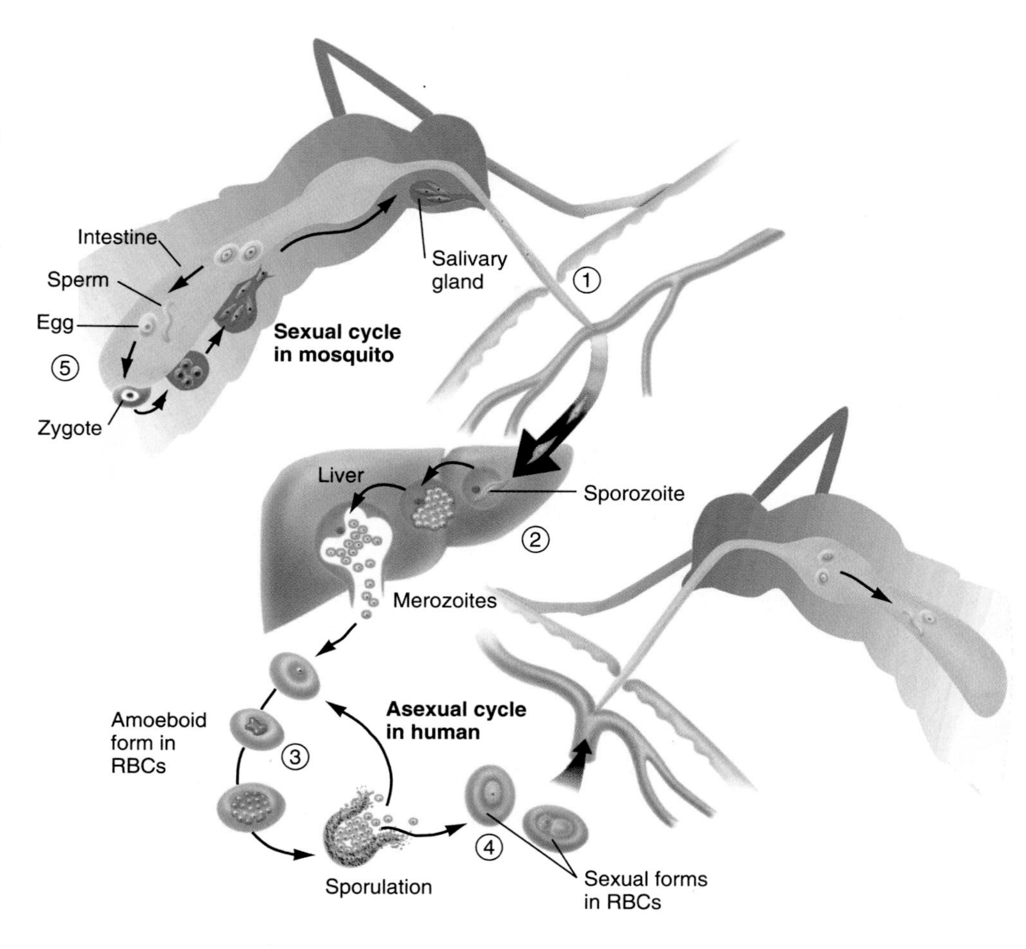

Figure 30.21

Ciliates include free-moving forms such as *Tetrahymena,* which propels itself with its cilia, and *Stylonychia,* which moves with a walking motion on its cirri. *Stentor* remains attached by its base but can free itself to move to new locations; *Vorticella* cell bodies are attached by long, thin stalks, which frequently contract quickly. *Podophrya* is a suctorian; its tentacles are used for sucking the cytoplasm out of the cells it preys on.

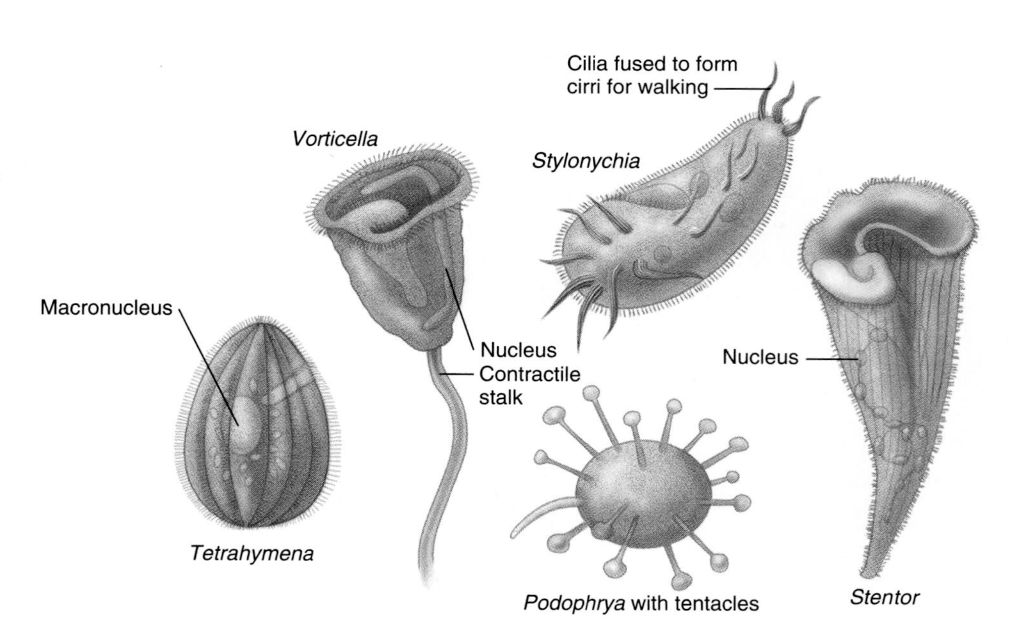

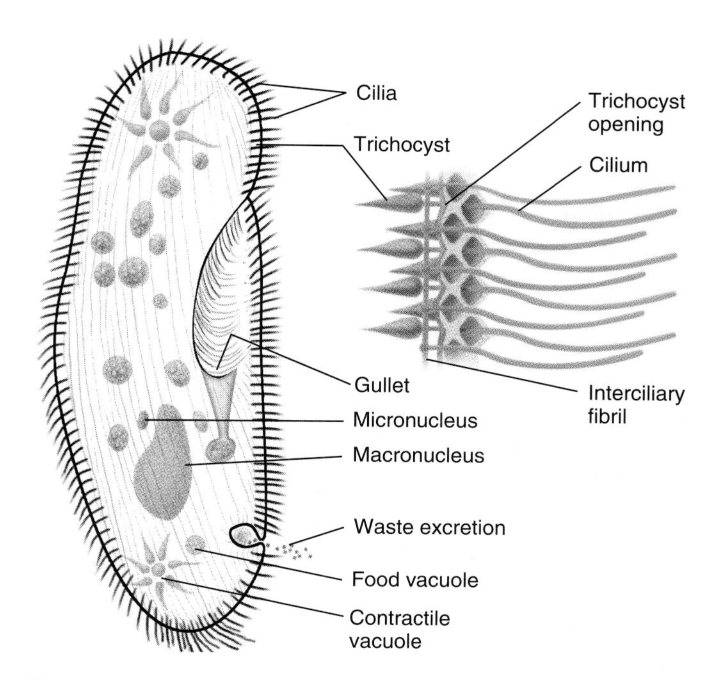

Figure 30.22

The ciliate *Paramecium* shows the network of fibrils that connect cilia and presumably coordinate their movement. Trichocysts lie between the ciliary bases.

of the flagella. Connecting fibrils coordinate the whole ciliary field, so the cilia beat together and the direction of movement can be changed quickly. Some ciliates also have internal contractile elements made of microtubules that allow them to wiggle or bounce about on long stalks, as *Vorticella* does. Scattered among the cilia on the surface of a cell are *trichocysts,* organelles similar to a whaler's harpoon with a sharp, sometimes poisoned barb on a long thread (Figure 30.22). Trichocysts shoot out of their retracted positions when stimulated, and they can be used for defense, for procuring food, or for anchoring the cell.

Paramecium, the representative ciliate shown in Figure 30.22, has two kinds of nuclei: One or more small *micronuclei* provide genetic continuity from generation to generation, and a large *macronucleus,* containing several times as much DNA as a micronucleus, provides the templates for protein synthesis and cellular regulation between divisions. When the cell divides, the macronucleus splits, and half goes to each daughter cell. While *Paramecium* and other ciliates generally reproduce by asexual cell division, they occasionally engage in **conjugation,** or sexual mating (Figure 30.23). Strains of paramecia that don't conjugate eventually die out. Each species has several mating types, and only cells of compatible types can conjugate.

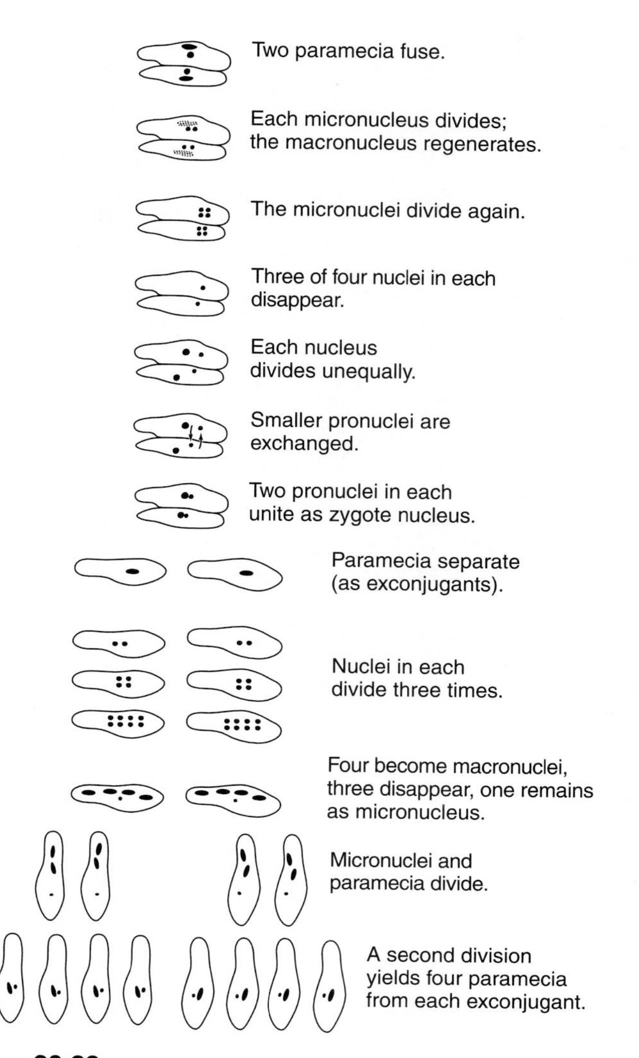

Figure 30.23

Conjugation in *Paramecium caudatum* begins when two compatible cells adhere to each other. The cells actually fuse while their macronuclei degenerate and disappear. Their micronuclei then undergo meiosis and one of each pair of daughter micronuclei migrates to the other cell. Micronuclei from opposite cells fuse to restore the diploid condition, so each cell incorporates genes from the other.

30.7 The slime molds include a variety of amoeboid saprobes.

Among the most enigmatic organisms are *slime molds,* found in damp soil, rotting wood, and decaying leaves and named for their glistening, slimy appearance (Figure 30.24*a*). Some species are white, but most are red, orange, or yellow. They fall into several groups that, while not very closely related, are lumped together in traditional classifications because they all feed and grow as amoeboid cells and then develop fruiting bodies that produce spores.

A **true slime mold** (phylum Myxomycota) grows vegetatively as a **plasmodium,** a large coenocytic mass (typically 5–8 cm across) that moves along slowly like a gigantic amoeba, phagocytizing organic particles as it goes. At some point, the diploid plasmodium stops moving and sends up fruiting bodies in which masses of haploid spores form by meiosis (Figure 30.24*b*). After falling to the ground, these spores germinate into flagellated gametes, which fuse in pairs to make new diploid vegetative cells. Then each cell begins to move about as an

(a)

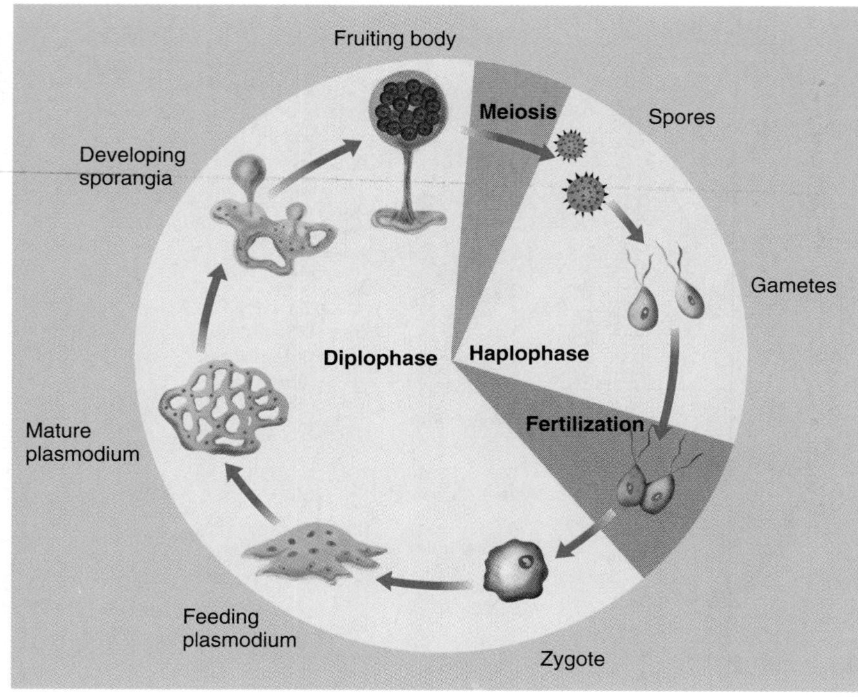

(b)

Figure 30.24

(a) Slime molds grow on woody substrates and many kinds of vegetation—here, on moss. *(b)* During the life cycle of a true slime mold, the zygote grows into a multinucleated coenocyte, which forms fruiting bodies where meiosis occurs.

amoeba again, feeding and undergoing repeated mitosis without cell division, so it grows into a new plasmodium.

Unrelated organisms with a similar life cycle are known as **cellular slime molds** (phylum Acrasiomycota). The best-studied type, *Dictyostelium,* consists of independent, uninucleate amoeboid cells that feed phagocytically as they move over the soil and divide repeatedly into more uninucleate cells (Figure 30.25). As their food supply diminishes, these cells stop feeding and become more rod-shaped. Some cells begin to secrete cyclic AMP (cAMP), an aggregation pheromone that induces more cells to aggregate and to start making their own cAMP. (cAMP also functions as an alarmone [see Section 18.6] and as an intracellular signal for growth factors and hormones [see Section 41.4].) Thus the cells recruit one another and stream into a central point where they form a **pseudoplasmodium,** so-called because the cells retain their cellular identities and do not form a coenocyte. The whole pseudoplasmodium may move, but it eventually grows a fruiting body in which spores form, and each spore can then develop into a new vegetative cell. There are no sexual events in this cycle, and all stages are apparently haploid.

C. Evolution among the basal eucaryotes

Knowing the main types of basal eucaryotes, we can now address some of the most interesting questions about their phylogeny. Keep in mind that, even though biologists have developed reasonable arguments for evolutionary relationships based on the available information, the evolutionary story is still subject to much uncertainty. Rather than looking for hard facts to memorize, think of this information as clues in a detective story that may never conclude. Because the first eucaryotic cells would be classified as protists, we begin by asking about the origin of the eucaryotic condition itself: a nucleus bounded by a double membrane, separate chromosomes, and mitotic nuclear division.

30.8 Eucaryotic cells and the mitotic apparatus arose through a series of complicated changes.

What were the major driving forces behind the evolution of eucaryotes? Biologists have speculated about the advantages of eucaryotic over procaryotic organization. One hypothesis is that the eucaryotic condition was a response to an oxidizing atmosphere, created by phototrophs that produce oxygen as a by-product. As useful as oxygen may be in metabolism, it can also be a poison that reacts with and destroys many biomolecules, including nucleic acids. The nucleus may have evolved as a place to sequester the precious genome and keep it safe from oxygen. However, this hypothesis does not explain how aerobic procaryotes are able to do so well. A second hypothesis is that the eucaryotic cell, with a mitotic apparatus, evolved as

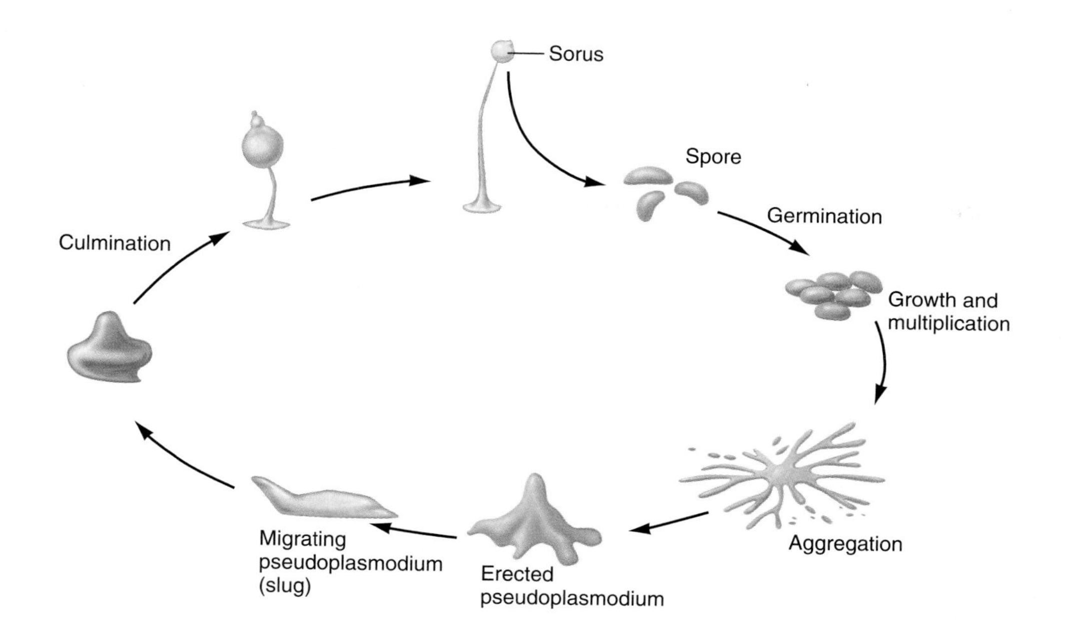

Figure 30.25

The life cycle of the cellular slime mold *Dictyostelium* has no sexual stage, and all forms are apparently haploid. Each spore releases an amoeboid cell, which moves over the substratum, growing and multiplying. Eventually, many cells aggregate into a pseudoplasmodium, which assumes a sluglike form. The slug settles down in one place and sends up a fruiting body made of a large sorus on top of a thin stalk. Cells in the sorus form spores again.

housing for a large genome and as a mechanism for regularly dividing such a genome among daughter cells. These features may have been necessary for the evolution of large cells capable of strong movements and for more complex activities than procaryotes engage in; ultimately, they opened the doorway to the evolution of plants and animals that require large genomes containing instructions for differentiation. Because mitochondria and chloroplasts provide efficient mechanisms for obtaining energy, more complex, energy-demanding processes are possible.

Figure 30.26 shows a likely scenario for the evolution of a procaryotic cell into a eucaryotic cell. Remember that procaryotic chromosomes are attached inside the cell membrane and that daughter chromosomes are separated by elongation of the whole cell (Figure 30.26*a*). Invagination of the cell membrane could form a nuclear envelope, with the chromosomes still attached to its inside. At first the nucleus probably separated daughter chromosomes by the bacterial mechanism—simply elongating the inner nuclear envelope and then pinching in two (Figure 30.26*b*). In a second stage, the cell acquired microtubules, which helped draw the nuclear envelope out toward the poles. Finally, the microtubules connected to the chromosomes directly and formed a spindle to pull daughter chromosomes toward the poles (Figure 30.26*c*).

Microtubules must have evolved early in the eucaryotic lineage. They are general organelles for motility and have become critical for separating sets of chromosomes precisely during mitosis. Some microtubules were incorporated into the basal body and flagella, thus opening up a new, rapid means of locomotion. Then, in some eucaryotes, the basal body apparently acquired a secondary function as a centriole that organizes the mitotic spindle.

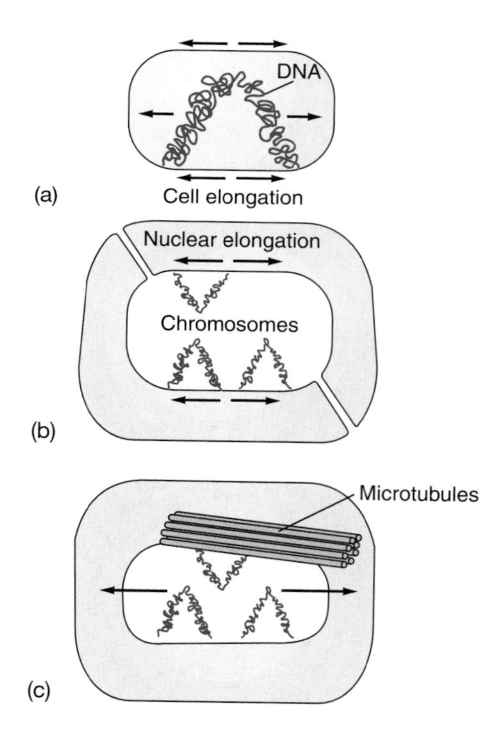

Figure 30.26

A nucleus and mitotic apparatus probably evolved in this way. *(a)* The genome of a procaryotic cell is a single long DNA molecule that is attached to the plasma membrane; its daughter molecules separate as the cell elongates. *(b)* The eucaryotic nucleus probably formed through invagination of the plasma membrane, and at first, the daughter chromosomes could still be separated by elongation of the nuclear envelope, even if the genome became divided into two or more chromosomes. *(c)* Later, microtubules came into play in helping separate the chromosomes and divide the nucleus, as shown in more detail in Figure 30.28.

30.9 Dinoflagellates and euglenoids have primitive types of nuclear structure and mitosis.

As eucaryotic cells evolved, mitosis—the process of dividing the nucleus and separating out sets of chromosomes—also evolved. We can trace the evolution of mitosis by studying some modern organisms that retain what appear to be remnants of early stages of the process. Of course, each clade evolved its mitotic apparatus independently of the others, and these organisms do not represent stages in a single process.

The discovery of primitive nuclear features and modes of mitosis among certain eucaryotes makes a fascinating story. During the 1960s, several investigators discovered the unusually primitive nuclear features of dinoflagellates; their chromosomes do not go through the typical cycle of uncoiling during interphase and condensing during mitosis. John Dodge showed that these chromosomes look more like the nuclear bodies of bacteria than like the chromosomes of other eucaryotes (Figure 30.27), and Peter Giesbrecht then found that these chromosomes lack the histones of other eucaryotic chromosomes.

Donna Kubai and Hans Ris showed that some dinoflagellates undergo mitosis as shown in Figure 30.28. The nuclear envelope does not break down; instead, the chromosomes are attached to its inner surface, and the whole nucleus divides by elongating with the aid and guidance of bundles of microtubules that invaginate it. The nucleus is pulled into two halves, like a drop of water that slowly stretches until it breaks. Meanwhile, each chromosome is pulled into a Y shape, then into a V, and finally into two separate bodies.

Dinoflagellates, then, resemble the hypothetical cell-within-a-cell stage of eucaryotic evolution. Their several chromosomes divide rather like those of bacteria, but with the aid of a new eucaryotic feature, microtubules. At this stage of evolution, the microtubules are still associated with the nuclear envelope in an unusual way and have not yet acquired connections with the chromosomes.

Although mitosis in other basal eucaryotes is quite varied, the process has apparently evolved through two kinds of changes (Figure 30.29): (1) Kinetochores were initially plaques in the nuclear envelope to which the chromosomes were bound, and sister kinetochores (and thus sister chromatids) were initially separated from each other by movements of the envelope itself. Later, the kinetochores became permanently attached to the chromosomes, and the nuclear envelope no longer had any role in mitosis. (2) The spindle was initially extranuclear. It contributed to the separation of chromosomes by helping the nuclear envelope elongate. Then some spindle elements began to separate the chromosomes by attaching directly to the kinetochores.

Primitive nuclei and mitosis also characterize the **euglenoid flagellates** (Figure 30.30); although their chromosomes are like those of other eucaryotes, their nuclei still divide by elongating and pinching in two, as the nuclei of dinoflagellates do. Euglenoids, however, are real phylogenetic enigmas. They are distinguished by several features: the AAA pathway of lysine synthesis, flagella with only a single row of filaments (unlike either the typical whiplash or tinsel flagellum), a gullet in which

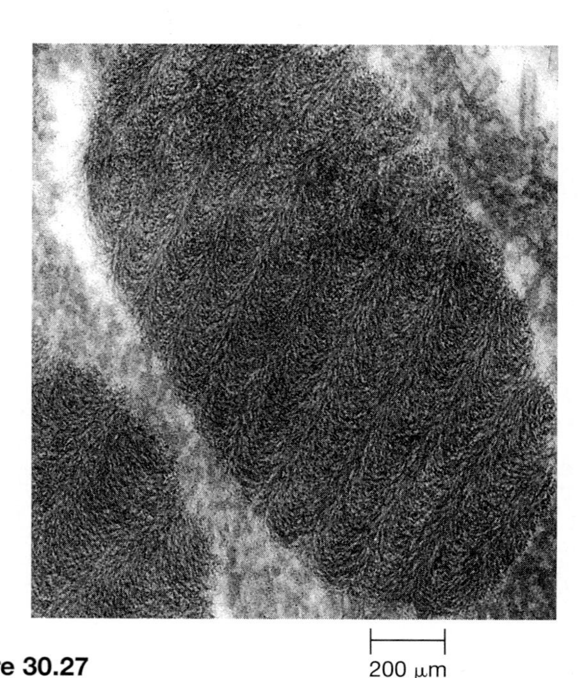

Figure 30.27

In an electron micrograph, chromosomes of the dinoflagellate *Prorocentrum* show a fine fibrillar structure that is reminiscent of a bacterial nuclear body, but is quite unlike the structure of other eucaryotic chromosomes.

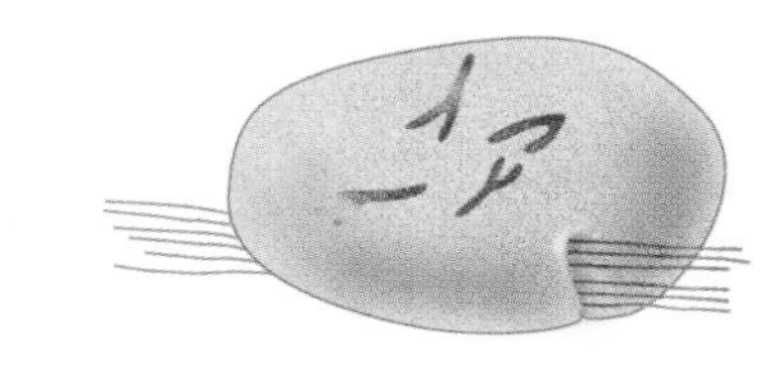

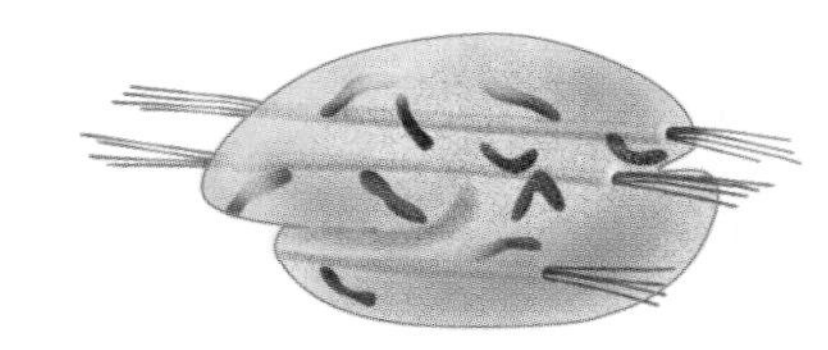

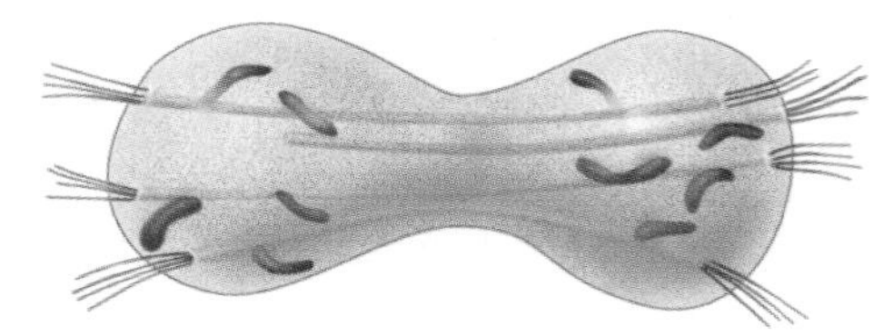

Figure 30.28

Nuclear division in a dinoflagellate occurs with the help of microtubules *(red lines),* but the nuclear envelope never breaks down. Channels through the nucleus that are filled with microtubules separate daughter chromosomes to opposite poles. The nuclear envelope then constricts to form two nuclei.

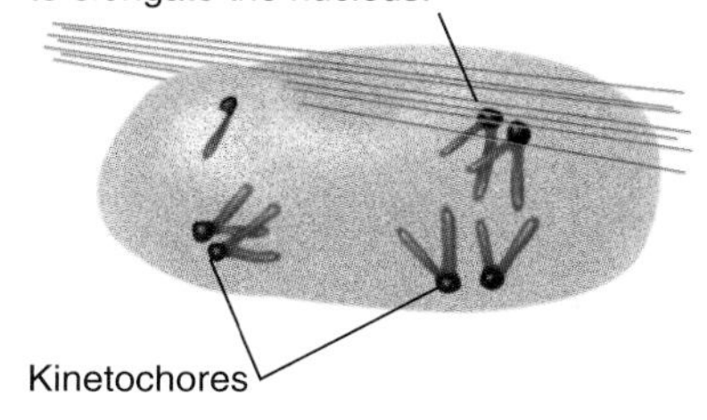

(1) Chromosomes are attached to the nuclear envelope at kinetochores and are separated by elongation of the envelope.

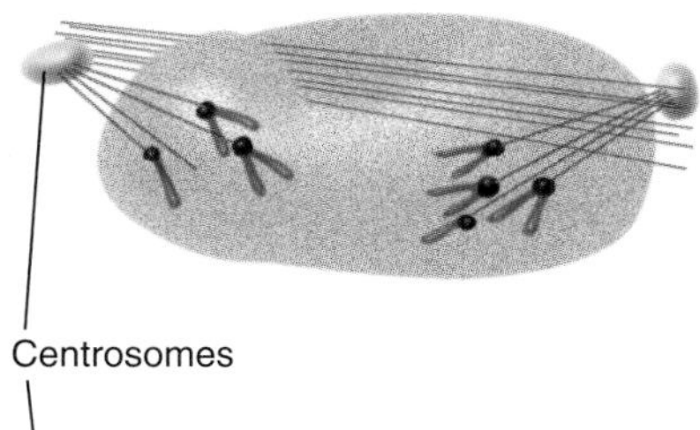

(2) Spindle invades the nuclear envelope, and some spindle fibers attach to kinetochores and help to separate the chromatids. Nuclear envelope still performs most separation of sister chromatids.

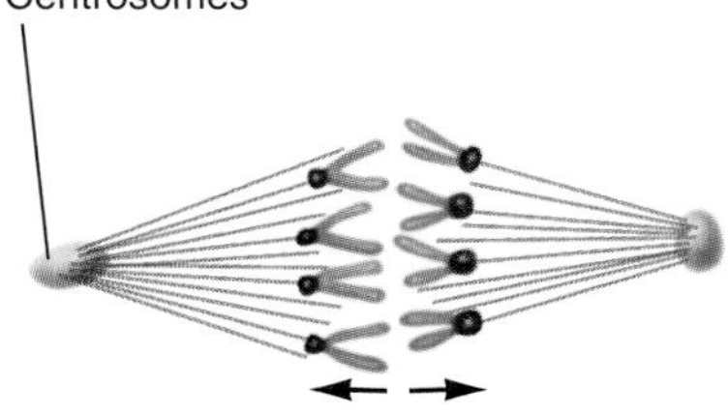

(3) Nuclear envelope disappears during mitosis. Kinetochores remain on chromosomes, and chromatids are separated entirely by spindle fibers.

Figure 30.29

The two principal changes in mitosis explained in the text can be seen in three identifiable stages in contemporary organisms: *(1)* mitosis in some dinoflagellates, as shown in Figure 30.28; *2)* mitosis in certain flagellates; *(3)* mitosis in plants and animals.

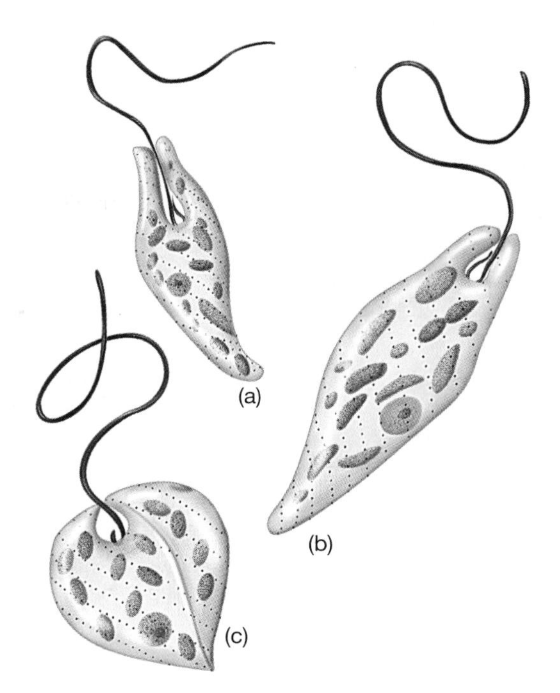

Figure 30.30

Some representative euglenoid flagellates include *(a) Astasia; (b) Euglena;* and *(c) Phacus.*

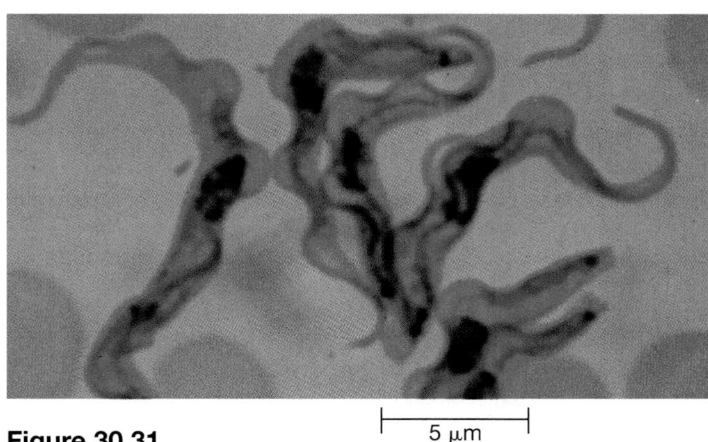

Figure 30.31

The trypanosomes *Trypanosoma gambiense* (here colored artificially) causes sleeping sickness. Each cell has a flagellum running along the edge of its undulating membrane.

they engulf bits of food and a brightly colored eyespot near the gullet that is apparently a light receptor, and a peculiar squirming motion in which the cell alternately bulges and narrows. Euglenoids have the chlorophylls *a* and *b* of the green algae and the storage materials of the chromist algae. Even assuming their organelles derived from various sources through endocytosis, as will be discussed later, it is hard to account for these creatures.

The euglenoids have another peculiar feature—a rod running through each flagellum parallel to the microtubules. This structure, as well as some other microtubule structures, shows their affinity with the **trypanosomes,** a group that includes such parasites as the causative agent of African sleeping sickness (Figure 30.31). The tsetse fly is the vector that carries these trypanosomes between humans and various animals. The bite of an infected fly transmits them first into the blood, where their continued growth produces fever and anemia, and eventually they invade the membranes around the brain and spinal cord, leading to unconsciousness and death.

30.10 Some eucaryotic organelles probably arose through endosymbiosis.

Few ideas in biology have captured the popular imagination more than the theory that some eucaryotic organelles originated through a process of symbiosis. As explained in Section 27.14, in *mutualism* (more loosely called *symbiosis*), two species live together in a cooperative way for their mutual benefit. Some features of eucaryotic organelles prompted the theory, championed largely by Lynn Margulis, that eucaryotic cells acquired mitochondria and chloroplasts by incorporating certain bacteria as **endosymbionts,** organisms that grow inside other cells (Figure 30.32*a*). For example, a number of contemporary protozoa and animals carry small algae as endosymbionts. According to this model, the primitive eucaryotic cell, developing a nucleus as outlined in Section 30.8, was fundamentally anaerobic and lived by fermentation. In fact, commonalities between eucaryotic and archaebacterial cells (Table 29.A, in Sidebar 29.1)

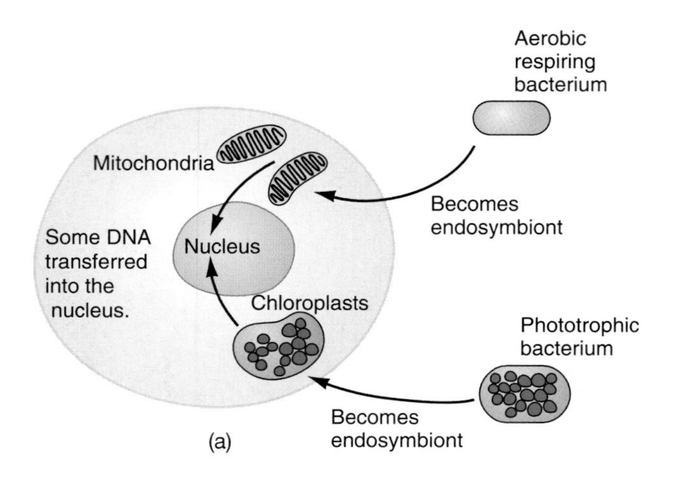

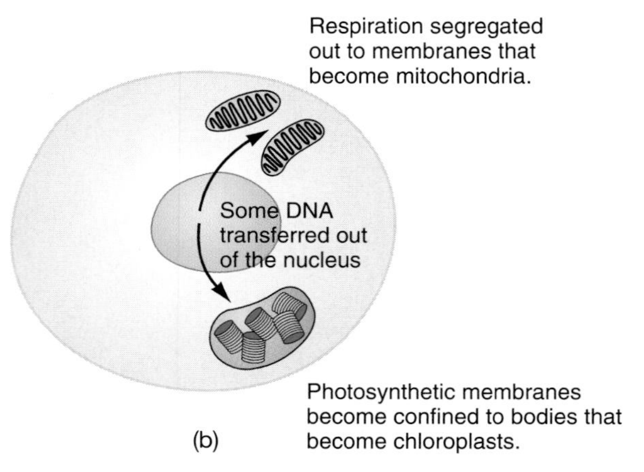

Figure 30.32

Two theories have been proposed to explain the origin of eucaryotic organelles. *(a)* The endosymbiotic theory posits that primitive eucaryotic cells incorporated certain bacteria as endosymbionts—not all at one time, and not all in each type of cell. These endosymbionts gradually lost their identities and became more highly integrated as mitochondria, as chloroplasts, and possibly as flagella, but they retained part of their own genomes and their apparatus for protein synthesis. *(b)* Alternatively, the organelles gradually evolved within primitive eucaryotes by moving some genes and their functions into different membrane-bound compartments.

suggest that eucaryotes arose from a procaryote of the archaebacterial type. Then some cells, in different clades, acquired endosymbionts that transformed their metabolism: a bacterium with an aerobic respiratory system that was the ancestor of the mitochondrion and photosynthetic bacteria that were the ancestors of various chloroplasts. Also, something like a spirochete may have brought the centriole and flagellum into the cell.

The main support for this hypothesis is that both mitochondria and chloroplasts have their own genetic apparatus. They contain small circular DNA molecules that encode some of their proteins, which are made in the organelle itself by its own ribosomes and tRNA molecules, and these components of the protein-synthesis apparatus are more like those of bacteria than like the cytoplasmic eucaryotic apparatus. Several investigators have used the sequences of ribosomal RNA from various organisms as an indicator of phylogeny, on the assumption that changes in these molecules should occur randomly and have little adaptive value. These studies and others have identified the probable ancestors of mitochondria and chloroplasts among particular groups of procaryotes.

An obvious strength of the model is that it explains why basal eucaryotes have such different chloroplasts. Because the cyanobacteria share several features with red algae—chlorophyll *a,* oxygen as a by-product of photosynthesis, phycobilin pigments in phycobilisomes—some ancient cyanobacterium was probably the ancestor of the red-algal chloroplast. The ancestors of green-algal chloroplasts might have been unusual bacteria such as *Prochlorion,* which is grass-green, possesses chlorophyll *a,* and also produces oxygen as a by-product. No one, however, has yet found a possible ancestor of the widespread brown or yellow-brown chloroplast of the chromist algae.

Although the endosymbiotic model is very plausible and explains a great deal, some biologists favor an alternative model (Figure 30.32*b*). Mitochondria could be derived just as well from internal changes in a cell. Mitochondrial DNA is similar in size to a typical plasmid, and the primitive eucaryotic cell might have acquired a plasmid, perhaps carrying genes for certain functions of the modern mitochondrion. The plasmid then exchanged some genes with the cellular genome as the mitochondrion formed around it. Neither theory is entirely supported by the facts about eucaryotic cell organization, and this kind of problem is hard to answer definitively, since it involves events from the remote past that cannot be recreated.

30.11 Can we derive a composite picture of early eucaryotic evolution?

Much of the information developed in this chapter has already led us to envision several lineages of evolution among the protists and the other eucaryotic kingdoms. Here we will summarize this information (Figure 30.33). It is important to see, however, that biologists' thinking about the phylogeny and systematics of eucaryotes is in a state of flux. The entire rRNA-based tree of protistan relationships, for instance, must be viewed as one useful working hypothesis, but many data are in conflict with it. Overall, our knowledge of relationships among the major groups of eucaryotes is quite uncertain and in need of substantial new data. Investigators are using molecular techniques to analyze the DNA, ribosomal RNA, and proteins of the problematic organisms now lumped together as protists, and new information from these sources will undoubtedly upset and clarify the phylogeny we outline here. Of course, that is only natural for science at its best.

According to the endosymbiotic model, various organelles were incorporated at different times, and events must have occurred quite independently in different clades. Consider first the acquisition of mitochondria and the eucaryotic flagellum. Some clades lack one, some lack the other. Cavalier-Smith has proposed a kingdom called Archezoa for what may be the most

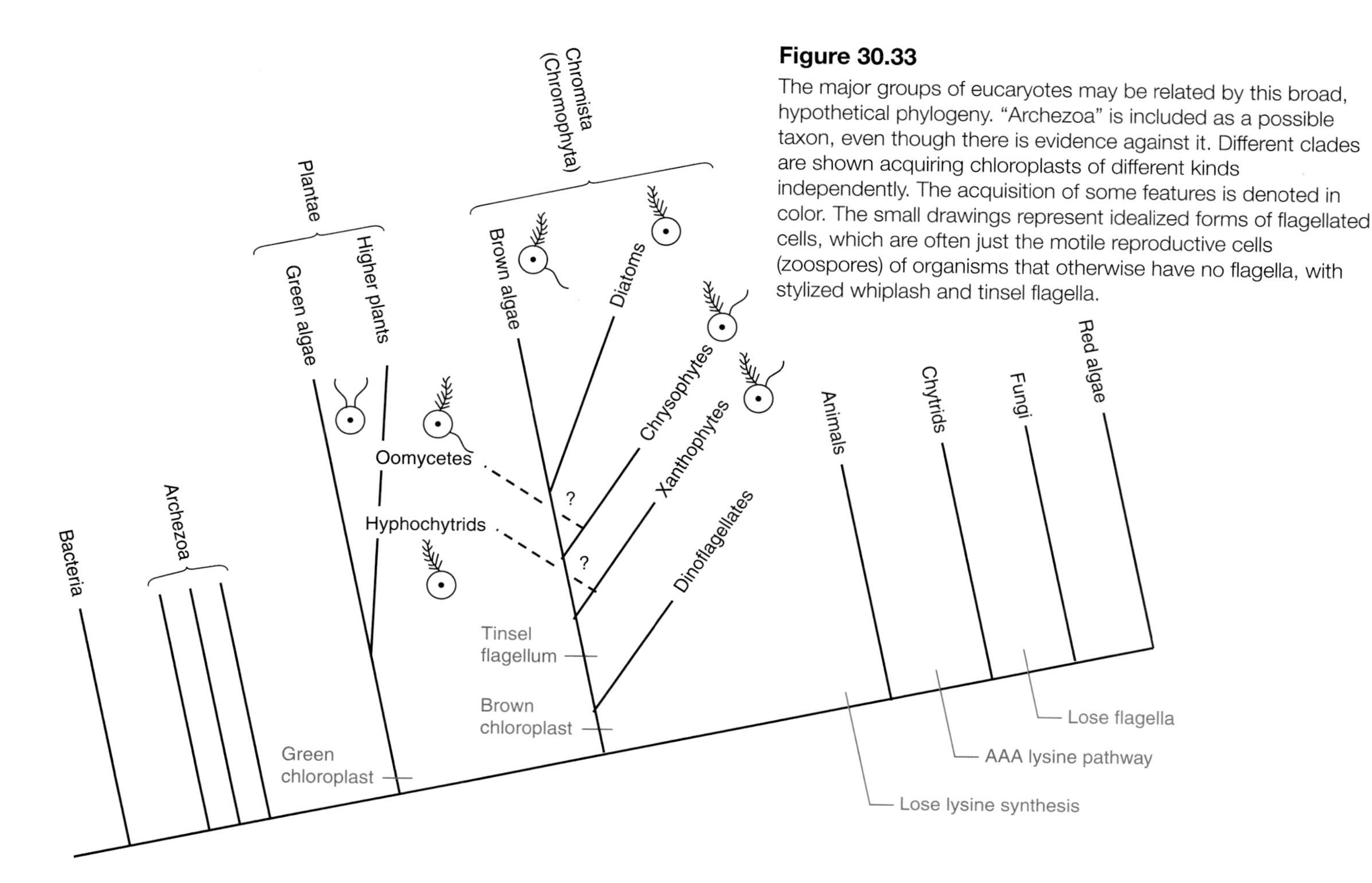

Figure 30.33

The major groups of eucaryotes may be related by this broad, hypothetical phylogeny. "Archezoa" is included as a possible taxon, even though there is evidence against it. Different clades are shown acquiring chloroplasts of different kinds independently. The acquisition of some features is denoted in color. The small drawings represent idealized forms of flagellated cells, which are often just the motile reproductive cells (zoospores) of organisms that otherwise have no flagella, with stylized whiplash and tinsel flagella.

primitive eucaryotes; they have small ribosomes, like those of bacteria, and may never have had mitochondria, Golgi membranes, and other organelles. Archezoa includes several puzzling organisms: the archamoebae, including species that can change from amoeboid to flagellate form; some parasites called microsporidians; and a group of flagellates called diplomonads. The latter include the freshwater organism *Giardia,* which is well known to hikers and travelers for causing a rather nasty intestinal upset (Figure 30.34). Diplomonads have the interesting feature of two separate nuclei, which gives them their name. The Archezoa concept is supported by rRNA evidence, but not by other evidence; there is good reason to think that these primitive eucaryotes evolved from ancestors that had mitochondria and lost these organelles in different ways.

The details of relationships among many of the protozoans has only limited interest for our general study of biology. However, a great deal of interest still resides in the puzzling relationships among three major groups: red algae, true fungi, and animals. Let's review some of the facts.

A body of fascinating evidence indicates that red algae and fungi are related. Both groups have no flagella or cilia and no trace of a centriole or basal body. In addition, many parallels in the life cycles of red algae and fungi argue for a close relationship

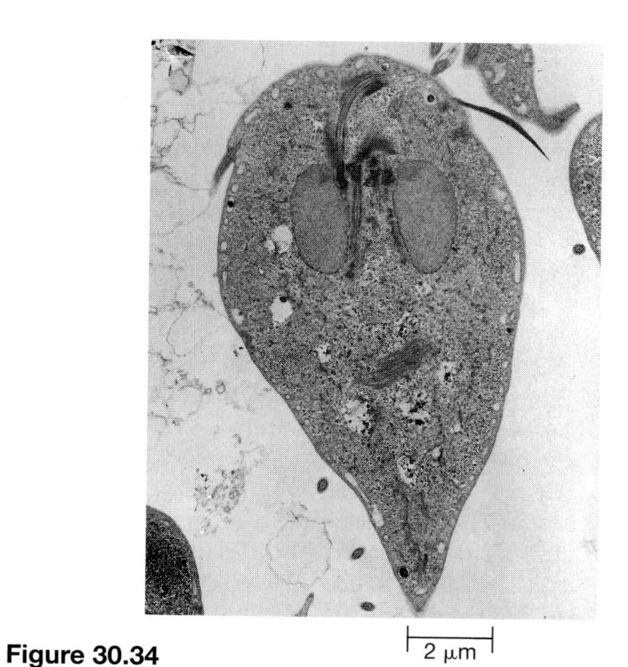

Figure 30.34

The diplomonad *Giardia,* shown in an electron micrograph, has two separate nuclei.

Figure 30.35

The life cycle of the chytrid *Allomyces* alternates between haploid and diploid thalli. Notice that these fungi are anisogamous—they reproduce with large female gametes which are fertilized by male gametes (zoospores) that swim with a posterior whiplash flagellum.

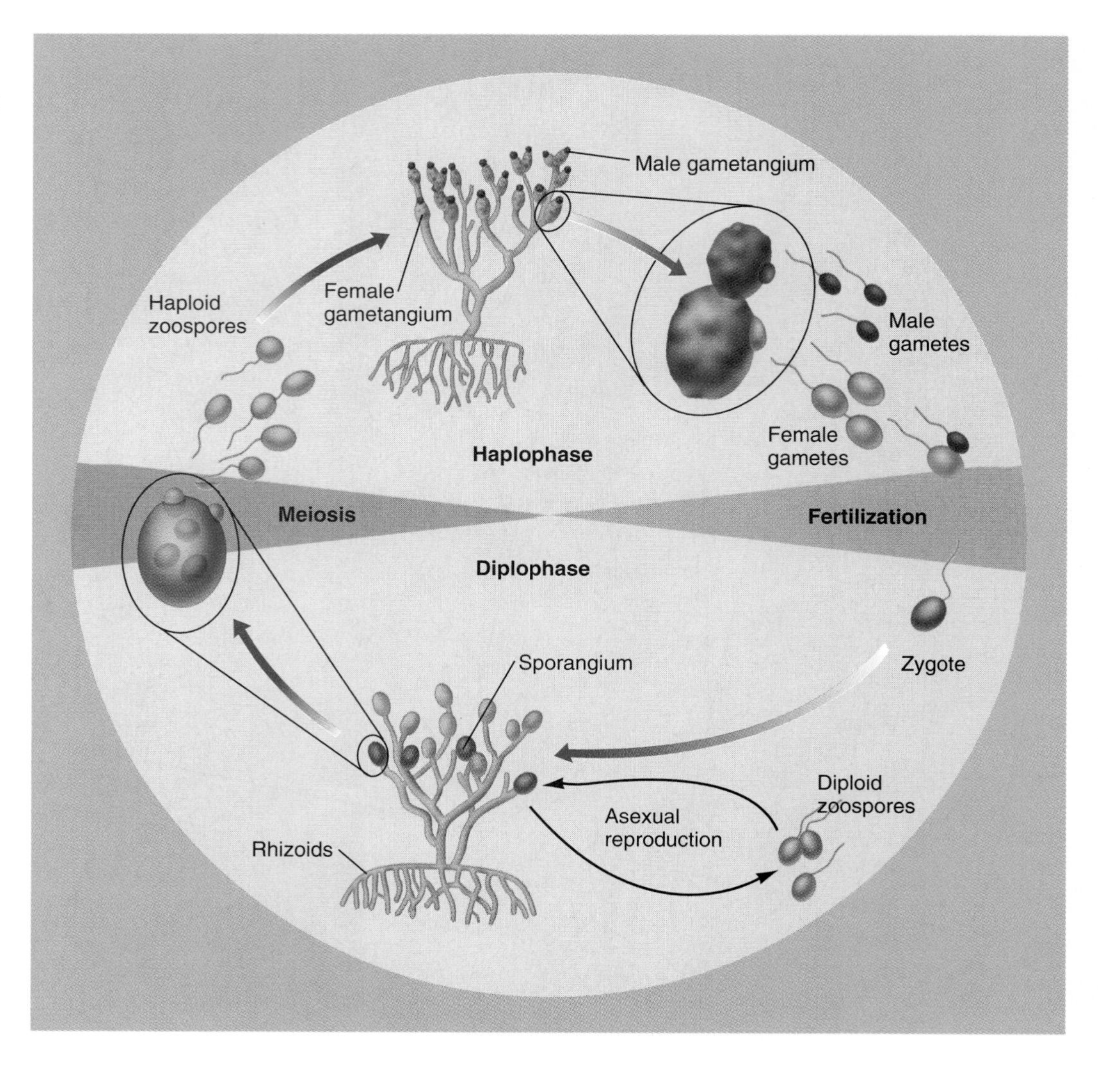

between them. Fungi such as rusts, for instance, have life cycles similar to the red-algal cycle shown in Figure 30.18, including stages in which a spore fuses with a long, tapering structure called a *trichogyne*. The migration of nuclei and the secondary pit formation in the red algae parallel the formation of connections in the main groups of fungi. Other biochemical features also indicate a close relationship; unfortunately, no one knows yet how red algae synthesize their lysine, but other phylogenetic guidelines predict that they use the AAA pathway as fungi do.

In 1993 Sandra Baldauf and Jeffrey Palmer presented evidence for a close relationship between fungi and animals, based on their own work and the research of several other biologists. For instance, it is remarkable that fungi and animals share a particular insert of 12 amino acids in a protein called EF-1α; it is almost impossible for them to have acquired this sequence independently. Baldauf and Palmer's analysis of other proteins supports the fungal-animal relationship.

But now we are left with a puzzle, particularly over a group of simple fungi called chytrids. Chytrids are not a prominent group, for most of them live obscurely as humble freshwater saprobes or as parasites on algae, plants, and even other fungi. We have not been calling them "true fungi" because they have flagella; they reproduce with motile spores powered by a single posterior flagellum (Figure 30.35). Yet they share other "true fungal" features, including chitin walls and the AAA pathway of lysine synthesis. We should also notice that animals, too, have reproductive cells (sperm) with a single posterior flagellum. Is that a meaningless coincidence?

We could understand the lack of flagella and centrioles among red algae and fungi if their clade is so ancient that it branched off the eucaryotic family tree before these structures evolved. Alternatively, they might have evolved from a flagellated ancestor and lost their flagella. Flagella have been lost in other groups, though they are usually retained in reproductive cells; fungi might have lost all their flagellated reproductive cells as they adapted to a terrestrial habitat, but why red algae would have lost them is a puzzle.

The most parsimonious resolution of this puzzle is shown in Figure 30.36. It assumes a good deal: that the red algae will turn out to use the AAA pathway and to have proteins similar to

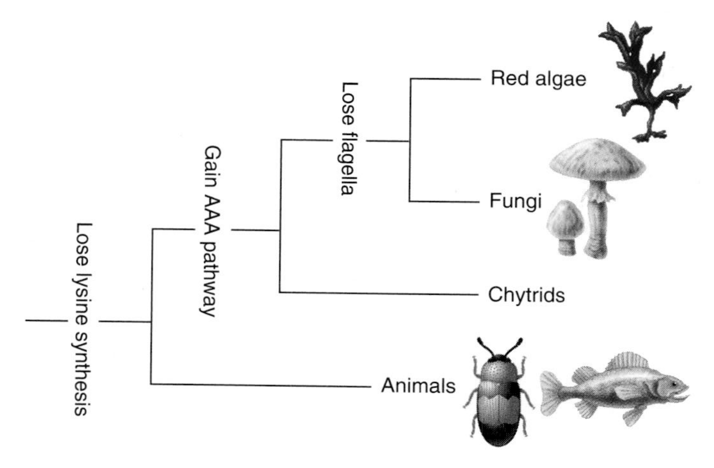

Figure 30.36

Here is one possible phylogeny that incorporates the available data about some important groups of organisms. The picture should not be taken too seriously, as new data could force its revision at any time.

those of fungi and animals. It may not be very satisfactory, since it entails loss of flagella instead of never having them to begin with. But let us remember that we are telling a detective story. The great detectives have always pieced together the best story possible on the basis of the evidence, revising it when they find new facts. We are not yet ready to assemble the suspects in the drawing room and resolve the plot.

Coda In thinking about the basically genetic nature of organisms, our attention is drawn to the difference between procaryotes and eucaryotes. Many eucaryotic cells contain about ten times as much DNA as a bacterial cell, and in some plants and animals, the ratio is about 1,000 times. Although much of that DNA does not encode useful information, even the simplest eucaryotes are more complex than procaryotes. It is important to remember that evolution is not progressive—that is, directed toward some "improvement." And yet, though procaryotes are very successful, occupying so many diverse niches, the course of evolution makes it appear that there is some advantage in being eucaryotic. At least, the evolutionary invention of the eucaryotic cell opened the door to the evolution of organisms with a vastly greater range of abilities. We tend to focus on the definition of being eucaryotic as having a distinct nucleus, but the more important factor may have been the evolution of mitosis, which allowed cells to acquire much larger genomes and propagate them regularly. Perhaps what we should celebrate is the invention of the tubulins, the subunits of microtubules, which have made possible mitosis and many other kinds of movement, and the invention of actin and myosin, which account for most other eucaryotic movements. Reflecting on the functions of these few proteins, we can see how they have allowed the protists to successfully occupy many niches. Millions of years after the simple protists evolved, the same proteins were exploited still more in the evolution of multicellular organisms, especially the animals, whose success depends largely on their ability to move.

Summary

1. Basal eucaryotes include three general types of organisms, classified according to their modes of nutrition: protozoa, molds, and algae. "Protozoa" is now an informal term for organisms that are rather like animals. Some eucaryotes are molds that are quite different from the true fungi placed in a kingdom of their own. Algae are simple photosynthetic organisms with a wide range of pigments.
2. Many protozoa and molds probably evolved from algae that lost their chloroplasts.
3. There have apparently been at least three major lines of evolution among the algae and their relatives: a green clade, which we place in the plant kingdom; a brown clade, forming a kingdom Chromista; and a clade including red algae and probably true fungi.
4. Several kinds of guidelines can be used to show relationships among the basal eucaryotes and sort them into major lines of evolution. Green-clade algae, which include the ancestors of higher plants, have chlorophylls *a* and *b*, single thylakoids in their chloroplasts, and whiplash flagella; they make α:1–4 glucose polymers. Brown-clade algae have chlorophylls *a* and *c*, thylakoids in bundles of three, and both whiplash and tinsel flagella; they make β:1–3 glucose polymers.
5. Metabolic features and pathways provide other clues to phylogeny. Lysine synthesis is especially useful, since the DAP pathway is associated with most organisms and the AAA pathway characterizes the true fungi and a few unusual groups of eucaryotes.
6. The chromist algae come in many shapes. The algae are colored by a distinctive series of carotenoids and include a wide variety of cell forms, including flagellates and amoebas. The diatoms have silica shells with complex, symmetrical sculpturing. The brown algae include large kelps, some of which have cellular differentiation and vascular tissue like the green plants. Brown algae also show a trend toward dominance of the diploid phase of the life cycle.
7. Some of the coenocytic algae were probably ancestors of the oomycetes, which are alga-like fungi with cellulose cell walls; the group includes prominent plant pathogens.
8. The various protozoans live mainly by engulfing food particles. Colorless flagellates are like flagellated algae that have lost their chloroplasts. The Sarcodina, which use pseudopods for moving and feeding, include a variety of amoebas and organisms such as heliozoans and radiolarians with beautiful silica skeletons; foraminiferans, which are abundant in the oceans, make multichambered shells of calcium carbonate. The sporozoans include many parasites, such as the one that causes malaria. Ciliates have both micro- and macronuclei, and their surfaces are covered with short cilia.
9. Slime molds are a heterogeneous assemblage of saprophytic organisms that move about like amoebas and produce spores in fruiting bodies. True slime molds form a coenocytic plasmodium; cellular slime molds form a similar amoeboid mass made of separate cells.
10. Eucaryotic cells arose through a series of complex changes. The eucaryotic nucleus probably evolved by invagination of the procaryotic cell membrane to enclose the chromosome(s), and at first the nucleus divided just as the bacterial nucleus does, by elongating and dividing in two. The mitotic apparatus evolved slowly.
11. Dinoflagellates retain some nuclear features that probably characterized the earliest eucaryotes. Their chromosomes lack

histones and are separated during cell division in an intranuclear process reminiscent of bacterial cell division. Other eucaryotes, such as euglenoids, have similar nuclear division mechanisms.

12. Mitosis probably evolved through two developments. Initially, the spindle, made of microtubules, was an extranuclear structure that helped elongate the nucleus; gradually, the nuclear envelope disappeared during mitosis, and the spindle became intranuclear. At first, chromosomes were attached to the nuclear envelope at kinetochores; the kinetochores lost their connection to the nuclear envelope, and the microtubules attached to them took on the role of separating the chromosomes.
13. Some eucaryotic organelles, especially mitochondria and chloroplasts, probably arose through a gradual incorporation of endosymbionts.
14. To sort out the basal eucaryotes, the brown clade is separated out as kingdom Chromista. Another group called Archezoa may contain primitive eucaryotes that lack mitochondria and some other eucaryotic organelles, although some data are not consistent with this idea. Red algae and true fungi may be a single clade, and there is also good evidence that fungi are closely related to animals. These relationships are now being sorted out using molecular evidence about ribosomal RNA and protein structures.

Key Terms

osmiotrophic 628
phagotrophic/holotrophic 628
saprobe/saprophyte 628
alga 629
apochromatic 629
tinsel flagellum 630
whiplash flagellum 630
golden-brown algae 632
yellow-green algae 632
coccoid 632
tetrasporine 632
siphonaceous 632
rhizopodial 632
diatom 633
dinoflagellate 633
oomycete 634
brown algae 634
thallus 634
parenchyma 634
red algae 635
sporozoa 636
ciliate 636
conjugation 639
true slime mold 639
plasmodium 639
cellular slime mold 640
pseudoplasmodium 640
euglenoid flagellate 642
trypanosome 643
endosymbiont 643

Multiple-Choice Questions

1. Among the basal eucaryotes, which are nonphotosynthetic, holotrophic, and generally unicellular?
   a. molds
   b. algae
   c. protozoans
   d. molds and algae
   e. molds, algae, and protozoans
2. Which lifestyle or form of nutrition is *not* found among the protozoans?
   a. osmiotrophy
   b. holotrophy
   c. phagotrophy
   d. parasitism
   e. chemoautotrophy
3. Which basal eucaryotes survive, grow, and reproduce using only the chemical energy of organic molecules?
   a. holotrophs
   b. osmiotrophs
   c. phototrophs
   d. *a* and *b*
   e. *a*, *b*, and *c*
4. Protozoans include all except
   a. amoebas.
   b. ciliates.
   c. molds.
   d. flagellates.
   e. parasites.
5. The most likely ancestors of many protozoans are
   a. animals.
   b. apochromatic algae.
   c. fungi.
   d. plants.
   e. all of these.
6. Which of these characteristics is found in all photosynthetic eucaryotes?
   a. chlorophyll *a*
   b. cellulose cell walls
   c. synthesis of starch or glycogen
   d. DAP pathway for lysine biosynthesis
   e. all of these
7. Golden-brown and yellow-green algae include organisms with all these characteristics except
   a. isogamous sexual reproduction as well as asexual reproduction.
   b. unicellular, filamentous and rhizopodial forms.
   c. uninucleate or multinucleate cells.
   d. photosynthetic and nonphotosynthetic forms.
   e. genetic recombination by conjugation.
8. The nontechnical name generally applied to photosynthetic basal eucaryotes is
   a. green-clade.
   b. brown-clade.
   c. algae.
   d. protozoans.
   e. protists.
9. The kingdom Chromista, or Chromophyta, includes
   a. members of the green-clade.
   b. photosynthetic procaryotes and eucaryotes.
   c. all protists plus unicellular green plants.
   d. all molds and fungi.
   e. eucaryotes within the brown clade.
10. A primitive form of eucaryotic cell division, believed to be the link between binary fission and modern processes of mitosis, occurs within the
    a. ciliates such as *Paramecium.*
    b. amoebas such as *Amoeba proteus.*
    c. dinoflagellates such as *Gonyaulax.*
    d. seaweeds such as *Laminaria.*
    e. water molds.

True-False Questions

Mark each statement true or false, and if false, restate it to make it true.

1. Protozoans are either osmiotrophs or phototrophs that live as saprobes.
2. Of all the chlorophylls, only chlorophyll *a* is found in all photosynthetic eucaryotes.
3. All photosynthetic eucaryotes produce glucose and polymerize it into the energy-rich storage carbohydrate called starch.
4. Although both green-clade and brown-clade algae contain chlorophyll *a*, the green-clade organisms also contain chlorophyll *b*, whereas the brown-clade organisms also have chlorophyll *c*.
5. Motility by means of pseudopodia is characteristic of some brown algae as well as some protozoans.

6. Protozoan amoebas are believed to be related to chromist amoebas and also to some of the flagellates.
7. Kinetochores were initially anchoring sites for cilia and flagella, and were located on the inside of the plasma membrane.
8. Endosymbiosis is the likely explanation for the evolution of the nuclear membrane.
9. The chromists range from small, single-celled organisms to large, plantlike kelps.
10. Many nonphotosynthetic protists, including the amoebas and unicellular flagellates, are believed to have evolved from apochromatic green algae.

Concept Questions

1. Explain the difference between osmiotrophy and holotrophy. How do both differ from phototrophy?
2. Distinguish the true slime molds from the cellular slime molds.
3. In the evolution of eucaryotes, which is believed to have come first—a membrane-enclosed nucleus or a mitotic spindle? Explain.
4. Outline the evidence for the endosymbiotic origin of mitochondria and chloroplasts.
5. Did the typical eucaryotic cells evolve all at once or in stages? Explain.

Additional Reading

Alexopoulos, Constantine J., and H. C. Bold. *Algae and Fungi.* Macmillan Co., New York, 1967.

Chapman, V. J. *The Algae.* Macmillan Co., London, and St. Martin's Press, New York, 1968.

Corliss, John O. *The Ciliated Protozoa.* Pergamon Press, New York, 1961.

Curtis, Helena. *The Marvelous Animals: An Introduction to the Protozoa.* The Natural History Press, Garden City (NY), 1968.

Donelson, John E., and Mervyn J. Turner. "How the Trypanosome Changes Its Coat." *Scientific American,* February 1985, p. 44. The parasite, which deprives much of Africa of meat and milk, survives in the bloodstream by evading the immune system. Its trick is to switch on new genes encoding new surface antigens.

Gall, Joseph G. (ed.). *The Molecular Biology of Ciliated Protozoa.* Academic Press, Orlando (FL), 1986.

Laybourn-Parry, Johanna. *A Functional Biology of Free-Living Protozoa.* University of California Press, Berkeley, 1984.

Leedale, Gordon F. *Euglenoid Flagellates.* Prentice-Hall, Englewood Cliffs (NJ), 1968.

Margulis, Lynn, et al. (eds.). *Handbook of Protoctista.* Jones and Bartlett Publishers, Boston, 1990.

Morris, Ian. *An Introduction to the Algae.* Hutchinson, London, 1967.

Round, F. E. *The Biology of the Algae.* St. Martin's Press, New York, 1973.

Tiffany, Lewis H. *Algae, the Grass of Many Waters,* 2d ed. C.C. Thomas, Springfield (IL), 1958.

Internet Resource

To further explore the content of this chapter, log on to the web site at:

http://www.mhhe.com/biosci/genbio/guttman/

31 The Fungi

Armillaria mellea, the Honey Mushroom, is prized by many mushroom-hunters, though it makes some people ill. It is actually a plant pathogen, and its phosphorescent mycelium glows in the dark.

Key Concepts

31.1 The kingdom Fungi is monophyletic and defined by a few characteristics.

31.2 Most fungi form mycelia, made of long, thin hyphae that grow from a spore.

31.3 Basidiomycetes include most of the common mushrooms.

31.4 Ascomycetes develop spores inside sacs.

31.5 Some deuteromycetes are harmful to humans, while others are beneficial.

31.6 Zygomycetes are molds that reproduce through zygospores.

31.7 Many fungi are involved in mutualistic relationships with other organisms.

Clean a pound of fresh mushrooms, remove the stems, and cut the caps into thick slices. Saute them in 2 tbsp. of butter until they are barely tender. Then transfer them to a well-buttered baking dish, arranging them in neat, overlapping rows. In a saucepan, combine a cup of sour cream, 2 tbsp. of flour, salt and pepper to taste. Simmer the mixture until heated, but do not let it boil. Spoon the sauce over the mushrooms. Top with 1/2 cup grated cheddar cheese, 1/2 cup grated monterey jack cheese, and a dash of cayenne pepper. Bake for 15 minutes at 350°. Serves 4.

Many mushrooms make wonderful additions to our cuisine, but this is only one context in which fungi enter our lives. Wise mushroom hunters know how to distinguish the delicious ones from their close relatives that can kill painfully. Similarly, some molds lend delightful tastes to cheeses, but others grow on our food and ruin it. Some fungi produce skin diseases, and others are terrible plant pests that can ruin crops. Fungi are well worth knowing about. Sometimes that knowledge literally makes the difference between life and death. Although the fungal kingdom embraces fewer than 80,000 named species, fungi are important both ecologically and in human affairs. We eat them and they eat us. They also eat (decompose) much of the detritus of our ecosystems. The study of fungi is **mycology,** and its practitioners are mycologists. Many amateur mycologists delight in the beauty and variety of mushrooms and their relatives, the challenge of identifying them and understanding their life histories, and the pleasant nuances some of them can add to our food. ■

31.1 The kingdom Fungi is monophyletic and defined by a few characteristics.

Fungi were once placed in the plant kingdom because they are superficially plantlike, but plants and fungi are only very distantly related. Plants are phototrophs, which obtain their energy from light, and fungi are osmiotrophs, which absorb nutrients through their cell membranes. Fungi grow as *saprobes* or *saprophytes* on decaying, dead organisms or as *parasites* on living organisms; the border between the two ways of life may be blurred.

Among the features all fungi share are cell walls made largely of chitin, no trace of flagella at any stage in their life cycle, and the use of the α-aminoadipic acid (AAA) pathway for lysine synthesis. By excluding organisms that do not share these features, the kingdom Fungi becomes an easily defined, monophyletic taxon. The kingdom includes three phyla: Basidiomycota, Ascomycota, and Zygomycota, along with organisms called trichomycetes, which we will not discuss. In addition, an artificial group called deuteromycetes, or imperfect fungi, includes species that do not reproduce sexually, so their affinities with the other fungi are obscure.

Besides the *true fungi* composing this kingdom, a heterogeneous assortment of "fungi" in the broader, looser sense of the word includes organisms that were once called "phycomycetes" or "algal fungi." Phylogenetic guidelines such as flagellation and cell wall structure indicate that these types represent at least two or three distinct clades and are not closely related. The oomycetes (water molds) and hyphochytrids are clearly members of the brown line of algal evolution, so we discuss these fungi with the kingdom Chromista in Section 30.4. The chytrids, which have a confusing combination of characteristics, are placed in kingdom Protista and discussed in Section 30.11.

31.2 Most fungi form mycelia, made of long, thin hyphae that grow from a spore.

A few fungi are small, rounded, unicellular organisms, including the yeasts, which are all unicellular and uninucleate. Many fungi take the form of long, threadlike cells called **hyphae** (sing., **hypha**), which grow into a branching network, a **mycelium** (Figure 31.1). The hyphae of some fungi are coenocytic, having many nuclei with little or no division of the cytoplasm into individual cells. In other fungi, the hyphae are divided by cross-walls and are actually multicellular. Hyphae have walls of *chitin*, a tough polysaccharide that also forms the arthropod exoskeleton. Fine hyphae, with their enormous ratio of surface area to volume, are well suited to the fungal ways of life, the lives of saprophytes and parasites. Hyphae make intimate contact with their sources of nourishment in soil, water, or the bodies of other organisms where they insinuate themselves. The enzymes they secrete digest food polymers into monomers, which the hyphae absorb. Fungi grow aerobically by respiration, although many yeasts are facultative anaerobes; they commonly grow on carbohydrates such as glucose and maltose, and they synthesize glycogen as an energy store. Fungi are very active metabolically, and their well-developed cytoplasmic streaming can carry proteins quickly from their sites of synthesis throughout the cytoplasm to the growing tips of hyphae. Hyphae sometimes grow so rapidly that they can be seen elongating under a microscope, and it isn't unusual for a clump of mushrooms to suddenly spring up overnight.

A generalized fungal life cycle (Figure 31.2) includes an extensive haplophase, a dicaryophase, and a very brief

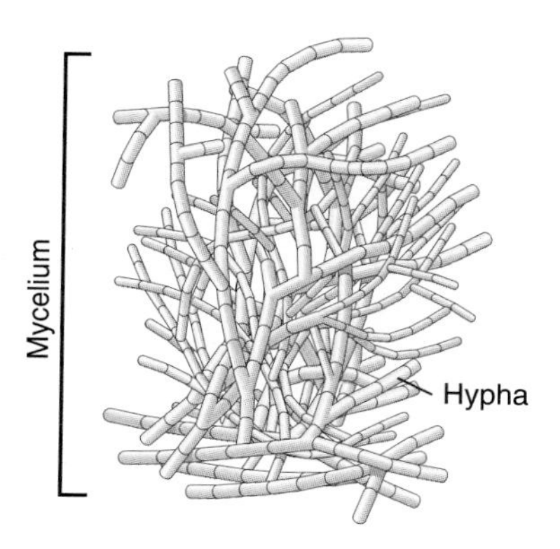

Figure 31.1
Typical fungi have the form of thin hyphae that form a mycelium. They may be coenocytic or multicellular.

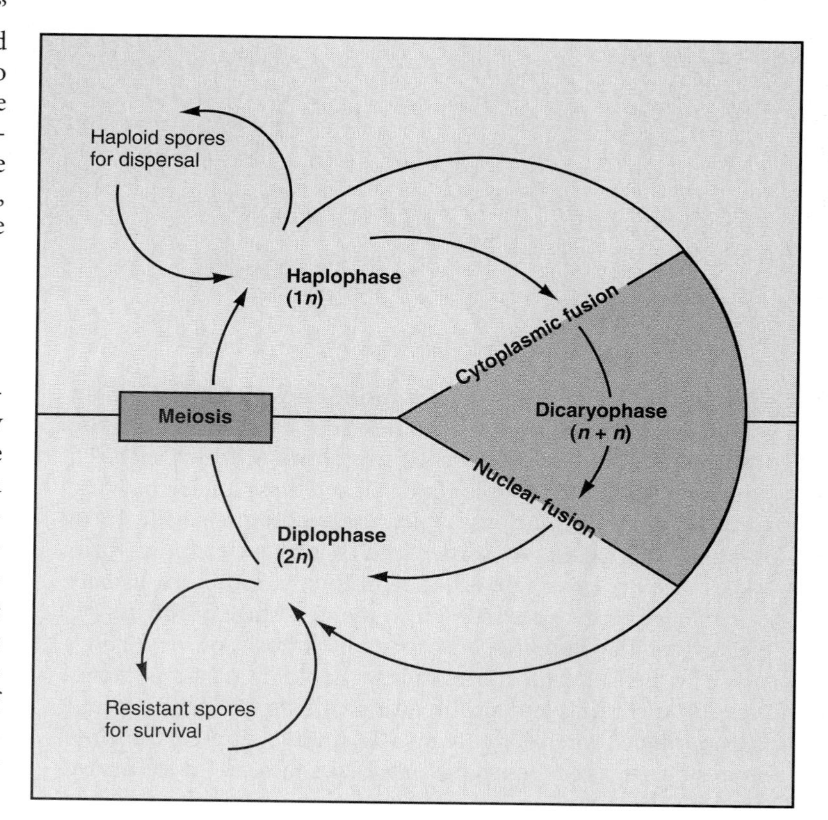

Figure 31.2
The most generalized fungal life cycle includes a dicaryophase, formed by cytoplasmic fusion of haploid cells, in which the haploid nuclei remain separate. The diplophase is only a single diploid cell or nucleus formed by the fusion of haploid nuclei. The diploid cell undergoes meiosis to produce haploid meiospores or haploid hyphae.

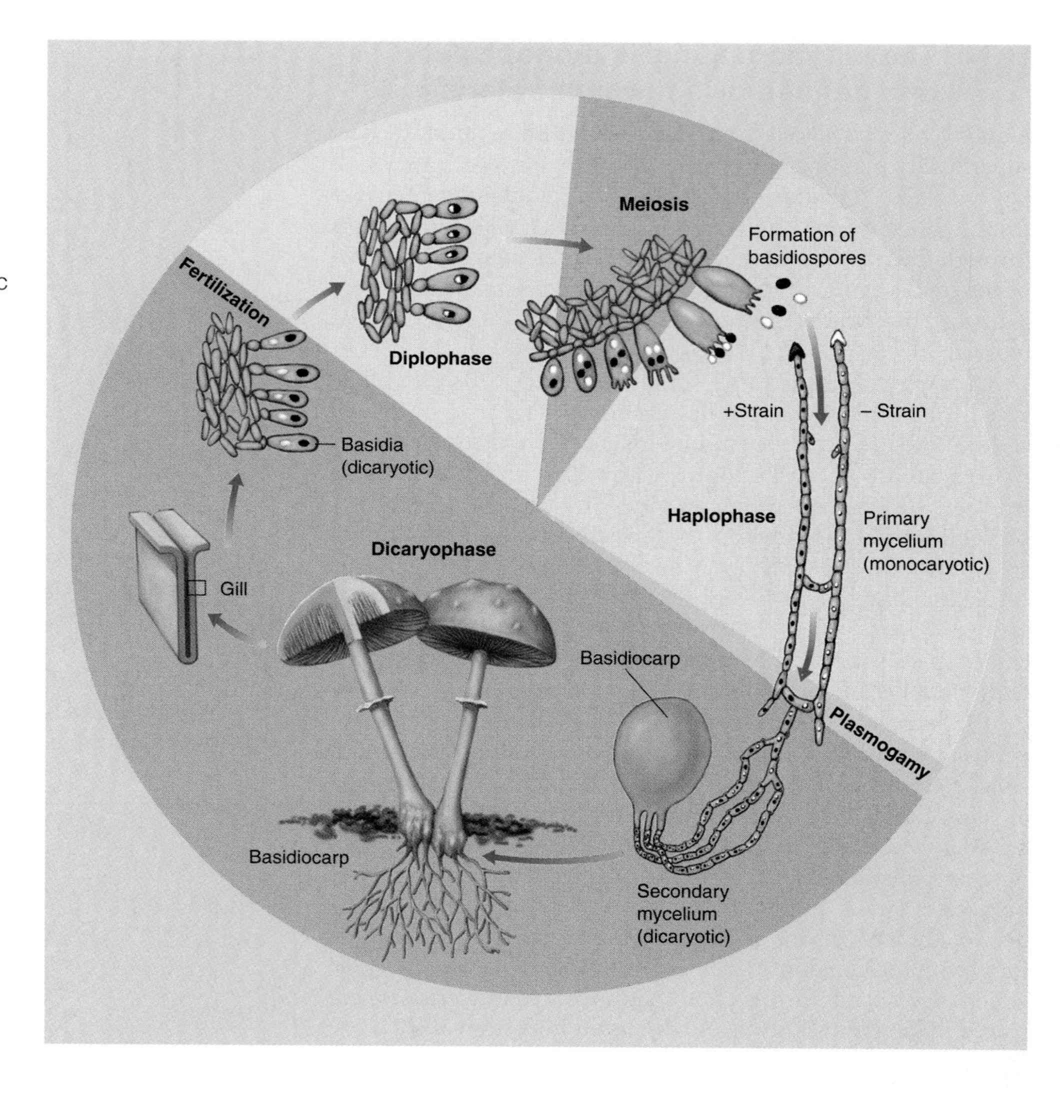

Figure 31.3
The life cycle of a basidiomycete is dominated by the dicaryophase, which typically takes the form of a mushroom. In some dicaryotic cells, the nuclei fuse to make a diploid nucleus, which undergoes meiosis to produce haploid basidiospores borne on club-shaped basidia. Basidiospores germinate into monocaryotic hyphae.

diplophase. Most fungi have a **monocaryotic mycelium** in which each cell has a single nucleus. Two monocaryotic mycelia join to form a **dicaryotic mycelium** in which each cell contains two separate haploid nuclei, and these nuclei may fuse to make a diploid nucleus, which undergoes meiosis to form haploid cells again. At some time in the cycle, fungi form *spores.* A spore is a single-celled reproductive structure, usually one that can grow (asexually) into the next stage of the life cycle. Spores may be *mitospores,* made by mitosis, or *meiospores,* made by meiosis. Mitospores (always haploid) are simply vehicles for spreading and proliferating. Meiospores are haploid cells produced when a zygote nucleus undergoes meiosis. Some spores develop very resistant coats that enable them to survive drought, heat, and other life-threatening conditions. Their small size makes spores excellent agents of dispersal; a bit of mold in your kitchen can release spores that will spread everywhere and start to grow on all the rest of your food if given a chance. Spores produced by mushrooms may be carried long distances by the wind. These terms will become more meaningful as we meet specific kinds of spores in the life cycles of various fungi.

31.3 Basidiomycetes include most of the common mushrooms.

To understand the fungi, we will examine the most obvious species, commonly known as mushrooms; they are classified as basidiomycetes or ascomycetes, depending on the way they form their spores, and we begin with the former.

All mushrooms are **fruiting bodies,** the spore-producing reproductive structures that arise from a fine dicaryotic mycelium buried in the substratum below, and most mushrooms are the fruiting bodies of basidiomycetes. The underground mycelium may spread for many square meters. Occasionally the popular press gets excited about a discovery of "the largest organism in the world." This is always an immense mycelium. A mycelium begins with a single **basidiospore,** with which basidiomycetes proliferate (Figure 31.3).

Figure 31.4

A gallery of basidiomycete mushrooms displays some of the most attractive species: *(a) Pleurotus cornucopiae,* horn of plenty; *(b) Dacromyces palmatus,* witches' butter; *(c) Ramaria araiospora,* var. *rubella,* carmine coral; *(d) Hygrophorus mineatus,* vermillion waxycap.

The variety of basidiomycetes is better illustrated than described (Figure 31.4). Under their caps, most mushrooms are divided into many thin partitions, or *gills,* radiating from a center; a few species have numerous pores instead. On the surfaces of these gills or pores, little club-shaped structures called **basidia** develop. In each basidium, two haploid nuclei fuse, and the resulting diploid nucleus immediately undergoes meiosis to make four haploid cells that develop into basidiospores—two of the plus (+) type and two of the minus (−) type. Notice that basidiospores are external, and they drop off as they mature. Their colors are an important aid in mushroom identification; if a whole mushroom cap is laid on a sheet of contrasting paper, the falling spores make a spore print with a distinctive color and pattern (Figure 31.5).

With the growing popular enthusiasm for collecting and eating mushrooms, it is important to understand that no simple

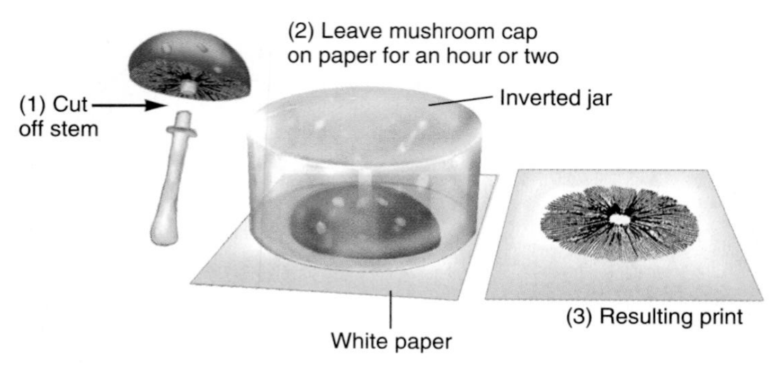

Figure 31.5

A spore print is made by allowing the spores from a mushroom cap to fall on white paper under a cover so they are not disturbed by air currents. The pattern and color of the spores are important in identification. Source: Clyde M. Christensen, *Common Edible Mushrooms,* 1943, University of Minnesota Press.

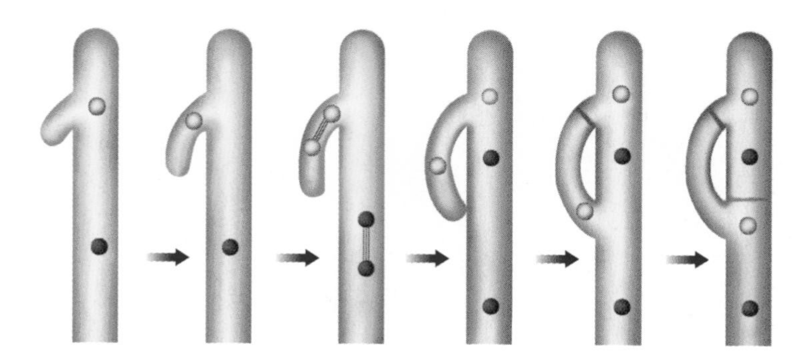

Figure 31.6

Clamp connections are made in dicaryotic hyphae through a series of nuclear divisions and migrations. This process ensures that each cell retains both types of nuclei.

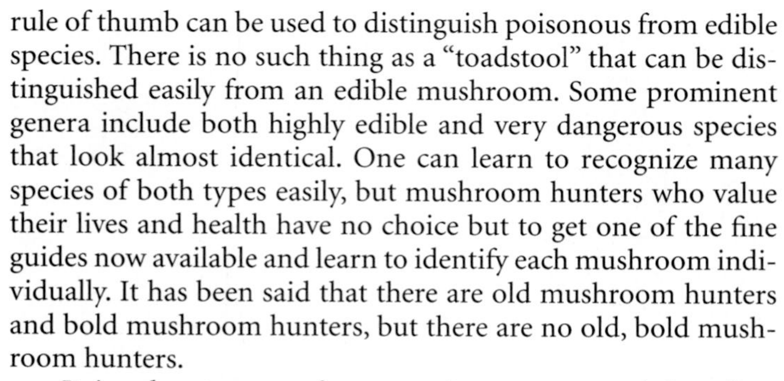

rule of thumb can be used to distinguish poisonous from edible species. There is no such thing as a "toadstool" that can be distinguished easily from an edible mushroom. Some prominent genera include both highly edible and very dangerous species that look almost identical. One can learn to recognize many species of both types easily, but mushroom hunters who value their lives and health have no choice but to get one of the fine guides now available and learn to identify each mushroom individually. It has been said that there are old mushroom hunters and bold mushroom hunters, but there are no old, bold mushroom hunters.

It is advantageous for a species to prevent inbreeding (breeding among closely related organisms) and to promote outbreeding, since this increases genetic diversity and enhances the species's ability to survive. Some basidiomycetes have complex genetic systems that determine mating type and prevent inbreeding; only certain types are able to mate with each other, like a society of nobles in which each family will only permit its offspring to marry into certain other families. For simplicity, consider a fungus that has only + and − types. Each haploid basidiospore grows into either a + or a − monocaryotic mycelium, which spreads out within the substratum. When monocaryotic mycelia of opposite types meet, cells from each mycelium unite, and they grow into a dicaryotic mycelium containing both types of haploid nuclei within each cell. This dicaryotic secondary mycelium eventually produces mushrooms.

Monocaryotic mycelia may be coenocytic at first, but they soon divide into many cells with single nuclei. In dicaryotic mycelia, the formation of a **clamp connection** at each division ensures that each cell will contain one nucleus of each type (Figure 31.6). These connections are made with nuclear movements like those of secondary pit formation in the red algae (see Section 30.5) and like another kind of nuclear movement we will meet in ascomycetes. The continuing reappearance of these related patterns of nuclear migration is one of the features that points so strongly to a phylogenetic relationship between true fungi and red algae.

Some basidiomycetes are plant parasites. The rusts and smuts are among the worst enemies of domestic crops. Wheat rusts have sometimes devastated crops and reduced people dependent on them to near starvation. A rust that infects coffee plants wiped out the extensive plantations on Sri Lanka in the mid-nineteenth century, while the industry has flourished in Latin America, where the parasite doesn't grow. Some rusts infect trees, and white-pine blister rust has been a serious scourge of commercial white pine forests. Wheat and other crops are sometimes infected by smuts, which get their name from the black, smutty patches of spores they form on infected plants. A careful comparison of the rust life cycle (see Figure 15.17) with a red algal cycle (see Figure 30.18) shows the close similarity between them, another indication that fungi and red algae are closely related.

Exercise 31.1 Compare the life cycles of the rust and the red alga, noting all the points of similarity. What point suggests a phylogenetic relationship between the two organisms?

The only basidiomycete known to be involved in a **mycosis,** a human disease caused by a fungus, was identified in 1975 as *Cryptococcus* (now renamed *Filobasidiella*) *neoformans.* It causes cryptococcosis, a generalized infection of the bloodstream that invades the central nervous system, lungs, and other organs.

31.4 Ascomycetes develop spores inside sacs.

Ascomycetes are known as sac fungi because their sexual reproductive spores form within closed sacs, or **asci** (sing., **ascus**). The simplest ascomycetes are the *yeasts.* Different types repro-

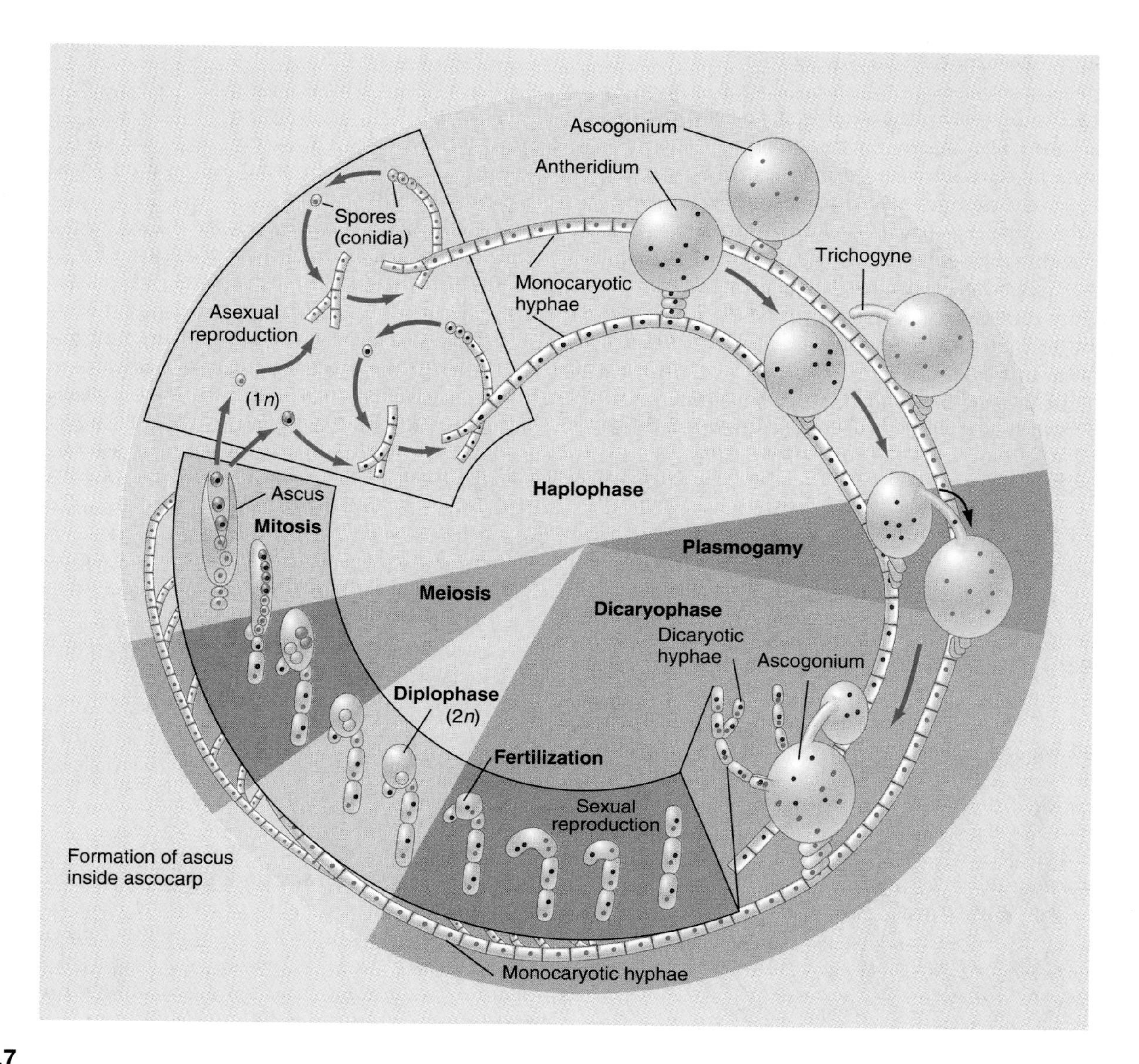

Figure 31.7

In the life cycle of an ascomycete, monocaryotic hyphae of opposite mating types develop gametangia: female ascogonia and male antheridia. Slender outgrowths called trichogynes grow from the ascogonia and fuse with the antheridia, followed by migration of male nuclei into the ascogonia. Dicaryotic hyphae containing both kinds of nuclei grow out of an ascogonium. A hook-shaped crozier forms at the tip of a dicaryotic hypha; two nuclei of opposite types fuse, and the diploid nucleus undergoes meiosis to produce ascospores. The fungus may also reproduce asexually by means of conidia.

duce asexually by either transverse fission or by budding. They also reproduce sexually by means of spores, like the common beer and bread yeast *Saccharomyces* whose cycle was illustrated in Figure 15.8. Two haploid yeast cells may fuse and form a diploid cell, which may continue to multiply vegetatively by mitosis. A diploid cell may also undergo meiosis, producing four haploid meiospores within the cell, which is an ascus. These spores eventually break out of the ascus and grow vegetatively. Yeasts are very simple fungi, but they are generally considered degenerate ascomycetes—that is, organisms that have evolved into a simple form from a more complex ancestor, rather than being primitively simple.

The mold *Neurospora,* long the darling of some geneticists, has a more typical life cycle (Figure 31.7). Its hyphae grow into visible, fluffy, red mycelia. In *Neurospora,* sexual reproduction only occurs between hyphae of opposite mating types, but each hypha generally produces both male and female gametangia, the structures that bear gametes or nuclei that function as gametes. The female gametangium, or **ascogonium,** is larger and often bears a narrow *trichogyne* (like red algal cells), while the male gametangium, or **antheridium,** is smaller. (There is some variation in detail among species; *N. crassa* does not produce antheridia, and an ascogonium fuses with a small reproductive cell called a *conidium.*) The fusion of male and female gametangia creates a dicaryotic cell containing both types of haploid nuclei. This cell grows into **ascogenous** (that is, ascus-generating) **hyphae,** from which fruiting bodies grow. An ascus develops where the tip of each hypha curls into a *crozier,* named after the similarly shaped shepherd's crook. Nuclei then migrate in the crozier in a manner reminiscent of the migration of nuclei during secondary pit formation in red algae and the formation of clamp connections in basidiomycetes. The two nuclei in the terminal cell of the crozier fuse and then undergo meiosis to make a series of meiospores, while the cell itself enlarges and becomes an ascus. The four products of meiosis typically go through another division, or even two, to make a series of 8 or 16 ascospores.

In *Neurospora* and some other ascomycetes, ascospores stay lined up inside the narrow ascus in exactly the order in which the nuclei divide, making these species very useful for genetic studies. Figure 31.8 shows *Neurospora* segregating genes for spore colors, and you can see how beautifully the spores in each ascus are grouped either 4-4 or 2-2-2-2. More advanced books on fungal genetics explain how these patterns of segregation can be used to map genes. Furthermore, fungi that segregate their spores in this way show different events during meiosis and allow geneticists to explore the mechanisms of recombination.

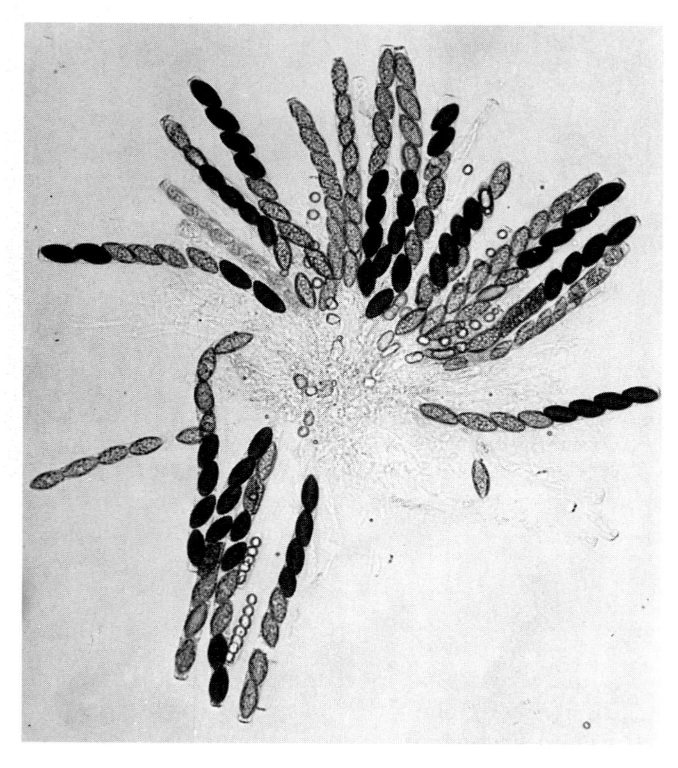

Figure 31.8

When two strains of *Neurospora* with different spore colors (black or white) are crossed, the resulting diploid cells produce asci in which the haploid spores are arranged in regular patterns (such as 4 black and 4 white, or 2, 2, 2, and 2). These patterns reflect the regular way genes segregate in meiosis, so *Neurospora* and similar ascomycetes have become favored genetic tools for studying meiosis and recombination.

In addition to these delicate molds, the ascomycetes include a number of mushrooms (Figure 31.9). Morels, for instance, bear their asci in the recesses of their spongy head; they are generally counted among the choicest and most delicious of mushrooms. Morels tend to grow in burned-over areas, and according to the mycologist Clyde Christensen, the Germans at one time set so many forest fires to encourage their growth that they had to be stopped by a special law. Most of the helvellas are also edible, and some are prized. Truffles, which grow underground, are famous delicacies. Many of the cup fungi that decorate the woods with brilliant bits of color are edible as well.

The ergots *(Claviceps)* have had a powerful effect on human society. Growing commonly on grains, especially rye, they produce alkaloids, including the infamous drug lysergic acid diethylamide (LSD). Animals that feed on ergotized grain suffer blindness, paralysis, and other nervous disorders, and every now and then outbreaks of ergotism occur in humans, who ingest the contaminated grain in bread. For some reason, there were a number of outbreaks during the Middle Ages, when the disease was known as holy fire or St. Anthony's fire. There is even good reason to believe that ergotism was responsible for the witch craze that struck Salem, Massachusetts, in 1692; although this episode was undoubtedly compounded by social factors, ergotism probably caused the strange behavior then identified as that of "witches." The last known outbreak occurred in 1951 in France, where a considerable amount of rye bread is made. A victim of ergotism suffers fever, convulsions, and psychotic delusions like those experienced by LSD users. Since the drug causes muscle contraction and constricts blood vessels, the extremities may lose their circulation, become blackened with gangrene, and drop off.

31.5 Some deuteromycetes are harmful to humans, while others are beneficial.

Many species of fungi are not known to reproduce sexually at all. They reproduce asexually by means of mitospores called **conidia** and are lumped together as an artificial phylum called Fungi Imperfecti, or Deuteromycota. This is a prime example of taxonomy for practical purposes. Most imperfect fungi are obviously ascomycetes without known sexual reproduction; most of the rest are rather easily identified as basidiomycetes or zygomycetes on the basis of vegetative features and are transferred to the appropriate phylum if sexual reproduction is discovered. Deuteromycetes include some notorious pathogens that cause infections of the skin, scalp, nails, and mucous membranes, generally known as ringworms or tineas. *Candida albicans,* a common pathogen of the mouth, vagina, and uterine cervix, is a typical opportunistic fungus (and not a yeast, as it is com-

(a) (b) (c) (d)

Figure 31.9
Ascomycete mushrooms may take quite unusual forms. Here are *(a) Morchella elata,* black morel; *(b) Helvella infula,* hooded helvella; *(c) Scutellinia scutellata,* eyelash cup; and *(d) Cordyceps militaris,* soldier grainy club.

monly called). In healthy people, the normal flora of other microorganisms in these places holds the pathogen in check, but *Candida* can take over rapidly in abnormal conditions such as disease (diabetes, for instance), obesity, vitamin deficiencies, or treatments with antibiotics or corticosteroids. On the other side of the economic coin, some deuteromycetes produce valuable antibiotics. Certain strains of *Penicillium* are used to make the antibiotic penicillin, while other types are used to produce cheeses such as Roquefort and Camembert. *Aspergillus* is grown industrially to produce citric acid and gluconic acid, which are used in many foods.

Some human mycoses caused by imperfect fungi include aspergillosis, a respiratory infection by *Aspergillus fumigatus,* histoplasmosis of the lymph nodes and other parts of the immune system (by *Emmonsiella capsulatum*), and several kinds of skin diseases. Ascomycetes also include serious plant parasites, like *Ceratocystis ulmi,* the cause of Dutch elm disease, and the powdery mildews that attack a host of fruits and vegetables.

One of America's most beloved trees, the chestnut, was virtually wiped out by chestnut blight early in this century. The responsible agent, *Endothia parasitica,* entered the country around 1900, and forty years later the trees were almost gone. *Endothia* invades a tree through small wounds in the bark, like those made by woodpeckers and small mammals, where it grows and digests the bark. Its mycelia grow in clumps that rupture the bark and form cavities just below it, and these cavities eventually fill with millions of spores. Then insects, squirrels, and birds such as woodpeckers pick up these spores on their feet and transmit the infection to other trees.

31.6 Zygomycetes are molds that reproduce through zygospores.

Most zygomycetes are saprobes that live in the soil on decaying organic matter, but a few are parasites of plants, insects, or small soil animals. Zygomycetes are relatively familiar fungi because they include such organisms as *Rhizopus,* a black bread mold, and *Mucor,* a soil saprobe that commonly grows on meats and vegetables. Since some of these organisms can grow at or just below freezing, they sometimes even appear on stored frozen foods.

The life cycle of *Rhizopus* is typical of zygomycetes (Figure 31.10). It grows asexually through aerial hyphae that spread over a surface. When a hypha touches the substratum, it sends out a cluster of rootlike rhizoids and another cluster of hyphae that grow into sporangiophores bearing reproductive structures called **sporangia** in which haploid mitospores develop. On falling to the surface, these spores germinate into new hyphae.

Rhizopus also reproduces sexually in a way that gives zygomycetes their name; they form a nonmotile zygote that is a key stage in reproduction. Each hypha is haploid and either a + or a − mating type. Hyphae of opposite types mutually induce each other to form gametangia, which grow together. Although gametangia are structures where gametes form, these fungi have no motile cells, and the gametes are simply haploid nuclei. The gametangia form a **zygosporangium** in which many + and − nuclei fuse by pairs to form zygotes, which go through meiosis. The zygote develops a heavy, thorny wall and eventually breaks open as haploid hyphae grow out of it.

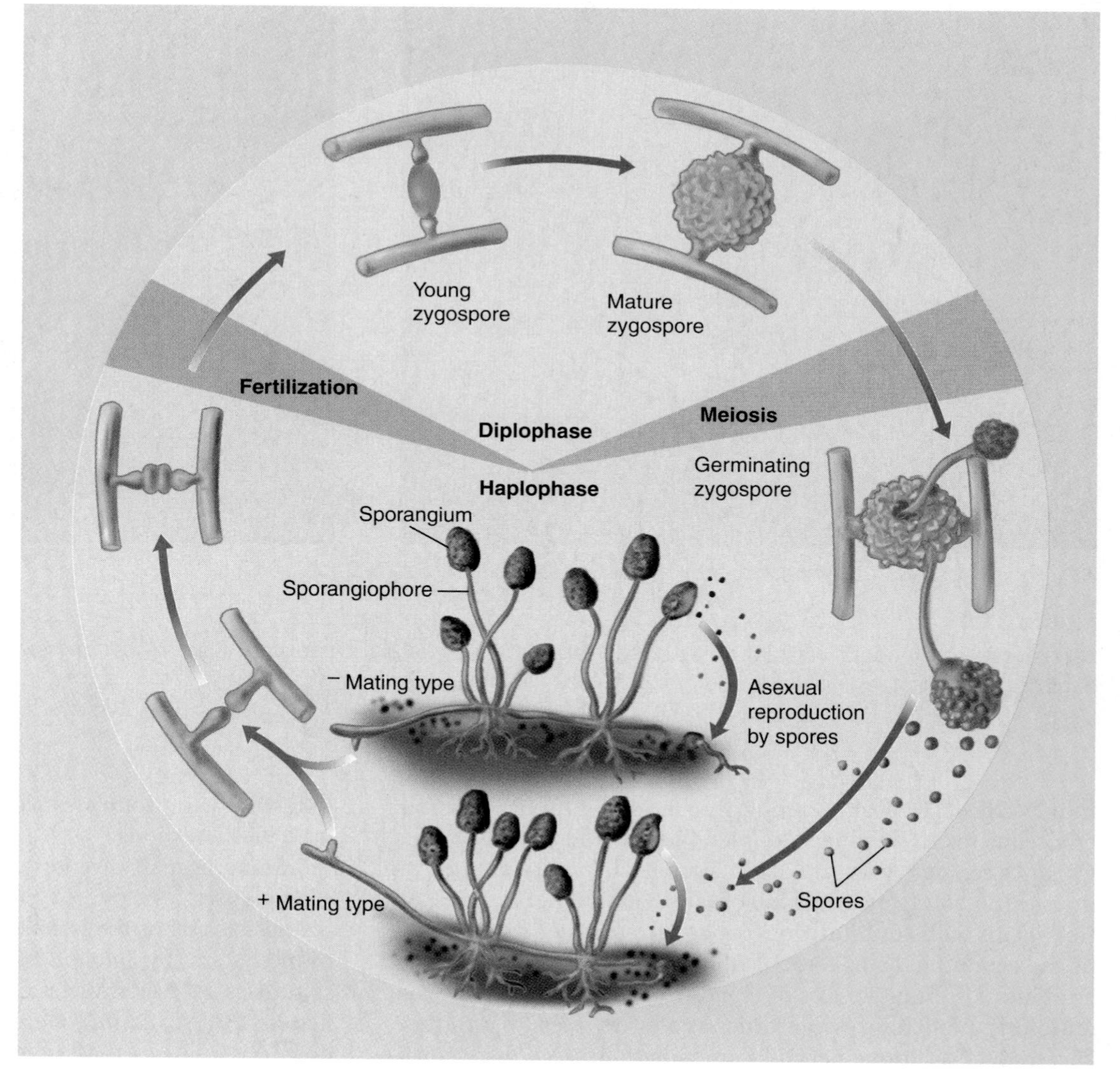

Figure 31.10
The life cycle of the zygomycete *Rhizopus* includes a brief diplophase and an extended haplophase. During the haplophase, the fungus reproduces asexually by means of spores produced in sporangia, shown in the micrograph.

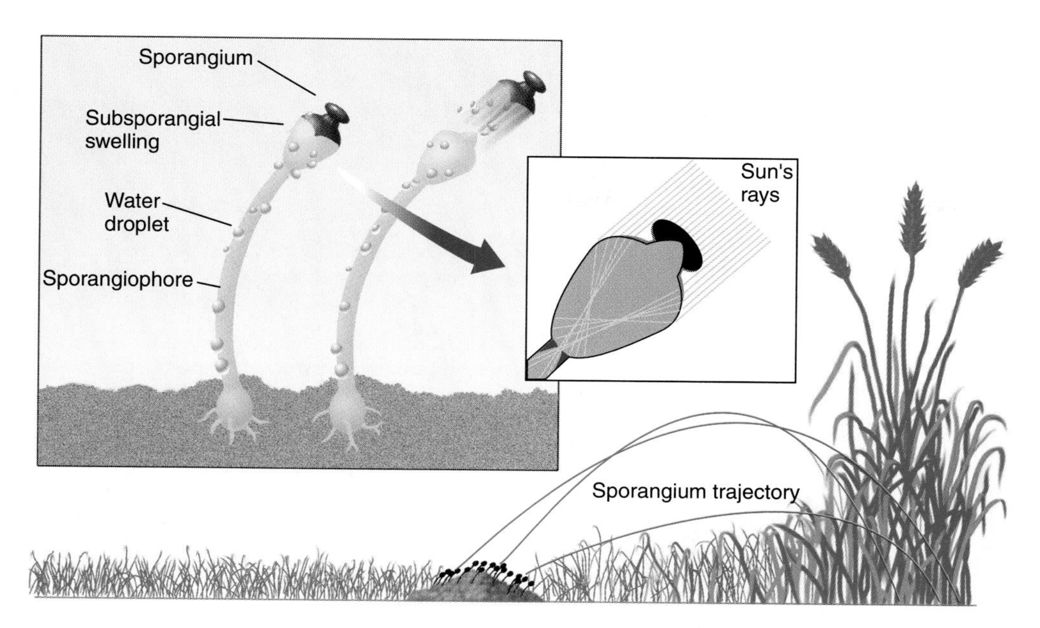

Figure 31.11

The zygomycete *Pilobolus* spreads its spores when a swelling beneath the sporangium is heated by the sun and bursts, shooting the sporangium as far as a few meters away.

Pilobolus is a fascinating little zygomycete that grows on the droppings of herbivores, forming sporangia much as *Rhizopus* does (Figure 31.11). As a *Pilobolus* sporangiophore matures, it develops a swelling just below the sporangium that focuses sunlight like a lens, directing growth toward the sunlight and also enhancing the turgor pressure in the swelling until it explodes and shoots the whole sporangium as far as 2.5 meters away on a jet of liquid. The sporangium is likely to adhere to a bit of vegetation by its sticky base, where it stays until the plant is eaten by an animal. The spores within it, being very resistant to digestion, are deposited in a bit of dung, where they repeat the whole cycle.

Some zygomycetes are *dimorphic (di-* = two; *morph* = form) and can grow as either hyphae or yeastlike cells; the yeastlike forms are implicated in a number of mycoses that are especially dangerous to people whose immune systems are suppressed. Other zygomycetes are used for nutritional enrichment. Normal soybean protein, for instance, contains an inhibitor of trypsin, one of the principal proteases of the intestine, so it is largely indigestible to humans. However, if soy protein is mixed with *Rhizopus oligosporus,* the fungus destroys the inhibitor and makes the protein more digestible and nutritious.

31.7 Many fungi are involved in mutualistic relationships with other organisms.

Lichens

Mutualisms are common in nature, but a remarkable number involve fungi. The best-known is the **lichen** symbiosis, in which a fungus (called the mycobiont) and an alga (the phycobiont) associate so intimately that they create a distinctive structure, a lichen thallus, that neither partner assumes by itself (Figure 31.12). The minority alga grows amid the moist fungus body and photosynthesizes, providing organic nutrients for both partners, while the fungus provides minerals and protection from drying and harsh light. The mycobiont and phycobiont are species in their own right, and some may grow independently, but many occur only in lichens. The lichen is sometimes considered a species of its own and named as such, but lichens are formally named for the mycobiont and classified on this basis. A single species of mycobiont may form lichens with several species of phycobionts.

The mycobiont always constitutes most of a lichen's volume; only about 7 percent is occupied by the phycobiont, which is always either a green alga or a cyanobacterium. Cyanobacterial phycobionts can fix atmospheric nitrogen, so these lichens are important in the nitrogen economies of their ecosystems. The lichen thallus is usually constructed in layers, including a dense protective layer and a distinct layer where the algae are scattered among the hyphae, and it grows in one of the three general forms shown in Figure 31.12: foliose, fruticose, or crustose. Lichens grow in virtually every habitat on earth, tolerating conditions no other organisms can endure such as dry bare rock and freezing polar environments. Lichens grow on living plants and on some animals, such as certain insects. A few species of lichens grow in freshwater, on ocean shores, and even submerged in the ocean. In the moist forests of the Pacific Northwest of North America, trees are shrouded in a covering of mosses and lichens. However, lichens have little or no ability to keep from drying, and they commonly dry out rapidly whenever their surroundings become dry, stay quite dry indefinitely, and rehydrate quickly when water returns.

Within the thallus, some hyphae hold the algal cells in place while other hyphae (haustoria) penetrate the algae, though always leaving the algal plasma membrane intact. As the phycobiont photosynthesizes, it exports over 90 percent of its carbohydrate product to the mycobiont—from cyanobacteria, mostly glucose, and from green algae, mostly a sugar alcohol such as sorbitol or ribitol. The mycobiont uses some of this

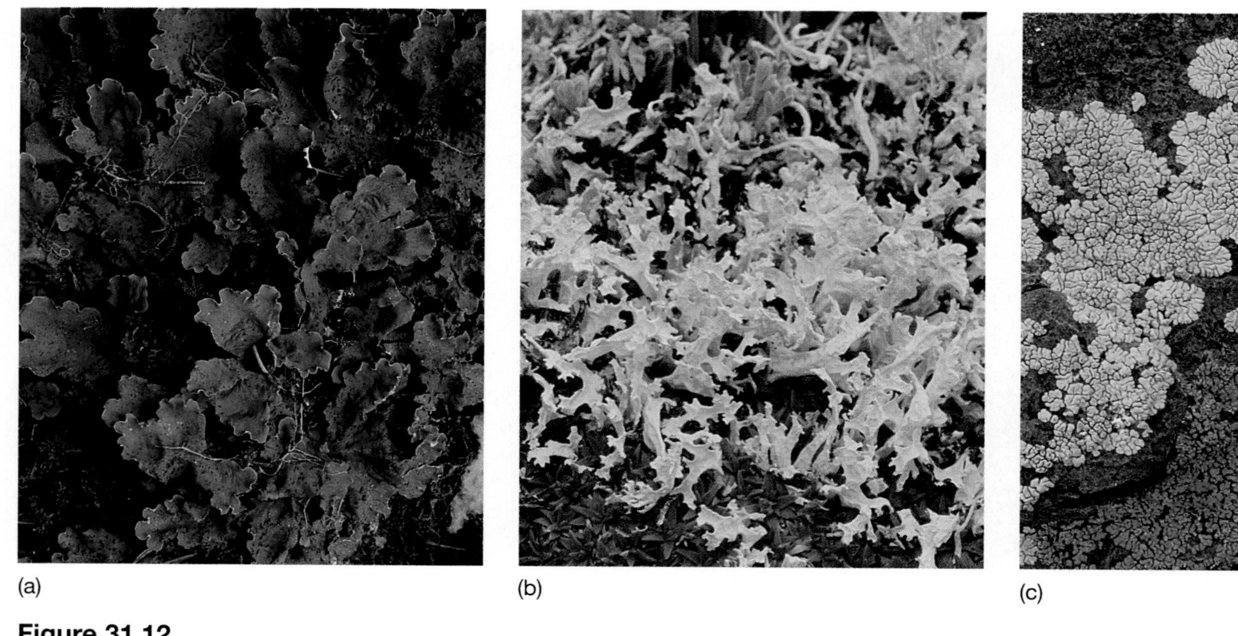

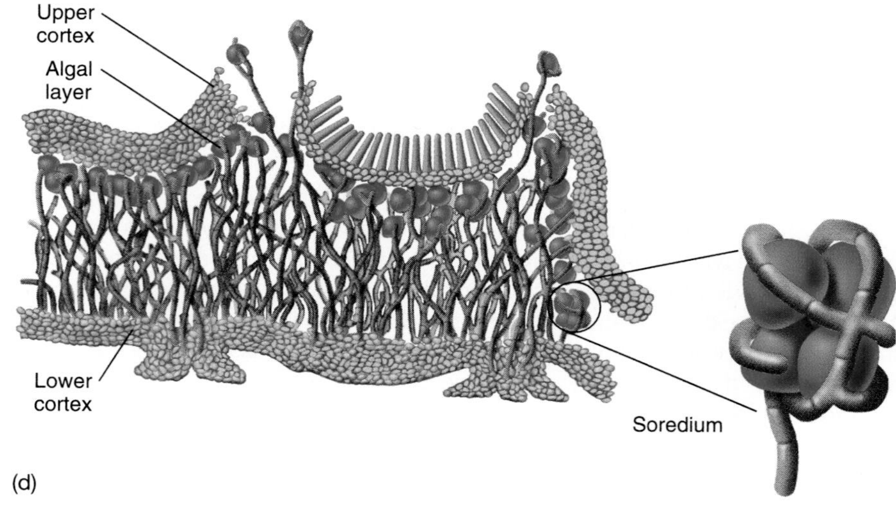

Figure 31.12

Lichens, which may be brightly colored, commonly grow in one of three forms. *(a)* Foliose lichens such as *Peltigera aphthosa* are flattened and divided into lobes. *(b)* Fruticose lichens, represented here by *Cetraria cucullata,* are shrubby or hairlike. *(c)* Crustose lichens have the form of thin crusts, often on bare rock. *Pleopsidium chlorophanum* is yellow, and *Caloplaca saxicola* is red-orange. *(d)* A thin section through a lichen shows the intimate association between algal cells and the fungal mycelium. A soredium, shown enlarged, is an asexual reproductive structure.

carbohydrate for growth but stores much of it as mannitol, which provides insurance against its harsh living conditions. This store is used during rehydration, and as a dry lichen becomes rehydrated, its respiration rate rises to a high level for several hours before falling back to the normal rate of a moist lichen. Lichens produce a variety of unusual compounds called lichen acids whose functions are unclear; some apparently are allomones, including antibiotics that inhibit the growth of bacteria that might decay the thallus, and others inhibit the growth of mosses, other lichens, and the seeds of plants. They may also discourage herbivores. Some may be light screens that protect the phycobiont, and others function in a mechanism in which the fungus regulates the rate of photosynthesis and the rate at which the alga exports carbohydrates.

Allomones, Section 27.9.

In contrast to the rapid growth of many fungal hyphae, lichens may grow very slowly, perhaps as little as 0.1 mm per year. Based on this growth rate, a few lichens are estimated to be thousands of years old. However, lichens are very sensitive to air pollutants, and like other sensitive organisms (for example, canaries carried by miners), they may serve as monitors of air quality. Lichens have disappeared from badly polluted urban areas and have returned later after the pollution has been reduced.

Other fungal mutualisms

A few marine fungi engage in mutualisms called mycophycobioses, which are rather the opposite of lichens in that the fungus becomes the minority partner with a much larger marine alga. The association never develops a new unique morphology, as lichens do, but appears to be a true mutualism in which the fungus obtains nutrients from the alga and in turn protects the alga from drying during exposure at low tide.

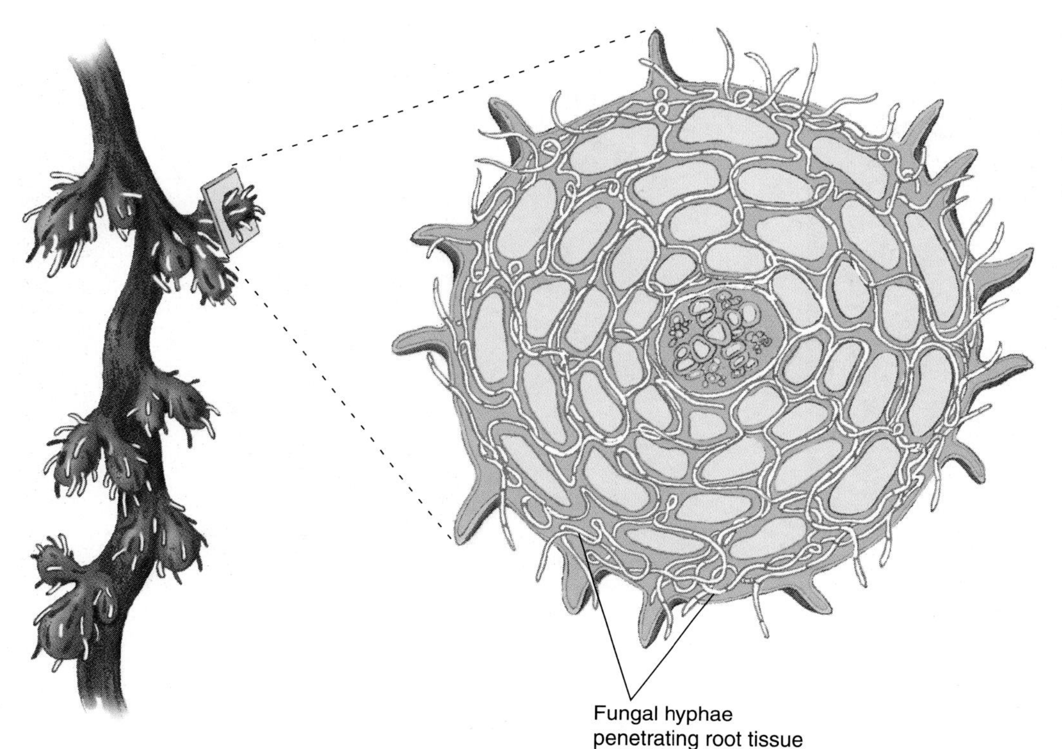

Figure 31.13
A mycorrhiza, shown here in a cross section through a root, develops when a fungal mycelium grows into the root tissue, through the continuous cell-wall system.

If you walk through a forest looking for mushrooms, you will find that many species are not distributed randomly, but rather, grow only around certain species of trees. The survival of trees often depends upon these **mycorrhizal associations** with specific fungi. An uncrowded tree can spread its roots broadly and thus create enough surface area to access all the water and nutrients it needs, but trees in a forest don't have this luxury. Their roots intertwine and compete with one another. Tree roots are commonly entwined with a rich mycelium that grows into the root tissue; the mycelium, being much finer than the roots, creates an enormous surface that absorbs nutrients from the soil and carries them into the plant (Figure 31.13).

Fungi also have a remarkable relationship with orchids. Most plant seeds are large enough to contain a store of oils and carbohydrates that supply food for the developing embryo, but orchid seeds are so small that they look like dust and are unable to contain food stores. When orchids became popular for their beauty and variety during the nineteenth century, horticulturists naturally tried to propagate them in greenhouses from seed—and failed. Then in 1904 the French botanist Nöel Bernard found that orchid seeds will germinate in the presence of certain fungi, and this method has been used ever since. The fungus provides an environment rich in carbohydrates and vitamins to nurture the young plant until it can grow its own root system and maintain itself; the fungus then forms a mycorrhizal association with the orchid and promotes its growth. Some species of orchids have become so dependent on a fungal association that they do not grow their own roots at all.

Certain insects, including the leaf-cutting ants shown in Figure 31.14, have formed remarkable mutualisms with fungi by essentially farming them. Ambrosia beetles are typical of a number of beetles that create extensive tunnels and galleries in wood. Because they are unable to digest the cellulose of the wood, they cultivate fungi that can. The beetles chew up the wood, making a substrate the fungi can grow on; the beetles then eat the fungi, keeping them from over-growing the insects' chambers. The beetles spread the fungi to new environments by packing fungal spores into a sack on the thorax and planting them as they colonize new trees. Remarkably, the fungi also produce ergosterol, which the beetles depend on for their development. Many termites also cultivate fungal gardens in their nests where, in some cases, the fungi digest cellulose into a soup the termites can eat.

Similar stories can be told about fungi and other kinds of insects. The story of scale insects could rival science fiction fantasies about life on other planets. One of these minute insects attaches itself to a plant, extends a proboscis into the plant's vascular system, and spends the rest of its life there drawing out the plant sap. Most scale insects develop a hard, chitinous shell that becomes cemented to the plant surface and protects the animal (Figure 31.15). But some species that live in tropical or subtropical regions have formed a remarkable association with a fungus, *Septobasidium.* When some females pick up the fungal spores, their bodies are invaded by a fine mycelium, and the insect becomes nothing more than a pump for transferring nutrients from the plant to the fungus. As the fungus grows, it leaves occasional small chambers that become occupied by other young female scale insects, some of whom remain uninfected and fertile. Believe it or not, the thick mat of mycelium covering these chambers is only attached on one side, so the female can

Figure 31.14
(a) Leaf-cutting ants cut small pieces of leaves, always of a particular species, and carry them back to their nests. They are sometimes called parasol ants because the leaf pieces they carry resemble parasols. *(b)* They chew up the leaves and deposit them in beds where particular species of fungi grow, fertilizing the beds with their feces and other kinds of waste, including dead ants. *(c)* The ant colony subsists entirely on these fungi.

raise it and extend her abdomen to be fertilized by passing males. She remains trapped inside the fungal chamber, where she consumes sap and lays eggs. Thus the whole arrangement is a three-way mutualism between sterile insects who pump plant sap, fertile insects who reproduce, and the fungus that grows on the sap while transmitting sap from one insect to another.

Coda Fungi share the role of decomposers with bacteria, and it should be clear that ecosystems would not operate for long without organisms that play this role. The detritus from other community members would quickly

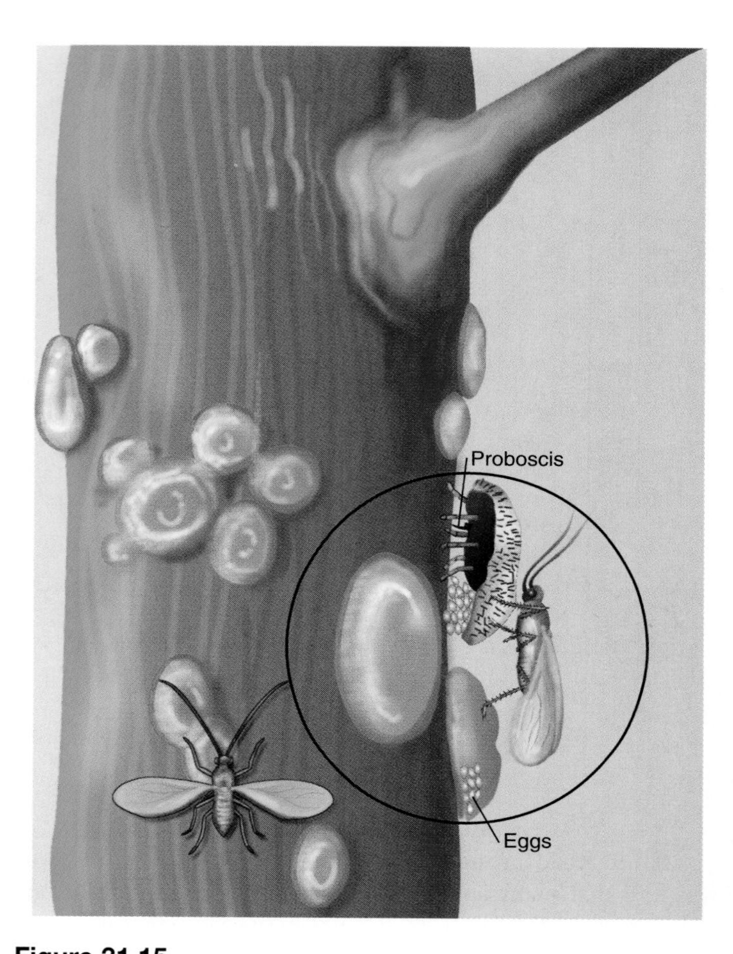

Figure 31.15
Scale insects are minute, flattened animals that stay attached to a plant surface, where they live on sap. Some scale insects form extensive mutualistic associations with fungi.

drown the ecosystem in waste if it were not removed and recycled, and the fungal growth form is ideal for that function. Slender, rapidly growing hyphae that secrete digestive enzymes can grow deep into the discarded parts and dead bodies of other organisms, reducing them to monomers to be recycled through the detritus food web. That a few fungi have taken to eating living organisms is only another illustration of the opportunism of evolution, for the saprobic fungus is only a step removed from the parasitic.

Is it remarkable that fungi engage in so many mutualistic associations with other organisms? Given the fungal mode of nutrition, such mutualism carries a certain inevitability. Fungi, being heterotrophs, require a source of organic compounds that must come from other organisms. The saprobic, decomposing way of life supplies these compounds, of course, but why not go directly to the source? By associating with algae in lichens and with vascular plants in mycorrhizas, fungi obtain the fruits of photosynthesis quite directly, and the fungi associated with scale insects are merely using the insects as pumps for plant sap. Given the decomposing abilities of fungi, perhaps it was inevitable that some animals, in turn, would use them as digestive agents for plant material. And given the interesting tastes of many fungi, perhaps it was inevitable that humans would establish another kind of mutualism with them—or is it merely predation on our part? If mushrooms had never evolved, maybe we would use some similar, delicately flavored vegetable to enhance our sauces, but the chefs on planets with fungi are probably making finer dishes than those that lack such interesting creatures.

Summary

1. The fungal kingdom includes heterotrophic saprobic (or parasitic) organisms. They typically consist of long hyphae that grow into a branching network called a mycelium, although some are quite simple cells.
2. The fungal kingdom includes only organisms that have no flagella at any stage, have chitin in their cell walls, and use the AAA pathway for lysine synthesis.
3. The fungal life cycle commonly includes extensive haploid and brief diploid phases that are joined by a dicaryophase in which distinct haploid nuclei occupy the same cytoplasm without fusing.
4. Fungi commonly form spores, which are single-celled reproductive structures. Spores may be made by mitosis or meiosis. Many have heavy walls and are resistant to environmental stresses such as drying and heating.
5. Mushrooms are fruiting bodies that develop from a dicaryotic mycelium buried in the substratum.
6. Most common mushrooms are basidiomycetes, which bear their spores on basidia on the gills or in the pores of their fruiting bodies.
7. Ascomycetes include yeasts and molds, as well as some mushrooms. They all develop spores inside sacs (asci).
8. Zygomycetes are molds that reproduce through zygospores, which form from hyphae of opposite mating types that extend toward one another and fuse.
9. Lichens are intimate mutualistic associations between a fungus and an alga. The fungus supplies water, minerals, and protection; the alga supplies organic nutrients through photosynthesis.
10. Mycorrhizas are common associations between a fungus and plant roots. The fungal mycelium grows into the roots and supplies the plant with additional water and nutrients.
11. Many insects form associations with fungi. Some ants and beetles cultivate particular species of fungi on which they feed.

Key Terms

mycology 650
hypha 651
mycelium 651
monocaryotic mycelium 652
dicaryotic mycelium 652
fruiting body 652
basidiospore 652
basidium 653
clamp connection 654
mycosis 654
ascus 654
ascogonium 655
antheridium 656
ascogenous hypha 656
conidia 656
sporangia 658
zygosporangium 658
lichen 659
mycorrhizal association 661

Multiple-Choice Questions

1. Which is *not* a general characteristic of all fungi?
   a. chitinous cell walls
   b. osmiotrophic nutrition
   c. all cells uninuclear
   d. lack of flagellated cells or structures
   e. AAA pathway for synthesizing the amino acid lysine
2. Among fungi, spores are produced
   a. by mitosis
   b. by meiosis.
   c. as a means of sexual reproduction.
   d. as a means of asexual reproduction.
   e. by all of the above.
3. Basidiospores are ______ cells that germinate to directly form ______.
   a. haploid; mycelia
   b. haploid; asci
   c. diploid; basidia
   d. diploid; mushrooms
   e. dicaryotic; mycelia
4. Which of the following is (are) not an ascomycete?
   a. rusts that parasitize plants
   b. truffles and morel mushrooms
   c. yeasts
   d. *Neurospora*
   e. the fungus that causes Dutch elm disease
5. With respect to the life cycle of a mushroom-producing basidiomycete, all of the following are dicaryotic *except*
   a. the mushroom itself.
   b. basidia.
   c. mycelium from which the mushroom grows.
   d. basidiospores.
   e. none of these.
6. In which structure(s) does genetic recombination occur?
   a. basidia
   b. ascus
   c. conidia
   d. *a* and *b,* but not *c*
   e. *a, b,* and *c*
7. All of the following are correct about asci, *except* they are
   a. formed from a dicaryotic cell.
   b. the site of meiosis.
   c. the site of mitosis.
   d. produced from the ascogonium.
   e. the place where conidia are produced.
8. In the sexual life cycle of an ascomycete, the ______ are haploid and not dicaryotic.
   a. conidia
   b. ascospores
   c. ascogonia
   d. ascogenous hyphae
   e. croziers
9. The role of fungi in mycorrhizal associations is to
   a. produce necessary nutrients.
   b. increase the absorptive surface for plant roots.
   c. decompose.
   d. synthesize antibiotics.
   e. provide a substitute for plant leaves.
10. What is the nutrient source for the fungal component of lichens?
    a. low-growing green plants
    b. nutrients dissolved in the rock inhabited by the lichen
    c. tree roots
    d. green algae
    e. insect symbiosis

True-False Questions

Mark each statement true or false, and if false, restate it to make it true.

1. The fundamental unit of a typical basidiomycete is the mycelium, which forms an intertwined mat called a hypha.
2. All members of the kingdom Fungi reproduce both sexually and asexually.
3. In the dicaryophase of a fungal life cycle, each cell has diploid nuclei containing DNA from both parental types.
4. A meiospore is a diploid cell that will undergo meiosis and produce four haploid spores.
5. In the gametangia of zygomycetes, meiosis produces haploid nuclei that fuse and grow into diploid hyphae.
6. Most mushrooms are basidiomycetes that produce haploid basidiospores in sacs under the mushroom cap.
7. The fruiting body of a basidiomycete is generally composed of a monocaryotic mycelium.
8. Yeasts can engage in asexual as well as sexual reproduction, and their growing stages can consist of either haploid cells or diploid cells.
9. Asci in ascomycetes and basidiospores in basidiomycetes both perform the same function.
10. Conidia are reproductive cells that function during the asexual life cycles of ascomycetes and imperfect fungi.

Concept Questions

1. How do the members of the kingdom Fungi differ from molds included in other kingdoms?
2. What are the principal commonalities in the life cycles of ascomycetes and basidiomycetes, and what are the principal differences between them?
3. Describe unique ascomycete and basidiomycete structures that are similar to a feature of red algal growth.
4. Why is the mycorrhizal association of plants and fungi considered mutualistic? How does each partner benefit from the association?
5. How does the sexual life cycle of zygomycetes differ from that of ascomycetes?

Additional Reading

Alexopoulos, Constantine J., Charles W. Mims, and Meredith Blackwell. *Introductory Mycology,* 4th ed. John Wiley, New York, 1996.

Burnett, John H. *Fundamentals of Mycology,* 2d ed. Edward Arnold, London, 1976.

Christensen, Clyde M. *Common Fleshy Fungi.* Burgess Publishing Co., Minneapolis, 1955.

———. *The Molds and Man: An Introduction to the Fungi,* 2d ed. University of Minnesota Press, Minneapolis, 1961.

Cooke, Roderic C. *The Biology of Symbiotic Fungi.* John Wiley, London and New York, 1977.

Deacon, J. W. *Introduction to Modern Mycology,* 2d ed. Blackwell Scientific Publications, Oxford and Boston, 1984.

Dix, Neville J., and John Webster. *Fungal Ecology.* Chapman & Hall, London and New York, 1995.

Lawrey, James D. *Biology of Lichenized Fungi.* Prager, New York, 1984.

McKnight, Kent H., and Vera B. McKnight. *A Field Guide to Mushrooms of North America.* Houghton Mifflin Co., Boston, 1987.

Moore-Landecker, Elizabeth. *Fundamentals of the Fungi,* 4th ed. Prentice-Hall, Upper Saddle River (NJ), 1996.

Internet Resource

To further explore the content of this chapter, log on to the web site at:

http://www.mhhe.com/biosci/genbio/guttman/

32

Introduction to Plants and Animals

Key Concepts

A. General Structural Considerations

32.1 Differences between plants and animals are correlated with their different modes of nutrition.

32.2 Some animals have become more like plants and have radial symmetry.

32.3 Plants and animals are built of specialized tissues and organs.

32.4 An organ is a complex of tissues serving a specific function.

32.5 Plants and animals may be either unitary or modular.

B. Water and Gas Relations

32.6 Plants and animals use water, carbon, and nitrogen in different ways.

32.7 Plant and animal cells exchange CO_2 and O_2 with their surroundings.

32.8 Plants and animals have two fluid compartments.

32.9 Plants and animals face comparable problems of water balance.

32.10 Plants and animals have vascular systems for distributing water and nutrients.

32.11 Reproduction in terrestrial organisms has generally become less dependent on water.

C. Considerations of Size

32.12 Surface/volume ratios determine many architectural features of organisms.

32.13 Biological architecture is often determined by mass/area restrictions.

32.14 An organism's metabolic rate is closely related to its size.

A lion cub rests in the grass of its home in Kenya.

We humans seem to empathize easily with other animals, especially those we use for pets or for labor, but it seems hard to empathize with plants, no matter how much we enjoy cultivating them in our homes and gardens. Animals move about, act in ways we can understand, and may even seem to have emotions like ours. Plants[1] stay in one place and

[1] When we refer to plants in this chapter, we mean only the multicellular organisms called vascular plants.

rarely do anything we could call acting or behaving, much less showing emotion! But are plants and animals really so different? After all, both perform many common functions, including:

- obtaining nutrients
- exchanging O_2 and CO_2
- maintaining the right balance of water and salts
- transporting materials to and from all tissues
- supporting themselves mechanically
- growing
- reproducing

Even though plants and animals have evolved to fill quite different ecological roles, both groups have faced the challenge of evolving multicellular bodies that perform many of the same functions. This is not surprising when you consider that the two clades arose from a common ancestry of protists that shared many basic mechanisms. Furthermore, biological materials have their own inherent limitations. Each type of molecule or structure can only accomplish so much, and it is likely that a limited number of possible structures can evolve to serve each function. So it is no wonder that plants and animals evolved many common ways to meet those challenges, and before we emphasize their differences in later chapters, we will explore those commonalities. Since we are comparing two distinct clades, the common structural features we will emphasize here are generally not *homologies* stemming from shared evolutionary ancestry but rather *analogies* that have arisen because of similar requirements. Generalizations of this kind are not always easy to make in biology. Since evolution is unpredictable, any rule may have exceptions, which we can address later, but in this chapter we will try to identify major themes of structure and function. ■

A. General structural considerations

32.1 Differences between plants and animals are correlated with their different modes of nutrition.

Plants and animals are primarily distinguished by their modes of nutrition, and this distinction dictates their very different forms (Table 32.1). Plants are phototrophs, which get their energy from light, and (with few exceptions) they are green because of the chlorophyll in their photosynthetic apparatus. They develop large, light-absorbing surfaces that take in CO_2 and release O_2. A plant grows as if one direction of the compass is about as good as another. A plant is **sessile,**[2] growing in one place and unable to move around. A typical plant has a central vertical column divided into a **root system** growing downward, which anchors the plant and secures water and nutrients, and a **shoot system** growing upward, made of stems and leaves that carry out photosynthesis and gas exchange (Figure 32.1). A **vascular system** of fine tubules carries water and nutrients between different parts of the plant. Since a plant can generally receive light and nutrients from all directions of the compass, it is largely indifferent to direction, and it radiates roots and shoots more or less symmetrically to all sides (Figure 32.2), though its roots will grow preferentially toward a source of water and its shoot toward a source of light. The resulting form contrasts sharply with that of most animals.

Animals do not photosynthesize. They are chemotrophs, which use the energy stored in organic molecules. They are also *holotrophic:* They engulf pieces of solid food or nutrient-rich liquids—often the bodies of plants or other animals—and digest them in an internal pouch or tube (Figure 32.3). Except for some rare and fascinating insectivorous species (see Figure 40.17), plants don't consume other organisms. To pursue the holotrophic life, an animal generally has to search for its food, rather than waiting for the food to come to it, so it must be **motile**—able to move from place to place and organized for getting food actively (chasing, climbing, grabbing, biting, sucking, ripping, trapping, or filtering). Since animals often prey on other motile animals, and are at risk of becoming food themselves, the ability to move fast is critical.

With the exception of some plantlike animals (see Section 32.2), these nutritional considerations have shaped animals with an emphasis on one direction, the direction of their movement. The vast majority of animals are based on a body plan that includes **cephalization** (*cephalon* = head), having a distinct head end with a concentration of special structures (Figure 32.4); this body plan evolved very early in the animal clade. Since the head end encounters new objects first, it must get as much information about the world as possible, as fast as possible. So the major sense organs are located there—eyes and ears, which monitor distant events via light and sound, and organs that detect chemicals, which indicate food or danger. These sense organs communicate quickly with the brain, a concentration of nervous tissue located in or near the head that processes incoming information and determines a course of action. The head-tail distinction is basic to animal anatomy, and structures are commonly described by referring to the head end (**anterior** or **cephalic**) or the tail end (**posterior** or **caudal**).

Much of an animal's body is taken up by the **digestive tract,** which processes food and absorbs its nutrients into the body. The head encounters food first. It has the sense organs needed to identify food and special structures for capturing and ingesting food—by sucking it into the mouth or drawing it in with special muscles, or by seizing and pushing it in with specialized limbs. The indigestible remainder (feces) passes out through the other end, the anus. (The so-called incomplete digestive tracts of some animals have a single opening that serves as both mouth and anus.) Metabolic wastes may also be excreted via this tract.

[2] In plant anatomy, "sessile" has another specialized use, meaning that an organ is not attached with a stalklike structure.

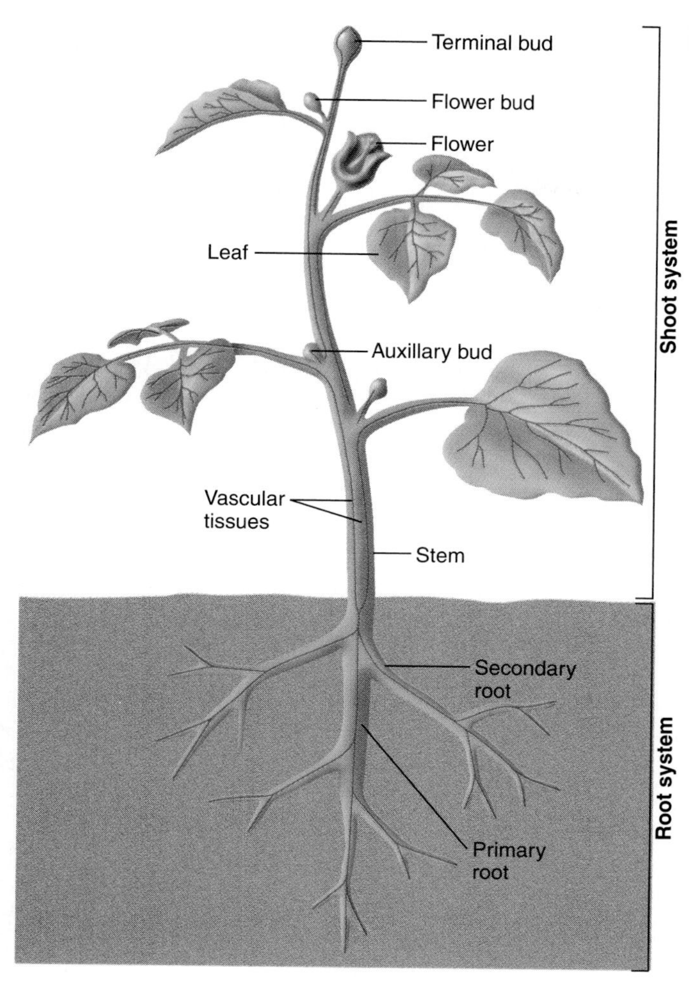

Figure 32.1

A vascular plant has a root system and a shoot system, centered on a vertical axis. Its vascular system consists of tubules that carry water from the roots upward and distribute nutrients from the shoot system throughout the plant.

Figure 32.2

A vascular plant *(viewed here from above)* grows more or less equally in all directions.

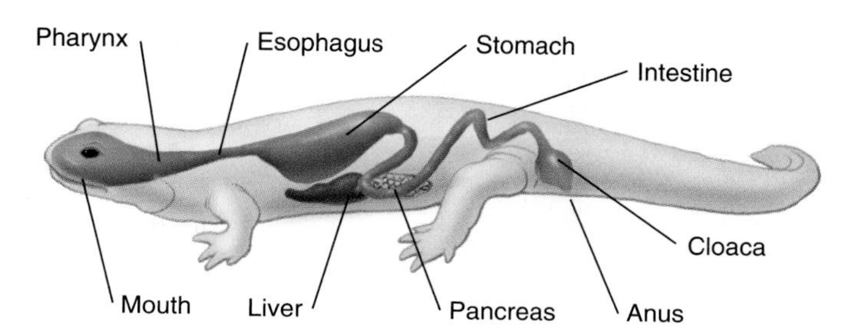

Figure 32.3

Most animals have a tubular digestive tract that conducts food from the mouth to the anus, with glands such as the pancreas and liver that secrete enzymes and other materials to effect and aid digestion.

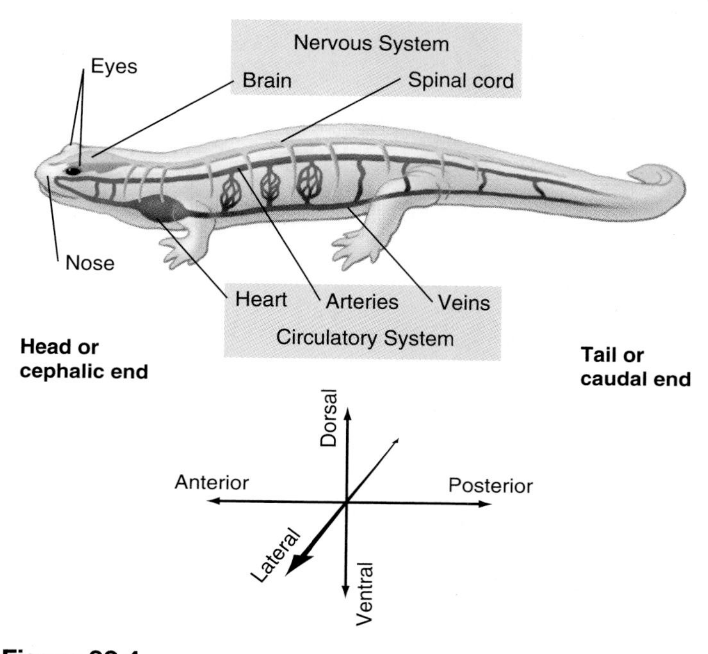

Figure 32.4

A typical animal is cephalized, with a brain and sense organs concentrated in the head. A circulatory, or cardiovascular, system with at least one heart distributes blood through the body. Notice the standard directions on the body. Anterior and posterior are the head and tail ends; dorsal and ventral sides are the back and belly; and the left and right sides are lateral.

Typical animals also have **bilateral symmetry,** which means the body can be divided into virtually identical mirror-image halves by a plane that passes through the middle of the animal (Figure 32.5). A bilaterally symmetrical animal moves forward by alternately moving its left and right appendages or by contracting sets of left and right muscles alternately to set up a wave of movement (Figure 32.6). Its legs, fins, flippers, or other appendages are arranged symmetrically, generally below the body and in contact with the ground. This gives the animal distinct upper **(dorsal)** and lower **(ventral)** surfaces (see Figure 32.4).

Table 32.1 Comparisons of Plants and Animals

Vascular Plants	Animals
The organism's source of energy determines its form:	
Autotrophic: Uses sunlight + air + water. Spreads absorbing surfaces *outside*. Leaves are generally thin for gas exchange.	Heterotrophic: Uses other organisms. Snares and ingests food; absorbing surfaces are *inside*. Ingestion permits eating now, digesting later; gives protection against other heterotrophs.
Metabolism:	
Uses CO_2 and produces O_2; uses inorganic nitrogen and may be limited by nitrogen source.	Uses O_2 and produces CO_2; gets excess nitrogen from its food and must excrete nitrogen.
Implications for motility and form:	
Generally sessile (stays in place). Thin, leafy "antennae" spread out to catch light and for gas exchange. Anchored with roots, a secure prop system that also supplies nutrients from soil and water. Supported by a "skeleton" of cells with heavy walls. Cannot obtain CO_2 without losing water; tends to lose a lot of water, so it needs extensive root system to absorb water.	Generally motile (moves). Has a cephalized body form with mouth, sensory organs, and brain in front:  Actively catches prey, evades predators, moves on to find new food. Therefore, needs a skeleton powered by a muscular system. Leads to centralization of functions: Thick body wall Heat regulation Large internal surfaces for absorbing nutrients and for gas exchange.
Distribution system:	
Xylem: 1-way water and solute transport. Phloem: Separate food transport system. Hormones (growth regulators): Slow distribution, induce long-term changes.	Circulation of blood: Suited for rapid exchange of food, O_2, and CO_2, and for hormone distribution. Hormones used for slow changes, but nerves for fast action.
These features determine embryology:	
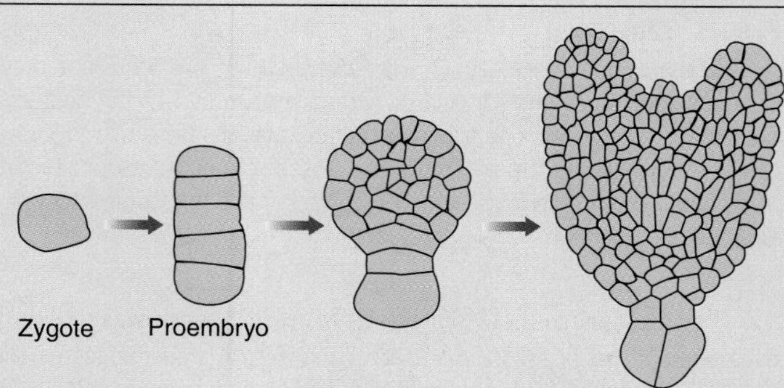 Development is typically modular and growth is indeterminate: Apical meristems (growth tissues) are as old as the plant. Meristems retain embryonic properties. Plant is drawn out between the meristems, and it matures toward them.	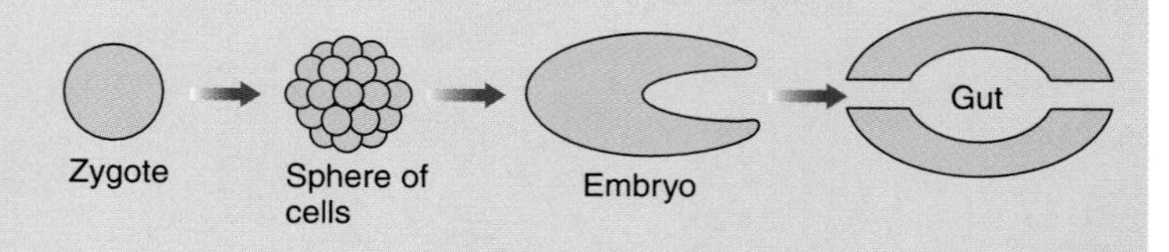 Development is typically unitary, and growth is determinate. An initial ball of cells forms tissues, which become determined early. Organism matures as a unit.

Figure 32.5
Most animals have bilateral symmetry. A line down the center of the body divides it into mirror-image left and right halves. However, the details of the body and all its internal organs are not necessarily symmetrical.

32.2 Some animals have become more like plants and have radial symmetry.

Although typical animals are motile, bilateralized, and cephalized, some animals have evolved into a sessile existence on the ocean floor. Instead of hunting for food, they remain in place and let the food come to them. Since they don't move in any one direction, all directions are equally important to them. Instead of a bilateral, cephalized form, they have **radial symmetry,** with their parts arranged regularly around a central axis (Figure 32.7). Having tentacles or other appendages around a central mouth, they wait for food to come into range and then suck it in or snatch it. These animals are so plantlike that some have been mistaken for plants.

32.3 Plants and animals are built of specialized tissues and organs.

We have been using the word "tissue" casually to refer to the cellular material of a plant or animal. Technically, though, a **tissue** is a mass of similar cells that are specialized for a particular function. Some leaf tissues are made of cells rich in chloroplasts that are specialized for photosynthesis; nervous tissue contains cells specialized for conducting nerve impulses; and muscle tissue contains cells that can exert a force by contracting. In both plants and animals, several kinds of tissue working together to

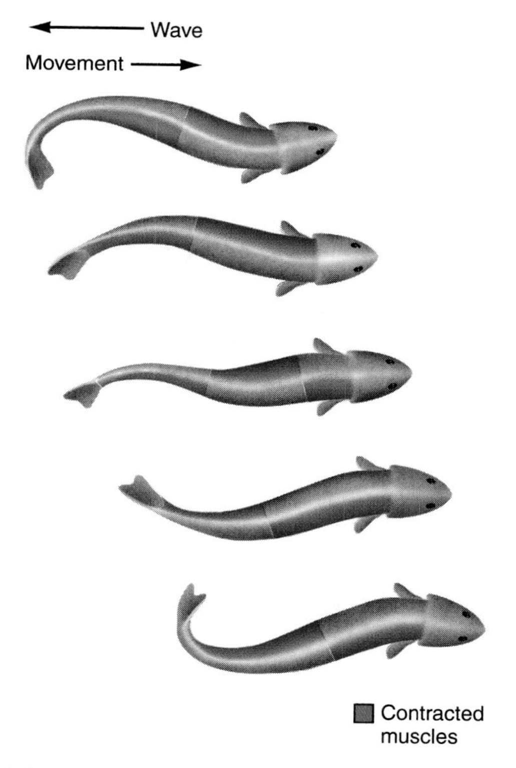

Figure 32.6
An animal such as a fish moves forward by means of opposing muscle pairs. Waves of contractions *(shown in red)* move from front to back on each side of the body, so the left and right sides contract alternately at each point, making the fish move forward. To move straight, it needs symmetrical left and right structures.

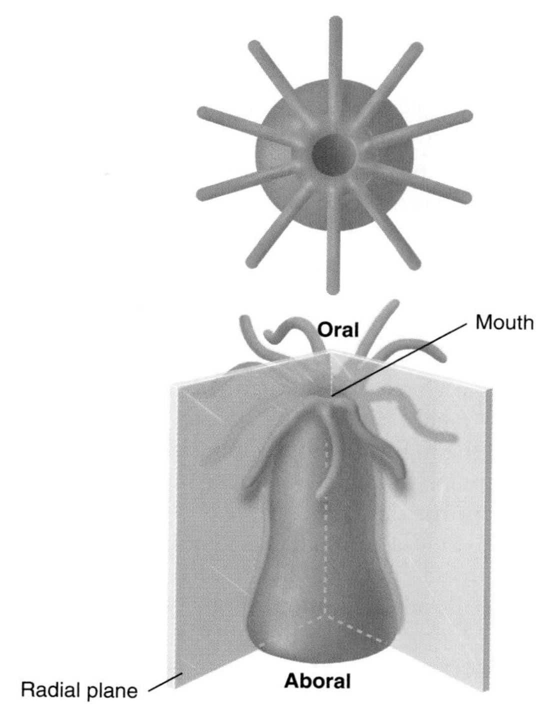

Figure 32.7
A sea anemone is an example of a nonmotile animal with radial symmetry. Instead of an anterior-posterior axis, it has an axis from the oral end at the opening of the digestive cavity to the aboral end. The animal reaches out in all directions to grasp food and convey it into the digestive cavity.

perform a specific function generally form an **organ,** such as a leaf, root, heart, kidney, or lung.

Multicellular organisms are traditionally said to have a hierarchy of different levels of organization: cells forming tissues, tissues forming organs, and organs forming organ systems. This viewpoint is actually an oversimplification and does not hold strictly true. Some cells are not functionally part of any tissue, for example, and some tissues are not functionally part of any organ. Nevertheless, the levels-of-organization concept is a useful way to start thinking about the structure of plants and animals.

As plants and animals evolved, they responded to certain common challenges by evolving similar structures, including analogous tissue types. In searching for commonalities among plants and animals, we find three general types of tissue common to both: surface tissues, connective or mechanical tissues, and bulk tissues. Keep in mind that, although similar, these tissue types are not identical in the two kingdoms because they may have quite different forms. However, their commonalities reinforce the general point of parallel evolution in the face of similar requirements.

Surface tissues

Surfaces are special places on any structure. The surfaces of structures in a multicellular organism often require special tissues to form a boundary, protect the structure, and transport materials in and out or prevent such transport. Plants and animals both have tissues specialized for these functions.

An animal **epithelium** is a layer of cells, generally only one cell thick, that covers an organ or the whole organism. The edges of epithelial cells fit together tightly, forming a solid barrier to the passage of water and other substances. The cells may be specialized for transporting certain materials across the epithelium. An epithelium alone can also comprise a large part of an organ; blood vessels and other tubes, for instance, are made of a rolled-up epithelium (called an *endothelium* because it forms the inside of a structure). Epithelial cells range in form from tall and narrow to flat (Figure 32.8). Layers of thin, scaly cells form a stratified squamous epithelium, as in the epidermis of the skin of most vertebrates; filled with the protein keratin, these cells make a hard, waterproof covering. The surface layers of the lining of the mouth are squamous epithelium; in biology laboratories, students often scrape out some of these cells for microscopic examination.

Mucosas are animal epithelia kept moist by a layer of mucus. Ciliated mucosas, bearing ciliated cells (Figure 32.9), cover surfaces such as the passages into the lungs of air-breathing vertebrates; their cilia continually move the mucus layer along in one direction, trapping dirt and microorganisms and carrying them away. Other mucosas secrete slippery fluids, such as the lubricant in the joints between bones, which is rich in protein and mucopolysaccharides.

Plants have **epidermal** layers, one cell thick, on the surfaces of stems, leaves, and growing roots. Epidermal cells, like animal epithelial cells, fit together tightly through interdigitations like the edges of jigsaw-puzzle pieces, sealing off the interior of the plant. On the cytoplasmic side of the cell wall, they produce a surface layer of *cuticle* a few micrometers thick. Cuticle consists of the waxy materials *cutin* (mostly on parts exposed to the air) and *suberin* (mostly on underground parts) embedded in wax. (Cutin and suberin, being polymers of modified fatty acids,

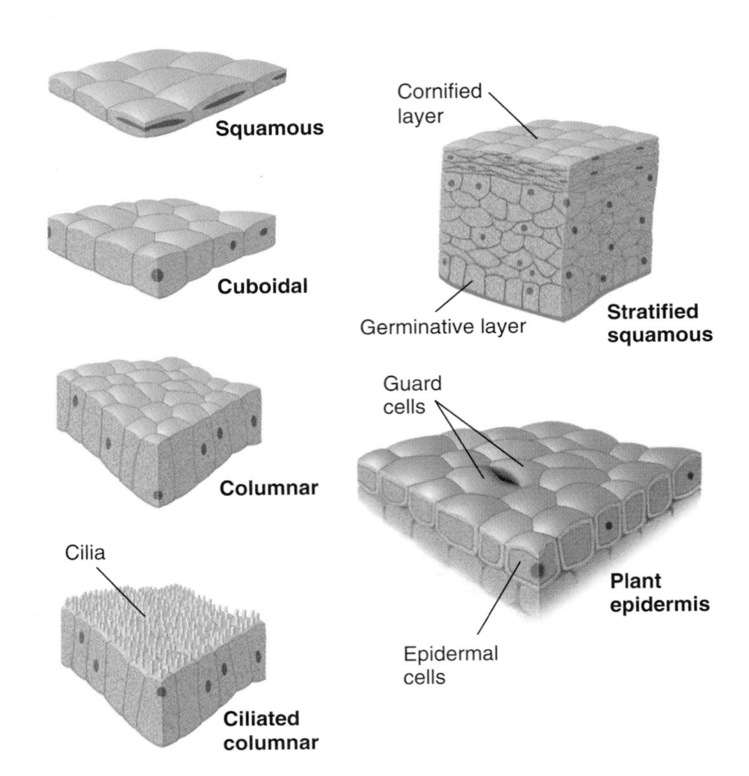

Figure 32.8
Epithelial or epidermal tissues consist of thin layers of cells shaped rather like bricks or tiles. These thin sheets form the outsides of structures, or they may be shaped into tubes or other inner surfaces.

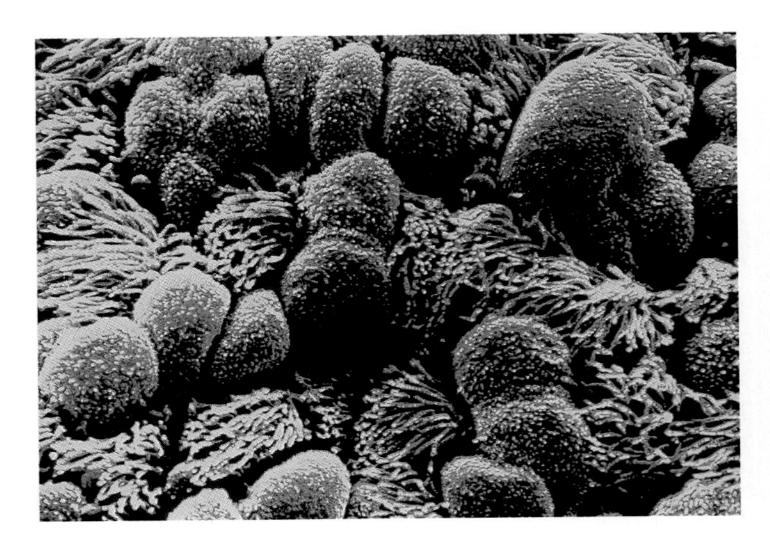

Figure 32.9
Animal epithelia often include ciliated cells. This scanning electron micrograph shows several cells covered with short, fingerlike cilia.

further illustrate the principle that large biomolecules are polymers.) Such thick, hydrophobic layers are very effective barriers to evaporation and to bacteria and other invaders. They are the sources of most commercial waxes, including the carnauba wax often used for polishes.

Connective or mechanical tissue

Plants and animals both need supports. Support is especially critical for terrestrial plants and animals, since the shoot systems of plants reach far above the ground, and most animals must hold their bodies off the ground. But even aquatic forms, whose weight is largely supported by the water they live in, are not just masses of soft cells. They need tissues specialized for binding other tissues together, for connecting and supporting their parts. Most animals form skeletons of bone or shell that serve primarily as frameworks for movement and secondarily for protection.

Plants and animals have both evolved tissues specialized for these functions. Although these tissues may be quite different in the two kingdoms, they share some remarkable features. The cells of all supporting tissues, such as cartilage, bone, and plant connective tissues, produce an extracellular *matrix* of tough, fibrous macromolecules, both proteins and polysaccharides (Figure 32.10). After forming this matrix, the cells themselves become relatively unimportant. Supporting plant tissues include cells with thickened walls called **collenchyma** and **sclerenchyma** (Figure 32.11). Sclerenchyma consists of long, heavy-walled cells called *fibers* and short, heavy-walled cells that form tubules. These tubules form half of a plant's vascular system, the half that conducts water throughout the plant in a tissue called *xylem.* Sclerenchymal cells lose their cytoplasm, and their walls remain to provide support. In animal connective tissues, fibroblast cells and other types synthesize a matrix of tough fibers that surrounds and connects other tissues.

Matrix materials are largely polysaccharide. In plants, the chief matrix polymer is *cellulose,* supplemented by proteins. In animals, the matrix is commonly the protein collagen combined with fibers of mucopolysaccharides; the mucopolysaccharide *chitin* is also an important structural matrix material in invertebrates. In both kingdoms, the matrix is often impregnated and strengthened with minerals such as calcium salts. An unmineralized matrix of collagen and mucopolysaccharide forms soft connective tissues such as ligaments, tendons, and soft invertebrate shells; the addition of calcium phosphate turns soft tissues into bone and hard shells.

Mucopolysaccharides, Section 4.2.

We take advantage of the strength of connective and supporting tissues. Our ancestors made many tools out of bone. Extensive xylem tissue forms *wood* (Greek, *xylon*), obviously one of the strongest supporting materials. We use the great strength and length of fibers by extracting them from plants to make textiles and ropes. A single fiber cell of flax, from which linen is made, may be 7 cm long, and the ramie plant, which is also used for textiles, has fibers 25 cm long.

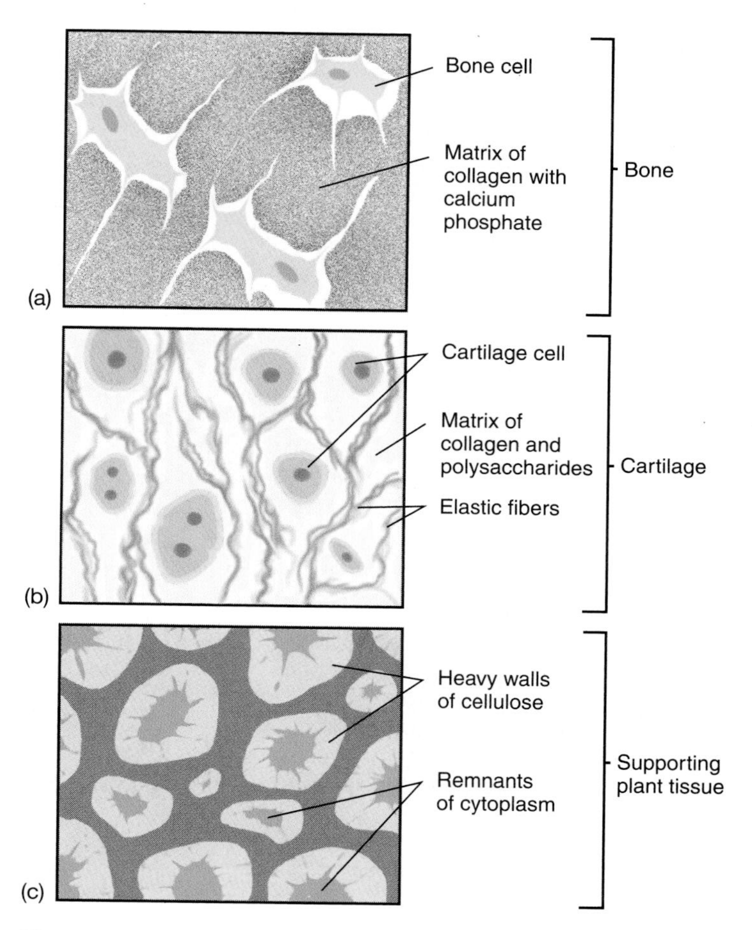

Figure 32.10

Connective and supporting tissues consist of cells embedded in a matrix of proteins and polysaccharides. *(a)* In cartilage and other soft animal tissues, the proteins are collagen and an elastic protein, elastin; the polysaccharides are mucopolysaccharides such as hyaluronic acid and chondroitin sulfate. *(b)* In bone, the matrix is hardened with crystals of calcium phosphate. *(c)* In plant connective tissue, the matrix consists largely of cellulose fibers in and between cell walls, while the cells themselves are often dead.

Bulk tissues

Parenchyma was originally a botanical term for cells that remain relatively undifferentiated and have thin walls (in contrast to most supporting cells; Figure 32.12*a, b*). When seen through a microscope, these cells look generally rectangular in shape, but we can think of them as bubbles in a froth, pressing against one another with surface tension reducing their surface area to a minimum. As the physicist Lord Kelvin showed in the nineteenth century, if the resulting cells are perfectly packed, they will have 14 faces, some hexagonal and some rectangular, a form called a tetrakaidecahedron (Figure 32.12*c*). When plant parenchymal cells are examined closely, they turn out to vary in numbers of sides (mostly 13, 14, or 15), with an average very close to 14. Cells in some animal tissues pack in the same way. Parenchymal cells may be somewhat specialized metabolically; many, for instance,

Figure 32.11

Plant tissues are supported by two kinds of specialized cells. *(a)* Collenchyma has regions of thickened cell wall but is still able to grow along with surrounding cells. *(b)* Sclerenchyma cells are mostly fibers and tubules that develop heavy cell walls, often followed by loss of the cytoplasm.

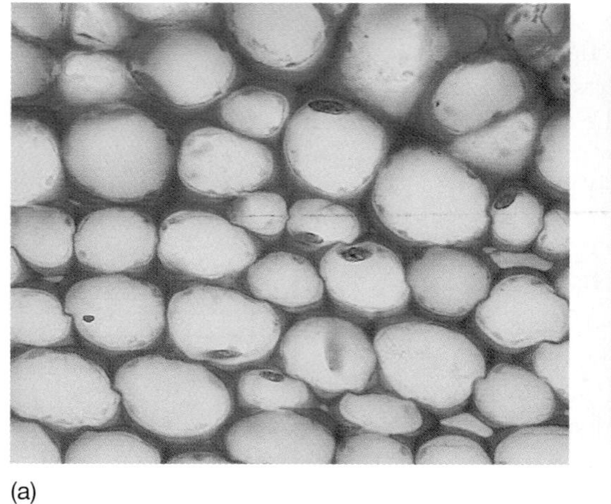

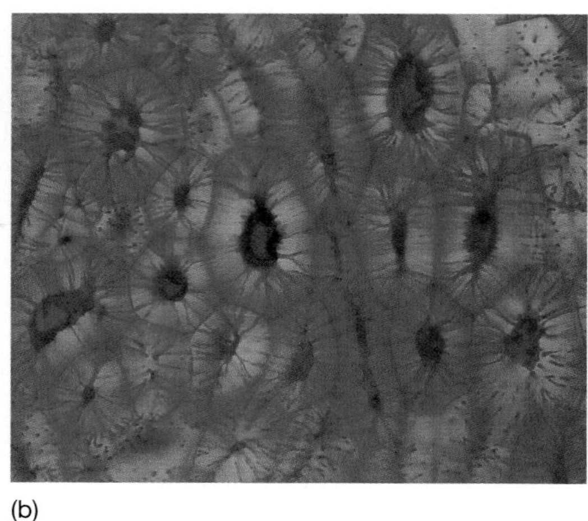

Figure 32.12

Parenchymal tissues from *(a)* a plant and *(b)* the liver of an animal consist of relatively undifferentiated cells that form the bulk of an organ. *(c)* If the cells of such a tissue are pressed against one another so as to fill the space, they assume the average shape of 14-sided solids with some rectangular and some hexagonal faces.

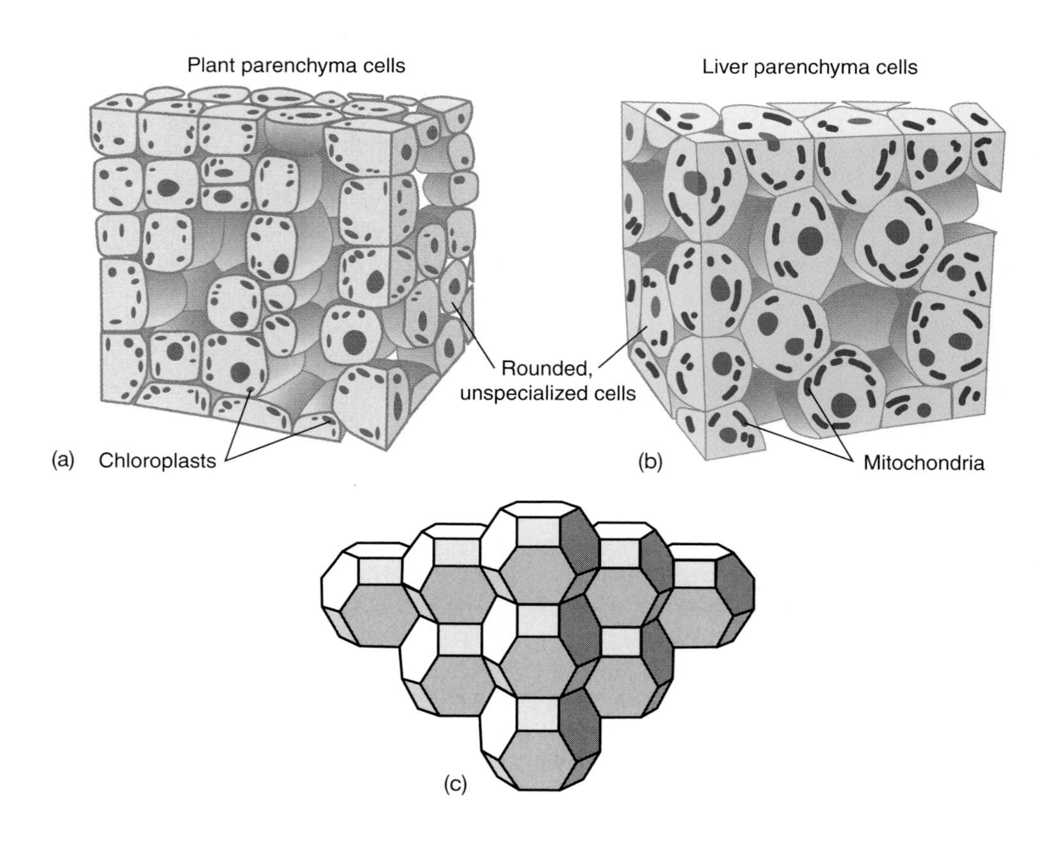

are rich in chloroplasts and are the primary sites of photosynthesis. Plant parenchyma forms the bulk of structures, such as the middle layers of a leaf and the body of a stem or root. The term is also useful for certain animal tissues made of quite uniform masses of cells, such as the body of the liver.

32.4 An organ is a complex of tissues serving a specific function.

In both plants and animals, several kinds of tissues combined to perform a specific function comprise an **organ,** such as a leaf, root, heart, kidney, or lung. Although organs vary enormously and serve many functions, they generally share several features (Figure 32.13):

1. An epithelial or epidermal surface. Animal organs may be wrapped in connective tissue sheets for additional support and binding to other structures.
2. A body made of either parenchyma, sometimes containing tubules of epithelium, or muscle (in animals).
3. Vascular elements that serve the organism's cells—xylem and phloem tubules in plants, arteries, veins, and capillaries in animals.
4. In animals, nerves to control the organ's activities.

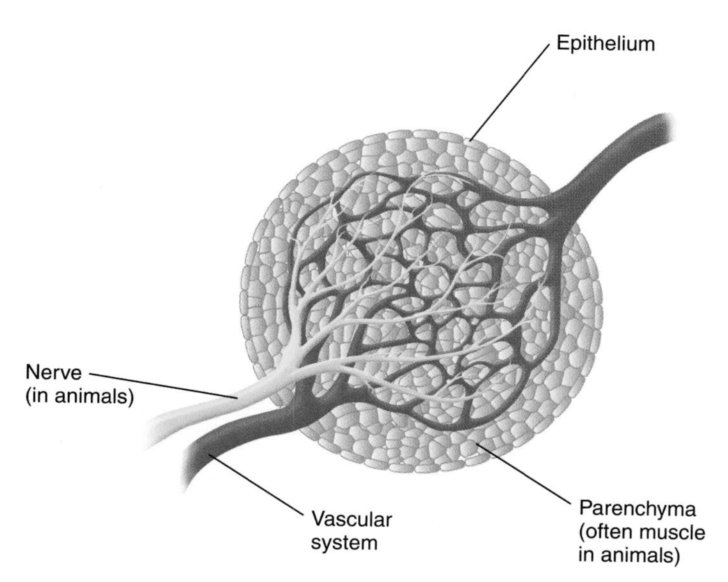

Figure 32.13

An organ often consists of a mass of parenchymal tissue enclosed in epidermis or epithelium, with vascular tissue that supplies its nutrients and removes its wastes. An animal organ may be made largely of muscle and always includes nervous tissue.

In animals we speak of organs as parts of an **organ system.** The brain, nerves, and sense organs, for instance, form a nervous system, and the various portions of the intestinal tract with associated glands make a digestive system. However, plants can hardly be divided into these systems. The few kinds of organs they possess are integrated functionally by the vascular tubules that connect them, and the whole plant is essentially equivalent to one organ system in an animal.

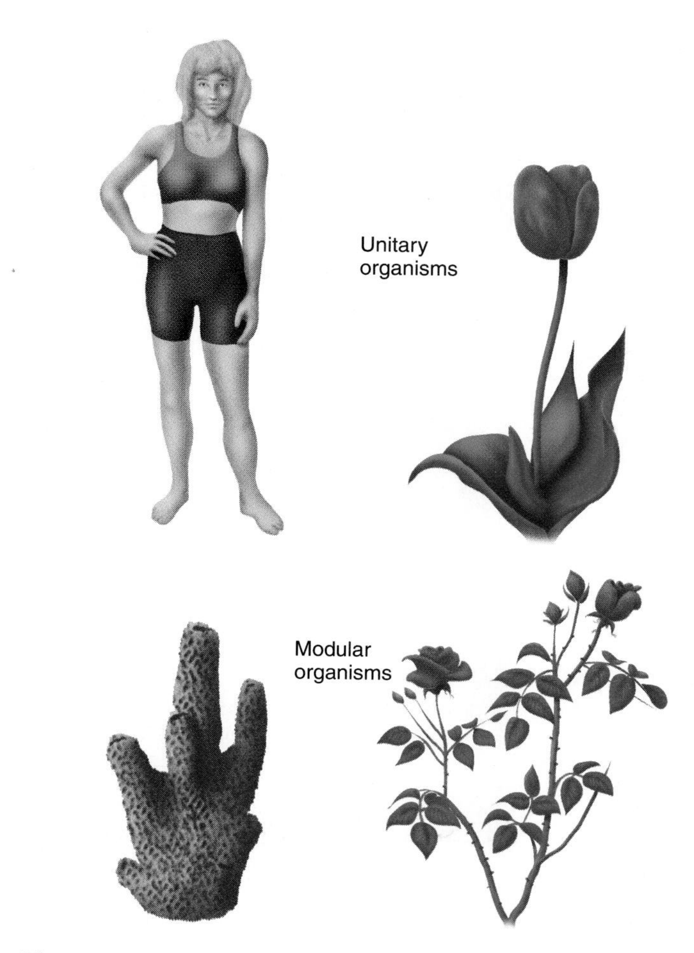

Figure 32.14

A human and a tulip represent unitary organisms; a sponge and a rosebush are typical modular organisms.

32.5 Plants and animals may be either unitary or modular.

Anatomy is such a stable, well-explored subject that discovering anything new about it seems unlikely. Yet recently some biologists, taking a fresh look at the organization of plants and animals, have recognized a fundamental distinction in organization: Both plants and animals can be either **unitary** or **modular** (Figure 32.14).

The distinction is based on an organism's form of growth and life cycle. A unitary organism is a distinct individual. In a human life cycle, a sperm and egg unite into a zygote that then grows into a multicellular adult, able to produce sperm or eggs to start the cycle again. Each person has two arms, two legs, and a single head with symmetrical features; barring accidents, these features will not change during its lifetime. A human has an individual name, and it will never sprout a smaller human so we would have to wonder which of the two ought to retain that name. Many plants are also unitary. A tulip sprouts from a bulb that was made by another tulip plant a year earlier, and it puts down a strong root, a circle of leaves, and one stem with a flower at the top.

In contrast, look at a sponge. A single sponge may begin as a zygote formed from a sperm and an egg, and at first it might be mistaken for a unitary animal, with a single, mouth-like orifice and a cavity through which water flows. However, as the sponge grows, it sprouts other units that look like the first one and remain joined, with interconnected water passages. It can grow indefinitely, making a *colony* with a variable number of modular units and no obvious limit on its size. The colony grows asexually, and there is little or no integration between the units. These are the characteristics of a modular organism.

Animals are most typically unitary and plants most typically modular. Grasses and strawberry plants, for instance, grow by sending out runners that periodically form new modules with roots and shoots. Rather than growing like a tulip, many plants grow in *shoot modules* consisting of a length of stem and a node where a leaf emerges with an axillary bud at its base (Figure 32.15). A plant organized in this way can grow without limit, simply by adding more modules. In a tree, the axillary buds of many modules grow into branches. Each branch can produce more branches, and thus the tree can continue to expand indefinitely. Figure 32.16 shows other examples of modular organization and growth in both plants and animals.

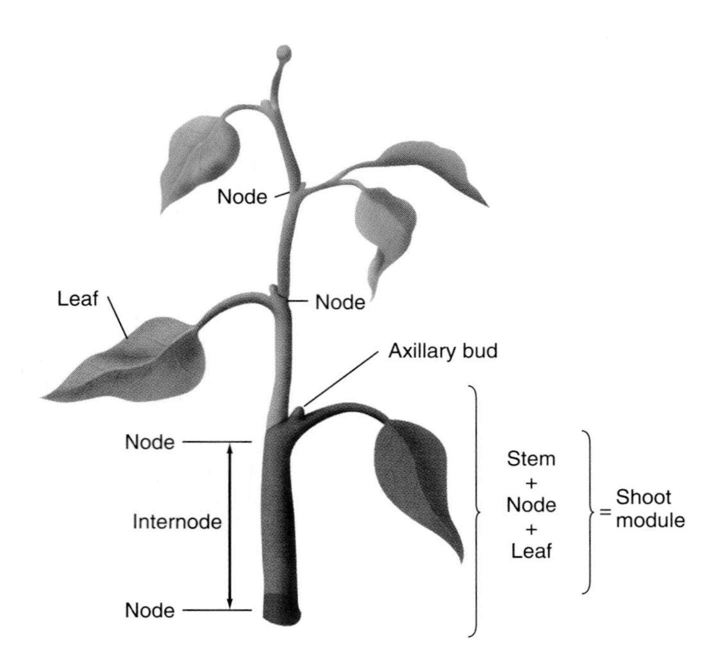

Figure 32.15

The growth unit of a modular plant is a shoot module. The node is the point at which a leaf or branch emerges; the internode is the space between nodes.

Unitary organisms tend to show **determinate growth,** growing to a certain size but no further (although some animals, such as fish, do continue to grow throughout their lives). We have definite expectations about the sizes of typical animals and many plants. Mice are small, rabbits bigger, elephants huge. Giant mice, tiny elephants, and gigantic rabbits are the creatures of fairy tales or bad dreams. Modular organisms, by contrast, exhibit **indeterminate growth;** each module tends to grow to a certain size, but the organism can add new modules without limit. There is nothing remarkable about seeing a small oak tree and a giant oak tree standing together, or a small sponge and a large sponge.

B. Water and gas relations

In spite of their fundamentally different modes of nutrition, plants and animals share several common metabolic requirements, including the need to supply all their cells with water and to maintain a balance of salts and other solutes in their fluids. Each cell is 70–80 percent water, and water is also the medium that transports other materials between cells: nutrients, wastes, and sometimes the critical gases CO_2 and oxygen.

32.6 Plants and animals use water, carbon, and nitrogen in different ways.

The balance of water, CO_2, and O_2 in an organism is strongly connected to the processes of photosynthesis and cellular respiration. In photosynthesis, water and CO_2 are consumed while O_2 is produced:

$$6\,CO_2 + 6\,H_2O \rightarrow 6\,O_2 + \underset{\text{Glucose}}{C_6H_{12}O_6}$$

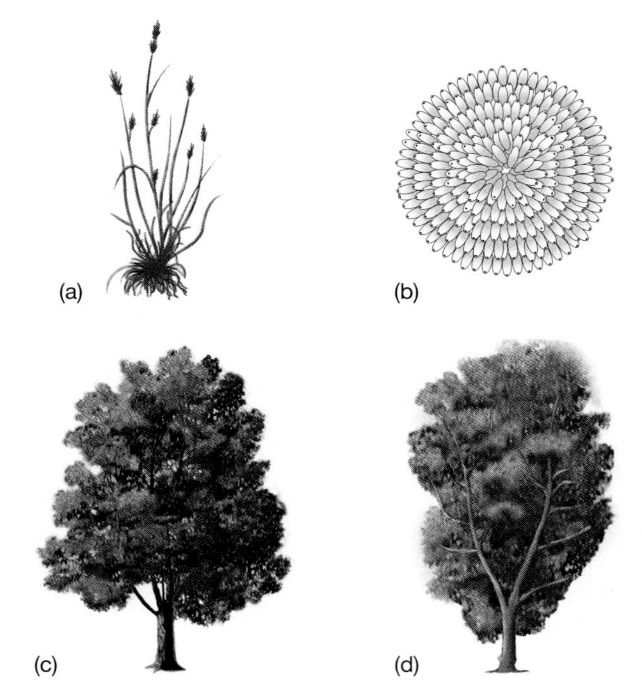

Figure 32.16

Some other plants and animals provide examples of modular organization. *(a)* Fescue grass grows from centers as a clump of stems and flowers, and *(b)* bryozoans are animals that grow in colonies, with each one developing asexually as a bud from an existing animal. *(c)* An oak tree and *(d)* a sea pen (a cnidarian, a type of primitive animal) both grow through multiple branching of small units.

In respiration, the reverse occurs:

$$6\,O_2 + C_6H_{12}O_6 \rightarrow 6\,CO_2 + 6\,H_2O$$

Since plants photosynthesize, it is tempting to think that only animals respire. That's wrong. Plant tissues also respire, even if they are also photosynthetic. Nonphotosynthetic parts, such as roots, can only get their energy by respiration. In balance, however, plants consume more CO_2 and water through photosynthesis than they produce through respiration, and overall they produce excess O_2. Animal metabolism consumes O_2 and produces excess CO_2 and water.

On the whole, water is a requirement to be supplied for plants and a waste to be eliminated by animals—yet a most precious waste for terrestrial animals, because they are continually challenged to get enough water.

A difference in nitrogen metabolism between plants and animals parallels a difference in the use of water, CO_2, and O_2. Plants, being autotrophs, synthesize their amino acids and other nitrogenous compounds, using ammonia or nitrate in the surrounding soil and water. (Amino acids and other nutrients are abundant in good soil, largely as a result of microbial activity, and plants can take up many of these substances, but this is not a significant factor in their overall nitrogen balance.) Animals, in contrast, get too much nitrogen by eating other organisms made largely of protein, and generally they have to excrete the excess amino groups. It can be hard for plants to obtain sufficient nitrogen (see Section 40.8) and hard for animals to eliminate it (see Chapter 46).

32.7 Plant and animal cells exchange CO_2 and O_2 with their surroundings.

The cells of both plants and animals must exchange gases with their surroundings, taking in CO_2 and eliminating oxygen for photosynthesis, and doing the opposite for cellular respiration. Small plants and animals can exchange gases entirely by diffusion alone, but diffusion is too slow to meet the metabolic needs of larger organisms. In many animals, exchanging gases between internal tissues and external surfaces is a principal function of the circulatory system, which we will describe later.

The gas-exchange systems of vascular plants show a remarkable parallel with those of arthropods such as spiders, millipedes, and insects. These animals have a system of fine, air-filled tubules called **tracheae** (sing., **trachea**) branching into still finer tubules between their cells. This arrangement brings internal tissues into quite direct contact with the atmosphere (Figure 32.17*a*). In vascular plants, an extensive **intercellular gas space** system of air-filled channels runs through tissues (Figure 32.17*b*), allowing oxygen and CO_2 to flow freely enough to support cellular respiration and photosynthesis. Gases exchange with the atmosphere at the boundaries of this space through **stomata,** openings in the leaves formed by a pair of guard cells:

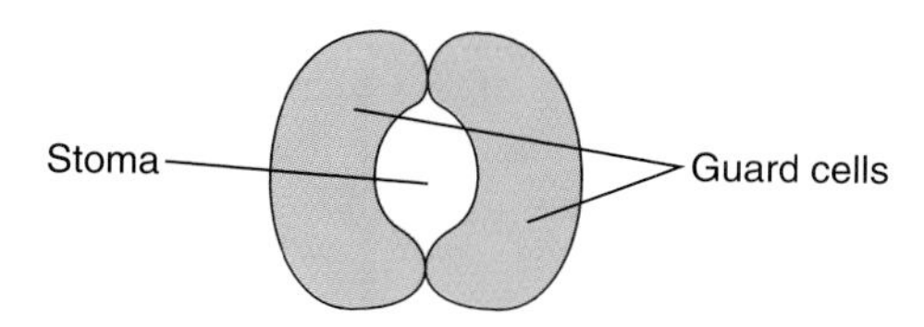

Guard cells can close off the stomata, and the openings of tracheae can also be closed. Both systems contain some water, which partially carries dissolved gases; however, excess water can block the system, and this is why the roots of some plants die when they are submerged.

32.8 Plants and animals have two fluid compartments.

Tissues, as seen in micrographs, look like masses of neat little cells fitted together like bricks. But the pictures are misleading. The cells themselves don't occupy all the space in any tissue. They are bathed in an **extracellular fluid** that has a different composition from their **intracellular fluid,** or cytosol, although both are mostly water containing many proteins, ions, nutrients, and metabolites. As we would expect, these two fluid compartments differ significantly in plants and animals.

In animal tissues, the extracellular fluid that occupies the spaces between cells is **interstitial fluid** (Figure 32.18). As the nineteenth-century physiologist Claude Bernard stressed, the interstitial fluid is a *milieu interieur,* an *internal environment* where all the cells live. A cell is most intimately in contact with this fluid—not with the blood or with the outside of the body. Materials move into and out of cells via the interstitial fluid, and much of an animal's regulatory activity revolves around maintaining the composition of this fluid. Interstitial fluid is distinct from blood in vertebrates, but many invertebrates have only a single extracellular fluid, which serves as both interstitial fluid and blood.

In vascular plants, large blocks of cells share a common intracellular space because they are interconnected through plasmodesmata (see Section 6.9). Neighboring cells can therefore exchange materials quite freely, without intervening

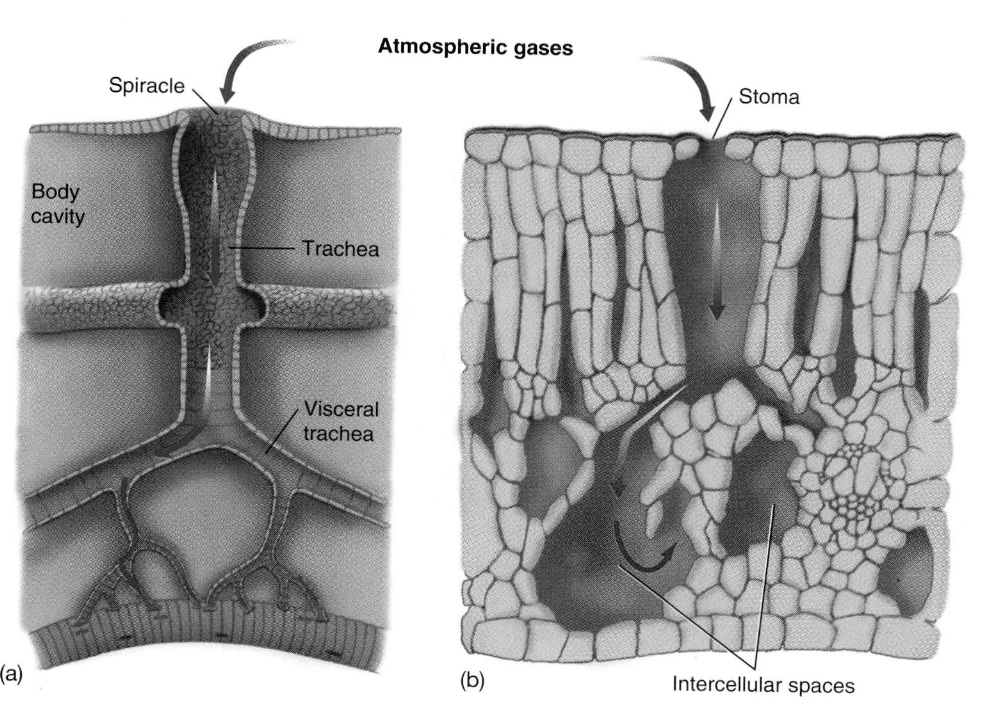

Figure 32.17
Insects and plants exchange gases between cells and the atmosphere in comparable ways. *(a)* Insects have tracheae, systems of fine tubules running deep into their tissues. *(b)* Plants have an intercellular gas space between their cells, sometimes occupying about as much space as the cells themselves.

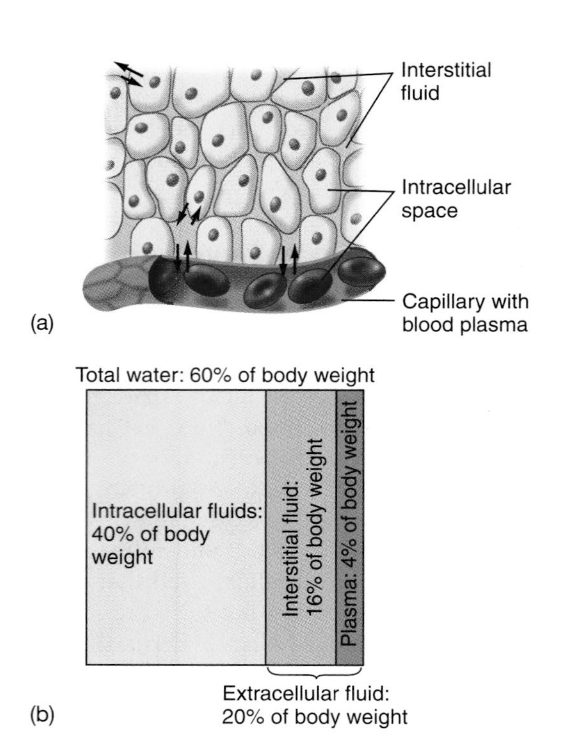

Figure 32.18

All plants and animals have at least two fluid compartments: intracellular fluid (cytosol) and extracellular fluid. *(a)* In vertebrates and many other animals, the extracellular fluid is divided into interstitial fluid bathing the cells and a distinct blood plasma confined to blood vessels, but many materials exchange freely between these two compartments. *(b)* The block shows the relative amounts of water in the three compartments of a human. Arthropods and some other invertebrates have only a single extracellular fluid that serves as both interstitial fluid and blood.

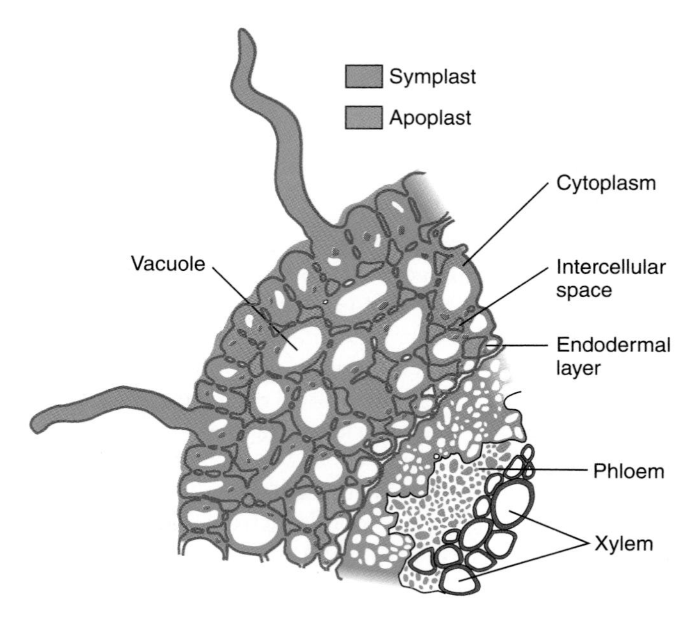

Figure 32.19

A cross section of a plant root showing the apoplast and the symplast. The apoplast consists of the cell walls and interstitial spaces through which water can move; notice that it is interrupted at the endodermal layer. The symplast consists of many cells with a common cytoplasm connected by plasmodesmata. Source: W. A. Jensen and L. G. Karalijian (Eds.), *Plant Biology Today, Advances and Challenges,* 1966, Wadsworth Publishing.

membranes. The cytoplasm of these interconnected cells comprises the **symplast** of the plant (Figure 32.19). The extracellular space, or **apoplast,** consists of the intercellular space and the fibrous walls of all the cells. Water moves freely through the apoplast, just as it soaks through a paper towel. Water enters both the apoplast and symplast of roots, which are in intimate contact with the water in the surrounding soil; it then moves through the symplast alone and into the rest of the plant.

32.9 Plants and animals face comparable problems of water balance.

Water is so integral to the structure and operation of organisms that water balance will be a continuing theme throughout our examination of both plant and animal physiology. Here we will only outline some general principles regarding water balance, leaving most of the specifics to the discussions of regulation in plants (Chapter 38) and in animals (Chapter 45). Multicellular organisms encountered several problems in their adaptation to different habitats—freshwater, marine, or terrestrial. Each habitat presents its peculiar challenges, and each one may have either quite constant or extremely variable conditions. Water may be scarce for two reasons: Either it simply does not exist in a particular habitat (aridity), or it is mixed with salts. So organisms face three kinds of problems: having enough water, having the right amounts of critical ions, and having an osmotic balance of water and ions (Figure 32.20).

We humans must have sufficient water all the time, or we will die. In other words, we maintain the *hydration* of our tissues within narrow limits, and so we are **homeohydric** organisms. Many other organisms, however, are **poikilohydric:** Their water content varies with environmental conditions. Remember that, according to the cell theory, a multicellular organism's activities are largely the activities of its individual cells; the organism's health depends upon having these cells sufficiently hydrated to carry out normal metabolism, but this sensitivity seems to vary enormously. Small changes in the volume of extracellular fluid can cause great damage in mammals, especially to the nervous system. Poikilohydric organisms, however, can withstand a certain degree of desiccation and still function normally when they are rehydrated.

The simplest organisms have little ability to regulate their hydration. Their only response to severe changes may be to go into a protected form, such as a spore that is resistant to dehydration. Some organisms endure hard times in a **cryptobiotic** state, a severely dehydrated condition in which they carry out little or no metabolism. Thus some mosses, ferns, and invertebrates can dry out and then "come alive," recovering their normal metabolism when hydrated again. Seeds are a cryptobiotic state for many plants, a state that is especially valuable for those adapted to extremely arid environments; their seeds lie dormant in the

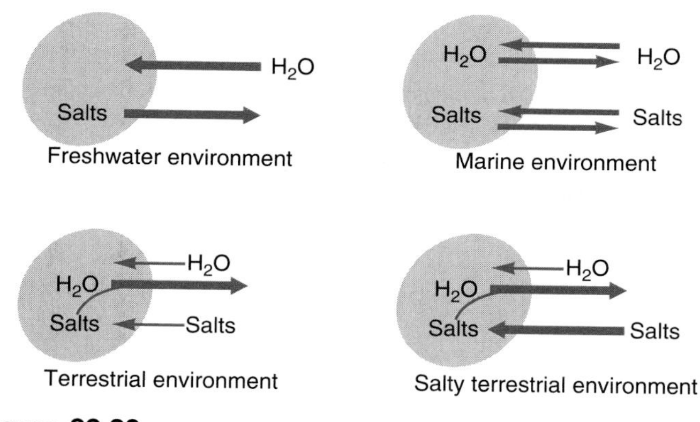

Figure 32.20

Organisms *(ellipses)* living in different habitats meet distinct challenges to their balance of water and salts. Arrows show the tendencies of water and salts to move in and out of the body; many organisms have special adaptations to prevent excessive movements of water or salts in one direction.

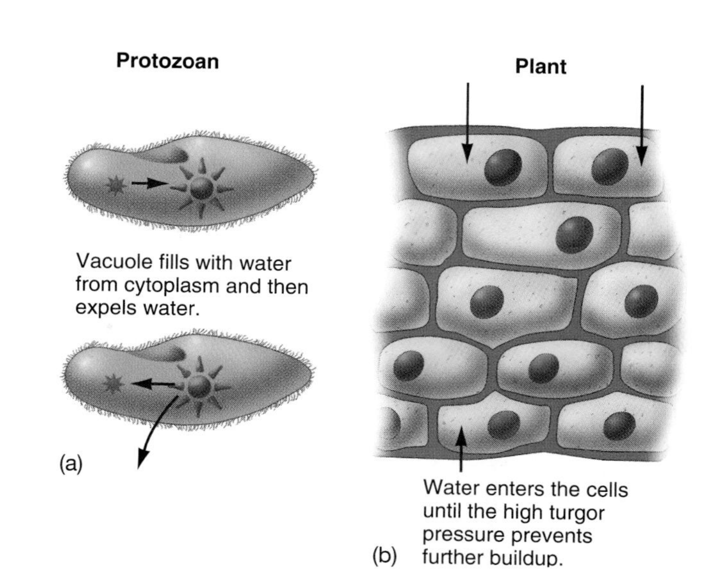

Figure 32.21

Cytoplasm, containing ions and dissolved organic molecules, is hypertonic to freshwater. Therefore, water always diffuses into the cells of freshwater organisms by osmosis. *(a)* Protozoa excrete excess water through contractile vacuoles. Animals have organs, such as kidneys, that remove excess water, along with wastes. *(b)* Algae and plant cells resist osmosis due to their strong cell walls, so they build up an internal turgor pressure that prevents further net diffusion.

soil until a brief wet period comes along, and then grow quickly into new plants that produce another generation of seeds.

Sporulation in bacteria, Section 15.6.
Sporulation in algae, Section 15.3.

Aquatic organisms may be adapted to salt water or to fresh water. We can trace the ways plants and animals fundamentally deal with water by looking back at their ancestry. The cytoplasm of all cells is approximately isotonic to seawater, though perhaps water with a lower concentration of salts than the modern oceans, indicating that early evolution occurred in such an environment. Some protists and chromists remain adapted to a marine environment, where they exchange water and ions with little or no regulatory problems. Some organisms then moved into freshwater habitats. But cytoplasm is hypertonic to freshwater, and therefore water constantly tends to diffuse into the cells of freshwater organisms by osmosis. Basal eucaryotes evolved two different ways to resist this influx and regulate their water content (Figure 32.21). The cells of algae are surrounded by strong walls, which allow a high pressure, **turgor pressure,** to build up inside until the cell comes to an equilibrium because the internal pressure stops the influx of more water. A cell with a significant turgor pressure is said to be **turgid**—bloated and stiff, with a definite shape, like a well-inflated tire or balloon. Green algae, in particular, are mostly freshwater organisms, and the more complex plants that evolved from them retained cell walls as a feature that figures importantly in their water relationships. It is the loss of water—hence the loss of turgor pressure and turgidity—that makes a plant *wilt,* becoming flabby and shapeless. A crisp plant is full of water, turgid, and healthy; a wilting plant is suffering from water stress to some degree, and its metabolism is correspondingly disrupted. Tolerance to this condition varies among poikilohydrics, but the plant may be getting dangerously close to death.

In contrast to the algae, protozoa living in freshwater evolved contractile vacuoles, also called water-expulsion vesicles, which accumulate and then flush out excess water. Much later, freshwater animals evolved still more complex organs such as kidneys for actively removing excess water from their body fluids, as we show in Chapter 46. However, to think realistically about the water relationships of animals, we must recognize that animals are "sea beasts," in the words of J.B.S. Haldane. The first animals evolved in seawater, and all animals owe their characteristics partly to their marine origins. Their cytoplasm, with its crucial concentration of various ions, is commonly isotonic to salt water. Many marine animals do not regulate the osmolarity of their extracellular fluid; they are in osmotic equilibrium with their surroundings and are said to be **osmoconformers.** However, any severe change in their extracellular fluid would create a problem of osmotic balance between their intra- and extracellular environments, so osmoconformers are generally confined to habitats where the ionic concentration remains quite constant, such as open seashores. Some animals, called **osmoregulators,** are able to regulate their extracellular fluids, at least to a degree, and they can live in changeable environments, such as estuaries.

As long as plants and animals remained aquatic, they had plenty of water—sometimes too much. However, those that invaded terrestrial environments faced the challenge of obtaining and retaining enough water. Terrestrial animals and plants easily lose water to the environment through evaporation, especially at surfaces where O_2 and CO_2 are exchanged with the atmosphere. This continuous loss of water vapor from an organism is called **transpiration.** (Transpiration is generally considered a plant process, but the term really covers the loss of water by evaporation from all organisms.) The loss can be enormous: A plant may transpire 99 percent of the water taken up by its roots. Balancing gas exchange with transpirational loss may be a problem for every terrestrial organism. Terrestrial plants obtain water through their roots; animals obtain theirs by drinking and eating. As terrestrial plants and animals lose water, however, they may also be losing salt. This problem has challenged plants to evolve special mechanisms for ionic regulation, while animals have adapted the excretory organs of aquatic animals for striking the required balance of water and salt retention, while eliminating wastes.

Plants and animals are particularly challenged by extremely salty conditions; they can adapt either by resisting the salt or living with it. One form of resistance is to evolve mechanisms to keep the salt out; the roots of some plants are adapted for excluding most salt. A second strategy is to allow salt to enter freely and evolve ways to excrete it. Halophilic plants—those adapted to high-salt conditions—may have glands in their leaves that constantly excrete concentrated salt solutions; marine animals such as reptiles and birds do the same with special glands. Or a plant may accumulate the salt into its leaves and then shed the leaves. A third strategy common to plants and animals is to let the salt enter their extracellular fluid, maintain good ion pumps to keep the salt ions out of their cells, and then increase the osmolarity of the cytosol with materials called organic osmolytes to prevent massive osmosis. Many marine algae and plants fill their cytosol with polyols, compounds rich in hydroxyl groups such as unusual sugars or sugars combined with glycerol; they are synthesized in direct proportion to the amount of salt in the external medium. Marine animals commonly use urea or trimethylamine oxide as osmolytes. Living with a lot of salt has been studied particularly in halophilic bacteria, which are unicellular organisms. Their proteins operate only in high salt concentrations and are rich in the acidic amino acids glutamate and aspartate, which bind Na^+ and K^+ ions. Their ribosomes and membranes are only stable in 3–4 M salt solutions, whereas most plants and animals have evolved proteins that operate in about 0.1 M salt solutions. These bacteria have adjusted their cellular organization to salt, and some plants and animals do the same. So the general rule is that halophilic plants do not exclude salt but have evolved to live with it internally.

In all our thinking about water relationships, we must recognize that water regulation is governed by an important general principle: *Water cannot be moved by itself, because cells have no water pumps.* Instead, cells transport ions across their membranes with ion pumps—specifically, Na^+ pumps or K^+ pumps—and water follows the ions passively, by osmosis, since cell membranes are permeable to water (Figure 32.22). This principle applies to many situations. Plants often change shape by means of K^+ pumps in motor cells; when K^+ ions are pumped inward, water follows, and the cells become turgid; when cells pump K^+ ions outward, they lose water and shrink slightly. Animals employ the principle to regulate their water and ion balances by having Na^+ and Cl^- pumps in their kidneys or other organs that move ions actively to regulate the passive movement of water. As we will show in specific instances, this principle is critical for understanding how plants and animals maintain a balance of water and salts.

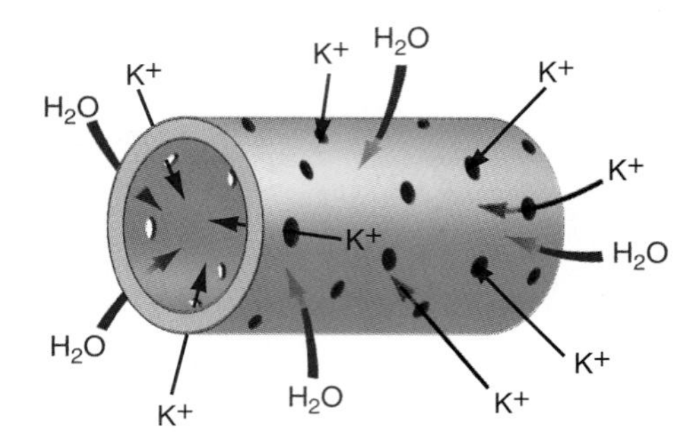

Figure 32.22

Ion pumps are used to change the osmotic balance of two compartments; here K^+-ion pumps move K^+ ions into a tubule. Then, since the barrier between the compartments is permeable to water, water follows the ions passively.

32.10 Plants and animals have vascular systems for distributing water and nutrients.

The most primitive plants are algae. Living only in water or in moist places on land, they exchange materials directly with their surroundings by diffusion. Mosses, some of the first plants to evolve into terrestrial niches, evolved from algae and also tend to occupy moist habitats where they can exchange materials much as algae do. For the most part, mosses are made of undifferentiated parenchymal cells, which pass water and the materials dissolved in it directly from one to another, although some mosses have simple vessels.

Most terrestrial plants, however, grow an extensive shoot system upward where photosynthesizing cells capture sunlight and form organic nutrients. Meanwhile, the root system descends into a moist environment where it can absorb water. Thus, a green plant is rather like a small ecosystem in itself, for it is built of cells that produce food for themselves and for the consuming cells elsewhere. The vascular system that communicates between the producing and consuming portions of the system consists of two kinds of tissue: **xylem** to carry water and dissolved minerals upward from the roots and **phloem** to carry sugars and other organic products of photosynthesis from their sources in the leaves to places where they are stored or used, in roots and rapidly growing regions (Figure 32.23a). The conducting elements of xylem are part of the apoplast, and those of phloem are part of the symplast.

The vascular system has allowed plants to adapt to terrestrial environments. Xylem tissue, with its thick-walled cells, conducts water to the tips of a plant and also provides a supporting skeleton. That support is necessary because, as plants grow heavier and reach to greater heights, turgor pressure and collenchymal cells cannot support them. (Mass isn't a problem for aquatic plants, even the largest, because the surrounding water supports them.) In larger terrestrial plants like trees, the extensive xylem (wood) allows growth to great heights; redwoods, for instance, grow to over 100 meters.

An animal, too, must keep its cells well supplied with water to carry nutrients and wash wastes away. The cells of small aquatic animals, like those of small plants, are served by diffusion alone, but once animals get past a critical size or become terrestrial, they must also have a vascular system. Whereas the water in a plant flows in one end and out the other, an animal continually recirculates its water, commonly with the help of a muscular, contractile *heart* that pumps the fluid inside a set of tubes (Figure 32.23*b*). An animal's vascular system is therefore called a **cardiovascular system** or **circulatory system** (Greek, *kardia* = heart). All vertebrates have a *closed* circulatory system in which the circulating fluid (blood) is confined to vessels, separate from the interstitial fluid that surrounds cells. In contrast, the vast majority of animal species—arthropods and molluscs—have an *open* circulatory system with only a single extracellular fluid that bathes the tissues and is also circulated in some confined vessels. Of course, all animals are open systems in a broader sense, and they continue to exchange water and other materials with the environment.

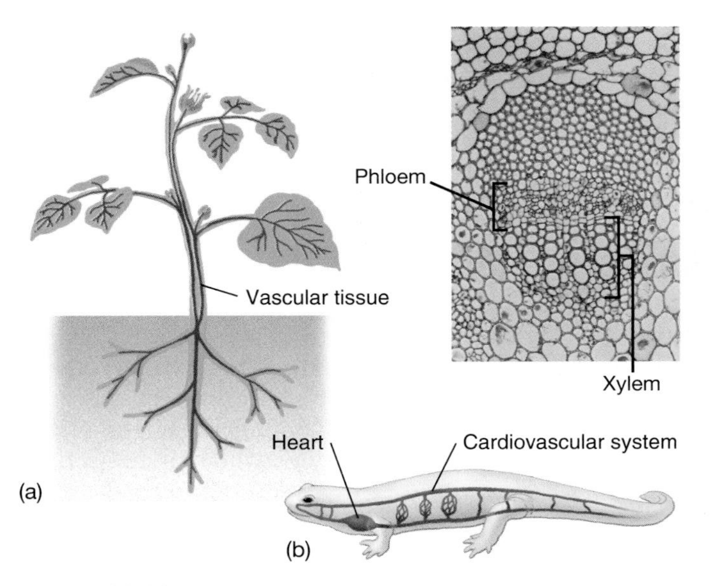

Figure 32.23

(a) A plant's vascular system consists of fine tubules—xylem that transports water from the roots upward and phloem that conducts nutrient-rich sap throughout the plant. *(b)* An animal's cardiovascular system consists of tubules (arteries and veins) that carry blood, with a heart to circulate the fluid in a circular path, either through a completely closed system or an open system.

32.11 Reproduction in terrestrial organisms has generally become less dependent on water.

Sexual reproduction began in aquatic environments and evolved into the common scenario of free-swimming sperm that must find the eggs of their species; the eggs may be released or held inside a receptacle (Figure 32.24). The earliest-evolved terrestrial plants also reproduce by means of motile sperm, but as plants became more independent of the water, they evolved ways to give up this method. Terrestrial plants did not achieve

Figure 32.24

(a) Aquatic animals such as sponges sometimes release vast numbers of sperm into the water; they must swim to the eggs of their species, and few will actually fertilize an egg. *(b)* Some plants release clouds of pollen, which is carried by the wind, independently of water; few pollen grains will ever engage in fertilization.

reproductive independence from water until at least 100 million years after the first land plants evolved; they did so by evolving remarkable reproductive devices, *pollen* and *seeds*, a story we will take up in Chapter 33.

As the most dominant animals have become independent of a watery environment, they have generally replaced external fertilization with internal fertilization, where sperm only have to swim a short distance in the fluids of the female reproductive tract to reach eggs. Insects, perhaps the most successful contemporary animals, use internal fertilization before laying the resulting zygotes outside. The reproductive success of reptiles and birds was linked to the development of eggs with a tough protective shell where the developing embryo lives on a supply of food, surrounded by water-filled membranes. In most mammals, the embryo is protected and nourished inside its mother's body.

C. Considerations of size

Organisms of different overall sizes lead different kinds of lives and are subject to different limitations. Biologists have documented many regular differences correlated with size, although the differences may be hard to explain. Small organisms, for instance, live faster and reproduce faster than large ones (Figure 32.25). Among birds and mammals, we find that the smaller animals have rapid heartbeats and the larger ones slower heartbeats; they live as if by a rule that each one gets only a certain number of beats in a lifetime. But there are no general rules that explain all these relationships.

On the graph in Figure 32.25, the logarithm of one variable is plotted against the logarithm of another, and the points fall along a line. A general equation for such a line, when the log of y is graphed against the log of x, is $\log y = \log k + a \log x$, and the slope of the line is a. Another form of this equation is $y = kx^a$. Any such relationship is said to be *allometric.* The number of allometric relationships between size and some other factor is remarkable, and we will discuss some of them in the next section, while generally avoiding equations that describe them.

32.12 Surface/volume ratios determine many architectural features of organisms.

Biological forms are shaped by elementary physical principles that are not unique to biology, and one of the most important is a purely geometrical principle: The surface/volume ratio of a structure decreases as the structure becomes larger. This principle constrains the form of structures wherever the activity

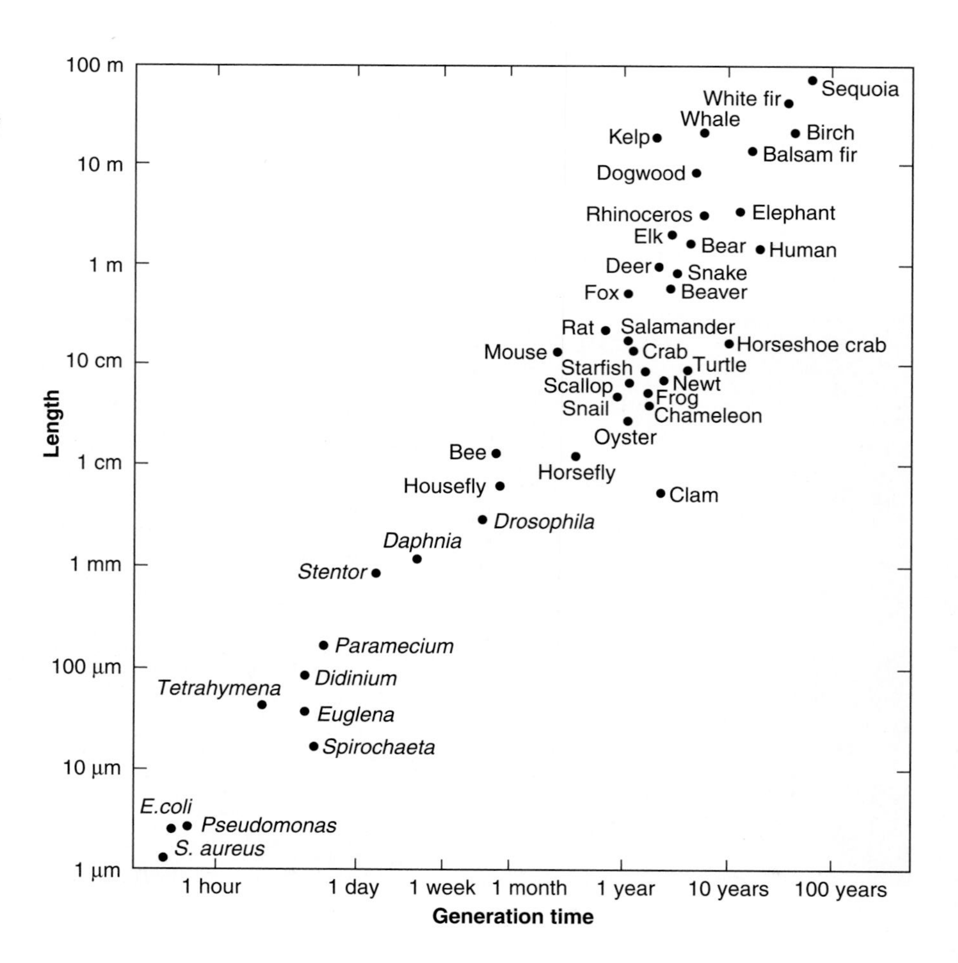

Figure 32.25
An organism's length is closely correlated with its *generation time:* the average time between the birth of an individual and the birth of its offspring or, for unicellular organisms, the time required for the cell to double its size and divide. Notice that both scales are logarithmic.

within a volume is limited by the transport of materials through the surface, including the small size of cells (see Section 6.3).

The rate at which a substance diffuses between two points is described by Fick's Law:

$$Q = DA(C_1 - C_2)/L$$

Q is the rate of diffusion. D is the coefficient of diffusion, which is a constant for each substance at a given temperature and medium; it is much higher for the volatile components of onions and garlic than for the oils you saute them in, so the appetizing odors of cooking quickly fill a house. D is also much higher in air than in water, and this fact alone has important implications for the abilities of terrestrial or aquatic organisms to obtain the nutrients they need. A is the cross-sectional area through which the substance is diffusing. C_1 and C_2 are the concentrations of the substance at the two points, and L is the distance between them. Therefore $(C_1 - C_2)L$ is a concentration gradient. If, for example, we use arbitrary concentrations of 10 units and 1 unit, diffusion will be 10 times slower if the difference of 9 units is spread over a distance of 10 cm than if it is concentrated in 1 cm.

Fick's Law gives us a foundation for thinking about biological structures that involve transport of materials. Rapid transport of a substance means giving it the largest possible surface area (A) to move through and the smallest possible distance (L) to diffuse through—in other words, reducing the volume through which it must move before reaching the surface. So wherever a process is occurring within a volume and it depends on a flow of material through a nearby area, the surface/volume ratio must be high enough for the flow to support the process. Consider, for instance, how a vascular system supplies nutrients to a tissue and removes its wastes. A single large tube clearly won't do the job. Although it can handle a large volume, its surface/volume ratio is so small that very little material can diffuse in and out of the fluid inside as it passes through the tissue (Figure 32.26). However, a vascular system divided into many small vessels can carry the same volume while providing a much larger surface in intimate contact with the region it serves, so materials can diffuse easily across the boundaries (Figure 32.26*b*).

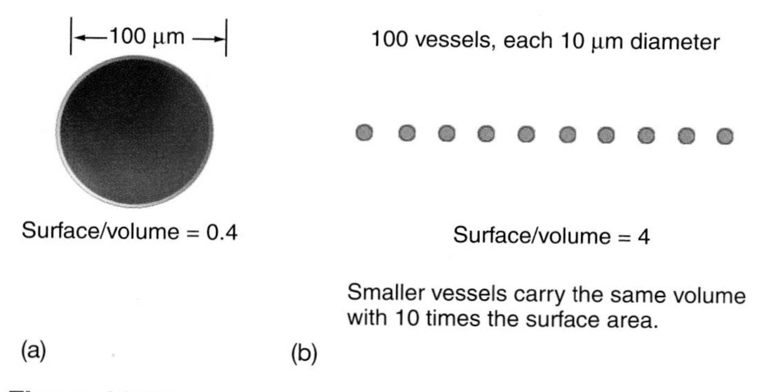

Figure 32.26

(a) A single large vessel, even with thin, permeable walls, could not supply materials to a tissue. *(b)* If the same volume of blood is carried by many small vessels with a high surface/volume ratio, materials can diffuse rapidly across their surfaces.

An animal's cardiovascular system is an example of a *branching* structure that subdivides into smaller and smaller branches to make vessels with a high surface/volume ratio. Plant roots are another familiar example; to absorb all the water and nutrients they need, plants develop extensive branching systems of fine roots. The surfaces of the smallest roots even project minute root hairs, further expanding the effective surface area of each cell. Figure 32.27 shows these and other adaptations dictated by surface/volume considerations.

Surface/volume considerations continually influence the larger shapes of organisms. Think, for instance, of the forms of plants, like succulents and cacti, that are adapted to very dry conditions. Instead of thin leaves with huge surfaces, they have barrel-like bodies with no leaves or with thick, knobby leaves, which maintain a low surface/volume ratio and minimize evaporation. We will show shortly how surface/volume factors in temperature regulation also influence the sizes and forms of animals.

Whenever we see a highly branched or convoluted biological structure, or one that is cut by small channels, we can be quite sure that its form has evolved because of a strong selective pressure to create a high surface/volume ratio. Be prepared to find such structures in many biological forms.

Exercise 32.1 The diffusion constant of a certain dye is 10^{-5} cm^2/sec, and it is diffusing between two points where its initial concentrations are 1 g/L and 0.1 g/L. Determine its diffusion rate in the following situations: *(a)* The dye can diffuse through an area of 1 cm^2 or 0.1 cm^2. *(b)* The distance between the two points can be 1 cm or 5 cm.

32.13 Biological architecture is often determined by mass/area restrictions.

The supporting materials of many plants and animals—wood, chitin, and bone—are among the strongest materials known, but every material has its breaking point. This fact dictates the relative sizes of many biological structures. An object's strength increases with its cross-sectional area, so if a concrete column 100 cm^2 in cross section can safely support a ton of material, a load of 10 tons demands a support with a cross section of $10 \times 100 = 1{,}000$ cm^2. But as an object increases in size, the cross-sectional area of its supports increases with the square of a dimension, and its volume—hence its mass—increases with the cube of a dimension. So a larger organism imposes a proportionally greater load on its supports. A small organism can therefore be organized quite differently from a large one because its dimensions impose different requirements.

Consider, for example, a mammal's skeletal supports (Figure 32.28). Its mass is borne on leg bones of a certain cross

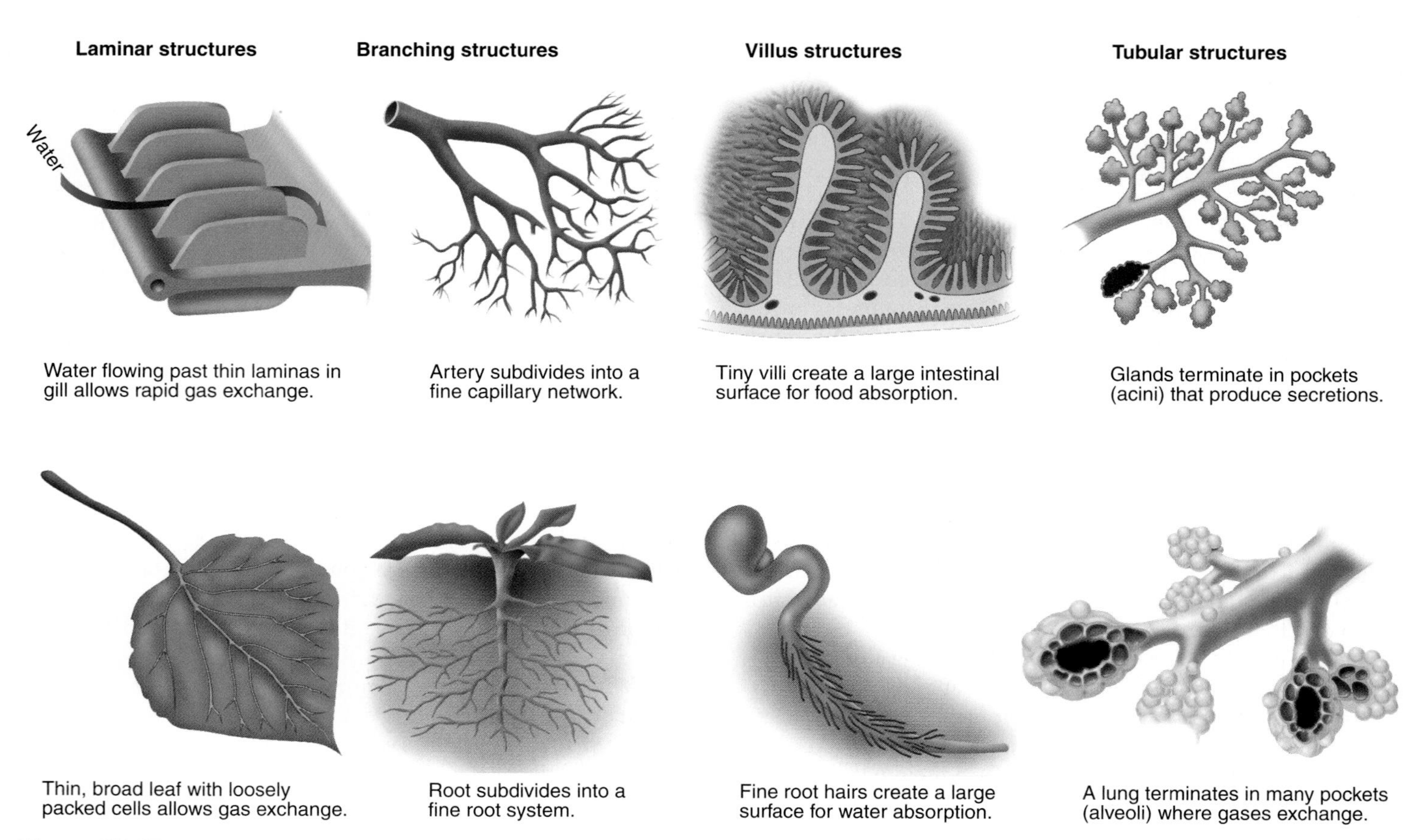

Figure 32.27

Various features create large surface areas across which materials can be exchanged.

section. A mouse can have delicate bones because they have to bear so little weight. An antelope bears its weight on hardy but still delicately shaped bones, and it can afford the luxury of a long neck to hold its head high and extend its sight range. A horse must be built more stoutly, with a shorter, heavier neck. And an elephant, which has to walk on massive legs to support its great weight, can hardly afford any neck at all. The heavier animals have relatively shorter and thicker legs. A giraffe is probably near the upper limit of height for an animal walking on stilt legs, for if it were much longer, the mass of its body would exceed the strength of its leg bones, and they would be crushed every time the animal took a step.

The same considerations constrain an animal's movements. The strength of a muscle is determined by its cross-sectional area, but the mass it pulls on increases with the cube of a dimension. A grasshopper can jump to a height of about a meter, nearly 100 times its own height. In a running high jump, a human can jump only to about his or her own height. Elephants don't jump.

The sizes and forms of plants are determined in much the same way. Herbaceous plants remain small and fragile, developing only minimal support from collenchyma and sclerenchyma. The limbs of a tree must grow thick to support the weight of their branches and foliage and perhaps a mass of fruit. Those that bear a lot of fruit have often evolved very flexible woods that bend under all that weight, for otherwise the branches could not grow strong enough without becoming a great deal thicker. The heights of trees must also be limited by the sheer mass of matter that must be borne at the base of the trunk. The huge redwoods may have reached the limits of height, for even though they also grow to great diameters, a taller tree might crush its base under its own weight. (They may also be limited by the height to which water can be drawn.)

Example 32.1 An animal stands on bones that are 1 cm in diameter at their narrowest point. If the animal's mass were doubled, how large would its bones have to be for the same proportional strength?

The force exerted on the bones is doubled, but the strength of the bone increases with its cross-sectional area, which depends on the square of the radius. So the radius of the bone should increase by the square root of 2, or 1.414 times.

Exercise 32.2 An animal stands on bones that are 1 cm in diameter at their narrowest point. If the animal's linear dimensions were doubled, how large would its bones have to be for the same proportional strength?

Exercise 32.3 In her classic paper about the creatures described in *Gulliver's Travels* (see Additional Reading), Florence Moog discusses several problems of transforming the human body from a height of 6 feet to Lilliputian height of 6 inches. Consider just two of these, assuming that all of the Lilliputians' linear dimensions are 1/12 those of a normal human. *(a)* The pitch of the voice varies with the square of the linear dimensions of the vocal cords. If the human voice is centered at middle C, 256 cycles/sec, could Gulliver have heard the Lilliputians speak, assuming the upper range of his hearing was 15,000 cycles/sec? *(b)* The cerebral cortex of the human brain contains about 14 billion cells; their complex interconnections are responsible for human intelligence. Allow the Lilliputians to have nerve cells only a quarter the size of ours and determine how many cortex cells they had. Comment on their probable intelligence.

32.14 An organism's metabolic rate is closely related to its size.

Every organism is adapted to living within a certain temperature range, a range in which its enzymes and other components are adapted to work best. This range may be very broad for some organisms and quite narrow for others, but there are always limits beyond which the organism grows poorly and may die. Aquatic organisms are generally not subjected to severe changes in temperature, since water changes temperature slowly. Some terrestrial organisms, however, live in habitats with extreme shifts in temperature, both daily and yearly, and they have had to evolve adaptations to meet these challenges. Terrestrial animals may migrate away from the cold, or they may pass the cold season in an egg or larval form; plants often go into dormant states during very cold (or dry) seasons.

Plants and most animals are **poikilothermic** (commonly said to be "cold-blooded") since their temperature changes with the temperature of the environment. Because they obtain most of their heat from outside—from the sun and their surroundings—they are said to be **ectothermic.** Poikilotherms regulate their temperature largely by their behavior—for example, by selecting suitable sunny places and exposing themselves to the sun. Plants may open and close their leaves and flowers on a daily cycle that matches the cycle of temperature changes. Opening the leaves exposes them to the sun, so they may absorb more heat and light; a butterfly does the same thing when it spreads its wings (Figure 32.29). Poikilothermic animals may be quite sluggish on a cold morning, but once warmed, they can move around actively and keep their temperatures within a suitable, narrow range.

Plants have the same kinds of temperature-regulation challenges as animals, but their mechanisms for regulating temperature are closely tied to the exchange of oxygen, carbon dioxide, and water vapor through their leaves. This situation is complex, and we discuss it in Section 38.7.

In some terrestrial vertebrates, adaptation to temperature takes on a new dimension. All organisms constantly produce heat metabolically. Birds and mammals are **endothermic** animals that use metabolic heat to maintain a relatively constant body temperature—in other words, to be **homeothermic** ("warm-blooded"). Homeotherms metabolize 3 to 10 times faster than poikilotherms of the same size, and they have extra insulation, including fat layers and feathers or hair, which conserves heat; their internal thermostats monitor the blood temperature and initiate various mechanisms to keep it constant. For instance, if a mammal's temperature begins to rise, it may start to pant or sweat, since the evaporation of water cools it. If its temperature falls, it is induced to generate heat by metabolizing faster and moving faster; shivering is a special heat-generating activity. Furthermore, the thermostat controls the overall metabolic rate of body tissues through hormonal

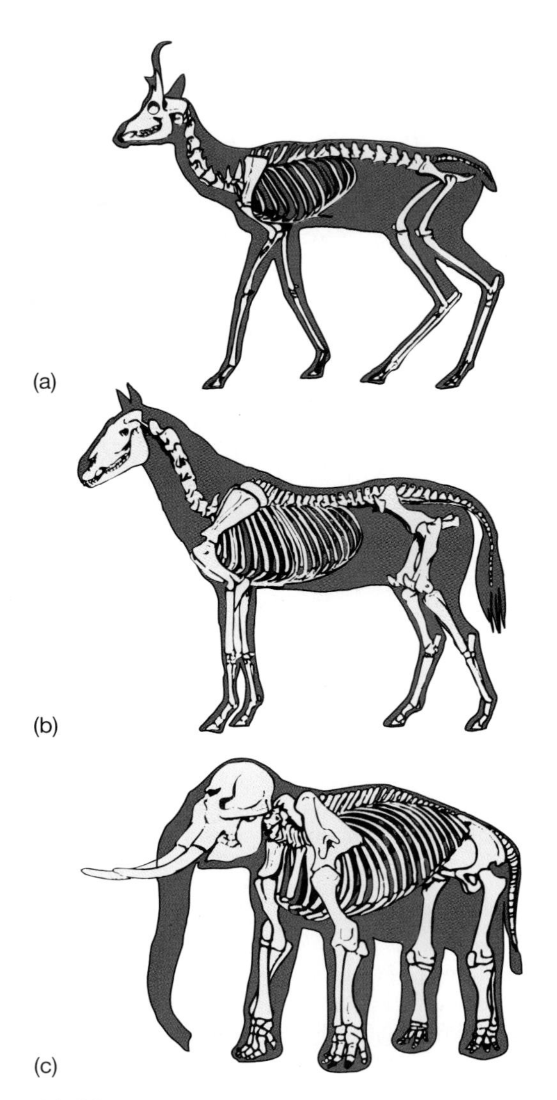

Figure 32.28
Three vertebrates drawn to identical heights show the sizes of bone needed to support their mass. The bones of the antelope *(a)* are rather delicate compared to the massive bones of the elephant *(c)*. The horse *(b)* has bones of intermediate size. The strength of a bone is proportional to its cross-sectional area.

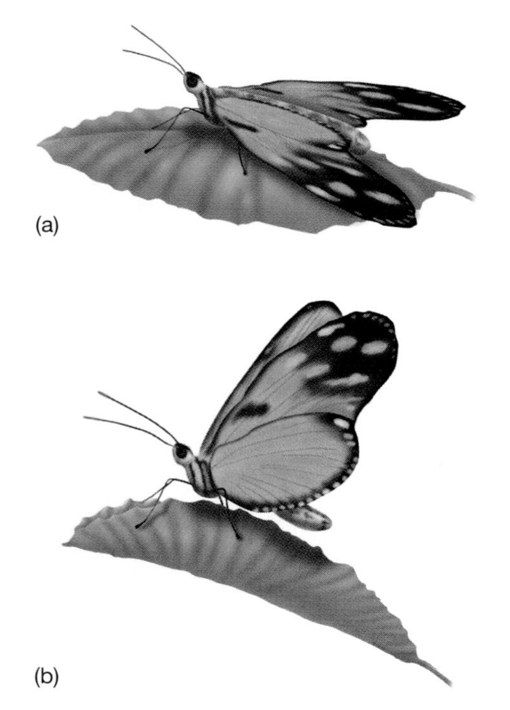

Figure 32.29

(a) A butterfly can increase its temperature by spreading its wings to make a large surface that absorbs sunlight, or *(b)* it can cool off by folding its wings. Plants that open and close their leaves achieve the same effect.

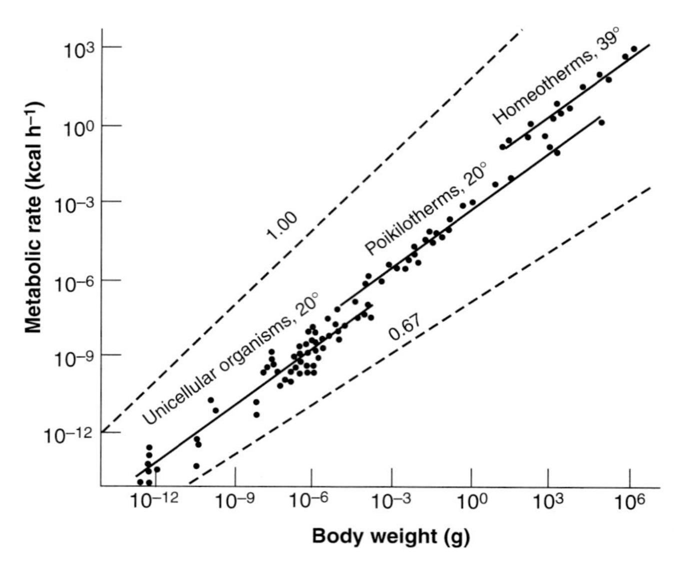

Figure 32.30

On logarithmic scales, the metabolic rates of different organisms are directly proportional to their masses. The slopes of the lines vary from about 0.65 to 0.80 and correspond to the exponent 0.75 in $m = kw^{0.75}$.

mechanisms. Mammals often have a special brown adipose tissue (brown because it is rich in mitochondria containing cytochrome oxidase) that metabolizes stored fat solely to release heat, without forming ATP.

Some organisms can reduce their activities to very low rates without apparently causing harm, at least for a while. For instance, many bacterial cultures are routinely stored for months in refrigerators where the cells remain viable but do not grow. After warming, their metabolic rates increase. The rate of a chemical process, which is all metabolism is, tends to increase by the same factor for all equal increments of temperature, and the ratio of its rate at one temperature compared to a temperature 10°C lower is called its Q_{10}. The Q_{10} of metabolism is about 2, so every increase of 10°C approximately doubles the metabolic rate. If an organism is metabolizing at a low rate m at 5°C, its metabolic rate will be about $2m$ at 15°C, $4m$ at 25°C, and $8m$ at 35°C. This is why a snake or a lizard, a poikilotherm that can hardly move when it is very cold, will acquire such marvelous vitality after basking in the sun.

Homeothermic animals have a **basal metabolic rate (BMR),** the minimum rate of metabolism needed to keep the animal alive, with its heart beating and its blood circulating, but not doing any extra work. The BMR is commonly measured on a resting, temporarily fasting animal. The average BMRs for humans are 1,800 kcal/day for a man and 1,700 kcal/day for a woman. Complex formulas can estimate the human BMR more precisely on the basis of such variables as age and stature. The BMR sets a lower limit on the amount of food a person should eat daily to maintain a constant body weight, but of course, we must eat more to fuel our activities.

The equivalent measure of metabolism for a poikilotherm is the **standard metabolic rate (SMR),** its metabolic rate while resting and fasting *at a given temperature.* The BMR and SMR are useful numbers, but measurements of an animal's metabolism while it is active are probably more meaningful, since activity commonly increases the metabolic rate by a factor of 10 or so.

Interestingly, an animal's metabolic rate (BMR or SMR) is related to its size in a simple way, as shown in Figure 32.30. Over an enormous range, metabolic rate (m) is related to weight (w) by $m = kw^{0.75}$. When metabolic rate is measured in watts (joules per second) and weight in kilograms, the value of k is 0.018 for unicellular organisms, 0.14 for poikilotherms, and 4.1 for homeotherms, since homeotherms must metabolize at high rates to generate the heat that keeps their temperatures constant. If the relationship between metabolic rate and weight were simply a function of surface area and volume, the exponent in this equation would be 2/3, since surface area changes with the square of a dimension and volume with the cube of a dimension. But because it is actually 3/4, biologists have devised various complicated theories to explain this relationship.

The metabolic rates for different homeotherms also vary over a wide range. A large animal naturally consumes more energy overall than a small one, but comparing the metabolic rates *per gram of body weight* shows that small animals

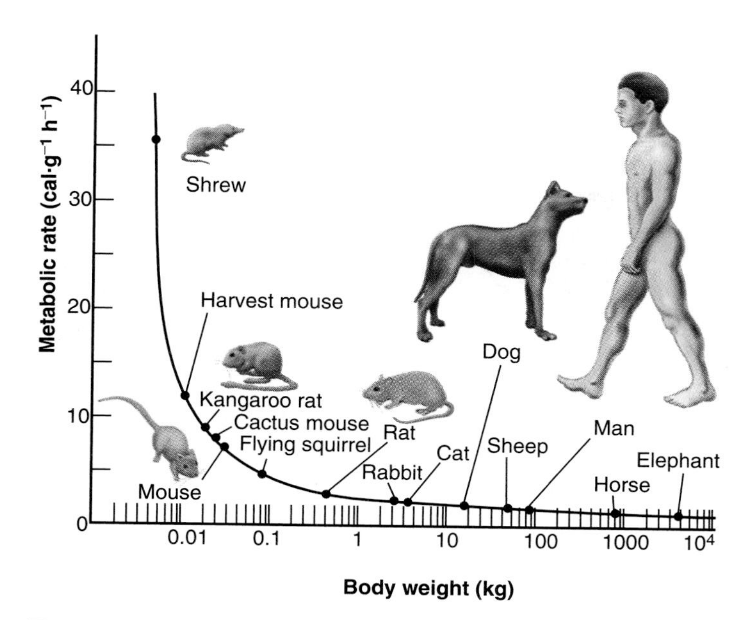

Figure 32.31

The metabolic rates of mammals *per gram of body weight* increase sharply as their weight decreases. This factor imposes a limit on the size of a homeothermic animal, so that a shrew is probably the smallest possible mammal.

metabolize much faster for their size than large ones do (Figure 32.31). The ratio of surface to volume is critically important in this regard. The great mass of a large endotherm produces a lot of heat, but with a small surface/volume ratio, it does not lose heat very fast. A small homeotherm, however, produces heat in a small volume but loses it through a relatively large surface (Figure 32.32). The small animal has to work hard just to stay warm, and its BMR is quite high. This, in fact, sets a lower limit on the size of an endothermic animal. The smallest endotherms are shrews and hummingbirds, and it is unlikely that animals any smaller than those could maintain a constant body temperature. There are also limits to metabolic rate and the rate at which an animal can eat; below a certain size, an animal wouldn't be able to eat fast enough to maintain itself, and it would use more energy in feeding than the food would provide.

Specific features of endothermic animals are also determined by surface/volume considerations. An old generalization called Bergmann's rule states that the average size of individuals tends to be smaller in warm climates and larger in colder climates. Another, called Allen's rule, says that animals living in colder climates tend to have shorter extremities, such as ears and legs, than those living in warmer conditions, because the former have less surface area for their volume and thus conserve heat better (Figure 32.33). Although other factors are involved, we see a similar regularity in plants; those of the taiga have needles or other very small leaves, while plants with very large leaves are common in tropical forests.

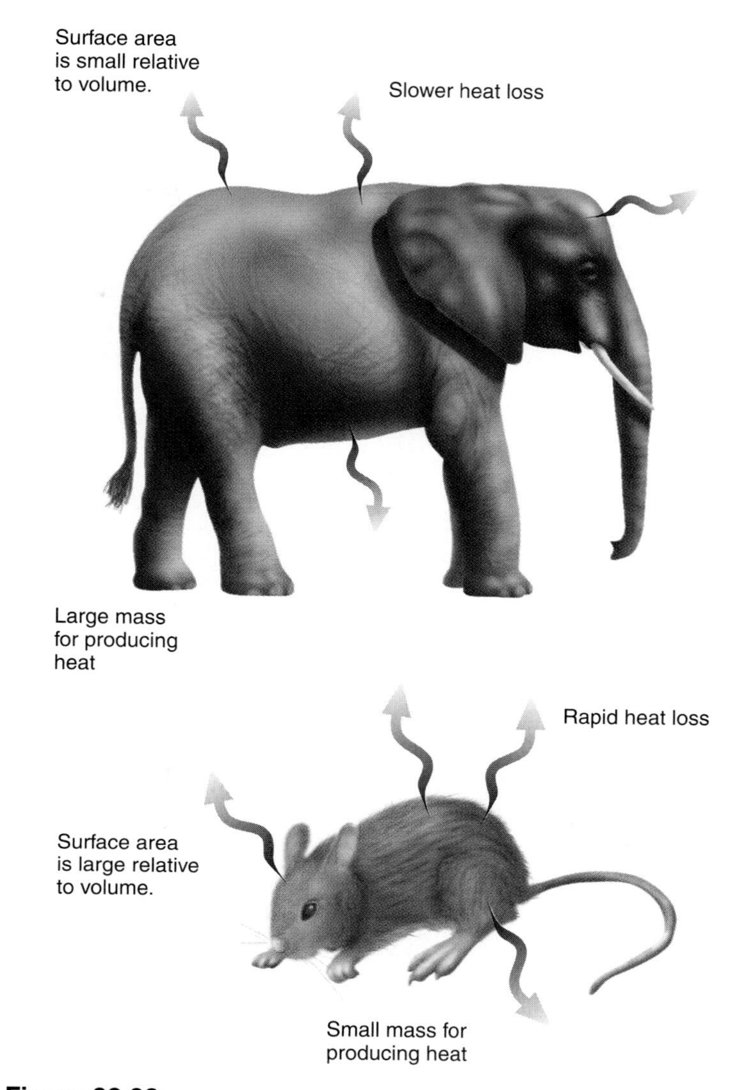

Figure 32.32

A large animal has a great mass of tissue that produces heat, and since its surface area is small relative to its mass, that heat is not lost very fast. But a small animal has only a small mass of tissue to produce heat, and its surface area is enormous relative to that mass, so it must metabolize rapidly to maintain a constant body temperature.

Exercise 32.4 If the human BMR is 1,800 kcal/day, what is the BMR of a Lilliputian? Compare the BMRs per kilogram of body weight.

Coda Once the random forces of evolution had produced large eucaryotic cells housing potentially very large genomes, the door was opened for the evolution of multicellular organisms of considerable complexity. The most obvious results are the multicellular plants and animals. In becoming large, and especially in invading the land, these organisms had to overcome various problems, challenges that

Figure 32.33
Animals of cold regions tend to have shorter extremities than similar, warm-climate animals. *(a)* Arctic hares *(Lepus arcticus)* live in far northern Canada and Greenland. *(b)* Whitetail jackrabbits *(L. townsendii)* are native to west-central United States and southern Canada. *(c)* Antelope jackrabbits *(L. alleni)* are native to Mexico and Arizona.

they met by exploiting their genetic potential. It is no wonder that organisms with a common genetic background evolved so many common responses to the same kinds of problems, despite their fundamentally different nutritional modes and different body organization.

Summary

1. Most species of organisms are plants or animals. These multicellular organisms represent two different ways of life—the phototrophic and the chemotrophic—and two quite different forms of construction.
2. In spite of their differences, plants and animals share major activities: obtaining nutrients, exchanging O_2 and CO_2, maintaining the right balance of water and salts, transporting materials to and from all tissues, supporting their structures, growing, and reproducing.
3. Plants are sessile (nonmotile), and they tend to grow in all directions around a central stem, with a crude radial symmetry. Terrestrial plants have a root system that grows downward to collect water and mineral nutrients and a shoot system that grows upward, supporting leaves in which photosynthesis occurs.
4. The growth pattern of plants requires support and water transport; both needs are fulfilled by xylem tissue, which is made of heavy-walled tubules that support the plant and transport water. Transport of dissolved organic nutrients from their sites of synthesis is accomplished by phloem tissue. Plants must maintain a stream of water upward through the xylem; water evaporates through stomata in the leaves in the process of transpiration.
5. A typical animal has a very different form to support its holotrophic nutrition (ingesting pieces of food). It becomes cephalized, with sense organs and a brain concentrated at its head end. This end also has a mouth through which food is taken in. Food moves through the intestinal tract to an anus at the other end, where wastes are excreted.
6. Some aquatic animals have become modified to live more like plants. They are radially symmetrical and generally remain attached in one place, where they reach out with various appendages to seize food.
7. All multicellular organisms (except sponges) are constructed of tissues, and these are generally organized into organs. The major tissue types are: *(a)* surface tissues, such as epithelium or epidermis, which form boundaries and sheets; *(b)* connective or supporting tissue, which supports and binds; and *(c)* bulk tissues, usually a parenchyma, which form the bulk of many organs.
8. An organ is a complex structure that serves a specific function. It is usually made of parenchyma or muscle (in animals), often with epithelial tubules, enclosed within epithelium, supported and bound together by connective tissue. It also contains vascular structures, and in animals, it is typically regulated by nerves.
9. Plants and animals may have either unitary or modular organization. A unitary organism has a limited number of substructures, generally grows to a definite size, and engages in sexual reproduction as a unit. A modular organism is constructed of an indefinite number of smaller modules, which grow independently and indeterminately, and may be able to reproduce independently and asexually.
10. Plants and animals both need water, but their metabolism imposes different requirements. Plants require water for photosynthesis; animals produce excess water through respiration, although plants also respire. Therefore, plants often need special adaptations for obtaining water, and animals often need special mechanisms for eliminating water.
11. Plants and animals both require vascular systems for distributing water and nutrients to all their cells, although very small organisms may be able to exchange materials directly with the environment without such a system. Plant vascular systems are made of xylem, which transports water and nutrients from the roots, and phloem, which transports water and dissolved organic nutrients from photosynthetic sources to other tissues.
12. Animals have cardiovascular systems, which include a muscular heart to force fluid through the system. Generally, the system circulates fluids, in contrast to plant systems, which maintain a one-way flow.
13. Multicellular organisms have considerable extracellular spaces, as well as intracellular ones, creating two fluid compartments. Animals have an extensive interstitial fluid that communicates with blood plasma. Plants have a symplast of cells connected by plasmodesmata and an apoplast of cell wall spaces. There is also an extensive intercellular gas space through which oxygen and CO_2 diffuse to support respiration and photosynthesis.
14. Plants and animals use carbon and nitrogen in different ways. Plants take in CO_2 and produce organic compounds; animals take in organic compounds and produce excess CO_2. Plants require nitrogen, generally as ammonia, for biosynthesis; animals usually obtain excess nitrogen as amino acids and must excrete the excess.
15. Both plant and animal cells must exchange CO_2 and O_2 with their surroundings. CO_2 is needed for photosynthesis and produced by respiration; O_2 is needed for respiration and produced by photosynthesis.
16. Reproduction in terrestrial organisms has generally become less dependent on water. More primitive organisms reproduce in water, with external fertilization of an egg by swimming sperm. More advanced plants developed pollen and seeds, as mechanisms for

reproducing without water. More advanced animals have internal fertilization rather than external.

17. The architecture of all organisms is largely determined by certain size limitations. The need for sufficient surface area to support the activities in a given volume causes many structures to grow with fine branches or as sheets with large surfaces. The ratio of volume to cross-sectional area, as in bones and muscles, sets limitations on the shapes and activities of many structures.

18. Organisms are commonly ectothermic, deriving most of their heat from their surroundings, so they are also poikilothermic, with temperatures determined by the environment. Some animals are endothermic, obtaining most of their heat internally from metabolism; they are therefore homeothermic and maintain a constant body temperature by regulating their production of heat.

19. Metabolic rate is closely related to an organism's size. The basal metabolic rate (for a homeotherm) or standard metabolic rate (for a poikilotherm) measures the minimum rate of metabolism needed to sustain life. This rate increases approximately with the 3/4 power of an organism's mass. The metabolic rate per unit of mass for a homeotherm is very high for small animals and decreases sharply for larger animals.

Key Terms

sessile 666
root system 666
shoot system 666
vascular system 666
motile 666
cephalization 666
anterior 666
cephalic 666
posterior 666
caudal 666
digestive tract 666
bilateral symmetry 667
dorsal 667
ventral 667
radial symmetry 669
tissue 669
organ 670
epithelium 670
epidermal 670
collenchyma 671
sclerenchyma 671
parenchyma 671
organ 672
organ system 673
unitary 673
modular 673
determinate growth 674
indeterminate growth 674
trachea 675
intercellular gas space 675
stoma 675
extracellular fluid 675
intracellular fluid 675
interstitial fluid 675
symplast 676
apoplast 676
homeohydric 676
poikilohydric 676
cryptobiotic 676
turgor pressure 677
turgid 677
osmoconformer 677
osmoregulator 677
transpiration 678
xylem 678
phloem 678
cardiovascular system 679
circulatory system 679
poikilothermic 683
ectothermic 683
endothermic 683
homeothermic 683
basal metabolic rate (BMR) 684
standard metabolic rate (SMR) 684

Multiple-Choice Questions

1. Which tissue has the primary function of protecting an organism from the entry of infecting microbes?
a. epithelium
b. connective tissue
c. parenchyma
d. vascular tissue
e. nervous tissue

2. A ciliated mucosa is an example of a(n) _____ tissue.
a. epithelial
b. connective or mechanical
c. parenchymal
d. nervous
e. vascular

3. Which features are common to mechanical or connective tissues?
a. an extracellular matrix
b. mineralization by calcium or other salts
c. large amounts of polysaccharide fibers
d. *a* and *b*, but not *c*
e. *a, b,* and *c*

4. Unitary organisms commonly have
a. indeterminate growth.
b. determinate growth.
c. large size.
d. constant body temperatures.
e. cephalization.

5. Where is interstitial fluid?
a. within plant and animal cells
b. within the ducts and chambers of vascular systems
c. between the cells in animal tissues
d. flowing through plasmodesmata
e. within the symplast

6. The primary soft matrix material in animals is made of
a. cellulose.
b. chitin.
c. collagen.
d. bone.
e. xylem.

7. Which of the following is *not* a mechanism for living in a salty environment?
a. membranes impermeable to salt
b. glands for excreting salt
c. osmolytes
d. osmosis
e. osmoconformity

8. If two animals have essentially the same structure but one is twice the length of the other, you would expect the mass of an organ in the larger one to be _____ times the mass of the organ in the smaller one.
a. 1.5
b. 2
c. 4
d. 6
e. 8

9. If a tree 3 meters high is supported by a trunk 10 cm in diameter, a trunk 40 cm in diameter will support a tree _____ meters high.
a. $4 \times 3 = 12$
b. $16 \times 3 = 48$
c. $\sqrt[3]{16} \times 3 = 7.6$
d. $\sqrt{16^3} \times 3 = 192$
e. $4\pi \times 3 = 37.7$

True-False Questions

Mark each statement true or false, and if false, restate it to make it true.

1. A cephalized animal has sense organs concentrated at its dorsal end.

2. Parenchyma, collenchyma, and sclerenchyma are supporting tissues of vascular plants.

3. One of the most important attributes of an epithelium or epidermis is being water-resistant (hydrophobic).

4. Cells that develop a high turgor pressure resist the influx of additional salt.
5. A poikilohydric organism is one whose temperature varies with changes in the environment.
6. Although plants excrete oxygen and animals excrete carbon dioxide, both plants and animals generally excrete a nitrogenous waste.
7. The apoplast of plants is analogous to the intercellular fluid of animals.
8. Ectothermic organisms are also homeothermic.
9. To conserve heat, plants and animals living in arctic regions would tend to be more compact and rounder than those living in tropical regions.
10. The smaller an animal, the higher it can jump.

Concept Questions

1. What types of tissues occur only in animals? Why have they not evolved in plants?
2. Explain why water is a metabolic waste product for animals but not for plants.
3. What sets the upper and lower limits of body size in endotherms?
4. Are the terms "ectothermic" and "poikilothermic" synonyms? What about the terms "endothermic" and "homeothermic"?
5. By considering cubes with different dimensions, show that the volume increases faster than the surface area.

Additional Reading

Bonner, John T. *Size and Cycle: An Essay on the Structure of Biology.* Princeton University Press, Princeton (NJ), 1965.

Heinrich, Bernd. "Thermoregulation in Winter Moths." *Scientific American,* March 1987, p. 104. Curiously lacking in highly specialized adaptations for the cold, certain nondescript moth species can nevertheless do what their relatives cannot: fly, feed, and mate at near-freezing temperatures.

McClanahan, Lon L., Rodolfo Ruibal, and Vaughan H. Shoemaker. "Frogs and Toads in Deserts." *Scientific American,* March 1994, p. 64. These amphibians have evolved many strategies for survival in extremely hot, dry climates.

McGowan, Christopher. *Diatoms to Dinosaurs: The Size and Scale of Living Things.* Island Press/Shearwater Books, Washington, DC, 1994.

McMahon, Thomas A., and John T. Bonner. *On Size and Life.* Scientific American Library, W. H. Freeman, New York, 1983.

Moog, Florence. "Gulliver Was a Bad Biologist." *Scientific American,* November 1948, p. 52.

Internet Resource

To further explore the content of this chapter, log on to the web site at:

http://www.mhhe.com/biosci/genbio/guttman/

33

Plants and the Evolution of Plant Reproduction

Key Concepts

33.1 Introducing the cast: A brief survey of the plants.
33.2 Green algae have evolved independently along several lines.
33.3 Methods of cell division among plants point to different lines of evolution.
33.4 Plant evolution has involved a modification of the haplodiplontic cycle.
33.5 Nontracheophytes include three divisions of closely related plants.
33.6 Some major trends have distinguished vascular plant evolution.
33.7 The first vascular plants were dichotomously branched cylinders.
33.8 Vascular plants probably evolved along three main lines.
33.9 Lycopsids developed cones made of fertile leaves, but were a side branch of evolution.
33.10 Equisetophytes have jointed stems.
33.11 Ferns were among the first plants to develop megaphyll leaves.
33.12 Seeds and pollen were a major innovation in reproduction.
33.13 Conifers are gymnosperms with relatively simple leaves.
33.14 Flowering plants have seeds enclosed in an ovary, which becomes a fruit.

A small pond in the Adirondack Mountains of New York in early autumn.

Only water striders break the mirror surface of the little woodland pond, but beneath the surface trillions of cells swim about, feeding among trillions of green and brownish algae. As they capture the light, the algae proliferate, turning the water greenish, providing food for a community of animals. But if hard times beset the pond—if the water turns too cold or becomes nitrogen deficient—those algae may stop growing and enter a cyst or spore stage until conditions improve.

The pond is shaded by the branches of surrounding trees, which flower and set seeds in the sunlight. Hard times may beset the trees too, but they have greater security than the algae. With their deep roots, their water transport system, and their defenses against drying, they can withstand considerable drought. The seeds they drop are also well protected, with heavy coats and stores of energy-rich food.

The path from the eucaryotic algae to the first trees took hundreds of millions of years. Green algae first appeared in the Bitter Springs sediment of Australia, dated at about one billion years ago. The first trees appeared in the Mississippian period, around 350 million years ago, and the first seed plant has been found in a plant from slightly older Devonian sediments. In this chapter, we will survey the plant kingdom and tell the story of plant evolution, a story that centers around changes in ways of reproduction, including the evolution of seed plants. ■

33.1 Introducing the cast: A brief survey of the plants.

The plant kingdom can be defined to either include or exclude the green algae, with good justifications for both alternatives. Many plant biologists place the green algae in the diverse kingdom of protists, but just as we reduce the heterogeneity of Protista by removing many organisms to a kingdom Chromista, we include the green algae among kingdom Plantae, thus giving this kingdom clear-cut features. Kingdom Plantae includes the green photosynthetic organisms with cell walls made of cellulose, food reserves of starch, and the photosynthetic pigments chlorophylls *a* and *b*, along with certain carotenoids. If they have any flagellated cells, the flagella are exclusively whiplash, rather than the tinsel flagella of brown-line algae (see Section 30.3). Biologists agree that the first organisms that are indisputably plants evolved from some kind of green algae, and the plant kingdom conceived in this way is probably monophyletic.

The systematics of plants, like so many other groups of organisms, is undergoing great changes as investigators obtain more information through molecular methods. As a framework for surveying the kingdom, we will use a slight modification of a classification devised in 1966 by three of the world's most respected botanists—Arthur Cronquist (in New York), Armen Takhtajan (in Leningrad), and Walter Zimmermann (in Tübingen). The plant kingdom is divided into two subkingdoms: thallophytes and embryophytes (Table 33.1). *Thallophytes* (subkingdom Thallo-

Table 33.1 An Outline of the Plant Kingdom

Rank				Taxon
Subkingdom 1.				Thallobionta: green algae and relatives. Unicellular, colonial, or having simple thalli; showing little or no cellular differentiation except for reproductive cells, and never forming an embryo that develops within a structure of the parent plant.
		Division[1] 1.		Prasinophyta: flagellate or coccoid algae
		Division 2.		Chlorophyta: grass-green algae
		Division 3.		Siphonophyta: tubelike, coenocytic (siphonaceous) green algae
		Division 4.		Zygnematophyta: conjugating green algae
		Division 5.		Charophyta: stoneworts. Macroscopic algae with precise segmentation into nodal and internodal segments, with branching whorls; anchored by rhizoids (simple rootlike structures)
Subkingdom 2.				Embryobionta. Plants in which the sporophyte begins its development attached to and dependent on the gametophyte, as an embryo.
	Infrakingdom 1.			Atracheophyta: plants without well-developed vascular tissues
		Division 1.		Bryophyta: mosses
		Division 2.		Hepatophyta: liverworts
		Division 3.		Anthocerophyta: hornworts
	Infrakingdom 2.			Tracheophyta: plants with distinct xylem and phloem tissue
		Division 4.		Rhyniophyta: primitive vascular plants
		Division 5.		Psilotophyta: unusual plants with primitive features
		Division 6.		Lycopodiophyta (Lycopsida): club mosses, with true roots, stems, leaves
		Division 7.		Equisetophyta: horsetails or sphenopsids
		Division 8.		Polypodiophyta: ferns
		Division 9.		Pinophyta: gymnosperms
			Subdivision 1.	Cycadophyta
			Class 1.	Pteridospermeae: seed ferns. Plants of the Carboniferous to Cretaceous periods having fernlike fronds with seeds
			Class 2.	Cycadatae: cycads. Like palm trees, with large frondlike leaves in a crown near the top
			Class 3.	Bennettitatae
			Subdivision 2.	Pinophyta
			Class 1.	Ginkgoatae: ginkgos
			Class 2.	Pinatae: conifers—pines, firs, yews, etc.
		Division 10.		Magnoliophyta: flowering plants (angiosperms)
			Class 1.	Magnoliatae: dicotyledons
			Class 2.	Liliatae: monocotyledons

[1] Botanists use the term "division" instead of "phylum."

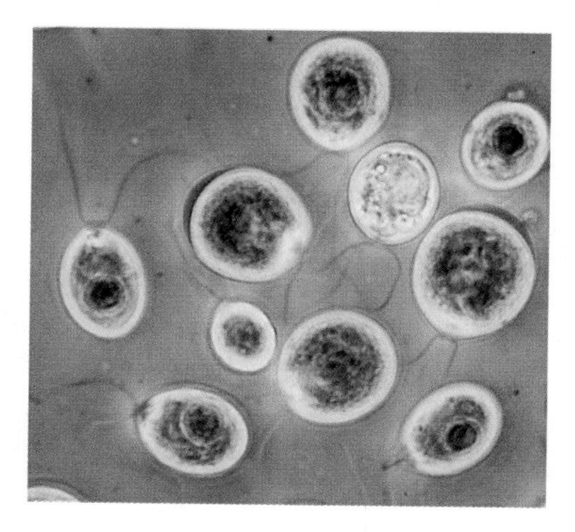

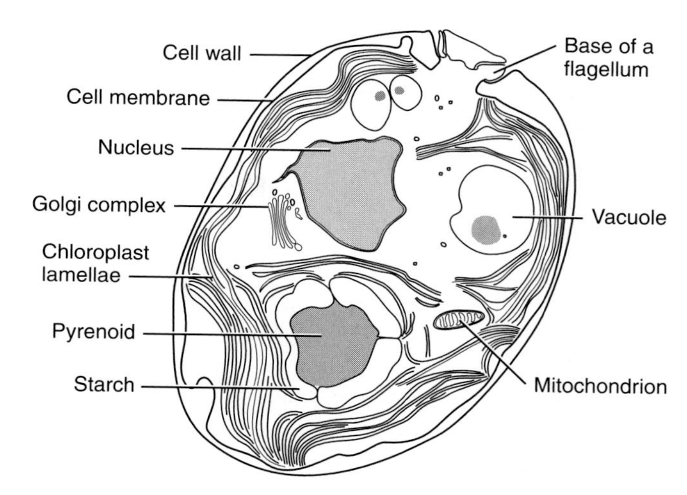

Figure 33.1

Chlamydomonas, a common unicellular flagellate, is representative of the simplest green algae.

bionta) include the green algae, which are single cells, colonies of a few cells, or simple thalli—that is, plants made of parenchyma-like cells with little or no differentiation. None of the algae have specialized organs such as roots, leaves, and stems. *Embryophytes* (subkingdom Embryobionta) include all the other plants, whose zygotes grow into multicellular embryos. Most of the embryophytes are terrestrial, and most have distinct roots and shoots.

Except for the mosses and their relatives, all embryophytes are vascular plants, with xylem and phloem for transport of materials and support, as outlined in Section 32.10. (The vascular plants are separated off as Tracheophyta.) The first vascular plants, known only from fossils, form one division (Rhyniophyta). Three types of common plants—club mosses, horsetails, and ferns—warrant divisions of their own, along with their abundant fossil ancestors.

The dominant plants all reproduce by means of seeds, and one of our major themes in this chapter will be the evolution of seeds and pollen as a means of reproduction independent of water. Seeds develop within structures called ovules, and most seed plants are divided into two large divisions on this basis: Gymnosperms (Pinophyta) have naked ovules (*gymnos* = naked), and angiosperms (Magnoliophyta) have ovules enclosed in an ovary (*angion* = vessel).

33.2 Green algae have evolved independently along several lines.

The 7,000 species of known green algae are wonderfully diverse in form. A unicellular flagellate such as *Chlamydomonas,* with a pair of flagella and a cup-shaped chloroplast, could be a model for a primitive green alga (Figure 33.1). Its life cycle was illustrated in Figure 15.7. At least three distinct types of algae have diverged from an ancestor of this kind. One trend in evolution has produced colonial flagellates, flagellated cells that remain together in regular packets after dividing, embedded in a gelatinous matrix. Such colonies range in size from small packets like *Gonium* and *Pandorina* to several thousand cells in *Volvox,* which gives the epithet "volvoccine" to these colonial flagellates (Figure 33.2).

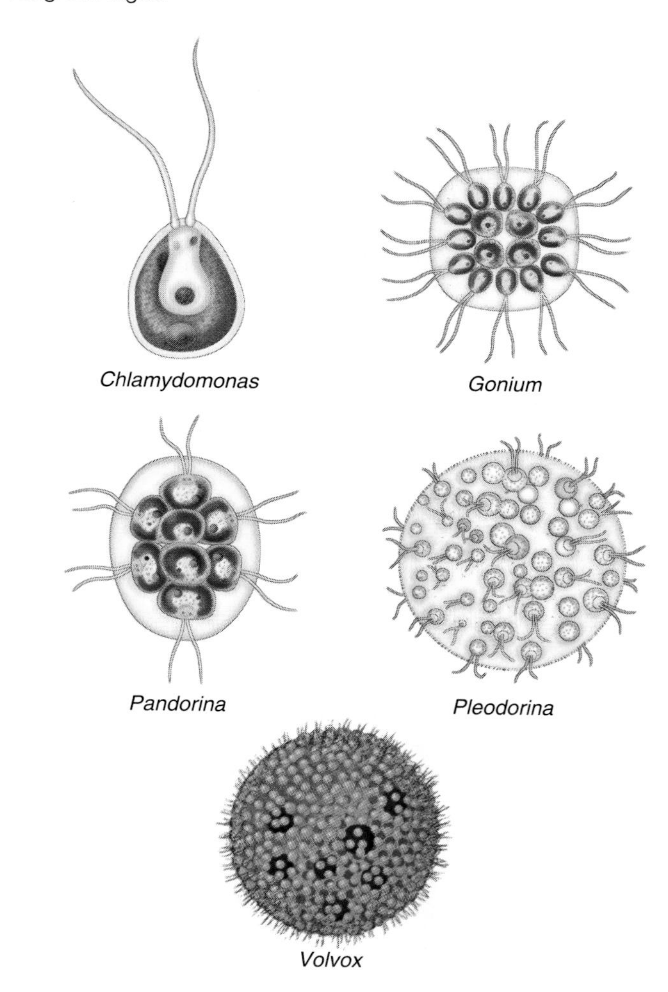

Figure 33.2

The volvoccine green algae retain their flagella and form clumps of cells or large colonies.

Figure 33.3

In the life cycle of the green alga *Volvox,* a diploid zygote (looking very much like a *Chlamydomonas* cell) develops flagella and then grows into a large, multicellular colony. The colony sometimes reproduces asexually by forming gonidia, which detach and grow into new colonies; it may also produce separate male and female colonies, which form sperm or ova by meiosis. Fertilization of an ovum by a sperm produces a new zygote.

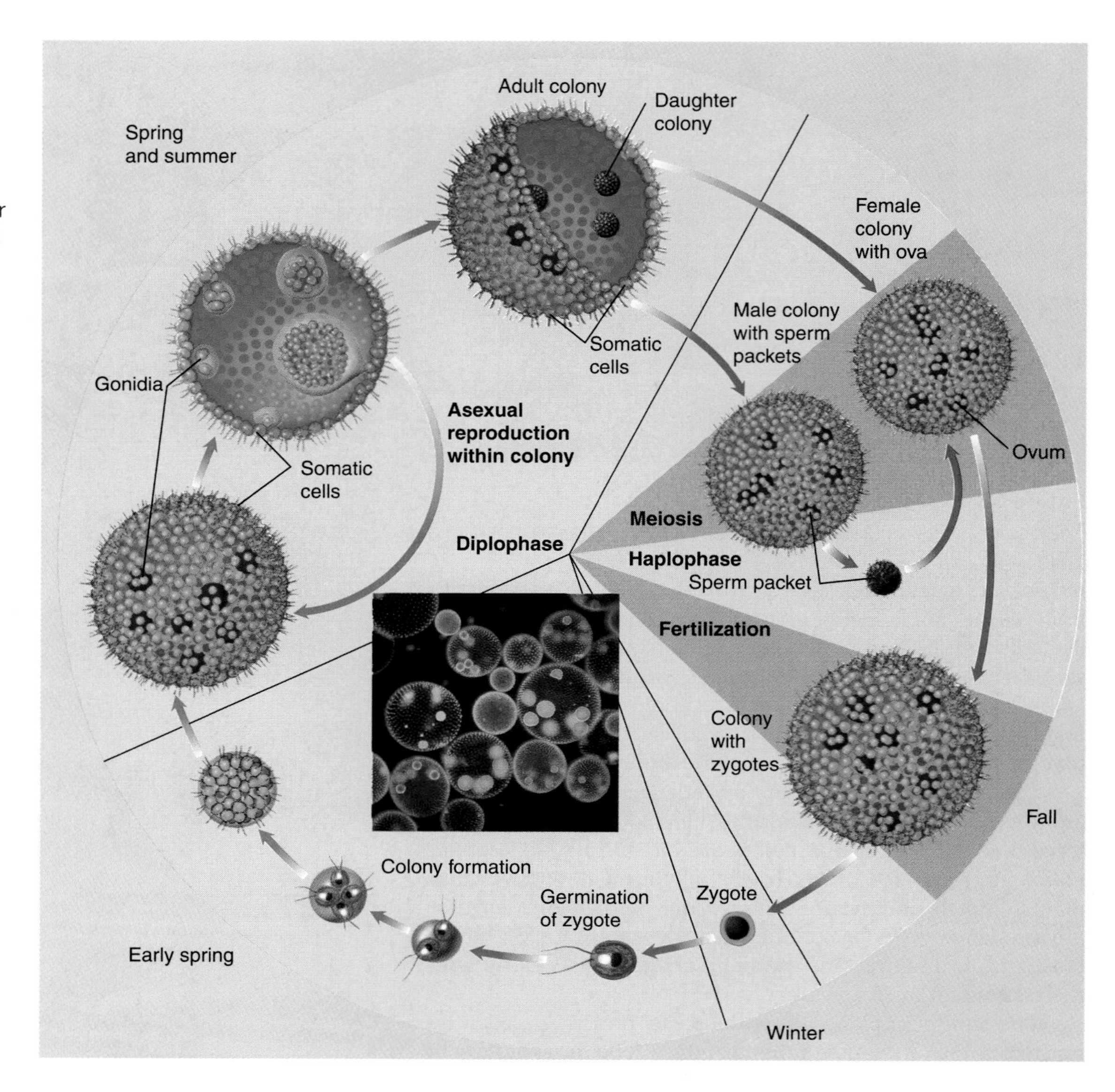

Volvox is a remarkable colonial organism whose cilia beat in coordination, whirling the whole ball through the water like a spectacular green crystal ornament.

Volvox aureus goes through an interesting life cycle over the course of a year (Figure 33.3). Most of the time it reproduces asexually: Individual clusters of cells within the colony divide repeatedly, and then the larger colony disintegrates and each cluster swims off as a new colony. But in the fall, some colonies develop into males, which form internal sperm packets, and others into females, which form internal ova. Fertilization inside a female colony produces a number of zygotes that develop heavy cell walls, enabling them to survive the coming winter. In the early spring, each zygote germinates into an independent cell, much like a *Chlamydomonas,* that divides and grows into a new colony.

A second trend in green algal evolution has produced tubelike, coenocytic species, which are also called siphonaceous because their continuous tubes are like siphons (Figure 33.4). These are mostly marine algae, some quite spectacular. *Codium* forms massive ropes of intertwined cells, and *Acetabularia,* with which Hämmerling demonstrated the importance of the cell nucleus, is uninucleate but has an umbrella shape with a long stalk.

Hämmerling's experiments, Section 13.1.

In a third evolutionary trend, called *tetrasporine,* the cells remain uninucleate but have lost their flagella and form sheets and filaments of uniform cells rather like parenchyma (Figure 33.5). *Ulva,* the "sea lettuce," has a broad, leaflike thallus made of nearly identical cubic cells packed into sheets one or two cells thick. A thallus is a body not differentiated into roots, stems and leaves. Other species, like *Ulothrix,* are long filaments. *Fritschiella,* perhaps the most complex tetrasporine alga, grows with horizontal filaments that send rhizoids (rootlike extensions) downward and other filaments upward. Although *Fritschiella* is aquatic, its resemblance to the more primitive terrestrial

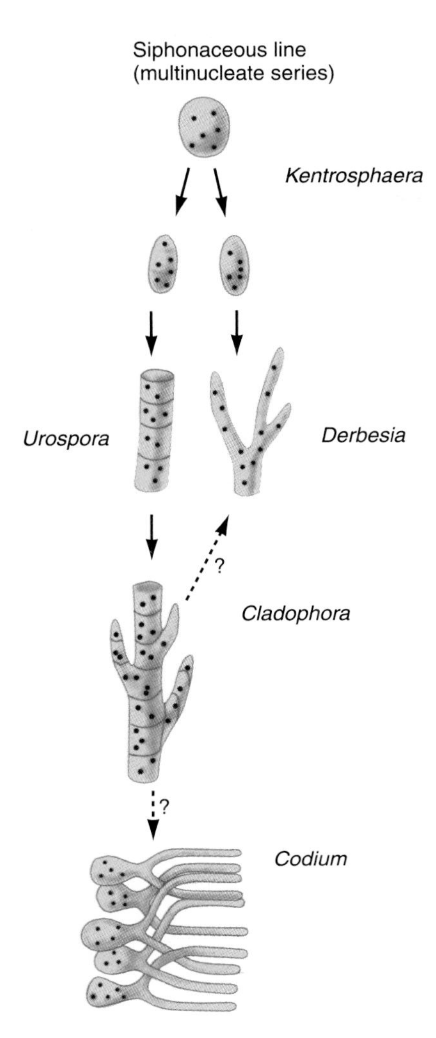

Figure 33.4
Species in the siphonaceous clade of green algae generally have elongated, tubelike cells, often having many nuclei within an undivided cytoplasm.

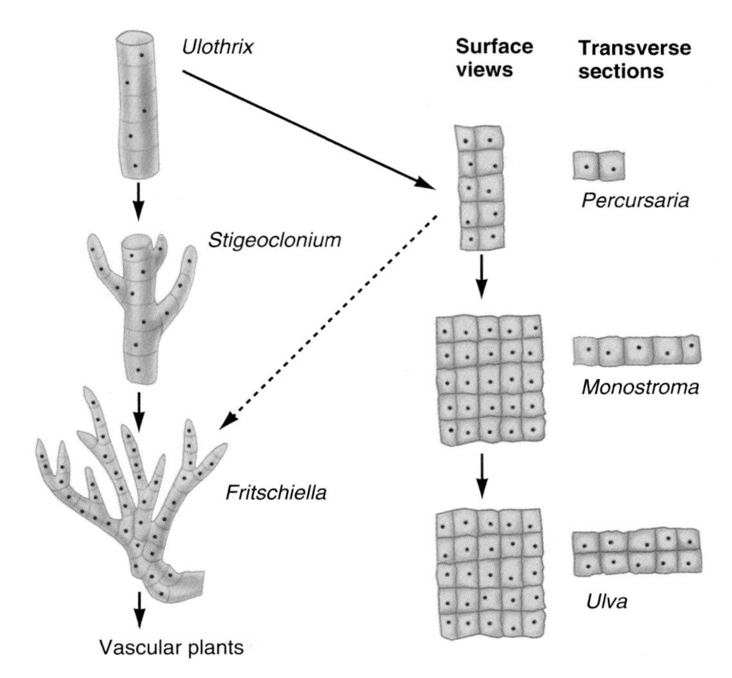

Figure 33.5
Species in the tetrasporine clade of green algae have nonflagellated cells with one nucleus each. They have the forms of filaments and sheets. Some algae with branched filaments, similar to the contemporary genus *Fritschiella,* are likely ancestors of the vascular plants.

embryophytes suggests that some early embryophytes evolved from algae like this.

The Charophyta, or stoneworts, are quite different from other thallophytes. They grow anchored to the substratum, generally in fresh water, extending a filament upward that is divided into nodes with projecting whorls of side filaments (Figure 33.6). The reproductive structures that develop at these nodes, and the biflagellate sperm produced in the male reproductive structures, are similar to those of mosses, leading some investigators to identify the charophytes as ancestors of mosses.

33.3 Methods of cell division among plants point to different lines of evolution.

It is quite clear that embryophyte plants evolved from some kind of green algae. But there are many types of algae. Which of them are the most likely ancestors of the more complex plants?

Figure 33.6
Stoneworts, division Charophyta, are complex algae with whorls of side branches arising from the nodes of a central filament. Their reproductive structures and other features suggest that they are ancestors of mosses.

Detailed studies of mitosis and cytokinesis among plants, especially by Jeremy Pickett-Heaps and his associates, have provided an improved perspective on plant evolution by showing that green algae have evolved two different modes of cell division. In most green algae, mitosis occurs in the classical way, except that the nuclear envelope remains intact and microtubules enter the nucleus through windows that open at the poles. Then an unusual structure, the *phycoplast,* forms along the plane of cytokinesis (Figure 33.7*a*). The phycoplast divides the cell in two, while drawing the widely separated anaphase nuclei closer together—so close that the telophase nuclei sometimes flatten against each other.

In contrast, cell division in other groups of green algae uses microtubules that assemble perpendicular to the plane of cytokinesis rather than within the plane. Many microtubules from the mitotic spindle persist after telophase to form a *phragmoplast* (Figure 33.7*b*). These microtubules guide Golgi-derived vesicles containing cell-wall material to the plane of cell division, where the vesicles fuse into a **cell plate.** The cell plate grows from the center outward to divide the cell in half. This mode of cell division occurs in only a few groups of green algae, such as the Zygnematales, *Fritschiella,* and the Charophyta. It is also used by the embryophytes, adding weight to the argument that algae of this kind were ancestral to the embryophytes.

33.4 Plant evolution has involved a modification of the haplodiplontic cycle.

It is time to outline the story that will unfold as we discuss the rest of the plant kingdom. Many plants have *haplodiplontic cycles,* in which both haploid and diploid stages grow vegetatively; this kind of sexual cycle is also called an *alternation of generations.* This cycle is characteristic of many brown algae (in the Chromista) and many green algae, for which *Ulva* will serve as an example; although all *Ulva* individuals look like large, green leaves, some are haploid and some are diploid. The life cycles of plants are described with the terms illustrated in Figure 33.8*a,* and the rest of this chapter will read like a foreign language unless you master these terms now. The haploid phase is the **gametophyte** because it produces gametes, which develop in organs called **gametangia;** an **antheridium** is a male, sperm-producing gametangium, and an **archegonium** is a female, egg-producing gametangium. After fertilization, the zygote develops into a diploid **sporophyte** plant, which eventually produces spores. Within sporophyte organs called **sporangia,** each meiotic division produces four meiospores, and each of these can grow into a new gametophyte.

All algae except charophytes have unicellular gametangia. In contrast, embryophytes have multicellular gametangia, and the cells that actually undergo meiosis are surrounded by a coat of protective cells. The embryo sporophyte develops within this coat. Embryophytes are also united by their mode of cell division, described in Section 33.3.

A diploid organism has certain advantages over a haploid, as reflected in the clear trend in plant evolution toward dominance of the sporophyte and reduction in the size of the gametophyte (see Chapter 15). At the same time, some plants have tended to develop into separate male and female individuals. The most primitive vascular plants were all **homosporous,** having only a single type of sporangium that produced one type of spore. Each spore developed into a gametophyte bearing both antheridia and archegonia. But some plants became **heterosporous,** able to produce two kinds of spores. Their sporophytes bear **megasporangia,** which produce a few relatively large megaspores, and **microsporangia,** which produce many small microspores (Figure 33.8*b*). Each megaspore then develops into a **megagametophyte** that eventually produces eggs, while each microspore develops into a **microgametophyte** that produces sperm. (So a megagametophyte is like a female and a microgametophyte like a male.) The gametophytes of the most primitive plants still have identifiable archegonia and antheridia, but in the more advanced groups of plants, the gametophytes are so reduced that these special organs hardly exist.

Exercise 33.1 After studying the terms introduced here, fill in this table without consulting the text.

Structure	1*n* or 2*n*	Derived From	Will Produce
Gametophyte			
Archegonium			
Egg			
Spore			
Antheridium			
Sporophyte			
Gametangium			
Sporangium			
Sperm			

33.5 Nontracheophytes include three divisions of closely related plants.

Find a rich, moist bed of moss and get down close to it (Figure 33.9*a*). The delicate, leafy forms of moss are easily overlooked, and you may need a magnifying glass to appreciate them fully, but with a little concentration, you could get lost in this minute world. By comparing the leaflets (they are not true leaves) of several mosses, you can see how diverse they really are. About 24,000 species of small plants are called nontracheophytes (Atracheophyta) because they lack the well-developed vascular tissue—xylem and phloem—of other plants. They are now classified in three divisions. Bryophyta includes about 14,000 species of the familiar leafy mosses; Hepatophyta includes some 9,000 species of liverworts, whose leathery, lobed forms are often found paving the surfaces of otherwise barren ground; and Anthocerophyta includes about 100 species of hornworts, which resemble liverworts. All these are commonly called bryophytes. Their lack of efficient vascular tissue for conducting water restricts the growth of nontracheophytes, and most

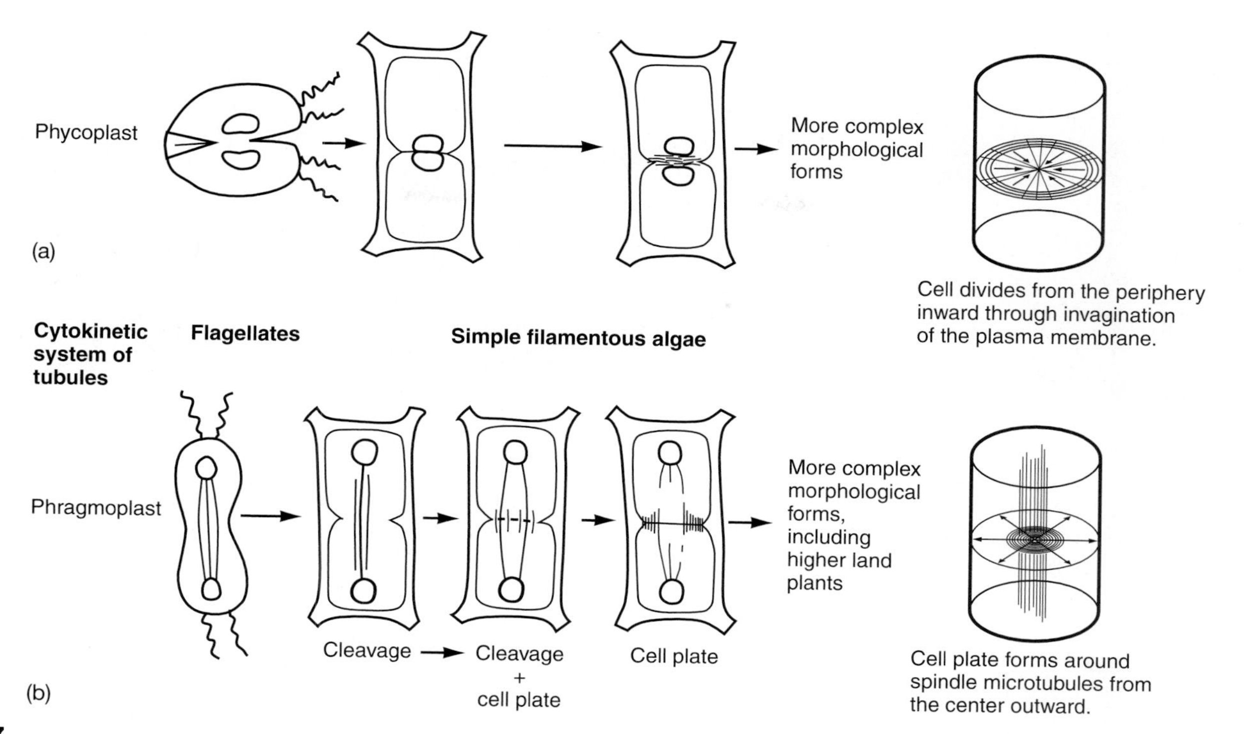

Figure 33.7

Green algae apparently evolved into two major clades, marked by different modes of cell division. Both modes evolved from flagellates to filamentous algae to more complex forms. *(a)* Most algae divide by means of a *phycoplast,* formed with microtubules parallel to the plane of division. The cell wall grows from the outside of the cell toward the center. *(b)* A few algae evolved division by means of a *phragmoplast,* a sheaf of microtubules perpendicular to the plane of division; in these algae, cleavage was supplemented and then replaced by formation of a cell plate, which grows from the center of the cell outward. Some algae in this clade were ancestral to embryophyte plants.

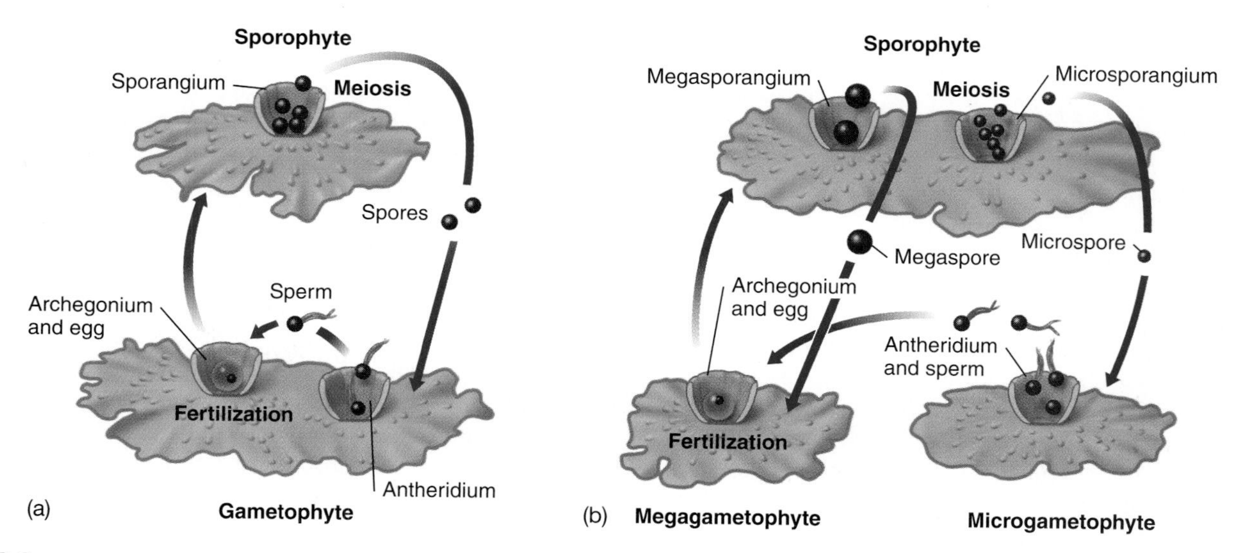

Figure 33.8

(a) The general plant life cycle entails an alternation between sporophyte and gametophyte stages. Note the terminology for the reproductive structures. *(b)* Many plants produce two types of spores—megaspores and microspores—which develop into distinct megagametophytes and microgametophytes.

(a)

(b)

(c)

Figure 33.9

(a) Mosses have green, leafy gametophytes. Their thin, ascending sporophytes *(b)* produce spores in the capsules at their tips. *(c)* Liverworts have rather heavy, fleshy gametophytes, and their distinctively shaped structures that bear gametangia separate them from mosses. The umbrellalike structures visible here bear archegonia.

grow only a few centimeters tall. In addition to having reproductive structures like vascular plants, they differ from algae by having protective epidermal tissue that resists drying and rootlike extensions called rhizoids for anchoring and taking up water from the soil or other supports.

Nontracheophytes grow most prolifically in cool, moist environments where they live on the soil, on the trunks and branches of trees, or on detritus such as fallen logs. Some species are adapted to wetlands. A few live in fresh water. Some are adapted to bare rock and are effective in breaking down rock into soil; these species may also grow on concrete structures. Bryophytes, on the whole, are prime examples of poikilohydric organisms that have a remarkable ability to adapt to exposed, dry places and to withstand drying for long times. We commonly find beds of dry moss, looking quite dead, in environments that receive rain irregularly. However, in contrast to vascular plants, they recover quickly with the first rain and start metabolizing normally.

Poikilohydric organisms, Section 32.9.

The moss shown in Figure 33.10 illustrates the bryophyte life cycle. The leafy green structures covering the ground are gametophytes. At the gametophyte tips, antheridia release sperm that swim in a film of water to the nearby archegonia, where they fertilize ova. As each zygote grows into a diploid embryo, the archegonium swells, and from it grows the sporophyte, usually a fine filament with a spore case (capsule) at its tip. The base of the sporophyte remains embedded in the gametophyte, from which it absorbs water and nutrients. However, the sporophyte has stomata—openings that permit gas exchange—and is a very active photosynthetic structure in most mosses, so it is only partially dependent on the gametophyte. The sporophytes eventually become brown, so a bed of moss becomes a green field sprouting thin, brown stalks. Within the capsule, cells divide meiotically to produce a mass of meiospores. Eventually the spores are released and fall to the ground, where each one can start to grow into a fine filament of cells, a **protonema.** From the protonema, a leafy gametophyte develops again.

These structures are quite simple. Although moss gametophytes may grow into beautifully sculptured forms, they are hardly more than filamentous green algae adapted to the land. A moss leaf is typically only a single layer of very similar parenchymal cells, but many species also have a thickened midrib that may contain elongated cells forming a primitive vascular system for conducting water and nutrients. (These tissues are analogous to the xylem and phloem of vascular plants, but have been called *hadrom* and *leptom.*) These cells transport some water internally; other mosses depend on transport of water externally, through capillary channels in the leaves. Mosses that obtain their water from below (from the soil) develop cuticle on their leaf surfaces to prevent dehydration; those that obtain their water through their leaves from above (from dew and rain) lack cuticle. Bryophytes anchor themselves with fine rhizoids, which can also collect water from the soil.

Some liverworts resemble mosses in having leafy, prostrate gametophytes, but the gametophytes of typical liverworts are flat, fleshy-looking plates. While the gametangia of mosses are hidden among their leaflets, liverworts develop two kinds of distinctive structures on their upper surfaces that bear gametangia; those that resemble umbrellas produce archegonia, and those with flat discs produce antheridia (Figure 33.9*c*). As in mosses, the motile sperm released by the antheridia swim in a film of water to fertilize eggs in the archegonia. The resulting sporophyte is a rather obscure structure, usually only a few millimeters long, anchored in the gametophyte and dependent on it. Within the capsule that develops at the end of the sporophyte, meiosis produces a new generation of haploid spores, which in some species are only released when the surrounding capsule rots. Among the spores are structures called *elaters,* elongated, dead cells whose cell walls have a helical thickening. Elaters of different species operate differently, but in one way or

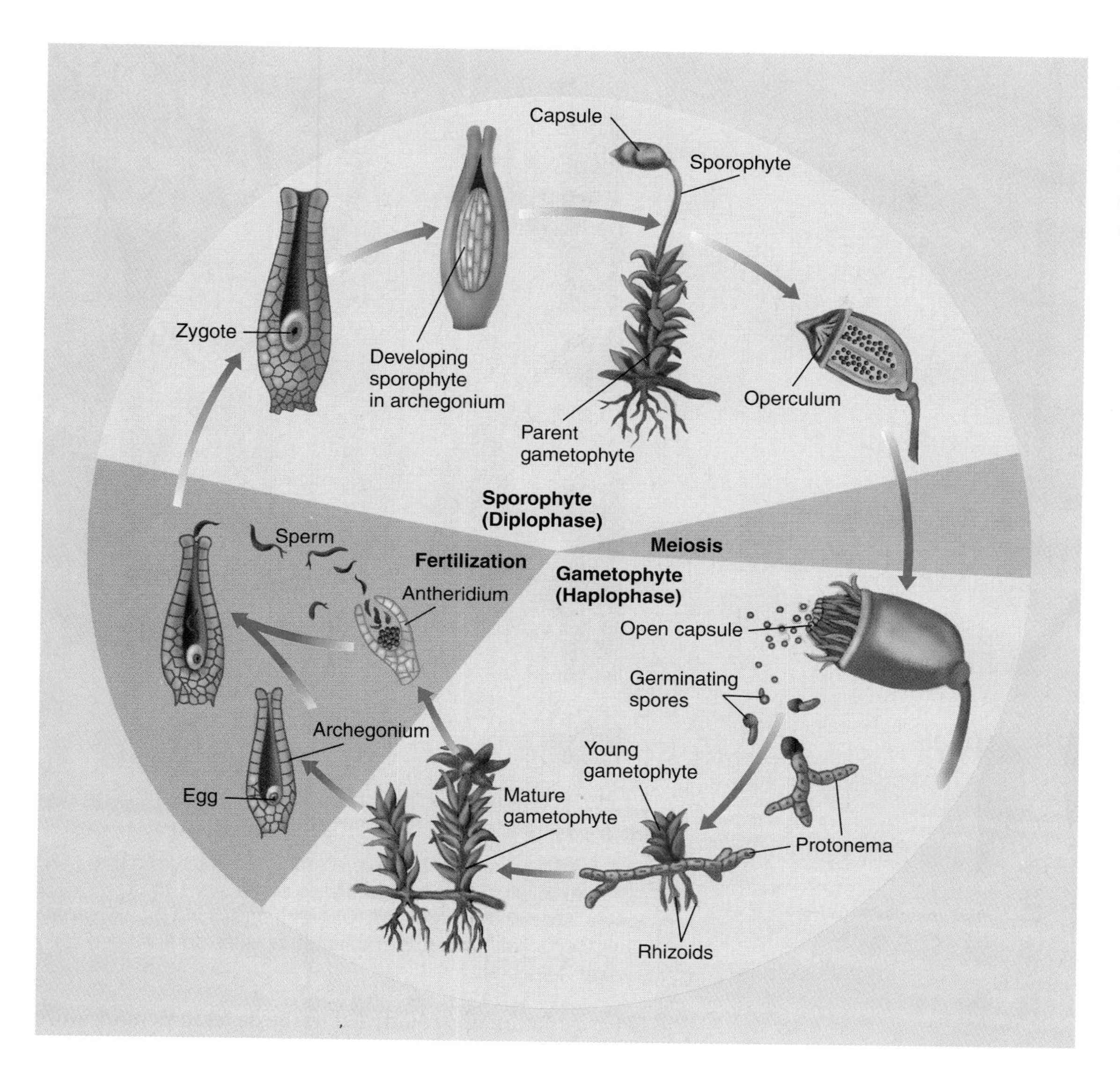

Figure 33.10
The life cycle of a moss is dominated by the green, leafy gametophyte (haplophase). The sporophyte (diplophase) consists only of the ascending stalk and capsule.

another their drying and hydration makes them act like tiny springs that disperse the spores in all directions. Some liverworts also reproduce asexually by forming lens-shaped structures called *gemmae* that can grow into new plants.

Hornworts include only about 100 species of plants, named for their tall, twisting sporophytes, which resemble little horns. Although they are similar to liverworts, their gametangia are embedded in the gametophyte tissue, not borne on specialized aerial stalks, and their sporophytes have stomata like those of mosses, which liverworts do not have. Each hornwort cell also has a single large chloroplast, in contrast to the many small chloroplasts of most other plants.

33.6 Some major trends have distinguished vascular plant evolution.

The remainder of the plant kingdom consists of plants with vascular tissue that provides support and transports water and nutrients. This feature allowed small, aquatic, herbaceous plants to evolve into large, terrestrial, woody ones. Although most species are not woody, trees and shrubs dominate the land. Several times during the evolution of plants, different groups developed tree forms soon after they arose, presumably because of the advantages of this form of growth: having a sturdy, durable structure that is not easily eaten, the ability to reach upward toward sunlight and to shade out competitors, and general dominance of the ecosystem. Vascular plant evolution is characterized by three major trends: changes in modes of reproduction (the dominant theme of this chapter), changes in vascular structure, and changes in leaf structure.

Evolution in vascular tissues

Xylem and phloem combine to form **vascular bundles.** The most primitive vascular plants have a core of vascular tissue, a **stele,** through the center of a stem. One major trend in evolution was the replacement of this structure by multiple vascular bundles separated by parenchyma. At the same time, the xylem has changed its structure. The most primitive xylem elements are long, spindle-shaped **tracheids** that fit together at their tapered ends. In later plants, these elements have tended to evolve

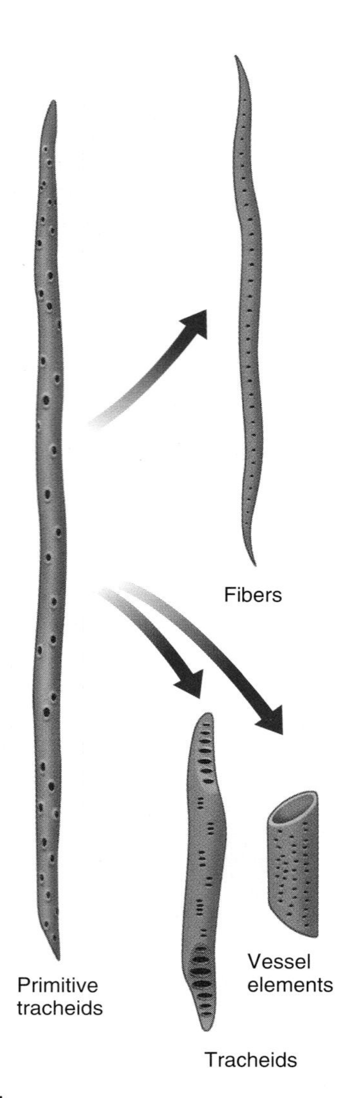

Figure 33.11
The long, tapered tracheids of primitive wood evolved into three types of cells in modern plants: fibers that provide support; vessel elements with flattened, open ends; and shorter tracheids that communicate through bordered pits on their tapered ends.

into three different types (Figure 33.11): shorter tracheids with flatter, less tapered ends, that communicate through bordered pits; **fibers** specialized for support; and **vessel elements** that are entirely open at their ends to form continuous **vessels.** The pits of neighboring tracheids, forming opposite each other, make pores that allow free movement of water.

Evolution in leaves

The earliest vascular plants carried out photosynthesis in their stems. But the evolution of larger plants required the evolution of large photosynthetic surfaces—leaves. The first leaves were **microphylls** ("small leaves") that are little more than extensions of the stem. Each one is served by a simple branch of a vascular bundle, and the branch does not interrupt the bundle (Figure 33.12*a*). This small vascular system is adequate for carrying materials in and out of leaves that are typically (but not always) small.

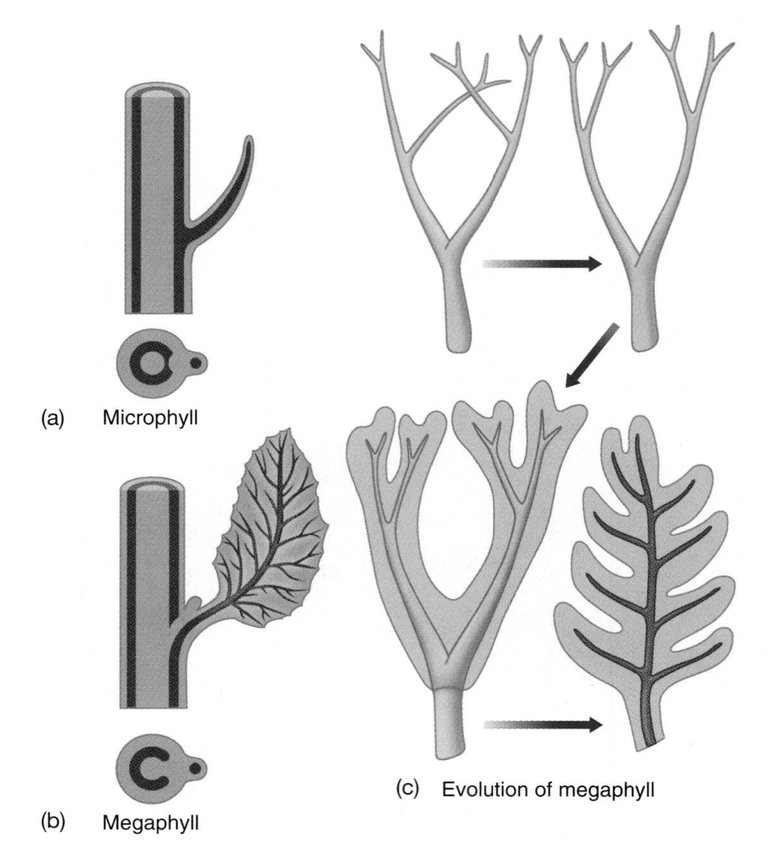

Figure 33.12
(a) Microphylls developed as simple lateral extensions from the stem, with no break in the vascular tissue of the stem. *(b,c)* Megaphylls apparently evolved as flattened, branching growths that were later filled in with parenchyma, and they interrupt the vascular tissue where they branch off.

Later evolution apparently led to the development of more extensive photosynthetic structures called **megaphylls** ("large leaves"). Each leaf of this type contains a small branch system, and there is a gap in the stele of the stem where the leaf joins it (Figure 33.12*b*). Megaphylls apparently did not evolve from microphylls. The stems of primitive vascular plants branched *dichotomously*—repeatedly forking into two branches:

There is good reason to think that megaphylls evolved in the sequence shown in Figure 33.12*c*, as some branching stems flattened out and parenchyma filled in between the vascular bundles. Later plant groups evolved other branching patterns of the veins in their leaves. Incidentally, the terms "microphyll" and "megaphyll" should not be taken literally to describe small and large leaves of modern plants; plants with leaves of both kinds have adapted to many ecological niches, with corresponding changes in the sizes of their leaves.

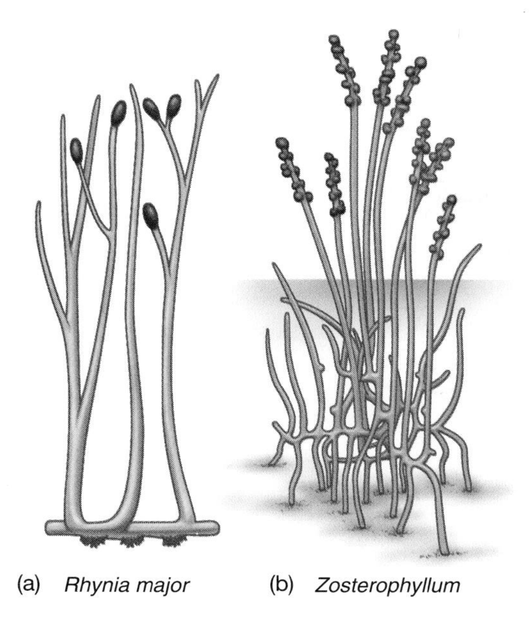

Figure 33.13
The earliest known vascular plants, from the Devonian, took two general forms. *(a) Rhynia* grew with dichotomous branching and terminal sporangia. *(b) Zosterophyllum* had H-shaped branching and lateral sporangia.

33.7 The first vascular plants were dichotomously branched cylinders.

Some Devonian sediments (375–400 million years ago) contain well-preserved fossils that are clearly primitive vascular plants. Although the plants were leafless and not divided into stems and roots, they had xylem elements. Reconstructions show two main patterns of branching and sporangia (Figure 33.13). The *Rhynia-Cooksonia* type had simple dichotomous branching with terminal sporangia, and the *Zosterophyllum* type grew in an H-shaped dichotomous pattern with lateral sporangia.

Whether or not these two types had a common origin—and there is no way to tell at this time—they continued to evolve in different directions. Harlan P. Banks has suggested that the *Zosterophyllum* group was ancestral to the lycopsids (see Section 33.9). The *Rhynia* types became more highly branched and developed clusters of sporangia at their tips, and they were probably ancestral to all the other vascular plants.

Two living genera, *Psilotum* and *Tmesipteris,* are placed in a separate division (Psilotophyta), but their actual taxonomic position is controversial (Figure 33.14). They have true embryos, and their sporophytes are larger than their gametophytes, but they lack true leaves and roots. Being leafless and homosporous, with a simple stele of xylem surrounded by phloem, they look like living remnants of the ancient regime of primitive vascular plants, but many botanists think they are more likely related to primitive ferns. *Psilotum,* a rather common tropical and subtropical plant, is commonly known as a whiskfern.

Figure 33.14
The modern psilotophytes *(a) Psilotum nudum* and *(b) Tmesipteris tannensis* are rather enigmatic plants whose true affinities are still being debated.

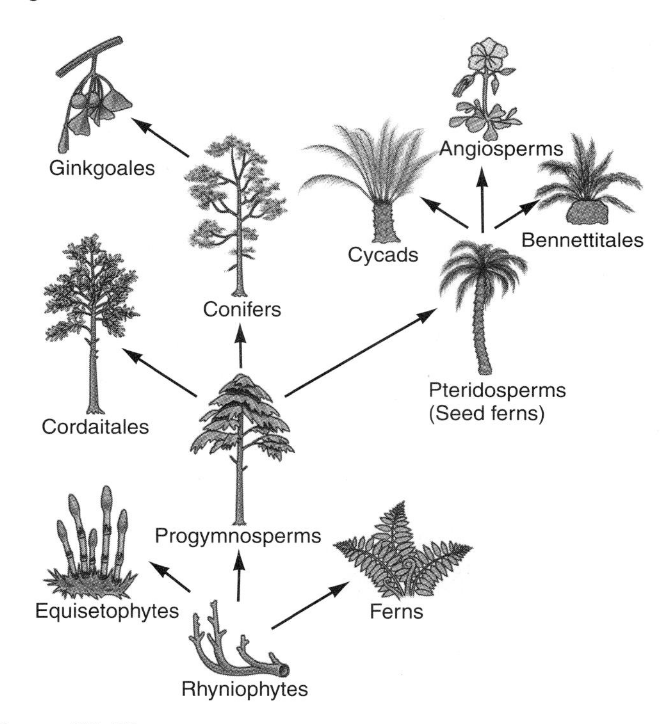

Figure 33.15
The major taxa of vascular plants probably evolved along these lines from primitive rhyniophytes; lycopsids appear to represent a more distantly related clade and are not shown. Notice the central positions of the progymnosperms and, later, the seed ferns.

33.8 Vascular plants probably evolved along three main lines.

Figure 33.15 summarizes the probable phylogeny of vascular plant groups, beginning with the rhyniophytes. One independent branch leads to the lycopsids, the club mosses, which are reminiscent of small pines and are the simplest plants that have true roots, leaves, and stems. The main line of evolution appears to have three branches: One leads to the equisetophytes, represented by modern horsetails; a second leads to the ferns; and a third leads to ancient plants called progymnosperms, which were apparently ancestral to all other plants.

Two major groups, pinophytes and cycadophytes, descended from progymnosperms. Pinophytes include conifers (such as pines, firs, and spruces) and some minor groups (cordaites and ginkgos). These plants have fairly hard wood made of small cells with heavy walls, and their small, simple leaves generally are dichotomously branched. Cycadophytes include fossil seed ferns and modern cycads, which have softer wood made of rather large cells with only moderately thick walls. Their leaves are large and frondlike, with a **pinnate** pattern of veins—a central vein with side branches:

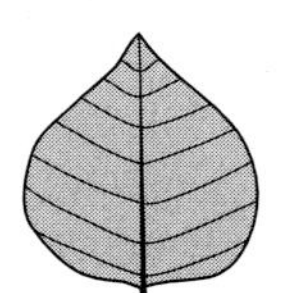

The flowering plants evolved from some ancestor in the cycadophyte branch, perhaps from seed ferns, but their origin is one of the real mysteries of plant evolution. No fossils have been found to connect them securely with any other plants.

33.9 Lycopsids developed cones made of fertile leaves, but were a side branch of evolution.

Lycopsids, such as the club moss *Lycopodium,* have microphylls arranged spirally around the stem (Figure 33.16). Here for the first time we see a phenomenon that occurred several times in the evolution of other plant groups: The sporangia that normally develop at the tips of branches become associated with microphylls, each leaf bearing a single sporangium on its upper surface to make a "fertile leaf." In *Lycopodium,* these structures aggregate at the tips of stems to form a structure resembling a pine cone. In the sporangium on each leaf, cells undergo meiosis to produce haploid spores. In contrast to *Lycopodium,* other

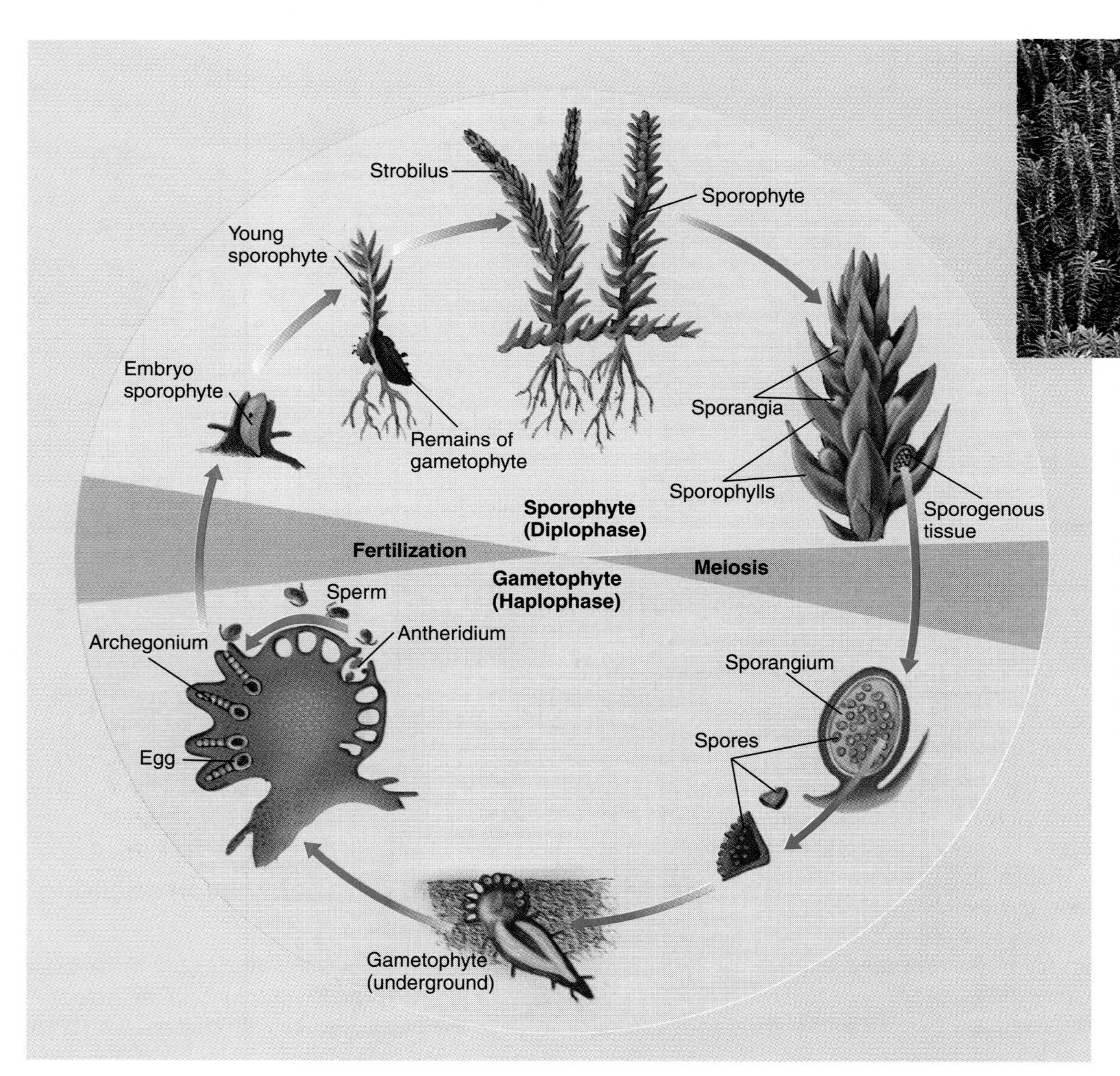

Figure 33.16
Club mosses *(Lycopodium)* resemble small pine trees. In contrast to mosses, their life cycle is dominated by the sporophyte (diplophase), and the gametophyte (haplophase) is only a small underground structure. Notice that fertilization still depends upon flagellated sperm, which must swim a short distance to the eggs.

Figure 33.17

(a) The tree lycopsid *Lepidodendron* from the Carboniferous period bore large sporangia and had bark with a characteristic diamond pattern. *(b)* Carboniferous forests, as reconstructed here in an exhibit at Chicago's Field Museum of Natural History, were dominated by tree lycopsids and included some insects of extraordinary size.

lycopsids such as *Selaginella* are heterosporous; they produce distinct megaspores and microspores, which grow into separate female and male gametophytes.

After its release, each *Lycopodium* spore grows into a gametophyte with a swollen underground part bearing rhizoids. Above ground, biflagellate sperm emerging from the antheridia swim to fertilize the single egg in each archegonium. Until it is large enough to live independently, the sporophyte that starts to develop in an archegonium lives parasitically on the gametophyte.

The living lycopsids are all small and herbaceous, but their ancestors in the Carboniferous period (Mississippian and Pennsylvanian) included trees like *Lepidodendron* that grew as large as some modern flowering trees, up to 35 meters high, with meter-thick trunks (Figure 33.17*a*). Their microphyll leaves were attached to the stems in spiral rows, leaving fossil branches covered with a characteristic diamond pattern. The extensive forests of that time dominated by lycopsid trees were often transformed into the great coal beds, rich in excellent fossils, that we now mine (Figure 33.17*b*).

33.10 Equisetophytes have jointed stems.

As the soil warms up each spring, the scruffy edges of many roads and paths often start to grow forests of green sticks reminiscent of asparagus spears (Figure 33.18). American pioneers found these surprisingly hard plants useful for cleaning out pots, and called them "scouring rushes." We know them as equisetophytes, or horsetails. Their hardness comes from having some cells whose walls are heavily impregnated with silica. And they are mostly stem. The central stem is made of a series of distinct joints (thus the alternative name *Arthrophyta* = jointed plants) that appear to be wedged together like the segments of a collapsible radio antenna. A cross section through the stem shows a central canal surrounded by multiple vascular bundles and two cylinders of smaller canals. Each of the lower joints is surrounded by a whorl of spindly stems and a ring of tiny microphyll leaves.

Figure 33.18

Horsetails, such as this Meadow Horsetail, *Equisetum pratense*, commonly grow in disturbed places such as the sides of roads and paths. They can be recognized by their central stem bearing whorls of fine leaves at several nodes.

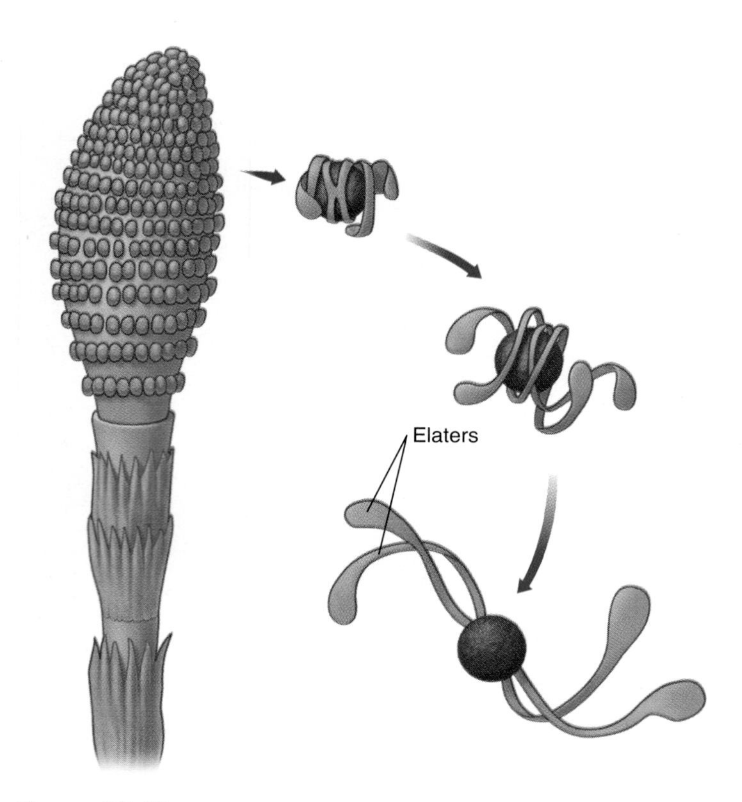

Figure 33.19

The cone at the tip of each *Equisetum* stem bears many spores whose elaters help them disperse. The spores tend to be blown by the wind, while the elaters are dry and extended, but when a spore reaches a moist habitat, the elaters curl up and the spore stays in place.

The life cycle of *Equisetum* is not particularly distinguished. The cone at the tip of each plant produces spores with a remarkable adaptation (Figure 33.19). Each spore bears four springy *elaters,* reminiscent of those in liverworts, that spread out when they are dry and curl up when wet, and this helps disperse the dry spores from the cone, but lets them lie still on moist soil. Each spore can grow into a small gametophyte that bears both antheridia, producing multiflagellated sperm, and archegonia. The zygote grows into a horizontal rhizome that produces rootlets and stems.

The equisetophytes were well represented by much larger plants long ago, especially in the Carboniferous period. *Calamites* trees grew to about 10 meters in height, and *Sphenophyllum* was a heavy herbaceous plant.

33.11 Ferns were among the first plants to develop megaphyll leaves.

Ferns do not bear seeds, and they require water for the fertilization stage of their life cycles, so they are mostly confined to moist habitats, especially in the tropics (Figure 33.20). They grace our moist temperate woods with their lacy forms; the few that occupy arid habitats rely on seasonal moisture for their reproduction requirements. Their large, distinctive leaves (called fronds) are megaphylls with pinnately branched veins instead of the primitive dichotomous branching pattern.

A fern's life cycle follows a by-now familiar pattern (Figure 33.21). The gametophyte is a small, heart-shaped structure where free-swimming sperm fertilize eggs. The resulting zygotes grow into sporophyte fronds, and fronds also grow from a rhizome that can grow along the ground. Fern sporangia are

Figure 33.20

Ferns are very graceful plants. The group includes *(a)* the Water Shamrock Fern, *Marsilea quadrifolia,* and *(b)* the Maidenhair Fern, *Adiantum pedatum.*

(a)

(b)

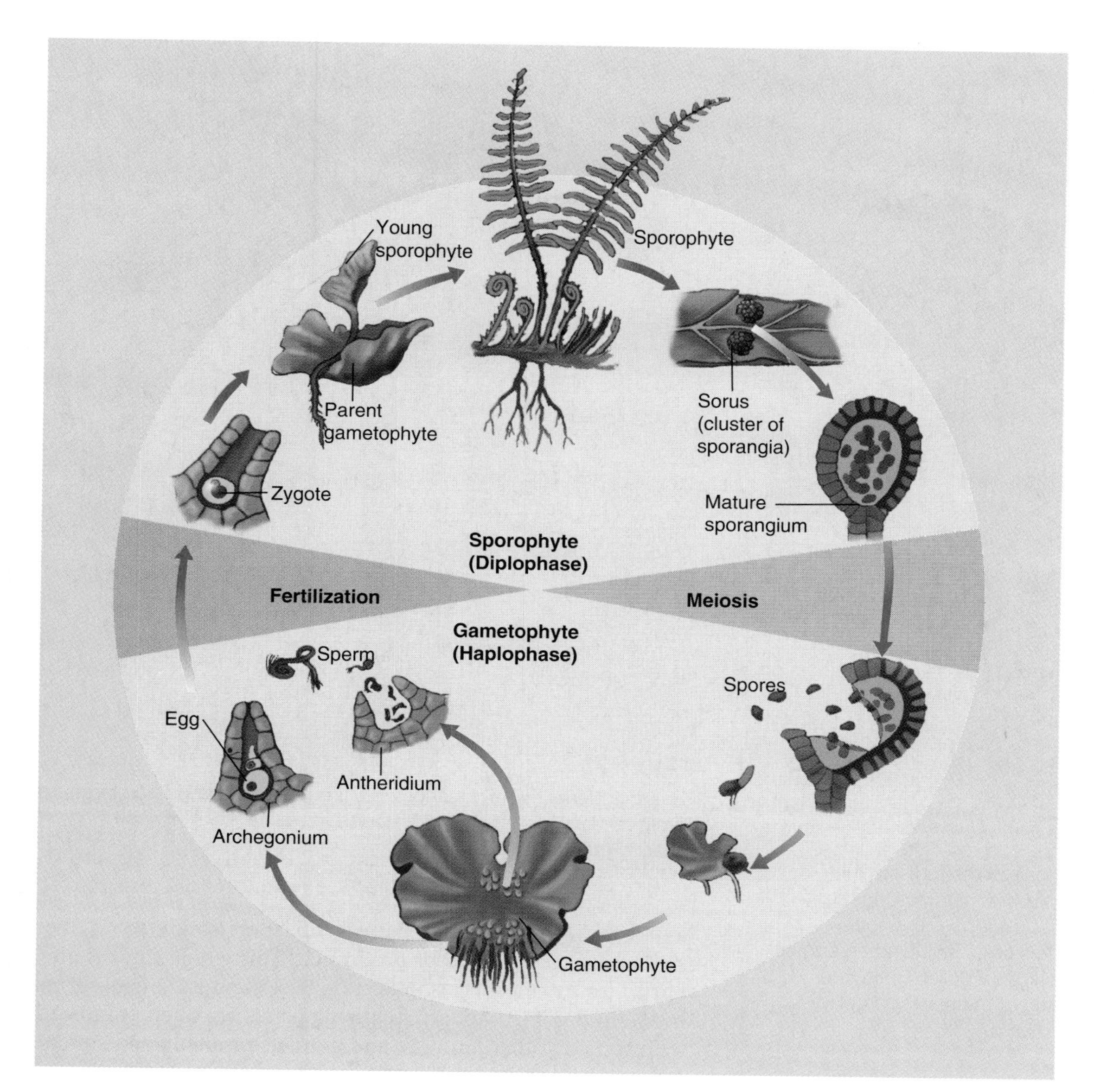

Figure 33.21
The life cycle of a fern is dominated by the sporophyte. Spores are formed in sporangia, clustered in the sori on the undersides of the leaves. Notice that fertilization still depends upon motile sperm.

sometimes attached to special shoots, but more often they are clustered into *sori* (sing., *sorus*) on the undersides of the fronds. When the mature spores turn brown, some home gardeners may mistake them for traces of a plant pathogen. Please don't try to cure your pet fern of its reproductive structures.

Even though the ancestors of modern-day ferns evolved megaphylls, other groups of advanced vascular plants did not evolve from ferns. The plants that dominate the land today evolved from ancestors called seed ferns that resembled ferns but were only distantly related to them, plants that flourished during the Carboniferous and Cretaceous periods.

33.12 Seeds and pollen were a major innovation in reproduction.

In adapting to terrestrial life, plants and animals had to become more independent of water. The more primitive vascular plants, whose motile sperm have to swim from the antheridia to the archegonia, are wedded to water for their reproduction. So there must have been considerable selective pressure for any reproductive mechanism that bypassed the motile sperm. Furthermore, an embryonic plant is in danger of drying out unless it stays in a moist place. The evolutionary solutions to these problems were **pollen** to carry the sperm and **seeds** to house and protect the embryo. These reproductive adaptations evolved independently at least twice.

Like non-seed plants, seed plants have spores within sporangia. Non-seed plants release those spores, but the essential feature of seed formation is that the megaspore is never released but produces a small gametophyte including an egg that is fertilized inside a megasporangium, where the resulting embryo remains (Figure 33.22). In seed plants, the number of megaspores in each sporangium is greatly reduced, ultimately to only one. The megasporangium itself is reduced to a thin covering (nucellus) around the megaspore, and this covering in turn is protected by one or two tough integuments that leave only one

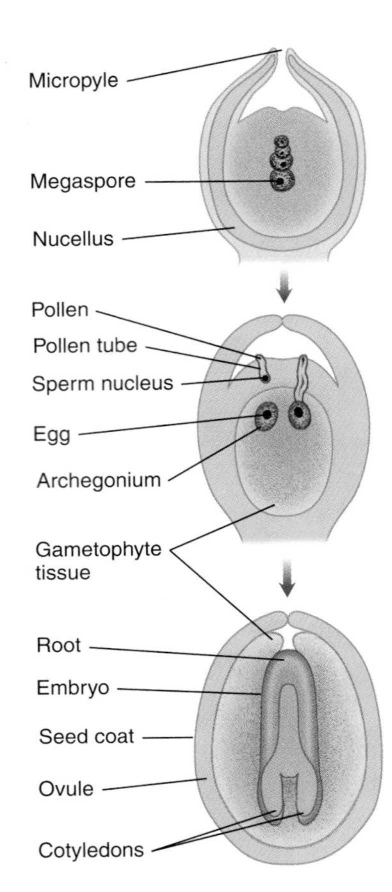

Figure 33.22
A seed develops from a megaspore (one of the four products of meiosis) that is retained with its megasporangium, called the nucellus. The megagametophyte that develops within the megasporangium consists of only a few cells, including an egg cell. When the egg is fertilized by a sperm from a pollen grain, the resulting diploid zygote is the beginning of the new sporophyte generation, and it develops into an embryo.

Figure 33.23
The seed ferns *(Medullosa)* of the Carboniferous period had large, finely divided leaves. They developed fertile leaves, each one bearing megasporangia.

opening where the sperm can get in. The whole structure—the integuments and everything within them—is an **ovule.** After fertilization, the ripened ovule containing a developing embryo is the *seed.* The seed has three distinct generations packed together: the embryo sporophyte within gametophyte tissue, surrounded by the old sporophyte.

A seed relieves the female half of the life cycle from dependence on water and protects the embryo plant against dehydration. Some seeds remain viable for very long times. However, it was the male half of the cycle, the sperm, that really depended on water for its motility in earlier plants, and it was even more critical for sperm to lose this dependence. This was done by a kind of heterochrony (see Section 24.11), which retards the development of the microspores so they don't become swimming sperm. Instead, the microspores of more advanced plants stay in the microsporangium; there each microspore becomes a **pollen grain** enclosing a microgametophyte made of only a few cells, rather than a small plant. Pollen grains, carried by the wind or by animals, stick to the female structures; each grain then grows a tube down to the ovule, where fertilization occurs, so no free-swimming sperm are ever produced. (This whole story is amplified for gymnosperms in Section 33.13 and for angiosperms in Section 33.14.) Modern flowering plants are often pollinated by insects or other animals, and animals probably had similar roles in the lives of much earlier plants.

Pollination and coevolution, Sidebar 24.1.

The evolution of seeds and pollen turned plant development on its head. Simpler vascular plants have a small, earthbound gametophyte that sprouts and temporarily nurtures a much larger sporophyte. But in a seed plant, nearly the entire plant is sporophyte, and the gametophyte is reduced to a few cells. The megagametophyte stays within the old sporangium, and the microgametophyte is reduced to a pollen grain. The megagametophyte develops into the embryo sac inside the ovule; after fertilization of the egg in the embryo sac, the ovule becomes the seed. These plants reverse the nutritional dependence of one generation on another; that is, in seed plants, the gametophyte is dependent on the sporophyte. A seed plant may not look particularly different from some of its ancestors ("a tree is a tree is a tree"), but the internal differences are profound.

The first known seed plants were seed ferns (pteridosperms) of Carboniferous-Jurassic age (Figure 33.23).

Figure 33.24
Modern cycads, *Cycas,* resemble palm trees, but they are more closely related to the seed ferns.

Because of their fronds, they were thought to be ferns until their seeds were discovered. These seeds were borne at the ends of certain fronds, on fertile leaves. Other fronds bore male sporangia that produced pollen grains. Since the ovules were naked and exposed on the tip of a frond, the seed that developed was also naked. All plants of this kind are therefore called **gymnosperms**—literally, "naked-seed" plants. Modern gymnosperms include the conifers and some related groups. This mode of seed formation did not change until the evolution of flowering plants, known as **angiosperms**—literally, "seeds in vessels," whose ovules develop inside a covering, an ovary.

The living plants most similar to seed ferns are cycads (Figure 33.24). Although they look like palms, their trunks bear a distinctive pattern of leaf traces, where leaves were attached, and their reproductive structures are cones, similar to the familiar cones of pines and firs. Modern cycads bear female cones, which produce seeds, and male cones, which produce pollen. A prominent group of Mesozoic-era plants, the Bennettitales, had cones that enclosed both male and female parts.

The tree ferns, cycads, and Bennettitales are cycadophytes. They so dominated the Mesozoic era that it could just as well be called the "Age of Cycads" as the "Age of Dinosaurs." But cycadophytes largely disappeared at the end of that era and were replaced by flowering plants, much as the great reptiles were replaced by mammals.

Exercise 33.2 What is the difference between a homosporous and a heterosporous plant?
Exercise 33.3 Only heterosporous plants have evolved into seed plants. Would it be possible for a homosporous plant to become a seed plant? If so, how would its ovule differ from that of an ovule as we know it?

Figure 33.25
Cordaites, of the Permian to Carboniferous periods, was one of the earliest conifers. It was characterized by its straplike leaves with dichotomous venation.

33.13 Conifers are gymnosperms with relatively simple leaves.

The Carboniferous and Permian forests where seed ferns flourished also contained trees now classified as Cordaitales (Figure 33.25). Standing 15–30 meters high, they bore long, strap-shaped leaves with primitive dichotomous venation. *Cordaites* also had separate male and female cones, and they are among the most primitive relatives of modern conifers: the pines, firs, spruces, cedars, and junipers (Figure 33.26).

The life cycle of a pine is typical of the conifers (Figure 33.27). The visible plant, of course, is a sporophyte, and it bears female (or ovulate) cones with megasporangia and male (or staminate) cones with microsporangia, both typically on the same plant. In a male cone, the microsporangia produce pollen grains, each containing only four haploid cells. One of these cells will eventually divide into a pair of sperm (actually, only sperm nuclei). A pollen grain, with its two winglike air sacs, floats on the air currents; jostling the ripe, pollen-laden cones of a pine tree in spring will release pollen in great yellow clouds (see Figure 32.24*b*).

A female cone has two ovules on each scale (shown in Figure 33.27). Four megaspores form in each ovule, but three of them degenerate, leaving the fourth to develop into a megagametophyte. This may be the slowest developmental process in the world. The meiotic divisions that produce the original four megaspores don't even begin until pollen grains arrive at the ovule. When the one megaspore does start to divide, after a delay of about a month, it spends about a year undergoing

Figure 33.26

Some familiar conifers include *(a)* spruce *(Picea)* and *(b)* White Pine *(Pinus strobus)*. Two conifers that are less familiar to North Americans are native to the Southern Hemisphere: *(c)* the Norfolk Island Pine and *(d)* the Monkey Puzzle tree, which belong to the same genus *(Araucaria)* in spite of their different appearance. Norfolk Island Pines, which are often grown as houseplants, usually show a remarkable pattern of branching: first two branches exactly opposite each other, then three branches spaced 120 degrees apart, then four spaced 90 degrees apart, and so on, adding one more branch at each level, with perfect symmetry.

repeated mitoses without cell division to make over 2,000 nuclei. Finally, cell walls form between these nuclei, and after another few months of leisurely growth, two or three archegonia develop in this mass of tissue.

At the entrance to the ovule, a pollen grain has been waiting all this time, with a pollen tube growing from it toward the gametophyte (in Figure 33.27) at the lethargic pace of the female cells. (The same very slow molecular clock may govern both processes.) Shortly before this tube reaches the megagametophyte, one of the four pollen nuclei divides into two sperm nuclei. One sperm nucleus fertilizes the single egg that has grown in one of the archegonia, and immediately afterward the zygote starts to develop into an embryo, with the beginnings of a root and shoot. Remember that each embryo lies in a seed, buried within the cone. The cone will generally start to dry and open, scattering its seeds as the wind catches their wings. But the cones of some species may stay closed and hold their seeds inside indefinitely, sometimes until the intense heat of a forest fire stimulates them to open and scatter their seeds over ground that needs reforesting. Pine trees are in no hurry.

Although conifers are called evergreens, their needles do not stay green forever; each one eventually drops off and is replaced by a fresh one. Aldo Leopold described the process:

> *Each species of pine has its own constitution, which prescribes a term of office for needles appropriate to its way of life. Thus the white pine retains its needles for a year and a half; the red and jackpines for two years and a half. Incoming needles take office in June, and outgoing needles write farewell addresses in October. All write the same thing, in the same tawny yellow ink,*

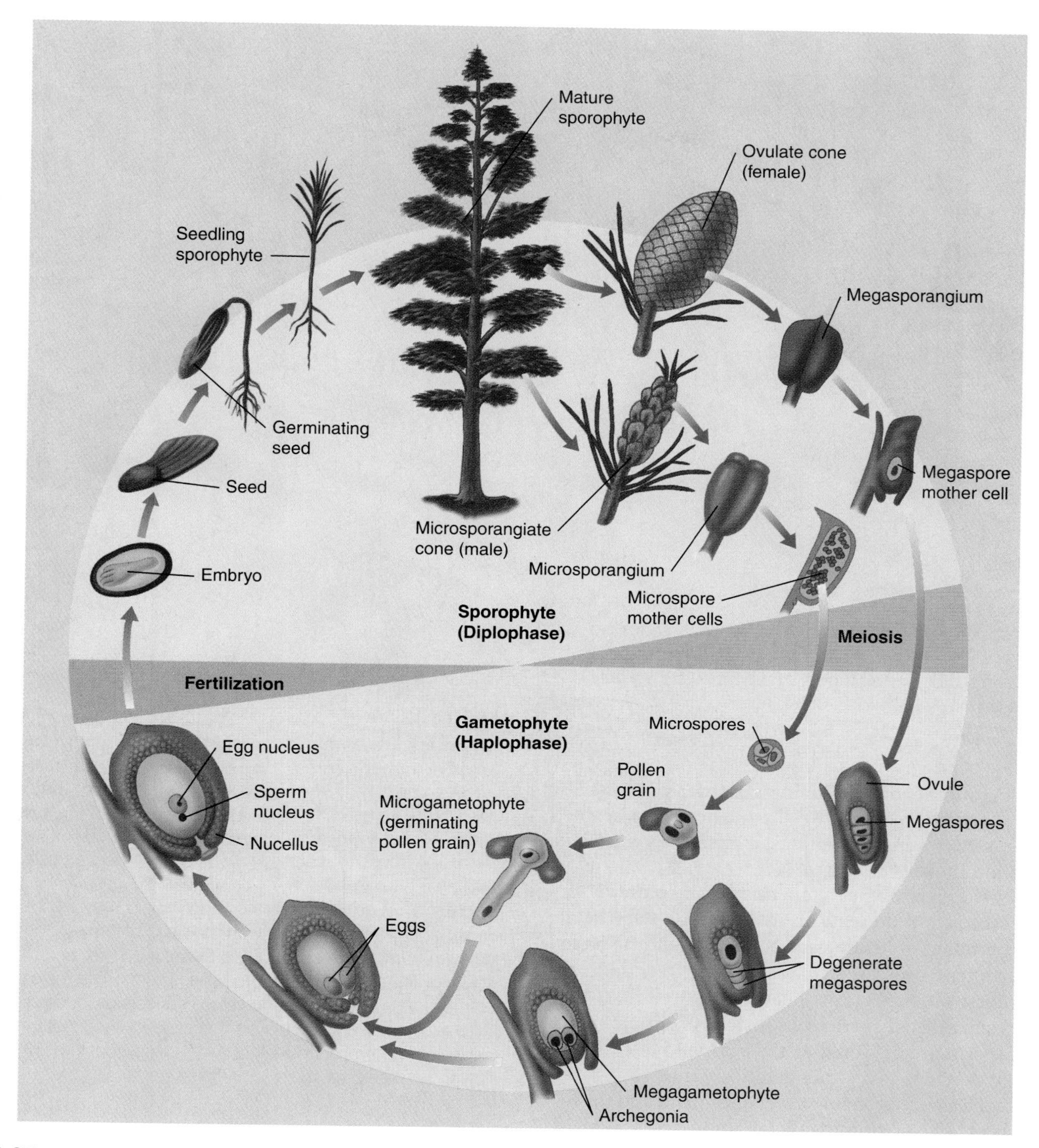

Figure 33.27

In the life cycle of a pine, the very small gametophytes are nutritionally dependent on the large sporophyte. Meiosis in microsporangia in male cones produces pollen. Meiosis in a megasporangium in each ovule on a female cone produces megaspores. One megaspore develops into a minute gametophyte, containing an egg, which is eventually fertilized by a sperm nucleus in a tube that grows from the pollen grain.

> *which by November turns brown. Then the needles fall, and are filed in the duff to enrich the wisdom of the stand. It is this accumulated wisdom that hushes the footsteps of whoever walks under pines.*[2]

[2] Aldo Leopold, *A Sand County Almanac* (Oxford University Press, 1949), p. 87. Reprinted by permission.

Needles look as if they should be microphylls, but they are actually highly reduced megaphylls, an adaptation to dry conditions as explained in Section 38.19. Furthermore, the growth habit that keeps the tree continuously green is an adaptation to life in the cold temperatures and short growing season of the far north or the high mountains, where it is difficult to grow a crop of new photosynthetic organs each year.

Figure 33.28

Leaves of the maidenhair tree, *Ginkgo biloba,* retain the primitive dichotomous venation pattern. Fruits borne on female plants are beautiful, but have the butyric acid odor of rancid butter, so only male plants are generally chosen as ornamentals.

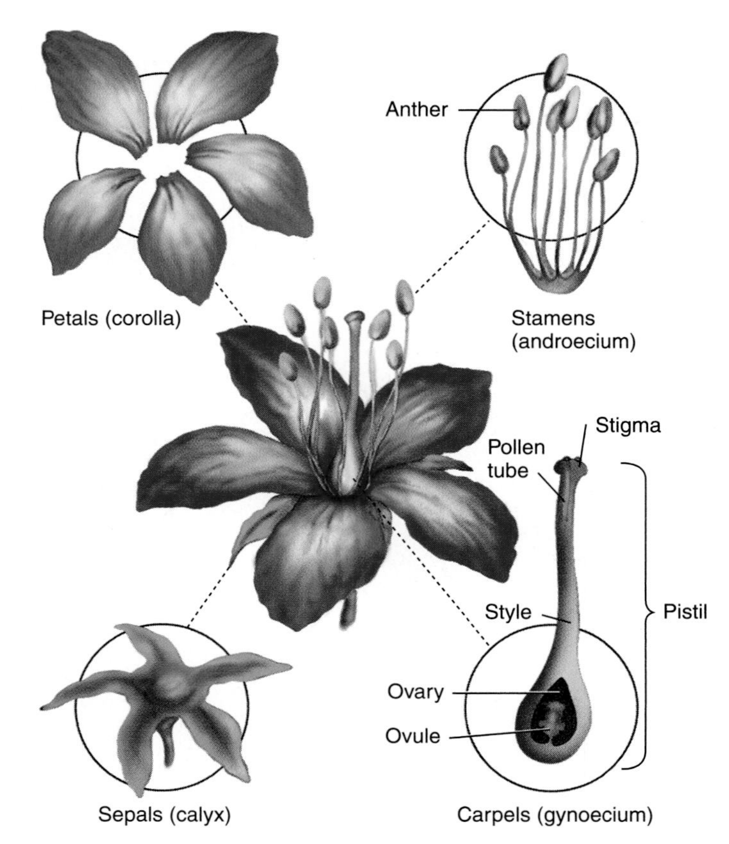

Figure 33.29

An idealized flower has a circle of brightly colored *petals,* which together form the *corolla,* surrounded by a smaller *calyx* made of green *sepals.* The calyx and corolla together form the *perianth* (literally, "around the flower"). The female part of the flower is a *pistil* with a closed *ovary* at its base containing the ovules, ready to be fertilized. The pistil is actually made of fertile leaves called *carpels* that bear ovules; collectively they make the *gynoecium* (literally, "woman's house") of the flower, and usually they are just fused together to make the pistil, with their bases forming the closed ovary. A narrow extension of the ovary, the *style,* ends in a *stigma,* a sticky or feathery tip that traps pollen grains. The male parts are *stamens* that end in *anthers,* where meiosis produces pollen. Stamens are also fertile leaves that collectively constitute the *androecium* ("man's house") outside the gynoecium (in a perfect flower that has both male and female parts).

One small group allied to the conifers is now represented by only a single species, *Ginkgo biloba,* the maidenhair tree. Two features make it interesting for botanists. First, the veins in its leaves are dichotomously branched (Figure 33.28), making it one of the few living species that retain this ancient pattern. Second, like the cycads, it represents an intermediate stage in the evolution of reproduction. Male trees produce air-borne pollen grains that germinate near the ovules of female trees and produce pollen tubes; the tubes grow near the egg cells, but they produce motile, flagellated sperm cells instead of simple sperm nuclei. These sperm are delivered to the eggs via pollen tubes, and a seed then develops as in other gymnosperms.

The Ginkgoales were probably never abundant, and the living ginkgo is kept alive predominantly as an ornamental plant. European scientists thought all ginkgoes were extinct until this species was discovered decorating the grounds of some Asian temples. The tree is popular because of its graceful form, but the developing seeds emit butyric acid, one of the obnoxious materials in rancid butter. For this reason, only male trees are generally selected as ornamentals.

33.14 Flowering plants have seeds enclosed in an ovary, which becomes a fruit.

The Magnoliophyta (or Anthophyta) are the most prominent of all modern plants, and they dominate modern ecosystems. The 250,000 species show an enormous range of size and variety, and include almost all the fruit and vegetable species we depend on for our diets.

The group is defined primarily by that distinctive reproductive structure, the **flower** (Figure 33.29). The female and male parts of a flower are actually fertile leaves, but they are so highly modified that only close study can identify them as such. Flower characteristics are often critical in identifying plants. Several flowers may be grouped in a certain way to make an **inflorescence,** and many flowers lack some of the structures of an idealized flower. A flower with both stamens and pistils is said to be *perfect;* one lacking either structure is *imperfect.* In some species, the female and male parts are separated into different imperfect flowers, either on a single plant (*monoecious,* or

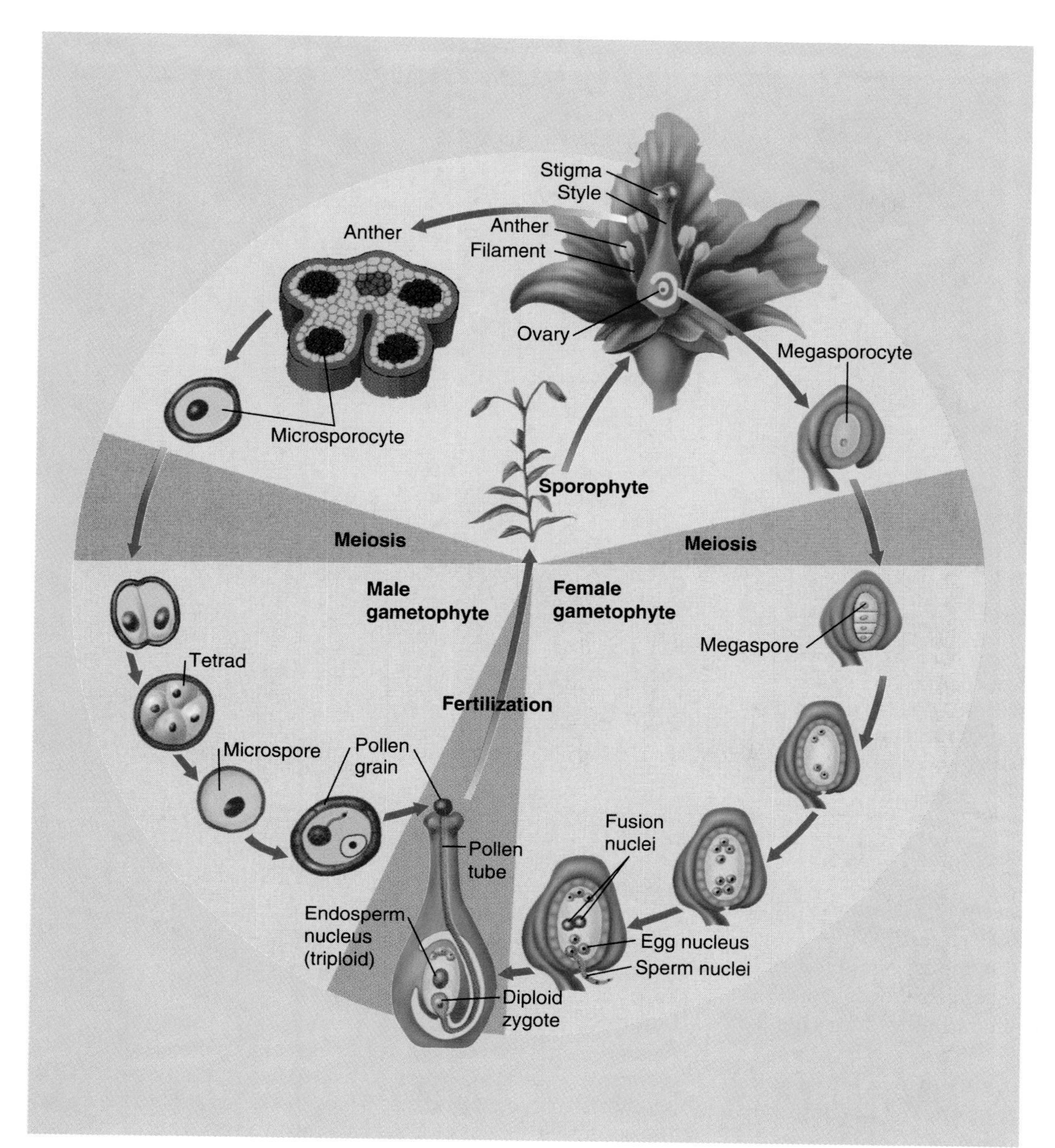

Figure 33.30

In the life cycle of an angiosperm, the male and female gametophytes are very small and nutritionally dependent on the large sporophyte (as in gymnosperms). Notice that the male half of the gametophyte phase is shown on the left and the female half on the right. Fertilization produces a diploid (sporophyte) embryo that grows into a new plant.

unisexual, species) or on separate plants (*dioecious,* or *bisexual,* species).

After fertilization, each fertilized ovule develops into a seed, and the ovary also ripens and develops into a **fruit,** which contains these seeds. The seeds are no longer naked, as in gymnosperms, but are protected by the ovary.

The events of the angiosperm life cycle should be easy to follow (Figure 33.30). The points worth commenting on happen in the flower itself. In the ovary, each ovule has one **megaspore mother cell,** which forms four megaspores through meiosis (Figure 33.31). Three of them usually degenerate, as in a gymnosperm, and the fourth divides a few times more to make an **embryo sac** (the megagametophyte), which in most species, has eight nuclei in seven cells. (This varies, and in some species all eight nuclei reside in a common cytoplasm.) One cell becomes the egg. The two polar nuclei in the large central cell fuse to make a diploid **fusion nucleus,** which will be used to make a food reserve tissue.

Each anther generally has four microsporangia where numerous microspore mother cells develop. Each mother cell goes through meiosis to make four microspores, and each microspore typically becomes a pollen grain. The pollen nucleus divides once more. After landing on a pistil of its own species, the pollen sprouts a pollen tube that burrows through the style toward an ovule (Figure 33.32). Then one nucleus divides again into two sperm nuclei. This division is reminiscent of gymnosperm pollen, where only one sperm nucleus participates in fertilization. In flowering plants, however, both nuclei are used in a unique **double fertilization.** When the pollen tube eventually breaks through the embryo sac, one sperm fertilizes the egg

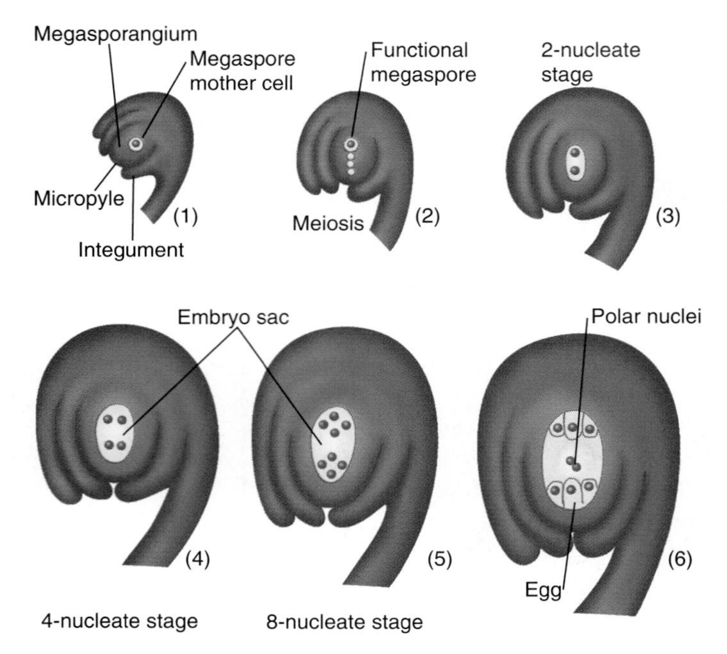

Figure 33.31

In an ovule, the (diploid) megaspore mother cell *(1)* undergoes meiosis *(2),* and three of the four products degenerate, leaving only one functional (haploid) megaspore. The megaspore nucleus divides once *(3),* these two nuclei divide again *(4),* and these four nuclei divide again *(5)* to make eight nuclei. One of them becomes the egg, and the two polar nuclei fuse to form a diploid fusion nucleus *(6).*

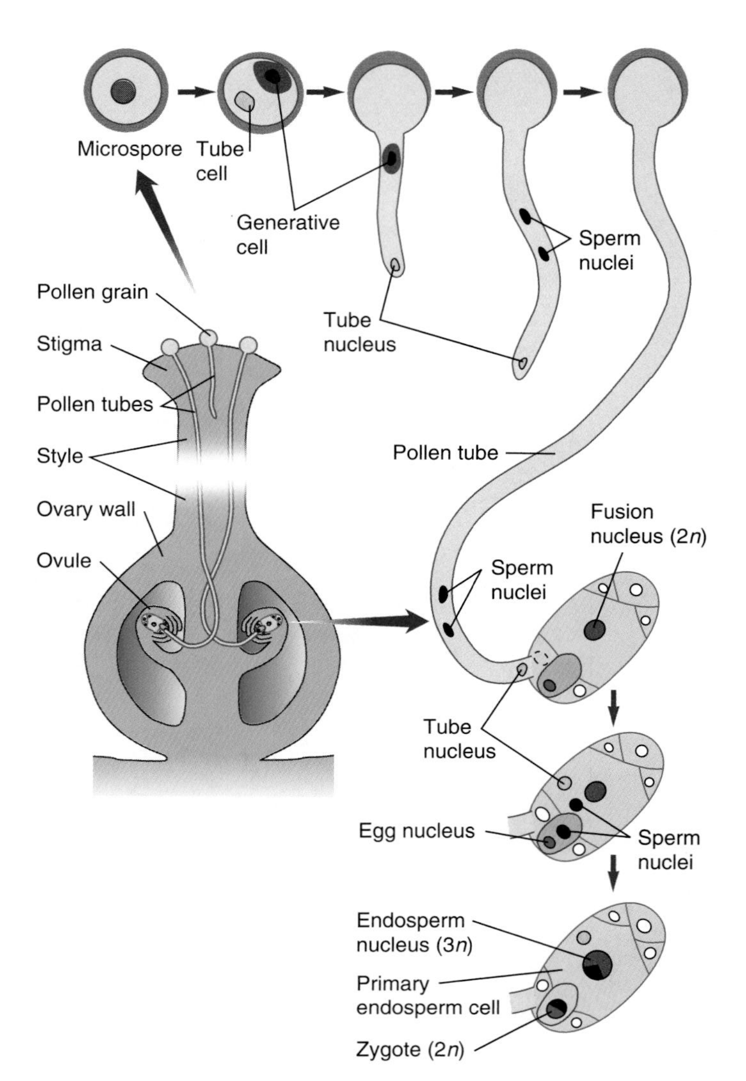

Figure 33.32

A pollen grain is a product of meiosis. Its nucleus divides into a *generative nucleus* and a *tube nucleus.* When a pollen grain sticks to the style of a flower, it produces a tube that grows toward the ovule. The generative nucleus divides again into two *sperm nuclei;* one of them fertilizes the egg to make a zygote, while the other combines with the fusion nucleus to make a triploid endosperm nucleus.

cell to form a zygote, and the other sperm fuses with the fusion nucleus to make a triploid ($3n$) **endosperm cell.** While the zygote develops into an embryo plant, this endosperm cell divides repeatedly to fill much of the embryo sac with endosperm, a tissue that contains a rich store of starch and other nutrients for the embryo.

Meanwhile, the embryo itself begins to develop (Figure 33.33). At the first division of the zygote, one cell develops into the embryo proper, and the other becomes a suspensor that attaches the embryo to the rest of the seed. The suspensor generally differentiates while the embryo grows longer and larger. Nutrients stored in the endosperm are usually digested and converted into food stores in the fleshy cotyledons, which make the bulk of the seed. When the seed matures, the embryo goes into a period of dormancy, an inactive state in which there is very little water to support metabolism. When a seed finds an appropriate environment, with moisture and the right temperature, it absorbs water, and the embryo then continues its growth. Here we leave the seed and embryo. We will continue the story of their development in Chapter 37.

Exercise 33.4 (a) What cell is the immediate precursor of a megaspore? Of a microspore? (b) Since megaspores and microspores are both haploid, do they combine to produce the diplophase?

Coda From humble unicellular beginnings, the plants have evolved into structures that dominate the landscape. Chromists in the form of kelps may dominate many marine vistas, but the green plants form the base of most ecosystems, both in their mass and in their role as the chief phototrophs of the earth, which bring energy into the biological community. Green algae support freshwater communities, and some of their descendants moved onto the land by evolving strong vascular tissue that transports water and supports the plant body. Some botanists have speculated that, ironically, terrestrial plants evolved as algae adapted to living more securely in the water; in habitats that dried up during the summer, some species survived not by going into cryptobiotic spore stages but by virtue of adaptations we now recognize in

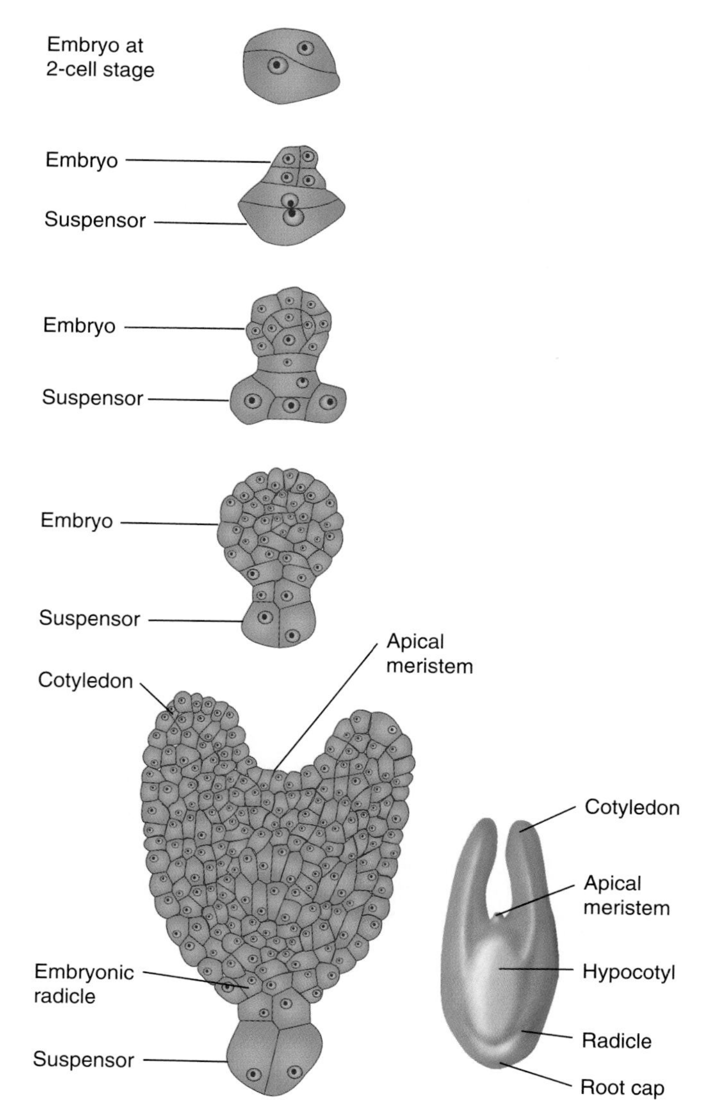

Figure 33.33

An angiosperm embryo first divides into two cells; one becomes the embryo proper, and the other becomes a suspensor, which is likened to the umbilical cord of a mammal because it conducts food reserves from the endosperm to the embryo. The embryo develops either one or two large cotyledons; the type of embryo shown here has two cotyledons, which store much of the food from the endosperm. The main embryonic axis extends between an embryonic root (radicle) and a second region of growth (apicle meristem) that will become the shoot.

embryophyte plants: rhizoids (later, roots) and tubelike cells for conducting water from the wet soil to remote parts of the plant, and some degree of waterproofing on exposed surfaces to prevent desiccation. We have seen that the defining features of tracheophytes, their efficient vascular systems, depend strongly on the virtues of cellulose and other hardy polymers that form the tough supporting walls of xylem cells. In fact, many important plant features depend on polymeric structure. The suberin and cutin used for waterproofing are polymers of fatty acids; lignin is a polymer of coniferyl alcohol and some of its relatives; and the walls of pollen grains contain an unusual polymer called sporopollenin. Plants also produce a remarkable variety of polysaccharides, including uncommon sugars such arabinose and xylose, as components of wood or fruits; since plants have evolved special enzymes to synthesize these molecules, they must have significant functions, which remain unknown to humans.

Of course, the conquest of the land has been largely an evolution in modes of reproduction. Even bryophytes form embryos that are held inside protective nonreproductive tissues, and all the embryophytes employ resistant spores at one stage or another. As we have repeatedly emphasized, the great change in reproduction came with the evolution of seeds and pollen, the structures that finally displaced water from its primitive role in reproduction and placed embryo sporophytes in secure, resistant enclosures where they might remain viable for years.

Summary

1. Plants are defined as phototrophic organisms with cellulose cell walls that store starch as a food reserve and have chlorophylls *a* and *b*.
2. The thallophytes are green algae that have evolved along several distinct lines, including colonial flagellates, coenocytic filaments, and parenchymal filaments and sheets. Some of the filamentous algae, which have both horizontal filaments and rhizoids, are probably like the ancestors of land plants.
3. The life cycles of more advanced plants include alternations between a haploid gametophyte and a diploid sporophyte. The gametophyte develops organs called gametangia where gametes are produced; male gametangia are antheridia, and female gametangia are archegonia. The zygote made by fertilization of an egg by a sperm develops into a sporophyte, which then forms sporangia where spores are formed. These spores develop into gametophytes.
4. Two trends in evolution are toward dominance of the sporophyte and increased prevalence of heterospory, the production of mega- and microspores that develop into separate female and male gametophyte plants.
5. The embryophytes are mostly terrestrial plants that have multicellular gametangia in which embryos develop. Their cells also divide by means of cell plate formation.
6. Bryophytes, including mosses and liverworts, are nonvascular plants with relatively simple, undifferentiated gametophyte bodies—green, leafy structures. They reproduce by means of sperm that swim in a film of water to the archegonia where they fertilize eggs; the resulting zygotes develop into sporophytes, which produce spores in sporangia at their tips.
7. Vascular plants evolved xylem and phloem tissues. The primitive xylem elements (tracheids) have evolved into shorter tracheids, long supporting fibers, and short vessel elements.
8. Primitive microphyll leaves have been replaced by megaphylls with dichotomous branching of their veins. The most primitive vascular plants had simple, dichotomously branched stems.
9. One branch of the plant family tree includes lycopsids, or club mosses, which developed fertile leaves, a common feature of other plants. Fertile leaves combine sporangia with leaves, and they commonly occur in clusters (cones).
10. Equisetophytes, which are now represented only by horsetails, have rigid stems heavily impregnated with silica and scale-like, nonfunctional microphyll leaves.

11. Ferns were among the first plants to develop megaphyll leaves; fern leaves have pinnate venation, a central vein with smaller veins arising from it. The fern gametophyte is very small; what most of us recognize as a fern is the sporophyte, which bears its spores in sporangia usually clustered in sori underneath the leaves.
12. Some plants evolved seeds and pollen as a mechanism for reproduction without the need for water-borne sperm. A seed develops from an ovule. The ovule consists of a megaspore enclosed in its megasporangium, where it undergoes meiosis to produce a few gametophyte cells. Among these haploid cells is an egg, which is fertilized by a sperm carried in the pollen tube of a pollen grain.
13. Gymnosperms bear their ovules on exposed scales that are clustered in cones. A tube grows from each pollen grain, and sperm migrate through it to fertilize the egg cells.
14. Angiosperms have flowers containing both stamens that produce pollen and a pistil that encloses ovules in an ovary. The pollen tube that grows toward an ovule develops two sperm, which engage in a double fertilization; one of them fertilizes the egg cell, and the other fertilizes the polar nucleus (fusion nucleus) to produce triploid endosperm tissue. Endosperm becomes a store of food to feed the developing embryo. After fertilization, the seeds develop within the ripened ovary, which becomes a fruit.

Key Terms

cell plate 694
gametophyte 694
gametangium 694
antheridium 694
archegonium 694
sporophyte 694
sporangium 694
homosporous 694
heterosporous 694
megasporangium 694
microsporangium 694
megagametophyte 694
microgametophyte 694
protonema 696
vascular bundle 697
stele 697
tracheid 697
fiber 698
vessel element 698
vessel 698
microphyll 698
megaphyll 698
pinnate 700
pollen 703
seed 703
ovule 704
pollen grain 704
gymnosperm 705
angiosperm 705
flower 708
inflorescence 708
fruit 709
megaspore mother cell 709
embryo sac 709
fusion nucleus 709
double fertilization 709
endosperm cell 710

Multiple-Choice Questions

1. Members of the plant kingdom have all these characteristics *except*
   a. tinsel flagella.
   b. starch as the storage carbohydrate.
   c. chlorophylls *a* and *b*.
   d. cell walls made of cellulose.
   e. autotrophy.
2. Plants whose bodies are thalli are
   a. algae.
   b. bryophytes.
   c. nonvascular.
   d. *a* and *b*, but not *c*.
   e. *a*, *b*, and *c*.
3. Which is true for ferns but not for bryophytes?
   a. presence of xylem and phloem
   b. ground-dwelling gametophyte plants
   c. photosynthetic gametophyte
   d. separate haploid and diploid structures
   e. sporophyte temporarily or permanently dependent on gametophyte
4. All of the following structural plans or forms are found among the green algae *except*
   a. coenocytic filaments.
   b. motility by means of pseudopodia.
   c. ciliated colonies.
   d. unicellular, flagellated organisms.
   e. flat sheets of parenchyma.
5. Which of the following are found among bryophytes (nontracheophyte embryophytes)?
   a. stomata
   b. flagellated sperm
   c. elaters
   d. *a* and *b*, but not *c*
   e. *a*, *b*, and *c*
6. Among vascular plants that do not produce seeds, meiosis occurs within
   a. archegonia.
   b. sporangia.
   c. megagametophytes.
   d. microgametophytes.
   e. antheridia.
7. Which pair of terms is mismatched?
   a. thallophytes—gametangia
   b. moss—nonvascular
   c. fern—seeds in ovules
   d. gymnosperms—naked seeds
   e. angiosperms—reduced mega- and microgametophytes
8. Modern ferns are characterized by all *except*
   a. megaphylls.
   b. pinnate branching of veins in leaves.
   c. seeds.
   d. sporangia on the underside of fronds.
   e. flagellated sperm that require water for fertilization.
9. Seeds contain
   a. an ovule containing a sporophyte embryo.
   b. a megasporangium.
   c. a fertilized meiospore.
   d. a fertilized megaspore.
   e. all of the above.
10. The fleshy part of an apple or peach consists of
   a. the embryo gametophyte containing a sporophyte.
   b. the embryo sporophyte containing a seed.
   c. an ovary containing gametophytes containing embryo sporophytes.
   d. a megaspore that has not yet undergone meiosis.
   e. a microspore within a microsporangium.

True-False Questions

Mark each statement true or false, and if false, restate it to make it true.

1. Gametes are produced in gametophytes that are part of a gametangial plant.
2. Antheridia and archegonia are spore-producing structures on the sporophyte plant.
3. Spore mother cells undergo meiosis and develop into haploid gametes.
4. Trees, including specimens such as maple, oak, or spruce, are the haploid sporophyte portion of a life cycle.

5. Both conifers and flowering plants are heterosporous, producing two kinds of gametophytes.
6. Within the plant kingdom, thallus formation is found only in the algae.
7. The first vascular plants had a central core of vascular tissue made of fibers and continuous vessels.
8. Primitive vascular plants had dichotomously branched stems, small leaves, and sporangia.
9. A pollen grain is really a microspore enclosing a microgametophyte.
10. Before the evolution of pollen and seeds, both the gametophyte and the embryo sporophyte stages of the plant life cycle were subject to the dangers of desiccation.

Concept Questions

1. What is the evidence that the Charophyta and/or *Fritschiella* may be ancestral to the embryophytes?
2. If mosses do not have true roots or leaves, how do they obtain nutrients (water and minerals)?
3. Contrast the most primitive arrangement of vascular tissue with the pattern seen in modern flowering trees. How does the modern arrangement better serve the needs of large, woody specimens?
4. Contrast the position of meiosis in the life cycle of modern plants with its position in sexually reproducing animals.
5. What parts of its life cycle are contained within each of the large cones on a pine tree?

Additional Reading

Attenborough, David. *The Private Life of Plants: A Natural History of Plant Behaviour.* Princeton University Press, Princeton (NJ), 1995.

Banks, Harlan P. *Evolution and Plants of the Past.* Wadsworth Publishing Co., Belmont (CA), 1970.

Barrett, Spencer C. H. "Mimicry in Plants." *Scientific American,* September 1987, p. 76. There are flowers that look like insects and weeds that masquerade as crop plants. Mimicry in plants results from natural selection: It attracts pollinators or deters predators.

Cobb, Boughton. *A Field Guide to the Ferns.* Houghton Mifflin Co., Boston, 1956.

Cox, Paul A. "Water-Pollinated Plants." *Scientific American,* October 1993. Some aquatic plants have evolved common strategies for pollination in the water.

Moore, R., W. D. Clark, and D. S. Vocopich. *Botany,* 2d ed. WCB/McGraw-Hill, Dubuque (IA), 1998.

Petrides, George A. *A Field Guide to Trees and Shrubs.* Houghton Mifflin Co., Boston, 1958.

Scagel, Robert F., et al. *Plant Diversity: An Evolutionary Approach.* Wadsworth Publishing Co., Belmont (CA), 1969.

———. *Nonvascular Plants: An Evolutionary Survey.* Wadsworth Publishing Co., Belmont (CA), 1982.

Watson, E. V. *The Structure and Life of Bryophytes,* 3d ed. Hutchinson & Co., London, 1971.

Internet Resource

To further explore the content of this chapter, log on to the web site at:

http://www.mhhe.com/biosci/genbio/guttman/

34 Animals I: General Features and the Lower Phyla

A trilobite fossil represents the rich variety of animals that appeared quite suddenly, by geological standards, in the Cambrian period.

Key Concepts

34.1 Animals are multicellular holotrophs that share many features.

34.2 Animals are divisible into several major taxa.

34.3 The origins of the metazoans are obscure.

34.4 Sponges are basically cellular aggregates without true tissues and with little integration.

34.5 Cnidarians are built around a central digestive cavity.

34.6 The flatworms are bilateral acoelomates.

34.7 Ribbon worms are acoelomates with more advanced animal features.

34.8 The evolution of a coelom enabled animals to occupy new niches.

34.9 Coeloms probably evolved several times through different paths.

34.10 Embryological features distinguish major groups of animals: protostomes and deuterostomes.

34.11 The pseudocoelomates share an unusual body structure.

A small but growing cadre of biologists would be tempted to sell their souls for the opportunity to go back and observe the earth between about 560 and 540 million years ago. Remarkable changes were going on at that time, changes we can only begin to fathom from the fossil record and from studies of living animals. The story biologists would love to complete begins something like this: Some colonial, flagellated, nonphotosynthetic protists started to evolve into flattened, ciliated creatures, the first animals. For a time, they were successful predators on their unicellular neighbors just by being large and able to swim fast. Then some of them developed two-layered bodies with an inner space, and—perhaps permanently, perhaps temporarily—folded their bodies into chambers where they trapped and digested food particles. They began to evolve what we now call *Hox* genes, genes that determine how bodies can change and differentiate. Within a short time, geologically speaking, these animals radiated into all the basic body forms we now see in modern animals, and probably many others that have since become extinct. They evolved muscles that permitted strong, rapid movements and mechanisms for communication between their cells to coordinate those movements. They evolved more complicated chambers for holding and digesting their food, with specializations for catching that food. They evolved fluid-filled hollows in their bodies that served as simple skeletons, allowing them to tap the rich organic debris of the ocean bottom. That is what seems to have happened. And although we know little more about *how* it happened, we will begin to look at some of the products of that evolution in this chapter and the next. ■

34.1 Animals are multicellular holotrophs that share many features.

Who are these creatures, the Animalia or Metazoa, that began their evolution during that short Precambrian time? Animals are multicellular holotrophs, meaning that they get their nutrients by engulfing food and digesting it. Most animals are diploid. They reproduce sexually with eggs and sperm that form a zygote, which then develops into an embryo and often into a **larva,** an immature developmental form that may be very different from the adult.

We will tell the story of the animals largely in relation to the course of their evolution, and we begin with some thoughts about the origins and early evolution of the animal kingdom in relation to ecology. Animals are best described as having a certain *body plan:* a specific architecture and anatomical arrangement that allows them to function as an integrated whole. Since animals are fundamentally holotrophs, their body plans must allow them to obtain and ingest food efficiently so as to occupy particular niches. Animals have evolved several distinctive ways of moving, obtaining their food, and thus occupying different adaptive zones.

Adaptive zones, Section 24.7.

Good arguments have been advanced to show that the kingdom Metazoa is monophyletic (see Figure 34.25). Nevertheless, the differences among the earliest-evolved metazoan phyla are so profound that it is hard to determine the nature of their common ancestor, a topic that has created profound disagreement among zoologists.

34.2 Animals are divisible into several major taxa.

Animals are so diverse that various authorities have divided them into between 20 and 40 phyla. We can make sense of this diversity by grouping these phyla in a series of larger taxa on the basis of the animals' distinctive body plans and embryology (Figure 34.1). Concepts 34.1 presents a brief review of the initial stages of animal development, as described at length in Chapter 20. The distinction between invertebrates and vertebrates, incidentally, is just a convenient bit of chauvinism. Were they able

Concepts 34.1 A Summary of Early Animal Development

A typical animal zygote divides (cleaves) into cells called *blastomeres.*

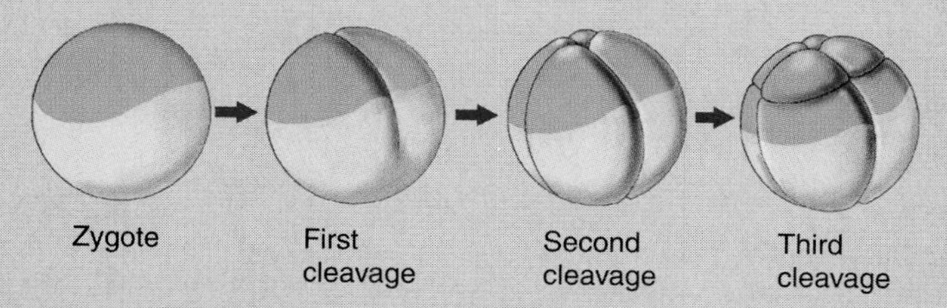

Soon the zygote becomes a hollow ball of cells, a *blastula,* enclosing a space, the *blastocoel:*

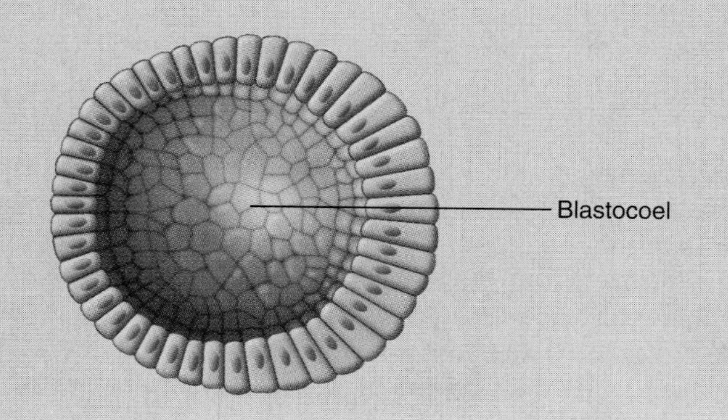

At one point on the blastula, an invagination, the *blastopore,* begins to form. Cells sweep inward through the blastopore to make a *gastrula,* forming an embryo with the beginnings of two tissue layers: *ectoderm* on the outside, which will eventually form surface tissues such as skin, and *endoderm* on the inside, which will eventually form the intestinal lining:

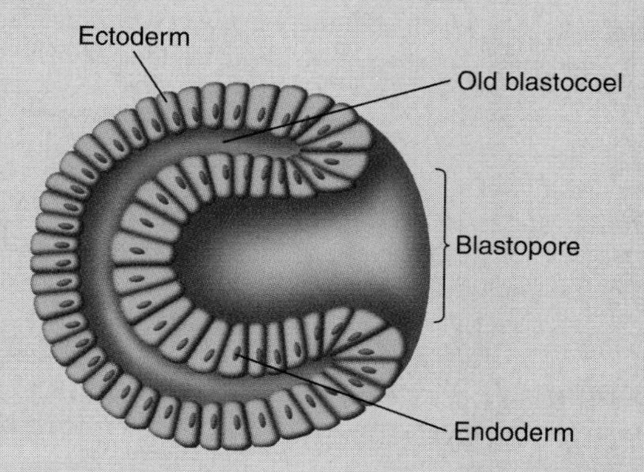

A middle layer of *mesoderm* then develops, which will make the bulk of internal organs and muscles. Thus a typical animal comes to have a tube-within-a-tube form, roughly identified with the three embryonic layers.

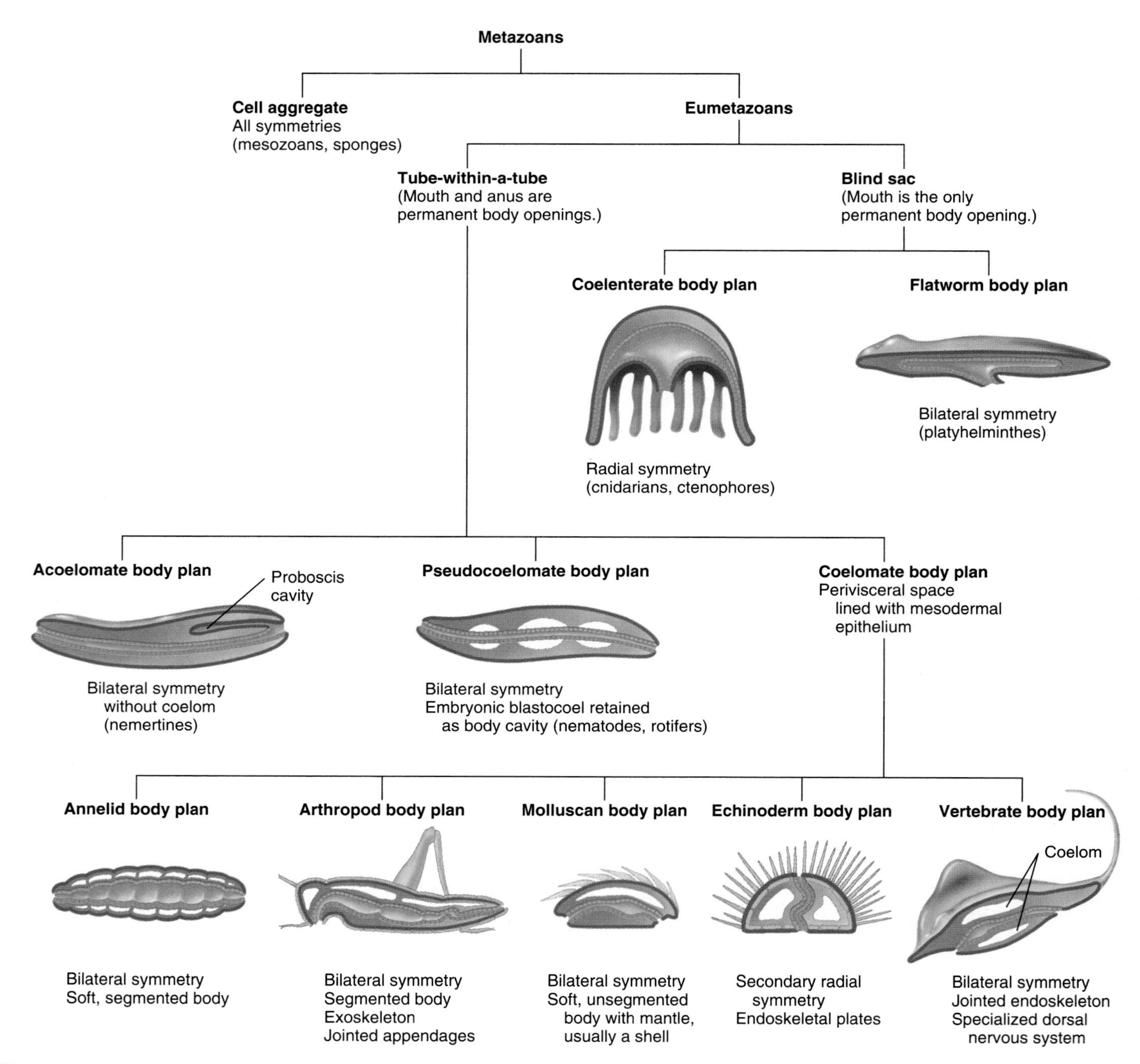

Figure 34.1

The principal animal phyla can be divided into categories on the basis of their body plans, as characterized by symmetry, arrangement of body cavities, segmentation, and skeletons. This arrangement correlates with the largest taxa in Table 34.1, except that it divides eumetazoans first by the form of their intestines instead of dividing them first into Radiata and Bilateria.

to do so, arthropods like insects and crabs might place their phylum in a special position and refer to everything else as "inarthropods."

Some primitive animals are set off in three subkingdoms, showing that they are not closely related to other animals (Table 34.1) A very peculiar group, the **Mesozoa,** are parasites made of one central (axial) cell surrounded by a few outer cells (Figure 34.2). Some experts think they are related to ciliates and place them among the protists (protozoa), but the majority believe they are a kind of degenerate flatworm. We will discuss animals in the subkingdom **Placozoa** in the next section in relationship to metazoan origins. The sponges and some of their extinct relatives, subkingdom **Parazoa,** are made of somewhat specialized cells that do not form true tissues and operate more independently than do the highly integrated systems of most other animals.

Table 34.1 Classification of the Animal Kingdom

Rank	Taxon
Subkingdom 1.	Mesozoa. Phylum Mesozoa
Subkingdom 2.	Placozoa. Phylum Placozoa
Subkingdom 3.	Parazoa. Phylum Porifera: sponges
Subkingdom 4.	Eumetozoa
Infrakingdom 1.	Radiata: with radial symmetry
	Phylum Cnidaria: coelenterates
	Phylum Ctenophora: comb jellies
Infrakingdom 2.	Bilateria: with bilateral symmetry or secondary radial symmetry
Branch 1.	Acoelomata: with no coelom
	Phylum Platyhelminthes: flatworms
	Phylum Nemertea: ribbon worms
Branch 2.	Pseudocoelomata: coelom derived from embryonic blastocoel with only partial mesodermal lining
	Phylum Rotifera: rotifers
	Phylum Nematoda: roundworms
	Phyla Gastrotricha, Kinorhyncha, Nematomorpha, Priapulida, Acanthocephala, Entoprocta, Gnathostomulida, Loricifera
Branch 3.	Coelomata: coelom completely lined with mesoderm
Subbranch 1.	Protostomia
	Phylum Annelida: segmented worms
	Phylum Arthropoda: crustaceans, arachnids, insects, centipedes, etc.
	Phylum Mollusca: snails, whelks, clams, oysters, squid, etc.
	Phyla Sipuncula, Echiura, Pogonophora, Vestimentaria
Subbranch 2.	Deuterostomia
	Phylum Echinodermata: sea stars, sea urchins, etc.
	Phylum Chaetognatha: arrow worms.
	Phyla Bryozoa, Brachiopoda, Phoronida: lophophorate animals
	Phylum Hemichordata
	Phylum Chordata: vertebrates and their relatives

Most animals constitute the subkingdom **Eumetazoa** ("true animals") and are divided into the infrakingdoms Radiata and Bilateria based on their body symmetry and number of tissue layers (see Table 34.1). Radiata are radially symmetrical animals such as corals, sea anemones, jellyfish, and their relatives. They are **diploblastic,** having only two layers of tissue with a poorly defined, jellylike material (mesoglea) in between, and they develop quite differently from other animals:

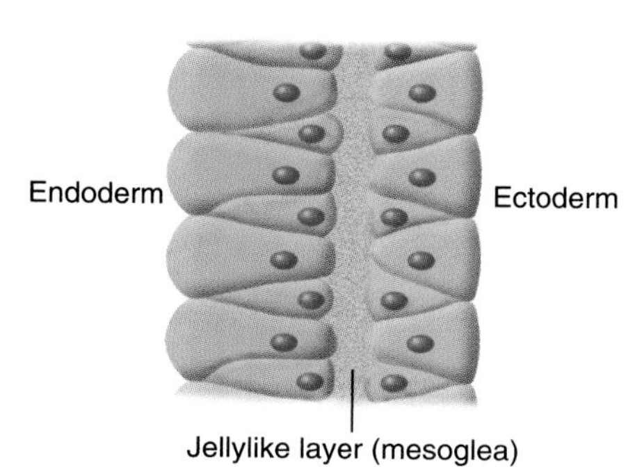

Bilateria include all other animals, which have a nominal bilateral symmetry, even though some prominent phyla include

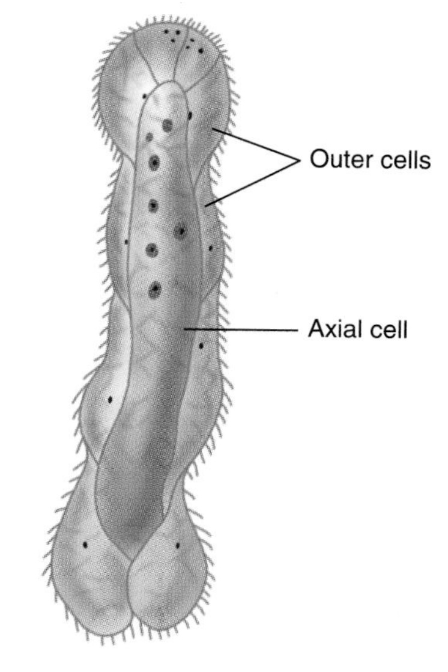

Figure 34.2
A mesozoan has a central axial cell surrounded by several outer cells.

animals that have secondarily acquired radial symmetry. All Bilateria are **triploblastic,** having three tissue layers. The Bilateria are then divided into major groups (here called branches) based on their possession of a **coelom,** a fluid-filled cavity between the body wall and intestine:

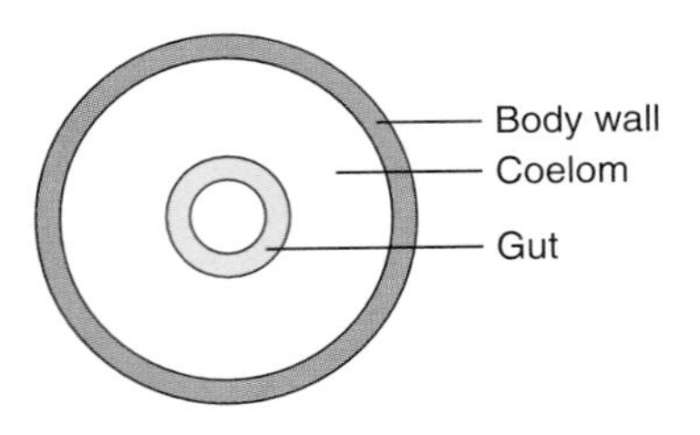

34.3 The origins of the metazoans are obscure.

The origin of the animal kingdom has been a matter of great controversy for over a century, and it is likely to remain one of those fascinating problems that have no definitive answer. But since most animals (eumetazoans) have flagellated or ciliated cells, especially in their larval stages, flagellated or ciliated protozoans are commonly seen as their ancestors.

Concepts 34.2 explains how an animal's embryonic development holds clues to the forms of its ancestors. Since a typical animal becomes a hollow blastula and then an invaginated gastrula, zoological theorists have commonly imagined an evolutionary parallel: an early hollow *blastea* followed in evolution by a *gastrea,* a hollow, cellular aggregate with an invagination that formed a simple digestive pouch. It is therefore interesting that the larval stage of the cnidarians is a flattened, hollow, two-layered *planula* and that sponges have a flat, one-layered larva. Most theorists now imagine that the ancestral metazoan was a flattened, colonial flagellate that evolved into a hollow, two-layered form and then perhaps into something like a gastrea. A century ago, Otto Bütschli proposed that the primitive eumetazoan was similar to a planula, which he called a *plakula* (Figure 34.3): a creeping organism with bilateral symmetry and two cell layers, that fed by ingesting food particles in its bottom layer. Such an organism might feed by raising itself up to form a temporary digestive chamber. Thus Bütschli provided a reason why a gastrulalike form might evolve. He also pointed out that F. E. Schulze had described an organism with just the right form and behavior, *Trichoplax adhaerens.* However, *Trichoplax* was generally considered to be just a cnidarian larva and was forgotten.

In 1974 Karl G. Grell rediscovered *Trichoplax* and demonstrated that it is not a larva, because it reproduces both sexually and asexually, and just as Bütschli had said, it forms a temporary digestive cavity. Since *Trichoplax* is not bilaterally symmetrical, a primitive animal of this kind could be

Concepts 34.2 Ontogeny (Sort of) Recapitulates Phylogeny

As some nineteenth-century zoologists compared the anatomy of animals with the anatomy of their embryos, they rallied under a motto that became a cliche of biology: "Ontogeny recapitulates phylogeny!" By this they meant that during an animal's development (ontogeny), it repeats (recapitulates) the structures of its ancestors (phylogeny). A mammal, for instance, evolved from a reptilian ancestor, which evolved from an amphibian, which evolved from a fish. So, according to this viewpoint, the early mammalian embryo looks like a fish, and then successively looks like an amphibian, a reptile, and finally a more generalized mammal.

This picture is only partly true, but it is true enough to be useful in thinking about both ontogeny and phylogeny. As a mammal develops, it first looks like a fish *embryo,* then like an amphibian *embryo,* then like a reptile *embryo* (see illustration). It makes sense that an organism should develop in this way. Since amphibians evolved by modification of fish anatomy, the developmental program in their genomes begins with instructions to develop into a fishlike form, but then the program changes to make them into amphibians. Similarly, reptiles which evolved from amphibians, carry developmental programs that make their embryos like those of amphibians up to a point, and then change into the pathway that makes them reptiles. In other words, every evolutionary change is built upon a developmental history, and this is why embryonic stages show the progression we see.

This view of development lies behind looking at larvae for clues to ancestors. This is what makes us think that the planula larva of cnidarians, for instance, must resemble the cnidarian ancestor.

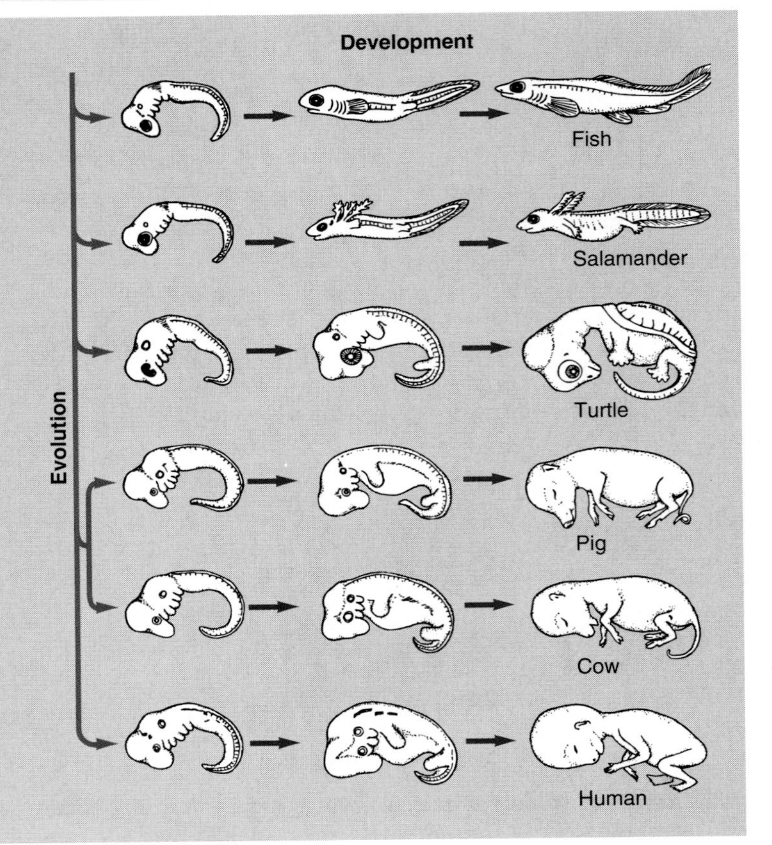

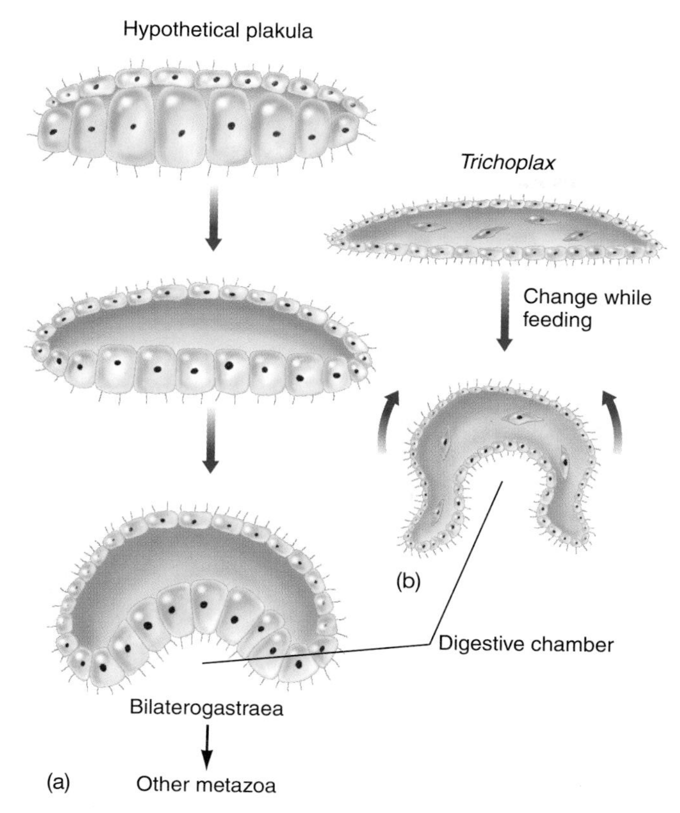

Figure 34.3
(a) A plakula, a hypothetical animal made of two layers of cells, might have been the ancestor of all other animals. An intermediate stage in evolution could have been a bilaterogastraea that could digest larger bits of food within its concavity. *(b) Trichoplax* is a real, contemporary animal of plakula form that feeds by ingesting food into its lower cells and temporarily forms a gastrula-like pocket similar to the hypothetical bilaterogastraea.

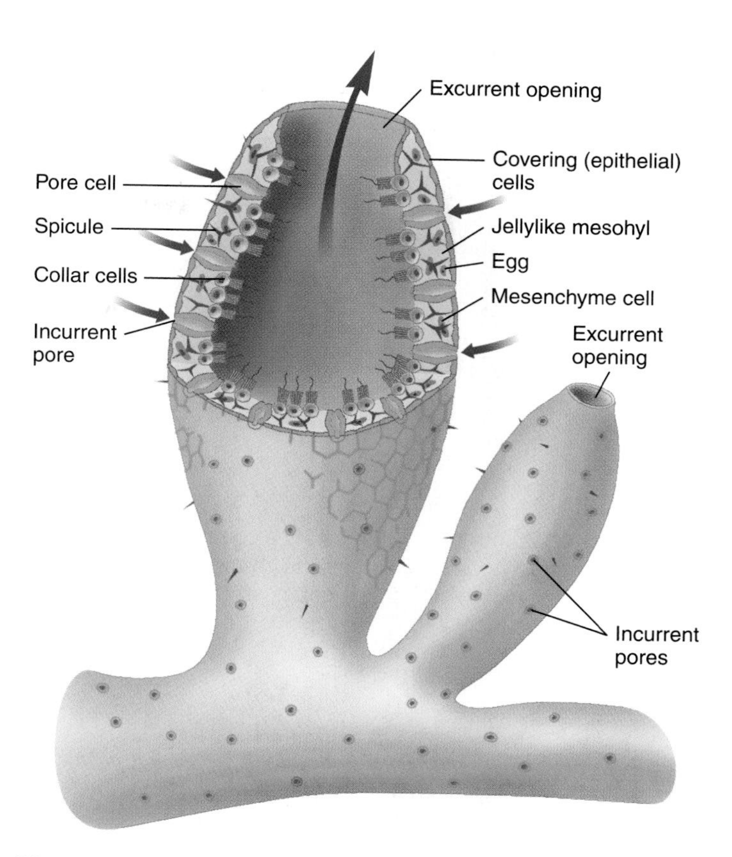

Figure 34.4
Sponges are animals that conduct a stream of water *(shown by arrows)* inward through incurrent pores and outward through the excurrent opening. Choanocytes in a sponge's interior trap food particles from the water.

ancestral to bilateral forms like flatworms, or to radial forms like cnidarians, or both. Thinking about animals of this kind seems to hold a solution to the old phylogenetic problem of the origin of the Metazoa.

Exercise 34.1 If you had a well-equipped molecular biology laboratory and live specimens of protists and primitive animals, such as ciliates, mesozoans, *Trichoplax,* and flatworms, what kind of experiments would you do to sort out the phylogenetic relationships among them?

34.4 Sponges are basically cellular aggregates without true tissues and with little integration.

Sponges are among the simplest animals, but their sessile habit and plantlike form led naturalists to think they were plants until the end of the eighteenth century; many skindivers make the same mistake today. The 10,000 species, ranging from a few millimeters to a meter or more in diameter, are sometimes dull gray and brown but more often brilliant orange, purple, or red. They are *modular* rather than unitary individuals, made of small units that sprout from one another asexually.

Sponges are filter feeders. Each unit pumps a stream of water inward from many *incurrent pores,* through a central cavity, and out by way of an excurrent opening, or osculum (Figure 34.4). This current is maintained by **collar cells,** or **choanocytes** (Figure 34.5), which move the water by the coordinated beating of their flagella. Each collar cell also traps dissolved nutrients and particles of food from the water with its collar and moves them down into its cell body to be engulfed and digested. Sponges are of three principal types, the more complex forms having a maze of canals and chambers lined with choanocytes (Figure 34.6). Though the water stream has very low pressure, the more complicated sponges can replace their entire water volume in a minute.

Although a few animals in other phyla also have choanocytes, they are a distinctive feature of sponges and are used to connect the sponges with their probable ancestors. Some colonial flagellates known as choanoflagellates and craspedomonads have virtually identical cells and are like very simple sponges (Figure 34.7). Although some zoologists

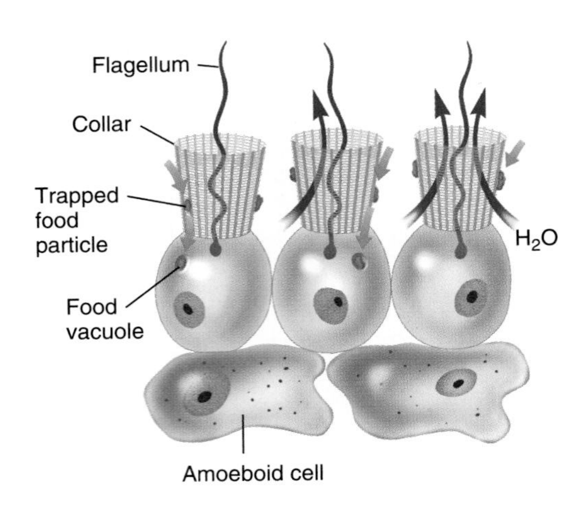

Figure 34.5

The collar of a choanocyte, or collar cell, is a comblike structure made of many fine cytoplasmic extensions (microvilli). The flagellum moves a stream of water through the collar, which traps food particles and moves them into food vacuoles for digestion.

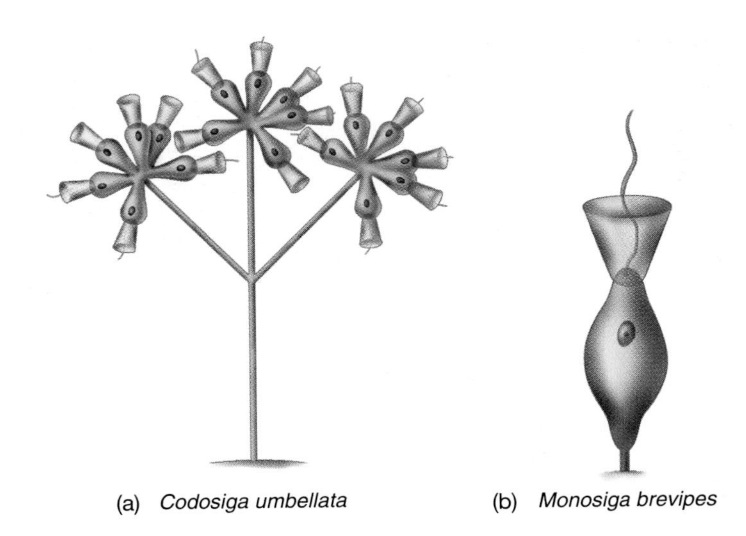

Figure 34.7

Some flagellates, including choanoflagellates such as *Codosiga (a)* and craspedomonads such as *Monosiga (b),* appear to be related to sponges.

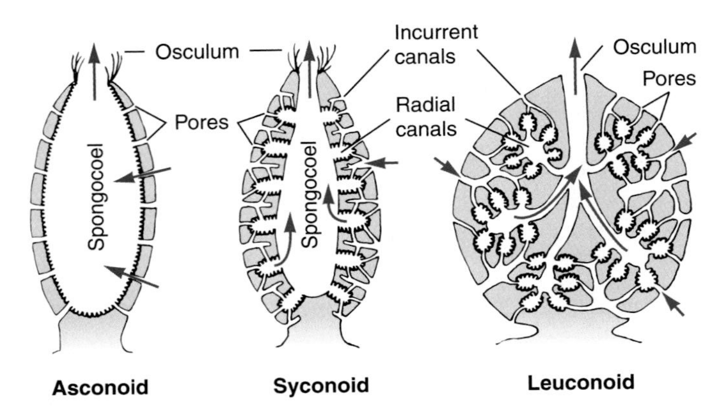

Figure 34.6

Sponges may be one of three principal types. Asconoids have only simple chambers lined by choanocytes; syconoids have passages lined by choanocytes between chambers; and leuconoids have choanocyte-lined chambers connected by passages.

Figure 34.8

Representative spicules taken from various sponges have regular geometric forms.

disagree on this point, sponges probably evolved from choanoflagellates.

Sponges are aggregates of cells without true tissues and little differentiation, but they are made of several cell types (see Figure 34.4). The surface is made of epithelial cells. In asconoid sponges, each incurrent pore is made by a pore cell *(porocyte).* The inner passages of syconoid and leuconoid sponges are surrounded by rings of contractile cells called *myocytes;* they can contract and close off the pore in response to a touch or to excessive sediment in the water. In contrast to eumetazoans, sponges don't coordinate their activities by nerves and hormones. The myocytes are in contact with one another and form an elementary integrative network. If one cell contracts, it stimulates others to do the same, so a wave of contraction spreads quickly.

All these cells are embedded in a jellylike middle layer (mesohyl) inhabited by very mobile cells called *amoebocytes,* which digest food particles and transfer nutrients from the collar cells to other cells. Amoebocytes are totipotent, undifferentiated cells that heal wounds by migrating to the injured place and developing into collar, epithelial, or other cells.

A sponge is supported by a skeleton made of hard crystals called *spicules* whose shape and composition are important taxonomic features (Figure 34.8). Calcareous sponges have spicules of calcium carbonate, the material of marble and

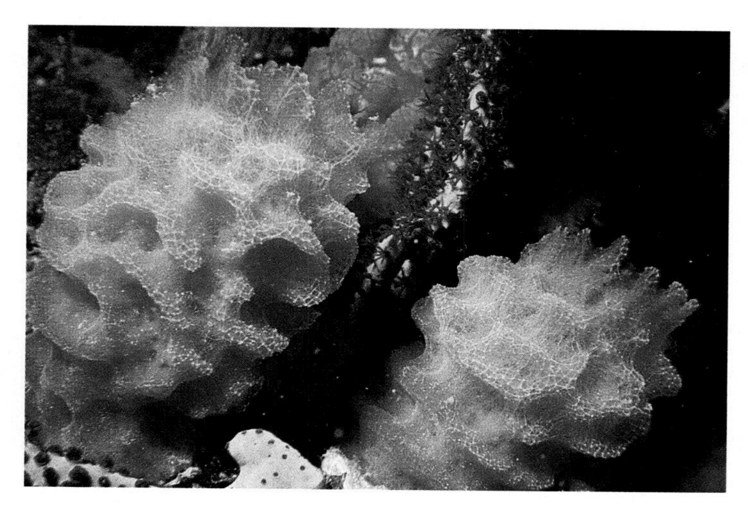

Figure 34.9

Glass sponges are deep-sea animals whose silica spicules form a delicate network.

limestone. The silica spicules of the hexactinellid, or glass, sponges are welded into a delicate, glassy network (Figure 34.9). Demosponges have siliceous spicules and a network of a fibrous protein, spongin, that is similar to collagen; from this group we get common household sponges, which are made by soaking dead sponges in shallow water until all the cellular material has decayed, leaving the spongin network behind. (Of course, most of the sponges sold today are just plastic and have nothing to do with real sponges.)

Sponges are diploid and reproduce sexually. Most species are hermaphroditic but produce sperm and eggs at different times. Amoeboid cells or choanocytes develop into both eggs and sperm, and the sperm are released into the water, sometimes in milky clouds. Collar cells trap these sperm and transfer them to eggs lying deeper in the sponge. The diploid zygote then develops into a hollow ball of cells that becomes ciliated and swims about freely for a while. After settling on the bottom, it turns itself inside out as ciliated cells migrate to the inside, where they develop into collar cells. Although this development is reminiscent of the formation of a blastula followed by gastrulation in eumetazoans, the two processes are really quite different.

Some sponges can reproduce asexually. A group of cells may simply break off from a sponge, attach somewhere else, and continue to grow. In adverse conditions, some sponges—particularly freshwater species—may form *gemmules,* which are special clusters of cells surrounded by a coating that makes them resistant to drying. When appropriate conditions return, gemmules develop into new sponges.

Exercise 34.2 A sponge moves water through itself with the flagella of its collar cells, and the water really moves in a stream; it is not just stirred around randomly. What does this imply about integration within the sponge?

34.5 Cnidarians are built around a central digestive cavity.

The **cnidarians** (*knides* = nettle, referring to their stinging cells) are a diverse group of primarily marine animals, with a few freshwater species. Some cnidarians show clear bilateral symmetry in their musculature, but most are radially symmetrical with one of two body types: either a sessile, vaselike **polyp** with its mouth pointed upward, or a floating, bowl-shaped **medusa** with its mouth pointed downward (Figure 34.10). Both have a **coelenteron,** or **gastrovascular cavity,** a central cavity that serves for both digestion and circulation. The animal draws water, containing oxygen and food, into the coelenteron through its single opening, digesting the food partly by enzymes secreted into the cavity and partly within the cells that line it, which then pass the nutrients along to other cells. The radial form is clearly adaptive for a sessile or a floating life, in which the animal receives food from all directions. Terms such as anterior and posterior do not apply to such a form; its main body axis extends from an **oral** surface at the mouth to the opposite, **aboral** surface.

Although polyps are sessile, they do not wait passively for food. The whole external surface, especially the tentacles around the mouth, is armed with cells called *cnidocytes,* which contain stinging structures called *nematocysts* that are used for various holding and clinging functions, including prey capture (Figure 34.11). Nematocysts may bear a sticky covering, barbs, poison, or all three. Once trapped—and often paralyzed—the food is pulled back by the tentacles and stuffed into the gastrovascular cavity to be digested.

Cnidarians have a definite endoderm and ectoderm, but the jellylike *mesoglea* in between is not really a mesoderm—it isn't highly organized and contains mostly scattered amoeboid cells. The ectoderm and endoderm are mixtures of cell types, including some contractile *epitheliomuscular cells* that form a surface like an epithelium and also contract like muscles. The actions of cnidarians are coordinated by a simple network of nerve cells, a clear advance over sponge organization that anticipates the complex nervous systems of other animals.

Cnidarians coordinate their movements with a simple *nerve net,* a loose network of nerve cells connecting the tentacles and mouth with the rest of the body, but they have no central brain or site where nerve cells are concentrated. Whereas the nerve cells of other animals conduct signals in only one direction, the cells of the cnidarian nervous system are unpolarized and can carry messages in any direction through the nerve net.

Cnidarians are divided into three classes, largely based on their body forms. The Anthozoa, including corals and sea anemones, have only polyp forms; the Scyphozoa, or jellyfish, have small polyps at an early stage of development, but their life cycle is dominated by medusas; and the Hydrozoa typically have both hydroids and medusas of comparable size. All cnidarians pass through a sexual stage in which they produce sperm and eggs, and the zygote they form develops into an interesting **planula larva,** a solid, ciliated mass of cells, as shown

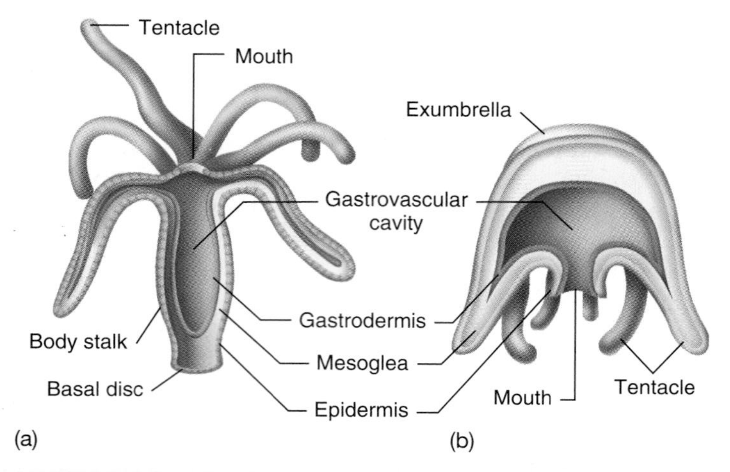

(c)

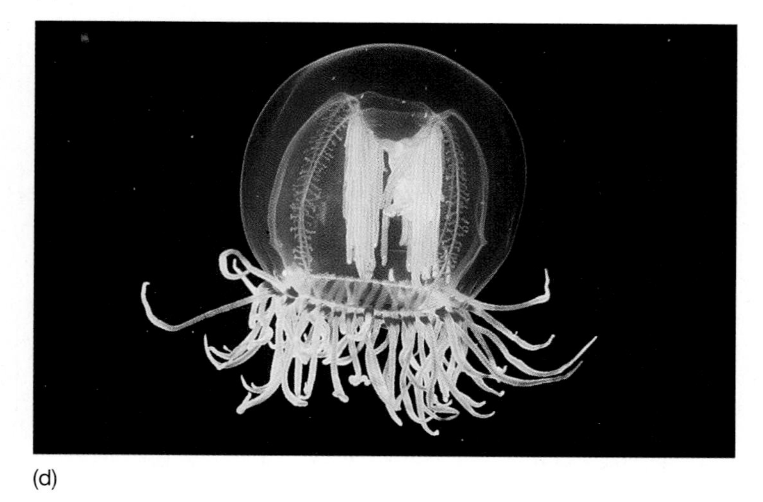
(d)

Figure 34.10
The body plan of a cnidarian is either *(a)* a polyp or *(b)* a medusa. In many species, these are alternating stages in a life cycle. *(c)* The jewel anemone *Corynactis* has a polyp form, and *(d)* the jellyfish *Polyorchis* has a medusa form.

in Figure 34.12. A planula eventually settles down, develops tentacles, and becomes a polyp. In anthozoans and *Hydra,* the polyp itself reproduces sexually, but in many hydrozoans, the polyp is an intermediate, colonial stage in a longer life cycle; the polyp stage buds into a colony, and some of its polyps bud off medusas, the animal's sexual stage. A planula larva presumably reflects the primitive flagellated or ciliated cnidarian ancestor.

The comb jellies, phylum **Ctenophora** ("ten-ah′for-uh"; the "c" is silent again), get their name from the eight ciliated, comblike plates (*cteno-* = comb; *-phore* = bear) they use for swimming (Figure 34.13). Though they look superficially like jellyfish, they belong to a separate phylum because they have biradial symmetry (generally radial, but with planes of bilateral symmetry), they lack cnidocytes, and those with tentacles have only two rather than a circle of them around the mouth. In fact, it has been argued that ctenophores show at least as much affinity with the flatworms as with the cnidarians, and they probably evolved from bilaterally symmetrical ancestors. We leave them near the cnidarians only because this is the traditional classification.

Some of the delicate, almost ghostly ctenophores are luminescent and will glow brilliantly when irritated. People who venture off for a midnight swim in the ocean in late summer are often rewarded by the flashing of ctenophores disturbed by the movement of the water (although swimmers are sometimes stimulating the luminescence of marine dinoflagellates, too).

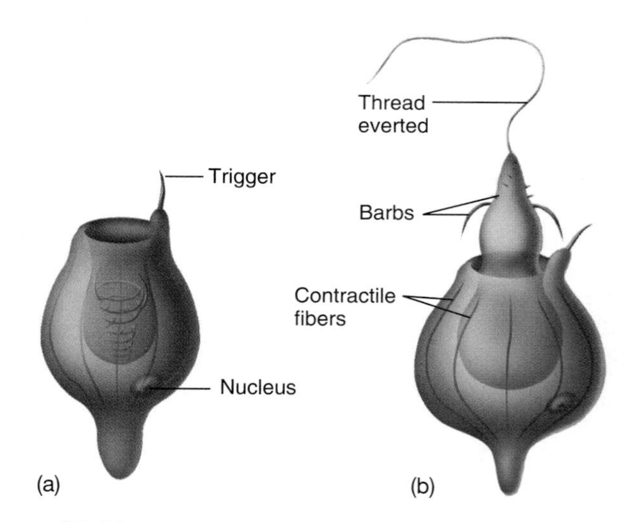

Figure 34.11
Cnidarians have a variety of nematocysts that are used for capturing prey and for protection. Each one consists of a poison-filled vacuole in a stinging cell (cnidoblast) with a long thread, which may penetrate prey with its barb, stick to the prey, or wrap around it. One type is shown here with the thread undischarged in its capsule *(a)* and in its discharged form *(b).*

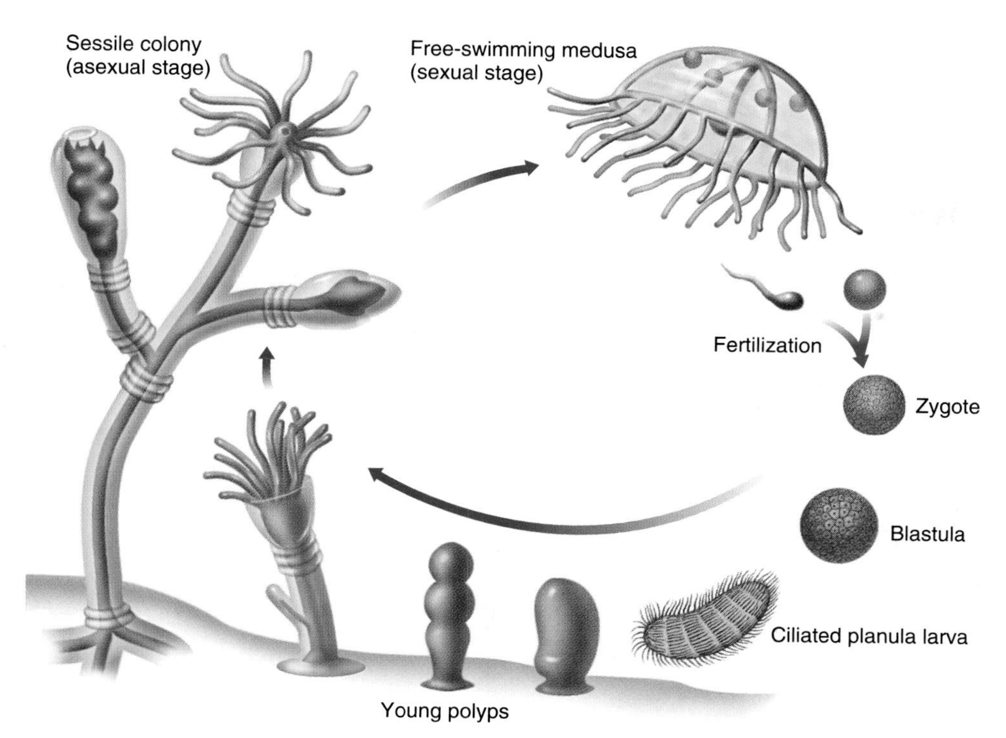

Figure 34.12
The hydroid *Obelia* shows a representative cnidarian life cycle. Notice particularly the ciliated planula larva.

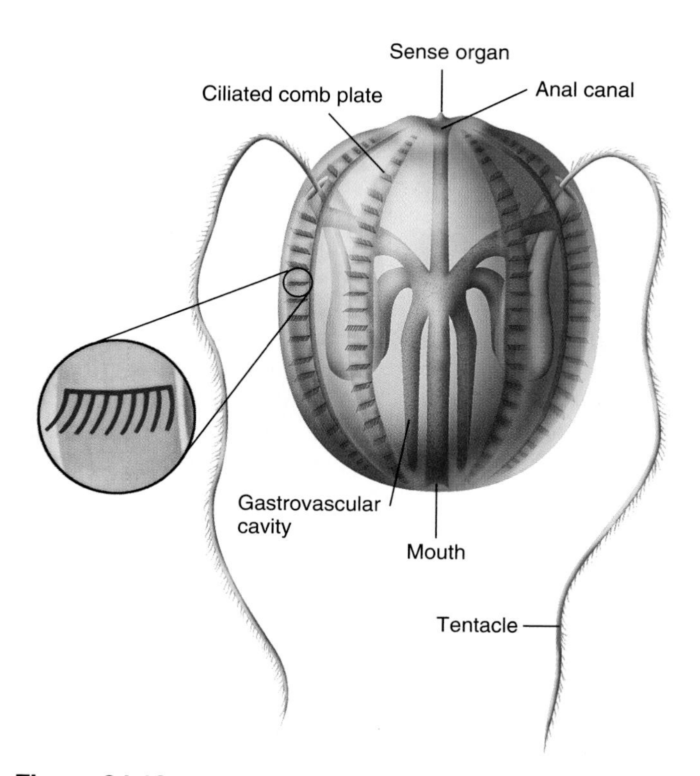

Figure 34.13
Pleurobrachia is a representative comb jelly of the phylum Ctenophora. Each comb is a short row of cilia that propel the animal.

34.6 The flatworms are bilateral acoelomates.

The flatworms, phylum **Platyhelminthes,** have bilateral symmetry and a definite head and tail, but no coelom; that is, they are **acoelomate** (*a-* = without). Although they have three germ layers, much of the mesoderm consists of undifferentiated, frequently amoeboid cells known as *mesenchyme.* In their basic body form, they are united with the cnidarians by having a digestive cavity with only a single opening—an *incomplete* digestive tract in contrast to the more usual animal intestine that has both a mouth and an anus. The three classes of flatworms include the free-living Turbellaria, the parasitic tapeworms (Cestoda), and the flukes (Trematoda).

Turbellarians, which are not modified by the peculiar demands of a parasitic life, live in both salt water and fresh water, with a few species occupying moist habitats on land. A representative turbellarian worm, a planarian, is only a few millimeters long (Figure 34.14). It has a mouth on the ventral side, between the head and the middle of the body, with a muscular **pharynx** ("throat") that in some species can be extended for feeding. The pharynx leads to the gut, a simple or multilobed sac. Planarians move on a secreted layer of lubricating mucus by contracting certain muscles and beating the cilia on their ventral surfaces. Since flatworms lack a circulatory system, metabolic wastes diffuse through the body surface (the integument) and through a series of tubes connected to external pores. Water is forced out through these pores by **flame cells,** hollow cells

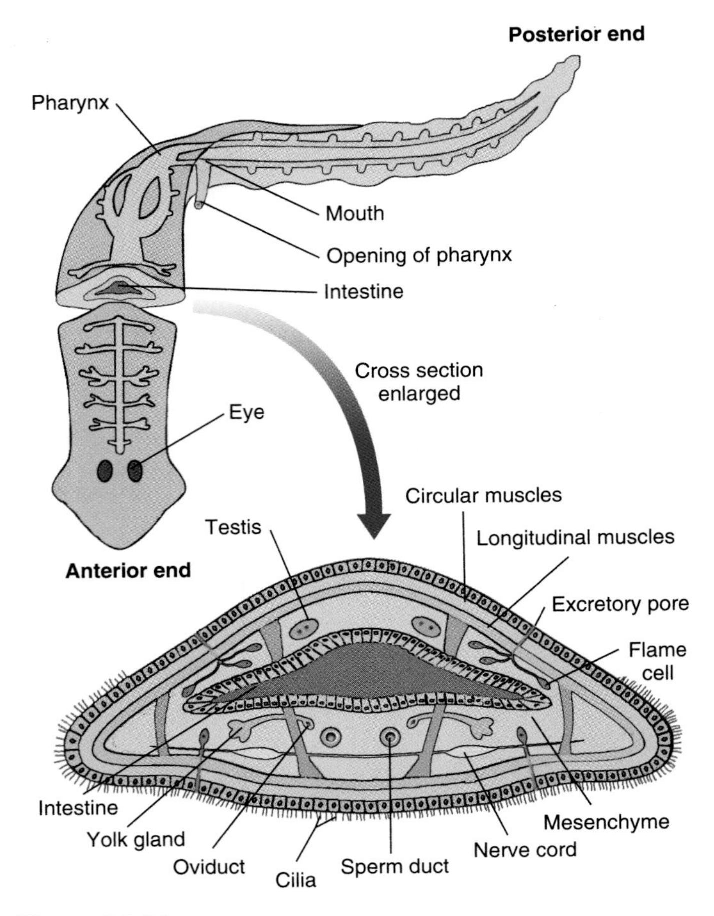

Figure 34.14

The planarian *Dugesia,* is a common turbellarian flatworm. Its highly branched gastrovascular cavity (intestine) has a high surface/volume ratio and occupies much of the body.

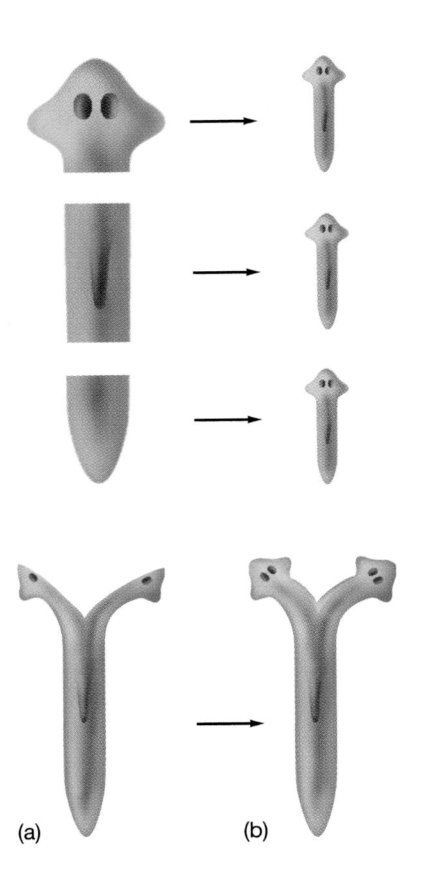

Figure 34.15

(a) If a planarian is cut into pieces, each piece will regenerate into a whole worm. *(b)* Even two-headed worms can be made in this way.

whose tufts of beating cilia resemble wavering flames. The rudimentary flatworm nervous system has two connected ganglia (sing., *ganglion* = a cluster of nerve cell bodies) in the head; extending from them toward the tail are two or more ventral nerve cords cross-connected in a ladder pattern. Many sense organs in the skin register touch, movement, taste, and light.

Turbellarians commonly reproduce sexually, and most species are hermaphroditic; each worm may have one or many testes, one or many ovaries, and a simple penis and vagina. Two individuals copulate and fertilize each other internally. In addition, these little animals are famous for their plasticity and their ability to form new parts from old by regeneration. Some species reproduce by budding or by fission, simply dividing into pieces that grow into new individuals. If a planarian is cut into several pieces, each one can become a whole new worm (Figure 34.15). Another aspect of this enormous plasticity is a starved worm's ability to digest its own body and grow smaller and smaller.

Flukes (trematodes; Figure 34.16) are parasites that usually attach by means of suckers to their hosts' internal organs, such as intestines, lungs, blood vessels, bladder, or liver. There the fluke feeds, often debilitating its host by robbing it of fluids and nutrients. Some flukes have complex life cycles (see Figure 15.18) that require several intermediate hosts, commonly snails and fish.

Tapeworms (cestodes) are also strictly parasitic, but having no digestive system or mouth, a worm simply attaches inside its host's intestine and absorbs nutrients through its epidermis. The worm's tough outer coat makes it impervious to its host's digestive enzymes. Its body is a series of continuously growing segments (proglottids) that remain attached to one another, forming the configuration that gives these worms their common name. Some worms may reach 10 meters in length (Figure 34.17). Cestodes are reproductive machines, and each proglottid is little more than a packet of sex organs where fertilization occurs and zygotes develop. As the proglottids grow and mature, they are pushed backward, and those at the end continually detach and are defecated by the host animal. The embryos develop when these segments are eaten by a new host. Tapeworms always parasitize vertebrates, but may require an intermediate invertebrate host in their life cycles.

Exercise 34.3 What kinds of surface structures would you expect an intestinal parasite to have to avoid being digested by its host's digestive enzymes?

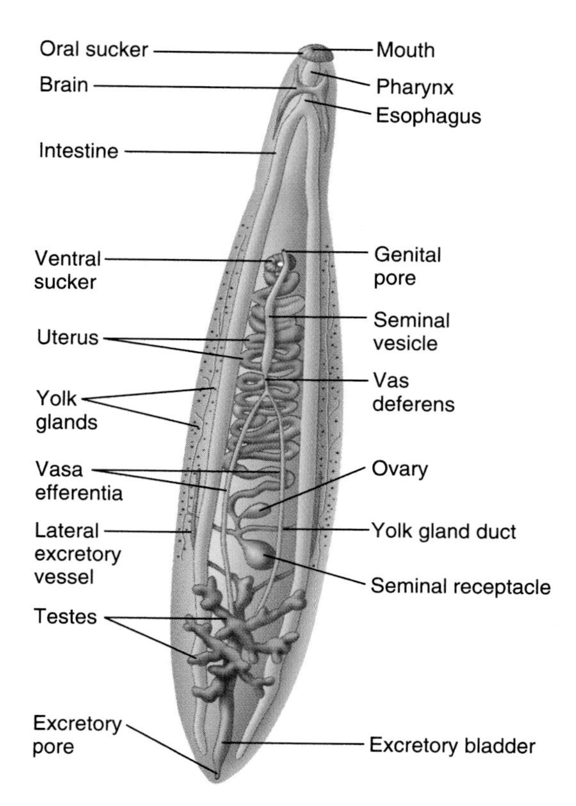

Figure 34.16

A Chinese liver fluke, *Opisthorchis sinensis,* is a typical trematode whose body is largely taken up by an intestine and reproductive organs.

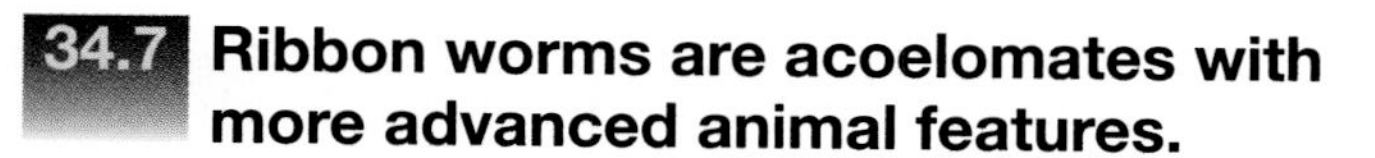

34.7 Ribbon worms are acoelomates with more advanced animal features.

The ribbon worms, phylum **Nemertea,** are mostly marine animals, usually only a few centimeters long, although species that grow 25–30 meters in length demonstrate the appropriateness of their name(Figure 34.18). The roughly 500 species resemble flatworms but differ in two important ways. First, in place of a digestive cavity with a single opening, they have a complete digestive tract with both a mouth and an anus. They take food in at one end, pass it through specialized digestive regions, and eliminate wastes at the other end. Second, they have a closed circulatory system with blood vessels, but no heart, and with blood cells containing hemoglobin. Three blood vessels—one dorsal and one on each side—absorb nutrients from the digestive tube and distribute them to other tissues.

Nemertea is also named phylum Rhynchocoela (= “hollow beak”), or the proboscis worms, because of their long, thin, anterior beak. The proboscis, a muscular tube kept inside a sheath, can be whipped out to a length sometimes approaching the length of the worm. The proboscis secretes a sticky mucus and in some species is tipped with a hard barb and poison glands. The worm lashes out with its proboscis to capture and kill prey, then withdraws the proboscis to pull the food into its mouth.

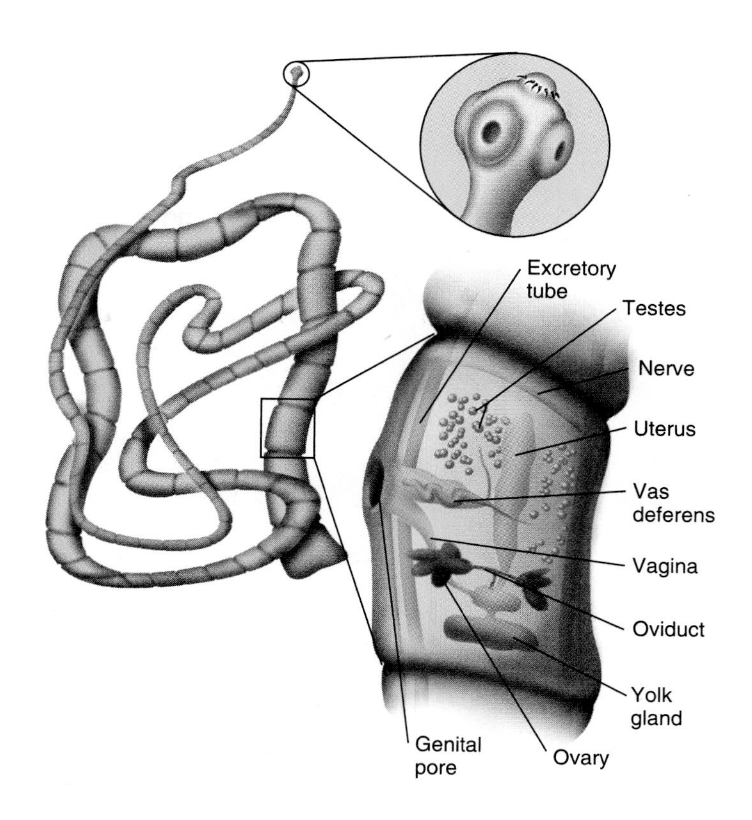

Figure 34.17

The pork tapeworm *Taenia solium* (class Cestoda) is a common parasite of humans, often transmitted from inadequately cooked pork. The worm anchors itself in the intestine by the hooks and suckers on its head. Segments (proglottids) mature as they are pushed back by the growth of new segments, and each one develops both male and female reproductive structures. Tapeworms can grow to astonishing length *(photo).*

(a)

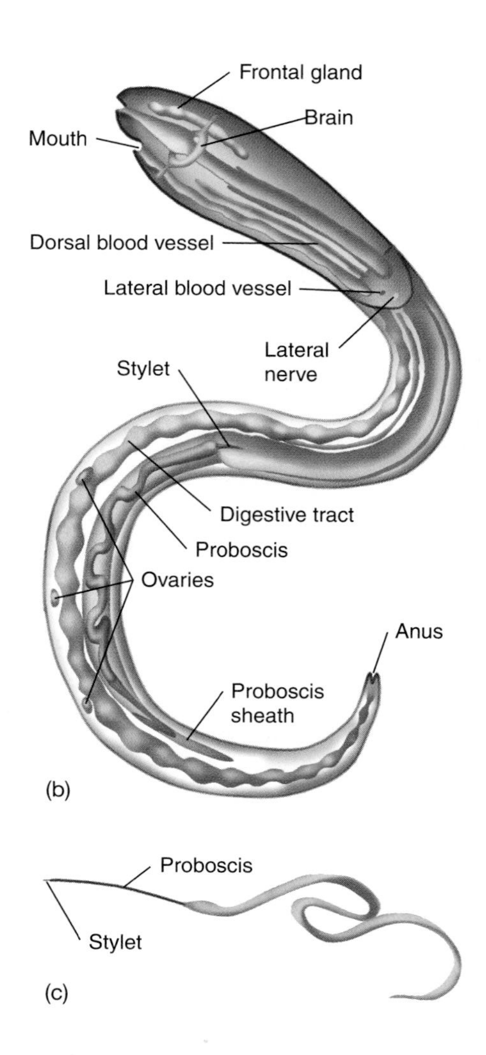

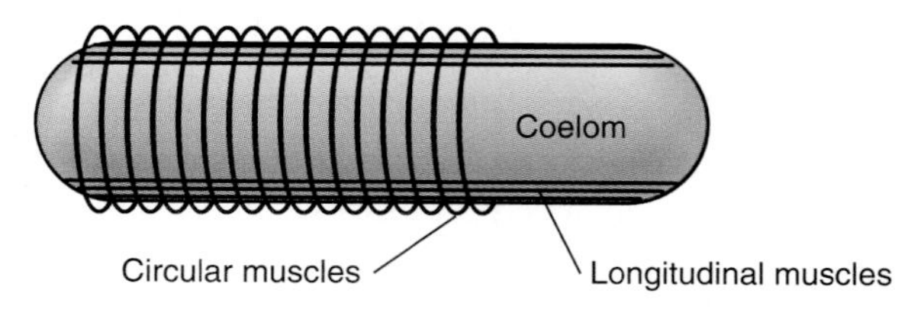

Figure 34.18

(a) Ribbon worms, phylum Nemertea, vary from 2 mm to 30 meters in length, averaging about 10 cm. They may be brightly and distinctively colored. *(b)* Internally, a worm has a complete intestine and a long proboscis that lies within a sheath. *(c)* The proboscis can be extended to seize prey and pull it in toward the mouth.

34.8 The evolution of a coelom enabled animals to occupy new niches.

We understand the evolution of animal structure partly in relation to their ecology. Each of the major developments in structure allowed a particular group of animals to move into a new adaptive zone. The story of animal evolution beyond the flatworm stage begins in an oceanic environment, and here we must distinguish two ways of living on the ocean bottom: either **epifaunal,** on the surface of the bottom, or **infaunal,** within the mud and sand of the bottom.

The animals that were beginning to evolve cephalized, bilaterally symmetrical bodies were probably something like the flattened, ciliated (or flagellated) blastea or gastrea forms discussed in Section 34.3. They fed by ingesting minute food particles (see Figure 34.3). Though largely epifaunal, these animals were in contact with the interstitial zone between the grains of sand and other debris, an area rich in food. Yet they could hardly enter this zone because they were too large (perhaps a millimeter or two in length) to squeeze between the sand grains and too weak to push most of them aside. The infaunal zone was thus wide open to any animal with both size and muscular strength.

The successors of these animals were small turbellarian flatworms, whose muscles allow them to flex and to move in new ways, but they were still primarily epifaunal, restricted to crawling over the ocean floor and probing its surface, because they lacked the power to burrow into it. The flatworms' limitation is that they are solid and have no coeloms. A coelom is most important as a device for locomotion. To be effective, muscles have to pull against a structure that offers resistance; most animals have a skeleton, either a bony endoskeleton inside the body or an exoskeleton of firm plates on the outside. In lieu of a solid skeleton like that, water confined inside a tight bag can form a stiff and nearly incompressible **hydrostatic skeleton.** A coelom forms this type of skeleton. In its most developed form, **circular muscles** running around the body and **longitudinal muscles** running lengthwise can squeeze against the water inside the coelom and therefore against each other:

Coelom
Circular muscles
Longitudinal muscles

Contraction of the circular muscles elongates the body and stretches the longitudinal muscles:

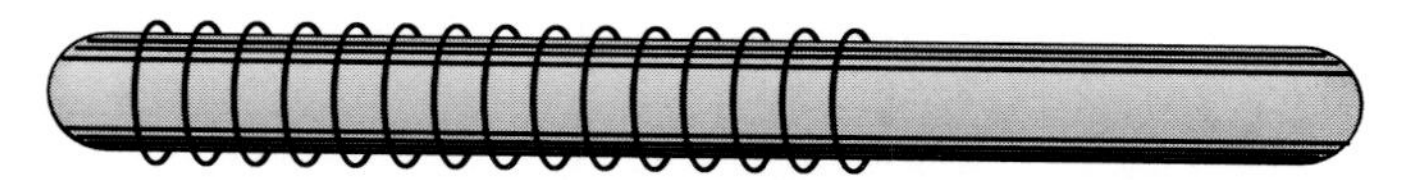

Contraction of the longitudinal muscles makes the body shorter and thicker:

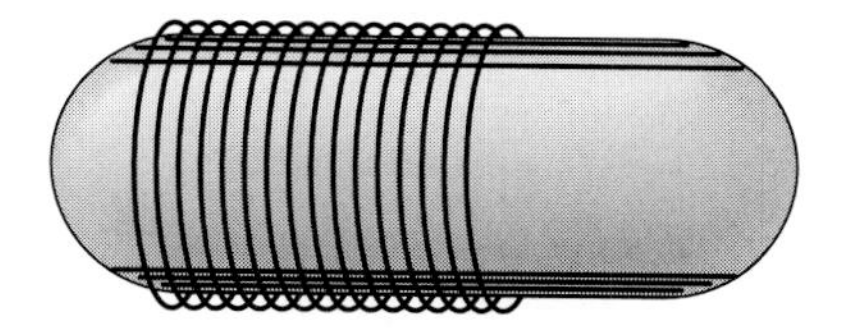

The first animals with a coelom surrounded by this arrangement of muscles could move efficiently, burrow into the sand and mud of the ocean bottom, and feed on the rich food buried there.

A coelom has other uses. It is a convenient reservoir where tissues can excrete[1] their metabolic wastes, especially the nitrogenous wastes and other substances that are by-products of the organic animal diet. Very small organisms can eliminate these substances easily just by diffusion through their surfaces, but larger animals require special mechanisms for excretion. Some time after coeloms evolved, larger animals developed circulatory systems that carry metabolic wastes to points of excretion. The coelom, however, serves excretion in organisms that have become too large to eliminate all their wastes through their epidermis and lack a circulatory system. Wastes can be held temporarily in the coelom, but some opening is needed for removing the wastes to the outside. Structures called nephridia (discussed in Section 34.9) evolved later for removing wastes from the coelomic fluid to the outside.

The coelom also gives the body as a whole, and the organs within it, more freedom of movement. Flatworms are quite solid, with their organs and tissues pressed together. As animals became larger and evolved circulatory systems, a space in which pumping hearts and other organs could move became essential.

The evolution of the hydrostatic skeleton also opened the door for a variety of animals, shown in Figure 34.19, to evolve a

[1]Excretion means eliminating metabolic wastes from the body. Materials to be used somewhere on or in the body are secreted.

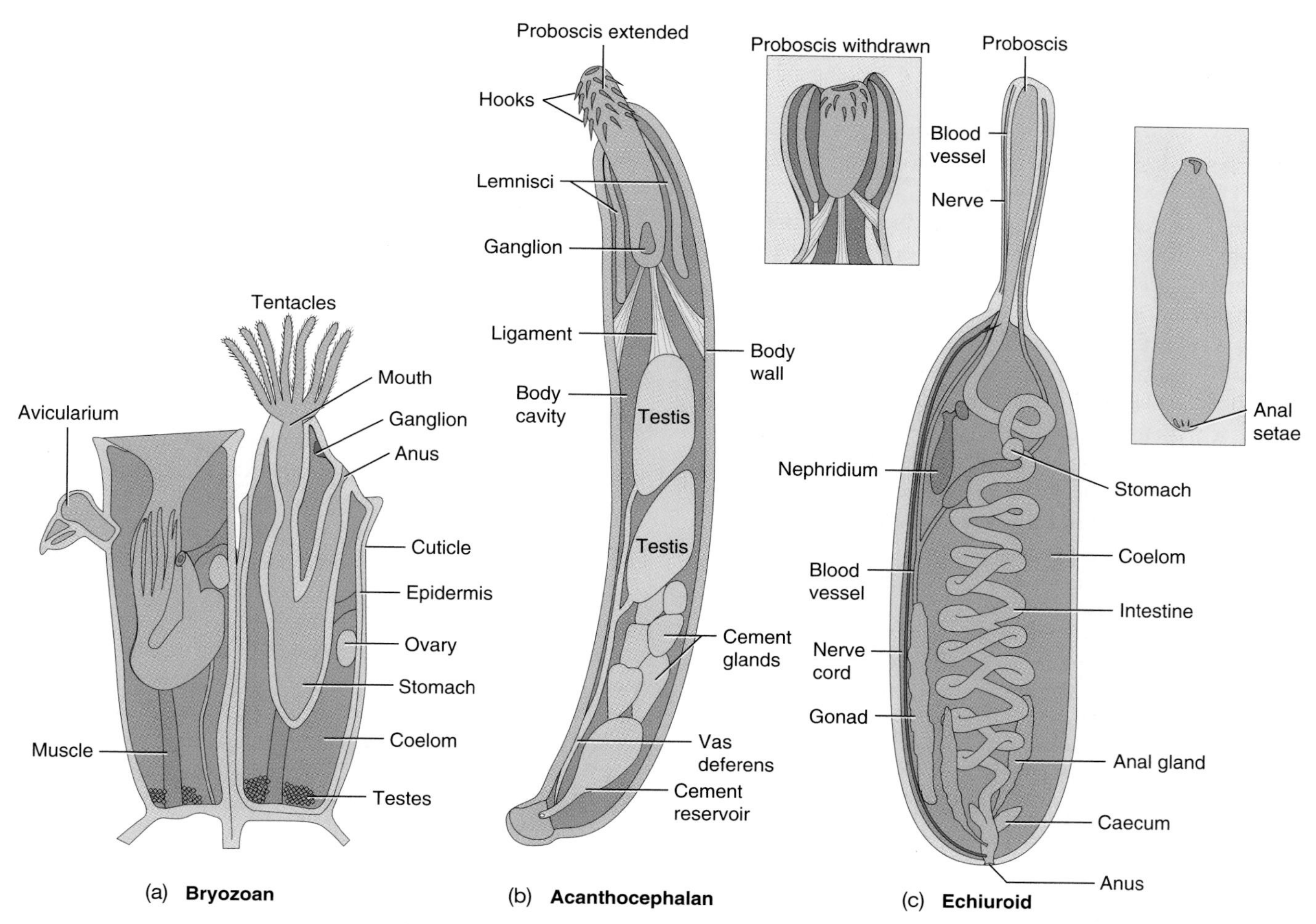

Figure 34.19

Several primitive wormlike animals, belonging to different phyla, are able to extend their bodies by squeezing on the coelom with circular muscles. In this way, a bryozoan *(a)* extends its tentacles and mouth, while an acanthocephalan *(b)* and echiuroid *(c)* extend their probosces for feeding and locomotion. The extensions are retracted by internal muscles.

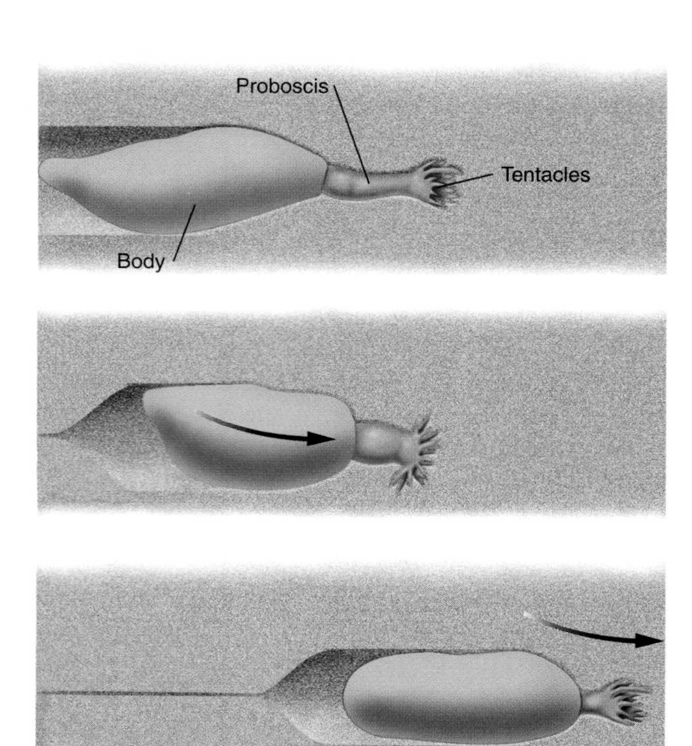

Figure 34.20

A worm such as *Sipunculus* moves by thrusting out its proboscis, anchoring it in the mud and sand, and then pulling its body forward.

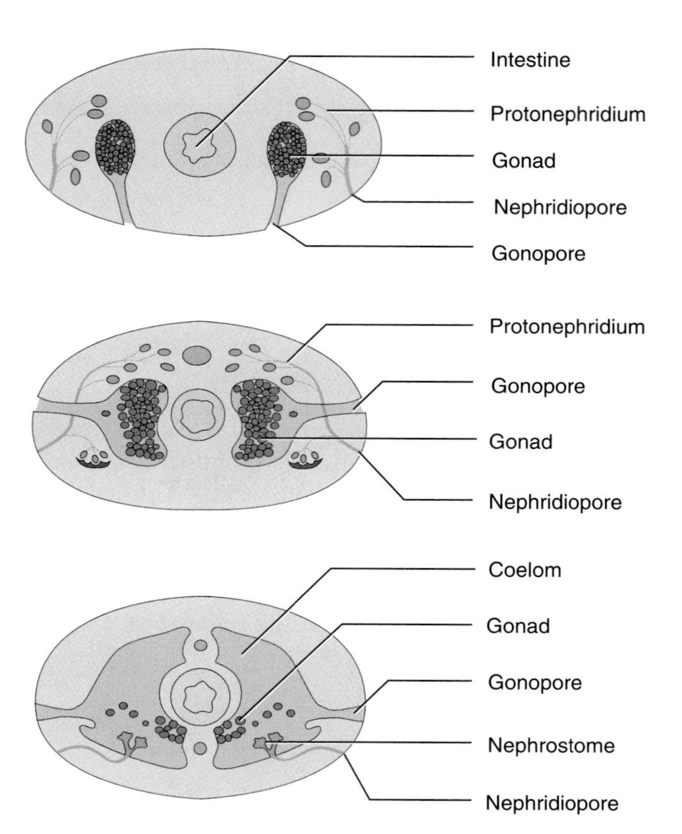

Figure 34.21

The gonocoel theory is one explanation for the origin of a coelom. In this scenario, the gonads were initially located in small cavities with external openings. These cavities became enlarged until they occupied much of the body space and had become coelomic cavities. Since one function of the coelom was to be a depository for wastes, the organs of excretion were changing at the same time; the protonephridia that excreted wastes from the acoelomate body changed into nephrostomes that removed wastes from the coelom.

kind of anterior end that can be extended by increasing pressure in the coelom. Notice that these head structures are always pushed outward by pressure from the inside and pulled inward again by the contraction of attached muscles. Generally this extendable unit is used for feeding, but it can have other uses as well. In the acanthocephalans (parasitic, spiny-headed worms), it is a barbed holdfast with which the animal attaches to its host. A sipunculid worm uses this proboscis for moving quickly through the mud: extending its head, anchoring, contracting the body, and extending the head again (Figure 34.20). (The ribbon worms, though technically acoelomate, do something similar with the extensible proboscis inside a cavity called a rhynchocoel, which may be homologous to a coelom.)

Exercise 34.4 Imagine you are a sipunculid worm. Which muscles do you contract to extend your proboscis? How do you retrieve the proboscis?

34.9 Coeloms probably evolved several times through different paths.

Because of the advantages listed in the previous section, coeloms have probably evolved independently at least three times in different branches of the animal family tree, as indicated by the fact that they develop in early embryos in various ways. Based on comparative studies of animal anatomy and embryology, many zoologists have speculated about the origin of the coelom, and we will briefly examine some of their ideas for the light they shed on characteristics of the various animal phyla.

The **gonocoel theory,** developed by Edwin S. Goodrich and other zoologists, provides one coherent explanation for the coelom of some animals (Figure 34.21). The theory proposes that the coelom developed through enlargement of the gonadal cavities of acoelomates, which open to the outside through ciliated pores. The external openings of these cavities became coelomoducts—that is, ducts for conducting the coelomic contents to the outside. The coelomoduct maintains its original function, providing passage for the gametes, in many animals. The gonocoel theory explains why gonads so often arise from the wall of the coelom and release their gametes into it. Even in mammals, the ovaries are swellings from the coelom wall that release eggs into the body cavity, to be swept down into the uterus through the ciliated funnel of the ovarian tube.

Ovulation and the ovarian tube, Figure 51.14.

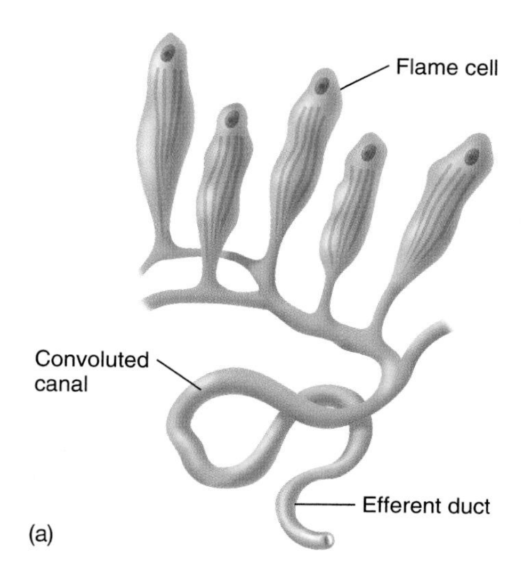

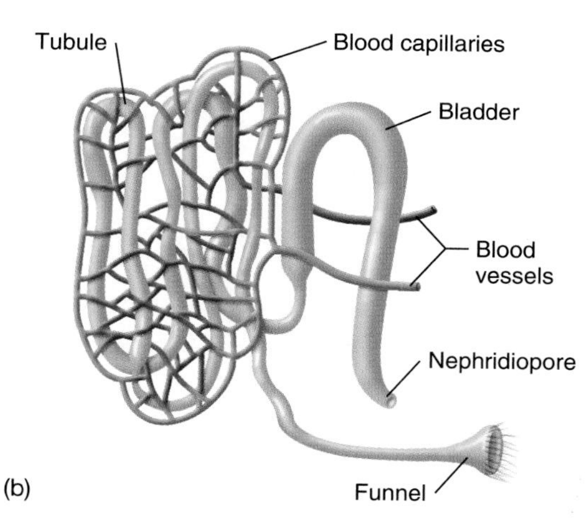

Figure 34.22

(a) A protonephridium is a primitive type of excretory structure, characteristic of acoelomates. Its flame cells remove wastes from the interstitial fluid and excrete them into a closed tube. *(b)* A metanephridium, as in an annelid worm, removes wastes from the circulation and from the coelom by means of its funnel.

A coelomoduct also provides an outlet for wastes that accumulate in the coelomic fluid, but this becomes the primary function of a second kind of tube, the **nephridium,** which eliminates wastes efficiently by forcing waste-laden fluid into one end of a channel. Goodrich distinguished the two types of tubules on the basis of their embryonic origins: A nephridium comes from the ectoderm and grows inward, while a coelomoduct comes from the mesoderm and grows outward. The flatworm protonephridium, as we noted earlier, ends in closed sacs containing flame cells (Figure 34.22*a*). More complex animals, such as annelid worms, have a comparable metanephridium with a funnel-shaped opening, or nephrostome, that conducts coelomic fluid into the tubule (Figure 34.22*b*).

An alternative proposal, the **enterocoel theory,** originated in 1877 with E. Ray Lankester. He observed that the coelom of many animals actually originates through outpocketings of the archenteron and proposed that here embryological development is recapitulating the evolution of the coelom. The enterocoel theory proposes that in some radially symmetrical ancestors of cnidarians, four pouches developed from the gut and became four coelomic pouches, later modified to three pouches. This theory explains why the animals to which it applies commonly have three-part coelomic cavities, as shown in Figure 34.23. The resulting animal then became bilaterally symmetrical and evolved a gut with both a mouth and an anus. In some animals, the anterior cavities might disappear while the

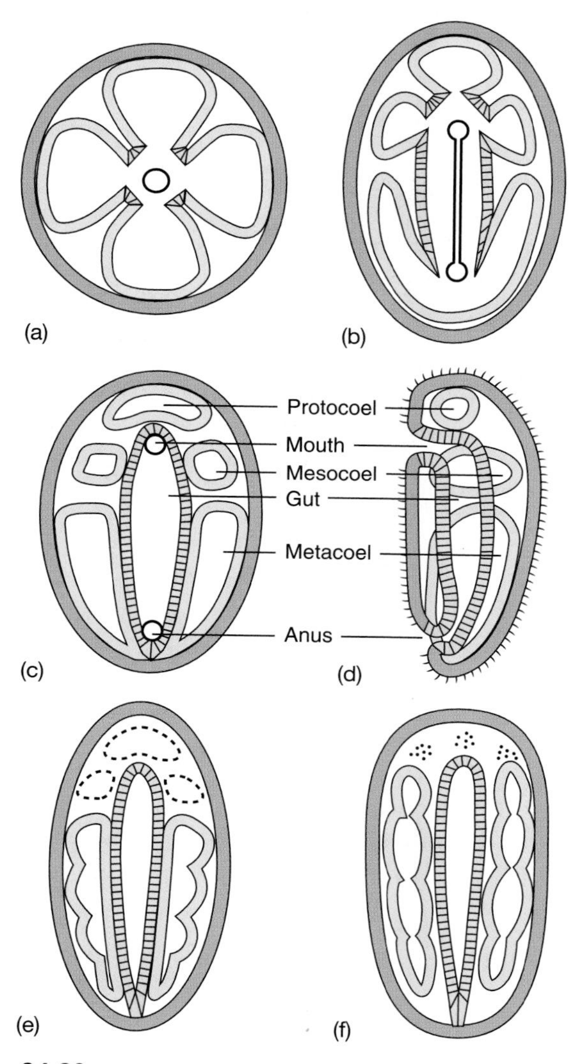

Figure 34.23

The enterocoel theory explains coelomic cavities as outpocketings from a primitive gut in an initially radially symmetrical animal *(a,b)*. This scheme accounts for the existence of three distinct coelomic cavities (protocoel, mesocoel, and metacoel) in some modern coelomates *(c)*. Meanwhile, the gut—which initially had only a single opening—elongates into a tube with separate mouth and anus, seen in side view in *(d)*. In some animals, the anterior cavities disappear, and the posterior cavities enlarge and become segmented *(e,f)*.

posterior cavities occupy most of the body and perhaps become divided into segments (for we will show later that segmentation is an important feature of most animals).

34.10 Embryological features distinguish major groups of animals: protostomes and deuterostomes.

We are now in a position to make some sense of the diversity of animals by focusing on their embryology in comparison with their adult body plans. Flatworms, we have already seen, have no coelom, and their bodies may be represented by three concentric germ layers:

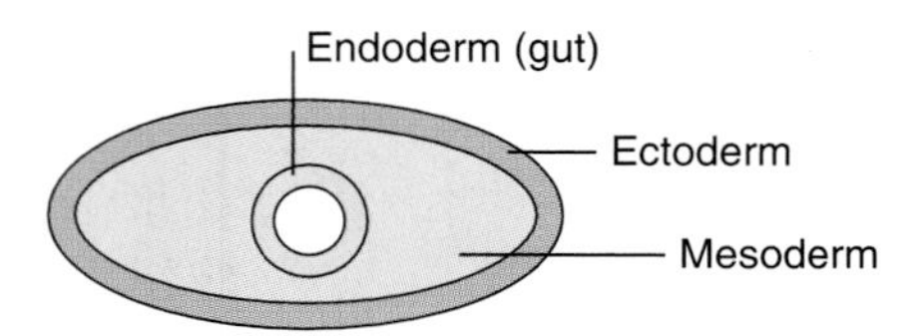

Most other eumetazoans are based on the tube-within-a-tube body plan, where the coelom occupies the space between the two tubes. This coelom clearly arises in at least three ways embryologically, almost certainly reflecting three distinct clades in animal evolution. The roundworms and their allies are distinguished as **pseudocoelomates** (*pseudo-* = false) because the coelom is formed from the former blastocoel. This space functions perfectly well as a coelom and is only called a "false coelom" because it has a layer of mesoderm only along one side:

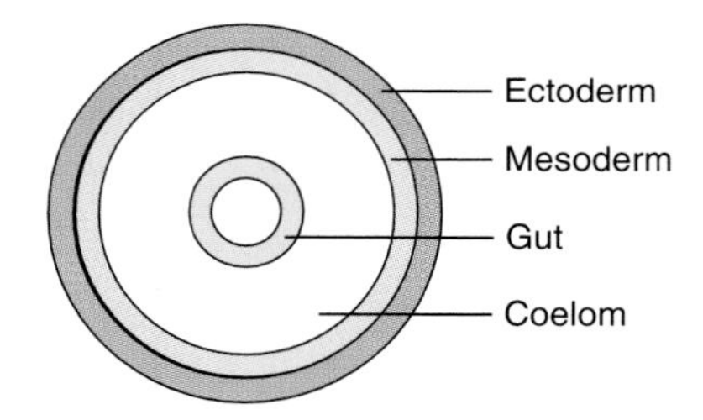

In contrast, all other bilaterians are **coelomate** and have a "true" coelom, which is completely lined by a layer of mesoderm called the **peritoneum:**

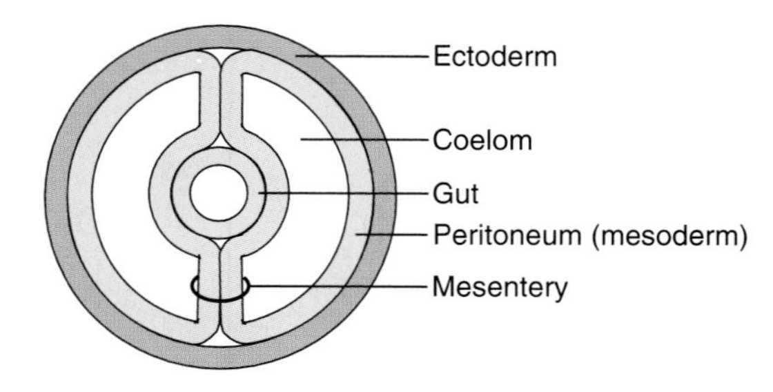

The peritoneum develops from a sheet of mesoderm on each side of the body, and where these sheets meet in the center, they form a **mesentery,** a double layer of tissue that surrounds and anchors the digestive tract and associated organs:

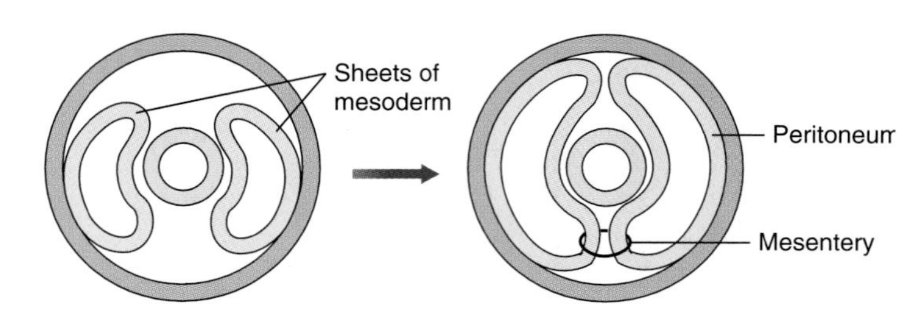

On the basis of comparative embryology, the coelomate animals are divided into two large groups: **Protostomia,** including molluscs, annelids, arthropods, and several minor phyla; and **Deuterostomia,** including chordates, echinoderms, and some other minor phyla (see Figure 34.25). These taxa are distinguished initially by the way the openings of the digestive tract arise. In protostomes, the mouth develops directly from the blastopore, and a second opening arises at the other end of the intestine for the anus. In deuterostomes, the anus develops from the blastopore, while the mouth arises as a new opening some distance away. Furthermore, protostome embryos divide by **spiral cleavage,** with each layer of early blastomeres twisted relative to the layer below it. Deuterostome embryos have **radial cleavage,** with the first blastomeres lined up in neat columns. As discussed in Chapter 21, protostomes generally show *mosaic* development, with the fate of each blastomere determined very early, while radial eggs tend to have *regulative* development, meaning that the blastomeres can change their fates until some later stage. Finally, many animals pass through a free-living larval stage in development; the protostome larva is typically a **trochophore,** and in those deuterostomes that have a larva, it is a **dipleurula** (Figure 34.24).

The origin of mesoderm correlates with these distinctions, too. In protostomes, the coelom is a **schizocoel.** In the spirally cleaving protostome embryo, as shown in Figure 21.3, a single blastomere called the *mesentoblast,* numbered 4d, produces the entire mesoderm; it develops into masses of cells that move into the blastocoel near the blastopore and then split (*schizo-* = to split)and spread out against the inside of the blastocoel:

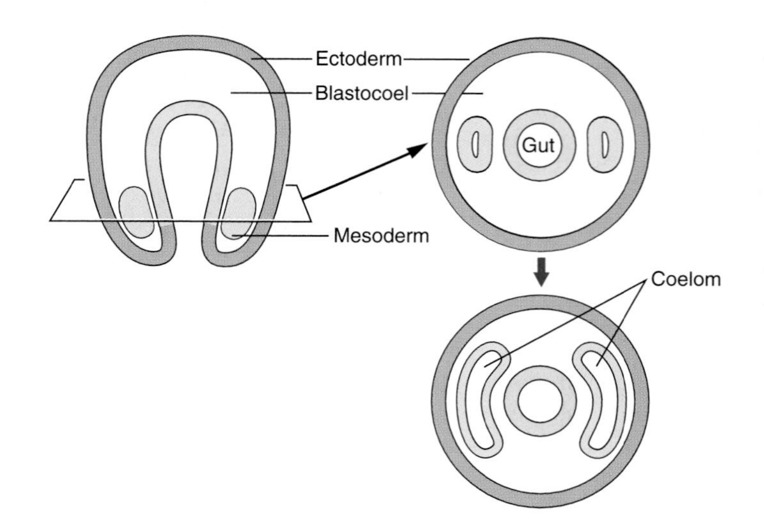

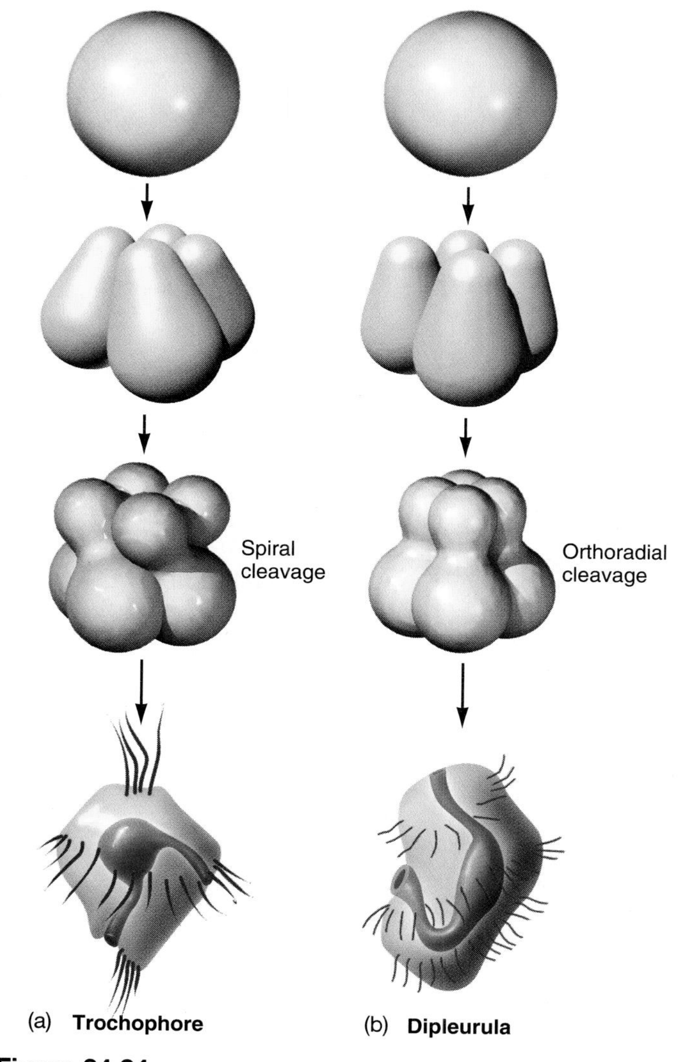

Figure 34.24

The coelomate phyla are divided on the basis of certain embryological features. *(a)* Annelid and mollusc embryos have spirally cleaving eggs that develop into trochophore larvae. *(b)* Echinoderm and primitive chordate embryos have radial cleavage and develop into dipleurula larvae.

In contrast, the coelom of deuterostomes is an **enterocoel,** which arises from pockets of cells that pouch out of the primitive gut and close into hollow masses:

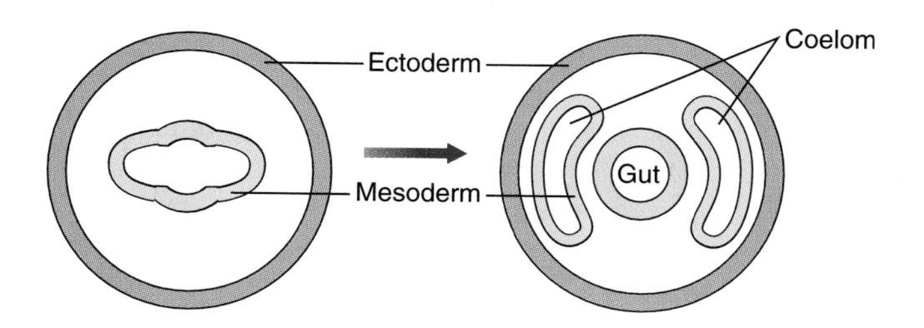

Some of the generalizations that define protostomes and deuterostomes have exceptions. Arthropods, for instance, do not have trochophore larvae, and their embryos divide in a way that cannot be described strictly as spiral cleavage, yet they are clearly closely related to animals with typical protostome features. Taking several characteristics together, though, indicates that coelomate animals belong to two main clades (Figure 34.25).

34.11 The pseudocoelomates share an unusual body structure.

Several groups of animals have bodies with a fluid-filled pseudocoelom, which develops from the blastocoel without a full lining of mesoderm. Two groups, the nematodes and the rotifers, deserve further attention.

Roundworms, or **nematodes,** are minute, hairlike, and extremely abundant creatures, and they are really quite round in cross section (Figure 34.26). Nematodes are among the most ubiquitous of animals. The 10,000 species so far described are probably only a small percentage of the number that exist, for they are found in every conceivable environment, and many are plant or animal parasites. Thousands may inhabit a handful of rich soil or a piece of rotting fruit. The most obscure roundworms that have come to our attention are specialized for living in the felt coasters placed under beer mugs in German taverns. It has been said that if everything in the biosphere disappeared except the nematodes, the former structures would remain, outlined in ghostly forms made of these fine, white worms.

A roundworm has a tough cuticle over a body wall made of unusual muscle cells arranged longitudinally. Because there are no circular muscles, roundworms cannot crawl and thus can be recognized by their characteristic thrashing motion. Separating the wall from the central intestine are a number of cells with very large vacuoles that all run together, creating a pseudocoelom that is not lined with mesoderm. Single dorsal and ventral nerve cords run along the body. Roundworms have an excretory system but no circulatory system, since the coelomic fluid itself helps distribute nutrients to the body tissues. Because of the simple structure of roundworms, one species *(Caenorhabditis elegans)* is now being used for extensive embryological studies (see Methods 21.1.)

Many nematodes are parasites of plants and animals. *Ascaris lumbricoides* is a common intestinal parasite of humans and pigs; the male worms are relatively small and slender, but some females may be 25 cm long and 6 mm wide (Figure 34.26*b*). An inhabitant this size in an intestine takes a considerable share of the incoming food, and dissection of a parasitized animal sometimes reveals a gut that is little more than a mass of worms. Hookworms are also common human parasites, especially where sanitation is poor and people go barefoot. An infected person or animal commonly defecates the extremely durable eggs of the roundworm parasite into the soil. A new host picks up the parasites, either via contaminated food or

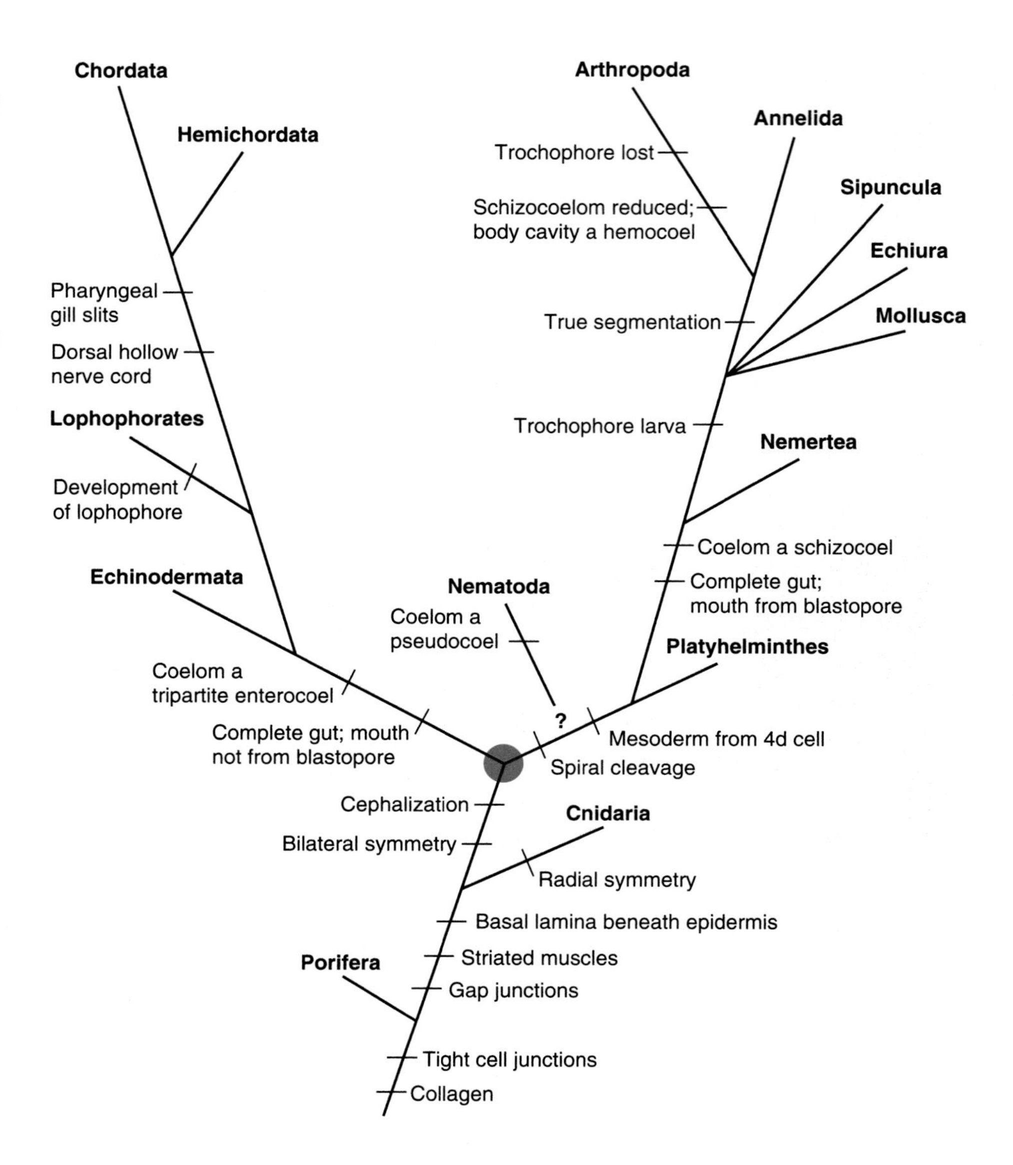

Figure 34.25
This diagram shows the most likely phylogeny of the animal kingdom, based on all the derived features of the various animal phyla. The most significant features are marked, including embryological features discussed in Chapter 20 and other characteristics discussed in Chapter 35. The branch point marked by a red circle divides most animals into two large clades.
(Modified from Brusca and Brusca, *Invertebrates* [Sinauer Associates, 1990].)

when a minute worm burrows into the skin of the feet and moves through the circulation to a favored place of residence, such as the lungs or intestine.

Trichinosis is a nematode disease that remains a problem long after it should have been eradicated by modern sanitation. In this infestation, minute worms, *Trichinella spiralis,* move out of the intestine, through the circulatory system, and into muscles, where they curl up and encyst (Figure 34.27). A new animal acquires the disease from infested meat that has not been adequately cooked. Rats and pigs commonly become infested and disseminate worms, partly through cannibalism. Humans generally get trichinosis from undercooked pork. Although the symptoms of infested people can be eased by drugs, there is no known cure. Proper cooking is the surest means of protection, and the whole cycle of reproduction can be stopped if potentially infested material, such as scraps from slaughterhouses, is cooked before being fed to pigs, a measure now required by law in many states.

Rotifers (phylum Rotifera) are abundant in ponds and streams. Often called "wheel animals" because the ciliated corona on their anterior ends resembles a couple of rotating wheels, rotifers are seldom more than 1–2 mm long, and are commonly only a tenth that size (Figure 34.28). They have an excretory system with flame cells and a small concentration of neurons that form a brain, but they are too small to need a circulatory system.

Some species of rotifers apparently consist exclusively of females that reproduce by parthenogenesis. In other species, small males appear only at certain times of the year. Most commonly, two kinds of summer eggs are laid, which develop parthenogenetically into either males or females; in the fall, heavy-walled winter eggs are produced that later develop only into females. These winter eggs are clearly a mechanism for resisting difficult times. In general, rotifers are good survivors, for when they start to dry out, many species encase themselves in an envelope of gelatin for protection against both dehydration and temperature stresses.

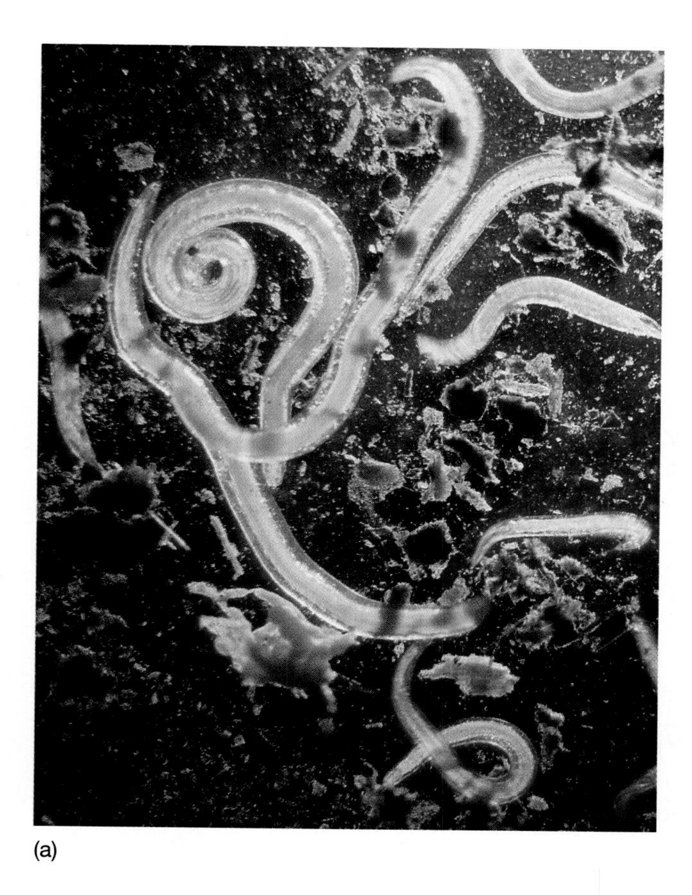

(a)

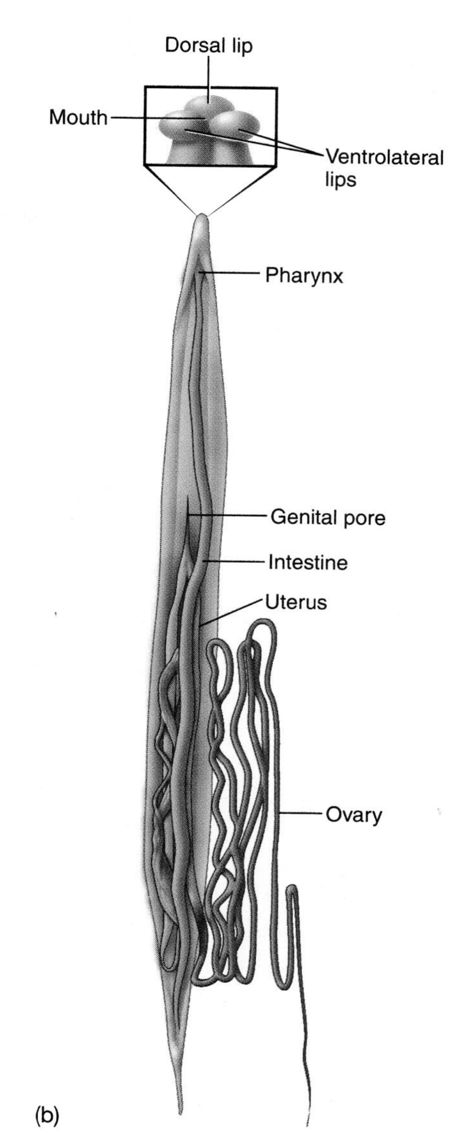

(b)

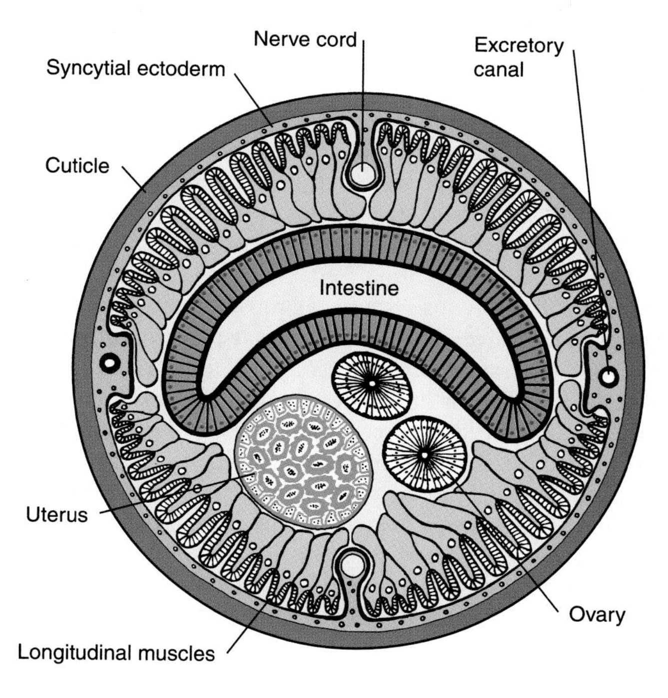

(c)

Figure 34.26

(a) Nematodes such as these small, soil-dwelling animals are simple roundworms. *(b)* The parasitic nematode *Ascaris lumbricoides,* which often inhabits human intestines, ranges from about 10 to 30 cm in length. *(c)* A cross section through the body of a female shows little more than a muscular wall, a large intestine, and reproductive organs.

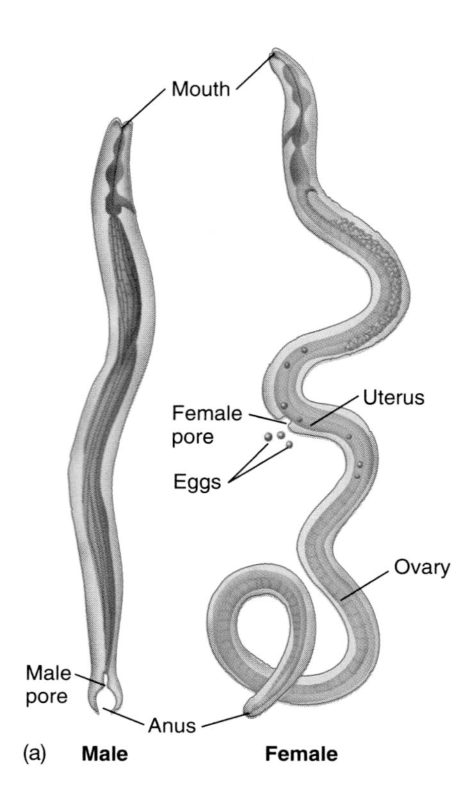

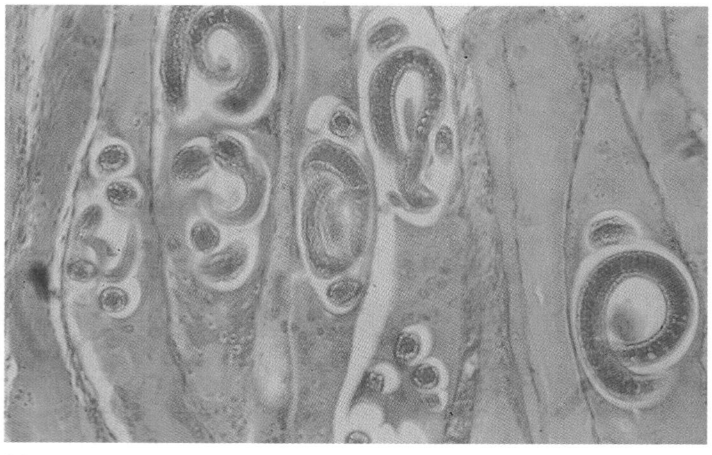

Figure 34.27

Trichina worms, *Trichinella spiralis,* are roundworms responsible for trichinosis, a painful, debilitating, and often fatal infestation of some animals, including humans. *(a)* Adult worms develop in the intestine. They produce millions or even billions of larvae, which migrate through the body and encyst in muscles *(b).*

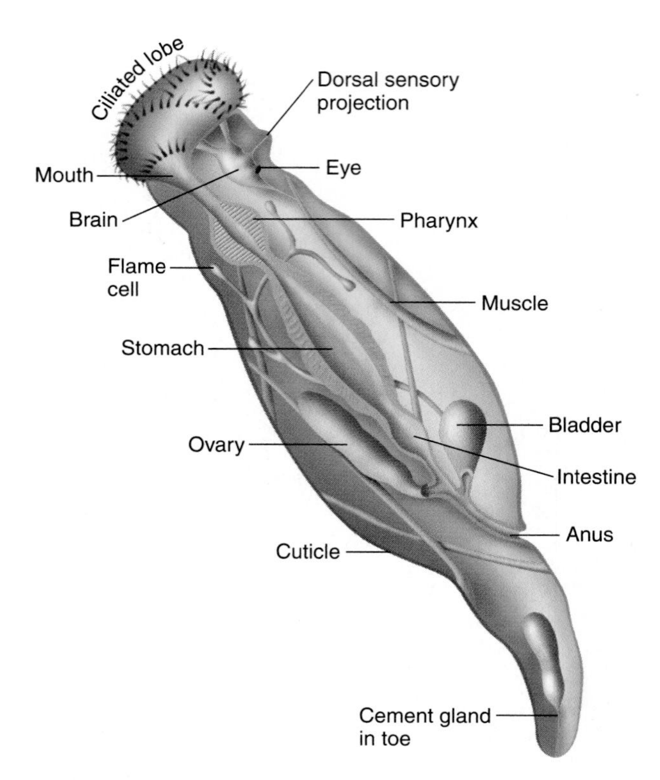

Figure 34.28

Rotifers are among the most common animals in freshwater habitats. They are pseudocoelomates characterized by a ciliated head organ that often looks like a pair of wheels turning. The animals move about freely or attach themselves to the substratum with cement produced in glands in their "toes."

Coda Holotrophy, the habit of engulfing other organisms and digesting them, evolved among protists such as ciliates, but it has been developed most strongly by the animals. Just as the plant kingdom has come to encompass most of the earth's producers, the animal kingdom has come to include its principal consumers. We can understand the major features of animal evolution as adaptations to becoming better, more efficient predators, with occasional side trips into parasitism. Animals are most characterized by movement, movement in pursuit of food or to escape becoming food; to this end, the actin-myosin complexes of primitive eucaryotes evolved into highly organized muscles controlled by rapid-transmission nervous systems. As movement became a primary factor in their lives, it became advantageous for animals to have skeletons as frameworks and supports for their muscles. They first evolved hydrostatic skeletons in the form of coeloms, not just once but at least three times. They evolved hard exoskeletons, which had the additional virtue of protecting their surfaces, and later the first vertebrates evolved endoskeletons made of bone. Animals evolved the protein collagen as a supporting material from the spicules of some sponges to the bones and other connective tissues of vertebrates.

Since animals take other organisms of all kinds into their bodies for food, they have to contend with a great variety of metabolic wastes. The coelom served as a convenient reservoir for waste disposal, and a variety of devices evolved later for filtering and cleansing the coelomic fluid and blood. Thus, in following animal evolution, we see how the forms of whole organisms are shaped by the related and overlapping functions of their separate components.

Summary

1. All animals are characterized by their mode of nutrition: They are multicellular organisms that ingest particulate or solid food. The typical animal has a complete digestive tract, with mouth and anus, running through an elongated body with bilateral symmetry.
2. Sponges are not closely related to other animals. They are modular organisms that pump water through their internal chambers, where collar cells trap small food particles. They have only a few other cell types, no nervous system, and are supported by a skeleton of spicules made of various materials.
3. Cnidarians, including corals, sea anemones, and jellyfish, have the forms of polyps or medusas. Both forms have a gastrovascular cavity in which food is digested. Each animal has only two tissue layers, with a jellylike mesoglea between.
4. The animals of some phyla (Platyhelminthes, Nemertea) have no coelom around the intestine. Others have a so-called pseudocoelom formed from the blastocoel with only a partial lining of mesoderm. The remaining phyla have a coelom, derived from the embryonic blastocoel and lined with mesoderm, although this is highly reduced in phyla such as Arthropoda.
5. Coelomate animals probably evolved because it was advantageous to move into an unused adaptive zone, the infaunal zone between grains of sand and debris on the ocean floor. A coelom, which is filled with fluid and surrounded by circular and longitudinal muscles, provides a firm foundation that allows free movement.
6. The coelom probably evolved independently in different groups of animals. The coelom of some animals may have arisen by enlargement of gonadal spaces, thus providing an internal fluid space with an external pore through which wastes can be eliminated. Animals have two types of tubes that can be used for elimination of wastes and conduction of gametes, the gonoduct and the nephridium. These are used in various combinations in the different animal phyla. Other animals probably evolved a coelom through outpocketings of the primitive gut in a radially symmetrical ancestor.
7. Flatworms include free-living types (turbellarians) and parasitic flukes and tapeworms.
8. The coelomate animals are divided into two large groups: protostomes and deuterostomes. In protostomes, the mouth develops from the embryo blastopore; these animals typically have eggs with spiral cleavage, their coeloms (schizocoels) arise through splitting of mesodermal masses, and many have a trochophore larva. In deuterostomes, the mouth arises from a separate opening in the embryo, the eggs have radial cleavage, and many have a dipleurula larva.
9. Pseudocoelomates include roundworms, which are very widely distributed, and rotifers, which are common inhabitants of fresh water.

Key Terms

larva 715
Mesozoa 716
Placozoa 716
Parazoa 716
Eumetazoa 717
diploblastic 717
triploblastic 718
coelom 718
collar cell/choanocyte 719
cnidarian 721
polyp 721
medusa 721
coelenteron/gastrovascular cavity 721
oral 721
aboral 721
planula larva 721
Ctenophora 722
Platyhelminthes 723
acoelomate 723
pharynx 723
flame cell 723
fluke 724
tapeworm 724
Nemertea 725
epifaunal 726
infaunal 726
hydrostatic skeleton 726
circular muscles 726
longitudinal muscles 726
gonocoel theory 728
nephridium 729
enterocoel theory 729
pseudocoelomate 730
coelomate 730
peritoneum 730
mesentery 730
Protostomia 730
Deuterostomia 730
spiral cleavage 730
radial cleavage 730
trochophore 730
dipleurula 730
schizocoel 730
enterocoel 731
roundworm/nematode 731
rotifer 732

Multiple-Choice Questions

1. Diploblastic animals
   a. are radially symmetrical.
   b. have a true coelom.
   c. are pseudocoelomate.
   d. have a mesoderm.
   e. are deuterostomes.
2. All triploblastic animals
   a. have a coelom.
   b. are protostomes.
   c. have a backbone.
   d. have fundamental (embryonic) bilateral symmetry.
   e. have all the above characteristics.
3. The coelom of most coelomates is
   a. lined on all sides by tissue derived from mesoderm.
   b. a closed cavity.
   c. bounded by a peritoneal membrane.
   d. present in protostomes and deuterostomes.
   e. all of the above.
4. All of the following are found in sponges *except*
   a. spicules.
   b. collar cells and amoebocytes.
   c. a well-developed digestive tube.
   d. sexual and/or asexual reproduction.
   e. multiple incurrent pores.
5. Platyhelminthes have all of the following *except*
   a. bilateral symmetry.
   b. a true coelom.
   c. three tissue layers.
   d. an incomplete digestive tube.
   e. flame cells for excretion.
6. Which of these characteristics does not belong with the rest?
   a. protostome
   b. pseudocoelom
   c. spiral cleavage
   d. mosaic development
   e. trochophore larvae
7. Chordates and echinoderms are grouped together because both
   a. are radially symmetrical.
   b. produce embryos in which the blastopore becomes the anus.
   c. include some species that have a pseudocoelom.
   d. have a peritoneum and mesenteries.
   e. are protostomes.

8. In protostomes, the mesoderm originates from
   a. a single ectodermal cell.
   b. a region of the ectoderm.
   c. two or three endodermal cells.
   d. the lining of the pseudocoelom.
   e. a single cell of the 16-cell embryo.
9. The human diseases known as hookworm and trichinosis are caused by parasitic
   a. platyhelminths.
   b. rotifers.
   c. nematodes.
   d. mesozoans.
   e. cnidarians.
10. Which phyla include both free-living and parasitic forms?
    a. flatworms
    b. roundworms
    c. ribbon worms
    d. *a* and *b*
    e. *a, b,* and *c*

True-False Questions

Mark each statement true or false, and if false, restate it to make it true.

1. Sponges, jellyfish, and flatworms are all classified as eumetazoans.
2. The two-layered planula larva is probably analogous to the zygote stage of eumetazoans.
3. Starfish and vertebrates are deuterostomes, whereas clams and insects are protostomes.
4. Cnidarians have surface cells called choanocytes that contain prey-capturing structures known as nematocysts.
5. Although ribbon worms resemble flatworms, they are considered more advanced because they have a complete digestive system and a closed circulatory system.
6. In all probability, the original function of the coelom was reproduction.
7. In most protostomes, the coelom develops from cells that migrate in from the skin or cuticle.
8. The coelom of most deuterostomes develops from outpocketings of the endoderm of the gut.
9. The peritoneum is formed from a double layer of mesodermal tissue called mesentery.
10. In coelomates, but not in pseudocoelomates, a peritoneum completely lines the coelomic cavity.

Concept Questions

1. Why do zoologists believe that the ancestral metazoan was a flattened, colonial organism?
2. Contrast the function of the central cavity of sponges with that of hydrozoans.
3. Why are embryonic and larval stages so important for determining evolutionary relationships?
4. What advantages does a complete digestive system have over an incomplete digestive system? In which phylum does a complete digestive system first appear?
5. Explain why a coelom is believed to have evolved more than once among animals.

Additional Reading

Borradaile, L. A., F. A. Potts, L. E. S. Eastham, and J. T. Saunders. *The Invertebrata.* Cambridge University Press, Cambridge, 1958.

Brusca, Richard C., and Gary J. Brusca. *Invertebrates.* Sinauer Associates, Inc., Sunderland (MA), 1990.

Buschbaum, Ralph. *Animals Without Backbones: An Introduction to the Invertebrates.* University of Chicago Press, Chicago, 1948.

Hotez, Peter J., and David I. Pritchard. "Hookworm Infection." *Scientific American,* June 1995, p. 42. How hookworms enter the body, and how they might be treated.

Levinton, Jeffrey S. "The Big Bang of Animal Evolution." *Scientific American,* November 1992, p. 52. Around 600 million years ago, all the basic body plans of modern animals arose. Why haven't there been any fundamentally new body plans since then?

McMenamin, Mark A. S. "The Emergence of Animals." *Scientific American,* April 1987, p. 94. The first adaptive radiation of animals produced all the major types we see today.

Wells, Martin. *Lower Animals.* World University Library, McGraw-Hill Book Co., New York, 1968.

Internet Resource

To further explore the content of this chapter, log on to the web site at:

http://www.mhhe.com/biosci/genbio/guttman/

Animals II: The Coelomates

35

Key Concepts

A. The Protostome Phyla

35.1 Segmentation is an important feature of many body plans.

35.2 Annelids have rather uniformly metamerized bodies.

35.3 *Peripatus* preserves some features of annelids and is similar to a forerunner of arthropods.

35.4 Arthropods have metameric bodies with jointed limbs.

35.5 Molluscs are unsegmented protostomes that usually form shells.

B. The Deuterostome Phyla

35.6 Animals in three small phyla feed with the aid of a lophophore.

35.7 Echinoderms are basically adapted to a sessile life.

35.8 Chordates share a notochord, dorsal nerve cord, and pharyngeal gill pouches.

35.9 Vertebrates are defined primarily by their backbone.

35.10 The first vertebrates were fishes.

35.11 Amphibians have made a partial transition to terrestrial life.

35.12 The evolution of the amniotic egg enabled reptiles to conquer the land.

35.13 Birds are essentially modified dinosaurs with feathers.

35.14 Mammals have body hair and suckle their young.

Gooseneck barnacles, with their white shells and black necks, resemble the common European Barnacle Goose.

As Europe entered the Renaissance, naturalists and adventurers began to write about the wonders they had seen or been told about. An often-repeated story described trees whose fruit was tiny geese that then dropped off and matured into full-sized geese. In the thirteenth century, Thomas de Cantimpré wrote,

> *This is a bird that grows from wood and that wood has many branches from which the birds sprout so that many of them hang from one tree. These birds are smaller than geese and have feet like ducks, but they are of black color. They hang from the tree by their beaks, also from the bark and the trunk. In time they fall into the sea and grow on the sea until they begin to fly.*

Later writers who were skeptical of this story were always reassured by one "authority" or another claiming to have seen the phenomenon with his own eyes. But in 1527, Boethius of Aberdeen not only expressed his disbelief in the story of the goose trees, but asserted instead that the trees that fall into the sea become riddled with worms that first have heads and feet, then develop wings and feathers, become small geese, and eventually mature into large geese. Naturalists of the seventeenth century gradually became more accurate, skeptical observers, so that around 1700, it finally became clear that the story of geese growing on trees was based on floating logs covered with gooseneck barnacles, *Lepas anatifera;* a casual observer might mistake their white shells and long, black necks for tiny forms of the white-bodied, black-necked bird known today as the Barnacle Goose, *Brenta bernicla.* The worms Boethius had interpreted as an earlier larval stage are molluscs called shipworms *(Teredo).*

It took a few centuries of improving observations for naturalists first to clearly separate plants from animals, and then to identify a mollusc, a vertebrate, and an arthropod as quite distinct species. In this chapter, we survey the dominant animals on earth, the coelomates, which chiefly belong to these three phyla. They comprise two large clades, the protostomes and deuterostomes. ■

A. The protostome phyla

Most species of modern coelomates are protostomes, animals in which the embryonic blastopore becomes the adult mouth. Some of the most prominent protostome groups have a developmental stage called a trochophore larva, which is shaped like a child's top and swims by means of a band of cilia around its middle (see Figure 34.24). Adult protostomes have ventral nervous systems, in contrast to the dorsal nervous systems of deuterostomes. Leaving aside some minor phyla, protostomes are divided into segmented animals and molluscs.

Because an introductory book like this is not the place to air all the complexities and difficulties discussed by the specialists in a subject, we will present a rather conventional view of animal phylogeny. Be aware, however, that many points of animal structure are really much more controversial than we indicate here. We cannot discuss all the controversies, but the Additional Reading section at the end of this chapter will open the door to some of them for students interested in exploring further.

35.1 Segmentation is an important feature of many body plans.

Most animals are built on a fundamental plan in which the body is divided into a series of segments called **metameres** or **somites,** each of which contains the most essential body parts.

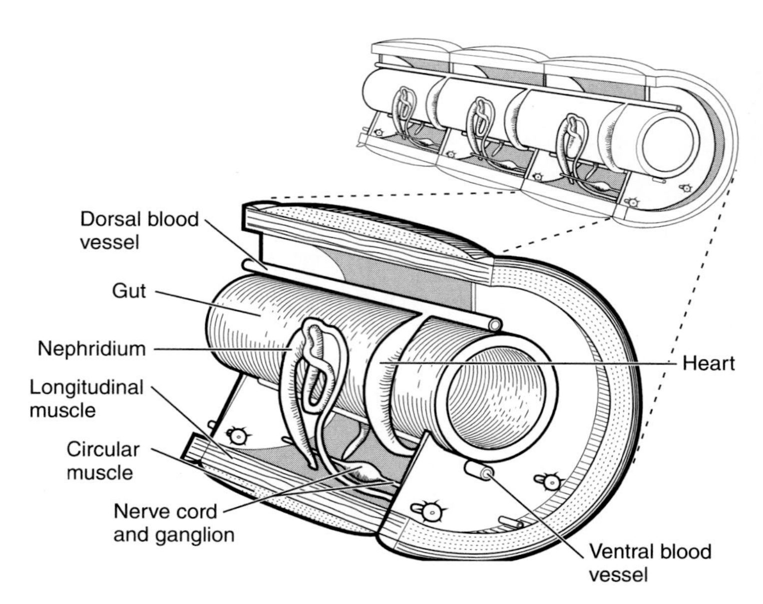

Figure 35.1
In a metameric animal, a single metamere containing organs and portions of organs is repeated many times. The uniform tubular animal may then be changed through specialization of some metameres.

The metameric design principle is similar to the principle of polymeric construction. Each metamere is a kind of "unit animal," and a whole animal is made by repeating this unit several times (Figure 35.1). Once an animal has evolved an embryological mechanism for forming one metamere with its organs, it is relatively simple to repeat that process many times in a chain. Note, however, that not all apparently segmented animals have true functional segmentation. For example, the "segments" of a tapeworm are just little bags of reproductive structures.

Principle of polymeric construction, Section 4.2.

As biologists assimilate the implications of research on *Hox* genes into their thinking about animal evolution, the idea of metamerism becomes more important and more problematic. Detailed studies of embryology show that the metamerism of the chordate body (phylum Chordata, including vertebrates) is quite different from that of segmented protostomes such as annelids and arthropods. Yet their sharing of *Hox* genes indicates that a fundamental plan for animal development was established long ago in the common ancestors of these groups; that plan is concerned, to a large extent, with defining body segments and making them develop into specialized structures. *Hox* genes exist even in nematodes, which are not metamerized and belong to an entirely separate clade, as shown by the form of their (pseudo)coeloms. We showed in Chapter 21 how developmental genes divide the body into regions, segments, and half-segments, and then impose an identity on each segment. We showed in Figure 24.29 how the evolution of homeotic genes has apparently changed the nearly identical segments of a primitive arthropod gradually into the more specialized segments of a flying insect. Thus it seems clear that much of the story of animal evolution is going to be told with reference to changes in these genes.

Homeotic genes and development, Section 21.11.

Metamerism appears to have evolved early in the protostome clade, and it dominates the structure of the annelids (segmented worms) and the arthropods (insects, scorpions, spiders, crabs, shrimp, and their relatives). The annelids and some arthropods retain the basic body plan, with relatively uniform segments. But in most arthropods, the structures of some segments have become modified for special functions; in particular, the simple walking appendages on each segment have changed into specialized appendages for swimming, grasping, or eating as well. We can understand the structure and evolution of the segmented animals by following the development of their metamerism.

A segmented coelomic cavity affords particular advantages for movement. Because an earthworm can extend or contract each of its segments independently of the others (Figure 35.2), it can easily extend its head end and then expand the head segments to anchor them into the surrounding soil while pulling the rest of the body forward. When we look at the segmented animals more closely, we will see other ways in which their structure allows them to move efficiently.

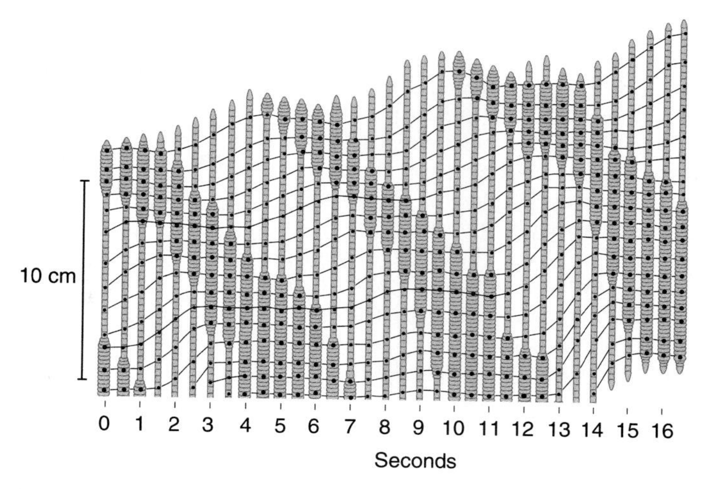

Figure 35.2

Because it is divided into segments, an earthworm can expand and contract regions of its body independently. It is therefore able to move by expanding part of its body to anchor itself to the surrounding soil and then pushing forward. A segment expands when the longitudinal muscles contract and becomes thin when the circular muscles contract, and the figure shows alternating waves of contraction of these two muscle sets moving along the worm's body. Source: Gray and Lissmann, *J. Exp. Biol.,* 15, 506–517, 1938.

35.2 Annelids have rather uniformly metamerized bodies.

The worms of the phylum Annelida (**annelids**) are constructed with a classical coelomate body plan—a tubular gut running through the middle of a coelom:

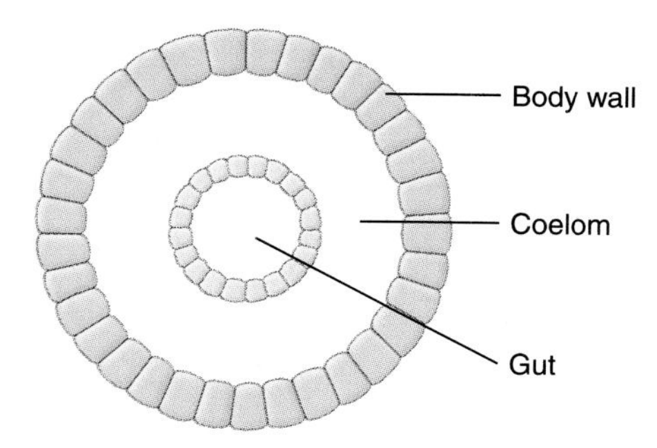

The familiar earthworm in Figure 35.3 shows this clearly. Annelids have a closed circulatory system powered by contractions of the dorsal aorta and a series of hearts. The blood exchanges CO_2 and oxygen with the surroundings as it moves through small vessels near the moist body surface. Annelids typically have a pair of metanephridia in each segment that remove wastes from the coelomic fluid. A ventral nerve cord emerges from the simple brain, a concentration of nervous tissue in the head end that receives signals from two sources: specialized sensory cells, which detect touch, light, and moisture, and chemoreceptors, which allow the animals to discriminate among various chemicals.

Earthworms represent the class Oligochaeta (*oligo-* = few; *chaete* = bristle or hair); the name refers to the bristles extending from their surface that help to anchor them in their surroundings. Many annelids in the second class, Polychaeta (*poly-* = many), show how metameric organization of the body muscles facilitates rapid movement. For instance, *Nereis* (Figure 35.4) is a free-living polychaete that can scurry over the ocean floor using its **parapodia,** the short, flat appendages on each segment:

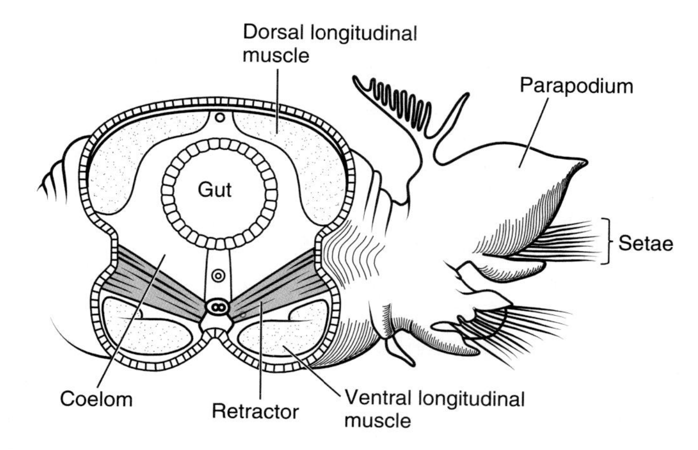

By contracting the longitudinal muscles in each segment, a polychaete can flex the dorsal and ventral parts of the

Figure 35.3

An earthworm, the most familiar annelid, has a classically simple coelomate body plan: a body wall with circular and longitudinal muscles (m. = muscle), a coelom, and a tubular intestinal tract running from its mouth to its anus. The animal is somewhat cephalized, with clusters of nerve cells (neurons) in ganglia at the anterior end connected to a ventral nerve cord. Nephridia are organs of excretion. The external bristles provide anchoring in the surrounding soil.

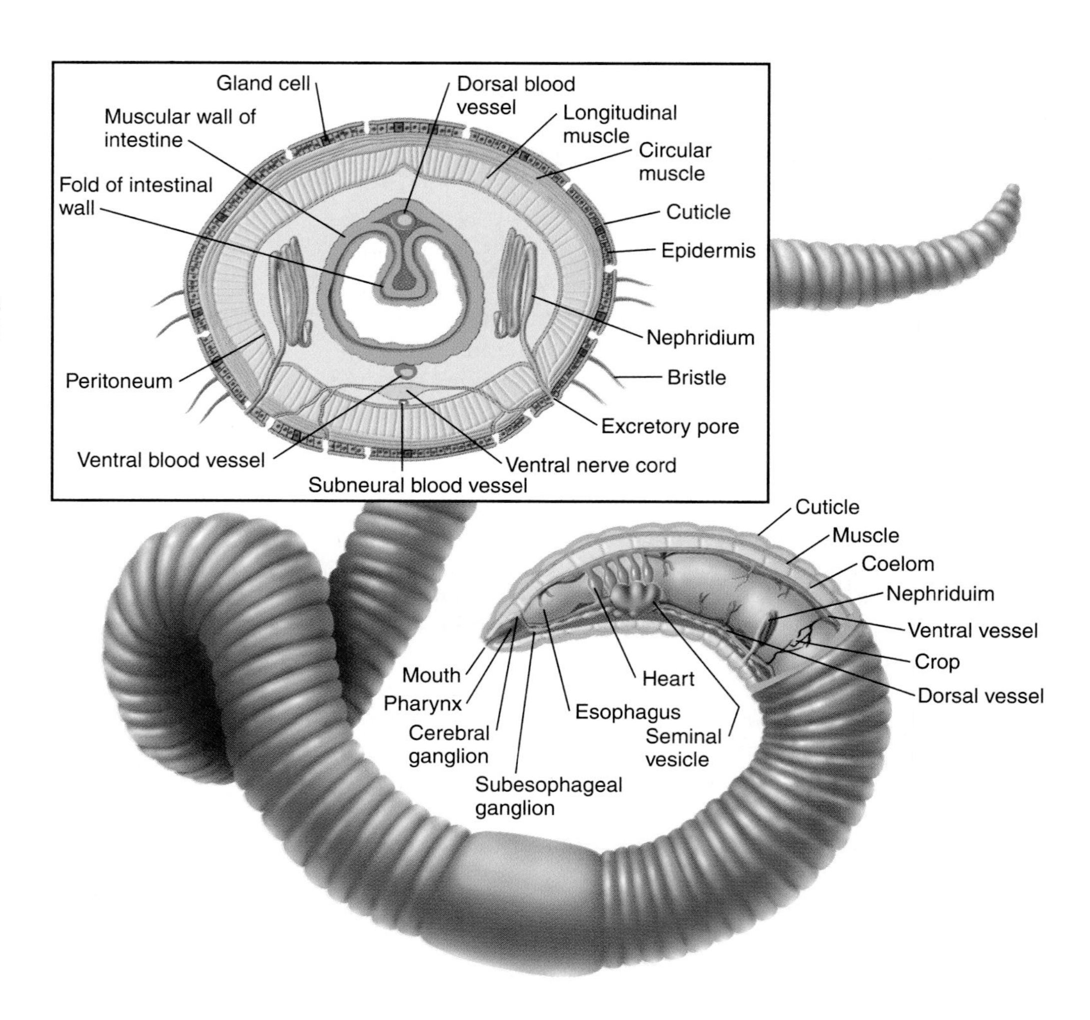

segment separately and thereby control the movement of its parapodia:

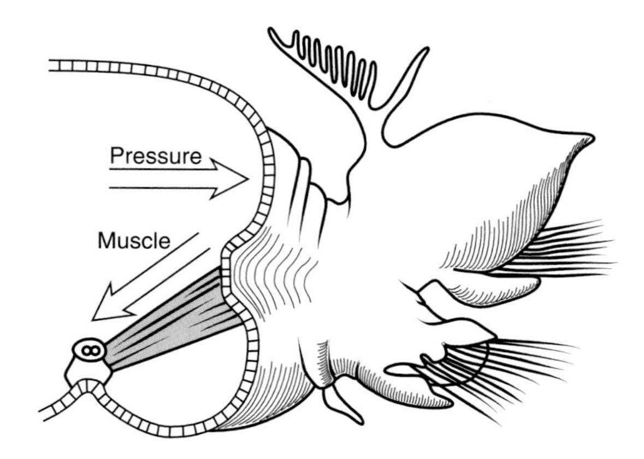

Each parapodium is extended by pressure on the coelomic fluid and withdrawn by muscles, reminiscent of the way the worms illustrated in Figure 34.19 move their specialized head segments. When swimming or running over the ocean bottom, a polychaete contracts the muscles on each side of its body in a wave running from the tail to the head as illustrated in Figure 35.5. Many arthropods use a similar swimming or walking motion.

Figure 35.4

The clam worm *Nereis virens* is a common marine polychaete worm.

In addition to swimming, a leech (class Hirudinea) moves by attaching its head to the surface, bringing its tail up to its head briefly, and then releasing the head end and pushing it

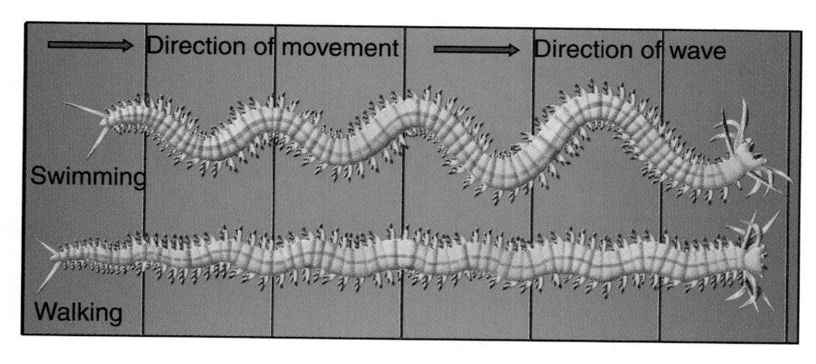

Figure 35.5

The swimming or walking motion of a polychaete illustrates a fundamental type of animal movement. The longitudinal muscles contract in a wave running from the animal's tail to its head, while it thrusts its parapodia outward to deflect the water or push against the substratum. The right side of each segment is held still for an instant while the left side is lifted and pivoted forward, and then the motion is reversed while the right side is pivoted forward. Source: Martin Wells, *Lower Animals,* 1968, McGraw-Hill Company.

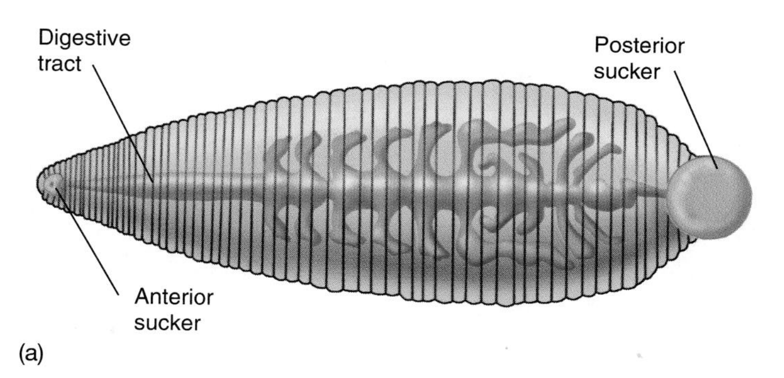

(a)

(b)

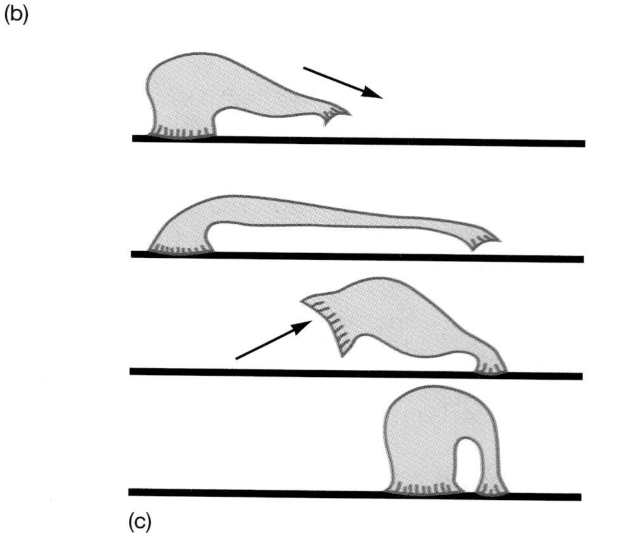

(c)

Figure 35.6

Most leeches are adapted for living on the blood of vertebrates, and a leech can store large amounts of blood in the side pouches of its intestine *(a).* The medicinal leech, *Hirudo medicinalis (b),* has been used to remove excess blood from some bruises or from the area around fingers or toes reattached surgically after an accident. *(c)* Leeches move with an "inch-worm" motion by attaching and releasing their suckers.

forward again (Figure 35.6). We see the same method of locomotion in insect larvae such as inchworms, the caterpillar larvae of certain moths, which seem to be measuring a leaf as they move over it.

35.3 *Peripatus* preserves some features of annelids and is similar to a forerunner of arthropods.

Hidden away under the debris in many tropical forests is a little creature called *Peripatus* (Figure 35.7*a*). A casual observer might dismiss it as just another large caterpillar, or a worm, or a millipede, but the essence of scientific discovery is to pay attention to details, and careful examination shows that *Peripatus* is not "just another" anything. *Peripatus* and its relatives in the phylum Onychophora look like highly modified annelids and share some features, especially in the appendages and circulatory system, with arthropods. Thus they appear to be a link between the two phyla.

Here we must reemphasize that no living organism can possibly be the ancestor of any other living organism. Onychophorans simply preserve the essential features of the annelid-like animal that presumably evolved into the ancestral arthropod. Annelid features, such as a closed circulatory system and muscles that do not extend into the appendages, are shared *primitive* characteristics; arthropod features, such as an open circulatory system and muscles that do extend into the appendages, are shared *derived* characteristics.

Peripatus is obviously metameric. Its gut and nervous system are like those of an annelid, and each of its segments bears a pair of coxal glands, used for excretion, that are similar to annelid nephridia. But there the resemblance ends. For one thing, each of the soft, cone-shaped appendages of a *Peripatus* (Figure 35.7*b*) ends in a tiny pair of claws that are quite different from the parapodia and bristles of annelids. Furthermore, instead of a closed circulatory system, like that of an annelid, *Peripatus* has an *open* circulatory system like that of arthropods, as we will illustrate later for a lobster (see Figure 35.10). Instead of two distinct fluids—blood confined to blood vessels and interstitial fluid surrounding the circulation—only one kind of fluid, **hemolymph,** serves as both blood and lymph. In *Peripatus,* a

(a)

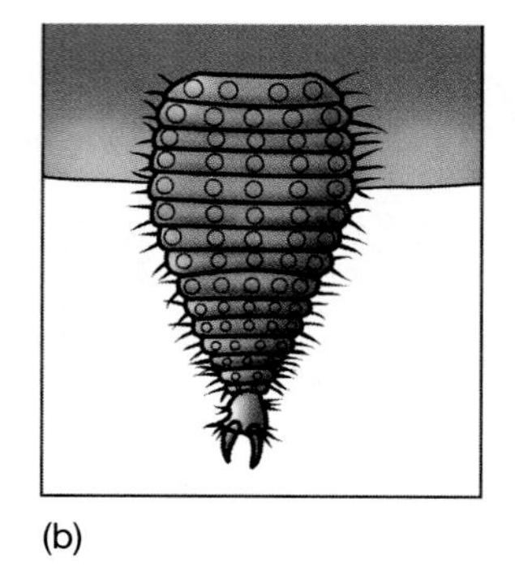

(b)

Figure 35.7

(a) Onychophorans such as this *Peripatus* have anatomical features of both annelids and arthropods. They resemble caterpillars, but have a simple gut and are not strongly cephalized. The body is covered with a protective cuticle that is molted periodically at times of growth, as in many arthropods. The legs, with a pair of terminal claws *(b),* move through a hydrostatic mechanism similar to that used by polychaete annelids.

dorsal heart pumps the hemolymph through a short vessel that is open at both ends, so it picks up hemolymph posteriorly, from a pair of openings in each body segment, and forces it back into the hemocoel at the anterior end.

35.4 Arthropods have metameric bodies with jointed limbs.

Arthropods, which constitute the largest of the animal phyla, are also metameric animals, and some retain uniform segments like those of annelids and onychophorans. More commonly, though, the segments are specialized in various ways. Arthropods are united by their segmented limbs (*arthro-* = jointed; *-pod* = foot), which are foreshadowed in the onychophoran leg, and specialization of segments most commonly takes the form of specialized limbs. In most arthropods, some limbs remain organs for locomotion while others are modified into sensitive antennae, mouthparts, or devices for grasping or swimming (Figure 35.8). Much of the arthropods' success in adapting to many distinct ways of life depends on the evolutionary modification of a basic, highly adaptable, jointed appendage. Arthropods are classified in part by the basic form of their appendages, which are either *uniramous,* with only a single branch at the end, or *biramous,* divided into a Y at the end.

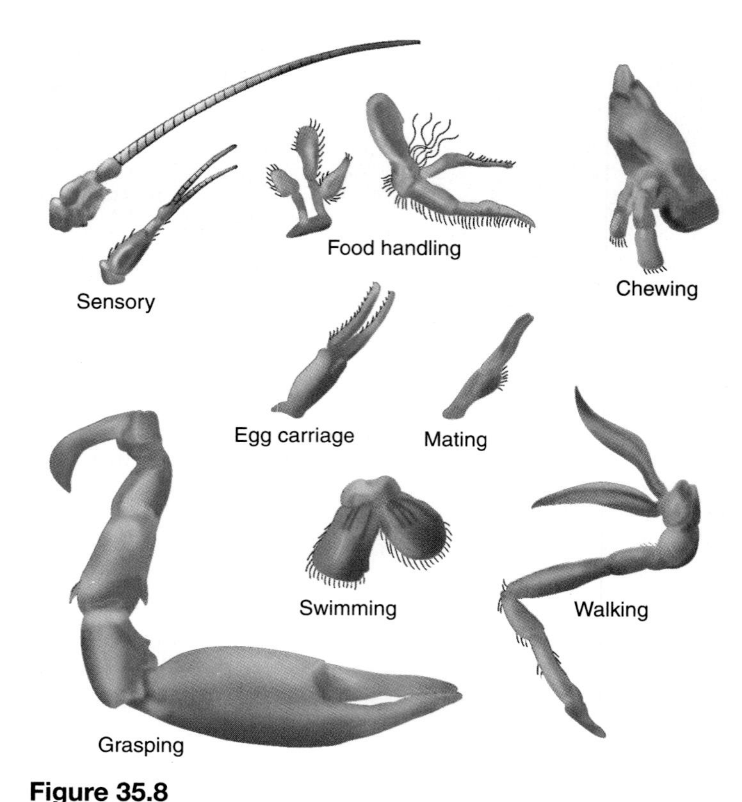

Figure 35.8

The jointed arthropod appendage can be modified into a great variety of useful structures in various groups. These are the appendages of a crayfish.

Arthropods also owe their success to their tough exoskeleton. Instead of moving with the hydrostatic skeleton of annelids, arthropods have muscles attached to the exoskeleton, allowing them to operate their many-functioned limbs. They can move these limbs independently in various ways and clamp their grasping limbs closed, sometimes with a powerful grip. The exoskeleton has permitted the most numerically successful of all animals, the insects, to evolve wings and to become one of only four groups of animals that can fly. The exoskeleton is made of the mucopolysaccharide **chitin.** The outer layer of exoskeleton is hardened into a rigid **cuticle** with cross-linked proteins, leaving a softer, flexible layer below. In crustaceans, the cuticle is hardened with calcium salts. The waterproof cuticle covers the whole body, and when topped with waxy lipids, it protects the animal not only against predators but also against excessive water loss or gain. Cuticle has allowed arthropods to become independent of the water and to move successfully into terrestrial habitats. However, cuticle has a disadvantage. If an arthropod is to grow, it must molt (shed its cuticle) periodically (a process called *ecdysis*), leaving it soft and vulnerable for a time. Since the exoskeleton extends to some parts of the digestive tract, even these are shed during ecdysis.

Exoskeleton and musculature, cuticle, insect flight, Chapter 44.

The **compound eye** has also contributed to the success of arthropods. One eye contains hundreds or even thousands of individual units called *ommatidia,* each with its own lens and light-sensitive cell, each forming part of a whole image (Figure 35.9). Compound eyes, while not well suited for forming detailed images, are exceptionally good for detecting motion. Some arthropods also have simple eyes (ocelli) that detect light but are unable to form images.

The open circulatory system of arthropods is illustrated by that of a lobster (Figure 35.10). Like *Peripatus,* one fluid, hemolymph, serves as both blood and interstitial fluid. The lobster heart pumps hemolymph into the tissues through arteries, but rather than branching into capillaries, the arteries open into **sinuses** from which all tissue cells receive their nutrients and where they deposit their wastes. The hemolymph circulates through an open body space, a **hemocoel** (which is distinct from the coelom, the coelom itself being reduced to a few small spaces). In lobsters and other aquatic arthropods, the hemolymph is forced back through feathery gills for gas exchange and then into the hemocoel again.

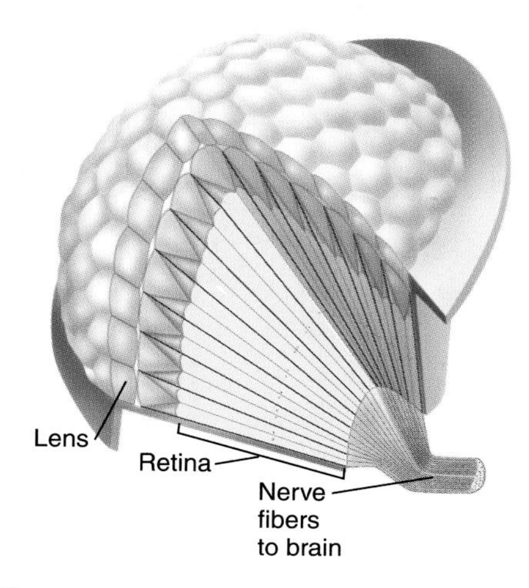

Figure 35.9

A compound eye is made of many separate units (ommatidia), each consisting of a lens that focuses light on a retina, from which nerve impulses are sent to the brain.

Instead of gills, most air-breathing arthropods use a system of fine **tracheae,** or air tubes, to carry air directly to their tissues (see Figure 32.17). Some spiders and other land arthropods, however, exchange gases with a **book lung** (Figure 35.11). This thin, folded diaphragm forms a large surface, like several pages of a book, with hemolymph on one side and air on the other. The air space opens to the atmosphere through a slit in the body, a spiracle. As hemolymph is pumped past the book lung, it exchanges gases with the atmosphere and then moves back into the hemolymph sinuses.

The arthropods are divided into four subphyla: Trilobitomorpha, Cheliceriformes, Uniramia, and Crustacea, based on the specializations of their appendages. The trilobites, all now extinct, had bodies divided into a central lobe and two lateral lobes, whence the name "tri-lobite." Their legs were relatively uniform and unspecialized (Figure 35.12). Most of the chelicerates are **arachnids**—spiders, scorpions, and their relatives (Figure 35.13); they have a pair of appendages called *pedipalps* that are modified for various functions and a second pair of appendages modified into *chelicerae,* like fangs or pincers. The chelicerate body is divided into an anterior **cephalothorax,** which generally bears mouthparts and four pairs of legs, and a

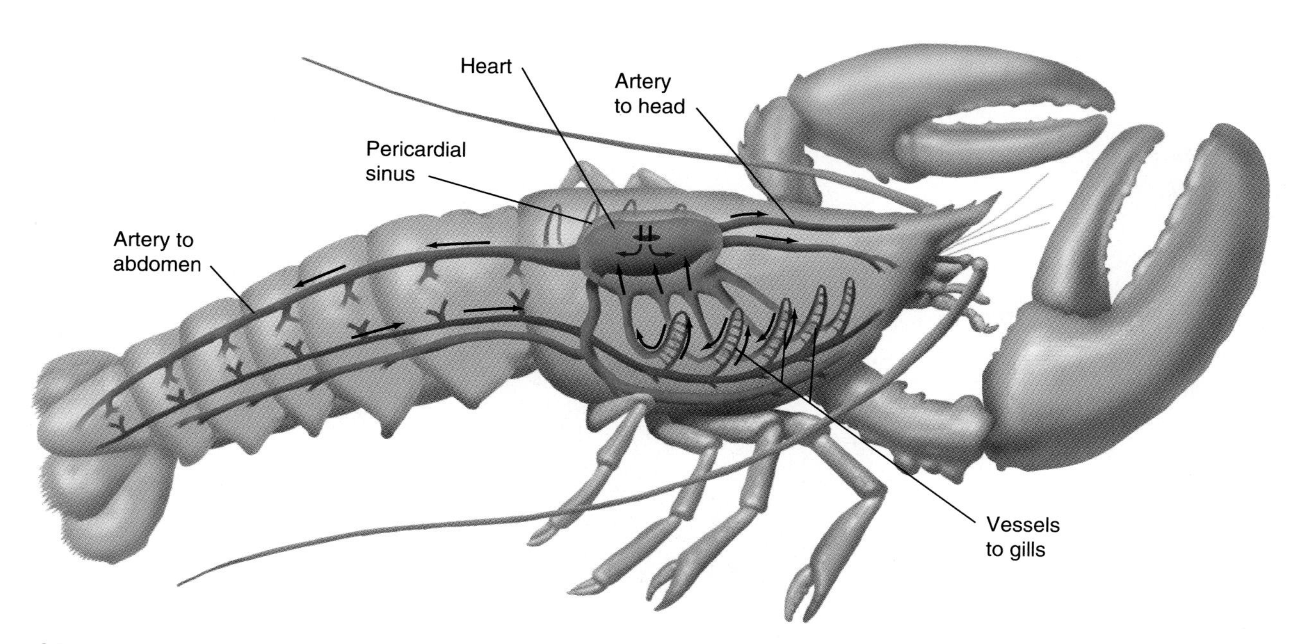

Figure 35.10

The open circulation of a lobster is typical of arthropods and similar to that of *Peripatus.*

posterior **abdomen** on which the legs are either lost or modified into sexual or respiratory structures.

Crustaceans include both macroscopic animals, such as barnacles, crabs, shrimp, and lobsters, and minute animals, such as water fleas and brine shrimp (Figure 35.14). They are characterized by a head made of five fused segments and a long trunk. The first few head appendages have been modified into one or two pairs of sensory **antennae,** a pair of strong **mandibles** suited for crushing or cutting food, and one or two pairs of **maxillae** that move food into the mouth. Crustacean appendages are biramous, and this feature separates them from the uniramians, which have similarly specialized head appendages.

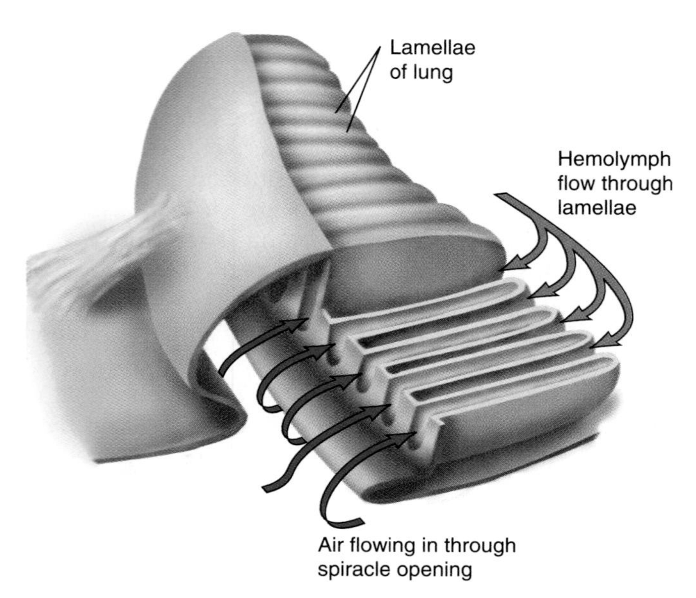

Figure 35.11

Many arachnids exchange gases through a book lung consisting of many thin plates (lamellae) in which hemolymph flowing on one side of each plate is separated by only a thin membrane from air on the other side.

Uniramians are divided into two classes: Myriapoda includes the many-legged centipedes (Chilopoda) and millipedes (Diplopoda); Hexapoda includes the familiar insects, with their six legs. The heads of all uniramians bear antennae, mandibles, and two pairs of maxillae. Myriapods have a long trunk; centipedes retain separate trunk segments, each with a pair of legs; and millipede segments fuse by pairs, so each unit has four legs. In insects, several segments fuse to make three regions—a **head** bearing sense organs and mouthparts, a **thorax** with three pairs of legs, and a posterior **abdomen** with no appendages:

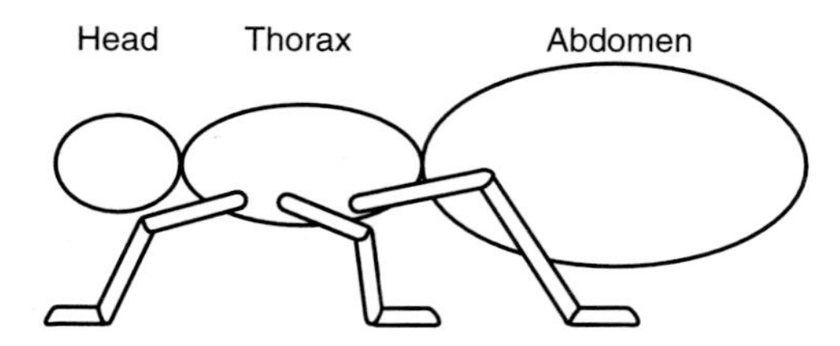

Some of these animals openly demonstrate their annelid affinities. Centipedes and millipedes are reminiscent of polychaetes running on land, and every caterpillar, with its uniform segmentation, is a reflection of its ancestors (Figure 35.15).

Arthropods retain the basic way of walking that first evolved in annelids, but the insects have improved and simplified it. A polychaete pivots each metamere back and forth while thrusting the attached parapodium out and swinging it forward. This action is limited because a parapodium has only retractor muscles and must be extended by pressure from the coelom. To use a hydrostatic skeleton in this way, the body must remain quite flexible, so pressure can be exerted independently

Figure 35.12

(a) A trilobite fossil shows the animal's relatively uniform metamerism. *(b)* The living animal had a series of rather undifferentiated legs and external gills.

(a)

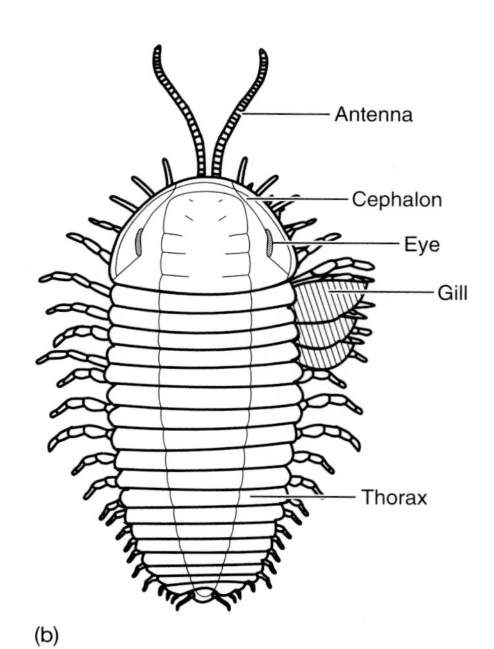

(b)

Figure 35.13

The class Arachnida includes eleven orders of terrestrial animals; these are representatives of the most common orders. The dimensions given are body lengths, not including legs.

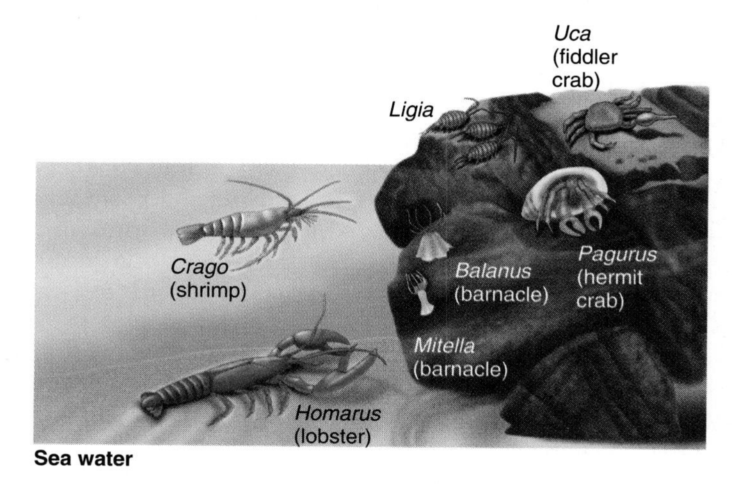

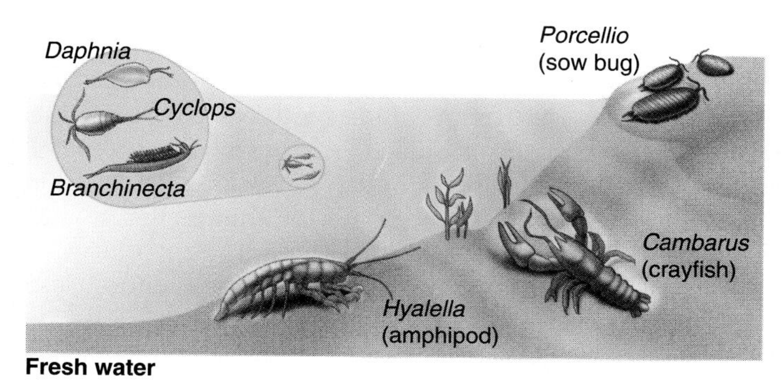

Figure 35.14

Crustaceans live in both sea water *(left)* and fresh water *(right),* as well as on land. *Daphnia, Cyclops,* and *Branchinecta* represent some orders of minute crustaceans that can hardly be seen except with a microscope.

Figure 35.15

The basic similarities among all uniramians are shown by *(a)* a millipede of the class Diplopoda and *(b)* a caterpillar, the larva of a Monarch butterfly of the class Insecta (Hexapoda).

(a)

(b)

in each segment, but it is much more efficient for the legs to push against a stiffened body, one that doesn't give with each step. Spiders, centipedes, and millipedes also lack extensor muscles for their legs, so they don't achieve the efficiency of insects, which use extensor muscles rather than by hydrostatic pressure in each of their six legs. Insects' legs are attached to a solid thorax and are extended by extensor muscles rather than by hydrostatic pressure. The thorax also makes an excellent foundation for wings, and some insects are superb fliers.

Figure 35.16 compares the walking motions of a millipede and an insect. The millipede moves with a wave of leg motion running from front to back, so at any instant alternating groups of legs are touching the ground or raised. An insect can move its legs slowly, individually, but when moving fast, it rests on a triangular base of three legs while moving the other three, thus gaining speed and stability. Other factors besides mobility have contributed to the incredible success of insects. Fully three-quarters of all known animal species are insects, and astonishingly, most of these are beetles. Figure 35.17 summarizes the most important orders of insects.

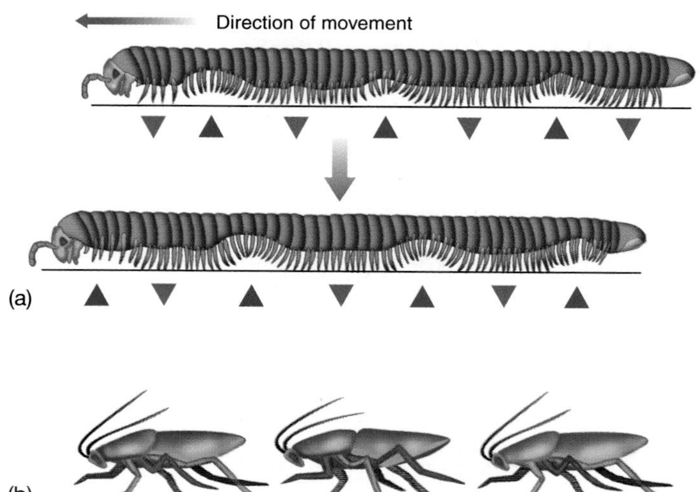

Figure 35.16

(a) A millipede moves with a wavelike pattern of leg motion, while *(b)* an insect supports itself on a triangular base of three legs while moving its other three legs.

35.5 Molluscs are unsegmented protostomes that usually form shells.

Molluscs are a familiar part of our world: Clams, oysters, and mussels are delicacies in our cuisines; the weirdly multilimbed octopuses add excitement to adventure stories; and we talk about a snail's pace and wonder how to keep slugs out of the garden. These animals are so varied that it isn't obvious that they all belong to the same phylum, Mollusca. Nor, since the phylum's name means "soft-bodied," is it obvious that some of its most prominent members are the "seashells" so prized for their beauty and durability—or rather, the animals that secrete these shells as their protective exoskeletons.

The basic molluscan body plan is shown by a chiton (class Polyplacophora or Amphineura), a bilaterally symmetrical animal covered with eight separate plates or valves (Figure 35.18). These valves make the animal appear metamerized on the outside, but internally there is no trace of segmentation. Instead, the body is divided dorsoventrally into three parts: A chiton rests on a remarkably large, muscular foot, bearing a mouth and sensory organs at its head end; above this is the visceral region where the organs of digestion, excretion, and reproduction are centered; and the mantle on top secretes the shell and encloses a cavity between the shell and the rest of the body.

The bilateral symmetry of molluscs is most obvious in clams, scallops, oysters, mussels, and other members of the class Bivalvia; each half of the shell is a valve, hence the name *bivalve.* However, the molluscan body plan is sharply modified in the *univalves,* molluscs that have a single valve with only one opening. To use this opening for both the mouth and anus, the animal's body is twisted, looping the intestine into a U and bringing the anus around to the head end. Figure 35.19 shows how this has been done in the class Gastropoda, composed of snails, whelks, cowries, and other animals that construct spiral shells. The body is also twisted in the tooth shells (Scaphopoda), which move through the sediment with only their tops protruding, and in the chambered nautilus of the class Cephalopoda. The body remains bilaterally symmetrical in chitons and molluscs without

Hemimetabola: Young are *nymphs* with compound eyes; metamorphosis is incomplete (gradual).

Odonata: Dragonflies

Large; body often brightly colored; filmy wings, not folded; large eyes; chewing mouthparts.

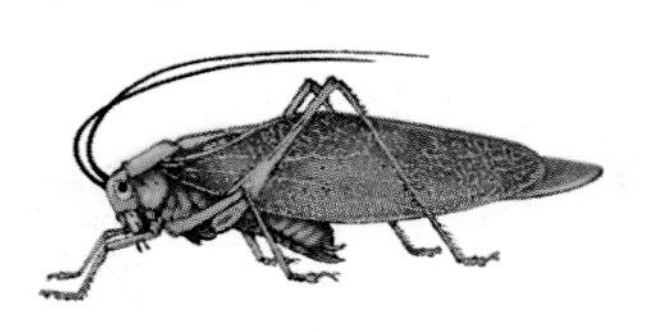

Orthoptera: Roaches, grasshoppers

Medium to large size; forewings leathery; hindwings thin; chewing mouthparts.

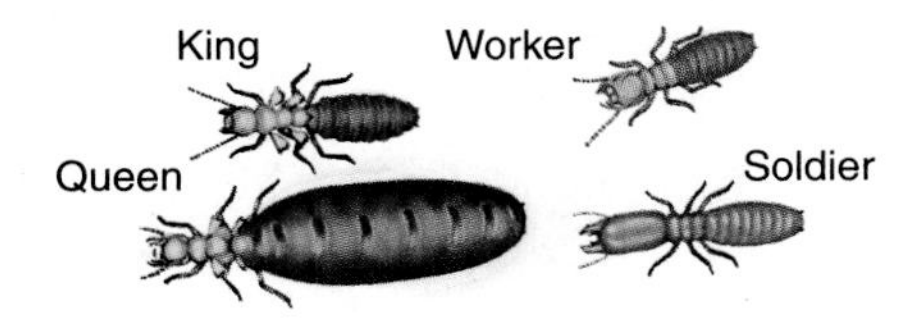

Isoptera: Termites

Body soft; chewing mouthparts; workers and soldiers wingless; sexual males and females with 2 pairs of membranous wings.

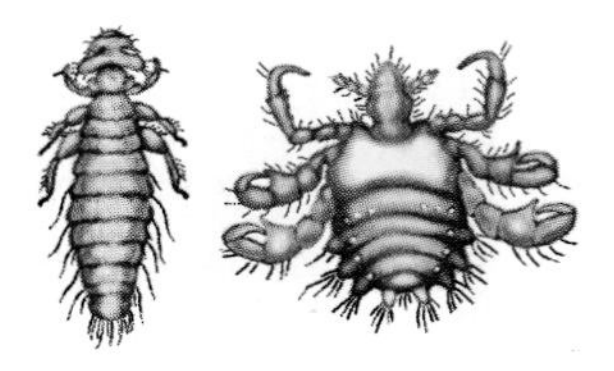

Mallophaga and **Anoplura:** Lice

Minute, flat, wingless; some Anoplura are sucking, others have chewing mouthparts.

Hemiptera: True bugs

Half-leathery forewings; filmy hindwings; piercing-sucking mouthparts; jointed beak far forward on head.

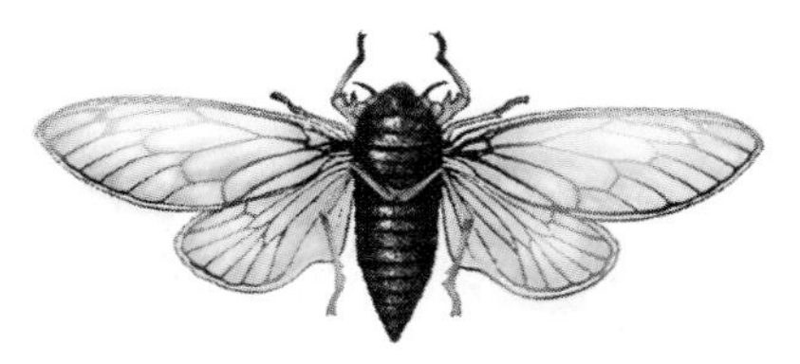

Homoptera: Cicadas, aphids, scale insects

Four or more membranous wings; piercing-sucking mouthparts; base of beak close to thorax.

Holometabola: Young are *larvae* with no compound eyes; metamorphosis is complete (complex).

Lepidoptera: Moths and butterflies

Four membranous wings covered with scales; chewing mouthparts in larvae, sucking in adult.

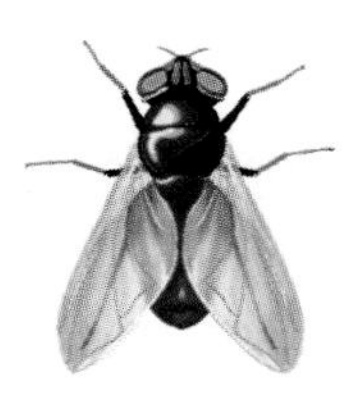

Diptera: True flies

One pair of wings; halteres (balancers) replace hindwings; piercing-sucking mouthparts, often forming a proboscis.

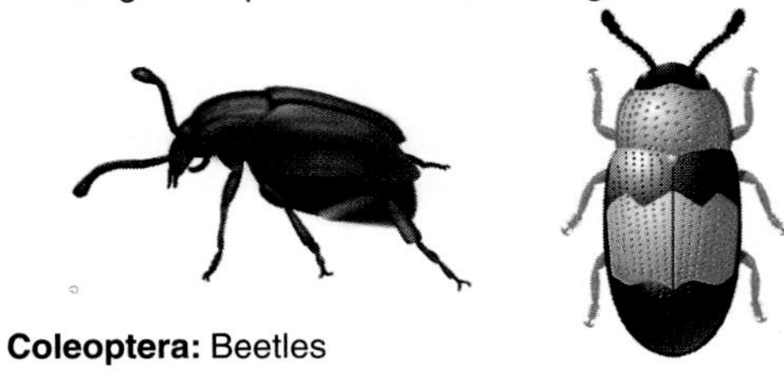

Coleoptera: Beetles

Hard, veinless forewings; filmy, folded hindwings; chewing mouthparts, some snoutlike.

Hymenoptera: Ants, wasps, bees

Four small, membranous wings interlocked in flight; chewing or chewing-lapping mouthparts.

Figure 35.17

The class Insecta is divided into two subclasses, one not shown (Apterygota) consisting of insects with no wings. The second subclass (Pterygota) is divided into two superorders. In Hemimetabola, the young insect grows gradually through stages that resemble the adult; Holometabola have complete metamorphosis through stages such as a larva and pupa.

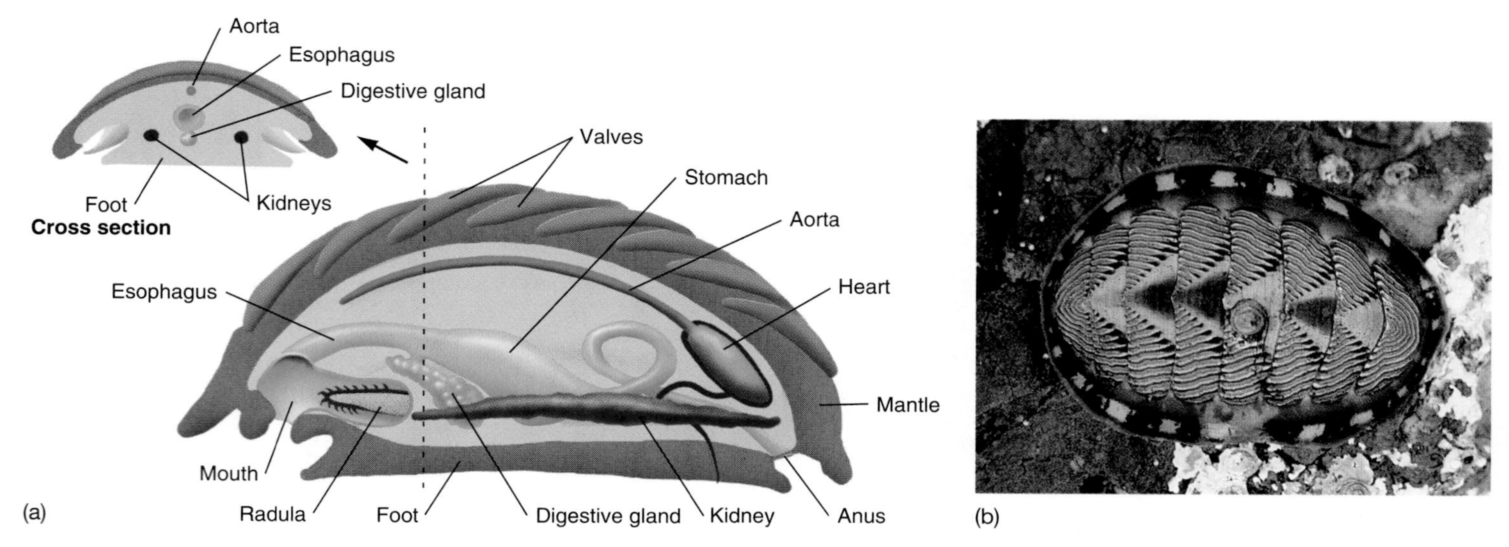

Figure 35.18

(a) A chiton, class Polyplacophora, is a relatively simple mollusc with bilateral symmetry as shown by the cross section. *(b)* The Sea Cradle chiton *Tonicella lineata* shows that the shell is made of eight valves.

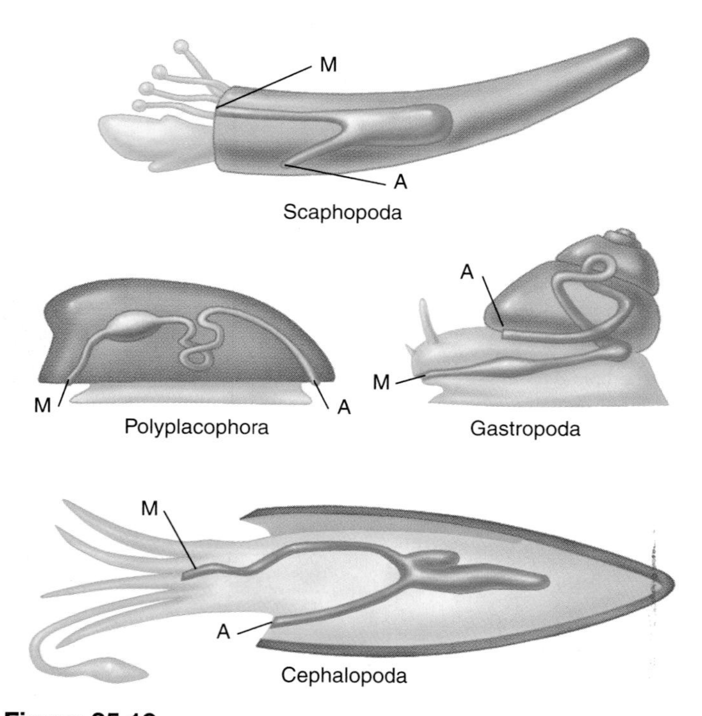

Figure 35.19

Body plans of the molluscan classes show how the intestinal tract has been twisted around to bring both the mouth (M) and anus (A) to the single valve opening, whereas the chitons (Polyplacophora) have the more typical animal intestinal tract with mouth and anus at opposite ends of the body.

shells, such as slugs among the gastropods and squid and octopuses among the cephalopods.

The molluscan circulatory system is open, like that of arthropods. The heart pumps hemolymph through a complex of arteries, and the hemolymph then empties into open sinuses in the tissues (Figure 35.20). Most molluscs are aquatic and use gills for respiration and nephridia for excretion. The simple digestive tract contains a large stomach and a digestive gland (called a liver). Some molluscs, such as gastropods and chitons, feed on large, solid food by rasping it with a specialized *radula,* a hard, toothed layer of chitin, which cuts and scrapes food into small fragments that are fed into the esophagus. Bivalves, in contrast, feed by pumping a stream of water through the chamber around their gills. While bringing oxygen to the gills, this stream carries a suspension of fine food particles, which are filtered by the gills and transported on a layer of mucus toward the mouth and into the esophagus. The molluscan nervous system consists of concentrations of neurons in ganglia at the head end. These ganglia are connected by a ring of nerves running around the esophagus and a pair of ventral nerve cords.

A mollusc's foot allows it to move rather slowly over a surface or to burrow through mud as the animal ingests fine-grained materials. The muscular foot sets up a wave of contraction running forward, thus moving the whole animal forward (Figure 35.21). In some of the bilaterally symmetrical gastropods that have no shells, such as nudibranchs, the foot becomes modified into what is virtually a wing, giving these animals a graceful swimming motion (Figure 35.22). Pteropods, or sea butterflies, crawl or float along on the ocean surface suspended on winglike extensions. Yet the masters of rapid movement are certainly cephalopods such as the squid that jet along by expelling a stream of water with a contraction of their muscular mantles.

The strong embryological indications of a close relationship among molluscs, annelids, and arthropods naturally raise the question of whether molluscs had metamerized ancestors. So it was very exciting when, in 1957, Henning Lemche reported a newly found mollusc, *Neopilina galatheae,* which had been collected from the depths of the ocean near Mexico by the Galathea expedition. *Neopilina* is just the kind of link with the

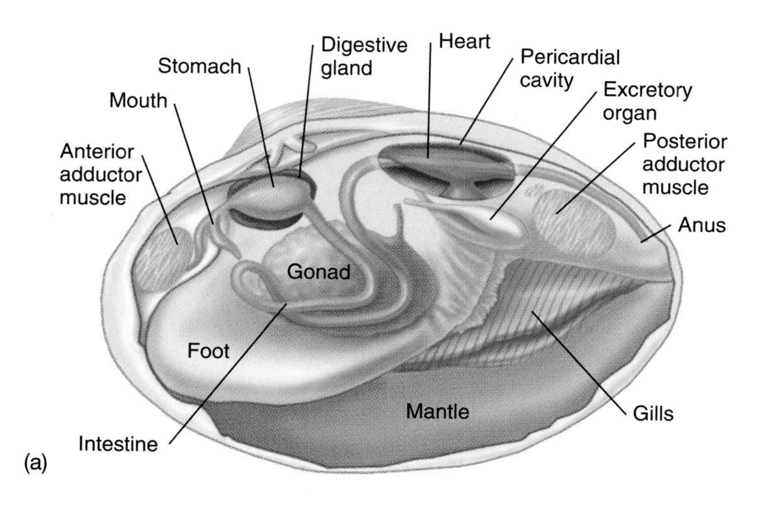

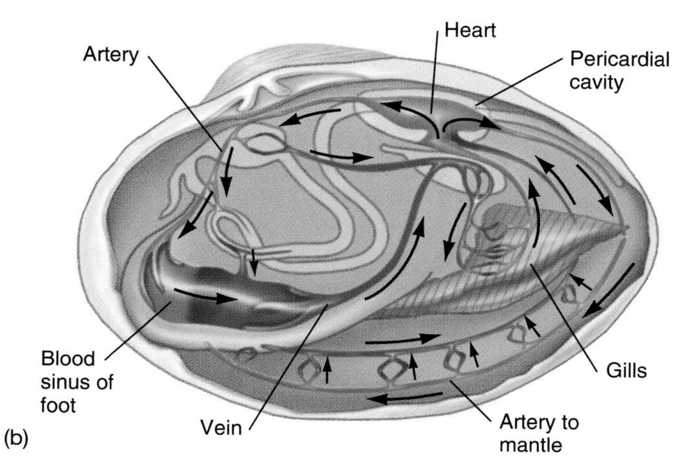

Figure 35.20
A clam, class Bivalvia, shows one set of gills and the foot dissected to reveal *(a)* the internal organs and *(b)* the open circulatory system.

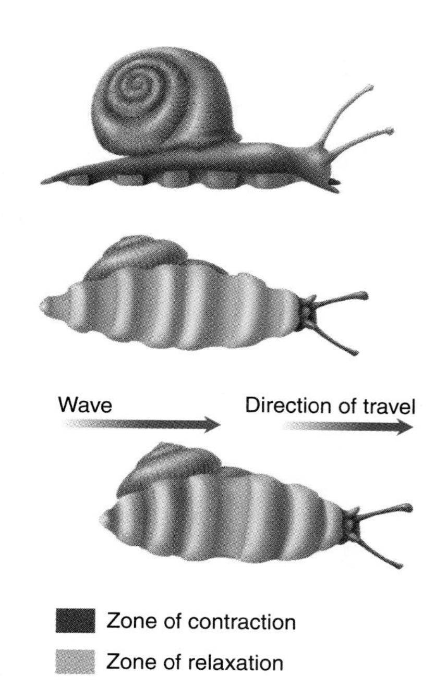

Figure 35.21
Molluscs typically move on a muscular foot that contracts in waves.

other protostome phyla that everyone had hoped to find, since it has five segments that each contain a pair of gills, a pair of nephridia, and other structures (Figure 35.23). It resembles a limpet—a kind of simple, noncoiled gastropod—but, along with some fossil relatives, it is placed in the class Monoplacophora. As zoologists have reexamined *Neopilina,* they have raised doubts about its apparent metamerism, pointing out that it is irregular and not strictly homologous with the segmentation of other protostomes. So the issue now remains in doubt. The molluscs most likely arose from the protostome clade sometime before the annelid-arthropod line, with its strong metamerism, evolved.

B. The deuterostome phyla

The second major group of coelomates, the deuterostomes, includes some wormlike animals as well as echinoderms and vertebrates. Remember that deuterostomes are defined by embryological features: The mouth does not develop from the blastopore, and the more primitive members have dipleurula larvae rather than trochophore larvae.

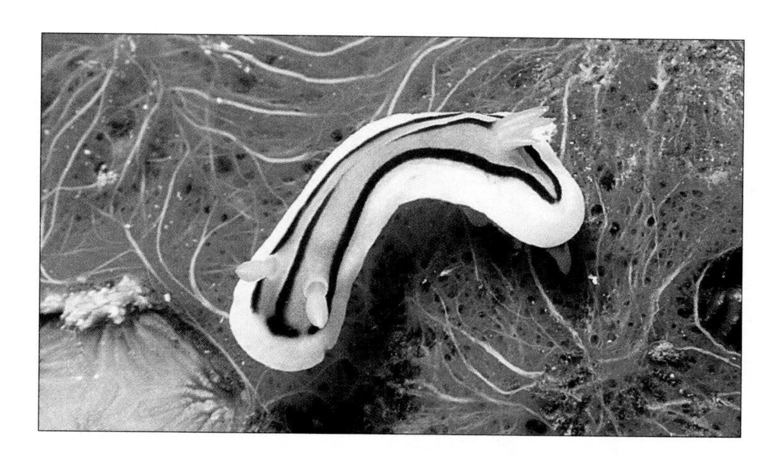

Figure 35.22
Nudibranchs are gastropods that have no shells and crawl along like slugs or swim on a broad undulating foot, which becomes rather like a wing. Some of them have incredibly bright colors.

35.6 Animals in three small phyla feed with the aid of a lophophore.

Three small phyla are united by a common body form: an intestine bent around into a U shape and a unique feeding device, a **lophophore,** consisting of a pair of spiral ridges fringed with many ciliated tentacles that sweep a stream of water toward the mouth. The members of all three phyla are relatively sessile animals that lie on the ocean floor or attach themselves to rocks or debris, where they draw in bits of food from the water. All are basically deuterostomes, except that the embryonic mouth develops from the blastopore as in protostomes.

The 15 species of **phoronids** (phylum Phoronidea) are wormlike animals that lie vertically in the ocean bottom with their lophophores exposed to collect food (Figure 35.24*a*). Animals of the other two phyla have similar internal anatomy, but each is enclosed in a hard shell of some kind. A **brachiopod** looks very much like a clam (Figure 35.24*b*), but its two valves are dorsal and ventral halves, whereas the valves of a clam are left and right halves. The lophophore occupies much of the space inside the shell. Brachiopods of many species are abundant in fossil beds, but they have now declined as a group to only about 200 living species. The **ectoprocts,** or bryozoa, are small animals, sometimes called "moss animals," that grow in colonies. Each animal looks like a phoronid inside a case (Figure 35.24*c*).

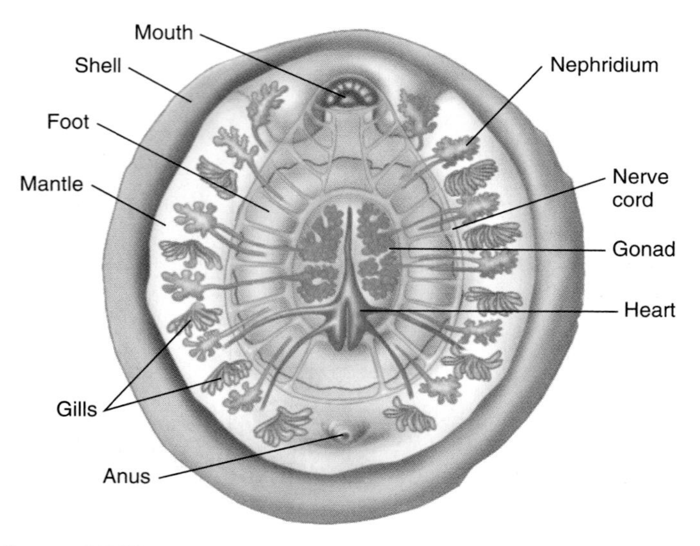

Figure 35.23
Neopilina has generally been considered a kind of living fossil mollusc that still shows metamerism; this ventral view shows a series of gills and nephridia. However, its metamerism has been questioned by new studies of its anatomy.

35.7 Echinoderms are basically adapted to a sessile life.

We have pointed out that animals that move from place to place are usually cephalized and bilaterally symmetrical while those that settle down to a sessile life become radially symmetrical. So it is puzzling, at first, that most members of the phylum Echinodermata **(echinoderms)**—sea stars, sea urchins, sand dollars, and their relatives—move about freely but have radial symmetry (Figure 35.25). The solution to this puzzle lies in the phylum's rich fossil history, which shows that ancestral echinoderms were adapted to a stationary life and so acquired radial symmetry; later, some echinoderms took to a motile life again, but retained this symmetry. The sessile echinoderms comprise a whole subphylum, now extinct except for a few crinoids (sea lilies) that remain attached to the ocean floor, looking like small trees. Throughout the phylum, fivefold symmetry is most

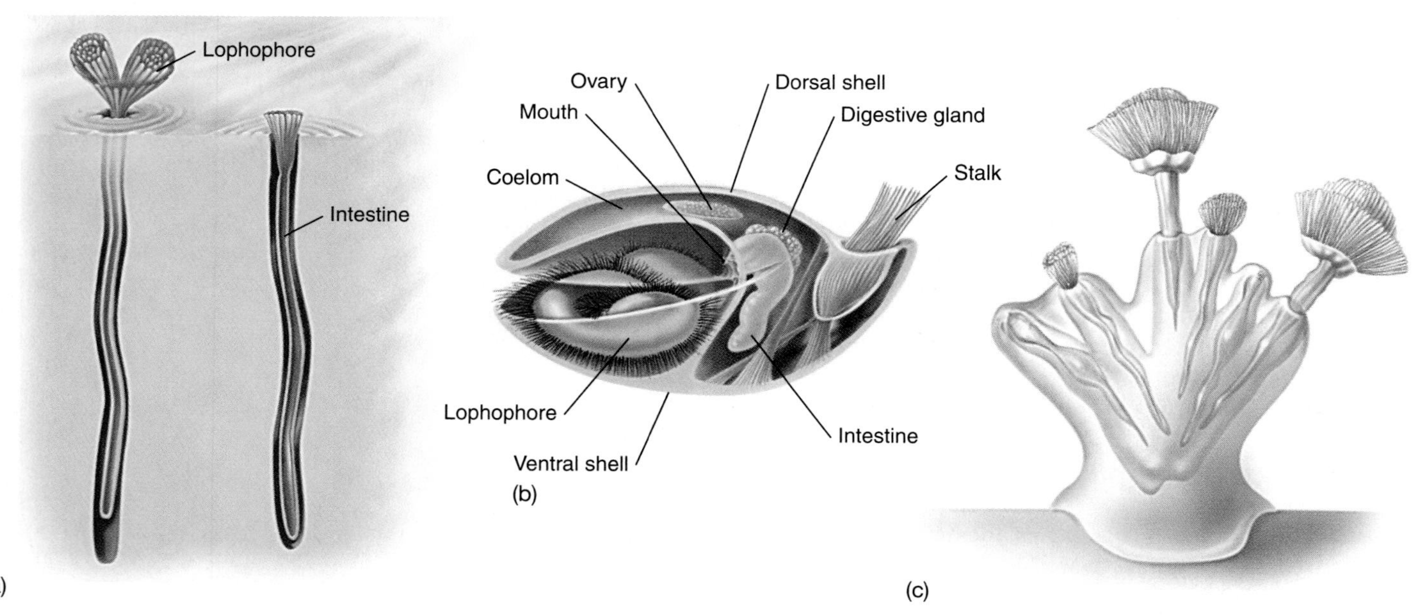

Figure 35.24
Animals in three phyla are united by their possession of a lophophore: *(a)* phoronids, extending mostly into the muddy ocean bottom; *(b)* a brachiopod; *(c)* an ectoproct.

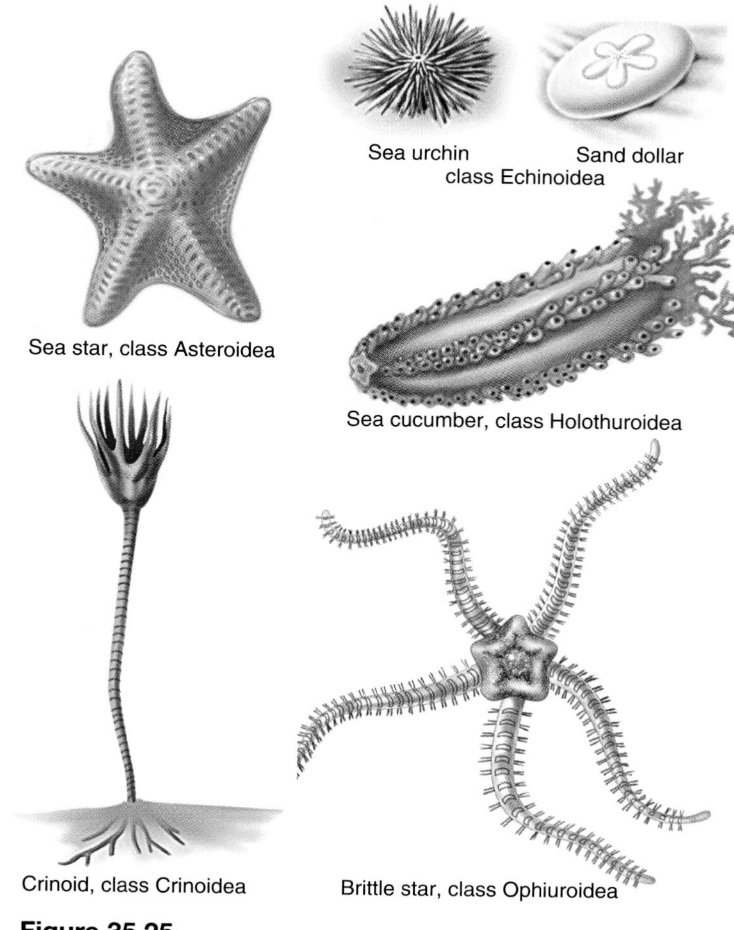

Figure 35.25
Echinoderms have spiny exoskeletons and usually have radial symmetry.

common, but some sea stars have many arms and not only in multiples of five. The sea cucumbers, Holothuroidea, have evolved much more of the cephalized, bilaterally symmetrical shape that characterizes other motile animals.

A sea star illustrates the general echinoderm body plan very well (Figure 35.26). Its arms radiate from a central disc, and since it has no head, the animal will move, sluggishly, in any direction. Its digestive tract runs from a mouth on the oral surface through a central stomach and out through an aboral anus, although the anus is believed to be nonfunctional in some species. The stomach is well supplied with digestive juices by large masses of *hepatic ceca* (see′kuh; sing., *cecum* = a pouch), glands that extend into the arms. A sea star has no respiratory, circulatory, or excretory systems; these functions are performed by fluid inside the coelomic cavity, which bathes all the internal organs. From a ring of nervous tissue surrounding the mouth, a nerve cord radiates down each arm, but echinoderms show no cephalization or brain formation.

Sea stars and other echinoderms move by means of a distinctive **water vascular system.** The dozens—sometimes hundreds—of tube feet on the underside of each arm are hollow, muscular cylinders that end in suckers. Each tube foot has a sac, or ampulla, at its base, and all the ampullae are connected to a radial canal leading to a central ring. This entire system is filled with water, and the animal controls its tube feet by controlling the flow of that water. By contracting, the ampullar sacs force water out into the feet, which stretch out and attach their suckers to a surface. Each foot then becomes firmly attached to the surface by external pressure (suction) when water is forced back into the ampulla. The animal can pull itself along by contracting the muscles near the base of each arm.

The water vascular system is also used in capturing prey. A sea star feeds on clams and other bivalves by straddling the edge of the mollusc opposite the hinge and persistently pulling with its many attached tube feet; meanwhile it secretes digestive

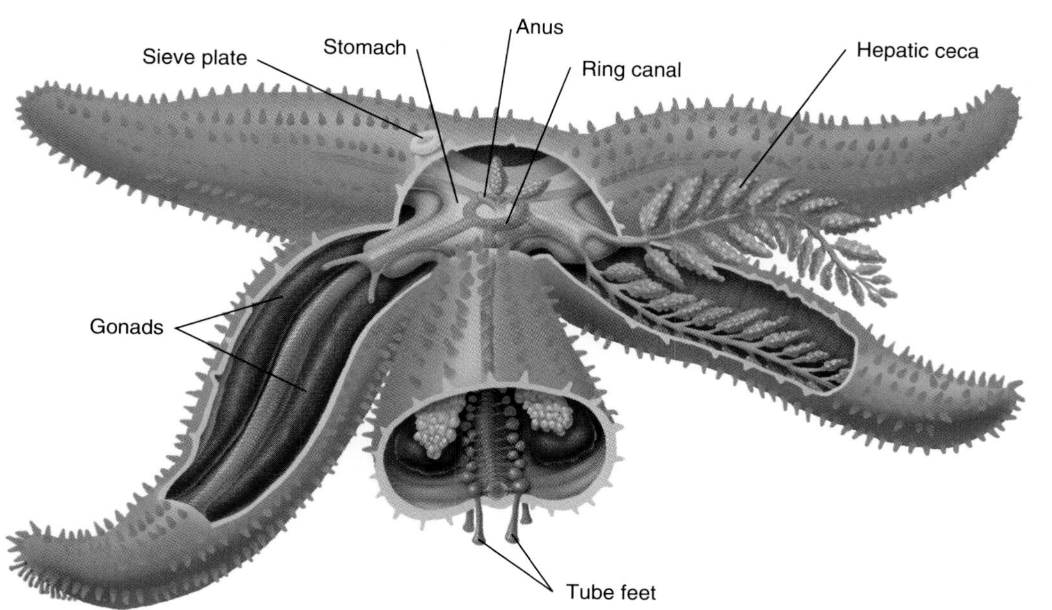

Figure 35.26
A sea star shows the structure of a representative echinoderm. The hepatic ceca produce digestive enzymes that pass into the central stomach. One arm is cut off to reveal the rows of tube feet, which are connected to the water vascular system that controls their movement.

enzymes, which begin to weaken the clam's muscles. Eventually a small slit opens between the clam's valves. The sea star actually turns its stomach inside-out, forcing it through its mouth and into the bivalve. Once inside, the stomach enzymes continue to digest the clam's tissues into a soup with a thick, clam-chowder consistency, which is then transferred into the sea star's hepatic ceca to be completely digested.

35.8 Chordates share a notochord, dorsal nerve cord, and pharyngeal gill pouches.

Leaving aside a few minor phyla, the other deuterostome coelomates are **chordates,** including the vertebrates. The 50,000 species in this phylum hardly rival the arthropods in numbers of species, but they certainly equal them in diversity and adaptability to a range of environments. Some zoologists prefer to make the chordate subphyla into separate phyla, but we unite them here because they share three important features (Figure 35.27), which always appear at some time during the life cycle but are not necessarily seen in adults:

- A **notochord,** a cartilaginous supporting rod, runs along the dorsal part of the body. It is always found in embryos, but in most vertebrates it is replaced during development by a backbone of bony or cartilaginous vertebrae.
- A **tubular dorsal nerve cord,** dorsal to the notochord, is formed during development by an infolding of the ectoderm. In vertebrates, the nerve cord eventually becomes encased and thus protected by the backbone.
- **Pharyngeal gill pouches** appear during embryonic development on both sides of the throat region, the **pharynx.** In the lower vertebrates, pouches grow outward to meet invaginations from the ectoderm, creating a series of slits that house the gills. Aquatic animals use these gills for respiration. The cartilaginous gill structures have no respiratory function in terrestrial animals, and they have become modified for other functions such as part of the jawbone and the three bones in the middle ear of mammals.

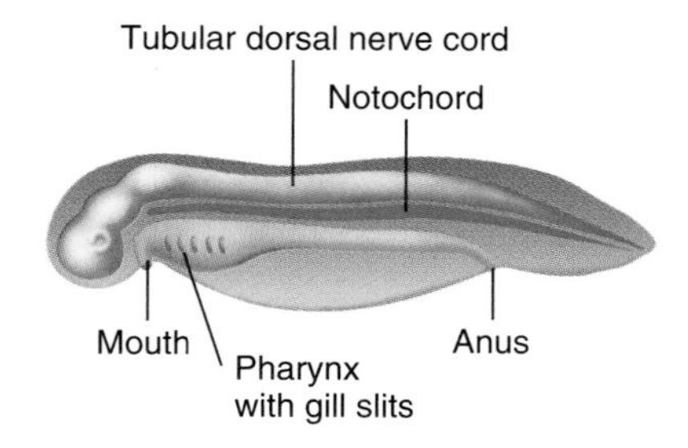

Figure 35.27

The characteristic features of a chordate are the tubular dorsal nerve cord, generally surrounded by a vertebral column, a dorsal notochord, and pharyngeal gill slits.

The chordates are divided into three subphyla. **Tunicates** are like leathery little bags that are either pelagic or attach to pilings, rocks, and seaweeds. They are also called sea squirts because a disturbed animal may contract and shoot streams of water from both of its siphons. Much of the body is occupied by a large pharynx with prominent gill slits (Figure 35.28*a*). Cilia in this pharyngeal basket continually move a stream of water through the animal, supplying it with oxygen and food particles. It would be hard to believe such a creature could be related to vertebrates if not for its larva (Figure 35.28*b*), which has all the chordate features, including a notochord in its tail, giving the subphylum the name Urochordata (*uro-* = tail).

The subphylum Cephalochordata contains the **amphioxus** or lancelet *(Branchiostoma),* which looks like a small fish and has the three chordate features as an adult (Figure 35.29). Significantly, amphioxus also shows clear metamerism, a feature that cannot be seen easily in other primitive deuterostomes. It is divided lengthwise into a series of muscle segments; its embryo has many gill slits, and a protonephridium for excretion is associated with each gill slit. Vertebrates, which comprise the third chordate subphylum, retain the same metamerism in internal structures. The rest of this chapter is devoted to them.

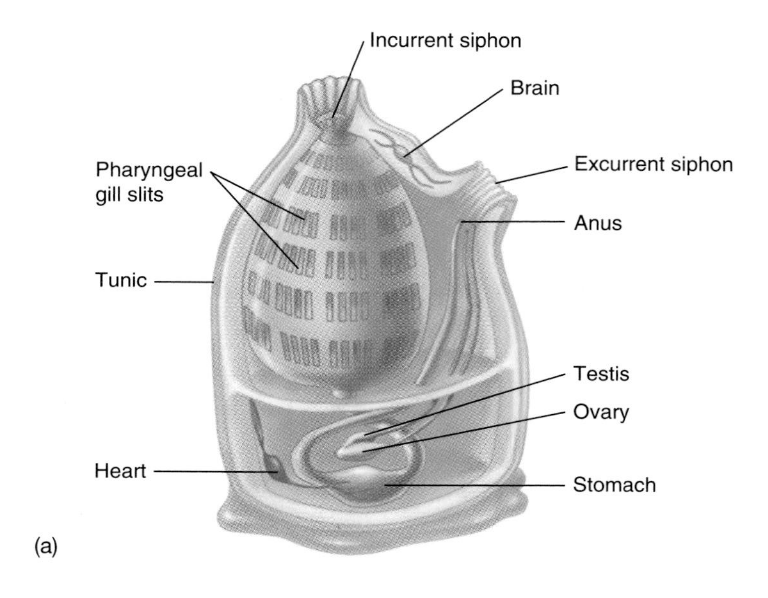

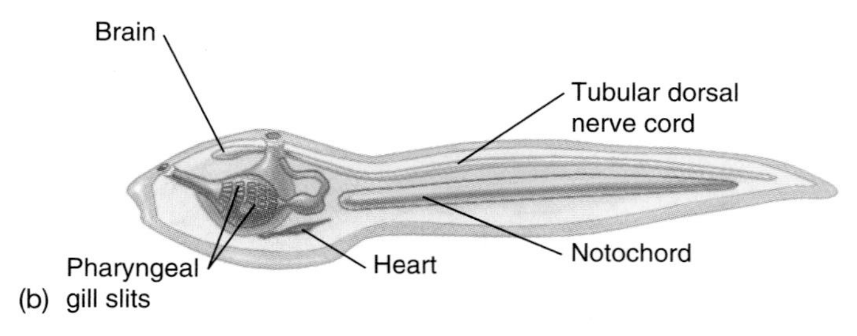

Figure 35.28

A mature tunicate *(a),* named for the tough, translucent "tunic" around the body, is usually a sessile organism that looks like a sac and is hardly recognizable as a chordate. However, its tailed larva *(b)* has typical chordate features.

35.9 Vertebrates are defined primarily by their backbone.

Animals in the subphylum Vertebrata (**vertebrates**) are set off from other chordates by several features. Most prominent, of course, is the endoskeleton of bone or cartilage, centering around a **vertebral column** (spine or backbone). Made of a series of separate **vertebrae** (showing internal metamerism again), a vertebral column combines flexibility with enough strength to support even a large body. Other vertebrate features include: (1) complex dorsal kidneys; (2) a tail (lost in some groups) extending behind the anus; (3) a closed circulatory system with a single, well-developed heart; (4) a brain at the anterior end of the spinal cord, with 10 or more pairs of cranial nerves; (5) a cranium (skull) protecting the brain; (6) paired sex organs in both males and females; and (7) two pairs of movable appendages—fins in the fishes, which evolved into legs in land vertebrates.

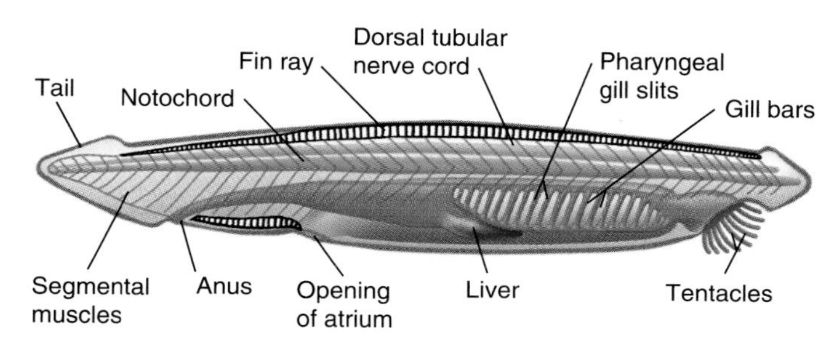

Figure 35.29

The lancelet, or amphioxus *(Branchiostoma),* has all three chordate features and is considered by many investigators to be a living representative of the ancestral chordate.

35.10 The first vertebrates were fishes.

Humans, being vertebrates themselves, have a special interest in the evolution of the phylum (Figure 35.30). The earliest vertebrates appeared during the Ordovician period, 500 million years ago. They were agnathans (*a-* = without; *-gnath* = jaw), small, jawless fishes up to about 20 cm long and also known as ostracoderms ("shell skin") because their bodies were covered with bony plates, most notably a head shield protecting the brain (Figure 35.31*a*). The first ostracoderms were marine bottom-dwellers that later moved into other marine and freshwater habitats. They fed on suspended food particles and could apparently draw in a stream of water with their pharyngeal muscles, rather than depending on the weak movements of cilia as their chordate ancestors had done. The only living agnathans are **cyclostomes,** known commonly as lampreys and hagfishes (Figure 35.31*b*); the adults have as many as 15 pairs of gill slits, and they lack the paired fins of all more advanced fishes. Furthermore, these animals have persistent notochords that are never replaced by a backbone, so they resemble larger versions of amphioxus. The cyclostome larva (ammocoete) is even more similar to amphioxus than is the adult, strongly suggesting that amphioxus is actually similar to the ancestors of vertebrates.

To understand the place of various types of fish, especially in relation to the later evolution of amphibians, it is helpful to see a classification (Table 35.1). You can pick the various groups out of this table as we discuss them.

The first jawed fishes, the acanthodians and placoderms, appear in upper Silurian and lower Devonian sediments, around 400 million years ago. Their ostracoderm ancestors, lacking jaws, were quite restricted to a small range of ecological niches, so the evolution of fishes with jaws around the mouth was an important innovation, one of the most critical in all of

Table 35.1 Classification of Fishes

Superclass 1.	Agnatha: fishes without jaws			
	Class 1.	Pteraspidomorphi: ostracoderms; Ordovician–Devonian		
	Class 2.	Cephalaspidomorphi: "head-shield" forms		
		Order 1.	Osteostracida	
		Order 2.	Anaspida	
		Order 3.	Cyclostomata: lampreys and hagfish	
Superclass 2.	Gnathostomata: vertebrates with jaws			
	Class 1.	Placodermi: early jawed fishes, mostly heavily armored		
	Class 2.	Chondrichthyes: cartilaginous fishes (sharks and rays)		
	Class 3.	Osteichthyes: bony fishes		
		Subclass 1.	Acanthodii: spiny fishes; Ordovician–Permian	
		Subclass 2.	Actinopterygii: ray fins; most modern fishes	
		Subclass 3:	Sarcopterygii: flesh fins	
			Order 1.	Crossopterygii: lobe fins
				Suborder 1. Rhipidistia: Devonian-Carboniferous. Probably ancestral to tetrapods
				Suborder 2. Coelocanthini
			Order 2. Dipnoi: Devonian–Recent; lungfishes	

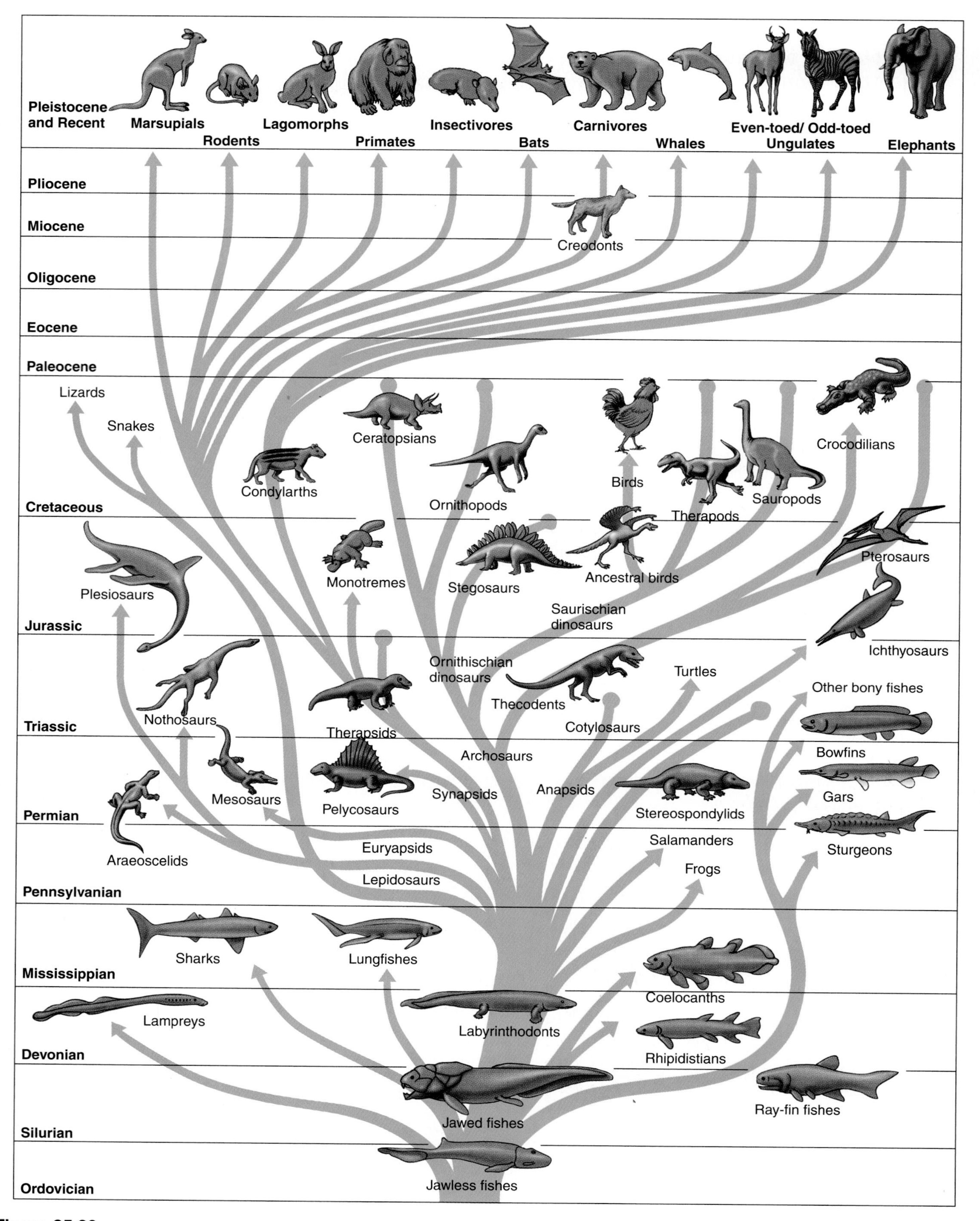

Figure 35.30

A phylogenetic tree shows the probable evolution of the major vertebrate groups.

vertebrate evolution. It opened the way to feeding on a greater variety of foods and occupying an enormous range of adaptive zones. In particular, jaws made possible predators that could bite into the flesh of other animals. This placed a premium on being able to swim quickly and maneuver well, and thus paired fins that accomplish this feat also took on high selective value. The story of jaw evolution from the cartilage of some gill arches is told in Sidebar 44.1.

The acanthodians retained the bony plate surfaces of ostracoderms and were probably ancestral to modern fishes, which are covered by small scales: the bony fishes (Osteichthyes) and the sharks and rays (Chondrichthyes) with cartilaginous skeletons. Because the embryos of other vertebrates first develop cartilaginous skeletons that are later replaced by bone, we are tempted to think that cartilaginous fishes are more primitive than bony fishes, but this isn't true: Sharks and rays apparently evolved from ancestors with bony skeletons.

Bony fishes have evolved an enormous variety of external forms, all based on the same fundamental anatomy—a highly metamerized series of vertebrae, ribs, and associated muscle segments (Figure 35.32). In later vertebrate evolution, it was primarily the skeleton and its musculature that changed; the circulation evolved as shown in Figure 45.7, while the rest of the internal organs changed very little.

The fins of most bony fishes are based on several nearly parallel rows of bony elements known as the ray-fin structure, but the lobe-fin fishes (Crossopterygii) have fins with a fleshy lobe containing a cluster of rather heavy basal bones. These are clearly homologous to the main bones of the tetrapod limb:

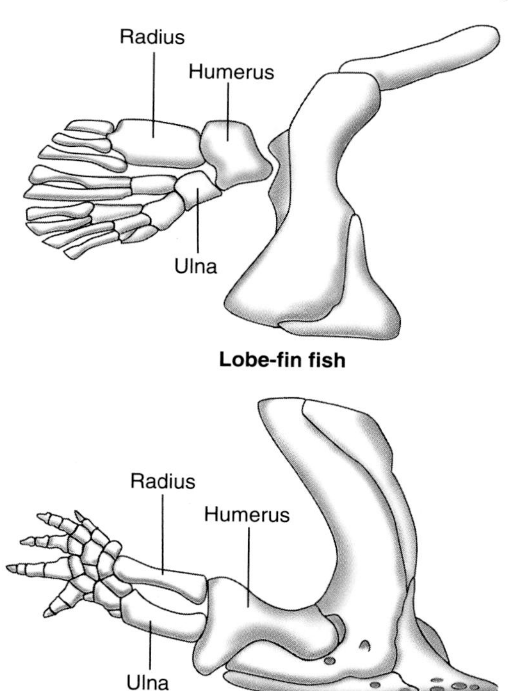

The lobe-fins emerged early from placoderm ancestors. One group of Devonian lobe-fins (rhipidistians) were the ancestors of amphibians, as shown by the details of their structure and fossil evidence that they had lungs and internal nostrils like amphibians. The lobe-fins are represented today only by the lungfishes and by a deep-sea fish, *Latimeria,* whose discovery off the east coast of Africa in 1938 provided a living member of the coelocanths (see′lo-kanths), a group of lobe-fin fishes thought to be extinct long ago. The ancestors of lungfishes (see Figure 46.2) were *not* on the main line leading to amphibians, but lungfish have evolved to depend on breathing air. In his fictionalized book *Kamongo,* the physiologist Homer W. Smith tells how his first specimens drowned in tubs that were too deep. With their much reduced gills, these fish can only live for short times below water; they are adapted to a habitat where they can lift their mouths out of the water periodically to fill their lungs with air.

Figure 35.31
The first vertebrates were jawless fish such as *(a)* the ostracoderm *Pterolepis,* which lived in streams during the Silurian period. *(b)* A lamprey, *Petromyzon,* is one of the few modern jawless fishes.

35.11 Amphibians have made a partial transition to terrestrial life.

The living amphibians include newts and salamanders, frogs and toads, and unusual legless creatures of the tropics called blindworms or caecilians. These are the remnants of a once much larger group that flourished from about the Mississippian to the Triassic periods, on the order of 350–200 million years ago. Although lungfish made a partial transition to living out of the water, amphibians were the real pioneers, the first to struggle onto land and become adapted to a life of breathing air while not constantly surrounded by water. The word "amphibian," meaning "both lives," refers to the animals' double life on land and in water. Their usual life cycle begins with eggs laid in water, which develop into aquatic larvae with external gills; in a development that recapitulates its evolution, the fishlike larva develops lungs and limbs, and becomes an adult. Many salamanders, however, grow directly from embryos into small versions of the adult and never pass through a larval stage. Adult salamanders are sometimes confined to an aquatic life, and even those that spend much of their time on land stay in moist places or live part-time in water.

Some of the totally aquatic amphibians, called axolotls, are salamanders that never achieve morphological adulthood and remain swimming larvae with gills all their lives (Figure 35.33). Since every species must reproduce, an axolotl becomes sexually

Figure 35.32

(a) A European perch, *Perca fluviatilis,* is a representative bony fish (class Osteichthyes). *(b)* A dissected view shows the segmented vertebral column (backbone) made of a series of vertebrae, each associated with one chevron-shaped block of trunk muscle. Notice that gas exchange occurs as water enters through the mouth and passes through the gills attached to the pharynx.

(a)

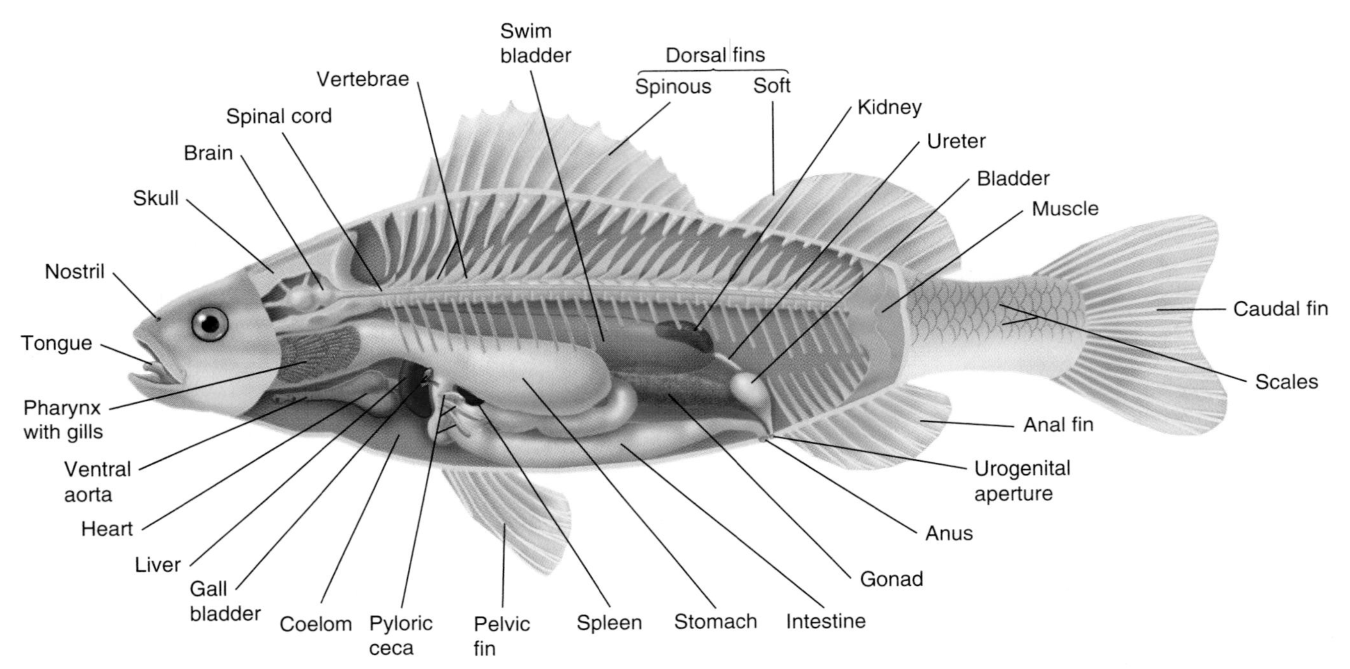

(b)

mature while still in a larval form, a phenomenon called *neoteny.*

The first known amphibians, from Devonian sediments, are known as labyrinthodonts because of a complex (labyrinthine) infolding of enamel in their teeth, a feature directly traceable to their crossopterygian ancestors. In adapting to terrestrial niches, labyrinthodonts evolved several new features. Fishes exchange gases through their gills. Although the moist skin of an amphibian is an effective organ of gaseous exchange, the amphibians were probably successful because they evolved efficient lungs. But they would have been confined to shallow-water habitats like lungfish unless they had also evolved limbs strong enough for effective movement on land. Fish obviously have no problems with water loss, but amphibians evolved a less permeable skin. The first amphibians (Embolomeri) occupied aquatic habitats where transpirational loss was

Figure 35.33
Axolotls are salamanders that never change morphologically into adults but still become sexually mature.

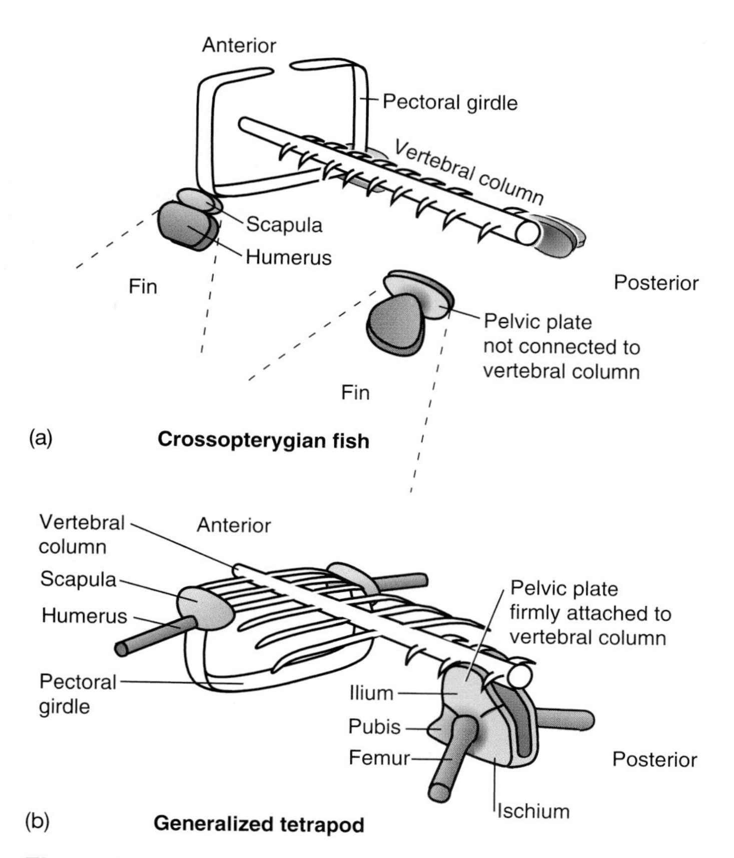

Figure 35.34
The pectoral *(a)* and pelvic girdles *(b)* have evolved to provide support for the limbs by anchoring them to the backbone.

not a great problem, but their successors in the Carboniferous period, the Rhachitomi, adapted to a terrestrial life with a tougher, less permeable skin. Perhaps the most formidable problem, however, was locomotion, since the fish body is adapted for movement by rhythmic undulations in a medium that supports it and tempers the pull of gravity. To support movement on land, the lobed fin changed from a flipper into an arm with its own powerful muscles. To overcome the pull of gravity, the weak girdles that anchor the fin bones of fishes had to be developed and strengthened. The resulting **pectoral** and **pelvic girdles,** which anchor and support the limbs, are major components of land vertebrates' skeletons (Figure 35.34). The Embolomeri had simple, fishlike vertebral columns that gave the body little support, and with their weak limbs supported by weak girdles, they were not very agile on land and probably spent most of their time in the water. But the Rhachitomi had more complex vertebral columns with interlocking vertebrae, thus providing more support. These amphibians, known broadly as temnospondyls, evolved into larger, stronger species, but they never became prominent terrestrial animals; all of the modern amphibians are quite small vertebrates. Primitive reptiles apparently evolved from amphibians soon after the latter had arisen, and the two types of animals were in competition from the beginning.

All these changes, of course, took time. Almost all the modern amphibians that have survived are quite small.

35.12 The evolution of the amniotic egg enabled reptiles to conquer the land.

Amphibians had terrestrial niches to themselves for about 150 million years, from the late Devonian to the Permian period, before the reptiles came to outstrip them and dominate the earth for about 200 million years. Indeed, the birds and mammals that evolved later are merely slightly altered reptiles, and technically, Reptilia is a paraphyletic taxon if it is not defined to include the birds and mammals. Modern reptiles include snakes, lizards, turtles, and a few large animals such as crocodilians and *Sphenodon,* an isolated, lizardlike New Zealander (Figure 35.35). Like the living amphibians, these reptiles are survivors, remnants of a group of often huge creatures that lasted until about 70 million years ago.

There are few fundamental differences in anatomy between the amphibians and the reptiles. The reptilian skull is deeper, but the most significant difference is in the structure of the reptilian girdles and limbs; reptiles have slimmer, more streamlined limbs, and the pectoral and pelvic girdles hold the limbs more directly downward, thus supporting the body firmly and allowing a more fluid, efficient motion:

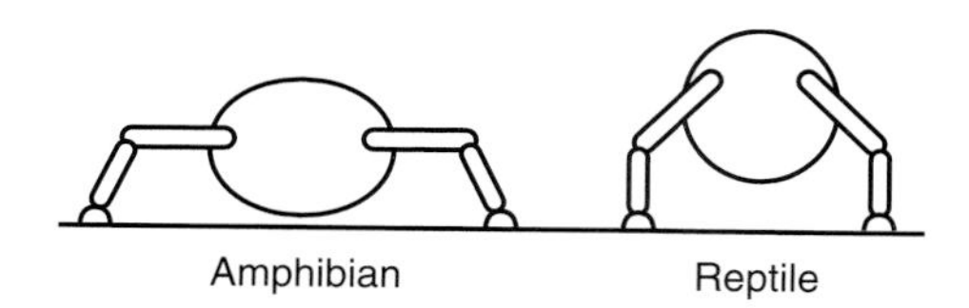

While amphibians are basically tied to the water, reptiles are not, freeing them to experience an adaptive radiation into many new niches. Reptiles became true land vertebrates in part by evolving more efficient lungs and a hard, scaly skin to withstand drying, but their most important adaptational development was the *amniotic egg* shown in Figure 20.9, in which an embryo can survive and develop on land. Within the protective shell of

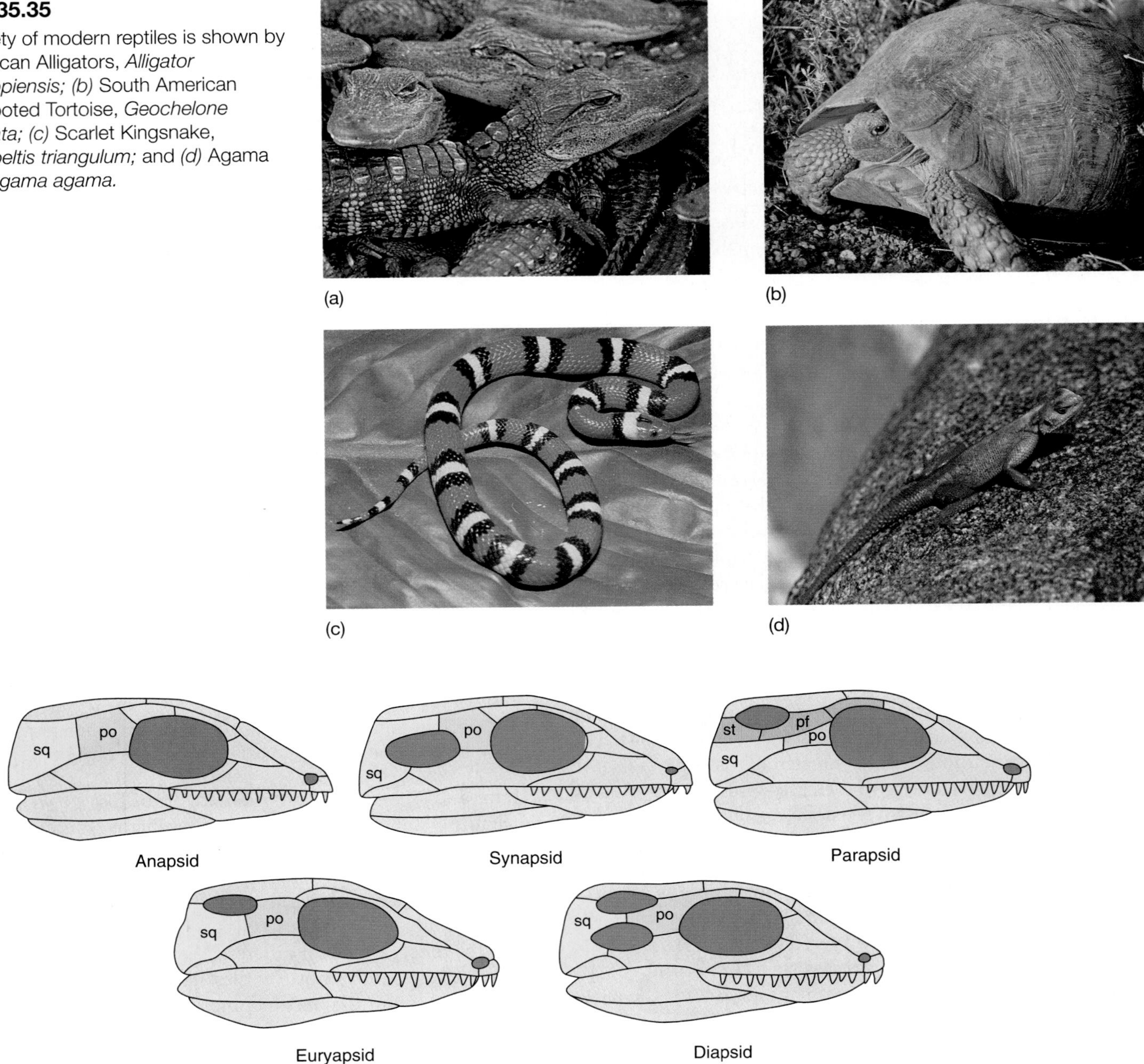

Figure 35.35
The variety of modern reptiles is shown by *(a)* American Alligators, *Alligator mississippiensis; (b)* South American Yellow-footed Tortoise, *Geochelone denticulata; (c)* Scarlet Kingsnake, *Lampropeltis triangulum;* and *(d)* Agama Lizard, *Agama agama.*

Figure 35.36
The subclasses of reptiles are defined by the anatomy of their skulls, which generally have openings behind the eye. Skull bones are abbreviated: *sq,* squamosal; *po,* postorbital; *st,* supratemporal; *pf,* postfrontal.

a reptile or bird egg, an embryo floats in the fluid of an amniotic sac, attached to a yolk sac that stores food, an allantois that collects wastes, and a chorion through which it respires.

The first animals with clear reptilian features appeared in the late Pennsylvanian and early Permian periods, about 300 million years ago, although the borderline between them and their amphibian ancestors is indistinct. Reptilian subclasses are defined by the presence or absence of extra openings in the side of the skull (Figure 35.36). The first reptiles had solid skulls (Anapsida). Turtles, with their specialized forms, are living anapsids. Over the next 150 million years, the other reptiles developed along several lines (Figure 35.37). Synapsids were the mammal-like reptiles, including therapsids that were actually ancestral to mammals. Parapsids included mesosaurs and ichthyosaurs, which occupied aquatic habitats. Diapsids became most prominent and include the Archosauria ("ruling reptiles"). Among them were pterosaurs, or flying reptiles, which became very successful and sometimes huge (*Pteranodon*

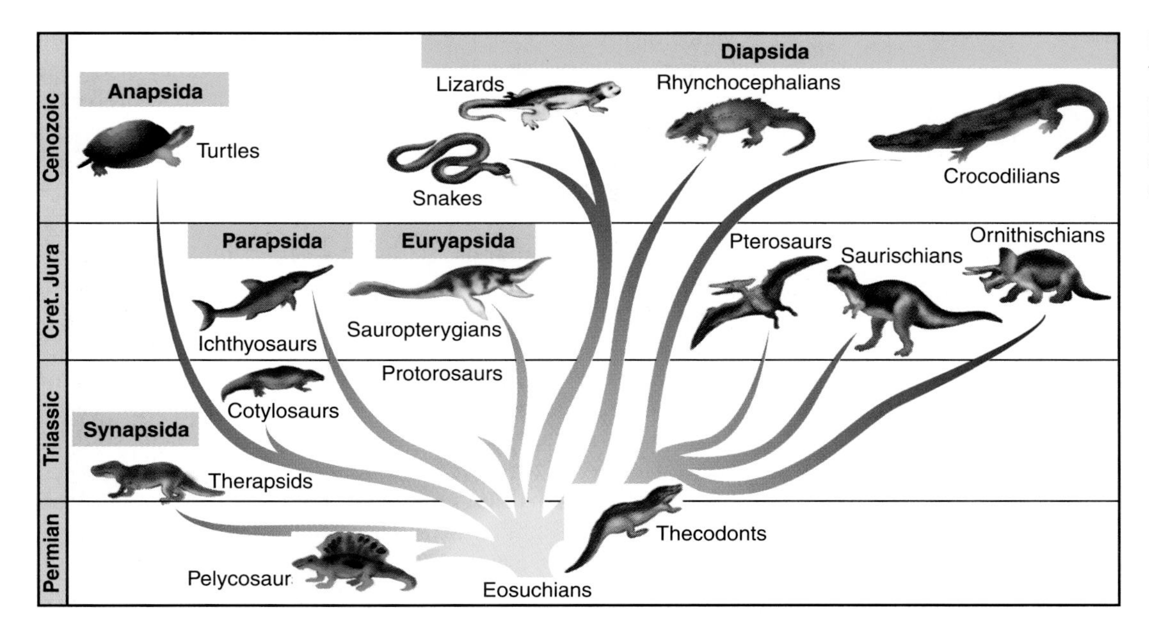

Figure 35.37
The major groups of reptiles probably evolved along these lines, starting with unspecialized diapsids of the Permian period called eosuchians.

had a wingspread of 8 meters) through the evolution of wings with a fundamentally different structure from those of birds, though based on the same vertebrate limb. The most prominent diapsids, however, were dinosaurs, which persisted all through the Mesozoic era and have thoroughly captured the human imagination with their often gigantic bones, eggs, and footprints. Their only apparent living relatives are the birds. Modern diapsids include crocodiles, alligators, lizards, and snakes.

The extinction of the pterosaurs and dinosaurs is still a mystery. Their demise coincided with a change in climate, perhaps one they could not adapt to or one that eliminated some of the vegetation on which they depended. It has also been suggested that they were wiped out by the spread of some new infectious disease. Another prominent theory postulates that a huge meteor known to have struck the earth around that time blackened the atmosphere with dust, killing off massive amounts of vegetation and, thus, the animals that depended on it. None of these ideas, however, accounts for the facts well enough to have become widely accepted, and the debate continues.

Figure 35.38
Archaeopteryx, the first known bird, had true feathers that were useful for insulation and perhaps also allowed the animal to form scoops with its wings for catching its prey. The colors used here are purely imaginary.

35.13 Birds are essentially modified dinosaurs with feathers.

The story of bird evolution from reptilian ancestors was given a new perspective in the mid-1970s through the work of Robert T. Bakker and John H. Ostrom. Bakker first became interested in the question of whether certain reptiles in the past were, in fact, ectothermic like modern reptiles or endothermic like modern birds and mammals. Since both birds and mammals evolved from reptiles, he wondered whether endothermy evolved in the birds and mammals themselves or in their reptilian ancestors. Bakker sought various kinds of evidence. For instance, the bones of ectothermic and endothermic animals are quite different in structure. Furthermore, an endothermic animal must eat a great deal more than an ectothermic one, since so much of its energy goes into keeping a constant temperature; for this reason, a given number of herbivores can support fewer endothermic carnivores than ectothermic carnivores. Bakker examined both the bone structures and the relative numbers of prey and predators in several fossil faunas, and his findings strongly suggest that dinosaurs were actually endothermic. At the same time, Ostrom reexamined some small dinosaurs and realized that they were very similar to *Archaeopteryx,* the first animal classified as a bird, except that they had no feathers (Figure 35.38).

These bits of evidence led Bakker and Ostrom to propose that birds and dinosaurs were not two separate branches from a common source, but rather, that birds evolved from dinosaurs and that dinosaurs themselves were already endothermic (Figure 35.39). Feathers, they contended, initially had adaptive

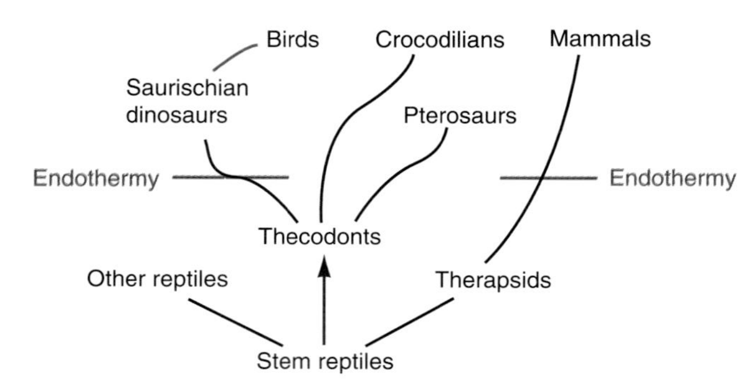

Figure 35.39

Following the work of Ostrom and Bakker, avian phylogeny has generally been reinterpreted to derive birds from one major group of dinosaurs. Fossil evidence strongly suggests that the dinosaurs were already endothermic (homeothermic). Endothermy also arose independently in the evolution of mammals.

value for insulation, not for flight, since insulation is important for a small endothermic animal. *Archaeopteryx* may also have used its feathered limbs as scoops for trapping small prey. Perhaps, having evolved feathers that served these functions, birds only later became gliding and then flying animals. (Some ornithologists disagree and derive birds from a different reptilian ancestor.)

Surely the demands of flight—being light and streamlined—selected the major anatomical characteristics of birds (Figure 35.40). Their bones are hollow and are supported by a triangulation of struts reminiscent of modern airplane wings. Their internal organs are intermeshed with air sacs connected to the lungs; heavy teeth are replaced by light bills; nitrogenous wastes are collected in a nearly solid state, as uric acid, so there is no need for a bladder; and the female has only one ovary, which remains small except during the breeding season. Birds have powerful flight muscles (the "white meat" of a chicken) solidly attached to a large keel on the sternum (breastbone).

To supply the oxygen needed for sustained flight, birds have an efficient cardiovascular system, with a four-chambered heart

Figure 35.40

A representative bird, a pigeon, shows general avian anatomy. The only part of the skeleton shown is the large sternum, or breastbone, to which the flight muscles are attached.

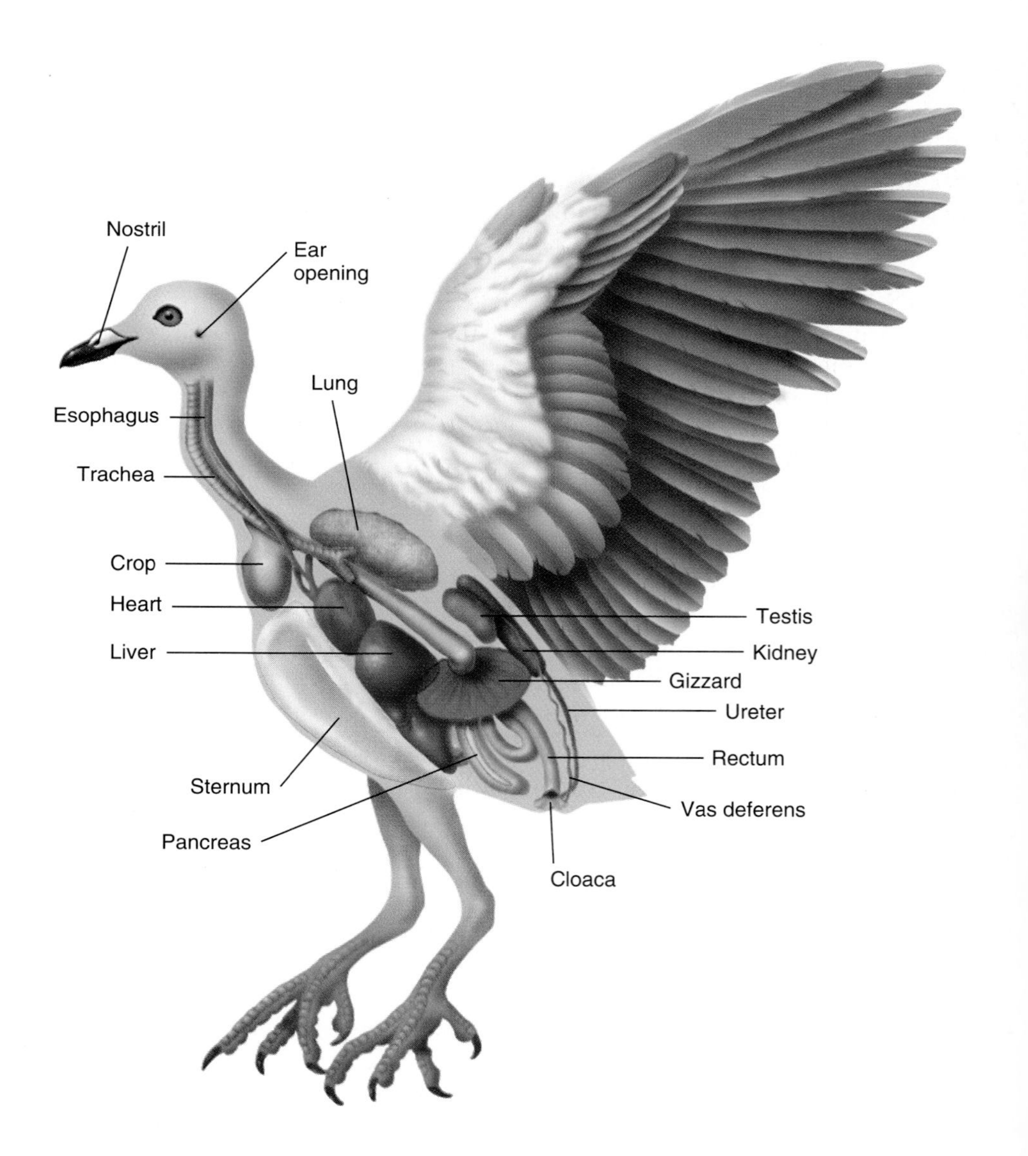

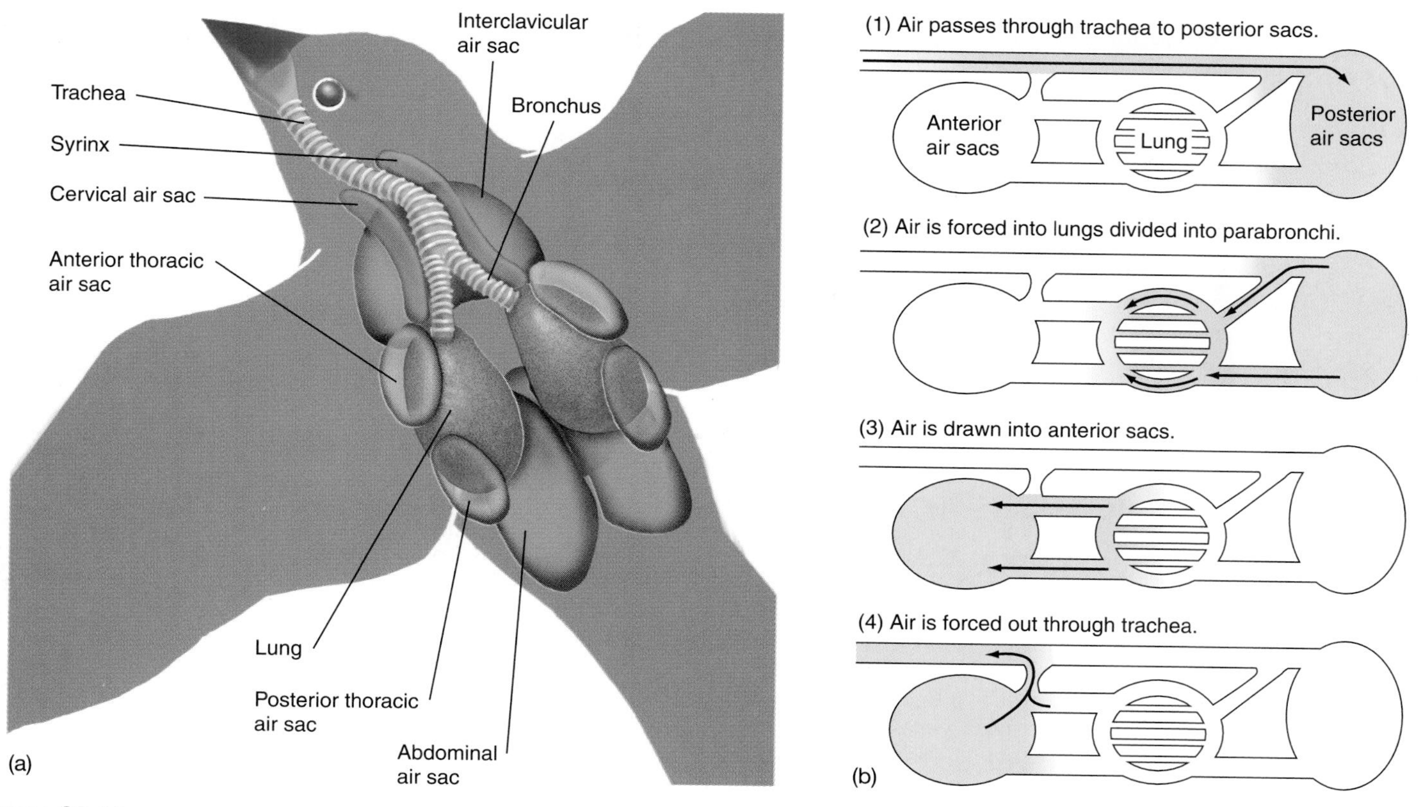

Figure 35.41

(a) Birds have a series of thin-walled air sacs connected to the lungs. *(b)* Gases do not exchange in the air sacs but rather in the lungs, which are divided into tubes called parabronchi. Inhaled air goes first to the posterior air sacs. They force the air through the lungs and on to anterior air sacs, from which it is expelled. Two breathing cycles are needed to move air completely through the system.

that can develop enough pressure to force blood quickly into all the tissues. The blood is oxygenated in lungs connected to a series of air sacs that function like bellows to blow air through the lungs. In mammalian lungs, air is drawn into a series of blind sacs and forced out again via the same route, making for some dead air space. In birds, the air passes in one direction, and almost continually, through the lungs (Figure 35.41). The insulation provided by their feathers conserves body heat and allows birds to live in cold climates that are inhospitable to reptiles.

About 9,000 species of birds are alive today. Perhaps because they are the most obvious vertebrates, perhaps because of their beauty and variety, they have attracted the attention of both amateur and professional ornithologists. The class Aves is divided into numerous orders and families, but most species are perching birds of the order Passeriformes, and most of these also are songbirds, which have membranes at the lower end of the trachea whose tension can be varied to make musical sounds of different pitches. Passeriforms include sparrows, larks, thrushes, warblers, and other small, familiar birds. The various orders are distinguished largely by specializations of their feet and bills (see Figure 47.4 for the latter). Thus the hawks, eagles, and falcons of the order Falconiformes and the owls of the order Strigiformes share adaptations for hunting and consuming small animals; the ducks, geese, and swans (Anatiformes) are excellent swimmers (and sometimes divers) that feed on water plants, fish, and molluscs; and the woodpeckers (Piciformes) have feet specialized for holding them in a vertical position on tree trunks while they use an apparatus of the bill and tongue for carving holes in trees and collecting insects.

35.14 Mammals have body hair and suckle their young.

Mammals are essentially a different kind of endothermic reptile, derived from therapsid reptiles of the Triassic period (see Figure 35.30). In mammals, the reptilian scales have been modified into hair, and females have **milk glands,** or **mammae,** with which they suckle their young. These characteristics, of course, are of no use in classifying fossil animals, so the strict definition of the class revolves around the maleus, incus, and stapes—the three tiny bones in the middle ear. All tetrapods have at least one of these bones, and an animal in which all three bones have shifted into their characteristic positions is classified as a mammal. In addition, mammals have specialized teeth—incisors for cutting and molars for grinding—instead of the rather undifferentiated teeth of reptiles. Mammals also have a distinctive ball joint at the base of the skull, so they can rotate their heads to survey the world in a way that reptiles cannot.

Table 35.2 Classification of Mammals

Subclass	Infraclass	Cohort	Order	Description
Subclass 1. Prototheria				
			Orders 1, 2, and 3.	All extinct
			Order 4. Monotremata:	egg-laying mammals
Subclass 2. Theria				
	Infraclass 1. Pantotheria:			All extinct
	Infraclass 2. Metatheria			
			Order 1. Marsupialia:	Marsupials
	Infraclass 3. Eutheria:			Placental mammals
		Cohort 1. Unguiculata		
			Order 1. Insectivora:	Insectivores
			Order 2. Chiroptera:	Bats
			Order 3. Edentata:	Armadillos, anteaters, sloths
			Order 4. Primates:	Monkeys, apes, humans
		Cohort 2. Glires		
			Order 1. Rodentia:	Rats and mice
			Order 2. Lagomorpha:	Rabbits
		Cohort 3. Mutica		
			Order 1. Cetacea:	Whales and dolphins
		Cohort 4. Ferungulata		
			Order 1. Carnivora:	Dogs, cats, bears, and allies
			Order 2. Tubulidentata:	Aardvarks
			Order 3. Proboscidea:	Elephants
			Order 4. Sirenia:	Sea-cows, manatees
			Order 5. Perissodactyla:	Horses, rhinoceros, tapirs
			Order 6. Artiodactyla:	Pigs, camels, deer, antelope, gazelles, goats, oxen

Figure 35.42

The Duck-billed Platypus, *Ornithorhynchus paradoxus,* is one of the most primitive living mammals. Although it lays eggs like a reptile, it is highly specialized for its way of life and not typical of the ancestral, reptilelike mammals.

The 4,500 species of modern mammals fall into a series of orders summarized in Table 35.2. They are grouped into two subclasses and three infraclasses, primarily on the basis of their modes of reproduction. In the **monotremes,** represented only by the duck-billed platypus (Figure 35.42) and the spiny anteater, or echidna, the females lay leathery-shelled eggs like those of turtles. In the **marsupials,** such as kangaroos, wombats, and opossums (Figure 35.43), the embryo develops in a uterus until it has enough musculature and coordination to move by itself; then, even though it is still very immature, it leaves the uterus, crawling up its mother's belly, with her guidance and the aid of special tracts of hair, into a pouch, or marsupium, where it completes its development attached to one of her mammary glands. In the **placental mammals** (Figure 35.44), the young develop inside the uterus attached to a *placenta,* an organ where nutrients and wastes are exchanged between the mother and the embryo's circulation. A young placental mammal develops to much greater maturity inside its mother's uterus than do the young of other mammals.

The mammal that holds the most interest for us is in the order Primates, family Hominidae, species *Homo sapiens.* We take up the story of this strange and fascinating creature in Chapter 36.

Coda The earth's fauna is dominated by arthropods and vertebrates. In searching for reasons for their success, it is instructive to consider their commonalities as well as their differences. Arthropods have open circulations, while vertebrates have closed circulations—perhaps proving only that both types of systems work. Arthropods exchange gases through gills, book lungs, or tracheae, systems that are often independent of the circulation, while vertebrates exchange gases via their circulation through gills, lungs, and skin.

Figure 35.43
Living marsupials are represented by *(a)* Red Kangaroos, *Macropus rufus; (b)* American Opossum, *Didelphis virginianus; (c)* Red-tailed Wambenger, *Phascogale calura.*

Figure 35.44
From over 4,000 species of placental mammals, four species will help to recall their variety: *(a)* Raccoon, *Procyon lotor; (b)* Atlantic Spotted Dolphin; *(c)* Polar Bear, *Thalarctos maritimus; (d)* Thomson's Gazelle, *Gazella thomsoni.*

Animals in both phyla are highly cephalized, with a strong concentration of nervous tissue and sense organs in the head; although they have no monopoly on this feature, they contrast strongly with other complex animals such as molluscs and echinoderms. Members of both phyla have strong musculature attached to solid skeletons; although arthropods have exoskeletons of chitinous plates, and vertebrates have endoskeletons of bone and cartilage, both structural systems support rapid movement, including flight (which evolved independently three times among vertebrates). Both types of animals are built on a metameric body plan, and we have seen that such a plan makes for independent movement of body parts and for appendages that may become highly specialized for different functions. Arthropods and virtually all vertebrates have evolved specialized mouth structures used primarily for feeding and defense. Many arthropods have chelicerae, which are used as fangs or pincers; others have strong mandibles for chewing, and many insects have sucking mouthparts. Except for a few species of primitive fishes, vertebrates have jaws equipped with teeth.

Thus success as an animal seems to lie in the combination of sensory acuteness and mobility, often with great muscular strength, as well as mechanisms for efficient feeding and protection. These features may evolve in animals with significantly different basic body plans, although we should not overlook the inherent value of metamerism. As we learn more about the developmental role of basic genes, especially *Hox* genes, it will be interesting to see just how such genes have been changed in evolution and what potentials for novelty they hold.

Summary

1. The protostome animals include the annelids and arthropods (segmented animals) and the molluscs.
2. Annelids have relatively simple, metamerized bodies with a ventral nerve cord and a closed circulatory system. Their external appendages remain relatively uniform and are used for locomotion or for moving a stream of water through the tubes in which some of them live.
3. Arthropods are characterized by jointed limbs and an exoskeleton made of chitin. They have an open circulatory system. While some retain the uniform metamerism of their ancestors, the appendages of most arthropods are modified for special functions, and their body segments are joined to make two or three large body regions.
4. The molluscs have an open circulatory system and soft bodies that are generally protected by calcareous shells. Their ancestral bilateral symmetry has been modified in some classes by twisting the body to fit inside this shell.
5. Three phyla of deuterostomes feed with the aid of a lophophore, a set of ciliated tentacles that directs a stream of water toward the mouth.
6. The major deuterostome animals are echinoderms and chordates. Echinoderms have radial symmetry associated with the sessile life led by their ancestors, and even those that have secondarily become motile retain this symmetry. They have a water vascular system made from a portion of the coelom, and their circulatory and excretory systems are highly reduced or absent.
7. Chordates are characterized by a dorsal nerve cord, a notochord generally overlaid with bone or cartilage, and a set of gill pouches. The ancestral chordate was probably metamerized, and this structure is still seen in lancelets and in fishes.
8. The primitive fishes were armored and jawless; jawed fishes were able to become active predators, and they evolved paired fins and muscles that allowed them to move rapidly.
9. The basic terrestrial vertebrates, the amphibia, evolved from lobe-finned fishes whose fins evolved into legs. These limbs had to be supported by the additional evolution of pectoral and pelvic girdles. Amphibians also evolved functioning lungs for breathing air, but they were still largely dependent on being in the water.
10. Reptiles became independent of an aquatic environment by evolving skin that withstands drying and amniotic eggs in which their embryos are enclosed in an aqueous environment. They radiated into a wide variety of niches and were the dominant vertebrates for many millions of years, until most of them mysteriously became extinct at the end of the Mesozoic era. Their major descendants include birds and mammals.
11. Birds apparently evolved from endothermic reptiles that evolved feathers for insulation; feathers later became devices for flight, and birds evolved other adaptations for light weight.
12. Mammals evolved internal fertilization and hair for insulation; all but two unusual species (the monotremes) bear live young, which are nourished for a time by milk produced by the females.

Key Terms

metamere 738
somite 738
annelid 739
parapodia 739
hemolymph 741
chitin 742
cuticle 742
compound eye 743
sinus 743
hemocoel 743
trachea 743
book lung 743
arachnid 743
cephalothorax 743
abdomen 744
crustacean 744
antenna 744
mandible 744
maxilla 744
head 744
thorax 744
abdomen 744
mollusc 746
lophophore 750
phoronid 750
brachiopod 750
ectoproct 750
echinoderm 750
water vascular system 751
chordate 752
notochord 752
tubular dorsal nerve cord 752
pharyngeal gill pouch 752
pharynx 752
tunicate 752
amphioxus 752
vertebrate 753
vertebral column 753
vertebra 753
cyclostome 753
pectoral girdle 757
pelvic girdle 757
milk glands/mammae 761
monotreme 762
marsupial 762
placental mammal 762

Multiple-Choice Questions

1. All of these animals are protostomes except
   a. clams.
   b. mosquitoes.
   c. earthworms.
   d. starfish.
   e. squid.
2. Which of these groups of coelomates is not segmented?
   a. annelids
   b. crustaceans
   c. chordates
   d. bivalve molluscs
   e. insects
3. In all segmented animals,
   a. all the segments are identical.
   b. each segment bears appendages that enable the animal to walk or swim.
   c. *Hox* gene activity during development produces segmental differences.
   d. all visceral organs are present in all the segments.
   e. each of the segments generates muscular movement.
4. Which of these structures is found in annelids?
   a. a dorsal nerve cord and a ventral aorta
   b. interior air-filled tubes that function for gas exchange
   c. a single kidney and a single heart
   d. a pseudocoelom
   e. none of the above
5. Leeches are members of the same phylum as
   a. earthworms.
   b. starfish.
   c. slugs.

d. squid.
e. shrimp.

6. Hemolymph differs from blood because only hemolymph
   a. is pigmented or contains pigmented cells.
   b. circulates within an open circulation.
   c. flows by diffusion and is not pumped.
   d. is made of chitin.
   e. is found in tracheae.
7. Which structures deliver oxygen to the cells of arthropods?
   a. external feathery gills
   b. book lungs
   c. tracheae
   d. hemolymph
   e. all of the above
8. Which is the most inclusive term?
   a. crustacean
   b. arachnid
   c. uniramian
   d. arthropod
   e. chilopod
9. Characteristics found in all chordates include all except
   a. bone.
   b. a notochord.
   c. a dorsal nerve cord.
   d. pharyngeal gill pouches.
   e. metamerism.
10. Both mammalian subclasses have all of the following structures except
   a. hair.
   b. three middle ear bones.
   c. milk glands.
   d. a placenta.
   e. teeth specialized for particular functions.

True-False Questions

Mark each statement true or false, and if false, restate it to make it true.

1. Of the major coelomate phyla, the only ones that are deuterostomes are echinoderms and vertebrates.
2. All segmented animals and even some that are not segmented share similar regulatory genes that control development of their body plans.
3. Locomotion by means of numerous parapodia is typical of all annelids.
4. Book lungs are found in crustaceans and molluscs, while external feathery gills are found in arachnids.
5. Although most modern fishes have movable jaws and paired fins, some primitive fishes lack both.
6. The first vertebrates to have sturdy paired appendages and lungs for breathing air were the amphibians.
7. All terrestrial vertebrates produce an amniotic egg.
8. Crossopterygians, the first known amphibians, are descended from labyrinthodont fishes.
9. While diapsid reptiles are the probable ancestors of birds, synapsid reptiles are ancestral to mammals.
10. The feathers of the ancestral bird Archaeopteryx were probably used for flight.

Concept Questions

1. Explain the relationship between *Hox* genes and metamerism.
2. What determines whether a circulatory system is "open" or "closed"?
3. Suppose a friend approaches you on the beach and tells you that he/she has a shelled invertebrate animal for you to identify. On your way to view the creature, you review in your mind all the aquatic animals having structures that might be called a shell. Which phyla are represented, and how do the shells differ from one phylum to another?
4. What is an *amniotic egg* and why is it an important evolutionary advance? What kinds of animals produce amniotic eggs?
5. What is homeothermy, and how many times has it probably evolved?

Additional Reading

Barrington, E. J. W. *Invertebrate Structure and Function,* 2d ed. Halsted-Wiley, New York, 1979.

Bond, C. E. *Biology of Fishes.* W. B. Saunders Co., Philadelphia, 1979.

Bourliere, Francois. *The Natural History of Mammals,* 3d. ed. Alfred A. Knopf, New York, 1964.

Colbert, Edwin H. *Evolution of the Vertebrates,* 3d ed. Wiley-Interscience, New York, 1980.

Diagram Group, The. *A Field Guide to Dinosaurs.* Avon Books, New York, 1983.

Dorst, Jean. *The Life of Birds.* Columbia University Press, New York, 1974.

Duellman, William E., and Linda Trueb. *Biology of Amphibians.* McGraw-Hill, New York, 1986.

Gill, Frank B. *Ornithology,* 2d ed. W. H. Freeman, San Francisco, 1990.

Gillott, Cedric. *Entomology.* Plenum Press, New York, 1980.

Linsenmaier, Walter. *Insects of the World.* McGraw-Hill, New York, 1972.

McLoughlin, John C. *Archosauria: A New Look at the Old Dinosaur.* Allen Lane, Penguin Books Ltd., London [The Viking Press, New York], 1979.

Purchon, R. D. *The Biology of the Mollusca.* Pergamon Press, New York, 1968.

Woota, Anthony. *Insects of the World.* Facts on File, New York, 1984.

Young, J. Z. *The Life of Vertebrates.* Oxford University Press, Oxford, England, 1950.

Internet Resource

To further explore the content of this chapter, log on to the web site at:

http://www.mhhe.com/biosci/genbio/guttman/

36 Human Origins and Evolution

Spencer Tracy and Fredric March played the roles of opposing lawyers in the film "Inherit the Wind," a dramatization of the Scopes trial over the teaching of evolution.

Key Concepts

36.1 Human evolution has been a troubling intellectual problem for over a century.

36.2 Primates are generalized mammals adapted to an arboreal life.

36.3 Who are the members of the order Primates?

36.4 Dryopithecine hominoids evolved in the Oligocene and Miocene epochs.

36.5 Australopithecines were small, erect hominids.

36.6 Human evolution is charted partially by cultural relics.

36.7 *Homo erectus* people created Lower-Paleolithic culture.

36.8 Neanderthal and later humans created Middle-Paleolithic culture.

36.9 Modern humans had finally displaced Neanderthals by 35,000 years ago.

36.10 Agriculture created a revolution in human life.

During the Miocene epoch, from about 25 million to 7 million years ago, the world's climate grew cooler and drier. Grasses, finding the climate hospitable, began to spread and create vast plains. This new food source invited the evolution of grazing mammals such as horses, antelope, and camel-like species. Mammals that had browsed leaves for much of their food now had the opportunity to tap new resources by evolving forms more dependent on ground plants and more suited to running than to climbing, and in the plains and savannas that stretched from East Africa through the Middle East to India, the first apes appeared. Among these apes were some that walked more erect and used their hands for manipulating objects, rather than just holding onto branches. They were the progenitors of a new way of life and of remarkable changes to occur on the earth's face, for we who assemble their bones and reckon their ages are their descendants. We tell the story of human evolution in this chapter, emphasizing earlier adaptations to an arboreal life and later adaptations to life on the ground. ■

36.1 Human evolution has been a troubling intellectual problem for over a century.

The importance of determining human origins did not escape Darwin and Wallace when they were preparing to publish their ideas about evolution. In December 1857, Darwin wrote Wallace, "I think I shall avoid the whole subject [of human origins] as so surrounded by prejudices, though I fully admit that it is the highest and most interesting problem for the naturalist." In fact, it was not until 1871, when he published *The Descent of Man,* that Darwin explicitly stated his ideas about human evolution. And immediately after the publication of *On the Origin of Species* in 1859, the intellectual public saw clearly that humankind fit integrally into the whole scheme of evolution that Darwin had laid before them. If natural selection was the force shaping plants and animals, humans could be no exception, and they must have evolved into their present form, from animal ancestors, over a long period of time.

Even though many people were immediately convinced that Darwin's thesis applied to humans, the conservative forces of society (known as "creationists") maintained just as strongly that the human species stood alone and had been created specially. The first, and most famous, confrontation between the evolutionists and creationists took place in Oxford, England, on June 30, 1860, at the annual meeting of the British Association for the Advancement of Science. Darwin wasn't present, but was represented by Thomas Henry Huxley. The creationists' spokesman was Bishop Samuel Wilberforce, who first delivered a brilliant oration attacking Darwin's thesis. Addressing his final remarks to Huxley, he begged to know, was it through his grandfather or his grandmother that Huxley claimed descent from a monkey?

As the Bishop sat down amid enthusiastic applause, Huxley took the platform. With great skill and logic he reviewed the scientific arguments in favor of evolutionary theory. At the conclusion, he turned to Wilberforce and replied to his gibe. Although his precise words were not recorded, he said in effect, "A man has no reason to be ashamed of having an ape for a grandfather or a grandmother. If I had the choice of an ancestor, whether it should be an ape or one who, having scholastic education, should use his great gifts to obscure the truth and mislead an untutored public, I would not hesitate for a moment to prefer the ape."

Wilberforce had led to his own downfall by violating the Victorian virtue of honesty. The subsequent roar of applause and laughter, largely from the several hundred students in the audience, gave the evolutionists a sense of victory. Yet the victory was momentary and hollow. More than 140 years later, the debate continues. Fundamentalist religious groups still petition school administrators and state legislatures to ban textbooks that advance the ideas of evolution, and they often demand equal time for creationist viewpoints.

Creationism, Sidebar 1.1.

We have no intention of continuing a fruitless debate here, since one basic premise of this book is that nothing in biology makes any sense except in the light of evolution. In this chapter, we will present the story of human evolution. Many details are lacking in the story, and it will probably always be an incomplete chapter in Earth's history. Nevertheless, the forces of evolution that operate on other organisms have shaped humanity just as inexorably, and they continue to do so today, however slowly.

36.2 Primates are generalized mammals adapted to an arboreal life.

The outline of human evolution is simple. Humans are members of the order Primates (Table 36.1), which includes the monkeys and the apes, such as gorillas, orangutans, and chimpanzees. We are most closely related to the apes of the family Pongidae and share with them ancestors that lived about 25–20 million years ago.

The primates evolved during the great adaptive radiation of mammals, around 70–65 million years ago, from Insectivora ("insect-eaters," though they eat more than just insects), an order that includes the moles and shrews. A modern group called tree shrews (order Scandentia) shares characteristics of both

Table 36.1 Classification of the Primates

Suborder Strepsirhini:	prosimians
Superfamily Lemuroidea:	lemurs
Superfamily Daubentonioidea:	aye-aye
Superfamily Lorisoidea:	lorises
Suborder Haplorhini:	simians
Superfamily Tarsioidea:	tarsiers
Superfamily Ceboidea:	New World monkeys
Family Callitrichidae:	marmosets and tamarins
Family Cebidae:	howlers, capuchins, other monkeys
Superfamily Cercopithecoidea:	Old World monkeys
Family Cercopithecidae:	langurs, guenons, baboons, macaques
Superfamily Hominoidea:	apes and humans
Family Hylobatidae:	gibbons
Family Pongidae:	orangutan, gorilla, chimpanzee
Family Hominidae:	humans and prehumans

Figure 36.1

Tree shrews, *Tupaia,* have some characteristics that show their affinities with primitive primates, including a somewhat opposable thumb that aids grasping, a large brain with good vision, and an arboreal lifestyle.

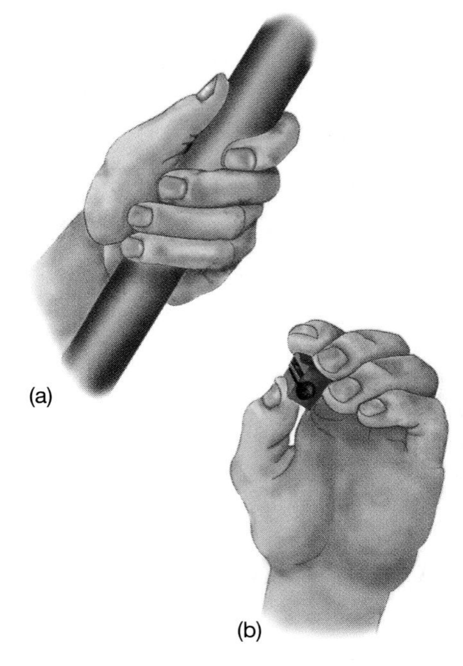

Figure 36.2

(a) The opposable thumb of primates gives them a power grip, using the flexor muscles of the arm. *(b)* Some primates, including humans, also have a delicate precision grip, using the small hand muscles.

insectivores and primates. The tree shrew *Tupaia* is a squirrel-sized Asiatic mammal with a long snout and tail (Figure 36.1). It is arboreal (tree-dwelling), while most of its insectivore relatives live on or under the ground. The first digit, or thumb, on each foot is set slightly apart from the other digits, which helps the animal grasp and climb. Its relatively large brain provides good vision but a poor sense of smell.

Modern tree shrews, which eat fruit as well as insects, illustrate how a primitive primate could have emerged as an insectivore that began to supplement its diet of insects, worms, and other small animals with fruit. Indeed, primates are generally frugivorous (fruit-eating) or omnivorous. Most primates are adapted to an arboreal life, specialized mostly for living on leaves and fruits among the small branches at the extremities of trees, where their major competitors are bats and frugivorous birds. A few species, such as humans, have switched back to life on the ground.

Many non-primate mammals have highly specialized extremities: Hoofed mammals are superbly adapted for running, but they can do little else with their hooves, while the marvelous flippers and fins of whales and seals make them excellent swimmers but restrict them to an aquatic life. Primates, however, are quite generalized mammals. Their five fingers retain primitive mammalian features but add an important new one, an **opposable thumb** that operates in a different plane from the other digits and can be pressed against the fingers to make a fine grasping tool. (Most apes also have an opposable big toe.) All primates can use their hands in a **power grip** for climbing and grasping tree branches (Figure 36.2*a*); more advanced primates, such as apes and humans, also have a more delicate **precision grip** for manipulating small objects (Figure 36.2*b*). Since chimpanzees and humans use their precision grip to make and use tools, it is a key factor in cultural evolution.

Primates typically have very long arms with which to reach to the extremities of trees and grasp fruit (another use for an opposable thumb). The primate arm also rotates very freely. Primates, especially humans, can rotate the two long bones of the forearm—the radius, on the thumb side, and the ulna—around each other to turn the forearm at least 180 degrees while the upper arm remains fixed (Figure 36.3). Additional flexibility comes from rotating the upper arm bone (humerus) in the shoulder socket. By watching a gibbon swinging from the bars in a zoo or a child on a jungle gym, you can see the value of these features for a tree-dweller.

The more advanced primates also have flat nails instead of claws. This leaves a bare, protected working surface at the ends of the digits, a surface rich in nerve endings that make it very sensitive and contribute to manual dexterity. Primates also lack the highly specialized teeth of other mammals, such as the prominent incisors of gnawing mammals and the prominent canines of carnivores. These features are presumably associated with primates' omnivorous eating habits.

The structure and placement of primates' eyes provide for excellent vision. In addition to rods, their retinas have cones for color vision and greater visual acuity, with a *fovea,* a central area about a millimeter square in humans, that is the point of clearest vision. The fovea contains only cones, not rods, and the lens of the eye focuses an image directly on it. Moreover, the other cellular layers that normally lie over the retina are displaced to the side in the fovea, so light that strikes the fovea doesn't have to pass through them. Locating the eyes at the front of the head, where they can focus on the same object from slightly different angles, affords stereoscopic vision and depth perception (Figure 36.4). Animals like horses and mice, whose eyes are directed to the side, have little or no depth perception; it is not essential for their ways of life. But excellent depth vision is obviously vital to an animal

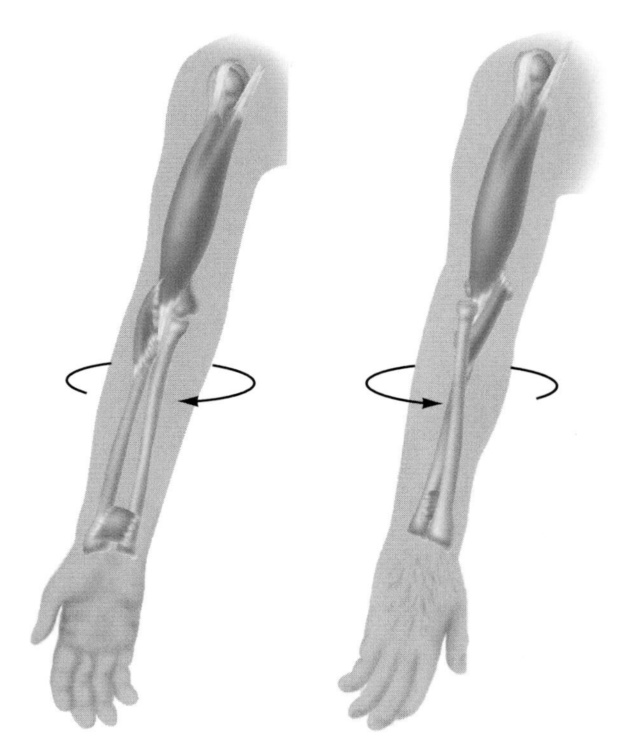

Figure 36.3

The primate forearm can be rotated at least half a turn at the elbow, and the upper arm also rotates quite freely within the shoulder joint, allowing the free movement needed for an arboreal life.

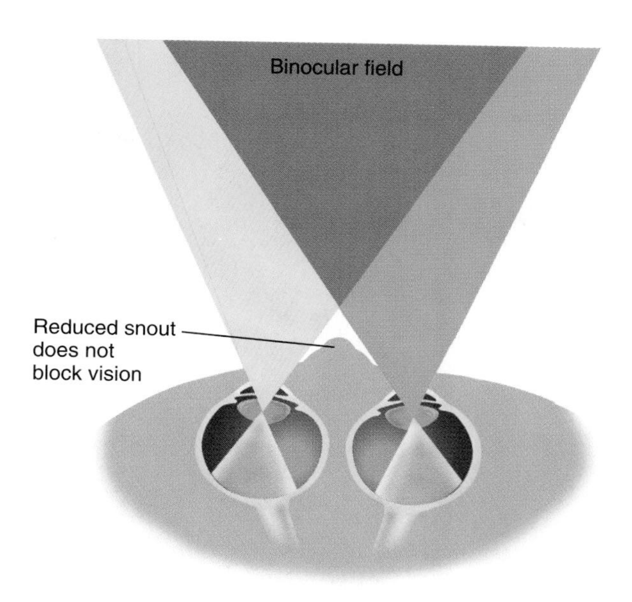

Figure 36.4

Primates' eyes are at the front of the head, with a greatly reduced snout, so the visual fields of the two eyes can overlap without interference. This provides depth perception (stereoscopic vision) by creating a binocular field where each object is seen from two slightly different angles.

moving quickly through a maze of branches. The portions of the primate brain that deal with the high-resolution information coming from the eyes are enlarged, as are other areas that control the sensitive hands and process the information coming from them. At the same time, the sense of smell (olfaction) is less useful to primates, and their olfactory brain centers are diminished.

A major trend in primate evolution has been toward more upright posture. Even monkeys sit in upright, human postures, and all higher primates hold their heads high, presumably a general adaptation for getting a better view of their surroundings. Upright posture not only affords a wider range of vision but also frees the sensitive and dexterous front limbs for carrying things, such as tools and babies.

Finally, primates have fewer offspring than do most other mammals, and their young are dependent on the parents longer. A relatively long, intense period of adult-child interaction is essential for transmitting cultural information and survival skills. This feature, of course, reaches its highest development in humans, where culture is transmitted via language throughout an extensive childhood.

36.3 Who are the members of the order Primates?

The order Primates is divided into two suborders (see Table 36.1). The Strepsirhini (literally, "twisted noses") includes the **prosimians** such as lorises and lemurs (Figure 36.5); these

Figure 36.5

Prosimians include *(a)* Slow Loris *(Nyctibus coucang)* from Africa, and *(b)* Ring-tailed Lemur *(Lemur catta)* from Madagascar.

Figure 36.6
The variety of simians is shown by *(a)* Philippine tarsier, *Tarsius syrichta; (b)* White-faced Gibbon, *Hylobates; (c)* Howler Monkey, *Alouatta,* a typical New World monkey; *(d)* Orangutan, *Pongo; (e)* Drill or Forest Baboon, *Mandrillus,* a representative Old World monkey; *(f)* Chimpanzee, *Pan troglodytes; (g)* Gorilla, *Gorilla gorilla.*

primitive primates have wet noses, which are associated with a sharp sense of smell, and their upper lips are immobile. Fossil evidence indicates that prosimians were quite abundant until about 40 million years ago, when they were probably displaced by the better-adapted monkeys that evolved at that time. Most prosimians now survive on the island of Madagascar, and the rest are reclusive, nocturnal denizens of the Asian tropics.

The **simians** (Figure 36.6) of the suborder Haplorhini ("simple noses") have dry noses, which are not so sensitive to smell, and mobile upper lips that permit a greater variety of facial expressions (see, for instance, Figure 50.7). Three families—gibbons, apes (chimpanzees, gorillas, and orangutans), and humans—make up the superfamily Hominoidea and are collectively called **hominoids.** Modern and fossil humans and prehumans of the family Hominidae are called **hominids.** Notice that the ending *-oid* identifies members of superfamilies (hominoid, tarsioid) and that *-id* identifies members of families (hominid, pongid).

Figure 36.7

(a) New World monkeys have relatively flat, platyrrhine faces with nostrils widely separated. *(b)* Old World monkeys have catarrhine faces with more prominent snouts and nostrils separated only by a thin septum.

Monkeys arose in the Oligocene epoch, around 40–30 million years ago, and diverged into New World and Old World clades. The modern New World monkeys of South and Central America (Ceboidea) include small marmosets and larger spider monkeys, howlers, and capuchins, all characterized by broad-nosed *(platyrrhine)* faces (Figure 36.7*a*). In association with their arboreal life, they have developed prehensile tails, giving them virtually a fifth hand to supplement their four excellent grasping limbs.

Information about plate tectonics and continental drift (see Section 25.11) has changed our understanding of primate evolution, although it has also created some mysteries. The great southern supercontinent, Gondwana, split about 110–100 million years ago into South America and Africa. Since the oldest known fossils of New World monkeys date from about 40 million years ago, the same time that all monkeys diverged from the ancestral primate line, they could not have occupied South America when the continental separation occurred. South America was relatively isolated for a long time. Its monkey fauna may have arisen independently from tarsioid primates, or it may have invaded the continent by island hopping, perhaps from North America by way of an archipelago.

The Old World monkeys include macaques, baboons, and mangabeys, which range throughout Africa and tropical Asia. They have *catarrhine* faces, with snouts and noses that point downward, and their tails are not prehensile. The Hominoidea arose from this branch of the primate tree, as shown by several features, including a common dental pattern. Anatomists describe the dental pattern of a mammal by listing the number of incisors, canines, premolars, and molars in that order. (The left and right halves of the jaw are mirror images, of course, and differences between the upper and lower jaws are described by one set of numbers over another.) New World monkeys have the pattern 2:1:3:3, as did the oldest primates; the formula of Old World monkeys, apes, and humans is 2:1:2:3. A paleontologist searching for evidence of primate evolution has the advantage that jawbones and teeth are among the most common finds in fossil beds, and they show diagnostic dental patterns and other features of hominoid evolution.

The Fayum beds of Egypt have yielded some remarkable simian fossils of Oligocene age, including the bones of *Parapithecus* and *Apidium,* rather small monkeys with the ancestral dental pattern. The late-Oligocene beds contain fossils of the first hominoids. *Aegyptopithecus zeuxis,* perhaps the oldest known hominoid, had teeth very much like those of a modern ape, with large molars increasing in size from front to back, elongated front premolars, and large canines. Fossils assigned to the genus *Propliopithecus* not only have the catarrhine dental pattern but also certain other dental features—such as incisors set vertically rather than jutting forward—that make them appear quite hominidlike. These Oligocene primates lived in tropical forests, were probably frugivorous or omnivorous, and had typical arboreal primate characteristics. The infamous "coming down out of the trees," which popular thought associates with human evolution, was several million years off. However, the animals that came down out of the trees were not humans but the ancestors of a large part of the hominoid branch.

36.4 Dryopithecine hominoids evolved in the Oligocene and Miocene epochs.

Apes are distinguished from monkeys by the lack of a tail, differences in tooth structure, and a more upright posture. The larger apes are too heavy to live primarily in trees as their ancestors did. (Modern gorillas are formidable in size. An adult male gorilla may reach 200 kg, a female 135 kg.) Even the smallest apes cannot walk on branches easily, so they move through the trees chiefly by **brachiation,** swinging from branch to branch with a movement made possible by their long arms and extremely mobile swivel joints (Figure 36.8). The largest apes, however, spend much of their time on the ground. They are clumsy walkers, since the form of the ape pelvis gives them a crouching, bent-over posture instead of the full upright posture of hominids (Figure 36.9). But they do walk on two legs. Chimpanzees and gorillas get around by knuckle-walking, putting a certain amount of weight on the knuckles of their hands from time to time, and this posture allows them to run while carrying things in their hands.

Figure 36.8

Gibbons are prime examples of primates that move by brachiation, swinging by the arms from one hold to another.

Sidebar 36.1 Evolving Populations and Species

Evolution rests on the variability in populations, and most individuals in a sexually reproducing population probably have unique genomes. This variation is obvious in humans, because we are sensitive to even little differences between individuals, but we tend not to see the variation in other creatures, in spite of warnings against typological thinking (see Section 2.3). The hominoid fossils we dig up are small samples from a large total, a few hundred parts of individuals out of many thousands that lived over millions of years. They come from variable, evolving populations. So we must not expect to find neat groups of fossils that can be labeled unequivocally as one species or another. Instead, we can expect these fossilized remnants of populations to show all their variations—one specimen with some features that were eliminated from later populations, another with features that became accentuated.

Since species living at the same time acquire isolating mechanisms and remain distinct, it is realistic to assign contemporary specimens to one species or another, using a biological or phylogenetic species concept. However, species that are portions of a lineage from different times must be defined morphologically and are basically conveniences of taxonomy; in Section 24.3 we used the term *paleospecies* for such groups. A complete series of fossils would show transition forms from one paleospecies into another, so we could never draw a line between them. We should see this more and more as specimens of human ancestors become more abundant.

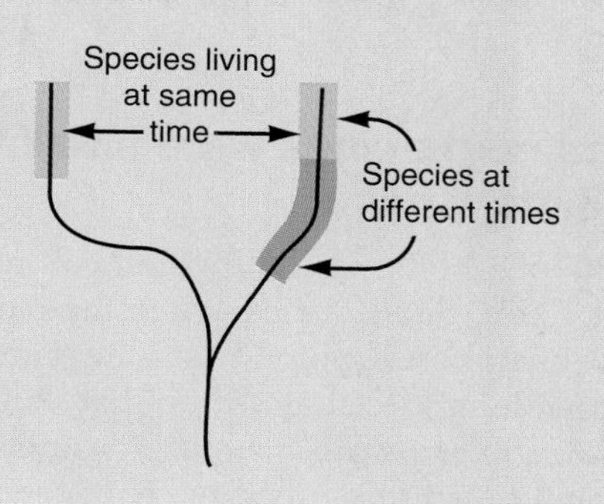

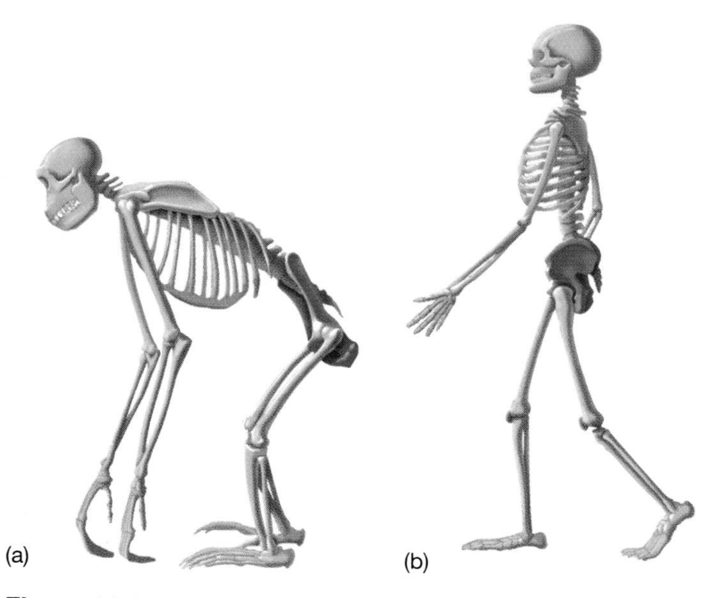

Figure 36.9
(a) The pelvis of an ape keeps the body bent forward, while *(b)* the human pelvis maintains an upright stance.

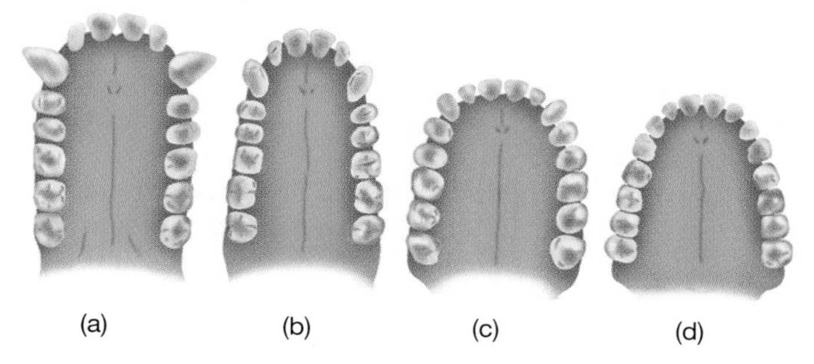

Figure 36.10
In contrast to a gorilla jaw *(a),* the jaw of dryopithecines *(b)* is rounded at the front, and the canines are much reduced. This trend in evolution continues with the australopithecines *(c)* and modern humans *(d).*

As the late-Oligocene climate grew cooler and drier, tropical forests in Africa and Asia gave way to subtropical forests and to extensive steppes (large grassy plains) and savannas (plains with scattered trees). During this period, horses, which are adapted for running in grasslands, radiated into several new species. In late-Oligocene and early-Miocene times, the first hominoids also moved into a new adaptive zone, living on the ground in the grasslands, out of competition with arboreal primates, replacing (or supplementing) their diet of fruit and leaves with tough foods such as nuts and roots. (Modern chimpanzees, while basically tropical forest animals, have a similarly varied diet, and some are almost as omnivorous as humans.) These were the **dryopithecine apes**—*Dryopithecus* and its relatives—whose fossils have been found over a vast area from southern Europe through East Africa, India, and China. Dryopithecines were rather generalized apes with monkeylike bodies. Some species have been tentatively identified as likely ancestors of various modern apes. (As we survey fossil primates, we should keep the ideas of Sidebar 36.1 in mind.)

Dryopithecines lost the huge, protruding canines of earlier simians, which were adapted to a chomping motion, and developed strong jaws with heavily enameled teeth, permitting a rotary grinding motion useful in chewing tough foods such as roots. In contrast to the rectangular jaw of apes, dryopithecines showed the beginning of a rounded hominid jaw, which became accentuated in the later australopithecines and in humans (Figure 36.10). In spite of their food habits, the fossilized teeth of later species such as *Ramapithecus* show less wear than do those of *Dryopithecus* and modern apes, suggesting that their permanent teeth emerged later, as in modern humans, and implying a longer period of development. The lack of wear may also mean that *Ramapithecus* used its hands a great deal, rather than tearing all its food with its teeth—one more indication of an upright posture.

The dryopithecines were gone by about 7 million years ago. The first definitive hominids, dating from just over 4 million years ago, walked fully upright. In the intervening period of about 3 million years are transitional species just now being discovered.

36.5 Australopithecines were small, erect hominids.

In 1924 the skull of a hominid child was discovered in a limestone quarry at Taung, South Africa. When the skull came into the hands of the anatomist Raymond Dart, he immediately recognized it as something special. Based on that skull, he defined a new species, *Australopithecus africanus* ("southern ape man from Africa"), and noted that it had many of the features of a bipedal hominid: the rounded shape, relatively large cranial capacity for its overall size, more rounded jaw, and attachment to the spine in a certain way. (A significant trend in hominid evolution is the movement of the opening at the base of the skull to a more central position, so the skull rests in a more balanced way on the vertebral column.)

Many other **australopithecine** fossils were discovered later, all clearly hominids but with the protruding jaws and heavy brow ridges that didn't disappear among hominids until much later. Those dated between about 3 and 1.5 million years ago represent two main groups—the *gracile* and the *robust* (Figure 36.11). The gracile forms, which retain the name *A. africanus,* were probably no taller than 1.2 meters and weighed 25–30 kg (Figure 36.12). Their jaws and teeth were similar to those of modern humans, except for their larger, sturdier molars, and the hinge between the jawbone and skull permitted a rotary chewing motion. Evidence from the pattern of dental wear and from animal bones associated with their bones indicates the gracile forms ate meat. Robert Broom and Louis and Mary Leakey later described robust forms, designated *A. robustus* and *A. boisei,* that were taller (up to 1.5 meters) and perhaps twice as heavy as the gracile form; they had more massive teeth and were probably vegetarians.

With the introduction of the australopithecines, we must pay attention to two hominid features. The first is brain size. Modern human brains range from about 1,200 to 1,800 cm^3 (cubic centimeters), with an average of about 1,400 cm^3. This large organ supports all the complex activities humans engage in, and its development from the australopithecine brain of 500–700 cm^3 was a major feature of human evolution. (The brains of australopithecines were about the same size as a modern gorilla's brain, but it is the ratio of brain size to body size that is correlated with intelligence, and the gorilla is a much larger animal.)

The second important adaptation in the australopithecine skeleton permitted **bipedalism,** the ability to walk erect on two legs. The arms of australopithecines may have been somewhat less mobile than ours, but their ankle bones are intermediate between human and typical ape forms, and seem suited for bearing weight in bipedal walking. The general structure of their thighs and hips seems well adapted for erect standing, walking, and running, but also for an arboreal life of climbing by grasping trunks and large branches (Figure 36.13). Their pelvic structure may have evolved initially as an adaptation for climbing trees but was just incidentally suited to bipedalism. Australopithecines probably divided their time between the ground and trees, perhaps sleeping in trees at night as modern chimpanzees do. Like chimps, they lived in small groups—extended families perhaps—of a few dozen individuals.

During the 1970s, Donald Johanson and his associates unearthed an extensive series of australopithecine fossils, 3.6–3.0 million years old, in the Afar Triangle (Figure 36.14), including the now-famous skeleton of a young female they named Lucy. Johanson and Timothy D. White established a new species, *Australopithecus afarensis,* to include these fossils and specimens that had been found to the south, at Laetoli, by Mary Leakey.

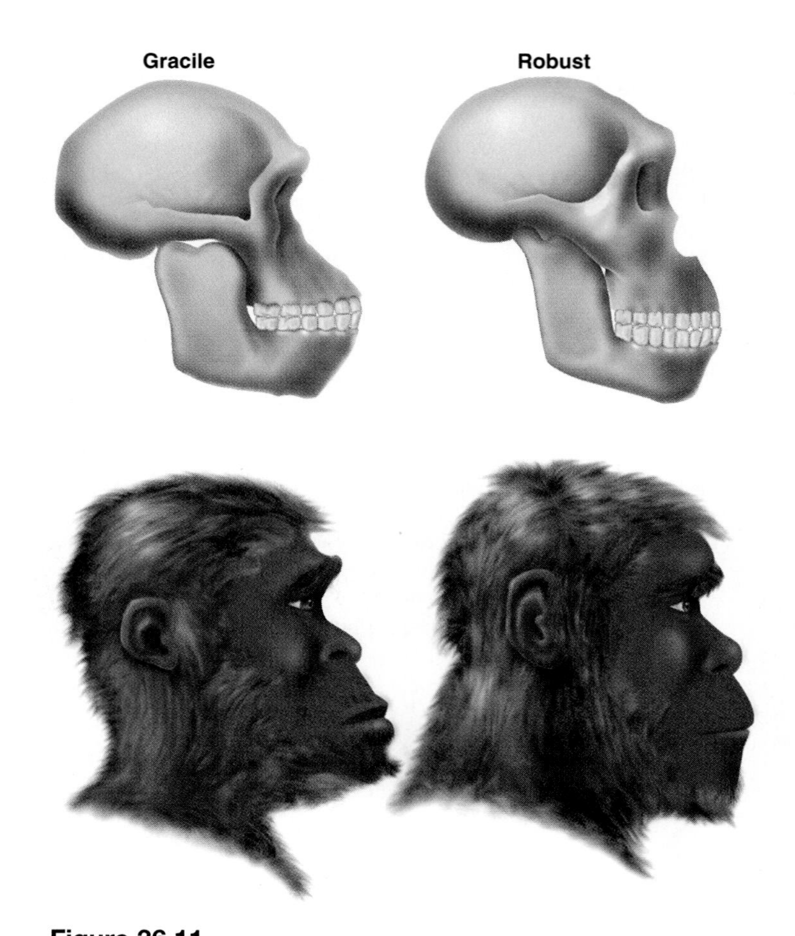

Figure 36.11
The fossilized skills of australopithecines have been used to reconstruct their probable appearance. The gracile types, designated *A. africanus,* were slim and stood about 1.2 meters tall. The robust types, *A. robustus,* were stocky and about 1.5 meters tall.

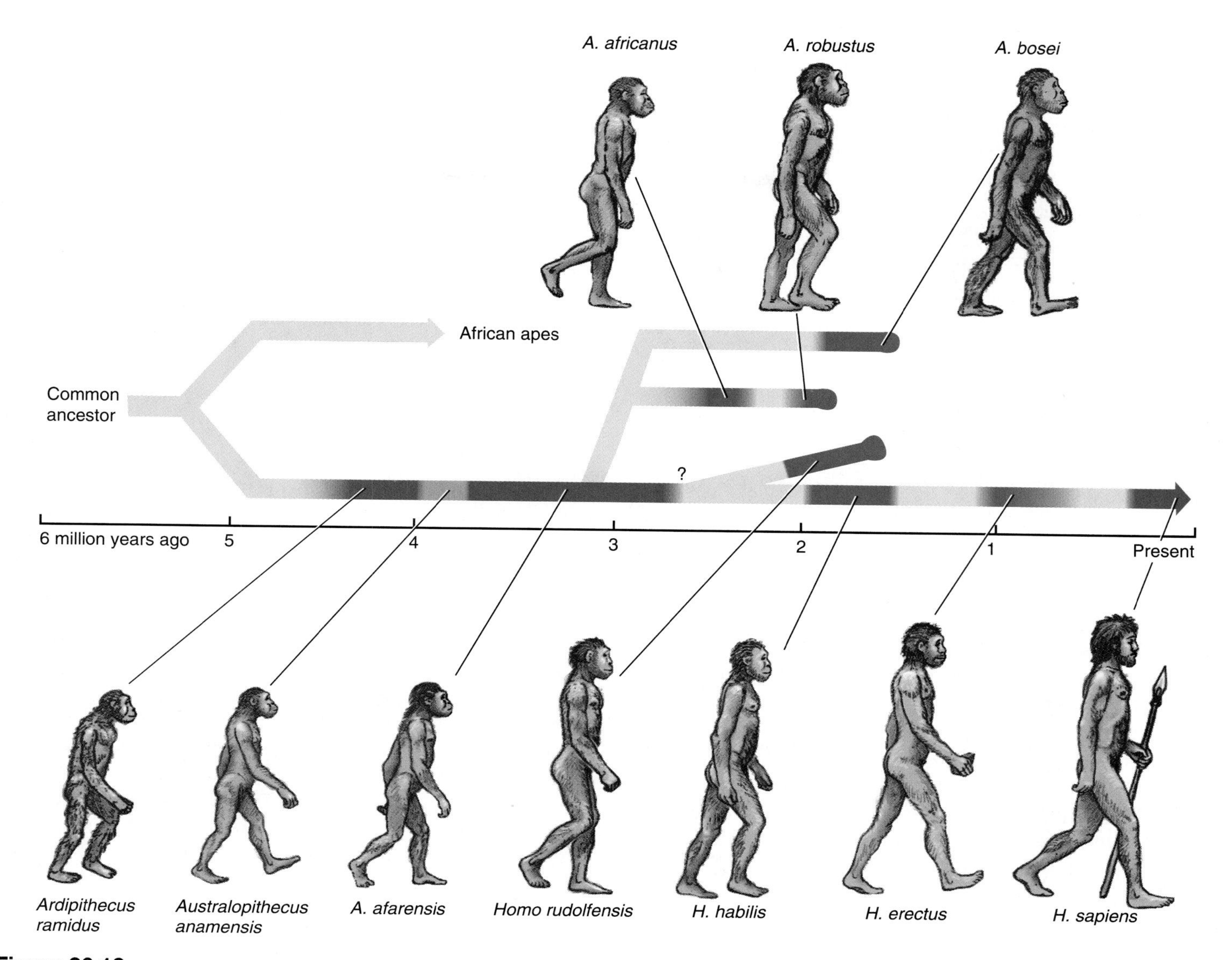

Figure 36.12

According to recent interpretations, hominids probably evolved along these lines.

A. afarensis is remarkable because it clearly shows that hominids were bipedal and stood fully upright almost four million years ago, even though their brains were still small (about 400–500 cm^3). As more evidence for bipedalism, Mary Leakey described a series of footprints in volcanic ash at Laetoli apparently made by two individuals of this species walking about 3.7 million years ago. The characteristics of *A. afarensis* undercut a formerly popular view: that large brains evolved along with erect posture, theoretically because they supported the development of tool use by hands that became free when they were no longer needed for walking. Bipedalism clearly evolved in hominids with rather small brains.

In 1992 White discovered fossils dated 4.4 million years ago that pushed the hominid line back still farther, virtually to the point where it branched off from the great apes. The species defined by these fossils, *Ardipithecus ramidus,* was described as "the most apelike hominid ancestor known." Although fossils

Figure 36.13

The erect pelvis of australopithecines allowed them to walk erect and also to easily grasp the trunks and large limbs of trees.

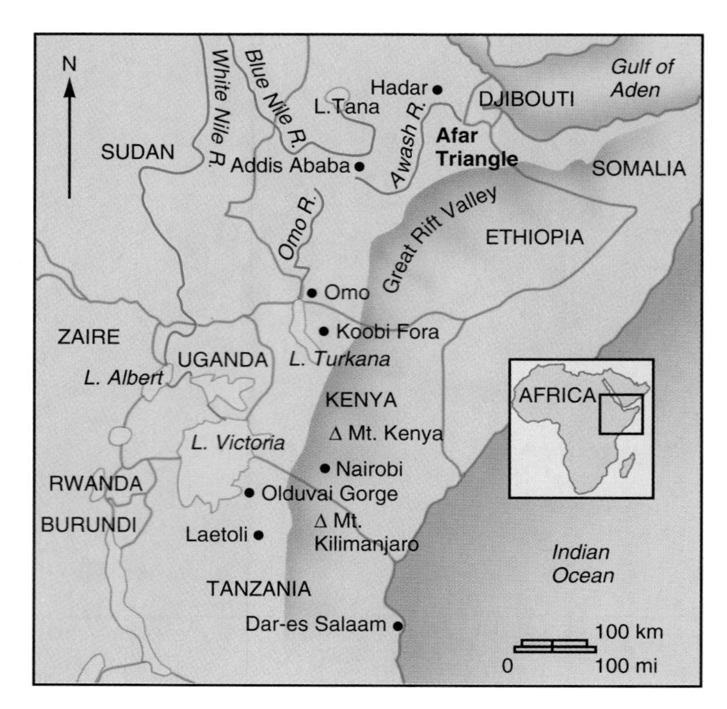

Figure 36.14

Early hominid fossils have come primarily from a few sites in the Great Rift Valley of East Africa, especially Olduvai Gorge, Laetoli, Hadar, and Koobi Fora.

are not yet available to show whether *ramidus* walked quite upright, the jaw and arm fragments are clearly those of a very early hominid that retains many features of a typical ape. The thin enamel covering on the teeth, for instance, suggests a diet like that of modern chimpanzees. In 1995 Alan Walker and Maeve Leakey found fossils dating from 4.2 million years ago that they described as a new species, *Australopithecus anamensis. Ardipithecus ramidus* and *A. anamensis* are more primitive and apelike than later hominids, but both were bipedal. These discoveries support the hominid phylogeny shown in Figure 36.12, with *A. afarensis* ancestral to both the later australopithecines and to the genus *Homo.* The hominid and ape clades diverged sometime between 7 and 5 million years ago. Although no anthropologist seriously talks about "the missing link" between apes and humans as many laymen do, some were willing to say, lightheartedly, that *A. ramidus* comes about as close as one could wish to being such a link.

36.6 Human evolution is charted partially by cultural relics.

Human evolution has been characterized by anatomical changes and by the development of an extensive culture, culminating in the use of language. But language—and most of culture in general—does not fossilize and cannot be dug out of the rocks. What we can dig up are tools. Until relatively recent times, tools were the sole remnants of hominid culture. Appearing first in strata laid down about 2 million years ago, they become more common, varied, and complex up to the present.

Figure 36.15

Simple pebble tools with minimal shaping are typical of the Olduvai culture.

Each collection of tools and other artifacts, along with the ways they are made, is called a *tradition* or an *industry,* and is named for a locality where typical remains are found. However, since we don't unearth hominid hands grasping those tools, we have to infer who the toolmakers were.

In Olduvai Gorge, south of Lake Victoria in Tanzania, Louis Leakey discovered primitive stone tools that represent the oldest known tradition, the Olduwan. As Figure 36.15 shows, they are little more than chipped pebbles, and the best evidence for their being tools is that they were found in a stratum where they would not have been deposited by purely geological forces, so they must have been worked as tools and then left to be uncovered millions of years later. At first it was assumed that these tools had been made by australopithecines, but this now seems unlikely. In 1961 Mary Leakey found bones of the species now called *Homo habilis* (literally, "handy man") at Olduvai. The Leakeys placed it in the genus *Homo,* for while its body appeared to be smaller than that of most australopithecines, it had astonishingly human teeth and a large brain capacity (650–700 cm^3). This find dates to about 1.75–2 million years ago, about the same age as the Olduwan tools, making *H. habilis* the most likely artificer of the Olduwan tradition. In 1972 Richard Leakey reported a new series of fossils from Koobi Fora, east of Lake Turkana (formerly Lake Rudolf) in Kenya, the prize fossil being a skull numbered KNM-ER (for Kenya National Museum-East Rudolf) 1470, which became something of a celebrity among human fossils. It was originally identified as a *H. habilis* skull, but its anatomy is very human, with a brain capacity of 800 cm^3. Furthermore, the KNM-ER 1470 bones were associated with artifacts that are definitely more advanced than those of the Olduwan tradition. Analysis of these and more recent finds in the Koobi Fora region now shows too much diversity among the fossils for a single species, and some fossils like KNM-ER 1470 have been assigned to the species *Homo rudolfensis. H. habilis* had a narrower face and a brain capacity of 500–700 cm^3, compared with 700–800 cm^3 for *H. rudolfensis.*

The current picture indicates two to four types of hominids coexisting in Africa for a period of about a million years—*Homo habilis* perhaps evolving into *H. rudolfensis,* and australopithecines evolving from the *africanus* to the *robustus* type. The two *Homo* species had larger brains than the australopithecines; the opening for the spinal cord is more centrally placed in their skulls; and their pelvises permitted an easier upright stride. With hands similar to those of modern humans, they were starting to develop manual dexterity. *H. habilis* and *H. rudolfensis* made primitive tools, but there is no evidence that any australopithecines made tools, perhaps indicating a difference in intelligence that explains why *Homo* eventually won out. The stage was set for the development of the next human species.

36.7 *Homo erectus* people created Lower-Paleolithic culture.

The robust australopithecines persisted until about 1.5–1 million years ago, about the time *H. habilis* was replaced by another species, *Homo erectus* (Figure 36.16). Eugene Dubois described the species in 1891 from specimens he had discovered in Java, and named it Java man or *Pithecanthropus erectus,* meaning "ape man that walks erect," describing features that were considered remarkable at the time. *H. erectus* people stood about 1.5 meters tall, with a skeleton much like ours, but a considerably more primitive skull. Their cranial capacity of 800–1,100 cm^3 approaches the modern human range, but they retained massive, primitive teeth and heavy eyebrow ridges. *H. erectus* was widely distributed throughout Europe, Africa, and the Near East. In 1985 Richard Leakey and his colleagues found a *H. erectus* skeleton in West Lake Turkana that they dated to 1.5 million years ago, and other skeletons assigned to this species have been dated back to 1.9 million years ago.

The *erectus* people left clear records of a developing culture. They were omnivorous hunters and gatherers. Although more primitive hominids undoubtedly hunted and ate meat, hunting apparently became a much more important activity for *H. erectus.* Large herds of animals, which must have provided plentiful food, roamed the extensive steppes and savannas of their Middle-Pleistocene world, about one million years ago. The campsites of these people contain vast bone remnants, not only from deer and antelope but from large, ferocious animals such as bears and elephants. They may have killed these animals with pointed wooden spears, which aren't preserved as well as stone tools, and they probably depended upon hunting techniques used by contemporary aborigines. A successful hunter can stealthily creep up on his prey, and bipedalism allows him to dog his prey at a steady trot for hours or even days, until it is too exhausted to resist. Primitive hunters may also have used the aboriginal American technique of stampeding herds of animals over a cliff to their death.

Although we have no indications that *erectus* people used stone tools for hunting, they left extensive collections of hand-axes that must have been used for more domestic purposes, such as fashioning wooden tools or cleaning and cutting the carcasses of animals. Their stone traditions define a long period, the Lower Paleolithic (Old Stone Age), that lasted until about 200,000 years ago. *H. erectus* is associated only with the oldest (Abbevillian) culture, characterized by the production of both core and flake tools (Figure 36.17). A **core tool** is made by chipping pieces off a large flint rock, at first just by hitting it with another rock. The large flakes that fall off the core can also be refined into scrapers or knives, known as **flake tools.** Crude as they are, these tools are clearly an improvement over those made by previous hominids because they are shaped on both sides, rather than just being chipped along one edge.

Like contemporary tribes of hunter-gatherers, the *erectus* people were probably nomads who roamed widely in small

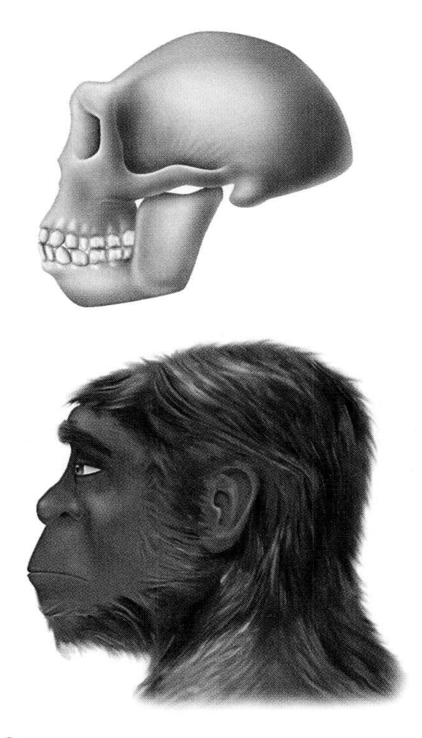

Figure 36.16

A *Homo erectus* skull has a larger cranial capacity than that of earlier hominids, but retains the receding chin line and has heavy brows. A reconstruction of the head based on such skulls shows the probable appearance of the living man.

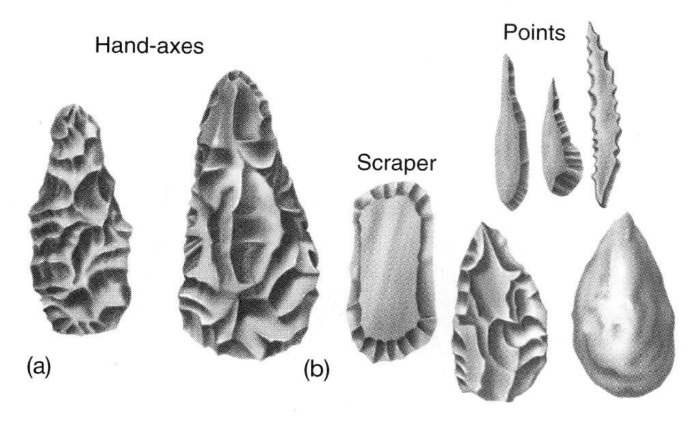

Figure 36.17

Some typical paleolithic stone tools include *(a)* hand-axes of Chellean age, which are core tools, and *(b)* scrapers and points of Aurignacian age, made from flakes split off a core.

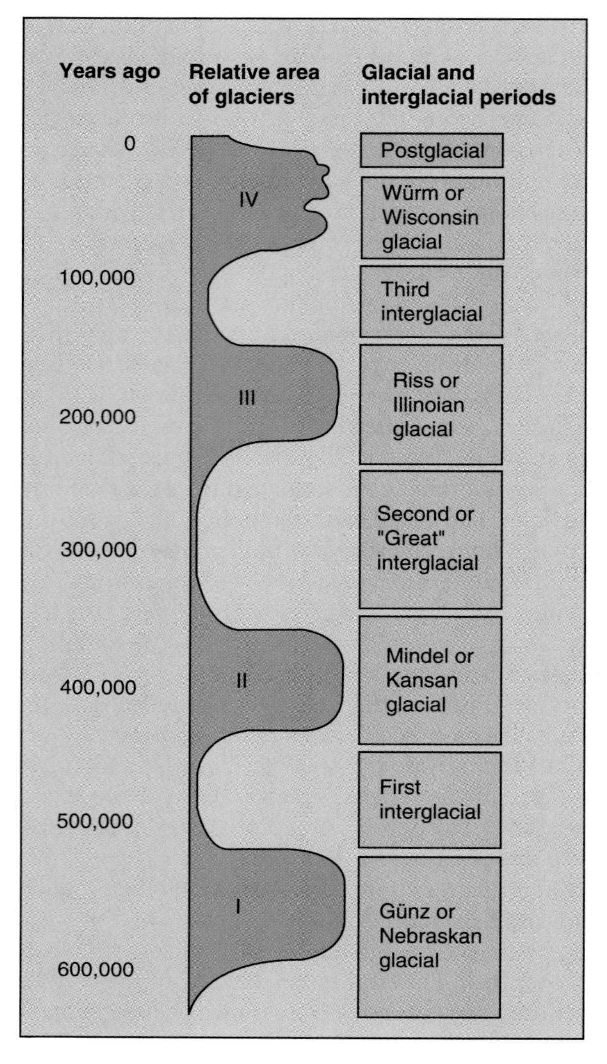

Figure 36.18

During the past 600,000 years, as humans have been evolving from *Homo erectus* into modern forms, the earth has experienced a series of glaciations (ice ages) separated by warmer interglacial periods.

bands and returned to the same base camps periodically. At some point, they also learned to use fire, for evidence from both Hungary and China clearly shows that they were cooking with fire a half million years ago.

The *erectus* humans lasted for nearly a million years, at least 40,000 generations, and during that time both they and their environment changed. At first, while the climate was still relatively warm and the people were quite unskilled, they probably went naked and lived in the open. But the Pleistocene climate was growing colder, and by 600,000 years ago a series of ice ages began, during which extensive areas of Eurasia and North America were covered by glaciers (Figure 36.18). The people of these times began to clothe themselves with animal skins, and some moved into caves. Their domestication of fire allowed them to heat those cold, damp spaces and to keep wild animals out.

The development of language is a key aspect of human evolution, since human culture could not have developed far without speech. The primate brain had been evolving linguistic capacities for a long time. Contemporary apes have considerably more linguistic ability than was once believed; gorillas are reported to communicate with 22 different sounds, and chimpanzees have an even larger natural vocabulary and are highly communicative. There is some dispute over experiments purporting to show that chimps can learn to use an artificial language as humans do, but at least they can communicate well, if not creatively.

Many factors in the lives of our ancestors put a premium on increased intelligence and on linguistic ability. Language offers a great selective advantage for a primitive people who are developing a social structure and hunting cooperatively, since social solidarity depends upon named classes of relations that cannot exist without language. The hierarchical relationships in families, for instance, can be passed from one generation to the next by language, whereas the dominance relationships of nonlinguistic primates must be reestablished every generation. Clues from artifacts and from skull morphology, which shows the size and shape of various brain areas, indicate that *H. erectus* probably used a kind of rudimentary language.

Culture became a new factor in natural selection that could shape behavior much faster and more effectively than purely genetic selection. At the same time, cultural requirements made it advantageous to have a larger brain for storing and processing information, so it is no wonder that in less than a million years of evolution the average human cranial capacity increased by at least 150 percent.

36.8 Neanderthal and later humans created Middle-Paleolithic culture.

By 250,000 years ago, the human skeleton as we now know it had been well established, and it was modified very little in further evolution. Between 250,000 and 150,000 years ago, *H. erectus* was replaced as the dominant hominid by people of our own species, *H. sapiens.* Various *sapiens* fossils are designated different subspecies. The earliest fossils define *H. sapiens neanderthalensis,* or Neanderthal humans, named after the Neander Valley in Germany where the first fossils were found. Averaging about 1.7 meters in height, the Neanderthals were similar to modern-day humans below the neck, though considerably more robust (heavier, sturdier, and stronger-boned). Like modern people, they had large cranial volumes, averaging about 1,400 cm^3. The major anatomical differences between Neanderthals and modern humans lie in facial features, since a classic Neanderthal skull has a markedly receding chin, heavy brow ridges, and a sloping forehead (Figure 36.19*a*). Modern humans, *H. sapiens sapiens,* have slight brow ridges, higher foreheads, and a more angular chin. Some anthropologists, however, regard the classical, heavy-browed western Neanderthals as a subgroup whose facial features may have been an adaptation to the cold, like the heavy brows and flat noses of contemporary Arctic peoples such as Eskimos. Over most of their range, from

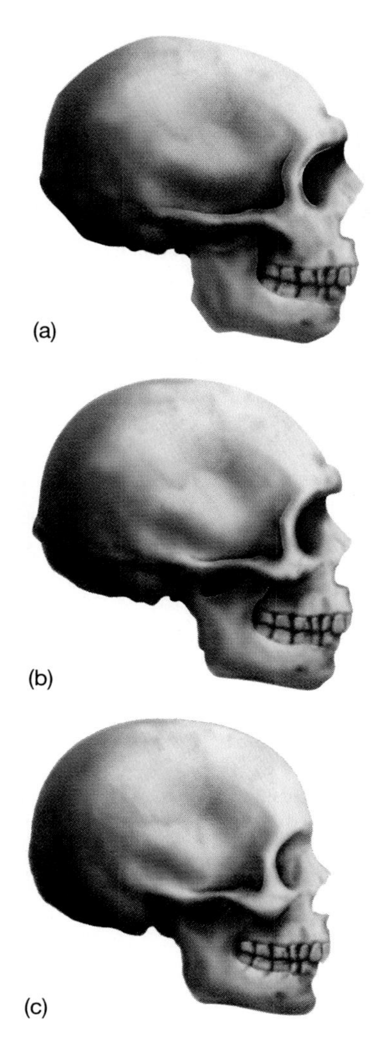

Figure 36.19

Neanderthal humans had receding chins, prominent cheeks, and fairly heavy brow ridges, but they were much more similar to modern humans than to *Homo erectus*. *(a)* A classic Neanderthal skull can be compared with *(b)* a progressive Neanderthal skull, which has more modern features, and with *(c)* a modern human skull.

western Europe through the Mediterranean, the Middle East, and East Africa, most Neanderthals were of the "progressive" type (Figure 36.19*b*), with skulls much more like ours.

The popular image of a Neanderthal as a near beast—stooped, shuffling, apelike, and stupid—was created by a few influential physical anthropologists, based largely on skeletons from western Europe. For some time, the Neanderthals were thought to have walked stooped over, rather than fully erect, due to misinterpretation of the type specimen from La Chapelle-au-Saints. When reexamined, it proved to be the skeleton of an old, arthritic man who clearly was not typical of his contemporaries.

During the period of Neanderthal dominance, from the third interglacial period until late in the fourth glacial period (roughly from 150,000 to 35,000 years ago), Eurasia was quite different from today. The present Mediterranean was two inland seas, completely isolated from the oceans, and a larger sea to the east encompassed what are now the Black and Caspian seas. As the glaciers advanced, burying northern Europe in a sheet of ice, humans shared the cool, temperate Mediterranean region with mammals that we now think of as typically African, such as elephants, hippopotamuses, and rhinoceroses, as well as extinct species like saber-toothed tigers, woolly mammoths, and woolly rhinoceroses.

Some time before 250,000 years ago, just as *erectus* humans were being replaced by Neanderthals, the Abbevillian tradition in toolmaking was being replaced by the Acheulian. Hand-axes that had been made by striking chips off a core were now made more carefully by striking a wooden tool held against the core to remove flakes. This method produced sharper, more symmetrical tools. The Acheulian industry was then replaced by the Mousterian, in which only the flakes removed from the core were delicately shaped into tools such as scrapers and spear points. The invention of a long wooden spear with a sharp stone tip must have been a major advance in hunting. Among Neanderthal artifacts are the bones of many animals. Neanderthal people possessed the knives to butcher animal carcasses and the fire to cook the meat. Due to the cold climate, fire was also needed for warmth, and the animal hides could be used for crude clothing.

The Neanderthals even left clear evidence that their thoughts went beyond mere survival. We have found sites where they carefully buried their dead; one of the bodies in Shanidar Cave in Iraq was buried with flowers, indicating some sort of thought or religious rite concerning life and death. Several caves contain the bones of cave bears, carefully arranged in a way that suggests their use in rituals, and some anthropologists believe a cult may have revolved around these animals.

36.9 Modern humans had finally displaced Neanderthals by 35,000 years ago.

Virtually all the skeletons dated later than about 35,000 years ago are those of *sapiens* people. The transition between Neanderthals and modern humans is now a subject of great controversy, and the evidence isn't complete enough to distinguish among the three models now defended by different students of human evolution. A *multiregional model* (Figure 36.20*a*) holds that the human races arose independently, in widely separated places, from the Far East through Europe and Africa, though some migration and intermixing probably occurred. On this model, the differences between human races are ancient and indicate that the people of each region underwent parallel evolution, gradually changing from *erectus* to Neanderthal types and then to modern *sapiens* types.

A second view, the *"out-of-Africa" model* (Figure 36.20*b*), postulates that modern *sapiens* people emerged from a single African source sometime around 100,000 years ago and spread through Europe and Asia, eventually replacing indigenous humans in each area. On this model, *H. erectus* was a distinct early species, and the Neanderthals might also have been a distinct species. A third view—the *hybridization model* shown in Figure 36.20*c*—suggests that different populations encountered one another and interbred to varying degrees in various localities, so modern humans carry complex intermixtures of ancient and modern features.

There is no resolution of these conflicting ideas yet. The out-of-Africa model was in vogue for a time and caught the popular imagination, especially when it was joined to the hypothesis that a single woman, a sort of "Mother Eve," might have been the ancestor of us all. This hypothesis was based on evidence from human mitochondrial DNA, which is very homogeneous; since mitochondria are inherited only from the mother, this evidence argued that we are all descended from a single woman or perhaps from a very small, homogeneous group. But other molecular evidence now makes this scenario unlikely; for instance, the great diversity of major histocompatibility genes is inconsistent with evolution from a very small, relatively recent ancestral group. The situation is compounded by better dating techniques, which have established that modern humans were living much earlier than 35,000 years ago at sites in western Asia, at the same time as Neanderthals. The two groups may have coexisted for tens of thousands of years, with no evidence of hybridization.

Regardless of the origins of modern humans, in the intervening 35,000 years, humans have become a truly cosmopolitan species, spread across the face of the earth, with only small genetic differences among contemporary races. Physical evolution in humans certainly has not stopped, but it is of secondary importance to cultural evolution, which has wrought remarkable changes in a very short time.

The first modern *sapiens* humans are known as Cro-Magnons, from the cave in southern France where their remains were found. In the Upper Paleolithic period, from about 35,000 to 8000 B.C., they created the foundations of our culture. With ever more refined methods, they learned to shape stones, and then bones, into excellent tools. We find spearheads and fishing hooks like those made by modern hunter-gatherers, and needles that indicate Cro-Magnons must have tailored clothing out of animal hides. Their level of intelligence and humanity probably equaled ours. They passed on their rituals and carefully tended their dead, painting the bodies with red ochre and burying them with various artifacts—evidence that they thought about an afterlife.

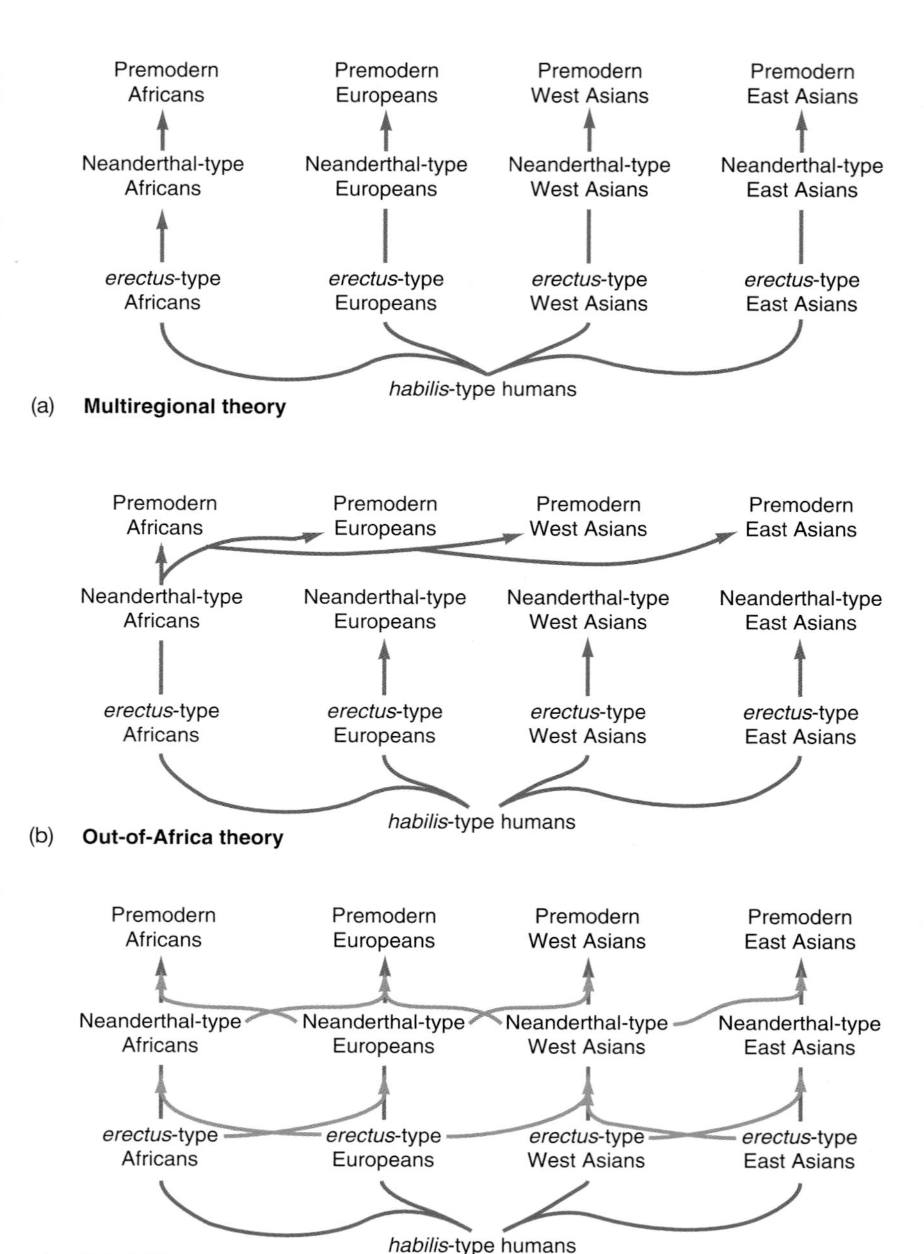

Figure 36.20

Anthropologists are still debating three models for the evolution of modern *sapiens* humans. *(a)* The multiregional theory proposes independent evolution from *erectus* to *sapiens* types in widely spaced locations. *(b)* The "out-of-Africa" theory proposes emergence of a single *sapiens* stock that spread out across the continents and displaced older populations. *(c)* Hybrid theories propose various degrees of hybridization between more ancient and more modern people in different places.

Cro-Magnon humans also expressed themselves through painting and sculpture. Some of their carvings rival our own in delicacy and realism, while others are clearly exaggerated representations of pregnant women that probably figured in fertility rites (Figure 36.21). (Fertility was important to primitive, disease-ridden people; their rate of infant and childhood mortality must have been enormous, and no more than 10 percent of them lived to the age of 40.) Paintings from at least 12,000 years ago in some French and Spanish caves still inspire awe and admiration. These people had learned to make paints from clays and animal fats colored with charcoal and metal

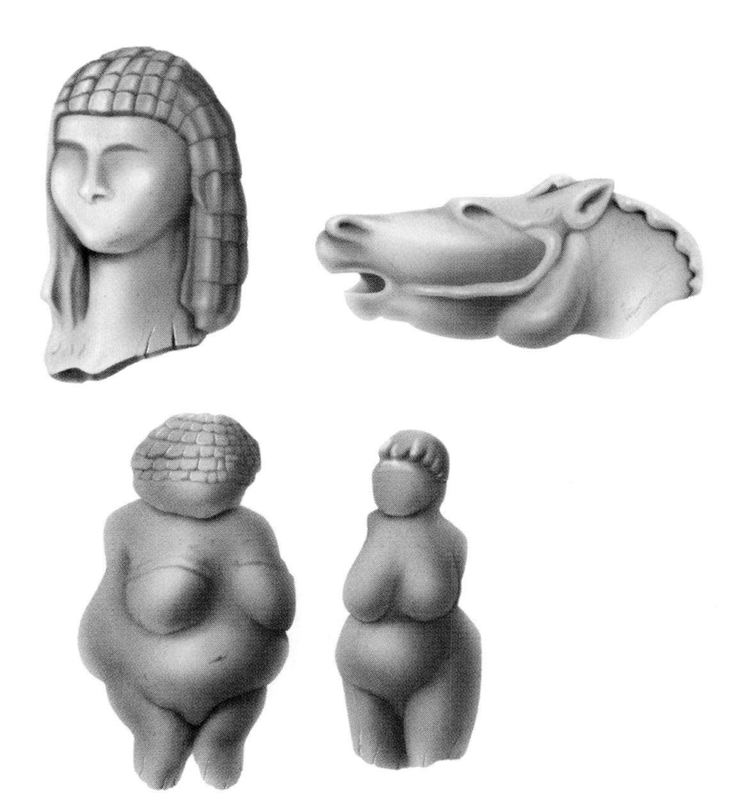

Figure 36.21

Typical small ivory and stone carvings of the Aurignacian age, about 35,000 years ago, show that the modern humans who emerged at that time were able to create art objects like those produced today.

oxides, and by daubing these paints on the cave walls or blowing them through a bone pipe, they created scenes that often portrayed the animals they hunted. Presumably these paintings held a magical or ritual significance and were meant to ensure successful hunting or the return of migrating herds, or to boast about past successes.

Contemporary with the Cro-Magnons in Europe, similar Upper Paleolithic cultures were thriving in Asia and Africa. All these people still lived by hunting and gathering, but by about 10,000 years ago many had started to settle down in permanent communities where food was abundant. At this time, there was a general warming of the earth, a retreat of the glaciers, and a transition to the *neothermal* conditions in which we now live. All ecosystems had to adapt to these changing conditions, and humans of this time were intimately entwined in their ecosystems. Northern European people, for instance, had long depended on herds of reindeer, but with warming conditions these herds had to adapt to new food plants and to the encroachment of forests into the European plains. Evidence also indicates that some very dry periods occurred about this time. All these factors forced people to find new food sources. Some tribes settled on lake and ocean shores where they created transitional cultures, known as Mesolithic, with efficient industries for catching fish and molluscs. While these European peoples continued to subsist by hunting, fishing, and gathering, others to the east were inventing a technique that would change the face of the earth: agriculture.

36.10 Agriculture created a revolution in human life.

Agriculture was another reaction to the neothermal transition. Traces of agricultural communities dating from about 8000 B.C. have been found in Asia, the Indian subcontinent, and the Middle East. Plant biologists can accurately locate early agricultural communities by comparing seeds from wild plants with those of domestic plants, and remnants of primitive domesticated plant forms are commonly found in ancient sites. For instance, matching wheat and oats with their ancestral wild grasses has confirmed that these plants were domesticated in the foothills of southwest Asia, from southern Turkey and Israel through Iran and Iraq. Wild goats and sheep from the same area were also domesticated. Other early developments in farming have been traced to the Andes of South America, where corn, tomatoes, potatoes, cocoa, and other economically important plants originated.

Evolution of wheat, Section 24.9.

These first agricultural communities are defined as Neolithic (New Stone Age), and we still live with the far-reaching consequences of the Neolithic revolution. Between 10,000 and 2,000 years ago, farming techniques spread rapidly. A dependable food supply made towns and cities possible and also led to a rapid population increase (see Figure 26.19). By some estimates, the world's population increased from about 5 million people 10,000 years ago to 130 million 2,000 years ago. The increase since then has been still faster, and again has been closely correlated with advances in agriculture.

The Neolithic revolution also ushered in our increasingly technological culture. Following the development of agriculture, people soon learned to smelt ores and to fashion metal objects. Lead and copper were used first, apparently as ornaments; we find jewelry dating from 4000 to 5000 B.C. The Bronze Age began around 3000 B.C. with the discovery that adding tin to copper created an alloy (bronze) from which superior weapons and tools could be made. Sometime shortly before 1000 B.C., the Philistines and other tribes were starting to make weapons of iron. But now we are starting a new story that belongs to the historians.

Coda Evolving the ability to live in trees seems to be good preparation for becoming intelligent. Arboreal success requires great manual dexterity, first-rate stereoscopic vision, and a nervous system capable of handling rapidly changing sensory inputs as an animal swings quickly through the trees. A sharp-eyed animal with the dexterity to grasp branches is preadapted for handling tools, looking at them closely, and thinking about ways to improve them. Intelligence and the use of language seem to be closely linked to these primate abilities. As the psycholinguist George Lakoff has shown, our language reflects our bodies' fundamental ways of engaging the world. We *grasp* ideas, *see* the solutions to problems, and *hold* opinions. Chapter 42 reveals that linguistic abilities are even stored in the regions of the brain associated with orienting in space and manipulating the world.

Primates are also the most social mammals. They develop highly structured families and societies, and they pass on a great deal of information culturally, through learning, rather than through their genes. Human societies are now passing through a critical period that will test the survival value of social structures, culture, and intelligence. Human history has been a story of enormous intergroup conflict, horrendous wars, and indescribable pain inflicted upon our fellow humans. While loving our brothers and sisters, we tend to hate and fear those who do not appear to share our genes. Now we must determine whether we can turn our intelligence toward creating truly humane societies characterized by harmony and equality. Our survival as a species depends on the outcome of this great biological experiment.

Summary

1. Primates are generalized mammals with several features that adapt them to an arboreal life: excellent sight, including binocular vision; long, mobile arms with prehensile hands; and sensitive, generalized digits capable of gripping with both power and precision.
2. Primates evolved from insectivore ancestors and are adapted to life in the extremities of trees, where they became fruit-eaters or, more generally, omnivores.
3. Prosimian primates include lemurs and lorises. The simians include tarsiers, Old-World and New-World monkeys, and hominoids (apes and humans).
4. Some primates of the Oligocene period (dryopithecines) became ground-dwelling apes that took advantage of the evolution of steppes and savannas. Their distinctive skeletons included a typical ape pelvis, which kept them bent over and unable to walk erect.
5. Australopithecines, which first appeared about 4 million years ago, were small hominids that were bipedal (walked erect). Some species persisted until about 1.5 million years ago, apparently as contemporaries of *Homo habilis,* an ancestor of modern humans.
6. Human evolution is charted partially by cultural relics. The first human artifacts were crudely chipped stone tools. These gradually became refined by later humans, who learned to shape smaller flakes of rock and, still later, bones and other softer material.
7. Between about 1 million and 250,000 years ago, *Homo erectus* people created Lower Paleolithic culture. Their cranial capacity was 800–1,100 cm^3, but they had heavy teeth and heavy eyebrow ridges.
8. Middle-Paleolithic culture was created by Neanderthal people and later humans between about 250,000 and 35,000 years ago. With cranial capacities of about 1,400 cm^3, the Neanderthals were very much like modern humans, except that they were more robust. Classic types had heavy eyebrow ridges and receding chins.
9. Modern humans had finally displaced Neanderthals by 35,000 years ago. Three alternative theories have been proposed to explain the origin of modern *Homo sapiens* people: separate ancient origins in different regions, an out-of-Africa model, and interbreeding between different populations in each region. However, present evidence is insufficient to determine which of these theories is most accurate.
10. Agriculture created a revolution in human life, apparently in response to changing climatic conditions. A major consequence of the agricultural revolution has been a huge, potentially disastrous increase in the human population.

Key Terms

opposable thumb 768
power grip 768
precision grip 768
prosimian 769
simian 771
hominoid 771
hominid 771
brachiation 772
dryopithecine ape 773
australopithecine 774
bipedalism 774
core tool 777
flake tool 777

Multiple-Choice Questions

1. Among primates and their ancestors, forward-facing eyes and three-dimensional vision are first seen in
   a. insectivores.
   b. prosimians.
   c. simians.
   d. hominoids.
   e. hominids.
2. Among primates, which have the sharpest sense of smell?
   a. lemurs
   b. tarsiers
   c. Old World monkeys
   d. New World monkeys
   e. apes
3. Which of the following are the most arboreal?
   a. pongids
   b. hominoids
   c. hominids
   d. Old World monkeys
   e. New World monkeys
4. Which is not a hominoid?
   a. orangutan
   b. monkey
   c. gibbon
   d. chimpanzee
   e. human
5. Street musicians called organ-grinders used to be accompanied by a small, entertaining simian companion with a long, prehensile tail. The organ-grinder is considered one of several hominids, while his companion is one of several
   a. Old World monkeys.
   b. New World monkeys.
   c. pongids.
   d. prosimians.
   e. hominoids.
6. Which of the following does *not* belong with the rest?
   a. ancestral to hominids
   b. 2:1:2:3 dental pattern
   c. Old World monkey
   d. platyrrhine face
   e. tail not prehensile
7. Which pair of terms is mismatched?
   a. dryopithecine—hominid
   b. australopithecine—hominid
   c. catarrhine—Old World monkey
   d. platyrrhine—New World monkey
   e. pongid—gorilla

8. Which primates were most equally at ease as bipedal walkers and as tree-dwelling climbers?
   a. gorillas
   b. dryopithecine apes
   c. australopithecine apes
   d. *Homo habilis*
   e. Cro-Magnon man
9. Which matched pair is *incorrect?*
   a. *Australopithecus africanus*—bipedal with gorilla-sized brain
   b. *Homo habilis*—simple tools and a brain nearly as large as modern human brain
   c. *Homo erectus*—Old Stone Age tools and fire
   d. *Homo sapiens neanderthalensis*—spears and knives
   e. *Homo sapiens sapiens*—guns and butter
10. Which of these is *not* a hominid?
   a. Neanderthal-type people
   b. *Homo sapiens sapiens*
   c. *Homo habilis*
   d. gorillas
   e. australopithecines

True-False Questions

Mark each statement true or false, and if false, restate it to make it true.

1. The precision grip is found in all primates, but the power grip is restricted to humans.
2. The pongids are extinct ancestors of humans.
3. Old World monkeys are more closely related to apes than to New World monkeys.
4. Knuckle-walking and brachiation are common characteristics of all monkeys.
5. Dryopithecine apes are believed to be ancestral to australopithecines.
6. Erect posture and bipedalism predated the large increase in brain size seen in the members of the genus *Homo.*
7. The primary distinctions between australopithecines and species within the genus *Homo* have to do with bipedalism.
8. All members of the genus *Homo* were probably capable of spoken language that was complex and symbolic.
9. The "out-of-Africa" model of human evolution is based on the hypothesis that modern *sapiens* people evolved in Africa before migrating to Eurasia.
10. The transition from hunting and gathering to agriculture first took place among the Neanderthals.

Concept Questions

1. Which characteristics of primates point to an arboreal life?
2. How do dryopithecines differ from australopithecines, and what is the relationship of these primates to modern apes such as chimps and gorillas?
3. What characteristics distinguish monkeys from apes?
4. What is the relationship between the development of an agricultural culture and the human population explosion?
5. Neanderthal fossils have been discovered in many locations in Europe and Asia. According to the "out-of-Africa" model, how closely related are these specimens, and how closely are they related to the modern humans in those locations?

Additional Reading

Hay, Richard L., and Mary D. Leakey. "The Fossil Footprints of Laetoli." *Scientific American,* February 1982, p. 50. The discovery of australopithecine footprints.

Jones, Steve, Robert Martin, and David Pilbeam. *The Cambridge Encyclopedia of Human Evolution.* Cambridge University Press, Cambridge and New York, 1992.

Klein, Jan, Naoyuki Takahata, and Francisco J. Ayala. "MHC Polymorphism and Human Origins." *Scientific American,* December 1993, p. 78. Presents the argument that the great variety of MHC proteins implies that the ancestral human population must have been large.

Molleson, Theya. "The Eloquent Bones of Abu Hureyra." *Scientific American,* August 1994. What skeletons reveal about human life and disease during Neolithic times.

Morgan, Elaine. *The Scars of Evolution.* Oxford University Press, New York, 1994. An exposition of the physiological and pathological aspects of human evolution.

Pilbeam, David. "The Descent of Hominoids and Hominids." *Scientific American,* March 1984, p. 84.

Shreeve, James. "The Neanderthal Peace." *Discover,* September 1995, p. 70. Fossils from Israel suggest that Neanderthal and modern-looking humans coexisted for up to 50,000 years with no morphological convergence, which means they never mated, perhaps because their mate-recognition systems were too different.

———"Erectus Rising." *Discover,* September 1994, p. 80. New dating techniques indicate that *Homo erectus* skulls from Java are 1.8 million years old, placing the human ancestor out of Africa a million years earlier than commonly thought and adding to the debate on single vs. simultaneous evolution of *Homo sapiens.*

Tattersall, Ian. "Madagascar's Lemurs." *Scientific American,* January 1993, p. 90. These remnants of primate evolution are disappearing fast as their habitat is destroyed. Unless hunting and deforestation stop, they may all become extinct.

van Lawick-Goodall, Jane. *In the Shadow of Man.* Houghton Mifflin Co., Boston, 1971. Jane Goodall's story of her studies on chimpanzees.

Internet Resource

To further explore the content of this chapter, log on to the web site at:

http://www.mhhe.com/biosci/genbio/guttman/

Plant Biology

37 Plant Structure and Development

The vegetables on display in a community Farmers' Market provide nutritious, colorful meals.

Key Concepts

A. Cell and Tissue Structure

37.1 The distinctive features of plant cells include walls, vacuoles, and plastids.

37.2 Plants are constructed of three basic tissue types.

37.3 Vascular tissues are constructed mostly of elongated conducting elements.

37.4 The epidermis is a plant's interface with its environment.

B. Development and Anatomy of the Plant

37.5 A seed contains an embryo plant in a dormant state.

37.6 The two types of flowering plants, monocots and dicots, have different organizations.

37.7 A vascular plant grows by elongation of a cylinder between two apical meristems.

37.8 Roots are built on a core of strong vascular tissue.

37.9 Stems grow from an apical meristem much as roots do.

37.10 The arrangement of vascular bundles gives the stem maximum strength.

37.11 Leaves are built of parenchyma with a network of vascular tissue.

37.12 Gymnosperm and dicot stems and roots increase in diameter through secondary growth.

37.13 Bark is a protective layer formed by secondary growth.

The fruits in the market didn't look very good today, so we're going to make our meal out of roots and stems. Doesn't sound like a meal for civilized people, you say? Perhaps—but you might be surprised at how tasty and nutritious they can be, since most roots and stems store food. For roots, we have some sweet potatoes, yams, and parsnips. For stems, here is asparagus and cabbage; yes, the cabbage looks like leaves, but it is really a very leafy stem. We can fill up on potatoes, which are also a stem growing underground. Turnips, beets, and carrots are very nutritious, and will add some color to the meal, but these underground storage parts are derived from both root and stem tissue, and it is hard to determine which parts come from which organ. For more flavor, we can add onions, radishes, and a bit of horseradish. On the whole, not a bad meal. And people living outside North America could add many other root and stem foods, including several wild plants that are commonly eaten but not cultivated.

Though vegetables like these are an ordinary part of our lives, we pay remarkably little attention to their structure. Most of us, being city folks, rarely think about how they grow and what the living plant looks like; many children would probably be surprised to learn that carrots and beets ever have green tops on them, and they could not identify the plants that will bear tomatoes. But somehow it seems only right that if we are going to eat them, we should know what we are eating.

All organisms share a unity of structure and metabolism. But in spite of the commonalities between plants and animals outlined in Chapter 32, plants have some unique physiological features, structures, and growth patterns. Since many plant parts grow constantly and their stages of development are inherent in the structure itself, we discuss their structure in the context of development, concentrating on the ecologically dominant gymnosperms and angiosperms introduced in Chapter 33. ■

A. Cell and tissue structure

37.1 The distinctive features of plant cells include walls, vacuoles, and plastids.

While adapting to their special ways of life, plants have evolved cells with several special features, most obviously a *cell wall* just outside the plasma membrane. The many plasmodesmata running through the cell walls link adjacent cytoplasms into a single, much larger cytoplasm. Plasmodesmata tend to be clustered in regions called *primary pit fields* (see Figure 6.17*a*). The walls of actively growing cells are relatively thin and pliable, whereas the walls of certain wood cells become thick and rigid, and may occupy over 90 percent of the cell volume. Cell walls develop in layers (Figure 37.1). First, at the end of mitosis, adjacent cells lay down and share a *middle lamella* composed largely of pectins, polymers of the sugar D-galacturonic acid; these are the pectins used to gel fruit preserves and jellies. Later, cells lay down cellulose inside the middle lamella to form a **primary cell wall.** Each plasma membrane contains rosettes made of six molecules of cellulose synthase (Figure 37.2); these remarkable structures, identified by T. H. Giddings, epitomize vectorial metabolism by drawing activated glucose molecules from the cytoplasm and extruding cellulose molecules on the cell surface. Each cellulose molecule, averaging a few micrometers long, is made of thousands of glucose residues. The cell lays down sheets of parallel microfibrils that are visible with an electron microscope, each made of 40–70 hydrogen-bonded cellulose molecules. Cells that become highly elongated, like some xylem elements, orient

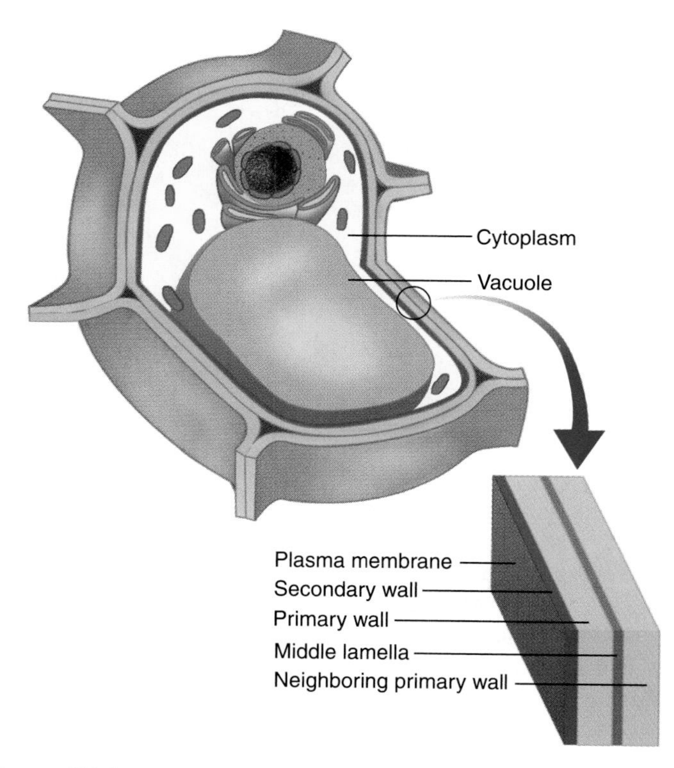

Figure 37.1

All plant cells have a primary wall made of proteins, cellulose, and other polysaccharides. Some cells have a secondary wall built inside the primary wall. Some cells have a secondary wall built inside the primary wall. The middle lamella is a thin region shared by neighboring cells. The plasma membrane that encloses the cytoplasm lies just inside the cell wall, but it is so thin that it is barely visible even in the enlarged view.

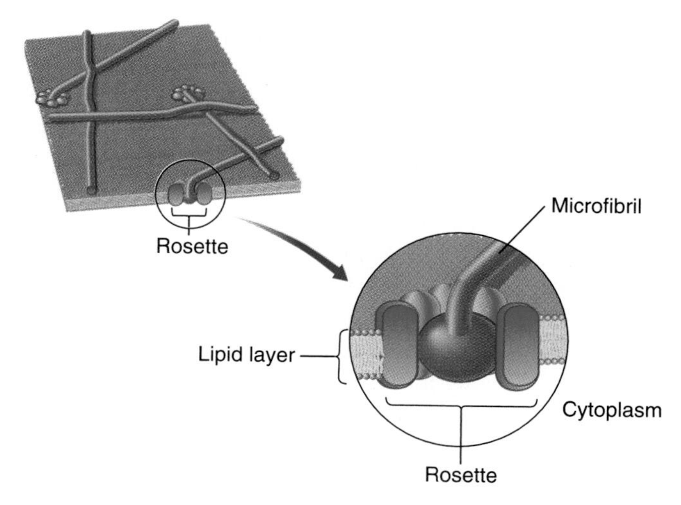

Figure 37.2

Cellulose-synthesizing rosettes, hexagonal complexes of cellulose synthase, form pores in the plasma membrane. They take glucose molecules from the cytoplasm and extrude cellulose molecules, aggregated into microfibrils, onto the cell surface, where they form the cell wall.

Figure 37.3

(a) In most plant cells, cellulose fibers are oriented almost perpendicularly to one another in successive layers, as shown in an electron micrograph *(c)*. In elongated cells *(b)*, many fibers are oriented around the cell, perpendicular to its long axis.

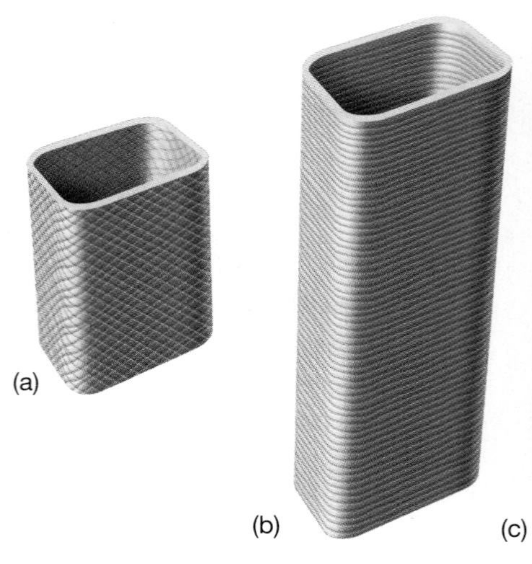

these microfibrils circularly, perpendicular to the long axis, while in cells that grow more uniformly, such as parenchyma, the fibrils run in all directions (Figure 37.3). Sheets of microfibrils are glued together through hydrogen bonding with pectins and other polysaccharides called hemicelluloses, and with special proteins, which are notable because they contain hydroxyproline, a modified amino acid found only here and in collagen, the main protein of animal connective tissue. The presence of hydroxyproline in these spots suggests that it is especially valuable in strong supporting materials.

Vectorial metabolism, Section 8.15.

Some cells have an additional **secondary cell wall** inside the primary wall (see Figure 37.1). Though made largely of the same materials as the primary wall, it is usually thicker and is strengthened with **lignin,** a material so tough it is sometimes called a "biological plastic." Lignin is a complex hydrophobic polymer of monomers such as coniferyl and coumaryl alcohols (Figure 37.4); it replaces the water in the cell wall, and a cell dies after becoming thoroughly lignified. (Lignin is very resistant to decay because few organisms make ligninases that can digest it.) Cell walls are also strengthened by incorporating ions such as calcium, which reinforces the point that these mineralized plant tissues, like the bones and shells of animals, are supports. The common structural features in plants and animals (complex polysaccharides, proteins with hydroxyproline, and Ca^{2+} ions), in spite of their long evolutionary separation, again show the close relationship between structure and function.

Virtually all living mature plant cells contain one or more large *vacuoles* that may take up most of the cell's volume, with the cytoplasm reduced to a thin layer around the cell periphery (Figure 37.5). A vacuole, with its unusual array of dissolved ions and molecules, is set off from the rest of the cell by a membrane, the *tonoplast.* In concert with the plasma membrane, the tonoplast controls ion concentrations in the cytosol. Wastes accumulate inside the vacuoles of some cells and may become so concentrated that they form distinctive crystals. The brilliant colors of flowers and leaves often come from anthocyanins,

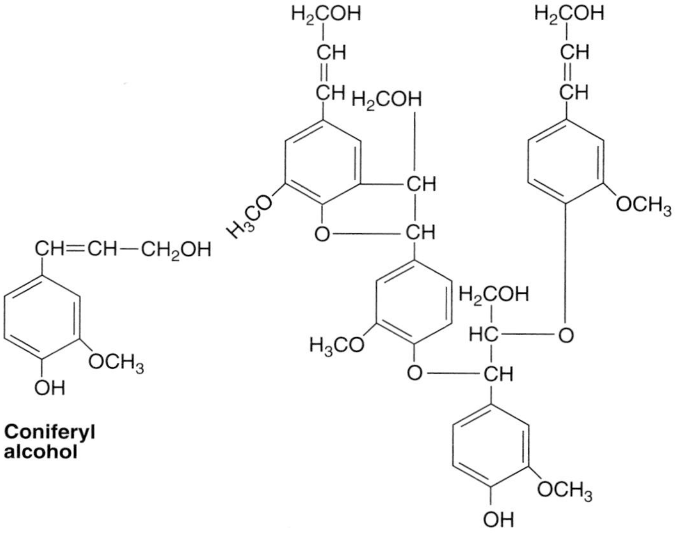

Coniferyl alcohol

Figure 37.4

Lignin is a tough, complex polymer made from monomers of coniferyl and related alcohols.

flavonoids, and other pigments deposited in these vacuoles. In a cell with a rigid wall, a vacuole can only expand at the expense of the cytoplasm. Before the cell wall solidifies, however, a vacuole can be an organelle of growth. Plants grow considerably due to elongation of cells after they have divided, rather than by cell division alone. By accumulating water in the vacuole, a cell builds up internal pressure, which can force it to elongate (Figure 37.6).

Mechanism of cell elongation, Section 39.2.

Chloroplasts, the sites of photosynthesis, are common in plant cells, along with related membrane-bounded organelles with varied functions. As a group, these organelles are called **plastids.** *Chromoplasts* filled with carotenoid pigments give color to fruits, such as the red of tomatoes and peppers, and to

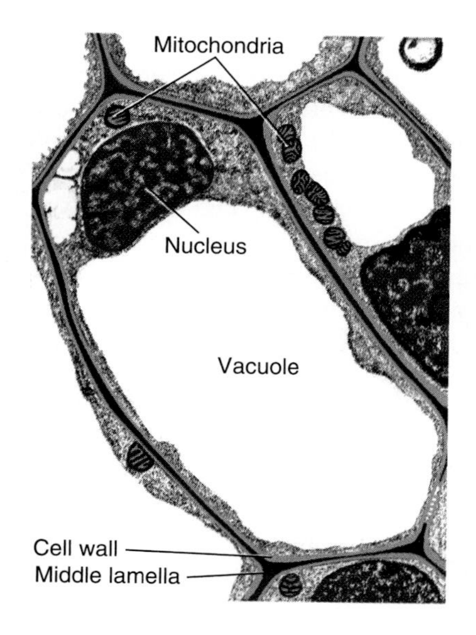

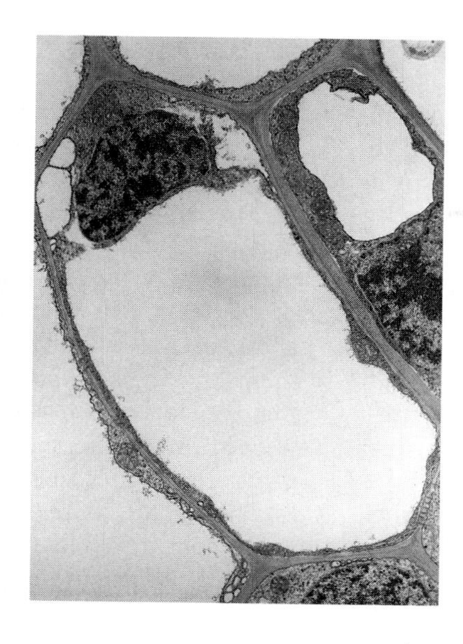

Figure 37.5

A typical plant cell has a large central vacuole that occupies much of its volume. The cytoplasm here occupies the periphery of the cell.

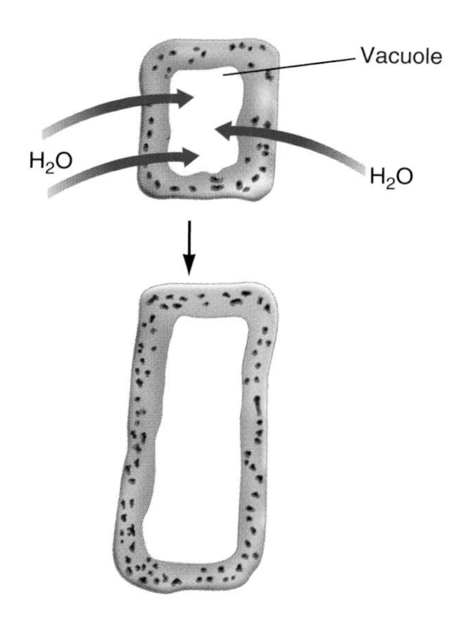

Figure 37.6

A cell may elongate through the expansion of its central vacuole if the fibrous structure in its cell wall is loosened in the right places.

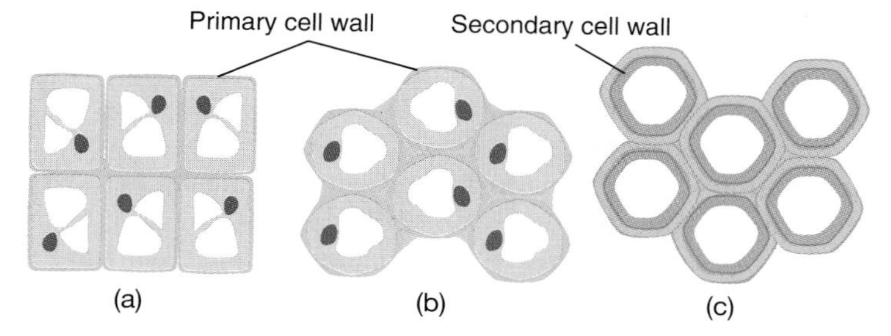

Figure 37.7

(a) Parenchyma has uniformly thin primary cell walls and retains its cytoplasm. *(b)* Collenchyma has thickened primary walls, sometimes thickened only where adjacent cells meet. *(c)* Sclerenchyma has regions of secondary cell wall built up inside the primary wall, and it generally loses its cytoplasm.

Exercise 37.1 Explain how the synthesis of cellulose fibrils is an example of vectorial metabolism.

some flowers. *Leucoplasts* are colorless bodies where starch or oils are synthesized. Plastids that accumulate large amounts of starch are called *amyloplasts,* and those located in the root cap have a role in detecting the force of gravity (see Section 39.4).

Plant cells contain typical eucaryotic organelles, like those of animal cells, except that they have no centrioles in their centrosomes. Their cytoplasm also contains various poorly characterized bodies such as *spherosomes,* spherical bodies about the size of mitochondria. Spherosomes that are rich in lipids are involved in lipid storage and transport; others are apparently equivalent to lysosomes. *Peroxisomes* are the sites where photorespiration occurs, and *glyoxysomes* are bodies where lipid reserves are converted into carbohydrates, as during the germination of seeds that store oil.

37.2 Plants are constructed of three basic tissue types.

As we outlined in Section 32.3, plants are made of three general types of tissue: parenchyma, collenchyma, and sclerenchyma (Figure 37.7). They are distinguished quite easily. Parenchyma cells have thin primary cell walls; collenchyma has thickened primary cell walls, often only in certain regions; and sclerenchyma has both primary and secondary cell walls. The most abundant tissue in an herbaceous plant is parenchyma, which takes many forms. *Chlorenchyma* is parenchyma with abundant chloroplasts, specialized for photosynthesis and production of the carbohydrates needed by the rest of the plant; chlorenchyma makes the bulk of most leaves. *Aerenchyma* is parenchyma characterized by extensive intercellular spaces, allowing the construction of large but lightweight tissues in which gases can exchange easily; parenchyma is especially important for gas exchange in aquatic plants. Some parenchyma is specialized for secretion, such as producing the nectar that attracts pollinators to flowers or secreting a gelatinous material around roots. The epidermis covering the outer surfaces of all plant organs is a kind of parenchyma. Storage parenchyma is abundant in seeds, fruits, and roots that store carbohydrates, proteins, and oils as food reserves. In addition, a plant reserves some parenchyma tissues as *meristems,* undifferentiated stem cells that retain the ability to grow and divide, as we discuss in Section 37.7.

Herbaceous plants are supported in large part by turgid parenchyma cells that are well filled with water, but they frequently need additional mechanical support; woody plants by definition are made of stronger tissue. The mechanical tissues of plants are collenchyma and sclerenchyma. Collenchyma retains its cytoplasm, and its thickened walls make it a strong, flexible support for growing stems and leaves. Its walls may be quite uniformly thickened by the deposition of cellulose layers, pectins, and water or thickened mostly at the angles where cells meet, giving it a distinctive pattern of adjoining triangles in cross section.

Sclerenchyma cells have thick, heavily lignified walls, and with few exceptions they die and lose their cytoplasm as they mature. Sclerenchyma is either conducting or nonconducting. Conducting sclerenchyma forms the xylem elements that carry water throughout the plant, as described in Section 37.3. Nonconducting sclerenchyma includes elongated *fibers* and more rounded *sclereids.* In Section 33.6, we showed how tracheids, the most primitive xylem elements, probably evolved into different forms, including long, narrow fibers that are specialized for support. Because of their elongated form, fibers are very flexible and form an essential part of woody tissues, mixed among the conducting elements. Sclereids, on the other hand, take on a variety of shapes, sometimes with many branches and extensions. They grow wherever tissues need support and protection with little flexibility, strengthening tough, largely inedible leaves and forming the outer coats of seeds and some fruits or seed pods. We perhaps know them best as the gritty particles in the flesh of pears.

Exercise 37.2 This observation is best done during a high wind, but you can also imagine the scene. Notice how much even very heavy trees bend with the wind; then explain the importance of fibers in the construction of wood.

37.3 Vascular tissues are constructed mostly of elongated conducting elements.

Parenchyma, collenchyma, and sclerenchyma are a plant's basic tissue types, but a plant is always made of more complex tissues in which the three types are mixed. To understand a plant's overall anatomy, we begin by examining its two types of conducting tissue, xylem and phloem. Both types of tissue also include parenchyma that stores food reserves and fibers that provide strength, but here we focus on the conducting elements.

Xylem is the tissue that conducts water and dissolved substances upward from the roots, through two kinds of tubes or tubelike structures called **tracheary elements. Tracheids** are elongated cells with no cytoplasm whose tapered ends fit together closely; where tracheid cells meet, water passes between them through bordered pits in the cell walls (Figure 37.8). Notice that in these pits material is still passing through the primary walls of the adjoining tracheids. **Vessels** are tubes formed by a series of cells called **vessel elements,** which have perforated end-walls and are laid end-to-end. Both kinds of tracheary elements, having lost their cytoplasm, are merely empty cell walls that are ideal for carrying water. In both types, the primary cell walls are strengthened by particular patterns of secondary wall formation, either in rings, in helices, or in circular bordered pits. Tracheids, however, retain their primary walls throughout, and where their tapered ends abut, water can only pass from one to the other through the primary walls within adjacent pits, making a *pit pair:*

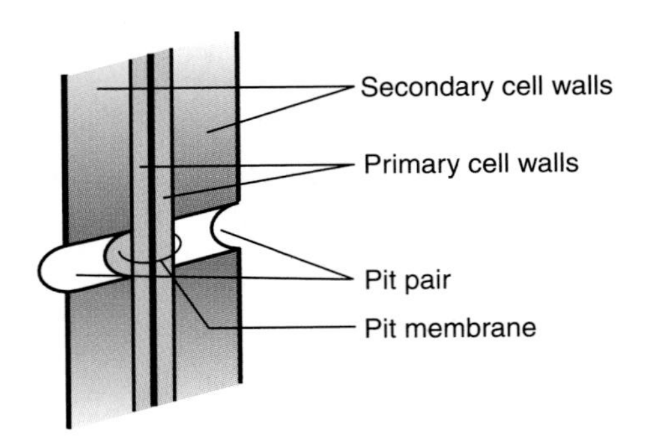

Vessels, on the other hand, have open, *perforated* ends only partially blocked by walls, allowing water to flow from one element to the next much more freely. A cross section through xylem tissue generally reveals a few large, rounded vessels alongside many smaller tracheids and some fibers. We will see later that the secondary xylem formed in woody plants also contains parenchyma that conducts substances radially—that is, in and out across the trunk, rather than up and down.

Phloem tissue distributes organic compounds manufactured in the leaves to the rest of the plant through **sieve elements: sieve-tube members** in angiosperms and **sieve cells** in other vascular plants (Figure 37.9). Sieve-tube members join end-to-end to form **sieve tubes;** two members meet and communicate at **sieve plates:**

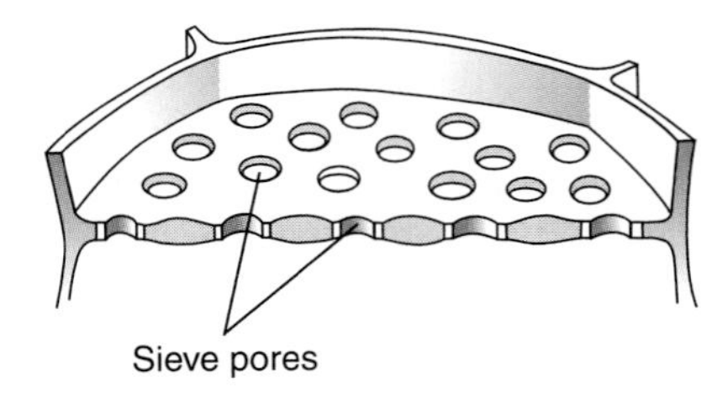

The sieve pores are modified, enlarged plasmodesmata concentrated in primary pit fields, making the cytoplasms of neighboring cells continuous. Sieve cells tend to be very long, often over

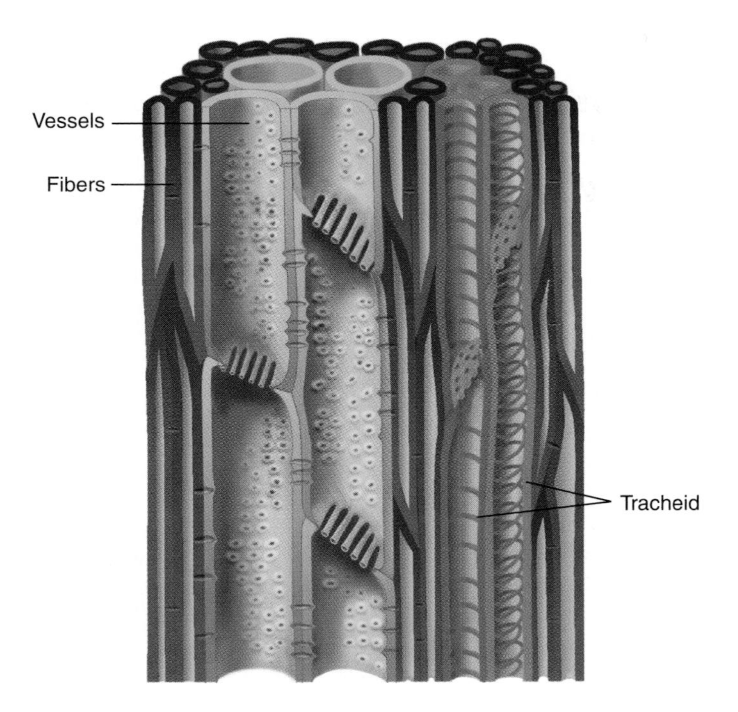

Figure 37.8

Xylem consists of vessels and tracheids that transport water axially. Tracheids are narrow and tapered; their elements communicate at pit pairs where only the primary walls of adjacent cells meet. Vessels are larger, often very large, and their elements meet at perforations to form open, continuous tubes that conduct water rapidly. Supporting fibers are interspersed between the other cells.

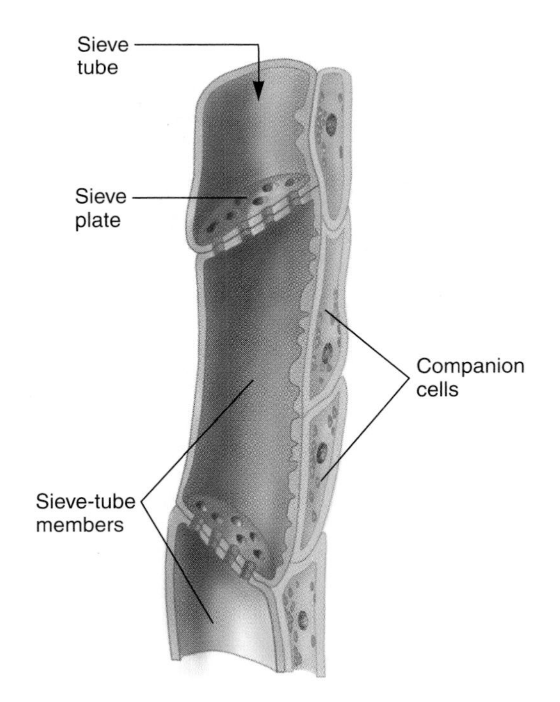

Figure 37.9

The major distinctive sieve elements of phloem are made by many cells (sieve-tube members) joined end-to-end so their cytoplasms are connected through the pores of sieve plates. Note the companion cells next to the sieve-tube members.

1 mm, and join one another at extremely long, slanted end-walls; sieve-tube members are much shorter (50–150 μm), often quite wide (up to 40 μm), and meet at transverse sieve plates. The evolution of these elements has paralleled that of the corresponding tracheary elements of xylem: the long, tapered sieve cells of earlier plants changing into the shorter, more tubelike sieve-tube members of angiosperms, just as tracheids changed into the tubelike vessel elements. In contrast to the dead, empty tracheary elements of xylem, the conducting cells of phloem retain their cytoplasm. As they mature, they lose their nuclei but stay attached to nucleated **companion cells** (albuminous cells in nonangiosperms), which were once assumed to perform maintenance functions for both themselves and the sieve-tube members. Now that the mechanism of phloem conduction seems reasonably well established, it is apparent that the primary function of companion cells is to lead materials into the sieve tubes.

Mechanism of phloem conduction, Section 38.4.

Exercise 37.3 Where tracheids meet to form continuous conducting elements, their walls are highly slanted, instead of meeting more or less perpendicular to their length. Explain the importance of this anatomy by considering how two tracheids communicate with each other.

Exercise 37.4 Vessel elements are generally quite short; tracheids tend to be very long, often more than a millimeter. What is the advantage of having such long tracheids?

37.4 The epidermis is a plant's interface with its environment.

The region of a plant that meets the outside world must have special properties. Internally, a plant has a certain degree of control over its activities, but it has little or no effect on the external world, and it must limit the impacts of that world. All plant structures are covered with a layer of **epidermis,** which protects underlying tissues from the outside. The bulk of epidermis is a single layer of cells with often distinctive shapes that fit together like paving blocks and create a surface that is almost impervious (Figure 37.10).

Water intake and conservation are chief functions of the epidermis. Root epidermis is generally the plant's source of water; this epidermis is highly permeable to water and even has ionic pumps for taking water up actively from the surrounding soil. The shoot epidermis, in contrast, must be waterproof to prevent evaporation; this already tight surface is covered with the lipid polymer cutin, and waxes are built up on top of the cutin or sometimes mixed with it. Therefore the shoot epidermis only allows any significant evaporation at its stomata, the openings we introduced in Section 32.7. We discuss the operation of stomata in Section 38.1.

Epidermis must also protect a plant against sunlight. This seems like a very odd notion at first, since plants obviously need sunlight as their source of energy. But the light they

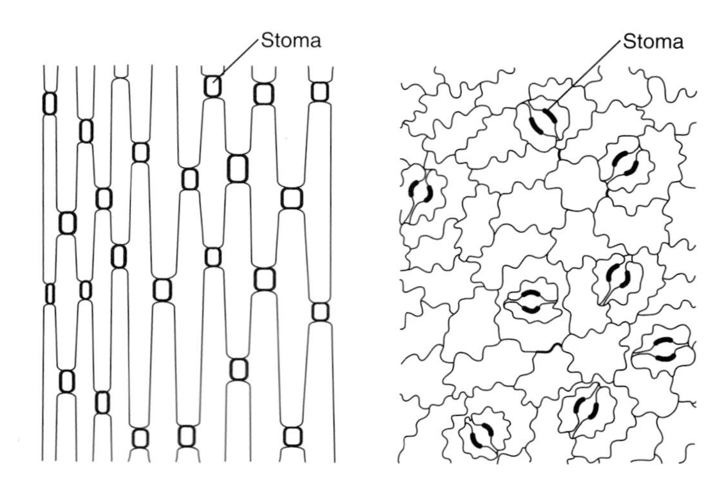

Figure 37.10

Epidermis consists of a single layer of cells that fit together tightly, often including stomata formed by a pair of guard cells. Here are two distinctive patterns as seen from the surface.

require is principally red and blue light in the visible range, the only wavelengths that the photosynthetic apparatus can absorb functionally; radiation of other wavelengths can be as damaging to a plant as to any other organism. Ultraviolet radiation can bleach chloroplasts, destroying their capacity to photosynthesize, and it can damage DNA. Infrared radiation can overheat the plant, inactivating its enzymes; we take up this problem of temperature regulation in Section 38.7. The epidermis, however, can reflect much of this damaging radiation and protect the tissues underneath. Cutin reflects light very well, including UV light.

Since every plant tissue is prospective food for some other organism, protection against viruses, bacteria, fungi, and animals such as insects is essential. Here again epidermis can play an important role. Cutin has special virtues, since no organism is known to have enzymes that digest it. Epidermis may protect against fungi and bacteria simply by virtue of its smoothness, which allows their spores to slide off; on the other hand, epidermis frequently develops various kinds of hairlike extensions, which inhibit the activities of insects. It may also be rich in secretory cells of various kinds that produce noxious substances that repel animal predators. Some plants combine the two kinds of defense in stinging hairs, like those of stinging nettles *(Urtica dioica);* each hair is made of a single elongated cell with a large vacuole filled with a toxin, arranged so an animal that brushes against the hair will break it off and release the toxin. Among the toxins found in such hairs are histamine, an agent of the inflammatory response in animals, and acetylcholine, a major neurotransmitter that initiates irritating stimuli in an animal's skin.

Finally, the epidermis may be the site of special glands, such as salt glands used by plants to excrete excess salt (see Section 38.10) or digestive glands used by carnivorous plants to digest insects and other animals (see Section 40.15).

B. Development and anatomy of the plant

37.5 A seed contains an embryo plant in a dormant state.

A seed houses an embryo plant waiting for the opportunity to grow (Figure 37.11). We left the seed in Chapter 33 at the point where it was ready to sprout, and now we take up its story again. As a seed matures inside its covering, usually a fruit, it loses moisture. Most mature seeds are less than 20 percent water. As reproductive devices that need to weather times with little or no water, most seeds can tolerate considerable desiccation, down to only a few percent water. Other seeds, however, can't tolerate desiccation at all and die if their moisture content falls below about 30 percent.

Some seeds lose their viability quickly; others retain it for long times. Occasionally, a well-preserved seed a few hundred years old will sprout. A mature, viable seed that has not been stimulated to develop is in a state of **dormancy,** a condition that ends with **germination,** when the embryo plant begins to grow. Growth begins with internal changes, and a seed is only said to have germinated when the embryonic root breaks through the outer layers. Various factors stimulate the end of dormancy in each species. For some, temperature is critical. If the plant lives in a temperate climate where it must endure a cold winter, the seeds only germinate after being exposed to a low temperature

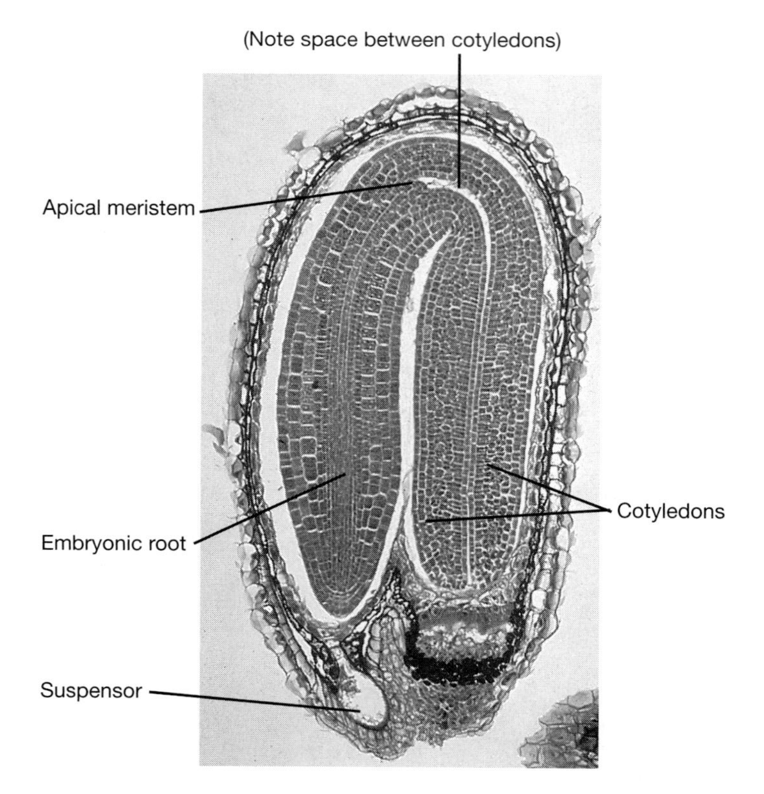

Figure 37.11

A seed contains an embryo plant. Virtually everything in the seed that is not embryo is endosperm, a storage tissue packed with nutrients on which the developing embryo can live.

(generally about 5°C) for a certain number of days, followed by a warmer temperature. Such seeds, in effect, are measuring the passing of winter and the beginning of spring.

Certain kinds of exposure to light may also end dormancy. Some seeds respond only to repeated exposures to light, others to a brief exposure to red light. This reaction depends on the phenomenon of photoperiodism, which we discuss in Chapter 39.

Seeds normally fall onto the soil, where they can absorb the water that is essential for the early stages of germination. During the first day or so after breaking dormancy, the seed rapidly absorbs (imbibes) water to about the level of normal tissue, roughly 60 percent, and its water content increases more slowly for some time afterward. The onset of germination is probably mostly expansion of the embryo tissues as they rehydrate. Then the seed ends its metabolic inertia, and the embryo within becomes a fully active plant. The enzymes that break down the oils, starches, and proteins in its food stores are activated, and these reserves support the embryo until its shoot emerges from the soil and develops leaves that can photosynthesize.

To examine the structure and early growth of a young plant, it is necessary to distinguish between the two classes of angiosperms.

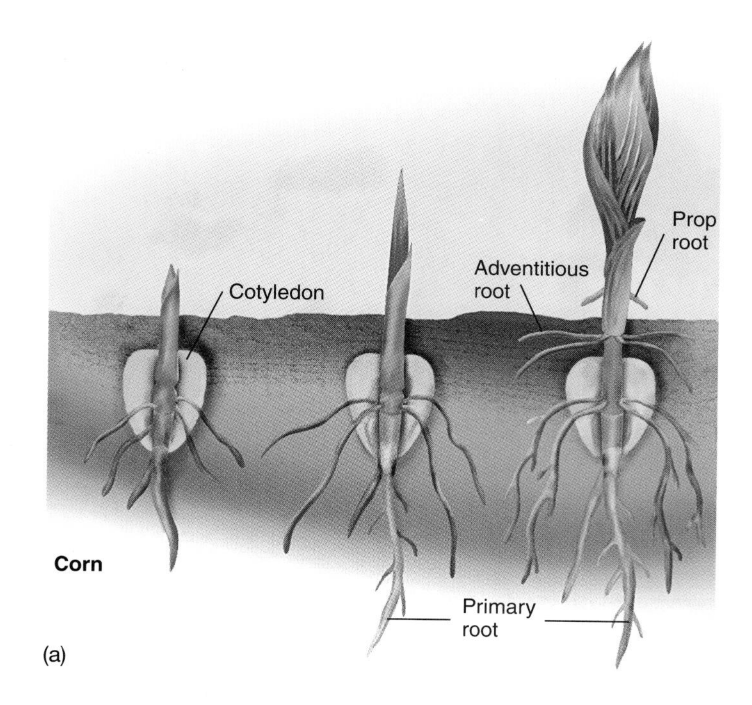

37.6 The two types of flowering plants, monocots and dicots, have different organizations.

Figure 37.12 contrasts the germination of a peanut seedling and a corn seedling. As the peanut embryo grows, the seed opens to reveal two **cotyledons,** fleshy extensions that remain attached to the infant stem and are pushed above ground with it. The corn seedling, however, bears only a single cotyledon. Although cotyledons can be photosynthetic and are sometimes called "seed leaves," they aren't really leaves. Their primary function is to store food that nourishes the young plant. They distinguish two classes of flowering plants: The **monocots** have a single cotyledon, and the **dicots** have two. Monocots (class Liliatae) include the grasses and, therefore, some of our most important food plants, like wheat, barley, and rye. Here, too, are the rushes, sedges, and cattails characteristic of marshes, as well as the tiny duckweeds that float on the surfaces of the water in marshes and ponds. The palms are monocots, as are flowers such as orchids, lilies, and irises. However, at least three-quarters of the flowering plants are dicots (class Magnoliatae), including our common deciduous trees and an enormous range of shrubs and plants with showy flowers. Figure 37.13 summarizes the principal differences in organization between the two classes.

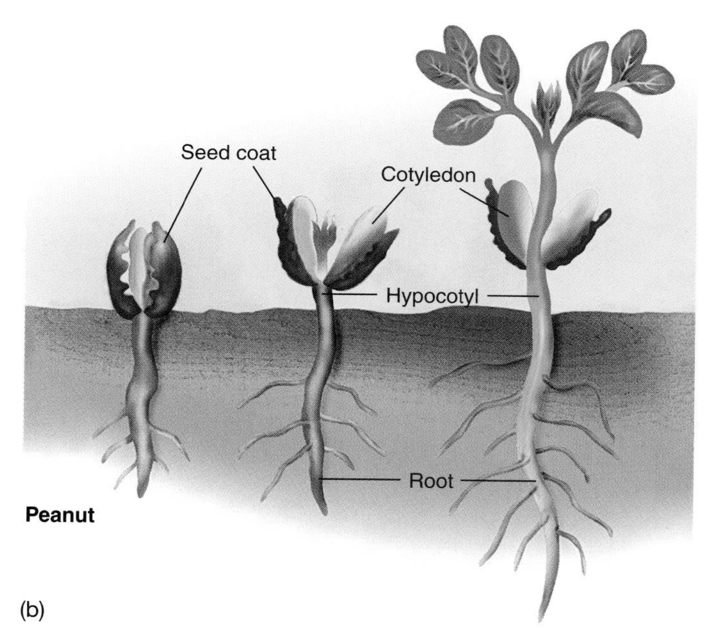

Figure 37.12

(a) Monocots, represented by a corn seedling, have a single cotyledon. *(b)* Dicots, represented by a peanut seedling, have two cotyledons.

37.7 A vascular plant grows by elongation of a cylinder between two apical meristems.

Notice that as an embryo unfolds (see Figure 33.33), it already has an embryonic root, or *radicle,* and an embryonic shoot divided into two parts that are named for their relationship to the cotyledons: the *hypocotyl* (below the cotyledon[s]) and the *epicotyl* (above the cotyledon[s]). As we discuss in Section 20.2, a principal difference between plant and animal development lies in the way their embryos establish an *axis of polarity.* Animals grow generally throughout their bodies, whereas plants grow only from localized growth regions called *meristems.* A plant is an elongated cylinder that grows primarily along an embryonic

Monocots Dicots

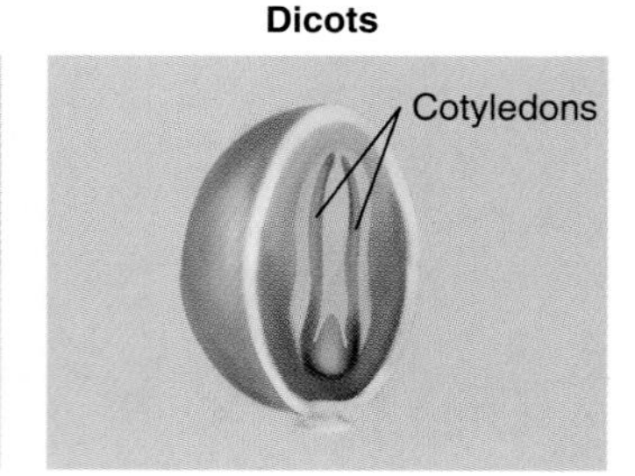

Embryos

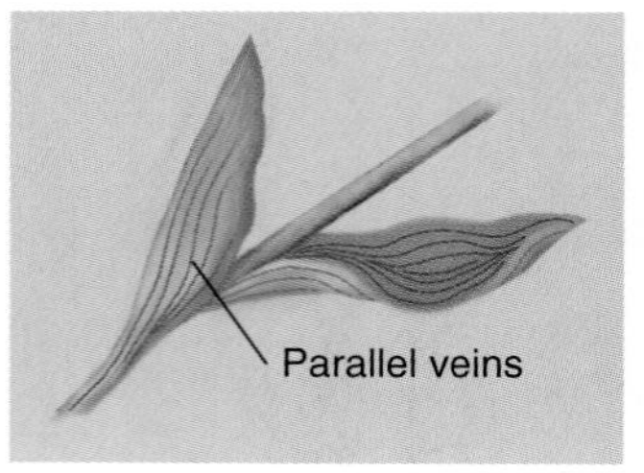

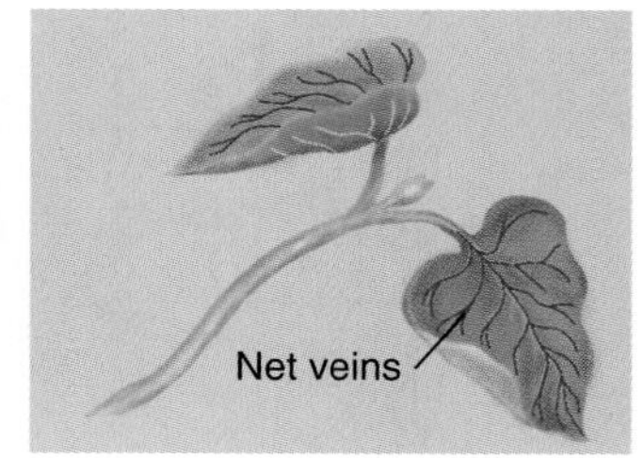

Leaves

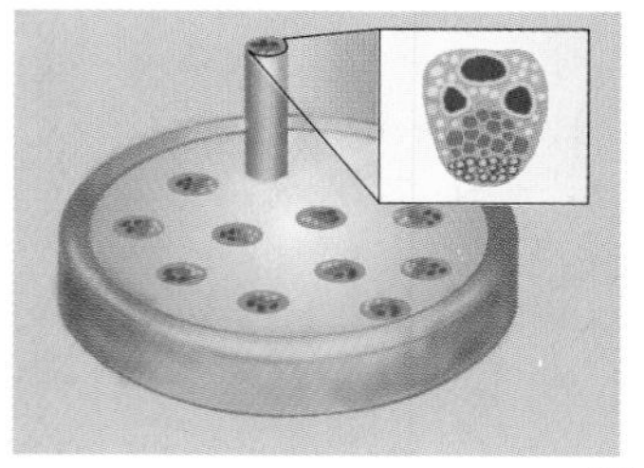

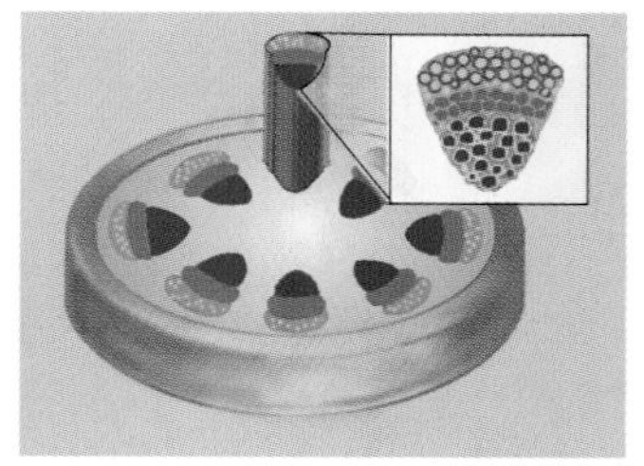

Stems

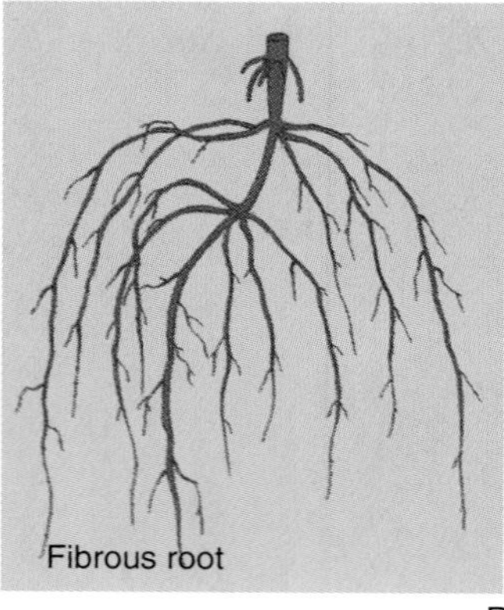

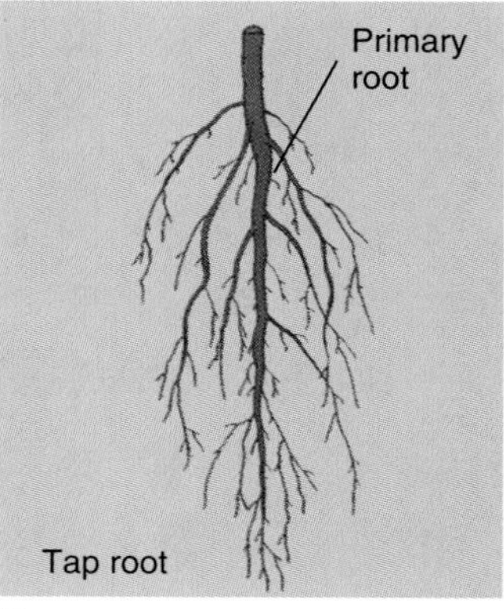

Roots

Figure 37.13

While monocots and dicots are defined by the number of cotyledons in a seed, the groups also differ in leaf structure, the arrangement of vascular bundles in the stem, and root structure.

axis stretched between two growing regions: a **shoot apical meristem** above and a **root apical meristem** below. These are the *primary meristems* already established in the embryo. Later a plant develops *secondary meristems* where particular organs develop, especially meristematic layers called **cambia** (sing., **cambium**) where stems, roots, or branches grow in diameter. But we will still consider a plant as primarily an elongating cylinder and its roots, stems, and leaves as elements of that elongated body that grow and transport materials along this axis, the **axial** direction.

Meristems, Section 21.5.

As meristematic cells divide, they follow the pattern shown in Figure 21.10, where one daughter cell becomes specialized while the other remains unspecialized and able to divide again. Meristematic cells produce different kinds of derivatives; a *protoderm* produces epidermis, and a *procambium* produces the primary vascular tissues (xylem and phloem). Notice that the shape of each organ depends strongly on how the division planes of the meristematic cells are oriented, as shown in Figure 20.12.

37.8 Roots are built on a core of strong vascular tissue.

The primary functions of roots are to anchor a plant in the soil, holding it upright, and to transfer water and minerals from the soil to the plant. Many perennials store food in their roots in the form of carbohydrates; the entire above-ground part of the plant may die back during the winter, leaving only the root to grow a new shoot system by expending its nutrient stores. Gardeners know to their sorrow how impossible it can be to eradicate pests like morning glory and dandelions if even a small piece of root is left in the ground. Roots are also the sole or primary sources of certain plant hormones.

You can predict the shape of the root system by noting whether a plant is a gymnosperm, a monocot, or a dicot. Most monocots have a highly branched *fibrous root system* made of many fibers of about the same diameter, whereas gymnosperms and dicots form one or more heavy *taproots* that sprout much smaller, secondary roots (Figure 37.13). Though we will concentrate here on typical roots that grow under the soil and beneath the shoot system of a plant, plants sometimes produce *adventitious roots,* those that arise from any organ that is not a root. Most commonly, generally in monocots, adventitious roots grow from the bottom of the stem and are called *prop roots* or *crown roots.*

The size of a root system is hard to predict, or even to imagine. The roots of a dicot, an alfalfa plant a meter or less in height, may extend more than 6 meters into the ground, and mesquite roots over 50 meters long have been found in mines in the southwestern United States. Graduate students in botany have occasionally been given the thankless task of counting roots and estimating the total size of the system; one such count showed that a ryegrass plant (a monocot) had 15 million rootlets with a total length of over 600 kilometers.

Plant roots are continuously strained as a plant is pulled this way and that by the wind and interactions with animals, and one structure that resists this tension well is a flexible cable with a strong center. A dicot root is built like this. A cross section through a mature dicot root shows that it is built in concentric layers (Figure 37.14). Its center is the **stele,** made of vas-

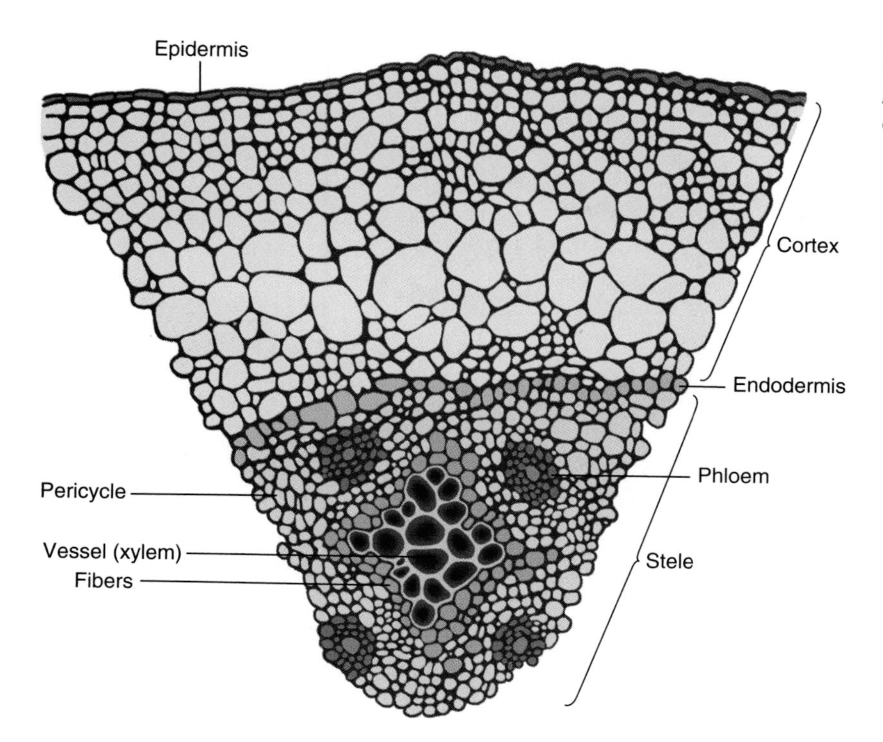

Figure 37.14
A cross section through a root shows the vascular cylinder, or stele, with surrounding cortex and epidermis.

cular tissue, which initially looks like a star in cross section. The core and points are made of tough xylem, with phloem between the points. Monocots may have parenchyma forming a **pith** within the stele. The stele is surrounded by a layer of parenchyma cells, the **pericycle,** a meristematic layer that produces secondary roots, which push outward through the outer layers of tissue. Eventually each rootlet develops its own vascular system, connected to the central vascular cylinder.

The bulk of the root surrounding the vascular cylinder is the **cortex,** made of rather loosely packed parenchyma cells with extensive intercellular gas spaces between them. **Endodermis,** the innermost layer of cortex just outside the pericycle, is made of specialized cells that regulate the transport of water and dissolved materials into the vascular cylinder. A layer of hypodermis sometimes marks the outside of the cortex, and a layer of epidermis surrounds the whole root.

Role of the endodermis in water transport, Section 38.2.

A mass of parenchyma cells forms a **root cap,** which covers and protects the root tip from damage as it pushes through the soil. Roots are notorious for growing in ways that require this protection. They snake through a soil made of rock particles, curling around some rocks and pushing others aside, insinuating themselves into the smallest cracks and slowly widening them with enormous force. Growing roots can generate enough force to destroy rock (or concrete, for that matter) by opening and widening cracks. The expendable cells of the root cap slough off as the root grows, and this loss, along with a slimy lubricant that the roots produce, eases the root's passage. Furthermore, the root cap, by a poorly understood mechanism, detects gravity and directs the growth of the root in response; the root cap is responsible for most roots growing downward.

Every root cell goes through three stages of growth—cell division, elongation, and differentiation—that roughly form three zones seen in a longitudinal (lengthwise) section through the root (Figure 37.15). As every cell grows older, it will successively reside in each zone. A cell originates in the root apical meristem immediately behind the root cap, a zone of rapidly dividing, undifferentiated cells that produce new root tissue as they proliferate. Meristematic cells here may go through a cell cycle in just 12 hours. In a zone behind the meristem, slightly older cells are elongating, growing somewhat wider and enormously longer—about 10 times the length of a meristematic cell. As we noted before, these maturing cells elongate through internal pressure from enlargement of their vacuoles, so their length and vacuolar structure set them off strikingly from younger cells. Finally, as the root cells achieve their full length, they differentiate into mature types such as epidermis, cortex, or vascular cylinder. The radial position of each cell determines its fate (shown in Figure 37.15). *Protoderm cells* on the outside will become epidermal tissue; the *ground meristem* inside this will become root cortex; and the central *procambium* will develop into xylem and phloem.

Epidermal cells in the zone of differentiation develop root hairs, which absorb much of the material that roots take up through their enormous surfaces (Figure 37.16); a square millimeter of root surface may sprout a few hundred hairs, multiplying the surface area 10–20 times. Furthermore, root hairs curl around soil particles, their mucilaginous outer layers merging so intimately with organic materials in the soil that plant and soil virtually become one. The root hair zone is no more than a few centimeters long, and each hair lasts for only a few

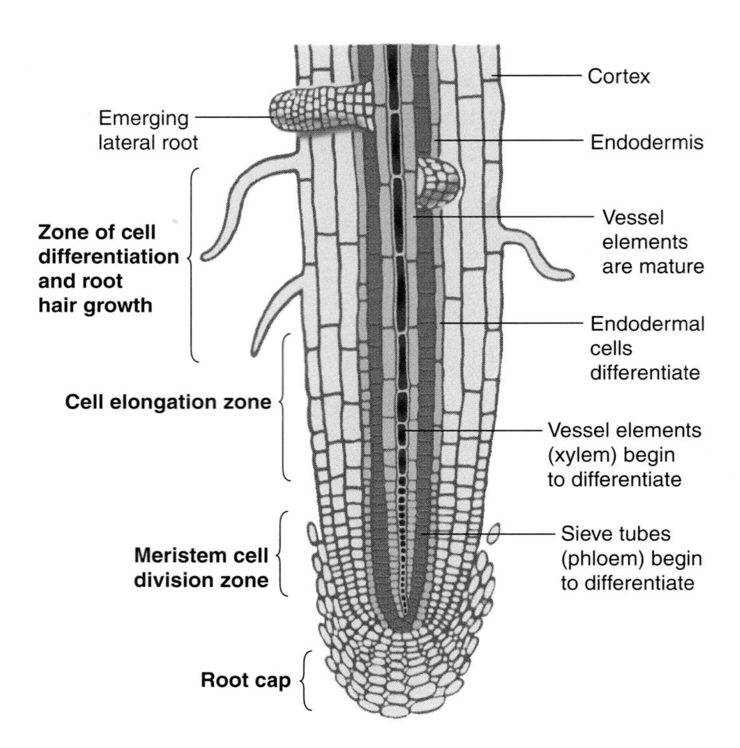

Figure 37.15

In longitudinal section through a root shows its cap and three zones of growth and cell differentiation.

Figure 37.16

The long, white root hairs are quite prominent on the young root of a radish.

days, or a few weeks at most, before it collapses and disappears. The older epidermal cells then develop a coating of cutin or suberin, and later grow bark, so they no longer take up nutrients. Many soil microorganisms will grow in the root tissue if they can, and these coverings on the older portions of the roots are protection against such invaders.

Root hairs are delicate structures that break off easily—one of the major problems in transplanting trees and other plants. Even though hairs that break off when the plant is lifted from the ground will be replaced in favorable conditions, it is important to protect the roots by keeping them surrounded with a good ball of soil. If this is done carefully, many of the hairs will not be broken at all.

37.9 Stems grow from an apical meristem much as roots do.

Shortly after an embryonic root appears, an embryonic shoot emerges from the germinating seed and grows upward to become the above-ground structures of the mature plant. The shoot grows very much as a root does. At its tip is a shoot apical meristem with its young, rapidly dividing cells pushing upward and leaving mature cells behind. The older cells gradually stop dividing; for a time they elongate and then differentiate, very much as root cells do, but the phases and zones are not quite as sharply marked.

At intervals, the upwardly growing shoot forms groups of meristematic cells (protoderm and procambium) at points called **nodes.** The lengths of stem between the nodes are **internodes.** Nodes are the sources of lateral outgrowths such as leaves, and in a mature plant the nodes will simply be identified as places where leaves are attached, the internodes as the spaces between. The patterns of nodes and internodes characterize different plant types. In some species, such as dandelions, the stem is very short, and leaves grow at the ground level, producing a rosette form, whereas the stems of species such as beans and sunflowers rise a considerable distance without sprouting leaves.

Leaves develop from meristematic tissues known as **leaf primordia,** which typically grow so fast that they outdistance the apical meristem and reach over it (Figure 37.17). Among the differentiating tissues in the young stem, long, dark strands of *provascular tissue* develop, leading from the stem up into the leaf primordia, and these gradually mature into xylem and phloem.

The stem grows longer through cell division at the apical meristem and elongation in the internodes. As the stem gets heavier, it develops collenchyma tissue, with its thickened cell walls, just inside the epidermis for support. Since its mechanical strength lies in thickened primary walls, collenchyma can continue to elongate and thus provide strength during growth. Eventually, as a region of the stem reaches its final length, fibers begin to differentiate in xylem and phloem and sometimes in the cortex; these tough, supportive elements cannot grow beyond a set length, and if they were to develop too early, they would stop the elongation of the stem.

Axillary buds that include dormant meristematic tissues grow between each leaf base and the stem. When a bud is activated, it grows into a side branch with an apical meristem that lays down new cells just as the main stem does. The growth of axillary buds is controlled by hormones, as we will see in Section 39.7.

Sidebar 37.1 Some Unusual, Specialized Stems

The stems of some plants, including a number of familiar species, take on unusual shapes and functions. *Rhizomes* are horizontal stems that grow at the ground surface, or just below it, in plants such as ginger and iris. They are frequently mistaken for roots, but they have visible nodes and internodes, and some have buds and leaf scales. Vertically growing *stolons* and horizontally growing *runners* are similar to rhizomes but have much longer internodes. The runners of strawberry plants spread along the ground and grow into new plants wherever buds develop. Some household plants develop runners that hang over the edge of the pot and root in a nearby pot if given the chance. In the Irish, or white, potato, stolons extend into the ground and accumulate carbohydrate deposits in their internodes. Eventually, the stolon dies, leaving the internode region swollen with starch; this is the potato that we harvest, and stems like this are called *tubers,* with nodes called "eyes."

Bulbs are really buds, and the major part of the bulb is made of leaves that contain stored food. A small stem grows at the lower end of the bulb, and the leaves surround it. (If you can stand to examine a longitudinal section of an onion, you can see this arrangement.) Adventitious roots grow out of the stem in such plants as English ivy and corn. *Corms,* such as in crocus and gladiolus, differ from bulbs in that they are mostly stem, with only a few thin leaves wrapped around the outside.

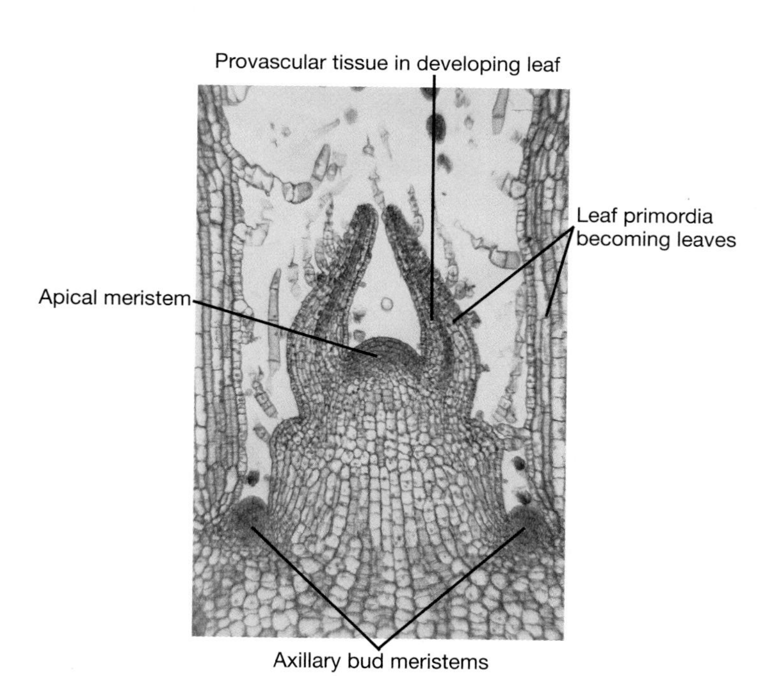

Figure 37.17

A section through the tip of a stem shows the apical meristem and the leaf primordia. The dark streaks of provascular tissue are procambium that is forming vascular bundles.

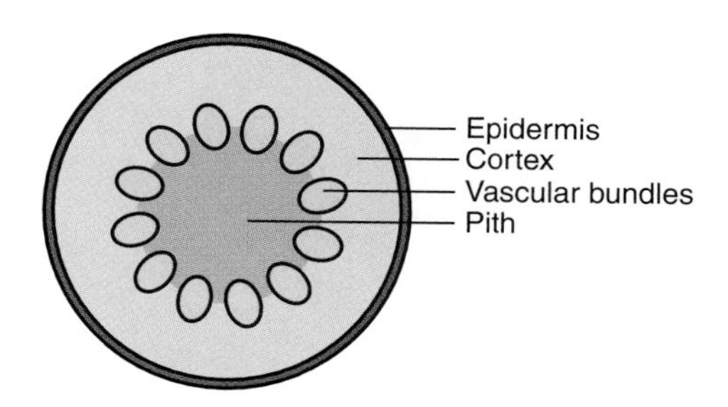

Figure 37.18

A cross section of a stem shows the general organization of tissues common to all vascular plants. It is primarily the arrangement of the vascular bundles that varies among major plant taxa.

37.10 The arrangement of vascular bundles gives the stem maximum strength.

A cross section through a stem (Figure 37.18) shows that it is made of concentric regions from the outside to the inside: epidermis, cortex, vascular bundles or cylinder, and pith, essentially the same as in a root. The only variation in this pattern, among all vascular plants, is in the arrangement of the vascular bundles. Sidebar 37.1 explores some unusual stems.

We discussed the epidermis layer in Section 37.4. The cortex, made primarily of parenchyma, may be quite a narrow region in some dicots, but in many herbaceous monocots, it comprises the bulk of the stem. Much of a stem is made of vascular tissues. Xylem and phloem are combined in **vascular bundles,** with the xylem of each bundle facing the center of the stem and the phloem facing outward. Since vascular bundles contain the strongest cells in the stem of an herbaceous plant, the plant's strength depends on their arrangement, just as a root derives its strength from a single, central vascular cylinder. Engineers have found two solutions to the problem of maximizing the strength of a support with a minimum of material, and the major taxa of seed plants just happen to have evolved precisely the same methods. The monocots employ the method used to strengthen concrete—their vascular bundles are dispersed through the stem like a series of separated steel rods (Figure 37.19*a*). The other engineering solution is to make the supporting material into a hollow tube, whose strength increases with its diameter. Gymnosperms and most dicots use this method, with their vascular bundles set in a circle (Figure 37.19*b*). In a dicot stem, the parenchyma ground tissue outside the circle is then known as *cortex* and that on the inside as *pith.*

The function of the vascular bundles, of course, is to serve the tissues of the stem and leaves, and eventually of branches arising from the stem. As the vascular bundles pass the nodes, subsidiary bundles called **leaf traces** leave them and feed into the leaves. When a plant grows branches, *branch traces* grow to serve them. The branching pattern is highly variable and sometimes rather complicated as axial bundles merge with one another or as a branch from one bundle leads into another; for

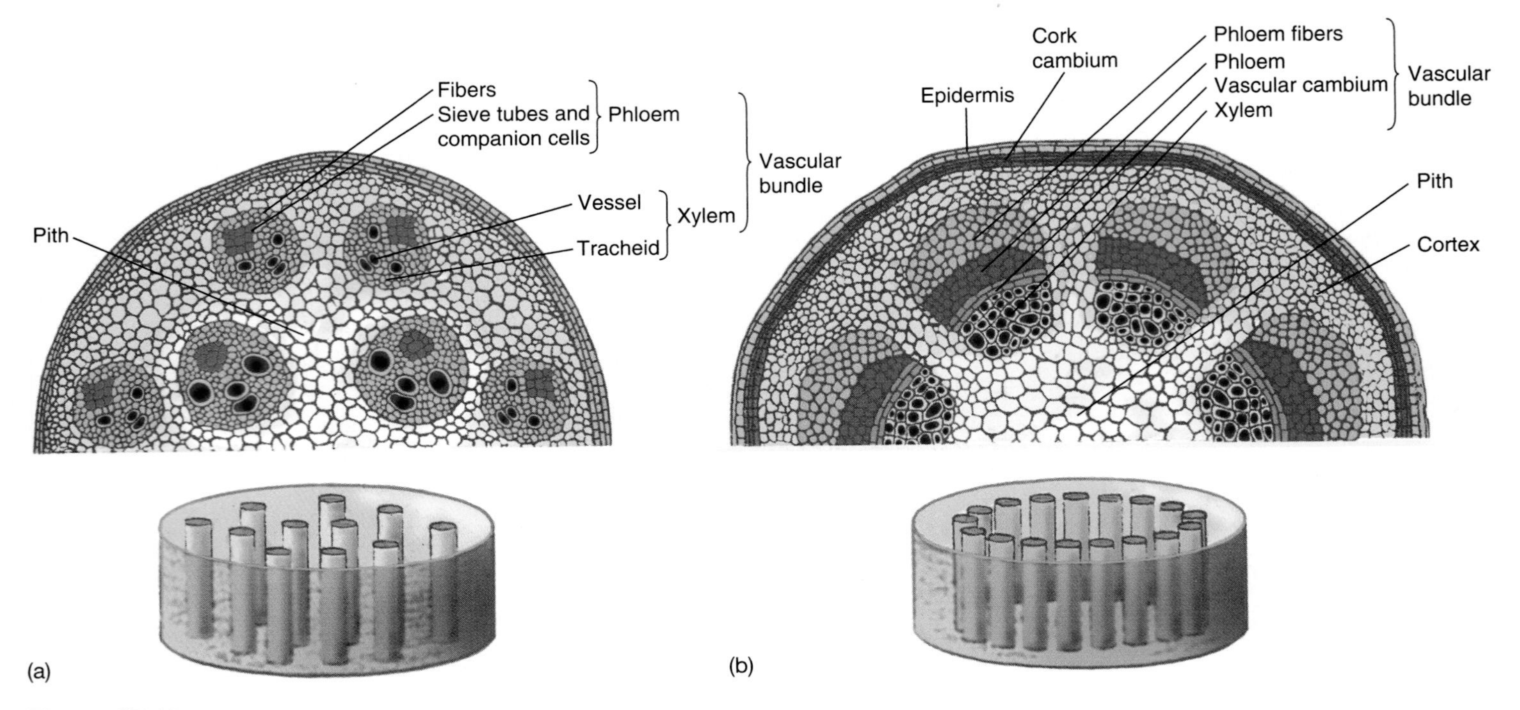

Figure 37.19

(a) A structure may be strengthened by embedding uniformly spaced strong rods, the way a monocot is strengthened with scattered vascular bundles. *(b)* A structure may also be strengthened with a cylinder, the wider the better, as the stem of a gymnosperm or dicot is strengthened with a ring of vascular bundles. As a woody stem grows, the bundles unite into a circle.

this introduction, the pattern shown in Figure 37.20 will serve as a good general model.

37.11 Leaves are built of parenchyma with a network of vascular tissue.

Leaves grow in an enormous range of shapes and sizes, from those of tiny water plants less than a millimeter wide to the huge Seychelles Island palm leaves that are six meters or more long. They may be needlelike, as in pines, or flat and circular, as in water lilies (Figure 37.21). Leaf characteristics are among the most important features used in identifying plants. The most obvious features are the shape of the **blade,** or **lamina,** the flattened part of the leaf, and the pattern of **veins,** which are vascular bundles.

Elongated conifer leaves, such as pine needles, have only one or two unbranched veins extending their whole length. Most monocots have a series of roughly parallel veins running from the base of the blade to its tip, whereas dicots have a complex branching or featherlike pattern in which smaller veins arise from larger ones. Veins converge at the base of the leaf, run through the **petiole,** the stalk that connects the leaf to the stem, and then join the leaf traces connecting to the stem's vascular bundles. The arrangement of leaves along the stem is also diagnostic in identifying plants. Leaves may branch off alternately along the stem, or spirally, or in opposite pairs. Some plants, such as ashes and hickories, have *compound leaves;* that is, the small blade that looks like a leaf is actually a *leaflet,* and the whole leaf is a symmetrical arrangement of leaflets along a central vascular strand. A true leaf can be identified by the axillary bud where its petiole joins the stem; when a leaf falls off in the process of abscission, it leaves a scar with a distinctive pattern of vascular bundles.

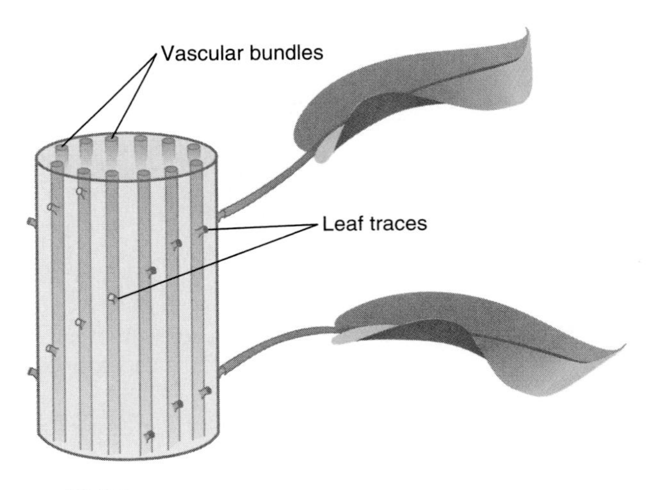

Figure 37.20

Vascular bundles run in the axial direction, all more or less parallel. They may connect with one another in complicated ways not shown here, and at the nodes of the stem some of them produce leaf traces that branch off to serve the leaves.

Figure 37.21

The leaves of seed plants take many forms. The veins that spread through the blade of the leaf and serve all its cells converge into a single vascular bundle running through the petiole into the stem. *(a)* The needles of fir are greatly reduced as an adaptation to cold, dry conditions. *(b)* Monocots such as lilies have parallel-veined leaves. Net-veined leaves may be *entire,* or made of a single blade without deep indentations as in elms *(c); lobed,* with indentations as in White Oak *(d);* or *compound,* divided into leaflets as in Butternut *(e).*

In spite of their different sizes and shapes, leaves all develop in much the same way from leaf primordia, and they all have surfaces of epidermal cells. Most of the leaf interior is **mesophyll,** a parenchyma tissue rich in chloroplasts where most photosynthesis occurs. Most dicot leaves are built with a *palisade mesophyll* next to the upper epidermis and a *spongy mesophyll* layer below (Figure 37.22). In leaves with both types of mesophyll, the chloroplasts are predominately in the palisade layer, where light from above strikes them most directly. Spongy mesophyll, as the name suggests, is an open tissue through which gases diffuse quite freely, giving palisade cells easy access to a supply of CO_2 and a route for water vapor and waste oxygen to diffuse away.

Veins branch and divide the leaf so completely that all mesophyll cells can exchange materials with some nearby vein by diffusion alone. The larger veins are often surrounded by a *bundle sheath* made of specialized parenchyma cells that control the flow of material into and out of the conducting vessels. The leaf mesophyll is thus connected, through a two-way system of xylem and phloem, with the rest of the plant.

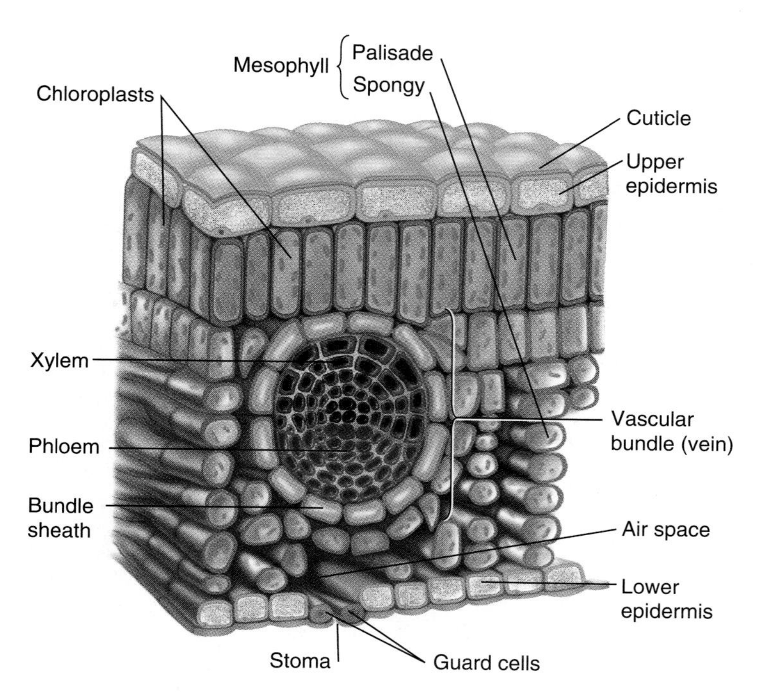

Figure 37.22

A typical leaf is bounded by epidermis above and below, protecting the mesophyll cells inside. Spongy mesophyll contains air spaces through which gases (CO_2, O_2, water vapor) can diffuse. These gases pass in and out through stomata, usually situated in the lower epidermis.

37.12 Gymnosperm and dicot stems and roots increase in diameter through secondary growth.

The lengthwise growth of stems and roots, originating in apical meristems, is *primary growth,* and the original xylem and phloem produced by differentiation of the procambium are also designated *primary.* In all gymnosperms and many dicots, a layer of permanently meristematic *vascular cambium* remains between the primary xylem and phloem. These plants grow into woody trees and shrubs as both their stems and roots enlarge in *secondary growth* through division of the vascular cambium (Figure 37.23). This mode of growth first appeared in the progymnosperms nearly 400 million years ago, and it has been the predominant mode of growth in their descendants ever since. Monocots and herbaceous dicots arose from clades that had secondary growth and then lost it.

We'll begin with the growth of stems. In stems capable of secondary growth, the vascular cambium divides in a tangential plane (Figure 37.24), increasing the stem's diameter by forming a *secondary xylem*—that is, wood—to the inside and a *secondary phloem* to the outside. At each division of a vascular cambium cell, one daughter cell remains a meristematic *initial,* and the other daughter cell becomes a xylem or phloem *mother cell.* These xylem and phloem cells continue to divide, elongate, and differentiate in much the same stages as in primary growth.

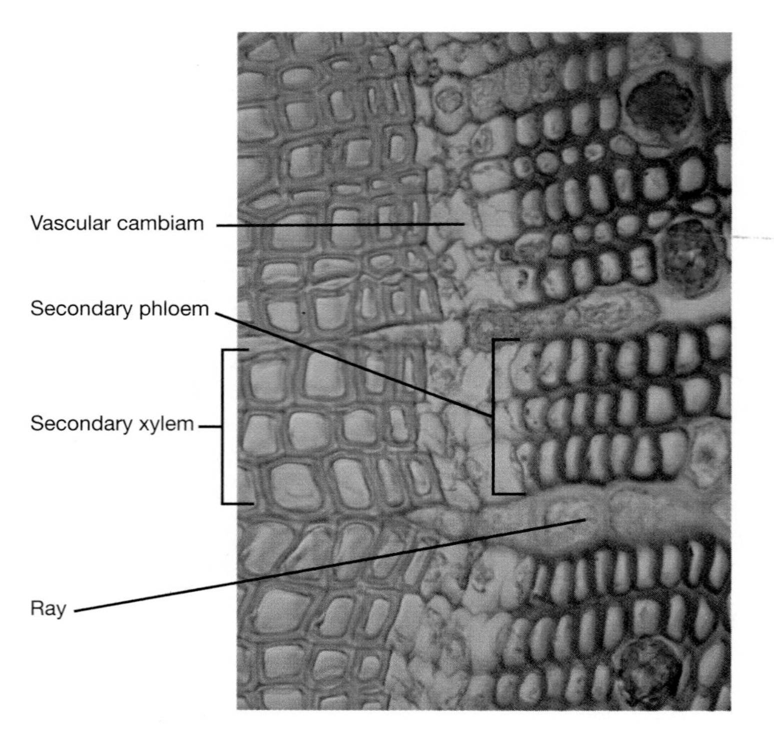

Figure 37.23

A meristem called vascular cambium lies between the xylem and phloem; it consists of a few layers of undifferentiated (meristematic) cells that can continue to grow and divide.

Both xylem and phloem remain a mixture of cell types. It should be clear, also, that as the diameter of the stem expands, xylem and phloem cells must grow so as to increase the circumference of the tissue as well. Some of the initials in the cambium therefore divide not only in the tangential direction but in the radial direction as well to create more initials:

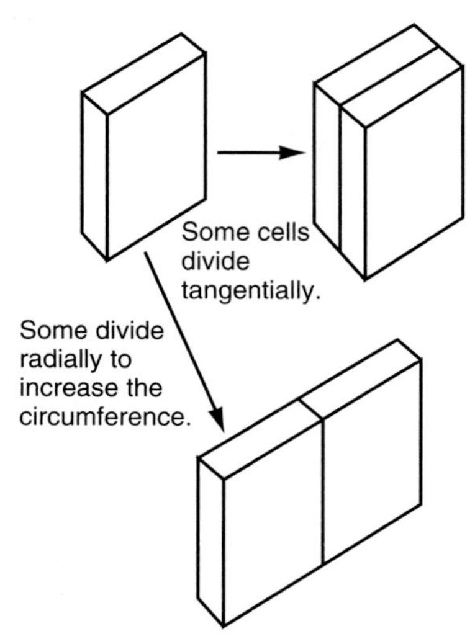

As secondary tissues grow, it becomes essential for the plant to conduct materials inward and outward in a radial direction, as well as in the axial direction. The vascular cambium therefore contains two types of initials. *Fusiform initials* elongate axially, and their daughter cells develop into the elongated elements of

Figure 37.24

When a vascular cambium cell divides, one daughter cell—called an initial—remains meristematic, and the other daughter cell starts to differentiate. Initials that divide toward the inside of the cambium produce xylem mother cells, and those that divide toward the outside produce phloem mother cells. These mother cells continue to divide to produce secondary xylem to the inside and secondary phloem to the outside.

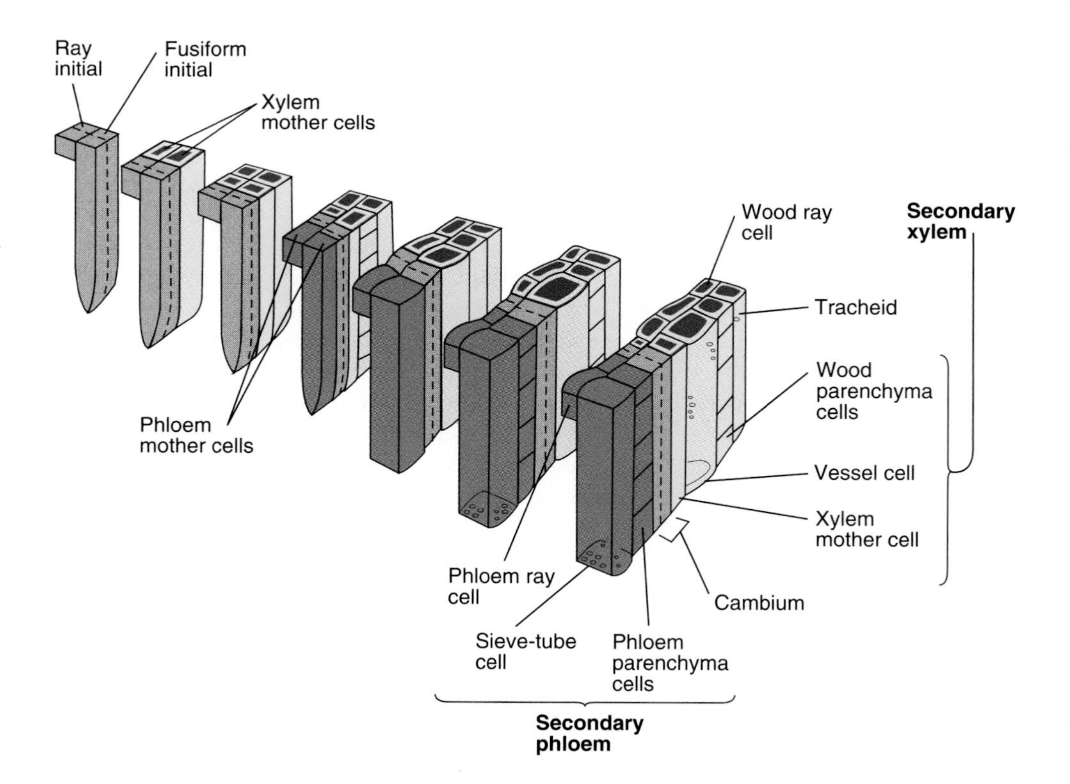

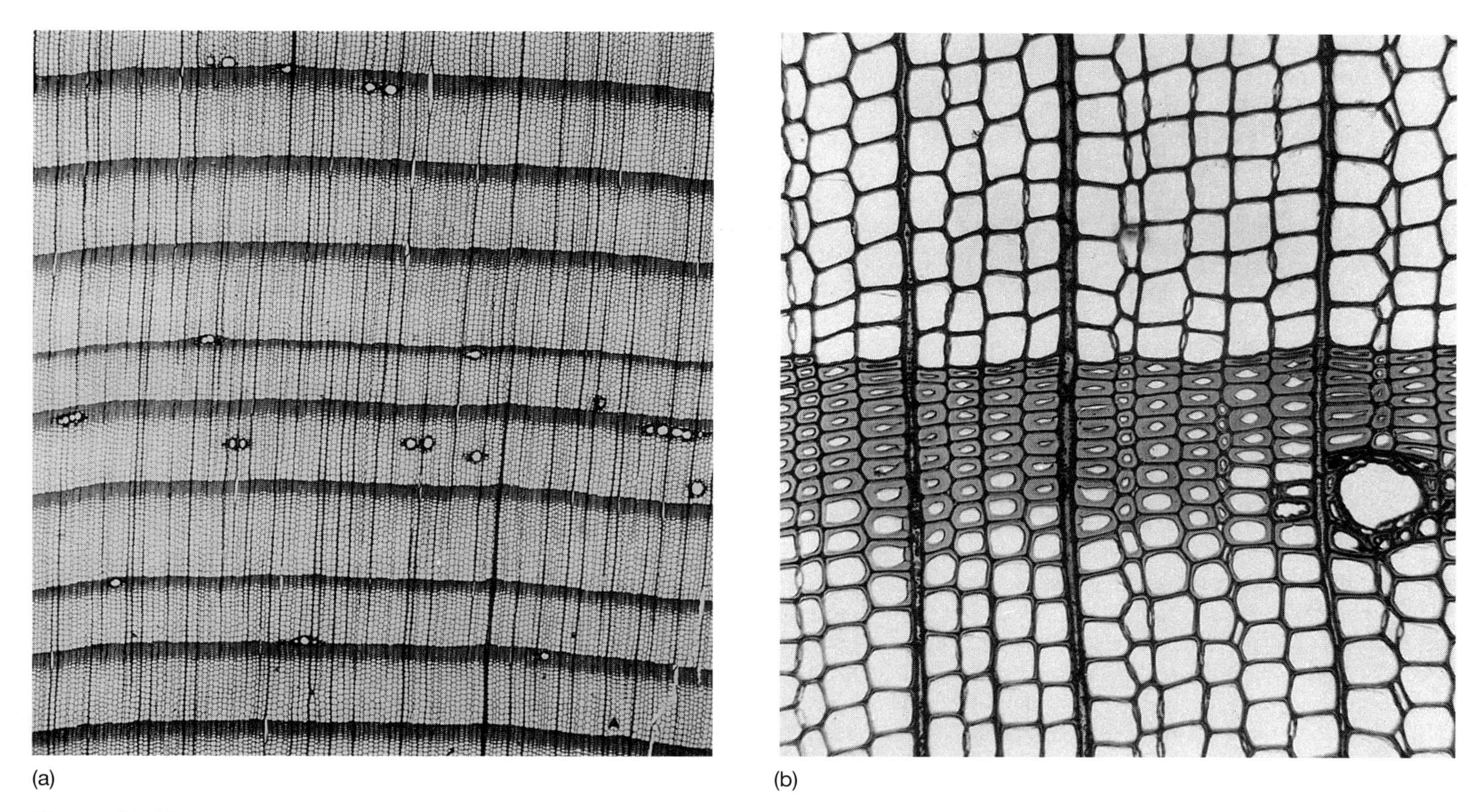

Figure 37.25

(a) Annual rings in wood reflect the temperature and availability of water during the growing season, so wide rings form during years with favorable conditions. Cells that grow late in the growing season are smaller and have heavy walls, making darker wood that marks the boundaries between rings. Differences in the thickness of cell walls are more obvious in an enlargement *(b).*

xylem and phloem: the tracheary and sieve elements, companion cells, and fibers. *Ray initials* develop into **rays** that run radially through secondary xylem and secondary phloem for radial transport. Most ray cells are parenchyma, and some store a lot of starch in amyloplasts; dicot rays in both xylem and phloem contain only parenchyma, but the xylem rays of some conifers contain a special type of tracheid. In this case, the axial tracheids will be in contact with one another and with radial tracheids at many points, and here there must be a precise alignment of the bordered pits to allow conduction in both directions.

The wood of gymnosperms and woody dicots is secondary xylem. In a temperate climate—as in most of North America—the cambium grows fastest in warm weather and nearly stops in winter. Each growing season leaves its own trace in the wood. Early growth in the warm, moist spring produces large xylem elements, but as the season progresses and water becomes scarcer, the new vessels become smaller; since they also tend to be dispersed among fibers and tracheids with thick walls, the wood produced later in the growing season is darker. This growth pattern leaves a series of *annual rings* in the wood (Figure 37.25), each representing a year's growth and recording the length and conditions of that growing season. In a cross section of an old tree, one can read back through history. The broad rings record good growing seasons, with warm weather, plenty of water, and little competition; narrow rings show less-than-ideal conditions. Archaeologists have used these records to date the remnants of wood found in ancient campsites, and ecologists can recreate the history of a forest from its trees.[1] However, some trees grow in such uniform conditions that they don't produce rings. The rings are more obscure in those growing near water, and some tropical trees, such as ebony, tend to grow in such constant conditions that their wood shows no rings.

The parenchyma and ray cells produced in each ring of new wood remain functional for some years, but eventually the oldest ones, at the center of the trunk, degenerate and become nonfunctional. Year by year, the older vessels and tracheids become inactive through cavitation—the breaking of the continuous water column in each tubule, as described in Section 38.3. This inner wood becomes *heartwood;* it turns dark from the deposition of dark-colored tannins and other materials, in contrast to the functional *sapwood* that continues to transport water. The

[1] For a beautifully written example, we highly recommend Aldo Leopold's essay "Good Oak" in his book *A Sand County Almanac,* Oxford University Press, 1949.

heartwood, of course, expands year by year. Although it is inactive, it continues to give the tree mechanical strength, and the materials deposited in it probably inhibit the growth of fungi and other microorganisms that could rot it and destroy the tree.

Secondary phloem grows at the same time as secondary xylem, and is rather simple in structure. Its rays are made only of parenchyma, and its axial elements are sieve elements and parenchyma. It contains storage cells that may be the principal sites for storage of carbohydrates, and it frequently bears all kinds of secretory cells. Here is where conifers produce the pitch that probably protects them against invading insects, and which humans harvest for turpentine and resin; here is also where rubber trees secrete the latex that is still harvested and made into natural rubber in some countries. But the fate of secondary phloem is much different from that of xylem. Generally, each layer of phloem functions for only one year, meaning that only the innermost layer is active. Instead of accumulating for the life of the plant, the older outer layers of phloem tend to be cracked, crushed, and sloughed off with the bark, as we will see next.

Secondary growth in woody roots is very similar to that in stems, and old roots, in cross section, can hardly be told from stems. The root vascular cambium also develops between xylem and phloem, initially following the indentations of the xylem cylinder. It then forms more secondary xylem in these indentations and soon fills them in to make a round vascular cylinder. A bark forms on the outside of the root, and in plants that grow for many years, the outer layers of bark are continually shed and replaced.

Exercise 37.5 The trunk of a tree has a vascular cambium 20 cm in diameter, and one of its branches has a cambium 1.0 cm in diameter. The trunk and branch both increase their diameters by 100 μm (0.1 mm). Calculate the increase in circumference ($C = \pi d$) in both cases. If a cambial cell is 20 μm wide, how many cells have to be added to each cambium to add the extra circumference? Which cambium has to divide more in the radial direction?

37.13 Bark is a protective layer formed by secondary growth.

Epidermis serves for protection in young plants and herbaceous plants, but in those that undergo secondary growth, it is replaced by thicker, tougher tissue called **periderm.** Periderm is a mixture of tissues formed when some cells of the epidermis, cortex, and secondary phloem secondarily acquire a renewed ability to divide and form a layer of **cork cambium,** or **phellogen.** This cambium divides outward to produce **cork,** or **phellem,** made of rather cubical cells that secrete the waxy polymer suberin as waterproofing and protection, and then die as they mature. The periderm is a combination of cork cambium and cork. The further combination of all the exterior tissues—periderm, secondary phloem, and bits of other primary tissues—forms **bark.**

Figure 37.26
The bark of White Birch makes it one of the most beautiful trees. It peels off in thin layers and has been used as paper. Its smooth texture makes its many thin lenticels quite obvious.

It takes only a little experience with trees, plus some judicious reflection on their functions, to realize the value of bark. Most obviously, it is mechanical protection. Old trees bear the scars of damage in their bark that might have been fatal to a less protected plant; given the ecological importance of fire, we can understand bark as a protective, insulating tissue that keeps inner tissues from cooking as a fire races through a forest. Many trees shed their bark continually, thus also shedding various pathogens or animals that might be growing there. On the other hand, bark keeps tissues just interior to it from obtaining CO_2 and eliminating O_2, so the periderm develops *lenticels,* patches of tissue with extensive intercellular gas spaces that allow this gas exchange to occur (Figure 37.26).

The first bark to form on a tree is made of the initial phellogen and various primary tissues, so it may look quite distinctive compared with later bark (Figure 37.27). The initial phellogen is generally replaced by another, more interior phellogen. We noted above that in the secondary phloem just interior to the outermost layers, only the newest innermost layer is functional. As the entire stem is growing larger with the formation of more xylem, the outermost layers of tissue are stretched, split, and crushed, and so in most plants each phellogen is replaced by another, more interior phellogen. The plant continues to grow new secondary phloems and new phellogens at more or less the same rate, and these tissues constitute an inner bark, which must be kept since it contains the only functional phloem. Tissues exterior to the phellogen become cut off from the vascular tissues inside, lose their ability to metabolize, and die; these tissues form an outer bark, which may continue to slough off. The outer bark can even be removed if the inner bark remains intact; cork oak trees can serve as a continuing source of cork because the outer bark is allowed to grow thick and then is carefully harvested for the manufacture of bottle stoppers, floor coverings, and other products.

Figure 37.27

The first bark of a tree may be quite different from that produced later. Pacific Madrone trees have a beautiful red bark, which eventually flakes off, in contrast to the inner yellowish bark.

Coda The first vascular plants evolved sometime in the Devonian Period around 400 million years ago. Consider the Devonian scene. Aquatic ecosystems were already rich and varied, but just beyond them lay a vast potential adaptive zone, the untouched space on land, if only . . . if only one could get around the water problem. The transition was accomplished by rather complex filamentous algae like modern charophytes, as discussed in Chapter 33, that gradually modified their simple cellular structures. They illustrate the plasticity and potential of cellular structure and the inventiveness of evolution. The word "inventiveness" may be surprising, but it is not unreasonable. We keep emphasizing that organisms have structures designed by a long evolution, and a good design should be inventive, even if it is done by an abstract force. The fossil record tells us essentially nothing about the details of the process, only the results. Some plants invaded the land in the form of bryophytes, the mosses and their relatives, which evolved some elongated cells that function as water-conducting elements; but mosses still are confined to moist areas close to the ground. Other plants developed elongated and strengthened cells that function as real vascular tissues, which not only became solutions to the water problem but to the problem of support as well. Now, 400 million years later, the result is a terrestrial flora that can support itself in some of the most challenging and inhospitable environments, and supports most of the animals as well. All honor to evolution! What a shame there are no prizes for the abstract forces of nature.

Summary

1. Plant cells have some distinctive features, including a cell wall made mostly of cellulose and, usually, a large central vacuole. Cell walls serve for support and strength. The vacuole builds up internal pressure, which forces a growing cell to elongate.
2. Plants are made of three basic tissue types: parenchyma, which retains its cytoplasm and has thin walls; collenchyma, which has cytoplasm and thickened walls; and sclerenchyma, which usually has no cytoplasm and develops secondary wall structures.
3. Xylem conducts water and dissolved minerals through tubules made of numerous tracheary elements of two kinds—tracheids and vessel elements—adjoined end-to-end. Phloem is distinguished by sieve cells or sieve tubes, made of sieve elements, which retain their cytoplasm and conduct streams of liquid through sieve plates where they join. Both xylem and phloem also contain supporting fibers and parenchyma.
4. Epidermis protects a plant against harmful radiation, invading microorganisms, and predators. It absorbs water in roots and retards evaporation in shoots. It is generally a single layer of tightly interlocking cells, interrupted by stomata.
5. Seeds contain embryo plants that break out of their dormant state when rehydrated.
6. Flowering plants are either monocots, whose embryos have a single cotyledon, or dicots, whose embryos have two cotyledons. These plants also differ in root structure, leaf venation, and arrangement of xylem and phloem elements in the stems.
7. A vascular plant is fundamentally a cylinder that grows along an axis between two growth centers, the root and shoot apical meristems.
8. Roots are built with a central column of tissue termed the stele, composed of xylem, phloem, and pith (in monocots), and bounded by a pericycle, the site of lateral root formation. The tissue outside the stele is cortex; its inner layer, the endodermis, controls the movement of water into the stele.
9. Roots grow at their tips. Root cells first divide, then elongate to perhaps ten times their original length, and finally differentiate into specific cell types. A cap over the root tip protects the growing root. Root hairs provide an enormous surface for absorbing water from the soil.
10. Stems, like roots, grow from apical meristems, and their cells also go through periods of elongation and differentiation. Lateral growth occurs at nodes that are separated by internodes.
11. Vascular bundles of xylem and phloem are arranged so as to give the stem maximum strength. In monocots, the bundles are scattered irregularly throughout the stem. In gymnosperms and dicots, the bundles are arranged in a circle, and they fuse into a ring as the stem grows wider. The vascular bundles produce leaf and branch traces, which feed organs growing to the side of the stem.
12. Leaves are built with outer layers of epidermis, coated with suberin and wax to reduce evaporation; between these are layers of mesophyll, typically (in dicots) a palisade and a spongy mesophyll, where most photosynthesis occurs. A network of vascular tissue supplies the mesophyll with nutrients.

13. Gymnosperms and woody dicot stems and roots increase in diameter through secondary growth. Between the xylem and phloem in each vascular bundle is a cambium layer made of meristematic tissue that continues to divide, thus producing secondary xylem (wood) and secondary phloem. Wood commonly grows in a pattern that produces annual rings. Only the newest layer of secondary phloem is generally active. Both xylem and phloem contain rays for conduction radially through the stem.
14. At the surface of the stem during secondary growth, epidermis is replaced by periderm, which consists of a cork cambium (phellogen) plus the cork (phellem) it produces. Bark is made of periderm plus the underlying secondary phloem. A tree continues to produce new secondary phloem and new phellogens.

Key Terms

primary cell wall 787
secondary cell wall 788
lignin 788
plastid 788
tracheary element 790
tracheid 790
vessel 790
vessel element 790
sieve element 790
sieve-tube member 790
sieve cell 790
sieve tube 790
sieve plate 790
companion cell 791
epidermis 791
dormancy 792
germination 792
cotyledon 793
monocot 793
dicot 793
shoot apical meristem 794
root apical meristem 794
cambium 794
axial 794
stele 794
pith 795
pericycle 795
cortex 795
endodermis 795
root cap 795
node 796
internode 796
leaf primordium 796
axillary bud 796
vascular bundle 797
leaf trace 797
blade 798
lamina 798
vein 798
petiole 798
mesophyll 799
ray 801
periderm 802
cork cambium/phellogen 802
cork/phellem 802
bark 802

Multiple-Choice Questions

1. All of the following are true about the primary cell wall *except*
   a. produced by vectorial metabolism
   b. primarily made of pectin and lignin
   c. porous in composition
   d. peripheral to the plasma membrane
   e. contains microfibrils of cellulose
2. Which is correct with respect to the tonoplast?
   a. can be a factor contributing to growth of the cell
   b. acts in a similar way to the Golgi body in animal cells
   c. important to photosynthesis because it regulates ionic balance in the chloroplasts
   d. surrounds plasma membrane and forms plasmodesmata
   e. synthesizes lignin and deposits it within the cell wall
3. Which of the following pairs is mismatched?
   a. tracheids—vessel elements
   b. sieve-tube members—companion cells
   c. parenchyma—secondary cell walls
   d. sclerenchyma—lignified cell walls
   e. parenchyma—chloroplasts
4. Shoot epidermis protects the underlying tissue from all of the following *except*
   a. evaporative water loss.
   b. lignification of cell walls.
   c. excessive salt accumulation.
   d. invasion by parasitic microorganisms.
   e. overheating.
5. Which one of the following is usually *not* important for germination?
   a. changes in light levels
   b. temperature changes
   c. increase in concentration of seed-eating predators
   d. rehydration of tissues within the seed
   e. imbibition of water
6. Which is the first part of the root to emerge in a newly germinated plant?
   a. epicotyl
   b. hypocotyl
   c. cotyledon
   d. radicle
   e. cambium
7. Which cells and/or tissues are found within the root?
   a. protoderm and ground meristem
   b. apical meristem and ground meristem
   c. pith and stele
   d. endodermis and pericycle
   e. all of the above
8. Bark includes
   a. primary xylem and primary phloem.
   b. secondary xylem and secondary phloem.
   c. secondary phloem, cork cambium, and periderm.
   d. vascular cambium, cork cambium, and pith.
   e. stele and endodermis.
9. All but _____ correctly matches a region of cell division with its descendent tissue.
   a. vascular cambium—periderm
   b. protoderm—epidermis
   c. ground meristem—cortex
   d. procambium—primary xylem
   e. cork cambium—cork
10. The most photosynthetically active cells are found in the
   a. spongy mesophyll.
   b. palisade mesophyll.
   c. periderm.
   d. epidermis.
   e. endodermis.

True-False Questions

Mark each statement true or false, and if false, restate it to make it true.

1. The enzymes that catalyze the conversion of activated glucose to cellulose are found within the primary cell wall.
2. The tonoplast surrounds the cytosol and regulates its ion concentrations.
3. Water passes from tracheid to tracheid through perforated end-walls.
4. Shoot epidermis is a waterproofing agent, but root epidermis assists in entry of water into the plant.
5. Periods of germination usually culminate in dormancy.
6. Some dicots are woody, and all woody plants are dicots.
7. Cells that have arisen by mitosis of root cap cells first enter the zone of elongation and then the zone of differentiation.

8. Root hairs are extensions of the pericycle that form in greatest number within the zone of elongation.
9. Vascular cambium gives rise to elements that will be incorporated into both the bark and the wood of a tree.
10. The vascular bundles of mature woody stems are scattered throughout the cortex, thereby strengthening the trunk.

Concept Questions

1. Explain the major differences between primary meristem, vascular cambium, and cork cambium.
2. How do taproots, fibrous roots, and adventitious roots differ?
3. Contrast the fate and function of secondary xylem with that of secondary phloem in the trunk of a woody dicot.
4. In a woody stem, which tissues derive from the primary apical meristems and which derive from secondary meristems?
5. What is meant by the term *initials?* Where are they located and what is their function?

Additional Reading

Carlquist, Sherwin J. *Comparative Plant Anatomy.* Holt, Rinehart and Winston, New York, 1961.

Duddington, C. L. *Evolution and Design in the Plant Kingdom.* Thomas Y. Crowell Co., New York, 1974.

Esau, Katherine. *Anatomy of Seed Plants.* John Wiley and Sons, New York, 1960.

Lee, Addison E., and Charles Heimsch. *Development and Structure of Plants, A Photographic Study.* Holt, Rinehart and Winston, New York, 1962.

Mandoli, Dina F., and Winslow R. Briggs. "Fiber Optics in Plants." *Scientific American,* August 1984, p. 90. The tissues of plant seedlings can guide light through distances as great as several centimeters, and plant cells may exploit "light pipes" to coordinate aspects of their physiology.

Mauseth, James D. *Plant Anatomy.* Benjamin/Cummings Publishing Co., Menlo Park (CA), 1988.

Metcalfe, Charles R., and L. Chalk. *Anatomy of the Dicotyledons: Leaves, Stem, and Wood in Relation to Taxonomy, with Notes on Economic Uses.* Oxford University Press, London and New York, 1957.

Internet Resource

To further explore the content of this chapter, log on to the web site at:

http://www.mhhe.com/biosci/genbio/guttman/

38 Translocation and Water Relations of Plants

Morning glories open their flowers in the morning and close them again in the evening.

Key Concepts

A. Transport Through Plants

38.1 Stomata regulate the flow of gases through the leaf.

38.2 Water and ions flow from the root epidermis into the xylem.

38.3 Several forces combine to move water into the shoot system.

38.4 Phloem sap moves by a combination of osmotic forces and specific pumps.

38.5 Many plant organs can move quickly through water exchanges in motor cells.

B. Water and Light Relationships of Plants

38.6 Plants have ways to balance photosynthesis, transpiration, and translocation for optimal activity.

38.7 Plants also tend to maintain their temperatures within optimal ranges.

38.8 Plants are adapted for growth in different light intensities.

38.9 Many plants have special adaptations for water shortage.

38.10 Xeromorphic characters are used for adaptation to high-salt conditions.

At dawn on what promises to be a fine spring day, we walk into the garden. All around us are the plants we have been nurturing so carefully. They sit still, quiet, apparently doing nothing. Only an occasional breeze makes the flower heads and branches bow and rise again. No other movement is to be seen, except for an occasional insect. But we cannot see the furious commotion inside all the green stems and brown branches around us. Though the early morning air is still cool and the light soft, every plant is already transporting water and minerals up from its roots through xylem tubules; by the end of a warm day, a plant may have carried ten to one hundred times its own weight in water into the atmosphere. As the sunlight intensifies, each plant manufactures organic compounds in its leaves and exports them through its phloem. Here is a patch of yellow hypericum, its flowers just emerging; a time-lapse film would show how its flower buds swell, burst, and spread their petals. But over here is a trellis of morning glories, and if we watch for only a few minutes we will actually see the deep purple flowers open themselves to the sun. Look at how the morning glory tendrils have wound themselves around the cedar bars, holding on like little springs. If insects begin to chew on the leaves of one of these plants, it will react quickly by starting to make noxious proteins to poison the invader. And throughout the day, each plant will balance the complex flows of water, oxygen, and carbon dioxide through its tissues, perhaps drooping or raising its leaves slightly to maintain an optimal temperature. A plant may not make as great a show of its enterprises as an animal does, but it is a quiet dynamo of activity. We will examine those activities later in this chapter. ■

A. Transport through plants

38.1 Stomata regulate the flow of gases through the leaf.

To understand most of a plant's activities, we have to keep in mind the general processes of photosynthesis and respiration. In the light, cells that contain chloroplasts carry out the photosynthetic process:

$$6\ CO_2 + 6\ H_2O \rightarrow 6\ O_2 + \underset{\text{glucose}}{C_6H_{12}O_6}$$

At the same time, all cells are engaged in respiration:

$$6\ O_2 + C_6H_{12}O_6 \rightarrow 6\ CO_2 + 6\ H_2O$$

These simple equations imply at least three essential plant activities:

- If photosynthesis is to be productive, a plant must maximize its intake of CO_2 and water.
- Sugar produced by the photosynthetic portions must be transported throughout the plant and perhaps stored.
- A plant needs a continuous flow of oxygen to support respiration.

Notice that we are talking about the transport of three gases—oxygen, CO_2, and water vapor, since the transport of liquid water is closely tied to the movements of its gaseous form. Gases move by diffusion. They enter and leave a vascular plant by diffusing through the **stomata** (sing., **stoma;** sometimes *stomate/stomates*) in the epidermis of its shoot system, mostly in the leaves (Figure 38.1). "Stoma" is sometimes used to mean only the space between guard cells and sometimes to include the guard cells themselves.

The epidermis of leaves and stems is a single layer of surface cells that secrete a protective cuticle of cutin and waxes. Stomata occur between cells and are usually concentrated on the lower surfaces of land plant leaves and in the upper epidermis of grasses and the leaves of floating water plants such as water lilies. The concentration of stomata in the leaves of different species, ranging from a few hundred to over a million per square centimeter, evolved as a long-term, adaptive response to the general availability of water in the environments they grow in. These minute openings in the leaf surface are surprisingly effective in promoting diffusion of gases in and out of the leaf. Diffusion from a surface is most rapid if the surface is not covered at all, but no plant could be constructed without

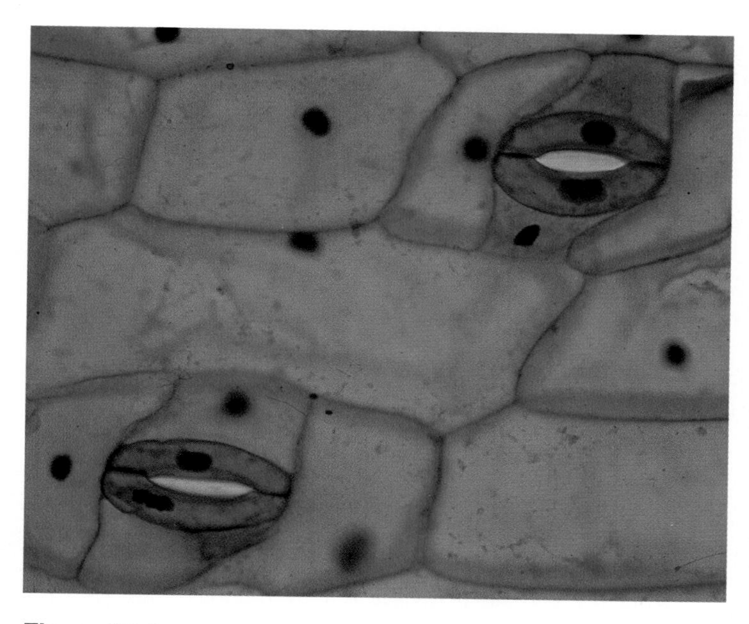

Figure 38.1
Stomata are formed by a pair of guard cells, which open or close the pore as they change shape.

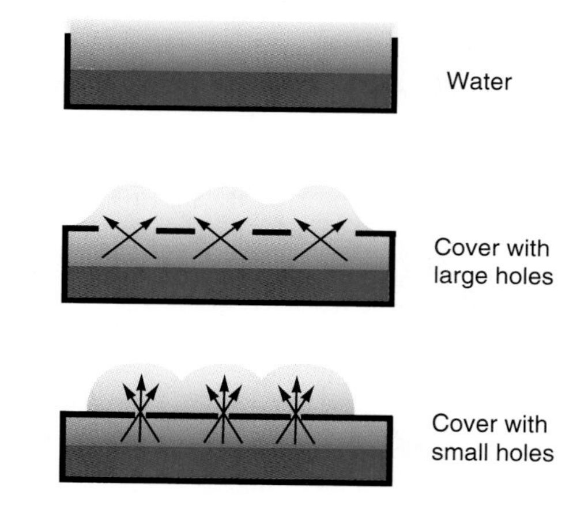

Figure 38.2
The diffusion of water vapor from an open tray *(top)* was compared to diffusion through a foil cover with either large or small holes. The density of color shows the concentration of water vapor above the surface, and the arrows indicate the paths that molecules take. Water vapor diffuses through large holes much as it does from an open tray, but small holes produce steep hemispherical concentration gradients, leading to much faster diffusion.

epidermis. A classical experiment to demonstrate the utility of small stomata compared diffusion through large and small openings (Figure 38.2) and showed that widely spaced small openings allow much faster diffusion. So the evolution of stomata again demonstrates how well natural selection is able to find optimal solutions to physical problems, given enough time.

Each stoma is formed by a pair of guard cells that can change shape to open or close the pore as their water content

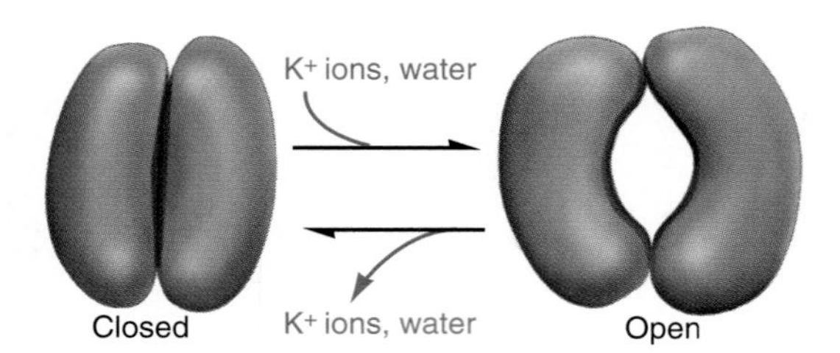

Figure 38.3

Stomata open and close when the guard cells around them either shrink or swell due to changes in their water content. The flux of water is controlled by changes in the K^+-ion concentration inside the cells.

changes. When the guard cells swell with water, the stoma opens; when they lose water, the stoma closes. Guard cells swell and shrink with the entry and loss of potassium (K^+) ions, illustrating the general principle that water movement is controlled by moving ions. When K^+ ions enter the cells, water follows quickly and the stoma opens. When the cells lose K^+ ions, water leaves and the stoma closes (Figure 38.3). Thus regulating the stomatal opening depends on controlling the influx and efflux of K^+ ions. This control process is not well understood yet, but K^+ flow is clearly related to the concentration of CO_2, which has a central role in leaf physiology.

Photosynthesis and respiration both go on in the interior of a leaf. Although both processes involve O_2 as well as CO_2, O_2 is much more abundant in the atmosphere than is CO_2, so the movement of CO_2 is much more critical. To maintain photosynthesis at a high level, a plant must be able to regulate the apertures of its stomata, the only portals where CO_2 can enter. If you could program the guard cells of a stoma like a computer to maximize photosynthesis, you would make the stoma close when CO_2 is abundant in the leaf and open to admit more whenever the CO_2 level falls. In fact, guard cells do seem to be sensitive to the CO_2 concentration inside the leaf. When light initiates photosynthesis, the internal CO_2 concentration falls, and K^+ ions that had been rather widely distributed through the leaf epidermis move into the guard cells. This causes the guard cells to absorb water and swell, so the stomata open and more CO_2 can enter. In the dark, when CO_2 is being produced by respiration but not consumed by photosynthesis, K^+ leaves the guard cells and the stomata close. Even in the dark, administering CO_2 to a leaf causes stomatal closure, while CO_2-free air produces stomatal opening.

The demand for CO_2 and oxygen must be balanced against **transpiration,** the enormous evaporative loss of water through the stomata. It is estimated that through transpiration some plants lose well over 90 percent of the water that enters their roots. The stomata therefore also respond to changes in the availability of water and are able to minimize the transpirational loss by closing when water is scarce. In Section 39.9, we show that a plant hormone, abscissic acid, also regulates stomatal opening to conserve water during times of stress. Furthermore, some plants have an endogenous stomatal rhythm: The stomata open and close on a regular daily cycle, even when all conditions, including water content, remain the same. The complications of balancing these factors are discussed in Section 38.6.

38.2 Water and ions flow from the root epidermis into the xylem.

The flow of water (and dissolved minerals) from the roots through the xylem supports transpiration. To supply all of a plant's tissues, the xylem must overcome the force of gravity and transport water to the topmost leaves, which may mean to the remarkable heights of some trees. The journey begins in the root, where water enters from the soil.

The epidermal and cortical cells of the root are joined into one unit, the symplast, through their plasmodesmata; the cell walls and intercellular spaces of the root cells constitute its apoplast. Water moves from the soil (called the soil solution) through the root cortex via both the apoplast and symplast (Figure 38.4). Water diffuses into the symplast through the epidermal cells, including the root hairs, as long as it is more concentrated in the soil solution than in the epidermal cytosol; this is probably always the case because root cells actively concentrate K^+ ions to a level several thousand times higher inside than outside. Once water and ions have crossed the membranes into the symplast, plasmodesmata provide conduits through the endodermis (around the vascular cylinder), so materials in the symplast have no additional cell membranes to cross.

Plasmodesmata, Section 6.9.
Symplast structure, Section 32.8.

The apoplast conducts water by **capillarity,** an effect seen in thin tubes (Figure 38.5). Because water molecules adhere to the wall of a tube by forming weak bonds with it, they tend to

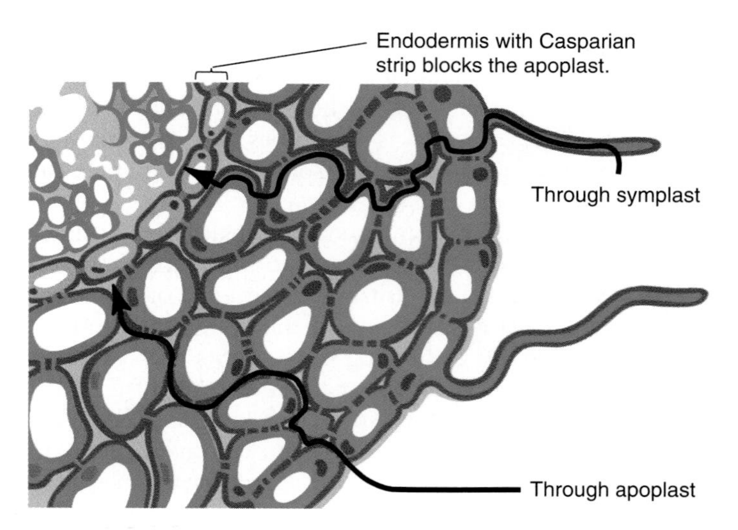

Figure 38.4

Water is transported partly through the apoplast and partly through the symplast of a plant.

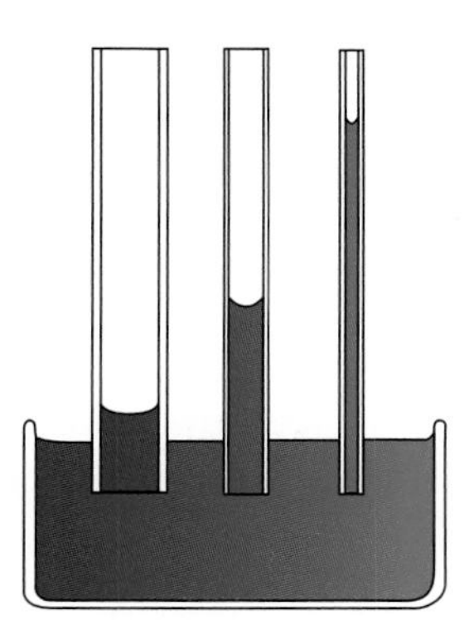

Figure 38.5

Water rises into thin tubes by capillarity, due to its adhesion to the walls of the tubes and strong cohesion between water molecules. The thinner the tube, the higher the water is drawn.

climb the wall slightly. Since they also cohere to one another through hydrogen bonds, those molecules that move up the wall pull others behind them, making a curved surface (meniscus). The thinner the tube, the higher the water can rise against gravity. Water seeps into the apoplast by capillarity among the fibers of cellulose and other cell-wall components, just as it does into a blotter or paper towel, infiltrating the thin channels between wood fibers. As it flows along, the water in the apoplast carries ions. However, the **Casparian strip,** a layer of waterproof suberin, completely blocks the intercellular spaces of the endodermis (Figure 38.6). This strip divides the root's apoplast into two regions, one outside the endodermis in the epidermis and cortex and one inside in the vascular cylinder. With the apoplastic route blocked, water and ions can only enter the vascular cylinder through the cytoplasm of endodermal cells themselves, so these cells control the entire flow of water and ions between the cortex and xylem. Endodermal cells actively transport K^+ ions into the vascular cylinder, and water accompanies them by osmosis (Figure 38.7). It is primarily here that essential ions are brought into the plant. The endodermis blocks the free diffusion of materials *out of* the vascular cylinder as well as *into* it; once transported across the endodermis and into the vascular cylinder, water and ions may diffuse out of the symplast, but they are trapped within the endodermis, and the xylem can transport them into the rest of the plant.

Because of the active transport of ions through the endodermis, an osmotic pressure known as **root pressure** builds up around the xylem, forcing water into the xylem vessels. You can demonstrate root pressure in a classical laboratory experiment by cutting the stem off a plant and fastening a glass tube to the portion remaining on the roots. Fluid will rise into the tube, sometimes to a height of a meter or so. Root pressures appear in most plants if there is adequate moisture in the soil and the humidity is high so that transpiration is low. They do not occur in conifers. Root pressures develop particularly at night, and early in the morning we sometimes see a resulting phenomenon called **guttation** on plants, especially grasses, where droplets of water have accumulated along the tips or edges of the leaves overnight.

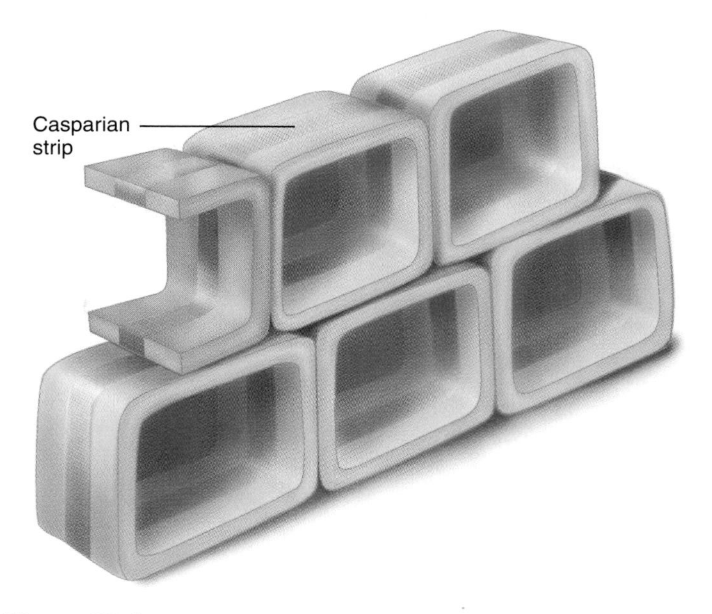

Figure 38.6

The Casparian strip consists of suberin that prevents water from moving around the endodermal cells in the root apoplast. Thus water must move into the vascular cylinder through the endodermal cells, which therefore control its rate of entry.

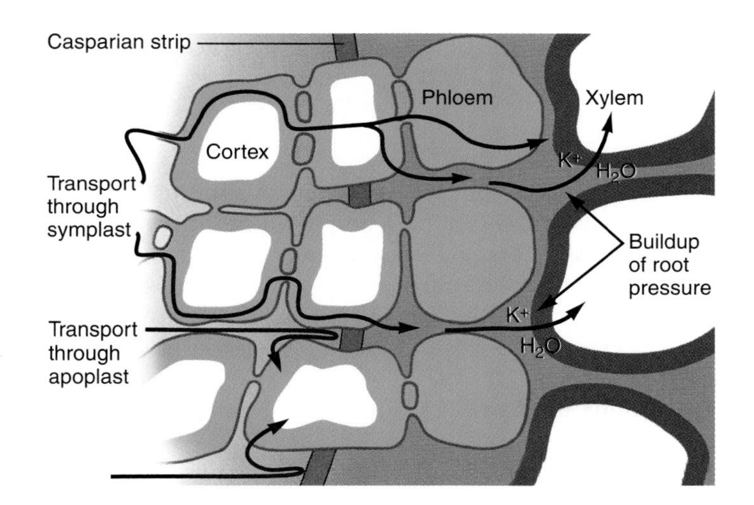

Figure 38.7

Roots exert a pressure on the xylem because endodermal cells transport K^+ and other ions into the region around the vascular cylinder, and water follows osmotically.

38.3 Several forces combine to move water into the shoot system.

Water and ions in the xylem must obviously rise the full height of the plant, which may be over 100 meters in some trees. Air pressure at sea level will only support a column of water about 10 meters high; root pressure alone might raise a column enough to supply the needs of short plants, but it is inadequate for transport in tall trees. How, then, is this accomplished?

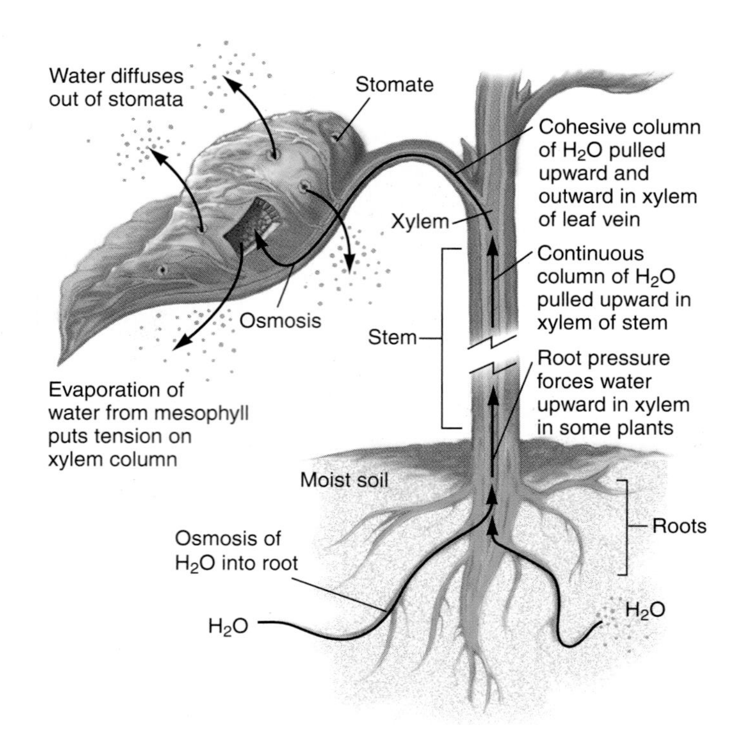

Figure 38.8

There are continuous columns of water in each xylem tube from the roots to the mesophyll layers of leaves, where the water is drawn off by transpiration. Due to cohesion of water, the whole column is drawn upward to replace molecules that are removed.

Water is raised to great heights in plants through a combination of capillarity, cohesion, and a pull on the xylem fluid from above. As long as the stomata are open, leaves continually transpire water, which moves from the mesophyll cells in the middle layer of a leaf into the air spaces and out through the stomata (Figure 38.8). As molecules evaporate from the most exposed mesophyll, they are replaced by the movement of water from one mesophyll cell to another, leading back to the smallest xylem columns in the leaves, which are branches of larger xylem columns. A xylem column contains an unbroken column of water, held together by cohesion between its molecules. So the movement of each water molecule into the mesophyll stretches the column slightly, creating a tension throughout the column that ultimately pulls more water into the roots. Thus, with root pressure from below and tension from above, the water that evaporates at the top is replaced by movement of the water column up through each xylem unit. The upward movement through a tree can be measured by heating the sap at one point and using a sensitive thermometer to determine when the heat reaches a point a short distance above. This method has shown that under optimal conditions, when transpiration is rapid, the xylem sap may be moving as fast as a meter per minute.

The diameter of the conducting elements in xylem varies considerably and has a powerful influence on their ability to transport water. Since water adheres to the walls of a tube, a layer of water adjacent to the walls is somewhat bound to the sides and is less mobile than water in the middle of the tube. The area of the tube increases with the square of its radius, so a slight increase in radius means a relatively large increase in the amount of unbound water that is free to move. Furthermore, the conductance (flow rate) in a vessel increases with the fourth power of its radius, so the flow rate increases enormously with only a small increase in radius, as you can prove to yourself in the following exercise.

Exercise 38.1 Consider vessels that are 10, 20, and 40 μm in diameter. If we call the flow rate in the smallest vessel 1 unit, and the conductance increases as r^4, what are the flow rates in the other two? (Take a moment to do this calculation.)

You can easily show that the relative flow rates in these vessels are 1, 16, and 256 units. (We meet this principle again in considering circulation in animals, in which the diameters of some blood vessels change slightly.) You could also determine that if these vessels were together in a vascular bundle, the 40-μm tube would carry 93.8 percent of the water, the 20-μm tube would carry 5.8 percent, and the smallest tube would carry only 0.4 percent. Considerations of flow rate alone would predict that plants will form large vessels for maximum flow. On the other hand, the water column in smaller vessels is stronger; the water column in a large vessel may break, or *cavitate,* if the combined strength of its hydrogen bonds at some point is less than the tension on the column, and a vessel with such a gap is useless. In narrow tubes, adhesion to the walls largely supports the column, so narrower tubes may be essential in carrying water to the tops of tall trees. Presumably a compromise between these considerations, acting through natural selection over long times, determines the actual sizes of vessels in angiosperms.

As we noted in Section 37.3, xylem conducts its water through two types of tracheary elements—either tracheids or vessels, but sometimes both. Vessels are common only in angiosperms. Among ferns, gymnosperms, and the more primitive vascular plants, tracheids are the rule, with only a few genera having vessels. Conduction in the two types of tracheary elements is quite different, since water moves from one tracheid to another by passing through pits covered by primary walls, whereas vessels conduct the water through continuous tubules. Tracheids therefore present considerable resistance to conduction, and they have tended to become quite long to reduce this resistance. If a plant 1 meter high has tracheids 100 μm long, its water must pass through 10,000 pit membranes, but if each element is 1 mm long, it must only pass through 1,000 of them, so the resistance to flow is considerably reduced.

Exercise 38.2 Explain why it is valuable for a plant to have vessels with different diameters.

38.4 Phloem sap moves by a combination of osmotic forces and specific pumps.

The function of phloem, the second half of a plant's vascular system, is **translocation**—distributing organic material produced in the leaves to the rest of the plant. Since considerable material may be stored in the roots and in parenchyma throughout the xylem and phloem, phloem may also conduct sugar from these reserves throughout the shoot. The phloem sap that moves through sieve tubes is quite a concentrated solution of organic compounds, sometimes as much as 27 percent (27 g of solute per 100 ml of solution). Although glucose is a primary product of photosynthesis, about 90 percent of the material dissolved in the sap is sucrose, a disaccharide of glucose and fructose; the remaining 10 percent consists of amino acids and other organic compounds. Phloem sap moves at rates of from 30 cm/hr to over 200 cm/hr in various plants—even 600 cm/hr in corn leaves. The sap carries along all kinds of material that cannot be transported by specific pumps, including virus particles.

A simple traditional rule of plant physiology is, "Xylem up, phloem down." This is correct for xylem but far too simple for phloem. The nutrients of the sap enter the phloem at **sources** (or **exporters**), primarily the leaves where most photosynthesis occurs, and are removed at **sinks** (or **importers**) where they are consumed or stored. When photosynthesizing leaves are exposed to radioactive CO_2, they make radioactive sugar that can be traced to show the movements of phloem sap. These studies have shown that traffic in the phloem follows a few simple rules:

1. A mature leaf is always a source. A growing leaf is a sink at first, switches over to being a source when it is about half grown, and never becomes a sink again.
2. An active sink is fed by its nearest source. Thus a growing fruit is fed by the nearest leaves.
3. Upper leaves feed the growing meristems of the top branches; leaves near the bottom of the plant feed the roots and lower parts of the stem; and leaves in between feed either or both, depending on the demands of the sinks.
4. If a source or sink is removed, the plant quickly compensates for the loss by changing its flow pattern.

So the movement in phloem is not always downward; the flow can go in any direction, depending on the locations of sources and sinks. Gravity is not a factor, since the forces generated in the phloem can overcome its pull.

Determining the mechanism of sap movement has been difficult. Although some fundamental principles are now known that explain translocation, many points still remain unclear. The most generally satisfactory explanation is the **pressure-flow model** (Figure 38.9*a*), which proposes that sucrose and other organic compounds are actively loaded into the phloem at sources and actively unloaded at sinks, with energy being expended at both ends. According to this model, osmotic pressure generated by the solute concentration in the sap drives the movement of phloem sap.

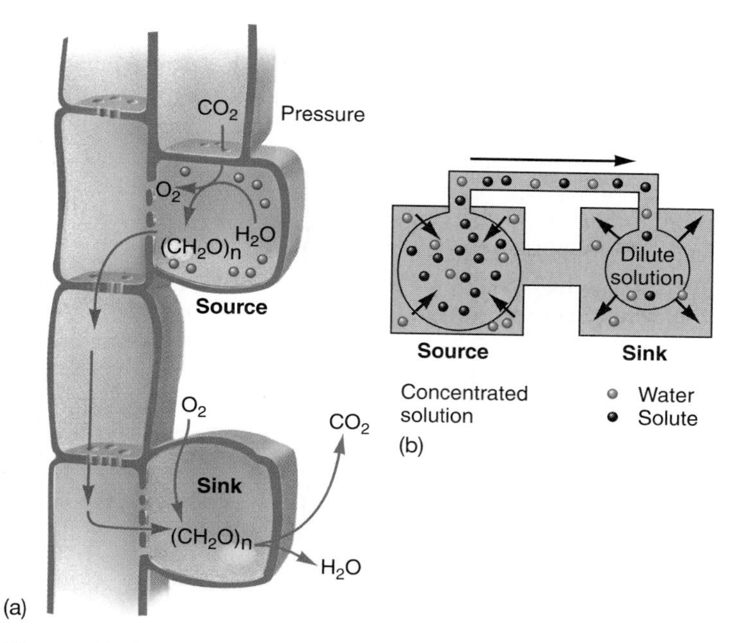

Figure 38.9

(a) The pressure-flow theory explains how materials are transported through the phloem from a source cell to a sink cell. *(b)* The theory is illustrated by a model consisting only of a tube with semipermeable membranes at its ends. If the concentration of sugar *(black dots)* is kept high at one end, the resulting osmotic pressure will create a constant flow of sugar, water, and other materials to the other end.

A simple demonstration of pressure-flow movement can be performed with glass tubing, as shown in Figure 38.9*b*. After the tube is filled with water, its ends are covered with semipermeable membranes and immersed in a water bath. Sucrose and a dye are put into one end. If the membranes are permeable to water but not to sucrose or dye molecules, osmosis will carry water in through the nearby membrane, creating a pressure that is transmitted through the tube to the membrane at the other end, where water will flow out. This flow carries the sucrose and dye marker along. This device is limited because sucrose must actually be added at one end and removed at the other, but the flow will go on indefinitely if this is done. The movement depends only on having a higher sucrose concentration at the source than at the sink.

Experimental evidence indicates that sucrose is actively loaded and unloaded. Bundle-sheath cells, sieve-tube elements, and, probably, companion cells do this by pumping hydrogen ions out of the cytosol, creating a hydrogen-ion gradient across their plasma membranes. A membrane carrier uses the energy of this gradient by symporting hydrogen ions and sucrose molecules together into the cell. Sucrose then moves into the sieve tubes through the plasmodesmata that join all these cells. At the sink, the process is reversed: A symport mechanism oriented the opposite way pumps sucrose out of the cells.

Symport mechanisms, Figure 8.25.

Investigators have had a hard time following the actual flow of sap through the phloem, principally because whenever a phloem cell is pierced with a fine tube to sample its contents, the pores connecting it to adjacent cells immediately seal off, and flow ceases. This response is probably a rapid reaction to injury, just as animal cells connected by gap junctions are sealed off from one another during an injury. Nevertheless, Tom E. Mittler solved the problem very neatly by taking advantage of the feeding mechanisms of aphids, the little insects that suck plant sap. The insect feeds by slipping a thin tube, or stylet, into the plant until it enters a sieve tube (Figure 38.10). Somehow the stylet doesn't set off the injury reaction, so sap continues to flow and the aphid takes its fill. (The excess sap, chemically altered in the aphid's digestive tract, comes out the other end of the aphid in droplets known as honeydew.) Mittler simply anaesthetized an aphid and then cut off its body with a sharp knife. The stylet, remaining in place, conducted sap out of the plant for several days, so the accumulating droplets could be sampled. These sampling experiments supported the pressure-flow model. The phloem sap is clearly under pressure, for sap oozes out of the phloem no matter where the stylets are. Furthermore, samples taken from stylets at various places in the plant show that the sap becomes less concentrated as it moves away from the source, although some of this change is likely due to dilution by water that moves into the phloem from the nearby xylem.

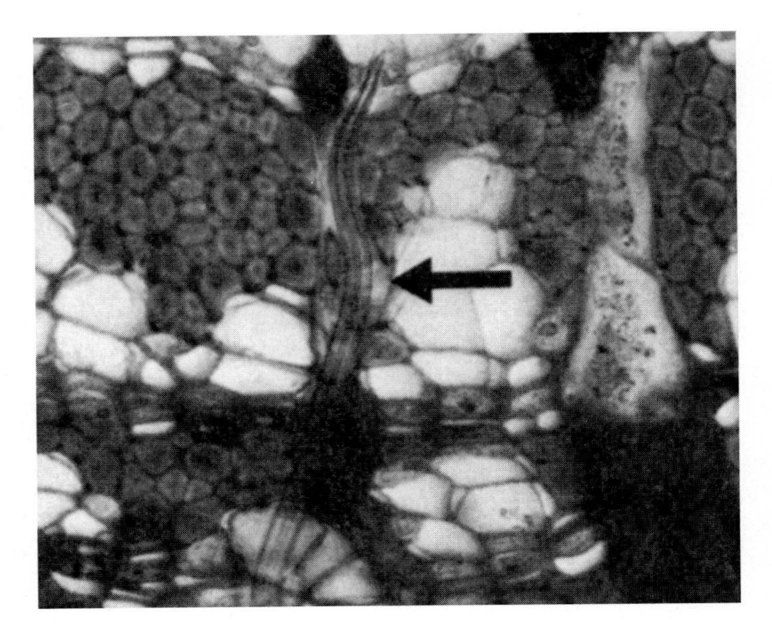

Figure 38.10
An aphid feeds by inserting its stylet *(arrow)* into a phloem tube.

Exercise 38.3 How much material does the phloem actually transport? This is measured by the specific mass transfer (SMT), which is typically 5–50 grams of material transported per hour per square centimeter of phloem. We can calculate some useful numbers from this information: *(a)* Suppose a phloem vessel has a square cross section, 20 μm on a side. How many vessels of this size are there per cm^2? (*Note:* 1 $cm^2 = 10^8$ μm^2.) *(b)* If 10 ml of sap is transported per cm^2 per hour, how many ml of sap moves through one vessel per hour? *(c)* Suppose the phloem sap has a sucrose concentration of 270 mg per ml. How fast must it flow through the phloem tube if the SMT is 10 grams per hour per cm^2?

Exercise 38.4 Why do orchardists remove some of the fruits from a tree early in the season if it starts to bear heavily?

38.5 Many plant organs can move quickly through water exchanges in motor cells.

The morning glory gets its name because its large, vaselike flowers open each morning to show their colors and then fold up again every night. The sensitive plant, *Mimosa pudica,* is famous for reacting to touch; even gently brushing a leafy branch makes it fold up and droop, as if it had been injured (Figure 38.11). Other mimosas are trees whose leaves fold up at night. Common three-leaved clovers open their leaflets in the day and close them at night, on a regular diurnal cycle. And when an insect wanders onto the open leaves of a Venus's-flytrap, the leaves close quite quickly, trapping the animal.

These are all examples of **nastic movements,** motions that occur in the same way each time because of the plant's anatomy. In contrast, **tropic movements** are directed by the stimulus that evokes them. For instance, plants generally show *phototropism,* growth in response to light, and *gravitropism,* growth upward or downward in response to gravity. Many plants grow around or along an object they touch *(thigmotropism),* such as a wall, a wire, or another plant, so as to hold onto as support. These tropic movements are basically directed growth, and we discuss them in the next chapter in relation to hormones that control growth.

Nastic movements are always produced by changes in a specialized structure that operates through a flow of water, and that is the reason for discussing them in this chapter. The opening and closing of stomata, in fact, is a kind of nastic movement effected by guard cells. In other cases, the effective organs are generally called **pulvini** (sing., **pulvinus**). Figure 38.12 shows the locations of pulvini on stems and leaves of a mimosa plant. A moderate stimulus first affects the tertiary pulvini at the bases of leaflets, making them fold up; with further stimulation, the secondary pulvini make whole leaves fold, and with stronger stimulation, the whole stem falls as the primary pulvinus at its base relaxes.

A typical pulvinus contains large parenchymal cells, which swell or shrink as they gain or lose water (Figure 38.13). These motor cells are the closest thing to muscles in the plant kingdom. The primary pulvinus of the mimosa stem, for instance, is poised to support the weight of the stem. Its extensor cells are full of water. When stimulated, the extensors quickly drain their water, so the branch falls. Later, and much more slowly, water is pumped back into the extensor cells (by means of ion pumps, of course), raising the stem again. Notice that lifting the branch

Figure 38.11
A sensitive mimosa *(a)* responds quickly to a touch by drooping its leaves *(b)*.

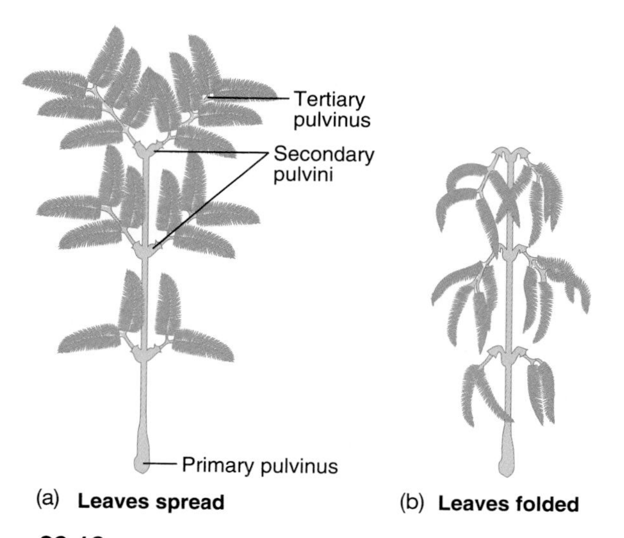

Figure 38.12
(a) The pulvini on a mimosa plant are arranged so as to hold the leaves in an open, spread position. *(b)* When stimulated, the pulvini lose water and let the leaves fold and droop.

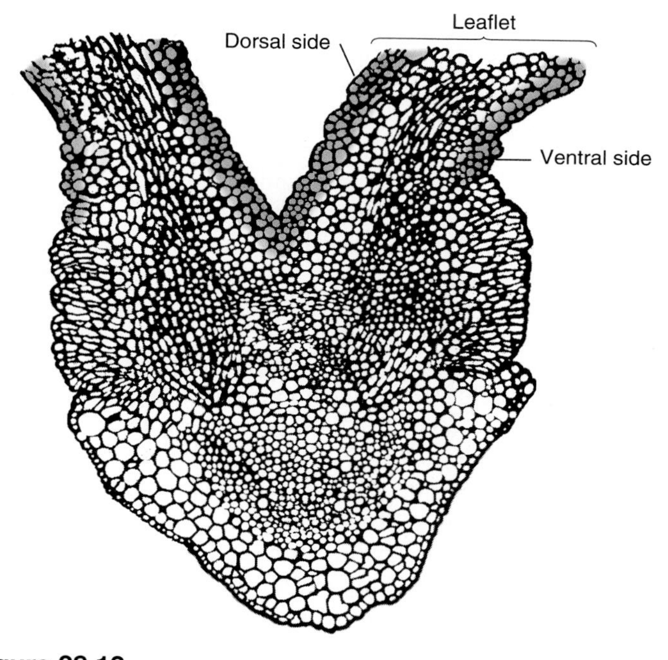

Figure 38.13
A pulvinus at the base of a pair of leaflets extends them when cells on the dorsal (upper) side swell and those on the ventral (lower) side shrink. The pulvinus raises the leaflets when the opposite happens.

requires energy; it falls of its own weight once its supports are removed.

In plants such as clover that open and close their leaves or leaflets on a daily cycle, the extensor and flexor cells alternately swell and relax. Experiments with several pulvini of this kind show that K^+ ions move into cells as they swell and out of them as they shrink—just like the guard cells of stomata. The same mechanism probably underlies all nastic movements.

What stimulates the movement of K^+ ions? Plants with a daily rhythm must be driven by an underlying clock mechanism, but such mechanisms are very poorly understood. However, we have a clearer idea of the important steps in nastic movements caused by touch. A mimosa leaf responds within about two seconds when it is touched almost anywhere, even though the pulvini may be located several centimeters from the site of stimulation. Surprisingly, parenchyma of the xylem and phloem can conduct slow signals like those in animal nerve cells (neurons), though they certainly cannot be called nerves and do not operate like neurons. (Animal neurons are elongated cells that communicate with each other through specific neurotransmitters.) Because of the properties of their plasma membranes, all cells maintain an electrical potential across the membrane, and in these plant cells the cytoplasm is about 200 millivolts more negative than the surrounding apoplast. Stimulation somehow induces an **action potential** in a cell (Figure 38.14), a transient change in the membrane potential explained in detail in Section 41.9. While neuronal action potentials last only a few milliseconds, those in plant cells may take 100 milliseconds to reach their peak, and the membrane potential does not return to its original low value for another 5–10 seconds. Nevertheless, these slow action potentials, relayed by a series of parenchymal cells, are responsible for triggering a response in motor cells.

Diurnal rhythms, Section 15.1.

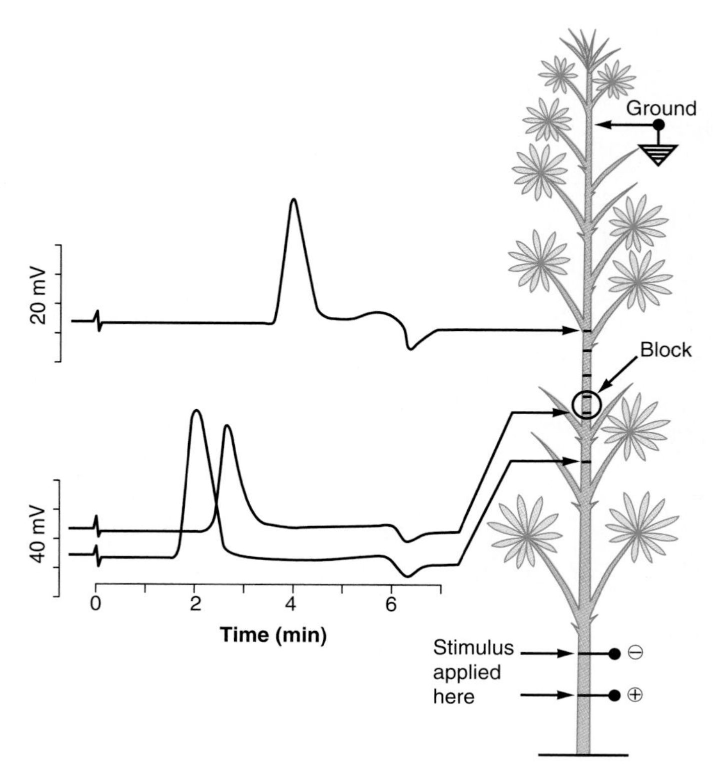

Figure 38.14

A signal can be carried by an action potential rather slowly through a series of parenchymal cells, as in the stem of *Lupinus augustifolius.* The cells have a high concentration of K^+ ions and a low concentration of Na^+ ions, with an electrochemical potential across the membrane. When the stem was stimulated near its base, action potentials were recorded a few minutes at points along the stem. Each action potential is recorded as a "spike," during which the membrane potential rises briefly and then slowly returns to its original value (note scales in millivolts). In this experiment, a block in the middle of the stem reduced the potential somewhat.

B. Water and light relationships of plants

38.6 Plants have ways to balance photosynthesis, transpiration, and translocation for optimal activity.

The principal plant processes of photosynthesis, transpiration, and translocation are interrelated in complicated ways. The challenge for a terrestrial plant is to perform them all at once and to balance these processes in the face of various environmental influences.

Let's consider photosynthesis and transpiration together. Photosynthesis requires equimolar amounts of CO_2 and water, and the elimination of the same molar amount of O_2. Water is supplied by the transpiration-driven flow in the xylem. CO_2 and O_2 exchange with the atmosphere through the stomata by diffusion, and the rate at which any material diffuses past a boundary, such as a stoma, depends on the difference in its concentration on both sides of the boundary: The greater the difference, the faster the material diffuses. Normally the CO_2 concentration in the atmosphere is only about 0.03 percent, while the concentration inside a leaf can be reduced to about 0.01 percent, so CO_2 diffusion can be driven by a concentration difference of 0.02 percent. On the other hand, the concentration of water vapor is about 1–2 percent in the atmosphere and 3–6 percent inside a leaf, so the difference in water vapor concentration is about 2–5 percent. Thus the concentration gradient of water vapor is about a hundred times that of CO_2, meaning that water can diffuse out of a leaf a hundred times faster than CO_2 can diffuse in when the stomata are open (Figure 38.15). Transpiration can therefore occur about a hundred times faster than photosynthesis. Since a plant only uses a small percentage of the water moving through it for metabolism, it must put up with an enormous transpirational loss, and water balance becomes one of its major problems.

A plant could improve its uptake of CO_2 through diffusion by lowering the CO_2 concentration in its leaves, but this is generally impossible because respiration keeps the CO_2 concentration at a minimum of 0.01 percent. Alternatively, a plant could achieve a better balance between photosynthesis and transpiration by closing its stomata somewhat. Since an open stoma passes water so much faster than it passes CO_2, closing the stomata should reduce transpiration without reducing photosynthesis proportionally. The CO_2-sensitivity of guard cells appears to be a mechanism for keeping an optimal CO_2 concentration in the leaf, but it is not clear whether plants generally use this mechanism to minimize the CO_2 concentration.

38.7 Plants also tend to maintain their temperatures within optimal ranges.

A plant's metabolism becomes even more complicated when we consider its overall energy balance and temperature. Every organism has an optimal temperature range, and a plant's heat balance is critical. It receives almost all its heat from radiation. The most obvious source is sunlight, but every object radiates energy in the infrared region of the spectrum. The warmth we feel from our surroundings, such as the walls and furnishings of a room, is mostly infrared radiation, and this is what every plant receives from its environment.

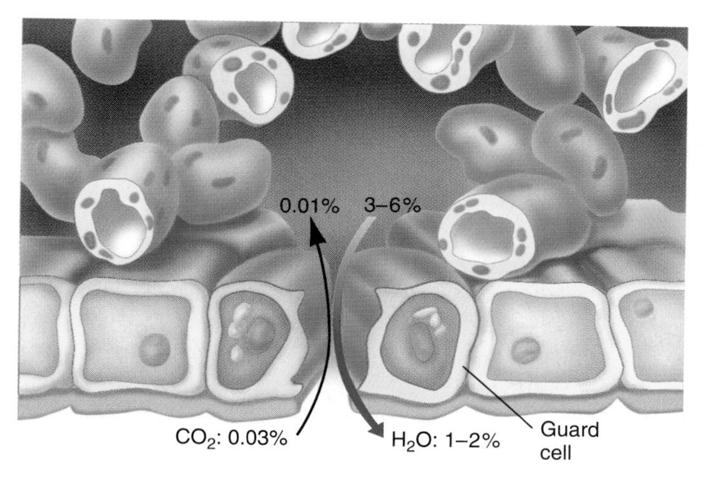

Figure 38.15
Water vapor and CO_2 establish independent concentration gradients in and around a leaf. The concentration differences determine the forces that tend to move each gas in or out of the leaf. These flows must be optimized so photosynthesis can occur as fast as possible with minimum loss of water by transpiration.

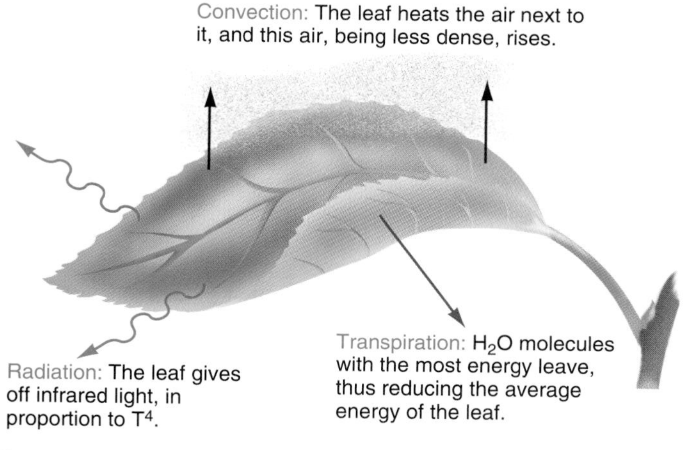

Figure 38.16
Heat is dissipated by radiation, by convection, or by transpiration.

Photosynthesis only uses a few percent of the radiant energy a plant absorbs; if the plant's temperature is to stay constant, the remainder must be dissipated by *radiation, convection,* or *transpiration* (Figure 38.16). Just as a plant receives infrared radiation from everything around it, so it radiates away some of its own energy. But radiation from any body rises in proportion to the fourth power of its temperature, so a leaf at 30°C radiates half again as much heat as one at 0°C. At typical growing temperatures, a leaf may reradiate 30–60 percent of the energy it absorbs. Convection transfers heat in bulk through air currents. A leaf, for instance, heats the air just above it, and layers of warm air, being less dense than cooler air, rise from the leaf, carrying heat away with them, as an appropriate optical arrangement shows (Figure 38.17). Every object is surrounded by a **boundary layer,** a layer of gas or liquid whose composition and temperature are influenced by the object (Figure 38.18). The leaf's boundary layer is a thin band of air that stays in contact with the leaf surface long enough to be warmed and then moves away. At the same time, the leaf takes CO_2 from this layer and adds O_2 and water by diffusion. This water loss is transpiration, of course, and it also carries heat away from the plant since a liquid is always cooled as some of it evaporates. The air currents that carry heat away by convection also help cool the leaf by increasing transpiration.

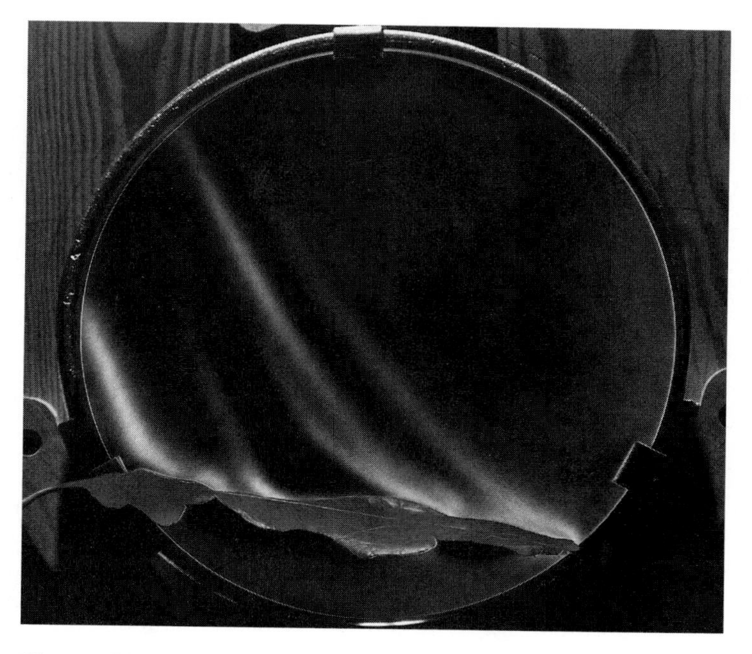

Figure 38.17
A photograph taken with Schlieren optics shows convection currents rising from a warm leaf. This optical system depends on the fact that as light passes through regions of different densities, it is deviated because the index of refraction of the medium changes with concentration. The system makes less dense air appear lighter in color. It reveals that a leaf heats the air above it, and layers of warm air, being lighter, move away from the leaf, carrying off heat and moisture.

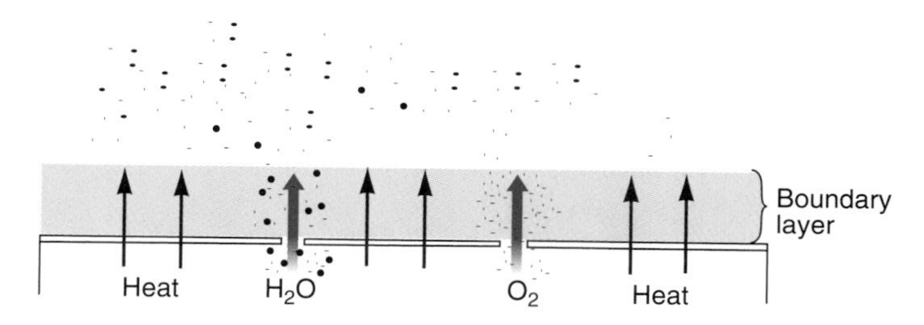

Figure 38.18
The boundary layer around an object consists of atoms or molecules that receive heat from the object. The composition of the boundary layer is also determined by exchanges of material with the object, such as exchanges through the stomata of a leaf.

Heat losses through convection and transpiration change in complicated ways with changing environmental factors. As the temperature rises, the ratio of transpiration to net radiation rises dramatically, so above 35°C, transpiration can actually keep a leaf cooler than the surrounding air. Above about 40°C, a plant may lose two to three times as much heat by transpiration as it absorbs by radiation. However, the rate of heat loss through convection falls sharply at higher temperatures.

Wind velocity has enormous influences on cooling. (Think of how much cooler even a slight breeze makes you feel on a hot day.) By sweeping air currents away from the leaf, the wind increases the rate of convection. Furthermore, moving air sweeps molecules out of the boundary layer, making this layer thinner as the wind velocity rises. The rate of transpiration is inversely proportional to the thickness of the boundary layer, so transpiration increases with increasing wind velocity.

It is hard to assess the combination of all these factors for any particular plant, although a number of investigators have tried to analyze them theoretically and have experimentally examined the rates of photosynthesis, transpiration, and other processes in various environments. The temperature of a leaf certainly rises as air temperature, humidity, and sunlight increase, but if the plant has enough water and can keep its stomata open, it can cool itself by maintaining a high rate of transpiration.

Since the Q_{10} of metabolism is about 2 (that is, the rate approximately doubles with every increase of 10°C), the rate of photosynthesis increases rapidly as a leaf grows warmer, and one might expect plants to be generally adapted to operate at quite high temperatures (say, 40–45°C). However, photorespiration, which competes with photosynthesis, increases much faster than photosynthesis as the temperature rises, so plants that use the common C_3 mechanism operate most efficiently at lower temperatures. Many plants that are adapted to higher temperatures and dry conditions do not photorespire; they use the alternative C_4 pathway of carbon fixation.

Photorespiration and the C4 pathway, Section 10.9.

The shape and orientation of a leaf are among the most important determinants of its heat balance and water balance. We noted in Section 38.1 that an array of small stomata allows rapid diffusion of gases from the leaf surface; these gases diffuse through the boundary layer, and the density and distribution of stomata in each species have undoubtedly been influenced by their effects on the water balance. The boundary layer is thinnest at the leaf edge, especially at the windward edge. A large leaf, with a low ratio of edge to surface, has a heavier boundary layer and is therefore cooled more slowly and loses water more slowly than a small leaf. Heat and water balance have probably been major determinants in the evolution of leaf shape, each shape representing an adaptation to a particular set of environmental conditions that affect the leaf temperature. The deeply indented leaves of oak and maple, with their relatively large ratios of edge to surface, represent one way of life, while the large leaves of catalpa represent another (Figure 38.19). Along with the evolution of a specific leaf shape, each species has evolved leaf enzyme systems that operate at the optimum temperature the plant can attain in its particular environment. Even the stiffness of the petiole and the angle at which it holds the blade are significant, for these factors determine how much sunlight and infrared radiation the leaf picks up.

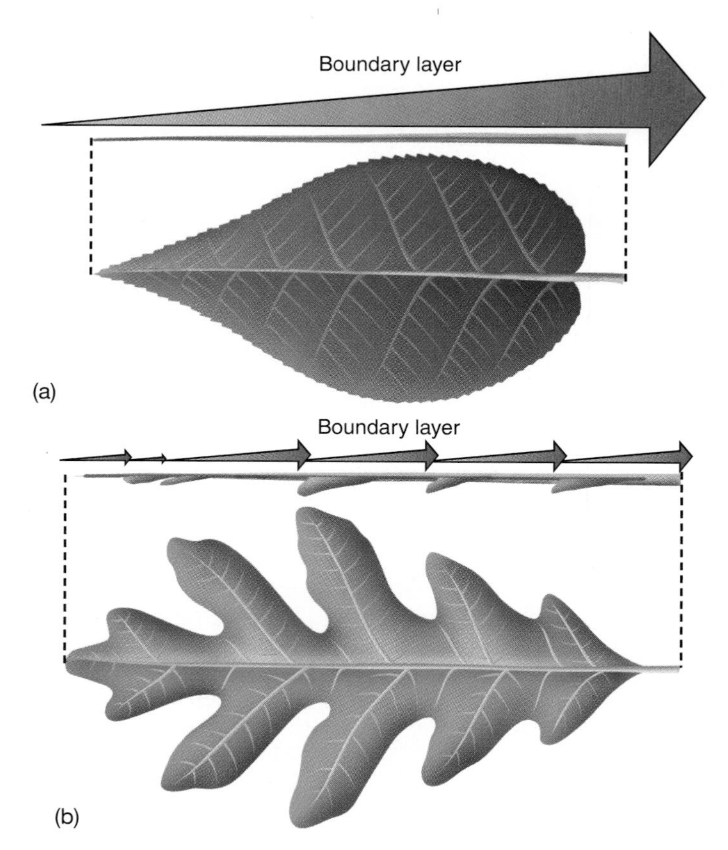

Figure 38.19
The temperature of a leaf is determined by a balance among the radiation it receives, the radiation coming from it, and convection carrying heat away from it. *(a)* A large leaf such as catalpa, with no indentations, has a low ratio of edge to surface, so its boundary layer will be heavier and it will retain more heat than a narrow leaf or one with deep indentations such as oak *(b),* which has a higher edge/surface ratio.

Notice, incidentally, that plants respond mechanically to a lack of water. If they are even slightly water deficient, the turgor pressure throughout the leaf tissue decreases, and the guard cells close the stomata, preventing any further water loss. As the water deficit becomes more severe, the leaves wilt and droop, so they are less exposed to radiation of all kinds and pick up less heat. This is not a healthy condition for a plant, but wilting is an appropriate adaptation to a condition that sometimes arises.

Turgor pressure, Section 32.9.
Wilting and the role of abscisic acid, Section 39.9.

38.8 Plants are adapted for growth in different light intensities.

Light conditions vary enormously. A plant growing in a field under the direct summer sun experiences very different conditions from one growing in the deep shade of a forest. Ernest Neal, whose study of woodland ecology provides some fine insights, noted that when the light in an open area exposed to the sun measured 39,000 lux (lumens/m^2), it averaged 2,600 lux in the forest shade and 200–500 lux under various shrub layers. Each plant species is adapted to a range of light, and we broadly distinguish **shade plants** from **sun plants.** A plant adapted to deep shade will outcompete others in its light range, and in brighter areas it will be outcompeted by other plants. (A few plants, classified as "intermediate," are adapted to partial shade.) Within its particular light habitat, each species has evolved a growth form that allows it to collect as much light as possible. Neal noted:

> *Ivy, with its creeping stems and leaves filling in all available space and hardly overlapping, covered a very large area of the woodland floor. Bluebell, with its narrow leaves radiating out in all directions, with little overlap, formed almost a complete circle of leaf surface. Primrose, with its rosette of leaves again covered all the ground in the neighourhood of the plants. This not only made use of all light available, but eliminated competition in the immediate vicinity.*

A leaf carries out photosynthesis and respiration simultaneously. The photosynthetic rate increases with light intensity, and the intensity at which CO_2 production just balances its consumption is called the **light compensation point.** However, as the temperature rises, the rate of respiration increases faster than that of photosynthesis, so the compensation point also increases with increasing temperature; that is, at higher temperatures a plant needs more light to keep up with respiration. As you might expect, the compensation points of shade plants are lower than those of sun plants, generally about one-third to one-fifth as much. Shade plants, in other words, require less light than sun plants to produce an excess of organic materials.

38.9 Many plants have special adaptations for water shortage.

One of the factors that may pose serious problems for plants is a shortage of water. The actual distribution of plants throughout the world owes much to this condition, for arid areas are widespread. Plants have evolved ingenious means to cope with dry environments.

The majority of plant species live in optimal environments with regard to water availability; they have enough water around their roots to ensure adequate turgor pressure in their cells and to provide a transpiration stream carrying minerals and nitrogen to their leaves. These plants can afford large losses of water through their stomata, even though some of their adaptations keep it to a minimum. Because they grow under intermediate moisture conditions, we call these plants **mesophytes** (*meso-* = in the middle).

Other plants, called **xerophytes,** have adapted to very dry conditions by evolving a set of **xeromorphic characteristics** (*xero-* = dry) (Figure 38.20). One xeromorphic adaptation is the crassulacean acid metabolism (CAM) mechanism of photosynthesis (see Section 10.9) that is common among succulent desert plants, especially cacti, and others that live in dry conditions. Their stomata stay closed during the heat of the day and open during the cooler, more humid night, thus conserving water. At night their leaves take up CO_2 and store it as the carboxyl groups of malic, isocitric, and other organic acids. Then during the day, CO_2 is recovered from these compounds and used for photosynthesis. (The related C_4 photosynthesis is more of an adaptation to high temperature than to scarce water.)

Other xeromorphic characteristics appear in leaves. Xerophytes commonly have thick layers of epidermis and heavy wax on the cuticle. Their cell walls may be especially rigid, thus preventing cell collapse during times of water deficiency, and they may have rubbery, shiny leaves that don't look wilted even when they are getting dry. (Cactus spines are the ultimate xeromorphic leaf adaptation.) Xeromorphic leaves often have surface hairs and stomata recessed in pits; both features reduce transpiration by trapping heavy boundary layers of humid air, which the wind cannot blow away easily. The leaves—usually needles—of conifers are an adaptation not only to the short growing seasons of biomes such as the taiga but also to the often-dry air there. A needle presents very little surface from which water can evaporate, and even that is covered by a thick, waterproof epidermis. The stomata of needles are also recessed, like those of plants that live in dry conditions, a feature that reduces the amount of water vapor carried away by air currents.

The leaves of some grasses roll and fold to expose fewer stomata to dry conditions. These automatic nastic movements depend on structures like pulvini. The upper surface of the leaf contains *hinge cells* (bulliform cells) that shrink as they get dry, thus folding the leaf. As the plant gets more water, the hinge cells swell and the leaf opens again.

Cacti, agave, and other succulents take advantage of what water they can get during the rare rainy periods. They may go through most of their growth and their entire reproductive cycles at such times. They also store water in specialized tissues in their stems and leaves as a reserve for long periods of drought.

38.10 Xeromorphic characters are used for adaptation to high-salt conditions.

Xeromorphic characteristics also allow plants to cope to some extent with excess salt in the soil or water where they grow, a situation that has much in common with lack of water. Water normally enters the roots of a plant by osmosis, because its concentration is higher in the soil solution around the roots than in the cytosol; salt water, however, reduces this difference in osmotic pressure, making it hard for the plant to take up water. Plants that grow in saline conditions, called **halophytes,** therefore face

(a)

(b)

(c)

Figure 38.20

Xeromorphic characteristics are adaptations to dry conditions. *(a)* A cactus has a heavy, wax-coated epidermis, and its leaves are reduced to spines. *(b)* A *Hakea* of the Australian desert has very small, thick leaves, and its seeds are enclosed in tough, drought-resistant pods. *(c)* A barrel cactus has thin spiny leaves and stores some reserve water within the tissues of its thick trunk.

much the same problems of water limitation that xerophytes face. Halophytes include representatives from many plant groups, including marine algae, as well as many protists and chromists. All have special adaptations. Considerable research is being directed toward understanding these adaptations in the hope of developing salt-resistant agricultural strains.

A salty environment threatens plants in two ways. First, salt surrounding the roots threatens to desiccate the plant by causing osmosis of water *out* of the roots rather than into them. Second, high salt concentrations threaten the plant's metabolic machinery by inhibiting most enzymes and disrupting membranes, so either the plant must exclude salt, or its enzymes and structures must be modified to resist it. Halophytes are distinguished from nonhalophytes by their ability to absorb a lot of sodium (Na^+) and other ions from the soil or water, and Na^+ is actually essential for the growth of some halophytes. In nonhalophytes, the principal cation of the cytoplasm is K^+; such plants don't require sodium and generally exclude it. We understand little about how enzymes and other structures in plants resist damage from salt. As we discuss in Chapter 32, halophilic bacteria have adaptations for living in high-salt concentrations, and halophytic plants probably use some of the same mechanisms. They also share salt-resisting mechanisms with animals, such as filling their cytosol with organic osmolytes.

Halophilic bacteria and organic osmolytes, Section 32.9.

Some plants accumulate excess sodium in their leaves and then drop the leaves to eliminate the salt. Many plants—not just halophytes—use this mechanism to rid themselves of excess minerals and organic matter. The common houseplant *Dieffenbachia* (dumb cane), for instance, produces excess oxalic acid, which is a poison. It stores the acid with calcium ions in its leaves as harmless, insoluble calcium oxalate, which is lost as older leaves fall off. Other plants do still better by having **salt glands** in their leaves that continually excrete salt by active transport. Salt crystals, sometimes visible as a grayish bloom, accumulate on the leaf surface, from which wind or rain then remove them. Mangroves, among the most successful halophytes, live with their roots submerged in salt water, and as many as 900 salt glands have been counted on a square centimeter of a mangrove leaf.

The salt problem frequently accompanies lack of water, for arid soils are often salty soils. Salt is of course found in soil near salt deposits, in salt marshes and estuaries, and near ocean shores, where it can be blown inland for miles. But salty soils are also a by-product of irrigation, since irrigation water contains small amounts of dissolved salts that remain when the water evaporates. This continuing deposition year after year finally produces *salinization* of the soil, a buildup of salt concentrations that inhibit plant growth. The British began extensive irrigation over a century ago in the Punjab area of India, which was once extraordinarily fertile, but now salting of the soil threatens the agricultural future of the region. The Central Valley of California, perhaps the world's most productive agricultural region, is now developing salt problems. The problem of salinization in agriculture is discussed further in Section 40.11.

Coda Considering the details of plant physiology brings home the similarities and differences between plants and animals. From a metabolic and cellular viewpoint, plants and animals are very similar, but evolution has led them into quite different ways of meeting common challenges. We tend to take humans as a model, to stress the importance of the cardiovascular system constantly circulating blood through all the tissues to supply them with oxygen and nutrients and to remove their wastes. A vascular plant, by contrast, engages in quite different flows of water. Only the movement through phloem can be said to be internal, and it is not a circulation, while xylem conducts a stream of water in at one end and out at the other. A plant has nothing like an excretory system (such as kidneys) for removing wastes. The oxygen produced by photosynthesis diffuses directly out of the photosynthetic tissues, and other wastes are sequestered and perhaps degraded in the central vacuoles.

Plants have evolved a remarkable range of mechanisms for meeting the constant environmental challenges to stay nourished, hydrated, and at an optimal temperature. Aside from elements derived mostly from the soil and water, a plant's principal source of nourishment is photosynthesis, and various plants have been able to adapt their metabolism for growing in at least a hundredfold range of light intensities. Some plants have adapted to being immersed in water and other to habitats that rarely see water. Perhaps most remarkably, they have evolved rather simple processes and structures that keep them in a temperature range favorable to metabolism and growth: by adjusting their rates of transpiration, by having leaves of various sizes and shapes, and by moving and orienting their organs so as to dissipate or absorb more energy. The fact that the world is generally green and that few areas are inhospitable to plants in spite of enormous variations in physical factors is testament to the adaptability and versatility of the basic plant design.

Summary

1. Plant physiology revolves around the basic processes of photosynthesis, which occurs in green parts of the shoot, and respiration, which occurs in all tissues. The basic functions of all plant structures are to support these two processes.
2. Leaves exchange gases with the atmosphere through stomata, which are pores in the leaf epidermis. Their size is regulated by pairs of guard cells in response to such factors as CO_2 concentration, water supply, and light intensity. Guard cells change their size through a mechanism that changes the concentration of K^+ ions.
3. Water and ions move from the root epidermis into the center of the root through both the apoplast and the symplast. The endodermis, whose cells are bound by a waterproof Casparian strip, regulates the passage of water into the xylem and prevents water loss back out of the vascular cylinder.
4. Water is drawn into the shoot system by a combination of forces: capillarity in the xylem, root pressure, and the cohesion of water molecules within the xylem. So as molecules are removed in the leaves, they are replaced by more molecules in the xylem vessels.
5. Xylem conducts water through tracheids and vessels of various diameters and lengths. Because the conductance of liquids in tubes increases with the fourth power of the radius, large vessels present the least resistance to flow and carry far more water than smaller tubes, but they are also most subject to cavitation. Tracheids impede flow at their pores, while vessels are open; therefore, tracheids tend to be quite long to minimize resistance.
6. Phloem sap moves by a combination of forces, as summarized by the pressure-flow theory. Concentrated sap is loaded at sources through specific pumps, and it moves from sources to sinks because of osmotic pressure.
7. Nastic movements are stereotyped actions of plants, such as the folding and unfolding of flowers and leaves, which are sometimes cyclical and sometimes evoked by stimuli. They are created by motor cells that swell or shrink, apparently as a result of K^+ ion movements. The signals that stimulate some nastic movements are carried by changes in membrane potential through specialized cells that act rather like the neurons of animals.
8. Stomatal openings must be properly regulated to balance the processes of photosynthesis and respiration with transpiration, which is important in maintaining a plant's temperature.

9. If plants are to maintain optimal rates of photosynthesis, they must maintain a high concentration of CO_2 in their leaves. They can do this only if their stomata are open, though by keeping their stomata open, they also allow a great deal of water vapor to escape through transpiration.
10. A plant's temperature is controlled by the balance of radiation, by convection, and by evaporation through transpiration. A plant must balance all these processes to maintain photosynthesis and respiration at the optimal levels.
11. Different species of plants are adapted to direct sunlight or to degrees of shade. Each species grows best at one light level, but those adapted to low light intensities have evolved growth forms that maximize their exposure to light.
12. Plants that are adapted to dry conditions have special features for withstanding water shortage, such as crassulacean acid metabolism, leaf hairs, depressed stomata, heavy wax cuticles on their shoot surfaces, forms that reduce their surface/volume ratios, and special behaviors.
13. The same kinds of adaptations that are valuable for conserving water are used for withstanding high-salt conditions. Plants adapted to high salt also may concentrate salt and excrete it through special glands.

Key Terms

stoma 807
transpiration 808
capillarity 808
Casparian strip 809
root pressure 809
guttation 809
translocation 811
source/exporter 811
sink/importer 811
pressure-flow model 811
nastic movement 812
tropic movement 812
pulvinus 812
action potential 814
boundary layer 815
shade plant 817
sun plant 817
light compensation point 817
mesophyte 817
xerophyte 817
xeromorphic characteristic 817
halophyte 817
salt gland 819

Multiple-Choice Questions

1. Changes in the shape of guard cells are regulated by
   a. CO_2.
   b. K^+.
   c. H_2O.
   d. abscisic acid.
   e. all of the above.
2. Stomata are part of the ______ tissue of plants.
   a. parenchymal
   b. endodermal
   c. epidermal
   d. xylem
   e. phloem
3. Which one of the following is *not* correct about root pressure?
   a. Moist soil increases root pressure.
   b. High humidity increases root pressure.
   c. High levels of transpiration increase root pressure.
   d. Root pressure increases at night.
   e. Root pressure can result in guttation.
4. The flow of water in the xylem depends on all of the following processes *except*
   a. root pressure.
   b. translocation of sugars.
   c. adhesion.
   d. cohesion.
   e. capillarity.
5. Which process would be most affected by cavitation in xylem vessels?
   a. transpiration
   b. capillary action
   c. ion transport into the endodermis
   d. osmosis
   e. development of root pressure
6. Which describes both guard cells and pulvini?
   a. part of the epidermis
   b. part of the endodermis
   c. primary regulators of transpiration
   d. regulated by transport of K^+
   e. Both are growth responses.
7. When mimosa leaves fold and droop, all but which of the following has caused the response?
   a. movement of K^+ ions out of cells
   b. slow conduction of membrane potential change
   c. change in size of guard cells
   d. flow of water out of extensor cells
   e. activity of pulvini
8. Transpiration occurs much more rapidly than photosynthesis because
   a. plants need more water than sugar.
   b. the relative concentrations inside and outside the leaf are greater for water vapor.
   c. CO_2 diffuses into the leaf more rapidly than water vapor moves out of the leaf.
   d. transpiration is powered by root pressure, and photosynthesis is not.
   e. transpiration is sensitive to temperature, and photosynthesis is not.
9. On hot, sunny days, plants gain heat mostly by
   a. transpiration.
   b. convection.
   c. radiation.
   d. photorespiration.
   e. photosynthesis.
10. Xeromorphic characteristics include all *except*
    a. C_4 metabolism.
    b. the CAM mechanism of photosynthesis.
    c. stomata open at night.
    d. a thick epidermis with a heavy wax covering.
    e. leaves that are reduced to spines or needles.

True-False Questions

Mark each statement true or false, and if false, restate it to make it true.

1. When K^+ enters guard cells, the change in the shape of these cells enables CO_2 to diffuse into the atmosphere.
2. The Casparian strip divides the apoplast into two separate regions.
3. The Casparian strip encircles the epidermis with a waterproof band that is rich in suberin.
4. The wider the xylem tube, the higher the flow rate of water through it.
5. Cavitation results from formation of small air bubbles within the phloem.
6. Water flows through xylem vessels more rapidly than through tracheids because of the resistance established by the end-walls of the tracheids.

7. The pressure-flow hypothesis predicts that if solutes were actively loaded into sieve tubes at a source and the surrounding cells were then made hypertonic to the sieve tube, bulk flow would continue.
8. A growing leaf may function as a phloem source or a phloem sink, depending on its stage of development.
9. On hot, windy days, plants primarily cool themselves by radiation.
10. Plants that live in salty soils tend to lose, rather than to gain, water by osmosis.

Concept Questions

1. List and explain the effect of the mechanisms that enable water to rise to the top of tall trees.
2. Explain the relationship between the Casparian strip and the development of root pressure.
3. Would you expect plants that grow in cool temperate regions to have leaves that are smaller than plants that grow in the tropics? Assume the plants have similarly shaped leaves and the environments differ only in temperature.
4. How do the processes of radiation, transpiration, and convection operate for heat gain and heat loss by plants? What effect does rising temperature have on each process individually?
5. Explain why sun-loving plants thrive in warm places while shade-loving plants grow best in cooler habitats.

Additional Reading

Bannister, P. *Introduction to Physiological Plant Ecology.* Blackwell Scientific Publications, Oxford, 1976.

Dennis, David T., and David H. Turpin (eds.). *Plant Physiology, Biochemistry, and Molecular Biology.* Longman Scientific & Technical, Essex (England) and John Wiley and Sons, New York, 1990.

Galston, Arthur W. *Life Processes of Plants.* Scientific American Library, New York, 1994.

Hopkins, William G. *Introduction to Plant Physiology.* John Wiley & Sons, New York, 1995.

Salisbury, Frank B., and Cleon W. Ross. *Plant Physiology,* 4th ed. Wadsworth Publishing Co., Belmont (CA), 1992.

Shigo, Alex L. "Compartmentalization of Decay in Trees." *Scientific American,* April 1985, p. 96. Animals heal, but trees compartmentalize. They endure a lifetime of injury and infection by setting boundaries that resist the spread of the invading microorganisms.

Taiz, Lincoln, and Eduardo Zeiger. *Plant Physiology.* Benjamin/Cummings Publishing Co., Redwood City (CA), 1991.

Internet Resource

To further explore the content of this chapter, log on to the web site at:

http://www.mhhe.com/biosci/genbio/guttman/

39 Plant Growth and Growth Regulators

A bed of pod seedlings lifts its shoots above the soil line.

Key Concepts

A. Plant Hormones

39.1 Plants often grow in response to the direction of light.

39.2 Auxin controls cell elongation.

39.3 Auxin controls several kinds of plant growth processes.

39.4 Gravitropic growth is influenced by auxin in still unknown ways.

39.5 Gibberellins enhance cell elongation and other processes.

39.6 Cytokinins regulate cell division and differentiation.

39.7 Hormones interact to control a plant's shape.

39.8 Ethylene promotes fruit ripening and other phenomena.

39.9 Ethylene also promotes the abscission of plant organs.

39.10 Systemin is a signal that induces the wound reaction in plants.

B. Photoperiodism

39.11 Flowering and other phenomena are regulated by the photoperiod.

39.12 The phytochrome system measures the photoperiod.

39.13 The phytochrome system controls many processes.

A pollen tube grows down to an ovary, a sperm joins an egg, and in a moment a new plant has been started. The new cell divides several times to form an embryo, drawing nourishment from the tissue around it. Soon it has formed a short root on one end, a shoot on the other. When dropped in fertile soil, it will continue to expand, its roots drawn downward, its shoot upward as they respond oppositely to the subtle pull of gravity. In response to genetic instructions, some cells will elongate and thicken into vessels while others remain small and begin to store up starch. Cells on one side of the shoot will grow slightly faster or longer, subtly bending it this way or that way as the sunlight directs an intricate interplay between the plant's hormones and its genes, shaping the branches, leaves, and flowers encoded in the genes. A leaf nibbled by a hungry insect will send out a chemical alarm to make itself and its neighbors distasteful, or poisonous. Plants, like animals, are mazes of wireless communication circuits, in which stimuli such as gravity, light, and the touch of a predator evoke the production of hormones that direct specific reactions.

In Chapters 20 and 21 we discuss growth and differentiation with an emphasis on animals; here we focus on those processes in plants. In many ways, we know more about growth and differentiation in plants than in animals, partly because plants are easier to handle in the laboratory. Plant growth is regulated internally by hormones but is also shaped by external influences such as light, gravity, temperature, and nutrients. Both types of factors can be manipulated in a laboratory and, increasingly, in the field, so laboratory results are now being applied in agriculture. Plant biology has advanced still more through molecular genetic methods, especially as a result of research with the little plant *Arabidopsis,* which has become the experimental equivalent of the nematode *Caenorhabditis* and the fruit fly *Drosophila* (see Methods 39.1). We will examine some of the insights into genetic regulation in plants that have come out of this research. ■

A. Plant hormones

Growth, as we noted in Chapter 11, means adding to the organic structure of a cell or organism. We see growth as an increase in mass, in the number of cells, and in the size of each cell. Although animal cells tend to remain about the same size throughout their lives, the roots and stems of plants grow largely by elongation of cells after cell division. If plant cells did not elongate, a 200-foot redwood would only attain a modest height of 40–50 feet.

Methods 39.1 *Arabidopsis* As an Experimental System

Molecular biology has thrived by using the bacterium *E. coli* and some of its viruses as an experimental system, and more recently by using yeast; genetics made great strides with the fruit fly *Drosophila;* and developmental animal biology is now progressing through work with the roundworm *Caenorhabditis elegans* (see Methods 21.1). Recently biologists have found that a tiny flowering plant, the mouse-ear cress *Arabidopsis thaliana,* is a wonderful laboratory organism for investigating the genetics, molecular biology, and physiology of plants, as well as other areas such as population biology and ecology. As information accumulates about these model systems, and as more people work with them, the boundaries between them break down. DNA can be moved from one system to another, so genes can be studied in the most suitable system. For example, using standard recombinant-DNA techniques, *Arabidopsis* DNA has been sequenced, replicated, and transcribed, and its genes have been isolated and transferred into bacteria, yeast, other plants, and even animals.

What makes *Arabidopsis* so special? First, it is a real flowering plant, with leaves, stems, and roots, which behaves like a typical plant, although it usually grows to only an inch or so in height. It is distributed worldwide and grows in diverse conditions. Because it is small, enormous numbers can be grown without elaborate facilities. Its flowers produce abundant seeds—at least 1,000 seeds per plant—from which new plants can be grown. Many *Arabidopsis* mutants have been selected, and it is easy to perform genetic experiments with them. Its life cycle, from seed to seed, is less than six weeks, a very short time for a plant.

Arabidopsis, of course, is a eucaryote with typical eucaryotic cells; it is easy to follow the movements of its chromosomes in mitosis and detect chromosomal abnormalities when they appear. A diploid genome has only ten chromosomes, made of an astonishingly small amount of DNA, only about 20 times as much as *E. coli,* whereas many eucaryotic genomes contain a thousand times as much as a bacterium.

Whole *Arabidopsis* plants can be grown in liquid or on solid nutrient media, and tissue can be removed and used to start *callus cultures,* lumps of proliferating undifferentiated tissue. Amazingly, the right concentrations of a few plant hormones will make these tissue lumps undergo differentiation in a controlled manner to form roots, leaves, stems, or even complete plants with all parts functioning normally.

By treating *Arabidopsis* tissues with enzymes that digest away their cell walls and allow the cells to be separated from one another, plant parts can be cultured as *protoplasts,* isolated cells with intact membranes but no walls. Now, with very little effort you can select a plant, perhaps a mutant, separate parts of it into individual cells, select a cell, and grow it in cell culture to produce vast numbers of identical cells. Some of these cells can be used to incorporate new DNA, from *Arabidopsis* or other organisms, so the genome can be tailored without limit. These altered cells can be removed and used to start a colony of cells for further study and manipulation, just as if you were working with *E. coli,* except that you are dealing with a plant cell. Another trick is also available: Add the right amounts of auxin and cytokinin to one of those cells and you have a whole plant again that will flower, produce seeds, and do all the things expected of a plant, but with its genome altered to test the effects of whatever genes you have added.

All aspects of plant growth are controlled by a small group of **plant hormones.** Only five classes of natural hormones are definitely known: *auxins, cytokinins, gibberellins, abscisic acid,* and *ethylene.* It is helpful to think of the first three as stimulating growth and the last two as inhibiting it, though they interact in complicated ways. In addition, many artificial materials called **plant growth regulators** have been synthesized that mimic natural hormones. Natural plant hormones are made in specific parts of a plant and are transported to other parts, where they are effective in very low concentrations. Unlike animal hormones, which regulate metabolic activities from moment to moment, plant hormones primarily regulate growth and differentiation.

39.1 Plants often grow in response to the direction of light.

The history of research on plant hormones starts with the work of Charles Darwin, as do so many important aspects of modern biology. Darwin and his son Francis devised several ingenious experiments to explore plant **tropisms,** movements directed toward or away from a stimulus. They were particularly interested in **gravitropism,** growth directed by gravity, and **phototropism,** growth directed by light. Roots, which grow downward (toward the source of gravitational attraction), are positively gravitropic, and stems, which grow upward, are negatively gravitropic. Stems are usually positively phototropic and grow toward a light source; the adaptive value of this growth habit is obvious, and we see it commonly in the way houseplants reach toward a window. In studying phototropism, the Darwins found that the tip of the stem plays a crucial role. They worked with *coleoptiles,* the protective sheaths through which monocot stems emerge from seeds, and found that coleoptiles will only bend toward the light if their tips are illuminated. The tips bent even if they were covered with clear glass caps, but not if the caps were blackened or covered with metal foil. Bending, however, doesn't occur at the tip itself but a few millimeters below it, due to elongation of the stem on the side *opposite* the light, pushing the tip toward the light (Figure 39.1).

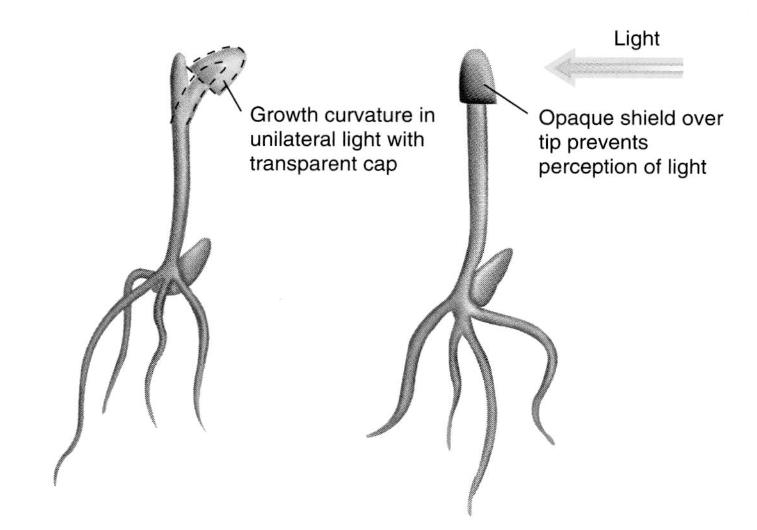

Figure 39.1
A seedling bends toward the light even if its tip is covered with a transparent cap, but not if the tip is covered with an opaque cap.

The Darwins published their work on phototropism in a book, *The Power of Movement in Plants,* in 1880. Although they postulated an "influence" transmitted from the tip to the affected region, they did not identify the "influence." It took some 50 years for other investigators to show that the effect is actually caused by **auxin,** or **indoleacetic acid (IAA).**

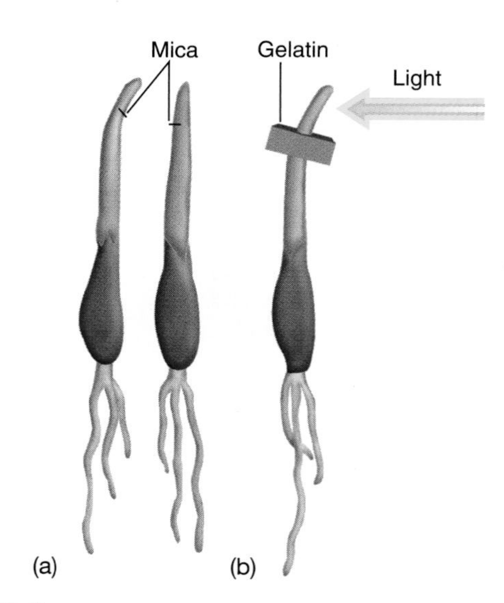

Figure 39.2
(a) Boysen-Jensen inserted slices of mica in coleoptiles illuminated from one side. If mica is inserted on the lighted side, the seedling still bends toward that side, but it does not bend if the mica is placed on the dark side, presumably because the slice blocks the flow of a hormone (auxin) from the tip. *(b)* The seedling bends even if it is sliced and a block of gelatin is inserted, showing that the movement of the effective material does not depend on an intact tissue structure.

39.2 Auxin controls cell elongation.

Between 1910 and 1913, the Danish botanist P. Boysen-Jensen extended the Darwins' observations in a series of experiments. Figure 39.2 shows how he made slices halfway through coleoptiles and inserted thin slices of mica, which is impervious to most chemicals. When inserted below the tip on the light side of an illuminated coleoptile, the mica slices had no effect on the bending reaction, but when placed on the dark side, they prevented bending. This observation can be understood if the mica slices are blocking the flow of material down the dark side of the shoot, where it would normally stimulate cell elongation.

Boysen-Jensen also showed that a coleoptile will bend normally if its tip is cut off and a block of gelatin is inserted between the tip and the rest of the coleoptile. This showed that the influence is probably due to a chemical diffusing through the block. Arpad Paal extended these observations by demonstrating that cutting out a notch of tissue on one side of a coleoptile will make it bend toward that side in the dark. This is understandable if the cut blocks the flow of a chemical that causes elongation from the tip. Paal also sliced off the coleoptile tip and replaced it off-center, causing the shoot to elongate on the side that continues to receive the growth-stimulating chemical from the tip.

In the late 1920s, Fritz Went, then working as a graduate student in his father's laboratory in the Netherlands, took up the problem. He saw the need for a quantitative assay of the substance causing bending and devised a test using oat coleoptiles that is still used today. Went cut off oat coleoptile tips and left them on small blocks of gelatin for a few hours to allow material to diffuse into the blocks. He then placed the blocks on one side of decapitated coleoptiles and found that they would bend, in the dark, away from that side, just as in Paal's experiment. The coleoptiles bend in proportion to the number of tips that have been placed on a block, providing an assay procedure for measuring the amount of this growth substance from any source (Figure 39.3). This experiment also clearly showed that a substance produced by a tip causes bending and excludes the possibility that the tip itself could have any residual influence.

Went found that the substance responsible for bending—now called auxin—was widespread in nature, but in very low concentrations. In 1934 Fritz Kogl and his associates found that human urine, of all things, is rich in auxin activity. From 40 gallons of urine obtained at a hospital, they purified 40 mg of crystalline material with extremely high activity and later isolated two similar compounds. These compounds all have auxin activity, but the material now known to be the major natural auxin is indoleacetic acid (IAA), a derivative of the amino acid tryptophan:

CH_2COOH

NH

Indoleacetic acid (IAA)

The phototropic effect of auxin can only be understood in a general way. It is obvious from the classical experiments that there is much more auxin on the dark side of a stem than on the light side. Light may inhibit its release, or stimulate its degradation, or even make it move laterally toward the dark side of the stem.

39.3 Auxin controls several kinds of plant growth processes.

The action of auxin partly explains the pattern of growth in a plant. An asymmetrical distribution of auxin, for example, produces asymmetrical growth, as in phototropism. Auxin is involved in most growth in plants and generally promotes cell elongation, not cell division (Figure 39.4). IAA promotes the elongation of cells in ordinary root and stem growth in the region directly behind a proliferating meristem by increasing the turgor pressure inside these cells while their wall structures are loosened. Within a few minutes after being applied to many plant tissues, IAA induces the transcription of a new class of genes. Among the newly induced proteins are some that pump H^+ ions out of the cell into the cell wall, probably directed

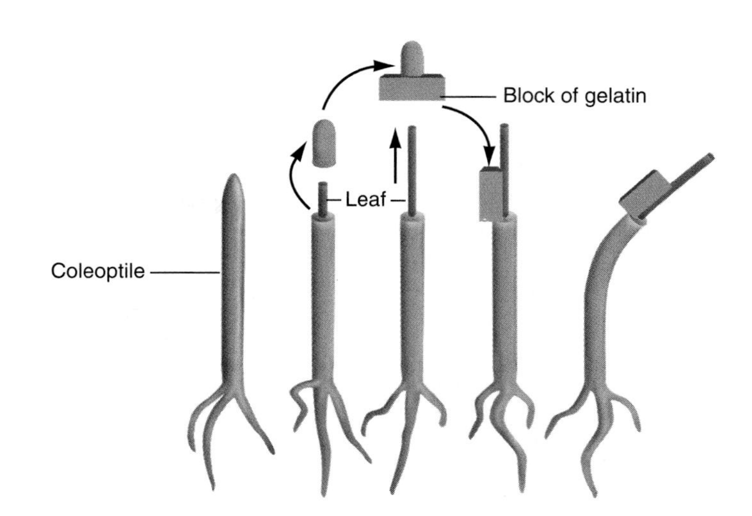

Figure 39.3
The tip of a coleoptile is cut off, and the leaf inside is pulled up. The tip is placed on a block of gelatin, which absorbs auxin and can then replace the tip as a growth inducer. The degree of bending that a gelatin block induces shows how much auxin it contains.

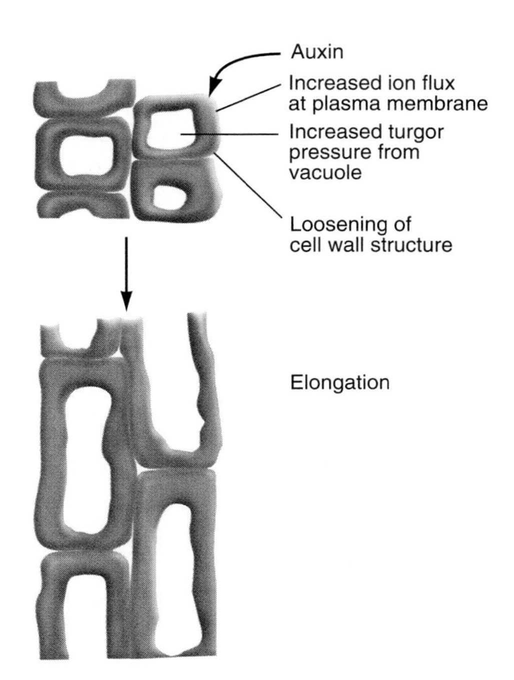

Figure 39.4
Auxin acts on cells of a specific age in the growing tips of roots and stems. It induces proteins to pump H^+ ions into the cell wall, thus loosening the wall structure. It induces pumping of K^+ ions into the central vacuole; water follows osmotically, building up an internal pressure that forces the cell to elongate.

ATPases of the cell membrane. This effect has inspired the **acid-growth hypothesis,** now well supported by experimental evidence, which postulates that reducing the pH in the cell wall loosens the wall structure, perhaps by activating enzymes that cut bonds between structural polysaccharides, so the wall can expand under increased turgor pressure. The situation is complicated, however, because auxin may have entirely different effects at different concentrations, in different plant organs, and even at different times of day and seasons of the year. Its action is also influenced by the concentrations of other hormones, with which it may work either antagonistically or in concert.

Auxin is transported from its sites of production, in the meristems of both stems and roots, to the region of cell elongation where it acts. The hormone is transported at the rate of about a centimeter per hour—too quickly for diffusion alone to be the motive force; auxin is actively transported between cells, rather than through the vascular system, as evidenced by the fact that it can move against a concentration gradient that its transport can be blocked by metabolic poisons. As with other hormones, the concentration of auxin is controlled by various enzymes: Some inactivate the hormone, others convert free auxin into storage complexes, and still others release the free hormone from storage.

Indoleacetic acid is the predominant auxin synthesized by plants, perhaps the only auxin in many plants. But organic chemists have ingeniously produced several important artificial compounds related to auxin. Some are used to promote root production in plants, particularly when cuttings are used for propagation. Auxin produced by a plant itself stimulates root development, but it can be enhanced by adding other auxins. A number of the synthetic auxins, such as indolebutyric acid (IBA) and naphthalene acetic acid (NAA), are even more effective than IAA in stimulating root development:

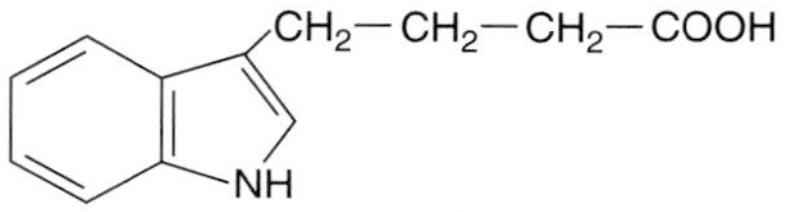

$CH_2—CH_2—CH_2—COOH$

NH

Indolebutyric acid (IBA)

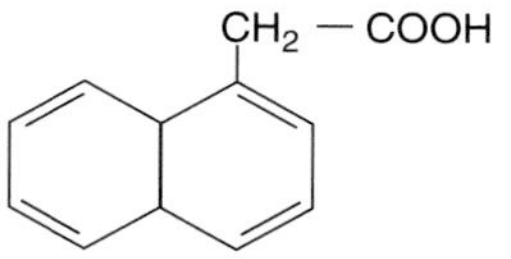

CH_2 — COOH

Naphthalene acetic acid (NAA)

Both are used commercially and are readily available.

Other synthetic auxins act as herbicides. The best known is 2,4-dichlorophenoxyacetic acid (2,4-D):

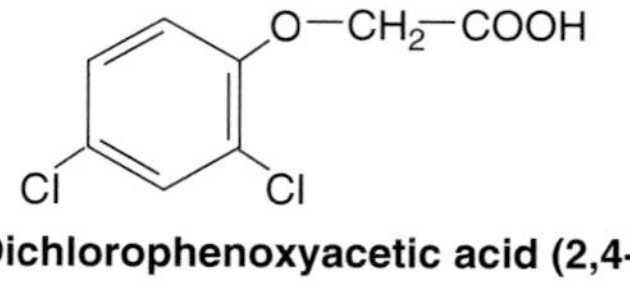

O—CH_2—COOH

Cl Cl

2,4-Dichlorophenoxyacetic acid (2,4-D)

Although 2,4-D promotes growth at extremely low concentrations, higher concentrations are lethal (because they promote production of ethylene, which we discuss later). Broadleaved dicots are more sensitive than monocots, so dicots can be eliminated preferentially by controlled use of 2,4-D. Thus dandelions can be eradicated from a lawn of monocot grass and broadleaved weeds from a crop of corn. On the other hand, 2,4-D has been implicated in harmful effects on reproduction in animals, and so its use has been greatly curtailed in recent years. Dozens of other synthetic growth regulators have been developed that play a major role in American agriculture and horticulture.

Effects of endocrine disruptors, Section 28.8.

39.4 Gravitropic growth is influenced by auxin in still unknown ways.

Auxin promotes gravitropism as well as phototropism. Gravity affects plants from the time they germinate, directing their roots to grow downward and their shoots to grow upward. If a young seedling is laid horizontally, its root will begin to bend downward while its stem bends upward. A plant responds the same way throughout its life. You have probably seen sizable trees in a forest that were obviously blown partway over years ago and have since grown upward again.

In trying to understand gravitropism, the first problem is to understand how a plant can *detect* gravity. Gravity doesn't automatically affect different sides of an organ differently, as light does. That is, light can affect one side of a stem more than the other, but gravity pulls on the top and bottom of a stem equally. An oat coleoptile assay shows that auxin is more concentrated in the lower parts of some cells or in the lower portion of a plant lying horizontally, as if it had been attracted by gravity, but gravity cannot sediment auxin molecules in the viscous fluid of a cell against the forces of diffusion.

Excellent evidence indicates that a plant detects the direction of gravity through cells called **statocytes** located in such places as root caps and the bundle sheath and endodermis of stems (Figure 39.5). A statocyte contains several **amyloplasts,** the plastids that store starch granules, which are dense enough to fall to the bottom of a statocyte and thus determine the downward direction. Several kinds of experimental evidence point to the role of amyloplasts in gravitropism. Their presence correlates with the ability of an organ to detect gravity, and the time it takes amyloplasts to fall in the cytoplasm is correlated with the time it takes an organ to detect the direction of gravity. Genetic evidence also shows that amyloplasts have a role in gravitropism. For instance, coleoptiles of a corn mutant with very small amyloplasts don't show the same upward curvature as normal coleoptiles when laid horizontally, and a mutant that doesn't store starch in certain amyloplasts has no gravitropic response. Unfortunately, other experiments give contrary results, and it is only clear that amyloplasts function in *some* gravitropic responses.

In spite of good evidence that auxin is effective in the gravitropic response in stems, this mechanism doesn't work at all in the root response, where a growth inhibitor, rather than a promoter, appears to be at work. The root cap is the gravitational detector for the root, and it appears to be the source of this inhibitor, which like auxin, must be more concentrated in the lower half of a horizontal root (Figure 39.6*a*). Experiments in blocking the flow of this inhibitor from the root cap confirm this mechanism; a block on one side of a vertical root makes the root curve away from that site (Figure 39.6*b*). Furthermore, chemical analysis shows that the

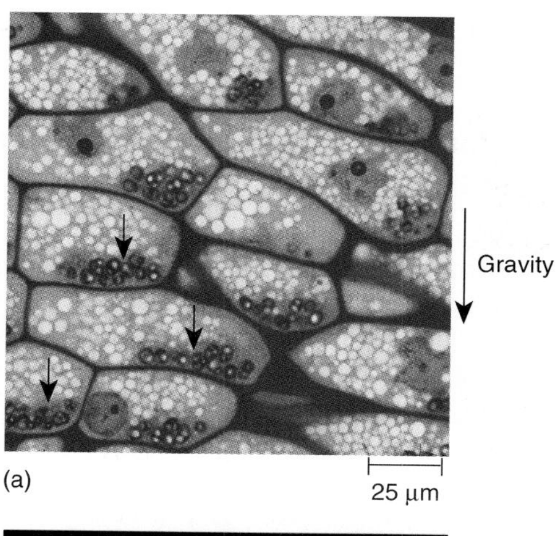

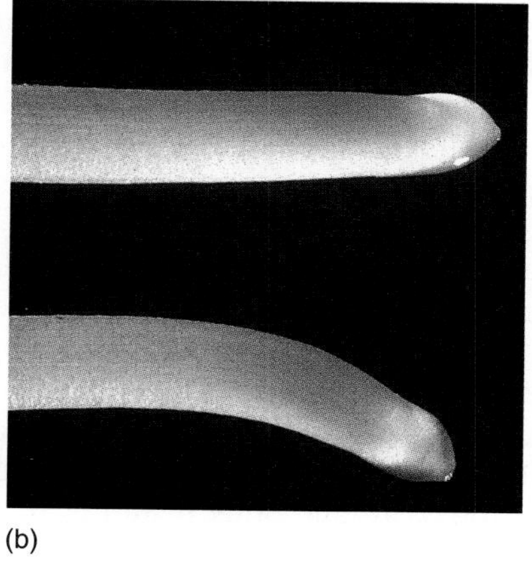

Figure 39.5

(a) A statocyte contains amyloplasts *(arrows)*, where starch is stored. These plastids are attracted by gravity, and the statocyte somehow responds to their orientation by sending out signals that direct growth in relation to the gravitational force *(b)*.

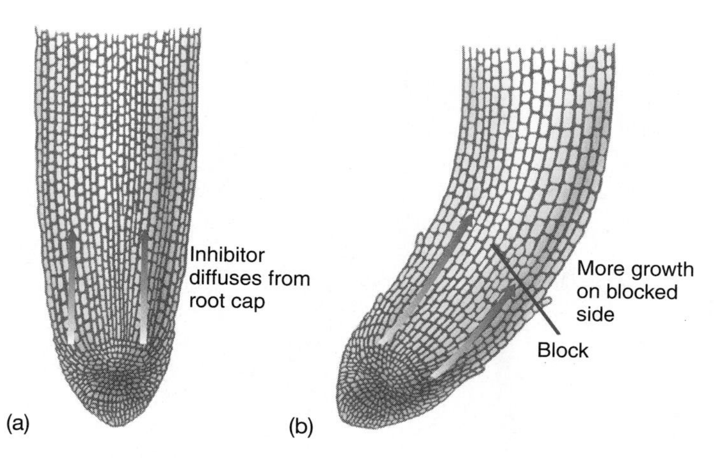

Figure 39.6

(a) The root cap directs the gravitropism of a root, apparently by producing a growth inhibitor. *(b)* If the flow from the cap is blocked on one side, that side grows more rapidly and the root turns away from it. The inhibitor is apparently concentrated in the lower part of a horizontal root, thus tending to make it curve downward.

inhibitor is likely to be abscisic acid, a plant hormone discussed in Section 39.9. Abscisic acid seems to be distributed in just the right way and to move in just the right way to have a gravitropic effect, though we do not know exactly how this happens.

There are still distinct gaps in these explanations of gravitropism. How is information about the direction of gravitation, once detected by amyloplasts or the root cap, translated into an asymmetrical distribution of auxin, or abscisic acid, or any other hormone? This and other questions remain unanswered, and until all the connections are made, we can hardly claim to understand the process.

39.5 Gibberellins enhance cell elongation and other processes.

Over 80 closely related compounds make up another class of plant hormones, the **gibberellins.** Early in this century, Japanese rice crops were threatened by foolish seedling disease (bakanae), an infection by the fungus *Gibberella fujikuroi.* Infected plants failed to set seed and developed long, straggly stems, so they fell over in the water and died (Figure 39.7). In 1928 the plant pathologist E. Kurosawa produced the same effect in healthy plants with an extract of the fungus. A few years later, the active agent was isolated and identified, and since then many related compounds have been identified that have the same effects on plant growth. All are called gibberellins or **gibberellic acid (GA):**

CH_3 H_3C COOH COOH CH_2

GA$_{12}$

O C=O HO OH CH_3 COOH CH_2

GA$_3$

Although gibberellins were first found in a fungus, they are widely distributed in plants and have been identified in algae and mosses as well as in angiosperms. They are normal hormones that stimulate growth in the shoot system but have little or no effect on roots. Gibberellins can stimulate growth far more dramatically than auxin can; for instance, cabbage plants treated with GAs grow to well over 2 meters in height! (They no longer look like cabbages because the internodes between the leaves become so elongated.)

The effects of gibberellins have been extensively studied by using strains of dwarf plants that have been of interest in horticulture for years. These plants have turned out to be homozygous for recessive mutations that block gibberellin production. Treating a dwarf corn seedling with a GA induces it to grow as tall as normal corn. This effect, which is also seen in dwarf strains of other plants, indicates that though the mutants cannot produce gibberellins, the targets of the hormone action are perfectly functional. These dwarf strains provide a ready assay for gibberellins, since they grow in proportion to the amount of hormone applied (up to some inherent limit, of course).

Gibberellins are normally transported through the phloem from their sources in the growing tips of young leaves or through

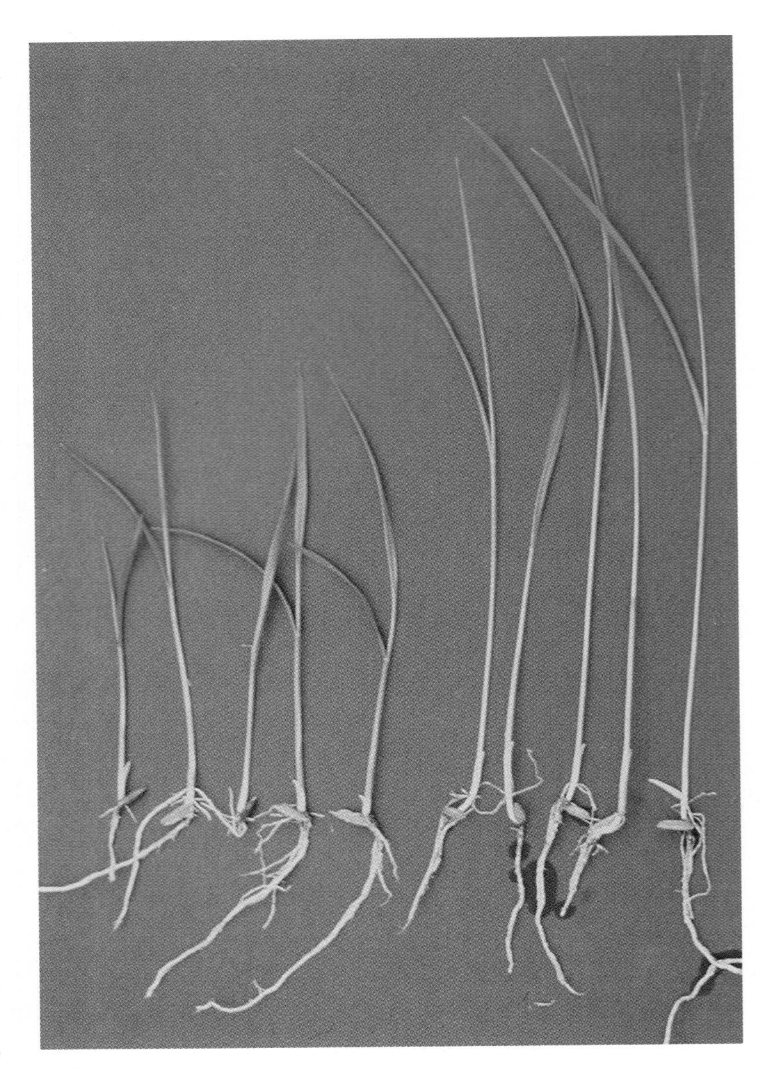

Figure 39.7
The fungus that causes foolish seedling disease in rice plants produces gibberellins that make the plants grow too fast. Infected plants, compared here with normal plants, become very spindly and fall over into the water.

the xylem from their sources in roots. They are especially concentrated in developing seeds. Like IAA, GAs promote cell elongation much more than cell division, but in addition to promoting shoot growth, they may also induce flowering, promote seed germination, break dormancy, and stimulate a number of other events. In some cases, they allow growth to occur at lower temperatures than normal. GAs are commonly used commercially, with particular success in increasing yields of seedless grapes and sugar cane.

GAs increase the transcription of messenger RNA, but it isn't clear if this is a general increase in all transcription or a specific induction of certain genes. They clearly do induce new enzymes during the germination of barley or corn seeds; gibberellins from the developing embryos diffuse to the aleurone layer surrounding the endosperm tissue where starch is stored, and they stimulate aleurone cells to synthesize several new enzymes, including hydrolytic enzymes that release food stored in the endosperm. For instance, α-amylase hydrolyzes starch in the endosperm into glucose, which is food for the young plant.

39.6 Cytokinins regulate cell division and differentiation.

Our knowledge of another class of plant hormones, the **cytokinins,** has come largely from the development of plant tissue culture techniques. Plant organs, small pieces of tissue, and individual cells can be grown in a nutrient medium, either in liquid or on a solid agar medium, using techniques similar to those for growing bacteria or animal cells in tissue culture. Some of the first experiments along these lines were designed to culture roots, and they produced a very interesting result: Roots grew very well for a while with sucrose as a carbon source, but then their growth stopped. It turned out that vitamins were needed to support continued growth—different combinations of vitamins for different species. Plants normally make their own vitamins, and these experiments showed that the vitamins are usually made in the leaves and other tissues and then transported to consuming tissues such as the roots.

In 1955 Carlos Miller, working in Folke Skoog's laboratory, tried to get tobacco pith tissue to grow in tissue cultures containing auxin. He found that the cells would grow, even to huge sizes as cells go, but would not divide. The nuclei of these giant cells continued to divide, but with no cytokinesis, each cell had many nuclei. Miller and his colleagues were able to induce normal cell division by adding extracts from yeast and other materials, thus showing that these extracts contained some kind of growth regulator. From these materials they eventually isolated a substance they named **kinetin,** which was later identified as a derivative of adenine:

HC—CH
HN—CH_2—C C
O
N N CH
H N N H

Kinetin

While developing tissue-culture techniques, Frederick C. Steward and others set out to find exactly what ingredients were needed to ensure growth. In addition to a number of vitamins and minerals, they found that a factor from coconut milk was essential for cell division. (Johannes van Overbeek had discovered the stimulatory effects of coconut milk on embryos in culture in 1942.) Other sources of this factor were identified later, and in 1964 *zeatin* was isolated from corn kernels:

H CH_3
HN—CH_2—C=C
CH_2OH
N N CH
H N N H

Zeatin

Figure 39.8
The balance of auxin and cytokinin determines whether a bit of tissue will grow into undifferentiated callus or develop into roots or shoots.

Zeatin is very similar in structure to kinetin, and all these compounds that promote cell division are known as cytokinins.

Cytokinins work cooperatively with auxin and do not promote cell division in its absence. Furthermore, the path of cell differentiation is controlled by the relative amounts of cytokinin and auxin, as the experiment shown in Figure 39.8 demonstrates. Pith tissue is removed from a tobacco plant and sterilely placed on the surface of nutrient agar medium lacking hormones. This control culture grows for a short time until the hormones carried along with the tissue are exhausted. The result is a small mass of new, undifferentiated tissue called **callus.** Varying amounts of cytokinin and auxin are added to other cultures of pith tissue. Low auxin concentrations and intermediate amounts of cytokinin support active cell proliferation, producing massive calluses of new, undifferentiated cells. Keeping the auxin level constant and increasing the cytokinin level inhibits callus formation and makes the shoot system of the plant develop. If the cytokinin level is reduced and the auxin level raised, a root system develops without any shoots. With care, one can adjust the concentrations of both regulators to just the level required to produce a perfectly normal plant with both root and shoot systems. Similar results have now been obtained with other plants.

This is a spectacular result. Merely adjusting the concentrations of two small organic molecules directs unspecialized cells either to proliferate and remain unspecialized or to develop along quite specific pathways. Apparently these two hormones can turn different sets of genes on and off, and this is remarkable in itself because it suggests that relatively simple mechanisms regulate the parts of the plant genome that are critical for differentiation. The comparable event in an animal—if it could be accomplished—would be to scrape off a bit of your skin tissue and grow it into carbon-copy clones of yourself in culture just by adding a few hormones.

Agricultural scientists are actively pursuing the practical outcomes of experiments like this. For instance, if a superior strain of agricultural or forest plant is found, large numbers of identical plants can be grown quickly, bypassing the long process of conventional plant breeding. After cells have been transformed with recombinant-DNA methods, they can be grown into new plants and tested for the desired traits. Agents such as polyethylene glycol cause cells in tissue culture to fuse and form new cells containing the genomes of both original cells. Then hormones can be used to induce these hybrid cells to grow into whole new plants with novel genetic potential. The possibilities are enormous.

Cytokinins do more than promote cell division and direct developmental pathways. They also delay aging and decay. For reasons not yet understood, the application of cytokinins keeps harvested plant products fresh. Lettuce, broccoli, and many other fruits and vegetables (as well as mushrooms, which are fungi) are treated with cytokinins to prolong their shelf life, and florists use cytokinins to keep cut flowers fresh.

In spite of all the phenomena in which cytokinins are involved, the molecular basis of their effects has eluded investigators. It is significant that cytokinins are adenine derivatives and that they also occur in cells as nucleotides. Cytokinins promote the transition from G_2 to mitosis by increasing the rate of protein synthesis, and evidence indicates that they do this by enhancing translation rather than transcription.

39.7 Hormones interact to control a plant's shape.

The forms of trees can be understood in terms of the distribution of hormones within them. The tall, stately conifers like pines and firs have a symmetrical conical shape, and many broadleaved trees grow with a distinctly pointed top. These shapes are the result of **apical dominance.** The growth of a stem tends to be dominated by its terminal bud, which suppresses the growth of other buds, especially the axillary buds that form between the stem and leaves (Figure 39.9). This inhibition is due to auxin, which is made in the terminal bud and spreads down through the stem. Auxin inhibits buds near the apex from growing into branches; as the tree grows taller, lower buds escape this inhibition and form branches, the oldest, lowest branches having grown for longer times (Figure 39.10). When a terminal bud is lost or cut off, nearby buds are relieved of their suppression. Horticulturists take advantage of this effect in pruning; a hedge, for instance, grows thicker as each terminal bud is cut off so the stem behind can branch out.

Cytokinins are essential for lateral stem growth. Increasing the cytokinin concentration overcomes the inhibition caused by auxin, and one reason the lower branches of a tree grow longer is that the lower buds are also within reach of cytokinin coming up from the roots. As with so many of the effects of plant hormones, the critical factor is the balance of auxin and cytokinin. Using this principle, horticulturists can regulate the height and bushiness of their plants without pruning, and wheat varieties can be bred with short stems to minimize wind damage.

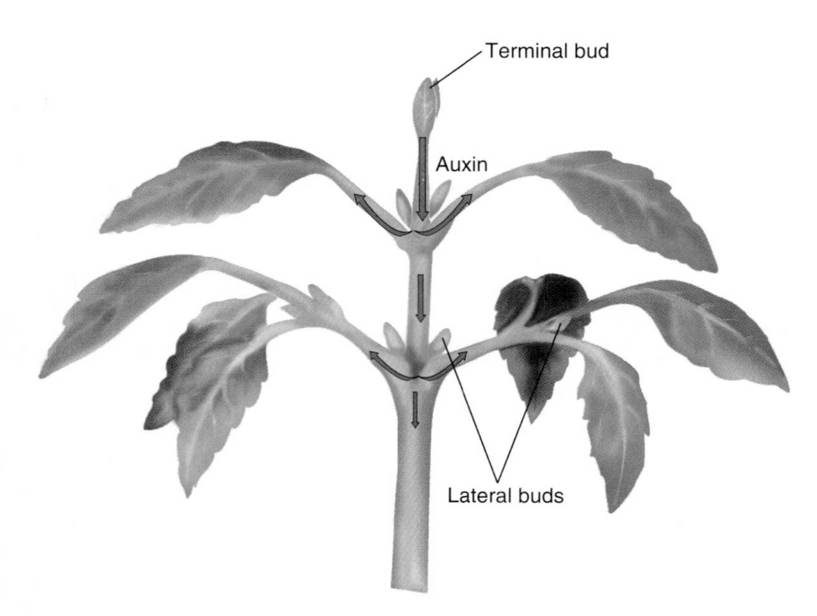

Figure 39.9
The terminal bud of a shoot produces auxin, which tends to inhibit the growth of other buds nearby. This is apical dominance.

Figure 39.10
Apical dominance is an important factor in determining the shape of a plant. A tree develops a conic shape because the high concentration of auxin at the top inhibits lateral growth. The inhibition decreases in proportion to the distance from the tip, so lower limbs grow longer. (The lower limbs have also been growing for longer times, of course.)

39.8 Ethylene promotes fruit ripening and other phenomena.

Perhaps the most amazing growth-regulating substance is ethylene ($H_2C{=}CH_2$). This gas is amazing because of the small, simple structure of the molecule, the extreme sensitivity of plants to it (as little as one part in 10 million parts of air may be effective), and its diverse effects.

Ethylene is synthesized by plant tissues and is also produced by human activities like burning natural gas and petroleum products. Although it has been used for centuries to ripen fruit, it wasn't until 1886 that a 17-year-old Russian scientist, Dimitry Neljubov, identified ethylene as the agent that has specific effects on the growth of pea seedlings. In 1934 the English biologist R. Gane provided evidence that ethylene is a natural product of plants and has a vital regulatory role. It used to be a common practice for orange and grapefruit growers to ripen their fruit in sealed rooms equipped with kerosene burners, and the ancient Chinese ripened fruit in the presence of burning incense. Combustion in both situations produced ethylene. Excess ethylene, though, can be detrimental to plants. Houses, and even greenhouses, heated with natural gas may prove inhospitable to growing plants in part because of the ethylene created as a by-product.

Ethylene is synthesized by many plant tissues, including seeds, flowers, fruit, leaves, and roots, and it does more than promote ripening. When seeds germinate, the stem pushes up through the soil toward the light, its movement facilitated by the formation of a **crook,** which is controlled by ethylene (Figure 39.11). As an advancing stem encounters resistance from the soil, ethylene production is stimulated in some unknown way, and the gas stimulates the crook to tighten so it can push through the soil particles more easily.

Auxin stimulates ethylene synthesis, and many of the direct effects of ethylene we now recognize were once thought to be

Figure 39.11
The crook shape of a growing stem allows it to push around barriers and through soil to reach the surface.

effects of auxin. Recall, for instance, that excess auxin inhibits growth. The reason, apparently, is that auxin stimulates ethylene production and ethylene inhibits cell elongation. In fact, ethylene induces a general expansion of cells in all directions rather than in the one preferred direction of stem growth. Ethylene may also be responsible for inhibiting root growth and the growth of axillary buds, which has been attributed to auxin. The different responses of shoots and roots to auxin might be explained this way too, by postulating that roots are more sensitive than stems to ethylene inhibition. Finally, ethylene is a classical inhibitor of gravitropism; seedlings respond to ethylene with the typical "triple response"—horizontal growth, inhibition of elongation, and swelling. In the presence of ethylene, the stem of a pea seedling never grows upward. This inhibition, too, was once attributed to auxin.

Ethylene synthesis begins with the amino acid methionine, and ACC oxidase is a key enzyme:

COOH–C(NH$_2$)–CH$_2$–CH$_2$–S–CH$_3$ (Methionine) $\xrightarrow{\text{ACC synthase}}$ HOOC–C(NH$_2$) in a ring with H_2C–CH_2 (Amino cyclopropane-1-carboxylic acid (ACC)) $\xrightarrow{\text{ACC oxidase}}$ CO_2 + HCN (Hydrogen cyanide) + $H_2C{=}CH_2$ (Ethylene)

Several factors, including ethylene itself, induce the synthesis of ACC oxidase, making a positive feedback system in which a little ethylene induces the production of much more. The pathway of ethylene action is being dissected through biochemical and genetic analysis, particularly using mutants of *Arabidopsis* and tomatoes. Ethylene first binds to a receptor protein associated with cell membranes; since the gas is quite soluble in cell membranes, it is probably able to enter the cytoplasm and bind to other receptors there. Some of these receptors appear to be protein kinases, so the first effect of ethylene is likely to be a kinase cascade (Figure 39.12), entailing the activation of some proteins and the inactivation of others. The final effects of ethylene require induction of a large set of new proteins; the promoters of the genes for these proteins include an ethylene-response element (ERE) with a characteristic repeating sequence (GCCGCC). A series of other proteins recognize the ERE and bind to it to regulate the responsive genes, and these proteins share sequence features with homeotic plant proteins, which are also known to be DNA-binding regulatory proteins. The ripening of fruit is an important response to ethylene, and studies in tomatoes particularly have shown how ripening is effected by certain newly induced enzymes (Figure 39.13).

Homeotic genes, Section 21.12.

The commercial use of ethylene for ripening fruit has reached impressive proportions, although agriculturists may not understand that ethylene is involved. For instance, the practice of putting unripe fruit in closed containers simply increases

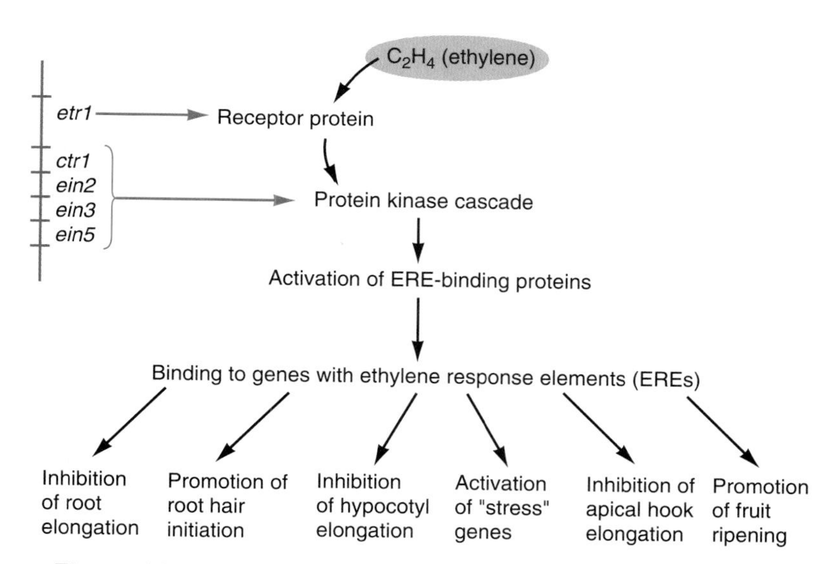

Figure 39.12

Ethylene acts through a complex series of events that have been analyzed largely with mutants; some *Arabidopsis* genes whose products are involved in the pathway are listed to the left. Ethylene binds to one or more receptor proteins which then start a kinase cascade. The cascade activates gene-regulation proteins that bind to ethylene response elements (EREs) associated with many genes.

the concentration of the gas, as does wrapping fruit individually in paper. A bit of ripe fruit will trigger ripening in nearby fruit on a massive scale because ethylene production is autocatalytic, so sometimes a few pieces of ripe fruit, often bananas, are placed with unripe fruit to hasten ripening. Ethylene is used in quite conscious and controlled ways in agriculture; bananas, mangos, citrus fruits, tomatoes, and many other fruits are picked before final ripening and then ripened when needed with ethylene.

Ethylene affects other plant processes besides ripening. It promotes seed germination and flower production, an effect that is exploited by pineapple and melon growers to stimulate flower formation. Ethylene also enhances the thickening of tree trunks, so nursery growers sometimes use it to make more stable ornamental trees. Silver ions specifically antagonize ethylene, perhaps by displacing a metal such as copper from the ethylene receptor. This effect is also used commercially—for instance, by spraying cut flowers with a silver compound to extend their vase life.

39.9 Ethylene also promotes the abscission of plant organs.

Questions about factors that stimulate or inhibit the growth of buds lead into questions about a plant's life cycle and the general matter of dormancy. After a cold winter with short days, a typical plant in a temperate climate renews its growth. Through the longer days of spring and a hot summer, its stems grow and it develops new leaves. In the fall, as the days grow shorter and colder again, it stops growing. Deciduous trees drop their

Figure 39.13

Ethylene induces the synthesis of several enzymes involved in the ripening of fruit, such as tomatoes. Pectins that constitute much of the green fruit are polygalacturonic acid, which is digested to soften the fruit, while other enzymes digest other polysaccharides. Chlorophyllase removes the green chlorophyll, and lycopene produces the red color of the ripe fruit. Other enzymes protect the fruit against bacteria and fungi.

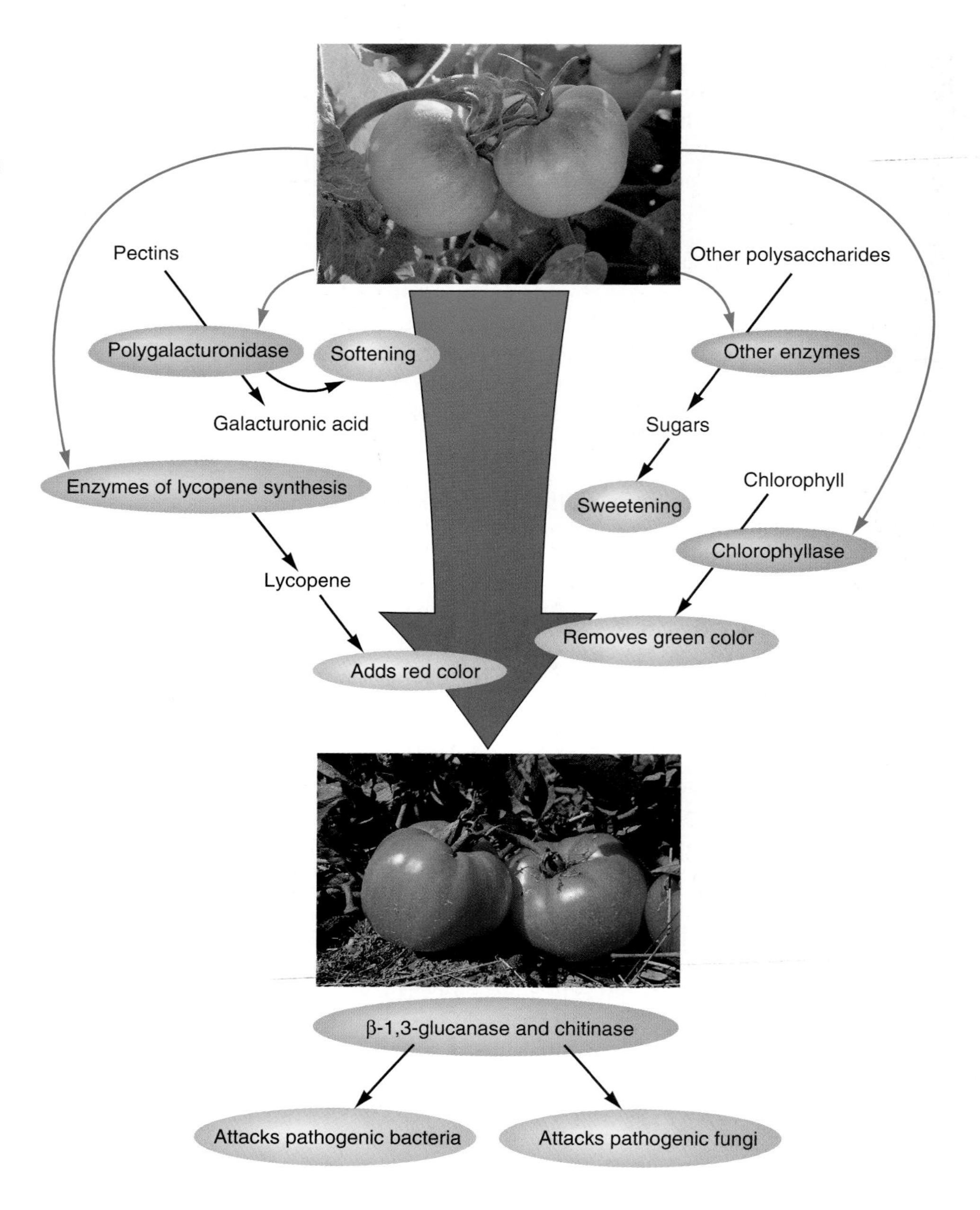

leaves, an adaptation that reduces water loss through transpiration; in the winter, less water is available to plants, the air is drier, and the ground freezes, so plants might die of desiccation if they retained their leaves. However, leaves, flowers, and fruits don't just break off. The plant separates them in the process of **abscission,** in which the vascular system is sealed off at appropriate points to prevent the loss of water and nutrients and to exclude bacteria, fungi, and other pathogens. An **abscission zone,** a layer of specialized cells, forms at the base of each part, and its cells die and become hardened by deposits of lignin and suberin, so by the time a leaf or fruit drops, its vascular system has been sealed off. The abscised part is hanging only by its epidermis and vascular bundles, and these slender connections are easily broken by a strong gust of wind.

In the course of research on abscission, a plant hormone called **abscisic acid (ABA)** was identified as the controlling agent:

H_3C CH_3 H CH_3 ... OH H ... CH ... COOH ... O ... CH_3

Abscisic acid (ABA)

Now, ironically, the compound that bears this name turns out to have nothing to do with abscission! ABA, which is synthesized mostly in chloroplasts, is a general inhibitor of many processes. But the abscission layer forms and hardens under the control of

IAA and ethylene. IAA inhibits the formation of the abscission layer, but when its production decreases due to shorter days, wounding, or other causes, ethylene production increases, and it is ethylene that stimulates cells in the abscission zone to harden.

Still, ABA has other roles in plant development. For instance, it induces tolerance to stress and aids in water conservation during wilting. When a plant loses water and begins to wilt, ABA production is enhanced, and ABA closes the stomata, among other effects. As soon as enough water returns to the plant, ABA is enzymatically degraded, and the guard cells return to normal.

Wilting and water conservation, Section 38.7.

Genetic regulation by ABA is understood better than regulation by most other plant hormones (Figure 39.14). ABA is known to regulate more than 70 genes, both positively and negatively. Those genes that have been sequenced share an ABA-responsive element—that is, a common DNA sequence associated with each gene that is recognized by regulatory proteins, which are now being identified. The entire regulatory pathway is complex and not known completely, but it fits a classic model of genetic regulation by a small ligand.

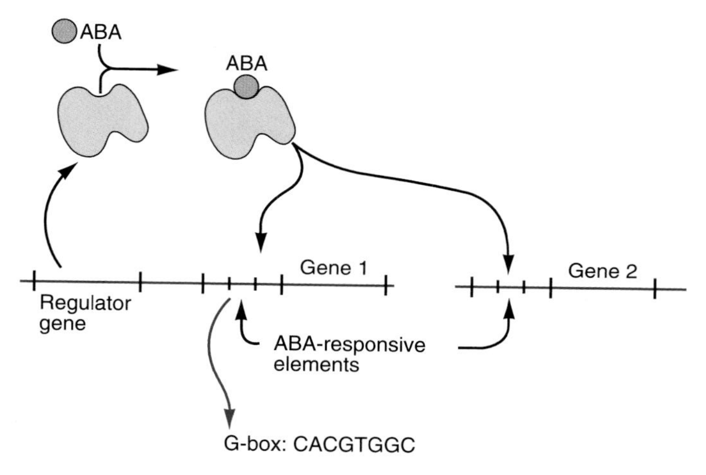

Figure 39.14
Many plant genes are regulated by abscisic acid (ABA). ABA binds to various regulatory proteins, which then bind to genes next to an ABA-responsive element. This is a sequence of 11 base pairs including a so-called G-box with the sequence CACGTGGC.

General model of genetic regulation, Section 18.3.

Figure 39.15 summarizes the effects of the various hormones and environmental stimuli.

39.10 Systemin is a signal that induces the wound reaction in plants.

So far, while discussing the differences between plants and animals, we have not mentioned defense. Animals, on the whole, seem better able to take care of themselves. Animals are able to move quickly, to attack or flee from a predator; if an herbivore starts to attack a plant, the plant seems to have no option except to be eaten. Yet it is unlikely that over hundreds of millions of years plants would not have been selected for other defenses. Indeed, some plants have sharp spines and barbs, often including toxins and irritants. Plants also have quite a rapid reaction, the **wound response,** to an attack by insects. They quickly accumulate defensive proteins, including protease inhibitors; as part of the digestive process, insects produce proteases, which digest the proteins in their food, but a protease inhibitor inactivates these proteins, leading to malnutrition, reduced growth, and perhaps death. Other enzymes that accumulate as part of the wound response have less obvious effects on the insect predators, but we may be sure they do the insects no good.

The signal ligand that induces the wound response is a peptide of 18 amino acids called **systemin.** It is cleaved off a protein, *prosystemin,* and carried through the phloem to all tissues near the point of injury. Within 60–90 minutes, systemin can be detected in leaves a few centimeters away from an injured leaf.

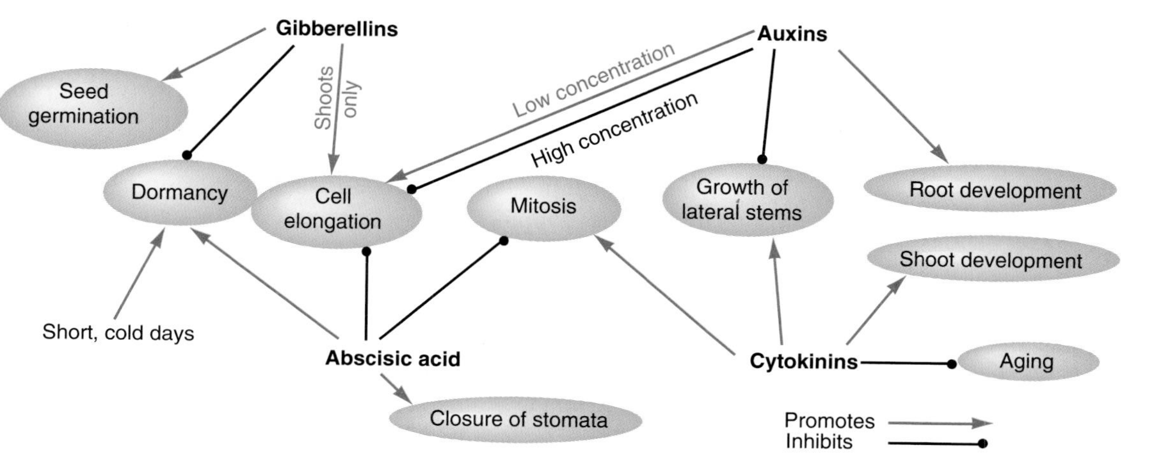

Figure 39.15
The four plant hormones (auxins, cytokinins, gibberellins, and abscisic acid) interact to promote and inhibit a number of processes, as summarized here.

Figure 39.16
Systemin signals initiation of the wound reaction. It is transported through phloem to target cells where it activates a pathway entailing the conversion of 18-carbon fatty acids. Jasmonic acid then induces transcription of several genes that encode defensive proteins.

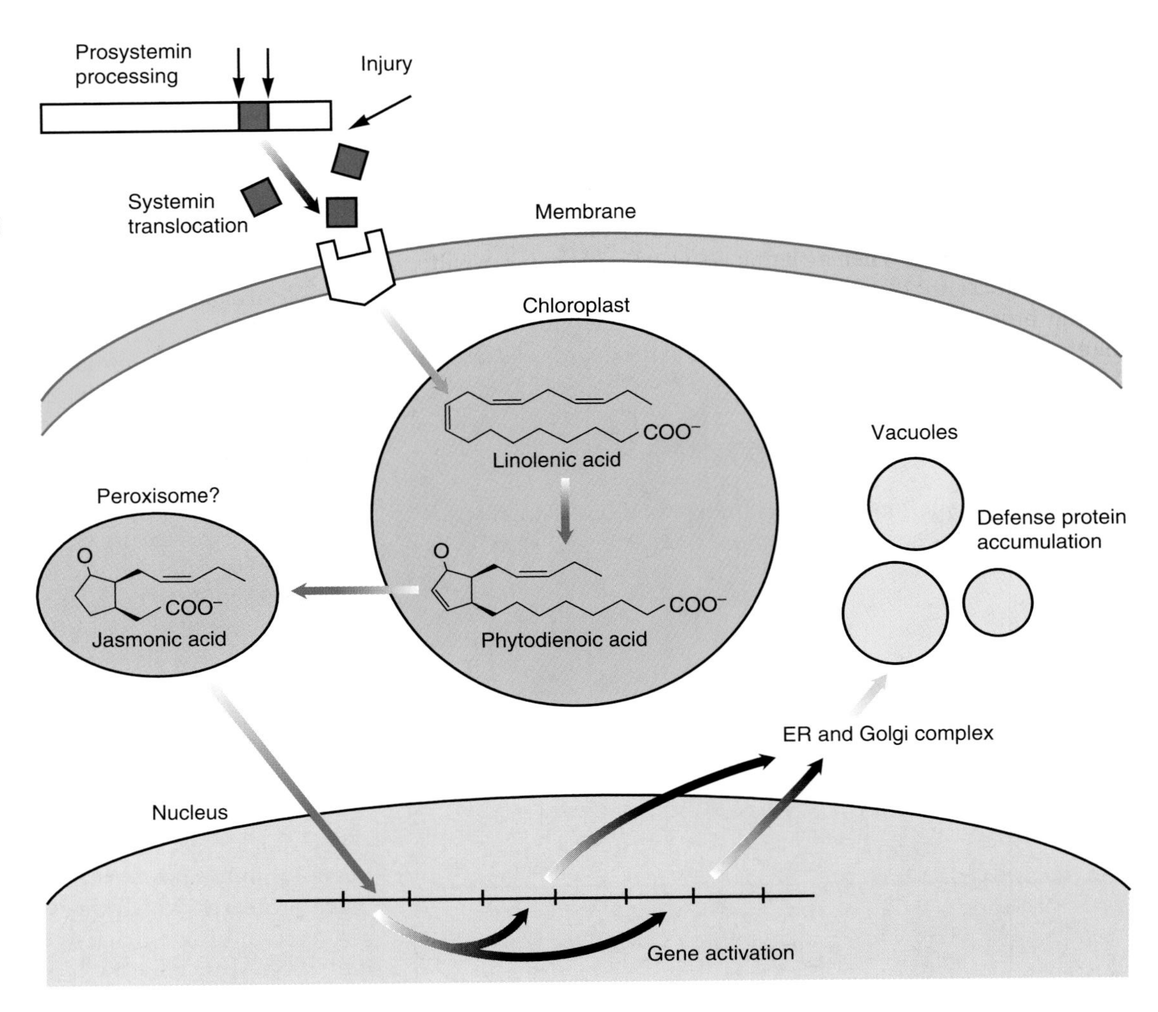

Systemin then induces genes for proteins of the wound response through a pathway that is interesting because of its similarity to animal pathways (Figure 39.16). The effective agents are 18-carbon fatty acids very similar to the prostaglandins and related effectors active in some animal pathways, especially in the inflammatory response.

Just what induces the production of systemin in response to a wound is still not clear. Abscisic acid may be necessary for the response, but it is apparently not the primary agent. Current research is showing that plants communicate internally through a variety of signals; systemin is just one of the best known at present.

B. Photoperiodism

The hormones described in Part A govern plant growth and development, but their release is controlled by environmental factors such as temperature and day length. So we now turn our attention to the fascinating questions of how light controls growth, development, and metabolism. The period of light during each day is the **photoperiod,** and an organism's response to it is **photoperiodism.** We will see that both the intensity and duration of light are significant in photoperiodic control.

39.11 Flowering and other phenomena are regulated by the photoperiod.

Year after year, plants develop and flower on a regular schedule. Snowdrops and crocuses bloom in early spring, ragweed and chrysanthemums flower in the fall, and other plants produce their flowers at the anticipated times. Extreme heat, cold, or unusual moisture conditions may shift the program slightly, but we know just about when to schedule a tulip or cherry blossom festival and when to walk through the woods looking for azaleas and dogwood. Furthermore, it is evident that any type of plant will flower earlier at a more southern location (in the Northern Hemisphere, that is; the reverse is true south of the equator). Spring beauties flower in Georgia before they do in New England, and poppies will brighten gardens in California before those in Washington.

How does a plant know when to bloom? The answer is that plants use the daily cycle of light and darkness as a calendar, a daily clock, and a map. In the Northern Hemisphere, days get longer (and nights shorter) as winter turns to summer, then shorten again as winter returns. The day length also varies with latitude; over a yearly cycle, day length varies little at the equator but changes between extremes of endless dark and endless light at the poles. Plants have evolved to use the cycles of light and dark as signals for flowering.

Photoperiodism was discovered in plants in 1918–20 by Wightman W. Garner and Harry A. Allard at the Plant Industry Station in Beltsville, Maryland. One of their tobacco strains, Maryland Mammoth, grew to great heights in test fields during the summer but would not flower unless it was brought into a greenhouse to continue growing in the winter. If, however, the same variety was raised from seed in a greenhouse in the fall, it grew to a normal height and flowered normally. After testing several factors, Garner and Allard finally speculated that the *duration* of light was most important. In testing this idea, they found that plants that wouldn't normally flower in Beltsville in the summer could be made to do so if they were subjected to artificially long nights—17 hours in one set of experiments, thus reducing the amount of daylight to seven hours. They obtained similar results with several varieties of soybeans.

It seems reasonable, at first, to think that the length of the photoperiod in a dark/light cycle determines flowering. In nature, the sum of night and day is always 24 hours, but in a laboratory cycles can be of any length, and experiments in changing the dark and light periods have revealed that the length of *darkness* is really the crucial factor. Unfortunately, the fact that plants actually react to the night length has been ignored in describing photoperiodism, and most species are classified as **short-day** or **long-day** plants. Yet these terms are very misleading. The critical factor is not the length of day or night but rather how each type of plant responds to a certain critical photoperiod. A short-day plant is not one that flowers only when the days become short but one that only flowers if subjected to photoperiods *shorter* than a certain critical day length. A long-day plant will only flower with photoperiods *longer* than a critical day length (Figure 39.17). But the critical day lengths for some long-day plants are shorter than those for some short-day plants.

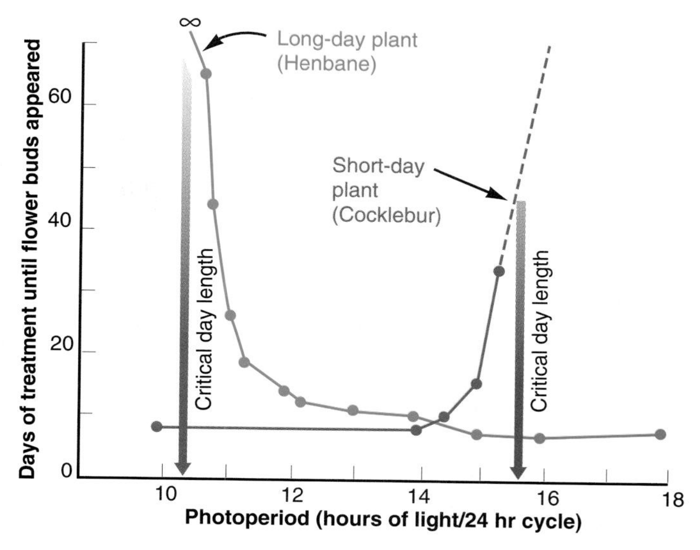

Figure 39.17

A long-day plant only flowers when the photoperiod exceeds a certain minimum critical length; it responds more strongly to longer photoperiods. A short-day plant only flowers when the photoperiod is less than a certain maximum critical length, and it responds more strongly to shorter photoperiods. Source: Ray, *The Living Plant*, 1968, Holt, Rinehart and Winston.

Exercise 39.1 Specimens of a certain plant that are not yet flowering are subjected to two light regimes, maintaining 24-hour total days. Both groups begin with 12 hours of light and 12 hours of darkness. Group 1 plants are given longer and longer dark periods, while Group 2 plants are given shorter and shorter dark periods. When the dark periods for Group 1 plants reach 14 hours, the plants begin to flower. Group 2 plants never flower. Is this species a long-day or a short-day plant?

Exercise 39.2 Specimens of a second species of plant that are not yet flowering are subjected to the same light regimes as those of the first species. Group 1 plants are given longer dark periods, while Group 2 plants are given shorter dark periods (starting with 12 hours of light and 12 of dark). When the dark periods for Group 2 plants reach 13 hours, the plants begin to flower. Group 1 plants never flower. Is this species a long-day or a short-day plant?

Exercise 39.3 The critical day length for the cocklebur *(Xanthium strumarium)*, a short-day plant, is 15.5 hours. The critical day length for henbane *(Hyoscyamus niger)*, a long-day plant, is about 11 hours. Explain how the critical time for a short-day plant can be longer than the time for a long-day plant.

Typical short-day plants will only flower if they are subjected to increasing dark periods of about 12–14 hours. These include strawberries, chrysanthemums, dahlias, goldenrod, sorghum, and violets. Long-day species, such as lettuce, spinach, wheat, potatoes, and larkspur, generally need shorter nights of perhaps 10–12 hours. Some plants flower only with short days that come after long days (as in the conditions of late summer or fall) or the reverse. A few plants flower either with short days or long days, but not with intermediate-length days. There are also day-neutral plants, whose flowering is not controlled by the light/dark cycle; these include dandelions, tomatoes, garden beans, snapdragons, cotton, sunflowers, and roses. Their behavior makes sense when you see that many of these species originated in the tropics.

In some species, control of flowering is very precise. This has been particularly well studied in agricultural crops where understanding photoperiodism is of major economic importance. In some varieties of soybeans, for instance, the difference between flowering and not flowering depends on less than a half hour difference in the photoperiod, which corresponds to 200 miles or less in latitude. Thus the northern growing limits of such plants may be determined more by photoperiod than by temperature.

Several activities besides flowering respond to the photoperiod. For instance, sprouting in potatoes is short-day regulated, while in strawberries, flowering is facilitated by short days and development of runners by long days. Dormancy and the breaking of dormancy are also regulated by both photoperiod and temperature. Just as seeds go through a period of dormancy, when their metabolism is suppressed and growth stops, a plant becomes dormant at the end of a growing season.

Abscisic acid is one factor that induces dormancy in buds, and it also suppresses the germination of seeds while they are still attached to a plant. Abscisic acid formation is apparently induced by the short days of autumn, so that even subjecting a plant to artificially short days during the summer can send it into dormancy. The cool fall is followed by a cold winter with very short days, and then lengthening days toward spring. Plants use these events as signals to break dormancy. Even in favorable temperature and light conditions, many plants won't break the dormancy of their buds and develop new leaves and flowers unless they have been previously exposed to a number of short, cold days, ideally a few degrees above freezing. In effect, plants have evolved a mechanism that anticipates the changing seasons and predicts that short, cold days will be followed by warmer days when they can grow. The induction of dormancy and abscission in the fall is a built-in prediction that hard times are about to follow.

Exercise 39.4 Explain why it makes sense that many plants native to the tropics are day-neutral plants.

Exercise 39.5 Gibberellins antagonize the effects of abscisic acid on dormancy. What effects would you expect if GAs are applied to dormant buds or to those that have not yet gone into dormancy?

Exercise 39.6 A species of plant only flowers when subjected to a series of short nights followed by longer nights. During what season will it bloom?

39.12 The phytochrome system measures the photoperiod.

We showed earlier that the length of darkness is most critical in photoperiodism. Subjecting plants to various light/dark cycles led to a critical discovery: Even a single burst of light, lasting only a minute, is enough to interrupt the dark period. A long dark period keeps a long-day (read: "short-night") plant from flowering, but a brief exposure to light during the dark period (a *night-break*) starts its flower development, as if the nights had suddenly been shortened as they are in spring. The same short burst will suppress flowering in a short-day (read: "long-night") plant. These results are very puzzling. For instance, an artificial 40-hour night will induce flowering in a short-day plant, but a night-break only 8 hours into this night will abolish the effect, even though the remaining 32 hours of darkness should certainly be enough to induce flowering. Such experiments reveal how much we still do not know about this system.

Insights into the mechanism of photoperiodism come from the observation that not all wavelengths of light are equally effective in establishing photoperiods. In long-day plants, a night-break burst of red light with a wavelength of 660 nm is most effective, suggesting that the light must be absorbed by a pigment with this absorption maximum. On the other hand, the effect will be reversed if a brief exposure to red light (or white light, which contains red) is followed within half an hour by a short exposure to far-red light, around 730 nm. A burst of red light will make a long-day plant flower, while far-red light will inhibit flowering. In fact, if exposed alternately to red and far-red light, the plant will respond to the last burst of light, whichever it is.

After investigating these mysterious phenomena for several years, Sterling B. Hendricks and Harry A. Borthwick postulated that plants have a light-receptor pigment, which they called **phytochrome,** that must exist in two forms: P_r, which absorbs red light, and P_{fr}, which absorbs far-red light. The molecule changes from one form into the other when it absorbs light. So when plants are illuminated with red light, as they are all day, the pigment is all converted from the P_r to the P_{fr} form. At night the P_{fr} form slowly changes spontaneously back to P_r. A burst of red light during the night converts the P_r back to P_{fr}. That is:

$$P_r \underset{\text{Far-red light or slow, spontaneous change}}{\overset{\text{Red light}}{\rightleftharpoons}} P_{fr}$$

Phytochrome was subsequently identified as a protein with just these properties. Its chromophore, the pigment that actually absorbs light, is a tetrapyrrole (Figure 39.18), an open tetrapyrrole similar to the bile pigments of animals and the

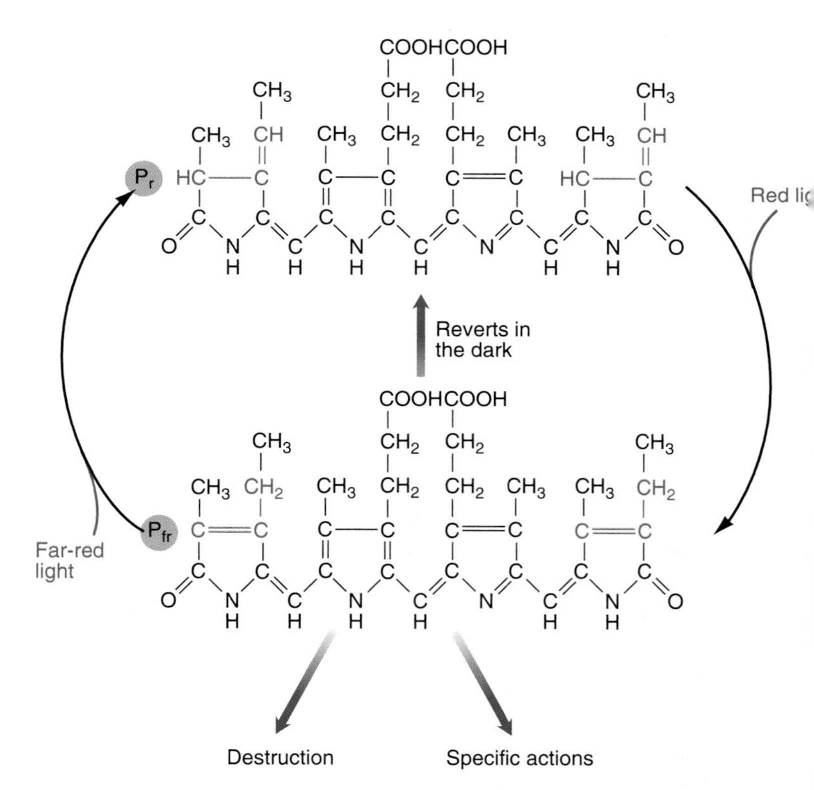

Figure 39.18

Phytochrome is a protein with a tetrapyrrole chromophore that can exist in two forms, and it is converted back and forth between the two by absorbing either red light or far-red light. P_{fr}, the active form of the protein, reverts to P_r in the dark and it is also destroyed by other cellular processes.

phycobilin pigments of cyanobacteria and red algae (see Chapter 26). Phytochrome responds to red or far-red light by shifting from one form to the other. P_r is metabolically inert; the active P_{fr} form has an exposed hydrophobic site on the protein. What are some of this protein's effects?

Tetrapyrroles, Concepts 4.2.
Phycobilin pigments, Section 29.4.

39.13 The phytochrome system controls many processes.

Phytochrome is a plant's principal light receptor. (Plants have other receptors that respond to blue light and to ultraviolet light, but relatively little is known about them, compared to phytochrome.) It allows plants to respond in many appropriate ways to the light that is so important to them: orienting both leaves and chloroplasts to receive more light or less, achieving growth and differentiation of chloroplasts, and synthesizing pigments such as the chlorophyll and carotenoids of leaves and the anthocyanins and flavonoids of flowers and fruits. Phytochrome also regulates more complex processes of growth and morphogenesis. You have probably turned over a rock or an old board and seen a small plant growing underneath, with a white, straggly stem and a few thin, yellowish leaves. Such plants are **etiolated** (Figure 39.19). Their growth form, with long internodes (the space between leaves) and tiny, chlorophyll-poor leaves, is adaptive for a plant in the dark that can't carry out photosynthesis and is most likely to survive by growing quickly into the light. Etiolation, being under the control of phytochrome, is eliminated by exposure to red or white light.

Figure 39.19
An etiolated seedling is characterized by a long, thin, yellow stem with small leaves.

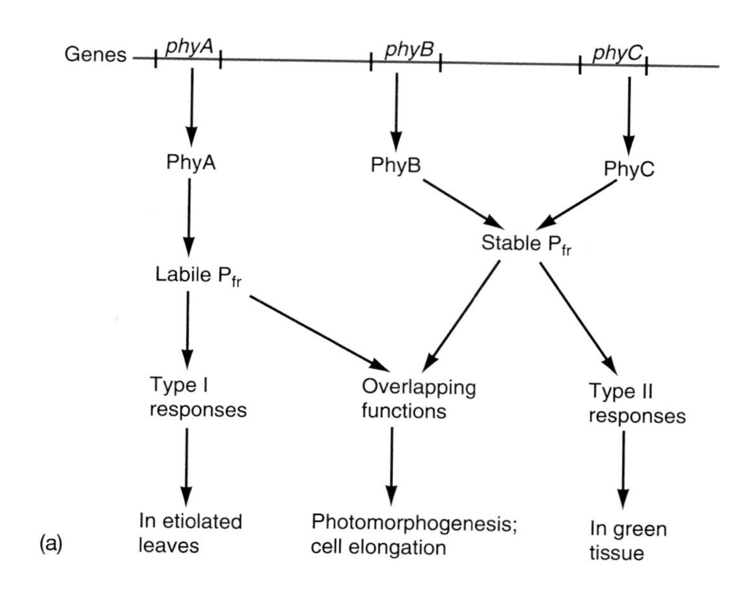

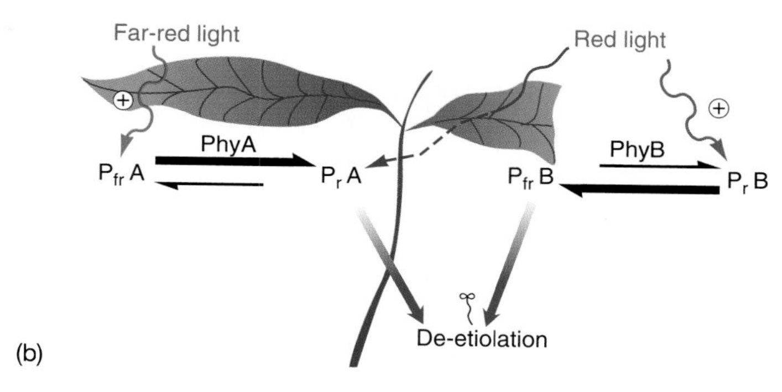

Figure 39.20
Plants adapted for growth in sunny areas grow away from shade through a reaction that depends on phytochromes. *(a)* Phytochromes A and B have different functions and stabilities. PhyA predominates in etiolated tissue and is the primary receptor of far-red light; PhyB predominates in green tissue and is the primary receptor of red light. (PhyC is not known very well but seems to function like PhyB.). *(b)* The shade of vegetation is enriched for far-red light, but the PhyA that receives it in etiolated seedlings is unstable and breaks down in continuous light. The stable PhyB in these shaded seedlings remains in the P_r form that maintains etiolation, so the seedling continues to elongate and grow away from the shade. When the plant grows into open sunlight, PhyB is converted into the P_{fr} form, which de-etiolates the plant.

Phytochrome has been identified in many plants, from angiosperms and gymnosperms to green algae; it is safe to say that it is found in all plants. In fact, plants have more than one species of phytochrome. *Arabidopsis* has five phytochrome genes, *phyA, B, C, D,* and *E;* the PhyA, PhyB, and PhyC proteins are well characterized and have different functions, as illustrated by the shade-avoidance reaction of some plants (Figure 39.20). PhyA is necessary for perception of continuous far-red light, and PhyB for perception of continuous red light. An emerging seedling in open sunlight, which

Figure 39.21

Phytochrome is a light receptor whose P_{fr} form acts through the standard signal-transduction pathway. The subscripts "i" and "a" refer to the inactive and active forms of proteins. CBP represents a number of calcium-binding proteins, including calcium-activated protein kinases. IP_3 is inositol triphosphate.

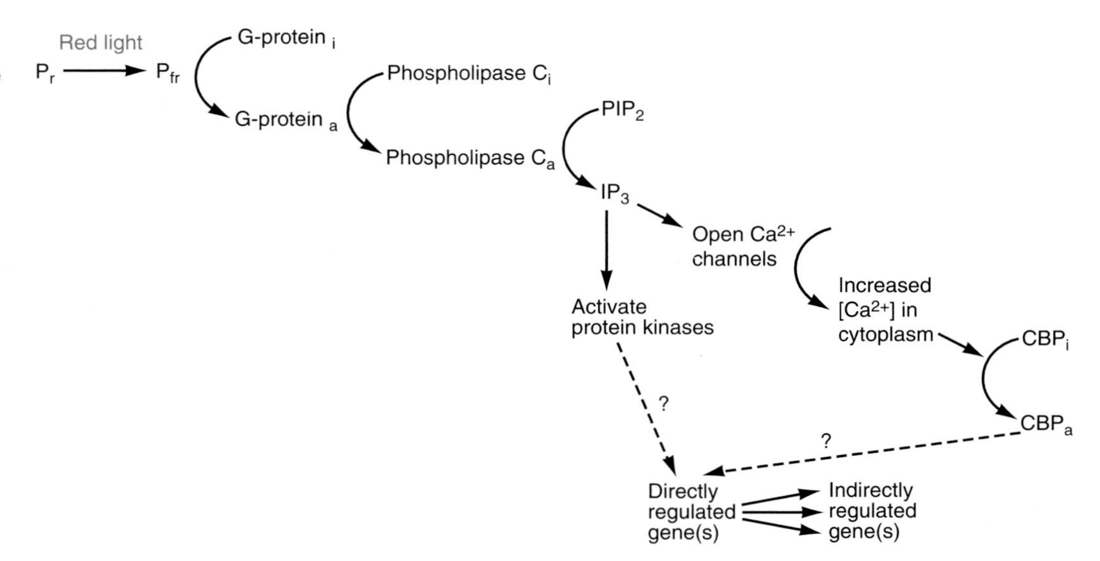

is rich in red light, is de-etiolated by the P_{fr} form of the stable PhyB. Because plants absorb red light in their chlorophyll, the shade of vegetation is enriched for far-red light; in an etiolated seedling emerging in the shade, PhyA absorbs this far-red light and de-etiolates the seedling. PhyA is labile, however, and is broken down under the influence of continuous light, so a shaded seedling is soon left with mostly PhyB, which induces the seedling to elongate so it tends to grow away from the shade.

Phytochrome is a receptor. It is linked to an effector system that employs the typical eucaryotic signal-transduction pathway of G-proteins and second messengers (Figure 39.21), though in plants, Ca^{2+} ions appear to be the principal second messenger, with some involvement of cyclic GMP and no involvement of cyclic AMP. At the end of this pathway, genes are clearly induced and repressed by regulatory proteins still being identified by means of mutations. Genes turned on by the P_{fr} form of phytochrome have a specific sequence, a light-responsive element, in their promoter regions. The clearest case of gene regulation by phytochrome is self-regulation of the *phyA* gene that encodes phytochrome A (Figure 39.22).

The effect of phytochrome is generally visualized with a kind of hourglass model: A process starts when the hourglass is turned over and is completed if there is enough time for all the sand to run through. Since the pigment reverts from P_{fr} to P_r in the dark, one can generally explain flowering in short-day plants if some process starts when a dark period begins, so if the dark lasts long enough, the process will go to completion, and flowering will begin. But this model is inadequate. For one thing, the conversion from P_{fr} to P_r in darkness happens in an hour or so, but the critical times are much longer. Other experiments show that flowering depends upon an internal rhythm, as shown in Figure 39.23. Various short-day plants will flower when exposed to dark periods of 10–15 hours. If the dark period is increased, however, the flowering response falls off and reappears in a rhythm. So the so-called critical day length is not

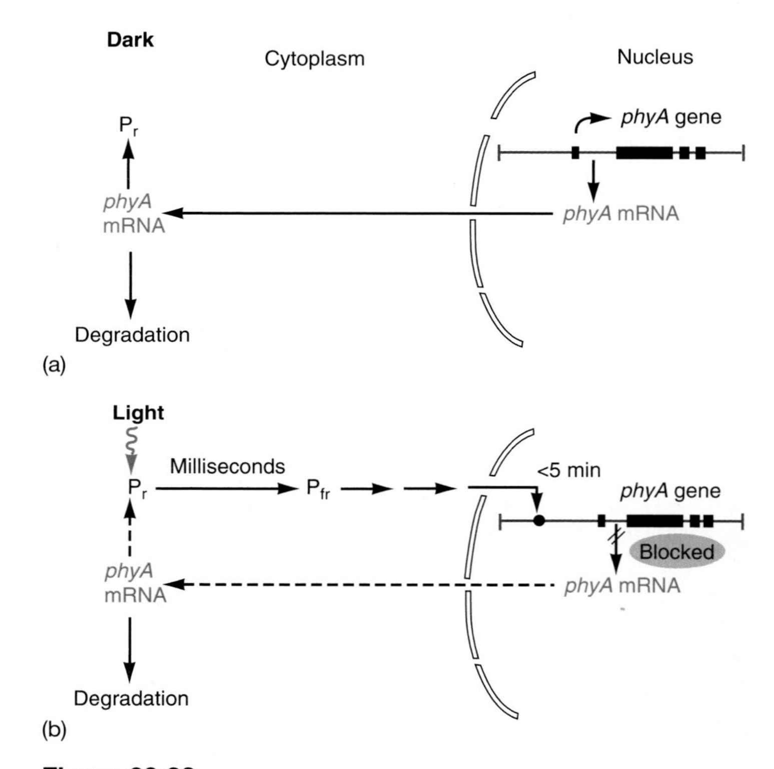

Figure 39.22

Phytochrome A regulates its own gene indirectly. *(a)* In the dark, the *phyA* gene is transcribed into mRNA, and phytochrome accumulates in the cytoplasm in its inactive, relatively stable P_r form. *(b)* Red light converts P_r into P_{fr} within milliseconds, and within minutes, transcription of the *phyA* gene is repressed. P_{fr} is slowly degraded.

a single time. The plant is evidently measuring light and dark periods with some kind of rhythmic mechanism. Unfortunately, even though biological rhythms are very common (see Chapter 15), they are among the most poorly understood biological phenomena. Therefore, after finding a clocklike

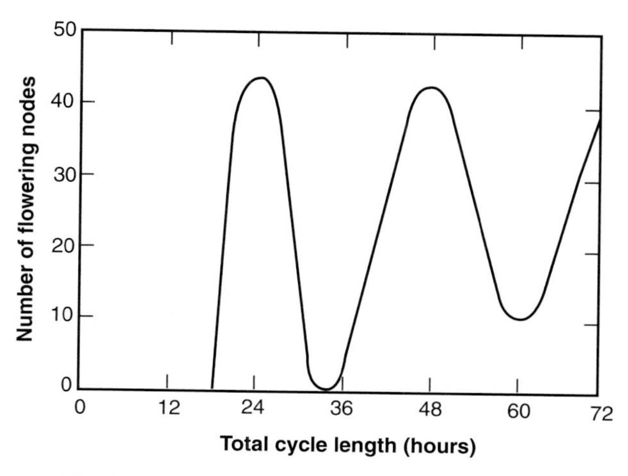

Figure 39.23

Experiments with several species of plants demonstrate that they respond to the photoperiod through a rhythmic mechanism. Plants are exposed to dark periods varying from 3 to 72 hours and are kept in continuous light at other times. Their flowering response shows a clear rhythm. Similar results are obtained by interrupting the dark period with red light after various times. Source: Salisbury, *The Flowering Process,* 1963, Pergamon Press.

mechanism at the base of flowering, we come up short, without a further explanation.

A flowering hormone, florigen, has been postulated, and if it exists, it might be triggered by the phytochrome reaction. But many plant physiologists suspect that there is no such hormone and that flowering is induced by the right combination of the hormones we have already discussed. There is also evidence that in certain cases hormones inhibit flowering, rather than promoting it, and that release of the inhibition, by eliminating the hormone(s), initiates flowering.

Much of the work designed to explore these problems has been done with cocklebur, a nearly indestructible and easily grown little plant. In one instructive experiment, cocklebur plants are grafted together so their vascular systems are continuous and so a substance made in one plant is transported to the others (Figure 39.24). If leaves of only one plant are exposed to a light/dark regimen that induces flowering, the plant bearing that leaf soon flowers; the other grafted plants then follow suit and flower in sequence. This experiment leads to two important conclusions: First, phytochrome detectors are located in leaves, and exposure of one cocklebur leaf is sufficient to set the photoperiod clock. Second, the substance (perhaps substances) that promotes flowering can be transported from plant to plant. Since actual grafting is necessary for the influence to spread, a gas such as ethylene cannot be the effective agent.

This experiment stands in sharp contrast to another designed to explore the breaking of dormancy, which demonstrates that all hormonal influences cannot be transmitted through the vascular system. In this case, only one branch of a lilac bush is exposed to the chill needed to break dormancy, and only that branch leafs out and flowers (Figure 39.25).

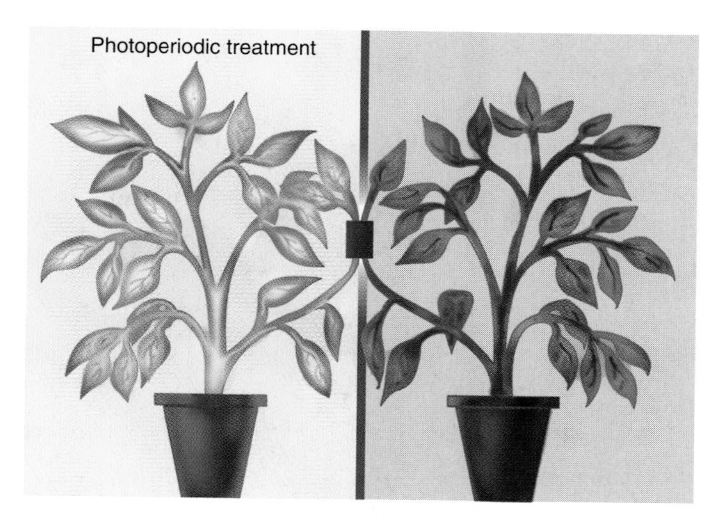

Figure 39.24

An experiment with grafted cocklebur plants demonstrates the existence of a flowering factor, since photoperiodic induction of one plant induces flowering in both.

Coda All organisms respond to their environments. The environment poses both dangers and opportunities, and the best-adapted organisms are able to avoid the former and take advantage of the latter. Most animals and many microorganisms can respond with rapid behaviors. But embryophyte plants cannot move as animals do, and their responses to the environment usually take the form of growth in particular ways and directions. Plant hormones direct this growth. Since the life of a plant revolves so strongly around photosynthesis and water relations, it is understandable that the shoot of a well-adapted plant grows upward and toward light, while its roots grow downward toward likely sources of water. Also, since firm support for the shoot system is advantageous, some plants naturally respond to the touch of a rigid structure with devices such as tendrils that wrap around supports.

Figure 39.25
Exposure of one branch of a lilac to low temperature breaks dormancy only in the flowers on that branch, thus demonstrating that a circulating factor is not involved.

Yet plants and animals use environmental cues and stimuli in quite similar ways to enhance reproduction. They both use the changing photoperiod as a calendar. Lengthening days are a sure sign of spring, a signal that the most favorable time for reproduction is at hand; shorter days signal the onset of autumn, the time to prepare for winter. For plants, that means shedding leaves that will soon be useless or a liability and dropping seeds that can overwinter until other stimuli signal their germination. The phytochrome system, which detects the photoperiod, also guides growth, leading plants in darkness or shade to extend themselves toward the light. As research reveals the details of these processes, we see how sequences of molecular interactions can transduce environmental factors into adaptive responses and how these molecular devices can be molded by natural selection.

Summary

1. All aspects of plant growth are controlled by five major classes of plant hormones: auxins, gibberellins, cytokinins, ethylene, and abscisic acid.
2. Phototropism of a stem occurs when a substance from the stem's tip causes growth on the side away from the light source. The agent has been identified as indoleacetic acid, or auxin, a hormone that promotes cell elongation.
3. Auxin is responsible for both phototropism (growth toward a light source) and gravitropism (growth in response to gravity). Auxin from terminal buds suppresses the development of lateral buds and creates the conical forms of many plants. Auxins have complex, multiple effects. Some synthetic auxins are herbicides.
4. The direction of gravity is apparently detected by amyloplasts in statocytes; even though auxin seems to promote gravitropism in stems, the mechanism is unknown.
5. Gibberellins promote stem elongation and other growth processes.
6. Cytokinins promote cell division. The balance of cytokinins and auxins regulates root or shoot development.
7. Ethylene is a potent promoter of fruit ripening. It induces and represses a large set of genes by means of specific regulatory proteins.
8. Abscission of leaves, flowers, and fruit occurs through the formation of an abscission layer that seals off a plant's vascular system. Abscisic acid was once thought to promote abscission, but now ethylene has turned out to be responsible.
9. Abscisic acid promotes tolerance to stress, induces dormancy of buds, and suppresses the germination of seeds. It appears to be a general inhibitor of growth.
10. Systemin is a peptide made in response to injury to a tissue, as from insects. It moves quickly to other tissues and induces the formation of proteins that inhibit or kill insects, such as protease inhibitors.
11. Flowering, dormancy, and some growth habits are regulated by the relative lengths of the light and dark periods in the daily cycle. Although the phenomenon is called photoperiodism, the critical factor is actually the length of the dark period.
12. Most plants are either short-day plants, which flower when the day is shorter than a certain critical time, or long-day plants, which flower when it is longer than a critical time. Some plants flower in response to intermediate day lengths, while others are not controlled by the photoperiod at all.
13. Photoperiodism depends on the phytochrome system. Phytochromes are protein with a tetrapyrrole pigment, and they exist in two forms. P_r absorbs red light and is thus converted into P_{fr}, which absorbs far-red light and is converted back to P_r. In the dark, P_{fr} reverts back to P_r. The P_{fr} forms of phytochromes act through the common signal-transduction pathway involving G-proteins, calcium ions, and protein kinases, resulting in the induction and repression of many genes. Phytochromes control many processes, including flowering, dormancy, and some kinds of growth, such as de-etiolation.

Key Terms

plant hormone 824
plant growth regulator 824
tropism 824
gravitropism 824
phototropism 824
auxin/indoleacetic acid (IAA) 824
acid-growth hypothesis 825
statocyte 826
amyloplast 826
gibberellin 827
gibberellic acid (GA) 827
cytokinin 828
kinetin 828
callus 829
apical dominance 829
crook 830
abscission 832
abscission zone 832
abscisic acid 832
wound response 833
systemin 833
photoperiod/photoperiodism 834
short-day plant 835
long-day plant 835
phytochrome 836
etiolate 837

Multiple-Choice Questions

1. Generally, indoleacetic acid, gibberellins, and cytokinins all
   a. stimulate growth.
   b. stimulate cell division.
   c. stimulate cell elongation.
   d. are also synthesized by mammals.
   e. are carried to their targets in the vascular system.
2. Which of the following reactions could be happening on the lighted side of a growing shoot?
   a. inhibition of auxin synthesis

b. inhibition of auxin release
c. increase in the rate of auxin breakdown
d. inhibition of auxin's activity within the cell
e. All of these could be effective.

3. The primary function of gibberellin is to
a. stimulate root growth.
b. inhibit root growth.
c. stimulate stem elongation.
d. inhibit stem elongation.
e. enable the stem to bend toward the light.

4. Which is correct about gibberellins but not auxins?
a. diffuse from cell to cell in the stem and root
b. most active in roots
c. increased concentration leads to more rapid cell division
d. highly concentrated in seeds
e. induce changes in protein synthesis

5. Which of these statements characterizes the production and action of gibberellins?
a. most highly concentrated on the dark side of the shoot
b. carried throughout the plant by the xylem and phloem
c. inactive until after dormancy has been broken and germination is underway
d. primarily affects gravitropism in contrast to phototropism
e. stimulates growth primarily by increasing the rate of cell division

6. Ethylene counteracts the effects of auxin by inhibiting
a. cell expansion.
b. apical dominance.
c. ripening.
d. abscission.
e. gravitropism.

7. Which of these hormones enhances or initiates protein synthesis?
a. auxin
b. cytokinins
c. ethylene
d. gibberellins
e. all of the above

8. Ethylene stimulates all of these activities *except*
a. ripening of fruit.
b. germination.
c. excision.
d. apical dominance.
e. horizontal growth.

9. Formation and hardening of the abscission zone involve all of the following *except*
a. abscisic acid.
b. ethylene.
c. lignin.
d. short days.
e. IAA.

10. Which of these hormones acts as a ligand that enters the nucleus and binds to DNA?
a. ethylene
b. ABA
c. IAA
d. gibberellin
e. cytokinin

True-False Questions

Mark each statement true or false, and if false, restate it to make it true.

1. The primary action of auxins and gibberellins is to increase the rate of cell division.
2. Although auxins and gibberellins both stimulate shoot growth, the gibberellins are far more potent than the auxins.
3. Plants detect the pull of gravity by means of cells called amyloplasts that are found in the root cap and endodermis of the stem.
4. In addition to stimulating growth, under certain conditions, auxins also induce flowering and seed germination.
5. Indoleacetic acid is more concentrated on the lighted side of a shoot than on the dark side.
6. To increase growth by means of cell division as well as cell elongation, both auxins and cytokinins must be present.
7. One site of abscisic acid synthesis is the vascular cells of the abscission zone.
8. If a required period of darkness is interrupted midway by a brief exposure to red light, short-day plants will begin to flower.
9. During the day, plants convert inactive phycobilin (P_{fr}) to its active form (P_r).
10. In *Arabidopsis,* phytochromes A, B, and C are the active forms, while D and E are the inactive forms.

Concept Questions

1. Systems of regulation generally include a stimulus, a receptor, and one or more levels of response. With respect to auxins and phototropism, identify each of these elements.
2. Compare and contrast the effect of indoleacetic acid, gibberellic acid, and kinetin.
3. While it is correct to label the effect of ethylene on its own synthesis as an example of positive feedback, is it technically correct to label the interaction of auxin and ethylene as a negative feedback?
4. Would you expect species that are native to arctic regions to be long-day, short-day, or day-neutral plants?
5. Explain the relationship between etiolation and phytochrome conversion.

Additional Reading

Addicott, F. T. (ed.). *Abscisic Acid.* Praeger, New York, 1983.

Devries, Peter (ed.). *Plant Hormones and Their Role in Plant Growth and Development.* M. Nijhoff, Hingham (MA), 1987.

Galston, A. W. *Life Processes in Plants.* W. H. Freeman and Co., New York, 1994.

Kendrick, R. E., and B. Frankland. *Phytochrome and Plant Growth.* Studies in Biology, no. 68, Edward Arnold, London, 1983.

Thimann, Kenneth V. *Hormone Action in the Whole Life of Plants.* University of Massachusetts Press, Amherst (MA), 1977.

Wilkins, M. B. *Plant Watching: How Plants Live, Feel, and Work.* Macmillan Publishing Co., London, 1988.

Internet Resource

To further explore the content of this chapter, log on to the web site at:

http://www.mhhe.com/biosci/genbio/guttman/

40 Plant Nutrition and the Practice of Agriculture

Orchid flowers often have fantastic forms, such as this hybrid *Oncidium* × *Miltonia*.

Key Concepts

A. Nutrient Requirements of Plants

40.1 Plants are similar in composition to other organisms.

40.2 Plants require several chemical elements for their metabolic activities.

40.3 Nutrient deficiencies can often be diagnosed by characteristic symptoms.

40.4 Most plants obtain their mineral nutrients from the soil.

B. Soils and Soil Processes

40.5 Soils are complicated ecosystems in which most plants grow.

40.6 Soil is formed by the breakdown and chemical alteration of rock.

40.7 Plant roots exchange many mineral nutrients with soil particles.

40.8 Nitrogen must be fixed for plants by procaryotes, which are often symbionts in roots.

40.9 Other transformations of nitrogen in the soil ecosystem are important for plant nutrition.

40.10 Plants are a source of reduced sulfur for the ecosystem.

C. Agricultural Practice

40.11 Agriculture, as it has been practiced, is an unsustainable process.

40.12 Ecological agriculture promotes sustainable food production practices.

40.13 Much agricultural research is directed toward improving the protein yield of crops.

D. The Ecology of Some Exceptional Plants

40.14 Epiphytes create a distinctive ecosystem that affects the nutrition of the plants they grow on.

40.15 Some plants have evolved parasitic and heterotrophic modes of metabolism.

Mr. Winter-Wedderburn, a passionate orchid fancier, buys a new orchid whose former owner died under mysterious circumstances: His corpse was found in a mangrove swamp, drained of blood, with one of the orchids crushed under his body. Winter-Wedderburn installs the new orchid in his steamy English greenhouse, where it grows into a vigorous plant bearing a web of gently waving aerial rootlets, much to the distress of the housekeeper who says they look "like fingers trying to get you." The plant then grows spikes of beautiful flowers, which open and release a powerful, sweet smell. Overwhelmed by the aroma, Winter-Wedderburn falls unconscious, and the roots descend upon him to drink his blood. He is only saved by the quick action of his housekeeper, who tears the roots from his face and throws a brick through the greenhouse glass, allowing the winter winds to blow the greenhouse clean and kill the tropical beast.

So runs H. G. Wells's story "The Flowering of the Strange Orchid." Many orchids do have that look about them, as if they might really be nightmarish monsters that drink the blood of animals deep in the jungle. Carnivorous plants that digest insects and other small animals are well known—why not vampire plants? The facts about plant nutrition, as we know them, are considerably less dramatic than Wells's story, and yet there really are plants whose modes of growth are bizarre in comparison with the tulips and chrysanthemums in our gardens. In this chapter we tell their story as well as the stories of more ordinary plants. ■

A. Nutrient requirements of plants

40.1 Plants are similar in composition to other organisms.

We commonly plant a slim twig and see it grow into a giant tree, or watch as tiny seeds sprout and eventually produce abundant vegetables for our tables. Where does all the new mass come from? Aristotle reflected the prevalent belief of his time that the substance of plants comes from the soil, an eminently sensible belief since plants commonly spring from the ground. This belief persisted until the beginnings of modern experimental science in the seventeenth century, when it was challenged by Jan Baptista van Helmont, a Belgian physician. He planted a willow seedling in a tub containing 91 kg of soil. After five years, the willow had grown into a tree weighing 77 kg, but only 60 grams of soil had disappeared from the tub. van Helmont attributed most of the willow's mass to the water he had added regularly. His conclusion was wrong, as we will see shortly, but a more realistic conception of plant growth and nutrition had to wait for the development of chemistry and physiology, when people became able to determine what elements actually form the structure of a plant.

Knowing the general chemical composition of organisms, we can make sense of their nutritional requirements and the sources of their components. We can analyze the composition of a plant by first comparing its mass before and after drying it (see Section 3.8). Plants, like all organisms, are largely water; herbaceous (nonwoody) plants are about 75–80 percent water, and even fresh wood is 40–65 percent water. Analysis of the *dry mass* of a plant—what remains after completely drying it—shows it is 90–95 percent organic matter: protein, nucleic acid, lipid, and polysaccharide. The inorganic minerals, which do come from the soil, make up only the remaining 5–10 percent. Thus carbon, oxygen, hydrogen, nitrogen, sulfur, and phosphorus, the ingredients of organic molecules, account for most of the dry mass of a plant. Water supplies most of the hydrogen and some of the oxygen for organic structure; it is also responsible for most of the increase in plant volume, since plants grow mainly by accumulating water in the central vacuoles of their cells. So van Helmont's conclusion was partly correct. However, 90–99 percent of the water a plant takes up merely passes through and is lost by transpiration. The bulk of the organic material of a plant, of course, comes from the CO_2 that is assimilated from the atmosphere through photosynthesis. Figure 40.1 summarizes the process of nutrition and materials flow through a plant.

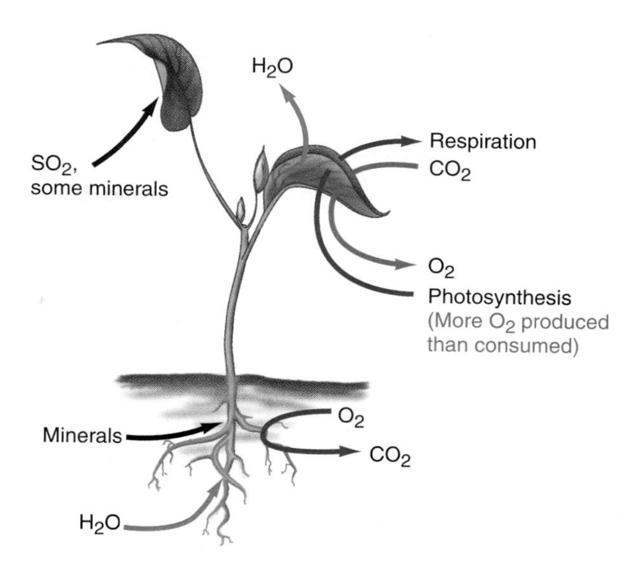

Figure 40.1
For a realistic view of plant nutrition, it is important to see that most of a plant's dry mass comes from CO_2 assimilated from the atmosphere through photosynthesis. A plant absorbs water and minerals through its roots, but can acquire some minerals and sulfur compounds through its leaves. Roots and other nonphotosynthetic parts also need access to O_2 for respiration, even though the plant is a net producer of O_2.

Exercise 40.1 Assume a reasonable value for the water content of van Helmont's tree. Suppose its dry mass was 95 percent organic material, and suppose all its inorganic dry mass came from the soil. Compare the mass that should have been removed from the soil with the mass he actually could attribute to the soil. What might account for the discrepancy?

40.2 Plants require several chemical elements for their metabolic activities.

We noted in Chapter 3 that all organisms require a specific, limited group of elements as nutrients. Since plants are similar in composition to all other organisms, their elemental nutritional requirements are not particularly different, but the details of their nutrition still warrant a closer look. Studies of plant

Table 40.1 Essential Plant Nutrients

Element	Fraction of Dry Weight	Major Uses
Macronutrients	***Percent***	
Carbon	45–50	In all organic molecules
Hydrogen	5–7	In all organic molecules
Oxygen	27–33	In most organic molecules; respiration
Nitrogen	8–10	In proteins, nucleic acids, coenzymes, hormones
Phosphorus	2–3	In nucleic acids and phospholipids; in coenzymes of energy metabolism; protein regulator
Sulfur	0.5	In proteins, coenzymes
Potassium	4–5	Major cation in cytosol; enzyme activator; water balance
Calcium	2–3	Cell wall component; maintains membrane structure; regulates cytoskeleton; second messenger in signal-transduction pathway
Magnesium	1	Major divalent cation in cytosol; in chlorophyll; enzyme activator
Micronutrients	***Parts per Million***	
Chlorine	500 ppm	Photosynthesis, ionic balance in cytosol
Iron	500 ppm	Enzyme activator; electron carrier of ETS
Manganese	250 ppm	Enzyme activator, especially Krebs cycle and amino acid biosynthesis enzymes
Boron	100 ppm	Cofactor in chlorophyll synthesis; may act in carbohydrate transport; not well understood
Zinc	100 ppm	Enzyme activator; auxin synthesis
Copper	30 ppm	Enzyme activator, especially redox reactions and lignin biosynthesis; electron carrier
Molybdenum	0.5 ppm	Enzyme activator, especially in nitrogen fixation and nitrate reduction
Nickel	??	Enzyme activator, nitrogen metabolism

nutrition have a very domestic focus. Just as we know a great deal about the food requirements of humans and a few domesticated animals, but very little about the millions of other species in natural ecosystems, our conception of essential plant nutrients derives from studies of food crops and a few ornamentals. Nevertheless, Table 40.1 lists the mineral elements known to be essential for plants and their functions. An element is considered an **essential nutrient** if a plant requires it to complete its life cycle, which usually means growing from a seed and producing another generation of seeds. Furthermore, the element cannot be replaceable by another element, and it must be required for a normal function—not just to cure an abnormality or relieve the toxic effect of another substance.

The essential nutrients in Table 40.1 are divided into two categories: the **macronutrients,** which plant tissues require in large amounts (at least 1 milligram per gram of dry mass), and the **micronutrients,** or **trace elements,** which are required in concentrations of less than 100 micrograms per gram of dry mass. The six main macronutrients—carbon, hydrogen, oxygen, nitrogen, sulfur, and phosphorus—are obviously the components of organic molecules. In addition to being a component of nucleic acids, phosphorus is central to energy metabolism in ATP and other nucleotides, and phosphate groups are used to regulate the actions of proteins.

Calcium, potassium, and magnesium are what we might call *inorganic* macronutrients. Calcium is a constituent of cell walls. As one of the principal second messengers in cells, it has an essential role in the common signal-transduction pathway that processes signals from hormones and other external ligands. Calcium also regulates the actions of the cytoskeleton and is essential for spindle formation in mitosis and meiosis. Potassium is the principal monovalent cation in plant tissues; it helps maintain the electrical neutrality of the cytoplasm by balancing the negative charges of ionized carboxyl groups ($-COO^-$) of proteins and organic acids, as well as the phosphate groups of nucleic acids. Potassium also has a role in moving water from one compartment to another; cells actively transport ions, most commonly K^+ ions, and water follows by osmosis. Magnesium has multiple functions, too—as the principal divalent cation of cytosol, as a cofactor of numerous enzymes, and as a constituent of chlorophyll.

The eight known micronutrients—chlorine, iron, manganese, boron, zinc, copper, molybdenum, and nickel—function in plants mainly as cofactors of enzymatic reactions (see Chapter 6). Iron is part of the heme group of cytochromes, the proteins that compose the bulk of electron transport systems in chloroplasts and mitochondria. Zinc, molybdenum, and nickel are required by certain enzymes, and with few exceptions each enzyme requires one particular element, which cannot be replaced by another. Plants need only minute quantities of these elements because they have catalytic functions and the enzymes they serve are not abundant. A typical dried plant may contain only one atom of nickel or molybdenum among several million atoms of hydrogen or carbon. Yet without those few nickel or molybdenum atoms, the plant cannot grow.

Plant physiologists have identified most of the essential elements by the hydroponic culturing technique (Figure 40.2). An element is considered essential if a plant grown hydroponically fails to grow, flower, or produce viable seed in the absence of that element. However, plants require some elements in such minute amounts that contaminating traces in the experimental apparatus have been enough to support plant functions. This was particularly a problem in early experiments before highly purified reagents were available, so several elements were wrongly assumed to be nonessential. For example, dust particles and water droplets in the air in laboratories near the ocean provided enough chloride for growing plants, and chlorine was not determined to be essential until fine air filters became available. A seed may contain enough of an essential trace element to supply the entire plant that grows from it and even part of the next generation. In spite of modern research facilities with highly purified water and reagents, new essential elements are rarely reported now. The list is not likely to be complete yet, especially since so few species have been studied. Just as some animals require nutrients that others don't need at all, some minerals are essential for certain plants but apparently not for others. But more sophisticated techniques may be required to find more essential elements.

Hydroponic methods have also been used commercially to a very limited extent. Plants grown this way are assured of good nutrition, since they can be given precisely measured nutrient solutions, but the method is limited by the relatively high costs of the lab and equipment compared with ordinary farming methods.

Exercise 40.2 A group of students are trying to determine whether chlorine is an essential nutrient for a certain species of plant. They have an immaculately clean modern lab and very pure reagents. Explain why it will make a difference if they handle the plants with their bare fingers or with disposable plastic gloves.

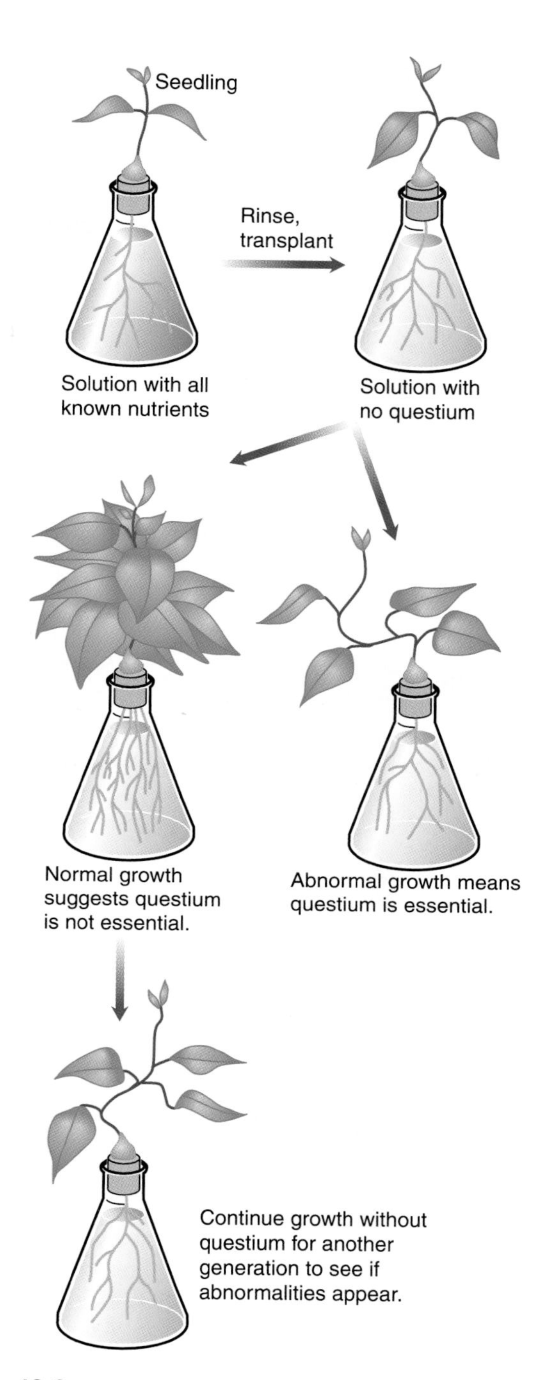

Figure 40.2
A simple hydroponic method is used to identify essential nutrients. A seedling is started in a full nutrient solution, rinsed thoroughly, and transferred to a solution lacking the imaginary element "questium." Failure to grow and reproduce is strong evidence that questium is an essential nutrient for this plant. If the plant grows and reproduces, questium is probably not essential, but it may be advisable to continue growth in the questium-free solution for at least another generation, since some nutrients are required in such minute amounts that the material in a seed may support the plant that grows from it.

40.3 Nutrient deficiencies can often be diagnosed by characteristic symptoms.

Horticulturists and agriculturists can diagnose many nutrient deficiencies in plants by noting characteristic **deficiency symptoms** that appear before a plant dies (Table 40.2; Figure 40.3). We can often understand these symptoms on the basis of the mineral's function. Since magnesium, for instance, is an essential component of chlorophyll, magnesium deficiency shows up as *chlorosis*, a yellowing of the leaves (Figure 40.3). Yet iron deficiency can also cause chlorosis, even though chlorophyll contains no iron, because one enzyme in the biosynthetic pathway of chlorophyll requires iron as a cofactor.

The symptoms of a mineral deficiency depend not only on the function of the element but also on its mobility in the plant. Some elements, such as nitrogen, may be redistributed within the plant while others, such as iron, are not. If a nutrient can move freely from one part of the plant to another, young, growing tissues are generally able to accumulate it; so if the nutrient is in short supply, deficiency symptoms will appear first in older tissues. Magnesium is a mobile element, so if a plant is

Table 40.2 Deficiency Symptions of Major Plant Nutrients

Element	Deficiency Symptoms
Boron	Tissues hard, brittle; stems rough and cracked; growing tips damaged; flowering inhibited; heart rot of root crops; poor legume nitrogen fixation
Calcium	Dieback of growing points; affects meristems (undeveloped terminal buds); stunted root growth; leaves curl
Chlorine	Young leaves blue-green color and shiny, then become chlorotic, necrotic, bronzed
Copper	Dieback of growing points; leaves chlorotic or blue-green in color, elongated; leaf margins curl or roll
Iron	Chlorosis of young leaves, but larger veins remaining green; short, slender stems
Magnesium	Marginal chlorosis and red, purple, or brown pigments in mature leaves first, with green venation
Manganese	Stunting, interveinal chlorosis in leaves; pale overall coloring; leaves malformed, with necrotic spots
Molybdenum	Interveinal chlorosis; pale, distorted, yellow leaves, with margins that curl or roll; stunting
Nitrogen	Chlorosis of leaves, beginning with lower leaves, retarded growth
Phosphorus	Retarded growth, blue-green or dark green color; red, purple, or brown pigments along veins, beginning on underside of leaf
Potassium	Dieback of growing points; blue-green or dark green color, necrotic spots; leaf margins necrotic ("scorched")
Sulfur	Chlorosis of whole plant, retarded growth
Zinc	"Little leaf," rosette formation; leaves necrotic, twisted, misshapen; late summer mottling of leaves

(a)

(b)

Figure 40.3

Characteristic coloring in plants often shows they are deficient in specific nutrients. *(a)* A grape leaf shows symptoms of potassium deficiency: necrosis and chlorosis around the leaf margin. *(b)* A rose leaf shows symptoms of magnesium deficiency: chlorosis around the margins of the leaf, with spots of dark pigments.

magnesium-deficient, chlorosis will show up first in its older leaves. On the other hand, deficiency symptoms of a nutrient that is relatively immobile will appear in young parts of the plant first because the older tissues used the limited amount of the element while they were growing, and they retain it. Since iron is an immobile element, iron deficiency will appear as chlorosis of the young leaves first, before any changes in the older leaves are visible. Thus, even though deficiencies of magnesium and iron produce similar symptoms, the difference in the age of affected organs may distinguish between them.

If a deficiency of a macronutrient is suspected, the diagnosis can sometimes be confirmed by analyzing the mineral content of the plant and soil. Deficiencies of nitrogen (N), potassium (P), and phosphorus (K)—the most common problems in domestic plants—are relatively easy to identify, and commercial fertilizers are commonly characterized by their N-P-K percentages. Shortages of micronutrients tend to only appear in geographically localized areas, due to differences in soil composition. Considering the minute amounts of some micronutrients that are needed, any such deficiency can be easily corrected.

Orchardists, for instance, have been able to cure a zinc deficiency in their fruit trees just by pounding a few galvanized (zinc-coated) nails into each tree trunk.

Having seen what minerals a plant requires, we can begin to ask where it normally obtains them, keeping in mind that most plants are *not* domesticated creatures nourished and watched over by farmers.

40.4 Most plants obtain their mineral nutrients from the soil.

Plants clearly obtain their carbon from the CO_2 in the atmosphere. Virtually all their other nutrients come from the soil. Roots with enormous surface areas grow down through the soil, extracting nutrient ions from the surrounding water through osmosis and active transport, as discussed in Section 38.2. We should note in passing that plants obtain the components of organic molecules from the environment in their most oxidized forms—carbon as carbon dioxide, hydrogen as water, nitrogen as nitrate, sulfur as sulfate, and phosphorus as phosphate. Except for phosphate, these elements must be reduced to the oxidation state of organic molecules, a process that requires considerable energy.

In addition to CO_2, plants are able to take up other compounds through their leaves. They remove some sulfur oxides from the atmosphere in this way. Farmers can also effectively fertilize certain plants by spraying their leaves with a nutrient solution. Plants can absorb more copper, manganese, and iron from such a foliar spray than from the soil. But the pH of the solution and the concentration of nutrient ions within it must be carefully adjusted, and the plants must be sprayed at the right time of day.

Plants that live in water are usually surrounded by a well-mixed solution containing the nutrients they need, but a terrestrial plant depends on local soil conditions that may be far from optimal. We noted in Figure 26.2 that the mineral composition of a soil can vary greatly even within a single square meter. If a plant starts to grow in nutrient-deficient soil, is it simply doomed? Not necessarily. Growth is a plant's way of moving to a better environment, and a plant can potentially reach sources of essential minerals by extending its root system or by extending runners and setting down new root systems. Through its roots, a plant may mine the soil so thoroughly that it depletes the local environment of essential materials. However, as its root system expands, a plant comes into contact with more and more soil potentially carrying additional nutrients.

As they grow, roots travel through a variable environment—encountering patches of different pH, with a patch of phosphate here where an organism has decayed, a patch of iron there where a nail has rusted. How do the roots respond to this variability? In contrast to an animal, which commonly ingests materials from its environment indiscriminately, plant roots show a degree of selectivity. To a large extent, they can match their uptake of minerals to the plant's nutritional needs, including elements that are present in the soil in very minute quantities. However, the mineral composition of plants often reflects the composition of the soil and water in which they grow. Plants growing on mine tailings, for instance, may contain gold or silver, and the nutritional value of fruits and vegetables can vary, depending on the composition of your garden soil. Occasionally a plant surprises us by concentrating unusual amounts of some material; for instance, locoweeds and poison vetch (both *Astragalus* species, Figure 40.4) accumulate a lot of selenium, so animals that eat them are killed or driven to wild behavior.

Figure 40.4
Locoweeds accumulate selenium, which has devastating effects on the nervous systems of animals that eat the plant.

To understand how plants obtain their nutrients from soil, we must understand soil structure and the nature of the soil ecosystem. This is our next task.

B. Soils and soil processes

40.5 Soils are complicated ecosystems in which most plants grow.

As the source of their water and mineral nutrients, soil is the external lifeblood of most plants. Soil gives terrestrial plants mechanical support and oxygen for their roots. Soils, like all ecosystems, have abiotic and biotic components. The abiotic part of a soil is basically rock fragments ranging in size from microscopic particles to large pebbles, whose composition depends on the parent rock they are derived from. As we show in Section 25.8, the structure of a terrestrial ecosystem is largely determined by a three-way interaction among soil, climate, and vegetation, so the soils of different regions vary enormously. Soils change constantly due to climatic influences such as heat and precipitation, and we will see that water is an especially potent force in forming and changing a soil. The living components in soil include plant roots and an astonishing number and variety of organisms, some whose activities supply essential nutrients and others that tend to damage or kill plants (Figure 40.5). A gram of topsoil may harbor a rich community of microorganisms and minute animals:

- 10^8–10^9 bacteria, including autotrophs that engage in nitrogen metabolism and many heterotrophs that decay organic material.
- 10^7–10^8 actinomycetes, filamentous procaryotes that are important in humus formation.
- 10^5–10^6 yeasts and molds involved in humus formation. Some produce antibiotics while others are plant pathogens; some molds are mycorrhizae associated with plant roots.
- 10^4–10^5 algae, protozoa, and rotifers, including some herbivores of the system.
- 10–100 nematodes; some are carnivores of the ecosystem, and others are plant parasites.

Soil microorganisms contribute to the cycling of carbon, nitrogen, sulfur, and other elements through the ecosystem, and the organic acids they produce also break down soil minerals and alter the soil's chemistry. Living on this microbial base are many other small animals, especially the detritivores shown in Figure 28.10. To a depth of only a few centimeters, each square meter of soil contains 10^3 to 10^5 insects, spiders, centipedes, slugs, snails, and other invertebrates, forming the usual complex food web. All these animals serve as food for variable numbers of vertebrate predators. Through their activities, all organisms change the physical and chemical properties of the soil.

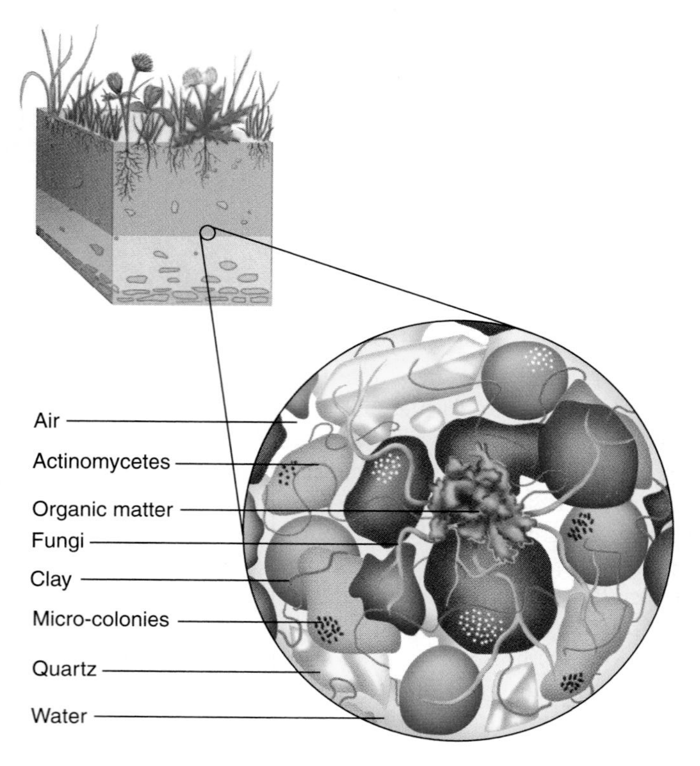

Figure 40.5
Soil particles create a habitat for a multitude of microorganisms, with a base of bacteria and fungi.

Typically, 30–300 earthworms per square meter mix and aerate the soil by their burrowing and add mucus that holds fine soil particles together. Finally, a soil contains from 50 to 3,000 kg per hectare of plant roots, which have enormous effects on soil chemistry as they remove water and minerals.

40.6 Soil is formed by the breakdown and chemical alteration of rock.

Soil is fundamentally highly weathered rock, broken into small fragments and altered chemically. So the type of soil in an area depends on the rock from which it formed and the climatic factors of heat and water acting on that rock. Soil formation, as outlined in Section 25.9, begins with the *mechanical weathering* of rocks into small fragments through repeated wetting, drying, and freezing. Water seeps into rock crevices, freezes, and fractures the rock. These processes are followed by *chemical weathering,* the chemical alteration of some rock components, particularly by dilute carbonic acid water, which is formed by the reaction of carbon dioxide with water:

$$CO_2 + H_2O \rightarrow H_2CO_3 \rightarrow H^+ + HCO_3^-$$

Soils begin with rocks that are mixtures of several minerals, commonly including quartz (silicon dioxide, SiO_2) and very complex silicates such as feldspar and mica. A typical feldspar has the formula $K_2 \cdot Al_2O_3 \cdot 6SiO_2$ and a typical mica has the formula $K_2Mg_6O \cdot 2FeO_3 \cdot Al_2O_3 \cdot 6SiO_2 \cdot 2H_2O$, or with sodium or calcium ions. (The centered dots here separate components of the mineral crystals.) Notice that these minerals contain essential plant nutrients, and if their other oxides are removed, they will liberate quartz. Carbonic acid rainwater filtering through the soil breaks down some mineral components and removes them; for example, potassium dissolves out as potassium carbonate, (K_2CO_3), leaving crystals of clay minerals such as kaolinite, ($Al_2O_3 \cdot 2SiO_2 \cdot 2H_2O$). The formation of clay minerals is critical. Notice that kaolinite has water molecules bound into its crystal structure; clay particles swell and shrink as they become hydrated and dehydrated, and they strongly determine the physical and chemical properties of soils. Each soil contains a particular mixture of clay minerals, depending on the parent rock.

Quartz, which is insoluble and not altered much chemically, falls loose as fragments of *silt* (between 2 μm and 20 μm in diameter) and *sand* (from 20 μm to a millimeter or two in diameter). The texture of topsoil depends in large part on the size of its mineral particles. Although water drains away from the larger spaces of the soil, it is held in smaller spaces by capillarity and by its adhesion to the hydrophilic surfaces of soil particles (Figure 40.6). Some of this water is so tightly bound that it is not available to plants, but the more mobile water constitutes the **soil solution** from which plants draw their water, with its dissolved nutrients. Each species of plant is adapted to a particular type of soil, which varies in texture, pH, and mineral composition. For agricultural purposes, the most fertile soils are **loams,** rich soils that contain a balance of sand, silt, and clay. (Methods 40.1 explains a simple way to estimate the mixture of

Methods 40.1 Estimating Soil Types

It is easy to judge a soil's general composition with a method explained by Marianne Sarrantonio. Pick up a handful of moist soil and squeeze it in your fist. Then open your hand. If the sample falls apart instead of remaining in a ball, your soil is sand. If you have a ball of soil, try to squeeze it upward with your thumb to form a ribbon. Form as long a ribbon as you can and measure it. If you cannot make a ribbon, your soil is loamy sand. Now add some water to the soil to make a soupy mud. Feel it with a finger of your other hand and decide whether it feels mostly gritty, mostly smooth, or equally smooth and gritty. Using this chart, you can now classify the soil.

	Feels Mostly Gritty	Feels Mostly Smooth	Feels Both Gritty and Smooth
Ribbon < 1 inch	Sandy loam	Silty loam	Loam
Ribbon 1–2 inches	Sandy clay loam	Silty clay loam	Silty clay
Ribbon > 2 inches	Sandy clay	Silty clay	Clay

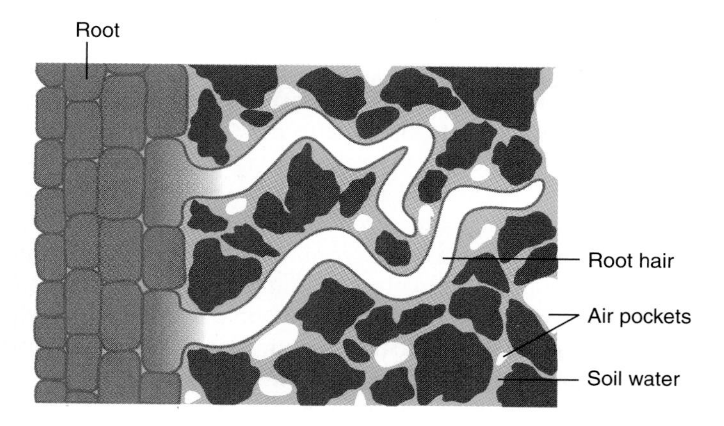

Figure 40.6

A healthy soil is made of fine particles of humus and rock, including clay minerals. It has an open, crumbly structure with much of the space between particles taken up by water, the soil solution from which roots draw their nutrients.

these components in a soil.) The fine particles of loams provide a large surface area for the adhesion of minerals and water; loams also have enough coarse particles to provide air spaces, allowing oxygen to get to the roots and support their cellular respiration. Roots suffocate in inadequately drained soils without air spaces, an environment that also favors the growth of molds that cause root-rot.

However, the texture of soil is determined less by the size of its mineral particles than by its *humus* content. Humus is a black or brown complex of decomposing organic matter formed by the actions of bacteria and fungi on all the detritus of the ecosystem. In humus, lignin and other woody compounds that are quite resistant to decay form **humic acids.** Although their structure isn't entirely known, they are acids that release hydrogen ions, and they combine with clay minerals to form **clay-humus complexes,** which are acidic particles with anionic (negative) surface groups. These complexes largely determine the soil's character. In a young soil, cations such as K^+, Na^+, Mg^{2+}, and Ca^{2+} initially released from the parent rock associate reversibly with the bound anions of the clay-humus complexes. Acidic precipitation that continues to filter through the soil carries H^+ ions, which tend to displace the mineral ions in a process called **cation exchange:**

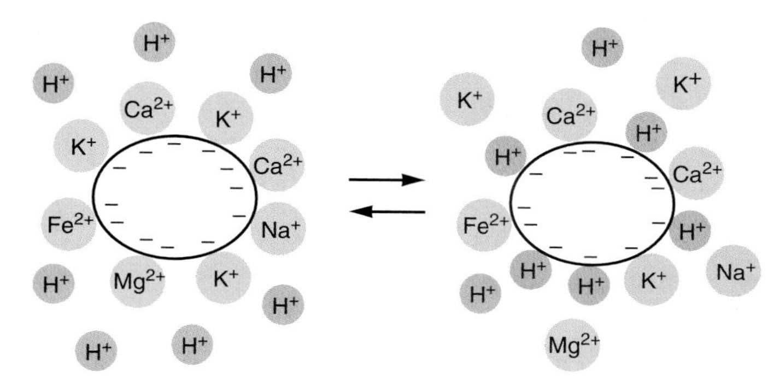

That is, H^+ ions trade places with bound mineral cations. A dynamic equilibrium will exist between bound mineral ions and H^+ ions. We think about this situation with the same logic used for all dynamic equilibria (see Section 5.2); the interactions of, say, H^+ and K^+ ions with the anionic sites are no different in principle from the binding of substrate and inhibitor molecules at the active site of an enzyme. At a high ratio of H^+ to K^+ ions, the H^+ ions will most likely be bound and the K^+ displaced.

Plants influence a soil's character largely by the types of carbon-rich materials they deposit as litter, which is transformed into humus. We have seen in Section 25.9 that the plants of different biomes are adapted to particular pH ranges and nutrient levels, which are reflected in their nutrient cycles. We have also seen that the soils of different biomes are characterized by various degrees of *leaching,* as precipitation percolating through the soil carries mineral ions away from the upper levels and deposits them in a lower level, creating characteristic layers, or *horizons,* as shown in Figure 25.34. The A horizon is defined as the uppermost zone from which minerals have been leached, and the B horizon as the lower zone where they accumulate. The parent rock from which soil is derived is the C horizon, a zone of little biological interest since roots are generally confined to the A horizon and sometimes reach into the B horizon, but rarely to the bedrock.

Now let's consider plant nutrition against this background.

40.7 Plant roots exchange many mineral nutrients with soil particles.

Plant roots growing in the soil create a special environment for themselves, a **rhizosphere** consisting of the particular microorganisms they nurture and the surrounding soil solution, soil particles, and clay-humus complexes (Figure 40.7). The roots determine the nature of the rhizosphere by secreting various organic acids into the soil solution and withdrawing its water and ions. Roots also secrete a slimy mucilage, a hydrated polysaccharide called *mucigel* that changes their surface properties. The mucigel forms the real interface between the cell surface and the soil solution; it may support the growth of microorganisms that are beneficial to the plant, and it may facilitate the uptake of ions. Many of the minerals essential for plant nutrition are the cations (K^+, Mg^{2+}, Ca^{2+}, and others) bound to clay-humus complexes, but the roots can only obtain them when they become free ions in the soil solution. They are freed by the same process of cation exchange involved in leaching. Now it should become clear that the pH of the soil is critical for plant growth. H^+ ions that can exchange with mineral ions are always present in the soil solution, but respiration from plant roots and soil organisms adds more CO_2, which makes the soil solution still more acid, thus releasing more mineral nutrients for plants (and incidentally selecting for the growth of particular microorganisms).

The fertility of a soil clearly depends on the slow release of minerals through cationic exchange at clay-humus complexes. A soil with little humus and clay loses these essential cationic nutrients through leaching, particularly during heavy rain or irrigation. Notice, however, that clay-humus complexes are only able to hold and exchange cations. Soils have no analogous exchange system for anions, and therefore anionic nutrients that are so important to plant nutrition, such as phosphate, nitrate, and sulfate, are easily leached to lower soil horizons.

Agriculture requires close attention to the pH of soil, which determines the availability of nutrient ions. Soil acidity also affects the chemical form of mineral nutrients. Plants may still be starving for an element that is abundant in a soil if it is bound too tightly to clay-humus complexes or is in a chemical form the roots cannot absorb. Farmers frequently reverse acidification of the soil by the practice of *liming,* applying calcium-rich materials (chalk, limestone, or ground-up mollusc shells) that release compounds such as calcium carbonate or calcium hydroxide. The added Ca^{2+} (a valuable macronutrient itself) releases H^+ ions from soil particles by ion exchange so it can be leached away, raising the soil pH. If a soil is too alkaline, sulfate can be added to lower the pH. However, balancing the pH of a soil is tricky, and it must be done to meet the requirements of each crop. Phosphorus, for example, is mostly available to plants in a very narrow pH range (Figure 40.8). Below pH 5.5, most macronutrients and some micronutrients are bound in insoluble forms; however, micronutrients such as iron, zinc, manganese, copper, and cobalt are unavailable at alkaline pH and only become soluble under more acidic conditions. On the other hand, toxic elements such as aluminum and lead become more of a problem at acid pH. On the whole, crops do best at slightly acidic conditions, around pH 6.3 to 6.8.

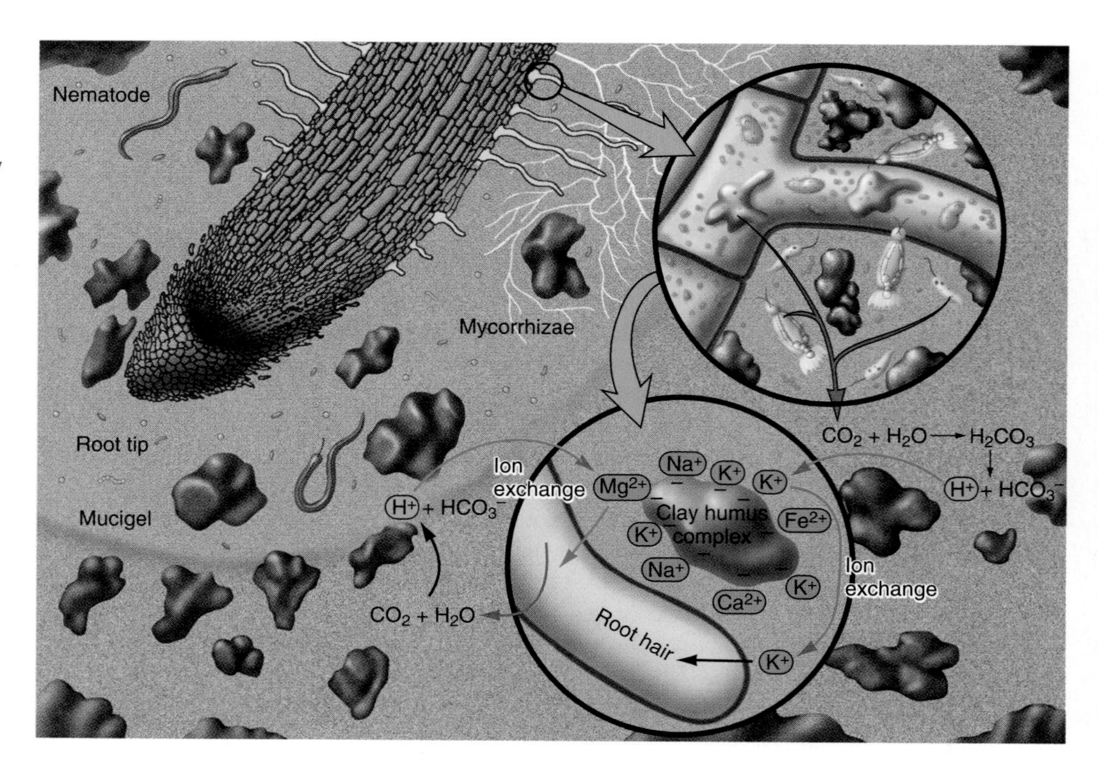

Figure 40.7

The rhizosphere is a region only a few millimeters in diameter where the root interacts with the surrounding soil and microorganisms. Its character is largely determined by the mucigel the root secretes, which nurtures certain organisms and inhibits others.

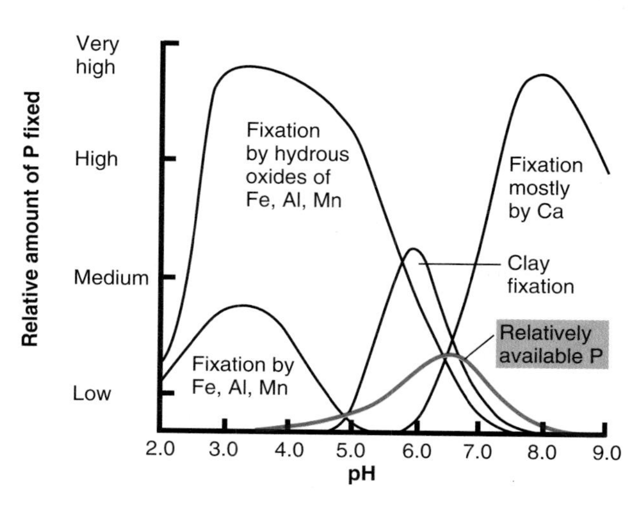

Figure 40.8

The availability of many mineral nutrients to plants changes with pH. Phosphorus, for example, is bound to other elements and compounds at very low or high pH, so it is most available between pH 6 and 7.

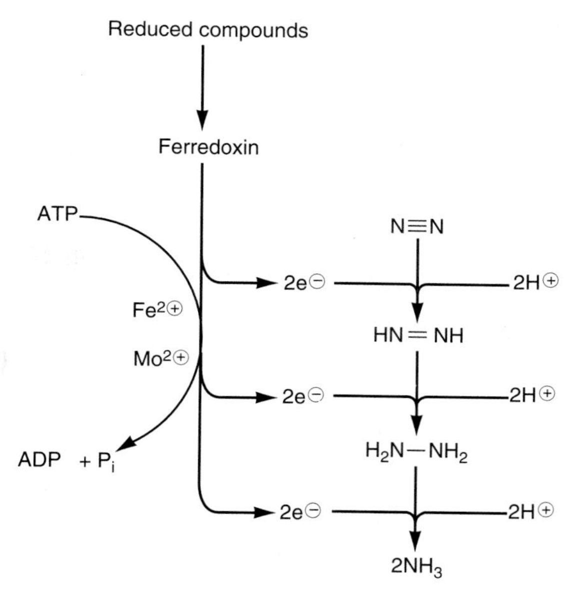

Figure 40.9

The nitrogenase system reduces molecular nitrogen to ammonia by adding three molecules of hydrogen to each N_2. The system uses energy from ATP and gets its hydrogen from ferredoxin (Fd).

40.8 Nitrogen must be fixed for plants by procaryotes, which are often symbionts in roots.

Photosynthesis supports most life on earth. By storing the energy of light in organic molecules, phototrophs supply the energy for the rest of an ecological community. Yet this process only fixes carbon, hydrogen, and oxygen in organic form. Where does a plant obtain the nitrogen that is such an integral component of proteins and nucleic acids and that may constitute 10 percent of a plant's dry weight? Even though four-fifths of the atmosphere is nitrogen gas (N_2, or dinitrogen), most organisms have no way to use this abundant resource. N_2 is very unreactive, and it takes a great deal of energy to convert it into one of the forms plants can use: nitrate ion (NO_3^-) or ammonia (NH_3). (In discussing nitrogen metabolism, we will generally refer to ammonia, knowing that ammonia dissolved in water will be in equilibrium with ammonium ion, NH_4^+). The process of reducing N_2 to NH_3 is called **nitrogen fixation.** Volcanic activity and lightning fix a little N_2, which is carried into the soil in rainwater. The fertilizer industry now fixes substantial amounts—tens of millions of metric tons per year—by a method called the Haber process. But most reduced nitrogen compounds are still made by nitrogen-fixing procaryotes, which convert about 9×10^{13} grams of atmospheric dinitrogen per year.

Nitrogen-fixing procaryotes have a **nitrogenase** enzyme system, which adds three hydrogen (H_2) molecules to the triple bond of a nitrogen molecule to make two ammonia molecules (Figure 40.9). The few known types of nitrogen-fixers make only a minute fraction of the earth's biomass, and yet the biosphere is as dependent upon them as on all the phototrophs that bring CO_2 and energy into the world's ecosystems.

Some nitrogen-fixers are free-living soil and water bacteria, particularly *Azotobacter* and cyanobacteria, but the most important nitrogen-fixation systems develop in mutualistic associations of procaryotes and certain plants or fungi; these mutualisms provide the bacteria with a place to grow and supply the eucaryote with ammonia. Some cyanobacteria fix nitrogen in mutualisms with fungi in lichens or with bryophytes, ferns, or cycads. The nitrogen metabolism of rice, a crop of enormous importance worldwide, depends strongly on the tropical water fern *Azolla,* which rice farmers cultivate in their paddies. Small pockets of cyanobacteria *(Anabaena)* grow between cells in the roots of *Azolla,* where they fix nitrogen. The growing rice then shades and kills the *Azolla,* which decomposes and fertilizes the rice plantation. More widespread mutualisms with nitrogen-fixers occur between certain angiosperms and two types of bacteria: the actinomycete *Frankia* or the small bacterium *Rhizobium.* (Some rhizobia are free-living soil organisms that do not fix nitrogen.) Both bacteria grow within root cells, where they induce the growth of round **root nodules** in which nitrogen is fixed (Figure 40.10). The filamentous *Frankia* forms **actinorrhizal associations** with plants such as alders, silverberries, and bayberries, while rhizobia associate with elms and primarily with legumes—clover, peas, alfalfa, soybeans, and many tropical shrubs and trees.

Mutualisms, Section 27.14.

Rhizobial associations have been studied extensively. Each plant requires a specific strain of rhizobia; the plant and the bacteria recognize each other by their cells' surface molecules. (In

Figure 40.10

Nitrogen fixation occurs in the root nodules of legumes and other plants that form mutualistic associations with *Rhizobium* bacteria.

some agricultural practices, seeds of a legume are soaked in a culture of bacteria to ensure that they find their specific *Rhizobium.*) Following recognition, the bacteria invade the root and form intimate associations with cortical root cells, stimulating them to divide and develop into nodules. There the two kinds of cells cooperate to create a nitrogenase system, producing ammonia from which both partners make amino acids. Hundreds of nodules may develop in a plant's root system, each containing between a million and a billion bacteria, and each supplied by a branch of the vascular cylinder so it receives water and sends nutrients—especially amino acids—to the rest of the plant. With age or the death of the plant, nodules eventually disintegrate, releasing bacteria that can spread to form new nodules.

Nitrogenase is exquisitely sensitive to molecular oxygen, which inactivates the enzyme. Traces of oxygen are bound up in the nodules by *leghemoglobin,* an iron-containing protein very similar to the hemoglobin that transports oxygen in the blood of many animals. Like animal hemoglobin, leghemoglobin binds oxygen reversibly, removing it from the nitrogenase environment. A functioning nodule may produce enough leghemoglobin to turn deep red in color. Remarkably, the plant cells produce the protein, the globin, since the gene is in their genome, while the bacteria make the heme group that binds the oxygen. Rhizobia are aerobes, and leghemoglobin also supplies oxygen for the intense respiration they must carry out to produce the ATP and reduced ferredoxin for nitrogen fixation.

Nitrogen-fixing legumes are used in **crop rotation.** A field is planted in alternate years with a nitrogen-demanding crop such as wheat and a legume such as clover or alfalfa. Instead of being harvested, the legume is generally plowed under and allowed to decompose as a "green manure." In the best of circumstances, root nodules fix so much nitrogen that they secrete excess ammonia, which further increases the fertility of the soil. Some plant biologists are trying recombinant-DNA technology to incorporate genes for the nitrogenase system into plants that can't support their own bacteria, as discussed in Section 40.13.

40.9 Other transformations of nitrogen in the soil ecosystem are important for plant nutrition.

A microbial soil ecosystem is involved in the further transformations of nitrogen compounds after nitrogen-fixing bacteria have produced ammonia and ammonium ions (Figure 40.11). Although ammonia is toxic to plants, they can take up and tolerate ammonium ions at low concentrations, especially at basic soil pH. Most plants, however, grow better with nitrate than with ammonium ions as a source of nitrogen, and they take up nitrate ions preferentially under more acidic conditions. A plant transports nitrate to its chloroplasts, where some of the energy

Figure 40.11

Bacteria in the soil ecosystem oxidize and reduce nitrogen compounds as they grow. Different plants use ammonium, nitrate, or both. Some synthesize amino acids in their roots and transport these upward; others carry nitrate to the chloroplasts where it is reduced to NH_3 and incorporated into amino acids. Plants and other organisms contribute to a "humus cycle," in which ammonia is recycled by ammonifying bacteria.

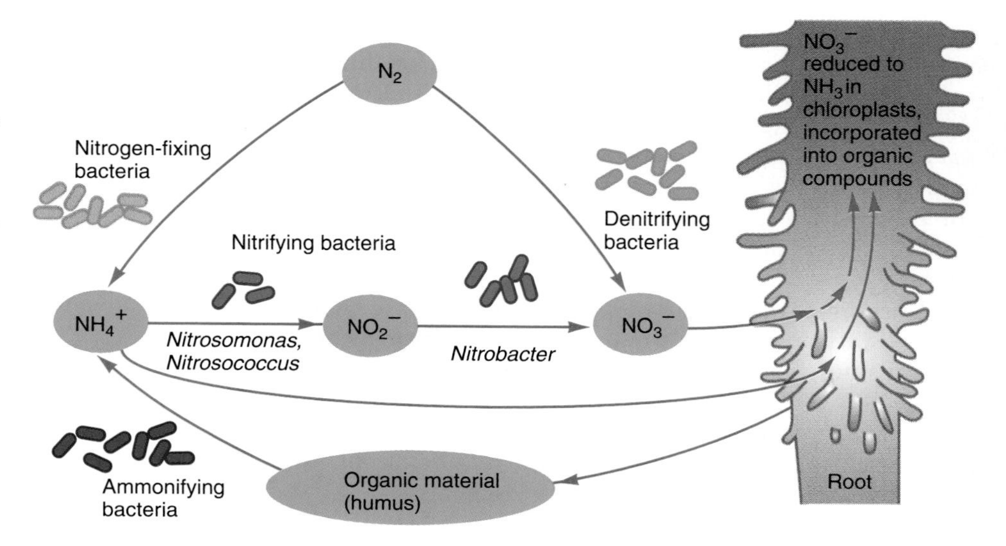

generated in the light-dependent reactions of photosynthesis is used to reduce it to ammonia for the synthesis of amino acids (see Section 10.8).

Soil bacteria oxidize ammonia to nitrate in the process known as **nitrification,** which has two stages: Bacteria of the genera *Nitrosomonas* and *Nitrosococcus* oxidize ammonia to nitrite ions (NO_2^-), and *Nitrobacter* bacteria oxidize nitrite to nitrate. These bacteria are all chemoautotrophs that carry out these oxidations as part of their energy metabolism. In growing as they do, they serve as an important ecological link by supplying nitrogen in a usable form to plants and therefore to the rest of an ecosystem.

Chemoautotrophic metabolism, Section 9.13.

A few bacteria create another link in the nitrogen cycle by carrying out the process of **denitrification** in which they return nitrogen to its atmospheric form, dinitrogen. Some common soil bacteria that normally grow aerobically, mostly species of the genera *Bacillus* and *Pseudomonas,* can grow anaerobically by using nitrate (NO_3^-) as a terminal electron acceptor in place of oxygen:

$$2\,NO_3^- + 10\,e^- + 12\,H^+ \rightarrow N_2 + 6\,H_2O$$

The nitrogen cycle, Section 28.5.

A major part of the nitrogen cycle is the conversion of the nitrogen to plant tissues to that of animal tissues. Animals can only acquire their nitrogen through the food web, so they are entirely dependent on the reactions of nitrogen fixation and nitrification that support plant growth.

40.10 Plants are a source of reduced sulfur for the ecosystem.

All organisms contain sulfur, mostly in the amino acids cysteine and methionine in their proteins. Sulfur is also a component of coenzymes such as thiamine, lipoic acid, and coenzyme A (see Sidebar 9.1). Plants can take sulfate ions (SO_4^{2-}) from the soil or water, reduce them to the sulfhydryl (–SH) level, and incorporate them into cysteine. Sulfur in this form is the precursor of the other sulfur-containing plant components. Sulfate reduction is analogous to the reduction of nitrate to ammonia, and animals depend upon the cysteine, methionine, and vitamins from plants, as they do for nitrogen. However, plants can reduce sulfate themselves and do not depend on bacteria.

In addition, many species of bacteria base their energy metabolism on the oxidation and reduction of sulfur compounds in their environment. *Desulfovibrio* and others carry out a sulfate respiration, reducing sulfate to sulfite and sulfide. The thiobacilli obtain their energy by oxidizing H_2S to sulfate, just as the nitrifiers oxidize ammonia to nitrate.

Sulfur oxidation and reduction in metabolism, Section 9.13.
The sulfur cycle, Section 28.5.

C. Agricultural practice

40.11 Agriculture, as it has been practiced, is an unsustainable process.

People are interested in plant nutrition primarily because we grow plants to use for our food. Agriculture was invented only around 10,000 years ago and has produced a much more dependable food supply than the hunting and gathering practices of earlier times, allowing the human population to grow by a factor of at least 25 times in the first 8,000 years. But at what cost to existing ecosystems?

Agriculture flourished in the Tigris-Euphrates valley that is now Iraq from about 5000 B.C. As the civilization of Mesopotamia was replaced by a series of others, wheat production decreased steadily because of **salinization,** the process in which the soil becomes increasingly salty as irrigation water evaporates year after year, leaving its dissolved salts behind. At first, farmers of the region were forced to grow more and more barley, a plant that is more tolerant of salty soil. But today, what was once a rich ecosystem is a desert. Much of western Asia, which was once covered with forests, now supports little more than goat-herding, the last stage of a degenerating agriculture. The Anasazi people of what is now New Mexico lived in a forest biome, but through their agricultural practices, they degraded it into a desert. Other stories of **desertification**—making deserts out of formerly rich ecosystems—can be told about other areas that have been virtually destroyed by standard agricultural practices and deforestation. Desertification continues intensively in Africa, especially at the southern edge of the Sahara Desert, around the Gobi Desert of Asia, on Asian hillsides, and in deforested regions of the Amazon basin.

In all natural ecosystems, mineral nutrients are recycled as organic material decomposes in the soil; the natural ecosystem is in a steady-state condition, with its mineral losses slowed by the structure of the ecosystem itself. But when a farmer harvests a crop, enormous amounts of essential elements are simply lost to the farm ecosystem. Agriculture depletes the mineral content of the soil.

Hubbard Brook studies on nutrient loss, Section 28.7.

In addition to salinization, agriculture commonly erodes the soil (Sidebar 40.1). A soil whose fertility has been built up over centuries can be destroyed in only a few years if mismanaged. In principle, soil is a renewable resource whose fertility can be preserved, so it should be able to sustain agricultural production for many human generations. Soil conservation measures are well known: terracing hillsides, using stands of woods as windbreaks between fields, planting and recycling certain crops as manures. But they have rarely been used consistently.

The modern practice of *intensive agriculture,* particularly as it has been developed by large industrial farms in the United States, treats soil like dirt, as several critics have charged. Instead of ecosystems to be nurtured, soils are considered containers to

Sidebar 40.1 The Soil and Food Situation: A Brief Overview

The human population is now about 5.5 billion, rapidly approaching 6 billion. (Do the math: Every second, there are three more people on Earth; a year is 3.15×10^7 seconds.) About 500 million people are starving or severely malnourished; another 300 million receive less than 90 percent of their daily food needs, and a billion or so are malnourished to somewhat lesser degrees. Yet one optimistic best estimate predicts the human population will double by 2030. How will we feed all these people?

Only 30 percent of the earth's surface, 14 billion hectares (ha), is land. Of this, only 3 billion hectares are potentially arable (farmable). Another 3 billion hectares are used for pasture and rangeland. Most of the 1.5 billion hectares that are readily usable are already in use. In an effort to find more land to farm, native peoples are slashing and burning forests and tilling the sides of hills and mountains, especially in India, the Philippines, and the Andes. But deforestation contributes to the general decline in the photosynthetic phase of the carbon cycle, and the land that has been cleared soon washes down the hillsides into streams, and into the sea.

Through intensive agriculture and exploitation, the world's food supplies have been increasing for a few decades, but those gains are now almost at an end. Production of grains (wheat, rice, corn, barley) had been increasing, but the supply per person has been declining since 1984 at the rate of about 1 percent per year. Production of soybeans, a major source of protein worldwide, has slowed since 1979 due to lack of cropland, and since soybeans are legumes that grow through their own nitrogen fixation, their production cannot be increased by using inorganic nitrogen fertilizers. Meanwhile, meat production has plateaued since 1987, and fishing, which has long supplied about a quarter of the world's protein, is at its peak. The great cod fisheries off the east coast of North America have been overfished and now must be restricted; the same fate will soon befall the huge fisheries of the north Pacific. Salmon stocks on the west coast of North America are being depleted by deforestation and the effects of cattle ranching, which lead to pollution of the streams where the salmon breed.

Meanwhile, poor soil-management practices have resulted in huge losses of the soil we depend on. The United Nations estimates that since 1945 one-fifth of all agricultural land has undergone moderate to extreme soil degradation, 85 percent of it from erosion. Our soil is blowing away and washing down rivers into the sea at an incredible rate; each year the world loses 1 percent of its topsoil through erosion and 6 million hectares of its most productive land, an area equal to that of Ireland. In Africa, 40 percent of the land is degraded, along with 30 percent in Asia, 20 percent in Latin America, 17 percent in Europe, and 4 percent in North America. In the past 200 years, the United States has lost one-third of its topsoil and 25–50 percent of its soil nitrogen. If humans are to be kept from massive starvation, we will have to become better stewards of the land.

be emptied and filled repeatedly. Intensive agriculture is largely **monoculture**—growing extensive fields of a single crop, such as one variety of corn or wheat, instead of smaller fields with several varieties. The method depends upon four largely destructive or futile practices: irrigation, application of inorganic fertilizers, use of pesticides, and genetic selection. Let's examine the consequences of these four practices.

Irrigation

Agriculture, especially intensive agriculture, requires huge amounts of water (Figure 40.12), and the availability of water is often the main limitation on plant growth. Many crops demand much more water than the natural vegetation that once grew on a plot of land, since the original vegetation was adapted to the area's climate and water resources. The lesson of history is that continued irrigation leads to salinization and the death of the soil. Irrigation also leaches mineral nutrients—especially anions—from the soil. Some agricultural programs use drip irrigation as an alternative to flooding fields, and this at least has the virtue of doing the damage more slowly. Furthermore, the water used for irrigation must come from limited resources. In the American midwest, water for agriculture has long been pumped out of the vast Oglalla aquifer, which once looked inexhaustible, but is now being drained much faster than it can refill from natural sources; in 20 years it may be too low to pump. Water is also being removed from rivers in competition with the demands of huge urban populations.

Some countries have undertaken massive irrigation projects to "transform the desert into a garden," but these enormous drains on water resources have more commonly done the opposite. The result has often been an ecological and social disaster. Many of the rivers in the southwestern United States that once supported vigorous ecosystems have been reduced to trickles by the diversion of water for irrigation. Lake Aral in the former Soviet Union—now Uzbekistan—is a monument to ecological folly. In the hope of improving the area's economy, the Soviet Union diverted the lake's water to nearby fields to support a huge cotton monoculture; as the lake level fell, the remaining water became increasingly saline, killing the fish that had once supported a fishing industry and supplied a great deal of food for the region. What was once the world's fourth largest lake is becoming a series of salty puddles. Salt from the dried lake blows over the cotton fields, killing the plants. Meanwhile, massive amounts of pesticides and herbicides were dumped on the cotton, with the result that people living in the region, now largely deprived of their food and livelihood, are also being deprived of their health by the chemical residues they are forced to eat and breathe. This experiment now costs about U.S.$10 billion per year, and it has been declared an ecological disaster beyond repair. An ecologist could have predicted the result beforehand.

Inorganic fertilization

The annual harvesting of crops removes the nutrients the plants took from the soil during growth. In growing a ton of wheat grain, the soil gives up 18.2 kg of nitrogen, 3.6 kg of phosphorus, and 4.1 kg of potassium. The soil becomes more impoverished every year unless fertilizers are used to replace these minerals. Because nitrogen, phosphorus, and potassium are the

Figure 40.12

Agriculture requires large amounts of water, which is commonly supplied by irrigation from rivers or lakes, or by pumping underground aquifers.

main mineral nutrients removed with each crop, they are the main nutrients that must be replaced. Instead of *organic fertilizers* such as compost, rotted plants, or manure, intensive agriculture uses *inorganic fertilizers,* commercially processed to contain certain N-P-K percentages. A 5-10-10 fertilizer, for example, contains 5 percent nitrogen, 10 percent phosphate, and 10 percent potassium. At present, American commercial agriculture uses 70 kg per year of nitrogen, phosphate, and potash fertilizer *for every person in the United States.* The production of inorganic fertilizers requires a great deal of energy, primarily from fossil fuels. As a result, we now expend 3 J of energy per J of food produced, and if the energy consumed in processing and transportation is included, we expend 10 J per J of food. In contrast, anthropologists estimate that hunter-gatherers living in their native ecosystems expend about 0.1 J per J of food.

Pesticides

Instead of promoting an ecological method of keeping insect pests under control, intensive agriculture uses massive amounts of pesticides—an expensive and futile process, since the insects and other organisms become resistant to the chemicals at least as fast as they can be invented. In 1948, when the chemical industry began massive production of pesticides that seemed to promise miracles, American farmers laid on 7 million kg of chemicals and lost 7 percent of their crop to insects; now they annually spread about 60 million kg and lose 13 percent. This process is obviously very good for the chemical industries; it is very bad for everyone else. The problem is exacerbated by monoculture; if a particularly destructive plant pathogen or insect appears, it can spread rapidly through huge fields of genetically identical plants, but smaller fields of diverse plants are generally more resistant.

An instructive story comes from Indonesia, where rice farmers, using intensive methods, had been trying to keep insects under control with vast amounts of pesticides. Then a major predator on the rice crop, the brown plant-hopper, became resistant to the pesticides at the same time its own major predators were being killed. In response, the Indonesian government banned the use of most pesticides and instituted the ecological pest-control methods, called integrative pest management, which are now keeping the insect problem within manageable limits.

Genetic selection

Just as intensive agriculture puts its faith in the technology of fertilizers and pesticides, it also depends upon being continually able to breed improved food plants. Its solution to the problem of salinization, for instance, is to develop plant varieties that require less water or can tolerate more salinity. Perhaps the most hopeful aspect of intensive agriculture, heralded since the 1960s, has been the so-called Green Revolution, which began with the breeding of high-yield grains. Increasing the mass of grain yielded by each unit of land has increased the food supply for many countries, but at great cost. Plants may have genomes for producing large grains, but to translate those genomes into plant material still takes water and nutrients, which do not come for free, and the Green Revolution has depended on the other destructive methods of intensive agriculture to achieve these improvements. In the Punjab region of India, which was intended to be converted "from a begging bowl to a bread bowl," farmers now contend with waterlogged deserts, weakened soils, and pest infestations. To supply the vast amounts of water required by high-yield crops, the water table is falling at the rate of 0.3–0.5 meter per year. The farmers themselves are in debt to pay the costs of this intensive agriculture, just as Korean farmers raising the "miracle rice" that was supposed to save the country have found their debt rising ten times faster than their income. In other countries, the Green Revolution has displaced small farmers from their land, creating large numbers of landless laborers who often cannot find jobs because of the mechanized methods of intensive agriculture. The general effects on local ecosystems and economies have been devastating, moving more wealth from the poor to the rich.

The overall result of intensive agriculture has included chemicalization of soil, displacing healthy soil ecosystems; compaction of soil from the use of heavy equipment and the elimination of humus; and massive rates of soil erosion (see Sidebar 40.1). Intensive methods are displacing ancient agricultures that were often quite well suited to their ecosystems and disrupting traditional societies and economies.

40.12 Ecological agriculture promotes sustainable food production practices.

If agriculture as standardly practiced is unsustainable, but billions of people still need to be fed, what is the alternative? Many people have sought, and are practicing, various versions of **ecological agriculture,** sometimes shortened to *ecoagriculture* and also known as *sustainable agriculture.* Its many forms share a few basic tenets, entailing social and economic, as well as ecological, ideas:

1. *Nature is capital.* Energy-intensive modes of conventional agriculture cannot be sustained. Agriculture must focus on recycling a finite supply of nutrients.
2. *Soil is the source of life.* Soil quality and nutrient balance are essential if agriculture is to have a future; human and animal health are directly related to the health of the soil.
3. *Feed the soil, not the plant.* Healthy plants, animals, and humans result from balanced, biologically active soil. Nutrients should sustain the soil ecosystem, not go directly into the plants.
4. *Diversify production systems.* Monoculture is overspecialized and environmentally unstable.
5. *Maintain independence.* Ecological agriculture promotes personal and community independence from energy-intensive production and distribution systems.

Ecological agriculture differs from the conventional in its emphasis on maintaining **tilth,** the health of the soil. This means maintaining a rich, fertile soil with an active biological community, rather than using inorganic fertilizers and pesticides, and it places great emphasis on maintaining the humus. A soil rich in humus is porous and crumbly, with large air spaces that are essential for respiration by plant roots; it acquires this texture partly from the gummy materials secreted by microorganisms growing there. By retaining water, up to 80–90 percent of its mass, humus makes a soil more drought-resistant. Humus holds mineral nutrients and returns them gradually to the soil as the organic matter decays; at the same time, it can bind toxic heavy metals such as lead, keeping them away from plants. Also, as we will see shortly, the interaction of clay-humus complexes with hydrogen ions in the soil gives humus a buffering action against extremes of pH. Because humus is dark in color, a humus-rich soil absorbs heat from sunlight more quickly than a poorer soil and warms up faster.

Buffers and buffering, Concepts 46.1.

Organic fertilizers are manure, compost, and other materials of biological origin that decompose in the soil. They become components of humus, providing all the advantages listed above, and as they decompose, they gradually release inorganic nutrients that roots can absorb. Gradual release is important because plants can only assimilate mineral nutrients in proportion to their general growth rate. Minerals in commercial fertilizers are available in large quantities immediately, and farmers may perceive them as superior because they provide an instantaneous supply of nutrients and can be formulated to meet the needs of a particular crop. But much of their mass cannot be *used* immediately; when applied to unhealthy soil lacking humus, the nutrients are leached away rather quickly by rainwater or irrigation. The major effect of commercial fertilizers may be to enter the groundwater and eventually pollute nearby freshwater ecosystems.

The use of inorganic fertilizers is an expensive addiction. A soil that has been fed inorganic fertilizer is like a helpless drug addict; the fertilizers kill off the natural bacterial populations that normally handle nitrogen metabolism, so the soil is helpless to cure itself and can only grow more crops by getting another "fix" of fertilizer—at the expense of becoming even more helpless.

The practice of ecological agriculture does not mean simply avoiding inorganic fertilizers and pesticides. Rather, it attempts to maintain a farm ecosystem as closely as possible to a natural ecosystem (Figure 40.13). This means crop diversity, as opposed to monoculture, and other ecologically sound practices such as crop rotation. It means maintaining farm animals as well as plants, so their manures can be used as fertilizers, and also remaining independent of the dominant, energy-intensive system of production and distribution. Ecoagricultural specialists have gone into many Third World communities to help the people develop agricultural systems appropriate to their local conditions and traditional ways of life, making them independent of expensive, destructive, high-technology methods.

Furthermore, ecological agriculture shares a broader ecological interest. As agriculture has moved toward huge monoculture agribusiness establishments, it has destroyed some of the natural ecosystems associated with traditional farms. In England, this has led to removing the hedgerows, shrubby areas between farms that have long supported many wild animals, at the rate of 3,000–5,000 km per year. One result is that the insect predators that formerly lived in the hedgerows are disappearing, leaving much larger insect populations to prey on the crops; in response, intensive agriculturists typically lay on more pesticides. But these pesticides cannot be confined to small farm areas; they affect all insect populations, with a subsequent disruption of ecosystems and a decline in the populations of birds and other animals (see Section 28.8).

Foods produced by organic and conventional methods have been compared by several criteria in well-controlled scientific studies. (Many of the studies have been conducted in Europe, which is generally more supportive of ecological practices than America.) Although differences such as appearance and taste are very subjective, criteria that can be evaluated objectively elicit strong preferences for foods grown organically. For instance, vegetables have exhibited better storage life when grown organically, apparently due to lower respiration rates and lower levels of the enzymes that produce spoilage. Compared to vegetables grown conventionally, those grown organically have higher levels of several desirable nutrients (protein, vitamin C, iron, potassium, and other mineral nutrients); they also have much lower levels of nitrate, which may promote human health since nitrate can be converted to nitrite, a mutagen and a carcinogen. Several studies on the nutrition and growth of animals have shown that they grow considerably better on organically produced food and show other advantages such as increased fertility in several species and better egg weights in chickens.

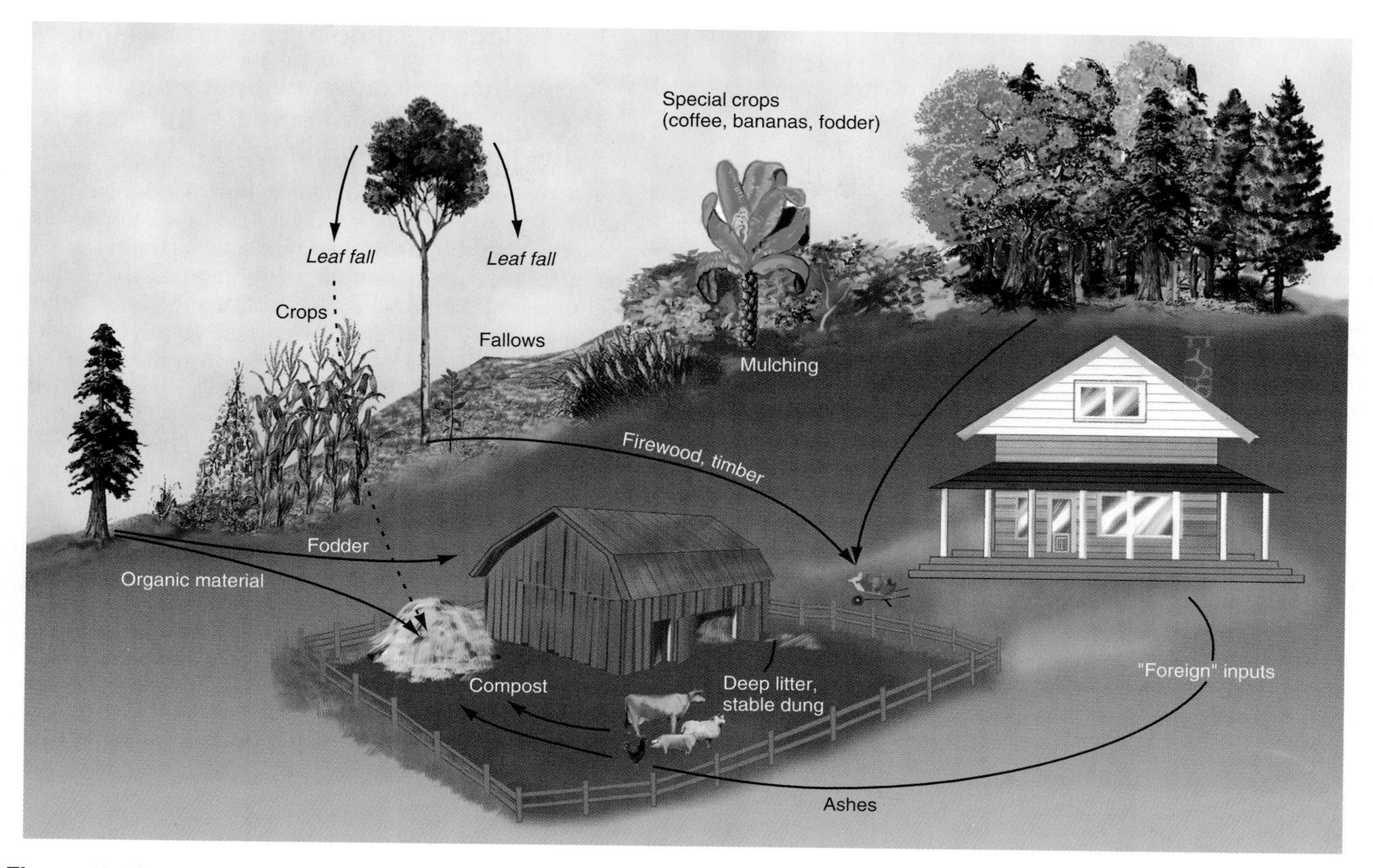

Figure 40.13

Ecological agriculture involves maintaining cycles of materials comparable to those in a natural ecosystem. A farm normally includes a small forest in which trees are cut on a sustainable basis for firewood and timber. Some fields are planted with crops on a rotating basis while others are allowed to lie fallow for a while to replenish the soil. Special crops appropriate to the climate may be grown for fodder or for sale. Farm animals are maintained by food from the land, and their manures are used as fertilizers.

40.13 Much agricultural research is directed toward improving the protein yield of crops.

Because plants are the only source of nitrogen for animals, plants' ability to make protein is an important factor in human nutrition. Malnutrition in humans is most often protein deficiency. The vast majority of the world's people today have a predominantly vegetarian diet and therefore rely mostly on plants for protein. But many plants contain relatively little protein, and what proteins they do have are often deficient in some of the essential amino acids that animals cannot synthesize for themselves. Some plant breeders have therefore sought to improve world health by developing plants with more and better proteins, and a great deal of agricultural research has been directed toward this goal. Indeed, protein-enriched strains of rice, corn, and wheat have been bred, but employing them to feed the malnourished people of the Third World has run into precisely the same difficulties that plague other agricultural projects. Proteins are nitrogen-rich compounds. These new types of grain therefore require large amounts of nitrogen-rich fertilizer, and this is the problem we have met before. Commercial inorganic fertilizers that can feed these plants are made through expensive, energy-demanding processes. (The energy, of course, generally comes from fossil fuels, adding to the larger energy problems of our civilization.) The fertilizers are expensive, and the countries most in need of the high-protein crops are the ones that can least afford to pay for them; if they do pay, they merely increase the personal debts of their farmers and their national debts to the rich nations.

If food plants could fix their own nitrogen, farmers would be able to raise much more food. At least two experimental avenues have been explored. Food plants might be modified so they could respond to rhizobia and form their own root nodules, or it might be possible to implant nitrogen-fixation (nitrogenase) genes directly into the plant tissue. The first approach requires identifying the factors essential for nodule formation in plants that do interact with rhizobia; the plant must be able to make essential proteins called nodulins as well as the protein part of leghemoglobin (see Section 40.8). Some investigators have managed to induce rhizobia to infect—and form root nodules in—nonlegumes, including rice and wheat. The grains have been able to fix nitrogen and assimilate some of it into their proteins. However, the methods used so far do not involve a stable genetic change in either the plants or the rhizobia.

The second possibility, transferring the nitrogenase system, would also be a big job; the *Nif* (nitrogen-fixation) system of the bacterium *Klebsiella pneumoniae* includes 17 genes in 8 operons. Assuming these genes could all be cloned into an appropriate vector (probably the Ti plasmid), several difficulties must be overcome. These are procaryotic genes, and their promoters would have to be replaced by eucaryotic promoters. The system would have to be protected from oxygen by leghemoglobin or a similar protein. And the plant would have to supply ferredoxin and a great deal of ATP—17 moles of ATP per mole of N_2 fixed. The drain on reducing power and energy might be too much for a plant and might wipe out any advantage it would gain from nitrogen fixation. Obviously, this research still has to progress from blackboard plans and laboratory experiments to field applications, but the prospect of food crops fixing their own nitrogen provides a strong incentive for the research.

Genetic engineering with plants, Section 18.13.

D. The ecology of some exceptional plants

40.14 Epiphytes create a distinctive ecosystem that affects the nutrition of the plants they grow on.

Typical higher plants are the kind of soil-dwellers we have been picturing all along, with their roots buried in soil and a shoot system extending above. Considering the opportunism of evolution, however, it should not be surprising that some plants take advantage of the great mass and surface area of trees, whose crowns create a world of their own. These **epiphytes** (*epi-* = upon; *-phyte* = plant) are plants that nourish themselves but grow on other plants called **phorophytes** (*phor-* = carry), usually on the branches or trunks of trees (Figure 40.14). Epiphytes are not parasites; the relationship is one of phoresis or commensalism (see Section 27.14). Epiphytes merely take advantage of a phorophyte's bulk and surfaces, particularly its height, since an epiphyte growing in the canopy of a forest (the treetop region) has access to sunlight that is not available on the forest floor. Some lichens, mosses, and ferns grow epiphytically, as do a number of flowering plants, especially orchids and some bromeliads of the pineapple family, Bromeliaceae. The epiphyte most familiar to North Americans is Spanish moss, *Tillandsia usneoides,* about which Peter Bernhardt has written,

> *It has the widest distribution of any* Tillandsia *species throughout the Americas. . . . It's all over the Caribbean and as*

(a)

(b)

Figure 40.14
Epiphytes are nonparasitic plants that grow entirely on other plants, most often on the limbs in the crown of a tree. *(a)* Spanish moss is very common in the American Southeast, while *(b)* many kinds of bromeliads grow on trees in the tropics.

far south as Chile and Argentina. This plant forms dense, gauzelike masses consisting of hundreds or thousands of individual plants. Each plant grows no more than three leaves before it flowers and dies. At one time an important cottage industry in the United States was milling the tiny plants and using them as a stuffing for upholstery. Today, Spanish moss is still regularly employed by horticulturists as a natural sunscreen in greenhouses and as a living humidifier to protect more delicate plants from desiccation.

Tillandsias produce seeds with a cluster of plumelike hairs that act as a parachute and easily become entangled in other vegetation, holding the seed in place until it can germinate. The vegetative plants of tillandias and other epiphytic bromeliads do not have functional roots, and they obtain water and mineral nutrients from rain and fog through their leaves, sometimes using special structures that are very efficient at absorbing water.

Although epiphytes are nutritionally independent of their phorophytes, some have been considered semiparasitic because they can take up nutrients from precipitation that would otherwise go to their larger hosts. Nearly half the organic material of the epiphytic zone is dead, composed of epiphytes that are decaying in place to form a *crown humus.* But epiphytes are rich in the nutrients they absorb from precipitation, and the mass of epiphytes in the crowns of trees may be several times the mass of the tree's foliage. Many epiphytes do, indeed, change the nutrient balance around themselves and may have a significant influence on the nitrogen balance of the ecosystem. But the nutritional relationship between epiphytes and phorophytes was given a new perspective when Nalini Nadkarni, while a graduate student at the University of Washington, climbed into forest canopies using modified mountain-climbing equipment and discovered that trees covered with epiphytes produce adventitious (extra) roots from their branches which spread out into the epiphyte layer and collect nutrients. So rather than robbing phorophytes of these nutrients, epiphytes are an additional source for the phorophytes to tap. By using radioisotopes, Nadkarni and Richard Primack showed that nutrients obtained through crown roots nourish areas of the phorophyte shoot quite near the root, whereas nutrients obtained from below-ground roots are broadly distributed; thus the above-ground roots are a source of supplementary nutrition that may be particularly important if nutrient uptake from the below-ground system is hindered.

We should note, in passing, that the ecology of forest canopies is barely known and offers enormous new challenges to the biological explorer. The crown humus is a real soil that supports its own community of microorganisms, earthworms, and insects. This community in turn supports a whole animal community, and Nadkarni has written,

Of the fifty-six bird species foraging in our sites, 60 percent hunted in epiphytes. . . . In nine out of ten visits, the ochraceous wren foraged in crown humus, and the purple-throated mountain-gem sipped nectar from epiphytic shrubs in the blueberry family. The most popular epiphyte, the woody shrub Norantea, *was a veritable smorgasbord for many bird species. Slate-throated redstarts picked insects off its foliage; silver-throated tanagers and emerald toucanets ate its red fruits; stripe-tailed hummingbirds and purple-throated mountain-gems sipped its nectar; and prong-billed barbets scavenged insects from its branches. . . . Birds gather epiphytic mosses and lichens to weave, pad or camouflage their nests, and they bathe in the pools of water that collect in bromeliads. White-faced monkeys pluck and peel back the leaves of tank bromeliads in search of insects. I have seen tree snakes slither along branches and pause at each pool—seeking to feast on a squatting frog or a bathing bird—then move on to the next pool like a teenager cruising fast-food restaurants on Main Street.*

40.15 Some plants have evolved parasitic and heterotrophic modes of metabolism.

Although the plant kingdom is defined as a clade of phototrophs, a few species have taken sidetracks in their evolution; instead of feeding themselves by photosynthesis, some plants have become parasites, and a few have even become partially heterotrophic.

Parasitic plants obtain their food directly from other living plants. If you are very lucky, some day you may discover a stark white plant emerging from the duff of a forest floor. You might mistake it for a rather unusual mushroom, but it is an Indian pipe, a flowering plant (Figure 40.15). Once thought to be a saprophyte (a decay-causing organism) that gets its nutrients from forest detritus, Indian pipe is now known to obtain its nutrients from nearby photosynthetic plants, with the aid of mycorrhizae (root fungi; see Section 31.6). Most of us are more familiar with the parasitic dodders and mistletoes. Dodders do

Figure 40.15
Indian pipe is a flowering plant that lives parasitically on the roots of other plants nearby.

Figure 40.16

Dodders are common parasites on other plants. The parasite wraps its tendrils the stems of its host and penetrates its vascular system to withdraw water and nutrients.

not carry out photosynthesis at all and draw all their nutrients from the plants they grow on (Figure 40.16). Although mistletoes are green and carry on some photosynthesis, they obtain water and mineral nutrients from the xylem sap of other plants by growing extensions called *haustoria.* Some mistletoes obtain photosynthetic products from their hosts.

Some of the world's most fascinating plants have evolved partly heterotrophic metabolism, so they augment their nitrogen and phosphorus supply by capturing and digesting insects. The 450 known species of plant carnivores are normally found in boggy regions with acidic soil; they include such well-known species as pitcher plants *(Sarracenia),* Venus's-flytrap *(Dionaea,)* and sundews *(Drosera)* (Figure 40.17). In discussing the flora of bogs (see Section 25.5), we noted that relatively little nitrogen is recycled in these acidic habitats because saprophytes generally require a more neutral pH to digest their substrates. Carnivory is therefore an adaptation for supplying a plant with nitrogen from animal proteins in this restricted environment. None of the carnivorous plants require insects for survival. They grow quite well without animal protein, but when insects are available, they grow faster and are a darker green because of the extra nitrogen and phosphorus, which is used partly to make chlorophyll.

The pitcher-shaped leaves of a pitcher plant collect rainwater. Insects wander into these leaves, possibly attracted by their brightly colored appearance or by some chemical attractant. The insect, however, finds it almost impossible to climb out over the stiff, downward-pointing hairs covering the leaves. Eventually it dies, falls into the water, and is digested by enzymes secreted by both the plant and the bacteria that live in the water.

Venus's-flytraps are fascinating little mechanical traps triggered by three hairs in the center of a leaf lobe. If a wandering

(a)

(b)

Figure 40.17

Some of the most common carnivorous plants are *(a)* the pitcher plant and *(b)* Venus's-flytrap, both of which capture and digest insects and other small animals.

insect touches one of these hairs, the two halves of the leaves close and imprison the insect. The leaves of sundews are covered with hairs that secrete a sticky liquid. An insect that touches a hair gets stuck, and then other hairs curve over the insect and stick to it. These plants, like a pitcher plant, secrete hydrolytic enzymes that digest the insect. The leaves then absorb the resulting monomers.

Coda The ecosystems that support all life on this planet, including human life, have been evolving for a long time. Those we see around us have grown while the human species was gradually evolving, and our ancestors evolved into certain niches within them. But humans have now come to exceed their original niches, to dominate their ecosystems, and often to engage in activities that destroy them. Agriculture is one of those activities. Some indigenous peoples, living close to and part of their ecosystems, have managed to maintain themselves and their environment very well for thousands of years, but the majority of humanity has been practicing unsustainable methods that quickly ruin the environment. If the human population were still small, the damage might be absorbed, but at just the time when the human population threatens to grow beyond capacity, damage to the ecosystems is also intensifying. If we are to adequately feed an estimated 10–12 billion people in the next few decades, agricultural methods will have to become ecologically based very quickly. But powerful economic interests may stand in the way of the necessary changes.

You will think we sing a gloomy song. *Homo sapiens* is, on the whole, an optimistic species. It has tended to march onward in the name of Progress, ever confident of success, ignoring realities. But if it does not start to face those realities quickly, our species may end in up massive social disruption and warfare as billions of people fight one another for pitifully little food. The readers of this book may be the most important people whose eyes can be opened to the facts.

Summary

1. Plants are similar in composition to other organisms, about 75–80 percent water. Their dry mass is 90–95 percent organic matter, composed of carbon, hydrogen, oxygen, nitrogen, phosphorus, and sulfur.
2. To grow and complete its life cycle, a plant requires a series of essential nutrients. Its macronutrients are the components of organic molecules plus potassium, magnesium, and calcium. Eight elements are micronutrients, or trace elements; they are primarily cofactors for certain enzymes. Essential nutrients are identified primarily through hydroponic culturing.
3. Plants often show characteristic deficiency symptoms, such as chlorosis (yellowing), if they lack certain nutrients. These symptoms can be used to diagnose and correct a deficiency.
4. Plants obtain their carbon from the CO_2 in the atmosphere. All other nutrients come from the soil, although plants can take up a few nutrients efficiently through their leaves. Root growth is a plant's way of moving to nutrient sources, and a plant is somewhat selective in the materials it absorbs through its roots.
5. Soils are rich ecosystems consisting of many kinds of microorganisms and animals mixed with rock fragments and humus, a brown or black material made of dead and decaying organic material. Soils are transformed by interactions among the soil community, the minerals, the water flowing through the system, and the humus.
6. Complex silicates such as mica and feldspar react with water from precipitation, which is a dilute carbonic acid solution. The minerals release cations such as K^+, Ca^{2+}, Mg^{2+}, and Fe^{2+}, leaving particles of clay minerals behind. Humic acids in the humus combine with the clay to form clay-humus complexes, which are acidic and bear anionic (negative) charges. The mineral cations tend to bind to these complexes.
7. In the process of cationic exchange, H^+ ions in the soil water exchange with mineral cations. These minerals are partly leached to lower levels in the soil, forming visible layers or horizons. Cation exchange also releases mineral nutrients that are taken up by plant roots. The pH of the soil solution is critical in making various minerals available or unavailable to roots.
8. A soil's texture is determined by its balance of sand, silt, and clay particles and by its humus content. A soil must not be too compact and must contain air spaces that allow oxygen to get to the roots to support their respiration.
9. Plants cannot use atmospheric nitrogen (N_2); they can only assimilate nitrogen from ammonia/ammonium or nitrate. Nitrogen is reduced to ammonia in the process of nitrogen fixation by certain procaryotes that can make nitrogenase. *Rhizobium* bacteria and the root cells of certain plants form a major nitrogen-fixing system in a mutualistic partnership. Nitrogen is also fixed by bacteria in other mutualisms and by some free-living soil and water bacteria.
10. Bacteria in the soil ecosystem carry out other conversions of nitrogen compounds. Nitrifying bacteria convert ammonia to nitrate, which is used by many plants in preference to ammonia. Denitrifying bacteria convert nitrate into N_2. The nitrogen in plants is the only nitrogen source for the animals in ecosystems.
11. Soil bacteria and plants also carry out interconversions of sulfur compounds. Plants can take up sulfate, reduce the sulfur to the sulfhydryl (–SH) level, and incorporate it into cysteine and other organic molecules. These compounds in plants are the only source of sulfur for the animals in ecosystems.
12. Agriculture, as it has been practiced, is usually destructive. Irrigation leads to salinization, the accumulation of salt in soils, and to desertification, the conversion of rich ecosystems into deserts. A great deal of topsoil is lost through soil erosion.
13. Modern intensive agriculture relies on irrigation, inorganic fertilizers, pesticides, and improved strains of food plants. Irrigation produces salinization and leaches away mineral nutrients. Inorganic fertilizers are expensive and wasteful; they eliminate the natural soil ecosystem and forsake all the benefits of humus. Pesticides have been futile, since insects and plant pathogens mutate to resistant states faster than pesticides can be invented. Raising improved crops still requires large amounts of water and nutrients, and promotion of their use has led to severe economic hardships.
14. Ecological agriculture seeks to grow food crops by developing soil ecosystems rich in humus and strong biological communities. It promotes small, integrated farming units where plant by-products and animal manures can be recycled into the soil. Ecoagriculture promotes practices that do not damage broader ecosystems and are appropriate to the local conditions and traditional lifestyles of people in developing countries, enabling them to be independent of expensive, destructive, high-technology methods.

15. Plant breeders are trying to develop plants with more and better proteins. Much research is directed toward getting rhizobia to grow with food plants or transferring genes for nitrogenase systems into those plants.
16. Epiphytes are nonparasitic plants that grow on other plants (phorophytes). Many of them create unique ecosystems and harbor rich communities in the canopies of trees. Phorophytes often grow extra roots that tap into these communities to obtain nutrients.
17. Some plants grow as parasites on others. A few hundred species are partially heterotrophic and supplement their nutrition by trapping and digesting insects and other small animals.

Key Terms

essential nutrient 844
macronutrient 844
micronutrient/trace element 844
deficiency symptom 845
soil solution 848
loam 848
humic acid 849
clay-humus complex 849
cation exchange 849
rhizosphere 850
nitrogen fixation 851
nitrogenase 851
root nodule 851
actinorrhizal association 851
crop rotation 852
nitrification 853
denitrification 853
salinization 853
desertification 853
monoculture 854
ecological agriculture 855
tilth 856
epiphyte 858
phorophyte 858

Multiple-Choice Questions

1. Chlorosis in mature leaves but not in young leaves often signals
   a. a deficiency in a mobile mineral.
   b. a deficiency in a nonmobile mineral.
   c. an imbalance in the ratio of O_2 to CO_2.
   d. death of meristems.
   e. lack of water.
2. Which are most abundant in soil?
   a. bacteria
   b. fungi
   c. algae
   d. earthworms
   e. invertebrate animals other than earthworms
3. Which component of loam is an organic product?
   a. clay
   b. sand
   c. silt
   d. humus
   e. none of the above
4. Why do farmers add calcium-rich lime to acid soil?
   a. to lower the pH by the addition of more cations
   b. to raise the pH by the addition of more anions
   c. to cause the exchange of Ca^{2+} ions for soil anions
   d. to cause the exchange of Ca^{2+} ions for H^+ ions in soil
   e. to increase the particle size of loam, which masks H^+ ions
5. In general, food plants grow best in soil that maintains a pH of
   a. 7.5–8.2.
   b. 6.8–7.5.
   c. 6.3–6.8.
   d. 5.5–6.3.
   e. below 5.5.
6. Which of these compounds contains nitrogen in a form that is most rapidly useful for plants?
   a. ammonia
   b. nitrites
   c. nitrates
   d. N_2
   e. amino acids
7. Which one of these statements is correct?
   a. Plants rely on bacteria for usable forms of nitrogen and sulfur.
   b. Animals rely on bacteria for usable forms of nitrogen and sulfur.
   c. Plants can reduce sulfate to sulfhydryl, but animals cannot.
   d. Sulfur and nitrogen are reduced and oxidized in the soil, without the intervention of living cells.
   e. Many plants obtain energy by oxidizing H_2S to sulfate.
8. All but one of the following statements correctly characterizes the benefits of using organic, rather than inorganic, fertilizer. Which of the following does not occur?
   a. The amount of humus in the soil increases.
   b. The need for irrigation systems increases.
   c. The loss of minerals by leaching decreases.
   d. The porosity of the soil increases.
   e. The diversity of soil microorganisms increases.
9. Crown roots and crown humus are products of
   a. the roots of phorophytes.
   b. low-growing plants surrounding phorophytes.
   c. the leaves of epiphytes.
   d. the interaction of epiphytes and phorophytes.
   e. crown plants.
10. Which one of the following statements is *not* correct?
   a. Some plants do not carry out the reactions of photosynthesis.
   b. Plants such as Venus's-flytrap require an animal source of nitrogen.
   c. Epiphytes such as Spanish moss do not kill their hosts.
   d. Plants that are carnivorous have adapted to acid soils.
   e. Some plants parasitize other plants; others parasitize animals.

True-False Questions

Mark each statement true or false, and if false, restate it to make it true.

1. The increase in the volume of a growing plant is attributable mostly to materials provided by the soil, while the increase in dry mass is attributable mostly to atmospheric carbon dioxide.
2. Essential minerals are generally absorbed in their most reduced form and must be converted to their elemental state for use in metabolism.
3. After incorporation into organic molecules, essential micronutrients generally function as structural components within plant cells.
4. Sand and silt, but not clay, are derived from quartz.
5. As a result of nitrogen fixation, atmospheric nitrogen is oxidized and converted into a form in which it can be assimilated by living cells.
6. Leghemoglobin is synthesized partly by cells of a leguminous plant and partly by the procaryotes living within its root nodules.
7. The process of oxidizing atmospheric nitrogen to nitrate is known as nitrogen fixation.
8. Plants absorb sulfur from the soil in the form of sulfhydryl compounds.
9. In general, epiphytes have a parasitic relationship with their host plants.
10. Ecological agricultural practices encourage the use of rapidly dissolving, high-nitrogen, inorganic fertilizers.

Concept Questions

1. Explain why symptoms of a nutritional deficiency are first visible in the leaves of a plant, rather than in other parts of the shoot system.
2. How does highly acidic rain limit soil fertility?

3. It is a clearly accepted principle that animal survival requires the photosynthetic activity of plants. What about plant survival? Do they depend on other living members of their ecosystem? Explain.
4. How do the processes of nitrogen fixation, nitrification, and denitrification differ? Which are carried out by procaryotes and which by eucaryotes?
5. What is meant by ecological agriculture? How does it differ from monoculture?

Additional Reading

Bernhardt, Peter. *Wily Violets and Underground Orchids.* William Morrow and Co., New York, 1989. Fascinating reading, with much information about epiphytes and parasites.

Brown, Lester R. *Who Will Feed China?* W. W. Norton and Co., New York, 1995. Having a fifth of the world's population, China exemplifies all the problems of the future in one nation.

Dennis, David T., and David H. Turpin (eds.). *Plant Physiology, Biochemistry, and Molecular Biology.* Longman Scientific & Technical, Essex (England); John Wiley, New York, 1990.

Eisenberg, Evan. "Back to Eden." *The Atlantic Monthly,* November 1989, p. 57. The story of Wes Jackson and his work on reforming agricultural practices.

Lampkin, Nicholas. *Organic Farming.* Farming Press, Ipswich (UK), 1992.

McMichael, A. J. *Planetary Overload.* Cambridge University Press, Cambridge and New York, 1993. A superb summary of the world situation, filled with references to sources, which we have used extensively in writing this chapter.

Pietropaolo, James, and Patricia Pietropaolo. *Carnivorous Plants of the World.* Timber Press, Portland (OR), 1986.

Salisbury, Frank B., and Cleon W. Ross. *Plant Physiology,* 4th ed. Wadsworth Publishing Co., Belmont (CA), 1992.

Internet Resource

To further study the content of this chapter, log on to the web site at:

http://www.mhhe.com/biosci/genbio/guttman/

Animal Biology

41 Integration and Control Systems

The French pianist Yvonne Loriod in concert in 1967.

Key Concepts

A. Hormones and General Features of Signaling Systems

41.1 Animals need control and communication systems to guide their actions.

41.2 Hormones carry signals from one part of the body to another through the blood.

41.3 Two peptide hormones, insulin and glucagon, regulate the glucose concentration in the blood of vertebrates.

41.4 Ligand receptors can act in four ways.

41.5 Animal cells communicate chemically in endocrine, paracrine, or synaptic modes.

B. Nervous Systems

41.6 The vertebrate nervous system illustrates the general properties of nervous systems.

41.7 A nervous system is made of neurons that communicate at synapses.

41.8 A nervous system collects and distributes information.

41.9 All cells maintain an electrical potential across their plasma membranes.

41.10 An increase in the Na^+ conductance of the plasma membrane generates an action potential.

41.11 A nerve impulse is an action potential that propagates itself along the axon.

41.12 A myelinated nerve can carry impulses rapidly.

41.13 Synaptic inputs determine a neuron's ability to fire impulses.

C. Connections Between Endocrine and Nervous Systems

41.14 The pituitary gland and hypothalamus form a center of regulation.

41.15 The autonomic nervous system controls many routine functions.

The pianist bows to the audience, sits down on the bench, lays her fingers on the keyboard, and looks up at the conductor. The tall, tuxedoed figure raises his hands, nods, gives a downbeat—and instantly the pianist's fingers begin to move dexterously across the keyboard, tapping out a rising arpeggio, moving into the opening theme of the concerto as the first notes from the strings rise behind her. Volleys of nerve impulses race back and forth between her brain and the arm muscles controlling her fingers, guiding movements too fast for thought, movements virtually stored in the memory of her muscles by years of practice. Hundreds of times a second, signals speed through a feedback loop between a dozen brain centers, ensuring that each fingertip strikes each key with precisely the energy and timing needed to transform the printed notes into a masterwork of expression. Meanwhile, her eyes dart constantly to the conductor's hands, taking in subtle cues to shape her timing to that of the orchestra behind her; as her ears pick up the music, this information feeds into her neural cycle, guiding and supporting her movements as soloist and orchestra together create a unified sonority. Her whole body is energized to its peak by an ancient system of hormones and nervous activity, once useful for escaping from lions and fighting baboons, now directed toward exciting every muscle to its peak of activity, draining the energy reserves in her liver and muscles, opening her pores to the cooling effects of sweat, heightening her sensitivity to the music around her and the subtle reaction of the audience. When the pianist rises to applause 20 minutes later, she will still be riding an adrenaline high, but soon after, she will rest, happily drained of energy.

The ability to give such a performance is perhaps the best definition of what it is to be human. Here we will begin to understand the complex of nervous and hormonal activity that makes it possible. ■

A. Hormones and general features of signaling systems

41.1 Animals need control and communication systems to guide their actions.

Animals are active creatures. With their *musculoskeletal system*—that is, muscles attached to a skeleton—they move around and encounter new situations. With a *digestive system,* they ingest and digest foods that vary unpredictably from moment to moment. Their *circulatory system* moves blood or hemolymph throughout the body; a *respiratory system* exchanges CO_2 and O_2 between the tissues and the environment, generally through the blood; and an *excretory system* maintains a constant composition of internal fluids and removes wastes. An *integumentary system* wraps the internal structures in a protective housing made of skin or other body covering. Many animals have cells that wander through the body, seeking out potential invading organisms and destroying them, and these cells may be part of a complex *immune system* that confers general protection against disease. To make use of their *reproductive systems,* animals generally have to find appropriate mates, recognize them as the same species, and go through a mating ritual. We discuss all these systems in the following chapters. But superimposed over all of them are *control and communication systems,* which we will discuss first because of their far-reaching effects.

We have seen that every organism collects and responds to critical information about its environment. In addition, as we emphasized in Chapter 32, a multicellular organism must continually monitor the internal environment where all its cells live. In reaction to this information, it maintains *homeostasis* by adjusting its enzyme systems, by turning genes on and off, by changing the concentrations of some substances, and sometimes by following a repertoire of behaviors. We showed in Chapter 11 that regulation always entails feedback loops using sensors that monitor a particular condition (such as temperature) and an effector that changes that condition. In an animal, the feedback loops become more extensive, complicated, and versatile than in other organisms, and the sensors and effectors are usually distinct cells—**sensor cells** and **effector cells**—or entire organs. Figure 41.1 illustrates this interactive process: An animal's sensors detect light, heat, mechanical pressures, and the concentrations of various molecules. Its effectors are *muscles* and *glands.* Between the sensors and effectors are two systems of *integrators,* or *communicators*—an endocrine (or hormonal) system and a nervous system. The two are closely connected; both carry messages between sensors and effectors, integrating information and creating regulatory circuits. In addition, a third system, the immune system, provides sensors of another kind that monitor foreign molecules and cells that may get into the body. It is becoming clear that the immune system is closely tied to the endocrine and nervous systems in ways that we do not entirely understand yet.

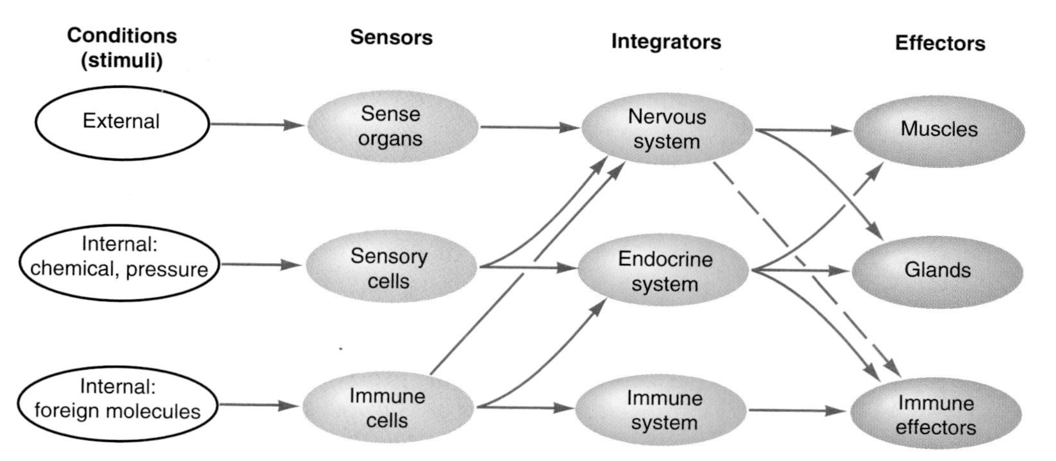

Figure 41.1

An animal's nervous, endocrine, and immune systems detect conditions in and around the body and respond to them. Sense organs sense external conditions through vision, hearing, touch, taste, and other senses; internal cells monitor chemical conditions and pressures; cells of the immune system detect foreign molecules. Each system includes sensors, integrators, and effectors.

Sensors in animals, Chapter 43.
The immune system, Chapter 48.

41.2 Hormones carry signals from one part of the body to another through the blood.

In 1902 William M. Bayliss and Ernest H. Starling were investigating digestion in dogs. Physiologists already knew that when the acidic contents of the stomach move into the small intestine during digestion, the pancreas neutralizes the acid by releasing alkaline fluid, rich in bicarbonate ions, into the intestine. Bayliss and Starling asked how the pancreas receives a signal that acid has reached the intestine. Is the signal sent through nerves or by other means? To answer this question, they simply cut the nerves between the small intestine and the pancreas and introduced acid into the small intestine; the pancreas still responded with the usual production of bicarbonate. This experiment showed that nerves could not be involved. Then they performed an ingenious test. After connecting the circulatory systems of two dogs so they shared a common blood supply, they injected acid into the small intestine of one dog, but both dogs secreted pancreatic fluid into their small intestines. The stimulating factor could only have been a chemical circulating in the blood.

Bayliss and Starling realized that the chemical must have been produced by cells in the small intestine, so they scraped off some intestinal lining, ground the tissue with sand, filtered it, and injected a bit of filtrate into a dog's vein; the result was a profuse secretion of material from the pancreas. They called the active chemical **secretin.** Bayliss and Starling also realized that many tissues were exposed to the secretin circulating throughout the body, but only specific cells of the pancreas responded—the **target cells** of secretin. Having thus shown that specific signals can be transmitted in an animal by a path other than the nervous system, they coined the general term *hormone* to describe the material that carries the signal (Greek, *hormaein,* = "to impel" or "to arouse to activity"). We use the same general concept of hormones today: *A hormone is a substance secreted into the circulation by certain cells that produces a response in certain other cells, the target cells.* It is important to see that hormonal communication entails two specificities: The hormone itself only affects target cells that are able to recognize and receive it, and the target cell has its own characteristic activity, so one hormone may stimulate two different cells to perform quite different processes.

Since Bayliss and Starling's work, many animal hormones have been identified and their actions explored. A hormone could be virtually any distinctive molecule, but animal hormones belong to four categories with different chemical structures and mechanisms of action:

- *Amino acid derivatives* sometimes function both as hormones and as signals in the nervous system called *neurotransmitters.*
- *Peptides* range in size from a few amino acids to many. Some of the amino acids may be modified forms of the familiar ones found in usual proteins.
- *Steroids* are lipids chemically related to cholesterol. In general, they enter their target cells and directly affect nucleic acid and protein synthesis.
- *Modified fatty acids* have special roles in the reproductive and immune systems.

41.3 Two peptide hormones, insulin and glucagon, regulate the glucose concentration in the blood of vertebrates.

The general action of hormones—peptide hormones, in this case—is nicely illustrated by the vertebrate hormones insulin and glucagon, which maintain a constant glucose concentration in the blood. Human blood plasma (the fluid part of blood) normally carries about 60–100 mg of glucose ("blood sugar") per 100 ml of plasma, and it must be kept in this range constantly, since glucose is the usual source of energy for all cells. The brain, in particular, depends on glucose. Too little glucose in the blood *(hypoglycemia)* produces weakness, dizziness, sweating, and even loss of consciousness. Too much glucose *(hyperglycemia)* does not have such immediate consequences, but eventually produces impaired circulation (as the blood virtually turns into syrup), damage to the eyes, kidneys, and other organs, and often death.

At each meal, a vertebrate may ingest a good deal of carbohydrate, much of which is broken down into glucose in its digestive tract. However, the concentration of glucose in blood plasma does not rise sharply after eating or fall low between meals because most of the glucose is polymerized into glycogen, or animal starch, stored in liver, muscle, and adipose (fat) cells, and released gradually. The blood glucose level is kept quite constant by a rather simple "glucostat," a classical negative feedback system that operates through insulin and glucagon. When the blood glucose level starts to rise after eating, insulin is secreted, promoting the buildup of glycogen stores. As glucose is depleted from the blood between meals, glucagon is secreted, promoting the breakdown of glycogen and the release of glucose into the blood:

$$\text{Glucose} \underset{\text{Glucagon}}{\overset{\text{Insulin}}{\rightleftharpoons}} \text{Glycogen}$$

(Note the confusing similarity of the words *glycogen,* the storage form of glucose, and *glucagon,* the hormone.)

Insulin and glucagon are synthesized and stored in the pancreas, an organ made of two kinds of anatomically and functionally different glands, exocrine glands and endocrine glands (Figure 41.2). Keep in mind that these glands exist in other organs and areas of the body as well; we use their role in the pancreas as one example of their function. **Exocrine glands** are pockets made of specialized epithelial cells; they produce materials such as digestive enzymes, sweat, tears, and oils that are released through ducts to the outside of the body or the intestinal tract. Exocrine cells of the pancreas produce bicarbonate ions and digestive enzymes used in the small intestine. **Endocrine glands** secrete hormones and release their secretions into the

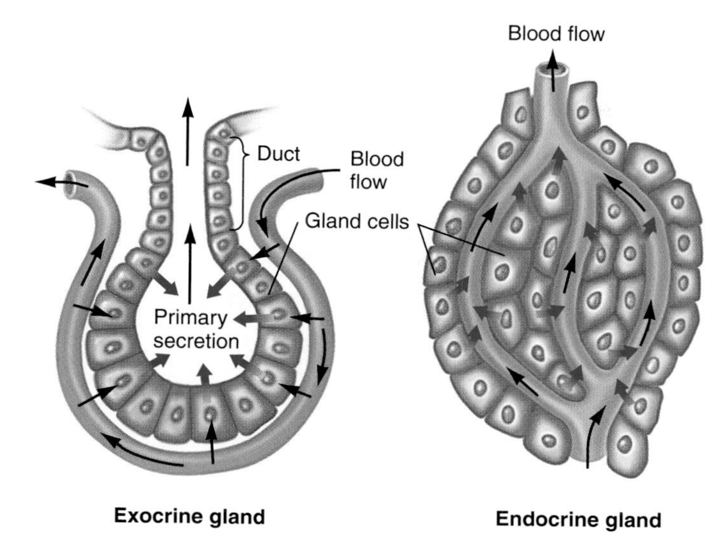

Figure 41.2

Exocrine glands release their products through a duct to a surface inside or outside the body. Endocrine glands release their products into the interstitial fluid and thus into the blood.

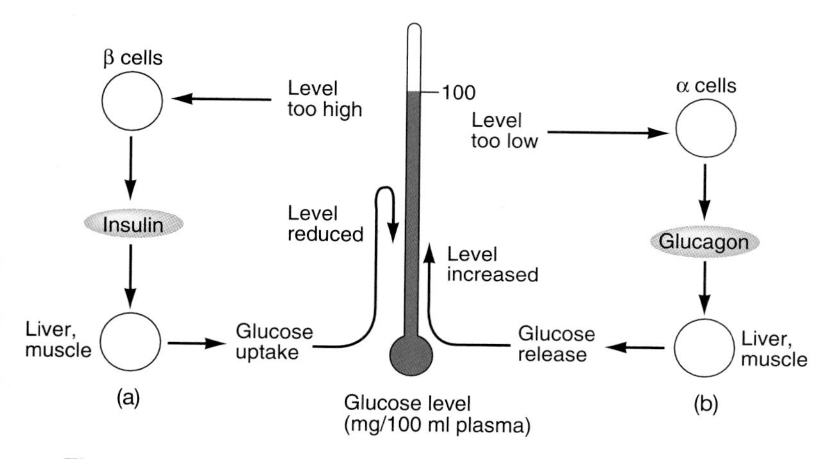

Figure 41.3

The plasma glucose concentration is regulated by a pair of hormones made by cells in the islets of Langerhans, scattered throughout the pancreas. *(a)* In response to a high level of glucose, β cells release insulin, which stimulates other cells to take up glucose. *(b)* In response to a low level of blood glucose, α cells release glucagon, which stimulates other cells to secrete glucose into the blood.

interstitial fluid, from where they pass into the blood. The endocrine regions of the pancreas are the islets of Langerhans, including α cells that synthesize glucagon and β cells that synthesize insulin. Both types of islet cells monitor the plasma glucose concentration. When this concentration rises above a set level, β cells sense the increase and release insulin; when the glucose concentration drops below a certain minimum level, α cells sense the decrease and release glucagon (Figure 41.3).

Target cells for glucagon or insulin have specific membrane receptors. Many types of cells come in contact with glucagon or insulin carried by the blood plasma but ignore these hormones.

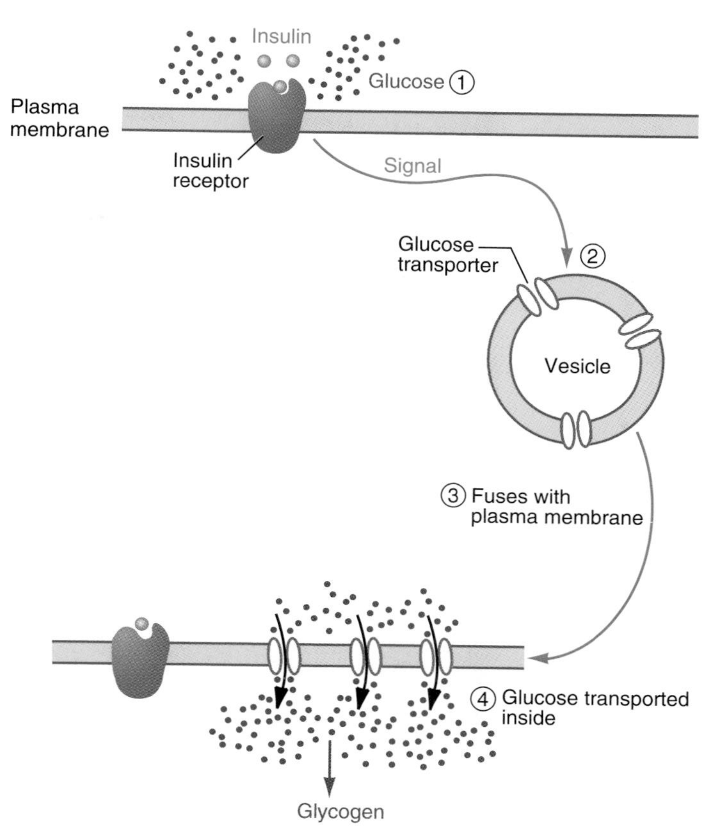

Figure 41.4

When insulin binds to insulin receptors *(1),* they send a signal to internal vesicles that carry glucose transport proteins *(2).* These vesicles move to the plasma membrane and fuse with it *(3),* so the transport proteins are able to move glucose into the cytoplasm *(4),* where it is converted into glycogen.

However, the plasma membranes of muscle and liver cells—the target cells of these hormones—carry specific receptor proteins capable of binding one hormone or the other. When glucagon receptors on a cell bind to glucagon, they change conformation and initiate a sequence of events that eventually results in the release of glucose from glycogen. Alternatively, when insulin binds to its plasma membrane receptor, it initiates another chain of events: Glucose transport proteins, sequestered in cytoplasmic vesicles, move to the cell membrane, where they transport glucose into the cytoplasm. The glucose is then metabolized or stored in the form of glycogen (Figure 41.4).

Exercise 41.1 You have discovered that a small cluster of cells in a mouse secretes a peptide of 12 amino acids. You are sure the peptide must be a hormone, but you have no idea of its function or what its target cells might be. How would you find its target cells? (*Hint:* Think radioactivity.)

41.4 Ligand receptors can act in four ways.

We have used the term **signal ligand** for hormones and other molecules that carry a chemical message. Despite the variety of signal ligands and their many effects, their receptors only act in

Figure 41.5

Four types of ligand receptors are known. *(a)* Many receptors activate G-proteins. *(b)* When activated by binding to their ligands, catalytic receptors become enzymes that can carry out new metabolic activities, perhaps activating other proteins. *(c)* Channel-linked receptors open channels and allow ions to flow in or out of the cell. *(d)* Lipid-soluble hormone receptors, such as steroid receptors, are intracellular, and they activate or repress genes directly.

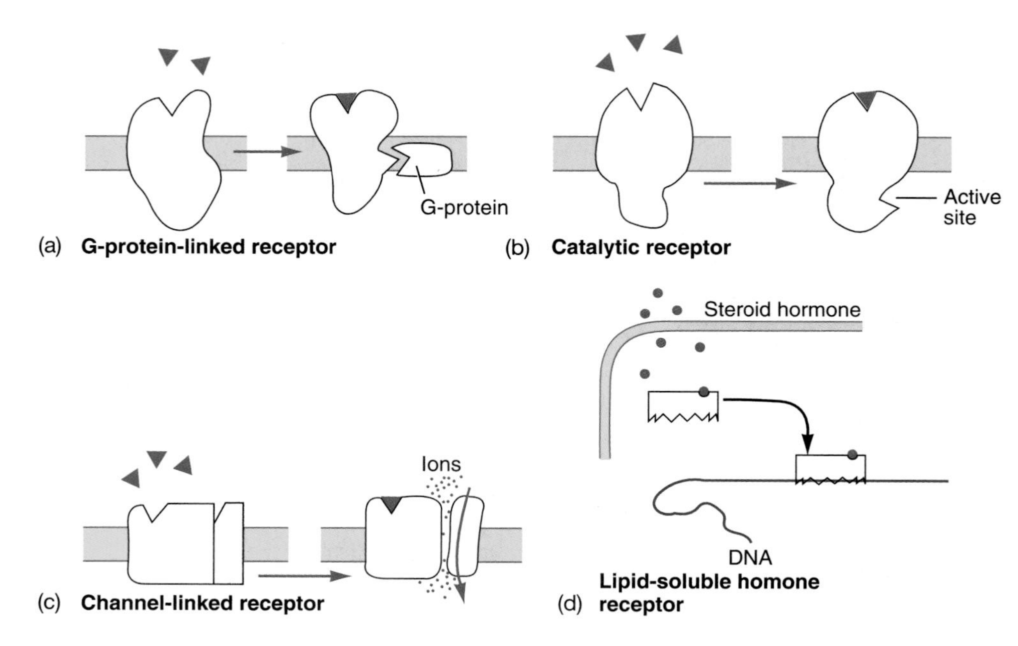

one of four ways, as shown in Figure 41.5. All receptors, of course, are proteins whose binding sites are specific for a particular ligand. Steroid- and thyroid-hormone receptors are intracellular. The others are surface receptors that bind ligands on the outside of the cell membrane and initiate specific internal changes.

G-protein-linked receptors

In Section 11.8, we describe the most common pathway of signal transduction in eucaryotic cells: through G-proteins, second messengers such as cyclic AMP or Ca^{2+} ions, and protein kinases. Now we add one additional mechanism, that explains how hormones can have such prodigious effects at low concentrations. The link between a small signal and a large effect is a **reaction cascade,** a kind of *biological amplifier:*

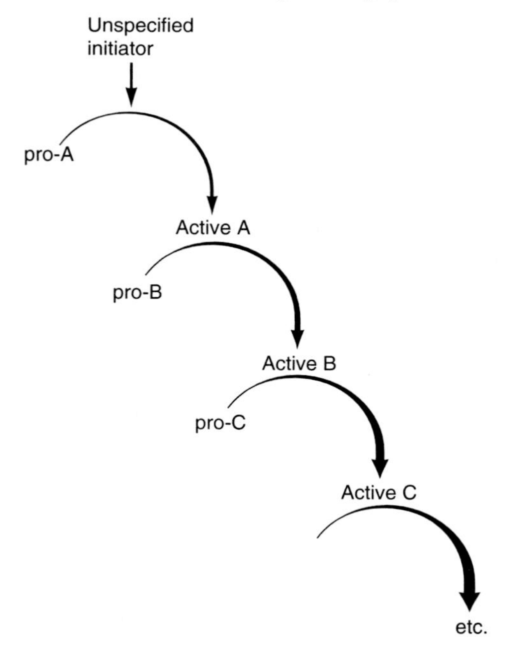

A cascade involves a series of inactive molecules, usually enzymes, that are activated in succession: A few molecules of the first activate several molecules of the second, which activate still more molecules of the third, and so on. If each step in a cascade only amplifies by a factor of 10 or so, a very small initial signal such as a few hormone molecules can quickly generate 10^4–10^5 molecules of the last component.

The cascade initiated by glucagon employs a series of protein kinases that activate one another through phosphorylation. This cascade causes the breakdown of glycogen to glucose (Figure 41.6). In some other cascades, the inactive proteins are **proenzymes** that have extra peptides blocking their active sites and are activated by removal of these peptides:

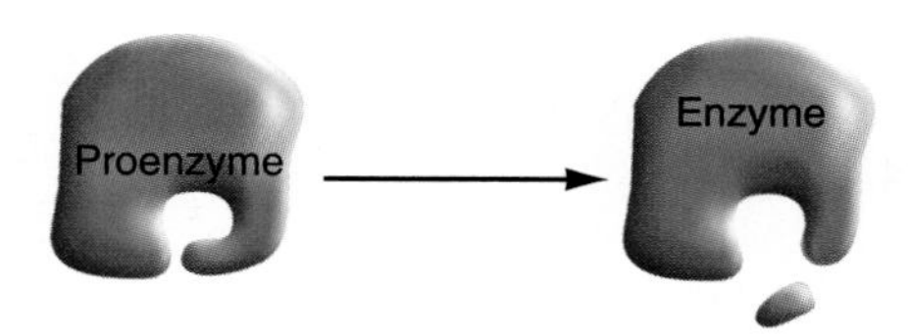

Catalytic receptors

Catalytic receptors are inactive membrane enzymes, usually protein kinases, that are activated when the ligand binds to them. These receptors also initiate reaction cascades.

Channel-linked receptors

Channel-linked receptors open ion-channel proteins, which conduct specific ions through membranes. These proteins are said to be **ligand-gated** because they only open when bound to an appropriate ligand, allowing ions to flow into or out of the cell. Ions always flow through channels passively, not by active transport. Channel-linked receptors are integral to the action of nerve cells, which operate through rapid changes in the permeability of the plasma membrane to specific ions.

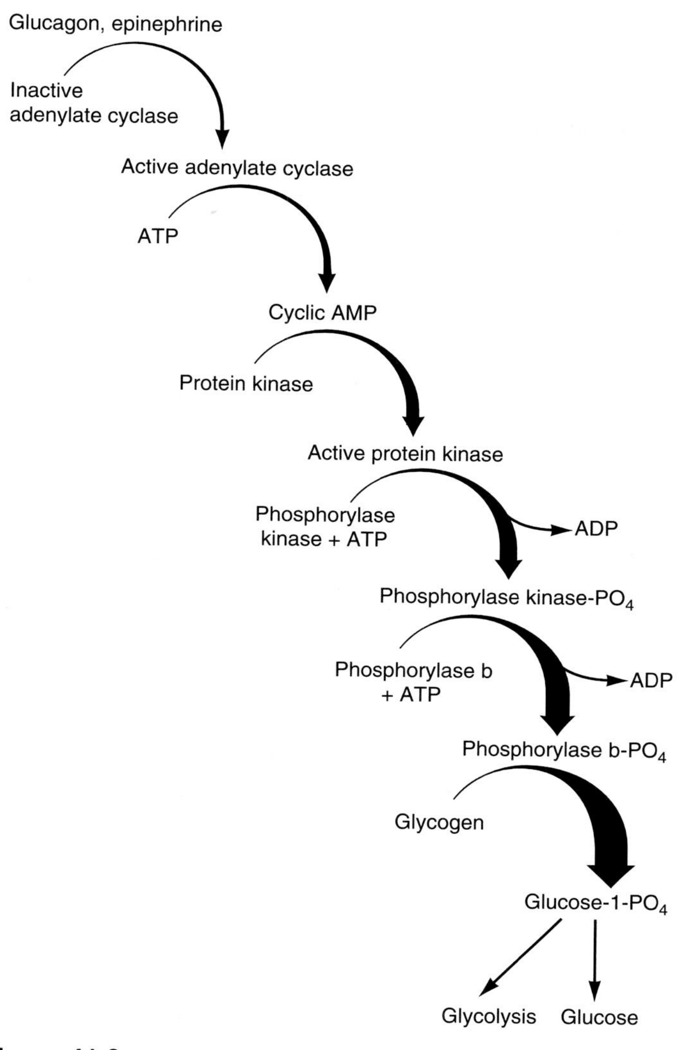

Figure 41.6

A cascade of enzyme activations initiated by the hormone glucagon finally leads to the breakdown of glycogen into glucose.

Lipid-soluble hormone receptors

Lipid-soluble hormones, including steroids and thyroid hormone, penetrate cell membranes and bind to specific intracellular protein receptors, which are always regulatory proteins for specific sets of genes. The female sex hormone estrogen, for instance, turns on one set of genes in a cell, while the salt-regulating hormone aldosterone, entering the same cell, binds to a different receptor and turns on different genes. At the same time, either hormone may enter cells of a different kind and bind to a different type of receptor, which regulates a different set of genes.

Exercise 41.2 A hormone is tagged with a radioisotope, and the radioactivity is found in the nuclei of its target cells. What general type of hormone is it (with respect to chemical structure)?

41.5 Animal cells communicate chemically in endocrine, paracrine, or synaptic modes.

Having outlined the features of communication through hormones, we can generalize about communication between the cells of a multicellular organism. Neighboring cells can communicate via gap junctions:

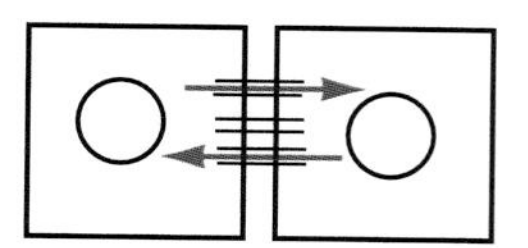

or via molecules bound to their surfaces:

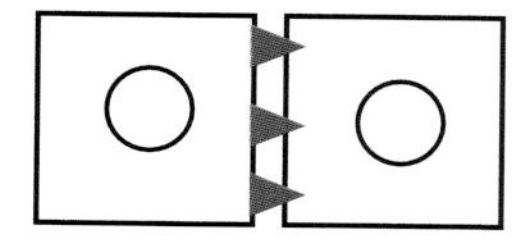

But cells that are not in direct contact use secreted signal ligands in any one of three types of communication: endocrine, synaptic, or paracrine (Figure 41.7).

Endocrine communication uses hormones, which are released by sensor cells into the bloodstream and are detected by effector cells that could be located anywhere in the body. Some endocrine cells just form small patches of tissue, but most of them are organized into distinct glands (Figure 41.8). Some are localized in the organs they regulate, just as specialized microcomputers are often placed in the instruments or appliances they control. A complex organism like a mammal requires many hormones to control its activities (Table 41.1). We will discuss the functions of most of these hormones in later chapters.

In **synaptic communication,** one cell signals another across sites called **synapses** where the cells are separated by only a narrow gap; the signal ligand is called a **neurotransmitter.** Synaptic transmission occurs only in nervous systems and the cells they control, which we explore later in this chapter.

In **paracrine communication,** cells regulate others very nearby in the same tissue or organ. Some cells communicate with themselves in **autocrine** mode by responding to the substances they secrete, but notice that in autocrine regulation, a cell is stimulating itself through an *external* pathway, which may be quite different from its internal regulatory pathways. The growth factors discussed in Section 20.8 are paracrine factors; they are all proteins that promote local cell growth and proliferation, and a long list of functionally related proteins called *lymphokines* are involved in the complicated processes of inflammation and immunity, as discussed in Chapter 48. One of the most interesting paracrine factors discovered in recent years is *nitric oxide* (NO), which is also a neurotransmitter in some parts of the nervous system. Nitric oxide and other small signal ligands that act paracrinally are called **local chemical mediators.** They are produced and degraded rapidly, causing local, short-lived effects.

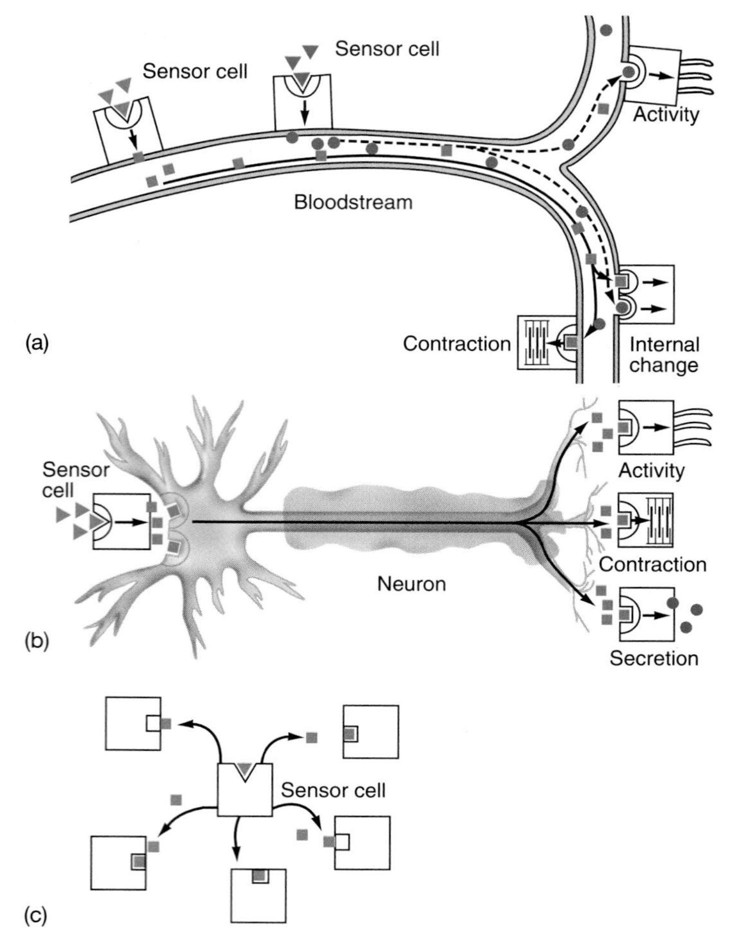

Figure 41.7

Cells communicate through signal ligands in three ways. *(a)* Endocrine communication uses hormones that are carried through the bloodstream. *(b)* The neurons that comprise a nervous system carry electrochemical signals to distant points, where they communicate synaptically by releasing neurotransmitters that diffuse the short distance across a synapse. *(c)* Paracrine communication uses local chemical mediators that diffuse to nearby cells.

The **prostaglandins** and their relatives are a family of fatty acid derivatives that act as local chemical mediators in many organs. Their precursor, arachidonic acid, is converted into either leukotrienes or prostaglandins by two enzymes:

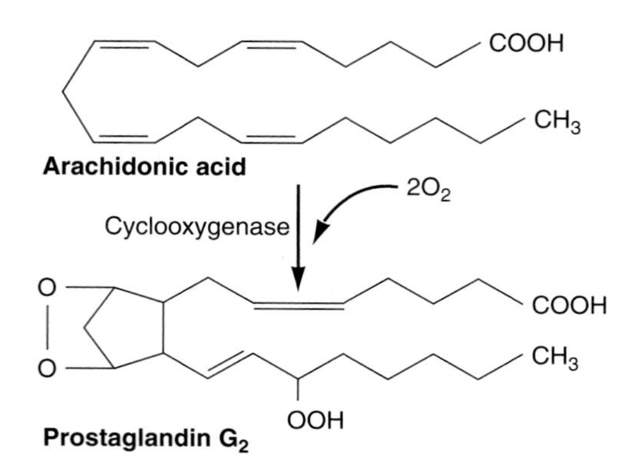

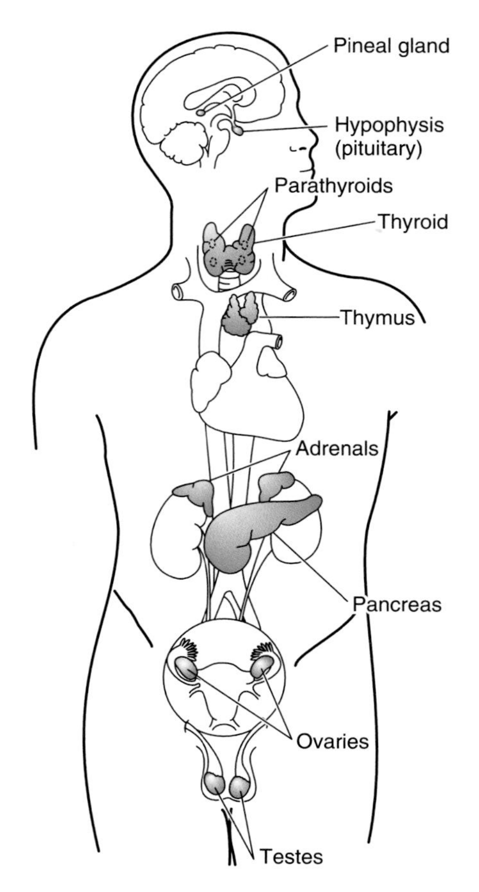

Figure 41.8

The principal endocrine glands are located at several points in the human body. This figure will serve as a reference map for glands to be discussed later.

Their primary effect is on smooth muscle. Prostaglandins were discovered in—and named for—secretions of the prostate gland that stimulate contractions in the smooth muscle of the uterus and oviduct, thus helping sperm reach an egg that may be moving down the oviduct to be fertilized. However, some prostaglandins stimulate smooth-muscle contraction, and others promote relaxation, so they may either constrict or relax blood vessels. Prostaglandins are important in the inflammatory process, which involves changes in the local circulation of a wounded area.

Naturally, all these signal systems are intimately connected and coordinated. Some cells are not strictly endocrine or strictly nervous, and often the same ligands serve as both hormones and neurotransmitters.

B. Nervous systems

All animals except sponges have nervous systems that allow them to respond quickly to changes in their environment, to behave in complicated ways, and to regulate themselves in general. A nervous system carries specific messages quickly through **neurons,** elongated cells that are strung in circuits from point to point. This system works in parallel with the endocrine

Table 41.1 Major Vertebrate Hormones

Gland	Hormone	Major Effects
Pituitary anterior lobe	Somatotropin (Growth hormone)	Stimulates growth; stimulates lipid hydrolysis, raises blood glucose
	Adrenocorticotropic hormone (ACTH)	Stimulates adrenal cortex
	Thyroid-stimulating hormone (TSH)	Stimulates thyroid gland
	Follicle-stimulating hormone (FSH)	Stimulates ovarian follicle, spermatogenesis
	Luteinizing hormone (LH)	Stimulates corpus luteum and ovulation in female, interstitial cells in male
	Prolactin	Stimulates milk production
Hypothalamus and posterior pituitary	Vasopressin (antidiuretic hormone, ADH)	Stimulates water reabsorption by kidney
	Oxytocin	Stimulates contraction of uterus, milk release
Pineal	Melatonin	Regulates gonadotropin production by anterior pituitary
Thyroid	Thyroxine, triiodothyronine	Stimulates respiration, helps regulate growth
	Calcitonin (thyrocalcitonin)	Stimulates uptake of calcium into bone
Parathyroids	Parathyroid hormone (PTH)	Stimulates release of calcium from bone, and other calcium regulatory systems
Adrenal cortex	Glucocorticoids (cortisone, cortisol)	Stimulate glucose synthesis from protein and lipids (glucogenesis)
	Mineralcorticoids (aldosterone, etc.)	Stimulate sodium and water retention by kidney; regulates angiotensin
Adrenal medulla	Epinephrine, norepinephrine	Elevates plasma glucose, "fight or flight" reaction; other metabolic functions
Ovaries	Estrogen	Develops secondary sexual characteristics, stimulates growth of uterine lining
	Progesterone	Stimulates growth of uterine lining, maintains pregnancy
Pancreas islet cells	Insulin	Reduces plasma glucose, stimulates glycogen storage
	Glucagon	Raises plasma glucose, stimulates glycogen breakdown

system, and the two are intimately linked in places. It is helpful to think of the nervous system as carrying signals rapidly to specific points while hormones carry broader, more diffuse signals more slowly to broader targets.

41.6 The vertebrate nervous system illustrates the general properties of nervous systems.

Because it is so familiar, the vertebrate nervous system, as illustrated by a human, is a convenient place to begin. The **central nervous system (CNS)** includes the **brain** and the **spinal cord,** which runs through the vertebral column. The **peripheral nervous system (PNS)** is made of *nerves,* paired right and left, that connect the CNS to effectors and receptors in the rest of the body (Figure 41.9*a*). The general terms "afferent," meaning *toward,* and "efferent," meaning *away from,* describe neurons in reference to the CNS. **Afferent neurons** carry messages from sense organs toward the CNS and are therefore also called **sensory neurons. Efferent neurons** carry messages from the CNS to the body's effectors, thus are also called **motor neurons:**

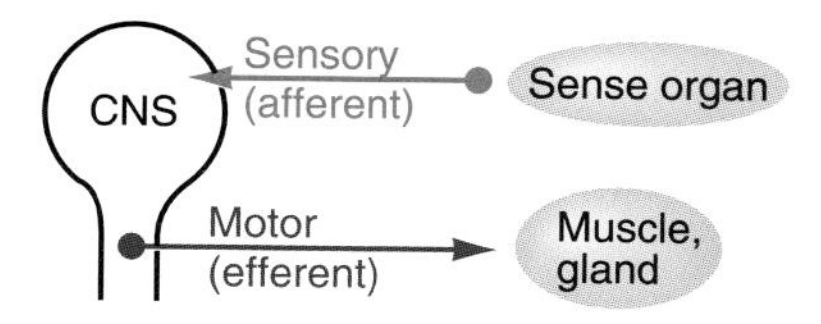

Twelve pairs of *cranial nerves* emerge from centers in the brain, and a pair of spinal nerves emerge at each level of the spinal cord, with a few exceptions, between the vertebrae (segments of the backbone).

The motor portion of the nervous system again has two parts. Regulatory functions are handled by the **autonomic**

Figure 41.9

(a) The human nervous system consists of a central portion (CNS: the brain and spinal cord) and a peripheral portion (PNS) consisting of nerves. *(b)* The nervous system is further subdivided anatomically and functionally. Nerves contain sensory, or afferent, neurons that carry signals toward the CNS, and motor, or efferent, neurons that carry signals away from the CNS to the effector organs. The motor part of the system is divided into a somatic system, which controls skeletal muscle (the muscles that move bones), and an autonomic system, which controls the cardiac muscle that forms the heart and the smooth muscle that forms internal organs such as the intestinal tract and the blood vessels. The two divisions of the autonomic nervous system are described in Section 41.15.

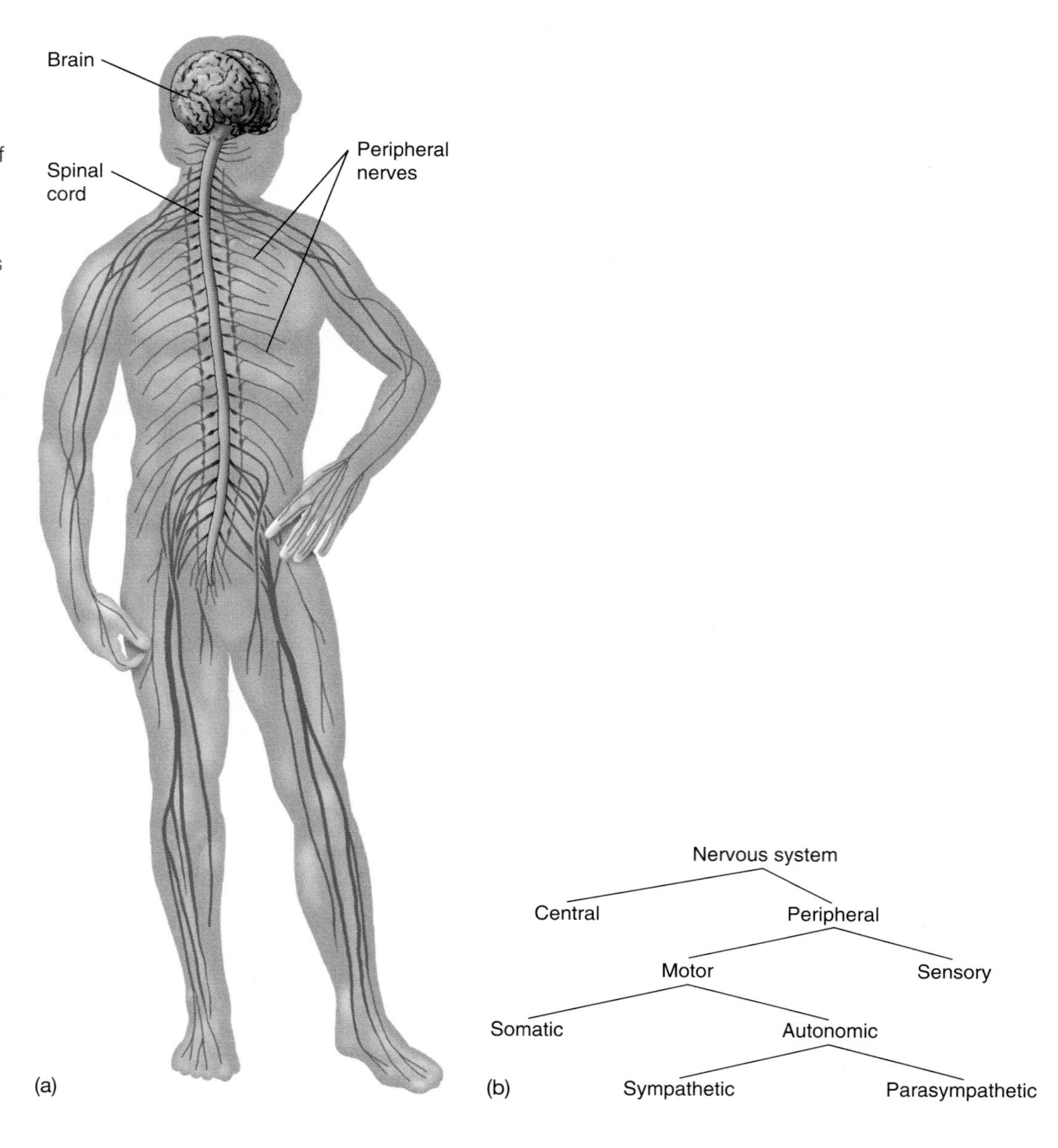

nervous system (ANS), which controls only glands and structures made of smooth or cardiac muscle such as the intestines and heart. This system both excites and inhibits effectors. The **somatic nervous system** controls the skeletal muscles, which move bones and parts of the skin; it is exclusively excitatory and controls general movement. Figure 41.9*b* diagrams the relationships between the various parts of the nervous system.

Please notice that "autonomic" is not "automatic." Our thinking about behavior is still plagued by an old, misguided conception that the autonomic system produces "involuntary" actions while the somatic nervous system is "voluntary," meaning that humans (and maybe other intelligent animals) have some conscious control over it. This distinction is a remnant of another misleading idea that a person has an automatically acting "body" and a consciously controlled "mind," which are only vaguely and mysteriously connected. In fact, all actions come from a complex of somatic and autonomic activities. For example, the actions we perceive as *behavior* are largely determined by the somatic system, but related actions such as crying and blushing are largely autonomic functions.

41.7 A nervous system is made of neurons that communicate at synapses.

A nervous system gathers information from many points in and around the body, processes and integrates this information, and sends instructions for action back to effector organs: muscles and glands. Through its sensors, an animal monitors a variety of environmental stimuli in the form of light, vibrations, pressure, or substances in its internal and external worlds. The nervous system, however, doesn't handle information in these forms; light itself is not transmitted along the optic nerve from the eye,

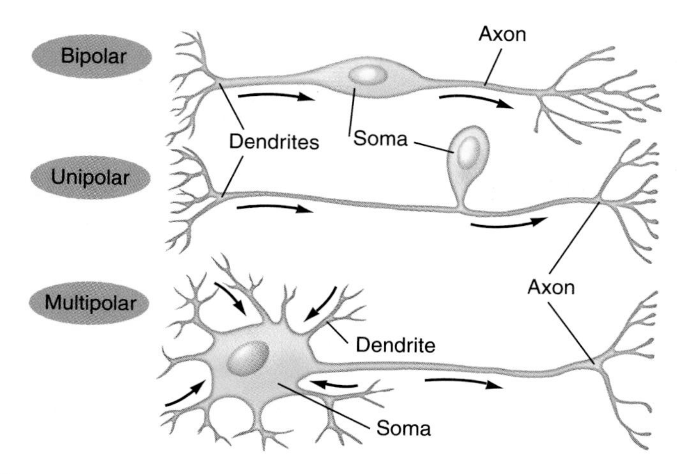

Figure 41.10

Neurons take many forms; the bipolar, unipolar, and multipolar types shown here are distinguished by the number of large extensions emerging from the cell body. The axon is a relatively large, uniform extension that does not branch near the cell body.

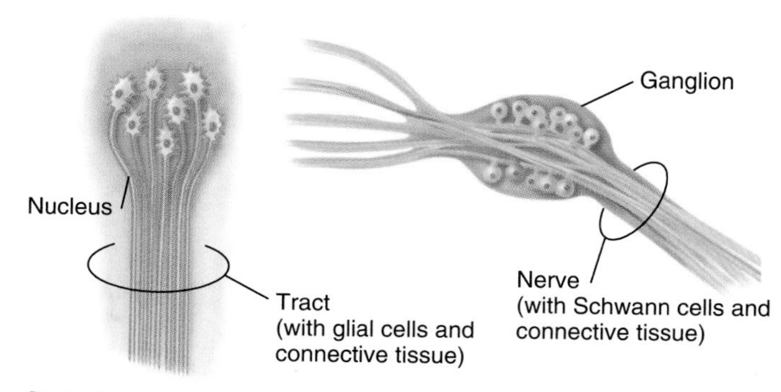

Figure 41.11

Comparable structures within the central nervous system and the peripheral nervous system have different names. Clusters of neuron cell bodies are called nuclei or ganglia; bunches of myelinated axons and dendrites are called tracts or nerves.

nor is sound pressure carried along the auditory nerve from the ear. The nervous system speaks a universal language of *nerve impulses,* waves of electrochemical change that travel rapidly in the membranes of neurons. External signals must be *transduced* into this language, and signals carried to effector organs must be transduced into their actions.

Signal transduction, Section 11.3.

Nervous tissue is made of several types of cells. A neuron, the type of cell that actually carries signals, has a large cell body, or **soma,** containing a nucleus and other typical eucaryotic organelles (Figure 41.10). **Dendrites** are small, branching extensions that carry signals toward the cell body. A longer, heavier extension, the **axon,** or **nerve fiber,** carries signals away from the cell body, perhaps branching to contact other neurons, muscle cells, or gland cells. An axon carries messages from the soma to these other cells; these messages are encoded in the language of nerve impulses by changing the rate at which a neuron "fires," or sends impulses, not by changing the size or properties of the impulses themselves. (Morse code, by analogy, carries information by changing the spacing or timing of "beeps," not by making them louder or softer.) Through these signals, the nervous system carries messages from sensory receptors and stimulates effectors into action.

The cell bodies of neurons with similar functions cluster together. Such a cluster is called a **nucleus** if it lies inside the CNS and a **ganglion,** which looks like a bulb or knob, in the PNS (Figure 41.11). A **tract** is a bundle of axons and dendrites bound together with connective tissue, lying within the CNS; a *nerve* is a comparable bundle of axons and dendrites outside the CNS.

A neuron communicates with another cell at a synapse, where information passes only one way: from the **presynaptic cell** (a neuron) to the **postsynaptic cell,** which could be another neuron or a gland or muscle cell (Figure 41.12). The axonal ending of a presynaptic cell is separated from a postsynaptic cell by a **synaptic cleft,** a space of only a few tens of nanometers. The presynaptic ending contains many small vesicles, 40–200 nm in diameter, full of neurotransmitter molecules. When a nerve impulse in the presynaptic cell reaches a synapse, it stimulates these vesicles to fuse with the cell membrane and release their contents into the synaptic cleft (Figure 41.12, *enlargement*). Diffusing quickly across the cleft, these ligands bind to receptors of ligand-gated channels on the postsynaptic plasma membrane, stimulating a response. If the postsynaptic cell is another neuron, its response is generally to change its firing rate.

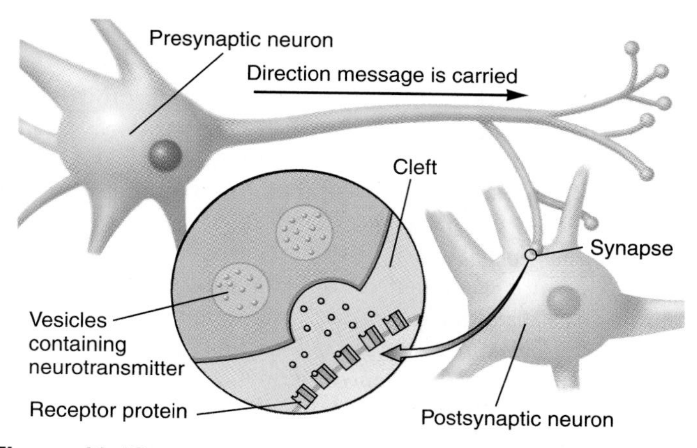

Figure 41.12

A signal passes across a synapse from the presynaptic to the postsynaptic cell. Neurotransmitters released at the tip of an axon (enlargement) diffuse across the synaptic cleft and stimulate the postsynaptic cell by binding to its receptor proteins. Information is therefore transmitted only one way across the synapse.

Exercise 41.3 Radio waves are just electromagnetic waves whose specific frequency identifies the transmitting station. A signal can be carried on a radio wave either by changing the size of the wave (amplitude modulation; AM radio) or by varying the frequency of the wave (frequency modulation; FM radio). Which method do neurons use?

Exercise 41.4 You are trying to investigate how a synapse operates by finding substances that inhibit it. Think of at least three processes that might be inhibited.

Exercise 41.5 A student talking to a neuroanatomist refers to "a nerve connecting two parts of the brain" and is corrected by the anatomist. What did the student say that was wrong?

41.8 A nervous system collects and distributes information.

A classical **reflex arc** nicely illustrates the general operation of a nervous system (Figure 41.13). A sensory neuron runs from a peripheral receptor (for example, a heat receptor on the hand) into the spinal cord. There it synapses with a short **interneuron,** a connecting neuron that, in turn, synapses with a motor neuron leading back to an effector (in this case, a group of arm muscles). Suppose your extended hand touches something hot. A receptor transduces information about that heat into a train of nerve impulses, which pass up the sensory neuron (the afferent fiber) to the spinal cord. The message is quickly transmitted across two synapses to the motor neuron, which sends signals through its efferent fiber to your arm muscles, causing them to contract and pull your hand away from the heat. This movement is a *reflex* action because it happens quickly, unconsciously, spontaneously, and in a relatively stereotyped way in response to a particular kind of stimulus.

Meanwhile, back at the spinal cord, the incoming signal is processed in other ways. The apparently simple reflex action is really more complicated than it may seem at first. The neurons in a reflex arc are modified by other fibers, both excitatory and inhibitory, that determine the extent and direction of the reflex action. Other muscles have to adjust to maintain your balance. Higher brain centers have to be notified about what is happening, so you can make other appropriate responses. Thus the incoming information passes through other synapses as well, to fibers running up and down the spinal cord. A fraction of a second after your arm pulls back, you feel the sensation of heat, indicating that the information has reached your brain. You can then start to take other actions—and you will remember the pain and where it came from.

The central nervous system contains a decision-making apparatus made of structures that process information. The reflex arc creates a simple, direct action in response to local conditions. The so-called lower brain centers gather information from single sources; higher brain centers bring together additional and more recent information: from the eyes, which are surveying the scene, from receptors in the muscles and ligaments about the posture and tension of each part of the body, and from all the other senses. Brain centers then integrate these ever-changing patterns of information with stored information from previous experiences to determine the appropriate direction, intensity, and timing of each reaction in the body.

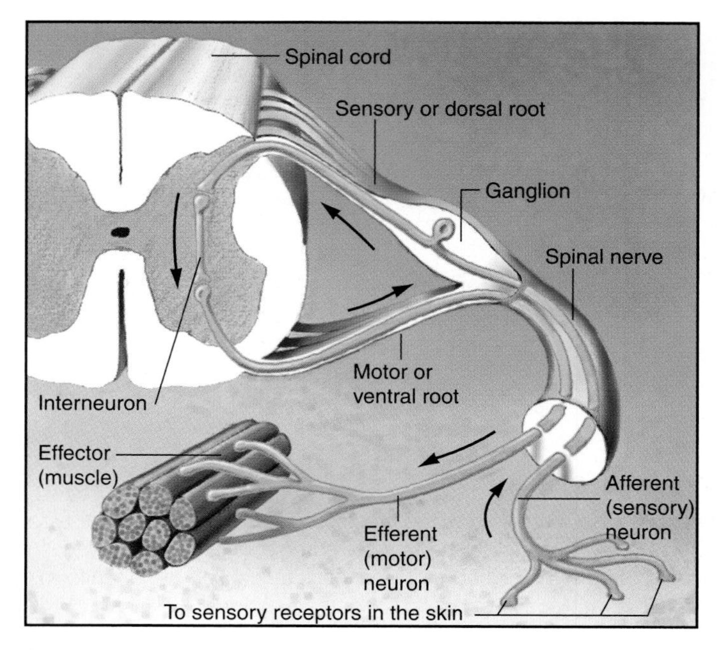

Figure 41.13

A reflex arc produces a rapid response to a stimulus. A sensory (afferent) neuron carries a signal from receptors into the spinal cord, where it synapses with an interneuron; the interneuron synapses with a motor (efferent) neuron leading back to an effector such as a muscle. Some reflex arcs do not include interneurons. The sensory neuron also synapses with other neurons that carry the signal to nearby points in the spinal cord and up the spinal cord to the brain.

Exercise 41.6 A dog has sustained a cut next to its spinal column. It does not respond to a stimulus to the skin that normally evokes a reflex, but stimulating the nearby region of the spinal cord evokes the normal muscular response. Did the cut sever the dorsal or the ventral branch of the spinal cord? (*Hint:* Examine Figure 41.13.)

41.9 All cells maintain an electrical potential across their plasma membranes.

Neurons operate through changes in their plasma membranes. Nerve impulses are changes in *voltage* or *electrical potential,* a measure of the force that tends to make ions or electrons move from one place to another. Remember that potential is a general term for a force that can do work; a *chemical potential,* for instance, exists wherever atoms or molecules are more concentrated in one spot than another, because they can do work as they diffuse from the higher to the lower concentration. A voltage exists wherever ions or electrons tend to move from place to place, and the magnitude of the voltage measures their potential for work; the 110-volt potential in your house wiring can do a lot more work than the 1.5-volt potential in a flashlight battery.

Electrical and chemical potentials, Section 7.8.

Because most neurons are too small to study easily, investigators have turned to some exceptionally large ones to understand the changes that take place in them. Much of our knowledge has come from work on the giant axons of squid (Figure 41.14), which can be 1 mm in diameter. Let's cut a section of a squid giant axon, tie off the ends, and place it in a solution (called Ringer's solution) containing ions in concentrations that match the interstitial fluid. We will connect the axon to an oscilloscope, which is just a voltmeter that directs a beam of electrons onto a screen to display how the voltage changes with time. It is set to measure voltage on a small scale from −100 to +100 mV (millivolts). To this instrument we attach two electrodes; one is a "ground" wire, lying in the Ringer's solution around the axon, and the other is a micropipette, a thin glass tube drawn out to a fine tip and filled with KCl solution, which we carefully push against the axon's plasma membrane (Figure 41.15). At first the voltage between the two electrodes is zero; there is no potential difference because the electrodes are in the same solution. But when we gently push the micropipette through the axon membrane, the voltage immediately moves to −70 mV, meaning that the inside of the axon is 70 mV negative with respect to the outside. This is the **membrane potential** of the giant axon. All cells have a membrane potential, although its magnitude varies from one type of cell to another. Why do cells have such a potential?

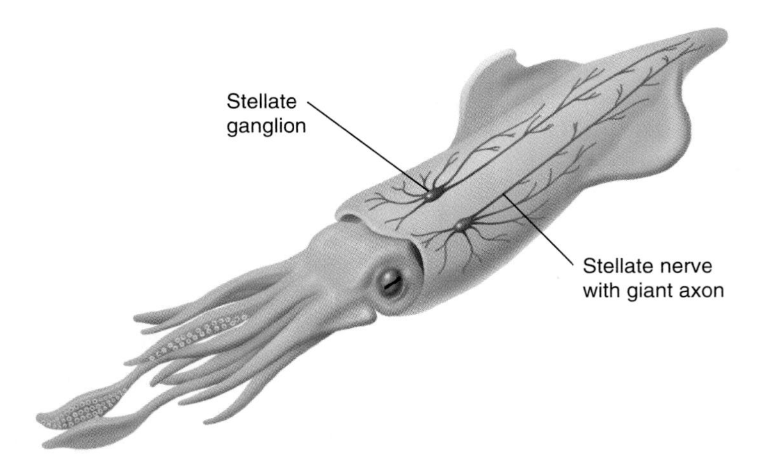

Figure 41.14
The arrangement of nerves in the muscular mantle of a squid facilitates rapid contraction of the mantle. The stellate nerves contain giant axons.

Because all eucaryotic cell membranes contain Na^+-K^+ ATPases that constantly pump K^+ ions inward and Na^+ ions outward, the axon cytoplasm (axoplasm) has a high concentration of K^+ ions and very little Na^+ and Cl^-; the surrounding interstitial fluid (replaced by Ringer's solution in this experiment) has just the opposite composition. The **conductance** of a membrane is a measure of its permeability to ions. The plasma membrane has a very low conductance for all inorganic ions, which move through it about 10^8–10^9 times slower than they do through water, but its conductance is about 30 times higher for K^+ ions than for Na^+ ions. Even though all ions contribute to the membrane potential, we can understand the properties of a resting neuron simply on the basis of K^+ ions.

In a squid axon, the internal concentration of K^+ ions is 410 mM, and the external concentration is only 22 mM. Because of this strong chemical potential, K^+ ions tend to diffuse out of the axoplasm:

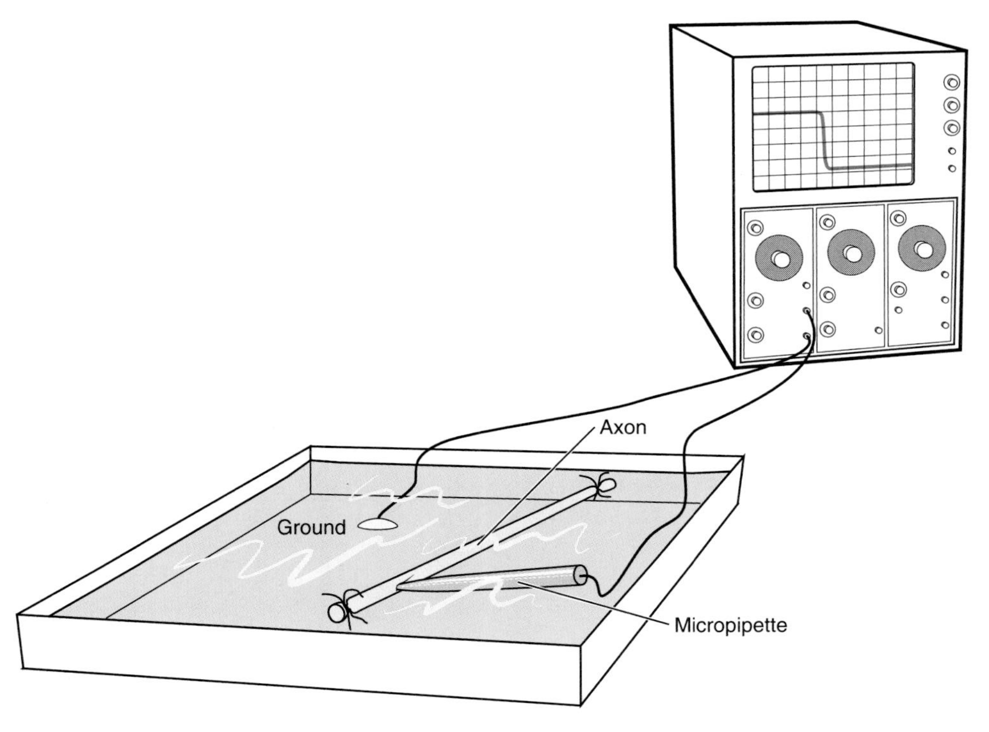

Figure 41.15
When a micropipette and a ground are both in the bath around the axon, there is no potential difference between them. When the micropipette is pushed inside the axon, the potential drops to −70 mV, the resting voltage of the axon membrane. (In reality, the axon is only about 1 mm in diameter; its size is exaggerated here for the sake of illustration.)

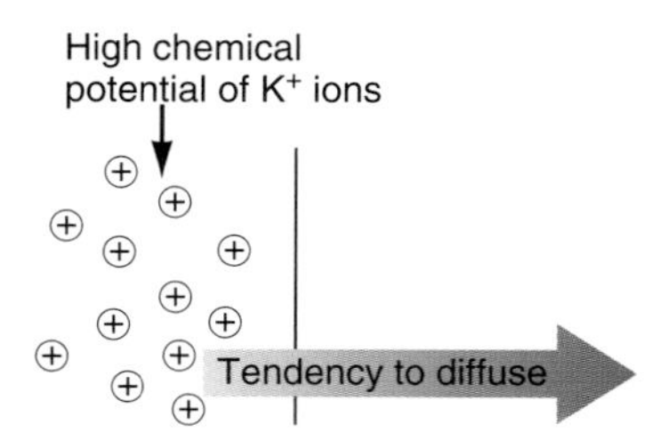

Then why don't they become equally distributed on both sides? Simply because each ion carries an electrical charge. As they start to diffuse outward, the ions create a charge imbalance, a voltage, with an excess of positive charge outside and negative charge inside:

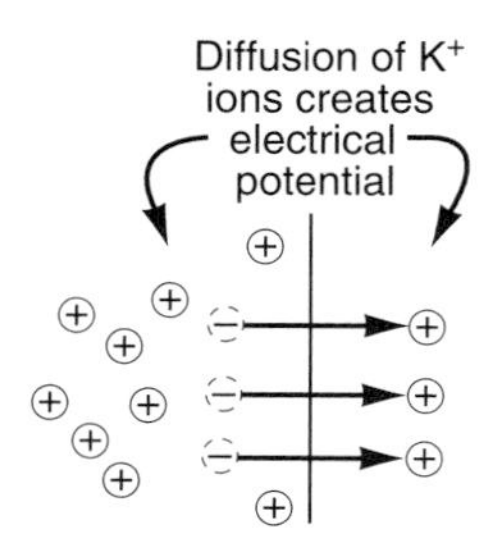

The cytoplasm also contains excess negative charges on proteins that can't move through the membrane, and all these negative charges tend to pull the K^+ ions back inside. Since the Na^+ conductance of the membrane is so low, Na^+ ions aren't free to move and restore the electrical balance. Therefore K^+ ions will only diffuse outward until their electrical potential (electrical attraction inward) just balances their chemical potential (the tendency to diffuse outward) and the ions come to an equilibrium distribution:

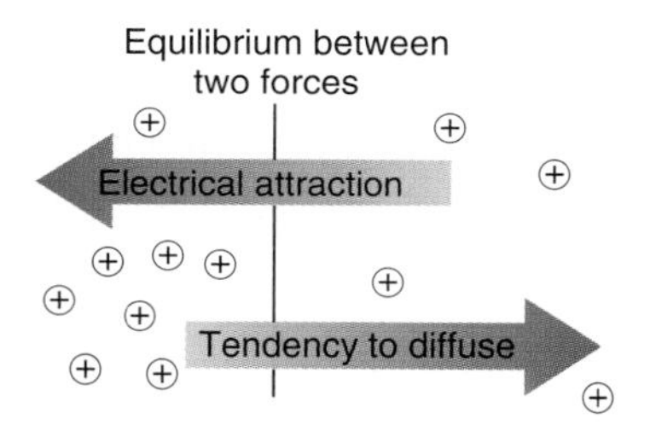

Electrical forces are so strong that this equilibrium occurs when only a small fraction of the K^+ ions are outside. At this point, there is an **electrochemical potential** across the membrane, an unequal distribution of ions that creates the resting potential of about −70 mV, with the inside negative. This kind of membrane potential is characteristic of all eucaryotic cells. Neurons are simply specialized to use the membrane potential to carry messages.

The terms used to describe voltages can be very confusing; we will try to ease the confusion by speaking a consistent language here. The membrane potential varies over the range on the oscilloscope, from −100 mV to around +60 mV; when we say the voltage is *lower* or *higher, increasing* or *decreasing,* we simply mean that it is changing along this scale, just like the temperature or any other variable that can take both negative and positive values. (We will never say that a voltage is getting larger or smaller, because this is ambiguous when the change is from one negative value to another.)

We will see next that, during a nerve impulse, neurons experience a transient flow of Na^+ ions inward and K^+ ions outward, so the potential is briefly reversed.

41.10 An increase in the Na^+ conductance of the plasma membrane generates an action potential.

Suppose we now stimulate the squid axon with a brief electric shock. The oscilloscope shows that in about a millisecond the membrane potential suddenly rises to +50 mV (indicating that the inside of the axon is now *positive* with respect to the outside) and falls again. This transient change in potential is an **action potential,** also called a "spike" because of its appearance on the oscilloscope screen:

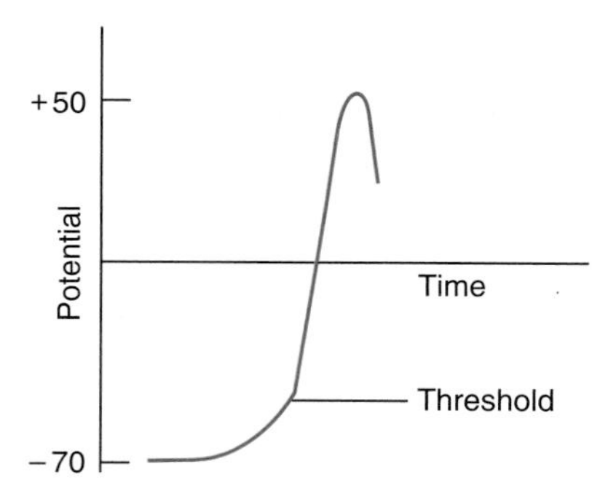

Instead of giving the axon a strong electric shock, suppose we just gradually increase the membrane voltage. As the voltage increases from −70 mV, nothing happens until the voltage reaches a point called a **threshold,** around −50 mV, where the axon suddenly responds with an action potential. An action potential is an all-or-none event: It doesn't occur until the potential reaches the threshold, and then it occurs completely. The change is characteristic of the axon itself and is always pretty much the same, no matter how it is triggered. After reaching its peak, the potential drops quickly to a low level, around −76 mV, and then rises to the resting potential:

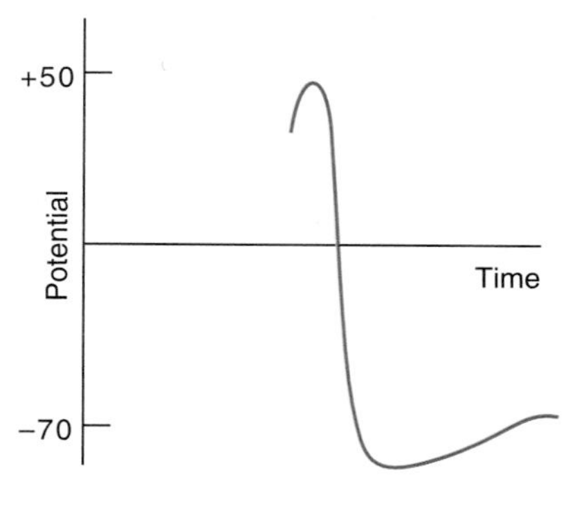

Why does this happen?

The classical experiments of Alan L. Hodgkin and Andrew F. Huxley around 1952 showed what is happening in the neuron; here we add some details that have been discovered since that time. The axon membrane contains **voltage-gated channels** for Na^+ and K^+ ions, transmembrane proteins that open their ion gates in response to the membrane voltage:

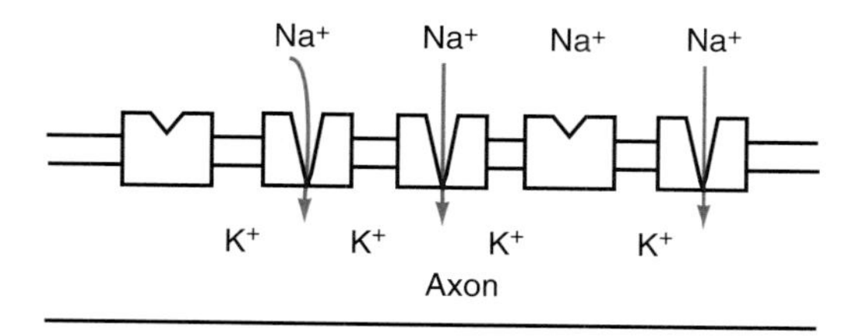

The Na^+ channels go through a positive feedback loop, the **Hodgkin cycle** (Figure 41.16). A slight increase in the membrane voltage at one point opens a few channels, increasing the Na^+ conductance, so some Na^+ ions (which are highly concentrated outside) move inward; these positive ions raise the voltage a little more, opening more channels, allowing more ions to enter, and so on. This positive feedback loop causes the membrane to react explosively, so Na^+ ions rush into the axon, making a local region inside the axon positive relative to the outside and creating the rising phase of the action potential. The Na^+ channels then close, making the membrane again impermeable to Na^+ ions, while voltage-gated K^+ channels open, and the outward flow of K^+ ions creates the falling phase of the action potential:

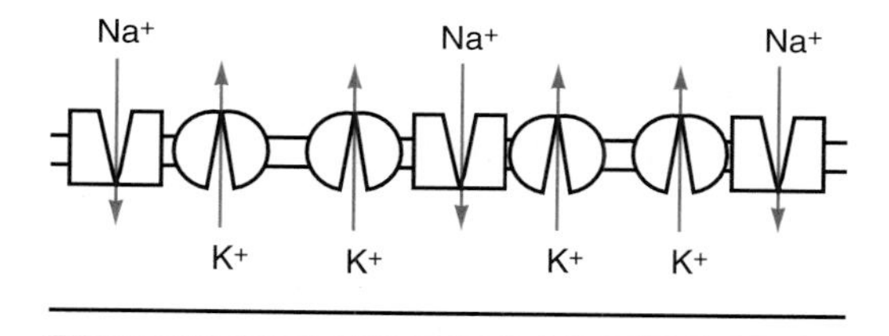

The local voltage falls below −70 mV because the K^+ channels are slow to close, but within a few milliseconds, the voltage returns to its resting value:

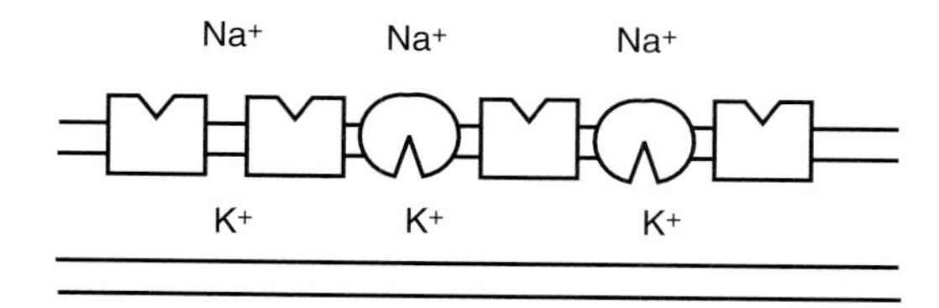

The overall composition of the cytosol actually changes very little during an action potential; very few ions, relative to the total inside the axon, move across the membrane to create this local potential change. The resting potential remains after many impulses because the Na^+-K^+ ATPases are continually exchanging Na^+ and K^+ ions, but so few ions pass through the membrane with each impulse that an axon can conduct many impulses even if these pumps are inhibited, as long as its membrane remains intact.

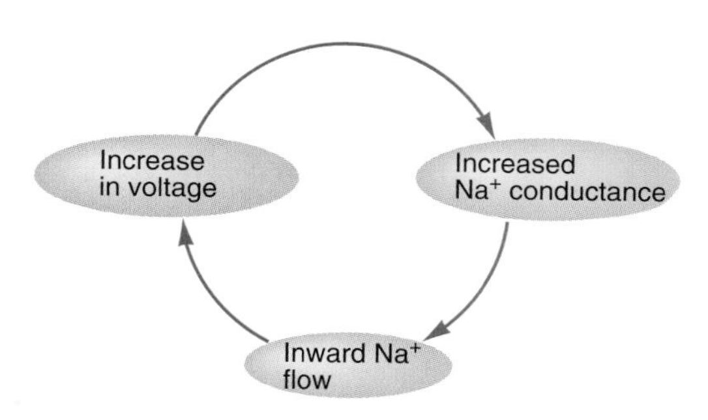

Figure 41.16

The Hodgkin cycle is a positive feedback loop. A small increase in voltage opens Na^+ gates, allowing some Na^+ ions to enter; this increases the voltage still more, opens more gates, and quickly generates an action potential.

Exercise 41.7 What is the difference between a ligand-gated channel and a voltage-gated channel?

41.11 A nerve impulse is an action potential that propagates itself along the axon.

Once an action potential is generated at one point on an axon, it becomes a nerve impulse that moves along the axon (Figure 41.17). The Na^+ ions that enter in one region of the axon create a local circuit: They repel other positive ions, which move to the side, raise the voltage in the neighboring region of the axon, increase the Na^+ conductance there, and thus generate an action potential in that area. In this way, activity in each region sets off the neighboring region, like people doing "the wave" at a baseball or football game, and this wave of activity is the nerve impulse. An impulse normally originates at the cell body and travels to the tips of the axon, but an axon stimulated in the middle can generate impulses that travel in both directions.

The events that create an action potential account for two important properties of a nerve impulse. First, it is *self-propagating:* Once created at any point, it keeps generating itself and moves all the way to the end of the axon. Second, it is *non-decremental:* Since the action potential at each point is all-or-nothing, the nerve impulse moves with undiminished amplitude along the axon.

With an appropriate electrical apparatus, we can set the voltage across an axon membrane to any desired level. If we raise the voltage above the normal resting value, the axon responds with a series of nerve impulses whose frequency increases as the voltage is set higher (Figure 41.18). This experiment shows one way a neuron encodes information: The frequency of a string of impulses reflects the strength of the stimulus that generates them. A nervous system apparently encodes information in other ways, but this is one of the most important.

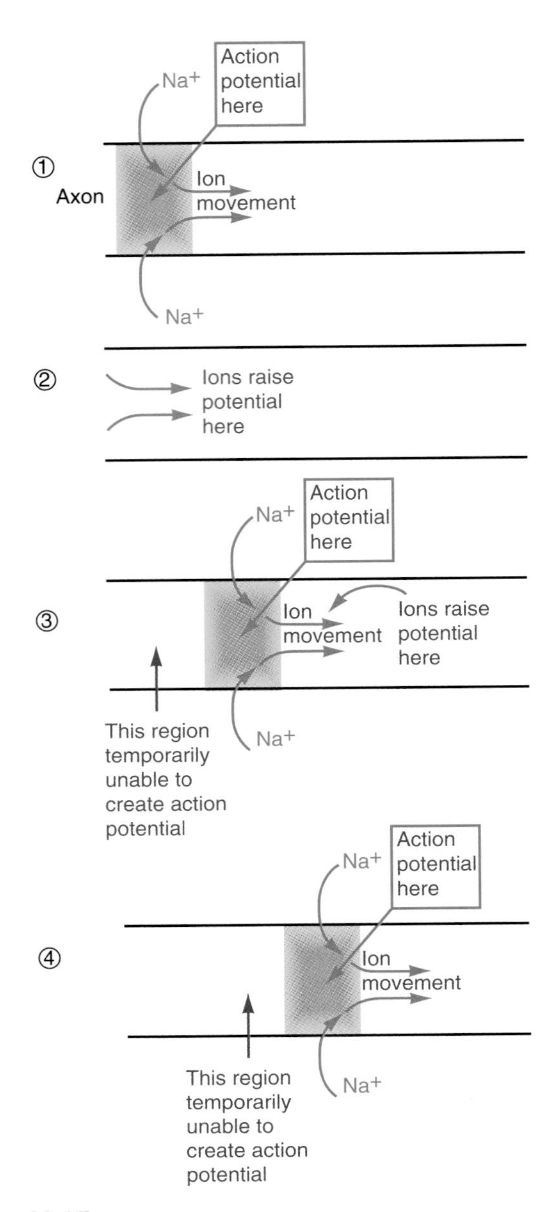

Figure 41.17

A nerve impulse moves along the axon. An action potential involving a local current of ions at one region *(1)* raises the concentration of cations in the adjacent region and raises the potential there *(2)*. This opens voltage-gated channels in the second region and induces a new Hodgkin cycle and action potential *(3)*. Thus the local current at each point induces an identical adjacent current, continually generating a new action potential, which keeps moving in one direction *(4)*.

Exercise 41.8 A membrane potential depends on the distribution of K^+ ions across a membrane. This distribution is largely maintained by the Na^+-K^+ ATPases, which are inhibited by the drug ouabain. A physiologist experimenting with a functioning squid axon gives the axon some ouabain, and it continues to function. Are you surprised? Why or why not?

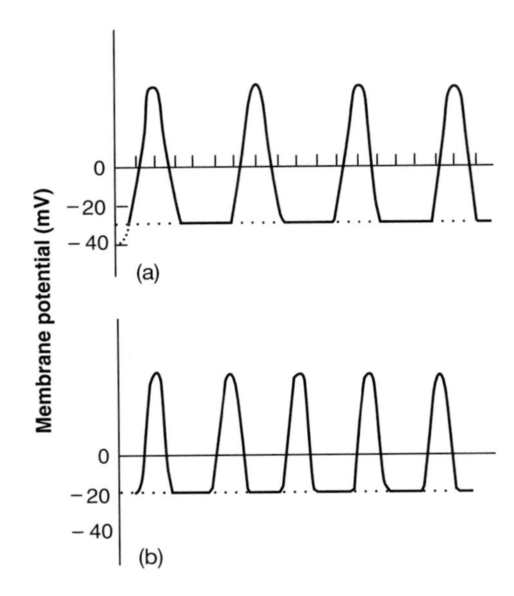

Figure 41.18

If we set the voltage across an axon membrane, raising the membrane potential above threshold, the axon fires a train of nerve impulses *(a)*. Applying a higher potential produces a higher frequency of impulses *(b)*. A neuron thus indicates the strength of the stimulus it is receiving by the frequency of its response.

41.12 A myelinated nerve can carry impulses rapidly.

Although neurons are the cells that carry signals within the nervous system, they comprise only about 10 percent of the cells in the mammalian nervous system. The rest are several kinds of **glial cells,** or **neuroglia,** that support, nourish, and protect the neurons. The glial cells of the vertebrate peripheral nervous system are **Schwann cells** that wrap around neurons (as shown in Figure 8.12) to make a **myelin sheath,** which insulates one axon from another. The myelin sheath also keeps ions from moving across the axon membrane by covering it closely. An axon completely covered with myelin sheath could not carry an impulse, but at small gaps between the Schwann cells, called the *nodes of Ranvier,* Na^+ and K^+ ions are free to move across the membrane through their channels. The nodes are spaced closely enough for an action potential at one node to generate an action potential at the next, so what was a local circuit of ionic movements in the unmyelinated axon becomes a much more extended circuit (Figure 41.19). Thus the nerve impulse skips from node to node in a process called **saltatory conduction** (Latin, *saltare* = to jump), which is faster than the continuous conduction in an unmyelinated axon.

The velocity of a nerve impulse not only increases with myelination but also with the diameter of the axon. C-type fibers, which are unmyelinated and range from less than 1 μm to 2 μm in diameter, only conduct at speeds of 0.7–2 meters per second. Large myelinated motor fibers 12–18 μm in diameter conduct impulses at 60–100 meters per second.

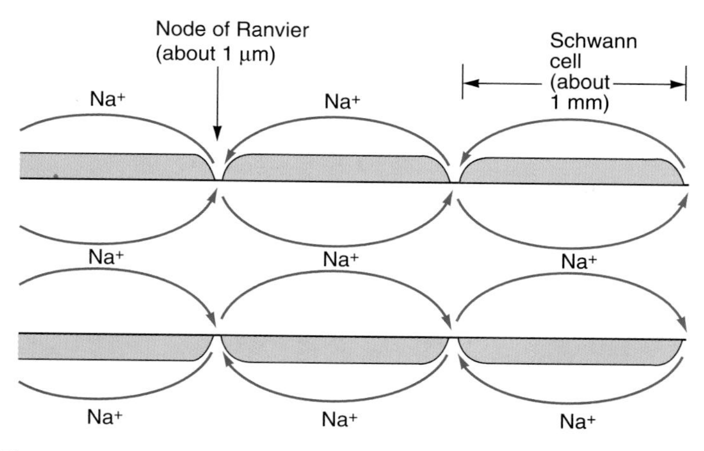

Figure 41.19

The myelin sheath around an axon is made of separate Schwann cells *(pink areas)* that leave short gaps (nodes) between themselves. The nerve impulse is conducted rapidly along the axon because the ionic current can jump (saltate) between the nodes.

Exercise 41.9 Multiple sclerosis is a disease in which myelin sheaths in areas of the CNS degenerate. Without knowing any clinical details, what would you expect the general effect to be?

41.13 Synaptic inputs determine a neuron's ability to fire impulses.

In a nervous system, nerve impulses are not triggered by an electrode in the middle of an axon but by synaptic inputs from other neurons or from receptor cells. The dendrites and soma of a typical neuron are studded with axon endings from other neurons (Figure 41.20), each providing some input. Every neuron is either **excitatory** or **inhibitory:** It tends to either increase or decrease the activity of the postsynaptic cells to which it is connected. With some exceptions, each neuron produces a single kind of neurotransmitter at its synapses, and both the neurotransmitter and synapse are also characteristically excitatory or inhibitory. So some synapses on a neuron are sending excitatory signals and some are sending inhibitory signals; the balance of these signals at any instant determines whether the neuron will fire. However, the plasma membrane of the dendrites and soma does not generate an action potential itself; rather, it allows the neuron to integrate signals coming to it.

Recall that, in a synapse, neurotransmitters from the presynaptic cell activate ligand-gated channels in the postsynaptic membrane. Ions moving through these channels change the membrane potential at that point on the cell. At an excitatory synapse, the neurotransmitter creates a local increase in the membrane potential, an **excitatory postsynaptic potential (EPSP).** At an inhibitory synapse, the neurotransmitter stimulates a local decrease, an **inhibitory postsynaptic potential (IPSP).** The plasma membrane of the dendrites and soma is different from that of the axon; its channels are ligand-gated, not

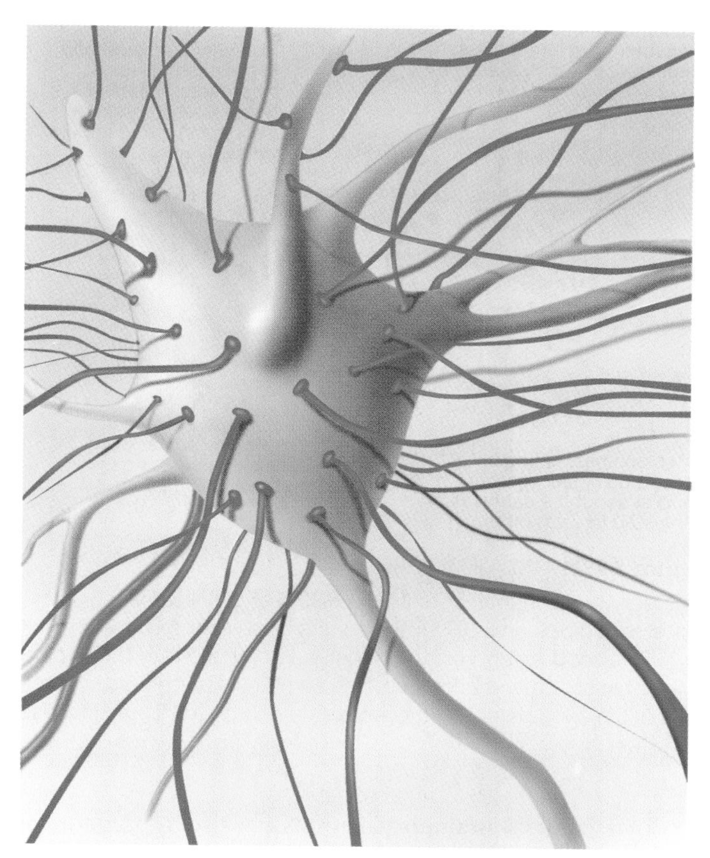

Figure 41.20

A single neuron is connected to many other neurons. It is studded with the endings of their axons, shown here as thin fibers ending in buttons (boutons). Each button forms a synapse where the neuron can receive an excitatory or inhibitory signal; the neuron summarizes these signals, and it may pass on other signals through its axon.

voltage-gated, so unlike potentials generated in an axon, EPSPs and IPSPs are *graded* potentials, not all-or-none: Their magnitudes are proportional to the strength of the incoming stimulus, which is probably determined by the amount of neurotransmitter released. Also, these potentials are conducted decrementally; they diminish as they spread across the membrane. EPSPs and IPSPs can add to, or subtract from, each other if they are generated close together in time and space (Figure 41.21). For instance, an EPSP and an IPSP of the same magnitude presumably cancel each other if they occur near each other and at about the same time.

A neuron acts like a little adding machine. From moment to moment the cell is summing up all the EPSPs and IPSPs being generated across its membrane. If the total potential at the base of the axon (the axon hillock) is above the threshold, it triggers a nerve impulse; if not, there is no impulse.

Excitatory and inhibitory neurons create a nervous system's logic. For instance, an inhibitory cell may form a negative feedback loop in a circuit, so once a cell has transmitted a signal, it is inhibited and will not fire again for a short time (Figure 41.22*a*); such a loop will sharpen a circuit's activity and make it

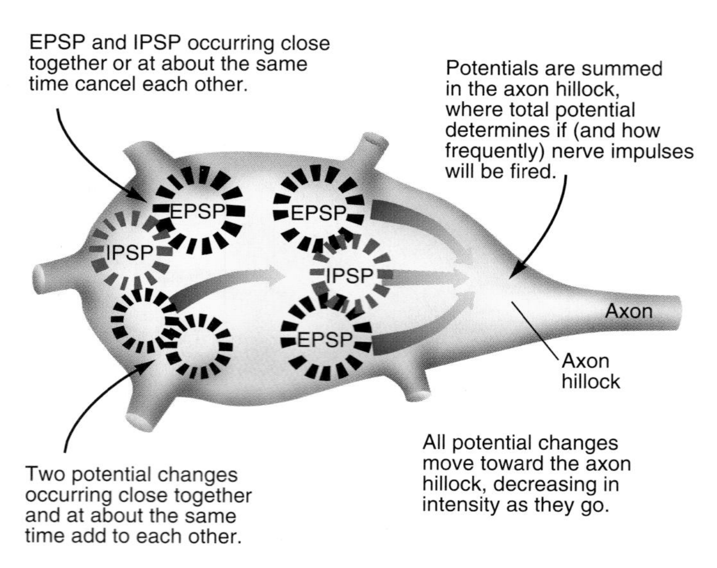

Figure 41.21

Excitatory postsynaptic potentials (EPSPs) and inhibitory postsynaptic potentials (IPSPs) are induced all over the neuron body. They add to, and subtract from, each other if they occur close together in space and/or in time. The sum of these potentials at the base of the axon determines whether the neuron will fire an impulse.

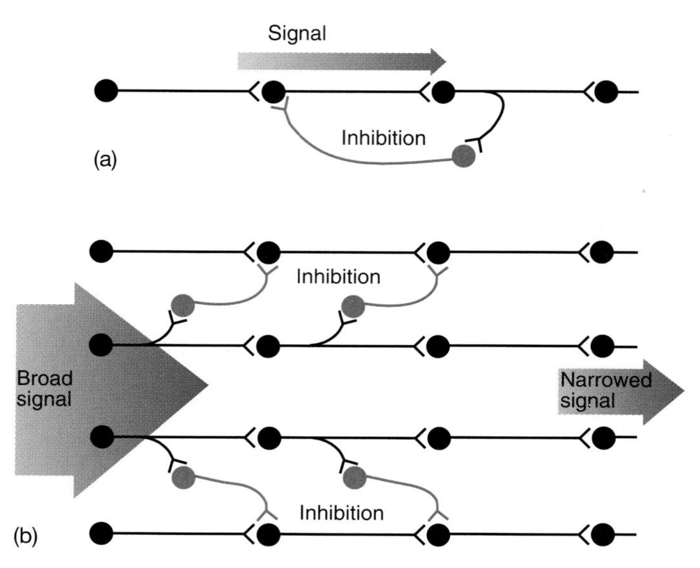

Figure 41.22

Inhibitory neurons can sharpen the activity of a nervous system in many ways. Two are shown here: *(a)* An inhibitory neuron can feed back to turn off a neuron that has already fired. *(b)* In a series of parallel neurons, inhibitory neurons reduce the activity of those on the outside, so the signal is narrowed and confined.

fire discrete bursts. Inhibitory neurons can also be used for **collateral inhibition** (Figure 41.22*b*). If several parallel neurons begin to carry a signal, those in the middle may inhibit the more peripheral ones through inhibitory interneurons, thus confining and sharpening the signal. In principle, all kinds of logical circuits can be formed by combinations of excitatory and inhibitory neurons.

Many neurons fire spontaneously and continually, without any stimulation by other neurons. It appears that their plasma membranes continually allow Na^+ ions to leak in slowly until they fire. Then their resting potential is restored, the potential slowly rises, and they fire again. Cells that fire spontaneously will continue to do so indefinitely, and spontaneously firing neurons are essential for some processes, such as the control of breathing. Other cells that are really modified muscle fibers fire spontaneously in the heart and are responsible for its intrinsic rhythm. In the brain, electrical activity continues even during rest, as shown by an electroencephalogram (EEG) recorded with electrodes connected to the scalp, giving a very crude representation of the brain's activity. The EEG would presumably continue even in the absence of all external stimuli. It is a record, in a sense, of the nervous system talking to itself.

Exercise 41.10 Section 11.B explains the logic of dissecting processes by means of drugs. Tetrodotoxin inhibits development of an action potential but not of an EPSP; curare has the opposite effect. Explain generally how this information can be used to explore neuron function.

C. Connections between endocrine and nervous systems

As we mentioned at the beginning of this chapter, the endocrine and nervous systems are intimately connected and coordinated, functioning together to keep all the other systems of the body properly regulated. In this section we illustrate that point with two prominent systems: the hypothalamus and pituitary gland, with their mixture of endocrine and nervous activities, and the autonomic nervous system, with its endocrine connections.

41.14 The pituitary gland and hypothalamus form a center of regulation.

The pituitary gland, or hypophysis, is a tiny organ just below the hypothalamus at the bottom of the brain, where it is well-protected by surrounding bone (see Figure 42.10). A human pituitary weighs only about half a gram. It is intimately connected to the hypothalamus, and together they constitute a center of regulation in vertebrates that controls most other endocrine glands and demonstrates some significant general points. Neurons in the hypothalamus sense such factors as temperature and ionic concentrations in the interstitial fluid, and they regulate these factors with nervous signals and signals to the cells of the pituitary.

The pituitary is made of two parts. The posterior pituitary (neurohypophysis) is an extension of the brain (Figure 41.23). This part secretes two small peptide hormones: *oxytocin,* which accelerates labor and childbirth, and *vasopressin,* which helps regulate blood pressure and the body's water content. These

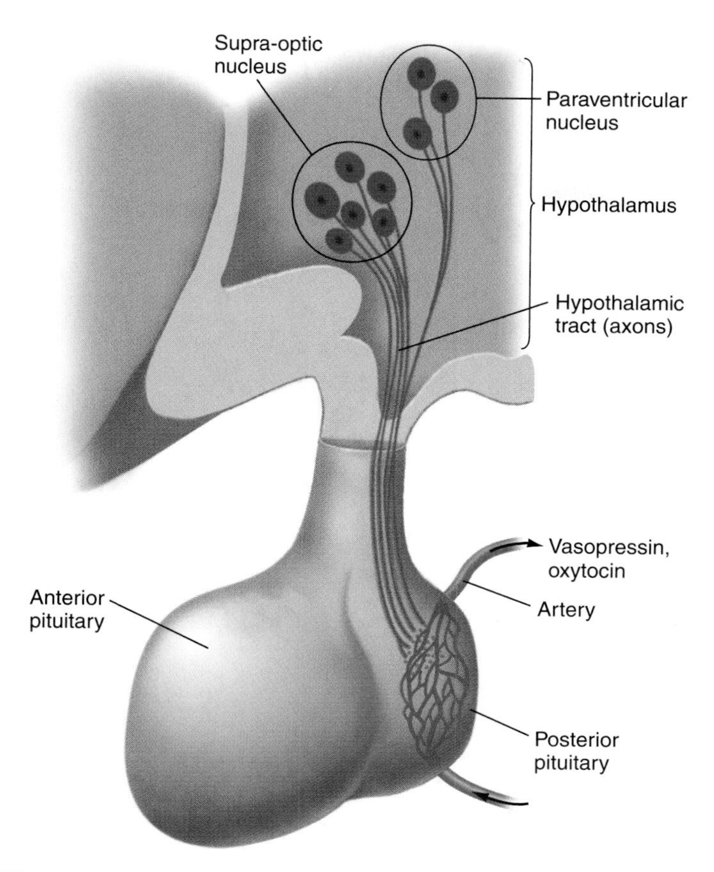

Figure 41.23

The posterior pituitary (neurohypophysis) is an extension of the hypothalamus above it. Neurosecretory cells (specialized neurons) in certain hypothalamic nuclei produce hormones, which are transported down the axons of these cells and stored in the posterior lobe, where they are released in response to neural stimulation.

hormones are not synthesized in the pituitary, however, but in neurons called **neurosecretory cells** whose cell bodies lie in the hypothalamus. Thus the posterior pituitary is basically an extension of the hypothalamus. Vasopressin and oxytocin made in the neuron cell bodies are packaged into vesicles that move slowly down the axons of these neurons, accumulating at their ends in the posterior pituitary. When stimulated by sensor cells in the hypothalamus and by complex neural circuits, these neurosecretory cells release their stored hormones into the bloodstream. Neurosecretory cells are just specialized neurons. Like other neurons, they store signal ligands in terminal vesicles and release them upon stimulation. In contrast to the neurotransmitters released by ordinary neurons, however, these signal ligands act on hormone receptors at distant sites rather than on synaptic receptors a few nanometers away. But you should now be able to see the basic similarity of these cells, regardless of their names.

The anterior pituitary gland (adenohypophysis) is quite different (Figure 41.24*a*). It develops as an outgrowth of the roof of the embryonic mouth rather than from neural tissue, and it produces seven peptide hormones (Figure 41.24*b*), including general regulators of cellular activity and others that regulate reproduction. Each of its specialized cells synthesizes one of these hormones under the control of equally specialized hypothalamic cells that produce **releasing hormones.** The hypothalamus communicates with the anterior pituitary by a very short portal[1] circulatory system, and releasing hormones move the few millimeters down this system to stimulate the appropriate pituitary cell. Some of the releasing hormones stimulate secretion by the pituitary cells, while others inhibit it. Thus the releasing hormones control the release of anterior-pituitary hormones, which may, in turn, stimulate the release of hormones by still other endocrine glands. Such complexity provides several points of regulation. Figure 41.25 shows how thyroxine and triiodothyronine, the thyroid hormones that stimulate metabolism throughout the body, feed back at two points to inhibit synthesis of the hormones that control it, so the level of thyroid hormones in the extracellular fluid can be adjusted continually. In Chapter 51, we'll see the still more intricate feedback system involving the sex hormones.

Exercise 41.11 Figure 41.26 shows a person with a *goiter,* an abnormal growth of the thyroid gland caused by inadequate iodine in the diet. The thyroid hormones contain iodine, and of course they cannot be synthesized in its absence. Examine the regulatory circuit of Figure 41.25, and explain why goiters develop.

41.15 The autonomic nervous system controls many routine functions.

A vertebrate can be in either of two behavioral states. It may be relaxed, going about its business calmly, digesting its food, storing up energy, and not alarmed by anything. This condition is maintained by the **parasympathetic** division of the autonomic nervous system (Figure 41.27), which keeps the eyes set for close vision, the heart pumping slowly, and the whole intestinal tract calmly churning away, digesting food. At other times, an animal may be alarmed by some threat. Then the **sympathetic** division takes over, and the animal goes into a "fight-or-flight" condition, literally ready to fight or run away. Its eyes quickly adjust to take in more light and focus on potential sources of danger in the distance; its heart pumps harder and faster, its air passages dilate, and its digestive activities stop. Everyone knows the feeling of this condition, the feeling of fear or anger.

In Section 47.7, we introduce the idea that the ANS has a third division, the *enteric nervous system,* which controls activities of the gastrointestinal tract. Here, however, we are only concerned with the fact that most organs in the body are stimulated to carry out opposite activities by the two classical divisions of the ANS. The general point is that a nervous system can carry different messages by having neurons that produce different neurotransmitters. Each neuron, together with its synapses, is

[1]A portal system is one that runs between two capillary beds. Most of the circulatory system connects a capillary bed with the heart.

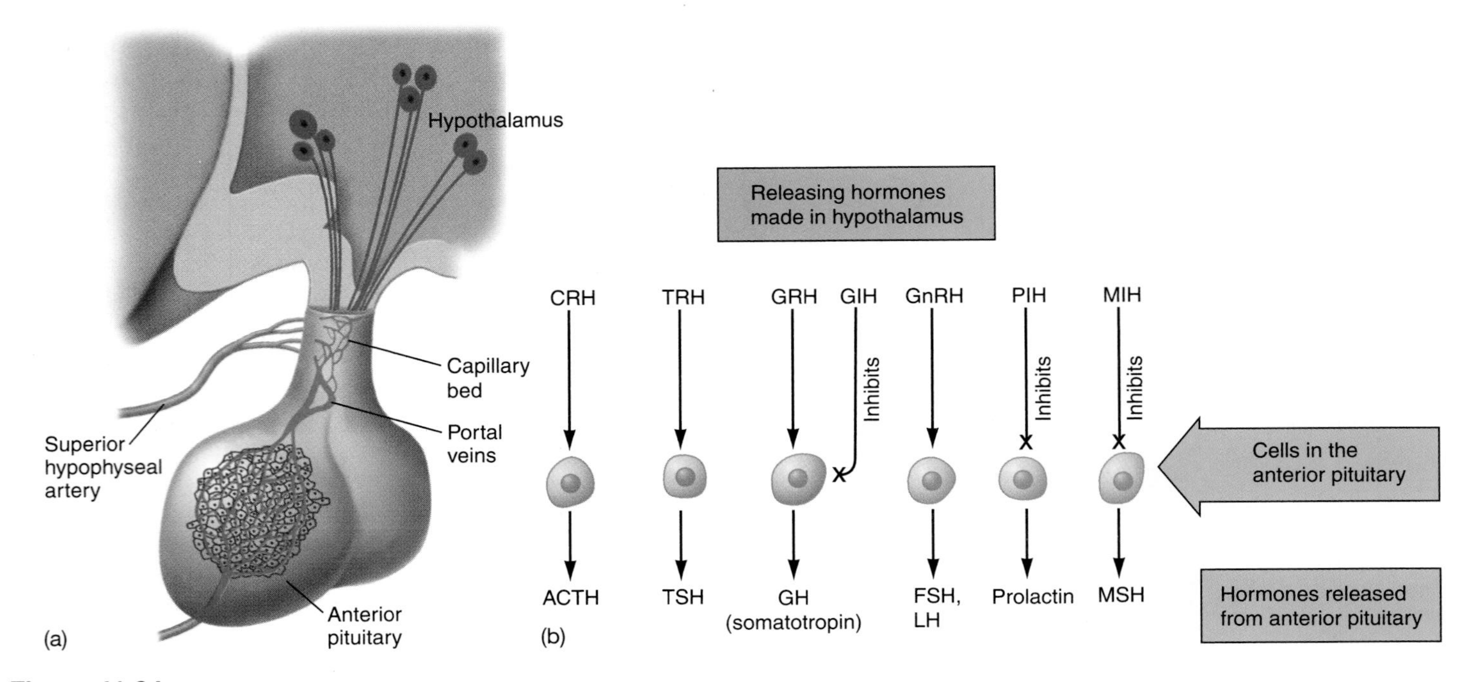

Figure 41.24

(a) The anterior pituitary (adenohypophysis) is connected to the hypothalamus by a short blood vessel, the hypothalamic-hypophyseal portal system. *(b)* In response to various stimuli, hypothalamic cells send releasing hormones (RHs) to the pituitary, where each of them stimulates (or sometimes inhibits) the release of another hormone whose target cells lie somewhere else. (Releasing hormones are: CRH, corticotropin RH; TRH, thyrotropin RH; GRH, growth hormone RH; GIH, growth hormone inhibiting hormone; GnRH, gonadotropin RH; PIH, prolactin inhibiting hormone; and MIH, melanocyte inhibiting hormone. Pituitary hormones are: ACTH, adrenocorticotropic hormone; TSH, thyroid stimulating hormone; GH, growth hormone; FSH, follicle stimulating hormone; LH, luteinizing hormone; prolactin; and MSH, melanocyte stimulating hormone.)

Figure 41.25

Production of thyroid hormones is controlled by a circuit with negative-feedback loops. In response to low skin temperature or a large meal, hypothalamic nuclei release thyrotropin releasing hormone (TRH) which stimulates release of thyrotropin (thyroid stimulating hormone: TSH). TSH stimulates growth of thyroid tissue and stimulates thyroid cells to release thyroid hormones (thyroxine and triiodothyronine), which stimulate cells of many tissues to metabolize faster and produce more heat. Thyroid hormones feed back at two points to inhibit TRH and TSH production.

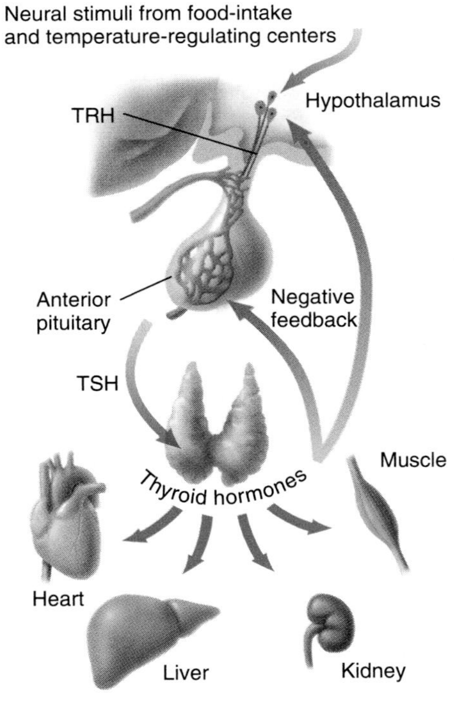

Figure 41.26

A goiter is an abnormal growth of the thyroid gland caused by a lack of iodine in the diet.

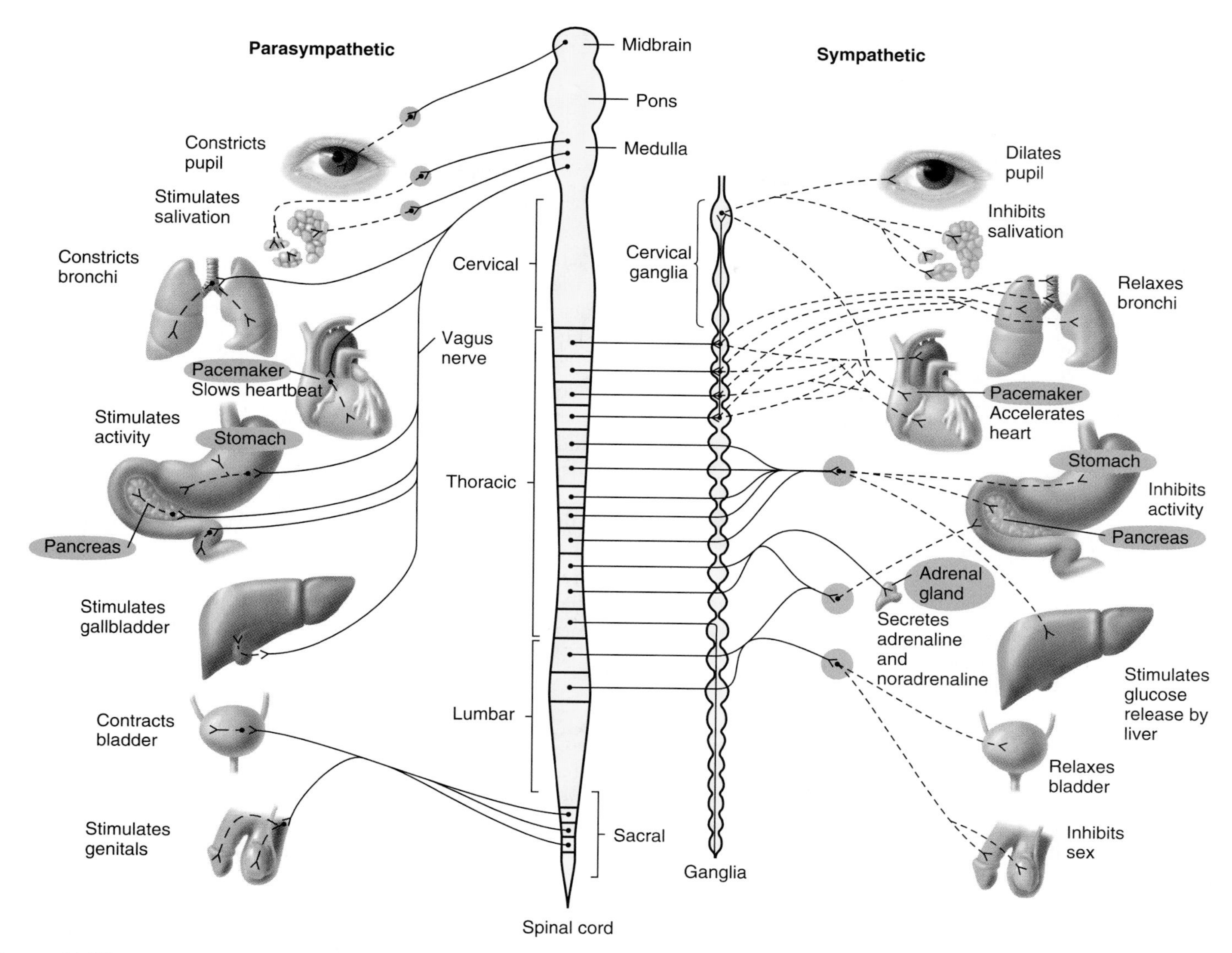

Figure 41.27

The autonomic nervous system has two parts. The parasympathetic division, operating through cranial (head) and sacral (sacrum) nerves, maintains placid behavior. The sympathetic division, operating through thoracic (chest) and some lumbar (lower back) nerves, excites the animal to emergency action. Note that in the sympathetic division most emerging fibers terminate at ganglia where they stimulate neurons running to the target organs, but in the parasympathetic division most emerging fibers run directly to the organs where they stimulate short neurons.

named for the single type of neurotransmitter it produces. **Adrenergic** neurons, for instance, produce the neurotransmitter noradrenaline (norepinephrine), and **cholinergic** neurons produce acetylcholine:

$$(HO)_2C_6H_3-CH(OH)-CH_2-NH_3^{\oplus}$$

Noradrenaline

$$H_3C-C(=O)-O-CH_2-CH_2-N^{\oplus}(CH_3)_3$$

Acetylcholine

The somatic and parasympathetic systems are built entirely of cholinergic neurons; the sympathetic system is built of both cholinergic and adrenergic neurons (Figure 41.28). These two signal ligands produce opposite autonomic effects on each organ, but whether each ligand excites or inhibits depends on the organ's signal-transduction mechanism. Thus noradrenaline stimulates pacemaker cells in the heart, accelerating its rate, but inhibits activity in the stomach and intestine, slowing digestion; acetylcholine inhibits heart pacemaker cells, slowing the heart, but stimulates the stomach and intestine, facilitating digestion.

A signal ligand may also have different effects because it binds to different receptors, the classic example being the binding of epinephrine to either α-adrenergic or β-adrenergic receptors. These receptors are distinguished experimentally by

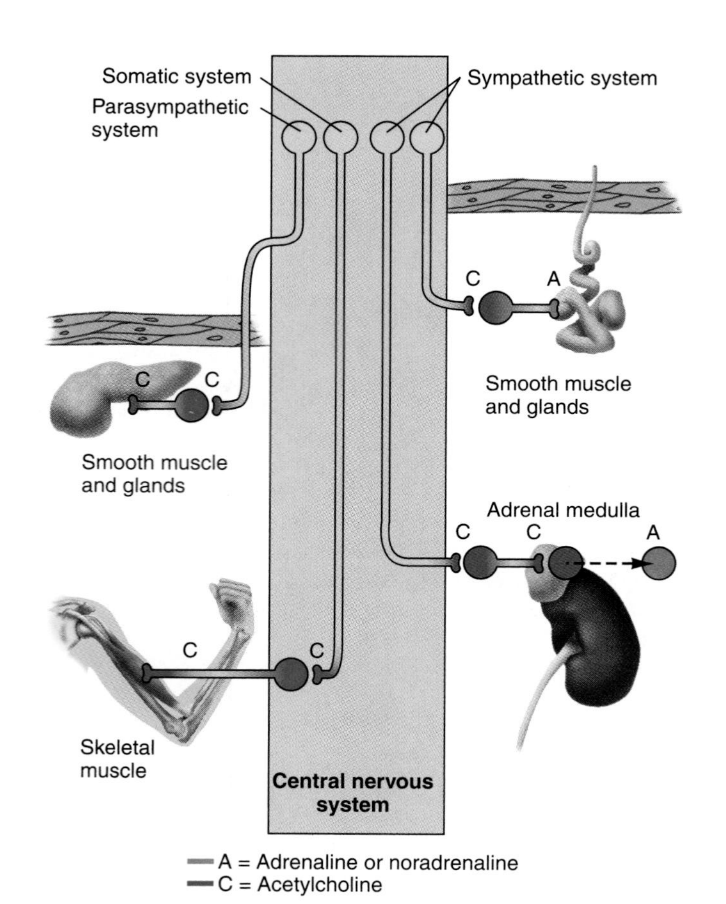

Figure 41.28

Neurons in the motor nervous system use distinctive neurotransmitters. Parasympathetic and somatic neurons are cholinergic (C). Sympathetic neurons from the CNS are cholinergic but eventually stimulate adrenergic (A) cells, either neurons or adrenal medulla cells.

their ability to bind drugs; α-receptors are blocked by phenoxybenzamine and β-receptors by propranolol. Functionally, an activated α-receptor initiates the reaction sequence leading to increased intracellular Ca^{2+} ions, whereas an activated β-receptor activates adenylate cyclase and produces an increase in cAMP. Tissue cells may have either type of receptor or both types; depending on the tissue, the two receptors may control quite different processes or may supplement each other by controlling the same process.

Distinguishing neurons by their neurotransmitters provides an insight into the close connections between the nervous system and certain hormones. Some sympathetic neurons run directly from the CNS to the **adrenal glands** lying on top of each kidney. The adrenals are really two glands together—a cortex on the outside and a medulla on the inside. Only the medulla figures in the system we're discussing here. Adrenal medulla cells are derived from the same embryonic cells (neural crest tissue) as cells of the ANS; although they have the form of gland cells rather than neurons, they behave like sympathetic neurons by releasing a mixture of noradrenaline (norepinephrine) and the structurally similar adrenaline (epinephrine):

$$(HO)_2C_6H_3-\underset{OH}{CH}-CH_2-\overset{+}{N}H_2-CH_3$$

Adrenaline

Activating the sympathetic division thus has two effects simultaneously: rapid but brief stimulation of glands and muscles by nerves that release noradrenaline directly and locally, and slower but more enduring stimulation through the blood by adrenaline and noradrenaline from the adrenal medullas. This twofold stimulation activates these organs quickly and then keeps them active for some time. Because the adrenal medulla and sympathetic system work together, it is useful to consider them a single **sympathoadrenal system.**

Like the association between the hypothalamus and the pituitary gland, the relationship between the adrenal medulla and the ANS demonstrates that nervous and endocrine systems are not sharply distinguished from each other. They are intimately connected. Their cells, whatever their form, are elements in communication systems that use fundamentally similar mechanisms.

Coda An organism without controls would not survive long. An organism that could not respond to changes in its environment would not survive long. Although we can only speculate about many events in evolution, sensory and regulatory proteins were probably evolving along with enzymes and proteins for other basic functions in the most primitive organisms. Sensory and control mechanisms depend—like so many other biological processes—on the binding of ligands to proteins; responding in an informational way to the presence of a ligand has great advantages and is not obviously more difficult to evolve than a metabolic pathway. We have seen that even unicellular organisms sense and respond to materials in their environment, sometimes by changing their metabolic systems, sometimes by moving away from a potential danger or toward a source of energy. Animals are multicellular organisms that have elaborated on those primitive abilities. For example, the G-protein-linked receptors that are so important in signaling in animals appear in simple eucaryotes such as the slime mold *Dictyostelium,* where they are involved in receiving the cyclic-AMP signal that induces aggregation of cells. Forerunners of these proteins even appear in related proteins in some bacteria. From this humble beginning, these signal-transduction proteins evolved into their many roles in the nervous, hormonal, and immune systems of animals. Molecules similar to some animal hormones, including insulin, have been identified in primitive eucaryotes; their functions in the simpler organisms are unknown, but they were available to be used informationally in the internal regulation of animals. Finally, some membrane functions are critical in signaling and control processes of animals. Neural signaling depends upon cells having a membrane potential and being able to change it rapidly. All cells probably have such a potential, and using it for signaling was another instance of evolution in animals elaborating on a primitive cellular characteristic.

Summary

1. A hormone is a compound produced by one type of cell and released into the blood or lymph, where it circulates to stimulate other cells.
2. The plasma glucose level is kept relatively constant by the peptide hormones insulin and glucagon, in a dual negative feedback system. When the plasma glucose level becomes too high, β cells in the pancreas release insulin, which stimulates liver, muscle, and other cells to take up excess glucose. When the glucose level becomes too low, α cells in the pancreas produce glucagon, which stimulates many cells to release glucose.
3. Exocrine glands are pockets of specialized epithelial cells that produce digestive enzymes and other materials, which are released to the outside of the body or the intestinal tract. Endocrine glands form around blood vessels and release their secretions directly into the interstitial fluid, from which it moves into the blood.
4. Ligands are always received by protein receptors, which are of four types: *(1)* G-protein-linked receptors ultimately produce second messengers. *(2)* Channel-linked receptors open channels and allow ions to flow in or out of the cell. *(3)* Catalytic receptors become activated as enzymes that can carry out new metabolic activities, perhaps activating other proteins. *(4)* Steroid and thyroid receptors, which are intracellular, regulate genes.
5. Reaction cascades are biological amplifiers, which begin with a small signal and produce larger and larger numbers of activated proteins to create a large end effect.
6. Animal cells may communicate via gap junctions, membrane-bound ligands, or secreted signal ligands. Communication via secreted ligands can be endocrine, using hormones carried through the bloodstream; paracrine, using local chemical mediators; or synaptic, using neurotransmitters between cells (chiefly neurons) that are in close contact across a synapse.
7. The main elements of a nervous system are neurons, which pass messages in one direction from a cell body along an axon to a synapse, where a second cell is stimulated. Transmission across the synapse is mediated by neurotransmitters, which may excite or inhibit the postsynaptic cell.
8. In vertebrates, the nervous system consists of a central portion (brain and spinal cord) and a peripheral portion that is made of both afferent (sensory) and efferent (motor) nerves. The motor portion consists of a somatic division, which controls skeletal muscles, and an autonomic division, which controls glands and other muscles. The autonomic nervous system has a sympathetic portion, which prepares an animal for emergency action, and a parasympathetic portion, which controls resting activities.
9. A reflex arc permits rapid, automatic reaction to stimuli, but the central nervous system processes and integrates information from many sources and sends out signals to maintain balanced action in the whole animal.
10. A neuron has an electrochemical potential across its membrane of about -70 mV, negative inside. It has a high internal concentration of K^+ ions generated by its Na^+-K^+ ATPases. A neuron membrane is only slightly permeable to K^+ ions and virtually impermeable to Na^+ or Cl^- ions. K^+ ions diffuse outward until their chemical potential is just balanced by an electrical potential, leaving the cell with an electrochemical potential.
11. When the membrane potential is raised above a threshold level, the axon generates an action potential, on the order of $+50$ mV, made chiefly because of a sudden influx of sodium ions followed by a compensating efflux of potassium ions. This excitation propagates itself along the axon as a nerve impulse. The frequency of a train of nerve impulses increases with the strength of the stimulus applied to the neuron.
12. Neurons carry messages in the form of nerve impulses along their axons. At the axon tips, the nerve impulse stimulates the release of neurotransmitters from presynaptic vesicles. These transmitters induce local excitatory or inhibitory postsynaptic potentials (EPSPs and IPSPs). A neuron acts like an adding machine; it sums up the excitatory and inhibitory signals that come to it both spatially and temporally to create a certain potential at the base of the axon, where nerve impulses normally originate. The neuron fires or does not fire at any instant accordingly. Many neurons spontaneously generate impulses at a low rate, and the main effect of stimulation is to increase this rate.
13. Many axons are surrounded by a myelin sheath made of Schwann cells. Impulses are transmitted rapidly along a myelinated neuron by saltatory conduction from one node to the next. The rate of transmission also increases with the diameter of the axon.
14. The pituitary gland is a major endocrine structure. Its two divisions illustrate two modes of hormone synthesis and control. In the posterior pituitary, hormones are made in neurosecretory cells with their bodies in the hypothalamus and are released upon neural stimulation. In the anterior pituitary, hormones are synthesized in specialized cells, and their release is stimulated or inhibited by releasing hormones made in the hypothalamus. This complex circuitry creates several points of control.
15. The autonomic nervous system has two parts: the parasympathetic, which maintains a calm state, and the sympathetic, which puts an animal into a "fight-or-flight" condition. The neurons in this system are specifically adrenergic (producing adrenaline) or cholinergic (producing acetylcholine). The adrenal medulla is also adrenergic tissue organized as a gland rather than as neurons.

Key Terms

sensor cell 867
effector cell 867
secretin 868
target cell 868
exocrine gland 868
endocrine gland 868
signal ligand 869
reaction cascade 870
proenzyme 870
ligand-gated 870
endocrine communication 871
synaptic communication 871
synapse 871
neurotransmitter 871
paracrine communication 871
autocrine 871
local chemical mediator 871
prostaglandin 872
neuron 872
central nervous system (CNS) 873
brain 873
spinal cord 873
peripheral nervous system (PNS) 873
sensory (afferent) neuron 873
motor (efferent) neuron 873
autonomic nervous system (ANS) 873
somatic nervous system 874
soma 875
dendrite 875
axon/nerve fiber 875
nucleus 875
ganglion 875
tract 875

presynaptic cell 875
postsynaptic cell 875
synaptic cleft 875
reflex arc 876
interneuron 876
membrane potential 877
conductance 877
electrochemical potential 878
action potential 878
threshold 878
voltage-gated channel 879
Hodgkin cycle 879
glial cell/neuroglia 880
Schwann cell 880
myelin sheath 880
saltatory conduction 880
excitatory neuron 881
inhibitory neuron 881
excitatory postsynaptic potential (EPSP) 881
inhibitory postsynaptic potential (IPSP) 881
collateral inhibition 882
neurosecretory cell 883
releasing hormone 883
parasympathetic 883
sympathetic 883
adrenergic 885
cholinergic 885
adrenal gland 886
sympathoadrenal system 886

Multiple-Choice Questions

1. In the Bayliss and Starling experiments, the stimulus for hormonal secretion was
   a. secretin.
   b. pancreatic fluid.
   c. bicarbonate.
   d. intestinal enzymes.
   e. stomach acid.
2. The molecular structure of hormones includes all but
   a. modified fatty acids.
   b. nucleic acids.
   c. peptides.
   d. steroids.
   e. amino acid derivatives.
3. Which of the following act neither as primary endocrine nor neural receptors?
   a. gap junction proteins
   b. G-linked membrane proteins
   c. catalytic protein kinases
   d. gene regulatory proteins
   e. channel proteins
4. Which are most likely to result in a cascade effect?
   a. receptors for steroid hormones
   b. protein kinases
   c. channel-linked receptors
   d. *a* and *b*, but not *c*
   e. *a, b,* and *c*
5. Molecular communication by means of gap junctions is classified as
   a. endocrine.
   b. paracrine.
   c. autocrine.
   d. synaptic.
   e. none of the above.
6. The significant difference between endocrine, paracrine, and autocrine communication is the
   a. chemical nature of the product.
   b. action elicited in the target.
   c. distance to the target.
   d. nature of the receptor.
   e. all of the above.
7. The brain is included in the
   a. PNS.
   b. CNS.
   c. ANS.
   d. afferent nervous system.
   e. all of the above.
8. Nerve impulses are effectively or functionally unidirectional because of
   a. dendrite responses.
   b. cell body metabolism.
   c. release of neurotransmitter at axon terminals.
   d. the computing function of the axon hillock.
   e. summation of IPSPs and EPSPs.
9. If you administer a single sub-threshold stimulus to a neuron, you should expect
   a. no response.
   b. a small action potential.
   c. a normal action potential.
   d. a large nerve impulse.
   e. an action potential that quickly dies out.
10. After an action potential reaches its peak and the potential falls, Na^+ ions ______ the cell primarily because of ______ .
   a. enter; Na^+-K^+ ATPases
   b. leave; Na^+-K^+ ATPases
   c. leave; open K^+ gates
   d. enter; open Na^+ gates
   e. leave; open Na^+ gates

True-False Questions

Mark each statement true or false, and if false, restate it to make it true.

1. Endocrine glands produce integrating hormones, while exocrine glands are sensors.
2. When blood sugar levels fall, the pancreas increases its secretion of secretin.
3. The primary target cells of insulin and glucagon are the β and α cells of the pancreas.
4. Neurotransmitters are paracrine secretions found in the nervous system.
5. The ANS regulates skeletal muscles, while efferent neurons of the somatic system regulate cardiac and smooth muscle.
6. Neurons are single cells that generally conduct unidirectionally; nerves carry nerve impulses bidirectionally because they contain collections of neuron processes.
7. A nerve is to a tract as a ganglion is to a nucleus.
8. An action potential begins with the opening of ligand-gated sodium ion channels.
9. In contrast to the all-or-none, nondecremental character of an action potential, postsynaptic potentials are variable and decremental.
10. The hormones of the anterior pituitary gland and adrenal cortex are really secretory products of neurosecretory cells.

Concept Questions

1. Explain how one hormone can elicit two different responses after binding with the same receptor in two different target cells and how one hormone can elicit two different effects within the same cell.
2. Distinguish endocrine from exocrine cells.
3. What are IPSPs and EPSPs, and how are both different from action potentials?
4. How does electrochemical conduction along axons differ from the flow of electricity along a wire?
5. Contrast the anatomical and functional relationships of each lobe of the pituitary gland with the hypothalamus.

Additional Reading

Carmichael, Stephen W., and Hans Winkler. "The Adrenal Chromaffin Cell." *Scientific American,* August 1985, p. 40. This cell synthesizes, stores, and secretes a complex mixture containing adrenaline, proteins, and peptides. Studies of these processes elucidate mechanisms relevant to other secretory cells, which include neurons.

Cheung, Wai Yiu. "Calmodulin." *Scientific American,* June 1982, p. 62. Calcium ions as intracellular messengers.

Nelson, Randy Joe. *An Introduction to Behavioral Endocrinology.* Sinauer Associates, Sunderland, MA, 1995. An introduction to the relationship between endocrine physiology and behavior, including information about the effects of drugs.

Internet Resource

To further explore the content of this chapter, log on to the web site at:

http://www.mhhe.com/biosci/genbio/guttman/

42

The Structure of Nervous Systems

In the film *Rain Man,* Dustin Hoffman (shown here with Tom Cruise) played the autistic savant Raymond.

Key Concepts

A. Invertebrate Nervous Systems and Behavior

42.1 Simple animals have netlike nervous systems.

42.2 More complex invertebrates have anterior ganglia and ventral nerve cords.

42.3 Behavior patterns can be related to the actions of specific cells and signal ligands.

B. Vertebrate Nervous Systems

42.4 The vertebrate brain develops from a tube.

42.5 Information is transmitted primarily along the anterior-posterior axis of the CNS.

42.6 The brain stem and diencephalon contain major control and relay centers.

42.7 The cerebrum exerts control over many lower brain centers.

42.8 Some functions are lateralized in the human cerebrum.

42.9 Motor activity is governed by complex feed-back mechanisms.

C. Some Aspects of Brain Chemistry

42.10 Many neurotransmitters are used for different pathways and functions.

42.11 Many peptides have roles in the nervous system.

42.12 Some neuropeptides are internal opiates.

42.13 Learning may involve changes in ion channel proteins, sometimes effected by second messengers.

Most of us follow a fairly typical learning routine from childhood on. We gradually learn to add and subtract, memorize the multiplication table, balance a checkbook, and determine how much change we are due at the grocery store, often feeling grateful for pocket calculators. We make childish drawings with stick figures, gradually refine them a bit, but rarely become good at drawing unless we take a course in which someone teaches us to see spaces properly. Some of us take up a musical instrument, play in the school band or orchestra, and eventually acquire enough facility to perform at parties. Yet certain exceptional people, mostly autistic people, can do one of these things superbly without progressing through the typical learning routine. Mathematical savants, like Raymond in the movie *Rain Man,* can instantly perform calculations that even computers struggle with. Artistic savants can make precise, graceful drawings of the most complicated scenes and buildings after just looking at them briefly. Musical savants can play the piano beautifully without a single lesson and can learn any piece of music by hearing it only once. Although the autistic are largely preoccupied with their own private worlds, these incredible talents show just what a human brain is capable of. They lead us to wish we could open a door in our own ordinary brains and instantly acquire the ability to make fine drawings, play music, or perform some other feat. Exceptional people, including those with unusual deficiencies in behavior or perception, also illustrate just how little we understand about the functioning of the 3-kg maze of neurons inside our heads. In this chapter, we will explore some of what modern biology knows about the basic structure of central nervous systems, with emphasis on the human brain and a few of its functions. ■

Figure 42.1

The nervous system of the cnidarian *Hydra* is a simple nerve net. Source: J. Hadzi, 1909, Univ. Wien Arb. Zool. Inst. 17.

A. Invertebrate nervous systems and behavior

42.1 Simple animals have netlike nervous systems.

Even though protozoans and sponges react to certain stimuli, real nervous systems only appear in eumetazoans. Cnidarians (corals, jellyfish, and sea anemones) have quite simple systems for responding to stimuli, with little of the nervous organization of most other animals. Sensory cells scattered over their surfaces detect touches and light, and send signals into the nervous system or communicate directly with effector cells. Their effectors, both epitheliomuscular cells and elongated muscle fibers, form thin muscular sheets that can contract to change the shape of the body. The cnidarian nervous system is a **nerve net** made of simple neurons that may have two, three, or several axonlike processes (Figure 42.1). These neurons conduct slow action potentials lasting about 20–50 milliseconds, in contrast to only 1–2 milliseconds in more complex animals; most synapses in the net act symmetrically and conduct signals in both directions. Nerve nets operate slowly because each signal passes through the whole network of many neurons instead of through a few neurons with long axons as in more advanced nervous systems.

The cnidarian nerve net can coordinate movements and stereotyped behaviors called *fixed-action patterns* (discussed extensively in Chapter 49). The freshwater cnidarian *Hydra,* for instance, feeds on small animals and bits of debris with directed feeding movements, including searching movements and contractions with the tentacles that carry food into its body cavity (Figure 42.2). The feeding pattern is always the same. Direct contact with prey stimulates discharge of the nematocysts (see Figure 34.11), which sting the prey and entangle it in their long

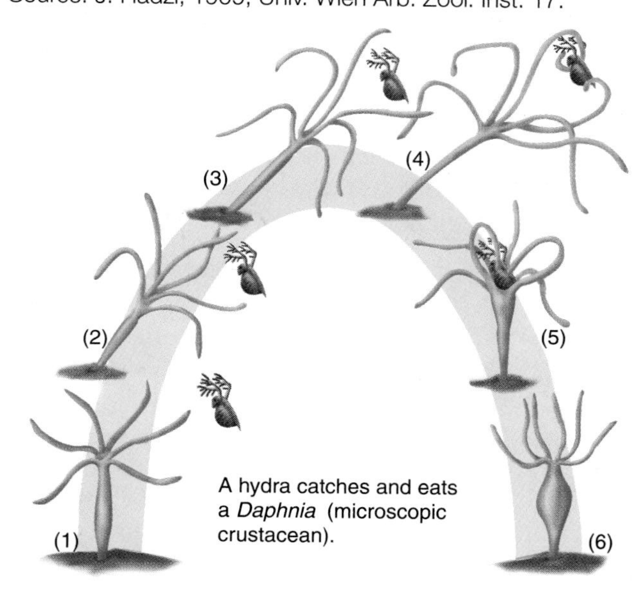

Figure 42.2

The feeding pattern of *Hydra* is a highly stereotyped fixed-action pattern that is determined by the structure of its nervous system. When a *Daphnia* comes near *(1),* the *Hydra's* tentacles reach up *(2),* and its body stretches out *(3).* As nematocysts sting the *Daphnia,* the tentacles grasp the prey *(4)* and pull it down toward the mouth, which rises to enclose it *(5).* The *Daphnia* is finally enclosed in the gastrovascular cavity *(6).*

fibers. Sensory cells signal the tentacles to grasp the prey, and as they contract and pull it down, the mouth opens to receive it. Then the tentacles relax and extend themselves again while the rim of the mouth glides up, encompasses the food, and closes. Circular contractions in the upper body force the food down

Concepts 42.1 A Vocabulary Review

These directional terms apply to a bilateral, cephalized animal.

Dorsal: back side
Ventral: underside
Anterior: head end
Posterior: tail end

When discussing the nervous system, it is essential to understand the following terms.

Neuron: A nerve cell that carries signals. A neuron has a *cell body,* or *soma,* where the nucleus and other organelles reside. *Dendrites* are usually short processes that carry signals toward the cell body, and the *axon* is a longer process that usually carries signals away from the cell body.

Glial cell: A cell that supports and protects neurons.

Ganglion: A cluster of neuron cell bodies outside the central nervous system (CNS) that may include synapses.

Nucleus: A bundle of neuron cell bodies inside the CNS.

Nerve: A bundle of axons and dendrites outside the CNS.

Tract: A bundle of axons and dendrites inside the CNS.

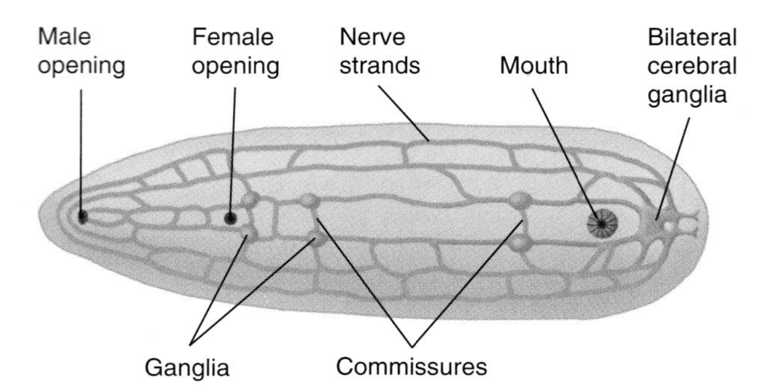

Figure 42.3
The more advanced flatworms have a simple central nervous system, with the beginnings of cephalization, as shown by ganglia concentrated at the head end. Several ganglia are also arranged along the body, with commissures between left and right halves, and connections joining ganglia along the length of the nerve strands.

into the central cavity where it is digested for a few hours, and then the indigestible remains are spewed back out through the mouth. When an appropriate bit of food appears near the animal's tentacles, the *Hydra* performs this sequence of movements smoothly and with good coordination; this behavior is a fixed-action pattern, determined entirely by the structure of the nerve net and the cells it controls. The behavior cannot be altered, although the tentacles are directed toward food particles wherever the food happens to be.

Some flatworms, the simplest bilateral animals, have nervous systems with features of the more cephalized animals that evolved later. In the simplest flatworms, the network has retained its external position just below the muscle layers, as in cnidarians. In more advanced flatworms, however, the nervous system as a whole evolved from the primitive peripheral location and moved deeper into the body, with neuron cell bodies concentrated in bilateral anterior ganglia that can be called a "brain" (Figure 42.3). These ganglia receive inputs from many sensory receptors, also concentrated at the anterior end. The nervous system has the form of a ladder, with left and right ganglia connected by tracts called **commissures.** In the evolution of later animal groups, the nervous system increasingly became formed of fibers—that is, axons—carrying signals for long distances, with fewer intervening cell bodies. Thus the primitive nerve net was gradually transformed into a central nervous system (CNS) with peripheral nerves. Later evolution led to the dominance of fibers that could rapidly conduct signals along the anterior-posterior axis to and from the anterior brain.

Concepts 42.1 reviews some of the basic terminology used to describe nervous systems.

42.2 More complex invertebrates have anterior ganglia and ventral nerve cords.

Most annelids, arthropods, and molluscs have well-developed central nervous systems with anterior ganglia, although primitive molluscs (chitons) have nerve nets, showing that the change from a nerve net to a CNS was made independently in different clades. Annelids have a brain in a somewhat dorsal position just above the esophagus, a subpharyngeal ganglion (below the esophagus), and a pair of ventral nerve cords with fused ganglia in each body segment (Figure 42.4). The nerve cords contain giant nerve fibers, like those of squid; such giant axons, which are quite common among invertebrates, conduct signals rapidly to produce fast escape reactions in response to threatening stimuli.

In arthropods, the body segments are specialized for different activities, and so are the segmental ganglia. One set of ganglia in the head controls mouthparts, thoracic ganglia control the walking legs, and abdominal ganglia control swimmerets or other small appendages. The brain is organized as in annelids, with some clear separation of functions, such as regions that process visual information.

Neurobiologists have selected certain molluscs with relatively simple nervous systems for detailed studies, following the precept that to understand complicated animals like humans we must first understand how simple systems operate. The large sea slug *Aplysia californica* has been a favorite subject since it was first used in 1941 by Angelique Arvanitaki, because its nervous system has only about 10^5 cells. (Contrast this with the human brain, which has about 10^{10} neurons and 10^{11} glial cells.) About 20,000

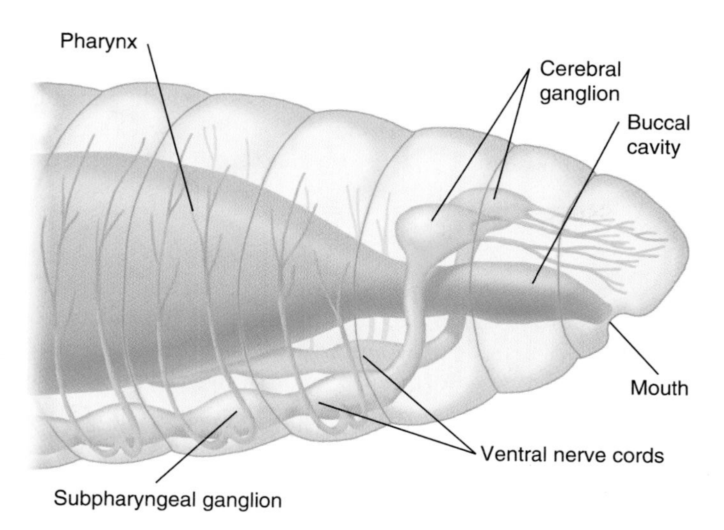

Figure 42.4

An annelid's nervous system shows distinct cephalization, with a brain consisting of large cerebral ganglia. Ventral nerve cords extend from the brain through the body, each cord composed of chains of ganglia joined by connectives.

Figure 42.5

The marine snail (or sea slug) *Aplysia* has become a favorite subject for neurophysiological research because it has a few ganglia that contain some very large cell bodies of neurons. The most prominent cells are all identifiable individually, making it possible to determine their connections and functions.

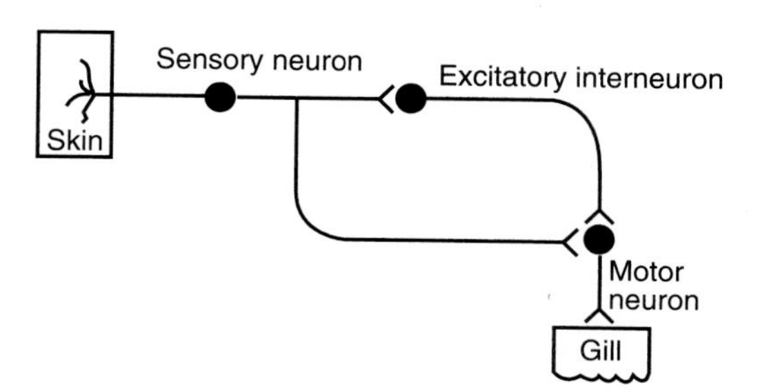

Figure 42.6

When an *Aplysia* is stimulated on the skin in the vicinity of the gills, it suddenly pulls the gills in. This gill-withdrawal reflex is controlled by a small set of neurons.

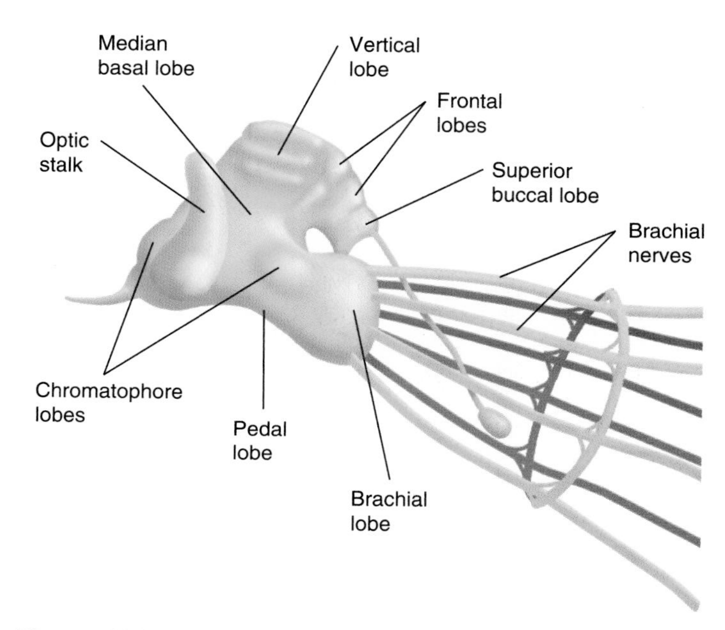

Figure 42.7

An octopus's brain has several lobes with distinct functions. We know that the vertical lobe is necessary for learning because its removal erases learned behaviors and prevents learning new ones. Source: From Boycott, "Learning in the Octopus" in *Scientific American,* 1965.

neurons in *Aplysia* are concentrated in four pairs of head ganglia and a large abdominal ganglion; some neuron cell bodies are giants, up to a millimeter in diameter, that can be identified individually because every slug has the same arrangement (Figure 42.5). Some major pathways have been worked out in this system. For instance, only a few neurons are responsible for a gill-withdrawal reflex initiated by sensory cells in the skin (Figure 42.6).

Other molluscs such as octopi and squid are interesting to neurobiologists for a different reason—not that their nervous systems are so simple but that they are complex enough for genuine learning. An octopus's brain, with about 10^8 neurons, is large enough to contain distinct centers for activities like respiration and changing color, and some of its areas are used for learning (Figure 42.7). By the judicious use of small electric shocks, for instance, an octopus can be conditioned to attack a crab only in the presence of certain sensory signals. After the octopus has learned this behavior, if the vertical lobe of the brain is removed, the animal forgets its training and can't be trained again with the standard schedule of one trial every two hours. However, if the operated animal is given a trial every few minutes, it retains its training for short times, showing that it still has short-term memory but not the long-term memory of true learning. (Short-term memory, which allows you to remember a phone number for a while after looking it up, is distinct from the long-term memory that allows you to remember childhood friends and phone numbers from the distant past.)

Exercise 42.1 Using a simple reflex network as a model, draw a network to accomplish this action: An animal's eye sees a certain stimulus (build that event into a single "black box") and signals a tentacle to withdraw quickly; the tentacle then extends again a few seconds later.

42.3 Behavior patterns can be related to the actions of specific cells and signal ligands.

Even the simplest fixed-action pattern must require complex interactions among several nervous system elements, muscles, and glands, but the underlying physiology can be worked out with the many techniques now available. Richard H. Scheller and his colleagues, for instance, have investigated the fixed-action pattern for egg-laying in *Aplysia,* using recombinant-DNA methods to clone the major genes involved in this system, sequence them, and study their functions. *Aplysia* lays long strings of more than a million eggs. When muscular contractions of the oviduct start to expel an egg string, the animal stops, grasps the string in its mouth, and begins to pull it out, winding it into a mass with glue produced by a mouth gland. Finally, with a forceful sweep of the head, it attaches the mass to a solid support.

Gene cloning, Section 17.13.

The entire action constitutes one fixed-action pattern. A group of neurons called bag cells produce an egg-laying hormone (ELH), a peptide of 37 amino acids that brings on the full pattern. ELH acts as a hormone on at least two target cells (Figure 42.8). ELH is made as part of a polyprotein, a long peptide that is later cut into functional pieces; the smaller peptides cut from the same polyprotein, called *bag cell factors,* also elicit parts of this fixed-action pattern. (Peptide neurotransmitters and hormones are commonly made as polyproteins.) This research has yet to explain the whole mechanism of the egg-laying pattern, but in principle, the mechanism can be worked

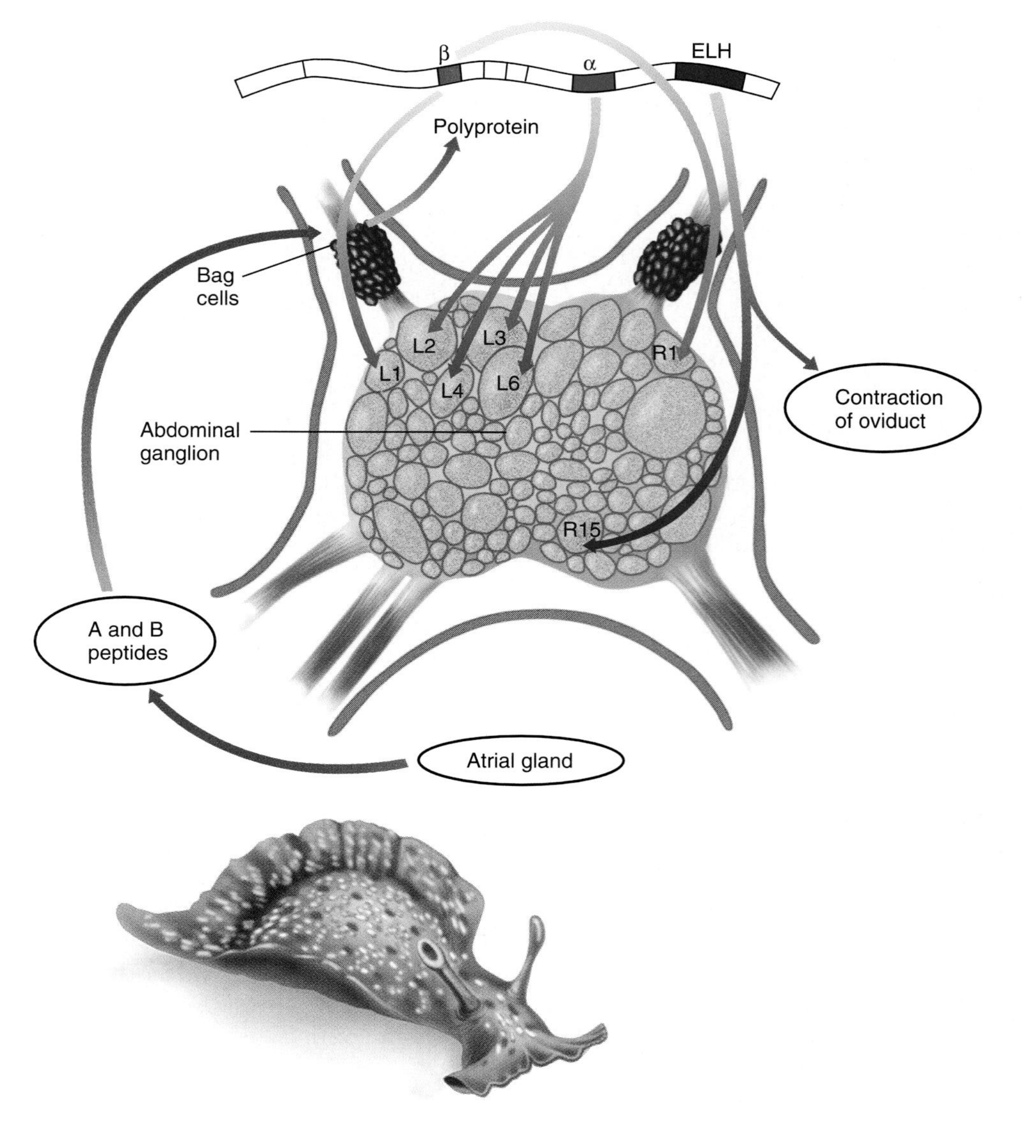

Figure 42.8
The egg-laying fixed-action pattern of the marine slug *Aplysia* is brought on largely by egg-laying hormone (ELH; *green*), acting in two ways. As a hormone, it stimulates the smooth muscle of the oviduct to contract and expel the egg string; as a neurotransmitter, it acts on at least one excitatory neuron, designated R15, in the abdominal ganglion. Meanwhile, the alpha bag cell factor (BCF; *blue*) inhibits neurons L2, L3, L4, and L6, while feeding back to stimulate the bag cells. The beta BCF *(red)* stimulates neurons L1 and R1, whose functions are unknown.

out by identifying the functions of neurons and signal ligands step by step.

Exercise 42.2 How can a single compound act as both a neurotransmitter and a hormone? Have we met this situation before in a vertebrate nervous system?

B. Vertebrate nervous systems

42.4 The vertebrate brain develops from a tube.

Invertebrates generally have ventral central nervous systems with one or two main nerve cords, but the vertebrate CNS is single, dorsal, and hollow. Early in the growth of a vertebrate embryo, the dorsal surface rolls into a tube that develops into the nervous system (see Section 20.2). The posterior half of the tube becomes the spinal cord, and the anterior half enlarges to form the brain (Figure 42.9). Failure of the neural tube to close properly at its anterior end results in a virtual lack of tissue in this region, called *anencephaly* (*an-* = without; *cephalo* = head) in humans; failure to close in a posterior region results in *spina bifida,* a condition in which the neural tube remains open and connected to the back. Early on, the brain divides into three regions: the **forebrain** (prosencephalon), **midbrain** (mesencephalon), and **hindbrain** (rhombencephalon). The hindbrain divides further into two regions: the metencephalon, which includes the **cerebellum** and the **pons,** and the myelencephalon, which develops into the **medulla oblongata** just anterior to the spinal cord. Together, the midbrain, pons, and medulla comprise the **brain stem.** Then two pockets grow out of the forebrain, twist a bit, and become the **cerebral hemispheres** (telencephalon), whose cortex is highly developed in mammals. The lower portion of the forebrain becomes the diencephalon, with the pituitary gland (hypophysis) below. These five regions and the structures they develop are recognizable in all vertebrates, but they vary in relative size and complexity.

Table 42.1 and Figure 42.10 summarize the main regions of a vertebrate brain.

No matter how much the parts of the brain may twist and enlarge, the brain remains a tube, containing four large, interconnected spaces, or **ventricles:** two lateral ventricles in the cerebrum, the third ventricle of the diencephalon, and the

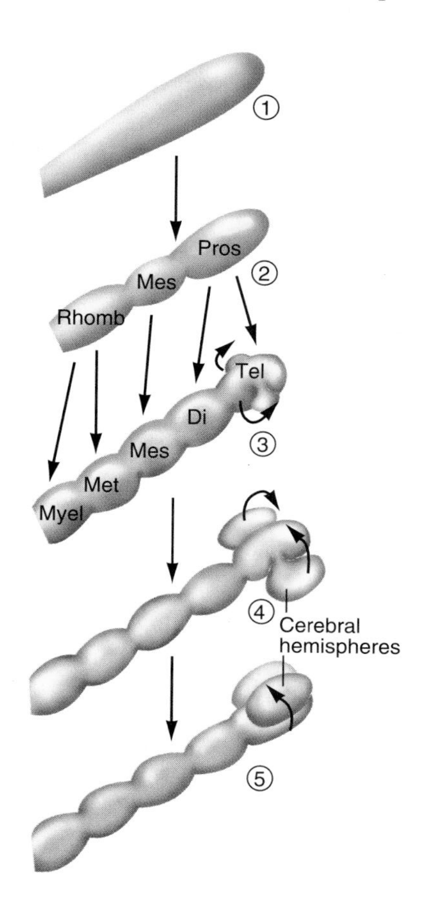

Figure 42.9
The vertebrate brain develops from the anterior end of the neural tube *(1).* Early in development, the brain divides into three main regions (prosencephalon, mesencephalon, rhombencephalon) *(2)* and then into five, adding telencephalon, diencephalon, metencephalon, and myelencephalon *(3).* The cerebral hemispheres emerge from the side of the telencephalon *(4)* and twist upward into their final positions *(5).*

Table 42.1 Major Regions of a Vertebrate Brain

Forebrain	Telencephalon	Cerebrum: Cortex and basal nuclei
	Diencephalon	Thalamus Hypothalamus Pineal body
Midbrain	Mesencephalon	Corpora quadrigemina
Hindbrain	Metencephalon	Cerebellum Pons
	Myelencephalon	Medulla oblongata

Figure 42.10

This view of the human brain shows its major structures as they would appear if some outer parts were made transparent. The hippocampus, fornix, and amygdala are three major basal nuclei of the forebrain.

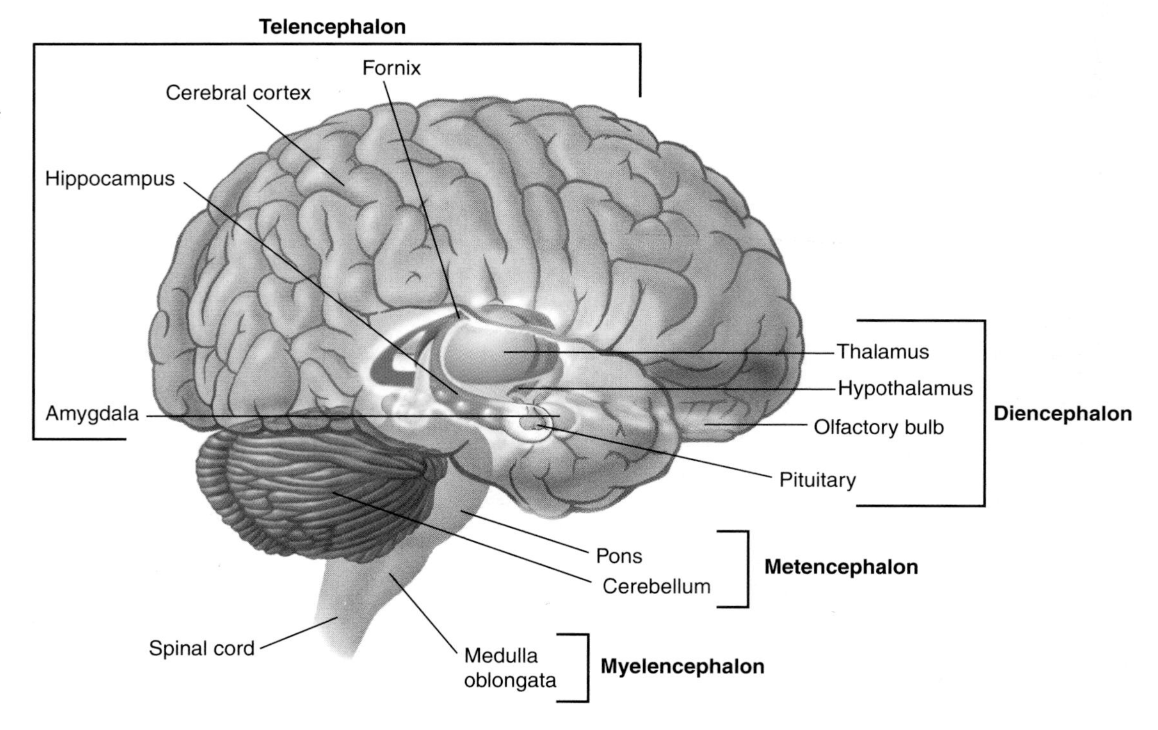

fourth ventricle of the hindbrain. The tube is filled with **cerebrospinal fluid (CSF),** similar to blood plasma. CSF is formed in the ventricles as a filtrate of blood by a tissue called the choroid plexus. The fluid circulates through the central canal of the spinal cord and outward from the lower part of the brain through small holes to the brain's surface. Eventually it returns to the bloodstream. If the circulation of CSF is obstructed during fetal development and the fluid accumulates inside, a fetus may develop hydrocephalus (water on the brain), resulting in severe deformation of the head.

The brain and spinal cord are wrapped in three layers of connective tissue called **meninges,** and CSF circulates in the space between two of them. Thus a fluid bath supports and protects the brain, inside and out. Because nervous tissue has the consistency of thick jelly, it needs this protection. Yet the brain is still easily damaged; it is no fable that boxers grow punch-drunk after years of having their brains knocked against the insides of their skulls.

42.5 Information is transmitted primarily along the anterior-posterior axis of the CNS.

Much informational traffic moves along the anterior-posterior axis of the CNS. Sensory information entering the spinal cord is transmitted through *ascending tracts,* toward the brain, while the brain transmits a lot of motor information through *descending tracts* to the muscles and glands. In evolution there has been a strong tendency for higher, more recently evolved centers to control lower, more primitive centers, making for additional anterior-posterior traffic between brain centers. When neurons originating in one center send their axons into a second center, we say that the first *projects* to the second.

A cross section through the spinal cord shows central, butterfly-shaped **gray matter** surrounded by **white matter** (Figure 42.11). White matter consists of tracts of myelinated axons and dendrites; its color comes from the density of myelin-sheath membranes. The myelin sheaths are made by **neuroglia,** supporting, insulating cells that are the equivalent, in the CNS, of Schwann cells in the peripheral nervous system. Gray matter consists mainly of cell bodies, dendrites, and unmyelinated axons. The organization of gray and white matter is modified extensively in the brain.

The white matter of the spinal cord contains numerous parallel tracts, both ascending and descending, that connect regions of the CNS. The names of these tracts simply show which parts of the CNS they connect, despite occasional fancy names such as "funiculus" or "lemniscus." Thus a corticospinal tract joins the cerebral cortex with the spinal cord, and a pontocerebellar tract runs from the pontine nucleus to the cerebellum. Given a list of brain centers, there is no mystery in these names. Curiously, tracts often cross over from one side to the other, so regions on the left side of the brain commonly control or receive sensory input from the right side of the body, and vice versa. Many commissures connecting right and left halves of each brain region keep them in communication.

42.6 The brain stem and diencephalon contain major control and relay centers.

As the spinal cord enters the skull, it becomes the medulla oblongata, the seat of nuclei that control many of the basic physiological functions mediated through the autonomic nervous system. A vasomotor center monitors blood pressure;

respiratory centers control breathing; and other nuclei control the heartbeat and some digestive activities. Major tracts connecting the spinal cord and higher brain centers run through the medulla, and seven of the twelve cranial nerves listed in Table 42.2 emerge from it. A **reticular formation,** extending from the medulla through the rest of the brain stem, has several important functions (Figure 42.12). Its ascending reticular activating system (ARAS) projects into the cerebrum, keeping

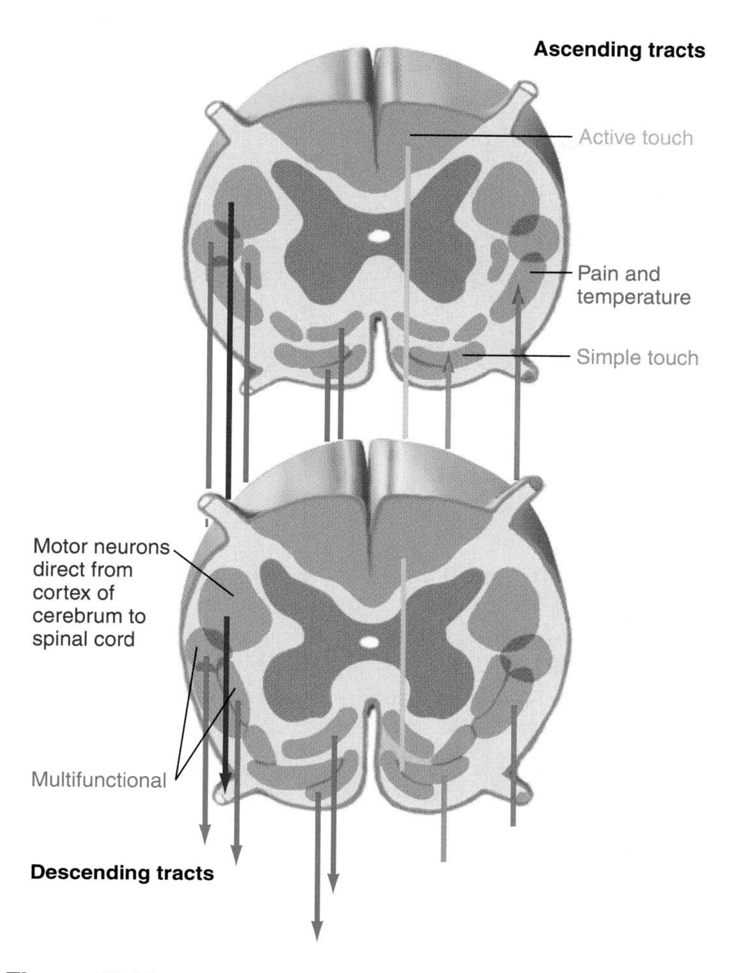

Figure 42.11
The spinal cord has a core of gray matter surrounded by white matter. The white matter contains major ascending tracts carrying sensory information, such as temperature, and descending tracts carrying motor signals to points in the body. Simple touch is the sensation of something touching the skin; active touch consists of sensations derived from manipulating objects actively.

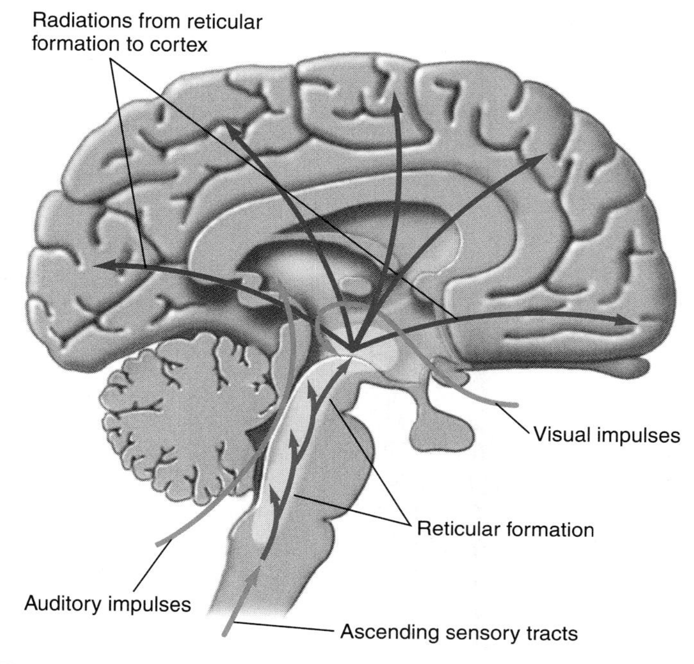

Figure 42.12
The reticular formation is a complex of nuclei and nerve fibers in the medulla, pons, and midbrain that radiates through the thalamus and hypothalamus into the cerebrum. It acts as an ascending reticular activating system that carries many kinds of sensory signals to the cerebral cortex and arouses it, so it is ready to receive other signals.

Table 42.2 Vertebrate Cranial Nerves

Nerve	Type	Site Innervated
1 Olfactory	Sensory	Nasal epithelium
2 Optic	Sensory	Retina of eye
3 Oculomotor	Motor	Four pairs of exterior eye muscles; internal eye muscles
4 Trochlear	Motor	One pair of exterior eye muscles
5 Trigeminal	Mixed	Head and face muscles; jaw muscles, lower jaw teeth
6 Abducens	Motor	One pair of exterior eye muscles
7 Facial	Mixed	Facial muscles, salivary glands, taste buds
8 Auditory	Sensory	Internal ear and balancing organs
9 Glossopharyngeal	Mixed	Tongue and pharynx; taste buds
10 Vagus	Mixed	Heart, intestines, etc. Lateral line organs in fish
11 Spinal accessory	Mixed	Accessory to vagus
12 Hypoglossal	Motor	Tongue muscles. Gill muscles in fish

Sidebar 42.1 Cycles, Depression, and the Pineal Gland

The seventeenth-century philosopher René Descartes believed the pineal body to be the seat of interactions between the mind and body because it is one of the few unpaired structures in the brain and is quite prominent in humans. Actually, its function is much more mundane. It appears to be a light receptor in ectothermic vertebrates, a role that seems absurd at first until you recognize that a substantial amount of light can penetrate the top of the skull in small animals with thin bones. The pineal forms a functional light receptor in certain reptiles and amphibians and receives some sensory input in all vertebrates. It is also a gland that produces the hormone **melatonin,** which controls photoperiodic phenomena such as those described in Section 15.1. In birds, for instance, more melatonin is made at night; it lowers their body temperature and puts them to sleep. Melatonin affects the body coloration of amphibians and reptiles, sometimes in a rhythm; a still-mysterious underlying clock creates the basic rhythm, which light signals keep correlated with the day-night cycle. The role of the pineal gland and melatonin in humans is not well understood. Though no light reaches the human pineal gland directly, many people experience real depression when the long nights and dark days of winter set in, a depression that can be reversed by treatments with high-intensity, full-spectrum lights. To this extent, the pineal gland may determine our moods which affect our thinking. Perhaps, metaphorically, Descartes was right.

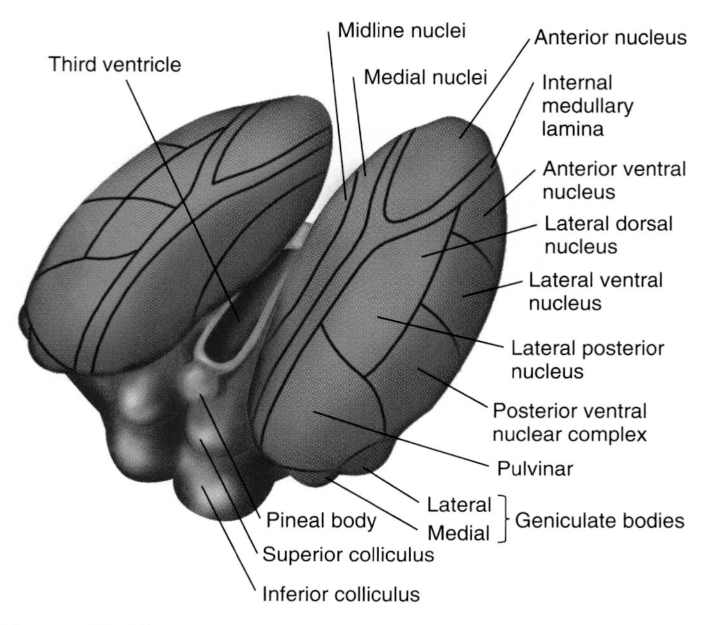

Figure 42.13

The diencephalon consists of many nuclei around the third ventricle: the thalamus *(blue)* around the sides of the ventricle and the hypothalamus, not visible here, below the thalamus around the ventral region of the ventricle. The pineal body emerges from the dorsal region. Notice also the geniculate bodies, which process visual and auditory information.

higher centers alert and awake; other brain stem centers override the ARAS during sleep, and activation of the ARAS awakens a sleeping animal. The ARAS also filters incoming stimuli, so parents living near a railroad line may sleep soundly through the regular nightly roar of trains, yet awaken at the first soft cry of their baby.

The diencephalon surrounds the third ventricle (Figure 42.13), with the pineal body arising from its roof. (Sidebar 42.1 discusses some of the functions of the pineal body.) All sensory pathways, other than those for smell, lead to nuclei of the **thalamus,** or two thalami that constitute the left and right walls of the third ventricle. Each thalamic nucleus coordinates and analyzes some kind of sensory information and relays the information to a region of the cerebral cortex. For instance, the lateral geniculate bodies relay visual information from the eyes, and the medial geniculate bodies relay auditory information. Below the third ventricle lie the *hypothalamus* and *pituitary gland,* discussed in Section 41.14. Hypothalamic nuclei govern temperature, eating, drinking, osmotic regulation, and reproductive and emotional behavior.

Exercise 42.3 Certain brain centers normally produce a regular rhythm of breathing. In mammals, fibers from the brain center descend to the breathing apparatus—diaphragm and rib muscles—through the spinal cord. If the brain stem of a mammal is cut above the medulla oblongata but below the pons, the rhythm continues but is irregular. If a cut is made below the medulla oblongata, there is no rhythm. What do these results tell us about the location and function of two breathing centers?

Exercise 42.4 In view of the general functions of the thalamus, what kind of defect would you expect an animal to exhibit most often from destruction of one of the thalamic nuclei?

Exercise 42.5 When one thalamic nucleus of a cat's brain is severely injured, the cat exhibits uncontrolled rage, continually baring its teeth and attacking or threatening to attack. What is the normal function of the nucleus?

42.7 The cerebrum exerts control over many lower brain centers.

The most significant change in the evolution of vertebrate brains has been the growing dominance of the cerebrum (Figure 42.14). In fishes, the cerebral hemispheres are small bulbs that process olfactory (smell) information. During the evolution of amphibians, this olfactory cortex was displaced somewhat by two new areas: the *archicortex,* used for association of information from different areas, and the *corpus striatum,* containing the nuclei for motor control in fishes and amphibians.

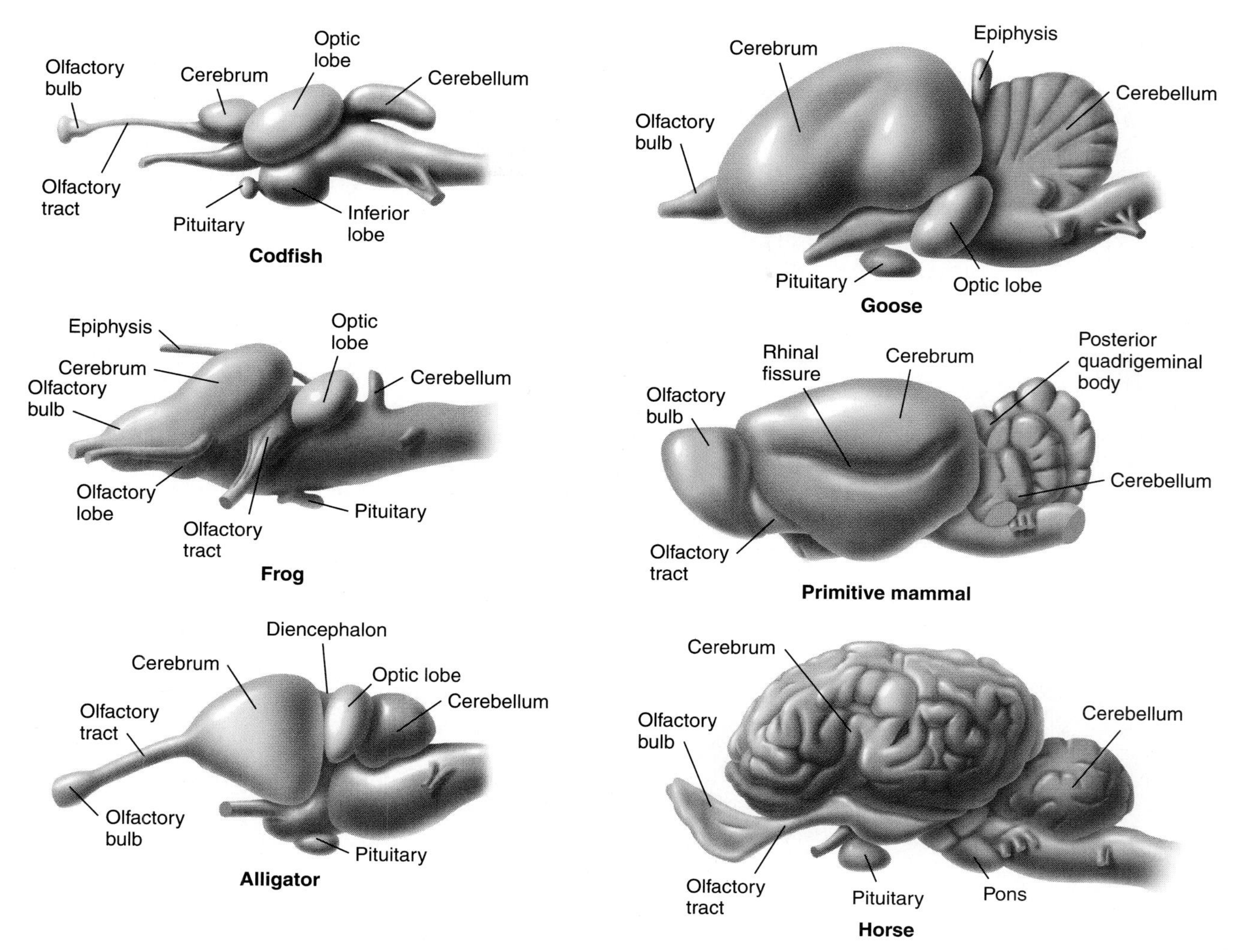

Figure 42.14

A comparison of the brains of several vertebrates shows some of the trends in evolution. Fishes and amphibians depend strongly on olfactory information, which is analyzed in the cerebrum. Birds and mammals have evolved an elaborate cerebral cortex with room to integrate olfactory, visual, and other information.

Two innovations occurred in reptiles. The corpus striatum was folded and pushed up inside the cerebrum as an isolated body, and a new area, the *neocortex,* appeared and was used to integrate inputs from all the senses. The neocortex eventually came to dominate the whole cerebrum. This transformation served the conversion from an olfactory to a visual world, since fishes and amphibians depend much more on olfactory information than birds and mammals do. In mammals, the neocortex began to expand and take over virtually the whole surface of the cerebrum. In the process, it compressed the other cortical areas, forcing the archicortex to fold up inside; this fold became the hippocampus of the mammalian brain, so named by anatomists who were struck by its resemblance to a sea horse (Greek, *hippos* = horse; *kampos* = sea monster). The mammalian neocortex became a center for integrating vast amounts of olfactory, visual, and other information.

The neocortex, the mammalian cerebral cortex, is large in all mammals, but primates have a particularly enlarged, complicated cerebral cortex with a large surface area. The human cerebral cortex is strongly sculpted into folds, or **gyri** (sing., *gyrus*), separated by grooves, or **sulci** (sing., *sulcus*). In fact, about two-thirds of the surface area of the whole cerebral cortex is buried and hidden in the sulci. Large, deep sulci called fissures divide each side into four lobes (frontal, parietal, temporal, and occipital), as shown in Figure 42.15. Smaller regions of the cortex are identified by numbers (Brodmann's areas) that show differences in cell structure.

During the 1950s, when it became possible to do delicate surgery on the brain, the neurosurgeon Wilder Penfield and his associates began to explore brain functions. Penfield's subjects were all undergoing brain surgery and were awake throughout the operation. Thus he delivered delicate electrical stimuli to points on the brain and was able to map the brain surface as the patients reported their experiences. Since the cerebrum has no pain receptors, patients felt no pain as the electrode was moved about. Penfield found that stimulating each point in the

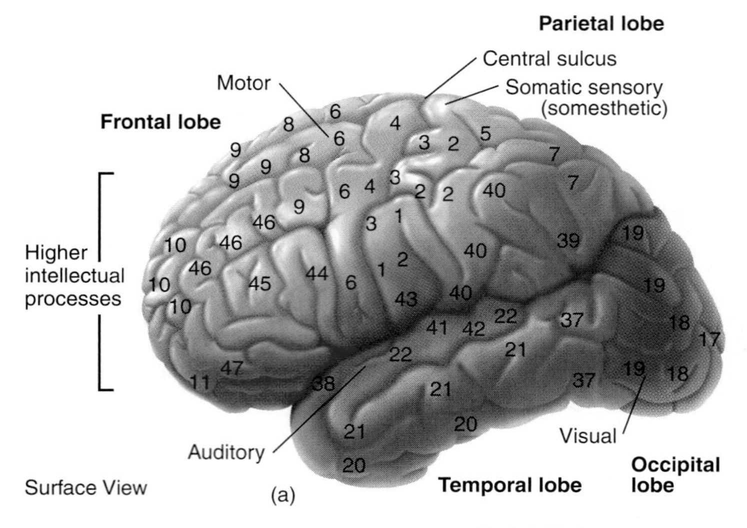

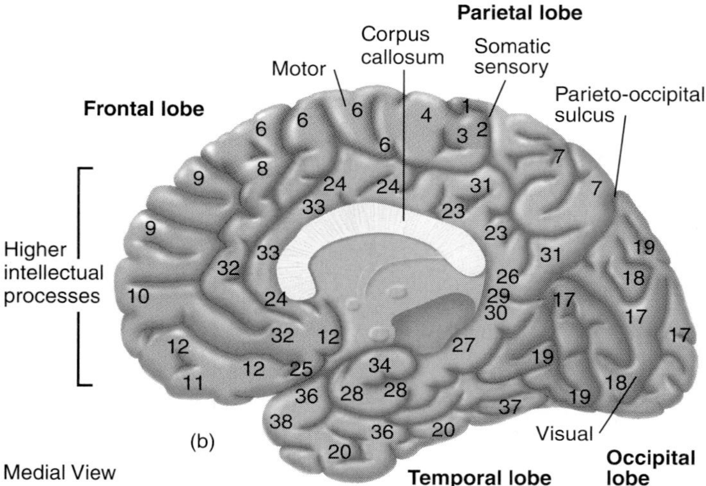

Figure 42.15

The human cerebral cortex, as viewed from the left side *(a)* and in a medial section *(b),* is divided into four principal lobes. The numbers identify Brodmann's areas, and the approximate limits of selected functional regions are shown in color.

postcentral gyrus (Brodmann's areas 1, 2, and 3) evoked the sensation that a particular point on the body was being touched; this is the **somesthetic** sense (*soma-* = body, *-esthetic* = of sense perception). The pattern of these sensations shows that the body surface is represented in a maplike way on the cerebral cortex (Figure 42.16*a*), with the order of body parts so well maintained that they can be illustrated by a homunculus, a miniature, distorted person. The hand and face take up enormous areas because they are dense with touch receptors, while the whole trunk of the body is represented by only a small area of cortex because it has few receptors. Each side of the cortex records from the opposite side of the body, because the ascending tracts cross over.

Penfield also discovered that stimulating the precentral gyrus (area 4), just in front of the central fissure, makes certain muscles twitch. This reaction allowed him to identify the region that controls each muscle group and to draw a map on this part of the cortex, known as the **motor cortex.** The motor map is very much like the somesthetic map (Figure 42.16*b*).

The occipital lobe, in areas 17, 18, and 19, contains centers of vision, discussed in Section 43.11. Sidebar 42.2 provides some insight into the function of the frontal lobes. The temporal lobe processes auditory information in areas 41 and 42. Nearby is Wernicke's area, a major speech center in humans (discussed in Section 42.8). Stimulating the temporal lobe can evoke strips of what Penfield first believed to be memories, but they are apparently random associations rather than recollections of actual events. Many areas are "silent," in that stimulating them doesn't produce any sensation or motor activity; these have been dubbed "association areas," on the assumption that they are places where information is integrated, or associated. Lesions in certain areas show that they are associated with memory, with emotions, or with specific abilities such as distinguishing left from right or performing mathematical operations. Because so many areas haven't been identified with specific functions, the myth has grown up that "we only use ten percent of our brains," a piece of folk wisdom that is repeated out of ignorance. Yet the great abilities of some autistic people, described in the introduction to this chapter, cause us to wonder if we all have some neural space that could be used for such talents.

The human brain is so complex, with so many distinct nuclei and tracts, that no brief course of study could provide an overall understanding of it. Indeed, the functions of many regions are just now being discovered. To provide some sense of how brain centers operate and interact, we will tell two stories in this chapter—about left and right specialization of the cerebrum and about motor activity (Sections 42.8 and 42.9). In Chapter 43, we will relate a story about a major sensory activity (vision).

Exercise 42.6 On a map of the brain, why are some large parts of the body represented in the motor cortex by small areas and some small parts represented by large areas?

42.8 Some functions are lateralized in the human cerebrum.

Few experiments have elicited greater interest than the "split-brain" research initiated by Roger W. Sperry and Robert E. Myers using cats, and continued by Sperry and Michael S. Gazzaniga using human subjects who had been operated on for epileptic disorders. Some kinds of epilepsy involve reverberations of impulses between the left and right cerebral hemispheres, and in these cases the patient can be cured by severing the **corpus callosum,** the major commissure connecting the hemispheres (visible in Figure 42.10). The main discovery—that each side of the brain is somewhat specialized—has provided some important insights into the brain's functional organization, but this work has been badly misunderstood and has generated some fuzzy-minded pseudoscience. The sober facts

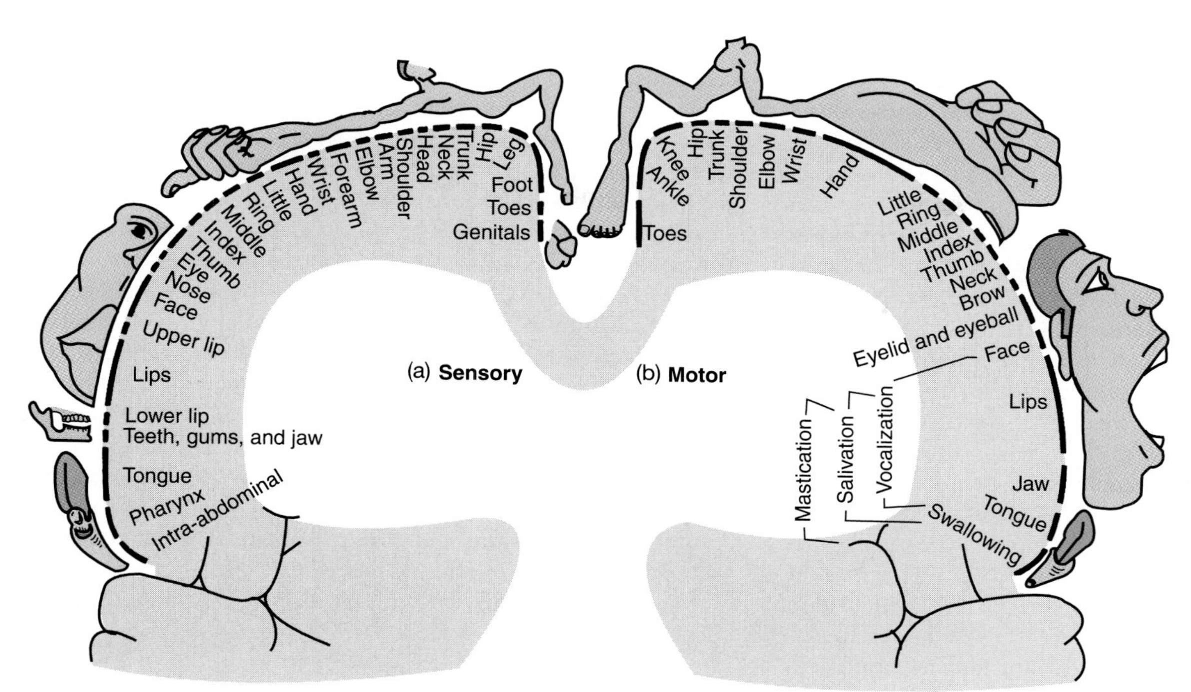

Figure 42.16

(a) A map of the sensory cortex shows which region receives sensations from each part of the body. *(b)* A similar picture of the motor cortex shows which region controls muscles in each part of the body.

Sidebar 42.2 A Remarkable Brain Experiment

On September 12, 1848, a young railroad foreman named Phineas Gage was packing blasting powder into a hole with a 13-pound, 3-foot-long iron bar when the powder exploded and drove the rod through his skull. The bar ran up through his left eye socket and neatly took out the left frontal lobe of his brain. Remarkably, Gage survived. He regained consciousness in a few minutes and was taken by oxcart to his hotel in a nearby town where Dr. John Harlow examined him two hours later. As chronicled by Dr. Harlow, the pipe was removed, leaving a massive wound that healed over the next two months. Hemorrhaging stopped, and Gage's condition gradually improved with episodes of delirium and occasional fetid discharges of brain tissue and other material. By the end of November, he was quite recovered.

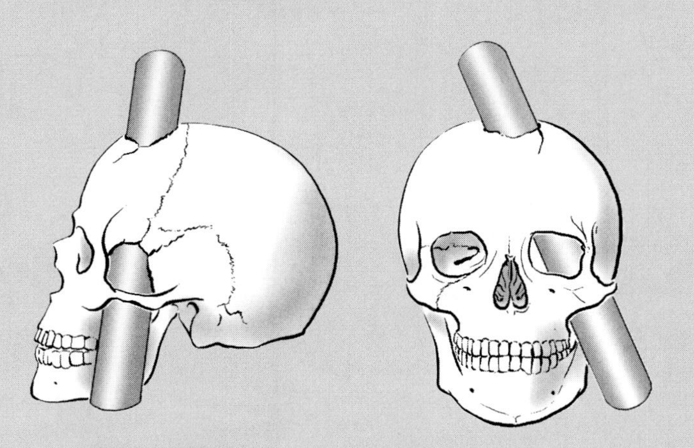

Gage's personality, however, was changed forever. He had been a dependable, well-liked, industrious workman. After the accident, he became "loud, profane, and impulsive." Dr. Harlow described him as "manifesting but little deference for his fellows, impatient of restraint or advice when it conflicts with desires, at times pertinaciously obstinate, yet capricious and vacillating, devising many plans of future operations, which are no sooner arranged than they are abandoned. His mind was radically changed, so that his friends and acquaintances said he was no longer Gage." The accident had inadvertently provided one of the first clues to the functions of the frontal lobes—controlling emotions and subtle features of personality. The frontal and prefrontal lobes are connected to a complex of brain structures called the limbic system that is involved with emotional behavior.

In the 1930s, investigators found that lesions in the prefrontal lobes could abolish aggressive behavior and reduce anxiety. This led Egas Moniz and Almeida Lima to use prefrontal lobotomy—slicing through the brain to separate the prefrontal lobes—as a cure for some types of mental illness, such as acute depression, anxiety, and uncontrolled aggression. The operation works on some patients but not on others. Since the 1950s, many drugs have been developed for effectively treating the mental and emotional problems that prefrontal lobotomies were once used for. Now the surgery is reserved for the few conditions for which pharmaceutical remedies are ineffective. Such surgery, of course, is based on a simplistic view of brain organization, and many physicians and psychologists have strongly condemned it; the brain structures associated with emotions are very complex, and soldiers who have suffered wounds to the prefrontal areas show the whole gamut of personality changes from euphoria to apathy to depression to psychopathology. A few centers in the limbic system correlate with specific emotional reactions such as anger and fear, but we do not expect to understand the whole range of emotions easily or soon.

remain exciting, even though they don't support the popular speculations.

The first split-brain experiments used cats whose optic pathways, as well as their corpora callosa, had been severed, so the right eye was connected only to the right side of the brain and the left eye only to the left side. These animals behaved as if they had two separate brains; if a cat was trained to respond to a stimulus presented to only one eye, the other side of the brain remained untrained and knew nothing about the stimulus. When these experiments are performed with human subjects, their visual systems remain intact, but different stimuli can be presented to each side of their brains with an apparatus that flashes a picture on only half of the retina (Figure 42.17). (The arrangement of visual pathways that makes this possible is shown in Figure 43.27.) Remarkably, these people report only what the left side of the brain sees. When shown a split picture of two faces, they describe the face presented in the right visual field, which is seen by the left side of the brain. However, at some level, the subjects know what the right side of the brain has seen. When the right side is shown pictures of nudes, for instance, the subjects may blush, giggle, and appear confused, even though they can't report what they see.

These patients behave this way because the speech centers are in the left side of the cerebral cortex, so the vocal apparatus can only report what the vision centers communicate to the left side. The right side, in contrast, carries the ability that Gazzaniga and Joseph LeDoux call **manipulospatiality.** Manipulospatiality is the ability to draw, arrange, construct, and otherwise manipulate objects so they are properly related to one another; it also involves the sense of **active touch,** in which we learn the size, shape, and orientation of an object by manipulating it—that is, obtaining information from an exchange of signals between the brain and the hands through a combination of sensory pathways and the active use of muscles. (Our sense of *simple touch* just tells us that an object with a certain texture is touching the skin somewhere.) Since the right motor cortex controls the left hand, a split-brained person who is given an object in the left hand can pick that object out of a collection (hidden behind a screen) with the same hand. The isolated right cortex cannot control speech, but it is superior at recognizing spatial relationships; so the left hand of a right-handed person, though clumsy, is better than the right hand at making simple drawings that show the arrangement of objects in space.

The split-brain experiments led some people to believe that the two halves of the human brain are really quite different functionally—two different "minds," as it were. The left half was characterized as linear, rational, and verbal; the right half as holistic and intuitive. This is the basis of a "pop" psychology in which people talk about being "right-brained" or "left-brained," as if we were all walking around with split brains. Knowing that the lateralization is simply a separation of linguistic from manipulospatial functions allows us to look at these popular misconceptions more soberly.

Manipulospatiality is highly developed in primates because primate evolution, as explored in Chapter 36, has been largely an adaptation to moving rapidly and skillfully through the trees, an activity that places a premium on manipulospatiality and excellent vision. In primates, including humans, manipulospatiality is vested in the inferior parietal lobules (IPL) of the neocortex (Figure 42.18*a*). The IPLs on both sides of the brain in nonhuman primates are active in spatial manipulations, but this arrangement was modified by the evolution of human language. As verbal abilities evolved in humans, the left IPL became specialized to control some of them, leaving manipulospatiality controlled by the right side. In normal people, of course, the two sides of the brain communicate with each other and integrate their information.

Speech is actually controlled by several areas in the left cortex (Figure 42.18*b*). One region was identified by Paul Broca over a century ago. Broca saw that damage to the left side of the

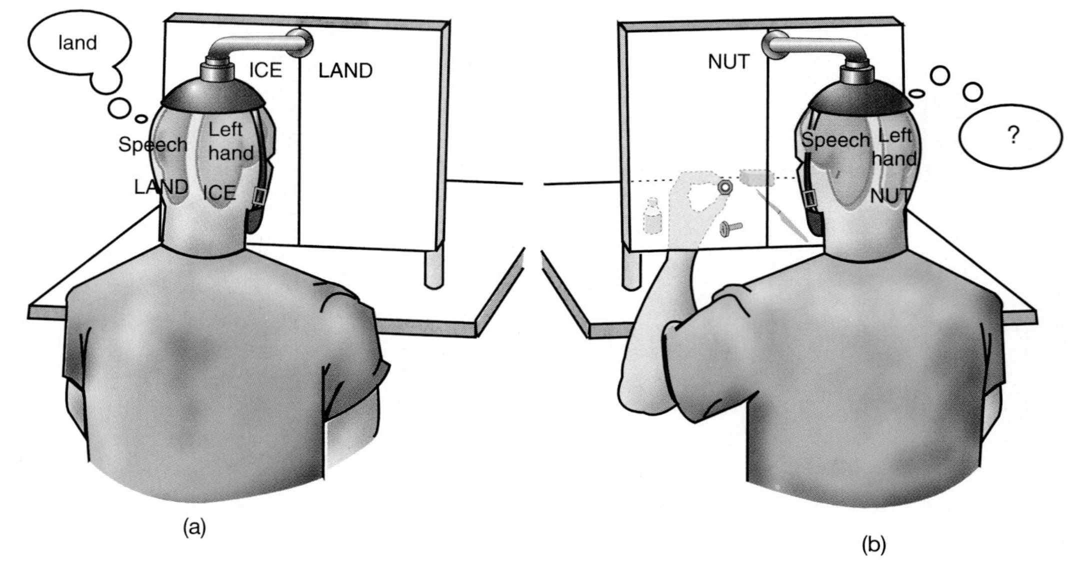

Figure 42.17
Lateralization of functions in the human brain can be tested in subjects who have had their cerebral hemispheres separated surgically. The subject's head is fixed in an apparatus, and images are flashed onto the screen so they fall in one or the other visual field. *(a)* If the word "iceland" is presented so that "ice" is in the left half of the field and "land" is in the right, subjects report they have only seen "land." *(b)* If the word "nut" is presented to the right hemisphere, the subject reports that he did not see anything, because the information cannot get to the speech centers in the left hemisphere. Nevertheless, with his left hand he is able to pick out a nut from a series of hidden objects.

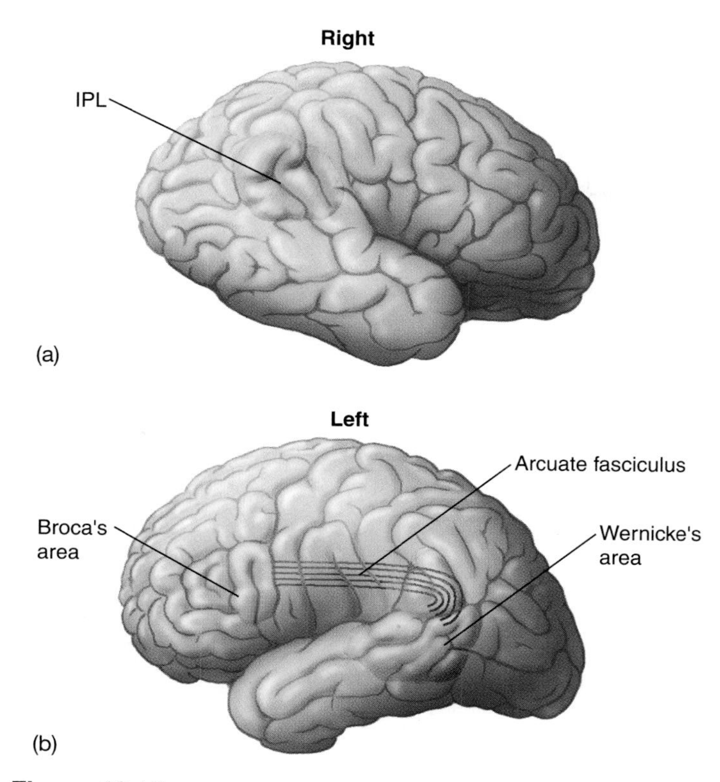

Figure 42.18

(a) In humans, the inferior parietal lobule (IPL) on the right side of the cerebral cortex is involved in manipulospatiality. *(b)* The corresponding center on the left side is close to, or identical with, Wernicke's area, a major speech center. Wernicke's area is apparently concerned with comprehension of words, Broca's area with articulation of speech sounds and with grammatical ability. These centers are connected by a tract called the arcuate fasciculus.

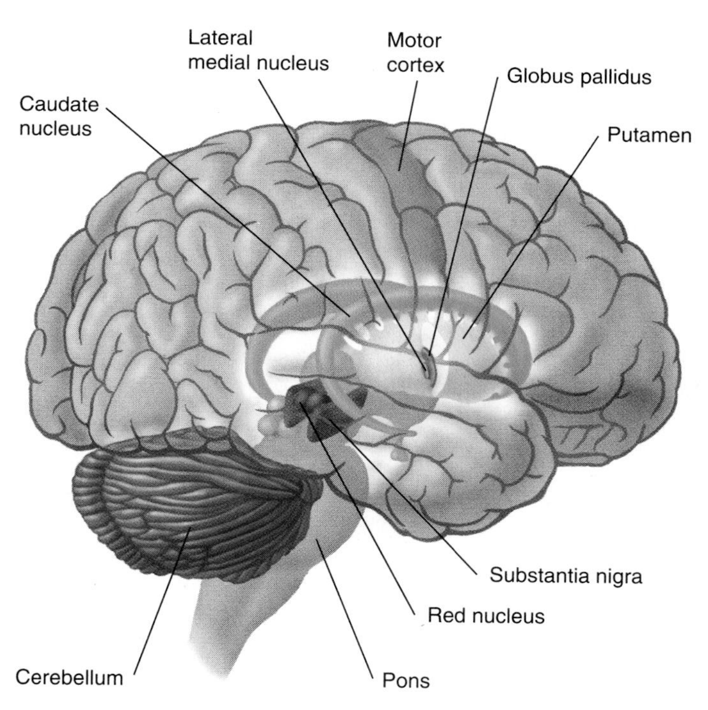

Figure 42.19

Motor activity is governed by the motor cortex of the cerebrum in concert with the cerebellum on the dorsal side of the medulla and a set of structures collectively called basal nuclei: caudate nucleus, lateral medial nucleus, globus pallidus, and putamen. The substantia nigra ("black substance") and red nucleus have key roles in regulating movements. The pons acts as a commissure (connecting link) between left and right sides of the cerebellum.

frontal lobe often produced an aphasia (speech disorder); this area is near the motor cortex that governs movements of the speech apparatus—the tongue, jaw, and throat. Damage in Broca's area does not simply paralyze the muscles used in speech; rather, the ability to control these muscles for speech is lost, along with some grammatical ability. Carl Wernicke identified a major speech area in the left temporal lobe, near the IPL, that governs speech comprehension and making proper connections with speech stimuli. If an experimental subject has to pronounce a word he hears or sees written, a signal containing the stimulus word passes through Wernicke's area, where it is apparently comprehended, and the proper speech signal goes on to Broca's area and to the speech apparatus. A defect in Wernicke's area may result in well-articulated, grammatical speech that has errors in sense and meaning.

42.9 Motor activity is governed by complex feedback mechanisms.

You sit at your desk reading this book, with a cup of coffee or tea in front of you. You reach for the cup, wrap your fingers around its handle, lift it to your lips, sip, and swallow as you set the cup back in its place, hardly thinking about your actions. Yet this routine action has employed neurons throughout much of your brain.

The very fact that you are able to sit up straight, with your muscles in a balance between the extremes of rigidity and relaxation, is remarkable. Your posture depends on the reticular formation (see Figure 42.12), which excites the entire musculature by continually sending out signals to increase muscle tone. Without this activation, you would be a limp bag of bones with no posture at all. An inhibitory segment of the same region tends to decrease muscle tone under control of higher brain centers. If all the connections to these centers are severed by slices above the brain stem, the excitatory portion of the reticular formation continues to increase muscle tone, resulting in a characteristic intense rigidity of the muscles (decerebrate rigidity).

Figure 42.19 shows the structures involved in drinking from a cup: principally the motor cortex of the cerebrum, the basal nuclei (formerly called basal ganglia), and the cerebellum. In nonmammalian vertebrates, the basal nuclei have primary control over motor activities, but in mammals this control has been transferred to the cerebral cortex. The basal nuclei still control simple stereotyped tasks and set muscle tone. Their destruction or impairment produces very strange behaviors, such as uncontrolled dancing movements or paralysis.

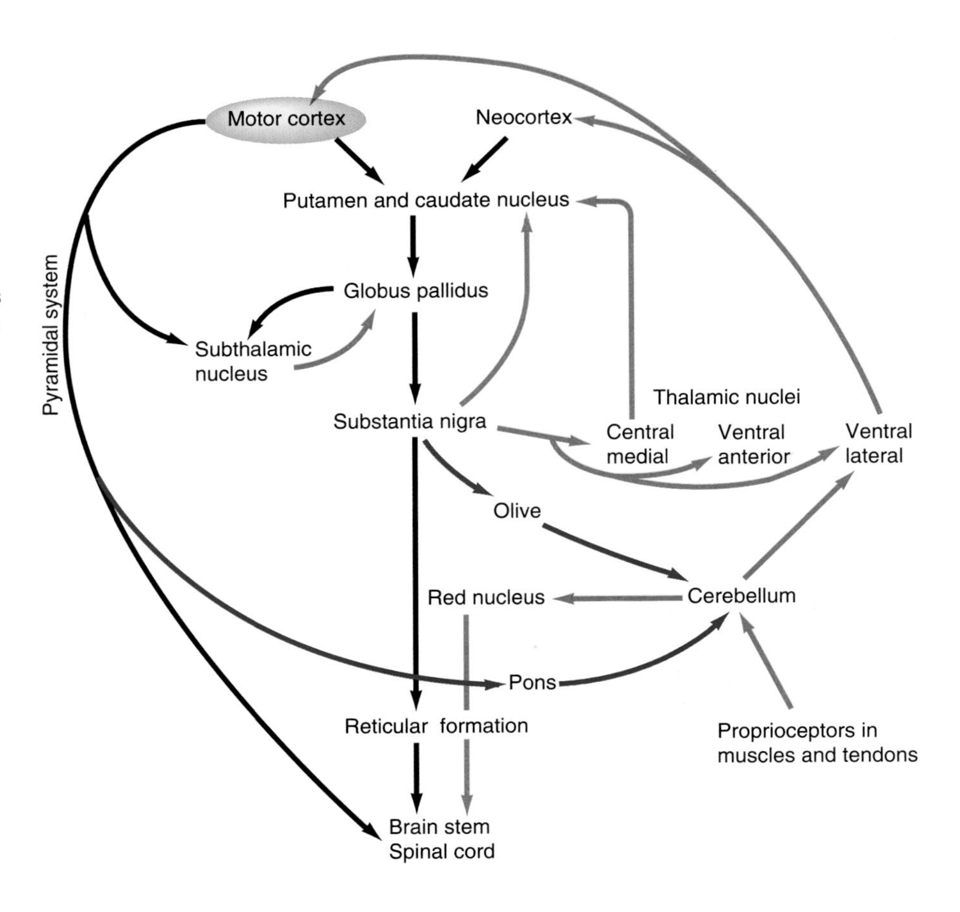

Figure 42.20

Signals that initiate a movement arise in the motor cortex *(blue ellipse).* One series of signals passes through the pyramidal motor system *(black arrows, left)* directly to synapses in the brain stem and spinal cord and through the extrapyramidal system *(center black arrows).* Simultaneously *(blue arrows),* information about the intended action is sent to the cerebellum from the pyramidal tract through the pons and from the extrapyramidal tract through the olive, a nucleus in the medulla. The cerebellum also receives information from receptors in the muscles and tendons about the actual position of the body. As it integrates this information, the cerebellum sends refining signals *(red arrows)* to the motor cortex through the thalamus and directly to the brain stem and spinal cord through the red nucleus. Other red arrows show pathways that regulate and refine the movements through feedback loops.

We don't know how the brain makes the decision to take a sip from a cup in the first place. The so-called association areas of the frontal lobes seem to be involved, and very early in the action, they transmit information to both the basal nuclei and to the lateral areas of the cerebellum. So the decision to act involves a "conference" between the basal nuclei, which will help initiate the action, and the cerebellum, which will regulate it. The action itself begins in the cerebral cortex, primarily in the motor cortex (area 4; refer back to Figure 42.15). Signals descend from the motor cortex through two general paths, the **pyramidal** and **extrapyramidal motor systems** (Figure 42.20). The pyramidal system, so named because its fibers originate in pyramid-shaped cells in the cortex, is responsible for precise, voluntary movements that can become quite skilled—all the movements we use in writing, typing, playing musical instruments, engaging in athletics, working with tools, and speaking. In this system, signals from the motor cortex descend through long axons all the way to the appropriate level of the spinal cord, where they synapse with interneurons that lead to motor neurons. So reaching for a cup must first involve signals to the large arm muscles. The pyramidal system controls the same set of motor neurons used in reflex actions, such as withdrawing your hand from a hot object, so the muscular movements are the same, whether they are initiated via a short reflex circuit or a long one involving the brain. As the pyramidal-tract fibers descend from the motor cortex, they cross over so the left cortex controls the right side of the body and vice versa. If you reach with your right hand, the signals are coming from the left motor cortex.

The basal nuclei are also involved very early. These nuclei control movement from one large area of space to another (in our example, the movement of your hand from somewhere near your body to a point on the desk), and they must do so on the basis of information about the present condition of the body. The corpus striatum is constantly receiving information about each part of the body, through numerous fibers from the sensory and motor cortices and from the nuclei of the thalamus. In addition, the corpus striatum receives input from the substantia nigra, a connection that is important in conditions such as Parkinson disease, as we will see later.

These direct pathways by themselves initiate movements, but the movement of your arm and hand require feedback regulation by the extrapyramidal system, to make the movements fine and precise. As signals are sent to the muscles via the pyramidal system, collateral signals are sent into the cerebellum; we may think of these as telling the cerebellum what movement is intended by the motor cortex. The cerebellum also receives sensory information about what has actually happened; it compares the two sets of signals and sends out correcting signals. You intend your fingers to wrap around the handle, and the cerebellum must signal each muscle to contract or relax just enough to accomplish this movement. One set of signals goes through the red nucleus and then directly back to the spinal motor neurons controlling the muscles. The other goes through the thalamus to the motor cortex, which corrects its activity and sends a new signal via the pyramidal system.

Enormous speed and dexterity can be incorporated into this system, as shown by the subtlety of a skilled crafts person's control over their fingers. To play a complex piece of music, a musician must practice an intricate pattern of movements until it becomes largely automatic. We can only imagine the accompanying series of signals moving through the motor system to make each finger strike in exactly the right position with the proper pressure and timing. Automatic skills of this kind are apparently encoded in the cerebellum. It is no wonder that people with cerebellar damage have great difficulty performing the simplest tasks well; instead of zeroing in on the cup, the arm of such a person may oscillate back and forth in what is called an *intention tremor,* as each attempt to correct the arm's position moves it farther in the wrong direction.

C. Some aspects of brain chemistry

42.10 Many neurotransmitters are used for different pathways and functions.

Neurons communicate with one another and with other cells by means of specific neurotransmitters (Figure 42.21). In the vertebrate CNS, each ligand is either excitatory or inhibitory, and each neural pathway operates largely or exclusively with a single neurotransmitter. Identifying these pathways challenges investigators. One method is to inject radioactively labeled neurotransmitter into a functioning brain, where it is taken up only by the neurons that use it. The brain is then sliced into thin sections that can be laid on photographic film to identify the labeled neurons by autoradiography (see Methods 10.1). Pathways are also elucidated by singling out neurons with fluorescence microscopy; in this process, neurotransmitters are made fluorescent, or fluorescent dyes are attached to antibodies against the enzymes that synthesize the neurotransmitters, so when a thin tissue slice is illuminated with ultraviolet light, the cells containing the neurotransmitter will fluoresce.

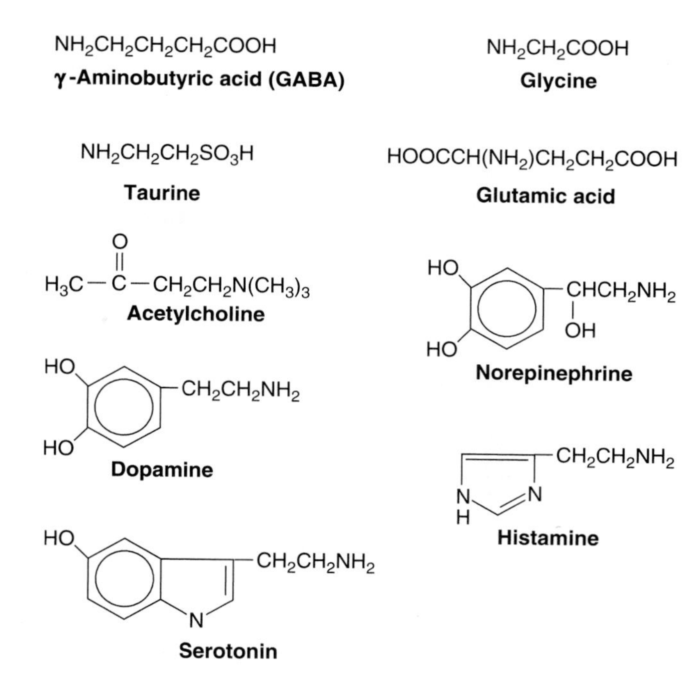

Figure 42.21
The common nonpeptide neurotransmitters are either monoamines or amino acids. They all have a positively charged nitrogen atom that binds to a complementary negative charge in their binding sites.

Use of antibodies in research, Methods 8.1.

After neurotransmitters have been released into a synaptic cleft by nerve impulses, they are rapidly removed, thus keeping the signals distinct. Presynaptic neurons withdraw transmitters from the synapses and recycle them, and they are degraded by enzymes on both pre- and postsynaptic membranes. The synapses use catecholamine, serotonin, or amino acid neurotransmitters, and the enzymes are monoamine oxidase and catechol-*O*-methyl transferase.

The importance of specific neurotransmitters is illustrated tragically by the degenerative diseases in humans that result from an inability to make these compounds. Huntington chorea, for instance, is an inherited disease that appears in middle age and is characterized by uncontrollable, spasmodic twisting and thrusting motions. These symptoms occur due to the degeneration of the corpus striatum, a center that normally controls relatively automatic movements through its inhibitory neurons, which use the inhibitory transmitter γ-aminobutyric acid (GABA). About one-third of the synapses in the CNS use GABA, and GABA deficiency is associated with Huntington chorea. The Purkinje cells of the cerebellum, which regulate motor activity by inhibiting other neurons in the cerebellum and brain stem, are critical GABAergic neurons.

Other tracts of the motor system use dopamine as a transmitter. Dopaminergic neurons arise in the substantia nigra and project into the corpus striatum (Figure 42.22). Parkinson disease, characterized by muscular paralysis and tremors, occurs with the death of cells in the substantia nigra. Patients have difficulty initiating movements and making slow, directed movements from one large region to another (such as moving your hand to the desktop and back to your mouth). These movements are normally controlled by the striatum cells, directed by the substantia nigra cells. Parkinson patients are treated with L-DOPA, a precursor of dopamine, which relaxes their muscles temporarily. Surgeons have experimented with implanting embryonic tissue that produces its own dopamine into the brains of Parkinson patients. Other patients have been helped by a delicate procedure that destroys the globus pallidus.

Even more tantalizing is an indication of the role of dopaminergic neurons in the limbic system, a complex of structures and pathways closely tied to emotional behaviors. Evidence indicates that the disorders known collectively as schizophrenia arise from faulty regulation of dopamine metabolism in some key pathways. Schizophrenic illnesses are characterized by extreme emotional withdrawal, loss of contact with reality, hallucinations, and deterioration in social behavior. These conditions can be controlled by drugs such as chlorpromazine and haloperidol that block dopamine receptors. On the other hand, amphetamine, whose effects mimic schizophrenia, enhances the release of dopamine from its synaptic vesicles and inhibits its uptake back into presynaptic neurons. Thus a contributing factor in schizophrenia could be an excess of dopamine. Compared to normal people, schizophrenics have lower levels of monoamine oxidase, which oxidizes dopamine, higher levels of dopamine, and more dopamine receptors, especially in their limbic systems. Schizophrenia has a clear genetic component, but it isn't inherited as a simple Mendelian characteristic; various chemical deficiencies can probably lead to similar symptoms.

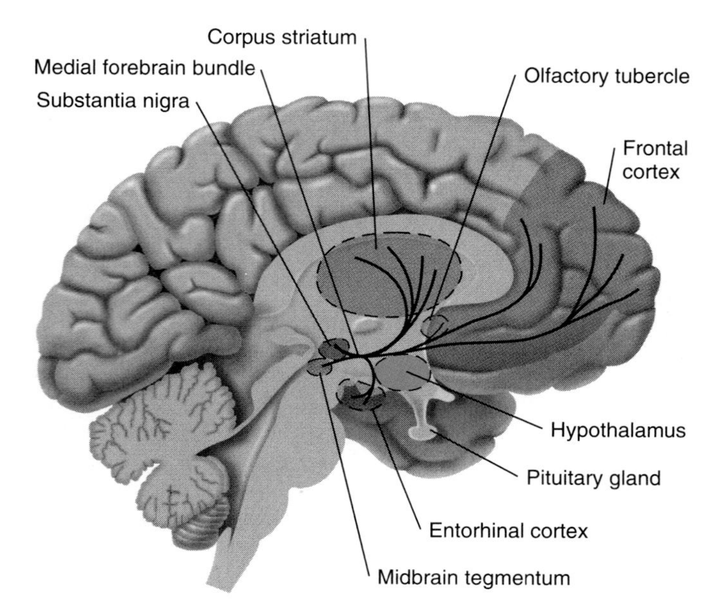

Figure 42.22
The principal dopaminergic pathways in the human brain center on the substantia nigra and basal nuclei, which are involved in motor activity.

42.11 Many peptides have roles in the nervous system.

An ever-growing list of small peptides have important roles in the nervous system. Known as **neuropeptides,** they include substances that have known functions elsewhere (Table 42.3). Cholecystokinin (CCK), for example, is a hormone in the digestive system (see Section 47.8). The cerebrum contains large amounts of a smaller peptide, CCK-8, made only of the critical eight amino acids required for CCK activity; CCK-8 acts as a neurotransmitter, and injecting it into the brains of some mammals causes them to stop eating, indicating that CCK-8 inhibits an appetite center. Angiotensin II, a hormone involved in regulating blood pressure, also occurs in parts of the thalamus and hypothalamus. Small amounts of angiotensin II, when injected into the third ventricle, induce an animal to drink and also cause secretion of vasopressin (ADH), another small peptide secreted by the hypothalamus through the posterior pituitary gland. The hypothalamus releases vasopressin if its interstitial fluid becomes more concentrated than normal (has a higher osmolarity than normal), and angiotensin II may be the neurotransmitter in the neurons that stimulate this event. These bits

Table 42.3 Structures of Principal Neuropeptides

Enkephalins	Tyr-Gly-Gly-Phe-Met Tyr-Gly-Gly-Phe-Leu
β-Endorphin	Tyr-Gly-Gly-Phe-Met-Thr-Ser-Glu-Lys-Ser-Gln-Thr-Pro-Leu-Val-Thr-Leu-Phe-Lys-Asn-Ala-Ile-Val-Lys-Asn-Ala-His-Lys-Lys-Gly-Glu
ACTH (corticotropin)	Ser-Tyr-Ser-Met-Glu-His-Phe-Arg-Tyr-Gly-Lys-Pro-Val-Gly-Lys-Lys-Arg-Arg-Pro-Val-Lys-Val-Tyr-Pro-Asp-Gly-Ala-Glu-Asp-Glu-Leu-Ala-Glu-Ala-Phe-Pro-Leu-Glu-Phe
Angiotensin II	Asp-Arg-Val-Tyr-Ile-His-Pro-Phe-NH_2
Oxytocin	Ile-Tyr-Cys Gln-Asn-Cys-Pro-Leu-Gly-NH_2
Vasopressin	Phe-Tyr-Cys Gln-Asn-Cys-Pro-Arg-Gly-NH_2
Thyrotropin releasing hormone (TRH)	(Pyro)Glu-His-Pro-NH_2
Gonadotropin releasing hormone (GnRH)	(Pyro)Glu-His-Trp-Ser-Tyr-Gly-Leu-Arg-Pro-Gly-NH_2
Tachykinins:	
Substance P	Arg-Pro-Lys-Pro-Gln-Gln-Phe-Phe-Gly-Leu-Met-NH_2
Substance K	His-Lys-Thr-Asp-Ser-Phe-Val-Gly-Leu-Met-NH_2
Neuromedin	Asp-Met-His-Asp-Phe-Phe-Val-Gly-Leu-Met-NH_2
Neurotensin	Glu-Leu-Tyr-Glu-Asn-Lys-Pro-Arg-Arg-Pro-Tyr-Ile-Leu
Somatostatin	Ala-Gly-Cys-Lys-Asn-Phe-Phe-Trp- Cys-Ser-Thr-Phe-Thr-Lys
Vasoactive intestinal polypeptide (VIP)	His-Ser-Asp-Ala-Val-Phe-Thr-Asp-Asn-Tyr-Thr-Arg-Leu-Arg-Lys-Gln-Met-Ala-Val-Lys-Lys-Tyr-Leu-Asn-Ser-Ile-Leu-Asn-NH_2
Cholecystokinin-8	Asp-Tyr-Met-Gly-Trp-Met-Asp-Phe-NH_2

of information show that neuropeptides may not simply act as broad-function neurotransmitters, such as norepinephrine, but may convey very specific messages.

Now it is becoming clear that neuropeptides may be of even greater importance outside the CNS. Vasoactive intestinal peptide (VIP), for instance, is found in many places in the body, especially in the enteric nervous system discussed in Section 47.7. Of perhaps greater interest, neuropeptides are being identified as mediators between the nervous and immune systems. Many of the lymphoid cells discussed at length in Chapter 48, such as lymphocytes and macrophages, have receptors for VIP, substance P, enkephalins, and other peptides, and these cells also produce many of the same peptides themselves. These connections are the subject of intensive research in many laboratories, but it is still too early to sort out their meaning. As we will see in Chapter 48, immune functions are strongly influenced both neurally and hormonally, and the many neuropeptides appear to be important in these connections.

Exercise 42.7 We will see in Chapter 46 that vasopressin causes the kidneys to reduce their output of urine. Given this information and the information in Section 42.11, rationalize the actions of angiotensin II.

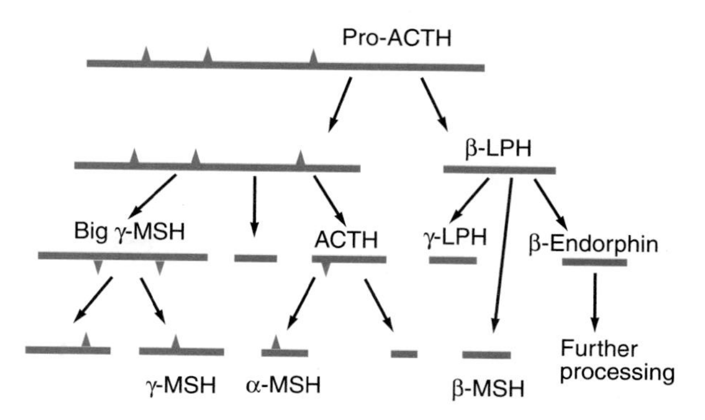

Figure 42.23
The protein pro-ACTH (pro-adrenocorticotropic hormone) is cleaved at several points to yield several neuropeptides, including functional ACTH and three types of melanocyte stimulating hormone (MSH). Points represent added sugar residues; LPH = lipotropin. ACTH induces secretion of hormones (glucocorticoids) from the adrenal cortex, particularly during stress. An excess of ACTH and glucocorticoids produces delusions of power and separation from reality, like one form of schizophrenia. In fishes and amphibians, MSH regulates melanocytes in the skin to lighten or darken skin color, but a function for MSH in mammals is not clear. Both ACTH and MSH promote learning in rats.

42.12 Some neuropeptides are internal opiates.

Studies of the neuropeptides called endorphins and enkephalins, which are natural analgesics (painkillers) in the nervous system, have generated great excitement. For hundreds of years, some of our most important analgesic drugs have been opiates, such as morphine, heroin, and codeine, all derived from the opium poppy. They are effective even in fishes. Opiates must bind to receptor proteins, but the vertebrate nervous system hasn't been evolving such proteins for hundreds of millions of years just on the off chance that some day someone would inject it with poppy juice. There must be natural ligands—now identified as the endorphins and enkephalins—that bind to opiate receptors, and the function of these receptors is to block the sensation of pain.

In 1973 Candace Pert and Solomon Snyder identified receptor proteins that bind labeled opiates and are concentrated in areas of the brain associated with pain perception. After it was shown that specific opiate receptors exist, several investigators began to search for natural opiate ligands in brain tissue, and in 1975 John Hughes and Hans Kosterlitz identified the first two of these compounds, which they called enkephalins. They are short peptides with morphinelike effects. A larger peptide, α-endorphin, also acts like an opiate and is concentrated in the pituitary gland. Like the egg-laying hormone of *Aplysia,* these small neuropeptides are cut out of polyproteins as shown in Figure 42.23.

β-endorphin produces analgesia and tranquility. Physicians who have experimented with small amounts of β-endorphin to treat neurological and psychiatric problems report moderate to high success. After receiving doses of the peptide, patients with depression, schizophrenia, and other disorders experience a release from their symptoms, improved moods, and more normal behavior. These experiments point toward a hopeful jfuture when many kinds of mental illness might be controlled chemically, using peptide hormones produced inexpensively from cloned genes.

Exercise 42.8 What is the functional value of making several hormones and neurotransmitters as segments of a polyprotein and then cutting it into functional peptides?

42.13 Learning may involve changes in ion channel proteins, sometimes effected by second messengers.

What happens in the nervous system when we learn—and how can we make it happen more easily and surely? These are some of the most interesting questions in neurophysiology, especially for students who are trying to learn a great deal in a short time. Although there are no good explanations yet for complex learning, some important insights have developed into the events of simple learning. Remember that the gill-withdrawal reflex of *Aplysia* uses a simple neural pathway (see Figure 42.6) from touch receptors on the siphon skin. Repeated stimulation of the skin, however, leads to *habituation*—that is, the animal stops responding. The effect is localized in the ends of sensory neurons (Figure 42.24), where the amount of neurotransmitter released depends on the activity of voltage-gated Ca^{2+}-ion channels.

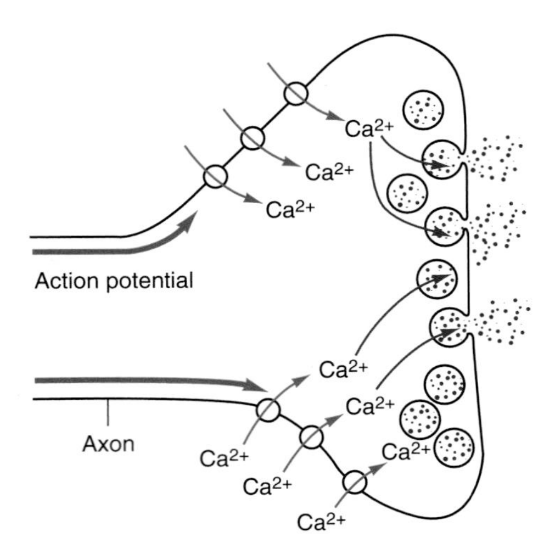

Figure 42.24
When an action potential reaches the terminus of an axon, it stimulates voltage-gated Ca^{2+}-ion channels to open, and this, in turn, stimulates the release of neurotransmitters. In habituation, the calcium-ion channels shut down.

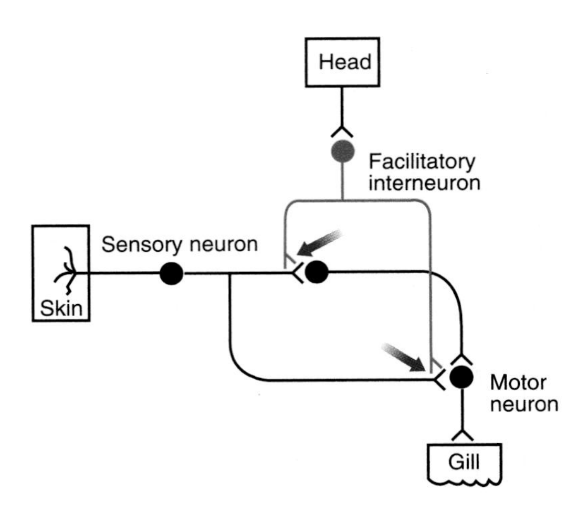

Figure 42.25
The gill-withdrawal reflex of *Aplysia* is accentuated by stimuli from the head through facilitatory interneurons *(red).* These neurons form axo-axonal synapses *(blue arrows)* on the sensory neurons, thus stimulating the motor neurons more strongly than usual.

With repeated stimulation, these channels become inactivated, causing less neurotransmitter to be released and thereby reducing the excitation of the postsynaptic motor neurons. This is a simple mechanism for the simplest kind of learning.

The opposite of habituation is *sensitization.* An animal may be sensitized to a noxious or harmful stimulus, so it pays more attention to it than usual. Sensitization in the gill-withdrawal reflex depends on facilitatory interneurons, which receive stimuli from the head; their axons end on the axon endings of the sensory neurons involved in the gill-withdrawal reflex (forming so-called axo-axonal synapses; Figure 42.25). Stimulation of these neurons facilitates the reflex by causing the sensory neurons to release even more neurotransmitter than usual on the motor neurons.

What are the facilitating neurons doing to these sensory neurons? They operate through that remarkable second messenger, cyclic AMP. The neurotransmitter released by the facilitatory neurons activates adenylate cyclase to make cAMP in the axon endings of the sensory neurons (Figure 42.26). Cyclic AMP activates protein kinases. Among the proteins phosphorylated by these kinases are certain membrane proteins, especially a K^+-channel protein that usually allows K^+ ions to leave the axon right after an action potential. Phosphorylation inactivates these channels, causing action potentials to last longer. Therefore more Ca^{2+} ions enter the axon through the voltage-gated Ca^{2+} channels, and more neurotransmitter is released into the synaptic cleft.

Although we have very little hard information about the basis for learning and memory in more complex animals, it is plausible that they involve the facilitation of certain neuronal circuits. When mammals are learning, a lot of activity occurs in the hippocampus, and much of it may involve simple changes of the kind that occur during facilitation in *Aplysia.* Long-term memory may also involve the formation of new connections

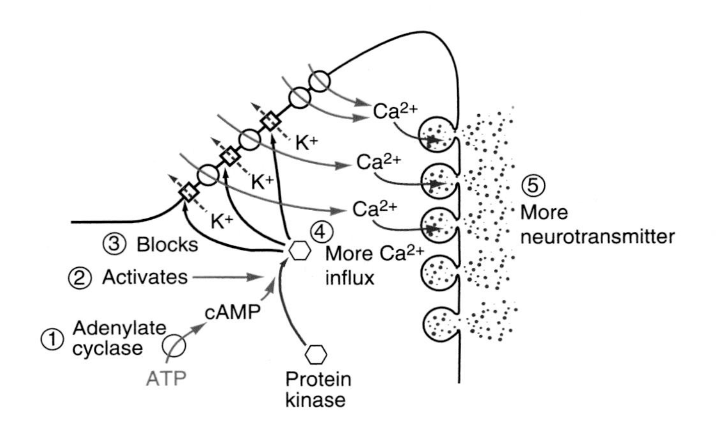

Figure 42.26
(1) Activation of adenylate cyclase in an axon ending produces cyclic AMP (cAMP), which activates protein kinases *(2).* The kinases phosphorylate a K^+-channel protein *(3);* this prolongs action potentials in the axon, allows more Ca^{2+} ions to enter the axon *(4),* and causes more neurotransmitter to be released to stimulate the motor neuron *(5).*

among neurons and perhaps changes in the supporting glial cells that surround the neurons. These are problems for future research. As for ways to make learning happen more easily, the answers presently lie in psychology, not in biology. The principal lesson is that learning happens actively, through the effort of writing, conversing, and otherwise playing with ideas and information aggressively; it happens much less effectively through passive activities such as listening to lectures.

Coda Nervous systems have evolved as centers that collect information about an animal's environment and internal conditions, integrate it, and create functional responses to changes. The resulting reactions and behaviors

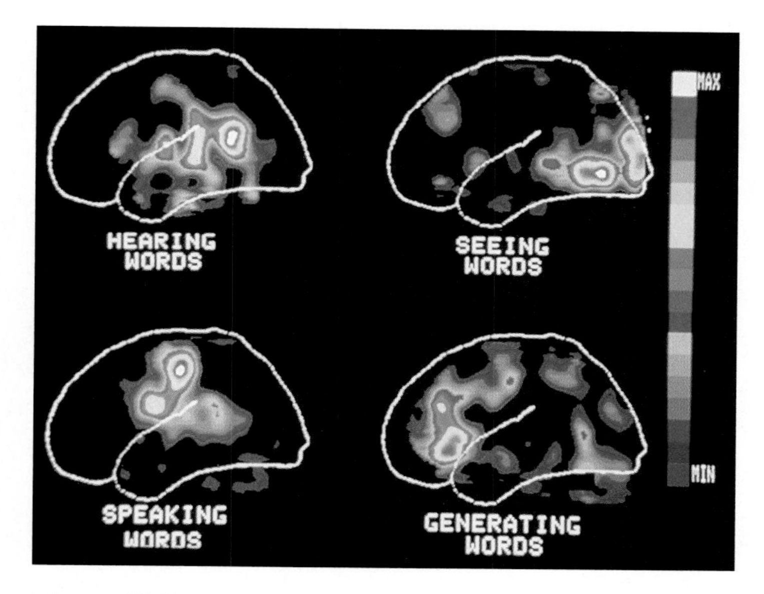

Figure 42.27

In positron-emission tomography (PET), radioisotopes that emit positrons (positive electrons) are injected into the circulatory system. When a positron collides with an electron, they annihilate each other and produce gamma rays, which are detected by a computerized analyzer. Artificially colored computer images show where the blood flow is highest.

maintain homeostasis more surely, make an animal more likely to survive, and increase its fitness. Even a simple reflex arc allows an animal to respond quickly and functionally to a stimulus, and some animals with minimal central nervous systems survive with behaviors of this kind alone. Evolution, however, is opportunistic. Opportunities constantly arise for evolving more complex, more adaptive actions and for developing senses that collect more sophisticated information about the world. Over time, various groups of animals have been able to take advantage of these genetic opportunities. The members of the animal kingdom have signed no pact agreeing to limit their "arms" to reflex actions, and the battles of competition have selected incredibly complex behaviors and the brain structures that underlie them. In later chapters, we explore some of the fascinating, complicated behavior patterns encoded in animals' central nervous systems.

Understanding brain functions is one of the greatest remaining challenges in biology. We know a little about what happens as sensory information is processed and how glands and muscles are activated in response to sensory inputs. Investigators continue to locate brain activities through new techniques such as *positron-emission tomography* (Figure 42.27), which shows changes in the pattern of blood flow and thus locates spots where neurons are carrying out specific functions. Yet we do not understand how even the simplest nervous system integrates all the information coming to it, much less how, in humans (and presumably in some other animals as well), the whole system formulates meaningful images of the world, becomes conscious of itself, and initiates actions on the basis of the events we call thoughts. We are currently far from answering these questions, but based on the progress made in answering other biological questions in the past few decades, there is good reason to be optimistic about the future in this field as well.

Summary

1. Cnidarians have only simple nerve nets that carry sensory information and can integrate their movements.
2. In flatworms, the nervous system evolved toward cephalization and lost the netlike form. Most invertebrate nervous systems have ganglia, often concentrated in the head, and ventral nerve cords connecting other ganglia.
3. The marine snail *Aplysia* has become a model system for neurophysiological research. In such a simple nervous system, it is becoming possible to identify the specific neurons and neurotransmitters that perform particular functions. By detailed analysis of relatively simple nervous systems, including recombinant-DNA methods, it is possible to show how an apparently complex behavior is produced by a series of specific cells and signal ligands.
4. The vertebrate nervous system develops from a tube of tissue and remains organized around a central canal filled with fluid. The brain is conventionally divided into fore-, mid-, and hindbrain regions. The forebrain develops into the cerebrum, which dominates the brain of mammals, and the diencephalon, including the thalamus and hypothalamus. The thalamus includes centers that relay signals to the cerebrum. The hypothalamus contains centers that control homeostatic functions, such as water balance. The hindbrain includes the cerebellum and the medulla oblongata, which contains the centers for several autonomically controlled functions.
5. The cerebrum, especially in mammals, receives and integrates signals from many lower brain centers. It contains areas that analyze sensory inputs and initiate motor activities. The cerebral cortex can be mapped into several functional areas. A strip of cortex at the front of the parietal lobe receives somesthetic signals, with each body part mapped onto a specific cortical area. Another strip of cortex just anterior to this region controls motor activities.
6. Experiments with humans whose brains have been severed through the corpus callosum show that some functions are confined to one cerebral hemisphere or the other. Centers that control speech are on the left side, while the right side has a major center that carries manipulospatial abilities—the abilities to manipulate and draw objects and to know how they are oriented in space.
7. Motor activities involve a pyramidal system, which sends signals directly to motor neurons in the spinal cord, and an extrapyramidal system, involving the cerebellum, basal nuclei, and other centers in the cerebrum. Movements are controlled through a series of feedback mechanisms primarily involving the cerebral cortex, cerebellum, and basal nuclei.
8. Specific brain tracts use distinctive neurotransmitters, some excitatory and some inhibitory. Certain diseases, including schizophrenias and motor disorders, are associated with excesses or deficiencies of some of these neurotransmitters or the neurons that employ them.
9. Many short peptides are now found to be neurotransmitters, including many that have other functions elsewhere in the body. The function of some is to block pain. The balance of these neurotransmitters may be critical in maintaining normal functions, and some abnormalities might be corrected by restoring that balance.
10. Long-term changes in the nervous system, including learning, result from activating or inhibiting ion channels, sometimes through cyclic AMP.

Key Terms

nerve net 891
commissure 892
forebrain 895
midbrain 895
hindbrain 895
cerebellum 895
pons 895
medulla oblongata 895
brain stem 895
cerebral hemispheres 895
ventricle 895
cerebrospinal fluid (CSF) 896
meninges 896
gray matter 896
white matter 896
neuroglia 896
reticular formation 897
melatonin 898
thalamus 898
gyrus 899
sulcus 899
somesthetic 900
motor cortex 900
corpus callosum 900
manipulospatiality 902
active touch 902
pyramidal/extrapyramidal motor systems 904
neuropeptide 906

Multiple-Choice Questions

1. Which of these phyla was the first to develop a brainlike structure composed of an anterior collection of neuron cell bodies?
   a. molluscs
   b. cnidarians
   c. flatworms
   d. annelid worms
   e. arthropods
2. Complex invertebrates such as annelids, arthropods, and molluscs have nervous systems that include all of these features except
   a. a central nervous system.
   b. a ventral nerve cord.
   c. the ability to form conditioned reflexes.
   d. the ability to learn.
   e. a nerve net.
3. The entire central nervous system of vertebrates
   a. is hollow.
   b. lies ventral to the gut.
   c. is formed from embryonic mesoderm.
   d. consists of a brain and nerves.
   e. is made up of the brain stem and spinal cord.
4. The hypothalamus is a part of the
   a. brain stem.
   b. myelencephalon.
   c. telencephalon.
   d. diencephalon.
   e. cerebellum.
5. Cerebrospinal fluid circulates
   a. in the central canal of the spinal cord.
   b. in the ventricles of the brain.
   c. peripheral to the brain and spinal cord.
   d. *a* and *b*, but not *c*.
   e. *a*, *b*, and *c*.
6. Neuroglia and Schwann cells both produce
   a. gray matter.
   b. myelin.
   c. CSF.
   d. neurotransmitters.
   e. meninges.
7. Which best describes the spatial relationship between the thalami, the hypothalamus, and the third ventricle?
   a. The ventricle is dorsal to the thalami and ventral to the hypothalamus.
   b. The thalami and hypothalamus are surrounded by the third ventricle.
   c. The third ventricle is surrounded by the thalami laterally and by the hypothalamus ventrally.
   d. All regions of the diencephalon are dorsal to the third ventricle.
   e. The thalami and hypothalamus are not near the third ventricle.
8. Which pair is mismatched?
   a. temporal lobe—hearing
   b. occipital lobe—smell
   c. frontal lobe—voluntary motor cortex
   d. parietal lobe—somesthetic cortex
   e. prefrontal lobe—expression of emotion
9. In the following list of statements, all are correct *except:*
   a. Some neuropeptides function as neurotransmitters.
   b. Neuropeptides have functions other than acting as neurotransmitters.
   c. Neuropeptides are generally cut from polyproteins.
   d. All neurotransmitters are neuropeptides.
   e. Neuropeptides are believed to create functional links between the nervous, endocrine, and immune systems.
10. Habituation is caused by
    a. repeating an activity until it becomes a habit.
    b. sensitization.
    c. activation of adenylate cyclase and production of cAMP.
    d. inactivation of voltage-gated calcium channels.
    e. muscle paralysis.

True-False Questions

Mark each statement true or false, and if false, restate it to make it true.

1. One general trend among multicellular animals has been the shift from a more central nervous system to a more peripheral system.
2. The entire central nervous system of vertebrates lies ventral to the gut.
3. Simple invertebrates, such as cnidarians, have nerve nets, whereas complex invertebrates and vertebrates have a central nervous system.
4. The ventricles of the brain are an extension of the central canal of the spinal cord.
5. The spinothalamic tract carries nerve impulses from the brain to the spinal cord.
6. The hypothalamus integrates sensory input, while the right and left thalami produce hormones that are secreted by the posterior pituitary gland.
7. All cranial nerves that are called mixed nerves carry both sensory and motor information.
8. The somesthetic cortex connects left and right cerebral hemispheres.
9. Parkinson disease results from death of cerebellar neurons that produce GABA.
10. CCK and VIP can bind to opiate receptors and act to mediate pain.

Concept Questions

1. Why would a nerve net be an inadequate system of communication in large, multicellular animals such as ourselves?
2. What is the origin and function of cerebrospinal fluid?
3. A young woman had been comatose for several years before she died of a respiratory infection. Autopsy of her brain revealed little or no damage to her cerebral hemispheres, but extensive damage to right and left thalami. Explain how thalamic damage could lead to a state of unresponsiveness.
4. Many pesticides inhibit the enzyme monoamine oxidase. Similar effects occur in both insects and humans, although insects, because they are much smaller, are affected more rapidly and severely. Why

should monoamine oxidase inhibitors have effects on the nervous system?

5. Suppose that a particular neurological disease results from oversecretion of one specific neurotransmitter. Describe three possible strategies that might relieve the symptoms of the disease.

Additional Reading

Caldwell, Mark. "Kernel of Fear." *Discover,* June, 1995, p. 96. Research has shown that the link between certain stimuli and fear responses, from common anxiety to post-traumatic stress syndrome, lies in a small brain structure called the amygdala.

Calvin, William H. *The Throwing Madonna: Essays on the Brain.* McGraw-Hill Book Co., New York, 1989.

———*Conversations with Neil's Brain: The Neural Nature of Thought and Language.* Addison-Wesley Publishing Co., Reading, MA, 1994.

Fine, Alan. "Transplantation in the Central Nervous System." *Scientific American,* August 1986, p. 52. Transplanted embryonic neurons can establish functional connections in the adult brain and spinal cord, long believed to be immutable in mammals. Such grafts might reverse damage from disease or injury.

Llinas, Rodolfo R. (ed.). *The Workings of the Brain: Development, Memory, and Perception:* Readings from *Scientific American Magazine.* W. H. Freeman, New York, 1990.

Mishkin, Mortimer, and Tim Appenzeller. "The Anatomy of Memory." *Scientific American,* June 1987, p. 80. An inquiry into the roots of human amnesia has shown how deep structures in the brain may interact with perceptual pathways in outer brain layers to transform sensory stimuli into memories.

Poggio, Tomaso, and Christof Koch. "Synapses that Compute Motion." *Scientific American,* May 1987, p. 46. How do nerve cells process the information they receive from the environment? Studies of cells in the eye that interpret movement may define a mechanism involved in many other neural operations.

Sacks, Oliver. *The Man Who Mistook His Wife for a Hat.* HarperCollins, New York, 1985.

——— *An Anthropologist on Mars.* Alfred A. Knopf, New York, 1995. Sacks tells wonderful stories about people with neurological disorders, the autistic, and others of interest to neurobiology.

Shreeve, James. "The Brain That Misplaced Its Body." *Discover,* May 1995, p. 82. The condition called anosognosia usually occurs in stroke victims who have suffered damage to the right parietal cortex.

Streit, Wolfgang J., and Carol A. Kincaid-Colton. "The Brain's Immune System." *Scientific American,* November 1995, p. 38. The role of microglia in protecting the brain against damage.

Internet Resource

To further explore the content of this chapter, log on to the web site at:

http://www.mhhe.com/biosci/genbio/guttman/

43 Sensory Receptors and Perception

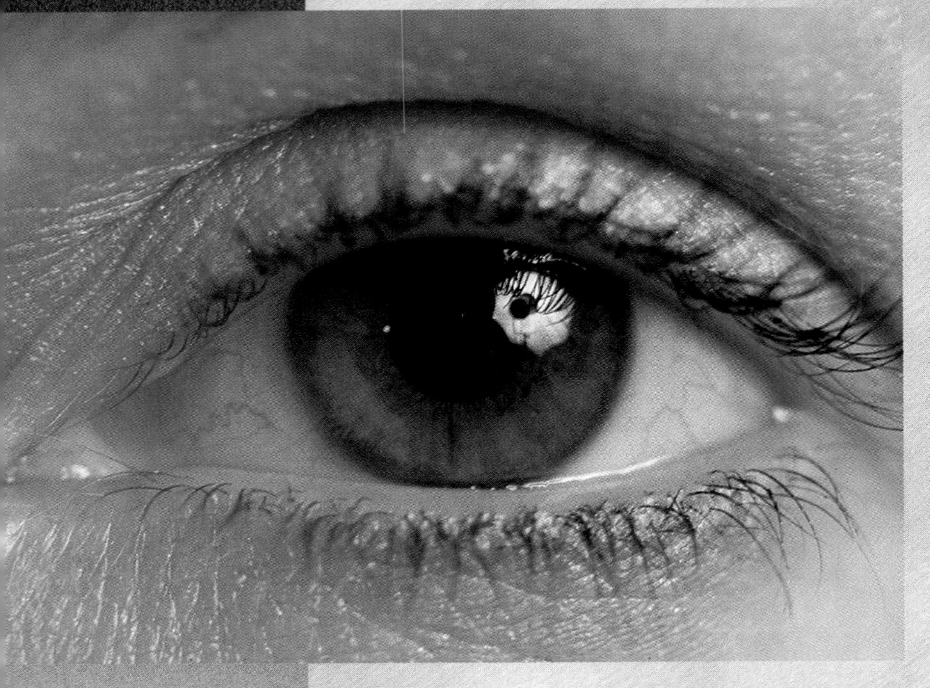

The eye is our principal window on the world. How closely does the image it produces correspond to reality?

Key Concepts

43.1 Receptors transduce, encode, and amplify.
43.2 Receptors respond to changes in the environment.
43.3 Chemoreceptors are basic and widely distributed.
43.4 Mechanoreceptors respond to tensions and pressures.
43.5 Many mechanoreceptors employ hair cells.
43.6 Hair cells in the ear detect sound vibrations.
43.7 Some thermoreceptors detect infrared radiation.
43.8 Photopigments in specialized membranes absorb light.
43.9 The vertebrate eye focuses light on the retina, a layer of receptor cells.
43.10 Analysis of information from visual receptors begins in the retina.
43.11 Vision depends on a hierarchy of cells in the visual cortex.

Each of us lives in a body closed off from the world by a tough skin. Like a prisoner in a cell with tiny windows, all we know about the world is what is revealed by a few specialized "receptors." These receptors have evolved to detect vibrations (sound), touch and pressure, heat or cold, a wide range of chemicals, and the part of the electromagnetic spectrum we call light. We wouldn't even know about the electromagnetic energy of longer and shorter wavelengths if people had not built instruments to detect it. Insects can see into the ultraviolet part of the spectrum, and they see the world differently. Some fishes can detect changes in local electrical fields that they create. We can't. Are we unaware of other forces and phenomena because our nervous systems never evolved receptors for them and we haven't yet devised instruments to find them? And how much does the structure of our sensory receptors distort the world we perceive? Questions like these, which have long troubled philosophers, should also trouble biologists.

Some things we perceive are subject to systematic distortions, as in the well-known visual illusions depicted in Figure 43.1. Having evolved in a world with certain regularities, the nervous system is designed to make certain assumptions. We have to distinguish *sensation,* the act of receiving raw sensory information at receptors, from *perception,* the act of converting

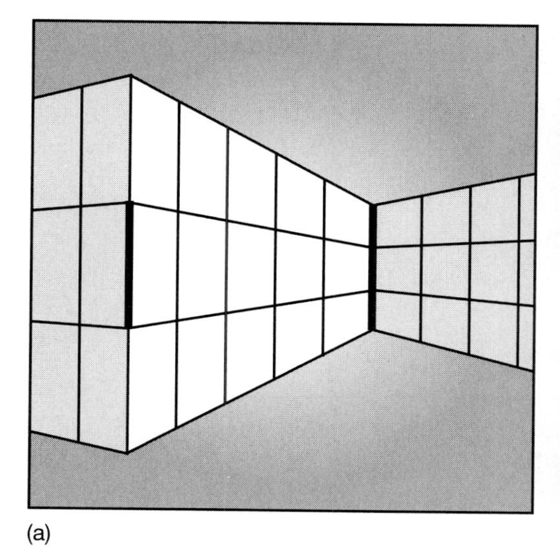

(a)

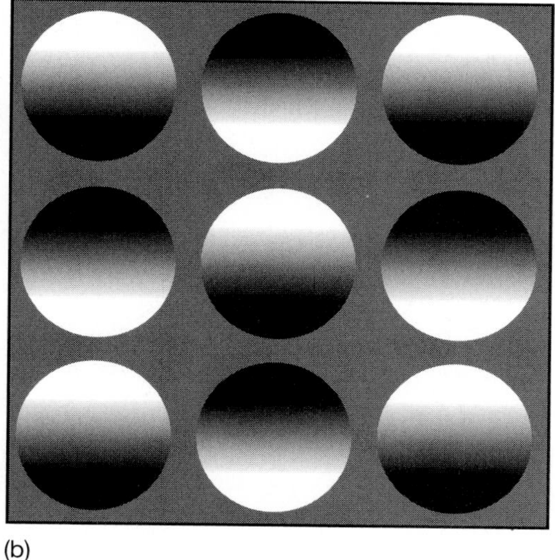

(b)

Figure 43.1

Two visual illusions reveal the kinds of assumptions about the structure of the world that are built into our sensory systems. *(a)* Objects that appear to be at a distance are interpreted as larger than those nearby; the two heavy lines are actually the same length. *(b)* Shading creates the illusion of objects being concave or convex because of the built-in assumption that light comes from above; rotate the page and notice how your perception of the figure changes. Source: Maya Pines, "Our Common Senses" in *Seeing, Hearing and Smelling the World,* Howard Hughes Medical Institute, 1995.

that information into conceptions of reality. In this chapter, we are primarily concerned with sensation, but we will also deal with some aspects of perception in discussing vision, where our knowledge is greatest. ■

43.1 Receptors transduce, encode, and amplify.

Animals have specialized receptors to detect information about their surroundings from several possible sources: **Chemoreceptors** respond to particular chemical structures, **photoreceptors** to light, **thermoreceptors** to heat (including infrared radiation), **mechanoreceptors** to mechanical changes such as touch, pressure, or sound waves, and **electroreceptors,** which only certain fishes possess, to changes in electric currents. Each receptor translates, or *transduces,* the stimulus into a change in membrane potential. At some point, this change is converted into a common language of nerve impulses, the only language the nervous system can interpret (Figure 43.2).

The same receptors are also distinguished by the *locations* of the stimuli they receive. **Exteroceptors** such as those in the eyes and ears respond to stimuli outside the body. **Interoceptors** monitor internal factors such as blood glucose concentration and blood pressure. **Proprioceptors** are mechanoreceptors that detect the orientation of the body—the positions of arms and legs or the tension in a muscle.

All receptors respond to stimuli by developing a **generator potential,** a rise in membrane potential similar to an EPSP. This is always a *graded* potential proportional to the energy of the stimulus impinging on the receptor. Some receptors are neurons themselves, and in these, the changes in membrane potential affect the frequency of action potentials the neurons send out along their axons (Figure 43.3). Other receptors are cells that synapse with neurons; their generator potentials determine the amount of neurotransmitter they release into the synaptic cleft, ultimately changing the frequency of action potentials produced by the postsynaptic cells. Thus, in both cases, the frequency of these impulses reflects the stimulus intensity, as reflected by the Law of Intensity Coding: *The frequency of nerve impulses from a receptor increases with the intensity of the stimulus.*

Stimuli may be very weak, but receptors and the cells they synapse with often amplify signals as they transduce them into a series of nerve impulses. Photoreceptors in the vertebrate eye, for example, can detect a single photon of light. The eye and brain, using their own stored energy, amplify this minute signal into nerve impulses with much greater energy.

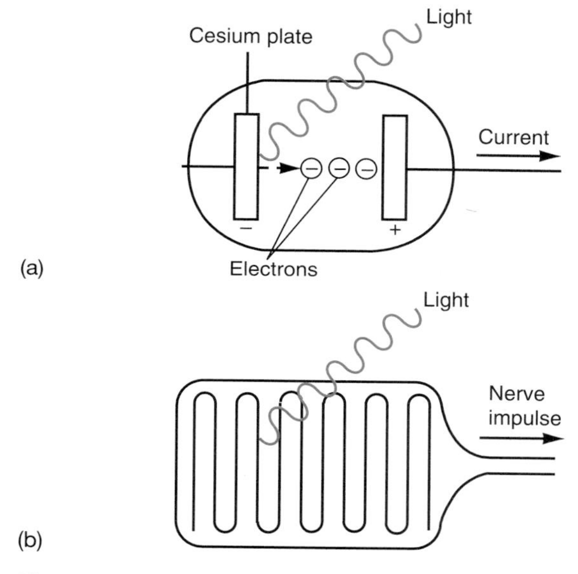

Figure 43.2

(a) A photocell absorbs light and converts some of its energy into an electric current. *(b)* A light-sensitive cell in an eye absorbs light and triggers nerve impulses in response. Both units are transducers.

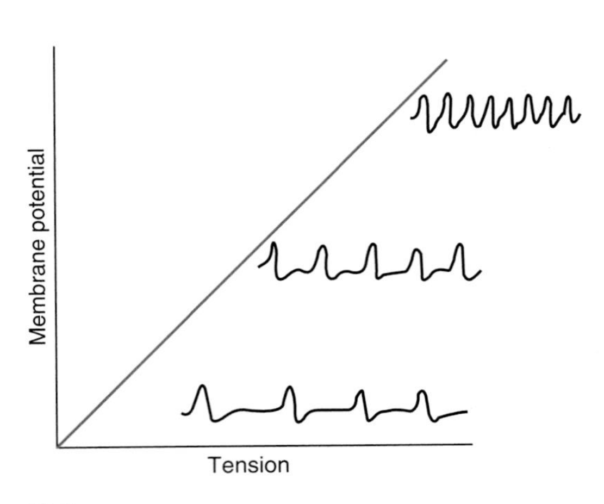

Figure 43.3
The potential generated by a tension receptor is proportional to the tension. The same relationship generally holds true for all kinds of receptors. The receptor sends out a series of nerve impulses, and their frequency increases with the potential on the receptor.

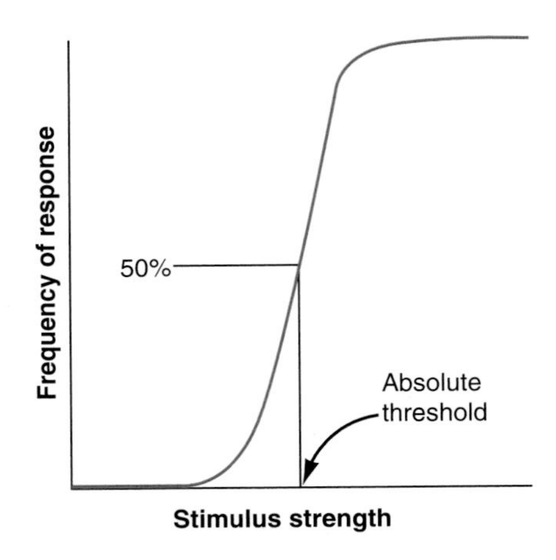

Figure 43.4
As the strength of a stimulus increases, it produces a response more often *(red curve).* The threshold is generally defined as the strength at which a response occurs half the time.

Exercise 43.1 *(a)* A fish has cells on its skin that detect light. Are they photoreceptors or exteroceptors? *(b)* You have cells in your arm that report its position and tension. Are they mechanoreceptors or proprioceptors?

Exercise 43.2 Some cells in your body monitor the level of Ca^{2+} ions in your blood; identify them with two kinds of designations.

Exercise 43.3 Cross your middle and index finger and then touch them at the crossing point with a pointer such as a pencil tip. How many points do you perceive? Why is it adaptive for your nervous system to be constructed in such a way that it can be deceived in this situation?

43.2 Receptors respond to changes in the environment.

A stimulus is a change in the environment that elicits a response, but not every change is a stimulus. Some touches are too light to feel, some sounds too soft to hear. Each sense organ has an **absolute threshold,** the minimum energy a stimulus must have, conventionally defined as the strength it must have to be detected half the time (Figure 43.4).

An animal is so inundated by possible stimuli that it must selectively ignore most of them. Its senses are set to take notice of a new stimulus, respond in some functional way, and then ignore the stimulus if it continues without changing. When you first put on your clothes in the morning, you feel their touch for a few seconds; later, you can make yourself aware of them if you want to, but you generally don't remain aware of their touch. Because your senses of touch must be ready to notice important new events, such as a touch that could mean danger, you ignore constant stimuli. Receptors of the **phasic** type produce action potentials for only a short time after being stimulated and then stop responding; this behavior is *sensory adaptation.* A **tonic** receptor, in contrast, continues to respond even to constant stimuli. If an animal no longer responds to stimuli arising from a tonic receptor, it is not undergoing adaptation but rather a more complex process of *habituation* through a change in the nervous system downstream of the receptor (see Figure 42.24).

How much *additional* stimulus strength is needed, over and above one already present, for a stimulus to be perceived? The **differential threshold** (ΔS) is the change in stimulus strength (S) that will be detected as a new stimulus (again, half the time). In 1834 Ernst Weber discovered a basic law of sensory physiology: Within limits, $\Delta S/S$ (the Weber fraction) is a constant for each sense. For example, if you are hefting similar weights, how much difference does there have to be between any two of them before you can detect it? The Weber fraction for this sense is about 1/50 (2 percent), so you can just tell the difference between a weight of 100 grams and one of 102 grams or 98 grams, because the difference is 2 grams out of 100, or 1 in 50. If you lift a weight of 300 grams, you would have to heft one of 306 grams to notice that it is heavier.

Exercise 43.4 The Weber fraction for the pitch of a sound at about 2,000 cycles is 1/333. What is the next highest frequency above 2,000 cycles that you could discriminate as a different pitch?

43.3 Chemoreceptors are basic and widely distributed.

Although we have often stressed the interactions between receptor proteins and ligands, external chemoreceptors deserve some special attention. Air-breathing animals like humans have distinct receptors and sensory pathways for two kinds of chemoreception: **olfaction,** or smell, and *taste.* Olfaction means the reception of many distinct molecules, while taste means the reception of a few general qualities such as sweetness or saltiness. What we call tasting is largely smelling, because some air

Sidebar 43.1 Human Pheromones and Little Pits in the Nostrils

Many mammals react strongly to chemicals produced by others of their species, which we have called *pheromones.* Some of the best-known pheromones convey a sexual message. A substance in the urine of male prairie voles prepares a female prairie vole for copulation: Her ovaries and uterus swell, she ovulates within two days, and is then ready to mate. Other female mammals react similarly to pheromones from the breath or skin of males, while pheromones from females raise the testosterone levels of males of the same species. However, the pheromones have little or no smell to the animals that react to them; instead, they stimulate a powerful reaction in a minute, little-known organ in the nose, the *vomeronasal organ (VNO).* It consists of two small pits in the nasal passage, one on each side of the central septum. Electrodes record strong signals from the VNO when it is exposed to the right pheromones, and animals that have had their VNOs destroyed do not react in their characteristic ways to their species's pheromones.

So much for mammals in general. But humans are mammals. Are our lives influenced by human pheromones, and do we have VNOs? In recent years, a few neurobiologists have become interested in these questions, and they now answer "yes" to both. David L. Berliner, for instance, has found that extracts from human skin cells, which have no odor, make people feel jovial, contented, confident, and social. So do humans have VNOs that detect these pheromones? For a long time it was believed that VNOs are present in human embryos but disappear in adults. Now, influenced by research on human pheromones, several anatomists have reexamined the question and found that we do have VNOs, one little pit on each side of the septum, about a half-inch up each nostril, and these pits contain recognizable, but unusual, neurons. Furthermore, recording nerve signals from the VNOs of volunteers show that they react powerfully to human pheromone extracts. The number of pheromones is still unknown. It is clear, however, that some produced by men are only detected by women, and vice versa.

Human pheromones probably account for the synchronization of menstrual cycles among women living together, and they are probably strong but subtle factors in people's attraction for one another. These discoveries raise important questions about the degree to which our behavior is influenced by chemical messages; they also raise ethical questions about producing pheromones commercially so people can use them to develop personal relationships.

normally rises from the back of the throat to the olfactory receptors while we chew food. That's why foods have little or no "taste" when you have a stuffy nose. Aquatic animals like fish also have different types of cells that correspond to taste and smell receptors and are connected to different brain centers. Fish have taste receptors all over their body surfaces, but concentrated on special organs such as the whiskers or barbels of catfish and their relatives.

Animals often have receptors specialized for certain pheromones, the chemical signals from others of their species, and stimulation of each receptor tends to induce a specific behavior pattern, such as moving toward the source of the pheromone. Steroids and other substances normally found in human sweat seem to be pheromones that have subtle effects on human behavior (Sidebar 43.1). More general chemoreceptors may also signal specific behaviors. For instance, sweetness indicates a source of energy, which is always desirable, and tends to attract animals and stimulate feeding; it is no wonder that modern sweets can present such a supernormal stimulus that people find it hard to resist them. Bitterness may warn of potentially harmful compounds to avoid, and sourness warns of acid—again, usually something to avoid. But the primitive functions of these taste receptors in ensuring the survival of our ancestors don't stop us from making and relishing foods with bitter or sour tastes. Some insects have specialized receptors for sugars and salt or for alkaloids made by plants as a warning to avoid those plants. At least one species of fly has a receptor for pure water, which detects the difference between water with and without dissolved ions.

Most mammalian taste receptors reside in taste buds on the thousands of lingual papillae on the tongue's surface (Figure 43.5*a*). The papillae tend to be specialized for each taste (Figure 43.5*b*), and each one may bear 200 or more taste buds, each made of several taste receptors. Fine microvilli extend from each receptor to the surface where they detect materials dissolved in the saliva. Only dissolved substances can be tasted. A compound is only detected when it binds to a receptor protein, and there is good evidence for the existence of distinct proteins that bind different kinds of ligands. Molecules that elicit one taste sensation, such as bitterness, presumably have common chemical features of shape and electrical charge that allow them all to bind to the same type of receptor protein.

Olfaction begins with receptor cells located in the olfactory mucosa—the surface layer of cells in the nose where odors are detected; these are bipolar neurons with cilia that form the sensory surface and axons that form the olfactory nerve leading to the brain (Figure 43.6). (Olfactory receptors are the only neurons known to regenerate; in humans, each one lasts for an average of 60 days and is then replaced by the differentiation of basal cells in the olfactory epithelium.) Our ability to distinguish many odors implies a much more complex system than the one that detects the four basic tastes. We don't know what the basic odor sensations are, and even our words for describing odors are most inadequate. Because odor is so important to certain industries, notably the perfume trade, there have been several attempts to devise a good vocabulary for describing an odor, but no one system is generally accepted. Several years ago, John A. Moore theorized that we have one type of chemoreceptor for each of seven basic odors, which he called camphoric, etherlike, floral, musky, pepperminty, pungent, and putrid. People who have a specific odor-blindness (anosmia), comparable to inherited color-blindness, provide evidence that specific

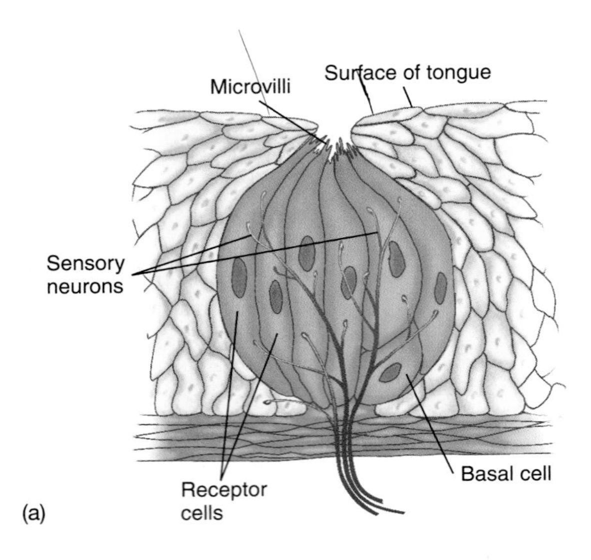

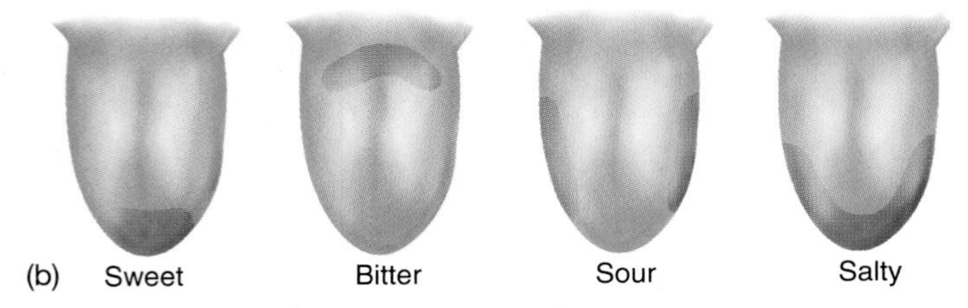

Figure 43.5

(a) A taste bud consists of receptor cells whose microvilli extend to the surface of the tongue. Receptor proteins reside on these microvilli. The ends of sensory neurons contact the receptor cells intimately. *(b)* The various taste modalities on the tongue are distributed in patches.

odor receptors exist. However, there appear to be many distinct receptors, not just seven. By searching for genes that encode proteins with certain common features of receptors, Richard Axel and Linda Buck have found evidence for at least 100 olfactory receptor proteins, which probably act through the common G-protein pathway. Current olfaction theory postulates that molecules of every *odorant*—a substance we can smell—bind stereospecifically to several proteins, each on a different olfactory receptor cell (Figure 43.7). Our sensation of a smell comes from the combination of signals from these receptors. The olfactory cortex, receiving their input, says, in effect, "Receptors 13, 20, 24, and 76 are all reporting, so this is cedarwood."

Exercise 43.5 Think about the structure of a small molecule and explain how it could bind to different kinds of receptors, as required by olfaction theory.

Exercise 43.6 Explain the argument that the existence of people with various anosmias shows that specific smell receptor proteins exist.

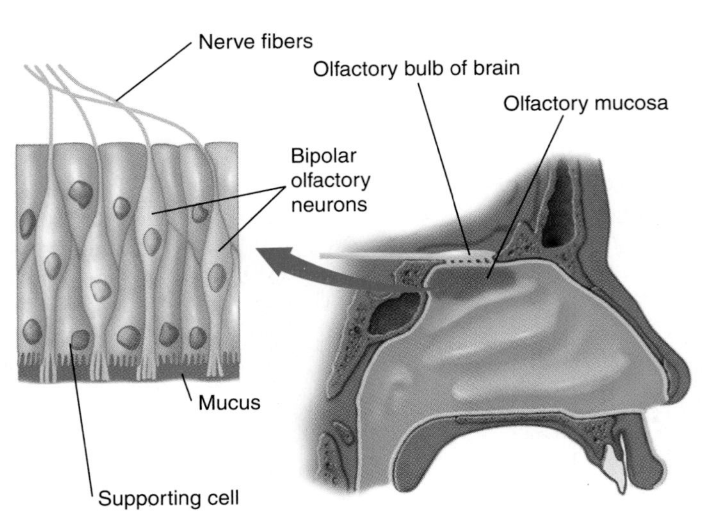

Figure 43.6

The sense of smell originates in bipolar neurons in the olfactory mucosa. These cells terminate in cilia that have the usual 9 + 2 microtubule structure at their bases but are not motile. Axons from these neurons project into the olfactory bulb.

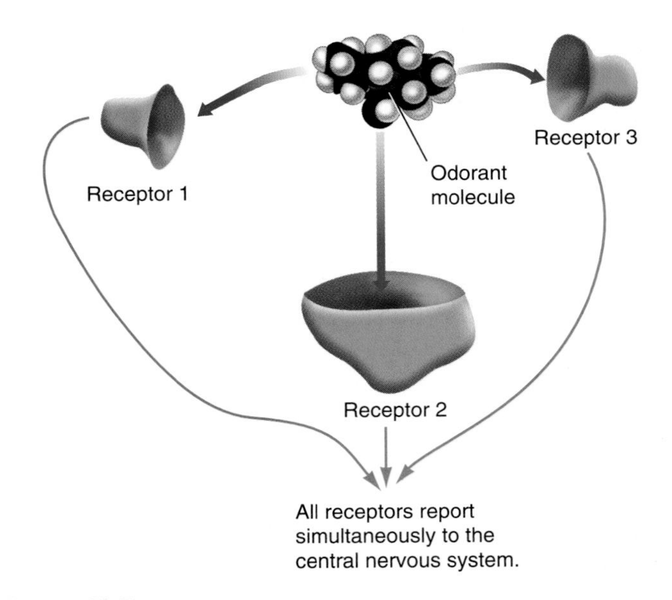

Figure 43.7

According to the stereochemical theory of odor reception, each odorant molecule excites cells by binding to stereospecific surface receptors. Different faces of a molecule have distinctive shapes, so the molecule can bind to, and stimulate, several receptors. Each receptor reports to the olfactory cortex, which combines their reports to form the sensation of a particular odor.

43.4 Mechanoreceptors respond to tensions and pressures.

Mechanoreceptors respond to movements. They are concentrated in the skin, especially in our hands and in particularly erotic areas such as the lips, nipples, and genitals. Some

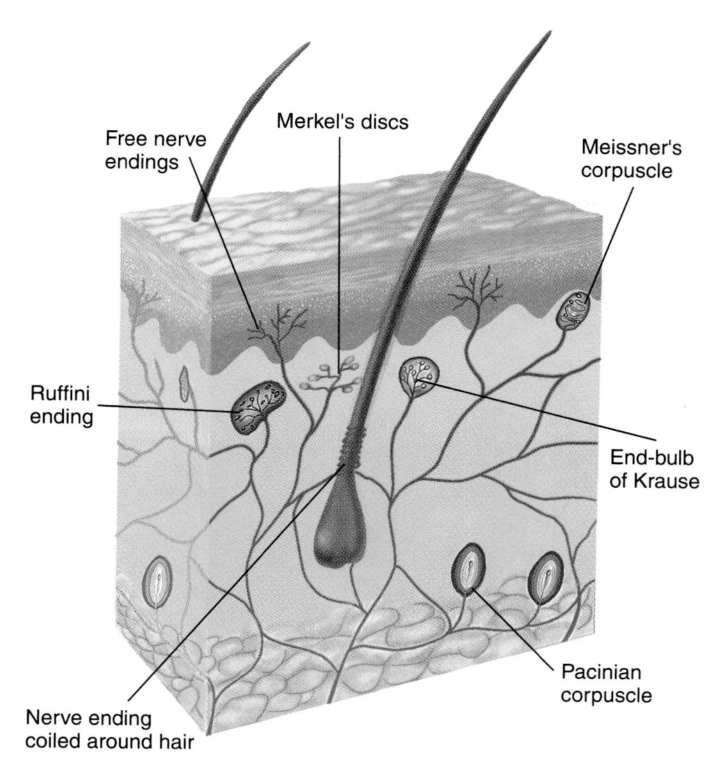

Figure 43.8
Hairs are sensitive detectors of movement, and nerve endings wrapped around bases form mechanoreceptors that detect deformations of the hair. The surrounding skin contains other special nerve endings, often named for the anatomists (Meissner, Ruffini, and others) who discovered them, which detect stimuli such as pressure and temperature.

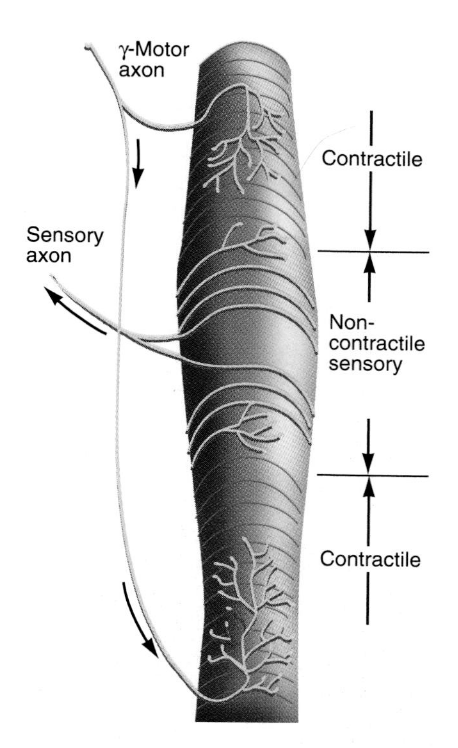

Figure 43.9
A muscle spindle consists of a specialized intrafusal muscle fiber that has contractile (striated) end regions and a non-contractile central region wrapped in the endings of sensory neurons. A neuron sends out signals when it is stretched, either by contraction of the striated portions of the spindle or by stretching of the surrounding muscle.

mechanoreceptors are just free nerve endings, the bare dendrites of sensory neurons, which may also detect heat and can produce the sensation of pain. Movements of hairs convey a lot of sensory information about events on or near the skin, because even the slight bending of a hair will stimulate nerve endings wrapped around its base and several other receptors located nearby (Figure 43.8). The skin is also rich in nerve endings surrounded by oddly shaped capsules of connective material, and many of these are mechanoreceptors. Pacinian corpuscles, for instance, are buried deep in the skin, especially near joints between bones, where they apparently detect deep touch and vibration, while Meissner's corpuscles and Merkel's discs detect lighter touches.

Proprioceptors are mechanoreceptors that provide essential information about ourselves (*proprius* = self) by constantly reporting on the positions and tensions of each part of the body. We depend on proprioception for balance and for knowing, without having to look, just how each muscle or finger is tensed and oriented. The Pacinian corpuscles just mentioned are an example of proprioreceptors. Information about muscle tension comes from *stretch receptors* called **muscle spindles** (Figure 43.9). Each spindle is built around distinctive muscle cells called *intrafusal fibers,* which are smaller than the extrafusal fibers that make the bulk of a muscle. The dendrites of a sensory neuron coil around the central section of each intrafusal fiber, and stretching the fiber sets off a series of nerve impulses. The more the fiber is stretched, the faster it fires, so a spindle continually reports the tension of the muscle around it.

Muscle spindles can set off **stretch (myotatic) reflexes,** such as the well-known knee-jerk reaction. When you sit with your leg hanging over the edge of a table, your leg muscles are set with a certain length and tension. Suddenly striking the knee tendon stretches spindles in the thigh muscles, causing sensory neurons in the muscle spindles to send signals quickly along sensory fibers and back through motor neurons to make the muscles contract, so your leg kicks (Figure 43.10). The kick allows the thigh muscles to contract enough to release the tension in their intrafusal fibers and stop them from signaling.

With this kind of circuitry, the CNS not only receives information about the tension in a muscle but can actually set any muscle to a certain length and tension. This is how the body maintains a posture. Consider, for example, how the CNS sets each muscle in your back to a certain length to keep you standing up straight. Using the kind of neural circuitry outlined in

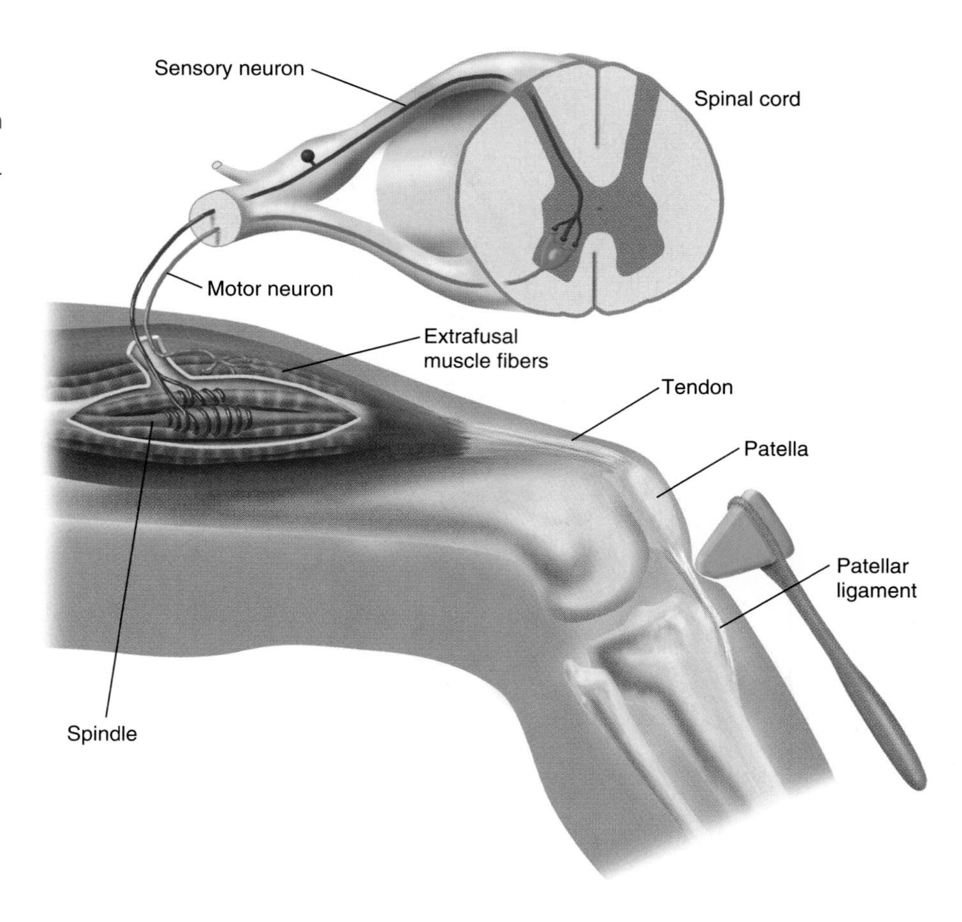

Figure 43.10
The knee-jerk reaction is a myotatic reflex. When a muscle is stretched suddenly, muscle spindles within it are stretched, and they send out signals through sensory nerve fibers that synapse directly with motor neurons. The motor neurons stimulate muscle contraction, bringing the muscle back to its original length. Thus the tension and length of each muscle can be set to achieve a certain posture.

Section 42.9, the CNS sets up a tension in each muscle by contracting the end (striated) portions of the intrafusal fibers. This causes each intrafusal fiber to initiate a stretch reflex, which contracts the whole muscle around it until the tension in the intrafusal fibers is relieved. Within seconds, the feedback circuits in the motor system will make the necessary fine adjustments to hold your body in the required position.

Exercise 43.7 When greeting someone, you extend your hand and arm, ready to shake hands. Describe the kind of signal exchange that occurs between the arm muscles and the central nervous system to keep your arm in that position.

43.5 Many mechanoreceptors employ hair cells.

Hair cells are widespread (but badly misnamed) mechanoreceptors built into several sense organs, including the vertebrate ear. The "hair" is no hair at all but rather a set of cilia that bend under slight pressure (Figure 43.11). A hair cell is in contact with a sensory neuron, and by releasing neurotransmitters into the synapse between them at a certain rate when the cilia are in a neutral position, it causes the neuron to send out a steady stream of nerve impulses. Then, by changing its rate of releasing neurotransmitter, the hair cell causes the neuron to fire at a higher frequency when the cilia are bent one direction and at a lower frequency when they are bent the other direction. Thus the angular position of the hair cell's cilia is encoded in the firing frequency of the sensory neuron.

A **statocyst** is a simple hair-cell device that functions as a gravity receptor in crustaceans and other invertebrates. It consists of a cavity lined by hair cells with a grain of sand (statolith) in the middle (Figure 43.12). As gravity pulls the grain downward, it presses against the hair cells beneath it, so by their arrangement and connection to the brain, they continually tell the animal which way is down. Proof that the position of the statolith tells the animal how to orient itself came from a clever experiment done in 1893 by Kreidl. He knew that when a shrimp molts, it sheds the lining of the statolith and the enclosed grain, and then showers itself with sand while its chambers are still open to replace the grains. Kreidl replaced the sand in the shrimp tank with bits of iron, which the shrimp then put into their statocysts. When he then held a strong magnet over the shrimp, the grains of iron were attracted upward, and the shrimp, now fooled into thinking that up was down, started swimming upside-down.

Fishes and amphibians use a **lateral line system,** with clusters of hair cells in a series of channels along the animal's side (Figure 43.13). These cells respond to small water displacements, especially to low-frequency vibrations in the water, so the lateral line system reports on events in the immediate environment, including vibrations from the movements of other animals.

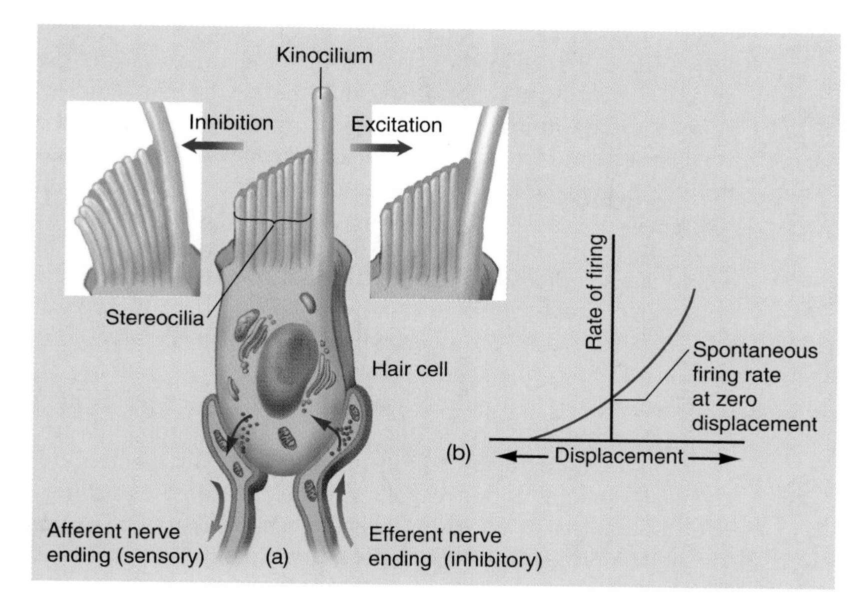

Figure 43.11
(a) A hair cell has a typical cilium (kinocilium) built on the familiar 9 + 2 pattern and several small extensions called stereocilia, which are thin microvilli with an actin core. *(b)* The spontaneous rate at which the cell fires is increased by moving the kinocilium one way and decreased by moving it the other way.

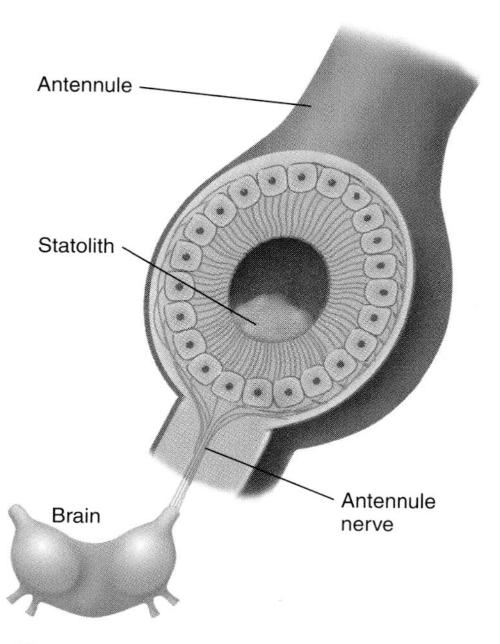

Figure 43.12
In the statocyst of a lobster (shown here much enlarged relative to the brain), a statolith rests on the ciliary extensions of many hair cells. The statolith stimulates different cells as its position changes with the animal's movements.

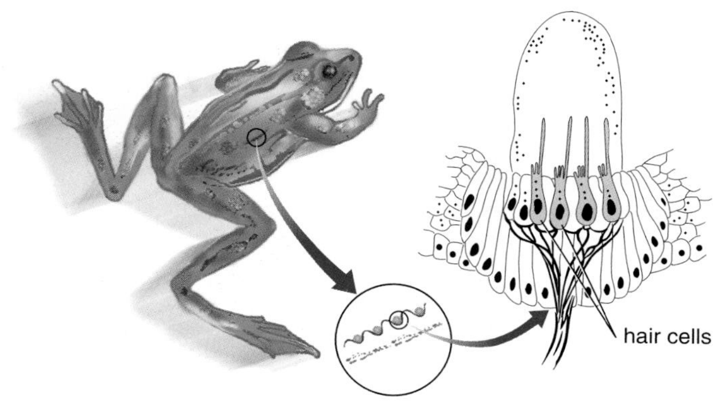

Figure 43.13
The lateral line system of a frog consists of channels that are open at some points to the surrounding water. Vibrations in the water stimulate the hair cells distributed along these channels. Source: Roger Eckert, et al., *Animal Physiology,* 3rd edition, 1995, W.H. Freeman.

The **labyrinth,** or **vestibular apparatus,** of the vertebrate inner ear (Figure 43.14) uses hair cells to detect acceleration—a change in the speed or direction of movement. We experience two forms of acceleration: from movement and from gravity; by detecting all accelerations, the labyrinth provides the information needed for orientation in equilibrium. The canals in the temporal bone of the skull are lined by a membrane, leaving a central duct filled with *endolymph,* a fluid similar to interstitial fluid. The cilia of hair cells in the utricle and saccule extend into a gelatinous *otolith membrane* covered by crystals of calcium carbonate known as *otoliths.* The otoliths and membrane affect the hair cells just as the statoliths of statocysts do. The layers of hair cells are arranged in two different planes, so the utricle detects mainly horizontal acceleration and the saccule mainly vertical acceleration, telling an animal how its head is oriented.

A second part of the labyrinth is a set of **semicircular canals** (more precisely, the ducts within them), three rings set at right angles to one another so they can detect rotational movements in the three dimensions of space. As hair cells in the canals are pushed one way or another by the movements of the endolymph, they detect accelerations, not steady movement. For example, you feel very little sensation of movement in a car moving steadily along a straight highway even at high speed, but if the car changes speed quickly or swerves to one side, the endolymph in your semicircular ducts is shifted in one direction,

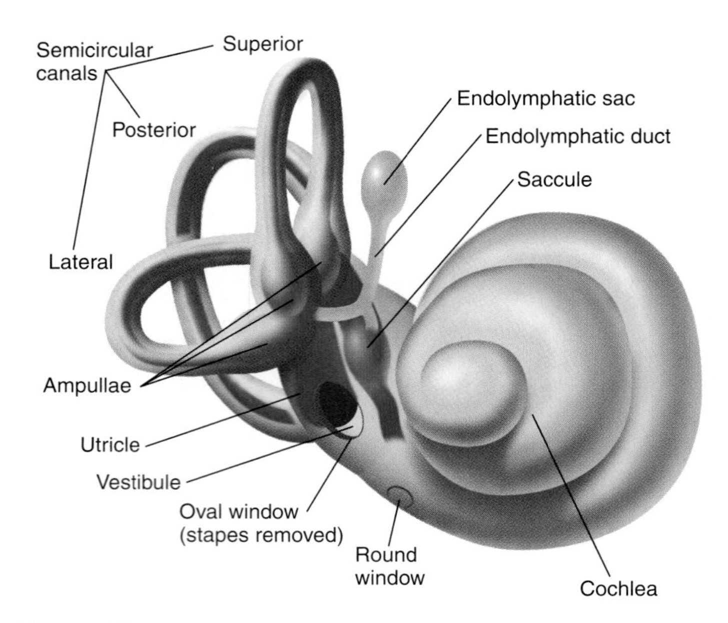

Figure 43.14

The inner ear of a mammal consists of a labyrinth, a maze of channels in the bone; this bony labyrinth is lined by a membrane, leaving an internal duct filled with endolymph *(blue)*. The labyrinth is divisible into two parts: the cochlea, which is the receptor of sound, and the vestibular apparatus, which is the receptor of acceleration and equilibrium. The vestibular apparatus includes three semicircular canals arranged at right angles to one another and two chambers at their bases, the utricle and saccule. Endolymph inside the vestibular apparatus moves as the head moves and stimulates hair cells located in the ampullae, utricle, and saccule.

giving you a sensation of movement as the fluid in each canal bends the cilia on the hair cells.

Exercise 43.8 When you are riding in a car, in addition to their effect on the labyrinth, accelerations throw your whole body this way and that, creating additional sensations of movement. How do you detect these motions?

43.6 Hair cells in the ear detect sound vibrations.

Hearing is the reception of vibrations, generally in air, which we call sound waves. The vertebrate ear is a device for delivering these waves to hair cells, which are eminently suited for detecting vibrations (Figure 43.15). Sound waves initially set up vibrations of the **tympanic membrane,** or **eardrum.** In mammals, these vibrations are then transmitted through the three small bones of the middle ear—hammer, anvil, and stirrup—that link the eardrum with the **oval window** of the **cochlea,** a snail-shaped cavity in the mastoid bone of the skull containing the innervated organ where vibrations are converted into nerve impulses. These bones also increase the strength of the vibrations they carry—that is, they exert more pressure. Pressure is force divided by the area on which the force acts, and the area of the eardrum is about 30 times that of the oval window, so each force impinging on the eardrum is concentrated to a smaller area and would be amplified 30 times if it were transmitted to the oval window without loss of energy. Some loss is inevitable, and so the sound pressure on the eardrum is amplified only about 22 times when it reaches the cochlea. (A lot of sound is also conducted directly through the skull; this is why your voice never sounds the same when you hear it through your ears from a recording as when you hear it through your skull as you speak.)

The cochlea is divided lengthwise into three fluid-filled canals: Vestibular and tympanic canals (scala vestibuli and scala tympani) contain the same fluid and join at the distal end of the cochlea; a smaller middle canal (scala media) contains a different fluid and partly encloses the ear's sensory tissue, the **organ of Corti.** A thin layer of tissue, the basilar membrane, separates the organ of Corti from the tympanic canal.

Experiments by Georg von Békésy showed that the organ of Corti operates on the principle that a small force applied to one side of a tightly stretched membrane will increase the tension in the membrane (Figure 43.15*d*). In the cochlea, the basilar membrane is under tension. The forces of vibration in the oval window are transmitted to the basilar membrane, which responds with a vibration of its own. These vibrations produce much larger shearing forces across the cilia of hair cells in the organ of Corti. (In mammals these are all stereocilia.) The hair cells then excite sensory neurons, which send signals through the auditory nerve. If the organ of Corti had been designed by a engineer, we would call it ingenious; since it has been shaped by random forces of evolution, we can at least say it is marvelously adapted to its function.

The basilar membrane responds to the pitch (frequency of vibration) of a sound, and at different points along the membrane to sounds of different pitches. As von Békésy described the situation, low-frequency sounds, up to about 60 Hz (hertz = cycles per second), cause the organ of Corti to send volleys of nerve impulses in synchrony with the rhythm of the vibration. The louder the sound, the more impulses are packed into each period of the vibration. Above 60 Hz, each pure tone makes the basilar membrane vibrate maximally at one point along its length. The part of the membrane closest to the oval window, being narrower and stiffer than the rest, responds to high-frequency sounds; the other end of the membrane vibrates in response to lower-frequency sounds. Thus the hair cells at each point along the organ of Corti send off nerve impulses reporting a different frequency. Since most sounds are complex mixtures of pitches, we hear them for what they are because the vibrations they set up in the eardrum are analyzed into their components by the cochlea, which sends its signals simultaneously to the brain where they are interpreted as one sound. The sense of pitch is sharpened by the phenomenon of collateral inhibition described in Section 41.13; the strongest signals, coming from a region of maximum stimulation, inhibit weaker

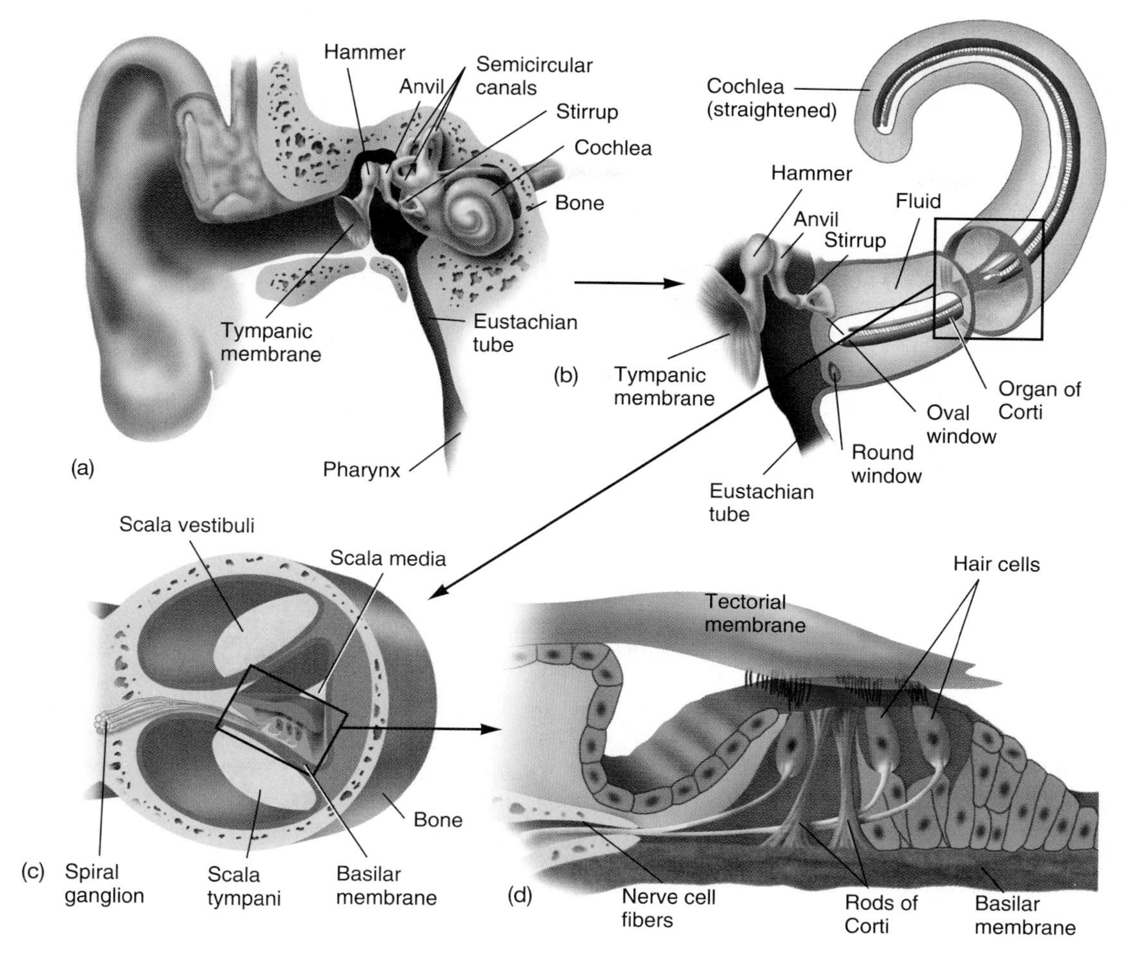

Figure 43.15
(a) The human ear has outer, middle, and inner regions. Sound waves in the outer ear produce vibrations of the tympanic membrane (eardrum). Its motion is transmitted through the middle ear by a chain of three bones: the hammer, anvil, and stirrup. The stirrup rides against the oval window of the spiral-shaped cochlea, the sensory organ of the inner ear. (Notice the position of the semicircular canals.) *(b,c)* The cochlea is filled with fluid (perilymph), so vibrations received at the oval window are converted into waves of pressure in the vestibular canal (scala vestibuli), continuing to the far end of the cochlea and back through the tympanic canal (scala tympani). Pressure in the tympanic canal is relieved by movement of the round window. *(d)* Sound vibrations are transduced into nerve impulses in the organ of Corti. Waves of pressure in the fluid of the scala tympani displace the tightly stretched basilar membrane, producing shearing movements of the stereocilia of hair cells. The hair cells report the pattern of their stimulation through signals in the auditory nerve fibers to the auditory centers of the cerebral cortex.

signals from nearby regions, thus making each signal sharper and more discrete.

Although the normal range of human hearing is from about 15 Hz to 20,000 Hz, young children may be able to detect sounds up to 40,000 Hz. The upper limit declines sharply with age in varying degrees. Some of the loss is due to a slight decline in the mobility of the middle-ear chain, but much of it is sensory deafness that depends strongly on the level of noise in one's environment. The inner-ear apparatus is highly sensitive to the kinds of loud noises that have become commonplace in our culture (jet planes, rock music, air hammers, motorcycles), and continuing exposure to such noise distorts and destroys the arrangements of delicate cilia in the organ of Corti (Figure 43.16). In contrast, even elderly natives of primitive habitats can still hear the soft movements of animals in the brush.

43.7 Some thermoreceptors detect infrared radiation.

Relatively little is known about thermoreceptors, which detect heat. Some mammalian thermoreceptors are just free nerve endings in the skin; others have distinctive shapes, like the end-bulb of Krause that detects cold and the Ruffini ending that detects warmth (see Figure 43.8). Thermoreceptors adapt rapidly; think of how quickly your body gets over the initial shock of a tub of hot water or a cold lake. Perhaps the most extensively analyzed heat receptors are the infrared detectors of snakes, particularly those of pit vipers, like rattlesnakes, that have heat-sensitive membranes in the pits along the sides of their heads (Figure 43.17). Since these receptors can detect very small differences in temperature, the snake uses them to accurately locate the warm body of its prey by orienting its head until it detects equal heat from both sides. Thermoreception serves a social purpose, too. Large numbers of rattlesnakes winter together in dens, all guided to the right place by going up the heat gradient to the warmest southern exposure.

43.8 Photopigments in specialized membranes absorb light.

The perception of light is probably understood better than any other sense. Most animals have photoreceptors. Invertebrates like earthworms and certain clams have photoreceptor cells scattered over their bodies that can only report the general

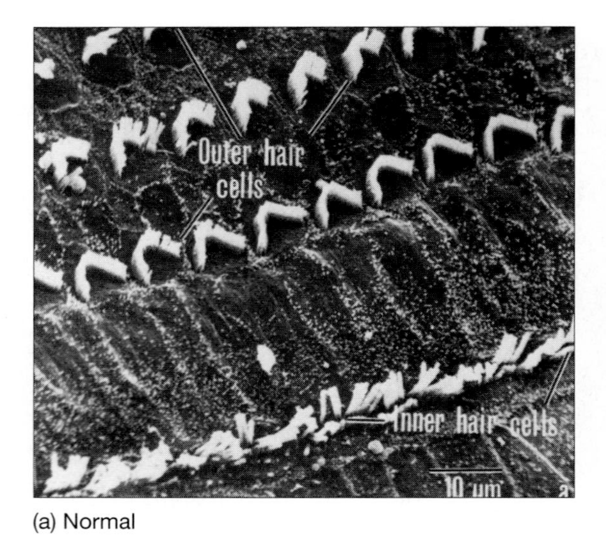

(a) Normal

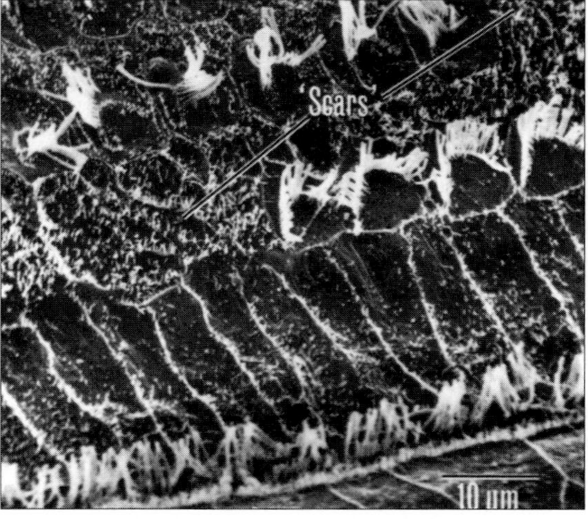

(b) Damaged

Figure 43.16
The neat, symmetrical arrays of hair cells in the organ of Corti are destroyed by continued exposure to loud noise.

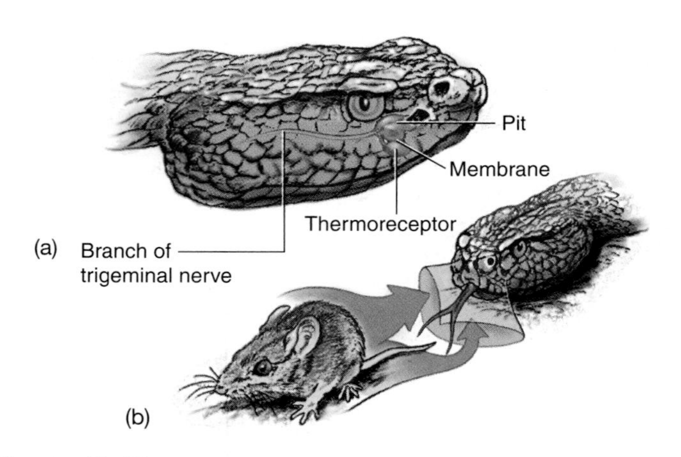

Figure 43.17
(a) The thermoreceptors of a rattlesnake, located in pits on the sides of the snout, can detect the heat produced by a nearby animal. *(b)* A snake can determine the animal's position by the relative intensity of heat recorded by its two receptors, just as we can locate the source of a sound by its relative intensity at our ears.

intensity of light. But most other animals have complex eyes with lenses that collect and focus light on clusters or sheets of photoreceptors. One of the more remarkable stories of parallel and convergent evolution concerns the independent evolution of eyes with similar structures in several phyla of animals (Figure 43.18).

Photoreceptors operate by means of **photopigments,** which are molecules that change structure when they absorb light, leading ultimately to a nerve impulse. Visual photopigments consist of a protein, *opsin,* bound to a *chromophore,* the small organic molecule that absorbs light. The main mammalian chromophore is the carotenoid *retinal,* a derivative of vitamin A. **Rhodopsin** (also called *visual purple*) is made of opsin bound to the colored 11-*cis*-retinal. When rhodopsin absorbs light, the retinal changes into all-*trans*-retinal, which comes off the opsin (and must be converted enzymatically back to its 11-*cis* form) (Figure 43.19). The opsin changes conformation into a colorless form and initiates the neural events that follow.

A photoreceptor in the dark maintains a membrane potential somewhat higher (closer to zero) than most neurons, due to a *dark current:* a continuous influx of Na^+ ions through Na^+ channels in its plasma membrane. These channels are kept open by cyclic GMP (cGMP), an internal second messenger similar to the cyclic AMP that plays a comparable regulatory role in so many other cells. Rhodopsin, in turn, is associated with G-proteins called *transducins;* when rhodopsin becomes bleached by light, the transducins activate an enzyme that converts cGMP to GMP. This reduces the dark current, and the membrane potential falls. The dark photoreceptor has been maintaining a certain level of an inhibitory neurotransmitter in its synapses with neurons; when its membrane potential falls, it reduces the rate of neurotransmitter secretion and thereby increases the firing rate of postsynaptic neurons.

Roles of cyclic AMP, Sections 11.8 and 41.4.

Exercise 43.9 Light stimulates neural signaling by photoreceptors by means of elements and changes that are virtually the opposite of those in the usual signal-transduction pathway. Compare the two pathways from this point of view.

Exercise 43.10 On a sunny day, you buy tickets to a movie and walk quickly into the dark theater. At first, you can hardly see anything, but over the next few minutes, you start to see the seats and other people clearly. What process alluded to in Section 43.8 accounts for most of this dark-adaptation?

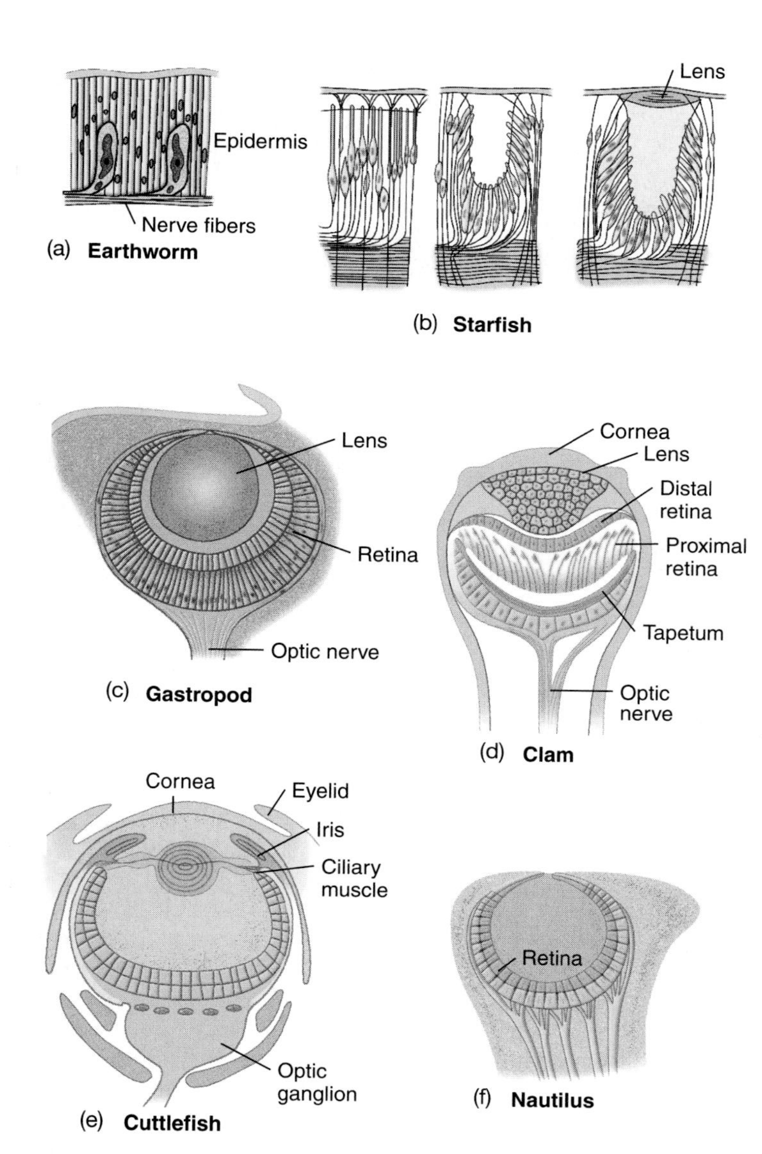

Figure 43.18

The eyes of several kinds of animals show convergent evolution: *(a)* light-sensitive cells of an earthworm; *(b)* photoreceptors of a starfish, including one with a rudimentary lens; *(c)* the eye of a gastropod; *(d)* the eye of a clam; *(e)* the eye of a cuttlefish; *(f)* the eye of a chambered nautilus, which uses a pinhole instead of a lens.

Photoreceptors are either *rhabdomeric* or *ciliary.* Rhabdomeric receptors, found in the eyes of molluscs and arthropods, have the outer segment of the cell membrane folded into a dense array of microvilli that contain the photopigment (Figure 43.20*a*). The ciliary receptors of a vertebrate eye are modified cilia, which show their ancestry by retaining a ninefold ring of microtubules (Figure 43.20*b* and *c*). Their photopigments reside in discs or in accordionlike foldings of a single membrane. A human eye contains two types of ciliary receptors, rods and cones.

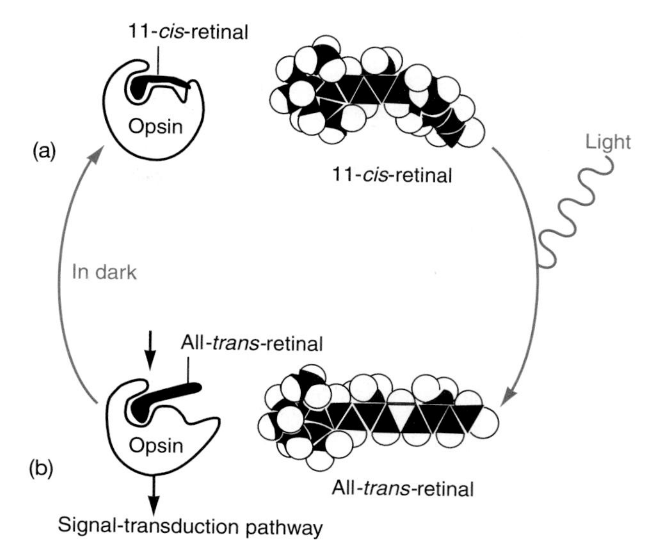

Figure 43.19

(a) The visual pigment rhodopsin consists of a protein (opsin) bound to the pigment 11-*cis*-retinal. *(b)* When this pigment absorbs light, it is converted to all-*trans*-retinal which has a different shape. This shift changes the conformation of the opsin. Through a signal-transduction pathway using G-proteins and cyclic GMP, the change in opsin changes the membrane potential of the receptor cell, leading to changes in the rate at which postsynaptic neurons fire nerve impulses.

Vertebrate **rods** are remarkably sensitive to dim light; under the best conditions, a single rhodopsin molecule absorbing a single photon can produce a measurable response in the rod membrane. Experiments suggest that people can reliably detect light flashes of only five photons. Yet the human eye can continue to detect forms and shapes over a 10^8-fold range of light intensities.

Cones, on the other hand, respond poorly to dim light. They are the receptors for color vision, and the human retina contains three types of cones that respond maximally to different wavelengths: to red-orange light at about 570 nm, to green light at 550 nm, or to blue-violet light at 440 nm (Figure 43.21). The color we see depends on the relative amounts of these colors that the eye detects. Since the cones respond poorly to dim light, we don't see colors well at night. Rods respond most strongly to blue-green light at about 500 nm, leading to a phenomenon called the *Purkinje effect* in which objects in this color range appear unusually bright in dim light. The use of blue-green paint for maximal visibility at night has only recently been adopted, and it will probably increase. A traditional red fire truck is quite visible during the day, but it tends to be very dark and hard to see at night.

Exercise 43.11 Look back at Figure 10.6 and then explain why people are advised to eat carrots to improve their vision.

Figure 43.20

These drawings compare three common types of visual receptors. *(a)* The rhabdomere of a horseshoe crab eye consists of vast numbers of parallel microvilli containing visual pigments on the surface membranes of a ring of retinular cells. *(b, c)* The rods and cones of vertebrate eyes are ciliary receptors; their outer segments are connected to the inner segments by stalks, which are modified cilia with the ninefold ring of microtubules but without the two central microtubules. Their visual pigments are built into an accordion-folded enlargement of the plasma membrane. In the outer segment of a rod, these membranes become detached as free discs.

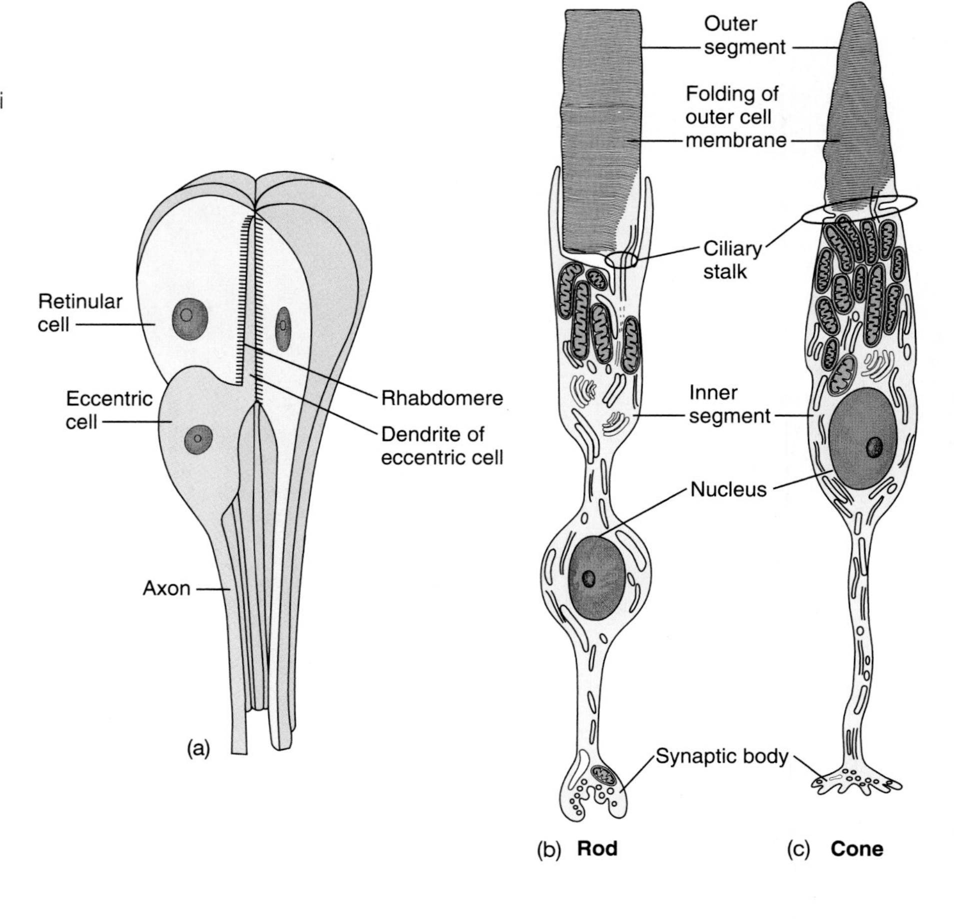

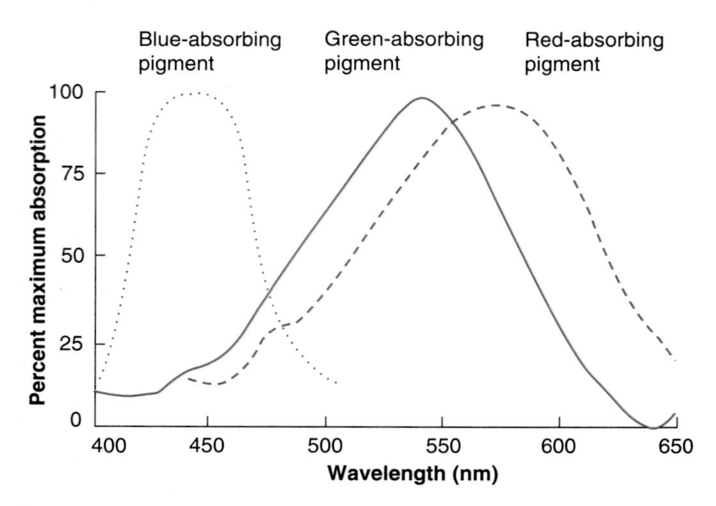

Figure 43.21

Color vision depends upon three types of cones that are maximally sensitive to three different wavelengths of light. The color a person sees at any point in the visual field depends on the relative stimulation of the cones, but probably not in a simple way.

43.9 The vertebrate eye focuses light on the retina, a layer of receptor cells.

The human eyeball (Figure 43.22*a*) is protected by a tough outer **sclera** that forms the transparent **cornea** in front. Just inside the sclera is the **choroid layer** containing blood vessels and a dark pigment that absorbs light, preventing internally reflected light from blurring the visual image. Just behind the cornea, the choroid extends inward to make the **iris,** the colored ring that gives an eye its characteristic color. The center of the iris is the **pupil,** a round hole that admits light to the body of the eye. Smooth muscles in the iris change the size of the pupil and therefore the amount of light admitted. The pupil is constricted by contraction of the muscles running circularly around the iris and dilated (enlarged) by contraction of those running radially. Changes in the ambient light intensity stimulate the *pupillary reflex,* a contraction or dilation of the iris that keeps the amount of light entering the eye relatively constant. Just behind the iris is the **lens,** made of elongated cells filled with proteins called crystallins. The eye's round shape is maintained by its strong sclera and by finely adjusted pressure on its internal fluids, a gelatinous *vitreous humor* filling most of the eyeball and an *aqueous humor* in front of the lens.

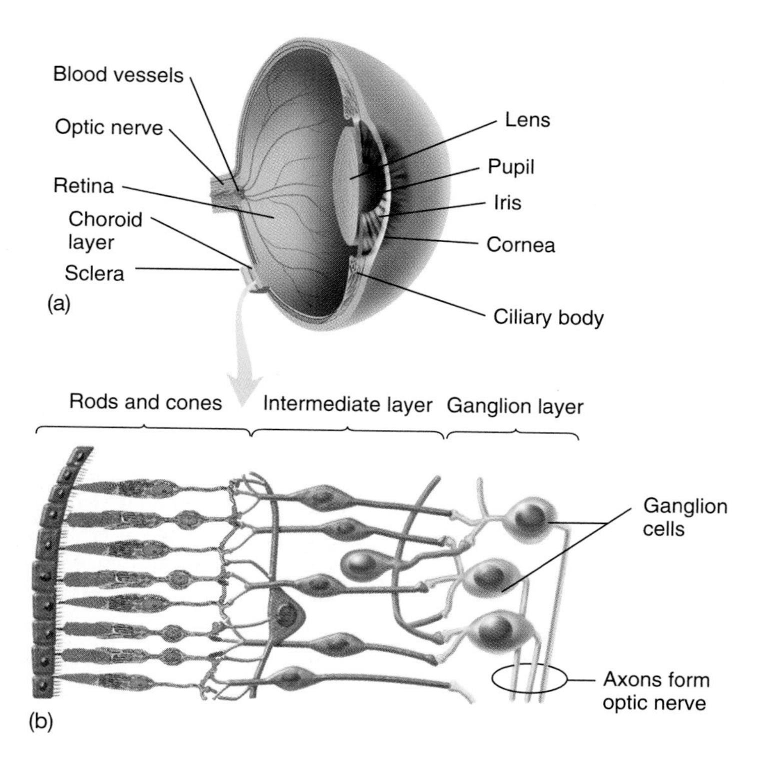

Figure 43.22

(a) A cross section through the human eye shows how light passes through the cornea (which focuses it partially), enters the pupil, and is focused on the retina by the lens. The size of the pupil is changed by expansion and contraction of the iris. The ciliary body and ciliary muscle exert pressure on the lens to change its shape. *(b)* The retina consists of three layers of cells, and light must pass through the ganglion layer and intermediate layer before it is absorbed by rods and cones. Axons of the ganglion cells form the optic nerve.

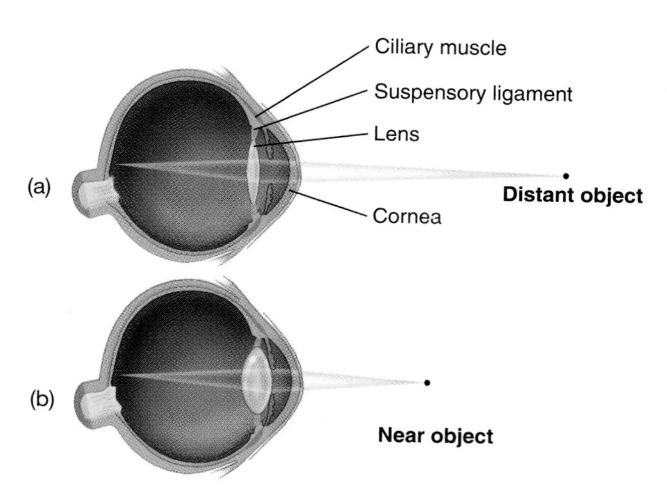

Figure 43.23

Light is focused partially by the cornea and more finely by the lens. The lens is made of cells largely filled with proteins (crystallins) and is very pliable. *(a)* When the ciliary muscle is relaxed, the lens has a flat shape suitable for viewing distant objects. *(b)* Contraction of the ciliary muscle exerts tension on the lens, so it bulges into a rounded shape to focus on close objects.

At the back of the eyeball lies the **retina,** made of three layers of cells. Rods and cones form the outside layer, farthest from the source of light, and synapse with an intermediate layer of cells that, in turn, synapse with cells of the ganglion layer (Figure 43.22*b*). Notice that light passes through two layers of cells (which are *not* stimulated) before reaching the photoreceptors. However, at one spot called the *fovea centralis* these overlying cells are displaced, so light does not have to pass through them before reaching the photoreceptors. Rods and cones are particularly dense in the fovea, and these features combine to make it a spot of particular visual acuity. The axons of the ganglion cells converge to form the **optic nerve** leading to the brain; the point where the optic nerve leaves the retina has no photoreceptors and consequently forms a *blind spot* in each eye.

The cornea and lens focus incoming light (Figure 43.23). Because light passing from one medium to another (as from air to glass) is bent, or refracted, the curved cornea acts as a fixed lens with just the right shape to focus parallel light rays onto the retina. Light rays from a source more than 5–6 meters away are essentially parallel, and an eye is best adapted to seeing objects at such distances, with the lens quite flat so as to produce minimal additional refraction of the incoming light. To view a closer object, the lens has to *accommodate*—become more rounded—to supplement refraction by the cornea and focus an image; radially oriented smooth muscles of the ciliary body contract, pulling the suspensory ligament into a smaller circle and letting the lens bulge out. Accommodation generally becomes more difficult with age, as the lens loses its flexibility and the fibers responsible for accommodation change their attachment to the lens; the condition is called *presbyopia,* and people entering their 50s commonly have to start wearing special glasses for reading.

Eyeballs commonly get out of shape. Nearsightedness (myopia) and farsightedness (hyperopia) are the result of eyeballs being either too long or too short, respectively, to focus light properly (Figure 43.24). The defects are corrected by glasses with concave or convex lenses.

This mechanical and optical mechanism is so much like the operation of a camera that it has become a cliche of elementary-school science to point out the parallels. The major difference is that an eye focuses by changing the shape of the lens, while a camera focuses by moving the lens nearer to or farther from the film. And yet a fish's eye focuses as a camera does, since it has no ciliary muscles.

The real mystery begins after light is focused on the retina. An image of the world is laid out over the back of the eye, and somehow the eye and brain together see that picture. Although the brain only knows patterns of nerve impulses, somehow it can recreate a version of the outside world. Next we will investigate what is known about this process.

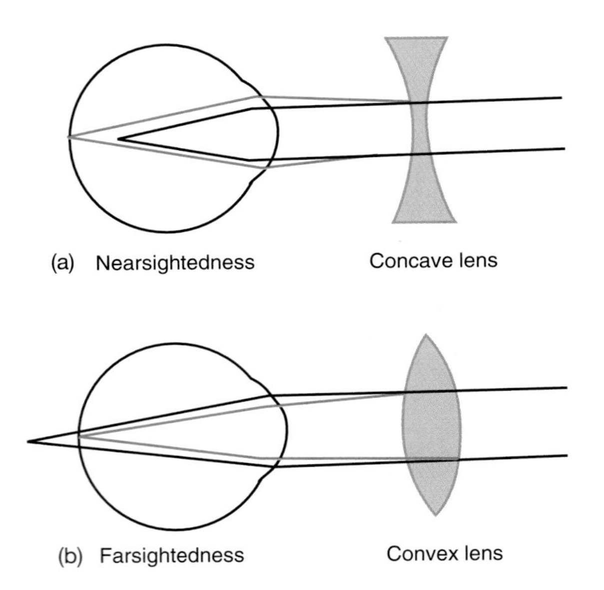

Figure 43.24

(a) When an eyeball is slightly elongated, light from distant sources is focused in front of the retina, making a fuzzy image (nearsightedness or myopia). This defect is corrected with a concave lens that spreads the light out slightly. *(b)* If an eyeball is too short, the image is formed behind the retina (farsightedness or presbyopia). This defect is corrected with a convex lens that concentrates the light somewhat.

Exercise 43.12 You walk out of a dark building into the sunlight. One part of your eyes reacts immediately, in a mechanical way. What happens, and what is its effect?

43.10 Analysis of information from visual receptors begins in the retina.

Light falling on the retina excites a pattern of rods and cones. But how can we actually see that pattern? We might imagine that the visual field is simply mapped onto the cerebral cortex, rather as points on the skin are mapped onto the somesthetic cortex (see Figure 42.16). The surface of the brain might be like a little television screen, with each cell wired to some cell of the retina, so each point of light coming into the eye would turn on a certain cell in the brain. The trouble with this model is that it doesn't answer the question, "Who is watching the screen?" Instead, it turns out that visual information is not just presented to the brain as a pattern of light and dark, but from the beginning, even in the retina itself, it is being analyzed into patterns and forms. At each higher level, these patterns are being synthesized into increasingly complex forms. Visual information is processed in the sequence:

Light → Rods and cones → Intermediate cells → Ganglion cells → *Via optic nerve* → Lateral geniculate nucleus of the thalamus → Visual cortex → Visual association cortex

The 10^8 rods and cones in each retina are connected to about 10^6 ganglion cells, whose axons form the optic nerve. But each ganglion cell isn't simply connected to 100 receptors so that it reports the average intensity of light reaching those 100 cells. Instead, the information from the receptors is processed through intermediate neurons so as to accentuate the pattern of light and dark falling across the retina.

The **visual field** of an eye is the entire space that the eye may be focused on at any time. Within the visual field, investigators map out the **receptive fields** of single neurons: the area or form within the visual field to which the neuron responds. We will see that the receptive field of each cell is a region with a distinctive shape; one cell might respond only when an animal is looking at a certain kind of line, another cell only when an animal is looking at a corner. To determine the receptive field of a cell, a tranquilized animal is positioned so it is looking at a blank screen on which images can be projected; then electrical responses are recorded from a microelectrode inserted in some cell in its visual system. If the cell responds with a change in membrane potential or firing frequency while a certain image is on the screen, this means the retina is being stimulated in just the right way to affect that cell.

In 1952 Stephen W. Kuffler reported that a cat's retina contains two types of ganglion cells. The receptive fields of both types have a circular center with a surrounding ring, but a cell with an **on-center field** fires when the center is illuminated and the surrounding ring is dark, whereas a cell with an **off-center field** fires when the center is dark and the ring is illuminated (Figure 43.25). The stimulus that makes a ganglion cell fire must reflect the way it is connected to a group of receptors, somewhat as shown in Figure 43.26. Every receptor actually communicates with many ganglion cells, so the ganglion cells' receptive fields overlap strongly. The ganglion cells then pass visual information on to brain centers.

Exercise 43.13 In some retinas, certain ganglion cells respond only to an image moving in one direction. Examine the hypothetical situation shown in the accompanying drawing and explain how the retinal cell will respond to a light moving toward the left and to one moving toward the right. Which direction of movement will make the cell fire?

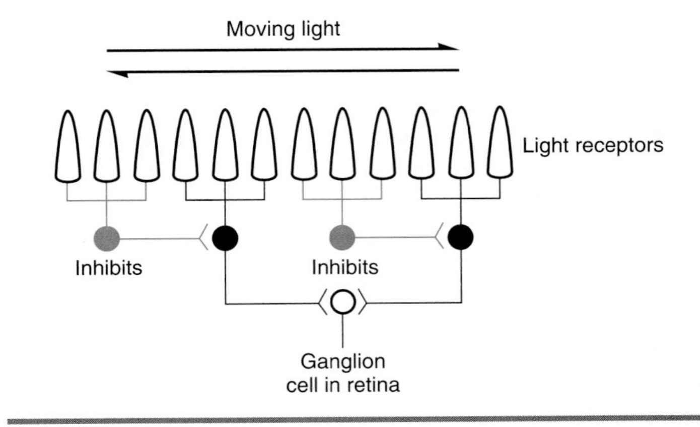

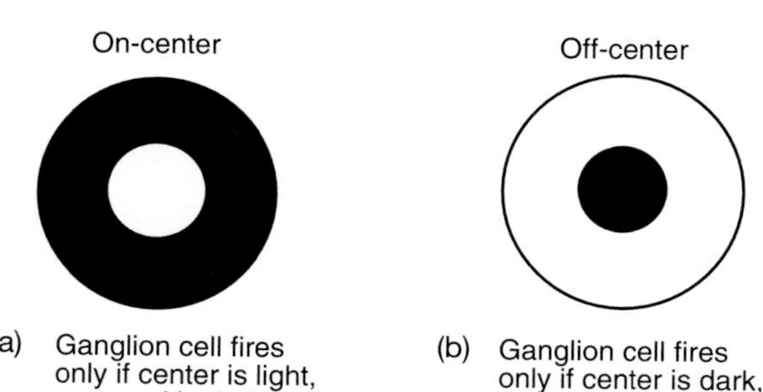

Figure 43.25
(a) A ganglion cell with an on-center field fires when the center is illuminated, provided the surrounding area is dark. *(b)* A cell with an off-center field fires when the center is dark, provided the surrounding area is light.

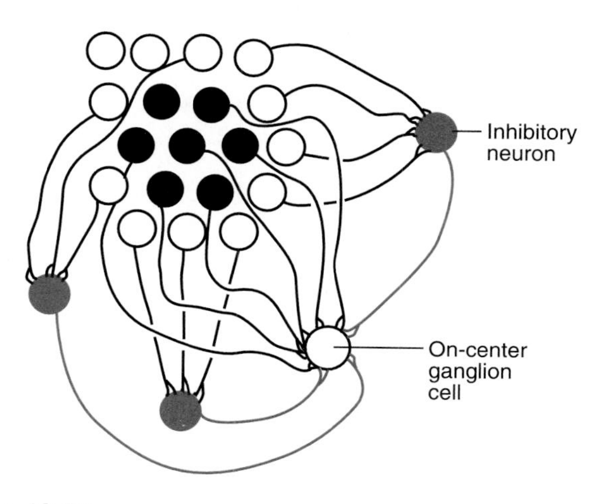

Figure 43.26
An on-center field could be created if a ganglion cell is excited by receptors *(black)* in a small area and by inhibitory neurons *(red)* from receptors surrounding the area. No attempt has been made here to show the actual neuronal connections.

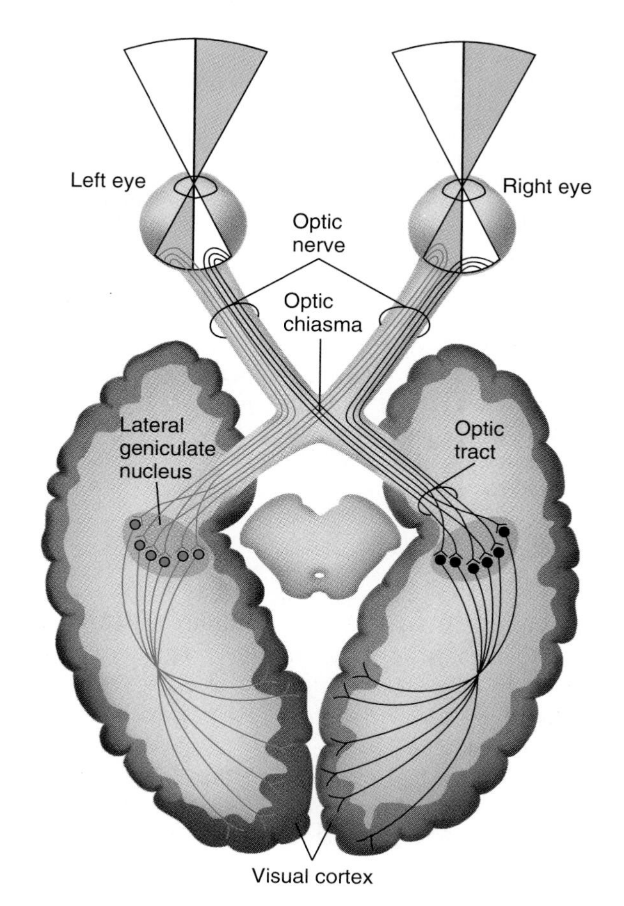

Figure 43.27
The human visual pathways, as seen here from above, partially cross in the optic chiasma, so visual fibers arising in the right side of each retina *(black lines)* project to the right visual cortex, and fibers arising from the left side of each retina *(green lines)* project to the left visual cortex.

43.11 Vision depends on a hierarchy of cells in the visual cortex.

In the main visual pathways of humans and other mammals, the optic nerves carry the axons of retinal ganglion cells through a crossroads, the **optic chiasma,** to the **lateral geniculate nuclei (LGN)** of the thalamus (Figure 43.27). Notice that the eyes send information from the left halves of both retinas to the left side of the brain and from the right halves to the right side. Each LGN therefore receives information about the same image from both eyes and somehow correlates this information, thus allowing us to have stereoscopic vision, which depends on information from two independent views of space. In mammals, the LGN cells project to the visual cortex, where their signals are analyzed by a complex series of cells.

Many years of careful, painstaking work by David H. Hubel and Torsten N. Wiesel has provided a detailed picture of some regions in the primate visual cortex. Using the experimental setup described in Section 43.10, with an animal watching a screen, they moved a light pattern across the screen to find the position and type of stimulus that produces maximum stimulation of each cell. The results are remarkable.

Area 17 of the visual cortex, where the LGN axons terminate, contains **simple cells** that respond to a straight line—a bright bar, an edge, a dark line on a lighter background—located in one part of the visual field with a particular orientation (Figure 43.28*a*). For instance, one cell might respond strongly to a line oriented at 40 degrees to the horizontal, another at 50 degrees, and so on. Each cell discriminates so well that changing the angle of the line by only 4 degrees (a very small angle) makes it ineffective as a stimulus for that cell. Collectively, the cortex cells respond to lines in every part of the visual field and with every possible orientation (about 4 degrees apart).

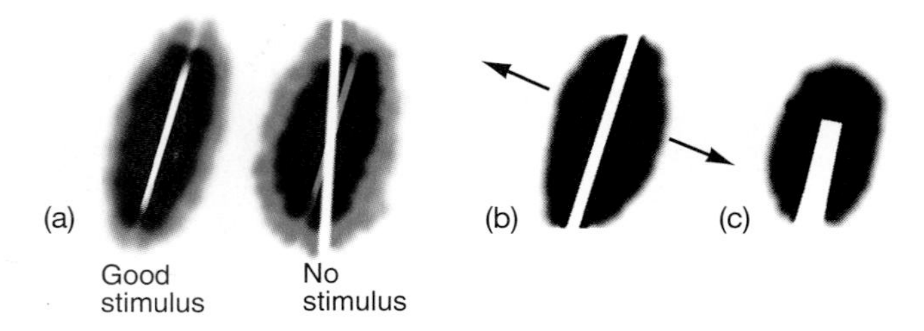

Figure 43.28

The receptive field of each cortical cell is a particular shape that stimulates that cell. *(a)* The field of a simple cell is a bar or edge at a certain angle in one place; a bar of light more than a few degrees away from that angle cannot stimulate the cell. *(b)* A complex cell responds to a bar or edge at some angle anywhere on the retina or moving across the retina, and *(c)* a hypercomplex cell responds only to a bar with a distinct end.

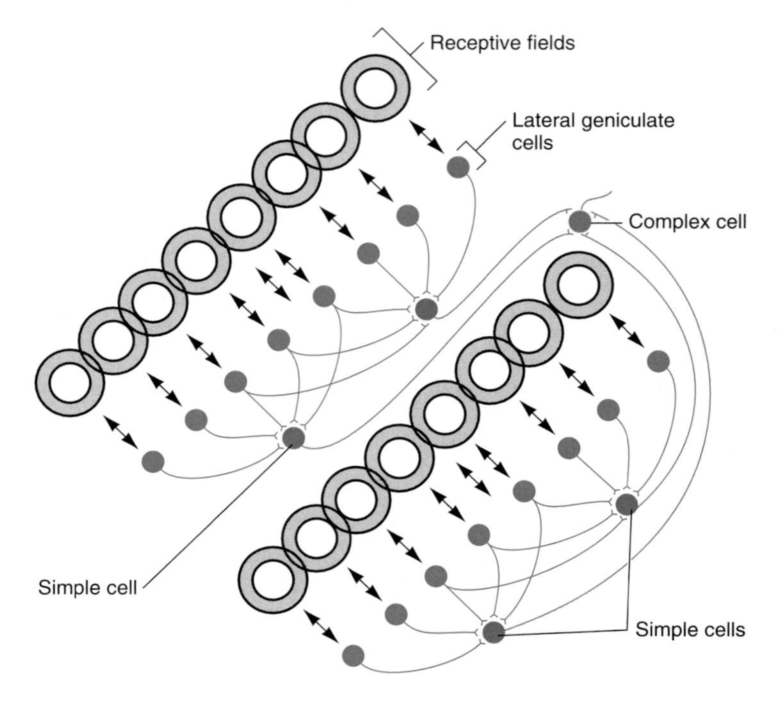

Figure 43.29

Cells at each level of the visual system have their own receptive fields and converge to form the receptive fields of higher-order cells. This wiring diagram indicates how cells can respond to a bar of light oriented at 45 degrees on the retina. The light simultaneously stimulates a line of on-center receptive fields, which report to a series of lateral geniculate cells *(green);* the latter are connected to simple cells, so when they all fire simultaneously, they stimulate a simple cell. Thus each simple cell fires when a bar of light falls on a part of a retina at a certain angle. Simple cells that respond to a bar at the same angle in different places are connected to a complex cell, which fires in response to a bar of light at one angle moving across the retina. Not shown are a series of inhibitory cells that refine this system by preventing cells from firing if specific conditions are *not* met.

The simple cells synapse with **complex cells** that also respond only to lines with specific orientations. But each complex cell responds to lines *located anywhere in a large part of the visual field* and to lines *moving through the field* with the right orientation (Figure 43.28*b*).

From area 17 of the cortex, connections go primarily to areas 18 and 19, where Hubel and Wiesel found **hypercomplex cells.** A hypercomplex cell responds maximally to a line in a particular orientation that has a definite end somewhere, or perhaps two definite ends (Figure 43.28*c*). Hypercomplex cells, in other words, respond to the length of a line as well as to its orientation. They also respond to corners with certain orientations. Other kinds of hypercomplex cells respond maximally to more complicated figures, such as a tonguelike bar of a certain length and width, but oriented at any angle and moving in any direction.

It isn't hard to see how neurons of the visual pathway can be connected to account for these receptive fields. Ganglion cells analyze the visual field into circular spots, and so do the LGN cells they synapse with. Figure 43.29 shows how a series of LGN cells whose receptive fields lie in a straight line could then be connected to stimulate one simple cell, *if and only if they are all stimulated at once.* A series of simple cells that all respond to lines with the same orientation could be connected so they will all stimulate a complex cell; it will fire if any of the simple cells are firing, especially if they fire one after another as a line sweeps across the visual field. Similarly, excitatory and inhibitory neurons connected to one another in a specific way could create a hypercomplex cell of any kind imaginable.

When a simple figure moves before our eyes, we see it because simple cells detect each of its lines, complex cells continue to detect them as they move, and hypercomplex cells define its corners and pick out its other features. Other systems of cells detect features of depth or distance and analyze colors. When all these cells pool their information in the cerebral cortex, we see an *object*—a car, or a house, or a table—instead of a mere collection of lines.

What kinds of cells process this visual information next? Are there cells that see increasingly complex forms, through *n* degrees of hypercomplexity, gradually combining their information to make greater sense out of items in the visual field? What is the end result? Is there finally one cell—or a very few—that fires whenever we see a green Volkswagen and another that is only fired at sight of a red Corvette? Visual processing could hardly be so specialized, since the evolving visual system could not anticipate what kinds of objects might appear in the world, but we all await answers to these questions from further investigations. Continuing work of the kind that Hubel, Wiesel, and others are pursuing should eventually provide a clearer understanding of how visual perception operates.

Exercise 43.14 You are looking at an object with the form of simple, straight lines in the shape of the letter H. Describe a sequence of increasingly complex cells that could analyze the visual information.

Exercise 43.15 Our eyes normally engage in small vibration movements, so an image constantly shifts to different parts of the retina. However, investigators have fitted people with a special contact lens that projects an image continuously onto one part of the retina; the subjects then report that parts of the image disappear and reappear at random. For instance, when viewing a letter H, the bar may disappear, then one of the vertical lines disappears, and so on. Why does this happen? Why doesn't it happen when we normally look at objects?

Coda We began this chapter by reflecting on the fact that our sensory receptors are our only link with the world, and we only know the world insofar as a few types of receptors allow us to detect certain substances or types of energy. This limitation is a continuing problem for philosophers concerned with epistemology, basic issues about the nature of knowledge. Continually wrestling with the meaning of this limitation, they wonder to what extent we can "know" the world at all, since it is filtered in this way. Yet our biological perspective provides an argument from evolutionary design: Our nervous systems apparently provide quite reliable knowledge of the external world because if it were not reliable—if it did not provide us with an accurate representation of reality—we would not be able to respond accurately to the dangers imposed by external factors. To put the case differently, our sensory receptors have been selected for accuracy because animals have relied on them for survival over millions of years. However, the existence of animals that can detect phenomena such as electrical fields suggests that the world may have features we are still unaware of.

People who don't believe in evolution often seize upon the evolution of receptors as an insurmountable difficulty. Their classical argument is that the vertebrate eye is an intricate structure that is only useful in its fully developed form, so it could not have evolved gradually through a series of imperfect, useless stages. But their argument rests on the false assumption that lesser eyes are useless, and the very existence of a variety of photoreceptors among invertebrates belies this assumption. Any cell that detects light has adaptive value. The photopigment retinal is a carotenoid of the kind common in unicellular phototrophs, so the basic light-absorbing apparatus is ancient. Primitive light-absorbing cells may be modified through many short steps into photoreceptors that provide valuable information to a nervous system, and many animals have taken these steps, in various directions. As the incredibly successful arthropods demonstrate, compound eyes have enormous value even though they are primarily good for detecting motion rather than forming images. Our growing knowledge of development should wallow us to provide perfectly coherent scenarios for the evolution of the most highly developed eyes from much simpler structures.

Summary

1. Known receptors can respond to chemical stimuli (chemoreceptors), light (photoreceptors), heat (thermoreceptors), mechanical stimuli (mechanoreceptors), and electrical fields (electroreceptors). Defined by the location of their stimuli, they may be interoceptors, which detect changes in the internal environment of the body; exteroceptors, which receive stimuli from outside; or proprioceptors, which report on the positions, tensions, and movements of body parts. Receptors encode information about the nature of stimuli and sometimes amplify the original signals.
2. Receptors transduce information about stimuli into the universal internal language of a series of nerve impulses. The frequency of the nerve impulses increases with the strength of the stimulus.
3. Receptors generally respond best to a change in a stimulus, rather than to a constant stimulus. The just-noticeable difference in a stimulus is always a characteristic constant fraction of the intensity of the stimulus (Weber's law).
4. Chemoreceptors bear receptor proteins specific for ligand molecules and are stimulated only by molecules with shapes that fit into their binding sites. The basic tastes—salty, sweet, bitter, and sour—signal the presence of general types of ligands, such as energy sources or potentially noxious materials. The sense of smell detects specific shape features of all kinds of molecules through a variety of receptor proteins in the olfactory mucosa.
5. Mechanoreceptors respond to movements. They are common in the skin, where they detect mechanical stimuli at the body surface, and in muscles and inner connective tissues, where they serve as proprioceptors to signal the orientation of body parts.
6. Muscle spindles are stretch receptors that signal the tension in a muscle and can be used to set its position and tension.
7. Several kinds of mechanoreceptors depend on hair cells, which change their firing rate as their terminal cilium is bent. Structures in which small grains rest on hair cells can detect gravity, and acceleration is signaled by hair cells that detect movements of the fluid in chambers.
8. The ear also depends on hair cells built into the organ of Corti in the cochlea. Sound waves that arrive at the eardrum are transmitted through a series of bones to the cochlea, where vibrations applied to the cochlea are converted into displacements of the basilar membrane. These cause disturbances of hair cells distributed along the organ of Corti, which send signals through the auditory nerve reporting vibrations of different frequencies, corresponding to their positions along the cochlea.
9. Photoreception depends upon receptor cells that carry photopigments made of a protein, opsin, bound to a carotenoid chromophore. When the carotenoid absorbs light, it and the protein change shape, initiating a change in membrane structure and potential that leads ultimately to nerve impulses.
10. Photoreceptors are either rhabdomeric or ciliary, depending on their structure. Vertebrate photoreceptors are either rods or cones; three types of cones, containing pigments sensitive to different wavelengths, discriminate color.
11. The mammalian eye consists of a cameralike chamber that focuses light, by means of its rounded cornea and lens, onto a layer of photoreceptors in the retina. These photoreceptors transfer their

information through intermediate cells to retinal ganglion cells, whose axons create the optic nerve.

12. Each cell in the visual system is stimulated by a particular type of stimulus in a certain position; these features define the receptive field of the cell. The receptive fields of retinal ganglion cells are circular; each cell is stimulated either by light in the center of the field and dark around it, or by the reverse.
13. The optic nerves of both eyes send information to the lateral geniculate bodies in the brain, which, in turn, send information to the visual cortex.
14. Visual cortex cells analyze information in the visual field into simple, particular shapes—primarily lines, edges, and corners—through several degrees of complexity. Simple cells detect lines or edges in a single position and orientation. Complex cells detect moving lines or lines in a wide range of positions. Various hypercomplex cells detect lines with specific lengths, corners, tongues, and other complex shapes. Other cells add features such as depth and color. It is not known how far this feature analysis extends.

Key Terms

chemoreceptor 913
photoreceptor 913
thermoreceptor 913
mechanoreceptor 913
electroreceptor 913
exteroceptor 913
interoceptor 913
proprioceptor 913
generator potential 913
absolute threshold 914
phasic 914
tonic 914
differential threshold 914
olfaction 914
muscle spindle 917
stretch (myotatic) reflex 917
hair cell 918
statocyst 918
lateral line system 918
labyrinth/vestibular apparatus 919
semicircular canals 919
tympanic membrane/eardrum 920
oval window 920
cochlea 920
organ of Corti 920
photopigment 922
rhodopsin 922
rod 923
cone 923
sclera 924
cornea 924
choroid layer 924
iris 924
pupil 924
lens 924
retina 925
optic nerve 925
visual field 926
receptive field 926
on-center field 926
off-center field 926
optic chiasma 927
lateral geniculate nuclei (LGN) 927
simple cell 927
complex cell 928
hypercomplex cell 928

Multiple-Choice Questions

1. Which of the following pairs of terms is a mismatch?
   a. interoceptor—pH of blood
   b. mechanoreceptor—hearing
   c. proprioceptor—blood pressure
   d. exteroceptor—vision
   e. chemoreceptor—taste
2. A generator potential is
   a. all-or-none.
   b. found only in receptors that are true neurons.
   c. similar to an action potential.
   d. similar to an EPSP.
   e. similar to an IPSP.
3. The various receptors exhibit all of the following activities *except*
   a. secretion of neurotransmitter.
   b. production of an action potential as a result of a generator potential.
   c. production of a generator potential as a result of stimulation.
   d. amplification of a stimulus.
   e. an all-or-none response to a stimulus.
4. Which of these sensory structures operates by stimulation of hair cells?
   a. vestibular apparatus
   b. lateral line system
   c. semicircular ducts
   d. cochlea
   e. all of the above
5. Which is *not* a characteristic or property of all receptors in the inner ear?
   a. Stimuli bend cilia on hair cells.
   b. Receptors function as sensory neurons.
   c. All are mechanoreceptors.
   d. All are contained within fluid-filled compartments.
   e. All result in perception of sound, balance, and acceleration.
6. Which structure includes all of the others?
   a. hair cells
   b. organ of Corti
   c. basilar membrane
   d. cochlea
   e. vestibular canal
7. Which structure includes all the others?
   a. vestibular apparatus
   b. utricle
   c. saccule
   d. otolith membrane
   e. semicircular canals
8. Which of the following are or contain muscle tissue?
   a. lens and cornea
   b. iris and choroid
   c. sclera and cornea
   d. retina, sclera, and choroid
   e. iris and ciliary body
9. Which of these statements correctly describes the activities and properties of the retinal photoreceptors?
   a. When light falls on rods and cones, they secrete more neurotransmitter to activate retinal neurons.
   b. Rhodopsin remains bleached when it is dark and becomes colored when exposed to light.
   c. When the rods are stimulated, sodium ions rush into the photoreceptors.
   d. The signal-transduction pathway of the photoreceptors includes transducins and GMP.
   e. Rhodopsin is composed of the protein retinal and a visual pigment called opsin.
10. Which of these are true neurons?
    a. rods
    b. cones
    c. ganglion cells
    d. *a* and *b*
    e. *a, b,* and *c*

True-False Questions

Mark each statement true or false, and if false, restate it to make it true.

1. Generator potentials of receptors are most similar to action potentials of axons.

2. The frequency of action potentials depends on the kind of initiating receptor.
3. The differential threshold of a receptor is equal to the minimum energy necessary to produce a response at least 50 percent of the time.
4. We can detect the molecular nature of a compound more accurately by olfaction than by taste.
5. All mammalian proprioceptors are mechanoreceptors, but not all mechanoreceptors are proprioceptors.
6. The most frequent cause of age-related deafness is loss of function in the middle ear.
7. Statocysts in invertebrates and otoliths in vertebrates serve similar sensory purposes.
8. The vertebrate eye is structurally most similar to eyes found in some annelids.
9. As the light falling on rod cells increases, secretion of neurotransmitter also increases, which leads to inhibition of the postsynaptic neuron.
10. To clearly see the distant horizon, the ciliary muscle contracts, enabling the lens to become thicker.

Concept Questions

1. How does habituation differ from sensory adaptation?
2. Briefly describe at least three differences in eliciting a response to an odor as opposed to a taste.
3. Contrast the activity of the olfactory epithelium with that of the vomeronasal organ.
4. Distinguish the types and functions of receptors in the mammalian inner ear.
5. In the area of the fovea centralis, many ganglion cells are stimulated by activity in a single receptor cell. In the remainder of the retina, each ganglion cell receives input from many receptor cells. How does this wiring difference affect vision?

Additional Reading

Ackerman, Diane. *A Natural History of the Senses.* Random House, New York, 1990.

Axel, Richard. "The Molecular Logic of Smell." *Scientific American,* October 1995, p. 130. Current research and theories of olfaction.

Barlow, Robert B., Jr. "What the Brain Tells the Eye." *Scientific American,* April 1990, p. 90. A circadian clock adjusts the sensitivity of the horseshoe crab's eyes.

Brownlee, Shannon. "The Route of Phantom Pain: Amputees Lead Scientists to a Startling New View of How The Brain Works." *U.S. News & World Report,* October 2, 1995. Phantom-limb pain is common among amputees and recent research shows that the pain results from defective brain signals.

Craig, A. D., and M. C. Bushnell. "The Thermal Grill Illusion: Unmasking the Burn of Cold Pain." *Science,* July 8, 1994, p. 252. In Thunberg's thermal grill illusion, first demonstrated in 1896, a sensation of strong, often painful heat is elicited by touching interlaced warm and cool bars to the skin. The basis for the illusion is explored here.

Dusenbery, David B. *Sensory Ecology: How Organisms Acquire and Respond to Information.* W. H. Freeman, New York, 1992. The comparative physiology and ecology of the senses.

Gallagher, Winifred. "Touch and Balance: A Tribute to the Forgotten Senses." *American Health,* January–February, 1990, p. 45. Physiological and psychological aspects of our senses of touch and balance.

Konishi, Mazakazu. "Listening with Two Ears." *Scientific American,* April 1993, p. 34. How barn owls locate their prey by sound.

Lewis, Ricki. "When Smell and Taste Go Awry." *FDA Consumer,* November 1991, p. 29. The senses of taste and smell can sometimes be distorted or stifled. These disorders may be a side effect of drug treatment. Since less than 5 percent of the population is afflicted, the problem has not warranted too much attention.

Moss, Cynthia F., and Sara J. Shettleworth (eds.). *Neuroethological Studies of Cognitive and Perceptual Processes.* Westview Press, Boulder (CO), 1996.

Suga, Nobuo. "Biosonar and Neural Computation in Bats." *Scientific American,* June 1990, p. 60. Bats extract remarkably detailed information about their surroundings from biosonar signals. Neurons in their auditory systems are highly specialized for performing this task.

Wolfe, Jeremy M. "Hidden Visual Processes." *Scientific American,* February 1983, p. 94. Vision is actually divided into subsystems.

Internet Resource

To further explore the content of this chapter, log on to the web site at:

http://www.mhhe.com/biosci/genbio/guttman/

44 Skeleton and Muscle

A Blackpoll Warbler in its yellow fall plumage prepares for its long migration to South America.

Key Concepts

As the September days grow shorter, a flock of Blackpoll Warblers, small yellow and gray birds, has been feeding in the forests of New England, gleaning insects from the trees, storing up fat. And one morning, heeding some unknown signal, they take off and head southeast into the Atlantic. For 85 hours, they remain aloft, moving out into the ocean until easterly winds start to blow them westward and they finally land, exhausted, on the northern coast of South America. They have burned off all their fat, using its energy to drive their wings up and down 15 times per second, about 4,600,000 times in all. Billions of protein fibers built into symmetrical arrays in their breasts have pulled on one another again and again to work a system of bony levers, moving their feathered wings in just the right pattern to resist gravity and propel their diminutive bodies over 8,000 kilometers to forests where they can survive through the northern winter.

Animals move. It is one of their most characteristic features. In this chapter, we will examine the muscles that create the forces of movement and the skeleton that supports them (the *musculoskeletal system*), with some emphasis on the neural signals that initiate and control movements. ■

A. Skeletons and connective tissues

44.1 A musculoskeletal system is a highly integrated array of elements.

Under their cloaks of skin, fur, feathers, scales, and chitinous plates, most animals hide one of the most marvelous systems ever devised by natural selection: a complex of powerful muscles attached to a firm skeleton made of braces and levers. Muscle was a very early animal innovation, and the bodies of most animals are largely made of muscle; the structure of muscle reaches a height of power and elegance among the most familiar animals, such as insects and mammals (Figure 44.1). Muscle can drive a cheetah to a speed of 70 mph when chasing its prey and a human to almost 28 mph; it has powered athletes to nearly 8 feet in the high jump and nearly 29 feet in the long jump. The flight muscles of insects make them some of the world's fastest and most efficient flying machines. And throughout the animal kingdom, muscles support all the complex behaviors of feeding, reproduction, and social activity.

Muscles work by contracting, and they can generate enormous forces and incredibly rapid movements. However, most of an animal's constituents—digestive tube, cardiovascular system, nervous system, and muscles themselves—are soft. They are, after all, mostly water. This softness creates a problem in locomotion for animals above a certain size. By a basic law of physics (Newton's Second Law of Motion), every force is accompanied by an equal force in the opposite direction. You can encounter this principle when trying to step from a boat onto a dock. The force of your leg propels the boat backward as the boat propels you forward; if the boat is small and not held in place, you can wind up in the water. Similarly, if an animal's muscles exert a force against other soft tissues that offer little or no resistance, the animal won't go far. It needs a firm support for the muscles to pull against—that is, a skeleton.

Besides muscle, the other components of the musculoskeletal system are **connective tissues,** tissues that bind and support other tissues. **Bone** is one of the principal connective tissues in a vertebrate; bones are connected to one another by flexible **ligaments.** Muscles are wrapped in thin sheets of connective tissue called **fascia,** and the fasciae converge into **tendons,** which connect the muscles to the skeleton. **Cartilage** forms several distinctive structures. We will discuss the structures of these and other connective tissues later in this chapter.

Paradoxically, the simplest animal skeleton is made of water itself, confined to a tight bag to make a stiff, nearly incompressible *hydrostatic skeleton.* We see in Section 34.8 how the evolution of a coelom in primitive animals provided them with a skeleton and allowed them to develop new ways of life. More complex skeletons are constructed of hard materials. Vertebrates have an **endoskeleton** made of internal bones and cartilage, to which muscles are attached by tendons (Figure 44.2*a*).

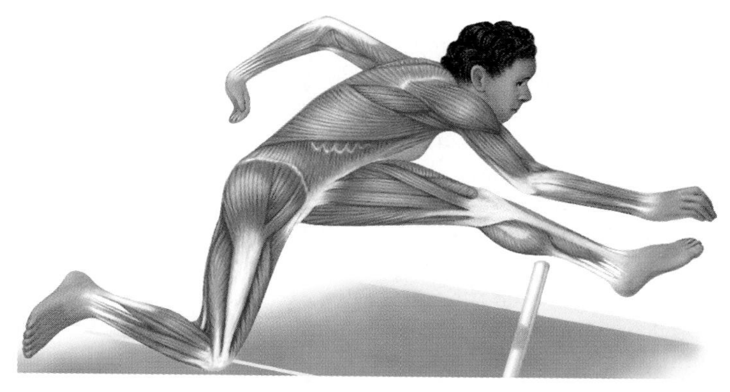

Figure 44.1
Athletic feats depend on a highly organized human musculature.

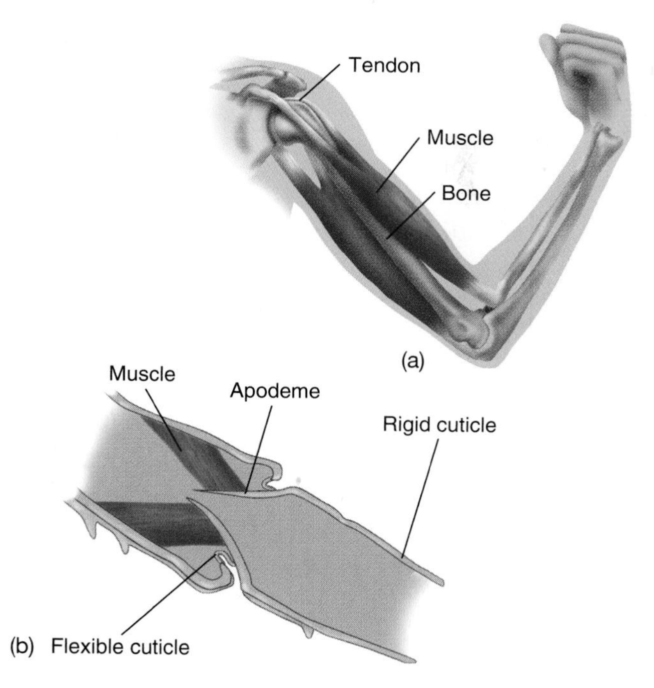

Figure 44.2
Hard skeletons may be internal or external. *(a)* Vertebrates have endoskeletons made of separate bones that are moved by muscles attached to their surfaces. *(b)* The exoskeletons of invertebrates such as arthropods are made of plates of chitin covering the body and moved by muscles attached inside.

Most invertebrates, such as molluscs and arthropods, have an **exoskeleton** made of external shells or plates, with the muscles attached on the inside. In arthropods, the rigid segments of a limb are connected by joints where the exoskeleton is soft and flexible (Figure 44.2*b*); muscles that span these joints pull neighboring segments together, rather like a vertebrate structure turned outside-in. Inward folds of cuticle make large surfaces for muscle attachment and are often extended into neighboring segments in long, internal plates called **apodemes.** Apodemes thus have the functions of tendons in vertebrate muscle.

A musculoskeletal system is basically a simple arrangement of levers and supports. A lever rests on a *fulcrum;* a force applied to one point moves a load (resistance) at a second point:

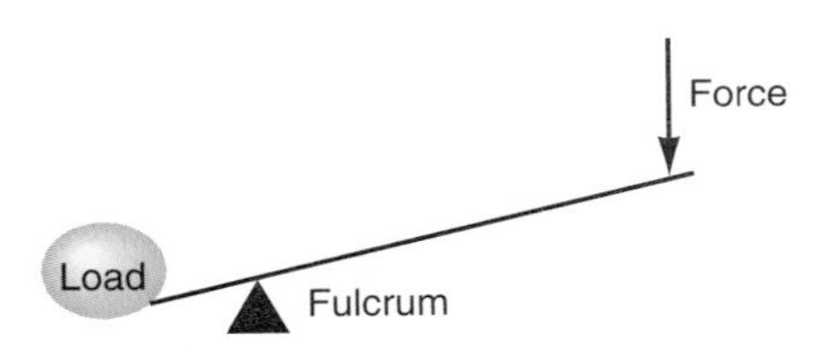

The distance between each force and the fulcrum is an *arm,* and the force times the length of its arm is called a *torque.* A simple lever such as a seesaw, with the fulcrum between the two forces, will balance if the torques are equal; the heavier of two children must sit closer to the center for balance:

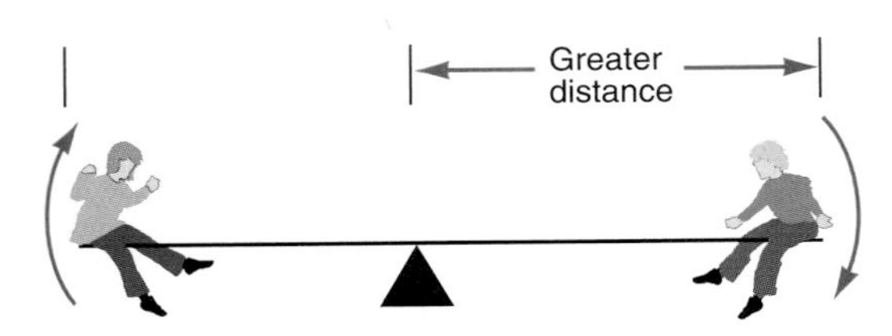

If a child leans backward, increasing her distance from the fulcrum, she exerts greater torque and goes down. We can use a long board or a tool such as a shovel to move and lift rocks and other heavy objects by placing the load close to the fulcrum and the applied force far from the fulcrum. The *mechanical advantage* of such a lever is equal to the ratio of the force arm to the resistance arm, so a small applied force can move a large load. For instance, a force of only 10 kg[1] acting through an arm of 1 meter just balances the resistance of 100 kg acting through 0.1 m:

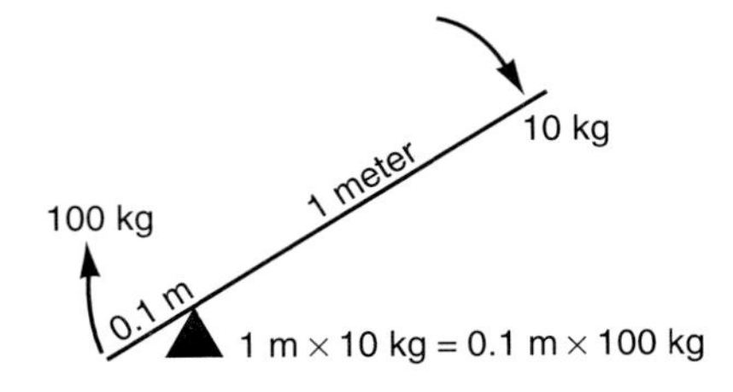

[1]Strictly speaking, a force is a mass times an acceleration and is measured in units called newtons (1 newton = 1 kg × 1 meter/sec^2. For discussion, it is sufficient to use units of weight, which is mass under the acceleration due to gravity on the earth.

The most common musculoskeletal arrangement places the applied force between the fulcrum and the load. Because muscles must be attached quite close to skeletal joints, a system may be working with a mechanical advantage of less than one—that is, a disadvantage. A biceps muscle attached only 5 cm from the elbow, for instance, can lift a 20-kg mass on the end of a 30-cm forearm, but the resisting mass has a sixfold (30/5 = 6) advantage, so 120 kg of force must be applied to raise it:

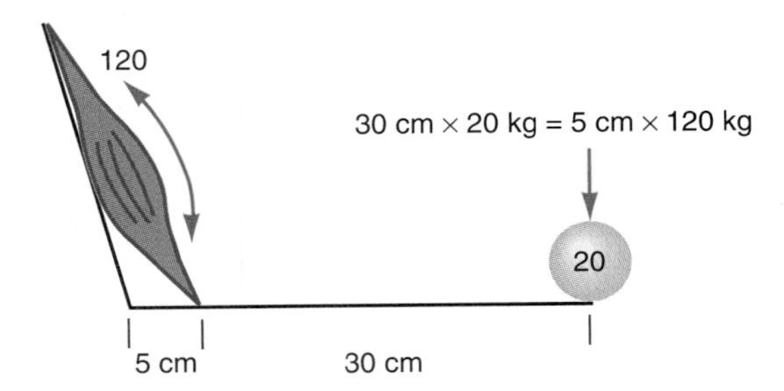

Some muscles run through a space or over another bone into a position where they pull the resistance more directly and have a better mechanical advantage; muscles that raise the lower leg, for example, pull across the kneecap (patella):

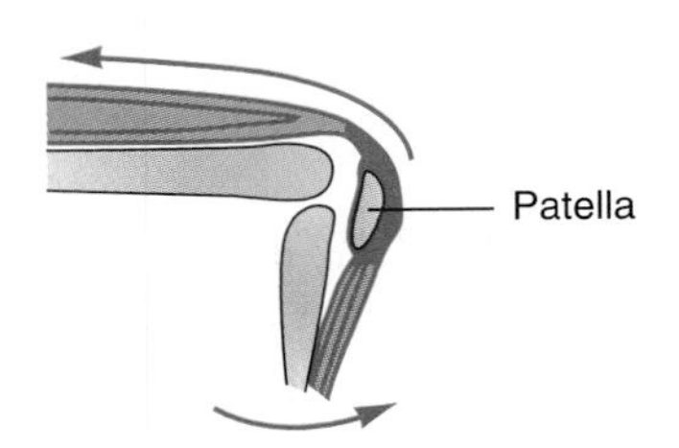

Work equals force *(F)* times the distance *(s)* through which the force acts, and force equals mass *(m)* times acceleration *(a);* therefore, work = $m \times a \times s$. A machine is generally designed to use a given amount of work (energy) for moving a large mass with a small acceleration or a small mass with a large acceleration, with a corresponding trade-off in speed. This has been obvious in the evolution of musculoskeletal systems. We have arms and legs with leverage systems that work with a mechanical disadvantage but move the ends of these limbs very rapidly. They are suitable for throwing a rock, a spear, or a baseball very fast; they are suitable for moving rapidly through the trees or over the ground in long strides, covering a long distance quickly. They are not suitable for lifting cars, and when this action became important, we were forced to invent machines with large mechanical advantages. Most muscle-bone leverage systems work with a mechanical disadvantage because, as one biologist put it, "We evolved to run away from the lion, not pick him up."

The very nature of muscle also requires that it be attached close to a joint, even though it must exert a strong force. Since a muscle is a bag of water packed with proteins, it is quite viscous, and if it contracts rapidly, a lot of energy is lost in simply overcoming this viscosity. A muscle attached far from the joint would have to contract rapidly through a long distance, so it is more efficient to attach it close to the joint, where it shortens less and can contract more slowly.

Exercise 44.1 A muscle is attached 5 cm from a knee joint; it pulls the lower half of a leg that is 40 cm long. If the leg lifts a load of 45 kg, how much force must the muscle exert?

Exercise 44.2 The leverage systems of vertebrate bones and muscles are easy to draw. Examine the structure of an arthropod joint in Figure 44.2*b* and draw a comparable picture to show how this can be reduced to a system of levers.

44.2 Connective tissue is built on a matrix of fibers.

The cells of all connective tissues are surrounded by an extensive extracellular *matrix,* which they secrete. This matrix is made mostly of fibers: Long fibers of collagen and crisscrossing reticular fibers give the tissue strength, and fibers of the rubbery protein elastin provide resilience. In cartilage, the matrix is rubbery; in bone, the matrix is supplemented by the addition of calcium phosphate crystals (technically, the mineral hydroxyapatite); and in other types of connective tissue, the matrix remains soft and jellylike.

Figure 44.3 shows four forms of connective tissues, which differ principally in the density and arrangement of their fibers and in the predominant fiber types. **Dense fibrous connective tissue,** the material of tendons, ligaments, and fascia, is mostly made of densely packed, parallel collagen fibers. **Elastic connective tissue,** which has relatively little collagen and much more elastin, forms layers; in large blood vessels, for instance, it stretches under the pressure of blood with each contraction of the heart and then rebounds elastically to keep the pressure more constant than it would be otherwise. **Loose connective tissue** is packed beneath the skin and around organs, protecting them from mechanical shocks. **Adipose tissue** has relatively few fibers, and most of the space is taken up by adipose cells specialized for storing fat.

Cartilage is another prominent type of connective tissue—the firm, rubbery material we commonly call "gristle." It is made by rounded cells called chondrocytes, which remain embedded in a tough intercellular matrix of collagen, elastin, and mucopolysaccharides that often has a blue-white color. Cartilage contains no blood vessels. It often covers the surfaces of bones within joints, forms discs within joints (as between the vertebrae), or forms whole structures such as the external ear or the larynx ("voice box") in the throat.

The predominant cells of connective tissue are **fibroblasts.** They arise from *mesenchyme,* a rather loose embryonic connective tissue, and when mature they secrete the collagen and other proteins of the matrix. Connective tissues are also home to leucocytes (white blood cells) that fight infections, as we see in Chapter 48.

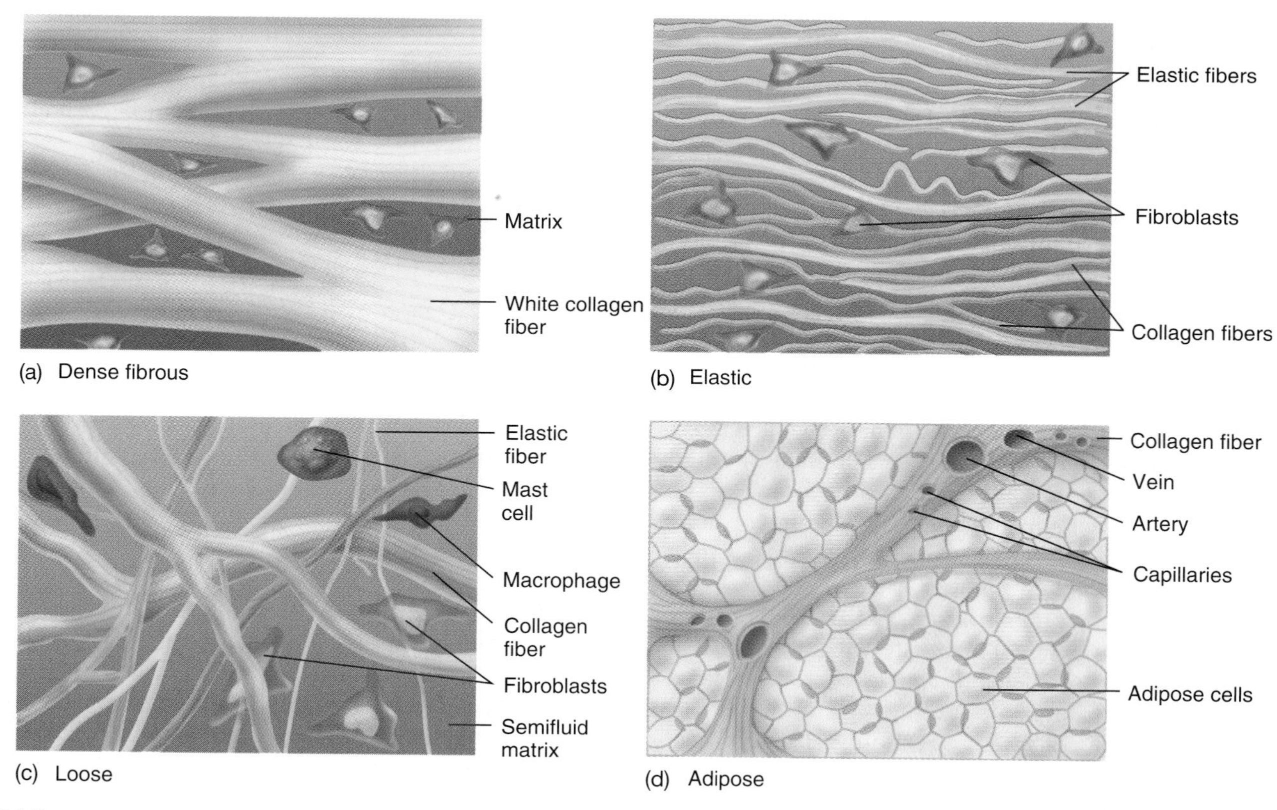

Figure 44.3
Connective tissues take several forms: *(a)* dense fibrous; *(b)* elastic; *(c)* loose; and *(d)* adipose.

Sidebar 44.1 The Evolution of Skeletal Elements

As one type of organism evolves into others that occupy different adaptive zones, it must obviously modify its structures for new functions; for instance, the flipperlike fins of some ancient fishes became gradually modified into the legs of amphibians. In the course of such a change, some existing structures may lose their functions, but the developmental paths that form them can be modified to produce different structures with totally new functions. Remember that structures are said to be *homologous* if they have a common embryological origin. By following the development of skeletal elements in different vertebrates, comparative anatomists have been able to determine that seemingly unrelated structures are actually homologous and have resulted from modifications of an ancient developmental path. One typical example describes the conversion of ancient gill arches into other structures in other vertebrates (see illustration).

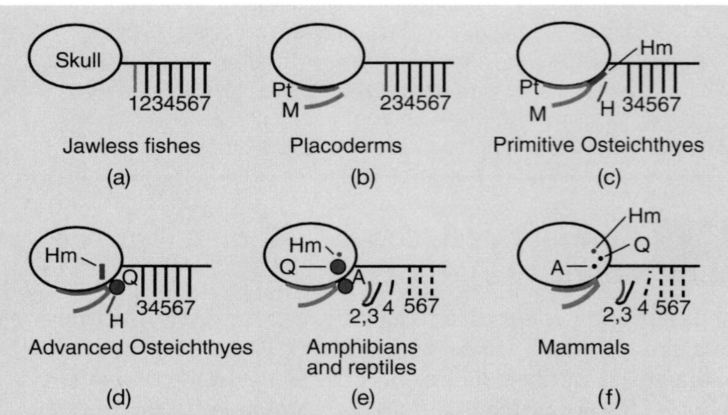

The gills of jawless fishes (agnathans) are supported on seven cartilaginous arches *(a)*. In fish that evolved jaws, the first two arches then changed into the cores of jaw components. In placoderms *(b)*, the first fishes with jaws, the first arch split into the pterygoquadrate (Pt) bone and Meckel's cartilage (M), the bases of the upper and lower jaws. As primitive Osteichthyes *(c)* evolved stronger jaws, the second arch split into hyomandibular cartilage (Hm), used as the support for the jaw, and hyoid cartilage (H). In advanced Osteichthyes *(d)*, part of the pterygoquadrate has separated off as the quadrate bone (Q), on which the jaw now moves. As amphibians and reptiles *(e)* evolved into land animals with no gills, the other gill arches became useless as gill supports but acquired new functions; arches 2, 3, and sometimes part of 4 were transformed into parts of the hyoid apparatus, the U-shaped bone in the neck that anchors muscles of the tongue and mouth. Also, the articular bone (A) split off from Meckel's cartilage, and the hyomandibular became a bone in the middle ear that transmits sound from the eardrum to the inner ear. In mammals *(f)*, sound is transmitted through the middle ear by three bones: the former quadrate (now the incus or anvil), the articular (now the malleus, or hammer), and the hyomandibular (now the stapes or stirrup). The hyoid is made only of arches 2 and 3; arches 4 and 5 become the thyroid cartilage, and arch 6 becomes the epiglottis.

44.3 A vertebrate skeleton is a series of bones connected at joints.

In spite of the great differences in body form among the vertebrates, their major skeletal features have been retained during evolution. Humans are unusual vertebrates because we are bipedal—we walk on two feet—and our whole musculoskeletal system has become modified for this way of moving. Nevertheless, the human skeleton makes a good point of departure for discussing vertebrate skeletons in general. The vertebrate skeleton can be divided into an **axial skeleton,** including the skull, vertebral column, sternum, and ribs, and an **appendicular skeleton,** made of the pectoral and pelvic girdles and limbs (Figure 44.4). Parts of the vertebrate skeleton have changed radically in the transition from fishes to terrestrial animals. Fishes differ from four-limbed vertebrates chiefly in the structure of their skulls and the way their appendicular skeletons are formed and attached to the axial skeleton. Sidebar 44.1 describes the transformations of some elements in the head and neck.

Bone is formed through the *ossification* of softer tissue—the deposition of hard calcium phosphate (hydroxyapatite) crystals—and it develops as either *endochondral bone* from cartilage or as *dermal bone* from skin. The most primitive vertebrates were jawless armored fishes (see Section 35.10), and the tough dermal plates of their armor became key elements of the skull, even though they were lost from most of the body. The remnants of these dermal plates are laid down in an embryo, and before they have all fused, one can see that the skull is made of many separate plates, some endochondral and some dermal (Figure 44.5). Several of these bones, which are easily discerned in an adult skull, were used to show homologous elements in Figure 2.10.

Bones connect to one another at joints, or articulations, of two kinds: *diarthroses,* in which there is fluid between the bones, and *synarthroses,* in which there is no fluid. Synarthroses include the joints where the skull bones meet along suture lines in the mature skull, as well as the *gomphoses* where the teeth fit into sockets in the mandible and skull; the teeth are held in place by collagen fibers connected to their roots, and though we are never aware of it, the roots are able to move slightly in their sockets as we chew. This slight movement takes up some of the impact from biting, pumps blood and lymph through the nearby circulation, and sends neural signals to the brain centers that control the muscles used in chewing, as a feedback circuit. Other synarthroses include the intervertebral discs of fibrous cartilage between the vertebrae, which allow the vertebral column to absorb shocks and bend a little; with these joints, the vertebral column is not very flexible, even if the spines of some acrobats and dancers seem to be more flexible than their arms and legs. (Anatomical authorities use different systems to classify the joints. An alternative system defines *synarthroses* as joints that allow essentially no movement, such as the sutures between skull bones; *amphiarthroses* as slightly movable joints with discs of cartilage, as between the vertebrae of the spinal

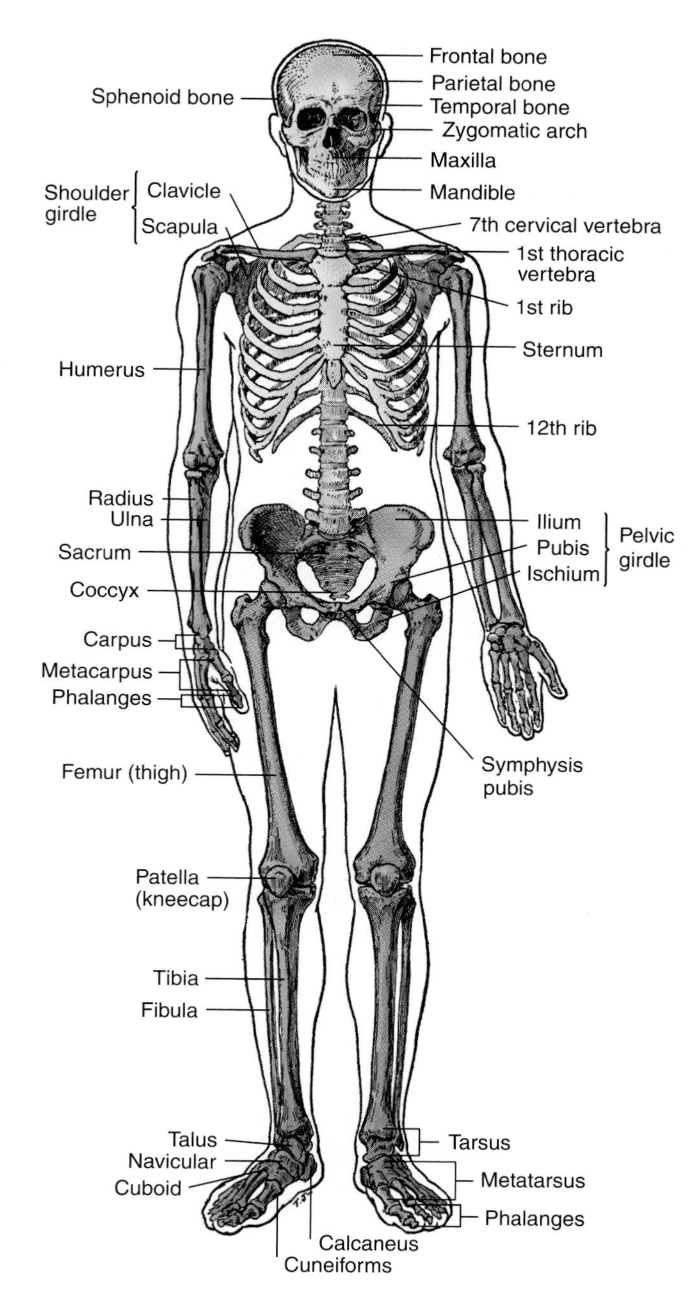

Figure 44.4

The human skeleton is based on an axial skeleton consisting of the skull, vertebral column, sternum, and ribs. To this is attached the appendicular skeleton *(blue),* which includes the pectoral (shoulder) and pelvic girdles and the limb bones.

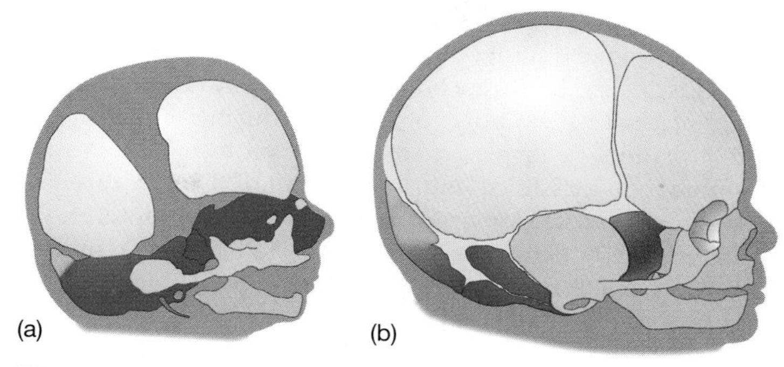

Figure 44.5

(a) An embryonic skull is made of several separate plates. They gradually grow together as the skull enlarges, and eventually fuse *(b).* Purple areas are endochrondral bone, and pale orange areas are dermal bone. Source: L.B. Arey, *Developmental Anatomy,* 6th edition, 1954, W.B. Saunders.

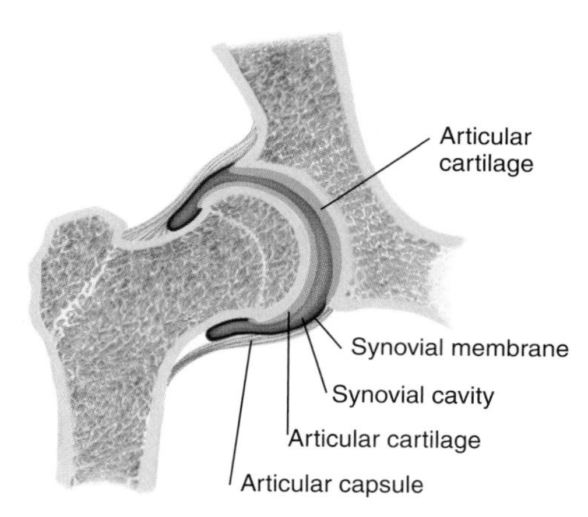

Figure 44.6

A diarthrosis is a freely movable joint between bones that encloses a fluid-filled cavity. The joint is enclosed in a capsule, and the opposing bone surfaces are covered by cartilages. The cavity is lined with a synovial membrane, which secretes a lubricating synovial fluid.

column; and *diarthroses* as freely movable joints, as between parts of the limbs.)

Diarthroses, found mostly in the appendicular skeleton, have a fluid-filled cavity between the ends of the bones, which are covered with cartilage (Figure 44.6). The cavity is encased in a fibrous capsule and lined by a **synovial membrane,** which secretes **synovial fluid** to lubricate movements of the bones. The joint can be shaped like a hinge, as in the knee, or like a ball and socket, as where the head of the femur joins the pelvis:

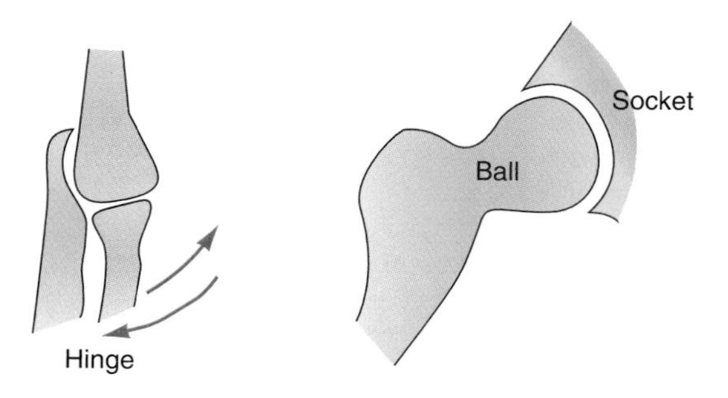

A series of ligaments surround each joint and sometimes run through it, holding the two bones together and restricting their movement. Ligaments also strengthen the joint so it can withstand the tremendous leverage exerted by attached muscles.

Muscles and bones are bound to rub against one another in some places as they move, possibly producing irritation and inflammation. Such places are usually protected by **bursas,** membranous sacs filled with synovial fluid. Due to excessive stress from overexertion or for unknown reasons, bursas are often invaded by calcium deposits, leading to bursitis, an inflammation characterized by pain, tenderness, and limitations on movement.

44.4 Bone is formed by the calcification of a soft model structure.

The paradox of bone is that the material needed to support movement also inhibits growth. A young animal with solid bones would be well supported and able to move, but it would soon outgrow its skeleton. Vertebrates resolve the paradox by forming bones that continually reshape themselves internally, so they can keep growing while providing support. The process requires a neat balance between **osteocytes,** the cells that secrete bone, and **osteoclasts,** the cells that remove it.

An adult bone's form is first laid down in a small model made of fibrous membranes and cartilage, and the model is gradually ossified—converted into bone—and calcified by the deposition of minerals. Fibroblasts deposit the long collagen fibers of the basic ground substance. Some fibroblasts then change into **osteoblasts,** which start to lay down calcium phosphate crystals along fibrous strands. A completely calcified strand is called a **trabecula.** As calcification proceeds, the trabeculae fuse into networks and create spongy (porous) bone surrounding marrow spaces. Figure 44.7 shows these processes at work in the growth of a long bone. The bone will continue to elongate as long as zones of growth remain between the centers of ossification at the ends and in the central marrow, but once the three centers of ossification fuse, the bone has achieved its adult size. In humans, this closure begins in the late teens in some bones and is not complete until about age 25.

As a bone grows and is reshaped, the original spongy bone in its outer regions is largely replaced by compact (hard) bone (Figure 44.8). By secreting a mineral structure around themselves, osteoblasts become trapped within spaces (lacunae) in the bone, with their cellular extensions reaching out into fine radiating canals; in this condition, the cells are properly called osteocytes. A cross section shows that the bone is organized into a series of **osteons,** each made of concentric layers around a central canal (Figure 44.9); blood and lymph vessels and nerves running through these canals serve the cells in each osteon. Compact bone is incredibly strong, yet light in weight. Bones have a tensile strength of 700–1,400 kg/cm^2 and can stand compression of 1,400–2,100 kg/cm^2, in the same range as aluminum and mild steel.

Exercise 44.3 Why do people stop growing taller by their early 20s?

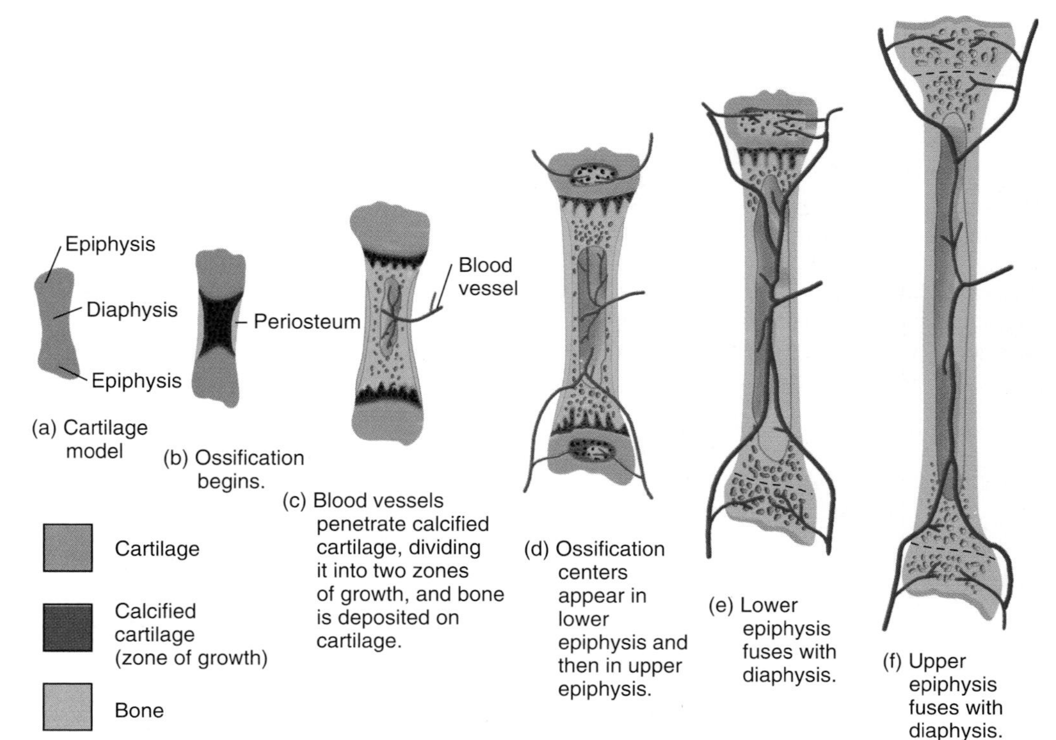

Figure 44.7
A long bone grows from a small model *(a)* made of fibrous membranes and cartilage, which has a shaft (diaphysis) between two enlarged ends (epiphyses). *(b)* Ossification starts beneath the layer of connective tissue (periosteum) surrounding the shaft, but at the same time the cartilage continues to grow below the epiphyses, so the bone continually enlarges. *(c)* Soon blood vessels penetrate the periosteum and infiltrate the developing bone; internal centers, which become bone marrow, start to open up. Some cells in the marrow become osteoblasts that start to build trabeculae of new bone, while other cells become osteoclasts that break down older bone, enabling the whole bone to expand. *(d)* Ossification centers appear in both epiphyses. Growth will continue as long as growth centers remain between these epiphyseal centers and the central marrow, but it stops when the epiphyses fuse with the diaphysis *(e, f)*.

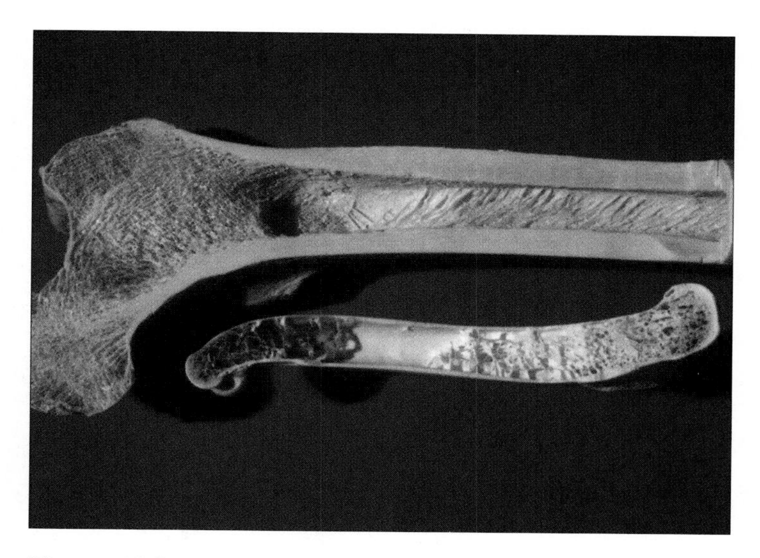

Figure 44.8

A mature bone has an outer layer of compact bone surrounding a body of spongy bone. The large bone is a human femur, the smaller one a turkey femur.

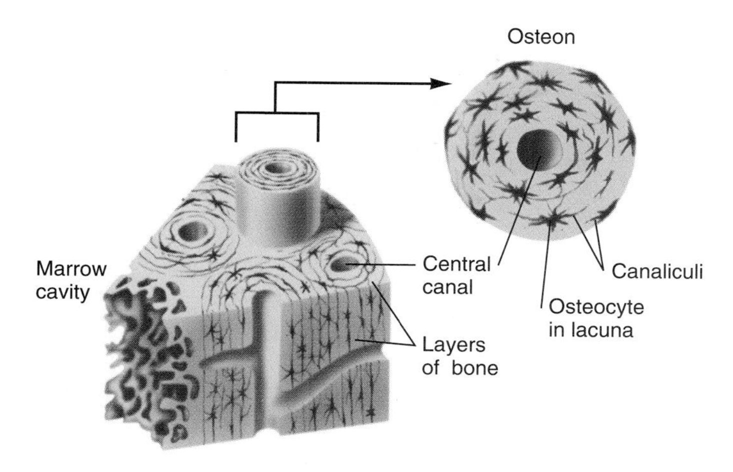

Figure 44.9

Compact bone is made of units called osteons, which consist of concentric layers of bone surrounding a central canal. Osteocytes remain trapped in spaces (lacunae) in the bone, and they are maintained by blood and lymph vessels running through the central canal. Extensions from neighboring osteocytes through the canaliculi meet and form gap junctions, so many osteocytes can communicate through a network.

B. Muscle

A skeleton is moved by an arrangement of opposing skeletal muscles.

Animal muscles are highly organized versions of the actin-myosin contractile systems that are so widespread in eucaryotic cells (see Chapter 11). The actin and myosin filaments of muscle exert forces on one another just as the less organized microfilaments of other cells do. Three major types of muscle have been identified (Figure 44.10). **Smooth muscle** is relatively simple, made of cells shaped like spindles (cylinders with tapered ends). This is the muscle of internal structures such as blood vessels and the digestive tract. **Skeletal muscle,** which moves the skeleton and parts of the skin (facial muscles, for instance), consists of long muscle fibers with a repeating pattern of bands (striations) that characterize it as *striated muscle.* **Cardiac muscle** forms the bulk of the vertebrate heart and is also made of long, striated fibers, fused with one another in a branching pattern. As we examine muscle structure, we will see that these prominent striations arise from a regular repeating alignment of the actin and myosin filaments.

Skeletal muscles work in opposing pairs. A muscle can only contract, so once it has pulled a bone in one direction, an opposing muscle must pull the bone in the opposite direction, simultaneously stretching the first muscle. Thus antagonistic muscles in the upper arm raise and extend the forearm, and another set of antagonistic muscles rotate the forearm back and

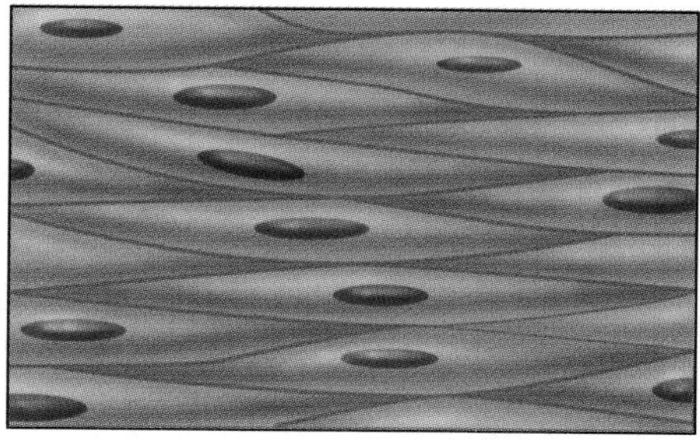

(a) Smooth

(b) Skeletal

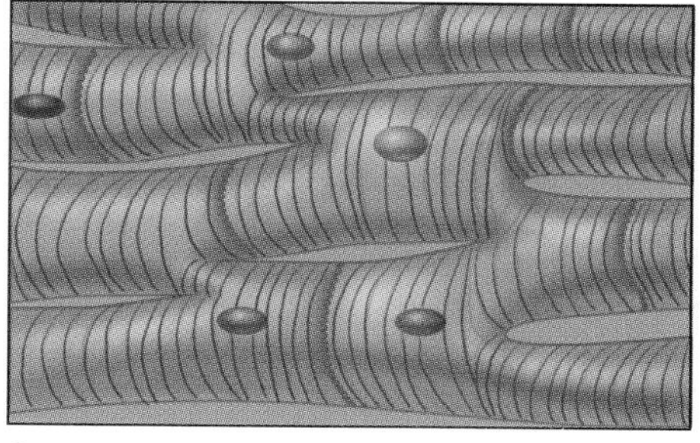

(c) Cardiac

Figure 44.10

Animals have three major types of muscle: *(a)* smooth, *(b)* skeletal, and *(c)* cardiac.

Figure 44.11
By itself, a muscle can only contract, but opposing pairs of muscles work together, each one extending its opponents, so by balancing each other, they effect smooth, controlled movements. Thus the brachialis and biceps work against the triceps to raise and lower the forearm *(a),* and the supinator and pronator muscles work against each other to rotate the forearm at the elbow *(b).*

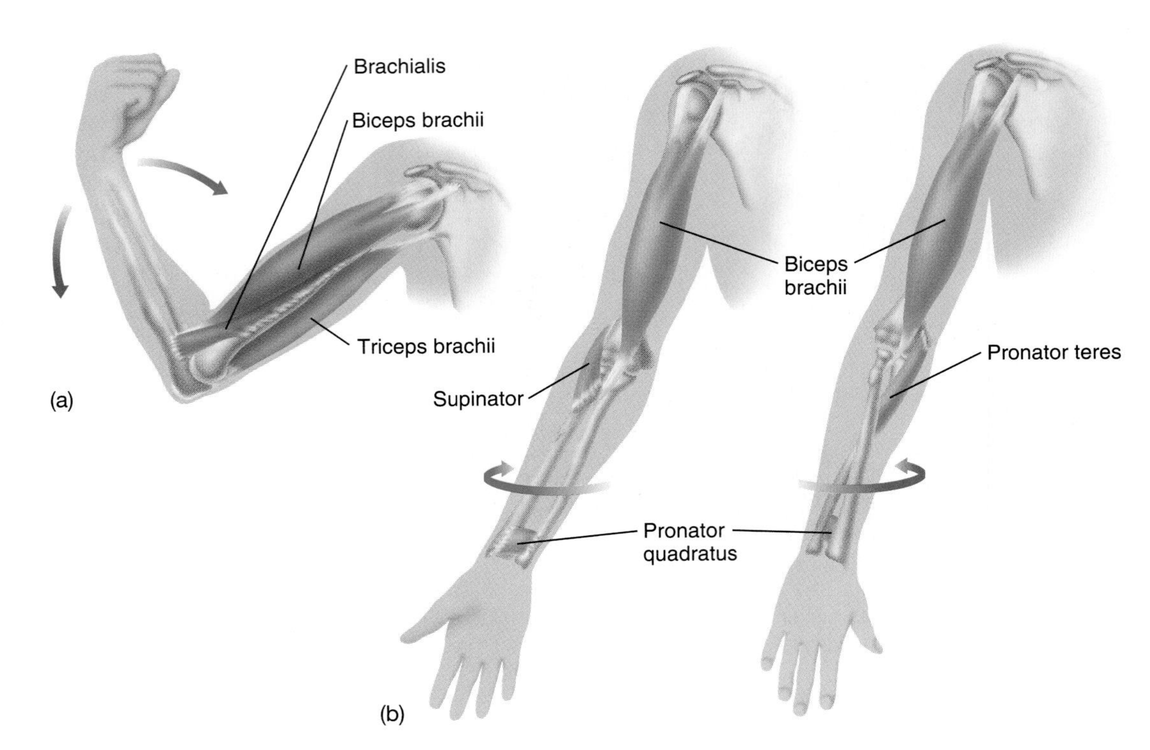

forth around the elbow joint (Figure 44.11). This arrangement allows muscles exerting opposing forces to cooperate and create smooth, delicate movements. Working in coordination, a system of muscles achieves what one alone could not achieve.

The fascia surrounding muscle bundles all converge into tendons at each end of the muscle. As a muscle contracts, the tension it develops is transmitted through the strong collagen fibers of the tendons, but the tendons and other connective tissue, being stretchable, create an elastic buffer that stretches as tension develops and recoils as it subsides. This arrangement creates the *series-elastic component* of muscle contraction. While most muscles attach directly to nearby bones, some have very long tendons attached to distant bones. The ability to locate a muscle so far from its attachment site allows some structures, such as a hand, to be quite streamlined. As Figure 44.12 shows, if all the muscles that move the fingers had to be in the hand itself, a hand would be huge and unwieldy; the muscles are actually located in the forearm, with only narrow tendons running through the hand.

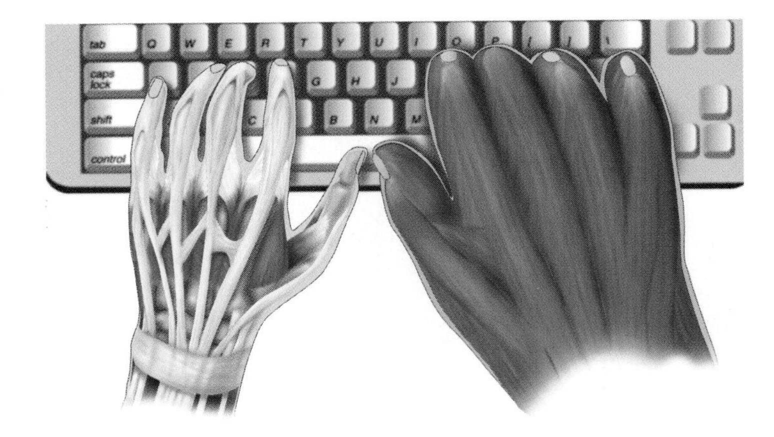

Figure 44.12
Long tendons attached to muscles in the forearm run through a hand. If each muscle had to be attached directly to the bone it moves, our hands would be very bulky and clumsy. Source: Hardin, *Biology: It's Principles and Implications,* 2nd edition, 1966, W.H. Freeman.

44.6 A muscle is a highly organized system of contractile proteins.

A whole skeletal muscle is divided into smaller and smaller bundles of tissue called *fascicles,* each wrapped in fascia (Figure 44.13). The cellular subunit of a muscle is a **muscle fiber,** a long cell with many nuclei called a *syncytium* (*syn-* = together; *-cyte* = cell), formed by the fusion of many uninucleate embryonic cells. Each fiber is as thick as an ordinary eucaryotic cell (10–100 μm) but may be millimeters or even centimeters long. It contains many **myofibrils,** which are bundles of contractile proteins called **myofilaments**—equivalent to microfilaments in other cells. Myofibrils are about 1 μm in diameter and so densely packed that other organelles, such as the mitochondria that supply energy for movement, are forced into spaces between them. Each myofibril is divided lengthwise into repeating structural units called **sarcomeres;** in striated muscles, the regular alignment of sarcomeres in adjacent myofibrils is what gives a muscle its characteristic appearance. A distinctive endoplasmic reticulum, called **sarcoplasmic reticulum (SR),** surrounds the myofibrils.

Beginning in the 1950s, an excellent combination of biochemistry and electron microscopy by Andrew F. Huxley and by Hugh Huxley and Jean Hanson showed that myofibrils

Figure 44.13

Striated muscle is organized in a hierarchy of structural units: muscle fibers, myofibrils, and myofilaments. Each myofibril is divided lengthwise into sections called sarcomeres. A sarcoplasmic reticulum encloses the myofibrils, and a system of transverse tubules (T tubules) extends from the plasma membrane (sarcolemma) deep into the fiber.

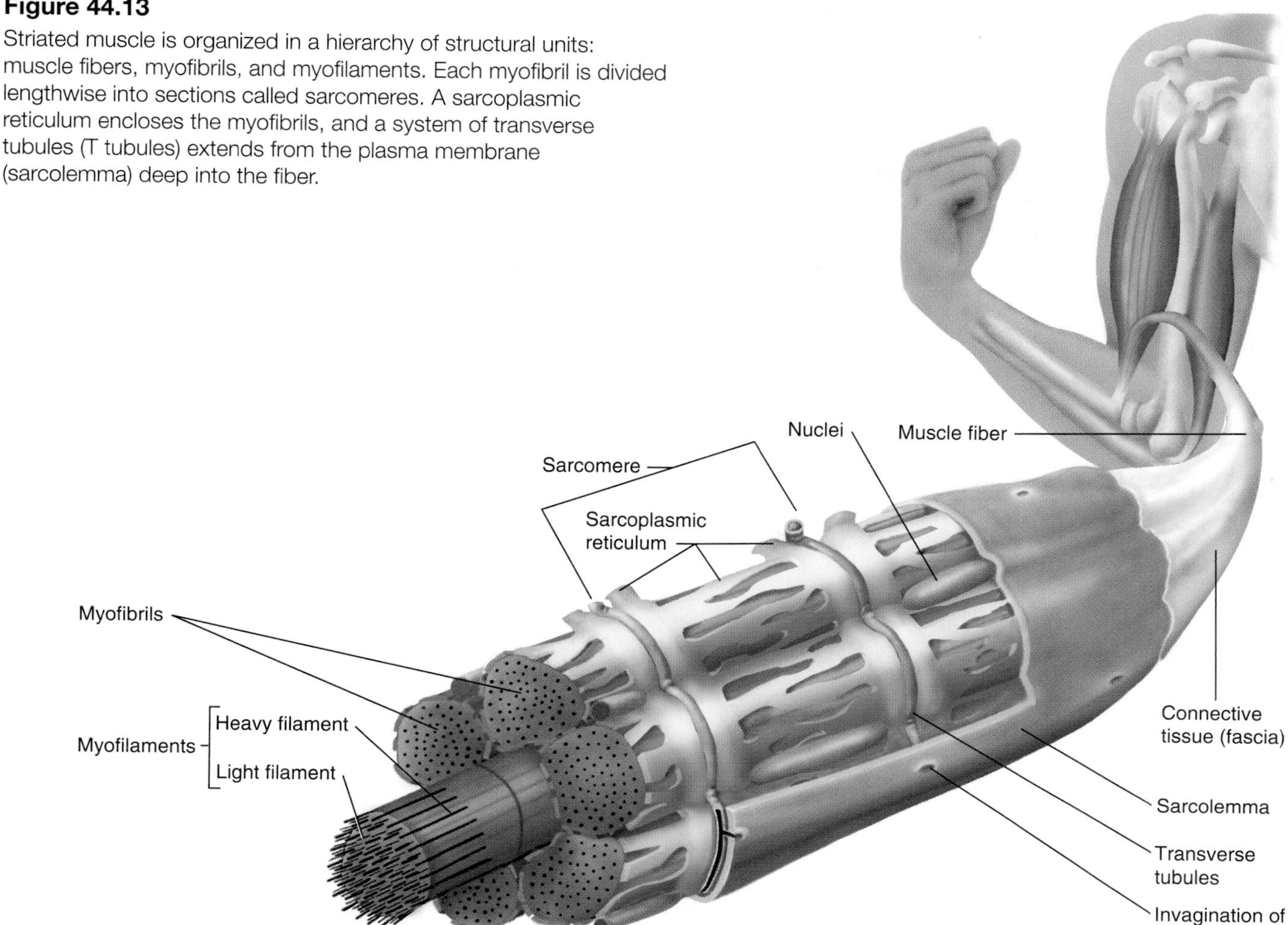

contain two kinds of myofilaments: *heavy filaments* of myosin and *light filaments* of actin (Figure 44.14), which in cross section are seen to be lined up in a most elegant hexagonal array (Figure 44.15). The actin filaments are held in this structure by their attachment on the Z lines, which are hexagonally symmetrical plates of the protein α-actinin. Each sarcomere, the unit between two Z lines, can be divided into several bands on the basis of its appearance under polarized light: Light I bands at the ends of the sarcomere contain actin filaments alone; the light H zone in the middle contains myosin filaments alone; and the dark A band is where actin and myosin filaments overlap.

Although muscles contract through the same actin-myosin interactions as in other cells (see Section 11.9), the larger, more highly organized muscle apparatus shows the process most clearly. The filaments don't change length. Instead, when a muscle contracts, the two types of filaments slide past each other. The globular myosin heads extending from each heavy filament attach to the actin filaments and pull on them a short distance. Through many repetitions of this action in a single contraction, each myosin filament pulls the neighboring actin filaments past itself. Since the ends of the myosin filaments pull in opposite directions, toward the center of the sarcomere, the myosin band pulls the actin bands and Z lines closer together, contracting the whole muscle fiber (Figure 44.16). Normally each sarcomere only contracts about 20 percent of its length (about 2 μm), but the combined contraction of thousands of sarcomeres in a series (and, of course, billions in the entire muscle) makes a whole muscle shorten visibly.

Contraction is powered by ATP, using the ATPase of the myosin heads. (Sidebar 44.2 describes a reserve energy source.) The energy of ATP goes into conformational changes in the head domains of the myosin, as they attach to the actin filaments, pull, release, and then reattach. After death, rigor mortis results from a lack of ATP, which is needed to break the cross-bridges between filaments.

Exercise 44.4 Suppose a muscle is 15 cm long and the myofibrils composing it have sarcomeres 2 μm long when stretched. How many sarcomeres would have to be placed in series in the longest part of the muscle? If all the sarcomeres contract 20 percent, how long will the muscle be when fully contracted?

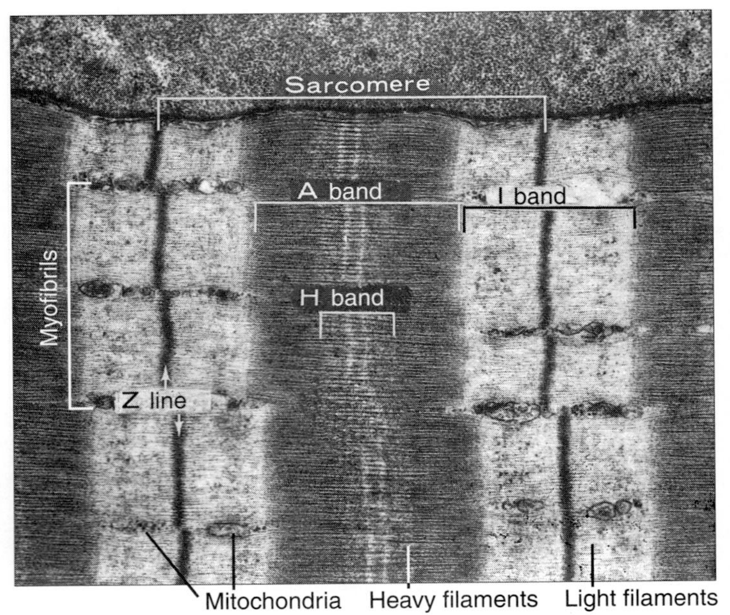

Figure 44.14

An electron micrograph of skeletal muscle shows five myofibrils, separated by densely packed mitochondria. Myofibrils are divided into sarcomeres by the Z lines. The I bands consist of light myofilaments of actin. In the A bands, actin myofilaments overlap heavy myosin myofilaments. The H bands include myosin myofilaments alone.

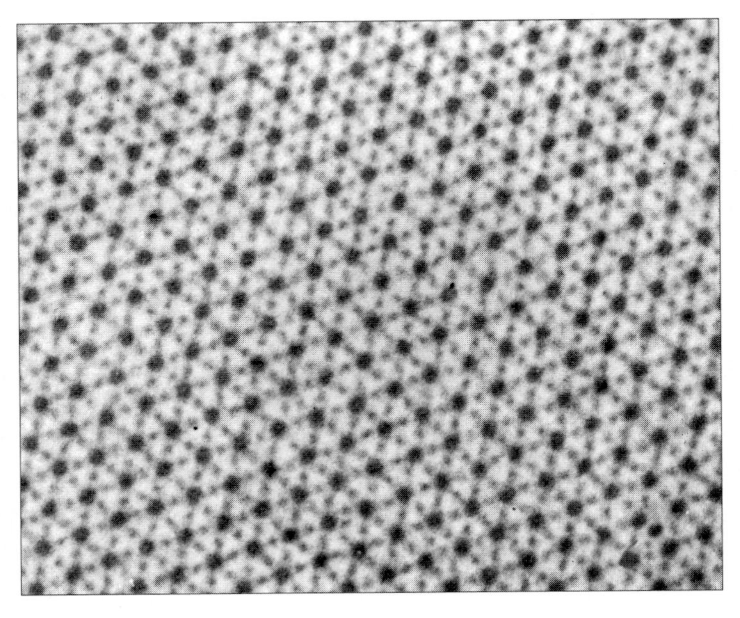

Figure 44.15

A cross section through a myofibril shows its hexagonal array of heavy myosin filaments, each surrounded by six light actin filaments. Notice the thin connections between the two.

Figure 44.16

A muscle contracts as the bundles of actin filaments are pulled closer together by sliding between the myosin filaments.

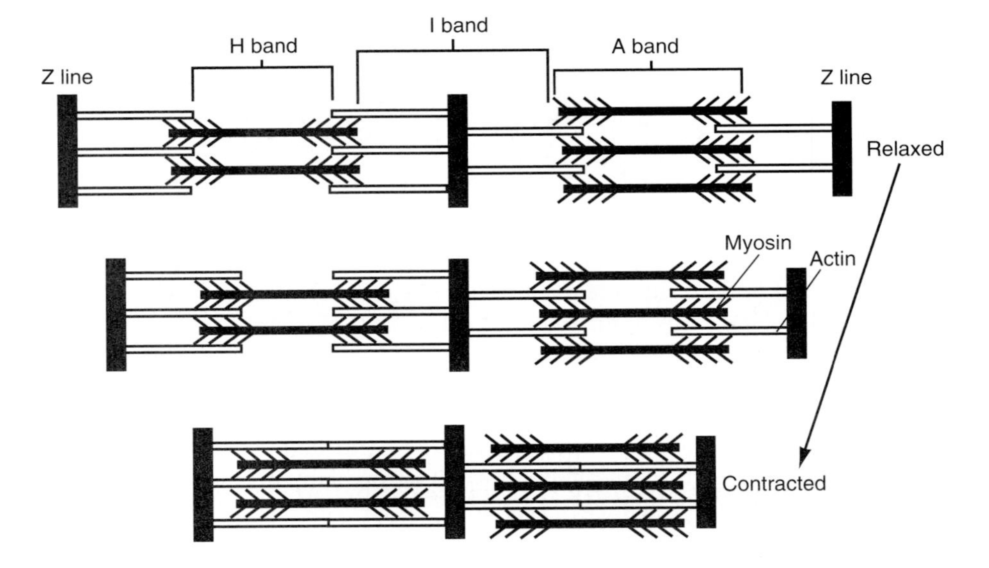

44.7 Muscle contraction is regulated by calcium ions.

Calcium ions are second messengers in muscle fibers. By controlling the interaction between actin and myosin, they control contraction itself. The control is exerted through two other proteins attached to actin: the calcium-binding protein **troponin** and the inhibitory protein **tropomyosin,** which prevents contraction by blocking the binding sites on actin to which the myosin heads attach (Figure 44.17). These two proteins interact with each other in a stereospecific complex that is sensitive to the Ca^{2+} concentration in the cytosol.[2] When Ca^{2+} enters the cytosol, it binds to troponin, changing the protein's conformation. Troponin then pushes tropomyosin aside, exposing the binding sites on actin where myosin interacts. This whole remarkable system depends on the properties and interactions of just four proteins—actin, myosin, troponin, and tropomyosin—

[2]The cytosol of a muscle fiber is often called *sarcoplasm,* and its plasma membrane is called the *sarcolemma,* but we will use the more general terms here.

Sidebar 44.2 Phosphocreatine

The energy source for muscle contraction is ATP, but during sustained muscle activity, ATP may be used faster than cellular respiration can create it. As an energy reserve, striated muscles build up a supply of *phosphocreatine* (creatine phosphate), a compound with a high enough phosphoryl transfer potential to convert ADP + P_i back to ATP. The molar concentration of phosphocreatine is over three times that of ATP, and it is built up while a muscle is not being worked hard.

The enzyme that transfers phosphoryl groups between ATP and phosphocreatine, *creatine phosphokinase*, has an important role in medical diagnosis. Skeletal and cardiac muscle have distinct isoenzymes (different forms of this enzyme). People with muscular dystrophy, a degenerative disease of skeletal muscle, are diagnosed in part by the elevated level of creatine phosphokinase in their blood. Similarly, people who have sustained a heart attack (technically, a myocardial infarction; see Chapter 45) have an increased level of the cardiac form of the enzyme in their blood plasma. The plasma level of this enzyme is used to measure the severity of the attack.

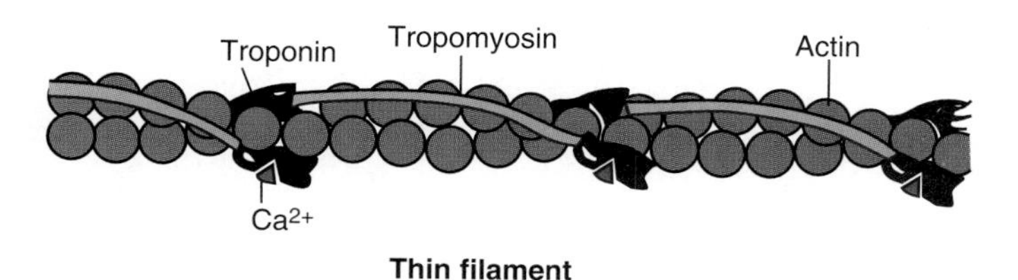

Figure 44.17
Contraction is controlled by two proteins: tropomyosin, which lies over the actin filaments, and troponin, which has a Ca^{2+}-binding site.

which interact with each other stereospecifically and change their shapes in response to Ca^{2+} and ATP.

Ca^{2+} ions act somewhat differently in smooth muscle contraction. These ions enter the cytoplasm through voltage-gated Ca^{2+} channels in the plasma membrane, rather than from a sarcoplasmic reticulum. Smooth muscle fibers have no troponin or tropomyosin. Ca^{2+} binds to the regulatory protein *calmodulin,* which is related to troponin, and calmodulin then activates a protein kinase. The kinase phosphorylates the myosin heads, allowing them to bind to actin and initiate contraction.

Calmodulin, Section 11.8.

Calcium ions regulate contraction, but what regulates calcium ions? Next we will examine the role of signals from the nervous system in this process.

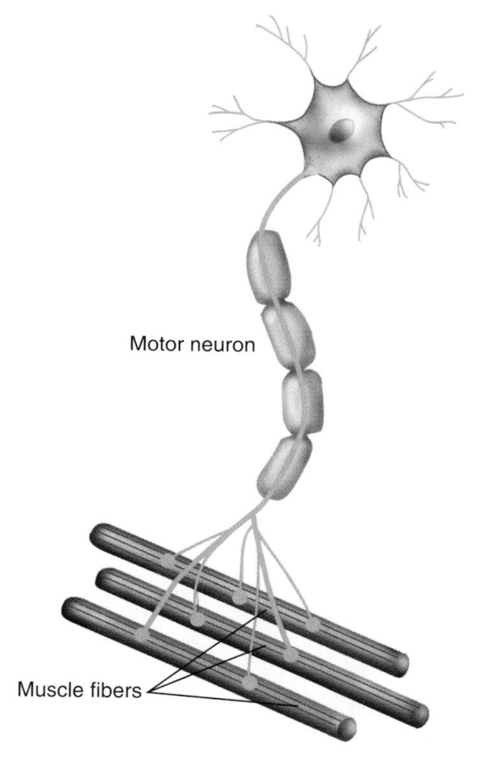

Figure 44.18
A motor unit consists of a motor neuron connected to one or more muscle fibers.

44.8 A neuron activates a muscle fiber by changing its membrane potential.

Activity in skeletal muscles is initiated and controlled by signals from the nervous system. A single motor neuron and all the muscle fibers it controls constitute a **motor unit** (Figure 44.18), the elementary unit of activity. The fibers of each motor unit contract in an all-or-none action; a muscle as a whole, however, contracts with increasing strength as more and more units are *recruited* to contract at one time, just as recruiting more people to pull on a rope increases the tension they exert. Motor units vary greatly in size. The fingers can be controlled very delicately and precisely because their muscles are divided into many motor units, each made of only a few muscle fibers and controlled independently. In contrast, back or chest muscles are divided into fewer units, each containing more muscle fibers. The motor cortex of the brain reflects this structure by allocating space in proportion to the number of motor units in each muscle (see Figure 42.16).

As a motor neuron's axon approaches a muscle, it branches out to make several contacts with each fiber it controls. The final intimate connection between a neuron and muscle occurs at a **neuromuscular synapse** (Figure 44.19), which is similar to a synapse between two neurons. Each ending of the motor neuron's axon contains vesicles of acetylcholine, and nerve impulses arriving at a neuromuscular synapse stimulate the release of this neurotransmitter into the synaptic cleft. Acetylcholine binds to

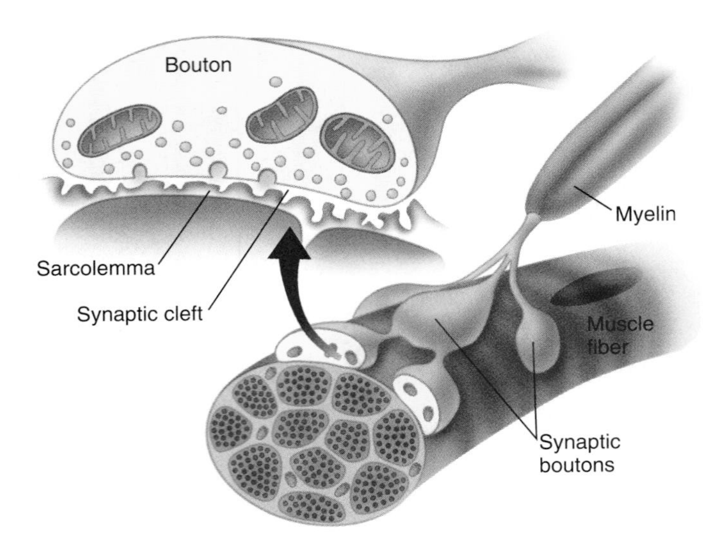

Figure 44.19
A neuromuscular synapse forms where a motor neuron makes contact with a muscle fiber. The bouton is the enlarged axon ending.

receptors in the **motor endplate,** the plasma membrane of the muscle fiber within the synapse; these receptors open ligand-gated channels for Na^+ ions, which flow into the motor endplate and increase the potential locally, creating an **endplate potential (EPP).** An EPP is almost the same as an excitatory postsynaptic potential at a synapse between neurons, but the EPP is much stronger because the endplate is much larger than other synapses. The EPP then initiates an action potential that spreads lengthwise along the plasma membrane of the muscle fiber, just as an action potential is conducted along an axon. This action potential, as we will show momentarily, activates the release of Ca^{2+} ions to produce muscle contraction by penetrating deep into the muscle fiber through transverse tubules.

44.9 A change in membrane potential triggers a release of calcium ions.

A muscle fiber may be 100 μm in diameter, but its myofibrils are only 1–2 μm thick and are buried inside. So a signal to contract—an action potential in the muscle fiber membrane—begins far from many of the myofibrils that actually do the work, and a system of **transverse tubules,** or **T tubules,** carries a signal from the membrane to the contractile elements (Figure 44.20). T tubules are deep invaginations of the plasma membrane that run through the muscle fiber close to the Z lines of each sarcomere, so the lumen of a T tubule is an extension of the extracellular space around the fiber. Consequently, an action potential initiated at a motor endplate propagates along the membrane of the muscle fiber and runs down the T tubules deep into the fiber's interior.

The sarcoplasmic reticulum (seen in Figure 44.20) around the myofibrils sequesters Ca^{2+} ions and keeps their concentration in the myofibrils very low (10^{-7} M or less). Though the reticulum membranes are in intimate contact with the T tubules at each end of a sarcomere, one membrane system does not open into the other. Yet action potentials moving down the T tubules stimulate the release of Ca^{2+} from the reticulum, increasing the Ca^{2+} level of the cytosol about a hundred times. Ca^{2+} ions then initiate contraction by binding to troponin, as outlined in Section 44.7.

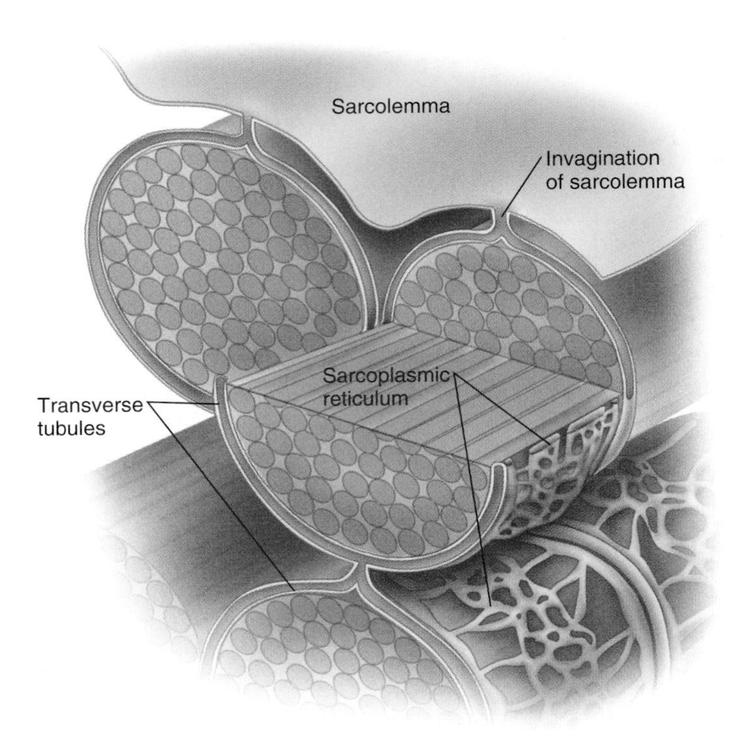

Figure 44.20
The T tubules are invaginations of the plasma membrane that run deep into each muscle fiber near the Z lines. They carry signals from the fiber surface to the sarcoplasmic reticulum surrounding the myofibrils, which stores Ca^{2+} ions that control muscle contraction.

In most skeletal muscles, a single impulse from a motor neuron triggers a single action potential in each muscle fiber in its motor unit, causing each fiber to respond with a **twitch,** a brief contraction that lasts for a few tenths of a second (Figure 44.21). Since the SR rapidly pumps Ca^{2+} ions back out of the cytosol, using its Ca^{2+}-linked ATPases, the contractile apparatus quickly stops working and the tension in the muscle subsides. Meanwhile, acetylcholinesterase on the motor endplate has been breaking down remaining molecules of acetylcholine, sharpening and defining each nervous signal. But if a series of nerve impulses arrive rapidly at a muscle fiber, they initiate contractions that build on one another and create a sustained state of contraction called **tetanus.** The Ca^{2+} concentration in the cytosol stays high, and the contractile apparatus keeps working.

Because the connective tissues of a muscle are elastic, the initial contractions stretch them and build up tension. Continued neural stimulation then builds tension in the muscle to higher and higher levels until the cross-bridges begin to slip and the contraction strength levels off. Further stimulation eventually produces fatigue, as the number of cross-bridges is reduced with a corresponding reduction in tension. However, the muscle fatigue we commonly feel after working a muscle hard is a complex phenomenon; it involves the ability of the SR to accumulate Ca^{2+} ions, the ability of the muscle fiber to develop an action potential, and overall changes in the fiber's metabolism, which varies with the fiber type, as described in the next section.

To summarize, the following series of events occurs in muscle contraction: An action potential in a motor neuron releases acetylcholine at a neuromuscular synapse. These signal ligands diffuse across the synapse and open ion gates in the motor endplate, creating an endplate potential. The EPP in turn elicits an action potential in the muscle fiber membrane, which is conducted along the length of the fiber and also conducted down the T tubules, stimulating the sarcoplasmic reticulum to release Ca^{2+} ions into the cytosol. Ca^{2+} ions bind to troponin molecules, which change conformation and shift the position of tropomyosin molecules on actin. Myosin is then able to interact with actin and begin contraction. Contraction uses the energy of ATP, which is hydrolyzed by ATPases in the heads of myosin molecules. As Ca^{2+}-linked ATPases in the sarcoplasmic reticulum remove Ca^{2+} ions from the vicinity of the contractile proteins, tropomyosin returns to its original position on actin, and the muscle fiber begins to relax.

Exercise 44.5 To determine how spaces in a muscle fiber are connected to one another, an electron-dense protein such as ferritin is placed just outside muscle fibers. It is useful because it cannot cross a cell membrane. After giving the ferritin molecules time to diffuse, the muscle is sectioned and examined with an electron microscope. Where would you expect to see the proteins? *(a)* In the T tubules? *(b)* In the sarcoplasmic reticulum? *(c)* In the cytoplasm?

44.10 Vertebrate striated muscles contain different types of fibers.

Skeletal muscle fibers are specialized for different functions and are distinguished by several features. **Tonic fibers** produce a steady contraction, and **phasic fibers** produce twitches that can build into tetanus. Fibers also differ in their speed, energy source, and color. Phasic fibers are either *slow-twitch* or *fast-twitch,* as determined by their myosin molecules, which can split ATP at different rates and therefore cycle cross-bridges with actin at different speeds. A fiber can metabolize oxidatively, using mitochondria to generate ATP, or it can use glycolysis alone, generating ATP even in the absence of oxygen. An oxidative fiber has many mitochondria and is full of myoglobin, which stores the oxygen needed for this kind of metabolism. Highly oxidative muscle is also highly vascularized, with a rich blood supply, and the combination of reddish myoglobin and vascularization makes these fibers *red;* a glycolytic fiber has few mitochondria, little myoglobin, and is less vascularized, so it is *white.* (We see this difference most often in the dark meat and white meat of a chicken.) Finally, fibers differ in their ability to resist fatigue. By these criteria, there are four types of fibers:

- *Tonic fibers* contract slowly, without twitching. Instead of developing action potentials, they depend on the endplate

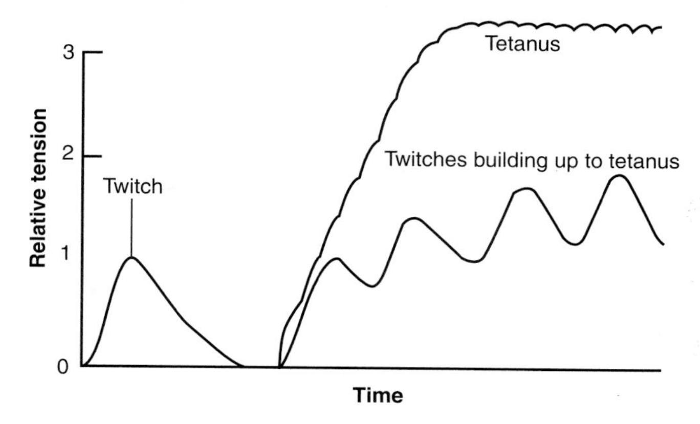

Figure 44.21
Tension develops in a muscle fiber in different ways, depending on the stimulation it receives. A single twitch develops a moderate tension, and a series of twitches can build somewhat higher tension. However, a rapid series of stimuli can produce tetanus, with a high, sustained tension.

potentials generated at each of the many synapses the motor neuron makes along a fiber's length. A single stimulus elicits only a small contraction, and a train of repeated stimuli produce more and more contractions, which build on one another. These are used as posture muscles by amphibians, reptiles, and birds, and in a few small mammalian muscles.
- *Oxidative slow phasic fibers* (red, or type I) contract with slow, all-or-nothing twitches. They combine oxidative metabolism with slow myosin-ATPases and are very resistant to fatigue. These fibers predominate in muscles like those in the trunk and hip, which maintain body posture by staying contracted for hours at a time.
- *Oxidative fast phasic fibers* (intermediate, or type IIA) contract quickly because of their rapid myosin-ATPases and are specialized for rapid, repetitive movement. They metabolize oxidatively and are very resistant to fatigue. These fibers predominate in arm and leg muscles, which produce rapid movements that are only sustained for short times.
- *Glycolytic fast phasic fibers* (white, or type IIB) are powerful fibers that contract quickly because of their rapid myosin-ATPases, but they use glycolytic metabolism and fatigue quite rapidly.

Sidebar 44.3 discusses how muscle-fiber types are related to exercise and athletic performance.

Exercise 44.6 *(a)* Domestic chickens and turkeys, which have little ability to fly, are unusual in having flight muscles made of white fibers. In contrast, what kinds of fibers should predominate in the flight muscles of birds that make long, sustained flights? *(b)* A Ruffed Grouse only flies a short distance when flushed; if a grouse is flushed three or four times successively, it can be picked up in the hand, exhausted. Explain why.

Sidebar 44.3 Exercise and Training

Knowing that our muscles are made of different kinds of fibers allows us to make sense of athletic performance and exercise regimes. Athletes who need to move quickly for short times, such as sprinters, should have an abundance of fast-twitch fibers in their leg muscles, whereas those who lift heavy weights should have mostly slow-twitch fibers. Then the question arises, are athletes born or made? Can an athlete train to develop muscles for the events she chooses, or must she choose events to match the muscle type she is born with? The standard answer has been that the ratio of fast- to slow-twitch fibers is genetically determined and cannot be changed by exercise. However, recent experiments show that lifting heavier loads during exercise favors the transformation of fast fibers to slow, and lifting lighter loads increases the percentages of fast fibers. It isn't clear just how effective such exercise can be. The best 100-meter sprinters have abundant fast-twitch fibers in their legs, and the best long-distance runners have mostly slow-twitch fibers. So, for greatest success, you should find out what you excel at and then develop the appropriate muscles further.

If exercise has a limited ability to change the number of fibers in a muscle, what are its physiological effects, and what is the best exercise regime? From the viewpoint of an athlete, there are two extremes of exercise regime, to build either strength or endurance. Those who depend on strength, such as power lifters, exercise with high weights but relatively few repetitions in each set; the result is to increase the size of each muscle fiber by the addition of new myofibrils, so it becomes stronger. Those who need greater endurance, such as swimmers, cyclists, and long-distance runners, exercise with low weights and many repetitions in each set; this training increases the vascularization of each muscle, so its oxidative slow-twitch fibers are better supplied with oxygen. Endurance regimes also build greater myoglobin concentrations and more mitochondria in all muscle fibers, so the fibers are able to generate more ATP, produce less lactic acid, and more efficiently use lipids for energy. The average person, and probably the majority of athletes, needs a combination of the two regimes, to increase both strength and endurance. For this purpose, medium weights with a medium number of repetitions are recommended to both increase the size of muscle fibers and increase vascularization. Trainers also emphasize that opposing muscle sets should be exercised equally for coordination and balance. In addition, exercise increases the pumping capacity of the heart; the heart of a well-conditioned athlete pumps a larger volume of blood with each contraction so the heart does not have to pump so frequently to supply the body with oxygen and nutrients.

44.11 Invertebrate musculoskeletal systems differ from vertebrate systems in several ways.

The exoskeletons of invertebrates are quite different from the skeletons of vertebrates, although cuticle, which makes up the exoskeletons of arthropods and some other invertebrates, shows some interesting similarities to bone at a molecular level (Figure 44.22). Cuticle is based on chitin, a tough polysaccharide comparable to the mucopolysaccharides of vertebrate connective tissue; chitin comprises from about 25 percent to 75 percent of the organic matter in cuticle, the remainder being protein. Chitin by itself is quite soft; cuticle remains soft in some places, but it is toughened, like bone, by the deposition of calcium carbonate.

Invertebrate and vertebrate muscles also differ in important ways. While both animal groups have striated muscle, most invertebrate muscles are tonic, rather than phasic; instead of contracting in an all-or-nothing twitch, each muscle fiber responds with a graded contraction whose force increases with greater stimulation. In arthropods, each fiber is innervated by both excitatory and inhibitory neurons (Figure 44.23), and each neuron makes several synapses on a fiber. However, an entire muscle is typically served by only a few motor neurons. So where the contraction of a vertebrate skeletal muscle is regulated by the number of muscle fibers recruited to contract simultaneously, contraction in an invertebrate muscle is regulated by a balance of graded excitatory and inhibitory stimuli to the same muscle fibers.

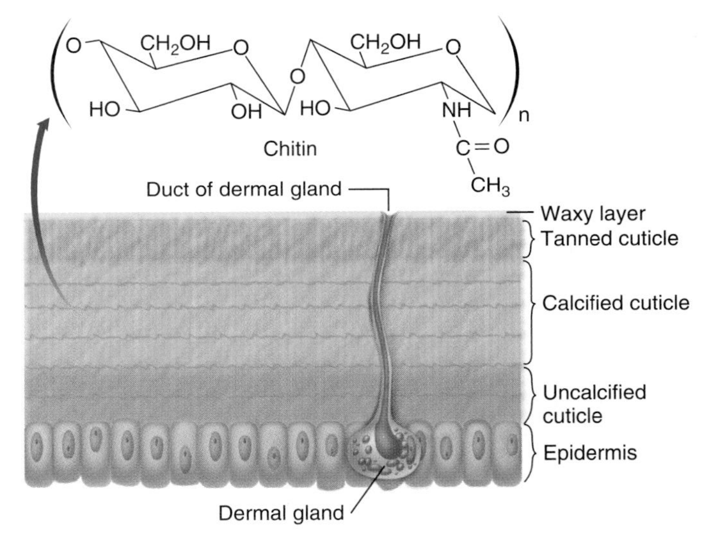

Figure 44.22
Arthropod cuticle consists of several layers secreted by the epidermis. It is mostly the fibrous mucopolysaccharide chitin supplemented by proteins. Its surface is hardened (tanned) by the cross-linking of proteins. An inner layer is mineralized by the deposition of calcium carbonate, just as calcium phosphate is deposited on collagen in bones.

Insect flight muscle, which has been studied extensively, is unique. An insect's wing may vibrate 300 to 1,000 times per second, far too fast for each muscle contraction to be stimulated by a nerve impulse. Instead, impulses in excitatory motor neurons initiate beating, and the subsequent stretching of each flight

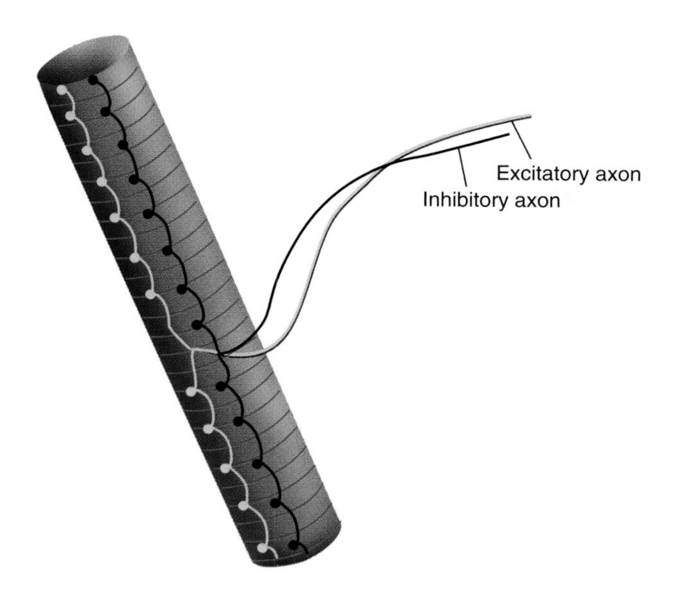

Figure 44.23

An arthropod muscle fiber is innervated by an excitatory and an inhibitory axon. Generally each fiber is innervated by several neurons of each kind, although axons from only two neurons are shown here.

muscle stimulates it to contract again and to maintain beating until being inhibited, as shown in Figure 44.24.

Coda The story of a planet lacking myosin and actin, or their functional equivalents, might make an interesting science fiction tale. If organisms had never evolved motor systems such as the myosin-actin complex, the world would certainly be a very different place. Would the planet be dominated even more by autotrophs, with the heterotrophs limited to fungi and to small, slow-moving creatures secreting external digestive enzymes? The ability to feed as animals depends on muscle for taking bits of food from the environment, pushing them into a digestive cavity, and moving indigestible materials back out. Even having a body large enough to feed in this way depends on muscle to power a circulatory system, for above a certain size a multicellular organism cannot sustain all its cells by diffusion alone. Indeed, this would be a quiet story, one lacking the action usually associated with exciting tales, since active, behaving creatures like humans would not exist. But could nervous systems exist? Could thinking exist? Could our counterparts be intelligent plants?

In fact, myosin and actin did evolve in primitive eucaryotes. The vast majority of species are animals, and the ability to move has become a hallmark of animals. The story of animal evolution is largely one of radiation into new adaptive zones, finding new places to live and new ways of moving and feeding. And so we have our complex world, with its complicated, animal-dominated ecology, depending to a large extent on the chance evolution of a pair of interacting proteins and on the ability of genetic systems to elaborate on their interactions.

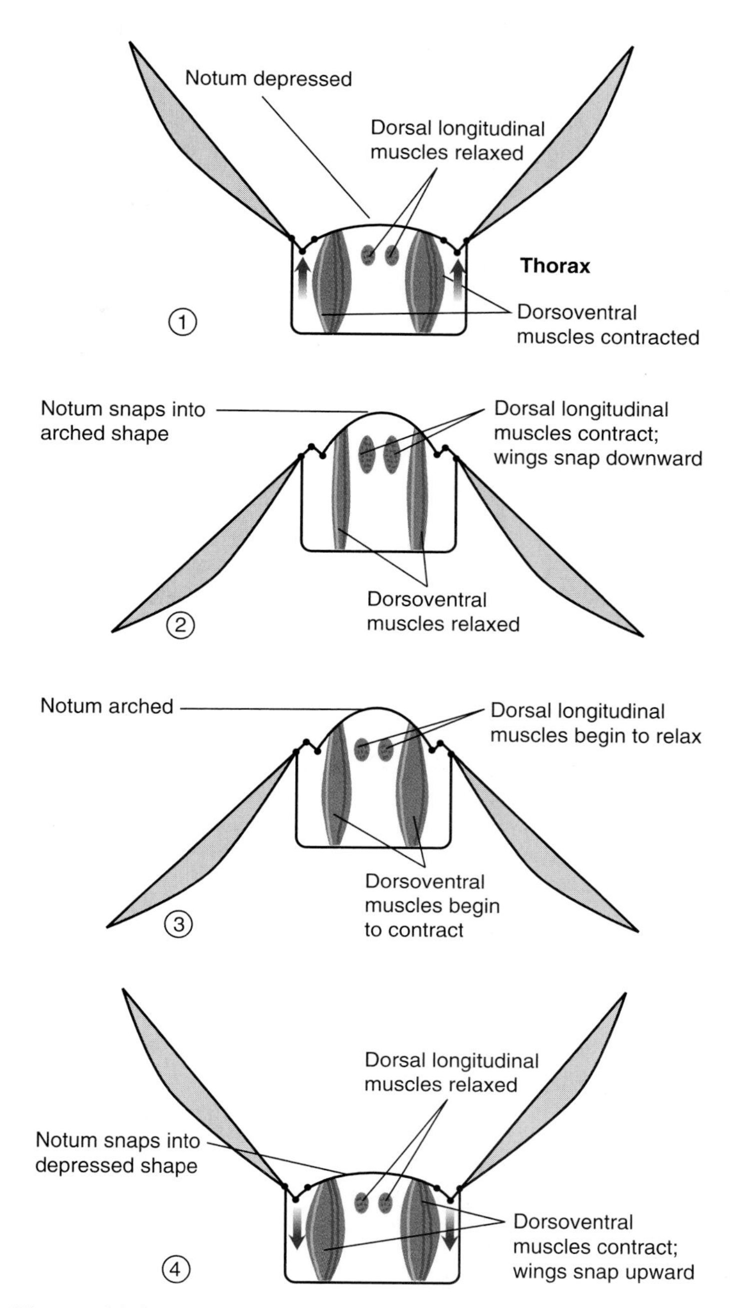

Figure 44.24

The upward and downward strokes of insect wings are created by antagonistic muscles, depressors running longitudinally and elevators dorsoventrally. The insect's wings are attached to the roof of the thorax (the notum), which is poised to spring back and forth between depressed and arched shapes, like the bottom of an oil can that snaps into one position or the other as it is pressed and released. The flight muscles snap the notum back and forth. *(1)* Start with the wings up and the notum depressed. *(2)* As the depressors (longitudinal muscles) contract, they suddenly snap the notum past its midpoint into its arched position, thus depressing the wings. This change relieves the depressor muscles so they can be pulled the other way, and extends the elevator (dorsoventral) muscles fully. But extending the elevator muscles immediately stimulates them to contract *(3),* so they snap the notum into its depressed shape and raise the wings again *(4).* This system sets up a resonance that drives the wings up and down.

Summary

1. Most animals have some kind of skeleton to support their movement: a hydrostatic skeleton, an endoskeleton, or an exoskeleton. The skeletal elements and their muscles can be analyzed as types of levers.
2. Connective tissue is built on an extensive matrix of extracellular fibers and other materials secreted by fibroblasts. These materials include collagen for strength, elastin for elasticity, and mineral crystals for rigidity.
3. A vertebrate skeleton is an articulated arrangement of bones. They are held together by ligaments at joints, or articulations, and are moved by muscles attached through tendons. Joints between bones may either be quite rigid or allow rather free motion. Movable joints are lubricated by synovial fluid and held together by ligaments.
4. Bone is formed by the calcification of a soft model structure.
5. A skeleton is moved by an arrangement of skeletal muscles.
6. A muscle is a highly organized system of contractile proteins. A myofibril is made of interdigitating filaments of myosin and actin. Contraction occurs when one set of fibers slides between the other, as the myosin fibers pull on the actin fibers. The globular heads of the myosins are ATPases, which change shape, using the energy of ATP, and attach to sites on the actin filaments. They then pull the actin filaments a short distance with each conformational change. Thus each bundle of myosin pulls two bundles of actin filaments closer together, thereby shortening the whole muscle.
7. Muscle contraction is regulated by calcium ions, which are stored in the sarcoplasmic reticulum. The interaction between actin and myosin is regulated by the proteins troponin and tropomyosin, which change their shape in response to an influx of calcium ions.
8. Each motor neuron controls a unit made of several skeletal muscle fibers. The neuron terminates at endplates on each muscle fiber and stimulates contraction of the fiber by eliciting an action potential in it. In vertebrates, neurons release acetylcholine, which opens ion channels in the plasma membrane of the endplate that admit Na^+ ions; this produces an endplate potential, leading to an action potential in the muscle fiber.
9. A muscle fiber action potential travels along the fiber membrane and down through a system of transverse (T) tubules into the heart of the fiber. There it stimulates the release of calcium ions from the sarcoplasmic reticulum, which surrounds the myofibrils.
10. Vertebrate skeletal muscle contains different types of fibers, which contract and fatigue at different rates and work oxidatively or glycolytically. These features make them suitable for different functions.
11. Invertebrate musculoskeletal systems show some major differences from vertebrate systems. In arthropods, the skeletons are plates made primarily of chitin, hardened by calcium carbonate. The muscle fibers are generally tonic, rather than phasic, so they produce graded contractions under the control of both excitatory and inhibitory neurons. Insect flight muscles are arranged to snap thoracic plates up and down. They contract too frequently to be controlled by neural signals; instead, they contract automatically when stretched to a certain length.

Key Terms

connective tissue 933
bone 933
ligament 933
fascia 933
tendon 933
cartilage 933
endoskeleton 933
exoskeleton 934
apodeme 934
dense fibrous connective tissue 935
elastic connective tissue 935
loose connective tissue 935
adipose tissue 935
fibroblast 935
axial skeleton 936
appendicular skeleton 936
synovial membrane 937
synovial fluid 937
bursa 938
osteocyte 938
osteoclast 938
osteoblast 938
trabecula 938
osteon 938
smooth muscle 939
skeletal muscle 939
cardiac muscle 939
muscle fiber 940
myofibril 940
myofilament 940
sarcomere 940
sarcoplasmic reticulum (SR) 940
troponin 942
tropomyosin 942
motor unit 943
neuromuscular synapse 943
motor endplate 944
endplate potential (EPP) 944
transverse (T) tubule 944
twitch 944
tetanus 944
tonic fiber 945
phasic fiber 945

Multiple-Choice Questions

1. All these structures are primarily composed of connective tissue *except*
   a. bone.
   b. muscle.
   c. tendon.
   d. ligament.
   e. fascia.
2. Which structural component is a part of invertebrate exoskeletons?
   a. cartilage
   b. bone
   c. apodemes
   d. tendons
   e. osteons
3. The biceps muscle is connected to the shoulder bone and forearm bone, and it raises the forearm. The fulcrum of this lever system is the
   a. shoulder.
   b. elbow.
   c. wrist.
   d. forearm.
   e. upper arm.
4. Dense connective tissue is the primary component of
   a. compact bone.
   b. trabeculae.
   c. cartilage.
   d. tendons.
   e. muscle.
5. The ankle is part of the ______ skeleton.
   a. exo-
   b. axial
   c. appendicular
   d. synovial
   e. chitinous
6. The elbow and knee both
   a. contain a joint cavity filled with synovial fluid.
   b. are diarthroses.
   c. are strengthened by ligaments.
   d. *a* and *b*, but not *c*.
   e. *a*, *b*, and *c*.
7. A fully grown long bone does not contain
   a. spongy bone.
   b. compact bone.
   c. bone marrow.

d. osteoblasts.
e. internal regions of cartilage.

8. Which of these structures completely contains all the others?
   a. myofibril
   b. sarcomere
   c. sarcoplasmic reticulum
   d. myofilament
   e. muscle fiber
9. The lowest concentration of mitochondria is found in ______ skeletal muscle fibers.
   a. oxidative slow phasic
   b. oxidative fast phasic
   c. glycolytic fast phasic
   d. tonic fibers
   e. *a* and *b*
10. Which of these physically shorten when a skeletal muscle fiber contracts?
   a. actin myofilaments
   b. myosin myofilaments
   c. tropomyosins
   d. troponins
   e. sarcomeres

True-False Questions

Mark each statement true or false, and if false, restate it to make it true.

1. Closely packed cells are the most abundant and functional constituent of connective tissue.
2. Bony matrix primarily consists of fibroblasts and bone cells.
3. Most bone-muscle lever systems in our body act to produce a mechanical advantage.
4. Structurally, a bursa is most like a tendon.
5. Bones grow larger and larger because osteoclasts deposit additional bony matrix.
6. Osteons are the basic structural units in compact bone but not in spongy bone.
7. Although we can push an object away and pull it toward ourselves, our muscles can only pull.
8. The fundamental unit of shortening in a skeletal muscle is the myofilament.
9. The energy for muscle contraction comes from the splitting of ATP by an enzyme located next to the calcium binding site of actin.
10. Troponin is a calcium binding protein, while tropomyosin blocks myosin binding sites.

Concept Questions

1. When a pair of muscles is called antagonistic, does that mean each is fighting the other?
2. What evidence can you find to show that the muscle fibers in one motor unit are scattered throughout the muscle rather than being grouped together?
3. Why is the sarcomere the basic unit of contraction?
4. Contrast the role of neuronal stimulation in vertebrate and invertebrate skeletal muscle.
5. Contrast the functions of the sarcoplasmic reticulum and the T-tubules during contraction of a skeletal muscle fiber.

Additional Reading

Caplan, Arnold I. "Cartilage." *Scientific American,* October 1984, p. 84. Cartilage holds spaces for tissues in the embryo, and then it cushions the body. Its fundamental properties of strength and resilience are now explained in terms of the tissue's molecular structure.

Junge, Douglas. *Nerve and Muscle Excitation,* 3d ed. Sinauer Associates, Sunderland MA, 1992. A general introduction to electrophysiology.

Pollack, Gerald H. (ed.). *Muscles and Molecules: Uncovering the Principles of Biological Motion.* Ebner & Sons, Seattle, 1990.

Internet Resource

To further explore the content of this chapter, log on to the web site at:

http://www.mhhe.com/biosci/genbio/guttman/

45 Circulation and Gas Exchange

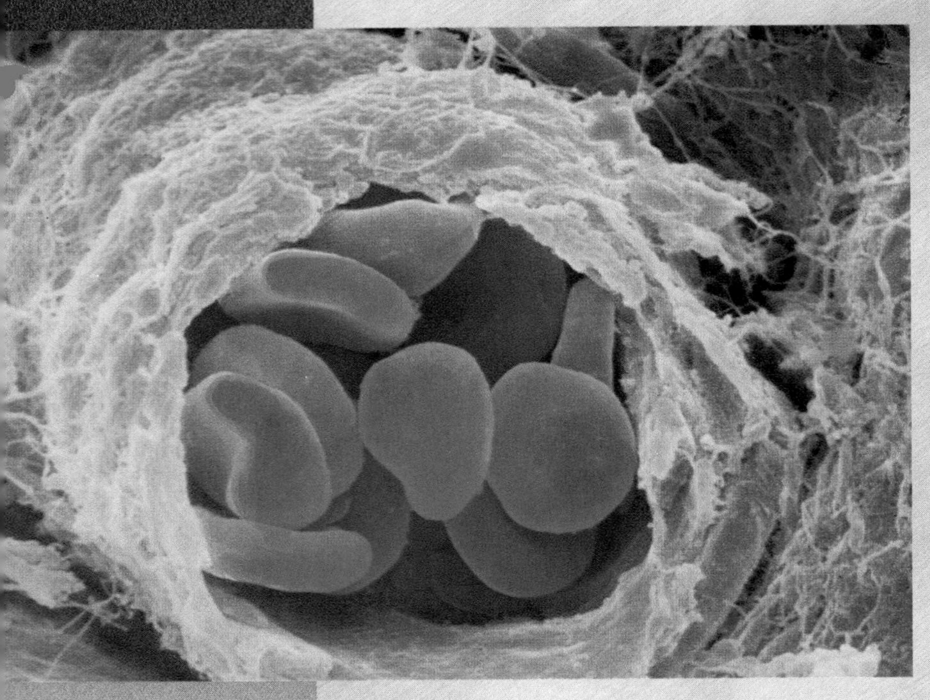

A scanning electron micrograph shows the cut end of a small blood vessel full of red blood cells (artificially colored).

Key Concepts

A. General Principles

45.1 The primary functions of a circulatory system are distribution and exchange.

45.2 Circulatory systems may be open or closed.

B. The Vertebrate Circulatory System

45.3 In a vertebrate circulatory system, a heart pumps blood successively through arteries, capillaries, and veins.

45.4 The vertebrate heart has an internal pacemaker.

45.5 Several regulatory factors combine to keep the heart functioning with quite constant blood pressure.

45.6 Blood is a tissue whose cells are embedded in a liquid plasma.

45.7 Blood plasma exchanges with interstitial fluid in a tissue.

45.8 The lymphatic system carries excess interstitial fluid.

45.9 Atherosclerosis and other disease processes can cause severe damage to the circulatory system.

C. Gas Exchange and Its Regulation

45.10 A circulatory system moves blood past a site of external gas exchange.

45.11 Blood and hemolymph transport oxygen into tissues.

45.12 Carbon dioxide is transported out of tissues.

45.13 An external circuit controls the breathing rhythm in mammals.

45.14 The nervous system regulates breathing in response to oxygen and hydrogen-ion levels.

It is thicker than water. About once a second, each heartbeat pumps it out of the heart into the body through a maze of tubes, carrying nutrients to every body cell, removing their wastes, keeping all the organs healthy. It is commonly considered a sign of life and the most precious of body fluids—so precious that we even store it in special banks. Our good health depends upon keeping its composition constant, with the correct balance of water and ions, and with sufficient sugar to supply all our cells with energy, while keeping the cholesterol concentrations low to avoid heart disease from clogged arteries.

We are talking about human blood, of course. Blood as a substance is precious to each life; blood as a concept is central to understanding animal biology. As we will see, every animal except the smallest aquatic forms requires a circulatory system to maintain all its cells—to supply their nutrients and remove their wastes. Blood also acts as an important medium of communication by carrying hormonal messages throughout the body, and as an agent of metabolism by carrying metabolites from one organ to another. We will focus in this chapter on blood and the circulatory system that distributes it, with emphasis on mammals, especially humans. ■

A. General principles

45.1 The primary functions of a circulatory system are distribution and exchange.

The activities of a circulatory system center largely on moving materials from one *compartment* to another, where a compartment may be an animal's environment, its bloodstream, its cells, or the fluid surrounding its cells. Each substance of importance in an animal's economy will move from a *source,* a compartment where the substance is abundant or is generated, to a *sink,* a compartment where it is consumed or dispersed. For instance, CO_2 will move from its source in cells through intermediate compartment to its sink in the atmosphere. In Section 32.12 we examined Fick's Law, which gives the rate of diffusion between points with concentrations C_1 and C_2, separated by a distance L:

$$Q = DA(C_1 - C_2)/L$$

We noted that this law implies the need for a high surface/volume ratio to support the transfer of a substance between compartments, since the diffusion rate is proportional to the surface area of a structure, and the rate at which the substance is generated or consumed is proportional to the volume of the structure.

Now compare the condition of a unicellular heterotroph in its aqueous medium with both a small aquatic animal and a liver cell in the human body. The unicellular organism is in direct contact with the source of its nutrients—organic food molecules, inorganic ions, oxygen—and a sink to deposit its wastes. Its needs are satisfied solely by diffusion of molecules across its boundary, although its plasma membrane supplements diffusion with some active transport systems that use energy. We speak of the two-way transfer of materials from one compartment to another as an *exchange,* even though the transfer does not necessarily mean trading one molecule for another as the word may imply. The tissues of many small aquatic animals are so close to the surrounding water that they, like the unicellular organism, can exchange materials directly with the water by diffusion alone (Figure 45.1).

A human liver cell, in contrast, is surrounded by other liver cells. The nutrients it needs for energy and biosynthesis are in a digestive system several centimeters away, while the nearest source of O_2 and sink for CO_2 is a few centimeters away in another direction. Its needs cannot possibly be supplied by diffusion. The liver cell only survives because humans have a circulatory system to carry nutrients from the digestive system and O_2 from the lungs, and to wash away its CO_2 and other wastes.

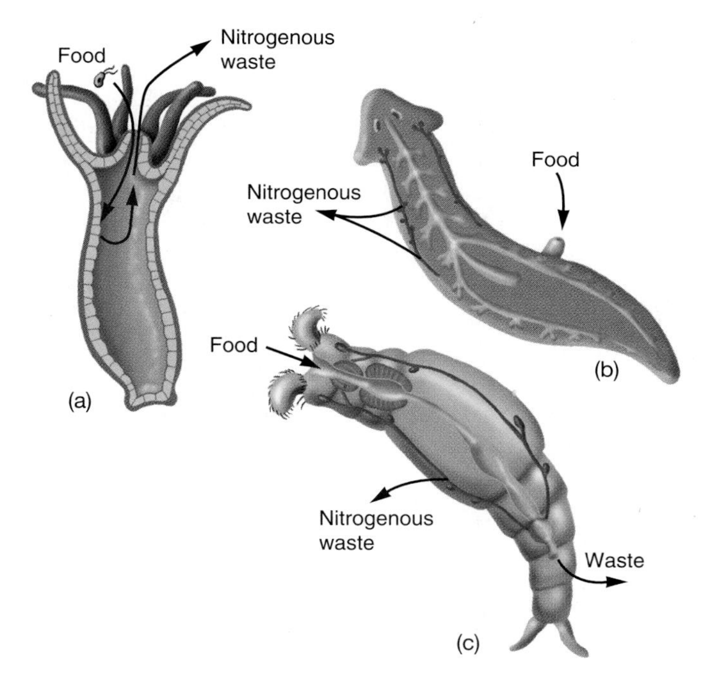

Figure 45.1
Small aquatic animals do not need circulatory or respiratory systems because their tissues are in such intimate contact with the surrounding water that gases can exchange with all cells by diffusion alone. Examples are *(a) Hydra; (b)* a flatworm, *Dugesia; (c)* a rotifer.

Not only is it far from the surfaces where it would have to exchange materials, but diffusion would have to occur through many layers of other cells and their membranes, with low diffusion coefficients.

One of the chief functions of circulation is to transport O_2 and CO_2 throughout the body. This **gas exchange** has commonly been called respiration, but recall from Chapter 9 that respiration in this sense is entirely different from cellular respiration, the oxidative metabolism that occurs inside cells. Because the two processes are so easily confused, we will generally use the term "gas exchange" and avoid "respiration" as a synonym. Nevertheless, O_2 and CO_2 are still called *respiratory gases,* and the brain centers that control breathing are still called respiratory centers. Gas exchange occurs in two steps, *external exchange,* in which the blood exchanges O_2 and CO_2 with the atmosphere as it moves through structures such as lungs, gills, and skin, and *internal exchange,* in which these gases exchange with the tissues:

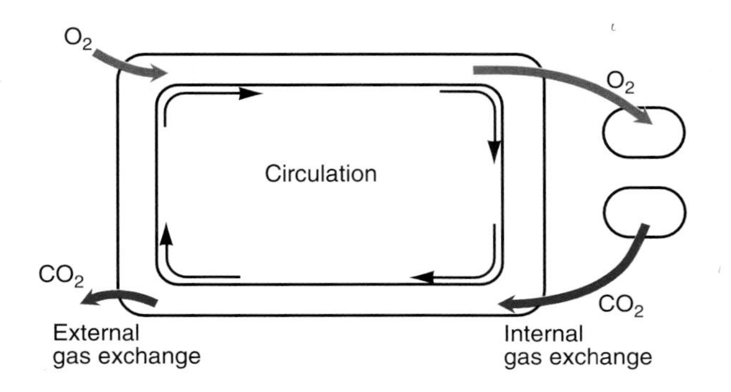

Blood also circulates through organs such as the kidneys where other wastes are eliminated (see Chapter 46), and it absorbs and distributes nutrients from the digestive tract (see Chapter 47).

Fick's Law has one further implication for biological activities: The medium through which diffusion occurs strongly influences the coefficient of diffusion, and this may impose a severe limitation on the structures and sizes of organisms. For instance, O_2 diffuses about 300,000 times faster through air than through water, and therefore animals that breathe air can obtain O_2 much more easily than those that obtain their oxygen from the surrounding water.

45.2 Circulatory systems may be open or closed.

Remember that an animal's water is divided into *intracellular* and *extracellular* compartments, inside and outside its cells (see Section 32.8). Vertebrates and other animals have a *closed* circulatory system confined to a series of tubes, whose extracellular compartment consists of **blood plasma,** the fluid part of blood, and **interstitial fluid,** the liquid in the spaces between cells in other tissues. But the majority of animal species, including arthropods and molluscs, have an *open* circulatory system and a single acellular fluid, **hemolymph,** instead of separate blood and interstitial fluid. Their hearts connect to a short, open blood vessel that pumps blood into all the body spaces; it returns to the heart through a sinus surrounding the heart and then through pores in the heart wall (Figure 45.2). Though open systems are somewhat less efficient in gas exchange than closed systems, arthropods can be quite large and active. Insects and some other terrestrial arthropods exchange gases through tracheae, which are independent of the circulation (see Section 32.7 and Figure 45.23*d*), so their less-efficient open circulatory system does not restrict their gas exchange and they can still metabolize quite rapidly.

By contrast, the circulatory system of vertebrates and many invertebrates is a series of closed blood vessels with at least one contractile heart to move the blood through them. An earthworm, for instance, has a closed circulatory system powered by the dorsal blood vessel and five pairs of hearts (Figure 45.3), they move the blood forward dorsally and downward into the

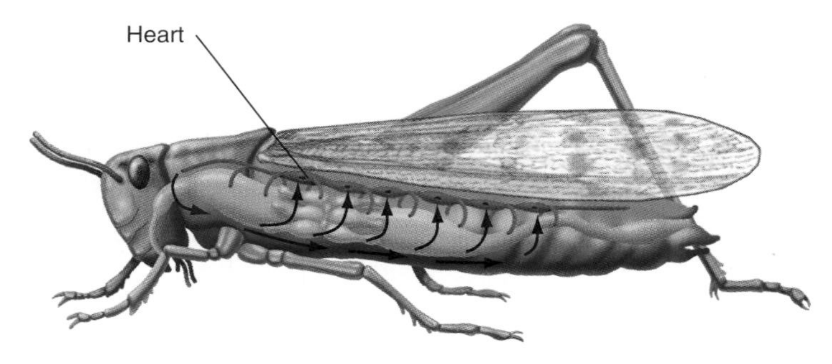

Figure 45.2

The open circulatory system of an insect has only a short blood vessel connected to the heart. Hemolymph flows out into the sinuses (cavities) in all tissues from the anterior end and is collected at the posterior end.

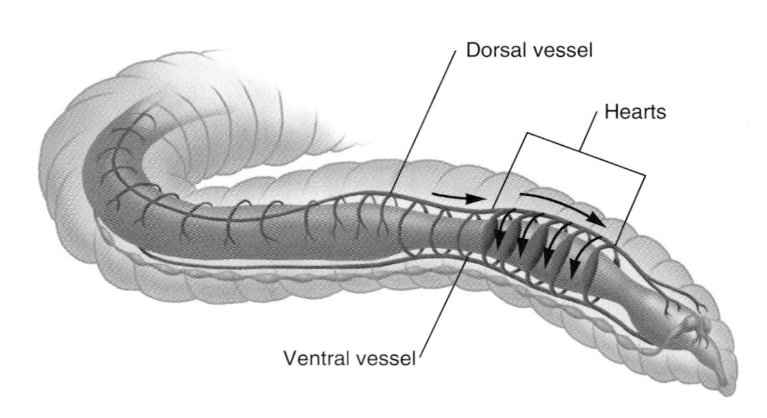

Figure 45.3

The circulatory system of an earthworm is simple and closed, powered by contractions of the dorsal blood vessel and five pairs of hearts.

ventral blood vessel, where it collects nutrients from the intestine and circulates them to the rest of the body. The blood circulates wholly inside this system of tubes and remains separated from the interstitial fluid. Instead of emptying into open sinuses, the blood stays confined in very fine vessels called **capillaries** whose walls are only the thickness of one thin epithelial cell; water, ions, and small molecules constantly diffuse back and forth between the blood inside the capillaries and the surrounding interstitial fluid. The capillaries in each tissue form a *capillary bed* in which each tissue cell is close enough to a capillary to be served by diffusion alone. Fine branches of the system also form capillary beds in the moist body surface, where external gas exchange occurs—O_2 moving inward, CO_2 outward—by simple diffusion through this large surface area. It must be clear that circulation is about transferring nutrients and oxygen in a path from outside—first to the blood, then to the interstitial fluid, and finally to the cells, while moving wastes in the reverse direction. The circulatory system is *not* in direct contact with tissue cells themselves.

B. The vertebrate circulatory system

45.3 In a vertebrate circulatory system, a heart pumps blood successively through arteries, capillaries, and veins.

All blood vessels in a vertebrate circulatory system leading away from the heart are **arteries,** and those leading toward it are **veins** (Figure 45.4). Both kinds of vessels are lined with a layer of flat epithelium, called *endothelium* (*endo-* = inside) because of its internal position. Veins and arteries differ mostly in their layers of smooth muscle. Heavily muscled arteries withstand pressure from blood being forced into them by the heart. Under endocrine and autonomic control, the smooth muscle in the walls of arteries can contract to constrict the arteries and supplement the pumping action of the heart; contracting the muscles of veins helps create enough pressure to return blood to the heart.

Blood leaves the heart through the **aorta,** the largest and toughest artery. Arteries diverging from the aorta fan out through the body, dividing into smaller and smaller vessels; the

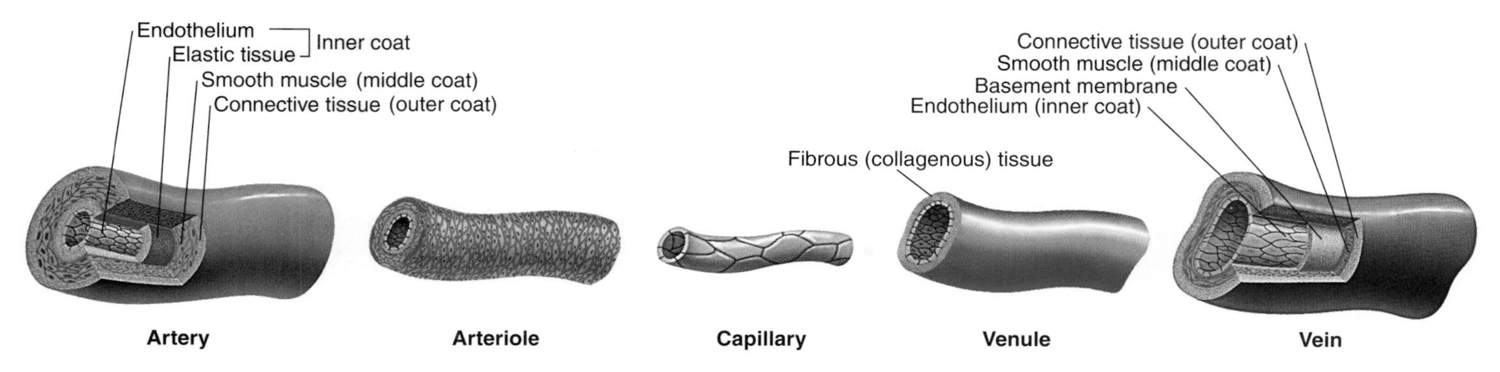

Figure 45.4

All blood vessels have an inner layer of endothelium, and capillaries are made of little more than endothelium, with no surrounding muscle. Arteries and veins both have a layer of connective tissue and an elastic layer around the endothelium, a middle layer of smooth muscle, which is much heavier in arteries, and an outer coat of connective tissue containing some smooth muscle cells. The arterial tree divides into smaller arteries, whose walls gradually become thinner; the smallest arteries, called arterioles, have only a single layer of smooth muscle around the endothelium, sometimes with collagen fibers and an elastic layer. Venules, the smallest veins, have a thin layer of collagen fibers; as they merge into larger and larger veins, they acquire more and more smooth muscle and become typical veins.

smallest arteries, called **arterioles,** lead to beds of **capillaries** made of endothelium alone, with no surrounding muscle:

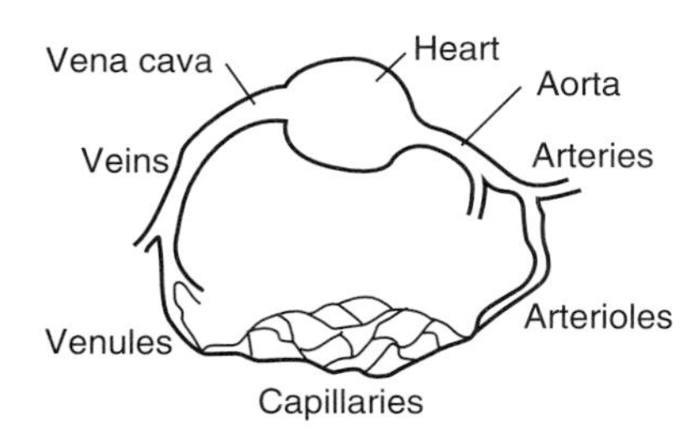

The total length of capillaries in a human has been estimated at 100,000 kilometers! The finest vessels leading out of a capillary bed are **venules,** which converge into a tree of increasingly larger veins, finally ending in the superior and inferior **venae cavae,** two large veins that return blood to the heart from above and below, respectively. Most veins contain a series of one-way valves, which open and close passively under pressure from the blood, keeping the blood flowing in one direction (Figure 45.5). Furthermore, the veins run through and between many muscles, which massage the veins as they contract and thus help force more blood back to the heart.

Let's begin the study of vertebrate circulation with the generally familiar mammalian system. A mammalian heart is best thought of as two pumps side by side, operating a *double-circulation system*—a **pulmonary circulation** through the lungs and a **systemic circulation** through the rest of the body:

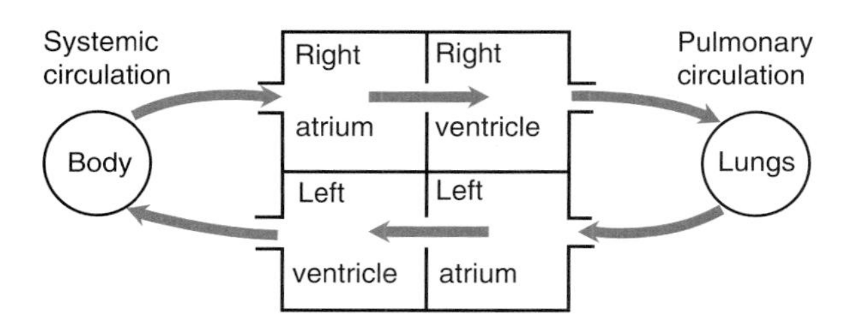

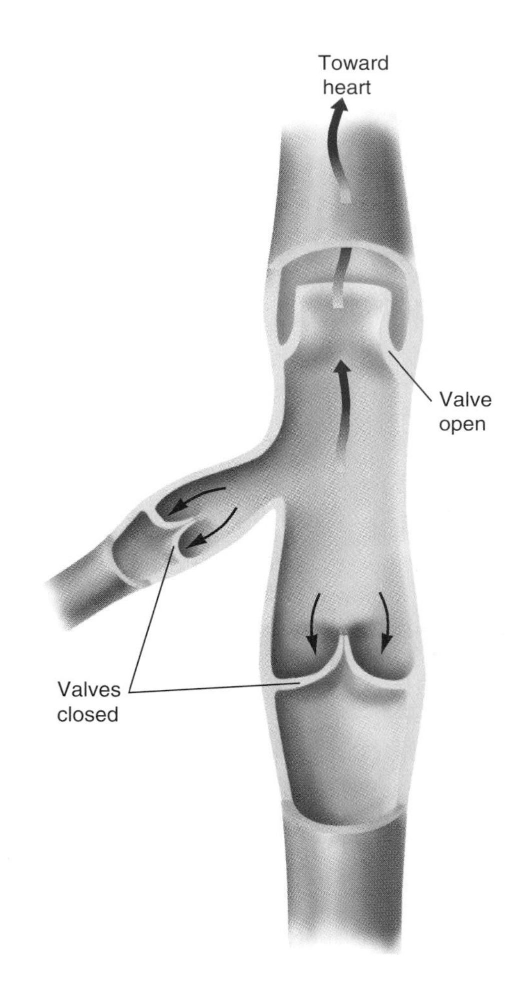

Figure 45.5

The valves in veins are shaped such that blood moving forward (toward the heart) opens each valve. As blood starts to move backward, it closes the valve, thus maintaining a one-way flow.

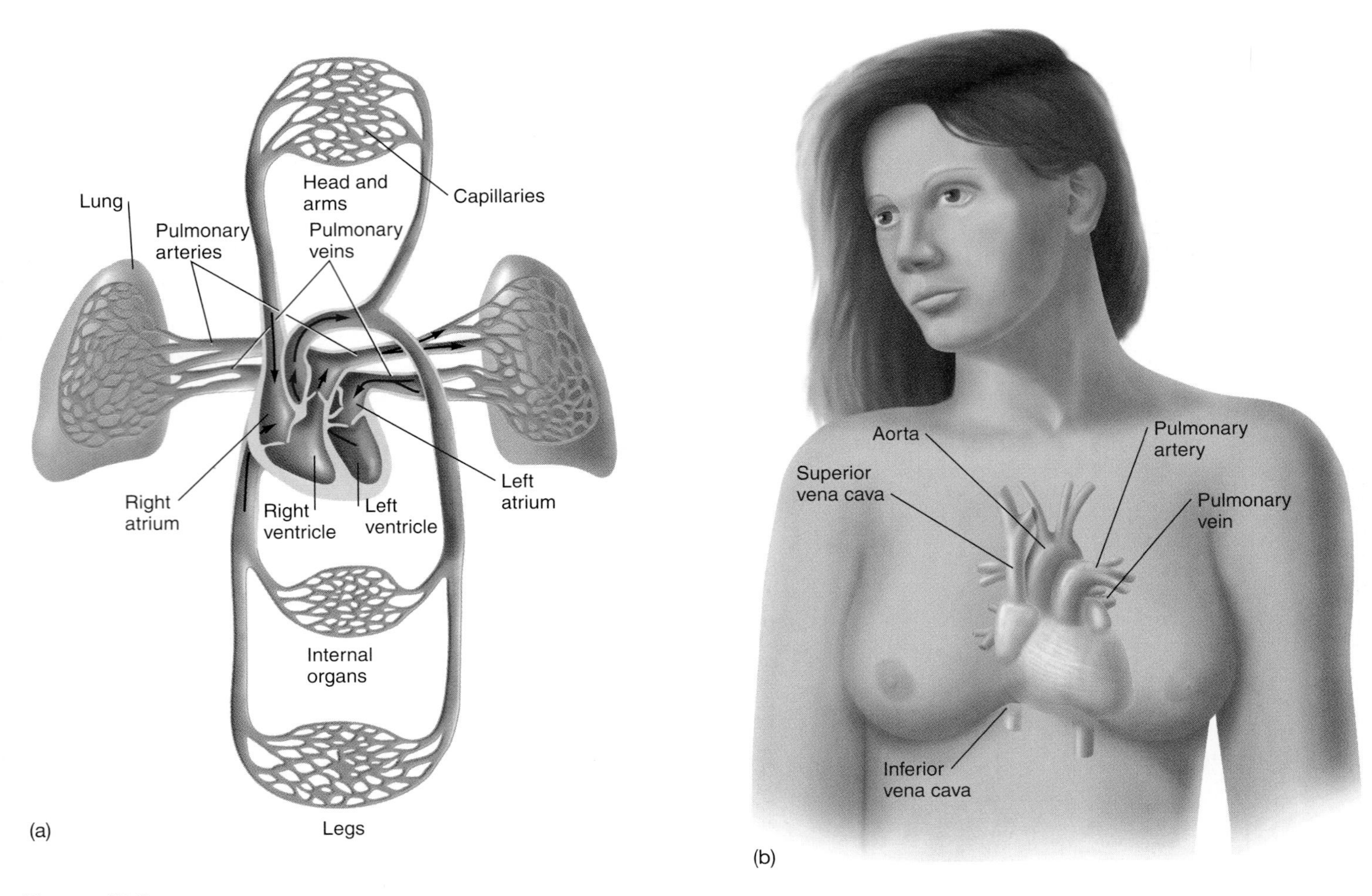

Figure 45.6

(a) Mammals have a double-circulation system, as shown diagrammatically with red representing oxygenated blood and blue representing deoxygenated blood. The *pulmonary circulation,* driven by the right side of the heart, moves blood through the lungs where it becomes oxygenated. The *systemic circulation,* driven by the left side of the heart, carries blood through the rest of the body tissues. *(b)* When the human heart is drawn in its proper position, we can see its major vessels in relation to other parts of the body.

The right side of the heart collects blood returning from the body in the right **atrium** and passes it into the right **ventricle** (Figure 45.6). The powerfully muscled ventricle drives the pulmonary circulation, forcing deoxygenated blood through the **pulmonary artery** to the lungs, where it flows into smaller arteries and a network of capillaries over the inner lung surfaces. Here the blood exchanges its gases with the atmosphere. Larger and larger veins then converge into the **pulmonary veins** that carry the oxygenated blood back to the left side of the heart. Blood returning from the lungs collects in the left atrium and moves into the left ventricle, which powers the systemic circulation by pumping blood into the aorta.

Vertebrate hearts have evolved through stages that are preserved in living animals (Figure 45.7). The heart of a fish, the earliest-evolved type of vertebrate, is a series of four relatively strong, muscular enlargements of a main blood vessel. It forces the blood ahead through the **gills,** for oxygenation, and then on to the rest of the body (Figure 45.8). This is a *single-circulation system* that can be drawn as one closed loop:

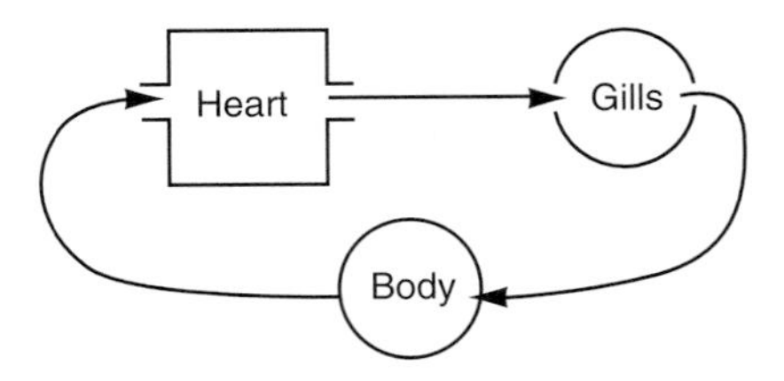

This system became modified into a double-circulation system in air-breathing animals with lungs. The heart of amphibians and most reptiles (turtles, snakes, and lizards; see Figure 45.7*b*) is at an intermediate stage—still a single tube with one ventricle, but with a second atrium that receives oxygenated blood from the lungs. Although oxygenated and deoxygenated blood pass through the ventricle together, they are actually quite well

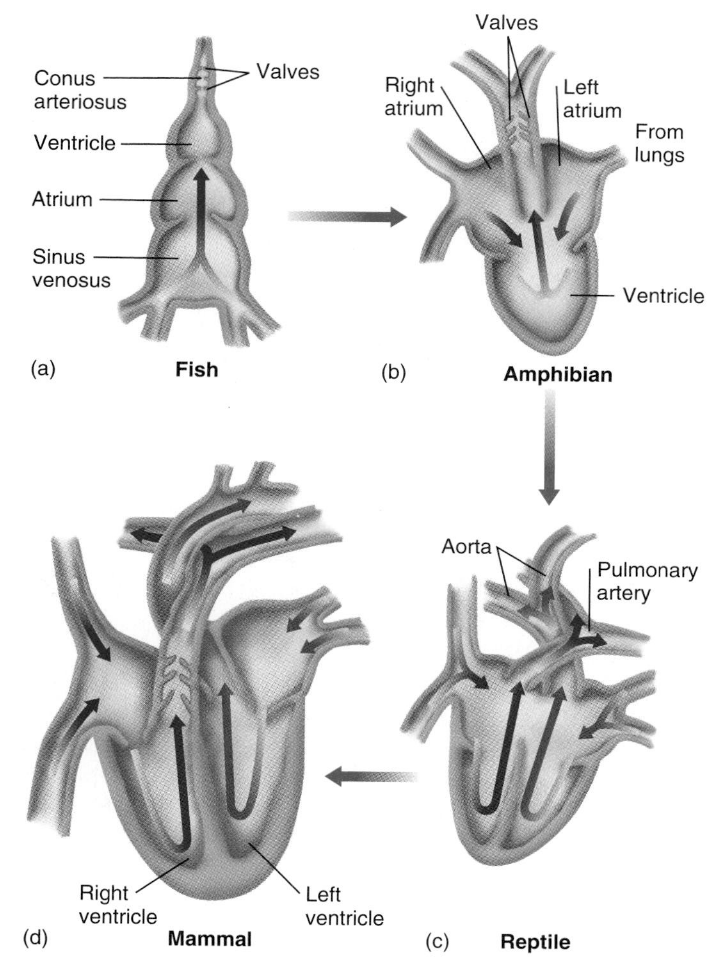

Figure 45.7

Comparing the hearts of four vertebrates shows how the circulatory system has evolved. *(a)* Fishes have a series of four chambers (sinus, venosus, atrium, ventricle, and conus arteriosus). This structure becomes twisted and reduced to three chambers, with two atria and a single ventricle, in amphibians and the more primitive reptiles *(b)*. In advanced reptiles *(c)*, the ventricle becomes partially divided, and birds and mammals *(d)* have two distinct ventricles, making a complete double-circulatory system.

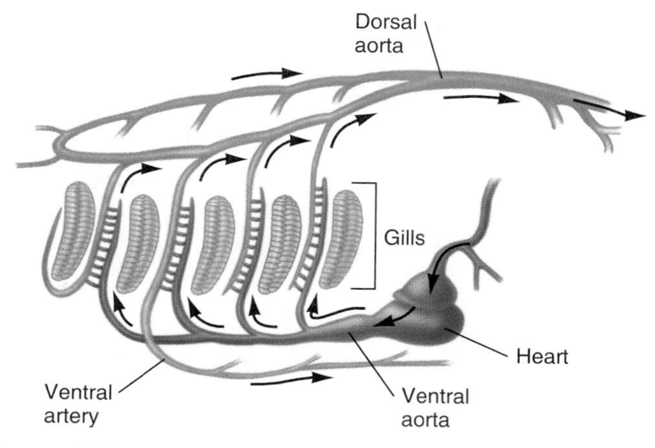

Figure 45.8

In a fish, the heart forces blood through a series of gills where it is oxygenated before moving through the rest of the body.

separated; the right atrium opens into the ventricle a little further forward than the left, so deoxygenated blood returning from the body enters the arterial cone first. There it meets an unusual spiral valve, which directs this first blood primarily to the pulmonary arteries, while the oxygenated blood that follows moves into the aorta and on to the rest of the body. In the heart of crocodilian reptiles, a septum through the middle of the ventricle tends to separate oxygenated from deoxygenated blood (see Figure 45.7*c*). Finally, birds and mammals have a complete septum and therefore a full double-circulation system (see Figure 45.7*d*).

Exercise 45.1 In the circulatory system of a fish, all the blood passes through capillary beds in the gills before moving on to the rest of the body. In mammals, blood moves into the tissues from arteries branching off the aorta. Compare the pressure behind blood moving through fish tissues with the pressure that moves blood through mammalian tissues.

45.4 The vertebrate heart has an internal pacemaker.

The heart maintains a regular *cardiac cycle,* a rhythm of contraction (**systole;** sis′to-lee) and relaxation (**diastole;** di-as′to-lee) of the ventricles. The atria also contract, toward the end of diastole, and relax, during systole. Strong heart valves, like those in veins, open and close passively as pressure changes within the heart chambers, keeping blood moving in one direction. Passages from the atria to the ventricles are closed by **atrioventricular (AV) valves:** a tricuspid valve made of three flaps on the right side and a bicuspid valve (or mitral valve) made of two flaps on the left side (Figure 45.9). These valves are connected through strong strands of connective tissue called *chordae tendineae* ("heart strings") to cone-shaped papillary muscles extending from the ventricle walls, which keep the valves from opening backwards into the atria when the ventricles contract. Let's follow the cycle, beginning with the heart in diastole, with both atria and ventricles relaxed. Pressure builds in the atria as they fill with returning blood. This pressure opens the AV valves and allows blood to flow into the ventricles. The atria then contract, forcing their contents into the ventricles, although it has been estimated that the ventricles are already about four-fifths full by this time; indeed, people with atrial damage who lack normal atrial contraction are not severely impaired except during exercise. At the onset of systole, pressure builds within the ventricles, closing the AV valves so blood cannot flow back into the atria and opening the **semilunar valves** at the origins of the

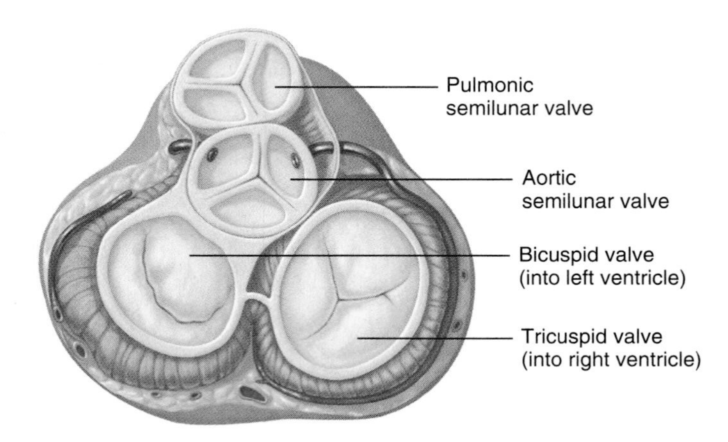

Figure 45.9
In the human heart, atrioventricular valves separate the atria from the ventricles; they are connected to papillary muscles, which keep them from opening backwards into the atria. Semilunar valves at the base of the pulmonary artery and aorta open during systole and close when the pressure in these arteries exceeds the pressure inside the ventricles.

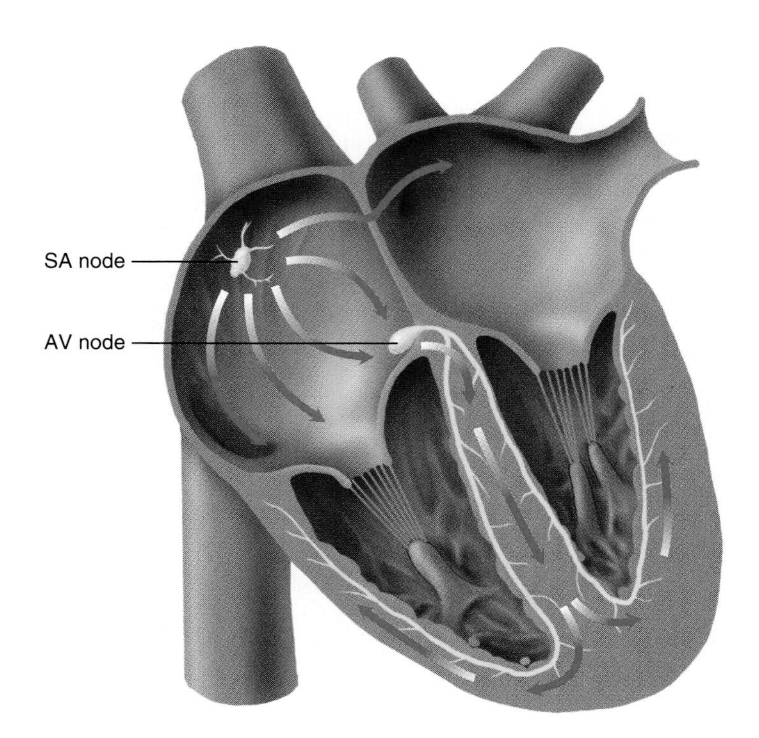

Figure 45.10
The heart's pace is set by the spontaneous activity of a small knot of specialized muscle tissue, the SA node, which sends out a signal that makes the atria contract. This signal then sets off the AV node, which sends out a second signal that passes around the ventricles and stimulates them to contract.

aorta and pulmonary artery; both AV and semilunar valves operate passively, opening under pressure from the contracting ventricles and then closing during diastole as pressure within the large arteries exceeds pressure within the ventricles.

Vertebrate heart muscle is a unique tissue. Though it is striated like skeletal muscle, cardiac muscle is made of cells that branch and join each other in a network connected by extensive gap junctions. Since gap junctions conduct signals rapidly from cell to cell, the entire interconnected tissue, called a **myocardium,** contracts as a unit when it is stimulated. In a mammalian heart, the two atria are formed by one sheet of myocardium and the two ventricles by another, so each pair of chambers contracts together.

Gap junctions, Sections 6.9 and 41.5.

About 100 years ago, Henry N. Martin demonstrated that an isolated mammalian heart, with no nervous connections, will continue to beat rhythmically as long as it is supplied with blood. The heart, in other words, has its own internal **pacemaker,** which stimulates contraction. This intrinsic pace is one of the wonders of vertebrate physiology. A turtle or frog heart, removed from the body and provided with an ionically balanced fluid (Ringer's solution) carrying oxygen, will keep beating for days. If individual embryonic heart cells are separated from one another in a tissue culture, each one establishes its own intrinsic rate of contraction, but as the cells grow together and reform gap junctions, they coordinate and beat together.

The pace of an intact mammalian heart is set by the **sinoatrial (SA) node,** a block of specialized muscle tissue only a few millimeters wide located in the right atrium (Figure 45.10). These modified muscle fibers fire with an intrinsic, spontaneous rhythm, about 75–80 times per minute in humans. An action potential generated by the SA cells is conducted to the atrial myocardium and spreads rapidly through the entire atrial muscle, contracting the atria in unison. Within a tenth of a second, this wave of excitation reaches a second control center, the **atrioventricular (AV) node,** located just above the right ventricle. The AV node and fibers leading from it are made of specialized myocardial cells. After a delay of about 70 milliseconds, which gives the atria time to empty, the AV node fires and sends an action potential through the *atrioventricular bundle,* located in the septum between the ventricles. This bundle divides into two bundles that pass through the bottoms of the ventricles and up their sides, and are continuous with muscle fibers (called *Purkinje fibers*) connected directly to the ventricular myocardium. Within 200 milliseconds, this series of fibers conducts the action potential from the AV node into the entire mass of ventricular tissue, stimulating the ventricles to contract in unison and to force blood into the pulmonary and systemic circulations.

Although the conduction system of the heart is highly innervated by the autonomic nervous system, neural signals merely increase or decrease its intrinsic rhythm. Parasympathetic stimulation produces a placid, relaxed condition in the whole animal and slows the heartbeat, while sympathetic stimulation prepares an animal for quick action by speeding up the heartbeat and increasing the strength of contraction.

Sidebar 45.1 Measuring Blood Pressure

Stephen Hales, an Anglican minister in Middlesex, England, was the first person known to have measured blood pressure. In 1711 he attached a pressure gauge to the carotid artery of a horse and noted that the pressure rose and fell with the animal's heartbeat. Today, there are far simpler devices for measuring blood pressure routinely.

What is a nurse or medical assistant doing when he places the sleeve of a blood-pressure device around your upper arm, pumps it up, and starts listening to your arm with a stethoscope? He is listening to the sounds—or silence—of blood flowing through your brachial artery, the main artery in the arm. Before the cuff is inflated, there is no sound because the blood flows through the artery smoothly, without turbulence. With the cuff completely inflated, there is no flow at all, and therefore no sound. The nurse then allows air to leak out of the cuff, all the while listening and watching a *sphygmomanometer,* a gauge showing the pressure on the cuff. The artery will stay closed as long as the cuff pressure is higher than the systolic pressure exerted by the heart; when the cuff pressure falls to the systolic pressure, a little blood will start to push through the artery at every contraction of the ventricles, but the flow will be turbulent, not smooth, because the artery closes during diastole and opens with a brief pulse of blood flow during systole. This turbulence produces a tapping noise called the *sounds of Korotkoff.* (They are *not* the lub-dup sounds of the heart itself.) The Korotkoff sounds continue at each systole as the cuff pressure falls, and the last sound will be heard when the cuff pressure is equal to the diastolic pressure, because at this point blood is flowing normally and smoothly through the brachial artery and it makes no sound.

Normal human blood pressure rises steadily until middle age, from about 90/60 torr at one year of age to 120/78–80 torr in young adulthood. Women typically have slightly lower pressures than men until about age 45–50, when their pressures become slightly higher for some reason. Still, from age 60 onward, a normal man's pressure should remain around 145–50/80 torr and a woman's no more than a few torr higher at systole. Diastolic pressure indicates the condition of the blood vessels, and so an elevated diastolic pressure is a cause for greater concern than an elevated systolic pressure.

Exercise 45.2 With a stethoscope, you can listen to the heart sounds, which are commonly represented by "lub-dup." The first sound, "lub," is made by the closing of the AV valves; the second sound, "dup," by the closing of the aortic valves. How are these sounds related to the times of systole and diastole?

45.5 Several regulatory factors combine to keep the heart functioning with quite constant blood pressure.

As blood is forced out of the heart during systole, it pushes against the arteries, so the blood pressure rises sharply (**systolic pressure**) and then falls again as the heart relaxes (**diastolic pressure**). A human heart, working at top speed, can force blood through the aorta as fast as 2 meters per second. To contain this gushing stream, the arteries have powerful, muscular walls whose smooth muscle layers expand with each surge of blood and then rebound elastically so as to partly smooth out the variation in blood pressure. Connective tissue in the arterial walls also contains elastic fibers, so it recoils elastically to each surge in blood pressure. Blood pressure is much greater in the largest arteries than in the smaller ones; when measured in an arm, a healthy person's pressure varies from about 120 torr (mm of mercury) at systole to about 80 torr at diastole, with men having slightly higher pressure than women (Sidebar 45.1). Blood pressure is an important indicator of health. It must be great enough to carry blood through all the narrow capillaries fast enough to support metabolism, but not so great as to damage blood vessels or overwork the heart. Blood pressure increases through stimulation by the sympathetic division of the ANS at times of stress, but many conditions raise the pressure chronically to abnormal levels.

Pressure, Concepts 5.1.

Blood flows through the circulatory system in vessels that vary in size—from the aorta, with an internal diameter of about a centimeter, to capillaries with diameters of only a few micrometers. Since blood pressure moves liquid through a series of tubes, we must first understand the factors of flow and resistance within a tube. Suppose you want to fill a child's wading pool with a garden hose as rapidly as possible and you have a choice of hose sizes. Which will deliver water faster—a narrow hose or a wider hose? A short hose or a longer hose? The flow rate of fluid in a tube, F, is proportional to $\Delta P/R$, where ΔP is the difference in pressure between the ends of the tube, and R is the resistance to flow. Resistance depends on two factors. First, it is proportional to the length of the tube, so doubling the length doubles the resistance; therefore a short hose will deliver water faster than a long one. Second, and more dramatic, resistance decreases in proportion to the fourth power of the tube's radius. So if equal pressures are exerted on a tube of radius r and on another tube of radius $2r$, the flow rate in the wider tube will be 16 times that in the smaller one, because $2^4 = 16$ (Figure 45.11). So even though garden hoses do not vary much in diameter, a slightly larger hose has much less resistance than a smaller one and will deliver water faster.

These considerations mean that even a slight change in the diameter of a blood vessel can have a great influence on the flow through it, as you can show for yourself with the following exercise.

Exercise 45.3 How will the flow through an artery be affected if its radius decreases by 10 percent, from r to $0.9r$, assuming pressure remains constant? (Do this calculation before reading further.)

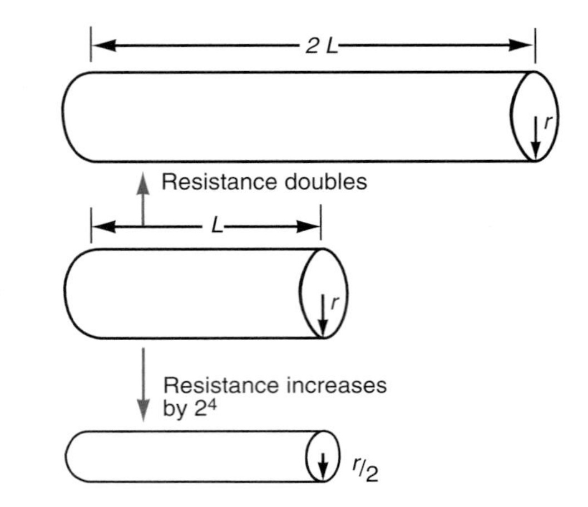

Figure 45.11

The resistance to flow through a tube increases in proportion to its length and with the fourth power of the tube's radius.

Since $(0.9)^4 = 0.655$, the flow rate will fall to about 65 percent of what it was. That can make a great difference. The diameters of arteries and veins can be changed by contraction or relaxation of their smooth muscles, under autonomic (sympathetic and parasympathetic) control. Contraction causes **vasoconstriction,** the narrowing of a vessel, and relaxation produces **vasodilation,** the widening of a vessel.

Returning now to the issue of blood pressure, we can see that arterial blood pressure depends on the push behind the blood and the resistance in front of it. The push is the **cardiac output,** the volume of blood expelled by the heart each minute. Large arteries, such as the aorta and those that branch from it directly, offer relatively little resistance to blood flow because of their size. Resistance is therefore almost all **peripheral resistance** from all the small blood vessels downstream, at the periphery of the circulation. In fact, the resistance is due almost entirely to the arterioles, since they are narrow vessels whose total cross-sectional area is small (Figure 45.12). The capillaries that branch off the arterioles have smaller diameters, but their total cross-sectional area is much larger, in keeping with their function of providing a large surface/volume ratio for the exchange of material with tissues.

With these factors in mind, we can begin to understand how blood pressure (BP) is kept to a narrow range by examining the relationship:

$$\text{BP} \propto \text{Cardiac output} \times \text{Peripheral resistance}$$

where $\propto$ means "is proportional to." First, we recognize that:

$$\text{Cardiac output} = \text{Cardiac rate} \times \text{Stroke volume}$$

That is, the heart can pump blood faster either by contracting faster (cardiac rate) or by pumping a larger volume at each stroke, or both. A 70-kg man, for example, has a blood volume of about 5 liters, and the heart circulates this entire volume each minute. At an average cardiac rate of 70 beats per minute, the average stroke volume is just over 70 ml. Regulation revolves

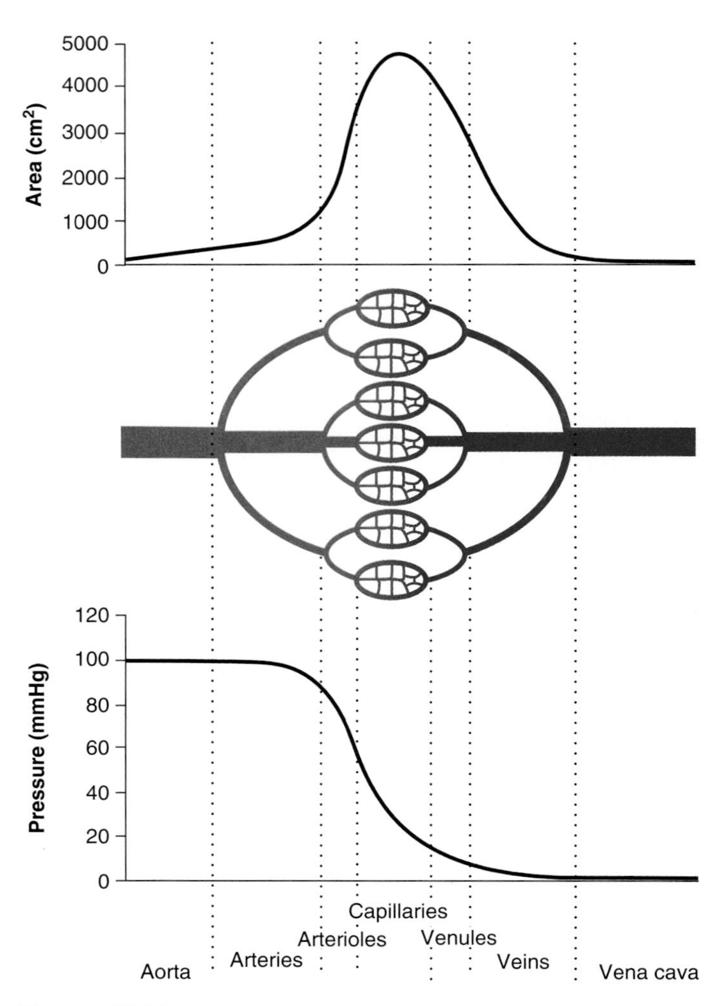

Figure 45.12

Blood pressure is high in the aorta and the large arteries branching from it because of pressure upstream from the heart and resistance downstream from the arterioles. Because the capillaries have such a large total cross-sectional area, they present relatively little resistance, and the pressure falls quickly within capillary beds. Pressure on blood returning through the veins is very low.

around making the three factors—cardiac rate, stroke volume, and peripheral resistance—compensate each other to keep blood pressure roughly constant, while adjusting the flow through all the tissues. To accomplish this feat, mammals (as exemplified by humans) exhibit a remarkable series of controls, as summarized in Figure 45.13. Regulation entails mechanisms intrinsic to the heart plus extrinsic control circuits.

Baroreceptor reflexes regulate the system extrinsically. Baroreceptors in the aortic arch and carotid sinus monitor BP and feed their information to centers in the medulla oblongata: cardiac control centers and vasomotor control centers. The *cardiac control centers* can change the heartbeat. We have seen that the heartrate is basically created by the internal SA and AV pacemakers. Their pace can be changed by both parasympathetic and sympathetic control. In addition, sympathetic stimulation can increase the contraction strength of the ventricles, thus

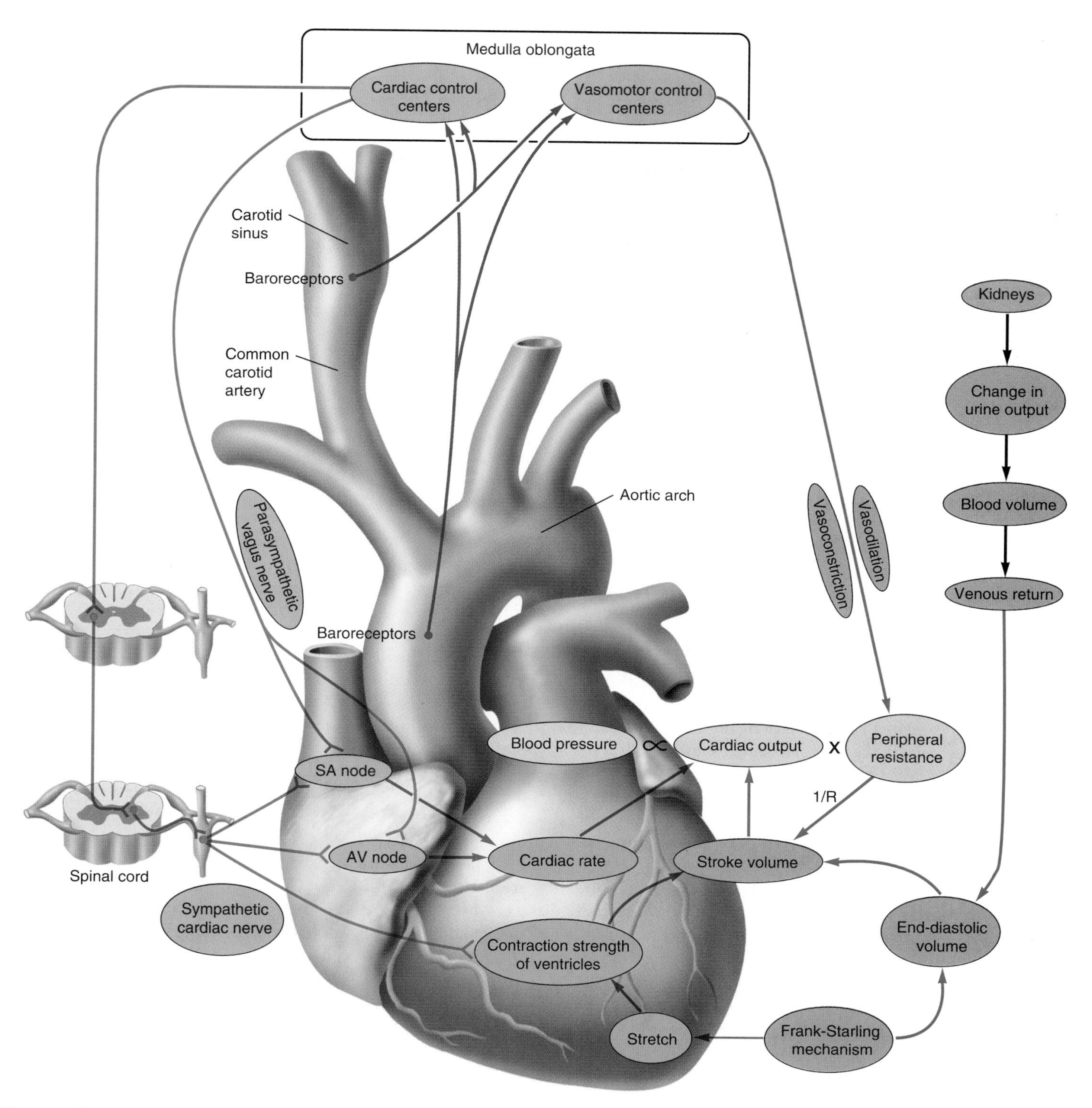

Figure 45.13
Many factors combine to keep the blood pressure within a limited range. The baroreceptor reflexes begin with pressure sensors (baroreceptors) in the aortic arch and carotid sinus, which report changes in blood pressure to centers in the medulla oblongata. Cardiac control centers can change the cardiac rate through their effects on the SA and AV nodes; sympathetic stimulation can also increase the contraction strength of the ventricles. Vasomotor control centers can change the peripheral resistance through vasodilation or vasoconstriction of arteries, especially arterioles. The stroke volume is also regulated intrinsically through the Frank-Starling mechanism, as explained in the text. The venous return rate to the right atrium depends on other factors, such as volume regulation by the kidneys.

Sidebar 45.2 Shock

A serious condition known as *shock* may be brought on by traumas, including severe psychological stress or physical injury, hemorrhage (excessive bleeding), and endotoxins produced by infecting microorganisms. Such a trauma immediately produces a sudden drop in blood pressure, which is detected by the aortic and carotid baroreceptors, initiating the baroreceptor reflex (see Section 45.5). In response (functionally, to raise the blood pressure), the cardiac output is increased while general vasoconstriction in the peripheral circulation reduces blood flow to the extremities, kidneys, and bowel. This has the effect of leaving more blood for the critical circulation to the brain and heart. The localized reduction in blood flow is called ischemia (see Section 45.9), and this initial reaction to the trauma is *ischemic (reversible) shock.* Although selective ischemia in nonessential tissues is necessary for immediate survival, it temporarily deprives those tissues of oxygen. If the shock reactions are not reversed (generally by transfusion of blood or a normal electrolyte solution), cells being starved for oxygen gradually begin to change their metabolism, and the system goes over into a state of stagnant shock.

Peripheral ischemia causes many cells to start metabolizing glucose anaerobically to lactic acid. Without aerobic respiration, they don't produce enough ATP to maintain their vital cellular ion pumps. Na^+ and water begin to enter the cells as K^+ and lactic acid start to leave, and the cells start to undergo irreversible changes leading to cell death. Acidosis, a decrease in blood pH, develops due to the excess lactic acid in the blood. Acidosis causes arteriolar sphincters to lose their tone, so they cannot stay constricted, while venous constriction continues, since venous sphincters function normally at a reduced pH. So blood is able to flow into the peripheral circulation, where it pools because of venous constriction. Increased hydrostatic pressure in these capillaries forces serum out of the circulation into the tissues, so they develop edema. These effects further reduce blood volume and pressure. The heart itself is damaged, reducing its ability to correct the situation. If this condition persists long enough, shock becomes *stagnant* (irreversible), and death follows.

increasing the stroke volume. The *vasomotor control centers,* through vasodilation and vasoconstriction of arterioles, can adjust the peripheral resistance. In fact, the flow into some capillary beds can be essentially cut off by contraction of selected arterioles—for instance, in the skin in very cold weather—producing a considerable increase in resistance and thus in blood pressure. Sidebar 45.2 explains the changes that occur during shock, a phenomenon that shows how closely the metabolism of individual cells is related to the health of the whole body.

The baroreceptor reflexes primarily affect the peripheral resistance and cardiac rate. Cardiac output is the product of cardiac rate and stroke volume; the latter is controlled by three factors: peripheral resistance, contraction strength of the ventricles, and end-diastolic volume. To understand these factors and their interaction, imagine the operation of the left ventricle as it contracts: As pressure builds inside, it forces blood into the aorta until the back pressure within the aorta exceeds the pressure in the ventricle. At this point, the aortic semilunar valve closes. Thus the stroke volume will be inversely proportional to the peripheral resistance, since the lower that resistance, the more blood the ventricle will deliver before its valve closes.

Peripheral resistance can be overcome to a degree by increasing the contraction strength of the ventricles; we have just seen that this strength can be increased extrinsically through sympathetic stimulation from the cardiac control center. The contraction strength is also regulated intrinsically through the **Frank-Starling mechanism.** Otto Frank and Ernest H. Starling discovered a fundamental law about contraction of the heart: *The strength of ventricular contraction is proportional to the* end-diastolic volume, *the volume of blood in the ventricles at the end of diastole, when systole is about to begin.* Figure 45.14 shows this relationship and explains it. Because the heart operates according to the Frank-Starling law, it can quickly adjust its output in response to a change in peripheral resistance. If peripheral resistance rises, the ventricle will deliver less of its blood, and more blood will remain in the ventricle at the end of systole. But during the next contraction, the ventricle will contain more blood, so it will contract harder, overcoming the peripheral resistance and raising the cardiac output.

Superimposed on these processes, the end-diastolic volume is affected by the *venous return*—the volume of blood delivered by the vena cava to the right atrium. Venous return depends on the total blood volume, which the kidneys largely regulate through changes in the rate of urine production by mechanisms discussed in Section 46.7. Venous return also depends on venous blood pressure. As we have already noted, this pressure is always low, on the order of 2 torr compared with an average arterial pressure of 90–100 torr, but it can be changed by at least two factors. The vasomotor control centers regulate vasoconstriction of veins as well as arteries, and sympathetic stimulation can thus increase the venous pressure. Furthermore, the veins run through and between many muscles, which massage the veins as they contract and thus help force more blood back to the heart.

The complexity of Figure 45.13 should be enough to show that all these factors interact in complicated ways, and we have barely touched upon the closely related processes acting through the kidneys. But despite all these regulatory mechanisms, blood pressure is often not well regulated. About 20 percent of American adults have *hypertension* (high blood pressure) for various reasons, including kidney disorders, arteriosclerosis, endocrine imbalances, poor dietary habits, and stresses associated with modern living. In so complicated a system, many things can go wrong, and they commonly do.

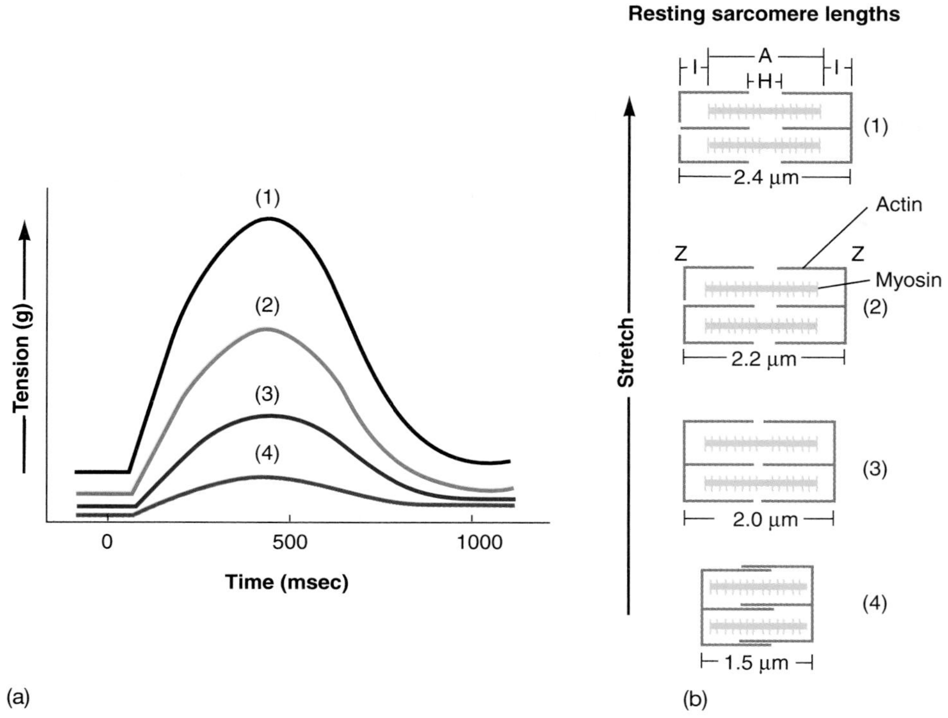

Figure 45.14
The Frank-Starling mechanism arises from the fact that the more heart muscle is stretched, the more forcefully it contracts. *(a)* Cardiac muscles stretched initially to four different lengths develop proportional tensions during contraction. Heart muscles will normally be stretched in proportion to the amount of blood in a chamber. *(b)* Drawings of actin-myosin relationships show why a stretched muscle develops more tension. In the most completely contracted muscle, actin filaments overlap in the middle of each sarcomere, and the muscle can develop little tension; the tension the muscle can develop increases as it is stretched out more, allowing more room for the actin to move along the myosin.

Exercise 45.4 Consider these two cases together: *(1)* Because of an accident on the freeway, all the traffic that had been moving in three lanes is forced to merge into one lane. Where will the "traffic pressure" be highest—before the point of the accident or after? Where will it be lowest? *(2)* The arterioles present a narrow constriction of the bloodstream compared with the capillaries downstream. Where will the blood pressure be high and where low—in the arteries or in the capillaries?

45.6 Blood is a tissue whose cells are embedded in a liquid plasma.

Blood, as Mephistopheles remarks in Goethe's *Faust,* is "a very special stuff." It is a tissue made of cells dispersed in a fluid matrix, the plasma.[1] If blood is centrifuged, the cells concentrate in the bottom of the tube, and one can measure the fraction of the blood they occupy when packed, the **hematocrit.** This amount is typically 45 percent in human males and a little less in females. Most of these cells are **red blood cells,** or **erythrocytes** (Figure 45.15), which are filled with hemoglobin—about 5 million cells per mm^3 of blood in humans. Only about 10,000 cells per mm^3 are **white blood cells,** or **leucocytes,** whose importance far outweighs their numbers, since they form a line of defense against infection, as we will see in Chapter 48. The plasma also carries small cell fragments, called **platelets,** that participate in blood clotting. The 5 liters of blood in a 70-kg person weigh about 5.6 kg (8 percent of body mass).

In adults, new red and white blood cells are constantly produced in bone marrow. The life of an erythrocyte is an excellent illustration of the continuous breakdown and resynthesis of body tissues. An adult human has about 3.1×10^{11} erythrocytes per kilogram of body weight. Thus a 70-kg person has 2×10^{13} erythrocytes (20 million million). Each cell circulates for about 120 days and is then captured and destroyed in the spleen and liver. That isn't simply an average time, as though some cells are destroyed when they are much younger and some when they are much older. Rather, some change that marks a cell for destruction seems to occur at just that age. Now, since every cell lives about 120 days, every day $1/120$ of the cells must be destroyed and an equivalent number made, so a 70-kg person makes 1.8×10^{11} new erythrocytes per day. That's 180 billion cells per day, or 2 million per second—every second, every day, for a lifetime!

Mammalian blood plasma is a salty solution containing a number of ions and other solutes (see Figure 45.15) and about 90 grams of protein per liter. Plasma proteins have many functions. They contain fibrinogen, which forms blood clots under the influence of other blood proteins; many proteins that transport lipids and mineral nutrients like iron; several protein hormones; and antibodies that recognize and attack invaders. Blood also carries a variety of organic compounds, including nonprotein hormones; glucose, the major energy source for all cells; metabolites in transit between points of specialized metabolism; and urea, the principal nitrogenous waste of mammals.

[1] Blood is usually classified as a connective tissue because its cells are surrounded by an extracellular matrix. This is a matter of tradition. The function of blood is obviously very different from that of connective tissues, and its matrix is a liquid, not a mass of strong fibers.

Figure 45.15

When whole human blood is centrifuged, the cells are compacted at the bottom of the tube, and one can see that the fraction they occupy, the hematocrit, is about 45 percent of the total. Standard staining methods reveal the major types of cells and other formed bodies in mammalian blood. Erythrocytes (red blood cells) carry O_2 and CO_2; leucocytes (white blood cells) have various roles in fighting infections. Without staining, the leucocytes are quite colorless. The table summarizes components of the plasma.

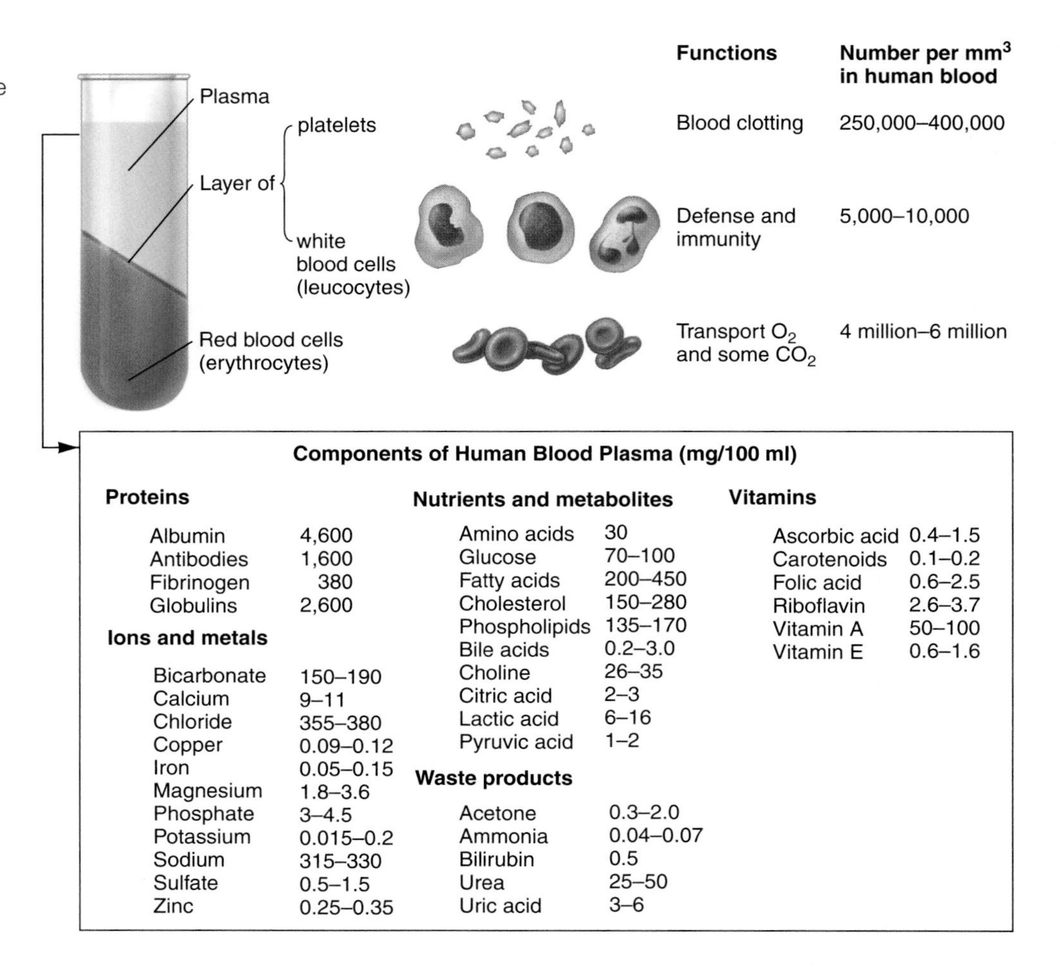

Components of Human Blood Plasma (mg/100 ml)

Proteins		Nutrients and metabolites		Vitamins	
Albumin	4,600	Amino acids	30	Ascorbic acid	0.4–1.5
Antibodies	1,600	Glucose	70–100	Carotenoids	0.1–0.2
Fibrinogen	380	Fatty acids	200–450	Folic acid	0.6–2.5
Globulins	2,600	Cholesterol	150–280	Riboflavin	2.6–3.7
Ions and metals		Phospholipids	135–170	Vitamin A	50–100
		Bile acids	0.2–3.0	Vitamin E	0.6–1.6
Bicarbonate	150–190	Choline	26–35		
Calcium	9–11	Citric acid	2–3		
Chloride	355–380	Lactic acid	6–16		
Copper	0.09–0.12	Pyruvic acid	1–2		
Iron	0.05–0.15	**Waste products**			
Magnesium	1.8–3.6				
Phosphate	3–4.5	Acetone	0.3–2.0		
Potassium	0.015–0.2	Ammonia	0.04–0.07		
Sodium	315–330	Bilirubin	0.5		
Sulfate	0.5–1.5	Urea	25–50		
Zinc	0.25–0.35	Uric acid	3–6		

Exercise 45.5 A liter is about a quart. To visualize how much blood an adult human contains, translate the 4–5 liters of blood into the volume of a group of soda bottles.

45.7 Blood plasma exchanges with interstitial fluid in a tissue.

Each tissue is served by a bed of fine capillaries whose high surface/volume ratio and exceedingly thin endothelial walls (Figure 45.16) permit rapid exchange of materials between the blood and tissue cells. **Thoroughfare channels** connect the arterioles leading into a capillary bed with the venules leading out of the bed (Figure 45.17). Blood flow through a capillary bed can be regulated by contraction and relaxation of the smooth muscle around arterioles and venules. In addition, **precapillary sphincters,** rings of smooth muscle surrounding each capillary where it leaves the thoroughfare channel, can contract and regulate the flow of blood into the capillary. We saw earlier how strongly a small change in the diameter of a blood vessel can affect the flow through it. Thoroughfare channels are always open, but the precapillary sphincters can reduce blood flow through the capillaries enormously.

The mechanism of vasodilation, incidentally, is now known to be a prime example of paracrine communication using that remarkable mediator nitric oxide (NO). When parasympathetic axons induce vasodilation, they are stimulating endothelial cells to produce NO, which then stimulates nearby smooth muscle cells to relax, using this mechanism:

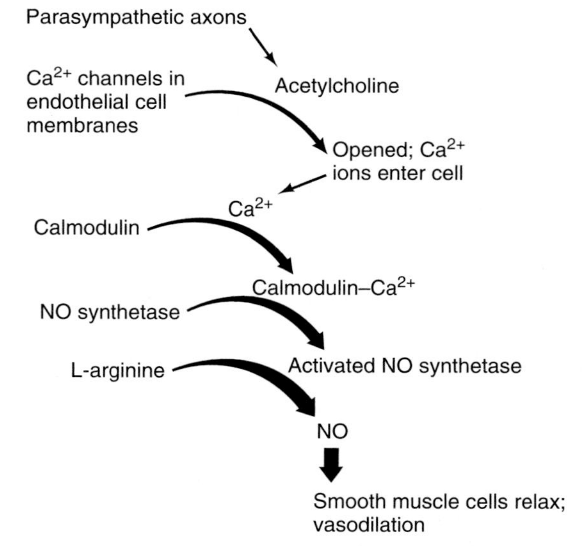

Nutrients, O_2, and other materials carried into each capillary bed move from the capillaries into the interstitial fluid

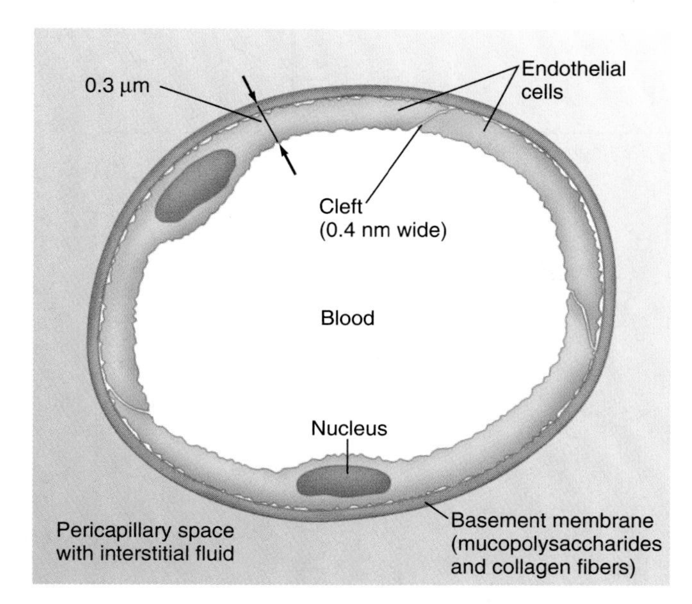

Figure 45.16

A cross section through a capillary shows it is made of thin endothelial cells that fit together like curved paving blocks. Materials diffuse easily over the short distances separating the blood from the surrounding interstitial fluid.

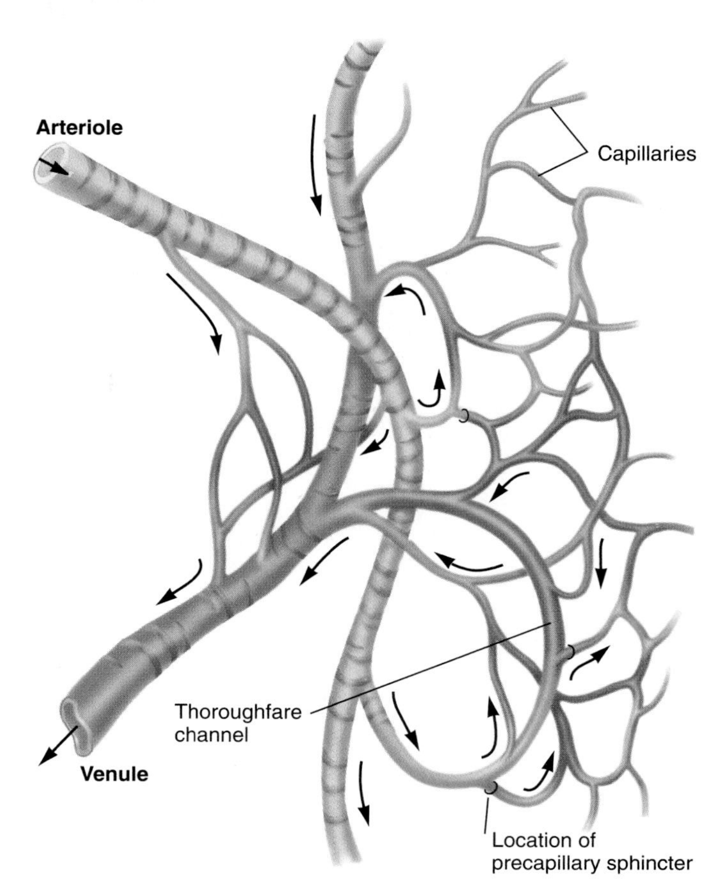

Figure 45.17

Blood flows through a capillary bed from arterioles through thoroughfare channels, which are always open, to venules. Where capillaries arise from the thoroughfare channels, they are surrounded by precapillary sphincters, rings of muscle that may close off the capillaries to severely reduce the flow through them.

and then into the cells of the tissue, while wastes from the tissue move back into the blood. How does this exchange occur if blood is confined to vessels with tight walls? Starling provided the general answer when he recognized that two forces can move fluid from one compartment to the other: *hydrostatic pressure,* the pressure of the fluid itself, and *osmotic pressure,* the tendency of water to diffuse across a membrane from a high water concentration to a low water concentration. Any imbalance between these forces will move fluid between the plasma and interstitial fluid in a capillary bed. Capillary walls act like very fine filters that restrict the movement of proteins but allow protein-free liquid to flow back and forth. This process is called **ultrafiltration.** Plasma and interstitial fluid are essentially the same fluid except for their proteins, and the balance of hydrostatic and osmotic forces determines whether this fluid moves in or out of the capillaries. As Figure 45.18 shows, at the arterial end of a capillary bed the hydrostatic pressure tending to move plasma outward is greater than the balance of osmotic pressures tending to resist its movement, so plasma flows out of the capillaries, carrying nutrients with it. At the venous end of the bed, the hydrostatic pressure has fallen, owing to resistance within the capillaries. The osmotic pressure around the capillaries remains the same, creating a net pressure that moves interstitial fluid into the capillaries, carrying wastes with it. The volume of fluid carried inward at the venous end is almost equal to the volume carried outward at the arterial end. In the next section we'll see what happens to the small excess of fluid that is not returned to the capillaries.

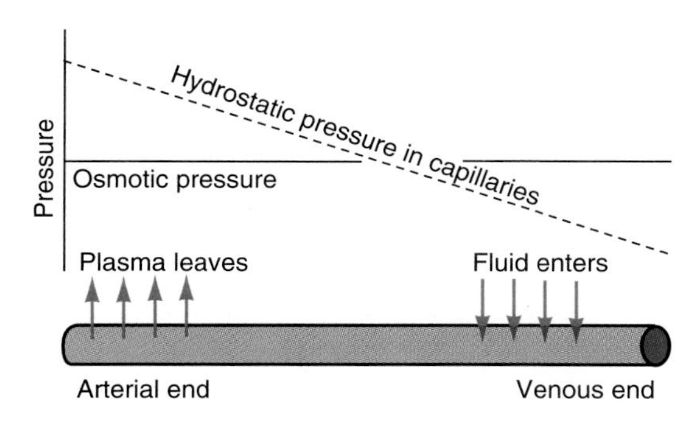

Figure 45.18

As blood moves through a capillary, it is subjected to two pressures: *(1)* hydrostatic pressure due to the force of the heart and the resistance within the vessels; *(2)* osmotic pressure due to different amounts of dissolved molecules in the blood and interstitial fluid. At the arterial end, hydrostatic pressure exceeds osmotic pressure, so plasma leaves the capillary. At the venous end, the hydrostatic pressure has fallen below the osmotic pressure, so interstitial fluid moves back into the capillary.

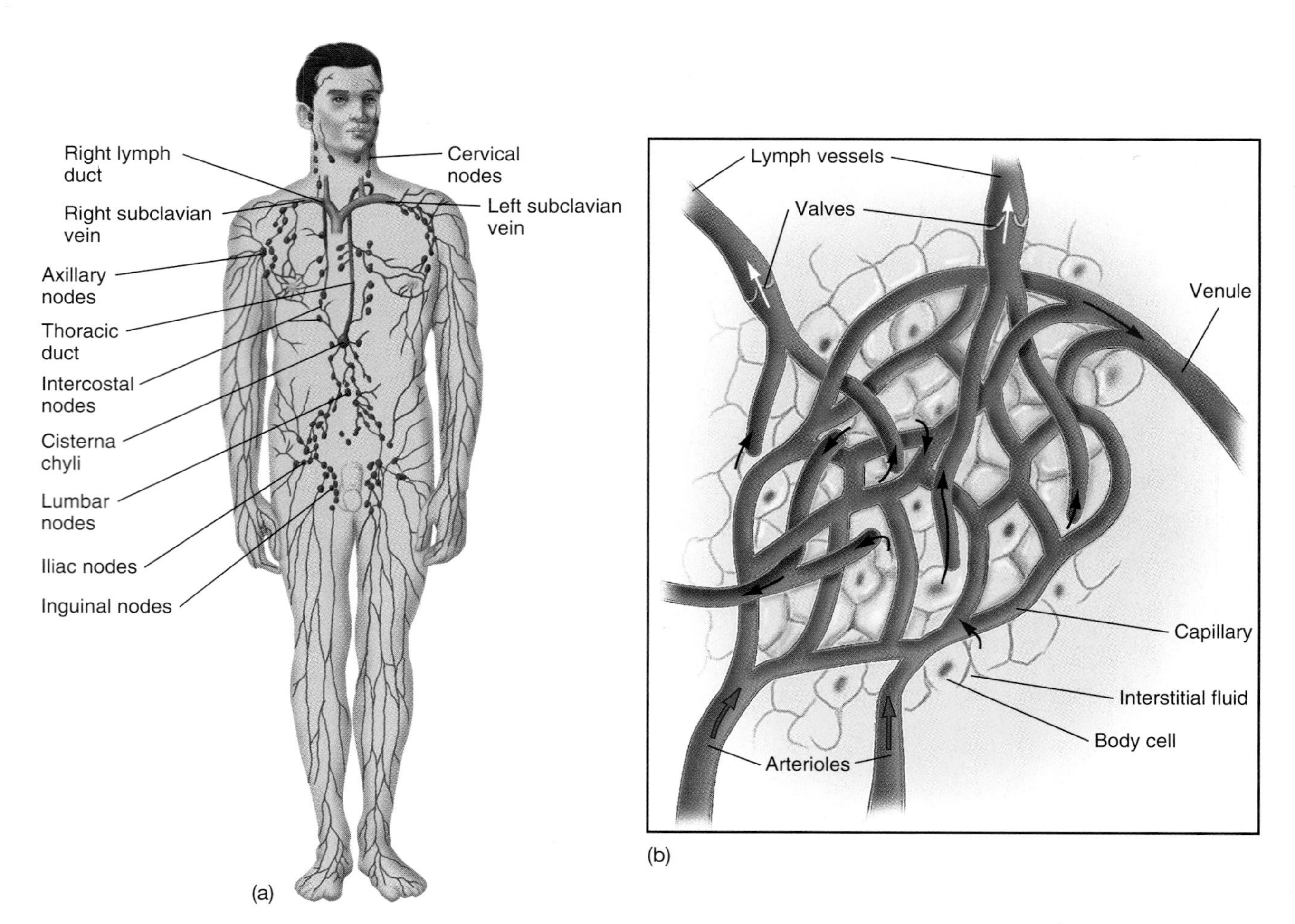

Figure 45.19

(a) The human lymphatic system is a tree of vessels that carries lymph from capillaries throughout the body into the subclavian veins. *(b)* Lymphatic vessels are interspersed with the vessels of a capillary bed.

45.8 The lymphatic system carries excess interstitial fluid.

A little of the plasma that leaves the capillaries doesn't return to them. This excess fluid, now called **lymph,** is picked up by the **lymphatic system,** which is made of auxiliary vessels leading back to the main circulation (Figure 45.19*a*). Lymph enters lymphatic capillaries running throughout intercellular spaces in almost all tissues. These are cul de sacs, so the lymph flows out of them in one direction into larger and larger vessels and eventually empties into veins in the base of the neck where it joins blood returning to the heart. The lymphatic flow is driven primarily by pressure from the organs the lymphatics pass through, supplemented by some contraction of smooth muscle around the larger lymphatic vessels, and lymph is kept moving in one direction by valves inside the vessels (Figure 45.19*b*).

If excess interstitial fluid could not get back into the circulation at all, the tissues would become swollen with fluid, a condition known as *edema.* Edema may result from high arterial blood pressure, obstruction of the venous circulation, kidney disease, and many other causes. In some people living in the tropics, lymph may fail to return to the circulation for a terrible reason: The lymphatics are blocked by minute nematode worms (filarias), which are transmitted by mosquitoes, and produce the condition known as elephantiasis (Figure 45.20).

Lymphatic capillaries are made only of endothelial cells with porous junctions and so are permeable to many materials, including proteins and potential pathogens such as viruses and bacteria that have invaded a tissue. Leucocytes move in and out of lymphatic vessels quite freely, as do cancer cells. At many points the lymph passes through **lymph nodes,** small nodules that filter the lymph and subject it to scrutiny by phagocytes, the protective white blood cells that form a staunch disease-fighting system (as discussed in Chapter 48).

45.9 Atherosclerosis and other disease processes can cause severe damage to the circulatory system.

A variety of disease processes, both infectious and functional, affect the circulatory system. The heart, for instance, may be damaged by infection with streptococcal bacteria that cause

Figure 45.20

People suffering from filariasis, an infection by minute nematode worms, develop elephantiasis due to blockage of their lymphatics by larval worms.

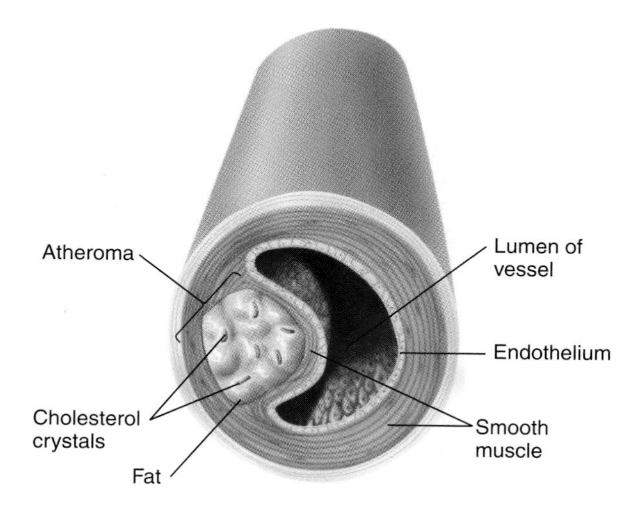

Figure 45.21

A plaque, or atheroma, is formed in the wall of a blood vessel when a smooth muscle cell begins to accumulate cholesterol and other fatty materials. It enlarges, becomes harder with the deposition of calcium, and blocks the vessel.

rheumatic fever, weakening heart muscle and damaging some of the heart valves. Infections are unpredictable and not easily avoided, but functional diseases are associated with certain living habits, many of which are avoidable. The principal risk factors in cardiovascular disease are a sedentary lifestyle with little exercise, a diet rich in fats, overeating, smoking, obesity, excessive use of alcohol, diabetes, stress, and a family history of heart disease.

The most common cardiovascular disease is **atherosclerosis,** the formation of fatty deposits called **plaques,** or **atheromas,** in the endothelial walls of arteries (Figure 45.21). While the process of plaque formation is still not well understood, it begins with some kind of damage to the endothelium. There is good evidence that a single smooth muscle cell is stimulated to start accumulating cholesterol and other fatty substances. Plaques enlarge through the interactions of many cells, including leucocytes that are involved in the process of acute inflammation discussed in Chapter 48. As they grow, plaques block the artery. Since the resistance to blood flow decreases with the fourth power of the vessel's diameter, it is evident that even a slight narrowing, called **stenosis,** may have severe effects on circulation. Stenosis of an artery results in **ischemia** (see Sidebar 45.2), decreased blood flow to a region of tissue and starvation for oxygen. The effect is most severe and deadly in the heart itself. Heart tissue is nourished by its own cardiac circulation, not by the massive volumes of blood moving through its chambers. Ischemia is very obvious in an experimental procedure where a cardiac artery is purposely tied off (Figure 45.22). If the ischemia is severe enough, the affected myocardial tissue undergoes a **myocardial infarction**—it literally dies; this event is experienced as a heart attack. Infarctions may be so minor that they are hardly noticed; sometimes a person only experiences fatigue, anxiety, and a vague discomfort. A serious heart attack, however, is characterized by extreme pain, called *angina pectoris,* often radiating into the left shoulder and arm, and difficulty in breathing. The prognosis for the victim depends, of course, on the extent of the infarcted region. Though this region forms scar tissue and no longer contracts, the heart may regain its normal function with rest. By changing his or her lifestyle to eliminate as many of the risk factors as possible, a heart patient may become healthier than before the attack.

Lipoproteins and plaque formation, Section 47.12.

Plaques are also dangerous because they stimulate the formation of blood clots, or *thrombi* (sing., *thrombus*). A thrombus in a cardiac artery can also produce a myocardial infarction, known in this case as a *coronary thrombosis.* A thrombus formed anywhere in the circulation may break loose, becoming an *embolus,* and travel to some other point. Lodged in a brain artery, the embolus blocks circulation to a portion of the brain, producing one kind of *stroke.* (A stroke may also result from hemorrhage in a part of the brain.) Since each part of the brain has its own distinctive function, strokes have various effects on behavior and mental functioning. Often they are extremely debilitating.

Mechanism of blood clot formation, Section 48.4.

In modern society, the risk factors listed in this section conspire to make cardiovascular disease the greatest single cause of death. However, most of these factors can be controlled. Keeping your weight below the recommended maximum for your height and exercising regularly are the two most important

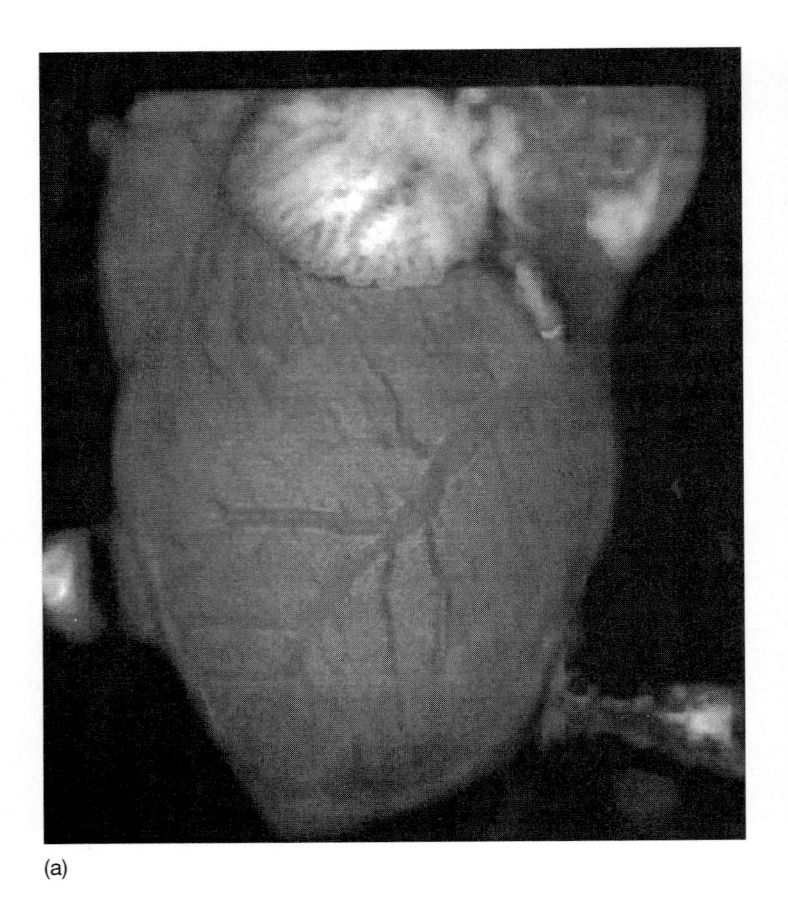

(a)

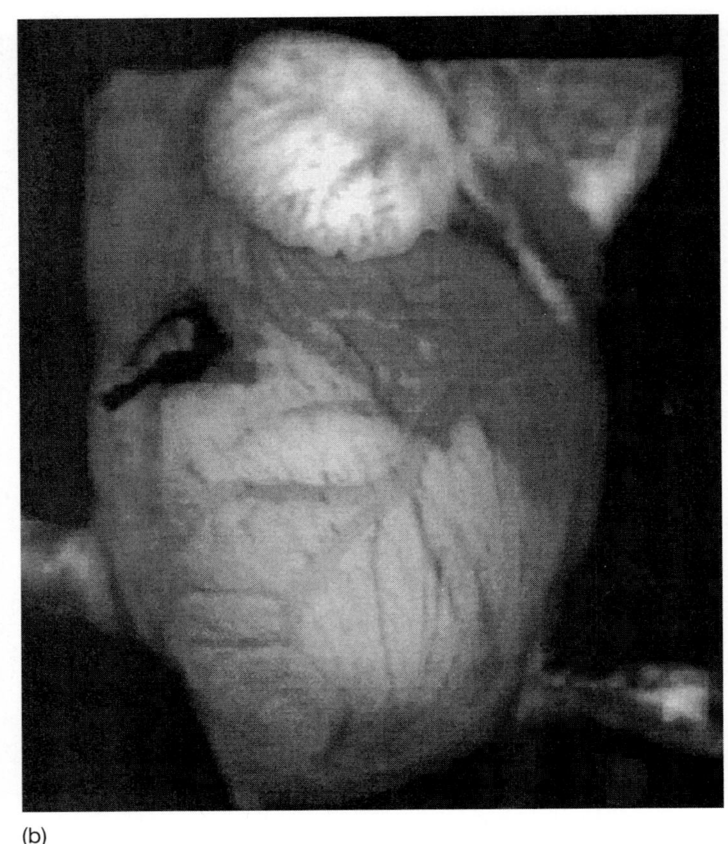

(b)

Figure 45.22

An experimental heart attack is induced in a rat heart. The heart is flashed with light that is absorbed by NADH in the heart tissue, which then fluoresces at a wavelength of 430 nm. The oxidized form, NAD^+, does not fluoresce. *(a)* Normal heart tissue appears quite dark because it is well oxygenated and its $NADH/NAD^+$ is primarily in the NAD^+ form. One cardiac artery is then tied off, and within half a minute *(b)* the region served by this artery appears as a bright, fluorescent patch, showing that its $NADH/NAD^+$ is now largely in the NADH form.

factors in maintaining cardiovascular health; even moderate regular exercise such as walking and climbing stairs (instead of driving and using elevators) decreases the chances of a cardiovascular accident considerably, in comparison with a sedentary life. Aerobic exercise, in which the body metabolizes aerobically at a high rate, is especially beneficial. Section 47.12 also discusses the development of atherosclerosis as it is mediated through blood lipoproteins and affected by exercise and diet.

C. Gas exchange and its regulation

45.10 A circulatory system moves blood past a site of external gas exchange.

Although very small animals can exchange respiratory gases through diffusion alone, animals greater than a millimeter or two in thickness require organs of external gas exchange, most commonly lungs, gills, skin, or the tracheal systems of arthropods (Figure 45.23). These organs are prime examples of structures that provide large surface/volume ratios (review Figure 32.27), which are essential because gas exchange occurs through diffusion alone; no proteins have ever evolved for active transport of respiratory gases. Organs for gas exchange are designed so that either air or water containing dissolved oxygen can move past a large surface where oxygen can diffuse down its concentration gradient into the blood, and CO_2 can diffuse down its concentration gradient in the opposite direction, out of the blood. Skin, as in an earthworm, is the most primitive surface for gas exchange and is generally replaced by more efficient organs in later evolution. Still, some aquatic animals with lungs, such as frogs, exchange a great deal of gas through their moist skin, and blood returning from the skin may be richer in oxygen than that returning from the lungs. Even humans exchange a little gas through the skin.

As a general rule, terrestrial animals draw air into inward-directed tubes and pockets, whereas aquatic animals circulate water around and through protruding gills, which may lie under the body surface as in molluscs, crustaceans, and fishes. This difference reflects the fact that air, being so much less viscous than water, can easily be drawn through tubes; yet there are exceptions to the rule, since some aquatic animals, such as sea cucumbers, draw water through a branched internal system (a

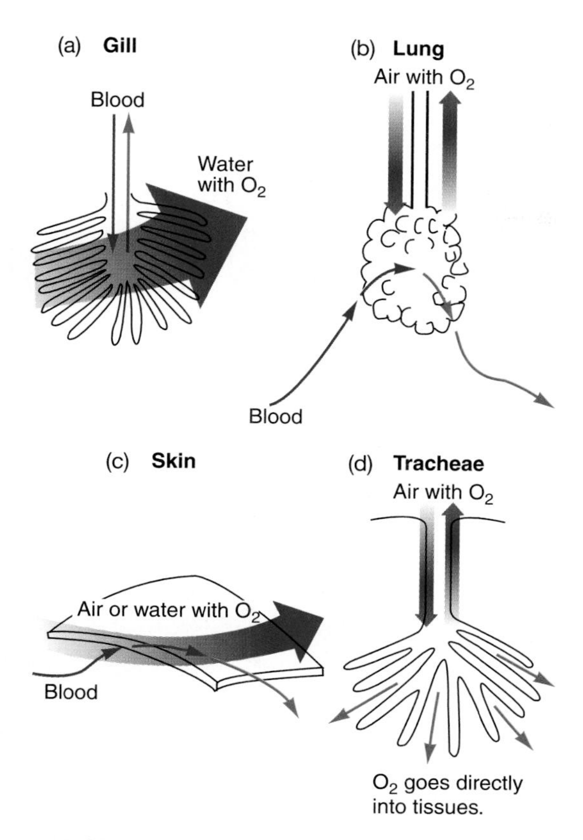

Figure 45.23

Several kinds of structures serve external gas exchange. *(a)* Most aquatic animals possess gills, which are fingers or plates extending outward from the body. *(b)* Many terrestrial animals draw air into internal pockets, such as the lungs of vertebrates. *(c)* In many animals, the skin alone provides a great deal of gas exchange. *(d)* Insects and other terrestrial arthropods have a system of tracheae that penetrates deep in the tissues.

respiratory tree), and some fish, remarkably, exchange a great deal of gas externally through their large intestinal surfaces. Fick's Law of Diffusion shows that a large surface is one factor that makes for rapid exchange. A second factor is maintaining the maximum possible concentration gradient across the exchange membrane, and we will see that the anatomy of some aquatic animals ensures this. Third, the path that respiratory gases must take during exchange can be made as short as possible, and this too is done by making the structures involved thin and the spaces between them narrow.

Gas-exchange structures linked to a circulatory system all depend on fluids moving past the exchange surface. On the inside, they are *perfused* by moving blood through the lungs, gills, or skin, and on the outside they are *ventilated* by moving the external air or water over the surface. Thus fresh blood constantly delivers more CO_2 and removes the incoming O_2, while the outside medium does the opposite. External gills are ventilated automatically by movements of the surrounding water, but animals with lungs or internal gills need a breathing mechanism.

The atmosphere contains 200 ml of oxygen per liter. Seawater carries only about 5 ml of oxygen per liter, and freshwater about 7 ml per liter, though the concentration in a stream that has been well-aerated by waterfalls approaches 10 ml per liter. So there has been considerable selective pressure during evolution to make gills efficient in extracting oxygen from the surrounding water. In fact, the gills of fishes can remove as much as 80 percent of the oxygen in the water flowing through them by using **countercurrent exchange,** described in Concepts 45.1. The direction of blood flow in the gill lamellae is opposite to the flow of water, thus maximizing the gradients of both O_2 and CO_2 concentration between the blood and the water (Figure 45.24). Remember that the rate of diffusion between two compartments is proportional to the concentration difference between them; by maximizing this difference, the gills very efficiently remove O_2 from the water and eliminate CO_2 from the blood. To maintain a ventilation stream through their gills, fish use a double-pump system: They close and contract their mouth cavity to create a positive pressure, while opening and closing the opercular flaps over the gills to create negative pressure.

Insects, the most abundant and diverse air-breathing invertebrates, use a system of **tracheae** (see Figures 45.23*d* and 32.17*a*) quite unlike anything else in the animal kingdom. These tubules open to the atmosphere through openings called *spiracles* in the insect's abdomen; the main tubes then divide into finer tracheoles and finally into many fine air capillaries, which carry air deep into all the insect's tissues. Gases diffuse through these tubes and, in larger insects, are forced back and forth by the contraction of body muscles. Thus respiratory gases exchange directly with the atmosphere without the need for a circulatory system, and since gases diffuse so much faster in air than in water, insects are able to maintain high metabolic rates. Indeed, the highly active flight muscles of some insects are richly supplied with air capillaries. However, insects are still limited to small sizes with relatively short tracheal tubes by the physical limitation discussed in Section 45.5: Flow through a tube meets resistance in proportion to its length and the fourth power of its diameter.

Insect flight muscle, Section 44.11.

In air-breathing vertebrates, air is drawn into the lungs through a large tube (also called a **trachea**), which branches off from the throat. The esophagus leading to the stomach begins in the same area, and the opening of the trachea is covered with a flap of flexible cartilage, the *epiglottis,* which closes during swallowing and prevents food from going into the trachea. The trachea branches into a respiratory tree: two smaller tubes, the primary **bronchi** (sing., *bronchus*), which then branch into a series of **bronchioles** of decreasing size that end in small pockets called **alveoli** (Figure 45.25). Each alveolus is a rounded sac whose walls are rich in capillaries where the air and blood are in such intimate contact that gases can easily exchange between them over an enormous surface. An adult human has 100–200 square meters of alveolar surface. (The area of a tennis court is about 250 square meters.)

Concepts 45.1 Countercurrent Exchange and Wonderful Nets

A countercurrent flow system is an elegantly simple device that can conserve heat, energy, or materials. For example, the pipes carrying cold water into a heater can be placed in contact with the pipes carrying hot water out, like this:

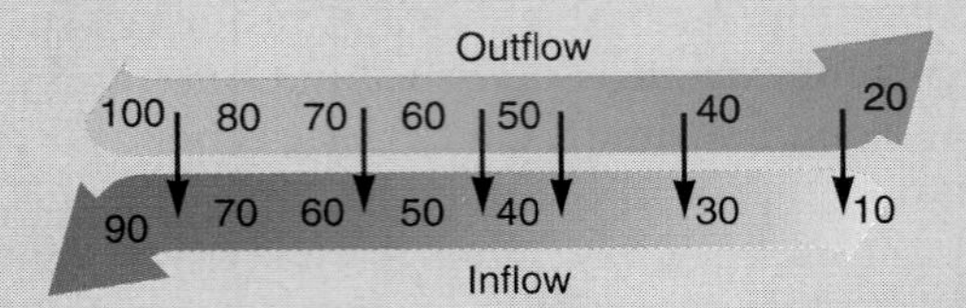

The cold water is preheated by the hot water, so some of the heat that would otherwise be lost from the hot pipe is carried back into the system and conserved.

Many animal structures use a similar device. For instance, an animal needs a continuous flow of blood into its extremities to maintain metabolism, but the extremities don't necessarily have to be warmed to the temperature of the body. A wading bird standing in cold water does not lose a great deal of heat through its legs because the arteries and veins in its legs, carrying blood in opposite directions, are interspersed in a network called a *rete mirabile,* or "wonderful net," so they are in intimate contact with one another:

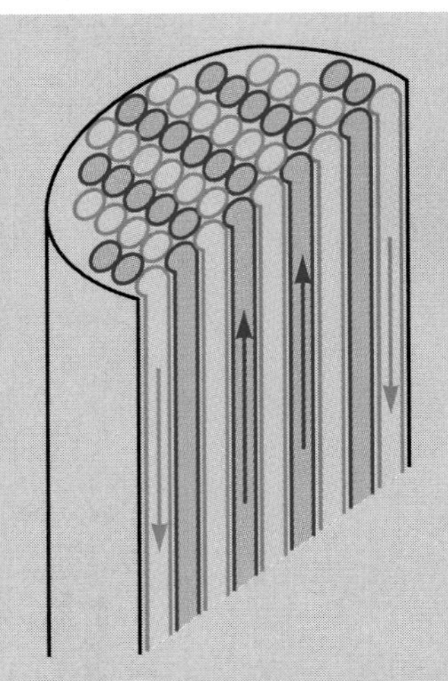

Heat is conserved as the warm arterial blood flowing outward gives up much of its heat to the cool venous blood returning. Some fish use a rete mirabile in their swim bladders, which must be kept filled with oxygen. The bladder loses very little oxygen because arteries carrying blood inward pick up oxygen from the veins leaving the bladder. Many other organs incorporate the countercurrent exchange principle and rete mirabile into their designs. We will see in Chapter 46 that is it particularly important in kidney function.

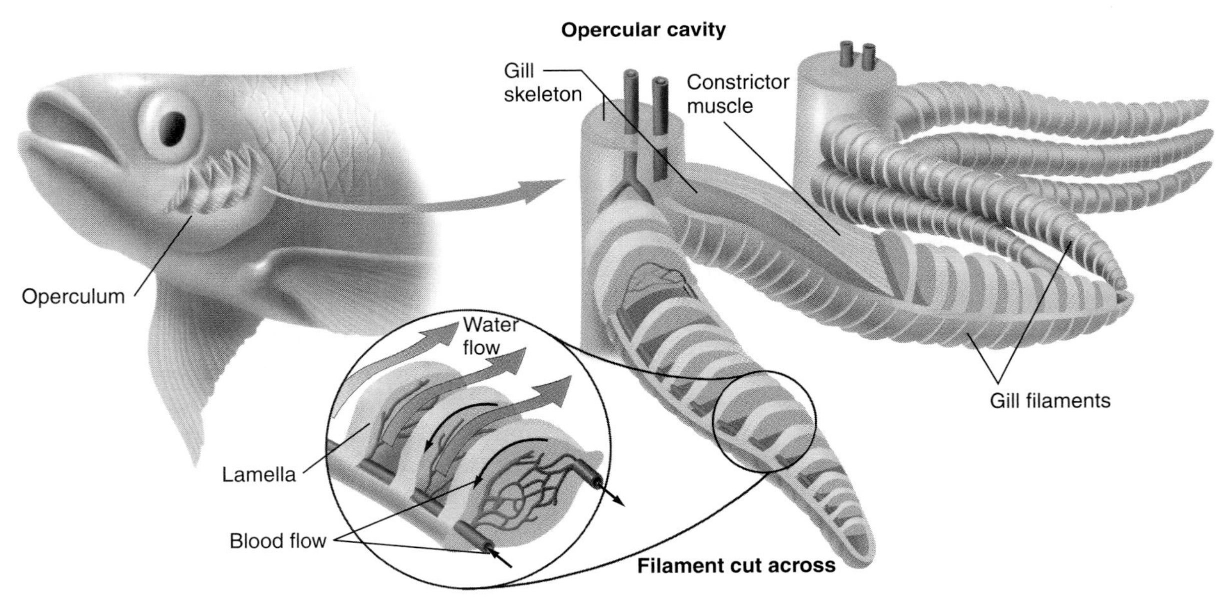

Figure 45.24
The gills of a fish lie under a gill cover (operculum). Each gill arch bears a stack of thin lamellae. Blood flows through each lamella in parallel capillaries, and this flow is oriented opposite to the flow of water through the gill. Thus the system acts by countercurrent distribution, maintaining the maximum gradient of oxygen concentration between the blood and water, so that a maximum of oxygen is transferred to the blood.

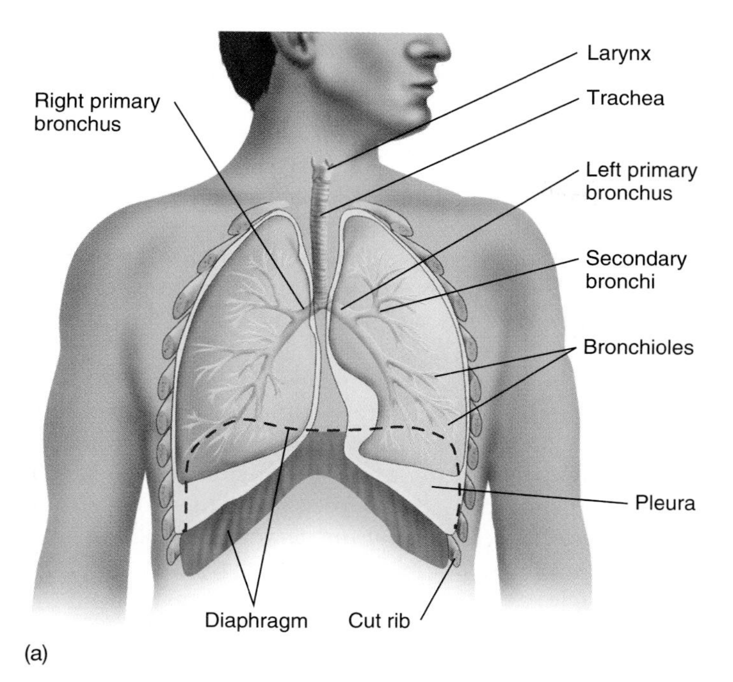

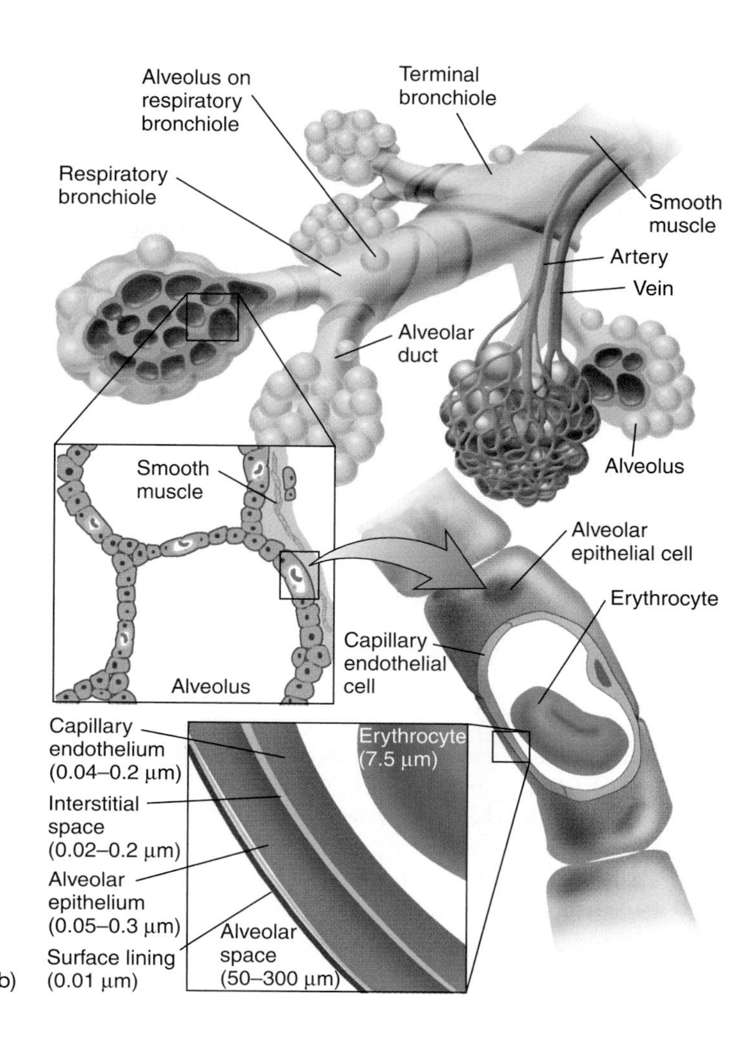

Figure 45.25

(a) Human lungs draw air in and out through a respiratory tree, which finally divides into small bronchioles. *(b)* At its end, each bronchiole leads to several alveoli where gas exchange occurs. A section through the bronchioles and alveoli shows that capillaries are interspersed among alveolar cells. A capillary is about the same diameter as the red blood cells that pass through it, so gases have only short distances to diffuse across the membranes of an erythrocyte, the capillary wall, and the lung surface.

The lungs of vertebrates lie inside the thoracic cavity, at the anterior end of the main body cavity. In mammals, the floor of the thoracic cavity is formed by the heavily muscled **diaphragm.** The lungs are ventilated by changing pressure in the thoracic cavity. Pleural membranes surround the lungs, dividing the thoracic cavity into right and left pleural cavities; there is a little liquid between the lungs and pleura, but no air. (*Pleurisy* is a painful inflammation of the pleura associated with pneumonia and other infections.) When the diaphragm is most relaxed, the thoracic cavity is smallest. During inhalation, or *inspiration,* the diaphragm contracts and pulls downward; at the same time, one set of muscles between the ribs (external intercostal muscles) contracts to move the rib cage upward and outward (Figure 45.26). These actions together expand the thoracic cavity. As it expands, the pressure within it becomes less than the air pressure in the respiratory tree, and air rushes into the lungs. Exhalation, or *expiration,* occurs passively as the rib muscles and diaphragm relax, the lungs recoil elastically to their less-inflated shape, and the walls of the chest cavity return to their former positions. Exhalation may be enhanced by contracting the opposite set of rib muscles (internal intercostal muscles) and the muscles of the abdominal cavity. (Try

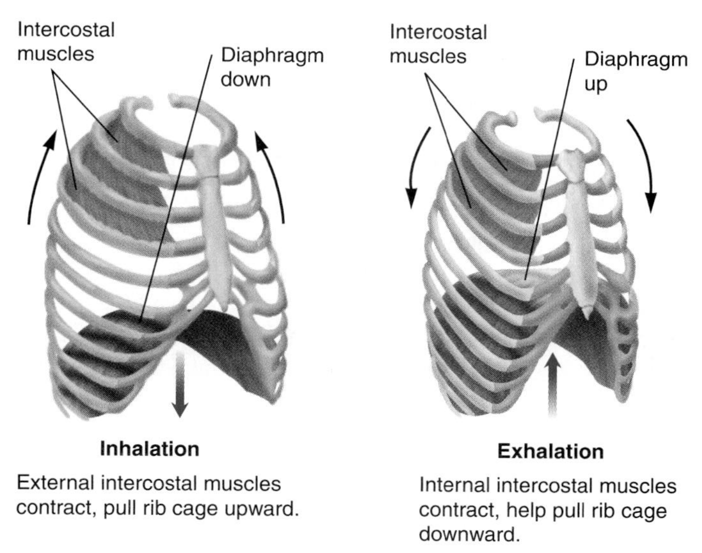

Figure 45.26

The thoracic cavity in which the lungs lie is lined by a mucous membrane, the pleura; the bottom of the thoracic cavity is the muscular diaphragm whose contraction makes the lungs inflate. Intercostal muscles between the ribs assist in breathing.

contracting your abdominal muscles and feel how that action forces air out of your lungs.)

The alveoli are covered by a thin layer of water, which creates an interesting physical problem that had to be overcome before the lung could function efficiently as an organ of gas exchange. Remember that the cohesiveness of water creates a surface tension at a water surface. The water in an alveolus is a mere film, so it has a high tension like that holding a soap bubble together. This tension creates a pressure directed toward the center of the bubble—or the alveolus—that is inversely proportional to the radius of the sphere:

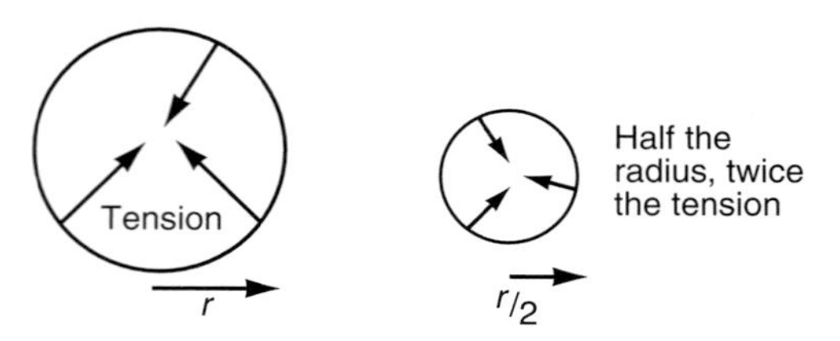

The smallest alveoli, therefore, are in danger of collapsing. Alveolar collapse is prevented, however, by a remarkably simple device: Certain cells in the alveolar walls produce a *surfactant* (short for *surface-active agent*), the phospholipid dipalmitoyl lecithin, which disrupts the film of water and reduces the surface tension. A newborn baby requires this lung surfactant to begin breathing, but surfactant synthesis begins late in fetal development; babies born before seven months of gestation usually develop *respiratory distress syndrome* and have great difficulty inflating their lungs.

Exercise 45.6 What happens to someone who sustains a puncture wound to the chest deep enough to penetrate a pleural cavity but not the lung? Assume the wound stays open and doesn't close up spontaneously.

45.11 Blood and hemolymph transport oxygen into tissues.

Most animals cannot go for long without oxygen, since their tissues die without it. (Refer to Sidebar 45.2 for the kinds of changes that occur in tissues when they are starved for oxygen.) Muscles and some other tissues are able to metabolize fermentatively for a while by oxidizing glucose to lactic acid, but they build up an oxygen debt that must be repaid eventually. Even animals in the deep ocean live aerobically, and we show in Section 25.3 how important it is for vertical ocean currents to carry oxygen into the ocean depths.

The small amount of O_2 dissolved in blood plasma or hemolymph is important in an animal's total oxygen balance, but by itself it is far from enough, so most animals depend on O_2-carrying blood proteins. Some molluscs and arthropods use the bluish copper protein hemocyanin. Many worms use the violet iron protein hemerythrin or the green chlorocrurin. But by far the most common O_2 carrier is hemoglobin. It is used by echinoderms, annelids, some arthropods and molluscs, and many minor animal groups, as well as by vertebrates. One kind of hemoglobin even has a role in the nitrogen-fixing nodules of certain plant roots (see Section 40.8). Some animals have hemoglobin dissolved in their plasma, but it is advantageous to confine it within red blood cells, as vertebrates do: More protein can be packed into the blood if it is concentrated in a streamlined body, a blood cell that slips through the capillaries, whereas blood with an equal amount of dissolved protein would be very viscous, too thick to circulate. (Some capillaries are so narrow that red blood cells are distorted as they slip through.)

At this point you should review Sections 5.9 and 5.10, where we examine the structure and function of myoglobin and hemoglobin. Remember that these two proteins, while very similar, have evolved to perform different functions, as reflected in the way they bind O_2. Both proteins bind more oxygen as the oxygen pressure increases, but the hemoglobin subunits interact cooperatively, so at low O_2 pressures, as in the tissues, hemoglobin releases O_2 more readily, thus making it the more efficient carrier.

An additional adaptation of hemoglobin is the **Bohr effect,** a shift in oxygen binding due to interactions with hydrogen ions and CO_2. As hemoglobin (Hb) releases its O_2, it picks up a H^+ ion:

$$HbO_2 + H^+ \rightarrow HHb + O_2$$

Hemoglobin is an allosteric protein whose conformation and activity are changed by binding certain small ligands (allosteric effectors). H^+ ions also act as allosteric effectors by accentuating the cooperative interactions among hemoglobin subunits. As the H^+ ion concentration increases, the hemoglobin-saturation curve shifts to the right, which means that hemoglobin releases more of its O_2 at low oxygen pressures (Figure 45.27). Now, where does hemoglobin meet such a high concentration of H^+ ions? In the tissues, of course, where metabolism produces acid wastes such as carbonic and lactic acids. The Bohr effect therefore forces hemoglobin to release even more of its O_2 where it is needed most.

Exercise 45.7 Here are dissociation curves for the hemoglobins of certain mammals:

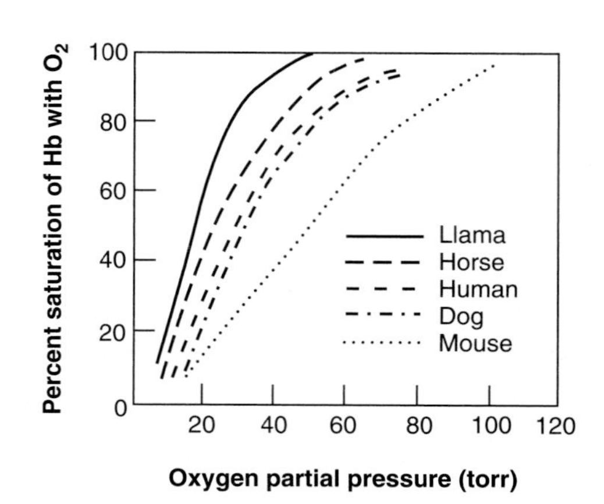

Llamas live high in the Andes, where low air pressure makes it difficult to get enough oxygen. Explain how the behavior of their hemoglobin is an adaptation to these conditions.

Exercise 45.8 The dissociation curve for hemoglobin also shifts to the right as the temperature increases, thus releasing more oxygen for a given oxygen pressure. Think of at least one condition in which your blood temperature could be increased locally and this feature of hemoglobin would then be advantageous.

Exercise 45.9 A typical adult human requires about 17 moles of O_2 per day just for basal metabolism. If each mole of gas occupies 22.4 liters, calculate how many liters of O_2 must be consumed per hour and how many milliliters per minute. Since air is about 20 percent oxygen, how many liters of air must be consumed?

Basal metabolism, Section 32.14.

Exercise 45.10 A normal person has 15 grams of hemoglobin per 100 ml of blood, and each gram can carry 1.34 ml of O_2. How much O_2 is carried per liter of blood in this way?

Exercise 45.11 The normal cardiac output is 5 liters/min. Suppose the hemoglobin, in passing through the tissues, releases only 25 percent of its O_2. How much will it supply per minute?

Exercise 45.12 If you have done the preceding exercises correctly, you have determined that a person requires about 265 ml of O_2 per minute, but the hemoglobin only supplies about 250 ml. Before you conclude that we are all starved for oxygen, consider that blood plasma can carry about 0.3 ml of dissolved O_2 per 100 ml. Determine whether this makes up the deficit.

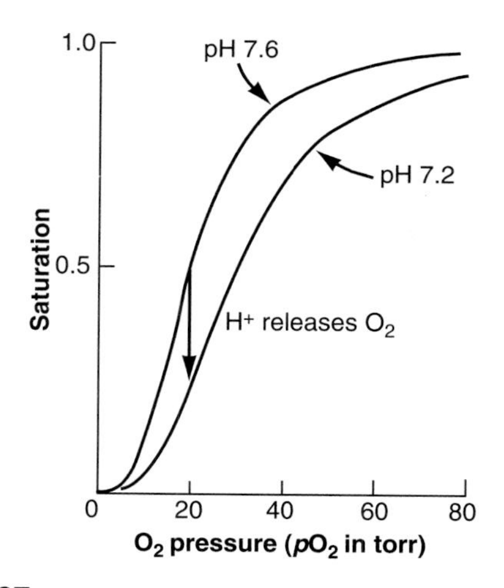

Figure 45.27

The Bohr effect is seen in saturation curves of hemoglobin at different pHs. The curve is sigmoid (S-shaped) because of cooperative interactions among the protein subunits, so where the oxygen pressure is low, in the tissues, hemoglobin tends to release its oxygen. H^+ ions induce more strongly cooperative behavior. The saturation curve shifts more strongly to the right as the pH decreases; therefore, in tissues, where the pH is lower, hemoglobin tends to release its O_2 more completely.

45.12 Carbon dioxide is transported out of tissues.

Removing carbon dioxide is one of the circulatory system's chief functions. As with O_2, a small amount of CO_2 dissolves in blood plasma or hemolymph and is carried this way, but not nearly enough to satisfy the needs of even a resting animal. In vertebrates, blood disposes of most CO_2 through the action of two proteins we have met in earlier chapters. First, *carbonic anhydrase,* an enzyme abundant in erythrocytes, converts CO_2 into bicarbonate ion:

$$CO_2 + H_2O \rightarrow H_2CO_3 \rightarrow HCO_3^- + H^+$$

Then *anion exchange protein* (see Section 8.10) in the erythrocyte membrane conducts a *chloride shift* by exchanging the bicarbonate for chloride ion:

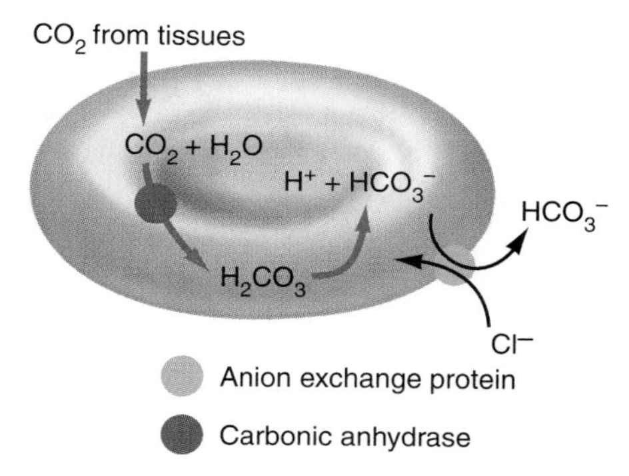

Bicarbonate, being very soluble, is carried to the gills or lungs, where the first processes are just reversed, producing CO_2 to be released:

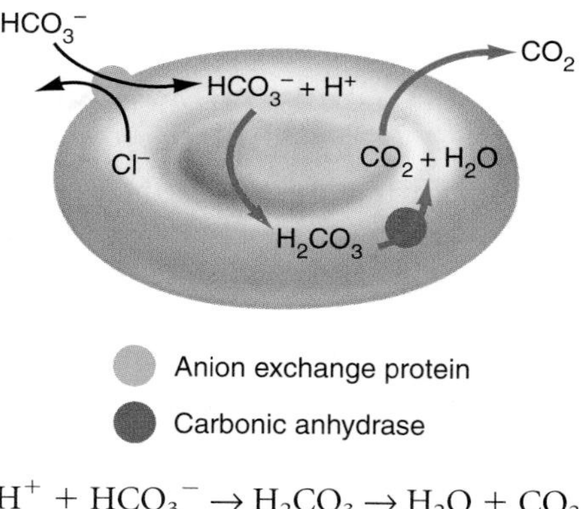

$$H^+ + HCO_3^- \rightarrow H_2CO_3 \rightarrow H_2O + CO_2$$

Notice that these transformations involve H^+ ions. This poses a potential problem because the pH of blood and intracellular fluids must be kept quite constant. The second equation shows that H^+ ions are eliminated when bicarbonate is converted back into CO_2 and water, which are both exhaled. As we'll see in Section 45.14, a reduced plasma pH stimulates faster, deeper breathing—an important homeostatic mechanism.

Hemoglobin plays a dual role in gas exchange. In addition to carrying oxygen into the tissues, it carries CO_2 away by binding about a quarter of the CO_2 covalently to its N-terminal amino group. Another aspect of the Bohr effect is that

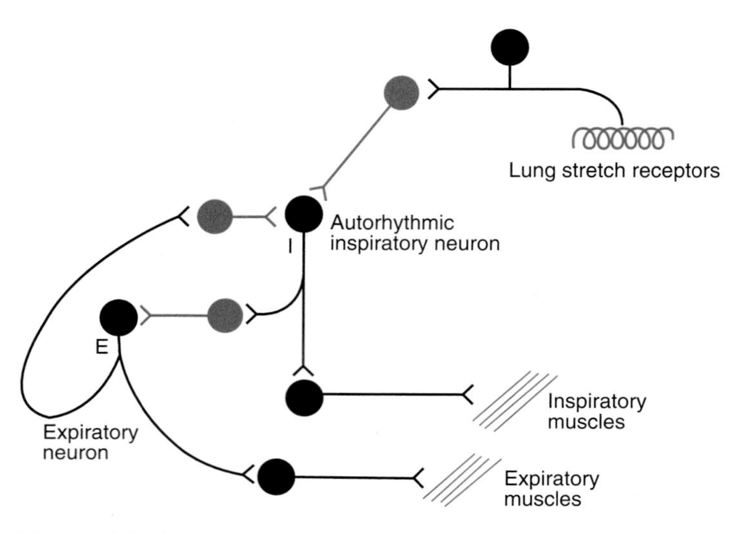

Figure 45.28

A neural circuit in the rhythmicity center of the medulla oblongata uses inspiratory and expiratory neurons. Inspiratory neurons (I) stimulate inspiration, and through inhibitory neurons *(red),* they inhibit expiratory neurons (E). After a short time, the inspiratory neurons tire and expiratory neurons begin to fire; they, in turn, inhibit the inspiratory neurons. The two sets are thus turned on and off rhythmically. In addition, stretch receptors in the lungs inhibit inspiratory neurons.

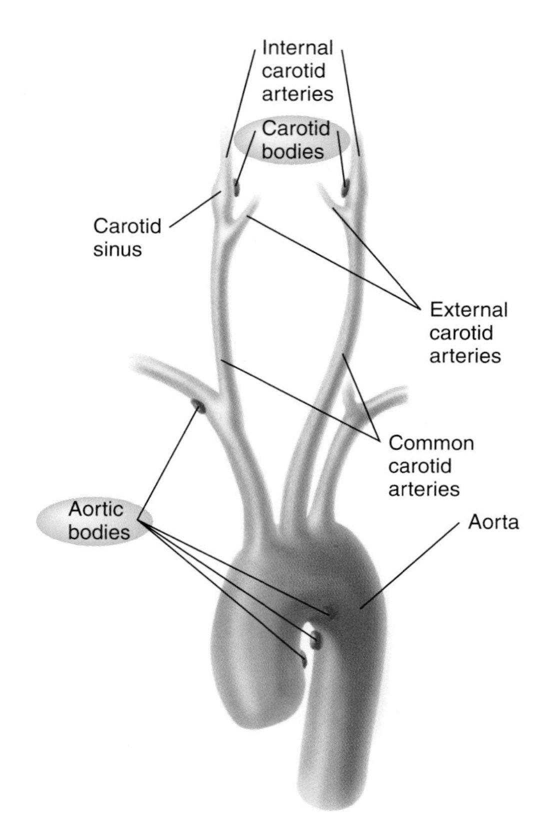

Figure 45.29

The carotid and aortic bodies, along with a center in the medulla oblongata, monitor the pH and O_2 concentration of blood and cerebrospinal fluid.

hemoglobin with CO_2 bound to it shifts its oxygen-carrying capacity to release more O_2, so picking up CO_2 helps make more O_2 available.

45.13 An external circuit controls the breathing rhythm in mammals.

We never have to think about breathing. It is totally involuntary. Mammals pump air in and out of their lungs in a steady rhythm created in a center of the medulla oblongata by an interesting circuit of excitatory and inhibitory neurons (Figure 45.28). Inspiratory neurons stimulate inspiration by contracting the diaphragm and external intercostal muscles, which expand the thoracic cavity; expiratory neurons inhibit the inspiratory neurons for a while. The two kinds of neurons play upon each other reciprocally through inhibitory neurons: When the inspiratory neurons fire, they keep the expiratory neurons quiet, and vice versa.

Stretch receptors in the lungs superimpose a second regulatory system. When stretched by lung expansion, these receptors send signals through the vagus nerve that inhibit the inspiratory neurons. This action is known as the Hering-Breuer reflex. The whole breathing apparatus is also subject to still higher brain centers, allowing some conscious control over the rate of breathing. As we will see next, the rate and depth of breathing are controlled automatically in response to metabolic conditions.

45.14 The nervous system regulates breathing in response to oxygen and hydrogen-ion levels.

Breathing normally keeps the concentrations of respiratory gases in the blood within certain limits. We saw earlier that the CO_2 concentration of blood plasma is closely related to its pH, so acidity in the blood can indicate the need for faster ventilation to eliminate CO_2. The ventilation rate is strongly regulated by signals from specialized sensory cells in contact with the blood that monitor its O_2 and CO_2 levels and its pH. The chemosensors responsible are located in two centers in mammals—the medulla oblongata and the **carotid bodies** and **aortic bodies** on major arteries (Figure 45.29). All three centers monitor the CO_2 concentration by responding to reduced pH; the medullary center senses the pH of cerebrospinal fluid. Experiments in which acid is added to the bloodstream show that the rate of breathing rises as the pH falls. In addition, the carotid bodies directly measure O_2 concentration.

Breathing is more sensitive to the CO_2 level of the blood than to O_2 (Figure 45.30). The classical experiments of John S. Haldane and John G. Priestley around 1905 clearly showed that ventilation becomes more rapid as the CO_2 concentration of

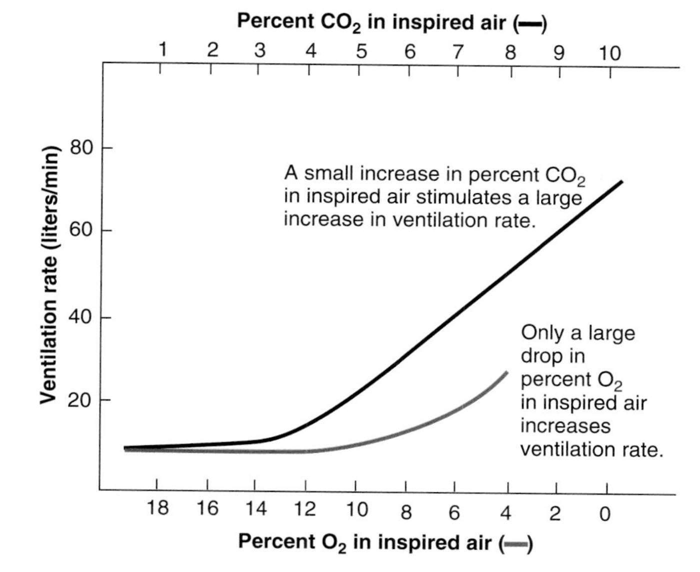

Figure 45.30

As the CO_2 concentration in inspired air increases, the rate of ventilation increases quite sharply *(black curve)*. The rate does not respond strongly to a decreased concentration of O_2 until the concentration becomes very low *(red curve)*.

inspired air increases. Haldane's son, J. B. S. Haldane, later recalled his work:

> *He started his work in Dundee. He collected samples of the very worst air that he could find. He went, between 12:30 and 4:30 in the morning, to the worst slums, taking samples of air from rooms where as many as eight people were sleeping in one bed. He went down the sewers also. Many years later I was with him in Dundee, and to remember his way above ground, he had to imagine himself back in the sewers below the streets.*

J. S. Haldane himself described the effects of CO_2 on breathing:

> *The effects on the breathing of air so vitiated attracted attention particularly. It was found that when the proportion of CO_2 in the air rose to about 3 percent and the oxygen fell simultaneously to about 17 percent (there being 20.03 percent of oxygen and 0.03 percent of CO_2 in fresh air) the breathing began to be noticeably increased. With further vitiation the increase in breathing became more and more marked until with about 6 percent of CO_2 and 13 percent of O_2 the panting was very great and caused much subsequent exhaustion.*

Although Haldane, in this passage, writes as if O_2 and CO_2 simply replace each other in the atmosphere, it must be clear that the concentrations of O_2 and CO_2 can be changed independently, to distinguish their effects, as shown in Figure 45.30.

Exercise 45.13 Draw a simple regulatory circuit summing up how various factors, receptors, and brain centers interact to change the rate of breathing in response to increasing production of CO_2.

Coda Considering the operation of a circulatory system brings out many of the themes of this book. Evolution pervades our discussion. The fact that animals display either open or closed systems reminds us of the opportunism of evolution, for very different systems—all quite functional—have evolved as different animal clades have followed the random genetic events that create different opportunities for survival. We also read an interesting evolutionary story about the changes that have occurred in vertebrate hearts and the associated circulation; comparative anatomists tell this story in even more detail than we have outlined here, for they can show how particular vessels leading to the gills in fishes became modified into the aorta and pulmonary circulation of later air-breathing vertebrates. These changes clearly adapt each species for its particular way of life. Years ago, the three-chambered amphibian heart was often pictured as a poorly adapted structure that has not yet achieved the efficiency of a four-chambered heart, but we can hardly believe that such inefficiency would have persisted among amphibians for 300 million years. In fact, amphibian hearts are wonderfully well adapted for directing oxygenated and deoxygenated blood to their proper destinations.

We have also seen repeatedly that molecular interactions are the basis for biological activity. The muscular contractions that power and regulate circulation are interactions of actin and myosin fibers, with their control proteins. We now know that vasodilation involves the paracrine mediator nitric oxide, so smooth muscle cells respond by virtue of their receptors that bind this simple ligand. Furthermore, circulation and respiration are constantly regulated by receptors that detect such factors as pH and the concentrations of CO_2 and O_2. These receptors contain specific proteins that recognize chemical components of the blood and interstitial fluid, sending signals back to the heart pacemakers and respiratory brain centers. We understand the operations of large physiological systems by keeping in mind the molecular interactions that underlie them.

Summary

1. Only the smallest animals, which can exchange materials with the surrounding water entirely by diffusion, do not need a circulatory system.
2. In a closed circulatory system, as in vertebrates, blood moves through a system of arteries and veins because of pressure generated by one or more hearts, and materials exchange in capillary beds between the blood and interstitial fluid. In contrast, arthropods, molluscs, and some other animals have open systems in which the heart pumps a single fluid, hemolymph, through open sinuses in the tissues.
3. The mammalian circulation is driven by the right side of the heart, which pumps deoxygenated blood to the lungs, and the left side of the heart, which receives oxygenated blood from the lungs and then pumps it through the arteries, capillaries, and veins of the systemic circulation.
4. The vertebrate circulatory system has evolved from a single loop in fishes into the double-circulation type in mammals and birds.
5. The heart contracts with a rhythm set by internal pacemaker tissue, the sinoatrial and atrioventricular nodes, which send signals through

modified muscle fibers to stimulate coordinated contractions of the myocardia. Its pace is merely increased or decreased by outside nervous control.

6. Blood pressure is proportional to the cardiac output times the peripheral resistance. The cardiac output is the cardiac rate times the stroke volume. These factors are regulated both extrinsically and intrinsically to keep the average blood pressure relatively constant. Extrinsic controls include baroreceptor reflexes mediated through the medulla oblongata and autonomic nerves. Intrinsic controls include the Frank-Starling mechanism: Stroke volume is proportional to the end-diastolic volume.
7. Blood is a tissue consisting of cells in plasma, a complex fluid containing salts and many proteins and metabolites. Blood cells of vertebrates are red cells that carry O_2 and white cells that play various roles in fighting infection.
8. In a capillary bed, blood flows through thoroughfare channels from arterioles to venules; the flow through capillaries can be strongly regulated by precapillary sphincters that reduce the diameter of a capillary opening.
9. In a capillary bed, blood exchanges with interstitial fluid because of the balance of hydrostatic pressure and osmotic pressure. It is ultrafiltered, so plasma proteins remain in the capillaries.
10. Interstitial fluid that is not picked up by the veins becomes lymph and is carried back to the circulation through an auxiliary set of vessels, the lymphatic system.
11. The circulatory system exchanges gases (O_2 and CO_2) with the environment through the skin or via gills or lungs. Gills extend into the water and extract a maximum of O_2 by maintaining a blood flow in each gill segment opposite to the flow of water.
12. Lungs are organs in which a highly branched tree of bronchi terminates in alveoli with a rich capillary network. Lungs are inflated when the diaphragm contracts and expands the thoracic cavity, supplemented by the contraction of intercostal muscles.
13. Oxygen is carried primarily by special proteins, such as hemoglobin; in vertebrates, hemoglobin is packed into red blood cells. Most vertebrate hemoglobin has four subunits, which interact cooperatively to accept a maximum of O_2 from outside and to release a maximum into the tissues.
14. Carbon dioxide is transported primarily as bicarbonate ion, which becomes CO_2 again in the lungs. Since the pH of tissues must be kept near 7.4, H^+ ions that are produced along with bicarbonate create a problem. These ions are largely carried by hemoglobin molecules that have given up their O_2. Deoxygenated Hb thus buffers excess H^+ ions and allows the pH of blood to remain quite constant.
15. Inspiratory and expiratory centers in the brain, which inhibit each other alternately, maintain the rhythm of breathing. The H^+ level of the plasma is detected by brain centers that increase the rate of breathing in response to an excess. Sensor cells in the carotid and aortic bodies also regulate breathing by monitoring the O_2 level and H^+-ion concentration in blood.

Key Terms

gas exchange 951
blood plasma 952
interstitial fluid 952
hemolymph 952
capillary 952
artery 952
vein 952
aorta 952
arteriole 953
capillary 953
venule 953
vena cava 953
pulmonary circulation 953
systemic circulation 953
atrium 954
ventricle 954
pulmonary artery 954
pulmonary vein 954
gill 954
systole 955
diastole 955
atrioventricular (AV) valve 955
semilunar valve 955
myocardium 956
pacemaker 956
sinoatrial (SA) node 956
atrioventricular (AV) node 956
systolic pressure 957
diastolic pressure 957
vasoconstriction 958
vasodilation 958
cardiac output 958
peripheral resistance 958
Frank-Starling mechanism 960
hematocrit 961
red blood cell/erythrocyte 961
white blood cell/leucocyte 961
platelet 961
thoroughfare channel 962
precapillary sphincter 962
ultrafiltration 963
lymph 964
lymphatic system 964
lymph node 964
atherosclerosis 965
plaque/atheroma 965
stenosis 965
ischemia 965
myocardial infarction 965
countercurrent exchange 967
tracheae (of insects) 967
trachea (of vertebrates) 967
bronchus 967
bronchiole 967
alveolus 967
diaphragm 969
Bohr effect 970
carotid body and aortic body 972

Multiple-Choice Questions

1. In vertebrates, all blood vessels include
   a. valves.
   b. endothelium.
   c. regions where exchange with interstitial fluid takes place.
   d. sphincters.
   e. one or more muscle layers.
2. By definition, all arteries carry
   a. oxygenated blood.
   b. deoxygenated blood.
   c. blood toward the heart.
   d. blood away from the heart.
   e. red, not blue, blood.
3. The semilunar valves open as a direct result of
   a. atrial diastole.
   b. atrial systole.
   c. ventricular diastole.
   d. ventricular systole.
   e. stimulation by the autonomic nervous system.
4. In mammals, oxygenated blood is found in the
   a. right side of the heart.
   b. left side of the heart.
   c. pulmonary artery.
   d. vena cava.
   e. none of the above.
5. Which is present in veins but not in arteries?
   a. valves
   b. pumps
   c. endothelium
   d. smooth muscle tissue
   e. sphincters
6. Sounds of Korotkoff are present only when pressure in the cuff
   a. exceeds systolic pressure.
   b. is less than diastolic pressure.
   c. is higher than diastolic pressure but lower than systolic pressure.
   d. is lower than diastolic pressure and higher than systolic pressure.
   e. is equal to atmospheric pressure.

7. Peripheral resistance is a measure of
   a. systolic blood pressure.
   b. the size of the heart.
   c. elasticity of capillaries.
   d. the diameter of arterioles.
   e. the heart rate.
8. Cardiac output is defined as the
   a. volume of blood pumped by the heart per stroke.
   b. volume of blood pumped by the heart per minute.
   c. energy expended by the heart per beat.
   d. level of oxygenation of left-ventricular blood.
   e. difference between blood pressure within the heart and in the radial artery.
9. Which of the following is regulated by all the others?
   a. cardiac output
   b. stroke volume
   c. peripheral resistance
   d. venous return
   e. heart rate
10. Cardiac ischemia can result from
   a. an embolus.
   b. stenosis.
   c. shock.
   d. coronary thrombosis.
   e. all of the above.

True-False Questions

Mark each statement true or false, and if false, restate it to make it true.

1. Blood leaving the fish heart is low in carbon dioxide, low in oxygen, and relatively low in nutrients.
2. The capillaries of the pulmonary circulation are the site of internal gas exchange.
3. In fishes, the gill membranes are the site of external gas exchange.
4. Blood leaving a capillary bed generally flows directly into an arteriole.
5. Nutrients and gases are exchanged between the blood and interstitial fluid throughout the arterial and venous systems.
6. The hydrostatic pressure of the blood is primarily a result of ventricular systole.
7. The primary function of lymph nodes is to pump the excess interstitial fluid through the lymphatic vessels and back to the bloodstream.
8. As the blood level of oxygen falls and that of carbon dioxide rises, the rate of ventilation falls.
9. The Bohr effect describes the exchange of chloride ions for bicarbonate ions within the lung capillaries.
10. Hemoglobin releases oxygen more rapidly in regions where the oxygen concentration is low and more slowly in regions where oxygen concentrations are high.

Concept Questions

1. What is the relationship between blood, interstitial fluid, and lymph? How are they different from one another?
2. Contrast the effect of the cardiac control centers and the vasomotor centers with respect to the nature and site of the cells that each stimulates.
3. When on duty, the guards at several royal palaces must stand at attention without moving so much as a leg muscle. Explain why fainting is common in such situations.
4. The skin is an important thermoregulator because it contains extensive superficial capillary beds, deeper thoroughfare channels, and precapillary sphincters. The activity in these vessels is visibly apparent because we flush when hot and become pale when cold. Explain which route is preferentially open to blood flow at high and at low temperatures.
5. With respect to fluid circulation and gas exchange, why is it unlikely that an insect as large as an elephant could evolve?

Additional Reading

Feder, Martin E., and Warren W. Burggren. "Skin Breathing in Vertebrates." *Scientific American,* November 1985, p. 126. Skin breathing can supplement or replace breathing through lungs or gills. Special adaptations of the skin and the circulatory system help regulate the cutaneous exchange of oxygen and carbon dioxide.

Fozzard, Harry A., et al. *The Heart and Cardiovascular System: Scientific Foundations,* 2d ed. Raven Press, New York, 1991. A general introduction, including diseases of the cardiovascular system.

Johansen, Kaj. "Aneurysms." *Scientific American,* July 1982, p. 110. How balloonlike dilatations of an artery wall are created.

Pool, Robert. "Heart Like a Wheel: A Relatively Simple Model of Electrical Activity in the Heart May Help Explain Sudden Coronary Deaths." *Science,* March 16, 1990, p. 1294.

Vogel, Steven. *Life's Devices.* Princeton University Press, Princeton, 1988.

Wood, Stephen C. (ed.). *Comparative Pulmonary Physiology: Current Concepts.* Marcel Dekker, New York, 1989. Current thinking about the comparative physiology of respiration.

Internet Resource

To further explore the content of this chapter, log on to the website at:

http://www.mhhe.com/biosci/genbio/guttman/

46 Excretion and Osmotic Balance

Birkenia elegans was a jawless fish of upper Silurian times, one of the armored ostracoderm fishes in which the first kidneys evolved.

Key Concepts

46.1 Animals excrete excess nitrogen as ammonia, urea, or uric acid.

46.2 Metabolic wastes are removed from body fluids as they pass through an excretory organ.

46.3 Some epithelia can pump ions and regulate the movement of water.

46.4 Insects use their Malpighian tubules and hindgut to produce a concentrated urine.

46.5 The vertebrate kidney is composed of nephrons.

46.6 The loop of Henle and the vasa recta form a countercurrent system that can produce a very concentrated urine.

46.7 Kidneys regulate blood pressure and osmolarity by changing their rate of water and Na^+ excretion.

46.8 Kidneys and lungs regulate the acidity of the blood.

46.9 Vertebrate kidneys have evolved to deal with different types of aqueous environments.

46.10 Many animals adapt to environmental demands through the chloride cells of their gills.

In some cool Ordovician sea, minute aquatic creatures began to invade the brackish waters around the mouths of rivers. Other animals already lived there of course, but these new types had a special advantage—tubules for draining their coeloms and removing the water that constantly threatened to invade their tissues, inflate their bodies, and rob them of vital salts. Year after year, these tubules became further refined, and some of the animals evolved a waterproof armor as well, which prevented a massive influx of water. With such adaptations, the animals became able to invade the still more hostile fresh waters of the rivers themselves, where they found abundant food and few competitors. As they evolved in the fast-flowing rivers, they became fine swimmers with sleek bodies and paired fins, and diverged into new forms with specialized niches. They had become fishes. Many of their descendants eventually returned to the sea. Others invaded shallow waters and the surrounding land. We, of course, are their descendants too.

The kidneys that our aquatic ancestors evolved are a precious and unique system in the animal world. Digestion, circulation, gas exchange, hormonal and neural control are old hat;

all animals beyond a certain size and complexity have evolved such mechanisms. But many—perhaps most—aquatic invertebrates simply live with the water and ions around them, conforming osmotically to their surroundings. Animals need a special mechanism for **osmoregulation,** to adjust the volume of water and the balance of dissolved ions in their body fluids. Once the kidney evolved to handle that problem, it also turned out to be highly adaptable for excreting nitrogenous wastes and clearing the blood of miscellaneous toxins. The kidney allowed vertebrates to become terrestrial and to escape total dependence on the water. Only the arthropods, with their radically different anatomy and physiology, rival the vertebrates for dominance on land. ■

46.1 Animals excrete excess nitrogen as ammonia, urea, or uric acid.

Although kidneys evolved primarily as osmoregulatory devices and only secondarily became organs for excreting nitrogenous wastes, the story of these wastes is so central to the whole picture that we must begin with it. Animals get excess nitrogen, mostly the amino groups of amino acids, from foods and from the continuous breakdown of cellular proteins. They use or store the carbon skeletons of amino acids and excrete the excess amino groups as ammonia, **urea,** or **uric acid,** depending on their way of life (Figure 46.1). An $-NH_2$ group removed from an amino acid easily becomes ammonia, NH_3, or an ammonium ion, NH_4^+. But since ammonia is toxic, this direct method of elimination is used only by aquatic animals that can flush vast amounts of water through their bodies and keep their internal ammonia concentration low. More commonly, amino groups are converted to urea, which is quite soluble in water and not very toxic. Terrestrial animals for whom water conservation is particularly important—including birds, reptiles, and most terrestrial invertebrates—excrete uric acid, which is nearly insoluble in water and can be excreted as practically a solid waste. Uric acid is the white material in bird droppings.

Each urea molecule, synthesized in a cycle of reactions in the vertebrate liver, effectively combines two ammonia molecules. Urea can accumulate to some extent without harm, and some animals retain urea for regulatory purposes, as we will show when discussing kidney function later in this chapter. Many aquatic animals excrete urea instead of ammonia.

Mammals normally produce some *urate* (the base form of uric acid) as part of purine metabolism, but some people produce excess urate that may be deposited as sodium urate crystals in the joints, where it causes inflammation and the excruciating pain of gout. Excess urate may also be deposited as kidney stones.

Some animals change their modes of nitrogen excretion during different periods in their lives. Lungfishes of Africa and South America, for instance, excrete ammonia as long as water is plentiful. They can survive the droughts that are common in their habitats by burrowing into the mud and forming a kind of cocoon (Figure 46.2). Then they start to make urea, since accumulated ammonia would kill them otherwise. Similarly, amphibians such as frogs typically begin their lives as aquatic larvae (tadpoles) that excrete ammonia, then start to make urea as they metamorphose into terrestrial adults. Their remarkable metabolism includes a molecular switch from one metabolic pathway to another, a switch thrown by a rising level of ammonia in their tissues.

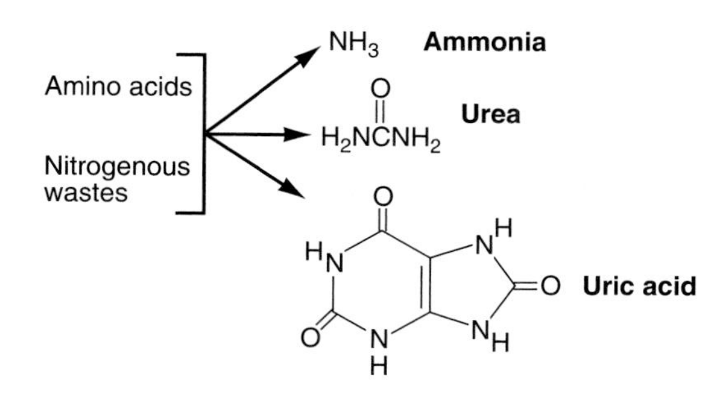

Figure 46.1
Animals excrete excess nitrogen in the form of ammonia, urea, or uric acid.

46.2 Metabolic wastes are removed from body fluids as they pass through an excretory organ.

In addition to excess nitrogen and the CO_2 resulting from respiration, animals must eliminate a variety of waste products, including metabolic by-products and many odd substances from their food. Plants make all kinds of unusual metabolites (such as pigments, hormones, and allomones), which may create interesting flavors but may also be quite toxic. Animals can't avoid eating these compounds, as well as many noxious materials made by bacteria and fungi. Enzymes in organs such as the liver and kidneys partly break down many of these substances, supposedly to nontoxic products (however, see Methods 20.1), but these products, too, have to be excreted. Specialized excretory organs remove all these substances from the body.

Allomones, Sections 11.7 and 27.9.

All excretory systems share a common structure and a three-phase general mechanism (Figure 46.3).

1. Extracellular fluid—blood plasma or hemolymph—is discharged into the lumen of a dead-end channel composed of a single layer of epithelial cells. If there is blood pressure on the fluid, it may be forced across in a process of *ultrafiltration* in which water, ions, and dissolved metabolites pass through the epithelium (similar to the movement of plasma through a capillary wall) while proteins remain behind. If there is little pressure, materials are actively *secreted* into the channel by transport proteins.
2. Fluid moves through the channel, and sometimes additional materials are secreted into it. Then carefully regulated

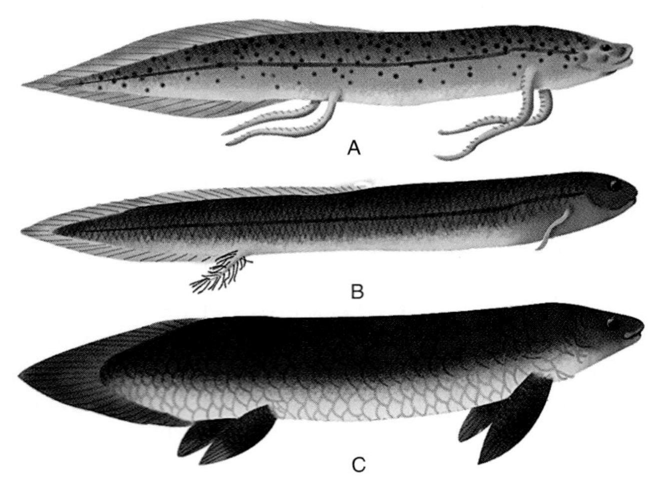

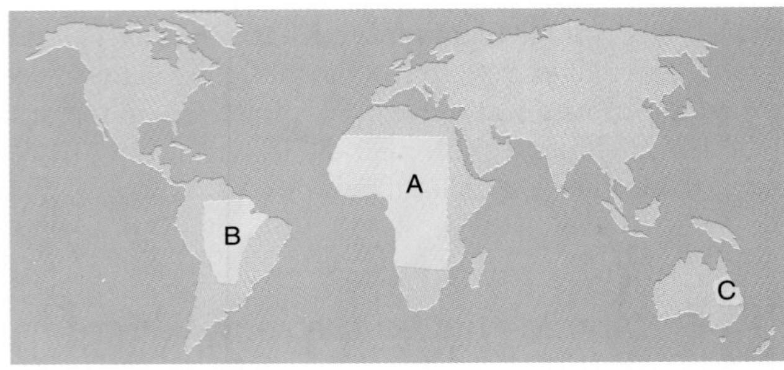

(a)

(b)

Figure 46.2

(a) The three contemporary species of lungfish live in Africa, South America, and Australia. *(b)* During droughts, African and South American lungfish encase themselves in cocoons of mud, where they undergo a change in excretory metabolism.

amounts of certain materials (such as glucose, water, and certain ions) are *reabsorbed* from the channel into the blood or hemolymph.

3. The remaining contents of the channel are excreted through an external pore.

These systems use a general principle of housecleaning: Carry everything out of the closet or garage and then bring back only those things you want to keep. Foreign and unwanted molecules are so diverse that animals couldn't possibly evolve proteins to recognize and remove all of them, so the simplest solution is to remove all the materials in the blood without discrimination and then bring back only what is needed. The same process gets rid of nitrogenous wastes, although urea is partially recovered in vertebrates.

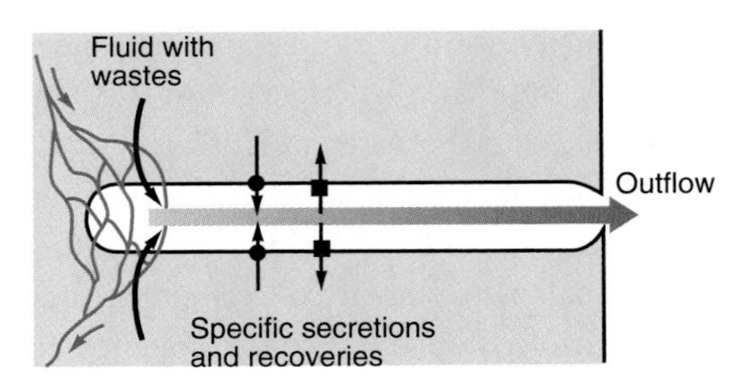

Figure 46.3

Excretory organs generally operate by ultrafiltration of blood plasma (or some other fluid) into a dead-end channel. As the filtrate moves along the channel, other materials may be added to it, and some substances are recovered. Whatever reaches the end of the channel is excreted from the body.

46.3 Some epithelia can pump ions and regulate the movement of water.

The devices animals use for eliminating wastes and maintaining the proper levels of water and ions all depend on the remarkable properties of certain epithelia, the sheets of cells that form the boundaries of many structures. Some epithelia are just protective surfaces, but others are specialized as glands for secreting enzymes (as in the stomach and intestine) or for moving ions and water; these are the epithelia that concern us here.

During the 1930s and '40s, the physiologists Ernst Huf and Hans Ussing investigated the properties of epithelia by clamping pieces of frog skin between two chambers that could be filled with different solutions. Then in 1947, Ussing used different sodium isotopes to demonstrate that Na^+ ions flow in one direction, from the outside surface of the skin to the inside, and he showed later that this flow can generate an electrical potential across the skin. This mechanism, commonly called a **sodium pump,** is essential in the life of a frog; as it sits in fresh water, a frog absorbs water by osmosis and excretes the excess in its urine. Because it tends to lose ions in the urine, the pump in its skin is needed to restore these ions. Different epithelia contain sodium pumps, or similar mechanisms that move other ions, in various animal organs. The question is, how does a sodium pump work?

The movement of Na^+ is clearly due to active transport, since it can be stopped by metabolic poisons such as cyanide and by specific poisons such as ouabain, which inhibits the Na^+-K^+ ATPase that transports K^+ ions into cells and Na^+ ions outward. Na^+-K^+ ATPases maintain the high concentration of K^+ ions in cytoplasm and are ultimately responsible for the voltage that cells maintain across their membranes. When

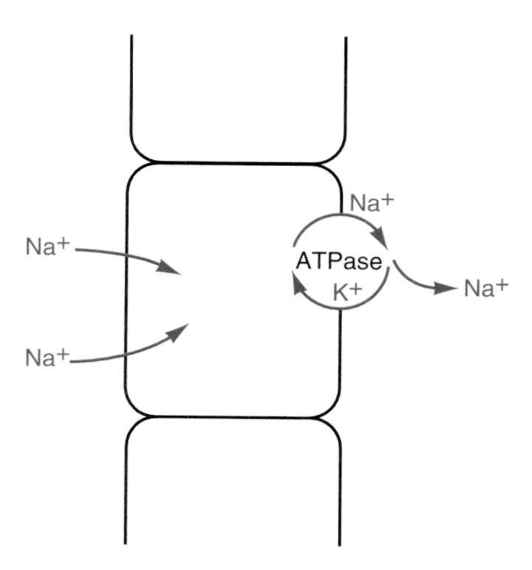

Figure 46.4

An epithelium acts as a sodium pump by having one surface that allows Na^+ ions to enter by facilitated diffusion and an opposite surface with Na^+-K^+ ATPases that move these ions outward.

integrated into certain epithelia, these transport proteins create epithelial sodium pumps. (Similar proteins can create pumps for H^+, K^+, HCO_3^-, and other ions.) In frog skin, the Na^+-K^+ ATPases are confined to the inner surface of the epithelium (Figure 46.4), and K^+ ions can leak out from this surface, creating the normal membrane voltage. The outer surface of the epithelium, however, has a very low K^+ conductance but relatively high conductance for Na^+ ions, which enter the cytosol through Na^+-channel proteins. This epithelium therefore moves Na^+ ions in only one direction—into the body.

Na^+-K^+ ATPases, Section 8.12.

Other epithelia specifically move water. Here it's important to remember a fundamental principle established in Section 32.9: There are no cellular pumps for water, so water is transported by moving ions and letting the water passively follow by osmosis. In water-pumping epithelia, deep clefts between the cells are closed at one end by tight junctions, which stop virtually all movement in that direction (Figure 46.5). The adjacent cells pump ions into the base of the cleft; water follows osmotically and then moves out of the open end of the cleft (along with a gradual movement of the ions, of course). It is now clear that, in cells like these, water moves through specific water channels in proteins called **aquaporins,** which make plasma membranes 100–200 times more permeable to water than pure phospholipid-cholesterol membranes.

Tight junctions, Section 6.9.

In the following descriptions of kidneys and other organs that regulate water and ions, we will discuss the movements of materials across specialized epithelia without further explanation. However, it is important to always keep the anatomy of these epithelia in mind. On one side is interstitial fluid and the microcirculation; on the other is either the outside of the animal or a passage leading to the outside, such as the gut or a kidney tubule. Substances will diffuse through the surrounding interstitial fluid between the epithelium and the circulation (Figure 46.6). When the epithelium is operating normally, each substance will reach a *steady-state concentration* in the

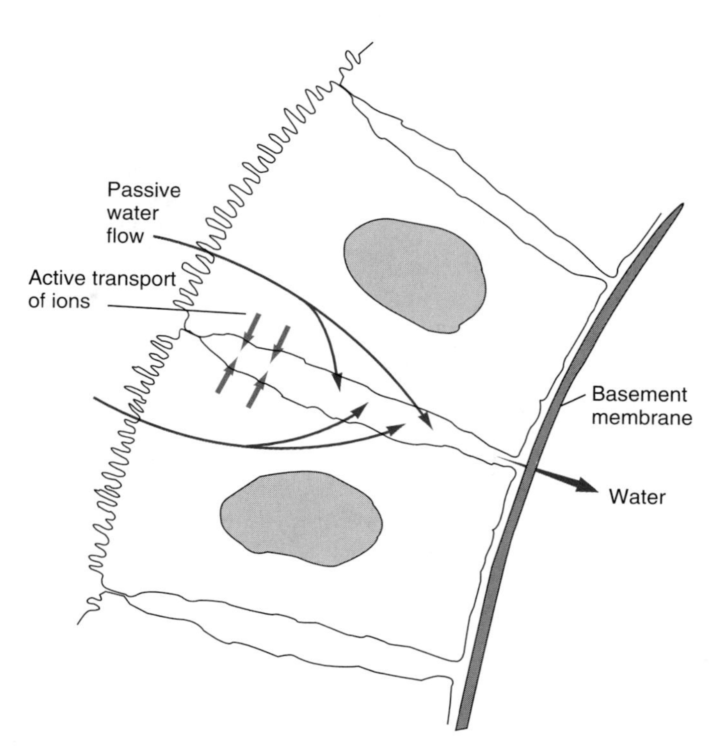

Figure 46.5

An epithelium that transports water uses the principle that water follows ions. The cells pump ions into the intercellular cleft. Water diffuses after the ions through aquaporins and then diffuses out of the cleft in the only direction open to it.

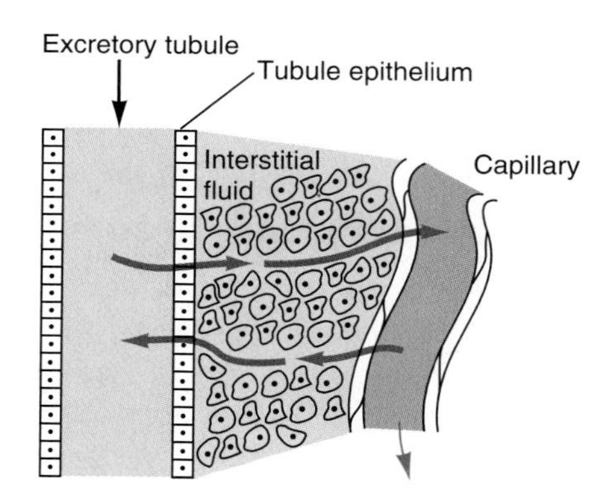

Figure 46.6

Ions and molecules move through interstitial fluid between the liquid inside an excretory tubule and the nearby capillaries. Each substance will reach a steady-state concentration in the interstitial fluid. If the epithelium of the tubule is transporting a substance passively, its movement depends on the relative concentrations inside the tubule and in the interstitial fluid.

epithelium and interstitial fluid. The operation of an excretory organ relies upon the establishment of these steady states.

Steady states, Section 11.3.

46.4 Insects use their Malpighian tubules and hindgut to produce a concentrated urine.

Excretion in insects illustrates the principles we have just developed. Insects have **Malpighian tubules,** numerous dead-end channels connected to the animal's hindgut (Figure 46.7), and their epithelia move materials between two compartments: the hemolymph and the lumen of the tubules and gut. The system uses a mechanism that we will see again in mammalian kidneys: creating a concentrated fluid around a tubule such that water is drawn out of the tubule osmotically, and the material inside becomes more concentrated. Since the hemolymph surrounding the tubules doesn't exert enough pressure to force wastes into the tubules, urine is actively secreted into them. The mechanism of secretion, while variable from one species to another, depends upon the principle of water movement by osmosis. The tubule epithelium secretes K^+ ions into the lumen of each tubule; Cl^- ions and water then follow, carrying along nitrogenous wastes—in insects, mostly uric acid. Uric acid may also be secreted into the tubule. As urine moves through the hindgut, potassium pumps in the gut wall transport K^+ ions into the hemolymph, raising the concentration of K^+ ions around the gut so water is drawn back out into the hemolymph while wastes stay behind. Water and ions are continually recycled through the gut, hemolymph, and tubules, while a very concentrated solution of wastes accumulates as urine in the gut.

46.5 The vertebrate kidney is composed of nephrons.

The vertebrate kidney is a remarkable device. A normal human kidney, for instance, produces 1–2 liters of urine per day, but it does so by filtering about 100 times that volume of blood plasma, removing its wastes, adjusting the concentrations of various materials, and returning 99 percent of the water to the blood.

The kidneys are paired, left and right (Figure 46.8). Each one contains about a million tubules called **nephrons,** which filter the blood and regulate its composition. Each nephron spans the kidney's outer **cortex** and inner **medulla.** (These general terms are also used to describe the outside and inside of other organs, such as the adrenal glands.) Each kidney receives blood directly off the aorta through a short, broad renal artery, so the pressure remains quite high. In humans, about one-fifth of the blood is passing through the kidneys at any instant, and since the entire blood volume of 4–5 liters circulates every minute, it takes only five minutes for the kidneys to filter the equivalent of all the blood in the body.

Urine formation begins with the ultrafiltration of blood plasma. Under the high blood pressure in the renal artery, plasma is forced from a knot of capillaries, a **glomerulus,** through the surrounding **Bowman's capsule** into the lumen of the nephron to form a **glomerular filtrate.** The glomerular capillaries are characterized by large pores, 20–50 nm in diameter, making them at least 100 times more permeable to plasma solutes than other capillaries. The glomerular filtrate must pass through these pores, through the basal lamina around the capillary endothelium, and through a barrier made by unique cells called **podocytes** (Figure 46.9). Podocytes cover the capillaries with interdigitating extensions that leave only very narrow slits through which the glomerular filtrate can pass. Although the pores in the glomerular capillaries are large enough to admit some proteins, very little protein passes into the glomerular filtrate, apparently due largely to the barrier imposed by the podocytes.

The rate at which the kidneys excrete any material depends initially on the **glomerular filtration rate (GFR),** the rate at which fluid is filtered through the glomeruli. The GFR depends on the same factors of osmotic and hydrostatic pressure that determine the exchange of plasma and interstitial fluid. Figure 46.10 shows that the hydrostatic pressure in a glomerulus, about 75 torr, is exerted against two back pressures: the hydrostatic pressure from the nephron and surrounding interstitial

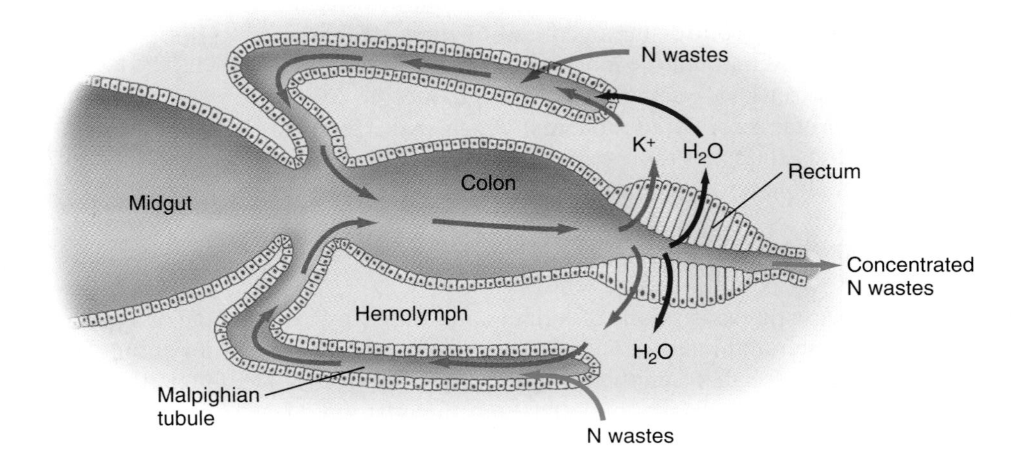

Figure 46.7

The Malpighian tubules of insects are blind extensions off the hindgut. (Only two of the many tubules are shown here.) K^+-ion pumps actively secrete K^+ into each tubule, with Cl^- ions, nitrogenous wastes, and water following passively. As the urine passes through the hindgut, it is concentrated by continual removal of K^+ ions and water into the hemolymph. Through the secretion of K^+ ions, the continuous circulation of water through the gut, hemolymph, tubules, and back out of the gut leaves behind a concentrated urine containing wastes.

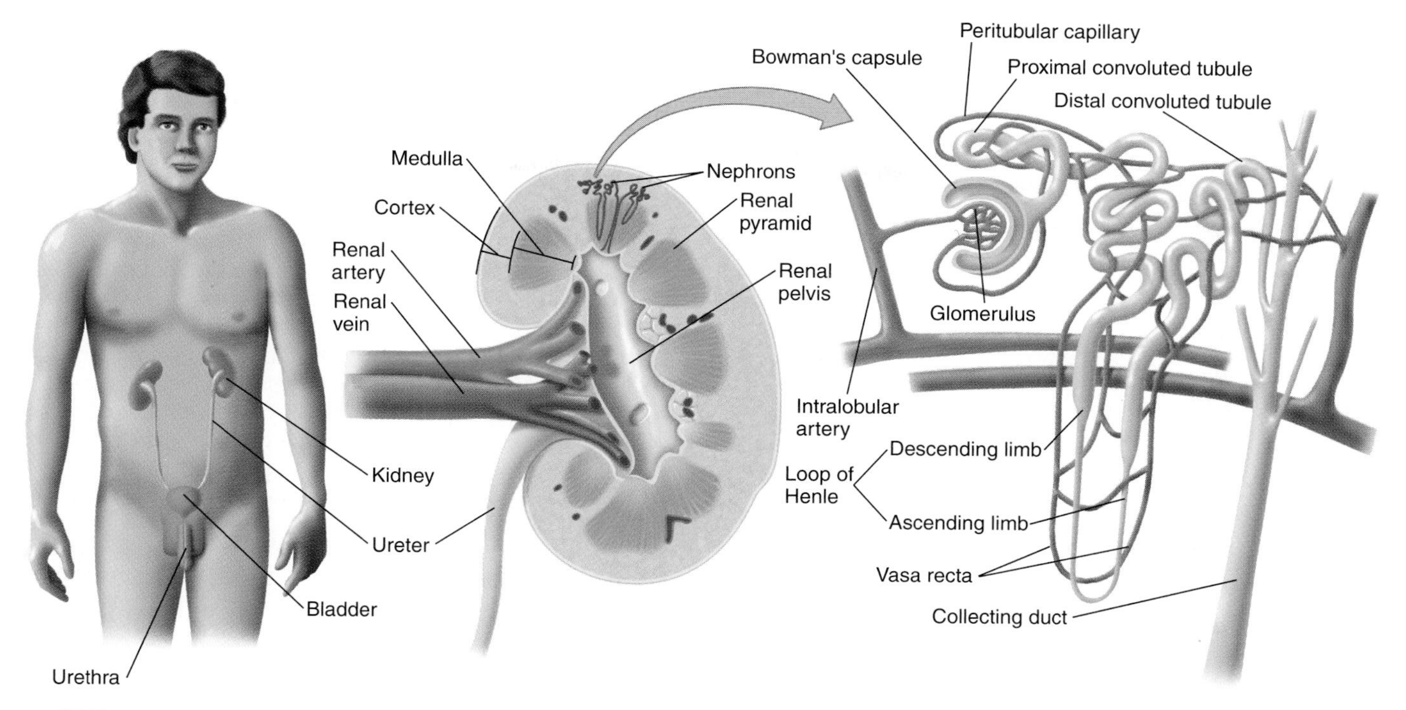

Figure 46.8

Each human kidney is made of about a million nephron units; here a single unit is accentuated. Each nephron extends from the outer cortex of the kidney into the renal medulla, which is divided into outer and inner regions; the latter contains the loops of Henle. Several nephrons discharge their contents into each collecting duct, and all the ducts carry urine into the inner renal pelvis. Urine collected in the renal pelvis then moves down the ureter into the urinary bladder.

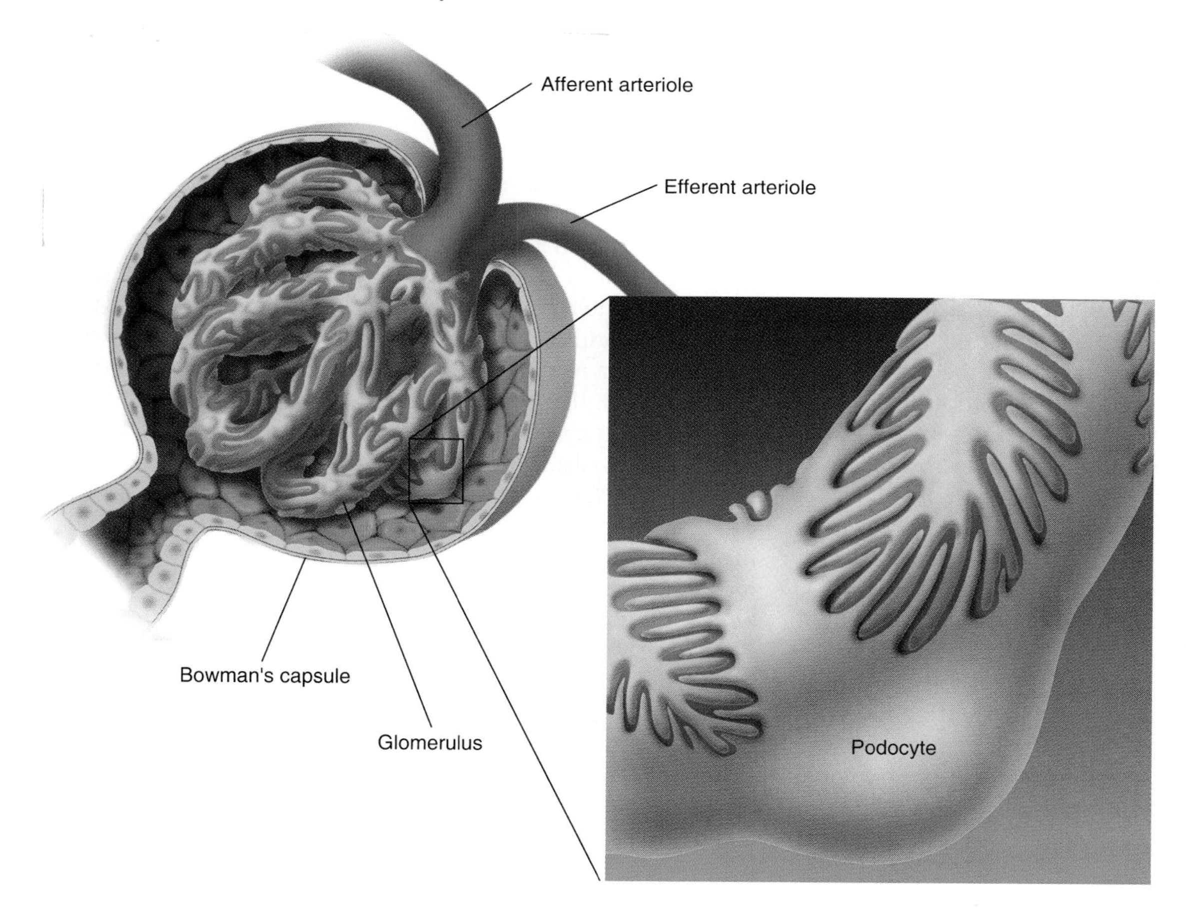

Figure 46.9

The glomerular capillaries are quite permeable, but they are covered with podocytes *(shown here in yellow)* whose fingerlike processes interdigitate, leaving only a narrow space *(red)* through which blood plasma is filtered.

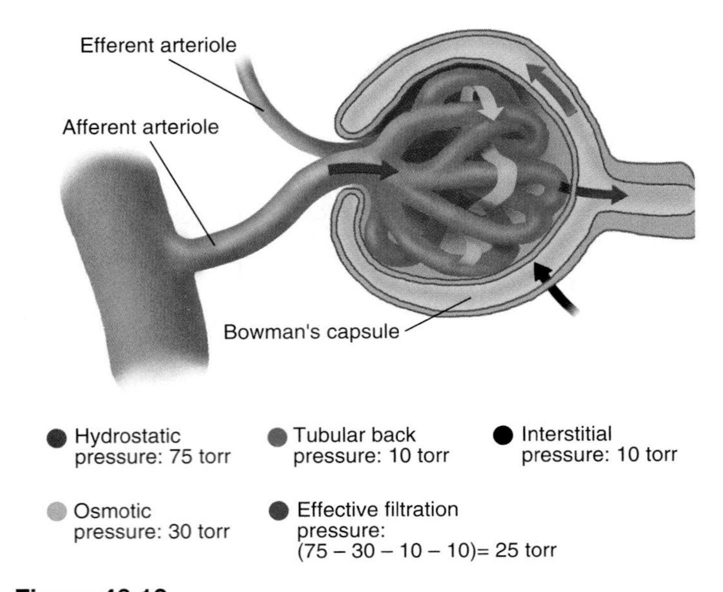

Figure 46.10
The balance of pressures around the glomerulus produces a net outward pressure of 25 torr.

fluid (20 torr) and the total osmotic pressure (30 torr). So the net pressure is $75 - 20 - 30 = 25$ torr outward. This pressure moves plasma from the glomerulus into the nephron. The GFR can be changed by increasing or decreasing blood pressure. For instance, vasoconstriction of the arterioles leading into the glomeruli increases their resistance (by the fourth power of the radius), reduces the flow rate through the glomeruli, and thus reduces the GFR.

Glomerular filtration is followed by tubular reabsorption of many substances, so the initial filtrate is quite different in composition from the urine that will finally be made from it. Since only about one percent of the filtrate will become urine, a nephron recovers virtually all the water passing through it, and carriers in the nephron walls recover all the glucose and amino acids, along with about half the urea. A nephron can actively secrete H^+ and K^+ ions and other materials into the filtrate, as needed for homeostasis. This combination of filtration, reabsorption, and secretion, with hormonal and neural controls superimposed, maintains the proper balance of each substance in the extracellular fluid.

One of a kidney's chief functions is to regulate the concentration of Na^+, the principal extracellular ion. Sodium pumps in the proximal convoluted tubule of the nephron remove two-thirds of the Na^+ in the glomerular filtrate. The flux of Na^+ ions carries along most of the Cl^- ions and about 75 percent of the water, because the epithelial cells in this part of the tubule are rich in aquaporins. We will see later in this chapter how regulatory systems that measure the body's water and ionic balance determine the amounts of additional water and Na^+ to be recovered from the filtrate or allowed to remain in the urine.

Exercise 46.1 The GFR in men is about 125 ml per minute. If a man has a blood volume of 5.5 liters and 55 percent of the volume is plasma, how long does it take for the whole plasma volume to be filtered?

Exercise 46.2 Measuring the GFR can be a problem, as shown by this situation: Suppose the urea concentration in the blood plasma of a mammal is 0.2 mg/ml and you determine that 10 mg of urea is filtered through its glomeruli per minute. *(a)* What GFR can you calculate from these data? *(b)* The problem is that *all* the urea is not removed from the blood as it passes through the glomerulus. Therefore, is the calculated GFR a minimum or maximum estimate of the true GFR? *(c)* If you knew that precisely half the urea is filtered from the blood, what would the GFR be? *(d)* A more challenging problem is to suggest a way to measure the actual GFR by modifying this general method.

46.6 The loop of Henle and the vasa recta form a countercurrent system that can produce a very concentrated urine.

Among vertebrates, only birds and mammals can produce a urine that is more concentrated than their interstitial fluid, and they are the only animals whose nephrons contain a **loop of Henle.** The form of the loop, with two thin, parallel tubes, suggests that it is a countercurrent exchange device, and indeed it is. But it operates in conjunction with a parallel network of capillaries known as the **vasa recta,** shown in Figure 46.8. The loop of Henle and vasa recta together create a strong *concentration gradient* of Na^+ ions, Cl^- ions, and urea throughout the interstitial fluid of the kidney, from a low concentration in the cortex to a high concentration in the medulla. Since we have to understand the activity of a kidney, or any similar device, as a steady-state flow of materials from one compartment to another, we start with this gradient already established and see how it is maintained. The vasa recta itself is a countercurrent exchanger (see Concepts 45.1). Like a simple mechanical device that exchanges heat between two pipes, the vasa recta and loop of Henle both exchange salt and water between their ascending and descending limbs, thus establishing concentration gradients within themselves (Figure 46.11). The walls of the vasa recta, in contrast to those of the nephron, are freely permeable to water, salt, and urea, so the concentration gradient in the blood extends to the surrounding interstitial fluid. As the blood moves through the vasa recta, the concentrations of salt and urea become equal in the blood and interstitial fluid. However, because of the high concentration of proteins in the plasma, water enters the blood by osmosis and is carried out, just as in any other capillary bed (see Section 45.7). Thus the vasa recta leaves a gradient of salt and urea in the interstitial fluid but constantly removes water from the region.

The loop of Henle is the more important mechanism in this system (Figure 46.12). Its operation depends on differences

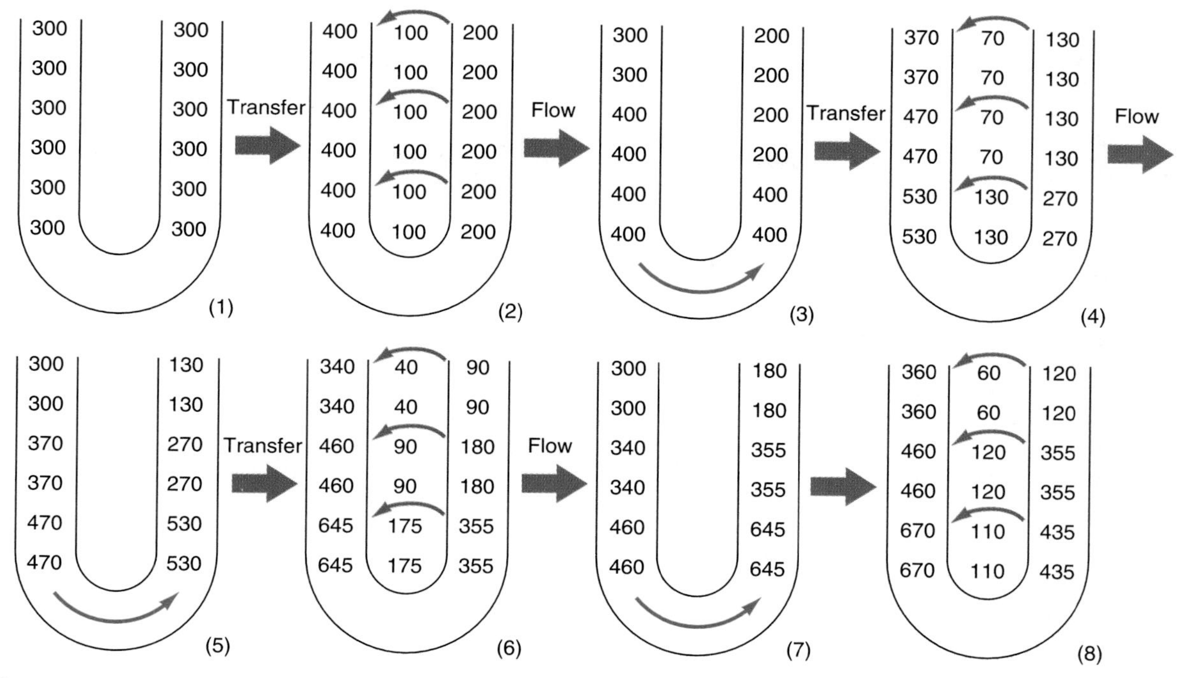

Figure 46.11

We can see how the vasa recta will establish a concentration gradient by operating it in stages, using some imaginary numbers. Suppose the system begins with blood at a normal concentration of 300 mOsm *(1)*. We arbitrarily transfer half of the salt in the ascending arm directly across to the descending arm *(2)*, let the blood flow a little, and stop *(3)*. We again transfer about a third of the salt across *(4)*, and let the blood flow a little *(5)*. Notice that with each transfer and flow, the gradient becomes stronger. Transferring half of the salt with each step is arbitrary, of course; if we transferred less, the gradient would just become established more slowly. If a gradient already exists, this system will maintain it.

in active transport and permeability in different sections of a nephron:

Thin descending limb: highly permeable to water, slightly to urea
Thin ascending limb: highly permeable to NaCl, slightly to urea
Thick ascending limb: actively transports NaCl outward (probably transports Cl^-, and Na^+ then follows)
Outer medullary collecting duct: highly permeable to water
Inner medullary collecting duct: variably permeable to water and highly permeable to urea.

Let's follow the filtrate as it moves through this region.

1. Filtrate enters the thin descending limb. The epithelium here contains aquaporins and is permeable to water. Because the surrounding interstitial fluid is hypertonic to the filtrate, water is drawn out of the filtrate osmotically. This water is carried out of the region by the vasa recta.
2. Filtrate moves to the ascending limb, where Na^+ and Cl^- ions move out—passively in the thin region and actively in the thick region. Most of the increased Na^+ and Cl^- concentration in the interstitial fluid comes from NaCl that leaves the thick region of the ascending limb, in the outer medulla.

Now notice that as the ascending limb extrudes more salt to the interstitial fluid, more water is drawn out of the filtrate both in the descending limb and as it rounds the bend of the loop and rises into the ascending limb. And the more concentrated the filtrate rising into the ascending limb, the more salt the ascending limb can extrude to the interstitial fluid, and the more water will be drawn out of the descending limb. This system thus creates a *positive feedback loop* that is able to draw more salt out of the filtrate and create a stronger concentration gradient than the vasa recta alone:

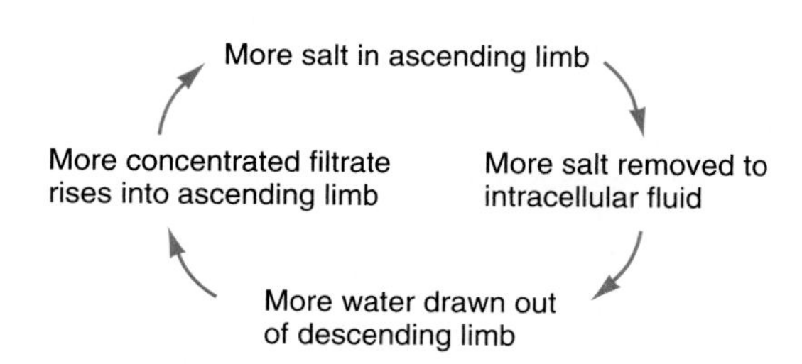

It is therefore called a *countercurrent multiplier,* rather than a countercurrent exchange device.

The importance of this system becomes clear if we follow the filtrate through the collecting duct. By this time, it has lost much of its water and has become quite a concentrated solution of urea, but it is still more dilute than the surrounding interstitial fluid. Water therefore diffuses passively out of the outer medullary collecting duct through aquaporins (aquaporin-CD). The inner medullary part of the collecting duct is permeable to urea and allows a great deal of urea to diffuse out, creating a high concentration of urea in the interstitial fluid of the inner medulla.

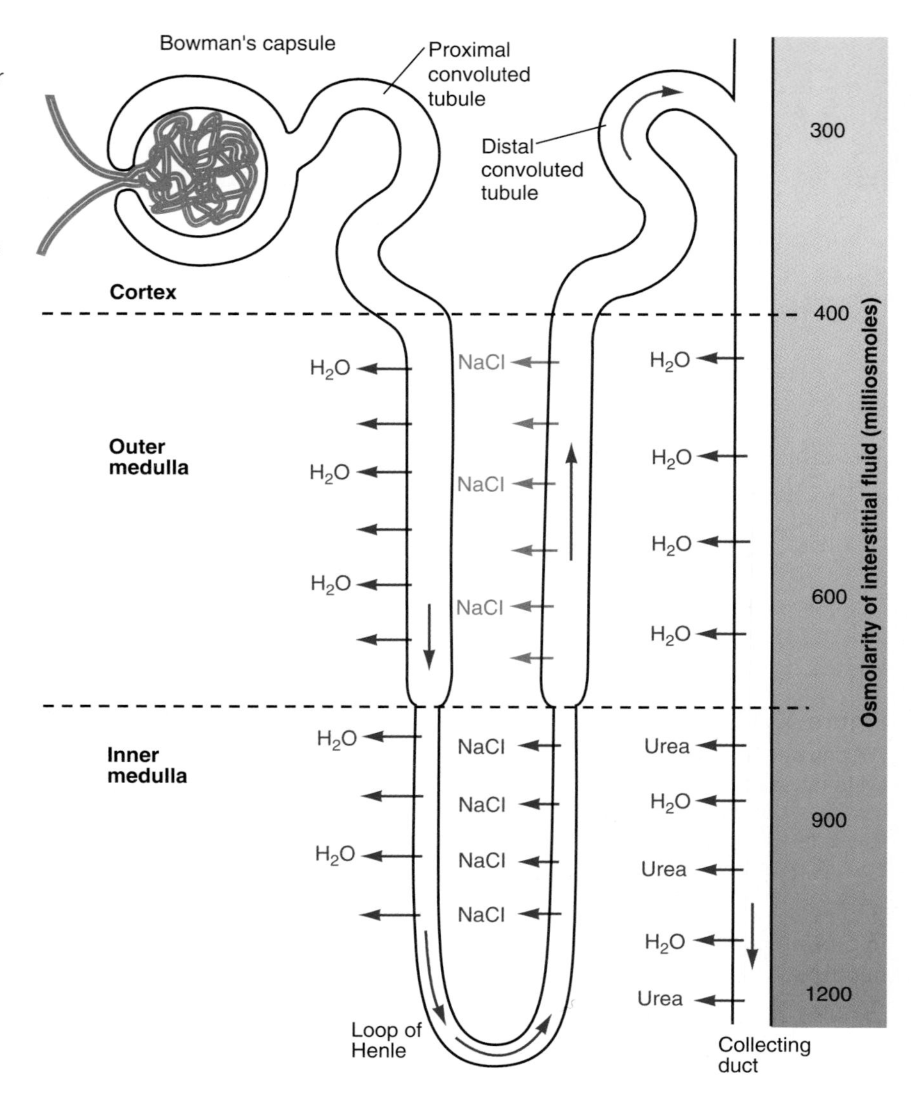

Figure 46.12

The loop of Henle operates as a countercurrent multiplier device. A steep concentration gradient of Na^+, Cl^-, and urea has already been established in the interstitial fluid *[reddish strip].* As filtrate enters the descending limb, it begins to lose water passively, becoming more concentrated. Moving up the thin ascending limb, NaCl begins to diffuse out, and it is actively pumped out in the thick ascending limb. Some additional Na^+, with Cl^-, is generally removed in the distal tubule. As the filtrate enters the collecting duct, it begins to lose more water; by this time, the withdrawal of NaCl and water has given the filtrate a high concentration of urea, so urea diffuses out in the inner region of the collecting duct, creating a major part of the concentration gradient in the interstitial fluid, and this gradient withdraws much of the water remaining in the filtrate.

This system maintains a strong concentration gradient of urea and ions in the interstitial fluid around the nephrons. So as glomerular filtrate passes through the collecting tubule, a great deal of water is drawn out of it, and the urine that accumulates in the medulla can be *hyperosmotic* to the blood and to the average interstitial fluid. We will see next that the aquaporin-CDs of the collecting duct are under hormonal control, so the permeability of the collecting duct to water can be regulated.

Exercise 46.3 How would you expect the loops of Henle in desert mammals, which must conserve water very efficiently, to compare with those in humans?

Exercise 46.4 We spread salt on icy streets and sidewalks because salt lowers the freezing point of water, in proportion to the salt concentration. How can investigators determine the concentration of materials in any region of a kidney by using this principle? (*Note:* It is possible to draw thin glass tubes out into very fine, sharp needles.)

46.7 Kidneys regulate blood pressure and osmolarity by changing their rate of water and Na^+ excretion.

The health of an animal's cells depends on maintaining specific ionic concentrations in its intracellular and extracellular fluids and an overall osmotic balance. Tissue cells are bathed in an interstitial fluid whose principal ion is Na^+, and its Na^+ ion concentration strongly influences the water flux in and out of cells. This flux, in turn, determines the intracellular and extracellular volumes. If an animal takes in too much Na^+ without enough water, the osmolarity of its extracellular fluid will increase, and water will flow out of its cells by osmosis; thus its intracellular volume will shrink, with possibly severe consequences. Shipwreck survivors who make the mistake of drinking sea water start to lose water from all their cells and become mentally deranged as their neurons lose normal function. They die if they cannot get fresh water soon. Regulating the Na^+ concentration of extracellular fluid is therefore critical.

The composition of extracellular fluid is regulated on the basis of its volume, which is mostly water, and its osmolarity, which is determined primarily by its Na^+ concentration. These factors are monitored by two kinds of interoceptors: **baroreceptors** that measure blood pressure and **osmoreceptors** that measure the osmolarity of the plasma and interstitial fluid. In response to signals from these receptors, the kidneys regulate the volume and osmolarity of extracellular fluid through two systems of hormones that affect the balance between filtration through the glomerulus and reabsorption from the filtrate. The glomerular filtration rate can be changed by adjusting the volume and pressure of blood flowing into the glomerulus. Most of the water and Na^+ is withdrawn from the glomerular filtrate in the proximal region of the nephron and the loop of Henle; as the remainder passes through the distal convoluted tubule and collecting duct, its water and Na^+ content are adjusted by three hormonal mechanisms that we will discuss next.

One regulatory circuit responds to changes in extracellular osmolarity through **antidiuretic hormone** (**ADH,** also called **vasopressin**) from the posterior pituitary gland (see Figure 41.14). (Diuresis means "producing urine," so antidiuresis means "producing less urine.") Osmoreceptors in the hypothalamus monitor the osmolarity of the interstitial fluid. Once they detect osmotic pressure above a set point, they signal the secretion of ADH, which targets the collecting duct endothelium of the nephron (Figure 46.13). Under the influence of ADH, the aquaporin-CDs of the endothelium open and let more water diffuse out of the glomerular filtrate. Thus water that would have been lost in the urine is retained and returned to the extracellular fluid.

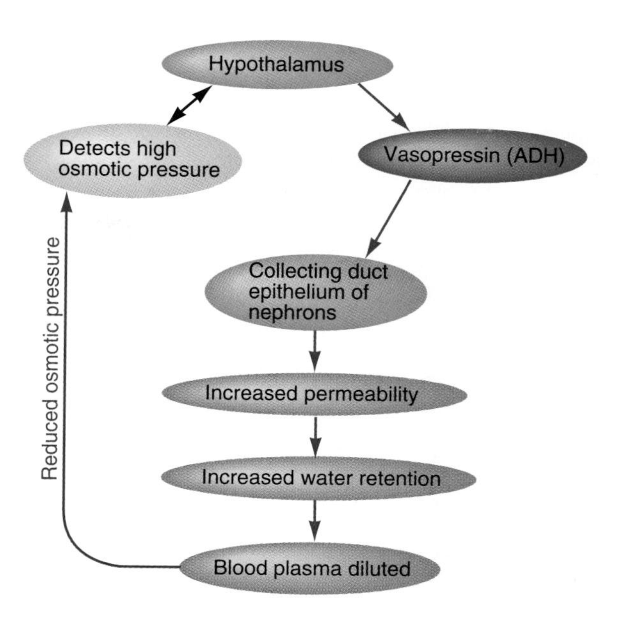

Figure 46.13

The circuit for release of vasopressin (antidiuretic hormone) begins with hypothalamic osmoreceptors. The targets of the hormone's action are in the collecting duct endothelium of the nephron.

ADH secretion is also influenced by blood pressure. Elevated blood pressure could result from elevated blood volume, and it could be maladaptive to retain water if the plasma volume is already too high. Baroreceptors located in the left atrium of the heart monitor blood pressure, and in response to elevated pressure they send neural signals to the hypothalamus, inhibiting ADH release. As a result, more water is excreted, reducing the blood volume.

A second regulatory pathway is the renin-angiotensin-aldosterone system, a complex circuit shown in Figure 46.14. In the kidney, the **juxtaglomerular apparatus** made of specialized cells next to each glomerulus (*juxta-* = next to) monitors blood pressure; if it detects a decrease in pressure, which could be the result of reduced blood volume, it secretes the enzyme *renin* (ree′nin). Already circulating in the blood is angiotensinogen, an inactive peptide hormone precursor made by the liver. Renin converts angiotensinogen to angiotensin I, which is then converted into the active hormone angiotensin II by converting enzymes in capillaries of the lung and other organs. Angiotensin II produces three principal effects. First, it causes general vasoconstriction in blood vessels throughout the circulatory system, thus raising the blood pressure. Second, it stimulates thirst centers in the hypothalamus, thus stimulating drinking, which can increase the blood volume. Third, it causes the adrenal cortex (the outer region of the adrenal glands) to release the steroid hormone **aldosterone,** whose target cells are in the distal convoluted tubule of the kidney. Aldosterone stimulates the epithelium of the distal convoluted tubule to reabsorb more Na^+ ions, which are followed osmotically by water. Thus the juxtaglomerular apparatus increases blood pressure both by the direct action of angiotensin II on the diameters of blood vessels and by stimulating the retention of Na^+ and water, raising the blood volume. Without aldosterone, a person could excrete 25 grams of sodium per day, but in its presence the kidney can absorb sodium so completely that hardly a trace is excreted.

Superimposed on these controls is a third hormone discovered only a few years ago. High blood pressure may be lowered by the active secretion of Na^+ ions, followed passively by water, a process called *natriuresis* (*natrium* = sodium), and cells in the atria of the heart produce a peptide hormone called **atrial natriuretic factor** (**ANF**) that promotes Na^+ and water excretion by the kidney, thus reducing the blood pressure. In addition, ANF antagonizes some actions of the other two hormonal systems, reducing the secretion of aldosterone and opposing some actions of angiotensin II by promoting vasodilation and thus increased glomerular filtration. In this way ANF helps fine-tune the regulation of blood pressure and volume.

Keep in mind that each endocrine factor—ADH, ANF, angiotensin II, and so on—is one element in a complex control system. The regulatory elements that affect the kidney must be added into all the controls on blood pressure discussed in Section 45.5. For example, the baroreceptor reflex partially regulates the GFR, because in response to a change in blood pressure, sympathetic or parasympathetic activation produces a slight vasoconstriction or vasodilation of arterioles leading to the glomerulus,

Figure 46.14

In response to decreased blood volume, the juxtaglomerular apparatus cells (JGA) secrete renin, which converts angiotensinogen into angiotensin I. Angiotensin I is converted into angiotensin II in the lungs and some other tissues. Angiotensin II stimulates release of aldosterone, which stimulates uptake of Na^+ from the renal tubule. Angiotensin II also causes general vasoconstriction, thus increasing blood pressure and reducing the GFR.

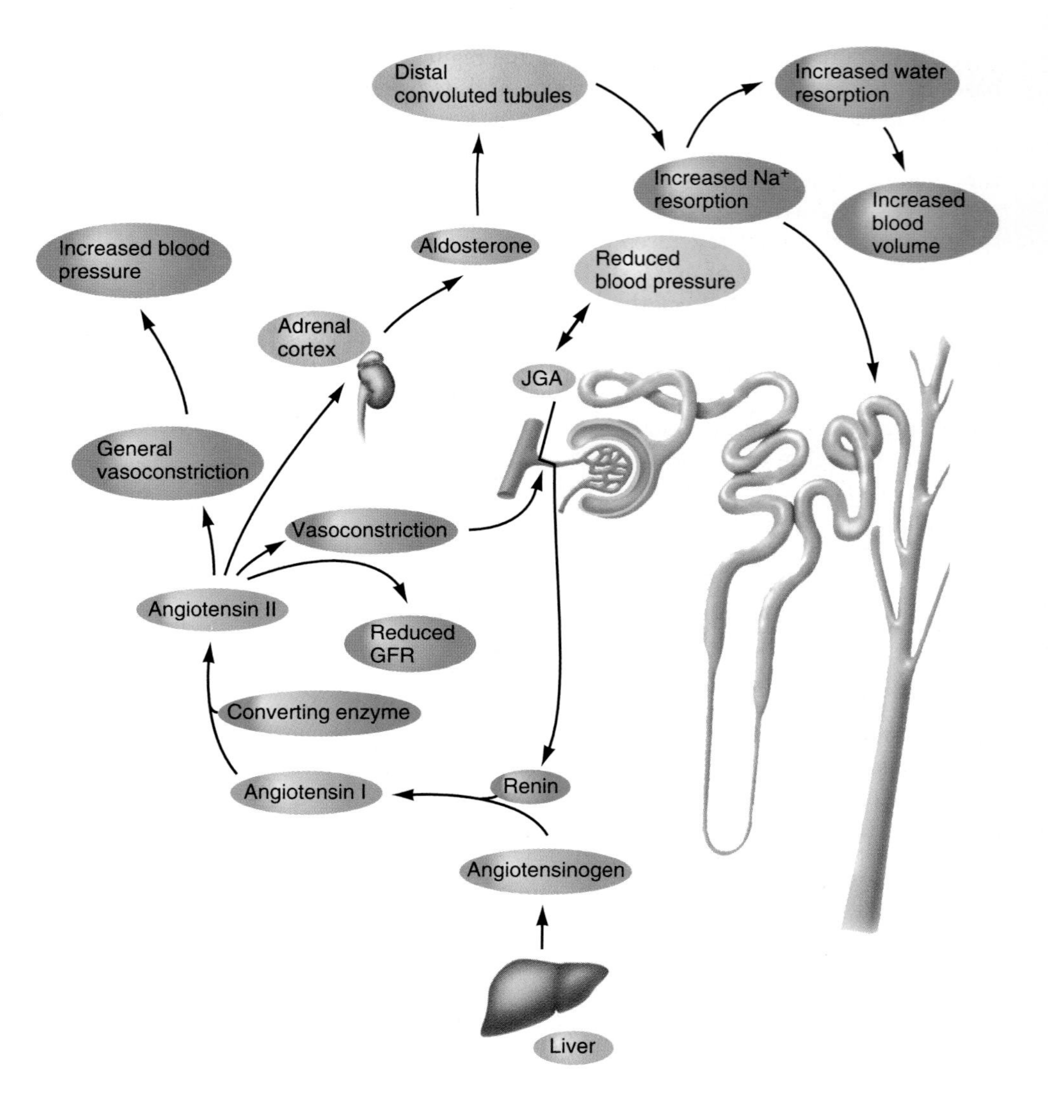

thus changing the blood flow and the GFR. The baroreceptor reflex also regulates blood pressure by changing the cardiac output, which in turn affects the GFR and other factors of kidney function. In reality, all these control systems are balanced against one another by the integrative activities of the nervous system.

46.8 Kidneys and lungs regulate the acidity of the blood.

The pH of blood and intracellular fluid is critical, since cells are adapted to narrow pH ranges. Animal cells operate near neutrality; for instance, human plasma is normally pH 7.4, with a slight excess of OH^- ions. Acidosis (pH less than 7.4) or alkalosis (pH greater than 7.4) can result from functional disorders and can, in turn, produce severe illness; a person cannot tolerate a pH less than 7.0 or more than 7.8.

H^+ ions in blood plasma come from three main sources: metabolic CO_2 that is converted to HCO_3^- and H^+ ions, lactic and other acids being transported through the blood, and sulfuric and phosphoric acids from the sulfide and phosphate groups released when proteins and nucleic acids are broken down. H^+ ions move in and out of the plasma too quickly for hormones to be of much use in maintaining a constant pH. Fortunately, it is easy to keep a solution at a constant pH by means of buffers (Concepts 46.1). Blood plasma is largely buffered by the HCO_3^- ion itself, with a ratio of H_2CO_3 to HCO_3^- that normally keeps the plasma at pH 7.4. If extra H^+ ions enter the blood, they associate with HCO_3^- to make H_2CO_3; if H^+ ions are removed, H_2CO_3 dissociates to HCO_3^- plus H^+ again. With this system, respiration alone partially controls the acidity of plasma because every CO_2 molecule removed through the lungs or gills also removes a H^+ ion. The respiratory control centers of mammals respond to increasing acidity in the blood by increasing the rate of breathing, thus eliminating H^+ ions faster (as discussed in Section 45.13).

Blood proteins also serve as buffers. Hemoglobin, for instance, takes care of a great deal of acid by picking up one H^+ ion for each O_2 it releases. (This is the mechanism of the Bohr effect, described in Section 45.11.) Additional H^+ ions combine at other points on the hemoglobin molecule, primarily at histidine residues that are stronger bases in the non-oxygen-binding conformation than in the binding conformation. These subtle

Concepts 46.1 Buffers

When working with chemicals, it is often important to carry out reactions in a medium that keeps the pH constant. A simple system called a **buffer** does this very well, and a buffering system in the blood keeps the blood pH relatively constant for optimal functioning. A buffer is made by combining a weak acid with its base. "Weak" means that it does not dissociate very strongly; for instance, acetic acid in water comes to an equilibrium between acid (associated) and base (dissociated) forms when only 0.0042 of the molecules are dissociated:

$$HA \rightleftarrows H^+ + A^-$$

where A stands for the acetate ion CH_3COO^-. If we combine acetic acid with one of its salts, sodium acetate, we can add more acetate ions without increasing the concentration of hydrogen ions:

$$NaA \rightleftarrows Na^+ + A^-$$

The pH of the resulting solution depends on the relative amounts of acetic acid and sodium acetate. So if some H^+ ions are added, they will combine with acetate ions to form acetic acid, maintaining the equilibrium with the same pH as before. If some H^+ ions are removed, enough acetic acid will dissociate to replace them. The buffer therefore resists changes in pH. Chemical theory makes it easy to select an appropriate buffer system (such as acetate, carbonate, or phosphate) and to calculate the ratio of acid to base that will yield a desired pH.

structural changes make hemoglobin well adapted for its function of carrying O_2 and CO_2. So without any need for more complex regulation, the pH of plasma is kept quite constant, provided large amounts of H^+ aren't added or removed. In that event, some regulation is required.

The nephrons can deal effectively with acidosis and alkalosis. They respond to acidosis by pumping H^+ or ammonium ions into the urine to acidify it (Figure 46.15*a*). In alkalosis, the kidneys can restore H^+ ions to the plasma by actively secreting K^+ ions into the glomerular filtrate (Figure 46.15*b*); the K^+ ions exchange with H^+ ions to maintain electrical neutrality, thus restoring H^+ ions to the extracellular fluid. Furthermore, nephrons normally reabsorb HCO_3^- from the filtrate, and they can change the rate of reabsorption. If there is too much HCO_3^- in the plasma, as occurs during alkalosis, it is removed by excreting $NaHCO_3$ in the urine.

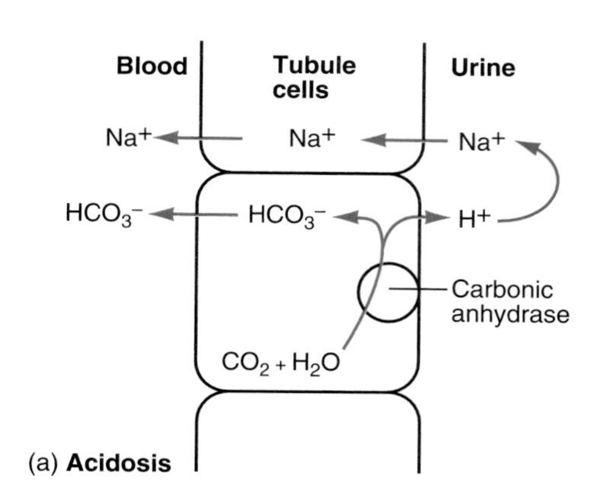

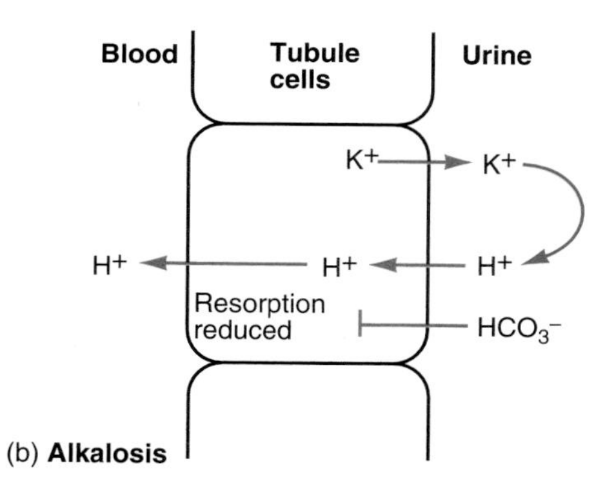

Figure 46.15
(a) In acidosis, excess H^+ ions are secreted primarily by the action of carbonic anhydrase in the nephron endothelium. The nephron can also secrete H^+ ions as ammonium. *(b)* In alkalosis, H^+ ions can be recovered from the glomerular filtrate in exchange for K^+ ions, and the tubule cells absorb less HCO_3^- ion from the filtrate.

Exercise 46.5 We said previously that every CO_2 molecule removed through the lungs or gills removes a H^+ ion. But the CO_2 molecule does not literally take a H^+ ion with it to the outside. Explain how removing CO_2 removes H^+ ions.

Exercise 46.6 Hyperventilation (excessively deep and rapid breathing) leads to light-headedness and fainting. What effect would hyperventilation have on the composition of blood?

Exercise 46.7 Consider the statement that some residues in hemoglobin are stronger bases in the non-oxygen-binding conformation than in the binding conformation. *(a)* Explain what a base is and what it means for a base to be stronger or weaker. *(b)* Explain when or where hemoglobin is in one conformation or the other, and how this characteristic is adaptive for buffering the blood.

46.9 Vertebrate kidneys have evolved to deal with different types of aqueous environments.

The first animals evolved in the oceans between 600 and 540 million years ago. By that time, the oceans were already quite salty, so animals became adapted to those conditions. The composition of animals' body fluids reflects the general composition of those ancient seas. Early animals could maintain a cytoplasm that was isotonic to sea water, so they had no need for rigid cell walls to keep from rupturing under the pressure of incoming water. This alone gave them great flexibility, both literally and figuratively.

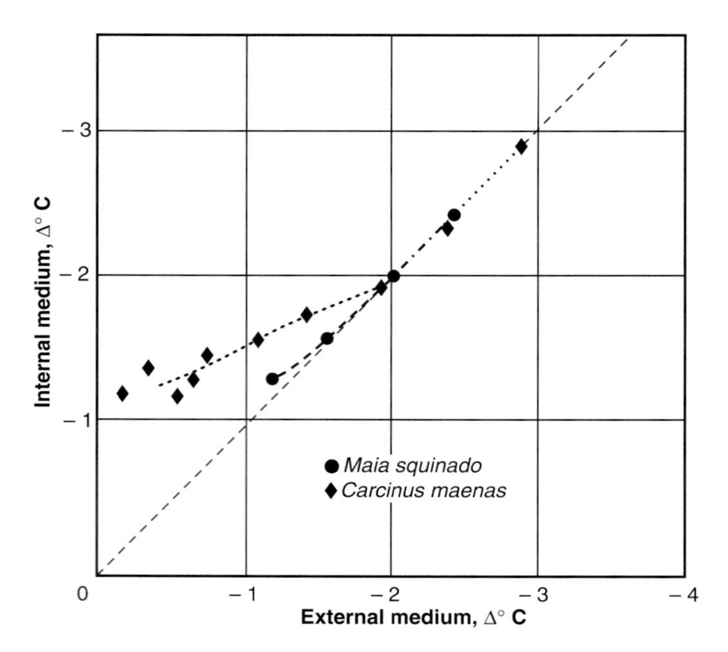

Figure 46.16

The concentrations of blood plasma in two species of crab are very similar to the concentrations of the sea water in which they live. The measurements are given in units of freezing point depression, which is proportional to the concentration of solutes in the liquid. Sources: Data from W.T.W. Potts and G. Perry, *Osmotic and Ionic Regulation in Animals,* 1964, Oxford; Pergamon Press Limited; and M. Duval, *annales de l'Institut Oceanographique,* tome II, fasc. 3 (1925) 232-407.

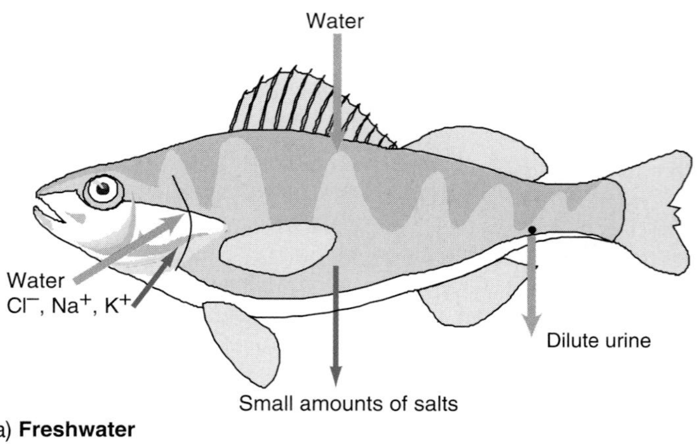

(a) **Freshwater**

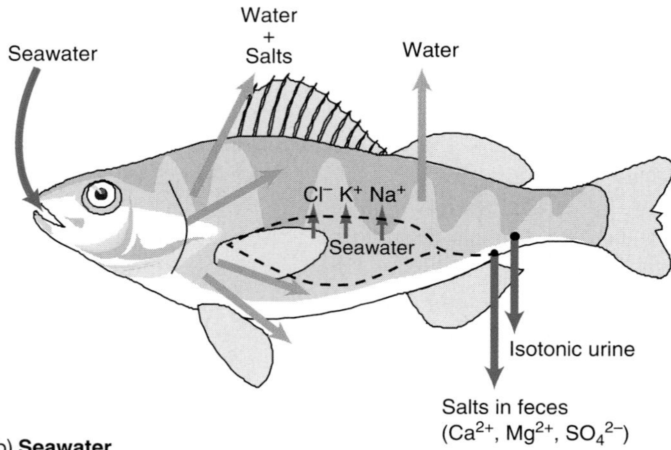

(b) **Seawater**

Figure 46.17

Freshwater and marine fishes face opposite osmoregulatory challenges. *(a)* Freshwater fish tend to take in excess water through their gills and skin; they excrete water in a very dilute (hypotonic) urine and actively take in extra salt through the chloride cells in their gills, as described in Section 46.10. *(b)* Marine fish drink sea water and also tend to become dehydrated by loss of water through their skin. They produce isotonic urine and excrete monovalent ions through their kidneys and the chloride cells in their gills, eliminating the divalent ions with their feces.

An animal surrounded by salty water can adopt one of two strategies for dealing with the osmotic concentration of its internal fluids: Either live with the salt or fight it. Many marine invertebrates are *osmoconformers* whose body fluids conform to the osmolarity of their surroundings (Figure 46.16). However, even though the overall ionic concentration of their fluids keeps them in osmotic balance with their surroundings, their specific ion pumps can maintain some ions at concentrations different from those of sea water. Other animals, in contrast, are *osmoregulators* whose kidneys or other organs regulate the composition of their body fluids and maintain an osmotic pressure quite different from that of their surroundings.

As we mentioned at the beginning of this chapter, the first vertebrates probably evolved in freshwater habitats during the Ordovician period, first in the brackish water in the mouths of rivers and then in the rivers themselves. As vertebrates have subsequently evolved to occupy different habitats, their kidneys and osmoregulatory mechanisms have changed correspondingly.

A freshwater fish is in constant danger of taking in too much water, while its extracellular salts tend to wash out (Figure 46.17*a*). The first fishes evolved kidneys that collect and excrete excess extracellular water; at first the kidney was simply a series of nephron units along the back of the coelom, one per segment, with an opening (a coelomostome) to drain fluid from the coelom (Figure 46.18). Such a kidney only appears today in some fish embryos, in adult hagfish (one of the few remaining jawless fishes), and in some adult teleosts (bony fish). Other fishes and amphibians have more efficient kidneys; their nephrons are associated with glomeruli for filtering water from the blood, at first rather loosely connected with a Bowman's capsule. With such kidneys, freshwater fish excrete very dilute, hypoosmotic urine. They replace the salts lost in their urine by absorbing extra salt through pumps in their gills, as we will see in Section 46.10.

In the Devonian period, long after fishes had evolved their basic form in fresh water, some of them moved back into marine habitats. Sea water, however, is hyperosmotic to the body fluids of fishes, so salt continually tends to enter their bodies while water tends to leave through the gills and other exposed membranes (Figure 46.17*b*). This problem was overcome first by sharks and rays (Chondrichthyes); instead of compensating

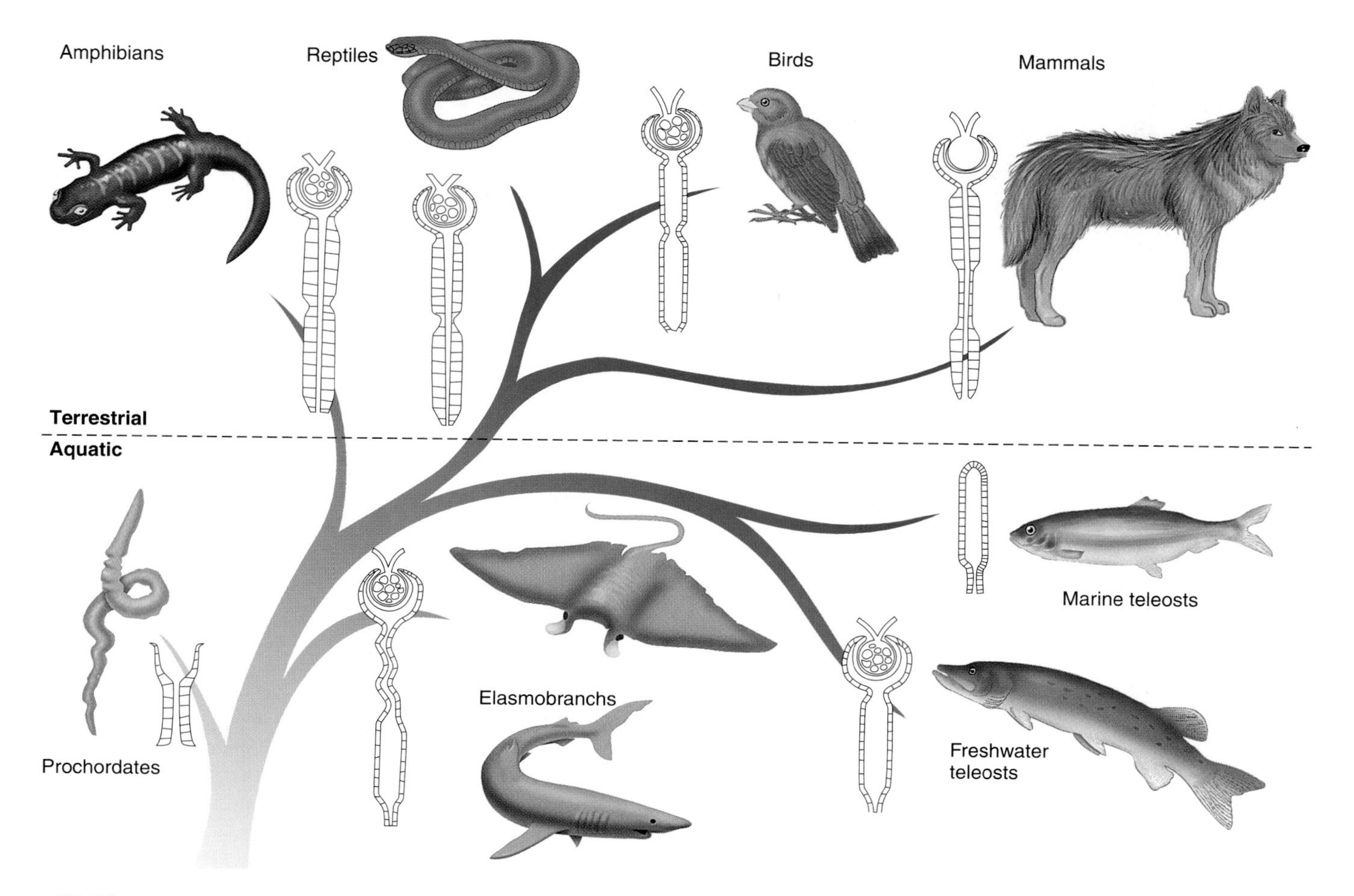

Figure 46.18

The evolution of vertebrate kidneys is shown for the principal groups, depicted with typical nephron units. Some prochordates have simple nephrons that drain fluid from the coelom. The first vertebrates evolved in freshwater habitats; their nephrons drained coelomic fluid, and they developed glomeruli associated with Bowman's capsules, as still seen in freshwater teleosts. Some fishes later moved into marine habitats. Marine elasmobranchs have kidneys similar to those of freshwater teleosts, but the nephrons of marine teleosts have lost the glomeruli and Bowman's capsules, so they lack a filtration mechanism and operate entirely by secretion and reabsorption. The nephrons of freshwater fishes served very well for amphibians and reptiles; in birds and mammals, the nephron became a structure for conserving water more efficiently with the evolution of the loop of Henle, represented by a thin region.

through a radical change in their kidneys, they evolved the marvelously simple alternative of accumulating urea in their extracellular fluid. This makes their fluids isosmotic with the sea water, or even somewhat hyperosmotic, so water still diffuses inward.

Sometime later, some teleosts also returned to the marine environment, adapting to it through significant changes in their kidneys. Marine teleosts take in a lot of water and have evolved kidneys that actively secrete the ions in this water, especially divalent and trivalent ions such as sulfate, Ca^{2+}, Mg^{2+}, and phosphate. Thus, instead of retaining salts and excreting water, the kidneys of marine fishes excrete salts and retain water. Again, they excrete some of the salt through their gills.

The kidneys of amphibians, which spend much of their time in the water, are very similar to those of their freshwater fish ancestors. The kidneys of reptiles are also similar. However, reptiles that occupy relatively dry terrestrial habitats are severely limited by their kidneys. One of their adaptations to life on land is to excrete urea or uric acid instead of ammonia, but they cannot conserve water nearly as well as birds and mammals do because their kidneys have no loop of Henle. Being unable to produce urine that is hypertonic to their interstitial fluid, they lose relatively more water than their avian and mammalian competitors. However, both amphibians and reptiles have mechanisms for reabsorbing water from both their urinary bladder and *cloaca,* a chamber where feces and urine are held before being expelled. Many reptiles also have salt glands (discussed in Section 46.10) and this combination of adaptations suits them very well to dry habitats. It is only in birds and mammals that the loop of Henle evolved, allowing some species to become very well adapted to dry conditions and sometimes to live exclusively on their own metabolic water—the water they produce as a by-product of metabolism (Figure 46.19).

Figure 46.19
Some animals, such as this kangaroo rat, have become adapted to extremely dry conditions by developing kidneys with very long loops of Henle, thus conserving a large part of the water they create metabolically.

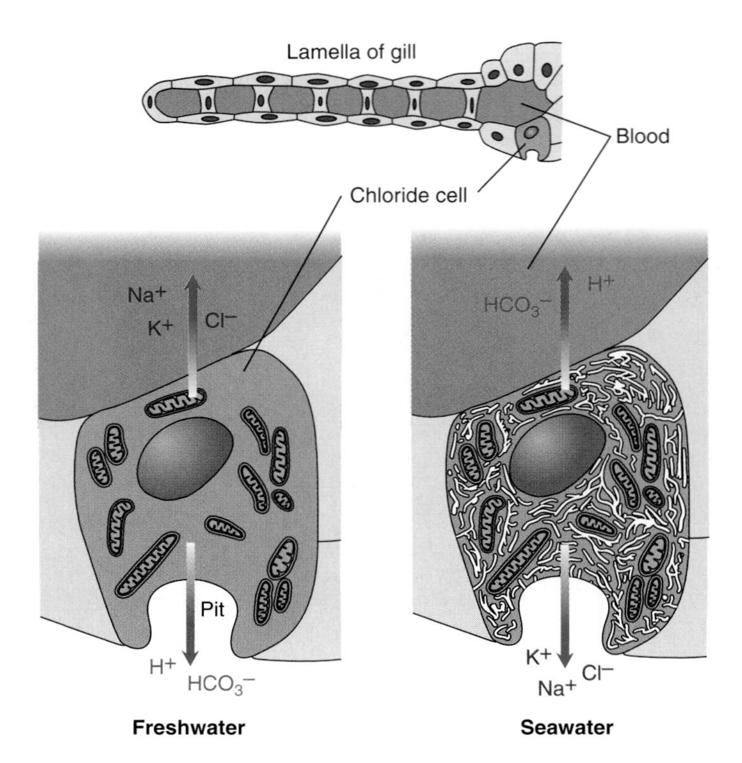

Figure 46.20
Chloride cells in the gills can transport ions either inward or outward. They adapt to the appropriate mode of action in response to external conditions (freshwater or seawater).

46.10 Many animals adapt to environmental demands through the chloride cells of their gills.

Aquatic animals rarely have the luxury of living in water with just the right concentrations of all the essential ions, and they use their ion pumps to continually adjust their internal conditions. Animals that migrate from one habitat to another especially depend on the flexibility and adaptability of these pump systems.

Fishes provide an excellent example of how ion pumps are used to meet environmental challenges. In addition to their kidneys, fishes have a unique osmoregulatory mechanism in the **chloride cells** of their gills which stand directly between the blood on one side and the open water on the other (Figure 46.20). Their inward sides are invaded by many dead-end channels that become even more numerous in fish adapted to fresh water. In freshwater fishes, these cells pump Na^+ and Cl^- ions into the blood with antiport mechanisms of the kind we have seen before, including the classical Na^+-K^+ exchange system and one that exchanges Na^+ for H^+. In marine fishes, the chloride cells pump in the opposite direction; the Na^+-K^+ exchange system removes excess Na^+ ions from the blood in exchange for K^+ ions. Since the intracellular concentration of K^+ is still about 10 times that in sea water, K^+ ions tend to continually leak out. Salmon and other fish that live part of their lives in fresh water and part in sea water can adapt to the different environments through a change in the pumps of their chloride cells. As they migrate into the sea, the increasing concentration of Na^+ ions induces a new exchange system and represses the old one.

Antiport mechanisms, Section 8.14.

Tube-nosed birds such as albatrosses and fulmars, which spend all their lives at sea, must drink sea water and eliminate the salt. They excrete massive amounts through a remarkable **salt gland** located in the corner of each eye, which creates an extremely concentrated brine that continuously runs down the beak (Figure 46.21). Operating on the countercurrent exchange principle (see Concepts 45.1), these glands use ion pumps that move Na^+ outward, with Cl^- following passively. Turtles, crocodilians, marine snakes, and lizards also have salt glands.

Coda Much of animal evolution could be described from the viewpoint of excretion and osmoregulation, and Homer Smith did write such a history of the vertebrates in his book *From Fish to Philosopher* (see Additional Reading). All multicellular organisms have problems of water balance, as discussed in Chapter 32, but the plant and animal solutions to those problems are very different from each other. Since animal cells have only thin plasma membranes, without the cell walls that help keep plant cells intact, animals must either remain in environments with ionic compositions similar to those of their cells or have osmoregulatory devices. Although many marine animals perform little osmoregulation, all animals need to excrete wastes, especially nitrogenous wastes.

We noted in Section 34.9 that animals have evolved two distinct kinds of ducts (coelomoducts and nephridia) that variously serve the functions of osmoregulation and excretion. In

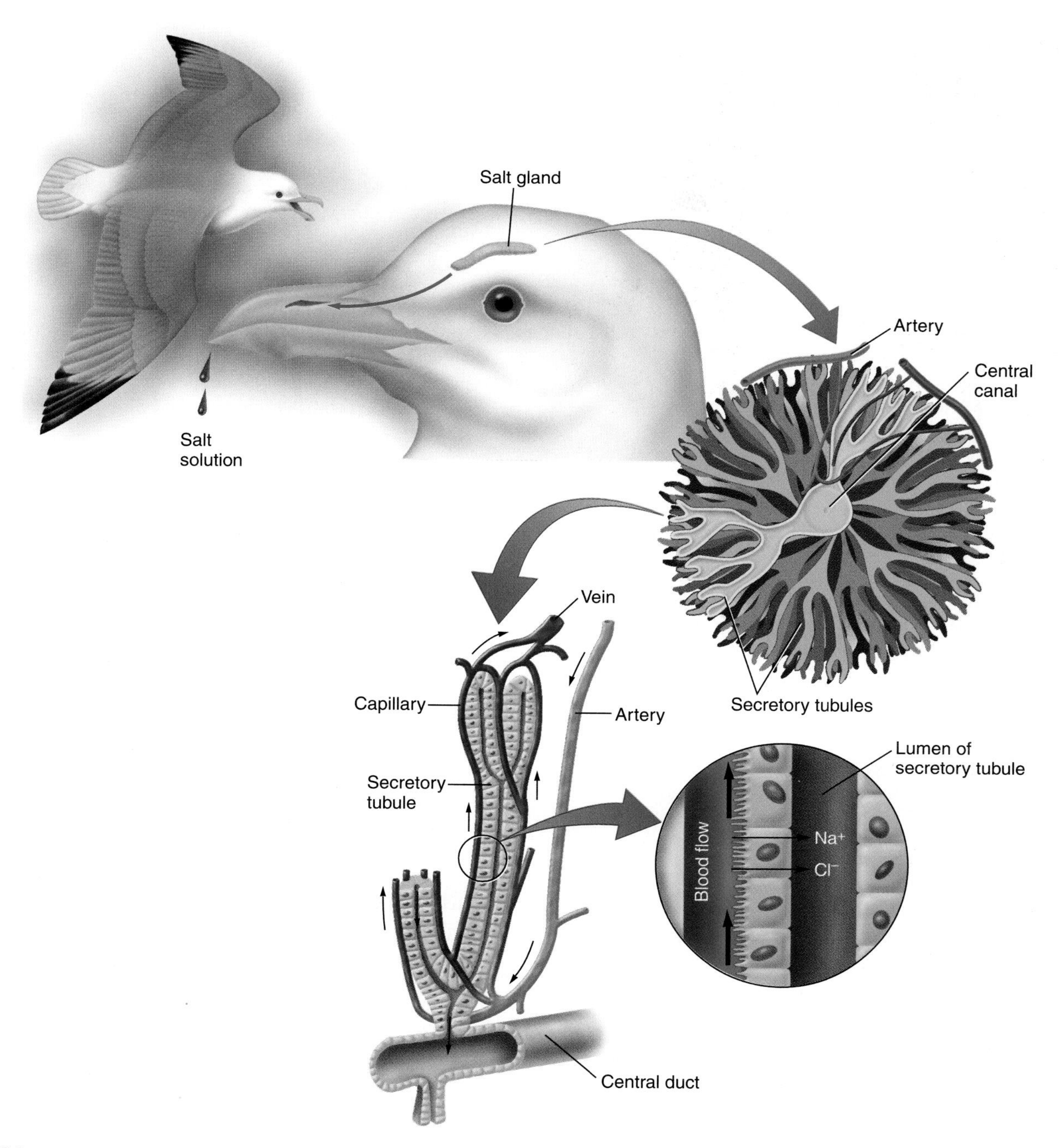

Figure 46.21

Some sea birds have salt glands that produce a highly concentrated salt solution. Blood flows in capillaries through an epithelium that secretes Na^+ ions into a central tubule, with Cl^- ions following. Because the blood flows in the opposite direction from the secretion in the tubule, salt secretion is enhanced by countercurrent exchange.

the course of their evolution, animals have transformed these ducts, singly or in combination, into a remarkable variety of devices, but all unified by the common mechanisms we discussed at the beginning of this chapter: filtration of fluids into a channel followed by specific secretions and recoveries as the fluid moves along the channel to the outside. These devices depend on a second common feature, an ion pump. The sodium pumps and other ion pumps built into the nephrons of vertebrate kidneys allow these animals to live in habitats as diverse as deserts and fresh water. Sodium pumps in the skin and elsewhere adapt fishes and amphibians to fresh water; fishes adapt to either sea water or fresh water with different versions of chloride cells in their gills; and marine birds and reptiles excrete salt with their special glands. Focusing on these functions emphasizes the general point that most evolutionary novelty arises from modifying, elaborating, and recombining previously evolved systems, rather than the evolutionary "invention" of totally novel structures. The success and adaptability of a group of organisms depends upon the happenstance of acquiring systems with great potential for further modification.

Summary

1. An excretory organ removes foreign molecules and wastes, including excess nitrogen, and it may also regulate the balance of water and salt ions. All excretory organs depend upon a dead-end channel made of epithelial cells; wastes are either forced into it by pressure or secreted into it, some materials are withdrawn, and the remainder is excreted. The kidney is an excretory organ designed to excrete foreign materials and nitrogenous wastes, and to balance water and ions.
2. Animals excrete excess nitrogen as ammonia (primarily in aquatic organisms), as urea, and as uric acid (primarily in terrestrial organisms that must conserve water). Some animals can switch from one mode to another in response to changes in their environments.
3. Epithelial cells create a sodium pump by having Na^+-K^+ ATPases on one cell surface and Na^+ channels on the other face. Some epithelia can also move water by pumping ions into clefts between the cells; water follows by diffusing through protein channels (aquaporins) and then diffusing out of the clefts.
4. In organs that regulate water and ions, there is a steady-state flow of each material through the surrounding interstitial fluid between an excretory tubule and the circulation.
5. Insects produce a concentrated urine by cycling material between their hindgut, hemolymph, and Malpighian tubules. The system is driven by active transport of K^+ ions.
6. A kidney unit, or nephron, is a long tubule closed at one end. It filters large amounts of plasma to make a glomerular filtrate and then selectively reabsorbs materials from it and secretes additional materials into it. Plasma is filtered across the glomerulus at a rate that depends on the total blood pressure, the resistance in arterioles of the kidney, and the back pressures from the nephron.
7. Sodium pumps in the proximal convoluted tubule remove most of the Na^+ ions from the filtrate, and most of the water follows. In general, water follows the salt ions that are removed from the filtrate. Some Na^+ pumps are regulated by aldosterone from the adrenal cortex, through a regulatory circuit involving renin, produced by the kidney.
8. In birds and mammals, the vasa recta, acting through countercurrent exchange, and the loop of Henle, acting by means of ion pumps, can produce a concentrated interstitial fluid that induces the absorption of extra water from the glomerular filtrate.
9. The kidneys regulate the volume and osmolarity of the blood in response to signals from baroreceptors that measure pressure and from osmoreceptors that measure osmolarity; the latter is primarily a function of the Na^+ ion concentration.
10. Two regulatory circuits respond to changes in blood pressure and osmolarity. One circuit responds to high osmolarity and releases vasopressin or antidiuretic hormone, which makes the collecting ducts of the nephron more permeable to water. The plasma Na^+ level and blood pressure are regulated by a feedback circuit involving angiotensin and aldosterone.
11. The rate of glomerular filtration depends on the flow rate through arterioles leading to the glomeruli and on the overall blood pressure. These factors are regulated partially by the baroreceptor reflex that effects vasoconstriction and vasodilation and also signals changes in the activity of the heart.
12. The pH of plasma is kept constant through buffering by plasma proteins and bicarbonate. The HCO_3^- content is regulated by the rate of breathing. The kidney has special mechanisms for eliminating or reabsorbing H^+ ions.
13. Animals can live either as osmoconformers, which change the concentration of their fluids to conform to their surroundings, or as osmoregulators, which maintain relatively constant concentrations of solute in spite of changes in their environment.
14. The ancestors of vertebrates evolved in the oceans, surrounded by salt water. Vertebrates probably evolved in brackish or fresh water, and they evolved kidneys to continually excrete the excess water that tends to diffuse into their tissues. Fishes that returned to the oceans later had to evolve different types of kidneys or other mechanisms to resist their tendency to continually lose water to their surroundings.
15. Animals that spend their lives in different salt concentrations can adjust their ion pumps to compensate for these changes. Fish use the chloride cells in their gills to pump ions inward or outward, depending on their environment. Some birds have special salt glands to excrete excess salt.

Key Terms

osmoregulation 977
urea 977
uric acid 977
sodium pump 978
aquaporin 979
Malpighian tubule 980
nephron 980
cortex 980
medulla 980
glomerulus 980
Bowman's capsule 980
glomerular filtrate 980
podocyte 980
glomerular filtration rate (GFR) 980
loop of Henle 982
vasa recta 982
baroreceptor 985
osmoreceptor 985
antidiuretic hormone (ADH)/vasopressin 985
juxtaglomerular apparatus 985
aldosterone 985
atrial natriuretic factor (ANF) 985
buffer 987
chloride cell 990
salt gland 990

Multiple-Choice Questions

1. Which statement is the most accurate about ecologically successful organisms that lack any mechanism of osmoregulation?
   a. They will soon die.
   b. They will soon become extinct.
   c. They can tolerate fluctuating internal concentrations of solutes.
   d. They must restrict their habitats to dry land.
   e. They produce no waste products.
2. Excretory mechanisms rely on all of the following *except*
   a. secretion.
   b. reabsorption.
   c. ultrafiltration.
   d. energy-requiring water pumps.
   e. energy-requiring ion pumps.
3. An aquaporin is a
   a. channel through which water moves by osmosis.
   b. channel through which water is pumped.
   c. protein that changes the thickness of fluid-mosaic membranes.
   d. protein that allows nonpolar solutes to cross cell membranes.
   e. protein channel that harnesses water movement to make ATP.
4. Which of the following structures is part of the vascular structure of the human kidney?
   a. Bowman's capsule
   b. cortex
   c. medulla

d. glomerulus
e. loop of Henle

5. All of the following except the ______ are part of a mammalian nephron.
   a. Bowman's capsule
   b. glomerulus
   c. proximal convoluted tubule
   d. ascending limb of loop of Henle
   e. descending limb of loop of Henle
6. Which term describes the flow from the vascular system into the nephron?
   a. ultrafiltration
   b. secretion
   c. tubular reabsorption
   d. *a* and *b*, but not *c*
   e. *a*, *b*, and *c*
7. Which term describes the flow from the nephron to the vascular system within the kidney?
   a. ultrafiltration
   b. filtration
   c. reabsorption
   d. secretion
   e. none of these
8. If the loop of Henle is working properly, it will do all of these activities *except*
   a. create a gradient of salt ions within the kidney medulla.
   b. increase the countercurrent effects of the vasa recta.
   c. enable further concentration of urine in the collecting ducts.
   d. prevent excessive water loss from the body.
   e. carry conserved fluid away from the nephron.
9. ADH is produced in the ______, and its target cells are in the ______.
   a. vasa recta; loop of Henle
   b. hypothalamus; collecting ducts
   c. anterior pituitary gland; podocytes
   d. Bowman's capsule; proximal convoluted tubule
   e. adrenal glands; liver
10. Which of the following is a secretory product of the kidney that assists in the regulation of renal function?
    a. ADH
    b. ANF
    c. renin
    d. adrenalin
    e. angiotensin II

True-False Questions

Mark each statement true or false, and if false, restate it to make it true.

1. Excretion of excess nitrogen in the form of ammonia is the method of choice for terrestrial animals.
2. Terrestrial animals excrete either ammonia or urea.
3. Ion/ATPase pumps use the energy obtained from ion flow to produce ATP.
4. In insects, metabolic wastes are excreted and water is retained by the action of a sodium ion pump in the Malpighian tubules.
5. Glomerular filtration pressure results in the flow of blood from the glomerulus into Bowman's capsule.
6. Podocytes restrict the flow of water from the glomerulus into the Bowman's capsule.
7. The urine output of a normal kidney represents about 20 percent of the amount of filtrate it processes.
8. An increase in fluid loss from the body could be caused by an increase in ANF or a decrease in angiotensin II.
9. Acidosis will stimulate the nephrons to pump hydrogen ions from the urine back into the nephron.
10. Freshwater fishes generally produce a highly concentrated urine, and they actively move salts out of their bodies by means of ion pumps in their gills.

Concept Questions

1. How does glomerular filtrate differ from urine?
2. Explain why an individual's glomerular filtration rate fluctuates throughout the day.
3. In a healthy individual, the capillaries that bring blood to the glomeruli are larger in diameter than those that drain blood from the glomeruli. What is the effect of this difference, and what would occur if the sizes were reversed?
4. Explain the activity of the juxtaglomerular apparatus. Does it act as an excretory, secretory, or sensory structure, or is it more like an effector, such as a muscle or a gland?
5. Examination of the kidneys of many freshwater and saltwater fishes shows a significant difference in glomerular size. Contrast the osmoregulatory challenge faced by these fishes and predict which would have the larger glomeruli.

Additional Reading

Dantzler, William H. *Comparative Physiology of the Vertebrate Kidney.* Springer-Verlag, Berlin and New York, 1989.

Guyton, Arthur C. "Blood Pressure Control—Special Role of the Kidneys and Body Fluids." *Science,* June 28, 1991, p. 1813. A research paper on the regulation of blood pressure.

Smith, Homer W. *Kamongo, or The Lungfish and the Padre.* The Viking Press, New York, 1949 [1932]. In the form of a novel, Smith tells the story of his research and philosophizes about evolution.

———. *From Fish to Philosopher: The Story of Our Internal Environment.* Little, Brown, and Co., New York, 1959. A popular story of evolution, focusing on the role of the kidney.

Internet Resource

To further explore the content of this chapter, log on to the web site at:

http://www.mhhe.com/biosci/genbio/guttman/

47 Digestion, Assimilation, and Nutrition

A group of African children illustrates the growing problem of feeding the world's hungry people. Although they are being fed with imported food, several of them still show the enlarged stomachs typical of kwashiorkor.

Key Concepts

A. Structure of the Digestive System

47.1 The digestive system of most animals is a long tube to which digestive enzymes are added.

47.2 Animals use many different mechanisms for feeding.

47.3 Food is often broken down mechanically into small pieces.

47.4 The mammalian GI tract consists of a series of compartments with specialized functions.

47.5 Carbohydrates are digested in the mouth and small intestine.

47.6 A series of enzymes digest proteins.

47.7 Neural and hormonal circuits regulate the secretion of materials involved in digestion.

47.8 Hormones from the small intestine stimulate pancreatic secretion.

47.9 Bile aids the digestion and absorption of lipids.

47.10 The small intestine has an enormous absorptive surface.

B. Absorption and Distribution of Nutrients

47.11 The products of digestion are distributed to all tissues.

47.12 Lipids are transported in lipid-protein complexes.

47.13 Metabolic patterns change during a fasting period.

47.14 The plasma glucose concentration is regulated by the action of several hormones.

47.15 Calcitonin and parathyroid hormone regulate the plasma calcium level.

47.16 Iron is transferred and stored by two proteins, transferrin and ferritin.

C. Aspects of Nutrition

47.17 All tissues are in a constant state of flux.

47.18 Animals must obtain eight of the twenty amino acids from their food.

47.19 Human diets have changed with civilization, generally for the worse.

47.20 The balance of nutritional factors is complex.

The visitor walks among the starving masses in disbelief, trying to control his emotions. Everywhere he sees walking human skeletons and tiny, wasted children with huge stomachs. "Kwashiorkor," he is told. "Protein starvation." The relief agencies fly in planeloads of grain and dole it out fairly, but there is not enough to end the misery, only enough to hold it in check for a while longer. The visitor knows that even if these people regain some semblance of health and recover an agricultural base, the children will never be normal; they have been starved for protein too long at a critical time in the development of their brains. And almost surely the whole story will be repeated soon elsewhere.

Observing this all-too-common socioeconomic problem raises many questions, including the central one of how a growing human population can be fed (see Sidebar 40.1). What nutrients does a person require to be adequately nourished? For example, protein has a special role in animal metabolism, because animals are unable to make several of its constituent amino acids; how much protein does a person need, and what kind? What happens to proteins and other substances after they are eaten? Fat also seems to be of great concern to people—that is, having enough, but not too much—and some fats seem to be much better for our health than others. In this chapter, we will address these and related questions. ■

A. Structure of the digestive system

47.1 The digestive system of most animals is a long tube to which digestive enzymes are added.

Animal cells metabolize much like the cells of other heterotrophs, but rather than getting their own nutrients individually, they depend on the animal's digestive system. Fundamentally, digestion is very simple. Food is made mostly of polymers, and digestion is just a matter of hydrolyzing food molecules into their monomeric units and then absorbing these into the body. The story becomes more complicated, of course, inside the digestive organs of a real animal, with all its regulatory mechanisms.

The most primitive animals, such as the hydra and flatworm in Figure 47.1, digest their food in a gastrovascular cavity, a blind pouch with a single opening that serves for both ingestion and the expulsion of undigested material. Cells lining the pouch absorb nutrients and digest food particles intracellularly. However, most animals have a digestive tube—a **gastrointestinal (GI) tract** or **alimentary canal**—running through the body from mouth to anus. As food passes through the tube, digestive enzymes secreted by glands in the intestinal walls and by accessory glands digest the food into its monomers. The monomers are then selectively absorbed through the walls of the tract into the circulatory system, and indigestible materials are expelled from the anus. It is important to realize that the lumen of the GI tract is really outside the body, in the same position as the hole in a doughnut:

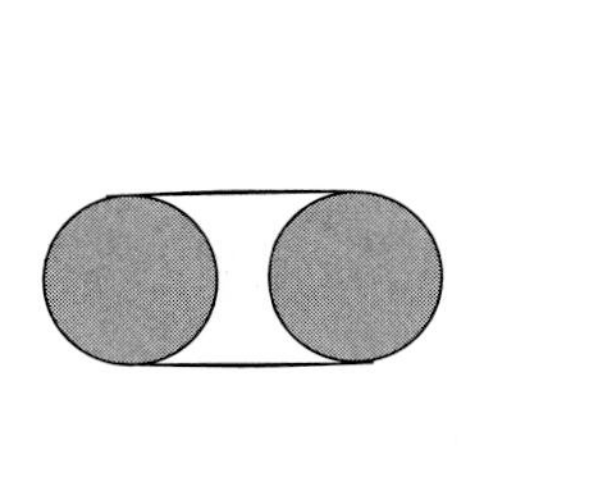

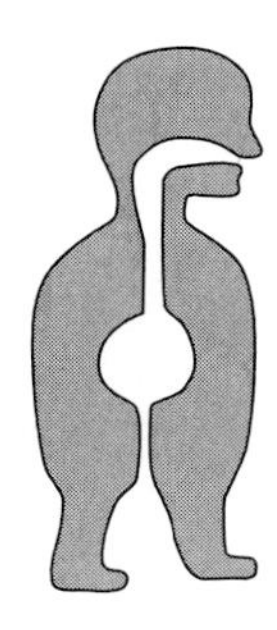

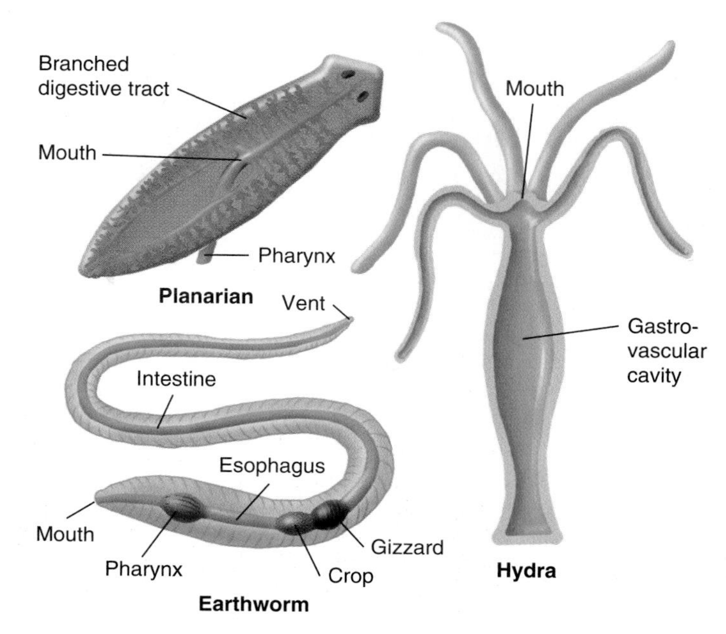

Figure 47.1

Cnidarians such as a hydra and flatworms such as the common planarian *Dugesia* have digestive cavities with only a single opening. However, most animals, as represented here by an earthworm, have complete digestive tracts.

Materials only enter the body proper when they move across the intestinal walls. Absorbed substances are transported throughout the body, to be metabolized for their energy or **assimilated**—that is, incorporated—into the body's structure. The whole process is regulated both hormonally and neurally.

Vertebrates digest their food with the secretions (enzymes and accessory materials) of three major glands—salivary glands, pancreas, and liver—and of gland cells not organized into organs that line portions of the digestive tract itself. Other types of animals have a variety of digestive glands, but we will focus on those in the vertebrate system.

47.2 Animals use many different mechanisms for feeding.

While an animal is feeding, it is acting as one link in a food web, and the way it chooses its food partly structures its ecosystem. Each species is adapted to consuming a certain limited range of foods that will supply all its essential nutrients.

Animals can feed in many ways. They eat anything that can provide nourishment, employing an amazing variety of structures to capture and ingest it. Some insects, such as grasshoppers and dragonflies, use their mandibles to chew solid food, while others have sucking mouthparts and live on the nutrient-rich liquids of other organisms (Figure 47.2). Mosquitoes, fleas, and lice suck animal juices, and cicadas, aphids, and bugs suck plant juices. You'll often see butterflies, moths, and bees extending their long probosces (feeding tubes) deep into flowers to extract the nutritious nectar, and many birds, such as hummingbirds, have also adopted this way of feeding.

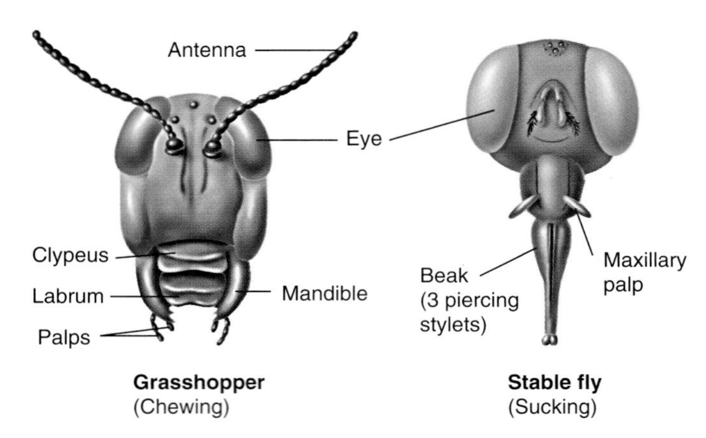

Figure 47.2
The mouthparts of insects are specialized for chewing or sucking.

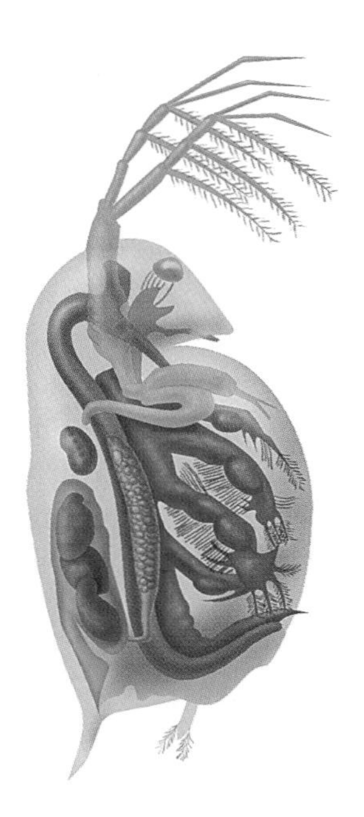

Figure 47.3
A filter feeder like the water flea *Daphnia* collects food in a comb of bristles.

Filter feeders strain small particles of food out of the water they live in. The little water flea *Daphnia,* for instance, has modified legs covered with a comb of bristles (Figure 47.3). As it swims along, tirelessly beating these legs, the combs trap small bits of food from the water. Another set of bristles then guides the trapped food into the animal's mouth. Clams and other bivalve molluscs continually move a stream of water through their gills, where food particles and dissolved nutrients are trapped in a layer of mucus. Beating cilia move the mucus and food across the gills and into the mouth.

Large animals can also trap their food by filtering. Some fish have extensive rakes on their gills that trap small crustaceans and other plankton. One large group of whales feeds by means of a baleen, or whalebone, which forms large brushes at the sides of the mouth. After swimming open-mouthed through a swarm of shrimp and other small crustaceans, the whale sieves out the food by closing its mouth and forcing the water through the baleen with its tongue.

The mouths of various animals are designed for biting, chewing, gobbling, tearing, or in some other way forcing whole organisms, or pieces of them, into the digestive system. Figure 47.4 shows a variety of bird bills that are specialized for certain foods. Obviously the size of a mouth determines the size of the food. This simple fact is ecologically important; it means that animals divide resources in part by acquiring feeding apparatuses of different sizes. Much of the food available in an ecosystem consists of small particles of many sizes, and correspondingly many kinds of small invertebrates move through this food supply, eating whatever is appropriate to their size and using genetically encoded feeding patterns that are perhaps modified by learning.

47.3 Food is often broken down mechanically into small pieces.

Animals commonly feed by tearing, grinding, and chewing. Since cooking tenderizes foods, we who live mainly on cooked food easily forget just how hard natural foods can be, but plant and animal structures are made of many strong connective molecules, including cellulose cell walls and collagen fibers. Although naturalists tell spectacular stories of animals that swallow others whole and digest them slowly, most animals tear food into pieces with their limbs or with grinding or chewing devices in their GI tracts. Plant cytoplasm is enclosed in largely indigestible cellulose walls, which must be broken down by extensive grinding, so animals that eat plant tissues spend considerable time chewing them. Breaking food down mechanically increases its surface/volume ratio, exposing it more thoroughly to enzymes for faster, more complete digestion.

Figure 47.5 illustrates the GI tract of a bird. Swallowed food enters the thin-walled crop and then passes on to the stomach and the muscular gizzard, where most digestion takes place. The crop is just a storage organ, which allows the animal to ingest large amounts at once whenever it has the opportunity and then to digest it later. This is a very useful feature, since food may not always be available and the threat of predators often makes it essential for animals to "eat and run." The gizzard, a crushing and grinding organ, breaks food apart mechanically, often aided by small stones (grit) that the animal swallows. (Such stones, called gastroliths, have been found with dinosaur skeletons, telling us that dinosaurs had similar structures and feeding habits.) The gizzard itself is powerful. The French naturalist Reaumur reported in 1752 that in 24 hours a turkey's gizzard could flatten and partially roll a tube of sheet iron that could only be dented under a force of 36 kilograms. Other animals have similar muscular organs in their intestines, often with hard internal ridges.

Vertebrates commonly break up their food with teeth, although some teeth serve other functions; for example, the

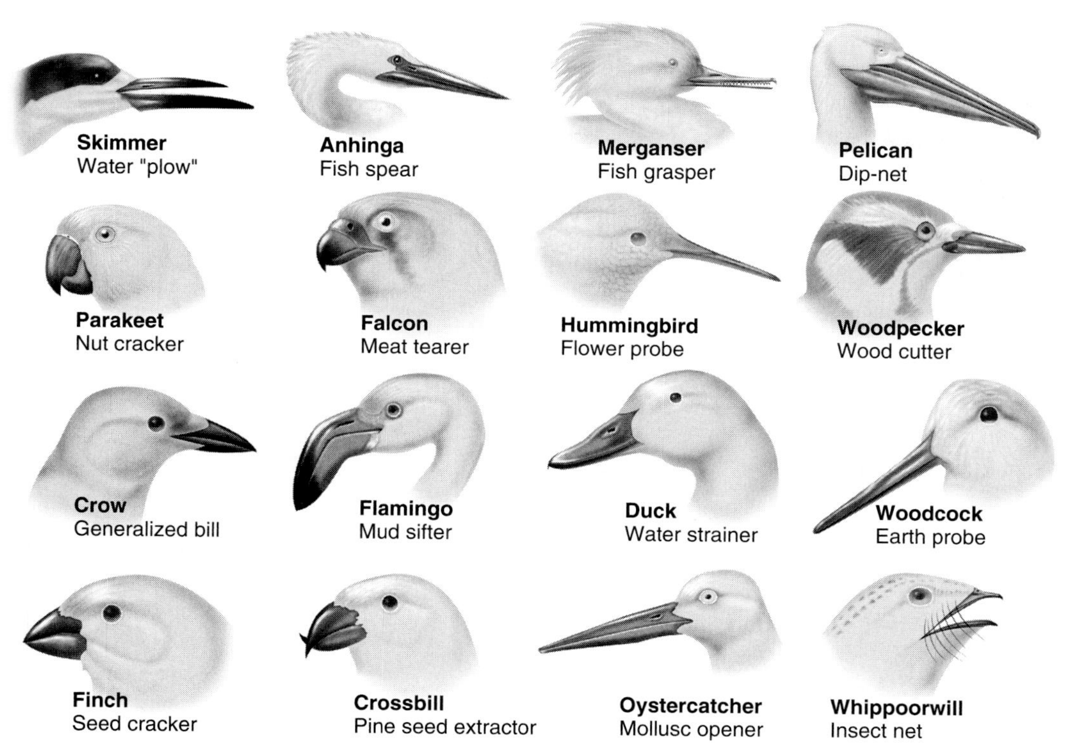

Figure 47.4
The specialized forms of birds' bills are adaptations for feeding in different ways on a wide variety of foods.

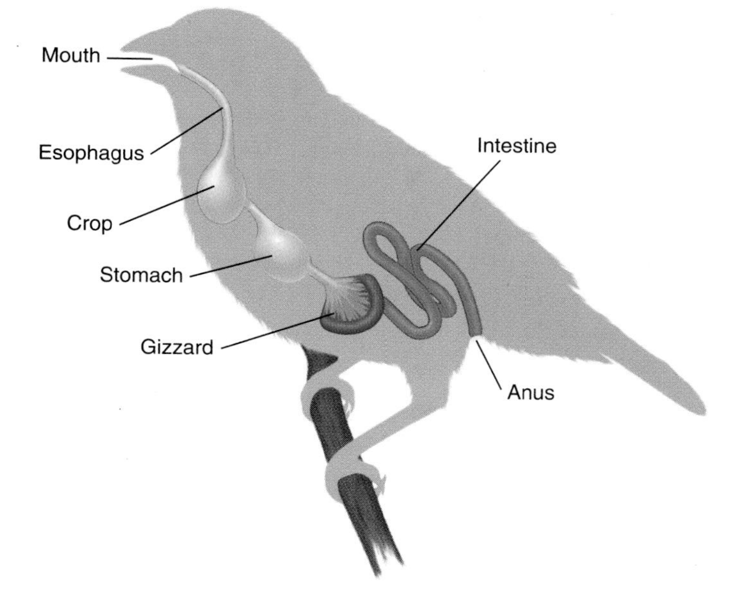

Figure 47.5
In birds, food is taken into the crop, which serves as a storage chamber. Digestion occurs in the stomach, which secretes digestive juices, and especially in the muscular gizzard, which is lined by horny plates that break food down mechanically. To aid this process, birds commonly swallow pieces of grit and small pebbles, which lodge in the gizzard.

inward-pointing teeth of many fishes and snakes are designed to trap a large prey animal so it can't pull out while muscles of the mouth and throat pull it in (Figure 47.6). Mammals have four kinds of teeth for cutting, tearing, and chewing. They cut food with their sharp, chisel-shaped *incisors* in front, tear it with their *canines* (especially well developed in carnivores), and grind and crush it with the flat, knurled surfaces of the *premolars* and *molars* along the sides of their jaws (Figure 47.7). To aid in chewing, the tongue moves food around, mixes it with the digestive enzymes in saliva, and shapes it into a round wad, or *bolus,* that can be swallowed. Some mammals, the ruminants, stuff food into the first pouch of their stomach (called the rumen) and then regurgitate and chew it, as a cud. So, like birds that stuff food into their crops, ruminants can swallow in haste and digest at leisure.

47.4 The mammalian GI tract consists of a series of compartments with specialized functions.

Before examining the digestion of particular foods, let's consider the general organization of the mammalian GI tract. This long tube is made of four tissue layers called, from inside to outside, *mucosa, submucosa, muscularis,* and *serosa* (Figure 47.8). Its inner lining, the mucosa, is made of columnar epithelial cells, which absorb digested food in some regions, and cells that secrete digestive enzymes, hormones, and the mucus that covers

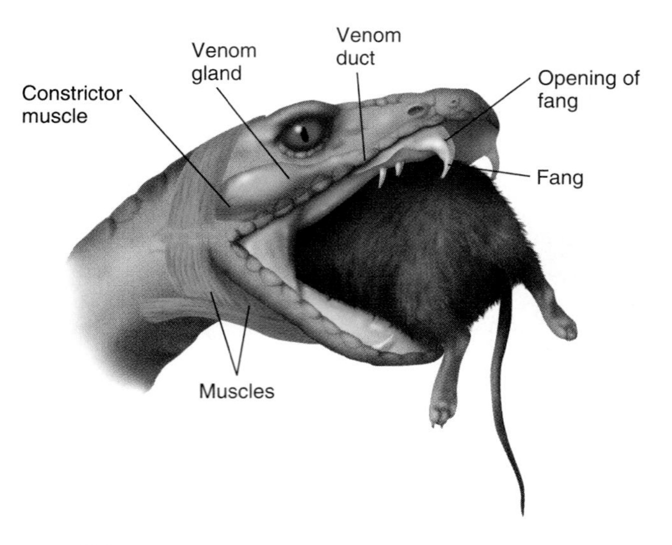

Figure 47.6

A snake has teeth that hold its prey, including fairly large animals, so it cannot escape. Muscles of the mouth and throat then gradually draw the food into the snake's intestine, where it is digested without being ground into smaller pieces.

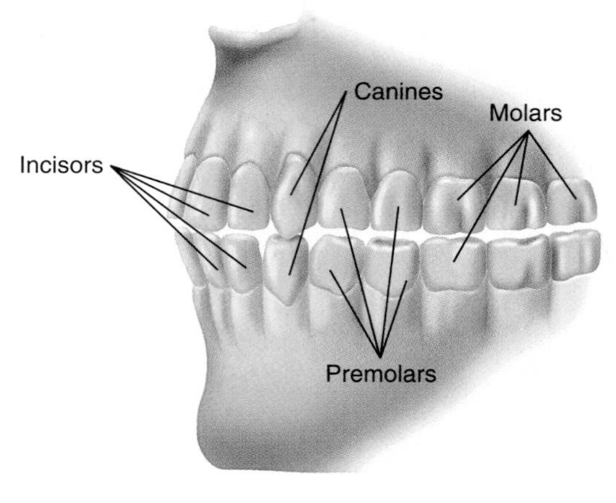

Figure 47.7

Humans, like other mammals, have four kinds of teeth: incisors for cutting, canines for tearing, and premolars and molars for grinding and crushing.

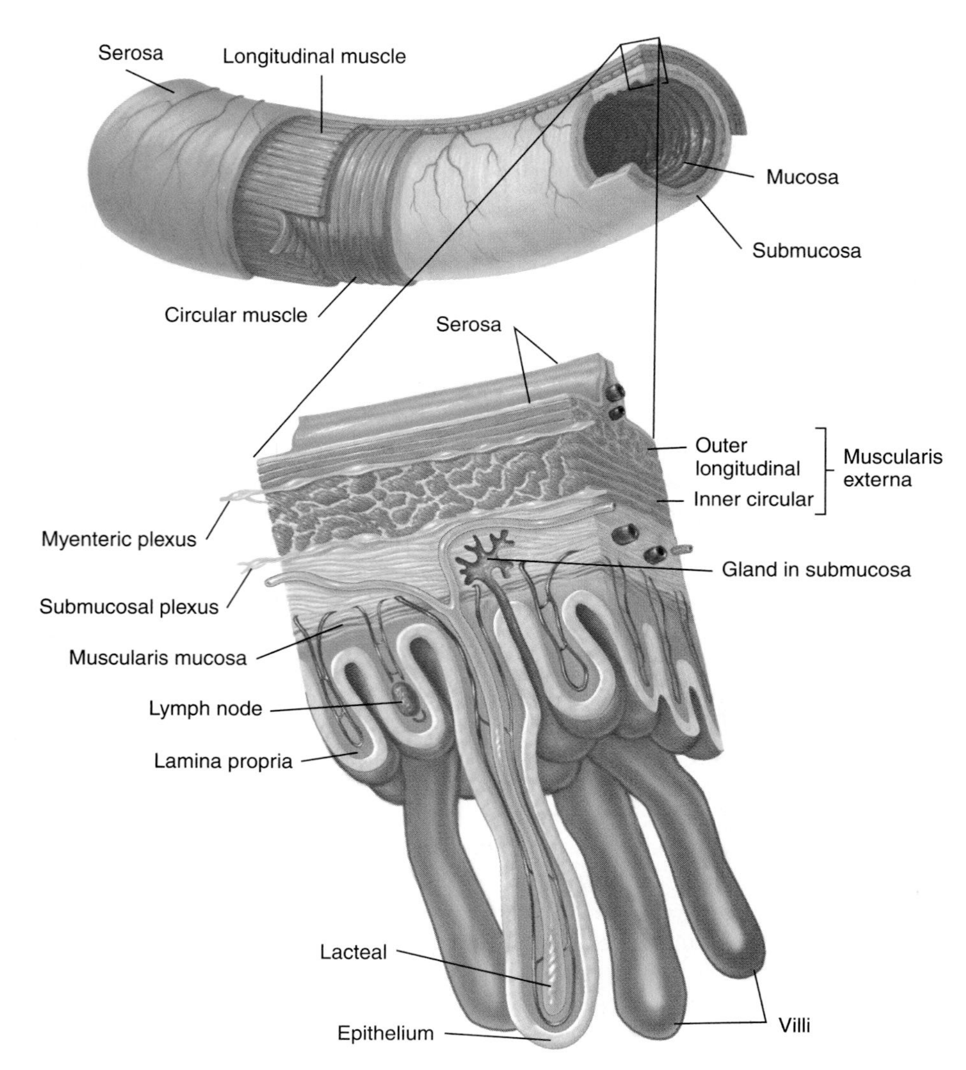

Figure 47.8

The intestinal tract is made of four concentric layers of tissue. The innermost *mucosa* secretes enzymes and other substances and absorbs food through its large surface. The mucosa consists of an inner columnar epithelium, the lamina propria, surrounded by a thin layer of smooth muscle, the muscularis mucosa. The *submucosa* is connective tissue rich in blood vessels. The *muscularis* includes an inner layer of circular muscles and an outer layer of longitudinal muscles; it also includes nervous tissue forming two plexuses that control the actions of the intestinal tract. The outer *serosa* is connective tissue covered by squamous epithelium.

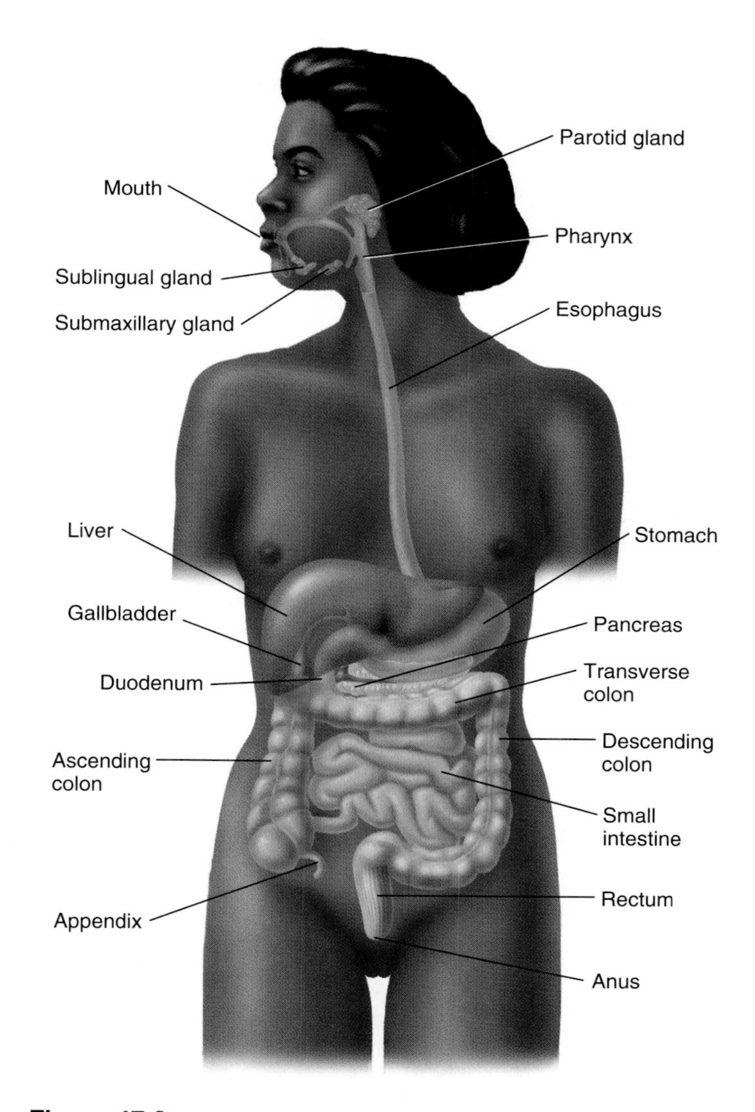

Figure 47.9

The human digestive system includes the mouth, esophagus, stomach, small intestine, and large intestine (colon). Glands that secrete digestive juices and enzymes include three pairs of salivary glands around the mouth as well as the pancreas, which fits into a curve between the small intestine and stomach. The liver also has important glandular functions and secretes bile, which is stored in the gallbladder. The small intestine is actually much longer than shown here.

most of the tract. The submucosa is a connective tissue layer that includes blood vessels and glands. It is surrounded by the muscularis made of smooth muscle in an inner circular layer and an outer longitudinal layer; the muscularis creates the various movements that aid digestion and propel materials through the tract. The whole tract is then surrounded by a serosa made of connective tissue and squamous epithelium.

Once food has been formed into a bolus, it is swallowed through the action of tongue and throat muscles and passes down the **esophagus** to the **stomach** (Figure 47.9). But a bolus doesn't just drop passively down the tube; it is pushed by **peristalsis,** a series of muscular contractions (Figure 47.10). Gravity has nothing to do with this action; one can actually swallow upside-down. The junction between the esophagus and stomach is normally closed by the lower esophageal sphincter, a circular ring of muscle that keeps the stomach contents from moving back into the esophagus. The pyloric sphincter at the lower end of the stomach closes it off from the small intestine.

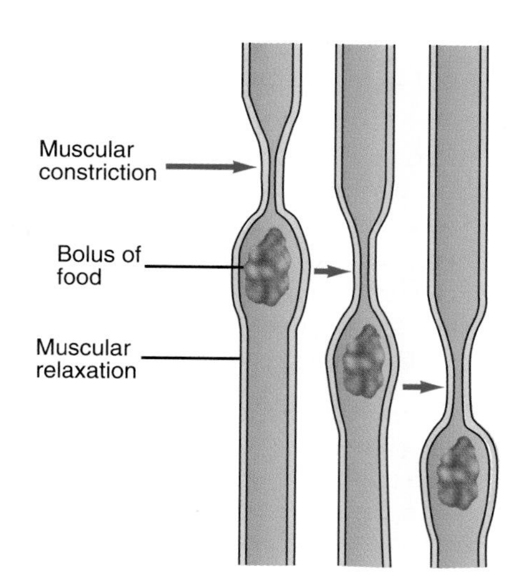

Figure 47.10

In peristalsis, a wave of muscular contractions *(blue arrow)* pushes a bolus of food ahead.

In the stomach, food is mixed with juices and enzymes to produce a thick, souplike **chyme** (pronounced "kime"). Mucus secreted by the mucosal stomach lining shields the surface against digesting itself. (When this protective mechanism fails, as it sometimes does in people under stress, the stomach does begin to digest itself and creates an ulcer, although infection with the bacterium *Helicobacter pylori* now appears to be the primary cause of ulcers.) The presence of food in the stomach stimulates peristaltic waves that run from one end to the other to assist digestion by mixing the chyme and breaking food down mechanically. Somewhat like the heart (see Section 45.4), the stomach has its own pacemaker and rate of contraction.

After an average of 30–60 minutes for digestion in the human stomach, the pyloric sphincter opens, and each stomach contraction squirts a bit of chyme into the **small intestine.** There it is mixed with additional digestive enzymes secreted by the pancreas and intestinal gland cells as well as bile from the liver, which aids in the digestion of lipids. Most digestion and absorption occurs in the small intestine.

The indigestible components of food pass from the small intestine into the **large intestine,** or **colon,** whose main function is to regulate the amount of water and ions absorbed into the bloodstream or left in the digestive system. The most common colon malfunctions are matters of water imbalance: diarrhea from material moving through the colon too quickly for enough water to be reabsorbed, or constipation from material moving so slowly that too much water is absorbed. People need to eat plenty of indigestible material (roughage such as cellulose and hemicellulose) whose bulk stimulates good peristalsis in the colon and prevents constipation. Eventually the contents of the colon reach the **rectum,** where they are compacted and stored for elimination. About half the solid mass of the feces consists of bacteria that inhabit the colon, called the "intestinal flora" because bacteria used to be considered plants. A rich bacterial population is essential for normal colon function and good nutrition, since these bacteria provide bulk and also supply some vitamins, especially vitamin K, which is essential for the liver to form certain blood-clotting proteins.

Now we will trace the digestion of different foodstuffs to see how a system of hormones regulates the process.

Exercise 47.1 In addition to peristalsis, the intestine engages in segmentation movements: constricting to divide the contents into segments, then constricting at different points to divide it into different segments, and so on, at a rate of 6 to 10 times a minute. Sketch what happens to the chyme in this process and explain its function:

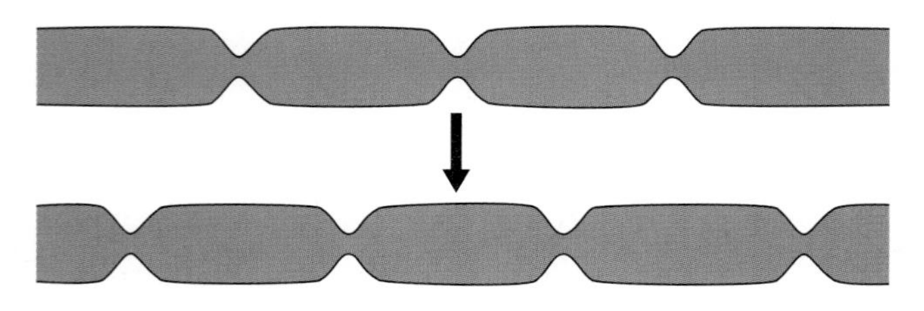

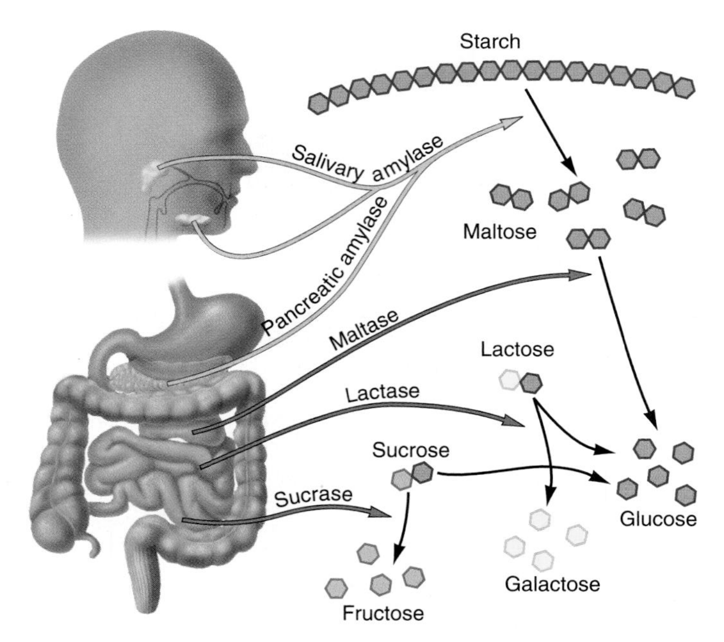

Figure 47.11
Amylase, secreted in saliva and in pancreatic juices, digests starch into short chains (dextrins) and into the disaccharide maltose. Intestinal maltase digests maltose into glucose. Lactase digests lactose into glucose plus galactose, and sucrase digests sucrose into glucose plus fructose.

47.5 Carbohydrates are digested in the mouth and small intestine.

Food commonly includes two glucose polymers—starch from plant cells and glycogen from animal cells; both are easily digested, beginning in the mouth where the food is mixed with **amylase** from the salivary glands (Figure 47.11). Amylase attacks the bonds between glucose residues, leaving maltose, a disaccharide of glucose, and short glucose chains known as dextrins. This digestive activity stops in the stomach, where amylase is inactivated, and begins again in the small intestine with another amylase secreted by the pancreas.

Glands in the mucosa of the small intestine also produce maltase, which digests maltose; sucrase, which digests sucrose; and lactase, which digests lactose. The simple 6-carbon sugars released by these enzymes are then absorbed by the small intestine. Children make lactase up to about age 4, and Caucasian people generally retain the ability throughout their lives. But many adults of other ethnic groups, particularly in Asia and Africa, cannot make it, so they become lactose intolerant; after eating milk and milk products, they experience cramps, gas, and diarrhea from undigested lactose.

Much of plant biomass consists of cellulose and other polysaccharides that our enzymes cannot digest. However, a few species of microorganisms have evolved *cellulases,* enzymes that can attack cellulose. Only by harboring these symbiotic microorganisms in their intestines can many animals get nourishment from plant foods. In fact, it is realistic to consider such animals to be bacteriovores, which live in part by digesting the bacteria supported by their food. Ruminants such as cattle, for instance, harbor cellulase-producing bacteria in their rumens, where their food goes through the first stages of digestion. Termites, which eat wood, have flagellated protozoans and various bacteria in their intestines that digest plant material (see Figure 27.22). Newly hatched termites eat some of their parents' feces to acquire their own intestinal microorganisms.

Exercise 47.2 When you eat fruit, you are ingesting many complex gums and other polysaccharides made of sugars such as rhamnose, arabinose, and xylose linked to one another in many ways. What happens to these compounds as they pass through your GI tract? Why?

Exercise 47.3 Some African tribes herd dairy cattle; others do not. Investigators have studied the presence of lactase in adults of several tribes. Which adults do you think retain the ability to produce lactase?

47.6 A series of enzymes digest proteins.

Protein digestion is a bit complicated because most proteolytic enzymes, or **proteases,** are remarkably specific and will only cut bonds next to certain amino acid residues. This fact is useful in determining amino acid sequences (see Section 4.8). Proteases from different glands supplement one another to digest proteins quite thoroughly (Figure 47.12). Protein digestion begins in the stomach. Pits in the stomach mucosa contain *chief cells* that secrete **pepsinogen** and *parietal cells* that secrete hydrochloric acid (Figure 47.13); the acid attacks many of the peptide linkages in proteins and creates the acidic conditions in

Concepts 47.1 Cell Types and Communication Circuits

As we describe the digestive system, we sometimes point to specific cell types, such as the chief and parietal cells. The purpose is not to give you additional facts to memorize but to emphasize the fact that animals are constructed of many types of cells, largely sensor and effector cells, each specialized for performing a certain task. Each cell type has specific receptor proteins for hormones, neurotransmitters, and other signal ligands. Since hormones circulate through the blood, they might affect any cell in the body, but these receptors ensure that each cell responds only to appropriate signals from hormones and other molecules or ions. In this way, an animal becomes a maze of highly efficient, specialized regulatory circuits. It is valuable to see an animal from this informational, functional viewpoint, as well as from a purely structural viewpoint.

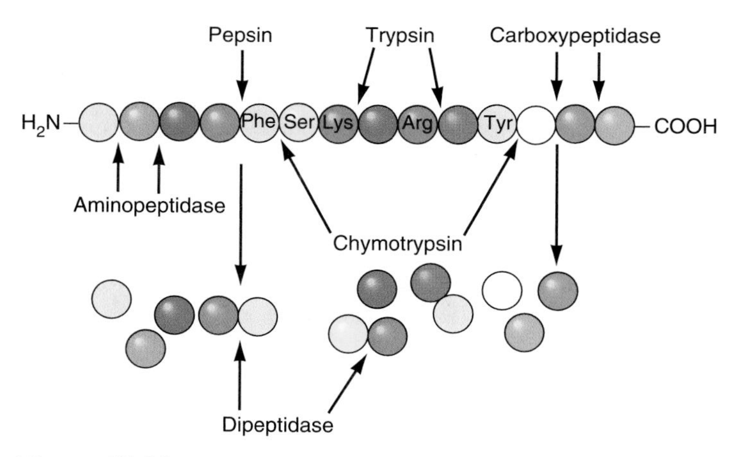

Figure 47.12
Each proteolytic enzyme attacks a specific kind of peptide linkage, next to certain types of amino acids; aminopeptidase and carboxypeptidase remove amino acids one at a time from the N-terminus or C-terminus of the peptide chain respectively. Here each circle represents one amino acid. In concert, these enzymes reduce proteins to amino acids, dipeptides, and tripeptides, all of which can be absorbed in the small intestine.

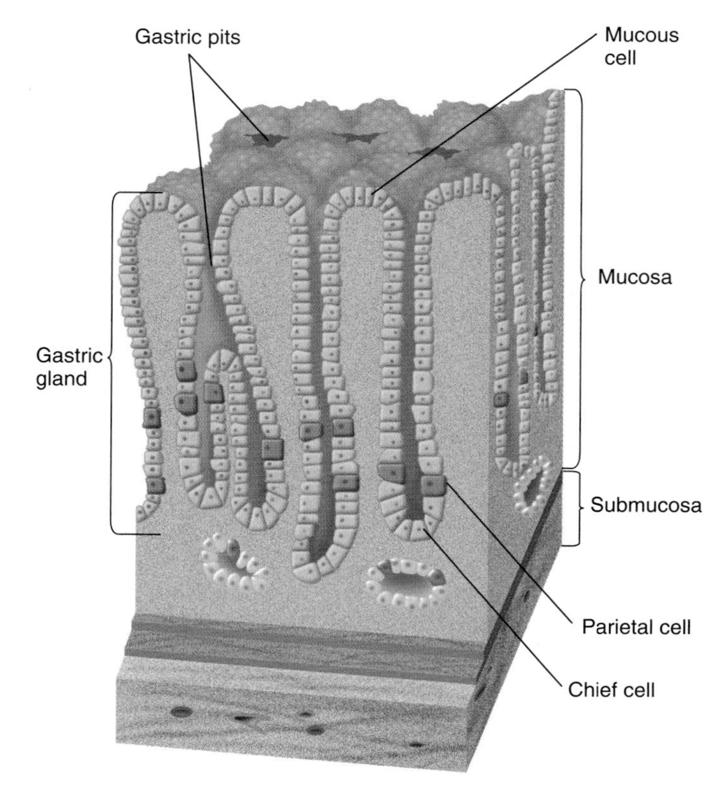

Figure 47.13
Pits in the stomach are made of three types of gland cells. Most of the stomach mucosa is made of mucous cells, which secrete mucus. Chief cells secrete pepsinogen. Parietal cells are marked by a series of cytoplasmic channels where H^+ and Cl^- ions are secreted.

which pepsin is most active. Concepts 47.1 addresses the variety of cell types in the digestive system.

Protein digestion continues in the small intestine through proteases secreted by the pancreas: trypsin, chymotrypsin, and enzymes that attack peptide chains from the N-terminus (aminopeptidase) and C-terminus (carboxypeptidase). (Notice that enzymes like trypsin and chymotrypsin are named in the old way, not with the ending *-ase.*) Enzymes in the mucosal walls of the small intestine split di- and tripeptides, although these short peptides—as well as free amino acids—are absorbed through the intestinal walls. The pancreas also secretes a bicarbonate solution that neutralizes the acidic contents of the stomach and creates the basic pH that is optimal for these enzymes.

Active proteases would digest one another and the cell that makes them. To keep this from happening, gland cells synthesize and store proteolytic enzymes in an inactive **proenzyme,** or **zymogen,** form. (The inactive form of a protein either has the prefix *pro-* or the suffix *-ogen,* and some proteins have been named both ways, by different scientists.) Chief cells in the stomach make pepsin as the proenzyme pepsinogen, which has stretches of amino acids covering its active site. Hydrogen ions from the HCl secreted in the stomach attack and remove the extra peptides from pepsinogen, converting it into pepsin. Trypsinogen, chymotrypsinogen, and procarboxypeptidase are proenzyme forms of other proteolytic enzymes.

Zymogen synthesis, Section 14.7.

The intestine and pancreas also secrete nucleases that digest DNA and RNA, and lipases that release fatty acids from triglycerides and phospholipids. All these enzymes conspire to reduce most foodstuffs to molecules small enough to be absorbed.

Exercise 47.4 Some self-styled nutritionists advise people to eat a certain food because it contains a certain enzyme presumably needed in metabolism. Considering what happens to food proteins in the GI tract, why can you ignore such advice?

47.7 Neural and hormonal circuits regulate the secretion of materials involved in digestion.

Generations of biologists and physicians have grown up knowing that the autonomic nervous system (ANS) has two divisions, sympathetic and parasympathetic. But as anatomists and physiologists have investigated the neural connections of the intestinal tract, they have started to see it as a system of its own, the **enteric nervous system (ENS),** which deserves status as a third division of the ANS.

With more neurons than the spinal cord, the ENS is centered in structures called **nerve plexuses,** best described as webs of neurons lying within the walls of the gut. The submucosal and myenteric plexuses extend along the length of the GI tract and regulate its activities. The submucosal plexus, lying within the submucosa, regulates the activity of glands and smooth muscle in the mucosa. The myenteric plexus, lying between the circular and longitudinal layers of muscle in the muscularis, provides most of the nerves to the gut and regulates gut movements such as peristalsis and segmentation. Cell bodies of enteric neurons lie within these plexuses. They receive inputs from both the sympathetic and parasympathetic systems, so the gut's activity is regulated like that of other organs during general sympathetic and parasympathetic stimulation. But even if nerves from both these divisions of the ANS are cut, the gut will continue to exhibit mobility and waves of contraction, showing that its activity is generated intrinsically. All the movements of the gut are due to local reflexes confined to the ENS. In other words, they are controlled by very short reflex arcs, in contrast to the long reflex arcs passing through the CNS. This means that the ENS, unlike the sympathetic and parasympathetic divisions of the ANS, includes sensory neurons and interneurons as well as motor neurons. In addition, a sequence of hormones superimposed on the nervous system carries signals between parts of the GI tract (Figure 47.14).

As neurotransmitters, the enteric neurons use acetylcholine, epinephrine, dopamine, ATP, serotonin, and several other compounds; in fact, though serotonin has important functions in the brain, most of it is in the gut.

The GI tract can be stimulated to activity just by seeing, smelling, or even thinking about food; these actions send signals through hypothalamic centers and the vagus nerve to start the secretion of digestive enzymes in the stomach. Then as food enters the stomach and small intestine, it stimulates glands to secrete digestive enzymes and juices of the right acidity in a functional sequence. Secretion is controlled internally through local ENS circuits and hormonal circuits.

Distension of the stomach by food triggers the release of hydrochloric acid and pepsin by the parietal and chief cells, thus starting the digestion of proteins. The resulting peptides and amino acids stimulate stomach endocrine cells to secrete **gastrin,** which in turn stimulates further secretion by the chief and parietal cells (Figure 47.15). This is a *positive feedback circuit:* The acid and pepsin produce more protein fragments, which stimulate more gastrin, which stimulates more gastric secretion. The accumulating acid eventually lowers the pH of the stomach so far that it inhibits gastrin secretion, thus terminating the positive feedback circuit. After digestion has proceeded in the stomach for a while, activity moves on to the small intestine.

Secretin

His•Ser•Asp•Gly•Thr•Phe•Thr•Ser•Glu•Leu•Ser•Arg•Leu•Arg•Ser•Ala•Arg•Leu•Glu•Arg•Leu•Leu•Glu•Gly•Leu•Val•NH_2

Glucagon

His•Ser•Gln•Gly•Thr•Phe•Thr•Ser•Asp•Tyr•Ser•Lys•Tyr•Leu•Asp•Ser•Arg•Arg•Ala•Glu•Asp•Phe•Val•Glu•Trp•Leu•Met•Asn•Thr

Gastrin

Phe•Asp•Met•Trp•Gly•Tyr•Ala•Glu•Leu•Trp•Pro•Gly•Glu

Gastrin activity

Phe•Asp•Met•Trp•Gly•Met•Tyr•Asp•Arg•Asp•Ser•20 more amino acids

CCK activity

Figure 47.14

Four peptide hormones that regulate digestion show similarities in their structures. Glucagon, which regulates plasma glucose, is similar to secretin. Colored boxes enclose the portions of gastrin and CCK that actually have the hormonal activity (presumably the parts that bind to target cell receptors). Since CCK shares part of the gastrin sequence, it has some gastrin-like activity.

Exercise 47.5 Notice in Figure 47.14 that the amino acid sequences of secretin and glucagon are very similar. In the context of evolution, what does this similarity mean?

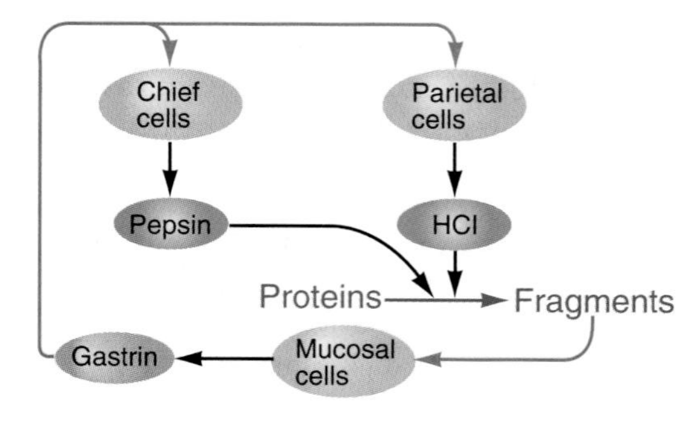

Figure 47.15

Gastrin, whose secretion is induced by proteins and peptides in the stomach, stimulates the release of pepsin (pepsinogen) and hydrochloric acid. Pepsin and acid produce more peptides, resulting in a positive feedback loop.

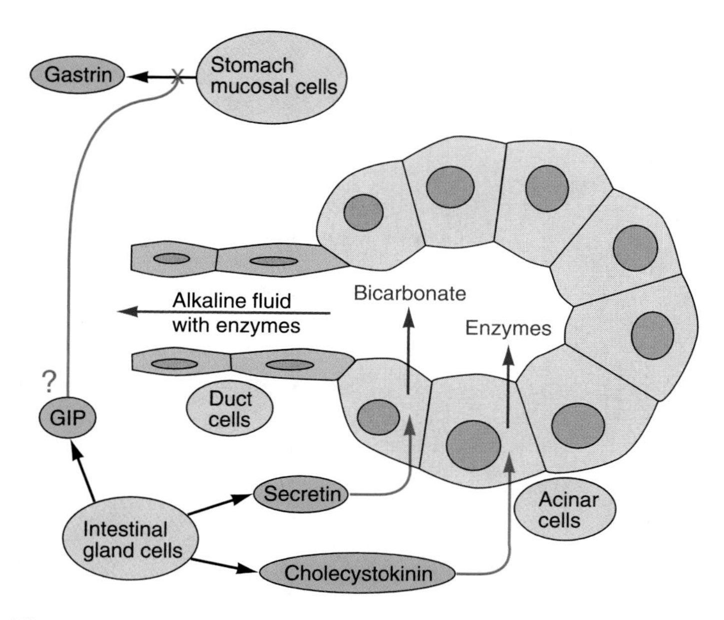

Figure 47.16

Secretin and cholecystokinin stimulate secretion of bicarbonate and enzymes, respectively, by the pancreas. Gastrin inhibitory peptide (GIP) feeds back to stop the activities of the stomach.

47.8 Hormones from the small intestine stimulate pancreatic secretion.

Digestion in the small intestine occurs in an alkaline environment created by a bicarbonate solution secreted by the pancreas and intestinal mucosa. Acidic chyme entering the intestine from the stomach triggers the release of **secretin** from intestinal gland cells, which stimulates the *duct cells* of the pancreas to release their alkaline fluid into the intestine (Figure 47.16), a story we tell in Chapter 41. Meanwhile, amino acids and lipids entering the small intestine stimulate other intestinal gland cells to secrete the hormone **cholecystokinin (CCK)**[1]**,** which then stimulates the *acinar cells* of the pancreas to secrete digestive enzymes. Thus two hormones carry signals to neutralize the acid chyme and add new enzymes to continue food digestion.

Meanwhile, another hormone, probably the *gastric inhibitory peptide* identified only a few years ago, feeds back to inhibit motility of the stomach and secretion by the chief and parietal cells. This negative feedback coordinates gastric and intestinal activity, ensuring that the stomach processes food slowly enough for the intestine to keep pace. Once the stomach has done its job and emptied its contents into the small intestine, it returns to its resting condition. For the next several hours, only the intestine is active, continually contracting in a peristaltic rhythm that mixes its contents. Amino acids, sugars, fatty acids, and other food molecules are absorbed across the walls of the small intestine. To complete this picture, we have only to add the action of the liver and its secretions.

[1]This peptide hormone has been called pancreozymin, because it stimulates secretion of enzymes from the pancreas, and cholecystokinin, because it stimulates the release of bile from the gallbladder. For a time, the two names were combined; physiologists have now settled on the single name cholecystokinin.

47.9 Bile aids the digestion and absorption of lipids.

One of the liver's many functions (Concepts 47.2) is to produce **bile,** a mixture of fatty materials (including cholesterol) and bile salts (cholate and taurocholate), that is essential for the assimilation of fat molecules in food. Bile salts act as detergent molecules, amphipathic molecules that *emulsify* fats—that is, they mix the fats into the surrounding water much as soaps and detergents do—and effectively dissolve fats in the aqueous contents of the intestine for digestion and absorption. Bile is synthesized in the liver, released into the bile duct, and stored in the **gallbladder** lying under the liver. Cholecystokinin, secreted by the small intestine, stimulates the smooth muscle walls of the gallbladder to contract and also relaxes a sphincter on the bile duct, allowing bile to flow into the small intestine. At the same time, secretin stimulates the liver to produce more bile.

Amphipathic molecules and detergents, Section 8.4.

As the small intestine absorbs lipids from food, the bile components are circulated back to the liver and reused in a cycle known as the **enterohepatic circulation** (Figure 47.17).

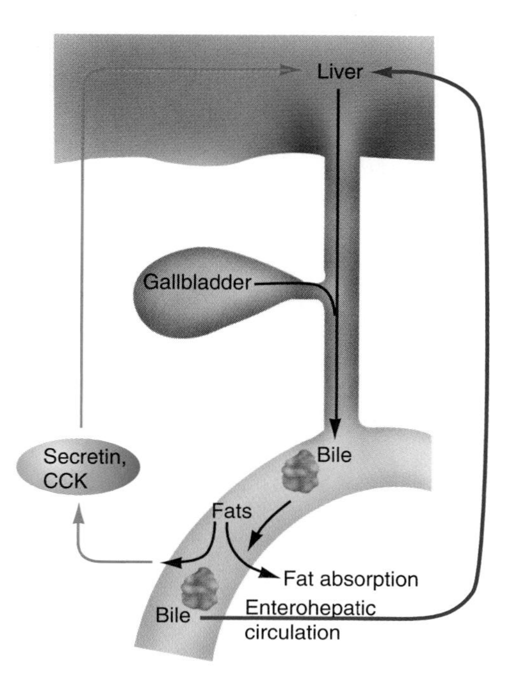

Figure 47.17

Secretion from the liver and gallbladder is stimulated by secretin and CCK. Bile components are recovered from the intestine by the liver and then secreted again, in an enterohepatic circulation.

Concepts 47.2 Summing Up Liver Functions

The liver is such a prominent organ in mammals, and plays so many roles in metabolism, that it is useful to sum up its functions.

Participates in Digestion and Assimilation

- Stores glycogen and releases glucose; converts lactate and other compounds to glucose and glycogen
- Secretes bile, which emulsifies fats in the small intestine
- Synthesizes lipoproteins that carry excess cholesterol; removes cholesterol from lipoproteins
- Converts free fatty acids into ketone bodies
- Stores iron as a complex with ferritin

Stores Lipid-soluble Vitamins

- Vitamin A for vision
- Vitamin D_3 for calcium metabolism; partly converts into the dihydro form
- Vitamin E, an antioxidant
- Vitamin K for synthesizing active forms of blood-clotting proteins

Synthesizes Proteins

- Albumin
- Clotting factors (prothrombin, fibrinogen, others)
- Plasma transport proteins
- Angiotensinogen, an inactive peptide hormone precursor
- Somatomedin, a hormone that targets cartilage cells, stimulates cell division and growth

Detoxifies and Eliminates Wastes

- Removes amino groups from amino acids and nucleic acids as ammonia, urea, uric acid
- Detoxifies drugs and potentially harmful substances in foods (such as allomones) with its P450 enzymes
- Contains part of the reticuloendothelial system, fixed macrophages lining the liver cavities that actively engulf foreign debris
- Destroys old red blood cells

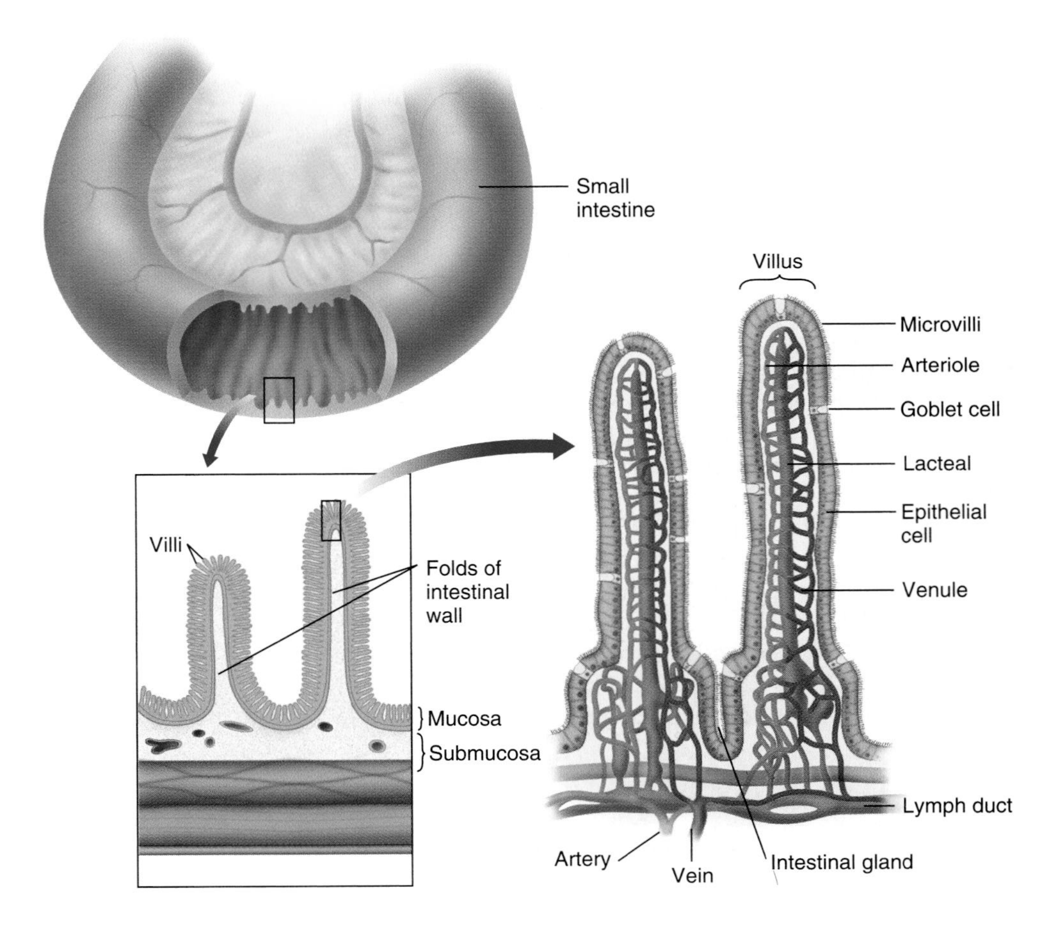

Figure 47.18
The structure of the small intestine, with its folds and villi, creates an extremely large surface area for food absorption.

Recycling conserves a lot of material that otherwise would have to be continually resynthesized.

47.10 The small intestine has an enormous absorptive surface.

The vertebrate GI tract, particularly the small intestine, clearly shows the importance of a large surface/volume ratio (Figure 47.18). Enormous numbers of small molecules and ions pass through the intestinal mucosa in a rather short time, and since the requisite transport systems are located in plasma membranes, the more plasma membrane, the faster uptake occurs. The human small intestine, a tube 2–3 cm in diameter and nearly 8 meters long, would have a lot of surface without any embellishment. Yet this area is increased by large folds, which are covered in turn by an estimated 4–5 million **villi,** tiny fingers about a millimeter long. Still finer projections, the **microvilli,** cover the epithelial cells of each villus (Figure 47.19). This huge surface, estimated at nearly 200 square meters, absorbs digested food. Each villus contains a bed of capillaries that collect most of the material absorbed through the mucosa and transport it into the circulatory system, to be carried throughout the body. Each villus also has a *lacteal,* an extension of the lymphatic system that is more permeable than the capillaries and absorbs lipids from the intestine.

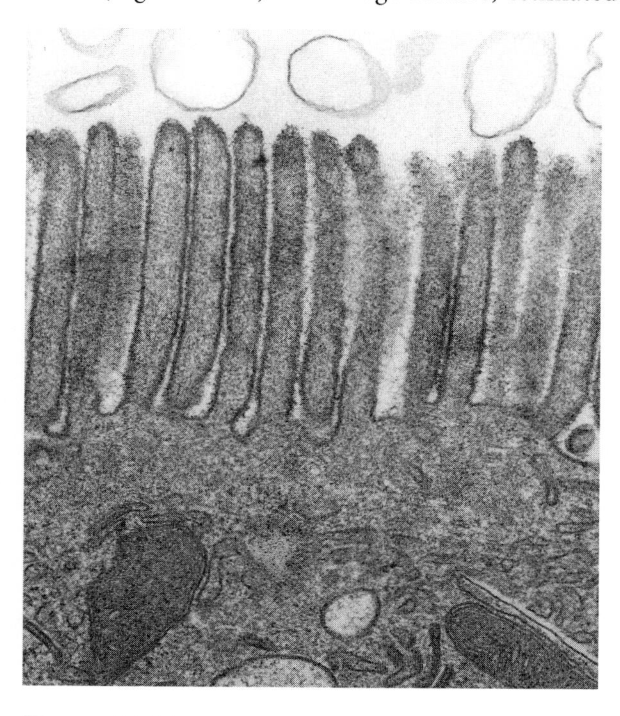

Figure 47.19
Microvilli along the border of an intestinal cell increase its absorptive surface enormously.

Notice that the microvilli are covered with an extensive glycocalyx, a fringe of oligosaccharides. This layer binds several of the intestinal enzymes, including maltase, sucrase, lactase, aminopeptidase, and an alkaline phosphatase.

Exercise 47.6 A football field is 100 yards long. A meter is approximately a yard. If you marked off a strip of a football field between goal lines, how wide would you have to make it to give it the same area as the small intestine?

B. Absorption and distribution of nutrients

47.11 The products of digestion are distributed to all tissues.

Small molecules transported through the intestinal surface enter the rich capillary beds inside its villi. Veins arising from the villi converge to form the **hepatic portal system** (Figure 47.20), which transports the absorbed food molecules to the liver. (A *portal* circulatory system is a venous system that runs between two capillary beds.)

During and after a meal, an animal is in the **absorptive phase** nutritionally, when it is digesting and absorbing food, so all cells usually have more than enough glucose. Excess glucose is first converted into glycogen and stored in muscles and liver cells. But cells have a limited capacity for storing glycogen, so considerable glucose is converted into triglycerides and stored as fat in adipose tissue. Table 47.1 shows the approximate energy stores in a 70-kg human. Notice that far more energy is stored as fat than as glycogen. This is efficient from the standpoint of mass alone, since the oxidation of 1 gram of fat yields about 9 kilocalories, while oxidation of a gram of carbohydrate yields only 4 kilocalories.

During the absorptive phase, all tissues can synthesize new proteins with the abundant amino acids derived from food proteins. Some excess amino acids are oxidized and used for energy,

Table 47.1 Energy Stores in a Normal 70-kg Human

Type of Material	Mass (kg)	Caloric Equivalent (kcal)
Neutral fats (adipose tissue)	15.0	141,000
Proteins (mainly muscle)	6.0	24,000
Glycogen (muscle and liver)	0.225	900
Circulating fuels (fatty acids, glucose, neutral fats, etc.)	0.023	100

Source: Data from A. L. Lehninger, *Biochemistry,* 2nd edition, 1975, Worth Publishers, New York.

Figure 47.20

The hepatic portal system carries material absorbed from the intestine to the liver, where much of it is removed before the blood goes on to the general circulation.

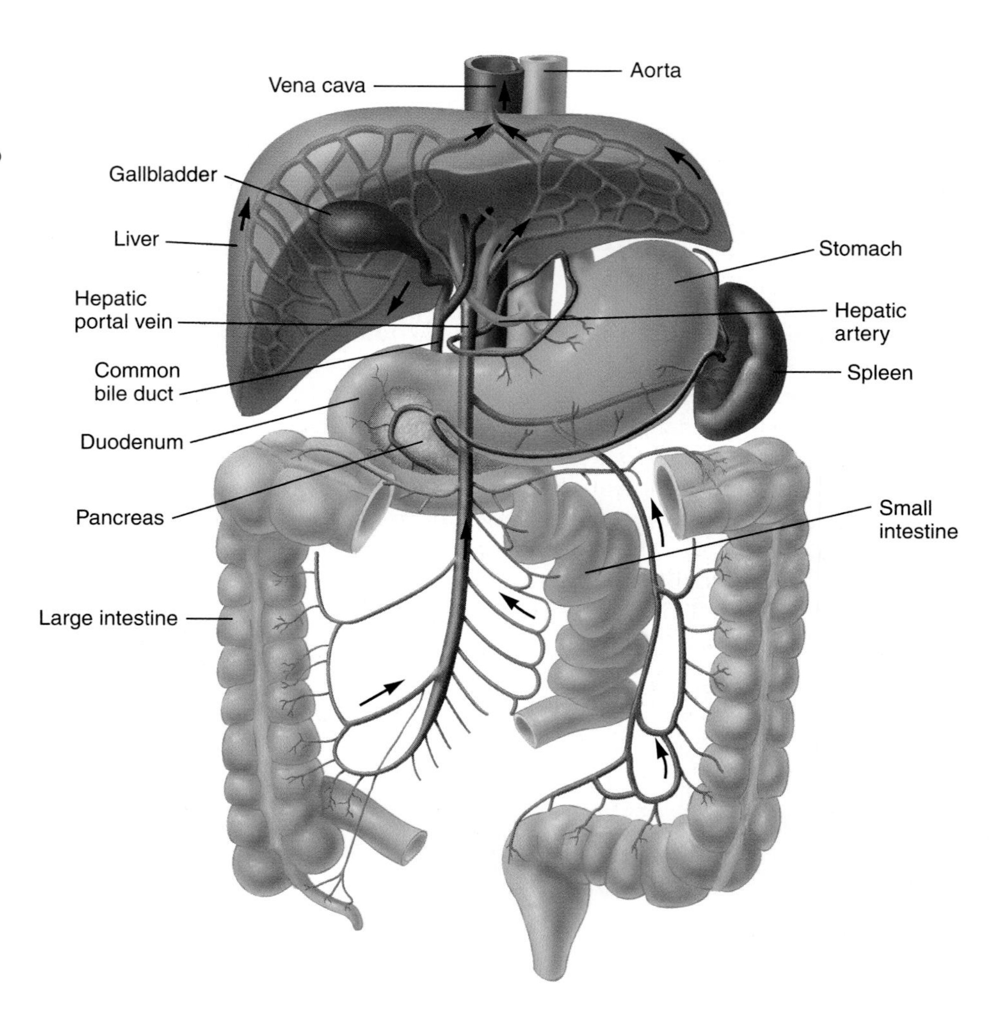

but most are converted to fatty acids and glycogen. This process too is largely carried out by the liver.

Catabolism of amino acids, Section 9.14

47.12 Lipids are transported in lipid-protein complexes.

In the intestine, neutral fats and phospholipids from foods are digested into fatty acids, glycerol, and the phospholipid bases. Intestinal epithelial cells that absorb these molecules resynthesize them into triglycerides for export to the circulation, to be carried to tissues where they are stored or metabolized. Lipids, being hydrophobic, can only be transported in blood plasma in soluble structures called **lipoproteins,** attached to carrier proteins that mask their hydrophobic regions. The largest lipoproteins synthesized in the intestinal mucosa are **chylomicrons,** balls of lipid and protein similar to a lipid bilayer with the hydrophobic ends of the lipids inside (Figure 47.21). Though too large to enter capillaries, chylomicrons are picked up by the lacteals in the intestinal villi and transported in the lymphatic system.

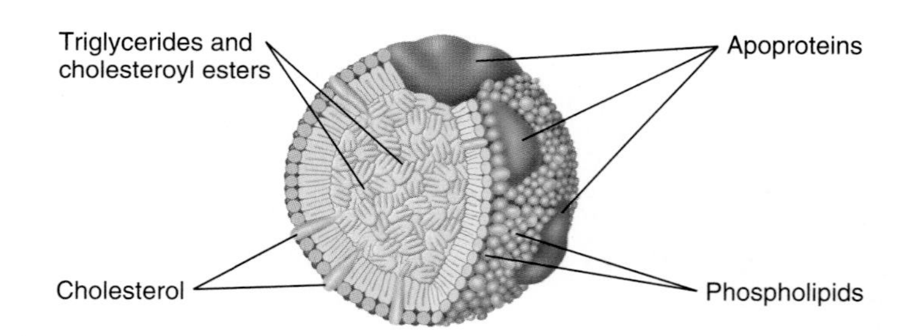

Figure 47.21

A chylomicron has the structure of a spherical lipid layer, about 0.1–0.5 μm in diameter. It is made of distinctive proteins, with triglycerides and cholesterol esters inside.

Other lipoproteins are identified as high-density (HDL), low-density (LDL), and very-low-density (VLDL), those with the most lipid being the least dense (Figure 47.22). (Remember that oil floats on water.) A lot of medical interest centers on the lipoproteins because of their role in cardiovascular diseases. *Atherosclerosis,* described in Section 45.9, is due to the abnormal deposition of cholesterol in lumpy plaques on the inner walls of blood vessels. Atherosclerosis is most common in people with

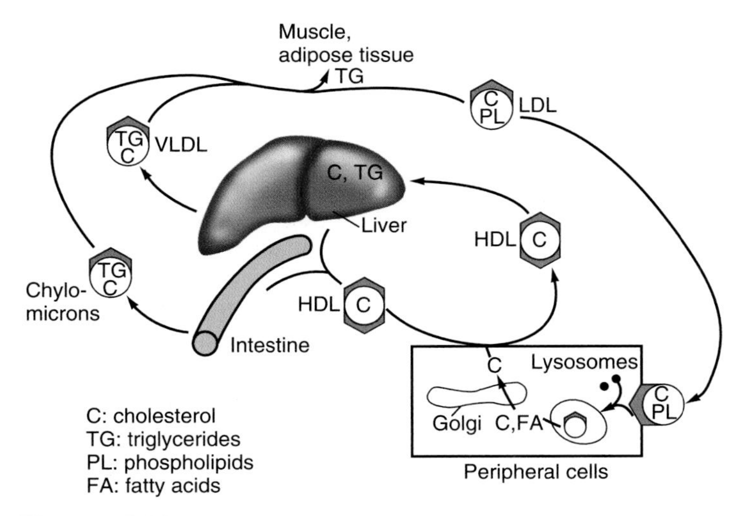

Figure 47.22

Four types of lipoproteins absorb and circulate lipids. Chylomicrons and very-low-density lipoproteins (VLDL) are both made in the liver; they carry triglycerides (TG) and some cholesterol (C) to muscle and adipose tissues, where the triglycerides are removed and either stored or oxidized for their energy. As their lipids are removed, VLDLs are converted into low-density lipoproteins (LDL), which carry cholesterol and phospholipids (PL) to other tissues that contain specific LDL receptors. The high-density lipoproteins (HDL) made in the liver and intestine contain considerable amounts of cholesterol, largely recovered from used chylomicrons and VLDL, and they transport cholesterol back to the liver, where it is converted into bile components.

high levels of LDL and less common in people with high levels of HDL. The HDL level is increased by aerobic exercise. Marathon runners have more HDL cholesterol than joggers, and joggers have more than people who don't exercise at all. HDL carries excess cholesterol to the liver, and the cholesterol in HDL is not deposited in arteries because they have no HDL receptors.

A diet high in fatty acids and cholesterol produces high LDL levels and is a major factor in cardiovascular disease. LDL levels are also highly elevated in people with *familial hypercholesterolemia,* an autosomal dominant condition; the allele responsible leads to reduced numbers of LDL receptors on liver cells, and consequently more cholesterol is deposited in blood vessels.

47.13 Metabolic patterns change during a fasting period.

Animals commonly go through periods when food is abundant, interspersed with times when they can find little or nothing to eat. Nevertheless, they must maintain adequate levels of energy sources in their blood, and regulate the levels of critical elements, such as calcium and iron. Well-adapted bodies have metabolic strategies for dealing with these differences by means of hormones that respond to different situations.

Having absorbed all the food from its intestine, an animal passes into the **postabsorptive phase** (or fasting phase) in which some processes that occur during the absorptive phase are reversed (Figure 47.23). Without any incoming glucose, glycogen stores in both muscle and liver cells start to break down. Glycogen stores, being small, supply energy for only a

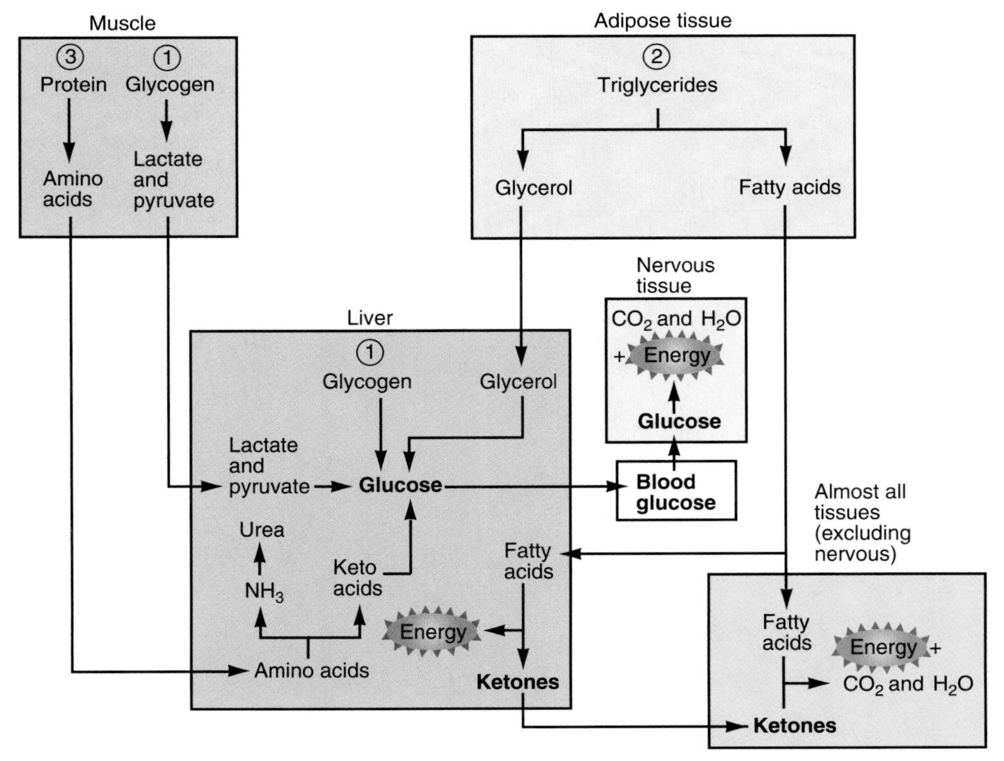

Figure 47.23

New paths of metabolism are employed during the postabsorptive phase. *(1)* Glycogen is used first, largely from stores in the liver. *(2)* Next, triglycerides stored in adipose tissue are converted into ketone bodies, which can be metabolized by many tissues, and into glycerol, which can be converted into glucose. The blood glucose level is maintained because the nervous system depends on it. *(3)* Proteins are also broken down into their amino acids, some of which are converted into glucose in the liver.

relatively short time. Then, under stimulation by glucagon and epinephrine, adipose tissue cells start to break down triglycerides, a process known as lipolysis. The resulting fatty acids are released into the bloodstream and carried to other tissues by *serum albumin,* which constitutes about half the protein in blood serum. Most tissues in the body then start to use the fatty acids for energy instead of glucose.

Fatty acids are oxidized to acetyl-CoA (see Section 9.12), which is usually oxidized further through the Krebs cycle. However, as cells continue to depend on fatty acids, they switch over and begin converting much of the acetyl-CoA into compounds called *ketone bodies:* acetone, acetoacetate, and β-hydroxybutyrate. The volatile acetone is discharged from the lungs, giving people who are fasting (or have uncontrolled diabetes) a characteristic odor on their breath; the other two ketone bodies are used by all tissues for energy. Even some neural tissue, which usually depends on glucose and cannot oxidize fatty acids, is able to switch over after prolonged glucose deprivation and use ketone bodies for about two-thirds of its energy.

An interesting evolutionary footnote comes from studies of the Sherpas of the high Tibetan mountains. In most people, cardiac muscle uses fatty acids as an energy source in preference to other nutrients, but the Sherpas' hearts preferentially use glucose, giving them greater endurance. Thus, even in the relatively short time since modern humans spread out across the world, particular groups have been able to evolve metabolic adaptations to the demands of their environments.

Humans generally think of fasting as a temporary time without food, but most animals can't predict when they will get more food. So after absorbing a meal, the animal body switches into a highly adaptive metabolic mode, preparing for the possibility that the next meal may be a long time in coming. Its metabolism slows down, and the body temperature of a homeotherm eventually falls. As lipid stores become depleted, some protein is broken down into its amino acids; those that can be converted into pyruvate are converted further into glucose, which is essential for maintaining the nervous system, and the other amino acids are catabolized for their energy. Protein breakdown isn't desirable, of course. It eventually means depletion of muscle proteins that are essential for quick action, and people in the last stages of starvation become blind and otherwise incapacitated. Still, a lot of protein can actually be catabolized without doing serious damage to the body.

47.14 The plasma glucose concentration is regulated by the action of several hormones.

Figure 47.24 summarizes some points we have already discussed and puts the regulation of the plasma glucose concentration into a broader context. The glucose pool in the plasma exchanges with liver and muscle glycogen stores, with fat, and with protein, and of course it is fed by incoming food. One additional factor in insulin metabolism is the level of dietary chromium. *Glucose tolerance factor,* a combination of chromium ion and the coenzyme NAD^+, is required for the action of insulin, and people living on highly refined modern diets may lack the chromium needed for normal carbohydrate metabolism.

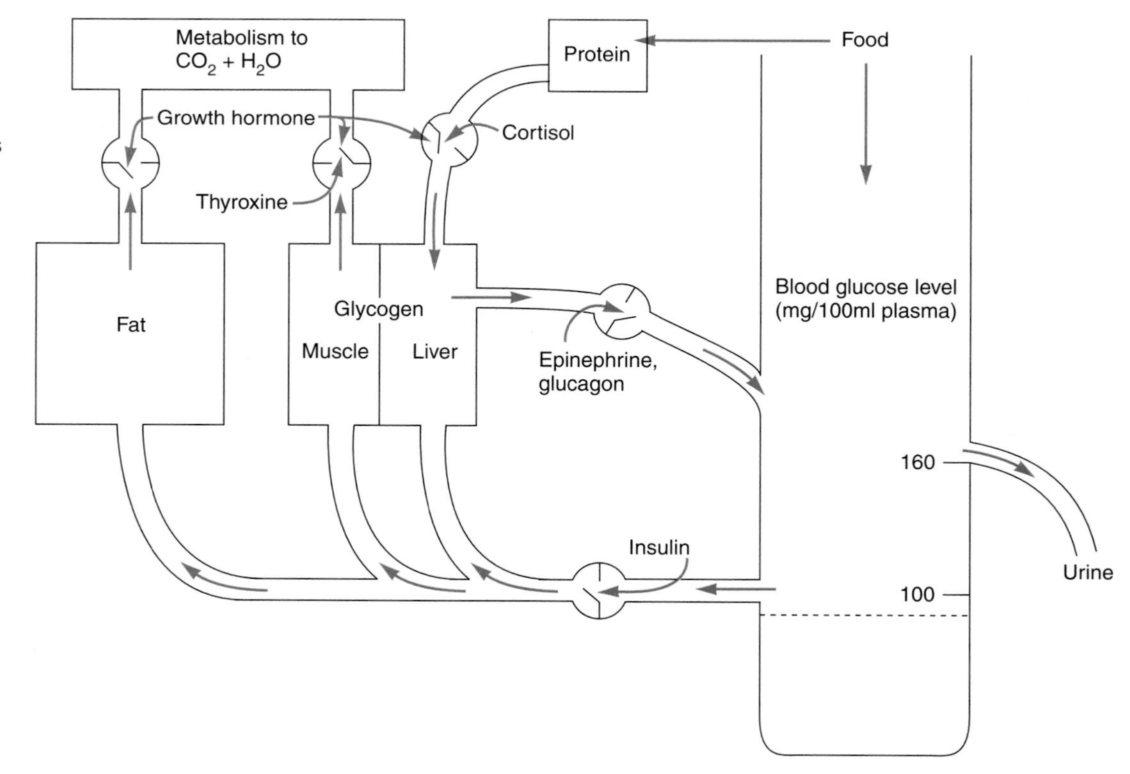

Figure 47.24
Stores of glycogen, fat, and protein, as well as incoming food, contribute to the plasma glucose level. Gates that control the various flows are regulated by several hormones: insulin, glucagon, epinephrine, thyroxine, and growth hormone. Insulin stimulates lipogenesis, the synthesis of lipid stores, as well as the storage of glucose in glycogen. Notice that when the glucose concentration becomes too high (more than 160 mg per 100 ml of plasma), it starts to spill over into the urine, producing one of the principal signs of diabetes.

Although insulin and glucagon are the major factors in regulating the plasma glucose level (see Section 41.3), several other hormones play important roles. Epinephrine (adrenaline) is released from both sympathetic neurons and the adrenal medulla under sympathetic stimulation (see Section 41.15). Cells in the hypothalamus of the brain monitor the plasma glucose level, and upon sensing too little glucose, they signal the release of epinephrine to accelerate the breakdown of glycogen in the liver. The same hypothalamic glucose receptors also signal the release of **growth hormone,** or **somatotropin,** from the anterior pituitary gland. Rather than stimulating glycogen breakdown, somatotropin has a kind of "anti-insulin" effect on all tissues except nervous tissue, inhibiting their uptake of glucose and thereby ensuring that it is spared for use by the brain. Somatotropin stimulates growth in all tissues that are capable of growth, and it inhibits the breakdown of proteins. Insulin stimulates cells to take up amino acids, as well as glucose, thus helping supply the raw materials for growth. Cortisol, one of the principal hormones of the adrenal cortex, has the opposite effect by promoting the breakdown of protein to amino acids and their conversion to glucose.

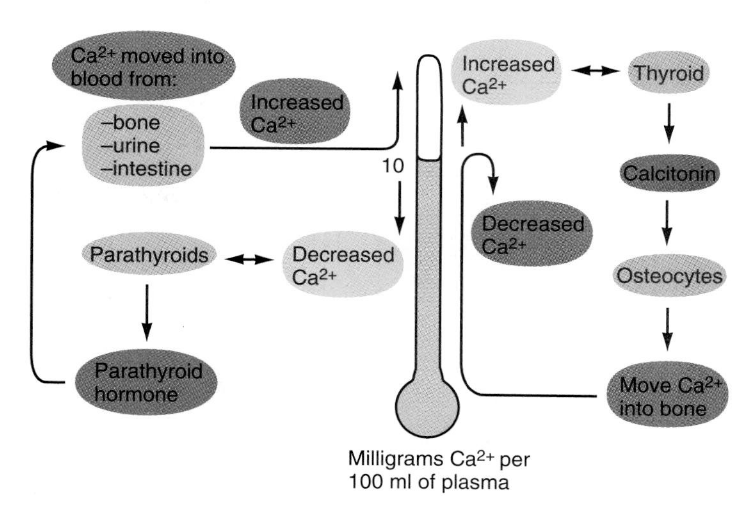

Figure 47.25

The blood calcium level is regulated by two hormones: calcitonin, which prevents the level from becoming too high, and parathyroid hormone, which prevents it from becoming too low.

47.15 Calcitonin and parathyroid hormone regulate the plasma calcium level.

Homeostatic mechanisms in mammals also maintain quite constant levels of various ions. The kidneys regulate the bulk of the ions in the body—Na^+, Cl^-, and K^+—and another pair of hormones, **calcitonin** and **parathyroid hormone (PTH),** keep the plasma calcium level quite constant. We have seen that Ca^{2+} ions are critical second messengers that regulate many processes, including muscle contraction. People who raise dogs, cats, and other domesticated animals sometimes see a nursing mother, who is excreting a lot of calcium in her milk, suddenly lie down and go into tetany, with her muscles twitching uncontrollably because they are so depleted of calcium.

About 99 percent of the body's calcium is built into bone, where most of it stays. But bone is actually a fluctuating reservoir of calcium (and phosphate), whose content can be raised and lowered; calcitonin and parathyroid hormone create a simple negative feedback system that keeps the plasma calcium pool at a constant level through their effects on bones, kidneys, and intestines (Figure 47.25). When the plasma calcium level rises above 10 mg per 100 ml, thyroid gland cells produce calcitonin (thyrocalcitonin), which stimulates osteoblasts, the bone-forming cells, to deposit more calcium phosphate as bone. Calcitonin also stimulates the kidney to excrete calcium and phosphate and the intestine to reduce the absorption of calcium ions. If, on the other hand, the plasma calcium level falls below 7 mg per 100 ml, the parathyroid glands produce parathyroid hormone, which produces the opposite effects: It stimulates osteoclasts, the cells that dissolve calcium phosphate in bone, to release more calcium into the blood. PTH also stimulates kidney pumps to reabsorb calcium so less calcium is excreted in urine, and it stimulates the intestine to absorb more calcium.

Calcium nutrition depends on a healthy exposure to sunlight. The absorption of calcium in the intestine requires a calcium-binding protein in the intestinal mucosa, and the synthesis of this protein is induced by an interesting hormone, dihydroxy-vitamin D_3. It is unusual to find a hormone that is also a vitamin, but D_3 can be obtained from many foods. D_3 is known as the sunshine vitamin, because skin cells synthesize it from cholesterol under the influence of ultraviolet light, and it is made into the dihydroxy form in the liver and kidneys (Figure 47.26). D_3 deficiency often manifests itself as the disease *rickets,* characterized by abnormal bone growth due to a lack of calcium. Rickets has been common in children with impoverished diets who often live in inner-city neighborhoods where large buildings block the sunlight they need.

Exercise 47.7 One of the two hormones that regulates the calcium level stimulates synthesis of the kidney enzyme that converts vitamin D_3 into its active form by adding one of the hydroxyl groups. Which hormone and why?

47.16 Iron is transferred and stored by two proteins, transferrin and ferritin.

Iron is an essential element for all organisms because it is an enzyme activator and a component of the electron-transport proteins, such as cytochromes. Animals that have hemoglobins, of course, require a lot of iron. An adult human body contains 3–5 grams of iron, about half of it in hemoglobin, a quarter in cytochromes and other iron-containing proteins, and the remaining quarter in storage. Iron circulates in the blood, from the intestine or from storage sites, combined with a protein appropriately named *transferrin* (Latin, *ferrum* = iron) by a

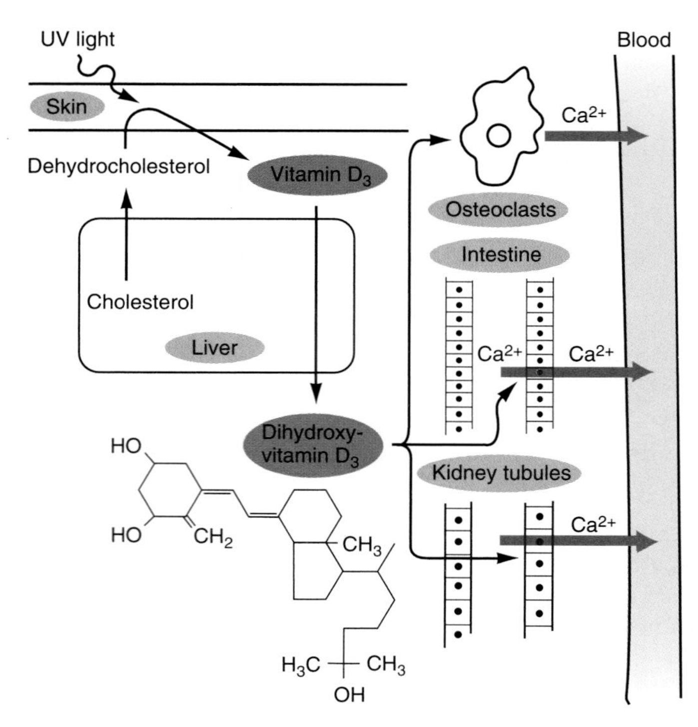

Figure 47.26

Dihydroxy-vitamin D_3 is needed for calcium absorption from the intestine. After being converted into its active form in the liver and kidneys, D_3 directly induces synthesis of the needed calcium-transport protein. D_3 also helps in the transport of calcium between the bones and blood plasma and in calcium retention by the kidneys.

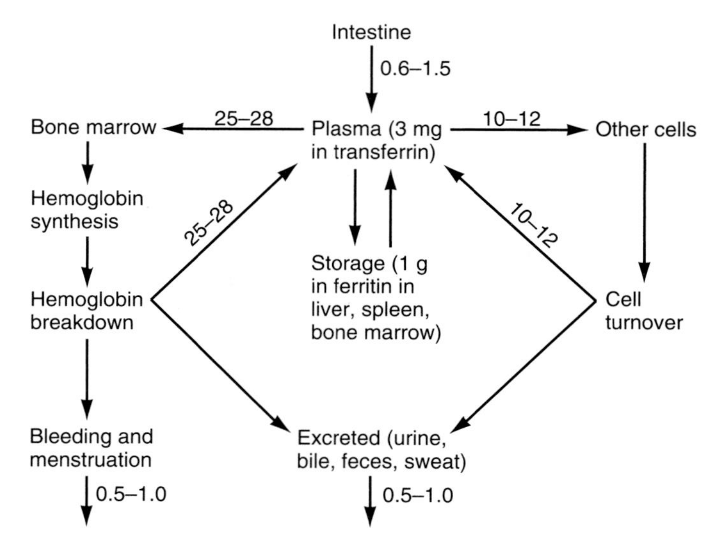

Figure 47.27

Iron moves between several compartments in the body; numbers are milligrams of iron transferred per day in an average adult human. The bulk of iron metabolism involves the hemoglobin in red blood cells, but a significant amount involves cell turnover generally, since iron is a component of the electron transport systems in mitochondria. Transferrin and ferritin are proteins that transport iron in the blood and store it in the liver, respectively.

punful biochemist (Figure 47.27). In this form, iron is carried to the bone marrow, the soft core of bone, where it is used by maturing erythrocytes. Iron is stored in the liver, spleen, and bone marrow as a complex with the protein *ferritin,* which makes large cubic crystals. Iron-ferritin complexes sometimes form large, membrane-bounded bodies called *siderosomes.*

Erythrocytes turn over rapidly (see Section 45.6). As old erythrocytes are digested by phagocytes, the heme groups in their hemoglobin are converted into green and orange pigments (biliverdin and bilirubin), which are removed by the liver and excreted into the intestines as components of bile. These pigments and others made from them are responsible for the green color of bile and the brown color of feces; they also contribute to the yellow color of urine, since they are partially excreted through the kidneys. In *jaundice,* a common symptom of liver disease, the skin and eyeballs acquire a distinct yellowish cast from excess bilirubin.

Erythrocyte destruction in humans releases 25–28 mg of iron per day, which is almost all recycled, so we lose no more than about 0.5–1 mg per day through urine, sweat, feces, and sloughed-off skin cells. Menstruating women lose an average of 0.5 mg more per day, although a few lose over 1 mg per day in this way. All the lost iron has to be replaced. An average diet supplies about 12–15 mg per day, of which only 5–10 percent is usually absorbed through the intestine. The rate of intestinal absorption is determined by the size of iron stores in the intestinal mucosa. If those stores decrease—as they do when there is a demand for iron elsewhere—more iron is absorbed to replenish them.

As iron is transferred between ferritin and transferrin, it is oxidized and reduced by two enzymes, one containing copper and one containing molybdenum. So normal iron metabolism requires these other trace elements, illustrating the subtle interactions among different nutrients.

C. Aspects of nutrition

47.17 All tissues are in a constant state of flux.

The tissues of all organisms are in a dynamic state, a state of constant turnover. This means that some molecules in the structure are continually being broken down and their atoms recycled or eliminated, while others are coming in to replace them. We can demonstrate this flux by labeling tissues with isotopic tracers such as ^{15}N and following the label over a period of time. The result is that cellular components always break down and are removed at characteristic rates, as shown in Table 47.2. These numbers are the average times for which a given material resides in a tissue; short times indicate rapid breakdown and resynthesis of some materials, while long times indicate slow turnover. Brain tissue, for example, turns over very slowly, and mature nerve cells are seldom if ever broken down and regenerated. Yet it has been estimated that 25 percent of the calcium in

Table 47.2 Turnover Times of Some Rat Tissues

Tissue	Residence Time (Days)
Liver	
Total protein	5–6
Glycogen	0.5–1
Lipids	1–2
Cholesterol	5–7
Muscle	
Total protein	30
Glycogen	0.5–1
Brain	
Neutral fats	10–15
Phospholipids	200
Cholesterol	> 100

Source: Data from A. L. Lehninger, *Biochemistry,* 2d edition, 1975, Worth Publishers, New York.

blood exchanges with the bone reservoirs every day. Overall, most of the cells in our bodies are being continually replaced by new ones, and their individual components are always turning over.

We are literally not the same people we were yesterday. Yet we retain a sense of continuity and an ongoing identity. An old man once claimed he owned an axe that had belonged to Abraham Lincoln. "'Course," he said, "it's had three new handles and two new heads since then." But we wouldn't think he was joking if he had said, "I was born 92 years ago in Kentucky and lived most of my life in Chicago." He has a continuous identity and is, in some sense, the same person, even though hardly an atom of his present body was there when he was a child. We understand that *dynamic identity,* identity in spite of dramatic changes in structure, is an aspect of being alive.

Exercise 47.8 This is a *thought experiment* (Gedankenexperiment). You don't actually perform it, just think about doing it. Then answer the questions at the end.

Weigh yourself on a very sensitive scale. Then perform each of these operations and weigh yourself again each time: eat, drink, urinate, defecate, sweat, sneeze, cough, vomit, inhale and exhale; get somewhat dehydrated and then drink some water to rehydrate yourself; cut yourself and bleed a little; get a haircut or shave; cry and let the tears flow; cut your fingernails and toenails; take a shower and rub off the surface skin; brush your teeth and floss well; have your teeth cleaned; menstruate or ejaculate; have your tonsils or appendix removed; go on an exercise program to build more muscle and lose several pounds of fat.

a. What is your *real* weight?
b. Which of the things you've lost or gained are really you?
c. What happens to the little bits of "you" that have been lost to the environment? Is the world full of little bits that were once you?

47.18 Animals must obtain eight of the twenty amino acids from their food.

Animals, in contrast to the rest of the biological world, have lost the ability to synthesize several *essential amino acids,* which must therefore be supplied in their food. For adult humans, these are methionine, threonine, leucine, isoleucine, valine, lysine, tryptophan, and phenylalanine, though cysteine and tyrosine can partly substitute for methionine and phenylalanine, respectively. In addition, histidine is essential for children and possibly for adults; infants also require arginine. Humans need only a few tenths of a gram to a little over one gram of each essential amino acid per day, but people beset by cultural, emotional, or economic restrictions may have difficulty getting the proper mixture if they depend on plant proteins. Wheat and other grains do not provide enough lysine and isoleucine; beans are deficient in tryptophan and methionine. Nevertheless, native diets have remarkable amino acid balance; without any knowledge of biochemistry, people have found combinations of foods that supplement one another's deficiencies, such as the corn and beans of some Latin Americans. Other cooking folkways, such as cooking in acidic solutions, promote the release of some amino acids from food. Central American natives have learned to prepare corn in highly alkaline solutions, which release much more of the available lysine and isoleucine and make corn a nutritious mainstay of their diet.

Aside from specific amino acid requirements, every animal needs enough bulk protein to maintain a proper **nitrogen balance** (Figure 47.28). A positive nitrogen balance means the body is taking in more nitrogen than it is excreting and there is some

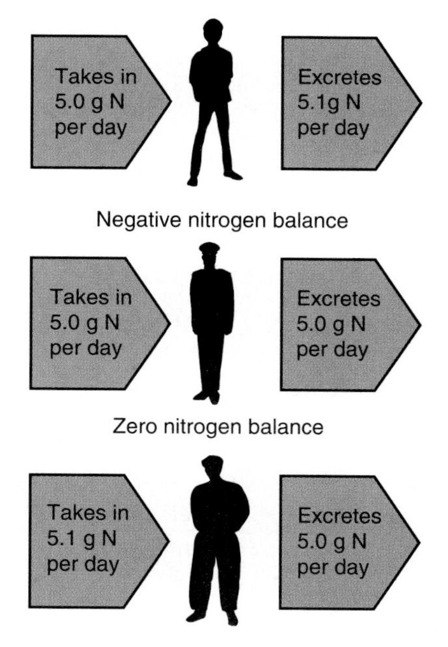

Figure 47.28
Nitrogen balance is a measure of the relative amounts of nitrogen entering and leaving an organism.

net synthesis of protein, the principal nitrogenous component of tissues. A negative nitrogen balance means more protein is being broken down and excreted than is being synthesized.

Most amino acids, either from food or from the breakdown of old tissue, are metabolized to glucose or fatty acids; their amino groups are removed and eliminated, primarily as urea through the urine. Unless those amino groups are replaced by new amino acids, there will not be enough for protein synthesis and the body will be in a negative nitrogen balance. A typical adult loses around 4.5 grams of nitrogen per day: 3 grams as urea, 1 gram through the feces, and a few tenths of a gram by sloughing skin and hair protein. This nitrogen must be replaced by consuming at least 28 grams of protein per day, since protein weighs 6.25 times the nitrogen it contains. In fact, the daily intake of protein recommended to maintain good health is even higher than this. One standard is 0.8 grams of protein per kilogram of body weight, or 56 grams of protein per day for a 70-kg person.

Kwashiorkor, an East African term, is a dietary protein deficiency characterized by edema, apathy, and modification of skin and hair. It is particularly prevalent in malnourished young children, who have characteristically bloated stomachs due to edema in the abdominal cavity. These children never recover completely even if their diets eventually improve. Certain effects are irreversible; since brain growth occurs in spurts, a protein-deficient child who can't make brain tissue at the right time will never be able to fill in the missing tissue and will never recover normal mental function.

Exercise 47.9 Suppose a piece of meat is 80 percent water and half the remainder is protein. How large a piece is needed to supply 10 grams of protein?

Exercise 47.10 Think about the process of protein synthesis. Imagine that the ribosomes in one of your growing cells are synthesizing protein, and suddenly they exhaust your supply of one essential amino acid. Describe the condition of the protein-synthesis apparatus.

47.19 Human diets have changed with civilization, generally for the worse.

We do not eat as our ancestors did. Yet we retain the metabolic apparatus that evolved over millions of years to process a certain diet, so in changing that diet through our civilized ways, we put great strains on our health. S. B. Eaton and M. Konner have estimated the content of human diets during three periods: the hunter-gatherer stage beginning a million or more years ago, the agriculturist stage beginning about 10,000 years ago, and the modern period of industrialized society starting about 200 years ago (Table 47.3).

People subsisting today on a typical European-American diet clearly obtain much less protein and much more fat than their remote ancestors did, and the fat is far more likely to be saturated than polyunsaturated (the P:S ratio in Table 47.3). Diets that contain more saturated fats are associated with hypercholesterolemia (high serum cholesterol) and atherosclerosis (see Section 47.12). Although we eat about the same amount of carbohydrate as our ancestors did, we now obtain a lot of it as simple sugar, which enters the bloodstream quickly, instead of complex starch, which enters more slowly. Our hormones have never evolved to handle the rapid changes in blood glucose associated with a modern diet, and excess refined sugar is probably one factor in the development of diabetes. Our consumption of dietary fiber has fallen, while our consumption of salt has risen sharply. Yet the fiber is essential for normal bowel function, while the excess salt strains renal mechanisms for ionic regulation and is correlated with the development of high blood pressure. The moral seems quite obvious.

Exercise 47.11 Given enough time, evolution would presumably solve the problems associated with a poor diet through adaptive changes in human metabolism. However, ignoring the long times required for evolutionary changes, explain why such adaptations are unlikely to occur in modern humans. (*Hint:* At what age do people generally die of the illnesses mentioned in the preceding section?)

47.20 The balance of nutritional factors is complex.

In response to people's growing awareness of the importance of nutrition, many self-styled experts now claim to know just what you should eat to stay healthy forever. Frequently they attribute human health to one simple factor or another: All we have to do is eat enough calcium or enough vitamin E or restrict our intake of saturated fats, and we will be a nation of supermen and superwomen. A good understanding of nutrition, however,

Table 47.3 Human Diets During Three Historical Periods

Nutrient	Hunter-Gatherer	Agriculturist	Modern
Protein (%)	30	10–15	15
Fat (%)	15	15	40
Carbohydrate (%	55	70–75	45
Simple (%)	5	5–10	20
Complex (%)	50	65–70	25
Dietary fiber, (g/day)	40	80–100	15–20
Salt (g/day)	1–3	3–5	10
P:S ratio	1.4	1.6	0.5

requires a specialized course of study, and even then a lot of unanswered questions remain. Nutrition is one of the youngest biological sciences. A great deal is yet to be learned, and nutritional factors are so complex that there are always surprises. For instance, we know something about calcium metabolism and about amino acid metabolism, but these separate bits of knowledge do not explain why a high intake of protein (approaching 100 grams per day) increases calcium excretion and produces a negative calcium balance, which cannot be made up no matter how much calcium is supplied in the diet. This means that Americans, who eat very large amounts of protein, gradually deplete their bones of calcium, and it partially explains why older people in our society have a high incidence of osteoporosis, a condition in which the bones have lost so much calcium that they are porous and badly weakened. Still, we don't know the metabolic connection between the two nutritional factors.

With all we know about protein metabolism, we still do not understand why people who take in less nitrogen per day require smaller amounts of the essential amino acids, yet this seems to be so. People who consume a total of 5 grams of nitrogen per day in their diets require about 50–400 mg of each amino acid, while those who receive 9 grams of nitrogen per day require three to four times as much.

Even as we learn about the role of glucose tolerance factor in sugar metabolism, we do not fully understand the need for chromium in the human diet and its relationship to the various lipoproteins—yet there is one. Why does β-carotene, taken as a dietary supplement, interfere with vitamin-E metabolism? We don't know. The point is that complex animals have complex nutritional requirements, and the road to personal health lies partially in understanding complex nutritional interactions. People who have studied these matters carefully give balanced, cautious advice, but it is essential to be skeptical of books on health, particularly those on nutrition; some of them can contain dangerous nonsense. In this chapter we have tried to lay a foundation that will help you separate the sense from the nonsense.

Coda Digestion is the essence of being an animal. We showed in Chapter 32 that the basic features of plants and animals stem from their fundamentally different modes of nutrition, and animals are holotrophs—heterotrophs that ingest and digest food. An animal's anatomy centers around its GI tract, a tube within the body-wall tube in all but the most primitive animals, and its embryology largely revolves around cellular movements that form that inner tube.

As holotrophs, animals have evolved a characteristic and specialized metabolism. Since their food always contains a mixture of amino acids, they could afford to lose the complicated metabolic pathways needed to synthesize several of the amino acids. They have elaborated and diversified the digestive enzymes of their ancestors, and have evolved ways to store nutrients, since their food intake may swing unpredictably between feast and famine.

As consumers, animals obviously have critical roles in the planet's ecology. This raises the question of whether a biosphere could evolve without consumers. Suppose phototrophs evolve on a planet. If decomposers do not evolve, the phototrophs will be buried in their own waste, so a stable biosphere appears to require at least producers plus decomposers. Since the first organisms to evolve out of a primordial soup must be heterotrophs, the ancestral decomposers are already present. But once organisms have evolved that can digest biomolecules of detritus, there is nothing to keep some from evolving into organisms that digest the biomolecules of living tissue, so parasites and predators seem inevitable. These speculations suggest that if organisms have evolved on other planets, their biota probably includes animals, very likely ones who live much as animals do on Earth.

Summary

1. A typical animal digestive system is a long tube with both a mouth and an anus. Along its length are specialized regions and accessory glands that secrete digestive enzymes. The walls of the intestine then absorb the products of digestion.
2. An animal eating its food is engaging in one of the principal activities that structure an ecosystem by consuming certain species and by preferring some over others.
3. Animals obtain their food in a variety of ways. Some suck juices from plants or animals; others are filter feeders that strain particulate and dissolved food from water. Most animals bite or tear their food. Because the size of an animal's mouth determines what it can eat, this is an important way in which the food resources in an ecosystem are divided.
4. Solid food is broken down mechanically before digestion, either by means of teeth or other similarly hard structures or by means of muscular pouches, such as the gizzard of birds.
5. In the human system, food that has been mixed by chewing is carried down the esophagus by waves of contraction called peristalsis. It is mixed in the stomach with digestive juices to make a thick chyme.
6. Food then passes into the intestine, where it receives bile from the liver and additional enzymes from the pancreas and the intestinal walls. Small molecules released by these enzymes are absorbed by the walls of the small intestine. Water is absorbed last, in the large intestine.
7. Carbohydrate digestion begins with salivary amylase, which attacks starch and glycogen. A similar amylase is made by the pancreas. The disaccharides sucrose, maltose, and lactose are digested by intestinal enzymes. Few organisms can digest cellulose.
8. Protein digestion begins with pepsin in the stomach and is continued by several other enzymes in the intestine, which attack peptide linkages between particular amino acid residues.
9. Many digestive enzymes are made in an inactive proenzyme form and are then activated by removal of small peptides.
10. The activities of the GI tract are regulated largely by a local enteric nervous system centered in two plexuses. This system regulates movement and secretion through local intrinsic circuits; longer extrinsic circuits through the central nervous system are superimposed on the enteric system.
11. Secretion of digestive enzymes and accessory fluids is controlled by a few simple hormonal circuits, starting in the stomach with a positive feedback loop that includes gastrin production and protein fragments produced by digestion.
12. Acids and specific food molecules entering the intestine stimulate the secretion of the hormones secretin and cholecystokinin, which in turn

stimulate the secretion of bicarbonate ions and enzymes, respectively, from the pancreas.

13. Bile from the liver helps to emulsify lipids, which are digested by other pancreatic enzymes.
14. The intestinal wall absorbs monomeric units (sometimes dimers and trimers) from the intestine through highly selective carriers. Its surface is increased manyfold by fine villi and by microvilli on the surfaces of the villi. Food absorbed there is carried by the hepatic portal circulation to the liver.
15. During the absorption phase, the sugars are used for metabolism or are stored as glycogen or lipids in the liver, muscles, and adipose tissue. Amino acids are used for protein synthesis or are converted into energy stores.
16. In the postabsorptive phase, stores of glycogen and lipid are broken down for their energy, and protein may also be catabolized as an energy source.
17. Some materials are carried by special plasma proteins. Lipids are distributed through the blood by means of carrier proteins (lipoproteins) and chylomicrons.
18. The plasma glucose level is kept relatively constant through exchanges with glycogen and fat stores, supplemented by protein. The system is regulated primarily by insulin and glucagon, supplemented by cortisol and growth hormone. Somatotropin spares glucose for use by the brain and indirectly stimulates growth. When the concentration of amino acids in plasma increases, insulin stimulates their uptake and somatotropin stimulates their use in protein synthesis.
19. The level of plasma calcium is maintained by two other hormones. Calcitonin reduces the calcium level by stimulating osteoblasts to remove calcium; parathyroid hormone raises the calcium level by stimulating osteoclasts to break down stores of calcium in bone. These hormones also affect calcium excretion in the kidney and calcium absorption by the intestine. Calcium uptake from the intestine depends on dihydroxyvitamin D_3.
20. Although large numbers of red blood cells, hence large amounts of iron, turn over every day, humans require only small amounts of new iron. Iron is stored and transported by the proteins ferritin and transferrin, respectively.
21. All tissues are in a constant state of flux, as their constituent molecules turn over at different characteristic rates. Red blood cells are destroyed after a lifetime of about 120 days, and their constituents are recycled or excreted. Yet each structure maintains its integrity in spite of this turnover.
22. Animals require several essential amino acids, which they cannot synthesize themselves, and they require new amino acids of all kinds to support tissue turnover.
23. The diet of people living in modern industrialized society differs significantly from the ancestral human diet. Human physiology evolved to regulate the ancestral diet, and various functional disease states are associated with the modern diet.
24. Minimum amounts of each nutrient can be recommended, but the interactions between dietary factors are complex and poorly understood. Dietary regimes should be based on a careful evaluation of this information, not on simplistic recommendations from self-styled experts.

Key Terms

gastrointestinal (GI) tract 995
alimentary canal 995
assimilate 995
filter feeder 996
esophagus 999
stomach 999
peristalsis 999
chyme 999
small intestine 999
large intestine 999
colon 999
rectum 1000
amylase 1000
protease 1000
pepsinogen 1000
proenzyme 1001
zymogen 1001
enteric nervous system (ENS) 1002
nerve plexus 1002
gastrin 1002
secretin 1003
cholecystokinin (CCK) 1003
bile 1003
gallbladder 1003
enterohepatic circulation 1003
villus 1005
microvillius 1005
hepatic portal system 1005
absorptive phase 1005
lipoprotein 1006
chylomicron 1006
postabsorptive phase 1007
growth hormone/somatotropin 1009
calcitonin 1009
parathyroid hormone (PTH) 1009
nitrogen balance 1011

Multiple-Choice Questions

1. Mechanical breakdown of food is one function of the
   a. crop.
   b. liver.
   c. gizzard.
   d. rumen.
   e. serosa.
2. Filter-feeding animals are generally
   a. large.
   b. aquatic.
   c. air-breathing.
   d. predatory.
   e. radially symmetrical.
3. Which of these mixtures is a noncellular secretory product?
   a. chyme
   b. mucus
   c. feces
   d. bolus
   e. rumen
4. All but _____ takes place within the stomach immediately after a meal.
   a. hydrolysis of proteins
   b. mechanical breakdown
   c. emulsification of lipids
   d. acidic sterilization of chyme
   e. stimulation of gastrin secretion
5. All are functions of the liver except
   a. bile synthesis.
   b. glycogen storage.
   c. secretion of digestive enzymes.
   d. deamination of amino acids.
   e. storage of lipid-soluble vitamins.
6. Which of the following is a hormone that is secreted by the organ that is also its target?
   a. amylase
   b. gastrin
   c. CCK
   d. secretin
   e. bile
7. Which of the following is a small intestinal hormone that targets the pancreas?
   a. secretin
   b. CCK
   c. gastric inhibitory peptide

d. *a* and *b*, but not *c*
e. *a*, *b*, and *c*

8. Bile is synthesized and secreted by the ______ and is stored in the ______.
a. liver; gallbladder
b. gallbladder; liver
c. liver; pancreas
d. pancreas; liver
e. liver; small intestine

9. The hormone ______ promotes lipolysis, whereas the hormone ______ promotes lipogenesis.
a. insulin; glucagon
b. glucagon; insulin
c. epinephrine; albumin
d. insulin; epinephrine
e. glucagon; epinephrine

10. Which endocrine organ is instrumental in regulating calcium but not glucose?
a. parathyroids
b. pancreas
c. thyroid
d. adrenal medulla
e. adrenal cortex

True-False Questions

Mark each statement true or false, and if false, restate it to make it true.

1. Molar teeth primarily function for biting and tearing.
2. Carbohydrates are hydrolyzed in all regions of the GI tract except the large intestine.
3. Parietal cells of the gastric mucosa secrete pepsin, whereas chief cells secrete hydrochloric acid.
4. Serum albumin functions as a transport protein to carry other proteins within the blood.
5. Lipogenesis, the formation of lipids from amino acids and monosaccharides, primarily occurs in the absorptive phase.
6. Iron, in metallic form, is transported by the blood to bone marrow, where iron-ferritin complexes are formed.
7. Glycogen stores in the liver and muscles are depleted during the postabsorptive phase and replenished during the absorptive phase.
8. Chromium deprivation is associated with elevated blood glucose levels because glucagon cannot work properly in the absence of chromium.
9. High-protein diets seem to be associated with osteoporosis.
10. Several amino acids are called "essential" because they are only obtained by synthesis in all body cells.

Concept Questions

1. Would you expect a dog to have a GI tract that was proportionately shorter or longer than that of a rabbit? Explain your answer.
2. Explain why natural selection has resulted in the secretion of protein-hydrolyzing enzymes in inactive form, whereas enzymes that digest carbohydrates are active when secreted.
3. Generally, digestive functions remain intact in patients whose spinal cords have been severed. Suggest a mechanism whereby digestion is controlled and regulated.
4. Explain the interactions of insulin, glucagon, epinephrine, cortisol, and somatotropin in the regulation of blood glucose levels.
5. Describe the feedback loops that regulate the level of blood calcium.

Additional Reading

Atkinson, Mark A., and Noel K. Maclaren. "What Causes Diabetes?" *Scientific American,* July 1990, p. 62. For insulin-dependent diabetes, the answer is an autoimmune ambush of the body's insulin-producing cells. Why the attack begins and persists is now becoming clear.

Baldwin, R.L. "Digestion and Metabolism of Ruminants." *BioScience,* April 1984, p. 244.

Bels, V. L., M. Chardon, and P. Vandewalle (eds.). *Biomechanics of Feeding in Vertebrates.* Springer-Verlag, Berlin and New York, 1994. A collection of papers about the comparative anatomy and mechanics of an apparently simple process that is not really so simple.

Block, Eric. "The Chemistry of Garlic and Onions." *Scientific American,* March 1985, p. 114. A number of curious sulfur compounds underlie the odor of garlic and the tears brought on by slicing an onion. The compounds also account for medical properties long ascribed to garlic and onions.

Cerami, Anthony, Helen Vlassara, and Michael Brownlee. "Glucose and Aging." *Scientific American,* May 1987, p. 90. Once considered biologically inert, the body's most abundant sugar can permanently alter some proteins. In doing so, it may contribute to age-associated declines in the functioning of cells and tissues.

Chivers, D. J., and P. Langer (eds.). *The Digestive System in Mammals: Food, Form, and Function.* Cambridge University Press, Cambridge and New York, 1994. Papers from a workshop about the physiology of the mammalian intestinal system.

Lawn, Richard M. "Lipoprotein(s) in Heart Disease." *Scientific American,* June 1992, p. 26. Some additional information about the structure of lipoproteins and their medical implications.

Radetsky, Peter. "Gut Thinking." *Discover,* May 1995, p. 76. Anthropologist Katherine Milton discovered that spider and howler monkeys in the same environment in Panama have very different brain sizes. She attributes this to the effects of their different diets on their evolution.

Sanderson, S. Laurie, and Richard Wassersug. "Suspension-feeding Vertebrates." *Scientific American,* March 1990, p. 96. Animals that filter their food out of the water can reap the abundance of plankton and grow in huge numbers or to enormous size.

Wilkinson, Gerald S. "Food Sharing in Vampire Bats." *Scientific American,* February 1990, p. 76. A buddy system among vampire bats ensures that food distribution among the bats is equitable.

Internet Resource

To further explore the content of this chapter, log on to the web site at:

http://www.mhhe.com/biosci/genbio/guttman/

48 Animal Defense Systems: Inflammation and Immunity

The miserable feeling of an infection is largely the result of the body's secondary defense systems working to eliminate a virus or microorganism that has eluded its primary defenses.

Key Concepts

A. Nonspecific Defenses and Inflammation

48.1 Animals have several nonspecific lines of defense against infection.

48.2 Acute inflammation, a general reaction to injury or infection, is another nonspecific line of defense.

48.3 Inflammation occurs through reaction cascades and chemical signals exchanged between blood cells.

48.4 Blood clotting and other cascades are essential components of inflammation.

B. The Immune System

48.5 The immune defense system supplements inflammation.

48.6 The immune response occurs in lymphoid organs and tissues.

48.7 All immunoglobulins share a common basic structure.

C. The Development of Immunity

48.8 Self is determined primarily by proteins of the major histocompatibility complex.

48.9 Each clone of differentiated lymphocytes makes antibodies with only a single type of antigen-binding site.

48.10 Immunoglobulins are encoded by several gene segments.

48.11 Immune and inflammatory processes are closely linked to the nervous system and endocrine controls.

D. The Dark Side of Immunity

48.12 The AIDS virus attacks the heart of the immune system.

48.13 An animal's immune system sometimes turns against it.

As you wake up in the morning, you become aware of a familiar, miserable feeling. An odd ache pervades your muscles. Your throat is scratchy, rolling your eyes brings on an ache deep behind them, and you probably have a fever. Your fondest wish is to go back to sleep, or maybe die. You have an infection. Some virus, probably, has slipped past your body's defenses and is now multiplying at your expense. Having weathered similar infections and recovered, you can be quite confident that you will survive this one too. As a reader of this book, you probably live in an industrialized country, most likely in North America, and so are likely to encounter pretty benign viruses. This one has circumvented some of your defenses, but you have others that will stop it from going further. In many other countries, though, you could be experiencing the first symptoms of a disease that would wrack your body terribly and eventually kill you.

The world is a tough, dangerous place, and one of the facts of life is that organisms die, not just from old age as their parts wear out, but from injuries and infections. Not only are animals in constant danger of bleeding to death from wounds, but viruses, bacteria, fungi, and worms besiege them and seize every opportunity to grow in their tissues. Over the eons, however, many animals have evolved complex defense systems to ward off dangers. We will discuss some of these defense mechanisms in this chapter, again concentrating on mammals, especially humans. ■

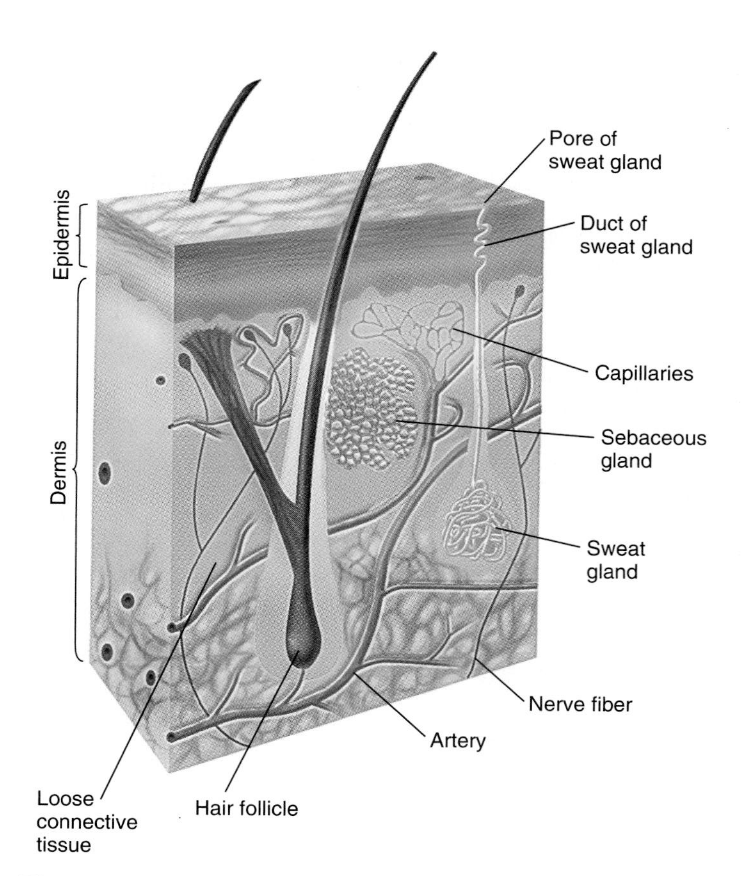

Figure 48.1

A cross section through human skin shows the structure of its sweat glands and sebaceous glands, which secrete a fatty sebum. Both secretions contain antimicrobial substances.

A. Nonspecific defenses and inflammation

48.1 Animals have several nonspecific lines of defense against infection.

Why aren't we all infected all the time?

The world is full of infectious agents that cause diseases at every opportunity. They live all over our bodies, on our skin, in our intestines, amd in our body openings. Even if we make peace with our own microbes, every outside social contact exposes us to new ones. It's a wonder we are ever well! But in fact, we fight infections through *nonspecific defenses* against all potential pathogens, especially one called the *acute inflammatory response,* and through *specific defenses* against particular organisms or viruses via the *immune system.* We will see that the immune system differs principally from nonspecific defenses in having a molecular memory that can react to foreign molecules and remember which ones it has seen. We begin with the nonspecific means of defense.

The first line of defense against infection is *mechanical* and *chemical.* Mammals are covered with **skin,** the body's largest organ and perhaps the least appreciated. Its water-resistant surface is made of tough, dead cells that are always being rubbed off and replaced by new cells growing from the epidermis below. Skin cells are sealed together by *tight junctions,* forming an impenetrable barrier to microorganisms that would find the moist, nutrient-rich tissue just below a delightful place to live. Burn victims who have lost large areas of skin are subject to infection as well as dehydration, since the skin also retains body fluids while excluding invaders. **Sebaceous glands,** or oil glands, produce an oily **sebum** rich in waxes, fats, free fatty acids, and cholesterol, which inhibits the growth of fungi and bacteria on the skin (Figure 48.1). Sebaceous glands are concentrated in folds, in crevices, and around body openings, where the skin is likely to be moist and there is danger of invasion by resident microorganisms. In general, infectious agents only get through the skin when it is broken.

Would-be invaders who enter by other routes meet different barriers. The ciliated epithelia of the nasal passages, mouth, and bronchi are covered with mucus that traps viruses and microorganisms, and the cilia sweep them outward or into the digestive tract, where they are digested by the acid and hydrolytic enzymes of the stomach. Mucous secretions and tears contain hydrolytic (digestive) enzymes that attack bacteria and viruses, such as lysozyme that hydrolyzes the murein of bacterial cell walls. The acidic mucous secretions of the vaginal canal are barriers to some invaders. Furthermore, coughing, sneezing, and vomiting reflexes expel noxious materials, including microorganisms.

Bacterial cell walls and mureins, Section 29.5.

Pathogens that surmount these barriers and invade a tissue meet a second line of nonspecific defense: They are attacked and eaten by an army of white blood cells broadly called **phagocytes** (*phago-* = to eat; *-cyte* = cell; thus, cells that eat) that patrol the lymphatic system, lungs, and other spaces (Figure 48.2). Phagocytes engulf foreign cells and viruses and digest them

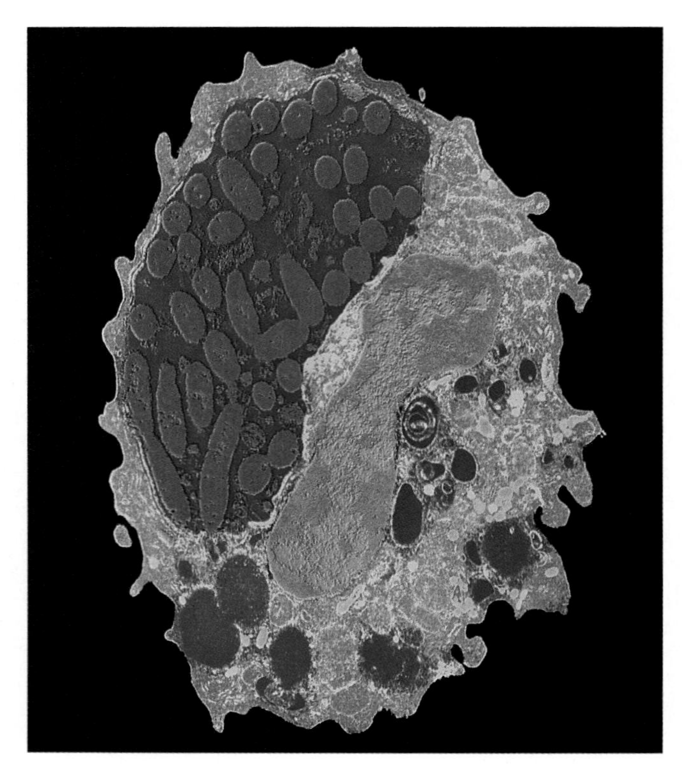

Figure 48.2

A phagocytic white blood cell (neutrophil), as seen by SEM, has a complicated surface with which it detects and engulfs bacteria.

with their lysosomal enzymes. They are particularly called into an area of infection during inflammation.

Skin, chemical secretions, and an army of phagocytes are usually enough to prevent infections, but if they fail, one of the body's most complex reactions, acute inflammation, is set in motion.

48.2 Acute inflammation, a general reaction to injury or infection, is another nonspecific line of defense.

As you are cutting up some vegetables for dinner, your knife slips and cuts your finger. Not a serious cut, but it stings and starts to bleed. You quickly wrap a tissue around it, and after a few minutes the bleeding stops, but you will be aware of the cut for several days. The area hurts, and soon it gets red, slightly swollen, and warm to the touch. You are observing a classical **acute inflammation** with its four cardinal signs as described by the Roman physician Celsus in the first century A.D. (Figure 48.3). You will also favor the finger for a few days and so experience the fifth sign, some loss of function, noted by the Greek physician Galen a century later. In making the cut, the knife has disrupted capillary beds, destroyed cells, and severed connective tissues. It has also forced a variety of bacteria into your tissues, where they could easily grow. Acute inflammation is designed to fight the invaders and restore the tissue. Although you will probably apply antiseptic and a small bandage to keep the wound clean, your body has its own protective mechanisms, which we will now explore.

Figure 48.3

The classical signs of acute inflammation are heat, redness, swelling, pain, and loss of function.

We hardly notice little cuts and scrapes. The inflammatory process heals them quickly. As a reaction to traumas such as cuts and burns, inflammation is usually a mechanism of **prophylaxis**—that is, it protects against disease (Greek, *pro-* = before; *phylax* = guard). But the system is a two-edged sword that cuts both ways. The mechanisms that so effectively prevent disease can sometimes turn against the animal they are designed to protect, especially by provoking a severe allergic reaction called **anaphylaxis** that entails labored breathing, shock, and even death (see Section 48.13). Even normal inflammation destroys some tissue to save the rest, and some symptoms of disease, such as pain and fever, are partially side effects of inflammation.

Shock, Sidebar 45.2.

An acute inflammation is basically a reaction in the microcirculation of the injured area. The arterioles dilate (expand), and the venules constrict, thus restricting blood flow so that blood accumulates in the region; this explains the redness and heat. (These reactions, called *vasodilation* and *vasoconstriction,* are discussed in Section 45.5). The capillaries become more permeable, so blood plasma exudes into the surrounding tissue, carrying many beneficial proteins. This explains the swelling. Some substances in the exudate affect nerve endings, causing pain. The junctions between capillary cells also loosen, and leucocytes (white blood cells) migrate into the tissue to fight invading microorganisms. So the observed changes are quite simple; the mechanisms underlying them, however, are marvelously complex, and it is a wonder that such complicated events can be going on inside us just to keep pathogens from taking over.

Inflammation depends on specialized leucocytes (Figure 48.4), which reside primarily in connective tissues and lymphoid tissues (see Figure 48.13) and also migrate through the lymphatics and blood. Human blood normally contains about 7,000 leucocytes per mm^3, the standard clinical measure. **Lymphocytes,** the key cells of the immune system, are relatively small. **Monocytes** are medium-sized phagocytes that may wander among tissue cells, in which case they become greatly enlarged and are called **macrophages.** The reticuloendothelial system is a series of macrophages that stay fixed in place, lining the sinuses (cavities) of lymph nodes, bone marrow, liver, and spleen, where they actively engulf foreign debris, removing bacteria and other particles.

Granulocytes are named for their distinctive cytoplasmic granules, which are actually lysosomes or vesicles of special materials; the cells are named for stains they take up that color these granules distinctively. By far the most abundant leucocytes—about half of the total—are **neutrophils,** which show a neutral reaction by taking up both basic and acidic dyes. **Basophils,** which take up basic dyes, secrete materials that are essential in inflammation; **mast cells** appear to be basophils that have left the circulation and reside in tissues where they may be needed to start an inflammation. **Eosinophils** take up the acidic dye eosin; they modulate inflammation by secreting substances that directly antagonize materials made by the basophils. Granulocytes are also called *polymorphonuclear leucocytes (PMNs)* because their nuclei take on many forms (*poly-* = many; *morph-* = form).

48.3 Inflammation occurs through reaction cascades and chemical signals exchanged between blood cells.

Acute inflammation begins when an injury stimulates two events more or less simultaneously, as shown in Figure 48.5. Injuries expose the blood to connective-tissue molecules it normally does not come in contact with, activating a plasma protein called *Hageman factor;* activated Hageman factor, in turn, initiates three reaction cascades. Remember that a cascade is a series of reactions that amplify a small initial stimulus into a large final effect (see Section 41.4). We'll come back to these cascades shortly. The injury also stimulates the mast

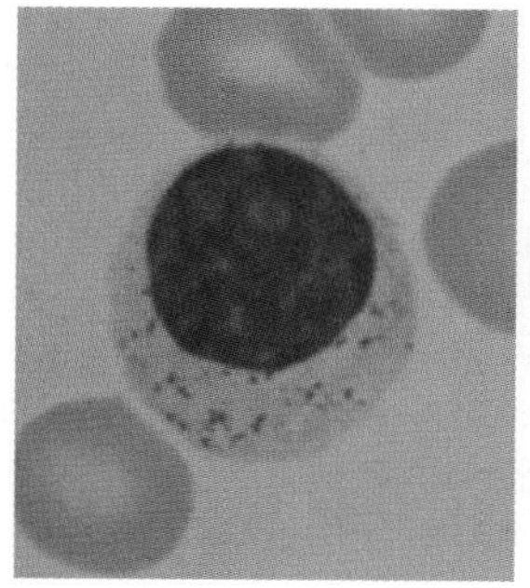

(a) Lymphocyte

(b) Monocyte

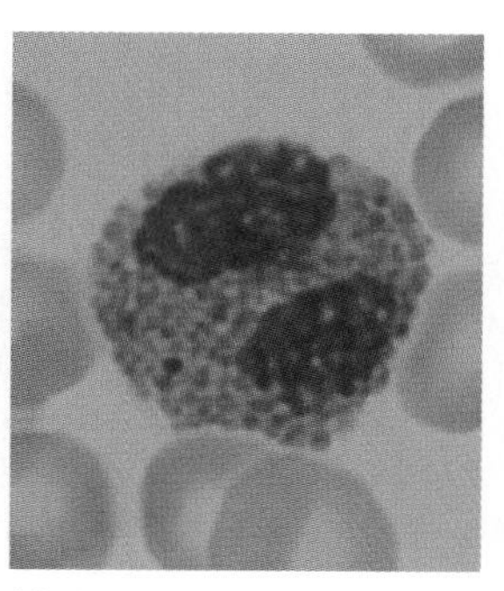

(c) Eosinophil

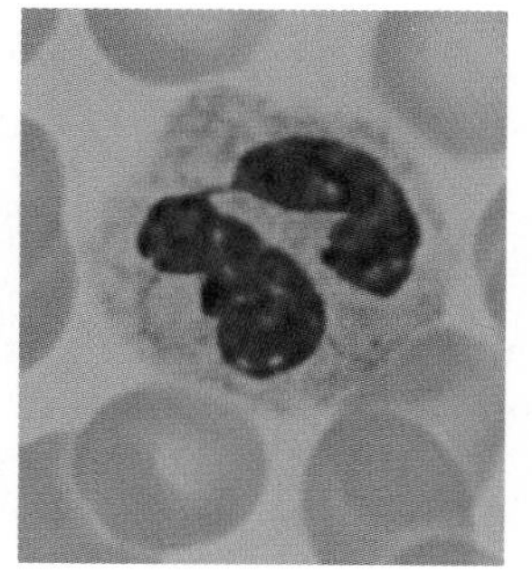

(d) Neutrophil

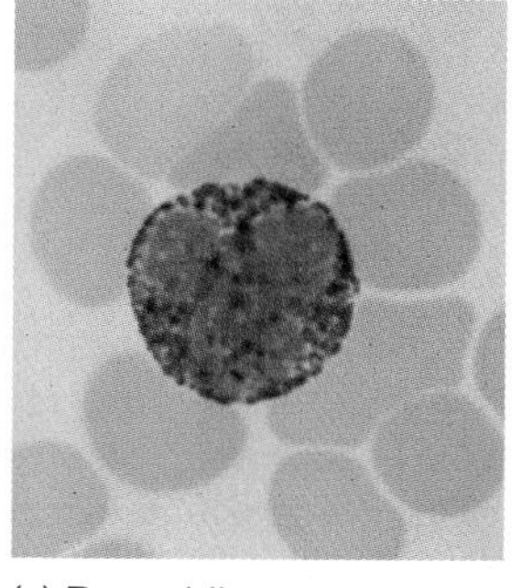

(e) Basophil

Figure 48.4

Three major types of leucocytes are now recognized. *(a)* Lymphocytes are small with ordinary, rounded nuclei. *(b)* Monocytes and macrophages have horseshoe-shaped nuclei. *(c,d,e)* Granulocytes (eosinophils, neutrophils, and basophils) have vesicles that stain distinctively and complex nuclei, usually with several lobes connected by thin strands.

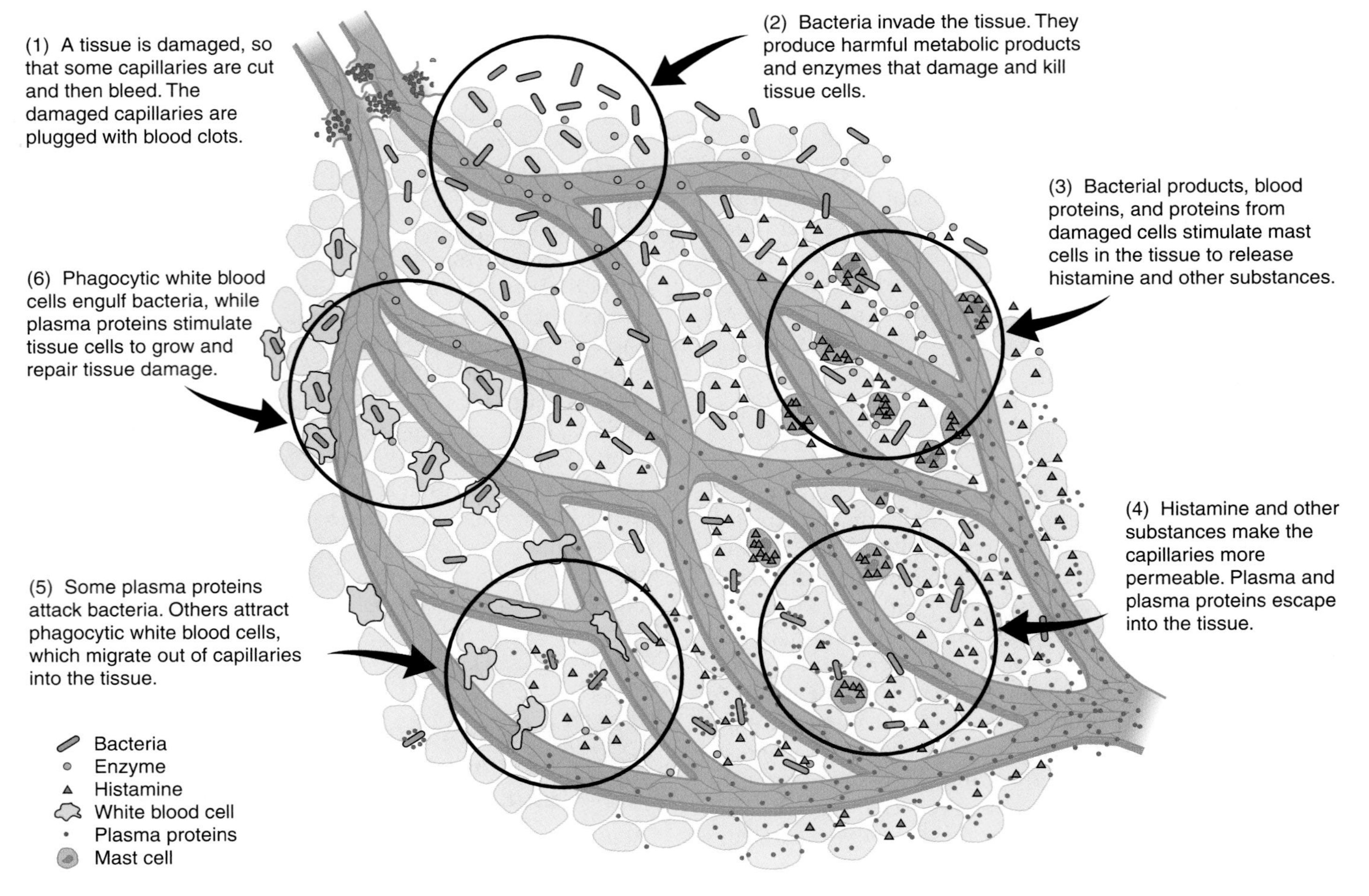

Figure 48.5

Inflammation begins with damage to the microcirculation, which generally allows microorganisms to invade. Blood usually clots quickly, but tissue damage and metabolic products from bacteria stimulate the release of histamine from mast cells in the tissue. Histamine and other factors produce changes in the microcirculation: vasoconstriction of venules and vasodilation of arterioles, increased permeability in capillaries, and loosening of the joints between capillary endothelial cells. Therefore, plasma carrying many kinds of proteins exudes into the tissue, and leucocytes migrate out of the capillaries into the tissue. The drawing shows the sequence of events.

cells in a tissue to expel the contents of their vesicles, mostly **histamine:**

N
HN
$CH_2CH_2NH_2$

Within minutes, the surrounding capillary bed reacts, since the smooth muscle and endothelial cells of the microcirculation have histamine receptors. Capillaries and arterioles dilate and become more permeable, while the venules constrict, and plasma with its beneficial proteins exudes into the tissue. As a result, leucocytes are attracted into the tissue by *chemotaxis,* guided by local chemicals; neutrophils arrive quickly, followed by macrophages and lymphocytes. They escape from the capillaries by squeezing through the joints between endothelial cells into the extracellular space (Figure 48.6) and start to clear the injured area of bacteria and other invaders.

During an inflammatory response, leucocytes communicate with one another through local chemical mediators (paracrine communication). Lymphocytes produce several proteins: a chemotactic factor that attracts monocytes into the tissue, where they develop into macrophages; another that inhibits their chemotaxis, so they remain in the tissue; and proteins that activate the lysosomes of macrophages, so they can aggressively phagocytize and digest invaders. Macrophages release the protein interleukin-1, which induces more release of histamine from mast cells, stimulates the breakdown of muscle proteins, thus inducing a feeling of weakness, and sets the body's thermostat (in the hypothalamus) higher, producing fever. So the miserable, aching feeling you get during an infection is a result of your body's attempt to fight the infection. Fever is another nonspecific defense, an adaptation for killing pathogens, since raising the temperature even a few degrees may inhibit the growth of pathogens just enough for the body's defenses to get the upper hand. Here, unfortunately, the defenses may turn

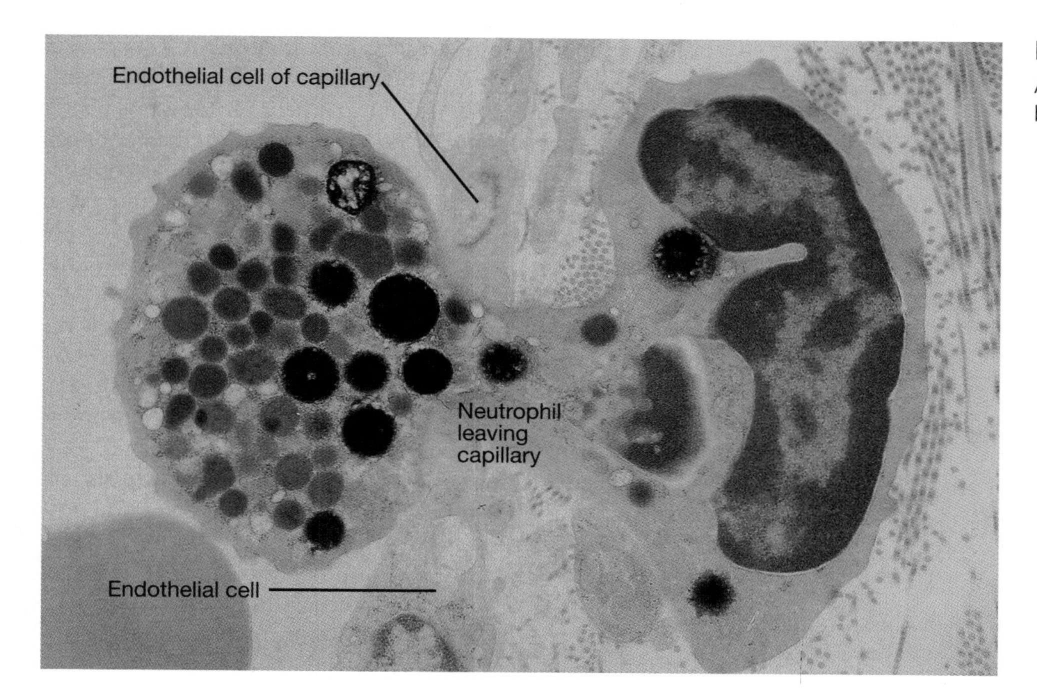

Figure 48.6
A neutrophil leaves a capillary by squeezing between the endothelial cells.

against the body, because a high, sustained fever can damage brain tissue.

Paracrine communication, Section 41.5.

The scene of the inflammation eventually looks like a battle zone, strewn with necrotic (dead and dying) cells, tissue debris, blood clots, partially digested bacteria, and bunches of leucocytes; this mass is called a purulent exudate—in short, *pus.* A localized mass of pus surrounded by inflamed tissue is an *abscess.* Tissue cells that haven't been injured directly may be killed by deprivation of oxygen and nutrients, due to the reduced circulation, and by lysosomal enzymes from phagocytes. However, some tissue debris enhances the activities of leucocytes and probably also induces tissue cells to grow and replace cells that have died. Soon fibroblasts start to proliferate and lay down new connective tissue fibers, and capillaries in the area start to grow into the injured region. Generally within a few days the wound is well on its way to being healed.

Exercise 48.1 To reduce the miserable symptoms of a cold, we often take antihistamines. What do they do, in molecular terms?

48.4 Blood clotting and other cascades are essential components of inflammation.

Now we can focus on the three cascades of inflammation. The kind of injury that starts an inflammatory reaction usually tears blood vessels, initiating an autonomic response in which arterioles leading into the injured area constrict. We saw in Section 45.5 that even a slight constriction reduces the rate of blood flow considerably. At the same time, the vital **blood clotting** reaction begins. A temporary plug made of **platelets** stops leakage from the vessel; platelets are small fragments (250,000–450,000 per mm^3 of human blood) of huge cells called megakaryocytes. Platelets swarm into an opening by chemotaxis and, being very sticky, adhere to the blood vessel lining and pack the opening.

Cascades as biological amplifiers, Section 41.4.

A stronger clot is soon made by the protein fibrin, which forms as the result of a long cascade of activations (Figure 48.7). Blood clotting begins with the activation of Hageman factor (also called factor XII); in the last steps of the cascade, prothrombin is converted into thrombin, and thrombin converts fibrinogen into fibrin. This long chain of clotting reactions, however, is vulnerable to mistakes, such as mutations; in fact, several types of hemophilia in humans are genetic deficiencies in one of the proteins. Classical hemophilia, like the kind discussed in Section 16.13 that was inherited from Queen Victoria, is a failure to produce factor X.

Blood drawn for laboratory analysis or transfusion is kept from clotting by small amounts of anticoagulants. Thrombin requires Ca^{2+} ions; some anticoagulants, such as citrate, chelate Ca^{2+} and thus prevent clotting. Other anticoagulants inactivate proteins in the cascade sequence; heparin, for instance, activates an antithrombin protein. The anticoagulant rat poison warfarin makes an animal bleed to death internally by antagonizing vitamin K, which is essential for converting prothrombin and other clotting proteins into an active form.

Chelation, Concepts 4.2.

While blood clotting is essential, it is also dangerous. A clot that breaks loose becomes an *embolus,* a floating plug that may

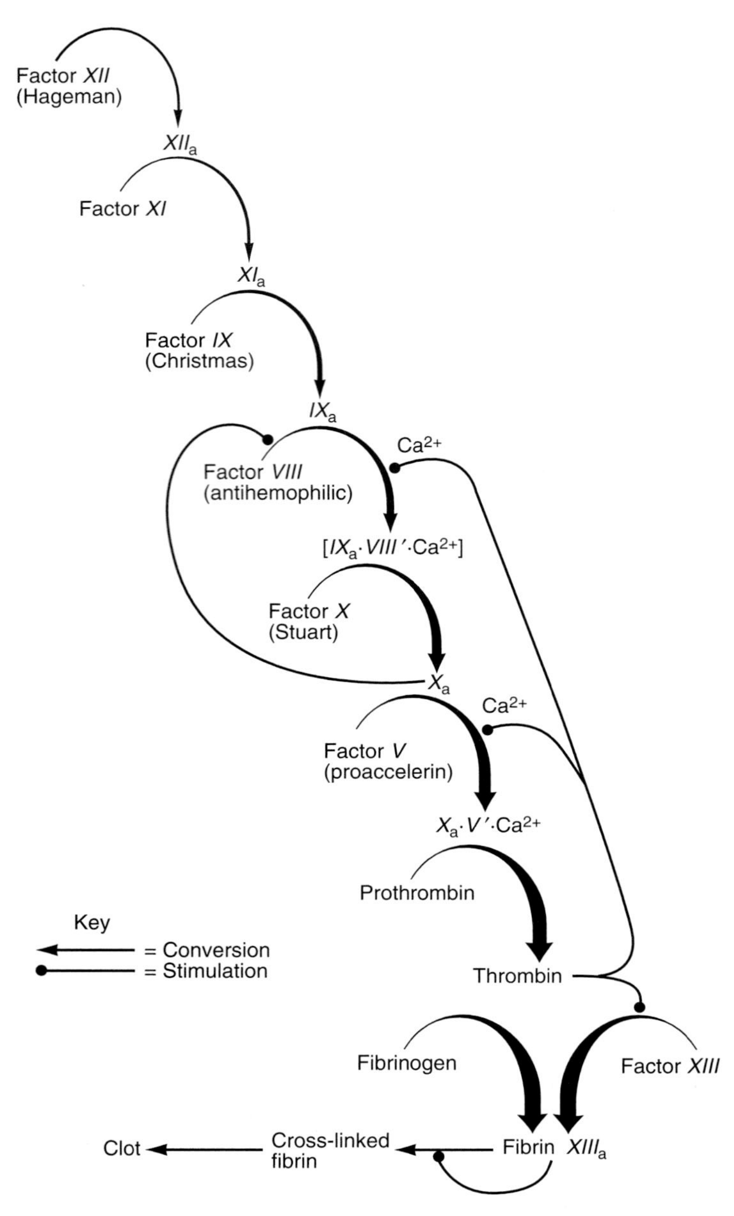

Figure 48.7

Blood clotting occurs through a long, complex cascade. Several of the proteins are named for patients in whom they were first found to be missing. Note that calcium ions are necessary for clotting and that some factors feed back positively at certain points to enhance the process.

lodge in a narrow vessel in the heart or brain, blocking blood flow and causing a heart attack (myocardial infarction) or stroke. Normal functioning depends on a delicate balance between clotting reactions and *fibrinolysis reactions* that prevent clotting and break up formed clots. In fibrinolysis, a tissue protein, plasminogen activator, converts the blood protein plasminogen into plasmin, which then digests fibrin. Some medical research has focused on using plasminogen activator to assure blood flow after a heart attack or to break up an embolus that has caused a stroke.

Hageman factor initiates a second inflammatory cascade in a series of about 20 plasma proteins known collectively as **complement** (Figure 48.8). Several complement proteins lyse bacteria by forming a *membrane attack complex* in the bacterial cell membrane, allowing the cytoplasm to leak out. Other complement proteins coat bacteria, making them more attractive to phagocytes. Two complement fragments (C3a and C5a) are anaphylatoxins, small peptides that can initiate and enhance the whole inflammatory response, and C5,6,7 is a chemotactic factor for neutrophils. (We will show later that complement is activated by other factors, including antibodies bound to bacteria.) On the other hand, after an ill-matched blood transfusion with red blood cells of the wrong type, complement proteins turn against us by lysing the transfused red blood cells.

A third, short cascade produces peptides called **kinins,** especially bradykinin (Figure 48.9*a*). Kinins, like histamine, increase vascular permeability and vasodilation, and they stimulate mast cells to release more histamine. Even small amounts of kinins produce an exquisite and excruciating pain by stimulating sensitive nerve endings, thus accounting for another symptom of inflammation. An active ingredient of one wasp venom (Figure 48.9*b*) is a kinin, and a Brazilian toad is poisonous to the touch because its skin secretes a phyllokinin. You can see how other animals are able to evolve protective strategies by subverting mammalian defenses.

Exercise 48.2 Find a place in the blood-clotting cascade where the product of a reaction feeds back to stimulate an earlier reaction. What kind of feedback loop is this? What is its function?

B. The immune system

48.5 The immune defense system supplements inflammation.

Mechanical and chemical barriers to infection and the reactions of inflammation are nonspecific mechanisms that protect against all kinds of infective agents. Vertebrates, particularly birds and mammals, back up these defenses with an **immune system,** which recognizes the unique chemical structures of viruses and foreign cells and inactivates them. Animals with an immune system are able to distinguish their own parts from something foreign—to distinguish "self" from "nonself"—and to destroy foreign materials that enter the body. As you will see, such a system is very effective and incredibly complex.

Knowledge of immunity began with the ancient observation that people who had recovered from a disease seemed immune to future episodes of the same disease. The English physician Edward Jenner observed that milkmaids who contracted the mild disease cowpox had clear skin because they were immune to the serious, disfiguring disease smallpox. He reasoned that infecting a person with cowpox could prevent smallpox, and in 1798 he tested his idea by inoculating an eight-year-old boy with material from a cowpox pustule and later challenging

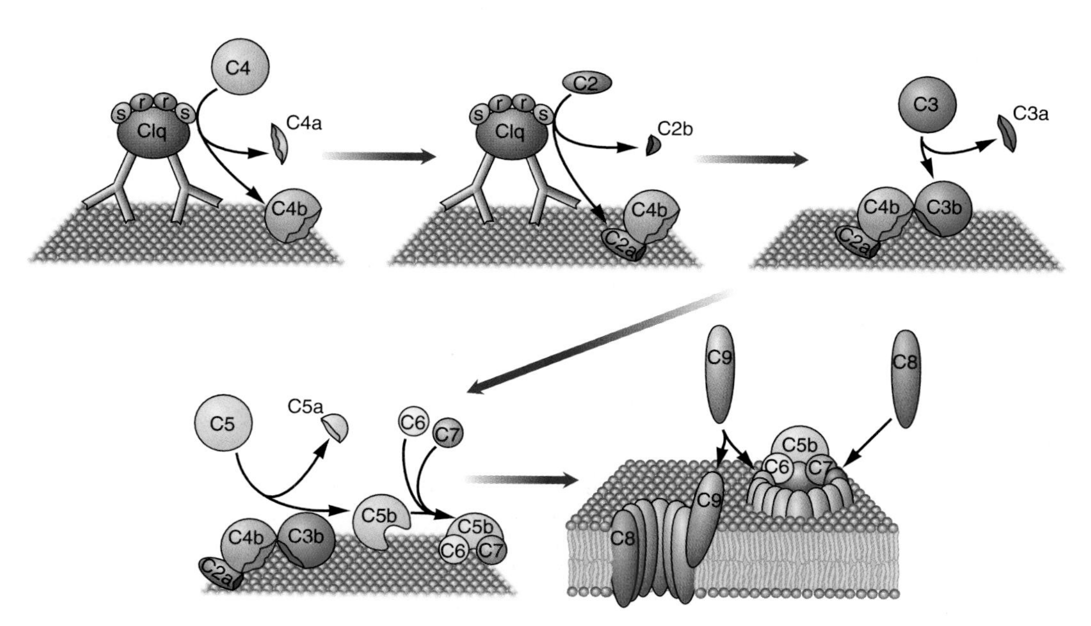

Figure 48.8
Complement proteins are activated in a cascade, initiated by various stimuli, including bacteria. The last proteins to be activated make membrane attack complexes in the bacterial cell membranes, which kill the cells.

him—that is, testing for a reaction—with material from a smallpox pustule. The boy did not contract smallpox. (With our modern sensitivity to medical testing on humans, we recognize that Jenner's experiment was ethically dubious and could have been very dangerous.) This technique, called **vaccination** (from *vacca* = cow) or **immunization,** became widely used. Nearly a century later, Louis Pasteur extended this method to fight fowl cholera, a disease of chickens. Pasteur accidentally discovered that an old culture of the cholera bacteria would confer immunity without causing the disease, and thus he developed the general technique of creating a **vaccine:** a preparation of weakened or heat-inactivated bacteria or viruses that can be injected to confer immunity to a specific disease. Children are now commonly immunized in infancy against a whole list of diseases, including measles, polio, whooping cough, and diphtheria.

To illustrate immunization we can inject a rabbit with diphtheria vaccine made from dead diphtheria bacteria (corynebacteria) that still have the chemical structures of live cells. After several days, we take a small blood sample from the rabbit and centrifuge it to collect the **serum,** the liquid part of blood lacking cells and clotting proteins:

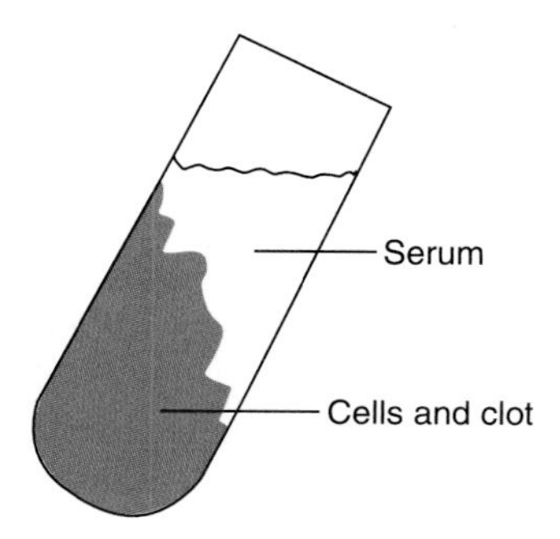

When corynebacteria are mixed with the serum, they form a fluffy complex; in contrast, bacteria mixed with serum of a nonimmunized rabbit just settle to the bottom of the tube. Because it gives this kind of reaction against the bacteria, the rabbit's serum is called an **immune serum,** or **antiserum.** The vaccine has induced the rabbit to form some new blood proteins called **antibodies;** the vaccine that induced their formation is an **antigen.** Antibody and antigen are defined together: An antigen is any substance capable of inducing an animal to form antibodies, and antibodies are proteins, formed in response to an antigen, that combine specifically with the antigen. Antibodies and antigens interact because they have complementary shapes and bind to each other just as a ligand binds to a receptor protein or a substrate binds to an enzyme. Therefore they form a network complex (Figure 48.10). The antibodies made against corynebacteria are very specific. They will not combine with any other kind of bacteria, or with any virus, or with any other substance. This specificity is the chief characteristic of antibodies. The immune system identifies an antigen as foreign, as nonself, because of its molecular shape, and in response produces antibody proteins that have sites with a complementary shape.

Most antigens are proteins or polysaccharides. For simplicity, we'll only use protein antigens as examples. To be antigenic, a molecule must have a minimum molecular weight of about

Arg • Phe • Pro • Ser • Phe • Gly • Pro • Pro • Arg
(a) **Bradykinin**

Arg • Phe • Pro • Ser • Phe • Gly • Pro • Pro • Arg • Gly
(b) **Wasp venom**

Figure 48.9
The structures of *(a)* bradykinin and *(b)* a component of wasp venom show that other animals can turn vertebrate defense systems against themselves by evolving the ability to produce some of their components.

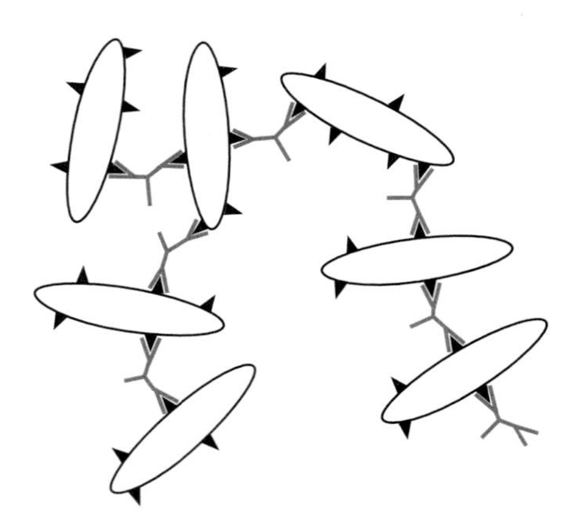

Figure 48.10

Antigens and antibodies have sites with complementary shapes, so they bind to one another to make large complexes. The Y-shaped antibodies are drawn much larger than their actual size relative to the bacteria.

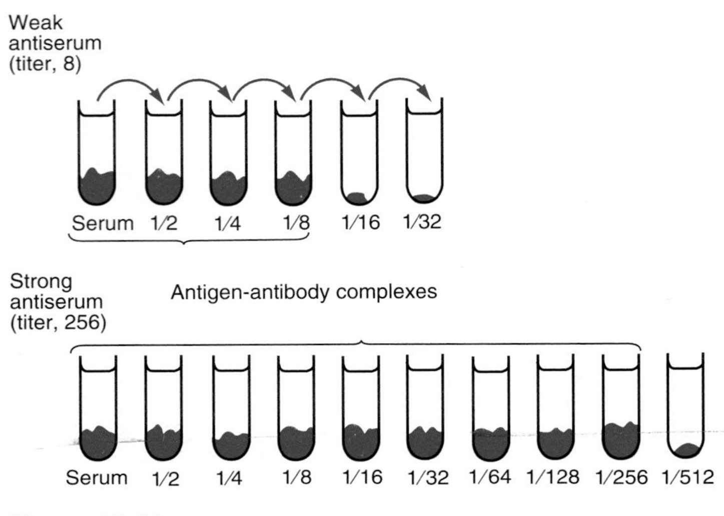

Figure 48.11

The strength (titer) of an antiserum is measured by diluting it several times by a factor of 2 and mixing each tube with antigens (in this case, bacteria). A strong antiserum shows a reaction with the antigens at a greater dilution than does a weak antiserum. The titer of the antiserum is measured by the highest dilution that still shows a reaction.

5,000 (varying with the type of compound), but small organic molecules can become antigens if they are bound to proteins or other macromolecules.

The **titer** of antibody in a serum is a measure of the antibody's concentration. (Titer is a measure of biological activity, not a physical measure such as milligrams per milliliter.) The titer of bacterial antiserum is measured by diluting serum serially (Figure 48.11) and adding the same amount of bacteria to each tube; the antigen-antibody complex only forms in the more concentrated tubes, and the titer of the serum is measured by the most diluted tube where the complex still forms. If we

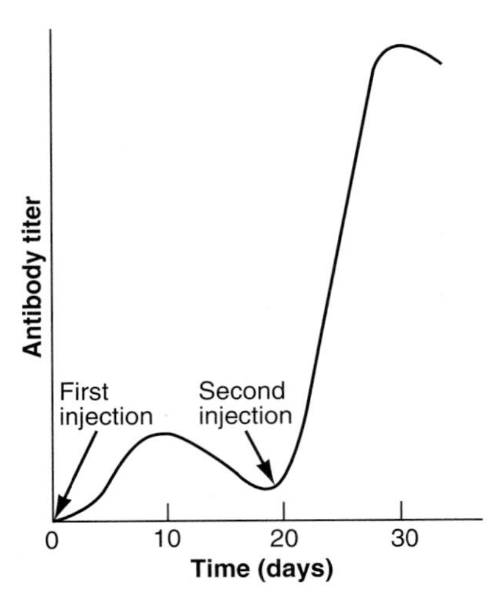

Figure 48.12

The primary response to an antigen is relatively weak, and the antibodies produced last for only a short time. A secondary response produces a much higher titer and long-lasting antibodies.

take samples of a rabbit's serum at various times after inoculation, we find that the titer of antibody in the rabbit's serum increases for a while and then falls (Figure 48.12); this reaction is the **primary response.** A second injection of vaccine elicits a **secondary response** with a much higher titer that may last for a very long time. An infant can be protected against common diseases for a lifetime by a series of immunizations. The fact that an antigen can elicit a secondary response, even long after it has elicited a primary response, shows that the immune system has *memory;* we will see later that its memory resides in certain specific cells.

Exercise 48.3 Two samples of antiserum against the same antigen are tested by diluting them serially by half several times. One produces a positive reaction to the sixth tube and the other to the tenth tube. What are the approximate relative titers of the two antisera?

48.6 The immune response occurs in lymphoid organs and tissues.

The immune system resides in scattered organs and patches of tissue where *immunocytes,* especially lymphocytes and macrophages, produce two closely related events, the *humoral* and the *cell-mediated* responses to foreign materials. Lymphocytes grow and mature in **primary lymphoid tissues** in the bone marrow and thymus gland (Figure 48.13). They differentiate into several types, which appear identical under the microscope but can be distinguished by their surface proteins. Precursor lymphocytes in bone marrow differentiate into **B-lymphocytes,** or **B-cells.** ("B" stands for "bone marrow" in mammals; in birds, B-cells are named for a small pouch off the intestine, the bursa of Fabricius

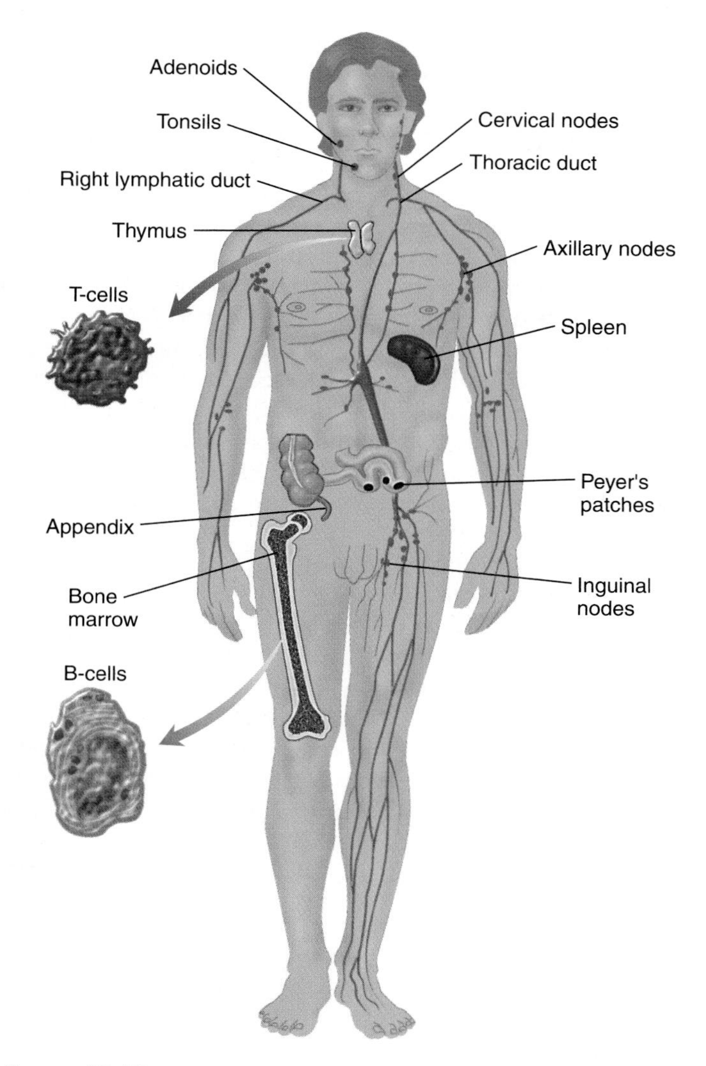

Figure 48.13

The immune system operates through several organs and tissues composed of lymphoid (lymphocyte-bearing) tissue. Lymphocytes mature in *primary lymphoid tissues:* Bone marrow is the source of B-cells, and the thymus gland is the source of T-cells. *Secondary lymphoid tissues,* such as the lymph nodes, spleen, and Peyer's patches, trap antigens and expose them to lymphocytes and other cells of the immune system.

[fa-bree′ shus], where they mature.) B-cells create **humoral**[1] **immunity** by producing soluble antibodies, which circulate in the blood and lymph and are secreted by mucous membranes; this system is primarily effective against bacteria, extracellular viruses, and foreign molecules in solution.

Lymphocytes become **T-lymphocytes,** or **T-cells,** as they are processed through the thymus gland. The human thymus, located in the upper chest, grows smaller as a person matures and ages. T-cells have one of two general roles. *T-helper cells* have a key role in the immune response. *T-killer (T-cytotoxic) cells* are responsible for **cell-mediated immunity;** their function is to destroy other cells and protect against fungi, parasitic worms, and intracellular virus infections. (Unfortunately, we know these cells best because they destroy transplanted tissues.) T-cytotoxic cells attack other cells by making holes in their membranes with proteins called *perforins* (similar to the pores made by complement proteins) through which toxic proteins and other substances are injected. These aren't nonspecific killer cells; they carry stereospecific recognition proteins similar to antibodies and only attack cells marked by the right matching antigens, as explained in Section 48.8. In addition, there are **natural killer (NK) cells,** lymphocytes with neither B nor T specificity that are able to kill cells that are not specifically marked. NK cells may be particularly important in the natural surveillance system that finds and destroys tumor cells.

In **secondary lymphoid tissues,** including the spleen, lymph nodes, tonsils, adenoids, and the Peyer's patches around the intestine (see Figure 48.13), immunocytes meet antigens and interact with each other to produce antibodies. As lymph nodes filter lymph from nearby tissues and vascular spaces, they trap many materials and inactivate those whose antigens identify them as foreign. With even a minor infection, you can often feel the lymph nodes that filter the region—under your arms, in your neck, in your groin, and elsewhere—becoming hard and sensitive as they undergo inflammatory reactions in response to foreign agents.

48.7 All immunoglobulins share a common basic structure.

Antibodies are Y-shaped molecules called **immunoglobulins** that all have the same general structure, two **light (L) chains** and two **heavy (H) chains** held together by disulfide (S–S) bonds:

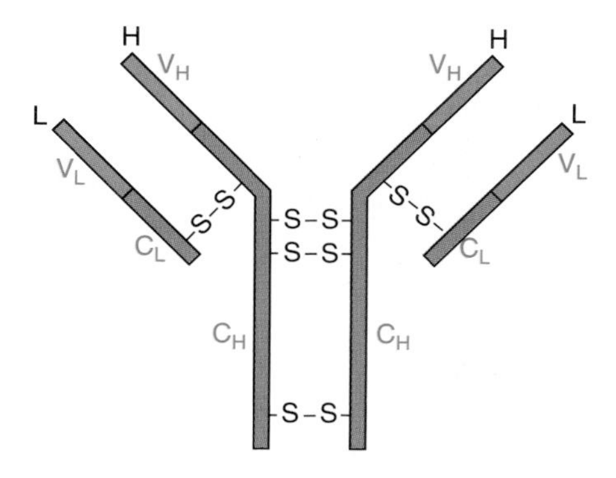

Comparisons of many immunoglobulins show that both L and H chains have a **constant (C) region,** which is the same or almost the same in all immunoglobulins of a single class from one species, and a **variable (V) region** unique to each type of antibody. In the two "arms" of each molecule, the V regions of H and L chains come together and create an **antigen-binding**

[1]"Humoral" refers to soluble proteins not bound to cells; ancient students of medicine talked about "humors," or fluids, that controlled a person's health and emotions. Modern language retains "good humor" and "bad humor," and gives us "humoral immunity."

site, which is stereospecific for an antigen with a complementary shape:

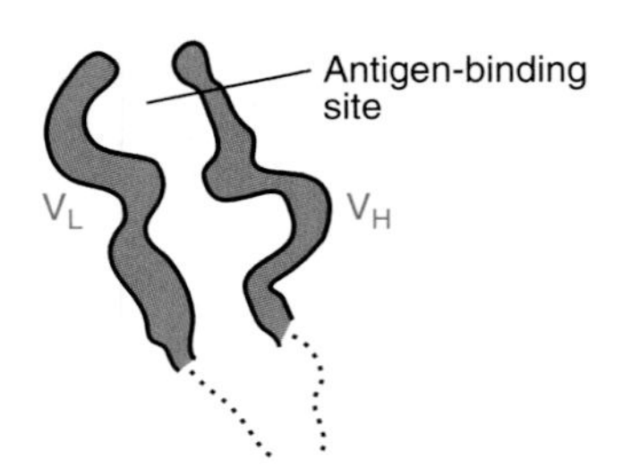

A typical bird or mammal can probably make millions of distinct immunoglobulins, each with a unique antigen-binding site, but it does not carry millions of genes to encode these proteins. Since each site is made from one variable heavy (V_H) chain and one variable light (V_L) chain, many different sites can be made by combining these chains randomly. If an animal could make only 100 kinds of V_H chains and another 100 kinds of V_L chains, it could make $100 \times 100 = 10{,}000$ different sites. With 1,000 chains of each kind, it could make a million sites ($10^3 \times 10^3 = 10^6$).

An immunoglobulin isn't directed against a whole antigenic protein, only a small, distinctively shaped portion called an **epitope** or **antigenic determinant** made of a few amino acids, perhaps with an attached chemical group:

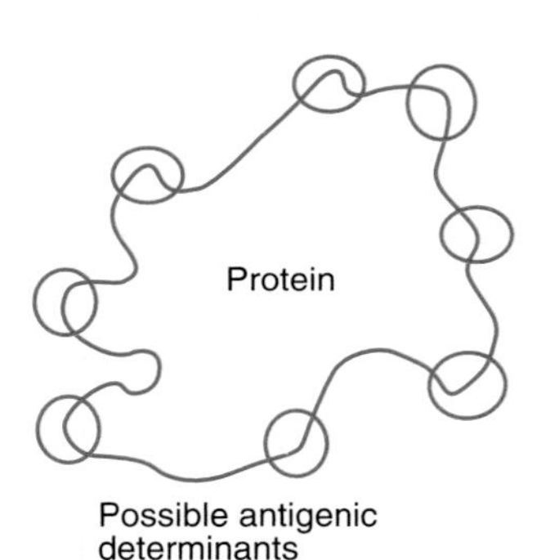

Antigens and antibodies form network complexes in which the two identical antigen-binding sites of each antibody bind to identical epitopes on different antigens:

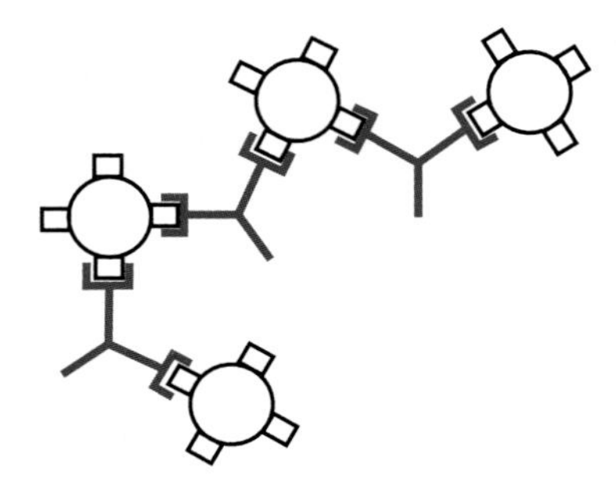

Blood serum and mucous secretions carry five classes of immunoglobulins (abbreviated Ig): IgA, IgD, IgE, IgG, and IgM. Immunoglobulin class is determined by the H chains; all five classes have the same type of L chain. IgA is secreted in mucus and in milk, making it one of the chief benefits of nursing a baby. (Notice, incidentally, that the immunity a baby obtains from its mother's antibodies is *passive immunity;* it only acquires *active immunity* when it makes its own antibodies.) IgA is the main antibody in the intestinal tract; we all secrete quite massive amounts of IgA every day as protection against the prospective pathogens that use this portal into the body. IgD is apparently a surface receptor on B-cells. IgE binds to mast cells and is involved in allergic reactions. IgG is the principal serum antibody and is also the only type of immunoglobulin that can cross the placenta and confer a mother's immunity on her developing fetus. The first antibodies made in response to a new antigen are mostly IgM.

Once antibodies have been released into the blood or secreted in mucus or milk, they bind to their complementary antigens and precipitate them into clumps that phagocytes can engulf and digest. The antigens may be bacteria, viruses, or large, soluble molecules such as the protein or polysaccharide toxins made by bacteria.

When the antigen-binding sites of an antibody (on the arms of the Y) bind to complementary epitopes, the constant region of the H chains (the base of the Y) changes shape and may be activated to perform other functions. For instance, activated IgM and IgG molecules will bind to the first component of complement and initiate that cascade. Antibodies directed against bacterial antigens will coat bacteria, and phagocytes have a special affinity for bacteria tagged with IgM, IgG, or the C3b complement protein. Bacteria tagged with antibodies are said to be **opsonized** (Greek, *opsonion* = relish) or made attractive to phagocytes:

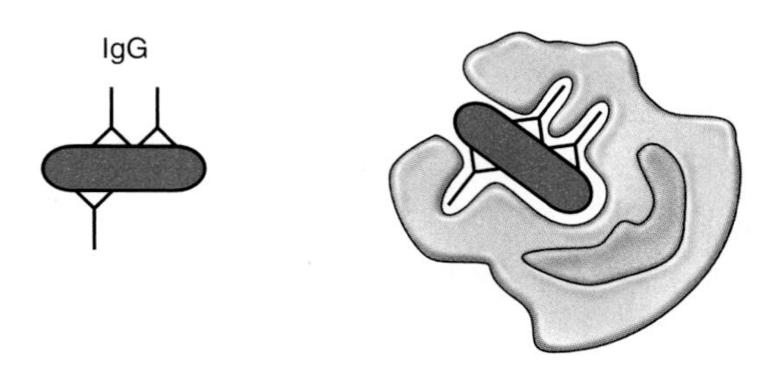

Exercise 48.4 If an antigen is a protein, one of its epitopes is a region made of a few amino acids. Sketch what you imagine an epitope of a polysaccharide might look like.

Exercise 48.5 Suppose a mouse protein bearing two identical epitopes is injected into a sheep. Draw the kind of complex the mouse antigens and sheep antibodies can form.

Exercise 48.6 Although only large molecules are antigenic, antibodies can also be made against small chemicals called *haptens* that are bound to macromolecules. Small molecular groups can be added to a benzene ring in *ortho-*, *meta-*, or *para-* positions relative to some other group:

NO_2 HO — *ortho*-nitrophenol

NO_2 HO — *meta*-nitrophenol

HO — NO_2 *para*-nitrophenol

A rabbit is immunized with a protein bound to *ortho*-nitrophenol; the antibodies it makes bind strongly to this antigen, but only weakly to the same protein with *meta*- or *para*-nitrophenol bound. What does this result show about the nature of epitopes and antibody specificity?

Exercise 48.7 In general, antibodies made against one strain of bacteria will only combine with those particular bacteria. Sometimes, however, other bacteria do react with the same antibodies, but more weakly. Suppose antibodies made against strain A react with 100 percent strength (by definition) against bacteria A, 50 percent against bacteria B, 45 percent against C, and 30 percent against D. *(a)* Explain why such cross reactions occur. *(b)* What do these results show about similarities among the different strains of bacteria?

C. The development of immunity

48.8 Self is determined primarily by proteins of the major histocompatibility complex.

We have said that the immune system is devoted to distinguishing self from nonself. *Self* is defined as anything the body normally doesn't make antibodies against and *nonself* as anything that elicits antibody formation. In addition to the many proteins and polysaccharides that incidentally mark the cells of different individuals and species, a series of special proteins appear to be designed only for making each animal's cells distinct. These proteins are encoded by 40–50 genes of the **major histocompatibility complex (MHC),** such as the H-2 genes of mice or the HLA genes of humans. The MHC was discovered in connection with attempts to transplant tissues and organs from one person to another. Grafts of skin from one region to another on the same person (autografts) or tissue transplants between closely related people (homografts) work reasonably well. Closely related people have similar or identical MHC genes and proteins, so one person's immune system may recognize the other person's cells as so nearly identical to its own that they are tolerated as "self." However, as the genetic difference between individuals increases, the likelihood of a successful graft decreases.

The MHC proteins mark cell surfaces, but they obviously did not evolve to frustrate surgeons who might come along millions of years later. Most MHC proteins belong to two classes that have distinct roles in immune responses (Figure 48.14). Class II proteins appear only on macrophages and other immunocytes that process antigens; they are involved in presenting epitopes to B-lymphocytes, as we will describe in the next section. Class I MHC proteins mark the surfaces of all nucleated cells. When complexed with antigens, they are what a killer T-cell recognizes as foreign. But how can an ordinary tissue cell have foreign antigens on it, and why would an animal have killer cells to destroy its own tissues? The cell being attacked is most likely infected by viruses, and this requires a word of explanation. As viruses proliferate within the cells of their host (Figure 48.15), some viral proteins appear in the plasma membrane, where they combine with the cell's class I MHC proteins and mark the cell as different (Figure 48.16). Killing infected cells before they can produce more viruses is the most effective way to fight a viral infection, and this is a primary function of T-killer cells. Unfortunately, T-killer cells also kill the cells of tissues that are being transplanted for therapeutic reasons if they carry recognizably foreign MHC proteins.

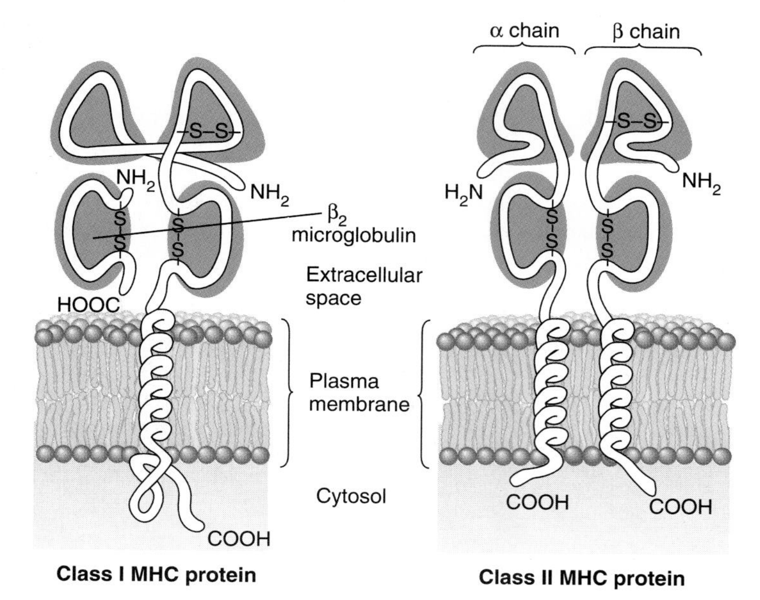

Figure 48.14
Class I and class II MHC proteins are similar in structure, but they have quite different functions.

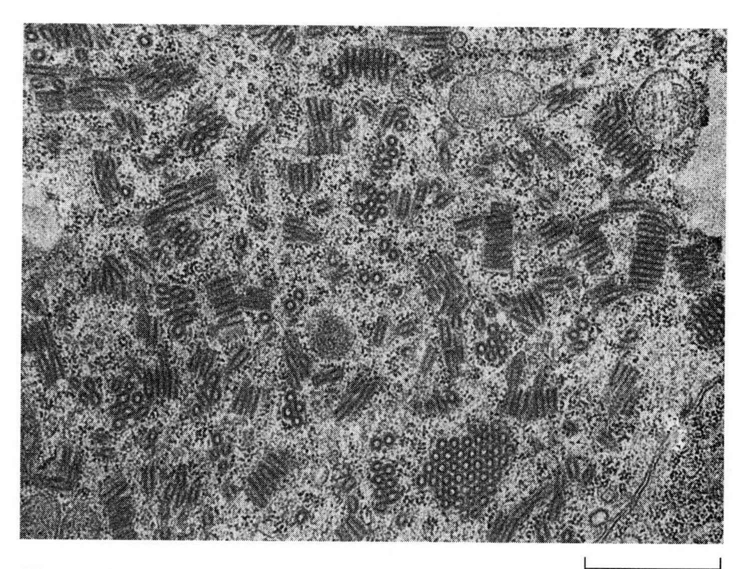

Figure 48.15
The cytoplasm of a cell infected by rabies virus is filled with new virus particles. The particles are bullet-shaped, so they appear round when cut in cross section and like short rods when cut lengthwise.

Concepts 48.1 An Antibody to You Is an Antigen to Me

Students sometimes get confused by discussions of immunology because "antigen" and "antibody" are relative terms defined by the animals involved in experimental and medical procedures rather than well-defined substances. To immunize people against a certain virus, we inject them with a viral vaccine; this is the antigen, and the immunoglobulins they produce are the antibodies. Suppose, however, we would like to learn more about those human antivirus antibodies (let's call them HAA). Since these immunoglobulins are distinctive proteins themselves, made by human cells, they can be antigens to another animal. For example, we could isolate these HAAs and inject them into a rabbit. The rabbit's immune system, recognizing them as nonself, will make its own distinctive antibodies against them—in other words, rabbit anti-HAA. (Actually, the rabbit will make a mixture of antibodies directed against all human immunoglobulins, but we single out these particular rabbit antibodies to make a point.) It is important to see that antibodies, or other proteins of the immune system, can be antigens in another context. The histocompatibility proteins, for instance, are like antibodies in that they can combine with foreign proteins, but they are also antigenic in the sense that other immune cells can recognize them as foreign proteins.

The distinctive surface properties of different kinds of lymphocytes can be used experimentally to separate and purify the various cells. Suppose we isolate some human T-cells and inject them into a rabbit to make anti-human-T-cell (anti-HTC) antibodies. (We will have to remove antibodies made against human cells in general, but that's a minor technical matter.) We then bind the anti-HTC to little beads of a resin (so their antigen-binding arms are free) and pack this material into a tube. If a mixture of human blood cells is allowed to flow slowly through the tube, the cells will interact with the bound rabbit antibodies. Only the T-cells will bind to the rabbit antibodies and remain bound to the resin, while all other cells will pass through the tube.

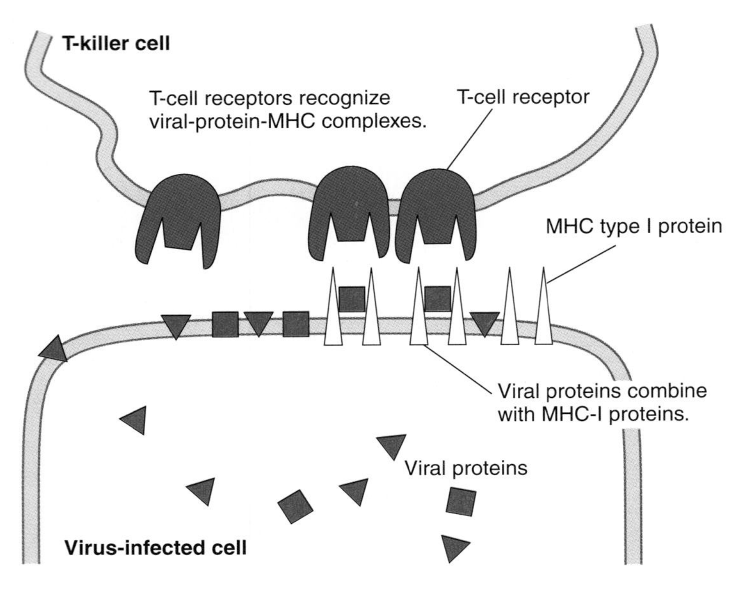

Figure 48.16

When a cell is infected by viruses, some of the specific viral proteins it produces move into the plasma membrane in preparation for the emergence of new virions. T-killer cells can recognize these viral proteins and kill the infected cell.

An animal's own cells might also appear to be foreign if they have become tumor cells. As we note in Chapter 20, a cancer cell acts rather like an undifferentiated embryonic cell and carries unusual proteins, which the immune system can recognize as foreign. These cells, too, ought to be killed for the animal's well-being, and suppression of tumors is another important function of the immune system. It only rarely fails us.

The MHC proteins are antigens like the ABO blood antigens (see Concepts 48.1 for help in thinking about this point), and the rules governing blood transfusions also apply to grafts. For instance, if a purebred mouse with H-2^a genes is mated to one with H-2^b genes, the F_1 hybrid has both *a* and *b* specificities. It will accept tissue from either parent, but each parent rejects tissues from the offspring because they also have antigens of the opposite type.

48.9 Each clone of differentiated lymphocytes makes antibodies with only a single type of antigen-binding site.

During its development, an immunologically competent animal makes an enormous number of lymphocytes, between 10^8 and 10^{12}. These cells are not identical. The B-lymphocytes differentiate into many distinct clones, each programmed (we say *committed*) to make antibodies with a single kind of antigen-binding site. (We will see how this happens in Section 48.10.) On its surface, each lymphocyte bears many immunoglobulins of the type it is committed to making. They are like an advertisement to the body: "I can make antibodies with this shape. Do we need antibodies like this?" If an antigen gets into the body, it *selects* those clones making complementary antibodies by stimulating some cells in each clone to proliferate still more and to make large amounts of their antibodies. This **clonal selection model** for the activation of antibody-making cells is well-confirmed (Figure 48.17). Its essential point is that clones of cells are formed during early development, and each kind of antigen only activates those clones that make complementary antibodies.

Let's return to the example of a rabbit encountering some corynebacteria (see Section 48.5). Macrophages phagocytize the bacteria. They save bacterial epitopes and combine them with MHC class II proteins on their surfaces. In a complex interaction with T-helper cells, they present these epitopes to those B-cells that bear complementary antigen-binding sites (shown in

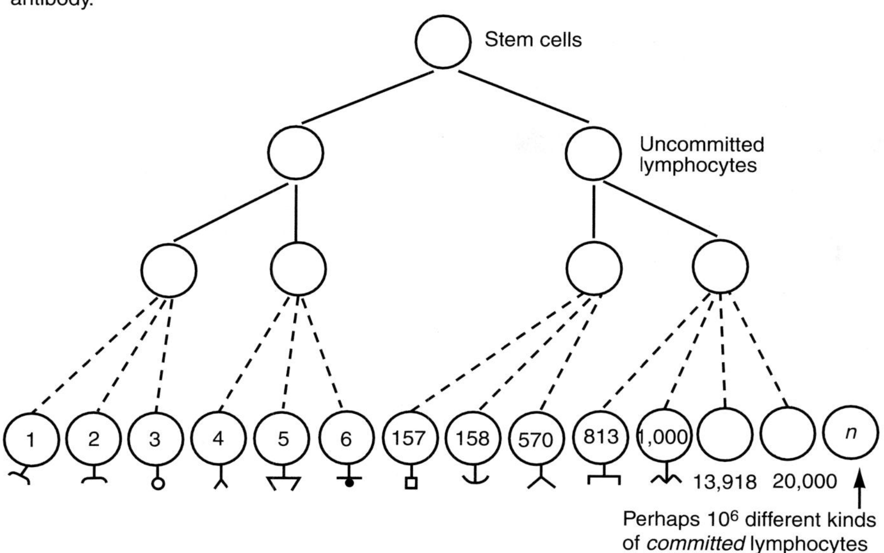

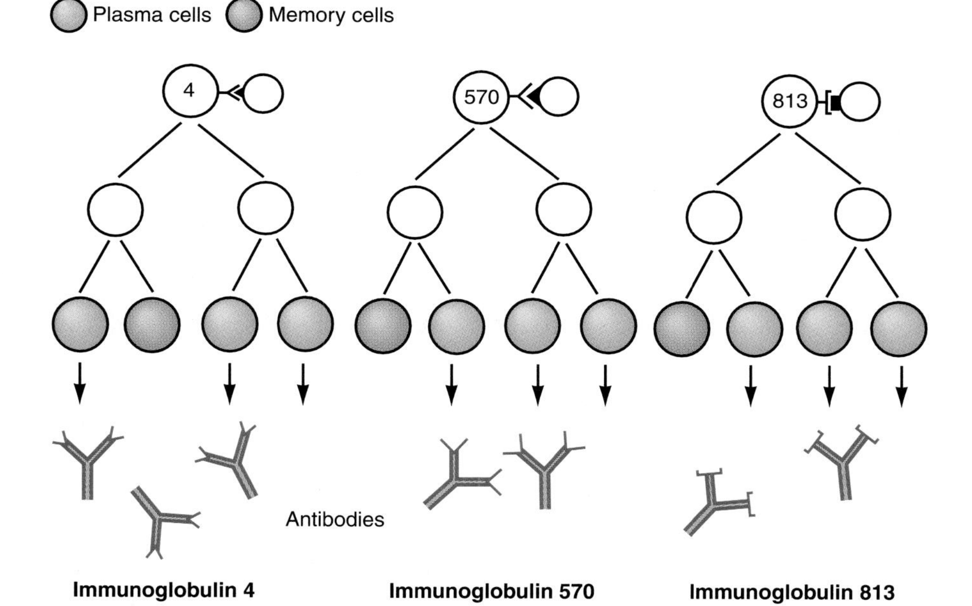

Figure 48.17

(1) Clonal selection begins with differentiating B-lymphocytes, which form many primary clones, each programmed to make antibodies with a single kind of antigen-binding site. Each lymphocyte bears some of these antibodies as distinctive surface receptors that recognize their complementary antigens. *(2)* When a new antigen appears (in this case, a new virus), it binds to those lymphocytes that make complementary antibodies, stimulating these cells to proliferate and differentiate into secondary clones, *plasma cells* and *memory cells (3).* Plasma cells are large antibody factories that produce antibodies for a brief period and then die *(4);* those plasma cells that have been selected by their complementary antigens produce large amounts of their antibodies. The memory cells in these clones are ready to respond even more strongly to a second encounter with the antigen in the future. T-lymphocytes that can respond to cell-bound antigens are selected to proliferate in much the same way as B-lymphocytes.

the lower portion of Figure 48.17). These B-cells are stimulated to proliferate, and many of them differentiate into **plasma cells** that produce large amounts of antibody. Plasma cells only live for a few days, and soon the titer of anti-corynebacteria antibodies in the serum declines. However, many lymphocytes differentiate into **memory cells** instead of becoming plasma cells. In a molecular sense, they remember their encounter with corynebacterial antigens, and when challenged with a second dose of these antigens, they respond rapidly with a secondary response. The secondary response is so much stronger than the primary response because there are so many memory cells, compared with the few initial committed lymphocytes, and the antibodies they make, mostly IgG, reach a much higher titer and may last for a long time. Still more memory cells are produced in the secondary response, so another encounter with the antigens will produce an even faster and stronger tertiary response.

These processes are governed by several paracrine factors called **interleukins** and **interferons.** The macrophages start to

Methods 48.1 Monoclonal Antibodies

Even a quite pure antigen elicits a mixture of antibodies made by several clones of lymphocytes. But it would be valuable to have antibodies with only a single type of antigen-binding site—made by a single clone of B-cells—that could be directed against a particular target. Such **monoclonal antibodies** can now be synthesized routinely. First we inject a mouse with the desired antigen to induce many of its B-lymphocytes to multiply. Then we remove the mouse's spleen and extract lymphocytes from it. These cells alone would soon die, but we make them "immortal" by hybridizing them with cancer cells; the reagent polyethylene glycol induces B-cells to fuse with multiple myeloma cells to make cells called *hybridomas.* Each hybridoma continues to multiply indefinitely while producing antibodies of a single type. We distribute these cells to small plastic chambers where they can grow, and during a few rounds of growth and testing, we select hybridomas producing antibodies that bind strongly to the desired antigen. The best of them are chosen for large-scale cultures.

Applications of monoclonal antibodies continue to develop. They can be used in research to isolate one kind of molecule, such as an enzyme, out of a mixture. Sensitive and specific medical tests, such as pregnancy tests, have been developed with them. And since monoclonal antibodies can be directed against specific cancer cells, they hold the potential for directed therapy; for instance, they can be coupled with cytotoxic drugs or proteins (perhaps perforins or complement proteins) to deliver these agents only to the cancer cells, avoiding the destruction of normal cells that accompanies many current therapies.

produce interleukin-1, which activates T-helper cells (Figure 48.18). Macrophages then combine with T-helper cells, which start to make several proteins:

- Interleukin-2 stimulates the growth of still more T-helper cells and of T-killer cells.
- Interleukins-4 and -5 stimulate the B-cells to grow.
- Interleukin-6 stimulates B-cells to become plasma cells and produce antibodies.
- Gamma interferon activates more macrophages and T-killer cells, so they can attack invading cells, and activates plasma cells to produce more antibody.

Gamma interferon is actually a rather nonspecific defense made in response to viral infection. Virtually all cells in the body can make some form of interferon if they become infected, and while these cells will die, the interferon blocks virus infection in other cells. Interferons are specific to each type of animal, not to the virus that induces them. Research with interferon has elicited great excitement because of its potential as an anticancer agent.

Exercise 48.8 Think about the structure of a protein. If a pure protein is injected into a competent animal, do you imagine that the animal makes antibodies with only a single kind of antigen-binding site or a mixture of several antibodies with different kinds of antigen-binding sites?

48.10 Immunoglobulins are encoded by several gene segments.

One of the marvelous aspects of immunity is that each primary clone of lymphocytes becomes committed to making only a single kind of antigen-binding site. Immunoglobulin chains are encoded by several *gene segments,* each encoding a portion of the whole polypeptide; as lymphocytes differentiate, these segments are recombined with a certain randomness, and some DNA is eliminated, leaving only a few DNA sequences to be transcribed into messenger RNA and translated into protein. This process is similar to ordinary eucaryotic processing to produce a protein encoded by a gene interrupted by introns, but there the intron sequences are removed from RNA transcripts to make the messenger RNA. In lymphocytes, the DNA is permanently rearranged to bring a few gene segments together. The process is probably unique in differentiation, because a permanent change in genes apparently doesn't happen in ordinary differentiation.

Mice and humans (the species that have been studied most) have three families of immunoglobulin genes (Figure 48.19): one for the heavy chains and one for each of the two classes of light chains, called kappa and lambda. Each family is a cluster of gene segments on one chromosome. For instance, an L chain is encoded by three groups of segments: V_L (for most of the V region), a joining region J (for the rest of the V region), and C_L (for the C region) (Figure 48.20). The gene family for the kappa L chain contains about 300 different V_L gene segments and five J segments. In every B-lymphocyte precursor cell, a recombination combines a randomly selected V_L segment with one J segment and with the C_L segment to make the sequence coding for a whole L chain. Some additional variations in sequence arise from other random events during recombination. At the same time, random recombination makes a sequence coding for an H chain in a similar (but slightly more complicated) way.

Because different gene segments are combined, by chance, in every differentiating B-cell, each animal acquires an army of lymphocytes programmed to produce antibodies with many distinct antigen-binding sites, and capable of fighting every kind of pathogen. Some embryonic lymphocytes, however, will make antibodies to the animal's own antigens—to its blood proteins, to the proteins and polysaccharides of its cell surfaces, and so on. During development, clones making such antibodies are discarded, and at this point the animal has made the clear and crucial distinction between self and nonself.

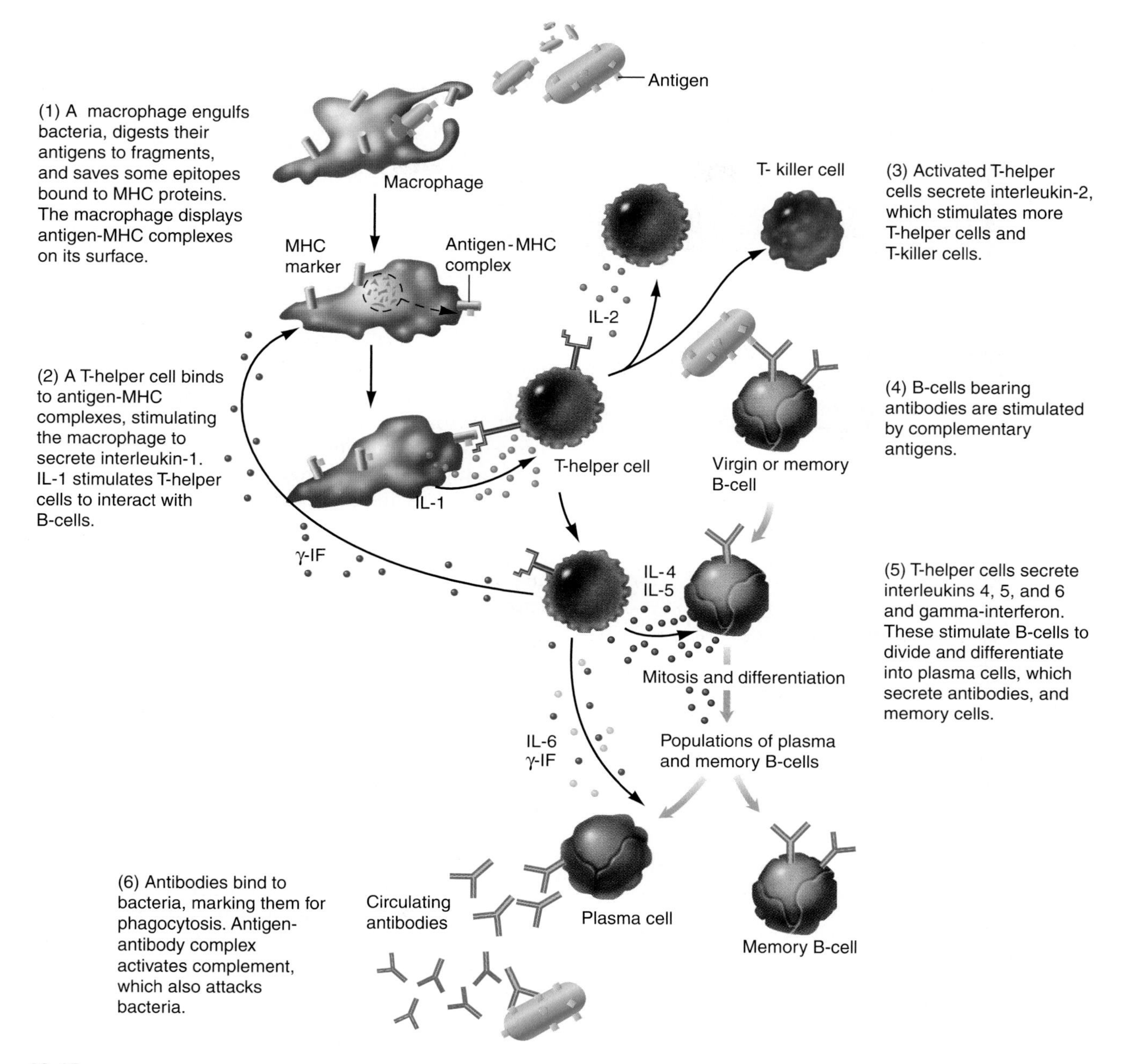

Figure 48.18

The production of antibodies requires a series of interactions among macrophages and lymphocytes, which stimulate one another through local cell mediators.

Exercise 48.9 For the L (kappa) chain, a mouse has 300 V gene segments and 5 J segments. How many possible kappa chains can it make, ignoring additional variations?

Exercise 48.10 For H chains, the mouse has 500 V segments; these are combined randomly with one of 12 D segments and then with one of 4 J segments to make a complete V region. How many possible H chains can it make, ignoring additional variations?

Exercise 48.11 Given these numbers for L and H chains, how many possible kinds of antigen-binding sites can the mouse make?

48.11 Immune and inflammatory processes are closely linked to the nervous system and endocrine controls.

It is easy to perceive the immune system as a lot of isolated cells only vaguely connected to the rest of the body, even though they are often localized in structures such as the lymphatic system. This viewpoint is gradually being replaced by the recognition that immunocytes are closely connected to the nervous system and are strongly influenced by hormonal controls; with this recognition, the immune system becomes another set of sensors and effectors on a par with traditionally recognized

Figure 48.19

Immunoglobulin genes form three clusters, located on different chromosomes.

Kappa family: $V_{\kappa 1}$ $V_{\kappa 2}$ $V_{\kappa 3}$ . . . C_{κ}

About 300 V-gene segments

Lambda family: $V_{\lambda 1}$ $V_{\lambda 2}$ $V_{\lambda 3}$ $V_{\lambda 4}$ . . . $C_{\lambda 1}$ $C_{\lambda 2}$ $C_{\lambda 3}$ $C_{\lambda 4}$

About 500 V-gene segments

Heavy family: V_{H1} V_{H2} V_{H3} . . . $C_{\mu 1}$ $C_{\gamma 4}$ $C_{\gamma 2}$ $C_{\gamma 1}$ $C_{\gamma 3}$ $C_{\alpha 2}$ $C_{\alpha 1}$ C_{δ} C_{ε}

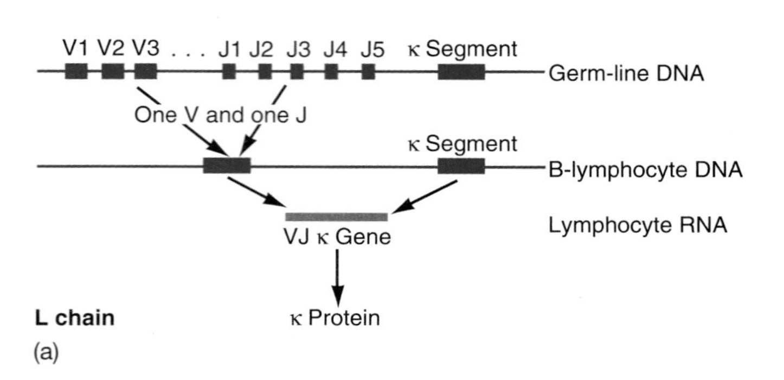

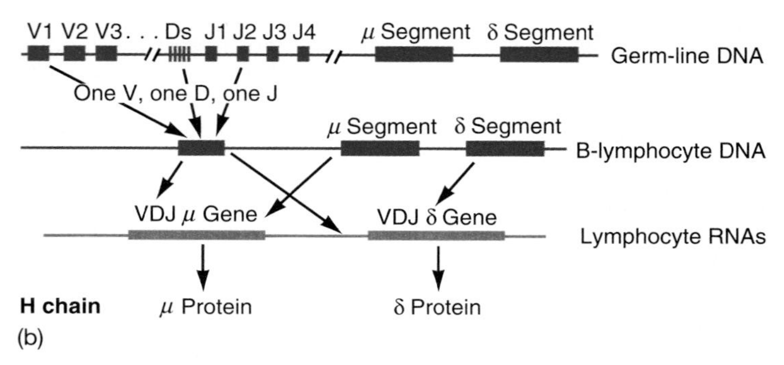

Figure 48.20

Functional immunoglobulin genes are created by recombination of gene segments. *(a)* An L (kappa) chain is encoded by one of about 300 V_L gene segments combined with one of five *J* segments and then with a C_L (κ) segment. The whole gene is made by recombination in the DNA and then by excision of intervening sequences from the messenger RNA to make a single *VJ*κ sequence. *(b)* An H (heavy) chain is made similarly, except that the variable region is coded by one of many V_H segments combined with a *D* and a *J* segment. This combination is then linked to a segment for a constant region, and it can be combined with a μ segment to make a complete μ chain for IgM and also with a δ segment to make a complete δ chain for IgD.

elements of the nervous system but specialized for dealing with a special class of stimuli—foreign invaders that might be pathogens.

The lore and literature of medicine has long included the knowledge that emotional factors are strongly linked to health. Psychosomatic medicine has shown how stresses and emotional traumas can make people sick, whereas people with generally stress-free, happy lives are more likely to stay well. Dentists know that the condition called acute necrotizing gingivitis, in which the normal oral flora becomes invasive and starts infections, is associated with stress. People who suffer from the autoimmune disease rheumatoid arthritis have been described as quiet, introverted, conscientious, conforming, self-sacrificing, limited in their ability to express anger, stubborn, rigid, and controlling. People who have experienced the death of a loved one and are feeling extreme grief are more susceptible than average to infectious disease and cancer. Studies have shown that the levels of natural killer cells, which are important in destroying cancer cells, are higher in patients who have good social support systems and positive, aggressive attitudes toward fighting the disease.

A growing science variously called *neuroimmunology* or *psychoneuroimmunology* has been elucidating the physical basis for some of these correlations. First, it is clear that the nervous system innervates key lymphoid tissues, including the thymus gland, bone marrow, spleen, lymph nodes, and probably other sites. It is particularly interesting that neuropeptides, such as vasoactive intestinal peptide, substance P, and neuropeptide Y, have been identified as the neurotransmitters in some of these sites; we pointed out in Section 42.11 that these neuropeptides may act as mediators between the nervous and immune systems. Furthermore, the various leucocytes not only have receptors for certain neuropeptides but also produce neuropeptides themselves. Though the data are too complicated to sort into a comprehensive picture, this implies two-way communication between immunocytes and the nervous system, as well as communication among immunocytes by means of neuropeptides. Several hormones have strong effects on immune functions; in general, growth hormone, insulin, and thyroxine enhance immune reactions, and steroid hormones such as glucocorticoids, androgens, and estrogens suppress these reactions. All this information suggests general ways in which immunocytes relate to the nervous and endocrine systems, and one clear regulatory circuit involving the adrenal cortex has been demonstrated (Figure 48.21).

Furthermore, some immune responses can be conditioned just like other behaviors; Robert Ader and Nicholas Cohen made this surprising discovery when Ader tried to condition rats to avoid drinking a saccharine solution by associating it with an immunosuppressive drug, cyclophosphamide, that induces nausea. As expected, the rats learned to avoid the saccharine solution, but those that drank the most saccharine also began to die at an unusually high rate. It turned out that they had

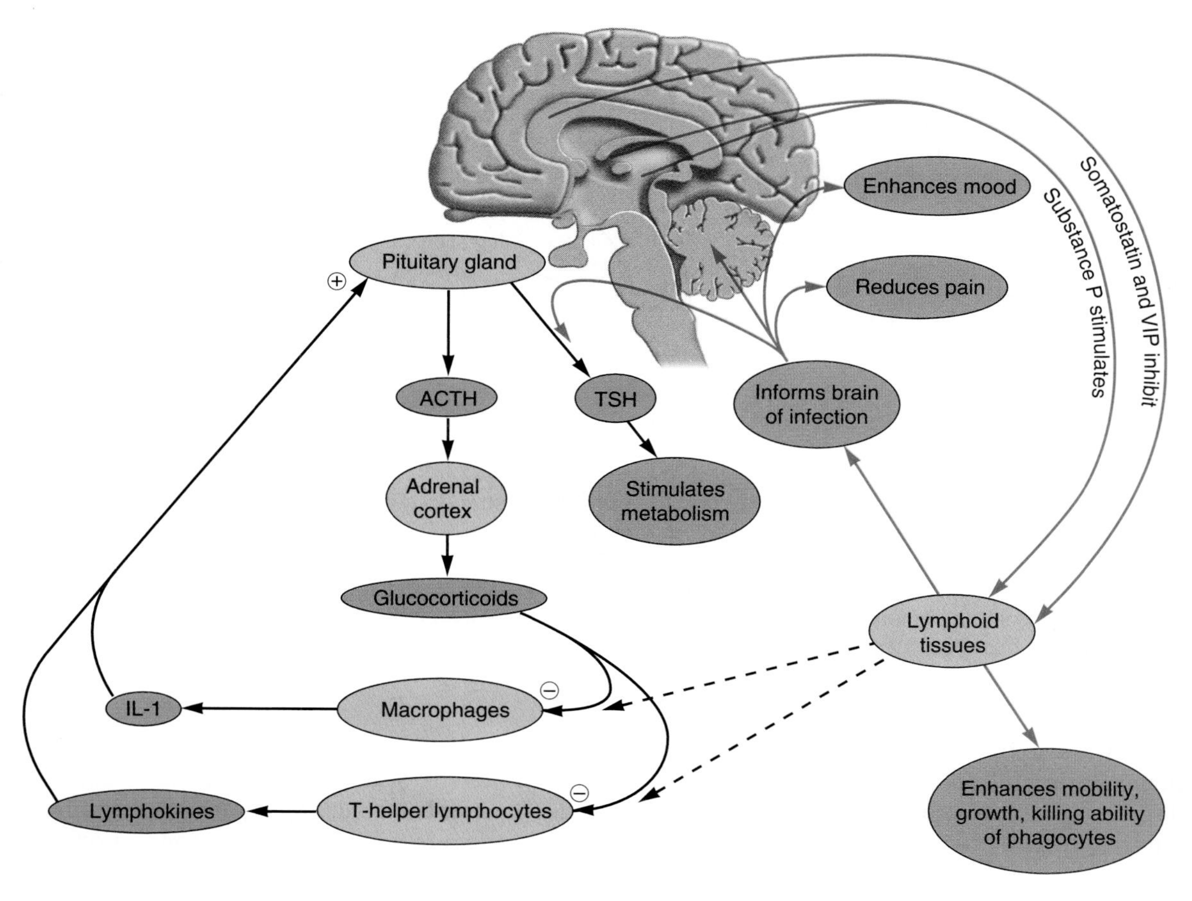

Figure 48.21
In one interaction between the immune and endocrine systems *(black arrows),* the interleukins (IL-1) and lymphokines produced by macrophages and T-helper lymphocytes stimulate the pituitary gland, which produces ACTH to stimulate the adrenal cortex to produce glucocorticoids; the latter inhibits the activities of the macrophages and lymphocytes. In interactions known less specifically *(red arrows),* the brain regulates activities of the lymphoid tissues through factors such as substance P, somatostatin, and vasoactive intestinal peptide (VIP), while the lymphoid tissues send signals to the brain that affect metabolism, mood, and perception of pain.

learned to associate the sweet solution not only with nausea but also with immunosuppression and were suppressing their immune systems as they drank the saccharine.

Conditioning and learning, Section 49.10.

A variety of nontraditional medical practices and pseudosciences, sometimes employing vague philosophies about "mind-body" connections, have used psychological and other practices to cure illness: massage techniques, yoga, visualization techniques, meditation, prayer, herbal medicines, dietary practices, and others that traditional physicians variously ignore or scoff at. While these practices have often depended on anecdotal evidence and lacked real scientific rigor, several practitioners have been employing them extensively in well-controlled studies, along with standard medical methods, to improve people's health and cure diseases. Cancer patients, for example, have been able to extend their lives and successfully fight their cancers by supplementing drug and radiation therapies with methods that teach them positive, aggressive attitudes toward their condition. As neuroimmunology advances, there will be a stronger basis for psychological and other treatments that emphasize maintaining a healthy immune system, rather than relying on medical practices to cure diseases after the body's natural defenses have failed.

D. The dark side of immunity

48.12 The AIDS virus attacks the heart of the immune system.

One of the most insidious infectious agents to emerge in a long time is the **human immunodeficiency virus (HIV),** the causative agent of **acquired immune deficiency syndrome (AIDS).** One student called it a "genius virus" because it destroys precisely the part of the immune system that is needed to fight it.

AIDS was recognized as a distinct syndrome in the late 1970s when several young men developed Kaposi's sarcoma, a rare cancer of the blood vessel linings that usually affects only elderly Italian or Jewish men. At the same time, several men with homosexual histories appeared who had immune deficiencies, characterized by opportunistic infections, depletion of T-helper cells, and sometimes Kaposi's sarcoma. The new syndrome was named AIDS by the Centers for Disease Control (CDC) in 1981. During its early years, the syndrome was found mostly among homosexual men, intravenous drug users, and recipients of frequent blood transfusions. However, there is nothing intrinsic to the disease that confines it to these populations; AIDS is epidemic among heterosexuals in Africa, and human sexual practices are spreading it quickly throughout the world.

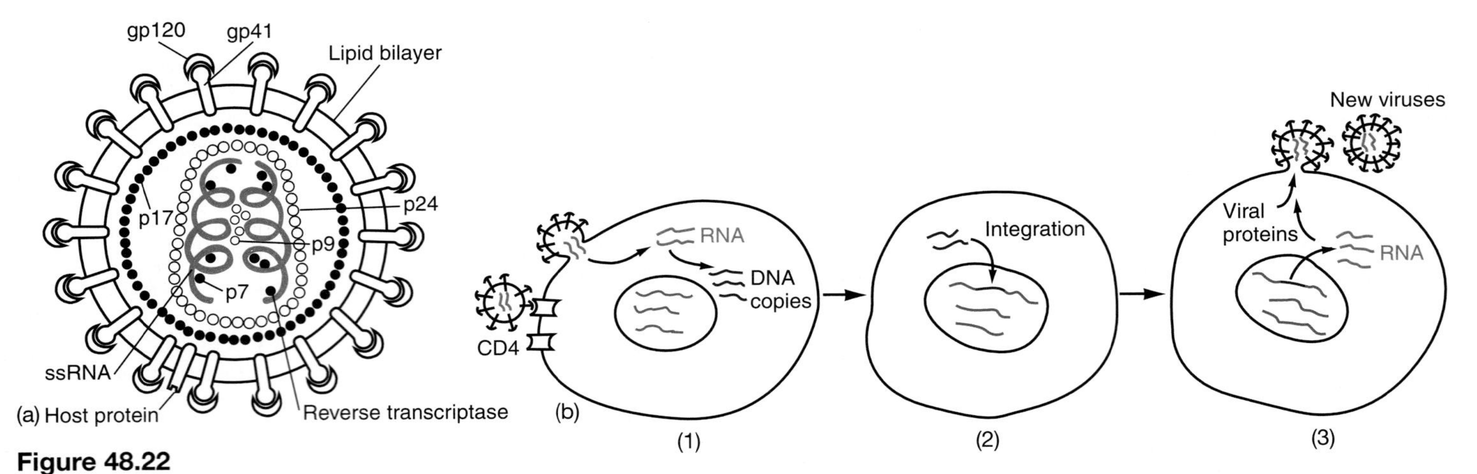

Figure 48.22

(a) An HIV virion is a complex structure; notice the location of the gp120 proteins and the single-stranded RNA genome (ssRNA). *(b)* HIV infects a cell through attachment of its gp120 protein to a CD4 receptor on some cell, usually a T-helper cell. *(1)* The outer membrane of the virion fuses with the cell membrane, and the inner layers of viral protein are removed by cellular enzymes, thus freeing the RNA genome *(red).* Reverse transcriptase packaged in the virion makes a DNA copy of the genome *(blue),* which integrates itself as a provirus somewhere in the cellular genome, just as a prophage integrates into a bacterial genome *(2).* The integrated genome may lie dormant, or the infected cell may produce many new viruses *(3).* Eventually the host becomes depleted of helper cells and loses its ability to resist infections.

The causative agent of AIDS was identified in 1982–83 by Luc Montagnier at the Pasteur Institute and by Robert Gallo at the National Institutes of Health. It is a retrovirus—an RNA virus that produces a DNA copy of its genome, which integrates into the chromosomes of its host (Figure 48.22). The principal surface protein of the virion, gp120, determines what cells the virus will attack; it has a high affinity for the CD4 protein found mostly on the surface of T-helper cells and, to some extent, on macrophages and other leucocytes. CD4 also occurs on some cells in the nervous system, which may account for the neurological effects of AIDS. gp120 binds to CD4, and the membrane of the virus fuses with the cell membrane, thus putting the viral core into the cytoplasm. Cellular enzymes remove the inner coats of the virus, releasing the RNA genome. Reverse transcriptase, packaged with the genome, then transcribes the RNA genome into a DNA copy, which integrates into a host chromosome.

The provirus DNA, integrated in a chromosome, may lie dormant. It now appears that cells containing the provirus start producing massive numbers of new viruses; for a while, the body is able to resist them, but T-helper cells are depleted over a period of years and the infected person loses the ability to fight off even quite ordinary infectious agents.

Selenium (Se) is an essential trace nutrient for many organisms, including humans. Selenium has gradually been implicated in several disease processes; it opposes the formation of cancers, including those caused by certain tumor viruses, and Se deficiency causes certain kinds of chronic damage to heart muscle (cardiomyopathy). We noted in Section 14.4 that Se is incorporated into certain proteins as an unusual amino acid, selenocysteine (Sec). Remember that Sec is encoded by UGA (which is usually a chain terminator) in a special RNA sequence that interacts with tRNAs carrying Sec, so Sec is incorporated if sufficient selenium is present. Selenocysteine is being found in a widening circle of viral, bacterial, and mammalian proteins, and several laboratories are now pursuing the connections between the actions of these proteins and the presence of selenium. Will Taylor and his associates have identified the Sec codon in HIV genes, including *nef,* the critical control gene associated with HIV pathogenesis. It is also clear that Se deficiency is an important factor in HIV pathogenesis and that HIV-infected people undergoing the transition to full-blown AIDS have a reduced level of selenium. These facts imply that a lack of Se leads to truncation of the Nef protein, and perhaps others, leading to a change in HIV growth. Although the story is incomplete, it emphasizes the subtlety of regulatory mechanisms and shows that obscure biochemical processes can have great implications for human health.

In spite of considerable effort, including intense recombinant-DNA work, no cure for AIDS has yet been discovered. The nucleoside analogue 3-azidodeoxy-thymidine (AZT) interferes with reverse transcription and effectively holds the disease symptoms in check for a time, but it is very expensive, has side effects, and does not provide a permanent cure. Purified CD4 protein will combine with virus particles and block their infection of new cells, but these proteins have short half-lives. Investigators are trying to engineer better blocking agents, but the prospect for anyone who becomes infected with AIDS is still very dark for the near future.

The only sensible course of action is to avoid infection. After some initial panic over casual association with AIDS patients, people now generally seem to feel assured that the disease is only transmitted by either direct contact with sexual fluids or through blood or blood products. This is a concern for health-care workers, a few of whom have been infected by the blood of infected patients during surgery or through infected hypodermic syringes. Also, the general public must realize that unprotected sexual contact is dangerous, especially since a contact with one person actually puts you into the entire network of people with whom he or she has been sexually connected in the past.

48.13 An animal's immune system sometimes turns against it.

We said earlier that inflammation and immunity are a two-edged sword, and sometimes the immune system can seriously damage the animal it is supposed to protect. Some of the most common adverse effects are **hypersensitivity** reactions, including **allergy.** Let's consider a couple of examples.

Anaphylaxis, a type of hypersensitivity, was discovered by Paul Portier and Charles Richet in 1902 when they injected a dog with a bit of sea anemone poison, hoping to induce immunity to the agent. But when they injected more poison, the dog went into anaphylactic shock and died. Portier and Richet then invented the term "anaphylaxis" to mean the opposite of prophylaxis (see Section 48.2). Anaphylactic reactions may be relatively mild, such as a little localized itching or hives, or quite generalized and severe. People going into anaphylactic shock have a skin rash or flushed skin, and may experience headache and itching. Then they start to develop respiratory distress—choking, wheezing, labored breathing—followed by vomiting and diarrhea, generalized shock, and even death. These reactions are all consequences of inflammation, beginning when mast cells are activated via an immunoglobulin, IgE (also called reagin), on their surface. When cell-bound IgE binds to its complementary antigen, it stimulates the release of histamine and initiates an inflammatory response. Histamine particularly affects so-called shock tissues: The epithelium of breathing passages disintegrates and starts to produce excessive mucus, while the smooth muscle of the respiratory tree contracts and inhibits breathing. An extreme reaction to histamine includes generalized edema, a drop in blood pressure, and shock. In Portier and Richet's experiment, the poison acted as an allergen and induced allergy rather than acting as an immunogen; the first injection sensitized the animal by eliciting the formation of complementary IgE molecules.

Twenty years later, C. Präusnitz and H. Kustner showed that sensitivity is transferrable, with a test that is still used for investigating allergy. Kustner broke out in hives when he came in contact with fish, so Präusnitz tried injecting a bit of Kustner's serum under his own skin. When he then injected fish extract into the same area, he experienced a strong, localized hive reaction—a wheal (a localized edema) and flare (a surrounding zone of reddening), which are typical inflammatory signs. The reaction in Präusnitz's skin resulted from Kustner's specific IgE that was complementary to fish antigens.

It is still not clear why 10–20 percent of humans have strong allergies to all kinds of substances. The tendency is certainly inherited, but the immune response is very complex

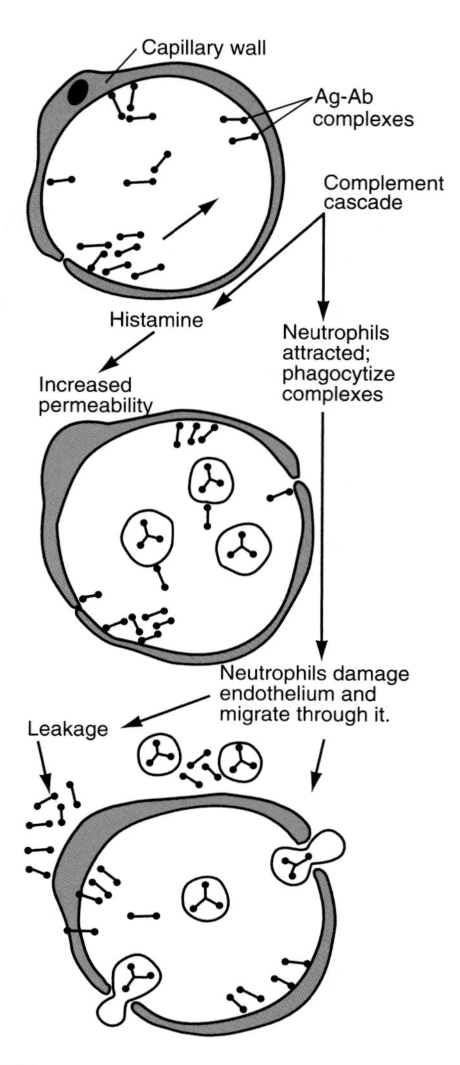

Figure 48.23
Acute glomerulonephritis is the result of antigen-antibody (Ag-Ab) complexes inducing inflammation around the glomeruli of the kidney. (The suffix *-itis* indicates an inflammation.)

genetically. One factor seems to be the persistence of specific IgE molecules in allergic people; nonallergic people eliminate these molecules after a few months.

Complex-mediated hypersensitivity, which results from antigen-antibody complexes forming in certain tissues, is quite a different kind of reaction. Although circulating antibodies—IgG and IgM—help eliminate antigens, small antigen-antibody complexes can pass through inflamed capillaries and initiate the blood-clotting and complement cascades. The blood-clotting reaction produces small clots (microthrombi) that block the circulation; the complement cascade produces anaphylatoxins that stimulate histamine release from mast cells, initiating a localized inflammation. This happens commonly in the kidneys, producing an acute glomerulonephritis (Figure 48.23).

Quite a different problem arises occasionally if the immune system reacts against normal body tissues, creating an

Figure 48.24

The proteins of the immunoglobulin superfamily share many features and homologies, indicating their common evolution from an ancestral protein. Like the immunoglobulins on B-cells, T-cell receptors make T-cells specific for antigens. Poly-Ig receptor transports IgA across mucosal membranes. N-CAM is a cell adhesion molecule discussed in Chapter 20. The various CD proteins mark cells of the immune system. Thy-1 is on T-cells and neurons. The Fc receptor of phagocytes recognizes IgG and IgM molecules bound to antigens. Other proteins in this superfamily have poorly known functions.

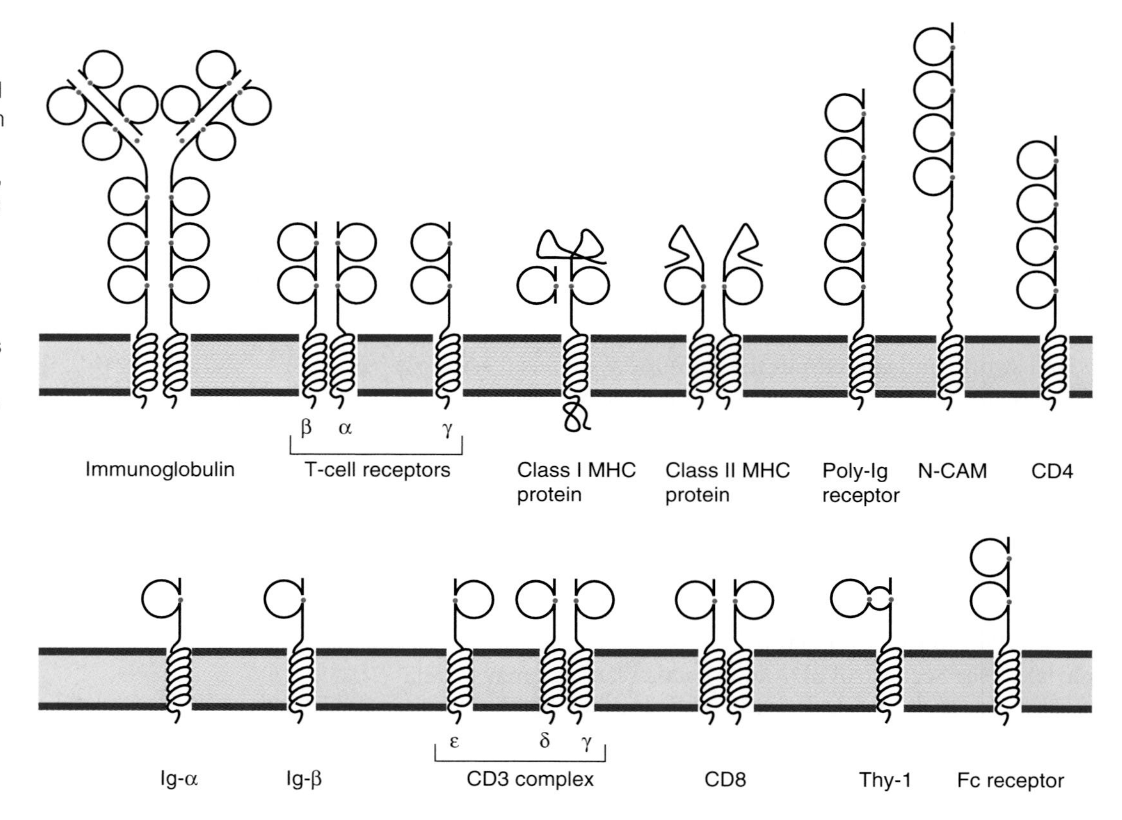

autoimmune disease. This is fundamentally a failure to distinguish self from nonself, and it happens when molecules normally hidden from the immune system become exposed to it. In *rheumatoid arthritis,* for instance, a viral infection or some comparable disturbance allows antibodies to form against collagen and other connective-tissue molecules; long-term stimulation produces excess IgG, which forms antigen-antibody complexes in the synovial membranes of joints, leading to local inflammation, with swelling and radiating pain. *Lupus erythematosus* is a slow-progressing disease characterized by lesions of the skin and internal organs and general wasting. In this disease, antibodies are apparently being made to intracellular proteins, such as those of the nuclear envelope. Some people with lupus develop symmetrical, butterfly-shaped patches across the middle of their faces, and these patches tend to grow hair. To the superstitious, their faces resemble wolves (*Lupus* means "wolf.") Furthermore, people with lupus are very sensitive to light and tend to restrict their outdoor excursions to night hours. It's not hard to see how the legends of werewolves arose in an age of superstition, based on the appearance and behavior of people suffering from lupus erythematosus.

Immunity is not a perfect system, although it has been evolving for many millions of years. Its incredible complexity, along with imperfect internal controls, can create serious side effects. We have shown some of the dark side of immunity to make another point: Disease processes are simply one reaction of ordinary tissue cells to foreign materials and their toxic products. We understand disease just as we understand the normal behavior of healthy organisms—in terms of interactions between specific cells and specific molecules.

Coda The theme of molecular interactions as the basis for biological processes reaches its height in discussing inflammation and immunity. A diagram of these processes, spread over a large sheet of paper, would be a maze of arrows indicating molecules and cells binding to each other, activating one another, forming molecular complexes, stimulating cells to produce still other molecules or inhibiting this production. We have not tried to produce such a diagram here because it would be too large, and perhaps rather intimidating, but one of its virtues is to inspire a sense of wonder at the beauty and complexity of processes occurring in our bodies. We are, in the biblical phrase, "fearfully and wonderfully made."

How could such complicated systems evolve? Part of the answer comes from closely examining the structures of the most important proteins involved in immunity (Figure 48.24). We can now see how similar and closely related they are; they are said to form a superfamily of proteins, since each drawing represents several variants that form families. We can now start to draw scenarios in which one primitive protein acquired selective value because it could bind to some pathogens and inhibit them. From that starting point, the now-familiar pattern of gene duplications and divergencies could begin to form other proteins with more specialized functions and even with great diversity. Since infections are among the most serious threats to an animal's survival, even a slight protection must have great selective value.

Summary

1. Animals' nonspecific defenses against infection include: mechanical barriers, such as skin; skin glands that produce antibacterial oils; ciliated epithelia that sweep away invading organisms; enzymes that destroy potential pathogens; and phagocytic leucocytes (white blood cells).
2. Acute inflammation, a general reaction to injury, is characterized by swelling, heat, pain, redness, and loss of function. Inflammation is induced by cuts and burns, and by potential pathogens; it eliminates the invaders and heals the damaged tissue.
3. Inflammation depends on a number of leucocytes: Lymphocytes mediate immune reactions; macrophages and neutrophils are phagocytes, and macrophages process antigens for presentation to lymphocytes. Granulocytes (polymorphonuclear leucocytes) include neutrophils, basophils (and mast cells), and eosinophils.
4. Acute inflammation is a reaction of the microcirculation to injury or infection. Inflammation begins with the release of histamine and other factors from mast cells in tissue. Histamine, binding to receptors on cells of the microcirculation, induces pooling of blood, vasodilation, and increased permeability of capillaries.
5. During inflammation, phagocytic cells (neutrophils and macrophages) are called into the injured area and are induced to phagocytize foreign cells. Lymphocytes are called into the area, where they produce antibodies to fight infection.
6. Inflammation ends with the cleanup of damaged tissue, formation of new capillary beds, and growth of new supporting tissue and new cells.
7. Three reaction cascades are activated in inflammation. The blood-clotting cascade stops bleeding, complement proteins attack invading bacteria, and kinins accelerate the inflammatory process.
8. The immune system is a defense superimposed over inflammation that recognizes specific molecular structures. It produces two kinds of responses, humoral and cell-mediated, which both recognize antigens (molecules foreign to the animal) and respond to them in ways that inactivate the antigens.
9. Antibodies are highly specific proteins made in response to antigens. Antigens must be large molecules with distinctive shapes. Each type of antibody is able to recognize and combine with only a small part of the antigen, an antigenic determinant or epitope. Antibodies and antigens bind to each other to form complexes.
10. The humoral immune response operates through B-lymphocytes, which produce antibodies that circlate in the blood and lymph or are secreted in mucus, as in the intestine. This type of response is primarily effective against bacteria, free viruses, and foreign proteins in solution.
11. T-lymphocytes mature in the thymus gland. T-cells include helper cells that are involved in the humoral immune response.
12. The cell-mediated response occurs principally through T-lymphocytes called T-killer or T-cytotoxic cells; they recognize foreign cells, or cells bearing foreign antigens, by means of antibody-like proteins on their surfaces that bind to cellular antigens. They destroy target cells by creating pores in their membranes. The cell-mediated response is primarily effective against intracellular viruses (whose antigens appear on the surfaces of cells they are growing in) and animal parasites.
13. All immunoglobulins have the same basic, symmetrical Y-shaped structure, made of two heavy chains and two light chains linked by disulfide bonds.
14. Both heavy and light chains consist of variable regions and constant regions. The variable regions of the two chains come together at the arms of the Y, where they form antigen-binding sites with a distinctive shape that are able to bind to antigens.
15. Tissue cells are marked by distinctive proteins of the major histocompatibility complex (MHC). Type I MHC proteins mark all nucleated cells; if such a cell is infected by viruses or mutates to become a tumor cell, it has distinctive surface antigens that combine with the MHC protein and are recognized by T-cytotoxic cells. Such cells are then destroyed. Type II MHC proteins appear on macrophages and other cells of the immune system. Macrophages process antigens and present antigenic determinants on their surfaces in complexes with these MHC proteins.
16. Each primary clone of differentiated B-lymphocytes makes antibodies with only a single shape of antigen-binding site. Lymphocytes become specialized during embryonic development. An antigen selects clones that produce complementary antibodies and stimulates them to grow into secondary clones and produce their antibodies.
17. When stimulated by antigen from macrophages, B-lymphocytes differentiate into plasma cells, which produce antibodies, and memory cells, which remember they have encountered an antigen and respond more rapidly and with more antibody when stimulated again.
18. The interactions among immune cells involve many specific proteins, including interleukins and interferons, that carry signals between cells.
19. Several gene segments carry the information for amino acid sequences of immunoglobulins. Each chain is made by random recombination of a variable segment: a J (joining) segment, a D segment (for heavy chains), and a constant segment. Random selection of these segments, and random events during recombination, produce antibodies with enormous numbers of different kinds of antigen-binding sites.
20. AIDS is caused by a human immunodeficiency virus (HIV), a retrovirus whose RNA genome makes a DNA copy, which integrates into the cellular genome. HIV has a strong affinity for the CD4 protein that is abundant on T-helper cells. The virus therefore infects and kills precisely those cells most necessary for making antibodies against it. There is no known cure at present.
21. Hypersensitivity reactions damage the host, primarily by inducing parts of the inflammatory response inappropriately—for instance, by forming antibody-antigen complexes that initiate the complement cascade and break down tissues.
22. Autoimmune diseases occur when the immune system reacts against self-antigens it normally does not see.

Key Terms

antigen 1023
titer 1024
primary response 1024
secondary response 1024
primary lymphoid tissue 1024
B-lymphocyte/B-cell 1024
humoral immunity 1025
T-lymphocyte/T-cell 1025
cell-mediated immunity 1025
natural killer (NK) cells 1025
secondary lymphoid tissue 1025
immunoglobulin 1025
light (L) chain 1025
heavy (H) chain 1025
constant (C) region 1025
variable (V) region 1025
antigen-binding site 1025
epitope/antigenic determinant 1026
opsonize 1026
major histocompatibility complex (MHC) 1027
clonal selection model 1028
plasma cell 1029
memory cell 1029
interleukin 1029
interferon 1029
monoclonal antibody 1030
human immunodeficiency virus (HIV) 1033
acquired immune deficiency syndrome (AIDS) 1033
hypersensitivity 1035
allergy 1035
autoimmune disease 1036

Multiple-Choice Questions

1. All of the following except ______ are nonspecific defense mechanisms.
   a. macrophages
   b. acute inflammation
   c. secretions of skin and mucous membranes
   d. production of complement
   e. production of antibodies
2. Leucocytes are found in the
   a. blood and lymph.
   b. lymph nodes.
   c. tissue fluid.
   d. connective tissue spaces.
   e. all of the above.
3. Which cell is most responsible for initiating an episode of acute inflammation?
   a. mast cell
   b. macrophage
   c. lymphocyte
   d. erythrocyte
   e. eosinophil
4. Which cell is not correctly matched with its secretion or function?
   a. T-lymphocyte—cell-mediated immunity
   b. macrophage—phagocytosis
   c. neutrophil—immunoglobulins
   d. platelet—clotting
   e. B-lymphocyte—humoral immunity
5. Which term most accurately describes the action of thrombin?
   a. hormone
   b. paracrine secretion
   c. anticoagulant
   d. enzyme
   e. forms substance of blood clot
6. Which of the following is not correct about complement?
   a. attracts neutrophils to site of injury
   b. a subset of lymphocytes
   c. forms membrane attack complexes that perforate bacterial surfaces
   d. enhances phagocytic activity of macrophages
   e. can initiate acute inflammation
7. Which is characteristic of antibodies?
   a. produced by mast cells
   b. released by splitting of megakaryocytes
   c. circulate in blood and secreted by mucous membranes
   d. produced only by cells that mature in lymph nodes
   e. can be a protein or a polysaccharide
8. All of the following except ______ are part of immunoglobulins.
   a. epitope
   b. light chain
   c. heavy chain
   d. antigen-binding site
   e. C and V regions
9. Antigenic determinants specifically bind to
   a. epitopes.
   b. antigen-binding sites.
   c. C-regions of the heavy and light chains.
   d. complement.
   e. B-lymphocytes.
10. The key concept of the clonal selection model is that
   a. each of us can form many different kinds of antibodies.
   b. particular antigens react with one of the five types of immunoglobulins.
   c. each lymphocyte produces only one kind of antibody.
   d. B-lymphocytes undergo mitosis to form a clone of plasma cells.
   e. specific epitopes initiate mitotic division of one small subset of lymphocytes.

True-False Questions

Mark each statement true or false, and if false, restate it to make it true.

1. In developmental terms, a monocyte is to a macrophage as a basophil is to a mast cell.
2. Interleukin-1 is released from neutrophils following stimulation by macrophages.
3. If administered immediately following a heart attack or stroke, tissue plasminogen activator may be effective in causing fibrinolysis.
4. The activity of an antiserum is quantified by determining the concentration of the protein antibody.
5. B-cells mature in bone marrow while T-cells mature in the thyroid gland.
6. Circulating antibodies are the primary defense against virus particles that have invaded cells.
7. Each antibody is made of two polypeptide chains and has one antigen-binding site.
8. Class I MHC proteins work in the humoral response, while class II MHC proteins function in cell-mediated immunity.
9. The determination between the proteins of self and nonself is subject to change until the age of maturity.
10. HIV is a DNA-containing virus that integrates within the genome of cytotoxic T-cells, thereby destroying the very cells that could be effective in controlling the virus.

Concept Questions

1. List the signs of acute inflammation and explain the immediate cause of each.
2. Contrast the location and functions of the primary lymphoid tissues and the secondary lymphoid tissues.
3. Keeping in mind that each antibody is a product of genes, how is it possible to produce more kinds of antibodies than we have genes to direct their synthesis?
4. Explain how one's own tumor cells or virally infected cells are targeted for recognition as nonself.

5. Explain the differences between the primary and secondary immune responses.

Additional Reading

Boon, Thiery. "Teaching the Immune System to Fight Cancer." *Scientific American,* March 1993, p. 32. The search for ways to induce the immune system to find antigens that it normally ignores on cancer cells and fight the disease.

Buisseret, Paul D. "Allergy." *Scientific American,* August 1982, p. 86. A process of immunity gone wrong.

Edelson, Richard L., and Joseph M. Fink. "The Immunologic Function of Skin." *Scientific American,* June 1985, p. 46. The body's largest organ is more than a passive protective covering; it is also an active element of the immune system. Specialized cells in the skin have interacting roles in the response to foreign invaders.

Goodfield, June. *An Imagined World: A Story of Scientific Discovery.* Harper & Row, New York, 1981. The author spent several months in the laboratory of an immunologist and tells the story of her research program.

Johnson, Howard M., Fuller W. Bazer, Brian E. Szente, and Michael A. Jarpe. "How Interferons Fight Disease." *Scientific American,* May 1994, p. 40. The role of these proteins in fighting viral infections and cancer.

Nowak, Martin A., and Andrew J. McMichael. "How HIV Defeats the Immune System." *Scientific American,* August 1995, p. 42. The process is a competition between the virus and the body's defenses.

Roitt, Ivan. *Essential Immunology,* 7th ed. Blackwell Scientific Publications, London, 1991.

Sapolsky, Robert M. "Stress in the Wild." *Scientific American,* January 1990, p. 116. Studies of free-ranging baboons help explain why humans can differ in their vulnerability to stress-related diseases.

Smith, Kendall A. "Interleukin-2." *Scientific American,* March 1990, p. 50. The first hormone of the immune system to be recognized, interleukin-2 helps the body mount a defense against microorganisms by triggering the multiplication of only those cells that attack an invader.

Internet Resource

To further explore the content of this chapter, log on to the web site at:

http://www.mhhe.com/biosci/genbio/guttman/

49 Fundamentals of Animal Behavior

A Great Blue Heron searches for food in a marsh.

Key Concepts

49.1 Ethology is the science of animal behavior.

49.2 A behavior is a strategy for survival.

49.3 Behavior patterns are mostly instinctive, but usually require maturation.

49.4 Simple, unlearned behaviors are fixed-action patterns.

49.5 Complex behaviors are sequences of fixed-action patterns.

49.6 Releasers are highly selected objects and events from the environment.

49.7 Behavioral patterns are often organized hierarchically.

49.8 Animals of the same species create ritualized behavior by mutually inducing fixed-action patterns in one another.

49.9 Many behavior patterns require the imprinting of specific environmental information.

49.10 Learning is a mechanism for rapid adaptation to new situations.

49.11 Behavior can be modified through both classical and operant conditioning.

49.12 Many patterns of behavior are traditions that have been learned.

49.13 Animals can use various cues to orient themselves.

In a cold, gray dawn, by the misty shore of a quiet inlet, a Great Blue Heron stands motionless in the shallow water, hunting for food. Its patience is remarkable, for the bird may stand as if frozen for many minutes, only occasionally slashing the water with its head to retrieve a fish. A careful observer, however, will see that the bird has quite a repertoire of behaviors. Standing upright is its most common activity and occasionally a bird will start to Walk Slowly. When it stops, it may Peer Over, extending its neck and pointing its bill straight down toward the water to get a binocular view of potential prey. Or, it may engage in Head Cocking, turning its head to the side so one eye points straight down and the other straight up. To stalk its prey, or when preparing to strike, the heron may go into a Crouched posture, with its legs bent and its body parallel to the ground, and sometimes it may Face Down, extending its neck, head, and bill straight downward, perhaps to probe something in the water.

Why does the heron act in these ways? Obviously to catch its food, but that's only a general answer. These behaviors are actually encoded in the genomes of the birds. Each bird has few options in its behavior, but they are good options, functional options, that have been selected over millions of years because they serve these birds well. Students of bird behavior commonly capitalize the names of these behaviors to emphasize that they are patterned, stereotyped actions observed in many herons again and again. Behaviors of this kind are the main subject of this chapter. ■

49.1 Ethology is the science of animal behavior.

Watching an animal such as a heron, we can catalog its repertoire of behaviors and see it shift from one type to another (Figure 49.1). What brings on each behavior? Is it responding to stimuli that set off each pattern? Then is the shift into a new behavior built into its nervous system genetically, or does it learn, as it matures, which activity is best? How much can a heron learn? Does it learn from its own experience, or is it taught some behaviors by its elders? To what extent is the bird acting consciously and intelligently?

Questions like these are what make **ethology,** the biological science of behavior, so fascinating. The popularity of the many television programs showing animals in their natural habitats testifies to the endless fascination that animal behavior holds for us. The trouble with many of these programs is their anthropomorphism, making it appear that animals have thoughts and feelings like humans. Certainly many animals are able to think, and many must also experience emotions similar to ours, although we have no reliable way of knowing this. Yet the behavior of most animals can only be understood if we see them as automatons whose genetic programs strictly determine their actions so that they can hardly even learn new behaviors, much less think.

Behavior is not easy to define; it includes all of an animal's overt, observable activities, including fighting and avoiding fights, searching for food and feeding, courtship and mating, and all the actions it uses for communication. Behavior is usually confined to muscular activity, but it includes the action of glands, too (salivating and sweating, for instance), and there is no sharp line between behavior and an animal's other activities. Most of what we will discuss in this chapter comes from ethological research, which was pioneered by Charles Whitman and Oskar Heinroth early in this century and then highly developed by the work of Nikolaas Tinbergen and Konrad Lorenz. Some information from classical psychology has already fit into this picture, and now deeper insights into the physical foundations of behavior are coming out of investigations into the simple

Figure 49.1

The Great Blue Heron, a common North American bird, has quite a repertoire of hunting and feeding behaviors, including *(a)* Standing Upright, *(b)* Peering Over, *(c)* Head Cocking, *(d)* Crouching, *(e)* Facing Down, and *(f)* Walking Slowly.

nervous systems of certain animals. Many of our examples will involve birds, vertebrates that are easily observed in their natural habitats and make excellent laboratory subjects.

49.2 A behavior is a strategy for survival.

Why does a heron behave as it does? After all, a bird could hunt for food in a lake or marsh in various ways. It could dive like a sea duck, or tip over and stretch its neck down like a pond duck. It could spin rapidly to stir up small organisms, as a phalarope does; it could fly back and forth low over the water, as gulls and terns do. Each behavior is a strategy in a game of life. Through natural selection, each species evolves strategies that work and make it successful. As with biological structures, a species discards a strategy if individuals that try to use that strategy reproduce poorly or die. Herons have been successful for millions of years by using their distinctive feeding patterns, and that is why present-day herons use them.

The game of life shifts continuously between solitaire and poker. Much of an organism's activity consists of internal physiological events, affecting no one but itself. In this way, it is playing solitaire, and if it finds a good strategy, so much the better for itself and the genes it carries. An animal whose genes give it a slightly improved kidney, the better to resist drought, may be able to move into an unoccupied desert niche. Success at solitaire hurts no one else, except that each surviving organism is a competitor for resources.

Behavior, however, is largely poker, and usually an individual wins only by diminishing the winnings of another. Here is where different strategies may compete, and where the strategy of one individual depends very much on what others do. The ultimate question is whether a particular behavioral pattern is an **evolutionarily stable strategy (ESS),** a behavior that an animal can maintain against possible counter-behaviors by other organisms. John Maynard Smith has analyzed these situations very neatly by means of game theory, and some of his simpler cases are very illuminating. Simple game theory, based on the work of the extraordinary mathematician John von Neumann, assigns win and loss values to a person's options in different situations. In one simple child's game, for example, two players simultaneously hold out one finger or two fingers; one wins a penny if the sum is even, and the other wins a penny if the sum is odd. This game is represented by the matrix:

	One	Two
One	1	−1
Two	−1	1

where "one" and "two" mean the number of fingers shown, and the numbers give the payoff to the player at the left side of the matrix. In this game, of course, the players have equal payoffs, and strategy lies in choosing a sequence and anticipating the opponent's moves.

Game theory can be applied to a population of animals with genetically determined behavior patterns as they come into conflict with one another while searching for food or other resources. Here, of course, the animals are not consciously making rational choices but rather are evolving behavioral patterns that might be successful. This kind of quantitative analysis lifts the study of behavior beyond the realm of anecdotes and helps identify the crucial factors involved and their interactions. The following hawk/dove exercise shows the kind of analysis that can predict the outcome of different behavioral strategies and helps us understand the role of behavior in evolution.

As with political stances on war, imagine that each animal can be either a *dove* (peaceful) or a *hawk* (aggressive). Suppose all initially have genes for dove behavior. When two individuals come into potential conflict, they may engage in some display or ritual conflict, and then one of them meekly gives up. On the average, each individual wins half its conflicts, and resources are shared equitably. In this contest, let's assign 50 points to the winner and −10 points to the loser for having wasted some time and energy. Since each individual wins half its conflicts, its average winnings are $\frac{1}{2}(50 - 10) = 20$ points. This pleasant utopian state seems stable, but it isn't. Eventually some dove gene is going to mutate into a hawk gene, and every animal who inherits such a gene will win every time it encounters a dove. Assuming that getting food determines who will survive and breed, life will be very good for hawks—that is, until there are so many hawks that they frequently encounter one another. When two hawks meet, they will fight until one is injured and gives up. Now the game is more serious (Figure 49.2). In addition to a potential 50 points for winning and −10 points for wasting time, a hawk is in danger of getting −100 points for serious injury. The potential gains and losses may be tallied up with a matrix representing the payoff to the individual at the left:

	Hawk	Dove
Hawk	−25	50
Dove	0	20

When a hawk meets a hawk, it can expect to win half its fights, so its payoff is the average of the 50 points gained for winning and the −100-point penalty for being injured—that is, $\frac{1}{2}(50 - 100) = -25$. On the other hand, each hawk easily wins 50 points when encountering a dove. Each dove loses immediately upon meeting a hawk, but it isn't injured and doesn't waste time in mock battle, so in encounters with other doves it still gains 20 points on average.

What will happen in this population? Life appears to have advantages for both hawks and doves, depending on the number of other hawks or doves they encounter. That, indeed, is the answer to the question: The advantage of hawkness or doveness depends on the makeup of the population, and both kinds of animals will have equal average payoffs (of 9.1 points) when 0.545 of them are hawks and 0.455 are doves. (Letting h and d be the proportions of hawks and doves, the payoff for a hawk is equal to the payoff for a dove when $-25h + 50d = 20d$; from this equation, if $h/d = 30/25$, then $h = 0.545$ and $d = 0.455$.) So the fraction of hawks in the population should increase until it reaches this equilibrium point and will then stay close to equilibrium. Using these particular numbers, one possible evolutionarily stable strategy would be for each individual to behave like a hawk 0.545 of the time and like a dove the rest of the time.

If we amplify this simple model to allow for other behavior patterns, as Maynard Smith does, it turns out that neither hawk nor dove is an evolutionarily stable strategy. For instance, a *bully* mutant could act like a hawk until challenged by another hawk and then run away; bully behavior is not an ESS either, because a population of bullies could be invaded by hawks, who would always win their fights with bullies.

The point of this exercise is that one *can* analyze behavioral patterns in this way quite accurately, even though a lot of research may be required to assign realistic numbers of points. The points assigned to interactions here measure the time and energy an animal gains or loses, and both can be critical. The average animal is probably living close to the edge; it has to be continually on the lookout for food and constantly wary of threats, so it can't afford to waste time in its search for food, and even a simple injury can prove fatal.

The ultimate payoff for an animal, of course, isn't just the energy to live for another hour or another day, but the ability to reproduce. Hawks initially increase in the dove population because they get more resources, faster and more easily, than doves do. An individual may not live very long, but the genome the individual is carrying can live indefinitely, as long as it creates viable strategies for itself—as long as the genome creates, in Richard Dawkins's words, a viable survival machine for itself. The survival machine is the whole organism, with both its solitaire strategies and its poker strategies.

Figure 49.2

Behavioral "hawks" and "doves" can interact in four ways, and they win or lose points (time, energy, etc.) for gaining resources, for wasting time, or for being injured. The overall balance of points determines the composition of the population.

49.3 Behavior patterns are mostly instinctive, but usually require maturation.

Nikolaas Tinbergen's 1951 book is entitled *The Study of Instinct*, but many biologists now try to avoid talking about "instinct" because the word has been misused, as if it were a real explanation ("Oh, it does that because it has an instinct for self-preservation"). *Instinct* is useful as a general term for genetically determined, species-specific patterns of behavior, as long as we realize that saying a behavior is instinctive explains little, no more than saying that a fruit fly has red eyes because it inherits genes for them from its parents. Instead, we want to know the mechanisms that underlie each behavior.

We will show that, for the vast majority of animals, the genetic programs governing physiological processes also govern overt behavior—their behavior is almost entirely instinctive. Because so much of human behavior is learned and our animal pets can learn, we tend to think of learning as the norm. But if learning is defined as modification of an animal's behavior on the basis of its experiences, then learning plays only a minor role in the behavior of most species. Still, the innate (instinctive) and learned aspects of a behavior may be closely intertwined, and separating them is a research problem. The problem is complicated by the question of *maturation.* Behaviors develop as an animal develops. A newborn mammal, for instance, must develop and mature before it can show any sexual behavior. Some behaviors develop without any need for practice or for environmental input, while in other cases specific stimuli may be needed. Squirrels and hamsters raised in isolation later show normal display and brood-raising behaviors. J. Grohmann raised some pigeons in narrow tubes so they were unable to use their wings and found that when released they could fly almost as well as pigeons who had always been able to move normally and practice their flying. On the other hand, some animals learn complex behaviors from their parents, and a broad range of behaviors require specific environmental inputs, called *imprinting* (see Section 49.9). Some behavior patterns are genetically encoded but require practice as the animal matures before they can be performed well. For instance, Martin J. Wells studied the maturation of hunting behavior in the cuttlefish *Sepia,* a relative of the squid and octopus (Figure 49.3). When a young cuttlefish sees a shrimp, it doesn't respond immediately but pauses for a long latent period of up to two minutes, while it does nothing. Then the cuttlefish's eyes fix on the shrimp; it turns, and finally attacks. The initial latent period declines rapidly with experience, and after three to five trials, a young cuttlefish will respond only 10 seconds after seeing the shrimp. The latent period never becomes much shorter than this. This change in behavior is considered maturation rather than learning because the rate of decline cannot be influenced—for instance, by rewarding the animal or by starving it beforehand.

In looking for the causes of behaviors, we often postulate that animals have certain *drives.* Clearly, no animal is disposed to display its entire repertoire of behavior all the time; sometimes it tends to eat or drink, other times to mate, and still other times to fight. To explain why an animal engages in one behavior rather than another at any time, it is common to invoke a specific drive that motivates each type of behavior. Thus hunger is a drive that motivates an animal to search for food, thirst

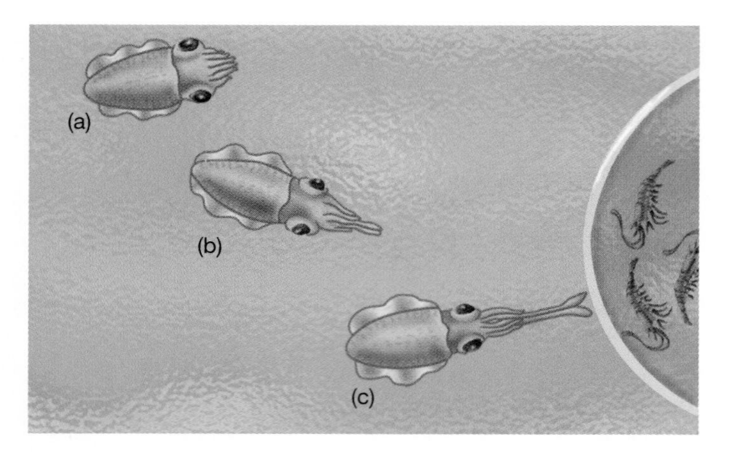

Figure 49.3
(a) A newborn cuttlefish sees the shrimp inside a glass tube. *(b)* It swims toward the prey, its eyes turned and fixated. *(c)* Finally it stabs at the shrimp with its tentacles. The behavior is executed faster as the cuttlefish matures, but it never changes.

motivates it to drink, and so on. It may make good sense to speak of drive if a physiological foundation for it can be identified, but like instinct, the concept of drive is useful only if we understand its limitations. It is really not very enlightening to explain sexual activity by saying, "Oh, that's just the sex drive at work."

Behavior associated with thirst and drinking makes a good example. Some behavior is certainly motivated by the feeling of thirst—a dryness of the palate; given access to water, a thirsty animal drinks its fill and then refuses to drink for a while. The drive it is responding to, however, is basically a condition in centers in the lateral regions of its hypothalamus, where high osmotic pressure in the interstitial fluid induces the release of antidiuretic hormone (ADH) and also promotes drinking behavior. Just injecting salt water into this area induces water-satiated goats to drink. The active neurons are cholinergic, and even injecting cholinergic agents will induce drinking in rats. But now the abstraction of a "drive" has been reduced to a simple physical condition in some regulatory centers, which maintain water balance by inducing both internal water-conservation actions and overt behavior. When these centers are activated, we may say that the drive to obtain water is high, but the very idea of a drive seems superfluous if we can speak, instead, about a physical condition.

Exercise 49.1 Give an example of the kind of experiment you would perform to determine whether a simple behavior pattern of some animal is learned or instinctive.

Exercise 49.2 In the 1930s, B. P. Wiesner and N. M. Sheard studied maternal behavior in rats, one aspect of which is retrieving any young who have been moved out of the nest. They found that injecting rats with gonadotropic hormone, which stimulates growth of the ovaries, will enhance retrieving behavior and even induce it in virgin females and in males. What does this result imply about the nature of the "maternal instinct" or "maternal drive"?

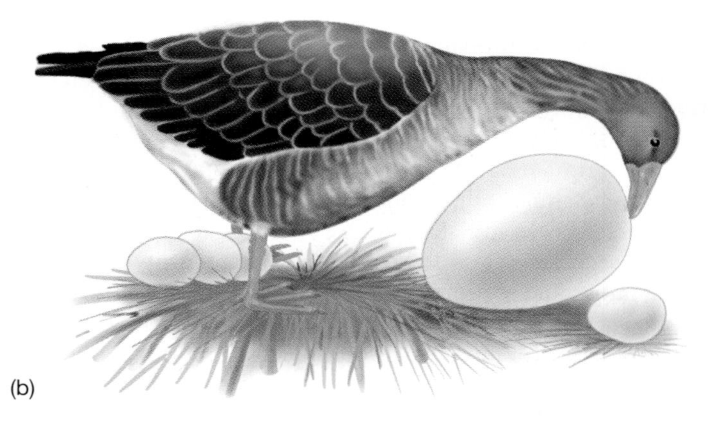

Figure 49.4
(a) A goose retrieves eggs that have rolled out of the nest with a stereotyped movement, guided by small taxic movements. *(b)* The goose is highly attracted to a large egg, which presents a supernormal stimulus for the action.

49.4 Simple, unlearned behaviors are fixed-action patterns.

Ethology rests on the concept of a **fixed-action pattern,** a highly stereotyped, genetically encoded action that an animal does not learn (although its performance commonly improves with practice). Learned behaviors, of course, can also be very stereotyped, so mere repetition in a standard way is no proof that a behavior is a fixed-action pattern.

An animal often refines a fixed-action pattern through directing or orienting movements called **taxes** (sing., **taxis**). The distinction between fixed-action patterns and taxes is illustrated nicely by Lorenz and Tinbergen's study of the Greylag Goose. Because goose nests are mere depressions in the ground, eggs sometimes roll out of them, but a goose simply retrieves stray eggs by extending its neck, or walking out if necessary, far enough to pull the egg back with the underside of its bill (Figure 49.4*a*). Although the egg tends to roll slightly to one side or the other, the goose adjusts its movements with taxes, moving its head one way or the other to keep the egg on course. It continues these actions until it has rolled the egg back into the nest. When Lorenz and Tinbergen removed an egg while a goose was rolling it back, the goose behaved as if the egg were still there, continuing to pull its neck all the way back until the entire movement was complete. Such behavior is called a fixed-action

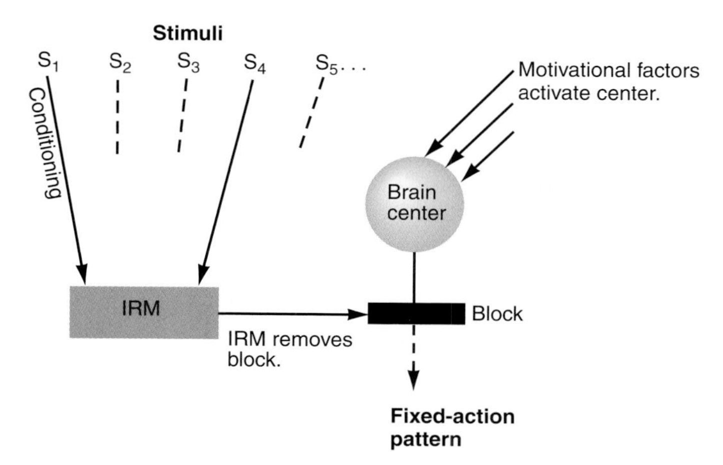

Figure 49.5

A general model for understanding behavior postulates that motivational factors prepare a brain center to produce a fixed-action pattern, but the action is blocked until brought on by certain stimuli. An IRM picks suitable stimuli out of the environment and removes the block. Through conditioning, other stimuli can take the place of the stimuli that normally activate the IRM.

pattern because it cannot be changed much by experience. However, the sham motion, in the absence of the egg, lacks all the side-to-side balancing movements. These taxic movements refine the fixed-action pattern, and presumably a goose learns to make them effectively through direct experience with eggs.

A fixed-action pattern is only performed at certain times, and any stimulus that sets it off is a **releaser,** sometimes called a **sign stimulus.** The releaser for egg-retrieving behavior is the sight of an egg near the nest; that stimulus sets off the appropriate fixed-action pattern, which can't be stopped until it is completed. Even a rock or a ball that looks sufficiently like an egg will set off the behavior.

To explain the connection between the releaser and the onset of the fixed-action pattern, Lorenz and Tinbergen postulated an **innate releasing mechanism (IRM).** An IRM is a stimulus filter. No animal can be sensitive to all possible stimuli; it has to ignore most of the world most of the time. So something has to selectively direct the animal's attention to a releasing stimulus and then initiate the appropriate fixed-action pattern. That is the role of the IRM. Experiments with extraordinary, prominent stimuli show how an IRM reacts to a releaser; for instance, a goose becomes obsessed with a huge egg (Figure 49.4*b*) because the IRM picks out this supernormal stimulus and sets the fixed-action pattern to work overtime.

The model in Figure 49.5, with a few additions, explains a great deal about animal behavior, at least in outline. Some IRMs must be completed by experience, and some fixed-action patterns can be modified by learning (conditioning). Also, an animal must be in a certain physiological state to activate an IRM, since its entire behavioral repertoire isn't constantly ready. Sexual behavior patterns, for instance, are only set off if the animal has the right sex hormones coursing through its blood.

49.5 Complex behaviors are sequences of fixed-action patterns.

Even if an animal had the ability to learn functional behaviors, it might have no opportunity to do so. Take the solitary digger wasps of the genus *Ammophila.* A female wasp emerges from her burrow in the spring, after developing there during the winter. Her parents died almost a year earlier, and she herself will only live for a few weeks. The survival of her species depends on the instructions for reproduction encoded in her genes and translated into the structure of her nervous system. Accordingly, she finds a male, recognizes him as one of her species, and mates with him. She then seeks a sandy, well-drained soil and starts to dig a burrow several centimeters deep with her front feet and mandibles, scattering the pebbles and sand grains so she doesn't build up a pile of sand at the entrance. With great vigor, accompanied by characteristic buzzing sounds, she excavates a tunnel and chamber where her offspring will grow. Then she emerges, closes the entrance temporarily with sand and pebbles, circles around the location, and flies off to find an appropriate prey animal.

Each type of solitary wasp has a narrow range of prey, and *Ammophila*'s victim is always a caterpillar. She may have to search for a long time and may only find one victim a day, but she knows exactly how to handle it. No matter how fiercely the caterpillar struggles, the wasp finds the right place to ram in her stinger and inject a paralyzing dose of venom. She then drags the caterpillar back to her burrow (Figure 49.6). (One observer reported on a wasp that worked for two hours to drag her prey nearly 100 meters.) Arriving at the burrow, she uncovers the entrance, leaves the caterpillar for a moment, and goes inside, as if for a final inspection. She comes back out, drags the caterpillar inside, and lays a single egg on it. The caterpillar, still alive, will serve as food for her young. The wasp then leaves the burrow forever, carefully closing the entrance with sand and pebbles, which she hammers in with a small pebble held in her mouth as a tool. Then she goes off to repeat the process.

This whole pattern of behavior is a stereotyped and virtually unbreakable chain of fixed-action patterns. The nineteenth-

Figure 49.6

At a burrow it has dug, an *Ammophila* wasp pauses with its prey, a caterpillar. It will lay an egg and raise one offspring that will feed on the prey.

century French naturalist Jean-Henri Fabre observed a similar pattern in the *Sphex* wasps, which hunt crickets. In one experiment, Fabre moved the cricket a few centimeters away from the burrow entrance while the wasp was inside making the final inspection. When the wasp emerged, she saw the cricket in the wrong position, moved it back to the threshold of the burrow, and went back in again. Again Fabre moved the cricket; again she moved it back and went inside. After 40 repetitions of this little game, Fabre grew tired and gave it up. The experiment demonstrates that the wasp has a fixed-action pattern that involves dragging the prey to the threshold of its burrow. The sight of the prey there, or the completion of this activity, or both, releases the next fixed-action pattern, which is to go into the burrow. Apparently the stimulus of seeing the prey lying in the right position when she emerges releases the next fixed-action pattern, dragging the prey inside. So if she comes out and doesn't see the prey where it should be, the chain of behavior is interrupted and she has to go back to an earlier stage to recreate that stimulus. In this species, the pattern is probably unchangeable. On the other hand, George and Ernest Peckham, earlier in this century, tried the same experiment with an American species of *Sphex* and found that after a few repetitions of the game the wasp did drag her prey directly into the burrow.

These wasps can apparently learn little or nothing to modify this chain of fixed-action patterns, yet they can learn other important things in connection with the process. In circling around her burrow before leaving it, the wasp acts as if she is memorizing the terrain, and she does recognize the burrow entrance on the basis of visual cues. Tinbergen and W. Kruyt showed this with the wasp *Philanthus triangulum* by surrounding one of its burrows with a ring of pine cones while the wasp was still inside (Figure 49.7). She emerged and flew over the area, and while she was off hunting her prey, they moved the cones about a meter away. Upon returning, she searched for the burrow opening within the ring. The wasp couldn't locate her burrow until the cones were put back into their original place, and the same thing happened during many repetitions of the experiment.

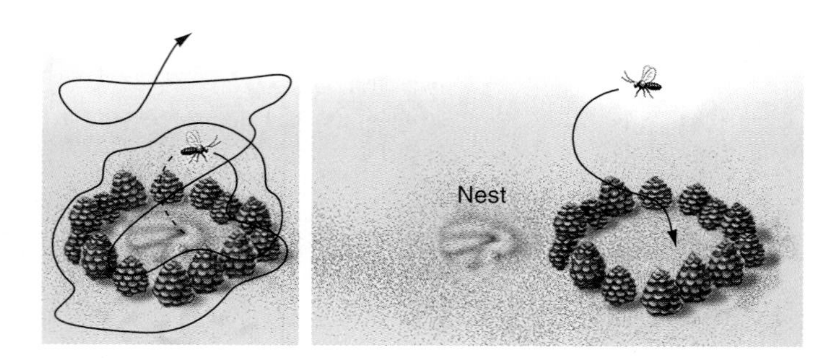

Figure 49.7

A *Philanthus* wasp finds its burrow by using visual cues. After leaving its nest, the wasp flies over the area to learn the location of the entrance, and it cannot find its burrow if the ring of pine cones is moved.

Occasionally an opportunity arises to study the genetic basis of a behavior. Two species of African lovebirds carry nesting material in different ways (Figure 49.8): Peach-faced Lovebirds cut short strips of material and carry several at a time by tucking them into the feathers of their backs, while Fischer's Lovebirds cut longer pieces and carry them, one at a time, in their bills. Hybrids between the two are confused. They cut strips of intermediate length and make clumsy attempts to tuck them into their back feathers, but generally fail. After some experience, they learn to carry pieces in their bills but still tend to make futile movements toward their backs. This case illustrates how separate species may evolve functional adaptations that become confused and ill-adapted in hybrids.

Exercise 49.3 W. C. Rothenbuhler studied hygienic behavior in honeybees. In hygienic strains, the workers will uncap a cell containing a dead bee and remove the corpse. Unhygienic strains don't do this. Hybrids between the two types are all unhygienic, so this trait seems to be dominant. Rothenbuhler crossed the hybrids to the recessive hygienic strain and studied 29 resulting F_2 colonies. Nine of the colonies uncapped cells containing dead larvae but did not remove the corpses. Six didn't uncap cells, but would remove the corpses if Rothenbuhler uncapped them. Eight would neither uncap nor remove corpses, and six were completely hygienic. In such a small experiment, these numbers are essentially equal for statistical purposes. *(a)* How many genes can you identify that shape this behavior, and what behavior do they specify? *(b)* What are the genotypes of the four strains? (Use whatever symbols you choose.)

49.6 Releasers are highly selected objects and events from the environment.

A releaser is a special bit of the environment selected out of a background of possible stimuli by an innate releasing mechanism. The nature of that stimulus is shown nicely by experiments with silvery little fish called sticklebacks.

During the breeding season, male sticklebacks have bright red bellies and display aggressive behavior toward one another. Their aggression is, in fact, a response to a red belly. Tinbergen presented various models to male sticklebacks (Figure 49.9) and found that they ignored a realistic model with no red but attacked various shapes with red undersides. A red-bellied model, in fact, is attacked most vigorously when presented in a head-down position, the species's characteristic threatening posture. Males actually get excited by any bit of red they see during the breeding season, such as a red flower petal that falls into the water, and Tinbergen described one male who even made a threatening display when a red mail truck passed the window of the laboratory. A releaser is most appropriately called a *sign stimulus* when it is a specific morphological feature that appears to have evolved only to trigger a behavior. It is clearly an advantage for animals to develop such features and to have a sensory system that notices them and releases appropriate behavior in

Figure 49.8
(a) Fischer's Lovebirds carry longer strips in their bills; *(b)* Peach-faced Lovebirds carry shorter strips in their back feathers. Both methods are genetically encoded. *(c)* Hybrids between them try, at first, to tuck strips into their back feathers but fail to make the proper movements. Eventually they learn to carry strips in their beaks, but still make sham motions toward their backs.

Figure 49.9
Tinbergen's models with red undersides provoked even stronger aggressive behavior among male sticklebacks than did a realistic model fish without red; the key stimulus is the red belly.

Figure 49.10
Young Blackbirds (*Turdus merula*) will gape when an adult bird appears on the nest, and even simple models made of circles that simulate the adult's shape release gaping behavior. The young birds react as if the smaller circle is the adult's head. In turn, the gaping mouth of a baby bird, bordered with yellow-green skin, releases feeding behavior in adults.

response. Even if they waste some time and energy responding to false signals, the system is generally adaptive and well designed.

The behavior of birds feeding their young also shows the importance of sign stimuli. The young sit in their nest waiting for a parent to return with food. When the adult appears, the chicks open their mouths and gape widely. Tinbergen showed that the releaser for this action can simply be a silhouette about the size of an adult bird that stands above the nest, moves, and has a "head" of the right size in proportion to the body (Figure 49.10). The sign stimulus that signals the adults to stuff food down the babies' throats is the open mouth, with the characteristic pattern of mouth color revealed when the babies gape; nestlings typically have a border of green or yellow skin along the edges of their beaks, and adults stuff food into the yellow diamond formed when the young gape. (An adult will stuff food into a yellow-bordered, diamond-shaped hole in a piece of cardboard.) Furthermore, the hungrier a bird is, the wider it gapes, and the adults respond most strongly to the mouth that is open widest. So the hungrier birds tend to be fed first, and the food is distributed according to need. In contrast, a baby bird that doesn't gape is a releaser for the adult to roll it out of the nest, for ordinarily such a bird would be dead.

Protective mimicry provides another example of releasers. A variety of animal features, such as shape, coloration, and characteristic movements, can be understood as sign stimuli that release certain behaviors in other species. For example, it is adaptive for many small animals to show alarm and flee when confronted by

Figure 49.11

The large eyelike spots on the wings of some moths are flashed in response to a light touch, and they provoke alarm behavior in small birds, presumably because they set off the same innate releasing mechanism as do the real eyes of owls and predatory mammals.

a pair of large eyes, like those of owls and predatory mammals. So it is also adaptive for insects to evolve large eyelike spots on their wings as a form of mimicry (Figure 49.11). This aspect of behavior is discussed at greater length in Section 27.16.

49.7 Behavioral patterns are often organized hierarchically.

Most animal behavior is directed toward some final act—drinking, eating, or mating—and is set off by releasers if the animal is in the proper physiological state. An animal, however, can rarely just get up and perform the final act. It first has to search for water or food or for a mate, and then perform other, preliminary activities. So behavior is commonly guided into narrower, more directed actions. An activity begins with **appetitive behavior,** wide, searching actions that are likely to bring the animal into contact with specific releasers that will set off the next phase of activity. A hungry hawk, for instance, cruises over a wide range, searching the ground below for some sign of suitable food, such as the movement of a small animal. This stimulus initiates another series of actions, including chasing and pouncing, that may result in a kill. If the prey gets away, the hawk goes back to an earlier stage. Eventually it kills some animal, using a characteristic behavior pattern, and eats it—the **consummatory behavior** that is the point of the whole activity. This final act is generally followed by a quiet period when the animal shows no more appetitive behavior and doesn't respond to stimuli that would normally guide its appetitive activities.

The mating of sticklebacks illustrates this pattern. In spring, increasing daylight activates hormones that induce appetitive behavior in the males. They start to migrate away from the deep water where they have been living all winter into shallower water where they nest. Each male establishes a territory, an act that is apparently stimulated by the sight of distinctive terrain and plants. If several males are put into a featureless aquarium, they stay together and show no reproductive behavior, but if a plant is put in one corner, one male will swim over and establish a territory there. So their migration is, in effect, a search for appropriate visual stimuli.

With his territory established, a male stickleback begins to change from his dull, brownish winter colors to his brilliant spring colors: red belly, blue back, and bright blue eyes. (This change in coloration is external and induced by the environment; is it a behavior? The lines defining behavior are blurred, not sharp.) As he continues into the mating ritual, these colors are accentuated. The stickleback chooses a nest site and makes a shallow pit by scooping out mouthfuls of sand. Over the pit he builds a pile of nest material, mostly threads of algae and pieces of plant, which he glues together with a sticky material secreted by his kidneys. Finally he wriggles through the mound to create a tunnel, and the nest is completed.

Any other male appearing in his territory during this time sets off aggressive display behavior. If the intruder flees, the owner chases him; if the intruder displays threateningly, the owner does the same. Meanwhile, the owner stands alert for any females that may come along, and they release a pattern of courtship behavior that we will describe shortly.

Here again the animal is led step by step, stimulus by stimulus, from behavior that is only broadly directed toward the goal to the final consummatory act of fertilization. To account for this pattern of activity, Tinbergen postulated a hierarchical organization of behavior, with a series of innate releasing mechanisms arranged so that each one leads to the next (Figure 49.12). Note that this pattern is a model of the functional relationships among behavior patterns, not of the physical structure of the nervous system.

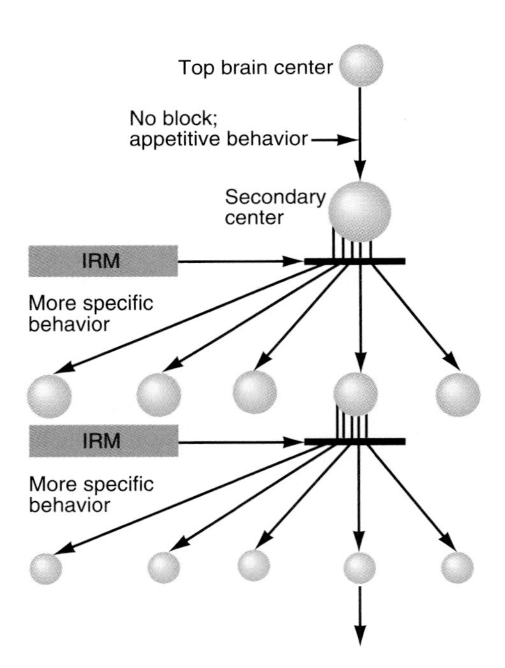

Figure 49.12

Behavior patterns are often arranged hierarchically, so that an IRM for a broad behavior pattern leads to narrower patterns. The chain is finally consummated in a behavior that satisfies a drive and ends the pattern.

49.8 Animals of the same species create ritualized behavior by mutually inducing fixed-action patterns in one another.

As the behavior of digger wasps shows (see Section 49.5), complex behavior patterns have evolved as chains of smaller fixed behaviors in which the completion of one act, with the resulting internal and external conditions, sets off the next act. Sometimes the sequence is made by two individuals who respond to each other.

In his studies of the stickleback, Tinbergen identified a pattern of courtship actions leading to fertilization (Figure 49.13). A male, having built a nest and defended a territory around it, waits until females appear. Sighting a school of females signals him to begin a zigzag dance of rapid back-and-forth movements. A female who is ready to lay her eggs responds to this dance and to his red belly by moving closer; she displays her belly, swollen with eggs, which releases his next behavior. He leads her toward the nest and shows her the entrance by sticking his head into the opening. While he lies on his side, she slips into the nest with her head and tail protruding. This behavior then signals him to butt the base of her tail with his snout; in response, she spawns (lays her eggs in the nest) and then swims off. Then he swims into the nest and milts (releases his sperm over the eggs) and stays there to guard the developing young. This whole ballet is a chain of stimuli and responses, each response being the stimulus that sets off the next event.

49.9 Many behavior patterns require the imprinting of specific environmental information.

Behavior patterns that develop fully with maturation or a bit of practice require no special input of information. In contrast, some fixed-action patterns seem to have a blank space in their innate releasing mechanisms, requiring an input of additional information through the process of **imprinting.**

Lorenz once put half of a clutch of Graylag Goose eggs into a laboratory incubator and left the other half in the field. When the eggs in the field hatched, they identified the first large

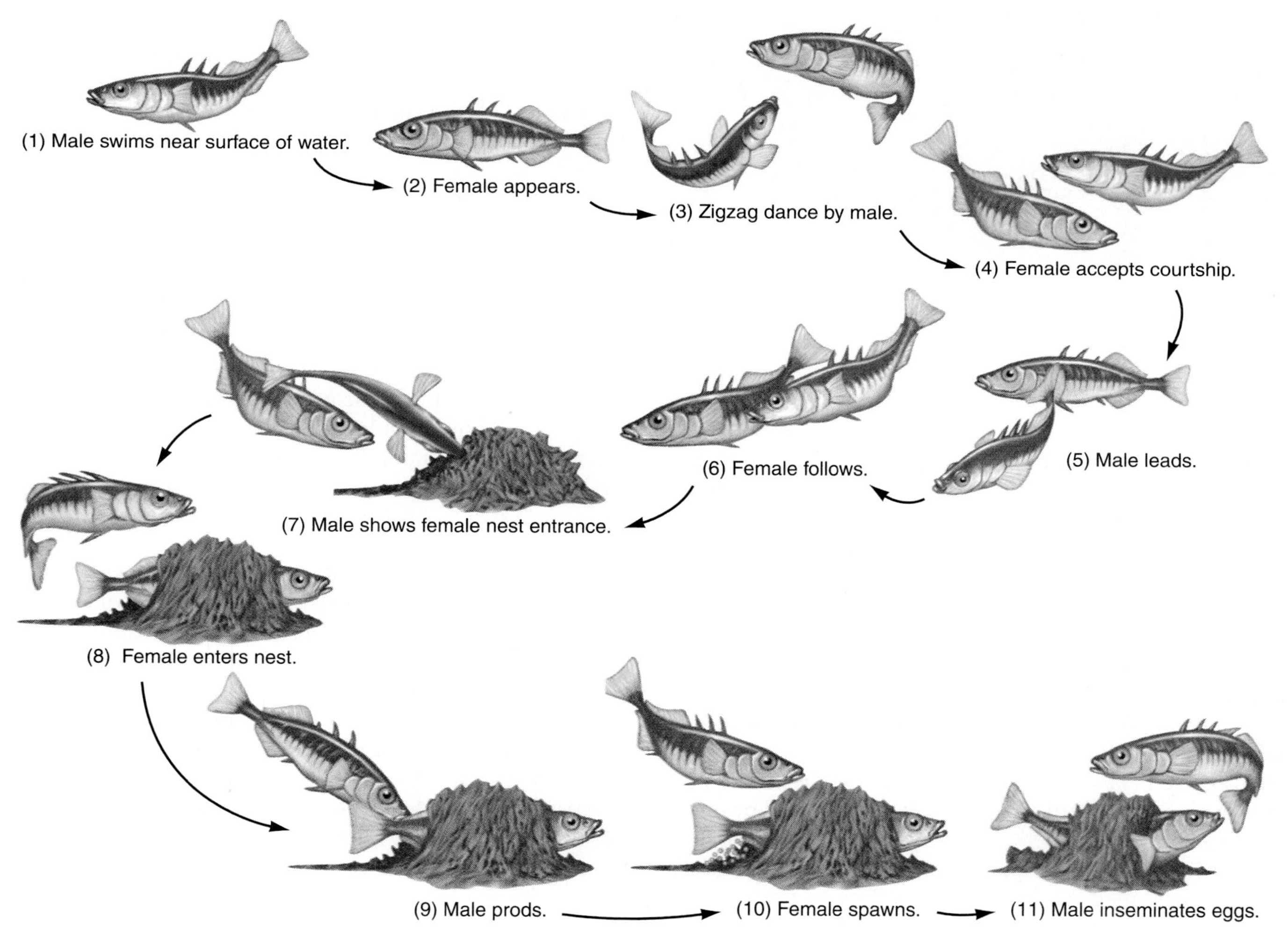

Figure 49.13
The mating behavior of sticklebacks is a series of fixed-action patterns in which each action by one fish releases a new action by the other fish.

Figure 49.14

Newly hatched goslings will imprint on any large moving object and follow it as if it is their mother.

Figure 49.15

A male White-crowned Sparrow sings a song that is largely encoded genetically but must be refined by experience with songs properly sung by other adult males.

moving object the young goslings saw as their mother, and they started to follow her around as good little goslings do. When the lab eggs hatched, the first large moving object the young birds saw was Lorenz, so they started to follow him around as if he were their mother (Figure 49.14). Lorenz called this phenomenon "imprinting" because each gosling's IRM for following behavior needs to get a personal imprint in a blank space: "Follow ____ around." A selective attention mechanism in each gosling picks out the first large moving thing it sees and writes its description in the blank. The object doesn't have to do anything but move; ducklings can be imprinted to a mechanical, moving duck decoy or even to a box that makes noise. This behavior really serves the geese very well, even if it can be subverted by human experimentation.

Imprinting underlies many other animal behaviors, such as the homing behavior of salmon. After hatching in freshwater streams, salmon migrate downstream to the ocean, where they grow and mature. After a few years, the adult fish migrate back upstream, almost unerringly finding the stream where they were hatched. This remarkable behavior has been documented by massive studies in which thousands of marked salmon are released into streams and the survivors are recaptured when they return. Salmon rarely home to the wrong stream. Studies by Arthur D. Hasler and his colleagues have shown that homing depends on the salmon's acute olfactory sense. Fish with destroyed olfactory mucosas or plugged nostrils can't detect odors and can no longer home into the right stream. Every stream has a distinctive odor, which gets imprinted into the fishes' homing mechanism early in development. Salmon clearly have a screening mechanism that recognizes one combination of chemicals and directs locomotion toward it.

Behavioral patterns that require imprinting are important in both reproduction and social coherence. Imprinting systems should evolve whenever the general outline of a needed behavior is clear and inheritable but the details are too unpredictable or complex to be genetically encoded. For example, there is survival value in having young birds stay close to their mothers, but a genetic system probably can't encode the appearance of one mother rather than another. Similarly, no genetic system could predict the smell of a particular stream.

The development of bird song shows the role of maturation and imprinting. Male birds use songs to establish their territories. The young of several species have an inherited ability to sing something approximating their species's song, which is fine-tuned through learning and experience. Peter Marler and M. Tamura studied the singing patterns of White-crowned Sparrows (Figure 49.15), whose various Pacific-Coast populations sing different dialects of the same song. They found a critical time in the life of a juvenile bird, up to the age of about three months, when it must be exposed to properly executed White-crowned Sparrow songs if it is to sing properly at maturity nearly a year later. These birds have a general pattern of their species's song already encoded, and they will sing this simple version later if they don't hear a proper version early on.

While they will learn any dialect of a White-crowned Sparrow song that is played to them, recordings of the songs of other species have no influence on their development, showing they have a selective mechanism that can pick out a pattern of sounds typical of their species.

These and other studies show that imprinting occurs when an animal is ready to learn a detail that completes a genetically determined behavior pattern. Imprinting must occur at a critical period in the animal's development, and if the proper stimulus isn't filled in during this period, the behavior remains incomplete.

49.10 Learning is a mechanism for rapid adaptation to new situations.

In addition to highly stereotyped, genetically encoded behaviors, many kinds of behaviors are clearly learned, at least in part. As we discussed in Section 49.3, learning means an enduring change in behavior as a result of experience, not simply because of maturation. The ability to learn obviously confers a selective advantage on animals by giving them a wider, more adaptive range of options than could be encoded by the genetic apparatus alone.

Think of the situation in terms of the length of a feedback loop. Natural selection shapes fixed-action patterns into successful, adaptive forms because individuals who engage in an adaptive behavior pass on their genes, so their offspring engage in the same behavior. But each genetic feedback loop takes a generation, and shaping only occurs over many generations. Learning, on the other hand, operates with a very short feedback loop that takes only seconds or minutes. An animal that engages in an adaptive behavior receives immediate confirmation of its success—it is rewarded in some way—and is inclined to repeat the behavior. By surviving a bad experience, an animal can learn in one trial not to do the same thing again. So an animal that can modify its behavior can experiment with better ways to get such rewards as food, water, sex, or freedom from pain.

The simplest learning is **habituation,** where an animal learns to ignore constant or persistent stimuli. This happens all the time, as when we ignore constant background noises in a room. Animals that normally react with alarm to certain signals—for instance, chickens reacting to the shape of a hawk—will become habituated to the stimulus if it is presented repeatedly without harming them, until they eventually ignore it altogether. One cellular basis for habituation is discussed in Section 42.13.

At the other extreme is complex learning which requires insight, concept formation, and—in humans—the use of language. Complex learning isn't limited to humans. In several experiments (supplemented by casual observations), animals are allowed to learn one rather complicated path to food; the path is then blocked, and the animals easily find another path, showing that they have an internal map of the area. Otto Köhler showed that some birds and mammals could develop an abstract conception of number. With appropriate rewards, for instance, they could be taught to take only a certain number of pieces of food. A parrot learned that if Köhler presented four light flashes or rang a bell four times, it was to take only four kernels of corn. A raven learned that if Köhler presented a sign with two marks on it, it should open only a food box with two marks on its lid.

Wolfgang Kohler, during the 1920s, performed some famous experiments that demonstrated chimpanzees' abilities to solve problems. He put them in situations where they could only reach a banana by using sticks and boxes in novel ways, such as putting two sticks together to make a longer one or piling boxes on top of one another. The animals showed the ability to use insight and to perceive how objects could be used as tools.

Jane Goodall showed that chimps, in their native habitats, commonly use tools to get food, for example by using a carefully prepared stick to fish termites from their nest. These and other observations show that many birds and mammals can learn a great deal. Learning has also been demonstrated in the octopus, an animal with a relatively large and complex brain (discussed in Section 42.2), and simple learning has been studied in some other large molluscs as well. Still, the ability to learn beyond simple habituation is the exception among animals.

We still understand little about what happens when an animal learns, but we will examine some aspects of simple learning in the following section.

Exercise 49.4 An ant colony recognizes an intruding ant as a foreigner by its odor, and often kills it. But if a stranger comes in while the colony is occupied with other business, it may go unnoticed and eventually be accepted as part of the colony. Provide one simple explanation.

49.11 Behavior can be modified through both classical and operant conditioning.

The most intensively studied learned behaviors are acquired through **conditioning,** a simple type of learning that strengthens the association between a stimulus and a response. **Classical,** or **respondent, conditioning** was described by the Russian physiologist Ivan Pavlov, who conditioned a dog to salivate and secrete gastric juices at the sound of a bell by always ringing the bell just before blowing meat powder into its mouth. After many trials, the sound of the bell became associated with food and would induce salivation by itself. In classical terminology, the food is an *unconditioned stimulus* and salivation an *unconditioned response,* but in our terminology they are a releaser and a fixed-action pattern (Figure 49.16*a*). After the bell becomes associated with salivation, its sound becomes a *conditioned stimulus* and salivation a *conditioned response.* Somehow a conditioned stimulus associated with a releaser takes on all the force of the releaser and can evoke the fixed-action pattern by itself.

Timing is important in conditioning. Learning occurs if the conditioned stimulus and the natural releaser are presented together or if the conditioned stimulus leads by a short interval.

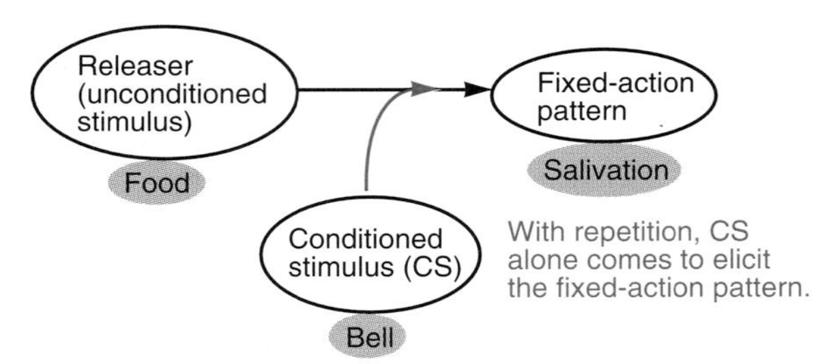

(a) **Classical conditioning**

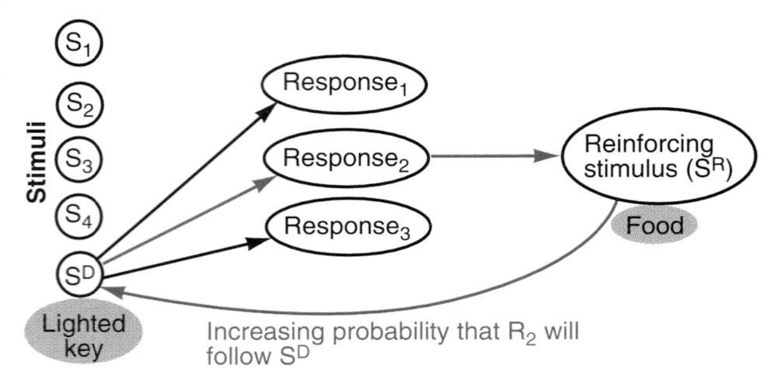

(b) **Operant conditioning**

Figure 49.16

In classical conditioning *(a),* an unconditioned stimulus releases a fixed-action pattern; an artificial, or conditioned, stimulus given simultaneously with the unconditioned stimulus then comes to release the fixed-action pattern. In operant conditioning *(b),* a response that occurs in the presence of a discriminative stimulus (S^D) is followed by a reinforcing stimulus (S^R). With repetition, the response comes to consistently follow S^D.

Figure 49.17

The Skinner box, invented by psychologist B. F. Skinner, provides a convenient environment for operant conditioning. Here, a pigeon pecks at the light and is reinforced with a bit of food.

However, no conditioning occurs if the conditioned stimulus follows the releaser.

Operant, or **instrumental, conditioning** is best explained by an example. A pigeon is placed in a specially designed environment (Figure 49.17), confronted with a lighted key and with a gated hole where it can get food or water. An untrained pigeon moves around and pecks at things more or less randomly, as pigeons normally do. Eventually it pecks at the lighted key, triggering the food hole to open so it can feed for a few seconds. After a time, the pigeon will peck at the light again, and again it will be fed. Soon pecking the light becomes associated with feeding, and the pigeon is pecking away furiously at the key.

In this situation, the pigeon is initially confronted by many stimuli, and the experimenter's task is to direct its attention to only one of them, the **discriminative stimulus** (in this case, the lighted key) (see Figure 49.16*b*). The experimenter wants to strengthen only one of the pigeon's natural behaviors, so this behavior is rewarded (in this case, with food). The reward **reinforces** the desired behavior of the animal—that is, increases the likelihood that the behavior will occur again. Interestingly, conditioned behavior is strengthened more by intermittent reinforcement than by successive reinforcements for each correct behavior. An animal that is only occasionally reinforced will work much harder for its reward than one that is consistently rewarded.

Both classical and operant conditioning are subject to **extinction:** If the reinforcement stops, the learned behavior diminishes in frequency and eventually dies out. Both types of performance increase with the number of reinforced trials and decrease with the number of nonreinforced trials. Both types, also, are subject to **stimulus generalization.** For instance, a dog classically conditioned to salivate to the sound of a bell with a tone of 1,000 cycles per second will also respond, but less strongly, to bells with higher or lower pitches. Similarly, a pigeon conditioned to peck vigorously at a key displaying a 550-nm light will also peck at light of longer or shorter wavelengths according to a very neat curve (Figure 49.18), pecking at the more distant colors with decreasing intensity.

Classical conditioning relies on fixed-action patterns that are elicited by specific stimuli; operant conditioning relies on behaviors that the animal engages in spontaneously and that are not elicited by external stimuli. Salivation, for example, is a reflex response to food (which explains why classical conditioning is sometimes called respondent conditioning). A pigeon, however, pecks all the time, regardless of the environment, so such activities are called operant behaviors because they are operations the animal performs on the environment, not responses to it.

To get at the foundations of learning, some biologists have turned to fruit flies, *Drosophila,* the classic genetic tool that has yielded so much information about other basic processes. After completing his classic genetic analysis of bacteriophage (see Chapter 17), Seymour Benzer set out to find *Drosophila* mutants with various behavioral defects, and several investigators have pursued this line of research. Fruit flies can be conditioned in a simple T-maze, a set of tubes where they have a choice of

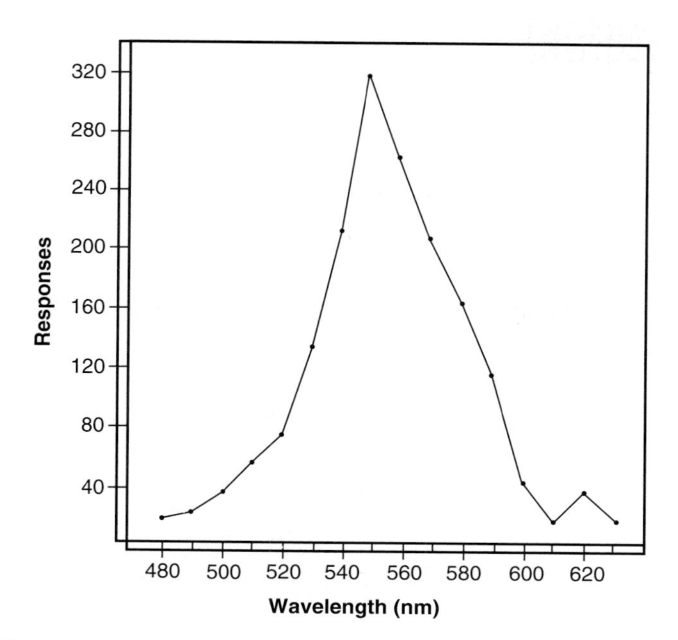

Figure 49.18

A pigeon is trained to peck at a high rate at a 550-nm light. When allowed to peck at lights of other colors (without reinforcement), it generalizes pecking behavior but pecks at lower rates the further the wavelength departs from 550 nm.

going to the left or the right, toward one of two chemicals with distinctive odors. They are given a mild electric shock in the presence of one chemical (the conditioned stimulus). After a few conditioning sessions, over 90 percent of the flies avoid the conditioned stimulus. In wild-type flies, conditioning lasts at least an hour, and with a longer training series, it can last for about a day. Several distinctive mutants—bearing such charming names as *dance, rutabaga,* and *amnesiac*—have been collected; they can be conditioned, although to lower levels than wild-type flies, but they "forget" rapidly—that is, the conditioning is extinguished very quickly. In a few mutants that have been analyzed, the defects have been localized in elements of the general signal-transduction pathway, which has been implicated in learning in *Aplysia* (see Section 42.13). For instance, rutabaga is defective in the Ca^{2+}-activation of adenylate cyclase (which makes cyclic AMP), and *dunce* is the gene for a phosphodiesterase that attacks cyclic AMP. So the same general mechanisms involved in response to hormones and other kinds of signal transduction are involved in learning.

General signal-transduction pathway, Section 11.8

Exercise 49.5 You have a radio that works irregularly. You start to fiddle with the controls, and occasionally it comes to life and works for a few seconds. After a while, you look at the clock and notice that you've been playing with the radio for over an hour, and it still doesn't work right. What has happened to you?

49.12 Many patterns of behavior are traditions that have been learned.

Not many years ago in many places in Africa, you could encounter a rather dull brown and white bird flying a few meters ahead of you, making a churring sound and fanning its tail to attract your attention. It might perch conspicuously in front of you, and as you approached, it would fly low to the ground and swoop ahead to another branch, still making its sound and waiting for you to follow. The bird would continue to lead you until, eventually, it alighted on a tree and sat quietly, waiting for you to take the next step. The bird would have brought you to a beehive. Waiting for you to break the hive open and take the honey, it would then fly down to feed on dead bees and pieces of wax.

The bird's generic name, *Indicator,* describes its behavior. Its popular name is honeyguide, and it is well known to the African natives, who once depended on it for their honey. The bird is even incorporated into their folklore. An African fable tells of a honeyguide that found an elephant's carcass, which he thought would make a nice home. After marking the elephant, he went off to find his friends and relatives, but while he was gone a mouse found the carcass and claimed it for his own. The two argued over it and then decided to let a bee settle the dispute. Despite the evidence of the honeyguide's mark, the bee decided in favor of the mouse. From that time on, bees and honeyguides have hated each other, and so the birds began to guide people to the bees' nests to destroy them.

Baboons as well as humans respond to the honeyguides' behavior, but their strongest ally is the Ratel, or Honeybadger (Figure 49.19). When the Zulu people of South Africa want some honey, they make a grunting sound that they claim helps attract the birds, and they say this is the same sound the Ratel makes. Humans probably learned to follow the birds by watching Ratels.

The guiding habit is clearly learned, on both sides, and when Herbert Friedmann wrote his definitive study of the honeyguides in 1955, the habit was apparently fading out. It had disappeared entirely in some areas, and in other places the birds were doing it less frequently. The extinction of this behavior has been associated with encroaching civilization on the honeyguides' habitats; local grocery stores have replaced bees' hives as a source of sweets, so humans are no longer reinforcing the

Figure 49.19

The honeyguide and the Ratel (Honeybadger) cooperate in a learned behavior that provides food for both of them.

Figure 49.20
An oystercatcher opens a mussel either by hammering at it to break the shell or by inserting its bill into the open siphon and prying the shell open.

honeyguides' behavior. The fact that only two nonhuman species have been able to fit their behavior patterns to that of the honeyguides shows that it isn't easy. Furthermore, even though all eleven species of honeyguides (Indicatoridae) regularly eat beeswax, only two have developed the guiding habit. The birds are evidently able to get plenty of wax by themselves without help from humans or other animals.

Anyone seeing the behaviors that animals have been taught to perform—in circuses, for example—by patient application of conditioning techniques shouldn't be surprised to find animals learning complex behaviors naturally, although it is hard to specify just how these complex behavior patterns develop in nature. Nevertheless, animals do have such traditions and can learn from one another, as another story will show.

Oystercatchers are large shorebirds that feed on clams and mussels. M. N. Norton-Griffiths found that oystercatchers can get past the tough shell of a mussel to the flesh inside in two ways (Figure 49.20): either by laying a mussel on the sand and hammering at it with their heavy bills to break the shell or by stabbing their bills into the open siphon of a mussel under water and cutting the adductor muscle that keeps the shell closed. Each bird is either a hammerer or a stabber, and at first glance these behaviors seem to be inherited, since young birds make appropriate movements of one kind or the other long before they develop any skill. However, by switching eggs between the nests of hammerers and stabbers, Norton-Griffiths showed that each behavior is actually learned, not genetically encoded. These are difficult skills, and it takes a young bird a long time to learn them. Where oystercatchers subsist on worms, the young stay with their parents for only 6–7 weeks, but where the birds live mostly on shellfish, the young stay with their parents for 18–26 weeks while becoming self-sufficient.

Other animal traditions, Section 50.12.

Figure 49.21
Worker ants follow a trail of pheromone laid down by each ant as it carries food.

Exercise 49.6 Write a scenario to explain how a honeyguide and some other animal could both learn to engage in honey-finding behavior.

49.13 Animals can use various cues to orient themselves.

Among the most remarkable behaviors is an animal's ability to orient itself in relation to important fixed points, such as a nest, breeding ground, or source of food. We have seen the digger wasp's use of visual cues to locate its burrow and the salmon's use of chemical cues to find its stream. Ants use a different kind of chemical signal, marking the trail to food with a pheromone (Figure 49.21). A worker ant returning from a food source touches the tip of its abdomen to the ground periodically and secretes a bit of trail substance. Other workers then follow this trail, and as they return with food, they leave more of the trail substance, so the track gets stronger. The pheromone is volatile—that is, it evaporates quickly—and since workers who don't find food lay no trail, the track disappears once the food source is used up. Furthermore, different species lay different pheromones, so their trails don't become confused even if they cross.

Migration has long been one of the great biological mysteries, now only partially clarified by modern research. Humpback Whales migrate thousands of miles each year from winter waters to breeding waters; bats that breed in New England migrate to caves in Connecticut, New York, Pennsylvania, and Vermont for the winter; and huge herds of Barren-grounds Caribou in North America migrate regularly across Canada and Alaska. While insects are usually not migratory, the Monarch butterflies of North America migrate every fall to one lone valley high in the mountains of Michocan, Mexico, and return in the spring, even though those that make the journey in one direction are several generations removed from those that make the return journey. Many birds migrate hundreds and even thousands of miles from their breeding grounds to a winter range, always following the same route and on such a regular schedule that observers can predict very closely when each species will pass through a region (Figure 49.22).

The motive force behind bird migration lies in day length, for birds change their activities in response to the photoperiod. As days lengthen during the spring, the birds' gonads grow in preparation for mating, and they lay down extra fat, because migration requires large energy stores; just before migration begins, as much as half of a bird's mass may be fat. Birds also display a migratory restlessness (German: *zugunruhe*), becoming quite active and nervous. Pioneering studies by William Rowan during the 1920s and 1930s showed that birds exposed to artificially lengthened days in the middle of the winter will go through these changes, and when released, will migrate in the direction they normally would take in the spring. Since then, other studies have confirmed that migratory behavior, as well as internal physiological changes, can be controlled by altering the photoperiod.

However, this does not explain why the birds follow the same route their ancestors took. Most young birds probably do not learn by following older birds who already know the route, since birds of different ages may leave at different times—often, the young birds first. Migration must have evolved independently in many species, and it is now clear that birds rely on a variety of directional cues. Some use landmarks; huge numbers of birds funnel through narrow points of land, such as the Straits of Gibraltar and Cape May, New Jersey, and some species clearly orient by flying along coastlines. An experiment by F. Bolle showed that Hooded Crows orient generally by direction; he captured these crows in the spring at Rossiten in eastern Germany, banded them, and released them at Flensburg in western Germany (Figure 49.23). The birds migrated in the same direction they had been taking before they were captured and ended up in Sweden instead of the Baltic countries.

Other species depend primarily on visual cues from the sky—the position of the sun and the stars—in combination with information from their internal clocks. Since the sun and stars change position as the earth rotates, birds that use these astronomical guides need a clock for correlating their position with the time. Experiments with homing pigeons have shown the importance of the clock. One group of pigeons was kept on

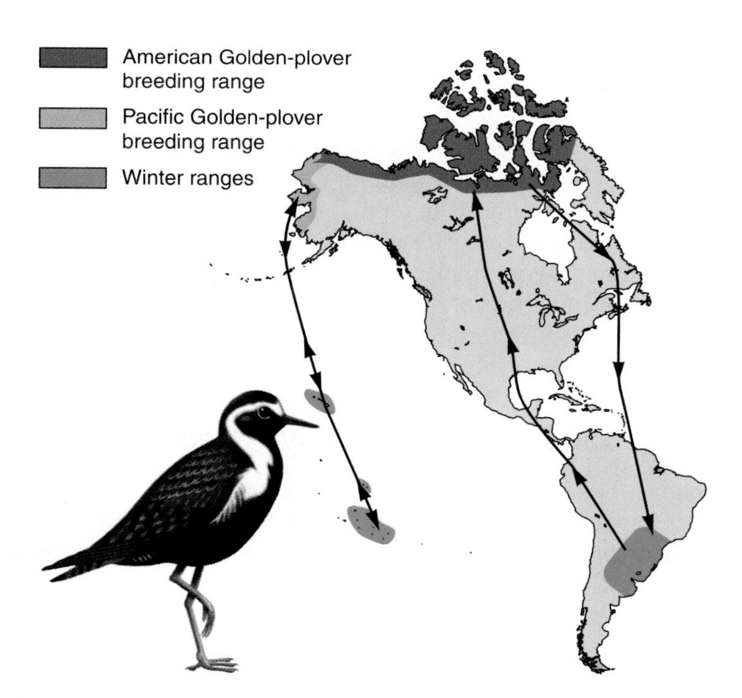

Figure 49.22
The American Golden-plover, one of the greatest bird migrants, covers about 8,000 miles from its Arctic breeding range to its South American winter range. The Pacific Golden-plover takes a similar route to Hawaii and the Marquesas Islands.

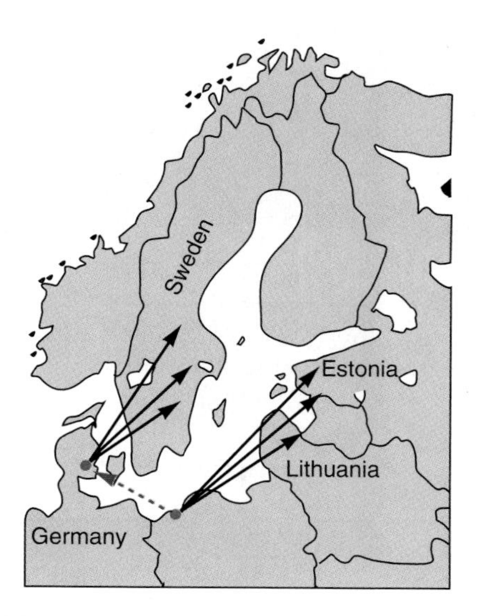

Figure 49.23
Hooded Crows that would normally migrate northeast from eastern Germany were taken farther west and released. They still migrated in the right compass direction and therefore wound up in the wrong place.

normal time, another group in a room with a light-dark cycle six hours (a quarter of a day) out of phase with normal time. When the birds were taken away from home and set free, those on normal time headed for home with their usual accuracy while the group with altered internal clocks were off by 90 degrees (a quarter of a circle) one way or the other, depending on the direction their clocks had been set for.

Several experiments have shown that migratory birds receive directional cues from the position of the sun. Gustav Kramer kept migratory birds in cages and observed their orientation as their zugunruhe increased. If they could see the sun, they began to move in the direction they would normally take in migration; when Kramer changed the apparent position of the sun with mirrors, the birds' orientation changed in a corresponding direction. Geoffrey Matthews showed that homing pigeons flew home directly from unfamiliar territory if they could see the sun, but on overcast days they had difficulty orienting.

Some birds also navigate by the stars, as shown in experiments by Franz and Eleanor Sauer and by Stephen T. Emlen. The Sauers experimented with Garden Warblers in a planetarium where they could see the star pattern and found that birds raised in a cage, who have never seen the sky and never migrated, still orient themselves in a direction that has been effectively calculated from the star pattern displayed overhead, on the basis of their internal clocks. A bird will also compensate for changes in the latitude and for the time of night. Emlen took Indigo Buntings that were ready to migrate south into a planetarium (Figure 49.24). The birds oriented toward the south as indicated by the stars, regardless of where true south was. Emlen also showed that migration depends on the birds' physiological state. By manipulating the photoperiod of some buntings, he brought one group into the normal spring condition and another into the normal fall condition simultaneously. When brought into a planetarium showing the spring sky, the spring group oriented toward the north and the fall group toward the south. In later experiments, Emlen showed that the birds are using certain constellations near the North Star, such as the Big Dipper, Draco, and Cassiopeia; young buntings learn these features through imprinting during their first month, and if they are not exposed to the sky during this time, they cannot orient themselves later.

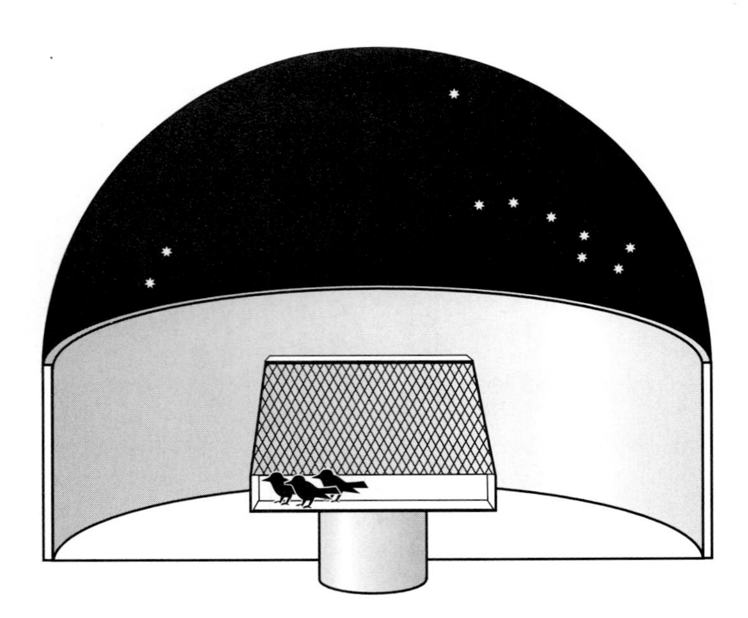

Figure 49.24

When Indigo Buntings in different physiological states are taken into a planetarium, they orient appropriately by the star map. Those that have been brought into breeding condition, as in the spring, try to migrate northward, and those that are in fall condition try to migrate southward.

In addition to all these navigational clues, many birds are sensitive to the earth's magnetic field and use it for orientation. William Keeton demonstrated that homing pigeons with small magnets attached have difficulty orienting on cloudy days. Charles Wolcott and Robert Green then fitted pigeons with caps containing electromagnetic coils; the magnetic fields created by these coils disoriented the birds under overcast skies. Reversing the electric current, which reversed the field, made the birds reverse their direction.

So birds have shown the opportunism of evolution by evolving migration patterns that use all the geographical information available to them, in various combinations. Each species has evolved a strategy that allows it to migrate successfully between its wintering and breeding grounds. Migratory animals of other kinds must use the same information.

Coda Since behavior is the essence of human life, perhaps it is anthropocentric and a bit romantic to consider behavior the most complicated expression of a genome. Nevertheless, it is a simple matter for a gene to encode a protein that helps structure a cell or catalyzes one step in metabolism; it is much more complicated for the protein to coordinate the activities of many cells in a plant or animal, perhaps in development or hormonal regulation; and still more complicated for the protein to make an animal move in an elaborate, functional pattern, shaping its environment and the behavior of other animals. What makes behavior possible? First, of course, an animal must be able to act; action lies mostly in muscle, and muscle is just a more organized arrangement of the actin and myosin that evolved very early in eucaryotic history. Second, an animal must be able to communicate internally, to signal muscles to contract in a certain sequence and pattern; this ability stems from membrane potentials and the ability to change them, which depends, in turn, on proteins that transport ions or simply channel their movements. Fundamentally, then, a complicated action still depends upon a series of molecular interactions, mostly between proteins. This perspective does not explain any particular behavior pattern, nor make it any less amazing, but analyzing behavior in general into its basic protein elements gives us a useful evolutionary perspective. Behavior, we can see, has simple foundations and ancient roots. The stickleback's courtship, the heron's foraging behavior, the pianist playing an elaborate concerto were all inherent in the flagellate's wriggling over a billion years ago. One aspect of animal evolution during that billion years or so has been a gradual genetic shaping and refining of protein interactions, finally forming nervous and muscular systems capable of the behavior we now wonder at—including the ability to have that sense of wonder.

Summary

1. Behavior includes primarily overt activities such as fighting, feeding, and mating, but also glandular activities such as sweating.
2. A behavior is a strategy for survival and is shaped by evolution like any feature of anatomy or metabolism. A behavioral pattern is an evolutionarily stable strategy if an animal can maintain it against possible counter-behaviors by other organisms.
3. Although the word "instinct" must be used with care, most behaviors of most animals are genetically encoded. Many behavior patterns will mature without practice or environmental input, while others require practice or specific information. A "drive," whose strength may change, may be postulated to explain why behavior occurs only in specific circumstances.
4. Behaviors can be understood as series of fixed-action patterns, stereotyped behaviors initiated by releasing stimuli.
5. Fixed-action patterns frequently follow one another in sequences, where the sign stimulus for one pattern is provided by completion of the previous pattern. Sometimes the sign stimulus is another individual's response to the previous action.
6. Some fixed-action patterns are always repeated without change as soon as an animal becomes mature enough to perform them; others require a period of practice before they can be performed.
7. Each fixed-action pattern is set in motion by a releaser, or sign stimulus, and it has also been postulated that innate releasing mechanisms select releasers out of the environment and remove the blocks to execution of the action. The stimuli that release a fixed-action pattern can be changed through learning.
8. Sequences of fixed-action patterns are often organized hierarchically. That is, the first pattern is a broad appetitive behavior in which a more specific releaser is sought, and this leads to more refined and directed behaviors, culminating in a consummatory act such as eating or copulating.
9. Many behavior patterns depend on the reciprocal activities of two animals. The behavior of one provides the sign stimulus for the next step in the other's behavior.
10. Some fixed-action patterns are incompletely specified by the animal's genetic program and must have the missing pieces of information supplied through experience; this is the process of imprinting.
11. Through learning, animals can rapidly adapt their behavior to new situations. Simple animals can modify their behavior very little, but some vertebrates are capable of learning relatively complex behavior patterns.
12. Two patterns of learning are classical conditioning and operant conditioning, which both depend on reinforcement of a new behavior through rewards that satisfy some drive. In classical conditioning, an animal learns to perform a fixed-action pattern in response to a stimulus that becomes associated with a natural releaser. In operant conditioning, an animal learns to modify a natural behavior pattern to perform tasks that are rewarded.
13. Intelligent animals have remarkable abilities to learn, and many complex behaviors are undoubtedly learned, often through imitation, and carried on in the species by tradition and culture rather than by genetic inheritance.
14. Many animals are capable of orienting themselves by means of cues, primarily those that are chemical or visual.
15. Among the most remarkable feats of orientation is bird migration. Birds navigate by means of an internal clock mechanism, which permits them to use visual cues related to the position of the sun and stars.

Key Terms

ethology 1041
evolutionarily stable strategy (ESS) 1042
fixed-action pattern 1044
taxis 1044
releaser 1045
sign stimulus 1045
innate releasing mechanism (IRM) 1045
appetitive behavior 1048
consummatory behavior 1048
imprinting 1049
habituation 1051
conditioning 1051
classical (respondent) conditioning 1051
operant (instrumental) conditioning 1052
discriminative stimulus 1052
reinforce 1052
extinction 1052
stimulus generalization 1052

Multiple-Choice Questions

1. Use the same hawk-dove matrix as in the chapter, but assume that in each hawk-hawk confrontation, one hawk was killed (−150 pts). How would the ratio of 54.5 percent hawks to 45.5 percent doves change?
   a. The proportion of hawks would increase to 100 percent.
   b. The proportion of doves would increase to 100 percent.
   c. The proportion of hawks would rise, but not to 100 percent.
   d. The proportion of doves would rise, but not to 100 percent.
   e. There would be no change in the ratio.
2. Instinctive behaviors include all of these features except
   a. species specificity.
   b. genetic regulation.
   c. specificity.
   d. effector organs such as muscles and glands.
   e. change as a result of experience.
3. A young (human) baby will grasp any solid, rodlike object pressed into the palm of its hand. This behavior is
   a. imprinting.
   b. operant conditioning.
   c. classical conditioning.
   d. a fixed-action pattern.
   e. a releaser.
4. During a critical period, young children assimilate the speech sounds (or even sign language movements) of their parents into universal human grammatical patterns, so language acquisition appears to be
   a. imprinting.
   b. operant conditioning.
   c. classical conditioning.
   d. a fixed-action pattern.
   e. a releaser.
5. Releasers constitute one form of
   a. instinctive behavior.
   b. taxic behavior.
   c. conditioning.
   d. sign stimuli.
   e. drive.
6. An innate releasing mechanism
   a. describes the end result of a fixed-action pattern.
   b. is any behavior that is produced by a drive.
   c. is a physiological filter that mediates response to a sign stimulus.
   d. includes all actions that result in movement toward food or mates.
   e. is none of these.

7. Some people wash their hands repeatedly, sometimes until they bleed profusely. Why is such stereotyped behavior not considered a fixed-action pattern?
   a. It is initially learned.
   b. It does not require releasers.
   c. It is not adaptive.
   d. It does not occur generally in the species.
   e. All of the above are true.
8. Which of the following is an adaptive form of mimicry?
   a. large eyespot on an insect's wings
   b. pigmented skin around mouths of nestlings
   c. red belly on male sticklebacks
   d. egg-laying behavior in wasps
   e. silhouetted shape of adult bird head
9. The evolution of a characteristic in one species that is a sign stimulus to another species may be an example of
   a. appetitive behavior.
   b. consummatory behavior.
   c. territoriality.
   d. protective mimicry.
   e. imprinting.
10. Complex, ritualized behaviors involving two adult individuals often can be analyzed as
   a. the results of imprinting.
   b. a temporal series of reciprocal releasers.
   c. appetitive and consummatory behavior.
   d. a series of taxes.
   e. maturation of sensory filters.

True-False Questions

Mark each question true or false, and if false, restate it to make it true.

1. Instinctive behaviors are only present in primitive organisms.
2. A drive is a motion-related instinct.
3. In rats, maternal behavior is apparently an instinctive response to the presence of a newborn.
4. In order to undertake and complete its complex egg-laying behavior, a female wasp must be presented with an array of specific releasers in a specific order.
5. During the breeding season, male sticklebacks become aggressive when they see anything shaped like another male fish.
6. A sign stimulus is a releaser that has no other function than to elicit a particular behavior.
7. Classical conditioning combines fixed-action patterns, behaviors that are subject to maturation, and a variable releaser.
8. Imprinting is the name given to the phenomenon whereby we ignore persistent, nonharmful stimuli such as the weight of our clothes or the background music in stores.
9. In classical conditioning, the unconditioned and conditioned stimuli are different, but the unconditioned response and the conditioned response are the same.
10. The function of operant conditioning is to maintain and enhance a particular response while substituting one stimulus for another.

Concept Questions

1. What distinguishes a behavior from a physiological activity?
2. Why is imprinting in baby birds not considered simply a matter of the maturation of fixed-action patterns?
3. Contrast classical and operant conditioning with respect to fixed-action patterns and releasers.
4. What evidence shows that the stabbing or hammering habit among oystercatchers is a learned behavior, not a fixed-action pattern or an imprinted behavior?
5. What evidence supports the hypothesis that navigational ability evolved independently in different species rather than just once?

Additional Reading

Archer, John. *Ethology and Human Development.* Harvester-Wheatsheaf, Lanham (MD), 1992. Applications of ethology to the psychology of human development.

Eibl-Eibesfeldt, I. *Ethology: The Biology of Behavior.* Holt, Rinehart and Winston, New York, 1970.

Friedmann, Herbert. *The Honey-Guides.* Smithsonian Institution, United States National Museum, Bulletin 208. Government Printing Office, Washington, DC, 1955.

Gould, James L., and Peter Marler. "Learning by Instinct." *Scientific American,* January 1987, p. 74. Though usually seen as opposites, learning and instinct are partners: The process of learning is often initiated and controlled by instinct.

Grier, James W., and Theodore Burk. *Biology of Animal Behavior,* 2d ed. Mosby-Year Book, St. Louis, 1992.

Gwinner, Eberhard. "Internal Rhythms in Bird Migration." *Scientific American,* April 1986, p. 84. Migratory birds have a clock that tells them when to begin and end their flight. It is based on rhythms with a period of about a year. Remarkably, the clock also helps the birds find their destinations.

Huber, Franz, and John Thorson. "Cricket Auditory Communication." *Scientific American,* December 1985, p. 60. The female's ability to recognize the male's calling song and to seek out the source of the song can be used to study how nervous system activity underlies animal behavior.

Krebs, J. R., and N. B. Davies. *An Introduction to Behavioural Ecology,* 3d ed. Blackwell Scientific Publications, Oxford and Boston, 1993.

Lorenz, Konrad Z. *King Solomon's Ring: New Light on Animal Ways.* Time Incorporated, New York, 1962 [1952].

Moore, Janice. "Parasites That Change the Behavior of Their Host." *Scientific American,* May 1984, p. 108. In doing so, they make the host more vulnerable to predation by their next host.

Narins, Peter M. "Frog Communication." *Scientific American,* August 1995, p. 62. Not a chorus; they are all trying to make themselves heard above the din.

Shettleworth, Sara J. "Memory in Food-hoarding Birds." *Scientific American,* March 1983, p. 102. How can birds remember where they have hidden food?

Internet Resource

To further explore the content of this chapter, log on to the web site at:

http://www.mhhe.com/biosci/genbio/guttman/

Social Behavior

50

Key Concepts

A. General Features of Social Behavior

50.1 Social behaviors entail a balance of advantages and disadvantages.

50.2 Societies may be organized through various types of bonds.

50.3 Societies are often structured by dominance hierarchies.

50.4 Social behavior requires communication among members of the group.

50.5 Territoriality is a strategy for dividing space and resources.

B. The Genetics of Altruism and Competition

50.6 The survival of a genome can be enhanced by the behavior of all individuals who share any of its genes.

50.7 Many birds mate and raise their young cooperatively.

50.8 In many species, individuals promote their own offspring and jealously destroy competitors.

C. Studies of Three Social Species

50.9 Honeybees exhibit an extreme division of labor among castes.

50.10 Honeybees communicate through dancing, sounds, and scents.

50.11 Herring Gull societies operate through fixed-action patterns.

50.12 Macaques have complex societies in which culture plays a large part.

Brazilians dressed for a festival show how humans, like many other animals, engage in complex social behaviors that promote coherence in groups.

At some time in the next few days, you will probably do something with a group. It may be a totally spontaneous "hanging-out together," dinner and a movie with a few friends, or a well-organized party or trip, but most people are drawn to others at least part of the time. You may also spend time with one other special person—a lover, a partner, a spouse. Sexual behavior and forming sexual bonds are also a part of social behavior. Though sexual behavior stems from different motivations and different stimuli, the group still offers the potential for reproduction, being the place where we commonly find our mates. The group also offers security and protection. We take comfort from doing what the group does, from the sense of belonging, including a feeling that these people will support and protect us. Our remote ancestors lived in small groups, in tribes, the human equivalent of primate bands, where survival quite literally depended on group action. We no longer need a little group to go off and kill a mastodon for dinner, but the need for the group remains.

Social behavior encompasses a wide range of interactions between animals of the same species. (Though interactions among individuals of different species may appear to be social—such as mammals drinking together at a water hole—these interspecific behaviors are best considered aspects of community structure, as described in Chapter 28.) And not even all behaviors that involve a collection of individuals of the same species are social, of course. In a swarm of houseflies feeding on a piece of rotting fruit or a group of moths congregating around a light, the animals are probably acting independently, oblivious to the others and unaffected by them. Social behavior is particularly difficult to define, but we must limit the term to situations in which members of the group interact with one another in specific, structured ways, as we will show in this chapter. ■

A. General features of social behavior

50.1 Social behaviors entail a balance of advantages and disadvantages.

Social behavior has important benefits to a species. Social interactions may increase an animal's ability to find food and other resources, shelter, space, mates, and conditions that promote reproduction. Furthermore, social behavior generally makes an animal safer. A flock of birds is an example of a *motion group.* It can confuse a predator by its dizzying weaving motion or by dispersing in all directions, leaving the predator unable to decide which one to chase (Figure 50.1). The collective eyes and ears of a social group are more sensitive than those of an individual, and a group can create a formidable wall of defense against predators.

Figure 50.1

A flock of birds may disperse when attacked, thus confusing a predator. A flock of sandpipers, like the one shown here, moves in unison, with sudden, unpredictable turns, making a poor target for a predator.

The potential benefits of social behavior are partly offset by disadvantages for an individual or a species. Although a group of animals may put up a better defense against predators, the group may also attract the attention of predators where a lone animal might have gone undetected. Also, predators often evolve their own behaviors in response to the social behavior of their prey. For instance, some hawks hunt only flocks of birds, not individuals, and tunas feed only on schools of other fish. Flock and herd behavior did not protect the now-extinct passenger pigeon or the once nearly extinct American bison from modern hunters.

Groups of animals are also subject to epidemics, episodes of disease that spread rapidly from one animal to another but often spare solitary individuals. Several studies have shown animals living alone to be healthier than their counterparts living in groups. Social organization also promotes conflicts between members of the same species, conflicts that consume time and energy and may be settled only by damaging fights or by establishing social hierarchies in which subordinate individuals are harassed and allowed less access to food and mates.

In spite of the costs, social organization is common in vertebrates, and a number of invertebrates show complex social interactions as well. Social behavior always has a strong genetic foundation, so where it has evolved, its advantages have apparently outweighed the disadvantages; otherwise, natural selection would have eliminated any tendency toward social interactions.

Exercise 50.1 On the planet Zexo, mammal-like creatures have evolved that occupy its one large, tropical continent. The intensity of ultraviolet light at the planet's surface is greater than on Earth, and the temperature and humidity there make it particularly easy for bacteria to pass back and forth between individual animals. What type of social behavior is likely to evolve under these conditions?

50.2 Societies may be organized through various types of bonds.

A *society* is a group of individuals that share a territory and common patterns of social behavior and are held together by *interpersonal bonds,* as illustrated by a baboon troop in Figure 50.2. A bond is a connectedness between individuals, as shown by their tendency to stay together and interact in certain ways. Parental bonds are shown by the tendency of parents and offspring to stay together—the parents feeding and protecting their offspring, the young seeking out a parent for protection or comfort. Sexual bonds are shown by a male and female staying together, perhaps feeding each other, grooming each other as many primates do, defending a territory and their offspring together.

A society may be a nuclear family or an extended family—parents, offspring, and other relatives held together by parental

Figure 50.2
A baboon troop is structured by several kinds of bonds. Males and females establish sexual bonds, of various durations. Mothers develop particularly strong bonds with their offspring. As juveniles mature, they form peer bonds with one another through play, which sometimes becomes quite rough; if a juvenile cries out from an injury, the adults step in and terminate the play, so juveniles acquire bonds to adults other than their parents. Gradually the bonds between peers of similar age are strengthened, and the adult-juvenile bonds grow weaker.

and familial bonds and to some extent by sexual bonds. It may include more than one family, held together both by sexual bonds and by weaker bonds between adults based on cooperation; when offspring of different families associate, they gain access to a wider variety of genes. Each individual belongs to social groups of different levels—the parental society, the sexually bonded society, and the group held together by nonfamilial social bonds—which overlap like a series of interlocking circles:

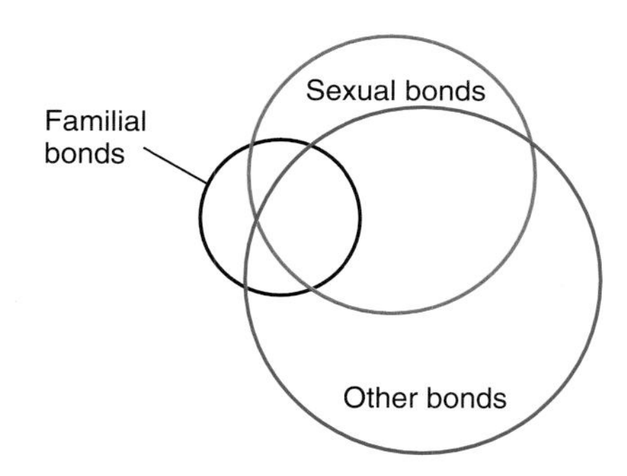

Exclusively male or exclusively female societies are structured by bonds of antagonism, cooperation, and the baboon equivalent of friendship. Adult males and females both form associations with others of the same sex who prefer to be together and who defend members of their group against other adults.

Societies frequently show division of labor. In a troop of baboons or monkeys (Figure 50.3), males protect the troop against predators, staying on its periphery as a shield, while females, who care for the young, remain in the protected interior of the group. Morphological specialization for division of labor is common among social insects, such as ants and termites (Figure 50.4). Males and females of the Huia, an extinct New Zealand bird (Figure 50.5), had differently shaped bills; the male used his shorter, straight beak to open up rotted logs, while the female used her curved beak to pick out insect larvae. Birds may show a sexual division of labor in building nests and rearing young, although often these duties are shared equally.

Exercise 50.2 Apply the bonds in a baboon society, as described in Figure 50.2, to the kinds of bonds that exist among people in a human society.

Exercise 50.3 Do humans exhibit division of labor in social groups? Do they have morphological specializations for such a division?

50.3 Societies are often structured by dominance hierarchies.

Chickens in a flock don't treat one another as equals. One chicken, the *alpha hen,* is dominant over all others in that she is able to peck them, but they do not peck her. Below her, the *beta hen* doesn't peck the alpha hen but may peck all the others, and so it goes down the hierarchy (or pecking order), each chicken pecking and being pecked, except for the last one, the *omega hen,* who only gets pecked. Once a hierarchy is established, however, relatively little pecking actually goes on except when new animals are added to the flock, in which case the pecking order must be reestablished, or when food is scarce and the dominant animals must assert their positions.

Dominance hierarchies, established through battles in which one animal brings others to submission, are common among many species. One male, the leader of the troop, is dominant over the other males. Rats in a cage soon develop dominant and submissive classes, and only the dominant ones mate.

Figure 50.3

A baboon troop is organized by space and division of labor. The structure differs somewhat from one troop to another, reflecting different traditions. In this troop, adult males stay on the edge of the troop, guarding against threats, while females and young stay in the middle.

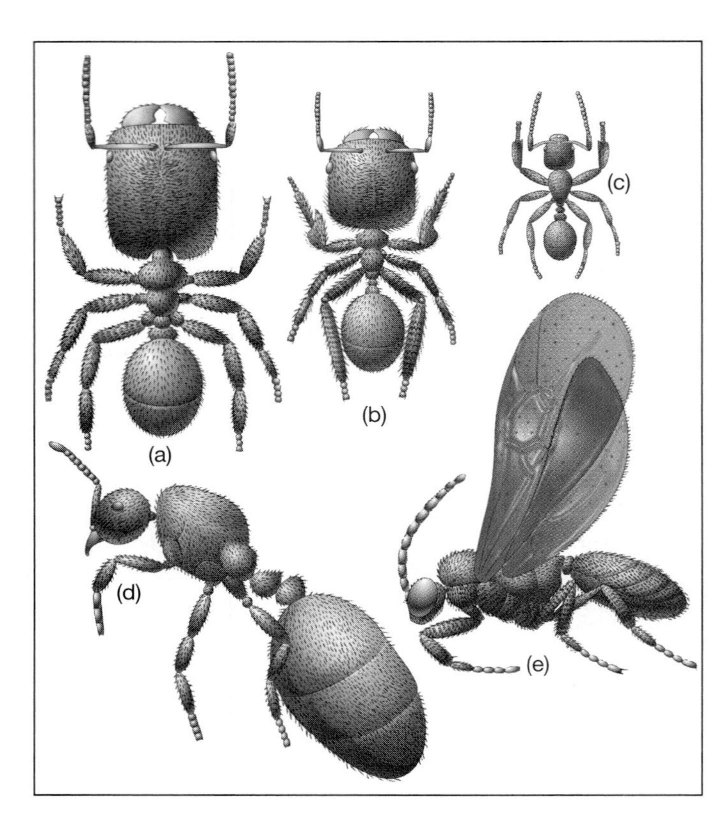

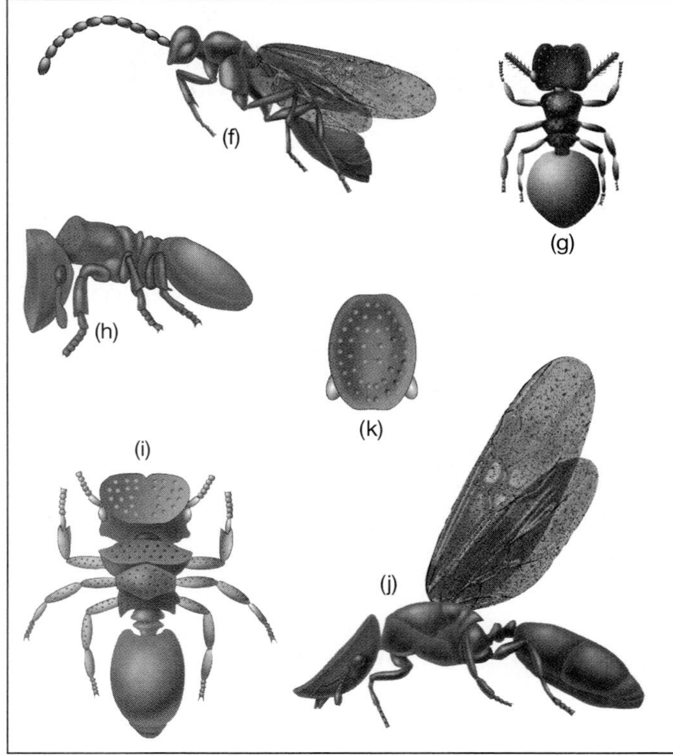

Figure 50.4

In some species of ants, a colony consists of several quite distinct castes with highly specialized forms. *Pheidole instabilis* has five castes: *(a)* soldier, *(b)* intermediate worker, *(c)* typical worker, *(d)* female, *(e)* male. In *Cryptocerus varians,* the males *(f)* and workers *(g)* are not particularly remarkable, but the soldiers shown at *(h)* and *(i)* and the females *(j)* have flattened heads specialized as door plugs; these individuals stand with their heads blocking the openings to the nest, excluding all but members of the colony. A soldier's head viewed from the front is at *(k).* Source: William M. Wheeler, *Ants: Their Structure, Development and Behavior,* 1910, Columbia University Press.

Physical combat is only one of many **agonistic** behaviors, those that involve hostile encounters, including threats and attacks, as well as submission, fleeing, and appeasement. In many encounters, symbolic displays take the place of actual fights. Thus a dominant wolf or dog can threaten with just a certain posture or facial expression, and a submissive wolf or dog acquiesces by exposing its vulnerable underside to a dominant animal (Figure 50.6).

Agonistic behavior patterns in several species of animals, including doves, mice, and chimpanzees, are clearly correlated with hormonal levels. Males higher in the hierarchy have higher serum levels of sex hormones (androgens). A castrated male has reduced aggressiveness, but he recovers it if he is injected with androgens. Even females can be made relatively aggressive if they are given androgens at a critical time during youth and then again when they reach maturity. While some investigators have found correlations between aggression and androgen levels in human males, these studies are open to some doubt.

Exercise 50.4 Do humans have dominance hierarchies? If so, give an example.

50.4 Social behavior requires communication among members of the group.

The members of a society must be able to communicate, to influence one another's behavior through specific signals. Social bonds can only be established through communication, and individuals use ritual signals to establish and maintain territories and to structure their societies into dominance hierarchies, in which each animal secures status for obtaining food, space, or access to mates.

Vocal communication is very common. Sounds in human language, the most sophisticated form of vocal communication, can convey subtle and abstract ideas, but vocalization is no less important to other species, even if it conveys more homely messages. A frog's croak is his love song. Birds use their calls to inform each other of territorial boundaries, to warn of danger, and to court mates. Humpback whales talk to each other with eerie songs that can be heard by others of their species more than a thousand kilometers away.

Bird song has been particularly well studied. Birds sing to establish and define their territories and to court and bond with mates. Song reaches a remarkable peak in African shrikes and other birds that live in dense foliage; a pair of birds can hardly see each other as they move through the bush, but they stay in contact by duetting: they may sing together in unison or sing one song by alternating notes, so perfectly timed that they sound like a single bird. In this way, they maintain a pair bond without visual contact. Species such as parrots, mynas, and mockingbirds have a well-known, extraordinary ability to imitate the songs of other birds and even to imitate human speech. Jerram L. Brown has suggested that this ability has evolved as a way for a pair of birds to share a distinctive, private song, so they can recognize each other more easily and share more intensive interactions.

Communication also takes many nonvocal forms. We have seen in Chapter 49 that insects lay down trails of pheromones to guide their comrades along a path. Wolves mark their territories with their urine, and their domesticated relatives use the same method to mark trees around their homes. (A classic story, perhaps apocryphal, tells of a naturalist who staked out a territory

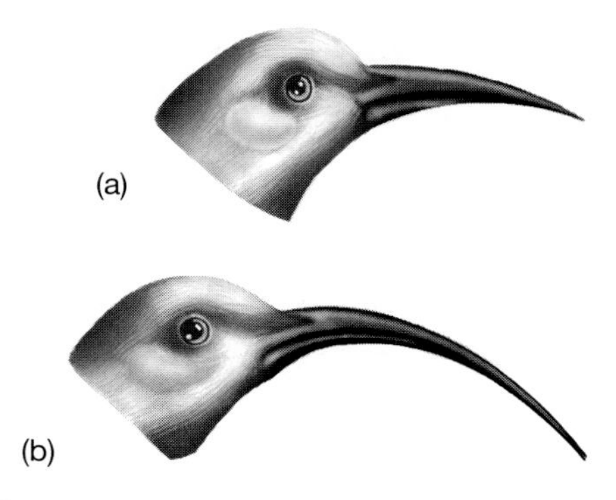

Figure 50.5
The now-extinct Huia showed sexual dimorphism, with specialization for cooperative feeding: *(a)* male, *(b)* female.

Figure 50.6
(a) A wolf threatens by standing erect and raising his hair and ears, thus making himself look larger. *(b)* One wolf appeases another by rolling over and displaying his soft belly.

Figure 50.7
Chimpanzees exhibit these characteristic facial expressions, according to Jane Goodall.

for himself in this way while studying the behavior of wolves, so they would avoid his camp.) A scouting honeybee performs a highly informative dance to tell the hive where it has found a favorable feeding spot. A dominant baboon keeps intruders at a distance by a mere flick of an eyebrow.

Many mammals communicate through facial expressions, often accompanied by vocalizations, that leave no doubt about their moods and intentions. Jane Goodall, who has devoted her professional life to studying chimpanzees, has identified the roles of their distinctive facial expressions (Figure 50.7). The compressed-lips face is used during aggression, and the play-face during play. Troops of chimpanzees commonly give loud "pant-hoot" calls to communicate over long distances, and they put on a distinctive face during calling. Frightened or excited animals show a full open grin, with the mouth wide open, and those who are less excited show a full closed grin, usually accompanied by high-pitched squeaking that may change to screaming or whimpering. Whimpering may be accompanied by a horizontal pout; whimpering and pouting are also shown by youngsters who have been attacked or by submissive members of the troop, perhaps as they beg for food or grooming from an individual with higher status.

Irenäus Eibl-Eibesfeldt and others have described apparently innate communication gestures in humans. Bowing is a universal sign of submission, and raising the arms is a sign of excitement and triumph. People everywhere greet acquaintances with a spontaneous eyebrow flash—a quick lift of the eyebrows, often accompanied by raising the head—at the moment of recognition (Figure 50.8). People of quite different cultures, including those who have been blind from birth, say "yes" by nodding the head and "no" by shaking it from side to side. These actions appear to be genetically encoded fixed-action patterns because they are universal in human societies and are performed spontaneously and unconsciously. However, there is no direct genetic evidence to that effect.

Fireflies signal to potential mates by flashing their abdominal lanterns, and where several species occur together, each species flashes in a distinctive pattern (Figure 50.9). These light displays can reach spectacular proportions. Thousands of male fireflies swarming in Thai mangrove trees display such intense, synchronized flashing that they instantly turn pitch dark into brilliant light. (Some predatory insects flash these patterns to attract fireflies of the species they prey upon.)

Sidebar 50.1 describes a unique method of communication between two different species.

Exercise 50.5 Describe some nonverbal ways humans communicate, and explain their social functions.

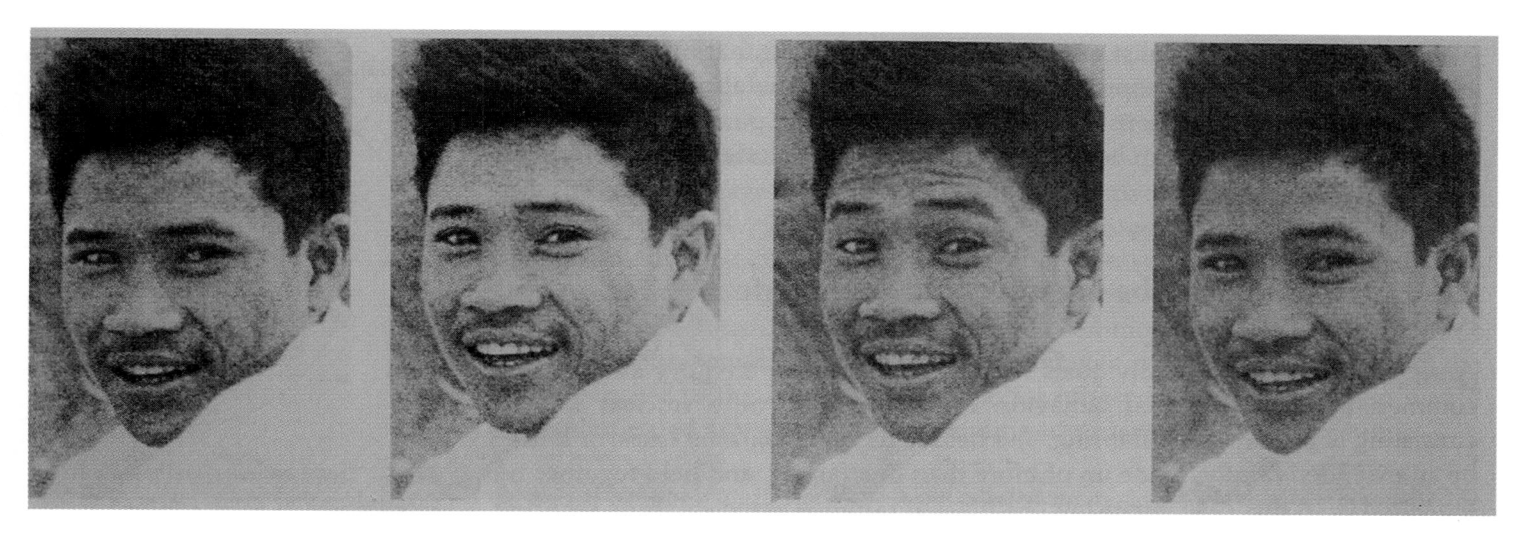

Figure 50.8
The eyebrow flash, seen in these film segments, is a universal sign of recognition among humans.

Sidebar 50.1 Singing Caterpillars and Ant Mutualisms

Although we can be quite sure that organisms sharing an ecosystem are involved in complex relationships, a lot of work is required to elucidate them. Stories of ecology are often about binary relationships between just two species, and ethologists generally describe the behavior of single species. However, Phillip DeVries and others have described some fascinating interactions that involve a unique mode of communication.

The caterpillars of several tropical lycaenoid butterflies have intimate relationships with ants and are said to be *myrmecophilous*—"ant loving" (*myrmeco-* = ant); various species illustrate the spectrum of symbiotic relationships (see Figure 27.16), from commensalism through mutualism to parasitism. DeVries has studied certain species with mutualistic associations; the caterpillars benefit from associating with ants because social wasps prey on the caterpillars, cut them into pieces, and carry the pieces back to their nests to feed their young, but ants are pugnacious insects that defend the caterpillar and fight off the wasps. Caterpillars associated with ants clearly survive longer than those separated from ants. In return, the ants get special nutrient secretions from the caterpillars. Both insects obtain nutrient-rich nectar from the plants they live on, but the ants prefer the caterpillars' secretions, which are richer in amino acids. When stroked by an ant, a caterpillar secretes droplets from special glands (ant organs), and ants can be seen eagerly lapping them up.

Since the association is so beneficial to both species, we might expect them to have a way to communicate, so that the ants can find the caterpillars easily. The caterpillars may produce pheromones, though none have been identified, but DeVries has shown that they use distinctive protrusions on their heads, called *vibratory papillae,* to create vibrations that are transmitted through the ground to attract ants. A papilla is remarkably similar to a guiro, a folk instrument made of a gourd with shallow grooves cut into one surface that is played by rubbing a stick over the surface; a caterpillar creates vibrations by rubbing the grooved surface of a papilla over tiny knobs (see photograph).

Of the many species of ants that live around lycaenoid caterpillars, the only species that interact with the caterpillars are those that also tend other types of secretion-producing insects such as aphids and bugs. The ants live on particular plants, which they protect from many herbivorous insects, but they allow certain insects to feed there and, in return, obtain their secretions. So this part of the ecosystem is structured by the mutualistic relationships of several species with ants.

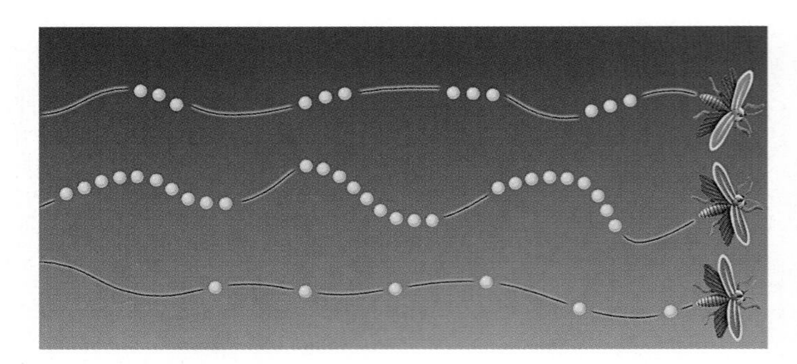

Figure 50.9
Various species of fireflies flash in different patterns, drawn out here as they would appear in sequence on a motion picture film.

Figure 50.10
Birds space themselves on a perch so as to maintain their personal spaces.

50.5 Territoriality is a strategy for dividing space and resources.

Territoriality is the behavior of claiming a space or an object, such as a nesthole, as one's own. It is characteristic of insects such as ants and dragonflies and some fish, lizards, and frogs, but it is most highly developed in birds and mammals. Some animals defend an entire home range—the area where they roam, feed, mate, and rear their young—while others defend only mating or nesting grounds.

Territoriality is related to, but different from, *personal spacing.* A territory is always a specific plot of land, an identified tree, or a claimed nesting place, whereas personal space moves along with the animal. You have probably seen personal spacing without recognizing it; a flock of birds sitting on a utility line or a pier will move about until they are spaced quite uniformly (Figure 50.10). When a new arrival alights, the birds all shift to preserve their distances. Social psychologists, anthropologists, and diplomats know how important personal space is to humans. Each culture has its own conception of the difference between intimate, personal, and social distance, and these unwritten rules must be respected. In one culture, a person feels insulted if you talk to him from a distance of more than a foot or two, while in another culture a person feels crowded and uncomfortable at the same distance.

The habit of defending an entire home range has been well studied in Palm Warblers. In May, when these birds return to Canada from their southern migration, each male immediately establishes a territory of some 1–2 hectares. He proclaims his rights by singing and by chasing away all other singing male Palm Warblers from his region. Once a bird has successfully laid claim to his territory, most aggressive behavior ceases. The owner merely advertises his proprietorship by sitting in a favorite tree and singing frequently, and most other birds respect this space. As in other species, a bird who intrudes on the territory of another senses that he is out of place and reacts submissively to attack by the owner, so territorial disputes are quickly settled with the retreat of the intruder. A female warbler will soon settle down with each male, and for the rest of the summer they will remain in their territory hunting insects, mating, and rearing young.

Unlike Palm Warblers, other animals defend only the territory where they mate; the California Sea Lion is an example. During the summer, bulls adopt territories in the Channel Islands, off the coast of southern California, for mating. They noisily engage in constant defense of their areas. Richard S. Peterson and George A. Bartholomew have described the ceremony that goes on at the boundaries:

> *[The bulls] rush toward each other, barking rapidly, with vibrissae [whiskers] extended anteriorly. Just before reaching the boundary between their territories they stop barking and fall on their chests, open their mouths widely, shake their heads rapidly from side to side, and weave their necks laterally at a rate much slower than the head shaking which is simultaneously going on. They then rear themselves to a maximum height, twist their heads sideways, and stare obliquely at each other. . . . This process is ritualized, and the animals do not touch each other; if they happen to be unusually close together they skillfully avoid contact. Sometimes during the boundary ceremony the two bulls partially cross the boundary and stand side by side, facing opposite directions, making feinting jabs at the chest and flippers of each other. The oblique stare is often the last component of the display prior to the bulls' separation, or prior to repetition of the entire head shaking, neck weaving sequence.*[1]

Sometimes, however, an intruder invades a bull's territory without noise or ceremony, and a fight erupts. The bulls lunge, bite, weave their heads, and rear up to bring their faces together.

[1] R. S. Peterson and G. A. Bartholomew, *The Natural History and Behavior of the California Sea Lion* (Stillwater, OK: American Society of Mammalogists, 1967), 25.

If the challenger wins, he takes over the territory, and the ritualized boundary ceremonies with neighbors are resumed. Fights like these are common, and few bulls are able to hold their mating territories for more than a couple of weeks.

Many birds establish territories for courting and mating. An individual's mating area is called a **court,** and a collection of courts is an **arena,** or a **lek.** A European shore bird, the Ruff, clears small circular courts a foot or two in diameter in a meadow and returns to the same area year after year. (One flock refused to abandon their arena even after a road was built through its middle.) A male Ruff struts and poses in his court while females (called reeves) look him over until one of them is sufficiently attracted to his actions to mate with him (Figure 50.11).

Figure 50.11
Male European Ruffs perform their courting ritual.

The bowerbirds of Australia and New Guinea build very elaborate courts (Figure 50.12); some species decorate them with "tokens of love" such as shafts of bamboo, poles of resin, beetle skeletons, snail shells, and lumps of charcoal. The Satin Bowerbird sets up two parallel rows of twigs and paints the inside of this hallway with chewed black bark. He then decorates it with trinkets and flowers, preferring blue and greenish-yellow shades. A female who is attracted to his work enters the bower and mates with him.

Some species develop group territories. For example, Black-tailed Prairie Dogs in the Black Hills of South Dakota and elsewhere build extensive towns. Each town is a massive system of tunnels and burrows that is divided into separate territories

Figure 50.12
A male bowerbird is shown with the bower it has constructed.

inhabited by subgroups, or coteries. While coterie members groom and play with one another and dig interconnecting burrows, intruders into their territory are barked at, chased, and bitten.

B. The genetics of altruism and competition

50.6 The survival of a genome can be enhanced by the behavior of all individuals who share any of its genes.

A baboon troop or any similar society holds some interesting puzzles. Some members of the troop, with low ranking in the dominance hierarchy, may never breed, or at least are likely to have fewer offspring than high-ranking individuals. Yet they will help protect and care for young baboons, and they will fight to protect the troop and may even die for it. What advantage does membership in the troop have for them? Why don't such individuals find mates and go off to live nonsocially, caring only for themselves and their own offspring?

The answer lies, in part, in taking the viewpoint of a genome rather than an organism. Samuel Butler, the English essayist and novelist, said that a chicken is just an egg's way of making another egg. That was an important, and rare, insight. Most of us take the chicken's point of view, which is just the opposite. But Butler was looking at the world as an intelligent genome would see it: The purpose of life is to make another generation of genomes, and the organism is just an instrument for doing that. A genome, after all, is the ultimate unit that can reproduce. An organism simply reproduces by following the program of its genome, and in this way it passes along copies of the genome. It is valuable to look at evolution from this point of view. In Richard Dawkins's metaphor, genomes (Dawkins says merely "genes") are *selfish*—their only "interest" is in replicating and surviving, while we organisms are "survival machines—robot vehicles blindly programmed to preserve the selfish molecules known as genes."

"Selfish genome" is just a metaphor, of course. Literal-minded people have criticized Dawkins for suggesting that genes can be selfish in the way your little cousin can be when asked to share a piece of cake. Genomes, or genes, aren't consciously selfish, any more than electrons are consciously negative or some quarks are consciously strange,[2] but they are selected from generation to generation by their survivability, and a genome that produces an inferior survival machine around itself simply will not survive. Those that do survive are, ipso facto, those that have acted selfishly by producing survival machines with superior characteristics.

We wonder why a baboon would lay down its life for a relative. When J. B. S. Haldane was asked if he would lay down his life for a brother, he said, "I would lay down my life for two brothers or eight cousins." Haldane was amplifying one of Charles Darwin's thoughts: Unselfish regard for the welfare of others, behavior we call *altruistic,* tends to ensure that one's own genome will be preserved in succeeding generations if the altruistic acts are extended to relatives. Suppose genome A is the ancestor of genomes B and C, which each carry half of A's genes. If B and C both survive and reproduce, A has effectively survived. It therefore makes sense for A to create a survival machine that tends to perpetuate both B and C; put more simply, the organism carrying genome A should take some care for the survival of its offspring. It also makes sense for the individuals carrying B and C to take care of each other. Organisms that share genomes can be expected to act as their brothers' and sisters' keepers.

Altruistic acts seldom extend beyond genetically related groups, and an animal that helps one of its relatives is actually helping to save copies of some of its own genes. W. D. Hamilton stated the idea more precisely in 1964 with the concept of **inclusive fitness.** As explained in Section 23.4, fitness is an indication of an organism's adaptation to its environment, as measured by reproductive success. Comparing organisms with similar genotypes, we assign higher fitness values to the genotypes that leave the greater numbers of offspring. The object of the evolutionary game, from the genome's viewpoint, is just to get as many copies of itself as possible into the next generation. Hamilton saw, however, that in this game an individual doesn't stand alone. A diploid individual carries half the genes of its parents, but its siblings all carry equivalent amounts, and on the average two siblings carry the equivalent of its own genes. A cousin carries an eighth of the individual's genes. So two surviving siblings or eight surviving cousins carry, on average, the equivalent of the individual's genes (Figure 50.13). This explains Haldane's remark.

Animals do behave altruistically. A dolphin, for instance, will buoy up a wounded comrade, keeping it at the surface so it doesn't drown; chimpanzees share some of the meat after killing a monkey; a bird will warn its flock of approaching danger, even though such action may call attention to itself. In experimental situations, rats and monkeys will perform a task that enables another animal to get a reward even when they themselves aren't rewarded. As we will show later, an animal in a social group frequently helps its relatives reproduce more effectively, even if this means giving up its own ability to reproduce; such behavior only makes sense if selection has favored the reproduction of genomes, regardless of the bodies they are in.

Jerram L. Brown has described inclusive fitness as follows. Let W, the average ordinary (classical) fitness of the individuals in a population, be 1. Then the fitness of each individual is 1 plus or minus fitness effects due to its particular genome:

$$W = 1 \pm \text{Fitness effects}$$

These fitness effects are of two kinds. *Personal effects* include all the usual adaptations for survival (good enzymes, strong muscles, protective coloration, and many other features); *parental*

[2]Physicists characterize quarks by several numbers, including one that measures a particle's "strangeness."

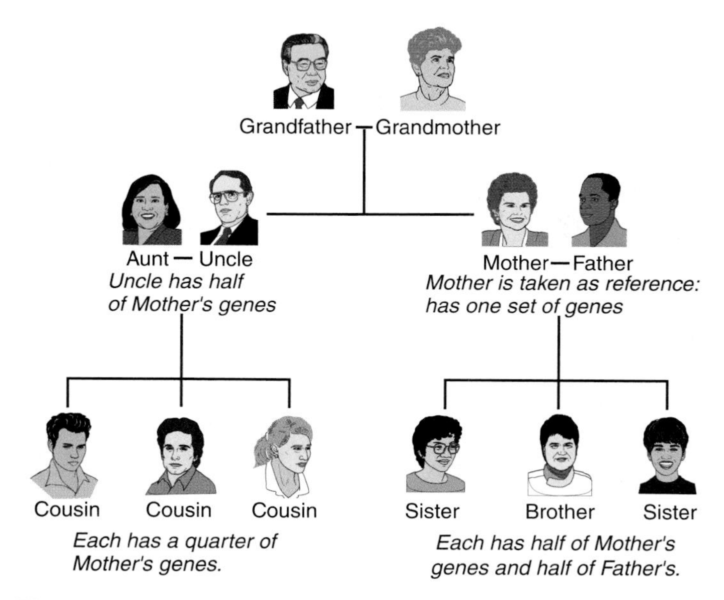

Figure 50.13

Taking the mother's genome as a reference, each child has half her genes (and half of the father's). The mother's brother has half her genes also, so each of his children has a quarter of her genes. Thus, each cousin shares an eighth of the genes of any one of the mother's children.

effects are those due to parental care, and perhaps care by grandparents, that increase the survival of offspring:

$$W = 1 \pm \text{Personal effects} + \text{Parental effects}$$

Defined in this way, parental effects refer only to effects on descendants:

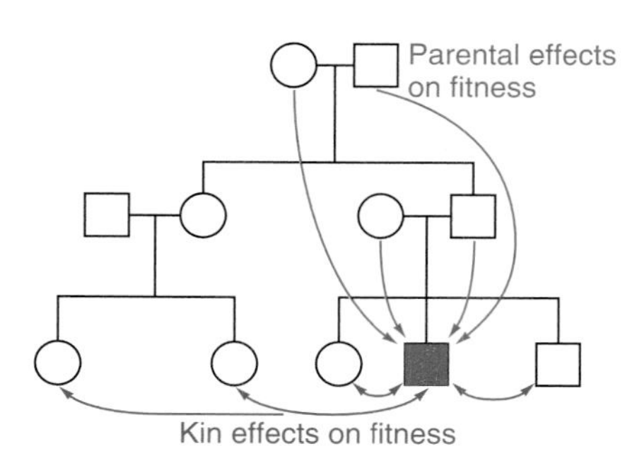

But Hamilton recognized that a genotype's fitness may be enhanced by an individual's effects on its nondescendant kin as well—its sisters, brothers, and cousins. An individual not only enhances its fitness by reproducing but also by supporting the reproduction of close relatives who share its genes. Inclusive fitness, E, may be defined as:

$$E = 1 \pm \text{Personal effects} + \text{Effects on all kin}$$

Because inclusive fitness encompasses effects on nondescendant kin, it is very different from classical fitness. An individual directly affects its descendants, those that actually incorporate some of its DNA; it indirectly affects nondescendant kin that haven't received any of its DNA but share copies of some of its genes.

If inclusive fitness is meaningful, social behavior should evolve as a way for organisms to look out for their relatives. Behavior patterns that promote the survival of other individuals shouldn't surprise us. This behavior is altruistic from the viewpoint of each organism, but from the viewpoint of the genomes and partial genomes they carry, it is just looking out for Number One.

Exercise 50.6 Suppose a woman has three children and all three live to have their own children. Has the woman's genome been perpetuated? Has every one of her genes been perpetuated? How many grandchildren must she have for the equivalent of her genome to be perpetuated for another generation?

Exercise 50.7 Mr. Jones has two children, Bill and Nancy. Bill has two children, Ed and Frank. Nancy has two children, Meg and Paul. All four children are in a boat, and the boat tips over. Meg is the only one who can swim well. If she can only save one other person, who should her genes tell her to save? Why?

50.7 Many birds mate and raise their young cooperatively.

During the last few years, it has become clear that altruistic behavior occurs in many species of birds with cooperative breeding systems in which the close relatives of a pair of birds help to raise the offspring instead of attempting to breed themselves. Various forms of this behavior have been documented in 222 of the 9,000 species of known birds. For instance, a colony of Acorn Woodpeckers—flashy, noisy birds of the American Southwest—is held together by a strong social structure that includes cooperative food gathering (Figure 50.14*a*). A group of birds establishes a granary, generally an old tree that they riddle with little cavities for storing acorns. By building up a storehouse during good times, they provide food for the group during lean times. The birds commonly breed cooperatively, a breeding group centering on a male-female pair or on a small group of males breeding with a small group of females. The males are generally brothers, but they may be father and sons; the females may be sisters or mother and daughters. Such a group shares a single nest cavity, and all members of the group, whether they are parents or not, become helpers who aid in feeding the young birds.

Interestingly, a certain amount of competition remains within such a group. This sometimes takes the form of egg tossing. Suppose female A is the first to lay an egg in the nest. Female B may then go into the nest, remove A's egg, and take it to a nearby tree, where she and the other females (including A) peck it open and eat its contents. Several of A's eggs may be

Figure 50.14
Cooperative breeding is common in birds. Two well-studied examples are *(a)* Acorn Woodpeckers and *(b)* Scrub Jays.

tossed, and this continues until B lays her first egg, for at this point neither bird can distinguish one egg from another. From this time on, the two females will lay a clutch of eggs together, and the group will cooperate in incubating them.

Cooperation in other birds takes somewhat different forms. Breeding groups of American Scrub Jays (Figure 50.14*b*) and African Green Woodhoopoes are based on a single breeding pair; related birds, often offspring of the previous year's brood, stay on as nonbreeders who help tend the new brood of young birds. This arrangement is only disrupted when one of the breeding birds dies; then its place is taken by a nonbreeder of the same sex. Opportunities for breeding are limited by territoriality, since there is only room for so many territories in any region. Sometimes a nonbreeder goes off in search of a territory where some other individual has died, but it has relatively little chance of finding one. On the whole, a nonbreeder therefore contributes more to the survival of its species (its genome) by staying at home and helping its relatives.

Cooperative breeding clearly yields greater overall success than breeding by isolated pairs. Birds that breed cooperatively always raise more offspring successfully than those that breed independently, sometimes twice as many. Cooperative breeding makes excellent sense from a genetic viewpoint, and even if a nonbreeding sister or brother has no offspring of its own, it is still helping to perpetuate the genome it shares with its siblings or parents.

50.8 In many species, individuals promote their own offspring and jealously destroy competitors.

Alongside reports of individuals recognizing their relatives and helping them breed, the literature of biology is rife with instances of individuals recognizing their nonrelatives and taking steps to prevent or nullify their breeding. African lions are a notorious case. A pride of lions consists of a dominant male or two with a harem of females. Occasionally a group of males from outside the pride attacks it, kills or chases away its males, and takes over (Figure 50.15). When this happens, the new males also kill all the cubs left by the previous males. By eliminating individuals that carry competing genes, they promote the transmission of their own, so this behavior is perpetuated genetically.

Many experimental studies on a variety of animals—including insects, frogs, birds, and mammals—show that they can distinguish relatives from nonrelatives and tend to associate with relatives. Recognition most likely depends on odor.

Figure 50.15
(a) A pride of lions consists of several females who mate with one male or a group of males. *(b)* Occasionally a male has to defend his position against intruding males. Successful intruders drive off the old males and kill the cubs they had sired. This behavior tends to perpetuate the genes of the new males and ensures that their future efforts will go toward the support of their own offspring—their own genes.

Genetically encoded enzymes produce chemical structures with definite odors, and there is good reason to think that among vertebrates the critical genes for odor-based recognition are included in the major histocompatibility complex. As explained in Section 48.8, these genes also create the distinctive cell-surface antigens that mark the individuality of tissues and obstruct tissue and organ transplantation.

In the following section we look in some detail at the lives of three social animals: honeybees, herring gulls, and Japanese macaques. These species have been extensively studied, and we rely on quotations from the work of those who have examined them in the field to give a sense of the animals' behavior and of the style, methods, and personal excitement of these investigators.

Figure 50.16
The three honeybee castes—queen, worker, and drone—have distinctive forms.

Exercise 50.8 Experiments have shown that women tend to prefer the odors of men with MHC complexes different from their own, thus promoting outbreeding (mating with nonrelatives) and a greater mixing of genes. However, pregnant women tend to prefer the odors of men with MHC complexes similar to their own—in effect, their own kin, who can be expected to protect them. What problem might a woman encounter if she falls in love and gets married while taking birth control pills, which mimic some conditions of pregnancy?

C. Studies of three social species

50.9 Honeybees exhibit an extreme division of labor among castes.

The social insects—wasps, bees, termites, and ants—have reached a pinnacle of social behavior that is "hard-wired" into their nervous systems: complex activities and interactions largely programmed in their genomes. We will focus on the common honeybee, *Apis mellifera.* Honeybees, in E. O. Wilson's terminology, are **eusocial** (truly social) because they have three characteristics: cooperation in caring for the brood, division of labor connected with reproduction, and an overlap of at least two generations, so offspring assist their parents.

A honeybee colony is made of three castes—queen, worker, and drone—marked by specialized anatomy and activities (Figure 50.16). The castes cooperate in the daily activities of the hive, which may have 20,000–80,000 workers, a few drones, and one queen. Drones are haploid males. Queens and workers are diploid females whose two sets of chromosomes result from ordinary fertilization; diploid zygotes can develop into either caste, depending on their early environment. Each bee develops in its own special type of chamber in waxy combs that bees make with architectural precision. Queen eggs develop in special queen chambers provided with royal jelly, a substance secreted by worker nurse bees. Worker eggs develop in other chambers without royal jelly. The developmental path of an egg can be changed by royal jelly in the first 72 hours after it is laid and put in an egg chamber, but thereafter its path of differentiation is irreversible.

Workers have definite duties determined by their age, not by their genotype or developmental history. During the first three weeks of their adult lives, they are "house bees" who nurse larvae and construct honey and brood chambers from wax secreted by their wax glands. When these glands shrink, the workers become "field bees" who gather pollen, nectar, and water outside the hive. Contrary to popular myth and the fables told to children, a worker bee doesn't work all the time; at least two-thirds of its day is spent resting or wandering around aimlessly. Although members of the colony communicate by movement and sound, communication and coordination are largely chemical, by means of pheromones, and depend largely on the habit of *trophollaxis:* Individuals regurgitate liquid food and pass it to one another. If a few individuals are given some radioactive syrup and the stomach contents of others are sampled over the next few hours or days, it becomes apparent that the radioactivity is quickly diffusing through the whole colony. Because of trophollaxis, the colony shares a kind of "community stomach." The stomach contents of individuals are quickly homogenized, and they all have a common nutritional state. If there is a lack of food, the whole colony feels it, and an individual who obtains food or water is obtaining it for the whole colony. And while they are sharing food, they are sharing information too, as we will see shortly.

A queen's only role in the colony is to lay eggs. She is much larger than a worker and can lay several thousand eggs a day, a mass greater than her own weight. A colony is restricted to one queen at a time because the reigning queen continually produces a pheromone called queen substance, *trans*-9-ketodecenoic acid, which the workers continually lick off her body, along with other pheromones, and share through trophollaxis. This queen substance inhibits the workers from producing royal jelly, which would stimulate the development of another queen egg. If the queen is killed, production of the pheromone naturally ceases, and the workers sense this lack within hours. Some of them begin to construct the special cells in which queens grow, stock them with royal jelly, and transfer some young zygotes into them. Production of the pheromone also ceases when a new queen is needed at the time a new colony breaks off from the old one.

The third caste consists of drones, haploid males whose only function is to provide the queen with sperm. Drones are produced parthenogenetically from unfertilized eggs (see Sidebar 15.1) and develop in special drone egg chambers, where they are fed by workers, at least as long as food is plentiful. But in times of shortage, drones are the first bees to be neglected, and their starving bodies accumulate on the floor of the hive until workers sweep them out.

A fully staffed hive of maximum size is ready to divide in two to create a new colony. As preparations begin for the colony to divide, the queen slackens production of *trans*-9-ketodecenoic acid, and the workers begin to construct special queen egg chambers, where the queen deposits one egg each. She also lays unfertilized eggs in drone egg chambers. Workers begin treating her aggressively, sometimes jumping up and down on her. Eventually they push her out of the hive, and she takes off in flight, secreting a second pheromone, *trans*-9-hydroxydecenoic acid, which excites a large swarm of workers to follow her. The swarm then alights on a tree branch or other perch while scouts go off in search of a permanent site for a new hive. When they find the right spot, they return and signal its location to the rest of the bees by means of a waggle dance—which we will describe shortly—and the swarm moves off to its new home.

Meanwhile, the remainder of the old colony prepares to install a new queen and to get her mated. The first virgin queen to emerge, just over two weeks after the egg is laid, searches out other queens as they emerge, making a quacking sound that the other queens answer. When two queens meet, a fight ensues that only ends when one has been killed. These fights continue until only one virgin queen is left. (Since the stingers of queens are unbarbed, they can sting with impunity again and again. A worker, however, can only sting once, since it cannot remove its barbed stinger from the victim.)

The lone queen is now ready to mate. Drones have hatched and are flying about, often mixing with drones of other colonies. Workers urge—push!—the young queen out of the hive, and she begins to secrete the pheromone that suppresses royal jelly formation, which also attracts the drones to her as she soars in a mating flight. One male finally catches her and copulates with her violently. The pair drops to the ground, and the queen can only free herself from him by ripping off his genitals, which are trapped within her. Leaving the dying male, the queen returns to the colony where workers remove the remaining male structures from her. The queen may make as many as a dozen such wedding flights, storing the sperm in a storage sac, or spermatheca. These are the only sperm she will have available for her several years of egg-laying. She finally settles down in the hive and begins producing the thousands of needed eggs.

One of the workers' most important activities is controlling the temperature in the hive. Honeybees originated in the tropics but have spread—or have been taken by humans—all over the world. They require a constant, elevated temperature for their activities. In summer, this temperature is almost always maintained within 0.05° of 35.0°C, slightly below human body temperature. Temperature control is tricky. As the temperature rises above the preferred 35°C, workers begin to circulate air with their wings. If the temperature continues to rise, workers bring water to the hive and sprinkle it on the brood combs. While some workers spread the water in thin films, others fan their wings to promote evaporative cooling.

In winter, the temperature remains elevated but isn't as precisely regulated. One hive was found to be 31°C inside when the outside temperature was −28°C. The hive is heated by metabolism, of course. It has been estimated that each bee produces about 0.1 cal/min at 10°C, so the thousands of bees in a hive have an impressive heat output. In winter, the bees clump together, blanketing the brood chambers. Bees on the outer surface of the cluster remain motionless and serve as an insulating layer, while those inside move constantly to generate heat, but they change places periodically so none remain on the outside too long. The colder it is, the more tightly the clump is packed.

Exercise 50.9 Naked mole rats of East Africa live in extensive underground burrows where most individuals serve a single queen, who produces most of the offspring. What other information do you need to determine if they are eusocial?

50.10 Honeybees communicate through dancing, sounds, and scents.

Perhaps the most amazing case of social communication in the animal kingdom, outside human language, is the dancing of the honeybee, first described about 50 years ago by the German ethologist Karl von Frisch. Pheromones and other signaling devices discovered since von Frisch's observations work in conjunction with dancing, but the dance itself conveys the essential information.

A foraging worker performs the dance to let the others know the location of a food source she has discovered. From his observations of bees in a glass-covered hive, von Frisch described the dance as follows:

> *To study the behavior of bees which have just discovered a rich source of food one may set out near the observation hive a glass dish filled with sugar-water. When a foraging worker comes to this feeding place she is marked with a colored spot while she is sucking up the sugar, so that we can recognize her later in the hive. After she has returned to the hive our marked bee is first seen to deliver most of the sugar-water to other bees. Then she begins to perform what I have called a round dance. On the same spot she turns around, once to the right, once to the left, repeating these circles again and again with great vigor. Often the dance is continued for half a minute or longer at the same spot. Frequently the dancer then moves to another spot on the honeycomb and repeats the round dance and afterwards ordinarily returns to the feeding place to gather more sugar. During the dance the bees near the dancer become greatly excited; they troop behind her as she circles, keeping their antennae close to her body. Suddenly one of them turns away and leaves the hive. Others do likewise, and soon some of these bees appear at the feeding place. After they have returned home they also dance, and the more bees there are dancing in the hive the more appear at the feeding place. It is clear that the dance inside the hive reports the existence of food. But it is not clear how the bees that have been aroused by the dance manage to find the feeding place.*[3]

Since it isn't natural for bees to feed at dishes of sugar-water, von Frisch placed sugar-water on various flowers. He found that the bees recruited by a worker who had fed on cyclamen, for example, would only return to cyclamen, ignoring other sources of food. Thus the foraging bee evidently carries back an odor of the flowers on her body:

> *The bees that troop after the returning forager as she dances on the honeycomb perceive the flower scent in two ways. By holding their antennae toward the dancer they smell the scent adhering to her body as a result of her contact with the flower. The upper surface of the bee's body has the ability to hold scents for long periods. Second, during pauses in the dance, the dancer feeds the bees that are following her by regurgitating a droplet of nectar from her honey stomach. This nectar was gathered from the bottom of the flower and is saturated with its characteristic scent.*[4]

The initial forager also returns to the food source and guides workers who follow her, using a special scent as a trail-maker.

Von Frisch then discovered that this round dance (Figure 50.17a) is used only when the food source is relatively close to the hive, within about 85 meters. When he placed food at greater distances, he found a new phenomenon:

> *When we now look into the observation hive we see a truly curious sight: all the bees marked at the 10-meter food source are performing round dances . . . but all the bees that have come from the more distant feeding place are dancing in quite a different manner. They perform what I have called a "wagging dance." They run a short distance in a straight line while wagging the abdomen very rapidly from side to side; then they make a complete 360-degree turn to the left, run straight ahead once more, turn to the right, and repeat this pattern over and over again [Figure 50.17*b*]. This wagging dance was one that I had observed many years before; but I had always taken it for the characteristic dance of bees bringing pollen to the hive, whereas now I saw that it was performed most vigorously by bees which were bringing in sugar solutions from the experimental feeding place at 300 meters.*[5]

Analysis of the waggle dance (as it is now called) shows that a worker conveys three kinds of information. First, the distance to the food source is indicated by the number of turns in the dance during a given time. Second, the vigor and duration of the dance indicate the richness of the source. Third, the direction of the food from the hive is given by using the sun as a reference. The dance is performed on a vertical honeycomb, and the direction of the sun from the hive is taken to be straight up (Figure 50.18). Then the angle between a vertical line and the direction of the waggle dance is equal to the angle between the sun's direction and the direction to the food. For instance, if the direction to the food is 30 degrees clockwise from the sun, the dancer will run upward 30 degrees to the right of vertical. Although the inside of the hive is dark, the other workers can follow the dancing bee's movements because she creates a subtle sound by vibrating her wings during the straight part of the dance.

[3]K. von Frisch, *Bees, Their Vision, Chemical Senses, and Language* (Ithaca, NY: Cornell University Press, 1950), 55–56.

[4]von Frisch, 60–61.

[5]von Frisch, 69–70.

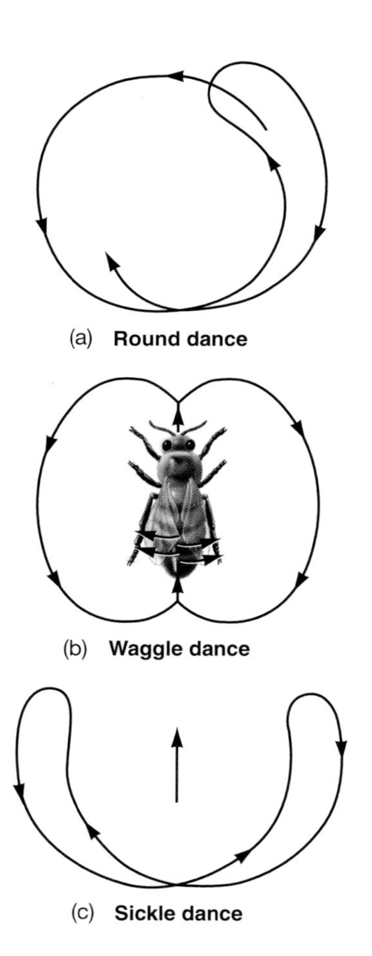

Figure 50.17
Honeybees may use three dances to indicate the location of food. *(a)* In the round dance, the worker takes a sweeping circular path. *(b)* During the waggle dance, the worker waggles her abdomen during the straight run, then turns left or right and returns to the starting point. *(c)* In the sickle dance, the worker moves back and forth in an arc, and the direction to a food source is shown by a line bisecting the arc.

There are several dialects of this bee language. Italian honeybees, for instance, perform a sickle dance (Figure 50.17*c*) when food is roughly 10–40 meters away; the direction to the food is shown by the arrow bisecting the opening of the sickle.

Exercise 50.10 Brothers and sisters each have half their father's genes and half their mother's, so the degree of genetic relatedness between them is ½. All the workers in a hive have a haploid father. (Assume they all have the same father, even though a queen may have sperm from several males.) What is the degree of genetic relatedness among all these workers? What does this imply about the degree to which they should engage in behavior that benefits the hive as a whole?

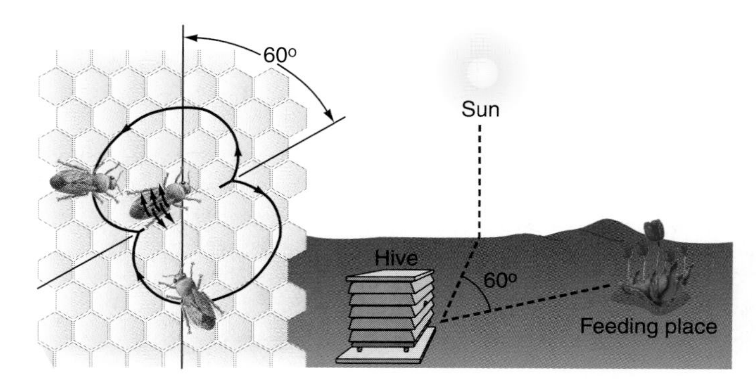

Figure 50.18
In a waggle dance, the vertical direction represents the sun's position. The angle between the sun's direction and the food source is indicated by the angle between the vertical and the straight run of the dance.

50.11 Herring Gull societies operate through fixed-action patterns.

Nikolaas Tinbergen, one of the world's most astute observers of animal behavior, made detailed studies of Herring Gull colonies. These birds illustrate some of the behavior patterns we have discussed. "All through autumn and winter," writes Tinbergen, "herring gulls live in flocks. They feed in flocks, migrate in flocks, sleep in flocks." When approached by an observer, one bird may utter an alarm call, a rhythmic "ga-ga-ga," and then fly off, followed immediately by the others.

In spring, the flock establishes its nesting grounds on sand dunes. The birds first gather in "clubs," where mating pairs form. Females initiate pair formation. An unmated female walks slowly around the male of her choice in a peculiar horizontal crouching attitude. The male responds either by strutting around and attacking other males or by walking away with the female. She then begins to beg for food with tossing movements of her head. The male responds by regurgitating some food, which she eats (Figure 50.19*a*). After a time, the two birds become bonded to each other and walk off to establish a territory somewhere. They build a nest cooperatively by collecting nest material and carrying it to the nesting site, where they mold it into a shallow pit and line it with grass and moss.

> *Coinciding with pair formation, nest building, courtship feeding, and copulation, another behaviour pattern has appeared, particularly in the male: fighting. Already in the club the male's aggressiveness can be so intense that it chases away all the gulls in the vicinity. Once established on the territory the male becomes entirely intolerant of trespassers. Each intruding male is attacked. Usually no genuine attack is made, threat alone is often sufficient to drive a stranger off. There are three types of threat. The mildest form is the "upright threat posture": the male stretches its neck, points its bill down, and sometimes lifts its wings [Figure 50.19*b*]. In this attitude it walks toward the stranger in a remarkably stiff way, all its muscles tense. A*

stronger expression of the same intention is "grass pulling" [Figure 50.19c]. The male walks up till quite close to the opponent, and all at once bends down and pecks furiously into the ground. It takes hold of some grass, or moss, or roots, and pulls it out. When male and female face a neighboring pair together, they show a third type of threat: "choking." They bend their heels, lower the breast, point their beaks down, and with lowered tongue bones, which give them a very curious facial expression, they make a series of incomplete pecking movements at the ground. This is accompanied by a rhythmic, hoarse sort of cooing call.[6]

Grass pulling is a kind of **displacement activity,** a totally irrelevant behavior that an animal appears to engage in as an outlet for frustration. Two males sometimes fight beak to beak at the edges of their territories, neither giving an inch; then they step aside and pull grass, as if they have built up a store of energy that must somehow be released.

The birds copulate once or twice a day, and when the eggs are laid, the pair take turns sitting on them. In Tinbergen's words:

Here again their cooperation is very impressive. They never leave the nest alone. While one is incubating, the other may be feeding miles away. When it comes back, the sitting bird waits until the newcomer walks up to the nest. This approach is accompanied by special movements and calls. Usually the long-drawn "mew call" is uttered, and often some nest material is carried. Then the sitting bird rises, and the other takes its place.

Soon after birth, the gull chick begins a begging activity. By pecking at the adult's bill, it stimulates the parent to regurgitate a bit of food, which it holds in its bill in front of the chick; the young gull is aroused to eat by the parent's "mew call" and by the sight of the food in the bill. The begging response of the newborn chick can be released by showing it a cardboard model of a gull's head, which must have one critical feature: a long, thin "bill" pointing downward. In fact, a model of the bill alone without the head will elicit the response. The bill must have one additional cue—a red spot near its tip like the one on the adult gull. No other colored spot will release pecking as frequently or as completely as a red one.

[6]N. Tinbergen, *Social Behavior in Animals* (London: Methuen and Co. Ltd., 1953), 5–6.

All these behaviors appear to be nothing more than series of fixed-action patterns released one after another—in Tinbergen's words, "rigid and immediate responses to internal and external stimuli." Both parent and chick play out the programs of their heredity. Their inflexible behavior continues only if the right cues are received on the right schedule. A chick's parents will defend it vigorously and feed it frequently as long as it moves and they can hear its calls. But the instant the chick no longer responds—if it dies or is killed—the parents eat it. The chick's failure to send the expected signals transforms it from an offspring to be nurtured into food.

50.12 Macaques have complex societies in which culture plays a large part.

The social behavior of bees and gulls depends almost entirely on species-specific fixed-action patterns, but our closest relatives, the nonhuman primates, are highly intelligent animals whose behavior can be greatly modified by experience. Unfortunately, these primates can be so much like people that observers tend to lose their objectivity and view their subjects as primitive and inferior humans or as cute little playthings. While these are certainly intelligent creatures with many similarities to humans, too often human emotions, intelligence, or motives have been wrongly attributed to them.

It has been difficult to study primate behavior for many reasons. Most primates live in forests and jungles in small bands

(a) (b) (c)

Figure 50.19

(a) A male Herring Gull assumes a characteristic posture when preparing to feed a female. *(b)* When threatening another male, a male Herring Gull assumes an upright threatening posture. *(c)* Grass pulling is a displacement activity that occurs during encounters between males.

that move quickly, and individuals are hard to identify. As native habitats are destroyed by human encroachment, many of these species have entered the endangered list; remaining gorillas, for example, are numbered in the hundreds. Many species have therefore been brought into captivity—in zoos, for instance—where some valuable behavioral studies have been done on them. Yet this imposes another problem for ethologists, a kind of biological uncertainty principle: Intervention by observers modifies the behavior of the observed animals. They simply behave differently in the wild than they do in captivity, where food is provided for them and predators are absent.

Nevertheless, several species of higher primates have been studied closely; here we look at Japanese Macaques, among the best-studied primates in the world. These monkeys are familiar to Westerners as the three animals that depict the Buddhist adage, "See no evil, hear no evil, speak no evil." They are about 60 cm tall, the males a bit larger than the females, with pinkish faces, short ears, shaggy hair, and short, furry tails (Figure 50.20). Living in bands of about 50–150 members, macaques are at home both in trees and on the ground, and they can swim. The lives of these highly intelligent animals are governed by subtle, highly adaptive social interactions. Some years ago, when it became clear that their population was decreasing with the destruction of their habitats by deforestation, macaque preserves were set up in Japan, and some colonies were established outside Japan as well, including several in the United States.

Macaque social life revolves around a dominance hierarchy with a few males at the top. The alpha male, the leader of the entire troop, reaches that position well after he has attained maturity, but oddly enough, neither aggressiveness nor physical size and strength are necessary. Rather, it is the social position of his mother that is important. This interesting matriarchal system emerges because mothers defend their offspring vigorously when play among the juveniles gets rough. Females establish a hierarchy during the times when mothers clash in defense of their young. Out of that hierarchy comes the ranking of males and the overall hierarchy of the troop, both male and female. In accordance with this method of establishing the hierarchy, the androgen levels of these males are not correlated with their hierarchical rank.

Figure 50.20
Japanese Macaques.

In most cases, observers can unambiguously rank all the individuals in the troop. Beneath the alpha male are five or six subleader males, and beneath them are the adult females of the troop, each with her position. At the bottom of the hierarchy are the remaining males, who live away from the center of the troop. Those at the top are seldom, if ever, attacked by other animals; with decreasing rank, the number of potential attacks increases, and the lowest animal is fair game for all. But an animal soon learns that for safety it should never provoke another higher in the pecking order, so a kind of restless peace prevails. Aggressive behavior is rather rare, and when it occurs it may be mostly symbolic.

The maternally established social hierarchy also determines a monkey's role in the troop. The alpha male, of course, is boss. He decides when a troop should move, and he directs its defense if the group is threatened. The male subleaders police the troop, with occasional help from the alpha male, moving quickly to stop fights. Males at the bottom of the social order are often referred to as peripheral males, for in the wild they remain at the edges of the troop, watching for predators and issuing warning cries when they perceive a threat. They also play a role in socializing and disciplining juvenile males, with whom they spend a lot of time. In turn, they are groomed by the juveniles. Peripheral males may break away from one troop to join another, thus spreading genes from troop to troop.

Mating is seasonal, confined to fall and winter. When mating season comes, the males parade on display with tails erect; they grab objects and wave them in the air, leap up and down, and beat their feet on the ground. They also begin to groom females, and eventually males and females form mating pairs. Although males initiate courtship behavior, it is the female who ultimately decides if, and with whom, she will mate. A male's position in the dominance hierarchy is surprisingly unimportant here, for a female may reject an alpha male and accept the advances of a low-ranking peripheral male. A male signals his intention to mate by placing his hand on the female's back; she may then allow him to mount or refuse him by walking away. After a short time—perhaps hours, sometimes a couple of days—the two stop interacting with each other and break up. They may form other pairs for mating, so an animal can have a number of partners during the mating season.

The young are born after a six-month gestation period, most often in April or May. Females raise the offspring and protect them until after puberty—three years for females and four for males. The two sexes behave quite differently when very young. Young males spend their time playing together in groups, while females spend much of their time grooming their mothers and other females. Close and permanent bonds form between mothers and daughters, but by a year or two after puberty, the males show little attachment to their mothers.

Much is rightly made of the transmissions of culture in human societies, where each generation passes on customs and knowledge as well as its genes. Cultural transmission isn't restricted to humans, as we show in Chapter 49, and it is generally important among all primates. Primates have fewer offspring than other mammals, so they can devote more attention to each one. While all baby mammals depend on their mothers for some time, young monkeys and apes depend on their mothers for much longer times than do other young mammals. During this prolonged interaction, the young learn from their parents. In this relationship between two highly intelligent animals—one older and more experienced, the other young and in need of survival skills—it is genetically advantageous for the elder to help ensure the survival of the youngster who carries half her genes. It would be dangerous for the younger animal to acquire all the needed skills through experience, for each learning experience is also potentially deadly. But the mother can teach her offspring survival skills while protecting them. In humans, it is during this period of acculturation that language is developed and much that is basic to survival gets passed on.

Two stories will illustrate the role of culture in the lives of the Japanese macaques. Starting in 1952, the monkeys on the island of Koshima were fed sweet potatoes by humans, and shortly after the food was introduced, a young female called Imo was seen washing the sand off her potatoes in the ocean. This behavior spread through the population, beginning with those who were closest to Imo, as one monkey imitated another. A decade later, the behavior had become a tradition; three-quarters of all the monkeys over the age of two were potato-washers, and many had even started to salt them by dipping them in seawater before taking each bite.

G. Gray Eaton and others at the Oregon Regional Primate Research Center observed another example of innovation in macaque culture. Following a heavy snow in 1971, a low-ranking male named Big X discovered how to roll a snowball. His first was 45 cm or so in diameter—impressive for a monkey who isn't much bigger than that himself. Other macaques took up the sport, and every winter since then, they have made snowballs, the juveniles playing with them and the adults just sitting on top of them. Like potato-washing by the Koshima macaques, snowball-making by the Oregon macaques is a learned behavior that has now been firmly incorporated into the culture of the troop.

Coda A common theme of science fiction stories has been the social superorganism, an integrated being whose subunits are individual animals tightly integrated into a single unit. After all, taking a hierarchical view of biological organization, cells are made of organelles, animals of cells, and societies of individual animals, so why shouldn't the next step in evolution be a superorganism made of separate animals? Indeed, eusocial animals seem to be a kind of intermediate step in such an evolution. The members of a colony share a genome more intimately than the members of other populations; they live together, work cooperatively toward a common end, share food and have virtually a common digestive system, and often sacrifice themselves for the good of the colony as readily as phagocytes in the blood sacrifice themselves by killing the bacteria in a cut. What more is needed to create a superorganism?

Summary

1. Social behavior, entailing specific interactions between members of a species, has evolved in many instances because its advantages, such as greater security from predators and more assured access to mates, outweigh its disadvantages, such as more rapid transmission of disease.
2. Societies are held together by interpersonal bonds, primarily parental-familial and sexual bonds. Some societies are organized through more general needs, such as division of labor, particularly in the social insects.
3. Agonistic behavior, which is common in social animals, may entail fighting or threatening to fight, as well as fleeing, submission, and appeasement. By means of such behaviors, social animals typically form dominance hierarchies in which each individual achieves a stable position—dominated by some, dominating others. Once established, such hierarchies create social stability and reduce actual conflicts.
4. From the viewpoint of a genome, the critical factor in reproduction is to form more copies of itself, and how these copies are distributed in organisms is less important. The organism is seen as a survival machine that protects and passes on the genome. Genomes tend to program organisms with behavior that ensures survival of copies of the genome. Behavior that helps relatives carrying part of an organism's genome increases its inclusive fitness, so selection favors such behavior. Therefore, altruistic behavior toward close relatives is to be expected.
5. Communication is a foundation for all social interactions, and signals can be carried vocally, visually, chemically, or through specific movements, such as the honeybee dances that provide directions to food sources.
6. Human language is special because of its range and subtlety, but other animals convey all kinds of information vocally, including warnings, threats, and invitations to mate. Many species convey such information through facial expressions, postures, light signals, and specific pheromones.
7. Social animals are frequently territorial. Certain species defend an entire home range where they breed and feed; others defend only a limited area where they reproduce. In this way, a species can allocate resources—including mates—without undue overcrowding and with controlled competition.
8. Many birds breed cooperatively. Sometimes a group of males and females share reproduction; sometimes a breeding pair is assisted by their kin, who are often offspring from the previous year. Cooperative breeding increases the inclusive fitness of all members of the group and allows more offspring to survive than does independent breeding.
9. In some societies, such as those of lions, individuals clearly promote their own genes and kill the offspring that carry potentially competing genes.
10. Honeybees and Herring Gulls illustrate societies based entirely on inherited fixed-action patterns. Primates, such as Japanese Macaques,

can modify their behavior through experience, and they have an established culture, in which innovations can be integrated, that probably governs much of their social structure.

Key Terms

agonistic 1062	lek 1067
territoriality 1066	inclusive fitness 1068
court 1067	eusocial 1071
arena 1067	displacement activity 1075

Multiple-Choice Questions

1. An animal that is part of a social group exhibits social behavior
   a. as a choice determined by environmental conditions.
   b. as a result of evolutionary selection of individual cost vs. individual benefit.
   c. to ensure survival of the group.
   d. *a* and *b*, but not *c*.
   e. *a*, *b*, and *c*.
2. Which of the following characterize eusocial animals?
   a. cooperation in caring for young
   b. offspring assist their parents
   c. division of labor connected with reproduction
   d. *a*, *b*, and *c*
   e. *a* and *b*, but not *c*
3. Dominance hierarchies in birds and mammals are generally characterized by
   a. frequently lethal agonistic behaviors.
   b. presence of both males and females within the ladder of dominance.
   c. permanence when new members arrive.
   d. hormonally caused status differences.
   e. inclusion of members of more than one species.
4. In all probability, animals most frequently communicate through
   a. vocalization.
   b. facial expression.
   c. physical appearance.
   d. limb gestures.
   e. chemical pheromones.
5. All of the following except ______ characterize territoriality.
   a. protected space
   b. space used for mating and nesting
   c. space used for feeding
   d. changing boundaries as the animal moves
   e. adaptive behavior for social species
6. A general synonym for a territorial mating area is a
   a. bower.
   b. arena.
   c. court.
   d. lek.
   e. range.
7. The word or phrase that comes closest in meaning to "selfish" in the phrase "selfish genome" is
   a. brutish in appearance or action.
   b. leaves more descendants than alternatives.
   c. mean-spirited.
   d. most capable of stealth.
   e. bullying.
8. In a beehive, which members have cells that undergo meiosis?
   a. queen only
   b. drone only
   c. workers only
   d. queen and drone, but not workers
   e. all members of the hive
9. When calculating an individual's inclusive fitness, it is important to know all of the following except
   a. its personal fitness.
   b. the total number of lineal descendants.
   c. the total number of nondescendant kin.
   d. the personal fitness of all descendant and nondescendant kin.
   e. its effects on survival of its kin.
10. If degrees of cooperation and altruism are completely tied to degrees of genetic relatedness, an Acorn Woodpecker should primarily help
    a. one of its half-brothers.
    b. a sibling.
    c. any male in the group.
    d. a bird born in the same year but in another nest.
    e. any member of the group because all are equally related.

True-False Questions

Mark each statement true or false, and if false, restate it to make it true.

1. The strongest social bonds are likely to be between members of the same sex.
2. Particular levels of testosterone, but not estrogen, appear to be correlated with formation of parent-child social bonds.
3. Social organization depends upon vocal communication.
4. It is possible for a grandchild to carry more than 50 percent of the genes from one particular grandparent.
5. The inclusive fitness of an individual depends upon all the benefits it receives from its relatives.
6. If two worker bees are full sisters, they share on average 75 percent of their genes, but two human sisters share on average only 50 percent of their genes.
7. The round dance of honeybees signals that workers have brought nectar to the hive, while the waggle dance signals that pollen has been brought.
8. If the members of a eusocial group engage in trophollaxis, they will all have equal motivation to seek new food sources if food becomes scarce.
9. While strong bonds form between members of mated pairs of Herring Gulls, a dominance hierarchy is established between the males within the flock.
10. The rigid dominance hierarchy formed by male Japanese Macaques more closely resembles a matriarchal monarchy than a "might-makes-right" society.

Concept Questions

1. Human males and females have evolved smaller differences in height and weight than other higher primates. How might this phenotypic characteristic affect the evolution of social bonds in human and other primate societies?
2. Many students of human behavior claim that we make relatively firm decisions about the character and intentions of another person within minutes of the first encounter. What does that suggest about human communication in general and about the kinds of bonds in the hierarchy of social groupings that are supported by verbal communication in particular?
3. In his book *The Selfish Gene,* Dawkins offers an alternative interpretation of a beehive: The hive is maintained for the benefit of the workers who farm the queen bee in order to increase the number of workers. Explain the logic behind this point of view.

4. Explain why the percent of shared genes is the same between full siblings (non-twins) as between a parent and child.
5. An equal degree of genetic relatedness exists between a parent and its child as between a child and its parent. Despite this equality, parents are much more likely to behave altruistically toward their children than vice versa. Explain.

Additional Reading

Barash, David. *The Whisperings Within.* Harper and Row, New York, 1979.

————. and Judith Eve Lipton. *The Caveman and the Bomb.* McGraw-Hill Book Co., New York, 1985.

Beauchamp, Gary K., Kunio Yamazaki, and Edward A. Boyse. "The Chemosensory Recognition of Genetic Individuality." *Scientific American,* July 1985, p. 86. Genes regulating immunologic function impart to individual mice a characteristic scent. A mouse can thus discriminate genetic differences among its potential mates by smell alone.

Blaustein, Andrew R., and Richard K. O'Hara. "Kin Recognition in Tadpoles." *Scientific American,* January 1986, p. 108. Tadpoles of the Cascades frog prefer to associate with siblings, which they distinguish from nonsiblings. The ability to recognize kin is not based on familiarity; it may have a genetic component.

Bonner, John Tyler. *The Evolution of Culture in Animals.* Princeton University Press, Princeton, (NJ), 1980.

Brown, Jerram L. *Helping and Communal Breeding in Birds: Ecology and Evolution.* Princeton University Press, Princeton, (NJ), 1987.

Clutton-Brock, T. H. and G. A. Parker, "Punishment in Animal Societies." *Nature,* January 19, 1995, p. 209

de Waal, Frans B. M. "Bonobo Sex and Society." *Scientific American,* March 1995, p. 58. Sexual behavior in the small chimpanzee, Bonobo, challenges ideas about the evolution of male supremacy in human society.

Ghiglieri, Michael P. "The Social Ecology of Chimpanzees." *Scientific American,* June 1985, p. 102. Wild chimpanzees have rarely been studied without the lure of food, which can distort their social relations. A study of unprovisioned apes shows their social structure is most similar to that of humans.

Greenspan, Ralph J. "Understanding the Genetic Construction of Behavior." *Scientific American,* April 1995, p. 74. Studies of behavior in *Drosophila* show that even simple behavior patterns are probably controlled by many genes.

Kirchner, Wolfgang H., and William F. Towne. "The Sensory Basis of the Honeybee's Dance Language. *Scientific American,* June 1994, p. 52. The bees may communicate through both their dance and sound.

Ligon, J. David, and Sandra H. Ligon. "The Cooperative Breeding Behavior of the Green Woodhoopoe." *Scientific American,* July 1982, p. 126. Further information about a cooperative breeder.

Pfennig, David W., and Paul W. Sherman. "Kin Recognition." *Scientific American,* June 1995, p. 68. Animals recognize their nearest relatives through chemical cues and other methods.

Sherman, Paul W., Jennifer U. M. Jarvis, and Stanton H. Braude. "Naked Mole Rats." *Scientific American,* August 1992, p. 72. These African rodents have a eusocial organization, in which only a few individuals breed, and nonreproductive workers attend a fertile queen.

Stacey, Peter B., and Walter D. Koenig. "Cooperative Breeding in the Acorn Woodpecker." *Scientific American,* August 1984, p. 114. Reproductive biology of a fascinating bird.

Tinbergen, N. *Social Behaviour in Animals, with Special Reference to Vertebrates.* Methuen & Co., Ltd., London, 1953.

————. *The Herring Gull's World: A Study of the Social Behaviour of Birds.* Basic Books, Inc., New York, 1960.

van Lawick-Goodall, Jane. *In the Shadow of Man.* Houghton Mifflin Co., Boston, 1971.

Internet Resource

To further explore the content of this chapter, log on to the web site at:

http://www.mhhe.com/biosci/genbio/guttman/

51 Sexual Behavior and Reproduction

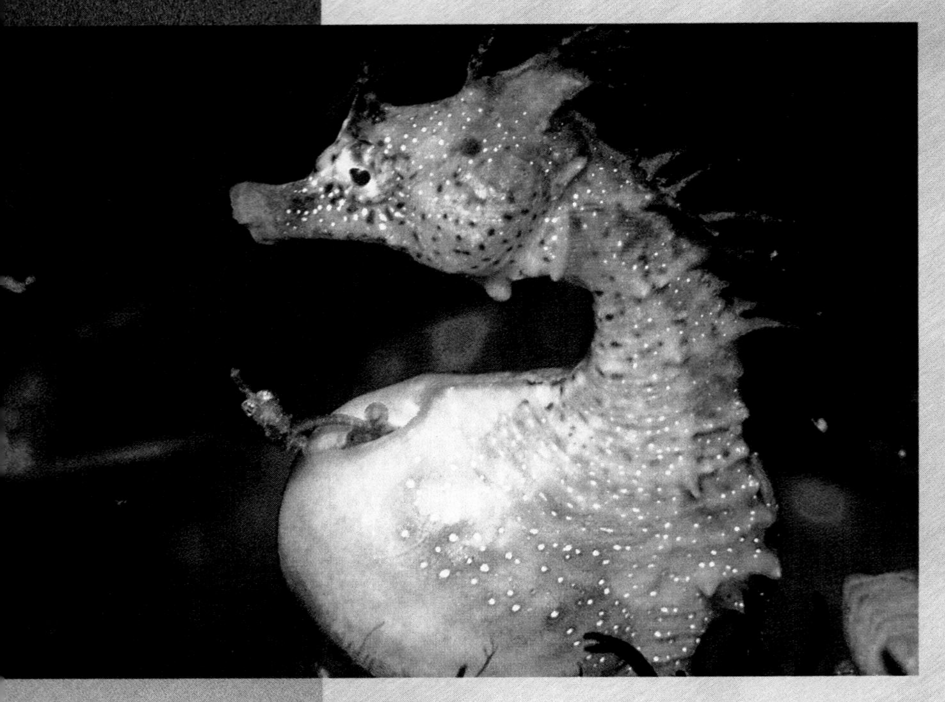

A male seahorse giving birth to the tiny offspring he has been incubating in his pouch.

Key Concepts

A. Sexual Attraction and Copulation

51.1 Gametes of the same species must be brought together for reproduction.

51.2 Complex mating rituals often precede copulation.

51.3 The reproductive systems of males and females are specialized for producing gametes and facilitating fertilization.

51.4 Copulation entails a regular sequence of acts in each species.

51.5 Sexual response in humans develops in four phases.

B. The Vertebrate Reproductive System and Its Regulation

51.6 Gonads produce steroid sex hormones under control of the anterior pituitary gland.

51.7 The gonadotropins stimulate spermatogenesis and androgen production in males.

51.8 An interplay of hormones controls the female reproductive cycle.

51.9 Other mammals ovulate on different schedules from those of humans.

C. Embryonic Development and Pregnancy

51.10 A mammalian embryo is attached to its mother through a placenta.

51.11 New hormonal pathways become active during pregnancy.

Reproduction in the oddly elongated pipefishes and horsefishes is especially interesting because the male generally incubates and protects the embryo fishes. During mating, the female transfers her eggs to a pouch on the male's body, where he fertilizes them and retains the zygotes. Seahorses have a bony crest called a coronet on the back of the head that makes a clicking noise as the head moves, and a mating pair make these sounds as part of their mating ritual. J. R. Norman wrote that mating in the Florida Pipefish follows an elaborate courtship. Male and female swim around in nearly vertical positions, with the head and shoulder region bent forward. "They then swim slowly past one another, their bodies come into contact, and the male demonstrates his affection by contortions of the body, caressing his mate with his snout. Just before the actual transfer, the male becomes violently excited, wriggles his body about in corkscrew fashion, and rubs the belly of the female with his snout." As they embrace, the female transfers her eggs by rapidly thrusting her protruding oviduct into the opening in the male's pouch. The pair repeats this ritual until the pouch is full, after which he remains quiescent and appears exhausted. In some species, a male may even receive the eggs of several females if there is room. As the embryos start to develop, the lining of the pouch thickens and acquires blood vessels in its folds and processes, bringing oxygen to the embryos, and it secretes nutrients for them. The male incubates the developing embryos for about three weeks and then gives birth by forcing them out. The young start to feed immediately but stay near the male and may even retreat into his pouch if threatened.

Each species of animal has its own way to bring male and female gametes together and reproduce. In this chapter, we concentrate on reproduction in mammals, especially humans. ■

A. Sexual attraction and copulation

51.1 Gametes of the same species must be brought together for reproduction.

To reproduce, male and female animals have to form pairs of the right type: one male and one female of the same species, of the right age, both physiologically ready to reproduce. Strong evolutionary pressures tend to make sexual organisms mate only with others of their species; as we explain in Section 24.6, species are adapted to slightly different niches, so hybrids between them, at least in animals, are usually not well adapted to either niche and are less viable than the inbred offspring of each species. Animals (and plants, to some extent) therefore tend to evolve all kinds of mechanisms that keep two individuals of different species from even trying to mate. Many of them are behavioral.

The most primitive mechanisms for bringing the correct gametes together are chemical. Sexual protists, algae, and fungi don't always have the equivalent of our male and female sexes. They may be best described as having **mating types,** which differ only in a few cell-surface structures, rather than in obvious morphological features. Compatible gametes of different mating types adhere to each other and fuse. These organisms generally release gametes into the surrounding water, where merely finding one another is a problem. So one mating type usually produces a **sex pheromone** that attracts gametes of the other type. Since the fungi are now known to be close relatives of the animals, let's use a fungal example: The large female gametes of the mold *Allomyces* produce the aptly named pheromone sirenin, which attracts the smaller, motile male gametes (Figure 51.1). (In the *Odyssey,* the sirens were creatures who wooed mariners to their destruction with song.) Another water mold, *Achyla bisexualis,* produces the sterol antheridiol, which stimulates antheridia (where male gametes form) to grow toward the oogonia (where female gametes reside). Sterols promote the formation of reproductive structures and control reproduction in animals and a number of fungi.

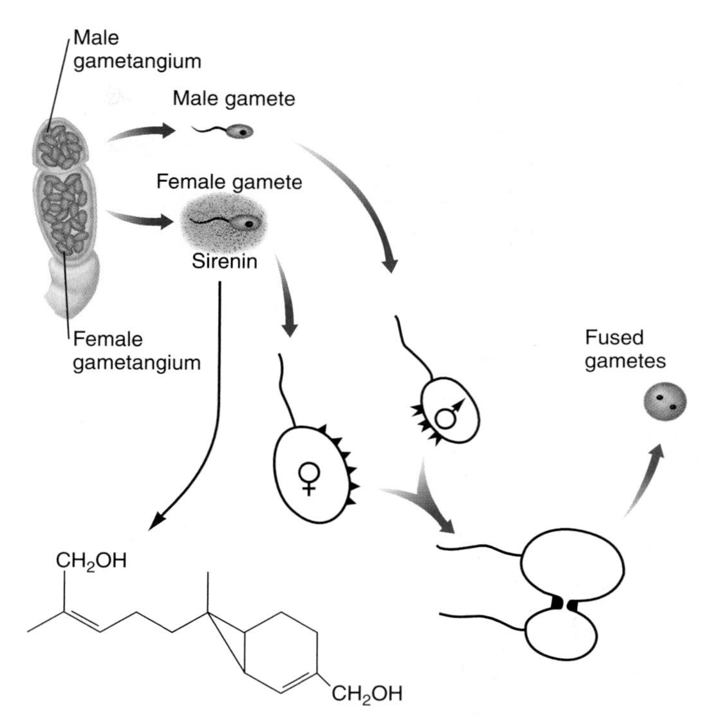

Figure 51.1
The male gametes of *Allomyces* are attracted to sirenin produced by the female gametes. The gametes fuse when their complementary surface structures join.

Sex pheromones in animals induce aggregation and complex behaviors. Male and female insects find each other because they are sensitive to tiny amounts of very volatile pheromones that diffuse over long distances. For instance, a female silkworm moth *(Bombyx mori)* produces less than 10 nanograms (10^{-8} gram) of bombykol, a distinctive alcohol:

$$CH_3-CH_2-CH_2-CH=CH-CH=CH-(CH_2)_8-CH_2OH$$

Only male moths of the same species are attracted to it, since the males of each species have receptor proteins on their antennae that bind the pheromones released by their females. Males are attracted from over 5 kilometers away. They are sensitive to as little as a few thousand molecules per cubic centimeter of air, and even a single molecule of pheromone can induce a moth to turn toward its source. First the males fly upwind and then up the chemical gradient as they approach the source, seeking the highest concentration until they reach the secreting female. (Studies done while keeping females inside gauze cages where they can't be seen have shown that males don't use visual cues in this search.)

Figure 51.2
A male Queen butterfly *(Danaus gilippus)* uses the hairpencils on its abdomen in its mating behavior.

Other species use pheromones to initiate complex sexual behaviors. Female Rhesus Macaque monkeys, for instance, produce copulin in their vaginal secretions when they are ready to mate, and its smell induces mating behavior in males. Male danaid butterflies, such as the Queen butterfly (Figure 51.2), have scent scales, or androconia, on their hind wings, with a pair of hairpencils resembling little brushes near the ends of their abdomens. After rubbing the hairpencils over his androconia, a male pursues a female in flight and rubs her antennae with his hairpencils. This induces her to alight, and the male continues to rub her antennae until she becomes quiet and they can copulate.

51.2 Complex mating rituals often precede copulation.

In the simplest organisms, gametes attracted to each other by sex pheromones come in contact and fuse. But in many animals, a mere chemical signal is not enough to ensure the pairing of a proper male and female; in addition, these species have evolved complicated rituals that lead to pairing and mating, rituals reminiscent of two spies or members of an arcane fraternal organization who must exchange elaborate secret handshakes and passwords before they can acknowledge each other. These rituals probably serve two functions: to identify the individuals as members of the same species and to identify high-quality mates.

Gary Nuechterlein and Deborah Buitron have described the ceremonies of the Western Grebe, a large North American water bird (Figure 51.3). Courtship begins with vocalizations, leading to the Rushing Ceremony:

> *When a male or female Western Grebe arrives on its breeding marsh in Manitoba, it must first find a suitable mate. During the first few weeks, a newly arrived male swims through many miles of channels through cattails, bulrushes or reeds, repeatedly advertising his availability by giving a series of loud calls in bouts of three to five. An interested female answers with a slightly shorter, higher-pitched series of calls.*
>
> *This behavior initiates a highly stereotyped sequence of courtship displays. Calling back and forth, the two birds cautiously approach one another, until they eventually catch sight of each other. They come to a standstill face to face, often only inches apart. Pointing their spear-like bills at one another, each bird stares into the other's brilliant red eyes and gives a fierce-sounding, ratchety call. Then one bird abruptly breaks the gaze to dip its bill into the water and shake its head vigorously from side to side. As the dip-shaking bird renews the staredown, the other bird does the same.*
>
> *Tension builds as the two grebes alternately ratchet-point and dip-shake. Almost simultaneously, the two birds make their next move. Turning at right angles, they emerge from the water and run rapidly across the surface side by side. In this famous rushing display, both birds propel themselves across the water's surface using only the powerful thrust from their lobed feet. They hold their partially opened wings stiffly at their sides, which may provide lift and balance as in a hydrofoil. After running 10 yards or so, the female and the male dive into the water with barely a ripple.*

The Rushing Ceremony then leads a male-female pair into the Weed Ceremony:

> *As the male Western Grebe emerges from the water, he turns back toward the female with his neck stretched high. After reestablishing eye contact, the pair alternately and silently dip their bills in the water and stretch their necks high again with a head shake. Each then gives a soft trill, and they both dive beneath the surface to search for water plants, materials similar to those used in nest-building. With the plants in their bills, they slowly and deliberately swim toward each other. As the birds come face to face, they stretch their necks upward, rise out of the water, and join their beaks together over their heads in a graceful dance.*
>
> *Finally, both grebes toss the plants away and begin to vigorously preen their back feathers. Following each of these deliberate bob-preens, the birds stretch their necks high. The preening becomes more synchronous, and soon their movements take on the likeness of a ballet, with each bird mirroring the image of its partner. As the preening subsides, the pair swims away side by side in the rather peculiar high-arch posture. If you are very close to them, you can hear them clicking softly to each other as they swim away from the center of courtship activities.*[1]

This ritual helps ensure that both birds are high-quality members of the same species, as if the genetically encoded logic of their nervous systems says, "If you know how to do the dance, then you must have all the right genes." By exchanging chemical and other signals, males and females of all species recognize each other as acceptable mates and prepare for copulation. Such rituals have evolved to ensure that fertilization occurs only between a sperm and an egg with compatible half-genomes and no obvious genetic defects.

[1]G. Neuchterlein and D. Buitron, "Water Dancers," *WildBird Magazine,* June 1994, 44–47.

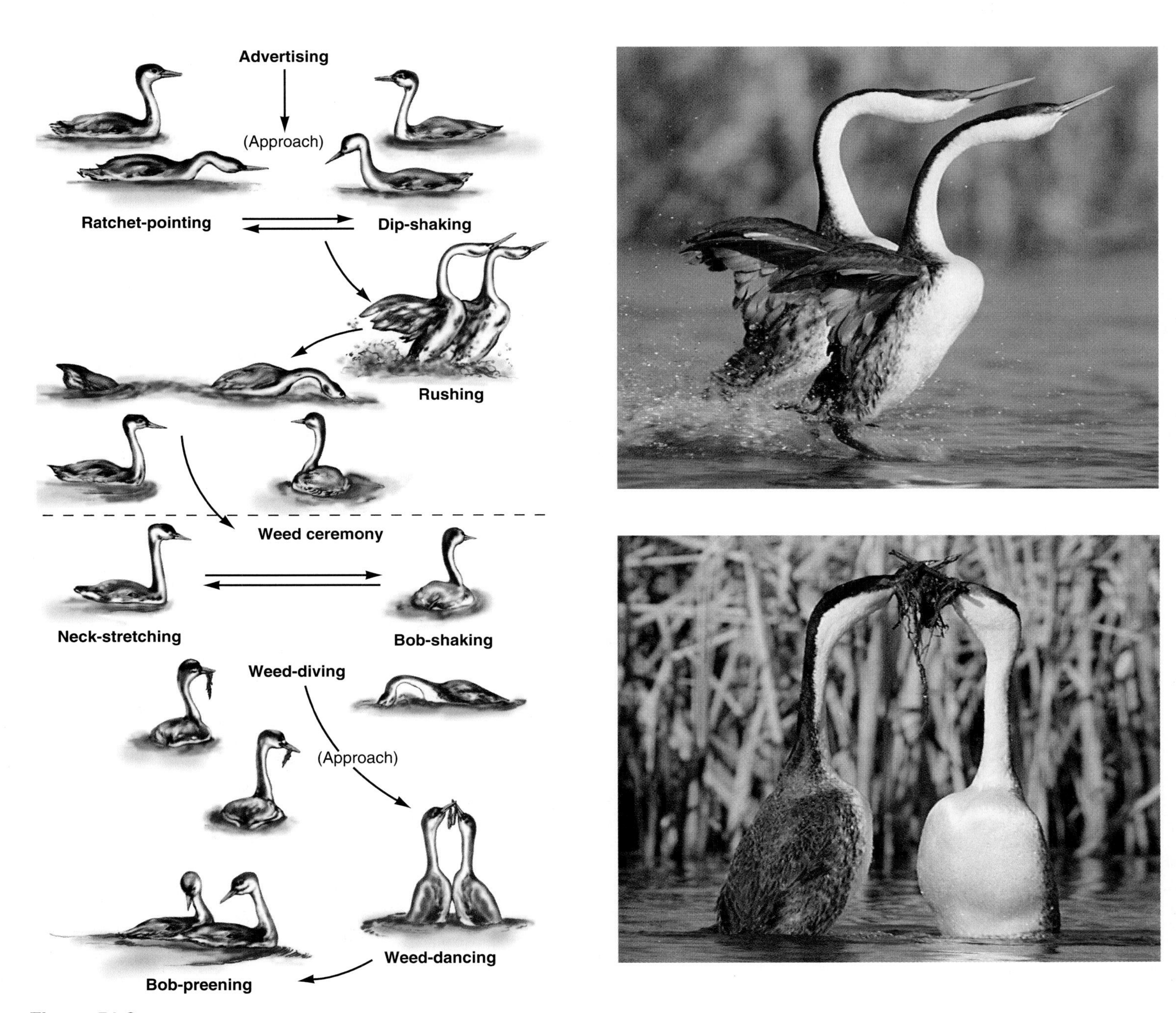

Figure 51.3

The mating ritual of the Western Grebe includes the spectacular Rushing Ceremony, in which the birds run across the surface of the water, and the Weed Ceremony, in which they dive together and emerge with water weeds on their bills.

51.3 The reproductive systems of males and females are specialized for producing gametes and facilitating fertilization.

Before continuing our discussion of mating behavior, we will review vertebrate genital structures, again using humans as a model since we will emphasize human reproduction in later sections.

The function of the male reproductive system is to deliver a mass of sperm to the vicinity of viable eggs (Figure 51.4). Sperm formation (spermatogenesis) occurs in about 100 meters of **seminiferous tubules** in each **testis.** Because spermatogenesis is inhibited at the core body temperature, the testes are suspended outside the body cavity in the **scrotum,** where their temperature is 2–4°C lower. The seminiferous tubules empty into collecting ducts that terminate in the **epididymis** on the surface of each testis, where sperm mature; under the influence of secretions from the walls of the epididymis, the sperm become motile, acquire resistance to pH and temperature changes, and become able to fertilize an egg. Each epididymis, in turn, leads into a **vas deferens.**

The vasa deferentia run up the spermatic cord, curve through the body cavity around the urinary bladder, and join the

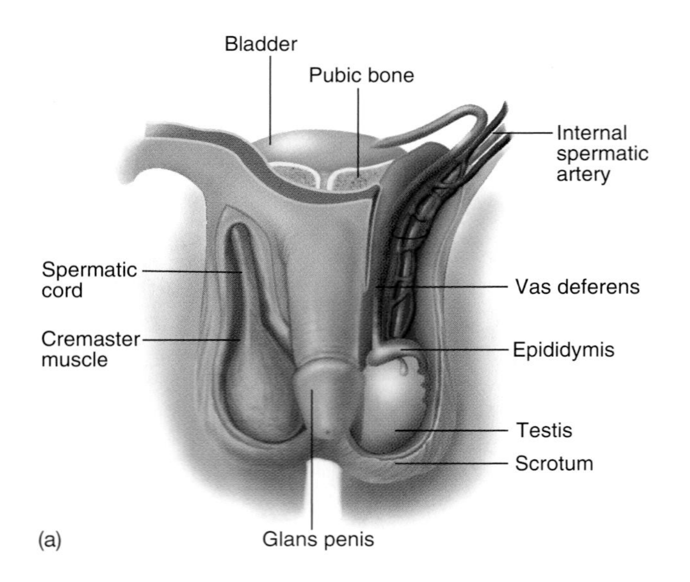

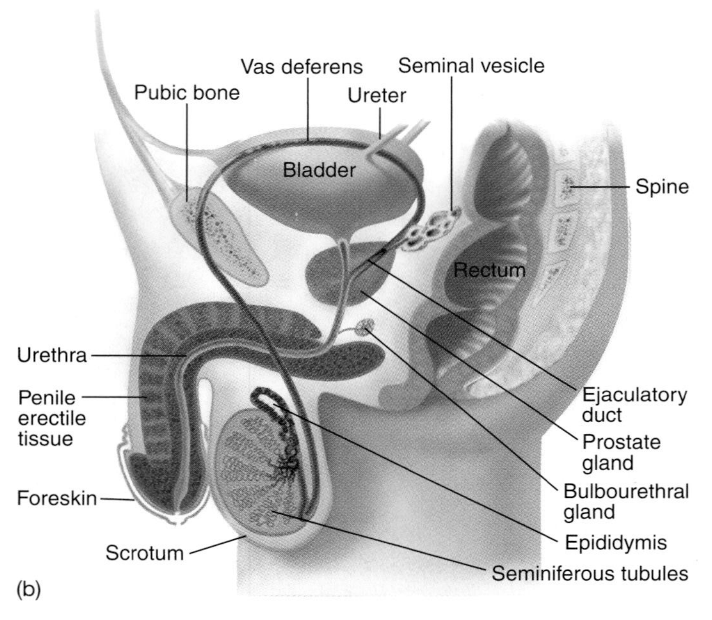

Figure 51.4

(a) A frontal view shows the external genitals of a human male, and *(b)* a section through the middle of the body shows the internal structures.

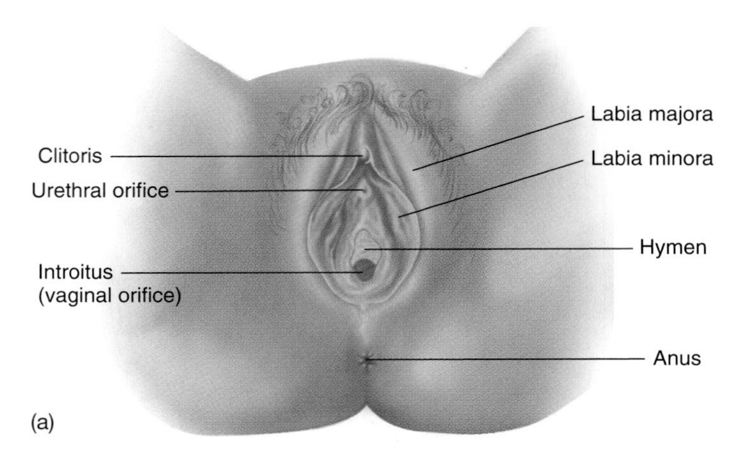

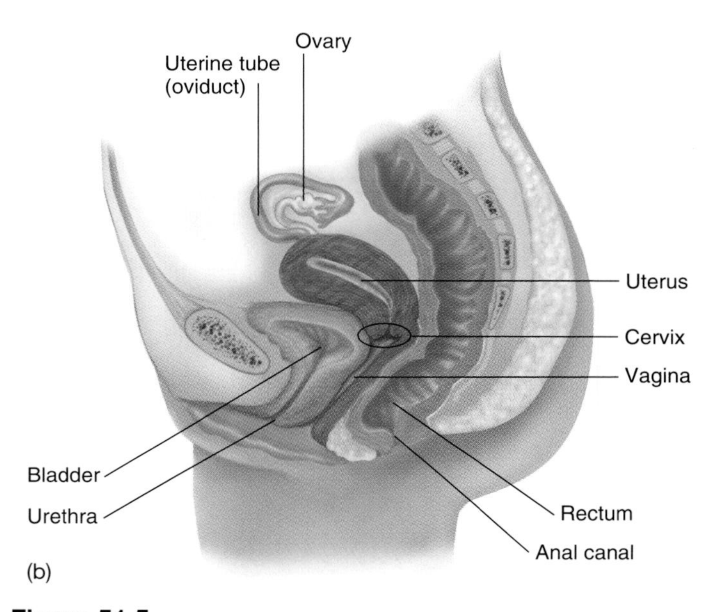

Figure 51.5

(a) A frontal view shows the external genitals of a human female, and *(b)* a section through the middle of the body shows the internal structures.

seminal vesicles. The volume of sperm-laden fluid from the testes is small compared to the total volume of material ejaculated. The seminal vesicles secrete about two-thirds of the ejaculate, a clear fluid containing mucus, amino acids, and fructose, the latter an energy source for the sperm that increases their motility in the female reproductive tract. Prostaglandins in the seminal fluid stimulate contraction of the uterus and uterine tubes, an action that helps propel the sperm up the uterus toward the oviduct. The ducts of each seminal vesicle and vas deferens fuse into **ejaculatory ducts** that run through the **prostate gland,** which contributes a quarter to a third of the **semen,** the sperm-laden fluid that is finally ejaculated. The prostate gland secretes a thin, milky, alkaline fluid that neutralizes the natural acidity of the vagina, extending the sperms' lives. The two ejaculatory ducts join the urethra coming out of the bladder, and this common urethra, carrying both urine and sperm, runs through the **penis.** The **bulbourethral gland** (Cowper's gland) just below the urethra produces a clear lubricating fluid upon sexual arousal. (Don't confuse the urethra with the ureter, one of the tubes that conducts urine from a kidney to the urinary bladder.)

The female reproductive system (Figure 51.5) produces ova, guides sperm to a place where fertilization can occur, and protects and nourishes a developing embryo. Eggs develop in the two **ovaries,** located on either side of the lower abdominal cavity, well protected from external shocks. Approximately once every 28 days, one ovary releases an ovum into the abdominal cavity, where fingerlike ciliated projections called fimbriae guide it into the funnel-like opening of the **oviduct**—also called a **uterine tube,** or **Fallopian tube.** Cilia lining the oviduct sweep the ovum down toward the muscular **uterus** (womb), where an embryo normally develops. If sperm are present, fertilization usually happens in the upper part of the oviduct. Rarely, however (in about 0.05 percent of human pregnancies), a zygote escapes into the abdominal cavity and starts to develop there, resulting in an ectopic pregnancy in which the embryo implants on the lining (peritoneum) of the abdominal cavity. This places both the fetus and the mother in jeopardy.

The uterus opens through a narrow passage, the **cervix,** into the **vagina,** a muscular tube that leads to the outside of the body. The **introitus,** or external opening of the vagina, is flanked by the **labia minora** and **labia majora;** the **clitoris,** the embryonic homolog of the male penis, lies anterior to the vaginal opening. In females who have not had intercourse, the introitus may be partially or completely covered by the **hymen,** a thin membrane that surrounds or bridges the introitus but rarely closes it completely. The opening of the urethra, which in females is completely separate from the reproductive tubes, lies between the clitoris and introitus. **Vestibular glands** (Bartholin's glands), homologs of the male Cowper's glands, also open into the introitus, but they secrete relatively little lubricant compared with glands in the vaginal walls.

A developing mammalian embryo begins with a set of *indifferent* genital structures, which can develop into either male or female genitals (Figure 51.6). Femaleness, however, is the body's default setting. Genitals will always develop along the female pathway unless they are given male-specific chemical signals during development, including adult sex hormones and factors present only at certain times during development. Under genetic and hormonal instructions, the primordia of the external genitals take one developmental path or the other: The glans region either diminishes into a clitoris or enlarges and becomes the glans penis; the labioscrotal swellings develop into either the labia majora or the scrotum; and the center of the region either becomes the vaginal opening or closes completely as the skin of the scrotum. Internal reproductive organs—including the ovaries, testes, oviducts, and vasa deferentia—develop from another set of indifferent primordia.

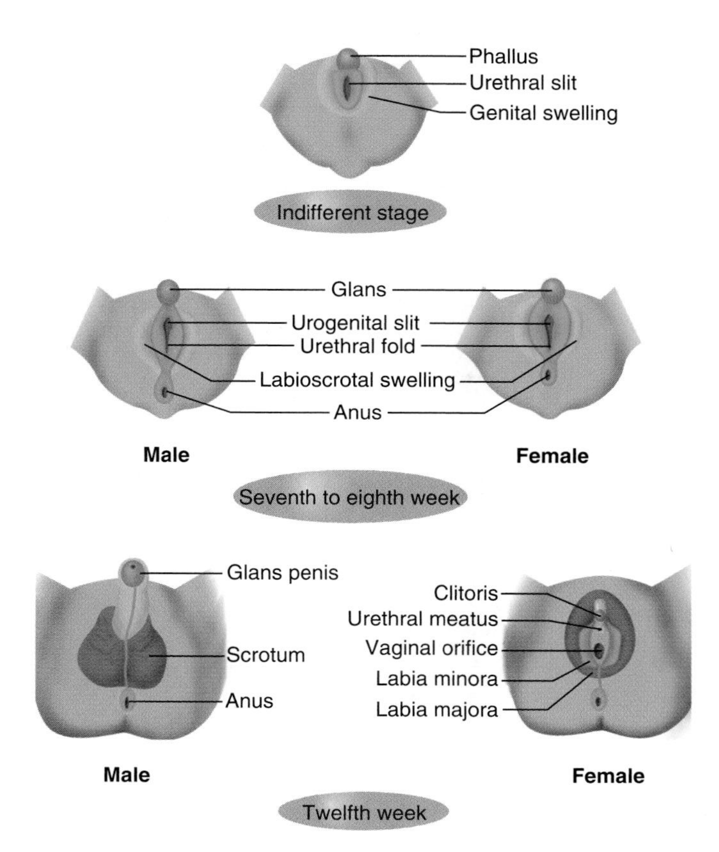

Figure 51.6
The external human genitals develop from neuter structures into either male or female forms. In the absence of male hormones, female structures always develop.

Exercise 51.1 Trace the path a human sperm takes from a testis to the outside, and explain the source and function of each new fluid it encounters along the way.

51.4 Copulation entails a regular sequence of acts in each species.

Fertilization in many animals occurs through the mixing of eggs and sperm in water, so there is no need for copulation (coitus or intercourse). But where copulation does occur, it is as varied as the animals that engage in it. Even though copulation is one of the most important events in the lives of mammals, including humans, sexual behavior has long been a neglected area of human biology, and reliable, detailed information about the physiology of human arousal and intercourse has only been available since about 1970. Before examining human sexual behavior in the following sections, we look briefly at copulation in mice, which appears to be a series of fixed-action patterns (Figure 51.7).

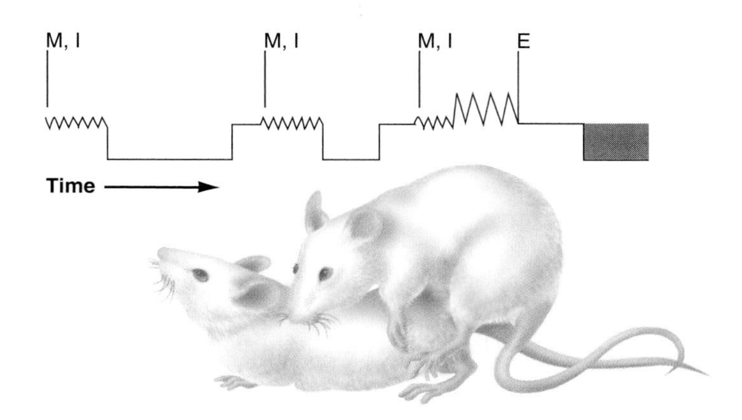

Figure 51.7
The pattern of copulation in mice is indicated by a time line. The male mounts many times (M), and intromission (I) usually follows, but not invariably. Thrusts by the male are indicated by jags in the line. In the final mount, the thrusts become very deep and are followed by ejaculation (E).

Sidebar 51.1 Variations on a Theme

Genital structures vary greatly, even among mammals. Some mammals, for example, have a duplex uterus with two cervixes (*a:* rabbit); one cervix and a small uterine body with two horns, (*b:* pig); a prominent uterine body (*c:* cat, dog); or a large, simplex uterine body with no horns (*d:* primates);

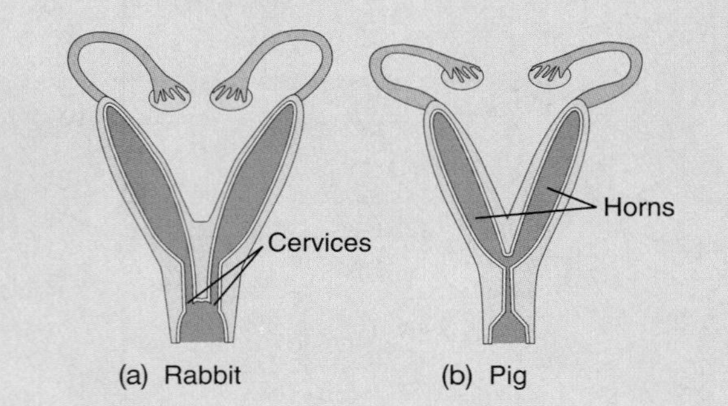

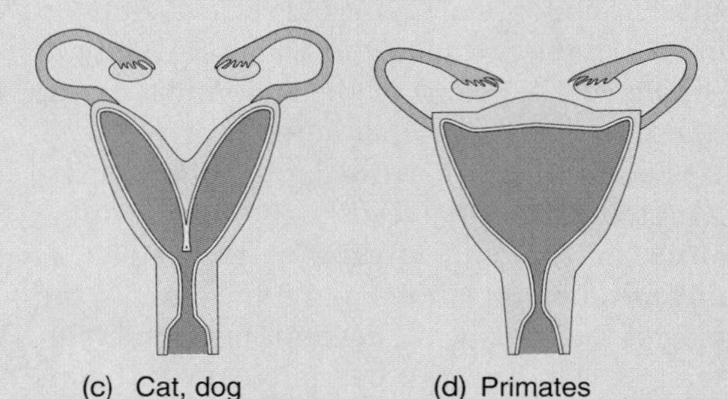

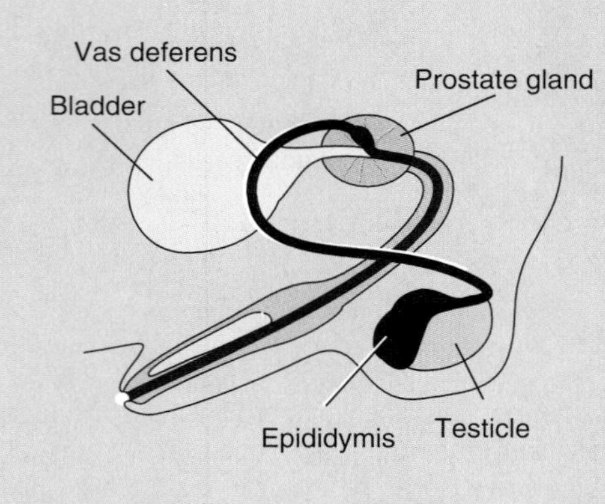

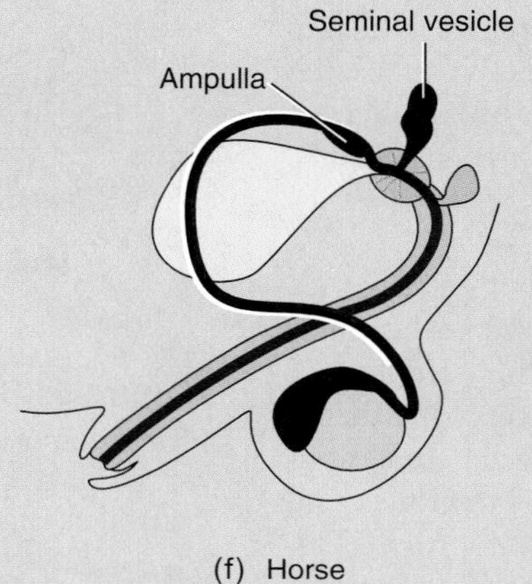

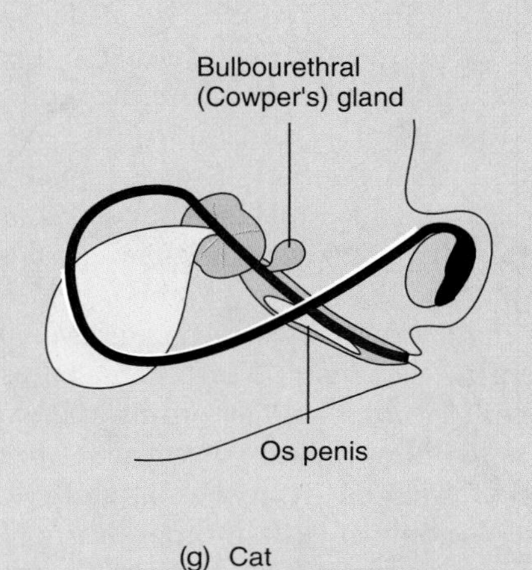

In most male mammals, the erectile part of the penis is enclosed and emerges only upon sexual arousal (*e:* dog; *f:* horse); a number of animals also have a special bone, the os penis, that helps support it (*g:* cat).

A fully receptive female mouse shows her readiness to mate by assuming a posture called **lordosis,** with her hindquarters raised and her tail to one side. A male approaches her, waits for a short time, and then tries to mount from behind. He begins a series of rapid, shallow pelvic thrusts, back and forth, while squeezing her sides with his forepaws and raising one of his hind legs high on her flanks. His movements typically result in **intromission,** the insertion of his penis into her vagina, and his thrusts then become somewhat slower and deeper. However, he usually thrusts only about five to ten times and then dismounts without ejaculating. After a short break, he starts again. He repeats this behavior several times; during the final mount, he thrusts faster and deeper until, at last, he raises himself on the female's back with all four limbs. He ejaculates, quivers, and finally falls off to the side, the female sometimes falling with him. The sequence completed, the two separate. The male will usually not be ready to repeat the pattern for a period ranging from an hour to a day.

T. E. McGill and his colleagues, studying copulation in several inbred strains of mice, found that each strain has a characteristic copulatory sequence, with considerable variation in such features as the time elapsing before the male ejaculates and the number of mounts in a sequence. By studying the expression of these characteristics in hybrids between the various strains, McGill's group showed that some features of the behavior are inherited in a simple way and could be determined by single genes, although most features are more complex.

51.5 Sexual response in humans develops in four phases.

Many cultural and psychological factors influence human sexual arousal and intercourse, but we will focus here on the biological aspects. These were long shrouded in misinformation, until 1948 when Alfred Kinsey published a pioneering study of sexual behavior in men, followed five years later by a parallel

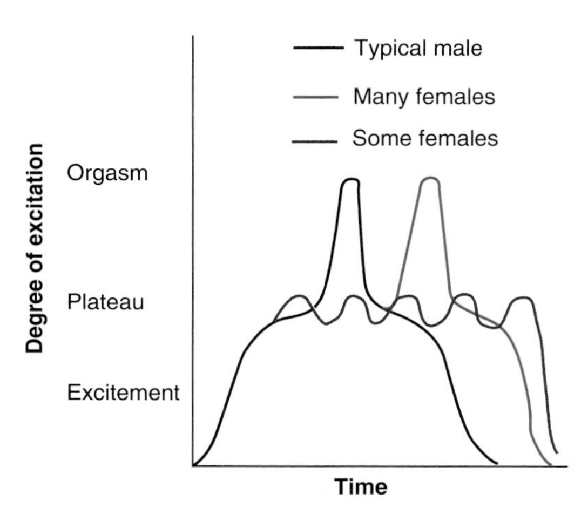

Figure 51.8

Sexual arousal and orgasm in men and women occurs in four phases, as shown by a graph in which the vertical axis is a rough measure of excitement and sexual tension. Both men and women go through excitement to a plateau level, which can last for varying times. Men typically then experience orgasm, followed by the decline of excitement in a resolution phase (not represented on the graph). Many women exhibit a similar pattern, though they may experience a series of orgasms instead of just one; some women experience a sustained orgasmic state for a considerable time before going into resolution.

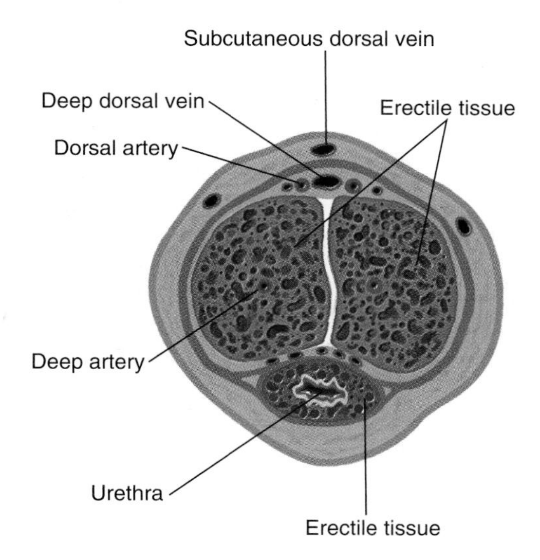

Figure 51.9

The circulation in the penis allows erectile tissues to become engorged with blood when the arteries dilate. The engorged tissues then squeeze the veins, reducing outflow so the tissue remains erect.

study of women. This work opened human sex to scientific analysis. Kinsey's work on sexual behavior was taken a step further with studies of sexual physiology by William Masters, Virginia Johnson, and others.

After studying hundreds of subjects, Masters and Johnson concluded that men and women have a common sexual arousal cycle that can be divided into four phases: excitement, plateau, orgasm, and resolution (Figure 51.8). An amazing variety of stimuli can initiate sexual excitement, or arousal. Visual cues are very effective, especially the sight of a prospective partner's secondary sexual characteristics such as a woman's breasts and hips or a man's chest and buttocks. People exchange both subtle and overt visual cues that may elicit arousal. Chemical stimuli are also important. Humans produce genuine sex pheromones that do not have odors (see Sidebar 43.1), and mature men and women also produce distinctive scents, especially from their sweat glands, that may excite members of the opposite sex.

Tactile (touch) stimuli soon come into play. Virtually any region of the body can be an erogenous zone for someone, but during foreplay the partners generally concentrate on the breasts, especially the nipples and surrounding areolae, the lips and mouth, and the genitals—the sensitive glans of the penis and the vaginal introitus and clitoris.

In men, the excitement phase is indicated by erection of the penis, which may occur within seconds after initial arousal. Erection results when three cylinders of spongy erectile tissue become engorged with blood (Figure 51.9). The arteries leading into them dilate, and the engorged tissues squeeze on the veins, reducing the outward flow of blood. A human penis increases in length from an average of 7–10 cm when flaccid to about 15 cm when erect; a shorter penis increases proportionately more than a longer one during erection, thus reducing the variation in penis sizes. During erection, the spermatic cord contracts and raises the testes. In women, the excitement phase is marked by secretion of a lubricating fluid from the vaginal walls rather uniformly, like perspiration. The nipples also become erect. As the labia minora and majora become engorged with blood, they swell, and the labia minora become bright pink. While the clitoris does not really achieve an erection like a penis, it also becomes engorged with blood and swells.

During the plateau phase, the man's penis enlarges more, his testes are elevated more, and his Cowper's glands begin to secrete their lubricating fluids. The woman's labia minora change from pink to red, and they and the labia majora swell to about twice their initial size. Her clitoris, rather paradoxically, remains engorged but becomes shorter and is virtually buried under the folds of tissue above it. Her breasts continue to become engorged with blood, and the nipples protrude further. All through the excitement and plateau phases, the walls of the vagina have been relaxing and expanding, but now—particularly in the lower third of the canal—the walls swell and narrow. These changes are essential for the occurrence of orgasm, since it is the swollen condition of the labia minora and the outer third of the vagina, which Masters and Johnson call the orgasmic platform, that allows the proper muscles to contract in orgasm.

Sexual excitement reaches its peak in the orgasmic phase. With intromission, the partners thrust rhythmically to maximize their mutual stimulation, and sexual tension builds toward orgasm. Orgasm entails a sudden release of tension that momentarily occupies the whole body but is concentrated in regular contractions of the perineal muscles between the legs just below the external genitals; these contractions, at intervals of about

0.8–1.0 seconds, increase in intensity and duration and then taper off. In women, the orgasmic platform contracts, while in mature men orgasm is normally accompanied by ejaculation, through contractions of the vasa deferentia, seminal vesicles, prostate gland, ejaculatory duct, and finally the penis. Orgasm is not the same as ejaculation; although coupled in mature men, the two events are produced by different sets of muscles, and preadolescent boys can experience orgasm without ejaculation.

During the resolution phase following orgasm, sexual excitement declines, and blood drains from engorged organs, returning the penis, clitoris, labia minora and majora, and breasts to their normal size. Orgasm is more variable in women than in men (see Figure 51.8). Men typically go directly into a refractory phase and are unresponsive to further stimulation. Many women, in contrast, can be restimulated to one or more additional orgasms, and some women experience a sustained series of orgasmic contractions in a fusion of the plateau and orgasmic phases. This is a prominent difference between men and women.

It hardly need be said that sex—especially orgasm—is intensely pleasurable. Mark Twain wrote, ". . . the human being . . . naturally places sexual intercourse far and away above all other joys. . . . The very thought of it excites him; opportunity sets him wild; in this state he will risk life, reputation, everything . . . to make good that opportunity and ride it to the overwhelming climax." Sexual stimuli activate certain pleasure centers in the mammalian brain; rats will work themselves to exhaustion if their behavior is reinforced by stimulation of these centers through implanted electrodes. The connection between coitus and pleasure probably evolved as a mechanism for rewarding what otherwise would be very unlikely behavior—behavior that cannot be explained by some vague "reproductive instinct." Pleasure centers evolved because those animals who had the right kinds of nervous connections, and stimulated them through copulation, just incidentally reproduced themselves.

Because sexual activity is intrinsically rewarding, self-stimulation of the genitals, or masturbation, is quite common. Masturbation is not just an inferior substitute for coitus, used only by those who cannot find a partner. The practice has been observed in monkeys and other mammals, and it is virtually universal among humans, even in sexually permissive societies. Masters and Johnson found that some of their volunteer subjects could achieve more intense orgasms through masturbation than through intercourse. The primary rewards of intercourse appear to be emotional while those of masturbation are physical.

B. The vertebrate reproductive system and its regulation

51.6 Gonads produce steroid sex hormones under control of the anterior pituitary gland.

In vertebrates, sexual development and the major sexual functions are regulated by hormonal communication between two centers of endocrine activity: the hypothalamus–anterior pituitary and the gonads. Gonads produce gametes and also secrete steroid hormones: **androgens,** such as testosterone, in males and **estrogens** in females (Figure 51.10). Gonadal tissues of males and females are regulated via very similar pathways (Figure 51.11) by two **gonadotropin** hormones from the anterior pituitary: **luteinizing hormone (LH),** also known in males as interstitial-cell stimulating hormone (ICSH), and **follicle-stimulating hormone (FSH),** named for its action in females. Recall that the anterior pituitary gland is controlled by the hypothalamus above it through releasing hormones. FSH and LH are both controlled by **gonadotropin-releasing hormone (GnRH),** and all three hormones are regulated by the usual negative feedback loop: An increased level of each steroid in the blood plasma inhibits GnRH secretion in the hypothalamus and inhibits production of either FSH or LH in the anterior pituitary. FSH stimulates the seminiferous tubules of the testes and other cells of the ovary to produce a regulatory hormone, inhibin, which feeds back to inhibit production of FSH. GnRH is also under complex neural control. For instance, in animals that mate only at certain times of the year, the entire system develops under control of neural signals that convey information about the photoperiod. During most of the year, a male bird has small testes and virtually no testosterone in his blood; as the days grow longer in spring, neural signals activate GnRH secretion, with subsequent growth of larger testes. At the same time, he will be stimulated to begin mating behavior and, in many species, to migrate.

51.7 The gonadotropins stimulate spermatogenesis and androgen production in males.

In males, LH stimulates the interstitial (Leydig) cells surrounding the tubules to produce testosterone and other androgens, and FSH stimulates the seminiferous tubules of the testes to produce sperm. As we described for the general system, testosterone feeds back at both the hypothalamus and pituitary to inhibit production of LH, thus reducing testosterone synthesis. Androgens promote growth of the genitals and the development of secondary sexual characteristics, such as the lion's mane, the rooster's comb, and the antlers of elk and deer (Figure 51.12). In humans, these characteristics include the growth of facial and body hair, broadening of the shoulders, and enlargement of the larynx (voice box), with its projecting Adam's apple and attendant deeper voice. Androgens stimulate protein synthesis and contribute to muscle development, making men, on the average, larger and stronger than women. Androgens also enhance libido (sexual drive) and stimulate the full range of emotional and sexual behavior in various animals.

Human males, lacking a yearly cycle, have significant amounts of sex hormones and well-developed testes year around. All male mammals, however, react to sexual stimulation—at least at the right time of year—with rapid production of LH and testosterone, even if the stimulus is nothing more than the sight of a female. Humans are no exception. Some years ago, the scientific journal *Nature* published a paper by an

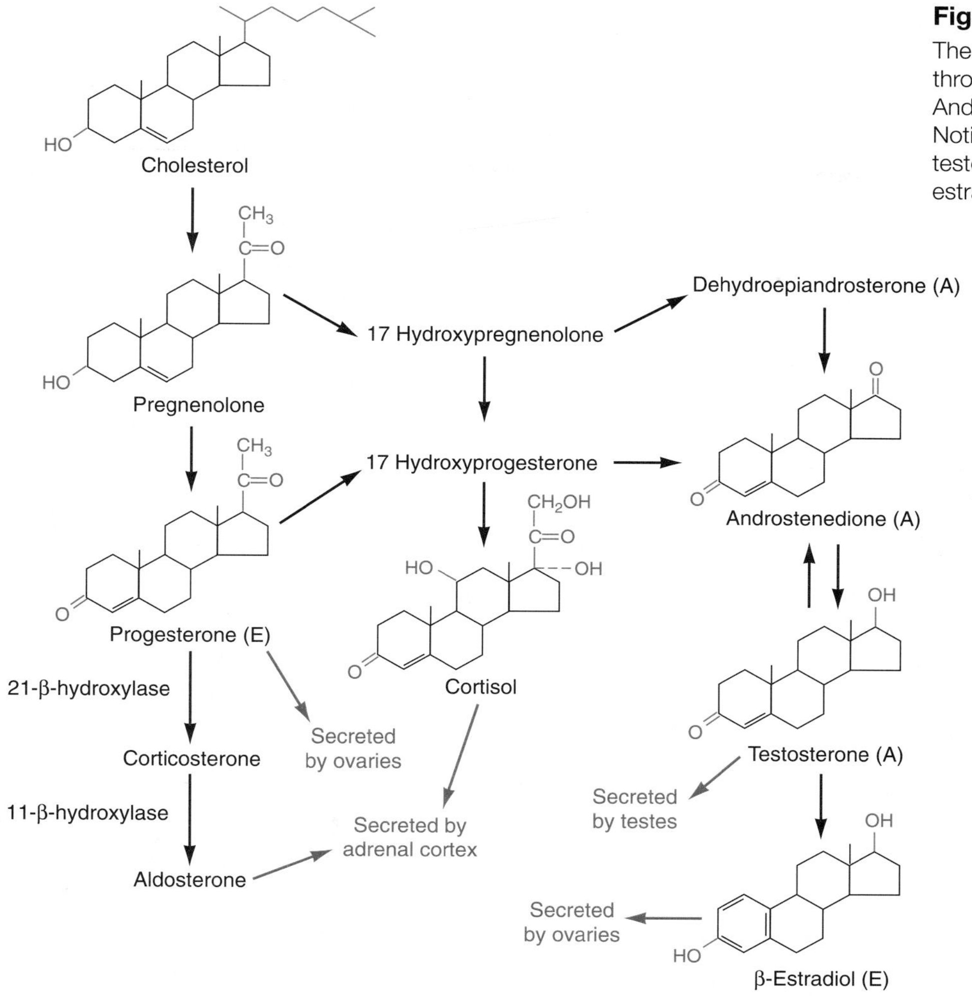

Figure 51.10
The major steroid hormones are synthesized through a common pathway in all tissues. Androgens are marked A and estrogens E. Notice that the principal androgen, testosterone, differs by only one step from estradiol, the principal estrogen.

anonymous scientist whose field trips frequently took him away from society. He noticed that at such times his beard seemed to grow more slowly, and so in the spirit of scientific investigation, he collected and weighed the bits of hair from his razor after shaving. He found that his beard did indeed grow faster when he was in the society of other people, particularly women—that there was a definite response to "specific female company," as he put it. His explanation was that women provide stimuli that elevate the testosterone level—and thus the rate of beard growth—through the hypothalamic circuit.

51.8 An interplay of hormones controls the female reproductive cycle.

Unlike the continuous secretion of hormones and the continuous production of sperm in males, the interplay of pituitary and ovarian hormones in females creates a regular hormonal cycle for the periodic production of ova (see Figure 51.11*c*). During the human female cycle, illustrated in Figure 51.13, three series of events are happening simultaneously:

- An ovum develops in an ovary and is released in mid-cycle; the site where it grows in the ovary also produces hormones.
- The lining of the uterus becomes thicker in preparation for receiving an embryo and then returns to its previous thickness if fertilization does not occur.
- An interplay of hormones between the pituitary gland and ovary controls the first two series of events.

Ovarian events

The ovaries house oocytes, which develop into ova in response to gonadotropins. A woman usually matures only one ovum at a time, selected apparently at random in either her left or right ovary. Each primary oocyte is still a diploid cell, but it is in the early stages of meiosis; in fact, it began meiosis while the woman herself was still a fetus inside her own mother, but then it stopped maturing until this time. The primary oocyte completes the first meiotic division and becomes a secondary oocyte going through the second division; this division generally is not completed until after fertilization, so a simple haploid ovum

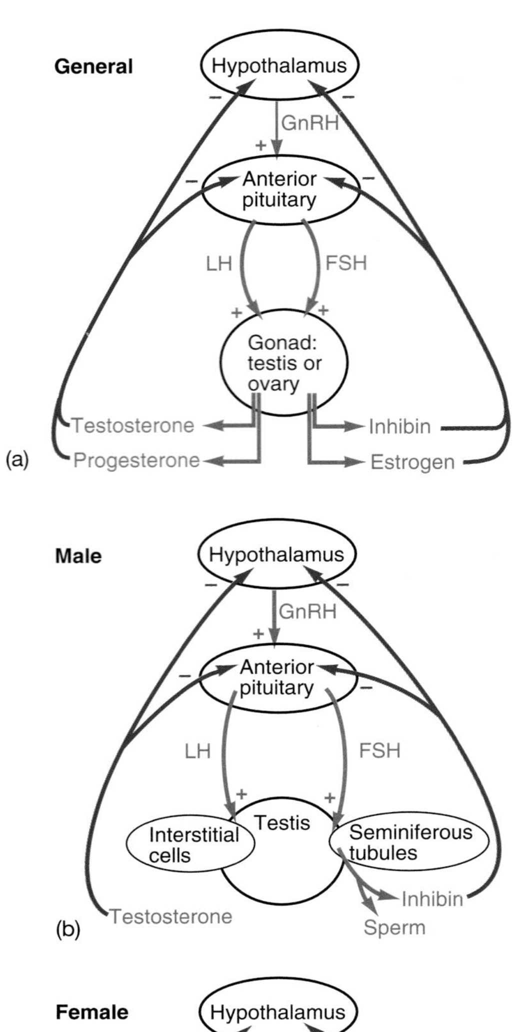

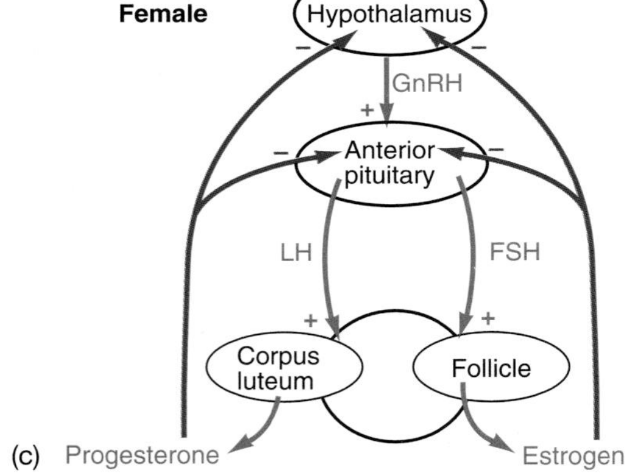

Figure 51.11

(a) The same general hormonal pathways control the gonads of males and females. Plus signs (+) indicate stimulations and minus signs (−) inhibitions. GnRH is gonadotropin-releasing hormone, LH is luteinizing hormone, and FSH is follicle-stimulating hormone. The general system takes particular forms in males *(b)* and females *(c)*.

equivalent to the male sperm never really exists, but the cell is commonly called an ovum or egg. As each oocyte begins to develop, a **follicle**—a small spherical sac made of other cells—grows around it. The follicle supports the growth of the oocyte until the time of **ovulation,** when it is released (Figure 51.14). The follicle's cavity then fills with blood and lymph for a while, until other cells displace these fluids, fill the space, and convert the follicle into a **corpus luteum** ("yellow body"). A developing follicle and corpus luteum both produce estrogens, primarily estradiol, which stimulate development of a woman's secondary sexual characteristics—enlarged breasts, wide hips, a characteristic pattern of body hair, and a general rounding of the body with subcutaneous fat. The corpus luteum also produces progesterone, which, in combination with estrogen, has a role in pregnancy.

Uterine events

A mature woman goes through a regular **menstrual cycle.** Early in the cycle, estrogen and progesterone from the ovary stimulate development of the mucous-membrane lining of the uterus, or **endometrium,** preparing it for implantation of an embryo if fertilization occurs. As the endometrium grows in thickness from about 1 mm to 7 mm, it develops a rich bed of blood vessels and many glands that secrete a mucus rich in glycogen. The thickened endometrium is maintained for several days, but if no embryo implants, it is sloughed off with the discharge of some blood in the process of **menstruation.** The first day of menstruation is counted as day one of the next cycle. Menstrual flow continues for about five days, while the endometrium becomes thinner. A woman has her first menstruation (menarche) at age 12–14 years. The age of menarche has been getting earlier in western societies for about a century, although it is apparently now stabilized. This trend has generally been explained by improvements in nutrition and general health.

Hormonal events and the whole cycle

The female reproductive cycle is divisible into two phases. With reference to the uterus, the first half of the cycle is the **proliferative phase,** since during this time estrogen secreted by the growing follicle stimulates proliferation of the endometrium. With reference to the ovary, the first half is the **follicular phase,** since a steadily increasing amount of FSH from the anterior pituitary stimulates one ovarian follicle to grow and produce estrogen. As the follicle grows, the low level of estrogen (and later progesterone) it produces at first feeds back to inhibit the synthesis of both LRH and LH, just as in men. A low level of estrogen inhibits the secretion of LH and FSH in response to GnRH and may also inhibit the secretion of GnRH. At the same time, inhibin secreted by the follicle feeds back to inhibit the secretion of FSH. We now understand this system a little better because investigators have found two distinct hypothalamic centers that produce GnRH; a *tonic center* produces GnRH more or less continually, thus keeping estrogen at a low level and supporting the secondary sexual characteristics, and a *surge center* produces GnRH only before ovulation.

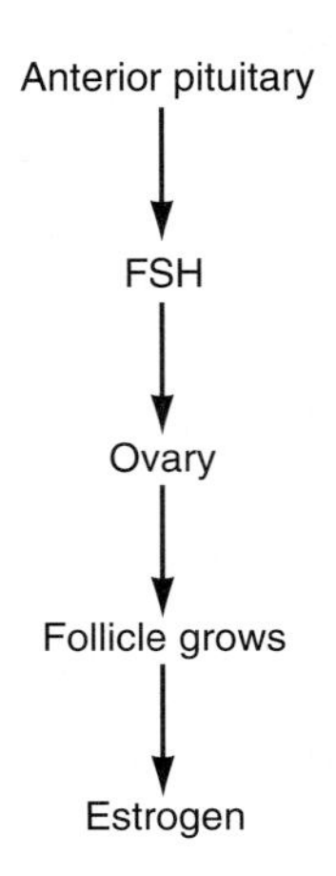

Figure 51.12

Secondary sexual characteristics of animals include the lion's mane, the rooster's comb, and the antlers of deer.

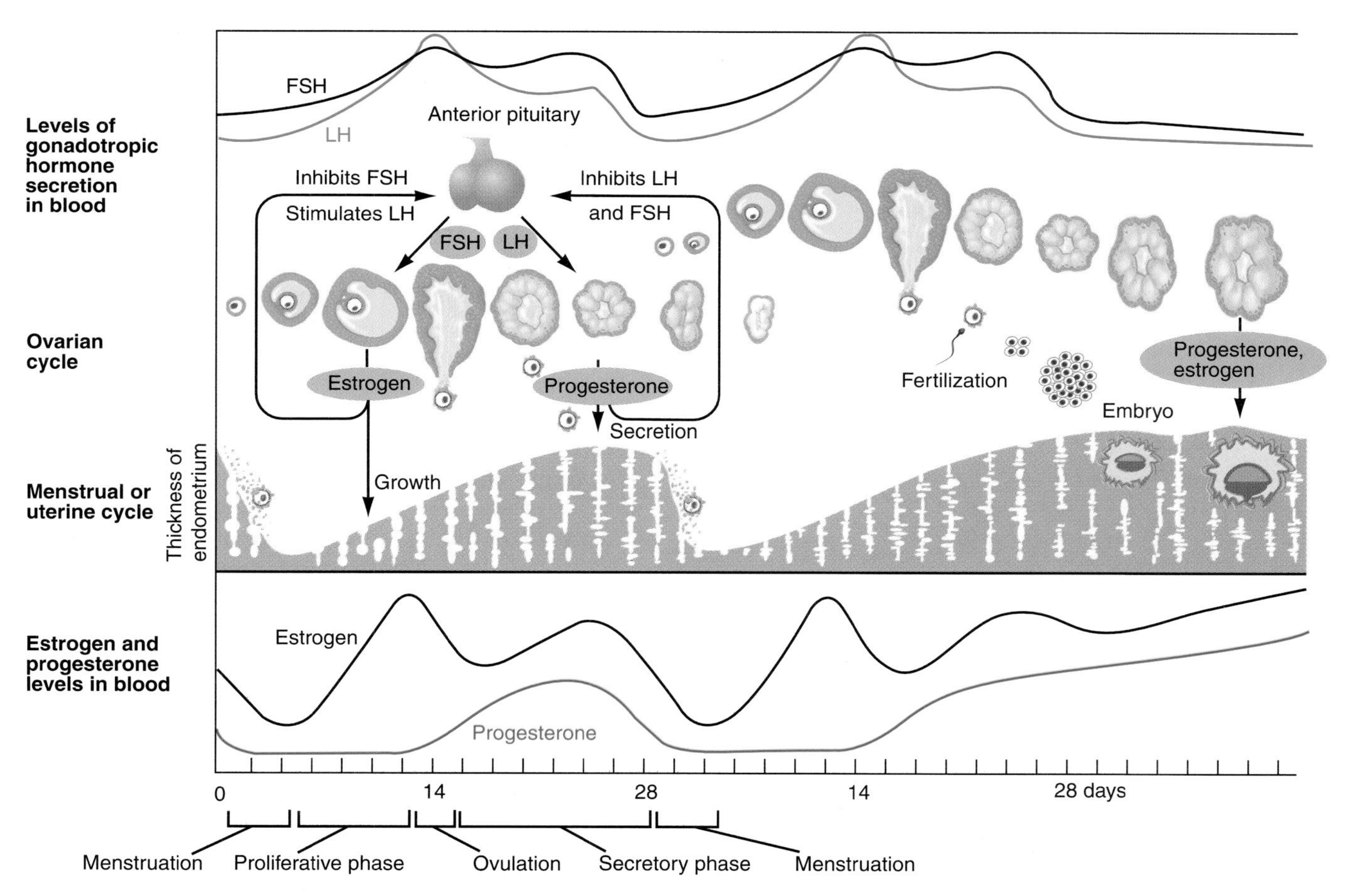

Figure 51.13

The ovarian and menstrual cycles depend on alternating dominance between two gonadotropic hormones and two steroid hormones. FSH stimulates growth of a follicle, which secretes estrogen. Estrogen stimulates growth of the uterine endometrium and feeds back to inhibit FSH and stimulate a surge of LH; LH, in turn, stimulates development of the corpus luteum. Finally progesterone, secreted by the corpus luteum, stimulates secretion by the endometrium. The steroids remain at high levels if an embryo becomes implanted; otherwise, their levels fall and menstruation begins.

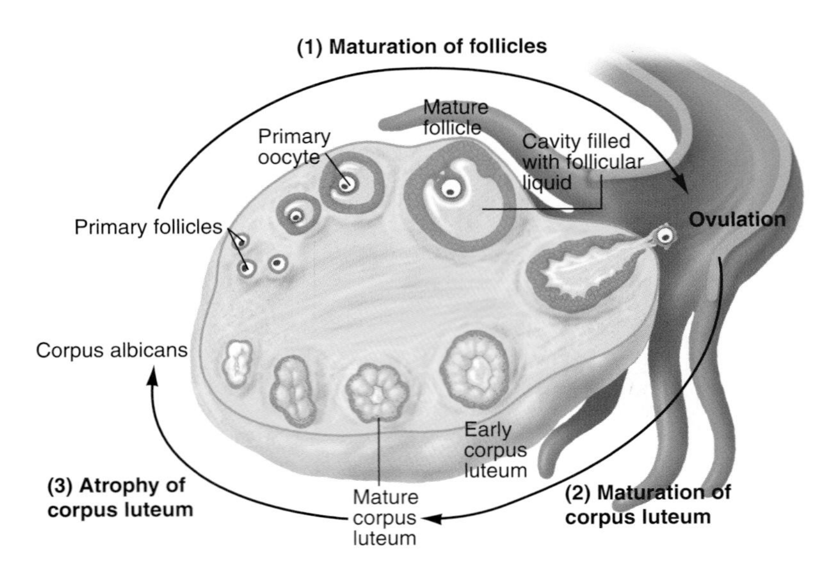

Figure 51.14

An ovarian follicle goes through three stages. First, a primary follicle enlarges into a mature follicle while the primary oocyte within it goes through meiosis and becomes a secondary oocyte. The oocyte is released during ovulation. Second, the empty follicle develops into a corpus luteum (yellow body), which is maintained for several days while it secretes steroid hormones. Third, if pregnancy does not occur, the corpus luteum atrophies into a corpus albicans (white body) and becomes inactive.

By around day 13 of the cycle, the follicle has enlarged into a bubble about a centimeter wide on the surface of the ovary. The endometrium is near its maximum thickness. As the follicle develops, it pours out estrogen, and around day 14 the high level of estrogen has a *positive* feedback effect on the surge center, causing it to release a surge of GnRH, which in turn stimulates the anterior pituitary to suddenly release its LH in an *LH surge.* About 12 hours later, the follicle responds by rupturing and releasing its ovum. Some women experience ovulation as a brief, sharp abdominal pain known as *mittelschmerz* (midpain), apparently due to irritation of the abdominal cavity by the blood being released.

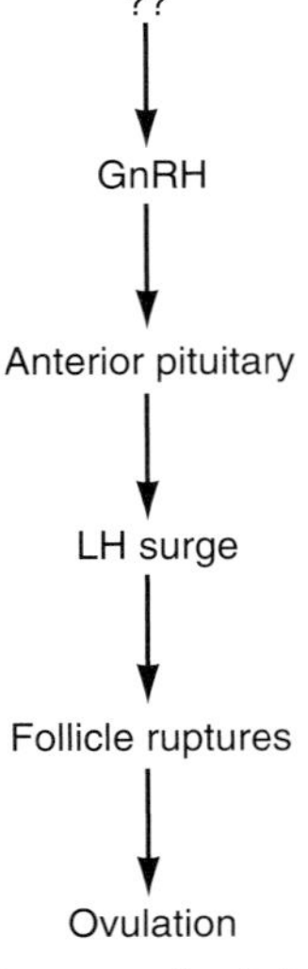

At this point, the cycle shifts into its **luteal phase,** as the follicle closes and becomes a corpus luteum. Under the influence of LH, the corpus luteum continues to secrete estrogen and progesterone, which maintain the endometrium and stimulate the growth of its glands and the secretion of their glycogen-rich mucus. This half of the cycle is therefore also called the **secretory phase.** Progesterone also exerts a powerful inhibitory effect on the hypothalamus, shutting down its production of GnRH. That in turn reduces the release of FSH and LH by the anterior pituitary gland. Lacking LH, the corpus luteum stops growing around day 23–24 and starts to degenerate. As it stops producing estrogen and progesterone, the endometrium responds in turn. Its blood flow decreases; it stops secreting and begins to shrink; and its surface layer degenerates and is sloughed off, marking the beginning of menstruation. The disappearance of the corpus luteum and its progesterone removes the inhibition on the hypothalamus, allowing it to resume secretion of GnRH and begin the next cycle.

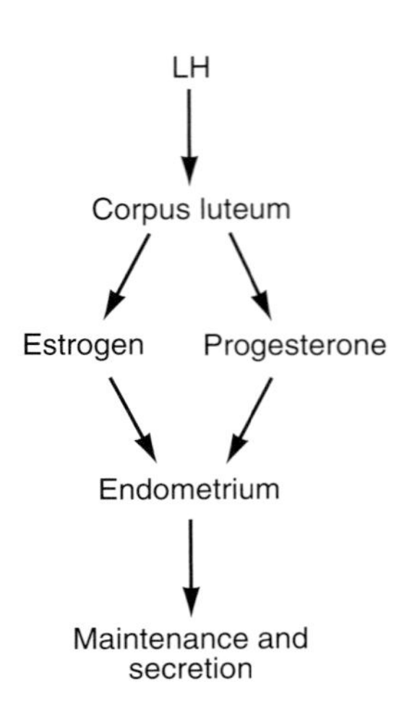

51.9 Other mammals ovulate on different schedules from those of humans.

In contrast to the menstrual cycles of humans (as well as apes and Old World monkeys), most female mammals have an **estrous cycle** in which they periodically come into **estrus** (or heat), and are receptive to copulation and capable of conceiving. Some mammals such as deer and foxes come into estrus only once a year. Female dogs are in estrus for 6–12 days about every six months, cows for about 24 hours every 18–21 days, and rats for a few hours about every four days. At the end of estrus, menstruation does not occur; the endometrium is simply reabsorbed by the uterus if no ova are fertilized. In most other respects, menstrual and estrous cycles are similar. Estrus is brought on by a high level of circulating estrogens, produced by the developing follicles and controlled by the hypothalamic-pituitary circuit.

The females of most species ovulate spontaneously during estrus, and since that is the only time they are receptive to mating, fertilization is likely to occur. But in some species—including cats, rabbits, and mink—fertilization is almost ensured because ovulation is not spontaneous and is only induced by the act of copulation itself. While LH triggers ovulation in all mammals, it rises and falls cyclically in spontaneous ovulators but is only released in response to copulation in induced ovulators. In species that come into estrus once a year, the secretion of LH is controlled by brain centers sensitive to such factors as increasing daylight.

C. Embryonic development and pregnancy

51.10 A mammalian embryo is attached to its mother through a placenta.

The function of the whole reproductive system, of course, is to bring together an egg and a sperm for fertilization. After ovulation, an oocyte is pushed down the oviduct by peristaltic contractions of its walls and by beating of the cilia that line it, while

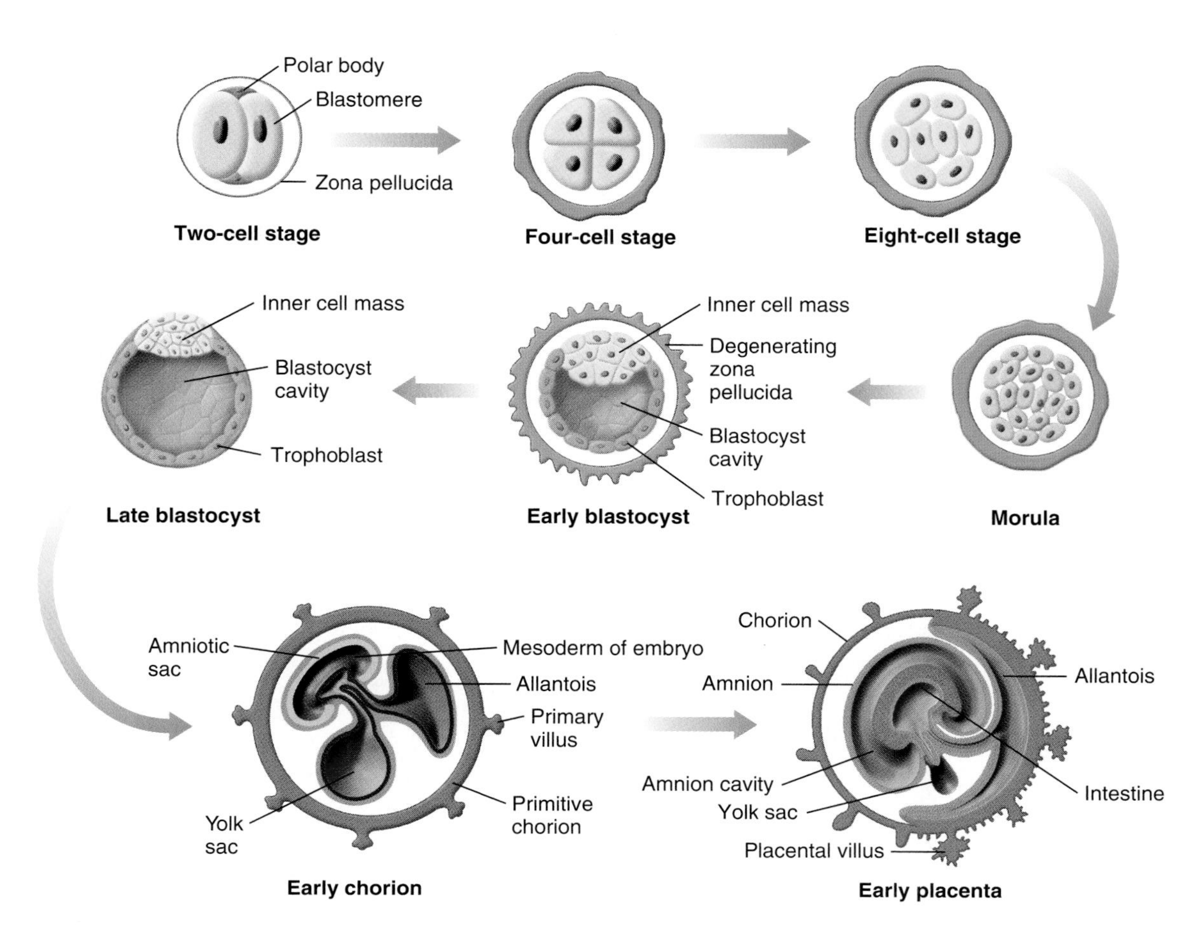

Figure 51.15
The mammalian embryo goes through the early stages of division into blastomeres; for a while it is surrounded by a membrane formed by follicle cells, the zona pellucida, which eventually degenerates. The embryo then develops into a blastocyst, consisting of an inner cell mass and trophoblast. The trophoblast enlarges to form the chorion, which combines with the allantois and begins to develop villi. The chorion and allantois will combine with the endometrium of the uterus to form the placenta.

sperm swim upward against this stream. Of the 3×10^8 sperm in an average human ejaculate, only a few hundred reach the oviduct and have a chance to fertilize the oocyte. If fertilization is to happen at all, it will occur in the oviduct within about 24 hours of ovulation, since the unfertilized oocyte won't live much longer. Sperm, with their limited energy reserves, also have only a short life span.

After fertilization, a mammalian embryo develops as described in Chapter 20. However, mammals are modified reptiles, and their development reflects this. A mammalian egg has hardly any yolk, and the reptilian membranes that were used for absorbing food from the yolk and for gas exchange are used in placental mammals (all mammals except monotremes and marsupials) to make an intimate connection with the mother's blood through which the embryo receives food and oxygen. While still enroute down the oviduct to the uterus, a mammalian zygote develops to the blastocyst stage, consisting of an *inner cell mass* from which the embryo itself develops and an outer layer of cells, the *trophoblast,* which grows into an extraembryonic membrane, the chorion (Figure 51.15). After a journey of two to four days, the blastocyst begins to implant itself in the endometrium of the uterus, where it will be nourished and continue to develop. Trophoblast cells begin to produce hydrolytic enzymes that digest the epithelial layer of the endometrium so the blastocyst can sink into it. By day 11–12, a human blastocyst is surrounded by a pool of maternal blood in the endometrium. The chorion combines with the allantois, which has grown out of the embryonic gut, to form a *chorioallantoic membrane;* its outer chorion layer develops thousands of fine projections called chorionic villi, which begin to create a large surface area for exchanging materials with the endometrium. The chorion and its villi eventually become highly vascularized (that is, full of blood vessels), and they join tissues from the uterine wall to form a large organ, the **placenta.** Although the maternal and embryonic circulations remain separate, the placenta develops a rich field of capillaries where the two bloodstreams are in such intimate contact that they can easily exchange nutrients, oxygen, carbon dioxide, and wastes (Figure 51.16). Cells and most macromolecules, however, do not move between maternal and fetal circulation; a major exception is immunoglobulin G, so the mother's circulating antibodies also confer immunity on the embryo. The growing embryo is eventually connected to its mother through an umbilical cord.

Figure 51.16

In a human placenta, villi from the embryonic chorion become highly vascularized and develop arterioles and venules that carry blood both away from the embryo and back into the embryonic circulation. Spiral arterioles in the uterine endometrium enlarge into intervillus spaces where maternal blood surrounds the chorionic villi. In these areas, the fetal and maternal circulations are closely intertwined but still separate, and materials can exchange between them.

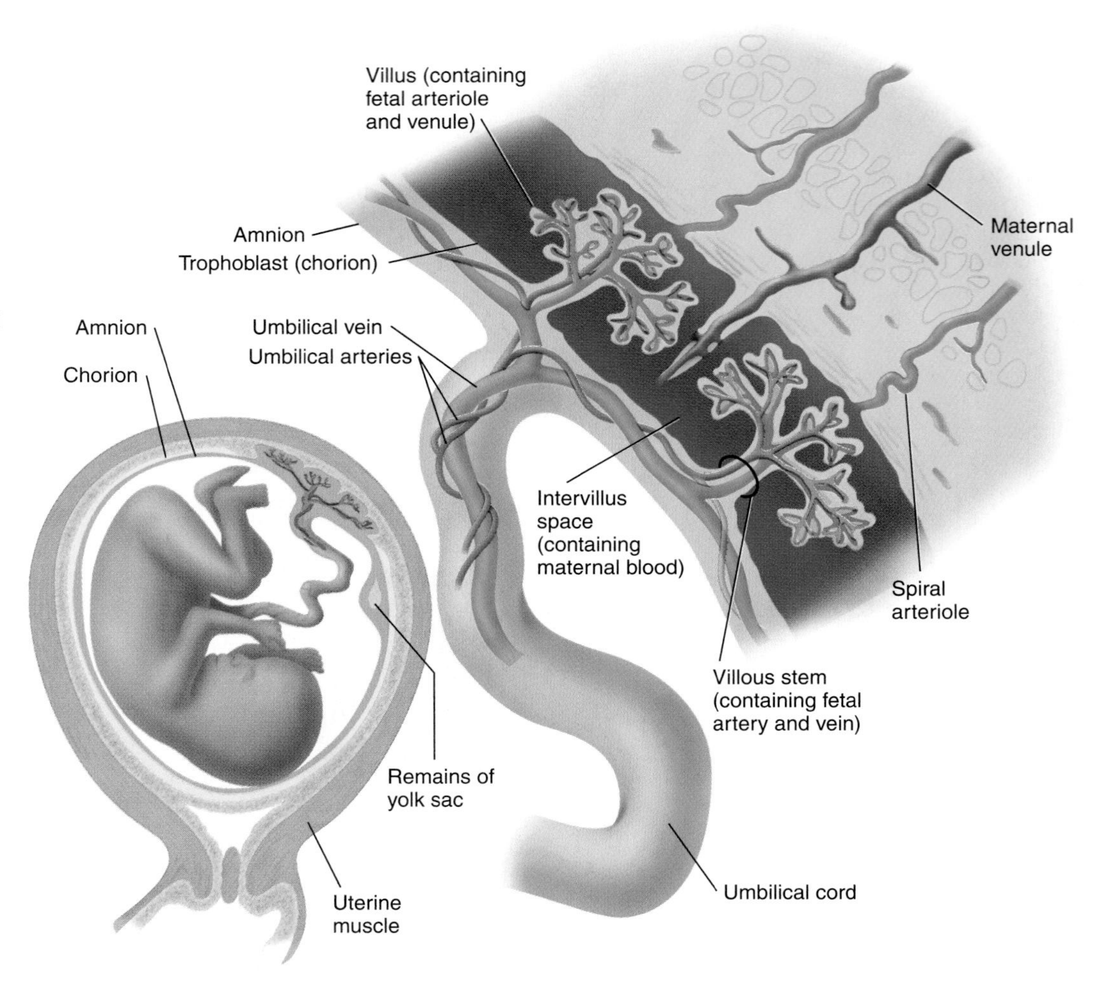

51.11 New hormonal pathways become active during pregnancy.

While an embryo is implanting in the endometrium, the corpus luteum produces progesterone and estrogen, which sustain the endometrium. As we have already noted, these two hormones (especially progesterone) also feed back to inhibit the hypothalamus and anterior pituitary gland, shutting down the release of FSH and LH and thus preventing the maturation of other ovarian follicles. Birth control pills contain a combination of synthetic estrogens and progesterones that create this condition artificially.

Since the hormones released by the corpus luteum block the secretion of FSH and LH, and since LH is needed to maintain the corpus luteum, you might expect the corpus luteum to degenerate during pregnancy as it does at the end of a menstrual cycle. But in fact, it remains viable and functional throughout the first three months of pregnancy (the first trimester) in humans because the embryonic chorion begins to secrete a hormone, **chorionic gonadotropin (CG),** an analog of LH, which sustains the corpus luteum. Thus the chorion and corpus luteum mutually sustain each other until the embryo is well-established, since the corpus luteum secretes estrogen and progesterone to support the endometrium. CG can be detected in a woman's urine by an immunological test, providing a convenient diagnosis of pregnancy. This is the basis for the simple color-change test kits now sold in drugstores. In humans, CG reaches a peak concentration near the end of the first trimester and quickly declines thereafter, signaling the degeneration of the corpus luteum. By then, however, the placenta is producing its own estrogens and progesterone, both of which maintain the endometrium.

The average duration of pregnancy in humans, from the time of fertilization to birth, is 266 days. Some major events during this time are outlined in Sidebar 51.2. Two hormones, **relaxin** and **oxytocin,** assist the birth process. Relaxin causes the ligaments and joints between bones in the pelvis to relax, expanding the birth canal. Relaxin is produced throughout pregnancy, first by the corpus luteum and then by the placenta, but its level builds to a maximum just before birth. The peptide hormone oxytocin, released by the posterior pituitary, stimulates birth by inducing stronger contractions of the

Sidebar 51.2 Human Pregnancy and Fetal Development

Janet Smith and her husband Sam are looking forward to their senior year in college, with big plans for graduate school. Two weeks after classes begin in early September, Janet misses her period, but she isn't terribly concerned because this has happened occasionally before in times of stress. However, when she misses her next period in mid-October, during midterm exams, she is worried and calls for a doctor's appointment. By the end of October, she has a reliable confirmation of her pregnancy. As a responsible mother she now takes on a new regimen due to biological imperatives. She will have to adjust her diet because 85 percent of her dietary calcium and iron goes to the embryo. She will also refrain from drinking alcoholic beverages because of the dangers of *fetal alcohol syndrome (FAS)*, the third most common type of mental retardation. FAS infants are much smaller than average, do not develop normally, and are mentally retarded, with an average IQ of about 70. They also have distinctive faces: small heads with a low bridge to the nose, a narrow upper lip, and an indistinct indentation of the area beneath the nose (the philtrum).

Before Janet even knows she is pregnant, the embryo has developed considerably. The first few weeks are spent in the basic embryonic changes outlined in Chapter 20, but mesodermal cells do not even appear until day 11. By day 14, both the amnion and chorion have developed, and a placenta is starting to form. By day 19, gastrulation has begun, and what will become the head is beginning to emerge from the flat embryonic disc. During the third and fourth weeks, the neural tube—the bare beginning of a nervous system—is forming, and the embryo develops many somites, the segmental tissues that will become skeleton and muscles. A 26-day embryo shows the beginning of eyes, a mandible (lower jaw), and a heart, but it is only about 3–4 mm long. At four weeks, the embryo has a liver and limb buds, which will become arms and legs. Incidentally, it was during this period of about 20–36

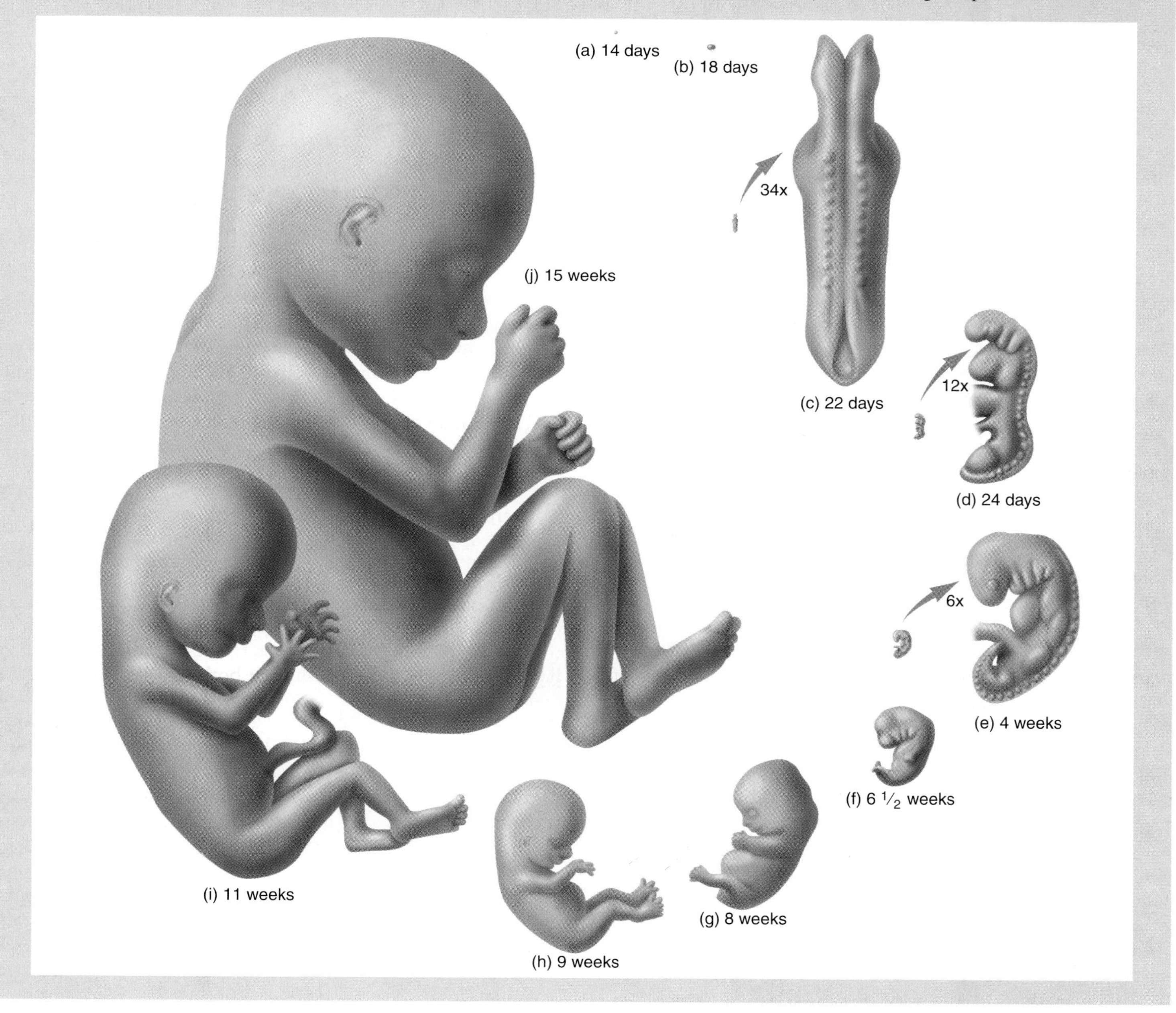

Sidebar 51.2 Continued

days after conception that the sedative *thalidomide* had its effect on limb development before the drug was withdrawn from the market in late 1961. Many women had used it during early pregnancy, and over 7,000 babies were born with a rare condition called *phocomelia,* in which the long bones of the arms and legs do not grow properly, so the infant is left with flipper-like appendages. Fortunately, thalidomide was kept off the American market by a skeptical investigator at the U.S. Food and Drug Administration.

By six weeks, the embryo is about 12 mm long; it has definite arms and legs, and its anatomy is dominated by a large head. By the eighth week, in early November, a neck is recognizable, the external genitals appear in their sexless condition, and connections are forming between the nervous system and muscles, producing some spontaneous movements. During this first trimester, especially the first five weeks of pregnancy, infection with *rubella* (German measles) is a particular danger. While the eyes, ears, and heart are developing, rubella infection can cause cataracts, deafness, and heart malformations in babies.

Beginning with the third month of development, the embryo is properly called a *fetus.* By 11–12 weeks—Thanksgiving time—it has started to look like a human. Its eyes have moved from the sides of the head to the front, and its sex can be determined externally. Nails are forming, and ossification centers appear in most bones. The lungs and brain have both acquired their proper general form, but are very primitive. From this point on, the fetus will grow without major changes in its external form.

While Janet is finishing the first semester and enjoying the Christmas break, the four-month-old fetus is developing a human-looking face, although it is broad and the eyes are still widely spaced. As second-semester classes begin in January, the body of the five-month fetus becomes covered with downy hair, called lanugo, and some head hairs appear. About this time, Janet begins to feel fetal movements. As she takes her midterm exams in March, during the seventh month, her blood volume is 30 percent greater than normal, and about 16 percent of her blood is in the uterus and placenta. A seven-month fetus has been described as resembling a dried-up old person, its red, wrinkled skin covered with a pasty mixture of dead cells and waxy secretions; its eyelids, which had been fused, are now open. At eight months, during April, the fetus is acquiring subcutaneous fat. Janet is determined to finish the semester. As she takes her final exams in early May, the baby she is carrying has lost the redness and wrinkling of its skin, and its body is acquiring the roundness of a full-term baby. A few days later, as other students are going through commencement ceremonies, Janet is giving birth, with Sam at her side.

uterine muscles (aided by prostaglandins). As explained in Section 11.3, contraction of the uterine muscles stimulates the production of more oxytocin, a prime example of positive feedback. However, these two hormones alone are not responsible for the advent of birth, since a number of complex nervous and hormonal events are taking place at this time.

One final event following birth is the production of milk in the mother's breasts. Mammary tissues grow throughout pregnancy in response to estrogens and progesterone; the placenta produces **chorionic somatomammotropin,** a hormone that stimulates the growth of milk glands and the lactiferous ducts leading to the nipple of the breast. However, milk production depends upon the hormone **prolactin** secreted from the anterior pituitary (Figure 51.17). Prolactin release, in turn, is usually inhibited by prolactin-inhibiting hormone (PIH) from the hypothalamus, and PIH is stimulated by estrogen. The estrogen and progesterone level is so high while the placenta is in place that it inhibits the actual formation of milk. So the breasts only begin to produce milk after the afterbirth (placenta and umbilical cord) is sloughed, when no more progesterone is released, PIH is no longer released, and prolactin production can begin. PIH synthesis is also stopped by neural signals stimulated by the action of the infant suckling at the breast and by the mother's sight of the infant. Prolactin, however, only stimulates production of milk and its release into the ducts. The same neural signals also stimulate the release of oxytocin, which in turn stimulates the *milk-ejection reflex,* commonly called "milk let-down." Oxytocin causes contraction of the lactiferous ducts, so milk can flow out through the nipple.

Reproduction is just one phase of the cycle of life that characterizes every organism. With the development and birth of a new individual, a new cycle has begun. The baby will grow and maintain itself through all the complex physiological mechanisms we have been exploring in this book. It will respond to stimuli and learn new patterns of behavior from the adults around it. Eventually it will mature and be attracted to another individual of the opposite sex. And the cycle will repeat once again.

Exercise 51.2 Oxytocin stimulates contraction of the uterine muscles as well as the lactiferous ducts of the breast. What additional benefit does a woman get by breast-feeding her baby?

Coda No matter how long an animal lives, all its complex physiological mechanisms have gone for nought if it fails to reproduce itself with viable offspring, or at least contribute to the reproduction of its close relatives (see Section 50.6). The most essential act of life is self-perpetuation. For asexual organisms, simply growing and dividing is the same as reproducing, but sexual reproduction makes the lives of animals a lot more interesting in ways that we have explored in this chapter. Sexual reproduction allows chromosomes to be reshuffled each generation, thus producing greater variation in a population and more genetic experimentation. Since the rate of evolution is proportional to the variance in a population (see Chapter 24), sex makes it more possible for a population to

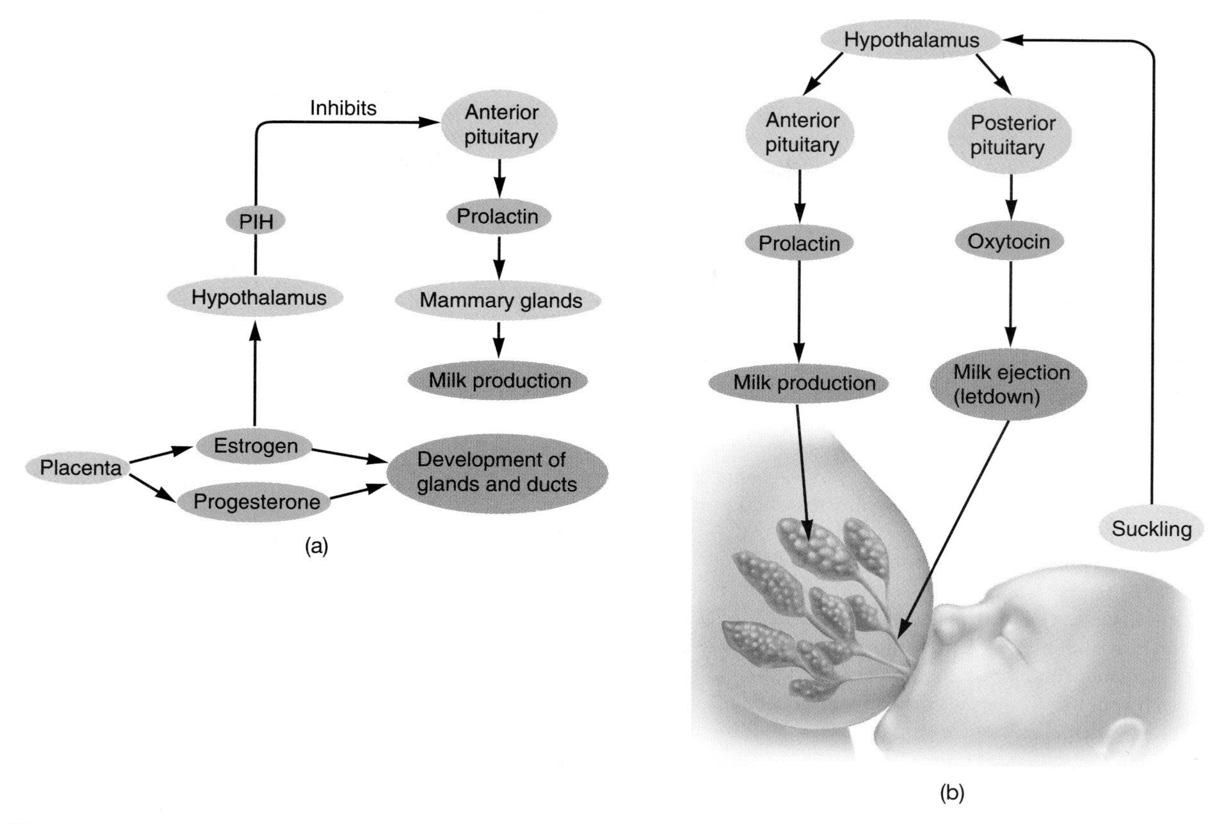

Figure 51.17

(a) During pregnancy, lactation is inhibited by a high level of circulating estrogen stimulating secretion of prolactin inhibiting factor (PIH), which prevents secretion of prolactin. *(b)* After birth, the placenta is no longer present to secrete estrogen, so PIH secretion stops. The baby's suckling stimulates secretion of prolactin, which stimulates milk production, and oxytocin, which stimulates contraction of the lactiferous ducts to eject the milk.

possess the kind of variations needed to survive unpredictable events.

It is no wonder that sexually reproducing animals have evolved signals that bring the sexes together, test their mutual fitness to reproduce, and eventually result in fertilization. The behaviors of some animals may strike us as fascinating, bizarre, or disgusting—but they all work. Each one is an independently evolved strategy that has been shaped and selected because it results in reproduction. We can view human characteristics and human behavior in the same light. The secondary sexual characteristics of men and women are intrinsically attracting; we do not have to learn them. We also know very well how to enhance those attractive features with paints and perfumes, with clothes draped provocatively, or even with an occasional surgical adjustment. In reality, we are playing with releasing stimuli matched to fixed-action patterns only slightly modified or held in check by social conventions. At the same time, we are not just seeking mates with whom to mix our gametes, but high-quality mates: women inclined to be nurturing and protective mothers, men inclined to be loyal mates and fathers and good providers. Humans are wise, we think, to recognize the barely conscious signals implanted in our nervous systems—what David Barash has called the "whisperings within"—and to recognize that these signals have led us and our ancestors through successful reproduction for millions of years.

Summary

1. In the sexual reproduction of simple organisms, gametes are often attracted to each other by simple chemical mechanisms, particularly by pheromones.
2. In insects and other animals, pheromones produced by one sex are recognized by the other by means of specific receptor proteins. Binding of the pheromone induces behavior that leads to mating.
3. In some species, particularly birds and mammals, pair formation and mating are preceded by a courting ritual consisting of a series of fixed-action patterns. Such rituals ensure that only high-quality individuals of the same species will mate.
4. The male genital system in mammals is designed to deliver an ejaculate of sperm from the testes combined with additional fluids from accessory glands. Sperm are carried from each testis by a vas deferens, which becomes the ejaculatory duct and then joins the urethra. Glands along this route add special fluids to the ejaculate.
5. The mammalian female genital system is designed to produce a mature egg, to receive the sperm, and to conduct them into the oviducts where fertilization can occur. Ova released from the ovaries

are swept down the oviducts toward the uterus. Fertilization normally occurs in the oviduct. The resulting embryo is held and nourished in the endometrial lining of the uterus.

6. Both the external and internal genital structures of mammals develop from embryological structures that are not yet differentiated into male or female. Under specific genetic and hormonal instructions, each region becomes a characteristic male or female part; the glans, for instance, becomes either the female clitoris or the male penis.
7. Copulation entails a ritualized series of fixed-action patterns that lead to intromission (insertion of the penis into the vagina) and ejaculation by the male.
8. Human sexual arousal can be divided into four phases. The excitement phase is brought on primarily by visual and olfactory stimuli, and heightened by the touching of sensitive areas. Excitement entails erection of the penis in men and erection of the nipples and engorgement of the clitoris and labia minora and majora in women, in preparation for intromission. In the plateau phase, these changes are accentuated, and the production of lubricating fluids increases. Orgasm is primarily a series of highly pleasurable muscular contractions, with release of sexual tension, accompanied by ejaculation in men. Men generally can have only one orgasm at a time, while women may have several orgasms or even a sustained series of orgasmic contractions. Both sexes then go into a resolution phase in which tissue engorgement and muscle tension rapidly decline.
9. The primary sex hormones are steroids made by the gonads. Testosterone is the principal androgen, or male hormone. It is made by interstitial cells of the testes under the influence of luteinizing hormone from the anterior pituitary gland. Testosterone stimulates the development of male secondary sexual characteristics and sexual behavior; in cooperation with follicle-stimulating hormone (FSH) from the pituitary, it stimulates spermatogenesis.
10. The female steroid hormones are estrogens and progesterone. Estrogens are made primarily by the ovarian follicles as they develop under the influence of FSH. They are responsible for female secondary sexual characteristics. Progesterone has various regulatory effects.
11. In females, hormones are made in a regular menstrual cycle. In the first half of the cycle, an ovarian follicle and ovum begin to develop because of FSH stimulation. The estrogen produced by the follicle stimulates growth of the endometrium, the lining of the uterus, which thickens in preparation for implantation of an embryo.
12. In a feedback cycle to the pituitary gland, estrogens stimulate the buildup of luteinizing hormone, which stimulates ovulation when it is suddenly released midway through the cycle.
13. The empty follicle grows into a corpus luteum that produces progesterone, which maintains the endometrium. In the absence of fertilization, the corpus luteum degenerates after several days, and the endometrium is sloughed off in menstruation.
14. An ovum is ejected from the ovary into the abdominal cavity close to the opening of the oviduct, which generally conducts it down toward the uterus through muscular contraction and the action of its ciliary lining. Fertilization typically occurs in the upper part of the tube, and the resulting zygote develops for two to four days while it descends to the uterus, where it implants in the endometrium.
15. The outer layer of the embryo develops into a chorionic membrane, which later grows into the placenta, an organ that makes intimate contact with the maternal blood supply in the endometrium and acts as a point of exchange between fetal and maternal blood.
16. The chorion produces a gonadotropin that maintains the corpus luteum; the corpus luteum produces progesterone, which maintains the endometrium and, thus, the chorion. Later in pregnancy, the placenta maintains itself with its own estrogens and progesterone.
17. Birth is aided by relaxin, a hormone that relaxes connective tissue in the pelvis, and is stimulated by oxytocin, a hormone that induces contraction of the uterine muscles. Estrogens stimulate growth of the breasts, and milk production is stimulated by prolactin from the anterior pituitary.

Key Terms

mating type 1081
sex pheromone 1081
seminiferous tubule 1083
testis 1083
scrotum 1083
epididymis 1083
vas deferens 1083
seminal vesicle 1084
ejaculatory duct 1084
prostate gland 1084
semen 1084
penis 1084
bulbourethral gland 1084
ovary 1084
oviduct/uterine tube/Fallopian tube 1085
uterus 1085
cervix 1085
vagina 1085
introitus 1085
labia minora 1085
labia majora 1085
clitoris 1085
hymen 1085
vestibular gland 1085
lordosis 1086
intromission 1086
androgen 1088
estrogen 1088
gonadotropin 1088
luteinizing hormone (LH) 1088
follicle-stimulating hormone (FSH) 1088
gonadotropin-releasing hormone (GnRH) 1088
follicle 1090
ovulation 1090
corpus luteum 1090
menstrual cycle 1090
endometrium 1090
menstruation 1090
proliferative phase 1090
follicular phase 1090
luteal phase 1092
secretory phase 1092
estrous cycle 1092
estrus 1092
placenta 1093
chorionic gonadotropin (CG) 1094
relaxin 1094
oxytocin 1094
chorionic somatomammotropin 1096
prolactin 1096

Multiple-Choice Questions

1. Which of these structures is within the mammalian testis?
   a. epididymis
   b. ejaculatory duct
   c. vas deferens
   d. seminal vesicle
   e. seminiferous tubule
2. All but ______ are a pair of sexually homologous structures.
   a. clitoris—glans penis
   b. ovary—testis
   c. labia majora—scrotum
   d. prostate gland—seminal vesicles
   e. vestibular glands—bulbourethral glands
3. Prostatic secretions function primarily to
   a. provide sperm with energy-rich nutrients.
   b. neutralize acidic vaginal secretions.
   c. stimulate sperm motility.
   d. stimulate the onset of sperm maturation.
   e. inhibit spermatogenesis.
4. Which hormones are included in the category of gonadotropins?
   a. estrogen and progesterone
   b. estrogen, progesterone, and testosterone

c. LH and FSH
d. LH, FSH, and inhibin
e. hypothalamic release factors

5. Gonadotropins have a negative feedback relationship with
a. estrogen.
b. testosterone.
c. progesterone.
d. inhibin.
e. all of the above.

6. The effect of GnRH on cells that produce FSH and LH is best described as
a. stimulatory.
b. inhibitory.
c. negative feedback.
d. positive feedback.
e. no relationship.

7. Which of the following would have the most immediate effect in raising FSH synthesis?
a. rise in GnRH
b. rise in progesterone
c. rise in LH
d. rise in estrogen
e. rise in testosterone

8. All but ______ is produced by gonadal cells.
a. testosterone
b. inhibin
c. progesterone
d. prolactin
e. estrogen

9. In some female mammals, corpora lutea do not degenerate completely, but simply turn into a spot of scar tissue. If you were to count these spots in one particular female, you could deduce the total number of her
a. offspring.
b. pregnancies.
c. follicles.
d. eggs ovulated.
e. reproductive cycles.

10. LH and CG both
a. are polypeptide gonadotropins.
b. maintain the corpus luteum and stimulate the secretion of progesterone.
c. are secreted by the ovary.
d. *a* and *b*, but not *c*.
e. *a*, *b*, and *c*.

True-False Questions

Mark each statement true or false, and if false, restate it to make it true.

1. In human males, both sperm and urine flow in the urethra and the ureters.
2. After release, unfertilized human eggs enter the abdominal coelom before arriving at the oviduct, whereas sperm are always contained within a tube or duct.
3. In humans, eggs are generally fertilized within the uterus.
4. In males, the primary effect of FSH is to stimulate testosterone-producing interstitial cells of the testes.
5. Male secondary sexual characteristics, such as antlers and larger muscle mass, result from the action of LH directly on the body cells involved.
6. At different times in the reproductive cycle, gonadotropins are produced and secreted by the pituitary, the ovaries, and the testes.
7. The ovarian cycle is divided into the proliferative phase, marked especially by rising estrogen levels, and the luteal phase, when progesterone levels rise dramatically.
8. The LH surge occurs because a low level of estrogen inhibits LH secretion but a high level stimulates GnRH and LH secretion.
9. LH provides the immediate hormonal trigger for lactation.
10. Mammals produce milk only when the levels of prolactin and estrogen are elevated.

Concept Questions

1. Contrast the sites of synthesis and activity of FSH and LH and their roles in females and males.
2. Identify the hormones required for lactation, and explain why milk is produced and released only after birth.
3. Explain why fertility rates are highest in mammals that are induced ovulators, slightly lower in spontaneous ovulators, and lower still in mammals (higher primates) that have menstrual cycles.
4. Identify the tissues that form the human placenta.
5. Explain why removal of the corpus luteum during the first trimester terminates the pregnancy, whereas removal during the second or third trimester has no effect.

Additional Reading

Alexander, Nancy J. "Future Contraceptives." *Scientific American*, September 1995, p. 104. A review of methods now being developed that may become common in the near future.

Boppre, Michael. "Sex, Drugs, and Butterflies." *Natural History*, January 1994, p. 26. Butterflies and moths emit chemicals from their odoriferous organs to attract the females of their species. This sexual chemical stimulation is a common mating strategy in many butterfly species.

Borgia, Gerald. "Sexual Selection in Bowerbirds." *Scientific American*, June 1986, p. 92. The bower, or mating site, of these extraordinary birds of Australia and New Guinea is the center of intense competition among males. The female's mating choice is based on its architectural adornment.

Catton, Chris, and James Gray. *Sex in Nature*. Facts on File Publications, New York, 1985.

Crews, David. "Animal Sexuality." *Scientific American*, January 1994, p. 96. A survey of different methods of sexual differentiation and a theory about the evolution of gender.

———. "Courtship in Unisexual Lizards: A Model for Brain Evolution." *Scientific American*, December 1987, p. 116. An all-female species of whiptail lizards presents a unique opportunity to test hypotheses regarding the nature and the evolution of sexual behavior.

Diamond, Jared. "Everything Else You Always Wanted to Know About Sex, but That We Were Afraid You'd Never Ask." *Discover*, April 1985, p. 70. Some aspects of human sexuality still can't be explained by scientists.

Greenspan, Ralph J. "Understanding the Genetic Construction of Behavior." *Scientific American*, April 1995, p. 72. Research on sexual behavior in fruit flies.

Newman, Jack. "How Breast Milk Protects Newborns." *Scientific American*, December 1995, p. 58. How the mother's immune system is extended into her child through her milk.

Short, R. V. "Breast Feeding." *Scientific American*, April 1984, p. 35. The relationship of its contraceptive effect to population growth.

Special Issue: "The Science of Sex." *Discover,* June 1992. A list of books on the biology, physiology, and evolution of sexual reproduction. Topics include puberty, the evolution of the orgasm, and pseudohermaphrodites.

Ulmann, Andrew, Georges Teutsch, and Daniel Philibert. "RU 486." *Scientific American,* June 1990, p. 42. This controversial drug is now used widely in France to terminate unwanted pregnancies, but the compound actually has many possible applications.

Wallace, Robert A. *How They Do It.* William Morrow and Company, Inc., New York, 1980. A light-hearted but scientifically serious survey of the way various animals mate.

Internet Resource

To further explore the content of this chapter, log on to the web site at:

http://www.mhhe.com/biosci/genbio/guttman/

Appendix A Exponential Notation and the Metric System

Exponential Notation

Because we have ten fingers that are convenient counting devices, the base of our number system is ten. We count from zero to nine, then make a mark to show that we have counted one ten, and start again until we count a second ten. If we do this ten times, we make another mark to indicate one hundred and start counting tens again. All the numbers of the counting system—a thousand, a million, and so on—can be expressed by multiplying tens a certain number of times, so it is most useful to express all numbers in this way, especially very large or very small numbers.

Remember how exponents are used: The expression $3 \times 3 = 9$ can also be written $3^2 = 9$ (read "three squared equals nine.") The superscript 2 is an *exponent* that shows the number of times 3 (the *base*) is used as a factor. Similarly, 3^3 (read "three cubed") is $3 \times 3 \times 3 = 27$, and 3^4 (read "three to the fourth power" or simply "three to the fourth") is 81. One reason this system is so convenient is that numbers expressed in this form can be multiplied and divided easily just by adding and subtracting exponents, as long as they all have the same base. For instance, $3^2 \times 3^3 = (3 \times 3) \times (3 \times 3 \times 3) = 3^5$, so we can do the multiplication simply by adding the exponents: $2 + 3 = 5$. Similarly, we divide numbers by subtracting exponents: $4^{10}/4^6 = 4^{10-6} = 4^4$. Numbers written with a negative exponent are simply the inverses of the same numbers with a positive exponent. Thus, since $2^3 = 8$, $2^{-3} = 1/8$. Since $5^2 = 25$, $5^{-2} = 1/25$. These numbers are multiplied and divided just like any others. Thus, $5^9 \times 5^{-3} = 5^{9-3} = 5^6$, and $3^{-3}/3^4 = 3^{-3-4} = 3^{-7}$. Be warned, though, that you must carefully obey the algebraic rules for adding and subtracting negative numbers. For instance, to calculate $10^4/10^{-8}$, you must think: $4 - (-8) = 4 + 8 = 12$.

Notice that $n^7 \times n^{-7} = n^{7-7} = n^0$. *But* $n^7 \times n^{-7}$ *is also equal to* $n^7 \times 1/n^7 = 1$. *And so any number to the zero power equals 1.*

All these rules naturally apply to numbers written as powers of 10 (exponential notation, sometimes called "scientific" notation). Ten $= 10^1$, $100 = 10^2$, $1{,}000 = 10^3$, and so on. It is useful to learn that the major named numbers have exponents that are multiples of three: A million is 10^6, a billion is 10^9, a trillion is 10^{12}, and so on. Notice that the exponent is just the number of zeros in the number when it is written out: $100{,}000 = 10^5$. Fractional numbers are also easily expressed this way. $0.1 = 10^{-1}$, $0.01 = 10^{-2}$, and $0.00001 = 10^{-5}$. Here the exponent shows the number of places (not zeros) to the right of the decimal point, including the first significant digit. (The significant digits in any number start with the first number on the left that is not zero and end with the last number on the right that is not zero. So in 0.00234500, the significant digits are 2345, unless those last zeros are claimed as exact measurements, in which case they are also significant.)

To handle calculations of any kind, it is useful to express all numbers as $n \times 10^x$, where n is a number between 1 and 10. Thus, $230 = 2.3 \times 10^2$, $4{,}600{,}000 = 4.6 \times 10^6$, and $0.00075 = 7.5 \times 10^{-4}$. If all the numbers used in a complex calculation are written in this exponential form, they can be multiplied and divided easily. The decimals—which are all about the same size—are simply handled by themselves, using a calculator or slide rule, and the exponents are added and subtracted by themselves. This system avoids mistakes in powers of 10—mistakes of putting the decimal point in the wrong place. For practice, try doing the following exercises.

Express in exponential form:

1. 2,400
2. 78,000,000
3. 1,470,000,000,000
4. 0.0423
5. 0.000001
6. 0.00024

Calculate:

7. $10^7 \times 10^3 =$
8. $10^8/10^6 =$
9. $10^{-4} \times 10^{-2} =$
10. $10^{-6} \times 10^3 =$
11. $10^7/10^{-2} =$
12. $10^{14} \times 10^{-17} =$
13. $4.2 \times 10^{-6} \times 1.7 \times 10^{-6} =$
14. $3.8 \times 10^5 \times 7.4 \times 10^{-2} =$

(See answers at the end of this appendix.)

The SI (Metric) System

The system of expressing everything in powers of 10 naturally leads to a system of measures based on powers of 10, the metric system; it is now formally the *International System* (le Système Internationale), established by an international agreement in 1960. The metric system has only a few basic units, including length, time, mass, temperature, and electric charge; more complex units are defined for derived quantities such as energy. Then a uniform system of multiples of ten or fractions of ten provides all the larger or smaller units one could need. We will illustrate the system with lengths, based on the unit of a meter (about 1.1 yards). The meter (abbreviated m) is the unit. For longer distances, 10 meters is one dekameter (dam), 100 meters is one hectometer (hm), and 1,000 meters is one kilometer (km). For astronomical applications, a million meters is a megameter (Mm), and still larger units are defined by the prefixes given in the table at the end of this section. Biologists have more use for fractions of a meter. A tenth of a meter (about 4 inches) is a decimeter (dm), a hundredth of a meter is a centimeter (cm), and a thousandth of a meter is a millimeter (mm). Many of the objects we deal with in this book are measured in still smaller units. A micrometer (μm) is 10^{-6} meters, and a nanometer (nm) is 10^{-9} meters. Again, there are smaller units whose prefixes are given in the table.

Using the metric system offers important advantages. All you have to learn is a simple series of prefixes that apply to all units. Thus, if you know that a millimeter is 10^{-3} meters, you also know that a milligram is 10^{-3} grams and a milliliter is 10^{-3} liters. Furthermore, if you learn the system in terms of powers of 10, you can easily determine how many of one unit are contained in another. For instance, knowing that a nanometer is 10^{-9} meters and a millimeter is 10^{-3} meters, you can see that there are 10^6 nanometers in one millimeter, because $10^{-3}/10^{-9} = 10^6$. People who learn this system thoroughly and play around with such conversions find science much easier to comprehend.

Metric Prefixes

Prefix	Abbreviation	Numerical Equivalent
atto-	a	10^{-18}
femto-	f	10^{-15}
pico-	p	10^{-12}
nano-	n	10^{-9}
micro-	μ	10^{-6}
milli-	m	10^{-3}
centi-	c	10^{-2}
deci-	d	10^{-1}
deka-	da	10
hecto-	h	10^2
kilo-	k	10^3
mega-	M	10^6
giga-	G	10^9
tera-	T	10^{12}
peta-	P	10^{15}
exa-	E	10^{18}

Answers

1. 2.4×10^3
2. 7.8×10^7
3. 1.47×10^{12}
4. 4.23×10^{-2}
5. 1×10^{-6}
6. 2.4×10^{-4}
7. 10^{10}
8. 10^2
9. 10^{-6}
10. 10^{-3}
11. 10^9
12. 10^{-3}
13. 7.14×10^{-12}
14. 2.812×10^4

Appendix B Answers to Exercises and Questions

Chapter 1

Exercises

1.1 Mostly carbon and hydrogen, with many chains of carbon atoms; many oxygen atoms, often connected by double bonds or in the combination –OH. Rings of carbon atoms are common.

1.2 Cells are small, boxlike, have distinct boundaries, perhaps with heavy walls, and contain other structures, especially a round nucleus.

1.3 The first argument is invalid. The second is formally invalid, in that getting the predicted result does not prove the hypothesis is true; however, in the logic of science, such a result supports the hypothesis and strengthens belief in it.

1.4 The spokesman is correct—we can't *prove* that radiation caused any of the illnesses. However, the correlation certainly suggests that the radiation has been a causal factor in the illnesses, and further study, taking account of other known correlations between radiation and specific disease processes, will probably strengthen the argument.

1.5 The critic has confused the age of a tree with the time during which it lived. One might as well argue that, since Thomas Jefferson died at age 83, there is no reason to assume the American Revolution took place more than 83 years ago.

1.6 It is advantageous for birds to have light bones, so generation after generation, the birds with strong, light bones have been more successful on the average and have had more offspring than birds with heavier bones. Therefore, birds have gradually been shaped into light-boned animals.

1.7 Dr. Brown's "hypothesis" is not a scientific idea because his postulated bions cannot be detected by ordinary physical means, so there is no way to test their existence. Furthermore, it violates the Principle of Parsimony by postulating mysterious entities to explain processes that other scientists are explaining adequately by means of quite ordinary physical concepts.

Multiple-Choice Questions

1.	a	**6.**	e
2.	e	**7.**	d
3.	c	**8.**	c
4.	c	**9.**	c
5.	b	**10.**	b

True-False Questions

1. False. Organic molecules contain *hydrogen and carbon.*
2. False. Causal explanations are essential in physical science *and equally important* in biology.
3. False. *It is important to not believe that* cosmic evolution was directed toward eventually producing the elements of living organisms so organic evolution could begin.
4. True.
5. False. Experiments bear out the consequences of my hypothesis, so the hypothesis *may* be correct.
6. True.
7. False. If someone claims only he and his disciples can observe a certain phenomenon, his claim is not scientific because it is not *intersubjectively testable.*
8. False. Enjoying a Mozart symphony is *a nonscientific* activity.
9. False. *Teleonomic explanations* are based on evolutionary theory.
10. True.

Concept Questions

1. Scientific wonder is characterized by questions about naturalistic cause and effect of a process or entity, while aesthetic wonder is characterized by a variety of intellectual and emotional responses to the existence of a process or entity. The human brain is complex enough to use the same sensory input to independently initiate both processes without detriment to either.
2. Hypotheses are causal explanations; you have made a generalization. Your statement does not attempt to explain the forces that cause the sun to appear to rise in the east and set in the west.
3. The record holder in the men's 100-meter dash is usually considered the fastest man on Earth. No one is particularly surprised if someone else later breaks that record, nor is the first man denounced as untrustworthy! Technological improvements, other new findings, new insights, world conditions, and many other factors can lead to refinements in experimentation, which in turn, can lead to a partial or complete change in an explanation.
4. A hypothesis is one explanation about the cause of an observed effect. A theory is an integrated collection of accepted hypotheses that explain the causes and mechanisms underlying a large-scale phenomenon. Thus the differences are both quantitative and qualitative.
5. The requirement that evidence must be intersubjectively testable helps to eliminate individual biases. Additionally, the understanding that hypotheses and theories may be modified or even discarded by additional evidence tends to negate temporal or cultural biases.

Chapter 2

Exercises

2.1 The principles require us to look for the simpler explanation, and evolution is a simpler, more parsimonious explanation than a series of creations.

2.2 A virus is not a cell or cells, it does not have its own mechanisms for obtaining energy, and it depends on a functioning organism for all the apparatus for reproducing itself.

2.3 The Malthusian principle says that populations tend to outgrow their food supply, so many individuals do not survive. Natural selection means those that do survive are the ones most fit for their particular environment.

2.4 Approximately: $60 \times 1.2 = 72$; $40 \times 0.8 = 32$.

2.5 Correct. There are differences in the reproductive success of individuals, and by definition the most successful are those that are selected.

2.6 Crabs with ordinary shells are removed from the population and cannot reproduce; crabs with any hint of a face on their shells are thrown back and reproduce. So with each generation, the breeding crabs have genomes that specify more and more perfect faces.

Multiple-Choice Questions

1.	a	**6.**	c
2.	b	**7.**	c
3.	e	**8.**	b
4.	d	**9.**	c
5.	b	**10.**	e

True-False Questions

1. False. *Fewer than one-half* of the species on Earth have been identified and named.
2. False. Phylogenetic trees attempt to classify organisms according to their *genetic relatedness.*
3. False. *The environment selects preexisting genetic variations* that enable some organisms to produce more offspring than others.
4. False. If members of two different species exhibit several *homologous* sets of structures, it is likely that they share a relatively recent common ancestor.
5. True.
6. False. Kettlewell's observations showed that as industrial soot accumulated in the environment, *the proportion of melanic moths increased because the light moths were the more visible prey.*
7. False. A molecule composed of 20 identical subunits carries *less information than* one composed of 20 different subunits.
8. False. Viruses contain *genetic information but no mechanism* to express that information.
9. False. When the proportion of individuals in a population with a particular adaptation increases or decreases, the population *is evolving.*
10. True.

Concept Questions

1. Evolution by means of natural selection is a slow, gradual process. Before Darwin's ideas could be taken seriously, geologists had to show that (*a*) the earth was much older than commonly believed; (*b*) after the appearance of life on Earth, there was no time when all living things were destroyed by environmental catastrophes; and (*c*) the relative depth of the rock strata in which fossils were found could be used as a prehistoric time line.

2. A population is a geographically circumscribed group of organisms of the same species. A community includes all populations living in a defined area. An ecosystem includes the biotic community as well as the nonliving elements of the environment, such as climate, altitude, soil depth, etc. A niche refers to the ecological lifestyle of a population. For example, the niche of dolphins would be that of a marine carnivore, while the niche of wild horses would be that of a grassland herbivore.
3. An *adaptation* is any morphological or physiological trait that enables an organism to survive in an ecosystem. The term *fitness* is used to compare relative numbers of offspring produced by members of a population who exhibit differences in structure or function. An individual with a higher fitness may often exhibit a high level of morphological adaptation. However, a structurally well-adapted but sterile individual has a fitness of zero and will not contribute its genes to the next generation.
4. If all organisms had more than sufficient access to food, space, and mates, no one variant would be selected above any other. The pace of natural selection would be slow to nil.
5. Although we would expect to see adaptations to similar niches, the appearance of the inhabitants and the mix of kinds of creatures might be quite different. Evolution proceeds by random, unpredictable events.

Chapter 3

Exercises

3.1 LiCl, LiBr, LiI, NaCl, NaBr, NaI, KCl, KBr, KI; $MgCl_2$, $MgBr_2$, MgI_2, $CaCl_2$, $CaBr_2$, CaI_2; Li_2O, Li_2S, Na_2O, Na_2S, MgO, MgS; and so on.
3.2 pH = 9; $[H^+] = 10^{-5}$
3.3 5.3
3.4 By adding water with a pH of 7, you can never get the pH below 7.
3.5 In real water, with hydrogen bonding, a molecule must have much more energy to break free than it would if there were no bonding.
3.6 The molecules with the most energy are the ones that escape through evaporation, so the average energy of the remaining molecules is reduced and therefore the water is colder.
3.7. (*a*) $H_2C{=}CH{-}CH(NH_2){-}CH_2{-}C(=O){-}O{-}H$

(*b*) $H_3C{-}HC(NH_2)CH_2CH_2SH$, $H_2NCH_2HC{=}CHNH_2$

Multiple-Choice Questions

1. b
2. c
3. d
4. c
5. a
6. e
7. e
8. e
9. e
10. b

True-False Questions

1. False. An atom that has one unfilled valence orbital may achieve a more stable configuration by *losing electrons and becoming a cation* or by *gaining electrons and becoming an anion.*
2. False. Since a mole is a unit of weight, one mole of any compound weighs *its particular molecular weight in grams.*
3. True.
4. False. Oxygen is *less electronegative* than fluorine because the oxygen nucleus has fewer protons than the fluorine nucleus.
5. False. A 1 M solution of HCl contains *6×10^{23} hydrogen ions and 6×10^{23} chloride ions.*
6. True.
7. True.
8. False. When a molecule of water dissociates to form one hydroxyl ion and one hydrogen ion, the hydroxyl ion is left with *the extra electron* that has been released from the hydrogen ion.
9. False. Different *isomers* of an organic compound will have different molecular shapes.
10. True.

Concept Questions

1. Water's high heat capacity enables it to absorb significant amounts of heat produced by the engine without much change in temperature. For the same reason, organisms can maintain relatively stable body temperatures in the face of environmental temperature fluctuations.
2. Pure water is both a weak acid and a weak base. It contains a small but equal concentration of hydrogen ions and hydroxyl ions. Because water can dissociate and release hydrogen ions, it qualifies as an acid. However, water can also become a hydronium ion by combining with a free proton, making water also a base.
3. Hydrocarbons are formed by linear chains, or ring-shaped arrangements, of carbon atoms that are covalently bonded to each other and to hydrogen atoms. Functional groups may contain elements in addition to carbon and hydrogen. When bonded to a hydrocarbon backbone, the resulting molecule may have specific new properties such as polarity, the ability to ionize, or the ability to form specific kinds of chemical bonds.
4. Hydrophobic molecules are nonpolar and cannot occupy a stable place within the lattice formed by hydrogen-bonded water molecules. Hydrophilic molecules are polar and can partly replace polar water molecules within the lattice.
5. Alcohols consist of a hydrocarbon backbone that is nonpolar plus one or more hydroxyl groups that are polar. In short-chain alcohols, the polar region occupies a proportionately greater part of the whole molecule than in long-chain alcohols.

Chapter 4

Exercises

4.1 180 Da
4.2 Homopolymer; $-CF_2CF_2CF_2CF_2CF_2CF_2-$...
4.3 $HS{-}CH_2CH_2{-}(C{=}O){-}S{-}CH_2CH_2{-}(C{=}O){-}$...
4.4 $0.2 \times 0.5 \times 10^{-12}g = 10^{-13}g$ of protein
4.5 6×10^{10} Da; 2×10^6 protein molecules
4.6 No; CO_2 is a reactant, not a catalyst.
4.7 30,250 Da
4.8 $H_2N{-}CH_2{-}C(=O){-}NH{-}CH(CH_2OH){-}C(=O)OH$

Gly-Ser

$H_2N{-}CH(CH_2OH){-}C(=O){-}NH{-}CH(CH_3){-}C(=O)OH$

Ser-Ala

4.9 ···Ⓟ—Sugar—Ⓟ—Sugar··· (each Sugar bearing a Base)

Nucleic acid

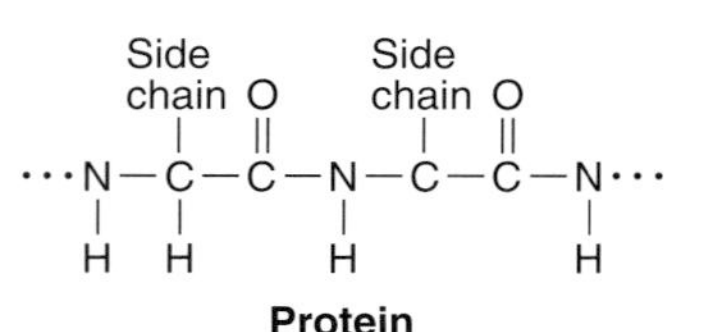

4.10 Water is added to the bond between monomers to restore the monomer structure. Proteins yield amino acids, polysaccharides yield sugars, nucleic acids yield nucleotides.
4.11 Since the Si–Si bond is weaker, molecules that form the organism's structure would be less stable.
4.12 Once the peptide linkages have formed, the amino and carboxyl groups no longer exist, so they can't ionize.
4.13 pH 2: $H_3N^+{-}CH(CH_3){-}C(=O)OH$; pH 11: $H_2N{-}CH(CH_3){-}C(=O)O^-$

4.14 About 10 amino acids; ACGGGSCELK
4.15 (*a*) 8,000 tripeptides; (*b*)20^{100}; 20^{300}

Multiple-Choice Questions

1. e
2. e
3. d
4. d
5. c
6. a
7. e
8. a
9. c
10. d

True-False Questions

1. True.
2. True.
3. False. The most abundant protein in many vertebrate animals is *collagen.*
4. False. The *amino and carboxyl groups* of amino acids participate in the formation of peptide linkages.
5. False. The higher the pH, the higher the concentration of *negatively charged amino acids.*

6. True.
7. True.
8. False. The three-dimensional configurations known as β-pleated sheets and α helices result from *internal hydrogen bonding within a polypeptide.*
9. False. Chlorophyll, hemoglobin, and myoglobin all contain tetrapyrroles. *Hemoglobin and myoglobin have an iron-containing porphyrin, while chlorophyll is a magnesium-containing tetrapyrrole.*
10. False. Although different proteins such as insulin and hemoglobin have different structures, *one kind of protein, such as insulin, made by different species will also vary slightly in structure.*

Concept Questions

1. Hydrolytic reactions are of primary importance during the digestion of a meal, as proteins, polysaccharides, nucleic acids, and fats are broken down to amino acids, monosaccharides, nucleotides, glycerol, and fatty acids. Dehydration synthesis reactions are of primary importance during growth and maintenance of the body.
2. If repeating glucose units form glycosidic linkages between the same series of carbon atoms (such as carbon 1 of one monomer and carbon 4 of the next), the chain will be unbranched. However, if a glucose monomer also forms an additional linkage using the hydroxyl group of still another carbon, the chain will be branched.
3. You would choose the product containing only L amino acids. Since living cells only use the L stereoisomer, you could conclude that the product containing the mixture was completely or partly synthetic.
4. Amino and carboxyl groups are the key components of peptide bonds, so these groups, along with the carbon, form the backbone of the molecule. The different R groups have different shapes and chemical properties, so they are key to the three-dimensional shape and the function of the polypeptide.
5. A multimer is a protein formed from two or more polypeptide chains, while a domain is a structurally and functionally distinct region that has a specific three-dimensional shape. Domains exist in proteins composed of one or more than one polypeptide chain.

Chapter 5

Exercises

5.1 He measured their speeds, determined the ratio of the two speeds, and assumed the speeds would double for every increase of 10°C.
5.2 He is talking about an enzyme; its function is to transfer something, and carbamoyl, whatever that is, may be what it transfers.
5.3 One oxidizes pyruvate, the other polymerizes DNA.
5.4 40 μmol/min
5.5 If the equilibrium constant is 10^3, the reaction proceeds strongly as written; if it is 10^{-3}, it proceeds strongly in the other direction.
5.6 c; an enzyme catalyzes the reaction in both directions.
5.7 b; the reaction is still occurring, but its forward and backward rates are the same.
5.8 The distilled water lacked trace minerals that the bacteria needed for growth.
5.9 (*a*) Saturation will follow a curve like Michaelis-Menten kinetics curve. (*b*) Each leaf will be vacant and occupied for random times. (*c*) The blue flies should be thought of as competitive inhibitors; the leaves will be occupied by one kind of fly or the other in proportion to their numbers.
5.10 Caproic acid is a competitive inhibitor.
5.11 In the mountain animal, the hemoglobin should saturate at a lower oxygen pressure.
5.12 Hb F has the higher affinity for oxygen, so oxygen is transferred from the mother's blood to the fetus's.

Multiple-Choice Questions

1. d
2. a
3. b
4. e
5. c
6. d
7. a
8. d
9. b
10. a

True-False Questions

1. False. *In general, weak bonds, such as hydrogen bonds,* link the active site of an enzyme to its substrate.
2. True.
3. False. If enzyme A forms tighter bonds to its substrate than enzyme B, the Michaelis constant of enzyme A will be *lower* than that for enzyme B.
4. True.
5. False. Competitive inhibitors *and irreversible inhibitors* bind to an enzyme's active site.
6. False. In terms of enzyme kinetics, the addition of a competitive inhibitor to an enzyme-catalyzed reaction will alter the V_{max} *of the reaction but not the turnover rate of the enzyme.*
7. False. At very high oxygen concentrations, myoglobin's saturation curve *is level.*
8. False. People who have only one gene for Hb S are resistant to malaria because *those red blood cells that sickle and are destroyed are also the site of infection.*
9. False. A person with two genes for Hb S will have anemia *and also be resistant to malaria.*
10. True.

Concept Questions

1. An energy barrier is the amount of energy required to raise the energy level of the average molecule in a collection of molecules to an activated state at which it can react. The activated state is an arrangement of atoms in the midst of changing from the reactant to the product.
2. First, because enzymes bind to their substrates in a precise way, the substrate molecules are no longer randomly distributed and are held in position for a reaction to occur. Second, the binding of an enzyme to its substrate distorts some chemical bonds in the substrate, thereby lowering the energy that must be added to break those bonds.
3. Human pathogens have evolved to flourish at normal human body temperature. A slightly increased body temperature can actually inhibit bacterial metabolism and reproduction.
4. A competitive inhibitor has a shape complementary to an enzyme's active site and weakly binds to the active site. An irreversible inhibitor forms strong bonds with the enzyme's active site and causes a permanent change in the shape of the active site so it cannot bind substrate.
5. Cooperative effect refers to the binding of oxygen to the heme group, hemoglobin's primary binding site. There are four heme groups per molecule of hemoglobin, and when oxygen has been bound to one of them, binding at the other three sites is facilitated. Allosteric regulation occurs when a ligand binds to a secondary site rather than an active site or a primary binding site, thus changing the shape of the primary site and the strength with which it binds oxygen.

Chapter 6

Exercises

6.1 The smaller room has a higher ratio of doors to people and can be emptied faster. Similarly, material can enter and leave a small cell faster than a large one.
6.2 The binoculars don't have sufficient power of resolution.
6.3

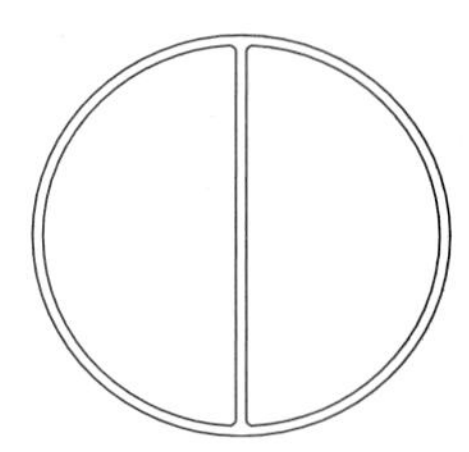

6.4 Ratio = 1,000
6.5 Eucaryote; the size of the cell doesn't matter.
6.6 The cells are making a lot of the enzymes, which are embedded in the membranes, so they have a lot of membranes.

Multiple-Choice Questions

1. a
2. d
3. d
4. a
5. b
6. c
7. d
8. b
9. c
10. e

True-False Questions

1. False. A rectangular cell that measures 2 × 4 × 8 μm has *four times the surface and eight times the volume* of a rectangular cell that measures 1 × 2 × 4 mm.
2. False. A microscope that uses infrared radiation would have *lower* resolving power than a microscope using ultraviolet radiation.
3. False. Every cell is surrounded by a *cell or plasma membrane* that regulates the inflow of nutrients and the outflow of metabolic wastes.
4. False. Phase-contrast microscopy is used to *visualize living cells.*
5. False. The *nucleoid* in procaryotic cells is functionally analogous to the nucleus in eucaryotic cells.
6. False. The elements of the cytoskeleton are primarily composed of *proteins.*
7. False. *Ribosomes on the cytoplasmic face* of the endoplasmic reticulum synthesize proteins.
8. True.
9. False. The glycocalyx is composed of *oligosaccharides* that provide the cell with cell-to-cell binding ability.
10. False. *Gap junctions* enable nutrients and ions to travel easily between neighboring cells.

Concept Questions

1. The higher resolving power of the electron microscope lets us see that adjacent but separate objects are indeed separate.

2. Tissues are killed, and the proteins are precipitated (fixed) before degenerative changes occur. They must be cut into thin slices to allow light to penetrate. Most cells are transparent, even though their organelles differ chemically from each other and from the cytosol. Stains react chemically with cellular structures and render them visible by increasing the contrast between structures.
3. Since the surface/volume ratio decreases as the volume increases, large cells cannot exchange materials with their surroundings fast enough to support and maintain metabolism. Smaller cells have proportionately more surface.
4. The rough endoplasmic reticulum is the primary site for protein synthesis; the smooth endoplasmic reticulum contains enzymes that synthesize lipids and detoxify a variety of metabolic poisons; and the Golgi complex processes and sorts internal cellular products, and ships external secretory products.
5. Gap junctions contain pores or channels of communication, enabling small molecules, ions, and/or electrochemical signals to travel from cell to cell. The protein pores can open or close in response to a variety of stimuli. Tight junctions are one kind of adhering junction; they are also made of proteins, but these proteins form a waterproof seal and "glue" between cells.

Chapter 7

Exercises

7.1 The energy of falling water is converted into the energy of a rotating generator that produces an electric current; the energy of the current is converted into the energy of a heated filament that produces light and a rotating motor armature, which increases the potential energy of the elevator.

7.2 447 kcal/mol

7.3 Ethane is more stable because energy is released when it is made from its elements.

7.4 Creatine phosphate can transfer a phosphoryl to ADP and glucose; ATP can transfer a phosphoryl to glucose; glucose 6-phosphoryl can't transfer its glucose to the others.

7.5 (*a*) Iron will not reduce magnesium; (*b*) $FADH_2$ will reduce pyruvate; (*c*) NADH will reduce oxygen; (*d*) nitrate will not reduce FAD.

Multiple-Choice Questions

1.	c	6.	d
2.	e	7.	e
3.	b	8.	b
4.	b	9.	e
5.	e	10.	d

True-False Questions

1. False. Heat is a form of *kinetic energy.*
2. True.
3. False. Chemical energy results from the *electromagnetic* force.
4. True.
5. False. Reactions that entail a *negative* change in free energy release energy to their surroundings.
6. True.
7. False. An exergonic reaction in which ΔG is -5 kcal/mol can drive an endergonic reaction in which ΔG is *less than* $+5$ *kcal/mol.*
8. False. Given the compounds A and AH_2, compound A contains *less energy than its reduced partner* AH_2.
9. False. NAD^+ is the *oxidized* form of NADH. Or: *NADPH is the reduced form of NADP.*
10. False. The formation of ATP from ADP and Pi is an example of *phosphorylation;* the formation of ADP and Pi from ATP is *hydrolysis.*

Concept Questions

1. Enzymes lower activation barriers, thereby speeding up the reactions they catalyze. Although an exergonic reaction is spontaneous, it might not occur at a rate that supports life. On the other hand, with no source of additional energy, endergonic reactions would not take place at all. These reactions are coupled to exergonic reactions in which ATP is hydrolyzed or a reduced coenzyme is oxidized.
2. Since endergonic processes are nonspontaneous, each endergonic process must be driven by the energy released by an exergonic (spontaneous) process. When specific chemical reactions or metabolic pathways are coupled, not only are they connected by the release of energy in a spontaneous reaction and use of energy in a nonspontaneous reaction, but also by the product(s) of one reaction serving as the substrate(s) of the other.
3. No. Enzymes lower the activation barrier so a reaction will proceed more rapidly, but they have no effect on the free energy change in the reaction.
4. Autotrophs are necessary to replenish the supply of energy available to do work, since the entropy of the environment tends to rise continuously. Heterotrophs, especially decomposers, are necessary to recycle matter.
5. ATP and NADH are both nucleotide-based compounds that act as energy shuttles and reaction couples. ATP and/or NADH formed in exergonic reactions drives endergonic reactions. ATP and NADH differ in the method by which they transfer energy. ATP works by phosphorylation, whereas NADH is a reducing agent.

Chapter 8

Exercises

8.1 As the atoms of mercury get hotter, they move faster and push each other harder as they collide, so the average distance between atoms increases and the mercury expands.

8.2 1 M glucose against 4 M glucose.

8.3 2 M NaCl against pure water.

8.4 Put the cells into solutions of different osmolarity and look for the one in which they tend to neither shrink nor swell.

8.5 Hypertonic: More; cell shrinks.
Hypotonic: Less; cell swells.

8.6 (*a*) Pentane quickly, because it is hydrophobic and lipid-soluble. (*b*) Malonate slowly, because it is charged. (*c*) ATP very slowly, because it is large and charged.

8.7 (*a*) Butane, because it is more hydrophobic. (*b*) Benzene, because it is not charged. (*c*) Glycine, because it is smaller.

8.8

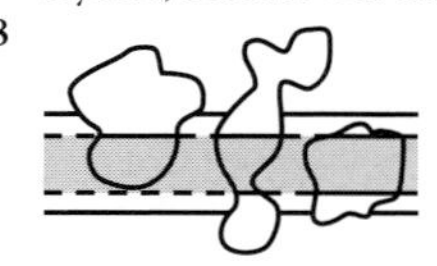

8.9 Just as a substrate binds to the active site of an enzyme, a ligand binds to the site of a carrier. In both cases, the number of proteins is limited, and the process is at its maximum when they are all occupied.

8.10 An enzyme located in the plasma membrane binds Q on the outer face of the membrane and simultaneously moves it through the membrane as it removes the amino group. Show the active site as a passage through the membrane.

Multiple-Choice Questions

1.	a	6.	d
2.	c	7.	e
3.	d	8.	d
4.	b	9.	b
5.	a	10.	c

True-False Questions

1. False. When a molecule diffuses, *either its overall movement* is directed down its concentration gradient or ttrajectory is random.
2. False. In an osmotic system containing two solutions that differ in osmolarity, water will flow toward the more *hyperosmotic* solution.
3. False. The cytosol of a cell has a *higher* osmolarity than pure water.
4. False. Cells placed in a hypertonic solution will *plasmolyze.* Or: Cells placed in a *hypotonic* solution will develop turgor pressure or burst.
5. True.
6. False. Substances that diffuse easily through the plasma membrane have a *high oil-water* partition coefficient.
7. True.
8. False. Permeases function as carriers during *facilitated diffusion.*
9. True.
10. False. When a symport protein transports two or more ligands, the ligands *bind simultaneously and are transported simultaneously.*

Concept Questions

1. A concentration gradient is a continuous change in the concentration of a diffusible substance from a region of relatively high concentration to a region of lower concentration.
2. Interior peripheral proteins anchor cytoskeletal proteins in the cytosol. Exterior glycoproteins act as receptors as well as recognition and binding sites. Transmembrane proteins may act enzymatically as transporters, channels, or pumps, as well as forming part of the physical structure of the membrane.
3. Active transport requires an energy source. To show that a solute is being actively transported, you would have to show that eliminating a specific energy source, such as ATP, will stop transport.
4. Channel proteins that are not uniformly open passageways are said to be gated. The gate is one or more protein domain(s) that may be opened by mechanical deformation, by a change in voltage across the membrane, or by the binding of a ligand to the gate. The ability to selectively open a channel can establish a concentration gradient across the membrane or allow the membrane to respond to a specific stimulus.
5. Primary active transport is an energy-dependent movement of one particular solute against its concentration gradient. Carrier proteins and ATP are used to develop and maintain the concentration gradient or the solute. Cotransport refers to

the movement of two or more solutes across the membrane, with the solutes moving in the same direction or in opposite directions. Once again, carrier proteins are activated when they bind their specific ligands. However, one or more of the solutes may travel down, rather than up, its concentration gradient. Secondary active transport means developing a work-producing concentration gradient that is then tapped to move a solute across the membrane.

Chapter 9

Exercises

9.1 30 ATPs × 11 kcal each = 330 kcal; 330/686 = 0.48.

9.2 With Complex IV inhibited, oxygen cannot oxidize the ETS, no ATP is made, NADH cannot be reoxidized, and the whole metabolic system backs up.

Multiple-Choice Questions

1. d
2. c
3. a
4. b
5. e
6. c
7. d
8. e
9. e
10. a

True-False Questions

1. False. In the overall equation of cellular respiration ($C_6H_{12}O_6 + 6O_2 \rightarrow 6H_2O + 6\ CO_2$), *both carbon and oxygen in CO_2 come from glucose.*
2. True.
3. False. Photosynthesis is primarily a process in which carbon is *reduced;* cellular respiration is a process in which carbon is *oxidized.*
4. False. After completion of glycolysis and the citric acid cycle, *the energy primarily is found in NADH, FADH, and ATP.*
5. False. *The reduction* of pyruvate to alcohol or lactate is known as fermentation.
6. False. The enzymes of the Krebs cycle are *contained within the matrix and inner membrane,* while the enzymes of the electron transfer system are found on the inner mitochondrial membrane.
7. True.
8. False. As electrons travel from one component of the electron transport system to the next, they *lose energy, which is used to produce the proton gradient.*
9. False. In oxidative metabolism, the final oxidizing agent is *inorganic,* whereas in fermentation it is *organic.*
10. False. As a result of the events of chemiosmosis, the oxygen we breathe is combined with *electrons and protons to form water.*

Concept Questions

1. The concentration of pyruvate would increase. The reactions of glycolysis would reverse, the glycolytic pathway would grind to a halt, and the Krebs cycle would stop.
2. The hydrogen atoms (or their protons and electrons) of glucose are used to reduce NAD^+ to NADH in glycolysis and the Krebs cycle and FAD to $FADH_2$ in the Krebs cycle. NADH and $FADH_2$ enter the ETS, where the protons are pumped across the membrane and the electrons are transferred through a series of electron carriers. Ultimately, the electrons and protons are combined with oxygen to form water.
3. Glycolysis and the Krebs cycle require NAD^+ and FAD, both of which are produced by the ETS. The ETS utilizes NADH and $FADH_2$, which are generated by glycolysis and the Krebs cycle.
4. The oxygen you inhale is used as the terminal electron acceptor in the ETS. The carbon dioxide you exhale is a waste product of the oxidation of pyruvate to acetyl-CoA and the Krebs cycle. The water in your exhaled breath is a waste product of the ETS.
5. The metabolic products of proteins, polysaccharides, and lipids all end up as metabolites of the pathway from pyruvate to acetyl-CoA and through the Krebs cycle. This pathway is a switching point between catabolic and anabolic reactions, where metabolites from each major class of biomolecules start to be converted into subunits of the others.

Chapter 10

Exercises

10.1 The membrane-metabolism principle.

10.2 It's quite easy to identify the main violet peak with Chl *a*, with a shoulder in the blue range from Chl *b* and carotenoids. The red peaks are clearly due to Chl *a* and Chl *b*.

10.3 Isolate chloroplasts and mix them with methylene blue. Irradiate the mixture with light and see if methylene blue becomes colorless with longer irradiation. As controls, see if methylene blue remains colored in the absence of light and if the chloroplasts are denatured or poisoned.

10.4 In separate experiments, incubate chloroplasts with labeled CO_2 or labeled H_2O. Collect the oxygen they produce and determine in which case the oxygen is labeled.

10.5 $NADP^+ + H^+ + 2e^- \rightarrow NADPH$. To activate 2 electrons, PS I must absorb 2 photons and PS II 2 more, or 4 photons per NADPH.

10.6 Increasing temperature favors C_4 plants, and as the CO_2 concentration increases, the crossover temperature increases, also favoring C_4 plants. However, an increasing CO_2 concentration may favor photosynthesis over respiration for C_3 plants. Conflicting factors make it hard to predict what will happen.

Multiple-Choice Questions

1. c
2. b
3. e
4. d
5. a
6. e
7. a
8. b
9. a
10. c

True-False Questions

1. False. The longer the wavelength of electromagnetic radiation, the *lower* its energy.
2. False. In the absence of other interactions, an electron that has been boosted to a higher energy level *will fall back to its ground state.*
3. True.
4. False. *Only photosystem I can carry out cyclic photophosphorylation, whereas photosystems I and II are engaged in noncyclic photophosphorylation.*
5. False. Since photosystem I absorbs light of longer wavelengths than photosystem II, *photosystem I absorbs quanta at a lower energy state.*
6. True.
7. False. During photosynthesis in bacterial autotrophs, reducing power comes from *compounds such as H_2S,* while energy comes from light.
8. True.
9. False. Several of the intermediate molecules of the Calvin cycle are the same or similar to intermediates in *glycolysis.*
10. True.

Concept Questions

1. Mitochondria include an outer and inner membrane, with much of the biochemistry taking place within the folded inner membrane and in the internal matrix. Although mitochondria contain cytochromes, they are not pigmented organelles. Chloroplasts are also enclosed by a double membrane, but they have a third internal thylakoid membrane folded into disc-shaped grana. The pigment chlorophyll is found within the thylakoid membranes, which function as the site of the light-dependent reactions. Because of the third membrane, chloroplasts have three spaces: an intermembrane space between the two outer membranes, an interior lumen within the thylakoid grana, and a stroma between the thylakoid membrane and the enclosing double membranes.
2. A proton-motive force is generated in the light-dependent reactions. During cyclic photophosphorylation, the proton gradient is generated by an ETS of photosystem I. In noncyclic photophosphorylation, the proton gradient is generated by an ETS between photosystem II and photosystem I.
3. Since 18 ATPs but only 12 NADPHs are used to produce one molecule of glucose, and since the noncyclic reactions produce equal amounts of ATP and NADPH, cyclic photophosphorylation can supply the additional required ATP.
4. The light-dependent reactions transduce electromagnetic energy of sunlight to chemical energy in ATP. In addition, the noncyclic light reactions generate reducing power in the form of NADPH. Using the energy and reducing power supplied by the light-dependent reactions, the light-independent reactions fix carbon into organic molecules, specifically sugar.
5. As long as a plant keeps its stomata open, oxygen from the light reaction can diffuse away, keeping its concentration within the leaf at a low level. Under those conditions, rubisco will effectively act as a carboxylase. Since 18 molecules of ATP are required to synthesize one sugar molecule by means of C_3 photosynthesis, whereas the C_4 reactions require 30 ATPs, the C_3 pathway is more efficient. However, if the stomata are kept closed, as is characteristic of plants during a hot, dry day, the concentration of oxygen will rise within the leaf. Rubisco will then act as an oxidase, and the fixation of carbon dioxide will fall, as will sugar production. Under those conditions, the C_4 pathway, despite its use of ATPs, becomes more efficient.

Chapter 11

Exercises

11.1 C of glucose will go into other organic compounds, along with N from NH_4^+: P will go into nucleic acids, ATP, and other nucleotides: S into proteins: Na, K, Cl will be ions in the cytosol.

11.2 It is more efficient to have a common pathway that produces all molecules with similar structures, rather than a separate pathway for each. Therefore, when metabolic pathways were evolving, organisms with the more efficient organization conserved their energy better, grew faster on available nutrients, and outcompeted those with less efficient organizations.

11.3 Eyes sense speed; when speed is slightly over 55, brain signals foot to ease pressure on the accelerator. When speed is slightly under 55, brain signals foot to increase pressure on the accelerator.

11.4 Glycerol level = controlled variable; liver cells = sensor and comparator; H = signal; fat cells = effector.

11.5 $K = [C]/[L][P]$. K is a constant by definition, and if [P] is constant, then the more ligand, the more complex. If the amount of ligand decreases, the amount of complex decreases.

11.6 If any organism had an enzyme later in the pathway regulated by feedback inhibition, it would waste energy and materials by carrying out unnecessary reactions. Therefore, organisms were selected that happened to have regulation at the most efficient points.

11.7 aa_1 and aa_2 should jointly regulate conversion of N to Q. aa_1 might regulate Q to R, and aa_2 might regulate Q to P. aa_3 should regulate N to U.

11.8 (*1*) Ligand binding causes a conformational change such that an active site is activated on the inside. (*2*) Ligand binding causes a conformational change such that a second protein binds to receptor and is activated.

11.9 Mitosis involves microtubules, and cytokinesis involves microfilaments.

11.10 Since their cilia are inactive, cilia on the epithelia of their respiratory tract cannot remove infecting microorganisms. Since sperm swim with flagella, males should be sterile.

11.11 Lipoprotein is carried into a lysosome; amino acid is digested out of the protein portion of the lipoprotein, passes through the lysosome membrane into the cytosol, and is used to synthesize a new protein. Lipid is digested out and passes into the membrane of the lysosome, eventually joining with other cell membranes.

Multiple-Choice Questions

1. a
2. e
3. d
4. d
5. c
6. d
7. a
8. a
9. e
10. a

True-False Questions

1. False. Growing cells need transaminase enzymes to convert keto acids from their central pathways *into amino acids.*
2. False. Bacteria cannot grow in a medium without sulfur because they need this element to synthesize *their proteins.*
3. False. An allosteric protein can bind two kinds of ligands *at two different binding sites,* where they have different effects on the protein's activity.
4. True.
5. False. End-product inhibition generally occurs when the metabolic product of a pathway inhibits one of the *initial* enzymes in the pathway.
6. False. Our sensations of taste and smell arise from the binding of *environmental ligands with allosteric receptors* in the plasma membrane of a receptor cell.
7. True.
8. False. Once activated, second messengers such as cAMP or Ca^{2+} ions act as *ligands that activate enzymes.*
9. False. The effects of certain inhibitors allow us to determine that *microfilaments* are the primary structures involved in cytoplasmic streaming, or cyclosis.
10. False. Materials brought into a cell *by endocytosis* end up in *lysosomes* where they are digested by *hydrolytic* enzymes.

Concept Questions

1. In a chemical equilibrium, there is a steady conversion of substrate to product and back again, which maintains overall unchanging concentrations of each. In a homeostatic steady state, products are continuously withdrawn—either to synthesize other compounds or to be catabolized. In order to maintain stable concentrations of products, new substrates must continuously be added.
2. Cycles of positive feedback continue to increase or decrease some variable until the system is disrupted by some fundamental change. For example, the positive feedback of childbirth culminates in birth. The positive feedback of an expanding fire ends when all combustible materials have burned. To achieve a steady state, an increase in the output of a system must lead to a decrease in the rate at which the product is produced.
3. Metabolic rate will increase as the food sensor in the brain compares the metabolic rate of the body with the amount of food. If the amount of food increases, the metabolic rate continues to increase. If food intake falls, the processes that increase metabolic rate are inhibited. If a process is inhibited by an increase in product, the cycle contains a negative step.
4. All are signaling or informational compounds. All are ligands that bind to allosteric receptors on the cell membrane or within the cell and trigger a cellular response. They differ primarily in the identity of, and distance between, the producing vs. responding cell. Alarmones are produced and bound within the same cell. Cells that produce hormones generally have different target cells within the same organism. Pheromones are chemical signals produced by one individual that act upon another individual of the same species.
5. MTOCs orient the position of microtubules and initiate their formation from tubulin dimers. Centrosomes and basal bodies are MTOCs, organizing the mitotic spindle and cilia or flagella, respectively. Molecular motors use the energy released by the hydrolysis of ATP to attach, pull on, and then release from tubulin. Dyneins are molecular motors that move from the plus to minus end of a microtubule; kinesins move from the minus to the plus end.

Chapter 12

Exercises

12.1 *trpX* *trpY*
$\rightarrow$ B $\rightarrow$ Trp

12.2 (*a*) A protease (carrier protein) for fructose. (*b*) Perfectly in line with "one gene, one polypeptide," but not "one gene, one enzyme."

12.3 $10^6 \times 10 \times 95 = 9.5 \times 10^8$ bacteria/ml

12.4 $3.1 — 10^{11}$ phage/ml

12.5 150 phage/cell

12.6 Phosphate has a negative charge; protein will have positive charges.

12.7 Negative phosphates on the inside would repel one another and would not make a stable molecule.

12.8 In this activity exercise, you should end up with two molecules identical to the original.

12.9 Information can only be encoded with two or more distinct elements, which don't exist in a homopolymer.

12.10 Pairs can specify 16 amino acids; triplets can specify 64.

Multiple-Choice Questions

1. b
2. c
3. e
4. d
5. b
6. e
7. c
8. e
9. a
10. e

True-False Questions

1. True.
2. False. If, in an environment that remains uniform, a mating between two flies carrying the same variation produces offspring that also exhibit the same variation, the cause of the variation *is most likely genetic.*
3. False. Wild-type *Neurospora* are prototrophs.
4. False. If a strain of *Neurospora crassa* is auxotrophic for arginine. *one or more of the enzymes in the arginine pathway is (are) not functioning properly.*
5. False. *Animal viruses, plant viruses, and bacteriophage are all infective and can only metabolize when inside a host cell.*
6. True.
7. False. *Adenine and guanine* are purine bases, while *thymine and cytosine* are pyrimidines.
8. False. DNA is said to have a 5' $\rightarrow$ 3' polarity, which means that *one phosphate group is bonded to the 5' carbon of the sugar, while another phosphate group is bonded to the 3' carbon of the sugar.*
9. True.
10. False. *Like most* other biological reactions, *DNA replication requires specific enzymes.*

Concept Questions

1. Compound Y is converted to compound Z, and Z is converted to X, in a series of enzymatic reactions. Strain 1 has a mutation in the gene that encodes the last enzyme in the pathway. Strain 2 has a mutation in the gene for the next-to-last enzyme, and strain 3 has a mutation in a gene that encodes an enzyme that acts earlier in the pathway.
2. A colony of bacteria is a clone of cells that result from the division of one bacterial cell. A plaque results when bacteriophage infect and destroy bacterial cells, creating an empty place in a lawn of bacteria.
3. Hershey and Chase relied on the fact that polypeptides contain carbon, hydrogen, oxygen, nitrogen, and sulfur while polynucleotides contain carbon, hydrogen, oxygen, nitrogen, and phosphorus. Since they were interested in determining which part of the virus (DNA or protein) entered the bacterial cell, they could exploit the chemical difference between the two molecules by labeling viruses with radioactive isotopes of either sulfur or phosphorus and seeing which isotope entered bacterial cells and which remained outside the cell.
4. Since bacterial viruses inject only their DNA core, not their protein heads, the presence of an intact virion ends at the time of infection. During the eclipse period, viral DNA causes the bacterial cell to copy the viral DNA and build viral protein, but whole virions do not start to reassemble for several minutes.

5. Watson and Crick's molecular models showed the space between the two backbones could be filled with one purine and one pyrimidine. Chargaff's data on relative quantities of bases indicated A = T and C = G. Thus there would not be room for adenine to pair with guanine (both purines), and a cytosine-thymine (both pyrimidines) pair would not be large enough to fill the space. Adenine would not pair with cytosine and guanine with thymine because the amino and keto groups that form the hydrogen bonds do not match properly.

Chapter 13

Exercises

13.1 After one round, half all heavy, and half all light. After two rounds, one-quarter all heavy, three-quarters light.

13.2 If replication were dispersive, single strands separated by heating would have various intermediate densities.

13.3 Each chromosome had one radioactive chromatid and one nonradioactive.

13.4 (3.8×10^6 nucleotide pairs)/(2×2400 sec) = 790 nucleotides/sec

13.5 The nuclear envelope breaks down and chromosomes condense, as if they were going to go through mitosis.

13.6 Activity exercise.

13.7 1 μm/min = 1 mm/1,000 min or about 1 cm/week

13.8 Units are 3 nm long, so about 330 units/μm; 13 units per turn of microtubule, so 4,290 units/min or 72 units/sec.

13.9 Number of units is irrelevant. (30,000 np)/(2×240 sec) = 62.5 np/sec

Multiple-Choice Questions

1. d	**6.** c
2. b	**7.** c
3. c	**8.** a
4. c	**9.** d
5. b	**10.** c

True-False Questions

1. False. Hämmerling's experiments with *Acetabularia* demonstrated *that the genetic program of a cell resides in its nucleus.*
2. False. It is possible to study the method of DNA replication in bacteria with density gradient centrifugation by using bacteria that have been growing and dividing in a medium *containing a heavy isotope of nitrogen.*
3. False. A replisome is composed of *several enzymes including DNA polymerase.*
4. True.
5. False. The energy needed to synthesize DNA is derived from the *breakage of phosphate bonds in nucleoside triphosphates.*
6. False. A eucaryotic chromosome contains *one chromatid* during all phases of the cell cycle except G_2 and the early stages of mitosis.
7. False. Barring the use of specific inhibitors, any cell that reaches G_2 in the cell cycle will undergo mitosis.
8. False. The eucaryotic cell cycle is regulated by the synthesis of *cyclin B* during interphase and its destruction during metaphase.
9. False. The primary function of a kinetochore is to *connect each chromatid to a spindle fiber.*
10. True.

Concept Questions

1. Nitrogen is a component of both protein and nucleic acid. Since eucaryotic chromosomes contain a much larger and more complex assortment of proteins than do procaryotic chromosomes, tracing heavy nitrogen through rounds of replication by means of density gradient studies would not differentiate between the protein and nucleic acid components of chromosomes. On the other hand, thymidine is incorporated into DNA but not proteins, so the use of a radioactive isotope and its detection by autoradiography would unambiguously follow DNA replication.
2. Helicase is an enzyme that enables the double helix of DNA to open, forming a replication fork. Single-strand binding proteins stabilize the single-stranded bubble of DNA. Primase is a polymerase that catalyzes the synthesis of a short primer. Primers are necessary because DNA polymerase cannot initiate the synthesis of a strand of DNA, but can only lengthen a preexisting strand.
3. There would be no need for Okazaki fragments since both strands would be synthesized continuously from one end of the double helix to the other. Primase would still be required, but only once, at the initiation of synthesis.
4. MPF is a complex protein made of two smaller proteins, cyclin B and Cdc2. MPF acts enzymatically as a protein kinase. It phosphorylates other proteins, such as histone H1, which enable the cell to perform activities necessary for mitosis, including chromosome coiling and nuclear membrane breakdown. MPF regulates the cell cycle by alternately forming and degrading as one of its components, cyclin B, is degraded at the end of metaphase. Without cyclin B, MPF is no longer complete and active, and the events that pushed the cell into mitosis cannot be sustained. As a result, the cell returns to interphase. As cyclin concentrations once again rise, MPF is remade, reactions necessary for mitosis begin, and the cell enters mitosis.
5. Bacteria have a single circular chromosome; eucaryotic cells have linear chromosomes. In bacteria, replication is initiated in one spot on the chromosome; eucaryotic chromosomes begin DNA replication at many sites simultaneously. Once the DNA strands have been unwound and separated at the initiation site, the bubble that is formed lengthens at both ends simultaneously as one replisome moves in one direction and the second replisome moves in the other direction. Bacterial replication is complete when both replisomes have circled and returned to the point where replication began. In eucaryotic cells, many replication bubbles are formed, and a pair of replisomes moves in opposite directions from each bubble. Replication is complete when a replisome moving east runs into the adjacent replisome that was moving west.

Chapter 14

Exercises

14.1 Activity exercise.

14.2 Activity exercise.

14.3 For 12 amino acids, a codon could be 2; for 80, it must be 4.

14.4 A–U–G | G–G–C | C–A–U– | G–C–A | A–G–C | C–U–U | U–A–G

14.5 Met–Gly–His–Ala–Ser–Leu–end; the code is unambiguous.

14.6 Several mutations are possible, such as: UGG (Trp) → UGA; UGG (Trp) → UAG; CAA (Gln) → UAA. In the middle of a gene, the mutant codon would terminate the protein prematurely.

14.7 TCG → CCG, Ser → Pro

14.8 A terminator ends synthesis of an RNA molecule; a termination codon ends a polypeptide chain.

14.9 900 nucleotides/55 nucleotides per sec = 16 sec; 300 amino acids/17 amino acids per sec = 18 sec (both answers rounded off).

14.10 They also synthesize A, B, and C, in the same order.

14.11 Galactose is added to the protein in the ER; fucose is added in the Golgi membranes.

Multiple-Choice Questions

1. a	**6.** e
2. e	**7.** a
3. d	**8.** c
4. a	**9.** e
5. e	**10.** a

True-False Questions

1. False. To qualify as a gene, a region of DNA must be *transcribed to RNA, and may be translated to a polypeptide.*
2. False. The promoter precisely aligns *RNA polymerase on DNA during transcription.*
3. False. Amino-acyl tRNA synthetase joins *an amino acid to its tRNA,* liberating pyrophosphate in the process.
4. False. Of the 64 possible codons, *60 code for amino acids, the methionine codon acts as the initiator,* and 3 are termination codons that signal the end of a gene.
5. True.
6. False. If one knows the sequence of *bases in DNA,* it is easy to deduce the order of *amino acids in a polypeptide.*
7. True.
8. False. The wobble hypothesis states that *one tRNA anticodon may base-pair with several mRNA codons.*
9. True.
10. False. In eucaryotes, *mature mRNA is* produced by removing the introns and splicing the exons together.

Concept Questions

1. The sequence of codons in the DNA of a gene, in the mRNA transcribed from it, and in the protein translated from the mRNA are all identical. A series of arrows relating these sequences will never cross one another. This means that the information in a gene is used in the most direct manner. The enzymes that conduct transcription and translation always move forward continuously, never jumping to a distant point.
2. Redundancy means that a particular amino acid may be encoded by more than one codon so there are more codons than amino acids. Wobble is a relationship between mRNA codons and tRNA anticodons: One anticodon may pair with more than one codon so there are more codons than anticodons.
3. The purpose of the technique is to localize each step of a multistep metabolic pathway. Thus the radioactive substrate must be given for only a short time so only one reaction step is tagged.

The quantity of nonradioactive chaser must be large so that any radioactive material that might not have yet been used is so dilute that the probability of its being used is low. The cell must be intact so that the normal geography of metabolism is maintained.

4. Present in all 4 test tubes: ^{32}P-labeled primer, dATP, dCTP, dGTP, dTTP, and DNA polymerase. Present in individual tubes: one variety of dideoxyribonucleotide, such that one tube contains ddATP, the second contains ddCTP, the third contains ddGTP, and the fourth contains ddTTP.
5. Genetic information is the sequence of bases in DNA that codes for the synthesis of all the proteins an organism needs to live. However, genetic information is not sufficient to functionally organize the proteins produced by gene action. Epigenetic information resides in the complex organization of the cellular components. Existing cellular structures assist in organizing newly produced membranes and organelles. Although both genetic and epigenetic information are handed down from one cell generation to the next, genetic information is circumscribed and generally limited to DNA and RNA. The nature of epigenetic information can change depending on the cell and the time of the life cycle. In the fertilized egg, the unequal distribution of many compounds constitutes epigenetic information, while in an older embryo, whether a cell is on the surface or in the interior constitutes another form of epigenetic information.

Chapter 15

Exercises

15.1 Answers will vary.

15.2 (*a*) Centromeres divide at anaphase in mitosis, only at second anaphase in meiosis. (*b*) Chromosomes move independently in mitosis and second meiotic division but form paired bivalents before first meiotic division. (*c*) Mitosis occurs in both haploid or diploid cells, meiosis only in diploid. (*d*) DNA replicates during interphase before both mitosis and meiosis but not during interphase between meiotic divisions.

15.3 Haploid cells could have both red, both blue, or the two combinations of one red and one blue.

15.4 For $n = 2$, either AB and ab or Ab and aB. For $n = 3$, ABC, abc; ABc, abC; Abc, aBC; or aBc, AbC.

15.5 In general, 2^n combinations. Each parent can make $2^{23} = 8.4 \times 10^6$ distinct gametes, so together they can make $2^{46} = 7 \times 10^{13}$ different children.

15.6 Because the zygote will become the embryo, it needs to retain the maximum cytoplasm possible, so the extra chromosome sets are removed with small amounts of cytoplasm.

15.7 Sperm and ovum, neither; ovary and testis, both; all others, mitosis only.

15.8 Distinct stages are adapted to different ecological niches, such as swimming from one animal to another or living in a particular organ; they have evolved different characteristics to fit into those niches.

Multiple-Choice Questions

1. e	6. e
2. e	7. a
3. d	8. c
4. a	9. b
5. b	10. d

True-False Questions

1. True.
2. False. *In both cycles,* meiosis occurs at the end of the diplophase, and in a haplontic cycle, meiosis occurs *at the end of the diplophase.*
3. False. Organisms with haplodiplontic cycles engage in fertilization *only after the haplophase.*
4. False. Cells replicate their DNA before going through mitosis, and those going through meiosis replicate their DNA *only before the first meiotic division.*
5. False. After meiotic anaphase II, each chromosome consists of one chromatid. Or: After meiotic anaphase I, each chromosome *still* consists of *two* chromatids.
6. True.
7. False. The cytoplasmic events in spermatogenesis and oogenesis *are quite different,* but the nuclear events *are similar.*
8. False. A flower ovary is to a human ovary as an anther is to a *testis.*
9. False. In males, the four sperm that result from a single meiotic event are *genetically different.*
10. False. If an organism has morphologically distinct stages in its life cycle, *each stage is using different genes from the same genome.*

Concept Questions

1. During prophase I of meiosis, synapsis occurs, chiasmata are formed, and crossing over takes place. None of these events occur in mitotic prophase. Another significant difference takes place at the start of anaphase. In meiosis, the centromeres do not divide, whereas during mitosis, they do.
2. All cells containing tetrads must be in meiosis I. Cells without tetrads but with an even number of chromosomes may be examples of mitosis or meiosis II. If each chromosome is clearly different in form from all others, the cell is in meiosis II. If each chromosome looks like one other, the cell may be dividing mitotically. Cells without tetrads and with an odd number of chromosomes are in meiosis II.
3. In haplontic life cycles, vegetative growth takes place when the cells are haploid. Cellular differentiation, but not meiosis, results in gametes that undergo fertilization. The spore that forms is diploid, and it enters meiosis to restore haploidy. In diplontic cycles, vegetative growth occurs in the diploid phase of the life cycle. Gametes are produced by meiosis, and fertilization reestablishes the diploid state.
4. Isogamous. If gametes are designated as + and −, it is likely that they are visually indistinguishable. Thus they are not oogamous.
5. Synapsis followed by crossing over shuffles the DNA between homologous chromosomes and can create chromosomes with different combinations of mutations. Genetic variability also results from the production of different combinations of homologs in the gametes of successive generations. This does not produce different genes, but rather different combinations of genes.

Chapter 16

Exercises

16.1 Activity exercise.

16.2 (*a*) p(12) = 1/36; (*b*) p(11) = 1/18; (*c*) p(7) = 1/6.

16.3 *TT* and *Tt* are tasters, *tt* is a nontaster.

16.4 The ratio is close to 3:1, so both parents are heterozygotes.

16.5 F_1 all wild-type; F_2 3/4 wild-type, 1/4 black.

16.6 Activity exercise.

16.7 She is heterozygous.

16.8 For $2^n = 4$, 4 kinds of gametes; for $2^n = 6$, 8 kinds; in general, 2^n kinds.

16.9 F_1 is all yellow, round; F_2 has the same 9:3:3:1 ratio as in the cross discussed in the text, showing that it makes no difference what combination of characteristics the parents have.

16.10 The woman is heterozygous, the man homozygous recessive, so half the children will have brown eyes, half blue.

16.11 3/4 brown hair, 1/4 blond. Therefore, 3/8 brown hair, brown eyes; 3/8 brown hair, blue eyes; 1/8 blond, brown eyes; 1/8 blond, blue eyes.

16.12 The usual ratio of 9:3:3:1 for wild-type: purple:ebony:purple, ebony.

16.13 $pr^+pr^+e^+e$

16.14 Activity exercise.

16.15 c^k, 40%; c^d, 20%; c^r, 5%; c^a, 0%.

16.16 The king determines the children's sex, not the queen.

16.17 Both are heterozygous for hair color; the mother is heterozygous for color-blindness.

16.18 (*a*) Heterozygous black. (*b*) The black and calico females must have had different fathers, one black and one yellow.

Multiple-Choice Questions

1. c	6. d
2. a	7. e
3. c	8. d
4. d	9. e
5. e	10. d

True-False Questions

1. True.
2. False. Independent assortment occurs *as a result of the independent arrangement of pairs of homologs during meiotic metaphase I.*
3. False. Segregation of alleles occurs as a result of *separation of homologous chromosomes following meiotic metaphase I.*
4. True.
5. False. If a particular gene is known to have more than two alleles, *only two are found in any individual.*
6. True.
7. False. If one observes 9:7 or 15:1 phenotypic ratios among F_2 members of standard genetic crosses, the traits under study probably result from the activity of *two pairs of alleles that interact epistatically.*
8. False. In a typical dihybrid cross (no linkage, true dominance, and recessiveness within each allelic pair) four *phenotypically* different classes will be formed.
9. False. In mammals, sex is chromosomally regulated so that the *females are homogametic, and the males are heterogametic.*

10. False. Since females carry two alleles for X-linked alleles while males have only one, recessive traits such as hemophilia and color blindness will be more frequently found in *males, since female heterozygotes will be phenotypically normal.*

Concept Questions

1. When used specifically to refer to a particular gene or allele, both terms refer to a particular region of DNA along a chromosome that governs the production of a particular polypeptide. The term *allele* refers to any single variant or mutant form of the region of DNA, while the term *gene* includes all the variants found in a population.
2. You hope it is true. With recessive alleles, what you see is what you get. If a spotted cow and a spotted bull are bred, it is likely that all of their offspring will be spotted.
3. Dominance refers to an expression relationship between alleles of one gene, whereas epistasis refers to an expression relationship between two separate genes.
4. Since $12 + 3 + 1 = 16$, a dihybrid cross ($AaBb \times AaBb$) is probable. If so, the phenotypic class that includes 12/16 of the offspring represents the sum of what would normally be the *A-B*-plus *A-bb* classes. If dominant allele *A* is present, neither *B* nor *b* can be expressed. *A* is therefore epistatic to either *B* or *b*.
5. Females inherit one X chromosome from each parent, so her father would be color blind and her mother would either be color blind or a carrier of the allele. Since the allele is relatively rare, one might assume the mother to be a carrier. However, the parents might have met at a club for color-blind people and been so compatible with their unique view of the world that they became a couple.

Chapter 17

Exercises

17.1 In third position of several codons, all four bases are equivalent; in several codons, the two purines or the two pyrimidines are equivalent.
17.2 Activity exercise.
17.3 *pyr tyr metB ara proC ala thi his proA val gal leu lac*, with the last marker linked to the first.
17.4 Cut both DNAs with Kpn; select cells by plating on kanamycin and replica plating onto spectinomycin.

Multiple-Choice Questions

1. b
2. a
3. e
4. d
5. c
6. d
7. a
8. e
9. d
10. b

True-False Questions

1. False. A DNA sequence leading from an initiation codon to a termination codon is called an open reading frame and *might be a gene.*
2. False. Most mutagens cause a mutation *by chemically changing bases* from DNA.
3. True.
4. False. The deletion or insertion of any number of nucleotide *pairs in a gene, other than a multiple of three,* results in a frameshift mutation.
5. True.
6. False. Both viruses and plasmids can exist as genetic elements inside cells *but only viruses can be independent* outside cells.
7. False. Episomes are genetic elements of bacteria that can only exist integrated into the bacterial chromosome *or as independent entities in the cytoplasm.*
8. False. When a prophage is excised from a bacterial chromosome, *it may take some* bacterial genes with it.
9. False. Each type of plasmid such as an R factor or virulence factor *is not restricted* to a particular species of bacterium and *commonly confers* its properties on other species of bacteria.
10. False. Restriction enzymes are *bacterial products that destroy viruses* by cutting up *viral DNA.*

Concept Questions

1. We can identify DNA sequences with initiation and termination codons that have the characteristics of genes, but we can only call them open reading frames (ORFs). Without specific experimental evidence, we cannot be sure how a DNA sequence is actually used and whether an ORF is a gene.
2. In recombination experiments, two different mutants are allowed to recombine their genomes to produce individuals with new genotypes; these experiments yield information about the frequency of recombination and are used for mapping mutations or genes. In complementation experiments, two mutants combine their genomes in the same cytoplasm so we can determine whether together they can produce a normal function that neither can produce alone; these experiments are used to establish the limits of genes by determining whether two mutations lie in the same gene.
3. The mutation rate for one gene is about 10^{-6} to 10^{-8} per generation. Even assuming the higher mutation rate, the probability that both mutations would revert to wild-type simultaneously in one cell is $(10^{-6})^2 = 10^{-12}$. Such an event is so improbable that we can dismiss it, and it is virtually impossible for it to occur in many cells simultaneously.
4. All three are relatively small genetic elements, pieces of double-stranded DNA that encode genes. Plasmids are circular DNA molecules that can exist independently in the cell, and some of them can transfer copies of themselves into other cells. Temperate phage are viruses, so they can exist as extracellular particles (virions). Episomes are elements that can exist either as independent molecules in the cytoplasm or in an integrated state in the bacterial chromosome; both plasmids and temperate phage can be episomes.
5. Antibiotics cannot change mutation rates because they are not mutagens. However, they act as selective agents that eliminate sensitive bacteria and only allow resistant bacteria to grow. In any case, most of the resistant bacteria we now find result from plasmids that carry genes for resistance and are transferred from cell to cell.

Chapter 18

Exercises

18.1

0.1	0.1	200	200
[given in table]			
100	0.1	200	100
0.1	0.1	0.1	0.1
0.1	0.1	0.1	0.1

18.2 C^- is dominant to C^+ because it encodes a functional protein.
18.3. #1 has a defective operator for *F-G-H.* #2 has a defective operator for *A-B.* #3 probably cannot make repressor protein.
18.4 It has a mutation in the site where CAP binds in the *lac* promoter region.
18.5 TTGACA
18.6 They have promoters whose sequences give them the right strength to be transcribed at appropriate rates.
18.7 The system depends on promoters of different types. The *reg* gene apparently changes the RNA polymerase so it recognizes promoters of late genes.
18.8 The essential feature of any circuit is that protein 1 (from gene 1) activates one or more new genes (2, 3, etc.). At least one of the latter genes then activates a third set of genes; meanwhile, one protein of the second set (2, 3, . . .) feeds back to inactivate gene 1; at each stage, one protein feeds ahead to activate new genes while another protein feeds back to inactivate previous genes.
18.9 Choose a sequence of amino acids encoded by a restricted set of codons, so the sequence of the probe will be as close as possible to the actual sequence in the gene.

Multiple-Choice Questions

1. d
2. a
3. d
4. e
5. b
6. a
7. a
8. e
9. b
10. d

True-False Questions

1. False. An operator can act only in the *cis position.*
2. False. The *lacI* gene can act in either the *cis* or *trans* position because it *produces a diffusible product.*
3. False. The three structural genes of the *lac* operon are regulated in a coordinated way because the *structural genes are transcribed as a single mRNA.*
4. True.
5. False. Alarmones are *ligands that bind to regulatory proteins* that alter the transcriptional activity of RNA polymerase.
6. False. Chromosomal regions that stain heavily are called heterochromatic, and are *indicative of a low level of transcription.*
7. False. In order to replicate DNA by PCR, one must combine a DNA primer, all four nucleoside triphosphates, *Taq polymerase,* and the target DNA in a thermal cycler.
8. True.
9. False. Transgenic plants have been constructed using a *plasmid* called Ti as the vector.
10. True.

Concept Questions

1. As long as glucose is present, the cAMP level in the cells remains low and the *lac* operon cannot be induced, so the cells only use glucose. When they run out of glucose, the cAMP level in the cells rises, the *lac* operon becomes induced, and the cells start to metabolize lactose.
2. Because procaryotes are unicellular organisms and must continuously adjust to environmental changes, one would expect th[illegible] have evolved primarily reversible pathways. The much more specialized cells of multicellular eucaryotes exhibit a division of labor. As a result, regulatory pathways for those tissue-specific proteins not produced in a differentiated cell type are likely to be irreversible.

3. All are protein-binding regions of DNA. Promoters lie immediately upstream of their particular gene or genes. Promoter regions bind to RNA polymerase, ensuring the correct orientation and start position for transcription. Enhancers and silencers lie more distant to the gene and can be either upstream or downstream. They bind to regulatory proteins, which in turn, bind to RNA polymerase. Enhancers enhance the binding between RNA polymerase and the promoter, while silencers inhibit the action of RNA polymerase.
4. While steroid hormones are characteristically found in animals, both work in a similar manner. After each binds to its specific receptor, the complex binds to a regulatory site on DNA, thereby altering the transcriptional activity of RNA polymerase.
5. Genomic DNA from eucaryotes contains introns and exons. Bacterial cells lack a mechanism for splicing out introns. Functional mRNA has already had the introns removed, so when it is used as a template for DNA, the DNA includes only exons.

Chapter 19

Exercises

19.1 (*a*)A; (*b*) A; (*c*) AB; (*d*) B; (*e*) B

19.2 (*a*) A, B, AB; (*b*) A, B, AB, O; (*c*) B, O; (*d*) A, B

19.3

A,O	A,B,O	A,B,AB	A,O
A,B,O	B,O	A,B,AB	B,O
A,B,AB	A,B,AB	A,B,AB	A,B
A,O	B,O	A,B	O

19.4 The man cannot be the father because he has no I^B allele.

19.5 The father has at least one allele for an antigen (Kell) which he has given to at least some of the children; the mother became immunized against this allele during an early pregnancy, and in a later pregnancy her antibodies react against the baby's blood cells, just as in the Rh case.

19.6 The daughter's one X chromosome carries an allele for color-blindness, which is expressed as it would be in a man.

19.7 The recessive allele is expressed in all cells in males and in patches of cells in heterozygous females.

Multiple-Choice Questions

1. d
2. e
3. e
4. d
5. e
6. e
7. a
8. b
9. c
10. d

True-False Questions

1. True.
2. False. Inherited traits that show up in each generation of an affected family are probably caused by *dominant* alleles.
3. False. The ABO blood group alleles are a good example of a *monogenic system involving three alleles.*
4. False. Rhesus antigen got its name when it was *found that both rhesus monkey and human blood contain an antigen that elicits antibody production in a rabbit.*
5. False. Each band in a chromosome stained with Giemsa stain or quinacrine mustard represents *several genes.*
6. True.
7. False. A Barr body can be formed by inactivation of *one X chromosome in a cell that has more than one.*
8. False. DNA fingerprinting relies on analyzing RFLPs found *in highly variable regions of DNA that are not located within structural genes.*
9. True.
10. False. *Because* the mutant CFTR protein is present and functions as a chloride channel in the plasma membrane of all body cells, the transfer and expression of the normal allele in lung cells *may alleviate respiratory symptoms but not other symptoms.*

Concept Questions

1. If an X-linked allele is rare, the trait appears in many more males than females. As the frequency of the allele rises in a population, the proportion of affected females will also rise. So pedigrees of affected families will start to look much more like those for an autosomal recessive trait.
2. Since embryonic cells randomly inactivate one or the other X chromosome, the same randomness is expressed in adult cell populations. One would expect a tiny proportion of female heterozygotes to have normal clotting times and an equally tiny proportion to exhibit the disease. The rest would fall somewhere along the spectrum between the two extremes.
3. Evidently the normal number of functional X chromosomes in humans is one, while the normal number for most of the other pairs is two. Males and XO females have only one X chromosome. Dosage compensation via Barr-body formation inactivates any additional X chromosomes in normal XX females or in aneuploids such as XXY males or XXX females.
4. Depending on the size and genetic homogeneity of a given population, it is statistically possible that two or more individuals have the same VNTR pattern. Thus a suspect with a matching VNTR pattern might be guilty or innocent. However, if the suspect's pattern does not match that prepared from the evidence, the suspect could not have committed the crime.
5. First, the genetic code is redundant, so knowing the correct order of amino acids does not unambiguously predict the correct order of nucleotides. Second, the amino acid sequence does not yield information about the number and arrangement of introns and exons within the DNA of the gene. Finally, knowing the amino acid sequence would not mitigate the difficulties encountered in finding a proper vector.

Chapter 20

Exercises

20.1 The outer surface of the blastula will become the inner surface facing the lumen of the archenteron.

20.2 The alternative is for the blastopore to become the mouth and for a second opening to break through for the anus.

20.3 Enlarge the vegetal hemisphere of the frog egg enormously to make all the yolk of the bird egg. Then the blastula cannot encompass all this yolk and must be only a small patch on its surface. The blastopore is converted into a central streak; gastrulation occurs through lateral migration of cells rather than migration into a central cavity. The gut tube only gradually closes as the yolk is absorbed.

20.4 A mass with P cells enclosing Q cells.

20.5 As two separate masses, not associated with each other.

20.6 Number of combinations = number of α types times number of β types.

20.7 They have contact inhibition of movement but not of growth.

20.8 Indicates that tumor cells have the surface properties of perpetually dividing cells.

20.9 Have cells in the middle elongate by growth of microtubules while cells toward the ends contract with microfilaments. The lengths of microtubules and degree of contraction by microfilaments can be graded as needed to get the right shape.

Multiple-Choice Questions

1. a
2. c
3. e
4. d
5. b
6. a
7. e
8. c
9. d
10. e

True-False Questions

1. False. The sea urchin oocyte is covered *first by its plasma membrane, then by the vitelline membrane, and finally by the jelly layer.*
2. False. The correct order of early embryonic stages is *morula, blastula, gastrula.*
3. False. During gastrulation, the blastocoel is *obliterated as the archenteron forms.*
4. True.
5. False. *Mesodermal cells tend to exhibit* the wandering activity typical of mesenchyme, whereas *ectodermal and endodermal cells generally form flat epithelial sheets.*
6. False. The extraembryonic membranes include the amnion, chorion, and allantoic membrane, *but not the vitelline membrane.*
7. True.
8. False. Adhesive proteins such as fibronectin and laminin are *proteins within the intercellular matrix that bind to other proteins located within or on the plasma membrane of cells.*
9. True.
10. False. Although many factors have been implicated in the development of cancer, *environmental factors* have been clearly shown to be the most significant.

Concept Questions

1. Morphogenesis is the attainment of three-dimensional shape by an organ, body region, or whole organism. Usually, many cells are involved in any process of morphogenesis in animals. Differentiation is a cellular process in which the cytosol and its constituents become more specialized in structure and function as a result of gene activity. Thus differentiation occurs at the level of an individual cell.
2. Since amphibian cleavage is holoblastic, the yolk is enclosed within cells that constitute the vegetal hemisphere. Cleavage in birds is meroblastic, and the yolk is not included within blastomeres. Thus the bird embryo does not have a yolky vegetal hemisphere and a nonyolky animal hemisphere.

3. Ectoderm, mesoderm, and endoderm are the primary embryonic tissue layers that develop after gastrulation in all vertebrates and almost all animals. Epiblast and hypoblast are formed prior to gastrulation from the blastodisc of amniote embryos. Endoderm and mesoderm form after cells of the epiblast have moved to the interior through the primitive streak.
4. If the neural crest cells lose their N-CAMs and begin to synthesize integrins that enable them to crawl on fibronectin and laminin fibers in the extracellular matrix, they will become mesenchymal. They settle down and form clumps when they once again produce CAMs. Different CAMs, substrate adhesion molecules, and cytoplasmic specializations could result from the varying stimuli from different groups of surrounding cells.
5. Yes and no. Although we certainly can become infected with a virus carrying an oncogene or inherit one or a pair of already-mutated oncogenic alleles, the development of cancer is generally a multistep process in which both alleles at several genetic loci must mutate. Inheritance and/or infection can certainly initiate or hasten the process. On the other hand, it is also possible for a zygote to carry the entire series of mutations needed for cancer, and the fetal environment might include mutagens to accelerate the process as well.

Chapter 21

Exercises

21.1 Failure to get normal development doesn't prove that the transplanted nuclei are unable to sustain it—that they are not totipotent. You may simply be using a bad technique. To show that the nuclei are defective in some way, you would have to get some positive result, such as demonstrating that they lack certain genes.

21.2 The cell is determined but not yet differentiated.

21.3 Regulative development.

21.4 First each cell runs through a routine for becoming a pluripotent stem cell. Then something selects a subroutine for each one, and it goes through the first stage of becoming a particular committed cell, such as a red blood cell; we might imagine that more specific subroutines are selected for it as it becomes more and more like the mature, specialized cell.

21.5 There are soluble growth factors carried through the blood.

21.6 (*a*) Instructive regulation; (*b*) permissive regulation.

21.7 As gene *E* is turned on, gene *C* will be transcribed, producing Regulator$_2$, which then turns off genes *A* and *B*. Regulator$_1$ will no longer be made, so increasingly genes *C* and *D* will be expressed and *A* and *B* will be repressed. This can be a general model for turning one set of genes on for a while and then replacing them by another set of genes.

21.8 The principle is that the arrangement of cells around the developing egg determines the distribution of materials in the egg cytoplasm and therefore its course of development. The arrangement of cells around the egg in this case determines dextral or sinistral coiling.

Multiple-Choice Questions

1. d	6. c
2. e	7. e
3. c	8. a
4. e	9. e
5. d	10. e

True-False Questions

1. True.
2. False. The cloning experiments performed by Briggs and King involved the transplantation of *nuclei* from older amphibian embryos into zygotes.
3. True.
4. False. Mosaic development and regulative development are two *ends of a spectrum seen in animal embryology, and all embryos develop somewhere along this spectrum.*
5. True.
6. False. If mesodermal tissue A always induces the formation of the same structure when grafted under ectodermal tissue anywhere on the body, then the ectodermal tissue must be competent everywhere, and the interaction is *instructive.*
7. False. As a *Drosophila* larva feeds and grows, *it produces imaginal discs from which adult structures will develop.*
8. False. In fruit flies, *the bicoid and nanos gene products activate the gap genes.*
9. False. There is *great* similarity in the DNA base sequence of homeotic genes from one organism to the next.
10. True.

Concept Questions

1. Determination occurs at the genetic level as genes are irreversibly turned on or off. Differentiation occurs at the cellular, and especially the cytoplasmic, level as a cell acquires a distinctive set of organelles that enable it to perform its specific functions. Thus differentiated cells are determined, but a determined cell may not yet have differentiated.
2. Although most cells that are capable of mitosis are relatively differentiated, stem cells are also capable of progressive determination to form two or more differentiated cell types. For example, blood stem cells give rise to several different blood cell lines, including red blood cells, lymphocytes, and other white blood cells. On the other hand, a skin cell or a pancreatic cell that is capable of division gives rise only to cells of the identical cell type.
3. During regulative development, the fate of a cell can be changed by influences from neighboring cells. Thus cells are subject to inductive forces. In mosaic development, each cell is preprogrammed by the determinants in its cytoplasm, and factors such as position or neighboring cells have little or no effect on the cell's developmental fate.
4. In Figure 21.6, presumptive epidermal and presumptive neural ectoderm were switched, while in Figure 21.12, the dorsal lip was transplanted. The dorsal lip acted as a second Spemann-Mangold organizer. The extra organizer developed into chordamesoderm, which induced the neighboring epidermal cells to produce an extra body axis. In Figure 21.6, each embryo contained only one organizer, so formed only one body axis.
5. Maternal *bicoid* and *nanos* genes are transcribed in different nurse cells that surround the egg during its maturation, and the mRNA so produced is translated into protein within the embryo. *Nanos* and *bicoid* gene products regulate the action of other sets of embryonic genes. The site of high Bicoid protein concentration is responsible for turning on sequential sets of genes that lead to head production. The site of concentrated Nanos protein acts similarly to produce the tail end of the organism.

Chapter 22

Exercises

22.1 *Phaecus graellsii*, because the earliest name has priority.

22.2 Kingdom, phylum, class, order, family, genus, species.

22.3 B D C A

Multiple-Choice Questions

1. c	6. e
2. c	7. d
3. e	8. e
4. d	9. a
5. d	10. e

True-False Questions

1. False. The evolution of an ancestral mammalian species into all the mammals we see today is an example of evolutionary *divergence.*
2. False. The difference in proportions of blood groups A, B, AB, and O among human populations is an example of *microevolution.*
3. True.
4. False. *Data on breeding patterns* are the most important evidence in defining species according to the biological species concept. Or: Phylogenetic data are the most important evidence in defining species according to the *phylogenetic* species concept.
5. True.
6. True.
7. False. Modern humans are included in a *monophyletic group.*
8. False. The lower the melting point of hybrid DNA, the *less* closely related are the parental species.
9. False. A binomial includes the genus name and the *trivial* name.
10. True.

Concept Questions

1. Phylogeny is the description of the evolutionary history of a species, whereas taxonomy is the identification and naming of groups of organisms. Although modern taxonomy is based on phylogenetic relationships, it is quite possible to base a taxonomy on other basic principles that need not even include the concept of evolution.
2. To reverse an evolutionary pathway, a reversal of environmental conditions as well as back-mutation of genes would have to occur simultaneously. Either one is a statistically improbable event. The probability of the simultaneous occurrence of both events would be the even less likely product of the probability of each event alone.

3. Evolution by means of natural selection is either adaptive in the short term or the result of chance events. Since the environment does not necessarily change linearly and progressively in the long term and since chance is not progressive, neither is evolutionary change.
4. Wheels on axles would be a shared primitive characteristic of all autos. Doors without running boards would be a shared derived trait of modern cars.
5. Essentially, the three concepts are historical responses to deficiencies in previous ideas. Each has incorporated data from advances in technology. The morphological species, the oldest of the three ideas of a species, was determined primarily by anatomical and embryological study. The biological species gives predominance to the species as an interbreeding, sexually reproducing unit, with all members sharing the same gene pool. A phylogenetic species, the newest of the three definitions of a species, relies less on information about interbreeding. The method uses morphological similarity and data from DNA studies to place organisms into the smallest morphological group that is also monophyletic.

Chapter 23

Exercises

23.1 Two reds, 0.01; two blonds, 0.81; mixed, $0.09 + 0.09 = 0.18$.

23.2 $p = 0.6, q = 0.4$

23.3 $p = 0.625, q = 0.375$

23.4 At equilibrium, $MM = 390$, $MN = 469$, $NN = 141$; the original population was not at equilibrium.

23.5 $p = 0.35, q = 0.65$. The population is not at equilibrium, since that would be $RR = 123$, $Rr = 455$, and $rr = 422$.

23.6 $q = 0.01$

23.7 In Japan, $q = 0.016$; in Ireland, $q = 0.065$.

23.8 $q = 0.316$, so $p = 0.684$. Since there are only 900 individuals instead of the expected 1,000, the population is reduced by 10%, and $s = 0.1$.

23.9 The species might become divided into breeding lines of different sizes, which could eventually become distinct species.

23.10 Heterozygotes don't always have an advantage over homozygotes, and heterozygosity is not the same as heterosis. It is hard to eliminate a recessive allele from a population just because selection only affects the minority of homozygous recessive individuals, whereas most of the recessive alleles are carried by heterozygotes, and this will be true even if they have the same fitness as homozygous dominants or even slightly reduced fitness.

23.11 It is advantageous for the finch population to carry genes for both large and small bills, since each of them has a higher fitness than the other at some time; selection will keep the species polymorphic.

Multiple-Choice Questions

1. c
2. d
3. e
4. b
5. e
6. a
7. b
8. d
9. e
10. d

True-False Questions

1. False. *In the absence of mutation and selection,* Mendelian populations do not undergo evolutionary change.
2. True.
3. False. If a population is at Hardy-Weinberg equilibrium for one locus, *it may be undergoing* evolution at any other locus.
4. False. If the frequency of a given allele in a population decreases by 5 percent in one generation, *the fitness of that allele is 0.95, or the coefficient of selection is 0.05.*
5. False. According to the Fundamental Theorem of Natural Selection, a species can change faster if its *genetic* variance increases.
6. False. A species occupying a homogeneous environment is *less* likely to be polymorphic than a species occupying a heterogeneous environment.
7. True.
8. False. If the rate of forward mutation (from A_1 to A_2) is twice the rate of back-mutation (from A_2 to A_1), eventually *an equilibrium between the two will result.*
9. False. Translocations occur when a segment of a chromosome breaks off and attaches to *a nonhomologous chromosome.*
10. True.

Concept Questions

1. You can determine the allelic frequencies with more certainty when the heterozygotes are distinctive, because in this case you can empirically determine the numbers of all three genotypes. When one allele shows complete dominance, you can only infer the frequency of heterozygotes by assuming a Hardy-Weinberg equilibrium for the locus, an assumption that may not be justified.
2. Probabilities only apply to relatively large sets of events. In a large population where individuals select their mates at random, and in the absence of mutation and selection, the allelic frequencies for a gene locus will remain constant because the possible matings between individuals of the three genotypes will occur in proportion to the allelic frequencies. But in a small population, only a few matings will occur each generation, and just by chance they may occur at frequencies quite different from those predicted by probability. For instance, a slight excess of matings involving homozygotes for one allele could significantly increase the frequency of that allele.
3. This is an example of polymorphism just because the gene has two alleles and there are three genotypes with distinctive phenotypes. Any population carrying both alleles has a genetic load because some individuals will die each generation for strictly genetic reasons—having a genotype with low fitness. However, in malarial areas, this is a case of balanced polymorphism because the heterozygotes are more fit than either homozygote, and the genetic load on the population is different because both homozygotes have reduced fitness.
4. Consider just three possible explanations. The mutation might create a minor change in a protein's structure, as in a region that does not affect its active or binding site. The mutant protein might have a partly different shape but still interact at its active or binding site just like other allelic forms. The mutant protein might produce a distinctive phenotype that makes no difference in the life of the organism, such as the M and N human blood types or the color morphs of Eastern Screech-owls.
5. The continental species will probably contain several or many populations, each of which is well adapted to its own particular habitat. Since each population will exhibit polymorphisms, the species as a whole will be highly polymorphic. Presumably, fewer habitats are present on an isolated island, leading to fewer distinct populations and a lower overall rate of polymorphism.

Chapter 24

Exercise

24.1 As in the model of Figure 24.16, each species apparently contained some individuals whose genomes gave them a tendency to hybridize with the other species. These individuals put all their gametes into a genetic dead-end because Koopman selected them out, so soon the species had no more genes that tended to produce interbreeding.

Multiple-Choice Questions

1. c
2. b
3. c
4. a
5. c
6. d
7. c
8. e
9. b
10. b

True-False Questions

1. False. *Adaptive radiation* means division of a parent species into two or more species occupying different adaptive zones.
2. False. According to the Red Queen hypothesis, every species is racing as fast as possible genetically to *just stay adequately adapted* to its ecological niche.
3. False. A species in a fairly stable environment is shaped largely by *stabilizing selection.*
4. False. Since the half-life of ^{14}C is 5,570 years, determining the ratio of ^{14}C to ^{12}C is an accurate dating method for fossil remains that are *less than about 10,000 years old.*
5. True.
6. False. Reproductive isolating mechanisms are likely to evolve more *rapidly* in isolated populations than in contiguous populations.
7. False. If a group of semispecies become sympatric by expanding their ranges and continue to interbreed with one another, we would probably classify them as *subspecies of a species.*
8. False. Two closely related, sympatric species usually occupy *distinct niches and are distinct morphologically.*
9. False. *In large populations,* directional selection is the mechanism that promotes speciation fastest. Or: In small populations, genetic drift is the mechanism that promotes speciation fastest.
10. False. If a hybrid plant is infertile but viable, it could become a new, fertile species through *allopolyploidy.*

Concept Questions

1. Although both adaptation and environmental change go on continually, environmental change is often random and generally more rapid than adaptive change.

2. Stabilizing selection is most likely to be found among stable, well-adapted species that live in stable environments. Directional and/or disruptive selection signals a changing environment or the appearance of new habitats. Whether selection becomes directional or disruptive depends on the magnitude, kind, and pace of environmental change.
3. Even though the two populations are sympatric, they will become more and more reproductively isolated, until they eventually become distinct species.
4. The biological species concept is based on the question of reproductive isolation between populations. To apply this test, the populations must be in contact during their breeding season, so we can determine whether members of the two populations successfully mate with or cross-fertilize one another. Thus, by definition, they must be at least partially sympatric.
5. Heterochrony means differential growth of an organism, or one of its parts, relative to another organism. During speciation, one individual might experience a mutation in a regulatory gene that changes the timing of its expression, thus causing one part of the organism to grow for a longer time (or a shorter time) than it did previously. Consequently, that part becomes larger (or smaller) than the corresponding part of organisms in a sister population; if this change is advantageous, or at least neutral, the mutation will spread through the population, making all its members morphologically distinct.

Chapter 25

Exercise

25.1 Evergreens, which retain their leaves and don't have to produce new leaves each year in the face of limited nutrients.

Multiple-Choice Questions

1. d	6. b
2. c	7. a
3. e	8. c
4. b	9. c
5. d	10. e

True-False Questions

1. True.
2. False. The overall pattern of air circulation constitutes the *climate,* but the day-to-day condition of the atmosphere at any one place constitutes the *weather.*
3. False. The primary producers in reef communities are *photosynthetic bacteria and algae.*
4. False. Within the ocean, the *intertidal zones and coral reefs* are richer in energy, nutrients, and organisms than the *open ocean habitats.*
5. True.
6. False. Eutrophic lakes are nutrient-rich and *oxygen-poor,* while oligotrophic lakes are nutrient-poor and *oxygen-rich.*
7. False. More species live in *tropical forests* than in any other biome.
8. False. Reindeer, polar bears, dwarfed shrubs, and permafrost are all characteristic of the *tundra.*
9. True.
10. False. Because they were once united in the continent of *Gondwana,* the fauna and flora of *Australia, Africa, and South America* show striking similarities.

Concept Questions

1. The reefs themselves are composed of the calcified skeletons of living and once-living coral animals. If they run alongside and close to a landmass, they are called fringing reefs. If they are farther from a landmass, the term barrier reef is appropriate. An atoll is a circular reef that encloses an island or sunken island. The heterotrophic inhabitants include the coral themselves, as well as annelids, molluscs, echinoderms, and fishes. The heterotrophs of reefs are supported by a variety of producers—mainly phototrophic algae and cyanobacteria.
2. Temperate and polar regions could not sustain freshwater communities. Ice would sink to the bottom and, if the climate were sufficiently cold, all of the contained water would freeze. All organisms would be frozen, and most would not survive the ordeal. Although this scenario would not occur in shallow water regions of the tropics and subtropics, there would be little or no spring or fall overturn, thereby depriving the water of its means of nutrient and oxygen refreshment.
3. The constant warmth in tropical rain forests enables rapid breakdown and recycling of dead plants and other organic litter into new above-ground plants. Thus the richness is mainly above the ground, while the soil is poor. In contrast, the nighttime and wintertime cold in temperate forests slows enzymatic recycling of forest-floor litter into above-ground structures, leaving a richer soil below ground.
4. Coral reefs and tropical rain forests contain the most species diversity, resulting in both cases from constant, warm temperature. The polar regions and open ocean environments are the poorest; however, temperature plays a dominant role on land, while light and turbulence are more important in the open ocean.
5. Convergent evolution is the process whereby relatively unrelated ancestral groups of organisms evolve some degree of phenotypic similarity because the environment presents each group with similar conditions and limitations. Since adaptation to each niche is associated with particular phenotypes, analogous structures will tend to develop in different ancestral species.

Chapter 26

Exercises

26.1 Assumes the marked individuals mix uniformly with the population; if they do not, the population may be larger than measured. Assumes marked individuals behave the same as others; if they do not, the measurement may be off in unpredictable ways.

26.2 If 10 percent of the beetles are marked, the whole population is 2,000. Since every population of insects has its predators and they are always being eaten, just knowing the identity of one predator give us no reason to change our estimate of the population size.

26.3 10,500

26.4 (*a*) and (*c*) are growing exponentially, (*b*) is not.

26.5 The larger the rodent population, the more likely it is that a hawk will be able to find and catch one.

26.6 Presumably a ratio of 5 to 1 in the two areas.

26.7 If 4 μl each, a bird spends 2 hrs/day feeding, or 8 kJ. If 6 μl each, a bird spends 1.33 hrs/day feeding, or 5.32 kJ. The saving is 2.68 kJ. But if defense takes 4.16 kJ, it isn't worthwhile in this case.

Multiple-Choice Questions

1. d	6. c
2. b	7. a
3. a	8. e
4. b	9. a
5. d	10. b

True-False Questions

1. False. The terms geographic range and habitat *cannot* be used interchangeably because *a single geographic range generally contains several or many habitats.*
2. False. Uniform distribution of a population within a habitat generally results from *negative interactions between members of the population.*
3. False. In mark-recapture experiments, the most critical value is the proportion of *recaptured individuals that are marked.*
4. True.
5. True.
6. False. Intraspecific competition can be either exploitative or *interfering.*
7. True.
8. False. The logistic curve *approaches the carrying capacity of its environment.*
9. False. The most successful populations are those that maximize the number of offspring *that survive to reproductive age.*
10. True.

Concept Questions

1. Yes, although it may be interspecific, intraspecific, or both. Competition results from limitation in quantity or quality of an essential environmental resource such as unpolluted space, nutrients, or mates. As long as the presumptive limiting factor is plentiful, populations will not be reproductively constrained. Once constraints exist, competition sets in, and natural selection ensues.
2. Suppose there are genetic mechanisms that maintain uniform population size. Since mutation is random, mutations will eventually occur that result in one or more individuals ignoring the selection process and out-reproducing the others. As a result, the genes for ignoring the group processes will increase, and group selection will decline.
3. The model of an ideal free distribution applies to organisms that are free to move until they find the optimum site or position for maximizing resources. Territoriality is found in species where the males (usually) compete, claim, and defend individual areas that serve as locations for mating, rearing of young, and/or feeding. Thus the ideal free distribution is a strategy for maximizing individual survival, while territoriality is a strategy for regulating population size. Both strategies could be used at different times by the same species. A species that was territorial during the mating season might freely distribute itself at other times.
4. Such populations will continue to grow as long as there are more individuals of prereproductive age than of reproductive age. Only after the prereproductive cohort moves through its reproductive years can growth be restrained.

5. By most structural measures, humans have all the attributes of K-selected, equilibrium species. We are large, long-lived, bear few young, and have a long immature period. However, human intelligence has given life to intellectual and technological revolutions that have pushed population growth toward the type seen in r-selected species.

Chapter 27

Exercises

27.1 Clusters of species that occur together repeatedly.

27.2 The snails show niche differentiation, with mouths adapted for eating food of different sizes. In this way, they divide the food resources so as to reduce competition with one another.

27.3 1.73 million kcal/acre

27.4 2.36 million kcal/acre. Clearly, not enough energy.

27.5 0.45×2.36 million $= 1.06$ million; 10.6 million

27.6 Briefly, if predators have learned to avoid a certain pattern or if avoidance of the pattern has become part of their genetically determined behavior, any animals that evolve to fit that pattern are less likely to be killed, and they will have a selective advantage.

Multiple-Choice Questions

1.	b	**6.**	d
2.	a	**7.**	e
3.	e	**8.**	b
4.	d	**9.**	c
5.	c	**10.**	c

True-False Questions

1. False. Tropical communities generally contain more individuals *and more individual species* overall than temperate communities.
2. False. A keystone species is the *species that plays the most critical role in a community.*
3. False. A species's *fundamental niche* is generally more inclusive than its *realized niche* because the fundamental niche does not take interspecific competitors into account.
4. True.
5. True.
6. False. *Both* niche differentiation *and* reproductive isolation are required for speciation.
7. False. Allomones function as *interspecific weapons.*
8. False. Production of antibiotics by some fungi and bacteria is an adaptation to *inhibit the growth of competing microorganisms.*
9. False. Predation is a *density-dependent* regulator for population size because as the prey population decreases, more refuge space becomes available to them, and their predators tend to hunt elsewhere.
10. True.

Concept Questions

1. Niche differentiation depends on the ability of each competing species to survive within a part of its former niche. This can occur either by the evolution of differentiating adaptations or by the ability to tolerate and thrive in more borderline conditions. Both occur more rapidly in species with high levels of genetic variability since there are more genetic possibilities to select from.
2. Both predation and parasitism are advantageous for one member of the interacting pair and harmful for the other. They differ in several other ways. First, predators generally act rapidly to kill their prey, while parasites may kill their host rapidly, slowly, or not at all. Second, predators are generally within the same size scale as their prey, while parasites are usually much smaller, often microscopic, in size. Third, parasites generally live on or within their host, getting both a place to live and nourishment; predators and their prey are generally both free-living within the larger community. Fourth, motility is implied in the predator-prey relationship. We commonly use the terms to refer to those animals that hunt and those that are hunted. Finally, many parasites are infectious and are passed from one host to the next, while predators usually are not.
3. It attempts to account for a fluctuating equilibrium between the size of the populations of predator and prey species based on negative feedback. As the number of predators increases relative to prey, less food is available for the predators, and some die of starvation. As the population of predators falls, more prey survive. This enhances the success of the predators, and their numbers rebound. The cycles of predators and prey are displaced relative to one another, since the predator cycle constitutes a response to the prey cycle.
4. Cryptic coloration is an adaptation whereby the prey avoids detection by blending into the background and becoming invisible. In contrast, aposematic coloration advertises the presence of the prey while also warning the predator of the ill effects of eating the prey. Both rely on different attributes of the predator. For example, cryptic coloration is more successful for predators that hunt visually rather than by smell. Aposematic coloration depends on the ability of the predator to learn by experience that particular colors and/or patterns are not tasty. Additionally, Batesian mimicry is more successful if the number of mimics is lower than the number of mimicked organisms.
5. Both kinds of mimicry utilize convergent evolutionary processes, and as a result share the same appearance as another unrelated species. Batesian mimics tell lies while Müllerian mimics tell the truth. The Batesian mimic is not dangerous, but looks dangerous. The Müllerian mimic both looks and is dangerous.

Chapter 28

Exercises

28.1 3.15×10^{21} J/yr

28.2 About 2×10^{17} g

28.3 About 0.67 percent

28.4 About 1,000 people

28.5 $40{,}000/250 = 160$ days

28.6 Tropical rain forest, 20.5 yr; temperate deciduous forest, 25 yr; desert, 7.8 yr.

Multiple-Choice Questions

1.	d	**6.**	c
2.	a	**7.**	a
3.	b	**8.**	e
4.	e	**9.**	b
5.	e	**10.**	d

True-False Questions

1. False. Matter cycles through the biosphere, while energy flows in a one-way path, entering as *light energy* and leaving as heat.
2. False. The biomass of the secondary carnivores in an ecosystem is *about 10 percent* of the biomass of the primary carnivores.
3. False. In the stages of succession, the ratio of photosynthesis to respiration *falls.*
4. True.
5. False. Bacteria as well as some industrial processes fix nitrogen by using *molecular nitrogen to generate ammonia.*
6. True.
7. True.
8. False. The necromass within a community is recycled by the cooperative activities of *bacteria, fungi, and animals.*
9. False. Older communities are characterized by *lower levels of productivity* as a result of an *increasingly woody biomass.*
10. True.

Concept Questions

1. Solar energy causes water to evaporate from the marine and freshwater basins; precipitation returns water to its basins. This cycle would occur regardless of living things. However, organisms effect changes in rate and distribution. Plants retard runoff, all organisms slow the overall rate of the cycle, evaporation is assisted by transpiration, metabolic uses divert water into organisms, and human activity currently seems to be at the root of an increase in global temperature, which may lead to glacial melting.
2. Molecular turnover is more rapid in aquatic communities. However, since aquatic ecosystems are more uniform in temperature and less subject to severe disruptions that would annihilate the resident species, species turnover is likely to be less rapid.
3. The early stages are likely to include a higher proportion of r-selected species. These organisms reproduce rapidly and exploit new environments. As the succession continues and the organisms get larger and live longer, the proportion of K-selected species increases.
4. Excess quantities of practically all water-soluble compounds are excreted and not stored. If lipid-soluble materials are applied to plants that are ingested by herbivores, they are likely to be retained, not excreted. Furthermore, they are not evenly distributed within the tissues of the herbivore, but rather are concentrated within the fatty tissue. When the herbivore is eaten by a carnivore, further concentration occurs.
5. After the forest is removed, transpiration rates fall and runoff increases. As a result, water does not percolate through the soil, new soil is not created, and the existing soil exits with the runoff. The surface that remains is rock hard and unsuitable for farming.

Chapter 29

Multiple-Choice Questions

1.	c	**6.**	a
2.	d	**7.**	c
3.	c	**8.**	e
4.	d	**9.**	e
5.	e	**10.**	c

True-False Questions

1. False. Haldane and Oparin proposed that the gases in Earth's original atmosphere were converted to an organic soup by the energy of *UV radiation and lightning.*

2. False. The first organisms were probably *unicellular heterotrophs.*
3. True.
4. False. Chemoheterotrophs synthesize sugar by ingesting *organic sources of carbon.*
5. True.
6. False. Like plants, all modern *cyanobacteria* contain and use chlorophyll *a.*
7. False. Bacterial cell walls are composed of *murein, which is made of amino acids and amino sugars.*
8. False. A protoplast is *a bacterial cell minus its cell wall.*
9. False. *Staphylococci* are clumps of spherical bacteria, and *streptococci* are chains of such cells.
10. False. Mycoplasmas are among the *smallest* bacteria and the actinomycetes among the *largest.*

Concept Questions

1. Nonpolar interfaces are essential to separate the interior of a protocell from the environment. Without a lipid-based membrane separating inner and outer watery solutions, cells could not have achieved the necessary concentration of macromolecules to support even a primitive metabolism. Furthermore, each genetic system must be enclosed in a cell so those with favorable genetic novelties can be selected, rather than sharing each novelty with other genetic systems.
2. Penicillin disrupts the construction of murein cell walls that are characteristic of bacteria. Since neither viruses nor eucaryotic cells have cell walls made of murein, penicillin has no effect on them.
3. The purpose of sterilization is to kill all microorganisms in or on an entity. The purpose of pasteurization is destruction of only those microorganisms that cause spoilage or are otherwise harmful. For example, cheese and yogurt may be made from pasteurized milk, but they contain microorganisms that give each product its characteristic flavor.
4. (*1*) The pathogen must be recovered from an affected organism. (*2*) It must be grown in pure culture. (*3*) When administered to healthy organisms, these organisms must contract the disease. (*4*) The pathogen must be recovered from these organisms. Stumbling blocks may accompany each of these steps. Small, intracellular pathogens, including some viruses, may be below the resolution of the available technology, thereby making conditions 1 and 4 difficult or impossible. For example, since it is not yet possible to isolate the HIV virus from an infected person, its presence is assumed if a person carries antibodies to the virus. Second, some healthy organisms may be genetically resistant to the pathogen, thereby making conditions 3 and 4 difficult to satisfy.
5. To survive, human pathogens must be adapted to a constant and relatively high body temperature. Since fishes, amphibians, and reptiles are not warm-blooded, it is doubtful that the bacteria could survive long enough to establish a reservoir of infection.

Chapter 30

Multiple-Choice Questions

1. c	**6.** a
2. e	**7.** e
3. d	**8.** c
4. c	**9.** e
5. b	**10.** c

True-False Questions

1. False. Protozoans are either *osmiotrophs or holotrophs that are free-living or parasitic.*
2. True.
3. False. All photosynthetic eucaryotes produce glucose, *but they polymerize it into a variety of carbohydrates, including starch, glycogen, and laminarin.*
4. True.
5. True.
6. True.
7. False. Kinetochores were initially anchoring sites for *chromosomes* and were located on the inside of the *nuclear membrane.*
8. False. Endosymbiosis is the likely explanation for the evolution of *chloroplasts and mitochondria.*
9. True.
10. False. Many nonphotosynthetic protists, including the amoebas and unicellular flagellates, are believed to have evolved from *apochromatic chromist algae.*

Concept Questions

1. Phototrophic organisms photosynthesize, using the sun as their source of energy. All the rest require energy in organic form. Osmiotrophs, either saprophytes or parasites, are fundamentally absorbers of monomeric-sized nutrients. Holotrophs ingest nutrients that are in larger packets than osmiotrophs. Most holotrophs use motility to catch their prey. Most osmiotrophs stay put or grow toward an energy source.
2. While both have amoeboid and plasmodial stages, the fundamental form for true slime molds is a coenocytic plasmodium containing diploid nuclei, whereas the primary form for cellular slime molds is the haploid amoeba. When cellular slime molds form a pseudoplasmodium, it is composed of many uninucleate cells.
3. The nucleus is believed to have preceded the evolution of microtubules. The nuclear membrane is believed to have evolved by invagination of the plasma membrane. Next, the transition from bacterial fission to mitosis probably went through stages in which chromosomes were attached to a nuclear membrane and distributed to daughter cells by an elongation of the nucleus, assisted by a microtubular apparatus. Then, as the genome enlarged, the microtubular spindle developed within the nucleus and acted as a mechanism to divide genetic material between two daughter cells in a more organized manner.
4. First, some modern protozoans and animals carry algal symbionts. Second, there are molecular similarities between archaebacterial and eucaryotic cells. Third, both mitochondria and chloroplasts have their own genetic apparatus, which they use for protein synthesis, and this genetic apparatus more closely resembles bacterial systems than eucaryotic systems.
5. Development was in stages. The evidence suggests that a membrane-enclosed nucleus arose first, followed by the shift in position of kinetochores from the nuclear membrane to the chromosomes. The mitotic spindle then became intranuclear, and mitotic division ensued. Some eucaryotes (i.e., plants and photosynthetic protists) evolved further by endosymbiosis with a photosynthetic procaryote as well as a mitochondrion-like procaryote, and possibly a spirochete-like procaryote. Fungi have mitochondria but not chloroplasts, cilia, or flagella. Animals have ciliated and flagellated cells and mitochondria but not chloroplasts. Although different eucaryotic clades could have lost organelles, there is more evidence to suggest that differences in endosymbiotic acquisition were more significant.

Chapter 31

Exercise

31.1 Basidiospores of the rust grow into separate + and − tissues that produce spermatia, which bind to receptive hyphae much as haploid spermatia produced by the male algal gametophyte attach to hyphae of the female algal gametophyte. The dicaryotic aecium phase of the rust is unique, but the two distinct diplophases of the rust (uredium and telium) are comparable to the diplophases of the alga (carposporophyte and tetrasporophyte).

Multiple-Choice Questions

1. c	**6.** d
2. e	**7.** e
3. a	**8.** b
4. a	**9.** b
5. d	**10.** d

True-False Questions

1. False. The fundamental units of a typical basidiomycete *are hyphae, which form* an intertwined mat called a *mycelium.*
2. False. All members of the kingdom Fungi, *except the imperfect fungi,* reproduce both sexually and asexually.
3. False. In the dicaryophase of a fungal life cycle, each cell has *two haploid nuclei, one of each parental type.*
4. False. A meiospore is a *haploid cell resulting from meiosis.*
5. False. In the gametangia of zygomycetes, *haploid nuclei fuse into diploid nuclei that undergo meiosis and grow into haploid hyphae.*
6. True.
7. False. The fruiting body of a basidiomycete is generally composed of a *dicaryotic* mycelium.
8. True.
9. False. Asci in ascomycetes and *basidia* in basidiomycetes both perform the same function.
10. True.

Concept Questions

1. The fungi have cell walls of chitin, no trace of flagella at any stage, and use the AAA pathway of lysine synthesis. Molds in the kingdom Chromista have cell walls of cellulose, have flagellated zygospores, and use the DAP pathway of lysine synthesis.
2. Members of the two phyla have similar life cycles in which monocaryotic hyphae fuse and grow into dicaryotic hyphae. Then some fruiting bodies form on the dicaryotic hyphae where the two nuclei in a cell fuse into a diploid nucleus, which then undergoes meiosis to produce haploid spores. The principal difference is that ascospores are produced in closed sacs (asci), while basiodiospores are produced on clubs (basidia) on exposed surfaces.
3. In the ascomycete crozier that develops into an ascus and in the clamp connections of dicaryotic basidiomycete hyphae, the formation of the cell and subsequent nuclear migration are very similar to the formation of secondary pit connections in red algae.

4. In exchange for increased absorption of water and minerals from the soil, the plant provides the fungus with photosynthetically produced nutrients.
5. In zygomycetes, hyphae of opposite mating types produce gametangia in which gametes are located—in this case, only haploid nuclei. The gametangia fuse, and many nuclei fuse into diploid zygotes; these zygotes then undergo meiosis, and the resulting haploid cells grow into haploid hyphae. In ascomycetes, hyphae of opposite mating types form gametangia designated ascogonia and antheridia. A trichogyne from the ascogonium fuses with the antheridium, allowing nuclei of opposite types to mix, but in this case dicaryotic hyphae emerge. These hyphae eventually develop asci in which nuclei fuse into diploid nuclei, which undergo meiosis and produce haploid ascospores.

Chapter 32

Exercises

32.1 (*a*) The dye will diffuse 10 times as fast through an area of 1 cm^2 as through 0.1 cm^2. (*b*) If the two points are 1 cm apart, the gradient will be 5 times greater than if they are 5 cm apart, and therefore diffusion will be 5 times as fast.

32.2 Doubling linear dimensions increases mass by a factor of 8; strength increases as d^2, so they would have to increase by $\sqrt{8} = 2.8$ times.

32.3 (*a*) Vocal pitch would increase to $12^2 \times 256 = 36{,}864$, way above the human hearing range. (*b*) $(1.4 \times 10^{10}$ cells$)/144 = 10^8$ cells, or 4×10^8, allowing for smaller size of the cells.

32.4 Using $w = 4.1m^{0.75}$, the Lilliputian BMR is 6.7 kcal/day. Per kilogram of body weight: humans, 25.7 kcal; Lilliputians, 167.5 kcal.

Multiple-Choice Questions

1. a
2. a
3. e
4. b
5. c
6. c
7. d
8. e
9. c

True-False Questions

1. False. A cephalized animal has sense organs concentrated at its *anterior* end.
2. False. Collenchyma and sclerenchyma are supporting tissues of vascular plants.
3. True.
4. False. Cells that develop a high turgor pressure resist the influx of additional *water.*
5. False. A poikilohydric organism is one whose *hydration* varies with changes in the environment.
6. False. Although plants excrete oxygen and animals excrete carbon dioxide, *only animals* generally excrete a nitrogenous waste.
7. False. The apoplast of plants is analogous to the *interstitial* fluid of animals.
8. False. Ectothermic organisms are also *poikilothermic.*
9. True.
10. False. The smaller an animal, the higher it can jump *relative to its size.*

Concept Questions

1. Animals contain muscle and nervous tissue, which are essential for organisms that often move quickly and must control their movements and other activities with rapidly transmitted signals. These tissues are not needed in plants, which rarely move except through growth and other slow processes.
2. Animals conduct respiration with oxygen as a terminal electron acceptor, producing water. Plants also respire, but they use water as a reactant in photosynthesis, so it is not a waste for them.
3. The lower limit of body size is set by their surface/volume ratio; the smallest animals need to eat almost continually to supply their energy needs. The upper limit is probably set by structural limitations on the strength of their constituents and by ecological limitations, since they must consume large amounts of food, even though their rate of consumption per unit of mass is minimal.
4. The terms are essentially synonymous in effect but not in fundamental meaning. "Poikilothermic" and "homeothermic" refer to the conditions of having quite variable or quite constant temperatures, respectively. "Ectothermic" and "endothermic" refer to the primary sources of heat—from the outside or the inside, respectively—reflecting the fact that the most obvious way for an organism to achieve homeothermy is by generating sufficient heat internally.
5. A one-unit cube has a volume of 1 cubic unit and a surface area of 6 square units; a 10-unit cube has a volume of 1,000 cubic units and a surface area of 600 square units. Similar calculations for cubes of various sizes will show easily that the volume increases much faster than the surface area.

Chapter 33

Exercises

33.1

Structure	*1n/2n*	*Derived From*	*Will Produce*
Gametophyte	1	Spore	Gametangia
Archegonium	1	Gametophyte	Egg
Egg	1	Archegonium	Zygote
Spore	1	Sporangium	Gametophyte
Antheridium	1	Gametophyte	Sperm
Sporophyte	2	Zygote	Sporangium
Gametangium	1	Gametophyte	Eggs or sperm
Sporangium	2	Sporophyte	Spores
Sperm	1	Antheridium	Zygote

33.2 A homosporous plant produces only one kind of spore, which develops into a gametophyte bearing both male and female gametangia. A heterosporous plant produces two kinds of spores; one of them develops into a macrogametophyte producing eggs, the other into a microgametophyte producing sperm.

33.3 Perhaps, but the ovule would contain a gametophyte that would produce both sperm and eggs, and it would probably fertilize itself unless some special mechanism evolved to prevent this. Since inbreeding produces genetic uniformity and outbreeding produces variability, such a plant probably would not compete well and probably would not survive.

33.4 (*a*) A megaspore mother cell; a microspore mother cell. (*b*) No, each spore first divides at least a couple of times mitotically to produce a small gametophyte, one cell of which becomes a gamete.

Multiple-Choice Questions

1. a
2. e
3. a
4. b
5. e
6. b
7. c
8. c
9. e
10. c

True-False Questions

1. False. Gametes are produced in *gametangia* that are part of a *gametophyte* plant.
2. False. Antheridia and archegonia are *gametangia* on the *gametophyte* plant.
3. False. Spore mother cells undergo meiosis and *develop into haploid spores that develop into gametophytes that produce haploid gametes.*
4. False. Trees, including specimens such as maple, oak, or spruce, are the *diploid* sporophyte portion of a life cycle.
5. True.
6. False. Within the plant kingdom, thallus formation is found *in algae and bryophytes.*
7. False. The first vascular plants had a central core of vascular tissue *made of separate spindle-shaped tracheids.*
8. False. Primitive vascular plants had dichotomously branched stems, *no leaves,* and sporangia.
9. True.
10. True.

Concept Questions

1. First, *Fritschiella* seems to exhibit a form of geotaxis and phototaxis because it produces downward-growing root-like rhizoids and upward-growing filaments. Second, unlike most green algae, these types produce a phragmoplast as part of the cytokinetic process, a structure that is also characteristic of the embryophytes. Third, like the embryophytes, the charophytes have multicellular gametangia.
2. Because the leaflike filaments are generally one cell thick, water from dew and rain can enter by osmosis. Some mosses have rootlike rhizoids where water and minerals are absorbed. Because the cells of the gametophyte contain chloroplasts, the haploid plant is phototrophic. The sporophytes also photosynthesize and obtain their nutrients from the gametophyte in which they are anchored.
3. The first vascular plants had a single central column of vascular tissue running through their stems, whereas modern angiosperm trees have many bundles of vascular tissue within a parenchyma. Additionally, the original xylem consisted only of separate spindle-shaped cells called tracheids, whereas modern xylem includes tracheids that fit together end-to-end, supporting fibers, and vessel elements. The tracheids communicate via pores, and the vessel elements unite to form unobstructed vessels. These changes result in much greater support for the shoot system of the plant as well as the more rapid upward flow of water and minerals in the vessels of the xylem.
4. In animals, meiosis occurs in diploid organs and results in haploid gametes—egg and sperm. In plants, meiosis occurs in diploid sporangia and results in haploid spores. These develop into gametophyte plants or at least gametophyte tissue. Gametes are mitotically produced by gametophytes.

5. The larger cones contain megasporangia; the small cones contain microsporangia. A scale on the larger, female cone contains two ovules, each initially containing four megaspores that have resulted from meiotic division of a diploid cell of the sporophyte. Three of the megaspores degenerate; the remaining megaspore divides by mitosis and develops into a megagametophyte containing about 2,000 cells. One or two archegonia develop within each megagametophyte, and each one produces an egg. After fertilization of the egg, the embryo sporophyte plus the megagametophyte plus its surrounding ovule constitute the seed.

Chapter 34

Exercises

34.1 Select certain proteins common to all, sequence them, compare their sequences. Compare sequences of distinctive genes, look for distinctive additions and deletions.

34.2 Implies some communication, coordination between regions of the spongocoel to create currents rather than mere mixing.

34.3 Structures very resistant to the particular pH of region where the parasite resides; protein sequences not easily digested by proteases characteristic of the region; mucus or other coverings resistant to digestion; also, production of enzymes that neutralize the host's digestive enzymes.

34.4 Extended by contraction of circular muscles around coelom; retracted by contraction of muscles attached inside the proboscis.

Multiple-Choice Questions

1.	a	**6.**	b
2.	d	**7.**	b
3.	e	**8.**	e
4.	c	**9.**	c
5.	b	**10.**	d

True-False Questions

1. False. Sponges, jellyfish, and flatworms are all classified as *animals.*
2. False. The two-layered planula larva is probably analogous to the *blastula* stage of eumetazoans.
3. True.
4. False. Cnidarians have surface cells called *cnidocytes* that contain prey-capturing structures known as nematocysts.
5. True.
6. False. In all probability, the original function of the coelom was *as a hydroskeleton.*
7. False. In most protostomes, the coelom develops from *a split within the mesoderm.*
8. True.
9. False. *Mesenteries are* formed from a double layer of mesodermal tissue *called a peritoneum.*
10. True.

Concept Questions

1. Sponges are more primitive than eumetazoans. They are colonial and have a flat larval stage. Cnidarians, among the most primitive metazoans, also have a flat, hollow planula larva. Finally, an organism named *Trichoplax* seems to be an intermediate form, resembling the hypothetical plakula. An organism such as *Trichoplax* could be ancestral to both radial and bilateral organisms. It forms a temporary digestive cavity suggestive of cnidarians.
2. Water flows into the central cavity of sponges through many incurrent pores and out through an excurrent pore. The cavity is really an internal extension of the surrounding environment from which each cell absorbs its nutrients and excretes its wastes. Although the central cavity does serve to circulate water, digestion occurs intracellularly. The gastrovascular cavity of hydrozoans has a single opening that serves as both mouth and anus. As in sponges, the cavity serves for circulation. However, the cells that line the cavity secrete enzymes into the cavity where digestion begins. Thus digestion is partly extracellular, and the cavity represents the beginning of a digestive system.
3. Since adult structures are constructed during embryonic stages, so too are the fundamental structural distinctions between phyla. Genes that regulate morphogenesis to create body patterns and body plans do so by altering the rate, length of time of activity, and/or sequence of activity of structural genes.
4. Aside from the aesthetics of separate openings for mouth and anus, the complete gut can specialize anatomically and physiologically. Specialized regions can take on one of the several functions of the system—ingestion, digestion, and absorption, among them. As a result, some aspects of digestion can become extracellular as enzymes are secreted into particular regions of the tube, and the process of digestion can become more efficient. With an incomplete gut, only part of digestion occurs in the gut cavity; digestion is completed in the cells that line the cavity.
5. Structures that are truly homologous can be seen to arise from the same embryonic tissues. In contrast, the coelom forms in different ways in embryos of different animal phyla. In protostomes, it forms as a schizocoel, by the splitting of a mesodermal mass. In deuterostomes, it forms as one or more outpocketings of the endodermal gut. In pseudocoelomates, it is a persistent blastocoel partly lined by mesoderm.

Chapter 35

Multiple-Choice Questions

1.	d	**6.**	b
2.	d	**7.**	e
3.	c	**8.**	d
4.	e	**9.**	a
5.	a	**10.**	d

True-False Questions

1. False. Of the major coelomate phyla, the only ones that are deuterostomes are echinoderms and *chordates.*
2. True.
3. False. Locomotion by means of numerous parapodia is typical of *polychaete annelids.*
4. False. Book lungs are found in *arachnids,* while external feathery gills are found in *crustaceans and molluscs.*
5. True.
6. False. The first vertebrates to have sturdy paired appendages and lungs for breathing air were the *lobe-fin fishes called crossopterygians.*
7. False. *All reptiles, birds, and mammals* produce an amniotic egg.
8. False. *Labyrinthodonts,* the first known amphibians, are descended from crossopterygian fishes.
9. True.
10. False. The feathers of the ancestral bird *Archaeopteryx* were probably used for *insulation.*

Concept Questions

1. The *Hox* genes act after other regulatory genes have determined the segmental organization of the embryo. Once the segments are established on an anterior-posterior axis, the *Hox* genes regulate the production of structures that are particular to each segment. In earthworms, most segments resemble each other closely, but in arthropods, different segments produce different kinds of appendages and show significant differences in structure.
2. In closed circulatory systems, the wall of a tubular vessel separates the circulating fluid (blood) from the tissue fluids. Systems are called open if, at least in particular regions, the circulating hemolymph is in direct contact with the tissues.
3. Arthropods have an exoskeleton composed of hard, chitinous plates. The exoskeleton extends over jointed legs as well as over the body of the animal. Bivalve molluscs, including clams and oysters, have a jointed pair of calcified plates as a shell. Some cephalopod molluscs, such as the chambered nautilus, have a spiral shell. Brachiopods have dorsal and ventral valves, or "shells", and echinoderms, such as sea urchins, are radially symmetric and covered with a thin casing studded with tube feet.
4. An amniotic egg contains several sac-like membranes produced by the embryo. These are, in turn, surrounded by shell membranes and sometimes a shell produced by the maternal organism. The embryonic sacs include a yolk sac, allantois, amnion, and chorion. Reptiles, birds, and mammals constitute the amniotes. Amniotic eggs enable an organism to live in a habitat that is far removed from a body of water, with an embryo that is contained and supported by the four embryonic sacs. The amnion encloses the embryo in a self-contained pond, the yolk sac digests the yolk, the allantois functions in respiration and waste disposal, and the chorion is protective.
5. Homeothermy is the ability to maintain an even body temperature by metabolic, as opposed to behavioral, means. Homeothermy evolved at least twice, once in mammals and once in birds or their reptilian ancestors.

Chapter 36

Multiple-Choice Questions

1.	b	**6.**	d
2.	a	**7.**	a
3.	e	**8.**	c
4.	b	**9.**	b
5.	b	**10.**	d

True-False Questions

1. False. The precision grip is found in *apes and humans,* but the power grip is *found in all primates.*
2. False. The pongids are *modern apes that coexist with humans.*
3. True.
4. False. Knuckle-walking and brachiation are *characteristic of apes.*
5. True.
6. True.
7. False. The primary distinctions between australopithecines and species within the genus *Homo* have to do with *tooth structure, brain size, and tool-making.*

8. False. *Homo erectus and their descendants* were probably capable of spoken language that was complex and symbolic.
9. True.
10. False. The transition from hunting and gathering to agriculture first took place among *modern Homo sapiens.*

Concept Questions

1. Excellent stereoscopic vision, opposable thumb, long arms, highly mobile shoulder joint, relatively small and light in weight (with a few exceptions), and few offspring.
2. Compared to australopithecines, dryopithecines were larger, not as fully bipedal, had a more posterior opening at the base of the skull so that the head could be carried in a more forward position, and had a smaller brain-to-body ratio. Presumably the dryopithecines were an offshoot of the main ape line. The main line led to modern chimps and gorillas, whereas the dryopithecines led directly or indirectly to australopithecines, which in turn were ancestral to humans.
3. Monkeys are generally smaller. Some have prehensile tails. They are generally arboreal and quadrupedal. Apes are usually larger, lack tails, are partly ground-dwelling, and tend toward bipedalism to some degree. The apes also have a higher brain size to body size ratio, more complex modes of communication, and a highly evolved social life.
4. Agricultural peoples have settlements. They can develop more highly structured defenses than can nomadic hunters and gatherers. They do not have to continuously transport their property or the juvenile members of the group, and barring climatic catastrophies, they have a regular food source. For all these reasons, mature females can give birth to children more frequently, and life expectancy lengthens.
5. The "out-of-Africa" model postulates the demise of any indigenous Neanderthal peoples, except for the ancestral African Neanderthal who migrated and evolved into all of the modern human races. Thus, in this view, widely distributed Neanderthal fossils were moderately related, having had as a common ancestor a *habilis*-type human. But, they were only distantly related (via a common *habilis*-type ancestor) to modern humans.

Chapter 37

Exercises

37.1 Glucose molecules enter the enzyme's active site from the cytosolic side and are polymerized into cellulose, which leaves on the exterior side.

37.2 As trees bend considerably in the wind, fibers parallel to the trunk axis are stretched, subjected to high tension. They must be flexible enough to bend and strong enough to not break under the tension and shearing forces.

37.3 Tracheids communicate through pores; tapered ends increase surface area, provide more area for pores, reduce resistance to flow from one tracheid to another.

37.4 Resistance to water flow is due primarily to diminished flow through pores at their ends; making a tracheid ten times longer than before removes 90 percent of the ends and reduces resistance.

37.5 Both increase their circumference by 314 μm, about 16 cells, but the branch increases by about 0.02 and the trunk by about 0.01 of the original diameter. So cells in the branch must divide twice as much in the circumferential direction to keep the cambium intact.

Multiple-Choice Questions

1. b	6. d
2. a	7. e
3. c	8. c
4. b	9. a
5. c	10. b

True-False Questions

1. False. The enzymes that catalyze the conversion of activated glucose to cellulose are found within the *plasma membrane.*
2. False. The tonoplast surrounds *vacuoles and regulates the ion concentration of the cytosol.*
3. False. Water passes from tracheid to tracheid through *small pits located where the walls of two tracheids overlap.*
4. True.
5. False. Periods of *dormancy* usually culminate in *germination.*
6. False. Some *woody plants are dicots or gymnosperms.*
7. False. Cells that have arisen by mitosis of *the root apical meristem* first enter the zone of elongation and then the zone of differentiation.
8. False. Root hairs are extensions of the *epidermis* that form in greatest number within the zone of *differentiation.*
9. True.
10. False. The vascular bundles of mature woody stems are *arranged in a fused ring,* thereby strengthening the trunk.

Concept Questions

1. The primary meristems are located at the apical ends of the plant. Cell division within primary meristems enables the root tip and shoot tip to elongate and to give rise to cells that differentiate within the root and shoot. Cambia are secondary meristems. The vascular cambium is found between xylem and phloem. The cork cambium is external to the secondary phloem of a woody plant. Cambia enable plants to increase in girth.
2. Adventitious roots are generally above-ground structures. They emerge from the shoot system, whereas taproots and fibrous roots emerge from an apical meristem below the shoot system and are generally below-ground structures. Fibrous roots are characteristic of monocots, taproots of conifers and dicots. Taproots have a single large cylinder, which may produce several small secondary roots. Fibrous roots consist of many, equal-sized rootlets that emanate from the pericycle.
3. Secondary xylem is wood; secondary phloem is included as part of the bark. The secondary xylem forms a structural (heartwood and sapwood) and functional (sapwood) part of the tree. The youngest layer of secondary phloem is conductive, but older rings of secondary phloem are sloughed off as secondary growth increases the girth of the tree.
4. The apical meristem gives rise to the protoderm, procambium, and ground meristem. In turn, protoderm differentiates into epidermis, procambium differentiates into primary vascular tissues, and ground meristem differentiates into cortex and pith. The secondary meristems constitute the vascular cambium and the cork cambium. Vascular cambium gives rise to secondary xylem and phloem. Cork cambium gives rise to the periderm.
5. Initials are mitotic stem cells in the vascular cambium. The fusiform initials give rise to axial secondary xylem and phloem. Vascular cambium also contains ray initials. In some plants, these act as storage sites in the stem; in other plants, the ray initials form vascular elements that conduct radially. Meristematic nodes along the stem mark the branch points where lateral stems grow. The tip of each lateral stem is formed by a central region of apical meristem surrounded by a lateral zone of rapidly dividing cells that give rise to leaf primordia. Within the leaf primordium, provascular tissues, as well as ground and protodermal meristems, grow and give rise to all three tissue systems of the leaf.

Chapter 38

Exercises

38.1 Ratio of 1 to 16 to 256 units.

38.2 Wide vessels permit rapid conductance of water, which is a prime consideration for many plants, but they are more likely to cavitate and become inactive. Narrow vessels conduct water more slowly, but they are more resistant to cavitation and thus serve as a kind of insurance to allow the plant to function even if large vessels have become inactive.

38.3 (*a*) $10^8\ \mu m^2/400\ \mu m^2 = 2.5 \times 10^5$ vessels. (*b*) 10 ml/2.5×10^5 vessels $= 4 \times 10^{-5}$ ml per hour. (*c*) If the concentration were 1 g/ml, 10 ml/hr would transport 10 g/hr; but the concentration is only 270 mg per ml, so the plant must transport $10/0.27 = 37$ ml/hr per cm^2.

38.4 The tree may not be able to supply enough organic nutrients to make all the fruits grow to the desired size and quality. If some fruits are removed, the phloem will adjust to supply nutrients only to the remaining fruits, so they develop fully.

Multiple-Choice Questions

1. e	6. d
2. c	7. c
3. c	8. b
4. b	9. c
5. a	10. a

True-False Questions

1. False. When K^+ enters guard cells, the change in the shape of these cells enables CO_2 to diffuse into the *plant.*
2. False. The Casparian strip divides the *symplast* into two separate regions.
3. False. The Casparian strip encircles the *endodermis.*
4. True.
5. False. Cavitation results from *breakage of the water column* within the *xylem.*
6. True.
7. False.The pressure-flow hypothesis predicts that if solutes were actively loaded into sieve tubes at a source and the surrounding cells were then made hypertonic to the sieve tube, bulk flow would *cease.*
8. True.
9. False. On hot, windy days, plants primarily cool themselves by *transpiration.*
10. True.

Concept Questions

1. Water rises to great heights as a result of root pressure and transpiration. Root pressure results from active transport of ions through the endodermis, which causes water to flow into the xylem vessels. Water rises in these vessels as water molecules adhere to the vessel walls and cohere to each other. Transpiration results in evaporation of water into the atmosphere. This creates a pull from above on the unbroken column of water beneath.
2. The waterproof nature of the Casparian strip prevents water from reaching the vascular bundles except by transport through the cytoplasm of endodermal cells. Since the cells of the endodermis actively transport ions, they effectively control the amount of water that will follow, producing an osmotic pressure that leads to root pressure.
3. No. Internal temperature is affected by the relative size of the boundary layer that surrounds the leaves. If shape is held constant, the smaller the leaf, the greater the ratio of surface to volume. The small leaf would lose heat and water faster than a similarly shaped larger leaf. All other conditions being equal, tropical plants would be expected to have smaller leaves than their temperate cousins.
4. Radiation is the primary method whereby plants gain heat from their surroundings; reradiation back into the environment is one primary method of heat loss. Transpirational evaporation and convection are both mechanisms for heat loss. Radiation and transpiration increase with increasing temperature, while heat loss by convection falls.
5. Metabolic efficiency for both types is optimal at the compensation point—i.e., the temperature at which CO_2 production by cell respiration equals CO_2 needs for photosynthesis. As temperature rises, the rate of respiration increases more than that of photosynthesis. Thus, in cool and sunny habitats, CO_2 is not produced as rapidly as it is needed. Conversely, the slower rates of photosynthesis in shady places are balanced by even slower rates of respiration, a condition that occurs only at relatively lower temperatures.

Chapter 39

Exercises

39.1 It is a short-day plant, because it only flowers when the nights become longer than 14 hours, which means the days have become shorter than 10 hours.

39.2 It is a long-day plant, because it only flowers when the dark period has been decreased to 13 hours, which means the light period has been increased to 11 hours.

39.3 A long-day plant flowers when the day is longer than a certain minimum; a short-day plant flowers when the day is shorter than a certain maximum. But those critical day lengths are not specified by these definitions.

39.4 In the tropics, day lengths change very little during the year, so plants are unlikely to use the day length as a signal for flowering.

39.5 GA will break the dormancy and start the buds growing again.

39.6 In late summer or early fall.

Multiple-Choice Questions

1. a
2. e
3. c
4. d
5. b
6. e
7. e
8. d
9. a
10. b

True-False Questions

1. False. The primary action of auxins and gibberellins is to *stimulate cell elongation.*
2. True.
3. False. Plants detect the pull of gravity by means of cells called *statocysts* that are found in the root cap and endodermis of the stem.
4. False. In addition to stimulating growth under certain conditions, *gibberellins* also induce flowering and seed germination.
5. False. Indoleacetic acid is more concentrated on the *dark* side of a shoot than on the *lighted* side.
6. True.
7. False. One site of abscisic acid synthesis is *in leaf chloroplasts.*
8. False. If a required period of darkness is interrupted midway by a brief exposure to red light, short-day plants will *be inhibited from flowering.*
9. False. During the day, plants convert *active phytochrome* (P_{fr}) *to its inactive form* (P_r).
10. False. In *Arabidopsis,* phytochromes *A, B, C, D, and E each have an active and inactive form.*

Concept Questions

1. The stimulus is sunlight. The shoot apical meristems are the receptors, producing auxins in response to increasing light levels. Cells on the dark side of the stem respond by initiating transcription of genes, which code for proteins that ultimately increase cellular turgor pressure and loosening of cell walls. This leads to elongation of cells on the dark side of the stem and a bending of the stem toward the light.
2. IAA and GA stimulate cell elongation; in the presence of IAA, kinetin stimulates cell division. The combination of IAA and kinetin also regulates differentiation of the root and shoot systems in ways that are determined by the relative concentrations of the two hormones. Kinetin also inhibits aging and decay.
3. No. In a negative feedback response, the product of the reaction system inhibits the reaction system itself. Auxin stimulates ethylene synthesis, and ethylene synthesis inhibits the cell-lengthening effect of auxin, but not the actual production of auxin.
4. The closer one gets to the poles, the greater the variation in day length between summer and winter. Given the short growing season and the fact that temperature plays a role in breaking dormancy, one would expect arctic species to be long-day plants—i.e., they would blossom when the nights shortened to the critical number of hours.
5. Etiolation occurs in the absence of light as plants grow rapidly to move their photosynthetic parts from the shade into the light. While in the shade, the inactive form of phytochrome is produced, etiolation continues, as does growth toward the light. When reaching sunlight, the inactive phytochrome is converted to its active form, etiolation ceases, and photosynthesis begins.

Chapter 40

Exercises

40.1 Assuming the tree was 75 percent water, its dry mass was 19.25 kg. If 95 percent of this was organic, the inorganic part was about 962 g. Van Helmont only found 60 g missing from the soil. We don't know where the rest came from, unless he was using mighty nutrient-rich water. We suspect sloppy technique.

40.2 Their skin carries minute amounts of salts, including chlorine, and if their bare skin touches a plant, it may transfer enough of the element to invalidate the experiment.

Multiple-Choice Questions

1. a
2. a
3. d
4. d
5. c
6. c
7. c
8. b
9. d
10. b

True-False Questions

1. True.
2. False. Essential minerals are generally absorbed in their most *oxidized form and converted to their reduced state* for use in metabolism.
3. False. After incorporation into organic molecules, essential micronutrients generally function as *metabolic regulators and enzyme cofactors.*
4. True.
5. False. As a result of nitrogen fixation, atmospheric nitrogen is *reduced* and converted into a form in which it can be assimilated by living cells.
6. True.
7. False. The process of *reducing* atmospheric nitrogen to *ammonia* is known as nitrogen fixation.
8. False. Plants absorb sulfur from the soil in the form of *sulfate* compounds.
9. False. In general, epiphytes have a *commensal* relationship with their host plants.
10. False. Ecological agricultural practices encourage the use of *slowly-dissolving organic humus.*

Concept Questions

1. Leaves consist largely of living cells that are active photosynthesizers. The rest of the shoot system has a mix of generally less active living cells and nonliving structural components.
2. A cation-exchange mechanism exists of the surface of clay-humus complexes in soil. Cations that are essential for plant nutrition are released by exchange with H^+ ions. If the exchange is too rapid, as would be the case with highly acid rainwater, essential cation release is too rapid. Under these conditions, the essential cations would be leached to soil levels that are too low to be useful for agricultural plants.
3. Although photosynthesis results in the fixation of carbon and hydrogen, nitrogen fixation requires metabolic activity by procaryotes. Without nitrogen fixation, proteins and nucleic acids could not be synthesized.
4. All are carried out by procaryotes and not by eucaryotes. The reduction of atmospheric N_2 to ammonia is called nitrogen fixation. The oxidation of ammonia to nitrite or nitrate is nitrification, and the conversion of nitrates back to N_2 is denitrification.

5. Ecological agriculture focuses on maintaining the fertility and structure of the soil by using organic fertilizers that decompose and release their nutrients slowly, by diversifying crops, and by avoiding energy-intensive farming methods. By contrast, monoculture means using energy-intensive farming practices to produce high, short-term yields of a single specialized crop.

Chapter 41

Exercises

41.1 Make some of the peptide labeled with a radioisotope and inject it into the bloodstreams of mice; then search for specific radioactive cells. Next, try to identify a protein on these cells to which the peptide binds specifically.

41.2 A lipid-soluble hormone, such as an estrogen.

41.3 Neurons are like FM radio; the nerve impulses have constant amplitude but vary in frequency.

41.4 Release of the neurotransmitter, binding of neurotransmitter to receptors, uptake or destruction of neurotransmitter after a signal has passed.

41.5 The connection between two brain centers is a tract, not a nerve.

41.6 The dorsal branch, where sensory fibers enter.

41.7 A ligand-gated channel opens when bound to a specific small molecule; a voltage-gated channel opens when the voltage across the membrane reaches a certain level.

41.8 At each action potential, the number of ions exchanged is a small fraction of the total number of ions responsible for the resting potential, so the neuron can fire many times even if the ATPases responsible for establishing the potential are poisoned.

41.9 Without myelin sheaths, nerve impulses will be transmitted slowly. (The scleroses are scars or hardenings that also interfere with transmission of impulses.)

41.10 In general, if an effect is abolished by curare but not by tetrodotoxin, it must result from an EPSP. If the effect is abolished by tetrodotoxin but not by curare, it must result from an action potential.

41.11 In the absence of iodine, thyroid hormones cannot be synthesized. These hormones feed back to inhibit release of TRH and thyrotropin, but in their absence thyrotropin continues to be released, stimulating growth of the thyroid tissues.

Multiple-Choice Questions

1.	e	**6.**	c
2.	b	**7.**	b
3.	a	**8.**	c
4.	b	**9.**	a
5.	e	**10.**	b

True-False Questions

1. False. Endocrine glands produce integrating hormones, while exocrine glands are *effectors.*
2. False. When blood sugar levels fall, the pancreas increases its secretion of *glucagon.*
3. False. The primary target cells of insulin and glucagon *are cells in the liver and muscles.*
4. True.
5. False. The ANS regulates *smooth and cardiac muscle,* while efferent neurons of the somatic system regulate *skeletal muscle.*
6. True.
7. True.
8. False. An action potential begins with the opening of *voltage-gated* sodium ion channels.
9. True.
10. False. The hormones of the *posterior pituitary and adrenal medulla* are really secretory products of neurosecretory cells.

Concept Questions

1. If the hormone can bind to two different receptors, each hormone-receptor complex may regulate different metabolic pathways or genes.
2. Endocrine cells secrete hormones into the surrounding extracellular fluid, and from there the secretion enters the bloodstream. It is transported throughout the body but affects only its target cells. Exocrine cells secrete materials through the duct onto a body surface such as the skin or the lining of the gut. Exocrine cells and multicellular exocrine glands are effector organs, while endocrine cells and endocrine glands are integrators.
3. IPSPs and EPSPs are graded, decremental changes in potential on postsynaptic membranes of dendrites and cell bodies. IPSPs are generated because ligand-gated channels decrease ion flow and thereby increase the potential across the membrane. EPSPs are generated when stimulatory neurotransmitters cause ligand-gated Na^+ channels to open, decreasing membrane potential. Action potentials are all-or-none, non-decremental changes in potential that occur along axons once the axon hillock reaches threshold.
4. Electricity results from the flow of electrons; nerve impulses result from the flow of ions, especially positively changed ions. Electrical conduction is decremental; nerve impulses along axons are nondecremental.
5. The posterior is an extension of the hypothalamus. Neurosecretory cells with cell bodies in the hypothalamus produce the hormones oxytocin and vasopressin. These hormones are released at the axon terminals of the neurosecretory cells, which lie within the posterior pituitary. Secretory cells wholly within the anterior pituitary produce hormones. The hypothalamus regulates the synthetic efforts of the anterior pituitary with a series of stimulatory and inhibitory releasing hormones. These enter a portal circulation in the hypothalamus and are carried in the blood to the anterior pituitary gland.

Chapter 42

Exercises

42.1 The "black box" stimulus entering the eye stimulates a single sensory neuron, which stimulates a single motor neuron that withdraws the tentacle. At the same time, let the sensory neuron stimulate a series of interneurons (thus creating a time-delay) whose final effect can be (1) to stimulate neurons that inhibit the motor neuron, (2) to stimulate other motor neurons that extend the tentacle, or both (1) and (2) together.

42.2 This is just like adrenaline (epinephrine), which acts as both a neurotransmitter and a hormone. The only requirement is to place receptors for the ligand on both neurons and other target cells.

42.3 The data suggest that a center in the medulla oblongata produces an irregular breathing rhythm, which is made regular by another center in the pons. A cut below the medulla oblongata interferes with signals to the breathing apparatus (the diaphragm); a cut above the medulla oblongata just eliminates the regularity.

42.4 Interference with a particular sensory function.

42.5 This is normally an inhibitory center that prevents expression of the fixed-action patterns of rage.

42.6 The size of a cortical area shows the number of controlling neurons. The small muscles controlling a small structure, such as a finger, may be subject to delicate controls by many neurons, while the large muscles of a large area of the body, such as the trunk, may be controlled much more coarsely by relatively few neurons.

42.7 This is a response to a need for additional water. It is functional for that need to be regulated via a single ligand, angiotensin II, that simultaneously induces a behavior (drinking water) and a physiological response (reducing urine output) that have the same effect.

42.8 If the peptides serve related functions, it is valuable to synthesize them all simultaneously in the same molar amounts. But then controls can be exerted on each enzyme that cuts the polyprotein to adjust the amounts of individual peptides.

Multiple-Choice Questions

1.	c	**6.**	b
2.	e	**7.**	c
3.	a	**8.**	b
4.	d	**9.**	d
5.	e	**10.**	d

True-False Questions

1. False. One general trend among multicellular animals has been the shift from a *peripheral to a more central* nervous system.
2. False. The entire central nervous sytem of vertebrates lies *dorsal* to the gut.
3. True.
4. True.
5. False. The spinothalamic tract carries nerve impulses from the *spinal cord to the thalamus in the brain.*
6. False. The *thalami integrate* sensory input, while the *hypothalamus produces* hormones that are secreted by the posterior pituitary gland.
7. True.
8. False. The *corpus callosum* connects left and right cerebral hemispheres.
9. False. Parkinson disease results from death of *neurons in the basal nuclei* that produce *dopamine.*
10. False. *Encephalins and endorphins* can bind to opiate receptors and act to mediate pain.

Concept Questions

1. First, conduction in a nerve net is slow because each stimulus travels throughout the entire net. Survival of large, multicellular animals depends on rapid and specific muscular responses, which in turn, depend on rapid, directed neural conduction. Second, the nerve net is essentially "hard-wired," but the nervous system needs flexibility in making connections to allow for learning.
2. CSF is a filtrate of blood produced by choroid plexuses that border the ventricles of the brain. It circulates both within and around the brain, between the brain and one of the meninges.

Since the CNS essentially floats within this inner pond, CSF insulates the CNS physically and thermally, and it also carries dissolved nutrients to neural tissue.

3. All sensory pathways, except for olfaction, terminate in the thalamus. Synaptic connections then feed from the thalami into the cerebral cortices, where conscious thought arises. Without thalamic input, the cortex has virtually no sensory information to stimulate motor responses.
4. Monoamine oxidase is an enzyme that destroys several neurotransmitters. Monoamine oxidase inhibitors prevent destruction of these neurotransmitters after they have activated the postsynaptic cell, making the affected synapses hyperactive.
5. Block some of the postsynaptic receptors. Block pathways in the presynaptic cell that produce the neurotransmitter. Increase the production of enzymes that destroy the neurotransmitter.

Chapter 43

Exercises

43.1 (*a*) Both photoreceptors and exteroceptors. (*b*) Both mechanoreceptors and proprioceptors.

43.2 They are chemoreceptors and interoceptors.

43.3 The regions of skin you are touching with one object are normally in places where they could only be touched by two distinct objects, so the nervous system ought to be programmed to report two objects.

43.4 $2000/333 = 6$, so you could detect a sound of 2006.

43.5 The molecule will fit into different proteins in different orientations by virtue of its ability to engage in different kinds of bonds.

43.6 Each receptor protein must be encoded by a distinct gene. A mutation in one gene will only inactivate the one protein and result in a very limited inability to smell certain compounds.

43.7 As you move your arm to approximately the desired position, the intrafusal fibers in the arm muscles initiate stretch reflexes, which contract the whole muscle until the tension in the intrafusal fibers is relieved. If the arm is then in the desired position, you leave it there; if not, you move it closer to the desired position, setting up another round of stretch reflexes and adjustments of muscle tensions, until the arm is finally where you want it.

43.8 Through proprioceptors that detect changes in tension and position in your muscles and joints.

43.9 In the usual signal-transduction pathway, a stimulus entails binding of a ligand to a membrane receptor, which activates a G-protein and causes production of cyclic AMP. The cell's resting potential is maintained without a flow of ions. In a photoreceptor, a membrane current is maintained by an influx of Na^+ ions through channels kept open by cyclic GMP. A light stimulus acts through G-proteins, converting cGMP to GMP and reducing the dark current.

43.10 Bleached photopigments must be gradually restored in an enzymatic process, which takes a few minutes.

43.11 Carrots contain carotenes, which can be converted into the carotenoid retinal for vision.

43.12 The pupillary reflex contracts your iris and closes the pupil.

43.13 If the light moves from right to left, it first stimulates excitatory cells that stimulate the receptors. But if it moves from left to right, it first stimulates inhibitory cells that prevent the receptors from firing.

43.14 If the H remains in one position in your visual field, the lines fall on two vertical rows of rods in the retina and one horizontal row. These rods stimulate series of off-center ganglion cells (because the H is black and its surroundings are white), which in turn stimulates two simple cells that see the vertical lines and another simple cell that sees the horizontal bar. If you move your eyes, a different group of simple cells is stimulated, but all these simple cells stimulate complex cells that continue to see the H as its position changes. Simultaneously, hypercomplex cells are being stimulated that detect the definite ends of the lines.

43.15 When visual cells are forced to fire continually as they "see" elements of the H, they tire and stop firing temporarily. Therefore, the part of the H each cell is registering will disappear for a few seconds. Normally this doesn't happen because the image is shifted continually from one set of cells to another.

Multiple-Choice Questions

1. c
2. d
3. e
4. e
5. b
6. d
7. a
8. e
9. d
10. c

True-False Questions

1. False. Generator potentials of receptors are most similar to *EPSPs of dendrites.*
2. False. The frequency of action potentials depends on the *strength of the stimulus.*
3. False. *The absolute threshold* of a receptor is equal to the minimum energy necessary to produce a response at least 50 percent of the time.
4. True.
5. True.
6. False. The most frequent cause of age-related deafness is loss of *cilia in the organ of Corti.*
7. True.
8. False. The vertebrate eye is structurally most similar to eyes found in some *cephalopod molluscs.*
9. False. As the light falling on rod cells increases, secretion of neurotransmitter *decreases,* which leads to *stimulation* of the postsynaptic neuron.
10. False. To clearly see the distant horizon, the ciliary muscle *relaxes,* enabling the lens to become *flatter.*

Concept Questions

1. Habituation takes place in the CNS; sensory adaptation takes place at the receptor. Although both are sensory filters that allow the organism to pay attention to meaningful and/or dangerous stimuli, habituation is associated with tonic receptors, while adaptation occurs with phasic receptors.
2. (*a*) Taste is a more generalized sense than smell. A particular kind of taste bud can be stimulated by a group of similar, but not identical, ligands. Olfactory receptors are more specific. (*b*) In humans, the olfactory epithelium contains cells that are simultaneously receptors and neurons; taste receptor cells synapse with neurons. (*c*) Olfactory neurons are able to regenerate; no other neurons, including those used for taste, can regenerate.
3. Although both receptors are found in nasal epithelium, the VNO is stimulated by pheromones, while the regular olfactory epithelium is stimulated by a wide variety of aromatic molecules produced by both animate and inanimate objects. Subjectively, the olfactory epithelium is associated with a sense of smell, while the VNO is not. Finally, the VNO apparently serves reproductive functions and behaviors primarily, whereas olfaction is not as functionally restricted.
4. The inner ear contains the cochlea and the labyrinth. The semicircular ducts, saccule, and utricle constitute the labyrinth. Although hair cells are the functional receptors in both the cochlea and labyrinth, they synapse with neurons that follow different routes into the brain, and they result in different perceptions. The stereocilia of the hair cells in the basilar membrane of the organ of Corti (cochlea) are activated by sound waves. Hair cells in the semicircular ducts are stimulated by dynamic motion of the head. In the saccule and utricle, hair cells are stimulated by static head position as otoliths cause deformation of the ciliated portion of the hair cells.
5. The fovea forms the region of clearest vision because light does not pass through a layer of overlying cells and because the rods and cones are very concentrated. Therefore a small change in receptor cell activation will produce a similar change in ganglion cell activation. Elsewhere on the retina, visual fields overlap much more because ganglion cells are connected to more overlapping fields of receptor cells.

Chapter 44

Exercises

44.1 $(40/5) \times 45$ kg $= 360$ kg

44.2

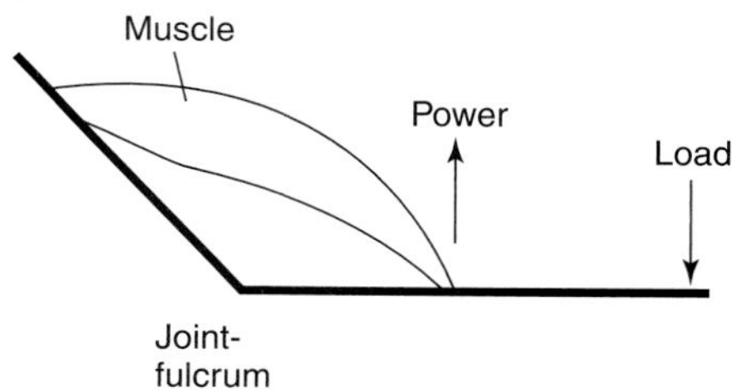

44.3 The centers of ossification in their long bones fuse, and the bones cannot grow longer.

44.4 15 cm $= 15 \times 10^4$ μm, so there are 7.5×10^4 sarcomeres in the longest fiber. If each sarcomere is shortened by 20 percent, the muscle would be 12 cm long.

44.5 Only in the T tubules; the ferritin cannot cross membranes to enter the sarcoplasmic reticulum or cytoplasm.

44.6 (*a*) Fast phasic oxidative fibers. (*b*) Its flight muscles are made of white fibers, which fatigue quickly.

Multiple-Choice Questions

1. b
2. c
3. b
4. d
5. c
6. e
7. e
8. e
9. c
10. e

True-False Questions

1. False. *The matrix* is the most abundant and functional constituent of connective tissue.
2. False. Bony matrix primarily consists of *hydroxyapatite.*
3. False. Most bone–muscle lever systems in our body act to produce a mechanical *disadvantage.*
4. False. Structurally, a bursa is most like *the synovial sac within synovial joints.*
5. False. Bones grow larger and larger because *osteoblasts* deposit additional bony matrix.
6. True.
7. True.
8. False. The fundamental unit of shortening in a skeletal muscle is the *sarcomere.*
9. False. The energy for muscle contraction comes from the splitting of ATP by an enzyme located *in the myosin heads.*
10. True.

Concept Questions

1. Antagonistic muscles work together to make movement smooth. For example, as you raise your forearm, your biceps contracts more and more while your triceps stretches more and more. Without antagonistic pairs of muscles, movements would be jerky.
2. If they were grouped, a gradual muscle contraction would begin in one section of the muscle and progress outward from there. The muscle would contract in a jerky fashion rather than smoothly as a whole.
3. The actin myofilaments are anchored at the Z lines; the Z lines form the boundaries of the sarcomeres. Moreover, although the muscle fiber contracts in an all-or-none fashion, each sarcomere contracts toward its own middle as actin and myosin filaments slide past each other.
4. Vertebrate skeletal muscle fibers are organized in motor units which contract in an all-or-none manner following stimulation. To increase the contraction of the entire muscle, the number of activated motor units is increased. Invertebrate skeletal muscles do not operate in an all-or-none fashion. Rather, they are innervated by stimulatory as well as inhibitory motor neurons and contract in a graded fashion that reflects the ratio of stimulation to inhibition.
5. The membranes that form the chambers of the sarcoplasmic reticulum pump calcium ions inward and then act as a storage depot for the mineral. The T tubules are extensions of the cell membrane that function in a manner similar to axons. After endplate potentials initiate an action potential, the action potential is propagated by the T tubules throughout the muscle fiber where it triggers the release of free calcium ions from the sarcoplasmic reticulum.

Chapter 45

Exercises

45.1 Blood pressure in the fish is quite low, because the blood passes through capillaries of the gills before moving on to the body tissues.

45.2 "Lub" occurs during diastole; "dup" occurs at the peak of systole.

45.3 Answer given and explained in the text.

45.4 The traffic pressure is high before the bottleneck and becomes quite low beyond it; the blood pressure is high before the capillaries and becomes quite low beyond them.

45.5 Activity exercise.

45.6 Since the chest cavity is no longer closed, air tends to enter the wound instead of the lungs when the diaphragm contracts.

45.7 The llama's hemoglobin has greater affinity for oxygen than the Hb of other animals and saturates at the relatively low oxygen pressure characteristic of high mountains.

45.8 During exercise, the muscles generate heat, and the higher temperature should supply more oxygen for their activity.

45.9 About 15.9 liter of O_2/hr or 265 ml/min; 1.32 liter of air/min.

45.10 201 ml of O_2/liter of blood.

45.11 5 liter blood/min $\times$ 201 ml O_2/liter blood $\times$ 0.25 = 250 ml O_2/min

45.12 3 ml of dissolved O_2 per liter blood $\times$ 5 liter blood/min = 15 ml O_2/min, which makes up the deficiency.

45.13. The circuit should show increased breathing rate as plasma pH falls, as CO_2 increases, and as O_2 decreases; it should also show which organs detect each of these factors.

Multiple-Choice Questions

1. b
2. d
3. d
4. b
5. a
6. c
7. d
8. b
9. a
10. e

True-False Questions

1. False. Blood leaving the fish heart is *high* in carbon dioxide, low in oxygen, and relatively low in nutrients.
2. False. The capillaries of the pulmonary circulation are the site of *external* gas exchange.
3. True.
4. False. Blood leaving a capillary bed generally flows directly into *a venule.*
5. False. Nutrients and gases are exchanged between the blood and interstitial fluid *at the level of the capillary beds.*
6. True.
7. False.The primary function of lymph nodes is to *filter the lymph as it moves* through the lymphatic vessels and back to the bloodstream.
8. False. As the blood level of oxygen falls and that of carbon dioxide rises, the rate of ventilation *rises.*
9. False. The *chloride shift* describes the exchange of chloride ions for bicarbonate ions within the lung capillaries.
10. True.

Concept Questions

1. Blood contains so-called "formed elements"—red blood cells, white blood cells, and platelets—in addition to the liquid portion, or plasma. Interstitial fluid is a filtrate of blood, similar to plasma in composition but lacking plasma proteins. Lymph is residual interstitial fluid, but the walls of lymph vessels are more porous than those of the blood vessels, and so plasma proteins as well as white blood cells, infecting bacteria, and viruses may be present.
2. The cardiac centers control the rate of firing of the SA and AV nodes, both of which are made of cardiac muscle cells specialized for conduction. The vasomotor centers regulate the precapillary sphincters in the walls of arterioles. These are made of smooth muscle.
3. If skeletal muscles cannot contract, blood will tend to pool at the lowest point in the body, in this case, the legs. Venous return to the heart will fall, as will cardiac output, and the brain and upper extremities will be deprived of oxygen. When a person faints, the head becomes the lowest, or one of the lowest, regions of the body, and blood will fall into the cerebral blood vessels. Fainting is thus a protective reflex that is triggered when the circulation to the brain drops.
4. When we are overheated, the thoroughfare channels are closed, blood flows through the more superficial capillary beds, and heat is lost to the surroundings. Because blood is close to the surface of the skin, lightly pigmented skin looks pink. When we are chilled, blood is shunted into thoroughfare channels. These are deeper than the capillaries, so the skin becomes pale, and heat is retained within the body.
5. Insects have an open circulation, which means the blood leaves enclosing vessels and enters large sinus chambers. With such a large available space to fill, blood pressure is low compared to that in a closed circulation. A low blood pressure would be insufficient to perfuse all the regions of a large creature. Tracheal air tubes are very efficient for gas exchange when they are short, but not when they are long. The larger the insect, the longer these tubes would have to be to reach all body cells.

Chapter 46

Exercises

46.1 About 24.2 min.

46.2 (*a*) 5 ml/min; (*b*) minimum estimate; (*c*) 10 ml/min; (*d*) use a substance that is known to be completely filtered.

46.3 The loops are longer, to create a stronger gradient.

46.4. Using fine needles, sample the glomerular filtrate, blood, and interstitial fluid at various points. Measure the freezing point of each sample; the freezing point depression is proportional to its concentration of solutes.

46.5 $H_2CO_3 \rightarrow H_2O + CO_2$, so removing CO_2 leaves a neutral water molecule in the place of a H^+ ion from dissociation of H_2CO_3.

46.6 Breathing too hard releases too much CO_2, removes too many H^+ ions, and produces alkalosis.

46.7 (*a*) A stronger base has more affinity for H^+ ions. (*b*) In the tissue capillary beds, Hb releases O_2, goes into its nonbinding conformation; stronger bases combine with H^+ more readily, so Hb is then able to pick up H^+ being produced in the tissues.

Multiple-Choice Questions

1. c
2. d
3. a
4. d
5. b
6. d
7. c
8. e
9. b
10. c

True-False Questions

1. False. Excretion of excess nitrogen in the form of ammonia is the method of choice for *certain aquatic animals.*
2. False. Terrestrial animals excrete either *urea or uric acid.*
3. False. Ion/ATPase pumps use the energy obtained from *the breakdown of ATP to pump ions.*
4. False. In insects, metabolic wastes are excreted and water is retained by the action of a *potassium ion pump.*

5. False. Glomerular filtration pressure results in the flow of *protein-free plasma* from the glomerulus into Bowman's capsule.
6. False. Podocytes restrict the flow of *plasma proteins* from the glomerulus into Bowman's capsule.
7. False. The urine output of a normal kidney represents about *1 percent* of the amount of filtrate it processes.
8. True.
9. False. Acidosis will stimulate the nephrons to pump hydrogen ions *from the nephron into the urine.*
10. False. Freshwater fishes generally produce a highly *dilute* urine, and they actively move salts *into* their bodies by means of ion pumps in their gills.

Concept Questions

1. Filtrate is essentially the same as the plasma minus its proteins. Urine results after filtrate has been subjected to the processes of reabsorption and secretion. Thus urine is a highly concentrated solution of wastes; most of the water, nutrients, and other useful compounds have been reabsorbed.
2. GFR will vary with blood pressure. When a person is at rest, blood pressure, and thus GFR, will be lower than during exercise.
3. As long as the entry is larger than the exit, hydrostatic pressure in the glomerulus will be elevated, and the process of filtration will be more effective. If the capillaries leaving the glomerulus had a larger diameter than the entering vessels, glomerular pressure would drop, as would GFR.
4. Since it monitors blood pressure, it is a sensory structure. However, unlike most sense organs, it does not respond by producing receptor potentials leading to action potentials. Instead it acts more like an effector, secreting the enzyme renin in response to lowered blood pressure.
5. Freshwater fishes have body fluids that are more concentrated than their environment. Consequently, they will tend to take on water. They have very large glomeruli, and they produce large volumes of filtrate and dilute urine. Saltwater fishes live in a hyperosmotic environment and will tend to lose water. The glomeruli of these fishes vary from very small to practically nonexistent. As a result, very little filtrate is produced, and very little urine exits the body.

Chapter 47

Exercises

47.1

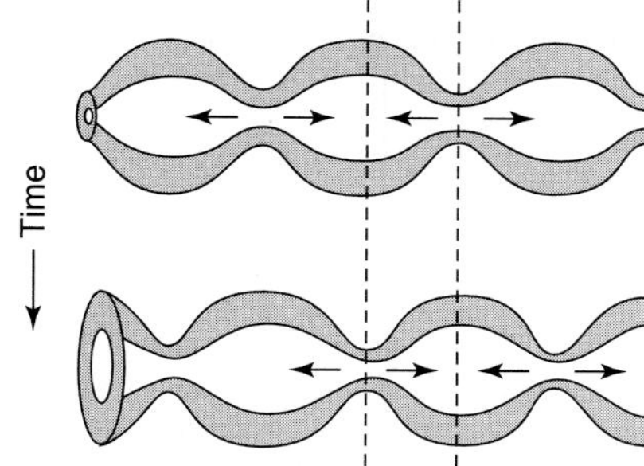

These segmentation movements divide and subdivide the intestinal contents, mixing them thoroughly (and bringing them in close contact with the intestinal walls).

47.2 Since we generally have no enzymes to cut polymers with these sugars, the molecules pass through our intestines intact and contribute to roughage.

47.3 People whose diet includes milk continue to make lactase; those who do not use milk do not make lactase.

47.4 Because the enzymes are just digested like all other proteins and have no other effects.

47.5 Secretin and glucagon must have evolved from the same ancestral protein.

47.6 About 2 meters wide.

47.7 PTH, because this increases uptake of Ca from the intestine.

47.8 Answers will vary.

47.9 100 g, or 3.5 oz.

47.10 Ribosomes are in the midst of synthesizing proteins, but are stopped, waiting for the required amino acid to be put into the next position.

Multiple-Choice Questions

1.	c	**6.**	b
2.	b	**7.**	d
3.	b	**8.**	a
4.	c	**9.**	b
5.	c	**10.**	a

True-False Questions

1. False. Molar teeth primarily function for *grinding.*
2. False. Carbohydrates are hydrolyzed in all regions of the GI tract except the *stomach and large intestine.*
3. False. Parietal cells of the gastric mucosa secrete *HCl,* whereas chief cells secrete *pepsinogen.*
4. False. Serum albumin functions as a transport protein to carry *fatty acids* within the blood.
5. True.
6. False. Iron, *complexed with transferrin,* is transported by the blood to bone marrow, where iron-ferritin complexes are formed.
7. True.
8. False. Chromium deprivation is associated with elevated blood glucose levels because *insulin* cannot work properly in the absence of chromium.
9. True.
10. False. Several amino acids are called "essential" because they are only obtained *from dietary sources.*

Concept Questions

1. The food eaten by herbivores, such as rabbits, is high in cellulose and requires a long intestine for digestion and absorption. Carnivores, such as dogs, generally have a short digestive tract, reflecting the higher protein and low cellulose content of their food.
2. Protein-hydrolyzing enzymes are capable of digesting other enzymes and the secretory cell itself. But since cellular membranes are primarily made of phospholipids and proteins, not polysaccharides such as starch that are digested in the intestine, the same danger does not exist for polysaccharides.
3. First, many of the control mechanisms are hormonal, not neural. Second, the enteric nervous system, plus input and output by the vagus cranial nerve, can support digestive functions even when the spinal reflexes are ineffective.
4. Glucagon, epinephrine, cortisol, and somatotropin elevate blood glucose levels, while insulin and epinephrine lower blood glucose levels.
5. Calcitonin, a thyroid secretion, lowers blood calcium levels, primarily by stimulating osteoblasts to increase the mineralization of bone. Parathyroid hormone, a parathyroid secretion, raises levels of blood calcium by stimulating bone breakdown by osteoclasts. Each of these hormones also works in the gut and kidney to raise or lower the absorption and retention of calcium in the gut and blood.

Chapter 48

Exercises

48.1 Antihistamines bind to histamine receptors and block them, so histamine cannot bind and have its usual effects.

48.2 Factor X feeds back to amplify conversion of factor VIII, and thrombin feeds back to amplify conversion of factors VIII and V. This is a positive feedback loop; positive feedback loops amplify and enhance processes.

48.3 About 16 times, since the stronger serum can be diluted by four factors of 2 beyond the weaker one, and $2^4 = 16$.

48.4

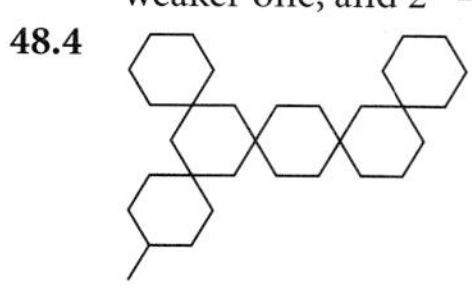

The epitope could be any comparable arrangement of hexagonal sugars.

48.5

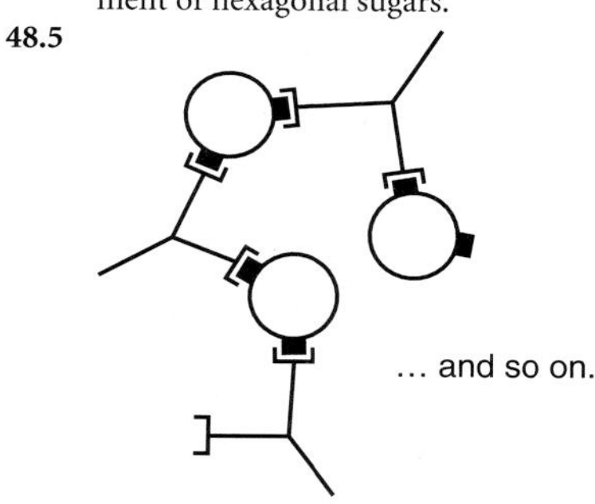

48.6 Epitopes are very small structures; antibodies made against one epitope fit its chemical structure very well, and even a slight change in chemical shape reduces the tightness of their binding.

48.7 (*a*) Cross reactions occur because other kinds of bacteria bear some of the same epitopes as strain A or other epitopes that are similar enough for antibodies to bind to them with reduced strength. (*b*) The numbers indicate the approximate degree of relatedness between strain A and the others.

48.8 Many different proteins with different specificities because the protein has many distinct epitopes on its surface.

48.9 $300 \times 5 = 1{,}500$

48.10 $500 \times 12 \times 4 = 24{,}000$

48.11. $1{,}500 \times 24{,}000 = 36{,}000{,}000$

Multiple-Choice Questions

1.	e	**6.**	b
2.	e	**7.**	c
3.	a	**8.**	a
4.	c	**9.**	b
5.	d	**10.**	e

True-False Questions

1. True.
2. False. Interleukin-1 is released from *macrophages* following stimulation by *mast cells.*
3. True.
4. False. The activity of antiserum is quantified by determining the *titer.*
5. False. B-cells mature in bone marrow, while T-cells mature in the *thymus.*
6. False. Circulating antibodies are the primary defense against *extracellular virus particles.*
7. False. Each antibody is made of *four polypeptide chains and two antigen-binding sites.*
8. False. Class I MHC proteins work in *cell-mediated immunity,* while class II MHC proteins function in *the humoral response.*
9. False. The determination between the proteins of self and nonself is *made during development.*
10. False. HIV is an *RNA-containing* virus that uses reverse transcriptase to make DNA, thereby destroying *T-helper cells.*

Concept Questions

1. Redness, swelling, warmth, and pain are the four primary signs. The immediate cause of all signs is local dilation of arterioles and local constriction of venules, which brings and traps excess blood in the area of the wound. Increases in capillary permeability enable plasma to enter the tissue spaces, and compounds such as kinins in the plasma stimulate sensory pain endings. All of these events result from histamine secretion from mast cells, interleukin-1 from macrophages, and other paracrine secretions produced by white blood cells.
2. The primary sites include the bone marrow and thymus. They are the locations where B- and T-lymphocytes mature. The secondary sites, including the spleen, lymph nodes, tonsils, adenoids, appendix, and Peyer's patches, are regions where immunocytes interact with antigens.
3. The variety of antibodies is far greater than the number of genes because of a mix-and-match process. If 1,000 or so gene segments code for light chains and an equal number code for heavy chains, and if a specific antigen-binding site includes one light and one heavy chain, the possibilities of unique combinations are in the millions.
4. When healthy cells are transformed into tumor cells or when they harbor a virus, proteins that are not normally present appear on the cell membrane. These combine with MHC I proteins that occur in the cell membrane of all nucleated cells, and the combination acts as an antigen.
5. The primary response results from an initial interaction between a small clone of B-lymphocytes and an epitope. These lymphocytes divide and differentiate into a small clone of plasma cells that secrete mostly IgM and a clone of memory cells. Subsequent encounters with the same antigen produce a more rapid, more powerful, and longer-lasting response as the memory cells rapidly multiply to form a much larger clone of plasma cells and also memory cells that synthesize IgG.

Chapter 49

Exercises

49.1 Raise young animals in isolation from older members of their species, and determine to what extent the behavior occurs in the absence of an opportunity to learn.

49.2 It means that "maternal instinct" is quite a vacuous term and that the animals are simply following behavior patterns induced by their biochemistry.

49.3 (*a*) Two genes, one for uncapping (symbols *U* and *u*) and one for cleaning out corpses (symbols *C* and *c*). (*b*) The four types are *UC, Uc, uC,* and *uc.*

49.4 This is an example of habituation.

49.5 You've been reinforced irregularly by the occasional bursts of the radio working, and this has conditioned you to keep fiddling with it.

49.6 One possible scenario: Suppose honeyguides display excited behavior when they find a hive, and this attracts the attention of a ratel, which opens the hive. The ratel starts to associate honeyguide behavior with honey; the honeyguide starts to associate the ratel with access to wax. The honeyguide encounters a ratel close to a hive, begins to act excited, attracts the ratel's attention, and leads it the short distance to the hive. From this point on, the two can interact through longer distances. Other members of each species then acquire the behavior through imitation.

Multiple-Choice Questions

1. d
2. e
3. d
4. a
5. d
6. c
7. e
8. a
9. d
10. b

True-False Questions

1. False. Instinctive behaviors *are present in all animals.*
2. False. A drive is *the motivation for instinctive behavior.*
3. False. In rats, maternal behavior is apparently an instinctive response to *a particular mix of gonadotropic hormones that is characteristic of postpartum females.*
4. True.
5. False. During the breeding season, male sticklebacks become aggressive when they see anything *colored red.*
6. True.
7. False. *Imprinting* combines fixed-action patterns, behaviors that are subject to maturation, and a variable releaser.
8. False. *Habituation* is the name given to the phenomenon whereby we ignore persistent, nonharmful stimuli such as the weight of our clothes or the background music in stores.
9. True.
10. False. The function of *classical conditioning* is to maintain and enhance a particular response while substituting one stimulus for another.

Concept Questions

1. Although the line is not finely drawn, behaviors are macroscopically observable and generally affect more than one individual. They are visible, complex outcomes of integrated sets of physiological responses.
2. Although the behavior of following is a fixed-action pattern, the entity to be followed may vary under experimental conditions.
3. In classical conditioning, a fixed-action pattern that is the usual response to one releaser is transferred to another, unusual releaser. In operant conditioning, the importance of one releaser within a range of releasers is enhanced by rewarding the animal when it pays attention to that releaser.
4. Hatchlings from eggs that have been exchanged between nests of hammerers and stabbers assume the behavior of their foster parents. This shows that the behavior is not a fixed-action pattern. Hatchlings in oystercatcher colonies that do not eat mussels do not assume either behavior, even in simpler form. This shows the behavior is not imprinted.
5. First, some cues rely on the activity of different structures and/or functions. For example, birds that use astronomical cues synchronize external events to an inner clock, whereas species that use landmarks do not. Those that navigate electromagnetically are able to sense electromagnetic fields. Birds that monitor the sun steer primarily by day, while those that use the stars steer by night. On the other hand, some elements appear to be common to all migratory species. For example, changing day length appears to be a stimulus that initiates migration in all migratory species and stimulates hormonal changes that regulate the physiology necessary for sustained activity.

Chapter 50

Exercises

50.1 Bacteria may have a high mutation rate due to the level of ultraviolet light; when combined with easy transmission of bacteria, epidemics could occur more easily in social groups. This factor might outweigh the advantages of strong social behavior and make these animals more solitary.

50.2 Humans form families maintained by sexual bonds between parents and parental bonds with their children. Children form bonds with one another both within a family and outside. Humans of all ages beyond infancy form friendship bonds with others. Larger societies are maintained by bonds of common interest and by bonds of loyalty to leaders at different levels.

50.3 Humans certainly show some division of labor based on factors such as intelligence and social rank. Traditional societies have gender-based division of labor, based in part on morphological differences between men and women, but this division tends to decrease in more modern societies.

50.4 Humans have many kinds of dominance hierarchies, mainly based on socioeconomic position. Obvious examples include the positions of people within business, political, or educational organizations.

50.5 Humans send many messages through body language and facial expressions. You can easily find examples.

50.6 Since her children carry three haploid sets of her genes, she has perpetuated the equivalent of her genome, but it is unlikely that they carry copies of all her genes. Since each grandchild carries a quarter of her genes, four grandchildren would carry the equivalent of her genome, but almost surely with multiple copies of some genes and none of other genes.

50.7 Paul, because he carries more of her genes than any cousin does.

50.8 While in a state mimicking pregnancy, she might prefer a man with an odor similar to hers and those of her family. But later, in a nonpregnant state, she might find herself married to a man who does not appeal to her strongly.

50.9 This is a eusocial species. It fits all the criteria. (See Additional Reading at the end of the chapter.)

50.10 Each worker carries half of her mother's genes and all of her father's genes, so their average genetic relatedness is 3/4. They should therefore be more inclined to behave in ways that benefit the group than animals that are less closely related.

Multiple-Choice Questions

1. b	**6.** c
2. d	**7.** b
3. d	**8.** a
4. e	**9.** d
5. d	**10.** b

True-False Questions

1. False. The strongest social bonds are likely to be between *parents and their children.*
2. False. Particular levels of testosterone, but not estrogen, appear to be correlated with formation of *dominance hierarchies.*
3. False. Social organization depends upon *vocal or nonvocal* communication.
4. True.
5. False. The inclusive fitness of an individual depends upon all the benefits it *bestows on its relatives.*
6. True.
7. False. The round dance of honeybees signals *a nearby food source,* while the waggle dance signals *a food source that is farther away.*
8. True.
9. False. While strong bonds form between members of mated pairs of Herring Gulls, *territorial aggression appears between* the males within the flock.
10. True.

Concept Questions

1. Bonds based on morphological specialization for division of labor are often determined by phenotypic differences between the sexes. The smaller the phenotypic differences, the less important will be divisions of labor based on physical strength.
2. If correct, most casual human communication is nonverbal. Verbal communication would then support more permanent bonds such as those between family members or members of closely bonded groups.
3. If one assumes that all workers in a hive had the same haploid father, the workers are more closely related to each other than any of them are related to the queen. Moreover, if each worker female could mate with one drone, her offspring would be less closely related to her than she is to her sisters. Thus the beehive should be viewed as a system that emphasizes altruism between workers and reproductive benefit to the workers.
4. The percent of shared genes in each case is 50 percent. A child receives 50 percent of its genes in the egg and 50 percent in the sperm. As a result of events of recombination during meiosis, siblings share 25 percent (on average) of their maternally derived genes and 25 percent of those that are paternally derived.
5. There are two good reasons. First, the child may require parental care for survival, so parental genes can only get to grandchildren if the child survives. Secondly, being younger, the child has a higher potential for reproduction.

Chapter 51

Exercises

51.1 Sperm is formed in the seminiferous tubules; resides in epididymis, where it matures and becomes motile; travels through vas deferens, picks up secretions from seminal vesicle that supply it with nutrients; continues through ejaculatory duct through prostate whose secretions neutralize the acidity of the vagina; then passes through urethra.

51.2 Contraction of the uterus reduces it to its nonpregnant size sooner.

Multiple-Choice Questions

1. e	**6.** a
2. d	**7.** a
3. b	**8.** d
4. c	**9.** d
5. e	**10.** d

True-False Questions

1. False. In human males, both sperm and urine flow into the urethra, *but only urine flows through the ureters.*
2. True.
3. False. In humans, eggs are generally fertilized within the *oviduct.*
4. False. In males, the primary effect of *LH* is to stimulate testosterone-producing interstitial cells of the testes.
5. False. Male secondary sexual characteristics, such as antlers and larger muscle mass, result from the action of *testosterone* directly on the body cells involved.
6. False. At different times in the reproductive cycle, gonadotropins are produced and secreted by the pituitary *and, in pregnant females, by the chorion.*
7. False. The ovarian cycle is divided into the *follicular* phase, marked especially by rising estrogen levels, and the luteal phase, when progesterone levels rise dramatically.
8. True.
9. False. LH provides the immediate hormonal trigger for *ovulation.*
10. False. Mammals produce milk only when the *level of prolactin is high and the level of estrogen is low.*

Concept Questions

1. Both are polypeptide hormones produced by the anterior pituitary. They are gonadotropins, which means their target cells are in the gonads. In males, LH stimulates the testosterone-producing interstitial cells in the testes, and in females it causes the mature follicle to rupture and become a progesterone-secreting corpus luteum. FSH promotes cellular growth and differentiation. In males, it promotes and supports the seminiferous tubules. In females, it causes growth and maturation of the follicle and oocyte.
2. Prolactin stimulates milk production, but only in mammary tissue that has been developing and proliferating under the influence of chorionic somatomammotropin. That is why milk is only produced after birth. Additionally, competent mammary tissue does not produce milk if the PIH level is high, as it is when the estrogen level is also high. Finally, oxytocin, a hypothalamic hormone, is produced at high levels during birth. It causes smooth muscle contraction in the uterus as well as in the mammary ducts, enabling milk to flow into the nipples.
3. The differences trace to the timing of copulation relative to ovulation. Induced ovulators, such as cats and rabbits, ovulate only if copulation has just occurred, ensuring that a mature sperm will encounter a mature oocyte. Spontaneous ovulators will copulate only during a small window of opportunity before and after ovulation, whereas copulation and ovulation are not synchronized at all in menstrual females.
4. The placenta is formed by a union of a region of the maternal endometrium and the fetal chorioallantoic membrane.
5. The corpus luteum secretes significant amounts of progesterone and smaller quantities of estrogen, both of which are necessary to maintain the uterine lining and keep uterine muscles from contracting. Although both hormones are also synthesized by placental tissue, there is insufficient production especially during early pregnancy. By the second trimester, the placenta is fully developed and its hormone production is in high gear.

Appendix C Summary Classification

This classification outline is simply intended to put the major groups of organisms into a context for an introduction to biology. It is highly modified from other current classifications, omitting many obscure minor groups and several levels of taxa that have been used to make fine distinctions among organisms. Students should also be aware that the endings of names vary a great deal between classifications. For instance, a taxon established for green algae has been variously listed as Chlorophyta, Chlorophyceae, Chlorobionta, and Chlorophycophyta in different systems. As the level of a taxon is raised and lowered, it may take on various standard endings such as -ales, -aceae, -atae, -ica, -ida, -ina, or -idae. Classifications are in a particularly great state of flux today as more molecular and cellular data are used as taxonomic guidelines.

Domain Archaea: archaebacteria

Domain Procarya: procaryotic organisms except for archaebacteria

Kingdom Monera

Division Gracilicutes: gram-negative bacteria
Division Firmicutes: gram-positive bacteria
Division Tenericutes: mycoplasmas

Domain Eucarya: organisms with eucaryotic cells

Kingdom Protista

Phylum Rhodophyta: red algae
Phylum Parabasalia: certain unicellular flagellates with characteristic parabasal bodies
Phylum Euglenozoa: euglenoid flagellates and relatives
Phylum Mycetozoa: slime molds
Class Myxogastrea: true slime molds
Class Dictyostelea: cellular slime molds
Phylum Choanozoa: choanoflagellates
Phylum Dinozoa: dinoflagellates and relatives
Phylum Apicomplexa: parasites (gregarines and coccidians) with a characteristic apical complex at one end of the cell
Phylum Ciliophora: ciliates
Phylum Heliozoa: amoeboid organisms with spherical bodies and no central capsule or shell
Phylum Radiolaria: amoeboid organisms with a central capsule and usually a skeleton
Phylum Rhizopoda: typical amoebas
Phylum Reticulosa: foraminifera and relatives

Kingdom Chromista

Phylum Cryptophyta: compressed unicellular flagellates with phycobilin pigments
Phylum Labyrinthulea: slime molds that glide along a cytoplasmic network
Phylum Xanthophyta: yellow-green algae
Phylum Chrysophyta: golden-brown algae
Phylum Phaeophyta: brown algae
Phylum Diatomea: diatoms
Phylum Pseudofungi: water molds, including oomycetes and hyphochytrids

Kingdom Plantae

Subkingdom Thallobionta: green algae and relatives. Plants with simple thalli, showing little or no cellular differentiation except for reproductive cells, and never forming an embryo that develops within a structure of the parent plant.
Division Prasinophyta: flagellate or coccoid algae
Division Chlorophyta: grass-green algae
Division Siphonophyta: tubelike, coenocytic (siphonaceous) green algae
Division Zygnematophyta: conjugating green algae
Division Charophyta: stoneworts. Macroscopic algae with precise segmentation into nodal and internodal segments.
Subkingdom Embryobionta: plants in which the sporophyte begins its development attached to and dependent on the gametophyte, as an embryo.
Division Rhyniophyta: primitive vascular plants
Division Bryophyta: mosses
Division Hepatophyta: liverworts
Division Anthocerophyta: hornworts
Division Psilotophyta: unusual plants with primitive features
Division Lycopodiophyta (Lycopsida): club mosses, with true roots, stems, leaves
Division Equisetophyta: horsetails
Division Polypodiophyta: ferns
Division Pinophyta: gymnosperms
Subdivision Cycadophyta
Class Pteridospermeae: seed ferns
Class Cycadatae: cycads
Class Bennettitatae
Subdivision Pinophyta
Class Ginkgoatae: ginkgos
Class Pinatae: conifers (pines, firs, yews, etc.)
Division Magnoliophyta: flowering plants (angiosperms)
Class Magnoliatae: dicotyledons
Class Liliatae: monocotyledons

Kingdom Fungi

Phylum Zygomycota: conjugating fungi. Bread molds, fly fungi, animal traps.
Phylum Ascomycota: fungi with spores borne in asci
Phylum Basidiomycota: fungi with spores borne on basidia
Phylum Deuteromycota: fungi imperfecti; fungi with no known sexual stage

Kingdom Animalia (Metazoa)

Subkingdom Mesozoa. Phylum Mesozoa
Subkingdom Placozoa. Phylum Placozoa
Subkingdom Parazoa. Phlum Porifera: sponges
Subkingdom Eumetazoa
Infrakingdom Radiata: with radial symmetry
Phylum Cnidaria: coelenterates
Phylum Ctenophora: comb jellies
Infrakingdom Bilateria: with bilateral symmetry or secondary radial symmetry
Branch Acoelomata: with no coelom
Phylum Platyhelminthes: flatworms
Phylum Nemertea: ribbon worms
Branch Pseudocoelomata: coelom derived from embryonic blastocoel with only partial mesodermal lining
Phylum Rotifera: rotifers
Phylum Nematoda: roundworms
Phyla Gastrotricha, Kinorhyncha, Nematomorpha, Priapulida, Acanthocephala, Entoprocta, Gnathostomulida, Loricifera
Branch Coelomata: coelom completely lined with mesoderm
Subbranch Protostomia
Phylum Annelida: segmented worms
Phylum Arthropoda: crustaceans, arachnids, insects, centipedes, etc.
Subphylum Trilobitomorpha: trilobites
Subphylum Chelicerata
Class Merostomata: horseshoe crabs
Class Arachnida: scorpions, mites, ticks, spiders
Subphylum Uniramia
Class Myriapoda: millipedes, centipedes
Class Insecta: insects
Subphylum Mandibulata (crustacea)
Class Branchiopoda: fairy shrimps, brine shrimps, cladocerans
Class Maxillopoda: ostracods, copepods, barnacles
Class Malacostraca: crabs, shrimp, lobsters, pill bugs
Phylum Mollusca: snails, whelks, clams, oysters, squid, etc.

Phylum Sipuncula, Echiura, Pogonophora, Vestimentaria
Subbranch Deuterostomia
Phylum Echinodermata: sea stars, sea urchins, and relatives
Phylum Chaetognatha: arrow worms
Phylum Ectoprocta: lophophorate animals in small cases
Phylum Brachiopoda: lophophorate enclosed in a bivalve shell
Phylum Phoronida: tubular lophophorate animals
Phylum Hemichordata
Phylum Chordata: vertebrates and their relatives
Subphylum Urochordata: tunicates
Subphylum Cephalochordata: lancelets
Subphylum Vertebrata: vertebrates
Superclass Agnatha: fishes without jaws
Superclass Gnathostomata: vertebrates with jaws
Class Placodermi
Class Chondrichthyes: cartilaginous fishes (sharks and rays)
Class Osteichthyes: bony fishes
Class Amphibia: frogs, toads, salamanders
Class Reptilia: reptiles
Class Aves: birds
Class Mammalia: mammals

Glossary

No attempt has been made here to include all terms used in this book, particularly the large number of structural terms and the variety of compounds and materials. Definitions of these terms, often accompanied by illustrations, are in the text and may be found by using the index. Instead, the glossary focuses on general terms, particularly adjectives and conceptual terms, on words that may be needed as background for the text, and on the most frequently used terms.

A

absorption spectrum a graph showing the relative amount of light absorbed by a compound as a function of wavelength (p. 199)

acetyl the portion of a molecule ($CH_3C{=}O$) made by removing the —OH group of acetic acid (CH_3COOH) (p. 180)

acid a compound that can donate (or release) a hydrogen ion (proton); compare with *base* (p. 51)

acidic having a pH less than 7; compare with *alkaline, basic* (p. 52)

actin a protein that forms microfilaments (a major part of the cytoskeleton) and is involved in cell shaping and movement (p. 226)

action potential a change in the membrane potential of a cell that is propagated along the length of the cell (p. 878)

action spectrum a graph that shows the relative effectiveness of light for driving some process (such as photosynthesis) as a function of wavelength (p. 201)

activator protein a genetic regulatory protein that initiates transcription of certain genes in response to binding a particular ligand (p. 354)

active site a site on an enzyme where catalysis takes place (p. 73)

active transport a process that uses energy to transport a substance across a cell membrane; compare with *facilitated diffusion* (p. 168)

adaptation 1. the evolutionary process in which organisms become fitted to a particular niche (p. 32); 2. a feature of an organism that improves its ability to survive and reproduce (p. 563)

adaptive radiation a process in which two or more species evolve simultaneously to occupy distinct adaptive zones (p. 494)

adaptive zone a particular mode of life characteristic of a species (p. 494)

adhesion the attraction between dissimilar substances, such as water and glass; compare with *cohesion* (p. 54)

adsorption the attachment of molecules or particles, such as viruses, to a surface; compare with *absorption,* taking materials into a volume rather than on the surface (p. 249)

aerobic 1. a condition in which oxygen is present (p. 186); 2. able to grow in the presence of oxygen, specifically by using oxygen in respiration; **aerobe,** an organism capable of growing aerobically (p. 614)

afferent directed toward a central system, such as the brain or heart; compare with *efferent* (p. 873)

affinity the tendency of one molecule to associate with another; also, a measure of that tendency (p. 95)

agonistic relating to aggressive or defensive social interactions between animals (p. 1062)

alarmone an intracellular signal ligand used to alert some metabolic system of a new situation to which it must respond (pp. 223, 357)

alcohol any of a class of organic compounds that bears one or more hydroxyl (—OH) groups (p. 62)

aldehyde an organic compound bearing a carbonyl group in a terminal position (p. 62)

alga (pl., **algae**) a photosynthetic organism, especially a eucaryote, consisting primarily of cells with little or no cellular differentiation (p. 629)

alkaline having a pH greater than 7 (p. 52)

allele a particular form of a gene with a specific nucleotide sequence and, generally, producing a distinctive phenotype (p. 313)

allelic frequency the number of copies of a particular allele in a population divided by the number of all other alleles at the given locus (p. 467)

allelochemic referring to chemical interactions among members of different species in an ecosystem (p. 563)

allomone a substance made by members of one species that kills or inhibits the activities of organisms of another species (p. 565)

allopatric living in different geographical areas; compare with *parapatric, sympatric* (p. 490)

allostery a mechanism in which the action of a protein is changed through the binding of ligands to two or more second distinct sites (p. 102)

alpha α 1. referring to a particular orientation of linkage between sugars and other molecules—contrasted with *beta* (p. 68); 2. the carbon atom in an organic acid to which the carboxyl group is attached (p. 73); 3. sometimes used simply as a name for a protein or a type of cell

alpha helix a common helical form assumed by polypeptide chains (pp. 78, 80)

amine a compound or molecule that bears an amino group (p. 62)

amino acid any organic compound of the general formula $R{-}CH(NH_2){-}COOH$, used especially as a monomer of proteins (p. 73)

amino group the group $-NH_2$ (p. 62)

amoeba a relatively formless unicellular organism that moves by means of pseudopods (p. 629)

amoeboid moving through a flow of cytoplasm, like an amoeba (p. 226)

amphipathic having an affinity for both polar and nonpolar molecules (p. 157)

anabolism biosynthesis; the phase of metabolism in which molecules are synthesized rather than being broken down (p. 139)

anaerobic 1. the absence of oxygen (p. 186); 2. able to live in the absence of oxygen; **anaerobe,** an organism that can live in the absence of oxygen (p. 614)

analog (also **analogue**) a compound that shares some features of another compound, such as shape or biological activity (pp. 1034, 1095)

analogous in comparative morphology, having the same general form but different evolutionary or embryonic origins (p. 25)

androgen a male sex hormone (p. 1088)

aneuploid having a set of chromosomes that is not an integral number of a basic haploid set (p. 299)

anion a negatively charged ion (p. 50)

anterior toward the front (head) end of an organism (p. 666)

antheridium in many protista, fungi, and plants, a male reproductive structure that produces sperm (pp. 694, 656)

antibiotic a substance that kills or inhibits the growth of a particular microorganism (p. 566)

antibody a protein made by some vertebrates that can recognize and bind to an antigen (p. 164)

anticodon a region of a transfer RNA molecule with a nucleotide sequence that recognizes a codon on messenger RNA (p. 277)

antigen a molecule that elicits the formation of specific antibodies and binds to an antibody (p. 164)

antiport a process in which two molecules or ions are simultaneously transported in opposite directions across a cell membrane (p. 170)

apical referring to the tip of a structure; compare with *basal* (p. 794)

aqueous referring to water (p. 51)

arboreal tree-dwelling (p. 768)

archegonium in many protista, fungi, and plants, the female reproductive structure that bears eggs (p. 694)

asexual reproduction reproduction that does not involve the union of two separate cells (p. 32)

assay a test for determining the presence or amount of some substance or activity
assimilate to incorporate into the structure of an organism (p. 216)
atomic mass the average mass of the atoms of an element; essentially, the total mass of protons and neutrons in the atom of the element (p. 46)
atomic number the number of protons in the atom of an element (p. 46)
ATP (adenosine triphosphate) a nucleotide of the base adenine with three phosphate groups that is a universal donor of free energy in organisms (p. 142)
autoradiography a process in which the presence of material is revealed by means of a tracer that is allowed to decay under a photographic emulsion, thus darkening the film wherever atoms of the tracer lie (p. 203)
autosome any chromosome other than a sex (such as X or Y) chromosome (p. 325)
autotrophy a nutritional mode in which an organism uses CO_2 as its only carbon source; compare with *heterotrophy*; **autotroph,** an organism that lives in this way (p. 143)
auxotroph a mutant organism that requires some additional substance (such as an amino acid) for its growth; compare with *prototroph* (p. 243)
axon the extension of a neuron that carries impulses away from the cell body (p. 875)

B

backcross a genetic cross between a hybrid organism and an organism of either parental strain (p. 498)
bacteriophage (also, **phage**) a virus that infects bacteria (p. 246)
bacterium generally, any procaryotic organism (pp. 116, 610)
basal referring to the base of a structure; compare with *apical* (p. 230)
basal body a structure at the base of a eucaryotic cilium or flagellum, consisting of nine triplets of microtubules (p. 230)
basal metabolic rate (BMR) the rate of metabolism required just to keep an animal alive (p. 684)
base a compound that can combine with a hydrogen ion; compare with *acid* (p. 51)
basic alkaline; having a pH greater than 7 (p. 52)
beta (β) 1. referring to a particular orientation of linkage between sugars and other molecules (p. 68); compare with *alpha*; 2. sometimes used to designate a particular protein subunit
beta pleated sheet see *pleated sheet*
bicarbonate the ion HCO_3^- (pp. 166, 206)
bilateral symmetry a structural pattern in which an object is divisible into mirror-image halves (p. 668)
bilayer a double layer of lipid molecules that forms the basis of a cell membrane (p. 160)
binary fission division of a cell into two essentially equal daughter cells (p. 215)
binding site a region of a protein where another molecule can bind (p. 101)
binomial system the standard system of biological nomenclature in which each species is given a generic name and a specific (trivial) name (p. 450)
biogeochemical cycle a series of transformations in the biosphere in which an element such as carbon, nitrogen, or sulfur is interconverted between various states in organisms and abiotic reservoirs (p. 585)
biogeographic realm one of the large geographic areas of the continents characterized by a distinctive flora and fauna (p. 531)
biomass the total mass of organisms in a particular place (p. 580)
biome a group of communities that share a distinctive type of vegetation, as determined by the climate (p. 509)
biomolecule a molecule characteristic of organisms, especially a polymer (p. 45)
biosphere that portion of the earth's surface in which organisms can exist; the sum total of all organisms and their habitats (p. 508)
biosynthesis the metabolic processes that build up biological structure; compare with *catabolism* (p. 138)
bond a specific attraction that holds atoms together; also, a particular attraction, in the general sense, that keeps animals associating with one another (p. 1061)
buffer a solution that maintains a constant pH when limited amounts of acid or base are added to it (p. 987)

C

C-terminal end the end of a polypeptide with a free carboxyl group; compare with N-terminal (p. 75)
calcareous containing calcium deposits, generally calcium carbonate (p. 720)
calorie the amount of heat required to raise the temperature of 1 gram of water 1°C (p. 54)
capillarity the tendency for a liquid to move into narrow spaces because of adhesion with the molecules of the space (p. 808)
carbohydrate a class of organic compounds that have multiple hydroxyl groups, generally with an aldehyde or ketone group; includes sugars, starches, and cellulose (p. 177)
carbon cycle the biogeochemical cycle in which carbon compounds are interconverted (p. 177)
carboxyl group the —COOH group that characterizes organic acids (p. 62)
cardiac referring to the heart
carnivorous flesh eating (p. 143)
carotenoid one of a number of yellow, orange, or red hydrocarbon pigments common in plants (p. 200)
carotid referring to structures (such as arteries) found in the neck
carrier protein a protein that transports a substance across a membrane (p. 167)
carrying capacity the upper limit on the size of a population set by the environment (p. 34)
catabolism the phase of metabolism in which molecules are broken down into smaller molecules, generally conserving some of their energy (p. 138)
catalyst an agent that changes the rate of a chemical reaction without being consumed by the reaction (p. 73)
cation a positively charged ion (p. 50)
caudal referring to the tail (p. 666)
cell a membrane-bounded structure that is the smallest unit capable of independent metabolism and self-reproduction (p. 5)
cell cortex a region of cytoplasm just inside the plasma membrane that is structured by elements of the cytoskeleton (p. 226)
cell culture a system of cells isolated from a multicellular organism growing in an artificial medium (p. 216)
cell cycle the regular series of events that cells engage in as they grow and multiply (p. 265)
cell wall a structure surrounding the plasma membrane in many organisms that holds the cell together and protects it (p. 110)
cellular respiration the metabolic process in which oxidizable substrates, usually organic, are oxidized and broken down to obtain their energy (p. 177)
cephalic referring to the head (p. 666)
cephalization the tendency in animal evolution for sensory and feeding structures to be concentrated in a head end (p. 666)
channel protein a membrane protein that allows the passage of specific ions and can be opened or closed by factors such as a voltage, a mechanical change, or a specific ligand (p. 170)
chemical reaction any process in which one or more chemical compounds are changed into other compounds (p. 51)
chemiosmotic coupling the mechanism in which ATP is formed from ADP and inorganic phosphate by using the energy stored in a proton gradient (p. 185)
chemoautotrophy a nutritional mode in which an organism obtains its energy by oxidizing reduced inorganic compounds and uses CO_2 as its sole carbon source; **chemoautotroph,** an organism that lives in this way (pp. 143, 189)
chemoheterotrophy a nutritional mode in which an organism obtains its energy by oxidizing reduced compounds and uses organic molecules as its sole carbon source; **chemoheterotroph,** an organism that lives in this way (p. 143)
chemoreceptor a protein (or system of proteins) that recognizes specific chemical structures (p. 222)
chemotrophy a nutritional mode in which energy is derived through the oxidation of reduced compounds; **chemotroph,** an organism that lives in this way; compare with *phototrophy* (p. 143)
chlorophyll a green pigment in phototrophs that captures light energy for use in photosynthesis (p. 200)
chloroplast a plastid in which photosynthesis occurs (p. 124)
chromatid one of the two identical halves of a chromosome found in eucaryotic cells after the time of DNA replication (p. 260)
chromosome the physical structure of a genome: primarily nucleic acid, typically with structural proteins (p. 116)
cilia (sing., **cilium**) the short, hairlike organelles on the surfaces of some eucaryotic cells, used for motility or for moving extracellular materials (p. 232)
ciliate a protist that bears cilia (p. 636)
clade a phylogenetic series including an ancestral species and all the other species derived from it (p. 457)

cleavage the division of a zygote into smaller cells (p. 400)
cline a gradual change in some characteristic, or in gene frequencies, across the range of a species (p. 450)
clone 1. a set of cells derived from one original cell by repeated division (p. 247); 2. a group of multicellular organisms derived from one organism by asexual reproduction; 3. an organism formed by introducing a nucleus from a developing organism into an enucleated egg (p. 424); 4. a system, usually a bacterium carrying a plasmid, in which a specific DNA fragment is propagated; 5. verb: to create a clone of type 3 or 4 (p. 347)
codon a set of (three) nucleotides that codes for a single amino acid (p. 275)
coelom a major cavity between the body wall and the gut of many animals, or various smaller cavities of the same kind (p. 718)
coenocytic consisting of an undivided cytoplasm containing several nuclei (p. 265)
coenzyme an organic compound that works in conjunction with an enzyme to catalyze a reaction but is not bound to the protein; compare with *prosthetic group* (p. 97)
coevolution the process in which two or more species mutually evolve adaptations to one another (p. 482)
cofactor a substance, especially a metal ion, that works in conjunction with an enzyme or other functional protein (p. 97)
cohesion the forces of attraction between identical molecules; **cohesive,** adjective (p. 54)
colinear describes molecules that encode pieces of information in identical linear sequences (p. 275)
colony 1. a group of organisms that are derived from one another asexually and live in close proximity or attached to one another (p. 247); 2. a similar group of closely related sexual organisms (p. 673)
commensalism an association between two species in which one benefits while the other neither benefits nor is damaged (p. 569)
community a complex of different organisms that live together and interact; the biological part of an **ecosystem,** a community of organisms plus their abiotic surroundings (p. 31)
compartmentalized divided into compartments in which distinct specialized processes occur (p. 118)
complementary said of two molecules whose shapes fit with each other (p. 46)
complementation test a procedure used to define genes experimentally. Two mutants are said to complement each other if they mutually supply genetic functions for each other. (p. 338)
complex a combination of molecules held together by weak (noncovalent) bonds (p. 98)
compound a substance made of a fixed combination of elements in a specific ratio (p. 46)
configuration the particular covalent arrangement of atoms in a molecule (p. 68)
conformation the particular orientation of the parts of a molecule relative to one another (p. 60)
conjugation a process in which two unicellular organisms exchange or transfer genetic material (pp. 341, 639)
consumer an organism in an ecosystem that lives by eating other organisms; compare with *producer, decomposer* (p. 143)
convection the transfer of heat by means of circulating gas or liquid; compare with *radiation* (p. 815)
convergent evolution evolution in which two or more species belonging to different clades acquire similar properties (p. 445)
corpus a general term meaning "body," designating a particular structure
cortex the outer region of an organ or body; compare with *medulla* (pp. 226, 795, 886)
cortical referring to the cortex
cotransport a process in which two substances are transported across a membrane together; compare with *antiport, symport* (p. 170)
coupled reaction a pair of chemical reactions that share a chemical species or in which a group is transferred from one molecule to another (p. 139)
covalent bond a bond between atoms formed by the sharing of one or more electron pairs (p. 48)
crossing over a process in which portions of homologous chromosomes are exchanged with each other (p. 302)
cyclosis the movement of cytoplasm in a cell, often in a circular pattern (p. 227)
cytochrome one of a number of proteins bearing heme groups that are used in electron transport systems (p. 185)
cytokinesis the process in which a cell divides into two daughter cells (p. 265)
cytoplasm the viscous contents of a cell, generally excluding the nucleus; compare with *cytosol, nucleoplasm* (p. 118)
cytoskeleton a complex of fibrous proteins, particularly microtubules and microfilaments, that holds a cell in shape (p. 226)
cytosol the aqueous portion of a cell, exclusive of organized bodies such as the nucleus, mitochondria, and various membranes (pp. 54, 118)

D

dalton an atomic mass unit; a unit of molecular weight equal to one-twelfth the mass of a carbon atom (essentially, the mass of a hydrogen atom) (p. 67)
daughter either of two structures or entities—such as cells, chromosomes, or species—derived from a common structure or entity. See also *sister.* (p. 241)
decomposer an organism in an ecosystem that degrades the remains of dead organisms into simpler materials (p. 143)
dehydration the removal of water
dehydration synthesis creation of a linkage between monomers by removal of a water molecule (p. 67)
dehydrogenation a chemical reaction in which a pair of hydrogen atoms is removed from a molecule (p. 147)
deletion a mutation in which a segment of nucleic acid is removed (p. 254)
deme a local population (p. 450)
denaturation the alteration of a material, especially a protein, through heating or chemical treatment, so it loses its characteristic activity (p. 85)
denitrification the conversion of nitrates and nitrites to nitrogen gas (p. 853)
density in physics, the ratio of the mass of an object to its volume; in population biology, the number of individuals per unit area (p. 541)
density-dependent factor a factor that increasingly affects the growth of a population as the population density increases (p. 545)
density-independent factor a factor whose effect on the growth of a population is independent of population density (p. 545)
deoxyribonucleic acid (DNA) a polymer of deoxyribonucleotides (p. 70)
deoxyribonucleotide a nucleotide based on the five-carbon sugar deoxyribose (p. 251)
detritivore an organism that consumes detritus (p. 509)
detritus the collective solid wastes of an ecosystem (p. 509)
differentiation the process in which the cells of an organism acquire distinct structures and functions (p. 398)
diffusion the spread of a material from a region of high concentration through random movements of its molecules (p. 154)
digestion the process of making food absorbable by breaking its large molecules (primarily polymers) into small molecules (p. 67)
dilation the enlargement or widening of a tube or opening (p. 958)
diploid having a double set of chromosomes, one from each parent (p. 295)
diplontic having a life cycle in which mitotic division occurs only during the diploid phase (p. 297)
dissociate to come apart into subunits, as a complex of a protein with a ligand coming apart into its components (p. 51)
distal away from some center or reference point; compare with *proximal*
domain a region of a protein that has a distinct structure and function (p. 85)
dominance 1. the phenotypic expression of only one of two alleles in a diploid organism (p. 313); 2. a relationship between two animals in which one behaves submissively toward the other in any kind of conflict (p. 1061)
dormancy a temporary suspension of growth, reproduction, or other activity (p. 831)
dorsal toward the back of a bilaterally symmetrical animal; compare with *ventral* (p. 668)
downstream the direction on a genome in which RNA transcription occurs; compare with *upstream* (p. 354)
drug a compound that has a relatively specific in vitro action because of its binding to a specific functional site (p. 226)

E

ecological niche see *niche*
ecological succession a process in which different types of ecosystems replace one another (p. 593)
ecology the branch of biology that deals with the relationships between organisms and their

environment, including populations of other species; also, a description of the relationships within a particular ecosystem (p. 31)

ecosystem the combination of a biological community and the inorganic world in which it lives (p. 31)

efferent directed away from a central structure, such as the brain or heart; compare with *afferent* (p. 873)

electrochemical potential a potential for doing work due to an unequal distribution of charged particles (p. 170)

electromagnetic radiation radiation that consists of simultaneous electrical and magnetic oscillations, including X rays and gamma rays; ultraviolet, visible, and infrared light; and microwaves and radio waves (pp. 113, 197)

electron a small basic particle of matter that has a negative charge (p. 46)

electron transport system (ETS) a chain of carrier molecules in a membrane that can transfer electrons from one to another and simultaneously conserve some of their energy in the form of a hydrogen-ion gradient (p. 179)

electrostatic referring to the attraction or repulsion of charged particles

element a substance that cannot be separated into simpler substances in a chemical reaction (p. 46)

endergonic a process in which the end products have greater free energy than the reactants; compare with *exergonic* (p. 136)

energy the ability to do work (p. 132)

enhancer a genomic control region that can enhance expression of a gene, usually distant from the controlled gene (p. 361)

entropy a measure of the degree of disorder in a physical system; also, the amount of energy that cannot be converted to useful work (p. 135)

enzyme a protein that catalyzes chemical reactions (p. 72)

equilibrium a state of balance between two or more processes in which the composition of a system remains unchanged; see also *steady state* (p. 93)

equilibrium constant a measure of the degree to which the components of a system tend to change in one direction or another (p. 94)

equilibrium species a species that tends to have a stable population density and to live in a steady-state condition with its surroundings; compare with *opportunistic species* (p. 551)

estrogen a female sex hormone (p. 1088)

eucaryotic (also **eukaryotic**) a type of cell with a true nucleus; compare with *procaryotic* (p. 116)

evolution a process in which the forms and overall genetic structures of organisms change with time (p. 5)

exchange a process in which two or more materials move in opposite directions across a barrier, not necessarily one molecule (or ion) for another (pp. 675, 951)

excretion the removal of wastes from an organism (p. 218)

exergonic a process in which the end products have less free energy than the reactants; compare with *endergonic* (p. 136)

exponential growth a mode of growth in which a population increases by a fixed proportion in each unit of time (p. 542)

expression the process in which information encoded in a genome is used (p. 332)

extinction the dying out of a species (p. 485)

F

facilitated diffusion the diffusion of materials across a membrane through combination with a specific carrier, without the need for additional energy; compare with *active transport* (p. 168)

facultative having different modes of activity, depending on conditions (p. 614)

feedback a mechanism in which part of the output from a system is used to change the activity of the system (p. 218)

fermentation an oxidative process in which the final electron (hydrogen) acceptor is an organic molecule (p. 186)

fitness a measure of the relative reproductive success of a particular genotype (pp. 34, 468)

fixed-action pattern a genetically encoded behavior pattern (p. 1044)

flagellum a long, whiplike organelle used by many cells for locomotion (p. 232)

food chain a sequence of organisms that eat one another (p. 143)

food web the complex of all the food chains in a community (p. 143)

free energy energy that is theoretically available to do work (p. 136)

functional disease a diseased condition resulting from the intrinsic deficiency or degeneration of a system rather than from infection; compare with *infectious disease* (p. 616)

functional group a characteristic group of atoms in an organic molecule (p. 62)

fundamental niche the entire niche that a species is capable of occupying in the absence of other organisms; compare with *realized niche* (p. 560)

fungus (pl., **fungi**) a heterotrophic organism that lives as a saprobe or parasite, whose body is either unicellular or consists of hyphae (p. 651)

G

gametangium in organisms such as fungi and algae, a structure in which gametes develop; compare with *antheridium, archegonium* (pp. 654, 694)

gamete one of the reproductive cells of a sexually reproducing organism (p. 295)

gametophyte the structure in a plant life cycle that produces gametes; compare with *sporophyte* (p. 297)

gastric referring to the stomach

gene a unit of inheritance, generally defined now as a portion of a nucleic acid that carries the information for a single polypeptide chain (p. 243)

gene flow the movement of genes from one population to others (p. 465)

gene pool the set of all the genes shared by the members of a population (p. 466)

genetic analysis a program of research in which the structure and function of a genome are analyzed by means of mutants (pp. 266, 332)

genetic information the information carried in the genome of an organism or virus (p. 28)

genetic map a representation of the structure of a genome as determined by genetic experiments (p. 332)

genome the structure of an organism that contains instructions for the organism's growth and reproduction (p. 28)

genotype a description of the particular set of genes that an organism carries, as contrasted with its appearance; compare with *phenotype* (p. 314)

genus a group of closely related species (p. 450)

geographic race a subspecies; a morphologically distinct portion of a species that occupies a particular geographic region (p. 450)

geographic range the area in which a particular species lives (p. 539)

germ cell any of the gametes (or their precursor cells) that might be used to produce a next generation through sexual reproduction; compare with *somatic cell* (p. 296)

glycolysis the process in which glucose is oxidized to pyruvic acid (p. 179)

glycoprotein a protein that bears sugars or oligosaccharides (p. 166)

gonad in animals, an organ (ovary or testis) that produces gametes (p. 296)

gradient a continuous change in space in a physical quantity, such as concentration or temperature (p. 154)

group transfer potential a measure (in terms of free energy) of a compound's ability to transfer a specific chemical group to another compound (p. 140)

H

habitat the place where a species normally lives (p. 509)

half life the time required for half of a substance, such as a radioactive isotope, to disappear (p. 484)

haploid having only a single set of chromosomes; compare with *diploid* (p. 295)

haplodiplontic a sexual cycle in which both haploid and diploid cells undergo mitosis (p. 297)

haplontic a sexual cycle in which only haploid cells undergo mitosis (p. 297)

heme a molecule made of four modified pyrrole rings with an iron atom in its center, used in various proteins (p. 84)

herbivore an animal that uses only plant material as food; **herbivorous,** adjective (p. 143)

heterokaryon a cell, especially in a fungus, that contains two types of genetically different nuclei (p. 162)

heterotrophy a nutritional mode in which an organism uses organic compounds as its carbon source; compare with *autotrophy* (p. 143)

heterozygous having two different alleles at a genetic locus in a diploid condition; compare with *homozygous* (p. 314)

homeostasis the tendency of an organism to maintain itself in a stable condition through compensatory adjustments to changing conditions (p. 218)

homeothermic maintaining a constant internal temperature independently of the environmental temperature; warm-blooded; compare with *poikilothermic* (p. 683)

homolog either of the two essentially identical copies of a chromosome in a diploid organism (p. 298)
homologous having a close identity of component parts and structures that provides evidence for a common evolutionary ancestry; also, derived from a common embryological precursor (p. 25)
homozygous having identical alleles at some gene locus in a diploid condition; compare with *heterozygous* (p. 313)
hormone a substance produced by some cells in a multicellular organism that has a specific influence on the activities of other cells (p. 223)
host a cell or organism that supports a parasitic or commensal organism or virus (pp. 246, 569)
humus the brown or black component of soil that consists of decaying organic material (p. 526)
hybrid the offspring of two sexually reproducing organisms of different varieties or genotypes (p. 312)
hydrocarbon an organic molecule made only of hydrogen and carbon (p. 58)
hydrogen bond a weak interaction between molecules formed by a hydrogen atom that creates a bridge between two more negative atoms (p. 53)
hydrologic cycle the cycle of transformations of water within the biosphere (p. 583)
hydrolysis the splitting of a bond by addition of water (p. 67)
hydrophilic a type of organic molecule or side chain that tends to dissolve in water and form bonds with water (p. 61)
hydrophobic a type of organic molecule or side chain that tends to associate with others like itself but not with water (p. 61)
hydrostatic pressure the pressure exerted uniformly in all directions in a fluid (pp. 726, 963)
hydroxyl group the group —OH (p. 62)
hyperosmotic having a higher osmotic pressure than a reference solution (p. 156)
hypertonic a solution in which a reference cell shrinks (p. 156)
hypha one of the individual filaments of a mycelium (p. 651)
hypoosmotic having a lower osmotic pressure than a reference solution (p. 156)
hypotonic a solution in which a reference cell swells (p. 156)

I

immune having resistance to infection by some agent; specifically, said of an animal that has antibodies against the agent (p. 1022)
induce 1. to activate a gene or set of genes to produce their products (p. 353); 2. in embryology, the process in which one tissue determines the developmental course of another; **inducer** (p. 429)
infectious disease disease caused by an organism or virus; compare with *functional disease* (p. 616)
information the specification of a limited set of possibilities (generally only one) out of a larger set (p. 30)
inhibitor a material that reduces the rate of a process (p. 97)
interspecific competition competition among the members of different species (p. 539)
interstitial fluid the fluid between the cells of an animal tissue (p. 675)
intracellular fluid cytosol; the fluid within a cell (p. 675)
intraspecific competition competition among the members of a single species (p. 539)
in vitro in a chemical apparatus; sometimes, in isolated cells or tissue; compare with *in vivo* (p. 278)
in vivo in an intact living organism (p. 278)
ion a charged atom or molecule (p. 49)
ionic bond an attraction between two ions (p. 50)
isosmotic having the same osmotic pressure as a reference solution (p. 156)
isotonic a solution in which a reference cell neither sinks nor swells (p. 156)
isotope a form of an element whose atoms have a distinctive number of neutrons (p. 47)

K

***K* selection** natural selection that tends to produce stable populations of large size; compare with *r* selection (p. 551)
kairomone a compound that tends to give a selective advantage to a species that receives it rather than to the one that produces it; compare with *allomone* (p. 566)
karyotype a description of the chromosome set of an organism (p. 298)
ketone an organic molecule with a carbonyl group at an internal position (p. 62)
kilocalorie a unit of 1,000 calories (p. 54)
kinetic energy the energy an object has by virtue of its motion (p. 132)
Krebs cycle a cycle of chemical reactions at the center of metabolism; also called **citric acid cycle** (p. 183)

L

labile readily undergoing chemical breakdown or denaturation; unstable
lamella a thin, platelike structure (p. 116)
larva a pre-adult form of an animal that has a distinctive form (p. 715)
lateral referring to the side (p. 668)
ligand a small molecule that can bind specifically to a protein (pp. 101, 219)
linkage group a set of genetic markers that are linked to one another (p. 323)
linkage map a description of the locations of mutations, genes, and other genetic elements in a genome (p. 321)
linked said of two or more genetic markers that tend to be inherited together because they reside on a single chromosome (p. 321)
lipid an organic molecule that dissolves readily in nonpolar solvents (p. 157)
locus the position where a gene (or other genetic unit) is located on a genetic map (p. 333)
lumen the inside of a hollow structure (p. 122)
lysis the destruction of a cell by rupturing its cell membrane; **lyse,** verb (p. 248)
lysogeny the condition of a cell harboring a virus in a provirus form (p. 343)

M

macromolecule a large molecule; see *polymer* (p. 67)
marker a mutation used to identify and locate a gene or other genetic element (pp. 323, 332)
medial toward the middle or midline of an object
medulla the central portion of a structure, contrasted with *cortex* (p. 886)
meiosis the process, in all sexually reproducing organisms, in which a diploid set of chromosomes is divided into one or more haploid sets (p. 296)
meiospore a haploid spore produced by meiosis (p. 651)
membrane a thin, pliable sheet; specifically: 1. any of the layers of lipid and protein that form the boundaries and many internal structures of cells (pp. 110, 154); 2. a thin tissue in many animals made of cells and fibers (p. 404)
messenger RNA (mRNA) an RNA transcript that carries the information for the structure of one or more proteins (p. 276)
metabolic pathway a sequence of chemical reactions catalyzed by a series of enzymes (p. 138)
metabolism the sum total, or any part, of the chemical reactions that occur in organisms (p. 13)
metabolite any of the molecules being transformed through a metabolic pathway (p. 138)
microfilament a thin filament composed of actin that is involved in shaping and moving eucaryotic cells (p. 226)
microorganism any organism that can only be seen by means of a microscope (p. 109)
microtubule a thin tubule made of tubulin that is involved in shaping and moving eucaryotic cells (p. 228)
mitochondrion an organelle in eucaryotic cells that carries out oxidative metabolism and produces ATP (p. 124)
mitosis the process, in eucaryotic cells, in which the nucleus divides and its chromosome complement is divided in two (p. 265)
mitospore a spore formed by mitosis (p. 651)
model 1. a conceptual scientific structure used to represent and explain a particular phenomenon (p. 28); 2. in evolutionary ecology, a species that a mimic species evolves to resemble (p. 575)
modular a mode of organization in which an organism is made of distinct subunits that grow, and sometimes reproduce, independently; compare with *unitary* (p. 673)
molar a measure based on a mole; specifically, a 1-molar solution contains a mole of a substance per liter (p. 52)
mole a quantity of a substance in grams numerically equal to its molecular (or atomic) weight; 6×10^{23} molecules (or atoms) of a substance (p. 51)
monomer one of the similar or identical molecules of which a polymer is made (p. 67)
monophyletic of a taxonomic unit derived from a single clade; compare with *paraphyletic, polyphyletic* (p. 453)
morph one of two or more distinct forms that coexist in a population and are not defined by age or sex (p. 446)
morphogenesis the process in which an organism develops its characteristic form (p. 398)
morphology the shape, color, and other general features of an organism; also, the science of the forms of organisms (p. 33)
mortality death; compare with *natality* (p. 543)
motile capable of moving from place to place; compare with *sessile* (p. 666)

multicellular consisting of several to many cells (p. 110)

multimer a complex made of two or more similar or identical protein molecules (protomers) (p. 84)

mutant a new form of an organism that arises as the result of a mutation (p. 242)

mutation a stable change in a genome that alters the information it carries or the expression of that information (pp. 34, 242)

mutualism a mutually beneficial association between two species (p. 571)

mycelium an interwoven mass of hyphae that forms the vegetative body of a fungus and of some filamentous bacteria (p. 651)

N

N-terminal end the end of a polypeptide with a free amino group; compare with *C-terminal end* (p. 75)

NAD^+ nicotinamide adenine dinucleotide, the major coenzyme used for dehydrogenation and hydrogenation reactions (p. 147)

nascent in the process of being formed, as a protein chain that is not yet completed (p. 278)

natality the formation of a new organism by birth, hatching, etc. (p. 543)

natural selection differential reproduction and survival of organisms in nature (p. 22)

neural referring to the nervous system

neurotransmitter a ligand used to carry signals across a synapse between two cells, especially between two neurons (p. 871)

neutron a neutral particle in the nuclei of most atoms (p. 46)

niche variously defined, but generally used to mean the total of an organism's way of life and the habitat in which it lives, including all factors that set a limit on that habitat (pp. 32, 560)

nitrogenous containing nitrogen (p. 69)

nonpolar of a compound or chemical group that does not have an electrical charge (p. 61)

normal distribution a distribution of values (such as size) around a mean in which most values fall close to the mean with decreasing numbers toward the extremes (pp. 315, 377)

nuclease an enzyme that hydrolyzes a nucleic acid (p. 346)

nucleic acid a polymer of nucleotides (p. 70)

nucleolus a dark-staining region of a eucaryotic nucleus in which ribosomes are synthesized (p. 120)

nucleoplasm the fluid contents of the nucleus (p. 118)

nucleoside a compound made of a pentose sugar and a nitrogenous base (p. 251)

nucleotide a nucleotide with a phosphate attached to the sugar (p. 69)

nucleus 1. the distinctive organelle of a eucaryotic cell, consisting of a membranous envelope in which the chromosomes reside (p. 116); 2. a cluster of neuron cell bodies within the central nervous system (p. 875); 3. the central body of an atom, made of protons and, usually, neutrons (p. 46)

nutrient any substance required for the growth and maintenance of an organism (p. 215)

O

obligate a way of life that an organism is obliged to follow; compare with *facultative* (p. 614)

olfaction the process of smelling; also, the sense of smell (p. 914)

omnivore an organism that eats a variety of foods, including plant and animal materials (p. 143)

ontogeny the course of embryonic development of a structure or organism (p. 718)

open reading frame (ORF) a DNA sequence that is potentially capable of encoding a protein but has not been shown empirically to be a gene (p. 332)

operator the site in an operon to which a regulatory protein binds (p. 354)

operon a set of genes that are regulated together and transcribed as a single messenger RNA unit (p. 354)

opportunistic (fugitive) species a species whose population density tends to change erratically, growing rapidly at times, decreasing suddenly at other times; compare with *equilibrium species* (p. 551)

orbital a region in an atom where electrons of a specific energy may reside (p. 48)

organelle a subcellular structure with a specific function that can be identified by microscopic examination (p. 118)

organic acid an organic molecule that bears a carboxyl group (p. 62)

organism a structure that is capable of synthesizing its own structure and generally of reproducing itself by obtaining energy and raw materials from its environment, following the program encoded in its genome (pp. 28–30)

osmolarity a measure of the concentration of a solution, taking into account all particles in the solution that can contribute to osmosis (p. 155)

osmoregulation the homeostatic regulation of osmotic balance in an organism (pp. 677, 977)

osmosis the process in which water tends to diffuse through a semipermeable membrane from a region of lower solute concentration to one of higher solute concentration (p. 155)

osmotic pressure a pressure that develops in a solution by virtue of osmosis from an adjacent solution (p. 155)

ovary an organ in which ova are produced (p. 296)

ovum (pl., **ova**) a female reproductive cell (p. 296)

oxidation a process in which electrons (or hydrogen atoms) are lost from a substance, or in which oxygen is added to it; **oxidize,** verb; compare with *reduction* (p. 144)

oxidative phosphorylation the process in which ATP is formed from ADP and inorganic phosphate in an electron transport system (p. 182)

oxidizing agent a substance that oxidizes another substance (p. 144)

P

parapatric living in adjacent geographic areas, not separated by physical barriers; compare with *allopatric* and *sympatric* (p. 490)

paraphyletic a taxonomic unit that includes some, but not all, of the descendants of a single species (p. 454)

parasitism a mode of life in which one organism lives at the expense of another, generally living on or in it and taking some resource from it, such as a share of its food; **brood parasitism,** one species using another species to care for its offspring (p. 570)

parthenogenesis reproduction in which eggs develop without fertilization (p. 297)

peptide a chain of amino acids joined by peptide linkages (p. 75)

peptide linkage the (NH)—(C=O) structure formed by joining an amino group to a carboxyl group with the removal of a water molecule (p. 75)

permeable allowing materials to pass through (p. 155)

permease a protein that facilitates the transport of a material across a cell membrane; a transport protein (p. 167)

pH a measure of the acidity or alkalinity of a solution; specifically, the negative logarithm of the molar concentration of hydrogen ions (p. 52)

phagocytosis the process in which a cell, especially a white blood cell, engulfs foreign material and digests it (p. 233)

phenotype a description of the appearance (or other manifestation) of an organism, as contrasted with its genotype (p. 314)

pheromone a substance produced by some individuals of a species that has a specific effect on other members (pp. 223, 547)

phosphate the ion PO_4^{3+} (p. 62)

phospholipid any of several lipids that contain a phosphate (p. 159)

phosphoryl the group PO_3^{3+}, with up to two hydrogen atoms attached, depending on the pH (p. 140)

phosphorylation the addition of a phosphoryl group to a compound (p. 142)

phosphoryl-group transfer potential a measure, in terms of free energy, of the ability of a phosphorylated compound to transfer its phosphoryl group to another substance (p. 141)

photoautotrophy a nutritional mode in which energy is obtained from light and carbon is obtained from CO_2; **photoautotroph,** an organism that lives in this way (p. 143)

photoheterotrophy a nutritional mode in which energy is obtained from light and carbon is obtained from organic compounds; **photoheterotroph,** an organism that lives in this way (p. 143)

photon a quantum (or "particle") of electromagnetic energy, such as light (p. 197)

photoperiod the length of a period of light, compared with that of the contiguous dark periods (p. 834)

photoperiodism the governance of behavior or activity by the photoperiod (p. 834)

photophosphorylation the synthesis of ATP driven by light energy (p. 202)

photosynthesis the synthesis of organic compounds through the use of energy captured from light (p. 143)

photosystem an organized system of pigments and electron carriers that can capture light energy and convert it into a stable chemical form (p. 202)

phototrophy a nutritional mode in which energy is obtained from light; **phototroph,** an organism that lives in this way; compare with *chemotrophy* (p. 143)

phylogeny the evolutionary history of a species or a group of species (p. 37)

pigment a chemical compound that absorbs visible light and is therefore colored (p. 197)

plasma membrane the limiting membrane of a cell (p. 110)

plasmid a small, independently replicating DNA molecule that resides in a cell in addition to the cellular genome (p. 341)

plastid any of several semi-autonomous plant organelles, such as chloroplasts, chromoplasts, and leucoplasts (p. 788)

pleated sheet a protein structure in which parallel or antiparallel peptide chains are hydrogen-bonded to one another (p. 78)

poikilothermic "cold blooded"; having a body temperature strongly influenced by the environment rather than a relatively constant temperature; compare with *homeothermic* (p. 683)

polar of a compound or chemical group that has (or can have) an electrical charge and thus interacts with other polar materials through electrostatic interactions; compare with *nonpolar* (p. 49)

polymer a macromolecule made by joining many similar or identical molecules (monomers) through similar or identical bonds (p. 67)

polymerase an enzyme that synthesizes polymers (pp. 263, 275)

polymerization the process of forming a polymer (p. 67)

polymorphic having two or more distinct forms within a single population that are not correlated with age or sex (p. 449)

polypeptide a polymer made of amino acids; a protein (p. 75)

polyphyletic said of a taxon derived from two or more ancestral lines (p. 453)

polyploid having more than two sets of chromosomes (p. 499)

population all the organisms of one species that live in one area (p. 27)

posterior toward the tail end of an organism, especially an animal; compare with *anterior* (p. 666)

potential a measure of a system's capacity to do work (p. 140)

potential energy energy that is stored in matter by virtue of its position (as in a chemical bond or in an object that can fall to a lower level), which could, in principle, be converted into work (p. 132)

precursor a chemical compound that can be transformed into some other compound, especially in a metabolic pathway

predation a mode of life in which food is obtained by catching and eating other organisms (p. 566)

primary consumer an herbivore (p. 143)

primary structure the sequence of monomers in a polymer, especially a protein (p. 77)

procaryotic (also **prokaryotic**) a type of cell that lacks a true nucleus bounded by a nuclear envelope (p. 116)

producer an organism that brings energy into an ecosystem, generally through photosynthesis; compare with *consumer, decomposer* (p. 143)

proliferate to grow by increasing in numbers; contrasted with growth by an increase in mass (p. 215)

promoter a region of a genome to which RNA polymerase binds and initiates transcription (p. 276)

prosthetic group a functional molecule bound to a protein; compare with *coenzyme* (p. 84)

protease an enzyme that digests a protein (p. 1000)

protein a functional polymer of amino acids (p. 73)

protomer one of the similar or identical protein subunits of a multimer (p. 84)

proton a heavy, positively charged, basic particle in the nuclei of all atoms (p. 46)

proton-motive force the electrochemical potential developed in a gradient of protons in an electron transport system (p. 185)

prototroph an organism that can synthesize all of its structure from simple nutrients; compare with *auxotroph* (p. 243)

proximal near the base or attachment point of an object; compare with *distal*

pseudopod a cytoplasmic extension, especially a relatively large, blunt one, of an amoeboid cell (p. 226)

pulmonary relating to the lungs

purine a nitrogenous base made of a 6-membered ring joined to a 5-membered ring; see *pyrimidine* (p. 251)

pyrimidine a nitrogenous base made of a ring of 6 atoms (p. 251)

pyrophosphate a compound ($H_4P_2O_7$) of two linked phosphates (p. 265)

Q

quantum a small unit of energy (p. 198)

quaternary structure protein structure resulting from the combination of two or more polypeptides (p. 84)

R

R group an abbreviation, in chemistry, for an unspecified atom or group of atoms attached to a molecule (p. 73)

***r* selection** natural selection that tends to develop species with rapid growth rates (p. 551)

race a subspecies (p. 450)

radial having parts arranged circularly around a center (p. 669)

radiation 1. energy transmitted as waves or particles, such as electromagnetic radiation; 2. the transmission of heat as infrared light, as contrasted with convection and conduction (p. 815)

radioisotope an isotope with unstable atoms that decay into a more stable form, with the release of radiation (p. 47)

reagent a chemical compound used in a certain context because of the particular reaction it engages in

realized niche the niche to which a species is restricted by competition with other species; compare with *fundamental niche* (p. 560)

receptor a protein, neuron, or other structure that is stimulated by some environmental factor or condition (p. 221)

recessive the quality of an allele that only produces a specific phenotype when homozygous (p. 313)

recognition a process in which the presence of some molecule has a specific effect through its binding to a biological molecule, generally a protein; **recognize,** verb (p. 222)

recombinant a genome composed of a combination of the alleles of two other genomes (p. 321)

recombination a process in which portions of two genomes are exchanged to make one with a new combination of alleles (p. 321)

reducing agent a substance that is able to reduce another substance (p. 143)

reduction a process in which electrons or hydrogen atoms are added to a substance; **reduce,** verb; compare with *oxidation* (p. 144)

reduction potential a measure of the relative ability of one substance to reduce another, the more negative number indicating the greatest potential (p. 146)

renal referring to the kidney

replication the process in which new copies of a genome are made (p. 241)

repressor a protein that regulates transcription of one or more genes by binding to the DNA at a specific site; also, a ligand that binds to such a protein to activate it; **repress,** verb (p. 354)

resolution the ability to distinguish two points as distinct objects rather than a single object (p. 112)

resource any object, substance, or place an organism requires that can be used up (p. 33)

respiration 1. breathing (external respiration); 2. oxidative metabolism in which an inorganic substance, usually oxygen, is used as the final electron (hydrogen) acceptor; compare with *fermentation* (pp. 177, 951)

resting (membrane) potential the difference in charge across the membrane of a cell, especially a neuron, that is not firing a nerve impulse (p. 877)

rhizoid a simple, rootlike structure of many fungi and plants (p. 692)

ribonucleic acid (RNA) a polymer of ribonucleotides (p. 70)

ribonucleotide a nucleotide whose sugar is ribose (p. 69)

ribose a five-carbon sugar, a principal component of nucleotides (p. 69)

ribosomal RNA any of several types of RNA built into the structure of a ribosome (p. 276)

ribosome a complex of protein and RNA that is the factory for protein synthesis (p. 120)

RNA polymerase an enzyme that synthesizes RNA (p. 275)

S

S period the phase of the eucaryotic cell cycle when DNA synthesis occurs (p. 265)

salt a neutral compound consisting of positive and negative ions (p. 51)

saprobe (also **saprophyte**) an organism that lives off of decaying dead organic matter (p. 628)

saturated 1. an organic molecule, as a fatty acid, with no double bonds carrying the maximum number of hydrogen atoms (p. 59); 2. said of a process or structure that is operating at its maximum rate (p. 95)

secondary consumer an organism that eats primary consumers; a carnivore (p. 143)

secondary structure the structure of a protein due to interactions among its imino and carbonyl groups;

most commonly an α helix or β-pleated sheet (p. 78)

seed the fertilized, ripened ovule of a seed plant (p. 703)

serum the liquid portion of blood from which cells and clotting proteins have been removed (p. 1023)

sessile immotile; not able to move about independently (p. 666)

sex chromosome a chromosome, generally designated X or Y, that is involved in the determination of sex; compare with *autosome* (p. 325)

sexual cycle a sequence of events in the life history of an organism consisting of a haploid phase, fertilization, a diploid phase, and meiosis (p. 295)

sexual reproduction reproduction that occurs through fusion of two gametes (p. 32)

signal ligand a molecule used only to carry information, as opposed to one that is metabolized (p. 223)

sink a place or compartment that absorbs or consumes a substance; compare with *source* (p. 811)

sister either of two structures or entities—such as chromosomes or species—derived from a single structure or entity; compare with *daughter* (p. 299)

site 1. a specific place on a molecule, especially a region of a protein to which other molecules can bind; 2. the position on a genetic map that represents a point mutation (p. 333)

somatic cell any body cell of an organism other than gametes and their precursors (p. 296)

source a place or compartment that produces a substance; compare with *sink* (p. 811)

speciation the process in which one species is divided into two or more (p. 484)

species 1. in sexual organisms, the most extensive population of organisms that are capable of interbreeding with one another; 2. in asexual organisms, and in many others for which there is insufficient information, a group of organisms with common ancestry and characteristics that are judged to be a relatively stable, homogeneous type (p. 447)

sperm a motile, haploid male reproductive cell (p. 296)

spindle a structure of microtubules formed during a typical mitosis or meiosis that separates chromosomes from one another (p. 267)

spore any of various single-celled reproductive structures that are able to develop into a vegetative structure in the reproductive cycle, especially those able to survive adverse conditions (p. 297)

sporophyte a structure in the life cycle of a plant that produces spores; compare with *gametophyte* (p. 298)

steady state a condition in which materials move through an open system at balanced rates such that the overall structure of the system does not change (p. 217)

stereospecificity the specific match between molecules based on their particular shapes (p. 101)

stimulus any event or aspect of the environment that influences the behavior of an organism (p. 221)

substrate a molecule upon which an enzyme acts (p. 73)

succession the process in which different types of ecosystems or communities replace one another (p. 593)

sugar an aldehyde or ketone with one or more hydroxyl groups (p. 62)

sulfhydryl group the group —SH (p. 62)

symbiosis any of the various modes of life in which two or more species live together, including phoresis, commensalism, mutualism, and parasitism; also used to mean mutualism alone (p. 569)

sympatric living in the same geographical region; compare with *allopatric, parapatric* (p. 490)

symport a process in which two molecules or ions are simultaneously transported across a membrane in the same direction (p. 170)

syncytium a cellular body with many nuclei in one cytoplasm formed by the fusion of several cells; sometimes, a coenocyte (p. 940)

system a portion of the universe chosen for description or study (p. 28)

T

target cell the cell on which a hormone acts (p. 868)

taxis movement or growth in response to a particular stimulus such as light or a chemical (p. 221)

taxon a taxonomic category such as species, genus, family, and so on (p. 450)

teleological an explanation that ascribes purpose or foresight to organisms that are unable to have an explicit purpose or foresight (p. 10)

teleonomic referring to the inherent purposefulness or functionality of organisms as a result of their evolution by natural selection (p. 10)

temperate 1. having a mild seasonal climate; 2. a type of virus that can multiply in a stable association with a host cell, rather than lysing the cell (p. 343)

template a molecule that serves to direct the synthesis of another molecule (p. 263)

termination codon a codon that terminates protein synthesis (p. 278)

terminator a signal in the genome that terminates RNA synthesis (p. 276)

territoriality a behavior pattern in which an animal establishes a living space and defends it against intruders (pp. 549, 1066)

tertiary structure the structure of a protein due to interactions among its amino acid side chains (p. 82)

thallus the body of some plants and algae in which there is little or no differentiation into stems, roots, and leaves (p. 634)

tissue a structure made of several to many cells of one or a few types (p. 669)

titer a measure of the amount, strength, or concentration of a substance or agent in terms of its biological activity rather than in terms of mass or number (pp. 246, 1024)

tracer a material used to follow the fate of some object or substance, such as a radioisotope used to follow the course of metabolism (p. 203)

transcript an RNA molecule made by transcription from DNA (p. 275)

transcription the process in which information is transferred from DNA to RNA (p. 274)

transducer a device that converts energy or information from one form to another ; **transduce,** verb (p. 218)

transduction 1. the conversion of energy or information from one form to another (p. 218); 2. the transfer of DNA from one cell to another by means of a virus (p. 346)

transfer RNA (tRNA) an RNA molecule that carries amino acids to their proper positions, as specified by a messenger RNA (p. 277)

transformation the conversion of a bacterium to a new genotype through the direct uptake of DNA (p. 245)

translation the process in which the information in RNA is converted into the amino acid sequence of a protein (p. 274)

translocation 1. the movement of molecules across a cell membrane (p. 171); 2. the movement of materials from one part of a plant to another through the phloem (p. 811); 3. a chromosomal rearrangement in which a segment of one chromosome becomes attached to another (p. 474)

transpiration the evaporation of water vapor from an organism (p. 678)

trophic referring to feeding or nutrition (p. 509)

trophic level the level of an organism in its food web (p. 509)

turgor the normal tense, distended condition of functioning cells, especially plant cells (p. 156)

turnover the process in which the components of an open system are continually replaced (p. 582)

U

ultraviolet light electromagnetic radiation with a wavelength longer than that of X rays but shorter than that of visible light, roughly the range of 200–350 nm (pp. 113, 198)

unicellular consisting of a single cell (p. 110)

unitary consisting of a single reproducing body, rather than of modules that function and reproduce independently; compare with *modular* (p. 673)

upstream in the direction along a genome opposite to the direction of RNA transcription; compare with *downstream* (p. 354)

V

vacuole a membrane-bounded cellular structure filled with fluid, characteristic of plant cells (p. 124)

vascular referring to the vessels or tubules that conduct fluids through a multicellular organism, such as veins, arteries, xylem, and phloem (p. 666)

vector 1. an organism that carries or transmits infectious agents (p. 618); 2. a plasmid or similar nucleic acid molecule capable of carrying a piece of cloned DNA (p. 347)

vegetative the condition of an organism that is growing normally, rather than forming spores or reproducing sexually; also, the comparable condition of a virus during intracellular multiplication (p. 297)

ventral toward the belly of an animal; compare with *dorsal* (p. 668)

ventricle 1. one of the chambers of a heart (p. 954); 2. one of the chambers of the vertebrate brain (p. 895)

vesicle any of various small, membrane-bounded cellular structures, often carrying materials from one region of a cell to another (p. 122)

viable capable of growth and metabolism; living (p. 246)

villus a slender, fingerlike process, especially one found on the inner intestinal wall (pp. 682, 1005)

virion a virus particle, consisting of a genome in a protective protein coat, sometimes with an additional membranous coat (p. 246)

virus a genetic system consisting of a nucleic acid genome (either DNA or RNA, but never both) enclosed at one stage in a protein coat to make a virion, and capable of multiplying only within a functioning cell (p. 30)

vitamin an organic compound that an organism requires in relatively small amounts, generally for use as a coenzyme, but one that it cannot synthesize for itself (p. 97)

voltage an electrical potential (pp. 140, 170)

W

wild type a common or typical variety of an organism chosen as standard, to which all mutant types are compared; **wild-type,** adjective (p. 242)

X

X-linked located on an X chromosome (p. 326)

xeromorphic of adaptations that allow an organism to live in dry (xeric) conditions (p. 817)

Z

zygote a cell formed as the result of fertilization (p. 296)

Credits

Photographs

Part Openers

1: © Tom McHugh/Photo Researchers, Inc., **2:** © Kim Taylor/Bruce Coleman, Inc., **3:** Courtesy of Warner Jenkinson Company, **4:** © K.G. Murti/Visuals Unlimited, **5:** © Bruce Berg/Visuals Unlimited, **6:** © Doug Martin/Photo Researchers, Inc., **7:** © Lynda Richardson/Peter Arnold, Inc., **8:** © Don W. Fawcett/Visuals Unlimited, **9:** © Budd Titlow/ Visuals Unlimited, **10:** © Randy Morse/Tom Stack & Associates, **11:** © Dr. F.C. Skvala/Peter Arnold, Inc., **12:** Courtesy of Elizabeth Kutter, **13:** © Camille Tokerud/Photo Researchers, Inc., **14:** © Jeff Greenberg/Visuals Unlimited, **15:** © Ronald Austing/Photo Researchers, Inc., **16:** © G.R. "Dick" Roberts, **17:** © K.G. Murti/Visuals Unlimited, **18:** © David M. Phillips/Visuals Unlimited, **19:** © SIU/Visuals Unlimited, **20:** © Petit Format/ Nestle/Photo Researchers, Inc., **21:** © CABISCO/ Visuals Unlimited, **22:** © Kjell Sandved/Photo Researchers, Inc., **23:** © G. Prance/Visuals Unlimited, **24:** © John Trager/Visuals Unlimited, **25:** Digital Stock/Landscapes, **26:** © E.R. Degginger/Bruce Coleman, Inc., **27:** © Tom Lyon, **28:** © Chris Johns/Tony Stone Images, **29:** © S. Lowry/Univ. Ulster/Tony Stone Images, **30:** © E.R. Degginger/ Photo Researchers, Inc., **31:** © Michael W. Beug, **32:** © Joe McDonald/Tom Stack & Associates, **33:** © Michael Gadomski/Photo Researchers, Inc., **34:** PhotoDisc/Details of Nature, **35:** © Fritz lking/Visuals Unlimited, **36:** © Springer/Corbis, **37:** Burton S. Guttman, **38:** © Bill Beatty/Visuals Unlimited, **39:** © D. Newman/Visuals Unlimited, **40:** © E.R. Degginger/Photo Researchers, Inc., **41:** © Hulton Deutsch Collection/Corbis, **42:** © Gamma Liaison, **43:** © SIU/Peter Arnold, Inc., **44:** © Thomas W. Martin/Photo Researchers, Inc., **45:** © P. Mota/S. Correr/SPL/Photo Researchers, Inc., **46:** © Martin Land/Science Photo Library/Photo Researchers, Inc., **47:** © Charles Cecil/Visuals Unlimited, **48:** © Peter Cade/Tony Stone Images, **49:** © Charlie Ott/Photo Researchers, Inc., **50:** © Carlos Sanuvo/Bruce Coleman, Inc., **51:** © Rudy H. Kuiter/Aquatic Photographics.

Chapter 1

Opener: © Tom McHugh/Photo Researchers, Inc., **1.1:** © Nigel Dennis/Photo Researchers, Inc., **1.2:** Burton S. Guttman, **1.4:** Courtesy of C.R. O'Dell, Rice University, NASA, **1.7:** © Kevin and Betty Collins/Visuals Unlimited, **1.8a:** © Tom Bean/Tony Stone Images, **1.8b:** © Sinclair Stammers/Photo Researchers, Inc., **1.9:** © M.D. Maser/Visuals Unlimited, **1.11:** © Kelvin Aitken/Peter Arnold, Inc., **1.12:** Burton S. Guttman.

Chapter 2

Opener: © Kim Taylor/Bruce Coleman, Inc., **2.3:** © Jeff Greenberg/Visuals Unlimited, **2.4:** Burton S. Guttman, **2.6:** © Mary Evans/Photo Researchers, Inc., **2.12a:** © Tim Davis/Photo Researchers, Inc., **2.12b:** © Charles E. Moher/Photo Researchers, Inc., **2.12c:** © John R. MacGregor/Peter Arnold, Inc., **2.13:** © Will Troyer/Visuals Unlimited, **2.14:** © Kenneth Eward/BioGrafx-Science Source/ Photo Researchers, Inc., **2.17:** © Michael Boyes/ Corbis, **2.19:** © George Herben Photo/Visuals Unlimited, **2.20a:** © Breck Kent/Animals Animals/ Earth Scenes, **2.20b:** © Breck Kent/Animals Animals Earth Scenes.

Chapter 3

Opener: Courtesy of Warner Jenkinson Company, **3.1:** Courtesy of IBM Corporation, Research Division/ Almaden Research Center, San Jose, CA., **3.4:** © Richard C. Walters/Visuals Unlimited, **3.5:** © R.G. Kessel-C.Y. Shih/Visuals Unlimited, **3.6:** © Biophoto Association/Photo Researchers, Inc., **3.7a:** © Bios/J. Douillet/Peter Arnold, Inc., **3.7b:** © Peter Johnson/Corbis, **3.8a:** © Doug Sokell/ Visuals Unlimited, **3.8b:** © Doug Sokell/Visuals Unlimited, **3.9a:** © W.A. Banaszewski/Visuals Unlimited, **3.9b:** © Richard Thom/Visuals Unlimited, **3.11:** Burton S. Guttman, **3.12:** © Corbis/ Robert Pickett.

Chapter 4

Opener: © K.G. Murti/Visuals Unlimited, **4.3:** © Steven Holt/Aigrette Photography, **4.5b:** Burton S. Guttman, **4.9:** Courtesy of Jerome Gross, **4.10:** © Kelvin Aitken/Peter Arnold, Inc.

Chapter 5

Opener: © Bruce Berg/Visuals Unlimited, **5.3:** Courtesy of Richard J. Feldman, **5.6:** © Hulton Deutsch Collection/Corbis, **5.9a:** © Science Source/ Photo Researchers, Inc., **5.9b:** © David M. Phillips/ Photo Researchers, Inc., **5.11:** © Terry Vine/Tony Stone Images.

Chapter 6

Opener: © Doug Martin/Photo Researchers, Inc., **6.1a:** © Kevin Collins/Visuals Unlimited, **6.1b:** National Library of Medicine, **6.2:** National Library of Medicine, **6.3a:** © O. Meckes/Nicole Ottawa/Photo Researchers, Inc., **6.3b:** © McGraw-Hill Higher Education/Kingsley Stern, photographer **6.5:** F.W. Sears Optics 3/e, **9.4(b):** © 1949 Addison - Wesley Publishing. Reprinted by permission of Addison - Wesley Longman, Inc., **6.6:** PhotoDisc/Family and Lifestyles Volume 15, **6.10:** © David M. Phillips/ Visuals Unlimited, **6.11:** © Keith Porter/Photo Researchers, Inc., **6.13:** © Andrew Syred/SPL/Photo Researchers, Inc., **6.14a,b:** Burton S. Guttman, **6.15a:** © Don W. Fawcett/Visuals Unlimited, **6.15b:** Burton S. Guttman, **6.16:** © K.G. Murti/ Visuals Unlimited, **6.18b:** Courtesy of Ron Milligan, **6.19b:** Burton S. Guttman, **6.20b:** Courtesy Don Gash, photo by Harold Parks, University of Kentucky, **6.21b:** Courtesy of Dr. Jeptha R. Hostetler, University of Kentucky, Department of Anatomy, **6.22b:** Courtesy of Dr. A.V. Grimstone/Cambridge University, **6.23b:** Burton S. Guttman, **6.24a:** © Ed Reschke/Peter Arnold, Inc., **6.24b:** © M. Eichelbeiger/Visuals Unlimited, **6.24c:** © M. Eichelberger/Visuals Unlimited, **6.25b:** Courtesy of Dr. Phillip W. Brandt, **6.26b:** Burton S. Guttman, **6.28a:** From Douglas E. Kelly, J. Cell Biol. 28, 51, (1966). Reproduced by copyright permission of The Rockefeller University Press, **6.28 b,c:** © Don W. Fawcett/Visuals Unlimited, **6.29b:** © Biophoto Association/Photo Researchers, Inc., **6.30b:** Courtesy of Dr. Antoinette Ryter, **6.31:** Burton S. Guttman.

Chapter 7

Opener: © Lynda Richardson/Peter Arnold, Inc., **7.4a:** © Jeffrey Howe/Visuals Unlimited, **7.4b:** Burton S. Guttman.

Chapter 8

Opener: © Don W. Fawcett/Visuals Unlimited, **8.1a-c:** Burton S. Guttman, **8.4:** © The McGraw Hill Companies, Bob Coyle, photographer, **8.6:** Burton S. Guttman, **8.14 a,b:** © Daniel Branton of Harvard University,

Chapter 9

Opener: © Budd Titlow/Visuals Unlimited, **9.1 a,b:** © Bill Robbins/Tony Stone Images, **9.3:** National Library of Medicine, **9.15:** © Philip Gould/Corbis, **9.17:** Courtesy of R.G. E. Murray/ The University of Western Ontario.

Chapter 10

Opener: © Randy Morse/Tom Stack & Associates, **10.1c:** Courtesy of Thomas Rost, Photo by Elliot Weier, **10.1d:** Courtesy of Lewis Shumway, **10.2 a,b:** © Dr. Samuel F. Conti, **10.2c:** Courtesy of Dr. Charles C. Remsen, **10.18:** © Max and Bea Hunn/Visuals Unlimited.

Chapter 11

Opener: © Dr. F.C. Skvala/Peter Arnold, Inc., **11.2:** © David M. Phillips/Visuals Unlimited, **11.11:** Courtesy Dr. Julius Adler, University of Wisconsin-Madison, **11.13 a, b:** Courtesy of Dr. Richard J. Feldman, **11.19:** Courtesy of Sandra L. Wolin, **11.20:** Burton S. Guttman, **11.23:** Courtesy of G. Albrecht-Buehler, Department of Cell Biology and Anatomy, Northwestern University, **11.27 (both):** Courtesy of Dr. Elias Lazarides and Dr. Jean Paul Revel, **11.29a:** Courtesy of Dr. A.V. Grimstone/ Cambridge University, **11.29b:** Courtesy of Dr. Keith Porter, **11.29c:** Burton S. Guttman, **11.30b:** Courtesy of Dr. Elton Stubblefield, **11.33a:** Burton S. Guttman, **11.33 b,c:** Courtesy Dr. Peter Satir, Albert Einstein College of Medicine, **11.35b:** © Mr. Bernard Schermetzler, University Extension/University of Wisconsin—Madison Archives, **11.37:** Courtesy Dr. Dorothy F. Bainton, University of California—San Francisco, **11.38 a-d:** © M.M. Perry and A.B. Gilbert "Cell Science."

Chapter 12

Opener: Courtesy of Elizabeth Kutter, **12.1:** Courtesy Dr. Mary M. Allen, Wellesley College, **12.7, 12.8 b, 12.9:** Burton S. Guttman, **12.12:** © Dr. Lee Simon, **12.15a:** Courtesy of Department of Biophysics, King's College London, photo by Maurice Wilkins, **12.15b:** © From "The Double Helix" by J.D. Watson, Atheneum Press, 1968, Courtesy of CSHC Archives.

Chapter 13

Opener: © Camille Tokerud/Photo Researchers, Inc., **13.1:** From "Mitotic and Meiotic Chromosomes of Amphiuma", by Grace M. Donnelly and Arnold H. Sparrow. Pages 91-98. Reprinted without change of paging from the Journal of Heredity, Washington, D.C., Vol. LVI, No. 3, May-June, 1965. © 1965 by the American Genetic Association, **13.2:** © L. Sims/Visuals Unlimited, **13.3:** Courtesy of Dr. David E. Pettijohn, University of Colorado-Health Sciences Center, **13.4a:** Courtesy of Dr. Charles Laird, University of Washington, Victoria Foe Department of Zoology, University of Washington, and Yean Chooi, Department of Biology, Indiana University, **Page 267(all):** Burton S. Guttman,

Chapter 14

Opener: © Jeff Greenberg/Visuals Unlimited, **14.9:** © Oscar L. Miller, Jr. /University of Virginia, Methods Box, **14.1:** © Roger Tully/Tony Stone Images, **14.20:** © Oscar L. Miller, Jr. and Barbara Beatty/ Biology Division/Oak Ridge National Lab.

Chapter 15

Opener: © Ronald Austing/Photo Researchers, Inc., **15.1b:** © Don Smetzer/Tony Stone Images, **15.2a:** © Andy Sacks/Tony Stone Images, **15.2b:** © Tony Stone Images, **15.2c:** © John Warden/Tony Stone Images, **15.3:** ©Alex Kerstitch/Visuals Unlimited, **15.4:** © S.J. Krasemann/Peter Arnold, Inc., **15.10 a, b:** © D.W. Fawcett/Photo Researchers, Inc., **Page 300, Page 301, Page 302:** From "Mitotic and Meiotic Chromosomes of Amphiuma," by Grace M. Donnelly and Ronald H. Sparrow. Pages 91-98. Reprinted without change of paging from the Journal of Heredity, Washington, D.C., Vol. VLI, No. 3, May-June, 1965. © 1965 by the American Genetic Association, **15.12:** © Fernand Ivaldi/Tony Stone Images, **15.15 b, c:** Courtesy of Dr. Edward M. Eddy,

Chapter 16

Opener: © G.R. "Dick" Roberts, **16.1:** © Jeff Greenberg/Photo Researchers, Inc., **16.14b:** © Biophoto Associates/Science Source/ Photo Researchers, Inc., **16.18a:** © The Bettmann Archive/Corbis.

Chapter 17

Opener: © K.G. Murti/Visuals Unlimited.

Chapter 18

Opener: © David M. Phillips/Visuals Unlimited, **18.21:** Courtesy Joseph G. Gall, Carnegie Institution, **18.22:** Burton S. Guttman, **18.23:** Courtesy of Jose Mariano Amabis, Dept. Biology, University of Sao Paulo, **18.32:** Courtesy of Dr. Keith Wood, **18.35:** © Fernand Ivaldi/Tony Stone, **18.36:** Courtesy Ralph Brinster.

Chapter 19

Opener: © SIU/Visuals Unlimited, **19.2:** Courtesy Prof. Linda Strausbaugh, Genetics Program, Dept. of Molecular and Cell Biology. Photo by P. Morenus, Univ. Relations, **19.8:** © Martha Cooper/Peter Arnold, Inc., **19.11a:** © L. Willatt/East Anghan Regional Genetics Services/Science Photo Library/Photo Researchers, Inc., **19.11b:** © Gary Parker/Science Photo Library/Photo Researchers, Inc., **19.13a:** © From "Medical Cytogenetics," by M. Bartalos and T.A. Barmski/Williams & Wilkins Co., **19.13b:** © Science VU/Valerie Lindgren/Visuals Unlimited, **19.14 a,b:** Courtesy of Dr. Kenneth L. Becker, from "Fertility and Sterility," 23:5668-78 1972, Williams & Wilkins, **19.15:** © Jacana/Photo Researchers, Inc., **19.17:** Courtesy of Dr. Thomas G. Brewster, **Sidebar 19.2:** Courtesy of Lifecodes.

Chapter 20

Opener: © Petit Format/Nestle/Photo Researchers, Inc., **20.2:** Courtesy of Dr. Edward M. Eddy, **20.3:** Courtesy of Dr. Edward M. Eddy, **20.13a:** © M. Eichelberger/Visuals Unlimited, **20.13b:** © D.P. Wilson/Eric and David Hosking/Photo Researchers, Inc.

Chapter 21

Opener: © CABISCO/Visuals Unlimited, **21.19b:** Courtesy Robinson, Douglas N., Cant, Kelly and Cooley, Lynn. 1994 Morphogenesis of Drosophila Ovarian Ring Canals. Development 120: 2015-2025, **21.22 b,c:** Courtesy Christiane Nusslein-Volhard/Max Plank Institute, **21.23a:** Courtesy Dr. Christiane Nusslein-Volhard/Max Planck Institute as published in "From Egg to Adult," © 1992 HHMI, **21.23 b-d:** Courtesy James Langeland, Stephen Paddock, and Sean Carroll as published in "From Egg to Adult," © 1992 HHMI, **21.25:** © Oliver Meckes/ Photo Researchers, Inc., **21.26:** Courtesy of Edward Lewis/California Institute of Technology.

Chapter 22

Opener: © Kjell Sandved/Photo Researchers, Inc., **22.3:** Burton S. Guttman, **22.4:** © James H. Carmichael, Jr./Photo Researchers, Inc.

Chapter 23

Opener: © G. Prance/Visuals Unlimited.

Chapter 24

Opener: © John Trager/Visuals Unlimited.

Chapter 25

Opener: Digital Stock/Landscapes, **25.6:** © Manfred Kage/Peter Arnold, Inc., **25.8:** © Science VU/WHOI/Visuals Unlimited, **25.9a:** © F. Stuart Westmorland/Corbis, **25.9b:** © Gary Braasch/Corbis, **25.10:** © Jim Zipp/Photo Researchers, Inc., **25.13:** © Brian Parker/Tom Stack & Associates, **25.14:** © Denise Tackett/Tom Stack & Associates, **25.17:** © Willard Clay/Tony Stone Images, **25.18:** © Patrica Caulfield/ Peter Arnold, Inc., **25.19:** © Gregory G. Dimijian, M.D./Photo Researchers, Inc., **25.23a:** Courtesy of Jack Longino, **25.24:** © Ken Lucas/Visuals Unlimited, **25.25a:** © Frank Oberle/Tony Stone Images, **25.26:** © Willard Clay/Tony Stone Images, **25.27:** © Larry Ulrich/Tony Stone Images, **25.28:** © Steve McCutcheon/Visuals Unlimited, **25.29a:** © Art Wolfe/Tony Stone Images, **25.29b:** © Kennan Ward/Corbis, **25.34:** © Bonnie Heidel/Visuals Unlimited.

Chapter 26

Opener: © E.R. Degginger/Bruce Coleman, Inc., **26.4:** © Doug Sokell/Visuals Unlimited, **26.5:** © Rod Planck/Photo Researchers, Inc., **26.12:** © Holt Studios/Photo Researchers, Inc.

Chapter 27

Opener: © Tom Lyon, **27.10b:** © Biophoto Association/Photo Researchers, Inc., **27.18:** © Mark Chester/Photo Researchers, Inc., **27.26:** © Ed Reschke/Peter Arnold, Inc., **27.27b:** © Glenn Oliver/ Visuals Unlimited, **27.28:** © Kjell B. Sandved/ Visuals Unlimited, **27.31a:** © Leroy Simon/Visuals Unlimited, **27.31b:** © A. Rider/Photo Researchers, Inc.

Chapter 28

Opener: © Chris Johns/Tony Stone Images, **28.12:** © Robert J. Ellison/Photo Researchers, Inc., **28.13:** © Brian Brake/Photo Researchers, Inc., **28.18a:** © Glenn M. Oliver/Visuals Unlimited, **28.18b:** © Jim Hughes/PhotoVenture/Visuals Unlimited.

Chapter 29

Opener: © S. Lowry/Univ. Ulster/Tony Stone Images, **29.5a, b, 29.6 a, b:** Courtesy Prof. Dr. Norbert Pfennig, **29.7a:** © Tom Adams/Visuals Unlimited, **29.7b:** © Michael Abbey/Visuals Unlimited, **29.7c:** © Dr. Dennis Kunkel/Phototake, **29.8:** Courtesy Dr. Mary M. Allen, Wellesley College, **29.9:** © Charles O'Rear, **29.12:** © T. J. Beveridge/ Visuals Unlimited, **29.13:** © Patricia L. Grilione/ Phototake, **29.14a:** Courtesy of J. Staley and J.P. Dalmasso, **29.14b:** © Richard L. Moore/Biological Photo Service., **29.15a:** © Dr. J. Burgress/Photo Researchers, Inc., **29.15b:** Courtesy of Dr. S. G. Bradley/American Society for Microbiology Journals Division, **29.16:** © David M. Phillips/Photo Researchers, Inc., **29.19:** © Jack M. Bostrack/Visuals Unlimited, **29.20 a, b:** © T.J. Beveridge/Visuals Unlimited, **29.23 a, b:** Courtesy Dr. Jeffrey C. Burnham and Dr. Samuel F. Conti, **29.26:** © Science Source/Photo Researchers, Inc., **29.27:** © Jeffrey Howe/Visuals Unlimited, **29.30, 29.31b, 29.32a:** Burton S. Guttman.

Chapter 30

Opener: © E.R. Degginger/Photo Researchers, Inc., **30.4 a,b:** Burton S. Guttman, **30.10:** © Biological Photo Service, **30.14:** © D.P. Wilson/Eric and David Hoskins/Photo Researchers, Inc., **30.15a:** © J. Robert Waaland/Biological Photo Service, **30.15b:** © Daniel W. Gotshall/Visuals Unlimited, **30.15c:** © J. Robert Waaland/Biological Photo Service, **30.16:** © Dr. Samuel F. Conti, **30.24a:** © Michael W. Beug, **30.27:** © Dr. John D. Dodge, **30.31:** © Arthur M. Siegelman/Visuals Unlimited, **30.34:** © E. White/ Visuals Unlimited.

Chapter 31

Opener: © Michael W. Beug, **31.4 a-d:** © Michael W. Beug, **31.8:** Courtesy of David Stadler, **31.9 a-d:** © Michael Beug, **31.10:** © Peter Arnold, Inc., **31.12 a-c:** Courtesy of Steven A. Trudell,

Chapter 32

Opener: © Joe McDonald/Tom Stack & Associates, **32.9:** © Thoresen/Science Source/Photo Researchers, Inc., **32.11a:** © Biophoto Associates/ Science Source/ Photo Researchers, Inc., **32.11b:** © Dennis Drenner/ Visuals Unlimited, **32.23a:** © George J. Wilder/ Science VU/Visuals Unlimited, **32.24a:** © Marty Snyderman/Visuals Unlimited, **32.24b:** © Gerald Corsi/Visuals Unlimited.

Chapter 33

Opener: © Michael P. Gadomski/Photo Researchers, Inc., **33.1:** © M.I. Walker/Photo Researchers, Inc., **33.3:** © Manfred Kage/Peter Arnold, Inc., **33.6:** © John D. Cunningham/Visuals Unlimited, **33.9a:** © Rod Planck/Tony Stone Images, **33.9b:** © Ed Pembleton, **33.9c:** © Glen Oliver/ Visuals Unlimited, **33.14a:** © Ed Pembleton, **33.14b:** © Walter Hodge/ Peter Arnold, Inc.,

33.16: © James Richardson/Visuals Unlimited, **33.17b:** Courtesy Field Museum of Natural History, Neg.# 75400C, **33.18:** © Photo Researchers, Inc., **33.20a:** © Pat Lynch/Photo Researchers, Inc., **33.20b:** © Rod Planck/Photo Researchers, Inc., **33.24:** © J.D. Cunningham/Visuals Unlimited, **33.26a,b:** © Ed Pembleton, **33.26c:** © R.C. Hermes/ Photo Researchers, Inc., **33.26d:** © George J. Wilder/ Visuals Unlimited.

Chapter 34

Opener: PhotoDisc/Details of Nature, **34.9:** © John C. Deitz/Photo Researchers, Inc., **34.10c:** © Fred Baverdam/Peter Arnold, Inc., **34.10d:** © James R. McCullough/Visuals Unlimited, **34.17:** © Manfred Kage/Peter Arnold, Inc., **34.18a:** © Norbert WU/ Peter Arnold, Inc., **34.26a:** © Bruce Rusell/Biomedia Assoc., **34.27b:** © G.L. Jensen/Visuals Unlimited.

Chapter 35

Opener: © Fritz lking/Visuals Unlimited, **35.4:** © Andrew J. Martinez/Photo Researchers, Inc., **35.6b:** © C.P. Hickman/Visuals Unlimited, **35.7a:** © CABISCO/Visuals Unlimited, **35.12a:** © Gary Retherford/Photo Researchers, Inc., **35.15a:** © Joseph T. Collins/Photo Researchers, Inc., **35.15b:** © Rod Planck/Tony Stone Worldwide, Ltd., **35.18b:** © Kjell B. Sandved/Visuals Unlimited, **35.22a:** © Marc Chamberlain/Tony Stone Images, Inc., **35.22b:** © Stephen Fink/Corbis, **35.32a:** © Jacana/ Photo Researchers, Inc., **35.33:** © Ken Lucas/Visuals Unlimited, **35.35a:** © Art Wolfe/Tony Stone Images, **35.35b:** © Hugh Sitto/Tony Stone Images, **35.35c:** © S.L. Collins and J.T. Collins/Photo Researchers, Inc., **35.35d:** © Wolfgang Kaehler/ Corbis, **35.42:** © Tom McHugh/Photo Researchers, Inc., **35.43a:** © Erwin C. "Bud" Nielsen/Visuals Unlimited, **35.43b:** © Leonard Lee Rue/Photo Researchers, Inc., **35.43c:** © Tom McHugh/Photo Researchers, Inc., **35.44a:** © John Warde/Tony Stone Worldwide, Inc., **35.44b:** © Chuck Davis/Tony Stone Images, Inc., **35.44c:** © Daniel J. Cox/Tony Stone Images, Inc., **35.44d:** © Corbis.

Chapter 36

Opener: © Springer/Corbis.

Chapter 37

Opener: Burton S. Guttman, **37.3c:** © CABISCO/ Visuals Unlimited, **37.5:** Burton S. Guttman, **37.11:** © BioPhoto Association/Photo Researchers, Inc., **37.16:** © Jerome Wexler/Photo Researchers, Inc., **37.17:** © Jack M. Bostrack/Visuals Unlimited, **37.23:** © Randy Moore/Visuals Unlimited, **37.25 a,b:** Burton S. Guttman, **37.26:** © J. Serrao/ Visuals Unlimited, **37.27:** Burton S. Guttman.

Chapter 38

Opener: © Bill Beatty/Visuals Unlimited, **38.1:** © Ed Reschke/Peter Arnold, Inc., **38.10:** Courtesy of Dr. P.B. Tomlinson, Photo by Martin H. Zimmerman, Harvard Forest, Harvard University , **38.11 a,b:** © Ed Pembleton, **38.17:** Courtesy of David M. Gates, The University of Michigan., **38.20a:** © Gary Braasch/Corbis, **38.20b:** © James Steinberg/Photo Researchers, Inc., **38.20c:** © D. Matherly/Visuals Unlimited.

Chapter 39

Opener: © D. Newman/Visuals Unlimited, **Methods 39.1:** © Dr. Jeremy Burgess/Science Photo Library/ Photo Researchers, Inc., **39.5(both):** © BioPhot, **39.7:** © From "Pictoral Guide to Rice Diseases", American Phytopathological Society, **39.11:** © David Newman/Visuals Unlimited, **39.13a:** © Norm Thomas/Photo Researchers, Inc., **39.13b:** © Michael P. Gadomski/Photo Researchers, Inc., **39.19:** © John D. Cunningham/Visuals Unlimited.

Chapter 40

Opener: © E.R. Degginger/Photo Researchers, Inc., **40.3 a, b:** © Holt Studios/Nigel Cattlin/Photo Researchers, Inc., **40.4:** © Derrick Ditchburn/Visuals Unlimited, **40.10:** © C.P. Vance/ Visuals Unlimited, **40.12:** © David R. Frazier/Photo Researchers, Inc., **40.14a:** © Eric Horan/Gamma Liaison International, **40.14b:** © Gary Retherford/ Photo Researchers, Inc., **40.15:** © Jeff Lere/Photo Researchers, Inc., **40.16:** © Michael Fogden/Bruce Coleman, Inc., **40.17a:** © John Gerlach/Visuals Unlimited, **40.17b:** © Scott Camazine/Photo Researchers, Inc.

Chapter 41

Opener: © Hulton Deutsch Collection/Corbis, **41.26:** © John Paul Kay/Peter Arnold, Inc.

Chapter 42

Opener: © Gamma Liaison, **42.5:** © Daniel Gotschall/Visuals Unlimited, **42.27:** Courtesy Marcus E. Raichle, M.D.

Chapter 43

Opener: © SIU/Peter Arnold, Inc.

Chapter 44

Opener: © Thomas W. Martin/Photo Researchers, Inc., **44.1:** © Lori Adamski Peek/Tony Stone Images, **44.8:** © Lester Bergman/Corbis, **44.14:** © D.W. Fawcett/Visuals Unlimited, **44.15:** Burton S. Guttman.

Chapter 45

Opener: © P. Mota/S. Correr/SPL/Photo Researchers, Inc., **45.20:** © SPL/Photo Researchers, Inc., **45.22a,b:** Courtesy of Dr. Clyde Barlow.

Chapter 46

Opener: © Martin Land/Science Photo Library/Photo Researchers, Inc., **46.2b:** © Alan Root/OSF/Animals Animals/Earth Scenes, **46.19:** © John Cancalosi/Peter Arnold, Inc.

Chapter 47

Opener: © Charles Cecil/Visuals Unlimited, **47.19:** Burton S. Guttman.

Chapter 48

Opener: © Peter Cade/Tony Stone Images, **48.2:** © A.B. Dowsett/SPL/Photo Researchers, Inc., **48.3:** Courtesy of National Library of Medicine, **48.4a:** © John D. Cunningham/Visuals Unlimited, **48.4 b-d:** © Ed Reschke/Peter Arnold, Inc., **48.4e:** © John D. Cunningham/Visuals Unlimited, **48.6:** © David M. Phillips/Visuals Unlimited, **48.15:** Courtesy of Dr. Klaus Hummeler, University of Pennsylvania.

Chapter 49

Opener: © Charlie Ott/Photo Researchers, Inc., **49.11:** © Bill Ivy/Tony Stone Worldwide, Ltd., **49.14:** © Nina Leen/Time Inc., **49.17:** Burton S. Guttman.

Chapter 50

Opener: © Carlos Sanuvo/Bruce Coleman, Inc., **50.1:** © Pat and Tom Leeson/Photo Researchers, Inc., **50.8:** © Dr. Irenaus Eibl-Eibesfeldt, **Sidebar 50.1:** Courtesy Philip J. DeVries, **50.10:** Burton S. Guttman, **50.11:** © Tom McHugh/Photo Researchers, Inc., **50.20:** © Akira Uchiyama/Photo Researchers, Inc.

Chapter 51

Opener: © Rudy H. Kuiter/Aquatic Photographics, **51.3 a,b:** © Gary Nuechierlein,

Line Art

Chapter 1

Sidebar 1.1: AIBS. 1995. Creationism is not science. *Bioscience,* 45:97. © 1995 American Institute of Biological Sciences. Reprinted by permission.

Chapter 2

Figure 2.1: Plate 49 from *A Field Guide to the Birds,* 2/e by Roger Tory Peterson. Copyright 1974 by Roger Tory Peterson. Reprinted with permission of Houghton-Mifflin Company. All rights reserved; **Figure 2.12:** Reprinted by permission of the publisher from *Prehistoric Life* by Percey E. Raymond, Cambridge, Mass.: Harvard University Press, Copyright © 1939, 1947 by the President and Fellows of Harvard College.

Chapter 4

Figure 4.14: Modified from a drawing by Jane S. Richardson. Used by permission of Jane S. Richardson.

Chapter 6

Figure 6B: Copyright © David S. Goodsell; **Figure 6.27:** Reproduced from *The Journal of Cell Biology,* 1965, vol. 24, No. 1, pp. 30-55 by copyright permission of The Rockefeller University Press.

Chapter 19

Figure 19.22: Reprinted by permission from J.M. Rommens, et al., *American Journal of Human Genetics,* 43:645-663. Copyright © 1988 by The University of Chicago. All rights reserved.

Chapter 23

Figure 23.12: © George V. Kelvin/Scientific American. **Figure 28.10:** Reprinted by permission from Begon, et al., *Ecology: Individuals, Populations and Communities,* 2nd edition. Copyright © 1990 Blackwell Scientific. Original data from M.J. Swift, et al., (1979) *Decomposition in Terrestrial Ecosystems,* Blackwell Science, Oxford.

Chapter 37

Figure 37.2: Reproduced from The Journal of Cell Biology, 1987, Vol. 84, p. 327 by copyright permission of The Rockefeller University Press.

Chapter 50

Figure 50.7: Illustrations from *In the Shadow of Man* by Jane Goodall. Copyright © 1971 by Hugo and Jane van Lawick-Goodall. Reprinted by permission of Houghton Mifflin Company. All rights reserved.

Index

A

B

C

D

E

F

G

H

I

J

K

L

M

N

O

P

Q

R

S

T

U

V

W

X

Y

Z

Index of Conceptual Themes

This book is tied together by four primary themes. This conceptual framework unifies all the specific facts and more limited principles developed in this textbook. This index is a sampling of where the themes occur within the text.

Organisms are genetic systems.

Organisms live in ecosystems, where they are adapted to particular ways of life and engage in complex interrelationships with other organisms and the environment.